Tolley's
Health and Safety
at Work
Handbook
2006

While every care has been taken to ensure the accuracy of this work, no responsibility for loss or damage occasioned to any person acting or refraining from action as a result of any statement in it can be accepted by the authors, editors or publishers.

# Tolley's Health and Safety at Work Handbook 2006

Eighteenth Edition

## Members of the LexisNexis Group worldwide

| | |
|---|---|
| United Kingdom | LexisNexis Butterworths, a Division of Reed Elsevier (UK) Ltd, Halsbury House, 35 Chancery Lane, LONDON, WC2A 1EL, and RSH, 1–3 Baxter's Place, Leith Walk EDINBURGH EH1 3AF |
| Argentina | LexisNexis Argentina, BUENOS AIRES |
| Australia | LexisNexis Butterworths, CHATSWOOD, New South Wales |
| Austria | LexisNexis Verlag ARD Orac GmbH & Co KG, VIENNA |
| Canada | LexisNexis Butterworths, MARKHAM, Ontario |
| Chile | LexisNexis Chile Ltda, SANTIAGO DE CHILE |
| Czech Republic | Nakladatelství Orac sro, PRAGUE |
| France | Editions du Juris-Classeur SA, PARIS |
| Germany | LexisNexis Deutschland GmbH, FRANKFURT and MUNSTER |
| Hong Kong | LexisNexis Butterworths, HONG KONG |
| Hungary | HVG-Orac, BUDAPEST |
| India | LexisNexis Butterworths, NEW DELHI |
| Ireland | Butterworths (Ireland) Ltd, DUBLIN |
| Italy | Giuffrè Editore, MILAN |
| Malaysia | Malayan Law Journal Sdn Bhd, KUALA LUMPUR |
| New Zealand | LexisNexis Butterworths, WELLINGTON |
| Poland | Wydawnictwo Prawnicze LexisNexis, WARSAW |
| Singapore | LexisNexis Butterworths, SINGAPORE |
| South Africa | LexisNexis Butterworths, Durban |
| Switzerland | Stämpfli Verlag AG, BERNE |
| USA | LexisNexis, DAYTON, Ohio |

First published in 1988

© Reed Elsevier (UK) Ltd 2005

All rights reserved. No part of this publication may be reproduced in any material form (including photocopying or storing it in any medium by electronic means and whether or not transiently or incidentally to some other use of this publication) without the written permission of the copyright owner except in accordance with the provisions of the Copyright, Designs and Patents Act 1988 or under the terms of a licence issued by the Copyright Licensing Agency Ltd, 90 Tottenham Court Road, London, England W1T 4LP. Applications for the copyright owner's written permission to reproduce any part of this publication should be addressed to the publisher.

Warning: The doing of an unauthorised act in relation to a copyright work may result in both a civil claim for damages and criminal prosecution.

Crown copyright material is reproduced with the permission of the Controller of HMSO and the Queen's Printer for Scotland. Any European material in this work which has been reproduced from EUR-lex, the official European Communities legislation website, is European Communities copyright.

A CIP Catalogue record for this book is available from the British Library.

ISBN 0 7545 2753 0

Typeset by Letterpart Ltd, Reigate, Surrey

Printed and bound in Great Britain by William Clowes Ltd, Beccles, Suffolk

Visit LexisNexis Butterworths at www.lexisnexis.co.uk

# List of Contributors

*Lawrence Bamber, BSc, DIS, MASSE, FIRM, FIOSH, RSP*
Managing Director, Risk Solutions International

*Mike Bateman, BSc, MIOSH, RSP,*
Independent Health and Safety Consultant

*Martin Bruffell, LLB,*
Partner, Berrymans Lace Mawer

*Jessica Burt, BA,*
Solicitor, CMS Cameron McKenna

*David Clarke, CIBiol,*
Food Safety Consultant

*Jill Cleaver, MSc, BSc (Hons),*
Senior Consultant, Hu-Tech Associates Ltd

*Nicola Coote, FIOSH, MIIRSM, MRSH, MCIPO, RSP,*
Director of Personnel, Health and Safety Consultants Ltd

*Dean Cross, BSc (Hons),*
Consultant, Sypol Limited

*Kitty Debenham, BA (Oxon),*
Solicitor, Simmons & Simmons

*Christopher Eskell, TD, LLB, MIMgt,*
Partner, Bond Pearce

*Janet Gaymer, MA (Oxon), LLM (Lond),*
Partner, Simmons & Simmons

*Margaret Hanson, BSc (Hons), MErgS,*
Senior Consultant, Hu-Tech Associates Ltd

*Hazel Harvey, MPhil, CCol, ASDC, MIOSH,*
Head of Professional Affairs, IOSH

*Dorothy Henderson, BA,*
Partner, Travers Smith Braithwaite

*Jo Hilliard, LLB,*
Solicitor, Cartwrights

*Nick Humphreys, LLB, LLM,*
Solicitor, Barrister, Richards Butler

*Ian Jerome, BSc,*
Senior Consultant, Education and Training, Loss Prevention Council

*List of Contributors*

*Raj Lakha, BA, MBA, MIIRSM, MIOSH, MInstD, AIMC,*
Director of Safety Solutions (UK) Ltd, Director of COASST International Ltd

*David Leckie, LLB,*
Barrister, Maclay Murray & Spens

*Deborah Lloyd, BSc, LLB, LLM,*
Solicitor, Ashurst Morris Crisp

*Subash Ludhra, BSc, RSP, FIOSH, Dip Env Man,*
Director Quality Assurance & Training, Initial Catering Services Ltd

*Peter Neild, MPhil,*
FCII, Berrymans Lace Mawer

*Andrea Oates, BSc,*
Health and Safety Researcher, Labour Research Department

*Stephen Perry, LLB,*
Solicitor, Bond Pearce

*George Pitblado, I Eng, MI Plant E, MIIRSM, MWM Soc,*
Partner, GMP Support Services Consultants

*Hani Raafat, BSc, MSc, PhD, CEng, MIMechE, FIOSH, RSP,*
Risk Management Consultant

*Hamish Robertson, BSc (Hons),*
Senior Consultant, Sypol Limited

*Mark Rutter, BSc, PhD,*
Environmental Advisor, CMS Cameron McKenna

*Kajal Sharma, BSc,*
Solicitor, CMS Cameron McKenna

*Dr R C Slade,*
College Safety Officer, King's College London

*Mary Spear, BA (Cantab),*
Solicitor, Cartwrights

*Rebecca Stewart, BA,*
Solicitor, Travers Smith Braithwaite

*Roger Tompsett, BSc(Eng), MIOA,*
Head of Acoustics, WS Atkins Noise & Vibration

*Mark Tyler, MA, LLM, MIOSH,*
Partner, CMS Cameron McKenna

*George Ventris,*
formerly secretary of the HSC's Construction Industry Advisory Committee (CONIAC) and a leader of HSE's Construction National Interest Group

*Jeffrey Wale, LLB,*
Partner, Berrymans Lace Mawer

*Lawrence Waterman, FIOSH, RSP, MBIOH, ROH,*
Chairman, Sypol Limited

*Natalie Wood, BSc (Eng),*
Solicitor, CMS Cameron McKenna

# Health and Safety at Work Editorial Board

The following are all members of the newly appointed Health and Safety Editorial Board for the handbook. Each member brings with them a wealth of experience and knowledge of health and safety matters. The benefits to be gained from introducing this board are to ensure that this publication continues to provide practical, authoritative coverage of health and safety law in all aspects of industrial and office workplaces.

## Lawrence Bamber

Lawrence Bamber is currently an Associate Consultant with Norwich Union Risk Services and is also a freelance OSH/ risk management consultant. He has worked in the OSH/ risk management field for over 30 years with General Accident, Stenhouse (Aon), as well as Norwich Union, and is well known for his training, consultancy and presentation skills. Lawrence is a past president of the Institution of Occupational Safety and Health (IOSH) and is the author of many papers and publications on OSH/ risk management.

## Mike Bateman

Mike Bateman has spent over 30 years in the health and safety field, firstly as a HSE inspector, and then managing health and safety departments in the shipbuilding and aluminium industries. Since 1991 he has been running his own consultancy, attracting both large and small clients from a wide range of occupational areas. Mike has considerable practical experience of risk assessment and is the author of *Tolley's Risk Assessment Handbook*.

## Professor Malcolm Harrington

Professor Malcolm Harrington, CBE, is Emiritus Professor of Occupational Medicine of the Faculty of Medicine and Dentistry, University of Birmingham. His major interests lie in epidemiology, occupational cancer, health care workers, and occupational health policy. In 1992, he was awarded the CBE, and in 2001 was awarded the lifetime achievement award by IOSH. During 1985–1996 Malcolm was Chairman of the Industrial Injuries Advisory Council. He is also the co-author of over 200 publications including research, review papers and chapters.

## Mark Tyler

Mark Tyler is head of the Health, Safety and Products team at CMS Cameron McKenna. He holds corporate membership of IOSH, and is a member of the CBI Health and Safety Panel. Mark is also a co-author of the books *Product Safety* and *Safer by Design*. He has acted in a number of major cases which include the Southall and Ladbroke Grove Rail Inquiries, the Organophosphate group litigation, and the South Kensington Legionnaires' Disease outbreak.

# Lawrence Waterman

Lawrence Waterman, FIOSH, RSP, MFOH, FRSA, has been a safety practitioner for over 25 years. Following a period in enforcement and in industry, he founded Sypol Limited, now one of the UK's leading health, safety and environmental consultancies. He is currently the Chairman of Sypol. Lawrence has a special interest in construction and in occupational health and hygiene which has been predominant in his career, but not to the exclusion of many other sectors and areas of work. He is a winner of the Gold Medal (jointly) for health and safety from the Institution of Civil Engineers and is Project Director of the Constructing Better Health occupational health pilot launched by the construction industry in October 2004. As of November 2004, Lawrence is President of IOSH.

# Contents

*List of Tables*   xxv
*List of Illustrations/Diagrams*   xxvii
*Abbreviations and References*   xxix

**Introduction** – *Lawrence Waterman*   Int/1

    The looseleaf   Int/1
    This Introduction is in five main parts   Int/1
    Drivers for health and safety performance   Int/1
    The changing world of work   Int/3
    Legal framework   Int/4
    The structure of UK health and safety law   Int/9
    Management of health and safety at work   Int/17

**Access, Traffic Routes and Vehicles** – *Nicola Coote*   A10/1

    Introduction   A10/1
    Statutory duties concerning workplace access and egress – Workplace
        (Health, Safety and Welfare) Regulations 1992 (SI 1992 No 3004)   A10/2
    Vehicles   A10/4
    Common law duty of care   A10/8
    Confined spaces   A10/8

**Accident Reporting and Investigation** – *Mark Tyler/Kajal Sharma*   A30/1

    Introduction   A30/1
    Reporting of Injuries, Diseases and Dangerous Occurrences Regulations
        1995 (RIDDOR) – (SI 1995 No 3163)   A30/1
    Relevant enforcing authority   A30/2
    Incident Reporting Centre   A30/2
    Persons responsible for notification and reporting   A30/2
    What is covered?   A30/4
    Major injuries or conditions   A30/4
    Injuries incapacitating for more than three consecutive days   A30/5
    Specified dangerous occurrences   A30/5
    Specified diseases   A30/5
    Duty to notify/report   A30/5
    Duty to keep records   A30/6
    Reporting the death of an employee   A30/7
    Road accidents   A30/7
    Duty to report gas incidents   A30/9
    Penalties   A30/9
    Obligations on employees   A30/9
    Objectives of accident reporting   A30/10
    Duty of disclosure of accident data   A30/10
    Accident Investigation   A30/12
    Appendix A
        List of Reportable Dangerous Occurrences   A30/19
    Appendix B
        List of Reportable Diseases   A30/31

*Contents*

| | |
|---|---|
| Appendix C | A30/42 |
| Appendix D | A30/43 |
| Appendix E | |
|     Health and Safety Executive Regional Offices | A30/44 |

**Asbestos** – *Lawrence Waterman*     A50/1

| | |
|---|---|
| Introduction | A50/1 |
| Historical exposures – manufacturing | A50/2 |
| Asbestos prohibitions | A50/3 |
| Asbestos removal | A50/3 |
| Exposure of building maintenance workers | A50/4 |
| Duties under the Control of Asbestos at Work Regulations 2002 | A50/4 |
| Asbestos building surveys | A50/10 |
| Asbestos waste | A50/11 |
| Asbestos works | A50/11 |
| Legislation, Approved Codes of Practice, and Guidance Notes | A50/12 |

**Business Continuity** – *Lawrence Bamber*     B80/1

| | |
|---|---|
| What is business continuity planning? | B80/1 |
| Business continuity planning in context | B80/1 |
| Why is business continuity planning of importance? | B80/2 |
| Business continuity drivers | B80/2 |
| Convincing the board – the business case | B80/4 |
| The business continuity model | B80/5 |
| Crisis communication and public relations | B80/14 |

**Compensation for Work Injuries/Diseases** – *David Leckie*     C60/1

| | |
|---|---|
| Introduction | C60/1 |
| Social security benefits | C60/1 |
| Industrial injuries benefit | C60/2 |
| Accident and personal injury provisions | C60/3 |
| Industrial injuries disablement benefit | C60/6 |
| Constant attendance allowance and exceptionally severe disablement allowance | C60/7 |
| Other sickness/disability benefits | C60/8 |
| Disability Tax Credit under Working Tax Credit | C60/8 |
| Payment | C60/8 |
| Damages for occupational injuries and diseases | C60/9 |
| Awards of damages and recovery of state benefits | C60/20 |
| Duties of the compensator | C60/20 |
| Complications | C60/21 |
| Appeals against certificates of recoverable benefits | C60/21 |
| Pneumoconiosis | C60/24 |

**Community Health and Safety** – *Mike Bateman*     C65/1

| | |
|---|---|
| Introduction | C65/1 |
| Relevant Legal Requirements | C65/1 |
| Neighbours and Passers-by | C65/4 |
| Trespassers on Work Premises | C65/7 |
| Customers and Users of Facilities and Services | C65/8 |
| Visiting Individuals and Groups | C65/11 |

| | |
|---|---|
| Volunteers | C65/14 |
| Major Public Events | C65/16 |
| Use of Contractors | C65/18 |
| Further Information | C65/19 |

**Confined Spaces** – *Mike Bateman* — C70/1

| | |
|---|---|
| Introduction | C70/1 |
| What the Regulations require | C70/1 |
| Other risks in confined spaces | C70/4 |
| Safe systems of work | C70/4 |
| Precautions for work in confined spaces | C70/10 |
| Emergency arrangements | C70/13 |

**Construction and Building Operations** – *George Ventris* — C80/1

| | |
|---|---|
| Introduction to construction work | C80/1 |
| Practical safety criteria | C80/1 |
| The Regulations | C80/4 |
| Construction (Design and Management) Regulations 1994 (SI 1994 No 3140) | C80/7 |
| Construction (Health, Safety and Welfare) Regulations 1996 (SI 1996 No 1592) | C80/14 |
| Confined spaces and harmful atmospheres | C80/24 |
| Work in compressed air on construction sites – Work in Compressed Air Regulations 1996 (SI 1996 No 1656) | C80/32 |
| Construction (Head Protection) Regulations 1989 (SI 1989 No 2209) | C80/36 |
| Composition of construction products | C80/39 |
| Protecting visitors and the public | C80/40 |
| Responsibility for contractors and subcontractors | C80/49 |
| Liability of employer/occupier in connection with contract work | C80/50 |

**Control of Major Accident Hazards** – *Mary Spear/Christopher Eskell* — C100/1

| | |
|---|---|
| Introduction | C100/1 |
| Control of Major Accident Hazards Regulations 1999 (SI 1999 No 743) | C100/1 |
| Radiation (Emergency Preparedness and Public Information) Regulations 2001 (SI 2001 No 2975) | C100/15 |
| Public Information for Radiation Emergencies Regulations 1992 (SI 1992 No 2997) | C100/16 |

**Dangerous Goods Systems** – *Hamish Robertson/Dean Cross* — D40/1

| | |
|---|---|
| Basic regulatory principles | D40/1 |
| The legal framework | D40/4 |
| General requirements | D40/10 |
| Specific requirements | D40/19 |

**Disaster and Emergency Management Systems (DEMS)** – *Raj Lakha* — D60/1

| | |
|---|---|
| Historical Development of Disaster and Emergency Management | D60/1 |
| Origin of Disaster and Emergency Management as a modern discipline | D60/2 |
| Definitions | D60/2 |
| The Need for Effective Disaster and Emergency Management | D60/6 |
| Disaster and Emergency Management Systems (DEMS) | D60/16 |

| | |
|---|---|
| External and Internal Factors | D60/17 |
| Establish a Disaster and Emergency Policy | D60/21 |
| Organise for Disasters and Emergencies | D60/24 |
| Disaster and Emergency Planning | D60/26 |
| Monitor the Disaster and Emergency Plan(s) | D60/41 |
| Audit and Review | D60/44 |
| Case studies | D60/45 |
| Conclusion | D60/46 |
| Sources of information | D60/46 |

**Display Screen Equipment** – *Mike Bateman* — D82/1

| | |
|---|---|
| Introduction | D82/1 |
| The Regulations summarised | D82/1 |
| DSE workstation assessments | D82/4 |
| Guidance on completing the DSE workstation self-assessment | D82/11 |
| Special situations | D82/13 |
| After the assessment | D82/14 |

**Dangerous Substances and Explosive Atmospheres** – *Hamish Robertson and Dean Cross* — D90/1

| | |
|---|---|
| Background | D90/1 |
| Introduction | D90/1 |
| The Regulations | D90/2 |
| Risk assessment | D90/5 |
| Management of risks | D90/7 |
| Classification of places with explosive atmospheres | D90/9 |
| Zoning | D90/11 |
| Equipment and protection systems | D90/12 |
| Arrangements to deal with incidents and emergencies | D90/13 |
| Provision of information, instruction and training | D90/14 |
| Identification of hazardous contents of pipes and containers | D90/14 |
| Exemption | D90/15 |
| Amendments to modernise petroleum and other legislation | D90/15 |
| Repeals and revocations of existing legislation | D90/16 |
| Enforcement | D90/17 |
| Regulatory impact assessment | D90/17 |
| Approved codes of practice | D90/18 |
| Sources and acknowledgement | D90/18 |

**Electricity** – *Chris Buck* — E30/1

| | |
|---|---|
| Introduction | E30/1 |
| Understanding electricity | E30/1 |
| Legal background and standards | E30/2 |
| Dangers of electricity | E30/3 |
| Fundamentals of controlling electrical risks | E30/6 |
| Competence | E30/7 |
| Design and construction of electrical equipment and systems | E30/8 |
| Maintenance | E30/12 |
| Safe systems of work | E30/13 |
| Special situations | E30/16 |
| Other relevant legislation | E30/19 |

*Contents*

| | |
|---|---|
| **Emissions into the Atmosphere** – *Roger Barrowcliffe* | E50/1 |
| Introduction | E50/1 |
| Regulatory position | E50/2 |
| Regulatory drivers | E50/2 |
| Key pollutants and issues | E50/5 |
| The European context | E50/9 |
| UK Regulations | E50/15 |
| | |
| **Employers' Liability Insurance** – *Christopher Eskell/Stephen Perry* | E130/1 |
| Introduction | E130/1 |
| Purpose of compulsory employers' liability insurance | E130/2 |
| General law relating to insurance contracts | E130/3 |
| Subrogation | E130/5 |
| Duty of employer to take out and maintain insurance | E130/5 |
| Issue, display and retention of certificates of insurance | E130/7 |
| Penalties | E130/8 |
| Cover provided by a typical policy | E130/8 |
| 'Prohibition' of certain conditions | E130/10 |
| Trade endorsements for certain types of work | E130/10 |
| Measure of risk and assessment of premium | E130/11 |
| Extension of cover | E130/11 |
| Insolvent employers' liability insurers | E130/11 |
| | |
| **Employment Protection** – *Kitty Debenham/Janet Gaymer* | E140/1 |
| Introduction | E140/1 |
| Sources of contractual terms | E140/2 |
| Employee employment protection rights | E140/8 |
| Enforcement of safety rules by the employer | E140/10 |
| Dismissal | E140/13 |
| Health and safety duties in relation to women at work | E140/18 |
| | |
| **Enforcement** – *Mark Tyler/Kajal Sharma* | E150/1 |
| Introduction | E150/1 |
| The role of the Health and Safety Commission and Executive | E150/2 |
| Enforcing authorities | E150/4 |
| 'Relevant statutory provisions' covered by the Health and Safety at Work etc. Act 1974 | E150/8 |
| Part A: Enforcement Powers of Inspectors | E150/8 |
| Part B: Offences and Penalties | E150/16 |
| | |
| **Environmental Management** – *Mark Rutter* | E160/1 |
| Introduction | E160/1 |
| Internal drivers | E160/2 |
| External drivers | E160/3 |
| Environmental management guidelines | E160/11 |
| Implementing environmental management | E160/14 |
| Benefits of environmental management | E160/19 |
| The future | E160/20 |

## Contents

**Ergonomics** – *Margaret Hanson/Jill Cleaver*    E170/1

    Introduction    E170/1
    The ergonomic approach    E170/2
    Designing for people    E170/4
    Designing the task    E170/7
    Workstation design    E170/11
    Physical work environment    E170/15
    Job design and work organisation    E170/18
    Musculoskeletal disorders    E170/20
    Ergonomics tools    E170/22
    Relevant legislation and guidance    E170/25

**Facilities Management – An Overview** – *Robert Greenfield*    F30/1

    Introduction    F30/1
    A good definition of facilities management    F30/1
    Facilities management    F30/1
    Hard and soft facilities management    F30/2
    Managing    F30/3
    Company approach to facilities management    F30/3
    In-house or outsourced facilities management?    F30/5
    Advantages to the company    F30/5
    Responsibilities of facilities management    F30/6
    Four most common health and safety issues for facilities management    F30/13
    How facilities management and building design can help a company    F30/16
    Recognition of facilities management    F30/16
    Innovations within the facilities management industry    F30/17
    The effects of safety, health and environmental legislation on facilities management    F30/18
    Summary of the facilities management    F30/20

**Fire Prevention and Control** – *Ian Jerome*    F50/1

    Elements of fire    F50/1
    Common causes of fires    F50/2
    Fire classification    F50/2
    Electrical fires    F50/3
    Fire extinction – active fire protection measures    F50/3
    Property risk    F50/4
    Passive and active fire protection    F50/4
    Fire procedures and portable equipment    F50/5
    Good 'housekeeping'    F50/7
    Pre-planning of fire prevention    F50/8
    Fire drill    F50/9
    Fires on construction sites    F50/10
    Fire and fire precautions – legislation    F50/12
    The current position – overview    F50/12
    Specific fire precautions regulations    F50/14
    Fire certification    F50/15
    Improvement notices    F50/20
    Prohibition notices    F50/22
    The Workplace Regulations    F50/23
    Fire Certificates (Special Premises) Regulations 1976 (SI 1976 No 2003)    F50/26
    Appeals relating to the issue of a fire certificate    F50/30

| | |
|---|---|
| Duty to keep fire certificate on the premises | F50/30 |
| Notification of alteration to premises | F50/30 |
| Licensed premises and fire safety | F50/30 |
| Residential and nursing homes | F50/32 |
| Crown premises | F50/32 |
| Civil liability at common law and under the Fires Prevention (Metropolis) Act 1774 for fire damage | F50/32 |
| Fire insurance | F50/33 |

**First-Aid** – *Subash Ludhra* — F70/1

| | |
|---|---|
| Introduction | F70/1 |
| The employer's duty to make provision for first-aid | F70/1 |
| The assessment of first-aid needs | F70/2 |
| Duty of the employer to inform employees of first-aid arrangements | F70/5 |
| First-aid and the self-employed | F70/5 |
| Number of first-aiders | F70/5 |
| Appointed persons | F70/6 |
| Records and record keeping | F70/7 |
| First-aid resources | F70/7 |
| Rooms designated as first-aid areas | F70/9 |
| Appendix A | |
|     Assessment of first-aid needs – checklist | F70/11 |
| Appendix B | |
|     First-aid competencies | F70/13 |

**Food Safety and Standards** – *David Clarke* — F90/1

| | |
|---|---|
| Introduction | F90/1 |
| Food Safety Act 1990 | F90/3 |
| Food premises regulations | F90/25 |
| Food hygiene regulations | F90/29 |
| Food Safety Manual – food safety policy and procedures | F90/43 |

**Gas Safety** – *Dr R C Slade* — G10/1

| | |
|---|---|
| Introduction | G10/1 |
| Gas supply – the Gas Act 1986, as amended | G10/2 |
| Gas supply management – the Gas Safety (Management) Regulations 1996 (SI 1996 No 551) | G10/3 |
| Rights of entry – the Gas Safety (Rights of Entry) Regulations 1996 (SI 1996 No 2535) | G10/4 |
| Pipelines – the Pipelines Safety Regulations 1996 (SI 1996 No 825) | G10/5 |
| Gas systems and appliances – the Gas Safety (Installation and Use) Regulations 1998 (SI 1998 No 2451) | G10/5 |
| Interface with other legislation | G10/14 |
| Summaries of relevant court cases | G10/15 |
| Gas safety requirements in factories | G10/18 |
| Pressure fluctuations | G10/18 |

**Harassment in the Workplace** – *Nick Humphreys* — H17/1

| | |
|---|---|
| Introduction | H17/1 |
| The European Commission Recommendation and Code of Practice | H17/2 |
| Definition | H17/3 |

*Contents*

| | |
|---|---|
| Examples of unlawful harassment | H17/7 |
| Victimisation | H17/10 |
| Employers' liability | H17/10 |
| Legal action and remedies | H17/14 |
| Preventing and handling claims | H17/20 |
| Future developments | H17/23 |

**Hazardous Substances in the Workplace** – *Mike Bateman* — H21/1

| | |
|---|---|
| Introduction | H21/1 |
| How hazardous substances harm the body | H21/1 |
| The COSHH Regulations summarised | H21/3 |
| What the Regulations require on assessments | H21/4 |
| Practical aspects of COSHH assessments | H21/10 |
| COSHH assessment records | H21/14 |
| Further requirements of the COSHH Regulations | H21/20 |
| Other important regulations | H21/28 |

**Joint Consultation in Safety – Safety Representatives, Safety Committees, Collective Agreements and Works Councils** – *Nick Humphreys* — J30/1

| | |
|---|---|
| Introduction | J30/1 |
| Regulatory framework | J30/2 |
| Consultation obligations for unionised employers | J30/3 |
| Safety committees | J30/10 |
| Non-unionised workforce – consultation obligations | J30/11 |
| Recourse for safety representatives | J30/13 |
| European developments in health and safety | J30/14 |

**Lifting Operations** – *Kevin Chicken and Nicola Coote* — L30/1

| | |
|---|---|
| Introduction | L30/1 |
| Statutory requirements relating to the manufacture of lifting equipment | L30/1 |
| Lifts Regulations 1997 (SI 1997 No 831) | L30/3 |
| Lifting operations and equipment failure | L30/18 |
| Hoists and lifts | L30/20 |
| Fork lift trucks | L30/21 |
| Patient/bath hoist | L30/22 |
| Vehicle lifting table | L30/23 |
| Lifting accessories | L30/23 |
| Shackles | L30/26 |
| Once only use accessories | L30/26 |
| Mobile lifting equipment | L30/26 |

**Lighting** – *Nicola Coote* — L50/1

| | |
|---|---|
| Introduction | L50/1 |
| Statutory lighting requirements | L50/1 |
| Sources of light | L50/3 |
| Standards of lighting or illuminance | L50/4 |
| Qualitative aspects of lighting and lighting design | L50/6 |

## Machinery Safety – *Hani Raafat* — M10/1

| | |
|---|---|
| Introduction | M10/1 |
| Legal requirements relating to machinery | M10/1 |
| Machinery safety – the risk based approach | M10/2 |
| The Provision and Use of Work Equipment Regulations 1998 (PUWER'98) | M10/2 |
| The Supply of Machinery (Safety) Regulations 1992 (as amended) | M10/12 |
| The Transposed Harmonised European Machinery Safety Standards | M10/17 |
| Machinery risk assessment | M10/18 |
| Machinery hazards identification | M10/23 |
| Options for machinery risk reduction | M10/26 |
| Types of safeguards and safety devices | M10/27 |
| Fixed guards | M10/28 |
| Safety devices | M10/34 |

## Managing Absence – *Lynda Macdonald* — M15/1

| | |
|---|---|
| Introduction | M15/1 |
| Managing long-term sickness absence | M15/1 |
| Managing short-term absences | M15/10 |

## Managing Health and Safety – *Mike Bateman/Lawrence Bamber* — M20/1

| | |
|---|---|
| Introduction | M20/1 |
| Legal requirements | M20/1 |
| The Management Regulations | M20/3 |
| Health and safety management systems | M20/9 |
| Policy | M20/11 |
| Organisation | M20/12 |
| Planning | M20/15 |
| Implementation | M20/17 |
| Monitoring | M20/18 |
| Audit | M20/19 |
| Review | M20/19 |
| Continual improvement | M20/20 |
| POPIMAR in action | M20/22 |

## Manual Handling – *Mike Bateman and Margaret Hanson* — M30/1

| | |
|---|---|
| Introduction | M30/1 |
| What the Regulations require | M30/1 |
| Risk of injury from manual handling | M30/3 |
| Avoiding or reducing risks | M30/9 |
| Planning and preparation | M30/15 |
| Making the assessment | M30/18 |
| After the assessment | M30/21 |

## Noise at Work – *Roger Tompsett* — N30/1

| | |
|---|---|
| Introduction | N30/1 |
| Scale of the problem and future developments | N30/1 |
| Hearing damage | N30/2 |
| The perception of sound | N30/2 |
| Noise indices | N30/3 |

*Contents*

| | |
|---|---|
| General legal requirements | N30/5 |
| Specific legal requirements | N30/7 |
| Compensation for occupational deafness | N30/8 |
| Action against employer at common law | N30/10 |
| General guidance on noise at work | N30/12 |
| Reducing noise in specific working environments | N30/13 |
| Workplace noise assessments | N30/14 |
| Noise reduction | N30/15 |
| Ear protection | N30/17 |

**Occupational Health and Diseases – An Overview** – *Martin Brufell, Peter Neild and Jeffrey Wale* — O10/1

| | |
|---|---|
| What is occupational health? | O10/1 |
| Why is it important? | O10/1 |
| Relationships between occupational health and safety | O10/2 |
| Prevention is better than cure | O10/3 |
| The occupational health team | O10/4 |
| Health records | O10/6 |
| Buying services and priorities | O10/7 |
| Initiatives and targets | O10/7 |
| Occupational diseases and disorders | O10/8 |
| Hand arm vibration syndrome (havs) and whole body vibration | O10/11 |
| Diseases and disorders of the eye | O10/13 |
| Diseases of the skin | O10/16 |
| Occupational respiratory diseases | O10/18 |
| Musculoskeletal disorders | O10/21 |
| Psychological disorders (stress, anxiety and depression) | O10/24 |
| Other occupational health concerns | O10/26 |

**Occupiers' Liability** – *Jessica Burt* — O30/1

| | |
|---|---|
| Introduction | O30/1 |
| Duties owed under the Occupiers' Liability Act 1957 | O30/1 |
| Duty owed to trespassers, at common law and under the Occupiers' Liability Act 1984 | O30/4 |
| Dangers to guard against | O30/6 |
| Waiver of duty and the Unfair Contract Terms Act 1977 (UCTA) | O30/8 |
| Risks willingly accepted – 'volenti non fit injuria' | O30/8 |
| Actions against factory occupiers | O30/10 |
| Occupier's duties under HSWA 1974 | O30/10 |

**Offshore Operations** – *Fred Osliff* — O70/1

| | |
|---|---|
| Introduction | O70/1 |
| Offshore Installations (Safety Case) Regulations 1992 (SI 1992 No 2885) | O70/7 |
| Offshore Installations and Wells (Design and Construction, etc.) Regulations 1996 (SI 1996 No 913) | O70/14 |
| Offshore Installations (Prevention of Fire and Explosion, and Emergency Response) Regulations 1995 (SI 1995 No 743) | O70/15 |
| Offshore Installations and Pipeline Works (Management and Administration) Regulations 1995 (SI 1995 No 738) | O70/18 |
| Hazardous operations with the potential to cause a major accident | O70/19 |
| Construction and use of submarine pipelines | O70/19 |

| | |
|---|---|
| Pipelines Safety Regulations 1996 (SI 1996 No 825) | O70/20 |
| The Diving at Work Regulations 1997 | O70/20 |
| Other offshore specific legislation | O70/22 |
| Offshore Installations (Logbooks and Registration of Death) Regulations 1972 | O70/22 |
| Offshore Installations (Inspectors and Casualties) Regulations 1973 | O70/22 |
| Offshore Installations (Safety Representatives and Safety Committees) Regulations 1989 | O70/24 |
| First-aid – the Offshore Installations and Pipeline Works (First-Aid) Regulations 1989 | O70/31 |
| The Offshore Electricity and Noise Regulations 1997 | O70/33 |
| Legislation for other hazardous offshore activities | O70/34 |
| Written procedures and training programmes for dealing with hazards | O70/35 |
| Hazards in offshore activities | O70/36 |

**Personal Protective Equipment** – *Nicola Coote* — P30/1

| | |
|---|---|
| Introduction | P30/1 |
| The Personal Protective Equipment at Work Regulations 1992 (SI 1992 No 2966 as amended by SI 2002 No 2174) | P30/1 |
| Work activities/processes requiring personal protective equipment | P30/2 |
| Statutory requirements in connection with personal protective equipment | P30/2 |
| Increased importance of uniform European standards | P30/5 |
| Main types of personal protection | P30/12 |
| Common law requirements | P30/20 |

**Pressure Systems** – *Kevin Chicken* — P70/1

| | |
|---|---|
| Introduction | P70/1 |
| Simple pressure vessels | P70/3 |
| Simple Pressure Vessels (Safety) Regulations (SPVSR) 1991 | P70/3 |
| EC Certificate of Adequacy | P70/3 |
| EC Type-Examination Certificate | P70/4 |
| Pressure Equipment Regulations 1999 (SI 1999 No 2001) and the EU Pressure Equipment Directive (97/23/EC) | P70/4 |
| Essential safety requirements | P70/4 |
| Conformity assessment procedure | P70/5 |
| Notified bodies | P70/6 |
| The Pressure Equipment (Amendment) Regulations 2002 (SI 2002 No 1267) | P70/6 |
| Pressure Systems Safety Regulations 2000 (SI 2000 No 128) | P70/6 |
| Systems and equipment covered by the regulations | P70/7 |
| Systems and equipment excepted from some or all the regulations | P70/8 |
| HSC approved code of practice and guidance | P70/10 |
| Interpretation | P70/10 |
| Design, construction and installation | P70/11 |
| Documentation and marking | P70/11 |
| Competent persons | P70/12 |
| Written scheme of examination | P70/13 |
| Operations, maintenance, modification and repair | P70/18 |
| Record keeping | P70/19 |
| Related health and safety legislation | P70/19 |

*Contents*

**Product Safety** – *Natalie Wood* — P90/1

    Introduction — P90/1
    Consumer products — P90/2
    General Product Safety Directive — P90/5
    Industrial products — P90/9
    Regulations made under the Health and Safety at Work etc Act 1974 — P90/10
    Criminal liability for breach of statutory duties — P90/10
    Civil liability for unsafe products – historical background — P90/13
    Consumer Protection Act 1987 — P90/14
    Contractual liability for substandard products — P90/21
    Unfair Contract Terms Act 1977 (UCTA) — P90/24
    Unfair Terms in Consumer Contracts Regulations 1999 (SI 1999 No 2083) — P90/25

**Radiation** – *Leslie Hawkins and Donald Bruce* — R10/1

    Non-ionising and ionising radiation — R10/1

**Rehabilitation** – *Leslie Hawkins* — R20/1

    What is Rehabilitation? — R20/1
    Why should we be concerned? — R20/1
    Managing sickness absence — R20/5
    Long term absence and rehabilitation — R20/7
    Using professional advice — R20/8

**Risk Assessment** – *Mike Bateman* — R30/1

    Introduction — R30/1
    HSWA 1974 requirements — R30/1
    The Management of Health and Safety at Work Regulations 1999 — R30/2
    Common regulations requiring risk assessment — R30/3
    Specialist regulations requiring risk assessment — R30/5
    Related health and safety concepts — R30/6
    Management Regulations requirements — R30/7
    Children and young persons — R30/11
    New or expectant mothers — R30/12
    Other vulnerable persons — R30/15
    Who should carry out the risk assessment? — R30/16
    Assessment units — R30/16
    Relevant sources of information — R30/17
    Consider who might be at risk — R30/18
    Identify the issues to be addressed — R30/19
    Variations in work practices — R30/19
    Checklist of possible risks — R30/20
    Making the risk assessment — R30/21
    Assessment records — R30/22
    After the assessment — R30/24
    Content of assessment records — R30/25
    Illustrative assessment records — R30/25

**Safe Systems of Work** – *Lawrence Bamber* — S30/1

    Introduction — S30/1
    Components of a safe system of work — S30/1

|  |  |
|---|---|
| Which type of safe system of work is appropriate for the level of risk? | S30/4 |
| Development of safe systems of work | S30/5 |
| Permit to work systems | S30/9 |
| Isolation procedures | S30/14 |
| Further reading | S30/15 |

## Statements of Health and Safety Policy – *Mike Bateman*  S70/1

|  |  |
|---|---|
| Introduction | S70/1 |
| The legal requirements | S70/1 |
| Content of the policy statement | S70/1 |

## Stress at Work – *Nicola Coote*  S110/1

|  |  |
|---|---|
| Introduction | S110/1 |
| Definition of stress | S110/1 |
| Understanding the stress response | S110/1 |
| Why manage stress? | S110/3 |
| Occupational stress risk factors | S110/5 |
| The work | S110/10 |
| Company structure and job organisation | S110/15 |
| Personal factors | S110/16 |
| Stress management techniques | S110/18 |

## Training and Competence in Occupational Safety and Health – *Hazel Harvey*  T70/1

|  |  |
|---|---|
| Introduction | T70/1 |
| Recognising competence | T70/2 |
| Legal obligations for training and competence | T70/3 |
| Setting national standards in health and safety | T70/6 |
| Education and training towards national standards of competence | T70/7 |
| Maintaining competence | T70/16 |
| Useful contacts | T70/17 |

## Ventilation – *George Pitblado*  V30/1

|  |  |
|---|---|
| Introduction | V30/1 |
| Pollution of the working environment and dangers of exposure to dust and fumes | V30/2 |
| Statutory requirements | V30/3 |
| Specific processes or industries | V30/4 |
| Control of airborne particulates | V30/6 |
| Respiratory protection | V30/8 |

## Vibration – *Roger Tompsett*  V50/1

|  |  |
|---|---|
| The scope of this chapter | V50/1 |
| An introduction to the effects of vibration on people | V50/1 |
| Vibration measurement | V50/3 |
| Causes and effects of hand-arm vibration exposure | V50/4 |
| Vibration-induced white finger as an occupational disease | V50/4 |
| Physical Agents (Vibration) Directive (2002/44/EC) and proposed Control of Vibration at Work Regulations | V50/6 |
| Advice and obligations in respect of hand-arm vibration | V50/7 |

| | |
|---|---|
| Personal protection against hand-arm vibration | V50/8 |
| Advice and obligations in respect of whole-body vibration | V50/9 |
| HSE guidance on hand-arm and whole-body vibration | V50/10 |

**Violence in the Workplace** – *Bill Fox and Charles Tolley* — V80/1

| | |
|---|---|
| Introduction | V80/1 |
| The nature and extent of the problem | V80/1 |
| Who is at risk? | V80/2 |
| Work-related violence – the legislation | V80/3 |
| Developing and implementing policy | V80/7 |
| Risk assessment for workplace violence | V80/12 |
| Reporting, recording and monitoring system | V80/18 |
| Risk reduction measures | V80/23 |
| Training | V80/27 |
| Incident management | V80/37 |
| Post-incident management | V80/41 |
| Sources of information and support | V80/46 |

**Vulnerable Persons** – *Mike Bateman* — V120/1

| | |
|---|---|
| Introduction | V120/1 |
| Relevant legislation | V120/1 |
| Children and young persons | V120/3 |
| New or expectant mothers | V120/11 |
| Lone workers | V120/21 |
| Disabled persons | V120/25 |
| Inexperienced workers | V120/28 |

**Work at Heights** – *George Ventris* — W90/1

| | |
|---|---|
| Introduction | W90/1 |
| Risk management of working at heights – employers' legal duties | W90/2 |
| The Work at Heights Regulations 2005 | W90/3 |

**Working Time** – *Dorothy Henderson/Rebecca Stewart* — W100/1

| | |
|---|---|
| Introduction | W100/1 |
| Definitions | W100/1 |
| Maximum weekly working time | W100/4 |
| Rest periods and breaks | W100/7 |
| Night work | W100/9 |
| Annual leave | W100/10 |
| Records | W100/14 |
| Excluded sectors | W100/15 |
| Following the adoption of five Directives aimed at extending the working time provisions to previously excluded sectors, this is no longer the case and most of the above sectors have gained some protection under the Regulations. The Directives are: | W100/15 |
| Enforcement | W100/19 |

**Workplaces – Health, Safety and Welfare** – *Andrea Oates* — W110/1

| | |
|---|---|
| Introduction | W110/1 |
| Workplace (Health, Safety and Welfare) Regulations 1992 (SI 1992 No 3004) | W110/2 |

| | |
|---|---|
| Safety signs at work – Health and Safety (Safety Signs and Signals) Regulations 1996 (SI 1996 No 341) | W110/15 |
| Disability Discrimination Act 1995 | W110/24 |
| Building Regulations 1991 (SI 1991 No 2768) | W110/25 |
| Factories Act 1961 | W110/28 |
| **Appendix 1** – Associations, organisations and departments connected with health and safety | A10/1 |
| **Appendix 2** – Current HSE Publications | A10/5 |
| **Appendix 3** – Relevant British Standards | A10/32 |
| **Appendix 4** – A Summary of Risk Assessment Requirements | A10/44 |
| **Appendix 5** – A Summary of Requirements for the Provision of Information | A10/57 |
| **Appendix 6** – A Summary of the Requirements for Training/Instruction and Consultation | A10/65 |
| *Table of Cases* | TC/1 |
| *Table of Statutes* | TS/1 |
| *Table of Statutory Instruments* | TSI/1 |
| *Index* | IND/1 |

# List of Tables

| Title | Page |
|---|---|
| Persons responsible for reporting accidents | A30/3 |
| Interviewing witnesses | A30/15 |
| Dangerous substances: qualifying quantities of named substances | C100/3 |
| Dangerous substances: qualifying quantities of categories of dangerous substances | C100/4 |
| Classification of Incidents, Accidents and Major Incident Types | D60/4 |
| Decisional Matrix | D60/29 |
| Monitoring Matrix | D60/41 |
| Provision of information under DSE Regulations | D82/4 |
| Display screen equipment workstation assessment | D82/7 |
| Display screen equipment workstation self-assessment checklist | D82/9 |
| Activities for which the HSE is the enforcing authority | E150/5 |
| Activities for which local authorities are the enforcing authorities | E150/7 |
| The ISO 14000 series | E160/13 |
| Guidance for allocating functions | E170/6 |
| Suitability of extinguishing types | F50/3 |
| 'Special premises' for which a fire certificate is required by the HSE | F50/27 |
| First-aid personnel | F70/6 |
| Assessment of first-aid needs | F70/12 |
| Safe working load for slings | L30/25 |
| Safe working load for slings – increased tension | L30/25 |
| Average illuminances and minimum measured illuminances for different types of work | L50/5 |
| Maximum ratios of illuminance for adjacent areas | L50/6 |
| Hazards identification checklist | M10/25 |
| Percentage degree of disablement in relation to hearing loss | N30/10 |
| Checklist of possible risks | R30/20 |
| Risk Assessment Matrix | R30/23 |
| Taylor-Pelmear Scale System | V50/5 |
| Checklist on work presenting increased risk for children and young persons | V120/9 |
| Aspects of pregnancy that may affect work | V120/17 |

# List of Illustrations/Diagrams

| Title | Page |
|---|---|
| Plan-Do-Check Act | Int/18 |
| Action to be taken by employers and others when accidents occur at work | A30/8 |
| Internal incident notification form | A30/11 |
| Elements of the investigation process | A30/13 |
| Prescribed form for reporting an injury or dangerous occurrence | A30/42 |
| Prescribed form for reporting a case of disease | A30/43 |
| Seven-step guide | A50/7 |
| Black on orange hazard symbols | D40/2 |
| UN classification system | D40/3 |
| Example of UN mark | D40/13 |
| Proper Shipping Name (PSN) and UN Number | D40/15 |
| Principles of thresholds | D40/18 |
| Truckmarking | D40/21 |
| Example of ADR vehicle labelling | D40/22 |
| Disaster and Emergency Management Systems (DEMS) | D60/16 |
| Major Incident Matrix | D60/27 |
| Ergonomic Approach | E170/3 |
| The 'S' Shaped Curve of the Spine | E170/5 |
| Zone of Convenient Reach | E170/13 |
| Viewing Angle | E170/14 |
| Body Map | E170/35 |
| Application for a fire certificate | F50/17 |
| Detention of food notice | F90/9 |
| Withdrawal of detention of food notice | F90/11 |
| Food condemnation warning notice | F90/13 |
| Improvement notice | F90/16 |
| Prohibition order | F90/18 |
| Emergency prohibition notice | F90/20 |
| Form of application for registration of food premises | F90/27 |
| Consulting employees | J30/5 |
| Example of an inspection record sheet | M10/5 |
| Example of a Declaration of Conformity | M10/16 |
| Structure of the Harmonised European Standards | M10/19 |
| The Risk Matrix | M10/21 |
| Risk assessment framework | M10/22 |
| Examples of mechanical hazards | M10/24 |
| Design/selection of safeguards | M10/28 |
| Example of a fixed guard | M10/29 |
| Fixed guard with adjustable element | M10/30 |
| Example of an automatic guard | M10/31 |
| Main types of electrical interlocking switches | M10/32 |
| Trapped key system for power interlocking | M10/34 |
| Example of a mechanical interlocking system | M10/34 |
| The costs of accidents | M20/2 |
| Systematic management cycle | M20/4 |
| Continual improvement loop | M20/10 |
| ILO model of continual improvement | M20/21 |
| Lifting and lowering | M30/5 |
| Handling while seated | M30/5 |

*List of Illustrations/Diagrams*

| | |
|---|---|
| Sign for informing that ear protectors must be worn (white on a circular blue background) | N30/7 |
| CE mark of conformity for personal protective equipment | P30/7 |
| Stress Response Curve | S110/2 |
| Examples of prohibitory signs | W110/20 |
| Examples of warning signs | W110/20 |
| Examples of mandatory signs | W110/22 |
| Examples of emergency escape or first-aid signs | W110/21 |
| Examples of fire-fighting signs | W110/22 |
| Hand signals | W110/22 |

# Abbreviations and References

Many abbreviations occur only in one section of the work and are set out in full there. The following is a list of abbreviations used more frequently or throughout the work.

**Legislation**

| | | |
|---|---|---|
| ACoP | = | Approved Code of Practice |
| BPR | = | Biocidal Products Regulations |
| CAWR | = | Control of Asbestos at Work Regulations |
| CDM | = | Construction (Design and Management) Regulations |
| CHIP | = | Chemicals (Hazard Information and Packaging for Supply) Regulations |
| CHSW | = | Construction (Health, Safety and Welfare) Regulations |
| CLAW | = | Control of Lead at Work Regulations |
| COMAH | = | Control of Major Accident Hazards Regulation |
| COSHH | = | Control of Substances Hazardous to Health Regulations |
| CPA | = | Consumer Protection Act |
| DSEAR | = | Dangerous Substances and Explosive Atmospheres Regulations |
| EAWR | = | Electricity at Work Regulations |
| ECHR | = | European Convention on Human Rights |
| EPA | = | Environmental Protection Act |
| EPCA | = | Employment Protection (Consolidation) Act |
| FA | = | Factories Act |
| FSA | = | Food Safety Act |
| HRA | = | Human Rights Act |
| HSWA | = | Health and Safety at Work etc Act |
| LOLER | = | Lifting Operations and Lifting Equipment Regulations |
| OLA | = | Occupiers' Liability Act |
| OSRPA | = | Offices, Shops and Railway Premises Act |
| NONS | = | Notification of New Substances Regulations |
| POMSTER | = | Placing on the Market and Supervision of Transfers of Explosives Regulations |
| PUWER | = | Provision and Use of Work Equipment Regulations |
| Reg | = | Regulation |
| RIDDOR | = | Reporting of Injuries, Diseases and Dangerous Occurrences Regulations |
| s | = | Section |
| Sch | = | Schedule |
| SI | = | Statutory Instrument |
| SRSC | = | Safety Representatives and Safety Committees Regulations |
| SR&O | = | Statutory Rule and Order |
| TURERA | = | Trade Union Reform and Employment Rights Act 1993 |
| UCTA | = | Unfair Contract Terms Act 1977 |

**Case citations**

| | | |
|---|---|---|
| AC | = | Appeal Cases |
| All ER | = | All England Reports |
| ALJR | = | Australian Law Journal Reports |
| Camp | = | Campbell Reports |
| CB, NS | = | Common Bench, New Series (ended 1865) |

*Abbreviations and References*

| | | |
|---|---|---|
| Ch | = | Chancery Reports |
| CL | = | Current Law |
| CLY | = | Current Law Year Book |
| CMLR | = | Community Law Reports |
| COD | = | Crown Office Digest |
| Con LR | = | Construction Law Reports |
| Cr App R | = | Criminal Appeal Reports |
| East | = | East's Term Reports |
| EG | = | Estates Gazette |
| Exch | = | Exchequer Reports |
| F & F | = | Foster & Finlayson (ended 1867) |
| H & C | = | Hurlstone & Norman |
| HSIB | = | Health and Safety Information Bulletin |
| ICR | = | Industrial Cases Reports |
| IRLR | = | Industrial Relations Law Reports |
| JP | = | Justice of the Peace and Local Government Review |
| KB/QB | = | Law Reports, King's (Queen's) Bench Division |
| KIR | = | Knight's Industrial Reports |
| LGR | = | Local Government Reports |
| LJKB | = | Law Journal Reports – New Series, Kings Bench |
| Lloyd's Rep | = | Lloyd's List Reports |
| M & W | = | Meeson & Welby |
| Med LR | = | Medical Law Reports |
| NLJ | = | New Law Journal |
| PIQR | = | Personal Injuries and Quantum Reports |
| RTR | = | Road Traffic Reports |
| SJ | = | Solicitors' Journal |
| SLT/SLT (Notes) | = | Scots Law Times/(Notes) |
| Taunt | = | Taunton Reports |
| TLR | = | Times Law Reports |
| WLR | = | Weekly Law Reports |

Legal terminology

| | | |
|---|---|---|
| ECJ | = | European Court of Justice |
| HL | = | House of Lords |
| CA | = | Court of Appeal |
| EAT | = | Employment Appeal Tribunal |
| IT | = | Industrial Tribunal |
| J | = | Mr Justice, a junior judge, normally sitting in a court of first instance |
| LJ | = | Lord Justice, a senior judge, normally sitting in a court of appeal |
| plaintiff | = | the person presenting a claim in a civil action |
| defendant | = | the person against whom a claim is brought in a civil action |
| appellant | = | the person bringing an appeal in a civil action |
| respondent | = | the person against whom an appeal is brought in a civil action |
| tort | = | a species of civil action for injury or damage where the remedy or redress is an award of unliquidated damages |
| volenti non fit injuria | = | 'to one who is willing no harm is done'; a complete defence in a civil action |
| novus actus interveniens | = | supervening act of third party |
| obiter dicta | = | words said by the way |
| ratio decidendi | | principle of a case |

*Abbreviations and References*

## Organisations

| | | |
|---|---|---|
| ACAS | = | Advisory, Conciliation and Arbitration Service |
| ACoP | = | Approved Code of Practice |
| ACTS | = | Advisory Committee on Toxic Substances |
| ASSE | = | American Society of Safety Engineers |
| BCC | = | British Chambers of Commerce |
| BHF | = | British Heart Foundation |
| BOHRF | = | British Occupational Health Research Foundation |
| BSC | = | British Safety Council |
| BSI | = | British Standards Institute |
| CBI | = | Confederation of British Industry |
| CCA | = | Centre for Corporate Accountability |
| CHAS | = | Chemical Health and Safety Assessment Scheme |
| CIA | = | Chemical Industries Association |
| CIEH | = | Chartered Institute of Environmental Health |
| CITB | = | Construction Industry Training Board |
| COIT | = | Central Office of the Industrial Tribunals |
| CORGI | = | Council for Registered Gas Installers |
| CPS | = | Crown Prosecution Service |
| CSCS | = | Construction Skills Certification Scheme |
| DEFRA | = | Department for Environment, Food and Rural Affairs |
| DWP | = | Department for Work and Pensions |
| EA | = | Environment Agency |
| EEF | = | Engineering Employers Federation |
| EMAS | = | Employment Medical Advisory Service |
| ENSHPO | = | European Network of Safety and Health Practitioners Organisation |
| EU | = | European Union |
| FPA | = | Fire Protection Association |
| HMIP | = | HM Inspectorate of Pollution |
| HMSO | = | Her Majesty's Stationery Office |
| HSC | = | Health and Safety Commission |
| HSE | = | Health and Safety Executive |
| ICOH | = | International Congress on Occupational Health |
| IEE | = | Institute of Electrical Engineers |
| IEMA | = | Institute of Environmental Management and Assessment |
| ILO | = | International Labour Office |
| INSHPO | = | International Network of Safety and Health Practitioners Organisations |
| IOSH | = | Institution of Occupational Safety and Health |
| JCT | = | Joint Contracts Tribunal |
| LSC | = | Learning and Skills Council |
| NEBOSH | = | National Examination Board in Occupational Safety and Health |
| NRPB | = | National Radiological Protection Board |
| OSHA | = | Occupational Safety and Health Administration (USA) |
| RoSPA | = | Royal Society for the Prevention of Accidents |
| RSSB | = | Rail Safety and Standards Board |
| SPA | = | Safety Passport Alliance |
| TUC | = | Trades Union Congress |
| WATCH | = | Working Group on the Assessment of Toxic Chemicals |

## Other terminology

| | | |
|---|---|---|
| ACM | = | Asbestos-containing Material |
| BRM | = | Business Risk Management |
| BS | = | British Standard |
| BSE | = | Bovine Spongiform Encephalopathy |

## Abbreviations and References

| | | |
|---|---|---|
| CCTV | = | Closed Circuit Television |
| CHAN | = | Chemical Hazards Alert Notice |
| CRI | = | Corporate Responsibility Index |
| CSR | = | Corporate Social Responsibility |
| DSE | = | Display Screen Equipment |
| EHO | = | Environmental Health Officer |
| ELCI (ELI) | = | Employer's Liability (Compulsory) Insurance |
| EMF | = | Electromagnetic Field |
| EMS | = | Environmental Management System |
| FM | = | Facilities Management |
| FMD | = | Foot and Mouth Disease |
| FOPS | = | Falling Object Protective Structure |
| HAVS | = | Hand-arm Vibration Syndrome |
| IR | = | Infra-red (radiation) or Ionising Radiation |
| LEV | = | Local Exhaust Ventilation |
| LEL | = | Lower Explosive Limit |
| LOCAE | = | List of Classified and Authorised Explosives |
| LPG | = | Liquefied Petroleum Gas |
| MDF | = | Medium Density Fibreboard |
| MDHS | = | Methods for the Determination of Hazardous Substances |
| MEL | = | Maximum Exposure Limit |
| MEWP | = | Mobile Elevating Work Platform |
| MPE | = | Maximum Permissible Exposure |
| MSDS | = | Material Safety Data Sheet |
| NOEM | = | New or Expectant Mother |
| NVQ | = | National Vocational Qualification |
| OEL | = | Occupational Exposure Limit |
| OES | = | Occupational Exposure Standard |
| OHSAS | = | Occupational Health and Safety Assessment Series |
| OHSMS | = | Occupational Health and Safety Management Systems |
| PAH | = | Polycyclic Aromatic Hydrocarbons |
| PAT | = | Portable Appliance Testing |
| PCB | = | Polychlorinated Biphenyls |
| PLI | = | Public Liability Insurance |
| PPE | = | Personal Protective Equipment |
| ppm | = | parts per million |
| PTW | = | Permit to Work |
| RHS | = | Revitalising Health and Safety |
| ROPS | = | Roll-over Protective Structure |
| RPE | = | Respiratory Protective System |
| RSI | = | Repetitive Strain Injury |
| RSP | = | Registered Safety Practitioner |
| SHT | = | Securing Health Together |
| SWL | = | Safe Working Load |
| UEL | = | Upper Explosive Limit |
| ULD | = | Upper Limb Disorder |
| UV | = | Ultra-violet (radiation) |
| VCM | = | Vinyl Chloride Monomer |
| VDU | = | Visual Display Unit |
| WRULD | = | Work-related Upper Limb Disorder |
| WSA | = | Workers' Safety Advisors |
| YP | = | Young Person |

# Introduction

## The looseleaf

1.　　When the first issue of the looseleaf was published, health and safety was still a matter for specialists – often professional safety advisers, HR directors and others who were required to 'deal with' health and safety on behalf of their organisations. Now, whether it is a debate on GM foods or the appropriateness of a particular train protection system, everyone is engaged in discussing 'risk'. In the boardroom this is linked to corporate governance and corporate social responsibility, on the shop floor it is associated with a radical rejection of a 'compensation culture' in favour of a desire to return from work healthy and uninjured.

This introduction sets the scene for health and safety and indicates ways in which the individual, focused chapters may be used to identify legal requirements and also characterise good practice with hints and tips on how this may be achieved and maintained.

## This Introduction is in five main parts

2.
- drivers for health and safety performance (this starts at PARA 3);
- the changing world of work (this starts at PARA 6);
- recent trends in UK health and safety law and its enforcement (this starts at PARA 7);
- the structure of UK health and safety law (starting at PARA 19);
- the management of health and safety at work (starting at PARA 25).

## Drivers for health and safety performance
### Why do organisations seek to improve their performance?

3.　　This is an issue that has been subject to increasing research in the UK and around the world in recent years. There is also growing evidence of distinct differences between the perception of small businesses and larger companies. It is also worth noting that definitions of smaller businesses vary from less than 10 and less than 100 employees, to the Department of Trade and Industry's (DTi) category of small to medium-sized enterprises (SMEs), at less than 250 employees. In small companies, the attitudes and beliefs of the owners/managers are a dominant force whilst in larger corporations and public bodies the social norms are subject to the broader forces of corporate governance. A combination of the following appear to be of greatest influence, but each overlaps with the other and may be present in different proportions and combinations:

- beliefs and attitudes of owners/managers in very small organisations;
- risk management and reputation risk;
- corporate governance;
- legal compliance.

4. *Introduction*

## Beliefs and attitudes of owner/managers in very small organisations

4.     In the UK, across the European Union and in the United States of America, more than half of all workers are employed in organisations of less than 10 people. This represents a growing trend as the large industries of the most developed countries continue to decline and the new economy comprises services which are often provided by new organisations and by the self-employed. Identifying what motivates the leaders of such organisations is therefore important to governments and enforcing authorities for health and safety. It appears that the level of awareness of specific health and safety requirements embodied in regulations, is often poor and almost wholly dependent on suppliers, customers and peers. The suppliers to small businesses of goods and services are regarded as a reliable source of information. For example the labels on containers of materials and the accompanying safety data sheets are used by small businesses to establish working practices without, typically, any reference to the requirements of the *Control of Substances Hazardous to Health Regulations 2002* (*SI 2002 No 2677* as amended by *SI 2003 No 978*). For some small businesses which supply major clients, those larger organisations increasingly impose operating conditions and provide support which incorporates health and safety, as part of the management of a secure supply chain*. The last significant element of advice and guidance is from the informal contacts with other owners/managers. (* For a study into supply chain management for health and safety, see '*Managing risk – Adding value: How big firms manage contractual relations to reduce risk*' (1998), HSE Books.)

Whatever their source of information as to what they should be doing and how to do it, the owners/managers of small businesses who do act in this area seem to have a clear set of reasons for getting health and safety right. These motivators are:

- the *focus on employees* who are so well known because of daily contact that injury or ill health is treated almost as if a family member were affected (and in many cases family members do work in the small organisation, and children may attend the workplace in the school holidays, etc);
- the *focus on satisfying customer*s and responding to guidance from suppliers, a general approach to 'doing the right thing';
- a conviction amongst a minority that looking after worker health and safety and similar matters contributes to a *productive, healthier, happier workforce* and may have benefits such as controlling insurance premiums.

## Why do larger organisations seek to improve their performance?

5.     Most large organisations have been on a learning curve since the post-war period. At first there was an emphasis on managing the pure risk of getting health and safety 'wrong', and the impact that this could have on people (injuries), products (damage) and property (losses such as by fire). By the 1970s this effort had broadened to encompass environmental issues, and also the concept of business recovery – and as a result events which could lead to business disruption were being investigated to both prevent their occurrence but also to respond effectively if they did occur. By the 1980s risk management had matured into a discipline which also encompassed such matters as brand management and corporate reputation. This does not mean that large companies have eliminated such risks. The impact on companies involved in rail crashes or other high-profile health and safety failures is a testament to the correct evaluation that these are significant matters for corporate value – but most have implemented systems which are designed to identify and control such risks. Thus, protecting the reputation of a company as an efficient and well-managed organisation is one of the most significant drivers to improving health and safety performance.

Following scandals in the City of London in the late 1980s (such as Polly Peck and the Bank of Credit and Commerce International (BCCI)), and stimulated by press reports of 'Fat Cats' receiving unreasonable remuneration from poorly performing public companies, a series of committees reported on remedial steps – Cadbury, Greenbury and Hampel. The latter called for a new approach to corporate governance which included a component of risk management at the highest level. The Turnbull Report, which requires organisations to establish a risk management strategy, was adopted by the London Stock Exchange obliging listed companies (plcs) to publish an annual risk management report as part of their annual reporting arrangements. This has prompted many large businesses to adopt formal plan-do-check-act management systems to achieve the level of assurance required for directors to sign such reports. The approach to risk management has also dovetailed with an increasing interest in corporate social responsibility and the pressures on companies to be seen to look after their staff, the environment and their neighbouring communities. In the public sector, similar developments have been encouraged by the Audit Commission, and in the voluntary sector by the Charities Commission. (See '*Worth the risk*: *An introduction to risk management and its benefits*' (2001), Audit Commission; '*Accounting reporting by charities: Statement of recommended practice* (2000), SORP, which is effectively a mandatory equivalent of the Turnbull requirements.) Across the public and private sectors, therefore, the development of corporate governance is linked to and integrated with risk management – all organisations are asked to identify what could blow them off course, resulting in a diminution of services, a loss of profitability, the development of an unsustainable approach to their work. Health and safety is an important part of the matrix of risks identified in any such exercise, and the management and control of the health and safety risks has become part of what the board of directors or trustees have to achieve and assure. It was against this background that a guide was developed and published by the Health and Safety Executive (HSE) on the duties of directors (see also recent trends at PARA 7).

The third pressure for health and safety performance is the necessity to achieve legal compliance. Most people, including those who lead organisations, are social beings who accept that the rules they need to live and work by represent the framework for a civilised life. Even if an individual driver occasionally breaks the speed limit, he or she is unlikely to wish for unlimited speeds to be agreed for the street on which they live. For small organisations, there is an often-expressed view that they are very unlikely to be visited by an inspector, and there may be a low awareness of both the statutory requirements and often a misperception of the level of risks associated with their work. It is in the large organisation with a nervous board of directors and a management commitment to legal compliance that represents a significant stimulus. Identifying what regulations need to be complied with, those which are specifically applicable to the mix of work being undertaken, and then assuring that compliance, is a further push towards formal management systems.

## The changing world of work

6. There is a wide range of evidence about the changes in the world of work. In the UK we have many more people working in call centres (about 4 per cent of the workforce) than in mining. This move towards service industries, as traditional manufacturing continues to decline in numbers of people employed, is matched by the increasing dominance of small businesses and the self-employed. Across the western world, it is more common for workers to be employed in companies with less than 10 staff than in companies with more than 250 employees. Flexible work patterns and the temporary nature of much employment, with hardly any young workers anticipating a 'job for life, has also altered the balance of health and safety issues with psychosocial risks giving rise to stress-related sickness absence overtaking musculo-skeletal injuries as the most common work-related illness. In the UK, the great increase in female employment and the pressure on an ageing workforce to stay at work longer is also changing the nature of the 'at risk' group which requires protection whilst at work. This listing of changes being experienced at work suggests that the law, which often moves slowly, the enforcement authorities, employers, employees and professionals safety advisers are each under pressure to adapt and change in turn. The looseleaf reflects these changes, and some themes are teased out in the remainder of this introduction.

7. *Introduction*

# Legal framework

## *Revitalising Health and Safety in the UK*

7.  Twenty-five years after the *Health and Safety at Work etc Act 1974* (*HSWA 1974*) came into force, reported accidents at work seemed to have been stuck on a plateau, albeit one lower than the levels reported when the Act was introduced. It was also recognised that ill health caused by work had not received enough attention. This prompted *Revitalising Health and Safety*, the Government's ten-year strategy statement published by the Health and Safety Commission (HSC) and launched by the Deputy Prime Minister in June 2000. The statement included targets for reducing incidence rates for fatal and major injury accidents and for work-related ill health by 10 per cent and 20 per cent respectively, and also for reducing the proportion of working days lost due to work-related injury and ill health.

The ten-point strategy for achieving these reductions included:

- positive engagement of small firms;
- motivation of employers through insurance incentives;
- cultivation of a more deeply engrained culture of self regulation;
- more partnerships with workers on health and safety issues;
- government leading by example, including promoting best practice through the supply chain;
- improved education on health and safety;
- introduction of a new occupational health strategy.

The last point was addressed in more detail in *Securing Health Together*, published by HSC a month later in July 2000. The target of healthy work, healthy at work and healthy for life was summarised in terms of workability. The occupational health strategy seeks, in addition to reducing the extent to which the workplace is a site for health harm, to improve access to the workplace for those who are ill or disabled by tackling rehabilitation and disability access with real vigour. Some three years after its launch, the strategy is being reworked to integrate with the Department of Work and Pensions (DWP) programme of health, work and recovery.

Since launching the two related documents there has been some limited progress. Pilot studies in occupational health have begun to provide the evidence required for creating an effective national occupational health support programme. The directors' guide on the responsibilities of the heads of companies was designed to embed health and safety at the highest levels of corporate life (see '*Directors' responsibilities for health and safety*' (2001), (IND(G)343), HSE). Although the strategy document refers to the significantly increased resources the Government has made available to HSC/HSE, since then there has been a squeeze on funding although the process is intended to be underpinned by enforcement. Proposals on corporate killing, and also the Government's stated intention to increase maximum fines, extend the range of offences which may be subject to custodial sentences and introduce other more innovative penalties have all been subject to delays, postponements and cancellations. (Note: at the time of writing the Corporate Manslaughter Bill has now been placed back in the parliamentary timetable.) However, HSE has started to 'name and shame' companies and individuals convicted of offences – the roll call can be accessed at: www.hse-databases.co.uk/prosecutions/

## *Global legal framework*

8.  The International Labour Organisation (ILO) and the World Health Organisation (WHO) are both agencies of the United Nations. Each has a set of objectives embodied in programmes of action which reflect the international consensus that workers have the right to

employment which does not damage their health or prejudice their safety. In the WHO *Occupational Health for All*, the statement in the introduction could not be more clear:

'According to the principles of the United Nations, WHO and ILO, every citizen of the world has a right to healthy and safe work and to a work environment that enables him or her to live a socially and economically productive life.'

[*Global Strategy on Occupational Health for All*, WHO Beijing Declaration 1994]

It is also worth noting that the founding definition of 'health' from WHO in 1948 is still that health is a state of complete physical, mental and social well-being and not merely the absence of disease or infirmity. This is, of course, much more positive than simply the prevention of harm, and can apply to the workplace as well. From the global perspective, each region can develop its own approach to health and safety at work. In many ways the most advanced region is the European Union. In addition, there are some health and safety regulations which derive directly from international agreements such as those which apply to marine safety, and to the transport of dangerous goods by road, rail and air.

## The role of the European Union

9.  For the past decade the European Union (EU) has provided the principal motor for changes in UK health and safety legislation. Article 118A in the Treaty of Rome 1957 gives health and safety prominence in the objectives of the EU. The Social Charter also contains a declaration on health and safety, although this has no legal force. Whilst the EU can issue its own regulations, it mainly operates through directives requiring Member States to pass their own legislation. The 'Framework Directive' adopted in 1989 as part of the creation of the Single Market contained many broad duties, including the requirement to assess risks and introduce appropriate control measures. Other directives have since been adopted or proposed, including many relating to technical standards and safety requirements for specific products. The proposed new EU Constitution is unlikely to do other than confirm the role of health and safety in the continued development towards an ever-closer union of nation states, soon to expand to 25 members.

## Implementation of EU directives

10.  The UK, along with other Member States, is required to implement EU directives by designated dates. The UK sits about halfway in the league-table for formal implementation of directives by the due date, and there is not at present an effective mechanism for evaluating implementation and enforcement. Implementation in the UK is normally overseen by HSC (and the HSENI in Northern Ireland), which as with its own regulatory initiatives, first circulates consultative documents incorporating the proposed regulations together with a related Approved Code of Practice (ACoP) and/or guidance to interested parties (trades unions, employers' organisations, professional bodies, local authorities etc). Following the consultation process, HSC submits a final version to the Secretary of State who then lays the regulations before Parliament. It is at this stage that a date for their coming into force is determined – usually that stipulated by the original EU directive.

## Current EU initiatives

11.  The process of EU directives and their local implementation peaked on 1 January 1993 when six sets of regulations came into operation on the same day. Since then, the pace of change has been much slower – a mixture of periodic updating of existing directives such as the Council Directive on Dangerous Substances which is amended every year, resulting in new CHIP Regulations covering the classification, labelling and packaging of dangerous substances, and new areas being covered, e.g. work is progressing on the Physical Agents Directive which will impact with new regulations on hand-arm vibration and whole body vibration. Few UK regulations are developed now, such as the proposed forthcoming regulations on improving arrangements for

## Balancing business with regulation

12.　　The latter stages of the Conservative Government of 1979–1997 saw a drive towards deregulation because of the perceived burden of legislation on business and particularly on small employers. The present Government was presumed by many to be automatically in favour of tougher health and safety laws and stricter enforcement. In practice, while HSC/HSE have been peripatetic between government departments with a period led by the Deputy Prime Minister and now a new home within DWP, the Government has displayed a high degree of reluctance to seek improvements in health and safety performance through new laws. As a result, when new regulations are considered, such as those for accident investigation (rejected in favour of new guidance) or the proposals for corporate manslaughter and for improved worker consultation – there is real pressure to establish that there will be direct benefits with little cost to modern British industry.

## Health and Safety Executive

13.　　HSE has adopted eight Priority Programmes which are reflected in several ways, ie priorities for publications, for general resources and, above all, priorities for visiting inspectors on enforcement activities:

- Falls from height.
- Workplace transport.
- Musculoskeletal disorders.
- Work-related stress.
- Slips trips and falls.
- Construction sector.
- Agriculture sector.
- Healthcare sector.

Meanwhile, the Policy Section of HSE has been re-organised into a series of sections which combine health with traditional safety matters (the Health Directorate is now defunct). The current corporate plan for health and safety within HSE is instructive, ie dealing with display screens, continuing back awareness, seeking to reduce stress (HSE is publishing new management standards for work organisations to reduce stress), targeting slips, trips and falls, improving the management of contractors and reducing work-related road risks.

## Other influences

14.　　Both the Government and HSC/HSE are subject to pressure from public opinion and lobby groups. The aftermath of rail accidents at Southall, Paddington (Ladbroke Grove) and Hatfield and the Cullen Inquiry, together with the continuing campaign following the death of Simon Jones in his first day at work at Shoreham Docks, have kept the issue of corporate accountability in the public eye.

Many civil actions for work-related injuries or ill health are supported by trades unions and these often bring issues into the public eye. Following the successful action by a social worker against Northumberland County Council, many more claims have been made for work-related stress (see *Walker v Northumberland County Council [1995] 1 All ER 737*). Compensation claims now total £2bn each year – 65 per cent through court claims and 35 per cent through industrial injuries

## Penalties and prosecutions

15. Penalties being imposed by the courts are steadily increasing. Great Western Trains pleaded guilty to a charge under *s 3* of the *HSWA 1974* following the Southall rail crash and were fined £1.5 million (plus £680,000 costs) – a record for a single offence under health and safety legislation. In September 1994 six people were killed and seven seriously injured when a pedestrian walkway at the port of Ramsgate collapsed. After a twenty-five day trial early in 1997, two Swedish companies which had constructed and designed the walkway were fined £750,000 and £250,000 respectively, Lloyds Register of Shipping (which inspected the walkway) was fined £500,000 and Port Ramsgate, the client, was fined £200,000. With costs added to the total fines of £1.7 million, the bill to the defendants for the criminal proceedings alone came to a total of just under £2.5 million. The collapse in October 1994 of a rail tunnel being constructed at Heathrow did not result in any injuries but was described in court as one of the 'biggest near misses' in years. When the prosecution came to court in 1999, Balfour Beatty Civil Engineering were fined £700,000 plus £100,000 costs, with the tunnelling sub-contractor, Geoconsult GES MBH, fined £500,000 plus £100,000 costs.

Fines in excess of £100,000 for health and safety offences have now become relatively commonplace although both the Government and HSE believe that some courts continue to impose relatively lenient punishments. Guidelines on appropriate levels of penalties for breaches of health and safety legislation were laid down by the Court of Appeal in *R v F Howe & Son (Engineering) Ltd [1999] 2 All ER* and endorsed by the same court in *R v Rollco Screw and Rivet Co Ltd [1999] IRLR 439*. The first custodial sentence for a health and safety offence, one of three months, was imposed on Roy Hill by Bristol Crown Court, whilst Paul Evans received a nine-month term of imprisonment from Birmingham Crown Court in September 1998. Both penalties resulted from carrying out work involving asbestos without the necessary licence.

A recent case heard by the Court of Appeal will have a major influence on the level of fines imposed for breaches of health and safety legislation. In the case of *R v Colthrop Board Ltd, 31 January 2002 (Unreported)*, Lord Justice Gibbs giving judgment stated that although the remarks in a previous case that financial penalties in excess of £500,000 were reserved for cases of major public disasters, 'what is important is that companies in the position of the appellant can expect to receive financial penalties on a scale of up to at least half a million pounds for serious defaults and proportionately lesser sums if the limitation upon means or some lesser blame justifies it'.

This particular case did not involve a fatality. An employee suffered serious crushing injuries when he became entangled in a machine. A health and safety inspector had previously raised concerns about the machine. The Crown Court judge had imposed a total fine of £350,000, comprising of a fine of £200,000 for contravention of the *HSWA 1974, s 2(1)* and a further fine of £150,000 for contravention of the *Provision and Use of Work Equipment Regulations 1998 (SI 1998 No 2306* as amended by *SI 2002 No 2174)*.

The company lodged an appeal and argued that the fines imposed were excessive and were inconsistent with the scale of penalties to be derived from precedents in other cases and in particular the Court of Appeal. It also argued that the wrong approach had been taken to assessing its financial means for the purposes of sentencing after its business had ceased. When deciding whether the Crown Court judge had imposed a fine that was too high, Lord Justice Gibbs referred to other authorities, in particular *R v Friskies Petcare (UK) Ltd [2000] 2 Cr App Rep (S) 401*. He commented: 'we would not wish the sum of £500,00 to appear to be set in stone or to provide any sort of maximum limit for such cases. On the contrary, we anticipate that as time goes on and awareness of the importance of safety increases, the courts will uphold the sums of that amount and even in excess of them in serious cases, whether or not they involve what could be described as major public disasters'.

16. *Introduction*

After the date of the accident the company had decided to stop trading. The sale of its assets realised £17 million. The Court of Appeal indicated that for a company that had ceased trading the asset value on the sale was the correct measure to use when assessing its means for sentencing purposes, and not its levels of profits while it had still traded (which would have resulted in lower assessment of means).

The Court of Appeal decided to reduce the total fine in this case to £200,000 (£100,000 for each offence). It took the view that despite the aggravating features, the company was of moderate size compared to others upon whom bigger fines had been imposed in the past. The case did not involve a fatality. The overall impact of the decision however will be more cases being referred to the Crown Courts for sentence and higher fines for corporate defendants.

## Risk assessments

16. Much recent UK legislation has included a requirement for some type of risk assessment. The concept was introduced in the early 1980s in regulations applying to asbestos and lead but it came to wider prominence as a core requirement of the *Control of Substances Hazardous to Health Regulations 1988* ('*COSHH*'). The group of Regulations which came into force on 1 January 1993 derived from EU directives continued this trend with four sets of regulations requiring an assessment of one kind or another. The most important of these is the general requirement for risk assessment contained in *Reg 3* of the *Management of Health and Safety at Work Regulations 1999* (*SI 1999 No 3242*) ('the *Management Regulations*'). Other important regulations requiring more specific types of risk assessment relate to noise, manual handling operations, personal protective equipment, display screen equipment and fire precautions.

In practice a less formal type of risk assessment was already required by health and safety legislation – an assessment of the level of risk, the adequacy of existing precautions and the costs of additional precautions was necessary in order to determine what was 'reasonably practicable'. Risk assessment lies at the heart of the original concept of self-regulation enshrined in the Robens Report 1992 – employers are required to demonstrate that they have identified relevant risks together with appropriate precautions and in most cases must have records available to prove this.

## Supply-chain pressure

17. Since the introduction of *HSWA 1974*, many court decisions, most notably those involving *Swan Hunter Shipbuilders ([1984] IRLR 93–116)* and *Associated Octel ([1996] 1 WLR 1543)*, have emphasised that employers often have responsibilities for the activities of employees of other organisations. The structure of *HSWA 1974* and much subsidiary legislation is such that responsibilities overlap between employers rather than being neatly apportioned between them. Employers must do more than simply not turn a blind eye to the obvious health and safety failings of those with whom they come into contact: they must often take a pro-active interest in the health and safety standards of others.

This principle is exemplified by the Ramsgate prosecution (see PARA 15). Despite having engaged specialists to carry out the design, construction and checking of the passenger walkway, Port Ramsgate was still convicted of an offence. Mr Justice Clark stated that 'the jury has found, in my judgement correctly, that an owner and operator of a port cannot simply sit back and do nothing and rely on others, however expert'.

In recent years a growing number of larger companies, local authorities and other public bodies have had increasingly formalised procedures for checking the health and safety standards of contractors wishing to work for them. This process has been accelerated by the demands of the *Construction (Design and Management) Regulations 1994 (SI 1994 No 3140)* ('*CDM*') which require clients to satisfy themselves (via their planning supervisors) that potential principal contractors are capable of dealing with the health and safety issues associated with projects. The Regulations also place responsibilities on principal contractors in respect of sub-contractors. Consequently contractors are frequently required to provide details of their health and safety policies and generic risk assessments together with risk assessments and/or method statements for specific projects or activities. Many clients also take an extremely hands-on approach in policing the work of contractors on their premises. This has been turned into an aim of *Revitalising*, to make the Government a leader in procurement. This also chimes with current business developments, in that the supply chain used to be a little-regarded operation but now improving performance in an era of outsourcing is often focused on contractors. (See '*Managing risk – Adding value*'; '*Transforming the supply chain*' (2002), The Conference Board, New York.)

## Recent changes in legislation

18.

- *Biocidal Products (Amendment) Regulations 2003 (SI 2003 No 429)*;
- *Chemicals (Hazard Information and Packaging for Supply) Regulations 2002 (SI 2002 No 1689)*;
- *Control of Asbestos at Work Regulations 2002 (SI 2002 No 2675)*;
- *Control of Lead at Work Regulations 2002 (SI 2002 No 2676)*;
- *Control of Substances Hazardous to Health Regulations 2002 (SI 2002 No 2677 as amended by SI 2003 No 978)*;
- *Dangerous Substances and Explosive Atmospheres Regulations 2002 (SI 2002 No 2776)*;
- *Electricity Safety, Quality and Continuity Regulations 2002 (SI 2002 No 2665)*;
- *Employment Act 2002*;
- *Flexible Working (Eligibility, Complaints and Remedies) Regulations 2002 (SI 2002 No 3236)*;
- *Flexible Working (Procedural Requirements) Regulations 2002 (SI 2002 No 3207)*;
- *Genetically Modified Organisms (Contained Uses) (Amendment) Regulations 2002 (SI 2002 No 63)*;
- *Health and Safety (Miscellaneous Amendments) Regulations 2002 (SI 2002 No 2174)*;
- *Health and Safety (Fees) Regulations 2003 (SI 2003 No 547)*;
- *Notification of New Substances (Amendment) Regulations 2002 (SI 2002 No 2176)*;
- *Offshore Safety (Miscellaneous Amendments) Regulations 2002 (SI 2002 No 2175)*;
- *Packaging, Labelling and Carriage of Radioactive Material by Rail Regulations 2002 (SI 2002 No 2099)*;
- *Personal Protective Equipment Regulations 2002 (SI 2002 No 1144 as amended by SI 2002 No 2174)*;
- *Working Time (Amendment) Regulations 2003 (SI 2003 No 1684)*.

## The structure of UK health and safety law

19. Health and safety in the workplace involves two different branches of the law – criminal law (dealt with below at PARA 20) and civil law (referred to in PARA 24).

### Criminal law

20. Criminal law is the process by which society, through the courts, punishes organisations or individuals for breaches of its rules. These rules, known as 'statutory duties', are comprised in Acts passed by Parliament (e.g. *HSWA 1974*) or regulations which are made by

21. *Introduction*

government ministers using powers given to them by virtue of Acts (e.g. the *Manual Handling Operations Regulations 1992* (*SI 1992 No 2793* as amended by *SI 2002 No 2174*)).

Cases involving breaches of criminal law may be brought before the courts by the enforcement authorities which, in the case of health and safety law, are HSE and local authorities via their environmental health departments (see ENFORCEMENT). Magistrates' courts hear the vast majority of health and safety prosecutions although more serious cases can be heard by the Crown Courts. The maximum fine which can be imposed by the magistrates' courts is currently £20,000 for breaches of certain sections of *HSWA 1974* and £5,000 for most other offences.

Where cases are heard by the Crown Court there is no limit on the fines which can be imposed. PARA 15 details some significant fines resulting from recent cases. There are also a limited number of health and safety offences which can result in prison sentences of up to two years. These include:

- contravention of licensing requirements (e.g. for asbestos removal);
- explosives-related offences;
- contravention of an improvement or prohibition notice;
- contravention of a court remedy order.

As in all criminal prosecutions the case must be proved 'beyond all reasonable doubt'. There is a right of appeal to the Court of Appeal (Criminal Division) and eventually to the House of Lords or even the European courts, although in practice very few health and safety cases go to appeal. A death involving work activities might result in manslaughter charges which could lead to more severe penalties, although hitherto such cases have been rare. (PARA 7 refers to recent developments in relation to 'corporate killing' etc. and also to likely extensions to the range of penalties available for health and safety offences.)

## The Health and Safety at Work etc Act 1974 (HSWA 1974)

21.     *HSWA 1974* is the most important Act of Parliament relating to health and safety. It applies to everyone 'at work' – employers, self-employed and employees (with the exception of domestic servants in private households). It also protects the general public who may be affected by work activities. HSE has published a booklet, '*A guide to the Health and Safety at Work etc Act 1974*' (L1).

Some of the key sections of the Act are listed below.

## Section 2 – Duties of employers

*HSWA 1974*, s 2(*1*) is the catch-all provision: 'It shall be the duty of every employer to ensure, so far as is reasonably practicable, the health, safety and welfare at work of all his employees.' See below for further discussion of the term 'reasonably practicable'.

*Section 2(2)* goes on to detail more specific requirements relating to:

- the provision and maintenance of plant and systems of work;
- the use, handling, storage and transport of articles and substances;
- the provision of information, instruction, training and supervision;
- places of work and means of access and egress;
- the working environment, facilities and welfare arrangements.

These are also qualified by the term 'reasonably practicable'.

*Section 2(3)* provides that an employer with five or more employees must prepare a written health and safety policy statement, together with the organisation and arrangements for carrying it out, and bring this to the notice of employees (see PARA 27).

## Section 3 – Duties to others

*HSWA 1974, s 3(1)* provides: 'It shall be the duty of every employer to conduct his undertaking in such a way as to ensure, so far as is reasonably practicable, that persons not in his employment who may be affected thereby are not exposed to risks to their health or safety.'

Employers thus have duties to contractors (and their employees), visitors, customers, members of the emergency services, neighbours, passers-by and the public at large. This may extend to include trespassers, particularly if it is 'reasonably foreseeable' that they could be endangered, for example where high-risk workplaces are left unfenced.

Individuals who are self-employed are placed under a similar duty and must also take care of themselves. (If they have employees, they must comply with *section 2*).

## Section 4 – Duties relating to premises

Under *HSWA 1974, s 4* persons in total or partial control of work premises (and plant or substances within them) must take 'reasonable' measures to ensure the health and safety of those who are not their employees. These responsibilities might be held by landlords or managing agents etc, even if they have no presence on the premises.

## Section 6 – Duties of manufacturers, suppliers etc.

Those who design, manufacture, import, supply, erect or install any article, plant, machinery, equipment or appliances for use at work, or who manufacture, import or supply any substance for use at work, have duties under *s 6*.

## Section 7 – Duties of employees

'It shall be the duty of every employee while at work:

(a) to take reasonable care for the health and safety of himself and of other persons who may be affected by his acts or omissions at work; and

(b) as regards any duty or requirement imposed on his employer or any other person by or under any of the relevant statutory provisions, to co-operate with him so far as is necessary to enable that duty or requirement to be complied with.'

[*HSWA 1974, s 7*]

Consequently employees must not do, or fail to do, anything which could endanger themselves or others. It should be noted that managers and supervisors also hold these duties as employees.

## Section 8 – Interference and misuse

'No person shall intentionally or recklessly interfere with or misuse anything provided in the interests of health, safety or welfare in pursuance of any of the relevant statutory provisions.'

[*HSWA 1974, s 8*]

## Section 9 – Duty not to charge

'No employer shall levy or permit to be levied on any employee of his any charge in respect of anything done or provided in pursuance of any specific requirement of the relevant statutory provisions.'

[*HSWA 1974, s 9*]

## Levels of duty

22.     Health and safety law contains different levels of duty:

### Absolute

Absolute requirements must be complied with whatever the practicalities of the situation or the economic burden.

### Practicable

The term 'practicable' means that measures must be possible in the light of current knowledge and invention.

### Reasonably practicable

This term is contained in the main sections of *HSWA 1974* and many important regulations. It requires the risk to be weighed against the costs necessary to avert it (including time and trouble as well as financial cost). If, compared with the costs involved, the risk is small then the precautions need not be taken – it should be noted that such a comparison should be made before any incident has occurred. The burden of proof, however, rests on the person with the duty (usually the employer) – they must prove why something was not reasonably practicable at a particular point in time. The duty holder's ability to meet the cost is not a factor to be taken into account.

In effect, considering what is 'reasonably practicable' requires that a risk assessment be carried out. The existence of a well-documented and carefully considered risk assessment would go a long way towards supporting a case on what was or was not reasonably practicable. Neither risks nor costs remain the same forever and what is practicable or reasonably practicable will change with time – hence the need to keep risk assessments up to date.

## Important regulations

23.
### Construction (Design and Management) Regulations 1994 (SI 1994 No 3140)

The broad definition of 'construction work' used in the Regulations means that they apply to many medium-sized engineering and maintenance projects as well as to traditional construction activities and all demolition work. The Regulations provide for specific duties to be carried out by the 'client' (who must appoint a 'planning supervisor') and by the 'principal contractor' (who may in some cases also be the client). Key requirements of the Regulations are for the development and implementation of a formal 'health and safety plan' and the creation of a 'health and safety file' for the project.

## Control of Substances Hazardous to Health Regulations 2002 (COSHH) (SI 2002 No 2677 as amended by SI 2003 No 978)

These Regulations require an assessment to be made of all substances hazardous to health in order to identify means of preventing or controlling exposure. There are also requirements for the proper use and maintenance of control measures and for workplace monitoring and health surveillance in certain circumstances.

## Electricity at Work Regulations 1989 (SI 1989 No 635)

These Regulations contain requirements relating to the construction and maintenance of all electrical systems and work activities on or near such systems. They apply to all electrical equipment, from a battery-operated torch to a high-voltage transmission line.

## Health and Safety (Consultation with Employees) Regulations 1996 (SI 1996 No 1513)

These Regulations extended the previous requirements (contained in the *Safety Representatives and Safety Committees Regulations 1977* (*SI 1977 No 500*) so that employers must now also consult workers not covered by trade union safety representatives.

## Health and Safety (Display Screen Equipment) Regulations 1992 (SI 1992 No 2792 as amended by SI 2002 No 2174)

Where there is significant use of display screen equipment (DSE), employers must assess DSE workstations and offer 'users' eye and eyesight tests (which may necessitate provision of spectacles for DSE work).

## Health and Safety (First-Aid) Regulations 1981 (SI 1981 No 917 as amended by SI 2002 No 2174)

Basic first-aid equipment controlled by an 'appointed person' must be provided for all workplaces. Higher risk activities or larger numbers of employees may require additional equipment and fully trained first-aiders.

## Health and Safety (Safety Signs and Signals) Regulations 1996 (SI 1996 No 341 as amended by SI 2002 No 2174)

These Regulations require safety signs to be provided, where appropriate, for risks which cannot adequately be controlled by other means. Signs must be of the prescribed design and colours.

## Health and Safety (Training for Employment) Regulations 1990 (SI 1990 No 1380)

Those receiving 'relevant training' (through training for employment schemes or work experience programmes) are treated as being 'at work' for the purposes of health and safety law. The provider of the 'relevant training' is deemed to be their employer – youth trainees and students on work experience placements therefore have the status of employees and must be protected accordingly.

23. *Introduction*

## Management of Health and Safety at Work Regulations 1999 (SI 1999 No 3242)

Employers and the self-employed are required to manage the health and safety aspects of their activities in a systematic and responsible way. The Regulations include requirements for risk assessment, the availability of competent health and safety advice and emergency procedures – several of these management issues are dealt with later in this Introduction (see PARA 25).

## Manual Handling Operations Regulations 1992 (SI 1992 No 2793 as amended by SI 2002 No 2174)

Manual handling operations involving risk of injury must either be avoided or be assessed by the employer with steps taken to reduce the risk, so far as is reasonably practicable.

## Noise at Work Regulations 1989 (SI 1989 No 1790)

Employers must carry out an assessment to determine the level of exposure to noise of their employees. The precautions required include noise reduction measures, provision of hearing protection and the establishment of hearing protection zones.

## Personal Protective Equipment at Work Regulations 1992 (SI 1992 No 2966 as amended by SI 2002 No 2174)

Employers must assess the personal protective equipment (PPE) needs created by their work activities, provide the necessary PPE, and take reasonable steps to ensure its use.

## Provision and Use of Work Equipment Regulations 1998 (PUWER 1998) (SI 1998 No 2306 as amended by SI 2002 No 2174)

These Regulations cover equipment safety, including the guarding of machinery. The definition of 'work equipment' also includes hand tools, vehicles, laboratory apparatus, lifting equipment, access equipment etc. The 1998 Regulations introduced additional requirements in respect of mobile work equipment and also replaced previous specific regulations relating to power presses, woodworking machines and abrasive wheels.

## Reporting of Injuries, Diseases and Dangerous Occurrences Regulations 1995 (RIDDOR) (SI 1995 No 3163)

Fatal accidents, major injuries (as defined) and dangerous occurrences (as defined) must be reported immediately to the enforcing authority. Accidents involving four or more days' absence must be reported in writing within seven days.

## Safety Representatives and Safety Committees Regulations 1977 (SI 1977 No 500)

Members of recognised trade unions may appoint safety representatives to represent them formally in consultations with their employer in respect of health and safety issues. The functions and rights of safety representatives are detailed in the Regulations. The employer must establish a safety committee if at least two representatives request this in writing.

## Workplace (Health, Safety and Welfare) Regulations 1992 (SI 1992 No 3004 as amended by SI 2002 No 2174)

Physical working conditions, safe access for pedestrians and vehicles, and welfare provisions are covered by these Regulations.

### Civil law

24. A civil action can be initiated by an employee who has suffered injury or damage to health caused by their work. This may be based upon the law of negligence, i.e. where the employer has been in breach of the duty of care which he owes to the employee. Being part of the common law, the law of negligence has evolved, and continues to evolve, by virtue of decisions in the courts – Parliament has had virtually no role to play in its development.

Civil actions may also be brought on the grounds of breach of statutory duty – it should be noted, however, that *HSWA 1974* and most of the provisions in the *Management of Health and Safety at Work Regulations 1999* (*SI 1999 No 3242*) do not confer a right of civil action, although the statutory duties owed by employers to employees under *HSWA 1974* have their equivalent obligations at common law.

### Duty of care

Every member of society is under a 'duty of care', i.e. to take reasonable care to avoid acts or omissions which they can reasonably foresee are likely to injure their neighbour (anyone who ought reasonably to have been kept in mind). What is 'reasonable' will depend upon the circumstances.

Employers owe a duty of care not only to employees but also to such people as contractors, visitors, customers, and people on neighbouring property. In the case of the duty of care owed by employers to employees, it includes the duty to provide:

- safe premises;
- a safe system of work;
- safe plant, equipment and tools; and
- safe fellow workers.

Occupiers of premises are under statutory duties comprised in the *Occupiers' Liability Acts* of 1957 and 1984 (see OCCUPIERS' LIABILITY), and those suffering injury because of a defect in a product may sue the producer or importer under the *Consumer Protection Act 1987* (see PRODUCT SAFETY).

### Vicarious liability

Employers are liable to persons injured by the wrongful acts of their employees, if such acts are committed in the course of their employment. Thus if an employee's careless driving of a forklift truck injures another employee (or a contractor or customer), the employer is likely to be liable. There is no vicarious liability if the act is not committed in the course of employment – thus the employer is not likely to be held liable if one employee assaults another.

### Civil procedure

Civil actions must commence within three years from the time of knowledge of the cause of action. In an action for negligence, this will be the date on which the plaintiff knew or should have known that there was a significant injury and that it was caused by the employer's negligence. The plaintiff must be prepared to prove his case in the courts, but in practice most cases are settled out of court following negotiations between the plaintiff's legal representatives and the employer's insurers or their representatives. The *Employers' Liability (Compulsory Insurance) Act 1969* requires employers to be insured against such actions (see EMPLOYERS' LIABILITY INSURANCE), although some public bodies, for example local authorities, are exempt from the provisions of this Act.

As the result of recommendations made by Lord Woolf in his '*Access to Justice*' report of 1996, the *Civil Procedure Rules 1998* introduced widespread changes to civil procedure on 26 April 1999, affecting the progress of civil claims from their commencement to their conclusion. The rules involve a 'pre-action protocol' and govern the conduct of litigation in a way that is intended to limit delay. In most cases a single expert, medical or non-medical, will be instructed rather than each party using separate experts. Even if the case goes to court, the expert's report will usually be in writing, with both parties able to ask written questions of the expert and to see the replies. The new arrangements include a fast track system for personal injury claims up to a value of £15,000.

## Damages

Damages are assessed under a number of headings including:

- loss of earnings (prior to trial);
- damage to clothing, property etc;
- pain and suffering (before and after trial);
- future loss of earnings;
- disfigurement;
- medical or nursing expenses; and
- inability to pursue personal or social interests or activities.

## Defences

The plaintiff must prove breach of a statutory duty or of the duty of care on a balance of probabilities. However, a number of defences are available to the employer, including:

- *Contributory negligence*

    The employer may claim that the injured person was careless or reckless, for example, that he ignored clear safety rules or disobeyed instructions. Accidental errors are distinguished from a failure to take reasonable care. Damages will be reduced by the percentage of contributory negligence established, which will vary with the facts of each case.

- *Injuries not reasonably foreseeable*

    The employer may claim that the injuries were beyond normal expectation or control (an act of God). In cases of noise-induced hearing damage, mesothelioma (an asbestos-related cancer) or vibration-induced white finger, the courts have established dates after which a reasonable employer should have been aware of the relevant risks and taken precautions.

- *Voluntary assumption of risk*

    If an employee consents to take risks as part of the job, the employer may escape liability. However, this defence (*volenti non fit injuria*) cannot be used for cases involving breach of statutory duty – no one can contract out of their statutory obligations or be deprived of statutory protection.

## Other civil actions

Other health and safety related situations may result in civil actions by employees. Employment protection legislation has recently been strengthened in relation to dismissals or redundancies resulting from health and safety activities (including refusal to work in situations of serious and imminent danger). Suspension or dismissal on maternity or medical grounds may also give a right of action. See EMPLOYMENT PROTECTION.

## Management of health and safety at work

25.     In 2003 the Institution of Occupational Safety and Health (IOSH) published its guidance to health and safety management systems, which stated in its introduction that:

'Work-related accidents and ill health can be prevented and wellbeing at work can be enhanced by organisations managing health and safety with the same degree of expertise and to the same standards as other core business activities.

['*Systems in focus*' (2003), IOSH, Leicester]

The continuing development and adoption of formal occupational health and safety management systems (OHSMS) is a reflection of this conviction – such systems include those developed by HSE ('*Successful health and safety management*' (1997), (HSG65), 2nd edition), by the ILO ('*Guidelines on occupational health and safety management systems*', (ILO-OSH 2001), ILO Geneva), and BSI ('*Guide to occupational health and safety management systems*', (BS8800:1996), BSI London). Each system exhibits differences when compared to the others, but there is a similarity to the core elements of Plan-Do-Check-Act.

## The costs of accidents and ill health

26.     Various studies have been published to show the cost of accidents and ill health at work. Recent estimates of 40 million days lost to UK industry, of costs in the order of £20 bn, have a general impact but not at the company level. That is why a lot of work is going into developing business cases for taking health and safety more seriously. What is clear is that in large organisations, days lost represents in excess of 2 per cent of total costs. Staff take time off because they have been harmed, and the 'culture' permits and even encourages this. Large companies bear the cost, as do the individuals and their families. In small companies there is less absence, mainly because workers do not take the time off for equivalent ill effects – but this still impacts on productivity. And for small businesses, the loss of an employee for an extended period can be catastrophic. The UK has 6 times the number of people on long-term disability as in France, partly because we do not have ready access to rehabilitation and general occupational health services. This all represents work-in-progress but there are some lessons which can be learned:

- Where studies have been carried out, there is evidence that good health is good business.
- When staff are away from work ill, a return-to-work interview can provide invaluable information to prevent a recurrence.
- Failure to keep proper absence records hamstrings the ability to intervene positively.
- A culture which looks after employees suffers less stress and less absence than one which does not. High staff turnover as well as absence can be a lot more expensive than staff care and good work organisation.

## Health and safety policies

27.     '*Successful health and safety management*' stated that 'accidents are caused by the absence of adequate management control' and stressed the importance of effective health and safety policies in establishing such control. *Section 2(3)* of the *HSWA 1974* requires employers to prepare in writing:

27. *Introduction*

Figure 1: Plan-Do-Check-Act

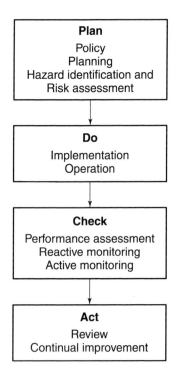

- a statement of their general policy with respect to the health and safety at work of their employees; and
- the organisation and arrangements for carrying out the policy.

It also requires the statement to be brought to the notice of all employees – employers with fewer than five employees are exempt from this requirement.

Policies are normally divided into three sections, to meet the three separate demands of *HSWA 1974*:

*(i) The statement of intent*

This involves a general statement of good intent, usually linked to a commitment to comply with relevant legislation. Many employers extend their policies so as to relate also to the health and safety of others affected by their activities. In order to demonstrate clearly that there is commitment at a high level, the statement should preferably be signed by the chairman, chief executive or someone in a similar position of seniority.

*(ii) Organisational responsibilities*

It is vitally important that the responsibilities for putting the good intentions into practice are clearly identified. In a small organisation this may be relatively simple but larger employers should identify the responsibilities held by those at different levels in the management structure.

Whilst reference to employees' responsibilities may be included, it should be emphasised that the law requires the employer's organisation to be detailed in writing. Types of responsibilities to be covered in the policy might include:

- making adequate resources available to implement the policy;
- setting health and safety objectives;
- developing suitable procedures and safe systems;
- delegating specific responsibilities to others;
- monitoring the effectiveness of others in carrying out their responsibilities;
- monitoring standards within the workplace; and
- feeding concerns up through the organisation.

## (iii) Arrangements

The policy need not contain all of the organisation's arrangements relating to health and safety but should contain information as to where they might be found, for example in a separate health and safety manual or within various procedural documents. Topics which may require detailed arrangements to be specified are:

- operational procedures relating to health and safety;
- training;
- personal protective equipment;
- health and safety inspection programmes;
- accident and incident investigation arrangements;
- fire and other emergency procedures;
- first aid;
- occupational health;
- control of contractors and visitors;
- consultation with employees; and
- audits of health and safety arrangements.

Employees must be aware of the policy and, in particular, must understand the arrangements which affect them and what their own responsibilities might be. They may be given their own copy (for example, within an employee handbook) or the policy might be displayed around the workplace. With regard to some arrangements detailed briefings may be necessary, for example as part of induction training.

Employers must revise their policies as often 'as may be appropriate'. Larger employers are likely to need to arrange for formal review and, where necessary, for revision to take place on a regular basis (e.g. by way of an ISO 9000 procedure). Dating of the policy document is an important part of this process.

## Sources of health and safety advice

28.     Within '*Successful health and safety management*', HSE emphasised the importance of establishing a positive health and safety culture within an organisation as a prerequisite of effective health and safety management. It referred to the 'four Cs' as key components in establishing such a culture: control, competence, communication and co-operation.

## 28. Introduction

While competence in health and safety matters is relevant throughout any workforce, it is particularly important at management levels. The *Management of Health and Safety at Work Regulations 1999* (*SI 1999 No 3242*) have taken this concept further by requiring (in *Reg 7*) every employer to appoint one or more competent persons to assist him in complying with the law. The ACoP accompanying the Regulations states that the size and type of resource required will be relative to the size of the organisation and the risks present in its activities. Full-time or part-time specialists may be appointed, or use may be made of external consultants, although the 1999 Management Regulations state a preference for employees. Smaller employers may appoint themselves, provided that they are competent – the ACoP refers to competence as comprising both the possession of theoretical knowledge and the capacity to put it into practice in the work situation. An awareness of the limits of one's own knowledge and capabilities is also important.

Health and safety training is available from many different sources. The following organisations either provide training themselves or oversee training through accredited training centres.

- National Examination Board in Occupational Safety and Health (NEBOSH)
  tel: 0116 288 8858
- Institution of Occupational Safety and Health (IOSH)
  tel: 0116 257 3100
- Chartered Institution of Environmental Health (CIEH)
  tel: 0207 928 6006
- Royal Society for the Prevention of Accidents (RoSPA)
  tel: 0121 248 2000
- British Safety Council (BSC)
  tel: 0208 741 1231

There are many independent consultants who can provide advice and assistance on health and safety matters. Consultants are listed in the Yellow Pages and the Institution of Occupational Safety and Health (see above) maintains a consultants' register. A free HSE leaflet, *'Need help on health and safety?'* (IND(G)322) provides guidance for employers on when and how to get advice on health and safety.

HSE itself can also be a valuable source of information and advice:

- HSE Books
  tel: 01787 881165
  fax: 01787 313995
  website: www.hse books.co.uk
  
  HSE has a huge range of priced publications and free leaflets, some of which are referred to elsewhere in this Introduction. *'The essentials of health and safety at work'* is a useful starting point for the small employer.
- HSE Infoline
  tel: 08701 545500
  
  Open Monday to Friday 8.30am to 5pm to provide information on workplace health and safety.
- HSE home page on the internet
  www.hse.gov.uk
  
  An online enquiry service can be accessed from the home page.

Of course, within this issue of *Tolley's Health and Safety at Work* you will find it easy to access much valuable information and practical advice, designed to keep you up-to-date with the law and all recent and relevant developments.

## Using this book

29.   In the following pages you will find a detailed compendium of information arranged in alphabetical order of subject, each prepared by an expert in the topic. It is possible to work through this material almost as a self-teaching aid in health and safety or to use each topic as a ready-reference when a question arises or a new company policy is being prepared in that subject area. The following is a general guide to the material.

*Access Traffic Routes and Vehicles* – this is one of HSE's current Priority Areas. If your workplace is visited by an inspector this is an aspect of your arrangements which is likely to be scrutinised. Many people are harmed from the poorly controlled and segregated movement of people and vehicles.

*Accident Reporting* – more than a guide to the legal requirements, it is worth noting that HSC considered introducing a new legal duty to investigate. In the Management Cycle (Plan-Do-Check-Act) the reactive monitoring of performance is a critical source of information for improvement. Proper accident investigation is a critical function.

*Asbestos* – the new legal regime for asbestos in workplace buildings is coming into force. For the material responsible for more work-related deaths over the past 100 years than any other, it repays close attention.

*Children and Young Persons* – children are recognised as particularly vulnerable due to their inexperience. Risk assessment and good communications and supervision are the key elements of good management.

*Compensation for Work Injuries/Diseases* – court cases lead to compensation of over £1.3bn each year (and rising), understanding what you need to do to comply with the Woolf requirements for handling claims is critical.

*Construction and Building Operations* – the UK's (and Europe's) most dangerous industry in terms of fatalities and people injured, the management of works requires close adherence to the rules to minimise the risks.

*Control of Major Accident Hazards* – the requirement to compile safety reports on major accident hazard sites and the annual revision of chemicals and quantities specified in the annex to the Seveso Directive mean that the regulatory regime needs to be carefully monitored and responded to by operators of such sites.

*Dangerous Substances and Explosive Atmospheres* – the European Union-derived safety regime for these materials defines the range of arrangements from workplace and equipment design to emergency procedures, which are vital for the safe management of such hazards.

*Dangerous Substances – Transportation* – the progressive tightening of transportation rules needs to be factored into an effective legal compliance programme.

*Disaster and Emergency Management Systems (DEMS)* – since the attacks including those on the New York World Trade Centre occurred, the world is perceived by many to be less stable and more dangerous, and this adds a further incentive to those seeking to develop effective disaster and emergency systems.

*Electricity* – the Risk Control System (RCS) concept is a structured approach to specifying the management of electrical risk. The concept is now routinely used to identify the agenda which will assist HSE inspectors in the assessment of management systems associated with electrical safety.

**29.** *Introduction*

*Emissions into the Atmosphere* – managing health and safety effectively in the modern context necessitates an awareness of the interface with environmental requirements, and this is a key aspect of these.

*Employees' Duties to their Employers and others* – the *Health and Safety at Work etc Act 1974* imposed most of the relevant duties on employers, but there are key requirements in the health and safety regime imposed upon employees which can be effectively communicated as part of training programmes.

*Employer's Duties to their Employees* – the focus of the health and safety regime is the protection of employees (although this is extended to other workers, and anyone else who may be affected by the work) and compliance with these duties is critical to legal compliance.

*Employers' Liability Insurance* – the cost, and sometimes difficulty of obtaining, employers' liability insurance, has been a major talking point over the past year including ministerial meetings on the subject. Understanding your rights and obligations was never more critical.

*Employment Protection* – a useful overview to sit alongside the health and safety law, focusing on the interface between the two legal regimes.

*Enforcement* – when 'an Inspector calls', what can you expect and what do you need to do.

*Environmental Management* – environmental breaches far outweigh, in numbers of prosecutions and in total fines levied, health and safety breaches. Understanding the legal framework is essential.

*Ergonomics* – estimates vary, but about a third of all work-related sickness absence is due to musculo-skeletal problems (one of HSE's eight priority programmes), and this is just one aspect that ergonomics addresses.

*Fire and Fire Precautions* – the proposed re-organisation of the Fire Service into the Fire and Emergency Service with emphasis on prevention is a reminder of the focus of workplace requirements.

*First-Aid* – one of the common factors in every workplace is the requirement to address this issue.

*Food Safety and Standards* – rarely out of the news, this remains one of the 'hot topics'.

*Gas Safety* – as prosecutions continue of landlords for failing to provide safe gas appliances, this remains an important topic for facilities managers.

*Harassment in the Workplace* – an element of the psychosocial issues which have come to the fore in recent years, and an example of the interface between health and safety and employment law.

*Hazardous Substances in the Workplace* – the new emphasis on occupational health has reinforced the necessity for dealing with hazardous substances effectively. Guidance on asthmagens, outbreaks of Legionnaires' Disease, are just some of the factors which need to be factored into an effective legal compliance programme.

*Joint Consultation in Safety* – *Safety Representatives, Safety Committees, Collective Agreements and Work Councils* – there are new plans to extend the legal rights for worker consultation, and in addition to an HSE initiative there are indications that the rules on Works Councils may be amended across the EU. The need to understand and implement current guidelines is clear.

*Lifting Machinery and Equipment* – the regulatory regime is clear, and the risks associated with lifting and handling operations necessitates careful compliance.

*Lighting* – one of the most significant causes of discomfort in indoor workplaces, and essential for safe working, the basic briefing will assist in providing advice on effective management.

*Machinery Safety* – one of the sets of Regulations introduced in 1992, the provision and use of work equipment has been updated since and remains a pillar of current effective safety management.

*Manual Handling* – estimates vary, but about a third of all work-related sickness absence is due to musculo-skeletal problems (one of HSE's eight priority programmes).

*Noise and Vibration* – with new regulations planned for both these agents, stimulated by the EU Physical Agents Directive, and many thousands of workers adversely affected by exposures, understanding the current regulatory regime is essential.

*Occupational Health and Diseases* – this is one of the major drives from the Government and from HSE, encouraging employers to tackle the high cost (in financial and human terms) of work-related ill health and disability of the workforce.

*Occupiers' Liability* – a useful briefing on the duties of a building occupier.

*Offices and Shops* – there remain requirements which reflect the way in which workplaces were categorised and the legal regime drawn up, and understanding these is an important part of defining a modern approach.

*Offshore Operations* – since the revolution in the legal regime which flowed from the Cullen Report into the loss of Piper Alpha, many specific regulatory requirements have been introduced and are subject to continual review and updating.

*Personal Protective Equipment* – this may be at the bottom of the hierarchy of control, but it still represents a major element of effective protection against risk.

*Pressure Systems* – the requirements to assess and draw up appropriate systems for maintenance, inspection and controlled work have achieved a significant reduction in incidents involving such systems despite their intrinsic hazardous nature.

*Product Safety* – the interface between product safety and health and safety is a critical area to understand.

*Risk Assessment* – this remains the 'motor' of modern health and safety, but it needs to be done well. Significant risks ignored can lead to accidents and deaths, and therefore a balance needs to be struck.

*Safe Systems of Work* – driven by risk assessment, and as part of an overall approach to management, safe systems are the heart of modern health and safety.

*Statements of Health and Safety Policy* – an opportunity to set the tone for an organisation, to create a 'Mission Statement' declaring a real commitment to improving performance.

*Stress at Work* – with the new Management Standards to minimise stress having been published by HSE, it is clear that reducing the adverse impact of excess stress within the workplace is both challenging and necessary.

*Training and Competence in Occupational Safety and Health* – this is a critical briefing on a topic which needs to be addressed by senior management. UK industry and commerce invests much in training, but unless it is effective the delegation of responsibility to junior staff will leave directors and senior managers vulnerable.

*Ventilation* – one of the key methods for the control of exposure to airborne contamination.

*Violence in the Workplace* – this a major issue in a variety of sectors from healthcare to banking, and as case law develops the duties of employers become ever more clear and defined to develop safe systems of work with appropriate emergency response, and robust report and investigation.

*Work at Heights* – falls from heights, one of HSE's eight priority programmes, continue to represent a significant cause of workplace fatalities, and new Regulations are imminent from an EU Directive.

*Working Time* – this continues to be a troublesome area, and a clear strategy based on understanding the legal requirements is essential.

**29.** *Introduction*

*Workplaces – Health, Safety and Welfare* – the management of the working environment represents one of the keys to effective improvements in health and safety performance.

# Access, Traffic Routes and Vehicles

## Introduction

A1001    With the regular daily flow of labour to and from the workplace and vehicles making deliveries and collecting items, access and egress points constitute potentially hazardous situations. For this reason there is a duty on employers and factory occupiers to 'provide and maintain' safe access to and egress from a place of work both under statute and at common law. As part and parcel of compliance with the *Building Regulations 1991 (SI 1991 No 2768)*. This includes access facilities for disabled workers and visitors (that is, persons who have difficulty walking or are wheelchair users, or, alternatively, have a hearing problem or impaired vision) (see further W11043 WORKPLACES – HEALTH, SAFETY AND WELFARE), as well as proper precautions to effect entry or exit to or from confined spaces, where a build-up of gas/combustible substances can be a real though not obvious danger (see A1022 below). Statutory requirements consist of general duties under the *Health and Safety at Work etc. Act 1974 (HSWA)*, which apply to all employers, and the more specific duties of the *Workplace (Health, Safety and Welfare) Regulations 1992 (SI 1992 No 3004)*.

The *Disability Discrimination Act 1995 (DDA 1995)* also relates to access of buildings and services. The Act defines a disabled person as someone with:

'a physical or mental impairment which has a substantial and long-term adverse effect on his ability to carry out normal day-to-day activities.'

From 1 October 2004 duties under *DDA 1995* affect all employers with fewer than 15 employees and anyone who provides a service to the public

The provisions, including those that require employers to consider making changes to the physical features of premises that they occupy, have been in force since December 1996. One way in which an employer might unlawfully discriminate against a disabled employee is by not making reasonable adjustments (without justification) and this includes adjusting the working to enable access and safe traffic routes. A Code of Practice – 'Elimination of discrimination in the field of employment against disabled persons or persons who have had a disability' gives general guidance on the main employment provisions of the Act. It is available from HMSO.

The term 'access' is a comprehensive one and refers to just about anything that can reasonably be regarded as means of entrance/exit to a workplace, even if it is not the usual method of access/egress. Unreasonable means of access/egress would not be included, such as a dangerous short-cut, particularly if management has drawn a worker's attention to the danger, though the fact that a worker is a trespasser has not prevented recovery of damages (*Westwood v The Post Office [1973] 3 AER 184*). Access to a fork lift truck qualified, for the purposes of the *Factories Act 1961, s 29(1)* (repealed as from 1 January 1996), the fork lift truck being a 'place' (*Gunnion v Roche Products Ltd, The Times, 4 November 1994*). The proper procedure is for the employer/factory occupier to designate points of access/egress for workers and see that they are safe, well-lit, maintained and (if necessary) manned and de-iced.

**A1002** *Access, Traffic Routes and Vehicles*

In the case of *Fildes v International Computers (1984) (Unreported)* an employee slipped on a patch of ice and injured his back whilst walking from his car in the car park to the factory gates. It was an agreed means of access.

Moreover, the statutory duties apply to access/egress points to any place where any employees have to work, and not merely their normal workplace.

This chapter summarises key statutory and common law duties in connection with:

(a)  access and egress;

(b)  vehicular traffic routes for internal traffic and deliveries;

(c)  work vehicles and delivery vehicles – with particular emphasis on potentially hazardous activities involving such vehicles; and

(d)  work in confined spaces, of necessity involving access and egress points.

## Statutory duties concerning workplace access and egress – Workplace (Health, Safety and Welfare) Regulations 1992 (SI 1992 No 3004)

**A1002** Statutory duties centre around:

(a)  the organisation of safe workplace transport systems;

(b)  the suitability of traffic routes for vehicles and pedestrians; and

(c)  the need to keep vehicles and pedestrians separate.

### Organisation of safe workplace transport systems

**A1003** Every workplace must (so far as is reasonably practicable) be so organised that pedestrians and vehicles can circulate in a safe manner. [*Reg 17(1)*].

### Suitability of traffic routes

**A1004** Traffic routes in a workplace must be suitable for the persons or vehicles using them, sufficient in number, in suitable positions and of sufficient size. [*Reg 17(2)*].

More particularly:

(a)  pedestrians or vehicles must be able to use traffic routes without endangering those at work;

(b)  there must be sufficient separation of traffic routes from doors, gates and pedestrian traffic routes, in the case of vehicles;

(c)  where vehicles and pedestrians use the same traffic routes, there must be sufficient space between them; and

(d)  where necessary, all traffic routes must be suitably indicated.

*[Reg 17(3)(4)]*.

Compliance with these statutory duties involves provision of safe access for:

(i)  vehicles, with attention being paid to design and layout of road systems, loading bays and parking spaces for employees and visitors; and

(ii)  pedestrians, so as to avoid their coming into contact with vehicles.

## Traffic routes for vehicles

**A1005** There should be sufficient traffic routes to allow vehicles to circulate safely and without difficulty. As for internal traffic, lines marked on roads/access routes in and between buildings should clearly indicate where vehicles are to pass e.g. fork lift trucks. Obstructions, such as limited headroom, are acceptable if clearly indicated. Temporary obstacles should be brought to the attention of drivers by warning signs or hazard cones or, alternatively, access prevented or restricted. Both internal and delivery traffic should be subject to sensible speed limits (e.g. 10 mph), which should be clearly displayed. Speed ramps (sleeping policemen), preceded by a warning sign or mark, are necessary, save for fork lift trucks, on workplace approaches. The traffic route should be wide enough to allow vehicles to pass and repass oncoming or parked traffic, and it may be advisable to introduce one way systems or parking restrictions. It may also be necessary to make allowance for pedestrian routes, where these interact with the vehicle traffic route. Traffic signs on roads, for example speed limit signs, must conform with those on public roads, whether or not the road is subject to the *Road Traffic Regulation Act 1984*.

## Checklist – safe traffic routes

**A1006** Safe traffic routes should:

(*a*) provide the safest route possible between places where vehicles have to call or deliver;

(*b*) be wide enough for the safe movement of the largest vehicle, including visiting vehicles (e.g. articulated lorries, ambulances etc.) and should allow vehicles to pass oncoming or parked vehicles safely. One way systems or parking restrictions are desirable;

(*c*) avoid vulnerable areas/items, such as fuel or chemical tanks or pipes, open or unprotected edges, and structures likely to collapse;

(*d*) incorporate safe areas for loading/unloading;

(*e*) avoid sharp or blind bends; if this is not possible, hazards should be indicated (e.g. blind corner);

(*f*) ensure that road/rail crossings are kept to a minimum and are clearly signed;

(*g*) ensure that entrances/gateways are wide enough; if necessary, to accommodate a second vehicle that may have stopped, without causing obstruction;

(*h*) set sensible speed limits, which are clearly signposted. Where necessary, ramps should be used to retard speed, and road humps or bollards to restrict the width of the road. These should be preceded by a warning sign or mark on the road;

(*i*) ensure that fork lift trucks should not have to pass over road humps, unless of a type capable of doing so;

(*j*) give prominent warning of limited headroom, both in advance and at an obstruction. Overhead electric cables or pipes containing flammable/hazardous chemicals should be shielded, i.e. using goal posts, height gauge posts or barriers;

(*k*) ensure that routes on open manoeuvring areas/yards are marked and signposted, and banksmen are employed to supervise the safe movement of vehicles;

(*l*) ensure that people at risk from exhaust fumes or material falling from vehicles are screened or protected; and

(*m*) restrict vehicle access where high-risk substances are stored (e.g. LPG) and where refuelling takes place.

(*n*) consider installation of refuge points (safe havens) where vehicles need to reverse into delivery areas or dead ends, or position barriers to prevent vehicles reversing from colliding into people.

## Traffic routes for pedestrians

### Checklist – safe traffic routes

**A1007** In the case of pedestrians, the main object of the traffic route is to prevent their coming into contact with vehicles. Safe traffic routes should:

(a) provide separate routes/pavements for pedestrians, to keep them away from vehicles;

(b) where necessary, provide suitable barriers/guard rails at entrances/exits and at the corners of buildings;

(c) where traffic routes are used by both pedestrians and vehicles, be wide enough to allow vehicles to pass pedestrians safely;

(d) where pedestrian and vehicle routes cross, provide appropriate crossing points. These should be clearly marked and signposted. If necessary, barriers or rails should be provided to prevent pedestrians crossing at dangerous points and to direct them to designated crossing points;

(e) where traffic volume is high, traffic lights, bridges or subways should be used to control movement and ensure a smooth, safe flow;

(f) where crowds use or are likely to use roadways, e.g. at the end of a shift, stop vehicles from using them at such times;

(g) provide separate vehicle and pedestrian doors in premises, with vision panels on all doors;

(h) provide high visibility clothing for people permitted in delivery areas (e.g. bright jackets/overalls);

(i) where the public has access (e.g. at a farm or factory shop), public access points should be as near as possible to shops and separate from work activities.

(See also W11014 WORKPLACES.)

## Vehicles

**A1008** Vehicles account for a high percentage of deaths and injuries at work. In 2003/4 38 people were killed by moving vehicles, 191 sustained a defined major injury and 1,975 workers had to take three or more days off work. Many of these casualties occur whilst vehicles are reversing, though activities such as loading and unloading, sheeting and unsheeting, as well as cleaning, can similarly lead to injuries, especially where employees are struck by a falling load or a fall from a height on, say, a tanker or HGV. So, too, climbing and descending ladders on tankers during delivery and 'dipping' at petrol forecourts can be hazardous, access onto vehicles and egress being as important as design and construction. Tipping, too, has its dangers, with tipping vehicles, tipping trailers and tankers overturning in considerable numbers. Also, sheeting and unsheeting operations have led to sheeters slipping or losing their grip or falling whilst walking on top of loads or in consequence of ropes breaking; absence of, or inadequate, training being an additional factor in injuries involving work vehicles. This part of the chapter considers the general statutory requirements relating to vehicles at work, precautions in connection with potentially hazardous operations involving vehicles, as well as providing a checklist for vehicle safety. Specific construction and use requirements are not considered.

### General statutory requirements

**A1009** Both work and private vehicles come within the parameters of health and safety at work. Regarding work vehicles, employers have the direct responsibilities of provision and maintenance generally under *HSWA, s 2*, and, more specifically, under the *Provision and Use of Work Equipment Regulations 1998 (SI 1998 No 2306)* (*PUWER*), vehicles qualifying as 'work equipment'. Regarding private vehicles, employers have much less control – at least, as far as design, construction and use are concerned – but should, nevertheless, endeavour to ensure regulated use via:

(a)  restricted routes and access;

(b)  provision of clearly signposted parking areas away from hazardous activities and operations; and

(c)  enforcement of speed limits.

## Work vehicles

A1010  Generally, work vehicles should be as safe, stable, efficient and roadworthy as private vehicles on public roads. As work equipment, they are subject to the controls of *PUWER*, which specifies provision, maintenance, access and safety provisions in the event of rolling or falling over, whilst employers must also ensure that drivers are suitably trained in conformity with the requirements of the *Management of Health and Safety at Work Regulations 1999* (*SI 1999 No 3242*).

### Provision

A1011  All employers must ensure that vehicles:

(a)  are constructed and adapted as to be suitable for its purpose;

(b)  when selected, caters for risks to the health and safety of persons where the vehicles are to be used; and

(c)  are only used for operations specified and under suitable conditions.

*[PUWER, Reg 5]*.

Compliance with these requirements on the part of operators of HGVs, fork lift trucks, dump trucks and mobile cranes presupposes conformity with the following checklist, namely:

(i)  a high level of stability;

(ii)  safe means of access and egress to and from the cab;

(iii)  suitable and effective service and parking brakes;

(iv)  windscreens with wipers and external mirrors giving optimum all-round visibility;

(v)  a horn, vehicle lights, reflectors, reversing lights, reversing alarms;

(vi)  suitable painting/markings so as to be conspicuous;

(vii)  provision of a seat and seat belts;

(viii)  guards on dangerous parts (e.g. power take-offs);

(ix)  driver protection to prevent injury from overturning, and from falling objects or materials; and

(x)  driver protection from adverse weather.

### Maintenance

A1012  All employers must ensure that work equipment (including vehicles) is maintained in an efficient state, in efficient working order and in good repair. *[PUWER, Reg 6]*. This combines the need for basic daily safety checks by the driver before using the vehicle, as well as preventive inspections and services carried out at regular intervals of time and/or mileage, in accordance with manufacturers recommendations. As regards basic daily safety checks, employers should provide drivers with a log book in which to record visual inspections undertaken and the findings of the following:

- brakes, tyres, steering, mirrors, windscreen washers and wipers, warning signals and specific safety systems (e.g. control interlocks)

Employers should see that drivers carry out the checks.

## Training of drivers

**A1013** All employers must:

(a) in entrusting tasks to employees, take into account their capabilities as regards health and safety; and

(b) ensure that employees are provided with adequate health and safety training on recruitment and exposure to new or increased risks.

*[Management of Health and Safety at Work Regulations 1999 (SI 1999 No 3242)].*

In order to conform with these requirements, employers should ensure that, for general purposes, drivers of work vehicles are over 17 and have passed their driving test or, in the case of drivers of HGVs, that they are over 21 and have passed the HGV test. Moreover, to ensure continued competence, or to accommodate new risks at work or a changing work environment, employers should provide safety updates on an on-going basis as well as refresher training, and require approved drivers to report any conviction for a driving offence, whether or not involving a company vehicle. One method for ensuring this is to require all drivers to submit their licence annually so that a copy can be held in their personnel file.

## Contractors and subcontractors

**A1014** Similar assurances (see A1013 above) should be obtained from drivers of contractors and subcontractors visiting an employer's workplace. If they are not forthcoming, permission to work on site or in-house should be refused until either the contractor's vehicles comply with statutory requirement and/or his drivers are adequately trained. Training of contractor's drivers would normally be undertaken by contractors themselves, though site or in-house hazards, routes to be used etc. should be communicated to contractors by employers or occupiers. Contractors should be left in no doubt of the penalties involved for failure to conform with safe working practices – a useful way of ensuring enforcement on the part of contractors and subcontractors is to issue a licence.

## Access to vehicles

**A1015** In addition to *Regulation 5* of *PUWER* (see A1011 above), employers (and others having control, to any extent, of workplaces (see OCCUPIERS' LIABILITY)), who operate/use vehicles, are subject to:

(a) the fall prevention requirements of *Regulation 13* of the *Workplace (Health, Safety and Welfare) Regulations 1992 (SI 1992 No 3004)* (see W9003 WORK AT HEIGHTS). As far as possible, compliance with this regulation would obviate the need for climbing on top of vehicles (by bottom-filling) and also require vehicle operators to ensure that loads are evenly distributed, packaged properly and secured in the interests of drivers going down slopes and up steep hills (see further A1018 below).

(b) the co-operation requirements of *Regulation 9* of the *Management of Health and Safety at Work Regulations 1999 (SI 1999 No 3242)*, specifying that where activities of different employers interact, different employers may need to co-operate with each other and co-ordinate preventive and protective measures. This regulation is particularly relevant for example, to tanker deliveries and 'dipping' at petrol forecourts as well as loading and/or unloading operations. For 'dipping' purposes or gaining top access to tankers, access should be by a ladder at the front or rear, such ladders being properly constructed, maintained and securely fixed; ideally, they should incline inwards towards the top. There should be a means of preventing people from falling whilst on top of the tanker. Failing this, employers of tanker drivers and forecourt owners should liaise on potential risks involved in tanker deliveries, e.g. the provision of suitable step-ladders on the part of the latter. As for carriage of goods and loading/unloading operations, consignors should ensure that goods are evenly distributed, properly packaged and secured.

## Potentially hazardous operations

A1016    The following activities and operations are potentially hazardous in connection with vehicles.

## Reversing

A1017    Approximately a quarter of all deaths at work are caused by reversing vehicles; in addition, negligent reversing can result in costly damage to premises, plant and goods. Where possible, workplace design should aspire to obviate the need for reversing by the incorporation of one-way traffic systems. Failing this, reversing areas should be clearly identified and marked, and non-essential personnel excluded from the area. Ideally, banksmen wearing high-visibility clothing should be in attendance to guide drivers through, and keep non-essential personnel and pedestrians away from, the reversing area. Refuge points, also known as safe havens, should be constructed where possible to enable an escape route or safe place for people to stay in the event of a vehicle reversing in an unsafe manner. Vehicles should be fitted with external side-mounted and rear-view mirrors – as, indeed, many now are. Closed-circuit television systems are also advisable for enabling drivers to see round 'blind spots' and corners.

## Loading and unloading

A1018    Because employees can be seriously injured by falling loads or overturning vehicles, loading/unloading should not be carried out:

(*a*)    near passing traffic, pedestrians and other employees;

(*b*)    where there is a possibility of contact with overhead electric cables;

(*c*)    on steep gradients;

(*d*)    unless the load is spread evenly (racking will assist load stability);

(*e*)    unless the vehicle has its brakes applied or is stabilised (similarly with trailers); or

(*f*)    with the driver in the cab.

(See also OFFICES AND SHOPS.)

## Tipping

A1019    Overturning of lorries and trailers is the main hazard associated with tipping. In order to minimise the potential for injuries, tipping operations should only occur:

(*a*)    after drivers have consulted with site operators and checked that loads are evenly distributed;

(*b*)    when non-essential personnel are not present;

(*c*)    on level and stable ground away from power lines and pipework; and

(d)   with the driver in the cab and the cab door closed.

Moreover, after discharge, drivers should ensure that the body of the vehicle is completely empty and should not drive the vehicle in an endeavour to free a stuck load.

### Giving unauthorised lifts

A1020   Giving unauthorised lifts in work vehicles is both a criminal offence and can lead to employers being involved in civil liability. Thus, 'every employer shall ensure that work equipment is used only for operations for which, and under conditions for which, it is suitable'. [*PUWER, Reg 5*].

Where a driver of a work vehicle gives employees and/or others unauthorised lifts, his employer could find himself prosecuted for breach of the above regulation, whilst the driver himself may be similarly prosecuted for breach of *HSWA, s 7*, as endangering co-employees and members of the public. In addition, although acting in an unauthorised manner and contrary to instructions, the employee may well involve his employer in vicarious liability for any subsequent injury to a co-employee and/or member of the public (see further *Rose v Plenty* at E11004 EMPLOYERS' DUTIES TO THEIR EMPLOYEES).

Where instructions to employees not to give unauthorised lifts are clearly displayed in a work vehicle, but an employee nevertheless gives an unauthorised lift and a co-employee or member of the public is injured or killed as a result of the employee's negligent driving, it can be argued that the employee has exceeded the scope of his employment and so the employer is absolved from liability. (In *Twine v Bean's Express [1946] 1 AER 202* a driver gave a lift to a third party who was killed in consequence of his negligent driving. There was a notice in the van prohibiting drivers from giving lifts. It was held that the employer was not liable, as the driver was acting outside the parameters of his employment when giving a lift. The injured passenger knew that the driver should not give lifts.)

Conversely, courts have taken the view that such conduct, on the part of drivers, does not circumscribe the scope of employment but rather constitutes performance of work in an unauthorised manner, so leaving the employer liable as the employee is doing what he is employed to do but doing it wrongly (see further *Rose v Plenty and Century Insurance Co v Northern Ireland Road Transport Board* at E11004 EMPLOYERS' DUTIES TO THEIR EMPLOYEES). Yet again, if the passenger, having seen the prohibition on unauthorised lifts in the vehicle, nevertheless accepted a lift and was injured, arguably, if an adult rather than a minor, he has agreed to run the risk of negligent injury and so will forfeit the right to compensation.

## Common law duty of care

A1021   At common law every employer owes all his employees a duty to provide and maintain safe means of access to and egress from places of work. Moreover, this duty extends to the workforce of another employer/contractor who happens to be working temporarily on the premises. Hence the common law duty covers all workplaces, out of doors as well as indoors, above ground or below, and extends to factories, mines, schools, universities, aircraft, ships, buses and even fire engines and appliances (*Cox v Angus [1981] ICR 683*, where a fireman injured in a cab was entitled to damages at common law).

## Confined spaces

A1022   Accidents and fatalities such as drowning, poisoning by fumes or gassing, have happened as a result of working in confined spaces. See, for instance, the case of *Baker v T E Hopkins & Son Ltd [1959] 3 AER 225* where a doctor was overcome by carbon monoxide fumes while going to rescue two workmen down a well – the defence of *volenti non fit injuria* failed).

Normal safe practice is a formalised permit to work system or checklist tailored to a particular task and requiring appropriate and sufficient personal protective equipment. Hazards typical of this sort of operation are:

(a) atmospheric hazards – oxygen deficiency, enrichment (see *R v Swan Hunter Shipbuilders Ltd* at C8060 CONSTRUCTION AND BUILDING OPERATIONS), toxic gases (e.g. carbon monoxide), explosive atmospheres (e.g. methane in sewers);

(b) physical hazards – low entry headroom or low working headroom, protruding pipes, wet surfaces underfoot as well as any electrical or mechanical hazards;

(c) chemical hazards – concentration of toxic gas can quickly build up, where there is a combination of chemical cleaning substances and restricted air flow or movement.

In order to combat this variety of hazards peculiar to work in confined spaces, use of both gas detection equipment and suitable personal protective equipment are a prerequisite, since entry/ exit paths are necessarily restricted.

Prior to entry, gas checks should test for (a) oxygen deficiency/enrichment, then (b) combustible gas and (c) toxic gas, by detection equipment being lowered into the space. This will determine the nature of personal protective equipment necessary. If gas is present in any quantity, the offending space should then be either naturally or mechanically ventilated. Where gas is present, entry should only take place in emergencies, subject to the correct respiratory protective equipment being worn. Assuming gas checks establish that there is no gaseous atmosphere, entry can then be made without use of respiratory equipment. Gas detection equipment should continue to be used whilst people are in the confined space so that any atmospheric change can subsequently be registered on the gas detection equipment. It is essential that an emergency plan is devised when gaining access to confined spaces, which includes effective two way communication between people inside and immediately outside the confined space, contact with emergency services and first aid, and fire prevention personnel on hand.

## *Statutory requirements*

A1023 It should be noted that the *Confined Spaces Regulations 1997* (*SI 1997 No 1713*) came into force on 28 January 1998. These repeal *Factories Act 1961, s 30*, and impose requirements and prohibitions with respect to the health and safety of persons carrying out work in confined spaces.

A 'confined space' is defined in *reg 1(2)* as 'any place, including any chamber, tank, vat, silo, pit, trench, pipe, sewer, flue, well or other similar space in which, by virtue of its enclosed nature, there arises a reasonably foreseeable specified risk'.

A 'specified risk' means a risk of:

(a) serious injury to any person at work arising from a fire or explosion;

(b) without prejudice to paragraph (a) –

　(i) the loss of consciousness of any person at work arising from an increase in body temperature;

　(ii) the loss of consciousness or asphyxiation of any person at work arising from gas, fume, vapour or the lack of oxygen;

(c) the drowning of any person at work arising from an increase in the level of a liquid; or

(d) the asphyxiation of any person at work arising from a free flowing solid or the inability to reach a respirable environment due to entrapment by a free flowing solid.

*Regulation 4* prohibits a person from entering a confined space to carry out work for any purpose where it is reasonably practicable to carry out the work by other means.

If, however, a person is required to work in a confined space, a risk assessment must be undertaken to comply with the requirements of the *Management of Health and Safety at Work Regulations 1999 (SI 1999 No 3242), Reg 3*. The risk assessment must be undertaken by a competent person and the outcome of the risk assessment process will then provide the basis for the development of a safe system of work (*Confined Spaces Regulations 1997, Reg 3*).

The risk assessment process should make use of all available information such as engineering drawings, working plans, soil or geological information and take into consideration factors such as the general condition of the confined space, work to be undertaken in the space to minimise hazards produced in the area, need for isolation of the space and the requirements for emergency rescue. In particular, information should be collected and assessed on the previous contents of the confined space, residues that still may be present, contamination that may arise from adjacent plant, processes, gas mains, surrounding soil, land or strata; oxygen level and physical dimensions of the space that may limit safe access and/or egress. The work to be undertaken should be assessed to determine if additional risks will be produced as a result of this work and systems developed to control these risks. All information collected should be recorded and a safe system of work developed for safe entry.

The main elements to consider when designing a safe system of work include the following:

(a)  supervision,
(b)  competence levels for personnel working in confined spaces,
(c)  communications,
(d)  testing/monitoring the atmosphere,
(e)  gas purging,
(f)  ventilation,
(g)  removal of residues,
(h)  isolation from gases, liquids and other flowing materials,
(i)  isolation from mechanical and electrical equipment,
(j)  selection and use of suitable equipment,
(k)  personal protective equipment (PPE) and respiratory protective equipment (RPE),
(l)  portable gas cylinders and internal combustion engines,
(m)  gas supplied by pipes and hoses,
(n)  access and egress,
(o)  fire prevention,
(p)  lighting,
(q)  static electricity,
(r)  smoking,
(s)  emergencies and rescue,
(t)  limited working time.

[*Confined Spaces Regulations 1997, Reg 4*].

*Regulation 6* provides for circumstances allowing the Health and Safety Executive to grant exemption certificates.

# Accident Reporting and Investigation

## Introduction

**A3001** Employers (both onshore and offshore [*Reporting of Injuries, Diseases and Dangerous Occurrences Regulations 1995* (*RIDDOR*) (*SI 1995 No 3163*), *Reg 12*]) and other 'responsible persons' (see A3003 below) who have control over employees and work premises are required to notify and report to the relevant enforcing authority (see A3004 below) the following specified events occurring at work:

(*a*) accidents causing injuries, fatal and non-fatal, including:

  (i) acts of non-consensual physical violence committed at work, and

  (ii) acts of suicide occurring on, or in the course of, the operation of a railway, tramway, trolley or guided transport system

[*RIDDOR* (*SI 1995 No 3163*), *Reg 2(1)*];

(*b*) occupational diseases; and

(*c*) dangerous occurrences, even where no injury results.

The duty to report applies not only in the case of incidents involving employees, but also to visitors, customers and members of the public killed or injured by work activities [*RIDDOR* (*SI 1995 No 3163*), *Reg 3(1)*].

Employees also have certain obligations to report accidents to their employers.

## Reporting of Injuries, Diseases and Dangerous Occurrences Regulations 1995 (RIDDOR) – (SI 1995 No 3163)

**A3002** The *Reporting of Injuries, Diseases and Dangerous Occurrences Regulations 1995* cover:

(*a*) reportable work injuries (see A3007 below);

(*b*) reportable occupational diseases (see Appendix B at A3036 below) – now 47 in all;

(*c*) reportable dangerous occurrences (see A3009 and Appendix A at A3035 below) – now 83 in all;

(*d*) road accidents involving work (see A3019 below); and

(*e*) gas incidents (see A3020 below).

Records must be kept by employers and other 'responsible persons' (see A3005) of such injuries, diseases and dangerous occurrences for a minimum of three years from the date they were made. In addition, employers must also keep an Accident Book (Form BI510).

Two reporting routes now exist:

**A3003** *Accident Reporting and Investigation*

- Injuries and dangerous occurrences are reportable to the relevant enforcing authority on Form F2508 and diseases on Form F2508A (See A3012 and Appendices C and D at A3037, A3038 below); or

- Reports can be made by telephone or electronically to the national Incident Contact Centre (ICC) (see A3004).

As for notification of major accident hazards (see C10009 CONTROL OF MAJOR ACCIDENT HAZARDS), notification and reporting under RIDDOR is sufficient for the purposes of the *Control of Major Accident Hazards Regulations 1999 (SI 1999 No 743), Reg 15(4)*.

## Relevant enforcing authority

**A3003** Health and safety law is separately enforced either by the Health and Safety Executive (HSE), or by local authorities, through their environmental health departments, depending on the nature of the business activity in question. Under *RIDDOR*, notifications and reports should be directed to the authority responsible for the premises where the reportable event occurs (or in connection with which the work causing the event is being carried out). Details of HSE offices are listed in Appendix E to this chapter.

## Incident Reporting Centre

**A3004** For all incidents occurring after 1 April 2001, employers and others who are subject to the reporting requirements, can comply with their obligations by contacting the Incident Contact Centre (ICC). This will eliminate the need to identify the local agency or other local enforcing authority. The ICC will then pass the report on to them. Reporting to the local HSE office or local enforcement authority by phone and on the current statutory forms is still an option and this information will be forwarded to the ICC.

Incidents may be reported to the ICC through the following channels:

- By telephone: Telephone 0845 300 9923 (Monday to Friday, 8.30 a.m. to 5.00 p.m.)
- By facsimile: Fax 0845 300 9924
- By email: riddor@gov.uk
- By internet: www.riddor.gov.uk
- By post: Incident Contact Centre, Caerphilly Business Park, Caerphilly CF83 3GG

In taking advantage of these arrangements, the employer or other reporting person will not have a record of a statutory RIDDOR form, so the ICC will send out confirmation copies of reports which should be checked for accuracy and retained.

## Persons responsible for notification and reporting

**A3005** The person generally responsible for reporting injury-causing accidents, deaths or diseases is the employer. Failing that, the person having control of the work premises or activity will be the responsible person. [*RIDDOR (SI 1995 No 3163), Reg 2(1)*]. In certain cases these normal rules are displaced and there are specifically designated 'responsible persons', e.g. in the case of mines, quarries, offshore installations, vehicles, diving operations and pipelines (see Table 1 below).

| Table 1 |
|---|
| Persons generally responsible for reporting accidents |

| | | |
|---|---|---|
| Death, major injury, over-3-day injury or specified occupational disease: | of an employee at work | that person's employer |
| | of a person receiving training for employment | the person whose undertaking makes immediate provision of the training |
| | of a self-employed person at work in premises under the control of someone else | the person for the time being having control of the premises in connection with the carrying on by him of any trade, business or undertaking |
| Specified major injury or condition, or over-3-day injury: | of a self-employed person at work in premises under his control | |
| Death, or specified major injury or condition: | of a person who is not himself at work (but is affected by the work of someone else), e.g. a member of the public, a shop customer, a resident of a nursing home | the person for the time being having control of the premises in connection with the carrying on by him of any trade, business or undertaking at which, or in connection with the work at which, the accident causing the injury happened |
| [*Reg 2(1)(b), (c)*]. | | |

| Persons responsible for reporting accidents in specific locations | |
|---|---|
| A mine | the mine manager |
| A quarry | the quarry owner |
| A closed tip | the owner of the mine or quarry with which the tip is associated |
| An offshore installation (except in the case of reportable diseases) | the duty holder (see O7027 OFFSHORE OPERATIONS) |
| A dangerous occurrence at a pipeline | the owner of the pipeline |
| A dangerous occurrence at a well | the appointed person or, failing that, the concession owner |
| A diving operation (except in the case of reportable diseases) | the diving contractor |
| A vehicle | the vehicle operator |
| [*Reg 2(1)(a)*]. | |

In situations where the responsible person is difficult to identify because of the shared control of a site or operations, arrangements should be made to determine who will deal with *RIDDOR* reporting in line with the duty to co-operate and co-ordinate under the *Management of Health and Safety at Work Regulations 1999 (SI 1999 No 3242), Reg 11.*

## What is covered?

**A3006** Covered by these Regulations are events involving:

(a) employees;

(b) self-employed persons;

(c) trainees ;

(d) any person, not an employee or trainee, on premises under the control of another, or who was otherwise involved in an accident (e.g. a visitor, customer, passenger or bystander) (see A3004 above).

## Major injuries or conditions

**A3007** Where any person dies or suffers a major injury as a result of, or in connection with, work, such an incident must be notified immediately and details formally reported. The person who dies or suffers injury need not be at work; it is enough if the death or injury arose from a work activity. For example, reporting requirements would apply to a shopper who fell and was injured on an escalator, so long as the injury was connected with the escalator (*Woking Borough Council v British Home Stores [1995] 93 LGR 396*); a member of the public overcome by fumes on a visit to a factory and who lost consciousness; a patient in a nursing home who fell over an electrical cable lying across the floor and was injured; or a pupil or student killed or injured in the course of his curricular work which was supervised by a lecturer or teacher.

Reportable major injuries and conditions are as follows:

(a) any fracture (other than to fingers, thumb or toes);

(b) any amputation;

(c) dislocation of shoulder, hip, knee or spine;

(d) loss of sight (whether temporary or permanent);

(e) a chemical or hot metal burn to the eye or any penetrating eye injury;

(f) any injury resulting from an electrical shock or electrical burn (including one caused by arcing or arcing products) leading to unconsciousness or requiring resuscitation or admittance to hospital for more than 24 hours;

(g) any other injury

  (i) leading to hypothermia, heat-induced illness or unconsciousness,

  (ii) requiring resuscitation, or

  (iii) requiring admittance to hospital for more than 24 hours;

(h) loss of consciousness caused by asphyxia or by exposure to a harmful substance or biological agent;

(j) either

  (i) acute illness requiring medical treatment, or

  (ii) loss of consciousness

  resulting from the absorption of any substance by inhalation, ingestion or through the skin;

(k) acute illness requiring medical treatment where there is reason to believe that this resulted from exposure to a biological agent or its toxins or infected material.

[*RIDDOR (SI 1995 No 3163), Reg 2(1), Sch 1*].

## Injuries incapacitating for more than three consecutive days

**A3008** Where a person at work is incapacitated for more than three consecutive days from their normal contractual work (excluding the day of the accident but including any days which would not have been working days) owing to an injury resulting from an accident at work (other than an injury reportable as a major injury listed in A3007 above), a report of the accident must be made direct to the ICC or sent in writing on Form F2508 to the enforcing authority as soon as is practicable and in any event within ten days of the accident. [*RIDDOR* (*SI 1995 No 3163*), *Reg 3(2)*].

## Specified dangerous occurrences

**A3009** 83 types of reportable dangerous occurrences are specified in *RIDDOR* (*SI 1995 No 3163*), *Sch 2*, ranging from general dangerous occurrences (e.g. the collapse of a building or structure, the explosion of a pressure vessel, and accidental releases of significant quantities of dangerous substances) to more specific dangerous occurrences in mines, quarries, transport systems and offshore installations (see Appendix A at A3035 below).

## Specified diseases

**A3010** Where a worker suffers from an occupational disease related to a particular activity or process (as specified in *Schedule 3*), a report must be sent to the enforcing authority. [*RIDDOR* (*SI 1995 No 3163*), *Reg 5(1)*].

This duty arises only when an employer has received information in writing from a registered medical practitioner diagnosing one of the reportable diseases. (In the case of the self-employed, the duty is triggered regardless of whether or not the information is given to him in writing.) Many of these diseases are those for which disablement benefit is ordinarily prescribed. (In respect of offshore workers, such diseases tend to be communicable.)

Moreover, the *Industrial Diseases* (*Notification*) *Act 1981* and the *Registration of Births and Deaths Regulations 1987* (*SI 1987 No 2088*), *Sch 2*, Form 14, require that particulars are to be included on the death certificate as to whether death might have been due to, or contributed to by, the deceased's employment. Such particulars are to be supplied by the doctor who attended the deceased during the last illness. (The full list of specified diseases is contained in Appendix B at A3036 below.)

## Duty to notify/report

### Duty to notify

**A3011** 'Responsible persons' must notify enforcing authorities (see A3003 and A3004 above) by the quickest means practicable (normally by telephone) of the following:

(*a*) death as a result of an accident arising out of, or in connection with, work;

(*b*) 'major injury' (see A3007 above) of a person at work as a result of an accident arising out of, or in connection with, work;

(*c*) injury suffered by a person not at work (e.g. a visitor, customer, client, passenger or bystander) as a result of an accident arising out of, or in connection with, work, where that person is taken from the accident site to a hospital for treatment;

(*d*) major injury suffered by a person not at work, as a result of an accident arising out of, or in connection with, work at a hospital;

(e) a dangerous occurrence (see A3009 above)

[*RIDDOR (SI 1995 No 3163), Reg 3(1)*];

(f) road injuries or deaths [*RIDDOR (SI 1995 No 3163), Reg 10(2)*] (see A3019 below); and

(g) gas incidents [*RIDDOR (SI 1995 No 3163), Reg 6(1)*] (see A3020 below).

## Duty to report

**A3012** Responsible persons must also formally report events causing death or major injury, dangerous occurrences and details associated with workers' occupational diseases. In cases of death, major injury and accidents leading to hospitalisation, the duty to report extends to visitors, bystanders and other non-employees. More particularly, within ten days of the incident, responsible persons must send a written report form to the relevant enforcing authority in relation to:

(a) all events which require notification (listed in (a) to (g) at A3011 above);

(b) the death of an employee if it occurs within a year following a reportable injury (whether or not reported under (a)) [*RIDDOR (SI 1995 No 3163), Reg 4*] (see A3018 below);

(c) incapacitation for work of a person at work for more than three consecutive days as a result of an injury caused by an accident at work [*RIDDOR (SI 1995 No 3163), Reg 3(2)*] (see A3008 above);

(d) specified occupational diseases relating to persons at work (see A3009 above and Appendix B Part I at A3036 below), and also specifically those suffered by workers on offshore installations (see Appendix B Part II at A3036 below) [*RIDDOR (SI 1995 No 3163), Reg 5(1)*], provided that, in both cases,

   (i) the responsible person has received a written statement by a doctor diagnosing the specified disease, in the case of an employee, or

   (ii) a self-employed person has been informed by a doctor that he is suffering from a specified disease.

[*RIDDOR (SI 1995 No 3163), Reg 5(2)*].

## Exceptions to notification and reporting

**A3013** There is no requirement to notify or report the injury or death of:

(a) a patient undergoing treatment in a hospital or a doctor's or dentist's surgery [*RIDDOR (SI 1995 No 3163), Reg 10(1)*]; or

(b) a member of the armed forces of the Crown [*RIDDOR (SI 1995 No 3163), Reg 10(3)*].

# Duty to keep records

**A3014** Records of injury-causing accidents, dangerous occurrences and specified diseases must be kept by responsible persons for at least three years. [*RIDDOR (SI 1995 No 3163), Reg 7(3)*].

## Injuries and dangerous occurrences

**A3015** In the case of injuries and dangerous occurrences, such records must contain:

(a) date and time of the accident or dangerous occurrence;

(b) if an accident is suffered by a person at work –

   (i) full name,

(ii) occupation, and

(iii) nature of the injury;

(c) in the event of an accident suffered by a person not at work –

(i) full name,

(ii) status (e.g. passenger, customer, visitor or bystander), and

(iii) nature of injury;

(d) place where the accident or dangerous occurrence happened;

(e) a brief description of the circumstances;

(f) the date that the event was first reported to the enforcing authority;

(g) the method by which the event was reported.

[*RIDDOR (SI 1995 No 3163), Sch 4, Part I*].

## Disease

A3016   In the case of specified diseases, such records must contain:

(a) date of diagnosis;

(b) name of the person affected;

(c) occupation of the person affected;

(d) name or nature of the disease;

(e) the date on which the disease was first reported to the enforcing authority; and

(f) the method by which disease was reported.

[*RIDDOR (SI 1995 No 3163), Sch 4, Part II*].

## Action to be taken by employers and others when accidents occur at work

A3017

## Reporting the death of an employee

A3018   Where an employee, as a result of an accident at work, has suffered a reportable injury/condition which is the cause of his death within one year of the date of the accident, the employer must inform the enforcing authority in writing of the death as soon as it comes to his attention, whether or not the accident has been reported under *Regulation 3*. [*RIDDOR (SI 1995 No 3163), Reg 4*].

## Road accidents

A3019   Road accident deaths and injuries are notifiable and reportable if the death or major injury is caused by or connected with:

(a) exposure to any substance conveyed by road;

(b) loading or unloading vehicles;

(c) construction, demolition, alteration or maintenance activities on public roads; or

**Figure 1**

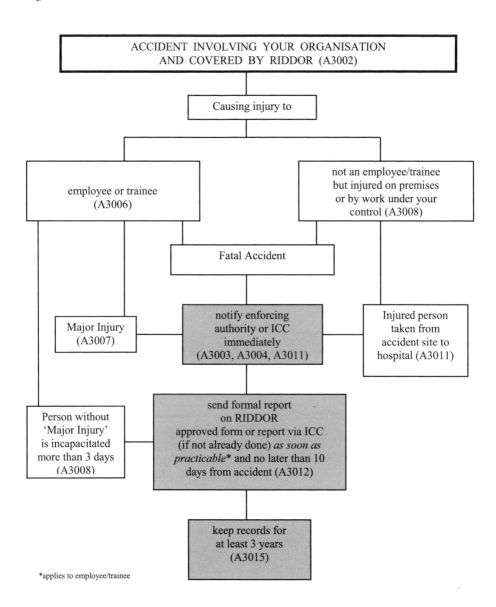

(d)   an accident involving a train.

[RIDDOR (SI 1995 No 3163), Reg 10(2)].

The person injured, whether fatally or not, may or may not be engaged in the above-mentioned activities. Thus, an employee struck by a passing vehicle or a motorist injured by falling scaffolding, are covered. In addition, certain dangerous occurrences on public highways and private roads are covered (see Appendix A at A3035 below).

## Duty to report gas incidents
### Gas suppliers
A3020  Where:

(*a*) a conveyor of flammable gas through a fixed pipe distribution system, or

(*b*) a filler, importer or supplier (not by way of the retail trade) of a refillable container containing liquefied petroleum gas

receives notification of any death, injury or condition (see A3012 above) which has arisen out of, or in connection with, the gas supplied, filled or imported, he must immediately notify the HSE and then send a report of the incident within 14 days. [*RIDDOR* (*SI 1995 No 3163*), *Reg 6(1)*].

### Gas fitters
A3021  Where an employer or self-employed person who is an approved gas fitter has information that a gas fitting, flue or ventilation is, or is likely to cause, death or major injury, he is required to report that fact to the HSE on a prescribed form within 14 days. [*RIDDOR* (*SI 1995 No 3163*), *Reg 6(2)*].

## Penalties
A3022  Contravention of any of the provisions of *RIDDOR* is an offence. The maximum penalty on summary conviction is a fine of £5,000. Conviction on indictment in the Crown Court carries an unlimited fine.

### Defence in proceedings for breach of Regulations
A3023  It is a defence under the Regulations for a person to prove that:

(*a*) he was not aware of the event requiring him to notify or send a report to the enforcing authority; and

(*b*) he had taken all reasonable steps to have all such events brought to his notice.

[*RIDDOR* (*SI 1995 No 3163*), *Reg 11*].

## Obligations on employees
A3024  All employees have duties to inform their employers of serious and immediate dangers and recognisable short comings in safety arrangements.

A3025  There are separate reporting requirements on employees by the *Social Security (Claims and Payments) Regulations 1979* (*SI 1979 No 628*), *Reg 24*. Although not actually bound in with the separate scheme of health and safety legislation, these reporting requirements complement those which employers have under *RIDDOR*, and there is an overlap in the record keeping requirements for both sets of provisions – see further below.

An accident to which the employees' reporting requirements apply is one 'in respect of which benefit may be payable ... '. [*Social Security (Claims and Payments) Regulations 1979* (*SI 1979 No 628*), *Reg 24(1)*]. Under *Regulation 24*, every 'employed earner' who suffers personal injury by an accident for the purposes of the Regulations is required by to give either *oral or written* notice to the employer in one of number of different ways. These ways include notice given to any supervisor or direct to a person designated by the employer. The method that has probably become the norm through custom and practice is by way of an entry of the appropriate particulars in a 'book kept specifically for these purposes' under the Regulations. These books are available from HSE Books as form BI 510. As well as containing sections for completing details of accidents etc. form BI 150 includes pages of instructions to employees and employers about their responsibilities. It is important only to use the latest (2003) version which has been updated so that it is compliant with the *Data Protection Act 1988*.

The entry in an Accident Book is to be made as soon as practicable after the happening of an accident by the employed earner or by some other person acting on his behalf. [*Social Security (Claims and Payments) Regulations 1979 (SI 1979 No 628), Reg 24(3)*]. Failure to so is an offence punishable by a fine. [*Social Security (Claims and Payments) Regulations 1979 (SI 1979 No 628), Reg 31*].

Under *Reg 25* of the 1979 Regulations, once an accident is reported by the employee, the employer must take reasonable steps to investigate the circumstances and, if there appear to be any discrepancies between the circumstances reported and the findings in these investigations, these should be recorded.

Where the accident in question is one to which *RIDDOR* applies there is, as well as the usual reporting requirements (see A3014 above), a requirement to keep a record of the accident. [*RIDDOR (SI 1995 No 3163, Reg 7*]. The HSE guide to *RIDDOR* provides (at paragraph 86) that an employer may choose to utilise the form BI 150 Accident Book as this record, and the format of the form includes space for the employer to initial the report as being one reportable under *RIDDOR*.

## Objectives of accident reporting

A3026    There should be an effective accident reporting and investigation system in all organisations. Accident reporting procedures should be clearly established in writing with individual reporting responsibilities specified. Staff should be trained in the system and disciplinary action may have to be taken where there is a failure to comply with it. Moreover, there is a case for all incidents, no matter how trivial they may seem and of the absence of any resulting injury being reported through the internal reporting procedures.

The capture of information about accidents and other incidents can be enhanced by the provision of a simple form used throughout the organisation.

## Duty of disclosure of accident data

A3027    Employers are under a duty to disclose accident data to works safety representatives and, in the course of litigation, to legal representatives of persons claiming damages for death or personal injury.

### Safety representatives

A3028    An employer must make available to safety representatives of both unionised and non-unionised workforces the information within the employer's knowledge necessary to enable them to fulfil their functions. [*Safety Representatives and Safety Committees Regulations 1977 (SI 1977 No 500), Reg 7(2)*; *Health and Safety (Consultation with Employees) Regulations 1996 (SI 1996 No 1513), Reg 5(1)*]. The Approved Code of Practice in association with these Regulations states that such information should include information which the employer keeps relating to the occurrence of any accident, dangerous occurrence or notifiable industrial disease and any associated statistical records. (Code of Practice: Safety Representatives and Safety Committees 1976, para 6(c)). See JOINT CONSULTATION – SAFETY REPRESENTATIVES AND SAFETY COMMITTEES.

Figure 2: Internal Incident Notification Form

| | | |
|---|---|---|
| | **INTERNAL INCIDENT NOTIFICATION FORM** | |
| Time/date: | 00.00 hrs | dd/mm/yy |
| Location: | | |
| Description of incident: | | |
| Person(s) injured: | | |
| Nature of injury: | | |
| Other persons involved: | | |
| First aid administered: | | |
| Accident Book completed: | Y/N | |
| Form completed by: | | |
| Date: | dd/mm/yy | |

## Legal representatives

**A3029** In the course of litigation (and sometimes before proceedings have actually begun) obligations may arise to give disclosure of relevant documents concerning an accident – even if they are confidential. Legal representatives of an injured party would expect reasonable access to *RIDDOR* information. The decision in *Waugh v British Railways Board [1979] 2 All ER 1169* established that where an employer seeks to withhold on grounds of privilege a report made following an accident, he can only do so if its dominant purpose is related to actual or potential hostile legal proceedings. In *Waugh* a report was commissioned for two purposes following the death of an employee: (*a*) to recommend improvements in safety measures, and (*b*) to gather material for the employer's defence. It was held that the report was not privileged.

## Accident Investigation

### Rationale for undertaking investigations

A3030   Unlike the highly prescriptive legal requirements for notification under *RIDDOR* (*SI 1995 No 3163*), there are no specific legal obligations requiring the investigation of accidents or the production of accident reports. The HSC consulted on proposals for a statutory duty to undertake investigations into accidents, diseases and dangerous occurrences in 2001, but the outcome was a decision to develop new guidelines with a view to encouraging the implementation of best practice voluntarily.

Nevertheless a number of statutory requirements demand some level of investigation is carried out, if only to determine whether it is necessary to undertake a review of the adequacy of existing risk assessments for the relevant activity in order to comply to with *Regulation 3(3)* of the *Management of Health and Safety at Work Regulations 1999* (*SI 1999 No 3242*) (see R3017).

Accident investigation does however does however have wider rationales. The main drivers for it are:

(*a*)   *learning lessons*: identifying and understanding the immediate and the underlying causes of accidents and identifying ways in which to prevent the recurrence of similar accidents.

(*b*)   *reassurance and explanation*: particularly the victim of an accident, but also others connected with it often need to be reassured that appropriate action is being taken or that a satisfactory explanation has been given for what has happened.

(*c*)   *monitoring of performance*: investigations produce data for the reactive monitoring of health and safety performance, measurement of results against an organisation's targets or for benchmarking, recording of essential information in databases ('corporate memory'), and the provision of management reports – ultimately to directors – on the management of operational risks.

(*d*)   *providing information to other interested parties*: the information is likely to be needed with which to brief insurers, safety representatives, also those who are responsible for dealing with media. Tthere is also an expectation on the part of health and safety inspectors as well that an accident investigation will be produced, and disclosed to them once completed, and an absence of a formal report may be taken as an indicator of management failure.

(*e*)   *disciplinary procedures*: where there is evidence of deliberate reckless failure to follow procedures an investigation may form part of the disciplinary process and evidence upon which the fairness of action perhaps ultimately resulting in dismissal might be judged.

(*f*)   *allocation of blame/liability*: this will be part and parcel of any investigation undertaken by enforcement agencies and insurers, any information may need to be obtained and analysed by the organisation involved in the accident in preparation for what are inevitably adversarial proceedings.

Accident investigation literature often tends to be focussed on the risk management and accident prevention benefits that can accrue from the process, while treating issues relating to blame as subsidiary to other objectives. This can result in a number of difficulties and potential conflicts, as it is difficult in reality for accidents to be investigated in an entirely blame-free context. Accident investigations can – if there is not careful consideration given to legal and disciplinary implications – prejudice not just the organisation but also the position of individual employees and managers who may be subject to actual or implied criticism. It should also be borne in mind that the HSE's internal work instructions for inspectors steer inspectors towards directing or using an organisation's investigation process in order to achieve enforcement-related goals which include providing '*an early insight into the duty holder's thoughts regarding cause and blame enabling any potential defence or mitigation to any subsequent proceedings to be identified*'.

It is not possible to produce a single system of investigation which ultimately resolves all the potential conflicts between the different rationales and needs for the investigation. However, the processes do need to be operated in a way which takes into account these conflicts and seeks to minimise them by appropriately involving different parts of the organisation, insurers and legal representatives. Particular attention needs to be given to identifying the information which is obtained for advice which is given in situations where legal privilege may apply and where the wider dissemination of that material might cause prejudice in future legal proceedings.

## *The main elements of investigations*

A3031     The following sections outline the main elements of an accident investigation process, and each stage of this process provides a framework within which more or less details and elaborate examination of relevant issues can be carried out depending on what is considered necessary or proportionate (see FIGURE 3). The process should not however be inflexible, and it may need to be adapted if for example there are parallel investigations taking place by enforcement authorities with a view to future prosecution.

Figure 3: Elements of the investigation process

It is possible to devise ad hoc arrangements for each investigation as it arises, but ideally (and essentially in large organisations where there is a steady occurrence of accidents) there should be defined procedures. Responsibilities should be designated, and there should be basic but clear criteria as to the types of occurrence which should be investigated more thoroughly than is necessary merely for the purposes of carrying out RIDDOR notifications. Thought needs to be given to how an investigation team is to be constituted, setting terms of reference, and determining what the reporting lines are during the process and not just for the delivery of the accident report. An agreed format for the contents of accident reports is desirable so that consistent and comprehensive results are produced. Those with responsibility for the investigation process may require training in some areas such as interviewing skills, accident causation principles and report writing.

An HSE investigation or claim for personal injury will raise liability issues and opportunities for internal or external legal advice should be built into the process at each stage.

It is important that the process does not exist in isolation of what is happening elsewhere in the organisation. There may need to be liaison not just with legal advisors, but with others who may be dealing with necessary actions following an accident such as HR managers, insurance managers and of course line managers responsible for the activity in which the accident has occurred who may need immediate guidance on whether (or how) to recommence or continue the activities.

## Collating evidence

**A3032** Every situation will be different, but typically the evidence will comprise some or all of the following:

*Conditions at the accident scene:*

- photographs and videos;
- sketches and plans;
- measurements;
- records of weather/environmental conditions;
- samples of substances;
- equipment or retained parts;
- condition of safety equipment/devices;
- first-aid records;
- information obtained from a reconstruction of the accident;
- lists of eye-witnesses;
- identities of others involved;
- notes of comments made immediately after the accident;
- internal memos and emails dealing with the accident and aftermath;
- statements taken by managers/others;
- other electronic records (e.g. time of emergency calls, or computerised process data).

*Safety documentation:*

- accident book entries;
- RIDDOR records;
- relevant risk assessments;
- method statements/safe systems of work;
- training materials;
- training records;
- equipment operating instructions;
- equipment logs/maintenance records;
- personnel records of those involved in the accident;

- reports on relevant previous incidents;
- relevant health and safety audits, management reports, consultants' advice;
- relevant emergency procedures.

The most difficult part of this stage of the process is likely to be obtaining statements promptly from witnesses, either because they are distressed or because they are unco-operative (which may be for a variety of reasons including concern that they might be criticised). It is nevertheless important to obtain witnesses' evidence as soon as possible after an accident. The accuracy of their information is reduced the more time elapses, and recall can be affected by *post-event enhancement* – the unconscious incorporation of ideas and perceptions based on discussing events with others in group setting and by *false confidence* – a hardening of a version of events from recalling and re-telling others or committing evidence to paper.

If witnesses are unco-operative or hostile it may serve no useful purpose to persevere in efforts to interview them. A period of delay can sometimes assist, in which it may be possible to resolve the underlying problems. Where the witness is an employee, ultimately the failure to co-operate with the employer may be treated as a disciplinary matter.

Another difficulty with interviewing witnesses can be that the HSE, police or other authorities may wish to restrict access to them – so that they can interview them first and take formal statements. There is no legal power to control witnesses in this way, but inspectors sometimes threaten employers with prosecution for obstructing them if they do not agree to postpone their interviews. It is therefore advisable to be open with the authorities about the interview process, and to offer inspectors or other officers an opportunity to attend interviews with witnesses about whom they have concerns.

| *Interviewing Witnesses* |  |
|---|---|
| When carrying out an investigation, it is always necessary to interview the people involved in the accident. The following procedures should be adopted. ||
| **Prepare for the interviews:** <br> • gather information (facts not hearsay) <br> • visit the accident scene <br> • allow adequate time <br> • minimise the risk of interruptions ||
| **The interview:** <br> • ask what happened <br> • do not interrupt; be a good listener <br> • do not ask leading questions <br> • do no make assumptions <br> • do not lead the interviewee; some people will tell you what they think you want to hear! <br> • be considerate, not authoritarian or judgemental <br> • ask key questions in order to tease out the facts concerning the accident <br> • emphasise that the purpose of the investigation is primarily to prevent a recurrence. It is not a whistleblowing, blame apportionment, fault-finding exercise. ||
| **Do:** <br> • put people at ease | **Don't:** <br> • interrogate |

**A3033** *Accident Reporting and Investigation*

| | |
|---|---|
| • listen | • interrupt |
| • be sympathetic | • get into an argument |
| • be firm | • point the finger |
| • be honest | • blame or find fault |
| • be tactful | • make promises |
| • be courteous | • describe what other witnesses have said |
| • be interested | • speculate about any facts or causes |

(Adapted from *Tolley's Workplace Accident Handbook* (2003) ISBN 0 7545 2023 4)

## Analysis of causes

**A3033** This stage of the process is concerned with two inter-related questions: 'what happened?' and 'why did it happen?' These apparently simple questions can be subjected to highly sophisticated analytical techniques. (See for example 'Root causes analysis: Literature review' – HSE Contract Research Report 325/2001 at website: www.hse.gov.uk/researcher/ crr_pdf/2001/crr01325.pdf, and 'Events and causal factors analysis' at website: www.tis.eh.doe.gov/analysis/trac/14/trac14.htm.) The following paragraphs summarise a basic approach, but those charged with accident investigations should have a fuller understanding of accident causation theory and knowledge of the background to *human factors* (see E17068).

Having collated the information about nature and circumstances of the accident the investigation team needs to decide what issues it needs to analyse. A useful starting point is often the organisation's standard template for accident investigation reports (see A3034 below) but not all the headings will necessarily be relevant in each case, and a degree of flexibility is needed so that the investigation does not become diverted by issues of low or no relevance.

Examples of the questions, which are typically explored, are:

- What was the chain of evens leading up to the accident?
- What planning of the activity had taken place?
- Were the risks known?
- Was the planning deficient?
- Were defective premises, equipment or maintenance a factor?
- Were the correct materials being used?
- Were those involved properly trained?
- Was the training adequate?
- Did people depart from instructions/procedures?
- Was supervision properly undertaken?
- Were the instructions/procedures adequate?
- Did organisational factors of the work play a role?, e.g. excessive hours, rushing to meet a deadline or target.
- Was there a failure to learn lessons from previous incidents or to follow-up previous warnings or complaints?
- Did the emergency procedures and provision of first-aid operate effectively?

It is not unusual for these or other questions to require additional evidence or for further queries to be raised with witnesses, and that the collating analysis stages may have to overlap.

Answers to these questions are used to inform the next stage of analysis which involves reaching an understanding as to the cause or causes of the accident, and determining whether the risk control measures that were in place are in need of improvement.

Causes of accidents can be complex and multi-factorial. It is often the unusual combination of a series of events which leads to injury or damage in what were hitherto routine workplace activities. Each event may be traced back to factors that preceded it, and so it has become common to analysis accidents in terms of three types of 'causes'.

1. 'Immediate' cause: an unsafe act (or omission) or an unsafe condition which leads directly to the injury or damage. (An example of an unsafe act is improper stacking of a load; an example of an unsafe condition is a machine with a missing guard). Occasionally the immediate cause of an accident may be a 'violation': a deliberate or reckless departure from correct procedures by an individual or group.

2. 'Underlying' cause: these precede the time of the accident and consist of management control factors (e.g. inadequate risk assessment or the absence of supervision), job factors (e.g. lack of competence or insufficient staffing), or environmental factors (such as working in harsh conditions).

3. 'Root' causes: a failing from which all the other causes have grown – usually a deficiency in the planning and organisation of management of health and safety.

Caution is needed in using this approach for a number of reasons. First, it may be difficult to separate and distinguish between several different causes as neatly in practice as it is in theory. Second, this approach does not give weight to the *gravity* of causal factors, which may be relevant in the wider employment or liability context. Third, and most importantly, there is a large element of subjectivity in determining what are the relevant underlying, and particular root, causes. The causal analysis therefore needs to be firmly based on the factual analysis, and the reasoning process explained at each stage by reference to the facts as they have found during the first stage of the analysis.

## Preparing the report

**A3034**  There are no hard and fast rules about the structure or content of accident investigation reports. The essential features however are:

- description of circumstances of the accident;
- consideration of control measures;
- conclusions as to causes;
- recommendations for remedial action.

A basic template which can be used for reports is as follows.

**Outline for Accident Investigation Report**

| Accident Investigation Report |
|---|
| 1. Time and place of accident |
| 2. Details of person(s) injured |
| 3. Description of events leading to accident and injury |
| 4. Working conditions/environment |

| | |
|---|---|
| 5. | Relevant equipment/substances |
| 6. | Planning and risk assessment |
| 7. | Adequacy of training |
| 8. | Instruction and supervision |
| 9. | Condition and use of PPE |
| 10. | Existence of similar risks elsewhere |
| 11. | Conclusions and cause(s) of accident |
| 12. | Recommendations for action |

Names of investigation team:
..............................................................................

Date of report: ......................................

Before finalising the report it is sometimes desirable to seek comments or corrections from those involved in the accident or connected with the background. Partly this is out of courtesy where individuals might perceive the conclusions as implied criticism, but it also provides a final opportunity to eliminate factual or analytical errors. It is also sensible that proposed recommendations for action are at least canvassed with the relevant managers who may become responsible for implementing them to ensure that they are not impractical or unrealistic. A draft may also be provided to the organisations' legal advisors to consider how it may affect liability issues and whether the conclusions are consistent with the findings of privileged investigations that have been, or are still being, carried out.

When the report is finalised there needs to be a clear and timely management plan of action to implement the recommendations.

The document itself will be disclosable in any subsequent legal proceedings and a copy may be requested by enforcing authorities. Delay or incompleteness in giving effect to the recommendations may expose the organisation to subsequent criticism.

# Appendix A
# List of Reportable Dangerous Occurrences
A3035
## PART I(A) – GENERAL
### Lifting machinery, etc.
1.  The collapse of, the overturning of, or the failure of any load-bearing part of any—
    (a)  lift or hoist;
    (b)  crane or derrick;
    (c)  mobile powered access platform;
    (d)  access cradle or window-cleaning cradle;
    (e)  excavator;
    (f)  pile-driving frame or rig having an overall height, when operating, of more than 7 metres; or
    (g)  fork lift truck.

### Pressure systems
2.  The failure of any closed vessel (including a boiler or boiler tube) or of any associated pipework, in which the internal pressure was above or below atmospheric pressure, where the failure has the potential to cause the death of any person.

### Freight containers
3.  (a)  The failure of any freight container in any of its load-bearing parts while it is being raised, lowered or suspended.
    (b)  In this paragraph, 'freight container' means a container as defined in regulation 2(1) of the Freight Containers (Safety Convention) Regulations 1984.

### Overhead electric lines
4.  Any unintentional incident in which plant or equipment either—
    (a)  comes into contact with an uninsulated overhead electric line in which the voltage exceeds 200 volts; or
    (b)  causes an electrical discharge from such an electric line by coming into close proximity to it.

### Electrical short circuit
5.  Electrical short circuit or overload attended by fire or explosion which results in the stoppage of the plant involved for more than 24 hours or which has the potential to cause the death of any person.

### Explosives
6.  (1)  Any of the following incidents involving explosives—
        (a)  the unintentional explosion or ignition of explosives other than one—

(i) caused by the unintentional discharge of a weapon where, apart from that unintentional discharge, the weapon and explosives functioned as they were designed to do; or

(ii) where a fail-safe device or safe system of work functioned so as to prevent any person from being injured in consequence of the explosion or ignition;

(b) a misfire (other than one at a mine or quarry or inside a well or one involving a weapon) except where a fail-safe device or safe system of work functioned so as to prevent any person from being endangered in consequence of the misfire;

(c) the failure of the shots in any demolition operation to cause the intended extent of collapse or direction of fall of a building or structure;

(d) the projection of material (other than at a quarry) beyond the boundary of the site on which the explosives are being used or beyond the danger zone in circumstances such that any person was or might have been injured thereby;

(e) any injury to a person (other than at a mine or quarry or one otherwise reportable under these Regulations) involving first-aid or medical treatment resulting from the explosion or discharge of any explosives or detonator.

(2) In this paragraph 'explosives' means any explosive of a type which would, were it being transported, be assigned to class 1 within the meaning of the Classification and Labelling of Explosives Regulations 1983 and 'danger zone' means the area from which persons have been excluded or forbidden to enter to avoid being endangered by any explosion or ignition of explosives.

## Biological agents

7. Any accident or incident which resulted or could have resulted in the release or escape of a biological agent likely to cause severe human infection or illness.

## Malfunction of radiation generators, etc.

8. (1) Any incident in which—

(a) the malfunction of a radiation generator or its ancillary equipment used in fixed or mobile industrial radiography, the irradiation of food or the processing of products by irradiation, which causes it to fail to de-energise at the end of the intended exposure period; or

(b) the malfunction of equipment used in fixed or mobile industrial radiography or gamma irradiation causes a radioactive source to fail to return to its safe position by the normal means at the end of the intended exposure period.

(2) In this paragraph 'radiation generator' means any electrical equipment emitting ionising radiation and containing components operating at a potential difference of more than 5 kV.

## Breathing apparatus

9. (1) Any incident in which breathing apparatus malfunctions—

(a) while in use, or

*Accident Reporting and Investigation*     A3035

      (b)    during testing immediately prior to use in such a way that had the malfunction occurred while the apparatus was in use it would have posed a danger to the health or safety of the user.

(2)    This paragraph shall not apply to breathing apparatus while it is being—

      (a)    used in a mine; or

      (b)    maintained or tested as part of a routine maintenance procedure.

## Diving operations

10.    Any of the following incidents in relation to a diving project—

      (a)    the failure or the endangering of—

           (i)    any lifting equipment associated with the diving operation, or

           (ii)    life support equipment, including control panels, hoses and breathing apparatus,

      which puts a diver at risk;

      (b)    any damage to, or endangering of, the dive platform, or any failure of the dive platform to remain on station, which puts a diver at risk;

      (c)    the trapping of a diver;

      (d)    any explosion in the vicinity of a diver; or

      (e)    any uncontrolled ascent or any omitted decompression which puts a diver at risk.

## Collapse of scaffolding

11.    The complete or partial collapse of—

      (a)    any scaffold which is—

           (i)    more than 5 metres in height which results in a substantial part of the scaffold falling or overturning; or

           (ii)    erected over or adjacent to water in circumstances such that there would be a risk of drowning to a person falling from the scaffold into the water; or

      (b)    the suspension arrangements (including any outrigger) of any slung or suspended scaffold which causes a working platform or cradle to fall.

## Train collisions

12.    Any unintended collision of a train with any other train or vehicle, other than one reportable under Part IV of this Schedule, which caused, or might have caused, the death of, or major injury to, any person.

## Wells

13.    Any of the following incidents in relation to a well (other than a well sunk for the purpose of the abstraction of water)—

      (a)    a blow-out (that is to say an uncontrolled flow of well-fluids from a well);

      (b)    the coming into operation of a blow-out prevention or diversion system to control a flow from a well where normal control procedures fail;

(c) the detection of hydrogen sulphide in the course of operations at a well or in samples of well-fluids from a well where the presence of hydrogen sulphide in the reservoir being drawn on by the well was not anticipated by the responsible person before that detection;

(d) the taking of precautionary measures additional to any contained in the original drilling programme following failure to maintain a planned minimum separation distance between wells drilled from a particular installation; or

(e) the mechanical failure of any safety critical element of a well (and for this purpose the safety critical element of a well is any part of a well whose failure would cause or contribute to, or whose purpose is to prevent or limit the effect of, the unintentional release of fluids from a well or a reservoir being drawn on by a well).

## Pipelines or pipeline works

14. The following incidents in respect of a pipeline or pipeline works—

(a) the uncontrolled or accidental escape of anything from, or inrush of anything into, a pipeline which has the potential to cause the death of, major injury or damage to the health of any person or which results in the pipeline being shut down for more than 24 hours;

(b) the unintentional ignition of anything in a pipeline or of anything which, immediately before it was ignited, was in a pipeline;

(c) any damage to any part of a pipeline which has the potential to cause the death of, major injury or damage to the health of any person or which results in the pipeline being shut down for more than 24 hours;

(d) any substantial and unintentional change in the position of a pipeline requiring immediate attention to safeguard the integrity or safety of a pipeline;

(e) any unintentional change in the subsoil or seabed in the vicinity of a pipeline which has the potential to affect the integrity or safety of a pipeline;

(f) any failure of any pipeline isolation device, equipment or system which has the potential to cause the death of, major injury or damage to the health of any person or which results in the pipeline being shut down for more than 24 hours; or

(g) any failure of equipment involved with pipeline works which has the potential to cause the death of, major injury or damage to the health of any person.

## Fairground equipment

15. The following incidents on fairground equipment in use or under test—

(a) the failure of any load-bearing part;

(b) the failure of any part designed to support or restrain passengers; or

(c) the derailment or the unintended collision of cars or trains.

## Carriage of dangerous substances by road

16. (1) Any incident involving a road tanker or tank container used for the carriage of dangerous goods in which—

(a) the road tanker or vehicle carrying the tank container overturns (including turning onto its side);

(b) the tank carrying the dangerous goods is seriously damaged;

(c) there is an uncontrolled release or escape of the dangerous goods being carried; or

(d) there is a fire involving the dangerous goods being carried.

17. (1) Any incident involving a vehicle used for the carriage of dangerous goods, other than a vehicle to which paragraph 16 applies, where there is –

(a) an uncontrolled release or escape of the dangerous goods being carried in such a quantity as to have the potential to cause the death of, or major injury to, any person; or

(b) a fire which involves the dangerous substance being carried.

17A. In paragraphs 16 and 17 above—

(a) 'road tanker' and 'tank container' have the same meanings as in Regulation 2(1) of the Carriage of Dangerous Goods (Classifications, Packaging and Labelling) and Use of Use of Transportable Pressure Receptacles Regulations 1996 ('the 1996 Regulations');

(b) 'carriage' has the same meaning as in Regulation 2(1) of the Carriage of Dangerous Goods by Road Regulations 1996; and

(c) 'dangerous goods' means any goods which fall within the definition of 'dangerous goods' in Regulation 2(1) of the 1996 Regulations, other than –

(i) explosives, or

(ii) radioactive material (other than that which is being carried in accordance with the condition specified in Schedules 1 to 4 of marginal 2704 to ADR),

and in this sub-paragraph 'ADR' has the meaning assigned to it by Regulation 2(1) of the 1996 Regulations.

## PART I(B) – DANGEROUS OCCURRENCES WHICH ARE REPORTABLE EXCEPT IN RELATION TO OFFSHORE WORKPLACES

### Collapse of building or structure

18. Any unintended collapse or partial collapse of—

(a) any building or structure (whether above or below ground) under construction, reconstruction, alteration or demolition which involves a fall of more than 5 tonnes of material;

(b) any floor or wall of any building (whether above or below ground) used as a place of work; or

(c) any false-work.

### Explosion or fire

19. An explosion or fire occurring in any plant or premises which results in the stoppage of that plant or as the case may be the suspension of normal work in those premises for more than 24 hours, where the explosion or fire was due to the ignition of any material.

### Escape of flammable substances

20. (1) The sudden, uncontrolled release—

(a) inside a building —
  (i) of 100 kilograms or more of a flammable liquid,
  (ii) of 10 kilograms or more of a flammable liquid at a temperature above its normal boiling point, or
  (iii) of 10 kilograms or more of a flammable gas; or
(b) in the open air, of 500 kilograms or more of any of the substances referred to in sub-paragraph (a) above.

(2) In this paragraph, 'flammable liquid' and 'flammable gas' mean respectively a liquid and a gas so classified in accordance with regulation 5(2), (3) or (5) of the Chemicals (Hazard Information and Packaging for Supply) Regulations 1994 (now revoked by SI 2002 No 1689).

## Escape of substances

21. The accidental release or escape of any substance in a quantity sufficient to cause the death, major injury or any other damage to the health of any person.

# PART II – DANGEROUS OCCURRENCES WHICH ARE REPORTABLE IN RELATION TO MINES

## Fire or ignition of gas

22. The ignition, below ground, of any gas (other than gas in a safety lamp) or of any dust.

23. The accidental ignition of any gas in part of a firedamp drainage system on the surface or in an exhauster house.

24. The outbreak of any fire below ground.

25. An incident where any person in consequence of any smoke or any other indication that a fire may have broken out below ground has been caused to leave any place pursuant to either Regulation 11(1) of the Coal and Other Mines (Fire and Rescue) Regulations 1956 or section 79 of the Mines and Quarries Act 1954.

26. The outbreak of any fire on the surface which endangers the operation of any winding or haulage apparatus installed at a shaft or unwalkable outlet or of any mechanically operated apparatus for producing ventilation below ground.

## Escape of gas

27. Any violent outburst of gas together with coal or other solid matter into the mine workings except when such outburst is caused intentionally.

## Failure of plant or equipment

28. The breakage of any rope, chain, coupling, balance rope, guide rope, suspension gear or other gear used for or in connection with the carrying of persons through any shaft or staple shaft.

29. The breakage or unintentional uncoupling of any rope, chain, coupling, rope tensioning system or other gear used for or in connection with the transport of persons below ground, or breakage of any belt, rope or other gear used for or in connection with a belt conveyor designated by the mine manager as a man-riding conveyor.

30. An incident where any conveyance being used for the carriage of persons is overwound; or any conveyance not being so used is overwound and becomes detached from its winding rope; or any conveyance operated by means of the friction of a rope on a winding sheave is brought to rest by the apparatus provided in the headframe of the shaft or in the part of the shaft below the lowest landing for the time being in use, being apparatus provided for bringing the conveyance to rest in the event of its being overwound.

31. The stoppage of any ventilating apparatus (other than an auxiliary fan) which causes a substantial reduction in ventilation of the mine lasting for a period exceeding 30 minutes, except when for the purpose of planned maintenance.

32. The collapse of any headframe, winding engine house, fan house or storage bunker.

## Breathing apparatus

33. At any mine an incident where—
    (a) breathing apparatus or a smoke helmet or other apparatus serving the same purpose or a self-rescuer, while being used, fails to function safely or develops a defect likely to affect its safe working; or
    (b) immediately after using and arising out of the use of breathing apparatus or a smoke helmet or other apparatus serving the same purpose or a self-rescuer, any person receives first-aid or medical treatment by reason of his unfitness or suspected unfitness at the mine.

## Injury by explosion of blasting material etc.

34. An incident in which any person suffers an injury (not being a major injury or one reportable under regulation 3(2)) which results from an explosion or discharge of any blasting material or device within the meaning of section 69(4) of the Mines and Quarries Act 1954 for which he receives first-aid or medical treatment at the mine.

## Use of emergency escape apparatus

35. An incident where any apparatus is used (other than for the purpose of training and practice) which has been provided at the mine in accordance with regulation 4 of the Mines (Safety of Exit) Regulations 1988 or where persons leave the mine when apparatus and equipment normally used by persons to leave the mine is unavailable.

## Inrush of gas or water

36. Any inrush of noxious or flammable gas from old workings.
37. Any inrush of water or material which flows when wet from any source.

## Insecure tip

38. Any movement of material or any fire or any other event which indicates that a tip to which Part I of the Mines and Quarries (Tips) Act 1969 applies, is or is likely to become insecure.

## Locomotives

39. Any incident where an underground locomotive when not used for testing purposes is brought to rest by means other than its safety circuit protective devices or normal service brakes.

## Falls of ground

40. Any fall of ground, not being part of the normal operations at a mine, which results from a failure of an underground support system and prevents persons travelling through the area affected by the fall or which otherwise exposes them to danger.

# PART III – DANGEROUS OCCURRENCES WHICH ARE REPORTABLE IN RELATION TO QUARRIES

## Collapse of storage bunkers

41. The collapse of any storage bunker.

## Sinking of craft

42. The sinking of any water-borne craft or hovercraft.

## Injuries

43. (1) An incident in which any person suffers an injury (not otherwise reportable under these Regulations) which results from an explosion or from the discharge of any explosives for which he receives first-aid or medical treatment at the quarry.

    (2) In this paragraph, 'explosives' has the same meaning as in regulation 2(1) of the Quarries Regulations 1999.

## Projection of substances outside quarry

44. Any incident in which any substance is ascertained to have been projected beyond a quarry boundary as a result of blasting operations in circumstances in which any person was or might have been endangered.

## Misfires

45. Any misfire, as defined by regulation 2(1) of the Quarries Regulations 1999.

## Insecure tips

46. Any event (including any movement of material or any fire) which indicates that a tip, to which the Quarries Regulations 1999 apply, is or is likely to become insecure.

## Movement of slopes or faces

47. Any movement or failure of an excavated slope or face which—

    (a) has the potential to cause the death of any person; or

    (b) adversely affects any building, contiguous land, transport system, footpath, public utility or service, watercourse, reservoir or area of public access.

## Explosions or fires in vehicles or plant

48. (1) Any explosion or fire occurring in any large vehicle or mobile plant which results in the stoppage of that vehicle or plant for more than 24 hours and which affects—

    (a) any place where persons normally work; or

(b) the route of egress from such a place.

(2) In this paragraph, 'large vehicle or mobile plant' means—

(a) a dump truck having a load capacity of at least 50 tonnes; or

(b) an excavator having a bucket capacity of at least 5 cubic metres.

## PART IV – DANGEROUS OCCURRENCES WHICH ARE REPORTABLE IN RESPECT OF RELEVANT TRANSPORT SYSTEMS

### Accidents to passenger trains

49. Any collision in which a passenger train collides with another train.

50. Any case where a passenger train or any part of such a train unintentionally leaves the rails.

### Accidents not involving passenger trains

51. Any collision between trains, other than one between a passenger train and another train, on a running line where any train sustains damage as a result of the collision, and any such collision in a siding which results in a running line being obstructed.

52. Any derailment, of a train other than a passenger train, on a running line, except a derailment which occurs during shunting operations and does not obstruct any other running line.

53. Any derailment, of a train other than a passenger train, in a siding which results in a running line being obstructed.

### Accidents involving any kind of train

54. Any case of a train striking a buffer stop, other than in a siding, where damage is caused to the train.

55. Any case of a train striking any cattle or horse, whether or not damage is caused to the train, or striking any other animal if, in consequence, damage (including damage to the windows of the driver's cab but excluding other damage consisting solely in the breakage of glass) is caused to the train necessitating immediate temporary or permanent repair.

56. Any case of a train on a running line striking or being struck by any object which causes damage (including damage to the windows of the driver's cab but excluding other damage consisting solely in the breakage of glass) necessitating immediate temporary or permanent repair or which might have been liable to derail the train.

57. Any case of a train, other than one on a railway, striking or being struck by a road vehicle.

58. Any case of a passenger train, or any other train not fitted with continuous self-applying brakes, becoming unintentionally divided.

59. (1) Any of the following classes of accident which occurs or is discovered whilst the train is on a running line—

(a) the failure of an axle;

(b) the failure of a wheel or tyre, including a tyre loose on its wheel;

(c) the failure of a rope or the fastenings thereof or of the winding plant or equipment involved in working an incline;

(d) any fire, severe electrical arcing or fusing in or on any part of a passenger train or a train carrying dangerous goods;

(e) in the case of any train other than a passenger train, any severe electrical arcing or fusing, or any fire which was extinguished by a fire-fighting service; or

(f) any other failure of any part of a train which is likely to cause an accident to that or any other train or to kill or injure any person.

(2) In this paragraph 'dangerous goods' have the meaning assigned to it in Regulation 2(1) of the Carriage of Dangerous Goods (Classification, Packaging and Labelling) and Use of Transportable Pressure Receptacles Regulations 1996 (SI 1996 No 2092).

## Accidents and incidents at level crossings

60. Any case of a train striking a road vehicle or gate at a level crossing.

61. Any case of a train running onto a level crossing when not authorised to do so.

62. A failure of the equipment at a level crossing which could endanger users of the road or path crossing the railway.

## Accidents involving the permanent way and other works on or connected with a relevant transport system

63. The failure of a rail in a running line or of a rack rail, which results in—

    (a) a complete fracture of the rail through its cross-section; or

    (b) in a piece becoming detached from the rail which necessitates an immediate stoppage of traffic or the immediate imposition of a speed restriction lower than that currently in force.

64. A buckle of a running line which necessitates an immediate stoppage of traffic or the immediate imposition of a speed restriction lower than that currently in force.

65. Any case of an aircraft or a vehicle of any kind landing on, running onto or coming to rest foul of the line, or damaging the line, which causes damage which obstructs the line or which damages any railway equipment at a level crossing.

66. The runaway of an escalator, lift or passenger conveyor.

67. Any fire or severe arcing or fusing which seriously affects the functioning of signalling equipment.

68. Any fire affecting the permanent way or works of a relevant transport system which necessitates the suspension of services over any line, or the closure of any part of a station or signal box or other premises, for a period—

    (a) in the case of a fire affecting any part of a relevant transport system below ground, of more than 30 minutes, and

    (b) in any other case, of more than 1 hour.

69. Any other fire which causes damage which has the potential to affect the running of a relevant transport system.

## Accidents involving failure of the works on or connected with a relevant transport system

70. (1) The following classes of accident where they are likely either to cause an accident to a train or to endanger any person—

(a) the failure of a tunnel, bridge, viaduct, culvert, station, or other structure or any part thereof including the fixed electrical equipment of an electrified relevant transport system;

(b) any failure in the signalling system which endangers or which has the potential to endanger the safe passage of trains other than a failure of a traffic light controlling the movement of vehicles on a road;

(c) a slip of a cutting or of an embankment;

(d) flooding of the permanent way;

(e) the striking of a bridge by a vessel or by a road vehicle or its load; or

(f) the failure of any other portion of the permanent way or works not specified above.

## Incidents of serious congestion

71. Any case where planned procedures or arrangements have been activated in order to control risks arising from an incident of undue passenger congestion at a station unless that congestion has been relieved within a period of time allowed for by those procedures or arrangements.

## Incidents of signals passed without authority

72. (1) Any case where a train, travelling on a running line or entering a running line from a siding, passes without authority a signal displaying a stop aspect unless—

(a) the stop aspect was not displayed in sufficient time for the driver to stop safely at the signal.

# PART V – DANGEROUS OCCURRENCES WHICH ARE REPORTABLE IN RESPECT OF AN OFFSHORE WORKPLACE

## Release of petroleum hydrocarbon

73. Any unintentional release of petroleum hydrocarbon on or from an offshore installation which—

(a) results in –

(i) a fire or explosion; or

(ii) the taking of action to prevent or limit the consequences of a potential fire or explosion; or

(b) has the potential to cause death or major injury to any person.

## Fire or explosion

74. Any fire or explosion at an offshore installation, other than one to which paragraph 73 above applies, which results in the stoppage of plant or the suspension of normal work.

## Release or escape of dangerous substances

75. The uncontrolled or unintentional release or escape of any substance (other than petroleum hydrocarbon) on or from an offshore installation which has the potential to cause the death of, major injury to or damage to the health of any person.

## Collapses

76. Any unintended collapse of any offshore installation or any unintended collapse of any part thereof or any plant thereon which jeopardises the overall structural integrity of the installation.

## Dangerous occurrences

77. Any of the following occurrences having the potential to cause death or major injury—

   (a) the failure of equipment required to maintain a floating offshore installation on station;

   (b) the dropping of any object on an offshore installation or on an attendant vessel or into the water adjacent to an installation or vessel; or

   (c) damage to or on an offshore installation caused by adverse weather conditions.

## Collisions

78. Any collision between a vessel or aircraft and an offshore installation which results in damage to the installation, the vessel or the aircraft.

79. Any occurrence with the potential for a collision between a vessel and an offshore installation where, had a collision occurred, it would have been liable to jeopardise the overall structural integrity of the offshore installation.

## Subsidence or collapse of seabed

80. Any subsidence or local collapse of the seabed likely to affect the foundations of an offshore installation or the overall structural integrity of an offshore installation.

## Loss of stability or buoyancy

81. Any incident involving loss of stability or buoyancy of a floating offshore installation.

## Evacuation

82. Any evacuation (other than one arising out of an incident reportable under any other provision of these Regulations) of an offshore installation, in whole or part, in the interests of safety.

## Falls into water

83. Any case of a person falling more than 2 metres into water (unless the fall results in death or injury required to be reported under sub-paragraphs (a)–(d) of regulation 3(1)).

[*The Reporting of Injuries, Diseases and Dangerous Occurrences Regulations 1995, Reg 2(1), Sch 2*].

# Appendix B
# List of Reportable Diseases

## PART I – OCCUPATIONAL DISEASES

| Column 1 | Column 2 |
|---|---|
| Diseases | Activities |
| *Conditions due to physical agents and the physical demands of work* | |
| 1. Inflammation, ulceration or malignant disease of the skin due to ionising radiation. | |
| 2. Malignant disease of the bones due to ionising radiation. | } Work with ionising radiation. |
| 3. Blood dyscrasia due to ionising radiation. | |
| 4. Cataract due to electromagnetic radiation. | Work involving exposure to electromagnetic radiation (including radiant heat). |
| 5. Decompression illness. | |
| 6. Barotrauma resulting in lung or other organ damage. | } Work involving breathing gases at increased pressure (including diving). |
| 7. Dysbaric osteonecrosis. | |
| 8. Cramp of the hand or forearm due to repetitive movements. | Work involving prolonged periods of handwriting, typing or other repetitive movements of the fingers, hand or arm. |
| 9. Subcutaneous cellulitis of the hand (beat hand). | Physically demanding work causing severe or prolonged friction or pressure on the hand. |
| 10. Bursitis or subcutaneous cellulitis arising at or about the knee due to severe or prolonged external friction or pressure at or about the knee (beat knee). | Physically demanding work causing severe or prolonged friction or pressure at or about the knee. |
| 11. Bursitis or subcutaneous cellulitis arising at or about the elbow due to severe or prolonged external friction or pressure at or about the elbow (beat elbow). | Physically demanding work causing severe or prolonged friction or pressure at or about the elbow. |
| 12. Traumatic inflammation of the tendons of the hand or forearm or of the associated tendon sheaths. | Physically demanding work, frequent or repeated movements, constrained postures or extremes of extension or flexion of the hand or wrist. |
| 13. Carpal tunnel syndrome. | Work involving the use of hand-held vibrating tools. |
| 14. Hand-arm vibration syndrome. | Work involving: |

|   |   |
|---|---|
|   | (a) the use of chain saws, brush cutters or hand-held or hand-fed circular saws in forestry or wood-working; |
|   | (b) the use of hand-held rotary tools in grinding material or in sanding or polishing metal; |
|   | (c) the holding of material being ground or metal being sanded or polished by rotary tools; |
|   | (d) the use of hand-held percussive metal-working tools or the holding of metal being worked upon by percussive tools in connection with riveting, caulking, chipping, hammering, fettling or swaging; |
|   | (e) the use of hand-held powered percussive drills or hand-held powered percussive hammers in mining, quarrying or demolition, or on roads or footpaths (including road construction); or |
|   | (f) the holding of material being worked upon by pounding machines in shoe manufacture. |
| *Infections due to biological agents* |   |
| 15. Anthrax. | (a) Work involving handling infected animals, their products or packaging containing infected material; or |
|   | (b) work on infected sites. |
| 16. Brucellosis. | Work involving contact with: |
|   | (a) animals or their carcasses (including any parts thereof) infected by brucella or the untreated products of same; or |
|   | (b) laboratory specimens or vaccines of or containing brucella. |
| 17. (a) Avian chlamydiosis. | Work involving contact with birds infected with chlamydia psittaci, or the remains or untreated products of such birds. |
| (b) Ovine chlamydiosis. | Work involving contact with sheep infected with chlamydia psittaci or the remains or untreated products of such sheep. |
| 18. Hepatitis. | Work involving contact with: |
|   | (a) human blood or human blood products; or |

*Accident Reporting and Investigation* **A3036**

| | | | |
|---|---|---|---|
| | | (b) | any source of viral hepatitis. |
| 19. | Legionellosis. | | Work on or near cooling systems which are located in the workplace and use water; or work on hot water service systems located in the workplace which are likely to be a source of contamination |
| 20. | Leptospirosis. | (a) | Work in places which are or are liable to be infested by rats, fieldmice, voles or other small mammals; |
| | | (b) | work at dog kennels or involving the care or handling of dogs; or |
| | | (c) | work involving contact with bovine animals or their meat products or pigs or their meat products. |
| 21. | Lyme disease. | | Work involving exposure to ticks (including in particular work by forestry workers, rangers, dairy farmers, game keepers and other persons engaged in countryside management). |
| 22. | Q fever. | | Work involving contact with animals, their remains or their untreated products. |
| 23. | Rabies. | | Work involving handling or contact with infected animals. |
| 24. | Streptococcus suis. | | Work involving contact with pigs infected with streptococcus suis, or with the carcasses, products or residues of pigs so affected. |
| 25. | Tetanus. | | Work involving contact with soil likely to be contaminated by animals. |
| 26. | Tuberculosis. | | Work with persons, animals, human or animal remains or any other material which might be a source of infection. |
| 27. | Any infection reliably attributable to the performance of the work specified in the entry opposite hereto. | | Work with micro-organisms; work with live or dead human beings in the course of providing any treatment or service or in conducting any investigation involving exposure to blood or body fluids; work with animals or any potentially infected material derived from any of the above. |

*Conditions due to substances*

| | | | |
|---|---|---|---|
| 28. | Poisonings by any of the following: | | Any activity. |
| | (a) | acrylamide monomer; | |
| | (b) | arsenic or one of its compounds; | |

|     |     |     |
| --- | --- | --- |
| | (c) | benzene or a homologue of benzene; |
| | (d) | beryllium or one of its compounds; |
| | (e) | cadmium or one of its compounds; |
| | (f) | carbon disulphide; |
| | (g) | diethylene dioxide (dioxan); |
| | (h) | ethylene oxide; |
| | (i) | lead or one of its compounds; |
| | (j) | manganese or one of its compounds; |
| | (k) | mercury or one of its compounds; |
| | (l) | methyl bromide; |
| | (m) | nitrochlorobenzene, or a nitro- or amino- or chloro-derivative of benzene or of a homologue of benzene; |
| | (n) | oxides of nitrogen; |
| | (o) | phosphorus or one of its compounds. |

29. Cancer of a bronchus or lung.

(a) Work in or about a building where nickel is produced by decomposition of a gaseous nickel compound or where any industrial process which is ancillary or incidental to that process is carried on; or

(b) work involving exposure to bis(chloromethyl) ether or any electrolytic chromium processes (excluding passivation) which involve hexavalent chromium compounds, chromate production or zinc chromate pigment manufacture.

30. Primary carcinoma of the lung where there is accompanying evidence of silicosis.

Any occupation in:

(a) glass manufacture;

(b) sandstone tunnelling or quarrying;

(c) the pottery industry;

(d) metal ore mining;

(e) slate quarrying or slate production;

|  |  |  |  |
|---|---|---|---|
|  |  | (f) | clay mining; |
|  |  | (g) | the use of siliceous materials as abrasives; |
|  |  | (h) | foundry work; |
|  |  | (i) | granite tunnelling or quarrying; or |
|  |  | (j) | stone cutting or masonry. |
| 31. | Cancer of the urinary tract. | \multicolumn{2}{l|}{1. Work involving exposure to any of the following substances:} |
|  |  | (a) | beta-naphthylamine or methylene-bis-orthochloroaniline; |
|  |  | (b) | diphenyl substituted by at least one nitro or primary amino group or by at least one nitro and primary amino group (including benzidine); |
|  |  | (c) | any of the substances mentioned in sub-paragraph (b) above if further ring substituted by halogeno, methyl or methoxy groups, but not by other groups; or |
|  |  | (d) | the salts of any of the substances mentioned in sub-paragraphs (a) to (c) above. |
|  |  | \multicolumn{2}{l|}{2. The manufacture of auramine or magenta.} |
| 32. | Bladder cancer. | \multicolumn{2}{l|}{Work involving exposure to aluminium smelting using the Soderberg process.} |
| 33. | Angiosarcoma of the liver. | (a) | Work in or about machinery or apparatus used for the polymerisation of vinyl chloride monomer, a process which, for the purposes of this sub-paragraph, comprises all operations up to and including the drying of the slurry produced by the polymerisation and the packaging of the dried product; or |
|  |  | (b) | work in a building or structure in which any part of the process referred to in the foregoing sub-paragraph takes place. |
| 34. | Peripheral neuropathy. | \multicolumn{2}{l|}{Work involving the use or handling of or exposure to the fumes of or vapour containing n-hexane or methyl n-butyl ketone.} |
| 35. | Chrome ulceration of: | \multicolumn{2}{l|}{Work involving exposure to chromic acid or to any other chromium compound.} |

|   |     |                               |                                                                                                                                                                                                                                                                                                                                                           |
|---|-----|-------------------------------|-----|
|   | (a) | the nose or throat; or        |     |
|   | (b) | the skin of the hands or forearm. |     |
| 36. | Folliculitis. | | |
| 37. | Acne. | | } Work involving exposure to mineral oil, tar, pitch or arsenic. |
| 38. | Skin cancer. | | |
| 39. | Pneumoconiosis (excluding asbestosis). | | |

1.(a) The mining, quarrying or working of silica rock or the working of dried quartzose sand, any dry deposit or residue of silica or any dry admixture containing such materials (including any activity in which any of the aforesaid operations are carried out incidentally to the mining or quarrying of other minerals or to the manufacture of articles containing crushed or ground silica rock); or

(b) the handling of any of the materials specified in the foregoing sub-paragraph in or incidentally to any of the operations mentioned therein or substantial exposure to the dust arising from such operations.

2. The breaking, crushing or grinding of flint, the working or handling of broken, crushed or ground flint or materials containing such flint or substantial exposure to the dust arising from any of such operations.

3. Sand blasting by means of compressed air with the use of quartzose sand or crushed silica rock or flint or substantial exposure to the dust arising from such sand blasting.

4. Work in a foundry or the performance of, or substantial exposure to the dust arising from, any of the following operations:

(a) the freeing of steel castings from adherent siliceous substance or;

(b) the freeing of metal castings from adherent siliceous substance:

(i) by blasting with an abrasive propelled by compressed air, steam or a wheel, or

(ii) by the use of power-driven tools.

|     |              |     |                                                                                                                                                                                                                         |
| --- | ------------ | --- | ----------------------------------------------------------------------------------------------------------------------------------------------------------------------------------------------------------------------- |
|     |              |     | 5. The manufacture of china or earthenware (including sanitary earthenware, electrical earthenware and earthenware tiles) and any activity involving substantial exposure to the dust arising therefrom.                |
|     |              |     | 6. The grinding of mineral graphite or substantial exposure to the dust arising from such grinding.                                                                                                                     |
|     |              |     | 7. The dressing of granite or any igneous rock by masons, the crushing of such materials or substantial exposure to the dust arising from such operations.                                                              |
|     |              |     | 8. The use or preparation for use of an abrasive wheel or substantial exposure to the dust arising therefrom.                                                                                                           |
|     |              | 9.(a) | Work underground in any mine in which one of the objects of the mining operations is the getting of any material;                                                                                             |
|     |              | (b) | the working or handling above ground at any coal or tin mine of any materials extracted therefrom or any operation incidental thereto;                                                                                  |
|     |              | (c) | the trimming of coal in any ship, barge, lighter, dock or harbour or at any wharf or quay; or                                                                                                                           |
|     |              | (d) | the sawing, splitting or dressing of slate or any operation incidental thereto.                                                                                                                                         |
|     |              |     | 10. The manufacture or work incidental to the manufacture of carbon electrodes by an industrial undertaking for use in the electrolytic extraction of aluminium from aluminium oxide and any activity involving substantial exposure to the dust therefrom. |
|     |              |     | 11. Boiler scaling or substantial exposure to the dust arising therefrom.                                                                                                                                               |
| 40. | Byssinosis.  |     | The spinning or manipulation of raw or waste cotton or flax or the weaving of cotton or flax, carried out in each case in a room in a factory, together with any other work carried out in such a room.                 |
| 41. | Mesothelioma. | (a) | The working or handling of asbestos or any admixture of asbestos;                                                                                                                                                       |
| 42. | Lung cancer. | (b) | the manufacture or repair of asbestos textiles or other articles containing or composed of asbestos;                                                                                                                    |

| | | | |
|---|---|---|---|
| 43. | Asbestosis. | (c) | the cleaning of any machinery or plant used in any of the foregoing operations and of any chambers, fixtures and appliances for the collection of asbestos dust; or |
| | | (d) | substantial exposure to the dust arising from any of the foregoing operations. |
| 44. | Cancer of the nasal cavity or associated air sinuses. | 1.(a) | Work in or about a building where wooden furniture is manufactured; |
| | | (b) | work in a building used for the manufacture of footwear or components of footwear made wholly or partly of leather or fibre board; or |
| | | (c) | work at a place used wholly or mainly for the repair of footwear made wholly or partly of leather or fibre board. |
| | | \-2. Work in or about a factory building where nickel is produced by decomposition of a gaseous nickel compound or in any process which is ancillary or incidental thereto. | |
| 45. | Occupational dermatitis. | | Work involving exposure to any of the following agents: |
| | | (a) | epoxy resin systems; |
| | | (b) | formaldehyde and its resins; |
| | | (c) | metalworking fluids; |
| | | (d) | chromate (hexavalent and derived from trivalent chromium); |
| | | (e) | cement, plaster or concrete; |
| | | (f) | acrylates and methacrylates; |
| | | (g) | colophony (rosin) and its modified products; |
| | | (h) | glutaraldehyde; |
| | | (i) | mercaptobenzothiazole, thiurams, substituted paraphenylene-diamines and related rubber processing chemicals; |
| | | (j) | biocides, anti-bacterials, preservatives or disinfectants; |
| | | (k) | organic solvents; |
| | | (l) | antibiotics and other pharmaceuticals and therapeutic agents; |

|     |                                              |     |                                                                                                                                                              |
| --- | -------------------------------------------- | --- | ------------------------------------------------------------------------------------------------------------------------------------------------------------ |
|     |                                              | (m) | strong acids, strong alkalis, strong solutions (e.g. brine) and oxidising agents including domestic bleach or reducing agents;                               |
|     |                                              | (n) | hairdressing products including in particular dyes, shampoos, bleaches and permanent waving solutions;                                                       |
|     |                                              | (o) | soaps and detergents;                                                                                                                                        |
|     |                                              | (p) | plants and plant-derived material including in particular the daffodil, tulip and chrysanthemum families, the parsley family (carrots, parsnips, parsley and celery), garlic and onion, hardwoods and the pine family; |
|     |                                              | (q) | fish, shell-fish or meat;                                                                                                                                    |
|     |                                              | (r) | sugar or flour; or                                                                                                                                           |
|     |                                              | (s) | any other known irritant or sensitising agent including in particular any chemical bearing the warning 'may cause sensitisation by skin contact' or 'irritating to the skin'. |
| 46. | Extrinsic alveolitis (including farmer's lung). | Exposure to moulds, fungal spores or heterologous proteins during work in: |
|     |                                              | (a) | agriculture, horticulture, forestry, cultivation of edible fungi or malt-working;                                                                            |
|     |                                              | (b) | loading, unloading or handling mouldy vegetable matter or edible fungi whilst same is being stored;                                                          |
|     |                                              | (c) | caring for or handling birds; or                                                                                                                             |
|     |                                              | (d) | handling bagasse.                                                                                                                                            |
| 47. | Occupational asthma.                         | Work involving exposure to any of the following agents: |
|     |                                              | (a) | isocyanates;                                                                                                                                                 |
|     |                                              | (b) | platinum salts;                                                                                                                                              |
|     |                                              | (c) | fumes or dust arising from the manufacture, transport or use of hardening agents (including epoxy resin curing agents) based on phthalic anhydride, tetrachlorophthalic anhydride, trimellitic anhydride or triethylene-tetramine; |

(d) fumes arising from the use of rosin as a soldering flux;

(e) proteolytic enzymes;

(f) animals including insects and other arthropods used for the purposes of research or education or in laboratories;

(g) dusts arising from the sowing, cultivation, harvesting, drying, handling, milling, transport or storage of barley, oats, rye, wheat or maize or the handling, milling, transport or storage of meal or flour made therefrom;

(h) antibiotics;

(i) cimetidine;

(j) wood dust;

(k) ispaghula;

(l) castor bean dust;

(m) ipecacuanha;

(n) azodicarbonamide;

(o) animals including insects and other arthropods (whether in their larval forms or not) used for the purposes of pest control or fruit cultivation or the larval forms of animals used for the purposes of research or education or in laboratories;

(p) glutaraldehyde;

(q) persulphate salts or henna;

(r) crustaceans or fish or products arising from these in the food processing industry;

(s) reactive dyes;

(t) soya bean;

(u) tea dust;

(v) green coffee bean dust;

(w) fumes from stainless steel welding;

(x) any other sensitising agent, including in particular any chemical bearing the warning 'may cause sensitisation by inhalation'.

## PART II – DISEASES ADDITIONALLY REPORTABLE IN RESPECT OF OFFSHORE WORKPLACES

48. Chickenpox.
49. Cholera.
50. Diphtheria.
51. Dysentery (amoebic or bacillary).
52. Acute encephalitis.
53. Erysipelas.
54. Food poisoning.
55. Legionellosis.
56. Malaria.
57. Measles.
58. Meningitis.
59. Meningococcal septicaemia (without meningitis).
60. Mumps.
61. Paratyphoid fever.
62. Plague.
63. Acute poliomyelitis.
64. Rabies.
65. Rubella.
66. Scarlet fever.
67. Tetanus.
68. Tuberculosis.
69. Typhoid fever.
70. Typhus.
71. Viral haemorrhagic fevers.
72. Viral hepatitis.

[*The Reporting of Injuries, Diseases and Dangerous Occurrences Regulations 1995, Reg 5(1)(2), Sch 3*].

A3037  *Accident Reporting and Investigation*

# Appendix C
## Prescribed Form for Reporting an Injury or Dangerous Occurrence
A3037

---

**HSE** Health & Safety Executive

Health and Safety at Work etc Act 1974
The Reporting of Injuries, Diseases and Dangerous Occurrences Regulations 1995

## Report of an injury or dangerous occurrence

**Filling in this form**
This form must be filled in by an employer or other responsible person.

### Part A
**About you**
1  What is your full name?

2  What is your job title?

3  What is your telephone number?

**About your organisation**
4  What is the name of your organisation?

5  What is its address and postcode?

6  What type of work does the organisation do?

### Part B
**About the incident**
1  On what date did the incident happen?

   / /

2  At what time did the incident happen?
   (Please use the 24-hour clock eg 0600)

3  Did the incident happen at the above address?
   Yes ☐ Go to question 4
   No ☐ Where did the incident happen?
   ☐ elsewhere in your organisation – give the name, address and postcode
   ☐ at someone else's premises – give the name, address and postcode
   ☐ in a public place – give details of where it happened

   If you do not know the postcode, what is the name of the local authority?

4  In which department, or where on the premises, did the incident happen?

### Part C
**About the injured person**
If you are reporting a dangerous occurrence, go to Part F.
If more than one person was injured in the same incident, please attach the details asked for in Part C and Part D for each injured person.

1  What is their full name?

2  What is their home address and postcode?

3  What is their home phone number?

4  How old are they?

5  Are they
   ☐ male?
   ☐ female?

6  What is their job title?

7  Was the injured person (tick only one box)
   ☐ one of your employees?
   ☐ on a training scheme? Give details:

   ☐ on work experience?
   ☐ employed by someone else? Give details of the employer:

   ☐ self-employed and at work?
   ☐ a member of the public?

### Part D
**About the injury**
1  What was the injury? (eg fracture, laceration)

2  What part of the body was injured?

F2508 (01/96)                                    Continued overleaf

A30/42

Accident Reporting and Investigation   A3038

# Appendix D
## Prescribed Form for Reporting a Case of Disease
A3038

3  Was the injury (tick the one box that applies)
- [ ] a fatality?
- [ ] a major injury or condition? (see accompanying notes)
- [ ] an injury to an employee or self-employed person which prevented them doing their normal work for more than 3 days?
- [ ] an injury to a member of the public which meant they had to be taken from the scene of the accident to a hospital for treatment?

4  Did the injured person (tick all the boxes that apply)
- [ ] become unconscious?
- [ ] need resuscitation?
- [ ] remain in hospital for more than 24 hours?
- [ ] none of the above.

## Part E
### About the kind of accident
Please tick the one box that best describes what happened, then go to Part G.

- [ ] Contact with moving machinery or material being machined
- [ ] Hit by a moving, flying or falling object
- [ ] Hit by a moving vehicle
- [ ] Hit something fixed or stationary
- [ ] Injured while handling, lifting or carrying
- [ ] Slipped, tripped or fell on the same level
- [ ] Fell from a height
      How high was the fall?
      _____ metres
- [ ] Trapped by something collapsing
- [ ] Drowned or asphyxiated
- [ ] Exposed to, or in contact with, a harmful substance
- [ ] Exposed to fire
- [ ] Exposed to an explosion
- [ ] Contact with electricity or an electrical discharge
- [ ] Injured by an animal
- [ ] Physically assaulted by a person
- [ ] Another kind of accident (describe it in Part G)

## Part F
### Dangerous occurrences
Enter the number of the dangerous occurrence you are reporting. (The numbers are given in the Regulations and in the notes which accompany this form.)

## Part G
### Describing what happened
Give as much detail as you can. For instance
- the name of any substance involved
- the name and type of any machine involved
- the events that led to the incident
- the part played by any people.

If it was a personal injury, give details of what the person was doing. Describe any action that has since been taken to prevent a similar incident. Use a separate piece of paper if you need to.

## Part H
### Your signature
Signature

Date
___ / ___ / ___

**Where to send the form**
Please send it to the Enforcing Authority for the place where it happened. If you do not know the Enforcing Authority, send it to the nearest HSE office.

| For official use | | | |
|---|---|---|---|
| Client number | Location number | Event number | |
|  |  |  | [ ] INV  REP [ ] Y [ ] N |

A30/43

# Appendix E
# Health and Safety Executive Regional Offices

## London and South East Division

Covers the counties of Kent, Surrey, East Sussex and West Sussex, and all London Boroughs.

St Dunstans House
201–211 Borough High Street
London SE1 1GZ
*Telephone*: 020 7556 2100
*Fax*: 020 7556 2200

3 East Grinstead House
London Road
East Grinstead RH19 1RR
*Telephone*: 01342 334 200
*Fax*: 01342 334 222

## Home Counties Division

Covers the counties of Bedfordshire, Berkshire, Buckinghamshire, Cambridgeshire, Dorset, Essex (except London Boroughs in Essex), Hampshire, Hertfordshire, Isle of Wight, Norfolk, Suffolk and Wiltshire.

14 Cardiff Road
Luton LU1 1PP
*Telephone*: 01582 444 200
*Fax*: 01582 444 320

Priestley House
Priestley Road
Basingstoke RG24 9NW
*Telephone*: 01256 404 000
*Fax*: 01256 404 100

39 Baddow Road
Chelmsford CM2 0HL
*Telephone*: 01245 706 200
*Fax*: 01245 706 222

## Midlands Division

Covers the counties of West Midlands, Leicestershire, Northamptonshire, Oxfordshire, Warwickshire, Derbyshire, Lincolnshire and Nottinghamshire.

McLaren Building
35 Dale End
Birmingham B4 7NP
*Telephone*: 0121 607 6200
*Fax*: 0121 607 6349

5th Floor
Belgrave House
1 Greyfriars
Northampton NN1 2BS
*Telephone*: 01604 738 300
*Fax*: 01604 738 333

1st Floor
The Pearson Building
55 Upper Parliament Street
Nottingham NG1 6AU
*Telephone*: 0115 971 2800
*Fax*: 0115 971 2802

## Yorkshire and North East Division

Covers the counties and unitary authorities of Hartlepool, Middlesbrough, Redcar and Cleveland, Stockton-on-Tees, Durham, Hull, North Lincolnshire, North East Lincolnshire, East Riding, York, North Yorkshire, Northumberland, West Yorkshire, Tyne and Wear and the Metropolitan Boroughs of Barnsley, Doncaster, Rotherham and Sheffield.

Marshalls Mill  
Marshall Street  
Leeds LS11 9YJ  
*Telephone*: 0113 283 4200  
*Fax*: 0113 283 4296

Sovereign House  
110 Queen Street  
Sheffield S1 2ES  
*Telephone*: 0114 291 2300  
*Fax*: 0114 291 2379

Arden House  
Regent Centre  
Regent Farm Road  
Gosforth  
Newcastle upon Tyne NE3 3JN  
*Telephone*: 0191 202 6200  
*Fax*: 0191 202 6300

## North West Division

Covers the counties of Cheshire, Cumbria, Greater Manchester, Lancashire and Merseyside.

Grove House  
Skerton Road  
Manchester M16 ORB  
*Telephone*: 0161 952 8200  
*Fax*: 0161 952 8222

Marshall House  
Ringway  
Preston PR1 2HS  
*Telephone*: *see* Manchestert  
*Fax*: 01772 836 222

## Wales and West Division

Covers Wales, and the unitary authorities of Cornwall, Devon, Somerset, North West Somerset, Bath and North East Somerset, Bristol, South Gloucestershire, Gloucestershire, Hereford and Worcester, Shropshire and Staffordshire.

Government Buildings  
Phase 1  
Ty Glas  
Llanishen  
Cardiff CF14 5SH  
*Telephone*: 029 2026 3000  
*Fax*: 029 2026 3120

Inter City House  
Mitchell Lane  
Victoria Street  
Bristol BS1 6AN  
*Telephone*: 0117 988 6000  
*Fax*: 0117 926 2998

The Marches House
Midway
Newcastle-under-Lyme
Staffordshire ST5 1DT
*Telephone*: 01782 602 300
*Fax*: 01782 602 400

## Scotland

Covers all the Scottish unitary authorities and island councils.

Belford House
59 Belford Road
Edinburgh EH4 3UE
*Telephone*: 0131 247 2000
*Fax*: 0131 247 2121

375 West George Street
Glasgow G2 4LW
*Telephone*: 0141 275 3000
*Fax*: 0141 276 3100

Offshore Safety Division
Lord Cullen House
Fraser Place
Aberdeen AB25 3UB
*Telephone*: 01224 252 500
*Fax*: 01224 252 662

# Asbestos

## Introduction

**A5001**   Asbestos is the generic name for a group of naturally occurring fibrous minerals, metallic silicates, which have a wide range of industrial applications. The most common forms of asbestos are chrysotile (white), amosite (brown) and crocidolite (blue); others are fibrous actinolite, fibrous anthophyllite and fibrous tremolite. Mined in Canada, South Africa, Russia and elsewhere, the material has been manufactured into products which use its characteristics of heat and chemical resistance. Despite its contribution to fire protection, asbestos is now regarded as one of the most significant causes of occupational disease over the past 125 years. The use of asbestos has been extensive, but as a result of prohibitions in the UK on new uses of asbestos-containing materials, products such as brake linings and other friction products are now unlikely to be found in most workplaces. However, its widespread application in materials used in construction has led to many buildings containing asbestos, and the current focus of asbestos controls is largely on the management of this legacy.

## Asbestos diseases

**A5002**   The diseases associated with asbestos only arise when asbestos fibres are inhaled and penetrate deep into the lung. Extremely small fibres are capable of bypassing the body's defence mechanisms to achieve this. These are therefore called respirable fibres and they are fibres with a diameter less than three micrometres. To illustrate how small these fibres are a human hair is approximately 50 micrometres in diameter. Asbestosis is a thickening of the wall of the alveoli, the tiny lung sacs where oxygen passes into the blood. Since the lung has a huge surface area one or two areas of such thickening would have little effect. However when millions of fibres are inhaled and reach the deep lung the reduction in gas exchange capacity caused by thickening of the alveoli walls becomes significant. The disease of asbestosis is recognised when the patient has a measurable decrease in lung capacity and even notices a shortness of breath. Signs of the disease can also be detected by x-ray. If exposure is extensive and prolonged the damage to the lungs can be severe and lead eventually to death. The progress of the disease is proportional to exposure, which in modern terms must be massive. It has occurred in industries where raw asbestos was handled in bulk, when raw asbestos was delivered in bales which were manually cut open and fed into hoppers for the manufacturing process. Workers often went home white with asbestos fibres adhering to their body and clothes. A condition associated with asbestosis was pleural plaques. This was areas of calcification or stiffening on the membrane on the outer surface of the lung. Asbestosis is a very similar disease to coal workers pneumoconiosis and since the level of exposure needed to develop these diseases no longer occurs in Britain today there are virtually no new cases appearing but workers exposed in the past are still alive, suffering from asbestosis and dying of it.

*Mesothelioma* is a cancer occurring on the outer membrane of the lung. It is normally a very rare disease and so its occurrence amongst asbestos workers was quickly recognised. Typically it is not detected in the early stages and therefore once diagnosed is advanced and usually progresses to death in months rather than years. However the latency period from initial exposure to onset of the disease appears to be long, in the region of decades. The risk of contracting mesothelioma is based on exposure. The greater the concentration of respirable asbestos fibres inhaled the greater is the risk of contracting the disease. In addition, the concentration of respirable asbestos fibres required to cause the disease is far lower than that associated with asbestosis. In the UK mesothelioma was at first associated with crocidolite or blue asbestos and the control limit (the maximum level to which workers could legally be exposed) for blue asbestos was tightened. Eventually it was recognised that all amphibole asbestos types were implicated. So amosite, brown asbestos, and crocidolite and the other amphibole asbestos types now have tighter control limits.

*Lung cancer* is now recognised as a disease associated with asbestos. The specific diagnosis of lung cancer being an asbestos-related disease is however occupational exposure. Since lung cancer occurring spontaneously, caused by smoking or caused by asbestos is indistinguishable it is the history of the patient which is used in the specific diagnosis. It is firmly established from epidemiology that there is an increased risk of lung cancer for smokers and that there is an increased risk of lung cancer for those exposed to asbestos. Indeed for smokers who are also exposed to asbestos it is estimated that the risk of lung cancer is multiplied rather than the two risks being additive.

When asbestos fibres are inhaled, they have a very long half life in the lungs. Their shape – long and thin – and robustness means that it is difficult for the usual lung clearance mechanisms to remove them from the lungs (in the manner in which dust particles are cleared), and they do not readily dissolve in lung fluid. Whilst the exact mechanisms which cause cancer remain obscure, the toughness and longevity of the fibres helps to explain why they are able to cause ill health many years after first exposure – although the abrasive and chemical effect of the fibres in very high lung concentrations can cause asbestosis on a shorter timescale this is largely an historical disease whilst the cancers continue to take their toll. If workers are exposed to airborne fibres, for example as a result of an incident in which asbestos has been inadvertently disturbed, there is no 'treatment' available to reduce the risk of future disease (except, of course to avoid inhaling cigarette smoke in future). However, it is worth noting that isolated exposures such as this represent a very (immeasurably) low level of risk. The significance of asbestos is that thousands of workers have been exposed on a daily basis to high levels, and that is why the last century saw an 'epidemic' of asbestos disease. This is not to argue that even single exposures to airborne asbestos should not be prevented, rather to encourage a recognition that the emphasis is on routine and continued exposure as this is the root of the problem.

## Historical exposures – manufacturing

A5003  In the 1950s and 1960s the predominant concern arising from exposure to asbestos was for manufacturing workers contracting asbestosis, following exposure to massive concentrations of respirable fibres. These workers were involved in manufacturing asbestos insulating boards and panels, asbestos cement products and friction products such as clutch plates, break shoes and gaskets. Anecdotally, such workers would refer to working in 'snow storms' of the material, which because the fibres are soft and silky did not create any immediate discomfort, as would be the case with glass fibre which is immediately irritating to the eyes, throat and skin. In order to control this risk, and against the background of a well-funded lobbying campaign by the asbestos mining and manufacturing interests (a campaign which still functions, as the Canadian government's challenge through the World Trade Organisation to the EU prohibitions on asbestos importation indicates) Asbestos Regulations were introduced imposing control limits on the maximum concentration of respirable asbestos fibres that workers could be exposed to. Where the concentrations could not be reduced below the control limits it was mandatory to provide suitable respirators. As the risk of lung cancer and mesothelioma was recognised and epidemiological data and analysis became available, the control limits for working with asbestos were gradually reduced. In addition, 'action levels' were adopted and defined in terms of concentrations of respirable airborne fibres measured over a twelve-week period. If the mean concentration of airborne fibres exceeded the action level then the area in the factory had to be designated an 'asbestos zone' and non-essential personnel excluded. If the airborne fibre concentration was likely to exceed the control limit then issuing and wearing respirators became mandatory.

In the *Control of Asbestos at Work Regulations 2002* (*SI 2002 No 2675*), the action levels are 72 fibre-hours per millilitre for chrysotile and 48 fibre-hours per millilitre for the other types of asbestos, such as amosite and crocidolite. This is equivalent to a continuous exposure to 0.15 fibres/ml for chrysotile and 0.1 fibres/ml for the other asbestos types. The control limits are 0.3 fibres/ml and 0.2 fibres/ml for other asbestos types, measured over four hours. There are also short-term control limits, measured over ten minutes, of 0.9 fibres/ml for chrysotile and 0.6 fibres/ml for other asbestos types. The Regulations also impose a general duty on all employers to assess the risk to the health of all employees before any work with asbestos commences and to put in place such control measures as are necessary to reduce the exposure to as low a level as reasonably practicable and, in any case, to below the control limit. Whilst these limits were introduced primarily for manufacturing, they did begin to have an effect on the management of asbestos installations in buildings, where everything from pipework lagging to floors, walls and ceilings could formed from asbestos products.

## Asbestos prohibitions

**A5004**     The *Asbestos* (*Prohibition*) *Regulations 1992* (*SI 1992 No 3067*), as amended, prohibit all imported materials to which asbestos has been added as part of a deliberate manufacturing process, but excludes naturally-occurring minerals which may contain traces of asbestos. Many producers and importers of mineral products carry out comprehensive testing to prevent materials containing significant quantities of asbestos getting into the supply chain. Where asbestos is found the material should not be sold unless the amount is trivially small. Even in those cases suppliers must inform their customers that trace quantities of asbestos may occasionally be found in their products. This will allow 'high-energy' users (for example traces of asbestos in blast-cleaning materials would release airborne fibres) to take appropriate precautions or consider alternative materials. In addition to prohibitions on the sale of asbestos products, where they may be discovered in store rooms or warehouses, they may not be used and must be disposed of safely. Whilst there is no prohibition on asbestos which is already in place, new uses are effectively banned.

## Asbestos removal

**A5005**     As the manufacture and use of asbestos declined, accelerated by the *Asbestos* (*Prohibition*) *Regulations 1992* (*SI 1992 No 3067*), the risk to the health of manufacturing workers also decreased. However, it was recognised that there was a new exposure group. These were the workers who were engaged to remove asbestos from buildings. Therefore, in the late 1980s, new Approved Codes of Practice (ACoPs) and Guidance Notes were introduced to specify how the *Control of Asbestos at Work Regulations 1987* (*SI 1987 No 2115*) (now revoked by *SI 2002 No 2675*) should be applied to the removal of asbestos from buildings. The *Asbestos* (*Licensing*) *Regulations 1983* (*SI 1983 No 1649*) were also introduced to provide the enforcing authorities with information on which companies were involved in removing asbestos from buildings, where and when the work was being carried out. Guidance Notes recommended that after asbestos was removed from an area the area should be visibly free from dust and debris and that the airborne fibre concentration measured in the area should be below 0.01 fibres/ml. This is the detection limit of the optical microscopy techniques used to measure airborne fibres and is referred to as the 'clearance indicator'. The clearance indicator is simply a measure of cleanliness of an area and is not a risk based limit such as the control limit. Following the introduction of the *Control of Asbestos at Work Regulations 2002* (*SI 2002 No 2675*), new editions of the Approved Codes of Practice were published, including 'Work with asbestos insulation, asbestos coating and asbestos insulating board', (Fourth edition, 2002). A full list of asbestos legislation, Approved Codes of Practice and Guidance Notes are appended at the end of this chapter (see A5030).

## Exposure of building maintenance workers

**A5006** As the *Asbestos (Licensing) Regulations 1983* (*SI 1983 No 1649*) were enforced, and the asbestos removal industry became more thoroughly regulated, the risk to the health of the workers in that industry began to be more effectively controlled. However, in the late 1990s it was recognised that there was still a very large group of workers whose exposure to asbestos was not adequately controlled by the existing asbestos control regime. Maintenance workers and those engaged in building refurbishment were still exposed to asbestos because many building owners did not know that asbestos was present and many employers of maintenance workers did not consider that they would be working with asbestos and so took no precautions. For these reasons, the *Control of Asbestos at Work Regulations 2002* (*SI 2002 No 2675*) were introduced. These provisions require building owners and occupiers to identify where asbestos is present in every building and have a documented plan to control the risks to health presented by that asbestos. The aim is to reduce over time the number of asbestos-related deaths in the UK. Currently, past exposure to asbestos leads to 3,000 deaths per year, and the death rate is likely to rise further for another 10 years or so and is unlikely to decrease quickly as the latency period of asbestos-related cancer is between 15 and 60 years. Recent research has indicated that asbestos-related deaths should peak in a few years and then fall as a direct result of the control measures adopted.

*Summary*: despite lobbying by asbestos interests, the evidence of the health risks associated with asbestos has led to a sequential establishment of a statutory regime of control. The protection was first designed for manufacturing workers, was extended to asbestos removers and latterly has addressed the management of buildings which contain some of the thousands of tonnes of asbestos used in construction. The death rate from the diseases caused by inhalation of asbestos fibres in the past is still rising, but is predicted to peak and then start to fall in a few years as a consequence of the control measures.

## Duties under the Control of Asbestos at Work Regulations 2002

**A5007** The *Control of Asbestos at Work Regulations 2002* (*SI 2002 No 2675*) ('*CAWR*') place duties on every employer in respect of his employees, and make it clear that these duties, so far as is reasonably practicable, are owed also to any other person, whether at work or not, who may be affected by the work activity undertaken by the employer. The exceptions are in connection with *Regulation 9* of the 2002 Regulations where the duty to inform, instruct and train does to apply to non-employees unless they are in the premises where the work is being carried out, and *Regulation 21* where the duty to carry out medical surveillance and maintain health records for exposed workers only applies to the employer's own employees. The *CAWR* apply to the self-employed, applying the duties of both employer and employee (*Regulation 3(2)*).

After four years of debate, consultation and review, the long-awaited amendments to the regulations have been enacted under the *CAWR*. Not only are employers subject to health and safety duties, and the existing asbestos control regime, in the usual way but the *CAWR* now also apply to all companies which own, occupy, manage or have any contractual responsibilities for maintenance or repair of non-domestic premises. The new duties, therefore, apply to all business, commercial, industrial, educational, healthcare, military, administrative, retail and leisure properties including hotels and guest houses. Even in the absence of agreement, all parties which control access or maintenance of a building could still be liable. Duty holders are under a duty either to:

- manage the risk from asbestos-containing materials ('ACMs'); or
- co-operate with whoever manages that risk.

### Affected premises

**A5008** The duty applies not only to all non-domestic premises, but also applies to the common parts of commercial buildings, housing developments and leasehold flats. Given the

## Identifying the duty holder

**A5009** Persons with a primary responsibility for a building, such as the owner, sole occupier or part occupier, have duties under the *CAWR* (*SI 2002 No 2675*). Those with any responsibility for the maintenance of premises by virtue of a contract or tenancy also have a legal duty to manage the risks from asbestos and comply with the *CAWR*. If no such formal agreement exists, the owner, sub-lessor or managing agent with any de facto control of a workplace premise, or any employer or self-employed person who occupies commercial premises in which persons work is subject to the same duty.

## Risk assessment

**A5010** The duty holder is required to make a suitable and sufficient risk assessment as to whether asbestos is, or is liable to be present in the premises. It should be noted that wherever a duty such as this applies, compliance requires that the duty holder ensures that it has been carried out. In this case, this means ensuring that a suitable and sufficient assessment has been conducted, documented and is available and used to inform decision-making, advise incoming occupiers, maintenance workers and others. Commissioning others to carry out such work is acceptable practice, so long as the persons employed are competent and allocate sufficient resource to carry out the work properly.

This includes taking into account the age of the premises, building plans and any other relevant information. In carrying out inspections it must be presumed that materials contain asbestos unless there is strong evidence to the contrary. At this stage it is necessary simply to identify whether asbestos may be present and, therefore, determine whether a more thorough investigation is required. It is not, however, sufficient to conclude that a building is free of asbestos simply because it was built after 1990. That conclusion can only be reached if there is documentary evidence from a designer or architect who specified that no asbestos was to be used in the construction of the building. The most likely asbestos materials to be found are:

- Sprayed asbestos and asbestos loose packing – typically used as fire breaks in ceiling voids and to protect steelwork, fire protection in ducts, firebreaks, etc.
- Moulded or preformed lagging – generally used in thermal insulation of pipes and boilers.
- Insulating board – fire protection, thermal insulation, panelling, partitioning, ducts.
- Ceiling tiles.
- Millboard and paper products – insulating electrical products and facing wood fibreboard.
- Woven products – boiler seals and old fire blankets.
- Cement products in flat or corrugated sheets.
- Textured coatings.
- Bitumen roofing materials.
- Vinyl and thermoplastic floor tiles.

## Records

**A5011** If asbestos is present, an up-to-date record of the location and condition of asbestos-containing materials ('ACMs') or presumed ACMs in the premises must be made. This is often referred to as an 'asbestos register'. The asbestos register should record where asbestos is present in the building the form in which it is present, e.g. asbestos insulating board, pipe lagging, ceiling tiles, asbestos cement sheets, and the condition of the material. The health risk depends on the inhalation of respirable fibres, so that the condition of the asbestos, as recorded in the asbestos register, should reflect the extent to which fibres may be released from the ACM during normal occupancy and during any maintenance work. Since maintenance work will be required at some stage, and it is necessary to either determine the type of asbestos present before such work commences or assume that it is not solely chrysotile, it is often considered appropriate to identify the type of asbestos present in the ACM and record this in the asbestos register. This requires samples of suspected ACM to be taken for laboratory analysis in accordance with the Methods for the Determination of Hazardous Substances – 'Asbestos in bulk materials' (MDHS 77). From November 2004, all analyses of asbestos samples may only be undertaken by an accredited laboratory. At present, there is some concern that there are insufficient accredited laboratories to cope with the huge demand for surveys which the CAWR (*SI 2002 No 2675*) are producing. Since the asbestos register is compiled and maintained in order to quickly identify where asbestos is present before planned or emergency work is undertaken the most appropriate format for the asbestos register should be considered. An electronic format on a computer server will facilitate the maintenance of the register and ensure that an up-to-date version is available at several locations. However where maintenance work may be undertaken where there is no access to a computer system, or at weekends and outside normal office hours, a paper-based register may be more effective. It is a legal requirement that the information is maintained in a form which can be readily accessed by those who need it.

## *Management plan*

**A5012** The compilation of an asbestos register is only one part of the new duties. There must also be a documented plan to manage the health risks presented by the presence of ACMs. Where the health risk during normal occupancy is low, the plan may be to maintain the ACM in place and label it. There should then be written procedures for planned maintenance work and procedures for emergency work, and staff familiar with their duties in implementation. These procedures would normally include calling in licensed asbestos contractors to remove the ACM under controlled conditions before the maintenance work proceeded. If the presence of asbestos in the building represents a risk to health under normal occupancy conditions or it is envisaged that emergency situations would involve the removal of ACM, then the asbestos management plan may be to immediately encapsulate or remove ACMs for particular areas. This work requires the involvement of contactors licensed under the *Asbestos (Licensing) Regulations 1983* (*SI 1983 No 1649*) working in accordance with the Approved Code of Practice 'Work with asbestos insulation, asbestos coating and asbestos insulating board (Fourth edition, 2002). Where asbestos removal operations are undertaken, it is good practice for the premises management to retain the services of a United Kingdom Accreditation Service (UKAS) accredited laboratory to both monitor and manage the asbestos removal work. The asbestos management plan should also contain the procedures for dealing with suspect ACM. Typically, such procedures would include the requirement to halt the work and call in a UKAS accredited laboratory to take samples and determine whether asbestos is present. The asbestos management plan should be reviewed regularly and the procedures and arrangements monitored to ensure that they are effective. It is important to test the procedures to ensure that they work. For instance, when routine maintenance is undertaken, checks that the tradesmen concerned were furnished with the appropriate information and that they followed the correct procedures should be made and recorded. In many commercial premises the effect of the *CAWR's* duty to manage may create more than one duty holder. Here it is necessary to examine the nature and extent of the repair and maintenance obligation owed by each party, or the level of control in determining their relative contributions to comply.

To ensure the effective compliance with the new duty to manage asbestos in non-domestic premises, there is a simple seven step programme to follow.

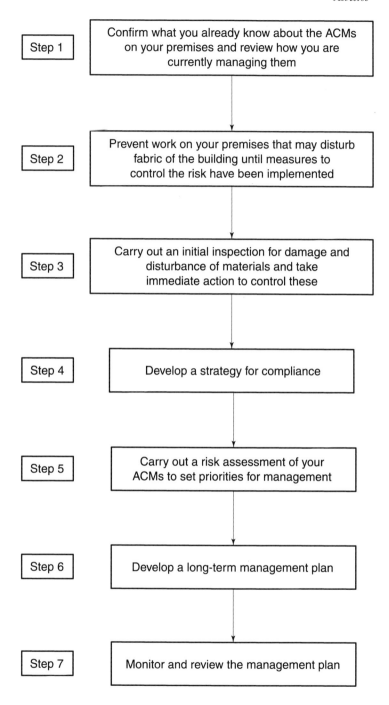

## Step 1: What ACMs do you know about and how are you managing them

A5013    Are there likely to be ACMs on your premises? If there are none, no further action is required. If ACMs are present, or there is uncertainty, a review of current management arrangements is required. Check that:

- Maintenance and building work is effectively controlled, with a system to prevent disturbance of asbestos and unknown material.

- Maintenance and building staff know the rules about ACMs, and how to work safely on or adjacent to them.

- The condition of known ACMs or materials presumed to contain asbestos is known, there is a system to check for damage and deterioration.

- Damaged and deteriorated ACMs are repaired, removed or isolated to prevent exposure to released fibres.

- There is a management plan for recording findings on ACMs, monitoring their condition and taking action as necessary.

- Everyone, including appointed contractors, knows their roles and responsibilities for the management of ACMs on the premises.

- A responsible person has been appointed, is competent and resourced (premises or facilities manager, or for small businesses the owner/manager). Access to competent advice from a qualified safety practitioner or hygienist particularly for buildings which are known or suspected to contain significant quantities of ACMs.

## Step 2: Control work on your premises

A5014    In every building, there needs to be a system which prevents anyone doing building or maintenance work, even running new telephone cables, without first checking whether ACMs may be disturbed (usually by reference to the Asbestos Register).

## Step 3: Carry out an initial inspection

A5015    After stopping new activities which could damage or disturb ACMs, it is essential to check on the condition of all existing known or suspected ACMs, and take immediate action to prevent exposure to airborne fibres where such damage is identified. This check is looking for *any* damage to building materials. If any damaged materials are identified, they can be tested to see if they are ACMs, and kept isolated until the results are available and the material removed or repaired if asbestos is confirmed.

## Step 4: Develop a strategy for compliance

A5016    After taking immediate control (steps 1–3), the remaining steps are designed to manage occupancy and building works. The first such step is to develop a formal strategy to manage the long-term risks:

- Commission a survey to establish the extent of ACMs.

- Document the management controls to minimise risk of damage to ACMs, and to identify and act on damage and deterioration effectively.

## Step 5: Risk assessment and priorities

A5017　The combination of the intrinsic potential of the material to release fibres *and* the likelihood of disturbance leading to that release which represents the core of the risk assessment.

## Step 6: Develop a long-term management plan

A5018　The plan should contain:

- How the location and condition of known or presumed ACMs is recorded.
- Priority assessments and an action plan.
- Decisions about management options and their rationale.
- Action timetable.
- Monitoring arrangements.
- Employees and their responsibilities.
- Contractor procedures including selection, appointment, information and monitoring.
- Training arrangements.
- Procedures and plan for implementation, including controlling work on the fabric of the building.
- Method for passing on information about ACMs to those who need it.
- Responsibility for the management plan and its updating.
- Procedure and timetable for review of the management plan.

## Step 7: Monitor and review the management plan

A5019　Implementation of the management plan is monitored such that checks are routinely carried out on the ACMs, records are kept up-to-date. Incidents and accidents require careful investigation and lessons learned should be used to improve the plan and its implementation. For most premises it is recommended that the plan is subject to at least a 6 monthly thorough review.

# Duties of third parties

A5020　The *CAWR* (*SI 2002 No 2675*) place any person under a duty to assist in the management of asbestos. Third parties will be required to co-operate with the duty holder so far as is necessary to enable the duty holder to comply with the *CAWR* by:

- making all information in relation to the premises available for inspection;
- providing them with information on the locations of asbestos, if known;
- allowing duty holders to identify asbestos or presumed asbestos for all parts of the premises repaired or maintained by the third party;
- assisting in the preparation and implementation of the asbestos management plan to control the health risk presented by ACMs;
- ensuring that everyone potentially at risk from ACMs receives information on the location and condition of the material so far as it is within the third party's control.

### CAWR contractual and tenancy agreements

**A5021** Organisations need to review the extent of their maintenance and repair obligations under any leases and be clear as to where the responsibility lies for asbestos control. Responsibility for dealing with any asbestos health risks, and therefore each party's contribution, should be considered in pre-contract lease negotiations and specific provisions should be made in the tenancy agreement. This will be especially important in determining who is responsible for common parts of the building where, for example, control over maintenance and repair activities in those parts could be shared.

### Minimum requirements

**A5022** It will be a defence to show that an employer has done everything in his power to prevent exposure to asbestos. He will need to be able to produce an up-to-date asbestos register in a suitable format and a documented asbestos management plan containing risk assessments, procedures and arrangements to manage the risk from exposure to asbestos during normal occupancy, planned maintenance and emergency situations. Evidence will also be required to demonstrate that the asbestos management plan has been implemented and that it is effective. The *CAWR* (*SI 2002 No 2675*) is accompanied by an ACoP 'The management of asbestos in non-domestic premises', which gives advice about ways to identify asbestos, the associated health risks and how to develop an asbestos management plan to manage those risks. The duty to manage asbestos took effect in May 2004 and employers should ensure familiarisation with the *CAWR* and to undertake surveys as soon as possible to locate whether there are any specific problem areas which should be dealt as a matter of urgency.

## Asbestos building surveys

**A5023** There are three types of asbestos surveys described in the Health and Safety Executive (HSE) publication MDHS 100 'Surveying, sampling and assessment of asbestos-containing materials'.

### Type 1 survey

**A5024** The Type I survey involves study of the building history, structure and plans to consider whether asbestos may be present. This is followed by a visual inspection to assess whether there are materials in place which have the appearance of ACMs. The output of a Type I survey may be the conclusion that the building is of a type which should not contain asbestos and that no suspect materials were seen, or that the building may contain asbestos and there were suspect materials which should be investigated. At this stage the building owner or occupier may assume that asbestos is present and draw up an asbestos management plan or commission a Type 2 survey.

### Type 2 survey: risk assessments

**A5025** A Type 2 survey involves a detailed inspection of the building with samples of suspect materials being taken for laboratory analysis. Such surveys should only be undertaken by organisations which hold UKAS accreditation for asbestos sampling with accreditation to ISO 17025 and comply with the provisions of EN45004 'General criteria for the operation of various types of bodies performing inspection'. After 20 November 2004 all organisations offering asbestos surveying services must hold this accreditation. The outcome of a Type 2 survey is a detailed report on the location of all ACMs within the building, in a form which can be readily understood and used by those who may be exposed. This usually means that the information is graphically displayed on a plan of the building. Photographs of the sampling site are also normally included to assist in the identification of the ACM position. The Type 2 survey report will also describe the nature of the material, e.g. asbestos insulating board, pipe lagging, ceiling tiles, asbestos rope, floor tiles, asbestos cement products etc. The report will also include the laboratory analysis of the type of asbestos present, e.g. amosite, crocidolite, chrysotile, etc. The condition of the ACM will be assessed in relation to the propensity for the release of respirable airborne fibres and so the consequent risk to health. This requires the judgement of a knowledgeable and experienced surveyor. This risk assessment will depend not only on the nature and condition of the ACM but also on its location and whether it is likely to be disturbed by normal occupancy or only during maintenance and refurbishment. This information is used to compile the asbestos register required in the *CAWR* (*SI 2002 No 2675*).

### Type 3 surveys: asbestos removal specifications

A5026   The outcome of a Type 3 survey is used as part of the specifications given to licensed asbestos removal contractors to tender for asbestos removal contracts prior to refurbishment or building demolition. The Type 3 survey report contains details of the location and type of all ACMs and the type of asbestos present. However, the report also specifies the amount of each ACM present. This allows contractors to decide how the ACM should be removed and the extent of the operation and, therefore, provide a method statement for removal and an estimate of costs.

## Asbestos waste

A5027   Most asbestos waste arising from asbestos removal operations, including asbestos cement is classified as a 'special waste' under the *Special Waste Regulations 1996* (*SI 1996 No 972*). Consequently, the waste must be transported in suitable containers and be adequately described in a consignment note in accordance with the *Environmental Protection (Duty of Care) Regulations 1991* (*SI 1991 No 2839*).

## Asbestos works

A5028   If work is undertaken on an ACM – to repair and encapsulate it, or to remove it, then there is a simple sequence which is required:

1. Identification of the type of asbestos or assumption that it is asbestos but not solely chrysotile.

2. Plan of work – detailing the nature and probably duration of the work, the location, the work methods and controls (based on the risk assessment) and the equipment to be used to protect and decontaminate the workers and others nearby. The plan should also detail emergency procedures.

3. Notification – if a licensed contractor is employed, he may have this requirement as a condition of license.

4. Information, instruction and training as required for those who may be affected by the works and/or may be required to supervise etc.

5. Once works are underway, the licensed operator is required to comply with a range of technical duties for the containment of asbestos, the labelling of the containment area, the protection of the workers engaged in ACM work and their decontamination, and the handling of waste. During the works the efficacy of the containment and other controls should be checked by air testing carried out by a body independent of the contractor, should be accredited by UKAS.

6. On completion of ACM work within a contained area, the premises must be assessed to determine that they are clean and fit for normal occupation. Such testing, to be carried out by a body independent of the contractor, should be accredited by UKAS.

## Licensing of asbestos works

**A5029** A licensing system for asbestos works concerning insulation or coating (that is, effectively not applying to cement products, thermoplastic floor tiles and other lower risk materials) was introduced in 1983 by the *Asbestos (Licensing) Regulations (SI 1983 No 1649*, as amended by *SI 1998 No 3233*). A licence is required, except for:

- work where any person carrying it out will not spend more than one hour on the work in seven consecutive days; and
- the total time spent by all persons on the work does not exceed two hours;

or

- the employer or self-employed person is the occupier of the premises on which the work is carried out; and
- he has given HSE 14 days' notice in advance, specifying the type of work to be carried out and the location address;

or

- the work consists solely of air monitoring or collecting samples.

Where the licence does not apply (above) the employer must still comply with the *CAWR (SI 2002 No 2675)*, including providing information, instruction and training, informing others, exercising effective control to minimise asbestos exposures and ensure that, as required, employees are under medical surveillance and health records are maintained.

# Legislation, Approved Codes of Practice, and Guidance Notes

**A5030** Legislation, Approved Codes of Practice and Guidance Notes dealing with asbestos for England and Wales include, but not exclusively, those listed below. Equivalent documentation exists for Scotland and Northern Ireland respectively.

All legislation, Approved Codes of Practice and Guidance Notes listed together with any subsequent amendments or revisions and any new relevant requirements should be considered before undertaking any work with asbestos or ACMs.

The following legislation, Approved Codes of Practice and Guidance Notes deal primarily with asbestos. Other legislation dealing with health and safety matters has not been listed here, although such legislation still applies to work with asbestos and should be considered at all times.

The following list was last revised in June 2004.

## Legislation

- *Health and Safety at Work etc. Act 1974*;
- *Environmental Protection Act 1990*;
- *Environment Act 1995*;
- *Consumer Safety Act 1978*;
- *Asbestos (Licensing) Regulations 1983 (SI 1983 No 1649)* and subsequent amendments (latest amendment *SI 1998 No 3233*);
- *Control of Asbestos at Work Regulations 2002 (SI 2002 No 2675)*;
- *Control of Asbestos in the Air Regulations 1990 (SI 1990 No 556)*;

- *Reporting of Injuries, Diseases and Dangerous Occurrence Regulations 1995 (SI 1995 No 3163)*;
- *Management of Health and Safety at Work Regulations 1999 (SI 1999 No 3242)*;
- *Workplace (Health, Safety and Welfare) Regulations 1992 (SI 1992 No 3004)*;
- *Personal Protective Equipment at Work Regulations 1992 (SI 1992 No 2966)*;
- *Provision and Use of Work Equipment Regulations 1998 (SI 1998 No 2306)*;
- *Construction (Design and Management) Regulations 1994 (SI 1994 No 3140)*;
- *Carriage of Dangerous Goods and Use of Transportable Pressure Equipment Regulations 2004 (SI 2004 No 568)*
- *Carriage of Dangerous Goods by Rail Regulations 1996 (SI 1996 No 2089)* (revoked in part by *SI 2004 No 568*);
- *Asbestos (Prohibitions) Regulations 1992 (SI 1992 No 3067)*;
- *Health and Safety (Emissions into the Atmosphere) Regulations 1983 (SI 1983 No 943)*;
- *Health and Safety (Fees) Regulations 2004 (SI 2004 No 456)*;
- *Environmental Protection (Duty of Care) Regulations 1991 (SI 1991 No 2839)*;
- *Waste Management Licensing Regulations 1994 (SI 1994 No 1056)*;
- *Controlled Waste Regulations 1992 (SI 1992 No 588)*;
- *Controlled Waste (Registration of Carriers and Seizure of Vehicles) Regulations 1991 (SI 1991 No 1624)*;
- *Environmental Protection (Prescribed Processes and Substances) Regulations 1991 (SI 1991 No 472)* (as amended);
- *Asbestos Products (Safety) Regulations 1985 (SI 1985 No 2042)*.

## Approved Codes of Practice

- L28 – 'Work with asbestos insulation, asbestos coating and asbestos insulating board' (Fourth edition, 2002);
- L27 – 'Work with asbestos which does not normally require a licence' (Fourth edition, 2002);
- L127 – 'The management of asbestos in non-domestic premises. Control of asbestos at Work Regulations 2002 (2002);
- 'A Guide to the Asbestos (Licensing) Regulations 1983' (1989);
- 'Waste management: The duty of care. A Code of Practice' (1991);
- 'Respiratory Protective Equipment: A practical guide for users' (1990);
- 'Respiratory Protective Equipment, legislative requirements and lists of HSE approved standards and type approved equipment' (Fourth edition, 1995).
- INDG288 – 'Selection of suitable respiratory protective equipment for work with asbestos' (1999).

## Health and Safety Executive Guidance Notes
### Environmental Hygiene Series
- EH 10 – 'Asbestos: Exposure limits and measurement of airborne dust concentrations' (revised 2001);
- EH 47 – 'The provision, use and maintenance of hygiene facilities for work with asbestos insulation, asbestos coating and asbestos insulating board' (revised 2002);
- EH 50 – 'Training operatives and supervisors for work with asbestos insulation and coatings' (March 1988);
- EH 51 – 'Enclosures provided for work with asbestos insulation, coatings and insulating board' (January 2000);
- EH 57 – 'The problems of asbestos removal at high temperatures' (December 1992).

### Medical Series
- MS 5 – 'Lung Function' (July 1977);
- MS 6 – 'Chest x-rays in dust diseases' (September 1980);
- MS 13 – 'Asbestos' (revised, April 1999).

### Methods for the Determination of Hazardous Substances Series
- MDHS 39/4 – 'Asbestos fibres in air, sampling and evaluation by Phase Contrast Microscopy (PCM) under the Control of Asbestos at Work Regulations' (November 1995), (Second Impression);
- MDHS 77 – 'Asbestos in bulk materials, sampling and identification by polarised light microscopy' (PLM) (June 1994);
- MDHS 100 – 'Surveying, sampling and assessment of asbestos-containing materials' (July 2001).

### Health and Safety Guidance Series
- HSG 189/1 – 'Controlled asbestos stripping techniques for work requiring a licence' (Second edition, 1999);
- HSG 189/2 – 'Working with asbestos cement' (Second edition, 1999)
- HSG 53 – 'The selection use and maintenance of respiratory protective equipment' (1998);
- HSG 210 – 'Asbestos essentials task manual: Task guidance sheets for the building maintenance and allied trades' (2001);
- HSG 213 – 'Introduction to asbestos essentials: Comprehensive guidance on working with asbestos in the building maintenance and allied trades' (2001);
- HSG 227 – 'A comprehensive guide to managing asbestos in premises' (2002).

# Business Continuity

## What is business continuity planning?

**B8001** Business continuity planning ('BCP') is not just about disaster recovery, crisis management, risk management or IT. It is a business risk management issue. It presents an opportunity to review the way an organisation performs its business processes, to improve procedures and practices, and increase resilience to interruption and loss.

To quote the Business Continuity Institute, the professional body for BCP:

> 'BCP is the act of anticipating incidents which will affect critical functions and processes for the organisation, and ensuring that it responds to any incident in a planned and rehearsed manner.'

The Turnbull Committee 'Guidance for Directors on Internal Controls' sets out an overall framework of best practice for business, based upon an assessment and control of its significant risks. For many companies, BCP will address some of these key risks and help them to achieve compliance.

Hence BCP as an activity has two primary objectives:

- minimise the risk of a disaster befalling the organisation; and
- maximise the ability of the organisation to recover from a disaster.

It differs from disaster recovery by recognising that approximately 80 per cent of disasters which organisations suffer are generated internally. As a result, BCP places more emphasis on managing out potential causes of disasters, than on recovering from those incidents which may not be avoided.

The BCP is therefore a combination of disaster/risk avoidance and the ability to recover.

## Business continuity planning in context

**B8002** Modern businesses cannot avoid all forms of corporate risk or potential damage. A realistic objective is to ensure the survival of an organisation by establishing a culture that will identify, assess and manage those risks that could cause it to suffer:

- inability to maintain customer services;
- damage to image, reputation or brand;
- failure to protect company assets;
- business control failure;
- failure to meet legal or regulatory requirements.

BCP therefore provides the strategic framework to achieve these objectives.

## Why is business continuity planning of importance?

**B8003** Very simply the answer to the above question is that disasters kill businesses.

In the UK, 80 per cent of companies who suffered a major disaster and did not have some form of BCP capability went into liquidation within eighteen months. A further 10 per cent suffered the same fate within five years.

Without BCP companies have only a one in ten chance of survival. Also, simply because:

- Customers expect continuity of supply in all circumstances.
- Shareholders expect directors/managers to be fully in control, and to be seen to be in control of any crisis.
- Employees and suppliers expect the business to protect their livelihoods.
- The company's reputation and brand is at risk without BCP.
- It is implicit in good corporate governance and demonstrates best practice in business management.

All the above needs to be taken into account against a business background in which:

- The pressures on business generally are changing and increasing.
- Technology is transforming and underpinning the business environment.
- Consolidation, restructuring, re-engineering, take-overs, acquisitions and mergers etc are now a fact of life.
- New risks and exposures are continually being created.

## Business continuity drivers

**B8004** The drivers or arguments that need to be understood by all involved in and concerned with BCP may be listed and grouped as follows:

- Commercial.
- Quality/Regulatory.
- Financial.
- Imported disaster.
- Public relations.

### Commercial drivers

**B8005** An organisation may operate in an intensively competitive market where any disruption to normal operation will be exploited by competitors. For example, when Perrier were forced into a total product withdrawal following a benzene contamination at source, the gaps on supermarket shelves were promptly filled by 30–40 competing brands.

The organisation may depend on a high degree of public confidence to generate/maintain sales, e.g. air transport industry. The issue for many organisations is not merely one of lost revenue, but also of lost customers, who may prove very difficult to get back. On occasions, it has been observed that companies who have suffered a disaster have spent three times their normal PR budget in an attempt to reassure customers that they are still around and worth doing business with.

Conversely, a visible and demonstrable BCP capability can be used to enhance and differentiate products and services.

## Quality and regulatory drivers

**B8006** If a company adheres to BS 7799 (the British Standard for information security management), there are specific requirements relating to the need for a BCP capability. If a company is BS EN ISO 9000 accredited, compliance with BS 7799 may well be an additional requirement.

Organisations who have implemented the Turnbull and Cadbury guidance on corporate governance will have agreed in principle to comply with all relevant ISO Standards. In addition, directors and officers have a duty to shareholders for the protection and wellbeing of their investments, hence the need for non-executive directors and an audit committee.

If the organisation operates under a licence agreement, a prolonged inability to operate could endanger the licence renewal. That inability to operate may also affect the ability to comply with statutory or regulatory requirements involving, for example, Inland Revenue, DTI, Stock Exchange, Customs and Excise, Companies House etc.

## Financial drivers

**B8007** Traditional financial measures of mitigating or cushioning against risk, such as insurance, are an important protective element within the overall organisational business risk management framework. Insurance on its own however has limitations which do not protect the complete business cycle.

Despite the fact that business interruption insurance is a fairly common component in most commercial insurance packages, cover is normally limited to two years from the event, and certain consequential losses, which may contain significant unquantifiable costs, will be excluded and hence non-recoverable.

For example, BI (business interruption) insurance does not cover the risk of lost customers, lost productivity or the attrition costs of key staff following a disaster. In some cases, this could be as much as 40 per cent of the total loss.

In some organisations, the disaster itself may be caused by, or made worse because of, loss of control over finances and cash flow, e.g. Barings, AIB.

## The imported disaster

**B8008** BCP is an approach which recognises that organisations do not operate in isolation of each other and that they rely on suppliers to continue to operate. Hence disasters elsewhere may have a knock-on effect within an organisation, and disasters within an organisation could adversely affect its customers.

With the concepts of single source supply and just-in-time delivery, there is an increasing need to undertake supplier audits to assess their BCP capability.

Also the risk of denial of access should be assessed, as in the case of the Manchester bomb explosion in 1996. Unaffected businesses were denied access by the emergency services for months after the explosion.

Natural disasters can also have the following knock-on effect:

Following the earthquake in Kobe, Japan, the world suffered a severe shortage of silicon chips (microprocessors). The price on the black market rose so dramatically, that it set off a spate of thefts worldwide. This led to any large office block thought to contain large numbers of PCs being hit.

## The public relations disaster

**B8009** Increasingly, companies strive to raise their profile in the market place and increase their visibility by promoting themselves as aggressively and publicly as possible. The public awareness they seek, and need, in order to sell their products and services is a window through which to view their operations.

Just as members of the public will be more aware of the products and services the company offer, so they will be more aware of any mistakes, accidents, disasters, poor decisions, poor performance etc. for which the company is ultimately responsible. The ill feeling that can be generated against the company by the general public can be so damaging that confidence in the company is totally undermined for a long period of time.

For example, Gerald Ratner jokingly referred to his products as 'cr*p' and later compounded the felony by suggesting that there was more value in a prawn sandwich than in some of his products. The ultimate consequence of his action was loss of sales and the subsequent break-up of his jewellery business. He lost control of his company and a number of employees lost their jobs through the resulting re-organisation.

Hoover's handling of the massive over-subscription to its 'Free Flights' promotion became a national scandal which caused the company to be pilloried almost weekly on prime time TV consumer programmes. Not only did several directors lose their jobs, but the Hoover crisis severely damaged the Hoover brand name so much that the company even considered changing its name away from one of the most powerful and successful brand names/images in history, hence the 'free flights fiasco!'

Thus, the ability of a company to avoid, or recover from, a disaster effectively can have a tremendous and positive PR impact, thus vastly improving stakeholders' (investors, public, customers, suppliers, employees etc.) confidence.

This area is covered in more detail in the section on crisis communication (B8038 below).

## Convincing the board – the business case

**B8010** A business case for BCP can be made by taking the following ten key points into account:

1. Find out how much time you have for your presentation and make sure you rehearse to make full use of what may be a limited time allocation.

2. Check the audience:
   - who will be there?
   - have non-executive directors previous experience of disaster and/or BCP?

3. Determine, in advance, if there are any regulatory influences that the board may be particularly sensitive to.

4. Achieve board buy-in/commitment to BCP by ensuring that they:
   - understand the negative impact of business interruption;
   - provide the necessary resources to reduce the risk of disaster;
   - appreciate the need to prepare and plan.

5. Stress the benefits of BCP:
   - peace of mind/sleep at night;
   - risk reduction.

6. Be prepared for the 'It won't happen here' syndrome. Have some statistics (not too many) and some local or industry examples that the board may be aware of.

7. Spell out that the board is ultimately responsible for business risk management. N.B. Turnbull/Corporate governance.

8. Avoid being drawn into a detailed discussion about how the BCP capability is to be developed.

9. Avoid being drawn into a discussion on actual or potential disasters, unless they are specifically relevant.

10. Summarise and reinforce the need for BCP by highlighting the unique selling points from the above list.

## The business continuity model

**B8011** The model is a phased and active process consisting of five main steps:

- *Stage 1: Understanding your business*

    Business impact analysis ('BIA') and risk evaluation/assessment tools are used to: identify the critical business processes within the business; evaluate recovery priorities; and assess risks which could lead to business interruption and/or damage to the organisation's reputation.

- *Stage 2: Continuity strategies*

    This involves determining the selection of alternative strategies available to mitigate loss, assessing the relative merits of these against the business environment; and deciding which are likely to be the most effective in protecting and maintaining the critical business processes.

- *Stage 3: Plan development*

    The development of a proactive BCP plan is designed to improve the risk profile by upgrading optional procedures and practices; introducing alternative business strategies; and using risk financing measures; including insurance.

- *Stage 4: Plan implementation*

    The introduction of BCP capability is achieved via education and awareness of all stakeholders, including employees, customers, suppliers and shareholders. The end result is to establish a BCP culture throughout the organisation. The BCPs should be widely communicated and published to all concerned so that everyone is aware of rules, responsibilities, accountabilities, timescales and KPIs (key performance indicator).

- *Stage 5: Plan testing and maintenance*

    The plan is not the capability. The plan is the means by which the capability is realised.

    Failure to plan is planning to fail!

    BCP is therefore not a 'one-off' exercise. There is a need for ongoing plan testing, maintenance, audit and change in the management of the BCP and its processes. Regular communication and feedback on the results of testing and maintenance are imperative so as to ensure continual improvement of the BCP capability.

The process of developing a BCP capability is a cyclic and evolutionary one in order to take account of the ever-changing organisation.

## Understanding your business (stage 1)

### Risk evaluation

**B8012** Risk evaluation is the process by which an organisation determines the pattern of risk that is unique to its operation.

Risk is used to describe the relationship between the impact and likelihood for a particular event or scenario:

Risk = Impact x Likelihood

Within the BCP framework, risk evaluation is the process which identifies those aspects of the company's operations, which are essential to the survival of the business, and assesses how well they are protected against serious disruption and/or potential disaster.

As in the field of occupational safety and health ('OSH'), risk evaluation is a combination of identification and assessment. Within BCP, there are three commonly used approaches to risk evaluation:

- scenario driven;
- component;
- critical business processes.

All approaches assume that all concerned fully understand what the company is in business for and what is vital for its survival.

### Scenario driven approach

**B8013** This approach involves the listing of all possible risk scenarios that can be thought of. Once the list has been compiled each scenario is considered in isolation with a view to planning specific counter-measures to tackle each risk scenario on an individual (and isolated) basis.

This approach has some serious limitations. It presupposes that risks exist in isolation and that they do not have a dynamic relationship with each other. This is clearly not the case. For example fire creates smoke, fumes and heat. The presence of smoke and fumes increases the risk of inhalation injury and may hinder evacuation or searching activities. If we then throw in the fact that the fumes are toxic or corrosive and also that they could adversely affect neighbouring sites, the single incidence of 'fire' has an exceptional number of associated risks.

Such an approach will have a high cost and also lead to complex planning implications.

### Component approach

**B8014** This approach examines each department or unit within the organisation in order to develop their individual risk profiles. Whilst this approach does eventually cover the whole organisation, it fails to prioritise amongst departments and hence does not determine those departments which are important/critical.

Also it sometimes misses linkages and dependencies between departments, e.g. the internal customer. The approach as a whole suffers from a lack of organisation-wide perspective, and hence it is difficult to establish which departments should be reinstated first.

### Critical business approach

**B8015** Business can be viewed as a set of linked processes. Some of these business processes, the critical ones, have a greater bearing on the core business or primary activity of the business. These are known as the critical business processes ('CBPs').

The fundamental question to be asked and understood is:

'What does my company actually do?'

Developing a model may help to clarify thinking and understanding in this regard. Reference to the company's vision, and mission statement and goals will also demonstrate where the organisation is focused. It is on mission critical activities – the critical business processes – where BCP should be focused.

There are four basic questions to be asked:

1. What is the business about?
2. When are we to achieve our goals?
3. Who is involved both internally and externally?
4. How are the goals to be achieved?

An organisation has many dependencies, both internally and externally, that support the mission critical business processes. These may include suppliers, customers, shareholders, key employees, IT systems, and manufacturing processes. It is important to identify these at an early stage and to involve people from the CBPs in the development of the BCP.

External influences on the CBPs may include government departments, regulators, competitors, trade bodies, and pressure groups. These must also be taken into account with the BCP framework.

In summary, the critical business approach:

- Needs a common understanding of the company's primary business objective – the focus or mission of the company. This is ultimately what the capability is being designed to protect and recover.

- Some BCPs are doomed to failure because the participants cannot agree to what the company's primary business objective actually is!

- Determines the CBPs by asking how important are they in achieving the business objective. Again agreement amongst the participants regarding the CBPs is vital in developing a relevant BCP.

- Develops a BCP capability to protect all identified CBPs whilst keeping focused on the primary business objective will prevent participants getting sidetracked or bogged down in detail.

## Business impact analysis

**B8016** Business impact analysis ('BIA') is the technique used to determine what the business stands to lose if its CBPs are disrupted.

BIA alerts the organisation to the likely cost of a business disruption and assists in the cost benefit analysis of risk reduction measures within the overall BCP capability. Indeed the approval and continuing support of your board for BCP may well hinge on the fact that the costs are justified in the light of potentially catastrophic business losses.

### Quantification of business impact

**B8017** It is important that all concerned in the BCP process understand that disasters are characterised by time and not event. What causes the disruption is less important to the survival of the organisation than how long the disruption lasts. BIA therefore should consider the impact of a disruption in any of the CBPs spread over time. Traditionally, the impact of a disruptive event was derived merely by determining how much revenue would be lost over the period during which the company was not operational:

Loss of revenue – *Out for 1 day* –> **Impact** = Annual Revenue/ No of working days/year

Loss of revenue  *Out for a year* –> **Impact** = Annual Revenue

On its own, however, loss of revenue does not give the full picture of the overall impact on the business. Other factors, which need to be taken into account, include:

- *Erosion of customer base*

    The organisation will lose customers if it cannot continue to service their requirements. Some will go simply because they perceive the service is unreliable and some will be forced to go for the sake of their own business continuity. Whatever the reason, it is generally estimated that it costs three times more to sell goods/services to a new customer than to an existing one. In the event of a prolonged disruption, the company could therefore have to treble its marketing, advertising, publicity and sales cost in order to restore its customer base.

- *Loss of key staff*

    Following a disaster, some companies have reported that up to 40 per cent of their workforce have left as a result. The loss of intellectual capital and the cost of recruitment can be considerable.

- *Loss of reputation*

    Most companies depend on their reputation as a reliable service provider to generate further sales. Also a great deal of resources, time and money, will have been spent over the years in cultivating a brand image linked to their reputation. Hence, any event which detracts from that image brings the company's reputation into question and not only jeopardises the previous investment but also seriously affects current and future sales. Such lack of confidence will also have a negative effect on the share price of publicly quoted companies.

- *Regulatory and contracted impact*

    The company may be subject to penalty clauses/payments if it cannot meet contracted obligations. It may also be barred from operating by an inability to comply with regulatory (e.g. prohibition notice requirements), reporting or licensing obligations.

- *Loss of credit-worthiness*

    The organisation may face increased costs of borrowing following a disruption to business, on the basis that it represents a greater investment risk to lenders as a result of the disaster.

## Business impact Analysis: outputs

**B8018**    BIA enables the organisation to focus on the CBPs and their negative impact on the business, rather than to conduct a global, risk-specific analysis. The process also takes time sensitivity into account, thus enabling the recovery objectives to be agreed.

It is essential to rate the impact of the loss of CBPs on the business. Risk rating may be qualitative: high, medium, low or quantitative, i.e. 1–9 rating scale. Where possible, financial values should be placed on the business impact.

It is also important to involve key personnel who work in the CBPs in the BIA process, especially as the CBPs are cross-functional/cross-departmental, and agreement must be reached on the relative ratings. Agreement from the board should then be obtained to ensure that the output from the BIA, i.e. the plan is put into place.

In summary therefore:

- Ensure involvement of appropriate functions/departments.

- Determine the impact on the business of the loss of CBPs.
- Apply risk ratings, including time dependencies.
- Obtain broad approval for BIA outputs, i.e. the plan.

## Continuity strategies (stage 2)

**B8019** Having identified those areas where the organisation is most at risk, a decision has to be made as to what approach is to be taken to protect the operation. With the introduction of the Turnbull Guidance on internal control (corporate governance) this decision must be taken at board level.

Many possibilities exist and it is likely that any strategy adopted will comprise a number of these approaches. Whichever are chosen, there are certain considerations to bear in mind, as indicated below:

- Do nothing (risk retention) – in some instances, the board may consider the risk to be commercially acceptable.
- Change or end the process (risk avoidance) – deciding to alter existing procedures have to be done, bearing in mind the organisation's key focus/mission.
- Insurance (risk transfer) – provides financial recompense/support following a loss but does not provide protection for brand, image and reputation.
- Loss mitigation (risk reduction) – tangible loss control programme to eliminate or reduce risk.
- Business continuity planning (risk reduction) – an approach that seeks to improve organisational resilience to interruption, allowing for the recovery of key business systems and processes within the recovery time frame objective whilst maintaining the organisation's critical business processes.

Any strategy must recognise the internal and external dependencies of the organisation and must be accepted by all members of management involved in the CBPs. Hence, in summary, strategy selection is as follows:

- Identify possible BCP strategies.
- Assess suitability of alternative strategies against the outputs of the BIA.
- Propose cost/benefit analyses of the various strategies.
- Present recommendations to the board for decision and approval.

## Developing a business continuity plan (stage 3)

**B8020** As previously stated:

- Failing to plan is planning to fail.
- The plan is not the capability.

The BCP is the means by which the capability is realised and can be viewed as the route which takes the organisation from its present state (crisis/disaster) to its desired state (normal operation).

Too often organisations derive a false sense of security from the mere presence of a disaster recovery or BCP and lose sight of the fact that the plan is passive. It has no inherent capability to assist the organisation in any way until it is used. Its value to the organisation can only be assessed by how well it enabled the business to either avoid or recover from a disaster-threatening scenario, i.e. after the event.

## The planning dilemma

**B8021**  *Question*: How does one include enough detail to provide meaningful instruction and guidance without including so much data that individuals or departments find the plan difficult to use?

There is no simple answer to this question since requirements will vary from business to business. However, via the outputs from the BIA, from the experience of individuals involved in the BCP process, and from regular testing and maintenance of the plan, a balance that suits the organisation will be reached through plan refinement.

## Plan components

**B8022**  The objective of creating a BCP capability is firstly to design/engineer out as much of the operational risk from the identified CPBs as possible, i.e. risk avoidance.

Thereafter the BCP should provide a structural means of recovering those lost processes within a given timeframe.

Thus the plan will have two main components:

- *Proactive* – range of cost-justified risk reduction measures, procedural, physical, behavioural, processes, i.e. engineering.
- *Reactive* – disaster recovery plan.

### Disaster recovery plan

**B8023**  The vast majority of disaster recovery plans will compose three stages:

1. *Emergency response – Immediate*

    This stage details the actions the company has predefined in immediate response to an emergency situation. Typically, it should include details of evacuation procedures, liaison with emergency services, how the BCP kicks in, notification procedures, damage assessment/ control/mitigation, deployment of first aid and other emergency resources.

2. *Fallback procedures – Short term*

    This stage details the process by which the company starts to cope with the aftermath of the disaster, i.e. the fallback position. This may involve a physical move to alternative premises, alternative working methods, or the commencement of contractual agreements with third parties to undertake one or more of the CBPs on behalf of the company. A list of potential suppliers of such services, together with contract details is a 'must' in this regard.

3. *Business resumption – Medium term*

    This stage details the process which determines when and how the organisation gets on the road back to normal operation. Typically it will contain logistical, commercial, and political criteria which will be used by the company to decide where the business will be resumed, which CBPs will be reinstated first and where resources will be allocated to get the operations back to normality as quickly as possible.

### Hierarchical planning structure

**B8024**  In order to co-ordinate the recovery process, particularly when some of the decision-makers are not located at the site of the disaster, many companies make use of hierarchical management planning structure:

- *The command and control team*

    This team is usually staffed by senior managers/decision-makers, supported by a business continuity professional. Their objective is to analyse the information being fed to them by the damage assessment team and to co-ordinate the overall response of the company to the disaster. It is their job to ensure that the recovery is on track; that they respond quickly and effectively to any developments in order to achieve restoration; to allocate emergency response resources; and to handle crisis communications both, internal and external.

- *The damage assessment team*

    The damage assessment team is generally a multi-disciplinary group able to convey accurate information from the disaster site to the command and control team. They are also typically responsible for on-site OSH issues, site security, access control, liaison with emergency services, providing the command and control team with regular updates on restoration progress, and general tactical decision-making.

- *Departmental team(s)*

    At some point within the BCP, those departments involved in CBPs will be instructed by the command and control team to incorporate their individual recovery plans. This may involve moving to an alternate location or developing an alternative way of working. Regular progress reports will be provided to the command and control team to enable them to keep track of the overall recovery. In this way, slippage against expected recovery timescales might be quickly identified, thus allowing additional resources to be allocated to that department, if available.

## Plan development process

**B8025**    The process for developing the BCP follows the criteria for project management methodology, increasingly employed by many organisations. It is essential that the BCP capability is developed in accordance with the following:

- Scope the development in terms of time and resources.
- Identify what skills are required to complete the recovery.
- Set milestones to demonstrate progress.
- Define reporting lines, roles, responsibilities, accountabilities.
- Establish a project tracking method.
- Set sign-off criteria.
- Move into maintenance phase.

## Plan implementation (stage 4)

**B8026**    The implementation process for the BCP requires the achievement of certain milestones in the following chronological order.

## Securing buy-in

**B8027**    Hopefully by this time, buy-in by the board will have been secured, using the techniques discussed above. It is vital, however, that this approval manifests itself in some way that allows it to be readily communicated throughout the organisation. A visible commitment to the BCP is required.

## Policy

**B8028** Having a clearly stated BCP policy is one good way of visibly demonstrating board level commitment. The policy can be issued under the signature of the MD/CEO as an endorsement of the BCP process. This then provides a mandate to those involved to get the BCP implemented throughout the organisation.

## Project authorisation

**B8029** The approval of the board should also include authorisation for the BCP project as a whole. This would generally be based on the information presented via the BIA, risk identification and risk evaluation exercises.

## Individual responsibilities

**B8030** All key individuals should, as part of their job, have clearly defined and agreed responsibilities, accountabilities, timescales, KPIs and action plans, so as to ensure smooth plan implementation. In this way individual personal performance is geared to successful BCP implementation and an ongoing BCP capability.

## Phased roll-out

**B8031** We have already determined that in order to provide the organisation with the most cost-beneficial BCP capability, there is a need to focus on the CBPs. In both the development and implementation stages, it is essential that the organisation stratifies its efforts to cover and protect those functions most directly involved in the core business (primary business objective) and the CBPs before moving on to less critical functions/areas.

This stratification will normally manifest itself as a phased roll-out of the BCP to the CBPs first, intermediate functions next, and non-business critical functions last, if at all.

## Specialist services

**B8032** Quite apart from the emergency services, not all the resources required to implement the plan will be available in-house or on-site.

The following list is by no means exhaustive but serves to illustrate further possible components of the BCP. It is imperative that contact details for all such support services are kept readily available (in at least two known locations) and up to date:

- Data recovery/IT back-up.
- Emergency telecommunications.
- Salvage/decontamination.
- Cleaning/restoration.
- Buildings/facilities management.
- Security.
- Counselling (post-traumatic stress disorder (PTSD)).

## Plan testing and maintenance (stage 5)

**B8033** Maintenance and testing are both vital components in ensuring that the BCP continues to support the disaster avoidance and recovery capability required by the organisation.

Maintenance is essential to avoid the plan becoming unusable because it no longer reflects the organisation's structure, vision, mission or business priorities.

The objective of testing is to provide a high degree of assurance that the plan will work when it is needed, and also to highlight and rectify any shortcomings in the plan so that the BCP process is as streamlined as possible.

Maintenance should be an ongoing process throughout the lifetime of the BCP. Its primary focus is to change the plan to continuously reflect changes within the organisation. New business lines or processes, new reporting lines, reorganisations, changes of personnel, location, telephone numbers, user IDs etc all need to be reflected in the plan.

It is strongly advisable to allocate responsibilities for plan maintenance and testing to a group of individuals and make it part of their job descriptions, CYOs (current year objectives) and KPIs.

## Testing strategies

### The desk check/audit

**B8034** Many plans fall into disrepair due to a lack of accuracy and, consequently, integrity in the detail of the plan itself. This situation seriously undermines the effectiveness of the plan and its ability to protect the organisation in a disaster situation. The desk check/audit is a type of testing which seeks to verify that the factual detail contained in the plan is both accurate and current. As well as confirming the factual correctness of the plan, it also provides an excellent indication of how well the maintenance function is being performed.

### The walkthrough test

**B8035** This test involves synthesising an incident or fictional set of circumstances, thereafter allowing participants to use the plan to resolve the described situation. Similar in concept to a war game, the rules of engagement are predefined and an external facilitator co-ordinates the introduction of extra information. The logic of the plan is tested conceptionally for robustness against a scenario designed to involve all sections of the plan.

### Component testing

**B8036** Component testing normally involves single functions or departments testing in isolation their own parts of the plan by enacting their emergency, fallback and resumption arrangements. The objective is to test the practicality of the plan and to derive an estimate of how effective the plan is in terms of logistics and recovery times. It has the merit of eventually covering the whole organisation but it does not adequately test the linkages between functions and departments, which may more fully represent the flow of activity within the organisation.

### Full simulation

**B8037** The organisation deliberately shuts down or denies access to CBPs, sometimes without warning, in order to get the most realistic feel for how groups and individuals react and perform in the event of a disaster.

Whilst this type of testing provides the highest possible degree of assurance that the plan works and the organisation can cope, it needs to be planned extremely carefully, otherwise it may well trigger off an alternative, real-life, disaster!

Normally a test of this type should not be undertaken until the organisation has successfully carried out the other types of test described with consistently favourable results.

## Crisis communication and public relations

**B8038** In a crisis, many decisions have to be taken within a very short timescale. Some of those decisions may determine whether the organisation fails or survives! This fact represents one of the most powerful arguments for BCP and hence should ensure that the organisation has a proactive, up-to-date plan to communicate with its internal and external audiences throughout the crisis.

Employees, suppliers, customers, shareholders, the public and the media all expect and need to be kept informed in order for the organisation to retain its reputation and to maintain confidence in the business. An inability to communicate with such audiences before, during and after a crisis can easily cause a public relations disaster, thereby compounding the original incident.

Examples of companies who did not, or chose not to, communicate include:

- Hoffman-La Roche: Seveso (July 1976).
- Eli Lilly: Opren Withdrawal (August 1982).
- Union Carbide: Bhopal (December 1984).
- Delta Airlines: Plane Crash (August 1985).
- Sandoz: Rhine Pollution ( November 1986).

In all cases, the share price fell within a short time of the disaster and did not get back up to the market average for at least a year after the event.

The converse of the situation faced by these companies is one where the organisation does communicate and manages to secure excellent PR from a difficult situation for example:

- British Midland: Kegworth Air Crash.
- Commercial Union: City IRA Bomb:

'We don't make a drama out of a crisis!'

In order to minimise the risk of negative PR and the damage it can do, many companies include a communications plan within their BCP. This also enables the company to be portrayed in a positive and responsible light.

### The communication plan

**B8039** The communications plan should ideally comprise three stages:

Stage 1: Pre-crisis – to minimise the risk of poor communications making the disaster worse

Stage 2: During crisis – to manage the volume and complexity of enquiries and information requirements

Stage 3: After crisis – to let shareholders know that we are back in business

### Pre-crisis measures

**B8040**

- Understanding which media are influential in your market and your location and developing a working relationship with them prior to any incident will tend to make them more sympathetic to your version of events in the reporting of a crisis.

- Press statements are an excellent way of ensuring that there is a single source of authorised information being disseminated to the media. They provide a consistent means of updating stockholders. Having them prepared in advance, for most possible scenarios, will save the company valuable time in a crisis.

- In the event of a disaster, which generally excites the interest of the media, it is likely that directors/senior managers will be approached for comment in an interview format. The physical environment and the circumstances of the interview often conspire to generate a highly charged atmosphere, where the untrained individual may fail to portray the organisation in a good light, hence missing the opportunity to use this very powerful medium of communication to good effect. Training and rehearsal can dramatically improve individual performance on camera and enhance the organisation's ability to harness the power of the media.

- It is therefore advisable for the company to identify and train not just directors and senior managers, but any individuals who have some detailed knowledge or experience of the organisation's critical business processes. They then become the company spokespersons if problems occur within their area of expertise.

- In the event of a disaster, organisations experience a phenomenon known as call deluge. Organisations can expect to receive up to ten times as many calls and enquiries following a disaster than on a normal day.

Under such a weight of demand, the risk of providing inaccurate or misleading information is vastly increased.

One way of managing call deluge is to set up a dedicated communications office to which are channelled all disaster-related calls. In some cases a separate crisis hotline telephone number should be given out and all received calls be dealt with by trained operators.

From this centre the enquiry can be satisfactorily answered and passed on to one of the company spokespersons, if considered necessary. This approach enables the company to maintain integrity and consistency in the information being released concerning the disaster and its ongoing effects and mitigation.

## *During crisis*

B8041

- The communications office swings into action. It is advisable to include the setting up of a dedicated communications office within the testing and maintenance section of the BCP.

- Spokespersons needed for comment and/or interview should be readily available throughout the duration of the crisis. Ideally, they should be on site, or near to the site, working in tandem with the command and control team.

- Press/media statements should be released at regular and frequent intervals to update them on damage mitigation and recovery progress.

- Key stakeholders (e.g. customers) should be updated at regular and frequent intervals with regard to the recovery process and the estimated time for the reinstatement of their service.

- As the situation develops, all staff need to be kept informed, particularly if they are required to participate in the recovery process. In any event, all staff should be aware of the likely timescales for the business getting back into full operation.

- As the recovery from the disaster takes place, a crisis event log should be kept to enable significant events, actions or milestones to be recorded with the purpose of learning from them in the aftermath of the disaster, and possibly reviewing and amending the BCP in the light of unique experiences.

## After the crisis

### B8042

- Use the crisis events log to review and analyse how well the organisation performed towards its agreed recovery targets. Does the BCP need amending in the light of unique experiences during actual recovery?

- Identify those aspects which worked well and those where there is room for improvement to the BCP, in order to make things work more smoothly if ever there is a next time!

- Publicise a summary of performance to all stakeholders in order to demonstrate that the BCP has actually worked in practice. This will breed confidence in the organisation and its ability to rise above and proactively manage any future potential disaster scenarios.

- And finally consider a dedicated marketing campaign to make capital of how the crisis was managed in a positive manner, and to share the lessons learnt with others operating in similar environments. This will again demonstrate the organisation's capability to cope before, during, and after any type of disaster or crisis.

# Compensation for Work Injuries/Diseases

## Introduction

**C6001** Compensation for work injuries, diseases and death is payable under the social security system and in the form of damages for civil wrongs (torts). This chapter examines the two systems and the interaction between them.

The social security system is a form of public insurance, funded by employers/employees and taxpayers, and benefit is payable irrespective of liability on the part of an employer, i.e. 'no fault'. However, connection with employment must be established. The social security system was created by legislation, such as the various Social Security Acts, the *Social Security Administration Act 1992*, the *Social Security Contributions and Benefits Act 1992* ('*SSCBA 1992*'), the *Statutory Sick Pay Act 1994*, the *Social Security (Incapacity for Work) Act 1994*, the *Social Security (Recoupment of Benefits) Act 1997*, and other legislation.

Liability in tort depends on proof of negligence or breach of statutory duty against an employer. Employers must be insured for such liability (see EMPLOYERS' LIABILITY INSURANCE). The current law relating to work injuries and diseases is to be found in a variety of Acts, such as the *Law Reform (Personal Injuries) Act 1948*, the *Employers' Liability (Compulsory Insurance) Act 1969*, the *Employers' Liability (Defective Equipment) Act 1969*, the *Damages Act 1996*, various health and safety regulations and a wide body of case law.

Some useful guides worth referring to are:

- *Guidelines for the assessment of general damages in personal injury cases* (2002), 6th edition (ISBN 0 19 925795 7), issued by the Judicial Studies Board, which is a publication which aims to unify judicial approaches to awards of damages. These guidelines are not a legal document and a full examination of the law applicable to each case is required.

- *Industrial injuries handbook for adjudicating medical officers*, was published in 1997. It sets out medical examination procedures and offers guidance on legislation, including discussion of case law.

## Social security benefits

**C6002** Under the *Social Security (Incapacity for Work) Act 1994*, and the Regulations made under that Act, the higher rate of incapacity benefit is not payable until after 364 days. The effect that this has had on industrial injuries benefits, including disablement benefit, is that industrial benefits continue to be payable after 90 days, but the benefits which are payable before and at the same time, that is, any national insurance benefits, are less for the relevant period. The higher adult dependant's allowance is delayed for 364 days when the long-term incapacity benefit increases begin. Claimants whose industrial injuries benefit is lower than the rate of incapacity for work benefit and who are thus entitled to claim this benefit are also affected. (See C6015–C6017 below).

Entitlement to industrial injuries benefits requires compliance with national insurance contribution conditions for incapacity benefit. The incapacity for work tests under the *Social Security*

(*Incapacity for Work*) *Act 1994* do not apply to industrial disablement benefits. The medical adjudication procedures for these benefits continue with some modifications as before. The incapacity for work tests will apply to sufferers from industrial disablement if they apply for incapacity benefit before the 90 qualifying days for industrial disablement benefits begin, or if they do not qualify for industrial disablement benefit under the percentage or other rules, or if incapacity benefit is payable at a higher rate than industrial disablement pension rates and incapacity benefit is chosen by the claimant.

Under the rules for claimants for incapacity benefit, the employee has to supply information and evidence of sickness and be prepared to submit to a medical examination to decide whether he is fit to work. If he has worked for more than eight weeks in the twenty-one weeks immediately preceding the first day of sickness, a test relevant to his incapacity to do work which he could reasonably be expected to do in the course of his occupation applies. This test criterion continues to the 197th day of incapacity. After that date, and for all claimants who do not qualify for the required period, an 'all work' test arises. Regulations provide for prescribed activities and a person's incapacity, by reason of a specific disease, or bodily or mental disablement, to perform these activities. These tests are detailed and are outside the scope of this publication. [*SSCBA 1992, ss 171(A)–(G)* as inserted by *Social Security* (*Incapacity for Work*) *Act 1994, ss 5–6*]. The *Social Security* (*Incapacity for Work*) (*General*) *Regulations 1995* (*SI 1995 No 311*) (as amended by the *Social Security* (*Incapacity for Work*) (*General*) *Regulations 1996* (*SI 1996 No 484*)), provide that certain people are deemed to be incapable of work, including those suffering from a severe condition as defined in the Regulations or those receiving certain regular treatment (such as chronic renal failure, hospital in-patients and those suffering from an infectious or contagious disease). 'Welfare to work beneficiaries' will also be treated as incapable of work in the circumstances prescribed by *Regulation 13A* of the 1995 Regulations.

A pregnant woman may also be deemed incapable of work if there is a serious risk to her health, or that of her unborn child, if she does not refrain from work, in the case of the 'own occupation' test; or in the case of the 'all work' test, if she does not refrain from work in any occupation. If she has no entitlement to maternity allowance or statutory maternity pay and the actual date of confinement has been certified, she is deemed to be incapable of work beginning with the first day of the sixth week before the expected week of confinement until the fourteenth day after the actual date of confinement and during this period she will therefore be entitled to claim incapacity benefit.

Whilst most actual work by a claimant disqualifies him from receiving this benefit on any day of such work, certain work does not stop benefit. [*Social Security* (*Incapacity for Work*) (*General*) *Regulations 1995* (*SI 1995 No 311*), *Reg 17*]. Earnings from such work must not exceed £67.50 per week and in most cases must be for less than 16 hours per week. The exempt work is work done on the advice of a doctor which:

- helps to improve, or to prevent or delay deterioration in the disease or bodily or mental disablement which causes that person's incapacity for work; or

- is part of the treatment programme undertaken as a hospital in-patient or out-patient under medical supervision; or

- is done when the claimant is attending a sheltered workshop for people with disabilities.

Voluntary work and duties as a member of a disability appeal tribunal or the Disability Living Allowance Advisory Board are also exempt work.

# Industrial injuries benefit

**C6003** The benefits which are available to new claimants for industrial injuries benefits are:

- industrial injuries disablement pension;

- constant attendance allowance; and

- exceptionally severe disablement allowance.

Disabled person's tax credit may also be claimed in some circumstances (see C6016 below). Some older benefits continue to be payable to recipients who were receiving them when they were otherwise abolished, or whose entitlement arose before the relevant dates. (For details of these older benefits and more detailed social security benefits information, see *Tolley's Social Security and State Benefits Handbook*.)

## Accident and personal injury provisions

**C6004** The employed earner must have suffered personal injury caused after 4 July 1948 by an accident arising out of and in the course of his employment, being employed earner's employment. [*SSCBA 1992, s 94*]. The *Reporting of Injuries, Diseases and Dangerous Occurrences Regulations 1995* (*SI 1995 No 3163*) contain detailed provisions which require reports to be made on prescribed forms to the Health and Safety Executive after the occurrence of an accident or on receipt of a report from a registered medical practitioner of his diagnosis of a prescribed disease. A self-employed person may arrange for this report to be sent by someone else. Records must be kept for three years containing prescribed details of accidents, or the date of diagnosis of the disease, the occupation of the person affected and the nature of the disease. For more information on these Regulations see ACCIDENT REPORTING.

Industrial accident and disease records may be kept for three years on:

- a B510 Accident Book;
- photocopies of completed form F2508; or
- computerised records.

### Personal injury caused by accident

**C6005** 'Personal injury' includes physical and mental impairment, a hurt to body or mind, which includes nervous disorders or shocks (R(I) 22/52; R(I) 22/59). The injury must have been caused by an accident. Although this is usually an unintended and unexpected occurrence, such as a fall, if a victim is injured by someone else, may also be considered to be an accident (*Trim Joint District School Board v Kelly [1914] AC 667*). A relevant accident may still have occurred where a series of accidents without separate definite times cause personal injury. An office worker was held to have suffered a series of accidents on each occasion she had been obliged to inhale her colleagues' tobacco smoke and this was held to have caused personal injury.

'Accident' must be distinguished from 'process', that is, bodily or mental derangement not ascribable to a particular event. Injuries to health caused by processes are not industrial injuries, unless they lead to prescribed industrial diseases (see C6010 below). 'There must come a time when the indefinite number of so-called accidents and the length of time over which they occur, take away the name of accident and substitute that of process' (*Roberts v Dorothea Slate Quarries Co Ltd (No 1) [1948] 2 All ER 201*). In *Chief Adjudication Officer v Faulds [2000] 1 WLR 1035*, the House of Lords, in disallowing a claim for damages for psychological injury suffered by a fireman, held there must be at least one identifiable accident that caused the injury. The fact that an employee might develop stress from a stressful occupation would not satisfy the definition of 'accident'.

### Accident arising out of and in the course of employment

**C6006** There is no need to show a cause for the relevant accident provided that the employee was working in the employer's premises at the time of the accident. An accident arising in the course of employment is presumed to have arisen out of that employment, in the absence of any evidence to the contrary.

What runs through all the case law is the common requirement giving rise to industrial injuries rights that the employee was doing something reasonably incidental to and within the scope of his employment, including extra-mural activities which the employee has agreed to do (R(I) 39/56). A male nurse who was injured in a football match watched by patients in the hospital grounds succeeded in a claim for benefit as this was reasonably incidental to and within the scope of his employment (R(I) 3/57), but a policeman who was injured whilst playing football for his force was unable to recover benefit despite the fact that his employers had encouraged him to play in the game (*R v National Insurance Commissioner, ex parte Michael [1977] 1 WLR 109*). Also, in *Faulkner v Chief Adjudication Officer [1994] PIQR P244*, a police officer who was injured whilst playing for a police football team was not entitled to industrial injuries benefit despite the benefit to the community resulting from his participation. He was not on duty at the time. Further guidance can be found in *Chief Adjudication Officer v Rhodes [1999] ICR 178* where it was held that the two main questions to be asked are:

- what are the employee's duties; and
- was he discharging them at the time of the accident.

## Supplementary rules

C6007    Five statutory provisions establish rules under which the employee is deemed to be acting in the course of his employment duties. If the occurrence falls within these rules, the employee will be covered by industrial injuries benefit. These provisions are as follows:

- Illegal employment – if the employee was not lawfully employed, or his employment was actually void because of some contravention of employment legislation. [*SSCBA 1992, s 97*].

- Acting in breach of regulations, or orders of the employer – if the employee is not acting outside his authority under his employment duties, and an accident occurs while the employee is doing something for the purposes of, and in connection with, the employer's business [*SSCBA 1992, s 98*], he will be covered. For instance, a kitchen porter hung up his apron to dry in a recess near to the ovens where he was forbidden to go. He was injured when he fell into a shallow pit. The hanging up of the apron was for the purposes of his employment, so the accident was deemed to have arisen out of, and in the course of, his employment duties (R(I) 6/55).

- Where a dock labourer was employed on loading a ship by the method of two slings, but he instead used a truck which he had not been authorised to use for this purpose, he was held to be acting in the course of his employment. (R(I) 1/70B). This contrasts with the earlier case of *R v D'Albuquerque, ex parte Bresnahan [1966] 1 Lloyd's Rep 69*, where a dock labourer was killed in an accident whilst driving a forklift truck, which he had no authority or permission to use to remove an obstruction. His widow was unable to recover industrial injuries benefit as her husband was held not to have been acting in the course of his employment.

- Travelling in an employer's transport – travel to and from work is not covered except where the employee is travelling in transport provided by the employer with his express or implied permission, whether or not the employee was bound to travel in this transport. [*SSCBA 1992, s 99*]. Outside this express provision, the employee will, in most cases, both be required to be on the employer's premises doing what he was authorised to do, unless his work takes him off the premises. A postman was able to recover benefit when he was bitten by a dog on the street, as his job required him to walk along streets (R(I) 10/57). If the journey is preparatory to the start of timed itinerant duties, for example, as a home help, there will be no entitlement to benefit in respect of injury sustained on the way to the first home, though if the employee has more discretion about his movements, he may be entitled to benefit on the way to his first call.

- An injury incurred while an employee is trying to prevent a danger to other people, or serious damage to property during an emergency. [*SSCBA 1992, s 100*].

- Accidents caused by another's misconduct, boisterousness or negligence (provided that the claimant did not directly induce or contribute to the accident by his own conduct), the behaviour of animals (including birds, fish and insects). If these cause an accident, or if a person is struck by lightning or by any object, respectively confer entitlement to industrial injuries benefit. [*SSCBA 1992, s 101*].

## Relevant employment

**C6008** 'Employed earner's employment' includes all persons who are gainfully employed in Great Britain under a contract of service, or as an office holder, and who are subject to income tax under Schedule E. Self-employed people and private contractors are thus not entitled to industrial injuries benefits.

Certain classes of person are expressly included for the purposes of industrial injuries benefits. They include unpaid apprentices, members of fire brigades (or other rescue brigades), first-aid, salvage or air raid precautions parties, inspectors of mines, special constables, certain off-shore oil and gas workers and certain mariners and air crew. Most trainees on Government training schemes are excluded from the scheme. [*SSCBA 1992, ss 2, 95*].

If an industrial accident occurs outside Great Britain, industrial injuries benefit has since 1 October 1986 been payable when the employee returns to Great Britain, provided the employer is paying UK national insurance contributions, or if the claimant is a voluntary worker overseas and is himself paying UK contributions. Accidents which occur, or prescribed diseases which develop in other EU countries, are covered by common rules. An employee who is entitled to make a claim for industrial benefit in another EU member state should make his claim for benefit in that state, regardless of the country which will actually pay the benefit.

Certain types of employment are specifically excluded from cover. [*SSCBA 1992, s 95; Employed Earners' Employments for Industrial Injuries Purposes Regulations 1975* (*SI 1975 No 467*), *Regs 2–7* and *Sch 1, 2*].

## Persons treated as employers

**C6009** The *Employed Earners' Employments for Industrial Injuries Purposes Regulations 1975* (*SI 1975 No 467*), *Sch 3* also provide for cases where certain people who may not have a contract with the employee, or who may be an agency employer, to be the relevant employer for the person who has suffered the industrial injury, or who has developed a prescribed disease. An agency that supplies an office cleaner, or a typist, will be the relevant employer. For casual employees of clubs, the club will be the relevant employer.

## Benefits for prescribed industrial diseases

**C6010** The rules for certain prescribed diseases (namely deafness, asthma and asbestos related diseases), differ in some respects from the provisions affecting prescribed diseases outlined below. (See C6055–C6059 below.) The different rules applicable to diseases resulting from exposure to asbestos at work are dealt with for industrial injuries purposes by a special medical board and there is a separate state scheme for statutory compensation where one of these diseases develops and there is no remedy against an employer, or entitlement to state benefits (see C6061 below). (See leaflets NI 12, NI 207 and NI 272 for help with making a claim.) To obtain the right to industrial injuries benefits for all other prescribed diseases, the claimant must show:

- that he is suffering from a prescribed disease;

- that the disease is prescribed for his particular occupation (where a disease is prescribed for a general activity, for example, contact with certain substances, he must clearly show that this was more than to a minimal extent); and
- that he contracted the disease through engaging in the particular occupation (there is a presumption that if the disease is prescribed for a particular occupation, the disease was caused by it, in the absence of evidence to the contrary).

(See also OCCUPATIONAL HEALTH AND DISEASES.)

Claims are made on Form BI 100B and there is a right to, and it is advisable to, claim immediately after the disease starts. The 90-day waiting period for receipt of benefit applies as for accidents, as do the percentage disabilities and aggregated assessment rules, except in the case of loss of faculty resulting from diffuse mesothelioma when entitlement begins on the first day of the claim.

Assessments are made by two doctors who will decide on the percentage disability and how long the disability will last. Benefit will then be payable for the period stated in the assessment, but if the doctors are not sure of the period, benefit will be paid for a while with a further review. If the disease recurs during that period, there will be no need to make a further claim, but if the condition has worsened, the assessment may be reviewed. If there is a further attack after the period of the assessment, a fresh claim will have to be made, which will be subject to a further 90-day waiting period.

# Industrial injuries disablement benefit
## Entitlement and assessment

C6011  A person is entitled to an industrial injuries disablement pension if:

- he suffers from a prescribed industrial disease (see C6010 above);
- he suffers as a result of the relevant accident, a loss of physical or mental faculty such that the assessed extent of the resulting disablement amounts to not less than 14 per cent [*SSCBA 1992, S 103, Sch 6*]. See leaflet NI6 (July 1999) (update);
- 90 days (excluding Sundays) have elapsed since the date of the accident or onset of the prescribed disease or injury.

An assessment of the percentage disablement up to 100 per cent will be made by an adjudicating medical practitioner who, in the case of accidents, looks at the claimant's physical and mental condition, comparing him in those respects with those of a normal person of the same age and sex. No other factors are relevant. The assessment may cover a fixed period/or the life of the claimant. The degrees of disablement are laid down in a scale so that, for example, loss of one hand is normally 60 per cent and loss of both hands 100 per cent. Disfigurement is included even if this causes no bodily handicap.

Where the claimant suffered from a pre-existing disability before the happening of the industrial accident at the onset of the industrial disease, benefit will only be payable in respect of the industrial accident or disease itself, and the medical adjudicators will compare the original disability with the industrial disability for this purpose. Where one or more disabilities result from industrial accidents or diseases, the level of resulting disability may be aggregated, but not so as to exceed the 100 per cent disability and its corresponding rate of benefit. [*Social Security (General Benefit) Regulations 1982 (SI 1982 No 1408), Reg 11*].

A list of initial application forms is supplied in NI 6 (July 1999) Industrial Injuries Disablement Benefit. Once a claim for industrial injuries benefit has been made, the Secretary of State will send Form BI/76 to the claimant. This form asks for full details of the accident or disease. The claimant must then submit to examination by at least two adjudicating medical practitioners and must agree to follow any appropriate medical treatment. If he fails to comply with these requirements, he will be disqualified from receiving this benefit for six months. A decision in writing will be sent to the claimant. It is possible to appeal a refusal of industrial benefit; details of how to appeal will be sent with the decision if it is negative.

This benefit can only be paid 90 days after the date of the accident excluding Sundays. [*SSCBA 1992, s 103*].

### Claims in accident cases

**C6012** Claims should be made on Form BI 100A obtainable from the post office, job centre, or from the Benefits Agency, by claimants disabled by the accident for nine weeks. Claims should be made within six months of the accident to avoid loss of benefit, as the usual period for which benefit will be backdated is three months. [*Social Security (Claims and Payments) Regulations 1987 (SI 1987 No 1968), Reg 19, Sch 4, para 3*]. It is possible to apply to the DSS for a declaration from an adjudicating officer that an accident is covered by the scheme if a claimant suspects that an accident at work may have lasting effects which may not become apparent until some time later. If a positive finding is made, this will bind the DSS on any later claim for benefit. [*Social Security Act 1998, s 29(4)*]. Form BI 95 should be used for making this application. (Claims in respect of prescribed diseases are mentioned at C6010 above; appeals at C6018 below.)

### Rate of industrial injuries disablement benefit

**C6013** Benefit is paid at one of two rates with the higher being paid to claimants aged over 18.

## Constant attendance allowance and exceptionally severe disablement allowance

**C6014** Constant attendance allowance is available if:

- the claimant is receiving industrial injuries disablement benefit based on 100 per cent disablement, or aggregated disablements that total 100 per cent or more;

- the claimant needs constant care and attention as a result of the effects of an industrial accident or disease.

Furthermore, if the carer of the claimant spends at least 35 hours a week looking after him and is of working age and is not earning more than £75 (before tax but after deduction of national insurance and other reasonable expenses) per week, invalid care allowance of £43.15 per week with allowances for the carer's own dependants (if any) may be granted to that person. Claims for industrial injuries benefit are made on Form BI 104. It is granted for a fixed period and may be renewed from time to time. There are four rates of payment:

- part-time (where full-time care is unnecessary);

- normal maximum rate (if the above conditions are fulfilled and full-time care is required);

- intermediate (if the claimant is exceptionally disabled and the degree of attendance required is greater, and the care necessary is greater than under the 'normal' classification, the benefit is limited to one and a half times the normal rate); and

- an exceptional rate (if the claimant is so exceptionally disabled as to be entirely dependent on full-time attendance for the necessities of life).

If the intermediate or exceptional rates are payable, an additional allowance (known as exceptionally severe disablement allowance) will be due if the condition is likely to be permanent. Constant attendance allowance may continue to be paid for up to four weeks if the claimant goes into hospital for free medical treatment.

## Other sickness/disability benefits

C6015 However, most applicants for industrial injuries benefit will, as their incapacity results from accidents or diseases contracted during their employment, be able to claim statutory sick pay or incapacity benefit up to the date of commencement of industrial injuries benefit as 90 days is less than 28 weeks. The rate of statutory sick pay is higher than that for incapacity benefit for 28 weeks when they are then paid at the same rate. Incapacity benefit rates in the first 28 weeks apply to the self-employed and to the unemployed, or those with too few contributions. For the purposes of claiming incapacity benefit before being entitled to claim a disablement benefit, it should be noted that the presumption that a claimant for benefit for industrial or prescribed diseases has satisfied the contribution requirements for incapacity benefit (as was the case with sickness benefit), has been removed. An age addition is added to incapacity benefit where the claimant is under certain age limits. Claims for incapacity benefit for the first time claimant should be made on Form BI 202. Where entitlement to industrial injuries benefit is below the incapacity benefit rates, then that benefit may be claimed to top up industrial injuries benefit to the current rate of incapacity benefit.

It is possible to claim extra for a spouse or person looking after children if the spouse is over 60, or, if younger, child benefit is being paid, and the claimant was maintaining the family to at least the extent of the dependency benefit being claimed. None of these restrictive rules apply to industrial injuries benefit claims.

Income support and other state benefits such as housing benefit and council tax benefit may be available if the claimant with or without dependants does not have sufficient to live on.

## Disability Tax Credit under Working Tax Credit

C6016 This benefit is for people aged 16 years or over who wish to work, but have a physical or mental disability which puts them at a disadvantage in securing a job under criteria set out in the regulations. The applicant must work for at least 16 hours a week to qualify and he receives a credit if he works for 30 hours or more a week. Self-employed people may qualify For a full analysis of the disability tax credit see: www.ircc.inrev.gov/taxcredits.

## Payment

C6017 For the financial year 2003/2004 the weekly benefit rates payable for Statutory sick pay, Incapacity benefit, Industrial injuries disablement benefit, Constant attendance allowance and Exceptionally severe disablement are set by the *Social Security Benefits Up-rating Order 2003* (*SI 2003 No 526*).

### Appeals

C6018 Appeals relating to all industrial injuries are made to an appeal tribunal and should be made within one month from the date that the decision maker sends the decision to the applicant.

### Change of circumstances and financial effects of receipt of benefit

C6019

- *Hospital*

    If a claimant enters hospital, industrial injuries disablement pension continues to be payable, as does exceptionally severe disablement allowance. Constant attendance allowance will stop after four weeks. Statutory sick pay will be reduced after six weeks.

- *Taxation*

    Industrial injuries disablement benefit, constant attendance allowance, exceptionally severe disablement allowance and disabled person's tax credit are not taxable. Incapacity benefit is taxable for all claimants who have claimed after 13 April 1995. Those who were in receipt of invalidity benefit before that date are not liable to tax on it. Statutory sick pay is taxable under the PAYE system.

### Benefit overlaps

**C6020** Industrial injuries benefits may be taken at the same time as incapacity for work benefit with no reduction. Disability living allowance and attendance allowance may not be received at the same time as constant attendance allowance, but if the claimant should receive a higher rate of either of those benefits, constant attendance allowance will be topped-up to bring it up to the higher benefit rate. (For recoupment of benefit after awards of damages see C6044 below.)

## Damages for occupational injuries and diseases

**C6021** When a person is injured or killed at work in circumstances indicating negligence or breach of duty on the part of an employer, he may be entitled to an award of damages. Damages are assessed by judges in accordance with previously decided cases; very exceptionally they may be assessed by a jury. Damages are also categorised as general and special damages, according to whether they reflect pre-trial or post-trial losses. Calculation of damages is often made by reference to Kemp and Kemp, *The quantum of damages*, and to the Judicial Studies Board's document, *Guidelines for the assessment of general damages in personal injury cases* (2002) (6th edition).

Damages normally take the form of a lump sum; however, 'structured settlements', whereby accident victims are paid a variable sum for the rest of their lives, are now a viable alternative [*Damages Act 1996, s 2*] (see C6025 below). This may well involve greater reliance on actuarial evidence and a rate of return of interest provided by index-linked government securities.

Legal aid is currently the subject of extensive Government reform. The Legal Services Commission (LSC) replaced the Legal Aid Board in April 2000. Legal aid is no longer available in negligence personal injury cases. It is now only available in non-negligence personal injury claims, i.e. where the injury has been intentionally as opposed to negligently caused. There is one further exception, legal aid may be available where the personal injury claim forms a small part of another claim for which legal aid is available. Legal aid is still available for clinical negligence. The Access to Justice Act 1999 introduced the conditional fee system whereby solicitors represent clients on a 'no win no fee' basis. In the absence of legal aid, such a system will be the way in which most personal injury actions are brought.

Accident Line, which is part of a scheme run by solicitors who are members of a specialist panel of personal injuries lawyers, provides a free half-hour consultation for claimants who have suffered personal injuries, including industrial accidents. The telephone number is 0500 192 939. In addition, the Association of Personal Injuries Lawyers has a helpful website at: www.apil.com.

### Basis of claim for damages for personal injury at work

**C6022** The basis of an award of damages is that an injured employee should be entitled to recoup the loss which he has suffered in consequence of the injury/disease at work. 'The broad general principle which should govern the assessment in cases such as this is that the court should award the injured party such a sum of money as will put him in the same position as he would have been in if he had not sustained the injuries' (per Earl Jowitt in *British Transport Commission v Gourley [1956] AC 185*). The *Damages Act 1996*, provides that periodical payments may be made in some cases.

It is not necessary that a particular injury be foreseeable, although it normally would be (see further *Smith v Leech Brain & Co Ltd [1962] 2 QB 405*). Moreover, if the original injury has made the claimant susceptible to further injury (which would not otherwise have happened), damages will be awarded in respect of such further injuries, unless the injuries were due to the negligence of the claimant himself (*Wieland v Cyril Lord Carpets Ltd [1969] 3 All ER 1006*). In this case, a woman, who had earlier injured her neck, was fitted with a surgical collar. She later fell on some stairs, injuring herself, because her bifocal glasses had been dislodged slightly by the surgical collar. It was held that damages were payable in respect of this later injury by the perpetrator of the original act of negligence. Conversely, where, in spite of having suffered an injury owing to an employer's negligence, an employee contracts a disease which has no causal connection with the earlier injury, and the subsequent illness prevents the worker from working, any damages awarded in respect of the injury will stop at that point, since the supervening illness would have prevented (and, indeed, has prevented) the worker from going on working.

**C6023** Listed below are the various losses for which the employee can expect to be compensated. Losses are classified as non-pecuniary and pecuniary.

- *Non-pecuniary losses*

The principal non-pecuniary losses are:

(i) pain and suffering prior to the trial;

(ii) disability and loss of amenity (i.e. faculty) before the trial;

(iii) pain and suffering in the future, whether permanent or temporary;

(iv) disability and loss of amenity in the future, whether permanent or temporary;

(v) bereavement.

(Damages for loss of expectation of life were abolished by the *Administration of Justice Act 1982, s 1(1)* (see further 'loss of amenity' at C6034 below).)

There are four main compensatable types of injury, namely:

(i) maximum severity injuries (or hopeless cases), e.g. irreversible brain damage, quadraplegia;

(ii) very serious injuries but not hopeless cases, e.g. severe head injuries/loss of sight in both eyes/injury to respiratory and/or excretory systems;

(iii) serious injuries, e.g. loss of arm, hand, leg;

(iv) less serious injuries, e.g. loss of a finger, thumb, toe etc.

There is a scale of rates applicable to the range of disabilities accompanying injury to workers but it is nowhere as precise as the scale for social security disablement benefit. Damages for maximum severity cases can vary from several hundred thousand pounds to millions of pounds. In *Biesheuval v Birrell [1999] PIQR Q40*, the High Court awarded total damages of £9,200,000 (see C6024 below) to a student who was almost completely paralysed in all four limbs after a car crash. In *Dashiell v Luttitt [2000] 3 QR 4* a settlement of £5,000,000 was reached for brain damage sustained by a child aged 14 at the time of a school minibus crash. In *Capocci v Bloomsbury Health Authority (21 January 2000)* the High Court awarded £2,275,000 damages to a 13-year-old boy who had been asphyxiated at birth and as a result suffered cerebral palsy and other severe physical and mental handicaps. Less serious injuries attract lower damages – for example, in *Williams v*

*Gloucestershire County Council (10 September 1999)* an out of court settlement of £3,639 was reached in a case where a 6-year-old lost the top of her little finger in an accident.

- *Pecuniary losses*

  These consist chiefly of:

  (i) loss of earnings prior to trial (i.e. special damages);

  (ii) expenses prior to the trial, e.g. medical expenses;

  (iii) loss of future earnings (see below);

  (iv) loss of earning capacity, i.e. the handicap on the open labour market following disability.

  In actions for pecuniary losses, employees can be required to disclose the general medical records of the whole of their medical history to the employer's medical advisers (*Dunn v British Coal Corporation [1993] ICR 591*).

## General and special damages
C6024

- *General damages*

  General damages are awarded for loss of future earnings, earning capacity and loss of amenity. They are, therefore, awarded in respect of both pecuniary and non-pecuniary losses. An award of general damages normally consists of:

  (i) damages for loss of future earnings;

  (ii) pain and suffering (before and after the trial); and

  (iii) loss of amenity (including disfigurement).

- *Special damages*

  Special damages are awarded for itemised expenses and loss of earnings incurred prior to the trial. Unlike general damages, this amount is normally agreed between the parties' solicitors. When making an award, judges normally specify separately awards for general and special damages. A statement of special damage must be served with the statement of claim which should suffice to give the defendant a fair idea of the case he has to answer. More detailed information must be supplied after the exchange of medical and expert reports.

*Example*

An example of the way in which damages awards are assessed and broken down is the case of *Biesheuval v Birrell* (referred to in C6023 above) where the High Court awarded £9,200,000 damages consisting of:

| | |
|---|---|
| Pain and suffering and loss of amenity | £137,000 |
| Interest on general damages | £6,617 |
| Past loss of earnings | £80,700 |
| Interest on past loss of earnings | £14,929 |
| Other special damages | £360,113 |
| Interest on special damages | £54,516 |
| Tax on interest | £41,215 |
| Loss of future earnings | £3,700,000 |
| Loss of pension rights | £67,491 |

Pain and suffering and loss of amenity £137,000
Initial capital expenditure £551,803
Recurring costs and future care £4,267,000

## Structured settlements

**C6025** A structured settlement is an agreement for settling a claim or action for damages on terms that the award is made wholly or partly in the form of periodic payments. Such settlements are expected to become more usual in the case of larger awards of damages. Under the *Damages Act 1996*, these periodic payments must be payable in the form of an annuity for life, or for a specified period, and may be held on trust for the claimant if that should be necessary. Provision may be added to the settlements for increases, percentages or adjustments where the court or claimant's advisers secure these variations in his interest. Structured settlements in favour of claimants may be made for the duration of their life. Knowledge of the claimant's special needs is thus vital for the structure to be successful – it should be recognised that structured settlements will not be suitable for all cases. If the claimant dies while in receipt of periodic payments, they pass under his estate. Structured settlements may also be made in awards of damages in respect of fatal accidents. Tax-free annuities are payable directly to the claimant by the Life Office.

Structured settlement awards made by the Criminal Injuries Compensation Authority are also tax free.

When agreeing a settlement (whether structured or not), it is better to agree whether payments are net of repayable benefits. If a settlement offer is silent as to repayable benefits, then a deduction will have to be made in respect of them, possibly with unplanned results for the claimant.

Structured settlements are advisable where brain damage makes the injured person at risk and suggestible to pressure from relatives. They are also useful where the injured person has little experience or interest in investment, or dislikes the possibility of becoming dependent on the State or relatives should funds run out. Disadvantages are that annuities only last for the lifetime of the injured person. There is also a loss of flexibility to deal with changed circumstances and of the better return gained with the skilful investment of a lump sum.

## Assessment of pecuniary losses

**C6026** Assessing pecuniary losses, i.e. loss of future earnings, can be a difficult process. As was authoritatively said, 'If (the claimant) had not been injured, he would have had the prospect of earning a continuing income, it may be, for many years, but there can be no certainty as to what would have happened. In many cases the amount of that income may be doubtful, even if he had remained in good health, and there is always the possibility that he might have died or suffered from some incapacity at any time. The loss which he has suffered between the date of the accident and the date of the trial [i.e. special damages (see above)] may be certain, but his prospective loss is not. Yet damages must be assessed as a lump sum once and for all [see 'Provisional awards' at C6040 below], not only in respect of loss accrued before the trial but also in respect of a prospective loss' (per Lord Reid in *British Transport Commission v Gourley [1956] AC 185*). Moreover, if, at the time of injury, a worker earns at a particular rate, it is presumed that this will remain the same. If, therefore, he wishes to claim more, he must show that his earnings were going to rise, for example, in line with a likely increase in productivity – a probable rise in national productivity is not enough.

When the court assesses loss of earnings, the claimant has to mitigate his loss by taking work if he can. In *Larby v Thurgood [1993] ICR 66*, the defendant applied to the court to dismiss an action brought by a fireman who was severely injured in a road traffic accident and who had taken employment as a driver earning £6,000 per annum, unless he agreed to be interviewed by an employment consultant who would then give expert evidence on whether the claimant could have obtained better paid employment. This application was refused. Evidence of whether the claimant could have obtained better paid employment depended partly on medical evidence of his capabilities and the present and future state of the job market where he lived, which could be established by an employment consultant. His general suitability for employment, his willingness and motivation, were matters of fact for the judge; thus expert opinion was not required for that purpose.

A claimant may be earning practically as much as he was before his accident, but may be more at risk of losing his present job, of not achieving expected promotion, or of disadvantages in the labour market. Damages may be claimed for such prospective losses and are known as *Smith v Manchester* damages (after *Smith v Manchester City Council (1974) 118 SJ 597*).

## Capitalisation of future losses

C6027 Loss of future earnings, often spanning many years ahead, is awarded normally as a once-and-for-all capital sum for the maintenance of the injured victim. The House of Lords considered the way in which lump sums should be calculated in the leading case of *Wells v Wells; Thomas v Brighton Health Authority [1999] 1 AC 345*. This also applies in Scotland (*McNulty v Marshall's Food Group Ltd [1999] SC 195*).

The award of damages is calculated on the basis of the present value of future losses – a sum less than the aggregate of prospective earnings because the final amount has to be discounted (or reduced) to give the present value of the future losses. Inflation is ignored when assessing future losses in the majority of cases (e.g. pension rights), since this was best left to prudent investment (*Lim Poh Choo v Camden and Islington Area Health Authority [1980] AC 174*). And where injury shortens the life of a worker, he can recover losses for the whole period for which he would have been working (net of income tax and social security contributions, which he would have had to pay), if his life had not been shortened by the accident. The present value of future losses can be gauged from actuarial or annuity tables. The net annual loss (based on rate of earnings at the time of trial) (the multiplicand) has to be multiplied by a suitable number of years (multiplier) which takes account of factors such as the claimant's life expectancy and the number of years that the disability or loss of earnings is expected to last. The multiplier normally ranges between 6 and 18, and is set out in actuarial tables. Both the multiplier and the multiplicand can vary for different periods; for example, where medical evidence shows that the need for care could increase or decrease over time – *Wells v Wells; Thomas v Brighton Health Authority [1998] 3 All ER 481*. In *McIlgrew v Devon County Council [1995] PIQR Q66*, the maximum multiplier of 18 was applied for permanent general losses, and the multiplier of 12 was applied to loss of earnings.

The multiplier is calculated on the assumption that the claimant will invest the lump sum prudently. Under *section 1(1)* of the *Damages Act 1996*, the Lord Chancellor may by order prescribe a rate of return which the courts must have regard to. Under this provision, the Lord Chancellor has set the rate of return at 2.5 per cent.

*Example*

In the case of a male worker, aged 30 at the date of trial, and earning £15,000 per year net, on a 2.5 per cent interest yield, the multiplier will be 22.80 (using Table 25 of the Ogden Tables). Hence, general damages will be about £342,000, assuming incapacity to work up to age 65.

In the case of a male worker, aged 50 at date of trial, earning £25,000 net, on a 2.5 per cent interest yield, the multiplier will be 12.06 (using Table 25 of the Ogden Tables). Hence, general damages will be about £301,500 assuming incapacity to work until age 65.

Deductions from awards under this head are made for the actual earnings of the injured claimant. Where the claimant takes a lighter, less well paid job, and thus suffers a loss of earnings, the courts have held that the fact that he gains more leisure through working shorter hours is not to be taken into account to reduce the amount of damages awarded for loss of earnings (*Potter v Arafa [1995] IRLR 316*).

## Institutional care and home care

**C6028** Expenses of medical treatment may be claimed, except where the victim is maintained at public expense in a hospital or a local authority financed nursing home, in which case any saving of income during his stay will be set off against the claim for pecuniary loss. Alterations to a home, purchase of a bungalow accessible to a wheelchair, adaptations to a car and equipment (such as lifting equipment), which are necessary for care, may be claimed. Nursing and care requiring constant or less attendance may be claimed whether or not the carer is a professional or voluntary carer. It was confirmed by the House of Lords in *Hunt v Severs [1994] AC 350*, that where an injured claimant is cared for by a voluntary carer, such as a member of his family, that damages could be recovered for this care, but the claimant should hold them in trust for the voluntary carer.

Where voluntary care is undertaken by relatives, compensation for the cost of this care is assessed as a percentage of the Crossroads rate agreed from time to time for community care by most local authorities; the actual percentage awarded being about two-thirds of that rate. A higher percentage of that rate will be awarded where the care being given is beyond the level of care normally provided by home helps (*Fairhurst v St Helens and Knowsley Health Authority [1995] PIQR 41*).

## Damages for lost years

**C6029** Damages may be payable up to retirement age for lost earnings resulting from the shortening of the claimant's life expectancy by reason of the injury or disease. Estimated costs of living expenses are deductible from these damages. Damages under this head may also be awarded to dependants if the victim has died.

## Non-pecuniary losses (loss of amenity)

**C6030** It is generally accepted by the courts that quantification of non-pecuniary losses is considerably more difficult than computing pecuniary losses. This becomes even more difficult where loss of sense of taste and smell are involved, or loss of reproductive or excretory organs. Unlike pecuniary losses, loss of amenity generally consists of two awards: an award for (i) actual loss of amenity and (ii) the impairment of the quality of life suffered in consequence (i.e the psychic loss).

Victims are generally conscious of their predicament, but in very serious cases they may not be, a distinction underlined in the leading case of *H West & Son Ltd v Shephard [1964] AC 326*. If a victim's injuries are of the maximum severity kind (e.g. tetraplegia) and he is conscious of his predicament, damages will be greater. However, where, as is often the case, the injuries shorten the life of an accident victim, damages for non-pecuniary losses will be reduced to take into account the fact of shortened life.

## Types of non-pecuniary losses recoverable

**C6031** The following are the non-pecuniary losses which are recoverable by way of damages:

- pain and suffering;
- loss of amenity;
- bereavement.

## Pain and suffering

**C6032** Pain and suffering refers principally to actual pain and suffering at the time of the injury and later. Since modern drugs can easily remove acute distress, actual pain and suffering is not likely to be great and so damages awarded will be relatively small.

Additionally, 'pain and suffering' includes 'mental distress' and related psychic conditions; more specifically (i) nervous shock, (ii) concomitant pain or illness following post-accident surgery and embarrassment or humiliation following disfigurement. Claustrophobia and fear are within the normal human emotional experience but even are not compensatable, unless amounting to a recognised psychiatric condition, such as post-traumatic stress syndrome (*Reilly v Merseyside Regional Health Authority [1995] 6 Med LR 246*).

In *Heil v Rankin and Another and joined appeals [2001] QB 272*, the Court of Appeal, reviewed the current awards for pain, suffering and loss of amenity. The court held that whilst payments under £10,000 should not increase, there would be a tapered increase for awards above that up to a maximum of 33 per cent for the highest level of damages.

### Nervous shock

**C6033** Nervous shock refers to actual and quantifiable damage to the nervous system, affecting nerves, glands and blood; and, although normally consequent upon earlier negligent physical injury, an action is nevertheless maintainable, even if shock is caused by property damage (*Attia v British Gas plc [1988] QB 304* where a house caught fire following a gas explosion. The claimant, who suffered nervous shock, was held entitled to damages).

Claimants fall into two categories, namely, (a) primary and (b) secondary victims, the former being directly involved in an accident and the latter are essentially spectators, bystanders or rescuers.

- *Primary victims*

    Primary victims can sue for damages for nervous shock/psychiatric injury, even if they have not suffered earlier physical injury (see *Page v Smith [1996] AC 155* in which the appellant, who was physically uninjured in a collision between his car and that of the respondent, had developed myalgic encephalomyelitis and chronic fatigue syndrome, which became permanent. It was held by the House of Lords that the respondent was liable for this condition in consequence of his negligent driving). Foreseeability of physical injury is sufficient to enable a claimant directly involved in an accident to recover damages for nervous shock. Thus, in the case of *Bourhill v Young [1943] AC 92* a pregnant woman, whilst getting off a tram, heard an accident some fifteen yards away between a motor cyclist and a car, in which the motor cyclist, driving negligently, was killed. In consequence, the claimant gave birth to a stillborn child. It was held on the facts of the case that the unknown motor cyclist could not have foreseen injury to the claimant who was unknown to him.

- *Secondary victims*

    As for secondary victims, defendants are taken to foresee the likelihood of nervous shock to rescuers attending to an injured person and to their close relatives, though the precise extent of nervous shock need not have been foreseen (*Brice v Brown [1984] 1 All ER 997*). Persons who witness distressing personal injuries, who are not related in either of these ways to the victim, are not only considered not to have been foreseen by the defendant, but are expected to be possessed of sufficient fortitude to be able to withstand the calamities of modern life. Only persons with a close tie with the victim or their rescuers who are within sight or sound of the accident or its immediate aftermath will be awarded damages for nervous shock as the law stands at the present. Husbands and wives will be presumed to have a sufficiently proximate tie whilst other relationships are considered on the evidence of the proximity of the relationship.

    In *Chadwick v British Railways Board [1967] 1 WLR 912*, following a serious railway accident, for which the defendant was held to be liable, a volunteer rescue worker suffered nervous shock and became psychoneurotic. As administratrix of the rescuer's estate, the claimant sued for nervous shock. It was held that (i) damages were recoverable for nervous shock, even though shock was not caused by fear for one's own safety or for that of one's children, (ii) the shock was foreseeable, and (iii) the defendant should have foreseen that volunteers might well offer to rescue and so owed them a duty of care.

The class of persons who can sue for damages for nervous shock is limited, depending on proximity of the claimant's relationship with the deceased or injured person (*McLoughlin v O'Brian [1983] 1 AC 410* where the claimant's husband and three children were involved in a serious road accident, owing to the defendant's negligence. One child was killed and the husband and other two children were badly injured. At the time of the accident, the claimant was two miles away at home, being told of the accident by a neighbour and taken to the hospital, where she saw the injured members of her family and heard that her daughter had been killed. In consequence of hearing and seeing the results of the road accident, the claimant suffered severe and recurrent shock. It was held that she was entitled to damages for nervous shock as she was present in the immediate aftermath). This approach was confirmed by the House of Lords in *Alcock v Chief Constable of South Yorkshire Police [1992] 1 AC 310*, where it was stated that the class of persons to whom this duty of care was owed as being sufficiently proximate, was not limited to particular relationships such as husband and wife or parent and child, but was based on ties of love and affection, the closeness of which would need to be proved in each case, except that of spouse or parent, when such closeness would be assumed. Similarly, in *Hinz v Berry [1970] 2 QB 40* the appellant left her husband and children in a lay-by while she crossed over the road to pick bluebells. The respondent negligently drove his car into the rear of the car of the appellant. The appellant heard the crash and later saw her husband and children lying severely injured, the former fatally. She became ill from nervous shock and successfully sued the respondent for damages. On this basis, an employee who suffers nervous shock as a result of witnessing the death of a co-employee at work, will be unlikely to be able to claim damages against his employer for nervous shock, as being a 'bystander who happens to be an employee', as distinct from an active participant in rescue (*Robertson v Forth Bridge Joint Board [1995] IRLR 251* where an employee was blown off the Forth Bridge in a high gale and fell to his death; a co-employee who watched this was unable to sue for damages for nervous shock).

In *Hunter v British Coal Corporation [1999] QB 140*, a claimant who was 30 metres away from the scene of a fatal accident and was told of the victim's death 15 minutes later was unable to recover damages for the psychiatric injury he suffered because he felt responsible for the accident. It was held that he was neither physically nor temporarily close enough to the accident to be a primary victim.

In any event, reasonable fortitude, on the part of the claimant, will be assumed, thereby disqualifying claims on the part of hypersensitive persons (see *McFarlane v EE Caledonia [1994] 2 All ER 1* where the owner of a rig did not owe a duty of reasonable care to avoid causing psychiatric injury to a crew member of a rescue vessel who witnessed horrific scenes at the Piper Alpha disaster).

In *White v Chief Constable of South Yorkshire Police [1999] 1 All ER 1*, the House of Lords considered the question of psychiatric injury to police officers on duty during the Hillsborough football stadium disaster when 95 spectators were crushed to death. It was held that police officers were not entitled to recover damages against the Chief Constable for psychiatric injury suffered as a result of assisting with the aftermath of a disaster, either as employees or as rescuers. An employee who suffered psychiatric injury in the course of his employment had to prove liability under the general rules of negligence, including the rules restricting the recovery of damages for psychiatric injury.

## Loss of amenity

**C6034** Loss of amenity is a loss, permanent or temporary, of a bodily or mental function, coupled with gradual deterioration in health, e.g. loss of finger, eye, hand etc. Traditionally, there are three kinds of loss of amenity, ranging from maximum severity injury (quadraplegias and irreversible brain damage), multiple injuries (very severe injuries but not hopeless cases) to less severe injuries (i.e. loss of sight, hearing etc.). Damages reflect the actual amenity loss rather than the concomitant psychic loss, at least in hopeless cases. Though, if a claimant is aware that his life has been shortened, he will be compensated for this loss. The *Administration of Justice Act 1982, s 1(1)* states:

'(i) damages are not recoverable in respect of loss of expectation of life caused to the injured person by the injuries; but

(ii) if the injured person's expectation of life has been reduced by injuries, there shall be taken into account any pain and suffering caused or likely to be caused by awareness that his expectation of life has been reduced.'

## Bereavement

**C6035** A statutory sum of £10,000 is awardable for bereavement by the *Fatal Accidents Act 1976, s 1*A (as amended by the *Administration of Justice Act 1982, s 3(1)* and the *Damages for Bereavement (Variation of Sum) Order 1990 (SI 1990 No 2575)*). This sum is awardable at the suit of husband or wife, or of parents provided the deceased was under eighteen at the date of death if the deceased was legitimate; or of the deceased's mother, if the deceased was illegitimate. (In *Doleman v Deakin (1990) 87 (13) LSG 43* it was held that where an injury was sustained before the deceased's eighteenth birthday, but the deceased actually died after his eighteenth birthday, bereavement damages were not recoverable by his parents.)

## Fatal injuries

**C6036** Death at work can give rise to two types of action for damages:

- damages in respect of death itself, payable under the *Fatal Accidents Act 1976* (as amended); and

- damages in respect of liability which an employer would have incurred had the employee lived; here the action is said to 'survive' for the benefit of the deceased worker's estate, payable under the *Law Reform (Miscellaneous Provisions) Act 1934*.

Actions of both kinds are, in practice, brought by the deceased's dependants, though actions of the second kind technically survive for the benefit of the deceased's estate. Moreover, previously paid state benefits are not deductible from damages for fatal injuries. [*Social Security Act 1989, s 22(4)(c)*]. Nor are insurance moneys payable on death deductible, e.g. life assurance moneys. [*Fatal Accident Act 1976, s 4*].

## Damages under the Fatal Accidents Act 1976

**C6037** 'If death is caused by any wrongful act, neglect or default which is such as would (if death had not ensued) have entitled the person injured to maintain an action and recover damages, the person who would have been liable if death had not ensued, shall be liable ... for damages ... '. [*Administration of Justice Act 1982, s 3(1)*].

Only dependants, which normally means the deceased's widow (or widower) and children and grandchildren can claim – the claim generally being brought by the bereaved spouse on behalf of him/herself and children. [*Administration of Justice Act 1982, s 1(2)*]. The basis of a successful claim is dependency, i.e. the claimant must show that he was, prior to the fatality, being maintained out of the income of the deceased. If, therefore, a widow had lived on her own private moneys prior to her husband's death, the claim will fail, as there is no dependency. The fact that both the deceased and his partner were at the time of the accident on state benefits is irrelevant to the question of loss in assessing damages – *Cox v Hockenhull [2000] 1 WLR 750*. Cointributory negligence on the part of the deceased will result in damages on the part of the dependants being reduced. [*Fatal Accidents Act 1976, s 5*].

## Survival of actions

**C6038** Actions for injury at work which the deceased worker might have had, had he lived, survive for the benefit of his estate, normally for the benefit of his widow. This is provided for in the *Law Reform (Miscellaneous Provisions) Act 1934*. Any damages paid or payable under one Act are 'set off' when damages are awarded under the other Act, as in practice actions in respect of deceased workers are brought simultaneously under both Acts.

## Assessment of damages in fatal injuries cases

**C6039** Damages in respect of a fatal injury are calculated by multiplying the net annual loss (i.e. earnings minus tax, social security contributions and deductions necessary for personal living (i.e. dependency)) by a suitable number of years' purchase. There is no deduction for things used jointly, such as a house or car. However, where a widow also works, this will reduce the dependency and she cannot claim a greater dependency in future on the ground that she and her deceased husband intended to have children.

It is possible to agree that fatal injury damages should be paid in the form of a structured settlement (see C6025 above).

Where both parents are dead as a result of negligence or a mother dies, dependency is assessed on the cost of supplying a nanny (*Watson v Willmott [1991] 1 QB 140*; *Cresswell v Eaton [1991] 1 WLR 1113*).

## Provisional awards

**C6040** Because medical prognosis can only estimate the chance of a victim's recovery, whether partial or total, or alternatively, deterioration or death, it is accepted that there is too much chance and uncertainty in the system of lump sum damages paid on a once-and-for-all basis. Serious deterioration denotes clear risk of deterioration beyond the norm that could be expected, ruling out pure speculation (*Willson v Ministry of Defence [1991] 1 All ER 638*). Similarly in the case of dependency awards under the Fatal Accidents Act 1976 it can never be known what the deceased's future would have been, yet courts are expected and called upon to make forecasts as to future income. To meet this problem it is provided that provisional awards may be made, e.g 'This section applies to an action for damages for personal injuries in which there is proved or admitted to be a chance that at some definite or indefinite time in the future the injured person will, as a result of the act or omission, which gave rise to the cause of action, develop some serious disease or suffer some serious deterioration in his physical or mental condition'. [*Administration of Justice Act 1982, s 6(1)*]. Moreover, 'Provision may be made by rules of the court for enabling the court to award the injured person:

- damages assessed on the assumption that the injured person will not develop the disease or suffer the deterioration in his condition; and

- further damages at a future date if he develops the disease or suffers the deterioration'.

[*Administration of Justice Act 1982, s 6(2)*].

A claim in respect of provisional damages must be included in the statement of claim to entitle the claimant to such damages. The disease or type of deterioration in respect of which any future applications may be made must be stated (the *Civil Procedure Rules 1998 (SI 1998 No 3132),Part 41 rule 2(2)*). The defendant may make a written offer if the statement of claim includes a claim for provisional damages, offering a specified sum on the basis that the claimant's condition will not deteriorate and agreeing to make an award of provisional damages in that sum.

## Interim awards

**C6041** In certain limited circumstances a claimant can apply to the court for an interim payment. This enables a claimant to recover part of the compensation to which he is entitled before the trial rather than waiting till the result of the trial is known – which may be some time away. This procedure is provided for in the *Civil Procedure Rules 1998* (*SI 1998 No 3132*), Part 25, but it only applies where the defendant is either:

- insured,
- a public authority, or
- a person whose resources are such as to enable him to make the interim payment.

Prior to making an interim payment a judge is under an obligation to take into account any effect the payment may have on whether there is a 'level playing field' for the hearing – see *Campbell v Mylchreest [1999] PIQR Q17*.

Compensation recovery applies to interim payments as well as to payments into court. Care needs to be taken when applying for interim payments to avoid putting the claimant at a disadvantage. Capital of over £8,000, which could include an interim award, removes entitlement to means tested benefits, particularly income support, and there are also reductions on a sliding scale in such benefits for any capital above £3,000. In the recent judgement of *Beattie v Department of Social Security [2001] EWCA Civ 498* the Court of Appeal held that payments from structured annuity funds constitutes 'income' for the purposes of the *Income Support* (*General*) *Regulations 1987* (*SI 1987 No 1967*).

## Payments into court

**C6042** A payment into court ('Part 36 payment') may be made in satisfaction of a claim even where liability is disputed. From the defendant's point of view, costs from the date of the Part 36 payment may be saved if the court does not order a higher payment of damages than that paid into court. The claimant may accept a Part 36 payment or Part 36 offer not less than 21 days before the start of the trial without needing the court's permission if he gives the defendant written notice of the acceptance not later than 21 days after the offer or payment was made. If the defendant's Part 36 offer or Part 36 payment is made less than 21 days before the start of the trial, or the claimant does not accept it within the specified period, then the court's permission is only required if liability for costs is not agreed between the parties. [*Civil Procedure Rules 1998* (*SI 1998 No 3132*), Part 36 rule 11].

The defendant may, instead of making a Part 36 payment, make a Part 36 offer (formerly known as a Calderbank letter) in which he sets out his terms for settling the action

## Interest on damages

**C6043** Damages constitute a judgment debt; such debt carries interest at 8 per cent (currently) up to date of payment. Courts have a discretion to award interest on any damages, total or partial, prior to date of payment (and this irrespective of whether part payment has already been made [*Administration of Justice Act 1982, s 15*]), though this does not apply in the case of damages for loss of earnings, since they are not yet due. Moreover, a claimant is entitled to interest at 2 per cent on damages relating to non-pecuniary losses (except bereavement), even though the actual damages themselves take into account inflation (*Wright v British Railways Board [1983] 2 AC 773*). Under *section 17* of the *Judgments Act 1838* (and *section 35A* of the *Supreme Court Act 1981* and *Schedule 1* to the *Administration of Justice Act 1982*), interest runs from the date of the damages judgment. Thus, where, as sometimes happens, there is a split trial, interest is payable from the date that the damages are quantified or recorded, rather than from the date (earlier) that liability is determined (*Thomas v Bunn, Wilson v Graham, Lea v British Aerospace plc [1991] 1 AC 362*). Moreover, interest at the recommended rate (of 8 per cent) is recoverable only after damages have been assessed, and not (earlier) when liability has been established (*Lindop v Goodwin Steel Castings Ltd, The Times, 19 June 1990*).

## Awards of damages and recovery of state benefits

**C6044** The *Social Security (Recovery of Benefits) Act 1997* and accompanying regulations made important changes to the rules for recoupment of benefit from compensation payments. One key change is that recoupment will not be taken from general damages for pain and suffering and for loss of amenity which it has been accepted should be paid in full. With respect to the other heads of compensation, namely loss of earnings, cost of care and compensation for loss of mobility, they are only to be subject to recoupment from specified benefits relevant to each of these heads of compensation. [*Social Security (Recovery of Benefits) Act 1997, s 8*].

## Duties of the compensator

**C6045** Before the compensator makes a compensation payment, he must apply to the Secretary of State for a certificate of recoverable benefits. [*Social Security (Recovery of Benefits) Act 1997, s 4*]. The Secretary of State must send a written acknowledgement of receipt of the application and must supply the certificate within four weeks of receipt of the application. [*Social Security (Recovery of Benefits) Act 1997, s 4*].

He must supply the following information with his application:

- full name and address of the injured person;
- his date of birth or national insurance number, if known;
- date of accident or injury when liability arose (or is alleged to have arisen);
- nature of the accident or disease;
- his payroll number (where known), if the injured person is employed under a contract of service and the period of five years during which benefits can be recouped includes a period prior to 1994 [*Social Security (Recovery of Benefits) Regulations 1997 (SI 1997 No 2205), Reg 5*]; and
- the amount of statutory sick pay paid to the injured person for five years since the date when liability first arose, as well as any statutory sick pay before 1994, if the compensator is also the injured person's employer. The causes of his incapacity for work must also be stated.

The certificate of recoverable benefits will show the benefits which have been paid.

The compensator must pay the sum certified within 14 days of the date following the date of issue of the certificate of recoverable benefits. [*Social Security (Recovery of Benefits) Act 1997, s 6*]. If the compensator makes a compensation payment without having applied for a certificate, or fails to pay within the prescribed fourteen days, the Secretary of State may issue a demand for payment immediately. A county court execution may be issued against the compensator to recover the sum as though under a court order. It is wise for the compensator to check the benefits required to be set off against the heads of compensation payment so that he is sure that the correct reduced compensation is paid to the injured person. Adjustments of recoupable benefits and the issue of fresh certificates are possible.

Where the compensator makes a reduced compensation payment to the injured person, he must inform him that the payment has been reduced. Statements that compensation has been reduced to nil must be made in a specific form. Once the compensator has paid the Secretary of State the correct compensation recovery amount and has made the statement as required, he is treated as having discharged his liability. [*Social Security (Recovery of Benefits) Act 1997, s 9*].

## Complications

### Contributory negligence

**C6046** Where damages have been reduced as the result of the claimant's contributory negligence, the reduction of compensation is ignored and recovery of benefits is set-off against the full compensation sum.

### Structured settlements

**C6047** The original sum agreed or awarded is subject to compensation recovery and for this purpose, the terms of the structured settlement are ignored and this original sum is treated as a single compensation payment. [*Social Security* (*Recovery of Benefits*) *Regulations 1997* (*SI 1997 No 2205*), *Reg 10*].

### Complex cases

**C6048** Where a lump sum payment has been made followed by a later lump sum, both payments are subject to recoupment of benefits where those benefits were recoupable. If the compensator has overpaid the Benefits Agency, he can seek a partial refund. [*Social Security* (*Recovery of Benefits*) *Regulations 1997* (*SI 1997 No 2205*), *Reg 9*].

### Information provisions

**C6049** Under *section 23* of the *Social Security* (*Recovery of Benefits*) *Act 1997*, anyone who is liable in respect of any accident, injury or disease must supply the Secretary of State with the following information within 14 days of the receipt of the claim against him:

- full name and address of the injured person;
- his date of birth or national insurance number, if known;
- date of accident or injury when liability arose (or is alleged to have arisen); and
- nature of the accident, or disease.

Where the injured person is employed under a contract of service, his employer should also supply the injured person's payroll number (if known) if the period of five years during which benefits may be recouped includes a period prior to 1994 and this is requested by the Secretary of State. [*Social Security* (*Recovery of Benefits*) *Regulations 1997* (*SI 1997 No 2205*), *Regs 3, 5* and *6*].

If the Secretary of State requests prescribed information from the injured person, it must be supplied within 14 days of the date of the request. This information includes details of the name and address of the person accused of the default which led to the accident, injury or disease, the name and address of the maker of any compensation claim and a list of the benefits received from the date of the claim. If statutory sick pay was received by the injured person, the name and address of the employer who has paid statutory sick pay during the five-year period from the date of the claim or before 6 April 1994.

## Appeals against certificates of recoverable benefits

**C6050** Appeals against the certificate of recoverable benefits must be in writing and made not less than three months after the compensation payment was made. The appeal should be made to the Compensation Recovery Unit for a hearing before a tribunal. Leave to appeal to a Commissioner against the decision of the tribunal should be made not later than three months after notice of the tribunal's decision.

## Treatment of deductible and non-deductible payments from awards of damages

**C6051** There are three well established exceptions to deductibility of financial gains:

- recovery under an insurance policy to which the claimant has contributed all or part of the premiums paid on the policy;
- retirement pensions;
- charitable or ex-gratia payments prompted by sympathy for the claimant's misfortune.

## Deductible financial gains

**C6052** The courts, applying the principles outlined above, have ruled that the following financial gains are deductible:

- tax rebates where the employee has been absent from work as a result of his injuries (*Hartley v Sandholme Iron Co [1975] QB 600*);
- domestic cost of living expenses (estimated) must be set off against the cost of care (*Lim Poh Choo v Camden and Islington Area Health Authority [1980] AC 174*);
- estimated living expenses must be set off against loss of earnings (*Lim Poh Choo*, above);
- payment from a job release scheme must be set off against loss of earnings (*Crawley v Mercer, The Times, 9 March 1984*);
- statutory sick pay must be set off against loss of earnings (*Palfrey v GLC [1985] ICR 437*);
- sick pay provided under an insurance policy must be set off against loss of earnings not paid as a lump sum (*Hussain v New Taplow Paper Mills [1988] AC 514*);
- health insurance payment under an occupational pension plan paid before retirement must be set off against loss of earnings where no separate premium had been paid by the employee who had paid contributions to the pension scheme (*Page v Sheerness Steel plc [1996] PIQR Q26*);
- reduced earnings allowance (not a disability benefit) must be set off against loss of earnings (*Flanagan v Watts Blake Bearne & Co plc [1992] PIQR Q144*);
- payments under *section 5* of the *Administration of Justice Act 1982* of any saving to the person who has sustained personal injuries through maintenance at the public expense must be set off against loss of earnings.

## Non-deductible financial gains against loss of wages

**C6053** The following financial gains are non-deductible:

- accident insurance payments under a personal insurance policy taken out by the employee (*Bradburn v Great Western Railway Co (1874) LR 10 Exch 1*) (contributory);
- lump-sum wage-related accident insurance payment under a personal accident group policy payable regardless of the fault of the employee (*McCamley v Cammell Laird Shipbuilders [1990] 1 WLR 963*) (benevolent);
- incapacity pension (from contributory insurance scheme) both before and after retirement age (*Longden v British Coal Corporation [1998] AC 653*). It should be noted that even though the incapacity pension was triggered by the accident, benefit flows from the prior contributions paid by the injured party (contributory);

- private retirement pensions (*Parry v Cleaver [1970] AC 1*; *Hewson v Downs [1970] 1 QB 73*; *Smoker v London Fire and Civil Defence Authority [1991] 2 AC 502*) (contributory or benevolent);

- redundancy payment unconnected with the accident or disease (*Mills v Hassal [1983] ICR 330*). Where the claimant was made redundant because he was unfit to take up employment in the same trade, however, the redundancy payment was deductible from damages for lost earnings (*Wilson v National Coal Board [1981] SLT 67*);

- ex-gratia payment by an employer (*Cunningham v Harrison [1973] QB 942*; *Bews v Scottish Hydro-Electric plc [1992] SLT 749*) (benevolent);

- ill health award and higher pension benefits provided by the employer (*Smoker v London Fire and Civil Defence Authority [1991] 2 AC 502*) (benevolent);

- moneys from a benevolent fund, paid through trustees (not directly to the injured person or dependant) in respect of injuries;

- charitable donations (but not where the tortfeasor is the donor).

## Non-deductible financial gains against cost of care
**C6054**

- Loss of board and lodging expenses awarded despite their being provided voluntarily by parents where the claimant had formerly paid such expenses herself (*Liffen v Watson [1940] 1 KB 556*).

## Social security and damages for prescribed diseases

**C6055** The claims for social security benefits are similar to those for personal injury. Claims for damages are also similar, with more latitude for late court applications because of recurrence of, or worsening of, industrial diseases. Lists of prescribed diseases are contained in social security leaflet NI 2. It is advisable always to refer to the latest edition of the leaflet, as the list of diseases is subject to amendment from time to time. There are special provisions applicable to asthma and deafness resulting from industrial processes, and also for pneumoconiosis and similar diseases.

## Occupational deafness (see DSS leaflet 207)

**C6056** If the claimant's deafness was caused by an accident at work, then to qualify for disablement benefit his average hearing loss must have been at least 50 decibels in both ears due to damage of the inner ear. The claimant's disablement must be at least 20 per cent or more for him to qualify for disablement benefit (total deafness being 100 per cent). The claimant must have worked for at least the five years immediately before making his claim, or for at least ten years, in one of the listed jobs set out in the current edition of the leaflet. A claimant who has been refused benefit because the rules were not complied with will have to wait for three more years when he may qualify if he has worked in one of the listed occupations for five years. Leaflet NI 196 'Social Security Benefits Rates' shows the rates payable, relevant to each percentage of loss, for a weekly disablement pension.

The claim will be decided by a Medical Board, whose members will also decide on reviews if the claimant believes that his deafness has worsened. They will inform the claimant in detail of their decision. Awards will be made for a period of five years, and there will be a review at the end of that period with another hearing test to determine whether benefit should continue, or be reduced.

Claims must be made on Form BI 100 (OD) which may be obtained free from the Department of the Environment HS ORI Level 4, Caxton House, London SW1H 9NA (tel: 020 7273 5248), or from a post office and some borough libraries.

## Asthma because of a job (see DSS leaflet NI 237)

**C6057**   As is well known, there has been an increase in the number of sufferers from asthma. To be able to claim disablement benefit because of asthma, a claimant must have worked for an employer for at least ten years and have been in contact with prescribed substances, and his disability must amount to at least 14 per cent. In the list of substances, item 23 includes 'any other sensitising agent encountered at work', so claims may be possible even if the offending sensitising substance is not yet on the list.

Claims must be made on Form BI 100 (OA) which may be obtained from the Department of the Environment (see address and telephone number above), or from a post office and some borough libraries.

Once a claim has been made, the claimant will be seen by a Medical Board whose members will decide whether he has asthma because of his job and, if so, how disabled he is by the asthma. They will also determine how long his disability is likely to last and the extent of his disability, and will then communicate all of this information to the claimant. If they consider that no change is likely, they may award benefit for life. If they expect that there will be changes, there will be a review at the end of the period for which benefit was awarded.

## Repetitive strain injuries

**C6058**   Repetitive strain injuries are included in the *Social Security (Industrial Injuries) (Prescribed Diseases) Regulations 1985 (SI 1985 No 967), Sch 1*, as item A4. (See OCCUPATIONAL HEALTH AND DISEASES.) The difficulties in establishing causation can be seen from the case of *Pickford v Imperial Chemical Industries [1998] 1 WLR 1189*, where the House of Lords reversed the decision of the Court of Appeal and upheld the trial judge's decision that a secretary was not entitled to damages for repetitive strain injury as she had failed to establish the cause of the injury and that the condition was not reasonably foreseeable. However, in many other cases damages for repetitive strain injury have been awarded – for example, *Ping v Esselte-Letraset Ltd [1992] PIQR P74* where nine claimants who worked at a printing factory were awarded damages ranging from £3,000 to £8,000. In *Fish v British Tissues (Sheffield County Court, 29 November 1994)*, damages of £57,482 were awarded to a factory packer for repetitive strain injuries to arms, hands and elbows.

# Pneumoconiosis

**C6059**   Pneumoconiosis is compensatable under different heads; first, as a ground for disablement benefit, under the *Social Security Contributions and Benefits Act 1992* and, secondly, by way of claim made under the *Pneumoconiosis etc. (Workers' Compensation) Act 1979* – the latter being in addition to any disablement benefit previously paid. See the Department of Environment guidance on the *Pneumoconiosis etc. (Workers' Compensation) Act 1979* on the DETR website: www.environment.detr.gov.uk//pneumo/index.htm.

## Pneumoconiosis/tuberculosis-pneumoconiosis/emphysema – disablement benefit

**C6060**   Where a person is suffering from pneumoconiosis accompanied by tuberculosis, then, for benefit purposes, tuberculosis is to be treated as pneumoconiosis. [*Social Security Contributions and Benefits Act 1992, s 110*]. This applies also to pneumoconiosis accompanied by emphysema or chronic bronchitis, provided that disablement from pneumoconiosis, or pneumoconiosis and tuberculosis, is assessed at, at least, 50 per cent. [*Social Security Contributions and Benefits Act 1992, s 110*]. However, a person suffering from byssinosis is not entitled to disablement, unless he is suffering from loss of faculty which is likely to be permanent. [*Social Security Contributions and Benefits Act 1992, s 110*].

Benefit may be payable if the claimant's disablement is assessed at 1 per cent or more, instead of the usual minimum of 14 per cent. If the assessment of disablement is 10 per cent or less (but at least 1 per cent), then the industrial injuries disablement pension will be payable at one-tenth of the 100 per cent rate.

## Pneumoconiosis – payment of compensation under the Pneumoconiosis etc. (Workers' Compensation) Act 1979

C6061    Claims made under the *Pneumoconiosis etc. (Workers' Compensation) Act 1979* are in addition to any disablement benefit paid or payable under the *Social Security (Industrial Injuries) (Prescribed Diseases) Regulations 1985 (SI 1985 No 967)*. Indeed, whereas the latter consists of periodical payments, the former resemble damages awarded against an employer at common law and/or for breach of statutory duty, except that under the above Act fault (or negligence) need not be proved. This statutory compensation is available only if the employee is unable to claim compensation from any of his former employers (for example, if the employer has become insolvent). There are no requirements that the employee must have worked in a particular industry to be able to claim compensation under the scheme. Claimants are entitled to benefits from the date of claim. There is no ninety-day waiting period under the scheme; it is sufficient if the claimant suffers from the prescribed disease. Coal miners are excluded from the scheme because they have their own statutory scheme. Claims made under the *Pneumoconiosis etc. (Workers' Compensation) Act 1979* must be made in the manner set out in the *Pneumoconiosis etc. (Workers' Compensation) (Determination of Claims) Regulations 1985 (SI 1985 No 1645)*. The claim must be made (except in the case of a 'specified disease' – see C6062 below) within 12 months from the date on which disablement benefit (under the Social Security Regulations) first become payable; or if the claim is by a dependant, within 12 months from the date of the deceased's death. [*Pneumoconiosis etc. (Workers' Compensation) (Determination of Claims) Regulations 1985 (SI 1985 No 1645), Reg 4(1), (2)*]. Awards for pneumoconiosis/byssinosis have slowly declined recently.

## Claims for specified diseases

C6062    Claims for a 'specified disease', i.e:

- pneumoconiosis, including (i) silicosis, (ii) asbestosis and kaolinosis;
- byssinosis (caused by cotton or flax dust);
- diffuse mesothelioma;
- primary carcinoma of the lung coupled with evidence of (i) asbestosis and/or (ii) bilateral diffuse pleural thickening;

[*Pneumoconiosis etc. (Workers' Compensation) (Specified Diseases) Order 1985 (SI 1985 No 2034)*],

must be made within 12 months from the date when disablement benefit first became payable, or, in the case of a dependant, within 12 months from the date of the deceased's death. [*Pneumoconiosis etc. (Workers' Compensation) (Determination of Claims) Regulations 1985 (SI 1985 No 1645), Reg 4(3)(4)*]. These time periods can be extended at the discretion of the Secretary of State. Moreover, where a person has already made a claim and has been refused payment, he can apply for a reconsideration of determination, on the ground that there has been a material change of circumstances since determination was made, or that determination was made in ignorance of, or based on, a mistake as to material fact.

# Community Health and Safety

## Introduction

**C6501** Since the *Health and Safety at Work etc. Act 1974* came into operation, every employer has had duties in respect of 'persons not in his employment', in addition to duties to his employees. These duties have come into closer focus through the risk assessment requirements contained in the Management of Health and Safety at Work Regulations 1999 and other sets of regulations.

The term 'community' embraces a variety of different categories, such as:

- Neighbours and passers-by
- Trespassers on work premises
- Customers and users of facilities and services
- Visiting individuals and groups
- Volunteers carrying out work

This chapter considers each of these categories in some detail – who is at risk, what types of risk they might be exposed to and the types of precautions likely to be necessary to protect them. In addition, it deals with major events that can involve several community categories and may be subject to the scrutiny of Safety Advisory Groups. The potential impact on the community of contractors used by businesses or hired in by event organisers is also considered.

## Relevant Legal Requirements

### Health and Safety at Work etc. Act 1974

**C6502** Section 3(1) of the Act places duties on employers in respect of members of the public and other non-employees:

> 'It shall be the duty of every employer to conduct his undertaking in such a way as to ensure, so far as is reasonably practicable, that persons not in his employment who may be affected thereby are not exposed to risks to their health or safety'.

As well as applying to various categories of members of the community (as summarised in C6501), these requirements also apply to others at work, such as contractors and their employees, agency staff, delivery personnel and members of the emergency services. It must also be noted that the duties are qualified by the phrase 'so far as is reasonably practicable' – namely, the level of risk must be considered together with the costs and practicalities of implementing related control measures.

The examples at C6503 below contain a variety of examples of successful prosecutions under Section 3 of HSWA in respect of risk or actual injury to members of the community.

## Examples of prosecutions under section 3 of the Health and Safety at Work etc. Act 1974

**C6503**

- Three contracting companies were fined a total of £80,000 after a member of the public was killed when she drove up the ramps of a low loader trailer during the hours of darkness. No lights were visible and there was no traffic management system in place.

- A manufacturing company was fined £15,000 when two private vehicles collided with a steel beam being carried by a side loading lift truck in the dark on an unlit service road accessible to the public.

- Two companies were fined a total of £75,000 when a motorcyclist was killed after colliding with the outrigger of a mobile crane being used at a shopping centre.

- A supermarket was fined £10,000 when an 11-year-old boy was knocked over by a reversing fork lift truck unloading goods from a lorry on a public road.

- A local authority was fined £80,000 after a reversing delivery lorry struck an elderly pedestrian who died from his injuries the following day.

- A plastics manufacturer was fined £70,000 after an outbreak of legionnaires disease in the surrounding area revealed poor management of their cooling tower.

- Test pressurisation of a railway tunnel resulted in a blow-out into playing fields above (fortunately at 4 a.m. when no-one was using the fields). The construction company involved was fined £700,000.

- A railway maintenance company was fined £200,000 after an eight-year-old girl was fatally electrocuted through coming into contact with a live third conductor rail. A robust system of fencing inspection was not in place in the area concerned.

- An electrical supply company was fined £10,000 for failing to prevent members of the public (including children) gaining access to effluent lagoons.

- A hospital was fined £25,000 after a fatal accident in which a vulnerable patient fell 3.6 metres from a window.

- A 72-year-old nursing home resident died after drinking a caustic substance in a water jug left unattended on an office desk. The owners of the home were fined £40,000.

- A local authority was fined £75,000 after a member of the public was killed when swimming into a submerged post in a boating and swimming lake.

- The company managing a sports centre was fined £15,000 after separate incidents in which a 13-year-old and an 8-year-old fell from a climbing wall.

- A disabled long-term hospital resident died after being severely scalded when put into an excessively hot bath. The hospital trust was fined £50,000.

- A manufacturing company was fined £10,000 for failing to prevent a child moving freely and unsupervised through their factory.

## The Management Regulations

**C6504** Regulation 3(1) of the *Management of Health and Safety at Work Regulations 1999* (*MHSWR*) (*SI 1999 No 3242*) states that:

> 'Every employer shall make a suitable and sufficient assessment of ... the risks to the health and safety of persons not in his employment arising out of or in connection with the conduct by him of his undertaking.'

Consequently risk assessments must take account of relevant members of the community, the risks that they might be exposed to and what precautions are necessary to control those risks to the extent required by the law – including the general requirements imposed the *Health and Safety at Work etc. Act 1974*.

*SI 1999 No 3242, Reg 5* also requires employers to take effective steps to implement the precautions identified through the risk assessment as being necessary. It states:

'Every employer shall make and give effect to such arrangements as are appropriate, having regard to the nature of his activities and the size of his undertaking, for the effective planning, organising, control, monitoring and review of the preventive and protective measures.'

Appropriate and effective management techniques must be used to ensure the protection of members of the community, as well as employees and others who are at work.

Employers with five or more employees must record both the significant findings of their risk assessments and their arrangements for implementing the related preventive and protective measures.

## Other Specific Requirements of Regulations

**C6505** Several sets of regulations contain requirements relating to the protection of members of the wider community and other non-employees. These include:

- *Control of Substances Hazardous to Health Regulations 2002 (COSHH) (SI 2002 No 2677)*

  *SI 2002 No 2677, Reg 3(1)* states that: 'Where a duty is placed by these Regulations on an employer in respect of his employees, he shall, so far as is reasonably practicable, be under a like duty in respect of any other person, whether at work or not, who may be affected by the work carried out by the employer ...'

  Whilst there are some exceptions to this requirement (e.g. in relation to health surveillance and training for non-employees), the main requirements of the *COSHH Regulations* apply in respect of members of the community. *COSHH* assessments must take account of risks to the community, particularly the ways in which substances are used or stored.

- *Control of Major Accident Hazards Regulations 1999 (COMAH) (SI 1999 No 743)*

  *COMAH* applies to a limited range of workplaces where specified quantities of high risk substances are stored or used, or may be generated e.g. through a chemical process. In developing their Major Accident Prevention Policy (MAPP), employers must consider safety, health and environmental risks both inside and outside their own premises. This includes the development of off-site emergency plans and the provision of information to the public.

- *Construction (Design and Management) Regulations 1994 (CDM) (SI 1994 No 3140)*

  These regulations contain a key requirement for a Health and Safety Plan to be prepared in respect of significant construction projects. This plan must take account of the wider community, as well as those carrying out construction work. Occupants of neighbouring property and passers-by must be considered, as must occupants of and visitors to premises where construction work is being carried out. Consultation began in 2005 with a view to revising these regulations during 2006, but the effective management of risk to the community is likely to be a core requirement of the replacement regulations.

- *Work at Height Regulations 2005 (SI 2005 No 735)*

  The regulations place duties on employers to prevent injuries to any person from falling objects. Members of the community must be protected, as well as those at work. These duties include preventing materials and objects from falling (e.g. through using toe boards, brick guards, tool belts and securing stored items), measures to prevent persons being struck by falling items (e.g. screens, nets etc.) and ensuring that materials or objects are not thrown or tipped from height in circumstances liable to cause injury.

## Occupiers Liability Acts 1957 and 1984

**C6506** The *Occupiers Liability Act 1957* established the liability of occupiers under civil law to exercise a duty of care in respect of lawful visitors to their property. The occupier's duty of care was extended by the *Occupiers Liability Act 1984* Act to also include trespassers, in respect of any injury suffered on the premises, either because of any danger due to the state of the premises or things done, or omitted to be done, in the following circumstances:

1   If he (the occupier) is aware of the danger or had reasonable grounds to believe that danger exists;

2   If he knows or has reasonable grounds to believe that a trespasser is in the vicinity of the danger concerned or that he may come into the vicinity of the danger; and

3   If the risk is one against which, in all the circumstances of the case, he may reasonably be expected to offer some protection.

The Act also states that 'no duty is owed … to any person in respect of risks willingly accepted by that person'.

Any case brought under this Act would be decided on its merits, but there are two important factors to consider:

- The Standard of Care Required

  A greater duty of care exists in relation to major risks (e.g. falls from significant height, presence of hazardous substances or deep water) than minor risks (e.g. uneven land or shallow water). The practicality of taking and maintaining precautions would also have to be taken into account. It may not be reasonable to expect to fence off a large relatively low-risk workplace or site in its entirety, especially in an area where vandalism was prevalent. However, it may be reasonable to fence off individual high-risk areas within the site.

- The Type of Trespasser

  Case law has already established that a greater duty of care exists in respect of child trespassers, although parents are expected to assume the prime responsibility for controlling the whereabouts of very young children. Occupiers must also be aware of the presence of items on sites that may attract children e.g. derelict machinery and opportunities to climb.

  The duty of care owed to adults will also vary. It will be much greater to those simply walking through a site (especially if straying accidentally from a public footpath) than to those indulging in illicit activities such as theft, vandalism or fly-tipping. However, even these latter categories cannot be completely ignored, particularly if the occupier is already aware that significant risk exists.

# Neighbours and Passers-by

## Who Must Be Taken Into Account?

**C6507** There are a number of categories of neighbours and passers-by who must be taken into account when carrying out risk assessments. Some of these will be at work, whilst others will be members of the wider community. They include:

- Persons within the workplace or site

  Those visiting or working within other parts of the premises, such as:

- other businesses;
- public buildings;
- community facilities; and
- voluntary groups.

- External Neighbours

  Occupants of property close to the workplace and those visiting them must be taken into account. In addition to the types of locations referred to above, this is also likely to include residential property.

- Passers-by

  Pedestrians, drivers and users of recreational areas must be considered – whether or not the roads, footpaths or areas that they are using are designated for public access.

## Risks to be considered

C6508    Neighbours and passers-by may be affected by risks even if they remain outside the workplace but more so if they are able to stray into work areas. The types of risks most likely to be of relevance are:

- Vehicle traffic

  Vehicles are likely to be delivering materials to or collecting them from most workplaces. Heavy items of mobile plant may also be in circulation, particularly on construction sites.

- Lifting operations

  Lifting operations may involve objects being lifted over occupied premises or over areas where pedestrians or vehicles circulate.

- Work at height

  Work at height will inevitably involve a potential risk to neighbours and passers-by from falling objects. There is also a risk of unauthorised persons using access equipment to reach high levels from which they might fall.

- Hazardous and dangerous substances

  Substances may create risks to neighbours and passers-by through 'routine' emissions or from more substantial leakages. Explosions within workplaces can also affect external areas. Waste substances or materials constitute a risk, particularly if left in unsecured areas.

- Work equipment

  Neighbours and passers-by may be at risk from contact with dangerous parts of equipment being used in areas to which they have access. They may also be hit by particles etc. thrown off from materials being worked on. Any tool, powered equipment or vehicle may be dangerous if it falls into the hands of unauthorised persons, particularly children.

- Dangerous activities

  There are a variety of activities which might present risks to neighbours or passers-by, even those who are some distance away, such as pressure testing, the use of radioactive substances and the use of explosives.

- Electrical equipment

  Uninsulated electrical cables and other unprotected electrical equipment are obvious risks to any unauthorised person who is able to gain access to them.

## Likely Precautions

**C6509** What precautions are necessary will be dependent upon the types of risks identified by the risk assessment, as described in C6508 above. For construction activities this will often be within the context of a CDM Health and Safety Plan. An HSE publication 'Protecting the public – your next move' (HSG 151) provides much practical guidance on precautions in respect of construction projects. Important precautions likely to be necessary for all types of work include:

- Security of the workplace

    The level of risk may justify the whole workplace or parts of it being made secure against access by neighbours or passers-by. This may involve the use of fencing or other barriers, with locked access gates or doors. Periodic checks will be necessary to ensure that these measures are still in good condition and proving effective. In some cases the presence of those carrying out the work (or a designated sentry) will be adequate to secure the work area against incursion from neighbours or passers-by.

- Traffic management

    Controlling the routes by which traffic enters and leaves work premises may be important in reducing the risks to neighbours and passers-by. It may be possible to avoid high-risk locations, e.g. shopping areas, school entrances. Direction of approach may have to be specified in order to leave an adequate turning circle for larger vehicles. The timing of vehicle traffic may have to be controlled to avoid times at which there will be large numbers of pedestrians or other vehicles in the vicinity. (There are detailed requirements governing work being carried out on the public highway itself.)

- Other traffic precautions

    Other precautions may be necessary to ensure the safe circulation of traffic where work activities are taking place, both on public and private roads. In dark conditions vehicles must possess appropriate lights, and adequate lighting must be provided for activities taking place on roads, e.g. loading or unloading of vehicles. Arrangements to clean roads of mud and other work-related debris may be necessary.

- Planning of lifting operations

    Timing may also be important in minimising the possible presence of pedestrians and vehicles in areas where lifting operations must be carried out. Loads can be lifted over buildings at times when they are unoccupied or when the removal of occupants causes minimum inconvenience. Measures may still be necessary to prevent any pedestrian or vehicle traffic which may be present from getting into risk areas.

- Storage locations

    Materials and equipment should be stored in locations where the risks they present to neighbours and passers-by are minimised. Where non-workers have regular access to workplaces, the security of storage facilities will be important.

- Planning of dangerous activities

    As with traffic movements and lifting operations, the timing of other potentially dangerous work activities should be planned to minimise the risks to others. Precautions in place to protect those carrying out the work should also protect neighbours and passers-by, but measures may still be necessary to keep them away from such activities.

- Emergency procedures

    Such procedures must take account of neighbours and passers-by:

    — how they will be notified of the emergency e.g. telephone messages, alarm sounds, loudspeaker announcements;

— how they should respond e.g. close all windows, evacuate to designated locations.
- Worker awareness

    It is essential that workers are made fully aware of how their workplace or work activities may create risks for others. In particular they must be made aware of precautions involving them, e.g. designated traffic routes, permitted times for access or other activities, the importance of securing work areas, equipment or materials and their role in emergency procedures.

- Signs and notices

    Signs and notices can play an important role in making workers aware of the need for precautions and also in reinforcing their awareness. Neighbours and passers-by can also be made aware of both risks and precautions, particularly the need to keep clear of areas where potentially dangerous activities are taking place.

## Trespassers on Work Premises

### Who must be taken into account?

**C6510** Employers have both statutory duties towards trespassers through the *Health and Safety at Work etc. Act 1974, s 3* and also a civil 'duty of care' (see C6506). These duties are much greater where the trespass is foreseeable, for example:

— pedestrians straying from a right of way;

— persons recovering lost property (such as footballs or golf balls);

— those recovering straying animals or children.

Potential trespass by children is a particular concern because of their lack of awareness of risks and their inability to read or understand warning notices. Children are also generally more inquisitive and adventurous than adults.

Deliberate trespass by adults, e.g. for theft, vandalism or other illicit purposes, presents particular difficulties. However, the actions of employers to protect their assets from such persons will often also protect this type of trespasser from any risks present.

### Risks to be considered

**C6511** A number of the risks to neighbours and passers-by mentioned in C6508 may also affect trespassers. However, risks of particular relevance in considering the need to protect trespassers include:

- Unsafe access
    - derelict buildings or structures (unsafe floors, staircases etc.);
    - fragile surfaces;
    - unstable ground or materials; and
    - falls from significant height (shafts, cliffs, scaffolds, towers, gantries, etc.).
- Water
    - deep water (including waste lagoons); and
    - fast-flowing water (both natural and artificial watercourses).
- Railway lines
    - moving trains; and

- power lines and live rails.
- Dangerous equipment or materials
  - abandoned equipment;
  - insecure items and those capable of toppling over; and
  - equipment capable of being used by unauthorised persons.
- Hazardous or dangerous substances
  - substances stored insecurely;
  - abandoned substances;
  - accessible waste products; and
  - toxic or asphyxiant atmospheres e.g. in sewers, pipes, conduits or other confined spaces.

### Likely precautions

C6512  Many of the precautions described in C6509 to protect neighbours and passers-by may also be appropriate to protect trespassers.

- Fencing and other security measures

  The standard of fencing, locks etc. necessary to prevent access by determined trespassers will generally need to be higher than to prevent more casual entry into work premises. Standards must also relate to the level of risk present within. Fencing and other enclosures may need to be reinforced by periodic security patrols, and in any case they must be inspected regularly to ensure they remain in a satisfactory condition.

- Signs and notices

  The use of signs and notices can be particularly important in making potential trespassers aware of risks within workplaces, particularly those which may not be readily apparent e.g. fragile surfaces, deep water, high voltage electrical equipment. The condition of signs and notices should also be subject to regular inspection.

- Secure storage

  Potentially dangerous materials and equipment should be stored so that they are not accessible to trespassers. The exact nature of the storage will depend upon the items stored, the storage location and the anticipated level of approach by trespassers. Storage facilities on an unattended site accessible to the public will need to be much more robust and secure than in an occupied workplace where any trespass is likely to be less frequent and less prolonged.

- Emergency equipment

  Lifelines or lifebelts may be necessary to safeguard trespassers (as well as workers and any legitimate passers-by) against risks from water. These, too, should be subject to periodic inspection.

## Customers and Users of Facilities and Services

### Who must be taken into account?

C6513  The need to protect other persons who may be present within workplaces was considered earlier in the chapter (see C6507). However, many workplaces involve direct contact by members of the community with work premises, work activities and people who are at work. Such members of the community can be described both as customers and as users of facilities and services.

- Customers

    Many businesses have customers who are members of the community:

    — shops, restaurants and other commercial premises;

    — transport undertakings (railways, buses, airlines, etc.);

    — the entertainment industry (including spectator sports and funfairs); and

    — those providing domestic installation and repair services (gas equipment, electrical work, construction work, gardening and similar activities).

- Facility and service users

    This includes members of the community who use facilities or services such as:

    — hotels and similar accommodation;

    — hospitals and care homes;

    — domiciliary care services;

    — sports facilities;

    — interactive attractions (e.g. at museums);

    — educational establishments (schools, colleges, universities); and

    — technical training services.

    Many other examples could be added to both of the above categories. (For brevity, all such persons will be referred to as 'customers' in the remainder of this chapter.)

## Risks to be considered

**C6514** Many risks to 'customers' are very similar to those to neighbours and passers-by – sections C6508 and C6509 deal with such risks and the related precautions. However, some risks relate particularly to the 'customers' themselves – who they are, why they are present in the workplace and what they are doing there. Consideration must be given to the following:

- Access within the premises

    Ways in which 'customers' will access the premises must be taken into account:

    — how many people will be present?

    — what routes will they follow?

    — are those access routes safe?

    — what other risks might be present on access routes?

- Equipment

    What equipment will 'customers' be expected (or have the opportunity) to use and what other equipment (used by workpeople) might they have close contact with?

- Activities

    What activities will 'customers' be involved in directly or come into close contact with? Some sporting and recreational activities (e.g. skiing, climbing or pole vaulting) involve an inevitable degree of risk, as do some types of technical training. Activities may also involve risks from hazardous or dangerous substances.

- Accidents and emergencies

    In some workplaces (such as the healthcare sector) there are specific obligations placed on employers to make first-aid provision for their 'customers', and in others there is often a general expectation that they will do so. Fire and other emergencies can affect 'customers' as well as those at work.

- Risks from products

    Risks to 'customers' from food products is a specialist subject in its own right. There may be risks to 'customers' from other materials, substances or equipment supplied to them by a business or simply present on display. This will be of particular importance in plant hire businesses and other establishments offering potentially high risk products.

- Capabilities of the 'customer'

    The capabilities of the 'customer' are important in considering the risks associated with all of the above. These may relate to their:

    — physical capabilities ('customers' may be elderly, disabled or have limited physical strength);

    — intellectual capabilities (these may be limited due to the ages of 'customers' or their mental condition);

    — technical capabilities (some activities may require a particular level of skill, knowledge or experience, as well as strength or intellectual ability); and

    — physical size (children in particular may be able to gain access to dangerous equipment or live electrical equipment through small apertures or may fall underneath guard rails).

## Likely precautions

C6515    The types of precautions necessary will need to relate to the types of 'customers' present and the risks it is foreseeable that they will be exposed to in the workplace. Precautions are likely to include:

- Maintenance of safe access

    Access routes must be maintained in a safe condition for the benefit of staff as well as 'customers'. However, particular consideration must be given to the numbers and types of 'customers' likely to be present. Special arrangements may be necessary where the elderly or those with disabilities (e.g. the partially sighted) are present on the premises.

- Selection and maintenance of equipment

    Where equipment is expected to be used by 'customers', it must be suitable for the purpose. It may need to be simpler to use and more robust than equipment intended for the use of workers. Arrangements must be in place for the maintenance of equipment, and some types of equipment will require formal training of intended users (see below).

- Screening of 'customers'

    In some situations it will be necessary to screen 'customers' in order to establish their physical, intellectual or technical capabilities. This is particularly important where they are likely to be involved in potentially risky activities (e.g. sporting or other outdoor pursuits) or using dangerous equipment. Such screening should establish the level of need for supervision, training etc. and any special individual arrangements. In some cases 'customers' may eventually be judged unsuitable to participate in the activity in question.

- Information and training

    'Customers' may need to be made aware of risks and related precautions through the use of signs and notices or via customer information packs, e.g. in hotels and similar accommodation. Some situations may justify the use of formal briefings of 'customers', particularly where they are likely to be involved in higher risk activities or where there are significant risks present in the workplace. Certain activities or types of equipment will necessitate 'customers' receiving formal training to ensure an adequate level of safety.

- Supervision and control

    An appropriate number of properly trained and informed staff must be available to ensure that 'customers' are adequately supervised. Whilst in some situations (e.g. the retail sector) this supervision will largely be passive, sporting and other outdoor activities will require more proactive forms of supervision, as will entertainment and other public events, through the presence of stewards. Those carrying out supervisory or stewarding duties must monitor the behaviour of their 'customers' and, in particular, restrict them from entering into higher risk areas of the workplace.

- Monitoring

    As well as monitoring 'customer' behaviour, there is a need to monitor the ongoing effectiveness of all arrangements for ensuring customer safety. In particular areas accessed by 'customers' and equipment used by them should be subjected to periodic formal inspections to ensure that they are still in satisfactory condition.

- Emergency arrangements

    In formulating emergency arrangements, account must be taken of 'customers'. Fire evacuation is an obvious case, but other foreseeable emergencies (e.g. chemical leaks or public disorder) may also need to be considered. 'Customers' must be informed about relevant emergency arrangements (see above regarding information and training). Staff also must be fully briefed on their role in respect of 'customers', which is particularly important where large numbers are present (e.g. spectator events, the educational and health care sectors and large retail outlets) and where 'customers' with disabilities may be present.

# Visiting Individuals and Groups

## Who must be taken into account?

C6516    Most of the immediately preceding sections (C6513–C6515) dealing with 'customers' and users of facilities and services is also of relevance to visitors. However, whilst dealing with 'customers' is an integral part of many work activities, other workplaces are less accessible to the community and may only have to deal with visitors now and again. Amongst such visitors are likely to be:

- Schoolchildren and other students

    Individuals may visit as part of work experience and similar programmes. By virtue of the Health and Safety (Training for Employment) Regulations 1990, schoolchildren on formal work experience programmes have the status of employees in respect of health and safety legislation.

- Student and community groups

    Some workplaces are willing to accept groups of visitors from schools and other educational establishments. Similarly, visits may be made by special interest groups or members of the local community.

- Informal individual visits

    Visits may be made by individual friends and relations of workers. Whilst employers will be aware of many such visits, others can pose more cause for concern, e.g. where a child is sent with a message for someone working outside normal working hours. During school holidays it is not uncommon for children to accompany parents driving vehicles in making deliveries to workplaces – sometimes even security staff may be unaware that anyone else is in the cab.

## Risks to be considered

C6517    Many of the risks referred to in C6514 in respect of 'customers' are also of relevance to visiting individuals and groups. Particular attention must be given to:

- Areas to be visited

    In order to gain benefit from their visit, it is likely that formal visitors will need to visit some areas of the workplace where risks may be present. A balance must be struck between the intended benefits and not exposing visitors to an unacceptable degree of risk.

    If informal visitors are not allowed to get beyond a reception area, a canteen or a security lodge, there will be little risk. However, granting informal visitors free access to working areas can introduce considerable risks to safety and also give rise to security issues.

- Activities to be carried out

    Work experience students and similar longer-term visitors are likely to be involved in carrying out actual work or projects related to work activities. A risk assessment is likely to be necessary in each case to identify the risks involved in the work or project and the precautions which must be taken, taking into account the capabilities of the individual (see below).

- Capabilities

    The physical, intellectual and technical capabilities of potential visitors are an important consideration. A visiting group of engineers or safety specialists can be expected to be much more risk aware than a group of schoolchildren. The same applies to individual visitors, whether visiting formally or informally. This is where the potential for children to be making unauthorised visits to workplaces should be of particular concern.

## Likely precautions

C6518    Inevitably the precautions necessary to protect visitors must relate to the areas they are visiting and activities they will be involved in.

- Restricted areas and routes

    Consideration must be given to which areas individual visitors are allowed to visit, particularly those to which they are to be granted unaccompanied access. In planning routes for visiting groups, risks on or close to the proposed route must be taken into account, but so also must the composition of the group – their physical capabilities, awareness of risk, degree of responsibility, the size of the group and the abilities of those in charge to supervise it effectively. In some situations it may be necessary to provide additional signage or barriers to ensure the group keeps to its intended route.

- Restricted activities

    Where individuals are making extended visits it must be made quite clear both to the visitor and those providing immediate supervision to them what the visitor is and is not permitted to do. These restrictions will relate to the capabilities of the individual and may be progressively removed as the individual acquires more knowledge and experience.

- Supervision

    The level of supervision necessary for a group will need to relate to its size and type, as well as the areas it will visit. Support may be necessary from teachers or others connected with the visitors in order to ensure effective supervision. Someone must be made responsible for the overall supervision of long-term individual visitors, who may also need to be accompanied when visiting high risk areas or carrying out high risk activities. Short-term visitors are likely to need to be accompanied at all times, except when visiting low risk areas – this will particularly be the case for any children. An area may need to be set aside for children to wait (under supervision) whilst their parents are making deliveries to the site.

- Information and instruction

    Visiting groups (and individuals) may need to be provided with information prior to their visit about such matters as:

    — the level of supervision they are expected to provide (for groups);

    — clothing and footwear to be worn; and

    — any special health and safety issues for individuals with specified disabilities e.g. lack of mobility, deafness, visual impairment, asthma.

    At the start of the visit further instruction and information will be necessary on topics such as:

    — risks the visitors may be exposed to;

    — behaviour during the visit;

    — restricted areas, routes and activities;

    — obeying instructions from visit guides/supervisors;

    — emergency arrangements;

    — use of Personal Protective Equipment (PPE) (see below); and

    — what to do if they have questions or concerns.

    During visits further emphasis may need to be given to the risks present or precautions necessary in individual areas. This might involve additional briefings from visit guides and/or use of signs and notices.

- Personal Protective Equipment (PPE)

    Some types of visitors are likely to be able to bring any PPE required for the visit, providing they are advised about this in advance. However, in other cases PPE will need to be made available by the organisation hosting the visit. This is relatively easy for individual visitors – long-term visitors will often, in effect, be treated as employees.

    Provision of PPE for visiting groups can pose practical problems, such as hygiene, availability of suitable sizes, need for adjustment, training in use, etc. Some types of PPE, e.g. safety helmets, high visibility clothing or disposable earplugs, are fairly simple to deal with. However, the issue of safety footwear or respiratory protection to visitors is much more problematical, and areas necessitating their use may need to be avoided. In some situations it may be acceptable for visitors to be instructed to wear sturdy leather footwear for their visit, namely no trainers, fashion shoes or sandals.

## Volunteers

### Who must be taken into account?

C6519  Volunteers carry out many types of work activities that might be carried out by paid staff in other locations. They are exposed to similar risks to workers but their capabilities may not be the same. Volunteers may have considerable knowledge and experience from previous work that they have carried out but this may not always be relevant to their voluntary work. Older volunteers may also have diminishing physical (and even mental) capabilities. There is often a fine line to be followed between ensuring the health and safety of the volunteer and putting them off carrying out the voluntary work completely.

There are many activities in which volunteers are involved:

- Regular volunteers

    Volunteers might work regularly in sectors such as:

    — charity shops and other retail outlets;

    — care work, including home visits;

    — providing transport e.g. for clients or meals on wheels;

    — drop-in centres;

    — coaching or officiating for sporting activities;

    — acting as guides e.g. at museums, historical properties;

    — maintaining equipment e.g. vintage vehicles, steam locomotives; and

    — carrying out environmental work e.g. in country parks or on trails.

- 'One-off' volunteers

    Many people also volunteer to work on a 'one-off' basis at major events, particularly those attracting large numbers, such as:

    — sporting or musical events;

    — carnivals; and

    — fund-raising events.

    Their duties on such occasions often put them in the front line in dealing with other members of the community in roles such as:

    — traffic control;

    — car parking;

    — entry control;

    — staffing registration or enquiry desks; and

    — crowd supervision.

### Risks to be considered

C6520  Given the wide range of activities that volunteers can become involved in, they are likely to be exposed to the same variety of risks as those who are at work – as described elsewhere in this publication. This section identifies some of the more common risks that must be taken into account during the risk assessment, including some which may tend to be overlooked. An HSE publication 'Charity and voluntary workers – a guide to health and safety at work' (HSG 192) provides further guidance on the subject.

- Travel

    Some volunteers need to travel about to carry out their work. Apart from the normal risks associated with road traffic, there may be risks to them in the areas they are travelling through or visiting. Such risks will be greater during hours of darkness, particularly for staff who must travel late at night.

- Entry into clients' homes

    Many people providing voluntary care services must enter clients' homes. There may be risks to their personal safety from the clients themselves, their relatives and friends and also from pets. In some cases risks may be due to difficult access (e.g. due to household clutter), defective equipment or dangerous electrical or gas installations.

- Other personal safety risks

    Volunteers will also be prone to personal attack where they are expected to restrict entry to events, handle cash or deal with problems involving the public. Risks will be much higher at certain types of event, with the potential level of alcohol consumption by members of the public often being a major factor.

- The work environment

    In some places where volunteers work it is possible that no-one will have been given specific responsibility for cleaning, tidying up or maintaining the premises. Consequently buildings themselves or electrical or gas installations may have deteriorated, possibly to a dangerous condition. Safe storage of equipment and the maintenance of safe access routes within premises will also merit particular attention.

- Equipment

    Voluntary organisations often receive donations of equipment that is no longer required elsewhere. Arrangements must be made to screen out any potentially dangerous equipment and suitable systems put in place to ensure all equipment is properly maintained and regularly inspected. Some types of equipment (e.g. powered machinery) may be dangerous if not used correctly, whilst other types (e.g. storage or display racking, trestle tables) may pose risks if incorrectly assembled. Equipment may also be dangerous if it is left accessible for use by unauthorised persons.

- Manual handling

    Some volunteers may be required to carry out significant amounts of manual handling in respect of goods, materials and equipment. The normal criteria for carrying out manual handling assessments for workers should be applied, namely account must be taken of the task, the load, the working environment and the capabilities of the individual (of particular relevance to many volunteers).

- Emergencies

    Volunteers must be capable of dealing with foreseeable emergencies both in respect of their own health and safety and also that of clients and other members of the community that they are working with. Volunteers may be responsible for assisting disabled clients in leaving buildings or shepherding members of the public away in the event of fires and similar emergencies.

## *Likely precautions*

**C6521** Some related precautions have already been identified in relation to the risks referred to in C6520 above, whilst others are detailed in other sections of this publication. Precautions of particular relevance to volunteers are identified below:

- Screening and assessment

  Volunteers should be screened in order to identify their existing knowledge and experience in relevant areas. Their physical and intellectual capabilities should also be assessed.

- Information and training

  It is likely that volunteers will need to be briefed on health and safety aspects of their intended work and to be trained in the use of certain equipment or the carrying out of some activities. This needs to be done with sensitivity – whilst volunteers will accept that they might require a certain level of information and training, they will react badly to spending time on subjects they regard as unnecessary or irrelevant. Sessions should be made as enjoyable as possible for the participants, with those who already possess relevant knowledge and experience being encouraged to share this with others.

- Restrictions

  Some activities or use of certain types of equipment may need to be restricted to volunteers who have undergone relevant training. In some cases volunteers may need to be accompanied by a competent paid member of staff.

- Communication

  Effective means of communication need to be established for volunteers working within the community (often remote from colleagues) or at major events. Use of radios, mobile phones and attack alarms may be necessary, depending on the situation. Peripatetic volunteers (e.g. those making home visits) should leave an intended itinerary back at base. There should be arrangements for them to log off on completing their duties and for following up any volunteers who fail to log off or do not keep in touch.

- Supervision and support

  As with paid workers, even very capable volunteers need to have someone that they can communicate with, particularly when problems arise. Support should be readily available to volunteers who need it, especially in emergencies, activated by a radio, telephone message or by the sounding of an alarm.

- Monitoring

  The working practices of volunteers should be monitored, as should also be the case for paid workers. However, a more diplomatic approach will need to be taken where corrective actions are necessary to modify the way that volunteers behave. Regular inspections of equipment and premises used by volunteers should take place and appropriate arrangements made for repair and maintenance work.

# Major Public Events

C6522    Major events can create potential health and safety risks for many of the community categories referred to already in the chapter:

— participants in the event ('customers');

— volunteer workers (stewards and performers); and

— neighbours and passers-by.

Not only must event organisers take these members of the community into account, but they must also consider contractors involved in the staging of the event and their potential impact on the community and on each other.

Many major events are now subject to Safety Advisory Groups (SAGs), at which the event is subject to scrutiny by relevant local authorities and the various emergency services. SAGs frequently ask organisers to prepare a properly documented risk assessment for their event so that a full event safety management plan can be developed.

## 'The Event Safety Guide'

**C6523** Public events can include music events (often requiring public entertainment licences), sporting events, carnivals, fun fairs, agricultural and county shows, firework displays, air shows, car rallies, religious gatherings and many other types of function.

'The Event Safety Guide' (HSG 195) was published by the HSE and is primarily aimed at major musical events, although it contains much of value to the organisers of other types of events. The booklet contains general information on the planning and management of events, together with the health and safety responsibilities of the various parties who may be involved.

## Risk Assessment

**C6524** Whilst some parts of the HSE publication 'The Event Safety Guide' deal solely with musical events, many of its other headings provide a checklist to use in carrying out a risk assessment for all types of public events. Topics include:

— venue and site design;
— fire safety;
— major incident planning;
— communication;
— crowd management;
— transport management;
— structures (e.g. stands, gantries, marquees, stages);
— barriers;
— electrical installations and lighting;
— food, drink and water;
— merchandising;
— amusements, attractions and promotional displays;
— sanitary facilities;
— waste management;
— special effects, fireworks and pyrotechnics;
— camping;
— facilities for people with special needs;
— medical, ambulance and first-aid management;
— information and welfare; and
— children

Of course, some of these topics will not be applicable to all events, whilst some events will necessitate risk topics being broken down in greater detail. For example, a running or cycling race taking place on public roads will need to take account of the direction of arrival and parking of competitors' vehicles and also the interface between competitors and other road users during the event itself. This in turn will require consideration of the race route, barriers, signage, possible road closures, race marshals and the possible need for direct police assistance.

## Use of Contractors

**C6525** Event organisers and others engaging contractors must consider the potential risks for members of the community posed by the contractors' equipment and activities. Contractors themselves have many duties under the *Health and Safety at Work etc. Act 1974* and associated legislation, but it has been established for many years that those engaging contractors also have legal responsibilities for risks that they create, both for those at work and for others who may be affected by their activities. Even where contractors already have good health and safety standards, they may not be fully aware of the risks associated with the environment that they are working in and the members of the community who may be present there.

Advance checks on contractors' health and safety standards should include topics such as:

— insurances (particularly public liability);

— health and safety policy;

— risk assessments and method statements;

— competence to provide the services required;

— the competence and training of staff; and

— availability of relevant licences, certificates, inspection records.

Once contractors have been engaged, they must be informed about precautions that they will be expected to take in their work. Precautions of particular relevance in ensuring that members of the community are adequately protected include:

— suitable arrangements for vehicle access and parking;

— maintenance of safe pedestrian access;

— segregation of work areas;

— adequate control of work equipment;

— safe storage and use of hazardous and dangerous substances; and

— suitable control of high risk activities.

Their attention must be drawn to relevant characteristics of members of the community who may be present in or around the workplace. This will be of particular importance in relation to customers and other users of buildings and facilities. Special precautions will be necessary in many cases to ensure the safety of children, the elderly and those with disabilities.

In respect of major public events, contractors must be informed about relevant sections of the event safety management plan. In addition to the matters referred to above, they will need to be told about:

— the layout of the event site;

— availability of key services and facilities;

— communications, e.g. with those managing the event and stewards; and

— emergency arrangements.

The performance of contractors should be monitored both at public events and in other work they carry out. Attention should be paid to their compliance with the health and safety standards set out in their own risk assessments and method statements, as well as to matters of importance in relation to the event or workplace concerned.

## Further Information

**C6526** Each of these publications contains details of further references – both within HSE publications and elsewhere. The HSE website is another potential source of information:

- 'Protecting the public – your next move' (HSG 151);
- 'Charity and voluntary workers – a guide to health and safety at work' (HSG 192); and
- 'The Event Safety Guide' (HSG 195).

# Confined Spaces

## Introduction

**C7001** The Confined Spaces Regulations 1997 contain the main legal requirements affecting work in confined spaces. Not only do they provide a definition of what is meant by a 'confined space' but they also define what 'specified risks' must be addressed in the context of the regulations. Both of these definitions are considered further below.

However, there are also other risks which may be present within confined spaces, which are outside the range of 'specified risks' and thus the application of the regulations (see C7007 below). These risks too must be considered during risk assessments and taken into account in establishing safe systems of work.

## What the Regulations require

**C7002** *Regulation 1* of the *Confined Spaces Regulations 1997* (*SI 1997 No 1713*) contains several definitions of which the most important are as follows:

'confined space' — means any place, including any chamber, tank, vat, silo, pit, trench, pipe, sewer, flue, well or other similar space in which, by virtue of its enclosed nature, there arises a reasonably foreseeable specified risk;

'free flowing solid' — means any substance consisting of solid particles and which is of, or is capable of being in, a flowing or running consistency, and includes flour, grain, sugar, sand or other similar material;

'specified risk' — means a risk of

(a) serious injury to any person at work arising from a fire or explosion;

(b) without prejudice to paragraph (a)–

   (i) the loss of consciousness of any person at work arising from an increase in body temperature;

   (ii) the loss of consciousness or asphyxiation of any person at work arising from gas, fume, vapour or the lack of oxygen;

(c) the drowning of any person at work arising from an increase in the level of liquid; or

(d) the asphyxiation of any person at work arising from a free flowing solid or the inability to reach a respirable environment due to entrapment by a free flowing solid;

'system of work' — includes the provision of suitable equipment which is in good working order.

The definition of 'confined space' contains a list of examples of what might constitute a confined space. The key words in the definition are in the latter part:

'... in which, by virtue of its enclosed nature, there arises a reasonably foreseeable specified risk;'.

A roof space in a building would not come within the definition in the normal course of events. However, should significant quantities of a volatile and highly flammable adhesive or paint be used in the roof space, it would become a confined space under the regulations. Similarly the regulations would be likely to apply if work was to be carried out there in extremely hot conditions.

The list of 'specified risks' can be paraphrased as meaning a risk of:

- serious injury from fire or explosion;
- loss of consciousness from increase in body temperature;
- loss of consciousness or asphyxiation from gas, fume, vapour or lack of oxygen;
- drowning from an increase in liquid level; or
- asphyxiation or entrapment by a free flowing solid.

## Persons with duties under the Regulations

**C7003**     *Regulation 3* of the *Confined Spaces Regulations 1997* (*SI 1997 No 1713*) places duties on both employers and the self-employed.

**Employers** have duties in respect of any work carried out by:

- their employees; and
- other persons (so far as is reasonably practicable), in relation to matters within their (the employers') control.

**Self-Employed Persons** have duties in respect of any work by:

- themselves; and
- other persons (so far as is reasonably practicable), in relation to matters within their (the self-employed persons') control.

This partially duplicates the position established by the *Health and Safety at Work etc. Act 1974* (*HSWA 1974*). In particular it places duties on employers who may bring in specialist contractors to work in confined spaces. Contractors and the self-employed also have duties towards each other.

## Preventing the need for entry

**C7004**     *Regulation 4 (1)* of the *Confined Spaces Regulations 1997* (*SI 1997 No 1713*) follows a pattern increasingly seen in health and safety regulations – a preference for avoiding the risk entirely. It states:

> 'No person at work shall enter a confined space to carry out work for any purpose unless it is not reasonably practicable to achieve that purpose without any such entry'

There are many ways in which entry into confined spaces can be avoided. These should be identified as part of a risk assessment which should determine what is reasonably practicable. Examples include:

- testing atmospheres within confined spaces using long probes;
- sampling or measuring contents using long tools, dip sticks or external instruments;
- cleaning or removing residues from outside by water jetting, steam or use of chemicals;
- integrating in-situ cleaning systems into the design of equipment (this also has hygiene and product quality advantages);

- clearing blockages using long tools or with the use of remotely-operated flails, vibrating devices or blasts of compressed air;
- providing sightglasses, openings etc. to see what is happening within vessels (internal glasses may need to be fitted with washers and/or wipers); and
- the use of CCTV cameras (additional internal lighting may be necessary).

Designers, manufacturers, importers and suppliers have duties under *section 6* of the *Health and Safety at Work etc. Act 1974* (*HSWA 1974*) to eliminate or minimise the need to enter confined spaces. Designers also have similar duties under *Regulation 13* of the *Construction* (*Design and Management*) *Regulations 1994* (*CDM*) (*SI 1994 No 3140*). Even where it is not reasonably practicable to eliminate the need for entry, the design must still facilitate the establishment of safe systems of work and the provision of emergency arrangements.

## Work in confined spaces

**C7005**  Where it is not reasonably practicable to prevent entry, *Regulation 4(2)* of the *Confined Spaces Regulations 1997* (*SI 1997 No 1713*) states that:

'Without prejudice to paragraph (1) above, so far as is reasonably practicable, no person at work shall enter or carry out work in or (other than as a result of an emergency) leave a confined space otherwise than in accordance with a system of work which, in relation to any relevant specified risks, renders that work safe and without risks to health'

Such safe systems of work are, of course, developed as a result of carrying out a risk assessment. Types of risks to be considered and the precautions likely to be necessary to control those risks are dealt with later in the chapter.

## Emergency arrangements

**C7006**  *Regulation 5* of the *Confined Spaces Regulations 1997* (*SI 1997 No 1713*) sets down requirements for emergency arrangements:

'(1) Without prejudice to regulation 4 of these Regulations, no person at work shall enter or carry out work in a confined space unless there have been prepared in respect of that confined space suitable and sufficient arrangements for the rescue of persons in the event of an emergency, whether or not arising out of a specified risk.

(2) Without prejudice to the generality of paragraph (1) above, the arrangements referred to in that paragraph shall not be suitable and sufficient unless–

  (a) they reduce, so far as is reasonably practicable, the risks to the health and safety of any person required to put the arrangements for rescue into operation; and

  (b) they require, where the need for resuscitation of any person is a likely consequence of a relevant specified risk, the provision and maintenance of such equipment as is necessary to enable resuscitation procedures to be carried out.

(3) Whenever there arises any circumstance to which the arrangements referred to in paragraph (1) above relate, these arrangements, or the relevant part or parts of those arrangements, shall immediately be put into operation.'

Several parts of this regulation require further emphasis:

- emergency arrangements must be in place for *any* type of emergency, not just those arising from specified risks – a worker experiencing a heart attack or a serious fall within the confined space may need to be rescued;
- risks to rescuers must be reduced, so far as is reasonably practicable;

- resuscitation equipment may need to be provided and maintained (a worker who has been asphyxiated is likely to need to be resuscitated quickly – well before the emergency services can get there); and
- emergency arrangements must be immediately put into operation should an emergency arise.

Practical guidance on emergency arrangements is provided later in the chapter at C7028 below.

## Other risks in confined spaces

**C7007** Many other risks may be present when working in confined spaces. Some of these risks may be greater because of the enclosed nature of the space, whilst rescue of accident victims is likely to be much more difficult. Such risks include:

- restricted openings for entry into and exit from the space;
- restricted access within the space;
- potential for falls from height or into water;
- slips, trips and falls at the same level;
- falling objects;
- lack of visibility within the space;
- mechanical or electrical equipment within the space;
- chemical, biological or dust hazards;
- excessive noise (damaging hearing or interfering with communication);
- excessive cold; and
- the remote location of the confined space.

Even though these other risks are not covered by the *Confined Spaces Regulations 1997* (*SI 1997 No 1713*), other specific regulations will apply e.g. the *Workplace (Health, Safety and Welfare) Regulations 1992* (*SI 1992 No 3004*), the *Work at Height Regulations 2005* (*SI 2005 No 735*), the *Electricity at Work Regulations 1989* (*SI 1989 No 635*), the *Provision and Use of Work Equipment Regulations 1998* (*PUWER*) (*SI 1998 No 2306*) and the *Control of Substances Hazardous to Health Regulations 2002* (*COSHH*) (*SI 2002 No 2677*).

The general requirements of *Health and Safety at Work etc. Act 1974* will also apply (*HSWA 1974, s2 (2)(a)* contains a similar requirement for a safe system of work) and the *Management of Health and Safety at Work Regulations 1999* (*SI 1999 No 3242*) still require a risk assessment to be carried out.

Figure 1 at C7009 below provides a checklist to aid in the identification of both 'specified risks' and other risks.

## Safe systems of work

**C7008** Both the specific requirements of the *Confined Spaces Regulations 1997* (*SI 1997 No 1713*) and the general requirements of *Health and Safety at Work etc. Act 1974* mean that work in confined spaces must be carried out in accordance with a safe system of work. The Approved Code of Practice (ACOP) accompanying the regulations (contained in HSE booklet L101 – 'Safe work in confined spaces') makes a very clear statement on how safe systems of work should be specified. It states:

'To be effective a safe system of work needs to be in writing. A safe system of work sets out the work to be done and the precautions to be taken. When written down it is a formal record that all foreseeable hazards and risks have been considered in advance. The safe procedure consists of all appropriate precautions taken in the correct sequence. In practice a safe system of work will only ever be as good as its implementation.'

## Permits to work

**C7009** The ACOP to the regulations goes on to describe a permit to work procedure as a means of recording the results of the risk assessment process and the key elements of the safe system of work. A correctly applied permit to work system should ensure that:

- people at work in the confined space are aware of the hazards involved;
- workers know the nature and extent of the work to be carried out;
- appropriate elements of a safe system of work are identified;
- there is a formal check that a safe system of work is in place before people enter or work in the confined space;
- suitable action is taken about other activities which may affect work in the confined space;
- any time limits on entry are specified;
- appropriate means of communication are established; and
- necessary respiratory protective equipment (RPE) and personal protective equipment (PPE) standards are identified.

Figure 2 (below) provides an example of a permit to work form for use in controlling entry into confined spaces. It identifies both the 'specified risks' defined in the regulations and a range of other risks. However, those intending to use this form should ensure that it addresses fully the risks likely to be present in the confined spaces that they are responsible for – adaptation of the form may well be necessary to deal adequately with local circumstances.

Figure 3 (below) is intended as an aide-memoire to accompany the permit to work form provided in Figure 2. It provides the permit issuer with a list of the precautions which may be necessary to control the risks involved in the work.

**Figure 1: Confined Spaces – Identification of Risks**

| DEPARTMENT | | |
|---|---|---|
| TASK/ LOCATION | | |
| SPECIFIED RISKS (Confined Spaces Regulations 1997) ✔ significant risk ? possible risk | | |
| | ✔ or ? | Notes |
| Serious injury from fire or explosion | | |
| Loss of consciousness – increase in body temperature | | |
| Loss of consciousness or asphyxiation due to: | | |
| — gas, fume or vapour | | |
| — lack of oxygen | | |
| Drowning – increase in level of liquid | | |
| Asphyxiation or entrapment – free flowing solid | | |
| **OTHER SIGNIFICANT RISKS** | | |
| Restricted entry/exit | | |

# C7009 Confined Spaces

| | | |
|---|---|---|
| Restricted access within | | |
| Falls from height/into water | | |
| Slips, trips, falls – same level | | |
| Falling objects | | |
| Lack of visibility | | |
| Mechanical equipment | | |
| Electrical | | |
| Chemical | | |
| Biological | | |
| Dust | | |
| Noise | | |
| Excessive cold | | |
| Remote location | | |
| Other (state) | | |
| Should access be controlled by a permit to work? | **YES/NO** | |
| Is a written task procedure required? | **YES/NO** | |
| Signature | Name | Date |

## Figure 2: Permit to Work – Entry into Confined Spaces

Confined Space            Purpose of Entry
Date                          Can the work be carried out from outside?          YES/NO

| RISKS IDENTIFIED | ✓ or ✗ | CONTROLS REQUIRED (inc. Tests) (If necessary, provide details on accompanying documents) |
|---|---|---|
| Fire or Explosion | | |
| Increase in Body Temperature | | |
| Gas, Fume, Vapour | | |
| Lack of Oxygen | | |
| Drowning | | |
| Free-flowing Solids | | |
| Restricted Entry/Exit/Access | | |
| Falls. Slips, trips | | |
| Falling Objects | | |
| Lack of visibility | | |
| Mechanical/Electrical | | |
| Chemical/BIological/Dust | | |
| Noise | | |
| Excessive Cold | | |
| Remote Location | | |
| Other (provide details) | | |

| | DETAILS |
|---|---|
| Communication Methods | |
| Emergency Equipment & Arrangements | |
| Person(s) covered Name(s) by this PTW Signature(s) | |

I confirm that the precautions detailed above have been/will be taken.
I authorise the persons named above to enter this confined space for the purpose stated.
They have been instructed on their roles and responsibilities.

| PERMIT VALID | FROM | | PERMIT CANCELLED AT | | |
|---|---|---|---|---|---|
| (Times) | TO | | (Time) | | (Date) |
| Signature | | (Permit Issuer) | Signature | | (Authorised Person) |

# Figure 3: Confined Spaces – Control Measures and Other Precautions

| RISKS | POSSIBLE CONTROLS |
|---|---|
| Fire or Explosion | Cleaning of residues. Atmospheric testing. Ongoing monitoring. |
| | Purging. Ventilation. Protected electrical equipment. |
| | Hot work restriction. No smoking. Fire fighting equipment. |
| Increase in Body Temperature | Delay before entry. Working time restrictions. Cooling fans. |
| Gas, Fume, Vapour | } Cleaning of residues. Atmospheric testing. Ongoing monitoring. |
| Lack of Oxygen (inc. presence of heavy gases nearby) | } Purging. Ventilation (natural or forced). Isolate incoming feeds. |
| | } Respiratory protection (inc. air-fed respirators or self-contained BA). |
| Drowning | } Isolate incoming feeds. Drain contents. |
| Free-flowing Solids | } Harness attached to secure lifeline. |
| Restricted Entry/Exit | Staff selection. Harness attached to secure lifeline. Instructions to staff. |
| Falls from height/into water | Harness attached to secure lifeline and inertia reel. |
| | Temporary scaffolding/platform. |
| Slips, trips, falls – same level | Clean up surfaces. Suitable footwear. |
| Falling Objects | Safety helmets. Protective screens. |
| Lack of visibility | Permanent or temporary lighting. Torches. BACK-UP ESSENTIAL. |
| Mechanical/Electrical | Secure isolation. |
| Chemical/Biological/Dust | Drain contents. Clean up. PPE e.g. gloves, body cover, eye protection. |
| | RPE – air-fed, self-contained BA, cartridge respirator, dust mask. |
| | Decontamination. |
| Noise | Hearing protection. Suitable communications (see below). |
| Excessive Cold | Suitable clothing and footwear. |
| Remote Location | Suitable communications (see below). Monitoring from base. |
| **KEY PRECAUTIONS** | **OPTIONS AVAILABLE TO CONSIDER** |
| Communications | Attendant outside. Word of mouth. Telephone. Portable phone. |
| | Radio. Air horn. Whistle. |

| | |
|---|---|
| | (Ensure that the equipment works in the work location.) |
| Emergency Arrangements (For all types of emergencies including falls, heart attacks etc.) | Attendant outside. Harness attached to secure lifeline. |
| | Recovery equipment e.g. winch (in position or available nearby). |
| | RPE/PPE worn by or readily available for rescuers. |
| | Resuscitation equipment. Stretcher. Other first aid equipment. |
| | Fire fighting equipment. |
| | Contact with emergency services/internal assistance. |
| | Medical staff on standby (high-risk situations only). |
| | Register of persons inside (large groups only). |
| | (Ensure relevant staff are trained in use of emergency equipment.) |

An essential ingredient of any permit to work system is that those issuing and cancelling the permit are competent for the purpose. Competence and training are dealt with at C7011 below.

The guidance contained in the HSE booklet on 'Safe work in confined spaces' (L101) does to a certain extent soften the previous statement about safe systems of work needing to be in writing. It states that a permit to work is unlikely to be needed where:

- assessed risks are low and can be controlled easily;
- the system of work is very simple; and
- other work activities cannot affect safe working in the confined space.

## *Task procedures*

**C7010** An alternative way of ensuring that a safe system of work is specified in writing may be to incorporate it into a formal 'task procedure" for carrying out the work. (The terms Job Safe Procedure' (JSP) or Standard Practice Instruction (SPI) are used by some to describe similar documents.) Such formal procedures are usually suitable where certain tasks are to be performed regularly within confined spaces. As with permits to work, they should result from a risk assessment of the task in question, and identify the risks which may be present and the various precautions which must be taken to control those risks to an adequate degree.

However, task procedures do not impose the same degree of formal checking that a safe system of work is actually in place in practice as that provided by the permit to work system. They will usually be more suitable for providing a point of reference for a well-trained in-house work team who are already familiar with the environment, the work involved and the risks it entails, as well as the task procedure itself. A permit to work is more likely to be necessary for controlling outsiders (such as contractors) or infrequently performed tasks.

## *Competence and training*

**C7011** Competence for work in confined spaces will be acquired through a combination of adequate training and relevant experience. Consideration should also be given to workers' physical and psychological suitability for work in confined spaces. Training should include:

- an awareness of the regulations and their requirements;

- an appreciation of the risks involved in work in confined spaces;
- the systems of work appropriate for different types of work;
- how to use relevant types of equipment e.g. for communications, access, testing, PPE; and
- emergency and rescue arrangements and equipment.

An increasing number of organisations can provide general training on work in confined spaces. However, this will usually need to be augmented by additional training and/or experience to ensure that workers are:

- familiar with the confined spaces they will work in and the risks involved;
- aware of the precautions necessary in their types of work; and
- capable of using the equipment required to ensure a safe system of work.

Where members of staff are required to issue and cancel permits to work for confined spaces they will need additional training, including an appreciation of the purpose of permits to work. They must be familiar both with the range of risks which may be present and the wide variety of precautions which may be necessary to ensure a safe system of work e.g. isolations, ventilation, gas testing, RPE, PPE etc. There should be a formal system in place for their training, testing and authorisation (as should be the case for all permit to work systems).

### Supervision

**C7012** All workers must be supervised effectively and this is particularly important for those working in confined spaces. Supervision will involve:

- identifying when work is to be carried out in confined spaces;
- arranging for necessary permits to work;
- ensuring that workers are aware of the risks involved and the precautions required;
- ensuring that they are capable of implementing the precautions and have all the equipment required;
- monitoring working practices (including the correct operation of the permit to work); and
- dealing with problems which may arise during the work.

For higher risk tasks the supervisor is likely to remain present whilst the work is carried out. In other cases it may be acceptable to visit periodically, although the supervisor should still be readily available if required.

## Precautions for work in confined spaces

**C7013** Figure 3 (see C7009 above) shows the range of precautions which may be necessary when working in confined spaces. Further detail is provided below on the types of precautions required to control 'specified risks'. Additional guidance is available in HSE booklet L101 'Safe work in confined spaces'. Other risks which may be present in confined spaces can be controlled using conventional techniques as set out in Figure 3.

### Communications

**C7014** Effective communications must be provided between those inside a confined space, between those inside and those outside, and also in order to summon help in an emergency. Communication could be by speech, rope tugs, telephone, radio etc. Consideration should be given to risks of using communication equipment in flammable or explosive atmospheres, to the difficulties of communicating when wearing respiratory protection equipment (RPE) and to the practical difficulties of using telephones or radios in some situations. Airhorns or whistles may provide a practical alternative for some environments. Communication equipment and methods should always be checked and tested at the work location before work starts.

## Testing and monitoring the atmosphere

C7015    Testing may be necessary in respect of hazardous gases, fumes or vapours (these may be hazardous to health, or flammable or explosive). The concentration of oxygen present in the atmosphere may also need to be checked. These test results will indicate possible needs for other precautions e.g. purging, ventilation, RPE. In some cases tests may need to continue to be carried out periodically or prior to each further entry. Constant monitoring of levels may also be necessary – some monitoring devices can give alarms if concentrations of hazardous substances rise above a pre-determined level (or if oxygen levels fall). Those carrying out testing must be competent for the task and test results should be retained.

## Gas purging

C7016    Flammable or toxic gases or vapours may need to be removed by purging with an inert gas (such as nitrogen) for flammable substances, or by an inert gas or air for toxic contaminants. Where inert gases are used, further purging with air is likely to be necessary before entry can take place without suitable RPE. Where purging has taken place, there is likely to be a need for further testing or monitoring of the atmosphere and the use of RPE may still be necessary. Consideration must also be given to the safe venting and disposal of the purged gases.

## Ventilation

C7017    Mechanical ventilation (or ventilation openings) may be necessary to provide fresh breathing air for those inside the confined space, to remove gas, fume or vapour produced by the work, or to remove low levels of emissions from residues within the space. Consideration should be given to the layout of the confined space and the location of any contaminants in order to ensure effective fresh air circulation. Suitable positions should be chosen for air inlets and outlets to avoid possible dead spots and also to prevent contaminated extract air from re-entering. Oxygen must never be used to 'sweeten' any atmosphere, particularly in confined spaces.

## Removal of contents and residues

C7018    Work may only be able to be undertaken safely once the contents of the confined space, together with any remaining residues, have been effectively removed. Methods for removing residues include mechanical means (e.g. scraping), steam, high-pressure water, or use of chemicals. In some cases the removal work may require entry to be made into the confined space, necessitating its own safe system of work. Testing or inspection after the work may be needed, in order to ensure residues have been removed effectively. Further cleaning may be required in some instances.

## Isolations

C7019    Frequently the system of work will require the effective isolation of feeds into the confined space (gases, liquids and other free flowing materials), or the isolation of mechanical and electrical equipment within the space. Such isolations are likely to need to be locked in order to remain effective – other precautions such as the removal of sections of supply pipe or the insertion of blanks may be required where a leaking valve could create a significant risk. Consideration may have to be given to the need to maintain essential services e.g. lighting, power or water for the work to be carried out whilst preventing undesirable risks e.g. re-energisation of process equipment or possible flooding.

## Electrical equipment

**C7020** Electrical equipment should be chosen in relation to risks within the confined space and the work to be carried out. It may need to be:

- suitable for use in a flammable or explosive atmosphere;
- suitable for use in damp conditions (or chemical exposure);
- low voltage because of electrocution risks (or at least protected by RCD's); or
- protected from mechanical damage or dropping.

## Gas equipment and internal combustion engines

**C7021** Use of gas cylinders within confined spaces should be avoided if possible. All gas equipment and pipelines should be checked for leaks before being taken into the confined space and removed to a well ventilated area at the end of every work period (this must include any gas cylinders). Where pipes and hoses cannot be removed they must be disconnected from the supply outside the confined space and their contents safely vented.

Petrol engines must not be used within confined spaces and use of diesel engines should be avoided whenever possible. Where the use of diesel engines or portable gas cylinders in confined spaces is unavoidable, adequate ventilation must be provided to ensure complete combustion takes place. Diesel engine exhausts must be vented well away from the confined space and downwind of any air intakes.

## Personal protective equipment (PPE) and respiratory protective equipment (RPE)

**C7022** Although the law expresses a preference for other types of control measures the use of PPE and RPE in confined space work will often be unavoidable. Wearing of harnesses with safety lines is particularly likely to be necessary as is the use of RPE providing higher levels of protection (self-contained breathing apparatus or air-fed respirators). Wearing PPE and RPE within a confined space may be an additional source of heat stress for workers.

## Access and egress

**C7023** Safe routes into and out of the confined space must be provided and maintained. These must also be suitable for emergency access or escape. Sizes of openings must allow for PPE and/or RPE that may need to be worn. Guidance on manhole sizes is provided in booklet L101 – 575mm diameter is the recommended minimum for new plant. Larger workers may find it physically difficult to get through small openings or to move around within the confined space. Signs prohibiting unauthorised entry may also be required.

## Fire prevention

**C7024** Precautions such as removal of residues, gas purging, ventilation or the control of ignition sources may be necessary as described earlier. Where flammable substances must be present in a confined space (e.g. paints, cleaning materials) they must be kept to a minimum. Provision of suitable fire fighting equipment will be necessary in such cases.

## Lighting

**C7025** Adequate and suitable lighting must be provided as for all workplaces. Suitable back-up lighting (e.g. torches) must be available in the event of lighting failure. See C7020 above for relevant criteria for the selection of electrical equipment.

### Static electricity, hot work and smoking

**C7026** Risks from static electricity or other sources of ignition, such as hot work, may need to be considered. Some types of equipment, clothing or activities (e.g. steam or water jetting) are prone to static build-up or discharge. Smoking should be prohibited in all confined spaces. The ACOP indicates it may be necessary to exclude smoking for 5 to 10 metres beyond the confined space.

### Limited working time and entry control

**C7027** Working within confined spaces may be physically demanding due to extremes of temperature or humidity, the difficulties of wearing RPE or PPE or cramped working conditions. Use of fans may be necessary to achieve a satisfactory thermal environment. Time limits for individuals to stay in the space and rest periods in between may need to be imposed. These may be enforced by a logging or tally system. Such systems could also be necessary in large confined spaces or for large work groups in order to provide a check as to who is inside at any point in time.

## Emergency arrangements

**C7028** Suitable arrangements must be prepared for emergency rescue from a confined space. This may be due to a 'specified risk' or 'other risk' associated with the confined space or for other types of emergency e.g. a worker inside suffering a heart attack or other medical condition. These arrangements must reduce risks to rescuers, so far as is reasonably practicable, and include the provision and maintenance of resuscitation equipment, where relevant due to 'specified risks'.

Emergency arrangements must be immediately put into operation should an emergency arise – a fall inside a confined space may allow time for medical and other assistance to be summoned, but a collapse due to gas, fume, vapour or lack of oxygen is likely to require immediate intervention by those on the spot e.g. rescue using a harness, lifeline entry into the space by rescuers using suitable respiratory protective equipment (RPE) and prompt use of resuscitation equipment.

Examples of emergency equipment and arrangements which may be necessary are:

- an attendant immediately outside the confined space;
- a harness attached to a secure lifeline worn by those inside;
- recovery equipment e.g. a tripod and winch (either already in position or available nearby, if required);
- suitable RPE or PPE for rescuers (either already worn by them or readily available);
- appropriate resuscitation equipment readily available;
- stretcher(s) and other appropriate first aid equipment near at hand;
- fire fighting equipment;
- means of contacting the emergency services or other sources of assistance;
- medical staff on standby nearby (only likely to be necessary for high risk situations);
- use of a register of persons inside the confined space (only necessary where large groups or large, complex confined spaces are involved); and
- arrangements to shut down nearby plant (which may be a source of noise or could increase the risks).

Relevant staff must be fully trained in the use of emergency equipment. This includes those working within confined spaces, acting as back-up attendants or intended to provide emergency assistance in other ways.

**C7029** *Confined Spaces*

The emergency arrangements required should be identified during the risk assessment process and should normally be recorded – either on the permit to work form or within the task procedure.

## Reference

C7029   L101 Safe work in confined spaces (HSE Books)

# Construction and Building Operations

## Introduction to construction work

**C8001** The main causes of accidents and ill-health in the construction industry are:

(a) *Falls* – particularly through fragile roofs and roof lights, from ladders and from scaffolds and other work;

(b) *Falling material and collapses*;

(c) *Electrical accidents*;

(d) *Mobile plant* – causing people to be struck by excavators, lift trucks, dumpers and other vehicles;

(e) *Exposure to dusts* – including asbestos, and other hazardous substances;

(f) *Lifting heavy and awkward loads*; and

(g) *Exposure to high noise levels and vibration*.

A recent Health and Safety Executive (HSE) campaign focused on the following health and safety aspects of construction work:

- Eliminating hazards and controlling risks through compliance with the *Construction (Design and Management) Regulations 1994 (SI 1994 No 3140)* (see C8015), for example, by eliminating the use of fragile roof lights in new and refurbished buildings;

- Controlling risks to scaffolders and to the public and other workers from falling material and scaffold collapses;

- Segregating people from vehicles and plant, and eliminating reversing and/or ensuring the use of suitable reversing aids;

- Ensuring appropriate precautions are taken during work by licensed contractors, including work with asbestos insulation board (see further ASBESTOS);

- Eliminating handling of heavy blocks over 20kg, or reducing the risk from single-person repetitive lifting of these blocks. HSE has recently published '*Backs for the future: Safe manual handling in construction*', which explains the principles for managing manual handling risks, provides an example of a manual handling risk assessment, and describes the role and lists the key points for designers, planning supervisors, principal contractors, contractors and manufacturers and suppliers. It also contains details of case studies where manual handling risks were identified and solutions to the problems found. (See further MANUAL HANDLING); and

- Eliminating or reducing risk from hand-arm vibration and noise at the design stage, and managing residual risk, particularly during scabbling, pile head removal by breaker and hand tunnelling. (See further NOISE AND VIBRATION)

## Practical safety criteria

**C8002** Good safety practice dictates the following.

## Ladders

**C8003**

(a) Only ladders in a sound condition should be used.

(b) The 'one out four up' rule should be strictly adhered to in all situations (i.e. that the vertical height from the ground to the ladder's point of rest should be four times the distance between the base of the vertical dimension and the foot of the ladder).

(c) Ladders should be securely fixed near to their upper resting place or, where this is impracticable, 'footed' by an individual or securely fixed at the base to prevent slipping.

(d) Ladders should be inspected on a regular basis and a record of such inspections maintained.

(See W9018 WORK AT HEIGHTS for statutory requirements.)

## Working platforms

**C8004**

(a) Working platforms should be adequately fenced by means of guard-rails and toe-boards.

(b) Platforms should be adequately covered with sound boards.

(c) Where mobile platforms are used, they should be stationed on a firm level base and, where possible, tied to the structure to prevent sideways movement. Wheel-locking devices should be provided and used.

(d) The following height to base ratios should be applied for all mobile working platforms:

outdoor work – 3:1; indoor work – 3.5:1.

(See W9009 WORK AT HEIGHTS for statutory requirements.)

## Materials

**C8005**

(a) Meticulous standards of housekeeping must be maintained on working platforms and other elevated working positions to prevent materials, tools and other items falling on to people working directly below. The correct positioning of toe-boards is most important here.

(b) Lifting operations should ensure correct hooking and slinging prior to raising, correct assembly of gin-wheels and a high degree of supervision.

(c) Catchment platforms or 'fans' should be installed to catch small items which may fall during construction, particularly where work is undertaken above a public thoroughfare.

## Excavations

**C8006**

(a) Trenches should be adequately timbered with regard to the depth and width of the trench, the nature of the surrounding ground and the load imposed by subsoil.

(b) Excavated ground and building materials should be stored well away from the verge of any excavation.

## Powered hand tools and machinery

**C8007**

(a) Electrically operated hand tools, such as drills, should comply with British Standard 2769: Series and, unless 'all insulated' or 'double insulated', must be effectively earthed.

(b) Portable tools and temporary lighting arrangements should operate through reduced voltages, using 110 volt mains isolation transformers with the secondary winding centre tapped to earth.

(c) Power take-offs, cooling fans, belt drives and other items of moving machinery should be securely fenced to prevent workers coming into contact with them. All machinery should comply with the *Provision and Use of Work Equipment Regulations 1998* (*SI 1998 No 2306*) (see further MACHINERY SAFETY AND ELECTRICITY). The HSE booklet, '*Electrical safety on construction sites*', provides more detailed guidance in this area.

(d) The health risks associated with vibration should be assessed when buying or hiring hand-held power tools, and existing power tools should also be checked (see further NOISE AND VIBRATION).

## Site transport

**C8008**

(a) Employees should not travel on site transport, such as dumper trucks, and notices should be affixed to such vehicles to that effect.

(b) All vehicle movement and tipping operations on site should be supervised by a person outside the driver's cab.

(c) Site vehicles should be subject to regular maintenance, particular attention being paid to braking and reversing systems.

(d) Only competent and trained drivers should be allowed to drive site vehicles.

(e) Site roadways should be maintained in a sound condition, free from mud, debris, obstructions and large puddles. The verges of the roadway should be clearly defined and adequate lighting provided, particularly at tipping points, reversing and turning areas.

(f) Site speed limits should be established and clearly marked with signs corresponding to speed limit signs on public roads.

## Demolition

**C8009**

(a) A pre-demolition survey should always be undertaken, making use of the original plans if available.

(b) Catching platforms should be installed not more than 6 m below the working level wherever there is a risk to the public.

(c) Employees should be provided with safety helmets incorporating chin straps, goggles, heavy duty gloves and safety boots with steel insoles. In certain cases, respiratory protection, safety belts or harnesses may also be necessary.

(d) Demolition should be undertaken, wherever possible, in the reverse order of erection.

(e) When using working platforms, all debris should be removed on a regular basis.

(f) Members of framed structures should be adequately supported and temporary props, bracing or guys installed to restrain remaining parts of the building.

(g) Employees should not work from the floor of a building which is currently being demolished.

(h) Where pulling arrangements, demolition ball, explosives or pusher arms are to be used, employees should be kept well away until these stages have been completed.

(i) Frequent inspections must be made of the demolition site to detect dangers which may have arisen following commencement of demolition.

## Fire

C8010

(a) All sources of ignition should be carefully controlled, e.g. welding activities, the use of blow lamps, gas or liquid fuel fired appliances.

(b) All flammable materials, including waste materials, should be carefully stored away from the main construction activity.

(c) All employees should be aware of the fire warning system, training sessions being undertaken according to need.

(d) There should be sufficient access for fire brigade appliances in the event of fire.

(e) There should be adequate space between buildings, e.g. site huts, canteen, etc.

(f) High-risk buildings should be separated from low-risk buildings.

(g) Controlled areas, where smoking and the use of naked lights are forbidden, should be established and suitably marked with warning signs.

(h) An adequate supply of water should be available for fire brigade appliances and on-site fire-fighting.

(i) Fire wardens should be appointed to undertake routine site inspections, together with the operation of a fire patrol, particularly at night and weekends.

(See also FIRE AND FIRE PRECAUTIONS.)

## Asbestos

C8011

(a) It should be assumed that buildings constructed or refurbished before the 1980s will contain asbestos-based materials.

(b) No work should be carried out which is likely to expose employees to asbestos unless an adequate assessment of exposure has been made.

(c) The area where work is to be carried out should be checked to identify the location, type and condition of any asbestos likely to be disturbed.

Two new guidance booklets from the HSE, '*Introduction to Asbestos Essentials*' (aimed at anyone who is liable to control or carry out maintenance work with asbestos-containing materials) and '*Asbestos Essentials Task Manual*' (aimed at workers), provide information on ensuring that building maintenance work involving asbestos-containing materials is carried out safely and in accordance with the law. (Also see further ASBESTOS.)

# The Regulations

C8012   The series of construction Regulations made under the *Factories Act 1961* referred to 'building operations' and 'works of engineering construction'. A major revision of these

Regulations gives a more comprehensive definition of 'construction work', which is based on a much broader concept and also includes the installation and removal of services which are part of a structure and the installation and removal of fixed plant where there is a risk of falling two metres or more (see also WORK AT HEIGHTS).

All construction activities are now contained in three sets of Regulations bearing the prefix 'Construction'.

- the *Construction (Head Protection) Regulations 1989* (*SI 1989 No 2209 as amended*) (see C8097 below);
- the *Construction (Design and Management) Regulations 1994* (*CDM*) (*SI 1994 No 3140*) (see C8015 below);
- the *Construction (Health, Safety and Welfare) Regulations 1996* (*CHSW*) (*SI 1996 No 1592*) (see C8031 below).

Generally, the definition of construction work covers the activities on all kinds of construction sites, from the smallest internal jobs to large-scale complex projects including:

- general building and construction work;
- refurbishment work;
- maintenance and repair work;
- engineering construction work;
- civil engineering work.

The exploration and extraction of mineral resources and the preparatory activities are excluded from the definition of construction work, which is extensively defined in the principal regulations. Reference to the full definitions should be made in any case of doubt or for litigation purposes. [*CDM* (*SI 1994 No 3140*), *Reg 2(1)*; *CHSW* (*SI 1996 No 1592*), *Reg 2(1)*].

The two principal sets of Regulations have different applications:

- *CDM* applies to construction sites only when the 'construction work' is notifiable to the HSE (see C8016) and where more than five people are likely to be employed at any one time;
- *CHSW* applies to all construction sites (that is, any place where the principal activity carried on is 'construction work') and when a person is at work on a construction site.

## Associated statutory requirements

**C8013**     The protection of the workforce who carry out construction work stems from the *HSWA 1974* and the *Management of Health and Safety at Work Regulations 1999* (*MHSWR*) (*SI 1999 No 3242*).

In particular, under the Act itself:

(*a*)   contractors (as employers) owe health and safety duties to their employees [*HSWA 1974, s 2*];

(*b*)   employees owe such duties to their employers (i.e. contractors/subcontractors) [*HSWA 1974, s 7*];

(*c*)   building owners (i.e contractors' employers) owe health and safety duties to subcontractors and their employees [*HSWA 1974, s 3(1)*] (also see *R v Associated Octel Co Ltd* at C8136);

(*d*)   self-employed contractors/workmen owe health and safety duties to other self-employed persons (e.g. building owners to main contractors), and to employees (of other organisations) [*HSWA 1974, s 3(2)*].

Self-styled 'labour-only' subcontractors have been held to be employees of a large main contractor and so were entitled to the protection of some of the *Construction Regulations* (*Ferguson v John Dawson & Partners (Contractors) Ltd [1976] IRLR 346* where a nominated subcontractor was liable to a 'self-employed labour-only subcontractor' for breach of *Reg 28(1)* of the *Construction (Working Places) Regulations 1966* (now repealed), requiring provision of guard-rails and toe-boards at working platforms and places);

(*e*) building owners and occupiers have duties to employees of contractors and subcontractors. [*HSWA 1974, s 4(1), (2)*]. Thus, *HSWA 1974, s 4* states that anyone 'having control to any extent' of premises, must take reasonable care in respect of persons working there, who are *not* employees. This duty can be subject to indemnity clauses in a lease and/or contract (see further OCCUPIERS' LIABILITY), it is strict and must be carried out so far as is reasonably practicable, that is, subject to cost-effective constraints;

(*f*) although ostensibly applicable to all workplaces, the *Workplace (Health, Safety and Welfare) Regulations 1992 (SI 1992 No 3004)* do *not* extend to building operations and works of engineering construction, unless some other activity is being carried on there (see further WORKPLACES – HEALTH, SAFETY AND WELFARE).

## Main Regulations relating to the construction industry

C8014   The most relevant legislation relating to construction work is listed below:

— *Construction (Design and Management) Regulations 1994 (SI 1994 No 3140);*

— *Construction (Health, Safety and Welfare) Regulations 1996 (SI 1996 No 1592);*

— *Management of Health and Safety at Work Regulations 1999 (SI 1999 No 3242);*

— *Provision and Use of Work Equipment Regulations 1998 (SI 1998 No 2306);*

— *Control of Substances Hazardous To Health Regulations 1999 (SI 1999 No 437);*

— *Control of Asbestos at Work Regulations 1987 (SI 1987 No 2115);*

— *Asbestos (Licensing) Regulations 1983 (SI 1983 No 1649);*

— *Manual Handling Operations Regulations 1992 (SI 1992 No 2793);*

— *Personal Protective Equipment Regulations 1992 (SI 1992 No 2966);*

— *Noise at Work Regulations 1989 (SI 1989 No 1790);*

— *Confined Spaces Regulations 1997 (SI 1997 No 1713);*

— *Work in Compressed Air Regulations 1996 (SI 1996 No 1656);*

— *Lifting Operations and Lifting Equipment Regulations 1998 (SI 1998 No 2307);*

— *Reporting of Injuries Diseases and Dangerous Occurrences Regulations 1995 (SI 1995 No 3163);*

— *Workplaces (Health, Safety and Welfare) Regulations 1992 (SI 1992 No 3004);*

— *Health and Safety (Enforcing Authority) Regulations 1998 (SI 1998 No 494);*

— *Safety Representatives and Safety Committees Regulations 1977 (SI 1977 No 500);*

— *Health and Safety (Consultation with Employees) Regulations 1996 (SI 1996 No 1513).*

## Construction (Design and Management) Regulations 1994 (SI 1994 No 3140)

**C8015** The *Construction (Design and Management) Regulations 1994* (*CDM*) (*SI 1994 No 3140*) implement (with minor exceptions) EU Directive 92/57/EEC on the minimum safety and health requirements at temporary or mobile construction sites. The Regulations are supported by an Approved Code of Practice, L54: *Managing construction for health and safety* (ISBN 0 7176 0792 5). This will be replaced by a new Approved Code of Practice (ACoP) on the Regulations, *Managing construction for health and safety*. The ACoP is expected to come into force (at the earliest) in summer 2002. The revision is intended to focus the duty holders on managing health and safety while avoiding unnecessary bureaucracy. It clarifies the duties involved in complying with the legislation and concentrates on the role of duty holders and particular topics including:

- Assessing competence and resources;
- Preparing health and safety plans; and
- Giving advice on the contents of the health and safety file.

Generally, the *CDM Regulations* apply to construction work (as defined) carried out on a construction site which is notifiable to the HSE, that is, a construction project which:

— is scheduled to last for more than 30 days; or

— will involve more than 500 man-days of work; or

— includes any demolition work regardless of the size or duration of the work; or

— involves five or more workers being on site at any one time.

The Regulations always apply when construction design work is involved.

### Notification of project

**C8016** Notification of a construction project (where notifiable) (see 'Exclusions' at C8017 below) should be given to HSE in writing (ideally) as soon as practicable after appointment of the planning supervisor, or (failing that), after the appointment of the principal contractor, but before construction work starts, specifying:

(*a*) date of forwarding;

(*b*) exact address of construction site;

(*c*) name and address of client(s);

(*d*) type of project;

(*e*) name and address of the planning supervisor;

(*f*) declaration of appointment by the planning supervisor;

(*g*) name and address of principal contractor;

(*h*) declaration of appointment of the principal contractor;

(*i*) date planned for start of the construction phase;

(*j*) planned duration of the construction phase;

(*k*) estimated maximum number of people at work on the construction site;

(*l*) planned number of contractors on construction site;

(m) name and address of any contractor(s) already chosen.

[CDM (SI 1994 No 3140), Reg 3, Sch 1].

## Exclusions

**C8017**  These Regulations are inapplicable (mainly) to construction work:

(a) where a client reasonably believes that
- (i) a project is not notifiable, and
- (ii) no more than four people are working at any one time (except for demolition and dismantling)

[CDM (SI 1994 No 3140), Reg 3(2), (3)];

(b) of a minor nature where the local authority is the enforcing authority [CDM (SI 1994 No 3140), Reg 3(4)];

(c) carried out for a domestic client, unless (as a result of agreement/arrangement with the developer):
- (i) land is transferred to the client,
- (ii) the developer undertakes to build on the land, or
- (iii) after construction, the land will incorporate premises to be occupied by the client

[CDM (SI 1994 No 3140), Reg 3(8)].

## Objectives of the CDM Regulations

**C8018**  The *CDM Regulations* introduce a control framework which requires the effective management of all stages of a construction project from conception, design, commissioning of work, its planning and execution, and in particular construction activities which are likely to pose significant risks to workers.

The overall aim of the *CDM Regulations* is to raise construction safety standards by improving co-ordination between various parties involved at both preparation and execution stage. More particularly, clients are required to appoint co-ordinators for preparation (planning supervisors) and execution stages (principal contractors), the co-ordinators being required to prepare and update the health and safety plan, which is the key to the operation of *CDM Regulations*, planning supervisors dealing mainly with designers and principal contractors dealing mainly with other contractors. The health and safety plan (see C8025 below) is initiated by the planning supervisor and forms part of tendering documentation. Before actual construction work begins, the plan should be scrutinised by the principal contractor in order to ensure that it is properly adjusted to contractors and site activities. Under *CDM (SI 1994 No 3140), Reg 6*, clients must ensure that adequate financial provision is made and adequate time allowed for completion of the project, with the main health and safety costs being assessed and included in tenders, the basis of this assessment being the health and safety plan. Under *CDM (SI 1994 No 3140), Reg 14*, planning supervisors must be competent to assess the project and advise the client, who must then set the budget and timescale and finance the project. As companies, individuals or partnerships, they will often be the lead member of the design team (e.g. architect or engineer); sometimes a main contractor may be a planning supervisor, sometimes even a client. But both planning supervisors and principal contractors should be appointed as early as possible during the planning stage of the project.

It is incumbent on clients to specify that contractors comply with the health and safety plan; in turn, contractors should price compliance, with clients requiring that contractors include in tender identifiable sums for the management of health and safety and to deal with specific hazards. For that reason, tenderers should advise clients of any hazards not identified in the plan which appear in their assessments. It is then for the planning supervisor to advise the client on adequacy of tender (and provision for health and safety), who should take this into account when awarding contracts. In this way, contracts should only be awarded to contractors/tenderers prepared to comply with health and safety requirements and standards.

## Duties and responsibilities

**C8019** Because the United Kingdom construction industry has characteristics which were not recognised by the EU Temporary or Mobile Construction Sites Directive, it was necessary to specify that the 'client' (a term which includes clients' agents and developers) must appoint other principals for the planning and the carrying out of the construction work within the scope of the Directive. These are the 'planning supervisor' and the 'principal contractor'.

The *CDM Regulations* give detailed 'job descriptions' for the various parties involved in the construction contract. It should be noted that the following duties imposed by the Regulations are additional to those imposed by the *Management of Health and Safety at Work Regulations 1999* (*SI 1999 No 3242*) (*MHSWR*) and other similar health and safety legislation.

### Clients

**C8020** Clients (including clients' agents) [*CDM (SI 1994 No 3140), Reg 4(1)*] must, prior to construction work, and in respect of each project:

(*a*) appoint

  (i) a competent planning supervisor [*CDM (SI 1994 No 3140), Reg 6(1), (3)*], and

  (ii) a competent principal contractor (who must be a contractor) [*CDM (SI 1994 No 3140), Reg 6(1)–(3)*].

  Such appointments can be terminated, changed or renewed (where necessary) to ensure that these roles are filled until construction is completed [*CDM (SI 1994 No 3140), Reg 6(5)*]. So long as competent to perform both roles, planning supervisors can also be principal contractors; and clients can be planning supervisors or principal contractors (or both) [*CDM (SI 1994 No 3140), Reg 6(6)*]. However, clients must not appoint planning supervisors, designers and contractors, unless satisfied as to their competence [*CDM (SI 1994 No 3140), Reg 8(1)–(3)*], and that all three have allocated, or will allocate, appropriate resources for the performance of their respective roles [*CDM (SI 1994 No 3140), Reg 9*];

(*b*) so far as is reasonably practicable, ensure that the construction phase of any project does not begin without preparation of a satisfactory health and safety plan [*CDM (SI 1994 No 3140), Reg 10*];

(*c*) as soon as is reasonably practicable but before commencement of work, ensure that the planning supervisor is provided with information relevant to the state of the premises on which construction work is to take place [*CDM (SI 1994 No 3140), Reg 11*];

(*d*) ensure that information in the health and safety file is kept available for inspection by any person, for the purposes of compliance with statutory requirements and prohibitions [*CDM (SI 1994 No 3140), Reg 12(1)*].

### Designers

**C8021** Designers must:

(1) advise clients as to their duties; and

(2) ensure that any design has regard to the need:

(a) to avoid foreseeable risks to the health and safety of any person involved in construction or cleaning work in or on the structure at any time, or anyone who may be affected by the work of such person (e.g. a member of the public);

(b) to combat at source risks to the health and safety of any person at work involved in construction, cleaning work or any person who may be affected by such work;

(c) to give priority to measures for protecting those involved in construction, cleaning and those who may be affected;

(d) to ensure that the design includes adequate information about any aspect of the project or structure of materials which might affect construction or cleaning workers or those who may be affected by their work; and

(e) to co-operate with the planning supervisor (and any other designer preparing a design in connection with the project) for the purposes of compliance with statutory requirements and prohibitions.

*[CDM (SI 1994 No 3140), Reg 13(2)].*

The Regulations concerning the definition of and duties of designers were amended by the *Construction (Design and Management) (Amendment) Regulations 2000 (SI 2000 No 2380).*

The amending Regulations were made as a result of a Court of Appeal decision in *R v Paul Wurth SA, The Times, 29 March 2000,* in which HSE had prosecuted Paul Wurth SA following a fatality during the installation of a conveyor at the British Steel Port Talbot works in 1997.

The HSE had alleged that the company had failed in its duties as designers under *Regulation 13(2)* of the 1994 Regulations. This places a duty on a designer to ensure that any design he prepares, and which he is aware will be used for the purposes of construction work, includes among its design considerations adequate regard to the need to avoid foreseeable risks to the health and safety of people as a result of construction work and subsequent cleaning work.

However, the Court of Appeal held that designers do not have responsibilities for designs which are prepared by others under their control, including their employees. This was contrary to the aim of the original provisions of the Regulations, and hence the need for the amendment.

*Regulations 2* (concerning the definition of 'designer'), *3* (regarding a person preparing a design), and *12* (concerning the client's duty to ensure the health and safety file is available for inspection) and *13* (concerning requirements on designers), have all been amended. The amended Regulations ensure that legal duties on designers to build safety into a design, apply not only to a design prepared by them personally, but also to a design prepared by an employee or other person under their control.

## *Planning supervisors*

**C8022** Planning supervisors must:

(a) ensure that the project is notified to the HSE (unless it is reasonably believed that the project is not notifiable (see 'Exclusions' at C8017 above));

(b) ensure that the design of a project includes

   (i) reference to health and safety management, and

   (ii) adequate information regarding structure and materials;

(c) ensure co-operation between designers, for the purposes of compliance with their duties as designers (under *CDM (SI 1994 No 3140), Reg 13);*

(d) be able to give adequate advice to

(i) any client/contractor regarding competence of personnel/allocation of resources, and

(ii) any client regarding competence of a contractor and a contractor's allocation of resources as well as the health and safety plan;

(e) ensure preparation of a health and safety file, containing

(i) information concerning aspects of the project, structure or materials that may affect health and safety, and

(ii) any other information which, foreseeably, will be necessary to ensure health and safety of persons involved in construction, cleaning and maintenance or demolition work;

(f) ensure that, on completion of the project, the health and safety file is delivered to the client

[CDM (SI 1994 No 3140), Reg 14]; and

(g) ensure that the health and safety plan is made available to every contractor before arrangements are made for them to manage or carry out construction work.

[CDM (SI 1994 No 3140), Reg 15(1), (2)].

## Principal contractors

**C8023** Principal contractors must:

(a) take reasonable steps to ensure co-operation between all contractors, for the purposes of compliance with statutory requirements and/or prohibitions;

(b) so far as is reasonably practicable, ensure that every contractor and every employee complies with the health and safety plan;

(c) take reasonable steps to ensure that only authorised persons are allowed where construction work is carried on;

(d) ensure that notification particulars (see C8016 above) are displayed prominently so that they can be read by construction personnel; and

(e) provide the planning supervisor promptly with any of the following information,

(i) which is in possession of the principal contractor, or which the latter could ascertain by making reasonable enquiries of a contractor,

(ii) which, reasonably, the planning supervisor would include in the health and safety file (in order to comply with his duties (under *Reg 14* above)), and

(iii) which is not in possession of the planning supervisor.

[CDM (SI 1994 No 3140), Reg 16(1)].

For the purposes of compliance with *Reg 16(1)*, principal contractors can

(i) give any necessary directions to any contractor, and

(ii) include in the health and safety plan rules for the management of construction work reasonably required for health and safety management; such rules being in writing and brought to the attention of those affected.

[CDM (SI 1994 No 3140), Reg 16(2), (3)];

(f) so far as is reasonably practicable, ensure that every contractor is provided with comprehensible information on health and safety risks to that contractor or to employees or other persons under that contractor's control.

(g) so far as is reasonably practicable, ensure that every contractor who is an employer provides his employees engaged in construction work with:

(i) information relating to

— health and safety risks,

— protective measures,

— procedures to be followed in imminent danger and in danger areas, and

— persons appointed to implement those procedures, and

(ii) health and safety training, both on recruitment and/or exposure to new/increased risks; such training to be repeated periodically

[*CDM* (*SI 1994 No 3140*), *Reg 17*];

(h) ensure that:

(i) employees/self-employed personnel are able to discuss, and offer advice on, matters foreseeably affecting their health and safety, and

(ii) there are arrangements for the co-ordination of employees' views (or their representatives')

[*CDM* (*SI 1994 No 3140*), *Reg 18*].

## Contractors

**C8024**  Contractors must:

(a) co-operate with the principal contractor, so that both can comply with their statutory duties;

(b) so far as is reasonably practicable, provide the principal contractor promptly with any information (including risk assessments for the purposes of the *Management of Health and Safety at Work Regulations 1999*) (*SI 1999 No 3242*), which might affect the health and safety of construction workers, or those who might be affected by construction work, or which might justify a review of the health and safety plan;

(c) comply with directions given by the principal contractor, for the purposes of compliance by contractors;

(d) comply with rules applicable to them in the health and safety plan;

(e) provide the principal contractor promptly with information relating to deaths, injuries, conditions and dangerous occurrences notifiable under the *Reporting of Injuries, Diseases and Dangerous Occurrences Regulations 1995* (*SI 1995 No 3163*) (see further ACCIDENT REPORTING);

(f) provide the principal contractor promptly with any information which:

(i) is in the possession of the contractor, or which he could ascertain by reasonable enquiries, and

(ii) it is reasonable to suppose the principal contractor would provide to the planning supervisor, for the purposes of inclusion in the health and safety file, which is not in the possession of the planning supervisor or the principal contractor

[*CDM* (*SI 1994 No 3140*), *Reg 19(1)*];

(g) not allow any employee to work on construction work, unless provided with:

(i) the name of the planning supervisor,

(ii) the name of the principal contractor, and

(iii) the contents of the health and safety plan relating to work being carried out by the employee.

Self-employed personnel must also be provided with this information [*Reg 19(2)–(4)*].

## Health and safety plan

**C8025** Before construction work starts, a duty is placed on the 'client' (as defined) to ensure that the appointed planning supervisor prepares a document known as the 'health and safety plan' which is to serve two purposes:

(*a*) during the pre-construction phase, to bring together the health and safety information obtained from the client and designers;

(*b*) during the construction phase, to include the principal contractor's health and safety policy and risk assessments.

The health and safety plan will continue to evolve and provide a focus for the co-ordination of health and safety matters as the construction work progresses.

The health and safety plan forms the basis of the health and safety management structure, being part of the tender documents. It should:

(*a*) indicate (in general terms) the approach to health and safety to be adopted by everyone (as per the *Management of Health and Safety at Work Regulations 1999 (SI 1999 No 3242)*);

(*b*) identify the main health and safety hazards likely to occur to employees, self-employed operatives and the general public – these having been specified earlier by the client and/or designers (e.g. work in compressed air);

(*c*) specify precautions to be taken; and

(*d*) require work to be done to recognised technical standards and in accordance with published guidance (which should be specified in the plan).

Before the commencement of work, the planning supervisor should acquaint both principal contractor and contractors with the health and safety plan, so that they can agree to it or to modifications in it, and then draw up tenders.

## The planning supervisor

**C8026** Planning supervisors should:

— prepare the health and safety plan;

— ensure that the plan forms part of the tender documentation;

— investigate significant differences in tender documents relating to the plan;

— assess the adequacy of sums specified in tenders vis-à-vis the plan;

— advise the client on the adequacy of tenders; and

— review the plan if the basis of the original advice changes.

## The principal contractor

**C8027** Principal contractors should ensure that the health and safety plan:

— translates into an intelligible working document for all those involved in the construction phase;

— incorporates arrangements submitted by individual contractors for the overall management of health and safety; and

— includes arrangements for compliance with the duties of the principal contractor (see C8022 above);

— specifies detailed arrangements for monitoring compliance with health and safety law;

— includes arrangements for assessing competence of subcontractors; and

— can be modified/updated in the light of the experience and information of contractors.

### Health and safety file

C8028    The health and safety file is a permanent record containing information about the particulars and arrangements relating to the design, methods and materials, maintenance and other information relating to the construction. In practice, the 'file' amounts to a manual to alert those who will be responsible for the structure after construction on safety matters which must be managed after handover. The manual should contain appropriate information regarding maintenance, repair, renovation and demolition.

### Prosecution – defence and civil liability
#### Defence

C8029    A defence is provided against prosecution where it can be shown that an employer or a self-employed person made all reasonable enquiries and reasonably believed either that the Regulations did not apply to the work in question or that he had been given the names of the planning supervisor and the principal contractor together with the relevant contents of the health and safety plan [*CDM* (*SI 1994 No 3140*), *Reg 19(5)*].

#### Civil liability

C8030    Generally, breach of the *CDM Regulations* is not actionable, except as regards:

(*a*)   the preparation of the health and safety plan, prior to construction work, under *Reg 10*; or

(*b*)   allowing only authorised personnel onto premises where construction work is going on, under *Reg 16(1)(c)*.

[*CDM* (*SI 1994 No 3140*), *Reg 21*].

## Construction (Health, Safety and Welfare) Regulations 1996 (SI 1996 No 1592)

C8031    The *Construction (Health, Safety and Welfare) Regulations 1996* (*CHSW*) (*SI 1996 No 1592*) promote the health and safety of everyone carrying out 'construction work' (as defined, and see also C8012 above) but the Regulations do not apply to workplaces on construction sites which are set aside for non-construction purposes. However, the *CHSW Regulations* give protection to other people who may be affected by the construction work.

A 'construction site' is defined in the Regulations as 'any place where the principal work activity being carried out is construction work' [*CHSW* (*SI 1996 No 1592*), *Reg 2(1)*].

For details of the requirements and standards of protection regarding falls from access equipment [*CHSW* (*SI 1996 No 1592*), *Reg 6*], see:

*Schedule 1:* Requirements for guardrails etc;

*Schedule 2:* Requirements for working platforms;

*Schedule 3:* Requirements for personal suspension equipment;

*Schedule 4:* Requirements for means of arresting falls;

*Schedule 5:* Requirements for ladders.

For details of the requirements for welfare facilities [*CHSW* (*SI 1996 No 1592*), *Reg 22*] see:

*Schedule 6:* Welfare facilities.

Training and site inspection by competent persons followed by preparation of a report to be kept on site and available for inspection by HSE inspectors are especially important requirements. In particular, anyone involved in construction work, which requires training, technical knowledge or experience for reduction of risk of injury, must possess such training, knowledge etc. or be under the supervision of a person with such knowledge. In addition, places of work on construction sites must only be used following inspection by competent persons and preparation of a report to the effect that such place is safe (see C8055 below).

These Regulations implement Annex IV of the Temporary or Mobile Construction Sites Directive, thereby extending to construction sites the health, safety and welfare requirements imposed on all other workplaces by the *Workplace* (*Health, Safety and Welfare*) *Regulations 1992* (*SI 1992 No 3004*). The Regulations revoke the *Construction Regulations 1961–1966* (with the exception of the *Construction* (*Lifting Operations*) *Regulations 1961* – which have been replaced by the *Lifting Operations and Lifting Equipment Regulations 1998* (*SI 1998 No 2307*).

These Regulations lay duties principally on employers, self-employed contractors and employees, and, to a lesser extent, on persons in control of construction sites. [*CHSW* (*SI 1996 No 1592*), *Regs 4, 22, 29(2)*]. All contractors, big or small, are covered. Breach of duty, followed by conviction, will be visited with criminal sanction (normally payment of a fine). There is also strict civil liability where breach of the Regulations results in injury or damage (see ENFORCEMENT and INTRODUCTION). Characteristically, 'I find it necessary to make some general observations about the interpretation of regulations of this kind. They are addressed to practical people skilled in the particular trade or industry, and their primary purpose is to prevent accidents by prescribing appropriate precautions. Any failure to take prescribed precautions is a criminal offence. The right to compensation, which arises when an accident is caused by a breach is a secondary matter. The Regulations supplement, but in no way supersede the ordinary common law obligations of an employer to care for the safety of his men, and they ought not to be expected to cover every possible kind of danger' (*Gill v Donald Humberstone & Co Ltd [1963] 3 AER 180*, as per Lord Reid).

## Duties under the regulations

**C8032** The following duties are laid on employers (and self-employed persons) in respect of employees involved in construction work and any persons under their control.

## Safe place of work

**C8033** Except in the case of a person making such place safe [*CHSW* (*SI 1996 No 1592*), *Reg 5(4)*]:

(a) every place of work must be made and kept safe and free from health risks for any person at work there, so far as is reasonably practicable [*CHSW* (*SI 1996 No 1592*), *Reg 5(2)*];

(b) so far as is reasonably practicable, suitable and sufficient safe access/egress, to/from, every place of work and any other place provided for use of a person at work, must be provided and properly maintained [*CHSW* (*SI 1996 No 1592*), *Reg 5(1)*];

(c) so far as is reasonably practicable, suitable and sufficient steps must be taken to deny access to any place not complying with (a) and (b) above [*CHSW* (*SI 1996 No 1592*), *Reg 5(3)*]; and

(d) so far as is reasonably practicable, every place of work must have sufficient working space and be so arranged as to be suitable for any person working or likely to work there [*CHSW (SI 1996 No 1592), Reg 5(5)*].

(For requirements relating to falls, fall-preventative equipment, scaffolds, personal suspension equipment, working platforms and ladders, set out in *Regs 6, 7* and *8*, see WORK AT HEIGHTS.)

## Stability of structures

**C8034** The Regulations require that:

(a) in order to prevent danger to any person, all practicable steps must be taken to ensure that any new or existing structure which may become unstable through construction work (including excavations) does not accidentally collapse [*CHSW (SI 1996 No 1592), Reg 9(1)*];

(b) no part of a structure must be so loaded as to make it unsafe [*CHSW (SI 1996 No 1592), Reg 9(2)*]; and

(c) any buttress, temporary support or structure (used to support a permanent structure) must be erected or dismantled under the surveillance of a competent person [*CHSW (SI 1996 No 1592), Reg 9(3)*].

## Demolition or dismantling

**C8035** Suitable and sufficient steps must be taken to ensure that demolition or dismantling of structures is planned and executed so as to prevent, so far as practicable, danger to any person. Planning and execution of demolition must be carried out under the supervision of a competent person [*CHSW (SI 1996 No 1592), Reg 10*].

## Explosives

**C8036** Explosive charges can only be used or fired if suitable and sufficient steps have been taken to ensure that no one is exposed to any risk of injury from the explosion or flying material [*CHSW (SI 1996 No 1592), Reg 11*].

## Excavations

**C8037** The Regulations require that:

(a) to prevent danger to any person, all practicable steps must be taken to ensure that new or existing excavations which are temporarily unstable due to the carrying out of construction work (including other excavation work) do not collapse accidentally [*CHSW (SI 1996 No 1592), Reg 12(1)*];

(b) so far as is reasonably practicable, suitable and sufficient steps must be taken to prevent any person being buried or trapped by a fall or dislodgement of material. [*CHSW (SI 1996 No 1592), Reg 12(2)*]. In particular, as early as practicable in the course of the work, the excavation must be sufficiently supported so as to prevent, so far as reasonably practicable, that fall or dislodgement of material [*CHSW (SI 1996 No 1592), Reg 12(3)*]; suitable and sufficient supporting equipment must be provided [*CHSW (SI 1996 No 1592), Reg 12(4)*]; and installation, alteration or dismantling must be carried out only under the supervision of a competent person [*CHSW (SI 1996 No 1592), Reg 12(5)*];

(c) suitable and sufficient steps must be taken to prevent any person, vehicle, plant and equipment, accumulation of earth or other material, from falling into an excavation [*CHSW (SI 1996 No 1592), Reg 12(6)*];

(d) where collapse of an excavation would endanger a person, no material, vehicle or plant and equipment must be placed or moved near any excavation [*CHSW (SI 1996 No 1592), Reg 12(7)*]; and

(e) no excavation work must be carried out unless suitable and sufficient steps have been taken to identify and, so far as reasonably practicable, to prevent any risk of injury arising from any underground cable or service [*CHSW (SI 1996 No 1592), Reg 12(8)*].

A HSE guidance booklet, '*Health and safety in excavations: Be safe and shore*', provides practical guidance on working in the ground without placing anyone at risk. It is intended for all those who may be involved with excavation works, e.g. clients, designers, planning supervisors, contractors, site managers and foremen. It aims to help identify the main failings which lead to injury and how to take the necessary measures, at the planning stage, to avoid them. It deals with:

— Hazards and control measures;

— Planning, design and management;

— Legal requirements; and

— Sources of further information.

It covers pipe and cable laying, manhole construction, foundations, small retaining walls, and other structures where earthworks are required.

## Cofferdams and caissons

**C8038** The Regulations require that:

(a) every cofferdam or caisson must be:

  (i) of suitable design and construction,

  (ii) of suitable and sound material,

  (iii) of sufficient strength and capacity for the purpose used, and

  (iv) properly maintained

  [*CHSW (SI 1996 No 1592), Reg 13(1)*]; and

(b) construction, installation, alteration or dismantling must be under the supervision of a competent person [*CHSW (SI 1996 No 1592), Reg 13(2)*].

## Prevention of drowning

**C8039** The Regulations require that where:

(a) after falling, any person is liable to fall into water (or other liquid) with a risk of drowning, suitable and sufficient steps must be taken to:

  (i) prevent, so far as is reasonably practicable, such person from falling,

  (ii) minimise the risk of drowning, and

  (iii) ensure the provision and maintenance of suitable rescue equipment

  [*CHSW (SI 1996 No 1592), Reg 14(1)*]; and

(b) where there are conveyances by water, there must be provision of safe transport to or from a place of work by water [*CHSW (SI 1996 No 1592), Reg 14(2)*]. Vessels used for conveyance purposes must be:

  (i) of suitable construction,

  (ii) properly maintained,

(iii) under the control of a competent person, and

(iv) not overcrowded or overloaded

[*CHSW (SI 1996 No 1592), Reg 14(3)*].

## Traffic routes

**C8040** Pedestrians and vehicles should be able to move safely and without health risks. To that end, traffic routes should be:

(*a*) suitable for persons and vehicles using them;

(*b*) sufficient in number;

(*c*) in suitable positions;

(*d*) of sufficient size; and

(*e*) indicated by suitable signs (see WORKPLACES – HEALTH, SAFETY AND WELFARE).

*[CHSW (SI 1996 No 1592), Reg 15(1), (2), (6)].*

These requirements are not satisfied unless:

(i) pedestrians or vehicles can use a traffic route without causing danger to the health or safety of persons near it;

(ii) any door or gate (intended for use by pedestrians) leading on to a traffic route for vehicles is so separated from that traffic route as to enable pedestrians to see any approaching vehicle or plant from a place of safety;

(iii) there is sufficient separation between vehicles and pedestrians to ensure safety, or if that is not reasonably practicable that:

(1) other means are provided for the protection of pedestrians, and

(2) effective arrangements are made for warning a person liable to be crushed or trapped by a vehicle of the vehicle's approach;

(iv) a loading bay has at least one exit point for exclusive use of pedestrians; and

(v) where it is unsafe for pedestrians to use any gate intended primarily for vehicles, one or more doors for pedestrians is provided in the immediate vicinity of such gate (such door(s) being clearly marked and obstruction free).

*[CHSW (SI 1996 No 1592), Reg 15(3)].*

Vehicles must not be driven on traffic routes unless, so far as is reasonably practicable, the route is free from obstruction and permits sufficient clearance [*CHSW (SI 1996 No 1592), Reg 15(4)*]. Where this is not reasonably practicable, drivers and persons riding on vehicles must be warned of any approaching obstruction or lack of clearance, e.g. by a sign. (For requirements relating to traffic routes generally, see ACCESS, TRAFFIC ROUTES AND VEHICLES.)

## Vehicles

**C8041** The Regulations require:

(*a*) unintended movement of vehicles (including mobile plant, locomotives and towed vehicles) must be either prevented or controlled [*CHSW (SI 1996 No 1592), Reg 17(1)*];

(*b*) where persons may be endangered by vehicle movement, the person in effective control of the vehicle must warn any person at work of the risk of injury [*CHSW (SI 1996 No 1592), Reg 17(2)*];

(c) construction work vehicles, when being driven, operated or towed, must be:

   (i) driven, operated or towed safely, and

   (ii) be so loaded as to be able to be driven, operated or towed safely

   [*CHSW* (*SI 1996 No 1592*), *Reg 17(3)*];

(d) no person must ride or be required or permitted to ride on any construction work vehicle other than in a safe place provided for that purpose [*CHSW* (*SI 1996 No 1592*), *Reg 17(4)*]. (For civil liability consequences of failure to do so, see ACCESS, TRAFFIC ROUTES AND VEHICLES.);

(e) no person must remain or be required or permitted to remain on a vehicle during the loading or unloading of loose material unless a safe place is provided and maintained [*CHSW* (*SI 1996 No 1592*), *Reg 17(5)*];

(f) excavating, handling and tipping vehicles must be prevented from:

   (i) falling into an excavation, pit or water, or

   (ii) overturning the edge of an embankment or earthwork

   [*CHSW* (*SI 1996 No 1592*), *Reg 17(6)*]; and

(g) in the case of rail vehicles, plant and equipment must be provided for replacing them on their track, or moving them, if derailed [*CHSW* (*SI 1996 No 1592*), *Reg 17(7)*].

The HSE publication, '*The safe use of vehicles on construction sites*', provides guidance for clients, designers, contractors, managers, and workers involved with construction transport. It provides guidance on how to establish safe workplaces for vehicle operations, vehicle selection, inspection and maintenance, safe driving and work practices, and managing construction transport.

## Doors and gates (not forming part of mobile plant and equipment)

**C8042** The Regulations require that:

(a) where necessary to prevent risk of injury, any door, gate, hatch (including temporary ones) – must incorporate (or be fitted with) suitable safety devices [*CHSW* (*SI 1996 No 1592*), *Reg 16(1)*]; and

(b) compliance with the Regulations presupposes that:

   (i) any sliding door, gate or hatch has a device to prevent it coming off its track during use,

   (ii) any upward opening door, gate or hatch has a device to prevent it falling back,

   (iii) any powered door, gate or hatch has suitable and effective features to prevent it trapping persons, and

   (iv) any powered door, gate or hatch can be operated manually unless it opens automatically if the power fails

   [*CHSW* (*SI 1996 No 1592*), *Reg 16(2)*].

## Prevention of fire risks

**C8043** The Regulations require that:

(a) so far as reasonably practicable, suitable and sufficient steps must be taken to prevent risk of injury from:

   (i) fire or explosion,

(ii) flooding, or

(iii) substances liable to cause asphyxiation

[*CHSW (SI 1996 No 1592), Reg 18*];

(b) if a work activity gives rise to risk of fire, such activity must not be carried out unless the worker is suitably instructed to prevent risk [*CHSW (SI 1996 No 1592), Reg 21(6)*];

(c) there must also be provision of:

(i) suitable and sufficient fire-fighting equipment, suitably located (e.g. in emergency routes),

(ii) suitable and sufficient fire detectors and alarm systems, suitably located (e.g. in emergency routes);

(iii) training of all persons on site in the correct use of appliances

[*CHSW (SI 1996 No 1592), Reg 21(5)*];

(d) fire-fighting equipment, detectors and alarm systems must be:

(i) properly maintained, examined and tested [*CHSW (SI 1996 No 1592), Reg 21(3)*],

(ii) indicated by suitable signs [*CHSW (SI 1996 No 1592), Reg 21(7)*], and

(iii) easily accessible, if not designed to come into use automatically [*CHSW (SI 1996 No 1592), Reg 21(4)*].

## Emergency routes and exits

**C8044** The Regulations require:

(a) a sufficient number of suitable emergency routes and exits, indicated by suitable signs (see WORKPLACES – HEALTH, SAFETY AND WELFARE), must be provided to enable any person to reach a place of safety quickly in the event of danger. [*CHSW (SI 1996 No 1592), Reg 19(1)*]. This should lead, as directly as possible, to an identified safe area [*CHSW (SI 1996 No 1592), Reg 19(2)*]; and

(b) emergency routes (and traffic routes or doors thereto) must be kept clear and obstruction free and, if necessary, be provided with emergency lighting [*CHSW (SI 1996 No 1592), Reg 19(3)*].

## Emergency procedures

**C8045** There must be prepared and implemented suitable and sufficient arrangements (to be tested by being put into effect at regular intervals) for dealing with any foreseeable emergency, including procedures for site evacuation. Moreover, persons likely to be affected must be acquainted with such arrangements. [*CHSW (SI 1996 No 1592), Reg 20(1), (3)*].

## Welfare facilities

**C8046** Duties in connection with the provision of welfare facilities on construction sites are laid principally on those in control of sites (i.e. occupiers, see generally C8131 below, and OCCUPIERS' LIABILITY). The fact that occupation (or control) of construction sites can be, and frequently is, shared between the building owner and contractor(s) means that both have obligations to see that welfare facilities are provided. In practice, actual provision would be made by the contractor(s). Hence, although overall duties are imposed on occupiers, employers (and self-employed persons) must ensure that workers under their control are provided with welfare facilities.

Facilities to be provided are:

(a) suitable and sufficient sanitary conveniences at readily accessible places [*CHSW* (*SI 1996 No 1592*), *Reg 22(3)*]. So far as is reasonably practicable, such conveniences must be:

   (i) adequately ventilated and lit,

   (ii) kept in a clean and orderly condition, with

   (iii) separate rooms containing sanitary conveniences provided for men and women (except where each convenience is in a separate room, the door of which can be secured from the inside)

   [*CHSW* (*SI 1996 No 1592*), *Sch 6*];

(b) suitable and sufficient washing facilities (including, where necessary, showers) at readily accessible places. So far as is reasonably practicable, washing facilities must:

   (i) be provided in the immediate vicinity of sanitary conveniences (except showers), and in the vicinity of changing rooms (whether or not provided elsewhere),

   (ii) include a supply of clean hot and cold (or warm) water (ideally running water), and soap, towels etc.

   Rooms containing washing facilities must be:

   (iii) sufficiently ventilated and lit,

   (iv) kept in a clean and orderly condition, and

   (v) must have separate washing facilities provided for men and women (except for washing hands, forearms and face only) unless provided in a room the door of which can be secured from the inside and the facilities in each such room are intended for use by only one person at a time

   [*CHSW* (*SI 1996 No 1592*), *Sch 6*];

(c) an adequate supply of wholesome drinking water at readily accessible places [*CHSW* (*SI 1996 No 1592*), *Reg 22(5)*]. So far as is reasonably practicable, every supply of drinking water must:

   (i) be conspicuously marked, and

   (ii) be provided with a sufficient number of suitable cups or other drinking vessels, unless the supply is from a jet

   [*CHSW* (*SI 1996 No 1592*), *Sch 6*];

(d) suitable and sufficient accommodation for:

   (i) accommodating the clothing of any person at work which is not worn during working hours, and

   (ii) special clothing worn by a person at work but which is not taken home

   [*CHSW* (*SI 1996 No 1592*), *Reg 22(6)*].

   So far as is reasonably practicable, clothing accommodation should include facilities for drying clothing [*CHSW* (*SI 1996 No 1592*), *Sch 6*];

(e) suitable and sufficient accommodation for changing clothing where:

   (i) a person has to wear special clothing at work, and

   (ii) that person cannot be expected to change elsewhere

   [*CHSW* (*SI 1996 No 1592*), *Reg 22(7)*].

(See *Post Office v Footitt [2000] IRLR 243* in W11026.)

Where necessary, facilities for changing clothing must be separate facilities for men and women [*CHSW (SI 1996 No 1592), Sch 6*];

(*f*)  suitable and sufficient rest facilities at readily accessible places [*Reg 22(8)*]. So far as reasonably practicable, rest facilities must include:

  (i)  suitable arrangements to protect non-smokers from discomfort caused by tobacco smoke, and where necessary

  (ii)  facilities for pregnant women or nursing mothers

[*CHSW (SI 1996 No 1592), Sch 6*].

(For welfare facilities generally, see WELFARE FACILITIES.)

## Fresh air

**C8047**  Every workplace on a construction site must, so far as reasonably practicable, have a supply of fresh or purified air so as to ensure safety and absence of health risks; and plant used for supply purposes must, where necessary, contain effective devices for giving visible or audible warning of failure. [*CHSW (SI 1996 No 1592), Reg 23*].

(For ventilation requirements generally, see VENTILATION.)

## Temperature

**C8048**  During working hours, so far as reasonably practicable, temperature at any indoor place of work must be reasonable, having regard to the purpose of the workplace. [*CHSW (SI 1996 No 1592), Reg 24(1)*].

(For temperature requirements generally, see WORKPLACES – HEALTH, SAFETY AND WELFARE.)

## Weather protection

**C8049**  Every place of work outdoors must, where necessary, be so arranged, so far as is reasonably practicable, as to provide protection from adverse weather. [*CHSW (SI 1996 No 1592), Reg 24(2)*].

## Lighting

**C8050**  Every place of work and traffic route must have suitable and sufficient lighting. In the case of artificial lighting where there would be a risk to a person's health or safety from failure of primary artificial lighting, suitable and sufficient secondary lighting must be provided. [*CHSW (SI 1996 No 1592), Reg 25(1), (3)*].

(For lighting requirements generally see LIGHTING.)

## Plant and equipment

**C8051**  The regulations require that, so far as is reasonably practicable, all plant and equipment used for construction work must be safe and without health risks, and be:

(*a*)  of good construction,

(*b*)  of suitable and sound materials,

(*c*)  of sufficient strength and suitability for its intended purpose, and (*d*) so used and maintained that it remains safe and without health risks

[*CHSW* (*SI 1996 No 1592*), *Reg 27*].

## Good order

**C8052** So far as reasonably practicable, every part of a construction site must be kept in good order and every place of work in a reasonable state of cleanliness. Perimeters should be identified by suitable signs (see WORKPLACES – HEALTH, SAFETY AND WELFARE), and sites arranged so that their extent is readily identifiable. Moreover, no timber material with projecting nails must be used in work where nails might be dangerous or allowed to remain in a place where nails could be a source of danger. [*CHSW* (*SI 1996 No 1592*), *Reg 26*].

## Training

**C8053** Any person who carries out construction work where training, technical knowledge or experience is necessary to reduce risk of injury, must either:

(*a*) possess such training, knowledge or experience, or

(*b*) be under the supervision of a person who so does.

[*CHSW* (*SI 1996 No 1592*), *Reg 28*].

## Inspection

**C8054** The Regulations require that in the case of the following places of work:

(*a*) working platforms or personal suspension equipment,

(*b*) excavations, and

(*c*) cofferdams or caissons

work can only be carried out if such place has been inspected by a competent person as follows.

(1) Working platforms and personal suspension equipment must be inspected by a competent person:

    (i) before being taken into use for the first time,

    (ii) after substantial addition, dismantling or other alteration,

    (iii) after any event likely to have affected their strength or stability, and

    (iv) at regular intervals not exceeding 7 days following the last inspection.

(2) Excavations must be inspected by a competent person:

    (i) before a person carries out work at the start of every shift,

    (ii) after any event likely to have affected the strength or stability of the excavation, and

    (iii) after an accidental fall of rock/earth.

(3) Cofferdams and caissons should be inspected by a competent person:

    (i) before the start of every shift, and

    (ii) after any event likely to have affected their strength or stability.

[*CHSW* (*SI 1996 No 1592*), *Reg 29(1), Sch 7*].

Where, following inspection, a place of work or plant and materials is not safe, this must be communicated to those in control and such place of work must not be used until defects have been remedied. [*CHSW* (*SI 1996 No 1592*), *Reg 29(3),(4)*].

In addition to the general duty to inspect places of work (above), in the case of scaffolds, excavations, cofferdams and caissons forming part of a place of work, employers must ensure that they are stable and of sound construction and that the requisite safeguards are in place before workers use such place of work for the first time. [*CHSW* (*SI 1996 No 1592*), *Reg 29(2)*].

## Reports

**C8055** Following inspection, a report must be prepared before the end of the working period in which the inspection was completed, and presented within 24 hours to the person on whose behalf it was carried out. [*CHSW* (*SI 1996 No 1592*), *Reg 30(1),(2)*]. The report (or a copy) must be kept at the site of such place of work and, after work there is completed, retained at the office of the person on whose behalf the inspection was made, for a minimum of three months. Such a report is available for inspection by an HSE inspector and, should he so require, extracts or copies must be sent to the inspector. [*CHSW* (*SI 1996 No 1592*), *Reg 30(3), (4)*].

Reports are not necessary in the following cases:

(i)  working platforms where persons are not liable to fall more than 2 metres; and

(ii) mobile towers (unless remaining erect for 7 days or more).

[*CHSW* (*SI 1996 No 1592*), *Reg 30(5),(6)*].

## Enforcement

**C8056** Penalties and defences in connection with the *Construction (Health, Safety and Welfare) Regulations 1996* are as for *HSWA 1974* (see ENFORCEMENT).

## Civil liability for breach of the regulations

**C8057** Breach of these Regulations (and the *Work in Compressed Air Regulations 1996* and the *Construction (Head Protection) Regulations 1989* – see C8081 and C8097 below), resulting in injury to an employee, will give rise to civil liability, since safety regulations, even if silent regarding civil liability, are actionable (see further INTRODUCTION). A short guide to these Regulations, '*A guide to the Construction (Health, Safety and Welfare) Regulations 1996*' is available free from HSE. This lists the series of further guidance.

# Confined spaces and harmful atmospheres

**C8058** The *Confined Spaces Regulations 1997* (*SI 1997 No 1713*) apply in all premises and work situations in Great Britain subject to the *HSWA 1974*, with the exception of diving operations and below ground in a mine.

The Regulations are supported by an Approved Code of Practice and guidance – *Safe work in confined spaces* (*L101*) – providing practical guidance with respect to the requirements of:

— the *Confined Spaces Regulations 1997* (*SI 1997 No 1713*) and the *HSWA 1974, ss 2, 4, 6* and *7*;

— the *Management of Health and Safety at Work Regulations 1999* (*SI 1999 No 3242*);

— the *Control of Substances Hazardous to Health Regulations 1999* (*SI 1999 No 1657*).

*Note:* the *COSHH Regulations* apply to all substances hazardous to health (other than lead or asbestos), such as toxic fume and injurious dust. The *Ionising Radiation Regulations 1999* (*SI 1999 No 3232*) may apply where radon gas can accumulate in confined spaces, such as sewers, and where industrial radiography is used to look at, for example, the integrity of welds in vessels.

— the *Personal Protective Equipment at Work Regulations 1992* (*SI 1992 No 2966*);
— the *Provision and Use of Work Equipment Regulations 1998* (*SI 1998 No 2306*).

## Entry into confined spaces

**C8059** A 'confined space' has two defining features. Firstly, it is a place which is substantially (though not always entirely) enclosed and, secondly, there will be a reasonably foreseeable risk of serious injury from hazardous substances or conditions within the space or nearby. [*SI 1997 No 1713, Reg 1(2)*].

Examples:

— ducts, vessels, culverts, tunnels, boreholes, bored piles, manholes, shafts, excavations, sumps, inspection pits, cofferdams, freight containers, building voids, some enclosed rooms (particularly plant rooms) and compartments within them, including some cellars, enclosures for the purpose of asbestos removal, and interiors of machines, plant or vehicles.

Some confined spaces are fairly easy to identify, for example closed tanks, vessels and sewers. Others are less obvious but may be equally dangerous, for example open-topped tanks and vats, closed and unventilated or inadequately ventilated rooms and silos, or constructions that become confined spaces during their manufacture. A confined space may not necessarily be enclosed on all sides.

## Respiratory protective equipment

**C8060** Where respiratory protective equipment (RPE) is provided or used in connection with confined space entry or for emergency or rescue, it should be suitable for the purpose for which it is intended, that is, correctly selected and matched both to the job and the wearer.

Where the intention is to provide emergency breathing apparatus to ensure safe egress or escape, or for self-rescue in case of emergency, the type commonly called an 'escape breathing apparatus' or 'self-rescuer' (escape set) may be suitable. These types are intended to allow time for the user to exit the hazard area. They are generally carried by the user or stationed inside the confined space, but are not used until needed. This equipment usually has a breathable supply of only short duration and provides limited protection to allow the user to move to a place of safety or refuge. This type of equipment is not suitable for normal work.

In some circumstances entry without the continuous wearing of breathing apparatus may be possible. Several conditions must be satisfied to allow such work including:

— a risk assessment must be done and a safe system of work in place including all required controls, and continuous ventilation;

— any airborne contamination must be of a generally non-toxic nature, or present in very low concentrations well below the relevant occupational exposure limits.

## Duties and responsibilities

**C8061** There have been several court cases involving accidents and fatalities in sewers and other confined spaces (see for instance *Baker v Hopkins & Son [1959] AER 225*). More recently, in 1998 at Cardiff Crown Court, a record fine was imposed on Neath Port Talbot Council following the deaths of two employees (*R v Neath Port Talbot Council*). The judge, John Prosser, said that the accident should never have happened; the dangers of toxic gases associated with sewer work were well known. The case demonstrated the need for employers to carry out, with strict care, the undertaking of such work and that the difficulty and danger must not be underestimated.

Duties to comply with the *Confined Spaces Regulations* (*SI 1997 No 1713*), are placed on:

— employers in respect of work carried out by their own employees and work carried out by any other person (for example, contractors) in so far as that work is to any extent under the employers' control [*SI 1997 No 1713, Reg 3(1)*]; and

— the self-employed in respect of their own work and work carried out by any other person in so far as that work is to any extent under the control of the self-employed [*SI 1997 No 1713, Reg 3(2)*].

## Duty to prevent entry

**C8062** The principal duty imposed on employers is to prevent entering or working inside a confined space where it is reasonably practicable to undertake the work by other means [*SI 1997 No 1713, Reg 4*].

The duty extends to others who are to any extent within the employers' control (such as contractors) and in many cases it will be necessary to modify working practices following a risk assessment of each requirement to enter a confined space.

Examples:

— modifying the confined space itself to avoid the need for entry, or to enable the work to be undertaken from outside;

— testing the atmosphere or sampling the contents of confined spaces from outside using appropriate long tools and probes.

Another case where it may be necessary to modify working practices is where employers or the self-employed have duties in relation to people at work who are not their employees – then the duty is to do what is 'reasonably practicable' in the circumstances. In many cases, the employer or self-employed will need to liaise and co-operate with other employers to agree the respective responsibilities in terms of the regulations and duties. It is also necessary to take all reasonably practicable steps to engage competent contractors.

## Associated duties and responsibilities

**C8063** In addition to the requirements of the *HSWA 1974*, other legislation imposes duties with regard to the design, construction and operation within confined spaces.

Some duties extend to erectors and installers of equipment and would include situations where plant and equipment unavoidably involved confined spaces. Where it is not possible to eliminate a confined space completely, procedures must be drawn up to minimise the need to enter such spaces both during normal use or working, and for cleaning and maintenance.

Regarding the type of PPE to be provided, this will depend on the identified hazards and the type of confined space. It may be necessary, for example, to include safety lines and harnesses, and suitable breathing apparatus.

Examples:

— the wearing of some respiratory protective equipment and personal protective equipment can contribute to heat stress;

— footwear and clothing may require insulating properties, e.g. to prevent softening of plastics that could lead to distortion of components such as visors, air hoses and crimped connections.

## Risk assessment – the development of a safe system of work

**C8064** The priority when carrying out a confined space risk assessment is to identify the measures needed so that entry into the confined space can be avoided. If it is not reasonably practicable to prevent work in a confined space the employer (or the self-employed) must assess the risks connected with persons entering or working in the space and also to others who could be affected by the work. The assessor(s) must understand the risks involved, be experienced and familiar with the relevant processes, plant and equipment and be competent to devise a safe system of working.

If, in the light of the risks identified, it cannot be considered reasonably practicable to carry out the work without entering the confined space, then it will be necessary to secure a safe system for working. The precautions required to create a safe system of work will depend on the nature of the confined space and the hazards identified during the risk assessment.

## Use of a permit-to-work procedure

**C8065** Not all work involving confined spaces requires the use of a permit-to-work system. For example, it is unlikely that a system would be needed where:

— the assessed risks are low and can be controlled easily; and
— the system of work is very simple; and
— it is known that other work activities being carried out cannot affect safe working in the confined space.

Although there is no set format for a permit system, it is often appropriate to include certain information relevant to all confined space working. In all cases, it is essential that a system is developed which ensures that:

(a) the people working in the confined space are aware of the hazards involved and the identity, nature and extent of the work to be carried out;

(b) there is a formal and methodological system of checks undertaken by competent people before the confined space is entered and which confirms that a safe system of work is in place;

(c) other people and their activities are not affected by the work or conditions in the confined space.

Isolation requirements, that is, the need to isolate the confined space to prevent dangers arising from outside, should also be included in the permit system. Permits are particularly appropriate if essential supplies and emergency services such as sprinkler systems, communications etc., are to be disconnected. The most effective isolation technique is to disconnect the confined space completely by removing a section of pipe or duct and fitting blanks. Other methods include the use of spectacle blinds and lockable valves.

## Workforce involvement

**C8066** Employees and their representatives should be consulted when assessing the risks connected with entering or working in a confined space. Particular attention is required where the work circumstances change frequently such as at construction sites or steel fabrications.

## Model or generic risk assessments

**C8067** Where a number of confined spaces (for example, sewers or manholes) are broadly the same in terms of the conditions and the activities being carried out, model risk assessments are permitted provided that the risks and measures to deal with them are the same. Any differences in particular cases which would alter the conclusions of the model risk assessment must be identified.

## Planning an entry into a confined space

**C8068** To satisfy the safe system requirement of the *Confined Spaces Regulations 1997* (*SI 1997 No 1713*), *Reg 4*, it is necessary to plan the work thoroughly and to organise various facilities and arrangements. For a large confined space and multiple entries, a logging or tally system may be necessary in order to check everyone in and out and to control duration of entry.

## Competence for confined space working

**C8069** The competent person carrying out the risk assessment for work in confined spaces will need to consider the suitability of individuals in view of the particular work to be done.

Examples:

— suitable build of individuals for exceptional constraints in the physical layout of the space (this may be necessary to protect both the individual and others who could be affected by the work to be done);

— medical fitness concerning claustrophobia or the wearing of breathing apparatus.

## Procedures and written instructions

**C8070** To be effective a safe system of work needs to be in writing – in the form of written instructions setting out the work to be done and the precautions to be taken. Each procedure should contain all appropriate precautions to be taken and in the correct sequence.

In particular, procedures for confined space working should include instructions and guidance for:

(*a*) *First aid* — the availability of appropriate first aid equipment for emergencies until professional medical help arrives.

(*b*) *First aiders* — the strategic positioning of trained personnel to deal with foreseeable injuries.

(*c*) *Limiting working time* — for example, when respiratory protective equipment is used, or when the work is to be carried out under extreme conditions of temperature and humidity.

(*d*) *Communications* — that is, the system of adequate arrangements to enable efficient communication between those working inside the confined space and others to summon help in case of emergency.

(*e*) *Engine driven equipment* — that is, the rules regarding the siting of such equipment which should be well away from the working area and downwind of any ventilator intakes.

(*f*) *Water surges* — especially the anticipation that sewers can be affected over long distances by water surges, for example following sudden heavy rainfall upstream of where the work is being carried out.

(*g*) *Toxic gas, fume or vapour* — procedures to ensure that work can be undertaken safely to include the availability of additional facilities and arrangements where residues may be trapped in sludge, scale or other deposits, brickwork, or behind loose linings, in liquid traps, joints in vessels, in pipe bends, or in other places where removal is difficult.

(*h*) *Testing/monitoring the atmosphere* — procedures for the regular testing for hazardous gas, fume or vapour or to check the concentration of oxygen before entry or re-entry into the confined space.

(*i*) *Gas purging* — the availability of suitable equipment to purge the gas or vapour from the confined space.

(*j*)   Ventilation requirements — the provision of suitable ventilation equipment to replace oxygen levels in the space, and to dilute and remove gas, fume or vapour produced by the work.

(*k*)   Lighting — procedures to ensure that the confined space is well lit by lighting equipment, including emergency lighting, which must be suitable for use in flammable or potentially explosive atmospheres.

(Generally all lighting to be used in confined spaces should be protected against knocks – for example, by a wire cage – and be waterproof. Where water is present in the space, suitable plug/socket connectors capable of withstanding wet or damp conditions should be used and protected by residual current devices (RCDs) suitable for protection against electric shock. The position of lighting may also be important, for example to give ample clearance for work or rescue to be carried out unobstructed.)

## Fire prevention and protection procedures

**C8071**   The presence of flammable substances and oxygen enrichment in a confined space creates a serious hazard to workers inside the space. There is also a risk of explosion from the ignition of airborne flammable contaminants. In addition, a fire or explosion can be caused by leaks from adjoining plant or processes and the use of unsuitable equipment.

*Note:* in the case of *R v Associated Octel Co Ltd [1996] 4 AER 846* (see C8136 below) a contractor was badly burned when an explosion occurred in the confined space (a chemical storage tank) he was working in. The principal cause of the accident was unsuitable lighting which broke and ignited some acetone solvent contained in an old emulsion bucket.

There are many fire precautions necessary for safe working in confined spaces; some of the more important of these are outlined below:

(*a*)   *Fire prevention measures* — procedures to ensure that no flammable or combustible materials are stored in confined spaces that have not been specifically created or allocated for that purpose. In any event, the quantity of the material should be kept to a minimum and stored in suitable fire-resistant containers.

(*b*)   *Fire protection and fire-fighting equipment* — procedures to ensure the availability of appropriate fire-fighting equipment where the risk of fire has been identified. In some situations, a sprinkler system may be appropriate.

(*c*)   *Smoking* — procedures to ensure the prohibition of all smoking within and around all confined spaces.

(*d*)   *Static electricity* — procedures to ensure that the build-up of static in a confined space is minimised. It may be necessary to obtain specialist advice regarding insulating characteristics (for example, most plastics), steam or water jetting equipment, clothing containing cotton or wool, flowing liquids or solids such as sand.

## Supervision and training

**C8072**   It is likely that the risk assessment will identify a level of risk requiring the appointment of a competent person to supervise the work and ensure that the precautions are adhered to. Competence for safe working in confined spaces requires adequate training – in addition, experience in the particular work involved is essential. Training standards must be appropriate to the task, and to the individuals' roles and responsibilities as indicated during the risk assessment.

## Emergency arrangements and procedures

**C8073** The arrangements for the rescue of persons in the event of an emergency must be suitable and sufficient and, where appropriate, include rescue and resuscitation equipment. The arrangements should be in place before any person enters or works in a confined space [*SI 1997 No 1713, Reg 5*].

The arrangements must cover any situation requiring the recovery of a person from a confined space, for example incapacitation following a fall.

## Size of openings to enable rescue from confined spaces

**C8074** Experience has shown that the minimum size of an opening to allow access with full rescue facilities including self-contained breathing apparatus is 575 mm diameter. This size should normally be used for new plant, although the openings for some confined spaces may need to be larger depending on the circumstances, for example to take account of a fully equipped employee, or the nature of the opening.

## Public emergency services

**C8075** In some circumstances, for example where there are prolonged operations in confined spaces and the risks justify it, there may be advantage in prior notification to the local emergency services before the work is undertaken. In all cases, however, arrangements must be in place for the rapid notification of the emergency services should an accident occur. On arrival, the emergency services should be given all known information about the conditions and risks of entering and/or leaving the confined space before a rescue is attempted.

## Training for emergencies and rescue

**C8076** To be suitable and sufficient the arrangements for training site personnel for rescue and resuscitation should include consideration of:

— rescue and resuscitation equipment;
— raising the alarm and rescue;
— safeguarding the rescuers;
— fire safety;
— control of plant;
— first aid.

Regular refresher training in the emergency procedures is essential and practice drills including emergency rescues will help to check that the size of openings and entry procedures are satisfactory. The risk assessment may indicate that at least one person, dedicated to the rescue role, should be stationed outside the confined space to keep those inside in constant direct visual sight.

All members of rescue parties should be trained in the operation of appropriate fire extinguishers which should be strategically located at the confined space. In some situations, a sprinkler system may be appropriate. In all cases, in the event of a fire the local fire service should be called in case the fire cannot be contained or extinguished by first-aid measures.

The training syllabus should include the following, where appropriate:

— the likely causes of an emergency;
— rescue techniques and the use of rescue equipment, for example breathing apparatus, lifelines, and where necessary a knowledge of its construction and how it works;

- the checking procedures to be followed when donning and using breathing apparatus;
- the checking of correct functioning and/or testing of emergency equipment (for immediate use and to enable specific periodic maintenance checks);
- identifying defects and dealing with malfunctions and failures of equipment during use;
- works, site or other local emergency procedures including the initiation of an emergency response;
- instruction on how to shut down relevant plant as appropriate (this knowledge would be required by anyone likely to perform a rescue);
- resuscitation procedures and, where appropriate, the correct use of relevant ancillary equipment and any resuscitation equipment provided (if intended to be operated by those receiving emergency rescue training);
- emergency first aid and the use of the first aid equipment provided;
- liaison with local emergency services in the event of an incident, providing relevant information about conditions and risks, and providing appropriate space and facilities to enable the emergency services to carry out their tasks.

## Rescue equipment

**C8077**   When safety harness and lines are provided, it is essential that proper facilities to secure the free end of the line are available. In most cases the line should be secured outside the entry to the confined space. Lifting equipment may be necessary and the harness should be of suitable construction, and made of suitable material to recognised standards capable of withstanding both the strain likely to be imposed, and attack from chemicals.

## Maintenance of safety and rescue equipment

**C8078**   All equipment provided or intended to be used for the purposes of securing the health and safety of people in connection with confined space entry or for emergency or rescue, should be maintained in an efficient state, in efficient working order and in good repair. This should include periodic examination and testing as necessary. Some types of equipment, for example breathing apparatus, should be inspected each time before use.

Atmospheric monitoring equipment – and special ventilating or other equipment provided or used in connection with confined space entry – needs to be properly maintained by competent persons. It should be examined thoroughly, and where necessary calibrated and checked at intervals in accordance with recommendations accompanying the equipment or, if these are not specified, at such intervals determined from the risk assessment.

Records of the examination and tests of equipment should normally be kept for at least five years. The records may be in any suitable format and may consist of a suitable summary of the reports. Records need to be kept readily available for inspection by the employees, their representatives, or by inspectors appointed by the relevant enforcing authority or by employment medical advisers.

## Equipment for use in explosive atmospheres

**C8079**   When selecting equipment for use in confined spaces where an explosive atmosphere may be present, the requirements of the EU-originated Regulations – *Equipment and Protective Systems Intended for Use in Potentially Explosive Atmospheres Regulations (SI 1996 No 192)* – must be complied with. These Regulations apply to 'equipment' and 'protective systems' intended for use in potentially explosive atmospheres. Some of the terms used in the Regulations are defined below:

— *equipment* means machines, apparatus, fixed or mobile devices, control components and instrumentation thereof and detection or prevention systems which, separately or jointly, are intended for the generation, transfer, storage, measurement, control and conversion of energy or the processing of material and which are capable of causing an explosion through their own potential sources of ignition;

— *protective systems* means design units which are intended to halt incipient explosions immediately and/or to limit the effective range of explosion flames and explosion pressures; protective systems may be integrated into equipment or separately placed on the market for use as autonomous systems;

— *devices* means safety devices, controlling devices and regulating devices intended for use outside potentially explosive atmospheres but required for or contributing to the safe functioning of equipment and protective systems with respect to the risks of explosion;

— *explosive atmosphere* means the mixture with air, under atmospheric conditions, of flammable substances in the form of gases, vapours, mists or dusts in which, after ignition has occurred, combustion spreads to the entire unburned mixture.

### Selection and use of equipment

C8080     All equipment must bear the approved CE mark properly fixed in accordance with the requirements of the 1996 Regulations.

Any equipment provided for use in a confined space needs to be suitable for the purpose. Where there is a risk of a flammable gas seeping into a confined space, which could be ignited by electrical sources (for example a portable hand lamp), specially protected electrical equipment must be used.

To be suitable the equipment should be selected on the basis of its intended use – proper earthing is essential to prevent static charge build-up; mechanical equipment may need to be secured against free rotation, as people may tread or lean on it.

## Work in compressed air on construction sites – Work in Compressed Air Regulations 1996 (SI 1996 No 1656)

C8081     Replacing the *Work in Compressed Air Special Regulations 1958*, the *Work in Compressed Air Regulations 1996*, which came into effect on 16 September 1996, reflect more modern decompression criteria, being more concerned with the long-term effects of rapid return to atmospheric pressure than the short-term effects which were addressed by earlier regulations. The new Regulations require principal contractors to appoint competent compressed air contractors, and the compressed air contractors to appoint contract medical advisers. In addition, greater provision is required in connection with fire prevention and protection measures (including, in particular, emergency means of escape and rescue). Duties are laid on principal contractors, employers (including the self-employed – tunnellers to whom these Regulations are substantially addressed are generally self-employed), and employees.

### Principal contractor's duties

C8082     The principal contractor must appoint a compressed air contractor in respect of work in compressed air, who must be competent. A compressed air contractor may be the principal contractor himself, if competent. [*SI 1996 No 1656, Reg 5(1), (2)*].

## Compressed air contractor's duties – notification

**C8083** The Regulations require that:

(a) the compressed air contractor must not allow work in compressed air to be carried out unless written notice has been forwarded to the HSE at least 14 days before commencement of work. Where this is not practicable, owing to an emergency, notice must be given as soon as practicable after the necessity for such work becomes known to the compressed air contractor, and, anyway, before work commences. [*SI 1996 No 1656, Reg 6(1), (2)*];

(b) no person must work in compressed air unless written notice is forwarded to:

　(i) the nearest suitably equipped hospital,

　(ii) the local ambulance service,

　(iii) the local fire service, and

　(iv) other establishments in the vicinity with an operable medical lock.

[*SI 1996 No 1656, Reg 6(3), (4)*];

(c) notification should be in writing and contain the following information:

　(i) the fact that work in compressed air is being undertaken,

　(ii) the location of the site,

　(iii) date of commencement and anticipated completion of work,

　(iv) name of compressed air contractor and a 24 hour contact telephone number,

　(v) name, address and telephone number of the contract medical adviser,

　(vi) intended pressure at which the work is to be undertaken,

　(vii) anticipated pattern of work (e.g. shifts), and

　(viii) number of workers likely to be in each shift

[*SI 1996 No 1656, Reg 6(4), Sch 1*].

## Competent persons

**C8084** The compressed air contractor must ensure that no person works in (or leaves) compressed air, except in accordance with a system of work which, so far as is reasonably practicable, is safe and without health risks. [*SI 1996 No 1656, Reg 7(1)*]. To this end, he must ensure that a sufficient number of 'competent persons' are immediately available on site to supervise execution of work in compressed air at all times and for up to 24 hours when work is being undertaken at or above a pressure of 0.7 bar. [*SI 1996 No 1656, Reg 7(2)*].

## Plant and equipment

**C8085** The compressed air contractor must ensure that all plant and equipment is:

(a) of a proper design and construction and of sufficient capacity;

(b) safe and without health risks and safely maintained; and

(c) where such plant and equipment is used for the purpose of containing air at a pressure greater than 0.15 bar, it is

　(i) examined and tested by a competent person and any faults rectified prior to use, and

　(ii) re-examined and re-tested after modification or alteration.

[*SI 1996 No 1656, Reg 8*].

## Compression and decompression procedures

**C8086** The compressed air contractor must ensure that:

(a) compression or decompression is only carried out as per procedures approved by the HSE;

(b) no worker is subjected to a pressure greater than 3.5 bar (except in emergencies);

(c) no worker is subjected to 'decanting' (i.e. rapid decompression in an airlock to atmospheric pressure followed promptly by rapid compression in an alternative airlock and subsequent decompression to atmospheric pressure – except in an emergency); and

(d) an adequate record is made of exposure in respect of times and pressures at which work in compressed air is carried out, and kept for a minimum of 40 years (including individual exposure records). Such records must be made available to the worker himself and his employer, the latter being required to keep the record for at least 40 years.

[SI 1996 No 1656, Reg 11].

## Provision and maintenance of adequate medical facilities

**C8087** The compressed air contractor must ensure provision and maintenance of adequate medical facilities (e.g. medical lock, recompression therapy) for the treatment of people working in compressed air and those who have worked in compressed air in the previous 24 hours. Where work is carried out at a pressure greater than 0.7 bar, facilities should include a medical lock; and where work is carried out at a pressure greater than 1.0 bar, a medical lock attendant should be present. [SI 1996 No 1656, Reg 12].

## Emergencies

**C8088** The compressed air contractor must ensure that no work in compressed air takes place in the absence of suitable and sufficient arrangements in the event of emergencies as follows:

(a) provision and maintenance of a sufficient number of suitable means of access;

(b) preparation of a suitable rescue plan which can be put into effect immediately (including the provision and maintenance of plant and equipment necessary to put the rescue plan into operation);

(c) provision and maintenance of suitable lighting;

(d) provision and maintenance of suitable means of raising the alarm; and

(e) in cases where an airlock is required, maintenance of the airlock, so that it is fit to receive persons in the event of emergency (with particular regard to air supply and temperature of the airlock).

[SI 1996 No 1656, Reg 13].

## Fire precautions

**C8089** The compressed air contractor must ensure provision of suitable and sufficient means for fighting fire and that any airlock or working chamber is maintained and operated so as to minimise the risk of fire, and must ensure the enforcement of the prohibition against smoking. [SI 1996 No 1656, Reg 14].

## Information, instruction and training

**C8090** The compressed air contractor must ensure provision of adequate information, instruction and training to employees, including, particularly, information relating to the risks arising from the work and the precautions to be observed. [*SI 1996 No 1656, Reg 15*].

## Fitness for work

**C8091** The compressed air contractor must ensure that no one works in compressed air where he has reason to believe that the worker is subject to a medical or physical condition likely to make him unfit or unsuitable for such work. [*SI 1996 No 1656, Reg 16*].

## Prohibition against alcohol and drugs

**C8092** The compressed air contractor must prohibit anyone from working in compressed air where he believes such worker to be under the influence of drink and/or drugs. [*SI 1996 No 1656, Reg 17*].

## Employees' duties

**C8093** Employees, including the self-employed, must:

(*a*) when required to do so, and at the cost of the employer (see OCCUPATIONAL HEALTH AND DISEASES), submit to medical surveillance procedures during working hours [*SI 1996 No 1656, Reg 10(6)*];

(*b*) avoid smoking or carrying smoking materials [*SI 1996 No 1656, Reg 14(2)*];

(*c*) avoid consumption of alcohol or drugs [*SI 1996 No 1656, Reg 17(2)*]; and

(*d*) wear a badge or label for 24 hours after leaving work in compressed air [*SI 1996 No 1656, Reg 19(2)*].

## Contract medical adviser

**C8094** Owing to the potentially serious dangers arising from pressure itself, or the construction work being done, appointment of a contract medical adviser is essential. Ideally, such person would be a doctor appointed by the HSE, who can carry out statutory medical examinations on compressed air workers on site. His principal role is to actively monitor incidence of decompression illness during work. In the event of decompression illness arising, the contract medical adviser should advise the medical lock attendant regarding appropriate treatment. Both the contract medical adviser and medical lock attendant are responsible for collation and maintenance of exposure records and completion of the worker's health and exposure record, and, on completion of the contract, assist the compressed air contractor in the preservation of formal health and exposure records for the statutory 40-year period.

## Competent persons

**C8095** The phrase 'competent persons' can refer to:

(*a*) the engineer in charge;

(*b*) the compressor attendants;

(*c*) the lock attendants;

(*d*) the medical lock attendants (for work in compressed air over 1.0 bar); and

(*e*) the contract medical adviser.

## Enforcement

**C8096** In any proceedings for an offence consisting of a contravention of *Reg 14(3)* or *17(3)* (compressed air contractor's duty to ensure compliance with prohibitions against smoking, alcohol or drugs), it is a defence for any person to prove that he took all reasonable precautions and exercised all due diligence to avoid the commission of the offence. [*SI 1996 No 1656, Reg 20*].

# Construction (Head Protection) Regulations 1989 (SI 1989 No 2209)

**C8097** Head injuries account for nearly one-third of all construction fatalities, but fell significantly after the introduction of the *Construction (Head Protection) Regulations 1989* (*SI 1989 No 2209*). These Regulations specify requirements for head protection during construction work, including offshore operations (see OFFSHORE OPERATIONS), but not diving operations at work. They place duties on employers, persons in control of construction sites, self-employed persons and employees regarding the wearing of head protection. The purpose of head protection is to prevent/mitigate head injury caused by: falling/swinging objects, e.g. materials and/or crane hooks; and striking the head against something, as where there is insufficient headroom. Circumstances where head injury is not reasonably foreseeable on construction sites are limited, but it probably would not be required on/in:

(a) sites where buildings are completed and there is no risk of falling materials/objects;

(b) site offices, cabins, toilets, canteens or mess rooms;

(c) cabs of vehicles, cranes etc.;

(d) work at ground level, e.g. road works.

## Duties of employers

**C8098** The following duties are laid on employers:

### (a) Provision/maintenance of head protection

Every employer (that is, main contractor, subcontractor etc.) must provide each employee, while at work on building/construction operations, with suitable head protection, and keep it maintained/replaced (as recommended by the manufacturer). [*SI 1989 No 2209, Reg 3(1), (2)*].

Moreover, head protection equipment must be kept in good condition and stored, when not in use, in a safe place, though not in direct sunlight or hot or humid conditions. It should be inspected regularly and have defective harness components replaced, and sweatbands regularly cleaned or replaced.

### (b) Ensuring head protection is worn

So far as reasonably practicable (for meaning, see E15039 ENFORCEMENT), every employer must ensure that each of his employees, whilst on construction work, wears suitable head protection, unless there is no foreseeable risk of injury to his head (other than by falling). [*SI 1989 No 2209, Reg 4(1)*].

Moreover, every employer (or employee) who has control (for meaning, see below) over any other person engaged in construction work, must ensure, so far as is reasonably practicable, that such persons wear suitable head protection, unless there is no foreseeable risk of injury to the head, other than by falling. [*SI 1989 No 2209, Reg 4(2)*].

## Persons in control of construction sites

**C8099** For the purposes of these Regulations, the following persons may be deemed to be 'in control' of construction sites:

(a)   main contractor;

(b)   managing contractor;

(c)   contractor bringing in subcontractors;

(d)   contract manager;

(e)   site manager;

(f)   subcontractor;

(g)   managers, including foremen, supervisors;

(h)   engineers and surveyors;

(i)   (sometimes) clients and architects with control over persons at work.

## Procedures and rule making

**C8100** Employers and others in control must:

(a)   identify when/where head protection should be worn;

(b)   inform site personnel procedurally when/where to wear head protection and post suitable safety signs to that effect;

(c)   provide adequate supervision;

(d)   check that head protection is, in fact, worn.

Supervision by those responsible for ensuring head protection is worn, is an on-going requirement, including monitoring helmet use at all times, starting early in the day and taking in arrivals on site.

Persons in control of construction works can (and should) make rules regulating the wearing of suitable head protection. Such rules must be in writing and be brought clearly to the attention of those involved. Such procedure is particularly useful to main/managing contractors on multi-contractor sites, and rules/regulations on head protection should form part of overall site safety procedures, such as construction phase safety plans in accordance with the *CDM Regulations* (see C8025 above).

## Wearing suitable head protection – duty of employees

**C8101** Employees must also make full and proper use of head protection and return it to the accommodation provided for it after use. [*SI 1989 No 2209, Reg 6 as amended by the Personal Protective Equipment at Work Regulations 1992, Sch 2, para 24*]. They must also comply with the rules and regulations made for the wearing of head protection mentioned in C8098 above (see also PERSONAL PROTECTIVE EQUIPMENT). All employees, provided with suitable head protection, must take reasonable care of it and report any loss of it or obvious defect in it, to the employer etc. [*SI 1989 No 2209, Reg 7*].

## Suitable head protection

**C8102** Suitable head protection refers to an industrial safety helmet conforming to British Standard BS EN 397: 1995 'Industrial Safety Helmets' – Specification for construction and performance (or an equivalent standard). For work in confined spaces, 'bump caps' to BS EN 812: 1998 are more suitable.

Suitability of head gear involves the following factors: (*a*) fit, (*b*) comfort, (*c*) compatibility with work to be done, and (*d*) user choice.

### (a) Fit

Head protection should be of an appropriate shell size for the person who is to wear it, and have an easily adjustable headband, nape and chin strap. The range of size adjustment should be sufficient to accommodate thermal liners in cold weather.

### (b) Comfort

Head gear should be as comfortable as possible, including:

(*a*) a flexible headband of adequate width and contoured vertically and horizontally to fit the forehead;

(*b*) an absorbent, easily cleanable or replaceable sweatband;

(*c*) textile cradle straps;

(*d*) chin straps (when fitted) which

   (i) fit round the ears,

   (ii) are compatible with any other personal protective equipment needed,

   (iii) are fitted with smooth, quick release buckles which do not dig into the skin,

   (iv) are made from non-irritant materials,

   (v) are capable of being stowed on the helmet when not in use.

### (c) Compatibility with work to be done

Head gear should not impede work to be done. For instance, an industrial safety helmet with little or no peak is functional for a surveyor taking measurements, using a theodolite or to allow unrestricted upward vision for a scaffold erector. If a job involves work in windy conditions, at heights, or repeated bending or constantly looking upwards, a secure retention system is necessary. Flexible headbands and Y-shaped chin straps can help to secure the helmet on the head. If other personal protective equipment, such as ear defenders or eye protectors, are required, the design must allow them to be worn safely and in comfort.

### (d) User choice

In order to avoid possibly unpleasant industrial relations consequences or a possible action for unfair dismissal (see further EMPLOYMENT PROTECTION), it is sensible and advisable to allow the user to participate in selection of head gear.

## Duties of self-employed personnel

**C8103** Every self-employed person involved in construction/building operations, must:

(i) provide himself with suitable head protection and maintain/replace it, whenever necessary [*SI 1989 No 2209, Reg 3(2)*];

(ii) ensure that any person over whom he has control, wears suitable head gear, unless there is no foreseeable risk of injury [*SI 1989 No 2209, Reg 4(2)*];

(iii) give directions to any other self-employed person regarding wearing of suitable head gear [*SI 1989 No 2209, Reg 5(4)*];

(iv) wear properly suitable head protection, unless there is no foreseeable risk of injury to the head, and make full and proper use of it and return it to the accommodation provided for it after use [*SI 1989 No 2209, Reg 6(2)–(4)*];

(v) where the presence of more than one risk to health or safety makes it necessary for him to wear or use simultaneously more than one item of personal protective equipment, see that such equipment is compatible and continues to be effective against the risk or risks in question [*Personal Protective Equipment at Work Regulations 1992 (SI 1992 No 2966), Reg 5(2)*].

## Exceptions

**C8104** The following categories of workers on construction sites are exempt from the Regulations:

(a) divers actually diving or preparing to dive;

(b) Sikhs wearing turbans on construction sites [*Employment Act 1989, ss 11, 12*], though the Regulations do apply to Sikhs not normally wearing turbans at work. (The probability is that this exemption, under the *Employment Act 1989, s 11*, is now subordinate to the requirements of the *Construction (Head Protection) Regulations 1989* and *HSWA 1974*, with the result that Sikhs working on construction sites will have to wear hard hats (*SS Dhanjal v British Steel plc (Case No 50740/91)*).)

## Visitors on site

**C8105** It is not necessary that visitors are provided with or, even less, wear head protection, under these Regulations. Nevertheless, in order to satisfy their general duty under the *Health and Safety at Work etc. Act 1974 (HSWA 1974), s 4* and additionally avoid any civil liability for injury at common law in an action for negligence (see further EMPLOYERS' DUTIES TO THEIR EMPLOYEES) and/or under the *Occupiers' Liability Act 1957* (see OCCUPIERS' LIABILITY), employers should provide visitors to the site with suitable head protection where there is a reasonably foreseeable likelihood of injury. This contention is further reinforced by the requirement for every employer to consider the risks inherent in his business which have the potential to harm his employees or any other persons who might be affected, and to take measures to remove or reduce such risks [*Management of Health and Safety at Work Regulations 1999 (SI 1999 No 3242), Reg 3*].

# Composition of construction products

**C8106** Products must be suitable for construction works and works of civil engineering and can then carry the 'CE' mark. To that end, when incorporated into design and building, construction products should satisfy the following criteria, namely,

(a) mechanical resistance and stability;

(b) safety in case of fire;

(c) hygiene, health and the environment;

(d) safety in use;

(e) protection against noise; and

(f) energy economy and heat retention.

[*Construction Products Regulations 1991* (SI 1991 No 1620), Reg 3, Sch 2 as amended by SI 1994 No 3051].

Manufacturers must show that their products conform to these specifications, if necessary, by submitting to third party testing (see also PERSONAL PROTECTIVE EQUIPMENT for products generally).

## Protecting visitors and the public

**C8107** The HSE have issued revised guidance, HS(G)151 *Protecting the public – your next move* (June 1997), which provides practical advice on the measures to be taken to minimise risks to the public and others not directly involved in construction activities. The advice is aimed at preventing accidents and ill health and, to a limited extent, at reducing incidents of nuisance. The guidance does not cover deliberate illegal trespass or forced entry on to sites by protest groups or those intent on criminal activity.

In addition, reference should be made to the following related statutory provisions which have aspects relating to public safety during construction or building operations:

### (a) Roads and streets

*Highways Act 1980:*

— *s 168* (building operations affecting public safety);

— *s 169* (the control of scaffolding on highways);

— *s 174* (the erection of barriers, signs and lighting etc.).

*New Roads and Street Works Act 1991:*

— *s 50* (lays down particular safety requirements for work in the street and, specifically, the measures to be taken to minimise inconvenience to the disabled).

*Environmental Protection Act 1990:*

— *s 79 as amended by the Noise and Statutory Nuisance Act 1993* (noise or vibration emitted from buildings and from or caused by a vehicle, machinery or equipment in a street).

### (b) Waste from building and demolition sites

Waste produced on construction sites is classed as controlled waste and as such must be controlled to comply with EU-based Directives:

*Environmental Protection Act 1990:*

— *ss 33–46* (deal with waste management and licensing control).

*Controlled Waste (Registration of Carriers and Seizure of Vehicles) Regulations 1991* (SI 1991 No 1624 as amended) (carriage of controlled waste by registered carriers only).

*Waste Management Licensing Regulations 1994* (SI 1994 No 1056 as amended) (registers, applications and waste regulation authorities for the recovery and disposal of waste).

*Special Waste Regulations 1996* (SI 1996 No 972 as amended) (hazardous properties of waste).

### Identifying hazards and evaluating risks

**C8108** Construction work is by its nature carried out by workers away from the home base and may not be open to direct management and supervision. The employer's general duty of care to ensure that employees are not put at risk by their work activities (*HSWA 1974, s 2*) still applies. In addition, if the work activities impinge on others, such as another organisation or members of the public, a further duty of care (*HSWA 1974, s 3*) applies.

The *HSWA 1974, s 4* relates to the control of premises, rather than the control of undertakings. Persons who control premises used by people who are at work, but who are not their employees, need to ensure, so far as is reasonably practicable, that the premises, access to them and plant and substances used on them are safe and free from risks to health and safety. Site occupiers therefore share a duty of care with contractors (as both are employers) to ensure that all reasonably practicable precautions are taken to safeguard their own employees, other persons on site and the public.

Under the *Occupiers' Liability Act 1957* and the *Occupiers' Liability Act 1984* a duty of care is imposed on occupiers of existing premises regarding visitors. The duty of care extends to children – it should be noted that a child is regarded as being at greater risk than an adult (see O3010 OCCUPIERS' LIABILITY). The 1984 Act further extends the duty of an occupier to people other than lawful visitors, such as trespassers, to ensure that they are not injured whilst on the premises. This may involve making unauthorised access more difficult or putting up suitable warning signs regarding hazards on site.

## Safety policies and written arrangements

**C8109** The *HSWA 1974, s 2(3)* requires written safety arrangements only with regard to employee safety. However, by virtue of the *MHSWR* (*SI 1999 No 3242*), *Reg 5* (health and safety arrangements) written arrangements are also required with regard to the protection of the public and other non-employees.

The *MHSWR* refer expressly to occupiers' responsibility to co-operate and co-ordinate arrangements (*SI 1999 No 3242*), *Regs 11* and *12*) and to provide information and training on risks and precautions (*SI 1999 No 3242, Regs 10* and *13*). These duties apply whether or not payment is involved, for example free surveys, estimates, measurements, maintenance and servicing under warranty, etc.

Measures to ensure the safety of visitors and members of the public must be one result of the risk assessment task as required by *MHSWR* (*SI 1999 No 3242*), *Reg 3* (see R3004 RISK ASSESSMENT). Chemical installations or other high risk manufacturing or chemical storage premises are also subject to specific duties towards the general public under the *Control of Major Accident Hazards Regulations 1999* (see CONTROL OF MAJOR ACCIDENT HAZARDS).

Companies with cooling towers on site are also subject to specific controls ultimately designed to protect the public. In *R v Board of Trustees of the Science Museum, The Times, March 15 1993, CA* the prosecution had alleged that members of the public outside the Science Museum had been exposed to risks to their health from *legionella pneumophila*, because of inadequate maintenance of the museum's air conditioning system. The basis of their case had been that it is sufficient for the prosecution to show that there has been a risk to health.

The Court of Appeal decided that the word 'risks' in *HSWA 1974, s 3(1)* implied the idea of potential danger. There was nothing in the subsection which narrowed this meaning. *HSWA 1974* should be interpreted so as to make it effective in its role of protecting public health and safety. *Section 3(1)* was intended to be an absolute prohibition, subject to the defence of reasonable practicability.

## Insurance cover

**C8110** The *Employers' Liability (Compulsory Insurance) Act 1969* states that an employer's legal liability for death, disease or bodily injury suffered by employees as a consequence of employment must be insured by the employer for their mutual protection under the duty of care owed by *HSWA 1974, s 2*. A copy of the current certificate of insurance (issued annually) must be displayed within all working premises. Other insurances, for example reflecting the risk of liability to non-employees under the duty of care imposed by *HSWA 1974, s 3*, are likely to be essential even if not compulsory.

## Site planning and layout

**C8111** Risk assessment should decide how the site perimeters will be defined, what type of barriers and fencing will be most effective and where they should be placed.

For most sites the perimeter will be the geographical area within which the construction work will be carried out. Determining the perimeter is an important aspect of managing public risk. It must always be recognised that site perimeters need to be changed as the work progresses.

Under the *CDM Regulations* (*SI 1994 No 3140*), the duties of the principal contractor are wide and varied. Importantly, *CDM* is excluded from use in civil proceedings, except for *Reg 10* which requires the client to ensure that an adequate health and safety plan has been prepared before the construction phase of the project commences, and *Reg 16(1)(c)* which requires the principal contractor to take appropriate steps to ensure that only authorised access is permitted to premises where construction work is continuing.

Similarly, where construction work is taking place on an occupied site, the client will impose existing security rules on everyone entering and leaving the site.

Under the *CDM Regulations*, other contractors involved in the construction work are required to co-operate with the principal contractor and comply with any directions or site rules. In addition, all contractors must provide appropriate information, including information relating to any injuries, diseases and dangerous occurrences.

## Employers' liability for the actions and safety of the public and non-employees

**C8112** Under *HSWA 1974*, both employers and the self-employed have duties not only to their own workpeople but also to outside contractors, workers employed by them and to members of the public – whether within or outside the workplace – who may be affected by work activities. Undertakings must be conducted in such a way as to ensure, so far as is reasonably practicable, that they do not expose people who are not their employees to risks to their health and safety. The duty extends to, for example, risks to the public outside the workplace from fire or explosion, from falls of unsafely erected scaffolding or from the release of harmful substances into the atmosphere.

In general, the standard of protection required for visitors and others within a construction site will be similar to that given to employees. There may, however, be a need to apply different criteria to achieve these standards when assessing the risks to members of the public. For example, it will be necessary to consider that certain people, such as the very young or disabled, may be more vulnerable than others and that people visiting or passing a workplace may have less knowledge of the potential hazards and of how to avoid them.

The responsibilities of employers and the self-employed to non-employees will in certain circumstances extend to people entering workplaces without permission. This is apart from any liability under common law towards trespassers. The duty under *HSWA 1974* to conduct the business in such a way as not to expose people to risks to health and safety implies taking certain precautions to deter people from unlawfully entering the workplace, for example by the provision of fences, barriers and notices warning of the danger. The duty towards 'unauthorised' people is qualified 'so far as is reasonably practicable', and on construction sites and other open-air workplaces, which have particular dangers as far as children are concerned, simply locking or guarding main doors and gates may not be adequate.

## Duties of all people

**C8113** *HSWA 1974* imposes one duty on all people, both people at work and members of the public, including children: this is not intentionally to interfere with or misuse anything that has been provided in the interests of health, safety or welfare, whether it has been provided for the protection of employees or other people. The purpose of the provision is clearly to protect things intended to ensure people's safety, including fire escapes and fire extinguishers, perimeter fencing, warning notices for particular hazards, protective clothing, guards on machinery and special containers for dangerous substances.

## Control measures

**C8114** The general duties of protection owed to visitors apply equally to the emergency services who should be given a plan or map of the premises together with information on specific high risk areas where high voltage or dangerous chemicals may be present. In addition, to comply with occupier's liability legislation the duty of care towards visitors includes contractors.

### (a) Visitors

**C8115** In general, visitors to a construction site should not be left unaccompanied and they should not, if possible, be taken into any hazardous areas. All visitors should be made to sign in on arrival and sign out on departure and, ideally, be given basic instructions on what to do in the event of an emergency. The main element of looking after visitors is to ensure that they are accompanied at all times, so that the host can lead them to safety in the event of fire or other emergency. Constant accompaniment of visitors will, of course, also improve security arrangements.

### (b) Contractors

**C8116** Under the *MHSWR* the occupier of the premises must ensure that contractors on site are provided with comprehensible information on:

— the risks to health and safety arising out of the activities on site; and

— the measures taken by the occupier to ensure compliance with statutory requirements.

*[SI 1999 No 3242, Reg 10].*

The ideal situation is where contractors can be provided with a completely separated area which can be designated as being under their control. Such an arrangement will normally only apply when the contractors are on site to undertake a major engineering or construction project; and it will only be successful if a contractor is actually given full control of the area – and the main site occupier and his or her employees only enter the area when authorised by the contractor. However, the main site occupier will retain the key responsibility for safety matters – as illustrated in the case of a south coast town council which employed contractors to remove part of a damaged pier. The council had accepted by far the lowest quote and did not discuss the system of work. During the demolition, there was a huge explosion which removed the derelict part of the pier but also caused considerable damage to cars and buildings on the seafront. The contractor and town council were held jointly responsible but the council suffered the larger fine because it failed to employ reputable contractors and did not request a method statement from them.

### (c) General public

**C8117** Measures to ensure the safety of the general public are usually more difficult and must be one result of the risk assessment task as required by *MHSWR (SI 1999 No 3242), Reg 3*. Members of the public are owed a duty of care under *HSWA 1974, ss 3 and 4*, to ensure that they are not put at risk by the employer's undertaking. For example, an employer engaged in construction work near a public place must ensure that risks are assessed and adequately controlled. The Act also imposes a duty on employees to co-operate with their employer on health and safety matters and not to do anything which puts others at risk.

As far as protecting the public is concerned, adequate control measures must be determined which should not rely on the use of protective equipment.

## Risk assessments

C8118    Risk assessments are an integral part of the *CDM Regulations* (*SI 1994 No 3140*) and the *Confined Spaces Regulations 1997* (*SI 1997 No 1713*). Additionally, with respect to visitors and the general public, employers may be liable to pay compensation to people injured on their premises under the terms of the *Occupiers' Liability Act 1957*.

When carrying out a risk assessment to safeguard members of the public it is necessary to adopt a very wide approach and consider all the possible hazards and subsequent risks. This would include, but not be limited to, compiling data and control measures for:

(*a*)   *chemical hazards* — e.g., mist, vapour, gas, smoke, dust, aerosol, fumes;

(*b*)   *physical hazards* — e.g., noise, temperature, lighting, vibration, radiation (ionising and non-ionising), pressure;

(*c*)   *biological hazards* — e.g., bacteria, parasites.

## Public vulnerability

C8119    Experience has shown that members of the public are particularly vulnerable to those hazards and risks which are not readily identifiable by the normal senses of sight, smell or hearing. The problem may be exacerbated by disabilities and sensory impairments as well as by physical and mental conditions. These conditions would have been assessed for everyone inside the confines of the site but the employer must address the specific needs of all people who may be affected by the work when the hazards may extend outside the site boundaries. The needs of children and the elderly must always be given top priority.

## Cooling towers

C8120    The public is particularly vulnerable to cooling towers, and employers must therefore ensure, so far as is reasonably practicable, that no one is put at risk from legionellosis as a result of work activities. Plant of this type includes hot and cold water services, air conditioning and industrial cooling systems, spas and whirlpool baths, humidifiers and air washers.

Designers, manufacturers, importers, suppliers and installers of such plant or water systems – and water treatment contractors – have a duty to ensure, so far as is reasonably practicable, that the plant or system is so designed and constructed that it will be without risks to health. Appropriate information must be provided to users, and tests carried out if required [*HSWA 1974, s 6*].

The use of this type of plant must be notified to the local authority under the *Notification of Cooling Towers and Evaporative Condensers Regulations 1992* (*SI 1992 No 2225*) and the requirements of the *COSHH Regulations* (*SI 1999 No 437*) must be complied with, particularly the provisions relating to:

— *risk assessment* – to include breakdowns, abnormal operation and the possibility of exposure of susceptible people (for example in hospitals) [*SI 1999 No 437, Reg 6*]. The assessment must be reviewed at least once every five years or if there is a change in plant or operation;

— *prevention or control of exposure* [*SI 1999 No 437, Regs 7, 8 and 9*]. Where potential exposure to infection cannot be prevented there must be a written control scheme to minimise exposure;

— health surveillance where appropriate [*SI 1999 No 437, Reg 11*];

— information, instruction and training [*SI 1999 No 437, Reg 12*].

(See further L8: *Legionnaires disease: The control of legionella bacteria in water systems* (ISBN 0 7176 1772 6), price £8.00, available from HSE books.

## Dusts and fibres

**C8121**   Dust in the form of particulates suspended in air is generated from a number of construction work activities, including, but not limited to:

— cutting bricks, blocks, tiles, slabs etc.;

— sawing wood;

— mixing cement, plasters etc.;

— grinding operations;

— blasting;

— demolition operations.

Fibres from asbestos demolitions are particularly harmful and strict precautions must be taken in case of an uncontrolled release of asbestos fibres from the workplace.

## Environmental safety aspects during asbestos removal

**C8022**   Under the *Control of Asbestos at Work Regulations 1987* (*SI 1987 No 2115*), Reg 3, employers have duties not only to their own employees but also to, for example:

(*a*)   visitors to the place where work with asbestos is being carried out;

(*b*)   the occupier's employees if the work is done in someone else's premises;

(*c*)   people in the neighbourhood who might be accidentally exposed to asbestos dust arising from the work.

Whenever two or more employers work with asbestos at the same time at one workplace they should co-operate in order to meet their separate responsibilities (for further detail see ASBESTOS).

Contractors undertaking work on materials containing asbestos were warned of the serious health hazards associated with such work when the *Control of Asbestos at Work Regulations 1987* came into force in 1988. The warning was reiterated in 1995 when a court case involving contractors cleaning an asbestos roof ruled that such contractors had a duty to take reasonable care and skill in the work including the taking of necessary precautions (*Barclays Bank plc v Fairclough Building Ltd (No 2) [1995] IRLR 605 CA*).

There have been numerous cases involving incidents during demolition and the removal of asbestos in the years following the above case and its warning. In a very significant case where the lives of children had been put at risk, HSE and the Environment Agency co-operated and raised a joint prosecution.

The case (*R v Rollco Screw and Rivet Co and Others [1998] HSE E198:98*) resulted in a defendant being jailed for nine months (the second custodial sentence for an asbestos offence). Five others and a company were ordered to pay fines and costs totalling £98,000. Birmingham Crown Court heard disturbing evidence of the casual attitude of the people carrying out the stripping of an asbestos roof and the subsequent disposal of blue, white and brown asbestos. The contractors were not licensed in accordance with the requirements of the *Asbestos (Licensing) Regulations 1983* (*SI 1983 No 1649*). The HSE inspector who conducted the investigation emphasised the importance of property owners and managers checking the qualifications of any person employed to carry out asbestos removal work. The inspector was especially critical of the clear attempt to gain financially from decisions taken which put the public at risk.

The QC for the Environment Agency gave evidence of nine contraventions of the *Environmental Protection Act 1990* (the keeping or disposal of controlled waste in such a way that pollution of the environment or harm to human health was likely) where approximately 300 bags of asbestos were dumped at different locations around the city. Some of the bags had been left open and others had burst, releasing asbestos fibres into the air. The court heard that children had been playing with loose asbestos material since it had been dumped recklessly and indiscriminately in a playground and in a supermarket car park as well as other locations.

Judge Charles Harris QC, in imposing the sentences, singled out one man for the custodial sentence and said that, unlike the others in the case, he knew the risks, he lied steadily and showed manifest dishonesty. He told him 'This was an act of the most astonishing criminal irresponsibility. You understood the nature of asbestos and yet you distributed it around Birmingham in places where people, including children, had easy access to it'. He described the co-defendants as being ignorant of the dangers of asbestos but their neglect had put others in serious danger.

## Fumes, mists and vapours

**C8123** Most dangers associated with fumes, mists and vapours from operations such as spreading adhesives, mixing and thinning paints and coatings and spraying are well known. In addition, care should be taken and appropriate precautions put in place for operations such as:

— cleaning operations;

— heat treatment processes;

— disturbance of sludge and scale from vessels;

— the release of gases from sewers and similar operations;

— emissions from extraction equipment, LEV systems and the venting of relief valves etc.;

— the use of aerosols;

— the use of pesticides.

Fumes from cutting, welding, soldering and brazing operations are particularly dangerous and require a thorough assessment.

## Radiations and hazardous waves

**C8124** The dangers associated from the use of equipment emitting radiations, radio-frequency waves and micro-waves are well documented and the precautions to be observed during the use of such equipment must be strictly adhered to at all times. Some operations where the public may be at risk from uncontrolled releases or discharges from the equipment or, in some cases, from natural causes are:

— infra-red and ultraviolet radiations during cutting and welding;

— ionising radiation from radiography;

— laser rays (for example the use of lasers during accurate alignment operations of machinery or structures);

— X-rays from high voltage sources (for example non-destructive testing operations);

— the presence of radon on some soils and rocks.

## Fire and smoke inhalation

**C8125** In addition to the dangers of fire spreading to areas and premises occupied by members of the public, it is necessary to consider the effects of smoke from burning refuse and discarded materials. Advice should be sought from appropriate specialists and approval granted before commencing any burning operations.

## Noise control on construction sites

**C8126** In addition to the *Noise at Work Regulations 1989* (*SI 1989 No 1790*) and HSE's supporting guidance to the Regulations (L108 – ISBN 0 7176 1511 1), there are other legislative requirements and codes designed to protect the public from noise on construction sites. Some of the most important are:

— *Control of Pollution Act 1974, s 71* – approval of codes and standards regarding noise control;

— the *Control of Noise* (*Codes of Practice for Construction and Open Sites*) *Orders 1984* (*SI 1984 No 1992*) *and 1987* (*SI 1987 No 1730*); and

— the *Construction Plant and Equipment* (*Harmonisation of Noise Emission Standards*) *Regulations 1988* (*SI 1988 No 361 as amended*) – noise from plant used in or about building or civil engineering operations.

## General construction operations

**C8127** The public are also at risk, but perhaps to a lesser degree, from normal construction work activities. Policies and procedures should be in place to cover contingencies and incidents arising from:

— electricity;
— excavations;
— explosives;
— flooding;
— lifting operations;
— materials handling;
— mechanical plant and general construction equipment;
— overhead working;
— piling;
— scaffolding;
— underground services;
— vehicles;
— vibration.

## References

**C8128** There are numerous publications available as reference documents which provide advice on the measures to be taken to protect visitors and members of the public during construction activities. In particular, the following HSE publications and Approved Codes of Practice (ACOPs) may be useful when formulating safety policies and procedures.

C8128  *Construction and Building Operations*

## Construction management

L111 A guide to the Control of Major Accident Hazard Regulations (ISBN 0 7176 1604 5);

HSR25 Memorandum of Guidance on the Electricity at Work Regulations 1989 (ISBN 0 11 883963 2);

L1 A guide to the Health and Safety at Work etc Act 1974 (ISBN 0 7176 0441 1);

L21 Management of health and safety at work (ISBN 0 7176 2488 9);

L24 Workplace health, safety and welfare: ACOP to the 1992 Regulations (ISBN 0 7176 0413 6);

L54 Managing construction for health and safety: ACOP to the *CDM Regulations 1994* (ISBN 0 7176 0792 5);

L55 Preventing asthma at work (ISBN 0 7176 0661 9);

L64 Safety signs and signals – guidance to the Health and Safety (Safety Signs and Signals) Regulations 1996 (ISBN 0 7176 0870 0);

L73 A guide to the Reporting of Injuries, Diseases and Dangerous Occurrences Regulations 1995 (ISBN 0 7176 1012 8).

## Dangerous substances

COP2 Control of lead at work: ACOP (ISBN 0 7176 1506 5);

L5 General COSHH ACOP, Carcinogens ACOP and Biological Agents ACOP, Control of Substances Hazardous to Health Regulations 1999 (ISBN 0 7176 1670 3);

L27 The control of asbestos at work: ACOP (ISBN 0 7176 1673 8);

L28 Work with asbestos insulation, asbestos coating and asbestos insulating board: ACOP (ISBN 0 1176 1674 6);

L62 Safety datasheets for substances and preparations dangerous for supply — guidance on regulation 6 of the Chemicals (Hazard Information and Packaging for Supply) Regulations 1994: ACOP (ISBN 0 7176 0859 X);

L86 Control of substances hazardous to health in fumigation operations: ACOP (ISBN 0 7176 1195 7).

## Explosives

L10 A guide to the Control of Explosives Regulations 1991 (ISBN 0 11 885670 7).

## Gas

COP20 Standards of training in safe gas installation: ACOP (ISBN 0 7176 0603 1);

L56 Safety in the installation and use of gas systems and appliances: The Gas Safety (Installations and Use) Regulations 1998 (ISBN 0 7176 1635 5);

L80 A guide to the Gas Safety (Management) Regulations 1996 (ISBN 0 7176 1159 0);

L81 Design, construction and installation of gas service pipes – ACOP to the Pipelines Safety Regulations 1996 (ISBN 0 7176 1172 8).

### LPG and petroleum spirit

**COP6** Plastic containers with nominal capacities up to 5 litres for petroleum spirit: Requirements for testing and marking or labelling (ISBN 0 11 883643 9).

### Pesticides

**L9** The safe use of pesticides for non-agricultural purposes. Control of Substances Hazardous to Health Regulations 1994: ACOP (ISBN 0 7176 0542 6).

### Pressure systems

**L122** Safety of pressure systems: Pressure Systems Safety Regulations 2000 (ISBN 0 7176 1767 X);

**L96** A guide to the Work in Compressed Air Regulations 1996 (ISBN 0 7176 1120 5).

### Radiations

**L121** Work with ionising radiation: Ionising Radiations Regulations 1999 (ISBN 0 7176 1746 7).

### Site vehicles

**L117** Rider-operated lift trucks: Operator training (ISBN 0 7176 2455 2).

## Responsibility for contractors and subcontractors

**C8129** Most industrial/commercial organisations delegate corporate functions and duties, placed on them by statute, regulation and common law, to contractors and subcontractors. This practice is particularly common in the construction industry, where a main contractor, in order the more competently and expeditiously to discharge his contractual obligations towards his employer (or builder owner), sublets performance of parts of the contract, e.g. steel erection, to specialists.

Significantly, this practice of subletting performance of parts of the entire contract is regarded as sufficiently important in the building industry to justify the existence of a Standard Form of Building Contract (the JCT Standard Form). This means that the rights/obligations of all interested parties, namely, the employer, main contractors and subcontractors, both nominated and domestic, are specified in a formal jointly witnessed contract, known as the Joint Contracts Tribunal (JCT). When a dispute arises between any of the interested parties, for example, who is liable for an injury to an employee of a subcontractor, reference is made to the Conditions of Contract (or Subcontract).

If necessary, such a dispute will be decided by arbitration, since the contract provides for independent arbitration machinery in the form of the RIBA (Royal Institute of British Architects). RIBA arbitration does not exclude jurisdiction of the courts but, in practice, that is often the result, since arbitration is quicker and cheaper. In other words, the building industry has its own quasi-judicial internal disputes machinery and procedures. This is preferable to no machinery at all, since all interested parties know where they stand – at least, that is the theory.

### Multiple occupation of construction sites – standard form work

**C8130** Where, as is normal on large construction sites, standard form (JCT) building work is being carried out, multiplicity of occupation (or control) is not uncommon. Here control will be shared among building owner(s), main contractor(s) and subcontractor(s). It has been decided that the legal nature of the relationship between a building owner and main contractor is that of licensor and licensee (*Hounslow London Borough Council v Twickenham Garden Developments Ltd [1970] 3 AER 326*).

This vests in the building owner some degree of control over the works e.g. if the contractor does not carry out and complete the works (in accordance with the requirements of Clause 2(1) of the JCT Standard Form Contract) the licence can, subject to certain exceptions, be terminated. Moreover, (given that some statutory duties can be modified) Clause 20(1) of the Conditions of Contract states that 'the contractor shall be liable for, and shall indemnify the employer against any expense, liability, loss or claim or proceedings whatsoever arising under any statute or at common law in respect of personal injury to or the death of any person whomsoever arising out of, or in the course of, or caused by the carrying out of the works, unless due to any act or neglect of the employer or of any person for whom the employer is responsible' (e.g. employee). The proviso to the clause 'unless due to any act or neglect of the employer' is limited solely to common law negligence and does not extend to statutory negligence.

In consequence, the building owner can insist that the main contractor(s) must take out insurance to meet that indemnity, and likewise the main contractor(s) can insist that the subcontractor(s) does the same (Clause 21.1 JCT Standard Form Contract). Thus, for the purposes of common law liability, based on occupation, persons injured on building sites can sue the building owner, main contractor and any subcontractor who may be responsible, the question of indemnity as between the liable parties being governed and determined by the terms of the JCT Contract.

## Liability of employer/occupier in connection with contract work

C8131  In practice, two sorts of situation give rise to liability:

(a) injuries/diseases, or the risk of them, to employees of the contractor as a result of working on the occupier's premises; or, alternatively, injuries or the risk of them to the employer's own workforce as a result of the employer failing to acquaint the contractor's workforce with dangers, thereby endangering his own employees (see *R v Swan Hunter Shipbuilders Ltd* below);

(b) injuries/damage to members of the public, or pollution or nuisance to neighbouring landowners.

Such liability can be both criminal and civil (often strict).

### Criminal liability

C8132  Criminal liability can arise under several statutes and at common law (e.g. where, owing to gross negligence, employers commit manslaughter – in practice, this is rare). Particularly relevant is the *HSWA 1974, ss 3(1), 4(2)*.

### Health and Safety at Work etc. Act 1974, s 3(1)

C8133  In particular, *HSWA 1974, s 3(1)* states: 'It shall be the duty of every employer to conduct his undertaking in such a way as to ensure, so far as is reasonably practicable, that persons not in his employment who may be affected thereby are not thereby exposed to risks to their health or safety'.

### The Swan Hunter case

C8134  An instructive case involving this section was *R v Swan Hunter Shipbuilders Ltd [1982] 1 AER 264*. During construction of a ship at a shipbuilder's yard, subcontractors, who had no contract with the shipbuilders, were working on the ship while it was being fitted out. The shipbuilders were aware that, because of use of oxygen hoses with fuel gases in welding, there was a risk of fire due to the atmosphere in confined and poorly ventilated spaces in the ship becoming oxygen enriched. In regard to that danger they had provided information and instruction for their own employees by way of a book of rules which stipulated that at the end of the day's work, all oxygen hoses should be returned from the lower decks to an open deck, or, where impracticable, the hoses should be disconnected at the cylinder or manifold. This rule book was *not* distributed to the subcontractor's employees working on the ship, and an employee of the subcontractor failed to disconnect the oxygen hose and oxygen was discharged during the night. In consequence, on the next morning, when a welder working in the lower deck lit his welding torch, a fierce fire escaped.

The shipbuilders were charged with failing to provide/maintain a safe system of work, contrary to *HSWA 1974, s 2(2)(a)*, and failing to provide such information/instruction as was necessary to ensure the health/safety of their employees, contrary to *HSWA 1974, s 2(2)(c)*.

The shipbuilders were also charged with failing to conduct their undertaking in such a way as to ensure that persons not in their employment, who might be affected thereby, were not exposed to risks to health and safety, contrary to *s 3(1)*.

On appeal against conviction on all three counts, it was held by the Court of Appeal that:

(i) duties imposed on an employer by *HSWA 1974, ss 2 and 3* followed the common law duty of care of the main contractors to co-ordinate operations at a place of work so as to ensure not only the safety of his own employees but also that of the subcontractor's employees. The main contractor had to prove, on a balance of probabilities, that it was not reasonably practicable for him to carry out the duties under *HSWA 1974, ss 2 and 3*;

(ii) the shipbuilders were under a duty, under *HSWA 1974, s 2(2)(a)*, to provide/maintain a safe system of work for the subcontractor's employees, so far as was reasonably practicable, and provide them with information/instruction so as to ensure their safety.

Accordingly, the main contractor, Swan Hunter Shipbuilders Ltd, was fined £3,000 and the subcontractor, Telemeters Ltd, £15,000 after eight men were trapped and killed on board HMS Glasgow.

Furthermore, if the main contractor fails, as in *Swan Hunter*, to comply with *HSWA 1974, s 3(1)*, in his duties towards subcontracted labour, it is likely that he would be in breach of his duty towards his own employees, under *s 2* of the 1974 Act. Thus, anyone who is responsible for co-ordinating work has to ensure that reasonable safety precautions are taken for the workmen of a contractor or subcontractor.

## The Rhone-Poulenc Rorer case

**C8135** Responsibility for contractors' work has been considered in several cases since the *Swan Hunter Shipbuilders* case resulting in further interpretation of *HSWA 1974, s 3* and the prosecution of employers. Decisions made in the courts have continued to emphasise the role of the main employer in organising and sharing responsibility for the safety of work carried out by contractors. One of these cases concerned a breach of the *Construction (Working Places) Regulations 1966, Reg 36(2)* (now revoked) in addition to *HSWA 1974, s 3* – the case of *R v Rhone-Poulenc Rorer Ltd [1996] ICR 1054* concerned the provision of suitable means for preventing a fall through fragile materials.

Counsel for the prosecution alleged that some sort of physical safety device is required to fulfil an employer's duty to prevent employees from falling through fragile material: neither a system of work based on instruction nor a Code of Practice will suffice.

The court heard that an employee of a subcontractor was instructed to repair a roof light at the company's factory in Dagenham. The company provided one of its own employees to supervise the work. The subcontractor was told not to climb on to the roof. He did so, and fell through the roof light on to a concrete floor 28 feet below and was killed. Rhone-Poulenc was charged under the 1974 Act and the 1966 Regulations. The company was fined £7,500 in respect of the former and £2,500 in respect of the latter, with £55,000 costs.

The company appealed and argued that the trial judge had misdirected the jury in telling them that it was necessary to prove that it had been impracticable to comply with the Regulations, not merely that it was not reasonable to do so.

The Court of Appeal rejected this argument. Wright J stated that the requirement under *Reg 36(2)* was absolute and that employers had a duty to provide such suitable means as might be necessary for preventing, so far as was reasonably practicable, persons from falling through fragile material. This meant that some sort of physical device, for example a safety harness, was required, where in the circumstances guard rails or covering could not be supplied. Falls through fragile material could not be prevented by the provision of a supervisor, a body of instructions, or a code of practice.

In relation to *HSWA 1974, s 3* it was held that because the occupier (Rhone-Poulenc Rorer Ltd) had been in breach of Regulations which applied to its own employees, it was also in breach of *s 3* so far as the contractor's employees were concerned.

## The Associated Octel case

C8136    It was 1996 before the House of Lords had to deal with the interpretation of *HSWA 1974, s 3*. In *R v Associated Octel Co Ltd [1996] 4 AER 846* it was decided that the duty placed on the employer of contractors extended to persons not in his employment.

In 1990, during the course of an annual shutdown for planned maintenance, a specialist contractor, RGP, was called in for the purposes of cleaning and repairing a tank at Octel's chlorine plant at Ellesmere Port. Octel had effectively approved the system of work by virtue of having issued a permit-to-work that required the contractor's employee to enter the tank, taking lighting with him, and grind the internal surfaces of the tank and clean residues with an acetone solvent. In the event the light broke causing an explosion in which the contractor's employee was badly injured.

The HSE inspector who investigated the incident was critical of the system of work and identified aspects of the operations which were unsafe. The points raised included:

— the acetone was carried in an old emulsion bucket;

— Octel provided an unsuitable lamp from its own stores;

— there was inadequate ventilation;

— the precautions listed on the permit-to-work referred only to the grinding work and did not specify adequate precautions relating to work in a confined space.

The inspector referred to the fact that the site was a major hazard site and under the control of the *Control of Industrial Major Accident Hazard Regulations 1984 (CIMAH) (SI 1984 No 1902 as amended)* (now replaced by the *Control of Major Accident Hazards Regulations 1999*) which required a safety case to be submitted and complied with. RGP had employees at Octel's site on a regular basis and had worked for Octel over a number of years. Previously, Octel had exercised a degree of control over RGP's work performance and had a high level of understanding of the dangers involved in tank cleaning operations.

The case argued in the Crown Court and the Court of Appeal was a complex one where the central issue concerned 'control' of the work of the contractor. Octel's counsel referred to the Robens Report of 1972, and cited, *inter alia*, as precedents *R v Board of Trustees of the Science Museum [1993] 3 AER 853, R v Swan Hunter Shipbuilders Ltd [1982] 1 AER 264, Mailer v Austin Rover Group Ltd [1989] 2 AER 1087* and *RMC Roadstone Products Ltd v Jester [1994] 4 AER 1037*. Octel argued consistently that there was no case to answer because it could not be shown that it was the conduct of Octel's undertaking that endangered the RGP employee. RGP is a specialist contractor and Octel claimed that it had no right to control the manner in which its independent contractor did its work.

Octel lost its argument before the trial judge, as well as before the Court of Appeal for different and complicated legal reasons. However, it was decided that the chemical business was Octel's undertaking, and the undertaking included having the tank cleaned, whether by its own employees or by contractors.

Lord Hoffmann delivered the judgment of the House of Lords, upholding the decisions of the Crown Court and the Court of Appeal. *The Times* (15 November 1996) reported the following points made by Lord Hoffmann:

— it was a question of fact in each case whether an activity which caused a risk to the health and safety of persons other than employees amounted to 'conduct of an undertaking';

— *section 3* of the 1974 Act was not concerned with vicarious liability, but imposed a duty upon the employer himself;

— if the employer engaged an independent contractor to do work forming part of the 'undertaking', then the employer had to stipulate whatever conditions were needed to avoid risks to health and safety;

— the question was simply whether the activity in question could be described as part of the employer's undertaking. Octel's undertaking was running a chemical plant and it was part of the conduct of that undertaking to have the factory cleaned by contractors;

— the tank was part of Octel's plant. The work formed part of a maintenance programme planned by the firm. The workers, although employed by an independent contractor, were almost permanently integrated into the firm's larger operations. In these circumstances, a properly instructed jury would undoubtedly have convicted.

## The Port Ramsgate ferry walkway case

**C8137** In this case a harbour operator, two foreign marine engineering companies (designers and builders) and Lloyd's Register of Shipping were all prosecuted and found guilty of failing to ensure the safety of passengers after the collapse of a walkway to a cross-channel ferry led to the deaths of six people and injured many more.

The case centred around the 'reasonably practicable' element of *HSWA 1974, s 3(1)* and concerned the design, construction, installation and the supervision and approval inspection of the walkway, contracted by the Port of Ramsgate Ltd. The accident occurred only four months after the walkway was commissioned because of the failure of a weld joining one of the feet to an axle. The walkway collapsed and fell some 30 feet on to a floating pontoon.

The QC who was prosecuting on behalf of the HSE gave evidence that the failure was the result of inaccurate stress calculations and also of inferior welding.

During the trial, many legal arguments took place as to who exactly was responsible. Port of Ramsgate Ltd was adamant that it had acted properly in placing contracts with appropriate, experienced contractors and that, by arranging for Lloyd's to manage the design, construction and installation, there was nothing else it could have done regarding the safety of its passengers. It argued strongly that there was no way of foreseeing the profound errors of judgment made by the various parties concerned.

The prosecution continued with its argument that there were measures that could have been taken to satisfy the 'reasonably practicable' element and that one of these might have been the inclusion of a simple fail-safe device such as safety chains.

The judge directed the jury to treat the question of the Port of Ramsgate's undertaking as a question of fact and left it for the jury to decide whether sufficient had been done to satisfy the grounds of 'reasonably practicability'. In the event, the jury found that all three companies were guilty of failing to satisfy the requirement. The judge imposed very high fines:

— Port of Ramsgate Ltd: £200,000;
— Lloyd's Register of Shipping: £500,000;
— Fartygsentreprenader AB and Others (Sweden): £1 million.

It should be noted that this was a highly complex case and its implications for further cases will be considerable.

Although the case was brought before the implementation of the *CDM Regulations*, it should be noted that these Regulations do not expressly require quality assurance checks to be carried out.

If the same type of case occurred offshore, the *Offshore Installations and Wells* (*Design and Construction*) *Regulations 1996* (*SI 1996 No 914*) would probably apply and in particular the requirement for 'verification' and the effect on the installation's safety case.

## Civil liability

**C8138** Normally employers are only liable for the negligent acts/omissions of their own employees or agents but there are certain important exceptions, such as where an employee of an occupier is injured by the contractor's negligence but overall control rests with the occupier.

### The McDermid Nash case

**C8139** An important case was heard in 1987 concerning a safe system of work, the delegation of the performance of that work and the competency of the person in charge of a situation where an individual, not being in the employ of that person, was injured.

During the case (*McDermid v Nash Dredging & Reclamation Co Ltd [1987] 2 AER 878*) several points were made regarding the legal responsibilities of the various people concerned when contracting an employee of another employer:

— the employer's safety duties towards his employees are owed personally to those employees – they cannot be got rid of by delegation;

— an employer who arranges for work to be done by another person, whether under contract or any other relationship, where that other will use employees of the employer, is vicariously liable for acts of negligence of the contractor or other person;

— liability arises not because the contractor or other person is, for the time being, the employer of the employee in the legal sense but merely from the fact that the usual employer has entrusted the safety of his employee to that other person;

— the employer still remains liable for the proper performance and the legal duties of his employee;

— this continued liability does not rely on any contract between the employer and the other party. It arises whenever an employer entrusts the performance of safety duties to another by asking that other party to perform work on his behalf using his employees.

The UK subsidiary of a Dutch company secured a contract to dredge a fjord in Sweden. It was to be an enterprise where some 100 of the UK company's employees were to work on board Dutch vessels under the joint captaincy of a UK and a Dutch captain, each working alternate shifts. Mr McDermid's job was to keep the deck tidy and to tie up and untie the tug from its moorings when it went alongside the dredger.

While the Dutch captain was in charge, Mr McDermid's leg got caught in one of the ship's hawsers and was so badly injured it had to be amputated. The evidence showed that the Dutch captain had not set up a safe system to be sure that Mr McDermid was free of the ropes before applying power to move the tug away from its mooring.

Because of the difficulties of suing a Dutch company while operating in Swedish territorial waters, Mr McDermid sought damages from his UK employer, even though at the time he was working under the control of a foreign captain who was clearly not an employee of the UK company.

The court decided that there was no need to prove even a temporary relationship of employer and employee between the Dutch company and Mr McDermid. Under the common law of vicarious liability, the employer's safety duties are owed to the employee personally. They cannot be got rid of by delegation. Further, when an employer puts his employee into the hands of another he is entrusting the performance of his own safety duties to that other but the employer remains responsible.

Taking all the factors into account, the court ruled that Mr McDermid's employer was vicariously liable for the negligence of the Dutch tug captain and stated:

> 'It is clear therefore that if an employer delegates to another person, whether an employee or not, his personal duty to take reasonable care for the safety of his employees, the employer is liable for injury caused through the negligence of that person because it is in the eyes of the law his own negligence.'

The employer's appeal to the House of Lords was dismissed, and it was held that an employer's duty to his employee to exercise reasonable care to ensure that the system of work provided for him is safe, is 'personal' and 'non-delegable'. It is no defence for the employer to show that he delegated its performance to a person whom he reasonably believed to be competent to perform it.

It was also stated in court that the negligence of the Dutch captain was not casual but central. It involved abandoning a safe system and operating in its place a manifestly unsafe system.

## Hazards left on highways

**C8140**  An occupier is not under a duty of care to check that a contractor removes hazards from a highway, and so is not liable for consequential injuries. In *Rowe v Herman and others, The Times, 9 June 1997*, the plaintiff (R) sought damages for injuries sustained by tripping over some metal plates left lying on the pavement by an independent contractor (L), which had been carrying out (and had finished) some building work for the occupier (H). H contended that he was not liable for L's negligence under the general principle that an employer is not liable for an independent contractor's negligence. There are two main exceptions to this basic principle – an employer is liable:

(i)  where the work commissioned involved extra hazardous acts; and

(ii) where danger was created by work on a highway.

The county court held that although H was not liable for the hazard whilst work was being carried out, he became liable under the principle in (ii) once L had finished the work. H appealed.

The Court of Appeal allowed H's appeal. Previous cases, where the employer was held liable, concerned obstructions to the highway being caused as a result of work being carried out under statutory powers, such obstructions arising directly from the work which the employer was required to do (being integral to it). In this case H was not obliged to carry out the building work, and it was not an integral part of the work to obstruct the footway (the plates had merely been laid to prevent lorries from damaging the pavement). It was clear that if the accident had occurred whilst L was carrying out the work, H would not have been liable; it made no sense to suddenly place H under such a duty once L had left the site. Following the general principle that H had no control over the manner in which L carried out its work, so too H had no control over the way that L cleared up.

# Control of Major Accident Hazards

## Introduction

**C10001** The *Control of Major Accident Hazards Regulations 1999* (*SI 1999 No 743*) implement the requirements of the 'Seveso II' Directive (96/82/EC) on the control of major accident hazards involving dangerous substances. Seveso II replaced the original Seveso Directive (82/501/EEC) which was implemented in Great Britain by the *Control of Industrial Major Accident Hazards Regulations 1984* (*CIMAH*) which in turn have been replaced by the *Control of Major Accident Hazards Regulations 1999* (*COMAH*). *COMAH* came into force on 1 April 1999. The provisions of Article 12 of Seveso II concerning land-use planning have been implemented in the *Planning* (*Control of Major Accident Hazards*) *Regulations 1999* (*SI 1999 No 981*).

The Health and Safety Executive (HSE) and the Environment Agency (in England and Wales) or the Scottish Protection Agency are jointly responsible as the competent authority (CA) for *COMAH*. The CA is the author of a report listing recent major accidents at industrial premises subject to *COMAH*. The report *'COMAH major accidents notified to the European Commission, England, Scotland and Wales 1999–2000'* details ten major accidents (one of which resulted in the death of an employee), and three near misses. The CA's concern at both the magnitude and frequency of these accidents is highlighted in the report together with its belief that a thorough implementation of *COMAH* will contribute significantly to future improvements in the safety of persons and the environment.

## Control of Major Accident Hazards Regulations 1999 (SI 1999 No 743)

**C10002** *COMAH* (*SI 1999 No 743*) gives effect to a safety regime for the prevention and mitigation of major accidents resulting from ultra-hazardous industrial activities. The emphasis is on controlling risks to both people and the environment through demonstrable safety management systems, which are integrated into the routine of business rather than dealt with as an add-on.

An occurrence is regarded as a major accident if:

— it results from uncontrolled developments (i.e. they are sudden, unexpected or unplanned) in the course of the operation of an establishment to which the Regulations apply; and

— it leads to serious danger to people or to the environment, on or off-site; and

— it involves one or more dangerous substances defined in the Regulations.

Major emissions, fires and explosions are the most typical major accidents.

The duties placed on operators by *COMAH* fall into two categories: 'lower tier' and 'top tier'. Lower tier duties fall upon all operators. Some operators are subject to additional, top-tier, duties.

An 'operator' is a person (including a company or partnership) who is in control of the operation of an establishment or installation. Where the establishment or installation is to be constructed or operated, the 'operator' is the person who proposes to control its operation – where that person is not known, then the operator is the person who has commissioned its design and construction.

As mentioned, the HSE and the Environment Agency (in England and Wales) or the Scottish Environment Protection Agency are jointly responsible as the competent authority (CA) for *COMAH*. A co-ordinated approach has been adopted with the HSE acting as the primary contact for operators. A charging regime has been introduced and the fees are payable by the operator to the HSE for work carried out by the HSE and the Environment Agency. Details of the current charging regime may be found on the HSE website at: www.hse.gov.uk. Under *COMAH (SI 1999 No 743), Reg 19*, the CA must organise an adequate system of inspections of all establishments, and then must prepare a report on the inspection.

The main HSE guidance to the Regulations is contained in *'Guide to the Control of Major Accident Hazards Regulations 1999'* (L111).

## Application

**C10003**   *COMAH (SI 1999 No 743)* applies to an *establishment* where:

(a)   dangerous substances are present; or

(b)   their presence is anticipated; or

(c)   it is reasonable to believe that they may be generated during the loss of control of an industrial chemical process.

'Loss of control' excludes expected, planned or permitted discharges.

'Industrial chemical process' means that premises with no such chemical process do not fall within the scope of the Regulations solely because of dangerous substances generated during an accident.

An 'establishment' means the whole area under the control of the same person where dangerous substances are present in one or more installations. Two or more areas under the control of the same person and separated only by a road, railway or inland waterway are to be treated as one whole area.

'Dangerous substances' are those which:

(a)   are named at the appropriate threshold (see Table 1); or

(b)   fall within a generic category at the appropriate threshold (see Table 2).

There are two threshold levels: lower tier (Column 2 of the Tables) and top tier (Column 3).

| Table 1 | | |
|---|---|---|
| Column 1 | Column 2 | Column 3 |
| Dangerous substances | Quantity in tonnes | |
| Ammonium nitrate (see note below) | 350 | 2,500 |
| Ammonium nitrate (see note below) | 1,250 | 5,000 |
| Arsenic pentoxide, arsenic (V) acid and/or salts | 1 | 2 |
| Arsenic trioxide, arsenious (III) acid and/or salts | 0.1 | 0.1 |
| Bromine | 20 | 100 |
| Chlorine | 10 | 25 |
| Nickel compounds in inhalable powder form (nickel monoxide, nickel dioxide, nickel sulphide, trinickel disulphide, dinickel trioxide) | 1 | 1 |

| | | |
|---|---|---|
| Ethyleneimine | 10 | 20 |
| Fluorine | 10 | 20 |
| Formaldehyde (concentration=>90%) | 5 | 50 |
| Hydrogen | 5 | 50 |
| Hydrogen chloride (liquefied gas) | 25 | 250 |
| Lead alkyls | 5 | 50 |
| Liquefied extremely flammable gases (including LPG) and natural gas (whether liquefied or not) | 50 | 200 |
| Acetylene | 5 | 50 |
| Ethylene oxide | 5 | 50 |
| Propylene oxide | 5 | 50 |
| Methanol | 500 | 5,000 |
| 4, 4-Methylenebis (2-chloraniline) and/or salts, in powder form | 0.01 | 0.01 |
| Methylisocyanate | 0.15 | 0.15 |
| Oxygen | 200 | 2,000 |
| Toluene diisocyanate | 10 | 100 |
| Carbonyl dichloride (phosgene) | 0.3 | 0.75 |
| Arsenic trihydride (arsine) | 0.2 | 1 |
| Phosphorus trihydride (phosphine) | 0.2 | 1 |
| Sulphur dichloride | 1 | 1 |
| Sulphur trioxide | 15 | 75 |
| Polychlorodibenzofurans and polychlorodibenzodioxins (including TCDD), calculated in TCDD equivalent | 0.001 | 0.001 |
| The following CARCINOGENS: | 0.001 | 0.001 |
| 4-Aminobiphenyl and/or its salts, Benzidine and/or salts, | | |
| Bis(chloromethyl) ether, Chloromethyl methyl ether, | | |
| Dimethylcarbamoyl chloride, Dimethylnitrosomine, | | |
| Hexamethylphosphoric triamide, 2-Naphthylamine and/or salts, | | |
| 1, 3 Propanesultone and 4-nitrodiphenyl | | |
| Automotive petrol and other petroleum spirit | 5,000 | 50,000 |

*[COMAH (SI 1999 No 743), Sch 1, Part 2].*

Ammonium nitrate:

(a)   The 350/2500 quantities apply to ammonium nitrate and ammonium nitrate compounds in which the nitrogen content as a result of the ammonium nitrate is more than 28 per cent by weight (except for those compounds referred to in (*b*) below) and to aqueous ammonium nitrate solutions in which the concentration of ammonium nitrate is more than 90 per cent by weight.

(*b*) The 1250/5000 quantities apply to simple ammonium-nitrate based fertilisers which conform with the requirements of the *Fertilisers Regulations 1991* (*SI 1991 No 2197*) and to composite fertilisers in which the nitrogen content as a result of the ammonium nitrate is more than 28 per cent in weight (a composite fertiliser contains ammonium nitrate with phosphate or potash, or phosphate and potash).

[*COMAH* (*SI 1999 No 743*), Notes 1 and 2 of *Part 2* of *Sch 1*].

Substances are classified according to the *Chemicals* (*Hazard Information and Packaging for Supply*) *Regulations 1994* (*SI 1994 No 3247*), *Reg 5* (now revoked and replaced by *SI 2002 No 1689, Reg 4*).

| Table 2 | | |
|---|---|---|
| Column 1 | Column 2 | Column 3 |
| Categories of dangerous substances | Quantity in tonnes | |
| 1. VERY TOXIC | 5 | 20 |
| 2. TOXIC | 50 | 200 |
| 3. OXIDISING | 50 | 200 |
| 4. EXPLOSIVE (see C10004 below) | 50 | 200 |
| 5. EXPLOSIVE (see C10004 below) | 10 | 50 |
| 6. FLAMMABLE (as defined in Sch 1 Part 3) | 5,000 | 50,000 |
| 7a. HIGHLY FLAMMABLE (as defined in Sch 1 Part 3) | 50 | 200 |
| 7b. HIGHLY FLAMMABLE liquids (as defined in Sch 1 Part 3) | 5,000 | 50,000 |
| 8. EXTREMELY FLAMMABLE (as defined in Sch 1 Part 3) | 10 | 50 |
| 9. DANGEROUS FOR THE ENVIRONMENT in combination with risk phrases: | | |
| R50: 'Very toxic to aquatic organisms' | 200 | 500 |
| R51: 'Toxic to aquatic organisms'; and R53: 'May cause long-term adverse effects in the aquatic environment' | 500 | 2,000 |
| 10. ANY CLASSIFICATION not covered by those given above in combination with risk phrases: | | |
| R14: 'Reacts violently with water' (including R14/15) | 100 | 500 |
| R29: 'in contact with water, liberates toxic gas' | 50 | 200 |

[*COMAH* (*SI 1999 No 743*), *Sch 1, Part 3*].

Where a substance or group of substances named in Table 1 also falls within a category in Table 2, the qualifying quantities set out in Table 1 must be used.

Dangerous substances present at an establishment only in quantities not exceeding 2 per cent of the relevant qualifying quantity are ignored for the purposes of calculating the total quantity present, provided that their location is such that they cannot initiate a major accident elsewhere on site.

## Explosives

**C10004**   Note 2 to *COMAH (SI 1999 No 743), Sch 1, Part 3*, defines an 'explosive' as:

(*a*)   (i)   a substance or preparation which creates the risk of an explosion by shock, friction, fire or other sources of ignition;

   (ii)   a pyrotechnic substance is a substance (or mixture of substances) designed to produce heat, light, sound, gas or smoke or a combination of such effects through non-detonating self-sustained exothermic chemical reactions; or

   (iii)   an explosive or pyrotechnic substance or preparation contained in objects;

(*b*)   a substance or preparation which creates extreme risks of explosion by shock, friction, fire or other sources of ignition.

## Aggregation

**C10005**   There is a rule for aggregation of dangerous substances set out in Note 4 to *COMAH (SI 1999 No 743), Sch 1, Part 3* . The rule will apply in the following circumstances:

(i)   for substances and preparations appearing in Table 1 at quantities less than their individual qualifying quantity present with substances having the same classification from Table 2, and the addition of substances and preparations with the same classification from Table 2;

(ii)   for the addition of categories 1, 2 and 9 present at an establishment together;

(iii)   for the addition of categories 3, 4, 5, 6, 7a, 7b and 8, present at an establishment together.

The sub-threshold quantities of named dangerous substances or categories of dangerous substances are expressed as partial fractions of the threshold quantities in column 3 of Table 1 or of Table 2, and added together. When the total exceeds 1, top-tier duties apply.

If the total is less than or equal to 1, the sub-threshold quantities of named dangerous substances or categories of dangerous substances are expressed as partial fractions of the threshold quantities in column 2 of Table 1 or of Table 2, and added together. If the total exceeds 1, lower tier duties apply.

Definitive guidance on aggregation can be found in HSE's 'Guide to the Control of Major Accident Hazards Regulations 1999'.

## Exclusions

**C10006**   *COMAH (SI 1999 No 743)* does not apply to:

— Ministry of Defence establishments;

— extractive industries exploring for, or exploiting, materials in mines and quarries;

— waste land-fill sites;

— transport related activities;

— substances at nuclear licensed sites which create a hazard from ionising radiation.

*COMAH does* apply to explosives and chemicals at nuclear installations.

## Lower tier duties

### General duty

**C10007** Under *COMAH (SI 1999 No 743)*, Reg 4, every operator is under a general duty to take *all measures necessary* to prevent major accidents and limit their consequences to persons and the environment.

Although the best practice is to completely eliminate a risk, the Regulations recognise that this is not always possible. Prevention should be considered in a hierarchy based on the principle of reducing risk to a level as low as is reasonably practicable, taking account of what is technically feasible and the balance between the costs and benefits of the measures taken. Operators must be able to demonstrate that they have adopted control measures which are adequate for the risks identified. Where hazards are high, high standards will be expected by the enforcement agencies to ensure that risks are acceptably low.

### Major accident prevention policy (MAPP)

**C10008** Under *COMAH (SI 1999 No 743)*, Reg 5(1), every operator must prepare and keep a document (MAPP) setting out his policy with respect to the prevention of major accidents.

The MAPP document must be in writing and must include sufficient particulars to demonstrate that the operator has established an appropriate safety management system. *COMAH (SI 1999 No 743), Sch 2*, sets out the principles to be taken into account when preparing a MAPP document and at *para 4* lists the following specific issues to be addressed by the safety management system:

(a) *organisation and personnel* – the roles and responsibilities of personnel involved in the management of major hazards at all levels in the organisation. The identification of training needs of such personnel and the provision of the training so identified. The involvement of employees and, where appropriate, sub-contractors;

(b) *identification and evaluation of major hazards* – adoption and implementation of procedures for systematically identifying major hazards arising from normal and abnormal operation and the assessment of their likelihood and severity;

(c) *operational control* – adoption and implementation of procedures and instructions for safe operation, including maintenance of plant, processes, equipment and temporary stoppages;

(d) *management of change* – adoption and implementation of procedures for planning modifications to, or the design of, new installations, processes or storage facilities;

(e) *planning for emergencies* – adoption and implementation of procedures to identify foreseeable emergencies by systematic analysis and to prepare, test and review emergency plans to respond to such emergencies;

(f) *monitoring performance* – adoption and implementation of procedures for the on-going assessment of compliance with the objectives set by the operator's major accident prevention policy and safety management system, and the mechanisms for investigation and taking corrective action in the case of non-compliance. The procedures should cover the operator's system for reporting major accidents or near misses, particularly those involving failure of protective measures, and their investigation and follow-up on the basis of lessons learnt;

(g) *audit and review* – adoption and implementation of procedures for periodic systematic assessment of the major accident prevention policy and the effectiveness and suitability of the safety management system; the documented review of performance of the policy and safety management system and its updating by senior management.

The MAPP is a concise but key document for operators which sets out the framework within which adequate identification, prevention/control and mitigation of major accident hazards is achieved. Its purpose is to compel operators to provide a statement of commitment to achieving high standards of major hazard control, together with an indication that there is a management system covering all the issues set out in (*a*)–(*g*) above. The guidance to the Regulations suggests that the essential questions which operators must ask themselves are:

— does the MAPP meet the requirements of the Regulations?

— will it deliver a high level of protection for people and the environment?

— are there management systems in place which achieve the objectives set out in the policy?

— are the policy, management systems, risk control systems and workplace precautions kept under review to ensure that they are implemented and that they are relevant?

Not only must the MAPP be kept up to date (*COMAH (SI 1999 No 743), Reg 5(4)*), but also the safety management system described in it must be put into operation (*COMAH (SI 1999 No 743), Reg 5(5)*).

## Notifications

**C10009**  Notification requirements are set out in *COMAH (SI 1999 No 743), Reg 6*. 'Notify' means notify in writing.

Within a reasonable period of time before the start of construction of an establishment and before the start of the operation of an establishment, the operator must send to the CA a notification containing the information which is specified in *COMAH (SI 1999 No 743), Sch 3*. There is no need for the notification sent before start-up to contain any information which has already been included in the notification sent before the start of construction, if that information is still valid.

The operator of an existing establishment was required to send a notification containing the specified information by 3 February 2000, unless a report had been sent to the HSE in accordance with the *Control of Industrial Major Accident Hazards Regulations 1984, Reg 7*.

The following information is specified in *Sch 3* for inclusion in a notification:

(*a*)   the name and address of the operator;

(*b*)   the address of the establishment concerned;

(*c*)   the name or position of the person in charge of the establishment;

(*d*)   information sufficient to identify the dangerous substances or category of dangerous substances present;

(*e*)   the quantity and physical form of the dangerous substances present;

(*f*)   a description of the activity or proposed activity of the installation concerned;

(*g*)   details of the elements of the immediate environment liable to cause a major accident or to aggravate the consequences thereof.

Notification is a continuing duty. *COMAH (SI 1999 No 743), Reg 6(4)* provides that an operator must notify the CA forthwith in the event of:

— any significant increase in the quantity of dangerous substances previously notified;

— any significant change in the nature or physical form of the dangerous substances previously notified, the processes employing them or any other information notified to the CA in respect of the establishment;

— *Regulation 7* ceasing to apply to the establishment as a result of a change in the quantity of dangerous substances present there; or

— permanent closure of an installation in the establishment.

Information that has been included in a safety report does not need to be notified.

## Duty to report a major accident

**C10010** Where a major accident has occurred at an establishment, the operator must immediately inform the CA of the accident [*COMAH (SI 1999 No 743), Reg 15(3)*].

This duty will be satisfied when the operator notifies a major accident to the HSE in accordance with the requirements of the *Reporting of Injuries, Diseases and Dangerous Occurrences Regulations 1995 (SI 1995 No 3163)*.

The CA must then conduct a thorough investigation into the accident [*COMAH (SI 1999 No 743), Reg 19*].

## Top-tier duties

### Safety report

**C10011** Operators of top-tier sites are required to produce a safety report – its key requirement is that operators must show that they have taken all necessary measures for the prevention of major accidents and for limiting the consequences to people and the environment of any that do occur.

Safety reports are required before construction as well as before start-up. The HSE publication *'Preparing safety reports'* (HSG 190), gives practical and comprehensive guidance to site operators in the preparation of a *COMAH* safety report. *COMAH (SI 1999 No 743), Reg 7(1)* requires that within a reasonable period of time before the start of construction of an establishment the operator must send to the CA a report:

— containing information which is sufficient for the purpose specified in *COMAH (SI 1999 No 743), Sch 4, Part 1, para 3(a)* (see below); and

— comprising at least such of the information specified in *COMAH (SI 1999 No 743), Sch 4, Part 2* (see below) as is relevant for that purpose.

Within a reasonable period of time before the start of the operation of an establishment, the operator must send to the CA a report containing information which is sufficient for the purposes specified in *Part 1 of Sch 4*, and comprising at least the information specified in *Part 2 of Sch 4* [*COMAH (SI 1999 No 743), Reg 7(5)*].

The report sent before start-up is not required to contain information already contained in the report sent before the start of construction.

An operator must ensure that neither construction of the establishment, nor its operation, is started until he has received the CA's conclusions on the report. The CA must communicate its conclusions within a reasonable period of time of receiving a safety report [*COMAH (SI 1999 No 743), Reg 17(1)(a)*].

The operator of an existing establishment must send to the CA a report meeting the requirements of *Parts 1* and *2* of *Sch 4*. Ex-CIMAH top-tier sites that submitted CIMAH safety reports in parts may continue to submit COMAH reports in parts. If a CIMAH report, or any part of it, was up for its three-yearly review before 3 February 2000, a COMAH report was required before that date or such later date (no later than 3 February 2001) as agreed by the CA. If the review would have fallen after that date a COMAH report was required by 3 February 2001. In any other case the safety report was required by 3 February 2002.

## PURPOSE OF SAFETY REPORTS

1. Demonstrating that a major accident prevention policy and a safety management system for implementing it have been put into effect in accordance with the information set out in *Sch 2;*

2. Demonstrating that major accident hazards have been identified and that the necessary measures have been taken to prevent such accidents and to limit their consequences to persons and the environment;

3. Demonstrating that adequate safety and reliability have been incorporated into the

    (a) design and construction, and

    (b) operation and maintenance,

    of any installation and equipment and infrastructure connected with its operation, and that they are linked to major accident hazards within the establishment;

4. Demonstrating that on-site emergency plans have been drawn up and supplying information to enable the off-site plan to be drawn up in order to take the necessary measures in the event of a major accident;

5. Providing sufficient information to the competent authority to enable decisions to be made in terms of the siting of new activities or developments around establishments.

[*COMAH (SI 1999 No 743), Sch 4, Part 1*]

## MINIMUM INFORMATION TO BE INCLUDED IN SAFETY REPORT

1. Information on the management system and on the organisation of the establishment with a view to major accident prevention.

    This information must contain the elements set out in *Sch 2.*

2. Presentation of the environment of the establishment:

    (a) a description of the site and its environment including the geographical location, meteorological, geographical, hydrographic conditions and, if necessary, its history;

    (b) identification of installations and other activities of the establishment which could present a major accident hazard;

    (c) a description of areas where a major accident may occur.

3. Description of installation:

    (a) a description of the main activities and products of the parts of the establishment which are important from the point of view of safety, sources of major accident risks and conditions under which such a major accident could happen, together with a description of proposed preventive measures;

    (b) description of processes, in particular the operating methods;

    (c) description of dangerous substances:

        (i) inventory of dangerous substances including –

            — the identification of dangerous substances: chemical name, the number allocated to the substance by the Chemicals Abstract

Service, name according to International Union of Pure and Applied Chemistry nomenclature;

— the maximum quantity of dangerous substances present;

(ii) physical, chemical, toxicological characteristics and indication of the hazards, both immediate and delayed, for people and the environment;

(iii) physical and chemical behaviour under normal conditions of use or under foreseeable accidental conditions.

4. Identification and accidental risks analysis and prevention methods:

(a) detailed description of the possible major accident scenarios and their probability or the conditions under which they occur including a summary of the events which may play a role in triggering each of these scenarios, the causes being internal or external to the installation;

(b) assessment of the extent and severity of the consequences of identified major accidents;

(c) description of technical parameters and equipment used for the safety of the installations.

5. Measures of protection and intervention to limit the consequences of an accident:

(a) description of the equipment installed in the plant to limit the consequences of major accidents;

(b) organisation of alert and intervention;

(c) description of mobilisable resources, internal or external;

(d) summary of elements described in sub-paragraphs (a), (b) and (c) necessary for drawing up the on-site emergency plan.

[COMAH (SI 1999 No 743), Sch 4, Part 2]

All or part of the information required to be included in a safety report can be so included by reference to information contained in another report or notification furnished by virtue of other statutory requirements. This should be done only where the information in the other document is up to date, and adequate in terms of scope and level of detail.

If an operator can demonstrate that particular dangerous substances are in a state incapable of creating a major accident hazard, the CA can limit the information required to be included in the safety report [COMAH (SI 1999 No 743), Reg 7(12)].

An operator must provide the CA with such further information as it may reasonably request in writing following its examination of the safety report [COMAH (SI 1999 No 743), Reg 7(13)].

## Review and revision of safety report

**C10012** Where a safety report has been sent to the CA, the operator must review it:

— at least every five years;

— whenever a review is necessary because of new facts or to take account of new technical knowledge about safety matters; and

— whenever the operator makes a change to the safety management system, which could have significant repercussions.

Where it is necessary to revise the report as a result of the review, the operator must carry out the revision immediately and inform the CA of the details [COMAH (SI 1999 No 743), Reg 8(1)].

An operator must inform the CA when he has reviewed the safety report but not revised it [*COMAH* (*SI 1999 No 743*), Reg 8(2)].

Ex-*CIMAH* top-tier sites that have submitted *CIMAH* safety reports in parts must review each part within five years from the time when that part was sent to the CA.

When modifications which could have significant repercussions are proposed (to the establishment or an installation in it, to the process carried on there or the nature or quantity of dangerous substances present there), the operator must review the safety report and where necessary revise it, in advance of any such modification. Details of any revision must be notified to the CA.

## On-site emergency plan

**C10013** The HSE publication '*Emergency planning for major accidents*' (HSG 191) is essential reading.

*COMAH* (*SI 1999 No 743*), Reg 9(1), requires operators of top-tier establishments to prepare an on-site emergency plan. The plan must be adequate to secure the objectives specified in *Sch 5*, *Part 1*, of the Regulations namely:

— containing and controlling incidents so as to minimise the effects, and to limit damage to persons, the environment and property;

— implementing the measures necessary to protect persons and the environment from the effects of major accidents;

— communicating the necessary information to the public and to the emergency services and authorities concerned in the area;

— providing for the restoration and clean-up of the environment following a major accident.

The plan must contain the following information:

— names or positions of persons authorised to set emergency procedures in motion and the person in charge of and co-ordinating the on-site mitigatory action;

— name or position of the person with responsibility for liaison with the local authority responsible for preparing the off-site emergency plan (see C10014 below);

— for foreseeable conditions or events which could be significant in bringing about a major accident, a description of the action which should be taken to control the conditions or events and to limit their consequences, including a description of the safety equipment and the resources available;

— arrangements for limiting the risks to persons on site including how warnings are to be given and the actions persons are expected to take on receipt of a warning;

— arrangements for providing early warning of the incident to the local authority responsible for setting the off-site emergency plan in motion, the type of information which should be contained in an initial warning and the arrangements for the provision of more detailed information as it becomes available;

— arrangements for training staff in the duties they will be expected to perform, and where necessary co-ordinating this with the emergency services;

— arrangements for providing assistance with off-site mitigatory action.

[*COMAH* (*SI 1999 No 743*), Sch 5, Part 2].

Ex-*CIMAH* top-tier sites that were subject to *CIMAH* requirements for the preparation of an on-site emergency plan were required to prepare a *COMAH* emergency plan by 3 February 2001. Any other existing establishment was required to prepare an on-site emergency plan by 3 February 2002. New establishments must prepare such a plan before start-up.

When preparing an on-site emergency plan, the operator must consult:

— employees at the establishment;
— the Environment Agency (or the Scottish Environment Protection Agency);
— the emergency services;
— the health authority for the area where the establishment is situated; and
— the local authority, unless it has been exempted from the requirement to prepare an off-site emergency plan.

## Off-site emergency plan

**C10014** The local authority for the area where a top-tier establishment is located must prepare an adequate emergency plan for dealing with off-site consequences of possible major accidents. As with the on-site plan, it should be in writing.

The objectives set out in *COMAH (SI 1999 No 743), Sch 5, Part 1*, (see C10013 above) also apply to off-site emergency plans.

The plan must contain the following information:

— names or positions of persons authorised to set emergency procedures in motion and of persons authorised to take charge of and co-ordinate off-site action;
— arrangements for receiving early warning of incidents, and alert and call-out procedures;
— arrangements for co-ordinating resources necessary to implement the off-site emergency plan;
— arrangements for providing assistance with on-site mitigatory action;
— arrangements for off-site mitigatory action;
— arrangements for providing the public with specific information relating to the accident and the behaviour which it should adopt;
— arrangements for the provision of information to the emergency services of other Member States in the event of a major accident with possible transboundary consequences.

An operator must supply the local authority with the information necessary for the authority's purposes, plus any additional information reasonably requested in writing by the local authority.

In preparing the off-site emergency plan, the local authority must consult:

— the operator;
— the emergency services;
— the CA;
— each health authority for the area in the vicinity of the establishment; and
— such members of the public as it deems appropriate.

*[COMAH (SI 1999 No 743), Reg 10(6)].*

In the light of the safety report, the CA may exempt a local authority from the requirement to prepare an off-site emergency plan in respect of an establishment [*COMAH (SI 1999 No 743), Reg 10(7)*].

## Reviewing, testing and implementing emergency plans

**C10015** *COMAH (SI 1999 No 743), Reg 11*, requires that emergency plans are reviewed and, where necessary, revised, at least every three years. Reviewing is a key process for addressing the adequacy and effectiveness of the components of the emergency plan – it should take into account:

— changes occurring in the establishment to which the plan relates;
— any changes in the emergency services relevant to the operation of the plan;
— advances in technical knowledge;
— knowledge gained as a result of major accidents either on-site or elsewhere; and
— lessons learned during the testing of emergency plans.

There is a new requirement to test emergency plans at least every three years. Such tests will assist in the assessment of the accuracy, completeness and practicability of the plan: if the test reveals any deficiencies, the relevant plan must be revised. Agreement should be reached beforehand between the operator, the emergency services and the local authority on the scale and nature of the emergency plan testing to be carried out.

Where there have been any modifications or significant changes to the establishment, operators should not wait for the three-year review before reviewing the adequacy and accuracy of the emergency planning arrangements.

When a major accident occurs, the operator and local authority are under a duty to implement the on-site and off-site emergency plans [*COMAH (SI 1999 No 743), Reg 12*].

## Local authority charges

**C10016** A local authority may charge the operator a fee for performing its functions under *COMAH (SI 1999 No 743), Regs 10* and *11*, i.e. for preparing, reviewing and testing off-site emergency plans.

The charges can only cover costs that have been reasonably incurred. If a local authority has contracted out some of the work to another organisation, the authority may recover the costs of the contract from the operator, provided that they are reasonable.

In presenting a fee to an operator, the local authority should provide an itemised, detailed statement of work done and costs incurred.

## Provision of information to the public

**C10017** The operator must supply information on safety measures to people within an area without their having to request it. The area is notified to the operator by the CA as being one in which people are liable to be affected by a major accident occurring at the establishment. The minimum information to be supplied to the public is specified in *COMAH (SI 1999 No 743), Sch 6*:

---

### INFORMATION TO BE SUPPLIED TO THE PUBLIC

— name of operator and address of the establishment;
— identification, by position held, of the person giving the information;
— confirmation that the establishment is subject to these Regulations and that the notification referred to in *Reg 6* or the safety report has been submitted to the competent authority;

- an explanation in simple terms of the activity or activities undertaken at the establishment;
- the common names or, in the case of dangerous substances covered by *Sch 1, Part 3*, the generic names or the general danger classification of the substances and preparations involved at the establishment which could give rise to a major accident, with an indication of their principal dangerous characteristics;
- general information relating to the nature of the major accident hazards, including their potential effects on the population and the environment;
- adequate information on how the population concerned will be warned and kept informed in the event of a major accident;
- adequate information on the actions the population concerned should take, and on the behaviour they should adopt, in the event of a major accident;
- confirmation that the operator is required to make adequate arrangements on site, in particular liaison with the emergency services, to deal with major accidents and to minimise their effects;
- a reference to the off-site emergency plan for the establishment. This should include advice to co-operate with any instructions or requests from the emergency services at the time of an accident;
- details of where further relevant information can be obtained, unless making that information available would be contrary to the interests of national security or personal confidentiality or would prejudice to an unreasonable degree the commercial interests of any person.

Under *COMAH (SI 1999 No 743), Sch 8*, the CA must maintain a public register which will include:

- the information included in the notifications submitted by the operators under *COMAH (SI 1999 No 743), Reg 6*;
- top-tier operators' safety reports;
- the CA's conclusions of its examination of safety reports.

## Powers to prohibit use

**C10018** The CA is required to prohibit the operation or bringing into operation of any establishment or installation or any part of it where the measures taken by the operator for the prevention and mitigation of major accidents are seriously deficient [*COMAH (SI 1999 No 743), Reg 18(1)*].

The CA may prohibit the operation or bringing into operation of any establishment or installation or any part of it if the operator has failed to submit any notification, safety report or other information required under the Regulations within the required time. Where the CA proposes to exercise its prohibitory powers, it must serve on the operator a notice giving reasons for the prohibition and specifying the date when it is to take effect. A notice may specify measures to be taken. The CA may, in writing, withdraw any notice.

## Enforcement

**C10019** The Regulations are treated as if they are health and safety regulations for the purpose of the *Health and Safety at Work etc Act 1974*. The provisions as to offences of the 1974

Act, ss 33–42, apply. A failure by the CA to discharge a duty under the Regulations is not an offence [COMAH (SI 1999 No 743), Reg 20(2)], although the remedy of judicial review is available.

## Radiation (Emergency Preparedness and Public Information) Regulations 2001 (SI 2001 No 2975)

**C10020** The *Radiation (Emergency Preparedness and Public Information) Regulations 2001 (SI 2001 No 2975)* revoke the *Public Information for Radiation Emergencies Regulations 1992 (SI 1992 No 2997)* (subject to savings). The 2001 Regulations came into force on 20 September 2001.

The Regulations implement the emergency planning aspects of Council Directive 96/29/Euratom, which lays down basic safety standards for the protection of health of workers and the general public against the dangers arising from ionising radiation. The Regulations impose requirements on operators of premises where radioactive substances are present (in quantities exceeding specified thresholds). They also impose requirements on carriers transporting radioactive substances (in quantities exceeding specified thresholds) by rail or conveying them through public places, with the exception of carriers conveying radioactive substances by rail, road, inland waterway, sea or air or by means of a pipeline or similar means.

Essential guidance is contained in the HSE publication '*Guide to the Radiation (Emergency Preparedness and Public Information) Regulations 2001*' (L126).

The Regulations:

(*a*) impose a duty on the operator and the carrier to make an assessment as to hazard identification and risk evaluation and, where the assessment reveals a radiation risk, to take all reasonably practicable steps to prevent a radiation accident of limit the consequences should such an accident occur [*SI 2001 No 2975, Reg 4*];

(*b*) impose a duty on the operator and carrier to send the HSE a report of an assessment containing specified matters at specified times and empower the HSE to require a detailed assessment of such further particulars as may reasonably require [*SI 2001 No 2975, Reg 6* and *Sch 5* and *6*];

(*c*) impose a duty on the operator and the carrier to make a further assessment following a major change to the work with the ionising radiation or within three years of the date of the last assessment, unless there has been no change of circumstances which would affect the last report of the assessment, and send the HSE a report of that further assessment [*SI 2001 No 2975, Regs 5* and *6*];

(*d*) where an assessment reveals a reasonably foreseeable radiation emergency arising, impose a duty on the operator or the carrier (as the case may be ) and, in the case of an operator, the local authority in whose area the premises in question are situated, to prepare emergency plans [*SI 2001 No 2975, Regs 7, 8* and *9* and *Sch 7* and *8*];

(*e*) require operators, carriers and local authorities to review, revise and test emergency plans at suitable intervals not exceeding three years [*SI 2001 No 2975, Reg 10*];

(*f*) make provision as to consultation and co-operation by operators, carriers, employers and local authorities [*SI 2001 No 2975, Reg 11*];

(*g*) make provision as to charging by local authorities for performing their functions under the Regulations in relation to emergency plans [*SI 2001 No 2975, Reg 12*];

(h) in the event of the occurrence of a radiation emergency or of an event which could reasonably be expected to lead to such an emergency, make provision as to the implementation of emergency plans, and in the event of the occurrence of a radiation emergency, the making of both provisional and final assessments as to the circumstances of the emergency [*SI 2001 No 2975, Reg 13*];

(i) where an emergency plan provides for the possibility of an employee receiving an emergency exposure, impose a duty on the employer to undertake specified arrangements for employees who may be subject to exposures, such as dose assessments, medical surveillance and the determination of appropriate dose levels and impose further duties on employers in the event that an emergency plan is implemented [*SI 2001 No 2975, Reg 14*];

(j) impose requirements on operators and carriers, where an operator or carrier carries out work with ionising radiation which could give rise to a reasonably foreseeable radiation emergency, and on local authorities, where there has been a radiation emergency in their area, to supply specified information to the public [*SI 2001 No 2975, Regs 16* and *17* and *Sch 9* and *10*].

## Public Information for Radiation Emergencies Regulations 1992 (SI 1992 No 2997)

C10021 The *Public Information for Radiation Emergencies Regulations 1992* (*SI 1992 No 2997*) were revoked by the *Radiation* (*Emergency Preparedness and Public Information*) *Regulations 2001* (*SI 2001 No 2975*). However, *Reg 3* of the 1992 Regulations continues in force to the extent that it applies in relation to the transport of radioactive substances by road, inland waterway, sea or air. Any other provisions of the 1992 Regulations continue in force so far as is necessary to give effect to *Reg 3*.

An employer (or self-employed person) must supply prior information where a radiation emergency is reasonably foreseeable from his undertaking in relation to the transport of radioactive substances by road, inland waterway, sea or air. He must:

(a) supply information to members of the public likely to be in the area in which they are liable to be affected by a radiation emergency arising from the undertaking with without their having to request it, concerning:

(i) basic facts about radioactivity and its effects on persons/environment;

(ii) the various types of radiation emergency covered and their consequences for people/environment;

(iii) emergency measures to alert, protect and assist people in the event of a radiation emergency;

(iv) action to be taken by people in the event of an emergency;

(v) authority/authorities responsible for implementing emergency measures.

(b) make the information publicly available; and

(c) update the information at least every three years.

[*Public Information for Radiation Emergencies Regulations 1992* (*SI 1992 No 2997*), *Reg 3* and *Sch 2*]

# Dangerous Goods Systems

## Basic regulatory principles

### The purpose of dangerous goods measures

D0401 Every day, organisations produce, package and transport dangerous goods. As well as industrial products, many household products also consist of 'common' dangerous goods used in numerous mundane applications. From the most common paints to the deadliest poisons, all these dangerous goods need to be carried, by sea, air, rail or, more often, by road.

In order to minimise risk, rules have been created to regulate this transport. The basic reason for transport control measures and systems is that *dangerous goods are dangerous*. This may seem to be an obvious statement, but it is necessary to remember that these products can cause serious harm or damage and that all reasonable precautions must be taken in order to control and minimise risks.

The aim of dangerous goods controls and regulations is to protect passengers, transport crews and equipment, property and the environment from any of the threats that may be posed by goods with the potential to cause danger within the transport systems.

### Fundamentals of control systems

D0402 Most goods that are consigned for transport pose no specific risk to the transport system and are therefore not covered by the controls present in the dangerous goods regulations. Bulky loads, which might present a manual handling hazard, are not classified as hazardous for transportation and are exempt from all dangerous goods regulations. The standard requirements of the *Road Traffic Acts* and others will still need to be followed.

However, certain types of goods may pose a direct hazard to the transport system or the environment within which it operates, and the carriage of these goods needs to be regulated to ensure suitable protection. Such items may pose many different types of hazards, from flammable liquids to radioactive substances and corrosive agents to compressed gases. The dangers that these goods present will be the same whether they are carried by road, rail sea or air so they can be referred to as multi-modal hazardous goods. Certain types of hazardous goods are only regarded as dangerous for specific modes of transport, such as magnetised materials which are dangerous goods for air transport but are not regulated for road transport.

Substances and preparations are classified as workplace hazards by the *Chemicals (Hazard Information and Packaging) for Supply Regulations 2002 (SI 2002 No 1689)* ('the *CHIP Regulations*') and are assigned certain categories of danger, such as toxic, corrosive, etc. These categories of danger are represented by black on orange hazard symbols, as indicated in FIGURE 1 below:

Substances classified as hazardous by the *CHIP Regulations* may also be classified as dangerous goods for transport, but this is not always the case. Some low to medium hazard materials are not considered to pose enough risk during transportation to warrant classification, the chance of exposure is lower during transport than use. Hazard classifications under the *CHIP Regulations* apply to materials that may show long-term chronic health effects as well as the results of acute exposure. A toxic paint that is hazardous to individuals who use it every day for a long period of time will not be classified as dangerous for transport if it poses no acute effects. However, some

Figure 1

materials with chronic effects, such as dichloromethane (methylene chloride) *are* classified as dangerous goods. An in depth knowledge of the dangerous goods regulations is necessary in order to determine whether specific items are governed by the regulations.

Dangerous goods may be grouped into two categories. A very small number of substances have been judged to be too dangerous to transport under any circumstances as no same method of containment is practicable. Such items include some unstabilised peroxides, particularly sensitive explosives and spontaneously reactive substances. These goods are prohibited for carriage, and transport not allowed on any occasion. Prohibited items are quite rare for road transport, but are much more common for air transport due to the severe consequences of any loss of containment. By far the majority of dangerous goods may be accepted for carriage by road, subject to compliance with the transport of dangerous goods regulations.

## *The principle of the United Nations dangerous goods classification system*

**D0403**     In order to maintain an international uniformity of approach to dangerous goods legislation, the United Nations has established guidelines on the regulation of dangerous goods. These guidelines are published as the 'Recommendations on the Transport of Dangerous Goods: Model Regulations' which is commonly referred to as the 'Orange Book'. These UN Recommendations have no legal force but are adopted worldwide as the basis of national legislation.

The Orange Book contains example procedural systems for the classification of types of danger, the identification of dangerous goods, containment system performance and use, documentation and training.

The UN classification system splits dangerous goods into nine hazard groups according to the kind of danger inherent to the item. Some of these classes are further broken down into smaller groupings, which are known as divisions. Sometimes items may show danger characteristics of more than one group. In this case, the material is assigned to the hazard class with the greatest risk, and the lesser risks are assigned as subsidiary hazard classes. Each class is allocated a diamond label to indicate the nature of the hazard on packaging and vehicles carrying dangerous goods. These are reproduced in FIGURE 2 below:

In addition to indicating the kind of danger presented by a substance the classification system also gives qualitative guidance as to how dangerous that substance is in relation to similar types of dangerous goods. This degree of danger is indicated by the 'packing group', which is defined below:

PG I = Greatest danger.

PG II = Medium danger.

PG III = Least danger.

**Figure 2**

The packing group is used to help determine the selection of acceptable packaging and threshold limits. The packing group system doesn't apply for Class 1,2,5.2, 6.2 or 7, as different specification criteria apply.

# The legal framework

## United Kingdom Carriage of Dangerous Goods Regulations

### Applicability of Dangerous Goods Regulations

D0404 Although the UN Recommendations laid out in the 'Orange Book' are used as the basis for dangerous goods legislation, they have no actual legal authority. National controls are governed by Acts and Regulations established by Parliament which establish duties under national law. International agreements to allow cross border movement of dangerous goods are also authorised by specific national laws.

Within Europe, the carriage of dangerous goods is covered by the European Agreement concerning the International Carriage of Dangerous Goods ('ADR'). This agreement is maintained by the United Nations Economic Commission for Europe ('UNECE'), not by the European Union ('EU'). ADR covers not only the EU, but stretches as far afield as North Africa and the Middle East. The only EU requirement for the transportation of dangerous goods is that Member States' national regulations follow the ADR Agreement.

### Interface of different modal controls and UK special concessions

D0405 While different modal regulations are all based upon UN Recommendations, separate specific requirements for road, rail, sea and air transport can cause the potential for conflict between various modes of transport. Differences between UK national requirements and the ADR requirements of international journeys can also lead to confusion about which requirements to follow. Luckily, the UK regulations generally follow the requirements of ADR and are flexible enough to accommodate such concerns.

UK regulations allow compliance with ADR for goods being transported within Britain as part of an international journey. A specific implication of international journeys ending or originating within the UK is that almost all such journeys will have to involve either travel by sea, air or channel tunnel. The requirements for sea or air transport, governed by the IMDG Code or IACO Technical Instructions are tougher than road transport requirements. To prevent having to label goods twice, compliance with UK national road regulations is not required as long as the goods have been correctly classified, packaged and labelled in accordance with applicable air and sea regulations.

### Structure of UK Carriage of Dangerous Goods Regulations

D0406 The previous series of legislative documents governing dangerous goods consignment and road transport in Britain have recently been replaced by the *Carriage of Dangerous Goods and Use of Transportable Pressure Equipment Regulations 2004* (*SI 2004 No 568*) ('*CDGTPE Regulations*') which came into force on May 10 2004.

These new Regulations act as one consolidated piece of legislation replacing the previous range of regulations. They also implement European Directive 2003/28/EC regarding the transport of dangerous goods by road and complete the implementation of Council Directive 1999/36/EC EC concerning transportable pressure equipment ('the Transportable Pressure Equipment Directive'). They directly reference the latest texts of the European Agreement concerning the International Carriage of Dangerous Goods by Road (ADR) (Current edition: 2003).

The main changes from the old rules include:

- New limited quantity and load thresholds for transporting dangerous goods.
- Packaging requirements now based on RID/ADR.
- The transport of diesel, gas oil and heating oil coming fully into scope.
- New reporting requirements.

The Regulations take account of national derogations agreed with the European Commission in respect of certain transport operations involving local journeys or the movement of dangerous goods in small quantities. They continue to use the mandatory Emergency Action Code system for placarding GB-registered vehicles on GB journeys instead of using the ADR placarding system. The Regulations will be supported by a number of letters of authorisation allowing time-limited relaxations to some requirements. Enforcement of the *CDGTPE Regulations* is undertaken by the Department of Transport, Local Government and the Regions rather than the Health and Safety Executive as used to be the case for the former regulations.

## Revocations

**D0407** The *CDGTPE Regulations* (*SI 2004 No 568*) completely revoke the following Regulations:

- the *Gas Cylinders (Pattern Approval) Regulations 1987* (*SI 1987 No 116*);
- the *Pressure Vessels (Verification) Regulations 1988* (*SI 1988 No 896*);
- the *Packaging of Explosives for Carriage Regulations 1991* (*SI 1991 No 2097*);
- the *Carriage of Dangerous Goods (Classification, Packaging and Labelling) and Use of Transportable Pressure Receptacles Regulations 1996* (*SI 1996 No 2092*);
- the *Carriage of Explosives by Road Regulations 1996* (*SI 1996 No 2093*);
- the *Carriage of Dangerous Goods by Road (Driver Training) Regulations 1996* (*SI 1996 No 2094*);
- the *Carriage of Dangerous Goods (Amendment) Regulations 1998* (*SI 1998 No 2885*);
- the *Carriage of Dangerous Goods (Amendment) Regulations 1999* (*SI 1999 No 303*);
- the *Transport of Dangerous Goods (Safety Advisers) Regulations 1999* (*SI 1999 No 257*);
- the *Transportable Pressure Vessels Regulations 2001* (*SI 2001 No 1426*);
- the *Carriage of Dangerous Goods and Transportable Pressure Vessels (Amendment) Regulations 2003* (*SI 2003 No 1431*).

The Regulations also partially revoke the following Regulation:

- the *Carriage of Dangerous Goods by Road Regulations 1996* (*SI 1996 No 2095*) ('*CDGRoad*').

    *Regulations 1* to *4*, *7* to *19*, *21* to *29* and the *Schedules* are revoked. The remaining regulations enpower certain approved documents to be used in conjunction with the new *CDGTPE Regulations* and enforce specific requirements regarding the unloading of petrol at petrol stations.

A number of legislative documents work in conjunction with the *CDGTPE Regulations* to provide complete legislative control of dangerous goods consignment and road transport in Britain. These are detailed below:

- the *Radioactive Material (Road Transport)Regulations 2002* (*SI 2002 No 1093*) ('*RAM-Road*').
- the *Radioactive Material (Road Transport)(Amendment) Regulations 2003* (*SI 2003 No 1867*) ('*RAMRoad Amendment*').

**D0408**  *Dangerous Goods Systems*

- the Dangerous Goods Emergency Action Code List ('DGEACL') (ISBN 0 11 341275 4).

These Regulations (except for *RAMRoad*) are empowered by the *Health and Safety at Work etc. Act 1974* and so follow the same format and type of presentation. Apart from items of Class 7 (additional measures covered by *RAMRoad*), the Regulations and supporting legal documents apply to all classes of dangerous goods. In each case, the appropriate regulation establishes the legal duties to ensure compliance and detailed schedules provide additional information to back up the regulations.

### The Approved List

**D0408** The former Approved Carriage List ('ACL') has been replaced by the ADR 'Approved List' listed in Table A of chapter 3.2 of ADR. This contains lists of all categories of dangerous goods authorised for road transport. If a substance is listed by name, or its hazardous properties described then it is subject to control and must be packaged, marked and labelled, its transport must be undertaken, the driver must be trained and a Dangerous Goods Safety Advisor ('DGSA') appointed in accordance with *CDGTPE* (*SI 2004 No 568*). Low quantities of dangerous goods in small receptacles or restricted load sixes limited in may be exempt from some or all of the above regulations under limited quantities concessions.

Each entry in the Approved List provides information to be used by the consignor to fulfil classification, identification, package selection, labelling and documentation completion duties. For example:

| UNNO | Proper Shipping Name | Class | CLS Code | Packing Group | Labels | Limited Quantities | Packaging | | | |
|---|---|---|---|---|---|---|---|---|---|---|
| | | | | | | | Packaging instructions | Special packing provisions | Special provisions | |
| 1106 | Amylamine | 3 | FC | II | 3 + 8 | LQ4 | P001 IBC02 | | MP19 | |

| UN Portable Tank | | ADR Tank | | Vehicle for tank carriage | Transport category | Special provisions for carriage | | | HIN |
|---|---|---|---|---|---|---|---|---|---|
| Instructions | Special provisions | Tank code | Special provisions | | | Bulk | Loading, unloading and handling | Operation | |
| T7 | TP1 | L4BH | TE1 | FL | 2 | Packages | | S2 S20 | 38 |

UNNO = United Nations Number
CLASS = Hazard Class
CLS Code = Hazard Classification Code
HIN = Hazard Identification Number

It is essential to check the implication of any classification codes in the table above as these can significantly affect consignor duties. These codes are described in chapter 3.2 of ADR. The carrier of the dangerous goods can use the Approved List to confirm the accuracy of the consignor's transport documentation, to determine the competence required by the driver carrying the load and to decide what vehicle markings are appropriate for the journey.

Controls of radioactive materials follow a significantly different pattern and refer to further sections of ADR and are discussed separately.

## Consignment duties

**D0409** Procedures for classification, identification, package selection and warning marks and labels are based upon the ADR system and are contained within *Part 2* of *CDGTPE* (*SI 2004 No 568*) (*Regulations 9* to *25*). A range of concessions for limited quantities are available, which are discussed later. Classification requires cross references to be made to the schedules in the back of the Regulations as well as to the relevant chapters within ADR, which provide detailed information on classification procedures.

Dangerous goods classification and identification requires use of both *CDGTPE* and the ACL.

For package selection decisions and marking and labelling duties, *CDGTPE* along with its schedules and both the Approved List and Requirements must be utilised.

## Carriage duties

**D0410** The consignor is required to create a 'transport document', which is handed to the carrier as an indication that they will probably have to comply with the remaining duties found in the *CDGTPE* (*SI 2004 No 568*) *Regulations*.

There are load size thresholds that may affect duties – if exceeded the vehicle operator and driver must comply with the duties laid down in the Regulations.

The Regulations also lay down requirements to:

- the suitability of vehicles for particular loads;
- loading;
- unloading and stowage procedures;
- the carriage of safety equipment;
- the need for safe working systems;
- emergency response actions; and
- parking and supervision duties.

## *The ADR Agreement*
### *Structure of the ADR Agreement*

**D0411** Volume I of this two volume document identifies hazard classes and allocates dangerous materials into these classes by UN number. It provides information to be used by the consignor to classify his hazardous goods for transportation in preparation for a journey. Volume II contains the duties to be followed by the carrier of the goods such as specific packing instructions, labelling requirements, package testing information, placarding and vehicle suitability/driver training arrangements.

Volume I comprises of an introduction, which is a brief summary of the duties, the legal text of the ADR Agreement and Annex A, the general provisions and provisions concerning dangerous substances and articles which outline the duties of the consignor of the goods.

The general provisions (chapter 1) detail the scope and availability of ADR, along with basic definitions, general obligations and transitional measures. Chapter 2 of Annex A then details general provisions and principles of hazard classification along with specific class provisions and approved test methods for classification and allocation of dangerous goods to a particular class and packing group. The final part (chapter 3) of volume I is a comprehensive dangerous goods list which details hazard class, packing group and other pertinent information for items falling into each specific UN number category. Its scope is the same as the ACL for UK regulations, but in addition includes details for explosives and radioactive materials (Classes 1 and 7).

Volume II begins with the concluding parts of chapter 3, detailing special provisions and exemptions for certain materials where appropriate. Chapter 4 describes packing and tank provisions, which covers general provisions along with special provisions for portable tanks, intermediate bulk containers ('IBCs') and other large vessels. Special provisions for explosives, gases, organic peroxides, self reactive substances and radioactive materials are also laid out. Chapter 5 describes consignment procedures including general provisions, marking and labelling requirements for packages, tanks and vehicles along with documentation requirements. Chapter 6 describes detailed requirements for the construction and testing of packages, IBC's, gas cylinders and tanks, etc and the resultant UN mark for approved packaging / receptacles. Chapter 7 details provisions concerning conditions of carriage, loading, unloading and handling of dangerous goods such as general conditions, mixed loading prohibitions, handling and stowage, cleaning vehicles after unloading and any additional provisions that may apply to specific items or goods. Annex B outlines provisions concerning transport equipment and transport operations and details requirements for vehicle crews, equipment, operation and documentation. Such matters as the provision of fire fighting equipment and spillage kits are covered, along with driver training requirements and other miscellaneous requirements to be complied with by the vehicle crew.

The ADR chapters are broken down into subchapters and subsections in a logical manner that allows easy reference to information throughout both volumes of the agreement.

## Transport of radioactive materials

### General principles of transport of radioactive material

**D0412** The International Atomic Energy Agency ('IAEA') is an agency of the United Nations which addresses the whole field of radioactive material control and not just the transportation of such substances.

The IAEA produces recommendations for the effective control of radioactive materials. The Transport Regulations are published in a collection of volumes entitled the Safety Series, and sixth in the series is the 'Regulations for the Safe Transport of Radioactive Material', referred to as Safety Series No. 6.

In 1996 'Regulations for the Safe Transport of Radioactive Material No. ST-1' was published in the IAEA Safety Series as a successor to Safety Series No. 6.

The ADR Agreement and the *RAMRoad Regulations* (*SI 2002 No 1093*) have adopted and adapted Safety Series No.6 and incorporated the information into the procedural duties of both systems, rendering the recommendations that they convey as legal duties. Therefore, the safety series books are not operational documents as the relevant directions from the series can be identified in the appropriate transport controls.

### UK Radioactive Transport Regulations

**D0413** The *Radioactive Material (Road Transport) Regulations 2002* (*SI 2002 No 1093*) ('*RAMRoad*'), control the movement of Class 7 items on road vehicles in Great Britain. *RAMRoad* is under the control of the Department of Transport, Local Government and the Regions, Radioactive Materials Transport Branch.

**D0414**  *Dangerous Goods Systems*

*RAMRoad* consists of 75 regulations and does not begin with an arrangement of regulations section. It also provides information on how to carry out the duties allocated through 14 schedules.

### The ADR radioactive material system

**D0414**     General consignor duties are outlined in chapter 1.7 of the ADR Agreement which provides the general requirements for transporting radioactive materials. Rather than the ECE which governs the requirements for all other classes of goods, the standards used are based on the IAEA Regulations for the Safe Transport of Radioactive Material (ST-1). These focus upon the need for the containment of radioactive contents, control of external radiation levels, prevention of criticality and heat damage. Chapter 6.4 of the ADR Agreement covers the requirements for the construction, testing and approval of packages and materials for containing radioactive materials.

In addition to radioactive and fissile properties, any subsidiary hazards of dangerous goods items such as flammability, corrosivity, etc, must be taken into account in the documentation, packaging, labelling, marking, placarding, stowage, segregation and carriage requirements to ensure compliance with all the requirements of ADR.

# General requirements

## Classification of dangerous goods

### Basic principles of classification

**D0415**     The purpose of classification is twofold; firstly, to determine which goods are dangerous during transport; secondly, to show what kind(s) of danger are to be found in a particular article or substance.

The United Nations Committee of Experts on the Transport of Dangerous Goods developed the system for testing and classification. Various modal and national experts have then modified this to produce the most appropriate controls for their method of transport.

All modes base their controls upon the UN system of 9 classes, follow the UN system of testing for the various classes and follow UN packaging group criteria.

The nine classes divide dangerous substances into the different types of hazard they present. Some of these classes are further broken down into smaller groupings, which are known as divisions.

| | Class/Division | Typical technical reference |
|---|---|---|
| 1 | Explosives | Pyrotechnic; non-detonating self-sustaining exothermic chemical reaction. |
| 2 | Gases | Vapour pressure above 300kPa at 50°C; gaseous at 20°C at 101.3kPa. |
| 3 | Flammable liquids | Initial boiling point; flashpoint. |
| 4.4 | Flammable solids | Burning time in seconds per 100mm; rate of detonation or deflagration; self-reactive. |
| 4.2 | Spontaneously combustible | Pyrophoric substance; self-heating substance. |
| 4.3 | Water reactive | Water reactive; rate of evolution of flammable gas; spontaneous ignition. |
| 5.1 | Oxidising substances | Increased burning rate. |

| 5.2 | Organic peroxide | Assigned only by competent authority. |
|---|---|---|
| 6.1 | Toxic | $LD_{50}$ oral; $LD_{50}$ dermal; $LC_{50}$ inhalation. |
| 6.2 | Infectious substances | Genetically modified; biological products; diagnostic specimens. |
| 7 | Radioactive materials | Specific activity more than 70kBq/kg. |
| 8 | Corrosive | Full thickness destruction of intact skin tissue. |
| 9 | Miscellaneous | Allocated by regulatory authority. |

The packing group indicates the degree of danger present by a substance, of which there are 3 levels.

| UN Packaging Group | Meaning |
|---|---|
| PG1 | HIGH danger |
| PG2 | MEDIUM danger |
| PG3 | LOW danger |

The packing group plays a wide role – from the determination of acceptable packaging and the definition of threshold limits, to stowage and segregation requirements. The packing group system does not apply to Class 1, 2, 5.2, 6.2 or 7, as different specification criteria and special packaging requirements apply.

## UK classification procedures

**D0416** Classification in the UK is based on the UN system of nine classes, the procedures for which are detailed in ADR. The ADR Approved List contains a large range of substances that have been previously classified and assigned packaging groups.

ADR allocates a substance that does not meet the criteria of any other hazard class, but is regarded as an 'aquatic pollutant', to Class 9 as an 'environmentally hazardous substance'. Solutions and mixtures with a concentration of 25 per cent or more of such a substance, should be also be classified as such.

## ADR classification procedures

**D0417** These largely conform to the UN system, including the hazard class system and the use of packing groups.

The ADR classification system differs from the former UK system in a few ways, as follows:

- Class 2 (gases) – the former separate divisions 2.1, 2.2 and 2.3 no longer exist; gases are split into nine groups according to their hazardous properties as follows:

    A   Asphyxiant
    O   Oxidising
    F   Flammable
    T   Toxic
    TF  Toxic and flammable
    TC  Toxic and corrosive
    TO  Toxic and oxidising

TFC Toxic, flammable and corrosive

TOC Toxic, oxidising and corrosive

- Class 3 (flammable liquids) – includes gas oil, diesel fume or heating oil (light), with a flashpoint in excess of 61°C.

The classification and packing group of a substance can be determined by looking up the name or UN number in the Approved List or alphabetically in Table B of chapter 3.2.

## Identification of dangerous goods
### Basic UN principles

**D0418**   In an emergency the emergency services, vehicle drivers, other transport staff and the general public need to know what dangers they face from certain substances. This creates a need for a product reference that can be quickly interpreted for an effective response.

The UN have devised 2 such references:

- Proper Shipping Names ('PSNs') – These are recommended names for a wide range of commonly moved substances, listed in the UN Orange Book. However, they are often complex chemical names and the need for the identification to be achieved on a worldwide basis led to the development of a second system.

- UN Numbers – These four digit identification numbers are directly linked to the PSN.

### UK/ADR identification

**D0419**   UK regulations are in line with the UN Recommendations and require the use of the same PSN and UN Numbers.

For consignment and transport, the PSN and corresponding UN number can be found in the ADR Approved List.

The Approved List is ordered by UN Number and may be cross referenced using the PSN.

A solution or mixture that contains a dangerous substance and one or more non-dangerous substances is identified using the PSN for the listed one, with the addition of the word 'solution' or 'mixture' e.g. methanol solution.

This does not apply where:

- The solution or mixture is identified in the Approved List.
- The entry in the Approved List only applies to the pure substance.
- The class, physical state or packing group is different to that of the pure substance.
- There is a significant change in the measures to be taken in an emergency.

If this occurs, then the solution or mixture is identified under the appropriate generic or not otherwise specified (n.o.s.) entry in the Approved List. The index contains a range of generic names that cover substances belonging to particular families (e.g. Alcohols n.o.s. and Ethers n.o.s.) and also involves more descriptive names that identify their danger (e.g. Flammable n.o.s. and Corrosive n.o.s.).

## Dangerous goods containment
### Principles of containment

**D0420**   All packages used for the transport of dangerous goods have to meet three criteria. The selected package must:

- meet the general requirements established by the regulation;
- be authorised to UN specification standards;
- meet the detailed directions within the applicable regulation for the packaging of the product.

All packages that meet these criteria are available for use to carry the product in the particular transport system.

This means that for most substances and modes of transport a wide range of package types and materials are available. However, in the worst-case scenario only one system of containment will be authorised for transport use.

UN specification sets the standards for packages used in the transportation of dangerous substances, in the UK and by ADR or RID. The UN system requires packaging design and materials to have proven their competence by passing practical tests. Firstly the packs are weathered to their most vulnerable condition, if this is likely to be a factor in their performance. They are then subjected to a number of physical tests such as being dropped, held in a stack and placed under pressure demands to simulate transport situations. Approved packages are given markings that must be clear, legible and readily visible.

Each of the regulatory systems has a basic set of standards that must be met by every package ever used. The requirements are wide ranging and largely overshadowed by the UN specifications, but there are some duties not repeated elsewhere such as the compatibility of the product and packaging material, the provision of ullage and the inspection of all packs before transport. Each system also carries a section detailing the significance of the packaging code, for example:

UN 1A1/X/550/095/GB/0103

UN = UN packaging symbol

1A1 = Type of packaging and material

X = Packing Group competence and maximum density of liquids

550 = Hydraulic pressure test result (kPa)

95 = Year of manufacture

GB = Authorising country

0103 = Packaging ID number

Figure 3: An example of a UN mark as it may appear on a fibreboard (cardboard) package

## UK/ADR selection

**D0421** Within the ADR Agreement, The General Packaging Conditions and the detail of UN specification are in chapter 4. Class specific provisions are detailed in Part 2 of ADR. The particular packaging requirements of each package type are identified in specific subchapters within Part 4.

Each packing instruction identified in the Approved List describes a range of packing options for substances depending upon their Packing Groups. These must be carefully examined in order to determine any volume or mass limits that may apply to a particular package type.

## Intermediate bulk container and tank systems

**D0422** Intermediate bulk containers ('IBCs') are designed for mechanical handling and have a general maximum capacity of $3m^3$ (3,000 litres). The selection procedures are similar to those for packages:

- they must meet a set of general requirements;
- be authorised to UN specification; and
- meet with the detailed directions in the applicable regulation for the product concerned.

IBCs that are acceptable packaging options in the ADR Agreement are listed in Volume II, chapter 4.1.

There are many types of tank used in the transport of dangerous goods, but they all are all built to tightly controlled criteria. A range of descriptions are used to identify the different types, unfortunately, terminology is not consistent across the various transport modes. In each modal control, definitions will be found indicating what kind of equipment is being referred to by any particular name.

The various controls offer two avenues of control of tank traffic:

- Identification of how tanks must be built and equipped to meet the demands of the regulation.
- Description of the type tank and ancillary equipment that must be used for the transport of a particular substance.

Chapter 4.2 to 4.5 of ADR contains constructional and equipment details for the various types of ADR approved tanks, which detail requirements for portable tanks, fixed tanks, demountable tanks, tank containers, tank swap bodies, reinforced plastic tanks, welded tanks and vacuum waste tanks.

## Marking, labelling and placarding

### Basic principles

**D0423** To ensure that everyone involved in handling or moving dangerous goods are aware of the hazards they present, information warning of the dangers are required to be clearly displayed on the outside of packages, tanks, vehicles, etc. This is also vital for rapid and effective action from the emergency response teams faced with dangerous goods incidents.

Terminology is not yet fully harmonised and a distinction is drawn between packages and transport units:

- *Package*: drums, jerricans, boxes, gas cylinders, IBCs, etc.
- *Transport unit*: road freight or road tanker vehicles, rail wagons or rail tanks, multi modal freight containers and tank containers.

Modal control requirements are all based on the UN Recommendations, the Orange Book, which identify the need for:

- Marking and labelling on packages.
- Placarding and marking of transport units.

### Marking and labelling

**D0424**   *Marking* on packages is the application of the complete Proper Shipping Name ('PSN') and UN Number, in a readily visible and legible location on the outside of all packages.

Figure 4

---

**Shipping Label for Fisher Scientific**                                S28555/   2

**DICHLOROMETHANE**

**UN 1593**

**Technical Name:**

**Fisher Item:**        D/1856/17
**2.5L DICHLOROMETHANE HPLC**

**Supplied by:**
Fisher Scientific UK Ltd., Bishop Meadow Road
Loughborough, Leics, LE11 5RG

| Chemical Class | – | 6.1 | Poisonous |
| Sub Class 1    | – |
| Sub Class 2    | – |

---

*Labelling* is the application of diamond shaped warning labels representing both the class of danger and any subsidiary risks. Class labels are distinguished by the appropriate class number in the bottom corner, with the upper half containing a pictorial symbol representing the danger. Text indicating the nature of the hazard e.g. 'Corrosive' may also be shown in the lower half – this is mandatory for Class 7, radioactive materials. Division numbers must also be shown on Class 1 and Class 5 labels.

The standard size for such labels is 100mm x 100mm, with a border line of 5mm. Symbols, text and numbers must be in black, except those containing the Class 8, corrosive, number and any associated text, which must be in white.

Limited quantity concessions of dangerous goods are not subject to the standard marking and labelling duties. The different modal regulations for these are not yet aligned and therefore the application of these varies depending on the type of journey.

## Placarding and marking

**D0425** Transport units must bear the appropriate placards and markings to warn of the dangers of the goods being transported. *Placards* are large versions of the warning labels required on packages, the size for which must be at least 250mm x 2,500mm with a border line of 12.5mm.

For certain journeys specific *markings* are required, the type, number and location of these vary depending on the transport mode and the nature of the journey – national or international. For example, the relevant UN number must be displayed on the outside of tank transport units. An elevated temperature mark is required for those units transporting liquids at temperatures of 100°C or more, or solids at 250°C or more. An emergency action code ('EAC'), which provides guidance for the emergency services in the event of an incident involving the load, along with other markings may also be needed for tank or bulk load transport, depending on the type of journey.

## UK and ADR requirements

**D0426** Every package must be marked with the PSN, UN Number and class risk label. The marks and labels required for certain products may be established through the relevant substance entries in the first six columns of the Approved List. Special packing provisions apply variations to the standard marking and labelling, these must be checked before determining the requirements of the package.

The ADR requirements for marking and labelling are the same ones applicable to rail (RID) journeys. These requirements are stated in chapter 5.2 'Marking and Danger Labels on Packages', in each hazard class section in Volume I. It requires every package to be marked with the PSN and labelled with the appropriate class and subsidiary risk labels. Labels are referred to by the label model number (e.g. label model no.3 is the label required for flammable liquids of Class 3) and are described in chapter 5.2.

## Documentation requirements

**D0427** Documentation relating to a dangerous goods shipment describes important aspects of the cargo and its containment as well as indicating its point of origin and eventual destination. This information is required to ensure that all individuals involved in the transport process are fully aware of the dangers of the material that they are handling and employ suitable equipment and procedures to minimise any risks involved. The documentation also provides vital information in a concise form to emergency personnel. Description of packages etc should be as clear as possible to enable rapid identification, and substance descriptions must conform exactly to the requirements of the regulations in order to minimise misunderstanding and errors during an emergency.

## UK transport documentation

**D0428** With the exception of goods carried under limited quantities exemptions, the consignor must provide each shipment with documentation detailing the designation of the goods (the proper shipping name, along with the designation 'waste' if applicable) and the class of each item of goods preceded by the word 'Class' or the classification. The UN Number, preceded by the letters 'UN' and the transport category must also be stated. The mass/volume of each package along with the number of packages, or the total mass/volume of each transport category along with the consignor and consignee names and addresses is also required. The consignor must also provide any other information necessary for the vehicle operator to comply with their transport documentation requirements. This information along with the emergency action code, the prescribed temperature and a container packing certificate, will provide emergency information to the driver and is referred to as 'instructions in writing'. This is most commonly done by the provision of a Tremcard. In addition, transport documentation for explosives must detail the total number of packages carried, the total *net* mass of explosives carried and the name and address of the vehicle operator.

Under the provisions of the *CDGTPE Regulations* (*SI 2004 No 568*), the driver of the vehicle must ensure that any documentation relevant to the load being carried is kept readily available on the vehicle, and must produce the documentation on request to any authorised person. The driver must also ensure that any documentation relating to dangerous goods *not* being carried is removed from the vehicle or is secured in a closed container and clearly marked to show that such information inside does not relate to the load currently on board. Once a journey is complete, the vehicle operator must maintain copies of transport documentation (but not emergency information) for three months.

## ADR documentation

**D0429** The ADR Agreement imposes specific duties relating to the provision and carriage of documentation when goods are moved by roads on international journeys. Similar to UK requirements the consignor, vehicle operator and driver all have duties under the provisions of ADR.

The consignor has a duty to provide a transport document (information on the goods consigned), a declaration of compliance with ADR, written emergency instructions (actions to be implemented in the event of an accident) and a container packing certificate (only needed if the goods are loaded onto a freight container exceeding $3m^3$ capacity) to the vehicle operator.

It is the vehicle operator's responsibility to ensure that the appropriate documents are carried. These could include:

- the transport document and declaration of compliance;
- a container packing certificate, a vehicle approval certificate;
- the driver training certificate;
- the written emergency instructions; and
- a journey authorisation permit.

It is the driver's duty to ensure that the relevant documents are kept readily available and that any irrelevant documents are kept separate and clearly identified to avoid confusion.

## Compliance thresholds/limited quantities

### Principles of thresholds

**D0430** Often dangerous goods are transported in such small receptacles that there is no significant risk of them causing a serious incident. Such items are subject to either simplified dangerous goods controls, or in certain circumstances are exempt from the scope of the regulations. This means that even if the substance is still dangerous, the quantity is so small that the package presents no particular danger for transport and is not considered to be a regulated package.

D0430  *Dangerous Goods Systems*

The definition of what is deemed to be a small receptacle varies greatly from class to class and packing group to packing group. Careful attention to detail is required to enable consignors to take advantage of limited quantity exemptions. Limited quantity concessions are available for shipments placed in combination packs, and there is usually a maximum weight specified for the package as well as a maximum size for the receptacles within the package.

Figure 5

Limited quantity – concessionary package. Full scope of Regulations does not apply

Not limited quantity – regulated load. Full scope of Regulations applies

The transport category of the dangerous goods determines the receptacle thresholds for limited quantity concessions. The transport categories are allocated on the basis of packing groups and class specific allocation in accordance with ADR 1.1.3.6 and entries for each specific material listed in column 15 of the Approved List:

| Transport category | | Maximum mass/ volume threshold per transport unit |
|---|---|---|
| 0 | Specific materials listed in Table 1.1.3.6.3 of ADR and in the Approved List | 0 |

| 1 | Packing Group I goods (unless otherwise specified) Specific materials listed in Table 1.1.3.6.3 of ADR and in the Approved List | 20 litre/kilogram |
|---|---|---|
| 2 | Packing Group II goods (unless otherwise specified) Specific materials listed in Table 1.1.3.6.3 of ADR and in the Approved List | 333 litre/kilogram |
| 3 | Packing Group III goods (unless otherwise specified) Specific materials listed in Table 1.1.3.6.3 of ADR and in the Approved List and any other dangerous goods not listed in other categories. | 1000 litre/kilogram |
| 4 | Specific materials listed in Table 1.1.3.6.3 of ADR and in the Approved List. These materials are effectively exempt from ADR controls | Unlimited |

## Exemptions and concessions

**D0431** Within ADR there are further concessions that confer exemption from the associated duties. To qualify for exemption, which extends to cover both the carrier of the goods as well as the consignor, both the receptacle and the package size maximum limits must be complied with. Details of such concessions are listed in column 7 of the Approved List. However, there are no concessions for Classes 1, 4.2, 6.2 or 7.

There are certain marking requirements applicable to any package that has limited quantity status: a package containing a single dangerous substance must display the appropriate PSN, preceded by the letters 'UN', a package containing dangerous substances with different PSNs must display the PSN for each of the substances, or the letter 'LQ'. Packages consigned as limited quantity and suitably marked are not subject to any other ADR duties.

## Training duties

**D0432**

- *Classification principles*

    If appropriate safety standards are to be established and maintained it is essential that adequate training be provided to all personnel involved in the consignment and transport of dangerous goods. All such training must be commensurate with individual responsibility and is a requirement of the UN Recommendations.

- *Regulatory training requirements*

    Formal training for drivers of road vehicles carrying dangerous goods is a requirement of both ADR and the *CDGTPE Regulations* (*SI 2004 No 568*). In addition there is a requirement on employers to provide adequate training for all employees loading, unloading and handling dangerous goods, and records to be kept by both the employer and employee training records must be kept for a minimum of 5 years.

# Specific requirements

## Vehicle requirements

**D0433** The vehicle and its ancillary equipment is an integral part of the containment system, which must be effective in order to allow the goods to be carried in a safe manner.

## UK and ADR vehicle and equipment requirements

**D0434** A range of requirements relating to the vehicle and its load carrying competence are laid down in the ADR Agreement (Annex B 'Provisions Concerning Transport Equipment and Transport Operations'). The information is presented at two levels: a set of duties that underpin all decision making on vehicle provision, equipment and use; and another set of directions related to each danger class and in some cases to specific item numbers within a class.

Annex B contains a number of general requirements regarding the physical competence of the vehicle. Chapter 8.5 deals with any extra provisions or variations particular to a class, or a certain substance within a class. The degree of supplementary information varies between classes and covers a range of issues. For example, Class 1 has the greatest level of detail, ranging from vehicle construction requirements, to bodywork and exhaust systems, where as Class 4 provisions cover aspects such as temperature control and the use of closed or sheeted vehicles. Chapters 8.5 and 9 should therefore always be checked to confirm any additional requirements.

The *CDGTPE Regulations* (*SI 2004 No 568*) require a check to be carried out on the vehicle, its equipment and the driver's documentation before loading the vehicle, to ensure that the journey can be undertaken in a compliant manner.

## Placarding and marking of different types of vehicle

**D0435** The operator and driver of the vehicle must ensure that any placards and markings are displayed as required during the journey, that they are kept in a clean and visible state and removed or covered when not in use.

## UK requirements

**D0436** Vehicle placarding and marking is the one major area where the new *CDGTPE Regulations* (*SI 2004 No 568*) differ substantially from ADR requirements. The former system has been retained as an authorised national concession by the UK and incorporated into the current legislation. For road journeys in the UK, combinations of placards and markings are required for different transport units, such as:

- Vehicles carrying packages (including IBCs) or freight containers loaded with packages.
- Road tanker vehicles (i.e. fixed tanks).
- Vehicles carrying tank containers.
- Vehicles carrying bulk loads.
- Vehicles carrying bulk loads in freight containers.

Some examples of UK vehicle labelling are given below in FIGURE 6.

*Placards* are the warning labels that represent the danger class of the load.

*Markings* consist of clearly visible rectangular orange-coloured plates, which can incorporate such items as the appropriate UN Number, the Emergency Action Code (EAC) and an emergency telephone number, where specialist advice may be obtained at any time during the carriage of the goods. Dangerous goods that are moved in tanks or as bulk loads may also display the appropriate class danger placard on a combined hazard-warning panel. Alternatively they can all be displayed separately according to the requirements under ADR.

*Regulation 55* and *Schedule 9* of *CGDTPE* (*SI 2004 No 568*) contain the requirements relating to the type, number and location of placards and markings for the transport of dangerous goods. The details for a particular load can be ascertained through the relevant entries in the Approved List, except for the EAC which may be found within the DGEACL. The EAC's effectively replace the ADR HIN codes for UK only journeys.

Figure 6

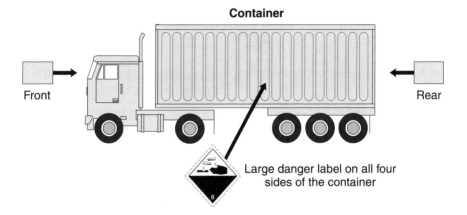

There are separate guidelines for those classes of dangerous substances not covered in *CDGTPE*. Those relevant to Class 7 items, are described in *Regulation 48* and *Schedule 6* of *RAMRoad* (*SI 2002 No 1093*).

## ADR requirements

**D0437** The requirements for ADR journeys and UK journeys are very closely aligned, in that all transport units must be placarded and marked.

Different placards and markings are required for the following transport units:

- Vehicles carrying packages (including IBCs).
- Vehicles carrying freight containers loaded with packages.
- Road tanker vehicles (i.e. fixed tanks).
- Vehicles carrying tank containers.
- Vehicles carrying a demountable tank(s).

- Vehicles carrying bulk loads.
- Vehicles carrying bulk loads in freight containers.
- Battery-vehicles (transport units with a manifold permanently attached to a frame, permanently fixed to the unit).

An example of ADR vehicle labelling is given below in FIGURE 7.

Figure 7

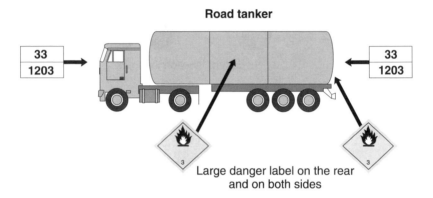

Large danger label on the rear and on both sides

*Placards* are the warning labels that represent the danger class of the load. In the ADR Agreement they are referred to as 'labels'.

*Markings* consist of clearly visible rectangular orange-coloured plates, which must be securely attached to the unit. If the goods are being transported in tanks or as bulk loads the markings must incorporate the appropriate SIN and hazard identification number ('HIN' or Kemler code), which identifies the dangers of the load being carried. This is explained in Appendix B.5, which also contains three tables for ascertaining the SIN and HIN details for particular loads and confirming the placarding requirements.

The nature, number and location of the placards and markings can be located in chapter 5.3, once the class, item number and danger letter of the substance being transported has been established.

## Regulated loads

**D0438** There are some concessions that ease the transport controls for vehicles that are carrying only small quantities of dangerous substances. These Limited Quantity ('LQ') exemptions and their conditions of applicability are detailed by reference to column 7 in the Approved List.

### Principles of concessions

**D0439** Some requirements of the *CDG Road Regulations* (*SI 1996 No 2095*) apply when any quantity of dangerous goods are being carried, e.g. to adequately segregate any toxic or infectious substances from food. The application of most requirements depends on whether the goods are being transported in a tank/bulk or in packages. If it is being transported in a tank or in bulk then the provisions of *CDG Road* apply. The Approved List provides full details of such requirements.

*Dangerous Goods Systems* **D0444**

## Filling, loading and stowage

### Filling and loading requirements

**D0440** The *CDGTPE Regulations* (*SI 2004 No 568*) contain duties to ensure the safe loading and unloading of vehicles.

A duty on all persons, to ensure that the manner in which goods are loaded, stowed or unloaded is not liable ' to create a significant risk or significantly increase any existing risk to the health and safety of any person'. This duty gives provisions for particular classes of dangerous goods, for example, where flammable goods are being transported by tank, the chassis must be properly earthed and flow rates controlled to prevent static discharges. It also provides more general health and safety requirements, such as prohibiting smoking in the vicinity of vehicles whilst loading and unloading and also cleaning tanks and freight containers after use.

ADR contains a number of provisions to ensure the safety of loading and unloading operations – which are in line. Class specific provisions can be found in chapter 2.

### Stowage requirements

**D0441** The *CDGTPE Regulations* (*SI 2004 No 568*) provide particular stowage requirements for certain dangerous goods. These requirements are detailed in chapter 7 of ADR and the subsections provide information on the various classes of dangerous goods, e.g. flammable solids must be loaded in a fashion that ensures free air circulation and the maintenance of a uniform temperature. A general duty to ensure that all goods are suitably stowed and secured on vehicles is also established.

### Segregation requirements

**D0442** From time to time a road traffic accident may occur, a tank valve may fail, a fire may break out. These incidents could result in two substances reacting dangerously if they came into direct contact with each other. Segregation ensures that acceptable levels of safety are maintained in the transport system, in case one of these events does occur.

Dangerous goods must not be packed together in the same outer packaging with dangerous or other goods if there is a possibility that they will react dangerously together to cause: combustion and/or the evolution of considerable heat; the evolution of flammable, toxic or asphyxiant gases; or the formation of corrosive or unstable substances. Toxic or infectious substances must not be carried on the same vehicle as food, unless it can be 'effectively separated'. It is therefore left to the operator to assess whether the loading arrangements are satisfactory. The circumstances under which the mixed packing of substances in a particular class is permitted with either other items from the same class, or with products from other classes, is clearly laid out in chapter 5.1 of ADR.

### Operational procedures

**D0443** The vehicle crew (driver, assistant(s), and attendant(s)) is the primary channel of safety during road journeys involving the carriage of dangerous goods. The working procedures, duties, training and instruction of the crew are both a safeguard against the likelihood of an incident occurring involving their hazardous load and a primary means of minimising the consequences of any such incident.

### UK duties and equipment

**D0444** The *CDGTPE Regulations* (*SI 2004 No 568*) contain various duties, many of which must be carried out by the vehicle crew.

The driver must:

- Ensure that any relevant transport documentation is kept readily available.
- Ensure that any relevant placards and markings required for the journey are displayed and removed when not relevant.
- If the load exceeds a specified limit then the special supervision and parking requirements must be satisfied.
- Ensure that any segregation, stowage or load safety instructions are followed.
- Ensure that the crew complies with any emergency information relating to the goods in the event of an accident and also ensure that the appropriate emergency services are contacted.
- Ensure that steps are taken to minimise the risk of fire or explosion.
- Ensure that the engine is shut off during loading and unloading.

The vehicle transporting dangerous goods must be equipped with a safety kit appropriate to the measures detailed in the 'emergency information', e.g. where toxic gases are being carried, the crew must be provided with suitable respiratory equipment.

The duties relating to the vehicle crew and its equipment are contained within Annex B of ADR.

Vehicle equipment requirements on dangerous goods journeys are:

- At least one wheel scotch.
- Two standing warning signs.
- A suitable warning vest or clothing.
- One intrinsically safe handlamp for each crewmember.
- Any other general or load-specific safety kit listed on the written 'emergency instructions'.

The duties laid down for the vehicle crew are contained in chapter 7.5 'Special Provisions Concerning Loading, Unloading and Handling'. These duties state that:

- If the load exceeds a specified limit then the vehicle must be supervised at all times unless it is in a secure factory or depot site.
- Crew members are the only personnel permitted to be carried on dangerous goods journeys.
- Crew members must know how to operate their fire extinguishers.
- Any relevant transport documentation is kept readily available.

## Emergency response requirements

**D0445** One function of the Dangerous Goods Safety Advisor ('DGSA') is to monitor the implementation of proper emergency procedures in the event of any accident or incident that may affect the safety during the transport of dangerous goods.

The *CDGTPE Regulations* (*SI 2004 No 568*) require that emergency information, usually in the form of a Tremcard, 'shall comprise details of the measures to be taken by the driver in the event of an accident or an emergency and other safety information regarding the goods being carried ... '. This information covers areas such as inherent dangers, measures to be taken if someone comes in to contact with the substance, measures to deal with a spillage and measures to be taken in the event of a fire. Additional requirements for specific types of material are given in chapter 7.5.11 of ADR.

## Environmental protection

**D0446** The *CDGTPE Regulations* (*SI 2004 No 568*) do not define a requirement for the Dangerous Goods Safety Advisor to be involved in environmental pollution activities. However, they should be aware of the relevant regulations or documents that deal with environmental measures.

The written emergency instructions should contain information on the potential for environmental damage and guidance to the immediate protective measures. In addition, those substances with known environmental effects should be labelled accordingly.

## Road transport training

**D0447** When carrying dangerous goods, special training and certification requirements apply to many drivers. Drivers must hold a vocational training certificate ('VTC'), issued by a competent authority established by *Part 3* of *CDGTPE* (*SI 2004 No 568*). These can only be obtained by attending a government approved training course and passing the examinations. There are some exemptions to these requirements and they are listed in the limited quantities exemptions. The training requirements of the ADR are similar to those of the UK requirements, in that the driver must hold a VTC in align with a European driving license.

# Disaster and Emergency Management Systems (DEMS)

## Historical Development of Disaster and Emergency Management

**D6001** The events which took place in New York on 11 September 2001 have marked a paradigm shift in the thinking, planning and perception of 'man-made disasters'. Throughout the 1990s the Federal Emergency Management Agency (FEMA) of the USA (a Competent Authority dealing with natural and man-made disasters) was being criticised by Congress and the media for being too resource consuming and inefficient. After '911' it was being hailed for its expertise in emergency response management. Suddenly 'corporate America' and the democratic world began to take more seriously the importance of 'Disaster and Emergency Management', 'Civil Defence Management', 'Civil Protection Management' and 'Business Continuity'. In the United Kingdom since June 2001 the 'civil contingencies' function has been transferred from the Home Office to the Cabinet Office, giving the Prime Minister's Office a greater strategic oversight of national 'Major Incidents'.

The UK experience broadly indicates that Disaster and Emergency Management (DEM) has evolved in three phases:

Disaster and Emergency Management (DEM) has evolved in three phases:

- *Phase 1: pre-CIMAH 1984 (Control of Industrial Major Accident Hazards Regulations 1984 (SI 1984 No 1902))*

    DEM was largely confined to local government, the emergency services, large organisations and civil protection agencies (environmental, military etc.). This phase is marked by large scale macro plans and detailed, sequential procedures to be followed in the event of a 'Disaster' or 'Emergency'. It is exemplified in the guidance prevalent on 'incidents' involving radioactivity, available from central government, the United Nations, Red Cross and others.

- *Phase 2: Liberalisation Phase*

    With the advent of decentralisation of industry and the growth of privatisation, organisations (large or otherwise) began to focus on 'procedures' and 'Business Continuity Planning'. British Telecom exemplified this with the establishment of its Disaster Recovery Unit and the offering of its expertise to customers at large. 'Major Incidents' such as Chernobyl, Piper Alpha, Kings Cross, The Marchioness and Challenger to name a few, which hit the world headlines in the 1980s also focused planners minds on effective DEM not just isolated procedures.

- *Phase 3: Holistic Phase*

    This has two dimensions, first the role of Europe. The impact of the *Framework and Daughter Directives*, as transposed into the '*6 Pack Regulations*' of 1992, introduced for the first time explicit and strict duties on organisations in general to plan for 'serious and imminent dangers'. The European Commission also began to take the risks of transnational 'Disasters' more seriously, as highlighted by the greater role given to the Civil

Protection Unit, in DGXI of the European Commission. Second, this phase sees domestic UK legislation becoming more cognisant of Disaster and Emergency issues (*Environment Act 1995* makes provisions for environmental emergencies; *Control of Major Accident Hazards Regulations 1999* (*COMAH*) (*SI 1999 No 743*) replaces *CIMAH 1984*; also the need for effective planning when carrying dangerous goods via road legislation, *Carriage of Dangerous Goods by Road Regulations 1996* (*SI 1996 No 2095*)).

The most noticeable index of how DEM has changed is the availability of information and templates on 'Disaster planning' and 'emergency planning' to organisations of all sizes. There is also a greater media focus on such issues, with the BBC for instance having dedicated web resources on 'Disasters'.

## Origin of Disaster and Emergency Management as a modern discipline

D6002 DEM is largely a fusion of four branches of knowledge:

- *Occupational Safety and Health (OSH)*

    This concerns 'internal' or work-based causes of systems failures, with major impact on life and property. It has been exemplified by Heinrich in his publication *Unsafe Acts and Unsafe Conditions* in the early part of the 1900s, Bird and Loftus with their 'management failing' explanations of accidents and incidents and Turner with his 'incubation' explanation of Man-Made Disasters in the 1970s. OSH academics have led the forefront in terms of analysing the branch and root causes of major industrial disasters.

- *Security Management*

    In the 1970s, security threats in the UK such as bomb explosions, terrorism and electronic surveillance failures have given insights to causation and the motivation behind man-made emergencies. Also, guidance from the Home Office as well as the emergency services has enabled a practical understanding of how to cope with emergency situations.

- *Business Management*

    The late 1980s saw a shift in the academic paradigm in economics and business management from 'static' or closed business planning – where businesses were told to make the assumption of *cetaris paribus*, that is assume all things are constant with the business acting as if it was the only one in the market place – to 'dynamic' or open planning. The latter sees uncertainty and risk being factored into decision making models. This influenced the development of 'business continuity management', developing strategies when the business faces major corporate uncertainty and crises as well as 'contingency planning'.

- *Insurance*

    The fourth significant influence comes from insurance and loss control. The occurrence of Disasters and Accidents has involved loss adjusters and actuarial personnel. The former have developed methods of analysing the basic and underlying causes of an event, whilst the latter have been developing statistical methods for calculating the chance of failure and the risk premiums needed to indemnify that failure.

    DEM uses both quantitative (statistics, quantified risk assessments, hazard analysis techniques, questionnaires, computer simulation etc.) and qualitative methods (inspections, audits, case studies etc.).

## Definitions

D6003 Stan Kaplan (*The Words of Risk Analysis*, 1997, Risk Analysis, Vol 17, No 4) stated that 50 per cent of the problems with communication are due to individuals using the same words with different meanings. The remaining 50 per cent are due to individuals using different words with the same meaning. In DEM this is a basic problem. Some authors have argued distinguishing definitions has no practical value and can be ' ... highly undesirable to try to control how others use them ...' (*Managing the Global Consequences of a Disaster,* Richard Read, Paper at the 2<sup>nd</sup> International Disaster and Emergency Readiness Conference, The Hague, October 1999, page 130 of IDER Papers). Nevertheless there is much confusion over the meaning given to core terms, so that basic definitions can assist in avoiding the Kaplan dilemma. Legislation or approved codes have not provided definitions of 'Disaster' or 'Emergency' for instance. *The New Oxford Dictionary of English*, OUP, (1999) provides the following primary meanings:

(a) Catastrophe: 'an event causing great and often sudden damage or suffering: a disaster', page 287.

(b) Crisis: 'a time of intense difficulty or danger', page 435.

(c) Disaster: 'a sudden event, such as an accident or a natural catastrophe, that causes great damage or loss of life', page 524.

A catastrophe and a crisis are types of Disaster, namely more severe.

(d) Emergency: ' a serious, unexpected and often dangerous situation requiring immediate action', page 603.

Whilst both Disasters and Emergencies can be sudden, the former has a macro, large scale impact whilst the latter requires an immediate response.

(e) Accident: '1. an unfortunate incident that happens unexpectedly and unintentionally, typically resulting in damage or injury,...2. an event that happens by chance or that is without apparent or deliberate cause', page 10.

It should be noted that the *COMAH Regulations (SI 1999 No 743)* have introduced the term 'Major Accident'; this is an event due to:

(i) unexpected, sudden, unplanned developments in the course of the operation;

(ii) leads to serious danger to people and the environment both on site at the place of work and off site;

(iii) the event involves at least one dangerous substance as defined by *COMAH*.

Given the potential for both human and property loss, and given the *COMAH* requirements for emergency On-Site and Off-site Plans, there seems to be little practical difference between a Major Accident and an Emergency. Both are *response based* concepts. However, if the legislators wished a Major Accident to be different from an Emergency then they would have either said so or implied so. One can regard a Major Accident as a type of Emergency situation.

(f) Incident: ' an event or occurrence', page 923.

The *Reporting of Injuries, Diseases and Dangerous Occurrences Regulations 1995 (RIDDOR) (SI 1995 No 3163)*, do not define an accident or incident, but gives a classification of the types.

(g) Major Incident

In *Dealing with Disaster* (Home Office, Third Edition currently being revised), a 'Major Incident' is the only term explicitly defined by the Home Office: A 'Major Incident' is any Emergency that requires the implementation of special arrangements by one or more of the emergency services, the NHS or the local authority for:

(i) the initial treatment, rescue and transport of a large number of casualties;

(ii) the involvement either directly or indirectly of large numbers of people;

(iii) the handling of a large number of enquiries likely to be generated both from the public and the news media, usually to the police;

(iv) the need for the large scale combined resources of two or more of the emergency services;

(v) the mobilisation and organisation of the emergency services and supporting organisations, e.g. local authority, to cater for the threat of death, serious injury or homelessness to a large number of people, page 43.

This definition is accepted by the police, fire service, local government and broadly the NHS (they also have a specific definition of Major Incident).

Thus the term 'Major Incident' is a broad phrase encompassing an array of events. 'Accident' has not been included as a type of Major Incident given the definition of the latter requiring 'major' mobilisation of human and physical resources which in most Accidents is not necessarily the case.

As an example:

(1) two trains missing each other would be an 'incident' (near miss);

(2) an employee or member of the public being injured on a train – this would be an 'Accident', irrespective of the type of injury or fatality (following *RIDDOR*);

(3) an event at a *COMAH* site where dangerous substances ignite causing damage to the plant, injury to personnel and emissions into the local community, would be a 'Major Accident' under *COMAH*;

(4) the immediate event after the collision and the response needed to the chaos – this would be an 'Emergency'. For example, the response by the emergency services to Ladbroke Grove;

(5) if two trains collide causing multiple fatalities and immediate property and environmental damage, that would be referred to as a 'Disaster'. For example Ladbroke Grove Rail Crash in 1999;

(6) if the collision, with the multiple fatalities and property damage is difficult to access, manage and control, this would be a 'crisis'. Ladbroke Grove fell short from being a 'crisis' in contrast to the Clapham Junction in 1988 where a triple train crash caused major access and logistical problems;

(7) if the event generated major environmental, public and social harm that has 'longer term' implications, over and above the immediate human and property loss, this would be a 'catastrophe' For example, Kings Cross Underground Fire (1987), Chernobyl (1986), Piper Alpha (1988) to name a few that had wider consequences over and above the immediate impact.

Events (1)–(7) would be Major Incidents.

Figure 1 summarises the essential differences between the above events.

**Figure 1: Classification of Incidents, Accidents and Major Incident types**

| | | | | 'MAJOR INCIDENTS' | | | | |
|---|---|---|---|---|---|---|---|---|
| | Event: | Incident | Accident | Major Accident | Emergency | Disaster | Crisis | Catastrophe |

| Characteristic: | | | | Types of emergency | | Types of disaster | | |
|---|---|---|---|---|---|---|---|---|
| 1. MPL | | Low | | | | | | Very High |
| 2. RISK: | | | | | | | | |
| | Severity | Near miss etc | | | | | | Multiple Fatalities |
| | Consequence | Minor | | | | | | Major |
| 3. NUMBERS | | 1–5 | | | | | | 100 + |
| 4. SOCIO-LEGAL IMPACT | | No change likely | | | | | | New Laws or Guidance |
| 5. TIME | | Short | | | | | | Longer Impact |
| 6. COST: | | | | | | | | |
| | Individual | Short Term | | | | | | Irreparable |
| | Social | None Usually | | | | | | Irreparable |
| | Environment | None Usually | | | | | | Irreparable |

Key:

MPL= Maximum Potential Loss (economic and property loss)

Risk= Severity x Consequence (Severity refers to the quantum of harm generated by the event whilst Consequence measures the scale of impact)

Numbers= The number of individuals affected

Socio-Legal Impact= The impact on social attitudes and legislation/guidance as a result of the event

Time= The length of the event

Cost= The loss suffered by the individual, society or nature

Such classifications are important, firstly from a philosophical perspective one needs to know how they differ, and secondly from a planning perspective as the resource allocation will accordingly differ. Thirdly, from a response perspective, the response to an incident differs from a Disaster, with the organisation needing to define and clarify when an event is an incident and not a Disaster.

# The Need for Effective Disaster and Emergency Management

**D6004** There are several reasons why DEM is needed: legal reasons (see D6005–D6022); insurance reasons (see D6023); corporate reasons (see D6024); societal reasons (see D6025); environmental reasons (see D6026); and humanitarian reasons (see D6027).

## Legal reasons

**D6005** There is an array of specific emergency-related legislation – much of which is currently under review, which may culminate in an Emergency Planning Bill. Existing legislation includes the following.

## The Civil Defence Act 1948

**D6006** This Act requires the UK Government to make arrangements to deal effectively to protect the population if and when a hostile attack takes place by a 'foreign power'. The first 'modern' *Civil Defence Act* was enacted in 1937, in response to threats from nationalists groups around the world during the era of the Empire. Under *s 1* of the *Civil Defence Act 1948*, a strict duty exists such that the responsible Minister considers the resources needed to protect civil society. Such resources include:

- training, equipment, management of suitable civil defence forces;
- providing the emergency services with training in civil defence;
- ensuring that suitable persons from the public can be trained to assist the civil defence forces/emergency services;
- ensuring that adequate commodities and rations are available if a civil defence situation arose;
- allowing alterations to be made to buildings or indeed temporary shelter to be constructed at short notice.

The Act is primarily a strategic control document such that Government has identified the source of the threat from a foreign power and has competent persons and suitable resources to protect its population. In addition the Act, highlights the links between national and local government and other competent authorities, the need for strategic planning, the importance of regular communication. The Act also provides a definition of 'civil defence' (in the broadest sense, protecting the public and the homeland).

## The Civil Defence (Grant) Regulations 1953 (SI 1953 No 1777)

**D6007** These Regulations were made under the *Civil Defence Act 1948, s 3*, and allow for grants to be paid to local authorities to assist with the costs of civil defence/war planning. The Regulations specify what are acceptable and legitimate costs that central Government will reimburse to a local authority when 'exceptional' events arise. Also a 'police authority' and/or 'fire and ambulance service' can make claims under the Regulations where specific and measurable resources are deployed for civil defence (which ordinarily would not have been).

It should be noted that these Regulations in particular have provided the impetus for local government in England and Wales to establish 'emergency planning units'. These units coordinate, control and oversee the effective preparation of Disaster Management and Emergency Response in a defined area. Their remit can range from natural catastrophe (e.g. floods) through to a population being exposed to chemical biological nuclear and radiological (CBNR) risks. The units' resources can be audited and expenses/costs/losses incurred measured.

## The Local Government Act 1972, s 138

**D6008** Section 138 of the *Local Government Act 1972* allows local authorities to make grants and loans available to those affected by Disasters and Emergencies. This could be to necessary business services, infrastructure organisations (electric, water etc) or it could be the public, again under 'exceptional conditions'.

## The Civil Protection in Peacetime Act 1986

**D6009** The *Civil Protection in Peacetime Act 1986* is 'An Act to enable local authorities to use their civil defence resources in connection with Emergencies and Disasters unconnected with any form of hostile attack by a foreign power' (Preamble Note to the Act). The 1986 Act allows local government to use civil defence resources to mitigate the effects of any Disaster whether it is natural or man-made, whether 'accidental' or 'internally hostile'. It can be noted that the greater the perceived threat of attack, the more likely local government is to deploy resources. In the 1990s as the perceived threat from the Eastern European block was diminishing, the deployment of resources also was reduced. Since '911' the resource allocation by local government to the identification of persons, social groups and technologies that can cause civil harm has increased.

## The Emergency Powers Act 1920

**D6010** For a 'state of emergency' to be proclaimed under the *Emergency Powers Act 1920* (as amended by the *Emergency Powers Act 1964*) there must be (or about to be) some interference in the 'essentials of life' e.g. food, water, fuel, light or transport. The proclamation of a state of emergency has a 'shelf life' of one month, although additional proclamations can be issued during or by the month end. It is important to note that Parliament must be notified by the Secretary of State. Regulations can be made under 'Order of Council' when a state of emergency exists. These Orders cannot impose compulsory military service or industrial conscription and neither can they prohibit the right to strike by recognised trade unions. The Regulations must be approved by both Houses of Parliament.

## The Civil Defence (General Local Authority Functions) Regulations 1993 (SI 1993 No 1812)

**D6011** These Regulations became law in August 1993. Local government is not under a duty to provide 'protected emergency centres' for war time preparation and planning. It was a requirement that county and regional councils have a Main and Standby Centre, whilst, district councils provide one Emergency Centre.

However, county and regional councils are required to:

- Create, assess, review, update and exercise Plans.
- Ensure county and district council staff/other relevant persons are trained.
- Action the Plans created (not just a 'paper exercise').
- To work with and consult other neighbouring authorities.

District councils are required to:

- Ensure that relevant information is given to the county council.
- Provide the necessary support in the making/revising of Plans.
- Assist where necessary in preparations.
- Assist in implementation of new Plans.
- To ensure staff are adequately trained.

Unitary authorities are also covered and have similar functions to county and regional councils.

It should be noted that Scotland, Wales and Northern Ireland have additional and specific legislation e.g. the *Civil Defence (Scotland) Regulations 2001 (SI 2001 No 139)*.

## The Health and Safety at Work etc Act 1974

D6012 The intentions of the *Health and Safety at Work etc Act 1974 (HSWA 1974)* are captured by s *1(1)(a)–(d)* which states that the Act is concerned with 'securing the health, safety and welfare of persons at work', protecting the public, effective control over dangerous substances/explosives and the effective control of emission of noxious substances into the atmosphere. Implicit at least is the intention that this Act will influence actions that contribute to Disaster and Emergency situations, whether industrial, chemical or environmental. The Act does give expressed powers to the Health and Safety Commission to investigate and to hold inquiries in relation to '… any accident, occurrence, situation …'. [*HSWA 1974, s 14(1)*].

## The Management of Health and Safety at Work Regulations 1999 (MHSWR) (SI 1999 No 3242)

D6013 These Regulations provide the main detail applicable to all organisations, for providing 'procedures for serious and imminent danger and for danger areas' [*MHSWR (SI 1999 No 3242), Reg 8*] and 'contacts with external services'. [*MHSWR (SI 1999 No 3242), Reg 9*]. In relation to industrial and chemical based incidents, Accidents and Major Incidents, the Regulations require the existence and implementation of safety procedures (internal) as well as links with emergency and para-emergency services (external). This contrasts with the 1992 and 1994 amended version of the same Regulations. *Regulation 1* (Citation, commencement and interpretation) does not define 'procedure for serious and imminent danger', although the ACOP L21 to the *MHSWR 1999 (SI 1999 No 3242)* exemplifies such situations, 'e.g. a fire, or for the police and emergency services an outbreak of public disorder' (page 20).

*Regulation 8(1)* says that a strict duty exists on all employers to:

(a) 'establish and where necessary give effect to appropriate procedures to be followed in the event of serious and imminent danger to persons at work in his undertaking'.

(b) 'nominate a sufficient number of competent persons to implement those procedures in so far as they relate to the evacuation from premises of persons at work in his undertaking'.

(c) 'ensure that none of his employees has access to any area occupied by him to which it is necessary to restrict access on grounds of health and safety unless the employee concerned has received adequate health and safety instruction'.

*Regulation 8(2)* says 'Without prejudice to the generality of *paragraph (1)(a)*, the procedures referred to in that sub-paragraph shall:

(a) so far as is practicable, require any persons at work who are exposed to serious and imminent danger to be informed of the nature of the hazard and of the steps taken or to be taken to protect them from it;

(b) enable the persons concerned (if necessary by taking appropriate steps in the absence of guidance or instruction and in the light of their knowledge and the technical means at their disposal) to stop work and immediately proceed to a place of safety in the event of their being exposed to serious, imminent and unavoidable danger; and

(c) save in exceptional cases for reasons duly substantiated (which cases and reasons shall be specified in those procedures), require the persons concerned to be prevented from resuming work in any situation where there is still a serious and imminent danger'.

*Regulation 8(3)* says 'A person shall be regarded as competent for the purposes of paragraph (1)(b) where he has sufficient training and experience or knowledge and other qualities to enable him properly to implement the evacuation procedures referred to in that sub-paragraph'.

*Regulation 8* can be summarised:

- Procedures need to be sequential, logical, documented and clear (clarity of procedures).
- Procedures need to be justified and authorised (legitimise procedures).
- A 'hierarchy of procedural control' seems to be advocated by the Regulation i.e:
  — give information to those potentially affected by serious and imminent dangers;
  — take actions or steps to protect such people;
  — stop work activity if necessary to reduce danger;
  — prevent the resuming of work if necessary until the danger has been reduced or eliminated.
- Appoint competent persons preferably from within the organisation who will assist in any evacuations. It is implied that the role and responsibility of such persons needs to be clearly demarcated.
- Generally prohibit access to dangerous areas (site management). If access to a 'danger area' is required i.e. a place which has an unacceptable level of risk but must be accessed by the employee, then appropriate measures must be taken as specified by other legislation (see below).

In addition, the Regulations imply:

- Risk Assessments will identify foreseeable events that may need to be covered by procedures. Such assessments may also identify 'additional risks' that need additional procedures. Thus the Risk Assessment, as shall be discussed below, is a vital tool to keep procedures in tune with current generic and specific risks.
- Procedures need to be dynamic – reflecting the fact that events can occur suddenly.
- There may be a need to co-ordinate procedures where workplaces are shared.
- Procedures should also reflect other legislative requirements (see below).

*Regulation 9* is a new addition to the *MHSWR (SI 1999 No 3242)*. *Regulation 9* reads ' Every employer shall ensure that any necessary contacts with external services are arranged, particularly as regards first-aid, emergency medical care and rescue work'. It can be inferred that 'necessary contacts with external services' does not only relate to the emergency services but the organisation needs to identify both private sector and voluntary organisations that can be called on for assistance. The organisation needs to identify contact names, addresses and contact numbers of such bodies and develop relations with them.

## The Control of Major Accident Hazards Regulations 1999 (COMAH) (SI 1999 No 743) (see C10002 for full details of COMAH 1999).

**D6014**  In summary:

- The General Rule

    After the Seveso Disaster in 1976, the European Union (EU) initiated an EU Directive called 'Seveso I'. Whilst no human was killed at Seveso the immense impact geographically and commercially in a key agricultural region meant the Italian Government and the European Commission were keen to ensure adequate safety and emergency systems were installed at 'high risk' production processes. At Seveso, the plant was manufacturing

TriChloroPhenol (TCP) and due to the requirement to switch off batch production during the weekend, the operator had not distilled the chemical. He had assumed the temperature of the substance would drop to a tolerable level until he returned on Monday. However, latent heat transfer resulted in an exothermic runaway reaction causing the change in chemistry of the substance to a dioxin. Seveso is referred to as a boiling liquid evaporating vapour explosion (BLEVE).

When an EU Directive is introduced, Member States are obliged to transpose this instrument into domestic legislation within four years. In 1984 the UK implemented the EU Directive as the *Control of Major Incident Hazards Regulations* (*CIMAH*). Likewise other Member States implemented the Directive. *CIMAH* was more of a 'why' document i.e. why a petro-chemical or manufacturing process needs to prevent a Seveso type Disaster. After Chernobyl and Bhopal, the EU realised it needed a more forceful document that showed the 'what' and the 'how' to prevent major industrial accidents. This resulted in the Seveso II Directive, which the UK transposed as the *Control of Major Accident Hazards Regulations 1999* (*COMAH*) (*SI 1999 No 743*). *COMAH* replaced *CIMAH* and indeed *COMAH* has been implemented in most of the 15 EU Member States and is used as a 'benchmark' globally (e.g. by the Government's of Brazil, India, and the Caribbean States etc).

*COMAH* applies to those establishments that keep on site any substance that is specified in *Sch 1* of *COMAH* (*SI 1999 No 743*). For example, keeping a certain amount of ammonium nitrate or oxygen on site. However, there are 'Top-Tier' and 'Lower-Tier' Threshold levels. If an establishment keeps an amount greater than or equal to an Upper Threshold Level, they are called Top-Tier. For instance, if an establishment keeps at least 2,500 tonnes of ammonium nitrate, then it is Top-Tier. If an establishment keeps a lower quantity than this, (greater than or equal to a lower threshold) then it is a Lower-Tier establishment. For example, the Lower-Tier for ammonium nitrate is 350 tonnes. Therefore, knowing the amount of substance being stocked is critical to knowing what tier the establishment falls under.

It should be noted that even if the quantity of substance being held is below the Threshold Level, the establishment could be still subject to *COMAH* so long as the specified substance could be produced, assuming there is loss of control of the chemical process. This implies dangerous substances are produced through reaction with the establishment being unable to control the production of such amounts. This could happen if there is production failure or through unexpected chemical reactions.

- Top-Tier establishments:

    (i) Must produce both On-Site and Off-Site Emergency Plans. The former deals with risks that occur on the establishment and is produced by the operator. The latter deal with risks to the immediate social and physical environment due to the On-Site Emergency. It is prepared by the local authority. Both Plans should link together and be co-ordinated.

    (ii) Top-tier establishments that exceed Top-Tier Thresholds are required to provide information to the public of known and foreseeable risks, as well as the controls adopted. They would also have to provide a Safety Report, which shows the actions taken by the operator to prevent Accidents and mitigate harm to the environment. The establishment must be able to prove that they are operating the plant in accordance with the safety report.

    (iii) Operators of existing *CIMAH* sites which became Top-Tier establishments on 1 April 1999 are given until February 2001 to prepare On-Site Emergency Plans and supply information to the local authority. Whilst other establishments that became Top-Tier sites on 1 April 1999 (e.g. those now encompassed by *COMAH* but not *CIMAH* (*SI 1984 No 1902*)), have until 3 February 2002 to prepare the On-Site

Emergency Plan and supply the information to the local authority. However, for those establishments that became top-tier after 1 April 1999, they have a duty to prepare the On-Site Emergency Plan before starting any operation.

(iv) The local authority has six months to prepare an Off-Site Emergency Plan. The six months can start from the time they receive all necessary information to prepare the Plan. However, the period can be extended to nine months. It is recommended that arrangements are made nevertheless by the operator until the Off-Site Plan is available.

(v) Emergency Plans should be reviewed and revised at least every three years.

- Lower-Tier establishments:

    (i) Whilst under no duty to produce Emergency Plans, they are required however to develop Emergency Arrangements in the Major Accident Prevention Policy (MAPP).

    (ii) The MAPP should specify safety and emergency management response systems in place in the event of a Major Accident and the procedures for identifying foreseeable emergency situations.

- *COMAH* does not apply to Ministry of Defence establishments, transport related activity, extractive industries exploiting/exploring in mines and quarries, waste land-fill sites nor to nuclear licensed sites that may have substances which generate ionising radiation. However, *COMAH* will apply to both chemicals and explosives at nuclear installations.

- A breach of *COMAH Regulations* carries the same penalties as those under *ss 33 – 42* of the *Health and Safety at Work etc Act 1974*, for example £20,000 fine per offence on summary conviction and unlimited fine and/or up to two years imprisonment on indictment.

## The Reporting of Injuries, Diseases and Dangerous Occurrences Regulations 1995 (RIDDOR) (SI 1995 No 3163)

**D6015** *RIDDOR* (*SI 1995 No 3163*) will apply to 'Accidents' through to 'Major Incidents'. *RIDDOR* is a reporting requirement that must be followed by the employer or 'responsible person' in notifying the enforcing authority by the quickest means practicable, when there is an event resulting in a reportable injury, reportable occupational disease, reportable dangerous occurrence, gas incidents, road incidents involving work and fatal or non-fatal injuries for example. In addition, work injuries lasting for three days or more are reported to the enforcing authority. (see A3001 ACCIDENT REPORTING for further details).

## The Health and Safety (First-Aid) Regulations 1981 (SI 1981 No 917)

**D6016** These Regulations do not explicitly mention first-aid Arrangements necessary in the event of an Emergency or Disaster. However, they will apply in all types of Accidents and Major Incidents as defined above. The Regulations are a general statement of best practice, of ensuring the existence of adequate medical equipment, competent and trained first-aiders or appointed persons and information on first-aid facilities and equipment to staff/others. The Regulations should be read as best practice to be followed in the event of an Accident or Major Incident.

**D6017** *Disaster and Emergency Management Systems (DEMS)*

## The Construction (Health, Safety and Welfare) Regulations 1996 (SI 1996 No 1592)

**D6017** *Regulation 20* requires employers to make emergency arrangements to cope with foreseeable emergency situations. It is primarily aimed at CDM sites where the Health and Safety Plans are operational. The Plans should factor in emergency arrangements and procedures in the event of a Major Incident on site.

## The Confined Spaces Regulations 1997 (SI 1997 No 1713)

**D6018** *Regulation 5* imposes duties on employers and others to make arrangements for emergencies in 'confined spaces'. In summary:

- 'Suitable and sufficient arrangements' need to be made for rescue of persons working in confined spaces, with no person accessing a confined space until such arrangements have been developed. [*SI 1997 No 1713, Reg 5(1)*].

- So far as is reasonably practicable, the risks to persons required to put the arrangements into operation must also be considered as should resuscitation equipment in the event of it being required. [*SI 1997 No 1713, Reg 5(2)*].

- There is also a duty to act, to operationalise the arrangements when an emergency results. [*SI 1997 No 1713, Reg 5(3)*].

Question: What are suitable and sufficient arrangements for rescue and resuscitation? This should include appropriate equipment, rescue procedures, warning systems that an emergency exists, fire safety systems, first-aid, control of access and egress, liaison with the emergency services, training and competence of rescuers etc.

## The Carriage of Dangerous Goods by Road Regulations 1996 (SI 1996 No 2095)

**D6019** A duty is imposed on the operator of a 'container', 'vehicle' or 'tank' to provide information to other operators engaged/contracted, regarding the handling of emergencies or accident situations. Such information includes:

- hazardness of the goods being carried and controls needed to make safe such hazards;

- actions required if a person is exposed/makes contact with the goods being carried;

- actions required to avert a fire and the equipment that should or should not be used to fight the fire;

- the handling of a breakage/spillage; and

- any additional information that should be given that can assist the operators or others.

## The Carriage of Explosives by Road Regulations 1996 (SI 1996 No 2093)

**D6020** The *Carriage of Explosives by Road Regulations 1996* (*SI 1996 No 2093*), *Reg 26* says that the operator is under a duty to ensure that in the event of an Emergency or Accident, people are protected from ignition as well as the security of the explosives [*Reg 26(3)*]. The Regulation also requires the operator to inform the HSE by the quickest practicable means if the emergency cannot be immediately controlled. [*Reg 26(2)*].

Duties are also placed on the driver of the vehicle that in the event of an Emergency or Accident, the emergency services and the operator are notified by the quickest practicable means [*Reg 26(1)*]. The driver must also ensure that people are safe that could be affected by any ignition and that the explosives are made secure. [*Reg 26(3)*].

## The Environment Act 1995

**D6021** Whilst the Act does not explicitly deal with DEM issues, its main objective is the mitigation or prevention of such events. For example, ss 14–18 creates the flood defence committees to co-ordinate the prediction, consequence and response needed in the event of floods.

## 'Legislating the Criminal Code: Involuntary Manslaughter', Law Commission (Law Com No 237)

**D6022** Disasters such as Piper Alpha, Kings Cross, Clapham Junction, Herald of Free Enterprise or Ladbroke Grove for example have seen multiple fatalities, allegations of managerial failure and negligence contributing to the Disasters. However, health and safety law has not permitted the successful prosecution and imprisonment of directors and senior managers because of the difficulty in establishing that the individuals in charge were 'the embodiment of the company'. All of the major Disasters have involved large and complex organisations with many management layers, so who is the 'embodiment of the company' or the 'directing mind'? The Law Commission has drafted an *Involuntary Homocide Bill 1995*, which is still being considered by the UK Government. The main intention of the Bill is stated in the preamble, to ' Create new offences of reckless killing, killing by gross carelessness and corporate killing to replace the offence of manslaughter in cases where death is caused without the intention of causing death or serious injury'. The intention is to have a codified system of law as regards these three offences. Thus:

(a)   *Reckless killing*

   This offence would be committed if the answers are 'yes' or 'affirmative' to all the following:

   (i)   a person through their conduct causes the death of another;

   (ii)  a person is aware that a risk exists and that their conduct whether causing that risk or acting on that risk will cause death or serious injury; and

   (iii) it is unreasonable for the person to take a risk, knowing the circumstances prevalent.

(b)   *Killing by gross carelessness*

   This offence would arise if the answers are 'yes' or 'affirmative' to all the following:

   (i)   a person by their conduct causes the death of another;

   (ii)  a reasonable person in the position of the person under question realises that there is a risk that the person's conduct will cause death or serious injury;

   (iii) the person 'is capable of appreciating' the risk at that time in question; and

   (iv)  either the person's conduct 'falls far below' what could be reasonably expected under those circumstances or, 'he or she intends by his conduct to cause some injury, or is aware of, and unreasonably takes, the risk that it may do so, *and* the conduct causing (or intended to cause) the injury constitutes an offence' (page 127/128). The former focuses on a breach of a duty whilst the second of this 'either or' option seems to require the elements of a crime, the *'action'* (action of 'unreasonably take' and 'conduct causing') and the *'intent'* ('intends by his conduct').

Both reckless killing and killing by gross carelessness are:

(1)   forms of 'individual manslaughter';

(2)   alternative verdicts to 'murder';

(3)   alternatives to each other for the court to consider;

(4) both applicable to workplace and other situations, for example be available to the court in cases of road deaths or serious injury (although the Law Commission do not recommend a change to the existing offence of 'causing death by bad driving');

(5) reckless killing would carry a maximum sentence of life imprisonment whilst the Commission do not recommend a maximum for killing by gross carelessness.

(c) *Corporate killing*

This contrasts to the two proposed individual manslaughter offences above. The Law Commission says on page 128/129:

(i) that there should be a special offence of corporate killing, broadly corresponding to the individual offence of killing by gross carelessness;

(ii) that (like the individual offence) the corporate offence should be committed only where the defendant's conduct in causing the death falls far below what could reasonably be expected;

(iii) that (unlike the individual offence) the corporate offence should not require that the risk be obvious, or that the defendant be capable of appreciating the risk; and

(iv) that for the purposes of the corporate offence, a death should be regarded as having been caused by the conduct of a corporation if it is caused by a failure, in the way in which the corporation's activities are managed or organised, to ensure the health and safety of persons employed in or affected by those activities.

Therefore:

(1) Corporate killing as an offence relies less on the 'intent' component and more on the actions/conduct and safety systems in place (or otherwise). There is also no reference to 'foreseeability of risk', with the Commission arguing it is difficult to apply this concept to a corporation vis-à-vis an individual.

(2) The Law Commission stress the role of 'management failure' by the corporation as a cause of death even though the 'immediate cause' could be the act or omission of some individual such as an employee.

(3) Corporate killing applies to 'incorporated bodies' only e.g. limited and public limited companies and would not apply to 'sole traders'. Thus, the offence of corporate killing can only be by an incorporated body.

(4) The Law Commission recommended that this offence should not require consent to bring a private prosecution for corporate killing.

(5) The Law Commission recommended that this offence be triable by indictment only.

(6) The Law Commission is suggesting that on conviction of corporate killing the court should be empowered (upon the suggestion of the prosecution or the HSE) to recommend improvements to management systems to ensure death or serious injury at work does not arise again.

(7) Assuming that the jury cannot convict under the proposed offences, it should be able to consider conviction under s 2 or 3 of the *Health and Safety at Work etc Act 1974*.

(d) *Unlawful act manslaughter*

The Law Commission recommends the abolition of unlawful act manslaughter as it currently stands. This refers to where a person causes death whilst committing a criminal act, with this act carrying a risk of *some* harm to the other person. The Commission says '.. we consider that it is wrong in principle that a person should be convicted for causing *death* when the gravest risk apparently inherent in his conduct was the risk of causing *some injury*'.

## Insurance reasons

D6023	There are also insurance based reasons for effective DEM:

- given the positive correlation between risk and premium, the existence of DEM indicates hazard and risk control, consequently the premium ought to be less;

- if insurers are not satisfied or dissatisfied with the DEM system in place then they may not insure the operation, in turn increasing corporate risk as well as reducing corporate credibility. It may also prohibit the operation from tendering for contracts.

## Corporate reasons

D6024

- The corporate experience of companies like P & O European Ferries (Dover) Ltd, indicates that proactive DEM systems would have saved the company considerable money, publicity and reputation. In March 1987, the *Herald of Free Enterprise* the roll-on roll-off car ferry left Zeebrugge for Dover, thereafter it sunk resulting in 187 deaths. The case against the company and the five senior corporate officers collapsed for reasons cited above ('embodiment of the company' and 'controlling mind', see D6022 above).

- Failures to have sound DEM systems can also have dire financial consequences as Disasters as disparate as Piper Alpha to The Challenger Space-shuttle indicate. In the former case, Occidental Petroleum were generating 10 per cent of all the UK's North Sea output from Piper Alpha. Lord Cullen, who led the inquiry said 'The safety policy and procedures were in place: the practice was deficient'. Piper Alpha showed that the company had ineffective emergency response procedures resulting in persons being trapped, dying of smoke inhalation or jumping into the cold North Sea (something the procedures prohibited but a significant number of those that did jump into the sea survived). Such a Disaster had consequences for Occidental's financial reputation, share value, ability to attract investment and growth potential. In contrast, in the case of The Challenger Disaster in January 1986, the o-rings that held the two segments of the rocket boosters, which carried the fuel to propel the Challenger into space, fell apart resulting in the Challenger exploding shortly after taking off. This shocked the USA public, resulting in delays and questions as to the viability of NASAs programmes. The failure of the o-rings was well known with the engineers down grading the risk from 'high to 'acceptable'.

## Societal reasons

D6025	Society expects its organisations to plan and prepare for the worst case scenarios and when this doesn't happen, society seeks the closure, forefeiture or expulsion of the organisation from society e.g. the legal and political costs to Union Carbide in India with its mis-management of its plant in Bhopal in India. The company has been banned from operating in India.

## Environmental reasons

D6026	Failure to prepare and plan for Disasters and Emergencies will also have environmental costs, no better illustrated than the Chernobyl Disaster in April 1986, with its impact not only on the population of Russia but as far as Wales in the United Kingdom. The aim was to test if the power reduction of a turbine generator would be sufficient given the use of some suitable voltage generator, to power an emergency core cooling system (ECCS) for a few minutes whilst the stand-by diesel generator became operational. The power fell to 7 per cent of full power with a benchmark of 20 per cent being critical for that make of RBMK reactor. Consequently, the reactor exploded with radioactive fall out polluting the physical environment.

## Humanitarian reasons

**D6027** The human cost for not planning for worst case outcomes is the most significant. All the major Disasters documented in the media resulted in multiple fatalities – of all demographic and socio-economic groups. After the Ladbroke Grove Disaster in October 1999, when two trains collided resulting in 31 fatalities and 160 being critically injured, the public and the media made a significant outcry for change. This culminated in John Prescott, the Secretary of State for the Environment, Transport and the Regions establishing a Committee headed by Lord Cullen (of the Piper Alpha Inquiry), to investigate 'safety culture' on the railways.

# Disaster and Emergency Management Systems (DEMS)

**D6028** A DEMS is outlined in Figure 2 below:

Figure 2: Disaster and Emergency Management Systems (DEMS)

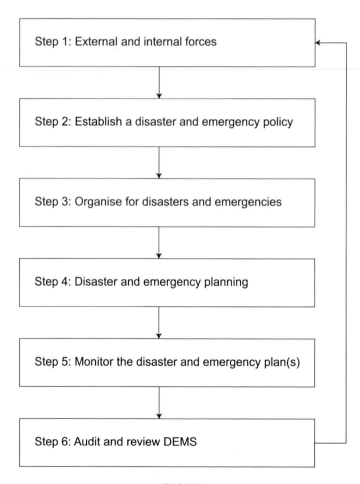

# External and Internal Factors

**D6029** The starting point with DEMS is understanding the variables that can influence or affect DEMS. These are both internal to the organisation and wider societal variables. It would be an erroneous assumption for the organisation to make, if it believed that Disaster and Emergency Planning can be in isolation of wider societal variables; these variables need to be understood and factored into any DEMS. The larger the organisation and the more hazardous its operation, the more it needs to provide specific detail.

## External Factors influencing DEMS

D6030

- The Natural Environment

    The organisation needs to identify physical variables that can influence its reaction to a Disaster or Emergency. The organisation needs to identify:

    (i) Seasonal weather conditions – including weather type(s) and temperature range. Effective rescue can be hampered by failing to know, note and record such variables. A simple chart, identifying in user-friendly terms on a monthly basis the weather conditions can assist the reader of any DEM Plans.

    (ii) Geography around the site – a description of the physical environment such as terrain, rivers or sea, soil type, geology as well as longitude and latitude positioning of the site. These should be brief descriptions unless the site is remote and could have difficulty being accessed in an Emergency.

    (iii) Time and distance – of the site from major emergency and accident centres, fire brigades, police etc. Both long and short routes need to be determined. Longitude and latitude co-ordinates should be identified.

- Societal Factors

    These include:

    (i) Demographics – the organisation needs to identify the age range, gender type, socio-economic structure of the immediate vicinity around the site. A Major Incident with impact on the local community requires the organisation to calculate risks to the community. *COMAH* Off-Site Emergency Plans indicates why planning is not just an On-Site 'in these four walls' activity.

    (ii) Social attitudes – the organisation needs to investigate briefly and be aware of attitudes (reactions and responses) of the local town or city and the country to Disasters and Emergencies. Union Carbide made the assumption that the population and authorities in Madhya Pradesh, where Bhopal is located, would react like the communities in the USA. The response rates, awareness levels and information available to different communities differs.

    (iii) Perception of risk – how does the local community view the site or operation? Is the risk 'tolerable' and 'acceptable' to them for having the operation in their community? Both Chernobyl and Bhopal highlight that economic necessity can alter the perception of risk in contrast to the statistical level of risk.

    (iv) History of community response – in brief the organisation needs to determine how many Major Incidents there have been in the past and how effectively have the community assisted (emergency services and volunteers). Such support will be critical in a Disaster.

(v) The built environment – a brief description of the urban environment, namely the street layout, urban concentration level, population density, access/egress to railways, motorways are useful. These issues could be covered by the inclusion of a map of the area.

- Government and political factors

  Relevant factors are:

  (i) The policy of national government to Disasters and Emergencies. Are they proactive in advising organisations, what information and guidance do they give? The organisation needs to identify its local or regional office of the HSE, the Environment Agency or a similar body and make contact with officials and seek early input into any planning.

  (ii) Local government – their Plans, information and guidance. DEM's need to be aware of any local government restrictions on managing Major Incidents. In the case of *COMAH* sites, they will need to involve the local authority in preparing Off-Site Emergency Plans.

  (iii) Committees and agencies – there are many legal and quasi-legal bodies created by legislation which have guidance, information and templates on response. For example the flooding committees mentioned above under the *Environment Act 1995*.

  (iv) Emergency and medical services – making contact, obtaining addresses/contact numbers and knowing the efficiency of the emergency services/accident and emergency are critical actions.

- Legal factors

  The organisation needs to know the legal constraints it has to operate within, not only to comply with the law but also to follow the best advice contained in legislation as a means of preventing Disaster and Emergency.

- Sources of information issues

  These include:

  (i) The local and national media – will not only report any Incidents but can be critical in relaying messages. Any planning requires the identification of local and national newspapers (their addresses/contact numbers), local/national radio details and any public sector media (through local government).

  (ii) Local library – they will in turn have considerable contact details and networks with other libraries so can be an efficient medium to relay urgent information. Making contact with the local librarian and identifying such contact details in any planing will be required.

  (iii) Business associations – Chambers of Commerce, Business Links, Training and Enterprise Councils need to be identified for the same reasons cited for local libraries. In the event of a Major Incident they can convey and provide information on a local rescue or occupational health organisation.

  (iv) Voluntary organisations – such as Red Cross, St. Johns Ambulance should be identified and noted.

  The organisation needs to build information networks locally and nationally. The experience of major Disasters such as Chernobyl or Kings Cross indicate that letting others know of the Incident is not a shameful or embarrassing matter – rather it can warn others and halt others from attending the site.

- Technological factors

    These concern what technology and equipment exists in the local community; what doesn't; lifting equipment, rescue, computers, monitoring equipment measuring equipment etc.

- Commercial factors

    These include:

    (i) Insurance – liaise with insurers from the beginning and ensure that all Plans are drawn to their attention and if possible approved by them. This can avoid difficulties of any claim that may arise due to a Disaster or Emergency as well as ensure that best advice from the insurer has been factored into the Plans.

    (ii) Customer and supplier response – larger organisations need to identify the responses from customers and suppliers, their willingness to work with the organisation in the event of a worst case scenario and possible alternatives. Suppliers need to be identified. Issues of customer convenience and loyalty also need to be addressed.

    (iii) Attitudes of the bank – liquidity issues will arise when a Major Incident strikes. Most organisations do not and cannot afford to make financial provisions for such eventualities. Therefore, if the operation is highly hazardous, establishing financial facilities before hand with the bank is necessary to ensure availability of liquid cash.

    (iv) Strength of the economic sector and economy – larger organisations that are significant players in a sector or the economy need to be cognisant of the 'multiplier effect' that damage to their operation can do to the community and suppliers, employees and others.

## *Internal Factors affecting DEMS*

**D6031**    DEMS also need to be constructed after accounting for various internal or organisational variables.

- Resource factors

Cash flow and budgetary planning whether annual or a longer period needs to account for resource availability for Major Incidents. This includes physical, human and financial resources. This is wider than banking facilities and encompasses equipment, trained personnel and contingency funds.

- Design and architecture

    Issues include – is the workplace physically/structurally capable of withholding a Major Incident? What are the main design and architectural risks? Also layout, access/egress, emergency routes, adequacy of space for vehicles etc.

- Corporate culture and practice

    The organisation needs to identify its own collective behaviour, attitude, strengths and weaknesses to cope with a Major Incident. Being a 'large organisation' does not mean it is a 'coping organisation'; there is also the delay in response that is associated with hierarchical structures. In this case, the organisation needs to consider small 'matrix' or project team cell in the hierarchy dedicated to incident response.

- Individual's perceived behaviour

    (a) The *Hale and Hale Model* is an attempt to explain how individuals internalise and digest perceived information of danger; make decisions/choices according to the cost/benefit associated with each decision/choice; and the actual actions they take as well as any reactions that result from their actions. In short, these five variables need to be understood in the organisation setting and a picture built up of the behavioural response of an individual.

(b)  The *Glendon & Hale Model* is a macro model of how the organisation (behaving like a system), being dynamic and fluid, with objectives and indeed limitations (systems boundary) can shape behaviour and in turn influence Human Error. Following Rasmussen, Human Error can be Skill Based (failing to perform an action correctly), Rule Based (have not learned the sequences to avoid harm) or Knowledge Based (breaching rules or best practice). If all three error types are committed then the danger level in the system is also greater. The model is a focus on how wider systems can contribute to Human Error and how that error can permeate into the organisation. The inference is the organisation needs to clearly define and communicate its intention and objectives and continuously monitor individual response.

(c)  *Reducing Error and Influencing Behaviour*, HSG 48, Health and Safety Executive, 1999 identifies the role of 'Human Error' and 'Human Factors' in Major Incident causation. The HSE says 'a human error is an action or decision which was *not intended*, which involved a deviation from an accepted standard, and which led to an undesirable outcome' (page 13). Following Rasmussen they classify four types:

(i)  *slips* (unintended action);

(ii)  *lapses* (short term memory failure) with slips and lapses being skill based;

(iii)  *mistakes* (incorrect decision) which are rule based; and

(iv)  *violations* (deliberate breach of rules) which are knowledge based.

Different types of Human Errors contribute in different ways to Major Incidents the HSE say and exemplify. Organisations need to identify from reported incidents the main types of Human Errors and why they are resulting and the negative harm generated. (Note that in the previous version of HSG 48 called *Human Factors in Industrial Safety*, five types of Human Errors were defined;

(A)  *misperception* (tunnel vision, excluding wider factors from one's senses e.g. the belief that smoking is safe on the underground or that a 'smouldering' is not a significant fire as in the case of Kings Cross in 1987);

(B)  *mistaken action* (doing something under the false belief it is correct e.g. the pilot switching off the good engine under the belief he was switching off the one with the fire, so both are off, hence the crash landing in the Kegworth Disaster);

(C)  *mistaken priority* (a clash of objectives, such as safety and finance as implied in the Herald of Free Enterprise Disaster in 1987);

(D)  *lapse of attention* (short term memory failure, not concentrating on a task e.g. turning on a valve under repair as in Piper Alpha in 1988); and

(E)  *wilfulness* (intentionally breaching rules e.g Lyme Bay Disaster when the principal director received a two year imprisonment for the death of teenagers at a leisure centre under his control)).

The HSE says that understanding Human Factors is a means of reducing Human Error potential. Human Factors is a combination of understanding the Person's behaviour, the Job they do (ergonomics) and the wider organisational system. Major Incidents can result if the organisation does not analyse and understand these three variables.

- Information systems at work

   The types of information systems, their effectiveness, and accuracy need to be identified. Thus telephone, fax, e-mail, cellular phone, telex etc need to be assessed for performance and efficacy during a worst case scenario.

## Establish a Disaster and Emergency Policy

D6032    The DEM Policy is a concise document that highlights the corporate intent to cope with and manage a Major Incident. It will have three parts: Statement of Policy for managing Major Incidents (see D6033); Arrangements for Major Incident management (see D6034); and Command and Control charge of Arrangements (see D6035).

### Statement of policy for managing major incidents

D6033    This should be a short (maximum 1 page) missionary and visionary statement covering the following:

- Senior management commitment to be responsible for the co-ordination of Major Incident response.
- To comply with the law, namely:
   - the protection of the health, safety and welfare of employees, visitors, the public and contractors;
   - to comply with the duties under *Regs 8* and *9* of the *Management of Health and Safety at Work Regulations 1999 (SI 1999 No 3242)* (see D6013);
   - to comply with any other legislation that may be applicable to the organisation.
- To make suitable arrangements to cope with a Major Incident and to be proactive and efficient in the implementation process.
- That this Statement applies to all levels of the organisation and all relevant sites in the country of jurisdiction.
- The commitment of human, physical and financial resources to prevent and manage Major Incidents.
- To consult with affected parties (employees, the local authority and others if needed).
- To review the Statement .
- To communicate the Statement.
- Signed and dated by the most senior corporate officer.

### The Arrangements for Major Incident management

D6034

- This refers to what the organisation has done, is doing and will do in the event of a Major Incident and *how* it will react in those circumstances. The Arrangements are a legal requirement under *Regs 5, 8* and *9* of the *Management of Health and Safety at Work Regulations 1999 (SI 1999 No 3242)*.
- Arrangements should be realistic and achievable. They should focus on major actions to be taken and issues to be addressed rather than being a 'shopping list'. The Arrangements will have to be verified (in particular for *COMAH* sites).

- Arrangements could be under the following headings with explanations under each. To repeat, the larger the organisation and the more complex the hazard facing it, the more detailed the Arrangements need to be. For example:

  (i) Medical assistance – including first-aid availability, first-aiders, links with accident and emergency at the medical centre, other specialists that could be called upon, rules on treating injured persons, specialised medical equipment and its availability etc.

  (ii) Facilities management – the location, site plans and accessibility to the main facilities (gas, electricity, water, substances etc.), rendering safe such facilities, availability of water supply in-house and within the perimeter of the site etc.

  (iii) Equipment to cope – identification of safety equipment available and/or accessible, location of such equipment, types (personal protective equipment, lifting, moving, working at height equipment etc.).

  (iv) Monitoring equipment – measuring, monitoring and recording devices needed, including basic items such as measuring tapes, paper, pens, tape recorders, intercom and loud-speakers.

  (v) Safe systems – procedures to access site, working safely by employees and contractors under Major Incident conditions (what can and cannot be done), hazard/risk assessments of dangers being confronted etc., risks to certain groups and procedures needed for rescue (disabled, young persons, children, pregnant women, elderly persons).

  (vi) Public safety – ensuring non access to Major Incident site by the public (in particular children, trespassers, the media and those with criminal intent), warning systems to the public etc.

  (vii) Contractor safety – guidance and information to contractors at the Major Incident on working safely.

  (viii) Information arrangements – the supply of information to staff, the media and others (insurers, enforcers) to inform them of the events. Where will the information be supplied from, when will it be done and updates?

  (ix) The media – managing the media, confining them to an area, handling pressure from them, what to say and what not to say etc.

  (x) Insurers/loss adjusters – notifying them and working with them at the earliest opportunity.

  (xi) Enforcers – notifying them of the Major Incident, working with them including several types e.g. HSE, Environment Agency (or Scottish equivalent) as well as local authority (environmental health, planning, building control for instance).

  (xii) Evidence and reporting arrangement – to cover strict rules on removal or evidence by employees or others, role and power of enforcers, incident reporting e.g under *RIDDOR (SI 1995 No 3163)* etc.

  (xiii) The emergency services – working with the police, fire, ambulance/NHS, and other specialists (Red Cross, Search & Rescue), rules of engagement, issues of information supply and communication with these services.

  (xiv) Specialist arrangements for specific Major Incidents such as bomb explosions – issues of contacting the police, ordnance disposal, access and egress, rescue and search, economic and human impact to name a few issues were most evident during the London Dockland and Manchester Bombings in the 1990s.

(xv) Human aspects – removing, storing and naming dead bodies or seriously injured persons during the incident. Informing the next-of-kin, issues of religious and cultural respect. Issues of counselling support and person-to-person support during the incident.

This is not an exhaustive list. The arrangements should not repeat those in the Safety Policy, rather the latter can be abbreviated and attached as a schedule to the above, so that the reader can have access succinctly to specific OSH arrangements such as fire safety, occupational health, safe systems at work, dangerous substances for instance.

## Command and Control Chart of Arrangements

D6035   This highlights who is responsible for the effective management of the Major Incident.

- It should be a graphical representation preferably in a hierarchical format, clearly delineating the division of labour between personnel in the organisation and the emergency services/others.

- The chart should display three broad levels of command and control, namely Strategic, Operational and Tactical. The first relates to the person(s) in overall charge of the Major Incident. Will this be the person who signed the Statement of Policy for Major Incidents or will it be another (disaster and emergency advisor, safety officer, others)? This person will make major decisions. The second relates to co-ordinators of teams. Operational level personnel need to have the above arrangements assigned to them in clear terms. The third refers to those at the front end of the Major Incident, for instance first-aiders.

  It is most important to note that internal Command and Control of Arrangements does not mean *overall* command and control of the Major Incident. This can (will be) vested with the appropriate emergency service, normally the police or the fire authority in the UK. In the event of any conflict of decisions, the external body such as the police will have the final veto. Therefore, the Chart and the Arrangements must reflect this variable.

- The chart should list on a separate page names/addresses/emergency phone, fax, e-mail, cellular numbers of those identified on the chart. It should also list the numbers for the emergency services as well as others (Red Cross, specialist search and rescue, loss adjusters, enforcing body).

- The chart should also clearly ratify a principle of command and control, as to who would be 'In-Charge 1', 'In-Charge 2', if the original person became unavailable.

- The chart and the list of numbers should also be accompanied by a set of 'rules of engagement' in short 'bullet points' to remind personnel of the importance of command and control e.g. safety, obedience, communication, accuracy, humanity for instance.

## Summary

D6036

- The three parts of the DEM Policy need to be in one document. Any detailed procedures can be separately documented ('Disaster and emergency procedures') and indeed could be an extensive source of information. However, unlike the Safety Policy and any accompanying Safety Manual, the same volume of information cannot apply to the DEM Policy. For obvious reasons it must be concise, clearly written, very practical and without complex cross-referencing.

- The DEM Policy must 'fit' with the Safety Policy, there can be no conflict so the safety officer and the DEM officer need to cross check and liaise. The DEM Policy must also fit with the broader corporate/business policy of the organisation.

- The DEM Policy must be proactive and reactive. The former concerned with preventing/mitigating loss and the latter concerned with managing the Major Incident when it does arise in a swift and least harmful manner.

- The DEM Policy needs to be reviewed 'regularly'. This could be when there is 'significant change' to the organisation, or as a part of an audit (semi-annual or annual).

- It must be remembered that the DEM Policy is a 'live' document so that it must be accessible and up-to-date.

- Although accessibility is important, the Policy should also have controlled circulation to core personnel only (for example those identified in the chart, the legal department). If the Policy was to be accessed by individuals wishing to harm the organisations, this will enable such persons to pre-empt and reduce the efficacy of the Policy.

- Finally and most crucially, the core contents of the Policy need to be communicated to all staff and others (contractors, temporary employees, possibly local authority).

## Organise for Disasters and Emergencies

D6037 Once the establishment has accounted for External and Internal factors and has produced a DEM Policy taking account of such factors, it is then necessary to ensure personnel and others are aware of the issues raised in the DEM policy. The '4' C's approach of HSG 65 provides a logical framework to generate this (*Successful Health & Safety Management*, HSE): communication (see D6038); co-operation (see D6039); competence (see D6040); and control (see D6041).

### Establish effective communication

D6038 Communication is a process of transmission, reception and feedback of information, whether that information is verbal, written, pictorial or intimated. Effective communication of the DEM Policy therefore is not a matter of circulating copies but requires the following:

(1) *Transmission*

- The whole policy should not be circulated as it will mean little to employees and others. Rather an abridged version, possibly in booklet format or as an addition to any OSH documentation supplied will make more sense. Such copies must be clear, user friendly, non-technical as possible, be aware of the end user's capabilities to digest the information, be logical/sequential in explanation and use pictorial representation as much as possible.

- Being aware of the audience is central to the effective communication of the DEM Policy. The audience is not one entity but will consist of:

    — direct employees;

    — temporary employees;

    — contractors;

    — the media;

    — the enforcers (*COMAH* sites);

    — the local authority (*COMAH* sites);

    — insurers;

    — emergency services (*COMAH* sites);

    — the public (*COMAH* sites).

This does not necessarily mean separate copies for each of these entities but the abridged copy will need to satisfy the needs of all such groups.

- Transmission should start from the board, through to departmental heads, and disseminated downwards and across.

(2) *Reception*

- What format will the end-user receive the abridged copy in (hard copy, electronic on disc, via e-mail, etc)?
- When will the copy be circulated – upon induction, upon training, ad hoc?

(3) *Feedback*

- Will the end-user have the opportunity to raise questions, make suggestions, be critical if they spot inconsistencies in the DEM Policy?
- There should also be 'tool box talks' and other general awareness programmes to inform individuals of the DEM Policy. This could be combined with general OSH programmes or wider personnel programmes, so that a holistic approach is presented.
- The importance of feedback is that the policy becomes owned by all individuals, which in turn is the single most important factor in successful pre-planning to prevent Major Incidents.

In general it may be useful to retain copies of the abridged and the full policy with other safety documentation in an 'in-house' company library. For smaller organisations, this could be one or two folders on a shelf through to a dedicated room for larger organisations. Second, the abridged copy could be pasted onto an intranet site.

## Co-operation

D6039    Co-operation is concerned with collaborating, working together to achieve the shared goal and objectives:

- Firstly, co-operation between strategic, operational and tactical level management is critical. This reflects the chart in the DEM Policy, as discussed above. This could be consolidated as a part of a broader corporate meeting or preferably dedicated time to cover OSH and fatal incident issues. This could be a semi-annual event, with a dedicated day allotted for all grades of management to interface. This is not the same as a board level discussion or a management discussion.
- To give responsibility to either the Safety Committee or the safety group to co-ordinate review, debate and assessment of the DEM Policy or to fuse this function within a broader business/corporate review committee. The former has advantages as it is safety dedicated whilst the latter would integrate DEM Policy issues into the wider business debate.
- Involvement of Safety Representatives/Representatives of Employee Safety (ROES), is both a legal requirement as well as inclusive safety management. These persons can be central 'nodes' in linking 'management' and the 'workers' together. In the UK, there has been an increasing realisation that trained safety representatives are a knowledgeable resource with many being trained to IOSH/NEBOSH standard (as with the AEEU (electrical engineers) trade union, with their National Academy of Safety and Health, NASH).
- Co-operation also needs to extend to contractors. The person responsible for interfacing with contractors needs to up-date them and make them aware of the DEM Policy and seek their support and suggestions. It may also be valuable to invite contractors to OSH/fatal incident awareness days or the general committee meetings as observers.

- Co-operation is also needed between the organisation and external agencies such as the local authority, insurers, enforcers, media etc. This can be achieved through providing abridged copies of minutes or a 'newsletter' (1–2 sides of A4) distributed semi-annually informing such bodies of the DEM Policy and any changes as well as other OSH issues. *COMAH* sites will have to demonstrate as a legal requirement that Plans and policies are up-to-date and effective.

## Competence

**D6040** Competence is a process of acquiring knowledge, skill and experience to enhance both individual and corporate response. Thus, competence is about enhancing and achieving standards (set by the organisation or others).

- Competence is important, so that certain key persons are trained to understand the DEMS process. The above issues and their link to OSH in particular require personnel that can assimilate, digest and convey the above issues. Training does not necessarily mean formal or academic training but can be vocational or in-house. Neither does it mean the organisation expending vast sums but can be a part of a wider in-house OSH awareness programme (e.g. 1 day per quarter of a year).
- For larger organisations, they may be able to recruit 'competent persons' to advice on OSH and fatal incident matters.
- In short, all employees need to brought up to a minimal standard. Piper Alpha showed that whilst Occidental Petroleum had detailed procedures, the employees generally did not fully understand them nor had they the minimal understanding of Major Incident evacuation. The organisation had failed to impart knowledge, skill and experience sufficient to cope with fires and explosions on off-shore sites.

## Control

**D6041** Control refers to establishing parameters, constraints and limits on the behaviour and action. This ensures that on the one hand an effective DEM policy exists and on the other, personnel will act and react in a co-ordinated and responsible manner. Controls can be achieved via for example:

- Contractual means – as a term of a contract of employment that instruction and direction on OSH and fatal incident matters must be followed by individuals.
- Corporate means – the organisation continuously makes individuals aware of following rules and best practice.
- Behavioural means – by establishing clear rules, training, supply of information, leading-through-example, showing top-level management commitment etc.
- Supervisory means – ensuring that supervisors monitor employee safety attitudes and risk perceptions.

The cumulative effect of the 4 C's should be a positive and proactive culture in which not only OSH issues but the DEM Policy issues are understood. Factors that can mitigate or prevent a positive and proactive culture developing include lack of management commitment, lack of awareness of requirements, poor attitudes to work and life, misperception of the risk facing the organisation, lack of resources or the unwillingness to commit resources, fatalistic beliefs etc.

# Disaster and Emergency Planning

**D6042** 'Planning' is a process of identifying a clear goal and objectives and pursuing the best means to achieve that goal/objectives. Although 'Disaster' and 'Emergency' are two separate

Disaster and Emergency Management Systems (DEMS)    D6044

but related terms, in the case of planning the two need to be viewed together. This is because in practice one cannot divorce the serious event (the Disaster) from the response to that serious event (Emergency).

Disaster and Emergency Planning can be viewed in three broad stages:

- stage 1: before the event; (see D6043)
- stage 2: factors to consider during the event; (see D6050)
- stage 3: after the event. (see D6051)

## Stage 1: before the event

**D6043**   Once the DEM Policy has been established and a culture created where the Policy has been understood and positively received, next one needs to be establish a 'state-of-preparedness', that is addressing issues, speculating on scenarios and developing support services when the Major Incident does strike. One can summarise this stage into three sections, with a special section on *COMAH* and the specific legal requirements for *COMAH* sites.

## The Risk Assessment of Major Incident potential and consequent Contingency Planning

**D6044**   What is the probability of the Major Incident resulting? What would be the severity? What type of Major Incident would it be? Figure 3 depicts a 'Major Incident Matrix':

Figure 3: Major Incident Matrix

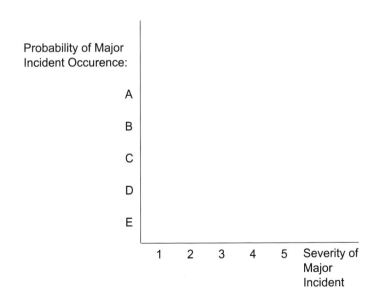

Where:
A=   Certainty of Occurrence (Probability = 1)
B=   Highly Probable

C= 0.5 Probability of Occurrence
D= Low Probability
E= Most Unlikely

And:

1= Serious Injuries
2= Multiple Fatalities
3= Multiple Fatalities and Serious Economic/Property Damage
4= Fatal Environmental, Bio-Sphere and Social Harm
5= Socio-Physical Catastrophe

The planning responses therefore will differ according to the cell one finds oneself in. For example, cell 1E requires the least complex state of preparedness in terms of resources, detail of planning and urgency. Whilst 5A is a major societal event that requires macro and holistic responses, in which case the organisation's sole efforts at planning are futile. The organisation needs to assess in which cell the Major Incidents it will confront, will mainly fall in and accordingly develop Plans. A conglomerate or an 'exposed operation' such as oil/gas refinery work may require planning at all 25 levels. This is 'Contingency Planning' – an analysis of the alternative outcomes and providing adequate responses to those outcomes.

During the concerns over the 'Millennium Bug', HSBC Bank provided some general advice on Contingency Planning (*Tackling the Millennium Bug: A final check on preparations including Contingency Planning*, HSBC, 1999). They suggested a 10-point strategy. 'A Contingency Plan is a carefully considered set of alternatives to your usual business processes' (page 18). This advice is generic and can be used more broadly for everyday planning, as follows.

(1) *Identify* – assume things will go wrong – focus on where the probability and severity would be of failure.

(2) *Prioritise* the processes into three categories:
   (a) Essential for business continuity.
   (b) Important for business continuity.
   (c) Non-essential for business continuity.

(3) *Analyse* e.g:
   (a) Identify the effects of risks on operation(s).
   (b) Assess 'domino effects'.
   (c) Grade risk using a suitable risk assessment approach.

(4) Develop solutions e.g:
   (a) For each operation develop a solution/control measure to the risk.
   (b) Encourage participation – involve suitable and critical stakeholders
   (c) Ensure new risks are not created.

(5) *Costs* – carry out a 'cost benefit analysis'.

(6) Formalise documents e.g:
   (a) Write and keep any assessments and inspections.
   (b) Top management involvement and 'joined up business'.
   (c) Develop an approach to implementation.
   (d) Regular training to be conducted to ensure improved competence.

(e) Identify and deploy resources needed (if available).

(f) Communications to be clear and simple – before, during and after.

(g) Contractual issues to be dealt with so that contractors, customers and insurers are kept informed.

(7) *Coordinate* – managers and team to ensure implementation of contingency plan.

(8) *Test* e.g. controlled testing of the contingency plan(s) (see below).

(9) *Update* e.g:

(a) Update the plan.

(b) Quality assurance and control of the plan.

(c) Accuracy – ensure that the organisation's lawyer and insurer has sight of the plan.

(10) Communicate e.g.

(a) Plan signed off by top management.

(b) Communicate salient features to employees.

Appendices can be included containing any relevant information.

## Decisional Planning

**D6045** Planning before the event also involves co-ordinating different levels of decision making. Figure 4 depicts a Decisional Matrix:

**Figure 4: Decisional Matrix**

|  | Production Level | 1. INPUT | 2. PROCESS | 3. OUTPUT |
|---|---|---|---|---|
| Management Level |  |  |  |  |
| A. STRATEGIC |  |  |  |  |
| B. OPERATIONAL |  |  |  |  |
| C. TACTICAL |  |  |  |  |

Where:

A= Strategic or Board/Shareholder/Controller Level (Gold Level)

B= Operational or Departmental Level (Silver Level)

C= Tactical or Factory/Office Level (Bronze Level)

And:

1= Inputs or those resources that make production possible, thus raw materials, people, information, machinery and financial resources.

2= Process or the manner in which the inputs are arranged or combined to enable production (method of production)

3= Output or the product or service that is generated.

Decisional Planning requires Major Incident risks to be identified for each cell. Essentially this is assessing where in the production process and at which management level can problems arise that can lead to Major Incidents. This matrix is not 'closed' but is 'open' and 'dynamic', which means

External Factors, which were discussed above as well as Internal Factors need to be accounted for. For example, in a commercial operation at board level, they need to consider:

- INPUTS: purchasing policy of any raw materials used (safety etc.), recruitment of stable staff, adequacy of resources to enable safe production, known and foreseeable risks of inputs used, reliability of supply, commitment to environmental safety and protection etc.
- PROCESS: external factors and their impact on production, production safety commitment, commitment to researching/investigating in the market for safer production processes etc.
- OUTPUT: boardroom commitment to quality assurance, environmental safety of the service and/or product produced, customer care and social responsibility etc.

The end result should be, in each cell major risks to production and their impact on a Major Incident needs to be identified. This need not be a major, time consuming exercise, rather it can be determined through a 'brain-storming' session or each management level fills out the matrix during their regular meeting and then it is jointly co-ordinated in a short (2–5 page) document.

## Testing

**D6046** This is considered to be a crucial aspect of pre-planning by all the emergency services in the UK. Testing is an objective rehearsal to examine the state of preparedness and to determine if the Policy and the Plan will perform as expected.

The Home Office guidance, *Dealing with Disaster* (1998, Third Edition) identifies Training and Exercising as types of Testing. Training is more personnel focused, aiming to assess how much the human resource knows about the Policy and Plan and means of enhancing the knowledge, skill and experience of that resource. Exercising is a broader approach, examining all aspects of the Policy and Plan, not just the human response but also physical and organisational capability to deal with the Major Incident.

## Training

**D6047** Carrying out a Training Needs Analysis (TNA) to determine the quantity (in terms of time) and quality of training needed is essential. It may be that personnel and others adequately understand the Policy and Plan, therefore the training response should be proportionate. Excessive training is a motivational threat to the interest and enthusiasm of the person as is under-training. The HSE in HSG 65, *Successful Health & Safety Management* suggest that the TNA consists of issues such as:

- Is the training 'necessary'? In other words, is there an alternative way of ensuring competence and awareness of the Policy and Plan, rather than just training? Could circulating the information be better? What about regular 'tool box talks'? It is important to weigh the costs and benefits of each option.
- Is it 'needed'? Will the training meet Personal, Job or Organisational needs thereby enhancing awareness and appreciation of the policy and plan ?
- What will be the intended learning objectives of the training? Will it be 'this is a course on the content of the DEM Policy and Plan'?
- What type of training will be offered? Class-room, on-site or both? What are the costs and benefits associated with each?
- Management of the training – where will the training be conducted? Who will deliver it? When will it be delivered – day or evening? What training aids will be provided – if any?
- How does one measure the effectiveness of such training ?

## Exercising

**D6048** The Home Office cite three types of exercising:

- Seminar Exercises

    — a broad, brain-storming session, assessing and analysing the efficacy of the Policy and Plan

    — seminars need to be 'inclusive', that is bring staff and others together in order to co-ordinate strategic, operational and tactical issues.

- Table-Top Exercises

    This is an attempt to identify visually using a model of the production or office site and surrounding areas, what types of problems could arise (access/egress, crowd control, logistics etc.).

- Live Exercises

    This is a rehearsal of the Major Incident, actively and pro-actively testing the responses of the individual, organisation and possibly the community (emergency services, the media etc.). London Underground, the railway sector commonly, the civil aviation sector, tend to annually assume that a Major Incident has occurred and the consequential conditions are created.

- Synthetic Simulation

    Whilst this is not mentioned in Home Office or other emergency service guidance, this fourth approach is becoming a most popular option. It applies the techniques used in flight simulation and military simulation, to Major Incidents. Thus, using computers one can model the site and operation and in 3-D format move around the screen. This then enables various eye-points around the site, achieved by moving the mouse. Some advanced systems enable 'computer generated entities' to be included into the data-set. For example, a collision can be simulated outside the production plant or the rate of noxious substance release modelled and calculated across the local community.

    There are costs and benefits associated with each of the four approaches. The need is for developing exercise budgets which enable all four to be used in proportion and to compliment each other.

    Training and exercising are both necessary to ensure that human resources understand the policy and plan, as well as to ensuring that problems can be identified and competent responses developed.

    It should be noted that the guidance to the *COMAH Regulations* (*SI 1999 No 743*) does not distinguish between 'training' or 'exercising'. It identifies drills, which test a specific aspect of the Emergency Plan (e.g. fire drills), seminar exercises, walk through exercises, which involves visiting the site, table-top exercises, control post exercises, which assess the physical and geographical posting of the personnel and emergency services during a Major Incident and live exercises.

## COMAH sites

**D6049** *COMAH* sites, in addition to the above, also need to ensure that On-Site and Off-Site Emergency Plans are compliant with guidance and are realistic and achievable.

(1) On-Site Emergency Plan

- A 'Site Main Controller' needs to be appointed. This is a strategic level person, who oversees the Major Incident (refer above to organisation chart under Policy). This person must:

- oversee and take control over the event;
- control the activities of site incident controller, that is operational and tactical level co-ordinators;
- must contact and confirm that the emergency services have been contacted;
- must confirm that On-Site Plan has been initiated;
- mobilise key personnel, identified in the policy and pre-planning;
- continuously review and assess the Major Incident;
- authorise the evacuation process;
- authorise closing down of plant and work equipment;
- ensure casualties are being cared for and notify their relatives, as well as listing missing persons;
- monitor weather conditions;
- liaise with the HSE, local authority or the Environment Agency;
- account for all personnel and contractors;
- traffic management, site access/egress etc. must be strategically controlled;
- record keeping of all decisions made, for later assessment and inquiry;
- provide for welfare facilities (food, clothing etc.) of personnel;
- liaise with the media;
- compliance with the law – non-removing of evidence etc;
- control of affected areas after the Major Incident.

- On-Site Emergency Control Centre (ECC)
  - This is a room or area near the Major Incident site, where the command and control decisions will be made.
  - The ECC will also co-ordinate liaison with the media and the public.
  - The ECC should contain: telecommunication equipment, cellular communication equipment, detailed site plans and maps, technical drawings of the operation and its shut-off points, list of critical/hazardous substances and waste on site, location of all safety equipment, location of fire safety equipment and access/egress points etc.

- The On-Site Emergency Plan must contain at least the following (*Sch 5, Part II* of *COMAH (SI 1999 No 743)*):
  - name/position of person that can activate the Emergency Plan and that person (unless being the same person) who will co-ordinate/manage and mitigate the impact of the Major Incident;
  - name/position of the person who will liaise with the local authority regarding the Off-Site Emergency Plan;
  - description of the foreseeable factors that can increase (or mitigate) the severity and consequence of the event, thereby either classifying it as a Major Accident or worse;
  - Arrangements for mitigating risks to persons on-site. Warning systems and rules that such persons must follow need to be listed;

- Arrangements for warning the local authority, so that they can activate the Off-Site Emergency Plan, as well as ensuring the local authority receives any other relevant information;
- Arrangements for training staff and effective liaison by them with the emergency services;
- Arrangements for providing assistance with 'off-site mitigating action' such as special equipment that off-site personnel can use, media liaison etc.

(2) Off-Site Emergency Plan

- *Schedule 5, Part III* of *COMAH* (*SI 1999 No 743*) says that this Plan must contain at least:
  - name/position of person that can activate the Emergency Plan and the person (unless the same person), who will co-ordinate the off-site actions. This should be senior, strategic level management that will co-ordinate links with the local authority and the public;
  - Arrangements for receiving 'early warning of incident'. This can be via on-site team, technology etc. In addition, there must be Arrangements for alerting others of the event and procedures for call-out of the emergency services, specialist assistance, other search and rescue bodies;
  - Arrangements for co-ordinating resources so that the Off-Site Emergency Plan can be effectively implemented. This should identify those organisations that can assist in the Off-Site Emergency, how each of these organisations will be alerted, how personnel of the organisation affected will recognise and identify the emergency services and specialist services and vice-versa, channels of communication between personnel and those other emergency services, specialist organisations, meeting place, off-site needs to be identified, for direct communication between personnel and others, access to the operation to use equipment;
  - Arrangements for providing on-site assistance, such as from the fire brigade;
  - Arrangements for off-site occurrences and the mitigating actions needed to protect the public and the environment;
  - Arrangements for providing information to the public and the appropriate behaviour expected from them. In addition, details of how the media could be utilised to disseminate information to the community and emergency organisations;
  - Arrangements for the provision of information to member states, when there is a 'reasonable likelihood' of impact on such neighbouring states.

- Off-Site Emergency Control Centre (ECC)
  - This parallels the on-site ECC and is at a safe distance from the site. It is a command and control centre for all off-site liaison.
  - It must also link up with the on-site ECC to ensure that Plans are in unison.

Thus, under *COMAH*, the On-Site and Off-Site Emergency Plans are required to reflect, account for and provide details of the Arrangements and Command and Control variables identified in stage 2 of the DEMS process. The difference is that under Emergency Plans, these variables are specific and must be practical, whilst in the Disaster and Emergency Policy of DEMS, these variables will also include wider issues not necessarily a legal requirement under COMAH but significant to Major Incident management. However, not all sites (majority of sites) are *COMAH* regulated.

## Stage 2: during the event
D6050

- Activate Disaster and Emergency Policy and pre-planning

  Major Incident response and management ought to be clear and effective if one has followed the previous stages in the DEMS; accounted for External and Internal factors and uncertainties, developed a Chain of Command as well as Arrangements to cope, Organised the operation to cope with a Major Incident and finally pre-planned if the event actually arises. In the main, activating the Arrangements made is the central action (refer above).

- Co-ordination

  This is the central action to ensuring efforts are in unison. This includes both liasing with external bodies such as the emergency services, media and internally at strategic, operational and tactical levels (as discussed above). The police will normally have overall command and control, so their guidance must be followed.

- Code of conduct

  A code of conduct is a set of 'golden rules' that staff and contractors need to follow if the Major Incident arises. This includes:

  (i)    following instruction from those with command and control responsibilities;

  (ii)   using self initiative;

  (iii)  facilities management (gas, electricity, water) – making safe, when and when not to switch on or off;

  (iv)   medical assistance – liaison with health service, first-aiders, when not to administer and when to;

  (v)    rules of evacuation;

  (vi)   site access and egress;

  (vii)  site security;

  (viii) work and rescue equipment – its safe use and logistics of use;

  (ix)   record keeping of the Major Incident (audio-visual, verbal and written);

  (x)    resources needed for effective Major Incident management;

  (xi)   media and public relations management;

  (xii)  issues of care and compassion of injured persons.

  The code of conduct is a reflection of the Disaster and Emergency Policy and the pre-planning. It should be reinforced verbally and in writing (1 page of A4) and reiterated to the Major Incident team before and during the Major Incident. Whilst this seems bureaucratic, it must be noted that if the core team (internal and external) fail to follow best practice and safe-guard themselves, this increases the risk factor of the Major Incident and could even lead to a double tragedy. A five minute or so reiteration is a minor time cost.

  *COMAH* sites will have to be aware of these issues also as well as activating the On-Site and Off-Site Emergency Plans.

## Stage 3: after the event

**D6051** Major Incident Planning does not cease as soon as the Major incident is physically over. There are continuous issues over time that need to be addressed (see below).

### I. Immediate term (immediately after the major incident and within a few days)

**D6052** The following paragraphs explain what steps should be taken.

#### Statutory investigation

**D6053** This will involve the HSE, local authority, Environment Agency/SEPA for instance. Under statute, these bodies have powers and a duty to investigate a Major Incident.

- The organisation investigated must fully co-operate with these bodies and afford any assistance they require.
- There must be no removal or tampering with anything at the Major Incident site (or Off-Site in *COMAH* cases). There may be forensic or other data gathering required by these bodies.
- There must be provision for all documentation and information to be provided to these bodies.

#### Business Continuity

**D6054** This has to be planned for even whilst the investigation by authorised inspectors is being carried out.

'Business Continuity' is a planning exercise to ensure critical facilities, processes and functions are operational and available during and immediately after the Major Incident thereby enabling the organisation to function commercially and socially.

There are many approaches to Business Continuity. Two are highlighted below – first, a 'generic approach', secondly the advice from the Business Continuity Institute (BCI). In addition, there are approaches from the Department of Trade and Industry, Disaster Recovery Institute of the USA, Australian National Audit Office as well as guidance contained in BS/ISO/ICE 17799:2000 (which is an information technology standard).

(1) Generic approach'

Business Continuity is a forecasting process of recovery, assessment and ensuring the adequacy of resources for the organisation to continue its operation.

    (*a*) Recovery

    Recovery is a state of regaining or salvaging assets that otherwise would have been permanently lost. In the event of a Major Incident, recovery phase needs to focus on:

        (i) Human resource recovery

            — ensuring personnel are safely evacuated;

            — others such as lawful visitors and trespassers are evacuated;

            — all persons in general are removed in the quickest and most practicable means.

        (ii) Information resource recovery

            — essential documents and software;

- private and confidential documents.

(iii) Physical resource recovery

- primary electrical and mechanical facilities such as power supply, water and gas services;
- essential work equipment, if possible and moveable.

(iv) Financial resource recovery

(v) Valuable assets recovery (if practicable)

The organisation also needs to rate the recovery potential – how possible is recovery? This could be rated from certain (1) through to not possible (5). This calculation requires different recovery responses; when the Major Incident strikes, those on the scene need to decide if the rating is 1 or say 5. Accordingly, the response will vary – if 5, then there is little purpose risking life and resources to salvage assets.

(b) Assessment

Upon the immediate recovery, the organisation must ask, how much damage (deterioration, infliction of harm or erosion of value) has been inflicted to the operation by the Major Incident? Damage Assessment is a three stage activity:

(i) Identify the type of damage

- human resource damage: physical and/or behavioural;
- informational resource damage: primary, secondary and tertiary documentation;
- physical resource damage: work equipment, property, facilities, environment;
- financial resource damage: cash, art, etc.

(ii) Identify severity of damage

This can be a qualitative scale which says 'low' or 'high' to the age or a quantitative scale – the damage is rated on some scale e.g. from 0 – 5. The severity needs to be forecasted for human, informational, physical and financial resources.

- Human resources could be scaled from serious injury (1) through to fatality and multiple fatality (5).
- Informational: from minor harm (1) through to destroyed (5).
- Physical: from reparable (1) to unsalvageable (5).
- Financial: from no impact on cash flow (1) to financial ruin (5).

(iii) Consequence of damage

How much does the damage affect the chance of the operation being resumed immediately or in the next few days? The longer it will take to resume, the worse has been the consequence. Consequence could be rated in quantitative terms also, such as a '1' for 'resume immediate' so the harm has been minimal through to say '5', or 'resumption will take months'. The consequence needs to be examined in each case:

- Human: how long will it take people to get back to work?

- Informational: how long will it take for manual and electronic systems to be operational?
- Physical: how long will it take for necessary facilities to be operational?
- Financial: how long will it take for cash flow to become positive or to access financial facilities?

Severity x Consequence will give an index of forecasted potential Loss, which can then be assessed as a spreadsheet over time. Loss adjustors use various detailed statistical models based upon such generic principles.

(c) Adequacy of resources

Once the organisation has forecasted the potential for recovery and hypothesised about damage assessment, next it needs to ensure that the organisation will have adequate resources to carry on operating, in light of the resource losses identified by the assessment. Adequacy needs to consider:

(i) Human resources:
- adequacy of competent and trained personnel at strategic, operational and tactical levels of the organisation;
- availability of key advisory support services (lawyer, accountant, etc.);
- if the Major Incident is classified as a 'Crisis', then there is a chance some of the key personnel are not available. Pre-planning therefore requires liaison with recruitment and selection specialists.

(ii) Physical resources:
- telephone, fax, e-mail, cellular connectivity and reliability;
- furniture and fittings;
- stationary;
- work equipment (including computers/type-writers, filing cabinets);
- working stock;
- working space;
- vehicles;
- safety equipment (personal protective equipment);
- etc.

(iii) Informational resources:
- legal documents (organisation's certificates of incorporation, insurance liability certificates etc.);
- personnel documents (PAYE, NIC, personnel records);
- financial documents (availability of bank books);
- sales/marketing documents, this is at the heart of the operation and there will be a need to develop databases, contact potential customers and re-establish commercial functionality.

(iv) Financial resources:
- adequacy of working capital.

(2) *Business Continuity Institute's approach*

In *Business Continuity Management: A Strategy for Business Survival* by the BCI, they say "BCM is not just about disaster recovery, crisis management, risk management or about IT. It is a business issue. It presents you with an opportunity to review the way your organisation performs its processes, to improve procedures and practices and increase resilience to interruption and loss. Business Continuity Management is the act of anticipating incidents which will affect critical functions and processes for the organisation and ensuring that it responds to any incident in a planned and rehearsed manner'.

Business Continuity Management (BCM) is the process of proactively identifying, anticipating, resourcing and planning to ensure that business operations can continue after the Disaster. In contrast, Business Continuity Planning (BCP) is a stage in the BCM process. It is the tactical and operational document that will be used when Disaster strikes. The BCI defines BCM as 'an approach that seeks to improve organisational resilience to interruption, allowing for the recovery of key business and systems processes within the recovery time frame objective, whilst maintaining the organisation's critical functions'.

The BCI present a 5 step approach to BCM.

- STEP 1: *'Understand your business'*

    Ensure that the main strengths, weaknesses, opportunities and threats are identified. Also ensure that key 'threats' and 'risks' are identified. Further, the support of the board/strategic management team (SMT) is necessary for the necessary resource allocations.

- STEP 2: *'Consider Continuity Strategies'* when the worst case scenario strikes e.g.

    (a) *Do nothing* – it maybe that no resource deployment or salvage strategy is required as the risks of significant loss are quite low.

    (b) *Changing or ending the process* – the board/SMT may consider that production or service processes/procedures need to be modified.

    (c) *Insurance* – should it be used? What are the costs?

    (d) *Loss Mitigation* – what actual/tangible actions can be taken to reduce risks?

    (e) *Business Continuity Planning* – is a more detailed plan needed given the severity and long term nature of the risk/threat?

- STEP 3: *'Develop the Response'*

    If BCP is needed then develop the plan. The BCI suggest the following sections to a BCP:

    — Section 1: General Introduction and Overview

        (i) Objectives

        (ii) Responsibilities

        (iii) Exercising

        (iv) Maintenance

    — Section 2: Plan Invocation

        (i) Disaster declaration

        (ii) Damage assessment

        (iii) Continuity actions and procedures

        (iv) Team organisation and responsibilities

(v) Emergency (Crisis) Operations Centre
- Section 3: Communications
    (i) Who should be informed
    (ii) Contacts
    (iii) Key messages
- Section 4: Suppliers
    (i) List of recovery suppliers
    (ii) Details of contract provision

- STEP 4: *'Establish a Continuity Culture'* – through education, training, information and awareness.
- STEP 5: *'Exercise, Plan Maintenance and Audit'*
    (*a*) Exercise the Plan frequently with participation of key stakeholders.
    (*b*) Plan Maintenance accounting for internal and external changes.
    (*c*) Use Auditing techniques for 'gap analysis' purposes.

## Additional matters to consider

**D6055** Also under the 'immediate term', in addition to the above, one needs to consider:

- Insurers/loss adjusters

Assuming that the insurance contract covers direct and consequential damage from a Major Incident (and not all will), the organisation needs to notify the insurer and ensure all paperwork is completed promptly. Most insurance contracts stipulate a time limit by which the paperwork has to be lodged with the insurer; the Major Incident will divert attention to other issues, so ensuring that a person is appointed to activate this insurance task is vital (this should be the organisation's lawyer).

The insurer in turn will notify their loss adjusters to investigate the basic and underlying causes of the event. Again, full disclosure and co-operation are implied insurance contractual requirements. Although, the organisation must check all documents that the loss adjusters completes, to ensure they are accurate and cover all aspects of the event.

- The police

In the suspicion that criminal neglect played a role in the event, then the police will need to interview all core board members, senior management and others.

- Building contractors

The organisation needs to plan for building contractors to visit the site and make it safe and secure. This may have to be done forthwith after the event, even if insurance issues have not been resolved. Therefore, adequacy of resources, as discussed above becomes vital.

- Visitors

Major incidents also lead to the public and media making visits. The arrangements made to handle such groups must extend to after the event.

- Counselling support

Counselling support to affected employees and possibly contractors is not just a personnel management requirement which shows 'caring management' but increasingly a legal duty of the organisation to provide such support. Issues of 'post traumatic stress', 'nervous shock' and 'bereavement' means that the organisation has common law obligations to offer medical and psychological support. This should involve a medical practitioner and an occupational nurse. The insurance policy can be extended before the event to cover the cost of such services.

## II. Short Term (a week onwards after the event)

D6056

- Investigation and inquiry

   This can be both a statutory inquiry (although most are called within days) and/or the in-house investigation of the event and lessons to be learnt. Issues to consider include:

   — Basic causes: was it a fire, bomb, an explosion or a natural peril?

   — Underlying causes: what led up to such a peril occurring? Examine the managerial, personnel, legal, technical, organisational and natural factors that could have caused the event.

   — Costs and losses involved.

   — Lessons for the future.

   — Did the Disaster and Emergency Plan operate as expected? Were there any failings? What improvements are required? There needs to be complete de-briefing and examination of the entire process involving internal personnel and external agencies.

   The organisation should weigh up the possibility of external persons carrying out this exercise or whether in-house staff are objective and dispassionate enough to assess what went wrong.

- Visit by enforcers

   The organisation also needs to be prepared for further visits from the enforcers and the possibility of statutory enforcement notices being served to either regulate or prohibit the activity. Multiple notices are possible, from health and safety officers, fire authority, Environment Agency/SEPA, building control or planning officers. This will affect the operation, production process and have economic implications. This ought not to occur if the organisation has taken due care to pre-plan for the Major Incident and had continuous safety monitoring of the operation. Notices will be served if there is a failure to make safe the site.

- Coping with speculation

   The public, the media and employees will speculate about causation and there is the risk of adverse publicity. Public relations is a central activity. For legal and moral reasons, it is best practice to disclose all known facts unless the statutory investigation prohibits otherwise.

## III. Medium term (a month plus)

D6057    The 'normalisation process' will begin. The organisation needs to carry on the operation, be prepared for further visits from enforcers, loss adjusters and re-assess its corporate/financial health.

## IV. Long term (six month onwards)

D6058

- Systems review

— A review of the impact of the Major Incident and whether the organisation is recovering from it.

— Impact on reputation.

— Legal threats.

— Any positive outcomes – learning from mistakes, improving technical know-how, wider industrial benefits from knowing the chain reaction of events etc.

### V. Longer term (one year onwards)

**D6059** The organisation's memory and experience needs to be included in:

- in-house training programmes;
- factored into systems and procedures;
- review of entire *espirit de corps* and corporate philosophy.

Disaster and Emergency Planning is an extensive exercise, being dynamic and accounting for a diverse array of phenomena as industrial, man-made, environmental, socio-technical, radiological and natural events.

## Monitor the Disaster and Emergency Plan(s)

**D6060** Monitoring is a process assessing and evaluating the value, efficiency and robustness of the Disaster and Emergency Plan(s). This involves looking at all three stages as discussed above (before, during and after the event stages) and not just the core document, 'the plan'. It is comprehensive and holistic in its questioning of the entire planning process.

Monitoring can be classified as proactive or reactive. The former attempts to identify problems with the plan(s) before the advent of a Major Incident. It is a case of continuously comparing the plan(s) with even minor incidents and loop-holes identified in any training/exercising sessions. Reactive monitoring occurs after a Major Incident or occurrences that could have led up-to a Major Incident, thereby reflecting back and assessing if the plan(s) need improvement. Both types are important.

Figure 5: Monitoring Matrix

| Stages in Planning | 1. Before the Major Incident | 2. During the Major Incident | 3. After the Major Incident |
|---|---|---|---|
| Monitoring Types | | | |
| A. PROACTIVE | • Risk Assessments | • Inspections | • Inquiries |
| | • Testing via Training or Exercises | • Live Interviews | • Systems Review |
| | • Major Incident Assessment | • Feedback | • Counselling Reports |
| | • Facilities Inspections | | |

| B. REACTIVE | • Incident Statistics | • Critical Assessments | • Brainstorming |
|---|---|---|---|
| | • Incident Reports | • Incident Levels | • Loss Assessments |
| | • Warnings | • Audio-Visual Assessment | • Enforcement |
| | • Notices | | |

The above cells are not strictly mutually exclusive; many techniques are both proactive and reactive. A third dimension is added in *COMAH* cases, that of, On-Site and Off-Site Emergency Plans (types of planning).

- Cell A1 (proactive before the Major Incident)
  - Risk Assessments and Hazard Analysis techniques will identify significant hazards and their risk level. Monitoring such risk is an index of danger, which in turn is a variable in the type and potential of the Major Incident, outlined in Figure 1. The HSE's *Five Steps approach* or their *Quantified Risk Assessment* methodology provides outlines of assessing risk. Hazard techniques include Hazard and Operability Studies, HAZANS, Fault and Event Tree Analysis etc.
  - Testing will identify any problems or concerns with the Disaster and emergency plan. For example, a live exercise or a synthetic simulation could identify factors the plan has not considered or which may not be practicable if the Major Incident was to arise. Thus, enabling questioning, critical appraisal and comments should not be perceived as a threat or being awkward with the plan(s) but can provide vital information.
  - Major Incident Assessment, which is a periodical overview of the plan(s), every quarter or semi-annual by both internal and external persons. This could identify areas of concern. This Assessment compares the plan(s) with the potential threat – can the former cope with the threat? Threats change as technology and know-how changes, so such Assessments become another vital source of information.
  - Facilities inspections of gas, electricity, water, building structure, equipment available/not available etc. can highlight issues of physical resourcing and adequacy of such resourcing.
- Cell B1 (reactive after the Major incident)
  - Incident statistics will show the type of incidences, the type of injury, when and where it occurred. This enables the organisation to hypothesise/build a picture of the potential and severity of a bigger incident. One cannot divorce occupational health and safety data from 'Disaster and Emergency Management'.
  - Analysing incident reports should enable issues of causation to be assessed. What type of occurrence could trigger a Major Incident? Identifying and developing a pattern of causes will enable one to assess if the plan(s) account for such causes.
  - Warnings from employees, contractors, enforcers, public and others of possible and serious problems are to be treated with seriousness. All such warnings are to be analysed and a common pattern and trend spotted.
  - Any notices served by enforcers will identify failings in the operation and the remedial actions required. These can be factored into the plan(s).
- Cell A2 (proactive during the Major Incident)
  - Inspections will be made even as the event occurs. Inspections can range from the stability of the structure through to how personnel behaved and coped. As these

inspections are made, the command and control team needs to evaluate if any aspect of the plan(s), which is a 'live document' need immediate changes.

— Live Interviews with internal personnel and emergency services' personnel will enable a continuous appraisal of any difficulty with procedures, arrangements and instructions that emanate from the plan(s). Again, these can lead to immediate changes to the plan(s).

— Feedback is a proactive technique of requesting regular, interval information on and off-site. This enables a picture to be constructed of what could happen next; trying to anticipate the next sequence and if the plan(s) can cope with it.

- Cell B2 (reactive during the Major Incident)

    — Critical Assessments are carried out after some unexpected occurrence, which causes uncertainty and may even threaten the efficacy of the plan(s). The Critical Assessment is by the command and control team as a whole. Why did this happen? Why did we not account for it in the plan(s)?

    — Incident levels – in particular if serious injury or fatalities are increasing, then at a moral or philosophical level one needs to ask if the plan(s) have been overwhelmed by reality. All forms of planning, including the statutory *COMAH* planning must not be viewed with rigidity. If the plan(s) are failing, it is better to re-appraise and re-plan. *COMAH (SI 1999 No 743)* does not overtly allow for this, although it stresses flexibility and continuous appraisal of the event. In such a case, there has to be quick and clear decision-making, with consequential command and control, as well as immediate communication of this 'alternative plan'. Training and exercising sessions need to factor in this dimension and equip people with decisional techniques.

    — Audio-visual assessments can be a dramatic means of understanding the actual event. This can be video or photographic footage shot by the incident personnel or from the emergency services. This enables monitoring of the extent, potential and actual threat from the Major Incident.

- A3 (proactive after the Major Incident)

    — Inquiries are proactive, even though the event has happened, the inquiry (whether internal or external) will identify strengths and weaknesses in the plan(s), which can lead to future improvements in planning.

    — Systems review, is an overhaul of the entire reaction and holistic experience of the organisation to the trauma of a Major Incident. This involves developing future coping strategies for personnel and issues of how well did the organisation respond? Were there adequate resources in place to cope?

    — Counselling reports will identify the experiences, perceptual and cognitive issues that affected personnel and others. This can provide probably the most significant information on behavioural response of the command and control team and those that were injured. Which in turn can be factored into training and exercise programmes, which itself will lead to personnel skill improvements.

- Cell B3 (reactive after the Major Incident)

    — Brainstorming is an open-ended, participative and indeed critical analysis of what went wrong and what was right with the plan(s). Brainstorming should also be inclusive, involving emergency services and possibly enforcers as well, so that their guidance is factored in.

— The loss adjusters report will be a vital document as to the chain reaction that lead to the event and the consequences that followed. For legal reasons, their findings may have to be applied before insurance cover is available.

— Enforcement notices and enforcers reports will contain recommendations, which need to be viewed as lessons for the future.

## Audit and Review

**D6061** The final stage of the DEMS process is Audit and Review. An audit is a comprehensive and holistic examination of the entire DEMS process (Figure 2). An Audit will identify stages in this process that need improvement. A Review is an act of 'zooming in' into that particular stage and carrying out those improvements.

### Major Incident Auditing

**D6062**

- Major Incident Auditing can be qualitative or quantitative. The former adopts a 'yes' or 'no' response format to questions. The latter asks the auditor to rate the issue being examined from say 0–5.

- The Audit must be comprehensive, assessing every aspect of the DEMS process. This means that the Audit will take time to be completed. The Audit is not some 'inspection' which is more random focusing upon a hazard rather than the complete system.

- Audits essentially benchmark (compare and contrast) performance. This can be against the DEMS process identified above or against legislation (e.g. *COMAH* (*SI 1999 No 743*)). The benchmarking could also be against another site or wider industry standards.

- Should Audits be carried out in-house or rely on external consultants? There are costs and benefits associated with both, with no definitive answer. The *Management of Health and Safety at Work Regulations 1999* (*SI 1999 No 3242*) in the UK, emphasises the need to develop and use in-house expertise in relation to general OSH issues, with a reliance on external specialists as a last resort. This may be interpreted as best practice for DEM.

- Audits can be annual or semi-annual. The more complex the operation and risk it poses, the greater the need for semi-annual audits.

- Audits should be proactive, that is learning from the weaknesses in the DEMS process and reducing or eliminating such weaknesses for the future.

- Finally, the results of the Audit need to be fed-back into the DEMS process and all affected persons informed of any changes and risk management issues arising.

- Audits are holistic (assess anything associated with Major Incidents), systemic (assess the entire DEMS process, i.e. the 'system') and systematic (that is logical and sequential in analysis).

### Review

**D6063**

(1) The Review of any specific problems needs to be actioned by the organisation. The consultant will identify the areas of concern and make recommendations but the final discussion and implementation lies with the organisation. This needs to be led by senior officials in the organisation.

(2) Reviews are by definition 'diagnostic', meaning the organisation needs to look at causation and cure of the failure in any part of the DEMS process.

(3) Budgeting both in time and resource terms is critical in the Review, as it will require management and external agency involvement.

(4) A Review can be carried out at the same time as an audit. A Review can also be a legal requirement, as with *Reg 11* of *COMAH* (*SI 1999 No 743*), which requires a Review and where necessary a revision of the On-Site and Off-Site Emergency Plans for Top-Tier establishments.

# Case studies

D6064    A summary of four Major Incidents follows (off-shore, rail underground, air aviation and land based chemical plant).

## Piper Alpha

D6065

- Offshore Disaster in 1988. 167 workers died in North Sea.
- Major explosion and fire on offshore platform. Piper Alpha involved four rigs linked together. Gas, crude oil and compressed gases were drilled.
- Piper Alpha involved various levels of work along the platform – mining at a certain level, accommodation above that and the helicopter pad at the very top.
- At 22.00 hours whilst 62 worked night shifts and 226 were on the platform, an explosion occurred then a fireball swept the platform. Thereafter a number of small explosions occurred.
- The water systems and emergency systems failed to respond.
- Three may day calls were sent out and personnel assembled at deck d.
- The radio and lighting systems then failed.
- At 22.20 p.m. there was a rupture of the gas riser on another rig connected to Piper Alpha.
- This was followed by explosions and ignition of gas and crude.
- At 22.50 p.m. another explosion occurred followed by the structure collapsing.
- The formal inquiry found both technical and organisational failings. The maintenance error that led to the leak was due to lack of training and poor maintenance procedures.
- There was a breakdown in the communication systems and the permit to work system at shift changeover.
- There were inefficient safety procedures.
- Human error therefore existed at various levels.

## Moorgate

D6066

- Underground rail accident in 1974.
- A passenger train carrying 300, overshot the platform at a speed to be around 30 – 40 mph.
- The warning light was knocked down on the track and hit hydraulic buffers.
- The lead car and hit the tunnel roof with the second carriage hitting this.
- 42 people were killed plus the driver and 74 injured.

- Driver was inexperienced and guard had little training.
- Brakes had to be checked visually.
- Driver should have slowed train down via manual controls.
- Safety systems were criticised – poor use of modern technology to slow down the train and the design and layout of impact equipment.

## Kegworth

D6067

- A Boeing 737 had taken off to Belfast. A known fault to the right hand engine had been logged and corrected.
- The air conditioning on a 737 is driven from the right hand engine in most cases.
- During flight, the pilot spotted vibrations, excess of smoke and fumes; he throttles the right engine back. The left engine throttles back automatically at the same time.
- It is said that there was a warning light showing fire in the right engine.
- The pilot seeks to land at the East Midlands Airport near Derby.
- The signal was incorrect and really the problem was with the left engine. The correct functioning had been turned off. Thus, there was one faulty engine and one switched off. The plane could have landed with one engine.
- The pilot landed on the M1 near Kegworth village.
- Kegworth demonstrates importance of cognitive factors and ergonomic design.

## Flixborough

D6068

- Chemical plant destroyed.
- A part (reaction vessel) is removed and in its place a bent pipe is inserted. The pipe is of inferior material and no problem assessment was carried out to assess the impact the pipe would have.
- The process was restarted and the pipe ruptured, releasing flammable vapour clouds that in turn ignited.
- Explosion resulted and other pipes were ruptured. Fires resulted in the complex.
- Poor maintenance and monitoring were prime factors in disaster.

## Conclusion

D6069     All organisations and societies need to prepare for worst case scenarios. DEMS provide a logical framework to understand the main stages in effective preparation. The larger the organisation, the more detailed and analytical the preparation needs to be. Finally, DEMS is a live and open system requiring continuous monitoring.

## Sources of information

D6070     These include:

1.   Legislation

- *COMAH* sites and indeed non-*COMAH* sites will find it useful in relation to *COMAH* (*SI 1999 No 743*) as a legal or managerial benchmark.

2. Guidance notes

    - *Emergency Planning for Major Accidents: Control of Major Accident Hazards Regulations 1999*, HSG 191 is joint Guidance from the HSE, Environment Agency and the Scottish equivalent, SEPA. This interprets *COMAH* in a user-friendly manner and recommends implementation approaches.
    - *Dealing with Disaster* by the Home Office (Third Edition) is a broad but useful outline of the managerial issues involved in planning. It is aimed at the Police, Fire Service and voluntary bodies.
    - *Planning for Major Incidents: The NHS Guidance*, published by the Department of Health, NHS Executive (1998 Edition) is similar in approach to the Home Office Guidance. This is also available at http://www.open.gov.uk/doh/epcu/epcu/index.htm
    - There are many other useful guidance documents that can be searched for at the following web sites:
        — http://www.open.gov.uk/hse

            This will enable one to search for specific guidance, case studies and documents.
        — epc.ho@gtnet.gov.uk

            The Home Office Emergency Planning College can be contacted at this web site where further assistance can be obtained.
        — http://www.environment-agency.gov.uk

            The Environment Agency will have details of specific natural environment or flood related guidance;

3. The European Commission

    DG XI is the department responsible for environment, nuclear safety and civil protection. It can be accessed via the EC web site http://www.europa.eu.int

4. International Agencies

    The United Nations web site is http://www.un.org and then search for 'humanitarian affairs'.

5. Professional bodies such as the:

    - International Institute of Risk & Safety Management, telephone 020 8741 0835.
    - Institution of Occupational Safety & Health, telephone 0116 257 3100.
    - Business Continuity Institute, telephone 0161 237 1007.
    - Fire Protection Association, telephone 020 7902 5306.
    - Loss Prevention Council, telephone 020 8207 2345.
    - Society of Industrial and Emergency Safety Officers, telephone 01642 816281.

Please note that since the reorganisation of 'civil contingencies' in the UK Government, the Home Office is no longer responsible for strategic 'Emergency Planning' rather it is the Cabinet Office. Consequently, the guidance and information sources will change and readers are requested to visit the Cabinet Office web site on a regular basis.

# Display Screen Equipment

## Introduction

**D8201** The *Health and Safety (Display Screen Equipment) Regulations 1992 (SI 1992 No 2792)* ('the *DSE Regulations*'), together with the other 'six pack' Regulations, came into operation on 1 January 1993. Their aim was to combat upper limb pains and discomfort, eye and eyesight effects, together with general fatigue and stress associated with work at display screen equipment ('DSE'). The Health and Safety Executive states in the guidance booklet to the Regulations (*Work with display screen equipment* (L26)) that they do not consider there are any radiation risks from DSE or special problems for pregnant women. The Regulations require assessments to be carried out to identify any health and safety risks at workstations used by 'users' or 'operators' (as defined). 'Users' have the right to an eye and eyesight test and any spectacles (or contact lenses) found to be necessary for their DSE work must be provided by their employer. Several minor amendments were made to the Regulations by the *Health and Safety (Miscellaneous Amendments) Regulations 2002 (SI 2002 2174)*. The latest HSE guidance booklet (L26) incorporates these amendments.

## The Regulations summarised

### Definitions

**D8202** *Regulation 1* of the *DSE Regulations (SI 1992 No 2792)* contains a number of important definitions which are summarised below:

- *Display screen equipment*

  Any alphanumeric or graphic display screen, regardless of the display process involved.

- *Workstation*

  An assembly comprising display screen equipment, any optional accessories, any disk drive, telephone, modem, printer, document holder, work chair, work desk, work surface or other peripheral item, and the immediate environment around.

- *User*

  An employee who habitually uses DSE as a significant part of his normal work.

- *Operator*

  A self-employed person who habitually uses DSE as a significant part of his normal work.

The HSE guidance states that the main factors to consider in determining whether a person is a 'user' or 'operator' are:

- use of DSE normally for continuous or near-continuous spells of an hour or more;
- use of DSE more or less daily;
- the need to transfer information quickly to or from the DSE;

Other factors are:

- high levels of attention and concentration are required;
- high dependence on DSE with little choice about using it;
- special training or skills are needed to use the DSE.

Detailed examples are given in the HSE guidance booklet providing pen-portraits of definite users, possible users and those who are definitely not users. Homeworkers, teleworkers and agency workers all come within the scope of the Regulations, if they fulfill the definition of user or operator – see D8213 below.

## Exclusions

**D8203** The *DSE Regulations* (*SI 1992 No 2792*) do not apply to:

- drivers' cabs or control cabs for vehicles or machinery;
- DSE on board a means of transport;
- DSE mainly intended for public use;
- portable systems not in prolonged use;
- calculators, cash registers or other equipment with small displays;
- window typewriters.

However, the *Health and Safety at Work etc. Act 1974* and regulations made under it, such as the *Workplace* (*Health, Safety and Welfare*) *Regulations 1992* (*SI 1992 No 3004*), still apply to the use of such equipment.

## Assessment of workstations and reduction of risk

**D8204** *Regulation 2* of the *DSE Regulations* (*SI 1992 No 2792*) requires employers to perform a 'suitable and sufficient' analysis of all workstations which are:

- used for their purposes by 'users';
- provided by them and used for their purpose by 'operators';

to assess the health and safety risks in consequence of that use. They must then reduce the risks identified to the lowest level reasonably practicable. As for other types of assessments, an assessment must be reviewed if there is reason to suspect it is no longer valid or there have been significant changes. Further guidance on assessment, including an assessment checklist (at D8216), is provided later in the chapter.

## Requirements for workstations

**D8205** *Regulation 3* of the *DSE Regulations* (*SI 1992 No 2792*) states that all workstations must meet the requirements laid down in the *Schedule* to the Regulations. The requirement must relate to a component present in the workstation concerned and be relevant in relation to the health, safety and welfare of workers. For example there is no need to provide a document holder (referred to in the *Schedule*) if there is little or no inputting from documents and some individuals with back complaints may benefit from a fixed back rest or a special chair without a back rest. The HSE guidance booklet provides further examples of where the requirements of the *Schedule* may not be appropriate.

## Daily work routine of users

**D8206** Employers are required by *Regulation 4* of the *DSE Regulations* (*SI 1992 No 2792*) to plan the activities of 'users' so that their DSE work is periodically interrupted by breaks or changes of activity. Breaks should be taken before the onset of fatigue and preferably away from the screen. It is best if users are given some discretion in planning their work and are able to arrange breaks informally rather than having formal breaks at regular intervals. Some software tools provide a means of ensuring that users take regular breaks but HSE guidance draws attention to the limitations of such software.

## Eyes and eyesight

**D8207**  *Regulation 5* of the *DSE Regulations* (*SI 1992 No 2792*) gives 'users' the right (at their employer's expense) to an appropriate eye and eyesight test by a competent person before becoming a 'user' and at regular intervals thereafter. (Some employers are re-testing at two yearly to five yearly intervals – in some cases varying the period depending on the age of the employee. Professional advice should be sought in cases of doubt). 'Users' are also entitled to tests on experiencing visual difficulties which may reasonably be considered to be caused by DSE work. Such tests are an entitlement for 'users' and are not compulsory.

Tests are normally carried out by opticians and involve a test of vision and an examination of the eye. Where companies have vision screening facilities, 'users' may opt for a screening test to see if a full eye test is needed. If the eye test shows a need for spectacles (other than the 'user's' normal spectacles) the basic cost of these must be met by the employer, but employees must pay for extras, e.g. designer frames or tinted lenses.

## Provision of training

**D8208**  'Users' are required by *Regulation 6* of the *DSE Regulations* (*SI 1992 No 2792*) to be provided with adequate health and safety training in the use of any workstation upon which they may be required to work. Training may also be required where workstations are substantially modified. The training should include:

- the causes of DSE-related problems, e.g. poor posture, screen reflections;
- the user's role in detecting and recognising risks;
- the importance of comfortable posture and postural change;
- equipment adjustment mechanisms, e.g. chairs, contrast, brightness;
- use and arrangement of workstation components;
- the need for regular screen cleaning;
- the need to take breaks and changes of activity;
- arrangements for reporting problems with workstations, or ill health symptoms;
- information about the Regulations (especially eyesight tests and breaks);
- the user's role in assessments.

The HSE has published a leaflet (*Working with VDUs* (INDG36)) which provides a useful reference for training purposes.

## Provision of information

**D8209**  *Regulation 7* of the *DSE Regulations* (*SI 1992 No 2792*) requires employers to ensure that operators and users at work within their undertaking are provided with adequate information. The table below shows the responsibility of the 'host' employer in this respect (and also gives a good guide to his responsibilities generally under the Regulations).

| The 'host' employer must inform on: | Regulation: | Own users: | Other users (i.e. agency staff): | Operators (self-employed): |
|---|---|---|---|---|
| DSE and workstation risks | | YES | YES | YES |
| Risk assessment and reduction measures | 2 and 3 | YES | YES | YES |
| Breaks and activity changes | 4 | YES | YES | NO |
| Eye and eyesight tests | 5 | YES | NO | NO |
| Initial training | 6(1) | YES | NO | NO |
| Training when workstation substantially modified | 6(2) | YES | YES | NO |

## DSE workstation assessments

### Decide who will carry out the assessments

**D8210** As with other types of risk assessments, DSE workstation assessments within an organisation may be carried out by an individual or by members of an assessment team. Those responsible for making assessments should have received appropriate training so that they are familiar with the requirements of the *DSE Regulations* (*SI 1992 No 2792*) and they should have the ability to :

- identify hazards (including less obvious ones) and assess risks from the workstation and the kind of DSE work being done;

- use additional sources of information or expertise as appropriate (recognising their own limitations);

- draw valid and reliable conclusions;

- make a clear record and communicate the findings to those who need to take action;

- recognise their own limitations so that further expertise can be utilised where necessary.

They may be health and safety specialists, IT managers, facilities managers or line managers, and there will often be benefits in involving employees' safety representatives in the assessment process.

### Identify the 'users'

**D8211** An important first step is to identify the users' (together with any 'operators') of DSE within the organisation. The definitions in *Regulation 1* of the *DSE Regulations* (*SI 1992 No 2792*) refer to habitual use of DSE. The HSE guidance booklet (L26) provides considerable advice on the factors which must be taken into account, of which time spent using DSE is the most significant. Some organisations have adopted a rule of thumb that anyone spending more than 50 per cent of their time in DSE work is a DSE 'user'. However, the HSE guidance indicates that a less simplistic approach should be taken.

The assessment checklist provided at D8216 includes reference to the other factors which the HSE states should be taken into account in deciding whether an individual is a 'user' or an 'operator'. It should be noted that employers have duties to assess workstations used for the purposes of their undertaking by *all* 'users' or 'operators'. This includes 'users' employed by others (e.g. agency-employed staff), 'operators' (e.g. self-employed draughtsmen or journalists), peripatetic staff (e.g. journalists, sales staff, careers advisors) and homeworkers or teleworkers (see D8213 below).

In practice, most employers do not find it too difficult to decide who their 'users' and 'operators' are. Many have taken the approach of assessing the workstations of those individuals where doubt exists and, if necessary, making a final decision then. More time and expense can often be wasted debating a few borderline cases than would be involved in including them in the definition.

## Decide on the assessment approach

**D8212** In some situations, especially smaller workplaces, it is practicable to carry out a personal assessment of each individual 'user's' or 'operator's' workstation, using a simple checklist (a completed example of such a checklist is provided at D8216). However, in larger organisations the number of DSE workstations may be so great that this approach becomes impracticable (particularly where there are regular office rearrangements taking place) or would require an unnecessarily large input of resources. In such cases the issuing of a self-assessment checklist to identified 'users' will often be more appropriate. A sample checklist, together with guidance to 'users' on its completion, is provided at D8216. The HSE guidance booklet (L26) provides a checklist covering similar topics. Providing that workers are given the necessary training and guidance in how to use such checklists, this approach is perfectly acceptable. However, it must be supported by other actions such as:

- an inspection of areas where DSE workstations are situated (evaluating general issues such as lighting, blinds, housekeeping, desk space and the standards of chairs and DSE hardware);
- providing employees with the option of an assessment of their individual workstation by a specialist;
- responding promptly to problems identified in the completed checklists.

It will often be appropriate to require staff to carry out a workstation self-assessment every year or two, and particularly after office rearrangements.

## Homeworkers and teleworkers

**D8213** Homeworkers and teleworkers (working away from their employer's premises) are subject to the *DSE Regulations* (*SI 1992 No 2792*), whether or not their workstation is provided by their employer. They will be subject to the normal risks from DSE work, some of which may be increased because of their social isolation, the absence of supervision and the practical difficulties of carrying out risk assessments. Risks can be reduced by:

- training such staff to carry out workstation self-assessments (see above);
- requiring them to carry out a self-assessment of their main workstation;
- encouraging them to carry out ad hoc assessments of temporary workstations, e.g. hotel rooms;
- emphasising the importance of ensuring good posture and taking adequate breaks;
- providing clear communication routes for reporting equipment defects and possible health problems;
- responding promptly and effectively to reports of defects or problems.

A free HSE guidance leaflet (*Homeworking: Guidance for employers and employees on health and safety* (INDG226)) provides general guidance on risk assessment in the home environment and the employer's responsibilities for home electrical systems and equipment.

## Observations at the workstation

**D8214** Where workstations are to be assessed on an individual basis the assessment checklist questions are intended to identify the principal factors that the assessor(s) need to look out for (as illustrated in the completed example).

Some of the more common problems identified are related to:

- Posture:
    - height of screen;
    - height of seat or position of backrest;
    - position of keyboard or keyboard technique;
    - need for footrest or document holder.
- Vision:
    - angle of screen;
    - position of lights or need for diffusers;
    - need for blinds to control sunlight;
    - adjustment of brightness or contrast controls.

## Discussions with 'users' and 'operators'

**D8215** One of the dangers of self-assessments is that a minority of employees will blame their DSE work for a variety of unrelated problems. An individual assessment provides an opportunity for a two-way dialogue between the assessor(s) and the 'user' and allows the assessor to evaluate whether stated problems are related to deficiencies in the workstation.

Common causes of problems are:

- Back, shoulders, neck:
    - the height or position of screen;
    - positioning of the seat, including the backrest;
    - the need for a footrest or document holder.
- Hands, wrists, arms:
    - position of keyboard or mouse;
    - keyboard technique.
- Tired eyes or headaches:
    - failure to take breaks or change activities;
    - reflected light (artificial or sunlight).
- Discussions with 'users' can also reveal other important pieces of information, e,g:
    - there are problems with sunlight at certain times of day or periods of the year;
    - the 'user' does not know how to adjust their chair or brightness/contrast controls;
    - the 'user's' chair is broken and incapable of being adjusted;
    - the 'user' has never been offered an eye test.

## Assessment records

**D8216** As for other types of assessment, there is no standard format for DSE workstation assessment records. The completed sample assessment checklist and the self-assessment checklist (both below), together with the guidance on its completion are offered as examples of record formats that have been found successful in practice.

The HSE guidance booklet states that records may be stored in electronic as well as paper form. Self-assessment checklists would particularly lend themselves to being completed and submitted electronically. No guidance is provided on how long records should be kept. Prudent employers may prefer to retain them indefinitely, bearing in mind that civil claims for alleged DSE-related conditions may be submitted many years after it is claimed the condition was first initiated.

| | | | |
|---|---|---|---|
| | Display Screen Equipment Workstation Assessment | | |
| | USER'S NAME: *Christine Jones* | | LOCATION: *Sales* |
| | FACTOR | | COMMENT |
| 1 | **WORK PATTERNS** | | |
| 1.1 | Most time spent per day at DSE | | *5 to 6 hours* |
| 1.2 | Average time per day at DSE | | *4 hours* |
| 1.3 | Number of days per week at DSE | | *5* |
| 1.4 | Longest spell without break | | *2 hours* |
| 1.5 | Can breaks be taken? | | *Yes – at Christine's discretion* |
| 1.6 | Concentration important? | | *Accuracy is important* |
| 1.7 | Speed of operation important? | | *Sometimes* |
| | 'USER' STATUS CONFIRMED | | *Yes* |
| 2 | **PROBLEMS EXPERIENCED** Has the user significant experience of problems with: | | |
| 2.1 | Back, shoulders or neck | | *Regular pains in shoulder and neck* |
| 2.2 | Hands wrists or arms | | *Occasional aching wrists* |
| 2.3 | Tired eyes or headaches | | *Sometimes* |
| 2.4 | Suitability of software | | *Suitable for all tasks* |
| 2.5 | Other problems | | *None* |
| 3 | **LIGHTING/ENVIRONMENT** | | |
| 3.1 | Artificial lighting: | Adequate to see documents | *Yes* |
| | | Any reflection or glare problems | *None – fittings recessed and diffused* |
| 3.2 | Sunlight: | Any reflection or glare problems | *On winter mornings from window behind* |
| | | Suitable blinds available (if necessary) | *Good vertical blinds provided* |
| 3.3 | Noise: | Hindering to communication | *No* |
| | | Distracting or stressful | *Occasionally distracting* |

**D8216** *Display Screen Equipment*

| 3.4 | Temperature and ventilation: | Satisfactory in summer and winter | *Office hot and stuffy in summer* |
|---|---|---|---|
| **4** | **SCREEN** | | |
| 4.1 | Set at suitable height | | *Screen height too low* |
| 4.2 | Stable image with clear characters | | *Good. Able to vary colours* |
| 4.3 | Brightness and contrast adjustable | | *Both have adjustable controls* |
| 4.4 | Swivels and tilts easily | | *Yes* |
| 4.5 | Cleaning materials available | | *In stationery cupboard* |
| **5** | **KEYBOARD** | | |
| 5.1 | Separate and tiltable | | *Yes* |
| 5.2 | Sufficient space in front | | *Too near edge of desk (wrists bent)* |
| 5.3 | Keys clearly visible | | *Yes* |
| **6** | **DESK AND CHAIR** | | |
| 6.1 | Desk size adequate | | *Satisfactory* |
| 6.2 | Sufficient leg room | | *Materials being stored in desk well* |
| 6.3 | Desk surface low reflectance | | *Yes – wooden surface* |
| 6.4 | Suitable document holder (if required) | | *Not provided* |
| 6.5 | Chair comfortable and stable | | *Yes* |
| 6.6 | Chair height adjustable | | *Yes* |
| 6.7 | Back adjustable (height and tilt) | | *Christine did not know how to adjust* |
| 6.8 | Footrest available (if required) | | *Not required* |

**OTHER COMMENTS**

*Christine was advised to take short breaks more regularly.*
*Some lengthy jobs without breaks seem to be the cause of her tired eyes and headaches.*
*Repositioning of the screen and provision of a document holder should overcome the shoulder and neck problems.*
*Christine was shown how to adjust her chair back.*

| No. | Actions required | Responsibility |
|---|---|---|
| 3.2 | Close blinds when sunlight bright. | *C Jones* |
| 3.4 | Free up office windows (seized up by paint). | *Facilities Manager* |
| 4.1 | Provide screen stand and keep top of screen at eye level. | *Office Manager/ C Jones* |
| 5.2 | Keep space in front of keyboard – wrists horizontal. | *C Jones* |
| 6.2 | Remove items from desk well. | *C Jones* |
| 6.4 | Provide document holder. | *Office Manager* |

| 6.7 | Ensure induction includes chair adjustment mechanisms. | *Training Department* |
|---|---|---|

| Assessor's name: | *B Wright* | Signature: | *B Wright* | Date: *24/5/04* |
|---|---|---|---|---|

**PROGRESS WITH ACTIONS**

All recommendations acted upon although Christine still needs to remember to take regular breaks on lengthy jobs. Pains in shoulder and neck have ceased and Christine only very occasionally experiences headaches and tired eyes.

*B. Wright 12/7/04*

Planned date for assessment review: July 2006

DISPLAY SCREEN EQUIPMENT
WORKSTATION

SELF ASSESSMENT
CHECKLIST

USER'S NAME: LOCATION:

| 1 | LIGHTING AND WORK ENVIRONMENT | COMMENTS |
|---|---|---|
| 1.1 | Is artificial lighting adequate? | |
| 1.2 | Does it cause any reflection or glare problems? | |
| 1.3 | Any reflection or glare problems from sunlight? | |
| 1.4 | Are suitable blinds available (if necessary)? | |
| 1.5 | Are temperature and ventilation satisfactory in summer and winter? | |
| 2 | SCREEN AND KEYBOARD | |
| 2.1 | Is your screen set at a suitable height? | |
| 2.2 | Stable image with clear characters? | |
| 2.3 | Brightness and contrast adjustable? | |
| 2.4 | Screen swivels and tilts easily? | |
| 2.5 | Cleaning materials available? | |
| 2.6 | Is your keyboard tiltable? | |
| 2.7 | Have you sufficient space in front of it? | |
| 3 | DESK AND CHAIR | |
| 3.1 | Is your desk size adequate? | |
| 3.2 | Is there sufficient leg room under it? | |

*Display Screen Equipment*

| | | |
|---|---|---|
| 3.3 | Do you have a suitable document holder (if required)? | |
| 3.4 | Is your chair comfortable and stable? | |
| 3.5 | Can you adjust your seat height? | |
| 3.6 | Can you adjust the height and tilt of your chair back? | |
| 3.7 | Do you have a footrest (if required)? | |
| 4 | HAVE YOU HAD SIGNIFICANT EXPERIENCE OF PROBLEMS WITH: | |
| 4.1 | Your back, shoulders or neck? | |
| 4.2 | Your hands, wrists or arms? | |
| 4.3 | Tired eyes or headaches? | |
| 4.4 | The suitability of the software you use? | |
| 4.5 | Other problems? | |
| ANY OTHER PROBLEMS OR COMMENTS? | | |
| WOULD YOU LIKE A FURTHER ASSESSMENT OF YOUR WORKSTATION? | | Yes/No |
| Signature: | Date: | |
| **COMPLETED FORMS MUST BE SENT TO THE HUMAN RESOURCES DEPARTMENT** | | |

# GUIDANCE ON COMPLETING THE DSE WORKSTATION SELF-ASSESSMENT

## Lighting and work environment

1.1 Artificial lighting should be adequate to see all the documents you work with.

1.2 Recessed lights with diffusers shouldn't cause problems. Lights suspended from ceilings might.

1.3 There may be problems in the early morning or afternoon, especially in winter when the sun is low.

1.4 Blinds provided should be effective in eliminating glare from the sun.

1.5 Strong sunlight may create significant thermal gain at times.

## Screen and keyboard

2.1 The top of your screen should normally be level with your eyes when you are sitting in a comfortable position.

2.2 There should be little or no flicker on your screen.

2.3 You should know where the brightness and contrast controls are.

2.4 The screen should swivel and tilt so that you can avoid reflections.

2.5 You should know where to get cleaning items for your screen (and keyboard, if necessary).

2.6 Small legs at the back of your keyboard should allow you to adjust its angle.

2.7 Space in front the keyboard allows you to keep your wrists horizontal and to rest your hands and wrists when not keying in.

## Desk and chair

3.1 Your desk should have sufficient space to allow you to have your screen and keyboard in a comfortable position and accommodate documents, document holder, phone etc.

3.2 There should be enough space under the desk to allow you to move your legs freely.

3.3 If you are inputting from documents, using a document holder helps avoid frequent neck movements.

3.4 Chairs with castors must have at least five (four is very unstable).

3.5 You should be able to adjust your seat height to work in a comfortable position (arms approximately horizontal and eyes level with the top of the screen).

3.6 The angle and height of your back support should be adjustable so that it provides a comfortable working position.

3.7 DSE users who are shorter may need a footrest to help them keep comfortable when sitting at the right height for their keyboard and screen (see 3.5).

## Possible problems

4.1 Problems with back, shoulders or neck might indicate your screen is at the wrong height, an incorrectly adjusted chair or need for a document holder.

4.2 Problems with hands, wrists or arms might indicate incorrect positioning of the keyboard or a poor keying technique. Hands should not be bent up at the wrist and a soft touch should be used on the keyboard, not overstretching the fingers.

4.3 Tired eyes or headaches could indicate problems with lighting, glare or reflections.

They may also indicate the need to take regular breaks away from the screen. Persistent problems might need an eye test – contact Human Resources to request one.

4.4 The software should be suitable for the work you have to do.

> IF YOU HAVE A PROBLEM DISCUSS IT WITH YOUR LINE MANAGER. REFER UNRESOLVED PROBLEMS OR REQUESTS FOR FURTHER ASSESSMENTS TO THE HUMAN RESOURCES DEPARTMENT.

# Special situations

**D8217** The HSE guidance booklet (L26) provides further guidance on several aspects of DSE work which have developed over the last decade as information technology has changed.

## Shared workstations

**D8218** 'Hot desking' arrangements mean that many workstations are shared, sometimes by workers of widely differing sizes. Assessments of such workstations should take into account aspects such as:

- whether chair adjustments can accommodate all the workers involved;
- the availability of footrests;
- adequacy of leg room for taller workers;
- arrangements for adjusting the heights of screens (e.g. stands).

In some situations the use of desks with adjustable heights may be appropriate.

## Work with portable DSE

**D8219** Work with laptop computers as well as mobile phones or personal organisers which can be used to compose, edit or view text is becoming increasingly common. Where mobile phones or personal organisers are used in this way for prolonged periods they are subject to the *DSE Regulations* (*SI 1992 No 2792*), as is all work with laptops.

Mobile phones or personal organisers should not generally be used for sufficiently long periods to require a workstation assessment but work involving laptops is likely to be of much longer duration. The revised HSE booklet (L26) contains detailed guidance on the selection and use of laptop computers.

The design of laptops is such that postural problems are much more likely to result and there will be a far greater need to ensure that sufficient breaks are taken. Where the laptop is used in the office, the risks can be reduced considerably by the use of a docking station and the HSE booklet also provides guidance on this. Alternative means of reducing risks include using laptops with detachable or height adjustable screens and a facility for attaching an external mouse, keyboard or keypad. Other risks associated with the use of portable equipment (e.g. manual handling issues and possible theft involving assault) should also be considered.

## Pointing devices

**D8220** Much work at DSE workstations involves the use of a mouse, trackball or similar device to move the cursor around the screen and carry out operations. The HSE booklet (L26) provides guidance on:

- *Choice of pointing devices*

    Suitability of the device for:
    - the environment (space, position, dust, vibration);
    - the individual (right or left handed, physical limitations, existing upper limb disorder);
    - the task (some devices are better than others in respect of speed or accuracy).

- *Use of pointing devices*

    Issues to be considered include:

- positioning (close to the midline of the user's body, not out to one side);
- work surface (particularly its height and degree of support for the arm);
- mousemats (smooth, large enough, without sharp edges);
- software settings (suitable for the individual user);
- task organisation (point device use mixed with other activities);
- training (how to set up and use devices);
- cleaning and maintenance.

The booklet also contains specific guidance on touch screens and speech interfaces.

# After the assessment

## Review and implementation of recommendations

**D8221** The general guidance provided in the chapter on RISK ASSESSMENT is equally applicable to recommendations made as a result of DSE workstation assessments. Some issues of general importance will need to be reviewed with others, particularly if there are significant cost implications or practical problems. Responsibility for implementing recommendations should be allocated clearly and all recommendations followed up. It should be noted that the sample assessment checklist includes a space which can be used at a follow-up of an assessment to describe 'Progress with actions'.

## Assessment review/re-assessments

**D8222** Changes to the layout of office accommodation often take place with bewildering rapidity. Some of these changes have significant implications for DSE workstations, others do not. To carry out a re-assessment every time the office layout is changed would be an inefficient use of resources. A better approach is to make both management and DSE 'users' aware that a review or reassessment should be requested from the assessor or assessment team whenever significant changes take place. The use of self-assessments is particularly relevant in such situations. This can be supported by observations by the assessor(s) of changes which have occurred or are in progress.

Because of the frequency of changes a periodic review of DSE workstation assessments would be beneficial. The HSE does not provide any guidance on this aspect. Reviews at two-yearly intervals may be appropriate for larger workplaces where changes are regularly taking place whilst five yearly intervals would probably be suitable for smaller workplaces where the layout of workstations is more static. Significant changes in DSE equipment, software, furniture, lighting etc. would obviously justify a review of relevant workstation assessments.

# Dangerous Substances and Explosive Atmospheres

## Background

**D0901** Until the *Dangerous Substances and Explosive Atmospheres Regulations 2002* (*SI 2002 No 2776*) ('*DSEAR*') came into force in December 2002, the regulation of substances that could give rise to the risks of fire and explosion had built up incrementally for different types of substances in different industries. There were separate regulations for highly flammable liquids and liquefied petroleum gases, celluloid in factories and workshops, dry cleaning, cinematograph film manufacture and petroleum. However, other substances and activities giving rise to these risks, such as flour dust and certain grinding operations, were not covered by health and safety regulations specific to the risks they created, and it was becoming increasingly impractical to legislate on a substance or activity specific basis as technology advanced.

The need for the UK government to implement two European Directives into UK law, i.e. the Chemical Agents Directive and the Explosive Atmospheres Directive, created the opportunity to modernise UK legislation and led to the enactment of the *DSEAR*. The Chemical Agents Directive ('CAD') requires employers to protect workers from fire, explosion and health risks arising from chemical agents present in the workplace. The Explosive Atmospheres Directive ('ATEX') requires employers to protect workers from the risks of fire and explosion caused by explosive atmospheres.

The health requirements of CAD have been implemented by changes to existing health legislation, mainly the *Control of Substances Hazardous to Health Regulations 2002* (*SI 2002 No 2677*) ('*COSHH*'), with some changes to legislation on asbestos and lead. On safety, it was decided that CAD and ATEX would be implemented together in one set of completely new regulations, i.e. *DSEAR*, dealing with all substances that could give rise to fires, explosive atmospheres, explosions and similar energetic events.

As the legislation would apply generically to all substances with these potential risks, it removed the need to have to separate regulations for particular substances or groups of substances, and covered substances for which specific regulation did not previously exist. As a result, around 20 pieces of old health and safety legislation could be repealed making a considerable simplification for duty holders and regulators alike. Significantly, the *DSEAR* have replaced the *Highly Flammable Liquids and Liquefied Petroleum Gases Regulations 1972* (*SI 1972 No 917*).

## Introduction

**D0902** The *DSEAR* (*SI 2002 No 2776*) follow the trend of modern goal setting regulation based on the concept of risk assessment and management by the employer or the person responsible for the risk. This allows the Regulations to deal generically with the risks of fire and explosion, where prescriptive requirements relate directly to the risks, and not to the characteristics of any particular substance or activity. The amount of prescriptive requirement is thereby minimised.

The onus in law is placed squarely on those responsible for the risks to assess the risks within their situation and to manage them safety. It requires duty holders to think for themselves about safety and to take the necessary action. In support of the Regulations and to assist duty holders, the Health and Safety Commission is publishing a series of Approved Codes of Practice giving practical advice on how to comply with specific aspects of the Regulations and additional guidance on good practice.

This chapter covers the requirements of the *DSEAR* in a form that should assist duty holders to understand easily what is needed to comply with the law. The first part of the chapter is concerned with:

- aspects relating to the application of the Regulations;
- typical activities;
- who the duty holders are;
- what workplaces are covered;
- what a dangerous substance is; and
- time limits for compliance with particular requirements.

The rest of the chapter then deals with the requirements under the Regulations, and more specifically:

- the need for a risk assessment and the principles of risk management, including the classification of places with explosive atmospheres;
- arrangements to deal with incidents and emergencies;
- provision of information, instruction and training and marking, and identification of containers.

Some examples and guidance are given on how to comply with the Regulations, but these should not be taken out of context.

The chapter will be useful to all employers and the self-employed where there is a possibility of dangerous substances or explosive atmospheres being present in their workplace. In practice, this will include a very wide range of businesses and public services, including manufacturing industries, service companies, some retailers, transport and shipping, garages, schools, universities and experimental research institutes, and local authorities. While all have a duty to comply, the manner of compliance can be commensurate with the risk. What is commensurate should be consistent with existing good practice and common sense, but there will no doubt be case law to refer to in time.

# The Regulations

## Scope

**D0903** The *DSEAR* (*SI 2002 No 2776*) set minimum requirements for the protection of persons from the risks of fire and explosion arising from dangerous substances and explosive atmospheres in the workplace. The Regulations assist duty holders to comply with the general requirement to manage risks from substances having these potential hazards under the *Management of Health and Safety at Work Regulations 1999* (*SI 1999 No 3242*). While the *DSEAR* protect against substances with the potential to cause fire and explosion, they stand alongside the *COSHH Regulations* (*SI 2002 No 2677*) which protect against substances hazardous to health.

Many substances are classified as dangerous under the criteria given in the *DSEAR*. These include petrol, liquefied petroleum gas, welding gases, paints, varnishes and certain types of combustible and explosive dusts produced in, for example, machining and sanding operations. The Regulations are therefore relevant to the many businesses in Great Britain at which these substances may be present.

## Typical activities

**D0904** The *DSEAR* (*SI 2002 No 2776*) apply to many types of activities commonly found at work in businesses and public services, of which the following are typical:

- Storage of petrol as a fuel for cars, motor boats, horticultural machinery.
- Use of flammable gases, such as acetylene for welding.
- Handling and storage of waste dusts in a range of manufacturing industries.
- Handling and storage of flammable wastes such as fuel oils.
- Hot work on tanks or drums that have contained flammable material.
- Work activities that could release naturally occurring methane.
- Mining or processing of coal producing dusts.
- Use of flammable solvents such as ether in pathology and school laboratories.
- Storage/display of flammable goods, such as paints, in the retail sector.
- Filling, storage and handling of aerosols with flammable propellants such as LPG.
- Transport of flammable liquids in containers around the workplace.
- Deliveries from road tankers, such as petrol or bulk powders.
- Chemical manufacture, processing and warehousing.
- Extraction and processing of petrochemicals – onshore and offshore.

The examples given in the list above are only representative of the very wide range of relevant activities. All employers and the self-employed must consider if the *DSEAR* could apply to them.

## General conditions for application

**D0905** The *DSEAR* (*SI 2002 No 2776*) apply whenever:

(a) There is work being carried out by an employer or self-employed person.

(b) A dangerous substance is present or liable to be present at the workplace.

(c) The dangerous substance presents a risk to the safety of persons (as opposed to a risk to health).

## Duty holders

**D0906** The responsibility for complying with the *DSEAR* (*SI 2002 No 2776*) falls on the employer, or the self-employed person, whichever is appropriate, who is responsible for the workplace where dangerous substances or explosive atmospheres may be present. [*DSEAR* (*SI 2002 No 2776*), *Reg 4*]. This is normally the owner, lessee or occupier of the workplace. (Note: references to the 'employer' throughout this chapter also apply to self-employed persons where appropriate.)

The duties extend not only to protecting employees but also to any other person, whether at work or not, who may be put at risk by dangerous substances and explosive atmospheres in the workplace. This includes:

- employees working for other employers;
- visitors to the site; and
- members of the public.

However, when making arrangements for dealing with accidents, incidents and emergencies and the provision of information, instruction and training, employers only have duties to persons who are normally at their workplace. [*DSEAR* (*SI 2002 No 2776*), *Reg 4(1)(b)*]. Further, their duties in law in respect of provision of suitable personal protective equipment and appropriate work clothing do not extend to persons who are not employees. [*DSEAR* (*SI 2002 No 2776*), *Reg 4(1)(a)*]. Responsible employers will, however, ensure that the number of persons who need to be in hazardous places is minimised and that all those that are in hazardous workplaces are suitably equipped.

Sometimes two or more employers will share the same workplace (whether permanently or temporarily) where an explosive atmosphere may occur. Here the employer responsible for the workplace (normally the owner or lessee) has the duty to co-ordinate the implementation of the measures to protect employees from the risk of an explosive atmosphere. The duty for implementing the measures relating to dangerous substances would be expected to lie with the employer responsible for those substances. [*DSEAR* (*SI 2002 No 2776*), *Reg 11*].

## Workplaces

**D0907** The *DSEAR* (*SI 2002 No 2776*) define 'workplace' widely as being any premises or part of premises used for or in connection with work. [*DSEAR* (*SI 2002 No 2776*), *Reg 2*]. It includes any place to which an employee has access to while at work, and covers any room, lobby, corridor, staircase or road or other place used for access or where facilities are provided. Public roads are not workplaces within the definition of the Regulations. Premises include:

- all industrial and commercial sites;
- land based installations;
- vehicles and vessels;
- shared buildings, private roads and paths on industrial estates and business parks;
- houses and other domestic premises where there is a work activity.

A few types of workplaces and activities are specifically exempt from some or all of the requirements of the *DSEAR* because there is other legislation fulfilling these requirements, for example, at offshore installations. [*DSEAR* (*SI 2002 No 2776*), *Reg 3*]. The Regulations do not in general apply to activities on board ships, except for construction, reconstruction, conversion and dry dock repairs carried out in Great Britain. The requirements in the *DSEAR* concerning zoning and co-ordination of safety in shared workplaces do not apply at workplaces in respect of the following substances, activities or types of equipment, where other controlling regulations exist:

(*a*) areas used directly for medical treatment;

(*b*) the use of gas appliances and fittings for cooking, heating, lighting etc;

(*c*) the manufacture, handling, use, storage and transport of explosives or chemically unstable substances;

(*d*) any activity at a mine quarry or borehole;

(*e*) land transport regulated by international agreements.

The *DSEAR* apply within Great Britain, and apply outside Great Britain to the extent defined by virtue of the *Health and Safety at Work etc Act 1974* (*Application outside Great Britain*) *Order 2001* (*SI 2001 No 2127*). [*DSEAR* (*SI 2002 No 2776*), *Reg 12*].

## Dangerous substances

**D0908** The *DSEAR* (*SI 2002 No 2776*) apply to any substance or preparation (or mixture of substances) with the potential to create a risk to persons from energetic events such as fire, explosions, or thermal runaway from exothermic reactions etc. The definition of 'dangerous' includes all substances that meet the criteria for being explosive, oxidising, extremely flammable, highly flammable or flammable according to the *Chemicals* (*Hazard Information and Packaging for Supply*) *Regulations* (*SI 2002 No 1689*) ('*CHIP*'). Substances classified as being explosive, oxidising, extremely flammable, highly flammable or flammable under the *CHIP Regulations* are therefore dangerous substances for the purposes of the *DSEAR*. [*DSEAR* (*SI 2002 No 2776*), *Reg 2*].

The definition of 'dangerous' also includes other substances that are not classified under the *CHIP Regulations*, but meet the criteria by virtue of the way they are used or are present at the workplace. For example, the *DSEAR* apply to substances that decompose or react exothermically when mixed with other substances. Peroxides, wood, flour and many other dusts involved with grinding or machining can, when mixed with air, ignite and explode. While diesel (or other high flash point oils) at ambient temperature are not classified as flammable under the *CHIP Regulations*, if they are heated and used at sufficiently high temperatures in a process that creates a fire risk, they may become a dangerous substance for the purposes of the *DSEAR*. [*DSEAR* (*SI 2002 No 2776*), *Reg 2*].

Thus, in order to determine if a substance present is dangerous, two steps are necessary:

(*a*) a check to find out if the substance has been classified under the *CHIP Regulations*;

(*b*) an assessment of whether the physical and chemical properties of the substance and the circumstances of its use in the workplace can create a safety risk to persons from a fire or explosion.

The *DSEAR* are only intended to protect persons from energetic events and the harmful physical effects from thermal radiation (burns), over-pressure (blast injuries) and oxygen depletion (asphyxiation) arising from fires and explosions. They implement the European Explosive Atmospheres Directive (99/92/EC) (ATEX 137) into UK law.

Many of the substances will, however, also create health risks and it should be noted that the *DSEAR* does not address these. Many solvents are toxic as well as being flammable. Health risks are dealt with by the *COSHH Regulations 2002* (*SI 2002 No 2677*), which have been amended to implement the health side of the European Chemical Agents Directive (98/24/EC).

## Commencement

**D0909** The *DSEAR* (*SI 2002 No 2776*) came fully into force on 30 June 2003, although certain transitional arrangements extend to 30 June 2006. [DSEAR (*SI 2002 No 2776*), *Regs 1* and *17*]. In particular, workplaces where explosive atmospheres may occur and that have been in use before 30 June 2003 have until 30 June 2006 to comply with the requirements of the *DSEAR* for zoning and co-ordination of safety in shared workplaces. However, workplaces that are used for the first time after 30 June 2003, and workplaces that are modified, reconstructed, extended or restructured after this date must comply with the zoning and co-ordination of safety requirements as soon as they come into use.

# Risk assessment

## Requirement and considerations

**D0910** When a dangerous substance is, or is liable to be present at the workplace, the employer is required to carry out a risk assessment before commencing any new work activity involving that substance. [*DSEAR* (*SI 2002 No 2776*), *Reg 5(1)*]. The risk assessment required by

**D0911**   *Dangerous Substances and Explosive Atmospheres*

the *DSEAR* is an identification and careful examination of the dangerous substance(s) that may be present, the work activities associated with the substance(s), and how the work activities might give rise to a risk of fire or explosion. The *DSEAR* require that the risk assessment considers certain aspects as follows [*DSEAR (SI 2002 No 2776), Reg 5(2)*]:

(*a*)   the hazardous properties of the substance;

(*b*)   information on safety or any safety data sheets that the supplier may provide;

(*c*)   the circumstances of the work including:

    (i)   the work processes and substances used and their possible interactions;

    (ii)   the amount of substance involved;

    (iii)   the risk when two or more dangerous substances are used in combination;

    (iv)   the arrangements for the safe handling, storage and transport of dangerous substances and waste containing dangerous substances;

(*d*)   activities such as maintenance where there is the potential for a high level of risk;

(*e*)   the effect of measures taken or to be taken to comply with the Regulations;

(*f*)   the likelihood that an explosive atmosphere will occur and persist;

(*g*)   the likelihood that ignition sources, including electrostatic discharges, will be present and become active and effective;

(*h*)   the scale of the anticipated effects of a fire or an explosion;

(*i*)   any places which are or can be connected via openings to places in which explosive atmospheres may occur;

(*j*)   such additional safety information as the employer may need to complete the risk assessment.

The requirements for a risk assessment and the management of risk are similar in principle to those required under the *COSHH Regulations (SI 2002 No 2677)*. A risk assessment may have already been carried out on dangerous substances under the *COSHH Regulations* if they were hazardous to health. However, because of the different nature of the hazards of fire or explosion, the risk assessment under the *DSEAR* has different considerations and must be treated separately. The management of dangerous substances will share aspects in common with the management of substances hazardous to health, but there will be individual elements of management required in each case.

The degree of detail in the risk assessment needs to be commensurate with the degree of risk. For a small workshop, a table showing the substances stored and the hazards that could be created (e.g. spillages) and the precautions in place (e.g. open windows when handling), with any recommended improvements (e.g. provision of an extra fire extinguisher) would probably suffice. For a large factory with many dangerous substances, a more sophisticated risk assessment would be expected that would analyse the threats, detail the protection systems and emergency procedures and the remedial actions in the event of an emergency.

A practical example of a fire risk assessment, and information about the risk assessment requirements contained in the *DSEAR* is given in *Tolley's Practical Risk Assessment Handbook*, 4th edition (November 2003).

## *Recording findings*

**D0911**   The purpose of the risk assessment is to enable employers to decide what measures are needed to eliminate, or reduce as far as reasonably practicable, the safety risks from dangerous substances. An employer with five or more employees is required to record in writing the significant findings of the risk assessment as soon as possible after the assessment is made, [*DSEAR* (*SI 2002 No 2776*), *Reg 5(4)*], including:

- the measures (technical and organisational) that have been taken or will be taken to eliminate and/or reduce the risk;
- sufficient information to show that the workplace and work equipment will be safe during operation and maintenance, including details of hazardous zones, co-ordination of safety in shared workplaces, arrangements to deal with accidents, incidents and emergencies and measures to inform, instruct and train employees.

After completing the risk assessment, the *DSEAR* require that the measures identified by the risk assessment are implemented before any new work commences. [*DSEAR* (*SI 2002 No 2776*), *Reg 5(5)*].

## Review

**D0912** The risk assessment must be kept up to date and the employer is required to review it regularly, [*DSEAR* (*SI 2002 No 2776*), *Reg 5(3)*], particularly if:

(*a*) there is reason to suspect that it is no longer valid, or

(*b*) there has been a significant change in the workplace, work processes or organisation of work.

When changes to the risk assessment are required as a result of the review, the risk assessment must be updated and any resulting safety measures implemented before work commences again. [*DSEAR* (*SI 2002 No 2776*), *Reg 5(5)*].

# Management of risks

## Safety principles

**D0913** Employers are required to ensure that the risks to the safety of employees and others from dangerous substances are either eliminated or reduced as far as reasonably practicable. [*DSEAR* (*SI 2002 No 2776*), *Reg 6(1)*]. Where it is not reasonably practicable to eliminate or reduce the risks, employers are required to take, so far as is reasonably practicable, measures to control the risks and measures to mitigate the detrimental consequences should a fire, explosion or similar event occur. The *DSEAR* therefore advocate the established safety management principles of elimination, reduction, control and mitigation of risk.

## Elimination and reduction of risk

**D0914** The avoidance of risk by replacement of the dangerous substance with another substance or process that eliminates or reduces the risk is preferable if possible. [*DSEAR* (*SI 2002 No 2776*), *Reg 6(2)*]. When it is not reasonably practicable to eliminate the risk, it may be possible to reduce the risk by replacing the dangerous substance with another that is less dangerous (e.g. by replacing a low flashpoint solvent with a high flashpoint one). Alternatively, it may be possible to redesign the process so as to reduce the quantities of dangerous substances involved.

In support of these measures, there are guides published by the Health and Safety Executive on 'Steps to successful substitution of hazardous substances' (HSG 110) which can be purchased from HSE Books) and by the Department of Trade and Industry on 'Process intensification' (which can be obtained from the DTI Publications Unit). When substituting other substances or redesigning the process, it is important to take care that the changes do not create new or increased safety or health risks from other sources.

## Control measures

**D0915** When dangerous substances are present, the *DSEAR (SI 2002 No 2776)* require employers to apply control measures, as far as reasonably practicable, consistent with the risk assessment and appropriate to the nature of the activity or operation [*DSEAR (SI 2002 No 2776), Regs 6(3) and 6(4)*], including the following in priority order:

(a) reducing the quantity of dangerous substance to a minimum (e.g. limit volume stored);

(b) avoiding or minimising releases (e.g. keep in sealed containers);

(c) controlling the release at source (e.g. ensure that valves and stoppers are used);

(d) preventing the formation of an explosive atmosphere (e.g. apply appropriate ventilation);

(e) collecting, containing and removing any releases to a safe place (e.g. fume extraction);

(f) avoiding ignition sources (e.g. sparks, naked flames etc);

(g) avoiding adverse conditions that could lead to danger (e.g. overheating, overpressure);

(h) keeping incompatible substances apart (e.g. by segregation).

## Mitigation

**D0916** When dangerous substances are present, the *DSEAR (SI 2002 No 2776)* require employers to apply mitigation measures, as far as reasonably practicable, consistent with the risk assessment and appropriate to the nature of the activity or operation [*DSEAR Regs 6(3) and 6(5)*], including the following:

(a) reducing the numbers of persons exposed to the risk (e.g. restricting access);

(b) providing plant that is explosion resistant (e.g. robust firmly mounted equipment);

(c) providing explosion suppression or relief equipment (e.g. pressure vents);

(d) taking measures to control or minimise the spread of fires or explosions (e.g. reducing combustible material, fire doors etc);

(e) providing suitable equipment for the protection of personnel (e.g. helmets, visors etc).

## General safety measures

**D0917** In addition to the specific measures relating to dangerous substances and their immediate vicinity, the *DSEAR (SI 2002 No 2776)* also specify a number of general safety measures that employers must take, as far as reasonably practicable. [*DSEAR (SI 2002 No 2776), Reg 6(8) and Sch 1*]. These measures relate to the workplace and work processes as a whole and include:

(a) Ensuring the workplace is designed, constructed and maintained so as to minimise risk (e.g. use of fire resistant materials, installation of appropriate blast walls etc).

(b) Providing work processes that are suitably designed, constructed, assembled, installed so as to reduce risk and ensuring that they are used properly and maintained in an efficient state, working order and good repair.

(c) Ensuring that equipment and protective systems meet the requirements for use in hazardous places (see below), and:

   (i) can be maintained in a safe state of operation independently of the rest of the plant in the event of a power failure;

   (ii) can be manually overridden when incorporated within automatic processes which deviate from the intended operating conditions;

(iii) can dissipate accumulated energy quickly and safely on operation of emergency shutdown;

(iv) contain measures to prevent confusion between connecting devices.

(d) Applying appropriate systems of work including written instructions, permits to work and other procedural systems of organising and controlling work.

(e) Identifying the hazardous contents of containers and pipes. [*DSEAR (SI 2002 No 2776), Reg 10*]. Many will already be marked or labelled under existing legislation. For those that are not, 'identification' may require labelling, marking or warning signs. It could include training, information or verbal instruction.

(f) Arranging for the safe handling, storage and transport of dangerous substances and waste containing dangerous substances. [*DSEAR (SI 2002 No 2776), Reg 6(6)*].

# Classification of places with explosive atmospheres
## Classification scheme

**D0918**  Gases, vapours, mists and dusts can all form explosive atmospheres with air. The source of the gases, vapours, mists and dusts may be a consequence of the natural state of the substance (e.g. hydrogen gas), or it may be a result of the means of storage or handling (as in the case of a flammable liquid), or as a result of a work activity (e.g. grinding or powder processing). Where there is any prospect of an explosive atmosphere, a hazardous area classification should be carried out as an integral part of the risk assessment, and for those areas classified as hazardous, *special precautions* over the sources of ignition are needed to prevent fires and explosions.

In workplaces where an explosive atmosphere may occur, employers must first classify the workplace into hazardous and non-hazardous places. [*DSEAR (SI 2002 No 2776), Reg 7(1)*]. The criteria for deciding whether a place is hazardous is whether an explosive atmosphere could occur in such quantities as to require *special precautions* to protect the health and safety of workers concerned within the meaning of the *DSEAR*. Similarly, a non-hazardous place is one where an explosive atmosphere is not expected to occur in such quantities as to require special precautions to protect the health and safety of workers concerned. [*DSEAR (SI 2002 No 2776), Sch 2 para 1*].

Thus, in the first instance, employers must assess the quantity of explosive atmosphere that could be present in a particular place. The term 'not expected to occur in such quantities' is deliberate and means that employers should also consider the likelihood or possibility of a release of an explosive atmosphere as well as the potential quantities of such a release. Places where it is credible that an explosive atmosphere could exist, even when this is not within normal operating limits should be classified as hazardous. If, on the other hand, a release is extremely unlikely to occur and/or if the quantities released are small, it may not be necessary to classify the area as hazardous.

As an example, a dangerous substance is being carried though a seamless pipe that has been properly installed, maintained and inspected. It is unlikely that the substance will be released. An explosive atmosphere would not be expected to occur from this source and the area surrounding the pipe would be non-hazardous. However, if the pipe contained flanged joints or fittings where there was the possibility of a leak or its condition was uncertain, a release would be a credible event and the area should be classified as hazardous.

With regard to quantity, a spillage from a small bottle of solvent used in a laboratory would release so little flammable vapour that no special precautions are needed other than general control of ignition sources (no smoking, for example) and cleaning and disposing of the spillage. The area would not be classified as hazardous. If, however, the vapour released also had a strong anaesthetic effect, and spillages were known to occur from time-to-time, then the area might be classified as hazardous. When deciding whether it is necessary to classify as 'hazardous' areas with small quantities of dangerous substances, the actual circumstances of use and any specific product guidance should be taken into account.

Where dangerous substances in small pre-packaged containers are stored or are on display for sale in retail premises, for example solvents or aerosols, the area would not normally need to be classified as hazardous. An exception to this might, however, be with storage in poorly ventilated basements. When such substances are held in large quantities, as in a warehouse, with the opportunity for a build up of vapours or a larger spillage, a hazardous classification would be expected. Procedures to control ignition sources and to clean up and dispose of any spillage/release would be needed.

In identifying hazardous and non-hazardous areas and assigning zones, the following matters should be considered:

- The hazardous properties of the dangerous substances involved.

- The amount of dangerous substances involved. The size of any potentially explosive atmospheres is, in part, related to the amount of dangerous substances present. Guidance is given in industry specific codes on the quantities of dangerous substances that may be safety stored. See, for example, the 'Code of Practice on the storage of full and empty LPG cylinders and cartridges' produced by the LPG Association.

- The work processes, and their interaction, including any cleaning, repair or maintenance activities that will be carried out.

- The temperatures and pressure of the dangerous substances. This will affect the nature and extent of any release. Some substances do not form explosive atmospheres unless they are heated (e.g. diesel oil), while others if released under pressure will form a fine mist that can explode, even if there is insufficient vapour.

- The containment system and controls to prevent liquids, gases, vapours or dusts escaping into the general atmosphere (e.g. trays, seals).

- The possibility of an explosive atmosphere forming in an enclosed plant or storage vessel. A release of a dangerous substance into an enclosed space always increases the risk of an explosive atmosphere forming.

- The ventilation (natural or a fan extract system) or any other measures designed to dilute sources of release and ensure that any explosive atmosphere does not persist for an extended time. Well-designed ventilation may prevent the need for any zoned area or reduce it so it has a negligible extent.

When considering the potential for dangerous atmospheres, it is important to consider all substances that may be present in the workplace. Waste products, residues, cleaning agents, materials used for maintenance and fuel could all be potentially dangerous. The possibility that combinations of substances could react to create an ignition source or an explosive atmosphere should not be overlooked.

The refuelling of cars or loading and unloading of petrol from tankers intended for use on public roads involve the introduction of potential sources of ignition in an area where a spill is possible. This would normally meet the criteria for being a hazardous area. In these circumstances, safety can be achieved by isolating power sources (turning off engines) where a transfer of fuel is taking place and making suitable checks before and after transfer and before and after vehicles are moved into or out of the refuelling area. Garage owners have a particular responsibility to monitor the forecourt and to ensure that these practices are implemented.

## Zoning

**D0919** For places classified as hazardous according to D0918 above, employers are required to further classify these places into zones on the basis of the frequency and duration of the explosive atmosphere occurring. [*DSEAR* (*SI 2002 No 2776*), *Reg 7(2)* and *Sch 2 para 2*]. The *DSEAR* give the following classification scheme for mixtures with air of dangerous substances (Zones 0, 1 and 2) and combustible dusts (Zones 20, 21, and 22). In decreasing order of risk:

- *Zone 0* – A place in which an explosive atmosphere consisting of a mixture with air of dangerous substances in the form of a gas, vapour or mist is present continuously or for long periods or frequently.

- *Zone 1* – A place in which an explosive atmosphere consisting of a mixture with air of dangerous substances in the form of a gas, vapour or mist is likely to occur in normal operation occasionally.

- *Zone 2* – A place in which an explosive atmosphere consisting of a mixture with air of dangerous substances in the form of a gas, vapour or mist is not likely to occur in normal operation, but if it does occur, will persist for a short period only.

- *Zone 20* – A place where an explosive atmosphere in the form of a combustible dust in air is present continuously or for long periods or frequently.

- *Zone 21* – A place where an explosive atmosphere in the form of a combustible dust in air is likely to occur in normal operation occasionally.

- *Zone 22* – A place where an explosive atmosphere in the form of a combustible dust in air is not likely to occur in normal operation, but if it does occur, will persist for a short period only.

Within this scheme, terms such as 'occasionally' or 'frequently' are not defined but are assumed to be that which a reasonable person would use. Normal operation is any situation when installations are used within their design parameters. Combustible dusts in the form of layers, deposits or heaps must be considered as a source of an explosive atmosphere and classified accordingly.

The basic principles of area classification are explained within the European standards, BS EN 60079/10 for gases and vapours and BS EN 61241/3 for dusts. These standards form a suitable basis for assessing the extent and type of the zone and can be used as a guide to complying with the requirements of the *DSEAR*. However, the guidance must be applied to the site-specific factors in order to determine the extent and type of zone in any particular case. It should be remembered that an explosive atmosphere may spread into areas away from the source of the hazard, for example through ducts, and these areas must also be included in the classification scheme.

Various organisations have published industry specific codes (e.g. Institute of Petroleum). Providing they are applied appropriately, they are valuable in encouraging a consistent interpretation of the requirements.

Area classification normally takes the form of drawings or plans identifying the hazardous areas and zones. Information about the dangerous substances that will be present, the work activities that have been considered, and other assumptions made will be given. These drawings or plans form part of the risk assessment record that the *DSEAR* require, and should be available for examination by those affected and regulators. They become particularly important when there is a change in the work activity and/or new equipment to be introduced into a zoned area.

Classification drawings or plans may need to be reviewed when maintenance is carried out. If the dangerous substances normally present have been removed it may be possible to treat the area as non-hazardous. Alternatively, if the maintenance creates a larger than normal risk of a release of a dangerous substance, for example if it is necessary to open a normally sealed pipe system or enter a tank where vapours could exist – a larger area may need to be treated as hazardous or the area may need to be to temporarily reclassified to a higher zone. It is not, however, normally necessary to create new classification drawings or plans for the duration of maintenance work, but the process needs to be carefully controlled.

## Equipment and protection systems

**D0920** Employers are required to take *special precautions* to ensure that equipment and protective systems in places classified as hazardous are selected so as to avoid sources of ignition from, for instance, sparks or flames. [*DSEAR* (*SI 2002 No 2776*), *Reg 7(2)* and *Sch 3*]. Specifically, equipment and protective systems must be selected on the basis of the requirements set out in the *Equipment and Protective Systems Intended for Use in Potentially Explosive Atmospheres Regulations 1996* (*SI 1996 No 192*), unless the risk assessment indicates otherwise. In particular, the following categories of equipment (as defined within these Regulations) must be used in the zones indicated, providing they are suitable:

- in Zone 0 or Zone 20 – category 1 equipment;
- in Zone 1 or Zone 21 – category 1 or 2 equipment;
- in Zone 2 or Zone 22 – category 1, 2 or 3 equipment.

Equipment already in use before 1 July 2003 can continue to be used indefinitely providing the risk assessment shows that it is safe to do so.

The *DSEAR* provide definitions of the terms 'equipment', 'protective systems', 'devices', 'components' and 'potentially explosive atmosphere'. This part of the *DSEAR* complements the *Equipment and Protective Systems Intended for Use in Potentially Explosive Atmospheres Regulations 1996*. For further information see the DTI website at: www. dti.gov.uk/strd/atex.htm.

Equipment built to the requirements of the 1996 Regulations can be identified as it will carry the explosion protection symbol 'Ex' in a hexagon, the equipment category (1,2,or 3) and the letter 'g' or 'd' depending on whether it is intended for use in gas or dust atmospheres. In many cases a temperature rating expressed as a 'T' marking will also be included. These indicate the limitations to safe use.

For most electrical equipment designed for use in explosive atmospheres there will be little change required, except in the details of the marking on the equipment. However, the 1996 Regulations also apply to mechanical equipment as a potential ignition source. A harmonised European standard for category 3 mechanical equipment is now available as BS EN 13463 Part 1.

### Warning signs on entry

**D0921** Employers are required to ensure that places classified as hazardous are marked at their points of entry with warning signs. [*DSEAR* (*SI 2002 No 2776*), *Reg 7(3)* and *Sch 4*]. The warning sign at entry into places where explosive atmosphere may occur is specified in Annex III of the European Council Directive 99/92/EC and is:

(*a*) triangular in shape; and

(*b*) contains black letters 'EX' on a yellow background with black edging (the yellow part taking up at least 50 per cent of the area of the sign).

### Verification of safety prior to first use

**D0922** Employers are required to ensure that before workplaces containing places classified as hazardous are used for the first time, they are confirmed as being safe (verified) by a person (independent or employed by an organisation) competent in the field of explosion protection. [*DSEAR* (*SI 2002 No 2776*), *Reg 7(4)*]. The person carrying out the verification must be competent to consider the particular risks at the workplace, the adequacy of the risk assessment, and the control and other measures put in place. As a result the person must have suitable experience, professional training or both.

## Work clothing

**D0923** Employers are required to ensure that appropriate work clothing that does not give rise to electrostatic discharges is provided to employees entering places classified as hazardous. [*DSEAR (SI 2002 No 2776), Reg 7(5)*]. Sparks from clothing could create a risk of igniting an explosive atmosphere. Advice should be sought from an industrial clothing supplier if the employer is in any doubt as to the suitability of any item.

## Commencement

**D0924** The requirements under the *DSEAR* come into effect at different times, depending on when the workplace is first used, according to the following table [*DSEAR (SI 2002 No 2776), Reg 7(6)*]:

| Workplace | When requirements must be met |
| --- | --- |
| Workplace already in use before July 2003 | Workplace must meet requirements by July 2006 |
| Workplace already in use before July 2003 but modified before July 2006 | Workplace must meet requirements from the time the modification takes place. |
| Workplace coming into use for the first time after 30 June 2003 | Workplace must meet requirements from the time it comes into use |

## Arrangements to deal with incidents and emergencies

**D0925** The *DSEAR* require employers to make arrangements to protect employees (and other persons who are at the workplace) in the event of incidents and accidents involving dangerous substances and explosive atmospheres. [*DSEAR (SI 2002 No 2776), Reg 8(1)*]. The provisions clarify what already needs to be done in relation to the safety management of dangerous substances and will not require any duties in addition to those already present in existing legislation. Specifically they build on existing requirements in *Regulation 8* of the *Management of Health and Safety at Work Regulations 1999 (SI 1999 No 3242)*.

The arrangements required are as follows:

(*a*)    Giving suitable warnings (including audible and visual alarms) where necessary to alert employees (and other persons who are at the workplace) immediately of a release of a dangerous substance or the presence an explosive atmosphere before explosive conditions are reached, so they may be withdrawn to a safe place.

(*b*)    Escape facilities (e.g. designated exit routes, stairways etc) identified as necessary by the risk assessment to ensure that employees (and other persons who are at the workplace) can, in the event of danger, leave the endangered place promptly and safety.

(*c*)    Warning and communication systems to enable an appropriate response to be made immediately when such an event occurs, including remedial actions and rescue operations to mitigate the effects and restore the situation to normal. [*DSEAR (SI 2002 No 2776), Reg 8(3)(a)*].

(*c*)    Emergency procedures to be followed in the event of an emergency, including defining responsibilities and actions to be taken.

(d) Providing equipment and clothing for essential personnel to deal with the incident. [*DSEAR (SI 2002 No 2776), Reg 8(3)*].

(e) Organising relevant safety drills at regular intervals designed to test the emergency procedures.

(f) The provision of appropriate first-aid facilities in a safe place.

(g) Making information available to employees (e.g. through a display at the workplace) on the warnings, escape facilities, emergency procedures, relevant work hazards and hazard identification (e.g. zoning and marking), and on specific hazards likely to arise at the time of an incident. [*DSEAR (SI 2002 No 2776), Reg 8(2)(b)*].

(h) Contacting the relevant accident and emergency services to let them know that information on emergency procedures is available (or providing them with any information that they consider necessary). [*DSEAR (SI 2002 No 2776), Reg 8(2)(a)*].

The scale and nature of these arrangements should be proportional to the risks, taking account of the frequency of such events and the possible consequences. When the results of the risk assessment show that because of the quantity of each dangerous substance at the workplace there is only a slight risk to employees and risk has been eliminated or reduced as far as reasonably practicable, specific emergency arrangements are not necessary. [*DSEAR (SI 2002 No 2776), Reg 8(4)*]. In this case, the general provisions for safety at the workplace will apply.

## Provision of information, instruction and training

**D0926**  Employers are required to provide employees and other people at the workplace who might be at risk from dangerous substances with suitable information, instruction and training on precautions and actions they need to take to safeguard themselves and others so as to reduce the possibility of an incident taking place. [*DSEAR (SI 2002 No 2776), Reg 9*]. This should include:

- names of the substances in use and the risks they present;
- access to any relevant data sheets;
- any specific legislative provisions concerning the hazardous properties of the substance;
- the significant findings of the risk assessment.

Other aspects may include procedures for correct handling, protective clothing and equipment to be used, and the use of safe work processes. Information, instruction and training should be adapted if significant changes in the type or methods of work occur. Existing health and safety legislation already covers much of this.

Information, instruction and training need only be provided to non-employees where it is required to ensure their safety. Where it is provided, it should be in proportion to the level of risk. In general, non-employees should not be exposed to such risks.

## Identification of hazardous contents of pipes and containers

**D0927**  Employers are required to identify pipes, vessels, tanks and other containers of dangerous substances in accordance with the legislation listed below where this is applicable. [*DSEAR (SI 2002 No 2776), Reg 10 and Sch 5*].

- the *Chemicals (Hazard Information and Packaging for Supply) Regulations 2002 (SI 2002 No 1689)*;
- the *Health and Safety (Safety Signs and Signals) Regulations 1996 (SI 1996 No 341)*;
- the *Good Laboratory Practice Regulations 1997 (SI 1997 No 654)*;

- the *Classification and Labelling of Explosives Regulations 1983* (*SI 1983 No 1140*);
- the *Carriage of Dangerous Goods by Rail Regulations 1996* (*SI 1996 No 2089*);
- the *Packaging, Labelling and Carriage of Radioactive Material by Rail Regulations 1996* (*SI 1996 No 2090*);
- the *Carriage of Dangerous Goods (Classification, Packaging and Labelling) and Use of Transportable Pressure Receptacles Regulations 1996* (*SI 1996 No 2092*);
- the *Carriage of Dangerous Goods by Road Regulations 1996* (*SI 1996 No 2095*);
- the *Carriage of Explosives by Road Regulations 1996* (*SI 1996 No 2093*);
- the *Radioactive Material (Road Transport) (Great Britain) Regulations 1996* (*SI 1996 No 1350*).

Where containers of dangerous substances are not marked in accordance with this legislation, for reasons of applicability, employers must ensure that the contents of those containers are clearly identifiable together with the nature of the associated hazards.

## Exemption

**D0928** The Health and Safety Executive (HSE) has the power to exempt any person or class of persons or any dangerous substance or class of substances from all or any of the requirements or prohibitions imposed by or under the *DSEAR*. [*DSEAR* (*SI 2002 No 2776*), *Reg 13*]. It is expected that any such exemption would only be used in exceptional circumstances. Such exemption is granted by the issue of a certificate in writing and may be subject to conditions and a limit of time and may be revoked at any time.

In exercising this power, the HSE must, having due regard to the circumstances of the case, be satisfied that the health and safety of persons likely to be affected will not be prejudiced. In addition, it must be satisfied that exemption will be compatible with the requirements of European Council Directives (98/24 and 99/92) relating to the protection of workers from risks from chemical agents and explosives, which take precedence.

The Secretary of State for Defence may, in the interests of national security, by a certificate in writing exempt from all or any of the requirements or prohibitions imposed by or under the *DSEAR* the following. [*DSEAR* (*SI 2002 No 2776*), *Reg 14*]:

- any of Her Majesty's Armed Forces;
- any visiting force and members attached to a headquarters;
- any person engaged in work involving dangerous substances, if that person is under the direct supervision or control of a representative of the Secretary of State for Defence.

Where any such exemption is granted, suitable arrangements must be made for the assessment of risk to safety created by the work and for adequately controlling the risk to the persons to whom the exemption relates. This power of exemption could apply at defence establishments and centres of defence related research.

## Amendments to modernise petroleum and other legislation

**D0929** Petroleum legislation is being modernised as part of the *DSEAR* (*SI 2002 No 2776*) by means of a number of amendments. Previously, the keeping of petrol in significant quantities was controlled by licences with conditions issued under the *Petroleum (Consolidation) Act 1928*. However, as petrol is a dangerous substance, the *DSEAR* apply to it and duplicate these controls. The *DSEAR* remove licensing requirements for holding petrol, except for petrol that is being kept for dispensing into vehicles (retail and non-retail) at, for example, garages. Accordingly the following Acts, Regulations and Orders are amended [*DSEAR (SI 2002 No 2776), Reg 15* and *Sch 6 Pts 1* and *2*]:

- the *Petroleum (Consolidation) Act 1928*;
- the *Petroleum Spirit (Motor Vehicles etc) Regulations 1929 (SI 1929 No 952)*;
- the *Petroleum (Liquid Methane) Order 1957 (SI 1957 No 859)*;
- the *Petroleum (Consolidation) Act 1928 (Enforcement) Regulations 1979 (SI 1979 No 427)*;
- the *Petroleum Spirit (Plastic Containers) Regulations 1982 (SI 1982 No 630)*.

In addition, the *DSEAR* amend a number of other pieces of legislation:

- the *Celluloid and Cinematograph Film Act 1922*;
- the *Dangerous Substances in Harbour Areas Regulations 1987 (SI 1987 No 37)*;
- the *Fire Precautions (Workplace) Regulations 1997 (SI 1997 No 1840)*;
- the *Fire Certificates (Special Premises) Regulations 1976 (SI 1976 No 2003)* ;
- the *Carriage of Dangerous Good by Road Regulations 1996 (SI 1996 No 2095)*.

It should be noted that the amendment of the *Fire Precautions (Workplace) Regulations 1997* changes the definition of a highly flammable liquid within *Schedule 1* of the 1997 Regulations (premises for which a fire certificate is required). The amendment to the *Carriage of Dangerous Good by Road Regulations 1996* prohibits the filling of fuel tanks for cars or other containers direct from a road tanker.

# Repeals and revocations of existing legislation

**D0930** The *DSEAR (SI 2002 No 2776)* repeal and revokes a large number of older Regulations and Orders relating to substances that now come under the general provisions of the 2002 Regulations. In this way the *DSEAR* are an important piece of modernising and reforming legislation. The following Acts and statutory instruments are completely repealed and revoked [*DSEAR (SI 2002 No 2776), Reg 16* and *Sch 7 Pts 1* and *2*]:

- the *Celluloid etc Factories and Workshops Regulations 1921 (SI 1921 No 1825)*;
- the *Manufacture of Cinematograph Film Regulations 1928 (SI 1928 No 82)*;
- the *Cinematograph Film Stripping Regulations 1939 (SI 1939 No 571)*;
- the *Petroleum (Carbide of Calcium) Order 1929 (SI 1929 No 992)*;
- the *Petroleum (Carbide of Calcium) Order 1947 (SI 1947 No 1442)*;
- the *Petroleum (Compressed Gases) Order 1930 (SI 1930 No 34)*;
- the *Magnesium (Grinding of Castings and other Articles) Special Regulations 1946 (SI 1946 No 2107)*;
- the *Dry Cleaning Special Regulations 1949 (SI 1949 No 2224)*;
- the *Dry Cleaning (Metrication) Regulations 1983 (SI 1983 No 977)*;
- the *Factories (Testing of Aircraft Engines and Accessories) Special Regulations 1952 (SI 1952 No 1689)*;
- the *Factories (Testing of Aircraft Engines and Accessories) (Metrication) Regulations 1983 (SI 1983 No 979)*;
- the *Highly Flammable Liquids and Liquefied Petroleum Gases Regulations 1972 (SI 1972 No 917)*;

- the *Abstract of Special Regulations* (*Highly Flammable Liquids and Liquefied Petroleum Gases*) *Order 1974* (*SI 1974 No 1587*).

All the Regulations above (except for the final two relate) to particular industries or activities of a specialised kind. The repeal and revocation of Regulations relating to highly flammable liquids and liquefied petroleum gases has wider effect. The more specific requirements of these Regulations are covered by the general goal setting provisions of the *DSEAR* but remain good practice. Extensive guidance on the safe management of flammable liquids and liquefied petroleum gases is available from the LPG Association.

For completeness, the *DSEAR* also repeal and revoke particular parts of the following legislation:

- the *Petroleum (Consolidation) Act 1928*;
- the *Factories Act 1961*;
- the *Shipbuilding and Ship Repairing Regulations 1960* (*SI 1960 No 1932*);
- the *Dangerous Substances in Harbour Areas Regulations 1987* (*SI 1987 No 37*);
- the *Carriage of Dangerous Goods by Road Regulations 1996* (*SI 1996 No 2095*);
- the *Carriage of Dangerous Goods* (*Classification, Packaging and Labelling*) *and Use of Transportable Pressure Receptacles Regulations 1996* (*SI 1996 No 2092*);
- the *Workplace (Health, Safety and Welfare) Regulations 1992* (*SI 1992 No 3004*).

## Enforcement

D0931  The application of the *DSEAR* (*SI 2002 No 2776*) will be enforced through existing powers by:

- the HSE or local authorities depending on the allocation of premises under the *Health and Safety (Enforcing Authority) Regulations 1998* (*SI 1998 No 494*). In the main, the HSE will enforce at industrial premises and local authorities (environmental health officers) elsewhere, e.g. in retail premises.
- fire brigades at most premises subject to the *DSEAR* in relation to general fire precautions such as means of escape;
- petroleum licensing authorities at retail petrol filling stations in relation to the storage and dispensing of petrol, LPG and other fuel subject to the 2002 Regulations.

## Regulatory impact assessment

D0932  The Health and Safety Commission has prepared a regulatory impact assessment (RIA) on the *DSEAR* (*SI 2002 No 2776*). This has been approved by ministers and published with the intention to inform on the public debate about the Regulations. The RIA describes the issues that have given rise to the need for the Regulations, compares various options, and identifies the costs and benefits.

The findings of the RIA showed that businesses complying with current health and safety legislation would expect little or no additional cost as a result of the *DSEAR*. Those not fully compliant may need to review and enhance their current arrangements, although most costs incurred would relate to more rigorous compliance with current legislation prompted by the Regulations. For example, many businesses with work processes involving potentially explosive atmospheres had not formally recorded zones where these could occur in their risk assessments. This and the need to mark zoned areas with a sign are now specific requirements of *DSEAR*.

## Approved codes of practice

**D0933** In support of the *DSEAR* (*SI 2002 No 2776*) and to assist duty holders in the interpretation and meeting the requirements of the Regulations, the following Approved Codes of Practice (ACoPs) are currently available or will be available shortly:

- DSEAR – Approved Code of Practice (L138);
- DSEAR – Design of plant equipment and workplaces (L134);
- DSEAR – Storage of dangerous substances (L135);
- DSEAR – Control and mitigation measures (L136);
- DSEAR – Safe maintenance, repair and cleaning procedures (L137);
- DSEAR – Unloading petrol from road tankers (L133);

These ACoPs are published for the Health and Safety Commission by HSE Books.

## Sources and acknowledgement

**D0934** In preparing this chapter, extensive use has been made of the wording of the *DSEAR* (*SI 2002 No 2776*) themselves and of the commentary provided by the Safety Policy Directorate of the HSE.

# Electricity

## Introduction

E3001   Electricity has become fundamental to our way of life – both at work and in the home. It provides an energy source to meet an ever-increasing range of needs and, used with care, is a good servant. However, treated with disrespect, it is a poor master and can result in serious injury or even death. The number of accidents at work involving electricity is small by comparison with the total for all causes (less than 0.5 per cent). However, it should be borne in mind that an electrical accident is very likely to result in serious injury or death. Recent published Health and Safety Executive (HSE) statistics show that in a typical year there are around 15 fatalities, 135 non-fatal major injuries and 450 other reportable injuries due to contact with electricity. Electricity accounts for around 6 per cent of all the fatalities, a significant number of which arise from construction activities.

## Understanding electricity

E3002   A basic knowledge of electrical principles is important to an understanding of the dangers of electricity and, more importantly, the control measures that need to be put in place to avoid these dangers and thereby prevent injury. The content of this section has been restricted to aiding understanding of the remainder of this chapter and a good textbook on electrical theory should be consulted for more extensive knowledge.

Because electricity is an unseen energy form its nature is often difficult to appreciate. A simple analogy is to compare it with the flow of water in a pipe, where the water flow is determined by the water pressure and resistance to the flow arising from the size of pipe. Fundamentally, the flow of electricity (electric current) arises from the movement of electrons in an electrical circuit. The amount of current (I) flowing in a circuit is determined by the electrical pressure or voltage (V) and the resistance (R) in the circuit. These three basic parameters are related by an equation known as Ohm's Law:

$I = V/R$

Thus varying the electrical pressure will alter the current in direct proportion whilst varying the resistance will vary the current in inverse proportion.

Electric current is measured in amperes or 'amps' for short (A), voltage in 'volts' (V) and resistance in ohms ($\Omega$). The range of values encountered for these basic parameters is very great, often from millions to millionths of the basic unit. Standard prefixes are used to denoted multiples and sub-multiples of the basic unit. For example, 1 mA is one thousandth of an amp whilst 1 m$\Omega$ is one thousandth of an ohm. Similarly, 1 kA is one thousand amps, 1 kV is one thousand volts and 1 M$\Omega$ is one million ohms.

In any electrical circuit, resistance is encountered in two forms, as continuity or conductor resistance and as insulation resistance. Continuity resistance is the conductor resistance measured end-to-end. This should be as low as possible in order not to unduly restrict the flow of current. Typical values of continuity resistance will usually be less than an ohm, and are often quoted in milliohms (m$\Omega$). Insulation resistance is the resistance between conductors and is therefore a measure of the quality of the electrical insulation surrounding the conductors. Because we do not want electricity to leak across or short-circuit from one conductor to another, or out of the circuit, the resistance value should be as high as possible. The insulation resistance of a circuit should therefore be a very large number of ohms, typically several million ohms (M$\Omega$).

There are two forms of current, alternating (a.c.) and direct (d.c.) and both are likely to be encountered in the workplace. Alternating current is the most common because this is the form of 'mains' electricity. Direct current is the form provided by batteries and is therefore encountered in workplaces where battery powered works vehicles are used. Certain specialist work processes also use d.c. obtained by conversion (rectification) of the a.c. mains supply. The process of rectification is not perfect and some oscillation generally appears on the d.c. output. This is referred to as ripple and has implications relating to electric shock. In the case of electrical circuits operating from the 'mains' the circuit resistance is more complicated because the oscillating waveform of the current brings into play additional characteristics, namely the magnetising and charging effects of the current. Suffice to say, the resistance of an a.c. circuit is referred to as its impedance (Z) which represents a fairly complex mathematical relationship involving the inductance and capacitance displayed by the circuit, in addition to its resistance. As with resistance, impedance is also measured in ohms.

The flow of an electric current represents power or energy. Electric power is the product of the current and voltage, measured in watts (W) or (kW), whilst electrical energy is the product of electric power and time, measured in watt seconds (Ws) or kilowatt hours (kWh).

## Legal background and standards

E3003   The fundamental piece of legislation regulating electrical safety at work is the *Electricity at Work Regulations 1989 (SI 1989 No 635)* ('*EWR*'). This chapter is mainly devoted to explaining the requirements of these Regulations but from E3029 onwards some information is included concerning other electrical legislation of relevance in specific circumstances.

The *EWR* comprise 33 individual regulations grouped into four parts:

- *Part I*: Introductory matters (*Regulations 1–3*);
- *Part II*: General technical provisions (*Regulations 4–16*);
- *Part III*: Provisions relating specifically to mines (*Regulations 17–28*);
- *Part IV*: Miscellaneous provisions (*Regulations 29–33*).

As these Regulations have been made under the *Health and Safety at Work etc. Act 1974* ('*HSWA*') they apply in the same circumstances as the parent Act, i.e. they are work-related rather than applying to specific categories of premises. Thus they apply to all types of work establishment and, like the parent Act, place responsibilities on employers, the self-employed and employees, in this case to avoid danger and prevent injury from electricity. The responsibilities of these duty holders are defined in *Regulation 3* and require action to be taken where the matter is under the control of the duty holder. An example of matters under the control of an employer would be the preparation of an appropriate policy relating to electrical safety and the organisation and arrangements for putting it into effect, such as for the maintenance of electrical equipment. An example of matching matters under the control of an employee would be to comply with those electrical maintenance procedures and use any tools and equipment, provided in connection with the work, in the correct manner. Some of the duties imposed by the Regulations are qualified, i.e. 'so far as is reasonably practicable', as in the parent Act; the others are absolute. In the case of a prosecution for a breach of an absolute duty, it is available to the employer to utilise the defence provided by *Regulation 29*. This states that it shall be a defence for any person to prove that he took all reasonable steps and exercised all due diligence to avoid the commission of that offence. Continuing with the example of electrical equipment maintenance, 'taking all reasonable steps' would involve establishing a maintenance policy and putting into place the organisation and arrangements for its implementation, whilst 'exercising all due diligence' would require on-going monitoring to ensure that the policy remained effective and that the maintenance was undertaken to the required standard.

The *EWR* apply to all electrical systems which, by definition (*Regulation 2*) means electrical equipment connected to a common source of energy. Electrical systems range from power stations and the national grid at one end of the spectrum to electrical installations within buildings and electrical equipment at the other. They require action to be taken only where there is danger to be avoided or injury prevented (see E3004).

When originally enacted, the *EWR* did not apply to offshore installations, i.e. outside territorial waters. However, one of the outcomes from the enquiry into the Piper-Alpha oil rig fire was to extend the application of the *EWR* to such installations through the introduction of the *Offshore Electricity and Noise Regulations 1997* (*SI 1997 No 1993*). Separate but virtually identical Regulations are in force in Northern Ireland, as the *Electricity at Work Regulations* (*Northern Ireland*) *1991*.

Modern health and safety legislation tends to be goal-setting and therefore non-prescriptive. The *EWR* are no exception to this principle! They are well-founded, laying down the fundamental principles for achieving good standards of electrical safety but do not prescribe the specific means for compliance. There is a wide range of guidance published by the HSE, as well as British, European and International Standards, to assist employers in determining how compliance is to be achieved. A key HSE publication is the Memorandum of Guidance on the Electricity at Work Regulations 1989 (HSE booklet HSR25 (ISBN 0 7176 1602 9)). This provides guidance of general application to workplaces. Because of the higher level of risk relating to the use of electricity in mines and quarries due to the nature of those environments, guidance in the form of Approved Codes of Practice (ACoPs) has been published specific to those places of work (e.g. The use of electricity in mines (HSE booklet L128) (ISBN 0 7176 2074 3); The use of electricity at quarries (HSE booklet COP35) (ISBN 0 11 885484 4)).

In the introduction to the HSE Memorandum of Guidance reference is made to one very important British Standard, BS 7671 – Requirements for electrical installations (ISBN 0 86341 373 0), otherwise known as the non-statutory Institution of Electrical Engineers (IEE) Wiring Regulations. This standard specifies requirements concerning the design, construction, maintenance, inspection and testing of low voltage electrical installations, i.e. installations operating at voltages up to 1,000 V a.c. or 1,500 V d.c. The HSE states that compliance with BS 7671 is likely to achieve compliance with the relevant aspects of the *EWR*. Furthermore, that an electrical installation installed to an earlier edition of the IEE Wiring Regulations, but current at that time, would not in itself mean that it did not comply with the *EWR*.

In should be borne in mind that other health and safety legislation of general application may also be of relevance to electrical safety. For example:

- the *Management of Health and Safety at Work Regulations 1999* (*SI 1999 No 3242*) which, among other things, impose a need for the assessment of electrical risks and the provision of training relevant to electrical work activities;

- the *Provision and Use of Work Equipment Regulations 1998* (*PUWER*) (*SI 1998 No 2306*), the requirements of which apply equally to electrical equipment, e.g. concerning the need for isolation from all energy sources prior to commencing work.

# Dangers of electricity

## General

**E3004**   *Regulation 2* of the *EWR* (*SI 1989 No 635*) gives the definitions of words used with special meaning in the Regulations. These include 'danger' and 'injury'. Danger is defined as risk of injury whilst injury means:

'... death or personal injury from electric shock, electric burn, electrical explosion or arcing, or from fire or explosion initiated by electrical energy, where any such death or injury is associated with the generation, provision, transmission, transformation, rectification, conversion, conduction, distribution, control, storage, measurement or use of electrical energy.'

The 1989 Regulations require action to be taken to avoid danger or, where appropriate, injury. The reason for distinguishing between danger and injury is that in some situations, e.g. live working, the electrical danger is not eliminated because of the presence of live parts and, therefore, precautions must be put in place to prevent injury from that danger. Ideally, injury is prevented by completely removing or eliminating the danger, e.g. as in the case of equipment isolation prior to undertaking maintenance work.

It is important to appreciate that injury does not always require physical contact to be made with a live conductor. At high voltages electricity has the capability of jumping an air gap (flashover), the flashover distance increasing as the voltage increases. Environmental conditions are also relevant. Damp or wet conditions will reduce the effectiveness of air as an insulator. For this reason the scope of the Regulations covers the prevention of injury from electrical danger arising from work near an electrical system or equipment as well as on it.

The Regulations do not, per se, make a distinction between high and low voltage systems because system voltage is not necessarily the only factor in determining danger. A 1.5 V torch battery would not be expected to present a shock or short-circuit hazard in a normal work environment but might provide sufficient energy to cause ignition in a potentially explosive atmosphere.

The definition of injury effectively encompasses the dangers of electricity which may be summarised as electric shock, electric burn, fire, arcing and explosion. These now will be briefly explained.

## *Electric shock*

**E3005**     Electric shock may arise either through direct or indirect contact with electricity. The former involves direct contact with a live part, e.g. an exposed live terminal or conductor, whilst the latter is a shock received from an exposed-conductive-part, e.g. the metal casing of electrical equipment, made live under a fault condition. An exposed-conductive-part is any conductive part of equipment which can be touched and which is not a live part but which may become live under fault conditions. The *EWR* (*SI 1989 No 635*) specify control measures to deal with each of these circumstances (*Regulations 7* and *8* respectively – see E3014 and E3015).

The severity of an electric shock depends on a number of factors, in particular the magnitude and duration of the shock current passing through the body. The current magnitude will depend on the contact voltage and resistance or, to be more strictly correct in the case of a mains supply, which is an alternating current (a.c.), the impedance (Z) of the shock path. Thus, in the case of contact with mains electricity the shock current is:

$I = V/Z$ (Ohms law)

Therefore, the outcome from an electric shock will be very dependent on the circumstances in which it is received, in particular the environment, since this will determine the shock path resistance or impedance. Indoors, standing on a floor covering acting as a good insulator, a hand-to-feet shock at mains voltage (230V nominal) may result in no more than a tingling sensation. Outdoors, however, with little insulation from earth, a shock at the same voltage would prove more severe, even with the possibility of being fatal. A hand-to-hand shock, where one hand makes contact with a live part whilst the other is simultaneously in contact with earthed metal, such as the earthed case of electrical equipment, represents an even higher risk because the shock path impedance is only that of the body. At mains voltage such a situation is quite likely to result in a fatality.

The International Electrotechnical Commission (IEC) has published information concerning the physiological effects of current flow through the human body (IEC report 479–1). This information is replicated in British Standard published document – BS PD 6519: Effects of current passing through the body (Parts 1 and 2). In summary the physiological effects of alternating current are:

| Around 0.5 mA | Threshold of perception |
| Up to 10 mA | Unpleasant but usually no harmful physiological effects |
| Above 10 mA | Increasing likelihood of muscular contractions and, at higher currents, possibility of respiratory system failure |
| 30 mA | Typical current trip rating for a residual current device (RCD) |
| Around 50 mA | Risk of ventricular fibrillation and cardiac arrest (increasing risk with increasing magnitude and duration of current flow) |
| 250 mA | Current drawn by a 60 W light bulb |

The above values serve only as a guide since the effect will vary from one person to another. With higher shock currents the effect on the body, in particular the probability of death, becomes increasingly dependent on how long the current flows for. Currents as low as 50mA (0.05A) can prove fatal, particularly if flowing through the body for several seconds. Higher currents will have a similar effect in a much shorter time.

Whilst a.c, per se, represents a high shock risk, contact with direct current (d.c.) is also hazardous and can disturb the normal heart rhythm. A direct current of approximately 3.75 times the a. c. value will have the same probability of inducing ventricular fibrillation. However, if ripple exists at more than 10 per cent of the nominal d.c. value due to imperfections in the process of rectification then it must be considered as hazardous as a.c.

## *Electric burn*

**E3006** Body tissue burns may result from the passage of electric current. The nature of such burns is that they tend to be deep seated and difficult to heal. Contact with high voltage is often characterised by body burn marks at the current entry and exit points, e.g. hand palm and soles of the feet in the case of a hand-to-feet shock path. Electric burns may also arise through exposure to high power sources of electro-magnetic radiation such as from radio transmission antennae or induction heating processes.

## *Fire*

**E3007** Fire may be initiated as a result of heat generation due to the overheating of cables or electrical equipment, which may occur for a number of reasons. Examples are:

- overloading due to insufficient current carrying capacity for the connected load;
- cables bunched or covered with thermal insulation, significantly reducing current carrying capacity;
- lack of, or incorrectly rated, excess current protection (e.g. fuses or circuit breakers) for circuits and equipment;
- blocked ventilation apertures in equipment enclosures, e.g. motors;
- leakage currents due to deterioration of insulation surrounding live parts.

Arcing arising from short-circuit flashover or sparks, generated within electrical equipment such as motors or switching devices, may provide an ignition source for adjacent flammable materials (solids, gases, vapours or dusts).

## Arcing

**E3008** Arcing or flashover will occur when short-circuiting or bridging between live parts at different voltages, e.g. between circuit conductors (phase to neutral or across phases in the case of a three-phase system) or from a live part to earth. As well as the risk of serious injury there is also the likelihood of damage to the equipment concerned. The severity of such a flashover will depend on the fault level at the point of short-circuit, i.e. the magnitude of the short-circuit current, which in turn will be determined by the upstream impedance in the electrical system back to the source of supply. Since this will be very low the prospective short-circuit current will be very high (Ohm's law again). In a factory situation prospective short-circuit currents as high as 10kA may well be encountered.

Short-circuit flashover may also result as a consequence of equipment failure, e.g. failure of a circuit protective device to disconnect the supply safely following a fault. Where failure has arisen because of a malfunction during a switching operation the operator is at risk of sustaining arc eye and severe burns.

## Explosion

**E3009** In the case of a severe short-circuit flashover there is often little to distinguish between the effects of arcing and an explosion. Where the short-circuit occurs in a restricted space, such as within a busbar chamber or inside switchgear, the explosive forces are likely to result in considerable damage to the equipment involved, e.g. rupture of the external housing or covers blown off. If the switchgear is of the oil-filled type this may lead to the ejection of burning oil, a further hazard for anyone with the misfortune to be close by.

Electricity presents special problems in potentially explosive atmospheres (PEAs), where an electric spark may act as an ignition source. For this reason particular attention must be given to the design and construction of electrical equipment intended for such environments – see E3027 below.

The assessment of electrical risks first requires the likelihood of electrical danger to be evaluated, following which the severity of injury can be predicted. The likelihood of danger will centre on the presence of hazards that may give rise to electric shock, burns, fire, etc. This will include situations where there may be exposed live parts, missing earth connections, incorrect fuse ratings, etc. The severity of the outcome will depend on factors such as the supply voltage, shock path and work environment in the case of electric shock and the fault level and bridging path in the case of short-circuit flashover. With electricity there is a very thin dividing line between the accident being survivable and death. It is therefore best to work on the basis of endeavouring to reduce the likelihood of an electrical accident to as near zero as possible, which is the objective of the *EWR (SI 1989 No 635)*.

# Fundamentals of controlling electrical risks

**E3010** The control measures aimed at preventing injury from electrical danger in general situations are prescribed in *Part II* of the *EWR (SI 1989 No 635)*, i.e. in *Regulations 4–16*. These cover:

| Regulation | Topic |
|---|---|
| 4 | Systems, work activities and protective equipment |
| 5 | Strength and capability of electrical equipment |
| 6 | Adverse or hazardous environments |
| 7 | Insulation, protection and placing of conductors |

| 8 | Earthing or other suitable precautions |
|---|---|
| 9 | Integrity of referenced conductors |
| 10 | Connections |
| 11 | Means for protecting from excess of current |
| 12 | Means for cutting off the supply and for isolation |
| 13 | Precautions for work on equipment made dead |
| 14 | Work on or near live conductors |
| 15 | Working space, access and lighting |
| 16 | Persons to be competent to prevent danger and injury |

*Regulation 4* of the *EWR* (*SI 1989 No 635*) is effectively an 'umbrella' requirement, covering everything relating to the life of an electrical system, which includes electrical equipment as well as fixed electrical installations. The Regulation is in four parts, requiring:

(1) the construction of electrical systems such as to prevent danger;

(2) the maintenance of electrical systems so as to prevent danger;

(3) all work activities to be carried out in such a manner as not to give rise to danger;

(4) any equipment provided under the Regulations for the purpose of protecting persons at work on or near electrical equipment to be suitable for use, maintained in a suitable condition and properly used.

The first three parts of the Regulation are qualified duties, i.e. 'so far as is reasonably practicable'.

The subsequent regulations in *Part II* cover three important aspects relating to electrical safety:

- the need for competence, which underlies the effective implementation of the *EWR* duties (*Regulation 16*);
- the design, construction and specification of the hardware, e.g. cables, switchgear circuit protection, etc (*Regulations 5–12* and *15*);
- safe systems of work (*Regulations 13* and *14*).

These aspects will now be considered in turn in more detail.

## Competence

**E3011** *Regulation 16* of the *EWR* (*SI 1989 No 635*) specifically requires that any person engaged in any work activity where technical knowledge or experience is necessary to prevent danger, or where appropriate injury, is competent to prevent danger and injury. Competence means the possession of such knowledge or experience as may be appropriate having regard to the nature of the work. If this is not the case, that person must be under an appropriate degree of supervision, again having regard to the nature of the work. This requirement for competence extends to non-electrical work activities in any situation where electrical danger may be present (see *EWR* (*SI 1989 No 635*), *Reg 4(3)*). The HSE has provided guidance concerning the scope of the phrase 'technical knowledge or experience'. This embraces:

- adequate knowledge of electricity;
- adequate experience of electrical work;
- adequate understanding of the system to be worked on and practical experience of that class of system;

- understanding of the hazards which may arise during the work and the precautions required to be taken;
- ability to recognise at all times whether it is safe for work to continue.

The above effectively provides a framework training specification. This can be used as a starting point for preparing a more detailed specification to suit a particular training need, e.g. to inspect and test portable electrical equipment. Thus, competence requires an understanding of the tasks to be performed, as well as of the equipment on which the work is to be undertaken, coupled with the necessary underpinning technical knowledge relating to electricity. As an example, an operative required to inspect and test portable electrical equipment would need a thorough knowledge of the different types and constructions of equipment likely to be encountered, in order to know what to look for when carrying out an inspection and what tests would be appropriate. An understanding of the test instruments to be used would also be required, in particular their functions and methods of use. The underpinning technical knowledge would enable meaningful measurements to be taken, properly recorded and the results correctly interpreted.

A higher level of competence is required for live work since the dangers of electricity remain present. It is necessary for those involved to understand fully the shock and short-circuit hazards in relation to the equipment being worked on and the work being undertaken, in particular possible shock and bridging paths that might arise during the course of the work. The fact that such hazards are always present whilst the work is in progress also must be fully appreciated.

Qualifications based on City & Guilds syllabuses exist for certain electrical activities, e.g. installation work, inspection and testing of fixed wiring and maintenance of electrical equipment, for which courses are offered by colleges and training organisations. Other routes are also available to acquire the necessary competence, such as on-the-job training linked to NVQs based on engineering occupational standards. Many electricians belong to one or more trade organisations such as the Electrical Contractors Association (ECA) for England and Wales or its Scottish equivalent (SELECT). The National Inspection Council for Electrical Installation Contracting (NICEIC) also exists, essentially as a consumer safety body, to maintain the standards of the electricians on its roll. However, membership of such organisations does not confer a particular level of electrical competence.

# Design and construction of electrical equipment and systems

## Strength and capability

E3012      *Regulation 5* of the *EWR* (*SI 1989 No 635*) requires that no electrical equipment shall be put into use where its strength and capability may be exceeded in such a way as may give rise to danger. This includes mechanical as well as electrical capability. Thus an overhead cable must have the mechanical strength not to break as well as the electrical capability to carry the required load current. Furthermore, electrical capability includes the capability to handle safely any prospective short-circuit currents that may flow under fault conditions.

## Adverse or hazardous environments

E3013      *Regulation 6* of the *EWR* (*SI 1989 No 635*) requires that electrical equipment shall be constructed or as necessary protected to prevent danger, so far as is reasonably practicable, from reasonably foreseeable adverse or hazardous environments, namely:

- mechanical damage;
- the effects of the weather, natural hazardous, temperature or pressure;
- the effects of wet, dirty, dusty or corrosive conditions;

- any flammable or explosive substance, including dusts, vapours or gases.

Whilst exposure to earthquakes in the UK might not be reasonably foreseeable, the siting of electrical equipment outdoors means that it is highly likely to be exposed to wet and possibly dirty, dusty or even corrosive conditions, such as salt spray in coastal locations. An appropriate construction specification would be needed, therefore, to accommodate this type of environment.

## *Protection from electric shock by direct contact*

E3014    Regulation 7 of the *EWR* (*SI 1989 No 635*) deals with the prevention of electric shock from direct contact with live conductors. It requires that all conductors in a system, which may give rise to danger, shall either:

(*a*)   be suitably covered with insulating material and as necessary protected so as to prevent, so far as is reasonably practicable, danger; or

(*b*)   have such precautions taken in respect of them (including, where appropriate, their being suitably placed) as will prevent, so far as is reasonably practicable, danger.

The normal means for preventing direct contact with live conductors is:

- electrical insulation (e.g. cable insulation);
- enclosure (e.g. busbar chamber);
- use of barrier fencing or screening (e.g. high voltage test bay).

In some circumstances the use of an extra-low voltage system (e.g. 12 V) may provide an adequate safeguard because at such a low voltage any accidental contact could not give rise to a serious electric shock. Nevertheless a short-circuit hazard may still exist, e.g. as with a 12 V high ampere-hour capacity battery.

The thickness of electrical insulation will depend on the conductor operating voltage. In some situations additional protection may be required to prevent damage to the electrical insulation. In industrial premises, steel wire armoured (swa) or mineral insulated copper sheathed (mics) cables are often used. Alternatively, cables may be placed in steel conduit or trunking, or placed out of reach on cable trays, to provide additional protection. Where enclosures are used it is important to ensure adequate protection against the ingress of foreign objects or fingers, in accordance with a British Standard known as the IP Code (BS EN 60529: Specification for degrees of protection provided by enclosures). Horizontal top surfaces of enclosures should have a degree of protection of at least IP4X (to prevent access by tools or wires greater than 1 mm in thickness) and, for all other surfaces, IP2X ( to prevent finger contact or the ingress of solid foreign objects of 12.5 mm diameter or greater up to 80 mm in length). Enclosure covers should be removable only by the use of a tool, i.e. secured using nuts/bolts and not wing nuts.

Barrier fencing is often used in test bays where there is a need to undertake live testing. Access gates must be interlocked with the test supply (e.g. key type) to ensure that entry cannot be gained whilst the supply is energised. In some circumstances it might be impracticable to enclose completely all live parts. Examples are:

- for functional reasons, e.g. the supply to electric overhead travelling cranes,
- railway traction supplies and overhead power distribution systems;
- in connection with operational requirements, e.g. fault finding and diagnostic live testing.

In such circumstances it is then necessary for other appropriate precautions to be implemented to minimise the risk of contact. Placing out of reach, coupled with the posting of warning notices, is a commonly adopted safeguard. However, it is important to recognise that occasions may arise where it becomes necessary to encroach within the prescribed safety clearances, in which case it is then necessary to implement a safe system of work based on isolation of the supply to remove the electrical danger at source.

## Protection from electric shock by indirect contact

**E3015** *Regulation 8* of the *EWR* (*SI 1989 No 635*) requires that precautions are taken, either by earthing or other suitable means, to prevent danger arising from any conductor (other than a circuit conductor) which may reasonably foreseeably become charged as a result of either the use of the system or a fault in the system. For example, an internal equipment fault may cause the metal case to be made live at the supply voltage, creating a shock risk. A range of techniques can be employed, either singly or in combination:

- earthing;
- equipotential bonding;
- double insulation;
- earth-free non-conducting environments;
- current/energy limitation;
- use of safe voltages;
- separated or isolated systems;
- connection to a common reference point on the system.

Earthing (for class I equipment) and double insulation (for class II equipment) are the techniques most commonly adopted. To achieve effective installation earthing it is most important that the earth fault loop impedance (the conductor continuity impedance/resistance from the supply source to the point of fault on the installation and back to source through the earth connection) is kept as low as possible in order that sufficient fault current will flow to enable the circuit protective device to operate to disconnect the fault current quickly. In buildings, connections (referred to as main equipotential bonding) are also made from the main earthing terminal to other services, as well as to the metal structure of the building if applicable. Further connections (supplementary bonding) are made between conductive parts in high shock risk situations (bathrooms, shower rooms, kitchens, etc.). This bonding serves a different role to earthing, namely to minimise the risk of voltage differences arising between exposed-conductive-parts during the time that the fault exists and before the supply is automatically disconnected.

In high risk situations a residual current device (RCD) may be used to supplement other protective measures, through very fast disconnection of the supply in the event of something going wrong (typically 20–40 ms). An RCD is a very sensitive device, capable of responding to the comparatively small currents typical of a shock situation. A common trip current rating is 30mA. Whilst an RCD will not prevent shock – a shock current must flow to create an out-of-balance situation, which is the basis of its operation – the shock should be survivable.

## Referenced conductors

**E3016** *Regulation 9* of the *EWR* (*SI 1989 No 635*) requires that nothing shall be placed in the neutral conductor of an electrical system to cause it to become open-circuited whilst the phase conductor(s) remain live.

This is for two main reasons:

(i) on modern distribution supply networks the neutral conductor is likely to be used also for earthing purposes, referred to as a protective multiple earthing (PME) system. An open-circuit in the neutral conductor therefore would result in loss of earthing;

(ii) the secondary (output) windings of three-phase LV distribution supply transformers are generally 'star' connected, with the star point earthed and serving as the system neutral. This provides a reference point for the phase voltages, to help maintain them within the prescribed statutory limits for supply. Loss of the neutral connection back to the supply transformer may result in undue voltage fluctuation depending on the loading on each phase.

Examples of prohibited devices are fuses and thyristors which cannot be relied upon not to give rise to danger by either becoming open-circuited or giving rise to a high conductor resistance. In the case of a three-phase four-pole switch the switch must break the neutral conductor last and make it first when operated. This situation does not arise with a triple-pole and neutral (TP&N) switch, where only the phase conductors are switched and the neutral remains connected.

## *Connections*

E3017    *Regulation 10* of the *EWR* (*SI 1989 No 635*) requires that all joints and connections should be mechanically and electrically suitable for use. This applies to both temporary and permanent connections and to connections in protective (earth) conductors as well as circuit conductors (phase and neutral). A broken phase connection could result in a short-circuit to a neutral conductor or to earth via an earthed metal equipment case. Failure of a joint or connection in a protective conductor could result in a loss of earthing and therefore loss of effective protection in the event of electric shock arising from indirect contact.

## *Excess current protection*

E3018    *Regulation 11* of the *EWR* (*SI 1989 No 635*) requires efficient means, suitably located, to be provided to protect every part of a system from excess current, as may be necessary to prevent danger. The abnormal conditions likely to give rise to excess current flow are:

- an overload, which can result in the overheating of cables or equipment and the possibility of fire;
- a short circuit, which can result in arcing or explosion, possibly accompanied by fire;
- an earth fault, giving rise to a shock risk through indirect contact.

In the case of electric shock protection it is important that the protective device (e.g. a fuse or circuit breaker) disconnects the supply as quickly as possible. BS 7671 specifies maximum disconnection times – 5 seconds for fixed equipment and 0.4 second for equipment supplied via socket outlets, although these disconnection times are reduced in high risk situations. To achieve the required disconnection time requires co-ordination between the circuit protective device (type and rating) and earth fault loop impedance. The higher the earth fault loop impedance, the lower the value of fault current that will flow through the protective device and therefore the longer the time that it will take to disconnect the supply. A further consideration is the fault rating of the device, i.e. the prospective short-circuit current that it is capable of interrupting. If the fault current exceeds the capability of the protective device it is likely that it will not disconnect the supply and may even blow up in the process, creating a second fault on the system.

## *Isolation*

E3019    *Regulation 12* of the *EWR* (*SI 1989 No 635*) interfaces with *Regulation 19* of *PUWER* (*SI 1998 No 2306*). It requires both a means for cutting off the supply as well as for isolation. In practice, switchgear often performs both functions. However, the fact that these are separately stated in this Regulation implies a difference. The distinction is made clear in the second part of the regulation, which explains the meaning of isolation in the electrical context. This amounts to a legal definition. The requirement for electrical isolation may be summed up as secure disconnection and separation from all supply sources. Key characteristics for compliance would be:

- adequate labelling, i.e. unambiguous and legible;

**E3020** *Electricity*

- adequate contact separation;
- means for securing the point of isolation, preferably by locking off.

Many accidents occur due to a failure to isolate plant correctly. Compliance with the requirements of *Regulation 12* is an important precursor to satisfying the requirements of the *Regulation 13* of the *EWR* (*SI 1989 No 635*), which is concerned with work on de-energised systems (see E3023 below).

## Working space, access and lighting

**E3020**  *Regulation 15* of the *EWR* (*SI 1989 No 635*) requires adequate working space, access and lighting. This is relevant to safe systems of work, particularly in the case of live work. For example, live testing within control panels for fault finding purposes requires accessibility, freedom of hand/arm movement, minimal risk of accidentally being pushed onto exposed live terminals and a lighting level sufficient to enable the live connections to be clearly observed, so that the testing can be undertaken safely. In terms of guidance for determining suitable accessibility, see Appendix 3 in the HSE Memorandum of Guidance, which reproduces a regulation from the former *Electricity (Factories Act) Special Regulations 1944* (*SI 1944 No 739*) (the 1944 Regulations extended the provisions of the original 1908 Regulations) which were revoked by the *EWR*. As a guide, a clear distance of the order of one metre should be allowed in front of fixed equipment where access is required for live diagnostic testing and other such work. With regard to the provision of adequate space requirements and accessibility for electrical switchgear in new buildings the provisions of the *Construction (Design and Management) Regulations 1994* (*SI 1994 No 3140*) are also now relevant. In other words, the designer must allow for adequate space provision in the initial building design, taking account of future operational and maintenance requirements.

# Maintenance

**E3021**  *Regulation 4(2)* of the *EWR* (*SI 1989 No 635*) states:

> 'As may be necessary to prevent danger, all systems shall be maintained so as to prevent, so far as is reasonably practicable, such danger.'

This legal requirement sets an objective, to ensure that sufficient maintenance is carried out on all electrical systems to prevent danger. The duty is a qualified one and therefore the extent and frequency of the maintenance required should be proportionate to the level of risk. By virtue of the definitions contained in *Regulation 2* of the *EWR* (*SI 1989 No 635*), the term 'system' embraces electrical equipment of all forms. In the case of equipment, examples of risk factors would include:

- age;
- operating voltage;
- environment of use (e.g. indoor or outdoor);
- type of equipment (hand held, portable, fixed, etc);
- class of construction for protection from electric shock (class I, II or III) (BS 2754: Memorandum. Construction of electrical equipment for protection against electric shock);
- nature of use (or even abuse).

The high end of the risk spectrum would be a hand-held class I mains powered tool used on outdoor construction work at different sites whilst the low end could be stationary class II IT equipment used in an office environment.

As a minimum, maintenance would include inspection to check for deterioration or damage and to verify safe condition for use by any user. However, some faults cannot be detected by inspection alone and therefore testing may also be required, e.g. to verify the effective earthing of exposed conductive parts such as a metallic equipment case.

The HSE has published guidance (Maintaining portable and transportable electrical equipment (HSG107) (ISBN 0 7176 0715 1) to assist duty holders in complying with the requirements of this Regulation. The IEE has published a code of practice which provides detailed guidance for both managers, with responsibilities for establishing policy and putting into place arrangements for electrical equipment maintenance, as well as staff undertaking inspection and maintenance tasks. It includes information relating to the different types and constructions of equipment as well as inspection criteria and test procedures (IEE Code of Practice for in-service inspection and testing of electrical equipment (ISBN 0 85296 776 4).

## Safe systems of work

### General

E3022   *Regulation 4(3)* of the *EWR (SI 1989 No 635)* imposes a general duty that all work activities, including work near an electrical system, shall be carried out in such a manner as not to give rise, so far as is reasonably practicable, to danger. The *EWR* provide for two basic approaches to achieving a safe system of work, covered respectively by *Regulations 13* and *14*:

(i) removal of any electrical danger by disconnection from all sources of supply and the implementation of a system of work aimed at verifying that the danger from electricity has been removed and remains removed for the duration of the work;

(ii) live working, recognising that the danger of electricity remains so that it is then necessary to implement a system of work aimed at preventing injury arising from the danger.

However, the second option is not an automatic choice since the intention to work live first must be properly justified, which forms the starting point for an assessment of electrical risks.

The HSE has published guidance specific to safe working practices involving electricity (Electricity at work – safe working practices (HSG85) (ISBN 0 7176 2164 2).

### Work on isolated equipment

E3023   A model safe system of work would require consideration of the following:

- isolation from all points of supply;
- earthing to discharge any residual electrical energy or to prevent the build up of an induced charge;
- proving dead at the point of work;
- demarcation of the safe zone of work;
- safety documentation.

*PUWER (SI 1998 No 2306)*, *Reg 19* places duties on employers in respect of isolation from all forms of energy, which therefore includes electrical energy. These duties require that:

- where appropriate, the work equipment shall be provided with suitable means to isolate it from all sources of energy;
- the means of isolation shall be clearly identifiable and readily accessible;
- appropriate measures shall be taken to ensure that re-connection of any energy source to work equipment does not expose any person using the equipment to any risk to his health or safety.

Electrical isolation is defined in *EWR* (*SI 1989 No 635*), *Reg 12* (also discussed at E3019 ABOVE) and means the secure disconnection and separation from all sources of electrical energy. The preferred means of securing points of isolation is by locking off, using special safety locks operated by unique keys.

Other safeguards include, where appropriate, the removal of fuses and the posting of warning notices. Where a number of different individuals or separate parties are working on the plant simultaneously a special locking device can be used to enable each party to apply its own safety lock, thereby maintaining personal control of the point of isolation for each person involved in the work.

Many accidents have arisen through incorrect plant isolation, coupled with a failure to prove dead. For example, this may be due to ambiguous, incorrect or missing circuit identification labelling or a failure to identify properly the isolation requirements appropriate to the intended work, particularly in the case of multiple supply sources. Confusion often arises concerning what has been made dead by the isolation of power and control circuits associated with factory production line equipment. For this reason, proving dead at the actual point of work is an essential element to achieving a safe system of work. Proprietary testers are available for this purpose and, because of the importance of using the right type of tester, the HSE has published a guidance note on the subject (Electrical test equipment for use by electricians (GS38) (ISBN 0 7176 0845 X)).

It must be recognised also that certain electrical equipment has the capability of retaining energy, in the form of an electrical charge, for a period of time following its isolation. This applies to plant containing capacitors, e.g. power factor correction equipment, or systems exhibiting capacitive effects, such as a long run of high voltage cable. Similarly, in the case of high voltage systems, a charge may be induced into an isolated circuit from an adjoining live circuit. Earthing is used to safeguard from static or induced charge.

Demarcation of the safe zone of work may be required to minimise the risk of the worker inadvertently seeking to gain access to adjacent live plant. In complex situations it is often useful to confirm the isolation, the equipment to be worked on and the work to be undertaken through the issue of a permit-to-work (PTW). The PTW also can serve to demarcate the safe zone of work. A PTW does not constitute the safe system of work but serves to strengthen communication and act as a control document to hold plant out of service until signed off by the person in charge of the work. The HSE recommends that a PTW should be used for all high voltage work (i.e. above 3kV) as well as for low voltage work involving multiple points of isolation. Where PTWs are adopted it is important that a management procedure exists to specify and control their use.

## Live work

**E3024** *Regulation 14* of the *EWR* (*SI 1989 No 635*) states that no person shall be engaged in any work activity on or so near any live conductor (other than one suitably covered with insulating material so as to prevent danger) that danger may arise unless:

(*a*)  it is unreasonable in all the circumstances for it to be dead; and

(*b*)  it is reasonable in all the circumstances for him to be at work on or near it whilst it is live.

Thus, justification centres around two issues. *First*, there must be good reason for the equipment to remain live whilst the work is carried out, e.g. the nature of certain electrical testing may require this to be undertaken live. *Second*, the equipment must be safe to work on whilst it remains live. This relates to the equipment design and construction, e.g. working space, extent of exposed live parts, shock and bridging paths (particularly involving any earthed metal case). *Regulation 14* applies to all work activities, including testing, i.e. there are no exemptions! The criterion is whether electrical danger is present or could arise during the work activity. It is also relevant to non-electrical activities, e.g. work in the vicinity of underground cables or overhead lines (discussed at E3028 below). The only exception to compliance with *Regulation 14* is where the live parts are suitably covered with insulating material and as necessary protected so as to prevent danger, thus instead satisfying the requirements of *Regulation 7* of the *EWR* (*SI 1989 No 635*) for the prevention of electric shock by direct contact.

Provided that live working can be justified in accordance with the first two parts of *Regulation 14*, the remaining part of that Regulation requires that suitable precautions shall be taken to prevent injury. This is an absolute duty. Thus, any accident occurring during the course of live work is prima facie evidence that the precautions taken were not suitable, i.e. in breach of *Regulation 14*.

Therefore, a safe system of work for live working must centre on the prevention of injury which, in turn, needs to recognise the types of injury and circumstances in which they might arise. The principal dangers are electric shock and short-circuit flashover. An understanding of shock and bridging paths, and how these might arise in the particular work activity, is therefore very important. Safeguards need to be devised and effectively implemented to prevent electric shock and short-circuit flashover.

A model safe system of work would require consideration of the following:

- capability of those undertaking the work;
- provision of appropriate information concerning the proposed work;
- use of suitable tools and other protective equipment;
- use of suitable instruments and test probes;
- consideration of the need for accompaniment;
- control of the work area.

There are important considerations concerning the competence of personnel intending to undertake live work, mentioned previously (see E3011 above). Additionally, adequate information needs to be provided concerning the work so that those directly involved can be satisfied that it is within their capability and experience.

Protective equipment must satisfy the requirements of *Regulation 4(4)* of the *EWR* (*SI 1989 No 635*), i.e. it must be:

- suitable for the use for which it is provided;
- maintained in a condition suitable for that use;
- properly used.

Suitability for use infers that management action has been taken to determine what protective equipment is needed and the circumstances in which it is to be used in contributing to the overall safe system of work. It is important that such equipment is of a suitable specification for the intended use, for which a number of British Standards exist:

- BS EN 60900: Hand tools for live working up to 1000 V a.c. and 1500 V d.c.
- BS EN 60903: Gloves and mitts of insulating material for live working.
- BS 921: Specification. Rubber mats for electrical purposes.

These same standards include information to assist in the implementation of appropriate in-service care regimes for the protective equipment. Correct use infers the provision to users of relevant information and training, particularly concerning the limitations of such equipment.

Whilst applied insulation may minimise the risk of electric shock, large areas of exposed metal, such as at tool working heads or test probe tips, may create a risk of short-circuit when used in a confined space with limited clearance between exposed live terminals or conductors.

The *EWR* requirements concerning protective equipment are supplemented by the *Personal Protective Equipment at Work Regulations 1992* (*PPE*) (*SI 1992 No 2966*). One important issue concerns compatibility between different items, e.g. insulated tools and gloves.

There is no automatic requirement for accompaniment in the case of live work. If it is considered necessary to have a second person present to render first aid in the event of an accident this implies that there could be doubt concerning the proposed system of work or its effective implementation. However, a second person may be necessary to enable safe working, e.g. by directly assisting with the job as 'another pair of hands' or to control the work area.

## Special situations

### Lightning protection systems

E3025    Although lightning is a natural phenomenon *Regulation 6* of the *EWR* (*SI 1989 No 635*) requires electrical equipment which may reasonably foreseeably be exposed to the effects of the weather, such as lightning, to be of such construction or as necessary protected as to prevent, so far as is reasonably practicable, danger arising from such exposure. In reality, it is not just electrical equipment that needs protection but buildings and the contents therein. A lightning strike onto a building may find its way to earth via the electrical installation. This risk can be minimised by careful attention to the design and installation of a lightning protection system for the building as a whole. Lightning protection is outside the scope of BS 7671 but is covered by another British Standard (BS 6651: Code of practice for protection of structures against lightning). The first step is to establish the need for lightning protection for if the risk of a strike is assessed to be low the expense may not be justified. The standard includes a methodology for determining the level of risk, taking as a starting point the ceraunic level (expected number of lightning flashes to the ground/$km^2$/year) for the site location. The computation then includes other factors having a bearing on the likelihood and consequences of a strike on the building in question, such as its location, construction, function and proximity of other taller structures. Lightning protection systems are designed on the 'Faraday cage' principle with the cage formed by lightning conductors running across the roof and down the outside of the exterior walls to earth electrodes. In the event of a strike, the theory is that the electrical discharge will find its way to earth without causing damage internally. Where lighting protection systems are installed it is important that these are regularly maintained, typically by an annual inspection and test. An inspection would include a check to verify that:

- all down conductors are in situ (these are usually formed of copper bar and therefore attractive to thieves);
- all lightning conductors are well secured to the building (in the event of a strike the high current creates strong electro-magnetic forces which can cause conductor whip, so pulling out poor fixings);
- all connections are sound, in particular the down conductor disconnecting links located near ground level to facilitate the measurement of earth electrode resistance;
- the building protection envelope has not been negated through the addition of extensions, chimneys, fume extraction vents, etc.

Testing requires measurement of the resistance of every earth electrode individually, the recommended maximum resistance being 10N $\Omega$, where N is the number of electrodes, subject to a maximum of 200 $\Omega$ where there are more than 20 of these.

### Static electricity

E3026    This is a common phenomenon. Most people have experienced the effects of static electricity as a short sharp shock when the body or another object appears to have become charged with electricity and discharge occurs as body contact, usually a hand, is made with the object.

Frequently occurring examples are alighting from a vehicle and opening a metal filing cabinet in an office. The mechanism surrounding the build up of static charge is not fully understood but involves interaction between different materials aided by particular atmospheric conditions. Certain work processes, involving the generation and movement of fine dusts or the flow of liquids through pipes and orifices, may give rise to the generation of static in work processes. The *EWR (SI 1989 No 635)*, in relating to electrical systems and associated equipment are not intended to apply to such situations. However, other legislation serves to ensure that appropriate safeguards are put in place. In particular, *PUWER (SI 1998 No 2306), Reg 12* requires that measures shall be taken to prevent or, where that is not reasonably practicable, to adequately control specified hazards associated with work equipment, such as the premature explosion of any article or substance produced, used or stored in it. Preventive measures might include redesigning the process or equipment or using different construction materials to prevent the build up of static. A common practice is to electrically bond together all conductive parts, to reduce the possibility of charge build up between different surfaces.

Some work processes depend on electrostatic, e.g. electrostatic paint spraying and chimney flue gas filters. In these cases *Regulation 6* of the *EWR (SI 1989 No 635)* is applicable and, in the case of electrostatic paint spraying, a British Standard gives specific guidance (BS 6742: Electrostatic painting and finishing equipment using flammable materials).

## *Electricity in potentially explosive atmospheres*

E3027　　The use of electricity in an area with a potentially explosive atmosphere (PEA) poses particular problems. Ideally, the risk should be minimised either by siting any electrical equipment outside the area or the work location designed and constructed so that natural ventilation will ensure that a PEA will never arise within it. In practice, the latter option may be difficult to confirm and attention will then need to be given to the design and construction of the equipment, taking account of the likelihood of a PEA arising, to achieve the required level of safety. A British Standard provides a classification for potentially explosive gas atmospheres based upon the likelihood and duration of a PEA arising (BS EN 60079: Electrical apparatus for potentially explosive gas atmospheres). This classification is given in the following table:

| Zone | Description |
| --- | --- |
| 0 | An area in which an explosive gas atmosphere is present continuously or for long periods of time |
| 1 | An area in which an explosive gas atmosphere is likely to occur in normal operation (i.e. 10–1,000 hours per annum) |
| 2 | An area in which an explosive gas atmosphere is not likely to occur in normal operation and, if it does occur, is likely to do so infrequently and will exist for a short period only (i.e. around 5–10 hours per annum) |

The types of protection employed in such PEAs are listed in the following table:

| Protection type | Title | Description |
| --- | --- | --- |
| d | Flameproof enclosure | Enclosure designed to withstand internal explosion of the flammable gas or vapour that may enter it and not to communicate this to the external explosive atmosphere |

| i | Intrinsically safe | Electrical energy within apparatus or system associated with the PEA restricted to a level below that which can cause ignition |
|---|---|---|
| p | Pressurisation/ dilution | Ingress of an explosive atmosphere to an area containing a source of ignition prevented by pressurisation with a protective gas and, where necessary, continuous dilution of atmosphere in an area containing a source of emission of gas to prevent build-up of an explosive atmosphere |
| e | Increased safety | Application of additional measures to apparatus to improve security against possibility of excessive temperature and recurrence of arcing/sparks throughout service life |
| N | Type of protection N | Application of protection so that, in normal service, the apparatus is not capable of igniting any surrounding explosive atmosphere and a fault capable of causing ignition is unlikely |
| s | Special protection | Appropriate to apparatus that, because of its nature, cannot comply with constructional or other requirements applicable to other protection types, but which can be shown to be suitable for use in prescribed situations |
| o | Oil immersion | Oil immersion so that any explosive atmosphere above or outside the enclosure will not be ignited |
| q | Powder/sand filled | Enclosure containing apparatus filled with granular material so that if arcing occurs it is not likely to ignite any surrounding flammable atmosphere |
| m | Encapsulation | Encapsulation with an appropriate compound of all parts that could cause ignition of an explosive atmosphere by sparking or heating |

Symbols relating to the method of protection are usually prefixed by 'EEx' so, for example, the marking on equipment for flameproof equipment would be 'EEx d'. Flameproof enclosures are characterised by their robustness of construction and older equipment may be marked FLP, a designation from an earlier standard.

Battery rooms are a common situation where a PEA may arise. The charging of lead-acid batteries, even those that are described as maintenance free, generates hydrogen gas. Lack of ventilation may allow the build up of a PEA since the lower explosive limit (LEL) is only 4 per cent. As hydrogen is lighter than air the PEA is likely to occur first at ceiling level where a light fitting could produce a spark to cause ignition. For this reason care must be given to the ventilation requirements for battery rooms, with vents at both high and low level to achieve a good level of natural ventilation. Additionally, light fittings of an appropriate type, should be wall-mounted. Because of the high number of accidents involving electric storage batteries the HSE has produced a leaflet concerning safe charging and use (Electric storage batteries: Safe charging and use (INDG139)).

A further British Standard provides guidance in respect of electrical equipment for use in the presence of flammable dusts (BS EN 50281: Electrical apparatus for use in the presence of combustible dust).

Two important European Community (EC) Directives, referred to as the ATEX Directives, have been introduced concerning equipment and installations in PEAs. One is an equipment directive relating to the design and manufacture of equipment for use in hazardous areas (EC Directive 94/9/EC (ATEX 100a)); the other is a users directive, relating to the use and operation of equipment in hazardous areas (EC Directive 99/92/EC (ATEX137)). They relate to non-electrical as well as to electrical equipment.

### Work near underground cables and overhead lines

E3028  By means of *Regulation 4(3)* the *EWR* (*SI 1989 No 635*) regulate work carried out in close proximity to live conductors, whether overhead lines or buried cables, in situations where there is electrical danger to be avoided. Many accidents involving inadvertent contact with live overhead lines arise during construction or agricultural activities. In the case of the former, examples are the movement or operation of cranes, dumpers, excavators and access equipment too close to a line and for the latter, the use of crop spraying or other irrigation equipment often poses particular problems. With high voltage overhead lines it is important to recognise that the electricity can flash across a gap and therefore actual contact does not have to be made for injury to occur. A safe system of work depends on establishing a safety clearance distance from the line and height clearances where plant is required to pass underneath. These measures are described in more detail in HSE guidance (Avoidance of danger from overhead electrical lines (GS6) (ISBN 0 7176 1348 8)) and, where there is any doubt, advice should be sought from the line owner. Excavation work accounts for many cable damages, some of which result in serious injury or even death. Again, it is essential to implement a safe system of work founded on:

- establishing the presence of any buried cables by reference to plans (electricity distribution companies are required to make such information available);
- verifying the position of cables by the use of a cable locating device;
- employing safe excavation techniques involving the digging of trail holes.

The HSE has published guidance on avoiding danger from all types of underground service, which includes electricity cables (Avoiding danger from underground services (HSG47) (ISBN 0 7176 1744 0)).

## Other relevant legislation

### The Electromagnetic Compatibility Regulations 1992 (SI 1992 No 2372)

E3029  These Regulations, with subsequent amendment, implement the European Union (EU) Directive on electromagnetic compatibility and are administered by the Department of Trade and Industry (DTI). They place duties on designers and constructors to ensure the electromagnetic compatibility of electrical equipment and systems in two ways, namely that it:

- does not give rise to excessive electromagnetic emissions, thereby causing interference with other equipment; and
- is not unduly affected by electromagnetic emissions originating from other equipment.

There are important safety implications concerning the remote operation of plant, such as an electric overhead travelling crane by wireless controller from ground level. Accidents have been known to happen because of unintended movement due to external signal influence.

### The Plugs and Sockets etc (Safety) Regulations 1994 (SI 1994 No 1768)

E3030  These Regulations were made under the *Consumer Safety Act 1978* and are administered by the DTI. They contain requirements for the safety of plugs, sockets, adaptors and fuse links designed for use at a voltage of not less than 200 V. For example, under these Regulations, the circuit conductor pins (L and N) of 13 A plugs are now required to be partially insulated.

## The Electrical Equipment (Safety) Regulations 1994 (SI 1994 No 3260)

E3031    These Regulations consolidate, with amendments, the *Low Voltage Electrical Equipment (Safety) Regulations 1989 (SI 1989 No 728)* which, in turn, implement the requirements of the EU Low Voltage Directive concerning the harmonisation of requirements across the EU relating to the design and construction of low voltage (i.e. up to 1000 V) electrical equipment. They are administered by the DTI and also include provisions relating to 'CE' marking. It is important to understand the limitations of such marking which does not signify product approval. The 'CE' mark is applied by the manufacturer to show that electrical equipment meets the relevant product standard(s). A disadvantage is that the scheme operates on the basis of self-certification by the manufacturer rather than using the services of an independent test house. A further problem is that counterfeit marks have been known to be applied to domestic type electrical products, such as plugs, originating from countries outside the EU. However, under the Regulations, the manufacturer is required to make available a written declaration of product conformity comprising:

- manufacturer's name and address (or details of authorised representative);
- a description of the electrical equipment;
- reference to the harmonised standards;
- where appropriate, references to the specifications with which conformity is declared;
- identification of the signatory who has been empowered to enter into commitments on behalf of the manufacturer or authorised representative;
- the last two digits of the year in which the CE mark was affixed.

## The Electricity Safety, Quality and Continuity Regulations 2002 (SI 2002 No 2665)

E3032    These Regulations are administered by the DTI and cover electricity supply systems and the provision of electricity supplies to consumers. Their prime purpose is to ensure public safety. Whilst mainly directed at those with responsibilities for public supply networks they will also be of interest to organisations operating private distribution systems and highway power supply networks, particularly if these encroach on public land. Duties are placed mainly on generators, distributors, suppliers and meter operators. Some companies may be undertaking more than one of these roles. Some responsibilities are placed also on consumers. The Regulations prescribe the nominal supply voltage and frequency and the statutory limits of variation for these. For standard low voltage supplies this is 400/230 V +10 per cent/–6 per cent, 50 Hz ±1 per cent. The Regulations specify important safety requirements concerning the design and construction of electricity networks, e.g. minimum ground clearances of overhead line conductors, the safeguarding of buried electricity cables and the protection of distribution equipment in substations, etc. Some of the requirements mirror those contained in the *EWR (SI 1989 No 635)*. The Regulations also specify fundamental requirements for safety concerning consumers' installations, prior to connection. These link in with the fundamental principles for the safety of electrical installations contained in chapter 13 of BS 7671. This is not only to ensure the safety of those using an installation but also to minimise the risk of the supply to other consumers being adversely affected, e.g. in terms of loss of supply or its quality, in consequence of how it is being used.

## The Building Regulations 2000 (SI 2000 No 2531) (as amended)

E3033    The Building Regulations for England and Wales have been amended to introduce a new *Part P*, which takes effect from 1 January 2005. The requirements of this part apply only to low voltage electrical installations in dwellings or the common parts of a building serving one or more dwellings, including outbuildings and gardens. Although not related to workplaces these requirements will be of interest to those organisations managing residential property, such as local authorities and housing associations. The Regulations cover the design, installation, inspection and testing of electrical installations and require:

- reasonable provision to be made in the design, installation, inspection and testing of electrical installations in order to protect persons from fire or injury; and
- sufficient information to be provided so that persons wishing to operate, maintain or alter an electrical installation can do so with reasonable safety.

The Regulations are supported by an approved document.

This new legislation has the effect of defining a legal standard for electrical installations in dwellings which, as with the *EWR* (*SI 1989 No 635*) in the case of work premises, can be met by fulfilling the requirements of BS 7671. With certain defined limited exceptions all work will need to be undertaken by a suitably competent person. Electricians will need to be registered under a *Part P* self-certification scheme. In Scotland, the Scottish Building Regulations already adopt BS 7671 as the appropriate standard concerning an electrical installation in a dwelling.

# Emissions into the Atmosphere

## Introduction

E5001 Pollution of the atmosphere we breathe is a subject which encompasses the whole range of distance scales and can be important over minutes or lifetimes. Regulation of emissions, therefore, has to account for all of these effects and, as a consequence, has to take a variety of forms. The concept of the different distance scales is illustrated in Figure 1 below:

Figure 1: Distance Scales

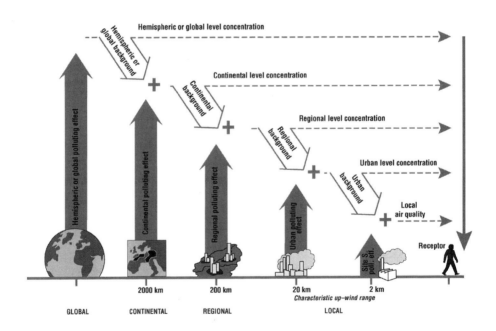

A 'receptor' (a human being, a population, an ecosystem) experiences air pollution as a sum of the contribution from sources which might be adjacent or at some other part of the planet. The extent to which the sources contribute might depend on the pollutant of concern. A pollutant such as sulphur dioxide is relatively short lived in the atmosphere and has the capacity to affect the lungs of people within tens of kilometres of the point of emission. At distances of hundreds of kilometres, however, the sulphur dioxide will have been transformed into sulphate, which may be absorbed into cloud water and be deposited on the ecosystems of another country. This is an example of trans-boundary pollution and popularly known as 'acid rain'.

Pollutants that are chemically more stable than sulphur dioxide have the potential to be dispersed throughout the globe. Some organic pollutants are an excellent example of this. Poly-chlorinated biphenyls were used widely in the middle of the twentieth century as a component part of electrical transformers. They are highly toxic and accumulate in body fats, so that animals (and humans) at the top of the food chain are most exposed. Hence, polar bears living in the Arctic Circle are far removed in distance and time from the original emission but still experience the impact.

Regulation of emissions from industry must recognise all these effects – from the large coal-fired power station with sulphur emissions to the small manufacturer using solvents. Where the potential exists for an air pollution impact, regulators will be seeking to restrict emissions. This chapter examines the means by which this goal is currently being achieved.

## Regulatory position

E5002    For air pollution, there are three main sources of legislation and regulation to which industry has to respond and which this chapter seeks to cover. These are the following:

- European Union directives on air quality and emissions;
- the Environment Agency and local authority implementation of Integrated Pollution Control (see E5059) and its successor, Integrated Pollution Prevention and Control (see E5027); and
- the Air Quality Strategy for England, Wales, Scotland and Northern Ireland, as prepared by the Department of the Environment, Transport and the Regions and the Scottish Executive.

In addition, emissions to atmosphere also encompass gases with the potential to contribute to global warming and stratospheric ozone depletion.

## Regulatory drivers

### Human health and toxicology

E5003    The effects of air pollution on human health have been recognised in the UK for more than 300 years. The often quoted example is the prohibition of coal burning in London in 1273, on the grounds that it was 'prejudicial to health'.

Coal burning has been the main cause of air pollution problems since then and has provoked most of the legislation aimed at improving air quality and human health. The infamous four day 'smog' of December 1952 was shown to have caused nearly 4,000 additional deaths through bronchitis and other diseases of the respiratory system. There was also an increase in heart disease.

It is less obvious now than in the 1950s that health is affected by exposure to air pollution, but epidemiologists can show statistically significant associations between high concentrations of certain pollutants and hospital admissions (and re-admissions), as well as mortality. Health effects include irritation and loss of organ functions (e.g. reduced lung capacity). The Committee on the Medical Effects of Air Pollutants estimates that 14,000–24,000 admissions a year are associated with short term air pollution, and that it causes between 12,000–24,000 premature deaths per year. The largest contribution to this total is made by inhalable particles, which, for example, may arise through combustion processes.

It is not only the 'classical' air pollutants that provoke concern with regard to their impact on human health through impairment of the respiratory and cardiovascular systems. Fears also exist about releases of substances with the potential for damage to other parts of the body through long term exposure at low levels. In particular, 'persistent organic pollutants' and heavy metals have the potential to cause human health effects through inhalation or entry into the food chain.

## The medical effects of air pollutants

### Nitrogen dioxide ($NO_2$)

**E5004** This gas can cause the inflammation of the airways when in high concentrations, and long-term exposure can affect lung function. This exposure can enhance the response that people have to allergens and is particularly harmful to asthmatics.

### Particulate matter ($PM_{10}$)

**E5005** $PM_{10}$ are particles in the air that have an aerodynamic diameter of less than 10μm. Particulate air pollution is associated with a range of health effects including effects to the respiratory and cardiovascular systems. $PM_{10}$ is responsible for causing premature deaths to those with pre-existing heart and lung disease. The smaller the particle, the greater the likelihood that it can reach the furthest part of the lung where the delicate tissues are.

### Sulphur dioxide ($SO_2$)

**E5006** This gas constricts the airways by stimulating the nerves in the lining of the respiratory tissues. Asthmatics are particularly vulnerable to this.

### Carbon monoxide

**E5007** This gas causes the formation of carboxyhaemoglobin. This reduces the capacity of the blood to carry oxygen round the body. Those with an existing disease connected to the heart or brain are particularly vulnerable.

### Benzene

**E5008** This gas is a genotoxic human carcinogen, and there is no safe level that has been defined. There is a risk of leukaemia when people are exposed to high levels.

### 1-3 butadiene

**E5009** This gas is also a genotoxic human carcinogen that can cause the induction of cancers of the lymphoid system, blood forming tissue, lymphomas and leukaemias.

### Ozone

**E5010** This gas can cause slight irritation to the eyes and nose and may exacerbate symptoms of those with respiratory conditions (e.g. asthma, pneumonia).

### Dioxins

**E5011** Short-term exposure may result in skin lesions and altered liver function. Long-term exposure may damage the immune system, the developing nervous system, the endocrine system and reproductive functions. One congener is a known human carcinogen, however it does not effect genetic material.

### Lead

**E5012** Lead causes problems in the synthesis of haemoglobin, it has effects on the kidneys, the gastrointestinal tract, joints and reproductive system, and the nervous system. The greatest cause for concern is the possible effect of lead on brain development in children. Food and water are the two main sources of ingestion for most people. Lead in air contributes to the lead levels in food through the deposition of dust and rain on crops and the soil.

## Polycyclic aromatic hydrocarbons ('PAHs')

E5013    Some PAHs are genotoxic human carcinogens, suspected of causing lung cancers and leading to the induction of other cancers.

## Acid deposition and damage to ecosystems

E5014    When atmospheric pollutants such as sulphur dioxide ($SO_2$), nitrogen oxide ($NO_2$), and ammonia ($NH_3$) are deposited at the surface, this is referred to as acidification. This can occur through direct 'dry' deposition of the gases or through 'wet' deposition after transformation into sulphates and nitrates and absorption into rain and cloud water. This has been recognised as a threat to ecosystems, resulting in major international research projects and negotiations to reduce emissions. It is usually referred to by the more popular term of 'acid rain'.

Pollutants can be transported over considerable distance, affecting the air quality and ecosystems in adjacent or distant countries. Such effects can extend over several thousands of kilometres. Deposition of excess acid can result in changes to the chemical composition of soil and surface water, in turn interfering with ecosystems.

## Consequences of acidification

E5015    Acidification has been a major problem in Europe with the highest deposition rates found in the highly populated and industrialised zone extending between Poland and the Netherlands. Within the next ten years, acid deposition in Europe is expected to decrease following the agreed reductions in European emissions of $SO_2$ and $NO_x$ (while ammonia emissions are not likely to change significantly).

Organisms in water are affected when the concentration of hydrogen ions increases (decreasing the pH of the water). As a result lake beds are covered in mosses, and there is an increase in the death of fish and other animals. Plants are affected by the changing chemistry and biology of the soil, altered by an increased amount of hydrogen ions. There has been a decline in the health of forests in central Europe because of acidification.

Modern forestry and agriculture can contribute to the pollution as well as suffer from it, through the use of fertilisers which release nitrogen and hydrogen ions into the air in the form of ammonia.

Deposition of excess nitrogen can lead to the eutrophication of fresh and marine waters.

Acidification in urban and industrial areas increases the deterioration of many buildings and construction materials. In sheltered positions, a black crust surface layer forms on calcareous materials, and the growth of lichen, mosses and algae is common. Historic monuments and buildings can be especially susceptible to damage.

Acidifying substances also contribute to climate change. Both $NO_2$ and $SO_2$ are radiatively active gases with a warming potential, although sulphate and nitrate aerosols result in cooling through reflection of sunlight in the upper part of the troposphere. Oxides of nitrogen are implicated strongly in the atmospheric chemistry of ozone formation, causing an excess of tropospheric ozone, and the depletion of stratospheric ozone.

## Nuisance

E5016    Air pollution can cause harm to the senses as well as harm to human health and the natural environment. Whilst not as damaging, odour and dust nuisance can result in costly disputes for industry and can be difficult to eliminate in many cases.

Another issue which can arise is the visual impact caused by plumes of water droplets, formed when a high moisture content plume enters the cool atmosphere and becomes visible through the condensation of its water vapour. The droplets then appear as a white plume, which may extend for some distance, depending on atmospheric conditions. Such a plume is harmless in terms of human health, but is disliked by regulatory authorities as it draws attention to the emission and can be a source of complaints.

At present, the UK regulatory climate discourages visible plumes, unlike Scandinavia and the USA for example, but the newer emphasis on energy efficiency may soon outweigh these concerns.

# Key pollutants and issues

## Introduction

E5017    The issues outlined above at E5003-E5016 are some of the main areas of concern which drive regulation aimed at reducing air pollution. The section below from E5018-E5024 provides background material on the main pollutants and their origin, as well as further material on the regional and global issues that result in industry being asked to emit less substances of a damaging nature to atmosphere.

## The UK priority pollutants

E5018    Emissions of pollutants have increased since the industrial revolution. Until quite recently, national air pollution control legislation was focussed on reducing air pollution from particular sectors of industry. Only since the late 1980s has this been added to by 'effects based' legislation. This approach takes as its start point an air quality standard or measure of harm in the environment and then policies are devised to enable this criterion to be attained.

This philosophy has reached its logical conclusion with the publication of the Air Quality Strategy (see below at E5064). This document targets eight priority pollutants and explains how the government will achieve air quality objectives relating to airborne concentrations of all of these eight pollutants.

## The priority pollutants

E5019

| Pollutant | Origin | Comments |
| --- | --- | --- |
| Sulphur dioxide ($SO_2$) | $SO_2$ dissolves in water to give an acidic solution (sulphuric acid). In the UK the main source of $SO_2$ is the combustion of fossil fuels which contain sulphur and the combustion of heavy oils. | At the beginning of the 20th century, $SO_2$ was emitted from the domestic sector, commercial and industrial premises, and power stations by the combustion of coal. The smogs of the 1950s prompted the *Clean Air Act 1956*. Cleaner fuels have since replaced coal in all sectors. However this use of coal is still significant in Northern Ireland and some other parts of the UK. |

| | | |
|---|---|---|
| Nitrogen dioxide ($NO_2$) | All combustion processes in air produce oxides of nitrogen. $NO_2$ and nitric oxide are oxides of nitrogen and together they are referred to as $NO_x$. In the UK 50% of $NO_x$ emissions are from road transport, 20% from the electricity industry, and other industrial and commercial activities account for 17%. In London, it is thought that 75% of $NO_x$ emissions come from road transport. | There is evidence that $NO_2$ emissions from motor vehicles may continue to increase because of the growth in private car use in recent years. Reductions will be slow even if vehicle numbers and distances travelled were to stop increasing. The Air Quality Strategy estimates that $NO_2$ from road transport will gradually decrease, levelling off in 2015, and that petrol cars make the greatest contribution to $NO_2$ transport emissions in the UK. |
| Carbon monoxide (CO) | This gas is formed by the incomplete combustion of carbon containing fuels. The main source is road transport, particularly petrol-engined vehicles. | |
| Lead | Lead is the most widely used non-ferrous metal. The largest use is for the manufacture of batteries. | The sale of leaded petrol was banned from 1 January 2000 in the UK. Its use has been phased out gradually since 1985. This has led to significant decreases in emissions in urban areas. |
| Ozone | Ozone is not directly emitted from any man-made source in any significant quantities. Instead it arises from chemical reactions in the atmosphere caused by sunlight. $NO_x$ and volatile organic compounds ('VOCs') are the most important precursors of elevated levels of ozone. Its production can also be stimulated by CO, methane, and other VOCs which come from natural sources. | |

| Particles (PM$_{10}$) | PM$_{10}$ are particles with an aerodynamic diameter of less than 10µm. This is emitted from road transport, stationary combustion sources, and other industrial processes. | Diesel vehicles emit particles at a greater rate than petrol vehicles and nationally, road transport contributed approximately 25% of PM$_{10}$ emissions in 1996. |
|---|---|---|
| Benzene | This gas is a VOC. In the UK the main sources are from the combustion and distribution of petrol. Diesel fuel is also a small source. | In recent years unleaded petrol consists of, on average, 2% by volume in the UK. |
| 1,3-Butadiene | There are trace amounts of this gas in the atmosphere from the combustion of petrol and other materials. It is used in industry for the production of rubber for tyres, however motor vehicles continue to be the dominant source. | |

[1] Airborne Particles Expert Group ('APEG') 1999 Source Appointment of Airborne Particulate Matter in the UK Department of the Environment Transport and the Regions, the Welsh Office and the Department of the Environment (Northern Ireland).

## Other important pollutants

**E5020**    The principal reason why the Air Quality Strategy is focussed on the eight priority pollutants is to protect the health of the population. It could be argued very easily that there are other pollutants which have the potential to harm health at sufficiently high concentrations. This might be especially true within the industrial sector where the possibility exists of localised 'hotspots' of concentrations of substances arising out of the particular industrial process in question. Examples are metals and individual organic compounds. For occupational hygiene and the indoor environment, the Health and Safety Executive publish safe limits for concentrations of several hundred substances. For the outdoor environment the same principles apply, but in most circumstances the concentrations are too low to cause a health effect.

In terms of the their abundance and toxicity, the following represents a list of pollutants which require regulation from industrial processes:

| | | | |
|---|---|---|---|
| • | cadmium | • | polycyclic aromatic hydrocarbons ('PAHs') |
| • | mercury | • | polychlorinated bipheyls |
| • | nickel | • | polychlorinated dibenzo-p-dioxins |
| • | arsenic | • | polychlorinated dibenzofurans |
| • | vanadium | • | hydrogen fluoride |
| • | hydrogen chloride | | |

## Trans-boundary pollution

**E5021** In the 1970s and 1980s much of the air pollution control policy was aimed at reducing the long range transport of acidic pollution across national boundaries. The reported decline in the health of trees and the loss of life in lakes from Scandinavia to eastern Europe had provoked a strong response from the public and politicians. This issue is still very much alive, although much has been done to reduce national emissions of sulphur and nitrogen, especially from large power plants burning coal and oil. The UK was once perceived as being the 'dirty man' of Europe in the context of sulphur dioxide emissions, but can now reasonably claim to have reduced national emissions substantially. The United Nations Economic Commission for Europe ('UNECE') Convention on long-range trans-boundary air pollution ('CLRTAP') is an important framework for environmental assessment and policy in Europe.

## Climate change

**E5022** The natural presence of 'radiatively-active' gases in the atmosphere is essential for life. These gases trap heat in the lower atmosphere, creating a greenhouse, and include the following:

- carbon dioxide;
- methane;
- nitrous oxide ($N_2O$);
- chlorofluorocarbons ('CFCs'); and
- ozone.

Increasing concentrations of these greenhouse gases enables more infra-red radiation to be absorbed in the lower atmosphere, upsetting the earth's radiation balance. This increase allows the troposphere to be warmed more significantly, and the upper layers to be cooled. Without any greenhouse effect at all, the earth would be uninhabitable for humans. In previous periods of geological history, the earth has been warmer and colder than at present. The issue today is the rate at which the climate appears to be changing.

Emissions of these greenhouse gases come mainly from anthropogenic activity, especially combustion of fuel and are expected to rise, doubling by 2030. As a result, the atmosphere is being gradually heated up; the trend of global mean surface temperatures indicates an increase of 0.45C over the past hundred years.

## Consequences

**E5023** Vegetation will be in greater competition as optimum conditions become scarce. This may cause shifts of plant and animal life. Extra warmth can trigger insect plagues and plant diseases. Both of these consequences affect agriculture, which is also directly affected by an annual rainfall decrease. All these negative changes to the land can expose vulnerable areas to land degradation.

## The depletion of stratospheric ozone

**E5024** The depletion of stratospheric ozone during the last few decades has been a global problem. Stratospheric ozone concentrations have been decreasing at an increasing rate over the last ten years. This trend is larger nearer the poles and is often referred to as the 'ozone hole'. The cause of this problem is the increase of industrially produced CFCs and halogens in the upper atmosphere, and these are expected to cause ozone depletion at least up to the year 2100.

## Consequences

E5025 The depletion of ozone means that certain wave lengths of ultra-violet solar radiation are able to penetrate through to the earth's surface, damaging human health and leading to an increase in the number of cases of skin cancer, as well as damaging ecosystems. This can also lead to a change in global circulation and climate, as the absorption of the radiation by ozone can lead to heat formation in the atmosphere.

# The European context
## The policy makers

E5026 In the world of air pollution, 'Europe' has several identities. There are two main bodies which drive legislation and protocols, namely the European Union and UNECE (see E5021 above). The latter body is composed of the EU member states, plus the countries of eastern Europe including Russia and the Ukraine. In all, it comprises 50 countries (and includes the USA and Canada.) Its chief role is to examine the consequences of trans-boundary pollution through CLRTAP (see E5021 above). The Convention states that all countries should;

> 'endeavour to limit and, as far as possible, gradually reduce and prevent air pollution, including long range trans-boundary pollution.'

In all, the Convention has been responsible for five protocols, as follows.

- *The 1985 Sulphur Protocol* required signatories to reduce national sulphur emissions by 30%, based on 1980 levels, on or before 1993. The UK was not a signatory to this Protocol. The Protocol was subsequently revised and signed in Oslo in June 1994 and came into force on 5 August 1998. This time the UK did ratify the Protocol (17 December 1996) and has agreed to reduce $SO_2$ emissions by 50% by 2000, 70% by 2005 and 80% by 2010, (again based on 1980 levels). This will require fuel switching and some advances in abatement technology. The government has drawn up a national plan to keep national emissions within these targets. In the main, these targets have most impact on the power generation and oil refining industries.

- *The 1988 Nitrogen Oxides Protocol*, signed in Sofia, came into effect in 1991 and sought to freeze emissions of $NO_x$ at their 1987 levels, with a target date of 1994. The UK ratified this Protocol in 1990. By 1996, UK emissions were 21% lower than in 1987.

- *The 1991 Volatile Organic Compounds Protocol* was signed in Geneva and came into force on 29 September 1997. The UK ratified the Protocol in June 1994. VOCs are defined as *'all organic compounds of anthropogenic nature, other than methane, that are capable of producing photochemical oxidants by reactions with nitrogen oxides in the presence of sunlight.'* The Protocol requires most parties to reduce overall VOC emissions by 30% on or before 1999, using a base year of 1988. Additionally, the protocol obliges signatories to control emissions from industries through new emission limits and to introduce abatement technologies as well as reduce solvent use and reduce VOC emissions through petrol distribution and refuelling.

- *The 1998 Heavy Metals Protocol* was signed in June 1998 at Aarhus. It requires emissions of cadmium, lead and mercury to be reduced below 1990 levels. Its aim is to reduce emissions through the use of stricter emissions limit values for industry and the use of best available technology (see E5027 below).

- *The Persistent Organic Pollutants Protocol* was also signed in June 1998 at Aarhus. It aims to phase out the production of and use of a defined list of substances, as well as imposing requirements to eliminate discharges, emissions and losses and to ensure safe disposal methods. The list covers 16 substances in three categories, (pesticides, industrial chemicals and by products or contaminants, e.g. dioxins.)

The effects of the Protocols on industry are not usually seen immediately or directly. Normally they take effect through the subsequent actions of national governments and regulatory agencies who propose action and legislation designed to achieve the aims of the Protocols. In this sense the European Commission is strongly interlinked with the CLRTAP Protocols. Not only is the Community a signatory in itself, but the Commission will frame directives so as to meet the aims of the Protocols.

The most active part of the Commission with regard to air quality legislation is Directorate General Environment ('DG Environment'). The units within this Directorate have direct responsibility for drafting new directives which set air quality standards and implement emissions reductions in key sectors. At present, their work has three main elements:

- The Auto-Oil II programme;
- National Emissions Ceilings;
- Daughter directives on air quality standards.

The DG Environment also has responsibility for some legislation relevant to industry and air pollution control, such as the implementation of Integrated Pollution Prevention and Control and the Waste Incineration Directive.

The work of the DG Environment in air pollution is characterised by a tension between the contribution from transport and from industry. The Auto-Oil programme of research and policy measures was established with the transport sector in mind. The feeling in the Commission now is that the transport 'problem' has been solved and more emphasis should return to the contribution from industry.

## Integrated pollution prevention and control ('IPPC') – an overview

E5027 The IPPC directive has been introduced into Europe (*Directive on Integrated Pollution Prevention Control 96/61/EC*), and is designed to prevent, reduce, and eliminate pollution at the source, through the careful use of natural resources. The directive is intended to help industries move towards greater environmental sustainability, and applies to the installation of the following activities:

- energy industries (power, oil and gas);
- the production and processing of metals (ferrous and non-ferrous);
- mineral industries (cement and glassworks);
- chemical industries (organic, inorganic, and pharmaceuticals);
- waste management (landfill sites and incinerators); and
- other (slaughter houses, food/milk processing, paper, tanneries, animal carcass disposal, pig/poultry units and organic solvent users).

IPPC covers emissions to air, land and water, as well as heat. IPPC goes further to cover noise, vibration, energy efficiency, environmental accidents, site protection, and many more processes.

> The IPPC directive requires that applications must show that installations are run in a way that prevents emissions, and must use the following principles:
> 
> - must apply the best available techniques ('BAT') to control emissions. Account must be taken of the relative costs and advantages of the available techniques;
> - waste is to be minimised and recycled where possible;
> - energy is to be conserved;

- accidents are to be prevented, and their environmental consequences to be limited; and

the site is to be returned to a satisfactory state after operations cease.

The overall objective for the directive is a high level of protection for the environment as a whole, and a system of permits has been set up to achieve this relating to:

- plant operating conditions;
- emission limits for certain substances to air, land and water; and
- annual reporting of pollutant releases.

When necessary, these permits will include conditions preventing emissions from crossing national boundaries. The emission limits and the operating conditions are based on BAT, taking into account the characteristics for installation, location, and local environmental conditions.

## BREF Notes

E5028   All permits for IPPC must be based on the concept of BAT, which is defined in Article 2 of the Directive. The licensing authorities need guidance as to which techniques are BAT. Consequently the European Commission is required to organise the exchange of information between industry, environmental organisations, and experts from the EU member states. This work is co-ordinated by the European IPPC Bureau. The Commission then publishes the results of the information exchange every three years in the form of BAT reference documents, known as BREF Notes. These are to be prepared for each industrial sector.

The BREF Notes form the basis of guidance issued in the UK. It has been proposed that the detailed guidance for a specific sector covering the full range of IPPC industries, which the information exchange intends to produce by 2001, will form a key input to the UK guidance. It has therefore been proposed that the UK guidance will not be produced until the appropriate BREF Note for the specific sector has been published. The BREFs are not guidance notes and so they need to be supplemented by domestic guidance.

The IPPC Bureau was set up in 1996 in Seville to control the information management system for the BREF Notes. There are four main bodies involved in the information exchange. These are the EC Directorate responsible for the environment (the DG Environment) which has overall control, the IPPC bureau, the information exchange forum (which is a committee of representatives from member states, industry and non-government organisations), and the technical working groups ('TWGs') which are made up of experts. The TWGs are where the information exchange really takes place. Information, BAT studies, and details on specific sectors and emission levels are provided by the experts, who meet at least twice for each industrial sector. The table below at E5029 shows the 'best guess' for the BREF Note completion for different industrial sectors.

## Best guess for BREF Note completion
E5029

| BREF Note: | Date: |
|---|---|
| Primary/Secondary Steel | Feb 2000 |
| Cement & Lime | Feb 2000 |
| Paper/Pulp | Sept 2000 |
| Tanneries | Sept 2000 |

| | |
|---|---|
| Ferrous Metal Processing | Sept 2000 |
| Non Ferrous Metal Production and Processing | Sept 2000 |
| Glass | Sept 2000 |
| Chloralkali | Sept 2000 |
| Textiles | Feb 2001 |
| Foundaries | Feb 2001 |
| Large Volume Organic | Sept 2001 |
| Livestock: Poultry | Sept 2001 |
| Batch Organics in Multi-purpose Plants | Sept 2001 |
| Refineries | Sept 2001 |
| Food & Milk | Sept 2002 |
| Ceramics | Sept 2002 |
| Large Volume Solid Inorganic | Sept 2002 |
| Large Volume Gas & Liquid Inorganic | Sept 2002 |
| Large Combustion Plant | Sept 2002 |
| Polymers | Sept 2003 |
| Slaughterhouses/Carcasses | Sept 2003 |
| Surface Treatment of Metals | Sept 2003 |
| Hazardous Waste Incineration | Sept 2003 |
| Municipal waste Incineration | Sept 2003 |
| Coating activities etc using organic solvents | Sept 2003 |
| Landfill | Sept 2004 |
| Waste disposal and recovery (other than landfill, incineration) | Sept 2004 |
| Speciality Inorganics | Sept 2004 |
| Organic Fine Chemicals | Sept 2004 |

## *The large combustion plant directive*

**E5030** The Large Combustion Plant Directive 1988 (*88/609/EEC*) commits member states to reductions in $SO_2$ and $NO_x$ from large fossil burning plants of $50MW_{th}$ or more. This Directive did not cover emissions from new solid fuel plants between 50–100MW. In 1994 an amendment *Directive (94/66)* was adopted to cover such a plant, setting an emission limit of 2000 mg m$_2$. Proposals covering emissions for smaller plants are currently under discussion.

In 1998 the EU published proposals for amending its 1988 Directive to account for the technical developments since adoption. It proposes setting emission limit values for $NO_x$ and $SO_2$ based on BAT. For a solid and liquid fuel plant of less than 100MW, an emission limit of 850 mg m $SO_2$ is proposed, and 200 mg m for those over 300MW. An emission limit of 400 mg m $NO_x$ is proposed for solid fuel plants with a capacity of less than 100MW–300MW. Large combustion plants over 50MW are required to comply with the IPPC Directive.

## Solvent emissions

**E5031** In March 1999 *Directive 99/13* was adopted on the limitations of emissions of VOCs, to come into force in 2001. This was due to the use of organic solvents in certain activities and installations. It aims to reduce emissions by 50% by 2010 from various activities including printing, coating, dry cleaning and shoe manufacturing.

## Fuel directives

**E5032** *Directive 93/12/EEC* limits the sulphur content of all gas oils (except aviation kerosene) to 0.2% by weight as from 1 October 1994, with a further reduction in the sulphur content of diesel fuel to 0.05% by weight as from 1 October 1996. The Directive also required member states to ensure that diesel fuel with a sulphur content of 0.05% or less becomes available from 1 October 1995. Thus enabling compliance with *EU Directive 91/542/EEC* on emissions from heavy duty vehicles. As part of its strategy to combat acidification, *Directive 99/32* adopted on 26 April 1999 amends *Directive 93/12/EC* to reduce the sulphur content of gas oil to 0.1% by weight from 1 January 2008.

## Framework directive on air quality assessment and management

**E5033** *Directive 96/62/EC* on ambient air quality assessment, the Air Quality Framework directive, provides a framework for the EU to use to set limit values for pollutants. This directive identifies twelve pollutants for which limit or target values will be set in daughter directives. These are:

- sulphur dioxide ($SO_2$);
- nitrogen dioxide ($NO_2$);
- particulate matter ($PM_{10}$);
- lead;
- carbon monoxide;
- benzene;
- ozone;
- polyaromatic hydrocarbons;
- cadium;
- arsenic;
- nickel; and
- mercury.

This directive is particularly important, as it prepares the ground for a number of significant daughter directives on air quality standards and control measures, as set out below.

## Daughter directives

**E5034**

### Daughter directive for limit values of $SO_2$, oxides of nitrogen, particulate matter, and lead in ambient air

The first Daughter Directive in 1998 established legally binding limit values for $SO_2$, $NO_2$ and $PM_{10}$, to be achieved by 1 January 2005 and 2010. This Directive was adopted in April 1999 and member states are required to implement it by July 2001.

## Daughter directive to reduce the sulphur content of liquid fuels

E5035 This daughter directive aims to reduce further the emissions of $SO_2$ resulting from the combustion of liquid fuels. For heavy fuel oil member states have to ensure that from 1 July 2003 the sulphur content is not over 1% by mass. For gas oil limits are 0.2% by mass by July 2000 and 0.1% by mass from 1 January 2008. Certain classes of fuel used by ships are excluded.

## Daughter directive on limit values for benzene and carbon monoxide in ambient air

E5036 In December of 1998 a proposal for a further daughter directive which sets limit values for benzene and carbon monoxide was published.

## Proposal for national emissions ceilings directive and ozone daughter directive

E5037 In June 1999 proposals were put forward for two further daughter directives. One was to set target values for ozone and the other emission ceilings for various pollutants. The latter is referred to as the *national emissions ceilings directive* ('NECD'). This directive sets ceilings for national emissions of $SO_2$, $NO_2$, $NH_3$ and VOCs to be achieved by 2010. The NECD is the main instrument for attaining air quality targets for ambient ozone, which are also included in the proposal.

## Proposal for a daughter directive on the control of emissions from heavy goods vehicles

E5038 This directive has been amended to introduce three stages of binding emission values.

## Other proposed daughter directives

E5039

- proposal for a daughter directive on the control of emissions from tractors;
- proposal for a daughter directive on consumer information fuel economy for new cars;
- proposal for a daughter directive to set tighter emission limits for large combustion plants;
- proposal for a daughter directive on setting EU-wide air quality standards for benzene and CO;
- proposal for a daughter directive on the reduction targets for $SO_2$ and $NO_2$ emissions; and
- proposal for a daughter directive on setting new reduction target for ozone emissions.

The EU also awaits a proposal for tighter controls on the quality of diesel and petrol fuels.

## Summary of directives

E5040

| Directive | Comment |
|---|---|
| *Air Quality* | |
| Directive 96/62/EEC | Ambient air quality assessment. |

| Daughter directives | Directive for limit values of $SO_2$, oxides of nitrogen, particulate matter, and lead in ambient air. |
|---|---|
| | Directive to reduce the sulphur content of liquid fuels. |
| | Directive on limit values for benzene and carbon monoxide in ambient air. |
| *Fuel directives* | |
| Directive 93/12/EEC | Limits the sulphur content of all gas oils. |
| Directive 91/542/EEC | Emissions from heavy duty vehicles. |
| Directive 99/32/EEC | Amends Directive 93/12/EC. |
| *Solvent emissions* | |
| Directive 99/13/EEC | Limitations of emissions of volatile organic compounds ('VOCs'). |
| *The large combustion plant directive* | |
| Directive 88/609/EEC | Commits member states to reductions in $SO_2$ and $NO_x$ |

# UK Regulations

## Introduction

E5041 Air pollution control for industry is dominated by the introduction of the IPPC Directive into UK law. This replaces the Integrated Pollution Control and Local Air Pollution Control ('LAPC') Regulations which have existed since 1991, which in turn replaced the old system of Best Practical Means Notes. Thus, the last decade has seen quite a transformation, overturning a system of controls and concepts which can be traced back to 1842.

## Definition of useful acronyms and terms
E5042

| IPC | Integrated Pollution Control |
|---|---|
| IPPC | Integrated Pollution Prevention Control |
| | IPPC is a regulatory system that employs an integrated approach to control the environmental impacts of certain industrial activities. It involves determining the appropriate controls for industry to protect the environment through a single permitting process. |
| LAPC | Local Air Pollution Control |
| BAT/BATNEEC | Best Available Technique /Best Available Technique Not Entailing Excessive Cost |

| | |
|---|---|
| | The Best Available Techniques are those that prevent or minimise pollution, and can be implemented effectively and are economically and technically viable while meeting the overall aims of the Directive. BATNEEC is a concept promoted by IPC and does not appear explicitly in IPPC legislation. |
| BPEO | Best Practical Environmental Option |
| Part A1 installations | The largest and most complex processes, regulated for releases to air, water and land by the Environment Agency in England and Wales. |
| Part A2 Installations | Processes with releases to air, water and land regulated by Local Authorities in England and Wales (but by SEPA in Scotland). |
| Part B installations | The smaller processes, with releases to air only regulated by Local Authorities in England and Wales (but by SEPA in Scotland). |

## Useful websites
E5043

- *Environment Agency:* http://www.environment-agency.gov.uk
- *Department of the Environment Transport and the Regions (DETR):* http://www.detr.gov.uk
- *European Commission:* http://europa.eu.int/comm/index.htm
- *The UK National Air Quality Information Archive:* http://www.aeat.co.uk/netcen/airqual/
- *European Integrated Pollution Prevention and Control Bureau:* http://eippcb.jrc.es/
  N.B.: BREF notes can be downloaded from this site.
- *Scottish Executive:* http://www.scotland.gov.uk/consultations/environment
- *Scottish Executive:* http://www.scotland.gov.uk/environment/airquality

## The regulators
### The environment agencies

E5044   The agencies were established by the *Environment Act 1995*, and formally took over their pollution control and other functions on 1 April 1996.

In England and Wales the Agency's pollution control responsibilities are as follows:

- authorisations, licences and consents for emissions to air, water and land. Monitoring compliance and enforcement, including prosecution permitting of installations and enforcement of Regulations under IPPC; and

- advice and guidance to industry and others on best environmental practice.

## Scottish Environment Protection Agency

**E5045**    With regard to the Scottish Environment Protection Agency ('SEPA'), Scotland has been divided into the west, east and north for operational purposes. The roles for SEPA are similar for the Environment Agency for England and Wales. In addition it regulates Part B processes for air pollution under the *Environmental Protection Act 1990, Part I*. In other words in England and Wales this function has not been transferred to the Agency and remains with local authorities. SEPA also controls the installations regulated for IPPC.

## Northern Ireland

**E5046**    The Department of Environment ('DoE') (Northern Ireland) has overall responsibility for air pollution control in Northern Ireland. It is introducing a system similar to the IPC and air pollution control regimes operating in the rest of the UK prior to July 2000.

Part A processes are regulated for IPC by the Industrial Pollution and Radiochemical Inspectorate within the DoE (Northern Ireland). Part B processes are also regulated by the same Inspectorate.

## *Local authorities*

**E5047**    Local Authorities have a wide range of responsibilities covering the whole spectrum of pollution control and environmental protection. The local authority associations and the Environment Agency have signed a Memorandum of Understanding (1997) covering those areas of environmental protection for which they have responsibility.

## ***The end of IPC and LAPC***

**E5048**    IPC should have been superseded by IPPC by 31 October 1999, as required by the Directive. In fact, it did not formally become part of legislation until August 2000. The consultation period was extended to April 2000 (with no less than five consultation papers being issued in all.) Implementation was delayed primarily because the regulatory structure was not in place to allow the environment agencies and local authorities to process applications for a permit. Whilst the transformation from IPC and LAPC to IPC is a straightforward one in principle, it has proved more difficult in practice to define the Regulations and to draft guidance to both industry and regulators on how they should be used.

In the UK the new system is known as Pollution Prevention Control ('PPC') as it includes provisions for both integrated (IPPC) and non-integrated (PPC) permits. In England, Wales and Scotland PPC has replaced the IPC and LAPC systems.

There are three main differences between IPC and IPPC:

(*a*)    IPC was concerned with releases of prescribed substances to air, water and land, whereas IPPC covers a wider range of environmental impacts.

(*b*)    IPPC covers many or more installations than IPC. The government estimates that 7,000 installations will be covered by IPPC, of which:

- (i)    2,000 were already covered by IPC;
- (ii)    2,500 were covered by LAPC;
- (iii)    1,000 are landfill sites;
- (iv)    1,000 are intensive pig and poultry farms;
- (v)    500 are food and drink factories.

*Emissions into the Atmosphere*

(c)  IPC has approximately 400 processes which are not covered by the IPPC Directive but which will be covered by the UK implementation of the IPPC regime.

The new format of IPPC, as proposed, has been superimposed on the existing system. The approach is designed to give the power to make regulation which would set out one consistent regime.

## The Pollution Prevention and Control Act 1999

E5049    *The Pollution Prevention and Control Act 1999* received Royal Assent on 27 July 1999. It is an enabling Act, paving the way for Regulations to be made which implement the IPPC Directive. As such, it is not inherently revealing – the Regulations contain the detail which affects industry.

From a legal standpoint, it provides for the repeal of *Environmental Protection Act 1990, Part I* which covers IPC and LAPC. The Act also makes provision for waste licences issued under the *Control of Pollution Act 1974* and which had expired to be treated as though they were still in force. Another area which the Act covers is regulation of off-shore oil installations.

## PPC (England and Wales) Regulations and the Draft PPC (Scotland) Regulations 2000

### Application for permit

E5050    A new installation cannot be brought into operation until an IPPC permit has been granted. Before the candidate makes an application they should check that they are actually required to make an application, and understand the requirements of the application. The application and associated Technical Guidance documents should also be consulted. It may be necessary to discuss the application with the Environment Agency before the application is prepared to help clarify important issues, and how to present the information that is required.

### Information on application

E5051    To make an application the candidate should complete the Environment Agency's or SEPA's, standard application form which can be obtained from the local Environment Agency or SEPA office, or from the Environment Agency website at http://www.environment-agency.gov.uk. (although the Scottish Draft Regulations state that an application for a permit to SEPA shall be acceptable in writing or in electronic format, they do not cite a web address to obtain the form.) The form should be submitted along with 10 copies (this requirement does not stand for SEPA applications) and all other additional information to the Environment Agency along with the appropriate fee. The application should include the following information:

(i)    full details of the operator, and the installation (address, national grid reference, map or plan of the installation), and the local authority in which the installation is situated;

(ii)   description of the installation and the activities that are to be carried out;

(iii)  the nature, quantities and sources of the likely emissions to the environment and their significant effects. (Part B installations release emissions to air only, and Part A installations release emissions to air, water and land);

(iv)  details of the technology that is to be used to prevent or reduce emissions; and

(v)   any proposals for monitoring emissions from the installation.

Applications for Part A installations must also include the following:

(i)    Measures that are to be taken for the prevention and recovery of waste that is generated by the installation when it is in operation.

(ii) Details on the raw materials, any substances, and energy which is to be used or generated by the operation of the installation.

(iii) A report on the current condition of the site, particularly identifying any substances that are around, in or under the site that may pose as a pollution risk in the future. The Agency is to produce guidance on the characterisation and the assessment of sites.

(iv) Relevant information must be included that has been obtained as a result of carrying out an Environmental Assessment in accordance with *EU Directive 85/337*.

In addressing the BATs the applicant should ensure that in the application the following issues are included:

(i) confirm compliance with any compulsory requirements e.g. other emission based Regulations/directives;

(ii) where the indicative requirements are clearly stated, confirm compliance or justify proposals by describing how they will meet the BAT criteria. In particular confirm that all the measures described are utilised and if any are not used, give justification why;

(iii) where there are no clear benchmarks or standards, describe how the proposals will meet the BAT criteria. In particular confirm that all the measures described are utilised and if any are not used, give justification why.

## Determination of an application

**E5052** Once the application has been submitted, the regulator then has four months from the date that it was received (or longer if agreed) in which to determine the application (see *figure 2 at* E5051 ABOVE). If the regulator requests additional information then the four months commences from when this is received. Commencing 14 days after submission, the applicant should publish details of the application made in one or more of the local newspapers in the area that the installation is located. Part A installations must also be published in the London Gazette. This advertisement must state where the site can be inspected and that any views on the application are to be sent to the Regulator within 28 days of the advertisement. Copies of all applications are put onto the public register. The applicant may request that certain information is excluded, the procedure for this is similar to the IPPC Application.

> Within 14 days of receiving the application, the regulator should distribute copies to all the statutory consultees:
> 
> - The Health and Safety Executive;
> - The Health Authority in which the installation is situated;
> - The Nature Conservancy Council for England/Scottish Natural Heritage;
> - Any other persons at the Secretary of State; and
> 
> The local authority in which the application is situated;
> 
> In Scotland, the harbour authority.

Part A installation applications should also be distributed to:

- Minister of Agriculture, Fisheries and Food;
- Sewerage undertaker, Harbour Authority as appropriate;
- The relevant planning authority (for applications involving waste management activity); and
- The Environment Agency (for A2 applications) or the local authority (for A1 applications).

**Figure 2: The Application Process**

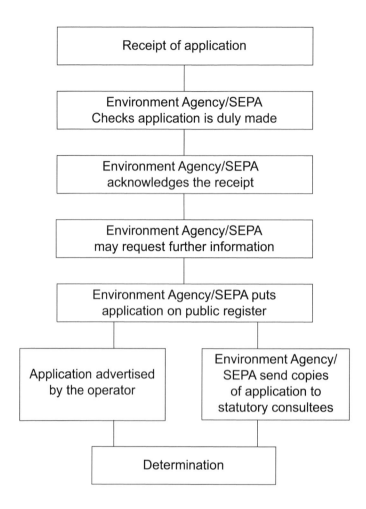

When the application includes off-site conditions then the Regulator should consult the owner, lessee, or occupier of the land and 28 days is allowed for representations.

The Secretary of State (or the Scottish Ministers) has (or have) the right to call upon any applications for determination. In these cases the regulator should inform the applicant of any representations that were received. Proceeding this the applicant and the regulator can request a hearing which should be made in writing within 21 days of being informed. These applications are then considered by the Secretary of State (or Scottish Ministers), who then directs the regulator as to whether a permit should be granted, and if so with what conditions.

When the Secretary of State (or the Scottish Ministers) is (or are) concerned that an installation will have significant negative effects on the environment of another Member State, then a copy of the permit application is sent to the Member State concerned to enable bi-lateral consultations to take place. Another Member State can also follow this procedure where they are likely to be significantly effected. This should be done as soon as the application is advertised. The four month period of determination does not begin until the consultations have been completed.

## Permits

**E5053**  One permit can cover more than one installation, as long as these installations are on the same site, and are operated by the same operator. Therefore a single application can be used if the operator meets these requirements. If this is done then a single submission should be made comprising of separate completed application forms for each installation, with appropriate cross referencing on common elements (e.g. site management and condition).

When determining an application, it must be taken into account whether the applicant will operate the installation in accordance with the permit and any conditions that are attached. These all ensure that the installation will be operated so that there is no significant pollution, and that all preventative measures are taken. It is implied that operators should use the BATs to prevent or reduce the emissions. For Part A installations, the conditions should also ensure:

(i) energy is used efficiently;

(ii) waste production is avoided;

(iii) measures are taken to prevent accidents;

(iv) transboundary pollution is minimised;

(v) protection of soil and groundwater, and appropriate management of waste; and

(vi) suitable emissions monitoring equipment is in place.

The Scottish Draft Regulations also refer to the following condition:

(vii) setting out the steps to be taken prior to the operation of the installation and after the final cessation of operations.

If a permit is granted then suitable conditions like the size of the emission limit values will be included for the pollutants that are likely to be emitted in significant quantities. These are based on the BATs and, in the case of Part A installations, will take into account pollution to other environmental media. These should ensure that there is no breach of the environmental quality standards that are set by the EU. The Environment Agency may also require that the installation is operated in specific ways, for example by referring to the proposals that were made in the permit application on any subsequent occasions.

When setting a condition which regulates a discharge to water from an A2 installation, the local authority must have regard for any notice from the Environment Agency which specifies an emission limit value, and they may not include a less stringent value. They may however include a stricter one.

For the aspects of the installation that are not regulated by specific conditions, the PPC Regulations impose an implied permit condition requiring the operator to use BAT to prevent and reduce emissions (often referred to as 'residual BAT').

The operator must comply with the conditions of the permit, submitting monitoring data of appropriate quality to demonstrate this. The Environment Agency may also carry out inspections taking various actions to enforce compliance. These include serving enforcement notices, suspending operation, bringing prosecutions and, in rare cases, revoking the permit.

If the operator receives a permit then they are responsible for the obligations arising under it (including the payment of fees) until the Environment Agency or SEPA accepts the transfer of the permit to another operator, accepts the surrender of the permit, or revokes the permit.

Permits are to be periodically reviewed, especially when the BATs make further reductions feasible (with no extra cost) in a situation where, for instance:

(*a*) there has been significant pollution from an installation and the emission limits need to be reviewed; or

(*b*) a change in technique is needed to operate safely.

This may lead to a variation of the permit conditions. Guidance notes published by the Environment Agency set out the normal review periods that are appropriate for installations in each sector. Powers are granted to the Scottish Ministers to issue guidance to SEPA in respect of carrying out any of their functions under the Regulations.

## Fees

**E5054** The Environment Agency and SEPA recover the costs of granting and ensuring compliance with environmental licences by charging the applicant. Charges are in accordance with the number of components that the process contains, and these charges are reviewed each year.

At the time of publication, no scale of fees had been published by the Environment Agency. They are expected to be slightly greater than those in effect for IPC, but not substantially so.

## Compliance

**E5055** Within IPPC an effective system of management is the key to ensuring that all the appropriate pollution prevention and control techniques are delivered reliably and on an integrated basis.

The Environment Agency strongly supports the operation of environmental management systems which will encourage the operator to look for continuous improvement of environmental performance.

## Low impact installations

**E5056** Unlike IPC, IPPC does not exempt installations from requiring a permit even though their emissions as a result of a prescribed activity are 'trivial'. Therefore it has been proposed that low impact installations have a simplified permit procedure.

For this procedure to apply, the operator would have to demonstrate, and the regulator would have to be satisfied, that the installation has only a low impact on the environment. The normal application and determination procedure still apply with permits for such installations including simplified conditions, such as:

(*a*)  operating the installation in accordance with the application, and monitoring its operation to ensure compliance;

(*b*)  reporting any releases not in accordance with the application;

(*c*)  advising if the changes are proposed to the operation of or if it is closed down or ceases to operate.

The definition of a low impact installation has been set out by the Agency, according to a set of principles relating to the quantity of emissions to air, water and land.

## PPC implementation in Scotland

**E5057** In Scotland, *all* installations are to be regulated by SEPA. Schedule 1 attached to the Scottish Regulations lists Part A activities to be regulated for IPPC and Part B activities for air pollution control only.

Applications are advertised in the Edinburgh Gazette, as distinct from the London Gazette.

The draft states that the operation of the Regulations will be monitored on an ongoing basis. There will also be a formal review of the operations of the new arrangements after five years.

## Northern Ireland

**E5058** The introduction of IPPC to Northern Ireland has not yet been considered formally. It is only recently that IPC has been adopted, through the *Industrial Pollution Control (Northern Ireland) Order (SI 2777)* on 26 November 1997. There are three tiers of control in the Northern Ireland version of IPC:

1. 'integrated central control': Part A processes with high pollution potential are regulated for IPC by the Industrial Pollution and Radiochemical Inspectorate ('IPRI') within the DOE (Northern Ireland);

2. 'restricted central control': Part B processes with the potential to cause serious air pollution are be regulated for air pollution control, through the IPRI:

3. 'local control': Part C processes with significant but lesser potential for air pollution are regulated by district councils.

## *IPC and its residual role*

**E5059** Although it is the clear intention that IPPC will replace IPC, there will be an inevitable period of transition (2000 – 2007), in which industrial processes will continue to be regulated by IPC (and LAPC in the case of Part B processes). The introduction of IPPC for existing processes will be accomplished through a timetable, conforming to the issue of BREF Notes. The Environment Agency and SEPA will seek to apply IPPC to industrial sectors one at a time, until all have an IPPC permit by 2007. So, for example, the cement and lime industry is the first to have to apply for IPPC permits, whereas organic fine chemicals are last on the timetable, as currently envisaged.

Until a permit for IPPC can be applied for and granted, it follows that industrial processes can only be regulated by IPC. The IPC Regulations and philosophy provides a good platform for the transition to IPPC, but there are differences. Not only does IPPC encompass more types of industrial activity, a greater number of processes overall and more environmental aspects than IPC, it also uses different terminology and has some important differences in approach. Some of these are listed below:

- an IPC licence is referred to as an 'authorisation' to operate, as distinct from a permit. The process of obtaining and maintaining an authorisation is similar to that described above for an IPPC permit.

- IPC applies to a site, typically. Its philosophy is one of preventing or minimising releases through an imaginary envelope at the site boundary to the external environment. IPPC relates to an installation, which could be an integral part of an overall industrial site, and takes a more holistic view of environmental issues.

- IPC made use of the BATNEEC concept. This has disappeared from IPPC legislation, although the BAT concept does incorporate the economic component.

Obtaining a new authorisation to operate under IPC should be an unlikely requirement, as new processes will be brought into the IPPC regime immediately – although how this will work in practice, given the absence of some of the key BREF Notes, is unclear.

Current guidance from the DETR is that existing processes can still apply for a 'variation' to their current IPC authorisation, provided that the changes are positive with respect to environmental impact. Should the change be a negative one, however, then an IPPC permit will have to be applied for.

The fees which applied to IPC for 1999/2000 are set out in the Environment Act Charging Scheme 1999.

## Application for fees for IPC
E5060

| Description | Fee | Notes |
|---|---|---|
| Application fee | £5,074 (per component) | |
| | (staged application £1,125 per day) | |
| Subsistence charge | £2,307 (per component) | This is payable annually, covering the cost of inspection, monitoring and enforcement. |
| Substantial variation fee | £1,699 | |

The cost of any routine monitoring is charged monthly in arrears to the operator, and reactive monitoring is covered by the subsistence charge. If the operator withdraws an application within 56 days of receipt from the Environment Agency then the fee will normally be refunded, however the Agency retains the right not to do so.

In August 1999 the Environment Agency published proposals for an interim charging scheme. This is to run from 31 October 1999 to 31 March 2001. During this time a new scheme will be developed for introduction from 1 April 2001. Installations with a low impact on the environment benefit from a reduction in charges. The final consultation of the Scottish regulations states that due to respondents making representations against the proposals issued separately by SEPA for an interim charging scheme for Part A installations, the Scottish Executive has asked SEPA to review the proposals. The interim charging scheme will need to have been approved by both the Treasury and the Scottish Ministers by the time the Regulations are in force.

## Public registers

E5061    Registers of IPC have been established containing information to be maintained by the enforcing authorities. The following documentation is to be placed in the register:

- application for authorisation, and information relating to the application requested by the enforcing authority;
- a copy of the advertisement relating to the application and representations made on response to the advertisement;
- authorisations and their conditions;
- written notice of transfer authorisation to another person;
- any notices issued, or notification of withdrawal;
- any documentation relating to an appeal;
- details of convictions in relation to carrying on of a process with or without authorisation;
- monitoring data collected by the enforcing authority or data that has been submitted by an operator as a condition of the authorisation, where this information has been omitted because of commercial confidentiality, the enforcing authority should put a statement on the register which confirms the compliance or otherwise with the relevant condition of the authorisation;
- any information given to the enforcing authority after 1 April 1996 in compliance with a condition of authorisation, variation, enforcement or prohibition notice;

- reports by the enforcing authority which relate to the environmental consequences of a prescribed process in the locality of premises where the prescribed process is carried on under authorisation; and

- directions from the Secretary of State.

Exclusion of data on the grounds of commercial confidentiality lasts for four years, and after that it is added to the register unless a further four year exclusion is applied for and agreed. The Local Authorities in England and Wales have copies of entries relating to IPC processes in their areas in the registers, and they also maintain registers for air pollution processes – Part B processes – under their control. In Scotland SEPA will have information in respect of Part B processes.

Applications for an authorisation or variation which have been advertised in a paper should be put on the register within 14 days of receipt by the enforcing authority.

Information is removed from the register in the following circumstances:

- withdrawal of application for authorisation prior to determination. All documents are to be removed not less than two months, but within three months after the date of the withdrawal;

- if a process ceases to be a prescribed process, all documents are to be removed not less than two months but within three months of the date on which the process ceases to be prescribed;

- monitoring information relating to a process – four years after entering in the register; and

- information relating to a process that has been superseded – four years after the later information has been entered in the register.

Public registers have proved to be a powerful tool in enforcing the Regulations, as the operators are exposed to scrutiny.

## The Clean Air Act 1993

**E5062** The IPPC legislation, and its predecessors IPC and LAPC, probably account for some 20,000 individual industrial processes. For those industrial processes which are not considered to be sufficiently polluting to warrant inclusion, it may be that other Regulations still apply to their activities. This is particularly true of processes which require the burning of fuel or waste. Historically, the *Clean Air Acts* of *1956* and *1968* were concerned very much with the control and reduction of smoke, largely through the burning of coal. These Acts, and other clean air legislation such as the *Control of Pollution Act 1974* and the *Control of Pollution (Amendment) Act 1989* have been consolidated in the *Clean Air Act 1993*.

There are three parts to this Act which are relevant:

1. *Part I: dark smoke*

   Prohibition of dark smoke from chimneys and from industrial or trade premises.

2. *Part II: smoke, grit, dust and fumes*

   The Regulations control the emissions of particles emitted from furnaces and boilers. Direction is also given on the height of chimneys.

3. *Part V: Information about air pollution*

   Regulations grant Local Authorities the power to provide information on emissions from specified premises, or to enter premises and make measurements.

## Statutory nuisance

**E5063** Air pollution control legislation has been rooted historically in the concept of nuisance to humans, rather than harm to the environment. The causes of such nuisance have traditionally been the burning of fuel or waste materials, giving rise to 'fumes' and 'smoke'. Odour is another form of nuisance.

The main legislation relating to nuisance is now in the *Environmental Protection Act 1990, Part III* as amended by the *Noise and Statutory Nuisance Act 1993*. It enables Local Authorities and individuals to secure the abatement of a statutory nuisance.

A nuisance is defined by the legislation, although only parts of the definition relate to air pollution. Others relate to noise and the state of premises which might be 'prejudicial to health'. In essence a nuisance can arise because of smoke or gases emitted from either private or business premises which are prejudicial to health or are a nuisance. A distinction is made between private dwellings and industrial premises in terms of the emissions. Private dwellings are exempt from causing a nuisance through emissions of 'dust, steam, smell or other effluvia'.

In effect, industrial premises currently regulated by IPC and LAPC are not likely to be prosecuted through the statutory nuisance legislation. Such action can only be taken with the consent of the Secretary of State. If a nuisance is reported, it is more likely that action will be taken by the appropriate regulatory authority.

The legislation also allows for exemption in the case of some specific cases, e.g. steam railway locomotives, recreational steam vessels and steam driven road vehicles.

## Air quality standards and guidelines

### Background

**E5064** For much of the twentieth century, UK air pollution control was exercised purely by reference to emissions, especially for industrial air pollution. Only in the 1990s did any concept of effects-based policies begin to emerge. This came about partly because European Community directives on air quality for pollutants such as $SO_2$ and $NO_2$ had been adopted, but also because the then Department of the Environment promoted such policies. In the acid rain debate, the effects were very much the start point for framing policies to reduce damage and the concept of a 'critical load' was used to define the extent of environmental damage. The trans-boundary issues were the dominant issue in air pollution policy debates in the 1980s, but subsequently interest in urban air quality and human health was re-awakened. No assessment of impacts in this topic can be made without reference to threshold concentrations at which health effects occur.

At present, it can be said with some justification that air quality standards are one of the primary regulatory influences on industry, transport and other emitters. For the UK, the main originators of air quality guidelines are the World Health Organisation and the DETR's Expert Panel on Air Quality Standards. These bodies are responsible for recommending guidelines on air quality that represent concentrations at which the most sensitive members of the population will be protected from the damaging effects of air pollutants. For most pollutants, it is possible to decide upon a threshold concentration below which no effects are discernible. For some pollutants, e.g. carcinogens, it is theoretically possible to see an effect right down to near zero concentrations.

A most important point to recognise is that air quality guidelines have no legal standing. They do provide the basis, however, for governments and the European Commission to develop standards. Usually, such standards reflect a degree of 'economic dilution' and the fact that it is rarely possible to achieve a target concentration for 100% of the time. It is more normal to permit a certain number of occasions on which a threshold value may be exceeded. For example, the UK Government in its 1997 version of the air quality strategy proposed that the 24 hour average value of inhalable particulate matter could be exceeded on several days in a year specifically to allow for national festivities, i.e. 'Bonfire Night'.

Different pollutants require different periods of time over which they should apply, depending on the health effect. Sulphur dioxide, for example, provokes a reaction in the human respiratory system over 10–15 minutes. Therefore, it is the short term peak fluctuations of $SO_2$ that need to be regulated. Lead, on the other hand, has an effect over a long period of time, as it accumulates in the human body. In this case, it is the annual average concentration of airborne lead which is the appropriate measure.

The multiplicity of ways in which air quality standards are expressed makes them rather confusing for those unfamiliar with the subject. In essence, however, they fall into two broad categories; those with averaging periods over a year and those with averaging periods over a short period of time, a day or less. The latter often have some expression of the number of occasions for which the threshold concentration may be exceeded. This is frequently described in terms of a 'percentile concentration'. For example, the most recent European Union Directive for $NO_2$ ambient concentrations specifies that the hourly concentration of 200 µg m should not be exceeded more than 18 times per year. A total of 18 hours represents 0.2% of the year. Hence, the standard can be expressed in an alternative manner as 200 µg m, as the 99.8 percentile concentration.

## World Health Organisation

E5065    The World Health Organisation ('WHO') formally produced a set of air quality guidelines for Europe in 1987. This was a much referenced document, despite having no legal status. The WHO began to revise these guidelines in the early 1990's and has produced a revision in draft form in 1995. Unfortunately, a formal publication is still awaited.

The most recent WHO recommended air quality guidelines are used as a basis for setting standards in EU directives, and were also taken into account by the UK Expert Panel on Air Quality Standard when making recommendations for UK air quality standards. The table below at E5066 shows WHO air quality guidelines. The table below at E5068 shows the standards and guidelines that the UK has adopted to meet with the Directive.

## WHO air quality guidelines
E5066

| *Substances* | *Time-weighted average* | *Averaging time* |
|---|---|---|
| nitrogen dioxide | 200 µg m$^{-3}$ | 1 hour |
|  | 40 µg m$^{-3}$ | annual |
| ozone | 120 µg m$^{-3}$ | 8 hours |
| sulphur dioxide | 500 µg m$^{-3}$ | 10 m10).ins |
|  | 125 µg m$^{-3}$ | 24 hour |
|  | 50 µg m$^{-3}$ | annual |
| carbon monoxide | 100 mg m$^{-3}$ | 15 mins |
|  | 60 mg m$^{-3}$ | 30 mins |
|  | 30 mg m$^{-3}$ | 1 hour |
|  | 10 mg m$^{-3}$ | 8 hours |
| lead | 0.5 µg m$^{-3}$ | annual |
| cadmium | 5x10$^{-3}$ µg m$^{-3}$ | annual |

| mercury | 1.0 µg m$^{-3}$ | annual |

In addition to the guidelines expressed as a threshold concentration, guidelines are also given as risk factors for exposure to carcinogens such as benzene and benzo(a)pyrene. The WHO also state that they are unable to set a threshold concentration as a guideline for exposure to inhalable particulate matter ($PM_{10}$).

## Air quality standards in the UK

**E5067** The most recent version of the Air Quality Strategy (see E5069 below), and the Regulations which have been passed subsequently, incorporate the EU directives on ambient air quality standards for the relevant pollutants.

These are summarised below at E5068.

## Summary of the objectives in the 2000 Air Quality Strategy

**E5068**

| Pollutant | Concentration | Measured as | Date to be achieved by |
|---|---|---|---|
| Benzene | 16.25 µg m$^{-3}$ | running annual mean | Dec 31, 2003 |
| 1,3-Butadiene | 2.25 µg m$^{-3}$ | running annual mean | Dec 31, 2003 |
| Carbon monoxide | 11.6 mg m$^{-3}$ | running 8 hour mean | Dec 31, 2003 |
| Lead | 0.5 µg m$^{-3}$ | annual mean | Dec 31, 2004 |
|  | 0.25 µg m$^{-3}$ | annual mean | Dec 31, 2008 |
| Nitrogen Oxide | 200 µg m$^{-3}$ | 1 hour mean not to be exceeded more than 18 times a year. | Dec 31, 2005 |
|  | 40 µg m$^{-3}$ | annual mean | Dec 31, 2005 |
| Particles ($PM_{10}$) | 50 µg m$^{-3}$ | 24 hour mean not to be exceeded more than 35 times a year. | Dec 31, 2004 |
|  | 40 µg m$^{-3}$ | annual mean | Dec 31, 2004 |
| Sulphur dioxide | 350 µg m$^{-3}$ | 1 hour mean not to be exceeded more than 24 times a year. | Dec 31, 2004 |
|  | 125 µg m$^{-3}$ | 24 hour mean not to be exceeded more than 3 times a year. | Dec 31, 2004 |
|  | 266 µg m$^{-3}$ | 15 minute mean not to be exceeded more than 35 times a year. | Dec 31, 2005 |

Notice that the UK standards have been expressed as 'objectives', with a target date by which these objectives are attained.

## The Air Quality Strategy ('AQS')

**E5069** The AQS for England, Wales, Scotland and Northern Ireland describes the policies by which air quality will be improved. As noted above, it gives objectives for eight priority pollutants, in terms of achieving compliance with air quality standards. These are now based on the implementation of the EU directives. The strategy describes the current and future concentrations of the eight pollutants, providing a framework to help everyone to identify what they can do to help reduce emissions.

The AQS fulfils a requirement under the *Environment Act 1995*, setting out policies for managing ambient air quality. A national air quality strategy was first published in March 1997, with the first review taking place in 1998. The government consulted on its revised proposals in January 1999. It is an intentional feature of the strategy that it should be reviewed and revised at intervals.

The AQS identifies the need to work at international, national and local level to achieve the objectives. This strategy is essential to anybody interested in achieving the objectives of the directives, not just those who are responsible for them. The strategy has five guiding principles:

- the best practicable protection to human health and the environment;
- the expert panel on AQS recommendations should be the basis for objectives except where an objective derives from an air quality daughter directive limit value based on WHO guidelines;
- it should allow the UK to comply with EU daughter directives, but allow for stricter national objectives for some pollutants;
- objectives should reflect the practicability of the measures needed to reduce pollutants, their costs and benefits and other social and economic factors; and
- objectives should take into account EU legislation and scientific advances.

## Local air quality management

**E5070** Local Authorities have a major role to play in delivering cleaner air. They are responsible for land-use planning and traffic management, and for controlling industrial pollution sources through the local air pollution control ('LAPC') regime, using action plans. Local Authorities have a long history of local environmental control. They have responsibilities in controlling:

- industrial pollution;
- local pollution hotspots under Local Air Quality Management ('LAQM'); and
- pollution from domestic sources.

In England and Wales the LAPC control certain industrial processes, in Scotland SEPA, and in Northern Ireland the Industrial Pollution and Radiochemical Inspectorate control the equivalent. Air quality has been given increased consideration more recently when they fulfil their strategic planning and transport roles. This has been a natural progression for Local Authorities.

Local Authorities are not obliged to prepare a local air quality strategy. The Environment Agency and other authorities are encouraged to work together by the government. All other authorities are encouraged to develop their own strategies, and are provided with advice. Under the new IPPC system most of the processes regulated under LAPC will be subject to control of emissions only, although some will be subject to integrated control. So Local Authorities will be responsible for some integrated permitting. The LAQM action plans are the major tool for tackling pollution hotspots often caused by road and transport.

*The Environment Act 1995, Part IV* requires Local Authorities in Scotland to conduct a review of the quality of air within their area. The *Air Quality (Scotland) Regulations 2000 (SI 2000 No 97)* prescribes the relevant period for the purpose of that review (*Regulation 3*) and sets out the air quality objectives to be achieved at the end of that period (*Regulation 4 and the Schedule*). The objectives are the same as those set out in the Air Quality Strategy for England, Scotland, Wales and Northern Ireland (SE 2000/3, January 2000) published by the Scottish Executive. Guidance entitled '*Review and Assessment: Pollutant Specific Guidance LAQM TGU*' has been issued in respect of the appropriate methodology to be used by Local Authorities in undertaking their air quality review and assessment. The Strategy and Guidance are available from the Scottish Executive, Air, Climate and Engineering Unit (Air Quality Team), Victoria Quay, Edinburgh, EH6 6QQ.

# Employers' Liability Insurance

## Introduction

E13001    Most employers carrying on business in Great Britain are under a statutory duty to take out insurance against claims for injuries/diseases brought against them by employees. When such a compulsory insurance policy is taken out the insurance company issues the employer with a certificate of insurance, and the employer must keep a copy of this displayed in a prominent position at his workplace, so that employees can see it. It is a criminal offence to fail to take out such insurance and/or to fail to display a certificate (see E13013–E13015 below); however, such a failure does not give rise to any civil liability on the part of a company director in England. In *Richardson v Pitt-Stanley [1995] 1 All ER 460* the plaintiff suffered a serious injury to his hand in an accident at work, and obtained judgment against his employer, a limited liability company, for breach of the *Factories Act 1961, s 14(1)* (failure to fence dangerous parts of machinery). Before damages were assessed, the company went into liquidation and there were no assets remaining to satisfy the plaintiff's judgment. The company had also failed to insure against liability for injury sustained by employees in the course of their employment, as required by the *Employers' Liability (Compulsory Insurance) Act 1969, s 1*. Section 5 of the *Employers' Liability (Compulsory Insurance) Act 1969* makes failure to insure a criminal offence. The plaintiff then sued the directors and secretary of the company who, he alleged, had committed an offence under *s 5*, claiming as damages a sum equal to the sum which he would have recovered against the company, had it been properly insured. His action failed. It was held that the *Employers' Liability (Compulsory Insurance) Act 1969* did not create a civil as well as criminal liability. In Scotland however the Sheriff Principal in *Quinn v McGinty 1999 SLT (Sh Ct) 27*, on similar facts did not follow the Court of Appeal in *Richardson v Pitt-Stanley*.

The above duties are contained in the *Employers' Liability (Compulsory Insurance) Act 1969* (referred to hereafter as the '*1969 Act*') and in the *Employers' Liability (Compulsory Insurance) Regulations 1998 (SI 1998 No 2573)* (referred to hereafter as the '*1998 Regulations*'). In addition, the requirements of the *1969 Act* extend to offshore installations but do not extend to injuries suffered by employees when carried on or in a vehicle, or entering or getting onto or alighting from a vehicle, where such injury is caused by, or arises out of use, by the employer, of a vehicle on the road. [*SI 1998 No 2573, Reg 9 and Sch 2, para 14*]. Such employees would normally be covered under the *Road Traffic Act 1988, s 145* as amended by the *Motor Vehicles (Compulsory Insurance) Regulations 1992 (SI 1992 No 3036)*. As from 1 July 1994, liability for injury to an employee whilst in a motor vehicle has been that of the employer's motor insurers.

Employees suffering from industrial diseases, including asbestos, will now find it easier to track down their employer's insurers thanks to a new Code of Practice launched in November 1999. The Code was drawn up by the Department of the Environment, Transport and the Regions, The Association of British Insurers, and the Non-Marine Association at Lloyds, and sets out the procedures and standards of service required of insurers. The Code does not have statutory authority and is a voluntary code that commits member insurers to make a thorough search when requested of all records which exist. The Code obliges insurers to keep records of all policies issued for 60 years. There is no charge levied for conducting a search. For the period 1 November 1999 to 31 October 2000, 1,062 enquiries were circulated to its member insurers with a 24.6 per cent success rate. Copies of the Code and enquiry forms can be obtained from the

Association of British Insurers, Employers Liability Enquiry Unit, 51 Gresham Street, London EC2V 7HQ, tel: 020 7216 7456; fax: 020 7367 8612.

## Purpose of compulsory employers' liability insurance

**E13002** The purpose of compulsory employers' liability insurance is to ensure that employers are covered for any legal liability to pay damages to employees who suffer bodily injury and/or disease during the course of employment and as a result of employment. It is the liability of the employer towards his employees which has to be covered; there is no question of compulsory insurance extending to employees, since employers are under no statutory or common law duty to insure employees against risk of injury, or even to advise on the desirability of insurance; it is their potential legal liability to employees which must be insured against (see E13008 below). Such liability is normally based on negligence, though not necessarily personal negligence on the part of the employer. Moreover, case law suggests that employers' liability is becoming stricter. The rule that employers must 'take their victims as they find them' underlines the need for long-tail cover because the employer may find himself liable for injuries/diseases which 'trigger off' or exacerbate existing conditions. (See further E11026 EMPLOYERS' DUTIES TO THEIR EMPLOYEES.)

An employers' liability policy is a legal liability policy. Hence, if there is no legal liability on the part of an employer, no insurance moneys will be paid out. Moreover, if the employee's action against the employer cannot succeed, the action for damages cannot be brought against an employer's insurer (*Bradley v Eagle Star Insurance Co Ltd [1989] 1 All ER 961* where the employer company had been wound up and dissolved before the employer's liability to the injured employee had been established) (see below for the transfer of an employer's indemnity policy to an employee). The effect of this decision has been reversed by the *Companies Act 1989, s 141*, amending the *Companies Act 1985* (*CA 1985*), *s 651* which allows the revival of a dissolved company within two years of its dissolution for the purpose of legal claims and, in personal injuries cases, the revival can take place at any time subject to the existing limitation of action rules contained in the *Limitation Act 1980*. For example, in the case of *Re Workvale Ltd (No 2) [1992] 2 All ER 627*, the court exercised its discretion under *section 33* of the *Limitation Act 1980* to allow a personal injuries claim to proceed after the three-year limitation period had expired. This meant that the company could also be revived under the provisions of *CA 1985, s 651(5)* and (6) (as amended). Thus, proceedings under the *Third Parties (Rights against Insurers) Act 1930, s 1(1)(b)* may be brought in this manner.

If the employer becomes bankrupt or if a company becomes insolvent, the employer's right to an indemnity from his insurers is transferred to the employee who may then keep the sums recovered with priority to his employer's creditors. This is only so if the employer has made his claim to this indemnity by trial, arbitration or agreement before he is made bankrupt or insolvent (*Bradley v Eagle Star Insurance Co Ltd [1989] 1 All ER 961*). The employee must also claim within the statutory limitation period from the date of his injury (see E13007 below for subrogation rights generally).

The policy protects an employer from third party claims; an employee as such is not covered since he normally incurs no liability. Although offering wide cover an employers' liability policy does not give cover to third party non-employees (e.g. independent contractors and members of the public). Such liability is covered by a public liability policy which, though advisable, is not compulsory.

This section examines:

— the general law relating to contracts of insurance (see E13003–E13006 below);

— the insurer's right of recovery (i.e. subrogation) (see E13007 below);

— the duty to take out employers' liability insurance (see E13008–E13012 below);

— issue and display of certificates of insurance (see E13013 below);

— penalties (see E13014, E13015 below);

— scope and cover of policy (see E13016–E13019 below);

— 'prohibition' of certain terms (see E13020 below);

— trade endorsement for certain types of work (see E13021, E13022 below).

— insolvent employers' liability insurers (see E13024 below).

## General law relating to insurance contracts

E13003    Insurance is a contract. When a person wishes to insure, for example, himself, his house, his liability towards his employees, valuable personal property or even loss of profits, he (the proposer) fills in a proposal form for insurance, at the same time making certain facts known to the insurer about what is to be insured. On the basis of the information disclosed in the proposal form, the insurer will decide whether to accept the risk or at what rate to fix the premium. If the insurer elects to accept the risk, a contract of insurance is then drawn up in the form of an insurance policy. (Incidentally, it seems to matter little whether the negotiations leading up to contract took place between the insured (proposer) and the insurance company or between the insured and a broker, since the broker is often regarded as the agent of one or the other, generally of the proposer (*Newsholme Brothers v Road Transport & General Insurance Co Ltd [1929] 2 KB 356)*.) However, a lot depends on the facts. If he is authorised to complete blank proposal forms, he may well be the agent of the insurer.

### Extent of duty of disclosure

E13004    A proposer must disclose to the insurer all material facts within his actual knowledge. This does not extend to disclosure of facts which he could not reasonably be expected to know. 'The duty is a duty to disclose, and you cannot disclose what you do not know. The obligation to disclose, therefore, necessarily depends on the knowledge you possess. This, however, must not be misunderstood. The proposer's opinion of the materiality of that knowledge is of no moment. If a reasonable man would have recognised that the knowledge in question was material to disclose, it is no excuse that you did not recognise it. But the question always is – Was the knowledge you possessed such that you ought to have disclosed it?' (*Joel v Law Union and Crown Insurance Co [1908] 2 KB 863* per Fletcher Moulton LJ).

The knowledge of those who represent the directing mind and will of a company and who control what it does, e.g. directors and officers, is likely to be identified as the company's knowledge whether or not those individuals are responsible for arranging the insurance cover in question (*PCW Syndicates v PCW Reinsurers [1996] 1 Lloyd's Rep 241*).

An element of consumer protection, in favour of insureds, was introduced into insurance contracts by the Statement of General Insurance Practice 1986, a form of self-regulation applicable to many but not to all insurers. This has consequences for the duty of disclosure, proposal forms (E13005 below), renewals and claims. In particular, with regard to the last element (claims), an insurer should not refuse to indemnify on the grounds of:

(*a*)   non-disclosure of a material fact which a policyholder could not reasonably be expected to have disclosed; or

(*b*)   misrepresentation (unless it is a deliberate non-disclosure of, or negligence regarding a material fact). Innocent misrepresentation is not a ground for avoidance of payment.

The trend towards greater consumer protection in (*inter alia*) insurance contracts is reflected in the *Unfair Terms in Consumer Contracts Regulations 1999 (SI 1999 No 2083)*. These Regulations apply in the case of 'standard form' (or non-individually negotiated) contracts. [*SI 1999 No 2083, Reg 5*]. A contractual term in such a contract shall:

(i) be regarded as 'unfair' if contrary to the requirement of 'good faith' it 'causes a significant imbalance in the parties' rights and obligations arising under the contract, to the insured's detriment' [SI 1999 No 2083, Reg 5(1), Sch 2];

(ii) always be regarded as having been individually negotiated where it has been drafted in advance and the consumer has not been able to influence the substance of the term [SI 1999 No 2083, Reg 5(2)];

(iii) require a written contract term to be expressed in plain, intelligible language. If there is doubt about the meaning of terminology, a construction in favour of the insured will prevail [SI 1999 No 2083, Reg 7]; and

(iv) where the insurer claims that a term was individually negotiated, he must prove it [SI 1999 No 2083, Reg 5(4)].

Complaints (other than ones which are frivolous or vexatious) or which are to be handled by a qualified body [SI 1999 No 2083, Sch 1] relating to 'unfair terms' in standard form contracts, are considered by the Director General of Fair Trading, who may prevent their continued use. [SI 1999 No 2083, Reg 12]. See also P9056 PRODUCT SAFETY.

## Filling in proposal form

**E13005** Generally only failure to make disclosure of relevant facts will allow an insurer subsequently to invalidate the policy and refuse to compensate for the loss. The test of whether a fact was or was not relevant is whether its omission would have influenced a prudent insurer in deciding whether to accept the risk, or at what rate to fix the premium.

The arm of *uberrima fides* (i.e. the utmost good faith) is a long one. If, when filling in a proposal form, a statement made by the proposer is at that time true, but is false in relation to other facts which are not stated, or becomes false before issue of the insurance policy, this entitles the insurer to refuse to indemnify. In *Condogianis v Guardian Assurance Co Ltd [1921] 2 AC 125* a proposal form for fire cover contained the following question: 'Has proponent ever been a claimant on a fire insurance company in respect of the property now proposed, or any other property? If so, state when and name of company'. The proposer answered 'Yes', '1917', 'Ocean'. This answer was literally true, since he had claimed against the Ocean Insurance Co in respect of a burning car. However, he had failed to say that in 1912 he had made another claim against another insurance company in respect of another burning car. It was held that the answer was not a true one and the policy was, therefore, invalidated.

## Loss mitigation

**E13006** There is an implied term in most insurance contracts that the insured will take all reasonable steps to mitigate loss caused by one or more of the insured perils. Thus, in the case of burglary cover of commercial premises, this could extend to provision of security patrols, the fitting of burglar alarm devices and guard dogs. In the case of employers' liability, it will extend to appointment or use of services of an accredited safety officer and/or occupational hygienist, either permanently or temporarily, particularly in light off the *Management of Health and Safety at Work Regulations 1999 (SI 1999 No 3242)*, to oversee, for example, application of the *COSHH Regulations 1999 (SI 1999 No 437)*. Again, in the case of fire cover, steps to mitigate the extent of the loss on the part of the insured, might well extend to regular visits by the local fire authority and/or advice on storage of products and materials by reputable risk management consultants. Indeed, it is compliance with this implied duty in insurance contracts that accounts for the growth of the practice of risk management, and good housekeeping on the part of more and more companies.

## Subrogation

**E13007**  Subrogation enables an insurer to make certain that the insured recovers no more than exact replacement of loss (i.e. indemnity). 'It [the doctrine of subrogation] was introduced in favour of the underwriters, in order to prevent their having to pay more than a full indemnity, not on the ground that the underwriters were sureties, for they are not so always, although their rights are sometimes similar to those of sureties, but in order to prevent the assured recovering more than a full indemnity' (*Castellain v Preston (1883) 11 QBD 380 per Brett LJ*). Subrogation does not extend to accident insurance moneys, whereby the insured (normally self-employed) is promised a fixed sum in the event of injury or illness (*Bradburn v Great Western Railway Co (1874) LR 10 Exch 1* where the appellant was injured whilst travelling on a train, owing to the negligence of the respondent. He had earlier bought personal accident insurance to cover him for the possibility of injury on the train. It was held that he was entitled to both damages for negligence *and* insurance moneys payable under the policy (see further COMPENSATION FOR WORK INJURIES/DISEASES)). The right of subrogation does not arise until the insurer has paid the insured in respect of his loss, and has been invoked infrequently in employers' liability cases. In *Morris v Ford Motor Co Ltd [1973] 2 All ER 1084* the Ford Motor Co had subcontracted cleaning at one of their plants to the X company, for which the appellant worked. Whilst engaged on this work at the plant, the appellant was injured owing to the negligence of an employee whilst driving a forklift truck. The appellant claimed damages from the respondent company for the negligence of their employee, on the grounds of vicarious liability. X company had, however, entered into a contract of indemnity with the respondent company, agreeing to indemnify the company for all losses or claims for injury arising out of the cleaning operations. Although accepting that they were bound by the terms of this contract of indemnity, the X company argued that they should be subrogated against the negligent Ford employee, on the ground that the employee had carried out his work negligently. It was held that the agreement by the British Insurance Association that they would not sue an employee of an insured employer in respect of injury caused to a co-employee, unless there was either (*a*) collusion and/or (*b*) wilful misconduct on the part of the employee, was binding and that the X company could not recoup its loss from the negligent employee.

## Duty of employer to take out and maintain insurance

**E13008**  'Every employer carrying on business in Great Britain shall insure, and maintain insurance against liability for bodily injury or disease sustained by his employees, and arising out of and in the course of their employment in Great Britain in that business.' [*Employers' Liability (Compulsory Insurance) Act 1969, s 1(1)*].

Such insurance must be provided under one or more 'approved policies'. An 'approved policy' is a policy of insurance not subject to any conditions or exceptions prohibited by regulations (see E13020 below). [*Employers' Liability (Compulsory Insurance) Act 1969, s 1(3)*]. This now includes insurance with an approved EU insurer. [*Insurance Companies (Amendment) Regulations 1992 (SI 1992 No 2890)*].

There is no duty under the *1969 Act* to warn or insure the employee against risks of employment outside Great Britain (*Reid v Rush Tompkins Group plc [1989] 3 All ER 228*) although the *1998 Regulations* require the employer to insure employees employed on or from offshore installations or associated structures – see E13009 below.

### Employees covered by the Act

**E13009**  Cover is required in respect of liability to employees who:

(*a*)   are ordinarily resident in Great Britain; or

(*b*)   though not ordinarily resident in Great Britain, are present in Great Britain in the course of employment here for a continuous period of not less than 14 days; or

(c) though not ordinarily resident in the United Kingdom, have been employed on or from an offshore installation or associated structure for a continuous period of not less than seven days.

[*Employers' Liability (Compulsory Insurance) Regulations 1998* (SI 1998 No 2573), Reg 1(2)].

## Employees not covered by the Act

**E13010** An employer is not required to insure against liability to an employee who is (*a*) a spouse, (*b*) father, (*c*) mother, (*d*) son, (*e*) daughter, (*f*) other close relative. [*Employers' Liability (Compulsory Insurance) Act 1969, s 2(2)(a)*]. Those who are not ordinarily resident in the UK are not covered by the Act except as above. Nor are employees working abroad covered. Such employees can sue under English law in limited circumstances (*Johnson v Coventry Churchill International Ltd [1992] 3 All ER 14* where an employee, working in Germany for an English manpower leasing company, was injured when he fell through a rotten plank. He was unable to sue his employer under German law; although he was working in Germany, it was held that England was the country with the most significant relationship with the claim because he had made the contract in England, his employers had covered him with personal liability insurance and he therefore expected them to compensate him through these insurers for any personal injury sustained in Germany).

## Degree of cover necessary

**E13011** The amount for which an employer is required to insure and maintain insurance is £5 million in respect of claims relating to any one or more of his employees, arising out of any one occurrence. [*Employers' Liability (Compulsory Insurance) Regulations 1998* (SI 1998 No 2573), Reg 3(1)].

Between 1 January 1972 (when the *Employer's Liability (Compulsory Insurance) Act 1969* came into force) and 1994, insurers, in practice, provided unlimited cover under employers' liability policies. As from 1 January 1995, as a result of payments made in respect of claims exceeding the amount of premiums received during the period 1989–1993, unlimited liability was withdrawn, but most insurers continued to offer a minimum of £10 million indemnity for onshore work. A consultative document issued by the Department of the Environment, Transport and the Regions entitled The Draft Employers' Liability (Compulsory Insurance) General Regulations [C4857 September 1997] (hereafter referred to as 'the 1997 consultative document') which preceded the *1998 Regulations* assumed that this practice would continue.

Where a company has subsidiaries, there will be sufficient compliance if a company insures/maintains insurance for itself *and* on behalf of its subsidiaries for £5 million in respect of claims affecting any one or more of its own employees and any one or more employees of its subsidiaries arising out of any one occurrence. [*Employers' Liability (Compulsory Insurance) Regulations 1998* (SI 1998 No 2573), Reg 3].

Insurers and the courts have interpreted the legislation to mean that all injuries resulting from one incident (e.g. an explosion) are treated as one occurrence, and each individual case of gradually occurring injury or disease is treated as an individual occurrence – the only exception being a situation where a sudden and immediate outbreak of a disease amongst the workforce is clearly attributable to an identifiable incident (e.g. the escape of a biological agent). The introduction to the 1997 consultative document suggested that this interpretation might be challenged and set out a possible alternative regulation to be used instead of what is now *Reg 3* of the *1998 Regulations* if clarification was felt necessary. This was not adopted and no new regulations are currently proposed and therefore it would appear that the Government are now satisfied that the position is clear.

## Exempted employers

E13012　　The following employers are exempt from the duty to take out and maintain insurance:

(a)　nationalised industries;

(b)　any body holding a Government department certificate that any claim which it cannot pay itself will be paid out of moneys provided by Parliament;

(c)　any Passenger Transport Executive and its subsidiaries, London Regional Transport and its subsidiaries;

(d)　statutory water undertakers and certain water boards;

(e)　the Commission for the New Towns;

(f)　health service bodies, National Health Service Trusts;

(g)　probation and after-care committees, magistrates' court committees, and any voluntary management committee of an approved bail or approved probation hostel;

(h)　governments of foreign states or commonwealth countries and some other specialised employers;

(j)　Railtrack Group plc and its subsidiaries (the exemption ceasing when it is no longer owned by the Crown);

(k)　the Qualifications & Curriculum Authority.

There are other types of employer specified in the Regulations, but these are the main exceptions. [*Employers' Liability Compulsory Insurance Act 1969, s 3; Employers' Liability (Compulsory Insurance) Regulations 1998 (SI 1998 No 2573), Sch 2*].

## Issue, display and retention of certificates of insurance

E13013　　The insurer must issue the employer with a certificate of insurance, which must be issued not later than 30 days after the date on which insurance was commenced or renewed. [*Employers' Liability (Compulsory Insurance) Act 1969, s 4(1); Employers' Liability (Compulsory Insurance) Regulations 1998 (SI 1998 No 2573), Reg 4*]. Where there are one or more contracts of insurance which jointly provide insurance cover of not less than £5 million, the certificate issued by any individual insurer must specify both the amount in excess of which insurance cover is provided by the individual policy, and the maximum amount of that cover. [*Employers' Liability (Compulsory Insurance) Regulations 1998 (SI 1998 No 2573), Reg 4(3)*].

A copy or copies of the certificate must be displayed at each place of business where there are any employees entitled to be covered by the insurance policy and the copy certificate(s) must be placed where employees can easily see and read it and be reasonably protected from being defaced or damaged. [*Employers' Liability (Compulsory Insurance) Regulations 1998 (SI 1998 No 2573), Reg 5*]. The exception is where an employee is employed on or from an offshore installation or associated structure, when the employer must produce, at the request of that employee and within ten days from such request, a copy of the certificate. [*Employers' Liability (Compulsory Insurance) Regulations 1998 (SI 1998 No 2573), Reg 5(4)*].

An employee must, if a notice has been served on him by the Health and Safety Executive, produce a copy of the policy to the officers specified in the notice and he must permit inspection of the policy by an inspector authorised by the Secretary of State to inspect the policy. [*Employers' Liability (Compulsory Insurance) Regulations 1998 (SI 1998 No 2573), Regs 7, 8*].

A change introduced by the *1998 Regulations* is that employers are now required by law to retain any certificate of employers' liability insurance (or a copy) for a period of 40 years beginning on the date on which the insurance to which it relates commences or is renewed [*Employers' Liability (Compulsory Insurance) Regulations 1998 (SI 1998 No 2573), Reg 4(4)*]. Companies may retain the copy in any eye-readable form in any one of the ways authorised by the *Companies Act 1985, ss 722 and 723* [*Employers' Liability (Compulsory Insurance) Regulations 1998 (SI 1998 No 2573), Reg 4(5)*].

## Penalties

### Failure to insure or maintain insurance

E13014  Failure by an employer to effect and maintain insurance for any day on which it is required is a criminal offence, carrying a maximum penalty on conviction of £2,500. [*Criminal Justice Act 1982, s 37(2)*].

### Failure to display a certificate of insurance

E13015  Failure on the part of an employer to display a certificate of insurance in a prominent position in the workplace is a criminal offence, carrying a maximum penalty on conviction of £1,000. [*Criminal Justice Act 1982, s 37(2)*].

In the 1997 consultative document, the Government suggests that penalties should be increased to become the same as those under the *Health and Safety at Work etc Act 1974* (*HSWA 1974*), i.e. fines of £20,000 in a magistrates' court and unlimited in the Crown Court. To implement this change will require primary legislation and currently it would appear that the Government has no plans to introduce such legislation.

## Cover provided by a typical policy

### Persons

E13016  Cover is limited to protection of employees. Independent contractors are not covered; liability to them should be covered by a public liability policy. Directors who are employed under a contract of employment are covered, but directors paid by fees who do not work full-time in the business are generally not regarded as 'employees'. Liability to them would normally be covered by a public liability policy. Similarly, since the judicial tendency is to construe 'labour-only' subcontractors in the construction industry as 'employees' (see CONSTRUCTION AND BUILDING OPERATIONS), employers' liability policies often contain the following endorsement: 'An employee shall also mean any labour master, and persons supplied by him, any person employed by labour-only subcontractors, any self-employed person, or any person hired from any public authority, company, firm or individual, while working for the insured in connection with the business'. The public liability policy should then be amended to exclude the insured's liability to 'employees' so designated.

In the 1997 consultative document, the Government points out that the issue as to what constitutes 'an employee' cannot be completely resolved without primary legislation, but proposes to issue guidance on interpretation, although it was proposed that guidance on interpretation would be issued. Such guidance has not been issued and therefore presumably the Government are now satisfied that the position is clear.

### Scope of cover

E13017  The policy provides for payment of:

(*a*)  costs and expenses of litigation, incurred with the insurer's consent, in defence of a claim against the insured (i.e. civil liability);

(b) solicitor's fees, incurred with the insurer's consent, for representation of the insured at proceedings in any court of summary jurisdiction (e.g. magistrates' court or Crown Court), coroner's inquest, or a fatal accident inquiry (i.e. criminal proceedings), arising out of an accident resulting in injury to an employee. It does *not* cover payment of a fine imposed by a criminal court.

The policy will often contain an excess negotiated between the insurer and employer, i.e. a provision that the employer pay the first £x of any claim. For the purposes of the *1969 Act*, any condition in a contract of insurance which requires a relevant employee to pay, or an insured employer to pay the relevant employee, the first amount of any claim or any aggregation of claims, is prohibited. Agreements will still be permitted which provide that the insurer will pay the claim in full and may then seek some reimbursement from the employer. [*Employers' Liability (Compulsory Insurance) Regulations 1998 (SI 1998 No 2573), Reg 2*].

## Geographical limits

E13018 Cover is normally limited to Great Britain, Northern Ireland, the Channel Islands and the Isle of Man, in respect of employees normally resident in any of the above, who sustain injury whilst working in those areas. Cover is also provided for such employees who are injured whilst temporarily working abroad, so long as the action for damages is brought in a court of law of Great Britain, Northern Ireland, the Channel Islands or the Isle of Man – though even this proviso is omitted from some policies.

Employers must also have employers' liability insurance in respect of employees who, though not ordinarily resident in the United Kingdom, have been employed on or from an offshore installation or associated structure for a continuous period of not less than seven days; or who, though not ordinarily resident in Great Britain, are present in Great Britain in the course of employment for not less than fourteen days. [*Employers' Liability (Compulsory Insurance) Regulations 1998 (SI 1998 No 2573), Reg 1(2)*].

## Conditions which must be satisfied

E13019

(a) Cover only relates to bodily injury or disease; it does not extend to employee's property. This latter cover is provided by an employers' public liability policy.

(b) Injury must arise out of and during the course of employment (see E11004 EMPLOYERS' DUTIES TO THEIR EMPLOYEES). If injury does not so arise, cover is normally provided by a public liability policy.

(c) Bodily injury must be caused during the period of insurance. Normally with injury-causing accidents there is no problem, since injury follows on from the accident almost immediately. Certain occupational diseases, however, may not manifest themselves until much later, e.g. asbestosis, mesothelioma, pneumoconiosis, deafness. Here legal liability takes place when the disease manifests itself, or is 'discovered'. Moreover, at least as far as occupational deafness is concerned, liability between employers can be apportioned, giving rise to contribution between insurers (see further NOISE AND VIBRATION).

(d) Claims must be notified by the insured to the insurer as soon as possible, or as stipulated by the policy.

The *Employers' Liability (Compulsory Insurance) Regulations 1998 (SI 1998 No 2573), Reg 2* does not fetter the freedom of underwriters to apply certain conditions in connection with intrinsically hazardous work; for instance, exclusion of liability for accidents arising out of demolition work, or in connection with use of explosives.

## 'Prohibition' of certain conditions

**E13020** All liability policies contain conditions with which the insured must comply if the insurer is to 'progress' his claim, e.g. notification of claims. Failure to comply with such condition(s) could jeopardise cover under the policy: the insured would be legally liable but without insurance protection. In the case of an employers' liability policy, an insurer might seek to avoid liability under the policy if the condition requiring the insured to take reasonable care to prevent injuries to employees, and/or comply with the provisions of any relevant statutes/statutory instruments (e.g. *HSWA 1974; Ionising Radiations Regulations 1999 (SI 1999 No 3232)*), or to keep records, was not complied with.

The object of the 1969 Act was to ensure that an employer who had a claim brought against him would be able to pay the employee any damages awarded. Regulations made under the Act, therefore, seek to prevent insurers from avoiding their liability by relying on breach of a policy condition, by way of 'prohibiting' certain conditions in policies taken out under the Act. More particularly, insurers cannot avoid liability in the following circumstances:

(a) some specified thing being done or being omitted to be done after the happening of the event giving rise to a claim (e.g. omission to notify the insurer of a claim within a stipulated time) [*Employers' Liability (Compulsory Insurance) Regulations 1998 (SI 1998 No 2573), Reg 2(1)(a)*];

(b) failure on the part of the policy-holder to take reasonable care to protect his employees against the risk of bodily injury or disease in the course of employment [*Employers' Liability (Compulsory Insurance) Regulations 1998 (SI 1998 No 2573, Reg 2(1)(b)*]. As to the meaning of 'reasonable care' or 'reasonable precaution' here, 'It is eminently reasonable for employers to entrust … tasks to a skilled and trusted foreman on whose competence they have every reason to rely'. (*Woolfall and Rimmer Ltd v Moyle and Another [1941] 3 All ER 304*). The prohibition is therefore, not broken by a negligent act on the part of a competent foreman selected by the employer. Where, however, an employer acted wilfully (in causing injury) and not merely negligently (though this would be rare), the insurer could presumably refuse to pay (*Hartley v Provincial Insurance Co Ltd [1957] Lloyd's Rep 121* where the insured employer had not taken steps to ensure that a stockbar was securely fenced for the purposes of the *Factories Act 1937, s 14(3)* in spite of repeated warnings from the factory inspector, with the result that an employee was scalped whilst working at a lathe. It was held that the insurer was justified in refusing to indemnify the employer who was in breach of statutory duty and so liable for damages). This was confirmed in *Aluminium Wire and Cable Co Ltd v Allstate Insurance Co Ltd [1985] 2 Lloyd's Rep 280*;

(c) failure on the part of the policy-holder to comply with statutory requirements for the protection of employees against the risk of injury [*Employers' Liability (Compulsory Insurance) Regulations 1998 (SI 1998 No 2573, Reg 2(1)(c)*] – the reasoning in *Hartley v Provincial Insurance Co Ltd* (see (b) above), that wilful breach may not be covered, probably applies here too;

(d) failure on the part of the policy-holder to keep specified records and make such information available to the insurer [*Employers' Liability (Compulsory Insurance) Regulations 1998 (SI 1998 No 2573, Reg 2(1)(d)*] (e.g. accident book or accounts relating to employees' wages and salaries (see ACCIDENT REPORTING));

(e) by means of the use of an excess in policies [*Employers' Liability (Compulsory Insurance) Regulations 1998 (SI 1998 No 2573, Reg 2(2)*] – see E13017 above

## Trade endorsements for certain types of work

**E13021** There are no policy exceptions to the standard employers' liability cover. Trade endorsements, however, are used frequently in underwriting employers' liability risks, and there is nothing in the *1969 Act* to prevent insurers from applying their normal underwriting principles and applying trade endorsements where they consider it necessary, i.e. they will amend their standard policy form to exclude certain risks. Thus, there may be specific exclusions of liability arising out of types of work, such as demolition, or the use of mechanically driven woodworking machinery, or work above certain heights, unless the appropriate rate of premium is paid. This does mean that there are still circumstances where an employee will not obtain compensation from his employer based on the employer's insurance cover.

## Measure of risk and assessment of premium

E13022    Certain trades or businesses are known to be more dangerous than others. For most trades or businesses insurers have their own rate for the risk, expressed as a rate per cent on wages (other than for clerical, managerial or non-manual employees for whom a very low rate applies). This rate is used as a guide and is altered upwards or downwards depending upon:

(*a*)    previous history of claims and cost of settlement;

(*b*)    size of wage roll;

(*c*)    whether certain risks are not to be covered, e.g. the premium will be lower if the insured elects to exclude from the policy certain risks, such as the use of power driven woodworking machinery;

(*d*)    the insured's attitude towards safety.

Many insurers survey premises with the object of improving the risk and minimising the incidence of accidents and diseases. This is an essential part of their service, and they often work in conjunction with the insured's own safety staff.

## Extension of cover

E13023    In addition to employers' liability insurance, it is becoming increasingly common for companies to buy insurance in respect of directors' personal liability. Indeed, in the United States, some directors refuse to take up appointments in the absence of such insurance being forthcoming.

## Insolvent employers' liability insurers

E13024    The problems that arise following the insolvency of employer's liability insurer gained particular prominence following the much publicised liquidation of Chester Street Holdings Limited ('Chester Street') on 9 January 2001. It was initially feared that individuals whose claims pre-dated 1972 (the date when employers' liability insurance became compulsory) would be at risk of not receiving compensation. The situation has, however, largely been clarified following the establishment of the Financial Services Compensation Scheme ('the FSCS') which was created under the *Financial Services and Markets Act 2000* and acts as a 'safety net' for customers of finance sector companies who are unable to play claims against authorised companies. The FSCS came into effect on 1 December 2001. Under the FSCS policy holders (save for any period of cover when the policy holder was a nationalised industry which is specifically excluded from the provisions of the scheme) are eligible for protection if they are insured by an authorised insurance company under a contract of insurance issued in the UK, Channel Islands or Isle of Man. The FSCS scheme pays 100 per cent compensation for post-1972 compulsory employers' liability insurance claims. If the employer still exists and is solvent then the employer initially pays the compensation to the claimant although is able to recoup 100 per cent of the compensation outlay from the FSCS. The FSCS pays 90 per cent compensation for pre-1972 claims providing that the employer is not still in existence and solvent. If however the employer is still in existence – the employer must meet its liability to the claimant and is not compensated by the FSCS.

**E13024** *Employers' Liability Insurance*

Special provisions apply to Chester Street for pre-1972 claims where:

- liability was established and quantified before insolvency on 9 January 2001, 90 per cent compensation is payable. Such claims are administered under the *Financial Services and Markets Act 2002* using the guidelines established by the Policyholders Protection Board;
- liability was agreed and quantified between 9 January and 30 November 2001. Such claims are subject to the requirements of a scheme set up by the Association of British Insurers who pay 90 per cent of the award.

Responsibility for claims handled by the Policyholder Protection Board (PPB) were transferred to the FSCS on 1 December 2001. PPB rules continue to apply to any claim arising from the insolvency of an insurance company where it occurred before 1 December 2001.

Further information can be obtained from FSCS, 7th Floor, Lloyds Chambers, 1 Portsoken Street, London E1 8BN, tel: 020 7892 7300.

# Employment Protection

## Introduction

E14001     The nature of employment protection is continually changing. At a domestic (national) level, the last decade has seen the extension of certain employment protection rights to workers as well as employees. For example, both the *Public Interest Disclosure Act 1998* and the *Working Time Regulations 1998* (*SI 1998 No 1833*) apply to 'workers' which is a wider category than persons categorised as employees under traditional English employment law analysis.

A number of new individual rights have been also introduced including the right to be accompanied at a disciplinary or grievance hearing and the right to request flexible working. Collective rights have also been enhanced with trade unions being given the right to statutory recognition in certain circumstances.

Europe has also continued to have a significant impact on the regulation of employment relations. Directives issued by Europe have required the introduction of legislation prohibiting discrimination on the grounds of sexual orientation and religion or belief, legislation requiring employers with more than 50 employees to set up national information and consultation procedures if requested by employees and will, in the future, require the prohibition of age discrimination. Further changes are also in the pipeline, perhaps the most important of which is the requirement for employers and employees to follow statutory dismissal, disciplinary and grievance procedures which came into force on 1 October 2004.

The relationship between an employer and an employee is a contractual one and as such must have all the elements of a legally binding contract to render it enforceable. In strict contractual terms an offer is made by the employer which is then accepted by the employee. As in the case of the offer, the acceptance may be oral, in writing, or by conduct, for example by the employee turning up for work. The consideration on the employer's part is the promise to pay wages and on the employee's part to provide his services for the employer. Once the employer's offer has been accepted, the contract comes into existence and both parties are bound by any terms contained within it (*Taylor v Furness, Withy & Co Ltd (1969) 6 KIR 488*).

The contractual analysis of the employment relationship is not entirely satisfactory in explaining the relationship between employee and employer. To fit the contract model, various elements comprising the reality of the employment relationship become part of the contract by implication.

An employment contract is unlike many other contracts, because many of the terms will not have been individually negotiated by the parties. The contract will contain the express terms that the parties have agreed – most commonly hours, pay, job description – and there will be a variety of other terms which will be implied into the contract from other sources and which the parties have not agreed. Many of these are relevant to health and safety. If any of the express or implied terms in the contract are breached, the innocent party will have certain remedies. The fact that various employee rights, particularly in relation to health and safety, are implied into the contractual terms and conditions is important for this reason.

In addition to terms implied into the contract by the common law, statute (such as the *Employment Rights Act 1996*) has created additional employment protection rights for employees, including some specific rights in relation to health and safety. These are in addition to detailed rights and duties arising from health and safety legislation and regulations which are discussed elsewhere.

An employer will often lay down health and safety rules and procedures. While the law allows an employer the ultimate sanction of dismissal as a method of ensuring that safety rules are observed, such dismissals should be lawful, that is generally with notice, and should be fair. Furthermore, statute has created specific protection from victimisation for employees who are protecting themselves or others against perceived health and safety risks. All of these provisions are the subject of this section.

As a specific health and safety protection measure, the *Health and Safety at Work etc Act 1974*, s 2 (*HSWA 1974*) lays a general duty on all employers to ensure, so far as is reasonably practicable (for the meaning of this expression, see E15039 ENFORCEMENT), the health, safety and welfare of all their employees. An employer is also under a duty to consult about health and safety matters. In addition, a number of codes of practice have been issued under the *HSWA 1974* by the Health and Safety Executive, for example relating to safety representatives and allowing them time off to train.

## Sources of contractual terms

### Express terms

E14002    These are the terms agreed by the parties themselves and may be oral or in writing. Normally the courts will uphold the express terms in the contract because these are what the parties have agreed. However, if the term is ambiguous the court may be called upon to interpret the ambiguity, for example what the parties meant by 'reasonable overtime'.

Generally the express terms cause no legal problems and the parties can insert such terms into the contract as they wish. There are, however, a number of restrictions which include:

(a)   An employer cannot restrict his liability for the death or personal injury of his employees caused by his negligence. Further, he can only restrict liability for damage to his employee's property if such a restriction is reasonable (*Unfair Contract Terms Act 1977*, s 2).

(b)   The terms in the contract cannot infringe discrimination legislation including:

- the *Equal Pay Act 1970* and the *Sex Discrimination Acts 1975* and *1986*;
- the *Race Relations Act 1976*;
- the *Disability Discrimination Act 1995*;
- the *Employment Equality (Sexual Orientation) Regulations 2003* (*SI 2003 No 1661*);
- the *Employment Equality (Religion or Belief) Regulations 2003* (*SI 2003 No 1660*).

(c)   The terms in the contract cannot infringe the *Part-time Workers (Prevention of Less Favourable Treatment) Regulations 2000* (*SI 2000 No 1551*). The Regulations provide that unless justified on objective grounds, a part-time worker has the right not to be treated less favourably than a comparable full-time worker on the ground that the worker is a part-timer in relation to the terms of the contract.

(d)   The terms in the contract cannot infringe the *Fixed-term Workers (Prevention of Less Favourable Treatment) Regulations 2002* (*SI 2002 No 2034*). The Regulations provide that unless justified on objective grounds, a fixed-term worker has the right not to be treated less favourably than a comparable full-time worker on the ground that the worker is a fixed-term worker.

(e)   The employer cannot have a notice provision which gives the employee less than the statutory minimum notice guaranteed by the *Employment Rights Act 1996*, s 86.

(f)   Some judges have suggested that any express terms regarding hours are subject to the employer's duty to ensure his employee's safety and must be read subject to this, so that a term requiring an employee to work 100 hours a week will not be enforceable (see for example Stuart-Smith LJ in *Johnstone v Bloomsbury Health Authority [1991] IRLR 118*). More specifically, the provisions of the *Working Time Regulations 1998* (*SI 1998 No 1833*) affect the contractual term in relation to working hours. The 1998 Regulations, save in the case of specified exemptions, set a maximum working week of 48 hours averaged over a 17 week reference period. In addition, the Regulations provide for an obligatory daily rest period, weekly rest, rest breaks, limits on night work and minimum annual leave and otherwise regulate working time. In *Barber v RJB Mining UK Ltd [1999] IRLR 308*, the court decided that the maximum imposed on weekly working time by the Regulations was part of the employees' contract. This decision gives some protection to employees who refuse to work beyond the statutorily stated maximum. It is possible under the 1998 Regulations for a worker to agree with his or her employer in writing to opt-out of the 48 hour working week, subject to complying with certain requirements.

(g) The terms in the contract cannot infringe the *Maternity and Parental Leave etc Regulations 1999* (*SI 1999 No 3312*), which entitle employees with one year's continuous service who have responsibility for a child born after 15 December 1999, to be absent from work for up to 13 weeks' parental leave.

(h) Confidentiality provisions should be subject to an employee's right to make a protected disclosure in accordance with the provisions set out in the *Employment Rights Act 1996*.

Whilst the *Sex Discrimination Acts 1975* and *1986* removed some of the restrictions on women and their employment generally and health and safety specifically, some restrictions/prohibitions on certain types of employment by women, in the interests of health and safety at work, still remain.

Under the *Control of Lead at Work Regulations 2002* (*SI 2002 No 2676*) an employer is prohibited from employing women of reproductive capacity or young people in particular activities relating to lead processes as follows:

(a) In the lead smelting and refining process:

  (i) handling, treating, sintering, smelting or refining any material containing 5 per cent or more of lead; or

  (ii) cleaning where any of the above activities have taken place.

(b) In the lead acid manufacturing process:

  (i) manipulating lead oxides;

  (ii) mixing or pasting;

  (iii) melting or casting;

  (iv) trimming, abrading or cutting of pasted plates; or

  (v) cleaning where any of the above activities have taken place.

## Common law implied duties – all contracts

**E14003** Both the employer and employee owe duties towards each other. These are duties implied into every contract of employment and should be distinguished from the implied terms discussed below which are implied into a particular individual contract. Although there are a number of different duties, three are of major importance in relation to health and safety:

(a) the duty on the part of the employee to obey lawful reasonable orders and to perform his work with reasonable care and skill; and

(b) the duty on the part of the employer to ensure his employee's safety.

The duty to obey lawful reasonable orders ensures that the employer's safety rules can be enforced and, as it is a contractual duty, breach will allow the employer to invoke certain sanctions against the employee, the ultimate of which may be dismissal.

The same is true of the duty to perform his work with reasonable care and skill. Should the employee be in breach of this duty and place his or others' safety at risk, the employer may impose sanctions against him including dismissal.

The imposition of the employer's duty is to complement the statutory provisions. Statutes such as the *HSWA 1974* provide sanctions against the employer should he fail to comply with the legislation or any regulations made thereunder. The common law duty provides the employee with a remedy should the duty be broken, either in the form of compensation if he is injured, or, potentially, with a claim of unfair dismissal. The employer's duty to ensure his employees' safety is one of the most important aspects of the employment relationship. At least one judge has argued that it is so important that any express term must be read subject to it (see *Johnstone v Bloomsbury Health Authority [1991] IRLR 118* at E14002 above). Breach of this duty can lead to the employee resigning and claiming constructive dismissal (see below). In *Walton & Morse v Dorrington [1997] IRLR 488*, an employee claimed that she had been constructively dismissed (unfairly) because her employer had breached the implied term of her contract of employment that it would provide, so far as reasonably practicable, a suitable working environment. The employee had been forced to work in a smoke-filled environment for a prolonged period of time and her employer did not take appropriate steps to redress the problem when she raised the issue. The Employment Appeal Tribunal agreed that she had been constructively dismissed because the employer had breached its duty to her. Further, the employer's duty to ensure his employee's safety has been held to cover stressful environments resulting in injury to the employee. In *Walker v Northumberland County Council [1995] IRLR 35*, the High Court held that an employer was liable for damages on the basis that they owed their employee a duty not to cause him psychiatric damage by the volume and/or character of work that he was required to undertake. In *Fraser v The State Hospitals Board for Scotland (11 January 2000) (2000 Rep LR 94)*, the Court of Session held that there was no reason to qualify an employer's duty to take reasonable care for the safety of employees so as to restrict the nature of the injury suffered to a physical one in circumstances where the employee claimed damages for psychological damage as a result of disciplinary measures. The claim failed on the basis of lack of foreseeability.

## *Common law implied terms – individual contracts*

E14004 The court may imply terms into the contract when a situation arises which was not anticipated by the parties at the time they negotiated the express terms. As such, the court is 'filling in the gaps' left by the parties' own negotiations. The courts use two tests to see if a term should be implied, (i) the 'business efficacy' test (*The Moorcock (1889) 14 PD 64*) or (ii) the 'officious bystander' or 'oh of course' test (*Shirlaw v Southern Foundries Ltd [1939] 2 KB 206*). Once the court has decided, by virtue of one of these tests, that a term should be implied, it will use the concept of reasonableness to decide the content of the term. Often this will involve looking at how the parties have worked the contract in the past. For example, if the contract does not contain a mobility clause, but the employee has always worked on different sites, the court will normally imply a mobility clause into the contract (*Courtaulds Northern Spinning Ltd v Sibson [1988] IRLR 305*). (See also *Aparau v Iceland Frozen Foods plc [1996] IRLR 119*, where the Employment Appeal Tribunal refused to imply a mobility clause on the basis that there were other ways of achieving the necessary flexibility.) In relation to dismissal, a term has been implied that, except in the case of summary dismissal, the employer will not terminate the employment contract while the employee is incapacitated where the effect would be to deprive the employee of permanent health insurance benefits (*Aspden v Webbs Poultry and Meat Group (Holdings) Ltd [1996] IRLR 521*). Terms can also be implied by the conduct of the parties or by custom and practice in a particular industry or area. The test for this is relatively difficult to fulfil – the term must be notorious and certain and, in effect, everyone in the industry/enterprise must know that it is part of the contract. Arguments based on custom and practice come into play in relation to issues such as statutory holidays and redundancy policies.

## Collective agreements

**E14005** Collective agreements are negotiated between an employer or employer's association and a trade union or unions. This means that they are not contracts between an employer and his individual employees because the employee was not one of the negotiating parties. Some terms of the collective agreement will be procedural and will govern the relationship between the employer and the union; some, on the other hand, will impact on the relationship between the employer and each individual employee, for example a collectively bargained pay increase. As the employee is not a party to the collective agreement, the only way he can enforce a term which is relevant to him is if the particular term from the collective agreement has become a term of his individual employment contract. Procedural provisions, policy and more general aspirations are not suitable for incorporation into an individual contract of employment. Incorporation is important because the collective agreement is not a legally binding contract between the employer and the union (*Trade Union and Labour Relations* (*Consolidation*) *Act 1992, s 179(1)*) and thus needs to be a term of an employment contract to make it legally enforceable.

The two main ways that a term from a collective agreement becomes a term of an employment contract is by express or implied incorporation. Until recently, implied incorporation was the most common and was complex. It generally required the employee to be a member of the union which negotiated the agreement, to have knowledge of the agreement and of the existence of the term, and to have conducted himself in such a way as to indicate that he accepted the term from the collective agreement as a term of his contract. A decision of the Employment Appeal Tribunal case, *Healy & Others v Corporation of London (24 June 1999) (unreported)*, illustrates that habitual acceptance of the benefits of a collective agreement does not, in itself, lead to the conclusion that the terms of that collective agreement have become contractually binding on an individual employee. There can be many reasons for an individual to accept the benefits of collective bargaining which do not amount to an acceptance that the underlying agreement forms part of his or her contract.

Express incorporation meant that the employee had expressly agreed (normally in his contract) that any term collectively agreed would become part of his contract. This used to be unusual, but with the change made to the statutory statement which must be given to all employees (see below) employees must be told of collective agreements which apply to them, and this has been held as expressly incorporating those agreements into the contract.

## Statutory statement of terms and conditions

**E14006** By the *Employment Rights Act 1996, s 1* every employee no later than two months after starting employment, must receive a statement of his basic terms and conditions. The statement must contain:

(*a*) the names of the employer and employee;

(*b*) the date the employment began;

(*c*) the date the employee's continuous employment began;

(*d*) the scale or rate of remuneration and how it is calculated;

(*e*) the intervals when remuneration is paid;

(*f*) terms and conditions relating to hours;

(*g*) terms and conditions relating to holidays;

(*h*) terms relating to sick pay (if any);

(*i*) terms and conditions relating to pensions;

(*j*) notice requirements;

(*k*) job description;

(*l*) title of the job;

(*m*) if the job is not permanent, the period of employment;

(*n*) place of work, or if various the address of the employer;

(*o*) any collective agreements which affect terms and conditions and, if the employer is not a party to the agreements, the persons with whom they were made;

(*p*) if the employee is required to work outside the UK for more than one month, the period he will be required to work, the currency in which he will be paid, any additional benefits paid to him and any terms and conditions relating to his return to the UK.

The terms in (*a*), (*b*), (*c*), (*d*), (*e*), (*f*), (*g*), (*k*), (*l*) and (*n*) must all be contained in a single document. In relation to pensions and sick pay the employer may refer the employee to a reasonably accessible document, and in respect of notice the employer can refer the employee to a reasonably accessible collective agreement or to the *Employment Rights Act 1996, s 86* which contains provisions relating to minimum notice periods.

In addition, if the employer employs more than twenty employees, he must give them written details of any disciplinary and grievance procedures which apply to them. If the employer employs fewer than twenty employees, he must let them know to which person they can take a grievance to – there is no requirement for him to give details of the disciplinary procedures. This exemption for small employers was removed from 1 October 2004.

There is no duty on an employer to give details of any disciplinary or grievance procedures relating to health and safety. Given the employer's duties under the *HSWA 1974, s 2*, however, and given the law relating to unfair dismissal, it is good industrial relations practice to ensure that all employees know of all the disciplinary procedures which could be invoked against them. The written statement must also provide (either by instalments or in one single document) details of whether the employment is contracted out of the State Second Pension.

## Works rules

**E14007** Works rules may or may not be part of the contract. If they are part of the contract and thus contractual terms, they can be altered only by mutual agreement, that is the employee must agree to any change. It is unusual, however, for such rules to be contractual – to be so, there would have to be some reference to them within the contract and an intention that they are terms of the contract. The more usual position with regard to the employer's rules was stated in *Secretary of State for Employment v ASLEF (No 2) [1972] 2 QB 455* where Lord Denning said that they were merely instructions from an employer to an employee. This means that they are non-contractual and the employer can alter the rules without the consent of the employees. The fact that they are not contractual does not mean that they cannot be enforced against an employee. All employees have a duty to obey lawful, reasonable orders (see E14003 above) and thus failing to comply with the rules will be a breach of this duty and therefore a breach of contract. The only requirement that the law stipulates is that the order must be lawful and reasonable and it is unlikely that an order to comply with any health and safety rules would infringe these requirements.

## Disciplinary and grievance procedures

**E14008** It has already been noted that the employer must give details of grievance procedures to all employees. Failing to do so could lead to the employee resigning and claiming constructive dismissal (*W A Goold (Pearmak) Ltd v McConnell [1995] IRLR 516*). In addition, if the employer employs more than 20 employees he must give details of the disciplinary grievance procedures to those employees. This requirement will also apply to those employers with less than 20 employees from 1 October 2004. Many employers adopt the ACAS Code of Practice on Disciplinary Procedures. This gives guidelines as to the sanctions which can be employed for breaches of the employer's rules. The Code incorporates the statutory right to be accompanied by a fellow employee or by a trade union representative during certain grievance and disciplinary proceedings under the *Employment Relations Act 1999, ss 10–15*. This right is enforceable in the employment tribunal and compensation is payable for any failure. It is worth noting that fellow employees are under no duty to perform the role of accompanying individual.

Subject to the above, employers can currently establish their own disciplinary procedures. Such procedures may become part of the contract. If, for example, the employer gives the employee a copy of the procedures with the contract, and the contract refers to the procedures and the employee signs for receipt of the contract and the procedures, it is likely that they will be contractual. Employers may prefer their disciplinary procedures not to be contractual. If the procedures are contractual, any employee will be able to claim that his or her contract has been breached if they are not followed. This possibility also applies to those employees who have been employed for less than the one year qualifying period required to bring a claim for unfair dismissal. In response to a claim of breach of contract, a court may award damages against the employer. These damages are based on an assessment of the time for which, if the procedure had been followed, the employee's employment would have continued.

From 1 October 2004, where an employer wishes to dismiss an employee for any reason or take action (other than suspension or the issue of warnings) on grounds of capability or conduct, the employer must follow a statutory dismissal and disciplinary procedure (DDP). A modified DDP applies to immediate 'gross misconduct' dismissals. If an employer fails to follow the statutory DDP, a tribunal will be able to increase the compensation awarded to an employee by up to 50 per cent, although this cannot take an award above the statutory cap (see below at E14022). It will be automatically unfair for an employer to dismiss an employee without completing the relevant DDP. Employees will receive a minimum of four-weeks' pay as compensation.

From 1 October 2004, an employee or ex-employee who has a grievance which could form the basis of a tribunal complaint will not be able to bring a tribunal claim based on that grievance unless they have raised the grievance with their employer under the statutory grievance procedure (GP). Having raised a grievance under the statutory GP, an employee must wait 28 days before lodging a tribunal application otherwise the application will be refused. If an employee attempts to lodge a tribunal claim without having raised the matter as a grievance and waited 28 days, the tribunal claim will be rejected but the employee will be automatically granted an extension of three months which is added to the original time limit of three months.

If an employee raises a grievance under the statutory GP and the employer fails to follow the statutory GP:

(*a*)   the admissibility criteria will not apply; and

(*b*)   the tribunal will be able to increase any award it makes by up to 50 per cent.

The principles underlying both the statutory disciplinary and grievance procedures are:

- There should be a letter setting out the problem.

- There should be time provided for the recipient to consider and understand it.

- The employer and employee should meet to discuss the problem. The employee has a right to be accompanied at this meeting.

- The employer's decision should be notified to the employee.

- An appeal should be held if the issue has not been settled.

ACAS has also issued a draft Revised Code of Practice on Disciplinary and Grievance Procedures which take into account the new statutory rules. The revised draft Code of Practice came into effect on 1 October 2004.

# Employee employment protection rights

## Right not to suffer a detriment in health and safety cases

E14009    By the *Employment Rights Act 1996, s 44* every employee has the right not to be subjected to a detriment, by any act or any failure to act, by his employer on the grounds that:

(a)   having been designated by the employer to carry out activities in connection with preventing or reducing risks to health and safety at work, the employee carried out (or proposed to carry out) any such activities;

(b)   being a representative of workers on matters of health and safety at work or a member of a safety committee –

(i)   in accordance with arrangements established under or by virtue of any enactment; or

(ii)  by reason of being acknowledged as such by the employer;

the employee performed (or proposed to perform) any functions as such a representative or a member of such committee;

(ba)  the employee took part (or proposed to take part) in consultation with the employer pursuant to the *Health and Safety (Consultation with Employees) Regulations 1996 (SI 1996 No 1513)* or in an election of representatives of employee safety within the meaning of those Regulations (whether as a candidate or otherwise);

(c)   being an employee at a place where –

(i)   there was no such representative or safety committee; or

(ii)  there was such a representative or safety committee but it was not reasonably practicable for the employee to raise the matter by those means;

he brought to his employer's attention, by reasonable means, circumstances connected with his work which he reasonably believed were harmful or potentially harmful to health or safety;

(d)   in circumstances of danger which the employee reasonably believed to be serious and imminent and which he could not reasonably have been expected to avert, he left (or proposed to leave) or (while the danger persisted) refused to return to his place of work or any dangerous part of his place of work; or

(e)   in circumstances of danger which the employee reasonably believed to be serious and imminent, he took (or proposed to take) appropriate steps to protect himself or other persons from the danger.

In considering whether the steps the employee took or proposed to take under (e) were reasonable, the court must have regard to all the circumstances including the employee's knowledge and the facilities and advice available to him (*Employment Rights Act 1996, s 44(2)*). In *Kerr v Nathan's Wastesavers Ltd (1995) IDS Brief 548*, however, the Employment Appeal Tribunal stressed that tribunals should not place too onerous a duty on the employee to make enquiries to determine if his belief is reasonable. Danger under (d) does not necessarily have to arise from the circumstances of the workplace, but can include the risk of attack by a fellow employee (*Harvest Press Ltd v McCaffrey [1999] IRLR 778*). Danger under (e) can include danger to others as well as the employee himself (*Mosiak v City Restaurants (UK) Ltd [1999] IRLR 180*). Various actions by the employer can constitute a detriment to the employee (such as disciplining the employee). Likewise, a failure to act on the part of the employer can also constitute a detriment (for example not sending the employee on a training course). Furthermore, the section is not restricted to the health and safety of the employee or his colleagues. In *Barton v Wandsworth Council (1995) IDS Brief 549* a tribunal ruled that an employee had been unlawfully disciplined when he voiced concerns over the safety of patients due to what he considered to be the lack of ability of newly introduced escorts. This shows that the legal protection is triggered in relation to any health and safety issue and includes cases where the employee voices concerns, and is not limited to only those circumstances where the employee commits more positive action. The *Employment Rights Act 1996, s 44(3)*, however, provides that an employee is not to be regarded as subjected to a detriment if the employer can show that the steps the employee took or proposed to take were so negligent that any reasonable employer would have treated him in the same manner. Furthermore, if the detriment suffered by the employee is dismissal, there is special protection under *s 100* (see below).

If the employee should suffer a detriment within the terms of *s 44* he may present a complaint to an employment tribunal (*Employment Rights Act 1996, s 48*). The complaint must be presented within three months of the act (or failure to act) complained of, or, if there is a series of acts, within three months of the date of the last act. The tribunal has a discretion to waive this time limit if it was not reasonably practicable for the employee to present his complaint in time. If the tribunal finds the complaint well founded, it must make a declaration to that effect and may make an award of compensation to the employee, the amount of compensation being what the tribunal regards as just and equitable in all the circumstances (*Employment Rights Act 1996, s 49(2)*). The amount of compensation shall take into account any expenses reasonably incurred by the employee in consequence of the employer's action and any loss of benefit caused by the employer's action. Compensation can be reduced because of the employee's contributory conduct.

The protection from being dismissed or subjected to a detriment on the health and safety grounds specified in *s 44* and *s 100* of the *Employment Rights Act 1996* has been reinforced by a new, more general protection for whistleblowers.

*The Public Interest Disclosure Act 1998* provides protection to workers who make disqualifying disclosures about health and safety matters, as well as about criminal acts, failure to comply with legal obligations, miscarriages of justice, damage to the environment and deliberate concealment of any of these matters. Under the Act, which inserts new sections into the *Employment Rights Act 1996*, a worker who makes a 'qualifying disclosure' which he reasonably believes shows one of these matters, may be protected against dismissal or being subjected to a detriment.

Disclosures are only protected if they are made to appropriate persons – which often means that the employer must to be approached in the first instance. There are other possibilities available under the Act: disclosure to a legal adviser, disclosure to a prescribed person (e.g. the FSA or the Commissioners of the Inland Revenue), and a more general category for disclosures made provided that certain specified conditions are met.

In terms of health and safety risks, protection for qualifying disclosures is not limited to cases of imminent or serious danger – it can apply where the health and safety of any individual has been, is being or is likely to be, endangered. In all cases the worker must have a reasonable belief and make the disclosure in good faith (except in the case of disclosure to a legal adviser).

If a disclosure is protected, and an employee is subjected to any detriment or dismissed as a result, it is unlawful. A dismissal in these circumstances is deemed to be automatically unfair and the tribunal is not required to consider whether or not the employer's actions were reasonable. There is no minimum qualifying period for entitlement to make an unfair dismissal claim for this reason, no upper age limit applies and there is no limit on the compensation available to whistleblowers who are unfairly dismissed because they have made a protected disclosure.

## *Dismissal on health and safety grounds*

**E14010** In addition to the normal protection against dismissal (see below), where an employee is dismissed (or selected for redundancy) and the reason or principal reason for the dismissal is one of the grounds listed in the *Employment Rights Act 1996, s 44*, the dismissal will be automatically unfair. The only defence available to an employer applies to a dismissal taken by the employee to protect himself or others from danger that the employee reasonably believed was serious and imminent (*Employment Rights Act 1996, ss 44(1)(e)* and *100(1)(e)*). The employer can escape a finding of unfair dismissal if he can show that the actions taken or proposed by the employee were so negligent that any reasonable employer would have dismissed the individual concerned. In respect of a dismissal falling within *s 100*, the normal qualifying period of employment does not apply nor does the upper age limit (*Employment Rights Act 1996, ss 108(3)(c) and 109(2)(c)*). Thus an employee who has only been employed for a few weeks or who is over the normal retirement age for the job can claim unfair dismissal for a breach of *s 100*. A dismissal which is not automatically unfair under *s 100* may nevertheless be unfair under the general reasonableness test under *s 98*.

There is no limit on the amount of compensation that may be awarded for an unfair dismissal on health and safety grounds.

## Dismissal for assertion of a statutory right

**E14011** By the *Employment Rights Act 1996, s 104*, an employee is deemed to be unfairly dismissed where the reason or principal reason for that dismissal was that the employee –

(*a*) brought proceedings against an employer to enforce a right of his which is a relevant statutory right, or

(*b*) alleged that the employer had infringed a right of his which is a relevant statutory right.

It is immaterial whether or not the employee has the right or whether or not the right has been infringed, as long as the employee made it clear to the employer what the right claimed to have been infringed was and the employee's claim is made in good faith. A statutory right for the purposes of the section is any right under the *Employment Rights Act 1996* in respect of which remedy for infringement is by way of complaint to an employment tribunal, a right under *s 86* of the 1996 Act (minimum notice requirements), rights in relation to trade union activities under the *Trade Union and Labour Relations* (*Consolidation*) *Act 1992* and the rights conferred by the *Working Time Regulations 1998* (*SI 1998 No 1833*).

This is an important right for employees. If, for example, after the employee has successfully claimed compensation from his employer for a breach of s *44* he is dismissed, the dismissal will be automatically unfair under *s 104*. Again, if the employer unlawfully demotes or suspends without pay as a disciplinary sanction for breach of health and safety rules, and after proceedings against him for an unlawful deduction from wages the employer dismisses the employee, this will be unfair under *s 104*. As with dismissal in health and safety cases under *s 100*, the normal qualifying period of employment does not apply – neither does the upper age limit.

# Enforcement of safety rules by the employer

## The rules

**E14012** Given the statutory duty on the employer, under the *HSWA 1974, s 2*, to have a written statement of health and safety policy, and the common law duty on the employer to ensure his employees' safety, the employer should lay down contractual health and safety rules, breach of which will lead to disciplinary action against the employee. These rules must be communicated to the employee and be clear and unambiguous so that the employee knows exactly what he can and cannot do.

The employer's disciplinary rules will often classify misconduct, e.g. as minor misconduct, serious misconduct and gross misconduct. It is unlikely that a tribunal would uphold as fair a dismissal for minor misconduct. It will underline the importance of health and safety rules if their breach is deemed to be serious or gross misconduct. The tribunal will, however, look at all the circumstances of the case – it does not automatically follow, therefore, if an employer has stated that a breach of a particular rule will be gross misconduct, that a tribunal will find a resultant dismissal fair.

## The procedures

**E14013** Once an employer has laid down his rules, he must ensure that he has adequate procedures (which should be non-contractual) to deal with a breach. The procedures used by an employer are scrutinised by a tribunal in any unfair dismissal claim and past cases indicate that many employers have lost such claims due to inadequate procedures. As discussed below, an employer in an unfair dismissal claim must show the tribunal that he acted reasonably. This concentrates on the fairness of the employer's actions and not on the fairness to the individual employee (*Polkey v A E Dayton Services Ltd [1987] IRLR 503*). This means that an employer cannot argue that a breach of procedures has made no difference to the final outcome and that he would have dismissed the employee even if he had adhered to his procedure. Breach of procedures themselves by an employer is likely to render a dismissal unfair regardless of which rule was broken.

Many employers adopt the ACAS procedures. Essentially any disciplinary procedure should contain three elements: an investigation, a hearing and an appeal and should observe the principles of natural justice.

### (a) Investigation

The law requires that the employer has a genuine belief in the employee's 'guilt', and that the belief is based on reasonable grounds after a reasonable investigation (*British Home Stores v Burchell [1978] IRLR 379*). If the employer suspends the employee during the investigation, this suspension should be with pay and in accordance with the disciplinary procedure. An investigation is important because it may reveal defects in the training of the employee, or reveal that the employee was not told of the rules, or that another employee was responsible for the breach. In all of these cases, disciplinary action against the suspended employee will be unfair. Any investigation should be as thorough as possible and documented. It should also take place as soon as possible since memories fade quickly and this is particularly important if other employees are to be questioned as witnesses. Likewise, taking too long to start an investigation may lead the employee to think that no action will be taken and to then discipline him may itself be unfair. No disciplinary action should be taken until a careful investigation has been concluded.

### (b) Hearing

Once the employer has investigated, he must conduct a hearing to make a decision as to the sanction, if any, he will impose. To act fairly, the employer must comply with the rules of a fair hearing. These are as follows.

(i) The employee must know the case against him to enable him to answer the complaint. This also means that the employee should be given sufficient time before the hearing with copies of relevant documents to enable him to prepare his case.

(ii) The employee should have an opportunity to put his side of the case, i.e. the employer should listen to the employee's side of the story and allow the employee to put forward any mitigating circumstances.

(iii) The employee must be allowed to be accompanied at the hearing by a fellow employee or a trade union representative of his choice (*Employment Relations Act 1999*).

(iv) The hearing should be unbiased, i.e. the person chairing the hearing should come to it with an open mind and not have prejudged the issue.

(v) The employee should be provided with an explanation as to why any sanctions are imposed.

(vi) The employee should be informed of his right to appeal (and the way in which he should go about it) to a higher level of management which has not been involved in the first hearing. If the employee fails to exercise his right of appeal, however, he will not have failed to mitigate his loss, if ultimately a tribunal finds that he has been unfairly dismissed and thus his compensation will not be reduced (*William Muir (Bond 9) Ltd v Lamb [1985] IRLR 95*). Failing to allow an employee to exercise a right of appeal will almost certainly render any dismissal unfair (*West Midlands Co-operative Society Ltd v Tipton [1986] IRLR 112*).

## (c) Appeal

In an unfair dismissal case a tribunal is required to consider the reasonableness of the employer's action taking into account the resources of the employer and the size of the employer's undertaking. This means that in the case of all but very small undertakings, the tribunal will expect the employer to have provided an appeal for the employee. All the rules of a fair hearing equally apply to an appeal. Only an appeal which is a complete rehearing of the case (rather than merely a review of the written notes of the disciplinary hearing) can rectify procedural flaws committed earlier on in the procedure (*Jones v Sainsbury's Supermarkets Ltd (2000) (unreported)*). An appeal, however, cannot endorse the sanction imposed by the earlier hearing for a different reason, unless the employee has had notice of the new reason and has been given an opportunity to put forward his argument in respect of it.

With effect from 1 October 2004, employers must also ensure that their procedure complies with the statutory dismissal and disciplinary procedure (see E14008 above).

## Sanctions other than dismissal

**E14014** There are a variety of sanctions apart from dismissal that an employer may impose. It is important however that the 'punishment fits the crime'. The imposition of too harsh a sanction may entitle the employee to resign and claim constructive dismissal (see below).

## (a) Warnings

The ACAS Code recommends three warnings in cases of normal misconduct before dismissing: the first oral, the second written and a final written warning stating that a repetition will result in dismissal. These are only guidelines, however, and it clearly depends on the circumstances of the case. A minor breach of a health and safety rule, for example, may justify a final written warning given the potential seriousness and consequences of breaches of such rules. The ACAS Code urges that, apart from gross misconduct, no employee should be dismissed for a first breach of discipline, although, again, breaches of health and safety rules have been held to be gross misconduct. The Code also recommends that warnings should remain on the employee's record for a definite period of time (six to twelve months). Once this time has expired, the warnings will be ignored when looking to see if the procedure has been followed in later cases of misconduct, but can be considered when the employer is looking at the employee's work record to decide what sanction to impose.

## (b) Fines or deductions

The employer must have contractual authority or the written permission of the employee before he can make a deduction from the employee's wages as a disciplinary sanction. Deducting without such authority is a breach of the *Employment Rights Act 1996, s 13* and gives the employee the right to sue for recovery in the employment tribunal. It will also lead to a potential constructive dismissal claim.

## (c) Suspension without pay

Any suspension without pay will have the same consequences as a fine or deduction if there is no contractual authority or written authorisation from the employee to impose such a sanction.

## (d) Demotion

Most demotions will involve a reduction in pay, and thus without written or contractual authority to demote the employer will be in breach of the *Employment Rights Act 1996, s 13* and liable to a constructive dismissal claim.

### Suitable alternative work

E14015   Where an employee is suspended from work on maternity grounds, the employer must offer available suitable alternative work. Alternative work will only be suitable if:

(*a*)  the work is of a kind which is both suitable in relation to the employee and appropriate for the employee to do in the circumstances; and

(*b*)  the terms and conditions applicable for performing the work are not substantially less favourable than corresponding terms and conditions applicable for performing the employee's usual work (*Employment Rights Act 1996, s 67*).

If an employer fails to offer suitable alternative work, the employee may bring a claim before an employment tribunal which can award 'just and equitable' compensation. Such complaint must normally be lodged within three months of the first day of the suspension (*Employment Rights Act 1996, s 70(4)*).

### Remuneration on suspension from work

E14016   An employee who is suspended on medical grounds is entitled to normal remuneration for up to 26 weeks. An employee who is suspended on maternity grounds, if no suitable alternative work is available, is entitled to normal remuneration for the duration of the suspension. However, in either case, if the employee unreasonably refuses an offer of suitable alternative work, no remuneration is payable for the period during which the offer applies. An employee may bring a complaint to an employment tribunal if an employer fails to pay the whole or any part of the remuneration to which the employee is entitled (*Employment Rights Act 1996, ss 64, 68, 70(1)*). (See *British Airways Ltd v Moore [2000] IRLR 296*, in which the Employment Appeal Tribunal upheld a purser's claim to a flying allowance on the basis that suitable work must be on terms and conditions not substantially less favourable.)

## Dismissal

E14017   Dismissal is the ultimate sanction that an employer can impose for breach of health and safety rules. All employees are protected against wrongful dismissal at common law, but, in addition, some employees have protection against unfair dismissal. The protection against unfair dismissal comes from statute (the *Employment Rights Act 1996*) and therefore the employee must satisfy any qualifying criteria laid down by the statute before he can claim. Given that the protection against wrongful and unfair dismissal rest alongside each other, an employee may claim for both, although he will not be compensated twice. Wrongful dismissal is based on a breach of contract by the employer and compensation will be in the form of damages for that breach – that is, the damage the employee has suffered because the employer did not comply with the contract. Unfair dismissal, on the other hand, is statute based and is not dependent on a breach of contract by the employer. Compensation for such dismissal is based on a formula within the statute (see E14022 below).

### Wrongful dismissal

E14018   A dismissal at common law is where the employer unilaterally terminates the employment relationship with or without notice. A wrongful dismissal is where the employer terminates the contract in breach, for example, by giving no notice or shorter notice than is required by the employee's contract and the employee's conduct does not justify this. An employer is entitled to dismiss without notice only if the employee has committed gross misconduct. In all other circumstances the employer must give contractual notice to end the relationship, or pay wages in lieu of notice. However, this does require qualification. Firstly, the law decides what is gross misconduct and not the employer. Just because the employer has stated that certain actions are gross misconduct does not mean that the law will regard it as such. Only very serious misconduct is regarded by the law as gross, such as refusing to obey lawful and reasonable orders, gross neglect, theft. Secondly, contractual notice periods are subject to the statutory minimum notice provisions contained in the *Employment Rights Act 1996, s 86*. Any attempt by the contract to give less than the statutory minimum notice is void. These periods apply to all employees who have been employed for one month or more and are:

(a) not less than one week if the employee has been employed for less than two years;

(b) after the employee has been employed for two years, one week for each year of service, up to a statutory maximum of 12 weeks.

Where an employer terminates the contract and pays the employee in lieu of notice, in the absence of an express right to do so, this will be a technical breach of contract. The employee can waive his right to notice or accept wages in lieu of notice. If the contract gives notice periods which are longer than the statutory minimum, the contractual notice prevails. Therefore, if the employer has an employee who has been employed for six years, and the employer sacks him with four weeks' notice, the employee can sue for a further two weeks' wages in the employment tribunal. Finally, if the employer fundamentally alters the terms of the employee's contract, without his consent, in reality the employer is terminating (repudiating) the original contract and substituting a new one. The employee should therefore be given the correct notice before the change comes into effect. A unilateral change by the employer to a fundamental term of the contract (followed by resignation by the employee) will amount to constructive dismissal (i.e. repudiation) and compensation in the form of damages for breach of contract. The employee must take all reasonable steps to mitigate his loss by seeking other employment.

## Unfair dismissal

### Dismissal

**E14019** While all employees are protected against wrongful dismissal, generally employees must be employed for one year or more before they gain protection against unfair dismissal. In certain circumstances, however, an employee is protected immediately and does not need a year of employment. One of these is dismissal on certain health and safety grounds discussed above.

Once an employee is protected against unfair dismissal the *Employment Rights Act 1996, s 95* recognises three situations which the law regards as dismissal. These are:

(a) the employer terminating the contract (with or without notice);

(b) a fixed term contract which expires and is not renewed;

(c) the employee resigning in circumstances in which he is entitled to do so without notice because of the employer's conduct – a constructive dismissal.

In the first situation the employer is unilaterally ending the relationship. Even if the employer gives the correct amount of notice so that the dismissal is lawful, it does not necessarily follow that the dismissal will be fair.

The second situation needs no explanation. If a fixed term contract has come to an end and is not renewed, this is, in effect, the employer deciding to end the relationship. As from 25 October 1999, it is no longer possible for employees to validly waive unfair dismissal rights in fixed term contracts (*Employment Relations Act 1999*).

The third situation, constructive dismissal, is much more complex. On the face of it the employee has resigned. However if the reason for his resignation is the employer's conduct, then the law treats the resignation as an employer termination. The action on the part of the employer which entitles the employee to resign and claim constructive dismissal is a repudiatory breach of contract. In other words, the employer has committed a breach which goes to the root of the contract and has, therefore, repudiated it. This means that not all breaches by the employer are constructive dismissals but that serious breaches may be. It is also important to recognise that, as discussed above, the terms of the contract may include those which have not been expressly agreed by the parties and therefore rules, disciplinary procedures, terms collectively bargained, and all the implied duties discussed in E14003 above, may all be contractual terms. Breach of the health and safety duties owed to all employees is likely to give rise to a constructive dismissal claim (*Day v T Pickles Farms Ltd [1999] IRLR 217*). In addition, the law requires as an implied term of the contract that both the employer and employee treat each other with mutual respect and do nothing to destroy the trust and confidence each has in the other. Breach of this duty may give rise to a constructive dismissal claim. In one case, a demotion imposed as a disciplinary sanction was held to be excessive by the Employment Appeal Tribunal. Its very excessiveness was a breach of the duty of mutual respect which entitled the employee to resign and claim constructive dismissal. It has also been held that an employer's disclosure in a reference on behalf of an employee of complaints against the employee, before first giving the individual an opportunity to explain, was a fundamental breach of the implied term of mutual trust and confidence which amounted to constructive (since the employee resigned) and unfair dismissal (*TSB Bank v Harris [2000] IRLR 157*). In *(1) Reed (2) Bull Information Systems Ltd v Stedman [1999] IRLR 299*, the Employment Appeal Tribunal held that in a case where the employer was aware of an employee's deteriorating health, and the employee concerned had complained to colleagues at work about harassment, it was encumbent on the employer to investigate and their failure to do so was enough to justify a finding of breach of trust and confidence and thus constructive dismissal.

Where an employee with one or more year's continuous service (unless one of the specified reasons apply which render a dismissal automatically unfair) is constructively dismissed, he or she will also have the right to claim unfair dismissal. Obviously, the employee must resign before he can make a claim for unfair dismissal. In the majority of cases the repudiatory breach by the employer is a fundamental alteration of the contractual terms (for example hours). In this situation, the employer still wishes to continue the relationship, albeit on different terms. The employee has two choices: he can resign or he can continue to work under the new terms. If the employee continues to work and accepts the changed terms, the contract is mutually varied and no action will lie, provided that the employee was given the correct notice before the change was implemented. If the employee resigns, however, he will have been dismissed. In *Walton & Morse v Dorrington [1997] IRLR 488*, the employee waited to find alternative employment before she resigned. The Employment Appeal Tribunal decided that, in her circumstances, this was a reasonable thing to have done and agreed that she had not accepted her employer's breach of its duty to her and had been constructively dismissed (see E14003 above).

## Reasons for dismissal

**E14020** Section 98 of the *Employment Rights Act 1996* gives five potentially fair reasons for dismissal. These are:

(*a*) capability or qualifications;

(*b*) conduct;

(*c*) redundancy;

(*d*) contravention of statute;

(*e*) some other substantial reason.

Dismissal on health and safety grounds could potentially fall within most of these reasons. It should, however, be remembered that, where an employee is dismissed in circumstances where continued employment involves a risk to the employee's health and safety, the employer may nevertheless face claims of unfair dismissal. Before terminating employment, an employer should consider all the circumstances of the case and assess the risk involved and take measures which are reasonably necessary to eliminate the risk.

Illness may make it unsafe to employ the employee; breach of health and safety rules will normally fall under misconduct; to continue to employ the employee may contravene health and safety legislation or it may be that the employer has had to reorganise his business on health and safety grounds and the employee is refusing to accept the change. This latter situation could be potentially fair under 'some other substantial reason'. (However, an employer may be liable under the *Disability Discrimination Act 1995*.)

## Reasonableness

**E14021**   Merely having a fair reason to dismiss does not mean that the dismissal is fair. Section 98(*4*) of the *Employment Rights Act 1996* requires the tribunal in any unfair dismissal case to consider whether the employer acted reasonably in all the circumstances (including the size and administrative resources of the employer's undertaking). This means that the tribunal will look at two things – (i) was the treatment of the employee procedurally fair, and (ii) was dismissal a reasonable sanction in relation to the employee's actions and all the circumstances of the case.

Procedures have already been discussed at E14013 above. If the employer has complied with his procedures, he will not be found to have acted procedurally unfairly unless the procedures themselves are unfair. This is unlikely if the employer is following the ACAS Code.

In respect of the fairness of the decision, the tribunal should consider whether the employer's decision to dismiss fell within the band of reasonable responses to the employee's conduct which a reasonable employer could adopt. The tribunal will look at three things – (i) has the employer acted consistently, (ii) has he taken the employee's past work record into account, and (iii) has the employer looked for alternative employment. The latter aspect is of major importance in relation to redundancy, incapability due to illness or dismissal because of a contravention of legislation, but will not usually be relevant in dismissals for misconduct. It is important to note in addition that special obligations apply to an employer in the case of a disabled employee within the meaning of the *Disability Discrimination Act 1995*.

When looking at consistency, the tribunal will look for evidence that the employer has treated the same misconduct the same way in the past. If the employer has treated past breaches of health and safety rules leniently it will be unfair to suddenly dismiss for the same breach, unless he has made it clear to the employees that his attitude has changed and breaches will be dealt with more severely in the future. Employees have to know the potential disciplinary consequences for breaches of the rules, and if the employer has never dismissed in the past he is misleading employees unless he tells them that things have changed. The law, however, only requires an employer to be consistent between cases which are the same. This is where a consideration of the employee's past work record is important. It is not inconsistent to give a long-standing employee with a clean record a final warning for a breach of health and safety rules and to dismiss another shorter-serving employee with a series of warnings behind him, as long as both employees know that the penalty for breach of the rules could be dismissal. The cases are not the same. It would, however, be unfair if both the employees had the same type of work record and length of service and only one was dismissed, and dismissal had never been imposed as a sanction for that type of breach in the past. In order for a misconduct dismissal to be fair, the employer must have had a reasonable belief in the guilt of the employee of the misconduct in question on the basis of a reasonable investigation.

## Remedies for unfair dismissal

### (a) Reinstatement

**E14022** The first remedy that the tribunal is required to consider is reinstatement of the employee. When doing so the tribunal must take into account whether the employee wishes to be reinstated, whether it is practicable for the employer to reinstate him and, if the employee's conduct contributed to or caused his dismissal, whether it is just to reinstate him. In order to resist an order for reinstatement, an employer must provide evidence to show that it is not practicable because the implied term of mutual trust and confidence between employer and employee has broken down (*IPC Magazines Ltd v Clements, EAT/456/99; Gentle & Ors v Perkins Group Ltd, EAT/670/99*). Reinstatement means that the employee must return to his old job with no loss of benefits. If reinstatement is ordered and the employer refuses to comply with the order, or only partially complies, compensation will be increased. Reinstatement, however, is rarely ordered by tribunals.

### (b) Re-engagement

If the tribunal does not consider that reinstatement is practicable, it must consider whether to order the employer to re-engage the employee. In making its decision the tribunal looks at the same factors as when it considers reinstatement. Re-engagement is an order requiring the employer to re-employ the employee on terms which are as favourable as those he enjoyed before his dismissal, but it does not require the employer to give the employee the same job back. Failure on the part of the employer to comply with an order of re-engagement will lead to increased compensation, although tribunals rarely make re-engagement orders.

### (c) Compensation

Compensation falls under a variety of different heads. In an unfair dismissal case the employee will receive:

**Basic award**: This is based on his age, years of service and salary –

(i) one and a half weeks' pay for each year of service over the age of 41;

(ii) one week's pay for each year of service between 41 and 22;

(iii) half a week's pay for each year of service below the age of 22.

This is subject to a statutory maximum of £270 a week, and a maximum of twenty years' service. Compensation is reduced by one-twelfth for each month the employee works during his 64th year. Where the employee is unfairly dismissed for health and safety reasons under the *Employment Rights Act 1996, s 100*, the minimum basic award is £3,600 (at present).

**Compensatory award**: This is payable in addition to the basic award to compensate the employee for loss of future earnings, benefits etc, which are in excess of the basic award. As with the basic award the compensatory award can be reduced for contributory conduct. The present maximum compensatory award is £55,000.

**Additional award**: If the employer fails to comply with a reinstatement or re-engagement order, the tribunal may make an additional award. This will be between 26 and 52 weeks' pay (subject to a maximum of £270 per week).

**Note**: There is no limit on:

- the compensatory award for employees who are dismissed for health and safety reasons (see E14010);
- awards of compensation in sex discrimination and equal pay cases (and interest can be included in such awards);

- the amount of the compensation that a tribunal may award in a case of discrimination under the *Disability Discrimination Act 1995*.

In cases of sex and disability discrimination, compensation may include compensation for injury to feelings.

# Health and safety duties in relation to women at work

## Sex discrimination

**E14023** Since health and safety issues may give rise to sex discrimination claims under the *Sex Discrimination Act 1975* (*SDA 1975*), it is appropriate to examine what particular considerations an employer needs to bear in mind in its relations with female employees.

The steps necessary to be taken by an employer, in order to comply with his duties under *HSWA 1974, s 2*, may differ for women.

New or expectant mothers are particularly vulnerable to adverse or indifferent working conditions. Indeed, most employers have probably taken measures to guard against risks to new and expectant mothers, in accordance with their general duties under *HSWA 1974, s 2*, and the *Management of Health and Safety at Work Regulations 1999* (*SI 1999 No 3242*). In addition, employers are required to protect new and expectant mothers in their employment from certain specified risks if it is reasonable to do so, and to carry out a risk assessment of such hazards. If the employer cannot avoid the risk(s), he must alter the working conditions of the employee concerned or the hours of work, offer suitable alternative work and, if no suitable alternative work is available, suspend the employee on full pay (see E14024 below).

Although *SDA 1975* prohibits discrimination on grounds of sex, *s 51(1)* provides that any action taken to comply with certain existing health and safety legislation (e.g. *HSWA 1974*) will *not* amount to unlawful discrimination. In *Page v Freight Hire (Tank Haulage) Ltd [1981] IRLR 13*, the complainant was an HGV driver. The employer, acting on the instructions of the manufacturer of the chemical dimethylformamide (DMF), refused to allow her to transport the chemical which was potentially harmful to women of child bearing age. She brought a complaint of sex discrimination. It was held that the fact that the discriminatory action was taken in the interests of safety did not of itself provide a defence to a complaint of unlawful discrimination. However, the employer was protected by *s 51(1)* of *SDA 1975* because the action taken was necessary to comply with the employer's duty under *HSWA 1974*.

## Pregnant workers, new and breastfeeding mothers

**E14024** *The Management of Health and Safety at Work Regulations 1999* (*SI 1999 No 3242*) impose a duty on employers to protect new or expectant mothers from any process or working conditions or certain physical, chemical and biological risks at work (see E14025 below). The phrase 'new or expectant mother' is defined as a worker who is pregnant, who has given birth within the previous six months, or who is breastfeeding. 'Given birth' is defined as having delivered a living child or, after 24 weeks of pregnancy, a stillborn child.

### Risk assessment

**E14025** The 1999 Regulations require employers to carry out an assessment of the specific risks posed to the health and safety of pregnant women and new mothers in the workplace and then to take steps to ensure that those risks are avoided. Risks include those to the unborn child or child of a woman who is still breastfeeding – not just risks to the mother.

An interesting development in relation to this requirement is the case of *Day v T Pickles Farms Ltd [1999] IRLR 217*, where the employee suffered nausea when pregnant as a result of the smell of food at her workplace. As a result of the nausea, she was unable to work and was eventually dismissed after a prolonged absence.

The Employment Appeal Tribunal found that she had not been constructively dismissed. However, the Employment Appeal Tribunal held that the obligation to carry out a risk assessment which considers possible risks to the health and safety of a pregnant female employee is relevant from the moment an employer employs a woman of childbearing age. The question of whether the applicant had been subjected to a detriment was remitted to the employment tribunal.

The Health and Safety Executive has published a booklet entitled '*New and expectant mothers at work – A guide for employers*'. The booklet provides guidance on what employers need to do to comply with the legislation. The booklet includes a list of the known risks to new and expectant mothers and suggests methods of avoidance.

The main risks to be avoided are as follows:

- Physical agents:
    - shocks/vibrations/movement (including travelling and other physical burdens),
    - handling of loads entailing risks,
    - noise,
    - non-ionising radiation,
    - extremes of heat and cold.
- Biological agents:
    - such as listeria, rubella and chicken pox virus, toxoplasma, cytomegalovirus, hepatitis B and HIV.
- Chemical agents:
    - such as mercury, antimiotic drugs, carbon monoxide, chemical agents of known and percutaneous absorption and chemicals listed under various Directives.
- Working conditions:
    - such as mining work and work with display screen equipment (VDUs).

Where a risk has been identified following the assessment, affected employees or their representatives should be informed of the risk and the preventive measures to be adopted. The assessment should be kept under review.

In particular, employers must consider removing the hazard or seek to prevent exposure to it. If a risk remains after preventive action has been taken, the employer must take the following course of action:

(i) temporarily adjust her working conditions or hours of work (*Management of Health and Safety at Work Regulations 1999* (*SI 1999 No 3242*), *Reg 16(2)*).

If it is not reasonable to do so or would not avoid the risk:

(ii) offer suitable alternative work (*Employment Rights Act 1996, s 67*).

If neither of the above options is viable:

(iii) suspend her on full pay for as long as necessary to protect her health and safety or that of her child (*Management of Health and Safety at Work Regulations 1999* (*SI 1999 No 3242*), *Regs 16(2)* and *16(3)*; *Employment Rights Act 1996, s 67*).

Appendix 1 of the booklet lists aspects of pregnancy such as morning sickness, varicose veins, increasing size etc. that may affect work and which employers may take into account in considering working arrangements for pregnant and breastfeeding workers. These are merely suggestions and not requirements of the law.

## Night work by new or expectant mother

E14026 Where a new or expectant mother works at night and has been issued with a certificate from a registered doctor or midwife stating that night work would affect her health and safety, the employer must first offer her suitable alternative daytime work, and suspend her as detailed at E14015 above if no suitable alternative employment can be found (*Management of Health and Safety at Work Regulations 1999* (*SI 1999 No 3242*), *Reg 17*).

## Notification

E14027 An employer is not required to alter a woman's working conditions or hours of work or suspend her from work under *Management of Health and Safety at Work Regulations 1999* (*SI 1999 No 3242*), *Reg 16(2)* or (*3*) until she notifies him in writing that she is pregnant, has given birth within the previous six months or is breastfeeding. Additionally, the suspension or amended working conditions do not have to be maintained if the employee fails to produce a medical certificate confirming her pregnancy in writing within a reasonable time if the employer requests her to do so. The same applies once the employer knows that the employee is no longer a new or expectant mother or if the employer cannot establish whether she remains so (*Management of Health and Safety at Work Regulations 1999* (*SI 1999 No 3242*), *Reg 18*). However, an employer has a general duty under *HSWA 1974* and the *Management of Health and Safety at Work Regulations 1999* to take steps to protect the health and safety of a new or expectant mother, even if she has not given written notification of her condition.

## Maternity leave

E14028 All pregnant workers have a right to 26 weeks' ordinary maternity leave, regardless of length of service and number of hours worked. Employees who have, at the beginning of the 14th week before the expected week of childbirth, being continuously employed for a period of not less than 26 weeks may be entitled to additional maternity leave. The ordinary maternity leave period starts from:

(*a*) the notified date of commencement; or

(*b*) the first day of absence because of pregnancy or childbirth after the beginning of the fourth week before the expected week of confinement; or

(*c*) the date of childbirth,

which ever is the earlier, and continues for 26 weeks or until the end of the compulsory leave period if later. Additional maternity leave continues until the end of the period of 26 weeks from the day after the last day or ordinary maternity leave.

If an employee is prohibited from working for a specified period after childbirth by virtue of a legislative requirement (e.g. under the *Public Health Act 1936, s 205*), her maternity leave period must continue until the expiry of that later period. The *Employment Rights Act 1996, s 72(1)*, provide that an employee entitled to maternity leave should not work or be permitted to work by her employer during the period of two weeks beginning with the date of childbirth.

# Enforcement

## Introduction

**E15001** Prior to the introduction of the *Health and Safety At Work etc Act 1974* ('*HSWA*'), the principal sanction against health and safety offences was prosecution. The HSWA provides enforcing authorities with a wider range of enforcement powers, which do not necessarily rely on prosecution for their efficacy; most notably the power to serve Improvement and Prohibition Notices. The HSE aims to achieve effective enforcement that is fair and proportionate to the risk. The HSE's current approach to enforcement stems from the Health and Safety Commissions' 'Strategy for Workplace Health and Safety in Great Britain to 2010 and beyond' launched in February 2004. The HSE Enforcement Guide is available on http://www.hse.gov.uk/enforce/enforcementguide/intro.htm. Whilst this guidance is informative on how HSE uses its discretion in making decision on enforcement it does not override the requirement for HSE to consider each case in accordance with the Code for Crown Prosecutors, the HSE Enforcement Policy Statement and the law.

The Lord Chancellor has stated in his budget speech on March 15 2005, that the recommendations laid out in a review of enforcement carried out by Philip Hampton, will be adopted in full. The report is based on a distinction between the 'policy costs' of regulation and the 'administrative costs'. Of particular note is the proposal that regulators apply a 'risk assessment' approach to inspection and enforcement such that the burden of enforcement falls most on those businesses deemed to be at highest risk, or resulting in greater severity from non-compliance, i.e. those with poor records of compliance and those in high risk businesses. The report further recommends a reduction in the number of enforcement agencies from 31 national regulatory bodies to seven with each new 'super' regulator, including HSE, being expanded and strengthened to cover the entire scope of environment, health and safety, food standards, consumer trading standards, animal health, agricultural inspection and rural and countryside issues.

Policy in England and Wales is that all deaths related to business activities are investigated for possible manslaughter charges. Corporate Manslaughter was first formally recognised as an offence following a ruling in the Zeebrugge case. In the last few years there has been a significant increase in the number of prosecutions of businesses and directors. The existing law on corporate manslaughter makes successful prosecution of large companies difficult. This is because for a company to be found guilty of manslaughter it is necessary to show gross negligence on the part of one or more of its senior executives who can be said to be the 'controlling mind' of the company. In large organisations with numerous tiers of management below board level the offence can be virtually impossible to prove. To date, there have been only six convictions of corporate defendants for this offence and all have been of small companies. A new offence of Corporate Manslaughter has been long awaited and in March 2005 a draft Bill was finally published for which consultation ends in June 2005. The draft Bill proposes a new specific offence of Corporate Manslaughter which focuses on the way in which activities in an organisation are managed or organised by its senior managers. The proposed offence requires a gross breach of a relevant duty of care by senior managers, which falls far below the standard reasonably expected in the circumstances resulting in a death.

**E15002** *Enforcement*

It is not just the employer who is liable to prosecution: employees and junior and middle management and even visitors to the workplace can also be prosecuted, either in tandem with the employer or alone. In the last year 25 employees were prosecuted of whom 19 were convicted. The average fine was £842.

Currently, prosecutions and other enforcement procedures are the responsibility of the appropriate 'enforcing authority' (see E15007–E15011 below). In addition, where an employee is injured or killed as a result of negligence or breach of a statutory requirement, a separate civil action may be brought against the employer for damages.

This section deals with the following aspects of enforcement.

— The role of the Health and Safety Commission and the Health and Safety Executive (see E15002–E15006).

— The 'enforcing authorities' (see E15007–E15011 below).

— The 'relevant statutory provisions', which can be enforced under *HSWA 1974* (see E15012 below).

Part A: Enforcement Powers of Inspectors'

— Improvement and prohibition notices (see E15013–E15018 below).

— Appeals against improvement and prohibition notices (see E15019, E15020 below).

— Grounds for appeal against a notice (see E15021–E15025 below).

— Inspectors' investigation powers (see E15026 below).

— Inspectors' powers of search and seizure (see E15027 below).

— Indemnification by enforcing authority (see E15028 below).

— Public register of notices (see E15029 below).

Part B: Offences and Penalties

— Prosecution for contravention of the relevant statutory provisions (see E15030, E15031 below).

— Main offences and penalties (see E15032–E15038 below).

— Offences committed by particular types of persons, including the Crown (see E15039–E15045 below).

— Sentencing guidelines (see E15043 below).

## The role of the Health and Safety Commission and Executive

**E15002** Under the *HSWA 1974* the administration and enforcement of health and safety law is carried out by the Health and Safety Commission (HSC) and its executive arm the Health and Safety Executive (HSE). The HSC consists of a chairman and not fewer than six, nor more than nine, other members. Of those members three represent employers, three employees and the remainder are from bodies such as local authorities and professional organisations. [*HSWA 1974*, s 10(2), (3)]. The HSE is the executive arm of the HSC and consists of a director general appointed by the HSE, a deputy director general and head of operations. Both the HSC and HSE are independent of any government department, but owe direct responsibilities to the relevant Secretary of State (the Secretary of State for the Environment, Transport and Regions).

## The functions of the HSC

**E15003** The functions of the HSC include promoting the general aims of the *HSWA 1974* and other health and safety legislation generally, drawing up proposals for regulations, replacing/updating existing law and regulations, preparing approved codes of practice and ensuring arrangements are in place for research and training. [*HSWA 1974, s 11(2)*].

The HSC is given consequential powers to make agreements with other government departments, established advisory committees etc for the purpose of discharging its functions. [*HSWA 1974, s 13*]. It may also delegate the exercise of certain of its functions to the HSE [*HSWA 1974, s 11(4)(a)*], for example, it may ask the HSE to carry out research or provide publicity or educational materials.

The HSC can also ask the HSE to conduct a special investigation and to report to them on their findings or, with the consent of the Secretary of State, direct that an inquiry be held into any accident, occurrence or matter. [*HSWA 1974, s 14*].

## Preparation by the HSC of draft regulations and issue of approved codes of practice

**E15004** One of the most important functions of the HSC is to prepare draft regulations for the Secretary of State, including new regulations seeking to implement EU directives relating to health and safety at work. This normally involves preparation of a consultative document and draft regulations which are considered by industry and employee bodies, trade associations and other interested bodies, such as local authorities and educational establishments. Draft regulations are submitted to the Secretary of State for approval after consultation. Although the power to make regulations rests with the Secretary of State, he has a duty to consult the HSC before doing so.

Such regulations can:

(*a*) repeal or modify any of the 'relevant statutory provisions' (see E15012 for the definition of this term);

(*b*) exclude or modify any of the 'existing statutory provisions' e.g. the *Employers' Health and Safety Policy Statements (Exception) Regulations 1975 (SI 1975 No 1584)*, exempt employers employing fewer than *five* employees from issuing a written company safety policy, as required by *HSWA 1974, s 2(3)*;

(*c*) make a specified authority responsible for the enforcement of any of the relevant statutory provisions.

[*HSWA 1974, s 15(3)(c)*].

Although the HSC does not have power to make regulations, it can approve and issue of codes of practice providing practical guidance on the requirements imposed by health and safety legislation or regulations. [*HSWA 1974, s 16*]. In all cases the HSC must consult with any appropriate government departments and obtain the consent of the Secretary of State before the issue of Approved Codes of Practice ('ACOPs').

## Functions of the Health and Safety Executive (HSE)

**E15005** The main function of the HSE is to enforce health and safety legislation unless responsibility lies with another enforcing body, such as a local authority. [*HSWA 1974, s 18(1)*]. The HSE's functions are principally performed by its inspectors and include carrying out routine inspections of premises, investigating workplace accidents, dangerous occurrences or cases of ill health and providing advice to companies and individuals on the legal requirements under health and safety legislation. The HSE also publish guidance documents (see E15006 below), provide an information service, carry out research and license or approve certain hazardous operations, such as nuclear site licensing and accepting offshore installation safety cases.

The *Health and Safety (Fees) Regulations 2004 (SI 2004 No 456)* also provide that the HSE may, in certain limited circumstances, charge for their services. In particular, fees are payable for (*inter alia*):

(*i*) an application for an explosives licence under *Part IX* of the *Dangerous Substances in Harbour Regulations 1987 (SI 1987 No 37)* as amended by *SI 2004 No 568*;

(*j*) certificates and approvals under the *Carriage of Dangerous Goods by Road (and Use of Transportable Pressure Equipment Regulations 2004 (SI 2004 No 568)*;

(*k*) an application under the *Radiation (Emergency Preparedness and Public Information) Regulations 2001 (SI 2001 No 2975)* as amended by *SI 2004 No568*;

(*l*) notifications and applications under the *Genetically Modified Organisms (Contained Use) Regulations 2000 (SI 2000 No 2831)*;

(*m*) notifications and applications under the *Notification of New Substances Regulations 1993 (SI 1993 No 3050)* as amended by *SI 2001 No 1055*;

(*n*) enforcement provisions in reation to offshore installations;

(*o*) an approval under the *Health and Safety (First-Aid) Regulations 1981 (SI 1981 No 917)* as amended *by SI 2002 No 2174*;

(*p*) an approval under the *Offshore Installations and Pipeline Works (First-Aid) Regulations 1989 (SI 1989 No 1671)* as amended by *SI 1995 No 738*;

(*q*) enforcement provisions in relation to a railway safety case prepared pursuant to the *Railways (Safety Case) Regulations 2000 (SI 2000 No 2688)* as amended by *SI 2001 No 3291*; and

(*r*) enforcement provisions in relation to gas safety cases prepared pursuant to the *Gas Safety (Management) Regulations 1996 (SI 1996 No 551)*.

## HSE publications

E15006    The HSE (through HSE Books) also publish a series of guidance notes/advisory literature for employers, local authorities, trade unions etc. on most aspects of health and safety of concern to industry. These are divided into five main areas i.e. Chemical Safety (CS), Environmental Hygiene (EH), General Series (GS), Medical Series (MS), and Plant or Machinery (PM).

In addition there are the Health and Safety (Guidance) Series (HS(G)); the Health and Safety (Regulations) Series (HS(R)); Legal Series (L); Best Practicable Means Leaflets (BPM); Emission Test Methods (ETM); Health and Safety Commission Leaflets (HSC); Health and Safety Executive Leaflets (HSE); Industry General Leaflets (IND(G)); Industry Safety Leaflets (IND(S)); similarly, Methods for the Determination of Hazardous Substances (MDHS); Toxicity Reviews (TR); Occasional Papers; and Agricultural Safety Leaflets (AS).

## Enforcing authorities

E15007    Whilst the HSE is the central body entrusted with the enforcement of health and safety legislation, in any given case enforcement powers rest with the body which is expressed by statute to be the 'enforcing authority'. Here the general rule is that the 'enforcing authority', in the case of industrial premises, is the HSE and, in the case of commercial premises within its area, the local authority (the enforcing authority in over a million premises), except that the HSE cannot enforce provisions in respect of its own premises, and similarly, local authorities' premises are inspected by the HSE. Each 'enforcing authority' is empowered to appoint suitably qualified persons as inspectors for the purpose of carrying into effect the 'relevant statutory provisions' within the authority's field of responsibility. [*HSWA 1974, s 19(1)*]. Inspectors so appointed can exercise any of the enforcement powers conferred by *HSWA 1974* (see E15013–E15018 below) and bring prosecutions (see E15030, E15031 below).

## The appropriate 'enforcing authority'

**E15008**  The general rule is that the HSE is the enforcing authority, except to the extent that:

(*a*)  regulations specify that the local authority is the enforcing authority instead; the regulations that so specify are the *Health and Safety (Enforcing Authority) Regulations 1998 (SI 1998 No 494)* or;

(*b*)  one of the 'relevant statutory provisions' specifies that some other body is responsible for the enforcement of a particular requirement.

[*HSWA 1974, s 18(1), (7)(a)*].

## Activities for which the HSE is the enforcing authority

**E15009**

> The HSE is specifically the enforcing authority in respect of the following activities (even though the main activity on the premises is one for which a local authority is usually the enforcing authority in accordance with the *Health and Safety (Enforcing Authority) Regulations 1998(SI 1998 No 494), Sch 1* (see E15010 below):
> 
> 1. Any activity in a mine or quarry;
> 2. Fairground activity;
> 3. Broadcasting, recording, filming, or video recording and any activity in premises occupied by a radio, television or film undertaking where such work is carried on;
> 4. The following work carried out by independent contractors:
>    (*a*)  certain construction work;
>    (*b*)  installation, maintenance or repair of gas systems or work in connection with a gas fitting;
>    (*c*)  installation, maintenance or repair of electricity systems;
>    (*d*)  most work with ionising radiations;
> 5. Use of ionising radiations for medical exposure;
> 6. Any activity in radiography premises where work with ionising radiations is carried out;
> 7. Agricultural activities, including agricultural shows (but excluding garden centres);
> 8. Any activity on board a sea-going ship;
> 9. Ski slope, ski lift, ski tow or cable car activities;
> 10. Fish, maggot and game breeding (but not in a zoo);
> 11. The operation of a railway;
> 12. Any activity in relation to a pipeline;

## E15009  Enforcement

13. Enforcement of *HSWA 1974*, *s 6* (duties of manufacturers/suppliers of industrial products).

*[Health and Safety (Enforcing Authority) Regulations 1998 (SI 1998 No 494), Reg 4(4)(b) and Sch 2].*

---

The HSE is the enforcing authority against the following, and for any premises they occupy, including parts of the premises occupied by others providing services for them. (This is so even though the main activity is listed in *Sch 1* of the 1998 Regulations, see E15010 below):

14. County councils;
15. Local authorities;
16. Parish or community councils;
17. Police authorities;
18. Fire authorities;
19. International HQ's and defence organisations and visiting forces;
20. United Kingdom Atomic Energy Authority (UKAEA);
21. The Crown (except where premises are occupied by the HSE itself).

*[Health and Safety (Enforcing Authority) Regulations 1998 (SI 1998 No 494), Reg 4(1)–(3)].*

---

The HSE is also the enforcing authority for the premises set out below, even if occupied by more than one occupier:

22. Airport land;
23. The Channel Tunnel system;
24. Offshore installations;
25. Building/construction sites;
26. University, polytechnic, college, school etc. campuses;
27. Hospitals;

*[Health and Safety at Work (Enforcing Authority) Regulations 1998 (SI 1998 No 494), Reg 3(5)].*

and for the following:

28. Common parts of domestic premises (*Reg 3(1)*);
29. Certain areas within an airport (*Reg 3(4)(b)*);
30. Common parts of railway stations/termini/goods yards (*Reg 3(6)*).

*Enforcement* E15010

## Activities for which local authorities are the enforcing authorities
E15010

Where the main activity carried on in non-domestic premises is one of the following, the local authority is the enforcing authority (i.e. the relevant county, district or borough council):

1. Sale or storage of goods for retail/wholesale distribution (including sale and fitting of motor car tyres, exhausts, windscreens or sunroofs), except:

   (*a*) at container depots where the main activity is the storage of goods which are of transit to or from dock premises, an airport or railway;

   (*b*) where the main activity is the sale or storage for wholesale distribution of dangerous substances;

   (*c*) where the main activity is the sale or storage of water or sewage or their by-products or natural or town gas.

2. Display or demonstration of goods at an exhibition, being offered or advertised for sale.

3. Office activities.

4. Catering services.

5. Provision of permanent or temporary residential accommodation, including sites for caravans or campers.

6. Consumer services provided in a shop, except:

   (*a*) dry cleaning;

   (*b*) radio/television repairs.

7. Cleaning (wet or dry) in coin-operated units in laundrettes etc.

8. Baths, saunas, solariums, massage parlours, premises for hair transplant, skin piercing, manicuring or other cosmetic services and therapeutic treatments, except where supervised by a doctor, dentist, physiotherapist, osteopath or chiropractor.

9. Practice or presentation of arts, sports, games, entertainment or other cultural/ recreational activities, save where the main activity is the exhibition of a cave to the public.

10. Hiring out of pleasure craft for use on inland waters.

11. Care, treatment, accommodation or exhibition of animals, birds or other creatures, except where the main activity is:

    (*a*) horse breeding/horse training at stables;

    (*b*) agricultural activity;

    (*c*) veterinary surgery.

12. Undertaking, but not embalming or coffin making.

13. Church worship/religious meetings.

14. Provision of care parking facilities within an airport.

15. Childcare, playgroup or nursery facilities.

[*Health and Safety (Enforcing Authority) Regulations 1998 (SI 1998 No 494), Reg 3(1) and Sch 1*].

E15011 *Enforcement*

### Transfer of responsibility between the HSE and local authorities

E15011  Enforcement can be transferred (though not in the case of Crown premises), by prior agreement, from the HSE to the local authority and vice versa. The Health and Safety Commission is also empowered to effect such a transfer, without the necessity of such agreement. In either case, parties who are affected by such transfer must be notified. [*Health and Safety (Enforcing Authority) Regulations 1998 (SI 1998 No 494), Reg 5*]. Transfer is effective even though the above procedure is not followed (i.e. the authority changes when the main activity changes) (*Hadley v Hancox (1987) 85 LGR 402*, decided under the previous Regulations).

Where there is uncertainty, these Regulations also allow responsibility to be assigned by the HSE and the local authority jointly to either body. [*Health and Safety (Enforcing Authority) Regulations 1998 (SI 1998 No 494), Reg 6(1)*].

## 'Relevant statutory provisions' covered by the Health and Safety at Work etc. Act 1974

E15012  The enforcement powers conferred by *HSWA 1974* extend to any of the 'relevant statutory provisions'. These comprise:

(*a*)  the provisions of *HSWA 1974, Part I* (i.e. ss 1–53);

(*b*)  any health and safety regulations passed under *HSWA 1974*, e.g. the *Ionising Radiations Regulations 1999 (SI 1999 No 3232)*, the *Management of Health and Safety at Work Regulations 1999 (SI 1999 No 3242)* as amended and the *Workplace (Health, Safety and Welfare) Regulations 1992 (SI 1992 No 3004)* as amended; and

(*c*)  the 'existing statutory provisions', i.e. all enactments specified in *HSWA 1974, Sch 1*, including any regulations etc. made under them, so long as they continue to have effect; that is, the *Explosives Acts 1875–1923*, the *Mines and Quarries Act 1954*, the *Factories Act 1961*, the *Public Health Act 1961*, the *Offices, Shops and Railway Premises Act 1963* and (by dint of the *Offshore Safety Act 1992*) the *Mineral Workings (Offshore Installations) Act 1971* [*HSWA 1974, s 53(1)*].

## Part A: Enforcement Powers of Inspectors

### Improvement and prohibition notices

E15013  It was recommended by the Robens Committee that 'inspectors' should have the power, without reference to the courts, to issue a formal improvement notice to an employer requiring him to remedy particular faults or to institute a specified programme of work within a stated time limit.' (Cmnd 5034, para 269). 'The improvement notice would be the inspector's main sanction. In addition, an alternative and stronger power should be available to the inspector for use where he considers the case for remedial action to be particularly serious. In such cases he should be able to issue a prohibition notice.' (Cmnd 5034, para 276). *HSWA 1974* put these recommendations into effect.

### Improvement notices

E15014  An inspector may serve an improvement notice if he is of the opinion that a person:

(*a*)  is contravening one or more of the 'relevant statutory provisions' (see E15012 above); or

(*b*)  has contravened one or more of those provisions in circumstances that make it likely that the contravention will continue or be repeated.

[*HSWA 1974, s 21*].

In the improvement notice the inspector must:

(i) state that he is of the opinion in (*a*) and (*b*) above; and

(ii) specify the provision(s) in his opinion contravened; and

(iii) give particulars of the reasons for his opinion; and

(iv) specify a period of time within which the person is required to remedy the contravention (or the matters occasioning such contravention).

[*HSWA 1974, s 21*].

The period specified in the notice within which the requirement must be carried out (see (iv) above) must be at least 21 days – this being the period within which an appeal may be lodged with an Employment tribunal (see E15019 below). [*HSWA 1974, s 21*]. In order to be validly served on a company, an improvement notice relating to the company's actions as an employer must be served at the registered office of the company, not elsewhere. Service at premises occupied by the company will only be valid if the notice relates to a contravention by the company in the capacity of occupier (*HSE v George Tancocks Garage (Exeter) [1993] COD 284*).

Although formal procedures requiring inspectors to give prior written notice of their intention to serve an improvement notice have been withdrawn, the HSC has announced (*Press Release C11:98, 31 March 1998*) that informal consultation between inspectors and employers regarding the issue of such notices will continue. Inspectors are expected to discuss with the employer concerned the alleged breaches of the law and any remedial action required in an attempt to resolve any points of difference before issuing a notice.

Failure to comply with an improvement notice can have serious penal consequences (see E15036 below).

## Prohibition notices

**E15015** If an inspector is of the opinion that, with regard to any activities to which *s 22(1)* applies (see below), the activities involve or will involve a risk of serious personal injury, he may serve on that person a notice (a prohibition notice). [*HSWA 1974, s 22(2)*].

It is incumbent on an inspector to show, on a balance of probabilities, that there is a risk to health and safety (*Readmans Ltd v Leeds CC [1992] COD 419* where an environmental health officer served a prohibition notice on the appellant regarding shopping trolleys with child seats on them, following an accident involving an eleven-month-old child. The appellant alleged that the Employment tribunal had wrongly placed the burden of proof on them, to show that the trolleys were not dangerous. It was held by the High Court (allowing the appeal), that it was for the inspector to prove that there was a health and/or safety risk).

It should be noted however, that a company's temporary cessation of an activity does not preclude an inspector from issuing a prohibition notice. In the case of *Railtrack v Smallwood [2001] EWHC Admin 78*, Mr Justice Sullivan considered an appeal by Railtrack following the decision of an Employment tribunal that a prohibition notice issued by HSE inspector, Mr Smallwood, following the Ladbroke Grove train crash was valid, notwithstanding that activities on the part of the track to which the notice applied, had ceased. Mr Justice Sullivan stated that:

'In my judgment, it would be very surprising, and a significant lacuna in the Act if an Inspector, in the aftermath of such a serious accident, when operations have been suspended precisely because of the gravity of the accident, was unable to issue a prohibition notice upon the basis of a present risk of serious personal injury'.

He went on to consider the intention behind section 22 and concluded that:

'Looking at the words of s 22(1) in this context, and bearing in mind in particular that s 22 must have been intended to confer powers upon Inspectors, not merely prior to, but also in the aftermath of, the most serious accidents, I am satisfied that "activities" are (still) being carried on for the purposes of s 22 if they have been temporarily interrupted or suspended as a result of a major accident. I do not consider that such an interpretation does any violence to the ordinary meaning of the words in s 22, provided they are considered in a realistic context'.

In his judgment Mr Justice Sullivan stated that all the surrounding circumstances must be considered, including the reason for the temporary inactivity:

'There may well be a distinction to be drawn on the facts between a factory that is inactive because the employer has gone bankrupt and dismissed all of his employees, and a factory that is inactive because it has closed for the summer holidays. If, however, the factory has just closed, and the workers had been sent home, because of a tragic accident, I do not consider that activities will have ceased for the purposes of s 22. The position may be different if, not as a result of an accident, but pursuant to a pre-planned closure programme, for example, to replace outdated and unsafe machinery, the factory is at a standstill for some weeks or months'.

'... I am satisfied that ... even though there are temporary interruptions when nothing is actually taking place on the ground, so "activities" will continue for the purposes of s 22 even though they have been interrupted as a result of a serious accident'.

Railtrack's appeal was subsequently dismissed.

Prohibition notices differ from improvement notices in two important ways:

(a) with prohibition notices, it is not necessary that an inspector believes that a provision of *HSWA 1974* or any other statutory provision is being or has been contravened;

(b) prohibition notices are served in *anticipation* of danger.

The *HSWA 1974*, s 22 applies where, in the inspector's opinion, there is a hazardous activity or state of affairs generally. It is irrelevant that the hazard or danger is not mentioned in *HSWA 1974*; it can exist by virtue of other legislation, or even in the absence of any relevant statutory duty. In this way notices are used to enforce the later statutory requirements of *HSWA 1974* and the earlier requirements of the *Factories Act 1961* and other protective occupational legislation.

A prohibition notice must:

(a) state that the inspector is of the opinion stated immediately above;

(b) specify the matters which create the risk in question;

(c) where there is actual or anticipatory breach of provisions and regulations, state that the inspector is of the opinion that this is so and give reasons;

(d) direct that the activities referred to in the notice must not be carried out on, by or under the control of the person on whom the notice is served, unless the matters referred to in (b) above have been remedied.

*[HSWA 1974, s 22(3)].*

Failure to comply with a prohibition notice can have serious penal consequences (see E15028 below).

## Differences between improvement and prohibition notices

E15016    Unlike an improvement notice, where time is allowed in which to correct a defect or offending state of affairs, a prohibition notice can take effect immediately.

A direction contained in a prohibition notice shall take effect:

(a) at the end of the period specified in the notice; or

(b) if the notice so declares, immediately.

[*HSWA 1974, s 22(4)* as substituted by *Consumer Protection Act 1987, Sch 3*].

Risk of injury need not be imminent, even if the notice is to take immediate effect (*Tesco Stores Ltd v Kippax COIT No 7605–6/90*).

An improvement notice gives a person upon whom it is served time to correct the defect or offending situation. A prohibition notice, which is a direction to stop the work activity in question rather than put it right, can take effect immediately on issue; alternatively, it may allow time for certain modifications to take place (i.e. deferred prohibition notice). Both types of notice will generally contain a schedule of work which the inspector will require to be carried out. If the nature of the work to be carried out is vague, the validity of the notice is not affected. If there is an appeal, an Employment tribunal may, within its powers to modify a notice, rephrase the schedule in more specific terms (*Chrysler (UK) Ltd v McCarthy [1978] ICR 939*).

### Effect of non-compliance with notice

**E15017** If, after expiry of the period specified in the notice, or in the event of an appeal, expiry of any additional time allowed for compliance by the Employment tribunal, an applicant does not comply with the notice or modified notice, he can be prosecuted. If convicted of contravening a prohibition notice, he may be imprisoned. [*HSWA 1974, s 33(1)(g), (2A)(b)*]. In *R v Kerr; R v Barker (1996) (unreported)* the directors of a company were each jailed for four months after allowing a machine which was subject to a prohibition notice – following an accident in which an employee lost an arm – to continue to be operated.

### Service of notice coupled with prosecution

**E15018** Where an inspector serves a notice, he may at the same time decide to prosecute for the substantive offence specified in the notice. The fact that a notice has been served is not relevant to the prosecution. Nevertheless, an inspector will not normally commence proceedings until after the expiry of 21 days, i.e. until he is satisfied that there is to be no appeal against the notice or until the Employment tribunal has heard the appeal and affirmed the notice, since it would be inconsistent if conviction by the magistrates were followed by cancellation of the notice by the Employment tribunal. The fact that an Employment tribunal has upheld a notice is not binding on a magistrates' court hearing a prosecution under the statutory provision of which the notice alleged a contravention; it is necessary for the prosecution to prove all the elements in the offence (see E15031 below).

Employment tribunals are mainly concerned with hearing unfair dismissal claims by employees'; only a tiny proportion of cases heard by them relate specifically to health and safety. Moreover, they are not empowered to determine breaches of criminal legislation.

### Appeals against improvement and prohibition notices

**E15019** A person on whom either type of notice is served may appeal to an Employment tribunal within 21 days from the date of service of the notice. The Employment tribunal may extend this time where it is satisfied, on application made in writing (either before or after expiry of the 21-day period), that it was not reasonably practicable for the appeal to be brought within the 21-day period. On appeal the Employment tribunal may either affirm or cancel the notice and, if it affirms it, may do so with modifications in the form of additions, omissions or amendments. [*HSWA 1974, ss 24(2), 82(1)(c); Employment Tribunals (Constitution and Rules of Procedure) Regulations 2001 (SI 2001 No 1171), Sch 5*].

### Effect of appeal

**E15020** Where an appeal is brought against a notice, the lodging of an appeal automatically suspends operation of an improvement notice, but a prohibition notice will continue to apply unless there is a direction to the contrary from the Employment tribunal. Thus:

(a) in the case of an improvement notice, the appeal has the effect of suspending the operation of the notice [*HSWA 1974, s 24(3)(a)*];

(b) in the case of a prohibition notice, the appeal only suspends the operation of the notice in the following circumstances:

   (i) if the Employment tribunal so directs, on the application of the appellant; and

   (ii) the suspension is then effective from the time when the Employment tribunal so directs.

[*HSWA 1974, s 24(3)*].

## Grounds for appeal

**E15021** The main grounds for appeal are:

(a) the inspector wrongly interpreted the law (see E15022 below);

(b) the inspector exceeded his powers, though not necessarily intentionally, under an Act or regulation (see E15023 below);

(c) breach of law is admitted but the proposed solution is not 'practicable' or not 'reasonably practicable', or that there was no 'best practicable means' other than that used where 'best practicable means' is also a defence to a charge of statutory nuisance (depending on the terminology of the particular statute) (see E15024 below);

(d) breach of law is admitted but the breach is so insignificant that the notice should be cancelled (see E15025 below).

The merits of appealing the issue or terms of an enforcement notice should always be considered carefully as, in practice, this may be the employer's only opportunity to dispute the reasonableness of the requirements imposed. Failure to comply with a notice is a strict liability offence and if a prosecution is brought it is not open to the employer to argue that he did everything 'reasonably practicable' to comply with the notice (*Deary v Mansion Hide Upholstery Ltd [1983] ICR 610*, see E15031 below).

The delay involved in lodging an appeal can have important practical consequences. Months might elapse before an appeal against a notice can be heard by the Employment tribunal. For this reason particularly, and in view of the fact that notice of appeal suspends operation of an improvement notice, some recipients opt for appealing, since at the time of service of notice they may not be in a position to meet the requirements of the notice; whereas, three or four months later, the position may have changed or a compromise arrangement reached with the inspector. If, however, the sole reason for appealing is to gain time and nothing else, the Employment tribunal is unlikely to be sympathetic and costs could be awarded against the unsuccessful appellant, though this is rare. [*Industrial Tribunals (Constitution and Rules of Procedure) Regulations 1993, Reg 8(4), Sch 4*, which states that: 'a Employment tribunal may make an Order that a party shall pay to another party either a specified sum in respect of the costs of or in connection with an appeal incurred by that other party or, in default of agreement, the taxed amount of those costs'.]

## Inspector's wrong interpretation of the law

**E15022** It is doubtful whether many cases have, or indeed would, succeed on this ground. Where regulations impose a strict duty (for example the duty under *Reg 6(1)* of the *Provision and Use of Work Equipment Regulations 1992 (SI 1992 No 2932)* to provide work equipment in good repair) there is no scope for argument by the employer. However, where the statute provides a defence, for example, it requires the duty to be carried out 'so far as reasonably practicable' (discussed more fully at E15031 below) or it provides for a due diligence defence, there is some scope to argue that the inspectors' interpretation of the law is incorrect. For example, in *Canterbury City Council v Howletts and Port Lympne Estates Limited (The Times, 13 December*

1996), a prohibition notice was served on Howletts Zoo following the death of a keeper while he was cleaning the tigers' enclosure. It was Howletts' policy to allow their animals to roam freely; the local authority argued that the zoo's keepers could have carried out their tasks in the tigers' enclosure with the animals secured. The High Court affirmed the Employment tribunal's decision to set aside the notice, holding that *HSWA 1974, s 2* was not intended to render illegal certain working practices simply because they were dangerous.

Employers should note that contesting a prohibition notice on the ground that there has been no legal contravention is pointless, since valid service of a prohibition notice does not depend on legal contravention (*Roberts v Day, COIT No 1/133, Case No 3053/77*).

## Inspector exceeded powers under statute

**E15023** It can happen that an inspector exceeds his powers under statute by reason of misinterpretation of the statute or regulation (*Deeley v Effer, COIT No 1/72, Case No 25354/77*). This case involved the requirement that 'all floors, steps, stairs, passages and gangways must, so far as is reasonably practicable, be kept free from obstruction and from any substance likely to cause persons to slip'. [*OSRPA 1963, s 16(1)*] (now replaced by equivalent duties under the *Workplace (Health, Safety and Welfare) Regulations 1992 (SI 1992 No 3004)*. The inspector considered that employees were endangered by baskets of wares in the shop entrance. The Employment tribunal ruled that the notice had to be cancelled, since the only persons endangered were members of the public, and *OSRPA 1963* was concerned with dangers to employees.

## Proposed solution not practicable

**E15024** The position in a case where, although breach of the law is admitted, the proposed solution is not considered practicable, depends upon the nature of the obligation. The duty may be strict, or have to be carried out so far as practicable or, alternatively, so far as reasonably practicable. In the first two situations cost of compliance is irrelevant; in the latter case, where a requirement has to be carried out 'so far as reasonably practicable', cost effectiveness is an important factor but has to be weighed against the risks to health and safety involved in failing to implement the remedial measures identified in the enforcement notice. Where there is a real danger of serious injury the cost of complying with the notice is unlikely to be decisive. Thus in a leading case, the appellant was served with an improvement notice requiring secure fencing on transmission machinery. An appeal was lodged on the ground that the proposed modifications were too costly (£1,900). It was argued that because of the intelligence and integrity of the operators a safety screen costing £200 would be adequate. The Employment tribunal dismissed the appeal: the risk justified the cost (*Belhaven Brewery Co Ltd v McLean [1975] IRLR 370*).

The cost of complying with a notice is likely to carry less weight in the case of a prohibition notice than an improvement notice, as there must be 'a risk of serious personal injury' for a prohibition notice to be served (*Nico Manufacturing Co Ltd v Hendry [1975] IRLR 225*, where the company argued that a prohibition notice in respect of the worn state of their power presses should be cancelled on the ground that it would result in a 'serious loss of production' and endanger the jobs of several employees'. The Employment tribunal dismissed this argument, having decided that using the machinery in its worn condition could cause serious danger to operators). Similarly, an undertaking by a company to take additional safety precautions against the risk of injury from unsafe plant until new equipment was installed was not sufficient (*Grovehurst Energy Ltd v Strawson (HM Inspector) COIT No 5035/90*).

Where cost is a factor this is not to be confused with the current financial position of the company. A company's financial position is irrelevant to the question whether an Employment tribunal should affirm an enforcement notice. Thus in *Harrison (Newcastle-under-Lyme) Ltd v Ramsay (HM Inspector) [1976] IRLR 135*, a notice requiring cleaning, preparation and painting of walls had to be complied with even though the company was on an economy drive.

E15025  *Enforcement*

Employment tribunals have power under *HSWA 1974, s 24* to alter or extend time limits attaching to improvement and prohibition notices (*D J M and A J Campion v Hughes (HM Inspector of Factories) [1975] IRLR 291*, where even though there was an imminent risk of serious personal injury, a further four months were allowed for the erection of fire escapes as it was not practicable to carry out the remedial works within the time limits set). Extensions of time in which to comply with notices are most commonly granted where the costs of the improvements and modifications required by the notice are significant.

## Breach of law is insignificant

E15025  Employment tribunals will rarely cancel a notice which concerns breach of an absolute duty where the breach is admitted but the appellant argues that the breach is trivial: *South Surbiton Co-operative Society Ltd v Wilcox [1975] IRLR 292*, where a notice had been issued in respect of a cracked wash-hand basin, being a breach of an absolute duty under the *Offices, Shops and Railway Premises Act 1963*. It was argued by the appellant that, in view of their excellent record of cleanliness, there was no need for officials to visit the premises. The appeal was dismissed.

## Inspectors investigation powers

E15026  Inspectors have wide ranging powers under *HSWA 1974* to investigate suspected health and safety offences. These include powers to:

(*a*) enter and search premises;

(*b*) direct that the premises or anything on them be left undisturbed for so long as is reasonably necessary for the purpose of the investigation;

(*c*) take measurements, photographs and recordings;

(*d*) take samples of articles or substances found in the premises and of the atmosphere in or in the vicinity of the premises;

(*e*) dismantle or test any article which appears to have caused or be likely to cause danger;

(*f*) detain items for testing or for use as evidence;

(*g*) interview any person;

(*h*) require the production and inspection of any documents and to take copies; and

(*i*) require the provision of facilities and assistance for the purpose of carrying out the investigation.

[*HSWA 1974, s 20*]

Under these powers inspectors may require interviewees to answer such questions as they think fit and sign a declaration that those answers are true. [*HSWA 1974, s 20(1)(j)*]. However, evidence given in this way is inadmissible in any proceedings subsequently taken against the person giving the statement or his or her spouse. [*HSWA 1974, s 20(7)*]. Where prosecution of an individual is contemplated the inspector will, therefore, usually exercise his evidence-gathering powers under the *Police and Criminal Evidence Act 1984* (*PACE*). Evidence given in this way is admissible against that person in later proceedings. Interviews conducted under *PACE* are subject to strict legal controls, for example, interviewees must be cautioned before the interview takes place, they have certain 'rights to silence' (although these were qualified by the *Criminal Justice and Public Order Act 1994*), and there are rules relating to the recording of the interview.

## Inspectors' powers of search and seizure in case of imminent danger

E15027    Where an inspector has reasonable cause to believe that there are on premises 'articles or substances ('substance' includes solids, liquids and gases – *HSWA 1974, s 53(1)*) which give rise to imminent risk of serious personal injury', he can:

(a)    seize them; and

(b)    cause them to be rendered harmless (by destruction or otherwise).

*[HSWA 1974, s 25(1)].*

Enforcing authorities are given similar powers regarding environmental pollution, under the *Environment Act 1995, ss 108, 109*. (See E5004 EMISSIONS INTO THE ATMOSPHERE.)

Before an article forming 'part of a batch of similar articles', or a substance is rendered harmless, an inspector must, if practicable, take a sample and give to a responsible person, at the premises where the article or substance was found, a portion which has been marked in such a way as to be identifiable. *[HSWA 1974, s 25(2)]*. After the article or substance has been rendered harmless, the inspector must sign a prepared report and give a copy of the report to:

(i)    a responsible person (e.g. safety officer); and

(ii)   the owner of the premises, unless he happens to be the 'responsible person'. (See A3005 ACCIDENT REPORTING for the meaning of this term.)

*[HSWA 1974, s 25(3)].*

Analogous powers are given to transport inspectors under the *Radioactive Material (Road Transport) Act 1991, s 5* in respect of vehicles carrying radioactive packages.

A customs officer may assist the enforcing authority or the inspector in his enforcement duties under the *HSWA 1974* by seizing any imported article or substance. He may then detain it for not more than 2 working days and may disclose information about it to the enforcing authorities or inspectors. *[HSWA 1974, ss 25A, 27A inserted by Consumer Protection Act 1987, Sch 3]*.

## Indemnification by enforcing authorities

E15028    Where an inspector has an action brought against him in respect of an act done in the execution or purported execution of any of the 'relevant statutory provisions' (see E15012 above), and is ordered to pay damages and costs (or expenses) in circumstances where he is not legally entitled to require the enforcing authority which appointed him to indemnify him, he may be able to take advantage of *HSWA 1974, s 26*. By virtue of that provision, the authority nevertheless has the power to indemnify the inspector against all or part of such damages where the authority is satisfied that the inspector honestly believed:

(a)    that the act complained of was within his powers; and

(b)    that his duty as an inspector required or entitled him to do it.

In practice there will be very few circumstances where an inspector is held liable for advice given or enforcement action taken as part of his statutory duties. The Court of Appeal has ruled (in *Harris v Evans and Another [1998] 3 All ER 522*) that an enforcing authority giving advice which leads to the issue of enforcement notices does not owe a duty of care to the owner of the premises affected by the notice and a claim for economic loss arising from such allegedly negligent advice cannot therefore arise. The court said that if enforcing authorities were to be exposed to liability in negligence at the suit of owners whose businesses are adversely affected by their decisions it would have a detrimental effect on the performance by inspectors of their statutory duties. The *HSWA 1974* contains its own statutory remedies against errors by inspectors and the court was not prepared to add to those measures. The court did, however, suggest that a possible exception might arise if a requirement imposed by the inspector introduced a new risk or danger which resulted in physical damage or economic loss.

## Public register of improvement and prohibition notices

E15029    Improvement (though not in the case of the *Fire Precautions Act 1971*) and prohibition notices relating to public safety matters have to be entered in a public register as follows:

(a)   within 14 days following the date on which notice is served in cases where there is no right of appeal;

(b)   within 14 days following the day on which the time limit expired, in cases where there is a right of appeal but no appeal has been lodged within the statutory 21 days.

(c)   within 14 days following the day when the appeal is disposed of, in cases where an appeal is brought.

*[Environment and Safety Information Act 1988, s 3].*

Notices which impose requirements or prohibitions solely for the protection of persons at work are not included in the register. *[Environment and Safety Information Act 1988, s 2(3)].*

In addition, registers must be kept of notices served by:

(i)    fire authorities, under the *Schedule* to the *Environment and Safety Information Act 1988* for the purpose of *s 10* of the *Fire Precautions Act 1971* (not improvement notices);

(ii)   local authorities, under the *Schedule* to the *Environment and Safety Information Act 1988* for the purpose of *s 10* of the *Safety of Sports Grounds Act 1975*;

(iii)  responsible authorities (as defined by the *Environment and Safety Information Act 1988, s 2(2)*) and the Minister of Agriculture, Fisheries and Food under *s 2* of the *Environment and Safety Information Act 1988* for the purpose of *s 19* of the *Food and Environment Protection Act 1985*;

(iv)  enforcing authorities, under *s 20* of the *Environmental Protection Act 1990*; and

(v)   enforcing authorities, under the *Radioactive Substances Act 1993* (see D1068 DANGEROUS SUBSTANCES I).

These registers are open to inspection by the public free of charge at reasonable hours and, on request and payment of a reasonable fee, copies can be obtained from the relevant authority. *[Environment and Safety Information Act 1988, s 1].* Such records can also be kept on computer.

# Part B: Offences and Penalties

## Prosecution for breach of the 'relevant statutory provisions'

E15030    Prosecutions can follow non-compliance with an improvement or prohibition notice, but equally inspectors' will sometimes prosecute without serving a notice. Service of notices remains the most usual method of enforcement. In 2003/04 approximately 11,000 enforcement notices were issued by the HSE compared with about 1,000 prosecutions for health and safety offences. Prosecutions are most commonly brought after a workplace accident or dangerous incident (such as a fire or explosion). Investigations by the enforcing authorities may take many months to complete and, in complex cases, it is not unusual for prosecutions to be commenced up to a year after the original incident. Prosecutions normally take place before the magistrates but there is increasing pressure for more serious cases, involving work-related fatalities or major injuries, to be prosecuted on indictment in the Crown Court, where increased penalties are available. The determining factor behind prosecution on indictment is the gravity of the offence.

Whilst the decision whether to prosecute, issue an enforcement notice, or simply to give advice lies within the discretion of the inspector concerned, the enforcing authorities aim to pursue a policy which is open, consistent and proportionate to the risks in deciding what enforcement action to take (see the HSC's *Enforcement Policy Statement* of *January 2002*, and the *HSC: A Strategy for Workplace Health and Safety in Great Britain too 2010 and Beyond*.

In practice the enforcing authorities usually only investigate the most serious workplace accidents; it is HSE policy to investigate all work-related deaths which are reported to HSE. The Health and Safety Statistics 2003/04, published by the HSE, shows an 8 per cent increase in the total number of prosecutions brought by the HSE in the period under review. The greatest number of cases brought was in the construction sector (635 offences prosecuted), followed closely by manufacturing (586 offences prosecuted). In the same period the HSE prosecuted a total of 17 Managers and Directors, of which eleven led to prosecutions.

## Burden of proof

**E15031** Throughout criminal law, the burden of proof of guilt is on the prosecution to show that the accused committed the particular offence (*Woolmington v DPP [1935] AC 463*). The burden is a great deal heavier than in civil law, requiring proof of guilt beyond a reasonable doubt as distinct from on a balance of probabilities. While not eliminating the need for the prosecution to establish general proof of guilt, *HSWA 1974, s 40* makes the task of the prosecution easier by transferring the onus of proof to the accused for one element of certain offences. *Section 40* states that in any proceedings for an offence consisting of a failure to comply with a duty or requirement to do something so far as is practicable, or so far as reasonably practicable, or to use the best practicable means to do something, the onus is on the accused to prove (as the case may be) that it was not practicable, or not reasonably practicable to do more than was in fact done to satisfy the duty or requirement, or that there was no better practicable means than was in fact used to satisfy the duty or requirement. However, *s 40* does not apply to an offence created by *s 33(1)(g)* of the *HSWA 1974* – failing to comply with an improvement notice.

In *Deary v Mansion Hide Upholstery Ltd [1983] ICR 610*, an improvement notice was served on the defendant company requiring it to provide the fire resistant storage for polyurethane foam. The company did not comply, nor did it appeal. It was irrelevant that the company had complied with the notice so far as 'reasonably practicable' as that was not a requirement of the offence charged. A similar burden of proof exists under the *Environmental Protection Act 1990* (see E5015 EMISSIONS INTO THE ATMOSPHERE).

Human rights legislation in the UK could have significant implications on the issue of the burden of proof in criminal cases. The case of *R v Lambert [2001] 3 WLR 206*, considered the effect of Article 6(2) of the European Convention on Human Rights ('ECHR') which requires that only an evidential burden can be placed on a defendant. In *R v Lambert* the defendant appealed against his conviction on the grounds that the trial judge's direction that the legal burden of proof was transferred to the defendant was contrary to Article 6(2). The appeal was dismissed by the Court of Appeal and the House of Lords on the grounds that the defendant could not rely on the *Human Rights Act 1998* ('*HRA 1998*') in relation to a conviction which predated the coming into force of the Act. However, in that case the defendant was challenging the trial judge's direction and not the authority which had brought the proceedings. Accordingly, the appeal brought by the defendant was not a proceeding by or instigated by a public authority so he was not entitled to claim retrospectivity under *s 22(4)* of the *HRA 1998*.

*R v Lambert* highlighted the potential for courts to review transfer of burden of proof cases. As a consequence, several cases have considered and challenged the legitimacy of the transfer of burden provisions under the *HSWA 1974, s 40*. The issue appears now to have been resolved in the case of *David Janway Davies v HSE [2002] WL 31676382*. In this case the appellant, Mr Davies, appealed against his conviction under *ss 3(1)* and *33(1) HSWA 1974*. In the court of first instance the judge ruled that *s 40 HSWA 1974* was compatible with the Convention and directed the jury that there was a legal burden on Mr Davies to prove, on the balance of probabilities, that it was not reasonably practicable to do more than he had in fact done. On appeal Lord Justice Tucker considered the construction of the HSWA and its compatibility with Article 6 and stated:

> 'The duty cast on the defendant is a "duty … to ensure as far as is reasonably practicable". It is a breach of this qualified duty which gives rise to the offence …
>
> This construction of the statute means that section 40 does make some inroad into the presumption of innocence. It does not follow however that it is incompatible with the Convention'.

Lord Justice Tuckey noted that the HSWA legislation is regulatory rather than prescriptive and that the objective of the legislation being to protect the public or broad segments of the public from potentially adverse affects of otherwise lawful activity. He also took into account the fact that there is no risk of imprisonment for any of the offences under *HSWA 1974*, ss 2–7 and *HSWA 1974, s 33(1)(a)*. He concluded that the imposition of a legal burden of proof in *s 40* was justified, necessary and proportionate for the following reasons:

(1) the purpose of the HSWA is to protect the health and safety of persons affected by activities referred to in *HSWA 1974, ss 2–6*;

(2) the reverse burden of proof takes into account that duty holders have chosen to operate in an activity which is regulated and therefore must be taken to accept the regulatory controls that go with it;

(3) the prosecution must first prove that the defendant owes the relevant duty of care and that the duty has been breached, before any reverse onus arises;

(4) the defence is flexible and does not restrict the away in which the defendant can show that he had done what was reasonably practicable;

(5) The reverse burden only applies to breaches under *HSWA 1974, ss 2–6*. It does not apply to *HSWA 1974, s 7* or to *HSWA 1974, s 37* implying that Parliament must have considered when the reverse onus was justified – due regard must be paid to its choice;

(6) the facts relied upon by the defence should not be difficult to prove since they will be within the knowledge of the defendant. In particularly complicated cases the relevant expertise may lie with the defendant or its state of the art supplier and therefore the prosecution would have considerable difficulties in assuming this burden of proof and enforcement could be virtually impossible if the defendant had only an evidential burden;

(7) the focus of the statutory scheme would be changed if all the defendant had to do was raise the defence to require the prosecution to disprove it; and

(8) where the reverse burden applies, the defendant does not face imprisonment.

## Main offences and penalties

E15032    Health and safety offences are either (*a*) triable summarily (i.e. without jury before the magistrates), or (*b*) triable summarily and on indictment (i.e. triable either way), or (*c*) triable only on indictment. Most health and safety offences, however, fall into categories (*a*) and (*b*). The main offences falling into these two categories are set out below.

## Summary only offences

E15033

(*a*) contravening a requirement imposed under *HSWA 1974, s 14* (power of the HSC to order an investigation);

(*b*) contravening a requirement imposed by an inspector under *HSWA 1974, s 20*;

(c)   preventing or attempting to prevent a person from appearing before an inspector, or from answering his questions;

(d)   intentionally obstructing an inspector or customs officer in the exercise of his powers;

(e)   falsely pretending to be an inspector.

## 'Either way' offences

**E15034**

(a)   failure to carry out one or more of the general duties of *HSWA 1974, ss 2–7*;

(b)   contravening either:

  (i)   *HSWA 1974, s 8* – intentionally or recklessly interfering with anything provided for safety;

  (ii)  *HSWA 1974, s 9* – levying payment for anything that an employer must by law provide in the interests of health and safety (e.g. personal protective clothing);

(c)   contravening any health and safety regulations;

(d)   contravening a requirement imposed by an inspector under *HSWA 1974, s 25* (power to seize and destroy articles and substances);

(e)   contravening a requirement of a prohibition or improvement notice;

(f)   intentionally or recklessly making false statements, where the statement is made:

  (i)   to comply with a requirement to furnish information; or

  (ii)  to obtain the issue of a document;

(g)   intentionally making a false entry in a register book, notice etc. which is required to be kept;

(h)   failing to comply with a remedial court order made under *HSWA 1974, s 42*.

*[HSWA 1974, s 33(1)].*

In England and Wales there is no time limit for bringing prosecutions for offences, except for offences tried summarily in the magistrates' courts – where the time limit is 6 months from the date the complaint was laid. [*Magistrates' Courts Act 1980, s 127(1)*]. In Scotland the 6-month time limit extends to 'either way' offences tried summarily. The period may be extended in the case of special reports, coroners' court hearings or in cases of death generally. [*HSWA 1974, s 34(1)*] There is no time limit for commencing hearings in the Crown Court.

## Summary trial or trial on indictment

**E15035**   Most offences triable either way are tried summarily. However an increasing number of serious offences are being referred to the Crown Court, which has increased sentencing powers, for trial on indictment. The Court of Appeal has advised magistrates to exercise caution in accepting jurisdiction in health and safety cases where the offence may require a penalty greater than they can impose or where death or serious injury has resulted from the offence (see *R v F Howe & Son (Engineering) Ltd* at E15043 below). The defendant may also refuse to consent to summary trial and opt for trial on indictment.

## Penalties for health and safety offences

**E15036**   Penalties tend to relate to the three main categories of offences characterising breach of health and safety legislation, namely:

(a) breaches of ss 2–6 of the *HSWA 1974* – serious offences:
   (i) summary conviction – a maximum £20,000 fine;
   (ii) conviction on indictment – an unlimited fine (but no imprisonment);
(b) breaches of improvement or prohibition orders, or orders under s 42 of the *HSWA 1974* to remedy the cause of the offence – serious offences:
   (i) summary conviction – a maximum £20,000 fine, or imprisonment for up to six months or both;
   (ii) conviction on indictment – an unlimited fine or imprisonment for up to two years or both;
(c) most other offences, including breaches of health and safety regulations:
   (i) summary conviction – a maximum fine of £5,000;
   (ii) conviction on indictment – an unlimited fine. For licence breaches and explosives offences, the court may also order up to two years imprisonment (see, for example, *R v Hill (1996) (unreported)*, where a demolition contractor was jailed for three months for breach of asbestos licensing regulations).

[*HSWA 1974, s 33(1A), (2)* and *(2A)* as inserted by the *Offshore Safety Act 1992*, and *HSWA 1974, s 33(3)*].

In December 1999, the Lord Chancellor announced that the Government intended to legislate to increase the penalties available for health and safety offences. A private members bill, the *Health and Safety at Work (Offences) Bill*, was first introduced in December 1999. The Bill has been reintroduced over the past two years for parliamentary sessions 2002/03 and 2003/04, but to date it has not been progressed. The Bill, if enacted, would raise the maximum fine which can be imposed in the Magistrates' Court for contraventions of requirements under health and safety regulations from the present £5,000 limit to match the £20,000 maximum that can be imposed for breaching the general duties under the *HSWA 1974, ss 2–7*. It would also make imprisonment an option for most health and safety offences in both the higher and lower courts.

Following the Court of Appeal decision in the case of *R v Colthorp Board Mill Ltd (January 2002) (unreported)*, an increase in the level of fines imposed for breaches of health and safety legislation is anticipated.

In his judgment, Lord Justice Gibbs referring to the authorities, in particular *R v F Howe & Son (Engineering) Ltd [1999] 2 All ER 249* and *R v Friskies Petcare UK Ltd (2000) 2 Cr App R (S) 401*, stated that although the remarks in a previous case that financial penalties in excess of £500,000 were reserved for cases of major public disaster,

'we would not wish the sum of £500,000 to appear to be set in stone or to provide any sort of maximum limit for such cases. On the contrary, we anticipate that as time goes on and awareness of the importance of safety increases, that courts will uphold sums of that amount and even in excess of them in serious cases, whether or not they involve what could be described as major public disasters'.

In the recent case of *R v Jarvis Facilities Ltd [2005] EWCA Crim 1409*, the Court of Appeal decided that where there is a significant public element, the court is entitled to take a more severe view of breaches of health and safety at work provisions. In this case the defendant, Jarvis Facilities Ltd had been required to carry out repair works to a set of points on a railway. Trained and qualified workers carried out the work but no specific method statement had been prepared. However, certain necessary processes had been omitted which resulted in the derailment of a freight train. The train had been travelling at 25 mph and remained upright and consequently there was no significant injury or damage. The defendant pleaded guilty to an offence contrary to *HSWA 1974, s 3(1)* but appealed against the £400,000 fine imposed by the court of first instance. The Court of Appeal concluded that public services cases would be likely to be treated more seriously than those where a breach is confined to a private sector, even where there is comparability between gravity of breach and economic strength of defendant. The Appeal court however considered the fine imposed as being excessive and accordingly reduced the original fine of £400,000 to a sum of £275,000 plus costs. The overall effect of these decisions is expected to result in more cases being referred to the Crown Courts for sentence and correspondingly higher fines for corporate defendants in particular where the offence involves a public element. The latest statistics from the HSE for the year 2003/04 show that the average fine per case has increased by 56 per cent from the previous year,

## Defences

E15037    Although no general defences are specified in the *Health and Safety at Work etc Act 1974*, some regulations passed under the Act carry the defence of 'due diligence' (for example, the *Control of Substances Hazardous to Health Regulations 2002* (*SI 2002 No 2677*) as amended by *SI 2003 No 978*. Currently, where a defence of 'due diligence' is available it is for the defendant to prove, on the balance of probabilities, that all appropriate steps to avoid the offence had been taken.

## Manslaughter

E15038    In 1996 the Law Commission proposed to reform the law on manslaughter with new statutory offences. Since then the Government has been looking carefully at this complex area of law and consulting with companies, business leaders and trade unions and after much delay a draft Bill for a new offence of Corporate Manslaughter was published in March 2005 which would apply in England and Wales only. The Bill contains an offence that is more limited in scope to previous government proposals in that there would be no special provisions relating to personal liability for directors or parent companies and it would be confined to corporations, regardless of where in the world it is incorporated (rather than all 'undertakings' as was previously proposed). It would also have limited application to government departments and other public bodies.

## The current position

E15038.1    For a company to be guilty of manslaughter under the current legislation, it is necessary to establish gross negligence on the part of one or more of its directors or other persons who can be classed as the 'controlling mind' of the company. In such cases, those persons actions and states of mind are 'identified' legally as those of the company.

In large companies where there are several tiers of management below board level it is almost impossible to show gross negligence on the part of one or more senior executives who can be said to be the 'controlling mind' of the company offence. Successful prosecutions to date have all been of small companies.

## The proposed new offence

E15038.2    An offence would be committed by an organisation when:

'the way in which any of the organisation's activities are managed or organised by its senior managers:

(*a*)    causes death; and

(*b*)    amounts to a gross breach of a relevant duty of care owed by the organisation to the deceased'.

Senior managers for the purposes of the offence are persons who play a significant role in either 'the making of decisions about how the whole or a substantial part of an organisation's activities are organised' or 'the actual managing or organising of the whole or substantial part of those activities'. The test for 'gross' breach would be whether the failure in question constituted 'conduct falling far below what can reasonably be expected of the organisation in the circumstances'. The Bill contains several factors to be taken into account when considering what this means, such as the failure to comply with relevant health and safety legislation or guidance, and whether the failure was serious, whether or not the senior managers knew or ought to have known of the failure or of the risk of serious harm associated with it and whether the organisation sought to profit from the failure.

The proposed new offence would be indictable and would attract unlimited fines. While, there is no provision for Crown immunity for the new offence, it is currently unsettled as to how fines would be applied to Crown bodies (which would in effect result in money being transferred between different public funds).

In cases of workplace death, manslaughter charges may also be brought if there is sufficient evidence. The decision to prosecute rests with the Crown Prosecution Service, not the enforcing authorities under health and safety legislation. Manslaughter convictions linked to breaches of safety legislation are rare but are becoming more common. There have been a series of individual prosecutions for manslaughter of directors or managers responsible for workplace deaths and, in 1994, the first successful prosecution for corporate manslaughter involved OLL Ltd, the activity centre responsible for organising the Lyme Bay canoeing trip in which four teenagers died. (See EMPLOYERS' DUTIES TO THEIR EMPLOYEES at E11014.)

## Offences committed by particular types of persons
### Corporate offences – delegation of duties to junior staff

**E15039** Companies cannot avoid liability for breach of general duties under *ss 2–6* of the *HSWA 1974* by arguing that the senior management and the 'directing mind' of the company had taken all reasonable precautions, and that responsibility for the offence lay with a more junior employee or agent who was at fault. The *HSWA 1974* generally imposes strict criminal liabilities on employers and others (subject to the employer being able to establish that all reasonably practicable precautions had been taken) and it is not open to corporate employers to seek to avoid liability by arguing that their general duties have been delegated to someone lower down the corporate tree. In *R v British Steel plc [1995] IRLR 310*, British Steel were prosecuted under *HSWA 1974, s 3* after a fatal accident to a subcontractor who was carrying out construction work under the supervision of a British Steel engineer. British Steel argued that it was not responsible under *s 3* for the actions of the supervising engineer as the engineer was not part of the 'directing mind' of the company and all reasonable precautions to ensure the safety of the work had been taken by senior management. The Court of Appeal dismissed this argument. A similar decision was reached in *R v Gateway Foodmarkets Ltd, [1997] IRLR 189*, which concerned a breach of *HSWA 1974, s 2* arising out of a fatal accident to an employee who fell through a trap door in the floor of a lift control room. The accident occurred while the store manager was manually attempting to rectify an electrical fault in the lift in accordance with a local practice which was not authorised by Gateway's head office. The Court of Appeal held that the failure at store manager level was attributable to the employer.

However, it does not follow that an employer will automatically be held criminally responsible for an isolated act of negligence by an employee performing work on its behalf. This is because it may still be possible for the employer to establish that it has done everything reasonably practicable in the conduct of its undertaking to ensure that employees and third parties are not exposed to risks to their health and safety by virtue of the way it has conducted its business (see *R v Nelson Group Services (Maintenance) Limited [1999] IRLR 646, Court of Appeal*).

The definition of 'reasonably practicable' was authoritatively laid down in *Edwards v National Coal Board [1949] 1 All ER 743* where it was said that: 'Reasonably practicable' is a narrower term than 'physically possible', and seems to imply that a computation must be made by the owner in which the quantum of risk is placed on one scale and the sacrifice involved in the measures necessary for averting the risk (whether in money, time or trouble) is placed in the other, and that, if it be known that there is a gross disproportion between them – the risk being insignificant in relation to the sacrifice – the defendants discharge the onus on them.'

In a series of decisions the Court of Appeal have concluded that the words 'so far as reasonably practicable' provide a limited defence to what are otherwise absolute obligations on employers and other duty holders (*R v British Steel Plc [1995] IRLR 310*, approved by the House of Lords in *R v Associated Octel Co Ltd [1996] 1 WLR 1543*). Although the circumstances where such a defence may be established are likely to be rare – it involves doing more than simply exercising reasonable care – the Court of Appeal in *R v Nelson Group Services (Maintenance) Limited* made clear that an isolated act of negligence by an employee performing work on behalf of the company does not preclude that employer from establishing a defence that it has done everything reasonably practicable. The court said:

> *'It is not necessary for the adequate protection of the public that the employer should be held criminally liable even for an isolated act of negligence by the employee performing the work. Such persons are themselves liable to criminal sanctions under the Act and under the Regulations. Moreover it is a sufficient obligation to place on the employer in order to protect the public to require the employer to show that everything reasonably practicable has been done to see that a person doing the work has the appropriate skill and instruction, has had laid down for him safe systems of doing the work, has been subject to adequate supervision, and has been provided with safe plant equipment for the proper performance of the work.'*

The question of what is reasonably practicable is a question of fact which must be determined in the light of the circumstances of each case. The burden of proving, on the balance of probabilities, that all reasonably practicable steps have been taken rests with the employer (see also E15024). In *R v Associated Octel* the Court considered whether the employer had done what was reasonably practicable. The Court stated:

> 'The question of what is reasonably practicable is a matter of fact and degree in each case. It will depend on a number of factors so far as concerns operations carried out by independent contractors: what is reasonably practicable for a large organisation employing safety officers or engineers contracting for the services of a small contractor on routine operations may differ markedly from what is reasonably practicable for a small shopkeeper employing a local builder on activities on which he has no expertise. The nature and gravity of the risk, the competence and experience of the workmen, the nature of the precautions to be taken are all relevant considerations'.

Similarly, where there is a defence of due diligence to the legislation allegedly breached, the company may be able to avoid liability if the members of senior management responsible for actual control of the company's operations have exercised due diligence and the failure occurs because of the actions of a junior member of staff: *Tesco Supermarkets Ltd v Nattrass [1972] AC 153*, a House of Lords decision on the wording of the *Trade Descriptions Act 1968*. The case was considered by the Court of Appeal in *R v British Steel plc [1994] IRLR 540* who distinguished it from the factual circumstances before them on the basis that the *Tesco* decision involved application of a due diligence defence which was not part of *HSWA 1974, s 3*.

## *Offences of directors or other officers of a company*

**E15040** Where an offence is committed by a body corporate, senior persons in the hierarchy of the company may also be individually liable. Thus, where the offence was committed with the consent or connivance of, or was attributable to any neglect on the part of a director or officer of the company, that person is himself guilty of an offence and liable to be punished accordingly. Those who may be so liable are:

(*a*)   any functional director;

(b) a manager (which does not include an *employee* in charge of a shop while the manager is away on a week's holiday (*R v Boal [1992] 1 QB 591*, concerning *s 23* of the *Fire Precautions Act 1971* – identical terminology to *HSWA 1974, s 37*));

(c) a company secretary;

(d) another similar officer of the company;

(e) anyone purporting to act as any of the above.

*[HSWA 1974, s 37(1)].*

It is not sufficient that the company through its 'directing mind' (its board of directors) has committed an offence – there must be some degree of personal culpability in the form of proof of consent, connivance or neglect by the individual concerned. Evidence of this sort can be difficult to obtain and prosecutions under *s 37(1)* have, in the past, been rarely compared with prosecutions of companies (although they are increasing in number). In 2003/04, the HSE prosecuted a total of 17 managers and directors, of which 11 led to conviction.

It is not sufficient that the company through its 'directing mind' (its board of directors) has committed an offence – there must be some degree of personal culpability in the form of proof of consent, connivance or neglect by the individual concerned. Evidence of this sort can be difficult to obtain and prosecutions under *HSWA 1974, s 37(1)* have, in the past, been rarely compared with prosecutions of companies (although they are increasing in number). In 2003/04, the HSE prosecuted a total of 17 managers and directors of which eleven led to conviction.

Directors, managers and company secretaries can be personally liable for ensuring that corporate safety duties are performed throughout the company (for example, a failure to maintain a safe system of work can give rise to personal liability). Liability may also arise as a result of a failure to perform an obligation placed on individuals by their employment contracts and job descriptions – for example, obligations imposed under a safety policy – not just in relation to duties imposed by law. In the case of *Armour v Skeen (Procurator Fiscal, Glasgow) [1977] IRLR 310*, an employee fell to his death whilst repairing a road bridge over the River Clyde. The appellant, who was the Director of Roads, was held to be under a duty to supervise the safety of council workmen. He had not prepared a written safety policy for roadwork, despite a written request that he do so, and was found to have breached *HSWA 1974, s 37(1)*.

Similar duties exist under the *Environmental Protection Act 1990, s 157* and the *Environment Act 1995, s 95(2)–(4)*. (See E5023 EMISSIONS INTO THE ATMOSPHERE.) Directors convicted of a breach of *HSWA 1974, s 37* may also be disqualified, for up to two years, from being a director of a company, under the provisions of the *Company Directors Disqualification Act 1986, s 2(1)* as having committed an indictable offence connected with (*inter alia*) the management of a company. In *R v Chapman (1992) (unreported)*, a director of a quarrying company was disqualified and fined £5,000 for contravening a prohibition notice on an unsafe quarry where there had been several fatalities and major injuries.

## Directors' insurance

For many years companies have been restricted by *section 310* of the *Companies Act 1985* (*CA 1985*) from 'immunising' their directors against liability by providing them with indemnities and similar arrangements. However, since 1990 companies have been able to buy insurance for their directors (usually known as Directors' and Officers', or D & O, insurance). Although the general restriction on companies providing indemnities to their directors was relaxed in certain respects by changes to *s 310* made on 6 April 2005, the ability for companies to purchase D & O insurance remains unchanged.

Such insurance may (subject to the terms of the policy), protect directors against claims made during the policy period for breach of contract, negligence, misrepresentation and negligent misstatement caused by the person in his capacity as a director. Some policies also cover breach of statutory duty, employment claims, misfeasance, disqualification proceedings, proceedings by regulatory authorities (such as the HSE or Financial Services Authority) and corporate manslaughter claims. Policies typically allow directors to recover their legal costs and expenses incurred in defending or settling such claims, but fines and other penalties are usually excluded.

Since 6 April companies have also been permitted (but not required) to indemnify their directors against certain liabilities (such as damages, costs and interest) awarded in a civil action brought by a third party such as an investor or creditor. In certain circumstances companies can also indemnify their directors against the costs of defending proceedings brought by such third parties, by the company itself or by a regulator. The existence of such an indemnity (but not its terms) will have to be disclosed in the company's annual report and accounts.

## Offences due to the act of another person

**E15042** *Section 36(1)* of *HSWA 1974* makes clear that although provision is separately made for the prosecution of less senior corporate staff, e.g. safety officers, works managers, this does not prevent a further prosecution against the company itself. The section states that where an offence under *HSWA 1974* is due to the act or default of some other person, then:

(*a*) that other person is guilty of an offence; and

(*b*) a second person can be charged and convicted, whether or not proceedings are taken against the first-mentioned person.

Where the enforcing authorities rely on *HSWA 1974, s 36*, this must be made clear to the defendant. In *West Cumberland By Products Ltd v DPP, The Times, 12 November 1987*, the conviction of a company operating a road haulage business for breach of regulations relating to the transport of dangerous substances was set aside as the offence charged related to the obligations of the driver of the vehicle and, in prosecuting the operating company, reliance was not placed on *HSWA 1974, s 36*.

## Sentencing Guidelines

**E15043** The level of fine imposed for health and safety offences will ultimately depend on the facts and circumstances of the case, including the gravity of the offence, whether the breach resulted in death or serious injury, and any mitigating evidence the defendant is able to put forward (including details of its means and ability to pay any fine imposed). In the past there have been wide variations in the sentences handed down by different courts for similar breaches of the legislation and there has also been concern at the general low level of fines imposed for health and safety offences. In *R v F Howe & Son (Engineering) Limited [1999] 2 All ER 249*, the Court of Appeal sought to address these concerns by laying down guidelines to assist magistrates and judges in sentencing health and safety offences. The case concerned an appeal by the company against fines totalling £48,000 and an order for costs of £7,500 imposed in respect of four health and safety offences arising from a fatal accident to one of the company's employees who was electrocuted while using an electric vacuum machine to clean a floor at the company's premises. In reducing the level of fine imposed to reflect the company's limited financial resources, the Court of Appeal laid down the following general guidelines:

---

### Sentencing Guidelines

General Principles

The level of fine should reflect:

- the gravity of the offence and the standard of the defendant's conduct;
- the degree of risk and extent of danger;

> - the extent of the breach – an isolated incident may attract a lower fine than a continuing unsafe state of affairs;
> - the defendant's resources and the effect of the fine on its business.
>
> ### Aggravating Factors
> - failure to heed warnings;
> - if the defendant deliberately flouts safety legislation for financial reasons;
> - if the offence results in a fatality.
>
> ### Mitigating factors
> - prompt admission of liability and guilty plea;
> - steps taken to remedy deficiencies;
> - a good safety record.

In applying these factors, the Court of Appeal made clear that every case needs to be considered on its own facts. It declined to lay down a tariff for particular offences or to link the level of fine directly to the defendant's turnover or net profit. However, it emphasised the importance of the defendant's means, as well as the gravity of the offence, in determining the appropriate level of fine, stating that this should be large enough to impress on both the management and shareholders of the defendant company the importance of providing a safe working environment. Although there might be cases where the offences are so serious that the defendant company ought not to be in business, in general the fine 'should not be so large as to emperil the earnings of employees or create a risk of bankruptcy'. In essence, the courts must answer two questions in determining the appropriate level of fine:

1. What financial penalty does the offence merit?
2. What penalty can the defendant reasonably be ordered to pay?

The Court of Appeal specifically made clear that the size and resources of the defendant company and its ability to provide safety measures or to employ in-house safety advisers are not a mitigating factor: the legislation imposes the same standard of care irrespective of the size of the organisation.

It also emphasised that if the offence results in a fatality that is a serious aggravating factor: 'the penalty should reflect public disquiet at the unnecessary loss of life.' In general, cases involving death or serious injury should be dealt with by the Crown Court, which has increased sentencing powers: the judgement in *Howe* made clear that 'magistrates should always think carefully before accepting jurisdiction' in such cases.

In *R v Rollco Screw & Rivet Co Ltd [1999] IRLR 439*, the Court of Appeal approved the principles laid down in *Howe* and indicated that the fine imposed should make clear that there is a personal responsibility on directors for their company's health and safety arrangements. It did, however, acknowledge that caution was necessary in the case of smaller companies, where the directors were also shareholders, to avoid imposing a fine which amounted to double punishment of the individuals concerned.

The court went on to suggest that, in appropriate circumstances, corporate defendants may be ordered to pay fines and costs by instalments over many years. In reducing Rollco's total payment period to 5 years and 7 months, the court held that there was no maximum period for payment of fines and costs. Although there might be good reason to limit the period of payment of fines and costs by a personal defendant, who may suffer anxiety because of his continuing financial obligations, the same considerations do not apply to a corporate defendant. The Court of Appeal indicated that, in proper circumstances, it might be appropriate to order payment of fines and costs by a corporate defendant over a 'substantially longer period' than would be appropriate in the case of an individual.

## Publication of convictions for health and safety offences and enforcement notices

**E15044** In an attempt to improve compliance with health and safety legislation, the HSE are also pursuing an active policy of naming companies that breach the legislation. Details of firms and individuals convicted of health and safety offences are published by the HSE in an annual report that can be accessed at their website: www.hse.gov.uk/action/content/off00–01.pdf. The HSE 'name and shame' website also publicises all prosecution cases taken by HSE since April 1999 which resulted in a conviction as well as, more recently, a register of improvement and prohibition notices issued by the HSE. This information can be accessed at: www.hse-databases.co.uk/notices/. In addition a local authority annual report names duty holders and individuals convicted of health and safety offences in Great Britain following investigations by their inspectors. This report can also be accessed on the HSE website.

## Position of the Crown

**E15045** The general duties of *HSWA 1974* bind the Crown. [*HSWA 1974, s 48(1)*]. (For the position under the *Factories Act 1961*, see W11031 WORKPLACES – HEALTH, SAFETY AND WELFARE.) However, improvement and prohibition notices cannot be served on the Crown, nor can the Crown be prosecuted [*HSWA 1974, s 48(1)*], although Crown employees' can be prosecuted for breaches of *HSWA 1974* [*HSWA 1974, s 48(2)*]. Non-statutory procedures are in place for the issue of Crown improvement and prohibition notices, and for the censure of Crown bodies in circumstances in which a prosecution would otherwise have been brought. In 2003/04, one censure, 14 improvement notices and two prohibition notices were issued against Crown bodies. Crown immunity is no longer enjoyed by health authorities, nor premises used by health authorities (defined as Crown premises) including hospitals (whether NHS hospitals or NHS trusts or private hospitals). [*National Health Service and Community Care Act 1990, s 60*]. Health authorities are also subject to the *Food Safety Act 1990*. Most Crown premises can be inspected by authorised officers in the same way as privately run concerns, though prosecution against the Crown is not possible. [*Food Safety Act 1990, s 54(2)*].

# Environmental Management

## Introduction

**E16001** The management of environmental performance is now well accepted as a critical issue for organisations in both the public and private sectors.

Broadly speaking, environmental management refers to the controls implemented by an organisation to minimise the adverse environmental impacts of its operations. Historically, environmental management tends to have been driven by a complex and interacting array of external pressures, to which business somewhat reluctantly responded. Nowadays, however, it is widely recognised that a positive policy towards environment issues can lead to opportunities to improve business performance in general.

As these benefits become more apparent, most large businesses have responded by becoming more proactive in their approach to environmental management. External pressures are still important influences, but increasingly they tend to shape the nature and scope of business' environmental management practices, rather than triggering them in the first place. The emergence of co-operative and constructive stakeholder dialogue as an element of corporate environmental management is an encouraging indication that business recognises the value of effective and proactive environmental management. Furthermore, the ongoing development of new and increasingly innovative approaches to environmental management reflects the fact that there is commercial value to be gained from continuous improvement.

Examples of sources of pressure on business to adopt more sustainable management practices include:

- corporate social responsibility (CSR) expectations;
- management information needs;
- employees;
- legislation;
- market mechanisms;
- the financial community;
- the supply chain;
- community and environmental groups;
- environmental crises;
- the business community;
- customers; and
- competitor initiatives.

## Internal drivers

### Corporate social responsibility (CSR)

E16002 The concept of CSR has gradually crept into mainstream business practice, reflecting the changed conditions in which business operates. Environmental management is an integral part of effective CSR. The removal of trade barriers, the subsequent growth and political influence of trans-national companies, the opening of previously restricted markets and a general reduction in corporate taxation rates have contributed to radical changes in the way business operates, including increasing the extent to which it controls its own performance. Consequently, the notion of CSR has also changed, with its increasing recognition by the media and its rise up the boardroom agenda. One reason for this has been a tendency for protest groups to target companies' wider impacts on global society as well as over environmental impact issues. In addition, society is looking less to government to control the social and environmental impacts of business activity, and instead is seeing business as being accountable for those impacts. Good corporate governance is no longer merely a reflection of responsible fiscal performance. As a result, the mandate of corporate directors and managers is expanding and they are recognising the need to operate in a more transparent and inclusive manner. Proactive environmental management is a key aspect of that.

In the UK, the Government updated its interpretation of CSR in 2004, describing it as essentially the business contribution to sustainable development. It defined CSR as the action companies take voluntarily over and above the minimum legal requirements for social and environmental performance. This will include environmental protection, as well as equal opportunities, employment terms and conditions and health and safety. It has also since 2002 appointed a Minister for CSR, and has a dedicated website containing information about what its various initiatives are.

It is also leading the way in the international arena after announcing its intentions to commit to international CSR by publishing a framework setting out the Government's approach to CSR at the global level. This framework, which includes the views of business and non-government organisations, is intended to enable responsible business practice for the environment by encouraging best practice and innovation. When implemented, it should bring together the relevant institutions to focus on maximising the positive contributions of UK business to international CSR development.

An initiative by Business in the Community (BITC), which consists of 700 member companies, is the incorporation of its index of corporate environmental engagement into a benchmark CSR index in 2003. Based on an overall score achieved for strategy and management practices in several social and environmental areas, BITC claims the index is the UK's leading benchmark of corporate environmental engagement. In 2004, the annual index, which comprised 176 firms, achieved an increase on 2003 in its overall average score from 72% to 79%. This index is gradually gaining popularity over the BITC environmental engagement index as a benchmark, as many leading companies are now integrating key environmental considerations with broader sustainability concerns into mainstream business management and reporting.

### Management information needs

E16003 Effective business management relies on timely and reliable information on the multitude of factors that influence it. As managers' understanding of the relationship between environmental performance and business performance increases, so too does their requirement for information pertaining to environmental performance. Such information helps to increase their control over those factors.

This reflects the growing acceptance of the sustainability or sustainable development concept, which recognises that long-term business success requires environmental, social and economic factors to be balanced. While there is no clear guidance or agreement on how such a balance should be achieved, it is clear that it must be based on appropriate information on all three primary elements. Thus the recognition of the relevance of environmental performance to business performance, the value of controlling it and the need for expanded management information is an increasingly important driver of environmental management practices. Structured environmental management systems not only provide a means of controlling environmental performance *per se*, but also allow more informed strategic and operational decisions to be made by management.

## Employees

E16004    Employees have a potentially strong influence over the environmental management practices of an organisation. The desire to minimise staff turnover means that companies are becoming more responsive to employee enquiries and suggestions regarding corporate environmental performance. Conversely, companies wishing to attract high calibre recruits are increasingly realising the importance of maintaining a strong and positive corporate image, which is often dependent on environmental performance, amongst a number of other things.

# External drivers

## Legislation

E16005    UK companies are influenced by a range of international treaties, conventions and protocols; European regulations and directives; and domestic legislation. The latter may be a tool for implementing European directives, or they may have been enacted independently of any requirement of the European Union (EU).

Enforcement of the various legal instruments within England and Wales is primarily the responsibility of the Environment Agency, although local authorities also play a role. In Scotland, responsibility lies with the Scottish Environment Protection Agency.

Good corporate environmental management requires a thorough understanding of the legal requirements imposed on a company, in addition to evidence that the company has made reasonable attempts to ensure ongoing compliance with them, either through technological, procedural and/or administrative mechanisms. Many published corporate environmental policies now commit to going 'beyond compliance'. That is, legislative compliance is increasingly being seen as a minimum standard.

Other initiatives are forcing companies that have not voluntarily responded, to give more systematic consideration to environmental management and performance in making strategic and operational business decisions.

The Company Law Review, launched by the Department of Trade and Industry (DTI) in March 1998 has been acknowledged as the most fundamental review of company law in 150 years.

Proposals in a White Paper, issued in 2002, and in a series of consultation papers, which followed, were intended to obligate around 1000 of the largest companies to publish an Operating and Financial Review (OFR), outlining the performance and future direction of the business. Environmental issues would need to be included in this. The last of these consultation exercises was an 'Exposure Draft' of a Reporting Standard on the OFR, issued by the Accounting Standards Board (ASB) in December 2004. After evaluating responses, the ASB will then publish a finalised Reporting Standard. Although quoted companies were required to prepare a statutory OFR for the first time for financial years beginning on or after 1 April 2005, neither the final Reporting Standard or Regulations giving legal effect to the proposals had been issued by that date.

The latest proposals involve a 'principles-based standard', which makes it clear that the OFR should reflect the directors' view of the business. It says that the information in the OFR should be useful to both investors and other users in gauging the strategies adopted by a company, and therefore should give an indication of the potential of those strategies of being successful. A basic framework for directors to apply in order to meet the requirements of the Regulations is also set out, although it is left to directors to consider how best to use this framework to structure the OFR. Future guidance will set out some illustrations and suggestions of specific content and related key performance indicators that might be included.

Directors will have to disclose information on financial and non-financial matters 'to the extent necessary' to allow investors to understand the company's development, performance and position. Among the topics to be considered for disclosure are environment, social and community issues. However, the draft does not contain details of the type of environmental information, if any, that might be disclosed, and states that directors believing their companies do not have a 'significant' environmental footprint do not need to disclose performance data. It is suggested that the OFR might apply where a company meets certain size criteria. At present this would include companies or groups with an estimated aggregate turnover in the region of £1,000 billion and to cover a quarter of corporate economic activity in the UK. It is suggested that the OFR might apply where a company meets certain size criteria. At present this would include companies or groups with an estimated aggregate turnover in the region of £1,000 billion and would cover a quarter of corporate economic activity in the UK.

*The Combined Code on Corporate Governance* and related guidance was issued by the Financial Reporting Council in July 2003. It replaced the *Combined Code Principles of Good Governance and Code of Best Practice*, published in June 1998, which was derived from the *Cadbury*, *Greenbury* and *Hampel* reports. The existing Listing Rules require listed companies to make a disclosure statement in two parts in relation to the Code. In the first part of the statement, the company has to report on how it applies the main and supporting principles in the Code. In the second part of the statement the company has either to confirm that it complies with the Code's provisions or, where it does not, to provide an explanation.

The nature of a company's internal controls depends on a proper identification and evaluation of the risks facing it, which could include environmental issues. Principle C.2 states that the company should maintain a sound system of internal controls to safeguard shareholders' investment and the company's assets. Such controls could include environmental issues. Under the provisions, the directors are required, at least annually, to conduct a review of the effectiveness of internal controls and to report to shareholders that they have done so. If the company does not have an internal audit function it should periodically review the need for one.

The European Commission's Green Paper on CSR, published in July 2001, defined CSR in terms of both company internal management and its impact on society and argued for a European framework for CSR. It covered areas such as social reporting and codes of practice and could eventually lead to legislation within this area. The Green Paper was followed in July 2002 by a Communication on a new strategy for CSR in which the European Commission defined CSR as 'company behaviour, which integrates social and environmental concerns into their operations over and above legal requirements'. Although it stated that CSR should not be mandatory, it did propose a new social and environmental role for business, and it also set up a European multi-stakeholder forum to establish principles for codes of conduct for CSR.

## *Market mechanisms*

While the command and control approach embodied in environmental legislation and regulations represents a significant pressure on business, policy makers have begun to examine new tools to encourage better management of environmental performance.

A range of measures is beginning to emerge designed to influence the economics of polluting activities. These so-called market-based or economic instruments impose costs on pollution-causing activities and provide incentives for companies to look for ways of minimising environmental damage. Such instruments include:

(a)  Climate Change Levy

The climate change levy ('CCL'), which came into effect in April 2001, is a tax on the business use of fossil fuels. It is designed to encourage energy conservation and a switch to cleaner fuels and renewable energy sources such as wind and solar power. All revenues are recycled back to business through a 0.3 per cent cut in employers' National Insurance contributions and additional support for energy-efficiency measures and energy-saving technologies.

(b) Landfill Tax

The Landfill Tax was introduced in October 1996 to encourage companies to reduce their volume of waste produced. The rate of tax paid by landfill operators from 1 April 2005 was £18 per tonne of biodegradable waste. It was confirmed in the 2003 Budget that the rate would increase by at least £3 per tonne per year until it reaches £35 a tonne by 2011 at the latest. The tax is intended to be revenue-neutral to business, with tax receipts going to fund the Landfill Tax Credit Scheme (LTCS). The LTCS allows landfill operators to support a wide range of non-profit environmental projects by giving them a 90 per cent tax credit against their donations. There were no plans to increase the £2 per tonne rate of landfill tax applying to inactive or inert waste in the near future.

(c) Aggregates Levy

An aggregates levy on virgin aggregates was launched in 2002. All the revenue raised is returned to business and the local communities affected by quarrying, through a 0.1 per cent cut in employers' National Insurance contributions and a new sustainability fund. In 2005, the rate was £1.60 per tonne, the same rate as when it began in April 2002. It is intended that the levy will help to ensure that the environmental impact of aggregate extraction is reflected in the price. This is aimed at encouraging more efficient use of aggregates and the development of alternatives including waste glass, tyres and recycled construction and demolition waste.

(d) Vehicle and Fuel Duty

Lower levels of duty on cleaner fuels and changes to company car taxation and vehicle excise duty are designed to reduce pollution, in particular carbon dioxide, from vehicles.

(e) Emissions Trading

Companies that reduce their emissions below a quota can sell the unused part of their quota to other firms, thus providing an incentive to improve emissions performance. Such a scheme is included in the Kyoto Protocol, which was adopted in 1997 and came into force in February 2005. It sets targets for cuts in greenhouse gas emissions by developed countries. Article 17 of the Protocol allows developed countries that reduce their emissions by more than their assigned target to gain credits, which can be sold to other developed countries. In the UK, a voluntary domestic emissions trading scheme for the six greenhouse gases included in the Kyoto Protocol began in April 2002. It was one of the first national greenhouse gases trading schemes in the world. EU wide emissions trading between companies started in January 2005 under a European Directive. This will continue until 2008, when the scheme will be modified during a second phase. It differs from the UK scheme in that it only covers, carbon dioxide, it is mandatory and is restricted to power generation companies and heavy industrial users of electricity.

## *The financial community*

E16007  The emerging realisation of the link between environmental performance and business performance has encouraged investors, shareholders and insurers to develop a direct interest in the environmental performance of companies. Investors are becoming increasingly concerned that companies that fail to manage their social and environmental exposure will suffer. The financial community is therefore now seeking information on how environmental issues will potentially impact on the long-term viability of the companies in which they have a commercial interest. In particular, they are concerned about the extent to which environmental risks are being controlled. Brand reputation accounts for an increasing proportion of stock market valuations. Therefore, they want reassurance that companies are not in breach of legal requirements with the consequent threats of fines, damage to reputation and the need for unanticipated expenditure. They need to be sure that assets, in the form of plant, equipment, property and brand value, against which they have lent money, are correctly valued. Raw materials, by-products and end products may need to be replaced, modified and/or discontinued, which may require provisions or contingent liabilities to be included in the corporate accounts. Such a situation may arise as a result of substances being phased out (for example, legal controls on CFC production), or it may reflect changing market attitudes such that the demand for less environmentally damaging products begins to decline. Additional research and development costs are likely to be associated with such changes.

Consideration of environmental risks and performance is now an established component of acquisitions, mergers, flotations, buyouts or divestments. Management of companies involved in any of these deals must be able to demonstrate that environmental liabilities do not constitute an unacceptable risk for investors or insurers.

Arguably the most striking example of progress is the rapid growth in socially responsible investment ('SRI'). The Dow Jones sustainability indexes and FTSE4Good are two examples of indexes designed to track stocks for the purpose of SRI. They include only those companies with the strongest records of corporate social and environmental performance. Morley, the UK Fund Management Company, has launched a sustainability matrix, which rates FTSE 100 companies according to social and environmental criteria. Business sustainability is rated A–E, depending on factors such as renewable energy use, healthcare and education. The aim of grading companies in this way is to provide clear and transparent analysis of their social and environmental policies and to encourage improvements, to protect and increase shareholder value and to continue to raise awareness of CSR.

Business in the Environment (BiE), the business led campaign for corporate environmental responsibility, publishes an annual index of corporate environmental engagement. The environmental index is acknowledged by the business community as having helped environmental engagement at board level. It has also shown consistently that most FTSE 100 companies are gaining competitive advantage by moving ahead of legislation in managing their environmental impacts. The 2005 report, renamed the Business in the Community Environment Index, concluded that companies that emphasise their environmental performance are viewed by the City as forward-looking and actively managing their reputation. It also highlighted the increasing numbers of companies paying greater attention to their indirect impacts, such as the environmental implications of lending and investment decisions.

Hermes, the independent fund manager has published 'The Hermes Principles', setting out ten investment principles to address what owners should expect from UK public companies and what these companies expect from their owners. The Hermes Principles are designed to encourage companies to communicate clearly the plans they are pursuing and the likely financial and wider consequences of those plans. Principles 9 and 10 deal with social, ethical and environmental issues. These principles call for companies to manage effectively relationships with their employees, suppliers and customers and say that they should behave ethically and have regard for the environment and society as a whole. In addition, they require that companies should support voluntary and statutory measures designed to minimise the externalisation of costs to the detriment of society.

The oil company ExxonMobil, subjected to a boycott campaign by environmental pressure groups because of it opposition to the Kyoto Protocol, has come under continual pressure from investors and some mainstream financial analysts concerned that the companies' environmental stance threats shareholder value. A number of resolutions have been lodged at annual shareholder meetings urging the company to improve its environment disclosure and performance.

A set of principles, known as the Equator Principles, has been developed for banks providing loans for large projects. They are intended to ensure that the projects they finance are developed in a socially responsible manner that reflects sound environmental management practices. The equator principles provide the framework for the issues that banks and borrowers need to be aware of. Most banks have adopted the principles, as they are useful in helping them document and manage their own risk exposure.

The UN Environment Programme has started to develop a set of global principles for investors to encourage them to investigate the risks and opportunities relating to environment and social issues of the companies they or their clients invest in. Due to conclude in November 2005, the proposed principles share a similarity with the banks' Equator Principles for project finance. The UK Government set up a similar initiative – the London Investment Principles – in conjunction with the Corporation of London as a contribution to the World Summit on Sustainable Development in Johannesburg in 2002. UNEP intends eventually to create a global framework involving governments as well as companies.

## Supply chain

E16008    Most businesses are both purchasers and suppliers of a range of goods and services. Introducing environmental criteria into procurement decision-making processes emphasises the importance of issues other than price and quality in purchasing goods or services. Examples of environmental criteria are selecting materials, components or products that were manufactured using relatively less energy than alternatives, or that require relatively less energy in operations. Another approach is to use environmental criteria as specifications for a particular material, component or product, such as conforming to an eco-label. The UK Government has introduced procedures for procurement under its *Greening Government* initiative, which outlines how environmental issues can be taken into account for all buying decisions. As part of a plan to ban hazardous chemicals from all their products, B&Q and Homebase have a policy of pressing suppliers to substitute chemical substances of concern with safer alternatives. This means that one company can directly influence the environmental performance of another. As a result, many companies are taking active steps to control their environmental performance throughout the supply chain. In its survey of 178 companies, which included FTSE100 and FTSE250 companies and sector leaders from the DowJones Sustainability Index, Business in the Community found that just over 25% of participants gave priority to ensuring that their procurement follows good environmental practice.

It is quite common for purchasing companies to require their suppliers to demonstrate ongoing compliance with formal environmental management standards such as ISO 14001 or EMAS, and to provide information on the environmental performance of the supplier and the products and services it provides. Regardless of the approach that is used, the criteria should be relevant to each procurement decision and should be designed to enable a business to directly influence the attributes and performance of its products at other stages of their life cycles.

The EU Eco-label award scheme has been in operation since 1993, when the first product groups were established, and was comprehensively revised in 2000. An Eco-label (the 'Flower symbol') is awarded to products possessing characteristics that contribute to improvements in environment protection. The main objective of the scheme is to encourage business to market greener products by providing information to allow consumers to make informed environmental choices when purchasing. There has been a steady increase in both the number of applications from manufacturers and the number of eco-labelled products marketed in the last few years. At the beginning of 2005 there were twenty-one different product groups, and more than 180 licences had been awarded for several hundred products. The eco-label scheme (as laid down in Regulation (EC) No 1980/2000) is part of a wider approach on Integrated Product Policy (IPP) within the EU Action Programme.

## Community and environment group pressure

E16009    The majority of public pressure on companies to improve their environmental performance is initiated by local communities that experience the direct effects of pollution. Most companies recognise the importance and value of working co-operatively with local communities. In fact many have established community liaison panels, which comprise representatives of the local community. Such panels interact with the company on a regular basis and provide input on environmental and other community issues.

Environmental pressure groups, often supported by a high level of legal and technical expertise, are also influential. Previously the relationship between environmental pressure groups and business was characterised by mutual mistrust and reactive criticism. The emergence of the concept of 'stakeholder engagement', whereby companies take a more inclusive approach to business management, means that environmental pressure groups can be expected to have more direct access to companies and management in the future. In fact their opinions are already actively sought by many organisations, which recognise the importance of constructive dialogue. While companies will not necessarily implement all suggestions made by external pressure groups, the trend towards more timely and constructive dialogue is likely to continue.

A good example of the potential for corporate reputation damage that could occur when a company fails to engage with environmental groups is Greenpeace's Stop Esso campaign. This campaign targeted Esso's parent company ExxonMobil worldwide, for its negative stance on climate change and lack of investment in renewable energy sources, by calling for a boycott of the company's products. Greenpeace claimed that by September 2002, its boycott campaign had resulted in around one million motorists, a quarter of the number of regular buyers of Esso petrol, boycotting its filling stations. Due largely to this campaign, Deutsche bank warned in October 2002 that ExxonMobil was being labelled as 'environmental enemy number one' and that this posed a significant risk to its business.

Similarly, BP has come under pressure from the Worldwide Fund for Nature (WWF), which sold its shares in the company in protest at its environment and employment policies in Alaska. Although the value of the shares was insignificant in financial terms, it is likely to impact on the environmental reputation of BP. This was followed by an announcement by one of the UK's leading ethical investment funds, Henderson Global Investors, that it intended to sell several million pounds worth of BP shares.

The evolution of the internet and other sophisticated communication media has significantly increased public awareness of and access to information about corporate environmental performance, as well as increasing the speed with which such information can be transferred and responded to. Under the EU Directive implementing the first objective of the Aarhus Convention concerning public access to information, the public have recently gained stronger rights to access information on companies' impacts on the environment. This also covers much health and safety information, which broadens the scope of the rules since many industrial projects can potentially affect public health. However, legally classified and commercially confidential information can still be excluded. The result is that pressure on companies to improve environmental performance has increased and has spread into the international arena. Nowadays, companies are finding it necessary to develop and adopt consistent environmental and social performance standards in all markets in which they operate, since their performance is increasingly subject to international scrutiny.

## Environmental crises

E16010    In many cases, high profile environmental crises are a catalyst for improved environmental management. There are two main types:

(a)    crises that are generated as a result of the actions of an individual company, the effects of which are generally experienced at a local scale;

(b) crises that are generated by collective action or by natural forces, the effects of which are often experienced at the national or international scale.

Individual companies that have been associated with environmentally damaging events such as oil spills generally find themselves exposed to intense pressure to improve their environmental management practices in the immediate future. The need to correct the damage to reputation caused by environmental crises is a further incentive to respond quickly. Obviously the costs of responding to pressures arising from catastrophic environmental incidents can be extremely high, and most companies seek to avoid those by incorporating environmental issues into their corporate risk management programmes.

Global or national environmental crises tend to emerge more gradually, and with considerably more debate about accountability and appropriate responses. Nevertheless, there are a number of examples of global environmental issues that have facilitated more systematic and intensive environmental management practices than may otherwise have occurred in the same time period. The most obvious examples are the depletion of the ozone layer and climate change. At a national level, issues such as water shortages, soil erosion and regional air pollution have triggered the adoption of improved environmental management practices on an extensive scale.

## *Business community*

E16011    Trade associations and business groups such as the Confederation of British Industry (CBI) and the International Chamber of Commerce (ICC) have played a leading and effective role in encouraging businesses to adopt environmental management practices. Increasingly the importance of doing this within a sustainable development framework is being accepted. Sustainable development is generally recognised as the inter-relationship and inter-dependence of the three core elements of economic, environmental and social consideration, although it is a concept that is open to wide interpretation. However, it is generally understood to mean achieving a better quality of life with effective environmental protection. The most widely used definition is contained in the 1987 United Nations Brundtland report to the World Commission on Environment and Development. This definition states that it is 'development which meets the needs of the present without compromising the ability of future generations to meet their own needs.'

There is a desire among employers and employees to improve the environment impacts of the workplace. A new website promoting partnership action in the workplace to help achieve sustainable development has been initiated by the Advisory Committee for Business and the Environment and the Trade Union Sustainable Development Advisory Committee. Sustainable Workplace (www.sustainableworkplace.co.uk) provides case studies of employer and union initiatives that have successfully improved organisations' environmental performance. It also contains links to other websites and organisations where business support, information on education for sustainable development, environmental management and reporting can be found, as well as the principles of sustainable development.

Thousands of companies worldwide have been helped by the ICC voluntary business initiative 'Business Charter for Sustainable Development'. Launched in 1991, as a tool to help companies tackle the challenges and opportunities of the environmental issues that emerged in the 1980s and early 1990s, the Charter has established a set of 16 principles to guide company strategies and operations towards sustainable development. Its principles for environmental management have provided a global alignment of business to common objectives and have helped thousands of companies worldwide establish the foundation on which to build their own integrated environmental management systems. The Charter highlights such areas as employee and customer education, research facilities and operations, contractors and suppliers and emergency preparedness. It requires organisations to support the transfer of technology, be open to concerns expressed by the public and employees and carry out regular environmental reviews and report progress.

**E16012** *Environmental Management*

In 1995 the World Industry Council on the Environment (WICE) and the Business Council for Sustainable Development (BCSD) merged to become the World Business Council for Sustainable Development (WBCSD). In 2003, it consisted of a coalition of some 165 international companies from 30 countries committed to the principles of sustainable development. WBCSD seeks to promote the effective implementation of these principles through a combination of advocacy, research, education, knowledge-sharing and policy development.

The World Business Council for Sustainable Development (WBCSD) is a coalition of 175 international companies from 35 countries and 20 major industrial sectors, committed to the principles of sustainable development. Its members include Aventis, Nokia, Lafarge, Renault, BP, Coca-Cola and Unilever. The WBCSD seeks to promote the effective implementation of these principles through a combination of advocacy, research, education, knowledge-sharing and policy development.

The Responsible Care programme is an example of an international initiative from a specific industry sector. The programme has been adopted by around 41 national chemical industry associations, including the Chemical Industries Association ('CIA') in the UK. All of those associations have made acceptance of Responsible Care requirements compulsory for individual member companies. Responsible Care is designed to promote continuous improvement, not only in environmental management, but also health and safety. The companies must also adopt a policy of openness by releasing information about their activities. Adherence to the principles and objectives of Responsible Care is a condition of membership of CIA.

Individual companies have also contributed to the development of improved environmental performance standards. As the relevance of good environmental management to overall business performance has become more apparent, progressive companies have voluntarily adopted a number of innovative and unique approaches to corporate environmental management. This has had the effect of constantly moving the frontiers of 'acceptable' environmental management practices and created substantial peer pressure which in turn has encouraged other companies to adopt similar or even more effective practices.

## *Customers*

**E16012** There is some evidence of the role of customers in the uptake by companies of a sound stance on environmental and other ethical issues. The Co-operative Bank is a well-established proponent of active ethical and environmental policies under which some customers are denied banking. It has attributed one fifth of its 2001–2002 pre-tax profits directly to these policies. Surveys for the bank showed that 14 per cent of account holders regarded ethical issues as the most important of a list of factors when choosing it, whilst 26 per cent viewed them as an important factor. The bank has increased its share of the current account market by 30 per cent since 1996. However, some competitors still feel little pressure from customers on these issues, preferring instead to work with industry to reduce environmental risks.

In the late 1980s and early 1990s some of the major retailers started to market 'green' products, including phosphate-free washing powders and biodegradable cleaning products, in response to consumer concern about high-risk chemicals. A new initiative was announced by Marks and Spencer in February 2005. It is to launch of a new range of bathroom, kitchen and carpet cleaning products, which list the ingredients and contain only 'natural' substances.

Further evidence of the importance of responsible environmental and social policies can be found in the Government's second annual CSR report. Published in May 2002, it claims that around half of consumers have identified that CSR is important when choosing to buy, and 20% will boycott or select products on these grounds.

# Environmental management guidelines

## Standards for environmental management

E16013   Various guidelines exist for responding to those many pressures to minimise damage to the environment and health. Effective environmental management, like quality management or financial management, requires *inter alia*:

(a)   the setting of objectives and performance measures;

(b)   the definition and allocation of responsibilities for implementing the various components of environmental management;

(c)   the measurement, monitoring and reporting of information on performance;

(d)   a process for ensuring feedback on systems and procedures so that the necessary changes can be actioned.

There are two main instruments that influence current approaches to environmental management. The first is the EU Eco-Management and Audit Scheme (EMAS) Regulation. The second is the International Standards Organisation Series of Environmental Management Standards.

## The Eco-Management and Audit Scheme ('EMAS') Regulation

E16014   This voluntary scheme came into force in July 1993 and has been open for participation by companies in all Member States since April 1995.

EMAS has three main aims:

(a)   the establishment and implementation of environmental policies, programmes and management systems.

(b)   the systematic, objective and periodic evaluation of these measures.

(c)   the provision of information to the public on environmental performance.

The Regulation sets out a number of elements of systematic environmental management. Sites that can demonstrate ongoing compliance with those requirements to an independent assessor have the right to be registered under EMAS.

Following the adoption of a company policy, an initial environmental review is made to identify the potential impacts of a site's operations, and an internal environmental protection system must be established. The system must include specific objectives for environmental performance and procedures for implementing them. The system, and the results of the initial environmental review, must be described in an initial environmental statement. The statement must be validated by an accredited external organisation before being submitted to nominated national authorities in individual Member States for registration of the site under the scheme.

There are a number of key points here:

(a)   the first stage in developing environmental management is to carry out a thorough review of impacts on the environment;

(b)   setting up an environmental management system is a prerequisite of registration under the scheme;

(c)   external validation of the environmental statement is intended to ensure consistency in environmental management systems;

(d)   the description of the environmental management system within the statement will be on the public record.

Once a site has been registered under the scheme, it will require regular audits to review the effectiveness of the environmental management system as well as giving information on environmental impacts of the site. Here it is sufficient to note that development of procedures for internal auditing is a crucial part of an environmental management system. In addition to the audit, the preparation and external validation of an environmental statement submitted to the competent authority for continued registration and made public, are elements of an on-going procedure.

In order to attract more registrations and make the scheme more competitive, EMAS was revised by Regulation (EC) No 761/2001, with the revisions taking effect in April 2001. The scheme is now open to all sectors of the economy, including financial companies, transport and local and public bodies, although industry still accounts for 70% of all EMAS registrations. In addition, registered organisations can use an official logo to publicise their participation in EMAS as well as gaining regulatory benefits. The new scheme also encourages more involvement by employees in implementation and strengthens the role of the environmental statement to improve the transparency of organisations and their stakeholders.

Twenty-five of Europe's top 100 firms have at least one EMAS registration. By February 2005, 3,084 organisations and 4,143 sites were registered to EMAS in Europe, with 61 organisations at 62 sites registered in the UK. Organisations appear to see less value in EMAS compared with ISO 14001, for which there were around 74,000 certified organisations worldwide, with nearly 5,500 of these located in the UK, as it does not extend beyond Europe and it is more exacting than ISO 14001.

## *International standards on environmental management (the ISO 14000 series)*

E16015    The International Standards Organisation (ISO) has developed a series of standards for various aspects of environmental management, which it keeps under review. All are designed to assist organisations with implementing more effective environmental management systems. Table 1 outlines the various standards and guidelines within the series.

The most high profile of these standards is ISO 14001, first published in June 1996, which sets out the characteristics for a certifiable environmental management system (EMS). ISO 14001 was based on the British Standard on Environmental Management Systems (BS 7750), although the latter has been superseded by the international standard.

The format of ISO 14001 reflects the procedures and manuals approach of the ISO 9000 quality management systems series. In practice this means that organisations that operate to the requirements of ISO 9000 can extend their management systems to incorporate the environmental management standard, although the existence of a certified quality management system is not a prerequisite for ISO 14001.

ISO 14001 requires an organisation to develop an environmental policy that provides the foundation for the rest of the system. The standard includes guidance on the development and implementation of other elements of an EMS, which is ultimately designed to allow an organisation to manage those environmental aspects over which it has control, and over which it can be expected to have an influence. ISO 14001 does not itself stipulate environmental performance criteria.

Revised versions of ISO 14001 and ISO 14004 were published in November 2004. The new ISO 14001 (ISO 14001:2004) is designed to be more closely aligned with ISO 9001, and to clarify a number of the requirements. ISO 14004:2004 is more consistent and compatible with ISO 14001:2004, to encourage their joint use, and to make it more accessible to SMEs. Although the new 14001 standard is very similar to the previous version, certain changes could affect existing EMSs. Organisations may need to review and amend their system to take account of the changes. The most significant impacts are likely to involve the scope, impacts, periodic evaluation of legal and other requirements, and corrective and preventive action under its EMS. As of May 2006, all

ISO 14001 certified organisations must be approved in accordance with ISO 14001:2004, when any certifications to the 1996 version will become invalid.

The changes to ISO 14001 are being phased in. From 15 November 2004 to 15 May 2005, audits and certification can be performed to either the old or new versions of the standard. During phase 2, from 16 May 2005 to 15 May 2006, no new certificates for IS014001:1996 can be issued. All new assessments and certificate renewals must be undertaken in accordance with ISO 14001:2004. By 16 May 2006 all 14001 all valid certificates must comply with the ISO14001:2004 standard. The British Standards Institution has resources available to help with the transition.

| Table 1: The ISO 14000 series | |
|---|---|
| ISO 14001: 2004 | Environmental management systems – Specifications with guidance for use |
| ISO 14004: 2004 | Environmental management systems – General guidelines on principles, systems and supporting techniques |
| ISO 14010: 1996 | Guidelines for environmental auditing – General principles |
| ISO 14011: 1996 | Guidelines for environmental auditing – Audit procedures – Auditing of environmental management systems |
| ISO 14012: 1996 | Guidelines for environmental auditing – Qualification criteria for environmental auditors |
| ISO 14040: 1997 | Environmental management – Life cycle assessment – Principles and framework |
| ISO 14041: 1998 | Environmental management – Life cycle assessment – Goal and scope definition and inventory analysis |
| ISO 14050: 1998 | Environmental management – Vocabulary |
| ISO/FDIS 14020 | Environmental labels and declarations – General principles |
| ISO/DIS 14021 | Environmental labels and declarations – Environmental labelling – Self-declared environmental claims – Terms and definitions |
| ISO/DIS 14024 | Environmental labels and declarations – Environmental labelling TYPE 1 – Guiding principles and procedures |
| ISO/FDIS 14031 | Environmental performance evaluation – Guidelines |
| ISO/DIS 14042 | Environmental management – Life cycle assessment – Impact assessment |

## *Differences between ISO 14001 and EMAS*

E16016    The key differences between ISO 14001 and EMAS are:

(*a*)    ISO 14001 can be applied on a company-wide basis, whereas EMAS registration is only granted at individual site level or for particular services;

(*b*)    ISO 14001 does not currently include a requirement for public reporting of environmental performance information, whereas sites registered under EMAS must produce an independently validated, publicly available environmental statement;

(*c*)    ISO 14001 does not require a register of environmental effects or legislation;

(d) the level of control of contractors and suppliers required in EMAS is not matched in ISO 14001, which stipulates only that required procedures are communicated to them.

The new revised EMAS allows integration of an existing ISO 14001 certification, to allow a smoother transition and avoid duplication when upgrading from ISO 14001 to EMAS.

# Implementing environmental management
## Practical requirements

E16017   Both EMAS and ISO 14001 set a pattern for companies wishing to develop environmental management systems. The key steps in implementing such systems involve:

(a) conducting an initial review, designed to establish the current situation with respect to legislative requirements, potential environmental impacts and existing environmental management controls;

(b) developing an environmental policy which will provide the basis of environmental management practices as well as informing day to day operational decisions;

(c) establishing specific objectives and performance improvement targets;

(d) developing a programme to implement the objectives and establish operational control over environmental performance;

(e) ensuring information systems are adequate to provide management with complete, reliable and timely information; and

(f) auditing the system to compare intended performance with actual performance.

## Review

E16018   In order to be able to actively manage its interactions with the environment, an organisation needs to understand the relationship between its business processes and its environmental performance. An environmental review should therefore be conducted, which clarifies how various business activities could potentially affect, or be affected by, the quality of different components of the environment (air, land, water and the use of natural resources). This will allow an organisation to understand which of its activities have the greatest potential impact on the environment. Usually, but not always, these will relate to procurement and/or manufacturing processes. It is also important to consider which elements of environmental management and performance have the greatest potential impact on business performance, for example, high profile environmental prosecutions can have a significant impact on corporate reputation and brand value.

The initial and subsequent review should also consider the current and likely future legislative requirements that the company must comply with. It should consider the overall organisational strategy, any other relevant corporate policies, customer specifications and community expectations that could influence environmental management practices. The review also offers the opportunity to establish a baseline of actual management organisation, systems and procedures, its compliance record and the range of initiatives already in place to improve performance.

Ideally, an environmental review should involve input from a range of stakeholders, both internal and external to the organisation. This promotes a wider perspective on potential environmental impacts, and ensures that the resulting policies and programmes to be developed by the organisation, reflects a comprehensive range of issues and risks. Consequently, the chances of unidentified and therefore uncontrolled risks emerging will be minimised.

## Policy

E16019    The results of the initial review should inform the development of a written environmental policy. The policy directs and underpins the remainder of the EMS, and represents a statement of intent with regard to environmental performance standards and priorities.

The policy should be endorsed by the highest level of management in the company, and should be communicated to all stakeholders.

## Objectives

E16020    It is important that the policy be supported by objectives that are both measurable and achievable. They should be cascaded throughout an organisation, so that at each level there are defined targets for each function to assist in the achievement of the objectives. A key part of ensuring continuous improvement in environmental performance is to review and update objectives in the light of progress and changing regulations and standards.

Objectives should be developed in conjunction with the groups and individuals who will have responsibility for achieving them. They should also be clearly linked to the overall business strategy and as far as possible with operational objectives. This ensures that environmental management is viewed as relevant and integral to business performance, rather than being seen as an isolated initiative.

The process of objective setting also needs to consider the most appropriate performance measures for tracking progress towards the ultimate objective. For example, if an objective is to reduce waste by 20 per cent over five years, a number of parameters could be used to reflect different aspects of the organisation's waste reduction efforts towards that, including volumes of waste recycled, efficiency with which raw materials are converted to product, and proportion of production staff that have received waste management training.

ISO 14031 provides useful guidance on the principles to be applied in the selection of appropriate environmental performance indicators. Performance in whichever parameters are selected should be measured regularly to enable corrective actions to be taken in a timely manner. As far as possible, performance measures and the achievement of quantified targets should be linked to existing appraisal systems for business units or individuals.

The Institution of Chemical Engineers (IChemE) has recently launched its own system to enable companies to quantify their sustainable development performance. Key environmental indicators used are resource use, emissions, effluent and waste. The system calculates the environmental burden of each product by using a weighted calculation of its acidification, global warming, damage to human health, ozone depletion and photochemical ozone formation potential. The National Chemicals Industries Association (CIA) has pledged to recommend that its members should use the system.

## Responsibilities

E16021    Allocation of responsibilities is vital for successful environmental management. Its implementation will typically involve changes in management systems and operations, training and awareness of personnel at all levels and in marketing and public relations.

A wide range of business functions will therefore need to be involved in developing and implementing environmental management systems. The commitment of senior personnel to introduce sound environmental management throughout the organisation, and to communicate it to all staff is vital.

Companies have adopted a range of organisational approaches as part of their environmental management systems. In some cases there is a single specialist function with responsibility for monitoring and auditing the system. An alternative is to have a central environmental function with only an advisory role, which can also undertake verification of the internal audits carried out by other divisions or departments. The approach needs to be adapted to the culture and structure of the organisation, but whatever system is adopted, there are a number of crucial elements:

(a) access to expertise in assessing environmental impacts and developing solutions;

(b) a degree of independence in the auditing function;

(c) a clear accountability for meeting environmental management objectives;

(d) adequate information systems to help those responsible for evaluating performance against objectives, to identify problem areas and to ensure that action is taken to solve them.

## Training and communications

E16022  Both ISO 14001 and EMAS include training and communications as a key requirement of the EMS. In particular, they focus on ensuring that employees at all levels, in addition to contractors and other business partners, are aware of company policy and objectives; of how their own work activities impact on the environment and the benefits of improved performance; what they need to do in their jobs to help meet the company's environmental objectives; and the risks to the organisation of failing to carry out standard operating procedures.

This can involve a significant investment for companies, but if integrated with existing training modules and reinforced regularly, many hours of essential training can be achieved. An important benefit of training is that it can be a fertile ground for new ideas to minimise adverse environmental impacts.

Communication with, for example, regulators, investors, public bodies and local communities is an important part of good environment management and requires a preparedness to be open, honest and informative. It also requires clear procedures for liaising with external groups.

## Operational controls

E16023  The operational control elements of an EMS define its scope and essentially set out a basis for effective day-to-day management of environmental performance. Typically, such controls would include:

(a) a register of relevant legislation and corporate policies that must be complied with;

(b) a plan of action for ensuring that the policy is met, objectives are achieved and environmental management is continuously improved. This sets out the various initiatives to be proactively implemented and milestones to be achieved, and could be considered a 'road map' for guiding environmental performance;

(c) a compilation of operating procedures which define the limits of acceptable and unacceptable practices within a company and which incorporate consideration of the environmental interactions identified in the review and the objectives and targets that were defined subsequently;

(d) an emergency response plan to be implemented in the event of a sudden, unexpected and potentially catastrophic event which could potentially influence a company's environmental performance in an adverse manner;

(e) a programme for monitoring, measuring, recording and reviewing environmental performance. This should incorporate a mechanism for regularly reporting back to senior management, since they are the key enablers of the EMS and because they retain ultimate responsibility of business performance.

These operational controls, and indeed all elements of an EMS, should be documented.

## Information management and public reporting

**E16024** The critical factor in an EMS is the quality and timeliness of conveying information to internal and external users. The importance of providing performance information to external users is increasing as expectations of greater transparency and CSR continue to grow. Such expectations have been supported in the UK by strong encouragement from Government for companies to voluntarily and publicly report on their environmental performance. The non-financial report tracking agency CorporateRegister claims that 82% of the UK's top 100 listed firms now publish stand-alone non-financial reports. While up to 1997 nearly all such reports were based on environment issues, it says that the majority are now broader and are focussed on CSR.

Regardless of whether they have adopted EMAS, ISO 14001, or neither, many companies have a statutory duty to report on some aspects of their environmental performance to demonstrate legislative compliance. Some of that information is publicly available. An increasing number of companies are choosing to provide information on their environmental performance, either in their annual report and accounts, or in a stand-alone document. In most cases, the information provided extends well beyond a demonstration of legislative compliance, and leans towards a general overview of all significant aspects of environmental management and performance. Furthermore, many companies, recognising the importance of sustainability, are beginning to measure and report on their performance in terms of social impact. The requirement for large quoted companies to produce an Operating and Financial Review (OFR) for financial years beginning on or after 1 April 2005 should raise the level of corporate environment and social disclosure.

As stakeholders and communication media become more sophisticated, and our understanding of environmental interactions increases, so information requirements become more complex. This means that management information systems must incorporate database management, modelling, measuring, monitoring and flexible reporting. This trend also means that environmental reports that have been produced solely as a means of improving public relations and to defuse external pressures are becoming less acceptable to many stakeholders. The main reasons for this appear to be that they are seldom generated in response to internal management information needs, which in turn means that they tend to focus on statements of management intent, qualitative claims and descriptive anecdotes rather than actual performance. They are therefore less likely to include detailed and verifiable information.

Some companies do recognise the value of maintaining a constructive and open dialogue with stakeholders, and published reports are being used as part of those efforts. The Association of Chartered Certified Accountants (ACCA) has held annual awards for environmental reporting for the last decade, over which time it has identified a steady increase in both the quantity and quality of environmental reports. Furthermore the initial tendency of large manufacturing companies to report has spread to small and medium sized enterprises, the public sector and service industries such as financial services. The structure of the awards was changed in 2001 to take account of the increasing awareness of the integrated environmental, social and economic impacts of business. Under the title 'The ACCA Awards for Sustainability Reporting', the new scheme includes three different award categories: the ACCA UK Environmental Reporting Awards, the ACCA Social Reporting Awards and a new category, the ACCA Sustainability Reporting Awards. In 2005, the ACCA also reviewed all reports for good practice in engaging with stakeholders.

Institutional investors needing reassurance that companies are aware of their environmental risks, are also driving companies into providing more environmental information. Morley Fund Management announced in 2001 that unless a FTSE 100 company publishes an annual environmental report, it would vote against the adoption of its annual reports and accounts. The Association of British Insurers (ABI) issued new investment guidelines in October 2001 after consultation with a wide range of fund managers, corporate executives and NGOs. These guidelines aimed to improve disclosure by companies of their approach to external social, ethical and environmental risks and single out environmental risks as among the most significant

CSR-related risks faced by companies. The guidelines set out the business case for CSR and recommend that CSR reporting should be integrated into annual financial reports to allow for independent verification. By setting out what institutional investors will expect to see disclosed in the annual reports of companies in which they hold stakes, the guidelines will increase transparency and allow shareholders to engage with companies where they consider significant risks have not been assessed adequately.

The introduction in July 2000 of a requirement for occupational pension funds to state the extent to which they consider environmental, social and ethical factors in investment decisions, has also increased the pressure for environmental reporting amongst companies. However, a survey of the SRI policies of the UK's top 100 occupational pension funds, published by Friends of the Earth in August 2001, concluded that although most funds now include SRI in their investment strategy, many impart little force on fund managers to ensure ethical standards. Furthermore, it found that most funds still have no means for monitoring whether trustees and managers are meeting their stated ethical policies.

While there is no unified standard for environmental reporting, a number of sources of guidance are available. Current 'best practice' environmental reporting in the UK is dictated by the United Nations Environment Program (UNEP)/SustainAbility benchmark study into corporate environmental reporting, the results of which have been published annually since 1996. The Global Reporting Initiative (GRI) was established in 1997 in an attempt to develop globally applicable guidelines for reporting on the economic, environmental and social performance of business, governments and NGOs. It incorporates the participation of corporations, NGOs, accountancy organisations, business associations and other stakeholders from around the world. The GRI's Sustainability Reporting Guidelines, released in June 2000, provide an expanded model for voluntary non-financial reporting and reflect the move towards broader sustainability reporting. New guidelines published in 2002, update and expand the original guidelines, which have been used by hundreds of firms worldwide. They include a core set of indicators, which, for the most part, are applicable to organisations in all sectors of commerce and include a list of expanded performance indicators aimed at placing a greater emphasis on social and economic issues, rather than just concentrating on environmental issues.

The guidelines also insist that statements on a company's vision and strategy regarding sustainable development, as well as specific company information on governance, should be addressed before a report can be described as in accordance with GRI guidelines. Guidelines are available for public authorities, the automotive, financial services, mining and metals, tour operators and telecommunications sectors. According to a survey for the UNEP carried out at the end of 2004, increasing numbers of companies are following the GRI guidelines, with UK firms' reports among the best.

Guidelines on corporate environmental reporting were issued and sent to the heads of the 350 largest firms in the UK by the Government in November 2001. They were developed in collaboration with consultants, academics and business, and aimed to set out how to produce an environmental report by drawing on existing reporting schemes, including the GRI, and outline ten basic elements that should feature in a report. The guidelines were accompanied by the results of a survey on the costs and benefits of environmental reporting across different sectors and company size. The survey was sent out to 109 companies, 38 of which were FTSE 100 companies. The content quality of the reports was found to be generally very high and the annual costs of environmental reporting ranged widely from £6,500 to £535,000, with an average cost per firm of £92,716. Four companies that published only on the Internet showed a significant reduction in reporting costs.

The WBCSD has produced its own guide to corporate sustainability reporting. It says that it is intended to help companies to identify the relevant information to be included in order to mitigate corporate reputational and financial risks and to gain competitive advantage. The guidance also contains recommendations on how to fulfil the information needs of the financial community. In addition, the WBCSD has set up a web-based portal to help companies report on their sustainable development activities. The portal provides examples of reporting practices from a variety of companies in a range of business sectors and is designed to generate ideas on what information to include in a report, rather than to prescribe rigid practices. At the end of 2004, the WBCSD published its report comparing the performance of its members with that of the wider market. It showed that companies taking sustainable development seriously consistently outperform the broader market.

The European Commission adopted a Recommendation in 2001 providing guidelines on the information relating to environmental expenditures, liabilities and risks that companies should publish in their annual accounts and reports. Among those proposed is experience in implementing environment protection measures. It also advocates that where relevant, this should contain quantitative measures in areas such as emissions and consumption of water, energy and materials. Unlike a Directive, however, a Recommendation is not binding. In fact, by the end of 2004, no EU Member State had fully implemented the Recommendation. Instead the emphasis has been on adhering to the Modernisation Directive on corporate financial accounts, passed in 2003, as a legal framework.

In an attempt to increase the credibility of published reports and identify opportunities for improving management information systems, a number of companies are seeking independent third party assurance on the reliability, completeness and likely accuracy of information contained in their reports. However, progress on developing standards for independent auditing of environmental reports has been relatively slow. One initiative is the global assurance standard for corporate public reporting on social, environmental and economic performance, launched by the British organisation Account Ability (AA) in March 2003. The AA 1000 Assurance Standard was developed after a two-year worldwide consultation involving hundreds of organisations, including the investment community, NGOs and business. It is openly accessible on a non-commercial basis and has been used by a number of leading companies in the UK. The Standard will also place new demands on external auditors and verifiers of CSR and environmental reports. They will be required to demonstrate independence and impartiality by publicly disclosing commercial relationships with their clients, as well as proving their competency and commenting where a report has omitted information that could be important to stakeholders. The standard is being supported by several organisations including the GRI.

## Auditing and review

E16025   Auditing of environmental management systems, whether conducted by internal or external parties, is a means of identifying potential risk areas and can assist in identifying actions and system improvements required to facilitate ongoing system and performance improvement.

# Benefits of environmental management
## Effective environmental management

E16026   Effective environmental management will involve changes across all business functions. It requires commitment from senior management and is likely to need additional human and financial resources initially. However, it can also offer significant benefits to businesses. These include:

(a)   avoidance of liability and risk. Good environmental management allows businesses to choose when and how to invest in better environmental performance, rather than reacting at the last minute to new legislation or consumer pressures. Unforeseen problems will be minimised, prosecution and litigation avoided;

(b)   gaining competitive advantage. A business with sound environmental management is more likely to make a good impact on its customers. The business will be better placed to identify and respond rapidly to opportunities for new products and services, to take advantage of 'green' markets and also respond to the increasing demand for information on supplier environmental performance;

(c) achievement of a better profile with investors, employees and the public. Increasingly, investors and their advisers are avoiding companies with a poor environmental record. The environmental performance of businesses is an increasing concern for existing staff and potential recruits. Some businesses are finding that a good environmental record helps to boost their public image;

(d) cost savings from better management of resources and reduction of wastes through attention to recovering, reusing and recycling; and reduced bills from more careful use of energy;

(e) an improved basis for corporate decision-making. Effective environmental management can provide valuable information to corporate decision-makers by expanding the basis of such decisions beyond financial considerations. Companies that understand the interactions between their business activities and environmental performance are in a strong position for integrating environmental management into their business, thereby incorporating key elements of the principles of sustainable development.

Effective environmental management can turn environmental issues from an area of threat and cost to one of profit and opportunity. As standards for environmental management systems are adopted and are applied widely, the question will increasingly become, as with quality management, can a company afford not to adopt environmental management? The external pressures on organisations to improve environmental performance are unlikely to abate. Environmental management systems can help companies respond to the pressures in a timely and cost-effective way.

# The future

E16027    Environmental management is now well established as an essential component of effective business management. Environmental management systems have been widely adopted and in most cases these have facilitated demonstrable and ongoing improvements in environmental performance. The involvement of a range of stakeholders in corporate environmental management is no longer the exception; the constructive contribution that they make is actively sought. The publication of environmental reports is also increasing and has become relatively common among leading companies; most include a mechanism for obtaining feedback from external stakeholders, and an increasing number are independently assured in a similar way to annual financial reports and accounts.

There is already a noticeable trend towards sustainability management, whereby companies are attempting to systematically balance economic, environmental and social considerations in business strategies and operations. The strong inter-relationships and inter-dependence between these three elements of sustainable development make it imperative for companies to move towards a more integrated approach to managing them.

The UK Strategy for Sustainable Development has four main objectives: social progress that meets the needs of everyone; prudent use of natural resources; effective protection of the environment; and maintenance of high and stable levels of economic growth. As a direct consequence of the 2002 Johannesburg World Summit on Sustainable Development, the Government announced the launch of a UK strategy for Sustainable Consumption and Production (SCP). Containing the steps to take forward commitments made in Johannesburg, it was designed as a framework for future action by Government and business. A key follow-up action highlighted in the SCP strategy document was a review of the range of Government sponsored programmes that aimed to deliver resource efficiency in business. The review examined the performance and effectiveness of 20 Government-funded initiatives offering support to business on resource efficiency. These included Envirowise, the Sustainable Technologies Initiative, the Waste and Resources Action Programme (WRAP) and the Carbon Trust.

The first set of indicators under the SCP strategy measuring the UK's progress on decoupling environmental impacts from rising economic growth were published in March 2005. They showed that while progress was made in some areas, for instance reduction in emissions of air pollutants and improving river quality, more action is needed in areas such as road transport. These indicators also form part of the Government's Sustainable Development Strategy, which is likely to result in a package of actions targeting producers and retailers to promote the development, production, purchase and use of 'better' products. In addition, and apart from legislation, pressure on business will increase due to initiatives from a number of organisations, including WBCSD, UNEP and the GRI consortium, seeking to implement their guidelines. An effective environment management system will allow business to respond rapidly to these pressures and demands.

Continued development of this approach will be greatly influenced by shareholders, market analysts and financial institutions recognising the value of non-financial performance information as a basis for evaluating management competence and predicting business performance. Consequently they can increasingly be expected to insist on changes to the information presented to them by companies, thereby generating a radical shift in the traditional business paradigm.

Therefore, while environmental management will remain a critical issue for business to address, it will increasingly become integrated with other aspects of business management, reflecting a growing acceptance of the importance of sustainability.

# Ergonomics

## Introduction

**E17001** Ergonomics is concerned with the fit between people and the things they use. People vary enormously in many attributes and abilities such as height, strength, visual ability, ability to handle information and so on. Ergonomics applies scientific information about human abilities, attributes and limitations to ensure that the tools or equipment used, tasks undertaken, workstation, environment and work organisation are designed to suit people. Ergonomics adopts a people-centred approach, aiming to fit the work to the person, rather than forcing the person to adapt to poorly designed equipment, furniture, environments, or work systems. Correct application of ergonomics will produce a work system which optimises human performance and minimises the risk to workers' health and safety. Ergonomics can be applied to any environment or system (work, leisure, travel, home etc) with which people interact.

The word 'ergonomics' comes from the Greek words 'ergos' meaning 'work' and 'nomos' meaning 'natural law'. Historically the term 'ergonomics' has been used in the UK and Europe, while 'human factors' has been used in North America, although this distinction is now blurring. Essentially the terms 'ergonomics' and 'human factors' are synonymous, although in some quarters a distinction has been drawn between physical workplace design (which is referred to as 'ergonomics') and system design and human behaviour (which is referred to as 'human factors'). This chapter does not make that distinction and the term 'ergonomics' is used to cover all aspects of the design of equipment, environment or system in relation to people.

## The benefits of applying ergonomics

**E17002** The benefits of ergonomically designed equipment, furniture, workplaces and tasks are as follows:

- *Improved safety* as people are less likely to make a mistake if equipment is designed taking account of abilities (e.g. if characters on displays are a suitable size and colour that can be read accurately).
- *Reduced ill health and sickness absence* as equipment and workstations are designed to fit people (e.g. seats are comfortable), and tasks are designed to take account of abilities (e.g. work rates and weights handled are within the person's capabilities).
- *Reduced fatigue and stress* as tools, equipment and systems are easier to use.
- *Increased efficiency, performance, productivity and quality of product* (e.g. if tools are designed to fit the hands of the users they will be easier to hold, users will be more comfortable, and may work more efficiently and make fewer mistakes).
- *Increased job satisfaction* as tasks are easier to perform.

Although people are highly adaptable and resourceful and are often able to use poorly designed equipment, systems etc, their use can lead to stress, errors, fatigue and injury. The consequences of not applying ergonomics can be enormous, potentially leading to human suffering through physical discomfort and disability, stress, and accidents. It may also lead to reduced efficiency and productivity.

Ill health resulting from poorly designed furniture, tasks, lifting and handling etc is also extremely costly. Musculoskeletal injuries are the most significant occupational ill health outcome in the UK. There were an estimated 1.2 million people reporting a work-related musculoskeletal disorder in 1995, resulting in approximately 9.9 million days of sickness absence. Manual handling injuries account for a significant proportion of these; more than a third of reportable accidents in 1996/7 were due to handling activities.

Not applying ergonomics can also be one factor that contributes to stress. Stress arises as a result of a mismatch between the demands of the job or situation and the abilities of the individual. Stress results in a significant amount of ill health, with an estimated 500,000 people reporting work related stress/depression, resulting in approximately 6.5 million working days lost in 1995.

Health effects of stress include headaches, indigestion, disturbed sleep and fatigue, changes in appetite, increased alcohol consumption, smoking or drug taking, loss of concentration, irritability, and loss of self esteem. Work-related stress may be caused by poor communication, lack of appropriate training, high workload, lack of control over work, inadequate feedback about work, or repetitive or boring work, i.e. if tasks, jobs and systems are not designed with full consideration of the users. Individuals vary in their response to these factors, but the employer has a responsibility to reduce the risks by designing work appropriately.

As well as increasing the potential for musculoskeletal injuries and increased stress, poor design which results in increased fatigue and error can contribute to accidents and system failures. Inadequate attention to human factors issues has been cited as a contributing element in many catastrophic failures (e.g. Ladbrook Grove rail disaster, Three Mile Island disaster, Kegworth air disaster – see DISASTER AND EMERGENCY MANAGEMENT SYSTEMS (DEMS)). These usually arise as a consequence of a series of errors which are a result of inappropriate design and management. The application of ergonomics can help to reduce these problems. More information on human error is contained in a HSE publication *Reducing error and influencing behaviour* (HS(G) 48) (see E17068 below).

## Core disciplines

E17003    Ergonomics adopts a multi-disciplinary approach to an issue and thus draws on a number of other key disciplines, using this integrated knowledge to obtain a holistic view of the work system. The core disciplines include anatomy (the structure of the body); physiology (the function and capabilities of the body); biomechanics (the effect of movement and forces on the body); psychology (the performance of the mind and mental capability including perception, memory, reasoning, concentration etc); and anthropometry (the physical dimensions of the body).

The following text outlines:

— an overview of the ergonomic approach (E17004);

— design guidelines and principles relating to different aspects of the system (E17005–E17043);

— musculoskeletal disorders (E17044–E17050);

— tools which can be used to assess work and assist with design (E17051–E17058);

— outline of legislation which promotes ergonomic design (E17059–E17068).

# The ergonomic approach

E17004    The ergonomic approach considers all aspects of the person's interaction with the task, workstation, environment and work system so that these can be designed to fit the abilities of the users and maximise the ease of use. The model below can be used to illustrate the framework of this approach.

**Figure 1: Ergonomic Approach**

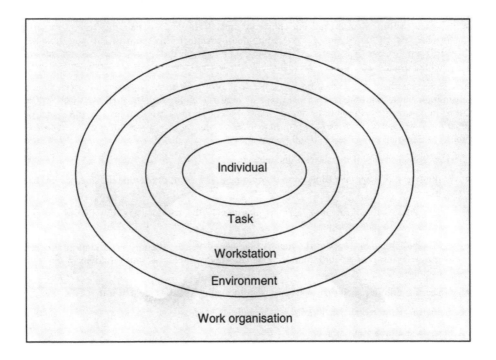

Ergonomics takes account of the physical and mental capabilities and needs of the potential users/ workers/ population (e.g. the size, strength and ability to handle information), before considering the task and the context that it will be undertaken in. By establishing the physical and mental capabilities of the user group, certain criteria or constraints can be established for the task. For example, most people can accurately remember up to seven numbers in short-term memory. A picking task requiring order numbers to be remembered should therefore be designed so that the order number is seven digits or less.

Likewise, the workstation should be designed in relation to the physical requirements of the person and the operational requirements of the task. For example, the users' height and reach will be relevant to specifying the dimensions of the workstation; the sequence of use of tools will be relevant to their arrangement on the workstation.

The work environment needs to be designed in relation to the person, task and workstation requirements. For example, the lighting and noise should be within acceptable limits for human work but also appropriate to the requirements of the task. (See also LIGHTING, and NOISE AND VIBRATION.)

The work system is the organisation within which these elements sit and which ultimately affect the whole work. This includes the way in which the person, task and workstation are organised, as well as the requirements and constraints that affect the way work is conducted. It should complement and not constrain the effectiveness of the person, task, workstation and environment interactions.

Ergonomic criterion for these elements are discussed in more detail in paragraphs E17005–E17043 below.

## Designing for people

**E17005** When adopting an ergonomic approach, the person is the central point of the design of any system. Work should be designed to optimise the person's ability to complete the work, i.e. so that it is within their mental and physical capabilities, and so that they can work in a comfortable and effective manner – the equipment, furniture, environment and task activities should all support this. It is therefore important to understand the abilities and characteristics of the potential users.

It is clear that there are wide differences between people in their physical and psychological capabilities. Understanding the normal range of human abilities and potential variations within this is important in being able to design appropriately. In most work, transport and domestic situations it will not be possible or appropriate to select who uses equipment or systems, and these should be designed to be useable by all potential users.

Physical characteristics of interest will include:

- the size and shape of people so that equipment and furniture can be designed to 'fit' the users;
- strength, so that loads and tasks are designed to be within users' capabilities without leading to fatigue or injury;
- visual abilities, in terms of the items to be viewed, taking account of differences between people in terms of colour blindness, long and short sightedness, and variations in this with age;
- hearing abilities, such that instructions and warnings can be clearly heard.

Psychological characteristics of interest will include:

- short and long term memory;
- uptake and processing of information;
- reaction time;
- motivation;
- concentration;
- perception.

These physical and psychological characteristics will vary between people and will depend on individual factors such as age, gender, training and skills, health and fitness, and previous injury or disability. For any individual these may also vary over time (e.g. visual ability tends to decrease with age).

## Posture

**E17006** In order to work comfortably and reduce the risk of musculoskeletal injury or discomfort, people should be able to adopt 'neutral' postures. A neutral posture is one in which the joints of the body are at about the midpoint of their comfortable range of movement and with the muscles relaxed (e.g. arms hanging by the side of the body). Deviations from this place a strain on the muscles and soft tissue.

The trunk is in a neutral posture when the natural 'S' curve of the spine is maintained – as when standing. The lower part of the spine (lumbar region) flattens when in a sitting posture due to rotation of the pelvis.

The wrist is in a neutral posture when the muscles are relaxed and the hand is in line with the forearm. The arm is in a neutral posture when relaxed by the side of the body.

Figure 2: The 'S' Shaped Curve of the Spine (S Pheasant Bodyspace: Anthropometry, ergonomics and the design of work (1996) p 69 fig. 4.1 reproduced with permission of Taylor and Francis)

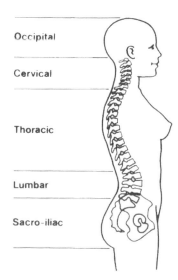

Depending on the degree of deviation and the duration of maintaining the posture, awkward postures (i.e. when the posture deviates from the neutral) and static postures (i.e. postures in which there is no movement for a period of time), can both lead to fatigue, discomfort, strain and possibly injury. Awkward postures such as bending, twisting and stretching, reaching the arms behind the shoulders, and reaching above shoulder height or below knee height should be avoided as far as possible; this can be achieved through careful positioning and adjustment of the equipment and furniture used. Introducing movement and changes in posture will help alleviate the discomfort which can arise from static postures.

## *Anthropometry*

E17007       Anthropometry (measurements of body dimensions) can be used to ensure that people will be able to reach, use and operate tools, equipment and workstations, by structuring and positioning items appropriately (e.g. so they are within easy reach, so they adjust sufficiently, so they allow or prevent access etc).

The size of the body varies with age, gender and ethnicity. Tables of anthropometric data are available which give data for many different body dimensions, representing different population groups (see S Pheasant *Peoplesize* (*Open Ergonomics*); *Anthropometrics: An introduction for schools and colleges* (1984) BSI p 7310). Variations in body size usually follow a normal Gaussian distribution, with most people falling within a central value for a given body dimension with fewer people tailing off at either side. It is common to design to take account of 90 per cent of the population, e.g. to design for those who are taller than the smallest 5 per cent of the population, and smaller than the tallest 5 per cent of the population (i.e. to design within the fifth to ninety-fifth percentile data). However, in some situations it is necessary to design for the extremes of the population. For example, the height of a door should take account of the tallest potential user; guarding on a machine should take account of the finger width of the smallest potential user.

## Population stereotypes

**E17008**    Population stereotypes are cues and expectations which provide information regarding the meaning, state or operation of items. We have certain expectations concerning colours, directions of movement, shapes, sounds, and abbreviations. For example, a red tap indicates that it supplies hot water; turning a control dial clockwise turns the power up. Designs that take account of population stereotypes, will result in increased accuracy and faster reaction times. Population stereotypes can vary by culture and often by industry.

Examples of these stereotypes are shown below.

| Population stereotypes in the UK |
| --- |
| *Colour* |
| ●Red = hot, danger, warning, stop |
| ●Green = safe, go |
| ●Yellow = caution |
| ●Blue = cold, information |
| *Shape of signs* |
| ●Round = enforceable |
| ●Triangle = caution/ attention |
| ●Square = advisory |
| *Direction* |
| ●Clockwise = on, tightens (screws), increases |
| ●Anti-clockwise = off, loosens, decreases |
| ●Upwards = switches off, increases |
| ●Downwards = switches on, decreases |

## Allocating tasks to people or machines

**E17009**    In most systems people interact with machines. When designing a system, tasks can be allocated to either the person or the machine. Decisions on whether a task should be undertaken by a person or by a machine will depend on an understanding of their relative capabilities. Some general guidelines on the relative capabilities of people and machines are shown below:

Figure 3: Guidance for allocating functions (adapted from Sanders and McCormick, *Human Factors in Engineering and Design* (1987) McGraw Hill)

| Person's Capabilities | Machine's Capabilities |
| --- | --- |
| ● Sensing stimuli – both at low levels and over background interference | ● Sensing stimuli outside range of human sensitivity |
| ● Recognising patterns of stimuli | ● Monitoring for prescribed events, especially infrequent ones |

| | |
|---|---|
| • Sensing unexpected occurrences | • Storing coded information quickly and in large quantities |
| • Decision making | • Retrieving coded information quickly and accurately |
| • Information recall | • Calculating |
| • Creating alternatives | • Exerting considerable force |
| • Interpreting subjective data | • High speed operations |
| • Adapting responses | • Repetitive tasks |
| | • Sustained accuracy and reliability |
| | • Counting or measuring |
| | • Simultaneous operations |

In allocating function between people and machines economic or social factors may also need to be considered beside performance. In some situations the 'best' performance may not be required and contact with people may be preferable to dealing with machines. Also, in order to optimise job design, retaining some 'interesting' tasks for people may be more important than mechanising.

## Designing the task

E17010    The task is the collection and sequence of activities and events that allow the work to be completed. The task should be designed to be within the capabilities of the person, but must take account of the constraints of the work process, environment or system. Mismatches between task requirements and individuals' capabilities increases the potential for human error.

Task design will influence the activities of the person e.g. the postures adopted, amount of force applied, repetition, and duration of the activities. These may have an influence on performance outcome or on health (e.g. musculoskeletal disorders). Issues to consider include:

- *Duration of the task* – long durations of work without a break may lead to reduced performance, and increased discomfort. Decline in concentration usually becomes evident after 30 minutes; physical discomfort increases if static postures are maintained or movements are repeated without a break.

- *Breaks and changes in activity* – infrequent breaks can lead to reduced performance (e.g. on a vigilance task such as inspection) and increased discomfort. Regular changes in posture and movement help reduce physical discomfort and injury. Prolonged periods looking at a display or screen without a break can also lead to visual discomfort. Varying the tasks that are undertaken will also help to reduce boredom.

- *Amount of repetition of movements required* – highly repetitive work which involves the same muscle groups and movements can lead to physical discomfort and boredom.

- *Frequency of the task* – frequently performed tasks may lead to boredom and increased physical discomfort (as the same movements are made). Conversely, there may be training and information needs for infrequently performed tasks.

- *Pacing* (if the person is required to work at a speed set by a machine or process) – this will dictate the speed of work and amount of recovery the person has between tasks or operations. This can have an impact on discomfort and performance. A comfortable work rate should be established.

- *Work rate* – excessively busy periods or quite periods can lead to stress and discomfort. There should not be benefits for completing the task early as this can lead to people rushing the task.
- *Complexity of the task* – a number of sub-tasks usually are required to complete the operation, and more than one operator may be required to assist in the operation. Tasks can be divided so that operators each undertake one sub-routine or a more complex combination of tasks. Jobs should be designed to facilitate development of skills, interest, variety and commitment. Several sub-tasks may be undertaken by one operator, but to reduce boredom and to prevent overload and the risk of musculoskeletal injury these should vary in the demands they place on the physical and psychological capabilities of the workers.
- *Awkward or static postures required* – these may be dictated by task requirements, or poor tool or workstation design. These can contribute to musculoskeletal disorders and should be avoided.
- *Force required to undertake the task* – application of frequent or excessive force, particularly in relation to the capabilities of the body part applying the force, can lead to discomfort. It may be appropriate to mechanise the task or use a tool or equipment to assist in generating force if required.
- *Sequence of use of tools/ equipment* – appropriate layout of equipment and tools will facilitate ease of use, so that, for example, those used most frequently are positioned closest to the user.
- *Levels of concentration and attention required* – the level of arousal (the state of consciousness) helps to maintain attention and affects performance. Low levels of arousal are found in undemanding jobs, where activities are either very mundane or very infrequent, e.g. production line packing, night security work. High levels of arousal can be found in stressful and demanding jobs, e.g. air traffic control and ticket inspection. Both high and low levels of arousal can lead to poor performance. Tasks should be designed to maintain the vigilance and arousal necessary for the task. Methods of facilitating concentration and vigilance include reducing the task time and introducing frequent breaks; providing music or background noise for routine tasks; and exaggerate the size or colour of the stimulus.
- *Information and training required for the task* – some tasks will require personnel to be trained in how to undertake them. For others, simple information or instructions may be adequate. The need for training and information will depend on the complexity of the task, and the experience and abilities of potential users.

Task design will be influenced by the equipment and workstation design, the organisation of the work, and the abilities of the individuals. Application of anthropometric and biomechanical knowledge will help ensure the physical elements of the task are within the physical capabilities of the person. Understanding the psychological abilities of the person such as their memory and concentration will help to design a task within limits acceptable to the person. The provision of any information or instruction should also account for the way in which people process information.

In general it is those who are undertaking the task who have the most knowledge about the needs of the work system. Involving workers in evaluation and re-design is an important component in ensuring the design meets their needs and those of the system, and that any changes made are accepted by workers.

## *Design of equipment*

E17011   Equipment should be designed and selected for those who will use it, taking account of the task to be completed and the conditions under which it is used.

## Hand tool design

**E17012**  Tools enable the use of the hand to be extended, e.g. to extend its ability to apply force, manipulate items, make precise movements and so on. Appropriate design of hand tools can facilitate the task, and reduce the risk of musculoskeletal injury.

Hand tools should be designed to allow neutral arm, wrist and finger postures in operation. The hand is in a neutral posture with the hand in line with the wrist, the thumb facing upwards and the palm facing inwards. This can be seen when the shoulder is relaxed and the arm is allowed to hang relaxed by the side of the body. The maximum grip strength can be applied when the wrist is in a neutral posture; deviations from this (ulnar or radial i.e. side to side, flexion or extension i.e. up or down) will reduce the amount of force that can be applied, with the hand able to apply the least force in a flexed posture. Working with the hand in a neutral posture also minimises the strain placed on joints and other soft tissue, which can be compressed or experience friction in awkward postures.

Angling either the work piece or the tool handle can help reduce the amount of wrist deviation required when using a hand tool; however, the potential for variation in orientation of the tool on the work piece may limit the benefit of this.

Key points in hand tool design are:

1. Handles should be long enough to fit the whole hand i.e. at least 100 mm, although a length of 120 mm is preferable.

2. In general, larger handles decrease the amount of muscular activity required when using the tool. A handle thickness of approximately 40mm is generally recommended. Slightly thicker handles may be beneficial when applying torque (e.g. for screwdrivers).

3. Tools with two handles which require a hand span (e.g. pliers, scissors) should have a span of approximately 60 mm. If the tool is used repetitively, an automatic spring opener will reduce the strain on the weaker finger extensor muscles (used to open the hand).

4. Ideally the handle surface should be compressible, textured and non-conductive. It should be free from sharp edges; avoid ridges or finger contouring on the handle as this can place pressure on the soft tissue of the hand. Avoid cold surfaces (e.g. metal), particularly if the tool is powered by compressed air.

5. An excessively smooth handle surface requires the user to grip the handle more tightly, increasing the amount of force required, and this should be avoided. If the hands are sweaty or if the user is wearing gloves they will also have to grip the handle more tightly.

6. It should be possible to use the tool in either hand; if this is not possible, specific tools for left handed workers should be provided.

7. Power-assistance – power assisted tools can greatly increase the speed of task completion, remove a large degree of force exertion from the operator and reduce some awkward postures such as rotating the wrist. However, the user may experience vibration from the tool. Pneumatically powered tools can blow cold air exhaust over the hands, increasing the risk of discomfort and musculoskeletal problems such as Hand Arm Vibration Syndrome (see OCCUPATIONAL HEALTH AND DISEASES). If power tools are used, ensure they are low-vibration and well maintained to reduce vibration transmitted to the hand and arm. Mounting power tools in a jig can reduce vibration transmission.

8. If force has to be applied through the tool, where possible use power tools rather than tools requiring manual application of force. However, power tools often weigh more than manually operated tools.

9. Weight – frequently used tools should weigh as little as possible; a maximum of 0.5kg is recommended. The distribution of weight in the tool should be even with the centre of

gravity as close to the hand as possible. Counterbalances may be required to support heavier tools or those with an off-set centre of gravity.

## Design of controls

E17013   Displays and controls are the mechanisms by which people and machines interact. The person gives instructions to the machine through the controls (e.g. knobs, switches, levers, buttons). Controls can be discrete, having a set number of conditions (e.g. on/ off/ standby); or continuous where the condition varies along a scale, (e.g. volume). Controls should be designed to facilitate the changes that they allow and the amount of effort required to operate them. The following factors should be considered:

- The amount of force required to operate the control (and the amount of resistance to prevent accidental operation). Fingers and hands should be used for quick, precise movements; arms and feet for operations requiring force.
- The size of the control should be appropriate for the force required, and body part required to activate it.
- The location of controls should facilitate their use. Hand-operated controls should be easily reached and grasped, between elbow and shoulder height. Controls should be sufficiently far apart to allow space for the fingertips. Sequence and frequency of use, and importance may also dictate the location of controls.
- Appropriate feedback should be provided to indicate activation of the control.
- Controls should be coded by controls by colour, shape, texture or size (to allow identification by touch).

## Design of displays

E17014   The state of the machine is relayed to the person through a display. Displays should be designed to enable users to easily and accurately assess information from the machine. The following principles should be followed:

1. Qualitative displays showing a small number of conditions should be used for discrete information (e.g. on/ off). Quantitative displays should be used to present numerical or continuous information, e.g. temperature, speed etc. Qualitative displays may be lights or words etc. Quantitative displays may be scales, counters etc.
2. Display scales should be clear, unobscured and concise. Scale intervals should increase left to right, bottom to top, preferably increasing in units of 10s, 100s, 1000s, etc. as appropriate.
3. Displays should conform to population stereotypes in terms of colours used (e.g. red = danger)
4. Ensure labelling, if used, is clear.
5. Limit the number of warning lights to avoid confusion and aid identification.
6. On dials with pointers, avoid parallax (the difference in scale reading depending on the viewing angle) by keeping the dial and pointer close together.
7. Text should be a suitable size and font to allow easy reading. If this is not appropriate, B and 8, O and 0 can be confused, and if resolution is not adequate, F and P can also be confused.

## Grouping of controls and displays

**E17015** The result of activating a control is often indicated in a display. The relationship between displays and controls should be clear; this can be achieved through location, arrangement, text, shapes, colours and responsiveness. Grouping of controls and displays can improve their association with their function and facilitate their ease of use. Controls and displays can be grouped according to sequence of use (e.g. start, run, finish) or by function (e.g. keeping all controls concerned with lights together) or by frequency of use, such that those most frequently used are within convenient reach and within the immediate viewing arc. Locating controls by importance of use can also aid their operation (e.g. emergency controls should be placed within convenient reach).

## Text

**E17016** Text used as information instruction or labelling should be clear and concise. Bold, italics, large fonts, underlining and colour can all be used to draw attention to or highlight information but should be used sparingly to preserve the meaning when they are used. Lower case should be used for phrases or sentences as UPPERCASE DECREASES THE DIFFERENTIATION BETWEEN LETTERS AND IS SLOWER TO READ.

Positive instructions should be used, not negative or double negatives. For example: 'When the alarm sounds turn the machine off' – not – 'When the alarm sounds do not leave the machine on'.

## Software design

**E17017** The field of Human Computer Interaction (HCI) is a specialist area; in this context suffice to say that software should be designed to make the use of a system intuitive and accessible (e.g. through the arrangement of icons, use of colours and text, structure of menus and commands).

# Workstation design

**E17018** The workstation (the area where the task is conducted) should be designed so that it is appropriate for the person using it. The workstation may be a desk or work surface, but may equally be at a conveyor belt or a driver's cab. Attention needs to be given to the physical dimensions of the workstation, and the arrangement of any necessary equipment/ tools/ components in it to allow good posture and acceptable movements.

The issues discussed below cover the main points that should be considered in the design of workstations.

## Worksurface height

**E17019** Many tasks will involve workers using a work surface. The height of the worksurface should permit a relaxed, upright posture and facilitate the use of equipment at the workstation. The surface height (whether designed for sitting or standing tasks) should be such as to avoid users stooping or reaching to its surface. In most situations the height of the worksurface is fixed. For seated tasks a height adjustable chair should be provided so that the working height can be set appropriately. A footrest may be required to support the feet when sitting at a comfortable height.

An appropriate worksurface height is dependent on the work that is done at that worksurface. As a general rule, the top of the worksurface should be level with the user's elbow height (the user may be sitting or standing, as required at the workstation). The height of the worksurface may need to be reduced to take account of the thickness of the item being worked on so that this is at an appropriate height.

There may be a range of users using the same workstation, and unless easy height adjustment is provided, it will not be possible to obtain the optimum height for all potential users. For a fixed height workstation some compromise will have to be made, based on the task requirements and the likely users.

The following working heights are recommended for different tasks, in relation to the user's height (S Pheasant *Ergonomics, work and health* (1991) Taylor and Francis).

| Task | Height |
| --- | --- |
| Manipulative, requiring force and precision | 50 mm – 100 mm BELOW elbow height |
| Delicate (including writing) | 50 mm – 100 mm ABOVE elbow height |
| Heavy – requiring downward pressure | 100 mm – 250 mm BELOW elbow height |
| Lifting and handling | Hands between knuckle and elbow height |
| Two-handed pushing/ pulling | Hands just below elbow height |
| Hand-operated controls | Located between elbow and shoulder height |

The variation in recommended working heights for different tasks arises from the aim to minimise the effort required to perform the tasks, by optimising the posture and muscle groups used.

Using this data, as an example, tasks such as component assembly where pneumatic screwdrivers are used should be conducted at between 100 mm and 250 mm below elbow height so that downward pressure can be applied. If the unit being assembled stands 50 mm above the worksurface, the worksurface will need to be 150 mm – 300 mm below elbow height. This assumes that all work at the workstation is of this nature. Workstations that are also used for writing tasks, for example, should be split-level, providing a writing surface at or just above elbow level. If a fixed height workstation and height adjustable chair are provided, users should adjust the height of the chair depending on the task. For example, VDU users who also write at their workstation will find it more comfortable to sit higher when they are keying and lower when they are writing.

Height adjustable workstations can be beneficial, particularly for standing tasks and where a range of different users may work at the same workstation. Where height adjustable equipment is provided users should be trained in how to adjust it and how to identify an appropriate working height.

## *Workstation characteristics*

E17020    Other characteristics recommended for the workstation include:

- The workstation should be a suitable size such that all equipment can be located and arranged conveniently for the person to complete the task.

- The workstation surface properties should not present a risk to the user – there should be no sharp edges or unprotected hot/ cold surfaces. A front edge of 90° can cause discomfort if the arms are rested or pivoted on it.

- Surrounding workstations or other items in the environment, e.g. columns or posts, should not constrain the user's posture and ability to get into and from the workstation easily.

- There should be sufficient legroom underneath the worksurface to allow the user to be able to sit comfortably, sufficiently close to the workstation, without their posture being constrained.

## Workstation layout

**E17021** To facilitate a good working posture, the equipment and items used to complete the tasks should be within convenient reach of the user, so they do not have to stretch or lean. The zone of convenient reach and the normal working area can be used to establish a suitable layout of the workstation.

### Zone of convenient reach

**E17022** The zone of convenient reach is defined by the area from the shoulder to the fingertips with the arm outstretched (upward, downward and to the sides), which can be reached without any undue exertion. Items required for the task (e.g. control buttons, handles, work surfaces, tools) should be placed within this zone. The extent of this area obviously depends on the length of the arm and its arc. The fifth percentile arm length (i.e. 95 per cent of the population will be able to reach items if placed within this area) is 735 mm for British men and 655 mm for British women when reaching forwards (S Pheasant *Peoplesize* (*Open Ergonomics*). If designing for both male and female users the shorter dimension (i.e. 655 mm) should be used.

**Figure 4: Zone of Convenient Reach (E Grandjean Fitting the Task to the Man (1988) p 51 fig. 42 reproduced with the permission of Taylor and Francis)**

### Normal working area

**E17023** The normal working area is defined by the comfortable sweep of the forearm with the elbow bent at 90° and the upper arm in line with the trunk. This provides an area that can be reached without extension of the arm at the shoulder and requiring no trunk movement. The extent of the area depends on the length of the arm from the elbow to the fingertips. The fifth percentile forearm length is 443 mm for men and 402 mm for women (S Pheasant *Peoplesize* (*Open Ergonomics*). Again, if designing for both male and female users the shorter dimension (i.e. 402 mm) should be used.

### Arrangement of items on workstation

**E17024** Frequently used items and critical items (e.g. emergency controls) should be placed in the normal working area. Occasionally used items can be sited within the zone of convenient reach. Items can be grouped by function or sequence of use, or importance.

Sufficient space should be provided at the workstation to allow a flexible arrangement of all the equipment that is required for the task. Items should be arranged so operators do not have to reach across their body e.g. the phone should be placed on the left hand side if it is answered with the left hand.

## Visual considerations

E17025    The following factors should be taken into consideration.

### Viewing angle

E17026    Items can be viewed both by movement of the eyes, and by movements of the head, neck and trunk.

Awkward neck and trunk postures may be required if the viewed item is low, high or to one side and this may cause discomfort. Items to be viewed should be positioned such that no twisting or bending of the neck, head or trunk is required to see them, and so that the eye muscles are in a neutral position.

The neutral position of the eye is generally taken as about 15° below the horizontal line of sight (i.e. 15° below level with the eyes). The eyes can comfortably move about 30–45° below the horizontal line of sight before the head must be inclined; an upward gaze of approximately 15° is achievable before the head must be tilted backwards to view further. An upward gaze is fatiguing when sustained for any length of time, as the muscles controlling upward movement of the eyes are not as strong as the muscles controlling downwards movement. The most comfortable eye position is an arc of 30° from the horizontal position of the eyes downward and this is the acceptable viewing angle.

Figure 5: Viewing Angle (S Pheasant Ergonomics, Work and Health (1991) reproduced with permission of Palgrave Macmillan)

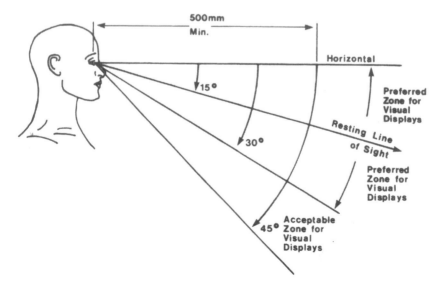

## Acceptable Viewing Angle

E17027    Items outside about a 60° arc in front of the person will involve neck twist to view. Items to be viewed should not be positioned outside this area.

### Viewing distance

E17028    The required viewing distance generally depends on the person's visual ability, the size of the item being viewed and the lighting levels. A viewing distance of between 500 mm and 750 mm is likely to be suitable for most items, but a shorter viewing distance may be required for fine work.

### Sitting versus standing

E17029    There are biomechanical benefits to both sitting and standing. Sitting can help to reduce the development of fatigue in the lower limbs, provided the chair provides adequate support and the feet are also supported. Tasks that require generally static postures (with limited or no trunk movement) will benefit from being undertaken while sitting. However, inappropriate seating and prolonged sitting can lead to discomfort and fatigue, and may prevent good posture (e.g. armrests may prevent the chair being brought sufficiently close to the workstation) (see WORKPLACES – HEALTH, SAFETY AND WELFARE W11008).

Standing is beneficial if the task requires a range of body movement, or handling of heavy items. However, prolonged standing, particularly with little walking, can result in workers experiencing tired or swollen legs and lower back discomfort. The *Workplace (Health, Safety and Welfare) Regulations 1992* (*SI 1992 No 3004*) require a suitable seat be provided for each person at work whose work (or a substantial part of it) can or must be done seated. Whether sitting or standing, workers should be able to vary their posture and should be able to take short, frequent breaks away from the workstation.

Sit/ stand stools can be used to allow work to be completed at standing height while removing some of the strain in the legs and back associated with standing.

Anti-fatigue matting can be used at standing workstations to cushion the feet, and this is especially beneficial on concrete floors as these are particularly fatiguing.

## Physical work environment

E17030    The work environment (the temperature, humidity, lighting and noise and general layout of the place of work) should also be designed to suit the person and the work. In general, the effects of the environment can impact on performance, subjective comfort, perception, attitudes, safety and health.

### Lighting

E17031    Appropriate levels of lighting are necessary to maintain acceptable levels of quality and performance, and to help avoid visual discomfort. Poor lighting can lead to visual fatigue (symptoms include red or sore eyes, blurred vision, headaches). Inappropriate lighting can also force users to adopt poor postures in order to view items more easily, and this can also lead to discomfort.

Issues to consider in designing visual environments are the brightness of the light sources, their position, the evenness of the light distribution, the contrast between the different work items, and the potential for glare/ reflections from other items in the work area.

The amount of illuminance required should be determined by the task demands e.g. visually demanding tasks such as inspection, or fine and precise work require a higher level of light than that required in walkways. Insufficient lighting can result in eye strain and discomfort, while too much light, or a bright light source within the visual field will cause glare and result in visual discomfort and reduced visual performance. Where light levels are already high it may be appropriate to improve the legibility of the source document or item being viewed (e.g. increasing size and contrast) rather than further increasing the lighting levels.

The location of lights will also effect the visual performance and comfort – light sources within the visual field may cause glare (which may be visually disabling or causes discomfort). Excessive differences in lighting levels within the work areas should be avoided. Surfaces of a highly reflective material should be covered or replaced to reduce glare. Directional lighting may be appropriate for some inspection tasks, but diffused light is generally more satisfactory as it reduces the amount of glare and shadows produced.

Light may come from a natural source (from windows) or an artificial source (electric lights). It is important to remember that lighting levels may vary by the time of day or year and therefore appropriate adjustments may need to be made e.g. provision of local lighting (i.e. lamps), or blinds.

The furniture and equipment in the room should be arranged so that the sources of light do not cause light to fall directly onto reflecting surfaces or into the eyes (e.g. sit at right angles to a window rather than facing or with the back to the window).

Colours should be chosen to provide contrast to aid detection but not to exaggerate glare or reflections. Colours with good contrast should be used for displays (e.g. black and white, blue and yellow etc). Where lighting levels are high, dark characters on a light background are generally easier to read.

In general it is appropriate to allow individuals to have control over the lighting levels in their area, as individuals vary in what they find comfortable. In addition, having control over the work environment can help reduce psychosocial stress.

Increased lighting levels may be required for those with poorer eyesight (e.g. older people), although this should be done with care as older people are generally more susceptible to glare.

For further details see LIGHTING.

Additional guidance is provided in the HSE Guidance Note *Lighting at Work* (HS(G) 38).

## *Noise*

**E17032**  Noise can have an effect both on hearing ability (hearing loss) and on performance. Too much or too little noise can adversely affect concentration. Where work is monotonous, noise can be introduced to provide a stimulus; but where noise is excessive it should be reduced to facilitate concentration. Noise levels should not exceed approximately 55dB(A) for tasks requiring concentration.

Noise can be used effectively as a warning or alarm, although excessive use of noise for warnings can be detrimental to their effectiveness. Hearing generally deteriorates with age; this, and the differences in hearing ability between people, should be taken into account when designing critical alarms.

There are several harmful effects of noise which include:

1. Physical damage to the eardrum and ossicles induced by excessively high noises e.g. explosives.

2. Hearing damage due to exposure to high levels of noise. Damage can be divided into temporary threshold shift (reversible) or noise-induced hearing loss (irreversible).

3. Annoyance and stress – which can lead to reduced concentration. Noises that are annoying are generally unexpected, infrequent, high frequency and those over which the listener has no control.
4. Hindering communication – which can lead to an increased rate of accidents as well as stress.

Where excessive noise is present (over 80dB(A)) methods to eliminate, reduce or control the risk (e.g. baffling and hearing protection) are required. Hearing protection should only be used as a last resort, with other methods of noise control being explored first, as hearing protection (muffs or plugs) is often found to be uncomfortable, incompatible with other forms of PPE, or the task requirements and may therefore not be worn.

For further information see NOISE AND VIBRATION.

## Vibration

E17033   Exposure to vibration is associated with musculoskeletal disorders, other health problems, and reduced performance. There are two main categories of vibration transmission to the body: whole body and hand/ arm (see N3027).

Whole body vibration usually occurs via the feet or seat when the person is working on/ in a vibrating environment (e.g. transport systems). Low frequencies of vibration (such as those from motor vehicles) can lead to chest pains, difficulties in breathing, back pain and impaired vision.

Hand-arm vibration affects only the hands and arms and can be experienced when using power tools, such as pneumatic screwdrivers and chainsaws. Exposure of the hands and arms to vibration is a known risk factor in the development of Hand Arm Vibration Syndrome (HAVS), a condition which affects the nerves and blood supply to the hand resulting in tingling or numbness, impairment or loss of function, and blanching of the fingers, particularly on exposure to the cold. It is particularly higher frequencies of vibration (such as those associated with hand-held tools) that have been associated with muscle, joint and bone disorders affecting the hand and arm. Transmission of vibration has been found to increase as the gripping force increases and where tight gloves are worn.

Vibration, as with noise, should be tackled at its source. Larger machines can be isolated or the vibrations insulated, e.g. by introducing better mountings on the machine. Regular maintenance of machines can also reduce vibration. Relocation of the machine and insulation of the source should be considered. Floors, seats and handgrips can all be fitted with damping material. Hand tools should be designed to reduce the vibration created, the weight and amount of force required, so that tools do not have to be gripped tightly. Where this cannot be achieved, exposure time should be limited, and the hands kept warm.

For further information see NOISE AND VIBRATION.

## Thermal environment

E17034   The internal body temperature needs to be maintained within a relatively narrow range for health, and the ambient conditions which facilitate this are also relatively narrow. Deviations from comfortable temperatures can lead to reduced performance, fatigue, discomfort, ill health (hypo and hyperthermia as the body temperature rises or drops below those necessary for health); in extreme circumstances this can be fatal.

Body temperature, and therefore comfort, is dependent on the ambient conditions (air temperature, humidity, any radiant heating, and air velocity), the activity being undertaken (and therefore the heat produced within the body), and the clothing worn. Individuals' response to the thermal environment vary and some people are more affected by high or low temperature than others; it is most satisfactory if employees are able to control the environment to suit themselves (e.g. opening windows etc), although this can be difficult in large, open plan offices.

As far as possible, environments should be designed and controlled to allow a comfortable working temperature (e.g. through heating, air conditioning etc). Where this is not possible (e.g. due to a work process or outdoor working), appropriate clothing can be used as a control measure, or in extreme situations it may be necessary to limit exposure time.

Low temperatures reduce manual dexterity and where this is required the hands should be kept warm (e.g. through appropriate gloves). Manual handling injuries are more likely in extreme thermal environments and account should be taken of this when planning tasks.

Guidelines for control of employees' work in hot/ humid/ cold environments are shown below.

1. Maintain temperatures between 20–24°C for sedentary work and 13–20°C for physical work.
2. Maintain humidity at between 40 and 70 per cent.
3. Reduce air velocity (draughts) to less than $0.1 ms^{-1}$ in moderate and cold environments. Increasing air velocity can be an effective control measure in hot environments, provided the air temperature is not close to or above body temperature when increased air velocity will increase the thermal load on the body.
4. Decrease physical work load in hot environments.
5. Schedule rest pauses or rotate personnel around more strenuous tasks in hot conditions.
6. In hot environments, allow rest periods to be taken in cooler environments
7. Schedule outdoor work to avoid periods of very high or low temperature.
8. Permit gradual acclimatisation to hot or cold environments (7–10 days)
9. Maintain hydration by consuming sufficient drinking water, particularly in hot environments.
10. Avoid long exposure periods in cold environments.
11. Avoid rest for long periods in cold environments.
12. Provide protection from wind and foul weather when working outside.

## Physical hazards

E17035    Other physical hazards may be present in the workplace (e.g. slipping, falling, tripping hazards). Ideally these should be eliminated at source, through appropriate design (e.g. removing the need for an operator to enter a potentially dangerous area). Where hazards are still present, guarding, safety barriers, good housekeeping, appropriate lighting, and good workplace layout will reduce the possibility of all types of accident. Raising awareness through education and training can be effective, although it should not be relied on as a first line of control. Personal protective equipment (PPE) should always be seen as the last resort control measure, due to issues related to fit, compatibility, impact on performance, loading on the body, etc which may mean that it is not worn correctly, or possibly at all.

# Job design and work organisation

E17036    The design and organisation of work can have a significant impact on health. The demands of the job should be matched to the skills and abilities of the individuals. Excessive demands or, conversely, under utilisation of skills can both lead to a decrease in performance. The design of the job or work task is important for obtaining the maximum performance from the person for the minimum effort, and also in terms of facilitating a level of job satisfaction. Factors to consider include job rotation, job enlargement and job enrichment, as well as job scheduling, job demands, and shift patterns.

## Job rotation

E17037  Job rotation, whereby people are moved from one task to another, allows changes in posture, movement, and mental demands. This may lead to the acquisition of additional skills and may help the person identify more with the completed product or service. Job rotation can also help to overcome boredom, particularly in repetitive jobs, and help to prevent fatigue, loss of concentration and deterioration in performance. To be effective as a control measure for musculoskeletal discomfort people should rotate to tasks that are significantly different in terms of postures, movements and forces required.

## Job enlargement

E17038  Job enlargement whereby the number and range of tasks undertaken by the individual is increased, allows them to become more flexible, develop other skills and have greater variety in their tasks.

## Job enrichment

E17039  Jobs can be enriched by incorporating motivating or growth factors, such as increased responsibility and involvement, opportunities for advancement and a greater sense of achievement. This should provide the workers with greater autonomy over the planning, execution and control of work.

## Work scheduling

E17040  Ideally workers should have control over how the work is scheduled. As far as possible work should be self-paced rather than machine-paced. Machine pacing may be too fast or too slow for an individual, both of which can lead to problems (lack of recovery time for muscles and soft tissue; frustration etc). Machine pacing has been implicated as being a contributory risk factor in upper limb disorders, as well as stress and stress-related illnesses.

Rest breaks help alleviate both mental and physical fatigue. Allowing the worker to select when and how often they take breaks is preferable to fixed breaks, although there is some evidence that people work until fatigue occurs rather than stopping before this point. Therefore, under some circumstances (e.g. tasks that require high concentration or those requiring physical effort) it may be preferable to give fixed rest breaks before the person is likely to become fatigued.

Breaks should be taken away from the workstation, not only because this allows for physical movement and a change of scene but also because it may facilitate reducing exposure to hazards such as noise. Short, frequent breaks are better than long, less frequent breaks.

## Shift-work

E17041  It is well recognised that shift-work affects health and safety at work. In particular, where shifts involve people changing their eating and sleeping patterns, symptoms such as irritability, depression, tiredness and gastrointestinal problems (e.g. indigestion, loss of appetite and constipation) can occur. These health problems are largely due to the disturbance of the body's physiological cycles. Although many shift-workers adapt to the disruptions, it is widely recognised that productivity levels are lower on night shifts than day shifts.

Night shifts should therefore be avoided if at all possible. Younger and older workers (those below 25 and above 50) may be more vulnerable to ill health effects from shift work. Those with digestive disorders, sleep problems or other health problems may find shift work aggravates these. Shift rotations should be short and followed by at least 24 hours' rest.

## Job support

**E17042** Appropriate design of jobs and the structure of the organisation and the way it operates are important design criteria. The following points are recommended for providing adequate support in a job.

- Provide opportunity for learning and problem solving, within the individual's competence.
- Provide opportunity for development and flexibility in ways that are relevant to the individual.
- Enable people to contribute to decisions affecting their jobs and their objectives.
- Ensure that the goals and other people's expectations are clear and provide a degree of challenge.
- Provide adequate resources (e.g. training, information, equipment, materials).
- Provide adequate supervision, support and assistance from contact with others.
- Provide feedback on performance and communication of objectives.

## Organisational support

**E17043** The following factors are considered to be necessary to provide an acceptable background against which people can work:

- Industrial relations policies and procedures should be agreed and understood and issues handled in accordance with these arrangements.
- Payment systems should be seen as fair and reflect the full contribution of individuals and groups.
- Other personnel policies and practices should be fair and adequate.
- Physical surroundings and the health and safety provision should be satisfactory.

# Musculoskeletal disorders

**E17044** The term 'musculoskeletal disorders' (MSDs) refers to problems affecting the muscles, tendons, ligaments, nerves, joints or other soft tissues of the body. These disorders most commonly affect the back, neck and upper limbs. Musculoskeletal disorders are the most common work related condition affecting the general population in Britain; they account for more than half of all self-reported occupational ill health (1.2 million cases in 1995). Of these, 44 per cent affected the back, 32 per cent the upper limbs or neck, and 9 per cent the lower limbs. An estimated 9.9 million working days were lost in Britain due to MSDs in 1995. The cost to employers of MSDs affecting the neck and upper limbs was estimated to be at least £200 million (HSE, 1995).

Musculoskeletal disorders give rise to clinical effects in the individual, i.e. symptoms (e.g. pain, numbness, tingling) and signs (e.g. changes in the appearance of the limb which may be identified on medical examination). They usually also result in some functional changes (e.g. reduced ability to use the part of the body affected and restrictions to movement or strength). They can also have a general effect, such that there may be a resultant reduction in general health or quality of life (e.g. through restriction of activities).

MSDs are frequently work related, but not necessarily caused by work. MSDs have been the subject of much civil litigation in recent years and a significant number of personal injury cases have been successfully brought against employers (See OCCUPATIONAL HEALTH AND DISEASES.). The employer's duty of care to their employees with respect to upper limb disorders is now well established in the civil courts. This civil law duty runs parallel to the employer's statutory responsibility under health and safety legislation.

There is currently no specific legislation relating to musculoskeletal disorders. However, the *Manual Handling Operation Regulations 1992* (*SI 1992 No 2793*), and the *Health and Safety (Display Screen Equipment) Regulations 1992* (*SI 1992 No 2792*) were introduced with the aim of reducing musculoskeletal disorders associated with these activities. There are also general duties under the *Health and Safety at Work etc Act 1974* and the *Management of Health and Safety at Work Regulations 1999* (*SI 1999 No 3242*). (For more details on legislation see E17059–E17068.)

Activities outside of work may also present similar types of risk of MSDs to those experienced in work activities. However, non-work activities are usually not as repetitive, forceful or prolonged as work tasks, and the individual is likely to have more control over whether the activities are undertaken and how long for.

## Back disorders

**E17045** Back discomfort is extremely common in adults. Over 80 per cent of adults will experience an episode of back pain during their life, but the vast majority of these (90 per cent) will recover within two weeks, although recurrence of back pain is common. Back pain is rarely caused by a single injury, but is generally cumulative in nature. However, a minor incident may be seen as the final straw which triggers discomfort.

Causes of back discomfort include manual handling activities; awkward postures such as twisting, leaning, bending, stretching and prolonged static postures; poor seat design (including car seats); low physical fitness; and exposure to whole body vibration.

## Upper limb disorders

**E17046** Upper limb disorders (ULDs) are a sub-category of MSDs and the term refers to injuries occurring in the upper limbs (fingers, hands, wrists, arms, shoulders and neck). It covers strains, sprains, injury and discomfort, including conditions such as tenosynovitis and carpal tunnel syndrome as well as non-specific disorders. The cause of some of these disorders can be related specifically to the work undertaken and become known as work-related upper limb disorders (WRULDs). These disorders are usually cumulative in nature, arising from a series of micro-traumas.

The main physical risk factors in the development of upper limb disorders are application of force, repetitive movements and awkward or static postures. It is usually the interaction of at least two of these risk factors that leads to disorders. Duration of exposure to these factors is obviously also significant, with long durations of exposure increasing the risk of injury.

1. *Force* – application of excessive force in relation to the capabilities of the upper limb muscle group, will place a strain on the muscles involved and may lead to injury. Different muscle groups are able to apply different levels of force (e.g. shoulder muscles can apply more force than finger muscles).

2. *Repetition* – frequent, repetitive movements require the same muscles to contract and relax over and over again. If this activity is very rapid or continues for a long period of time the muscle can become fatigued and this can lead to discomfort.

3. *Posture* – poor postures (both awkward postures and static postures) can lead to injury and discomfort by placing the muscles and joints under unnecessary strain.

4. *Vibration* – exposure to hand/arm vibration is a recognised risk factor in the development of Hand Arm Vibration Syndrome (HAVS).

WRULDs are not confined to particular activities or industries, but are widespread throughout the workforce. Some jobs have particularly been associated with WRULDs, and these involve recognised risk factors (force, repetition, awkward/ static postures, and long durations of exposure to these factors). Jobs such as assembly line work, construction work, garment machinists, meat and poultry processors, and display screen equipment work (particularly intensive data entry tasks) pose a risk of WRULDs, but this list of tasks is not exhaustive and is provided for illustration only.

Common ULDs are listed in O1047.

### Lower limb discomfort

E17047  Lower limb disorders are less common and often less debilitating than other musculoskeletal disorders. However, lower limb discomfort can be caused by prolonged standing, particularly on concrete floors; inappropriate (particularly hard soled), heavy or ill-fitting footwear; operation of foot controls; and a lack of adequate foot support if sitting.

### Psychosocial risk factors

E17048  As well as the physical factors that can lead to musculoskeletal disorders, there is strong evidence of the role of work related psychosocial factors (i.e. the worker's psychological response to work and workplace conditions) in the development of these disorders. Relevant factors include the design, organisation and management of work, the context of the work (overall social environment) and the content of the work (the specific impact of job factors).

Specifically, repetitive, monotonous tasks, excessive or undemanding workloads, lack of control over the task or organisation of the workplace, working in isolation, poor communication, and lack of involvement in decision making have been implicated as psychosocial risk factors.

It is thought that many of the effects of these psychosocial factors occur via stress-related processes which result in biochemical and physiological changes which can result in discomfort. Some can also have a direct impact on working practices and behaviours, e.g. work pressure may mean workers do not adjust the workstation to suit themselves at the start of the shift, or may forego rest breaks.

### Individual differences

E17049  For biological reasons some people may be more likely to develop a musculoskeletal injury. These include new employees who may need time to develop the necessary skills/ rate of work; those returning from holiday or sickness absence; older/ younger workers; new/ expectant mothers; those with particular health conditions. Account should also be taken of differences in body size which may require awkward postures/ reaches.

### Prevention of WRMSDs

E17050  The greatest risk reduction benefits will be achieved by tackling both physical and psychosocial risk factors in the workplace. These factors are best identified and tackled through consultation with the workforce. Appropriate design of tasks, workstations, tools as outlined in E17005–E17043 can help to prevent injury to the musculoskeletal system, by ensuring work is designed to be within the physical capabilities of people. Tasks or equipment may need to be modified for those returning to work following an absence related to a work related musculoskeletal disorder.

## Ergonomics tools

E17051  The following techniques can be used to ensure thorough and comprehensive investigation of all parts of the work so that these can be designed appropriately.

## Risk assessments

**E17052** Ergonomic risk assessments can identify the risk of injury, accident or reduced performance due to ergonomic deficiencies in the work or tasks. Any assessment should be systematic and comprehensive, accounting for all elements of the work including irregular activities. Checklists can be used to ensure all elements within the task are covered, i.e. the person, the work task, the workstation, the work environment and work system. One example of a simple ergonomics checklist is provided in J Dul and Weerdmeester *Ergonomics for beginners* (2001) Taylor and Francis. The particular questions that may be asked may relate specifically to the work situation present.

All hazards should be identified and recorded and an assessment of the risk posed by the hazard should be made. All hazards must be addressed, eliminated where possible or reduced as far as practicable.

A risk assessment checklist concerning the risks of manual handling is provided in the *Manual Handling Operations Regulations 1992* (*SI 1992 No 2793*). A list of factors to consider in the assessment of the risk of injury through DSE work is included in the *Health and Safety* (*Display Screen Equipment*) *Regulations 1992* (*SI 1992 No 2792*), although this is not presented as a risk assessment checklist. The revised HSE guidance on *Upper limb disorders in the workplace* (HS(G) 60 rev) contains a thorough risk assessment checklist for considering the risks associated with tasks.

Several checklists and assessment tools exist to facilitate in the assessment of working postures and risk and include:

- the Rapid Upper Limb Assessment tool (RULA)

    See L McAtamney and E N Corlett 'RULA: A survey method for the investigation of work-related upper limb disorders' (1993) *Applied Ergonomics* 24(2) pp 1991–1999;

- the Quick Exposure Checklist (QEC)

    See G Li and P Buckle 'Evaluating change in exposure to risk for musculoskeletal disorders: A practical tool' (1999) HSE Contract Research Report 251/1999;

- the Rapid Entire Body Assessment tool (REBA)

    See S Hignett and L McAtamney 'Rapid Entire Body Assessment (REBA)' (2000) *Applied Ergonomics* 31, pages 201–205.

See also RISK ASSESSMENT.

## Task analysis

**E17053** A task analysis provides information about the sequence and inter-relationship of activities undertaken in a task. It can be used to help in the evaluation of existing tasks, or the design of new ones, by identifying the demands on the individual, connections between activities, unplanned or unusual activities. From this tasks can be allocated to different individuals (or the decision may be taken to mechanise), difficulties with equipment, tasks, workstations can be identified and equipment and training needs can be identified.

## Workflow analysis

**E17054** In a workflow analysis the movements of an individual at the workstation (e.g. to the equipment that is used) or within a work area are assessed. This helps to identify the pattern of activities and the routes within the workplace, to ensure they are optimised. Equipment can be arranged to ensure that items used together or in sequence are placed appropriately.

## Frequency analysis

E17055 Information on the frequency with which tools and equipment are used can be important in positioning items or controls. The frequency of use of tools or adoption of postures can be recorded regularly using a frequency count.

## User trials

E17055 Mock ups of prototype equipment or workstations, and trials of new furniture can be useful ways of assessing the suitability of new equipment. A representative sample of the user population should be used to evaluate the design. Rating scales can be used to allow people to score their opinion of a new item in terms of comfort, ease of use (clarity, adjustment etc) and so on. Rating scales should have a mid-point (neutral); either 5 or 7 point rating scales are generally adequately discriminating.

## Data collection

E17056 Questionnaires and (formal or informal) interviews can also be useful means of collecting information and obtaining users' views of equipment, tasks or workplaces. Open (free response) or closed (series of options) questions can be used depending on the information that is required. Anonymity may be required in some situations (e.g. asking about health issues).

Discomfort surveys can be conducted to gain an understanding of the extent of any discomfort or disorders that are experienced by the workforce. In these, individuals report any discomfort they experience at the end of their shift, according to the part of the body affected, and rate its severity (e.g. on a scale of 1–5). The responses of individuals from a work area can be collated, and this combined data can be used to identify in which parts of the body operators are experiencing discomfort and also to identify any trends in relation to the work tasks or work areas. This information can also be used as a baseline against which the benefit of any interventions can be evaluated. It is a useful tool to use in relation to risk assessment as it can assist in the interpretation of risks, and prioritisation of risk reduction measures. An example of a body map is shown in figure 6 below.

Alternative ways of using the body map are for groups of employees to apply stickers to a body map to indicate any discomfort experience, thus building up a pattern of discomfort.

Accurate assessment of tasks requires good observational skills and sufficient time to assess the task adequately thoroughly. Observers should ensure that a representative sample of the workforce or tasks undertaken are observed. Irregularly undertaken tasks (e.g. maintenance) should not be overlooked in assessments. In some tasks, movements may be rapid, or only last for a short period. It can be useful therefore to collect video material of tasks to allow subsequent analysis.

## Using existing data

E17057 In some cases data will be available which may indicate ergonomic issues. Productivity, accident and ill health data may all be used to identify any particular issues, and the extent of the problem. An accident book may also provide useful information on problems experienced. Data can be analysed by work area or type of injury in order to help prioritise areas for risk reduction measures.

Regular communication (e.g. through safety meetings) can also provide a useful route for identifying problem tasks or equipment.

## Involving employees

E17058 A number of people can contribute to the design of the workplace or task. These may include planners, procurers, human resources as well as those who undertake the task itself.

**Figure 6: Body Map**

Involvement of the relevant people in any proposed changes is important to ensure that it is acceptable. In particular, involvement of those who undertake the task in risk assessment, redesign and evaluation is a key element in ensuring that the task is thoroughly understood, any redesign is suitable, and that any new equipment or changes are acceptable to the users. This can be achieved for example through questionnaires, discussions, and trials of new equipment.

## Relevant legislation and guidance

E17059　　Ergonomics is mentioned in several pieces of legislation, although there is no single specific piece of legislation concerning this subject. Requirements for ergonomic design (explicit or implicit) in legislation are summarised below. Specific pieces of legislation are discussed in more detail elsewhere in this publication. Selected relevant HSE guidance is also summarised.

### Health and Safety at Work etc Act 1974

E17060　　Under this Act employers have a general duty to ensure (as far as is reasonably practicable) the health, safety and welfare at work of their employees. This includes the provision of machinery and equipment that is without risks to health, the duty to keep workplaces in a safe condition and without risks to health, and to provide the information, instruction, training and supervision necessary to ensure employees' health and safety at work.

## Management of Health and Safety at Work Regulations 1999 (SI 1999 No 3232)

E17061     These Regulations require that any risks to health from work have to be assessed; employers must take appropriate preventative and protective measures, and employees must be informed of the risks and preventative measures taken.

Further information on these Regulations is contained in RISK ASSESSMENT.

## Manual Handling Operations Regulations 1992 (SI 1992 No 2793)

E17062     These Regulations take an ergonomic approach to handling tasks. They apply to all manual handling tasks, and place a duty on employers to avoid hazardous manual handling activities as far as is reasonably practicable. Where this is not practical employers should assess the risk of injury, and then take appropriate steps to reduce the risk of injury as far as possible through appropriate design of the task, load, and environment, taking account of the capabilities of the individual.

Further information on these Regulations is contained in MANUAL HANDLING.

## Health and Safety (Display Screen Equipment) Regulations 1992 (SI 1992 No 2792)

E17063     Particular ergonomic issues relate to work at display screen equipment (DSE) i.e. VDUs. The *Health and Safety (Display Screen Equipment) Regulations 1992* set out particular duties on the employer to assess the health and safety risks associated with DSE use, and to reduce the risks identified. The health risks particularly associated with DSE use include musculoskeletal disorders, eyestrain and stress. Specific requirements for the display screen equipment, workstation, software and task design are set out in the *Schedule* to these Regulations. These requirements ensure that it is possible to make certain adjustments to the equipment and furniture. It is advisable as part of the assessment to ensure that equipment is positioned in such a way for the user as to prevent awkward postures (e.g. twist to view screen), and that the adjustments are suitable for the user (e.g. seat height adjustment is adequate).

Further information on the DSE Regulations is contained in OFFICES AND SHOPS O5005.

Laptops computers pose a particular problem as it is not possible to independently adjust the distance between the screen and the keyboard, and this may result in visual and musculoskeletal discomfort. Laptops may be used in environments that do not allow adjustment to the workstation (e.g. trains, hotel rooms). There are also manual handling issues related to the transport of laptops, and can be concern over security and theft. These risks should be considered in a risk assessment.

Although there is currently no guidance from HSE on laptop use, it would be good practice to:

- Provide a docking station for the laptop when office based. Alternatively, provide a separate keyboard and mouse, and view the laptop screen (preferable); or a separate screen and use the laptop keyboard. If viewed, the height of the laptop screen may need to be raised so it is a suitable height.

- Provide information on adapting a work area to facilitate a comfortable posture (e.g. sitting in the passenger seat when working in a vehicle, using pillows to raise the height of the chair in a hotel).

- Encourage good working practices (e.g. regular breaks are particularly important because of the potential postural problems arising from laptop use).
- Provide a rucksack, or suitable carrying case, for easy transport of laptops. Rucksacks enable the weight to be carried close to the spine, reducing the amount of asymmetric loading and resultant stress on the spine and muscles that can occur if the bag is carried on one shoulder.
- Provide additional power supplies at regularly used locations, so these do not have to be carried.

If users are required to work at home, suitable furniture and equipment should be provided, and a risk assessment should be undertaken to ensure the workstation is suitable and the user can achieve a comfortable posture.

## Provision and Use of Work Equipment Regulations 1992 (SI 1992 No 2932)

E17064  The Regulations require employers to ensure that work equipment is suitable for the purpose and safe to use for the work so it does not pose any health and safety risk. General duties include: 'When selecting work equipment, employers should take account of ergonomic risks ... Operation of the equipment should not place undue strain on the user. Operators should not be expected to exert undue force or stretch or reach beyond their normal strength or physical reach limitation to carry out a task. This is particularly important for highly repetitive work.'

Further information on these Regulations is contained in MACHINERY SAFETY M1014.

## Personal Protective Equipment at Work Regulations 1992 (SI 1992 No 2966)

E17065  These Regulations place a duty on employers to ensure that suitable personal protective equipment (PPE) is provided to employees who may be exposed to a risk to their health and safety while at work, in circumstances where such risks cannot be adequately controlled by other means. PPE should take into account the ergonomic requirements of the wearer, and be capable of fitting them correctly.

PPE is designed to protect the wearer from a hazard; however, it may introduce other risks, such as reduced vision and hearing which may restrict ability to detect warnings. PPE may also restrict movement or the person's ability to perform a task (e.g. through reduced dexterity when wearing gloves). Some chemical protective clothing (particularly water vapour impermeable garments) can contribute to heat strain as the wearer has limited potential to evaporate sweat from the body. In selecting PPE the compatibility of different forms of PPE should be considered: it is often difficult to wear a hard hat with hearing defenders, or a chemical protective suit with a hard hat, although some integrated forms of PPE are available.

## Reporting of Injuries, Diseases and Dangerous Occurrences Regulations 1995 (RIDDOR) (SI 1995 No 3163)

E17066  Under the RIDDOR Regulations certain work-related accidents, diseases and dangerous occurrences have to be reported to the enforcing authorities. In terms of musculoskeletal disorders (which may arise due to poor ergonomics and inappropriate design), RIDDOR specifies that 'cramps in the hands or forearm due to repetitive movements' must be reported if they arise from 'work involving prolonged periods of handwriting, typing or other repetitive movements of the fingers, hands or arms.'

Further information on these Regulations is contained in ACCIDENT REPORTING A3002.

## Upper limb disorders in the workplace (HS(G) 60 (rev))

E17067    This revised guidance on upper limb disorders in the workplace provides a very useful and comprehensive approach to tackling these issues. It models a management approach, outlining seven stages in addressing these issues. These seven stages are:

1. Understand the issues and commit to action on ULDs.
2. Create the right organisational environment.
3. Assess the risks of ULDs in the workplace.
4. Reduce the risk of ULDs.
5. Educate and inform the workforce ULDs.
6. Manage any episodes of ULDs.
7. Carry out regular checks on the programme effectiveness.

These stages are not necessarily sequential, and different stages may well interact, such as educating the workforce concerning ULDs may be part of creating the right organisational environment where these issues are taken seriously and tackled supportively.

The guidance also includes a two stage risk assessment checklist. The first stage is a screening tool to help identify tasks where there may be a risk of injury and to assist in prioritising assessments. It contains five sections asking a small number of questions concerning any signs and symptoms of ULDs, repetition, working postures, application of force and exposure to vibration. More detailed risk assessment worksheets are also included. These are divided into eight sections concerning repetition, posture of the fingers, hands and wrist, posture of the arms and shoulders, posture of the head and neck, force, working environment, psychosocial factors, and individual differences. The risk assessment form encourages the identification of appropriate risk reduction measures for the risks identified.

The guidance also contains useful suggestions for reducing the risk of injury, medical aspects of ULDs and case studies illustrating how organisations have successfully tackled these issues.

## Reducing error and influencing behaviour (HS(G) 48)

E17068    This useful guidance promotes the consideration of human factors as a key factor in effective health and safety management. It provides practical advice on identifying, assessing and controlling risks arising from humans' interaction with the working environment.

# Facilities Management – An Overview

## Introduction

F3001   *Facilities management? – that's the department that moved us to our new premises – the building manager – the guy that negotiated the contract with the cleaning company – it's a helpdesk number that you ring if anything goes wrong.*

These are just a few examples of the perception that people have of facilities management and the services that may be provided in this relatively young but rapidly emerging industry, which is now recognised by government, commerce and researchers, as providing a business edge within today's highly competitive world. Facilities management today is a highly developed strategic discipline which provides a well managed and utilised built environment at best value for the company.

## A good definition of facilities management

F3002   The definition of facilities management is best described by the following two statements:

'the integrated approach to the operation, maintenance, improvement and adaptation of the buildings and infrastructure of an organisation in order to create an environment that most effectively supports the core business of that organisation.'

And:

'the co-ordination of specialist disciplines and the physical workplace with the staff and work processes of the organisation to create the optimum working environment for staff.'

An important element of both of these definitions is the integrated management approach. Even ten years ago there is no doubt that various departments in an average company would have been managing themselves independently of each other. Tasks and activities would have been duplicated due to the missing element of an integrated approach and this is what makes facilities management such an important element of an company's overall strategy.

## Facilities management

F3003   The modern company has to support a diverse range of activities from its core business through to the provision of activities such as maintenance and cleaning, which enables a company to concentrate on its core business. However, the company may not have the expertise or the desire to manage such non-core functions and may not be supported by the most ideal environment. Facilities management is able to effectively manage the non-core activities and use them to support the core business and add value.

The built environment and all associated assets are a major expense for a company and is something which must suit the needs of the core business but also be flexible enough to accommodate change. Being such an expense it is important that all steps are taken to protect the assets for the future. Facilities management, therefore, may cover the management of a very broad spectrum of activities and may include the following:

- maintenance;
- property management;
- energy management;
- health and safety management;
- human resources management;
- financial management;
- change management;
- domestic services e.g. cleaning, catering, security, postroom, and reprographics;
- utility procurement; and
- lifecycle costing.

An illustration of good facilities management can be seen in the following example.

> We all use buildings on a daily basis and to some extent take them for granted, but have you ever stopped and thought about just what goes on behind the scene to provide the lighting, heating, cooling, ventilation, catering, postroom, reprographics, grounds maintenance and all the other functions that go to make your time within a building, as comfortable as possible? I thought not, but then who would and yet when you consider it we use buildings all the time. When we are comfortable and are able to undertake our intended activity without problem then we do not give the environment within the building a second thought. It is not until the building or something within the building goes wrong that we can become uncomfortable and that's when we as humans experience some discomfort, and start to give our environment some thought.
>
> Imagine the discomfort of not being able to open your windows properly at home in the middle of summer on an extremely hot day. Your whole family will suffer from the heat during the day, which will cause fatigue and possible heat stroke if you are not careful and they will not be able to sleep properly at night, which will lead to tiredness and a lack of concentration the following day and so on. Imagine equating this up to a company employing a few hundred people in a building of modern design with no opening windows that relies upon mechanical ventilation and air-conditioning equipment that continually enables cooled air to be fed into the building. What if the ventilation or air-conditioning equipment behind the scenes, broke down. The staff would soon find themselves in a very uncomfortable hot and stuffy environment, which from the health and safety aspect, would rapidly become untenable and would inevitably lead to closure and loss of production. Imagine the cost to the company.

## Hard and soft facilities management

F3004    The activities associated with facilities management and the support of the core business is normally split into the following two discipline categories.

*Hard FM* – which includes the provision of all activities involved in the provision of maintenance services, refurbishment, minor and/or additional works of the actual structure of the built environment. In addition the provision of utilities and procurement could also fall into this category.

*Soft FM* – includes provision of all the remaining activities such as security, cleaning, catering, help desk, postroom, reprographics and reception duties.

## Managing

F3005　　Changes in society and within the workplace have seen the previous ten years experience major upheaval. The technology available to all companies is now staggering and affects staff in a variety of ways. Modern buildings are designed and operated in a very high-tech manner which makes full use of a vast amount of computer soft and hard ware. The same applies for manufacturing processes, which tend to be very automated and which rely on a massive amount of unseen infrastructure to support this process. The same is true for people management, due to the extensive range of well tested management techniques. During this period of change, which has seen technology and management techniques grow, the concept of facilities management has also grown. Gone are the days of the caretaker running the old school boilers, ordering paper towels and a few cleaning products whilst running his team of cleaners. Today's companies demand an integrated approach to the provision of buildings and their management, and this responsibility would fall to a specialist facilities management team, depending on the size of the task. Failure to embrace an effective facilities management strategy for some companies could lead to failure.

The person, or in the larger company, the team, responsible for the ventilation and air-conditioning equipment in the example above would be a facilities manager or team who could have been responsible for the design concept of the building, the selection and procurement of the ventilation and air-conditioning equipment, the maintenance and repair of the equipment and even the procurement of the energy actually powering the equipment. As can be seen facilities managers have a very difficult and responsible role, as the impact upon the business of a company, if they get it wrong, can be considerable. Most companies increasingly seem to outsource everything to a facilities management company, with the exception of their core business, which means that the facility manager is being expected to manage an even more diverse range of activities.

If a company is to place facilities management at the forefront of its strategy, then the lead facilities manager will be required to be highly skilled and, ideally, have a seat on the board of directors. If a board placement is not possible then there should be a communications line directly up to a strategic level. This is important as the company that has a facilities manager or a facilities management team will probably have embarked upon a high level facilities management strategy, which would have input to the management decisions regarding all aspects concerning the built environment.

## Company approach to facilities management

F3006　　There are a number of ways in which a company could approach the introduction of facilities management and this could depend upon it size, location, type of buildings, core business and objectives, as an example. Any approach must be tailored to the individual company's needs.

It is important that the company is able to effectively distinguish between its core activities, e.g. production of engineering components or provision of a service and its non-core activities, e.g. security, cleaning or maintenance. If the decision is made within the facilities management strategy to outsource, then it is the non-core activities that should be managed in this way, allowing the company to concentrate their efforts on their core business.

In order to manage the facilities of a company effectively a facilities strategy needs to be developed. The company must decide where it wants to go and how it is going to get there, and its mission statement and objectives must be clear to enable the development of the facilities management strategy. This strategy must be suitably developed to help cost efficiency. Scenario planning and business simulation may be useful in formulating a strategy. In this technique a series of 'what if?' questions are applied to produce a table of consequences enabling the organisation to form clearer decisions.

When determining objectives and strategy it is essential to build in flexibility and to have alternatives, should problems be encountered. Survival could depend upon the ability of a company to react to change quickly.

However the strategy is developed, the key is to take a systematic approach.

## Facilities strategy

F3007     *Analysis* of the company's present and future needs. To look closely at objectives, policies, resources, processes, and procedures. It is important to know where the company stands in the market place and, if available, some idea as to its customer's perception. What sort of image is the company producing now and is this in alignment with its desired objectives? What needs to be changed? Where could improvements be made? A full understanding of how the company's resources and assets are being utilised needs to be made. It is important to understand all the business processes surrounding the company's core business to enable a facilities management strategist to establish where improvements can be made, so as to add value.

*Development of Solutions* to make up the facilities management strategy can then be made by careful interpretation of the data and resulting information gained within the analysis. It is essential that solutions add flexibility, efficiency and are innovative e.g. by use of modern technology for communication of information. It can be useful to assemble a team who are capable of looking out side the 'box'. New ideas are crucial and whatever decisions are made they must be able to convince the senior management or board of the company that they will work, and ideally should be financially justified within any resulting report. Consider the use of a risk register to record the various scenarios and be sure to consider positives as well as negatives in this process. Any changes must ultimately be viewed as being cost effective and adding efficiency and value instead of being a cost burden.

*Implementation of the solutions* has to be carefully planned as without the use of effective 'change management' techniques will result in resistance to any changes being introduced and will probably result in staff problems taking up management time, thereby reducing efficiency. It is important to communicate with staff by way of group meetings and on a one-to-one basis to provide information and to allay any fears that they may have to proposed changes. Also listen to feedback from staff who may have some interesting ideas of their own. It is essential that the implementation plan is developed and includes time lines for each task with key milestones, performance measurement, proper ownership and responsibilities. It is important to ensure that the implementation plan includes all activities and elements of the company that will be affected by the change.

A typical facilities management strategy should be capable of the effective delivery of the following:

(*a*)   in-house/outsourced services – catering, cleaning, maintenance, security etc.;

(*b*)   specifications for required accommodation standards (RAS);

(*c*)   procurement strategy;

(*d*)   IT strategy;

(*e*)   human resources strategy;

(*f*)   provision of change management;

(*g*)   service level agreements (SLA);

(*h*)   performance measurement;

(*i*)   financial planning;

(*j*)   accommodation/real estate strategy; and

(k)  business systems and procedures.

## In-house or outsourced facilities management?

**F3008**  The decision taken is dependent upon the facilities management strategy of each individual company and may even be a combination of the following two options.

*In-house* – this is the retention of facilities management services within the company itself and with the expertise and management skills of providing effective services from within. This will detract from the core business activities but may suit company needs.

*Outsourced* – most companies consider that facilities management is such a specialist skill that they would prefer to outsource, as all the problems and issues of running their building are left with the total facilities management provider or a contractor, and allows the company to concentrate on its core business. The facilities company, who will in effect be a contractor, will have to be managed, but the added advantage is that facilities will be provided at a guaranteed price for the contract period which makes for easier budgeting. Consideration must also be given to the cost associated with the management of the facilities management contract to ensure that the agreed service is being delivered and that the service provider is complying with their health and safety duties.

The outsourced activities would be placed with a service provider, which would be one of the following:

- managing agent;
- managing contractor; or
- total FM provider.

## Advantages to the company

**F3009**  The advantages of using proper integrated facilities management within a company are vast and are essential to, in some cases, its very survival. Gone are the days where the various departments virtually managed themselves by doing their own thing using practices which had been handed down over the years. In such situations it was often found that tasks were carried out throughout the company with little or no communication, and certainly, no best practice. Today the facilities management department is responsible for advising a company's board and will most probably have a role in setting the overall business strategy for the future. It is strongly believed that the facilities management role in the future will also encompass human resources, information technology and eventually all the financial elements of running a large company. Such a transition would provide a totally joined-up management approach towards all elements of the business.

The facilities management integrated approach can add tremendous value to a company by identifying needs in all areas and looking at ways to satisfy such needs at a strategic level by innovation. For example, a department may need to bring in a temporary worker for short-term data entry. However, when viewed at strategic level it may be the case that other departments have administrative pressures at certain times and are also bringing in temporary staff. A view could be taken to either increase the head count of staff and use the new member of staff as a float to be used filling administrative requirements, or if this were not possible, the facilities management team could agree on more time for the temporary worker to act as the float in various departments and could therefore negotiate a better price.

By giving consideration to the needs of the entire company for the supply of an item or a service and grouping the procurement of goods and services together will result in added purchasing power to obtain the best deal from the suppliers and manufacturers. This concept could be taken further by establishing when, for example, a manufacturer is at their quietist. If a company's production run can coincide with a manufacturing quiet period, a more competitive rate could be negotiated with the added advantage to the manufacturer of not having to close the plant temporarily or be forced to lay off staff, thereby potentially losing skilled labour.

## Main advantages

F30010

- **Cost certainty** – by inclusion within the facilities management contract of a detailed set of specifications and service level agreements, the desired services can be delivered at an agreed price aiding budgeting within the company. By careful identification of the needs of the company, specifications may be tailored and delivered in a more efficient way adding value.

- **Reduced cost** – by the integrated management of all services, more efficient procurement, utilisation of company assets and staff management, huge savings can be realised which will lead to increased profits and will ensure competitiveness and the long-term survival of the company. The claim that the facilities management department is an overhead, which perhaps costs a company more than it could ever save, can be refuted. It has been proven that good strategic facilities management can improve the efficiency of a company's systems, thereby reducing running costs and assisting in the delivery of the company core objectives.

- **Service delivery** – by careful facilities management contract negation and management of company expectations, services can be delivered to an agreed level which, if subject to the contract, may be measured periodically to ensure long-term delivery to the required standard. A system of continual improvement could form part of the initial contract and this could be effected by both parties taking a proactive approach developing innovations in working practices.

- **Improved communications** – will ensue following the integrated management of all services through one department. Issues can be quickly resolved or brought to senior management for a decision. Having all information collated in one department will ensure wide-ranging, accurate and meaningful information can be utilised in the decision making process.

# Responsibilities of facilities management

F3011    As demonstrated earlier, the skills expected of the facilities manager are diverse and complex and are defined by the British Institute of Facilities Management (BIFM) competencies as 20 broad categories of know-how which have been broken down into numerous individual elements. An overview of a range of competencies is shown as follows.

## Development of a maintenance policy

F3012    Maintenance of any building or process plant is crucial as without this equipment the company's environment will cease to function efficiently. One aspect of facilities management is to protect the assets of a company and to ensure they are utilised to maximum effect. The facilities management department would be responsible for the development and implementation of the maintenance policy for the company. This will depend on the culture of the company, the type of work carried out by the company, the age of the building, and type of plant equipment. For example, it would not be advisable to adopt a breakdown maintenance policy if the company was reliant on one piece of very complicated plant equipment which was business critical. Therefore, the development and implementation of the maintenance policy for the company is important in the provision of services to enable the company to function, but could also save money in the long term through the by selection of the most suitable policy.

Before any decision is made on the selection of a maintenance policy, each asset must be entered in an asset register and a condition survey completed. This will provide the facilities manager with information, which can be used when considering the policy to adopt. The type of company and business processes conducted will be taken into account during this decision making process, as will the available budget.

Companies always try to cut costs and maintenance is often seen as an area costing too much that a few more per cent taken off will hardly make any difference. It is for the facilities manager to demonstrate that cost cutting in this area could be counterproductive, leading to unreliability of plant equipment and general dilapidation.

It is important to make the correct choice of maintenance policy with current and future trends being considered, together with the long-term effects that a particular maintenance policy could have upon the company's assets and future capital expenditure. By adopting a reduced maintenance policy now, of say, reduced frequency, or perhaps even a breakdown policy, which will no doubt have a detrimental effect upon the assets in the future and will perhaps reduce the life expectancy of the asset. This will either result in expensive maintenance in the future or complete replacement and therefore capital outlay before the expected life of the asset.

There are three maintenance policies or a combination of the three that may be implemented, as follows.

### *Planned preventative maintenance (PPM)*

F3013    In this policy, a maintenance schedule is developed for every item on the asset register and may include periodic inspection of the condition of assets, items of plant equipment being adjusted, and filters replaced or cleaned as appropriate. This is initially the most expensive form of maintenance, but will prevent unexpected breakdowns, will ensure that equipment runs efficiently and will prolong equipment life. This policy would normally be used to maintain any 'business critical' assets within the company.

### *Planned corrective maintenance*

F3014    This is where the assets are inspected periodically but are not touched in anyway unless they are running very inefficiently or having an impact upon the running of other assets, or perhaps where there is a high risk of the asset failing in the near future. This maintenance policy is cheaper than planned preventative maintenance but will make it difficult to predict breakdown and an asset could run so far outside its design tolerance that long-term damage is inflicted, thus reducing the life expectancy of the asset. This is very much a hands off approach.

### *Breakdown maintenance*

F3015    Assets are neglected and only touched when they fail. This is therefore purely a reactive system and by far the least expensive policy in the short term, but it's use makes it difficult to predict breakdowns. This policy does not protect assets for the future; dilapidation of equipment could prove costly.

Before deciding on the maintenance policy to be adopted the main aims and objectives of maintenance within the company need to be considered, some of which could include the following:

1. upkeep of the company's image;
2. to enable the built environment to fulfil its function;
3. compliance with legislation to provide a safe and healthy environment;
4. ensure plant runs efficiently to provide a comfortable environment;

5. minimum cost rather than reliability;
6. reliability of plant equipment; and
7. creation of a forward maintenance register to allow funds to be amortised for lifecycle replacement of equipment.

## Life cycle costing

**F3016** It is important to consider the building environment, it's condition and likely maintenance requirements for the future. The facilities management team can assist the company by using life cycle costing, also known as whole life costing. Life cycle costing provides the facilities manager with a cost profile which can be broken down into three basic elements:

- acquisition cost;
- operational cost; and
- disposal cost.

This technique may be applied to services as well as actual assets.

All components, whether they are the outer protective layer of a building, a heating pump, or even a light bulb, will have been designed with a life expectancy which the manufacturers will quote. Therefore, a whole building or section may be designed to provide a certain number of years' service before requiring total replacement, subject to the manufacturer's ongoing maintenance being completed throughout its expected life. The cost element will also be considered as a cheaper product may have to be included in the design which will require replacement earlier. However, the ongoing maintenance costs may be far higher than a more expensive product and this will also have to be taken into consideration. This highlights the fine balance a designer will have make to achieve the client's requirements and deliver within a prescribed budget. The compromise is that the client has to accept that although their initial capital expenditure is reduced at the construction phase, at some point in the future there will be an ongoing maintenance cost and further expenditure to replace the component.

There are a number of computer software programs available to assist a facilities management team to create an asset register for the building. Such programs may include everything within the building from the very fabric of the construction to the fixed items of plant equipment, general office furniture and carpets. By inputting full details and the condition of each asset a lifecycle capital expenditure programme can be produced to provide information indicating when assets should be replaced and guidance costs to assist in budgeting.

## Customer service

**F3017** One of the difficulties encountered when services are outsourced is the management of customer expectations which comes from the desire to have everything work all the time, or for items to be delivered exactly on time. A key driver in the decision making process to outsource, may have been to save costs and it may be that the customer (e.g. the client company staff) could be expecting a gold service when in fact the company is only now paying for a silver service. It is essential to manager customer expectations and to avoid confusion it is advisable to ensure that staff within the client company are provided with clear information as to the level of service that they should expect.

The facilities manager will be able to measure customer service by use of measures which should be agreed within the initial contract. Such measures are normally referred to as key performance Indicator's (KPI's) and will provide an ongoing indication of performance.

The KPI's which could be utilised for a typical facilities management help desk may be defined under the following three headings.

*Timeliness* – needs to be defined in a very precise way, but most probably that the help desk telephone is answered within five rings or perhaps within 15 seconds.

*Appropriateness* – when a help desk operator answers a call, was the customer's request understood or was the most appropriate action taken? Was the customer's expectation completely met or did the operator fail to establish a solution to the problem?

*Accuracy* – most probably the only acceptable standard in this category is 100%. In the example above, did the help desk operator enter the correct information into the system or were there inaccuracies that affected the resolution of the customer's problem?

The measuring and collating of this information is important to the facilities manager because it confirms whether or not the appropriate service is being delivered. With continuous performance measurement in this area it can be ensured that the service is delivered at the agreed level. This may be taken a stage further by publishing the service measures, or KPI's, to demonstrate how the help desk is performing and to encourage staff to provide suggestions on how service delivery may be improved. The performance may also be compared with those of help desks in other companies, provided the KPI measures are the same.

Service delivery is a key driver within the facilities management industry and is taken very seriously. In the example above, it would be expected that the help desk undertake its own surveys by phoning back a percentage of customers at random and asking them to complete a survey form to provide feedback, which is then collated and measured.

## Environmental and energy management

**F3018** Much emphasis is being placed on environmental protection and everybody has a duty to ensure that they use energy as efficiently as possible by carrying out basic tasks such as switching off lighting in areas that are not in use or lowering the temperature on a thermostat by a degree or so. Simple things that can be easily performed in everyday life, whether at home or in the workplace, and yet they can make such a difference. Expand this to all of the staff and processes contained within an average working environment and it can then be demonstrated that a company can make a huge difference. Energy consumption within a company burns fossil fuels, which adds to the greenhouse gas effect, which in turn creates global warming by retaining heat within the earth's atmosphere. Organic waste dumped in landfill sites also produces greenhouse gasses, which again adds to the problem.

Facilities management can apply their skills not only to comply with legislation concerning environmental issues, but also to save the company costs on energy bills. A number of companies have also developed a corporate and social responsibility strategy which would probably be part of the facilities management function to ensure that there is compliance with this as well.

The facilities management department would be able to assist energy management within a company at both a strategic and operational level. Typically, part of the environmental strategy would be for a company to agree upon its energy management strategy, which would normally consist of the following.

*Strategic*

- policy statement;
- objectives;
- roles and responsibilities;
- process of review;
- arrangements for the following:
    - heating – space temperatures, heating periods, and
    - cooling – air conditioning policy – natural/mechanical ventilation.

It is important that all staff are made aware of the environmental and energy policy, together with full details of the practical ways in which staff can really make a difference. For the policy to be effective a clear message must be provided from the top down in the company, with directors and senior managers setting a good example and being supported by the facilities management team whose overall responsibility would be to manage energy issues on a day-to-day basis, as follows.

*Operational*

- housekeeping issues;
  - switch off lighting,
  - doors and windows closed in winter,
  - switch off equipment not being used,
  - use water wisely,
  - wear sensible clothing for weather conditions,
- engineering issues;
  - install or replace existing lamps with energy efficient equipment,
  - ensure external lighting controlled by photo cell or time clock,
  - installation of waterless urinals in gents' toilets,
  - installation of internal lighting control by motion detectors (PIR),
  - use of improved ventilation rather than air-conditioning, and
  - installation of window film and or blinds to control solar gain.

The facilities management team would have a direct effect on the environmental impact of the company and would be able to reduce any adverse impact by careful management of the following service functions:

(a) products are procured from suppliers with environmental policies;

(b) all wood used is from sustainably managed sources;

(c) where possible, water is recycled in production or other processes;

(d) only use paints with a low volume of volatile organic compounds (VOC's);

(e) procure products made from recycled materials;

(f) recycle paper, cardboard, toner and printer cartridges, and mobile phones;

(g) manage ordering of marketing publications and leaflets to reduce waste;

(h) ensure that waste from refurbishment projects is recycled; and

(i) procure energy from renewable energy sources.

These are some of the activities that could be undertaken by a skilled facilities management team, all of which will ensure compliance with current legislation and above all will realise considerable savings in the company's energy consumption and resulting expenditure on energy bills.

## Acquisition and disposal of buildings

F3019    There are very few companies with buildings in their estate who could say that their buildings are completely suitable for their operation. Compromises are often made as companies continually change and even if a building was designed specifically for a company, and even if there was a certain amount of flexibility built into the design to cope with future needs, there is no doubt that by the time the building was finally ready for occupation, the company would have changed.

The facilities management team can use their skills in advising a company on the acquisition and disposal of buildings by taking an integrated approach, in which they would consider the company's objectives and current and future needs. If the company required more space in which to conduct its operations, then areas for consideration could be made, as follows.

- Is this temporary or permanent?
- What has created the need?
- Can some staff be allowed to work from home with a laptop?
- Is it possible to move a department into a satellite office?
- Is it financially viable to take short-term space nearby?
- Should the entire company relocate?
- What is the current lease situation on the present building?
- Is there a lease break or could the building be sub-let?
- What location would suit the company?
- Cost of any necessary refurbishment to new location?
- Maintenance and running costs?

There are so many areas to be considered that usually three options would be selected, each providing a range of advantages and disadvantages, although ultimately, senior management would have to compromise.

Similar skills would have to be applied to the disposal of a building. It could be that only part of a building is no longer required, so consideration could be given to establish whether that particular area could be sub-let, or if there is a lack of a lease break, then it may be feasible to sub-let the entire building. If the building is no longer required but has to be retained due to the lease, then consideration has to be given to reducing building services to an absolute minimum whilst preserving the assets for future final disposal.

## Space planning

**F3020** Further to the acquisition and disposal of buildings (at F3019 above) it can also be demonstrated that, due to the constantly changing company, space within a building will not be utilised to its full potential and facilities management can be called upon to find an appropriate solution.

The building occupied by a company is a large overhead. Reduced overheads and profits will be increased, thus ensuring survival in the marketplace. If space planning is undertaken effectively then the company will further benefit from improved production created by better flow of work and processes through the building. Staff will be working more efficiently as they will have a better environment in which to work and with better management control, defining space requirements and reducing clutter by good housekeeping. A well-planned and managed workplace will enhance the company's image, which will provide a good impression to visiting potential clients.

## Project management

**F3021** A company will often have many projects running concurrently, such as the development and introduction of a new information technology system, to the construction of an extension of a building. In most companies a facilities manager would normally manage small to medium sized projects, which would place further demands on the skills base required and workload. Often, it is unclear how to define what a project actually is. Each project will have its own unique set of co-ordinated activities being undertaken by an individual or a team brought together for this specific purpose. There will be defined start and finish points with objectives, time constraints and pre-set cost and performance parameters.

A facilities manager would be aware, from the media, of the many larger projects that have failed elsewhere, so facilities managers must use their experience to prevent any similar failure from occurring by ensuring that the following criteria are managed at the outset of the project:

1. agreeing a detailed project specification;
2. timescales set are realistic;
3. selection of appropriate staff in the team;
4. management of customer expectations; and
5. management of change when a project is implemented.

Successful projects are those that have the intended project sanctioned at a senior level within a company, containing a detailed project brief outlining the desired objective. It is important that time is spent defining a project brief, as this can save much effort and frustration at a later stage.

A typical project brief could include the following:

(a) goals and objectives – how will the project be deemed as being successful;
(b) scope of the project – remember to include areas out of scope;
(c) risks that will affect the project; and
(d) assumptions made.

Usually, a facilities manager will have to prepare a business case to put to senior management in order to obtain the desired budget. Part of the document would have to define cost, time limitations and benefits as well as operational risks of undertaking, or not undertaking the project.

During the project management of, say, a new IT system, the risks to a company in the changeover maybe considered too high until the new system has been proven. Both old and new systems could be run in parallel with the facilities management team analysing the performance of the new system until such a time as data is transferred completely and the old system discontinued.

## Contract management

F3022   This is a very complex skill, but one in which it is essential that the facilities management team succeed, as a poorly defined contract will lead to poor performance and eventually a complete breakdown in communications between the two parties.

The contract management process itself is one that enables both parties to agree a contract that enables them to deliver the objectives required. It also involves building a good relationship between the parties in order to be able to work proactively. It is unhelpful having a relationship built around an adversarial approach, as the aim is not only to achieve delivery of the agreed services or products, but also to achieve value for money by looking at innovations in delivery and in some cases aiming for continuous improvement.

There are two types of contract that a facilities management team would normally be expected to deal with, which are: service delivery and construction contracts.

Service delivery contracts are generally divided into three areas of:

1. service delivery management – to ensure that this is being delivered at the agreed level of performance;
2. relationship management – to ensure that the relationship between the parties is maintained; and

3. contract administration – to handle changes to the contract and to manage the more formal aspects.

The construction contract is very complex and the contract strategy will depend upon what suits the individual projects itself, and may be influenced by pricing and risk transfer. In this instance, the facilities manager will require much experience of the construction industry.

## Change management

F3023 Everything covered so far will involve an element of change. Humans tend to be creatures of habit and will often resist any changes, partly due to a fear or uncertainty of the unknown, created by a lack of knowledge of the intentions of the company and the reasoning behind the change.

It is important for the facilities management team to manage changes within a company, which is best carried out by effectively communicating with the staff affected. Such communication should be done at the earliest opportunity to reduce the risk of rumours escalating across the workforce.

The first stage in the process is to analyse what effect the intended changes will have on staff, so that a chart can be prepared to detail the likely resistance to change. Management can have in place a series of methods to alleviate any areas of resistance, so ensuring the effective transition.

The second stage will be to hold regular staff meetings to keep the workforce informed of the changes and the drivers behind such changes, e.g. the introduction of a new management information system for improved efficiency in reporting. It is most likely that the staff most affected will need to be met on a one-to-one basis.

The final stage before the go live date will be for the company to agree and produce new procedures that can be communicated to staff through training.

# Four most common health and safety issues for facilities management

## Asbestos

F3024 Asbestos is likely to be a major project for a facilities manager to deal with, especially in older buildings of pre-1981 construction, which potentially could contain large quantities of asbestos containing materials (ACM's). Buildings of pre-1999 construction can contain limited amounts of ACM's, which have also been found in recent new constructions where old gaskets have been fitted in pipe work.

The *Control of Asbestos at Work Regulations 2002* (*SI 2002 No 2675*), sometimes referred to as the duty to manage regulations, impose a number of requirements upon the 'duty holder' who has been defined as the party who has 'control' of the building or part of a building.

The duties imposed on the duty holder may be summarised as follows.

- It is essential that, unless already completed, a competent asbestos surveyor must take reasonable steps to locate ACM's within a building and to assess their condition.
- There is a further requirement to presume that materials contain asbestos unless there is strong evidence to prove that they do not. Instances where it could be presumed that there could be ACM's present is in areas where access was not gained, such as in lift shafts (fire compartmentalisation/suppression), locked areas, electrical switch gear (electrical flash protection) or in areas where samples are unable to be taken such as in listed buildings, where such damage would be prohibited.
- The findings of the asbestos survey must be maintained in a written record indicating the location and condition of known and presumed ACM's. This would normally be in the form of an asbestos register which should indicate the areas where known ACM's were

located, together with non-asbestos samples and areas of no-access. The findings, ideally, would also be supported by the use of photographs and annotated drawings indicating locations where samples were taken.

- The final requirement is to produce an active asbestos management plan which must be kept up to date at all times. This plan should indicate how known asbestos is to be managed e.g. removal, repair and encapsulation and periodic monitoring as well as persons responsible for management, emergency procedures and a record of all activities undertaken on the ACM's.

It is essential that ACM's are managed safely for the protection of all persons likely to come into contact with them and there is no doubt that this task would fall to the facilities management team.

## Work at height

**F3025** Facilities managers will have to pay considerable attention to working at height. Working at height may appear to be relatively simple, from changing a lamp using a small pair of stepladders, to a major refurbishment project being undertaken from scaffolding. However, in each of these examples the inherent risk of falling from height may be high unless control measures are in place to mitigate such risk. It is important that the facilities manager is aware of the risks and is able to ensure that persons working at height are working in a safe manner and that they not only protect themselves, but also others who may be nearby.

The new *Work at Height Regulations 2005* (*SI 2005 No 735*) came into force on 6 April 2005 and apply to all work at height where there is a risk of a fall liable to cause personal injury. The old two-metre rule has been replaced by the new regulations which apply to all heights where there is a risk of personal harm. As in the *Control of Asbestos at Work Regulations 2002* (*SI 2002 No 2675*) there are duties placed on dutyholders who are defined as employers, the self-employed and any person who controls the work of others e.g. a facilities manager.

Duty holders are required to ensure the following.

(*a*) Avoid work at height if at all possible.

(*b*) All work at height is planned and organised in a systematic way. Consider how long the work will take and if it can be successfully carried out with one hand from a ladder. Is there some other activity in the area that may affect the work?

(*c*) All persons working at height are competent. Ensure that if contractors are carrying out the work that their health and safety has been evaluated and that the operatives have been suitably trained.

(*d*) Risks from working at height have been assessed and that suitable control measures and work equipment are in place before work commences.

(*e*) Risks from working on fragile surfaces have been assessed and are properly controlled. Consider fitting moving bridges and gantries to span fragile surfaces requiring some form of maintenance.

(*f*) All equipment used for working at height is marked with a serial number or colour coding and is properly maintained and inspected. Operatives should also undertake their own inspection prior to using the equipment.

The facilities manager must ensure that there is an effective safe system of work in place such as a permit to work system for evaluating the competency of operatives, checking their risk assessments and method statements, checking that equipment to be used is suitable and has been maintained. Once work has commenced the facilities manager should supervise the work to ensure that the operatives are carrying out the work in the manner described in the risk assessment/method statement and that there are no other unforeseen hazards.

## Water quality

**F3026** A number of deaths from legionnaires' disease has been widely reported in the media over recent years, in spite of the introduction of the Code of Practice 'The Prevention or Control of Legionellosis' (including legionnaires' disease) L8 rev and its associated guidance document HS(G) 70. These deaths have resulted from poor maintenance of cooling towers and evaporative condensers.

An essential component of the above is to ensure that a water risk assessment (WRA) is undertaken for the company's building and to make recommendation to reduce the risk to the lowest possible. Within this document there will be an organisational chart indicating persons responsible for ensuring that water services are properly maintained. In a small to medium sized company the facilities manager will have responsibility for ensuring that all recommendation in the WRA are completed and that maintenance of the water system is carried out in accordance with the maintenance scheme shown in the WRA.

The following should be considered when dealing with the water services of a building:

1. ensure that a WRA has been carried out by a competent person;
2. ensure that all recommendations made in the WRA are completed and records maintained;
3. ensure that the maintenance regime shown in the WRA is undertaken at the prescribed intervals by a competent person;
4. ensure that all water samples taken are tested by a UKAS accredited laboratory;
5. ensure that records are maintained for all maintenance and testing; and
6. ensure that the WRA is reviewed on a regular basis or after any alterations have been made to the water system.

If the facilities manager adheres in essence to the above guidance and has maintained good, accurate and auditable records then the incidence of legionalla should be avoided.

## Management of contractors

**F3027** Any contractors working on site pose one of the most notable health and safety risks. Some senior managers share the mistaken belief that they are able to transfer any Health and Safety risk by contracting work out. In fact any such risk is shared because the company on whose premises work is undertaken also has a duty under health and safety legislation to ensure that persons not in their employ are working safely. It is highly probable that the facilities manager will be the person responsible for employing such contractors and will have to manage them whilst they are on site.

Before employing contractors, facilities management should ensure that the contractors have been thoroughly vetted by way of a health and safety evaluation questionnaire. Such a questionnaire can be designed to verify their areas of competency with points such as accident record, insurance, procedures and risk assessment systems being a minimum. Some companies keep an 'approved contractors' list which shows all contractors who have been subjected to such an evaluation process. A word of warning, however, facilities management should check to ensure that the contractor has been approved for the type of skill required for the work activity to be undertaken. It could be that the contractor has been added to the approved list for some other activity.

Once facilities management is satisfied that the contractor is competent to undertake the allotted work activity, facilities management can allow them to conduct a site survey before holding a meeting to agree how the work should proceed. At this stage the contractor should be preparing their method statements and risk assessments while the facilities manager should inform the contractor of any known hazards connected with the site and should also provide rules or contractor's guidance notes. The next stage would be for the facilities manager to review the method statements and risk assessments to ensure that the work is to be undertaken safely and will not impact upon any other activity on site.

When the contractor arrives on site the facilities manager must be satisfied that the contractor has all the necessary control measures in place to protect not only the contractors but also company staff and any one else who may be affected by the work activity, such as members of the public or visitors. It is important that work is undertaken as agreed and is recorded in the method statements and risk assessments with the facilities manager providing as much supervision as necessary. Provided there are good communications between both parties and with good preparation, any work should proceed smoothly. However, unforeseen circumstances can prevent work progress, in which case it is important that any changes in work procedure must be agreed and the facilities manager must ensure that any such changes will still result in safe working.

## How facilities management and building design can help a company

F3028    The way in which facilities are managed within the work environment is important in creating the desired corporate image, whilst concurrently achieving its corporate objectives. It is pointless creating the desired corporate image if the efficiency of the company is undermined by bad building design or its location.

It is not uncommon to see many fine buildings, which are difficult to maintain with lights or air conditioning cassette units located high up in large atriums of buildings. In these instances the changing light bulbs or cleaning and replacing filters suddenly becomes a major part of the maintenance procedure as safe access becomes a more significant priority with, for example, the constructing scaffolding or mobile towers in order to undertake such a relatively simple task. It would be ideal if an experienced and skilled facilities manager were involved at the design stage, as they could consider the maintenance and operation of the building in a practical way. This situation has now improved as designers have duties, under the *Construction (Design and Management) Regulations 1994 (CDM) (SI 1994 No 3140)*, where they have to consider the construction, maintenance and demolition of the building. However, it is still a problem in some areas.

A facilities manager is expected to be multi-skilled in a wide range of subjects, so that they are best able to advise a company how to create a safe and efficient workplace which provides a comfortable and positive atmosphere for all its staff.

Most visitors will quickly form a first impression of a company by noticing the building location, its design and its general condition. How a company's logo is displayed is all part of a carefully constructed corporate package. A skilled facilities manager with clever use of colours, furniture design and layout will display the company logo in such a way as to create the desired synergy and corporate image in order to help achieve the company's objectives.

## Recognition of facilities management

F3029    Facilities management of built environments is gaining greater recognition and credibility from a wide range of government departments, such as the Health and Safety Executive (HSE), the Office of the Deputy Prime Minister and the Sector Skills Council due, in part, to facility managers working in all types of companies who use a wide variety of skills. Consequently, facilities management is being recognised as an industry in its own right and facility managers are being used to provide feedback for government consultation documents, speak at conferences and attend innovation workshops. In fact, the facilities management industry has progressed to such an extent that it now employs more people than those working in the construction industry.

BIFM is working closely with the Office of the Deputy Prime Minister (ODPM) on a range of issues concerning energy and environmental management of the built environment and has produced a paper on Approved Document L2 of the *Building Regulations 2000* (*SI 2000 No 2531*). In addition, BIFM also has its own accredited qualification based around 20 core competencies, which further demonstrates the very diverse nature of facilities management.

Asset Skills, the Sector Skills Council for the places in which we live and work, has now created a facilities management sector board which is comprised of representatives from the facilities management industry and BIFM. The intention of the board is to:

- raise the standards of learning and professionalism;
- develop a National Vocational Qualification (NVQ) for facilities management;
- develop a research programme for facilities management (current research is being undertaken by Asset Skills Council, local councils and university estates management departments); and
- work closely with the facilities management industry and large companies to develop facilities management skills and best practice.

The Royal Institution of Chartered Surveyors (RICS) now has a facilities management faculty and a qualification of Chartered Facilities Management Surveyor. The institute is now actively undertaking research into best practice and is actively publishing facilities management articles on their website.

HSE recognise that the facilities management industry has considerable experience, retains broad skills and as a result holds regular update meetings with BIFM to discuss current trends and innovations. HSE is also contributing to a work at height guide designed for use by facilities managers. BIFM is also involved in the provision of feedback concerning government consultation documents by the collation of responses from specialists within its membership.

Industry is also recognising facilities management as an essential element of their strategy to ensure that they remain competitive in the market place. Increasingly, more companies are joining BIFM as corporate members and are actively involved in providing input into the institute and also to putting forward representation on committees across the United Kingdom. Some of the larger blue-chip companies run their own facilities management training and, through their considerable experience, are laying down the foundations of best practice, which in some cases are being adopted throughout the facilities management industry.

Universities are offering degrees in facilities management with a number of the more traditional degrees e.g. architecture and surveying including a number of modules consisting of facilities management disciplines. The same applies to health and safety qualifications where a number of course providers are delivering the requisite course content but are focusing this towards facilities management using practical examples to demonstrate the operational implications within the facilities environment.

## Innovations within the facilities management industry

F3030    New products and methods of working are being developed constantly as manufacturing companies have seen the opportunities to be realised within facilities management, particularly within:

- information technology software manufacture;
- new construction products and materials;
- cleaning;
- security;
- environmental; and

- energy management.

The development of telecommunications and software operating on the world-wide-web has provided an incredible choice of management systems to assist the facilities management industry. The large impact of such choice has led to the development of the 'virtual office'. A company has the opportunity to have staff based at home working from a laptop and phone line. Such working practices has advantages for both parties and can save the company costs in office space and staff equipment.

Within the cleaning industry, there have been considerable advances in cleaning products which are safe and easy to use and which have such a tolerance that they can be used by an unskilled operator and still provide a good finish. In other areas of cleaning, soap dispensers in toilets are being designed with large reservoirs to reduce the number of times they need to be refilled, helping reduce costs whilst dispensing soap over a longer period.

Floor cleaning equipment has progressed beyond large floor polishers with trailing leads and their inherent problems of trip, electrical and manual handling hazards. Modern machines tend to be smaller, lighter and are powered by maintenance-free batteries, thus eliminating the need for trailing cables. New flooring products manufactured from revolutionary materials are currently entering the marketplace and tend to be virtually maintenance-free. Flooring, especially in reception areas, can require much maintenance yet still look worn very quickly. Worn flooring may create a poor corporate image and as a result modern floor coverings are generally very durable and require little maintenance other than periodic vacuuming.

Technological advances have helped so-called 'intelligent buildings' realise much reduced running costs which have helped save energy and lower the environmental impact caused by the company. An intelligent building can detect when there is a lack of movement i.e. staff in an area, and can automatically switch off lighting and reduce the amount of heating and ventilation generated. Facilities and maintenance managers are being assisted by new hardware which can collate information from the central building management system (BMS) and transmit this information remotely to a handheld wireless computer such as a Blackberry, enabling facilities management real-time information. Recent developments in such building management software enable facilities management to pre-set parameters within the building and as soon as any readings extend outside those parameters a warning is sent to the Blackberry, warning the facilities manager, regardless of his location. This allows a facility manager to be proactive and take corrective action, hopefully before a problem arises.

The security industry has also seen much innovation with the introduction of new camera and software technology that can transmit pictures from remote, unmanned sites. The equipment runs in a hibernating mode until movement is detected, whereupon pictures and sound are transmitted back to a centrally manned site where they are then recorded. A mobile security patrol or the police may be called to deal with the situation. This equipment eliminates the need for a security guard to be exposed to lone working in a building and reduces security costs for the company.

Facilities management help desks have benefited from help desk software systems and the introduction of operative vehicle tracking modules, which not only show the location of the nearest operative, but their skills and competency. The helpdesk operative can locate and send the closest operative with the requisite skills for the work activity. Using wireless Blackberry hand-held computers jobs can be sent to operatives electronically and once completed the operative is then able to close the job down in the system remotely which means that this data within the help desk system is live.

# The effects of safety, health and environmental legislation on facilities management

F3031   A strong interrelationship exists between facilities management, safety, health and environmental (SHE) due to a large amount of compliance and legislation. Even just keeping abreast with ever changing legislation places a facilities manager under considerable pressure by not only having to acquaint themselves with new legislation, but by knowing also how to apply it to the practical aspect of how and when it will impact upon a company and the budget required for any remedial works, training and implementation. HSE has recognised that frequent changes in legislation has caused problems and are now due to release legislative changes twice-a-year in April and October.

Facilities management is becoming even more demanding in the field of SHE, as most small to medium sized companies will be unable to justify employing a SHE specialist or buying-in such advice from a consultant. A facilities manager is now expected to add health and safety to the ever-increasing range of facilities management skills. BIFM has recently reviewed its full range of facilities management competencies and has considerably expanded the section concerning SHE legislation to include the working environment, environmental protection, energy efficiency, development of health and safety policy and risk management.

As previously mentioned, building design is critical and if facilities management have some input at the design stage of a new build or refurbishment, then hopefully they can ensure that the building can be operated and maintained safely. A good, experienced facilities manager can make compromises when required and will take a proactive and flexible approach towards practical solutions to enable the safe operation of the building.

Every aspect of the facilities management role in the workplace, whether it is people management, or managing equipment, will at some stage be subject to an aspect of health and safety legislation. Staff within a company have to be managed to ensure that they have a safe and healthy place in which to work and that the actual way in which they work is also safe and will not endanger others. The facilities management role can influence the culture of a company in an effort to ensure that health and safety is viewed as a positive element of its business activities with benefits, rather than one that uses up valuable resources and impedes operations. It has often been noticed that a health and safety management system can generally identify operational issues before they become a problem, so helping to create a more efficient workplace.

Conversely, a building and all it's associated assets, will be subject to legislation throughout it's entire lifecycle, from design, through to its final demolition or disposal, thus making it essential that a facilities manager is kept fully aware of such legislation. This maybe demonstrated by consideration of the legislation and best practice compliance concerning for example a passenger lift within a building at F3032–F3034 below. However this three stage lifecycle model could be applied to any asset within the built environment.

## Design

F3032     *Construction (Design and Management) Regulations 1994 (CDM) (SI 1994 No 3140)* – place duties on the designer of the lift to ensure that the equipment is safe to construct, use, maintain and finally de-commission.

*Provision and Use of Work Equipment Regulations 1998 (PUWER) (SI 1998 No 2306)* – place duties on the manufacturers and suppliers to ensure that the equipment is suitable for the purpose for which is has been designed and manufactured, and is safe for use and maintenance.

## Construction

F3033     *Construction (Design and Management) Regulations 1994 (CDM) (SI 1994 No 3140)* – place duties on the client and principal contractor to ensure that the construction work is undertaken in a safe manner and is a safe system of work. There is also a requirement to keep a health and safety file, which is a record of all materials and equipment used and the methods of construction. This file must be kept in the building throughout its entire life as this contains valuable information.

## Operation

**F3034**    *Health and Safety at Work etc Act 1974* – places a duty on the owner/operator to ensure that the lift is safe.

*Lifting Operations and Lifting Equipment Regulations 1998 (SI 1998 No 2307)* – place duties upon owners and operators of lifts to ensure that they are subjected to preventative maintenance and also a thorough examination at regular intervals.

*Construction (Design and Management) Regulations 1994 (CDM) (SI 1994 No 3140* – ensure that the lift is demolished in a safe manner and information from the health and safety file would be used to assist in this process.

There are also environmental issues for a facilities manager to consider and increasingly more companies are developing and implementing environmental management systems (EMS) based upon the ISO 14001 model. Consequently, the suppliers to such companies are also expected to have an EMS.

Approved Document L2 of the *Building Regulations 2000 (SI 2000 No 2531)* deals with the conservation of fuel and power and therefore as a result has consequences for facilities managers as this requires a building logbook be maintained for all new or refurbished buildings. The logbook is designed to provide details of the plant equipment installed, it's controls and maintenance in order that energy consumption can be monitored and controlled. The object of such logbooks is to improve the efficiency of buildings by reducing energy costs and the emission of $CO_2$ gas into the atmosphere.

# Summary of the facilities management

**F3035**    Facilities management is gaining considerable recognition as an essential industry and efforts are being made to advertise facilities management within universities and colleges as a worthwhile and challenging career. It will be sometime before careers advisers are recommending facilities management to school leavers, but this should change in due course.

There is collaboration between BIFM in the UK and various facilities management institutes world-wide including Mainland Europe, the United States and Australia. These institutes have an agreement to share best practice and research, which will provide a powerful resource and will be invaluable to companies.

Future innovation and resultant advancements in facilities management will lead to improved efficiency, improved working conditions for staff, more added value and cost efficiencies for companies, reduced energy consumption and will reduce the impact to the environment.

# Fire Prevention and Control

## Elements of fire

**F5001** Fire is a chemical reaction resulting in heat and light. For a fire to occur the following need to be combined in the correct proportions:

- oxygen (supplied by the air around us which contains about 21 per cent oxygen);

- fuel (combustible or flammable substances either solids, liquids or gases);

- source of heat energy (ignition sources such as open flame, hot surfaces, overheated electrical components).

For a fire to burn the fuel must be in a gaseous or vapour form. This means that solids and some liquids such as oils will need to be heated up until sufficient vapour is given off to burn. The form that the fuel takes will also influence the ease with which it ignites and the speed with which it burns. Generally, finely divided materials are easier to ignite and burn more quickly than fuels in a solid form.

Once initiated, a fire will continue to burn as a result of the reinvestment of energy during the process provided sufficient fuel vapour, heat and oxygen are available and the process is not interrupted.

The 'triangle' of fire symbolises the fire process. Each side of the triangle represents fuel, heat or oxygen. Take any one side away and the fire will be extinguished or more importantly, stop all three sides from coming together and a fire can be prevented.

The first part of this chapter (F5001–F5029) considers generally the practical aspects of fire prevention and control. It should be borne in mind that *special risks* involving flammable or toxic liquids, metal fires or other hazards should be separately assessed for loss prevention and control techniques. For precautions against fire hazards from flammable liquids, see the HSE Guidance documents HS(G)51 and HS(G)176. Control of these hazards is among the areas dealt with under the *Dangerous Substances and Explosives Atmospheres Regulations 2002* (*SI 2002 No 2776*).

It is the responsibility of management to consider how safe is safe: that is, to balance the costs of improvement against the financial consequences of fire. Considerable improvement can often be made immediately at little or no cost. Other recommendations which may require a financial appraisal must be related to loss effect values. In certain cases, however, due to high loss effect, special protection may be needed almost regardless of cost.

Modern developments in fire prevention and protection can now provide a solution to most risk management problems within economic acceptability. It must be pointed out, however, that it is a waste of time and money installing protective equipment unless it is designed to be functional, the purpose of such equipment is understood and accepted by all personnel and the equipment is adequately inspected and maintained. The reasons for providing such equipment should, therefore, be fully covered in any fire-training course. Fire routines should also be amended as necessary to ensure that full advantage is taken of any new measures implemented.

## Common causes of fires

**F5002** The following, in no particular order of significance, are the most common causes of fire in the workplace:

- wilful fire raising and arson;
- careless disposal of cigarettes, matches;
- combustible material left near to sources of heat;
- accumulation of easily ignitable rubbish or paper;
- inadvertence on the part of contractors or maintenance workers, usually involving hot work;
- electrical equipment left on inadvertently when not in use;
- misuse of portable heaters;
- obstructing ventilation of heaters, machinery or office equipment;
- inadequate cleaning of work areas;
- inadequate supervision of cooking activities.

## Fire classification

**F5003** There are five classes of fire which are related to the fuel involved and the method of extinction, as follows.

- *Class A*

    This relates to fires generally involving solid organic materials, such as coal, wood, paper and natural fibres, in which the combustion takes place with the formation of glowing embers. Extinction is achieved through the application of water in jet or spray form. Water extinguishes the fire by removing or limiting the heat by cooling.

- *Class B*

    This relates to fires involving:

    (i) liquids, which can be separated into those liquids which mix with water, e.g. acetone, acetic acid and methanol; and those which do not mix with water, e.g. waxes, fats, petrol and hydrocarbon solvents; and

    (ii) liquefiable solids, e.g. animal fats, solid waxes, certain plastics.

    Foam, carbon dioxide and dry powder can be used on all these types of fire. However, some types of foam break down on contact with water-miscible liquids and thus special alcohol-resistant foam is needed for large volumes of such liquids. Water must not be used on fats, petrol, etc. With foam, carbon dioxide and dry powder, extinction is principally achieved by limiting or removing the oxygen by smothering.

- *Class C*

    This relates to fires involving gases and should only be extinguished by shutting off the gas supply if it is safe to do so. Burning gas should not be extinguished as to do so without isolating the supply may result in a build-up of unburnt gas which might explode.

- *Class D*

    This relates to fires involving combustible metals, such as aluminium or magnesium. These fires burn with very high temperatures, and their extinction can only be achieved by the use of special powders. These powders form a crust over the surface isolating the burning metal from the surroundings. It is extremely dangerous to use water on burning metals. Special training is required where the use of extinguishers on combustible metals is concerned.

- *Class F*

    This relates to fires involving cooking oils and fats. Extinguishers using special wet chemical extinguishing agents, foam or dry powder can be assessed for their effectiveness for this type of fire using test fires defined in BS 7937: 2000. Extinction is achieved primarily by limiting or removing the oxygen by smothering and in some cases by a degree of cooling. It is extremely dangerous to use water on fires involving cooking oils and fats.

The table below classifies fires which can be controlled by portable fire appliances (see also BS EN 2 and BS EN 3).

**Suitability of extinguishing types**

| Class of fire | Description | Appropriate extinguisher |
|---|---|---|
| A | Solid materials, usually organic, with glowing embers | Water (foam, dry powder or CO2 will work but may be less effective than water due to a lack of cooling properties). |
| B | Liquids and liquefiable solids: | Foam, CO2, dry powder |
|   | • miscible with water e.g. acetone, methanol | Alcohol-resistant foam, CO2, dry powder and the skilled use of water spray |
|   | • immiscible with water e.g. petrol, benzene, fats, waxes | Foam, dry powder, CO2 |
| F | • cooking oils and fats | Type F, special wet chemical extinguisher |

## Electrical fires

F5004    Fires involving electrical apparatus must always be tackled by first isolating the electricity supply and then by the use of carbon dioxide or dry powder, both of which are non-conducting extinguishing mediums. Dry powder and carbon dioxide extinguish the fire by limiting or removing the oxygen by smothering.

EU Regulations, which apply in all EU member countries including the UK, prohibit the use of halon as an extinguishant in most applications after 31 December 2003. Portable extinguishers and fixed systems using halon are required, by these regulations, to have been decommissioned by this date.

## Fire extinction – active fire protection measures

F5005    Extinction of a fire is achieved by one or more of the following:

- *starvation* – this is achieved through a reduction in the concentration of the fuel. It can be effected by:
    (i) removing the fuel from the fire;
    (ii) isolating the fire from the fuel source; and
    (iii) reducing the bulk or quantity of fuel present;
- *smothering* – this brings about a reduction in the concentration of oxygen available to support combustion. It is achieved by preventing the inward flow of more oxygen to the fire, or by adding an inert gas to the burning mixture;
- *cooling* – this is the most common means of fire-fighting, using water. The addition of water to a fire results in vaporisation of some of the water to steam, which means that a substantial proportion of the heat is not being returned to the fuel to maintain combustion. Eventually, insufficient heat is added to the fuel and continuous ignition ceases. Water in spray form is more efficient for this purpose as the spray droplets absorb heat more rapidly than water in the form of a jet.

## Property risk

**F5006** Fire safety precautions needed for the protection of life are dealt with by legislation, as detailed in F5030–F5077.

In order to ensure the survival of a business in the event of fire, property protection must be considered and a business fire risk assessment carried out. The involvement of a company's insurer is essential as insurers have considerable experience in this field. The business fire risk assessment follows similar stages to the life safety risk assessment (see F5053) but assesses the importance of each area to the function of the business and how vulnerable they are to fire.

For example, an area where essential records or documents are stored is likely to have a serious effect should a fire occur. Essential equipment, plant or stock which, if destroyed or severely damaged by fire, might be difficult to replace or have a serious effect on production would require special consideration, and often high fire protection requirements to minimise such an effect.

These risks should be determined by management; they need to be identified, considered and evaluated. A report should be produced by each departmental head, outlining areas which may require special consideration. Such a report should also include protection of essential drawings, records and other essential documents.

A typical area of high loss effect would be the communication equipment room. The loss of this equipment could have a serious and immediate effect upon communications generally. Fire separation (to keep a fire out) is therefore considered essential, and automatic fire suppression by a self-contained extinguishing system should be strongly recommended.

Essential data should be duplicated and the copy stored in a safe area which preferably is off-site. Paper records can be converted to electronic format and similarly stored.

## Passive and active fire protection

**F5007** Passive fire protection is where part of the structure of a building is inherently fire resistant. Most buildings are divided into fire-resisting compartments. These serve two functions. One is to limit the spread of fire and can be a property protection measure as well as life safety. The other is to protect escape routes by making the escape route a fire-resisting compartment. All internal fire escape stairways are fire-resisting compartments.

In day-to-day work these compartments are most obvious where doorways pass through the compartment walls. The doors in these openings are nearly always self-closing fire resisting door sets. It is essential that these doors are not obstructed and are allowed to self-close freely at all times. Any glazing in such doors must also be fire resistant and if damaged must be repaired to the appropriate standard. (Guidance on the most common types of fire door is given in BS 8214: 1990 *Code of practice for fire door assemblies with non-metallic leaves.*)

The breaching of fire resistant walls with services such as pipes, ducts and cables should be avoided. Where there is no alternative, it is important that all the openings are suitably protected to the same rating as the fire wall. Ducts should be fitted with fire dampers while openings around services should be suitably sealed or fire stopped. (Guidance on the fire safety aspects of the design and construction of air handling duct work is given in BS 5588-9: 1999 *Fire precautions in the design, construction and use of buildings: Code of practice for ventilation and air conditioning ductwork.*)

Active fire protection involves systems that are activated when a fire occurs, for example automatic fire detection or automatic sprinkler systems. As such systems are only required in an emergency it is essential that they are designed, installed, tested, inspected and maintained in accordance with acceptable standards and good practice.

## Fire procedures and portable equipment

### Fire procedures

F5008    The need for effective and easily understood fire procedures cannot be over-emphasised. It may be necessary to provide a fire procedure manual, so arranged that it can be used for overall fire defence arrangements, and sectioned for use in individual departments or for special risks.

It is essential that three separate procedures are considered:

- procedure during normal working hours;
- procedure during restricted manning on shifts;
- procedure when only security staff are on the premises.

All procedures should take into consideration absence of personnel due to sickness, leave, etc. The fire brigade should be called immediately if any fire occurs, irrespective of the size of the fire. Any delay in calling the fire brigade must be added to the delay before the fire brigade's actual arrival, which will be related to the traffic conditions or the local appliances already attending another fire.

A person should be given the responsibility for ensuring that pre-planned action is carried out when a fire occurs. Large fires often result from a delayed call, which may be due not to delayed discovery but to wrong action being taken in the early stages following discovery of a fire. A pre-planned fire routine is essential for fire safety. The fire brigade, when called, should be met on arrival by a designated person available to guide them directly to the area of the fire. It is essential that all fire routines, when finalised, be made known to the fire brigade.

### Fire equipment

F5009    There have been a number of cases where a person using an extinguisher has been seriously injured. Investigations have shown that either the wrong type of extinguisher was supplied or the operator had no training in the correct use of the extinguisher. The need for training staff cannot be over-emphasised, especially in areas of special risk, oil dipping tanks, furnace areas, highly flammable liquids, gas or cylinder fires etc.

The following recommendations are given in order to allow an evaluation of an existing problem and may need to be related to process risks:

- it is essential that persons be trained in the use of extinguishers, especially in areas where special risks require a specific type of extinguisher to be provided;

F5010    *Fire Prevention and Control*

- any person employed to work, who is requested to deal with a fire, should be clearly instructed that at no time should that person jeopardise their own safety or the safety of others;
- persons who may be wearing overalls contaminated with oil, grease, paint or solvents should not be instructed to attack a fire. Such contaminated materials may vaporise due to heat from the fire, and ignite.

## Types of fire extinguisher

F5010    The type of extinguisher provided should be suitable for the risk involved, adequately maintained and appropriate records kept of all inspections, tests etc. All fire extinguishers should be fitted on wall brackets or located in purpose built floor stands. It has been found that if this is not done, extinguishers are removed or knocked over and damaged. Extinguishers should be sited near exits or on the route to an exit.

## Water extinguishers

F5011    This type of extinguisher is suitable for ordinary combustible fires, for example wood and paper, but is not suitable for flammable liquid fires. Spray type water extinguishers are recommended. Water extinguishers should be labelled 'not to be used on fires involving live electricity'.

## Foam extinguishers

F5012    These are suitable for small liquid spill fires or small oil tank fires where it is possible for the foam to form a blanket over the surface of the flammable liquids involved. Foam extinguishers may not extinguish a flammable liquid fire on a vertical plane. Alcohols miscible with water, when on fire, will break down ordinary foam and should be considered a special risk. Alcohol resistant foams are available.

## Dry powder extinguishers

F5013    This type of extinguisher will deal effectively with flammable liquid fires and is recommended, as it is capable of quick knock-down of a fire. The size of the extinguisher is important and it must be capable of dealing effectively with the possible size of the spill fire which may occur, with some extinguishant in reserve. The recommended minimum size is a 9 kg trigger-controlled extinguisher. (Dry powder extinguishers are also safe on fires involving electrical equipment.) Some dry powders have been tested and listed as suitable for use on normal combustible material (Class A) fires.

## BCF extinguishers

F5014    The extinguishing medium is a halon. Manufacture of this class of chemicals is no longer permitted due to its adverse effect on the environment. New extinguishers of this type are no longer available although some old ones may still be in service. Recharging of existing extinguishers is no longer practised and replacement by dry powder or carbon dioxide extinguishers is recommended. EU Regulations prohibit the use of halon as an extinguishant in most applications (aircraft and military applications will be exempt) by the end of 2003. Any halon extinguishers in normal service should be decommissioned and disposed of safely by 31 December 2003.

## Carbon dioxide extinguishers

**F5015** For fires involving electrical equipment, carbon dioxide extinguishers are recommended. Carbon dioxide ($CO_2$) extinguishers are quite heavy and may be at high pressure. A minimum size of 2 kg is recommended. $CO_2$ is not recommended for flammable liquid fires, except for small fires. Training in the use of $CO_2$ extinguishers is essential. Particular care should be taken not to hold the discharge horn on a $CO_2$ extinguisher as the surface gets very cold and can cause cold burns to unprotected skin. As carbon dioxide is an asphxyiant it should not be used in confined unventilated spaces.

## Colour coding and distribution of portable fire extinguishers

**F5016** All newly certified fire extinguishers for use throughout the EU are coloured red in compliance with BS EN 3. Manufacturers are allowed under BS 7863 to affix different coloured areas on or above the operation instructions label. Existing extinguishers (complying with BS 5423) need not be replaced until they have served their useful life. The colour coded panels help identify the extinguisher type for example red for water, blue for dry powder and black for $CO_2$.

Extinguishers are distributed in accordance with BS 5306–8: 2000 *Code of practice for selection and installation of portable fire extinguishers*, based on an extinguishers rating (capability) rather than by the type and size of the unit.

## Fire alarms in 'certificated' premises

**F5017** A manually operated fire alarm system is required in any 'certificated' premises and in some cases an automatic fire detection system will be required (see F5039). Both these needs will be indicated on the fire certificate. The need for a fire detection and alarm system may also result from a fire risk assessment. The systems should comply with the requirements of BS 5839–Part1: 2002 *Fire detection and alarm systems for buildings: Code of practice for system design, installation, commissioning and maintenance.*

Fire alarm systems should be tested once a week

# Good 'housekeeping'

**F5018** The need for good 'housekeeping' cannot be over-emphasised. Poor housekeeping is the greatest single cause of fire. A carelessly discarded cigarette end, especially into a container of combustible waste or amongst combustible storage, often results in fire. The risk is higher in an area which is infrequently used. The following are essential guidelines:

- smoking should, preferably, be prohibited but where smoking is permitted, suitable deep metal ashtrays should be provided. Ashtrays should not be emptied into combustible waste unless the waste is to be removed from the building immediately. It is recommended that smoking ceases before close of work so that if smouldering occurs this will be detected before staff leave the premises;
- combustible waste and contaminated rags should be kept in separate metal bins with close fitting metal lids;
- cleaners should, preferably, be employed in the evenings when work ceases. This will ensure that combustible rubbish is removed from the building to a place of safety before the premises are left unoccupied;
- rubbish should not be kept in the building overnight, or stored in close proximity to the building;
- smoking should be prohibited in places which are infrequently used, e.g. stationery stores, oil stores, or telecommunications intake rooms. Suitable 'smoking prohibited' notices

should be displayed throughout such areas – notices should comply with BS 5499 and the *Health and Safety (Safety Signs and Signals) Regulations 1996 (SI 1996 No 341)*;

- where 'no smoking' is enforced due to legal requirements (for example, areas where flammable liquids are used or stored) or in areas of high risk or high loss effect, it is recommended that the notice read 'Smoking prohibited – dismissal offence';
- materials should not be stored on cupboard tops, and all filing cabinets should be properly closed, and locked if possible, at the end of the day.

## Pre-planning of fire prevention

**F5019**  A pre-planned approach to fire prevention and control is essential. Fire spreads extremely fast, and the temperature can rise to 1,000°C in only one minute. Smoke can be flammable and toxic. The essential factor is to re-evaluate the risk, identify areas of high loss effect or high risk, and plan accordingly to meet requirements and legal responsibilities.

### Means of escape

**F5020**  Means of escape routes consist of two main stages:

- travel within rooms;
- travel from rooms to the entrance to a stairway or corridor protected against the effects of fire or directly to a final exit where a protected route is not provided.

    In particular, for offices:

    (i) total travel distance between any point in a building and the nearest final exit or access to a protected route should not exceed:

    — 25 m, if there is only one exit, or

    — 45 m, if more than one exit;

    (ii) two or more exits are necessary:

    — from a room in which more than 60 people work, or

    — if any point in the room is more than 12 m from the nearest exit;

    (iii) minimum widths of escape routes (doors, corridors, stairs):

    — doors from rooms should have a minimum width of 750 mm for up to 100 people and 1.05 m for up to 200 people;

    — main corridors should be at least 1.05 m wide;

    — office corridors should be divided by fire-resisting doors if they are longer than 45 m;

    (iv) stairways should be at least 1.05 m wide and in a fire-resistant enclosure;

    (v) one stairway may be acceptable in a building of up to four storeys only. Depending on the floor area the stairs will need to be protected against the affects of fire including a lobby approach;

    (vi) escape doors should never be locked. If, for security reasons, they have to be secured against external entry, panic bolts or similar fastenings (complying with BS EN 1125) should be fitted;

    (vii) fire exit notices should be affixed to or above fire escape doors;

(viii) corridors and stairways, which are a means of escape and form part of a protected route should have a half-hour fire resistance and preferably be constructed from brick or concrete with a non-combustible surface;

(ix) fire alarms should be audible throughout the building. In large or multi-storey buildings such alarms will normally be electrically operated with suitable backup power supplies. In very small buildings a bell, gong or air horn may be sufficient;

(x) it should not normally be necessary for a person to travel more than 30 m to the nearest alarm point.

In any certificated premises the fire certificate must be available on demand. The following are essential:

- all doors affording means of escape in case of fire should be maintained easily and readily available for use at all times that persons are on the premises;

- all doors not in continuous use, affording a means of escape in case of fire, should be clearly indicated;

- sliding doors should also clearly indicate the direction of opening and, where possible, not be used on escape routes;

- doors should be adequately maintained and should not be locked or fastened in such a way that they cannot be easily and immediately opened by persons leaving the premises. Moreover, all gangways and escape routes should be kept clear at all times.

## *Unsatisfactory means of escape*

F5021  The following are unsatisfactory means of escape and should not be used in the event of fire:

- lifts;
- portable ladders;
- spiral staircases;
- escalators;
- lowering lines.

## Fire drill

F5022  As well as being a statutory requirement in procedures relating to 'events of serious and imminent danger' (under the *Management of Health and Safety at Work Regulations 1999 (SI 1999 No 3242), Reg 8(1)*) – as a matter of good housekeeping (in addition to being probably required as well by the fire certificate (see F5039, 'Contents of a fire certificate')), employers should acquaint the workforce with the arrangements for fire drill. This consists of putting up a notice in a prominent place stating the action employees should take on:

- hearing the alarm, or
- discovering the fire.

Ideally, employees should receive regular fire drills, even though normal working is interrupted. Indeed, fire alarms should be sounded weekly so that employees may familiarise themselves with the sound, and evacuation drills should be carried out at least annually. Trained employees should be designated as fire wardens and carry out head counts on evacuation, as well as acting as last man out and generally advising and shepherding the public. In addition, selected employees should be trained in the proper use of fire extinguishers. Moreover, periodical visits by the local fire authority should be encouraged by employers, since this provides a valuable source of practical information on fire fighting, fire protection and training.

## Typical fire drill notice

F5023

> When the fire alarm sounds:
> 1. Switch off electrical equipment and leave room, closing doors behind you.
> 2. Walk quickly along escape route to open air.
> 3. Do not use lifts.
> 4. Report to fire warden at assembly point.
> 5. Do not re-enter building.
>
> When you discover a fire:
> 1. Raise alarm (normally by operating a break glass call point).
> 2. Leave the room, closing doors behind you.
> 3. Leave the building by escape route.
> 4. Report to fire warden at assembly point.
> 5. Do not re-enter building.

## Fires on construction sites

F5024   Each year there are numerous fires, of a major kind, on construction sites and in buildings undergoing refurbishment. For that reason the *Joint Code of Practice on the Protection from Fire of Construction Sites and Buildings Undergoing Renovation*' (published by the Fire Protection Association) proposes that the main contractor should appoint a *site fire safety co-ordinator*, responsible for assessing the degree of fire risk and for formulating and regularly updating the *site fire safety plan*; and he should liaise with the co-ordinator for the design phase (see F5025). The site fire safety plan should detail:

- organisation of and responsibilities for fire safety;
- general site precautions, fire detection and warning alarms;
- requirements for a hot work permit system;
- site accommodation;
- fire escape and communications system (including evacuation plan and procedures for calling the fire brigade);
- fire brigade access, facilities and co-ordination;
- fire drill and training;
- effective security measures to minimise the risk of arson;
- materials storage and waste control system.

## Role of site fire safety co-ordinator

F5025   The site fire safety co-ordinator must:

- ensure that all procedures, precautionary measures and safety standards (as specified in the site fire safety plan) are clearly understood and complied with by all those on the project site;
- ensure establishment of hot work permit systems;
- carry out weekly checks of fire fighting equipment and test all alarm and detection devices;
- conduct weekly inspections of escape routes, fire brigade access, fire fighting facilities and work areas;
- liaise with local fire brigade for site inspections;
- liaise with security personnel;
- keep a written record of all checks, inspections, tests and fire drill procedures;
- monitor arrangements/procedures for calling the fire brigade;
- during the alarm, oversee safe evacuation of site, ensuring that all staff/visitors report to assembly points;
- promote a safe working environment.

## Emergency procedures

F5026    The following emergency procedures should be implemented, where necessary:

- establish a means of warning of fire, e.g. handbells, whistles etc.
- display written emergency procedures in prominent locations and give copies to all employees;
- maintain clear access to site and buildings;
- alert security personnel to unlock gates/doors in the event of an alarm;
- install clear signs in prominent positions, indicating locations of fire access routes, escape routes and positions of dry riser inlets and fire extinguishers.

## Designing out fire

F5027    Construction works should be designed and sequenced to accommodate:

- permanent fire escape stairs, including compartment walls;
- fire compartments in buildings under construction, including installation of fire doors;
- fire protective materials to structural steelwork;
- planned fire fighting shafts duly commissioned and maintained;
- lightning conductors;
- automatic fire detection systems;
- automatic sprinkler and other fixed fire fighting installations.

Moreover, adequate water supplies should be available and hydrants suitably marked and kept clear of obstruction.

## Other fire precautions on site

F5028    Portable fire extinguishers can represent the difference between a conflagration and a fire kept under control. Therefore, personnel should be trained in the use of portable fire fighting equipment and adequate numbers of suitable types of portable extinguishers should be available. They should be located in conspicuous positions near exits on each floor. In the open, they should be 500 mm above ground bearing the sign 'Fire Point' and be protected from both work activities and adverse weather conditions. In addition, all mechanically-propelled site plant should carry an appropriate fire extinguisher, and extinguishers, hydrants and fire protection equipment should be maintained and regularly inspected by the site fire safety co-ordinator.

Plant on construction sites also constitutes a potential danger. All internal combustion engines of powered equipment, therefore, should be positioned in the open air or in a well-ventilated non-combustible enclosure. They should be separated from working areas and sited so that exhaust pipes/gases are kept clear of combustible materials. Moreover, fuel tanks should not be filled whilst engines are running and compressors should be housed singly away from other plant in separate enclosures.

### Consequences of failure to comply with code

F5029    Non-compliance with the provisions of this code could well result in insurance ceasing to be available or being withdrawn, thereby constituting a breach of a Standard Form contract (see C8141 CONSTRUCTION AND BUILDING OPERATIONS). Where fire damage is caused to property by the negligence of employees of a subcontractor, then, in accordance with Clause 6.2 of the JCT Contract, to the effect that the contractor is liable for 'injury or damage to property', the contractor is liable and not the employer (or building owner); even though, under Clause 6.3B, the employer is required to insure against loss or damage to existing structures, to the work in progress, and to all unfixed materials and goods intended for, delivered to or placed on the works (*National Trust v Haden Young Ltd, The Times, 11 August 1994*). The practical implementation of fire precautions on construction sites forms part of the overall construction health and safety plan (see C8033 CONSTRUCTION AND BUILDING OPERATIONS).

## Fire and fire precautions – legislation

F5030    Fire safety legislation in the UK is concerned with the health and safety of the occupant's buildings. This will in most cases also include those, such as firefighters, who may need to enter a building on fire as part of their professional duties. There are two main sets of regulations that control this area: (*a*) the imposition of controls on the design and construction of buildings and (*b*) requirements leading to the safe management of buildings that are in use.

The provision of fire safety in new buildings relates to the physical provisions at the time of construction and are controlled through the Building Regulations in England and Wales, by Building Standards (Scotland) Regulations in Scotland and Building Regulations (Northern Ireland).

Buildings in use are currently controlled by a range of statutes relating to fire certification, workplace fire risk assessment and licensing of premises for consumption of alcohol, entertainment and similar licensed activities. These statutes generally relate to physical provisions and management requirements.

This section deals, almost exclusively, with the safe management of the workplace. The principle legislation relating to fire safety in the workplace is the *Fire Precautions Act 1971*. This Act is concerned with the fire safety provisions and management of designated premises. Designated premises include factories, offices, shops and railway premises as well as hotels and boarding houses. In addition to the 1971 Act the *Fire Precautions (Workplace) Regulations 1997 (SI 1997 No 1840)* as amended by the *Fire Precautions (Workplace) (Amendment) Regulations 1999 (SI 1999 No 1877)* extend control of fire safety to nearly all other workplaces.

## The current position – overview

F5031    Currently statute law relating to fire and fire precautions is extensive and complex. The legislation includes:

- the *Fire Precautions Act 1971* ('*FPA 1971*') (and regulations and orders made thereunder);
- the *Fire Safety and Safety of Places of Sport Act 1987*;
- the *Fire Precautions (Workplace) Regulations 1997 (SI 1997 No 1840)* as amended by the *Fire Precautions (Workplace) (Amendment) Regulations 1999 (SI 1999 No 1877)*;
- the *Health and Safety at Work etc Act 1974* ('*HSWA 1974*') (in the form of the *Fire Certificates (Special Premises) Regulations 1976 (SI 1976 No 2003)* (see F5060 below));
- the Petroleum Acts (and regulations made thereunder);
- the *Public Health Acts 1936–1961*;
- the *Building Act 1984* and the Building Regulations made thereunder;
- the *Fire Services Act 1947* and the *Fires Prevention (Metropolis) Act 1774*;
- as well as certain regulations, made under the *Factories Act 1961*.

This complex situation has been recognised by the UK Government and it is proposed to rationalise UK fire safety legislation using a regulatory reform order under the *Regulatory Reform Act 2001*. Consultation has been carried out during 2003 and it is expected that the Reform Act will find its way onto the statute books during 2004.

Many of the existing regulations will be rationalised. At least 50 items of legislation have been identified in the consultation document as needing repeal or amendment. The *Petroleum Spirit (Consolidation) Act 1928* which imposes requirements on storage and transportation of petroleum spirit is being reviewed and has been amended by the *Dangerous Substances and Explosive Atmospheres Regulations 2002 (SI 2002 No 2776)* ('*DSEAR*'), but is expected to remain for the time being and not be changed by the reform order. The fire safety aspects of the *Building Act 1984* and the current *Building Regulations 2000 (SI 2000 No 2531)* made under that Act are also expected to be affected but only in the areas that refer to the *FPA 1971*.

Much of the regulation concerned with dangerous and hazardous materials under the *HSWA 1974* has already been rationalised under the *DSEAR*. This includes the repeal of the *Highly Flammable Liquids and Liquefied Petroleum Gases Regulations 1972 (SI 1972 No 917)*.

At the time of writing, the exact content of the proposed Regulatory Reform Order has not been confirmed. However it is expected that the *FPA 1971* will be repealed in favour of risk assessment based legislation similar to the existing *Fire Precautions (Workplace) Regulations* of 1997 and 1999.

In view of this, many fire authorities are reluctant to issue new fire certificates under the *FPA 1971* and are concentrating on ensuring that adequate fire risk assessments have been carried out under the *Fire Precautions (Workplace) Regulations* for those premises that they consider high risk such as schools, hospitals and hotels. During all or part of the currency of this chapter the *FPA 1971* will still be in place. This Act imposes requirements relating to fire precautions upon occupiers of premises (whether or not employers) where there is a fire risk.

Readers should also be aware of the Fire and Rescue Services Bill published by the Office of the Deputy Prime Minister on 12 January 2004.

The Bill replaces the *Fire Services Act 1947* with a new statutory framework, which is designed to drive forward the Government's plans for a more preventative and risk-based approach, as set out in the White Paper, *Our Fire and Rescue Service*. It is expected to become law later this year (2004).

Under the Bill, the community fire safety work, which most fire and rescue authorities currently undertake, will be placed on a statutory footing. The wider role of the service beyond firefighting is also recognised in the Bill, which will create a duty on the service to respond to non-fire incidents, such as road traffic accidents, major flooding and terrorist incidents.

The Bill also places the Fire and Rescue National Framework (which calls on fire authorities to set up regional management boards) on a statutory footing. And it gives the Secretary of State powers to procure equipment and services for fire authorities and to direct authorities during specific emergencies – a move designed to ensure a co-ordinated and strategic response.

The Bill also:

- imposes duties on authorities and water undertakers to ensure an adequate supply of water for firefighting activities;
- updates existing powers to allow fire authorities to enter into reinforcement schemes with other authorities to provide mutual assistance;
- maintains the existing ability for fire authorities to charge for particular services. The Bill will continue to exclude the possibility of charging for firefighting;
- includes reserve powers to set up new bodies to negotiate pay and conditions of service. These powers could be used if the current review of National Joint Council arrangements does not result in satisfactory new mechanisms;
- devolves responsibility for the fire and rescue service in Wales to the National Assembly for Wales;
- abolishes the Central Fire Brigades Advisory Council, which has now been replaced with the more effective and flexible Practitioners' and Business and Community forums.

## Specific fire precautions regulations

F5032    Until the placement of the Regulatory Reform Order (See F5031), the main pieces of legislation relating to fire safety in workplaces in the UK are the *Fire Precautions (Workplace) Regulations 1997* (*SI 1997 No 1840*) as amended by the *Fire Precautions (Workplace) (Amendment) Regulations 1999* (*SI 1999 No 1877*), jointly referred to as 'the Workplace Regulations', and the *FPA 1971*. The key to operation of the *FPA 1971* is certification – that is, if premises were to be put to certain uses, they would require a valid fire certificate. The Workplace Regulations require all employers to conduct a fire risk assessment of the workplaces for which they are responsible. However, application must still be made for a fire certificate for premises designated under the *FPA 1971*, i.e. hotels and boarding houses, offices, factories and railway premises.

Similarly until the Regulatory Reform Order is in place fire certificates will also be required for 'special premises' (see below.)

The *FPA 1971* puts responsibilities on the owners or occupiers of properties whereas the responsibilities under the Workplace Regulations fall upon the employer.

Thus virtually all places of employment are now caught by the Workplace Regulations.

The Workplace Regulations do not allow the contents of a fire certificate to be such that people complying with the certificate will contravene these Regulations. The Workplace Regulations allow fire authorities to modify certificates accordingly.

Similarly, safety certificates issued under the *Safety of Sports Grounds Act 1975* or the *Fire Safety and Safety of Places of Sport Act 1987* are not allowed to require anyone to do anything that would cause them to contravene the Workplace Regulations.

The principal Regulations made under the *Fire Precautions Act 1971, s 12* (the regulation-enabling section) are:

- the *Fire Precautions (Factories, Offices, Shops and Railway Premises) Order 1989* (*SI 1989 No 76*);
- the *Fire Precautions (Hotels and Boarding Houses) Order 1972* (*SIs 1972 Nos 238, 382*);
- the *Fire Certificates (Special Premises) Regulations 1976* (*SI 1976 No 2003*); and
- the *Fire Precautions (Workplace) Regulations 1997* (*SI 1997 No 1840*).

(For offences and penalties in connection with the 'fire regulations', see F5040 below.)

Fire safety signs must be provided and maintained where indicated by a risk assessment to ensure that occupants leave safely in event of fire. [*Health and Safety (Safety Signs and Signals) Regulations 1996 (SI 1996 No 341), Reg 4*].

## Fire certification

**F5033** The prime tool of the *FPA 1971* was the requirement for 'designated' premises to be 'fire-certificated' – normally by the local fire authority, though in the case of exceptionally hazardous industrial premises (i.e. special premises) by the Health and Safety Executive (HSE). To date, two designation orders have been made affecting factories, offices, shops and railway premises *and* hotels and boarding houses (see further F5034 below). A fire certificate must be produced on demand to a fire officer, for inspection purposes, or a copy of it. [*FPA 1971, s 19(1)(c)*]. (For offences/penalties, see F5040 below.)

Fire certificates specify:

- use/uses of premises;
- means of escape in case of fire;
- how means of escape can be safely and effectively used;
- alarms and fire warning systems; and
- fire-fighting apparatus to be provided in the building.

Moreover, at its discretion, the fire authority can, additionally, impose requirements relating to:

- maintenance of means of escape and fire-fighting equipment;
- staff training; and
- restrictions on the number of people within the building.

Prior to the issue of a fire certificate, premises must be inspected by the fire authority, though failure to do so or inadequate inspection or advice cannot involve the fire authority in liability for negligence (see F5035 below).

### Fire certification – what you must do

**F5034** If your premises are a factory, office, shop or railway premises and fall under any of the following categories you must apply to your local fire authority for a fire certificate. (A shop includes restaurants. pubs and 'take away' outlets as these are considered as retail premises.)

(*a*)   (i)   where more than 20 people are at work, or

     (ii)   more than 10 people are at work elsewhere than on the ground floor,

     (in shops, factories, offices and railway premises);

(*b*)   in buildings in multiple occupation containing two or more individual factory, office, shop or railway premises, when the aggregate of people at work exceeds the same totals;

(*c*)   in factories where explosive or highly flammable materials are stored, or used in or under the premises, unless, in the opinion of the fire authority, there is no serious risk to employees.

[*Fire Precautions (Factories, Offices, Shops and Railway Premises) Order 1989 (SI 1989 No 76), Reg 5*].

In light of the amendment of the *FPA 1971* by the *Fire Safety and Safety of Places of Sport Act 1987*, some relaxation of earlier requirements is now possible (see F5041 below).

## Applications for local authority fire certificate

**F5035** Applications for a fire certificate (see F5034 above) must be made to the local fire authority on the correct form (see F5036 below), obtainable from each fire authority. Plans may be required and the premises will be inspected before issue of a certificate. If the fire authority is not satisfied as to existing arrangements, it will specify the steps to be taken before a certificate is issued, and notify the occupier or owner that they will not issue a certificate until such steps are taken within the specified time. [*FPA 1971*, s 5]. (Hospitals, NHS trusts and private hospitals must conform to the requirements of FIRECODE, which is a suite of documents produced by NHS Estates, making recommendations on fire safety in hospitals.) For the position of hospitals, see generally E15045 ENFORCEMENT and F5049 below.)

In view of the expected and imminent repeal of the *FPA 1971* many fire authorities are reluctant to issue new fire certificates under the *FPA 1971* and are concentrating on ensuring that adequate fire risk assessments have been carried out, under the Workplace Regulations.

## Form for application for fire certificate

**F5036** Applications for a fire certificate must be made on the form prescribed by the *Fire Precautions (Application for Certificate) Regulations 1989* (*SI 1989 No 77*).

## Who should apply?

**F5037** Application should normally be made by the occupier in the case of factories, offices and shops. In the following cases however it must be made by the owner or owners:

- premises consisting of part of a building, all parts of which are owned by the same person (i.e. multi-occupancy/single ownership situations);
- premises consisting of part of a building, the different parts being owned by different persons (i.e. multi-occupancy/plural ownership situations).

[*FPA 1971, s 5, Sch 2 Part II*].

## Categories of fire risk premises

**F5038** Where there is in the opinion of the fire authority a serious risk to persons from fire on premises, unless steps are taken to minimise that risk, and the fire authority thinks that a particular use of premises should be either prohibited or restricted, it may apply to the court for an order prohibiting or restricting that use of premises until remedial steps are taken. [*FPA 1971, s 10(2)*]. This requirement does not impose a duty on a fire authority to advise an occupier not to reopen; hence, if he reopens and suffers loss from fire, the fire authority will not be liable

The Workplace Regulations 1997 amend the *FPA 1971, ss 10 to 10B*, to include tents and other movable structures and places of work in the open air.

## Contents of a fire certificate

**F5039** A fire certificate specifies:

- the particular use or uses of premises which it covers;
- the means of escape in the case of fire (as per plan);
- the means for securing that the means of escape can be safely and effectively used at all relevant times (e.g. direction signs/emergency lighting/fire or smoke stop doors);
- the means for fighting fire for use by persons in the building;
- the means for giving warnings in the case of fire;

*Fire Prevention and Control*     F5039

## FIRE PRECAUTIONS ACT 1971, s.5

### APPLICATION FOR A FIRE CERTIFICATE

| For Official Use Only |

\* In the case of Crown premises, substitute H.M. Inspector of Fire Services.

To the Chief Executive of the Fire Authority\*

I hereby apply for a fire certificate in respect of the premises of which details are given below. I make the application as, or on behalf of, the occupier/owner of the premises.

Signature ...................................................................

Name: Mr/Mrs/Miss ......................................................
                  (in block capitals)

If signing on behalf of a company or some other person, state the capacity in which

signing ..................................................................................................................................

Address ..................................................................................................................................

Telephone number ..................................................... Date ........................

To be completed by the Applicant:-
1. Postal address of the premises ........................................................................

..................................................................................................................................

..................................................................................................................................

..................................................................................................................................

2. Name and address of the owner of the premises

Name ....................................................................................................................

Address ................................................................................................................

..................................................................................................................................

..................................................................................................................................

..................................................................................................................................

(In the case of premises in plural ownership the names and addresses of all owners should be given.)

3. Details of the premises
(If the fire certificate is to cover the use of two or more sets of premises in the same building, details of each set of premises should be given on a separate sheet.)

   (a) Name of occupier ........................................................................................

      (and any trading name, if different) ...............................................................

      ..................................................................................................................

(b) Use(s) to which premises put ..................................................................................

(c) Floor(s) in building on which premises situated (e.g. basement(s), ground floor, first floor etc.) ..................................................................................................................
..............................................................................................................................

(d) Number of persons employed to work in the premises ...............................................

(e) Maximum number of persons at work or it is proposed will work in the premises at any one time (including employees, self-employed persons and trainees)–

   (i) below the ground floor of the building ...............................................................

   (ii) on the ground floor of the building ....................................................................

   (iii) on the first floor of the building .......................................................................

   (iv) in the whole of the premises ............................................................................

(f) Maximum number of persons other than persons at work likely to be in the premises at any one time .....................................................................................................

(g) Number of persons (including staff, guests and other residents) for whom sleeping accommodation is provided in the premises–

   (i) below the ground floor of the building ...............................................................

   (ii) above the first floor of the building ..................................................................

   (iii) in the whole of the premises ............................................................................

**4.** If the premises consist of part only of a building, the uses to which the other parts of the building are put (on a floor by floor basis):

..............................................................................................................................
..............................................................................................................................
..............................................................................................................................

**5.** (a) Total number of floors (excluding basements) in the building in which the premises are situated ..........................................................................................................

(b) Total number of basements in that building ...............................................................

**6.** Approximate date of construction of the premises ........................................................

**7.** Nature and quantity of any explosive or highly flammable materials stored or used in or under the premises

| Materials | Maximum quantity stored | Method of storage | Maximum quantity liable to be exposed at any one time |
|---|---|---|---|
|  |  |  |  |

(Continue on a separate sheet if necessary)

8. Details of fire-fighting equipment available for use in the premises

| Nature of equipment | Number Provided | Where installed | Is the equipment regularly maintained? |
|---|---|---|---|
| (a) Hosereels | | | Yes/No |
| (b) Portable fire extinguishers | | | Yes/No |
| (c) Others | | | Yes/No |
| (specify types e.g. sand/water buckets, fire blanket) | | | |

(Continue on a separate sheet if necessary)

- in the case of any factory, particulars as to any explosives or highly flammable materials stored or used on the premises.

[*FPA 1971, s 6(1)*].

In addition, a fire certificate may require:

- maintenance of the means of escape and their freedom from obstruction;
- maintenance of other fire precautions set out in the certificate;
- training of employees on the premises as to what to do in the event of fire and keeping of suitable records of such training;
- limitation of number of persons who at any one time may be on the premises;
- any other relevant fire precautions.

[*FPA 1971, s 6(2)*].

## Offences/penalties for breach of the Fire Precautions Act 1971

**F5040** Failure to have/exhibit a valid fire certificate, or breach of any condition(s) in such certificate is an offence, committed by the holder, carrying:

- on summary conviction, a maximum fine at level 3 on the standard scale*; and
- on conviction on indictment, an indefinite fine or up to two years' imprisonment (or both).

[*FPA 1971, s 7(1), (5)* as amended by the *Criminal Justice Act 1991, s 17(1)*].

An occupier also commits an offence if the fire authority or inspectorate is not informed of a proposed structural or material alteration. [*FPA 1971, s 8*]. A breach carries the same penalties as for breach of *s 7* (above). Other offences include obstructing a fire inspector [*FPA 1971, s 19(6)*] and forging/falsifying a fire certificate [*FPA 1971, s 22(1)*].

*The standard scale of fines is defined in the *Criminal Justice Act 1982* and may be increased from time to time.

## Exemptions from the need to have a fire certificate

**F5041** A fire authority can grant exemption from certification requirements in the case of 'designated use' premises Exemption from certification requirements can be granted, either on application for a fire certificate, or at any time during the currency of a fire certificate. The usual criterion for exemption is that the fire authority deems the premises concerned to be low risk.

It is not necessary formally to apply for exemption as the fire authority will offer an exemption, if they see fit after inspecting the premises. Any exemption certificate must specify the greatest number of persons who can safely be in the premises at any one time. Such exemptions can be withdrawn by the fire authority without an inspection but must give notice of withdrawal.

## Change of conditions affecting premises for which exemption is granted

F5042    If, while an exemption is in force an occupier proposes to carry out material changes in the premises, the occupier must inform the fire authority of the proposed changes. Not to do so is to commit an offence. This applies where it is proposed:

- to make an extension of, or structural alteration to the premises which would affect the means of escape from the premises; or
- to make an alteration in the internal arrangement of the premises, or in the furniture or equipment, which would affect the means of escape from the premises; or
- to keep explosive or highly flammable materials under, in or on the premises, in a quantity or aggregate quantity greater than the prescribed maximum; or
- to make use of the premises which involves there being a greater number of persons on the premises than specified in the exemption certificate.

## Offences/penalties – exempted premises

F5043    The penalties for being found guilty of committing the offences (explained in F5042 above) are liable to the same penalties as described in F5040.

## Duties on occupiers of exempted premises

F5044    Premises which are exempt from fire certification must be provided with:

- means of escape in the case of fire; and
- means of fighting fire.

A code of practice has been published ('*Code of practice for fire precautions in factories, offices, shops and railway premises not required to have a fire certificate*' (1993), Home Office and Scottish Home and Health Department, ISBN 0 11 340904 4). This means that an occupier who followed the recommendations of the code will normally be considered to have complied with these requirements.

## Definition of 'escape'

F5045    In relation to premises 'escape' means escape from them to some place of safety beyond the building.

# Improvement notices

F5046    Use of improvement notices, which have proved effective in general health and safety law for upgrading standards of health and safety at the workplace, has been duplicated in fire precautions law. Thus, where a fire authority is of the opinion that the duty to provide:

- means of escape; and

- means of fire-fighting;

has been breached, they can serve on the occupier an improvement notice, specifying, particularly by reference to a code of practice (see F5044 above), what measures are necessary to remedy the breach. They can also require the occupier to carry out this remedial work within three weeks. The occupier has a right to appeal. [*FPA 1971, s 9D*]. Service of such notice need not be recorded in a public register (see further E15029 ENFORCEMENT).

## Building Regulations

**F5047** Where the *Building Regulations 2000* (*SI 2000 No 2531* as amended by *SI 2001 No 3335, SI 2002 No 440 and SI 2002 No 2871*) apply to premises, with respect to requirements as to means of escape in case of fire, the fire authority cannot serve an improvement notice requiring structural or other alterations. However, a notice can be served if the fire authority is satisfied that the means of escape in case of fire are inadequate, by reason of matters/circumstances of which particulars were not required by the Regulations. [*FPA 1971, s 9D(3)*].

The *Building Regulations 2000* (as amended) include requirements to make 'appropriate' provisions for the early warning of fire and 'reasonable facilities to assist firefighters in the protection of life'. This brings these Regulations into line with the Workplace Regulations 1997. The amendments are reflected in the accompanying approved document, i.e. '*Approved Document B: Fire Safety*' (2000 edition) that should be read in conjunction with '*Approved Document B: Fire safety (2002 Amendments)*' available from the Stationary Office or the Office of the Deputy Prime Minister, Building Regulations Department.

The amendments incorporate changes and discussion on the replacement of references to British Standards by harmonised BS EN standards, the application of the CE marking to building products and the implications of the Construction Products Directive.

The following requirements are specified in respect of new premises:

- The building must be designed/constructed so that there are *means of escape* in case of fire, to a place of safety outside.

*Internal fire spread*

- To inhibit internal fire spread, internal linings must:

    (i) adequately resist flame spread over surfaces;

    (ii) if ignited, have a reasonable rate of heat release.

- The building must be designed and constructed so that, in the event of fire, its stability will be maintained for a reasonable period; and a common wall should be able to resist fire spread between the buildings.

- The building must be designed and constructed so that unseen fire/smoke spread within concealed spaces in its fabric and structure, is inhibited.

*External fire spread*

- External walls shall adequately resist fire spread over walls and from one building to another.

- A roof should be able to resist fire spread over the roof and from one building to another.

- The building must be designed and constructed to provide facilities to firefighters and enable fire appliances to gain access.

*Means of warning and escape*

- The building shall be designed and constructed so that there are appropriate provisions for the early warning of fire, and appropriate means of escape in case of fire from the building to a place of safety outside the building capable of being safely and effectively used at all material times.

*Access and facilities for the fire service*

- The building shall be designed and constructed so as to provide reasonable facilities to assist firefighters in the protection of life and reasonable provision made within the site of the building to enable fire appliances to gain access.

### Appeals against improvement notices

F5048   An appeal must be lodged against an improvement notice within 21 days from the date of service. Moreover, unlike prohibition notices, the effect of an appeal is to suspend operation of the improvement notice. Presumably, a ground of appeal would be that occupiers of 'low risk' premises have achieved satisfactory fire safety standards by means other than those specified in a code of practice.

If the appeal fails, the occupier must carry out the remedial work specified in the notice. Failure to do so carries with it similar penalties as explained in F5040 above.

## Prohibition notices

F5049   In places of work generally, health and safety inspectors can serve prohibition notices requiring a hazardous activity to cease in cases where there is thought to be a serious risk of personal injury. In the case of fire hazards where there is thought to be a serious risk of injury to persons from fire, the fire authority is in a similar way empowered to serve on the occupier a prohibition notice. The effect of such a notice (which need not affect all the premises) will be to prohibit use of the premises or activity until the risk is removed. Prohibition notices can be placed on premises that are:

- providing sleeping accommodation;
- providing treatment/care;
- for the purposes of entertainment, recreation or instruction, or for a club, society or association;
- for teaching, training or research;
- providing access to members of the public, whether for payment or otherwise;
- places of work.

This includes hospitals, factories and places of public worship, but not private dwellings. [*Fire Safety and Safety of Places of Sport Act 1987, s 13*]. Thus, a fire authority could restrict/prohibit the use of any part of a hospital, factory or place of religious worship, presenting a serious fire risk to persons – premises which are also subject to the *Building Act 1984* (see W11044 WORKPLACES – HEALTH, SAFETY AND WELFARE). In particular, after consultation with the fire authority, if a local authority is not satisfied with the means of escape in case of fire, it can serve a notice requiring the owner to carry out remedial work in residential premises of all kinds, including inns, hotels and nursing homes and certain commercial premises with sleeping accommodation above them, whether a fire certificate is in force or not. [*Building Act 1984, s 72*].

The fire authority is most likely to serve prohibition notices in cases where it considers that means of escape are inadequate or could be improved. As with prohibition notices served generally in respect of workplaces, it can be immediate or deferred. Occupiers must lodge an appeal within 21 days.

Entry of such notices must appear in a public register. [*Environment and Safety Information Act 1988, ss 1, 3* and *Sch*].

### Offences/penalties

**F5050** A person found guilty of contravening a prohibition notice is liable to the same penalties as described in F5040 above.

## The Workplace Regulations

### Premises to which the Workplace Regulations apply

**F5051** By way of implementation of the EU Framework and Workplace Directives in relation to fire safety, the *Fire Precautions (Workplace) Regulations 1997 (SI 1997 No 1840)* pursuant to the *FPA 1971, s 12*, extend fire precautions requirements to most places of work. This includes some workplaces currently designated and certificated under the *Factories Act 1961* and the *Offices, Shops and Railway Premises Act 1963*.

The *Management of Health and Safety at Work Regulations 1999 (SI 1999 No 3242)* amend some aspects of the Workplace Regulations, mostly where references to the Management Regulations 1992 have been changed. Also *Regulation 3* of the Management Regulations 1999 requires employers, among other things, to carry out a fire risk assessment to identify measures needed to comply with the provisions of the *Fire Precautions (Workplace) Regulations 1997* (as amended).

The *Fire Precautions (Workplace) (Amendment) Regulations 1999 (SI 1999 No 1877)* make amendments to the Workplace Regulations 1997. The main changes are to extend the scope of the Regulations by removing some of the excepted categories of workplace that appeared in the Workplace Regulations 1997. These amendments have been brought in to address concerns expressed by the EU that the 1997 Regulations did not fully implement the Framework and Workplace Directives (see above) and to resolve the resulting legislative overlaps.

### Workplaces to which the Workplace Regulations do not apply

**F5052** These include the following:

- Workplaces used only by the self-employed.
- Private dwellings.
- Mine shafts and mine galleries, other than surface buildings.
- Construction sites (any workplace to which the *Construction (Health, Safety and Welfare) Regulations 1996 (SI 1996 No 1592)* apply).
- Ships within the meaning of the *Docks Regulations 1988 (SI 1988 No 1655)* (including those under construction or repair by persons other than the crew).
- Means of transport used outside the workplace and workplaces which are in or on a means of transport.
- Agricultural or forestry land situated away from the undertaking's main buildings.
- Offshore installations (workplaces to which the *Offshore Installations and Pipeline Works (Management and Administration) Regulations 1995 (SI 1995 No 738)* apply).

*Fire Prevention and Control*

[*Fire Precautions (Workplace) Regulations 1997 (SI 1997 No 1840)*, Reg 3(5) as amended by *Fire Precautions (Workplace) (Amendment) Regulations 1999 (SI 1999 No 1877)*, Reg 5(c)].

All other workplaces, including those subject to certification under the *FPA 1971* are covered by the Workplace Amendment Regulations 1999.

## Fire risk assessment

**F5053** The Workplace Regulations require all employees to assess the risk the fire poses to their employees and others who may be present in the workplace.

The fire risk assessment may involve six stages:

*Stage 1* – identifying the fire hazards;

*Stage 2* – identifying the people at risk;

*Stage 3* – removing or reducing the hazards;

*Stage 4* – assigning a risk category;

*Stage 5* – deciding if existing arrangements are satisfactory or need improvement;

*Stage 6* – recording the findings. This is a statutory requirement if more than five people are employed.

Further, detailed guidance is available in '*The Fire Protection Association Library of Fire Safety, Volume 5 – Fire risk management in the workplace*' published by the Fire Protection Association and in '*Fire Safety. An employer's guide*' published by the Stationary Office and HSE books.

Where there are five or more employees, the fire risk assessment must be recorded and retained.

The assessment should be reviewed periodically and when any significant changes occur to, for example, layout, management or staffing.

## Fire-fighting and fire detection

**F5054** Where necessary (whether due to the features of a workplace, the activity carried on there, any hazard present there or any other relevant circumstances, i.e. as a result of a fire risk assessment) in order to safeguard the safety of employees in case of fire:

(a) a workplace must, to the extent that is appropriate, be equipped with appropriate fire-fighting equipment and with fire detectors and alarms; and

(b) any non-automatic fire-fighting equipment so provided must be easily accessible, simple to use and indicated by signs,

and for the purposes of sub-paragraph (a) what is appropriate is to be determined by the dimensions and use of the building housing the workplace, the equipment it contains, the physical and chemical properties of the substances likely to be present and the maximum number of people that may be present at any one time i.e. as a result of a fire risk assessment. [*SI 1997 No 1840* (as amended), *Reg 4(1)*].

Where necessary in order to safeguard the safety of his employees in case of fire, an employer must:

- take measures for fire-fighting in the workplace, adapted to the nature of the activities carried on there and the size of his undertaking and of the workplace concerned and taking into account persons other than his employees who may be present;

- nominate employees to implement those measures and ensure that the number of such employees, their training and the equipment available to them are adequate, taking into account the size of, and the specific hazards involved in, the workplace concerned; and

- arrange any necessary contacts with external emergency services, particularly as regards rescue work and fire-fighting. [*SI 1997 No 1840* (as amended), *Reg 4(2)*].

## Emergency routes and exits

**F5055** Routes to emergency exits from a workplace and the exits themselves must be kept clear at all times, where this is necessary for safeguarding the safety of employees in case of fire. [*SI 1997 No 1840* (as amended), *Reg 5(1)*].

The following requirements must be complied with in respect of a workplace where necessary (whether due to the features of the workplace, the activity carried on there, any hazard present there or any other relevant circumstances) in order to safeguard the safety of employees in case of fire:

- emergency routes and exits must lead as directly as possible to a place of safety;
- in the event of danger, it must be possible for employees to evacuate the workplace quickly and as safely as possible;
- the number, distribution and dimensions of emergency routes and exits must be adequate having regard to the use, equipment and dimensions of the workplace and the maximum number of persons that may be present there at any one time;
- emergency doors must open in the direction of escape;
- sliding or revolving doors must not be used for exits specifically intended as emergency exits;
- emergency doors must not be so locked or fastened that they cannot be easily and immediately opened by any person who may require to use them in an emergency;
- emergency routes and exits must be indicated by signs; and
- emergency routes and exits requiring illumination must be provided with emergency lighting of adequate intensity in the case of failure of their normal lighting. [*SI 1997 No 1840* (as amended), *Reg 5(2)*].

## Maintenance

**F5056** The workplace and any equipment and devices provided in respect of the workplace under *Regs 4* and *5* of the 1997 Regulations, must be subject to a suitable system of maintenance and be maintained in an efficient state, in efficient working order and in good repair, where this is necessary for safeguarding the safety of employees in case of fire. [*SI 1997 No 1840* (as amended), *Reg 6*].

## Enforcement

**F5057** The duty of enforcing the workplace fire precautions legislation falls on fire authorities, who may appoint inspectors for this purpose. [*SI 1997 No 1840* (as amended), *Reg 10*].

A fire authority can issue an enforcement notice in respect of a serious breach. The enforcement notice must notify the person who is served with it that the fire authority is of the opinion that he is in breach of the workplace fire precautions legislation and that such breach is putting one or more employees at serious risk. The notice must also:

- specify the steps required to remedy the breach;
- require those steps to be taken within a given time; and

- provide details of the appeals procedure relating to enforcement notices.

[*SI 1997 No 1840* (as amended), *Reg 13*].

## Offences and penalties

**F5058** A person is guilty of an offence if:

- being under a requirement to do so, he fails to comply with any provision of the workplace fire precautions legislation;
- that failure places one or more employees at serious risk (i.e. subject to a risk of death or serious injury which is likely to materialise) in case of fire; and
- that failure is intentional or is due to his being reckless as to whether he complies or not.

[*SI 1997 No 1840* (as amended), *Reg 11(1)*].

Any person guilty of an offence under *Reg 11(1)* is liable:

- on summary conviction, to a fine; or
- on conviction on indictment, to a fine, or to imprisonment for a term not exceeding two years, or both.

[*SI 1997 No 1840* (as amended), *Reg 11(2)*].

A person is not guilty of an offence under *Reg 11(1)* in respect of any failure to comply with the workplace fire precautions legislation which is subject to an enforcement notice. [*SI 1997 No 1840* (as amended), *Reg 11(3)*].

## Application to the Crown

**F5059** The workplace fire precautions regulations do not extend to premises used solely by the armed forces, but otherwise apply to premises which are Crown occupied and Crown owned. *Section 10* of the *FPA 1971* only binds the Crown in so far as it applies to premises and workplaces owned by the Crown but not occupied by the Crown. [*SI 1997 No 1840* (as amended), *Reg 18*].

# Fire Certificates (Special Premises) Regulations 1976 (SI 1976 No 2003)

**F5060** It is expected that these Regulations will be revoked by the Regulatory Reform Order referred to in F5031. Until this Order enters the statute book the *Fire Certificates (Special Premises) Regulations 1976 (SI 1976 No 2003)* remain in force. Under these Regulations, in the case of premises containing hazardous materials or processes (i.e. special premises), a fire certificate must be obtained from HSE. This applies even if only a small number of persons are employed there. Exemption from this requirement may be granted where the Regulations are inappropriate or not reasonably practicable.

When a certificate has been issued by HSE the occupier of those premises must post a notice in those premises, stating:

- that the certificate has been issued; and
- the places where it (or a copy) can be inspected easily by any person who might be affected by its provisions; and
- the date of the posting of the notice.

[*Fire Certificates (Special Premises) Regulations 1976, Reg 5(5), (6)*].

These conditions do not override those applicable to a licence for the storage of petroleum spirit under the *Petroleum Spirit (Consolidation) Act 1928* apart from the special provision for harbours.

## 'Special premises' for which a fire certificate is required from HSE

**F5061** The 'special premises' for which a fire certificate is required from HSE are set out for reference in the table below.

| | 'Special premises' for which a fire certificate is required by HSE |
|---|---|
| 1 | Any premises at which are carried on any manufacturing processes in which the total quantity of any highly flammable liquid under pressure greater than atmospheric pressure and above its boiling point at atmospheric pressure may exceed 50 tonnes. |
| 2 | Any premises at which is carried on the manufacturing of expanded cellular plastics and at which the quantities manufactured are normally of, or in excess of, 50 tonnes per week. |
| 3 | Any premises at which there is stored, or there are facilities provided for the storage of, liquefied petroleum gas in quantities of, or in excess of, 100 tonnes except where the liquefied petroleum gas is kept for use at the premises either as a fuel, or for the production of an atmosphere for the heat-treatment of metals. |
| 4 | Any premises at which there is stored, or there are facilities provided for the storage of, liquefied natural gas in quantities of, or in excess of, 100 tonnes except where the liquefied natural gas is kept solely for use at the premises as a fuel. |
| 5 | Any premises at which there is stored, or there are facilities provided for the storage of, any liquefied flammable gas consisting predominantly of methyl acetylene in quantities of, or in excess of, 100 tonnes except where the liquefied flammable gas is kept solely for use at the premises as a fuel. |
| 6 | Any premises at which oxygen is manufactured and at which there are stored, or there are facilities provided for the storage of, quantities of liquid oxygen of, or in excess of, 135 tonnes. |
| 7 | Any premises at which there are stored, or there are facilities provided for the storage of, quantities of chlorine of, or in excess of, 50 tonnes except when the chlorine is kept solely for the purpose of water purification. |
| 8 | Any premises at which artificial fertilisers are manufactured and at which there are stored, or there are facilities provided for the storage of, quantities of ammonia of, or in excess of, 250 tonnes. |
| 9 | Any premises at which there are in process, manufacture, use or storage at any one time, or there are facilities provided for such processing, manufacture, use or storage of, quantities of any of the materials listed below in, or in excess of, the quantities specified —<br><br>Phosgene — 5 tonnes<br>Ethylene oxide — 20 tonnes<br>Carbon disulphide — 50 tonnes<br>Acrylonitrile — 50 tonnes<br>Hydrogen cyanide — 50 tonnes<br>Ethylene — 100 tonnes |

| | | |
|---|---|---|
| | Propylene | 100 tonnes |
| | Any highly flammable liquid not otherwise specified | 4,000 tonnes |
| 10 | Explosives, factories or magazines which are required to be licensed under the Explosives Act 1875. | |
| 11 | Any building on the surface at any mine within the meaning of the Mines and Quarries Act 1954. | |
| 12 | Any premises in which there is comprised — | |
| | (a) any undertaking on a site for which a licence is required in accordance with section 1 of the Nuclear Installations Act 1965 or for which a permit is required in accordance with section 2 of that Act; or | |
| | (b) any undertaking which would, except for the fact that it is carried on by the United Kingdom Atomic Energy Authority, or by, or on behalf of, the Crown, be required to have a licence or permit in accordance with the provisions mentioned in sub-paragraph (a) above. | |
| 13 | Any premises containing any machine or apparatus in which charged particles can be accelerated by the equivalent of a voltage of not less than 50 megavolts except where the premises are used as a hospital. | |
| 14 | Premises to which Regulation 26 of the Ionising Radiations Regulations 1985 (SI 1985 No 1333) applies. | |
| 15 | Any building, or part of a building, which either — | |
| | (a) is constructed for temporary occupation for the purposes of building operations or works of engineering construction; or | |
| | (b) is in existence at the first commencement there of any further such operations or works | |
| | and which is used for any process of work ancillary to any such operations or works (but see F5062 below). | |
| [*Fire Certificates (Special Premises) Regulations 1976, Sch 1 Part I*]. | | |

## *Application of Fire Certificates (Special Premises) Regulations to temporary buildings used for building operations or construction work*

**F5062** A fire certificate is required in the case of buildings constructed for temporary occupation for the purposes of building operations or works of engineering construction (see CONSTRUCTION AND BUILDING OPERATIONS). This is a requirement of the *Fire Certificates (Special Premises) Regulations 1976 (SI 1976 No 2003)*, and application for such a certificate should be made to HSE. Buildings that are already in existence for such purposes when such operations or works begin are included in this requirement. An exemption is available, however, if:

- fewer than 20 persons are employed at any one time, or fewer than 10 elsewhere than on the ground floor; and

- the nine conditions set out in *Sch 1 Pt II* of the 1976 Regulations are satisfied.

[*Fire Certificates (Special Premises) Regulations 1976 (SI 1976 No 2003, Reg 3(1), Sch 1, para 15*].

In all events suitable means of escape in case of fire, adequate fire-fighting equipment, exit doorways that can be easily and immediately opened from inside, unobstructed passageways as well as distinctive and conspicuous marking of fire exits must be provided.

## Other dangerous processes

**F5063** Until the *DSEAR* (*SI 2002 No 2776*) came into force, fire prevention measures for certain particularly dangerous processes were controlled by specific regulations. Most of these have been repealed or revoked and hazards are now controlled by the requirements of the *DSEAR*. Some of the regulations replaced by *DSEAR* are:

(a) the *Celluloid (Manufacture, etc.) Regulations 1921* (*SR & O 1921 No 1825*) – applying to the manufacture, manipulation and storage of celluloid and the disposal of celluloid waste;

(b) the *Manufacture of Cinematograph Film Regulations 1928* (*SR & O 1928 No 82*) – applicable to the manufacture, repair, manipulation or use of cinematograph film;

(c) the *Cinematograph Film Stripping Regulations 1939* (*SR & O 1939 No 571*) – applicable to the stripping, drying or storing of cinematograph film;

(d) the *Highly Flammable Liquids and Liquefied Petroleum Gases Regulations 1972* (*SI 1972 No 917*) – applicable to premises containing highly flammable liquids and liquefied petroleum gases;

(e) the *Magnesium (Grinding of Castings and Other Articles) Special Regulations 1946* (*SR & O 1946 No 2017*) – prohibition on smoking, open lights and fires;

(f) the *Factories (Testing of Aircraft Engines & Accessories) Special Regulations 1952* (*SI 1952 No 1689*) – applicable to the leakage or escape of petroleum spirit;

Some regulations with specific fire safety requirements that are still on the statute book are given below. However most of these are expected to be amended or replaced under the Regulatory Reform Order (see F5031 above).

- the *Offshore Installations (Prevention of Fire and Explosion, and Emergency Response) Regulations 1995* (*SI 1995 No 743*) – to prevent and minimise the effects of fire and explosion on offshore installation;

- the *Construction (Health, Safety and Welfare) Regulations 1996* (*SI 1996 No 1592*), Reg 21; and

- the *Work in Compressed Air Regulations 1996* (*SI 1996 No 1656*), Reg 14.

- the *Electricity at Work Regulations 1989* (*SI 1989 No 635*), Reg 6(d) – electrical equipment which may reasonably foreseeably be exposed to any flammable or explosive substance, must be constructed or protected so as to prevent danger from exposure;

- the *Dangerous Substances in Harbour Areas Regulations 1987* (*SI 1987 No 37*) – applicable to risks of fire and explosion in harbour areas. A fire certificate is not required for these premises.

## Application for fire certificate for 'special premises'

**F5064** In order to obtain a fire certificate for 'special premises' (see F5061 for definition), the occupier or owner (see F5037) must apply to the relevant HSE inspectorate, e.g. the factory inspectorate in respect of factory premises. No form is prescribed but certain particulars must be provided, including:

- the address and description of the premises;
- the nature of the processes carried on or to be carried on there;
- nature and approximate quantities of any explosive or highly flammable substance kept or to be kept on the premises;
- the maximum number of persons likely to be present on the premises;
- the name and address of the occupier.

Plans may be required to be deposited and premises will be inspected, and the occupier may well have to make improvements to the fire precautions before a certificate is issued. [*Fire Certificates (Special Premises) Regulations 1976 (SI 1976 No 2003, Reg 4*].

## Appeals relating to the issue of a fire certificate

**F5065** An appeal may be made to the magistrates' court by an applicant for a fire certificate in relation to:

- a requirement specified by a fire authority or HSE; or
- the fire authority or HSE refusing to issue a certificate; or
- the contents of a certificate.

An appeal must be lodged within 21 days of the date of notice from the fire authority. [*Fire Certificates (Special Premises) Regulations 1976 (SI 1976 No 2003, Reg 12)*].

## Duty to keep fire certificate on the premises

**F5066** Every fire certificate must be kept on the premises to which it relates (and preferably displayed), as long as it is in force. [*FPA 1971, s 6(8)*]. Failure to comply with this subsection is an offence. [*FPA 1971, s 7(6)*].

## Notification of alteration to premises

**F5067** Any proposed structural alterations or material internal alteration to premises for which a fire certificate has been issued, must first be notified to the local fire authority or inspectorate (as appropriate). Similarly, any proposed material alteration to equipment in the premises must be so notified. Not to do so is to commit an offence (see F5040 above). The premises may be inspected by staff from the relevant enforcement agency at any reasonable time while the certificate is in force. [*FPA 1971, s 8(1), (2)*]. A material alteration is one that may affect means of escape or other fire safety provisions such as removing a hose reel or changing the direction of an escape route.

### Penalties

**F5068** A person found guilty of contravening the notification requirements is liable to the same penalties as described in F5040 above.

## Licensed premises and fire safety

**F5069** Some premises must have a current licence in order to operate. A licence will be refused or not renewed by the local magistrates if the local fire authority is not satisfied as to the fire precautions necessary in the premises. If people are employed on these premises the Workplace Regulations will also apply (see F5051). Many of these licensing regulations are expected to be repealed or revoked under the Regulatory Reform Order (see F5031).

### Cinemas

**F5070** Safety in cinemas is controlled by the *Cinematograph (Safety) Regulations 1955* (*SI 1955 No 1129*) (as subsequently amended). Cinemas must be provided with:

- adequate, clearly marked exits, so placed as to afford safe means of exit;
- doors which are easily and fully openable outwards;

- passages and stairways kept free from obstruction;

[*Cinematograph (Safety) Regulations 1955 (SI 1955 No 1129), Reg 2*];
- suitable and properly maintained fire appliances;
- proper instruction of licensee and staff on fire precautions;
- treatment of curtains so that they will not readily catch fire;
- use of non-flammable substances for cleaning film or projectors;

[*Cinematograph (Safety) Regulations 1955 (SI 1955 No 1129), Reg 5*];
- prohibition on smoking in certain parts of the premises [*Cinematograph (Safety) Regulations 1955 (SI 1955 No 1129), Reg 6*];
- appropriate siting of heating appliances [*Cinematograph (Safety) Regulations 1955 (SI 1955 No 1129, Reg 24*].

## Theatres

**F5071** Under *s 12(1)* of the *Theatres Act 1968* premises used for the public performance of a play must be licensed. The conditions for obtaining or having a licence renewed include:

- compliance with rules relating to safety of persons in the theatre, and particularly, staff fire drills, provision of fire-fighting equipment, maintenance of a safety curtain and communication with the fire service;
- gangways and seating correctly arranged and free from obstruction, doors and exits, marking and method of opening, lighting arrangements;
- scenery and draperies must be non-flammable and there must be controls over smoking and overcrowding.

## Gaming houses (casinos, bingo halls etc.)

**F5072** The issue or retention of a licence to operate depends *inter alia* on compliance with fire requirements [*Gaming Act 1968*].

## Premises for music, dancing etc.

**F5073** The issue or retention of a licence to operate premises for public music or entertainment depends on compliance with the fire requirements. [*Local Government (Miscellaneous Provisions) Act 1982, Sch 1*].

Similarly, in the case of premises used for private music/dancing, e.g. dancing schools, there must be compliance with the fire requirements. [*Private Places of Entertainment (Licensing) Act 1967*].

## Schools

**F5074** In the case of local authority controlled schools, including special schools, the 'health and safety of their occupants, and in particular, their safe escape in the event of fire, must be reasonably assured', with particular reference to the design, construction, limitation of surface flame spread and fire resistance of structure and materials therein. [*Standards for School Premises Regulations 1972 (SI 1972 No 2051)*].

As far as a school is a place of work, the *Fire Precautions (Workplace) Regulations 1997 (SI 1997 No 1840)* also apply.

### Children's and community homes

**F5075** Both these local authority controlled establishments must carry out fire drills and practices and, in addition, consult with the fire authorities. [*Children's Homes Regulations 1991* (*SI 1991 No 1506*); *Community Homes Regulations 1972* (*SI 1972 No 319*)].

As far as a children's or community home is a place of work, the *Fire Precautions (Workplace) Regulations 1997* (*SI 1997 No 1840*) also apply.

## Residential and nursing homes

**F5076** Similar requirements apply in the case of:

- residential homes [*National Assistance (Conduct of Homes) Regulations 1962* (*SI 1962 No 2000*)]; and

- nursing homes [*Nursing Homes and Mental Nursing Homes Regulations 1981* (*SI 1981 No 932*) (as amended)].

In particular, satisfactory arrangements must be made for the evacuation of patients and staff in the event of fire.

As far as a residential or nursing home is a place of work, the *Fire Precautions (Workplace) Regulations 1997* (*SI 1997 No 1840*) also apply.

## Crown premises

**F5077** As with other health and safety duties and regulations, generally speaking, statutory fire duties and fire regulations apply to the Crown. [*FPA 1971, s 40; Fire Precautions (Workplace) Regulations 1997* (*SI 1997 No 1840* as amended), *Reg 18*]. However owing to Crown immunity in law, proceedings cannot be enforced against the Crown (see further E15045 ENFORCEMENT). This has the effect that Crown premises, that is, government buildings such as the Treasury and the Foreign Office as well as royal palaces, are required to be fire-certificated but, if they fail to apply for certification, they cannot, like other occupiers, be prosecuted. Secondly, failure to acquire fire-certificate status could prejudice the safety of fire-fighters called upon to combat fires in Crown premises. Moreover, fire certificates are not required in (*a*) prisons, (*b*) special hospitals for the mentally incapacitated and (*c*) premises occupied exclusively by the armed forces. [*FPA 1971, s 40(2)*].

## Civil liability at common law and under the Fires Prevention (Metropolis) Act 1774 for fire damage

**F5078** Only *sections 83* and *86* of the *Fires Prevention (Metropolis) Act 1774* still exist. The Law Commission has proposed that *section 83* be repealed and the Reform Order is expected to remove *section 86*. Until these proposals become law these sections are still in force and there effect is summarised as follows.

Both sections only apply to England and Wales. *Section 83* allows an insurance company to insist that, if there is any hint of fraud, the money paid out by it in respect to fire damage can only be used to rebuild or repair the damaged property. *Section 86* prevents any action being taken against a person who owns property that accidentally burns down.

### Liability of occupier

**F5079** An occupier of premises where fire breaks out can be liable to:

- lawful visitors to the premises injured by the fire or falling debris (and is also liable to unlawful visitors, i.e. trespassers, as the principle of 'common humanity', enunciated in *Herrington v British Railways Board [1972] 1 All ER 749* applies as does the *Occupiers' Liability Act 1984*, see OCCUPIERS' LIABILITY);
- firemen injured during fire-fighting operations; and
- adjoining occupiers.

In order to ensure therefore that occupiers and others involved may minimise their liability, insurance cover, though not compulsory, is highly desirable. The basic principles relating to fire cover are considered below (see F5080).

## Fire insurance

**F5080** Insurance is intended to provide the insured with cover in the event of an accident that is not predictable. This applies to fire insurance and is reflected in the terms of insurance policies. Thus if a fire occurs in a premises where it is thought that the insured might be involved in deliberately starting the fire the insurance company may refuse to pay out under the policy. However if the insured is innocent of any involvement, even if the fire was started deliberately, the insurer will pay out as arson is not predictable in the context of this issue. Where a high fire risk has been identified and adequate measures are not taken to manage the risk there may be a case for non-payment on the grounds of negligence. In practice insurance plays a very important role in the management of many risks in industry, including fire, and adequate appropriate fire cover is considered an essential part of modern business risk management.

# First-Aid

## Introduction

**F7001** People can and do suffer injury or fall ill at work. This may or may not be as a result of work related activity. However, it is important that they receive immediate attention.

The *Health and Safety (First-Aid) Regulations 1981* (*SI 1981 No 917*), which came into operation on 1 July 1982, require employers to have facilities for the provision of first-aid in their place of work.

Medical treatment should be provided at the scene promptly, efficiently and effectively before the arrival of any medical teams that may have been called. First-aid can save lives and can prevent minor injuries from becoming major ones. Employers are responsible for making arrangements for the immediate management of any illness or injury suffered by a person at work. First-aid at work covers the management of first-aid in the workplace – it does not include treating ill or injured people at work with medicines.

However, the Regulations do not prevent specially trained staff taking action beyond the initial management of the injured or ill at work.

## The employer's duty to make provision for first-aid

**F7002** Employers must provide or ensure that there is equipment and facilities provided that are adequate and appropriate for rendering first-aid if any of their employees are injured or become ill at work. Employers must also ensure that there are an adequate and appropriate number of suitable persons who:

(*a*) are trained and have such qualifications as the Health and Safety Executive may approve for the time being; and

(*b*) have additional training, if any, as may be appropriate in the circumstances.

Where such a suitable person is absent in temporary and exceptional circumstances, an employer can appoint a person, or ensure that a person is appointed:

(*a*) to take responsibility for first-aid in situations relating to an injured or ill employee who needs help from a medical practitioner or nurse;

(*b*) to ensure that equipment and facilities are adequate and appropriate in the circumstances.

With regard to any period of absence of the first-aider, consideration must be given to:

(*a*) the nature of the undertaking;

(*b*) the number of employees at work; and

(*c*) the location of the establishment.

## The assessment of first-aid needs

**F7003**   The employer should assess his first-aid needs and requirements as appropriate to the circumstances of his workplace. The employer's principal aim must be the reduction of the effects of injury and illness at the place of work. Adequate and appropriate first-aid personnel and facilities should be available for rendering assistance to persons with common injuries or illnesses and those likely to arise from specific hazards at work. Similarly, there must be adequate facilities for summoning an ambulance or other professional medical help.

The extent of first-aid provision in a particular workplace that an employer must make depends upon the circumstances of that workplace. There are no fixed levels of first-aid – the employer must assess what personnel and facilities are appropriate and adequate. Employers with access to advice from occupational health services may wish to use these resources for the purposes of conducting such assessment, and then take the advice given as to what first-aid provision would be deemed appropriate.

In workplaces that employ:

- qualified medical doctors registered with the General Medical Council; or
- nurses whose names are registered in Part 1, 2, 10 or 11 of the Single Professional Register maintained by the United Kingdom Central Council for Nursing, Midwifery and Health Visiting,

the employer may consider that there is no need to appoint first-aiders.

There is no requirement for the results of the first-aid risk assessment to be recorded in writing, although it may nevertheless be a useful exercise for the employer – for he may subsequently be asked to demonstrate that the first-aid provision is adequate and appropriate for the workplace.

When assessing first-aid needs, employers must consider the following:

— the hazards and risks in the workplace;
— the size of the organisation;
— the history of accidents in the organisation;
— the nature and distribution of the workforce;
— the distance from the workplace to emergency medical services;
— travelling, distant and lone workers' needs and requirements;
— employees working on shared or multi-occupied sites;
— annual leave and other absences of first-aiders and appointed persons.

### The hazards and risks in the workplace

**F7004**   *The Management of Health and Safety at Work Regulations 1999* (*SI 1999 No 3242*) require employers to make a suitable and sufficient assessment of the risks to health and safety at work of their employees. The assessment must be designed to identify the measures required for controlling or preventing any risks to the workforce: highlighting what types of accidents or injuries are most likely to occur will help employers address such key questions as the appropriate nature, quantity and location of first-aid personnel and facilities.

Where the risk assessment conducted by an employer identifies a low risk to health and safety, employers may only need to provide (i) a first-aid container clearly identified and suitably stocked, and (ii) an appointed person to look after first-aid arrangements and resources and to take control in emergencies.

Where risks to health and safety are greater, employers may need to ensure the following:

- the provision of an adequate number of first-aiders so that first-aid can be given immediately;
- the training of first-aiders to deal with specific risks or hazards;
- informing the local emergency services in writing of the risks and hazards on the site where hazardous substances or processes are in use;
- the provision of a first-aid room(s).

Employers will need to consider the different risks in each part of their company or organisation. Where an organisation occupies a large building with different processes being performed in different parts of the premises, each area's risks must be assessed separately. It would not be appropriate to conduct a generic assessment of needs to cover a variety of activities – the parts of the building with higher risks will need greater first-aid provision than those with lower risk.

## The size of the company

**F7005** In general, the amount of first-aid provision that is required will increase according to the number of employees involved. Employers should be aware, however, that in some organisations there may be few employees but the risks to their health and safety might be high – and, as a result, their first-aid needs will also be high.

## The history of incidents and accidents

**F7006** When assessing first-aid needs, employers might find it useful to collate data on accidents that have occurred in the past, and then analyse, for example, the numbers and types of accidents, their frequency and consequences. Organisations with large premises should refer to such information when determining the first-aid equipment, facilities and personnel that are required to cover specific areas.

## The character and dispersion of the workforce

**F7007** The employer should bear in mind that the size of the premises can affect the time it might take a first-aider to reach an incident. If there are a number of buildings on the site, or the building in question comprises several storeys, the most suitable arrangement might be for each building or floor to be provided with its own first-aiders. Where employees work in shifts or in self-contained areas, first-aid arrangements may need to be tailored to reflect that fact.

Employees who are potentially at higher risk, such as trainees, young workers or people with disabilities, will need to be given special consideration.

## The distance from the workplace to emergency medical services

**F7008** Where a workplace is far from emergency medical services, it may be necessary to make special arrangements for ensuring that appropriate transport can be provided for taking an injured person to the emergency medical services.

In every case where the place of work is remote, the very least that an employer should do is to give written details to the local emergency services of the layout of the workplace, plus any other relevant information, such as information on specific hazards.

## The needs and requirements of travelling, distant and lone workers

**F7009** Employers are responsible for meeting the needs of their employees whilst they are working away from the main company premises.

When assessing the needs of staff who travel long distances or who are constantly mobile, consideration should be given to the question of whether they ought to be provided with a personal first-aid kit.

Organisations with staff working in remote areas must make special arrangements for those employees in respect of communications, special training and emergency transport.

## Personnel on sites that are shared or multi-occupied

**F7010** Employers with personnel working on shared or multi-occupied sites can agree to have one employer on the site who is solely responsible for providing first-aid cover for all the workers. It is strongly recommended that this agreement is written to avoid confusion and misunderstandings between the employers. It will highlight the risks and hazards of each company on the site, and will make sure that the shared provision is suitable and sufficient. After the employers have agreed the arrangement, the personnel must be informed accordingly.

When employees are contracted out to other companies their employer must ensure they have access to first-aid facilities and equipment. The user employer bears the responsibility for providing such facilities and equipment.

## First-aiders on annual leave or absent from the workplace

**F7011** Adequate provision of first-aid must be available at all times. Employers should therefore ensure that the arrangements they make for the provision of first-aid at the workplace are adequate to cover for any annual leave of their first-aiders or appointed persons. Such arrangements must also be able to cover for any unplanned or unusual absences from the workplace of first-aiders or appointed persons.

## Other important points

**F7012** Employers are not obliged by these regulations to provide first-aid facilities for members of the public. Although the regulations are aimed at employees, including trainees, many organisations like health authorities, schools and colleges, places of entertainment, fairgrounds and shops do make first-aid provision for persons other than their employees. Other legislation deals with public safety and first-aid facilities for the public – for example, the *Road Traffic Act 1960* regulates first-aid provision on buses and coaches.

Where employers extend their first-aid provision to cover more than merely their employees, the provision for the employees must not be diminished and should not fall below the standard required by these regulations.

The compulsory element of employers' liability insurance does not cover litigation resulting from first-aid given to non-employees. It is advised that employers check their public liability insurance policy on this issue.

## Reassessing first-aid needs

**F7013** In order to ensure that first-aid provision continues to be adequate and appropriate, employers should from time to time review the first-aid needs in the workplace, particularly when changes have been made to working practices.

## Duty of the employer to inform employees of first-aid arrangements

**F7014** Employers are under a duty to inform employees of the arrangements that have been made for first-aid in their workplace. This may be achieved by:

- Distributing guidance to all employees which highlights the key issues in the first-aid arrangements, such as listing the names of all first-aiders and describing where first-aid resources are located.
- Nominating key employees to ensure that the guidance is kept up to date and is distributed to all staff, and to act also as information officers for first-aid in the workplace.
- Internal memos can be used as a method of keeping the personnel informed of any changes in the first-aid arrangements.
- Displaying announcements up on notice boards informing employees of the first-aid arrangements and of any changes to those arrangements.
- Providing new employees with the information as part of their induction package.

   Any person with a reading or language problem must be given the information in a way that they can understand.

## First-aid and the self-employed

**F7015** The self-employed should provide, or ensure that there is provided, such equipment, if any, as is appropriate in the circumstances to enable them to render first-aid to themselves whilst at work. The self-employed who work in low-risk areas, for example at home, are required merely to make first-aid provision appropriate to a domestic environment.

When self-employed people work together on the same site, they are each responsible for their own first-aid arrangements. If they wish to collaborate on first-aid provision, they may agree a joint arrangement to cover all personnel on that particular site.

## Number of first-aiders

**F7016** Sufficient numbers of first-aiders should be located strategically on the premises to allow for the administration of first-aid quickly when the occasion arises. The assessment of first-aid needs may have helped to highlight the extent to which there is a need for first-aiders. The table below gives suggested numbers of first-aiders or appointed persons who should be available at the workplace. The suggested numbers are not a legal requirement – they are merely for guidance.

There are no hard and fast rules on numbers – employers will have to make a judgement, taking into account all the circumstances of their organisation. If the company is a long way from a medical facility or there are shift workers on site or the premises cover a large area, the numbers of first-aid personnel set out below may not be sufficient – the employer may have to make provision for a greater number of first-aiders to be on site.

### Selection of first-aiders

**F7017** First-aiders must be reliable and of good disposition. Not only should they have good communication skills, but they must also possess the ability and aptitude for acquiring new knowledge and skills and must be able to handle physical and stressful emergency incidents and procedures. Their position in the company should be such that they are able to leave their place of work immediately to respond to an emergency.

**F7018** *First-Aid*

| Table 1<br>First-aid personnel | | |
|---|---|---|
| *Category of risk* | *Numbers employed at any location* | *Suggested number of first-aid personnel* |
| **Lower risk** – e.g. shops, offices, libraries | Fewer than 50<br>50–100<br>More than 100 | At least one appointed person<br>At least one first-aider<br>One additional first-aider for every 100 employed |
| **Medium risk** – e.g. light engineering and assembly work, food processing, warehousing | Fewer than 20<br>20–100<br>More than 100 | At least one appointed person<br>At least one first-aider for every 50 employed (or part thereof)<br>One additional first-aider for every 100 employed |
| **Higher risk** – e.g. most construction, slaughterhouse, chemical manufacturer, extensive work with dangerous machinery or sharp instruments | Fewer than 5<br>5–50<br>More than 50<br>Where there are hazards for which additional first-aid skills are necessary | At least one appointed person<br>At least one first-aider<br>One additional first-aider for every 50 employed<br>In addition, at least one first-aider trained in the specific emergency action |

## The training and qualifications of first-aid personnel

**F7018** Individuals nominated to be first-aiders must complete a programme of competence-based training in first-aid at work run by an organisation approved by the HSE. See APPENDIX B for a range of first-aid competencies which make up a basic curriculum. Organisations that are contracted to train first-aid personnel may be notified of any particular risks or hazards in a workplace so that the first-aid course that is provided can be tailored to include the risks specific to that workplace.

Additional special training may be undertaken to deal with unusual risks and hazards. This will enable the first-aider to be competent in dealing with such risks. Such special training may be separate from the basic course or an extension to it but does not need approval by the HSE. The first-aid certificate awarded may be endorsed to verify that special training has been received.

It is important for employers to understand that first-aid at work certificates are valid for three years. Employers wishing to arrange for re-testing of competence training must do so before the original first-aid certificate expires. If the certificate of any first-aider expires, he or she will have to attend a full course of training to regain their first-aid at work certificate. Employers can arrange for re-testing of competence training up to three months before the expiry date of the certificate – the new first-aid at work certificate will then run from the date of the expiry of the old certificate. It is advisable for employers to keep a record of first-aiders in the company, together with their certification dates, in order to assist them in organising refresher training. Employers should develop a programme of knowledge and skills training for their first-aiders to enable them to be updated on new skills and to make them aware of suitable sources of first-aid information, such as occupational health services and training organisations qualified by the HSE to conduct first-aid at work training.

## Appointed persons

**F7019** An appointed person is an individual who takes charge of first-aid arrangements for the company including looking after the facilities and equipment and calling the emergency services when required. The appointed person is allocated these duties when it is found through the first-aid needs assessment that a first-aider is not necessary. The appointed person is the minimum requirement an employer can have in the workplace. Clearly, even if the company is considered a low health and safety risk and, in the opinion of the employer, a first-aider is unnecessary, an accident or illness still may occur – therefore somebody should be nominated to call the emergency services, if required.

Appointed persons are not first-aiders – and therefore they should not be called upon to administer first-aid if they have not received the relevant training. However, employers may consider it prudent to send appointed persons on first-aid training courses. Such a course normally last for four hours and includes the following topics:

- action necessary in an emergency;
- cardio-pulmonary resuscitation;
- first-aid treatment for the unconscious casualty;
- first-aid treatment for the bleeding or the wounded.
- HSE approval is not required for this training.

The only time an appointed person can replace a first-aider is when the first-aider is absent, due to circumstances which are temporary, unforeseen and exceptional. Appointed persons cannot replace first-aiders who are on annual leave. If the first-aid assessment has identified a requirement for first-aiders, they should be available whenever there is a need for them in the place of work.

## Records and record keeping

F7020    It is considered good practice to keep records of incidents which required the attendance of a first-aider and treatment of an injured person. It is advisable for smaller companies to have one record book but for larger organisations this may not be practicable.

The data entered in the record book should include the following:

- the date, time and place of the incident;
- the injured person's name and job title;
- a description of the injury or illness and of the first-aid treatment administered;
- details of where the injured person went after the incident, i.e. hospital, home or back to work;
- the name and signature of the first-aider or person who dealt with the incident.

This information may be collated, to help the employer improve the environment with regard to health and safety in the workplace. It could be used to help determine future first-aid needs assessment and will be helpful for insurance and investigative purposes. The statutory accident book is not the same as the record book, but they may be combined.

## First-aid resources

F7021    Having completed the first-aid needs requirements assessment, the employer must provide the resources, i.e. the equipment, facilities and materials, that will be needed to ensure that an appropriate level of cover is available to the employees at all relevant times. First-aid equipment, suitably marked and obtainable, must be made available at specific sites in the workplace.

## First-aid containers

**F7022** First-aid equipment must be suitably stocked and contained in a properly identifiable container. One first-aid container with sufficient quantity of first-aid materials must be made available for each worksite – this is the minimum level of first-aid equipment. Larger premises, for example, will require the provision of more than one container.

First-aid containers should be easily accessible and, where possible, near hand-washing facilities. The containers should be used only for first-aid equipment. Tablets and medications should not be kept in them. The first-aid materials within the containers should be protected from damp and dust.

Having completed the first-aid needs assessment, the employer will have a good idea as to what first-aid materials should be stocked in the first-aid containers. If there is no specific risk in the workplace, a minimum stock of first-aid materials would normally comprise the following (there is no mandatory list):

- a leaflet giving guidance on first-aid (for example, HSE leaflet *Basic advice on first aid at work*);
- 20 individually wrapped sterile adhesive dressings (assorted sizes), appropriate to the type of work;
- two sterile eye pads;
- four individually wrapped triangular bandages (preferably sterile);
- six safety pins;
- six medium-sized individually wrapped sterile unmedicated wound dressings – approximately 12cm x 12cm;
- two large sterile individually wrapped unmedicated wound dressings – approximately 18cm x 18cm;
- one pair of disposable gloves.

This list is a suggestion only – other equivalent materials will be deemed acceptable.

An examination of the first-aid kits should be conducted frequently. Stocks should be replenished as soon as possible after use, and ample back-up supplies should be kept on the company premises. Any first-aid materials found to be out of date should be carefully discarded.

All first-aid containers should have a white cross on a green background as identification.

## Additional first-aid resources

**F7023** If the results of the assessment suggest a need for additional resources such as scissors, adhesive tape, disposable aprons, or individually wrapped moist wipes, they can be kept in the first-aid container if space allows. Otherwise they may be kept in a different container as long as they are ready for use if required.

If the assessment highlights the need for such items as protective equipment, they must be securely stored next to first-aid containers or in first-aid rooms or in the hazard area itself. Only persons who have been trained to use these items may be allowed to use them.

If there is a need for eye irrigation, and mains tap water is unavailable, at least a litre of sterile water or sterile normal saline solution (0.9%) in sealed, disposable containers should be provided. If the seal is broken, the containers should be disposed of and not reused. Such containers should also be disposed of when their expiry date has been passed.

## First-aid kits for travelling

**F7024** First-aid kits for travelling may contain the following items:

- a leaflet giving general guidance on first-aid (for example, HSE leaflet *Basic advice on first aid at work*);
- six individually wrapped sterile adhesive dressings;
- one large sterile unmedicated dressing – approximately 18cm x 18cm;
- two triangular bandages;
- two safety pins;
- individually wrapped moist cleansing wipes;
- one pair of disposable gloves.

This list is a suggestion only – it is not mandatory. However, the kit must be regularly inspected and topped up from a back-up store at the home site.

## Rooms designated as first-aid areas

**F7025** A suitable room should be made available for first-aid purposes where the first-aid needs assessment found such a room to be necessary. Such room(s) should have sufficient first-aid resources, be easily accessible to stretchers and be easily identifiable, and where possible should be used only for administering first-aid.

First-aid rooms are normally necessary in organisations operating within high-risk industries. Therefore they would be deemed to be necessary on, chemical, ship building and large construction sites or on large sites remote from medical services. A person should be made responsible for the first-aid room.

On the door of the first-aid room a list of the names and telephone extensions of all the first-aiders should be displayed, together with details as to how and where they may be contacted on site.

First-aid rooms should:

- have enough space to hold a couch with space in the room for people to work, a desk, a chair and any other resources found necessary (see below);
- where possible, be near an access point in the event that a person needs to be taken to hospital;
- have heating, lighting, and ventilation;
- have surfaces that can be easily washed;
- be kept clean and tidy; and
- be available and ready for use whenever employees are in the workplace.

The following is a list of resources that may be found in a first-aid room:

- a record book for logging incidents where first-aid has been administered;
- a telephone;
- a storage area for storing first-aid materials;
- a bed/couch with waterproof protection and clean pillows and blankets;
- a chair;

- a foot operated refuse bin with disposable yellow clinical waste bags or some receptacle suitable for the safe disposal of clinical waste;
- a sink that has hot and cold running water – also drinking water and disposable cups;
- soap and some form of disposable paper towel.

If the designated first-aid room has to be shared with the working processes of the company, the employer must consider the implications of the room being needed in an emergency and whether the working processes in that room could be stopped immediately. Can the equipment in the room be removed in an emergency so as not to interfere with any administration of first-aid? Can the first-aid resources and equipment be stored in such a place as to be available quickly when necessary? Lastly, the room must be appropriately identified and, where necessary, be signposted by white lettering or symbols on a green background.

# Appendix A
# Assessment of first-aid needs – checklist

The minimum first-aid provision for each worksite is:

- a suitably stocked first-aid container;
- an appointed person to take charge of first-aid arrangements;
- information for employees on first-aid arrangements.

*This checklist will help you assess whether you need to make any additional provisions.*

|    | Aspects to consider | Impact on first-aid provision |
|----|---------------------|-------------------------------|
| 1. | What are the risks of injury and ill health arising from the work as identified in your risk assessment? | If the risks are significant you may need to employ first-aiders. |
| 2. | Are there any specific risks, such as: | You will need to consider: |
|    | • hazardous substances; | • specific training for first-aiders; |
|    | • dangerous tools; | • extra first-aid equipment; |
|    | • dangerous machinery; | • precise siting of first-aid equipment; |
|    | • dangerous loads or animals? | • informing emergency services; |
|    |                         | • a first-aid room. |
| 3. | Are there parts of your establishment where different levels of risk can be identified (e.g. a university with research labs)? | You will probably need to make different levels of provision in different parts of the establishment. |
| 4. | Are large numbers of people employed on site? | You may need to employ first-aiders to deal with the higher probability of an accident. |
| 5. | What is your record of accidents and cases of ill- | You may need to: |
|    | health? What type are they and where did they happen? | • locate your provision in certain areas; |
|    |                         | • review the contents of the first-aid box. |
| 6. | Are there inexperienced workers on site, or | You will need to consider: |
|    | employees with disabilities or special health problems? | • special equipment; |
|    |                         | • local siting of equipment. |
| 7. | Are the premises spread out, e.g. are there several buildings on the site or multi-floor buildings? | You will need to consider provision in each building or on several floors. |

**F7025** *First-Aid*

| | | |
|---|---|---|
| 8. | Is there shiftwork or out-of-hours working? | Remember that there needs to be first-aid provision at all times people are at work. |
| 9. | Is your workplace remote from emergency medical services? | You will need to: <br>• inform local medical services of your location; <br>• consider special arrangements with the emergency services. |
| 10. | Do you have employees who travel a lot or work alone? | You will need to: <br>• consider issuing personal first-aid kits and training staff in their use; <br>• consider issuing personal communicators to employees. |
| 11. | Do any of your employees work at sites occupied by other employers? | You will need to make arrangements with the other site occupiers. |
| 12. | Do you have any work experience trainees? | Remember that your first-aid provision must cover them. |
| 13. | Do members of the public visit your premises? | You have no legal responsibilities for non-employees, but HSE strongly recommends that you include them in your first-aid provision. |
| 14. | Do you have employees with reading or language difficulties? | You will need to make special arrangements to give them first-aid information. |

Do not forget to allow for leave or absences of first-aiders and appointed persons. First-aid personnel must be available at all times when people are at work.

# Appendix B
# First-aid competencies

Employees who have successfully completed first-aid training must be able to apply the following competencies:

(a) to act safely, promptly and effectively when an incident occurs at work;

(b) to administer cardio-pulmonary resuscitation (CPR) promptly and effectively;

(c) to administer first-aid safely, promptly and effectively to a casualty who is unconscious;

(d) to administer first-aid safely, promptly and effectively to a casualty who is wounded or bleeding;

(e) to administer first-aid safely, promptly and effectively to a casualty who:

— is burnt or scalded;

— has an injury to bones, muscles or joints;

— is in shock;

— has an eye injury;

— is suffering from poisoning;

— has been overcome by gas or fumes;

(f) the transportation of the casualty safely in the workplace;

(g) the recognition and management of common major illnesses;

(h) the recognition and management of minor illnesses;

(i) the management of records and the provision of written information to medical staff if required.

First-aiders must also know and understand the following elements of first-aid at work:

(a) the legal requirements relating to the provision of first-aid at work;

(b) first-aider responsibilities and procedures in an emergency;

(c) safety and hygiene in first-aid procedures;

(d) how to use first-aid equipment provided in the workplace.

# Food Safety and Standards

## Introduction

**F9001** Legislation has governed the sale of food for centuries. In Europe we can trace it back at least to the Middle Ages, and the ancient Hebrew food laws, found notably in the Book of Deuteronomy, show that it goes back even further. In the UK, the trade guilds were instrumental in the introduction of legislation aimed at stopping the adulteration of a wide variety of foods and the adulteration of tea, coffee and bread was prohibited by legislation as long ago as the early-mid 18th century.

Two themes have existed from the start:

- *Food Safety* – the protection of the health and well being of anyone eating food; and

- *Food Standards* – the control of composition and adulteration of food for the prevention of fraud.

UK legislation is structured principally around the *Food Safety Act 1990* supported by a whole raft of more detailed regulations. Increasingly over the past 30 years directives and regulations from the European Union have determined UK legislation – and the last five years in particular have seen an almost total revision of food safety and consumer protection rules. Devolution within the UK has seen the responsibility for food policy shifted to the Scottish parliament and regional assemblies, with the resultant introduction of a number of local initiatives in areas such as nutrition but little impact, to date, in respect of traditional food safety.

## Food safety

**F9002** The principal legislation in the UK is the *Food Safety Act 1990*, to which significant amendments have been introduced since January 2005 in order to align the technical details with the EU General Food Law Regulation 178/2002/EC. In addition, fundamental changes to the UK food hygiene legislation will be introduced with effect from January 2006.

EU Regulation 178/2002/EC introduced new food safety requirements, new traceability requirements, and also measures to ensure effective product recall/withdrawals and notification to competent authorities. Although as an EU Regulation it was directly applicable in Member States, it was still necessary to amend the basic UK Act to introduce the new provisions and to ensure conformity with the EU law.

The *Food Safety Act 1990 (Amendment) Regulations 2004 (SI 2004 No 2990)* align the definition of 'food' in the *Food Safety Act 1990* with the definition in the General Food Law Regulation (EC) 178/2002/EC, and also make some minor amendments to the *Food Safety Act 1990* in respect of public consultation requirements, to remove duplication with directly applicable provisions in Regulation *178/2002/EC*.

The *General Food Regulations 2004 (SI 2004 No 3279)* introduce consequential amendments relating to definitions of safety, in particular to *ss 7* and *8* which provide the necessary enforcement powers in respect of the new food safety requirements.

Since the inception of the European Economic Community (EEC), now the European Union (EU), European 'hygiene' regulations have developed on the basis of specific measures for specific food sectors, the so-called 'vertical' regulations. A full complement of these was already in place by the 1990s covering the processing of most foods of animal origin such as meat, fish, eggs, milk and related products, and even honey. These 'vertical' regulations did not in general cover businesses selling food direct to the ultimate consumer, for example catering or retail businesses. The 'horizontal' Food Hygiene Directive 93/43/EEC had preciously established general principles for food hygiene and set hygiene standards for those businesses not covered by 'vertical' legislation such as food factories processing non-animal products and also retail or catering outlets.

However, the EU has recently completed a radical overhaul of all food hygiene legislation, consolidating and recasting both the general requirements and those relating to animal-derived products into two principal Regulations (*Regulation 852/2004 on the Hygiene of Foodstuffs* and *Regulation 853/2004 laying down Specific Hygiene Rules for Products of Animal Origin*). The requirements of these are in the final stages of transposition into UK legislation and will come into effect in January 2006, as the *Food Hygiene Regulations 2005* (*SI 2005 No 2059*). Each of the devolved regions will introduce their own set of hygiene legislation but, for the purposes of this summary, they will be grouped under the generic description *Food Hygiene Regulations 2005*. With the exception of the temperature control requirements in Scotland, the provisions are technically virtually identical (the reader is however advised to consult the original text appropriate to his region when the individual regulations have been finalised in late 2005). These Regulations are wide-ranging and will subsume/revoke numerous existing provisions, including the *Food Safety (General Food Hygiene) Regulations 1995* (*SI 1995 No 1763*) ('the General Food Hygiene Regulations'), the *Food Safety (Temperature Control) Regulations 1995* (*SI 1995 No 2200*) ('the Temperature Control Regulations') and the *Food Premises (Registration) Regulations 1991* (*SI 1991 No 2825*).

The following account will be largely restricted to the provisions of the *Food Safety Act 1990* (as amended) and the new *Food Hygiene Regulations 2005* (*SI 2005 No 2059*).

## Food standards

**F9003** Food standards are governed by a complex range of regulations. Despite its name, the *Food Safety Act 1990* is also the primary legislation covering most aspects of food standards and a number of regulations have been made under it. Over the recent past there has been a significant change in emphasis. Whereas previously there was a tendency to prescribe compositional standards for foods, many of these have disappeared and relatively little of this legacy remains. The approach proved insufficiently flexible in a climate of rapid innovation and product development, and particularly when exposed to the range of competing products traded within the EU Internal Market. The emphasis therefore shifted to allow more diversity of composition, but supported by informative labelling. The *Food Labelling Regulations 1996* (*SI 1996 No 1499*) play an important role in providing a substantial amount of information to consumers, including nutritional information in many cases, about the composition and properties of individual foods and also how they should be safely stored and prepared. The *Food Labelling (Amendment) Regulations 1998* (*SI 1998 No 1398*) took this even further with a specific requirement that pre-packed foods must carry a 'quantitative ingredient declaration' or 'QUID'. This dictates that the label must carry a declaration of the percentage quantity of characterising or emphasised ingredients of any food. Food standards and labelling may also be enforced through the more general *Trades Descriptions Act 1968*.

## Administrative regulations

**F9004** Finally, there are more general regulations made under the *Food Safety Act 1990*. These may regulate the law enforcement process, setting criteria for inspectors, food analysts or examiners. A new regime of Official Control legislation will be introduced from January 2006 to implement EU Regulations 854/2004, covering animal-derived products, and 882/2004, covering all aspects of other food and feed controls including imported products. Current UK legislation on the registration of premises and temperature control will become part of the new *Food Hygiene Regulations 2005* (*SI 2005 No 2059*) from that date. In addition, many other pieces of consumer protection legislation are applicable to food; notably the *Weights and Measures Act 1985*, which controls the way in which food is sold by weight or other measures, the *Consumer Protection Act 1987*, which establishes a civil law right of redress for death, or injury, caused by using defective consumer goods (the so called 'product liability' provisions). and the *Prices Act 1974* and the various Price Marking Orders laid under it. Detailed guidelines on the application of these pieces of legislation can be obtained from the DTI.

## Policy and enforcement

**F9005**  Since 2000, government policy has been derived principally from the Food Standards Agency (FSA). This is a non-ministerial government department that was created in the wake of BSE and other food 'crises' to take all of the policy functions from the Department of Health (DH) and former Ministry of Agriculture, Fisheries and Food (MAFF) into an organisation which operates at arm's length from the government and has a primary role of protecting the consumers' interests. The FSA is controlled by a board of directors and operates through a chief executive and officials, the majority of whom joined the FSA from similar roles in MAFF or DH. Complementary FSA structures exist in Scotland, Wales and Northern Ireland. The FSA was created and empowered by the *Food Standards Act 1999*.

Complex arrangements are in place for the enforcement of food legislation. Traditionally local authorities have had the responsibility although central government officials have particular responsibilities in 'upstream' parts of the food chain. This is particularly true of 'meat hygiene' in abattoirs and cutting plants. This role was taken away from local authorities to a central Meat Hygiene Service (MHS) with the MHS itself becoming a branch of the FSA from April 2000. The FSA has also developed a role of local authority monitoring and, increasingly, plays a direct role in enforcement issues that cross local authority borders such as food fraud and contamination.

Most food enforcement at the production and retail level falls to two groups of local government officers:

- Environmental Health Officers (EHOs) have responsibility for food safety legislation. They are generally employed at district council (town hall) level.

- Trading Standards Officers (TSOs) take control of food standards issues, including weights and measures. Generally they are employed at county council level.

The distinction between district and county councils has become increasingly blurred with much local government organised on a unitary basis. This includes most English metropolitan areas and all authorities in Scotland. In these cases TSO and EHO functions are usually combined in the same department.

## Food Safety Act 1990

**F9006**  The *Food Safety Act 1990*, as amended, describes the offences that may be committed if food legislation is contravened and also explains the defences that may be used if charges are instigated. It sets up the mechanisms to enforce the laws and fixes penalties that may be levied. The Act also enables the government to make regulations that include more detailed food safety and consumer protection measures.

The Act is divided into four principle sections:

(i) Part I: Definitions and responsibilities for enforcement;

(ii) Part II: Main provisions;

(iii) Part III: Administrative and enforcement issues such as powers of entry; and

(iv) Part IV: Miscellaneous arrangements notably the power to issue codes of practice.

EU Regulation 178/2002/EC (the 'General Food Law Regulation') came into force on 21 February 2002, with additional key provisions applied from 1 January 2005. These include new food safety requirements, new traceability requirements, and also measures to ensure effective product recall/withdrawals and notification to competent authorities. Although, as an EU Regulation, 178/2002/EC is directly applicable in Member States, consequential changes to domestic legislation were necessary, to introduce new enforcement provisions, and to ensure conformity with this EU law.

The *Food Safety Act 1990 (Amendment) Regulations 2004 (SI 2004 No 2990)* have made minor changes to the definition of 'food' in the *Food Safety Act 1990* to align it with the new EU definition and have also made some minor amendments to the Act itself in respect of public consultation requirements, and to remove duplication with the directly applicable provisions in Regulation 178/2002/EC. The *General Food Regulations 2004 (SI 2004 No 3279)* provide enforcement powers in respect of the new food safety requirements introduced under Regulation 178/2002/EC, designate competent authorities, specify enforcement authorities, make provision for offences and penalties, and introduce some consequential amendments to the *Food Safety Act 1990*, in particular *ss* 7 and 8.

The following is a brief summary of the main provisions contained within the Act.

## Section 1: definitions

**F9007** Under the Food Safety Act 1990, as amended, 'food' has a wide meaning and includes drink, chewing gum and any substance, including water, intentionally incorporated into the food during its manufacture, preparation or treatment. It includes water after the point of compliance as defined in Article 6 of Directive 98/83/EC (Water Quality for Human Consumption) and without prejudice to the requirements of Directives 80/778/EEC (Natural Mineral Waters) and 98/83/EC.

'Food' does not include live animals, birds or fish (although shellfish that are eaten raw and alive such as oysters and similar are classed as 'food'). Neither does 'food' include animal feed, residues and contaminants, nor controlled drugs that might be taken orally, medicines or cosmetics that are covered under separate Community legislation.

One key difference is that unlicensed medicinal products are excluded from the definition of 'food' if they are medicinal products within the meaning of the Medicines Directive 2001/83/EC. However, certain 'borderline' medicinal products (which do not fall within the definitions of Directive 2001/83/EC) will now be included in the new definition of 'food'. The Medicines and Healthcare Products Regulatory Agency (MHRA) will determine whether a given product is a 'medicinal product' or a food on a case by case basis, having regard to the overall presentation and function of the product.

The revised definition of 'food' automatically applies to other legislation in related areas that uses the Food Safety Act definition, for example the *Food Standards Act 1999* and the *Food and Environment Protection Act 1985*, as well as to Regulations and Orders made under these Acts.

The scope of the *Food Safety Act 1990* is very wide ranging and covers all commercial businesses, ranging from farmers, manufacturers, wholesalers, distributors, retailers to businesses such as canteens, clubs, schools, hospitals, care homes and so on, whether or not they are run for profit. Government establishments that once had 'crown immunity' are treated no differently to other food businesses. Charity events that sell food are also subject to the safety provisions of the Act but may be exempt from certain administrative requirements.

The Act thus covers any premises from an abattoir to a retail superstore, from a street vendor to a five star hotel. It includes any place (including premises used only occasionally for a food business such as a village hall), any vehicle, mobile stalls and temporary structures.

The *Food Safety Act 1990, s 2* provides an extended meaning to the term 'sale' of food. The meaning of 'sale' of food is very wide (and now encompasses the European concept of 'placing on the market'). The Act does not apply only to sales where money changes hands and the business is run for profit. It also covers food given as a prize or as a reward by way of business promotion or entertainment. Entertainment includes social gatherings, exhibitions, games, sport and so on. Thus, if food, which turns out to be unfit is given as a prize for a darts competition, it could be subject to an action under the 1990 Act. Similarly, a food business cannot avoid its obligations under the Act by giving away food with other non-food items for which payment is accepted.

The *Food Safety Act 1990, s 3* establishes a presumption that food is intended for human consumption. If a food business has any food or substance capable of being used in the preparation of food in its possession, the Act presumes that the intention is to sell it, unless it can be proved otherwise. Where food raw materials, ingredients, additives or finished products are on trade premises but are either not ready for consumption or have been rejected, it is prudent to keep them in separate rooms, areas or batches clearly identified as being 'not for human consumption'. For example, food past its 'use by' date must be segregated from food that is for sale and should be marked clearly otherwise an EHO or TSO could presume that it was for sale.

The *Food Safety Act 1990, ss 4–6*, defines the roles and responsibilities of local and central government and the officials authorised to act on their behalf. A formal Code of Practice (No 1) was made under s 40 of the Act, to offer guidance to non-metropolitan county and district councils, on these roles. This Code will be replaced imminently by a new, comprehensive Food Law Code of Practice, which will encompass enforcement of all aspects of the revised set of food hygiene rules and the requirements of Regulation 882/2004 on Official Food and Feed Controls.

## Part II: main provisions of the Act

**F9008**   The main provisions of the *Food Safety Act 1990* fall within *ss 7–22*. The offences are divided into two types: food safety and consumer protection which each breakdown into two separate offences:

(i) *Food Safety*

- *Section 7*: Rendering food injurious to health.
- *Section 8*: Selling food not complying with food safety requirements.

(ii) *Consumer Protection*

- *Section 14*: Food not of the nature, substance or quality demanded.
- *Section 15*: Falsely presenting or describing food.

## Section 7: rendering food injurious to health

**F9009**   It is an offence to do anything intentional that would make food harmful to anyone that eats it. Even if one did not know that it would have that effect, an offence would still have been committed. For example subjecting food to poor temperature control that allows bacteria to multiply could render food injurious to health. An example of a more deliberate offence would be the addition to, or removal from, food of substances or components so as to render the food injurious to health.

*Section 7* has been amended by the *General Food Regulations 2004 (SI 2004 No 3279)* to require regard to be had not only to the probable immediate and/or short-term and/or long-term effects of that food on the health of a person consuming it, but also on subsequent generations. It also requires the probable cumulative toxic effects and particular health sensitivities of a specific category of consumers, where the food is intended for that category of consumers, to be taken into account. Food would be injurious to health if it were likely to cause immediate harm to anyone that ate it, for example botulism in yoghurt. It would also be injurious to health if the harm were cumulative over a long period, for example fungal toxins in cereal products. There is no Court precedence to say whether an adverse allergenic reaction to a food (e.g. anaphylactic shock in response to undeclared peanut) would fall under this section of the Act or remain under the more general provisions of the *Consumer Protection Act 1987*.

## Section 8: selling food not complying with food safety requirements

**F9010** Whilst *s 7* deals with the person who renders food injurious to health, the *Food Safety Act 1990*, *s 8* goes on to make it an offence for him or anyone else in the food supply chain to sell food which 'fails to comply with food safety requirements' as defined in the EU Regulation.

Food would be deemed to be unsafe if it was considered to be either 'injurious to health' or 'unfit for human consumption'. The concept of 'injurious to health' relates to safety, but the concept of 'unfit' relates to unacceptability. Unfit food is not necessarily unsafe in the ordinary sense of the word (e.g. sour milk and certain mouldy foods).

In determining whether any food is 'unsafe', consideration would be given to the normal use of the food and its handling at each stage of the chain, and to any information provided to the consumer, such as labelling, to help him avoid specific adverse health effects.

In respect of whether a food is 'injurious to health', regard would be had not only to the probable immediate and/or short-term and/or long-term effects of the food on the health of someone who had eaten it, but also on subsequent generations and to any probable cumulative toxic effects. The particular health sensitivities of a specific category of consumers would also be relevant if the food is specifically intended for them. However, where a food is not specifically intended for such a group, the fact that it may harm someone in that group does not automatically mean that it is 'unsafe' for general consumption.

Food would be 'unfit' (i.e. unacceptable) for consumption if it was putrid or contained serious foreign material, for example a dead mouse. This section gives wide scope to prosecute someone for selling food that fails to meet a customer's expectations. For example, if the food is mouldy or contains a rusty nail or there are excessive antibiotic residues in meat.

(The *Food Safety Act 1990*, *ss 9–13* provide enforcement procedures for food safety offences. These sections are dealt with from F9014–F9019.)

## Section 14: selling food not of the nature, substance or quality demanded

**F9011** It is an offence to supply to the prejudice of the purchaser food which is not of the nature, substance or quality demanded. The purchaser in this case does not have to be a customer in a retail store or catering outlet. One company, large or small may purchase from another and an offence is committed if the food is inferior in nature or substance or quality to that which they demanded. Once again this allows the law to be applied at any point of the food chain.

The purchaser does not have to buy the food for his or her own use to be prejudiced. They may intend to give it to someone else, for example members of their family.

The three offences created by this section are separate and a charge will be brought under one of the three according to the circumstances. These offences have provided a flexible and far reaching means of preventing the adulteration, contamination and mis-description of food and have been the foundation of food law in the United Kingdom for over 125 years. The wide application of this section has been the subject of many cases in law, and a detailed consideration of them is outside the scope of this brief overview.

If a purchaser asks for cod and gets coley, or beef mince contains a mixture of lamb or chicken, it would be not of the 'nature' demanded.

If a purchaser asked for diet cola and was served regular cola, or expected a sheep's milk cheese and it was made from cow's milk, it would be not of the 'quality' demanded.

Food that was not of the quality or substance demanded often formed the basis for complaints of mouldy food or foreign material. Such complaints are likely to be taken under the *Food Safety Act 1990, s 8* (i.e. that the food is so contaminated that it would be unreasonable to expect it to be eaten).

More recently, the above provision has been used when a susceptible customer has requested information about peanuts or similar ingredients to which they may have a severe allergic reaction. If, notwithstanding such a request, the consumer is sold food that triggers a reaction, cases have been brought on the grounds that the food was not of the 'substance' demanded.

## Section 15: falsely presenting or describing food

**F9012** The *Food Safety Act 1990, s 15* provides the principal protection for consumers in respect of food composition and description. It prohibits false and misleading descriptions of food by way of labelling and advertising and is backed up with very detailed requirements in the *Food Labelling Regulations 1996* (*SI 1996 No 1499*) and other 'vertical' regulations that impose specific, additional labelling and compositional requirements on particular foods. It is also an offence if material that is technically accurate is presented in such a way as to mislead the consumer. The scope of this section includes misleading pictorial representation. For some offences, especially labelling by caterers on menus or chalkboards, TSOs may invoke the more general requirements under the *Trades Descriptions Act 1968, s 1*.

## Enforcement

**F9013** The *Food Safety Act 1990* gives enforcement officers strong powers, namely:

- To enter food premises to investigate possible offences.
- To inspect food.
- To detain or seize suspect food.
- To take action that requires a business to put things right.
- If all else fails:
  - to prosecute,
  - to prohibit the use of premises or equipment,
  - to bar an individual from working in a food business.

Under the *Food Safety Act 1990, ss 32* and *33*, enforcement officers have rights of access at any reasonable time and an offence will be committed if they are obstructed. They must be given reasonable information and assistance. They can inspect premises, processes and records. If records are kept on a computer, they are entitled to have access to them. They can take samples, take photographs and even videos, and (under specific circumstances, e.g. where there is reason to believe an offence has been committed) copy records. If they are refused access to any of these, they can apply to a magistrate for a warrant. They must give 24 hours notice to enter private houses used in connection with a food business.

Of course, none of these requirements to co-operate with officers and give them access to information cancel the basic right to avoid self-incrimination. In the extreme situation, one has the right to remain silent and to consult a solicitor.

Officers themselves commit an offence if they reveal any trade secrets learned in the course of official duties.

The *Food Safety Act 1990*, ss 29–31 allow officers to procure samples and control the sampling and testing procedures. A section 40 code of practice (no. 7) covers 'Sampling for Analysis or Examination'. Samples that are taken carelessly may give unreliable results and may be not be acceptable as evidence in court.

The *Food Safety Act 1990*, ss 9–13 provide enforcement officers with a series of enforcement measures to control food or food businesses that present a health risk.

## Section 9: detention and seizure of food

**F9014** If an officer suspects that food is unfit or fails to meet food safety requirements, he can issue a detention notice. This requires the person in charge of the food to keep it in a specified place and not to use it for human consumption pending investigation. The detention notice must be written on a prescribed form, *Detention of Food (Prescribed Forms) Regulations 1990* (*SI 1990 No 2614*). It is an offence to ignore or contravene a detention notice even if the business may believe that it is unjustified. The officer has 21 days to complete his investigations but must clearly act quickly where the food is highly perishable. If he concludes that the food was actually safe, he must withdraw the notice (using another prescribed form) and restore the food to the person in charge.

On the other hand, the EHO may conclude that the food is unsafe. He can seize it and take it before a justice of the peace (JP). The EHO may reach this conclusion as a result of tests performed during the 21 days after serving the detention notice. But in some circumstances, he may be confident that it is unsafe from the beginning and seize the food immediately without using a detention notice. Once again there is a form that must be used and the EHO must state exactly what is happening and when. After seizure the matter should be taken before the JP within two days, quicker in the case of perishable food.

The hearing before the JP may not only decide the fate of the batch of food but may also be a forerunner to other legal proceedings. One is allowed to make representations at this hearing and to call witnesses. It is probably advisable to take specific legal advice from a solicitor when the first detention notice is issued. It is wise to have legal representation at a hearing following seizure.

If the JP decides that the food is unsafe he can order it to be destroyed and he may order the offender to cover the costs of its disposal. If the seized food is only part of a larger batch the court will presume that the whole batch fails unless evidence is provided to the contrary. Disposal must follow the requirements of the relevant waste disposal legislation, using licensed waste disposal contractors, and must also be in accordance with the relevant provisions of Regulation (EC) No 1774/2002 on 'health rules concerning animal by-products not intended for human consumption', implemented by the *Animal By-Products Regulations 2003* (*SI 2003 No 1482*).

If it turns out that the EHO was wrong, the offender may be entitled to compensation. If the food has deteriorated as a result they may be entitled to compensation equal to the loss in value. If a reasonable sum cannot be agreed with the authority there is an arbitration procedure.

## Section 10: improvement notices

**F9015** *Section 10 provides* EHOs with a range of powers to deal with unsatisfactory premises from informal advice through to prosecution and emergency prohibition. If an authorised officer has reasonable grounds for believing that a business does not comply with hygiene or processing regulations, he can issue a formal improvement notice demanding that it puts matters right. This provides a quicker and cheaper remedy than taking the business to court. An improvement notice should not be ignored. It is an offence to fail to comply with a notice and either matters must be put right as indicated on the notice or an appeal lodged against it.

Authority

FORM 1

## Food Safety Act 1990 - Section 9

## DETENTION OF FOOD NOTICE

Reference Number:

1. To: ..........................................................................................................
   Of: ..........................................................................................................

2. Food to which this notice applies:
   Description: ..........................................................................................................
   Quantity: ..........................................................................................................
   Identification marks: ..........................................................................................................

3. *THIS FOOD IS NOT TO BE USED FOR HUMAN CONSUMPTION.*
   In my opinion, the food does not comply with food safety requirements because: ..........................................................................................................
   ..........................................................................................................
   ..........................................................................................................

4. The food must not be removed from:
   ..........................................................................................................
   ..........................................................................................................
   *unless it is moved to:
   ..........................................................................................................
   ..........................................................................................................
   (*Officer to delete if not applicable)

5. Within 21 days, either this notice will be withdrawn and the food released, or the food will be seized to be dealt with by a justice of the peace, or in Scotland a sheriff or magistrate, who may condemn it.

   Signed: .................................................................... Authorised Officer
   Name in capitals: ..........................................................................................................
   Date: ..........................................................................................................
   Address: ..........................................................................................................
   ..........................................................................................................
   ..........................................................................................................
   Tel: ........................... Fax:-..............................................................

   > *Please read the notes overleaf carefully. If you are not sure of your rights or the implications of this notice, you may want to seek legal advice.*

**NOTES**

1. The food described in this notice has been detained pending official investigation.

2. The food must not be used for human consumption until it is released by the officer.

3. The food must remain where it is. If it is moved, it may only be moved to the place stated in paragraph 4 of the notice.

4. If for some reason you need to move the food after receiving this notice, you should contact the officer.

5. Within 21 days the officer must tell you if the notice is being withdrawn or if he is seizing the food for it to be dealt with by a justice of the peace, or in Scotland a sheriff or magistrate, who may condemn it.

6. *COMPENSATION:* If this notice is withdrawn and the food released for human consumption, then you may be entitled to compensation from the authority. This compensation will be payable for any loss in value of the food resulting from the effect of the notice.

*WARNING*

*FAILURE TO COMPLY KNOWINGLY WITH THIS NOTICE IS AN OFFENCE*

*Offenders will be liable:*

   on summary conviction, to a fine of up to £2000
   and/or 6 months in prison,

   *or*

   on conviction on indictment, to an unlimited fine and/or
   up to 2 years in prison.

*Food Safety and Standards* **F9015**

Authority: _____   FORM 2

## Food Safety Act 1990—Section 9

# WITHDRAWAL OF DETENTION OF FOOD NOTICE

1. To: ........................................................................................................
   Of: ........................................................................................................
   ..............................................................................................................

2. Detention Notice Number ..................... dated ........................ and served on you on ................................. (date) is now withdrawn. The food described in paragraph 3 below can now be used for human consumption.

3. Food released for human consumption:
   Description: ...........................................................................................
   Quantity: ...............................................................................................
   Identification marks: ..............................................................................

   Signed: ............................................................................... Authorised Officer
   Name in capitals: ..................................................................................
   Date: ....................................................................................................
   Address: ................................................................................................
   ..............................................................................................................
   ..............................................................................................................
   Tel: ............................ Fax ............................................................

   > *Please read the notes overleaf carefully. If you are not sure of your rights or the implications of this notice, you may want to seek legal advice.*

**NOTES**

1. The food described in this notice has now been released for human consumption.

2. If this notice does not relate to all of the food originally detained, then the rest has been seized under section 9(3)(b) of the Food Safety Act 1990.

3. *COMPENSATION:* If you can show that any of the food now released for human consumption has lost value, you may be entitled to compensation from the authority. Compensation will be payable for any loss in value resulting from the effect of the notice.

*Food Safety and Standards* **F9015**

Authority:_____ FORM 3

## Food Safety Act 1990—Section 9

## FOOD CONDEMNATION WARNING NOTICE

Reference Number:

1. To: ..................................................................................................
   Of: ..................................................................................................
   ..................................................................................................

2. This Notice applies to the following food which has been seized by an officer of this authority:
   Description ..................................................................................................
   Quantity ..................................................................................................
   Identification marks ..................................................................................................

3. *IT IS MY INTENTION TO APPLY TO A JUSTICE OF THE PEACE, OR IN SCOTLAND A SHERIFF OR MAGISTRATE, AT*
   ..................................................................................................
   *ON ..................... (DATE) AT ............... AM/PM FOR THE ABOVE FOOD TO BE CONDEMNED.*
   because ..................................................................................................

4. As the person in charge of the food, you are entitled to attend and to bring witnesses.

5. A copy of this notice has also been given to:
   ..................................................................................................
   ..................................................................................................
   who may also attend and bring witnesses.

   Signed: ................................................................... Authorised Officer
   Name in capitals: ..................................................................................................
   Date: ..................................................................................................
   Address: ..................................................................................................
   ..................................................................................................
   ..................................................................................................
   Tel: ........................... Fax: ..................................................................

> *Please read the notes overleaf carefully. If you are not sure of your rights or the implications of this notice, you may want to seek legal advice.*

**NOTES**

1. You are being warned that the Authority will be applying to a justice of the peace, or in Scotland a sheriff or magistrate, for the food that has already been seized to be condemned.

2. The justice of the peace, or in Scotland the sheriff or magistrate, will listen to the authority's case that the food fails to comply with food safety requirements and should be condemned. You may say why it should not be condemned.

3. You may bring your own evidence and witnesses to challenge the views of the authority and you may be represented by a lawyer.

4. You are not being charged with an offence. The hearing is only to decide whether the food complies with food safety requirements. But the court may order the food to be condemned. However you may be prosecuted for offences under the Food Safety Act 1990.

5. *EXPENSES:* If the justice of the peace, or in Scotland the sheriff or magistrate, orders the food to be condemned, then the owner of the food will have to pay reasonable expenses for it to be destroyed or disposed of.

6. *COMPENSATION:* If the justice of the peace, or in Scotland the sheriff or magistrate, does not condemn the food, the owner of the food may be entitled to compensation from the authority for any loss in its value as a result of the action taken by the authority.

An improvement notice must contain four elements:

(i) the reasons for believing that the requirements are not being complied with;
(ii) the ways in which regulations are being breached;
(iii) the measure which should be taken to put things right;
(iv) the time allowed for putting things right (this cannot be less than 14 days).

The business can choose to put things right in a different way to that suggested by the EHO provided that the EHO is satisfied that it will have the required effect.

If a party does not agree with an improvement notice, the *Food Safety Act 1990, s 37* gives the right to appeal to a magistrate within one month of its issue or within the time specified on the notice if shorter than one month. The EHO must inform the party about their rights to appeal and provide the name and address of the court when he serves the notice.

The notice may contain a number of points but the whole notice does not have to be appealed. The proprietor may appeal just one point or simply ask for a time extension. The notice is effectively suspended during the appeal, but it is best to let the authority, as well as the court, know about the appeal. The *Food Safety Act 1990, s 39* allows the magistrate to modify or cancel any part of the notice.

With effect from January 2006, the provisions for Hygiene Improvement Notices will fall under the new *Food Hygiene Regulations 2005, (SI 2005 No 2059)*, Reg 8.

## Section 11: prohibition orders

**F9016** If a proprietor is convicted of food safety or processing offences, including failing to comply with an Improvement Notice, the authority can also apply for a prohibition order. If the court is satisfied that any processing, or the construction or state of the premises, or use of equipment, poses a risk to public health, it must issue a prohibition order to the proprietor or any manager of the business. The order can deal with any of one three issues depending upon the actual risk:

- to prohibit a process or treatment,
- to deal with the *structure* of the premises or the use of equipment,
- to deal with a problem that relates to the *condition* of the premises or equipment.

In either of the last two cases, the order may prohibit the use of the equipment or the premises.

Any prohibition order must be served on the proprietor of the business. If it relates to equipment, it will also be fixed to the equipment and if it relates to the premises it will be fixed prominently to the premises. It is automatically an offence to knowingly breach a prohibition order.

As soon as a party believes that they have put matters right and that the health risk no longer exists, they can apply to the EHO to have the prohibition lifted. The EHO must respond by reaching a decision within a fortnight and if they believe that things are satisfactory, issue a certificate within another three days. The certificate will state that enough has been done to ensure that there is no longer an unacceptable risk to public health. If the authority disagrees and refuses to issue a certificate they must give their reasons and the business can appeal to a magistrate to have the order lifted. The authority must give information on the right to appeal, and who to contact as well as the time scale (the appeal must be made within one month).

With effect from January 2006, the provisions for Hygiene Prohibition Orders will fall under the new *Food Hygiene Regulations 2005, Reg 9 (SI 2005 No 2059)*.

F9016  *Food Safety and Standards*

Authority: .................................................................................... FORM 1

**Food Safety Act 1990—Section 10**
**IMPROVEMENT NOTICE**

Reference Number:

1. To: .................................................................... (Proprietor of the food business)

    At: ..............................................................................................................

    ..................................................................................................................

    .................................................................... (Address of proprietor)

2. In my opinion the:

    ..................................................................................................................

    [Office to insert matters which do not comply with the Regulations]

    in connection with your food business ..........................................................

    .................................................................... (Name of business)

    at ................................................................................................................

    .................................................................... (Address of business)

    do/does* not meet the requirements of ........................................................

    of the .......................................... Regulations

    because:

    ..................................................................................................................

    [* Officer to delete as appropriate]

3. In my opinion, the following measures are needed for you to comply with these Regulations: ..........................................................................................

    ..................................................................................................................

    ..................................................................................................................

4. These measures or measures that will achieve the same effect must be taken by: .......................... (date)

5. It is an offence not to comply with this improvement notice by the date stated.

    Signed: .................................................................... Authorised Officer

    Name in capitals: ..........................................................

    Date: ..........................................................

    Address: ..........................................................

    ..................................................................................................................

    Tel: .......................................... Fax: ..........................................

> Please read the notes overleaf carefully.
> If you are not sure of your rights or the implications of this notice, you may want to seek legal advice.

*Food Safety and Standards* **F9016**

## NOTES

1. In the opinion of the officer you are not complying with the Regulations under Part II of the Food Safety Act 1990 described in paragraph 2 of the notice. The work needed in the officer's opinion to put matters right is described and it must be finished by the date set.

2. You are responsible for ensuring that the work is carried out within the period specified, which must be at least 14 days.

3. *You have a right to carry out work that will achieve the same effect as that described in the notice.* If you think that there is another equally effective way of complying with the law, you should first discuss it with the officer.

**YOUR RIGHT OF APPEAL**

4. If you disagree with all or part of this notice, you can appeal to the magistrates' court, or in Scotland to the sheriff. You must appeal within one calendar month of the date of the notice or the period ending with the date stated in paragraph 4 of the notice, whichever ends earlier.

5. If you decide to appeal, the time set out in the notice is suspended and you do not have to carry out the work described until the appeal is heard. *However, if you are not complying with the Regulations mentioned in the notice, you may still be prosecuted for failure to comply with those Regulations.*

6. When the appeal is heard, the magistrates' court, or in Scotland the sheriff, may confirm, cancel or vary the notice.

WARNING

*FAILURE TO COMPLY WITH THIS NOTICE IS AN OFFENCE*

*Offenders are liable to be fined and/or imprisoned for up to 2 years.*

F9016  *Food Safety and Standards*

# PROHIBITION ORDER

(Food Safety Act 1990, s. 11)

Magistrates' Court

(Code)

Date :
Accused : WHEREAS
of :

being the proprietor of a food business carried on at premises at

has today been convicted by this court of an offence under regulations to which section 11 of the Food Safety Act 1990 applies [by virtue of section 10(3)(b) of the said Act], namely:-

Decision : AND WHEREAS the Court is satisfied that there exists a risk of injury to health by reason of

[the use for the purposes of the said food business of a certain [process] [treatment], namely

 ]

[the [construction] [state or condition] of the premises at

used for the purposes of the said food business]

[the [use for the purposes of the said food business] the [state or condition] of certain equipment [used for the purposes of the said food business], namely

 ]

and it is ORDERED that

Order : [the use of the said [process][treatment] for the purposes of the business] [the use of the said [premises] [equipment] for the purposes of [the business] [any other food business of the same class or description, namely

 ]
[any food business]]
[and that the participation by the accused in the management of any food business [of the said class or description]
is prohibited.

[By Order of the Court]

[*Justice of the Peace* ] [*Justices' Clerk* ]

Delete any words within square brackets which do not apply.

FS 46    Copyright forms are reproduced by permission of the publishers, Shaw & Sons Ltd., Shaway House, Crayford, Kent, DA1 4BZ, from whom copies may be purchased.

F90/18

## Personal prohibition under section 11

**F9017** The court can also prohibit a person from running or managing a food business. An application to the court to lift a personal prohibition cannot be made within six months of its imposition and if an appeal is unsuccessful, it cannot be appealed again for a further three months.

## Section 12: emergency prohibition notices and orders

**F9018** Whereas *s 11* prohibitions are imposed only by a court following conviction, *s 12* permits, in an emergency, an EHO to act on his or her own authority, and with immediate effect. Where the authority is satisfied that a business presents an imminent risk to health, it can serve an emergency prohibition notice without prior reference to a court. Emergency prohibition can apply to the whole premises or a specific part. The notice will be fixed in a prominent place and anyone removing it or deliberately ignoring it is guilty of an offence.]

The officer must take the matter before a magistrates' court within three days of serving the notice. At least one day before it goes to court the EHO must serve a notice on the proprietor of the business telling him of the court hearing. If an order is not issued by the court or if no application for one is made within the three day period, the emergency prohibition notice lapses. Once again, if a business finds itself faced with action of this kind, it is advisable to seek specific legal advice.

If the court agrees that the prohibition actions were justified, it will make an emergency prohibition order which replaces the emergency prohibition notice. The arrangements for lifting the order are the same as a *s 11* order, above. If the court does not uphold an emergency prohibition, the proprietor may seek compensation for loss or damages. An emergency prohibition order cannot be made against an individual. Prohibition of a person can only be made under *s 11* following a conviction.

With effect from January 2006, the provisions for Hygiene Emergency Prohibition Notices and Orders will fall under the new *Food Hygiene Regulations 2005, regulation 10 (SI 2005 No 2059)*.

## Section 13: emergency control orders

**F9019** *Section 13* empowers the Secretary of State or the Food Standards Agency to issue emergency control orders where it appears that commercial food-related operations involve, or may involve, imminent risk of injury to health. In such circumstances either of the above may prohibit such operations. It is an offence to knowingly contravene an emergency control order. If an emergency control order is invoked, there is no right of appeal and no compensation arrangements.

It may be that an unsafe batch of food has already been distributed around the country. Emergency prohibition of the manufacturing site would stop further production but it would not control the risk from food already in the system. The *Food Safety Act 1990, s 13* allows an emergency control order to require all steps to be taken that will remove the threat. This is a wide-ranging power that will only be used in exceptional circumstances. Of course in most cases, any business told that it has received unsafe food from a supplier would stop selling it simply to protect their own reputation.

Voluntary arrangements are in place to co-ordinate food hazards and issue warnings and organise recalls when food is distributed more widely than the local area in which it is produced. These warnings are circulated by electronic mail to EHOs around the country. Where possible, trade organisations and the media are also informed. The Food Standards Agency is demonstrating a greater commitment to freedom of information than previous departments. Nowadays food hazard warnings and recalls are better publicised, not least through the FSA website at: www.foodstandards.gov.uk.

Authority: ..................................................................  FORM 2

**Food Safety Act 1990–Section 12**
**EMERGENCY PROHIBITION NOTICE**

Reference Number:

1. To: .................................................................. (Proprietor of the food business)

    At: ..................................................................

    ..................................................................

    .................................................. (Address of proprietor)

2*  I am satisfied that: ..................................................................

    ..................................................................

    ..................................................................

    at ..................................................................

    ----------------------------------------------------- (Address of business)

    POSES AN IMMINENT RISK OF INJURY TO HEALTH because:

    ..................................................................

    ..................................................................

    ..................................................................

    (*See Note 1 overleaf)

3.  YOU MUST NOT USE IT FOR THE PURPOSE OF THIS/ANY/THIS OR ANY SIMILAR* FOOD BUSINESS.

    [* Officer to delete as appropriate]

    Signed: .................................................................. Authorised Officer

    Name in capitals: ..................................................................

    Date: ..................................................................

    Address: ..................................................................

    ..................................................................

    Tel: .................................................. Fax: ..................................................

> Please read the notes overleaf carefully.
> If you are not sure of your rights or the
> implications of this notice, you may want to
> seek legal advice.

**NOTES**

1. *When you receive this notice you must IMMEDIATELY stop using the premises, process, treatment or equipment described by the officer in paragraph 2 of the notice and located at the address stated.*

2. *Within 3 days of service of this notice, the authority must apply to a magistrates' court, or in Scotland to a sheriff, for an order confirming the prohibition. You will be told the date of the hearing which you are entitled to attend and at which you may call witnesses if you wish.*

3. *If you believe that you have acted to remove the imminent risk of injury to health, you should apply in writing to the authority for a certificate which would allow you to use the premises, process, treatment or equipment again. You can do this even if the court hearing has not taken place.*

4. *You are not allowed to use the premises, process, treatment or equipment for the purpose specified in paragraph 3 of the notice (see section 11(3) of the Food Safety Act 1990) until (a) a court decides you may do so; (b) the authority issues you with a certificate as in paragraph 3 above; (c) 3 days have passed since the service of the notice and the authority has not applied to the court as in paragraph 2 above; or (d) the authority abandons the application.*

5. *A copy of this notice must, by law, be fixed on the premises or equipment which is not to be used. It is an offence (under section 1 of the Criminal Damage Act 1971 or, in Scotland, section 78 of the Criminal Justice (Scotland) Act 1980) to deface it.*

6. *COMPENSATION: If the authority does not apply to the magistrates' court, or in Scotland to the sheriff, for an order confirming its action within 3 days of the date of service of this notice, you will be entitled to compensation for any losses you have suffered because you could not use the premises, process, treatment or equipment because you were complying with this notice. You will also be entitled to such compensation if the magistrates' court, or in Scotland the sheriff, decide at the hearing that the authority's action was wrong.*

**WARNING**

*ANY ONE WHO KNOWINGLY CONTRAVENES THIS NOTICE IS GUILTY OF AN OFFENCE*

*Offenders are liable to be fined and/or imprisoned for up to 2 years.*

## Remedial Action Notices

A new type of enforcement notice will be introduced under the *Food Hygiene Regulations 2005, Regulation 11* (*SI 2005 No 2059*). With effect from January 2006, in the case of an establishment subject to approval under Article 4(2) of Regulation 853/2004 (animal products hygiene legislation), an authorised officer may serve a **Remedial Action Notice** where any of the requirements of the Hygiene Regulations is being breached or inspection under the Hygiene Regulations is being hampered. This may:

- prohibit the use of any equipment or any part of the establishment (and must specify the reason and the action needed to remedy it);
- impose conditions upon or prohibit the carrying out of any process;
- require the rate of operation to be reduced to such extent as to ensure hygienic operation, or to be stopped completely;
- require the detention of any animal or food of animal origin for sampling and examination.

A remedial action notice must be served as soon as practicable and state why it is being served. The notice must be withdrawn by a further written notice as soon as the authorised officer is satisfied that the necessary action has been taken.

Failure to comply with a remedial action notice is an offence, although the right to appeal to a magistrate within one month is granted.

## Sections 16 to 18: power to make regulations

**F9020** The *Food Safety Act 1990, ss 16–18* and *26* give the Secretary of State very wide-ranging powers to make regulations and orders for general food safety and consumer protection; for special provisions for particular foods; for the registration and licensing of food premises; for food hygiene training; and for facilities for sampling and the giving of information. For example:

- To prohibit specified substances in foods.
- To prohibit a certain food process.
- To require hygienic conditions in commercial food premises.
- To dictate microbiological standards for foods.
- To make standards on food composition.
- To specify requirements for food labelling.
- To restrict and control claims made in the advertising of food.

They also provide the power to ban from sale food originating from potentially diseased sources, such powers were used to prohibit the sale of certain parts of beef cattle for human food on the basis that they may have been suffering from BSE.

Much of the detail of food law is in the regulations made under *s 16, which also contains a requirement for consultation prior to the introduction of individual Statutory Instruments.*

*Section 17* allows the government to bring EU provisions into UK law. *Section 18* regulations are intended to cover very particular circumstances such as genetically modified and other novel foods, although in practice these are now harmonised at EU level and the UK has only very limited room in which to act independently. For example, all aspects of the approval, labelling and traceability of foods and food ingredients derived from GM crops are controlled under EU Regulations 1829 and 1830/2003.

## Section 19: registration and licensing

**F9021** The *Food Safety Act 1990, s 19* permits the government to bring in regulations demanding licensing or registration of certain food businesses.

Registration and licensing confer significantly different powers on enforcement authorities. Registration is simply an administrative exercise. It is designed primarily to let enforcement agencies know what food businesses are operating in their area. It also provides some basic information on the size of the business and the type of food that they produce. No conditions are attached to registration, the local authority cannot refuse to register a business, there is no charge, and it does not have to be renewed periodically.

Licensing (or 'prior approval') is quite different. It is a control measure. Precise licensing criteria will be specified and there would normally be prior inspection and approval before a new business is allowed to open. There is often a charge for a licence, which must be renewed periodically. The enforcement authority will usually have the right to withdraw a licence if standards fall and the business would have to close.

But at present, licensing only applies to a those businesses subject to the requirements of the EU 'vertical', animal product legislation, such as slaughter houses, cutting plants and dairy operations. Most other types of food businesses are subject only to registration.

The *Food Premises (Registration) Regulations 1991 (SI 1991 No 2828)* (as amended by *SI 1993 No 2022* and *SI 1997 No 723*) were enacted under *s 19* but will effectively be revoked with effect from January 2006 by the introduction of the *Food Hygiene Regulations 2005*, as will the current regime of licensing retail butchers' shops.

## Defences against charges under the Food Safety Act 1990

**F9022** The following paragraphs explain the defences available under the *Food Safety Act 1990*.

### Section 21: defence of due diligence

**F9023** The so called 'due diligence' defence first appeared within food safety legislation when the *Food Safety Act 1990* was passed. Previously, the provision had been tried and tested within a number of other pieces of legislation after it first appeared just over 100 years before in the *Merchandise Marks Act 1887*. Indeed legislation that included a due diligence defence, such as the *Trades Descriptions Act 1968*, already overlapped with food legislation. So although the defence was new to food law there were numerous precedents to signal the implications of the due diligence provision.

The *Food Safety Act 1990* creates offences of strict liability. The prosecution is not required to show that the offence was committed intentionally. However the due diligence defence is intended to balance the protection of the consumer against the right of traders not to be convicted for an offence that they have taken all reasonable care to avoid committing. The intention of the due diligence defence is to encourage traders to take proper responsibility for their products and processes.

The *Food Safety Act 1990, s 21* states:

> 'in any proceedings for an offence under any of the preceding paragraphs of this Part ... it shall be ... a defence for the person charged to prove that he took all reasonable precautions and exercised all due diligence to avoid the commission of the offence by him or any person under his control.'

The onus is on the business to establish its defence on the balance of probabilities. Part of the due diligence defence may be to establish that someone else was at fault. If that is what is intended, the prosecutor must be informed at least seven days before the hearing, or if a business has been in court already, within one month of that appearance. Under the 1990 Act, reliance upon a supplier's warranty alone is not possible, although a warranty may be part of a due diligence defence.

It is impossible to give precise advice on what any business would need to do to have a due diligence defence. Cases that have been decided by the courts serve only to demonstrate that each case is decided on its own facts and may also be determined by the size and resources otherwise available to a business. What is certain is that doing nothing will never provide a due diligence defence. Positive steps must be taken if a defence is to succeed.

The defence has two distinct parts:

- Taking *reasonable precautions* involves setting up a system of control. The business operator must consider the risks that threaten the operation and take *all reasonable precautions* that may be expected of a business of the size and type.

- Exercising *due diligence* means that those precautions are taken on a continuous basis.

*All* reasonable precautions and *all* due diligence are needed. The courts have decided consistently that if a precaution could reasonably have been taken but was not taken then the defence would not succeed.

The test is 'what is reasonable?' A business will be expected to take greater precautions with high risk, ready-to-eat foods than are needed for boiled sweets or biscuits. What is reasonable for a large-scale food business may not be reasonable for a smaller enterprise.

The business must ensure that the scope of 'due diligence' covers all aspects of food law requirements. It is not adequate to have excellent control of food safety hazards if there are no precautions in place to ensure that the composition and labelling of food complies with the regulations.

Documentation is important. Unless the precautions are written down, it will be difficult to persuade the EHO or the court that a proper system is in place. The same is true of records of any checks made to demonstrate that the system is followed diligently. The burden of proof lies with the business to establish that the defence is satisfied.

A good 'HACCP' plan (see below) may be helpful in showing that a systematic approach to identifying the food safety precautions needed in the operation has been adopted. Cross-reference to industry guidelines or codes of practice may show that the system has a sound basis. Documentation from the hazard analysis will add to the evidence that precautions have been taken. Records such as specifications, cleaning schedules, training programmes and correspondence with suppliers or customers may all play a part in establishing the defence. An organisation chart and job descriptions that identify roles and responsibilities will also be useful evidence of the precautions taken.

Not all food businesses have the same defence. Some traders within the retail and catering end of the food supply chain may use a simpler defence of 'deemed due diligence' from the *Food Safety Act 1990, s 21(3)* and *(4)*. Even then 'deemed due diligence' is not available against the *s 7* offence of rendering food injurious to health. Manufacturers and importers must satisfy the full defence. Larger businesses may be expected to take more precautions than smaller businesses. Those selling 'own label' packs may have greater responsibility for upstream activity than those selling manufacturers' 'branded' products.

## Section 35: penalties under the Food Safety Act 1990

**F9024** The courts decide penalties on the merits of individual cases but they are limited by maximum figures included in the *Food Safety Act 1990, s 35*. For most offences the crown court can send offenders to prison for up to two years and/or impose an unlimited fine. Even the magistrates' court, where most cases are heard, can set fines of up to £5,000 per offence and up to six months imprisonment. For the two food safety offences in *ss 7* and *8* and the first of the consumer protection offences (*s 14*), the fine may be up to £20,000. In Scotland the sheriff may impose equivalent penalties.

Each set of regulations made under the Act will have its own level of penalties that will not exceed the levels mentioned here.

### Section 40: codes of practice

**F9025** The *Food Safety Act 1990, s 40* allows for the development of codes of practice. *Section 40* codes of practice are primarily intended to instruct and inform enforcement officers and related bodies or personnel, for example food analysts or examiners. In carrying out their duties, enforcement professionals must have regard for *s 40* codes of practice. Any failure to follow a code requirement might seriously jeopardise their ability to bring a successful prosecution. Twenty such codes have been made under *s 40*. Some have undergone several revisions. The subjects of early codes are general matters, such as the demarcation of responsibility for enforcing different parts of the Act or general inspection procedures. Later ones provide more specific information on the enforcement of particular regulations, for example the Dairy Hygiene Regulations.

The most significant of the general codes is Code of Practice Number 9 on food hygiene inspections. It contains detailed advice to environmental health officers on how to conduct such inspections. It includes a 'risk rating' system whereby inspectors can prioritise different food businesses and determine the necessary inspection frequency. Codes also give guidance on such issues as the powers of inspecting officers and the communication of findings to the business. The FSA is currently finalising a radical revision and consolidation of the *s 40* codes related to all aspects of enforcement into a single document, leaving only those covering meat, fish and dairy products.

*Section 40* codes should not be confused with 'industry guides' that are promoted by the EU General Food Hygiene Directive and the General Food Hygiene Regulations. These 'guides' provide more detailed practical explanation of the implications of the particular regulations in a food business sector. There are also numerous other guides and codes of practice published by industry, government, and other organisations.

# Food premises regulations

**F9026** Current requirements for the registration of food premises will be subsumed into revised legislation with effect from January 2006. The *Food Premises (Registration) Regulations 1991* (as amended by *SI 1993 No 2022* and *SI 1997 No 723*) will be replaced by the requirement to comply with Article 6(2) of EU Regulation 852/2004, as enacted by the *Food Hygiene Regulations 2005* (*SI 2005 No 2059*).

It is intended that the existing requirements will, where possible, form the basis of new requirements, mainly through the registration of establishments with the local authority. One change that will be introduced will relate to wholesale markets, where in future, action will be possible against a single, infringing unit rather than seeking the closure of an entire market operation. Existing registered premises will not have to reregister.

The following paragraphs cover the principal requirements.

**F9027** Food businesses must notify the local authority of all food premises under their control that carry out any production, processing and distribution of food (including retail and catering premises), unless they are already required by other legislation to be licensed (i.e. approved under 'vertical' legislation relating to handling products of animal origin). The legislation requires the notification of individual food *premises* and not food businesses. Thus, premises that may be used only occasionally by food businesses, even by different businesses, (e.g.

a church hall, village hall or scout hut used from time to time for the sale of food) may have to be registered by the owner of the premises, depending on the frequency of use. The criterion is that they are used by commercial food businesses for five or more days (which do not have to be consecutive) in any period of five weeks.

Similarly, wholesale markets or other premises used by more than one food business must also be registered. The market operator or the person in charge of the premises is responsible. If mobile food premises are used they may have to be registered. These can range from a handcart in a market to a forty-foot trailer used for hospitality. Privately-owned, moveable premises used in a market must be registered even though the market is registered by the operator. (If stalls provided by the market are used, registration is not necessary). If mobile food premises are operated, the premises at which they are normally kept must be registered with the food authority local to the base.

Staff restaurants must be registered. Contractors and clients should have a clear agreement as to who will action this.

New premises must notify the authorities at least 28 days before opening. This is designed to allow an EHO the opportunity to look at a business before it opens. However, if the EHO does not take that opportunity, it is not necessary to wait for a visit or for permission from the EHO before opening, provided that the necessary registration form has been sent. To register, one must supply a range of information on a registration form obtained from the local EHO department, which is normally contacted via the Town Hall. The local authority may prefer their version of the form to be used instead. In case of doubt, it is best to keep a note of when the form was sent. If possible check that it has been received. Sending the form by fax may be helpful in establishing the transmission date.

Food business operators must also ensure that the authority always has up-to-date information on premises, including any significant change in food-related activities (e.g. change of product handled) and any closure of an existing establishment. The authority must be notified of the change within 28 days. Many of the other pieces of information on the application form may also change, for example the name of the manager, the phone number, the number of employees and so on but these changes do not need to be notified. Failing to register, giving false or incomplete information or failing to notify any of the changes mentioned above, can trigger prosecution proceedings.

Some food premises must by law be licensed, in which case additional notification/registration is not required. These premises are generally those (other than retail and catering) that handle raw/unprocessed products of animal origin, which fall into a number of categories under EU Regulation (EC) No 853/2004, and are approved ('licensed') by the competent authority, following at least one on-site visit.

## Premises licensed or registered under other measures
F9028

### Establishments handling raw products of animal origin

The Meat Hygiene Service is responsible for the approval of meat-handling establishments where Regulation 853/2004 requires control by an official veterinarian. These will include slaughterhouses, game handling establishments, cutting plants and any of these in which minced meat, meat preparations, mechanically separated meat, or meat products are also produced.

Local Authorities are responsible for the approval of establishments where control does not fall to an official veterinarian. Such 'product-specific establishments' will be producing any, or any combination, of the following: minced meat, meat preparations, mechanically separated meat, meat products, live bivalve molluscs, fishery products, dairy products, egg products, gelatine and collagen, and will include cold stores and certain wholesale markets.

*Food Safety and Standards* **F9028**

## FORM OF APPLICATION FOR REGISTRATION OF FOOD PREMISES

1. Address of premises ..........................................................................................
   (or address at which movable
   premises are kept)  Post code .........................................

2. Name of food business............................................  Telephone no: ...........................
   (trading name)

3. Type of premises Please tick ALL the boxes that apply

   | | |
   |---|---|
   | Farm/smallholding ☐ | Staff restaurant/canteen/kitchen ☐ |
   | Food/manufacturing/processing ☐ | Catering ☐ |
   | Slaughterer ☐ | Hospital/residential home/school ☐ |
   | Packer ☐ | Hotel/pub/guest house ☐ |
   | Importer ☐ | Private house used for a food business ☐ |
   | wholesale/cash and carry ☐ | Premises used by a number of businesses ☐ |
   | Distribution/warehousing ☐ | Moveable premises ☐ |
   | Retailer ☐ | |
   | Market ☐ | Other: please give details ........................... |
   | Restaurant/cafe/snack bar ☐ | ......................................................... |

4. Does your business handle or involve any of the following? Please tick ALL the boxes that apply

   | | |
   |---|---|
   | Chilled foods ☐ | Alcoholic drinks ☐ |
   | Frozen foods ☐ | Canning ☐ |
   | Fruit and vegetables ☐ | Vacuum packing ☐ |
   | Fish/fish products ☐ | Bottling and other packing ☐ |
   | Fresh/frozen meat ☐ | Table meals/snacks ☐ |
   | Fresh/frozen poultry ☐ | Takeaway food ☐ |
   | Meat products or delicatessen ☐ | Accommodation ☐ |
   | Dairy products ☐ | Delivery service ☐ |
   | Eggs ☐ | Chilled food storage ☐ |
   | Bakery ☐ | Bulk storage ☐ |
   | Sandwiches ☐ | Use of private water supply ☐ |
   | Confectionery ☐ | Other: please give details ........................... |
   | Ice cream ☐ | |

5. Are vehicles or ships used for transporting food kept at or used from the premises?  Yes/No
   
   Are vehicles, stalls or ships used for preparing or selling food, kept at or used from the premises?  Yes/No
   
   Number of vehicles/stalls/ships kept at or used from the premises, and used for preparing, selling or transporting food.
   5 or less ☐   6–10 ☐
   11-50 ☐   51 plus ☐

6. Name(s) of proprietor(s) of food business .................................................................
   Address of business head office or registered office ...................................................
   if different from address of premises
   ........................................................  Post code ................................................

7. Name of manager if different from proprietor ............................................................

8. If this is a new business......................
   Date you intend to open

9. If this is a seasonal business ..........................
   Period during which you intend to be open each year

10. Number of people engaged in food business
    Count part-timer(s) (25 hrs per week or less) as one-half
    
    0–10 ☐   11–50 ☐   51 plus ☐ (Please tick one box)

**It is an offence to give false or incomplete information**

*The completed form should be sent to:*

[                    ]

[                    ]

Signature ...............................................................
Date ......................................................................
Name ....................................................................
(BLOCK CAPITALS)
Position in company/business...........................

### Establishments producing 'composite products'

Regulation 853/2004 does not apply to food comprising mixtures of plant origin and processed products of animal origin. However, establishments producing such 'composite products' which are also engaged in activities which would otherwise require approval will need to be approved in relation to those activities. Otherwise, Regulation 852/2004 will apply to the part of the establishment where the composite products are assembled. Only the notification process is required in respect of establishments engaged solely in assembling products originating from approved establishments to create 'composite products'.

### Premises used only occasionally

**F9029**

- Premises used for less than five days (not necessarily consecutive) in five consecutive weeks.

### Premises and areas exempt unless used for retail sales

The following are examples of 'premises', areas and locations which have not themselves been subject to notification/licensing under the food hygiene legislation (although farms, smallholdings etc may be subject to measures under other legislation). In general, however, the products obtained from these become subject to food hygiene controls as soon as they are harvested, slaughtered, caught etc if they are destined for sale to third parties.

**F9030**

- Where game is killed in sport (grouse moors).
- Where fish is taken for food (but not processed).
- Where crops are harvested, cleaned, stored or packed. However, if the crops are put into the final consumer pack they must register.
- Where honey is harvested.
- Where eggs are produced or packed.
- Livestock farms and markets.
- Shellfish harvesting areas.
- Places where no food is kept, for example an administrative office or head quarters of a food business.

### Some domestic premises

**F9031** Some **domestic** premises that might technically be deemed to be food premises are also exempt in the following situations:

- Childminders caring for no more than six children (*SI 1993 No 2022*).
- People preparing foods for sale in WI Country Markets Ltd. (*SI 1997 No 723*).
- Where the resident is not the owner of the food business (for example a mobile food vehicle is sometimes kept at the premises of the driver). If the premises are used for the purpose of for example peeling shrimps or prawns they must be registered.
- Premises used for the production and sale of honey.
- Where crops are produced, cleaned, packed and sold whether wholesale or retail.

- Premises which provide bed and breakfast accommodation in not more than three bedrooms.

## Some vehicles
**F9032**
- Private cars.
- Aircraft.
- Ships unless permanently moored or used for pleasure excursions in inland or coastal waters.
- Food vehicles normally based outside the UK.
- Vehicles or stalls kept at premises that are themselves registered or exempt.
- Market stalls owned by the market controller.
- Tents, marquees and awnings.

## Other exemptions
**F9033**
- Places where the main activity has nothing to do with food but where light refreshments such as biscuits and drinks are served to customers without charge (for example hairdressers' salons).
- Where food is sold only through vending machines.
- Places run by voluntary or charitable organisations and used only by those organisations provided that no food is stored on the premises except tea, coffee, dry biscuits etc. For example some village or church halls, but not if they are used more than five days in five weeks by commercial caterers.
- Crown premises exempted for security reasons.
- Places supplying food and drink in religious ceremonies.
- Stores of food kept for an emergency or national disaster.

# Food hygiene regulations

**F9034** The EU has recently completed a fundamental overhaul of all food hygiene legislation, consolidating and recasting both the general requirements and those relating to animal-derived products into two principal Regulations (Regulation 852/2004 on the Hygiene of Foodstuffs and Regulation 853/2004 laying down Specific Hygiene Rules for Products of Animal Origin). The requirements of these are in the final stages of transposition into UK legislation and will come into effect in January 2006, as the *Food Hygiene Regulations* (*England*) *2005* (*SI 2005 No 2059*). Similar, parallel provisions will be introduced by the devolved administrations. These UK Regulations are wide-ranging and will subsume/revoke numerous existing provisions, including the *Food Safety* (*General Food Hygiene*) *Regulations 1995* (*SI 1995 No 1763*) ('the General Food Hygiene Regulations'), the *Food Safety* (*Temperature Control*) *Regulations 1995* (*SI 1995 No 2200*) ('the Temperature Control Regulations') and the *Food Premises* (*Registration*) *Regulations 1991* (*SI 1991 No 2825*).

Two further EU regulations have also been introduced to define the requirements and responsibilities for official controls by the authorities, Regulation 854/2004 laying down specific rules for

**F9035**  *Food Safety and Standards*

the organisation of official controls on products of animal origin intended for human consumption and Regulation 882/2004 on official controls performed to ensure the verification of compliance with feed and food law, animal health and animal welfare rules.

The following account will concentrate on the *Food Hygiene Regulations 2005* (*SI 2005 No 2059*) that stem from the Regulations 852/2004 and 853/2004.

It will also cover the associated temperature control requirements that are implemented by the *Food Hygiene Regulations 2005* (*SI 2005 No 2059*), *Reg 30 and Sch 4*, which largely re-enact the *Food Safety (Temperature Control) Regulations 1995* (*SI 1995 No 2200*) ('the Temperature Control Regulations'). These requirements apply to all food businesses except those specifically covered by EU Regulation 853/2004 or to food operations carried out on a ship or aircraft.

A detailed account of the hygiene requirements within the vertical Regulation 853/2004 is beyond the scope of this publication.

## Food Hygiene Regulations 2005

**F9035**  The requirements of the *Food Hygiene Regulations 2005* (*SI 2005 No 2059*) are based on the HACCP principles established by the Codex Alimentarius of the World Health Organisation. The significant difference is that the hazard analysis requirement within the 2005 Regulations is less formally structured than Codex and there is no overt requirement for documentation or records. The legislation also recognizes that the application of HACCP principles to primary production is not generally feasible but, nevertheless, encourages their use as far as possible.

For the purposes of the legislation, 'hygiene' is defined as the measures necessary to ensure fitness for human consumption of a foodstuff taking into account its intended use. The concepts of 'fitness' and food safety are now defined in the general food legislation, described above. There is no reference to 'wholesomeness', as was the case under previous legislation.

Since the UK legislation is implementing directly-applicable EU Regulations, it is necessary to refer to these in order to establish the precise requirements. Regulation 852/2004 (Article 4, Part A of Annex I and the 12 chapters of Annex II) sets out the general principles to be followed.

Article 5 requires food business operators to put in place, implement and maintain a permanent procedure or procedures based on the HACCP principles.

### Hazard analysis

**F9036**  EU Regulation 852/2004 specifies seven elements that must be included in the process of identifying steps in the activities of the business that are 'critical' to ensuring that food is safe by identifying and/or establishing:

- hazards that must be prevented, eliminated or reduced to acceptable levels;
- critical control points ('CCPs') at the step(s) at which control is essential to prevent, eliminate or reduce a hazard to acceptable levels;
- critical limits at CCPs to separate acceptability from unacceptability for the prevention, elimination or reduction of identified hazards;
- effective monitoring procedures at CCPs;
- corrective actions that need to be taken when monitoring indicates that a CCP is not under control;
- procedures, to be carried out regularly, to verify that the measures are working effectively;
- documents and records (appropriate to the nature and size of the food business) to demonstrate the effective application of the measures.

HACCP: A powerful and sophisticated management tool that can be used by businesses to develop a food control plan. Full and formal HACCP requires quite detailed expert knowledge and training. A significant amount of literature has built up over the years. It serves well those larger businesses that employ staff with technical training. Smaller food businesses tend to find HACCP literature to be too full of jargon and inaccessible.

Clear guidance has been developed by central government and international agencies for some specific food sectors. Critical control points (CCP) for microbiological food poisoning hazards will tend to focus on a small number of areas.

COOKING: In most processes there is likely to be a risk of pathogens in raw ingredients, so cooking is frequently a CCP. Adequate cooking temperatures will be the control measure, and target temperatures should be established. These can be monitored either by checking the food temperature directly or monitoring the process. For example using a cooking time and temperature that you know will achieve a satisfactory temperature.

CONTAMINATION: Process steps at which ready to eat food may become contaminated are also likely to be CCPs. Contamination could come from contact with raw food ('cross contamination'), or from contact with contaminated equipment or personnel. Separation of raw and ready to eat foods, and cleaning and disinfection routines will be appropriate control measures. In a good hazard analysis system these procedures will be specified objectively to allow them to be monitored. Written cleaning schedules will help. Nowadays there are also rapid test systems that will assess effectiveness of cleaning.

TIME AND TEMPERATURE CONTROLS: Cooling of food after cooking is often a CCP, as are storage times and temperatures of ready to eat foods. In most cases, controls will be expressed as a combination of time and temperature. The critical limits may be just a few hours if the food is kept at warm room temperature, or many days if it is kept in good refrigeration. In this area technology is providing increasingly sophisticated monitoring options include automatic alarms and real-time data monitoring. The temperature control regulations also intervene by prescribing certain targets.

## Documentation and records

F9037   A full, Codex-defined HACCP system would be fully documented. Any monitoring or verification procedures would be recorded and records kept. There is no explicit requirement for either of these in the new legislation. Most businesses would, nevertheless, be advised to keep some concise and succinct documentation. Without it, it may not be easy to demonstrate compliance with the legislation. It will certainly be difficult to muster a convincing due diligence defence if that should become necessary.

## Hygiene 'pre-requisites'

F9038   Codex Alimentarius, the definitive guide to HACCP, says that any HACCP system must be supported by basic hygiene controls. These have become known as the pre-requisites to HACCP. The *Food Hygiene Regulations 2005* (details taken from EU Regulation 852/2004) take a similar approach by establishing certain requirements for structures and services. They cover the following subjects within twelve chapters:

Chapter I: General requirements for food premises.

Chapter II: Specific requirements for rooms where food is prepared.

Chapter III: Movable or temporary premises, etc.

Chapter IV: Transport.

Chapter V: Equipment.

Chapter VI: Food waste.

Chapter VII: Water Supply.

Chapter VIII: Personal hygiene.

Chapter IX: Protection of food from contamination.

Chapter X: Wrapping and packaging

Chapter XI: Heat treatment

Chapter XII: Training

## Chapter I: General requirements for food premises, equipment and facilities

**F9039** Food premises must be kept clean, and in good repair and condition. The layout, design, construction, and size must permit good hygiene practice and be easy to clean and/or disinfect and should protect food against external sources of contamination such as pests, dirt, condensation etc. Cleaning agents and disinfectants are not to be stored in areas where food is handled.

Where necessary, suitable temperature-controlled handling and storage conditions should be available, with sufficient capacity for maintaining foodstuffs at appropriate temperatures and designed to allow those temperatures to be monitored and, where necessary, recorded.

Adequate sanitary and hand-washing facilities must be available and lavatories must not lead directly into food handling rooms. Sanitary conveniences are to have adequate natural or mechanical ventilation. Washbasins must have hot and cold (or better still mixed) running water and materials for cleaning and drying hands. Where necessary there must be separate facilities for washing food and hands. Drainage must be suitable and there must be adequate changing facilities.

Premises must have suitable natural or mechanical ventilation. Ventilation systems must be accessible for cleaning, for example to give easy access to filters, and adequate natural and/or artificial lighting.

## Chapter II: Specific requirements in food rooms

**F9040** Food rooms should have surface finishes that are in good repair, easy to clean and, where necessary, disinfected. This would, for instance, apply to wall, floor and equipment finishes. The rooms should also have adequate facilities for the storage and removal of food waste. Of course, every food premises must be kept clean. How they are cleaned and how often, will vary. For example it will be different for a manufacturer of ready-to-eat meals than for a bakery selling bread.

There must be facilities, including hot and cold water, for cleaning and where necessary disinfecting tools and equipment. There must also be facilities for washing food wherever this is needed.

## Chapter III: Temporary and occasional food businesses and vending machines

**F9041** Most of the requirements apply equally to food businesses trading from temporary or occasional locations like marquees or stalls. For reasons of practicability, some requirements are slightly modified.

## Chapter IV: Transport of food

**F9042** The design of containers and vehicles must allow cleaning and disinfection. Businesses must keep them clean and in good order to prevent contamination and place food in them so as to minimise risk of contamination. Precautions must be taken if containers or vehicles are used for different foods or, in particular, for both food and non-food products. You should also separate different products to protect against the risk of contamination, and clean vehicles or containers effectively between loads. In general, bulk foodstuffs in liquid, granular or powder form should be transported in receptacles and/or containers/tankers reserved for the transport of foodstuffs.

## Chapter V: Equipment

**F9043** All equipment and surfaces that come into contact with food must be well constructed and kept clean. One interpretation of this requirement is that wooden cutting boards are not deemed suitable for use with ready to eat foods. If necessary to ensure the safety of the food, the food room should be equipped with appropriate temperature etc recording systems.

## Chapter VI: Food waste

**F9044** Food and other waste must not accumulate in food rooms any more than is necessary for the proper functioning of the food business. Containers for waste must be kept in good condition and easy to clean and disinfect. Waste storage containers should be lidded to keep out pests. The waste storage area must be designed so that it can be easily cleaned and prevent pests gaining access. There should be arrangements for the frequent removal of refuse and the area should be kept clean.

All waste must be disposed in accordance with relevant legislation, using licensed waste disposal contractors and, in the case of products containing animal-derived ingredients, must also be in accordance with the relevant provisions of Regulation (EC) No 1774/2002 on 'health rules concerning animal by-products not intended for human consumption', implemented by the *Animal By-Products Regulations 2003 (SI 2003 No 1482)*.

## Chapter VII: Water supply

**F9045** There must be an adequate supply of potable (drinking) water. Ice must be made from potable water (there are special rules for ice and water used for cooling/washing fish. Recycled water used in processing or as an ingredient must be of potable standard, unless it can be shown not to affect the wholesomeness of the food.

Water used for cooling cans etc after thermal treatment must not be a source of contamination. Non-potable water may be used for fire control, steam production, refrigeration and other similar purposes, but must be in a separate duly identified system and not connect with potable water systems.

Steam used directly in contact with food is not to contain any substance that presents a hazard to health or is likely to contaminate the food.

## Chapter VIII: Personal hygiene

**F9046** Anyone who works in a food handling area must maintain a high degree of personal cleanliness. The way in which they work must also be clean and hygienic. Food handlers must wear clean and, where appropriate, protective clothes. Anyone whose work involves handling food should:

- follow good personal hygiene practices;
- wash their hands routinely when handling food;

- never smoke in food handling areas;
- be excluded from food handling if carrying an infection that may be transmitted through food if there is a risk of contamination of the food.

In this case, the obligation on the business proprietor to take action is supplemented by an obligation on every employee contained within the *Food Hygiene Regulations 2005 (SI 2005 No 2059)*. Personnel working in food handling areas are obliged to report any illness (like diarrhoea or vomiting, infected wounds, skin infections) immediately to the proprietor of the business. If a manager receives such notification he or she may have to exclude them from food handling areas. Such action should be taken urgently. If there is any doubt about the need to exclude, it is best to seek urgent medical advice or consult the local council EHO.

These personal hygiene provisions are wide ranging. They apply not only to food handlers, but anyone working in food handling areas. There have been well-documented cases of cleaners contaminating working surfaces or equipment with pathogens that are later transmitted to food.

## Chapter IX: Preventing food contamination

**F9047** Food and ingredients must be stored in conditions that prevent harmful deterioration and protected against contamination that may make them unfit for human consumption or a health hazard. For example, raw poultry must not be allowed to contaminate ready-to-eat foods.

Products likely to support the growth of pathogenic micro-organisms or toxin formation must be kept at appropriate temperatures (more details about temperature controls and management are given below at **F9050**). Frozen materials must be thawed in a way that similarly minimises the risk of micro-organisms growth or toxin formation. Any run-off liquid from the thawing process that may present a health risk must be adequately drained.

Adequate procedures to control pests and to prevent access by domestic animals must be in place. Hazardous and/or inedible substances, including animal feed, must be adequately labelled and stored in separate, secure containers.

## Chapter X: Wrapping and packaging

**F9047.1** Materials used for wrapping and packaging must be stored and used in a way which ensures that they are not to be a source of contamination. Where appropriate and, in particular, in the case of cans and glass jars, the integrity of the container's construction and its cleanliness must be checked prior to use.

Wrapping and packaging material re-used for foodstuffs must be easy to clean and, where necessary, disinfect.

## Chapter XI: Heat processing

**F9047.2** Specific requirements apply to foods processed and sold in hermetically-sealed containers to ensure safe processing and to avoid defective containers or re-contamination.

## Chapter XII: Training and supervising food handlers

**F9048** Whereas eleven of the chapters deal with structural and physical pre-requisites, the final chapter deals with the higher-level issue of competence of personnel to do their job. This provision is more subtle than it is often given credit for. The provision, carried forward from previous legislation, is still subject to occasional criticism that it is too lenient on training requirements but, in fact, it requires three complementary elements, instruction, training and supervision.

Food business operators must ensure that food handlers are supervised and instructed and/or trained in food hygiene matters commensurate with their work activity and that those responsible for the development and maintenance of the HACCP-based procedure or for the operation of relevant guides have received adequate training in HACCP principles. In practice, therefore, food handlers must be instructed on how to do their particular jobs properly and supervised to make sure that they follow instructions. The requirements will differ according to the nature of the business and the job of the individual staff member.

Businesses must also comply with any national legal requirements concerning training programmes for persons working in certain food sectors, e.g. the meat industry.

## Industry guides to good hygiene practice

**F9049** EU Regulation 852/2004 retains the concept of voluntary industry guides to good hygiene practice. These provide more detailed guidance on complying with the Regulations as they relate to specific industry sectors. They are usually produced by trade associations and recognised by the Food Standards Agency. Importantly, enforcement officers are obliged to have regard for them when examining how businesses are operating and industry demonstrated compliance is likely to contribute towards a potential defence of 'due diligence' should the need arise. Numerous guides have now been published in the UK. Sectors covered include:

- Vending.
- Catering.
- Retail.
- Baking.
- Wholesale.
- Markets and Fairs.
- Fresh Produce.
- Flour Milling.

## Temperature Control Requirements [formerly Food Safety (Temperature Control) Regulations 1995]

**F9050** Growth of pathogenic bacteria in food will significantly increase the risk of food poisoning. Indeed growth of any micro-organisms in food will compromise its wholesomeness. Temperature controls at certain food holding or processing steps will be CCPs in most HACCP plans. The importance of good control of food temperatures is emphasised by the fact that some controls are prescribed in regulations.

The *Food Safety (Temperature Control) Regulations 1995* (*SI 1995 No 2200*) will be revoked but the *Food Hygiene Regulations 2005* (*SI 2005 No 2059*) *Schedule 4* will effectively re-enact the principle provisions requiring food business proprietors to observe certain temperatures during the holding of food if this is necessary to prevent a risk to health. (In Scotland, the requirements for boiling and holding temperatures for gelatine, prior to use, have been dropped but the requirement to reheat food to 82°C will be retained.) The temperature requirements relate only to matters of food safety. For example, hard cheese that may go mouldy if kept at room temperature is not covered because it would not support the growth of pathogens. The temperature control requirements in the *Food Hygiene Regulations 2005* (*SI 2005 No 2059*) *Schedule 4* do not apply to foods and establishments that fall under EU Regulation 853/2004 which sets more specific rules for most foods of animal origin during processing and handling prior to the retail or catering outlet.

The legislation makes a fairly simple issue extraordinarily complex. Furthermore, Scotland does not have the same regulations as the rest of the UK. (EU Regulation 852/2004 does not set specific temperature controls and, thus, detailed decisions are made nationally.)

The following table contains a very brief summary of the requirements. The Regulations prescribe temperatures for some but not all steps. The Regulations in Scotland also have similar gaps but not always at the same step. The industry guides, particularly the catering guide, fill in the gaps with recommendations of good practice.

|  | Rest of the UK | Scotland | Recommended good practice targets |
|---|---|---|---|
| Chill store | 8°C maximum | Not specified | 5°C |
| Cook | Not specified | Not specified | 70°C for 2 minutes or 75°C minimum |
| Hot hold | 63°C | 63°C | 63°C min |
| Cool | Not specified | Not specified | Below 10°C in 4 hours |
| Reheat | Not specified | 82°C | 70°C for 2 minutes or 75°C minimum |

Frozen storage is not covered by the Food Hygiene Regulations and will not be a food safety CCP. Other legislation and best practice indicate a frozen storage temperature of –18°C for most foods but somewhat lower (–23°C for ice cream).

## *Relationship between time and temperature*

**F9051** Almost invariably the control of micro-organisms in food depends upon a combination of time and temperature. Temperature alone does not have an absolute effect. For example, destruction of micro-organisms can be effected in a very short time (seconds) at temperatures above 100°C or in a couple of minutes at around 70°C. A similar thermal destruction can be achieved even at temperatures as low as 60°C but exposure for around 45 minutes will be necessary. Similarly in chilled storage, food can remain safe and wholesome for many days if the temperature can be kept close to freezing point at around –1°C. However, it will have a much shorter life at higher storage temperatures around 10°C.

Of course one should note that the above remarks are generalisations and not all micro-organisms will react in the same way. Some organisms show greater resistance to heating; some are able to grow at storage temperatures as low as 3°C whereas others will show little growth a temperatures cooler than 15°C. But the general point that control is a function of time and temperature remains true.

Broadly speaking, this relationship is recognised in the Regulations. In some instances, time is actually prescribed in the temperature control regulations. An example is the so called 'four-hour rule' for display of food at a temperature warmer than 8°C. In other cases times are implicit in the temperature control regulations, such as the necessity to cool food rapidly after heating. Actual parameters for compliance are outlined in the industry guides. For other situations there is an inter-relationship with other regulations. For example the storage life of pre-packed, microbiologically perishable food must be displayed under the requirements of the *Food Labelling Regulations 1996*. A food must be labelled with an indication of its 'minimum durability' that takes the form of a 'use by' date and, in the case where the life of the product in the unopened pack has been extended, for example by the use of a modified atmosphere, the safe life after opening must also be displayed. There is an onus on the producer to ascertain a safe shelf life having regard for the nature of the food and its likely storage conditions and to use this as the basis of the date mark.

## Temperature controls – England, Wales and N. Ireland

(N.B. Requirements for Scotland are described in detail below, see F9063)

F9052 The regulations governing temperature controls were largely new provisions introduced and refined in the early 1990s, culminating in the *Food Safety (Temperature Control) Regulations 1995* (*SI 1995 No 2200*), and are continued, largely unchanged, under the new hygiene legislation. The structure of the Regulations in England, Wales and N. Ireland has the main requirement that chilled storage should be at 8°C or cooler, followed by a series of exemptions for particular circumstances. The general requirement for temperature control under *Reg 30, Sch 4* of the *Food Hygiene Regulations 2005* (*SI 2005 No 2059*) comes from the EU Regulation 852/2004, Article 4.3 and it is generally assumed that compliance with the other more specific regulations will deliver compliance. The requirements are outlined in the table below.

| Provision within the Food Hygiene Regulations | Requirement | Comment |
| --- | --- | --- |
| Reg 30, Sch 4 | A general requirement for all food to be kept under temperature control if that is needed to keep it safe. | Applies to raw materials and foods in preparation. Limited periods outside temperature control are permitted for certain practicalities. No temperature is specified. A combination of time and temperature will be important. |
| Sch 4, para 2.1 | Chilled food must be kept at 8°C or cooler. | Applies only to foods that would become unsafe. |
| Sch 4, para 3 | Various cold foods are exempt: shelf stable, canned foods, raw materials, cheeses during ripening, and others where there is no risk to health. | Soft cheeses once ripe, and perishable food from opened cans must be kept below 8°C. |
| Sch 4, para 4 | Manufacturers may recommend higher storage temperature/shorter storage life (provided that safety is verified). | Caterers must use the food within the 'use by' date indicated. |
| Sch 4, para 5.1 | Cold food on display or for service can be warmer than 8°C. A maximum of 4 hours is allowed. | Any item of food can be displayed outside temperature control only once. The burden of proof is on the caterer. |
| Sch 4, para 5.2 | Tolerances outside of temperature control are also allowed during transfer or preparation of food, and defrost or breakdown of equipment. | No time/temperature limits specified. Both should be minimized consistently with food safety. |
| Sch 4, para 6,7 | Hot food should be kept at 63°C or hotter. | Food may be kept at a temperature cooler than 63°C for maximum 2 hours if it is for service or on display. |

|  | Food must be cooled quickly after heating or preparation. | No limits are specified – must be consistent with food safety. |

Note that all temperatures specified are food temperatures not the air temperature of refrigerators, vehicles or hot cabinets.

## Chilled storage and foods included in the scope of the regulations

**F9053** The principal rule for chilled storage is contained within the *Food Hygiene Regulations 2005* (*SI 2005 No 2059*), *Sch 4.2:*

'Subject to sub-paragraph (2) and paragraph 3, any person who keeps any food–

(*a*) which is likely to support the growth of pathogenic micro-organisms or the formation of toxins; and

(*b*) with respect to which any commercial operation is being carried out, at or in food premises at a temperature above 8 degrees C shall be guilty of an offence.'

*Schedule 4.2* only applies to food that will support the growth of pathogenic micro-organisms. Such foods must be kept at 8°C or cooler. The types of food that will be subject to temperature control are indicated in the following table.

It is often good practice to keep foods at temperatures cooler than 8°C either to preserve quality or to allow longer-term storage and to allow a margin of error below the legal standard. Industry guides suggest that one should aim for a target food temperature of 5°C. This is especially important in cabinet fridges. There can be significant temperature rises during frequent door opening.

| *Food type* | *Comment* |
| --- | --- |
| Cooked meats and fish, meat and fish products. | Includes prepared meals, meat pies, pates, potted meats, quiches and similar dishes based on fish. |
| Cooked meats in cans that have been pasteurised rather than fully sterilised. | Typically large catering packs of ham or cured shoulder. |
| Cooked vegetable dishes. | Includes cereals, rice and pulses. Some cooked vegetables or dessert recipes may have sufficiently high sugar content* (possibly combined with other factors like acidity) to prevent the growth of pathogenic bacteria. These will not be subject to mandatory temperature control. |
| Any cooked dish containing egg or cheese. | Includes flans, pastries etc. |
| Prepared salads and dressings. | Includes mayonnaise and prepared salads with mayonnaise or any other style of dressing. Some salads or dressings may have a formulation (especially the level of acidity**) that is adequate to prevent growth of pathogens. |
| Soft cheeses/mould ripened cheeses (after ripening). | Cheeses will include Camembert, Brie, Stilton, Roquefort, Danish Blue and any similar style of cheese. |

| Smoked or cured fish, and raw scombroid fish. | For example smoked salmon, smoked trout, smoked mackerel etc. Also raw tuna, mackerel and other scombroid fish. |
|---|---|
| Any sandwiches whose fillings include any of the foods listed in this Table. | |
| Low acid** desserts and cream products. | Includes dairy desserts, fromage frais and cream cakes. Some artificial cream may be 'ambient stable' due to low water activity and/or high sugar. Any product that does not support the growth of pathogenic micro-organisms does not have to be kept below 8°C. It may be necessary to get clarification from suppliers. |
| Fresh pasta and uncooked or partly cooked pasta and dough products. | Includes unbaked pies and sausage rolls, unbaked pizzas and fresh pasta. |
| Smoked or cured meats which are not ambient stable. | Salami, parma hams and other fermented meats will not be subject to temperature controls if they are ambient shelf stable. |

*Technically Aw (water activity) is the key criterion.

**Technically, pH 4.5 (or more acid) is the critical limit.

## Mail order foods

**F9054** There is a controversial exemption from the specific 8°C requirement if the food is being conveyed by post or by a private or common carrier to the ultimate consumer. The sender still has a responsibility for the safety of the food but the specific 8°C temperature does not apply. This exemption for mail order foods cannot apply to supplies to caterers or retailers neither of whom meet the definition of 'ultimate consumers'. (N.B. there are no specific provisions relating to mail order in Scotland.)

## Short 'shelf life' products

**F9055** Some perishable foods are allowed to be kept at ambient temperatures for the duration of their shelf life with no risk to health. This may be because they are intended to be kept for only very short periods (sandwiches). Bakery products like fresh pies, pasties, custard tarts can also be kept without refrigeration for limited periods.

## Canned foods and similar

**F9056** Sterilised cans or similar packs do not have to be stored under temperature control until the hermetically sealed pack is opened. After that, perishable food must be kept chilled (for example corned beef, beans, canned fish, and dairy products).

However, some canned meats are not fully sterilised (e.g. large catering packs of ham or pork shoulder) and must be kept chilled even before the can is opened.

High acid canned or preserved foods (some fruit, tomatoes, etc.) do not have to be kept chilled after opening for safety reasons. However, it is advisable to remove these from the can for chilled storage after opening. It will inhibit mould growth and avoid any chemical reaction with the metal can body.

## Raw food for further processing

**F9057** Raw food that is intended for further processing does not have to be kept at 8°C or cooler, providing the further processing will render the food fit for consumption. Thus it is not against the Regulations to keep raw meat, poultry and fish out of the fridge. Of course, for quality reasons it is best kept in the fridge. Processed foods must be kept at 8°C or cooler, even if they are to be heated again.

Raw meat or fish that is intended to be eaten without further processing (for example beef for steak tartare, or fish for sushi) will not be exempted by this provision. It must be kept chilled. Raw scombroid fish (tuna, mackerel, etc.) will also not be exempted by this provision. The 'scombrotoxin' is heat stable and processing will not render contaminated food fit for consumption. Scombroid fish must be kept at 8°C or cooler.

## Variations from 8°C storage

**F9058** The *Food Hygiene Regulations 2005* (*SI 2005 No 2059*) *Reg 30, Sch 4.4* allow the producer of a food to recommend that it can be kept safely at a temperature above 8°C. He must label the food clearly with the recommended storage temperature and the safe shelf life at that temperature, or convey the information by other written form. He must also have good scientific evidence that the food is safe at that temperature. Little use has been made of this provision.

## The four-hour rule

**F9059** The 'four-hour rule' allows cold food to be kept above 8°C when it is on display. This is crucial. Without it, food could not be served on a buffet without refrigeration. The time that the food is on display must be controlled. The maximum time allowed is four hours. Most catering operations and many retailers depend upon this exemption for at least some of their operation.

Only one such period of display is allowed no matter how short. For example, if a dish of food is placed on display for one hour at the end of a service period, it cannot be displayed for a further three hours above 8°C at the next service. Food uneaten at the end of a display period does not have to be discarded provided that it is still fit for consumption. The food must be cooled to 8°C or cooler and kept at that temperature until it can be used safely.

The burden of proof is on the business, which must be able to demonstrate that the time limit is observed. Good management of food displays will be important. The amount of food on display must be kept to a minimum consistent with the pattern of trade.

There needs to be systems to help keep to the time limit and to demonstrate that they are being adhered to. These may include the labelling of dishes to indicate when they went onto display. Avoid topping-up of bulk displays of food. Food at the bottom of the dish will remain on display for much longer than four hours if topping-up is allowed.

## Exemptions for other contingencies

**F9060** The Regulations allow for the fact that food may rise above 8°C for limited periods of time in unavoidable circumstances such as:

- transfers to or from vehicles;
- during handling or preparation;
- during defrost of equipment;
- during temporary breakdown of equipment.

The Regulations do not put specific figures on the length of time allowed or how warm the food might become. This tolerance is allowed as a defence and the burden of proof will be on the business to show that:

- the food was unavoidably above 8°C for one of the reasons allowed;
- that it was above 8°C for only a limited period;
- that the break in temperature control was consistent with food safety.

The acceptable limits will obviously depend upon the combination of time and temperature. Under normal circumstances, a single period of up to two hours is unlikely to be questioned.

All transfers of food must be organised so that exposure to warm ambient temperatures is reduced and that rises in food temperature are kept to a minimum. For example, deliveries should be put away quickly and chilled food moved first, then frozen, then grocery.

If food from a cash and carry warehouse is being collected, insulated bags or boxes should be used for any chilled foods to which the Regulations apply. The chilled food should be taken straight back to the outlet and quickly put them into chilled storage.

Equipment breakdown should be avoided by ensuring planned and regular maintenance.

## Hot food

**F9061** Hot food must be kept at 63°C or hotter if this is necessary for safety. This will apply to the same types of food described in the earlier table (see F9053 above). For example, the rule does not apply to hot bread or doughnuts.

Food must be kept at 63°C or hotter, whether:

- it is in the kitchen or bakery awaiting service or dispatch;
- or in transit to a serving point, no matter how near or far;
- or actually on display in the serving area.

Some tolerance or limited exemptions are allowed for practical handling reasons. As similar to the four-hour rule for chilled food, hot food can be kept for service or display at less than 63°C for one period of up to two hours.

Again one must be able to show that:

- the food was for service or on display for sale;
- it had not been kept for more than two hours;
- it had only had one such period.

There is a further general exemption if a well-founded scientific assessment can show that storage and safety at below 63°C does not present any risk to health.

Generally there should be less problems with meeting the 63°C target in hot display equipment than encountered in achieving 8°C in 'chilled' display units. Hot display units are not designed to heat or cook foods from cold. They should be used for holding only not heating.

## Cooling

**F9062** Previous legal requirements for food that becomes warmer than 8°C during processing to be cooled again to 8°C 'as quickly as possible' have not been re-enacted and the necessary practices will therefore fall under 'good practice' and the general requirements to supply safe food. Over the years a wide variety of cooling times have been recommended by different sources. Cooling times as short as one-and-a-half hours are specified in UK Department of Health guidelines on cooking chilled foods and published in the early 1980s. Many EHOs have regarded

**F9063** *Food Safety and Standards*

this figure as definitive. More recent literature from the USA recommends a cooling time of exactly ten times that long. The UK industry guides recommend that cooling between 60°C and 10°C should be accomplished in a maximum of four hours. Recent UK research suggests that this may be conservative.

Note that rapid cooling is not only necessary for food that has been heated. Food that has become warm during processing should also be returned to chilled storage below 8°C as quickly as possible after the process is completed.

## *Temperature controls – Scotland*

**F9063** The Scottish Regulations have remained largely unchanged for many years. Some would say that they are now anachronistic. No chilled storage temperature is specified, yet the 82°C requirement for re-heating pre-cooked food appears unnecessarily high and unsupported by the science (although there is an exemption from this if it can be shown to be impossible to achieve without a deterioration in the quality in the product). However, the current requirements will continue largely unchanged under the 2005 regulations, the only change being the deletion of requirements to boil gelatin or to hold it above 71°C for 30 minutes prior to use.

| Provision within the Food Hygiene Regulations | Requirement | Comment |
| --- | --- | --- |
| Schedule 4 | A general requirement for all food to be kept under temperature control if that is needed to keep it safe. Additionally it includes the need for rapid cooling. | Applies to raw materials and foods in preparation. Limited periods outside temperature control are permitted for certain practicalities. No temperature is specified. A combination of time and temperature will be important. |
| Sch 4, para 2.1(a) | Cold food must be kept in a cool place or refrigerator. | No temperature is specified. |
| Sch 4, para 2.1(b) | Hot food must be kept at 63°C or hotter. | There are exemptions for food on display, during preparation, etc. |
| Sch 4, para 3.1 | Food that is reheated must reach 82°C. | Exemption in place if 82°C is detrimental to food quality. |
| | | Unused glaze must be chilled quickly and kept in a refrigerator. |

Again, note that all temperatures specified are food temperatures not the air temperature of refrigerators, vehicles or hot cabinets.

## Cool storage

**F9064** The *Food Hygiene Regulations 2005*, (*SI 2005 No 2059*) *Reg 30, Sch 4* will apply to any food that may support the growth of food poisoning organisms within the 'shelf life' for which the food will be kept.

The types of food that will be subject to the provision are the same as those that fall under *Reg 30, Sch 4* of the Regulations (applicable to England, Wales and N. Ireland, described above, see F9052–F9062).

If cold, this food must be kept either in a refrigerator or a cool well-ventilated place. The Scottish requirements do not specify a temperature but in practice, it is advisable to follow the same rules and recommendations that apply in the rest of the UK. In any case, the hazard analysis requirement of the General Food Hygiene Regulations does apply in Scotland, and thus appropriate controls at CCPs will need to be established.

### Hot food

**F9065** The target for hot food is the same as England, Wales and N. Ireland, 63°C or hotter.

### Cooling

**F9066** The Scottish requirements (*Sch 4 para 2(c)*) demand food to be cooled under hygienic conditions as quickly as possible to a temperature which would not result in a risk to health. This requirement is considered to have the same effect as *Sch 4, para 4* in the legislation applicable to England, Wales and N. Ireland and, in both cases, to fall under the general requirement for safe processing and handling of food.

### Exemptions during practical handling

**F9067** In certain circumstances food does not have to be kept cold or above 63°C:

- if it is undergoing preparation for sale;
- if it is exposed for sale or it has already been sold;
- if it is being cooled;
- if it is available to consumers for sale;
- if it is shelf stable.

Again these Regulations have remained unchanged for many years and are less detailed than those that apply in the rest of the UK.

### Reheating of food

**F9068** In Scotland only, food that has been heated and is being reheated must be raised to a temperature of 82°C or hotter. Fortunately this is qualified by a clause to say that this temperature is not necessary if it is detrimental to the quality of the food. This should provide the opportunity for most businesses to reheat food to the perfectly adequate and scientifically verified temperature of around 70–72°C.

### Gelatine

**F9069** The previous rules requiring gelatine to be boiled or held at 71°C or hotter for at least 30 minutes have not been retained.

# Food Safety Manual – food safety policy and procedures

**F9070** Both the *Food Safety Act 1990* and the Regulations made under it provide an obligation on proprietors of food businesses to operate within the law. In addition, the *Food Safety Act 1990, s 21* provides the defence of 'due diligence'. One way for a business to demonstrate its commitment to operating safely and within the law is to document its management system, including the relevant policies and procedures. This documentation may form a significant part of a due diligence defence. The policy and procedures will illustrate the precautions that should be taken by the business and its staff in their day-to-day operations. The procedures will normally require the keeping of various records, and these will in turn help to demonstrate that the *'reasonable precautions'* are being followed *'with all due diligence'*.

It is often convenient to keep these policies and procedures together in a 'Food Safety Manual'. The following provides a brief outline of the contents of a Food Safety Manual for a typical food business.

## Policy and procedures

F9071 The 'policy' will be a statement of intent by the business. A Food Safety Manual may begin with a fairly broad statement of policy. Later, there may be more specific policy statements linked to particular outcomes, for example a policy to take all reasonable precautions to minimise the contamination of food with foreign material.

Procedures will detail the ways in which the business intends to put the policies into effect. A typical layout of any 'Quality Manual' is to begin each section or subject heading with a statement of policy, and to follow that with the procedures intended to deliver it. Procedures are often referred to by the acronym SOPs – 'Standard Operating Procedures'.

## General policy and organisation

F9072 A Food Safety Manual would normally begin with a top-level statement of the business's policy with regard to food safety. Recognising the wider scope of 'food safety' within the meaning of the 1990 Act, it may be advisable to add a statement about commitment to true and accurate labelling of food and adherence to compositional requirements.

The statement should be signed by the proprietor of the business, who is the person with ultimate legal responsibility. In a large group this may be the group chief executive.

The early part of the manual should also illustrate the organisational structures and management hierarchy that has responsibility for implementing the policy and procedures. There may be technical support services, (e.g. pest control, testing laboratories) either from inside or outside the business. Their roles and responsibilities can also be identified within the organisational structures in this part of the manual.

## Hazard Analysis (HACCP)

F9073 Today, any food safety policy must be centred on a hazard analysis approach. The Food Safety Manual should include a policy statement to that effect. This will be followed by procedures to describe how the business will go about the hazard analysis and record its outcomes and implementation.

### Pre-requisites to Hazard Analysis

F9074 A modern approach to HACCP will focus on a very small number of process steps that are truly critical to the safety of the food. These are the steps that *must* be controlled (and monitored) to ensure that the food is safe. But this control must operate against a background of good hygiene practice. The so-called 'pre-requisites' to HACCP must all be in place. The Food Safety Manual should include policies and procedures for all of these.

- Design – structures and layout

    The Food Safety Manual should include a policy that the design, structure and layout will promote hygienic operation. The procedures should detail specific arrangements to achieve that. This section should also cover points such as the services, especially water and ventilation; provision for washing of hands, equipment and food; and removal of waste.

- Equipment specification

    The policy should be that the business will only acquire and use equipment that is fit for its purpose. The focus should extend beyond just cleanability alone. In many cases, the ability of equipment to do the job can be even more important. For example, can the refrigeration equipment keep food at a specified temperature in operational conditions?

- Maintenance

    A policy for maintenance should address both the premises in general and the equipment. Procedures should detail the arrangements for routine and emergency maintenance. There should be a record of any maintenance whether routine or emergency.

- Cleaning and disinfection

    Detailed cleaning schedules for sections of the building or particular pieces of equipment should be available. Cleaning should be recorded.

- Pest Control

    Employing a pest control contractor is only one small part of the pest control procedures. The prevention of pest access, the integrity of pest proofing, cleaning and the need for vigilance should also be included in the procedures. Records would normally include any sightings of or evidence of pests, positions of bait stations, visits by the contractor together with observations and action taken, and even details such as the changing of UV tubes on electronic flying insect killers.

- Personal hygiene

    The business should have a very clear policy on personal hygiene. The arrangements will include a very clear statement of the dress code and personal hygiene requirements for all staff working in or passing through food areas. There must also be very particular policies and procedures to deal with staff suffering illness that may potentially be food borne.

- Food purchasing

    It may be easy to state a policy that the business should purchase only from reputable suppliers. It may be more difficult to deliver such a policy, especially for smaller businesses that do not have the resources to inspect their suppliers. It may be possible to rely on third party inspection systems that have been developed during the late 1990s; especially those operated by accredited inspection or certification bodies.

    Supply specifications may also be drafted. Such specifications will usually cover food 'quality' in its broadest sense. They are the ideal route through which to state food safety conditions such as delivery temperature, shelf life remaining after delivery, packaging and so on.

- Storage

    Most businesses will have documented procedures for the storage of various categories of food as well as for stock checks to ensure proper storage temperature and stock rotation.

- Food handling and preparation

    The general policy should be to manage all steps in food handling and preparation in order to minimise contamination with any hazardous material or pathogenic micro-organisms. There should also be precautions to minimise the opportunities for growth of any micro-organisms that might get into the food. The procedures will be unique to the individual business, its particular range of foods and preparation methods.

- Foreign material

    The policy should require that all precautions are taken to minimise the risk of foreign material contamination, including chemical hazards such as cleaning materials. In general terms, procedures will cover three points:

    (*a*) Excluding or controlling potential sources of contamination;

    (*b*) Protecting food from contamination;

    (*c*) Taking any complaints seriously, tracking down what caused them and stopping them from happening again.

## Allergy (Anaphylaxis)

**F9075** A responsible business will have a policy to protect susceptible customers from exposure to foods to which they may be allergic. In general terms procedures will rely upon:

(*a*) Awareness amongst staff of the types of food that may cause reaction.

(*b*) Labelling of the foods that include such ingredients.

(*c*) Careful use of these ingredients in the kitchen to avoid 'cross-contamination' to other dishes.

(*d*) An ability to supply reliable information about all ingredients on request from guests.

There is also new legislation requiring the labelling of foods, the *Food Labelling (Amendment) (England) (No. 2) Regulations (SI 2004 No. 2824)*, implementing EU Directive 2003/89/EC, which requires the known presence of any of a defined list of serious allergens to be declared on the label of pre-packed foods. The detailed requirements of this legislation is outside the scope of this overview of food safety legislation.

## Vegetarians and other dietary preferences

**F9076** Similarly, most food businesses today recognise the need to cater for customers with particular dietary preferences, notably vegetarians. There should be a policy to be able to supply vegetarian dishes properly formulated and accurately labelled.

## Consumer concerns

**F9077** From time to time particular issues will come to the forefront of consumer consciousness. For example, the irradiation of food, genetically modified ingredients or animal welfare issues such as veal production, foie gras and intensive rearing of poultry. The Food Safety Manual could include a statement of the policy on whichever of these issues that one thinks might be relevant. Procedures will be needed to deliver them. In the case of food and food ingredients derived from genetically modified organisms, comprehensive legislation is now in place to require the traceability and labelling of all such materials throughout the food supply chain, based on EU Regulations 1929/2003 and 1830/2003.

## Labelling and composition of foods and 'fair-trading'

**F9078** In view of the fact that the *Food Safety Act 1990* deals not only with health and hygiene, and in the light of growing interest in these areas, it may be considered desirable to include policies and procedures on fair-trade topics such as food composition and labelling.

## Product recall

**F9079** Businesses involved in central production and distribution of food that will have several days of 'shelf life' (if not more) should have contingency plans for product recall by ensuring that products are labelled in such a way that individual batches can be identified and records kept to show where/when it might have been delivered.

## Complaints

**F9080** Complaints are an important source of information. A business should have a policy and procedures to deal with them. These will normally include keeping a record of every complaint and making every attempt to track down its cause.

## Training, instruction and supervision

**F9081** Cutting across all of the other policies and procedures will be a policy on staff training, instruction and supervision. Typically a business will keep a personal training record for all members of staff.

## *Summary*

**F9082** Modern businesses recognise the value of a properly documented management system. A Food Safety Manual can provide the focus of such a system. The design of a Food Safety Manual must be a fine balance between conflicting objectives. On the one hand, it must be reasonably comprehensive if it is to stand up as part of a 'due diligence' defence. On the other hand, it must be sufficiently concise to be usable as an effective working tool within the business.

Copyright forms are reproduced by permission of the publishers, Shaw & Sons Ltd, Shaway House, Crayford, Kent DA1 4BZ, from whom copies may be purchased.

# Gas Safety

## Introduction

**G1001** Not surprisingly, explosions caused by escaping gas have been responsible for large-scale damage to property and considerable personal injury, including death. Although escape of metered gas or gas in bulk holders might possibly have attracted civil liability under the rule in *Rylands v Fletcher* (see further F5054 FIRE AND FIRE PRECAUTIONS), statutory authority to perform a public utility (i.e. supply gas) used to constitute a defence in such proceedings, coupled with the fact that negligence had to be proved in order to establish liability for personal injury (*Read v J Lyons & Co Ltd [1947] AC 156*). This situation has now changed in favour of imposition of strict product liability for injury/damage caused by escape of gas and incomplete or inefficient combustion causing carbon monoxide poisoning.

According to research undertaken by the Consumer's Association, over half of the appliances surveyed in dwellings had not been serviced in accordance with the requirements of the Gas Safety Regulations. As some well-publicised court cases have demonstrated, landlords and those they employ to deal with gas installations face heavy fines and possibly imprisonment for breach of the regulations (see G1026 BELOW).

Changes have also been made to the gas industry itself, stemming from the Government's policy of creating competition in the gas supply field. These changes have resulted in the introduction of new and revised legislation, namely:

- *Gas Act 1986*

    This has been considerably modified and extended by the *Gas Act 1995*.

- *Gas Safety (Installation and Use) Regulations 1998 (SI 1998 No 2451)*

    These Regulations were made under the authority of the *Health and Safety at Work etc Act 1974*.

- *Gas Safety (Management) Regulations 1996 (SI 1996 No 551)*

    These Regulations were made under the authority of the *Health and Safety at Work etc Act 1974*.

- *Pipelines Safety Regulations 1996 (SI 1996 No 825)*

    These Regulations were made under the authority of the *Health and Safety at Work etc Act 1974*.

- *Gas Safety (Rights of Entry) Regulations 1996 (SI 1996 No 2535)*

    These Regulations were made under the authority of the *Gas Act 1986* (as amended).

In practice, many of the legislative changes were introduced for the purposes of increasing responsibilities for safety of gas processors, gas transporters and gas suppliers and are therefore outside the scope of this handbook. This chapter concentrates on the legislation which deals with the knowledge required by landlords, managing agents, health and safety managers, employers and other responsible persons who must ensure that gas appliances and fittings are installed safely and checked by a competent person every twelve months.

It should be noted that although the legislation discussed in this chapter is current at the date of publication, the Health and Safety Executive (HSE) is at present considering changes that could affect the *Gas Act 1986* (as amended) and some of the regulations referred to in this section. The proposed changes are outlined in a HSE document entitled *'Fundamental Review of Gas Safety Regime Proposals for Change'*, which is the result of consultation with interested parties throughout the gas industry.

Gas is defined in the *Gas Act 1986* (as amended) and the *Gas Safety (Installation and Use) Regulations 1998* (*SI 1998 No 2451*) to include methane, ethane, propane, butane, hydrogen and carbon monoxide mixtures of any two or more of these gases together with inert gases or other non-flammable gases; or combustible mixtures of one or more of these gases and air. However, the definition does not include gas consisting wholly or mainly of hydrogen when used in non-domestic premises.

## Gas supply – the Gas Act 1986, as amended

**G1002**   The *Gas Act 1986* (as amended) requires the Secretary of State for the Environment to establish a Gas Consumers' Council and appoint an officer – the Director General of Gas Supply – to perform the functions relating to the supply of gas as set out in *Part I* of the Act.

As far as safety in the supply of gas is concerned, the Secretary of State and the Director each have a principal duty to exercise the functions assigned to them under Part I for the protection of the public from dangers arising from the conveyance or from the use of gas conveyed through pipes.

The following persons are defined in the *Gas Act 1986* as having duties and responsibilities in respect of the safe supply of gas:

(a) Domestic customer

A domestic customer is a person who is supplied by a gas supplier with gas conveyed to particular premises at a rate which is reasonably expected not to exceed 2,500 therms a year.

(b) Owner

In relation to any premises or other property, 'owner' includes a lessee, and cognate expressions are construed accordingly.

(c) Relevant authority

— in relation to dangers arising from the conveyance of gas by a public gas transporter, or from the use of gas conveyed by such a transporter, 'relevant authority' means that transporter; and

— in relation to dangers arising from the conveyance of gas by a person other than a public gas transporter, or from the use of gas conveyed by such a person, 'relevant authority' means the Secretary of State for the Environment.

With regard to safety, the *Gas Act 1986* (as amended) provides for:

— licensing and general duties [*Gas Act 1986, ss 4A, 7–10* and *23*];

— security [*Gas Act 1986, s 11*];

— safety regulations regarding rights of entry for inspection of connected systems and equipment and the making safe and investigation of gas escapes [*Gas Act 1986, ss 18* and *18A*];

— pipeline capacity [*Gas Act 1986, ss 21, 22, 22A*];

— standards of performance [*Gas Act 1986, s 33* as amended by the *Competition and Service (Utilities) Act 1992*].

## Gas supply management – the Gas Safety (Management) Regulations 1996 (SI 1996 No 551)

G1003  Principally, the *Gas Safety (Management) Regulations 1996 (SI 1996 No 551)* require an appointed person – the 'network emergency co-ordinator' – to prepare a safety case for submission to and acceptance by HSE before a gas supplier can convey gas in a specified gas network [*SI 1996 No 551, Reg 3*].

### The safety case

G1004  The safety case document must contain:

— the name and address of the person preparing the safety case (the duty holder);
— a description of the operation intended to be undertaken by the duty holder;
— a general description of the plant, premises and interconnecting pipes;
— technical specifications;
— operation and maintenance procedures;
— a statement of the significant findings of the risk assessment carried out pursuant to the *Management of Health and Safety at Work Regulations 1999 (SI 1999 No 3242), Reg 3*;
— particulars to demonstrate the adequacy of the duty holder's management system to ensure the health and safety of his employees and of others (in respect of matters within his control);
— particulars to demonstrate adequacy in the dissemination of safety information;
— particulars to demonstrate the adequacy of audit and reporting arrangements;
— particulars to demonstrate the adequacy of arrangements for compliance with the duty of co-operation [*Gas Safety (Management) Regulations 1996 (SI 1996 No 551), Reg 6*];
— particulars to demonstrate the adequacy of arrangements for dealing with gas escapes and investigation [*Gas Safety (Management) Regulations 1996 (SI 1996 No 551), Reg 7*];
— particulars to demonstrate compliance with the content and characteristics of the gas to be conveyed in the network [*Gas Safety (Management) Regulations 1996 (SI 1996 No 551), Reg 8*];
— particulars to demonstrate the adequacy of the arrangements to minimise the risk of supply emergency;
— particulars of the emergency procedures [*Gas Safety (Management) Regulations 1996 (SI 1996 No 551), Reg 3(1)* and *Sch 1*].

### Duties of compliance and co-operation

G1005  The duty holder must ensure that the procedures and arrangements described in the safety case and any revision of it are followed. In addition, a duty of co-operation is placed upon specified persons including:

— a person conveying gas in the network;
— an emergency service provider;
— the network emergency co-ordinator in relation to a person conveying gas;

- a person conveying gas in pipes which are not part of a network;
- the holder of a licence issued under the *Gas Act 1986, s 7*;
- the person in control of a gas production or processing facility [*Gas Safety (Management) Regulations 1996 (SI 1996 No 551), Regs 5 and 6*].

The *Gas Safety (Management) Regulations 1996 (SI 1996 No 551), Reg 7* deals with the duties and responsibilities of gas suppliers, gas transporters and 'responsible persons' regarding actions to be taken in the event of gas escape incidents. All incidents must be investigated including those resulting in an accumulation of carbon monoxide gas from incomplete combustion of a gas fitting.

Anyone discovering or suspecting a gas leak must notify British Gas plc immediately by telephone. British Gas are obliged to provide a continuously manned telephone service in Great Britain. The person appointed 'responsible person' for the premises must take all reasonable steps to shut off the gas supply.

The reporting of gas incidents to the HSE is laid down in the *Reporting of Injuries, Diseases and Dangerous Occurrences Regulations 1995 (SI 1995 No 3163) (RIDDOR)* in conjunction with HSE Form F2508G.

For further information regarding compliance with these Regulations see the HSE publication L80: '*A guide to the Gas Safety (Management) Regulations 1996*' (ISBN 0 7176 1159 0).

# Rights of entry – the Gas Safety (Rights of Entry) Regulations 1996 (SI 1996 No 2535)

**G1006** Where an officer authorised by a public gas transporter has reasonable cause to suspect an escape of gas into or from premises supplied with gas, he is empowered to enter the premises and to take any steps necessary to avert danger to life or property. [*Gas Safety (Rights of Entry) Regulations (SI 1996 No 2535), Reg 4*].

## Inspection, testing and disconnection

**G1007** On production of an authenticated document, an authorised person must be allowed to enter premises in which there is a service pipe connected to a gas main for the purpose of inspecting any gas fitting, flue or means of ventilation. Fittings and any part of a gas system may be disconnected and sealed off when it is necessary to do so for the purpose of averting danger to life or property.

It is incumbent on the authorised officer carrying out any disconnection or sealing off activities to give a written statement to the consumer within five days. The notice must contain:

- the nature of the defect;
- the nature of the danger in question;
- the grounds and the manner regarding the appeal procedure available to the consumer.

Prominent and conspicuous notices must also be affixed at appropriate points of the gas system regarding the consequences of any unauthorised reconnection to the gas supply. [*Gas Safety (Rights of Entry) Regulations (SI 1996 No 2535), Regs 5–8*].

## Prohibition of reconnection

**G1008** It is an offence for any person, except with the consent of the relevant authority, to reconnect any fitting or part of a gas system. This provision is qualified by the term 'knows or has reason to believe that it has been so disconnected'. [*Gas Safety (Rights of Entry) Regulations (SI 1996 No 2535), Reg 9*].

## Pipelines – the Pipelines Safety Regulations 1996 (SI 1996 No 825)

**G1009** The *Pipelines Safety Regulations 1996* (*SI 1996 No 825*) apply to all pipelines in Great Britain, both on and offshore, with the following exceptions:

— pipelines wholly within premises;
— pipelines contained wholly within caravan sites;
— pipelines used as part of a railway infrastructure;
— pipelines which convey water.

For the purposes of the Regulations, a pipeline for supplying gas to premises is deemed not to include anything downstream of an emergency control valve, that is, a valve for shutting off the supply of gas in an emergency, being a valve intended for use by a consumer of gas. [*SI 1996 No 825, Reg 3(4)*].

The Regulations complement the *Gas Safety* (*Management*) *Regulations 1996* (*SI 1996 No 551*) and include:

— the definition of a pipeline [*SI 1996 No 825, Reg 3*];
— the general duties for all pipelines [*SI 1996 No 825, Regs 5–14*];
— the need for co-operation among pipeline operators [*SI 1996 No 825, Reg 17*];
— arrangements to prevent damage to pipelines [*SI 1996 No 825, Reg 16*];
— the description of a dangerous fluid [*SI 1996 No 825, Reg 18*];
— notification requirements [*SI 1996 No 825, Regs 20–22*];
— the major accident prevention document [*SI 1996 No 825, Reg 23*];
— the arrangements for emergency plans and procedures [*SI 1996 No 825, Regs 24–26*].

Generally, the Regulations place emphasis on 'major accidents' and the preparation of emergency plans by local authorities. As with the *Gas Safety* (*Management*) *Regulations 1996* (*SI 1996 No 551*), the detail relates to the specification and characteristics of gas and the design of the pipes to convey it. The design of gas service pipelines is specifically addressed in the HSE Approved Code of Practice, L81: '*Design, construction and installation of gas service pipes*' (1996) (ISBN 0 7176 1172 8). For further information on and guidance to the *Pipelines Safety Regulations 1996*, see the HSE Publication L82: '*A guide to the Pipelines Safety Regulations 1996*' (ISBN 0 7176 1182 5).

## Gas systems and appliances – the Gas Safety (Installation and Use) Regulations 1998 (SI 1998 No 2451)

**G1010** The *Gas Safety* (*Installation and Use*) *Regulations 1998* (*SI 1998 No 2451*) are supported by the HSE Approved Code of Practice and Guide, L56: '*The Gas Safety* (*Installation and Use*) *Regulations 1998*'.

The Regulations aim at protecting the gas-consuming public and cover natural gas, liquefied petroleum gas (LPG), landfill gas, coke, oven gas, and methane from coal mines when these products are 'used' by means of a gas appliance. Such appliances must be designed for use by a gas-consumer for heating, lighting and cooking. However, a gas appliance does not include a portable or mobile appliance supplied with gas from a cylinder except when such an appliance is under the control of an employer or self-employed person at a place of work. Similarly, the Regulations do not cover gas appliances in domestic premises where a tenant is entitled to remove such appliances from the premises.

In addition to the general provisions governing the safe installation of gas appliances and associated equipment by HSE approved and competent persons, the 1998 Regulations lay down specific duties for landlords.

Except in the case of 'escape of gas' (*SI 1998 No 2451, Reg 37*), and certain types of valves to control pressure fluctuations, the Regulations do not apply to the supply of gas when used in connection with:

— bunsen burners in an educational establishment [*SI 1998 No 2451, Reg 2*];

— mines or quarries [see *Mines and Quarries Act 1954*];

— factories [see *Factories Act 1961* (*FA61*), *s 175*] or electrical stations [*FA61, s 123*], institutions [*FA61, s 124*], docks [*FA61, s 125*] or ships [*FA61, s 126*];

— agricultural premises;

— temporary installations used in connection with any construction work within the meaning of the *Construction (Design and Management) Regulations 1994* (*SI 1994 No 3140*);

— premises used for the testing of gas fittings; or

— premises used for the treatment of sewage.

*Note*: The Regulations would apply in relation to the above premises (or parts of those premises) if used for domestic or residential purposes or as sleeping accommodation, but not to hired touring caravans. [*SI 1998 No 2451, Reg 2*].

Generally, gas must be used in order to come within the scope of the Regulations. Therefore, the venting of waste gas from coal mines and landfill sites is excepted unless the gas is collected and intended for use. Gas used as motive power or from grain drying is also not covered by the Regulations. Service pipes (that is, the pipes for distributing gas to premises from a distribution main and the outlet of the first emergency control downstream of the main) remain the property of the gas supplier or gas transporter. At least 24 hours' notice should be given to the relevant organisation before any work can be performed on a service pipe. [*L56, paras 2–11*].

'Work' in relation to a gas fitting is defined in the Regulations as including any of the following activities carried out by any person, whether an employee or not, that is to say:

(*a*) installing or reconnecting the fitting;

(*b*) maintaining, servicing, permanently adjusting, disconnecting, repairing, altering or renewing the fitting or purging it of air or gas;

(*c*) where the fitting is not readily movable, changing its position; and

(*d*) removing the fitting.

The Approved Code of Practice, L56: '*The Gas Safety (Installation and Use) Regulations 1998*', provides further guidance on the meaning of work and states that 'work' for the purposes of these Regulations also includes do-it-yourself activities, work undertaken for friends and work for which there is no expectation of reward or gain, such as charitable work. [*L56, para 12*].

## *Duties and responsibilities*

**G1011** Duties are placed on a wide range of persons associated with domestic premises and on persons concerned with the supply of gas to those premises and to industrial premises which have accommodation facilities.

Most of the legal requirements are 'absolute', i.e. they are not qualified by 'as far as is practicable' or 'so far as is reasonably practicable'; in all cases the duty must be satisfied to avoid committing an offence. Persons affected by the Regulations include:

— *individuals.* These include householders and other members of the general public;

— *responsible person.* This is the occupier of the premises or, where there is no occupier or the occupier is away, the owner of the premises or any person with authority for the time being to take appropriate action in relation to any gas fittings;

— *landlord.* In England and Wales the landlord is defined as follows:

   (i) where the relevant premises are occupied under a lease, the person for the time being entitled to the reversion expectant on that lease or who, apart from any statutory tenancy, would be entitled to possession of the premises; and

   (ii) where the relevant premises are occupied under a licence, the licensor, save where the licensor is himself a tenant in respect of those premises.

   In Scotland, the landlord is the person for the time being entitled to the landlord's interest under a lease;

— *tenant.* In England and Wales the tenant is defined as follows:

   (i) where the relevant premises are so occupied under a lease, the person for the time being entitled to the term of that lease; and

   (ii) where the relevant premises are so occupied under a licence, the licensee.

   In Scotland, the tenant is the person for the time being entitled to the tenant's interest under a lease.

Duties are imposed on employers and the self-employed to ensure that all persons carrying out work in relation to gas fittings are competent to do so and are members of HSE approved organisations. [*Gas Safety (Installation and Use) Regulations 1998 (SI 1998 No 2451), Reg 3*].

Persons connected with work activities include:

— *supplier.* In relation to gas, a 'supplier' means:

   (i) a person who supplies gas to any premises through a primary meter; or

   (ii) a person who provides a supply of gas to a consumer by means of the filling or refilling of a storage container designed to be filled with gas at the place where it is connected for use, whether or not such container is or remains the property of the supplier; or

   (iii) a person who provides gas in refillable cylinders for use by a consumer whether or not such cylinders are filled, or refilled, directly by that person and whether or not such cylinders are or remain the property of that person.

   *Note:* A retailer is not a supplier when he sells a brand of gas other than his own.

— *transporter.* This is defined as meaning a person, other than a supplier, who conveys gas through a distribution main.

— *gas installer.* This is any person who installs, services, maintains, removes or repairs gas fittings whether he is an employee, self-employed or working on his own behalf (for example, a do-it-yourself activity).

## *Escape of gas*

**G1012** It is the duty of the responsible person for the premises to take immediate action to shut off the gas supply to the affected appliance or fitting and notify the gas supplier (or the nominated gas emergency call-out office if different from the supplier). All reasonable steps must be taken to prevent further escapes of gas, and the gas supplier must stop the leak within twelve hours from the time of notification. This may entail cutting off the gas supply to the premises. [*Gas Safety (Installation and Use) Regulations 1998* (*SI 1998 No 2451*), *Reg 37(1)–(3)*].

An escape of gas also includes an emission of carbon monoxide resulting from incomplete combustion in a gas fitting. However, the legal duties regarding the action to be taken by the gas supplier are limited to making safe and advising of the need for immediate action by a competent person to examine, and if necessary carry out repairs, to the faulty fitting or appliance. [*Gas Safety (Installation and Use) Regulations 1998* (*SI 1998 No 2451*), *Reg 37(8)*].

## *Competent persons and quality control*

**G1013** No work is permitted to be carried out on a gas fitting or a gas storage vessel except by a competent person. [*Gas Safety (Installation and Use) Regulations 1998* (*SI 1998 No 2451*), *Reg 3(1)*].

Where work to a gas fitting is to any extent under their control – or is to be carried out at any place of work under their control – employers (and self-employed persons) must ensure that the person undertaking such work is registered with an HSE-approved body such as the Council of Registered Gas Installers (CORGI). [*Gas Safety (Installation and Use) Regulations 1998* (*SI 1998 No 2451*), *Reg 4*].

A gas fitting must not be installed unless every part of it is of good construction and sound material, of adequate strength and size to secure safety and of a type appropriate for the gas with which it is to be used. [*Gas Safety (Installation and Use) Regulations 1998* (*SI 1998 No 2451*), *Reg 5(1)*].

## *Competent persons – qualifications and supervision*

**G1014** To achieve competence in safe installation, a person's training must include knowledge of purging, commissioning, testing, servicing, maintenance, repair, disconnection, modification and dismantling of gas systems, fittings and appliances. In addition, a sound knowledge of combustion and its technology is essential, including:

— properties of fuel gases,

— combustion,

— flame characteristics,

— control and measurement of fuel gases,

— gas pressure and flow,

— construction and operation of burners, and

— operation of flues and ventilation.

To reach the approval standard required by *Gas Safety (Installation and Use) Regulations 1998* (*SI 1998 No 2451*), *Reg 3*, gas installers and gas fitters should know:

(*a*) where/how gas pipes/fittings (including valves, meters, governors and gas appliances) should be safely installed;

(*b*) how to site/install a gas system safely, with reference to safe ventilation and flues;

(*c*) associated electrical work (e.g. appropriate electrical power supply circuits, that is, overcurrent and shock protection from electrical circuits, earthing and bonding);

(*d*) electrical controls appropriate to the system being installed/maintained/repaired;

(*e*) when/how to check the whole system adequately before it is commissioned;

(*f*)   how to commission the system, leaving it safe for use.

In addition, they should know how to recognise and test for conditions that might cause danger and what remedial action to take, as well as being able to show consumers how to use any equipment they have installed or modified, including how to shut off the gas supply in an emergency. They should also alert customers to the significance of inadequate ventilation and gas leaks and the need for regular maintenance/servicing.

To meet the criteria for HSE approval, individual gas fitting operatives must be assessed (or reassessed) by a certification body accredited by the United Kingdom Accreditation Service. The scheme requires every registered gas fitter or gas installer to possess a certificate, which consumers can ask to see.

## Materials and workmanship

**G1015**   It is incumbent on gas installers to acquaint themselves with the appropriate standards about gas fittings and to ensure that the fittings they use meet those standards. Most gas appliances are subject to the *Gas Appliances (Safety) Regulations 1992 (SI 1992 No 711)* and therefore should carry the CE mark or an appropriate European/British Standard.

It is an offence to carry out any work in relation to a gas fitting or gas storage vessel other than in accordance with the appropriate standards and in such a way as to prevent danger to any person. [*Gas Safety (Installation and Use) Regulations 1998 (SI 1998 No 2451), Reg 5(3)*]. Gas pipes and pipe fittings installed in a building must be metallic or of a type constructed in an encased metallic sheath and installed, so far as is reasonably practicable, to prevent the escape of gas into the building if the pipe should fail. Pipes or pipe fittings made from lead or lead alloy must not be used. [*Gas Safety (Installation and Use) Regulations 1998 (SI 1998 No 2451), Reg 5(2)*].

## General safety precautions

**G1016**   A general duty is imposed on all persons in connection with work associated with gas fittings to prevent a release of gas unless steps are taken which ensure the safety of any person [*SI 1998 No 2451, Reg 6(1)*].

The *Gas Safety (Installation and Use) Regulations 1998 (SI 1998 No 2451), Regs 6(2)–(6)* provide that the following precautions must be observed when carrying out work activities:

— a gas fitting must not be left unattended unless every complete gasway has been sealed with an appropriate fitting;
— a disconnected gas fitting must be sealed at every outlet;
— smoking or the use of any source of ignition is prohibited near exposed gasways;
— it is prohibited to use any source of ignition when searching for escapes of gas;
— any work in relation to a gas fitting which might affect the tightness of the installation must be tested immediately for gas tightness.

With regard to gas storage vessels it is an offence for any person intentionally or recklessly to interfere with a vessel, or otherwise do anything which might affect it so that the subsequent use of that vessel might cause a danger to any person. [*SI 1998 No 2451, Reg 6(9)*]. In addition:

— gas storage vessels must only be installed where they can be used, filled or refilled without causing danger to any person [*SI 1998 No 2451, Reg 6(7)*];
— gas storage vessels, or appliances fuelled by LPG which have an automatic ignition device or a pilot light, must not be installed in cellars or basements [*SI 1998 No 2451, Reg 6(8)*];
— methane gas must not be stored in domestic premises [*SI 1998 No 2451, Reg 6(10)*].

## Gas appliances

**G1017** Precautions to be observed regarding gas appliances apply to everyone, not just gas installers. Generally, it is an offence for the occupier, owner or other responsible person to permit a gas appliance to be used if at any time he knows, or has reason to suspect, that the appliance is unsafe or that it cannot be used without constituting a danger to any person. [*SI 1998 No 2451, Reg 34*]. It is also an offence for anyone to carry out work in relation to a gas appliance which indicates that it no longer complies with approved safety standards. [*SI 1998 No 2451, Reg 26(7), (8)*]. The *Gas Safety (Installation and Use) Regulations 1998* (*SI 1998 No 2451*), Reg 27 imposes similar safety requirements on persons who install or connect gas appliances to flues (see G1019 BELOW).

All gas appliances must be installed in a manner which permits ready access for operation, inspection and maintenance. [*SI 1998 No 2451, Reg 28*]. Manufacturer's instructions regarding the appliance must be left with the owner or occupier of the premises after its installation. [*SI 1998 No 2451, Reg 29*]. If the appliance is of the type where it is designed to operate in a suspended position, the installation pipework and other associated fittings must be constructed and installed as to be able to support the weight of the appliance safely. [*SI 1998 No 2451, Reg 31*].

During installation it is essential to ensure that all gas appliances are:

— connected (in the case of a flued domestic gas appliance) to a gas supply system by a permanently fixed rigid pipe [*SI 1998 No 2451, Reg 26(2)*];

— installed with a means of shutting off the supply of gas to the appliance unless it is not reasonably practicable to do so [*SI 1998 No 2451, Reg 26(6)*].

With the exception of the direct disconnection of the gas supply from a gas appliance, or the purging of gas or air which does not adversely affect the safety of that appliance, all commissioning operations and any other work performed on a gas appliance must be immediately examined to ensure the effectiveness of flues, and checks must be made with regard to:

— the supply of combustion air;

— the operating pressure or heat input; and

— the operation of the appliance to ensure its safe functioning.

[*SI 1998 No 2451, Reg 26(9)*].

Any defects must be rectified and reported to the appropriate responsible person as soon as is practicable; the defect must be reported to the appropriate gas supply organisation if a responsible person or the owner is not available. [*SI 1998 No 2451, Reg 26(9)*].

## Room-sealed appliances

**G1018** A room-sealed appliance is an appliance whose combustion system is sealed from the room in which the appliance is located and which obtains air for combustion from a ventilated uninhabited space within the premises or directly from the open air outside the premises. The products of combustion must be vented safely to open air outside the premises.

A room-sealed appliance must be used in the following rooms:

— a bathroom or a shower room [*SI 1998 No 2451, Reg 30(1)*];

— in respect of a gas fire, gas space heater or gas water heater (including instantaneous water heaters) of more than 14 kilowatt gross heat input, in a room used or intended to be used as sleeping accommodation [*SI 1998 No 2451, Reg 30(2)*].

Gas heating appliances which have a gross heat input rating of less than 14 kilowatt or an instantaneous water heater may incorporate an alternative arrangement, i.e. an approved safety control designed to shut down the appliance before a build-up of dangerous combustion products can occur [*Reg 30(3)*].

For the purposes of the *Gas Safety (Installation and Use) Regulations 1998 (SI 1998 No 2451)*, *Reg 30(1)–(3)*, a room also includes:

— a cupboard or compartment within such a room; or

— a cupboard, compartment or space adjacent to such a room if there is an air vent from the cupboard, compartment or space into such a room [*SI 1998 No 2451, Reg 30(4)*].

## Flues and dampers

**G1019** A flue is a passage for conveying the products of combustion from a gas appliance to the external atmosphere and includes the internal ducts of the appliance. Generally, it is prohibited to install a flue other than in a safe position, and, in the case of a power-operated flue, it must prevent the operation of the appliance should the draught fail [*Gas Safety (Installation and Use) Regulations 1998 (SI 1998 No 2451), Reg 27(4), (5)*].

A flue must be suitable and in a proper condition for the operation of the appliance to which it is fitted. It is prohibited to install a flue pipe so that it enters a brick or masonry chimney in such a manner that the seal cannot be inspected. Similarly, an appliance must not be connected to a flue surrounded by an enclosure, unless the enclosure is sealed to prevent spillage into any room or internal space other than where the appliance is installed. [*Gas Safety (Installation and Use) Regulations 1998 (SI 1998 No 2451), SI 1998 No 2451, Reg 27(1)–(3)*].

Manually operated flue dampers must not be fitted to serve a domestic gas appliance, and, similarly, it is prohibited to install a domestic gas appliance to a flue which incorporates a manually operated damper unless the damper is permanently fixed in the open position.

In the case of automatic dampers, the damper must be interlocked with the gas supply so that the appliance cannot be operated unless the damper is open. A check must be carried out immediately after installation to verify that the appliance and the damper can be operated safely together and without danger to any person.

## Gas fittings

**G1020** Gas fittings are those parts of apparatus and appliances designed for domestic consumers of gas for heating, lighting, cooking or other approved purposes for which gas can be used (but not for the purpose of an industrial process occurring on industrial premises), namely:

— pipework;

— valves; and

— regulators, meters and associated fittings.

[*Gas Safety (Installation and Use) Regulations 1998 (SI 1998 No 2451), SI 1998 No 2451, Reg 2*].

## Emergency controls

**G1021** An emergency control is a valve for use by a consumer of gas for shutting off the supply of gas in an emergency. Emergency controls must be appropriately positioned with adequate access, and there must be a prominent notice or other indicator showing whether the control is open or shut. A notice must also be posted either next or near the emergency control indicating the procedure to be followed in the event of an escape of gas. [*Gas Safety (Installation and Use) Regulations 1998 (SI 1998 No 2451), SI 1998 No 2451, Reg 9*].

## Meters and regulators

**G1022** The *Gas Safety (Installation and Use) Regulations 1998 (SI 1998 No 2451)*, Reg 12 provides that gas meters must be installed where they are readily accessible for inspection and maintenance. A meter must not be so placed as to adversely affect a means of escape or where there is a risk of damage to it from electrical apparatus. After installation, meters and other other associated fittings should be tested for gastightness and then purged so as to remove safely all air and gas other than the gas to be supplied. [*SI 1998 No 2451, Reg 12(6)*].

Meters must be of sound construction so that, in the event of fire, gas cannot escape from them and, where a meter is housed in an outdoor meter box, the box must be so designed as to prevent gas entering the premises or any cavity wall, i.e. any escaping gas must disperse to the air outside. Combustible materials must not be kept inside meter boxes. Where a meter is housed in a box or compound that includes a lock, a suitably labelled key must be provided to the consumer. [*SI 1998 No 2451, Reg 13*].

The *Gas Safety (Installation and Use) Regulations 1998 (SI 1998 No 2451)*, Regs 14–17 prescribe detailed precautions to be observed during the installation of the various types of meters. Most of these are for compliance by gas installers and should be included in work procedures. In essence, a meter must not be installed in service pipework unless:

(a) there is a regulator to control the gas pressure [*SI 1998 No 2451, Reg 14(1)(a),(b)*];

(b) a relief valve or seal is fitted which is capable of venting safely [*SI 1998 No 2451, Reg 14(1)(c)*]; and

(c) the meter contains a prominent notice (in permanent form) specifying the procedure in the event of an escape of gas [*SI 1998 No 2451, Reg 15*].

Similar requirements are imposed regarding gas supplies from storage vessels and re-fillable cylinders. [*SI 1998 No 2451, Reg 14(2)–(4)*].

Where gas is supplied from a primary meter to a secondary meter, a line diagram must be provided and prominently displayed showing the configuration of all meters, installation pipework and emergency controls. It is incumbent on any person who changes the configuration to amend the diagram accordingly. [*SI 1998 No 2451, Reg 17*].

## Installation pipework

**G1023** For the purposes of this chapter, 'installation pipework' covers any pipework for conveying gas to the premises from a distribution main, including pipework which connects meters or emergency control valves to a gas appliance, and any shut-off devices at the inlet to the appliance. The term 'service pipe' is used to describe any pipe connecting the distribution main with the outlet of the first emergency control downstream from the distribution main. Similarly, 'service pipework' are those pipes which supply gas from a gas storage vessel. All service pipes normally remain the property of the gas supplier or transporter and their permission must be obtained before any work associated with such pipes can be performed.

Installation pipework must not be installed:

— where it cannot safely be used, having regard to other pipes, pipe supports, drains, sewers, cables, conduits and electrical apparatus;

— in or through any floor or wall, or under any building, unless adequate protection is provided against failure caused by the movement of these structures;

— in any shaft, duct or void, unless adequately ventilated;

— in a way which would impair the structure of a building or impair the fire resistance of any part of its structure;

— where deposit of liquid or solid matter is likely to occur, unless a suitable vessel for the reception and removal of the deposit is provided. (It should be noted that such clogging precautions do not normally need to be taken in respect of natural gas or LPG as these are 'dry' gases.)

[*Gas Safety (Installation and Use) Regulations 1998* (*SI 1998 No 2451*), *Regs 18–21*].

Following work on installation pipework, the pipes must immediately be tested for gastightness and, if necessary, a protective coating applied; followed, where gas is being supplied, by satisfactory purging to remove all unnecessary air and gas. Except in the case of domestic premises, the parts of installation pipework which are accessible to inspection must be permanently marked as being gas pipes. [*Gas Safety (Installation and Use) Regulations 1998* (*SI 1998 No 2451*), *Regs 22, 23*].

## Testing and maintenance requirements

**G1024** According to the *Gas Safety (Installation and Use) Regulations 1998* (*SI 1998 No 2451*), *Reg 33(1)*, a gas appliance must be tested when gas is supplied to premises to verify that it is gastight and to ensure that:

(a) the appliance has been installed in accordance with the requirements of the 1998 Regulations;

(b) the operating pressure is as recommended by the manufacturer;

(c) the appliance has been installed with due regard to any manufacturer's instructions accompanying the appliance; and

(d) all gas safety controls are in proper working order.

If adjustments are necessary in order to comply with (a) to (d) above, but cannot be carried out, the appliance must be disconnected or sealed off with an appropriate fitting. [*SI 1998 No 2451, Reg 33(2)*].

A general duty is imposed on employers and self-employed persons to ensure that any gas appliance, installation pipework or flue in places of work under their control is maintained in a safe condition so as to prevent risk of injury to any person. [*SI 1998 No 2451, Reg 35*].

A similar duty is imposed on landlords of premises occupied under a lease or a licence to ensure that relevant gas fittings and associated flues are maintained in a safe condition. [*SI 1998 No 2451, Reg 36(2)*]. A relevant gas fitting includes any gas appliance (other than an appliance which the tenant is entitled to remove from the premises) or installation (not service) pipework installed in such premises.

It is the responsibility of landlords to ensure that each gas appliance and flue is checked by a competent person (HSE approved person) at least every 12 months. [*SI 1998 No 2451, Reg 36(3)*]. Such safety checks shall include, but not be limited to, the following:

(a) the effectiveness of any flue;

(b) the supply of combustion air;

(c) the operating pressure and/or heat input of the appliance;

(d) the operation of the appliance to ensure its safe functioning.

Nothing done or agreed to be done by a tenant can be considered as discharging the duties of a landlord in respect of maintenance except in so far as it relates to access to the appliance or flue for the purposes of carrying out such maintenance or checking activities. [*SI 1998 No 2451, Reg 36(10)*].

**G1025** *Gas Safety*

In addition, it is the duty of landlords to ensure that appliances and relevant fittings are not installed in any room occupied as sleeping accommodation such as to cause a contravention of *Reg 30(2)* or *(3)* (see G1018 above).

### Inspection and maintenance records

**G1025** The *Gas Safety (Installation and Use) Regulations 1998 (SI 1998 No 2451)* mark a significant increase in the obligations now placed upon landlords.

A landlord is under a duty to retain a record of each inspection of any gas appliance or flue for a period of two years from the date of the inspection. The record must include:

— the date on which the appliance or flue is checked;
— the address of the premises at which the appliance or flue is installed;
— the name and address of the landlord of the premises at which the appliance or flue is installed;
— a description of, and the location of, each appliance and flue that has been checked together with details of defects and any remedial action taken.

The person who carries out the safety check must enter on the record his name and his particulars of registration. Then he should sign the record confirming that he has examined the following:

— the effectiveness of any flue;
— the supply of combustion air;
— the operating pressure and/or heat input of the appliance;
— the operation of the appliance to ensure its safe functioning.

[*SI 1998 No 2451, Reg 36(3)*].

Within 28 days of the date of the check, the landlord must ensure that a copy of the record is given to each tenant of the premises to which the record relates. [*SI 1998 No 2451, Reg 36(6)(a)*]. In addition, before any new tenant moves in, the landlord must provide any such tenant with a copy of the record – however, where the tenant's right to occupation is for a period not exceeding 28 days, a copy of the record may instead be prominently displayed within the premises. [*SI 1998 No 2451, Reg 36(6)(b)*]. A copy of the inspection record given to a tenant under *Reg 36(6)(b)* need not include a copy of the signature of the person who carried out the inspection, provided that it includes a statement that the tenant is entitled to have another copy, containing a copy of such signature, on request to the landlord at an address specified in the statement. Where the tenant makes such a request, the landlord must provide the tenant with such a copy of the record as soon as is practicable. [*SI 1998 No 2451, Reg 36(8)*].

The *Gas Safety (Installation and Use) Regulations 1998 (SI 1998 No 2451), Reg 36(7)* provides that where any room occupied, or about to be occupied, by a tenant does not contain any gas appliance, the landlord may – instead of giving the tenant a copy of the inspection record in accordance with *Reg 36(6)* (see above) – prominently display a copy of the record within the premises, together with a statement endorsed upon it that the tenant is entitled to have his own copy of the record on request to the landlord at an address specified in the statement. Where the tenant makes such a request, the landlord must provide the tenant with a copy of the record as soon as is practicable.

## Interface with other legislation

**G1026** Although wide-ranging in their scope the *Gas Safety (Installation and Use) Regulations 1998 (SI 1998 No 2451)* are not concerned directly with product safety. Certain premises, gas fittings and uses are also subject to exceptions. However, in many of these instances where exceptions are made similar requirements for gas safety are to be found in the *Health and Safety at Work etc Act 1974* and supporting legislation. The most important of these are:

- *Gas Appliances (Safety) Regulations 1995 (SI 1995 No 1629)*

    These Regulations were made under the authority of the *Consumer Protection Act 1987*. They implement European directives that require gas appliances and fittings to conform with essential safety requirements and to be safe when used normally.

- *Management of Health and Safety at Work Regulations 1999 (SI 1999 No 3242)* (see also RISK ASSESSMENT)

    These Regulations were made under the *Health and Safety at Work etc Act 1974*. They require employers and the self-employed to undertake a suitable and sufficient assessment of the risks to the health and safety of their employees and others who may be affected by the work that they do. These Regulations further require employers (including those who install or supply gas or gas appliances or who undertake gas maintenance work) to ensure controls are in place and kept under review.

- *Provision and Use of Work Equipment Regulations 1998 (SI 1998 No 2306)* (see also MACHINERY SAFETY)

    These Regulations were made under the authority of the *Health and Safety at Work etc Act 1974*. They impose health and safety requirements on work equipment including any gas appliance, apparatus, fitting or tool used or provided for use at work. Equipment must be suitable for its intended use, be maintained and inspected and users provided with appropriate health and safety information and where necessary written instructions.

- *Pressure Systems Safety Regulations 2000 (SI 2000 No 128)*

    These Regulations were made under the authority of the *Health and Safety at Work etc Act 1974*. Their objective is the prevention of serious injury from the failure of a pressure system or its components, including those that contain gas in excess of 0.5 bar above atmospheric pressure and pipelines used to convey gas at 2 bar above atmospheric pressure. Under these Regulations duties are placed on those who design, manufacture, use and maintain gas systems overlapping with similar duties under the *Gas Safety (Installation and Use) Regulations 1998 (SI 1998 No 2451)*.

- *Building Regulations 1991 (SI 1991 No 2768)* (see also WORKPLACES – HEALTH, SAFETY AND WELFARE)

    These Regulations (as amended in 1994) were made under the authority of the *Building Act 1984*. They are supported by non-mandatory approved documents that provide practical guidance on for example the safe installation of gas appliances used for heating – Approved Document J is concerned with air supply to gas heaters and discharge of the products of combustion to the open air.

## Summaries of relevant court cases

G1027   The importance of the provisions relating to the correct installation and regular maintenance aspects of gas appliances, even in the earlier Gas Safety Regulations, is demonstrated in the following summaries of recent court cases.

### Unsafe gas fittings

G1028   A Nottingham landlord narrowly escaped a charge of manslaughter when he was tried and convicted of carrying out unsafe work practices which resulted in the deaths of two people.

The court heard of makeshift repairs to a gas boiler which then leaked carbon monoxide into an adjacent flat. The gas supplier (British Gas) gave evidence of having previously made the boiler inoperable because of its unsafe condition but it was reconnected by the landlord without any authority to do so, in December 1992. The landlord was charged under the *Gas Safety (Installation and Use) Regulations 1984* and the *Health and Safety at Work etc Act 1974, s 3(2)*. He was accused of failing in his duty as a self-employed person to conduct his undertaking so as not to put himself or others at risk and of failing to carry out work in relation to a gas fitting in a proper and workmanlike manner.

The judge made it clear that if the legislation had carried a penalty of imprisonment, he would have imposed it. The seriousness of the case could only be reflected by a heavy fine and he sentenced the landlord with a fine of £32,000 with £24,800 costs.

[*Leicester Crown Court, June 1996*].

*Note:* The *Gas Safety (Installation and Use) Regulations 1994* introduced a wider scope of control and increased the penalties for contravention to an unlimited fine and a maximum of two years' imprisonment.

## Local authority (as landlord) guilty of bad gas safety management

G1029    A local authority found itself in court for nine breaches of the *Gas Safety (Installation and Use) Regulations 1994* (failing to carry out safety checks and to keep records) and for failing to ensure the safety of tenants, contrary to the *Health and Safety at Work etc Act 1974, s 3(1)*. The case was brought after the discovery of more than 150 flues which were left disconnected after refurbishment work on a London council estate; and on other estates, annual safety checks had not been carried out.

The magistrate commented that there had been serious risk to hundreds of householders and imposed fines on the council of £27,000 for the breaches of the *Gas Safety (Installation and Use) Regulations 1994* and £17,000 for the offences which contravened the *Health and Safety at Work etc Act 1974*.

[*Clerkenwell Magistrates' Court, January 1997*].

*Note:* The HSE pointed out that local authorities were the largest landlords in the country and that failure to manage in this case had put many tenants at risk of carbon monoxide poisoning and fire and explosion.

## Gas explosion results in three separate prosecutions

G1030    A major gas explosion at a twenty two-storey tower block resulted in the prosecution of a borough council, an installation company and a self employed electrician. The explosion blew off the roof of a boiler house, only minor injuries were sustained but there was considerable property damage from flying debris. There had been serious risk to hundreds of householders.

An investigation revealed that new gas burners and gas pressure booster pumps had been inadequately installed and not commissioned correctly.

The borough council was fined £75,000.00 for failing to manage the project.

The installation company was fined £10,200.00 for responsibility for the inadequate installation and management of the project.

The electrician was fined £3,000.00 for failing to install and commission the boilers correctly, not being competent, and not registered with CORGI.

[*Middlesex Guildhall Crown Court, September 1999*].

### Student death – landlord and gas fitter imprisoned

**G1031**   A landlord and an unregistered gas fitter were charged and convicted of manslaughter following the death of a student who inhaled carbon monoxide fumes from an inappropriate type of gas boiler which was not correctly ventilated.

The investigation into the incident by the HSE originated with the *Gas Safety (Installation and Use) Regulations 1994* which required gas fitters to be registered with the Council of Registered Gas Installers (CORGI). The charge of manslaughter was brought because of the seriousness of the offence: it had resulted in the death of one student and the hospitalisation of four other tenants who had been affected by poisonous fumes.

Both the landlord and the gas fitter pleaded guilty to manslaughter: the landlord was judged to be most responsible and was imprisoned for two years. The gas fitter was imprisoned for fifteen months.

[*Stafford Crown Court, December 1997*].

*Note:* Broadly, the law of manslaughter is part of the law of homicide. It is less serious than murder with the sentence at the discretion of the court. For health and safety purposes, manslaughter may be tentatively defined as killing without the intention to kill or to cause serious bodily harm. This case was the second health and safety manslaughter charge to result in imprisonment.

### Gas fitter jailed – chimney flue capped with concrete caused death

**G1032**   This case involved an unregistered gas fitter who failed to notice that the flue into which he had connected a gas fire was capped with a concrete slab.

The fitter, the landlord and the landlord's son were charged with manslaughter following the death of a tenant in a bedsit. The prosecuting counsel told the court that the landlord of the bedsit, and his son who was looking after the property on behalf of his father, were also responsible for the death. He gave evidence of negligence in 20 other properties where failure to maintain gas appliances properly could have had similar tragic consequences.

In the case in question the judge said that the landlord and his son had been lulled into a false sense of security by a gas safety certificate provided by the fitter. He imposed the following sentences:

Gas fitter: 12 months' imprisonment, six of which were suspended;

Landlord and his son: 9 months' imprisonment suspended for 18 months – the landlord's son was also ordered to pay £16,000 costs.

[*Norwich Crown Court, April 1998*].

### Unsafe gas appliance – landlord guilty

**G1033**   A landlord was charged with:

— failing to maintain gas appliances in a safe condition, contrary to the *Gas Safety (Installation and Use) Regulations 1994, Reg 35A*; and with

— providing false information in connection with his failure to comply with an improvement notice, contrary to the *Health and Safety at Work etc Act 1974, ss 20* and *21*.

Two tenants who suffered dizziness and headaches informed the gas emergency company which then inspected the gas boiler and declared it unsafe. The boiler was switched off and a label was attached, warning that it should not be used.

G1034    *Gas Safety*

The landlord instructed an unregistered gas fitter to repair the boiler; he failed to check for carbon monoxide leaks after switching the boiler back on. Two days later the two tenants collapsed after inhaling poisonous fumes and were hospitalised.

The HSE found that both the boiler and a gas cooker in the same property were leaking dangerously high levels of carbon monoxide gas. The landlord gave false information about the identity of the fitter who had carried out the work and switched the boiler back on.

The Crown Court judge said that the landlord was very lucky not to be facing manslaughter charges and imposed fines and costs totalling £24,000.

[*Cardiff Crown Court, May 1998*].

## Gas safety requirements in factories

G1034    Where part of plant contains explosive/flammable gas under pressure greater than atmospheric, that part must not be opened unless:

(a)    before the fastening of any joint of any pipe, connected with that part of the plant, any flow of gas has been stopped by a stop-valve;

(b)    before such fastening is removed, all practicable steps have been taken to reduce the gas pressure in the pipe or part of the plant, to atmospheric pressure; and

(c)    if such fastening is loosened or removed, inflammable gas is prevented from entering the pipe/part of the plant, until the fastening has been secured or securely replaced.

No hot work must be permitted on any plant or vessel which has contained gas until all practicable steps have been taken to remove the gas and any fumes arising from it, or to render it non-explosive or non-flammable. Similarly, and if any plant or vessel has been heated, no gas shall be allowed to enter it until the metal has cooled sufficiently to prevent any risk of ignition.

[*Factories Act 1961, s 31(3), (4)*].

## Pressure fluctuations

G1035    The *Gas Safety (Installation and Use) Regulations 1998 (SI 1998 No 2451), Reg 38*, provides that where gas is used in plant which is liable to produce pressure fluctuations in the gas supply, such as may cause danger to other consumers, the person responsible for the plant must ensure that any directions given to him by the gas transporter to prevent such danger are complied with.

If it is intended to use compressed air or any gaseous substance in connection with the consumption of gas, at least 14 days' written notice must be given to the gas transporter.

Any device fitted to prevent pressure fluctuations or to prevent the admission of a gaseous substance into the gas supply must be adequately maintained.

# Harassment in the Workplace

## Introduction

H1701 It is a common misconception that there exists in English law a particular cause of action for unlawful harassment in the workplace (whether on grounds of sex, race or disability). However, the law relating to harassment is but a specific example of the general law relating to discrimination in the workplace and, as such, the main pieces of legislation dealing with the subject of harassment are the *Sex Discrimination Act 1975* (*SDA 1975*), the *Race Relations Act 1976* (*RRA 1976*) and the *Disability Discrimination Act 1995* (*DDA 1995*) (together, 'the Discrimination Acts'). This said, although harassment is a particular example of discrimination, the area is of such significant importance that in the context of sex discrimination, the European Commission has been moved to publish a Recommendation (O.J.1992, L49/1) and annexed Code of Practice on the protection of the dignity of men and women at work (see H1703).

## History

H1702 The first significant decision on the issue of whether harassment amounted to discrimination for the purposes of the *SDA 1975* was *Porcelli v Strathclyde Regional Council [1986] IRLR 134*. Here, the applicant was a female laboratory assistant working at a school alongside two male colleagues. The male assistants did not like the employee and they subjected her to a campaign of harassment in the form of brushing against her in the workplace and making sexually suggestive remarks to her and about her in order to make her leave her job. The campaign was a 'success' in that it did indeed force the applicant to apply for a transfer to another school. Having done so, the applicant then brought a claim against her employer claiming that the actions of her former colleagues amounted to direct sex discrimination for the purposes of the *SDA 1975*, s *1(1)(a)*, in that she had been subjected to a detriment. The employer tried to defend the claim by calling in evidence the male employees responsible for the course of harassment. During their evidence they ventured to suggest that the conduct that had been meted out to the applicant would also have been displayed towards a hypothetical male comparator employee that they did not like and, consequently, the employer could not have committed sex discrimination against the applicant since the treatment was gender neutral.

The Scottish Court of Session refused to accept the defence. Lord Emslie stated of sexual harassment that it is:

'... a particularly degrading and unacceptable form of treatment which it must be taken to have been the intention of Parliament to restrain'.

The court likened the use of sexual harassment as a 'sexual sword' which had been 'unsheathed and used because the victim was a woman'. It added that, although an equally disliked male employee may well have been subjected to a campaign of harassment, the harassment would not have had as its cutting edge the gender of the victim. The court then went on to give general guidance as to what harassment is. It stated that sexual harassment amounted to:

'... [u]nwelcome acts which involve physical contact which, if proved, would also amount to offences at common law, such as assault or indecent assault; and also conduct falling short of such physical acts which can be fairly described as sexual harassment'.

The law has moved on since the *Porcelli* decision, most notably in the re-statement by the Employment Appeal Tribunal (EAT) in the case of *(1) Reed (2) Bull Information Systems Ltd v Stedman [1999] IRLR 299* (see below at H1710).

As will be seen in the course of this chapter, various defences to claims of harassment have been developed, and, as stated above, the European Commission was moved to set down the Recommendation and Code of Practice following a report commissioned by it from Michael Rubenstein in 1988.

## The European Commission Recommendation and Code of Practice

**H1703** As stated at H1701 above, the European Commission was prompted into drafting a Recommendation and Code of Practice dealing with the issue of sexual harassment following a report undertaken for it in 1988 (*The Dignity of Women at work: A report on the problem of sexual harassment in the Member States of the European Communities, 1988*). The Recommendation was made on 27 November 1991 and the Code was published in the Official Journal of the European Communities, 4 February 1992.

### Content of the Recommendation

**H1704** Art 1 of the Recommendation requires Member States to:

'... take action to promote awareness that conduct of a sexual nature, or other conduct based on sex affecting the dignity of women and men at work, including conduct of superiors and colleagues, is unacceptable if:

(a) such conducted is unwanted, unreasonable and offensive to the recipient;

(b) a person's rejection of or submission to, such conduct on the part of employers or workers (including superiors or colleagues) is used explicitly or implicitly as a basis for a decision which affects that person's access to vocational training, access to employment, continued employment, promotion, salary or any other employment decisions; and/or

(c) such conduct creates an intimidating, hostile or humiliating working environment for the recipient,

and that such conduct may, in certain circumstances, be contrary to the principle of equal treatment within the meaning of Articles 3, 4 and 5 of Directive 76/207/EEC [the Equal Treatment Directive]'.

The Recommendation is designed to draw to the attention of employers the nature of sexual harassment and to give broad outlines as to the ways in which harassment occurs together with providing an indication of the effect of such conduct (i.e. to highlight that it may be actionable as a form of discrimination).

### Content of the Code of Practice

**H1705** The Code of Practice goes on to provide clear guidance to employers as to:

(a) the nature of sexual harassment including the effect of harassment on victims;

(b) a definition of harassment. This amounts to: '... unwanted conduct of a sexual nature, or other conduct based on sex affecting the dignity of women and men at work. This can include unwelcome physical, verbal or non-verbal conduct';

(c) the legal implications for employers; and

(d) the steps employers should take to prevent the spread of harassment in the workplace.

The Code goes beyond the position as it currently stands at English law. For example, it expressly states that sexual harassment is capable of occurring to homosexuals who are harassed in the workplace on account of their sexual orientation. Yet, the Court of Appeal has stated that harassment of a person on account of their sexual orientation is not *per se* discriminatory within the meaning of the *SDA 1975* (per *Smith v Gardner Merchant Limited [1998] IRLR 520*, below at H1709).

Although the EAT in Scotland held that harassment on grounds of sexual orientation can amount to discrimination (*MacDonald v Ministry of Defence [2000] IRLR 748*), this decision was reversed by the Court of Session on appeal (*[2001] IRLR 431*).

There is no denying that employers should familiarise themselves with the contents of the Code. It amounts to a sensible extension to good human resources practice and, even though it may go beyond the requirements of national sex discrimination law, it should be remembered that harassment which is not currently actionable in terms of sex discrimination may, in any event, amount to conduct which could allow employees to claim that their contracts of employment have been repudiated by their employer thereby allowing them to claim unfair dismissal.

### Legal effect of the Recommendation and Code of Practice

**H1706** The Recommendation and Code are not law in this country and it still remains to be seen what the Government will do to adopt the Code on a formal basis. As regards employers, they should be aware that the Code is readily accepted by employment tribunals as evidence of good practice.

## Definition

### The Discrimination Acts

**H1707** The starting point in connection with harassment is the Discrimination Acts. Harassment on grounds of sex or race is direct discrimination, i.e. it is treatment meted out to the recipient simply because that person is a woman or a man or because of the race of a person. Disability discrimination is discrimination that is meted out to a person on account of that person's disability and which cannot be objectively justified. It seems unlikely that there could be any justification for harassment meted out to a person on the grounds of their disability.

The provisions of the Discrimination Acts dealing with harassment are contained within the *SDA 1975*, s 6(2)(b), the *RRA 1976*, s 4(2)(b), and the *DDA 1995*, which provide that it is unlawful to discriminate against a person employed at an establishment in Great Britain on grounds of sex, race or disability (as the case may be).

Effectively, for a cause of action to arise in relation to harassment, a three stage test has to be satisfied, this being that:

(a) less favourable treatment has been shown to an applicant;

(b) the proximate cause of the differential treatment is the applicant's gender, race or disability ('prohibited grounds'); and

(c) the differential treatment has subjected the victim to a detriment.

### Less favourable treatment

**H1708** In dealing with the question of whether or not sexual harassment amounted to less favourable treatment, the Court of Session in *Porcelli v Strathclyde Regional Council [1986] IRLR 134* stated that:

'In a case of alleged sexual harassment, the correct primary question is not "was there sexual harassment?", a phrase which is not found in the statute, but "was the applicant less favourably treated on the ground of her sex than a man would have been treated?" If that question is answered in the affirmative, there is discrimination within the meaning of the Sex Discrimination Act'.

The Discrimination Acts do, however, have their limitations. In the first instance, it is entirely conceivable that situations could arise where although an employer treats an employee in a manner that seems to amount to a *prima facie* case of harassment, it may do so in circumstances that would not amount to less favourable treatment. For example, for the purposes of the *SDA 1975*, there can be circumstances arising from conduct that could equally affront a male employee. A good illustration is the case of *Stewart v Cleveland Guest (Engineering) Ltd [1994] IRLR 440*. Here, the applicant was a female engineer who frequently had to make visits to the factory floor in order to carry out her duties. She was upset at the display of posters of nude women that adorned the factory floor area and complained to the employer's management. She asked for the posters to be removed. Her requests were ignored and the employee resigned and claimed that she had been constructively dismissed. She alleged that she had been subjected to less favourable treatment on the grounds of her sex which had ultimately led to her constructive dismissal. An employment tribunal and, on appeal, the EAT, found that the display of such posters was gender neutral since a male employee might also have been equally upset at the posters. In the circumstances, it was not possible to say that the employee had been treated less favourably than a male comparator.

Another example is the case of *Balgobin v London Borough of Tower Hamlets [1987] IRLR 401*. This case concerned complaints by female employees who claimed that they had been subjected to the detriment of being required to work with their alleged harasser after an investigation into the allegations of harassment was found to be inconclusive. Again, the EAT refused to accept that the employee had been treated less favourably on grounds of sex since it was possible that if a male employee had been harassed, the employer would have treated the male in a similar way. This is not to say that all cases will be treated in this fashion. Indeed, the EAT held in *Driskel v Peninsula Business Services [2000] IRLR 151* that, where a male supervisor engages in vulgar behaviour with a junior male colleague, it does not follow that such behaviour will provide a defence if the supervisor attempts to engage in similar behaviour with a female subordinate since behaviour of this nature where directed at a female subordinate is more likely to be intimidatory.

Finally, in the context of sex discrimination, it is conceivable that a harasser may be bi-sexual and therefore a defence could exist to the effect that the harasser would indeed treat a man in a similar fashion.

This is not to say that employers can use these lines of defence to totally escape liability. It needs to be remembered that although treatment might not be less favourable for the purposes of the Discrimination Acts, it may still amount to a breach of the implied duty in a contract of employment to maintain the trust and confidence of an employee and the harassment and may therefore be a precursor to an unfair dismissal claim.

In spite of the above, it seems that an argument still exists in the context of sexual harassment to the effect that, since sexual harassment is gender specific (in much the same way as pregnancy), there is no requirement for a comparison to be made as to whether a male employee would have been treated in the same way. Indeed, the EAT accepted this view in *British Telecommunications Ltd v Williams [1997] IRLR 668*. The EAT even went so far as to hold that it would be no defence for an employer to state that a male employee would have been treated in a similar fashion. Unfortunately, some of the credibility in the *Williams* case is lost in that it made no reference to the decision in *Stewart*. There must also be some doubt as to whether the argument can survive following the decision of the European Court of Justice in *Grant v South-West Trains Ltd [1998] CMLR 993* (dealing with an employer's policy of not providing travel concessions to partners in same sex relationships) which held that, provided a policy or treatment is applied equally to men and women, there can be no discrimination.

## Differential treatment is on prohibited grounds of sex

**H1709** In order for unlawful discrimination to occur, it is necessary that a claimant be treated less favourably than an actual or hypothetical male comparator. To this end, in the context of sex discrimination, one of the defences that has been raised is to argue that harassment against a female employee would have occurred equally to a male employee. The case of *Porcelli v Strathclyde Regional Council [1986] IRLR 134* (above) shows the limits of this line of defence. The Court of Session stated of the comparison between male and female employees:

> 'If any material part of unfavourable treatment to which a woman is subjected includes a significant element of a sexual character to which a man would not be vulnerable, the treatment is on grounds of the woman's sex within the meaning of $s\ 1(1)(a)$'.

It will be recalled that the defence raised by the employer in this case was that a hypothetical male employee would have been treated in a similar fashion. The defence failed and the Court of Session commented upon the nature of the treatment that had been received by the applicant in the following terms.

> 'The [employment] tribunal ... concentrated on the unpleasantness of the treatment meted out to the respondent and compared it with the unpleasantness of the treatment which they considered would have been meted out to a man whom his colleagues had disliked as much as her colleagues had disliked the respondent. That was a question of fact; but the question which had to be asked after that conclusion had been reached was "Was the treatment, or any part of it, which was meted out to the respondent less favourable, than that which would have been meted out to the notional man because the respondent was a woman?".'

An example of a decision that was resolved in a contrary fashion was the case of *Smith v Gardner Merchant Limited [1998] IRLR 520*. The facts of this case were that the applicant was a male homosexual who had been dismissed for gross misconduct. He brought a claim for sex discrimination, one of the limbs of which was an allegation that he had been harassed by a female employee who had asked a series of offensive questions about his sexuality, made various accusations to him along the lines that gay people had all sorts of diseases and stated to him that gay people who spread the HIV virus should be put on an island. Mr Smith claimed that he had been treated less favourably on account of his gender. The Court of Appeal took the view that although there may have been less favourable treatment, it was on account of Mr Smith's sexual orientation rather than his gender. Accordingly, the relevant grounds of comparison that would need to be made were whether a female homosexual with the same personal characteristics of Mr Smith would have been subjected to similar or differential treatment. Whether or not such a decision is consistent with Arts 8 and 14 contained within the *Human Rights Act 1998, Sch 1*, (which recognises a right to respect for a person's private life and prohibits discrimination in relation to that right on any ground) remains to be confirmed. The Scottish EAT in *MacDonald v Ministry of Defence [2000] IRLR 748* held that the *SDA 1975* must be construed in the light of Arts 8 and 14. Since sex discrimination for the purposes of Art 14 has been held to include discrimination on the grounds of sexual orientation (in *Salgueiro da Silva Mouta v Portugal, 33290/96, 21 December 1999*) the EAT concluded that, in line with its requirement to interpret the *SDA 1975* consistently with the *Human Rights Act 1998*, discrimination on grounds of sexual orientation must be unlawful. The Court of Session reversed the decision on appeal (*[2001] IRLR 431*) and in *Pearce v Governing Body of Mayfield Secondary School [2001] IRLR 669*, the EAT and the Court of Appeal in England accepted the orthodoxy of Smith. (Note: Ms Pearce has been given leave to appeal to the House of Lords.) From a common sense point of view, there does seem to be some contradiction in the law to the extent that it does not protect homosexuals from harassment as to their sexuality and yet does protect transsexuals against treatment meted out to them on the grounds of gender reassignment (as occurred in *Chessington World of Adventures Ltd v Reed [1997] IRLR 556* — see also now the *Sex Discrimination (Gender Reassignment) Regulations 1999 (SI 1999 No 1102)* outlawing discrimination on the grounds of gender reassignment). It remains to be seen whether the Government will pass similar legislation in relation to homosexuals following its defeat in the European Court of Human Rights in the case of *Smith & Grady v UK [1999] IRLR 734* (involving the right of homosexuals to serve in the UK armed forces).

In the context of disability discrimination, there is no requirement of an able bodied comparator (*Clark v Novacold Limited [1999] IRLR 318*) and it is simply enough for harassment to be meted out to a complainant on grounds of the complainant's disability.

## Detriment

H1710   The final hurdle to be crossed in order to bring a successful claim is to show that the employee has suffered a detriment. As to what is capable of amounting to a detriment, the starting point is again the decision in *Porcelli v Strathclyde Regional Council [1986] IRLR 134*. Here, the Court of Session stated that a detriment, in its statutory context, simply means a disadvantage. The point has also been considered in *Barclays Bank plc v Kapur [1989] ICR 753* where the Court of Appeal held that, since detriment means a disadvantage, it would cover almost any discriminatory conduct by an employer against an employee. The Court of Appeal did, however, limit the scope of this statement in *Barclays Bank plc v Kapur (No 2) [1995] IRLR 87*, where it was held that an unjustified sense of grievance did not amount to a detriment.

The concept of detriment was re-examined in the case of *Reed and Bull Information Systems v Stedman [1999] IRLR 299*. The facts of the case were that the claimant had resigned claiming that the behaviour of her manager amounted to sexual harassment and therefore discrimination. The employee cited a list of fifteen alleged incidents that she considered amounted to a course of harassment. The Employment Tribunal concentrated on four of the allegations, these being such matters as telling the employee that sex was a beneficial form of exercise, telling dirty jokes to colleagues in her presence, making a pretence of looking up the employee's skirt and then laughing when she angrily left the room and stating to her when she was listening to a trial presentation that 'You're going to love me so much for my presentation so that when I finish you will be screaming out for more and you will want to rip my clothes off'. The tribunal added that, of themselves, each of the allegations was insufficient to amount to sexual harassment. However, if the claimant were able to prove the incidents, they would amount to a course of conduct that would itself be sufficient to amount to harassment. The employment tribunal went on to find that the acts had occurred and that the employer and the manager had therefore discriminated against the applicant. The employer and manager appealed to the EAT. The EAT dismissed the appeal agreeing that the actions by the manager, although not meant to be sexually harassing, were indeed discriminatory. It added:

> 'It seems to us important to stress at the outset that 'sexual harassment' is not defined by the [*SDA 1975*]. It is a colloquial expression which describes one form of discrimination in the workplace made unlawful by *s 6* of the *Sex Discrimination Act 1975*. Because it is not a precise or defined phrase, its use, without regard to s 6, can lead to confusion. Under *s 6* it is unlawful to subject a person to a "detriment" on the grounds of their sex. Sexual harassment is a shorthand for describing a type of detriment. The word detriment is not further defined and its scope is to be defined by the fact-finding tribunal on a common-sense basis by reference to the facts of the particular case. The question in each case is whether the alleged victim has been subjected to a detriment and, second, was it on the grounds of sex'.

Consequently, in order to determine whether or not conduct of a sexual nature is sexual harassment and therefore a 'detriment' within the meaning of the *SDA 1975*, it is necessary to consider whether it is a disadvantage which is itself determined from the particular circumstances of the case.

### Detriment — subjective or objective determination?

H1711   The EAT in *Reed and Bull Information Systems* (above at H1710) also discussed the question as to whether a 'detriment' was something that was to be determined subjectively from the perspective of an applicant to a tribunal or objectively by the tribunal. The EAT dealt with the issue by posing a series of questions that would typically be asked in harassment cases, these being as follows.

(a) If a woman regards words or conduct of a sexual nature as harassment to which many women would not take exception or regard as harassment, has her claim been made out?

(b) If a man does not appreciate that his words or conduct are unwelcome, has her claim been proved?

(c) Is a 'one-off' act (words or conduct) sufficient to constitute harassment?

The tribunal answered the questions that it posed by stating as to the first of the questions that it is up to the victim of harassment to determine what she finds unwelcome or offensive. Simply because a tribunal does not necessarily find the words or conduct offensive will not lead to a complaint being dismissed. Indeed, the EAT stated that there is a range of factual situations that are capable of amounting to harassment. Inevitably, once a victim of harassment has received unwelcome attention from a person it is possible that further attention that would not ordinarily attract comment may take on a different context in the light of previous unwelcome attentions. Consequently, it is up to the victim to determine what words or conduct she is unhappy with and for a tribunal to determine in the totality of the facts whether those words or conduct could reasonably be viewed as offensive and unwelcome.

As to the second question, the EAT acknowledged that it is trite law that the motive and intention of the discriminator are irrelevant to the question of whether or not discrimination has occurred.

Finally, as to the third question, the EAT stated that the fundamental question to be addressed by a tribunal was whether the applicant was subjected to a detriment on account of her sex. It went on to add that this was not a question that was answered by counting up the number of incidents that had occurred. In fact, it was answered by looking at whether an incident or series of them damaged the working environment and constituted a barrier to sexual equality in the workplace which constituted a detriment.

Further guidance on the issue of detriment was given by the EAT in *Driskel v Peninsula Business Services Limited [2000] IRLR 151*. The case concerned allegations of a series of acts committed by the line manager of the female applicant that was alleged to amount to harassment on grounds of sex. The EAT held that, where there are a series of acts, tribunals should focus upon the cumulative effect of the acts rather than whether the acts in themselves are trivial in nature. Further, where a male supervisor engages in vulgar behaviour with male colleagues, it does not follow that a female employee subjected to vulgar behaviour is not treated less favourably since it is more likely that a female subordinate will find such behaviour to be intimidatory and likely to undermine her dignity.

Ultimately, the view of the EAT was that the test as to whether a detriment has been suffered focuses upon an applicant's subjective determination of whether conduct to which she is subjected is unwelcome and that determination has then to be judged by a tribunal which must ask itself whether that is a reasonable position for the applicant to take.

## Examples of unlawful harassment

**H1712** As was stated in the case of *Reed and Bull Information Systems v Stedman [1999] IRLR 299*, sexual harassment is a shorthand way of expressing the legal requirement in the *SDA 1975, s 6(2)* of a 'detriment'. The case also recognises that the type of situations in which sexual harassment can occur are legion. However, of the cases that have been reported to date they can broadly be categorised as:

(a) course of conduct cases;

(b) single incident cases; and

(c) failure to investigate or protect cases.

## Course of conduct cases

**H1713** As the heading suggests, the hallmark of this genus of cases is that an employee is subjected to repeated harassment in the course of her employment. The following are reported examples of such cases.

(a) *Porcelli v Strathclyde Regional Council [1986] IRLR 134.* The applicant was a female laboratory assistant and was the victim of a campaign of harassment by fellow male employees to make her leave her job. The male employees would frequently brush past the applicant in a sexually suggestive manner and direct comments of a sexual nature at her. The Court of Session held that the applicant had suffered a detriment in having to put up with the conduct which had clear sexual overtones and to which a similarly disliked hypothetical male comparator would not have been subjected.

(b) *Tower Boot Company Limited v Jones [1995] IRLR 529.* The applicant was a black employee who was subjected to such acts as being whipped with a leather belt, 'branded' with a red-hot screwdriver and subjected to racist name calling. The Court of Appeal held that, although employees were not employed to harass other employees, the acts had been committed during the course of the employees' employment and therefore the employer was liable for the discrimination.

(c) *Chessington World of Adventures Ltd v Reed [1997] IRLR 556.* The applicant was a biological male and was undergoing gender reassignment from male to female. The applicant was subjected to a prolonged campaign of ostracism and harassment (which the employer was found to have done nothing to investigate or curb) having as its proximate cause the therapy that the applicant was undertaking. In the circumstances, the employer was held to have subjected the applicant to a detriment.

(d) *Reed and Bull Information Systems v Stedman [1999] IRLR 299.* The applicant was a junior secretary responsible to the marketing manager of the employer. The applicant catalogued fifteen separate incidents ranging from the telling of dirty jokes by the manager to other colleagues in her presence to the pretence of looking up her skirt in the office smoking room. The EAT upheld the findings of the employment tribunal that, although the incidents by themselves would have been insufficient to amount to a detriment, collectively they amounted to a course of conduct of innuendo and general sexist behaviour amounting to harassment and therefore a detriment for the purposes of the *SDA 1975*.

(e) *Driskel v Peninsula Business Systems [2000] IRLR 151.* The applicant was a female employee who complained that she had been subjected to sexual banter and comments by the head of her department and, on the day before an interview for a promotion, she was told by the manager that she should wear a short skirt and a see-through blouse showing plenty of cleavage. The applicant left work and refused to return unless the manager was moved to another position. The employee's request was refused and she was subsequently dismissed. The EAT held that the dismissal was on grounds of the applicant's sex.

## Single incident cases

**H1714** The hallmark of 'single incident' cases is that they are effectively a 'bolt from the blue', yet comply with the requirements in *Reed and Bull Information Systems v Stedman [1999] IRLR 299* in that they provide a barrier to sexual equality in the workplace.

(a) *Bracebridge Engineering Ltd v Darby [1990] IRLR 3.* The applicant was physically manhandled into the office of the works manager by both a chargehand and the works manager and then indecently assaulted. She was warned not to complain because no one would believe her. The applicant did subsequently complain but the complaint was then not properly investigated by the employer. It was held by the EAT that the assault was sufficiently serious to amount to a detriment.

(b) *Insitu Cleaning Co Ltd v Heads [1995] IRLR 4.* The applicant was a cleaning supervisor who attended at a company meeting. Also present was the son of one of the directors of the company who greeted her by saying 'Hiya, big tits'. The applicant found the remarks particularly distressing as the director's son was approximately half her age. The EAT held that the remark was capable of subjecting the applicant to a detriment and rejected the employer's contention that the remark was of the same effect as a statement made to a bald male employee about his head.

(c) *Chief Constable of the Lincolnshire Police v Stubbs [1999] IRLR 81.* The applicant was a Detective Constable in the Lincolnshire Constabulary. She attended an off duty drink with some colleagues in a public house one evening after work. During the course of the evening a male colleague pulled up a stool next to her, flicked her hair and re-arranged her collar so as to give the impression that there was a relationship between himself and the applicant. The applicant found this attention to be distressing and moved away from him. On another occasion at a colleagues leaving party that she had attended with her boyfriend, she was accosted on her way to the toilet by the same officer who stated to her 'Fucking hell, you look worth one. Maybe I shouldn't say that it would be worth some money'. The applicant found the remark to be humiliating and brought proceedings for sex discrimination. The EAT upheld the decision of the employment tribunal that the statement amounted to a detriment.

## Failures to investigate or protect

**H1715** Where an employee brings an allegation of sexual harassment against another employee, the employer has a duty to carry out a proper investigation into the complaint and to take such further steps as are necessary in the light of the findings of the investigation. A failure to carry out such a proper investigation is capable of amounting to a detriment in its own right. The following cases are examples of alleged failures to investigate or protect employees.

(a) *Balgobin v London Borough of Tower Hamlets [1987] IRLR 401.* The employees were female cleaners working in a local authority hostel. They alleged that a male cook at the hostel had sexually harassed them. The employer suspended the cook and carried out a full investigation into the allegations that resulted in a failure to find that the complaints of the applicants had been proved. Consequently, the cook was allowed to return to work and the cleaners were required to carry on working with him The employees complained that they were subjected to the detriment of having to work with the cook. The EAT held that there had been no discrimination since it was possible that a male cleaner to whom homosexual advances had been made could have been treated in a similar fashion by the employer.

(b) *Bracebridge Engineering Ltd v Darby [1990] IRLR 3.* (See H1714 for the facts.) The assault that took place on the employee was coupled with a failure to properly investigate the incident on the part of the employer who, through its personnel manager, simply accepted that it was the word of two employees against one and that there was insufficient evidence to warrant further investigation. The EAT held that where serious allegations of sexual harassment are made, employers have a duty to properly investigate them.

(c) *Burton v De Vere Hotels [1996] IRLR 596.* The case took place under the provisions of both the *SDA 1975* and the identical provisions of the *RRA 1976*. The case was brought by a waitress at a hotel and concerned the detriment of suffering sexual and racial harassment from a third party not connected to the employer but whom over the employer was able to exercise control (the case concerned the applicant being subjected to sexist and racial abuse masquerading as comedy from a well known comedian at a hotel function. The applicant was humiliated in front of the audience by the comedian). Although the employer apologised to the employee for the actions of the comedian, the employee lodged proceedings for sexual and racial harassment and succeeded in the claim. The EAT held that, where an employer was in a position to exercise control over a third party so that discrimination would either not occur, or the extent of it would be reduced, the employer would be subjecting an applicant to a detriment by failing to take the steps necessary to control the employee.

(d) *Chessington World of Adventures Ltd v Reed [1997] IRLR 556*. (See above at H1713.) Added to the campaign of harassment that was suffered by the employee was a failure to curb the harassment when complaints were made by the applicant. This was held to be capable of amounting to a detriment since the employer was plainly able to stop the harassment.

(e) *Waters v Commissioner of Police for the Metropolis (No 2) [2000] IRLR 720*. The employee was allegedly subjected to bullying and a failure to investigate her complaint of rape. The House of Lords held that an employee owes a duty of care to investigate allegations of rape, bullying and harassment.

## Victimisation

H1716    Victimisation is a further category of discrimination that exists under the Discrimination Acts. Victimisation exists under the *SDA 1975, s 4*, the *RRA 1976, s 2*, and the *DDA 1995, s 55*.

Victimisation occurs where a person treats another person less favourably that he treats or would treat other persons and does so for an inadmissible reason. The inadmissible reasons are set out in the *SDA 1975, s 4(1)*, the *RRA 1976, s 2(1)*, and the *DDA 1995, s 55(2)*, respectively.

The victimisation provisions are aimed at preventing an employer from curtailing the right of an employee to legitimately protect his position opposite the employer. The leading case is the House of Lords decision in *Nagarajan v London Regional Transport [1999] IRLR 572* (decided under the provisions of the *RRA 1976*). Their Lordships held, by a majority (and in the face of a powerful dissenting speech by Lord Browne-Wilkinson), that victimisation occurs where less favourable treatment is meted out to a complainant because the victimised person had done an act protected under the provisions of the Act. Their Lordships went on to add that whether the discriminator intended to engage in an act of discrimination towards the victim was irrelevant. Accordingly, it is sufficient for a victimisation claim to arise that an employer is aware of the existence of the fact that an employee has a protected right under the Discrimination Acts and then unconsciously treats the employee less favourably than he would treat another not having a protected right.

However, in *Ledeatte v Tower Hamlets London Borough Council (26 June 2000)*, the EAT confirmed that there must be a causal link between the act complained of and the protected act.

### Remedies

H1717    The remedies that are available to an applicant successfully bringing a victimisation complaint are the same as those arising in relation to sex discrimination — see H1725.

## Employers' liability

### Vicarious liability

H1718    The legal problem that occurs in connection with harassment is that, often, it is not the employer that directly subjects a victim of harassment to the course of conduct; rather it is a fellow employee who is responsible. The employer may then be held responsible for the employee's actions on the grounds of vicarious liability.

The problem that arises in legal theory with vicarious liability is that employers do not employ staff to harass other members of the workforce. Therefore, it is necessary to determine what amounts to 'the course of employment' in order to see when an employer will be liable for the harassment of its employees by other members of staff.

## The course of employment

**H1719** The first case to take as a line of defence the fact that a harasser was not acting in the course of his employment was *Tower Boot Co Ltd v Jones [1997] IRLR 168*. The case concerned proceedings brought under the provisions of the *RRA 1976* in respect of racial harassment. The applicant had been subjected to a particularly vicious campaign of name-calling and physical assaults. An Employment Tribunal found that this constituted racial harassment and that the employer was vicariously liable for the acts of the employees engaged in the course of conduct. The employer appealed to the EAT which found for the employer on the ground that the common law test for vicarious liability had not been satisfied. The EAT took the view that for this test to be met the aggressor employees had to be engaged in doing the business of the employer, albeit in an improper manner, i.e. the act had to be a different mode of doing what the aggressor employees were employed to do. The EAT then held that on the facts, the actions of the aggressor employees could not 'by any stretch of the imagination' be found to be an improper method of doing what the aggressors were properly employed to do.

The EAT decision was widely condemned on the grounds that the more serious that harassment was, the less likely that an employer would be found to be vicariously liable for it. Indeed, the logical conclusion to the EAT's decision was that, since people were not employed to harass other members of staff, employers would never be liable, a surprising result indeed. Unsurprisingly, the decision was appealed to the Court of Appeal which readily found for the employee. In doing so, the court held that the words, 'in the course of his employment' in the *RRA 1976, s 32(1)*, were not subject to the common law rules in the Salmond test which would artificially restrict the natural everyday sense that every layman would understand of the wording in *s 32(1)*. Consequently, harassment by an aggressor employee did not have to be connected with acts authorised to be done as part of his work. In so deciding, the Court of Appeal attacked the reasoning of the EAT, stating that its interpretation of the section creating liability:

> '... cut across the whole legislative scheme and underlying policy which was to deter racial and sexual harassment through a widening of the net of responsibility beyond the guilty employees themselves by making all employers additionally liable for such harassment'.

## Requirement of causal link with employment

**H1720** It is not every act of harassment that is committed by an employee that will create a liability on the part of the employer. There must still be a causal link with employment. A good example of this point is the case of *Waters v Commissioner of Police of the Metropolis [1997] IRLR 589*. In this case, the complaint was one of sexual harassment in the form of an alleged sexual assault by a male police officer on a female police officer in her section house room after they had returned from a late night walk together. Given that the attack took place outside the normal working hours of the applicant, the employment tribunal refused to follow the decision in the *Tower Boot* case and held that the applicant was in the same position as if the alleged assailant had been a total stranger to the employer. The alleged assailant did not live in the section house, both officers were both off duty and there was nothing linking the alleged assailant to the employer other than the fact that he was a police officer. The Court of Appeal ultimately upheld the decision of the tribunal on appeal.

This said, the decision in *Waters* does not provide an absolute defence to harassment undertaken after employees leave work for the day. Whether there is a causal link between the harassment and the employment, thereby attaching liability to the employer, will be a question of fact in each particular situation. In the subsequent proceedings brought by Ms Waters in respect of bullying and failure to investigate her complaint, the House of Lords held that an employer owes a duty of care to investigate allegations of rape, bullying and harassment (*Waters v Commissioner of Police for the Metropolis (No 2) [2000] IRLR 720*). An example of this point is *Chief Constable of the Lincolnshire Police v Stubbs [1999] IRLR 81*. The complainant had been the subject of distressing personal comments of a sexual nature from a fellow police officer made at a colleague's leaving party (see above at H1714). The Employment Tribunal at first instance held that since the party was an organised leaving party, the incidents were 'connected' to work and the workplace. It stated that the incident:

> '... would not have happened but for [WPC Stubbs'] work. Work-related social functions are an extension of employment and we can see no reason to restrict the course of employment to purely what goes on in the workplace'.

The EAT emphasised, however, that it would have been different 'had the discriminatory acts occurred during a chance meeting' between the employee and the harasser outside of the workplace.

The decisions give a wide degree of discretion to tribunals on the facts of a particular case to decide whether harassment is committed in the course of employment. If a tribunal finds that it is not committed in such circumstances, the effect is that, subject to *Waters (No 2)*, whilst an employee may be able to sue the harasser in tort, the remedy that is provided may well be empty if the harasser does not have the means to pay for his wrongdoing.

## *Employer's defence*

**H1721** A defence is provided to employers in relation to allegations that they are vicariously liable under, respectively, the *SDA 1975, s 41(3)* and the *RRA 1976, s 32*. The defence provides that:

> 'In proceedings brought under this Act against any person in respect of an act alleged to have been done by an employee of his it shall be a defence for that person to prove that he took such steps as were reasonably practicable to prevent the employee from doing that act, or from doing in the course of his employment acts of that description'.

A similar, although not identical, provision exists under the *DDA 1995, s 58(5)*.

The problem for employers with the defence is that, whilst it will provide a defence in a situation where an employer has a full procedure dealing with harassment (including such matters as monitoring, complaints handling and disciplinary proceedings), the provision requires that the employer must pay more than mere lip service to the existence of these matters and must also be able to show that its employees were aware of the procedures and, where relevant, that the employer had fully implemented the procedures in practice. A good example of a case where the defence succeeded was *Balgobin v London Borough of Tower Hamlets [1987] IRLR 401* (see H1715 above). Here, the alleged harasser was suspended following the initial allegations, his conduct was fully investigated and it was found that there was insufficient evidence against the cook to justify the claims. The EAT was moved to comment that it was difficult to see what further steps the employer could have taken to avoid the alleged discrimination.

Ultimately, if an employer has gone to the expense of having policies and procedures drafted then they should not be kept in the office safe. Employees should be left under no illusions as to what will happen in cases of harassment.

## *Liability of employees*

**H1722** Although an employer may be vicariously liable for acts of harassment, primary responsibility for such action will remain with the employee who has committed the discrimination and, to this end, the *SDA 1975, s 42*, and the *RRA 1976, s 33*, provide as follows.

> '(1) A person who knowingly aids another person to do an act made unlawful by this Act shall be treated for the purposes of this Act as himself doing an unlawful act of the like description.

(2) For the purposes of subsection (1) an employee or agent for whose act the employer or principal is liable under section [*SDA 1975, s 41* or *RRA 1976, s 32*] (or would be so liable but for section [*41(3)* or *32(3) respectively*]) shall be deemed to aid the doing of the act by the employer or principal.'

The scope of the provision was given a liberal interpretation in *AM v WC & SPV [1999] IRLR 410*. Here, the EAT held that an Employment Tribunal had erred in not allowing a police officer to bring a complaint of harassment against another individual officer. The EAT considered that the construction of *SDA 1975, s 41(1)*, *(3)*, and *SDA, s 42*, was to make an employer vicariously liable for acts of discrimination committed by staff (*SDA 1975, s 41(1)*), to provide conscientious employers with a defence for the actions of staff where the actions are effectively a 'bolt from the blue' (*SDA 1975, s 41(3)*) and to allow an employee who has committed acts of discrimination against another to be joined as a 'partner' to proceedings commenced against an employer (*SDA 1975, s 42*). The EAT added that even where a defence did exist for an employer under *s 41(3)*, there was nothing in the SDA 1975 that prevented the victim of discrimination from launching proceedings against a discriminator employee alone.

The defence has also been considered by the House of Lords in *Anyanwu v South Bank Student Union [2001] IRLR 305*, where it was held that the words 'knowingly aids' means that the person providing aid must have given some kind of assistance to the person committing the discriminatory act which helps the discriminator to do the act. The amount or value of the help is of no importance. All that is needed is an act of some kind, done knowingly, which helps the discriminator to do the unlawful act.

The provisions under *DDA 1995, s 58(1)* and *(5)* are similar, although not identical, to those under the *SDA 1975* and *RRA 1976*.

## Narrowing the extent of employers' liability

H1723   Where an employee is able to establish that he or she has been the victim of unlawful harassment, there can be circumstances where the damages that would otherwise be awarded to a victim of such discrimination will be dramatically reduced. In the context of sexual harassment, this will be where the victim's conduct shows that she would not have suffered a serious detriment as a consequence of the harassment. For example, in *Snowball v Gardner Merchant Ltd [1987] IRLR 397*, the applicant complained that she had been sexually harassed by her manager. The allegations were denied and the applicant was subjected to questioning at the tribunal hearing as to her attitude to matters of sexual behaviour so as to show that even if she had been harassed, her feelings had not been injured. In particular, evidence was called as to the fact that she referred to her bed as her 'play pen' and that she had stated that she slept between black satin sheets. On appeal, the EAT held that the evidence had been correctly admitted so as to show that the applicant was not likely to be offended by a degree of familiarity having a sexual connotation.

Likewise, in *Wileman v Minilec Engineering Ltd [1988] IRLR 144*, the EAT refused to overturn an award of damages in the sum of £50 that had been made for injury to feelings. In this case, the applicant had made a complaint of sexual harassment against one of the directors of her employer. The award had been set at this figure on account of the fact that the victim had worn revealing clothes and it made it inevitable that comment would be passed.

With respect to the EAT, these cases, whilst reflecting that harassment can still occur in cases where a complainant has a high degree of confidence in sexual banter, unfortunately appear to send out the wrong message in that they can be construed as being as a 'harasser's charter' and do not seem to recognise that a person may be particularly upset by the unwelcome attentions of a specific person. It is worth noting that both of the decisions pre-date the EC Code of Practice and the enactment of the *Human Rights Act 1998* and it will be interesting to see if the effect of the decisions is in any way curbed at a future date by these matters.

# Legal action and remedies

**H1724** Where a person unlawfully harasses another, the consequences of such action can be extremely far-reaching and potentially spread across the full range of actions in both civil and criminal law.

## Unlawful discrimination

### Elements of the tort

**H1725** The most obvious form of action that a victim of harassment may consider is direct sex and race discrimination under *SDA 1975, s 1(1)(a)*, and *RRA 1976, s 1(1)(a)*, respectively. Liability arises because a person engaged in harassing another subjects that other to a detriment (*SDA 1975, s 6(2)(c)*, and *RRA 1976, s 4(2)(c)*). This head of liability is fully covered at H1703 above and the paragraphs that follow.

Disability discrimination may also occur under *DDA 1995, s 5(1)*, where the employer treats the employee less favourably on account of the employee's disability and the employer cannot show the less favourable treatment to be justified on disability neutral grounds.

### Remedies

**H1726** The remedies that are open to a victim of unlawful harassment are:

(a) a declaration;

(b) recommendation; or

(c) compensation.

A declaration is a formal acknowledgement from an employment tribunal that discrimination in the form of harassment has occurred. It will be granted in every successful case brought before a tribunal.

Recommendations are used where an applicant to an employment tribunal remains in the employment of her employer. The purpose of a recommendation is for a tribunal to give guidance to an employer relating to its business for the purposes of eliminating discrimination.

However, the main remedy that victims of harassment will seek is compensation for the loss that has been suffered as a result of the harassment including, particularly, injury to feelings.

## Unfair dismissal

**H1727** Unfair dismissal can arise as a claim where an employer fails to control unlawful harassment in the workplace or fails to properly investigate complaints of unlawful harassment. Such failures may and probably will also act to undermine the employee's trust and confidence in the employer. The result of such actions is that an employee may be able to claim that the employer has constructively dismissed the employee, thereby amounting to a dismissal for the purposes of the *Employment Rights Act 1996, s 95(1)(c)*. Indeed, where a unlawful harassment claim has been brought on the back of an employee's departure from his or her employment, most cases will also include a claim for unfair dismissal as well.

### Proving constructive dismissal

**H1728** The first hurdle that an employee has to overcome is to prove that he or she has been dismissed for the purposes of the *Employment Rights Act 1996 (ERA 1996), s 95(1)(c)*. It is trite law that for an employee to show that the employee has been constructively dismissed the acts of his or her employer must show that the employer, by its conduct, evinced an intention not to be bound by the terms and conditions of the contract of employment (see, to this effect, *Western*

*Excavating (ECC) Limited v Sharp [1978] IRLR 27*). The problem that an employee may face in this regard is that if an employer has in place a full anti-harassment policy that is properly implemented, the employee may fail to establish that the employer has acted in breach. For an example of this occurring in practice see *Balgobin v London Borough of Tower Hamlets [1987] IRLR 401* (above at H1715).

Likewise, an employee may also fail to show that an employer has breached the implied term of trust and confidence where the act is a serious one off act of harassment that the employer was unable to predict would happen and where the employer properly deals with the matter immediately upon a complaint being made. To the extent that the employer does not properly investigate a complaint in such circumstances it is almost inevitable that the employer will breach the contract of employment (see *Bracebridge Engineering Ltd v Darby [1990] IRLR 3*).

It should be noted that conduct by an employer towards an employee that may amount to a constructive dismissal for the purposes of *ERA 1996, s 95(1)(c)*, has been held by the EAT in *Commissioner of Police for the Metropolis v Harley [2001] IRLR 190* not to entitle an employee to resign and claim that he/she has been dismissed for the purposes of *DDA 1995, s 4(2)(d)*, because there is no statutory concept of constructive dismissal defined in the *DDA 1995*. The decision runs contrary to the decision of the EAT in *Derby Special Fabrication Limited v Burton [2001] IRLR 69* on the identical wording in the *RRA 1976*. Dismissal is defined by *SDA 1975, s 82(1A)*, as to include a constructive dismissal. With respect to the EAT, it is arguable that the decision in *Harley* is wrong in law given that the dismissal provision is common to all three of the Discrimination Acts. In any event, even if the EAT in *Harley* is correct in holding that constructive dismissal does not arise under the *DDA 1995*, conduct by an employer towards an employee that would amount to a constructive dismissal for the purposes of *ERA 1996, s 95(1)(c)*, may amount to a 'detriment' for the purposes of *DDA 1995, s 4(2)(d)*.

Finally, it will be recalled, that, following, *Smith v Gardner Merchant Limited [1998] IRLR 520*, sexual harassment on the grounds of sexual orientation is difficult prove since the applicant will have to show that the person subjecting her to harassment would also treat a homosexual man in like fashion. This does not release an employer from an obligation to investigate since such conduct on the part of fellow employees, if left unchecked by the employer, may breach the implied duty of the employer to maintain the employee's trust and confidence.

## Remedies

**H1729** The remedies that can be provided for unfair dismissal are those set out in *ERA 1986, ss 113–118*, namely:

(a)  an order for reinstatement;

(b)  an order for re-engagement; or

(c)  an order for compensation.

The practical consequence of an employee claiming that he or she has been constructively dismissed as a result of an employer's breach of the implied duty of trust and confidence is that a tribunal is likely to be unwilling to make an order of reinstatement or re-engagement and therefore the reality is that most cases will provide for compensation to be awarded. The downside to claiming unfair dismissal as opposed to bringing a claim for unlawful discrimination is that, whereas, the compensation for sex discrimination is both unlimited as regards the amount that can be awarded and can contain an award for injury to feelings, the compensation that is available for unfair dismissal is subject to a cap of £52,600 and can include an award for the manner of dismissal (following *Johnson v Unisys Ltd [2001] IRLR 279*).

## Actions in tort

**H1730** Broadly speaking, torts are the category of actionable civil wrongs not including claims for breach of contract or those arising in equity. Acts of unlawful harassment, depending upon the particular characteristics of the harassment, are capable of amounting to a variety of different torts.

## Assault and battery

**H1731** Assault and battery are forms of trespass to the person. They are separate torts and occur where a person fears that either he or she will receive or has been the subject of unlawful physical contact. As was stated of the torts in *Collins v Wilcock [1984] 3 All ER 374*:

> 'An assault is an act which causes another person to apprehend the infliction of immediate, unlawful, force on his person; a battery is the actual infliction of unlawful force on another person'.

For the tort of assault to be committed it is irrelevant that the victim is not in fear of harm. Consequently, it is possible for an assault to occur where a victim of harassment is threatened with the possibility that he or she will receive unwanted physical contact. All that must exist for the tort to be committed is that the victim must have a reasonable apprehension that she will imminently receive unwanted contact.

Further, for either tort, the use of the word 'force' in the formulation of the tort does not mean physical violence. It is not necessary that the victim be subjected to a full-scale attack (as occurred for example in *Bracebridge Engineering Ltd v Darby [1990] IRLR 3* and *Tower Boot Company Limited v Jones [1995] IRLR 529*). It is possible for the torts to be committed where a person brushes past another suggestively (as in *Porcelli v Strathclyde Regional Council [1986] IRLR 134*) or even kisses another without consent. The torts therefore cover the full range of possible forms of contact related harassment from physical injury to unwanted molestation and are capable of providing a cause of action even though no actual physical harm has been occasioned to the victim of the particular tort.

## Negligence

**H1732** While negligence may imply to the layman that an act committed against a person has been committed carelessly, it is perfectly possible for the tort to be committed intentionally by a person. The classic phrasing of the tort is derived from the case of *Donoghue v Stevenson [1932] AC 562*, where Lord Atkin stated of the elements of the tort:

> 'You must take reasonable care to avoid acts or omissions which you can reasonably foresee would be likely to injure your neighbour ... [who are] persons who are so closely and directly affected by my act that I ought reasonably to have them in contemplation as being so affected when I am directing my mind to the acts or omissions which are called in question'.

The test thus framed ensures that liability will be created where there is reasonable foresight of harm being inflicted upon persons foreseeably affected by an actor's conduct. Consequently, the conduct can be intentional or reckless as well as careless.

Where it is the case that an employee harasses a fellow employee, the harasser and victim are likely to be within the relationship of proximity required by the law (it being recognised that fellow employees owe a duty of care to each other).

Once it has been established that a duty is owed, three other factors become relevant these being whether harm has been sustained by a victim, whether the magnitude of the harm suffered by the victim is far greater than the actor intended and whether an employer has a defence if an action is brought for vicarious liability.

(a) *Requirement for loss.* For the tort of negligence to be committed, the victim must suffer loss or damage flowing from the negligent act. An obvious form that is likely to be sustained in the case of harassment is in the form of stress related illnesses that can occur from prolonged harassment (nervous shock). It is well recognised in law that where a person suffers from medically treatable stress related illnesses that are caused by the negligence of an employer, it is possible to recover damages for the injuries so inflicted (see to this end the non-harassment case of *Walker v Northumberland County Council [1995] ICR 702*).

(b) *'Eggshell-skulls'.* The extent of the injury suffered, provided that a victim could foreseeably have suffered a particular form of harm (e.g. distress) from the commission of the unlawful act, it is irrelevant that the final extent of the harm would not necessarily have been foreseen. This principle is known as the 'eggshell-skull' rule. The law recognises that it is possible for 'eggshell-personalities' to exist (see, for example, *Malcolm v Broadbent [1970] 3 All ER 508*). It is irrelevant that another female would not have suffered nervous shock as a consequence of harassment being meted out to her provided that the hypothetical female would have been distressed by it.

(c) *Vicarious liability.* Finally, there is the question as to whether it is possible for an employer to be sued for harm that is occasioned in negligence from an act of harassment. Whilst the case of *Chief Constable of the Lincolnshire Police v Stubbs [1999] IRLR 81* recognised that employers could become vicariously liable for the acts of employees for the tort of sex discrimination (due to the provisions of *s 41(1), SDA 1975*), it will also be recalled that in the earlier case of *Tower Boot Co Ltd v Jones [1997] IRLR 168* (dealing with racial harassment), it had been recognised by the EAT that, since, at common law, employers can only be liable for the misperformance of an act that an employee is performed to do, and, since an employer does not employ people to harass others, it should not be possible at common law for an employer to be vicariously liable for the harassment of an employee. Whilst the decision was overturned on appeal, the appeal only dealt with the creation of vicarious liability under the anti-discrimination legislation. In the circumstances, it is conceivable that an employer would still have a defence to a claim for negligence. However, where the claim is framed on the grounds that the employer was aware of the harassment yet failed to act when called upon to do so and harm was occasioned by the employer's failure to act, vicarious liability is not an issue since what is being complained of is the employer's primary failure.

Given that it is possible to recover damages for injury to feelings occasioned as a consequence of unlawful discrimination, it may well be the case that an employee will not want to pursue a claim for negligence. This said, it is not possible to obtain an injunction in respect of unlawful harassment under the Discrimination Acts.

## Remedies in tort

**H1733** Where an employee commits an actionable tort outside of harassment (the remedies for which are provided under the *Protection from Harassment Act 1997* — see below at H1734), two remedies will commonly be available to the victim. These will be:

(a) damages (which is available as of right for a tort); and

(b) an injunction (where damages are not adequate).

In tort, the correct measure of damages is that the victim should be given such a sum by way of compensation as would put the victim in the position as if the tort had not been committed (*Livingstone v Rawyards Coal Co (1880) 5 App Cas 25*). As to how much this will be is a question of fact in each case and, potentially, in the case of a course of harassment that induces nervous shock preventing an employee from working, could run into hundreds of thousands of pounds.

However, the problem incurred with the common law solution of awarding damages to the victim of a tort in the context of unlawful harassment is that the victim may not be properly compensated for on-going acts. In the circumstances, damages may not be adequate in such a situation and the better remedy may be to sue for an injunction.

Further, the use of an injunction to stop on-going harassment may be of more practical benefit to a currently distressed employee than would a future award of damages. This is because injunctions can be applied for and obtained relatively speedily as an intermediate remedy pending a full liability trial.

The problem with pursuing this remedy is that, assuming that the injunction is applied for whilst both the harasser and the victim are still in the employment of their employer, the victim's faith in his or her employer may well have become exhausted (by the employer's failure to take action in relation to the employee). Indeed, it would be surprising if this were not the case. Consequently, in many cases, the better course of action may well be to leave employment on the grounds of constructive dismissal and sue for compensation under the relevant Act.

## Protection from Harassment Act

**H1734** An offence and tort of harassment was created by the *Protection From Harassment Act 1997*. The legislation was passed to protect victims of 'stalking' but there is no reason why its provisions cannot be used to protect victims of sexual harassment in the workplace. Harassment is not specifically defined by the 1997 Act, although *s 7* provides that it must amount to a course of conduct causing alarm or distress to a person.

The phrase 'a course of conduct' is defined to be conduct on two or more occasions and 'conduct' includes words (*Protection From Harassment Act 1997, s 7(4)*).

It is possible to obtain an injunction against a harasser under *s 3* of the Act and also to claim damages.

## Breach of contract

**H1735** Where an employee harasses a fellow employee, the harasser will almost certainly breach one or more of the terms of his or her employment contract. In the first instance, most disciplinary procedures provide that harassment of fellow employees on grounds of sex, race or disability amounts to gross misconduct. Given that many contracts of employment provide that a disciplinary procedure is contractual, harassment per se may provide grounds for summary dismissal.

Even if an employer fails to provide a contractual disciplinary procedure, an employee may well be in breach of several of the implied terms under his contract of employment if he or she is engaged in unlawfully harassing a fellow employee. This is because the employee will owe a duty of care to the employer that requires the employee to take reasonable care for fellow employees. Deliberate unlawful harassment will breach this term.

Further, employees must maintain the trust and confidence of the employer during their employment. Unlawful harassment of fellow employees will undermine a victim's confidence and would provide grounds for a conduct related dismissal of a person committing acts of harassment.

## Remedies

**H1736** As with remedies provided for torts, the remedies for breach of contract will be either damages or an injunction.

The law places a different emphasis on the method of computing damages for breach of contract to that provided in relation to torts. Practically though, in the context of a contract of employment, the final outcome will usually be the same. In the law of contract, the victim is to be placed so far as possible in the position as if the contract had been properly performed by the employer (*Robinson v Harman (1848) 1 Exch 850*). The amount payable will vary according to the loss that has been suffered by a victim. Ordinarily, if the employee leaves his/her employment claiming constructive dismissal, the employee will be able to claim damages for loss of contractual benefits during her notice period (subject to the employee's duty to mitigate). However, it seems that in spite of the decision of the House of Lords in *Malik v BCCI [1997] IRLR 462*, following the subsequent decision of the House of Lords in *Johnson v Unisys Ltd [2001] IRLR 279*, it is not possible to recover damages for injury to feelings occasioned as a result of the humiliating manner of a dismissal.

This said, if the claim is framed by an employee as one going to the employer's duty of care to the employee (due to stress related illness being induced by harassment and the employer's unreasonable failure to prevent the cause of the stress) it is well established that an employee will be able to claim damages for the illness so caused (see, to this effect, *Walker v Northumberland County Council [1995] ICR 702*).

See H1733 for a discussion on injunctions.

## Criminal action

**H1737** The conduct of a person who sexually harasses another may also amount to one or more of the following crimes:

(*a*) battery;

(*b*) harassment under the *Protection from Harassment Act 1997* (see above at H1734);

(*c*) harassment under the *Public Order Act 1986*;

(*d*) racially aggravated harassment under the *Crime and Disorder Act 1998*.

## Battery

**H1738** Battery is a common law offence. The elements of the offences are that the harasser intentionally and unlawfully inflicts personal violence upon the victim. In *Collins v Wilcock [1984] 3 All ER 374* the Court of Appeal held that there was a general exception to the crime which embraced all forms of physical contact generally acceptable. However, it is clear from this that if the victim makes known to his/her harasser the fact that the victim does not welcome physical contact from him, the harasser will commit the offence simply by touching the victim.

The offence can only be tried summarily.

## Harassment under the Protection from Harassment Act 1997

**H1739** The tortious aspects of the *Protection from Harassment Act 1997* are considered at H1734 above. The 1997 Act also criminalises a course of conduct pursued by a person who 'knows or ought to know' that the conduct amounts to harassment. *Section 1(2)* of the Act provides that a person will be deemed to have knowledge where a reasonable person in possession of the same information as the harasser would think that the course of conduct amounted to harassment.

The offence is one that is triable either way. If the offence is tried on indictment, the maximum penalty is five years imprisonment or a fine or both. If tried summarily, the maximum penalty is a term of imprisonment not exceeding six months, a fine subject to the statutory maximum or both.

## Harassment under the Public Order Act 1986

**H1740** The offence exists under the *Public Order Act 1986, s 4A* and criminalises threatening, abusive or insulting words of behaviour thereby causing harassment, alarm or distress. The offence can be committed in a public or private place (other than a dwelling). Consequently, sexual harassment in the workplace is capable of being covered by the Act.

The offence can only be tried summarily and a fine on level 3 of the standard scale can be levied upon conviction.

### Racially aggravated crimes under the Crime and Disorder Act 1998

**H1741** The offence of racially aggravated harassment arises under the *Crime and Disorder Act 1998* (*CDA 1998*), s 32. The offence is committed where a person commits an offence against another for the purposes of the *Protection from Harassment Act 1997* and the offence is racially aggravated (as defined by *CDA 1998, s 28*).

The offence in either case is triable either way. In the event that a person is convicted on indictment of racially aggravated harassment under *CDA 1998, s 32(1)(a)*, the person can be sentenced to imprisonment for a term of up to two years, a fine or to both, or for putting people in fear of violence under *CDA 1998, s 32(1)(b)*, the person can be sentenced to imprisonment for a term of up to seven years, a fine or to both.

Where a person is summarily convicted of racially aggravated harassment under *CDA 1998, s 32(1)(a)*, the person can be sentenced to imprisonment for a term of up to six months, or a fine not exceeding the statutory maximum or to both, or for putting people in fear of violence under *CDA 1998, s 32(1)(b)*, the person can be sentenced to imprisonment for a term of up to six months or a fine not exceeding the statutory maximum or to both.

## Preventing and handling claims

### Harassment policy

**H1742** The Equal Opportunities Commission (EOC) recommends the creation and use an of equal opportunity policy by employers. In particular, the EOC's Code of Practice states:

> 'The primary responsibility at law rests with each employer to ensure that there is no unlawful discrimination. It is important, however, that measures to eliminate discrimination or promote equality of opportunity should be understood and supported by all employees. Employers are therefore recommended to involve their employees in each opportunity policies (para 4).
>
> An equal opportunities policy will ensure the effective use of human resources in the best interests of both the organisation and its employees. It is a commitment by an employer to the development and use of employment procedures and practices which do not discriminate on grounds of sex or marriage and which provide genuine equality of opportunity for all employees. The detail of the policy will vary according to size of the organisation (para 34)'.

In order for the policy to be effective and properly used, it must have the full support of senior managers within the business and must be used fairly and properly on each occasion that its provisions become relevant. So as to achieve this, it is important that the policy is clear and made known to all employees (para 35). A harassment policy is an essential feature of a proper equal opportunities policy and should exhibit the following characteristics:

(a) a complaints procedure;

(b) confidentiality;

(c) a method for informal action to be taken; and

(d) a method for formal action to be taken.

### Complaints procedure

**H1743** The hallmark of an effective policy is that provision is made for a complaints procedure that both encourages justifiable complaints of harassment to be made and protects the maker of the complaint. By way of guidance, the European Commission ('EC') Code of Practice states in this regard (at section B paragraph (iii)) that a formal complaints procedure is needed so as to:

'... give employees confidence that the organisation will take allegations of sexual harassment seriously'.

The EC Code also states that such a procedure should be used where employees feel that informal resolution of a dispute is inappropriate or has not worked.

In order for the complaints procedure to be effective, the EC Code recommends that employees should know clearly to whom to make a complaint and should also be provided with an alternative source of person to whom to complain in the event that the primary source is inappropriate (e.g. because that person is the alleged harasser).

The EC Code adds that it is good practice to allow complaints to be made to a person of the same sex as the victim.

## Confidentiality

**H1744** In order for complaints to be made effectively, it is necessary that some measure of confidentiality be maintained in the complaint making procedure. The problem that arises here is that there is an inevitable trade-off between the right of the victim of sexual harassment to be protected against recrimination and the right of an alleged harasser to know the case that he/she is meeting for disciplinary purposes. Indeed, to the extent that complaints are made against a person who is not told where, when and whom he/she is alleged to have harassed and the alleged harasser is then disciplined as a result of the complaint, there is a chance that the employer will be acting in breach of the duty to maintain the alleged harasser's trust and confidence.

To this end, the requirement for confidentiality exists to the extent that although a harasser may have the right to know the full nature of the complaint that he/she has to meet, the victim should be protected from recrimination in the workplace by the employer ensuring that all details surrounding the complaint are suppressed from other employees in the workplace save to the extent that it is necessary to involve them to either prove or disprove the allegations of harassment.

Further, the requirement of confidentiality also extends to ensuring that the allegations, to the extent that they are made in good faith, are not subsequently used against the complainant for any other purposes connected with the complainant's employment (thereby tying in with the duty of the employer to ensure that there is no victimisation occasioned to the complainant — see *Nagarajan v London Regional Transport [1999] IRLR 572* dealing with the issue of victimisation).

## Informal action

**H1745** Most victims of harassment simply want the unwanted attention or conduct towards them to stop. Whilst the employer's policy should leave employees in no doubt as to the fact that deliberate harassment will amount to a case of gross misconduct being brought against an employee, it may be the case that the form of the harassment is innocent in nature and can be remedied by informal action. As the EC Code points out, it may be the case that the victim of harassment can simply explain to the harasser that the conduct that he/she is receiving is unwelcome and makes the victim feel uncomfortable in the workplace. The EC Code goes on to state that where the victim finds it embarrassing or uncomfortable to address the harasser, it may be possible for a fellow employee (in the case of a larger organisation possibly a confidential counsellor) to make the requests to the harasser.

This said, in some cases it simply will not be possible or appropriate for the matter to be dealt with informally and where this is the case, the only possible resolution will be through formal channels.

## Formal action

**H1746** In the event that formal action is required to end harassment, both harassers and their victims should be made fully aware as to the procedures that an employer will adopt in relation to a claim of harassment. The EC Code provides an illustration as to the type of features that should be included in a policy. This will include providing details as to:

(a) how investigations are to be carried out;

(b) the time frame for investigations (which should be as short as possible for the benefit of both the victim of actual harassment and the alleged harasser);

(c) whether there is to be a period of suspension whilst a claim is investigated; and

(d) the penalty for an employee who is found to have engaged in harassment

The EC Code concludes that it may be necessary to require employees to be transferred to different duties or sites (in some cases even where a case is not proven against an alleged harasser so as to prevent the employees being forced to work with each other against their will). Indeed, where they are required to do so in the face of an unproved allegation, unless the matter has been thoroughly investigated by the employer, it may be a recipe for litigation as the case of *Balgobin v London Borough of Tower Hamlets [1987] IRLR 401* illustrates.

## Training

**H1747** Another of the statements made by the EC Code is that training of the workforce is essential in order to ensure that harassment does not occur in the first place. The EC Code specifically focuses on training for supervisory staff and managers. The recommendations for training include:

(a) identifying the factors contributing to a working environment free of sexual harassment;

(b) familiarisation of employees with the employer's harassment policy;

(c) specialist training of those employees required to implement the harassment policy; and

(d) training at induction days for new employees as to the employers policy on harassment.

## Monitoring and review

**H1748** The acid test of a company's equal opportunity policy is not how it looks on paper but whether and how it is implemented and monitored in practice. As to monitoring, much will depend upon the facts relating to particular employers. In the case of smaller employers, it may simply require the employer to be aware of any recommendations that are proposed by bodies such as the EOC or the Chartered Institute of Personnel and Development. However, where employers are larger, the monitoring function may require a full comparative analysis of complaints made in order to determine what matters need to be implemented with regard to training and whether any changes need to be made to the employer's policy for the purposes of such matters as investigating complaints and tightening up the policy generally.

In short, once the policy is adopted, it must be kept up to date with regard to what is acceptable within the workplace.

The Commission for Racial Equality's 'Code of Practice for the Elimination of Racial Discrimination and the Promotion of Equality of Opportunity in Employment (1983)' and the Secretary of State for Education and Employment's 'Code of Practice for the Elimination of Discrimination in the Field of Employment Against Disabled Persons or Persons Who Have a Disability (1996)' provide evidence of good practice and guidance for employers in relation to racial and disability discrimination.

## Future developments

### The EOC white paper — 'Equality in the 21st Century: A New Approach'

**H1749** The White Paper *Equality in the 21st Century: A New Approach* sets out what needs to be done by legislation to promote equal opportunities for men and women'. It provides a full review of the current legislation under the *SDA 1975* and the *Equal Pay Act 1970* and also examines the impact of the EOC Code of Practice in relation to harassment. The White Paper ultimately concludes that a new, unifying Act dealing with all forms of sex discrimination needs to be enacted in order for the Government to fully tackle sex discrimination.

### Recommendations in relation to sexual harassment

**H1750** Paragraph 22 of the White Paper specifically deals with the issue of sexual harassment. It recommends that the proposed new single Act should specifically prohibit sexual harassment and adopt the EOC Code definition of harassment for these purposes.

The White Paper goes on to add that the sexual harassment should be extended to cover all areas covered by the Act and not just employment.

It remains to be seen whether the White Paper will be adopted especially in the light of the fact that the Government has announced that it does not propose to enact any further changes to employment law in the lifetime of this current Parliament.

### EU proposals

**H1751** The European Union (EU) has recently proposed new legislation for outlawing sexual harassment in the workplace across Member States.

If a new law is agreed by the European Parliament the EU will for the first time have a clear definition of sexual harassment. This will be helpful for UK employers and courts who will no longer have to rely on case law and judicial interpretation to deal with the basic principles.

The law will reinforce protection for employees, both male and female, who complain about discrimination, and will acknowledge the right of women to return to their jobs after maternity leave.

The proposed new law, which amends the original 1976 directive on sexual harassment and which must be implemented by 2005, does not alter the law in the UK drastically. Much of the proposed law consolidates established UK practice. However, it is likely to have a greater impact on other EU countries. While France and Belgium have laws explicitly banning sexual harassment, most other EU Member States have only partial legislation – and Portugal and Greece have not developed this area of the law.

The proposed law defines harassment:

> 'Sexual harassment shall be deemed to be discrimination on the grounds of sex at the workplace when an unwanted conduct related to sex takes place with the purposes or effect of affecting the dignity of a person and/ or creating an intimidating, hostile or offensive or disturbing environment, in particular if a person's rejection of, or submission to, such conduct is a basis for a decision which affects that person.'

It has always been advisable for employers to have specific policies on equal opportunities and sexual harassment. However, with the proposed developments it seems that employers may now need to be more proactive in their approach and attitude. The proposed new law requires employers to take preventative measures against all forms of discrimination, especially sexual harassment, as well as introducing what are referred to as 'enterprise level equality plans' which are to be made available to workers.

# Hazardous Substances in the Workplace

## Introduction

H2101   For most employers, the main requirements relating to hazardous substances in the workplace are those contained in the *Control of Substances Hazardous to Health Regulations SI 2002 (SI 2002 No 2677)* ('*COSHH Regulations*'). This chapter is therefore mainly concerned with the requirements under those Regulations and the assessments they require.

The *COSHH Regulations* were first introduced in 1988 but have been subject to many changes and amendments since. Currently the *Control of Substances Hazardous to Health Regulations 2002* are in force, but these were amended in 2003 (by *SI 2003 No 978*) and further amendments are in the pipeline.

However, there are many other legal requirements affecting what would generally be regarded as hazardous substances, even though some of these do not fall within the definition of 'substance hazardous to health' contained in the *COSHH Regulations*. Relevant regulations include:

- *Chemicals (Hazard Information and Packaging for Supply) Regulations 2002 (SI 2002 No 1689)* (commonly known as '*CHIP*');
- *Control of Asbestos at Work Regulations 2002 (SI 2002 No 2675)*;
- *Control of Lead at Work Regulations 2002 (SI 2002 No 2676)*;
- *Control of Major Accident Hazards Regulations 1999 (SI 1999 No 743)* (known as '*COMAH*');
- *Dangerous Substances and Explosive Atmospheres Regulations 2002 (SI 2002 No 2776)* (known as '*DSEAR*')
- *Dangerous Substances (Notification and Marking of Sites) Regulations 1990 (SI 1990 No 304)*;
- *Ionising Radiations Regulations 1999 (SI 1999 No 3232)*.

The requirements of these regulations are summarised at the end of this chapter and some of them are dealt with more fully in other chapters.

## How hazardous substances harm the body
### Routes of entry

H2102   Hazardous substances can be present in the workplace in a variety of forms – solids, dusts, fumes, smoke, liquids, vapour, mists, aerosols and gases. In deciding what risks are posed by these substances, three potential entry routes into the body must be considered:

- *Inhalation* – the hazardous substance may cause damage directly to the respiratory tract or the lungs. Inhalation also allows substances to enter the bloodstream via the lungs and thus affect other parts of the body.

- *Ingestion* – accidental ingestion of hazardous substances is always possible, particularly if containers are not correctly labelled. Eating, drinking and smoking in the workplace introduce the risk of inadvertently ingesting small quantities of hazardous substances.
- *The skin* – substances can damage the skin (or eyes) through their corrosive or irritant effects. Some solvents can enter the body by absorption through the skin. Damage is more likely when there are cracks or cuts in the skin.

## Occupational ill health

**H2103** The adverse effects of hazardous substances are many and varied and are a field of study in their own right. Many types of occupational diseases must be reported to the enforcing authorities under the requirements of the *Reporting of Injuries, Diseases and Dangerous Occurrences Regulations 1995 (SI 1995 No 3163)* ('*RIDDOR*') and some are eligible for social security benefits under government schemes. Employees suffering from occupational ill health can of course bring civil actions for damages against their employers, former employers or others they consider responsible for their condition.

Some examples of occupational ill health caused by hazardous substances are:

- Respiratory problems
    - *Pneumoconiosis* – e.g. asbestosis, silicosis, byssinosis, siderosis.
    - *Respiratory irritation* – caused by inhaling acid or alkali gases or mists.
    - *Asthma* – sensitisation of the respiratory system which may be caused by many substances, e.g. isocyanates, flour, grain, hay, animal fur, wood dusts.
    - *Respiratory cancers* – e.g. those caused by certain chemicals, asbestos, pitch or tar.
    - *Metal fume fever* – a flu-like condition caused by inhaling zinc fumes.
- Poisoning

    Acute (short-term) or chronic (long-term) poisoning may be caused by:
    - *Metals and their compounds* – particularly lead, manganese, mercury, beryllium, cadmium.
    - *Organic chemicals* – which may affect the nervous system, the liver, the kidneys or the gastro-intestinal system.
    - *Inorganic chemicals* – as above plus possible asphyxiant effects such as those caused by carbon monoxide.
- Skin conditions
    - *Dermatitis*:
        - (i) caused by primary irritants where skin tissue is damaged by acids or alkalis or natural oils are removed by solvents, detergents etc.
        - (ii) due to the effects of sensitising agents, e.g. isocyanates, solvents, some foodstuffs.
    - *Skin cancer* – caused by contact with pitch, tar, soot, mineral oils etc.
- Biological problems

    From contact with animals, birds, fish (including their carcasses and products) or with micro-organisms from other sources, e.g:
    - *Livestock diseases* – e.g. anthrax or brucellosis.
    - *Allergic Alveolitis* – caused by inhaling mould or fungal spores present in grain etc.

- *Viral Hepatitis* – usually due to contact with blood or blood products.
- *Legionnaires' Disease* – caused by inhaling airborne water droplets containing the legionella bacteria (which can occur extensively in water at certain temperatures).
- *Leptospirosis* – Weil's disease, caused by contact with urine from small mammals, particularly rats.

Information on the harmful effects associated with individual products may be available on packaging but certainly should be contained in a material safety data sheet provided by the manufacturer or supplier.

## The COSHH Regulations summarised

### Substances hazardous to health

H2104   The definition of 'substance hazardous to health' is contained in the *COSHH Regulations 2002* (*SI 2002 No 2677*), *Reg 2(1)* and includes:

- substances designated as very toxic, toxic, corrosive, harmful or irritant under product labelling legislation (the *CHIP Regulations*) – these substances should be clearly identified by the standard orange and black symbols required by the *CHIP Regulations*;
- substances for which the Health and Safety Commission has approved a Maximum Exposure Limit or an Occupational Exposure Standard;
- biological agents (micro-organisms, cell cultures or human endoparasites);
- other dust of any kind, when present at a concentration in air equal or greater to 10 mg/m$^3$ (inhalable dust) or 4 mg/m$^3$ (respirable dust) as a time weighted average over an 8-hour period;
- any other substance creating a risk to health because of its chemical or toxicological properties and the way it is used or is present at the workplace;

Given this widely drawn definition, in any cases of doubt it is always advisable to treat the Regulations as applying.

### Duties under the Regulations

H2105   The *COSHH Regulations* place duties on employers in relation to their employees and also on employees themselves. [*SI 2002 No 2677, Reg 8*]. The self-employed have duties as if they were both employer and employee. *Regulation 3(1)* extends the employer's duties 'so far as is reasonably practicable' to 'any other person, whether at work or not, who may be affected by the work carried on'. Consideration must be given to:

- Others at work in the workplace, e.g. contractors, visitors or co-tenants.
- Visitors.
- Members of the public particularly in public places and buildings.
- Customers and service users.

Some of these (e.g. children) cannot be expected to behave in the same way as employees or contractors. An early prosecution under the *COSHH Regulations* was of a doctor's surgery when a young child gained access to an insecure cleaner's cupboard and drank the contents of a bottle of carbolic acid, suffering serious ill effects.

**H2106** *Hazardous Substances in the Workplace*

The main requirements of the Regulations are summarised below. Some will be explained in more detail later. Full details of the Regulations together with associated Approved Codes of Practice ('ACoPs') and guidance are contained in a single Health and Safety Executive ('HSE') booklet – *Control of substances hazardous to health* (L5).

- *Prohibitions relating to certain substances* [*SI 2002 No 2677, Reg 4*] – the manufacture, use, importation and supply of certain specified substances is prohibited (full details are contained in *Schedule 2* to the Regulations).

- *Application of Regulations 6 to 13* [*SI 2002 No 2677, Reg 5*] – exceptions are made from the application of these Regulations where other more specific regulations are in place. These relate to respirable dust in coal mines, lead, asbestos, the radioactive, explosive or flammable properties of substances, substances at high or low temperatures or high pressure and substances administered in the course of medical treatment.

- *Assessment* [*SI 2002 No 2677, Reg 6*] – an assessment of health risks must be made to identify steps necessary to comply with the Regulations and these steps must be implemented.

- *Prevention or control of exposure* [*SI 2002 No 2677, Reg 7*] – this must be achieved preferably through elimination or substitution or alternatively the provision of adequate controls, e.g. enclosure, LEV, ventilation, systems of work, personal protective equipment ('PPE'). This regulation requires a 'hierarchical' approach to be taken – this is explained in H2107 below.

- *Use of control measures etc.* [*SI 2002 No 2677, Reg 8*] – employees are required to make 'full and proper use of control measures, PPE etc.', and employers must 'take all reasonable steps' to ensure they do.

- *Maintenance, examination and testing of control measures* [*SI 2002 No 2677, Reg 9*] – control measures, including PPE, must be maintained in an efficient state. Controls must also be examined and tested periodically.

- *Monitoring exposure at the workplace* [*SI 2002 No 2677, Reg 10*] – monitoring (e.g. through occupational hygiene surveys) may be necessary in order to ensure adequate control or to protect employees' health.

- *Health surveillance* [*SI 2002 No 2677, Reg 11*] – surveillance of employees' health may be required in some circumstances.

- *Information, instruction and training* [*SI 2002 No 2677, Reg 12*] – must be provided for persons who may be exposed to hazardous substances (and also for those carrying out COSHH assessments).

- *Arrangements to deal with accidents, incidents and emergencies* [*SI 2002 No 2677, Reg 13*] – this new regulation, introduced in 2002, requires emergency procedures to be established and information on emergency arrangements to be provided.

# What the Regulations require on assessments
## Assessing the risk to health

**H2106** The wording of *Regulation 6* of the *COSHH Regulations 2002* (*SI 2002 No 2677*) was changed considerably from previous versions, with *sub-paragraph* (*2*) detailing many specific factors which must be taken into account during the assessment. The Regulation now states:

'(1) An employer shall not carry on any work which is liable to expose any employees to any substance hazardous to health unless he has—

(a) made a suitable and sufficient assessment of the risk created by that work to the health of those employees and of the steps that need to be taken to meet the requirements of these Regulations; and

(b) implemented the steps referred to in sub-paragraph (a).

(2) The risk assessment shall include consideration of—

(a) the hazardous properties of the substance;

(b) information on health effects provided by the supplier, including information contained in any relevant safety data sheet;

(c) the level, type and duration of exposure;

(d) the circumstances of the work, including the amount of the substance involved;

(e) activities, such as maintenance, where there is the potential for a high level of exposure;

(f) any relevant occupational exposure standard, maximum exposure limit or similar occupational exposure limit;

(g) the effect of preventive and control measures which have been or will be taken in accordance with regulation 7;

(h) the results of relevant health surveillance;

(i) the results of monitoring of exposure in accordance with regulation 10;

(j) in circumstances where the work will involve exposure to more than one substance hazardous to health, the risk presented by exposure to such substances in combination;

(k) the approved classification of any biological agent; and

(l) such additional information as the employer may need in order to complete the risk assessment.

(3) The risk assessment shall be reviewed regularly and forthwith if—

(a) there is reason to suspect that the risk assessment is no longer valid;

(b) there has been a significant change in the work to which the risk assessment relates; or

(c) the results of any monitoring carried out in accordance with regulation 10 show it to be necessary,

and where, as a result of the review, changes to the risk assessment are required, those changes shall be made.

(4) Where the employer employs 5 or more employees, he shall record—

(a) the significant findings of the risk assessment as soon as practicable after the risk assessment is made; and

(b) the steps which he has taken to meet the requirements of regulation 7.'

In carrying out assessments, it should be noted that *Regulation 3(1)* also requires employees to take account, so far as is reasonably practicable, of others who may be affected by the work, e.g. visitors, contractors, customers, passers-by and members of the emergency services.

## Prevention or control of exposure

**H2107** The *COSHH Regulations 2002* (*SI 2002 No 2677*), *Reg 6(1)(a)* state that the assessment must take into account the steps needed to meet the requirements of the Regulations. *Regulation 7* of the *COSHH Regulations 2002* sets out a clear hierarchy of measures which must be taken to prevent or control exposure to hazardous substances. It states:

'(1) Every employer shall ensure that the exposure of his employees to substances hazardous to health is either prevented or, where this is not reasonably practicable, adequately controlled.

(2) In complying with his duty of prevention under paragraph (1), substitution shall by preference be undertaken, whereby the employer shall avoid, so far as is reasonably practicable, the use of a substance hazardous to health at the workplace by replacing it with a substance or process which, under the conditions of its use, either eliminates or reduces the risk to the health of the employees.

(3) Where it is not reasonably practicable to prevent exposure to a substance hazardous to health, the employer shall comply with his duty of control under paragraph (1) by applying protection measures appropriate to the activity and consistent with the risk assessment, including in order of priority –

　(a) the design and use of appropriate work processes, systems and engineering controls and the provision and use of suitable work equipment and materials;

　(b) the control of exposure at source, including adequate ventilation systems and appropriate organisational measures; and

　(c) where adequate control of exposure cannot be achieved by other means, the provision of suitable personal protective equipment in addition to the measures required by sub-paragraphs (a) and (b).'

Thus a clear order of preference is established:

1. Prevention of exposure (by elimination or substitution).
2. Appropriate work processes, systems and engineering controls (e.g. enclosure) and suitable work equipment and materials.
3. Adequate control at source (by ventilation or organisational measures).
4. Adequate control through PPE.

(*Sub-paragraph (4)* of *Regulation 7* contains further details of what the measures referred to in *paragraph (3)* must include – these are referred to in H2109.)

## Prevention of exposure

**H2108** The first preference should always be to prevent exposure to hazardous substances. This could be achieved by:

- Changing work methods, e.g. adopting ultrasonic techniques or high-pressure water jets for cleaning purposes rather than using solvents.

- Using non-hazardous (or less hazardous) alternatives, e.g. replacing solvent-based paints or inks by water-based ones (or ones utilising less hazardous solvents).

- Using the same substances but in less-hazardous forms, e.g. substituting powdered materials by granules or pellets, using more dilute concentrations of hazardous substances.

- Modifying processes, e.g. eliminating the production of hazardous by-products, emissions or waste by changing process parameters such as temperature or pressure.

Exposure must be prevented 'so far as is reasonably practicable' – a definition of this qualifying phrase is provided in the chapter on RISK ASSESSMENT. The level of risk will be determined by the type and severity of the hazards associated with a substance and its circumstances of use.

Considerations weighing against the prevention of exposure being reasonably practicable might be the costs associated with using alternative methods or materials, the detrimental effect of alternatives on the product or additional risks which may be introduced by the alternatives.

For example:

- ultrasonic cleaning may not achieve the required standards of cleanliness;
- use of high pressure water can involve additional risks;
- water-based paints may result in increased problems from corrosion;
- alternative solvents may be more flammable.

However, much progress has been made in recent years in the substitution of hazardous materials. The HSE booklet *Seven steps to successful substitution of hazardous substances* provides useful guidance.

## Control measures, other than PPE

**H2109** The *COSHH Regulations 2002* (*SI 2002 No 2677*), *Reg 7(4)* states that the control measures required by *paragraph (3)* shall include:

'(a) arrangements for the safe handling, storage and transport of substances hazardous to health, and of waste containing such substances, at the workplace;

(b) the adoption of suitable maintenance procedures;

(c) reducing, to the minimum required for the work concerned—

   (i) the number of employees subject to exposure,

   (ii) the level and duration of exposure, and

   (iii) the quantity of substances hazardous to health present at the workplace;

(d) the control of the working environment, including appropriate general ventilation; and

(e) appropriate hygiene measures including adequate washing facilities.'

Examples of such control measures include:

- Enclosed or covered process vessels.
- Enclosed transfer systems, e.g. use of pumps or conveyors, rather than open transfer.
- Use of closed and clearly labelled containers.
- Plant, processes and systems of work which minimise the generation of, or suppress and contain, spills, leaks, dust, fumes and vapours. (This often involves a combination of partial enclosure and the use of local exhaust ventilation ('LEV')).
- Adequate levels of general ventilation.
- Appropriate maintenance procedures.
- Minimising the quantities of hazardous substances stored and used in workplaces.
- Restricting the numbers of persons in areas where hazardous substances are used.
- Prohibiting eating, drinking and smoking in areas of potential contamination.
- Providing showers in addition to normal washing facilities.

**H2110** *Hazardous Substances in the Workplace*

- Regular cleaning of walls and other surfaces.
- Using signs to indicate areas of potential contamination.
- Safe storage, handling and disposal of hazardous substances (including hazardous waste).

*Paragraphs* (5) and (6) of *Regulation 7* set out measures required to control exposure to carcinogens and biological agents – these are dealt with later in H2113.

## Control using PPE

**H2110** Control of exposure using PPE should be the last option, where prevention or adequate control by other means are not reasonably practicable. This might be the case where:

- the extent of use of hazardous substances is very small;
- adequate control by other means is impracticable from a technical viewpoint;
- control by other means is excessively expensive or difficult in relation to the level of risk;
- PPE is used as a temporary control measure pending the implementation of adequate control by other means;
- employees are dealing with emergency situations;
- infrequent maintenance activities are being carried out.

PPE necessary to control risks from hazardous substances might be:

- respiratory protective equipment (RPE);
- protective clothing;
- hand or arm protection (usually gloves);
- eye protection;
- protective footwear.

The *COSHH Regulations 2002* (*SI 2002 No 2677*), Reg 7(9) state that:

> 'Personal Protective Equipment provided by an employer in accordance with this regulation shall be suitable for the purpose and shall –
>
> (a) comply with any provision in the Personal Protective Equipment Regulations 2000 which is applicable to that item of personal protective equipment; or
>
> (b) in the case of respiratory protective equipment, where no provision referred to in sub-paragraph (a) applies, be of a type approved or shall conform to a standard approved, in either case, by the Executive.'

HSE publishes detailed guidance on PPE generally and on RPE in particular.

## Adequate control

**H2111** The *COSHH Regulations 2002* (*SI 2002 No 2677*), Reg 7(7), (8) and (11) define what is meant by the term 'adequate control' in respect of inhalation risks from hazardous substances:

> '(7) Without prejudice to the generality of paragraph (1), where there is exposure to a substance for which a maximum exposure limit has been approved, control of exposure shall, so far as the inhalation of that substance is concerned, only be treated as being adequate if the level of exposure is reduced so far as is reasonably practicable and in any case below the maximum exposure limit.

(8) Without prejudice to the generality of paragraph (1), where there is exposure to a substance for which an occupational exposure standard has been approved, control of exposure shall, so far as the inhalation of that substance is concerned, only be treated as being adequate if –

   (a) that occupational exposure standard is not exceeded; or

   (b) where that occupational exposure standard is exceeded, the employer identifies the reasons for the standard being exceeded and takes appropriate action to remedy the situation as soon as is reasonably practicable.

(11) In this regulation, "adequate" means adequate having regard only to the nature of the substance and the nature and degree of exposure to substances hazardous to health and "adequately" shall be construed accordingly.'

## MEL approved

**H2112** Where a Maximum Exposure Limit ('MEL') has been approved, the level of exposure must be reduced so far as is reasonably practicable as well as always being kept below the MEL at all times. More flexibility is given in respect of substances where an Occupational Exposure Standard ('OES') has been approved. If the OES is exceeded the employer must identify the reasons and remedy the situation as soon as is reasonably practicable. Listings of MELs and OESs are published regularly by HSE.

In some cases it will be relatively easy to make a judgement as to whether exposure is controlled to levels well within the MEL or OES but in other situations it may be necessary to carry out an occupational hygiene survey or even arrange for routine or periodic monitoring of exposure.

## Carcinogens and biological agents

**H2113** The *COSHH Regulations 2002 (SI 2002 No 2677)*, *Reg 7(5)* and *(6)* set out a number of specific measures which must be applied where it is not reasonably practicable to prevent exposure to carcinogens and biological agents. (These are in addition to the measures required by *paragraph (3)* of the regulation.)

Measures specified for controlling carcinogens are:

- totally enclosing process and handling systems (unless not reasonably practicable);
- prohibition of eating, drinking and smoking in areas that might be contaminated;
- cleaning of floors, walls and other surfaces regularly and whenever necessary;
- designating areas of potential contamination by warning signs;
- storing, handling and disposing of carcinogens safely, using closed and clearly labelled containers.

Appendix 1 to the HSE ACoP booklet provides further detail on the control of carcinogenic substances whilst Annex 2 of the booklet contains a background note on occupational cancer.

Measures for controlling biological agents are:

- use of warning signs (a biohazard sign is shown in *Schedule 3* of the Regulations);
- specifying appropriate decontamination and disinfection procedures;
- safe collection, storage and disposal of contaminated waste;
- testing for biological agents outside confined areas;
- specifying work and transportation procedures;
- where appropriate, making available effective vaccines;

- instituting appropriate hygiene measures;
- control and containment measures where human patients or animals are known (or suspected to be) infected with a Group 3 or 4 biological agent.

The *COSHH Regulations 2002 (SI 2002 No 2677), Reg 7(10)* require work with biological agents to be in accordance with *Schedule 3* to the Regulations which is contained in the HSE COSHH ACoP booklet (L5). This sets out several detailed requirements, including standards for containment measures.

## Practical aspects of COSHH assessments

### The requirements summarised

H2114   COSHH assessments must involve the following:

- identification of risks to the health of employees or others;
- consideration of whether it is reasonably practicable to prevent exposure to hazardous substances creating risks (by elimination or substitution);
- if prevention is not reasonably practicable, identification of the measures necessary to achieve adequate control of exposure, as required by the *COSHH Regulations 2002 (SI 2002 No 2677), Reg 7* (these control measures are described at H2109 AND H2110);
- identification of other measures necessary to comply with the *COSHH Regulations 2002 (SI 2002 No 2677), Regs 8–13*, e.g:
    - use of control measures;
    - maintenance, examination and test of control measures etc;
    - monitoring of exposure at the workplace;
    - health surveillance;
    - provision of information, instruction and training;
    - arrangements for accidents, incidents and emergencies.

    (These are described more fully at H2127–H2135 BELOW).

The assessment must be 'suitable and sufficient', and its complexity will vary considerably dependent on the circumstances. In some cases it may only be necessary to study relevant safety data sheets and decide whether existing practices are sufficient to achieve adequate control. However, for more complex processes and variable activities involving greater numbers of hazardous substances, the assessment must be much more detailed. The planning, preparation and implementation of a COSHH assessment will be very similar to other types of risk assessment and much of the guidance given in the chapter on RISK ASSESSMENT will be valid here.

### Who should carry out COSHH assessments?

H2115   *Regulation 7* of the *Management of Health and Safety at Work Regulations 1999 (SI 1999 No 3242)* requires employers to appoint a competent person or persons to assist them in complying with their duties, which will include carrying out COSHH assessments. *Regulation 12(4)* of the *COSHH Regulations 2002 (SI 2002 No 2677)* also states:

> 'Every employer shall ensure that any person (whether or not his employee) who carries out any work in connection with the employer's duties under these Regulations has suitable and sufficient information, instruction and training.'

The degree of knowledge and experience required will depend upon the circumstances. An assessment in a workplace containing a few simple uses of hazardous substances may be carried out by someone without any specialist qualifications or experience but with the capability to understand and apply the principles contained in this chapter. (However, such a person may need access to someone with greater health and safety knowledge and a more detailed understanding of the *COSHH Regulations*.) Some qualified and experienced individuals may be capable of carrying out detailed assessments of relatively high-risk situations themselves, although they are likely to need to consult others during the assessment process.

Higher risk situations, with much more complex and significant use of hazardous substances are more likely to require multi-disciplinary teams, some of whom have specialist knowledge and experience. Possible candidates for such a team might be:

- health and safety officers or managers;
- occupational hygienists;
- occupational health nurses;
- physicians with occupational health experience;
- chemical or process engineers;
- maintenance or ventilation engineers.

## *How should the assessments be organised?*

**H2116** In larger workplaces it is often best to divide them into manageable assessment units which could be based on:

- Departments or sections.
- Buildings or rooms.
- Process lines.
- Activities or services.

Where an assessment team is used, those individuals most suited for a particular situation can be selected to assess that unit.

## *Gathering information together*

**H2117** Prior to the assessment, information should be gathered together, particularly about which hazardous substances are present and what hazards are associated with them. Substances to be taken into account include:

- Raw materials used in processes.
- The contents of stores and cupboards.
- Substances produced by processes:
    — intermediate compounds;
    — products;
    — waste and by-products;
    — emissions from the process.
- Substances used in maintenance and cleaning work.
- Buildings and the work environment:
    — surface treatments;
    — pollution or contaminants;

- — bird droppings, animal faeces etc;
- — possible legionella sources.
- Substances involved in the activities of others, e.g. contractors.

Potential sources of information about the hazards associated with these substances include:

- Manufacturers' or suppliers' safety data sheets.
- Information provided on containers or packaging (including the symbols required by the *CHIP Regulations* (*SI 2002 No 1689*).
- HSE publications and the HSE website.
- Other reference books and technical literature.
- Direct enquiries to specialists or suppliers.

This information will primarily concern the hazards associated with the substance but is also likely to contain recommendations on control methods. However, the substance's actual circumstances of use may constitute a greater or lesser degree of risk than the manufacturer or supplier anticipated – it is for the assessor(s) to determine what control measures are necessary in each situation.

Other useful information may be available from previous assessments or surveys. Even if previous COSHH assessments were inadequate or are now out of date, they may still contain information of value in the assessment process. The results of previous occupational hygiene surveys (e.g. dust, gas or vapour concentrations) are also likely to be useful as will the collective results of any health surveillance work which has been carried out.

Details of existing control measures may be contained in operating procedures, health and safety rule books or listings of PPE requirements or may be referred to within training programmes. Specification information may be available for some control measures, e.g. design flow rates for ventilation equipment or performance standards for PPE such as respiratory protection or gloves. (Some of this information may also be sought during the assessment itself or prior to the preparation of the assessment record.)

## *Observations in the workplace*

**H2118** Time spent in making COSHH assessments can be reduced and made more productive by good planning and preparation. Nevertheless, observations of workplaces and work activities must still be an integral part of the COSHH assessment process.

Sufficient observations must be made to arrive at a conclusion on the level of risk and the adequacy of control measures for all the hazardous substances in the workplace – be they raw materials, products, by-products, waste, emissions etc. This should include the manner and extent to which employees (and others) are exposed and an evaluation of the equipment and working practices intended to achieve their control.

Aspects to be considered include:

- Are specified procedures being followed?
- Does local exhaust ventilation or general ventilation appear effective?
- Is PPE being used correctly?
- Is there evidence of leakage, spillage or dust accumulation?
- What equipment is available for cleaning?
- Are dusts, fumes or strong smells evident in the atmosphere?
- Are there any restrictions relating to eating, drinking or smoking?

- What arrangements are there for storage, cleaning or maintenance of PPE?
- Are there suitable arrangements for washing, showering or changing clothing?
- Are special arrangements for laundering clothing required?
- What arrangements are in place for storage and disposal of waste?
- Is there potential for major spillages or leaks?
- What emergency containment equipment or PPE is available?

## Discussions with staff

H2119    It is also important to talk to those working with hazardous substances, their safety representatives and those managing or supervising their work. Questions might relate to working practices, awareness of risks or precautions, the effectiveness of precautions or experience of problems. These might include:

- Why tasks are done in particular ways?
- Are the present work practices typical?
- How effective is the LEV/general ventilation?
- What happens if it ever breaks down?
- What types of PPE are required for the work?
- Are there any problems with the PPE?
- Have workers experienced any health problems?
- How often is the workplace cleaned?
- What methods are used for cleaning?
- How is waste disposed of?
- Is there any possibility of leaks occurring or emergencies arising?
- What would happen in such situations?
- What are the arrangements for washing/cleaning/maintaining PPE?

## Further assistance

H2120    Guidance on COSHH assessments is available in the HSE booklet *A step by step guide to COSHH assessment*. The HSE's Advisory Committee on Toxic Substances has also developed *COSHH essentials* which provides generic risk assessments for a wide range of substances. It is available in printed form via HSE Books but can also be accessed electronically via the HSE website. It contains advice on control measures for common tasks, e.g. mixing, weighing, spray painting.

## Further investigations

H2121    Assessment in the workplace may reveal the need for further tests on the effectiveness of control measures before the assessment can be concluded. The tests may be in the form of occupational hygiene surveys – usually to measure the airborne concentrations of dust, fume, vapour or gas and compare them with the relevant MEL or OES. Alternatively there may be a need to measure the effectiveness of local exhaust ventilation or general ventilation.

Further investigations may also be necessary into matters such as:

- hazards associated with substances which were not identified during the preparatory phase of the assessment;
- whether the specifications for PPE found in use are adequate for the exposure involved;
- the feasibility of using alternative work methods (e.g. preventing exposure) or alternative methods of control to those currently in place.

# COSHH assessment records

## Preparation of assessment records

H2122  Much of the guidance on note taking and the preparation of assessment records contained in the chapter on RISK ASSESSMENT is relevant to COSHH assessments. Illustrative examples of completed assessment records are provided at H2124 below, although it should be stressed that no single record format will automatically cater for all types of workplaces or work activities. The essential components which must be included in COSHH assessment records are:

- the work activities involving risks from hazardous substances;
- information as to the hazardous substances involved (and their form, e.g. liquid, powder, dust etc.);
- control measures which are (or should be) in place;
- improvements identified as being necessary.

The COSHH ACoP does state that assessments need not be recorded in the simplest and most obvious cases which can be easily repeated and explained at any time. However, it suggests that in most cases assessments will need to be recorded and kept readily accessible to those who may need to know the results. The amount of information should be proportionate to the risks posed by the work. In many cases it may be acceptable to state that in the circumstances of use there is little or no risk and that no more detailed assessment is necessary. The two illustrative COSHH assessment records (below) demonstrate the contrasting levels of detail appropriate for different situations. The COSHH ACoP provides further details of what might need to be included in more comprehensive assessment records. Employees or their representatives should be informed of the results of COSHH assessments.

## Review and implementation of recommendations

H2123  Once the assessment has been completed and the record has been prepared it is important that recommendations are reviewed, implemented and followed up. The review process is likely to involve people who were not involved in the original assessments – senior managers, or staff with relevant technical expertise. Whilst the assessment team may be prepared to make changes to their assessment findings or recommendations, they should not allow themselves to be pressurised into making changes they find unacceptable. If senior management choose not to implement COSHH assessment recommendations then they must accept the responsibility for their actions.

Responsibilities for implementing each recommendation within a designated timescale must be clearly allocated to individuals, possibly as part of an overall action plan. Each of these recommendations must then be followed up to ensure that they have actually been carried out and also to check that additional risks have not been created inadvertently. The assessment record should then be annotated to indicate that the recommendation has been completed (alternatively a new record could be prepared reflecting the improved situation).

## Example assessment records
H2124

Acorn Estate Agents, Newtown

## Risk Assessment

| Reference number: 6 | Risk topic/issue: Hazardous substances | Sheet 1 of 1 |
|---|---|---|
| **Risks identified** | **Precautions in place** | **Recommended improvements** |
| The substances listed below could present risks to both staff and clients: | | |
| Office materials | Office supplies are kept in a cupboard in a part of the office not normally accessible to clients. | |
| With warning symbols | Generally substances are only used in very small quantities for short periods presenting no significant risk. | |
| Old correction fluid (harmful) | | |
| New correction fluid (flammable) | | |
| Spray adhesive (harmful) | If the spray adhesive is used for more than a couple of minutes, a nearby window is opened which provides adequate ventilation. | |
| Without warning symbols | | |
| Photocopier and printer toners | The photocopier is used in well-ventilated areas and there are no noticeable ozone smells, even on long copying runs. | Provide disposable gloves for cleaning significant spillages of photocopier or printer toner. |
| Various felt tip pens (solvent based) | | |
| Cleaning materials | The cleaner's cupboard is kept locked except when substances are being removed. | |
| With warning symbols | Suitable gloves are provided (and worn) for handling the irritant and corrosive substances in concentrated form. | |
| Thick bleach (Irritant) | | Investigate replacing the thick bleach by a more dilute solution. |
| Polish stripper (Irritant) | | |

# Hazardous Substances in the Workplace

| Acid descaler (Corrosive) | The cleaner is aware of the risks of mixing bleach with other substances e.g. the acid descaler. | | Ensure the relief cleaner is also made aware of these risks. |
|---|---|---|---|
| **Without warning symbols** <br> Furniture polish <br> Floor polish <br> Window and glass cleaner | | | |
| **Signature(s)** K Stephenson, R Lewis | **Name(s)** K Stephenson, R Lewis | | Date 4/11/98 |
| **Dates for** | **Recommendation follow up:** December 1998 | | **Next routine review:** November 2003 |

**Rainbow Products**

## COSHH Assessment

| | |
|---|---|
| Assessment unit | Mixing Hall |
| Activities | Manufacture of solvent-based paints and other surface treatments. Solvents are pumped into the mixing vessels from an external tank farm. Solid constituents and some liquid components are charged into the mixing vessels from platforms above. After mixing, the products are pumped directly to filling stations in a neighbouring building. |
| Substances used (Data sheet file references) | 4, 7, 14, 23, 49, 50, 51, 60, 73, 80, 92, 106, 117. Detailed formulations for each product are available from the Quality Control Department. |
| Main risks | Dust from solid constituents during charging.<br><br>Solvent vapours:   – escaping from the charging hatches;<br>                  – from minor leaks at valves and pipe connections;<br>                  – from liquid components during charging.<br><br>Entry into mixing vessels for cleaning or maintenance purposes. |
| Those at risk | Mixing Hall production employees.<br>Maintenance staff when working in the area.<br>Contractors involved in vessel cleaning or maintenance work |

| | |
|---|---|
| Controls and other precautions in place | Good general ventilation (specification 20 air changes per hour).<br>Annual surveys of solvent vapours show levels well below all OESs.<br>Hoods with LEV over each vessel charging position. These are examined and tested annually by the Maintenance Department.<br>Exposure Monitoring Surveys (Aug 1996, May 1999) show production employees as well within OESs for all dusts and solvents.<br>Disposable dust masks are available for vessel charging, although their use is not compulsory.<br>A portable vacuum cleaning unit with suitable filter is available when required.<br>Maintenance carry out an annual physical inspection of all mixer vessels and pipelines.<br>A weigh station is provided in a fume cupboard for weighing out smaller quantities of solid constituents (this also is examined and tested annually).<br>Entry into vessels is controlled by the permit to work system.<br>All employees are subject to the company's annual health screening programme. |
| Observations | Discarded empty paper sacks were strewn on several loading platforms.<br>Liquid component transfer containers had been left open on platforms 1 and 4 (still containing residual materials).<br>The ventilation at charging stations 2 and 4 appeared inadequate.<br>There were dust accumulations on all loading platforms.<br>A brush appears to have been used to sweep up dust on some platforms.<br>A pipe flange below mixer 3 had developed a small leak. |
| Recommendations | 1. Improve the sack disposal containers on all loading platforms.<br>2. Remind staff of the importance of disposing of sacks correctly.<br>3. Remind staff that liquid transfer containers should have their lids replaced after use.<br>4. Rectify the ventilation at charging stations 2 and 4.<br>5. Introduce simply weekly checks on the ventilation at all charging stations.<br>6. Introduce weekly cleaning for all loading platforms.<br>7. Remind staff of the importance of using vacuum methods for cleaning.<br>8. Investigate obtaining vacuum cleaning units which can be lifted onto the platforms more easily. |

| | |
|---|---|
| 9. Repair the leaking flange below mixer 3.<br>10. Increase the frequency of pipeline inspections to six monthly.<br>11. Develop a standard procedure for entering vessels for cleaning or maintenance purposes (linked to the permit to work system).<br>12. Investigate alternative cleaning methods avoiding the need for entry, e.g. immersion in solvent over weekend periods. | |
| **Signature(s)** A Storm, P Gold | **Date** 2 July 2001 |
| **Recommendation follow up:** October 2001 | |
| **Assessment Review:** July 2003 | |

## Ongoing reviews of assessments

**H2125** The *COSHH Regulations 2002* (*SI 2002 No 2677*), *Reg 6(3)* require COSHH assessments to:

'be reviewed regularly and forthwith if—

(a) there is reason to suspect the risk assessment is no longer valid;

(b) there has been a significant change in the work to which the risk assessment relates; or

(c) the results of any monitoring carried out in accordance with regulation 10 show it to be necessary.'

- *Assessments no longer valid*

    Assessments might be shown to be no longer valid because of:

    — new information received about health risks, e.g. information from suppliers, revised HSE guidance, changes in the MEL or OES;

    — results from inspections or thorough examinations or tests (*Regulation 9*), e.g. indicating fundamental flaws in engineering controls;

    — regular reports or complaints about defects in control arrangements;

    — the results from workplace exposure monitoring (*Regulation 10*), e.g. showing the OES or MEL is regularly being exceeded (or approached);

    — results from health surveillance (*Regulation 11*), e.g. demonstrating an unsatisfactory or deteriorating position;

    — a confirmed case of an occupational disease.

- *Significant changes*

    Changes necessitating a review of an assessment might involve:

    — the types of substances used, their form or their source;

    — equipment used in the process or activity (including control measures);

    — altered methods of work or operational procedures;

    — variations in volume, rate or type of production;

    — staffing levels and related practical difficulties or pressures.

- *Regular review*

    The periods elapsing between reviews should relate to the degree of risk involved and the nature of the work itself. The COSHH ACoP previously stated that assessments should be reviewed at least every five years, but no longer contains such a recommendation. However, this would seem to be a reasonable maximum period to use for practical purposes.

    A review would not necessarily require a revision of the assessment – it may conclude that existing controls are still adequate despite changed circumstances. However, where changes are shown to be required, *paragraph* (*3*) of *Regulation 6* requires that these be implemented.

# Further requirements of the COSHH Regulations

**H2126** Some recommendations from COSHH assessments could relate to the ongoing maintenance of effective control measures and many of these aspects are covered by further specific requirements of the *COSHH Regulations*.

## Use of control measures

**H2127** The *COSHH Regulations 2002 (SI 2002 No 2677), Reg 8* place duties on both employers and employees in respect of the proper use of control measures, including PPE.

'(1) Every employer who provides any control measure, other thing or facility in accordance with these Regulations shall take all reasonable steps to ensure that it is properly used or applied as the case may be.

(2) Every employee shall make full and proper use of any control measure, other thing or facility provided in accordance with these Regulations and where relevant shall—

    (a) take all reasonable steps to ensure it is returned after use to any accommodation provided for it; and,

    (b) if he discovers any defect therein, report it forthwith to his employer.'

Employers have duties under the *Management of Health and Safety at Work Regulations 1999 (SI 1999 No 3242)* to monitor all of their health and safety arrangements. Areas where monitoring of COSHH control measures are likely to be necessary are:

- correct use of LEV equipment;
- compliance with specified systems of work;
- compliance with PPE requirements;
- storage and maintenance of PPE;
- compliance with requirements relating to eating, drinking or smoking;
- condition of washing and showering facilities and personal hygiene standards;
- whether defects are being reported by employees.

Where employees are unwilling to use control measures properly, employers should consider the use of disciplinary action, particularly in relation to persistent offenders.

## Maintenance, examination and testing of control measures

**H2128** The *COSHH Regulations 2002 (SI 2002 No 2677), Reg 9* contain both general and specific requirements for the maintenance of control measures:

'(1) Every employer who provides any control measure to meet the requirements of regulation 7 shall ensure that, where relevant, it is maintained in an efficient state, in efficient working order, in good repair and in a clean condition.

(2) Where engineering controls are provided to meet the requirements of regulation 7, the employer shall ensure that thorough examination and testing of those controls is carried out—

    (a) in the case of local exhaust ventilation plant, at least once every 14 months, or for local exhaust ventilation plant used in conjunction with a process specified in Column 1 of Schedule 4, at not more that the interval specified in the corresponding entry in Column 2 of that Schedule; or

    (b) in any other case, at suitable intervals.

(3) Where respiratory protective equipment (other than disposable respiratory protective equipment) is provided to meet the requirements of regulation 7, the employer shall ensure that thorough examination and, where appropriate, testing of that equipment is carried out at suitable intervals.

(4) Every employer shall keep a suitable record of the examinations and tests carried out in pursuance of paragraphs (2) and (3) and of repairs carried out as a result of those

examinations and tests, and that record or a suitable summary thereof shall be kept available for at least 5 years from the date on which it was made.'

*Paragraphs* (5), (6) and (7) contain requirements specifically relating to PPE – these are examined later.

## General maintenance of controls

**H2129** The COSHH ACoP states that, where possible, all engineering control measures should receive a visual check at least once every week. Such checks may simply confirm that there are no apparent leaks from vessels or pipes and that LEV or cleaning equipment appear to be in working order. No records of such checks need to be kept, although it is good practice (and prudent) to do so.

The *COSHH Regulations 2002* (*SI 2002 No 2677*), *Reg 9(2)* require thorough examinations and tests of engineering controls. Requirements relating to LEV are reviewed below at H2130 but for other engineering controls such examinations and tests must be 'at suitable intervals', and suitable records must be kept for at least five years. The nature of examinations and tests will depend upon the engineering control involved and the potential consequences of its deterioration or failure. Examples might involve:

- detailed visual inspections of tanks and pipelines;
- inspections and non-destructive testing of critical process vessels;
- testing of detectors and alarm systems;
- planned maintenance of general ventilation equipment;
- checks on filters in vacuum cleaning equipment.

Persons carrying out maintenance, examinations and testing must be competent for the purpose in accordance with the *COSHH Regulations 2002* (*SI 2002 No 2677*), *Reg 12(4)*.

## Local exhaust ventilation (LEV) plant

**H2130** Most LEV systems must be thoroughly examined and tested at least once every 14 months, although *Schedule 4* of the *COSHH Regulations 2002* (*SI 2002 No 2677*), requires increased frequencies for LEV used in conjunction with a handful of specified processes. Dependent upon the design and purpose of the LEV concerned, the examination and test might involve visual inspection, air flow or static pressure measurements or visual checks of efficiency using smoke generators or dust lamps. In some cases air sampling to confirm efficiency levels, filter integrity tests or checks on air flow sensors, may be appropriate. Detailed guidance is available in an HSE booklet *The maintenance, examination and testing of local exhaust ventilation*.

## Personal Protective Equipment (PPE)

**H2131** All types of PPE are subject to the general maintenance requirements contained in the *COSHH Regulations 2002* (*SI 2002 No 2677*), *Reg 9(1)* whilst *Regulation 9(3)* contains specific requirements relating to non-disposable respiratory protective equipment ('RPE').

Further requirements in respect of PPE are contained in *Regulation 9(5)*, (6) and (7) of the 2002 Regulations that state:

'(5) Every employer shall ensure that personal protective equipment, including protective clothing , is:

    (a) properly stored in a well-defined place;

    (b) checked at suitable intervals; and

(c) when discovered to be defective, repaired or replaced before further use.

(6) Personal protective equipment which may be contaminated by a substance hazardous to health shall be removed on leaving the working area and kept apart from uncontaminated clothing and equipment.

(7) The employer shall ensure that the equipment referred to in paragraph (6) is subsequently decontaminated and cleaned or, if necessary, destroyed.'

Some types of PPE can easily be seen to be defective by the user whilst in other cases (e.g. for gloves or clothing providing protection against strongly corrosive chemicals) it may be appropriate to introduce more formalised inspection systems.

For non-disposable RPE the COSHH ACoP states that thorough examinations and, where appropriate, tests should be made *at least* once every month, although it suggests that for RPE used less frequently, periods up to three months are acceptable. Alternatively, examination and testing prior to next use may be more appropriate. An HSE booklet *The selection, use and maintenance of respiratory protective equipment* provides detailed guidance on the subject.

## Monitoring exposure at the workplace

**H2132** Reference was made earlier in the chapter at H2111 and H2121 to the possible need to carry out air testing in the workplace as part of the COSHH assessment process in order to determine the adequacy of control measures. Such testing may also be necessary in order to ensure that adequate control continues to be maintained and a requirement for this is contained in the *COSHH Regulations 2002 (SI 2002 No 2677), Reg 10*:

'(1) Where the risk assessment indicates that—

(a) it is requisite for ensuring the maintenance of adequate control of the exposure of employees to substances hazardous to health; or

(b) it is otherwise requisite for protecting the health of employees,

the employer shall ensure that the exposure of employees to substances hazardous to health is monitored in accordance with a suitable procedure.

(2) Paragraph (1) shall not apply where the employer is able to demonstrate by another method of evaluation that the requirements of regulation 7(1) have been complied with.

(3) The monitoring referred to in paragraph (1) shall take place—

(a) at regular intervals; and

(b) when any change occurs which may affect that exposure.

(4) Where a substance or process is specified in Column 1 of Schedule 5, monitoring shall be carried out at least at the frequency specified in the corresponding entry in Column 2 of that Schedule.

(5) The employer shall ensure that a suitable record of any monitoring carried out for the purpose of this regulation is made and maintained and that record or a suitable summary thereof is kept available—

(a) where the record is representative of the personal exposures of identifiable employees, for at least 40 years; or

(b) in any other case, for at least 5 years,

from the date of the last entry made in it.

(6) Where an employee is required by regulation 11 to be under health surveillance, an individual record of any monitoring carried out in accordance with this regulation shall be made, maintained and kept in respect of that employee.

(7) The employer shall—
- (a) on reasonable notice being given, allow an employee access to his personal monitoring record;
- (b) provide the Executive with copies of such monitoring records as the Executive may require; and
- (c) if he ceases to trade, notify the Executive forthwith in writing and make available to the Executive all monitoring records kept by him.'

Schedule 5 to the *COSHH Regulations 2002* (*SI 2002 No 2677*) automatically requires monitoring to be carried out in processes involving vinyl chloride monomer and electrolytic chromium plating.

A number of techniques are available for monitoring air quality in the workplace. Most of these involve the use of chemical indicator tubes, direct reading instruments, or sampling pumps and filter heads. An HSE booklet *Monitoring strategies for toxic substances* provides further guidance on the subject.

## Health surveillance

H2133 The *COSHH Regulations 2002* (*SI 2002 No 2677*), *Reg 11(1)* and (*2*) contain the main requirements relating to the need for health surveillance.

'(1) Where it is appropriate for the protection of the health of his employees who are, or are liable to be, exposed to a substance hazardous to health, the employer shall ensure that such employees are under suitable health surveillance.

(2) Health surveillance shall be treated as being appropriate where—
- (a) the employee is exposed to one of the substances specified in Column 1 of Schedule 6 and is engaged in a process specified in Column 2 of that Schedule, and there is a reasonable likelihood that an identifiable disease or adverse health effect will result from that exposure; or
- (b) the exposure of the employee to a substance hazardous to health is such that—
  - (i) an identifiable disease or adverse health effect may be related to the exposure,
  - (ii) there is a reasonable likelihood that the disease or effect may occur under the particular conditions of his work, and
  - (iii) there are valid techniques for detecting indications of the disease or the effect,

and the technique of investigation is of low risk to the employee.

The remaining parts of *Regulation 10* (*paragraphs (3)–(11)*) relate to the manner in which health surveillance is conducted and used, together with the maintenance of and access to surveillance records.

The decision as to whether health surveillance is appropriate to protect the health of employees is one that would normally be taken at the time of a COSHH assessment or during its subsequent review. However, for those processes and substances specified in *Schedule 6* to the Regulations, surveillance must be carried out.

Normally health surveillance programmes would be initiated and carried out under the overall supervision of a registered medical practitioner, and preferably one with relevant occupational health experience. However, the surveillance itself may be carried out by an occupational health nurse, a technician or a responsible member of staff, providing that individual was competent for the purpose.

There are many different procedures available for health surveillance including:

- Biological monitoring, e.g. tests of blood, urine or exhaled air.
- Biological effect monitoring, e.g. through lung function testing.
- Medical surveillance, e.g. through physical examinations.
- Interviews about possible symptoms, e.g. skin abnormalities or shortness of breath.

For health surveillance to be 'appropriate' there must firstly be a significant enough risk to justify it and there must also be a valid technique for detecting indications of related occupational diseases or ill-health effects. HSE provide specialist guidance in their booklet *Health surveillance at work*. Records of health surveillance must contain information specified in the COSHH ACoP and must be retained for at least 40 years.

## Information, instruction and training

**H2134** The *COSHH Regulations 2002* (*SI 2002 No 2677*), *Reg 12* contain requirements relating to information, instruction and training for persons who may be exposed to substances hazardous to health.

'(1) Every employer who undertakes work which is liable to expose an employee to a substances hazardous to health shall provide that employee with suitable and sufficient information, instruction and training.

(2) Without prejudice to the generality of paragraph (1), the information, instruction and training provided under that paragraph shall include—

    (a) details of the substances hazardous to health to which the employee is liable to be exposed including—

        (i) the names of those substances and the risk which they present to health,

        (ii) any relevant occupational exposure standard, maximum exposure limit or similar occupational exposure limit,

        (iii) access to any relevant safety data sheet, and

        (iv) other legislative provisions which concern the hazardous properties of those substances;

    (b) the significant findings of the risk assessment;

    (c) the appropriate precautions and actions to be taken by the employee in order to safeguard himself and other employees at the workplace;

    (d) the results of any monitoring of exposure in accordance with regulation 10 and, in particular, in the case of any substance hazardous to health for which a maximum exposure limit has been approved, the employee or his representatives shall be informed forthwith, if the results of such monitoring show that the maximum exposure limit is exceeded;

    (e) the collective results of any health surveillance undertaken in accordance with regulation 11 in a form calculated to prevent those results from being identified as relating to a particular person; and

    (f) where employees are working with a Group 4 biological agent or material that may contain such an agent, the provision of written instructions and, if appropriate, the display of notices which outline the procedures for handling such an agent or material.

(3) The information, instruction and training required by paragraph (1) shall be—

(a) adapted to take account of significant changes in the type of work carried out or methods of work used by the employer; and

(b) provided in a manner appropriate to the level, type and duration of exposure identified by the risk assessment.

(4) Every employer shall ensure that any person (whether or not his employee) who carries out work in connection with the employer's duties under these Regulations has suitable and sufficient information, instruction and training.

(5) Where containers and pipes for substances hazardous to health used at work are not marked in accordance with any relevant legislation listed in Schedule 7, the employer shall, without prejudice to any derogations provided for in that legislation, ensure that the contents of those containers and pipes, together with the nature of those contents and any associated hazards, are clearly identifiable.'

The general requirements of *paragraph (1)* match those found in various other regulations but the contents of *paragraph (2)* are much more prescriptive than the requirements of previous versions of the *COSHH Regulations*. Whilst it is important that workers are aware of the substances they are exposed to and the risks that they present, only a minority are likely to comprehend fully the significance of OESs and MELs, or understand all of the contents of a typical safety data sheet. However, a good awareness of the risks will mean employees are more likely to take the appropriate precautions. Instruction and training will be particularly relevant in relation to:

- awareness of safe systems of work;
- correct use of LEV equipment;
- use, adjustment and maintenance of PPE (especially RPE);
- the importance of good personal hygiene standards;
- requirements relating to eating, drinking and smoking;
- emergency procedures;
- arrangements for cleaning and the disposal of waste;
- contents of containers and pipes (see *Regulation 12(5)* above).

Employers must also inform employees or their representatives of the results of workplace exposure monitoring forthwith if the MEL is shown to have been exceeded. Employees must also be informed of the *collective* results of health surveillance, e.g. the average results within a department or on a particular shift from biological monitoring, or the numbers of employees referred for further investigation following a skin inspection.

The type of information, instruction and training provided must be appropriate for the level, type and exposure involved and adapted to take account of significant changes.

Any person carrying out work on the employer's behalf (whether or not an employee) is required by *paragraph (4)* to have the necessary information, instruction and training. Thus a consultant carrying out a COSHH assessment or an occupational hygienist conducting exposure monitoring must be verified by the employer as being competent for the purpose and be provided with the information necessary to carry out their work effectively.

## Accidents, incidents and emergencies

**H2135** *Regulation 13* was introduced in the *COSHH Regulations 2002 (SI 2002 No 2677)* and requires employers to ensure that arrangements are in place to deal with accidents, incidents and emergencies related to hazardous substances. (This is in addition to the general duty to have procedures 'in the event of serious and imminent danger' contained in *Regulation 8* of the

*Management of Health and Safety at Work Regulations 1999 (SI 1999 No 3242.)* Based on HSE guidance in the COSHH ACoP booklet (L5) such emergencies might include:

- a serious process fire posing a serious risk to health;
- a serious spillage or leakage of a hazardous substance;
- a failure to contain biological or carcinogenic agents;
- a failure that could lead (or has led) to a sudden release of chemicals;
- a threatened significant exposure over an OES or MEL, e.g. due to a failure of LEV or other controls.

The regulation is rather lengthy and prescriptive in its requirements which are summarised below.

Arrangements should include:

- emergency procedures, such as:
  — appropriate first aid facilities;
  – relevant safety drills (tested at regular intervals);
- providing information on emergency arrangements:
  — including details of work hazards and emergencies likely to arise;
  — made available to relevant accident and emergency services;
  — displayed at the workplace, if appropriate;
- suitable warning and other communications:
  — to enable an appropriate response, including remedial actions and rescue operations.

In the event of an accident, incident or emergency, the employer must:

- take immediate steps to:
  — mitigate its effects;
  — restore the situation to normal;
  — inform employees who may be affected;
- ensure only essential persons are permitted in the affected area and that they are provided with:
  — appropriate PPE;
  — any necessary safety equipment and plant;
- in the case of a serious biological incident, inform employees or their representatives, as soon as practicable of:
  — the causes of the incident or accident;
  — the measures taken or being taken to rectify the situation.

*Regulation 13(4)* states that such emergency arrangements are not required where the risk assessment shows that because of the quantities of hazardous substances the risks are slight and control measures are sufficient to control that risk. (However, this 'exception' does not apply in the case of carcinogens or biological agents.)

*Regulation 13(5)* requires employees to report possible releases of a biological agent forthwith to their employer (or another employee with specific responsibility).

HSE guidance in the ACoP booklet states that whether or not arrangements are required under *Regulation 13* is a matter for judgement, based on the potential size and severity of accidents and emergencies which may occur. Many incidents will be capable of being dealt with by the control measures required by the *COSHH Regulations 2002* (*SI 2002 No 2677*), Reg 7. Even if an emergency does occur, the response should be proportionate – a small leak would not necessarily justify an evacuation of the workplace. The ACoP booklet provides further details of what emergency procedures might need to include, such as:

- the identity, location and quantities of hazardous substances present;
- foreseeable types of accidents, incidents or emergencies;
- special arrangements for emergencies not covered by general procedures;
- emergency equipment and PPE, and who is authorised to use it;
- first-aid facilities;
- emergency management responsibilities e.g. emergency controllers;
- how employees should respond to incidents;
- clear up and disposal arrangements;
- arrangements for regular drills or practice;
- dealing with special needs of disabled employees.

## Other important regulations

H2136    The introduction to this chapter referred to several other legal requirements relating to hazardous substances in the workplace and these are summarised below.

### Chemicals (Hazard Information and Packaging for Supply) Regulations 2002

H2137    The *Chemicals (Hazard Information and Packaging for Supply) Regulations 2002* (*SI 2002 No 1689*) (commonly known as 'CHIP') impose duties on suppliers (and importers into the EU) of substances classified as dangerous for supply. They must provide recipients with safety data sheets containing a number of specified types of information, including details on hazards; first aid, fire fighting and accidental release measures; handling and storage; exposure controls and personal protection.

Substances must be labelled according to the categories of danger they present, i.e:

- Physico-chemical, e.g. explosive, oxidising, highly flammable.
- Health, e.g. very toxic, toxic, harmful, corrosive, irritant.
- Environmental, e.g. dangerous to the environment.

This will involve the use of the standard orange and black symbols, usually accompanied by one of a number of specified risk phrases or safety phrases.

Suitable packaging must be provided for substances which:

- is designed and constructed to prevent escape or spillage of the controls;
- will not be adversely affected by its contents;
- where fitted with a replaceable closure, can be re-closed without contents escaping.

Like the *COSHH Regulations*, the *CHIP Regulations* are constantly subject to amendment and updating and employers directly affected by their detailed technical requirements should consult HSE either directly or via their publications or website.

## Control of Asbestos at Work Regulations 2002

H2138 Earlier regulations on asbestos predated the original *COSHH Regulations* and thus provided a framework on which the *COSHH Regulations* were based. They continue to follow a similar pattern, involving an assessment of work which exposes employees to asbestos and the preparation of a suitable written plan of work which must prevent or reduce exposure to asbestos. Many other requirements mirror COSHH requirements, e.g. information, instruction and training; use of control measures; maintenance of control measures; requirements for air monitoring (plus standards for air testing and analysis); and health records and medical surveillance.

There are also a number of other specific requirements including those relating to the notification of work with asbestos (to HSE), to the storage, distribution and labelling of raw asbestos and asbestos waste and to the supply of products containing asbestos (which in both cases must carry specified labels).

Significant work with asbestos insulation, asbestos coating and asbestos insulating board usually requires a licence from HSE under the *Asbestos (Licensing) Regulations 1983 (SI 1983 No 1649)*. However, minor work with such materials, or work with lower risk asbestos-containing materials (asbestos cement or materials of bitumen, plastic, resins or rubber which contain asbestos) can be carried out without requiring a licence. Both licensed and unlicensed work must comply with the requirements of the *Control of Asbestos at Work Regulations 2002 (SI 2002 No 2675)* and HSE provide detailed practical guidance on how this should be achieved.

## 'Duty to manage asbestos in non-domestic premises'

H2139 This extremely important new requirement is contained in *Regulation 4* of the *Control of Asbestos at Work Regulations 2002 (SI 2002 No 2675)*. It will have a major impact on many employers as asbestos-containing materials ('ACMs'). (See also APPENDIX in ASBESTOS.)

ACMs were commonly used as building materials in the 1950s, 1960s and 1970s and the use of asbestos-containing building products was not finally banned until 1999.

*Regulation 4* does not come into operation until 21 May 2004 but employers should prepare early to meet its requirements. HSE has already published two key booklets – an ACoP booklet *The management of asbestos in non-domestic premises* and *A comprehensive guide to managing asbestos in premises* which contains much practical guidance. They also have a free guidance leaflet on the subject.

The 'duty holder' under this regulation will depend on the terms of any lease or contract or upon who is in 'control' of the premises. This may be the owner or leaseholder, or the employer occupying the premises, or it could be a combined responsibility (the HSE ACoP booklet provides guidance on the subject). Owners have responsibility for non-domestic premises which are unoccupied.

Other persons are required to co-operate with duty holders under the regulation. This would mean occupiers providing owners with relevant information when they vacate premises but would also require contractors, suppliers and architects involved in constructing or refurbishing premises, to provide duty holders with information about materials used.

Duty holders under *Regulation 4* must:

- take reasonable steps to find ACMs and check their condition;
- presume materials contain asbestos unless there is strong evidence to the contrary;
- keep a written record of the location and condition of actual and presumed ACMs and keep it up to date;
- assess the risk of people being exposed to asbestos;

- prepare and put into effect a plan to manage that risk.

The duty holder must ensure that anyone carrying out work as a result of the regulation is competent for the task in hand. If a separate organisation or individual is appointed, they must:

- have adequate training and experience;
- demonstrate independence, impartiality and integrity;
- have an adequate quality management system;
- carry out any survey work in accordance with recommended guidance (HSE guidance booklet MDHS 100).

MDHS 100 specifies three types of survey which can be carried out. Organisations can be accredited under ISO 17020 to undertake ACM surveys and individual surveyors under EN 45013.

## *Finding ACMs and assessing their condition*

H2140   Information about possible ACMs should be gathered together from sources such as:

- plans and specifications;
- information from builders, architects and material suppliers;
- previous asbestos survey work.

A survey of the premises must then be carried out and materials categorised as:

- not containing asbestos materials (some are obviously wood, glass, metal, brick, stone etc.);
- materials known to be ACMs;
- materials presumed to be ACMs, until there is evidence to the contrary.

Some materials may be inaccessible, e.g. covered by wallpaper, contained within walls or fire doors or in positions which could be dangerous to access such as roof voids. In some cases it may be appropriate to make further investigations as to their composition or it may be acceptable just to presume that they are ACMs for the time being (see H2141 below). Some areas may only require a relatively straightforward inspection whilst others may justify a detailed survey including sampling of suspected ACM.

The locations and condition of actual and presumed ACMs must be clearly recorded on drawings or sketches as part of the assessment record. The use of photographs may be particularly useful.

HSE guidance sets out in detail what should be included in the assessment record.

## *Risk management plan*

H2141   The duty holder must prepare a written plan for managing the risks from ACMs and this must include measures such as:

- clear identification of actual and presumed ACMs (by signs, colour coding etc.);
- restriction of access to or maintenance work in ACM areas (through locked doors, use of permits to work or other control systems);
- analysis of suspect material prior to its disturbance;
- treatment of ACMs which may release fibres (sealing or safe removal);
- protection of ACMs from damage (e.g. from vehicles, trolleys);

- regular monitoring of ACMs to check for deterioration, disturbance or damage (including damage by vandalism);
- ensuring information about actual and presumed ACMs is readily available to those who might disturb it (and the emergency services).

Both the risk management plan and the ACM assessment must be reviewed if there is reason to suspect they are no longer valid or there are significant changes.

## Control of Major Accident Hazard Regulations 1999 (COMAH)

H2142 The *Control of Major Accident Hazard Regulations 1999 (SI 1999 No 743)* only apply to sites containing specified quantities of dangerous substances. They require the preparation of both on-site and off-site emergency plans with the objectives of:

- containing and controlling incidents so as to minimise their effects, and to limit damage to persons, the environment and property;
- implementing the measures necessary to protect persons and the environment from the effects of major accidents;
- communicating the necessary information to the public and to the emergency services and authorities concerned in the area;
- providing for the restoration and clean-up of the environment following a major accident.

Such plans must be prepared utilising risk assessment techniques. HSE have a number of booklets providing guidance on the methodology to be followed and the parameters to be taken into account.

## Control of Lead at Work Regulations 2002

H2143 Like asbestos, previous regulations on lead preceded the original *COSHH Regulations* and the *Control of Lead at Work Regulations 2002 (SI 2002 No 2676)* continue to follow similar principles. The Regulations together with an ACoP and related guidance are contained in an HSE booklet *The control of lead at work* which is supported by other HSE guidance material.

## Dangerous Substances and Explosive Atmospheres Regulations 2002 (DSEAR)

H2144 The *DSEAR Regulations (SI 2002 No 2776)* are concerned with protecting against risks from fire and explosion. They replaced several other regulations including the *Highly Flammable Liquids and Liquified Petroleum Gases Regulations 1972 (SI 1972 No 917)* and (apart from the dispensing of petrol from pumps) the requirement for licensing under the *Petroleum (Consolidation) Act 1928*.

### Dangerous substances

H2145 The *DSEAR Regulations (SI 2002 No 2776)* contain a detailed definition of 'dangerous substances' including:

- explosive, oxidising, extremely flammable, highly flammable or flammable substances or preparations (whether or not classified under *CHIP Regulations*), e.g. petrol, solvents, paints, LPG;
- other substances or preparations creating risks due to their physico-chemical or chemical properties;
- potentially explosive dusts e.g. flour, sugar, custard, pitch, wood.

## Explosive atmosphere

**H2146** This is defined as 'a mixture, under atmospheric conditions of air and one or more dangerous substances in the form of gases, vapours, mists or dusts, in which, after ignition has occurred, combustion spreads to the entire unburned mixture'.

The *DSEAR Regulations (SI 2002 No 2776)* require employers to:

- Carry out a risk assessment of work activities involving dangerous substances [*SI 2002 No 2776, Reg 5*].

- Eliminate or reduce risks as far as is reasonably practicable [*SI 2002 No 2776, Reg 6*].

- Classify places where explosive atmospheres may occur in zones, and ensure that equipment and protective systems in these places meet appropriate standards, marking zones with signs, where necessary [*SI 2002 No 2776, Reg 7*] (this duty is being phased in)

- Provide equipment and procedures to deal with accidents and emergencies [*SI 2002 No 2776, Reg 8*].

- Provide employees with suitable and sufficient information, instruction and training [*SI 2002 No 2776, Reg 9*].

*Regulation 6* requires control measures to be implemented according to a specified order of priority which can be summarised as:

- Reduce the quantity of dangerous substances to a minimum.

- Avoid or minimise releases of dangerous substances.

- Control releases at source.

- Prevent the formation of an explosive atmosphere (including providing appropriate ventilation).

- Collect, contain and remove any releases to a safe place or otherwise render them safe.

- Avoid the presence of ignition sources (including electrostatic discharges).

- Avoid adverse conditions that could lead to danger (e.g. by temperature or other controls).

- Segregation of incompatible dangerous substances.

The regulation also requires measures to mitigate the detrimental effects of a fire or explosion (or other harmful physical effects). These include minimising the number of employees exposed and the provision of explosion pressure relief arrangements or explosion suppression equipment.

## Dangerous Substances (Notification and Marking of Sites) Regulations 1990

**H2147** The *Dangerous Substances (Notification and Marking of Sites) Regulations 1990 (SI 1990 No 304)* are primarily for the benefit of the fire service and other emergency services. Both the fire authority and HSE must be provided with notification of dates when it is anticipated that a total quantity of 25 tonnes of 'dangerous substances' will be present on the site. (The definition of 'dangerous substances' is contained in the Regulations.) Further notifications must be made where significant changes occur in substances present, including cessation or reductions. Sites must also be marked with standard signs as specified in *Schedule 3* to the Regulations. A booklet providing detailed guidance on the regulations is available from HSE.

## Ionising Radiation Regulations 1999

**H2148** *Regulation 7* of the *Ionising Radiation Regulations 1999* (*SI 1999 No 3232*) require employers to carry out a risk assessment before commencing any new activity involving work with ionising radiation. The assessment must be:

' ... sufficient to demonstrate that—

(a) all hazards with the potential to cause a radiation accident have been identified; and

(b) the nature and magnitude of the risks to employers and other persons arising from those hazards have been evaluated.'

Where such radiation risks are identified, all reasonably practicable steps must be taken to:

- prevent any such accident;
- limit its consequences should such an accident occur;
- provide employees with necessary information, instruction, training and equipment necessary to restrict their exposure.

The Regulations contain many detailed requirements on the prevention of radiation accidents and related control measures. An HSE booklet contains the Regulations and an ACoP and guidance on the regulations. Much more guidance is available from HSE on specific aspects of radiation safety.

# Joint Consultation in Safety – Safety Representatives, Safety Committees, Collective Agreements and Works Councils

## Introduction

**J3001** The provision of 'information, consultation and participation' and of 'health protection and safety at the workplace' are outlined as two of the 'fundamental social rights of workers' in accordance with the Community Charter of Fundamental Rights of 1989. Combined with Article 138 of the Consolidated Version of the Treaty Establishing the European Community (i.e. Treaty of Rome 1957), the Health and Safety Directive 1989 (resulting in 'the six pack' collection of regulations) and the United Kingdom's signing of the Social Chapter in 1998, the EU has provided great stimuli for the introduction of extensive obligations to consult on health and safety issues.

At national level, there has already been change to the obligations owed by employers to their workers. The *Working Time Regulations 1998* (*SI 1998 No 1833*) ('*Working Time Regulations*') were introduced in October 1998 and have subsequently been amended.

Furthermore, the obligations owed under the European Works Councils Directive 1994 have been transposed into national law under the *Transnational Information and Consultation of Employees Regulations 1999* (*SI 1999 No 3323*) ('*Works Councils Regulations*').

Most recently, there has been the 'Mini Works Councils Directive' (Directive 2002/19/EC). This is to be introduced on a phased basis into all workplaces with 50 or more workers (in this regard, the Government has issued a consultation document, '*High performance workplaces: Informing and consulting employees*' and draft regulations, and consultation ends on 7 November 2003). Consultation with the workforce is likely to become a central issue for industrial relations and health and safety over the next few years.

The Health and Safety Commission ('HSC') is also currently seeking to promote change. In 1999, it published the consultative document '*Employee consultation and involvement in health and safety*', which made a number of recommendations in relation to health and safety consultation. In particular, it recommended:

- giving trade union safety representatives the legal power to issue their employer with provisional improvement notices requiring the employer to rectify stated defects within a fixed period of time;

- extending the unfair dismissal protection provided to 'whistleblowers' to health and safety representatives where the representatives recommend that workers leave their place of work due to circumstances giving rise to a serious, imminent and unavoidable danger at the workplace;

- extending access to specific information to all employees of an employer;

**J3002** *Joint Consultation in Safety – Safety Representatives, Safety Committees, Collective Agreements and Works Councils*

- the introduction of specially trained union safety representatives whose function is to undertake workplace inspections at the workplaces of different employers where their members work; and

- extending protection against victimisation to safety representatives.

HSC is proposing to publish a further consultation document on proposed legislation in this area in the early part of 2004.

This chapter looks at the existing domestic legislation that places an employer under an obligation to consult with its workforce.

## Regulatory framework

**J3002** The legal requirements for consultation are more convoluted than might be expected, principally because there are two groups of workers who are treated as distinct for consultation purposes. Until 1996 only those employers who recognised a trade union for any collective bargaining purpose were obliged to consult the workforce through safety representatives. The *Safety Representatives and Safety Committees Regulations 1977* (*SI 1977 No 500*) (*as amended*) ('*Safety Representatives Regulations*'), which were made under the provisions of the *Health and Safety at Work etc Act 1974* ('*HSWA 1974*'), s 2(6), came into effect in 1978 and introduced the right for recognised trade unions to appoint safety representatives. The *Safety Representatives Regulations* were amended in 1993 – by the *Management of Health and Safety at Work Regulations 1992* (*SI 1992 No 2051*) (now superceded by the *Management of Health and Safety at Work Regulations 1999* (*SI 1999 No 3242*)) – to extend employers' duties to consult and provide facilities for safety representatives.

The duty to consult was extended on 1 October 1996 by virtue of the *Health and Safety (Consultation with Employees) Regulations 1996* (*SI 1996 No 1513*) ('*Consultation with Employees Regulations*'), which have subsequently been amended. The Regulations were introduced as a 'top up' to the *Safety Representatives Regulations*, extending the obligation upon employers to consult all of their employees about health and safety measures. The *Consultation with Employees Regulations* expanded the obligation beyond just trade union appointed representatives. When introduced, the Regulations addressed the general reduction of union recognition over the preceding five years. However, the more recent trend in workforces, driven by the twin forces of Europe and the current Labour Government, is for greater worker participation in the employer's undertaking and the current framework facilitates this whether or not a union is the conduit for worker consultation.

The obligation of consultation, its enforcement and the details of the role and functions of safety representatives and representatives of employee safety, whether under the *Safety Representatives Regulations* or the *Consultation with Employees Regulations*, are almost identical but, for the sake of clarity, are dealt with separately below. The primary distinction is that different obligations apply depending upon whether the affected workers are unionised or not. The respective Regulations also cover persons working in host employers' undertakings.

There are specific supplemental provisions which apply to offshore installations, which are governed by the *Offshore Installations (Safety Representatives and Safety Committees) Regulations 1989* (*SI 1989 No 971*). Likewise, in the case of the education sector, HSC has published the 'Safety Representatives' Charter' ('the Charter'). The Charter has been developed to:

- Promote and emphasise the rights, roles and functions of safety representatives and to actively promote the involvement of safety representatives in the education sector's efforts to improve health and safety.

- Motivate employers, safety representatives and employees to work in partnership to develop a positive safety culture throughout the education sector.

- Raise awareness amongst employers of the important contribution of safety representatives towards the development of such a culture.

- Encourage employers to demonstrate their full commitment towards consulting and involving safety representatives in matters of health, safety and welfare.

- Increase the participation of education sector employees and their safety representatives in health and safety activities.

- Contribute towards an improved health and safety performance which aims to reduce accidents and ill health in the education sector.

In particular, the Charter sets out the legal obligations that employers in the education sector owe under the *Safety Representatives Regulations* and the *Consultation with Employees Regulations*.

Whilst it is not legislation itself, the Charter is evidence of best practice within the education sector.

Although the *Consultation with Employees Regulations* and the *Safety Representatives Regulations* are based upon good industrial relations practice, there is still the possibility of a prosecution by the Health and Safety Executive ('HSE') inspectors of employers who fail to consult their workforce on health and safety issues. There is substantial overlap, however, with employment protection legislation and the obligations in respect of consultation – as can be seen from the protective rights which are conferred upon safety representatives, such as the right not to be victimised or subjected to detriment for health and safety activities, together with the consequential right to present a complaint to an employment tribunal if they are dismissed or suffer a detriment as a result of carrying out their duties.

The *Consultation with Employees Regulations* also extend these rights to the armed forces. However, armed forces representatives are to be appointed rather than elected, and no paid time off is available. The Regulations do not apply to sea-going ships.

The Working Time Directive 1994 has resulted in the *Working Time Regulations 1998 (SI 1998 No 1833)* (subsequently amended) under national law. These Regulations allow collective modification of the night working requirements in *Regulation 6*, the daily rest provisions in *Regulation 10*, the weekly rest provisions in *Regulation 11*, the rest break provisions in *Regulation 12* and also allow modification of the averaging period for calculating weekly working time under *Regulation 4* where the same is either for objective, technical or organisational reasons. The method of modification is either by collective agreement in the case of a unionised workforce or by workforce agreement (as defined in the *Working Time Regulations 1998 (SI 1998 No 1833), Sch 1*) in the case of non-unionised employees.

In relation to pan-European employers, obligations are owed to workers under the European Works Councils Directive 1994. The United Kingdom adopted the Directive on 15 December 1997 and provisions were enacted into national law by the *Works Councils Regulations (SI 1999 No 3323)* and came into force on 15 January 2000. Whilst the *Works Councils Regulations* do not specifically encompass health and safety obligations, these issues are within the remit of a European Works Council.

## Consultation obligations for unionised employers

**J3003** Under the *Safety Representatives Regulations (SI 1977 No 500)*, a trade union has the right to appoint an individual to represent the workforce in consultations with the employer on all matters concerning health and safety at work, and to carry out periodic inspections of the workplace for hazards. Every employer has a duty to consult such union-appointed safety representatives on health and safety arrangements (and, if they so request him, to establish a safety committee to review the arrangements – see J3017 below). [*HSWA 1974, s 21*].

**J3004** *Joint Consultation in Safety – Safety Representatives, Safety Committees, Collective Agreements and Works Councils*

## General duty

**J3004** The general duty of the employer under the *Safety Representatives Regulations (SI 1977 No 500)* is to 'consult with safety representatives with regard to both the making and maintaining of arrangements that will enable the employer and its workforce to co-operate in promoting and developing health and safety at work, and monitoring its effectiveness'. General guidance has been issued by HSC in the form of the Codes of Practice '*Safety Representatives and Safety Committees*' and '*Time Off for the Training of Safety Representatives*' to which regard should be had generally.

## Appointment of safety representatives

**J3005** The right of appointment of safety representatives was, until 1 October 1996, restricted to independent trade unions who are recognised by employers for collective bargaining purposes. The *Consultation with Employees Regulations (SI 1996 No 1513)* extended this to duly elected representatives of employee safety, as detailed above.

The terms 'independent' and 'recognised' are defined in the *Safety Representatives Regulations (SI 1977 No 500)* and follow the definitions laid down in the *Trade Union and Labour Relations (Consolidation) Act 1992, ss 5* and *178(3)* respectively. The *Safety Representatives Regulations* make no provision for dealing with disputes which may arise over questions of independence or recognition (this is dealt with in the *Trade Union and Labour Relations (Consolidation) Act 1992, ss 6* and *8* – the amendments that have been made to the *Trade Union and Labour Relations (Consolidation) Act 1992* by the *Employment Relations Act 1999* in relation to recognition of unions do not help as *Sch A1* is confined to recognition disputes concerning pay, hours and holiday).

Safety representatives must be representatives of recognised independent trade unions, and it is up to each union to decide on its arrangements for the appointment or election of its safety representatives. [*Safety Representatives Regulations (SI 1977 No 500), Reg 3*]. Employers are not involved in this matter, except that they must be informed in writing of the names of the safety representatives appointed and of the group(s) of employees they represent. [*Safety Representatives Regulations (SI 1977 No 500), Reg 3(2)*].

The *Safety Representatives Regulations (SI 1977 No 500), Reg 8*, state that safety representatives must be employees except in the cases of members of the Musicians' Union and actors' Equity. In addition, where reasonably practicable, safety representatives should have at least two years' employment with their present employer or two years' experience in similar employment. The HSC guidance notes advise that it is not reasonably practicable for safety representatives to have two years' experience, or employment elsewhere, where:

- the employer is newly established,
- the workplace is newly established,
- the work is of short duration, or
- there is high labour turnover.

The same general guidance is followed for employee safety representatives under the *Consultation with Employees Regulations (SI 1996 No 1513)*.

## Number of representatives for workforce

**J3006** The *Safety Representatives Regulations (SI 1977 No 500)* do not lay down the number of safety representatives that unions are permitted to appoint for each workplace. This is a matter for unions themselves to decide, having regard to the number of workers involved and the hazards to which they are exposed. HSE's view is that each safety representative should be regarded as responsible for the interests of a defined group of workers. This approach has not been found to conflict with existing workplace trade union organisation based on defined groups

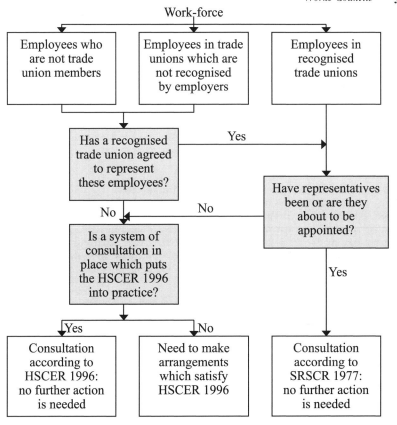

of workers. The size of these groups varies from union to union and from workplace to workplace. While normally each workplace area or constituency would need only one safety representative, additional safety representatives are sometimes required where workers are exposed to numerous or particularly severe hazards; where workers are distributed over a wide geographical area or over a variety of workplace locations; and where workers are employed on shiftwork.

## Role of safety representatives

**J3007** The *Safety Representatives Regulations (SI 1977 No 500), Reg 4(1)* (as amended by the *Management of Health and Safety at Work Regulations 1992 (SI 1992 No 2051)* – now superseded by the *Management of Health and Safety at Work Regulations 1999 (SI 1999 No 3242)*) lists a number of detailed functions for safety representatives:

(a)  to investigate potential hazards and causes of accidents at the workplace;

(b)  to investigate employee complaints concerning health etc. at work;

(c)  to make representations to the employer on matters arising out of (a) and (b) and on general matters affecting the health etc. of the employees at the workplace;

(d)  to carry out the following inspections (and see J3008 below):

- of the workplace (after giving reasonable written notice to the employer – see *Safety Representatives Regulations, Reg 5*);

- of the relevant area after a reportable accident or dangerous occurrence (see accident reporting) or if a reportable disease is contracted, if it is safe to do so and in the interests of the employees represented (see *Safety Representatives Regulations (SI 1977 No 500), Reg 6*);

- of documents relevant to the workplace or the employees represented which the employer is required to keep (see *Safety Representatives Regulations (SI 1977 No 500), Reg 7*) – reasonable notice must be given to the employer;

(e) to represent the employees they were appointed to represent in consultations with HSE inspectors, and to receive information from them (see J3016 below);

(f) to attend meetings of safety committees.

These functions are interrelated and are to be implemented proactively rather than reactively. Safety representatives should not just represent their members' interests when accidents or near-misses occur or at the time of periodic inspections, but should carry out their obligations on a continuing day-to-day basis. The *Safety Representatives Regulations* make this obligation clear by stating that safety representatives have the functions of investigating potential hazards and members' complaints *before* accidents, as well as investigating dangerous occurrences and the causes of accidents *after* they have occurred. These functions are only assumed when the employer has been notified in writing by the trade union or workforce of the identity of the representative.

Thus safety representatives may possibly be closely involved not only in the technical aspects of health, safety and welfare matters at work, but also in those areas which could be described as quasi-legal. In other words, they may become involved in the interpretation and clarification of terminology in the *Safety Representatives Regulations*, as well as in discussion and negotiation with employers as to how and when the regulations may be applied. This would often happen in committee meetings.

Under the *Safety Representatives Regulations (SI 1977 No 500), Reg 4A(1)* (introduced by the *Management of Health and Safety at Work Regulations 1992*), the subjects on which consultation 'in good time' between employers and safety representatives should take place are:

- the introduction of any new measure at a workplace which may substantially affect health and safety;

- arrangements for appointing competent persons to assist the employer with health and safety and implementing procedures for serious and imminent risk;

- any health and safety information the employer is required to provide; and

- the planning and organisation of health and safety training and health and safety implications of the introduction (or planning) of new technology.

The safety representative's terms of reference are, therefore, broad, and exceed the traditional 'accident prevention' area. For example, the *Safety Representatives Regulations (SI 1977 No 500), Reg 4(1)* empowers safety representatives to investigate 'potential hazards' and to take up issues which affect standards of health, safety and welfare at work. In practice it is becoming clear that four broad areas are now engaging the attention of safety representatives and safety committees:

- health;
- safety;
- environment;
- welfare.

These four broad areas effectively mean that safety representatives can, and indeed often do, examine standards relating, for example, to noise, dust, heating, lighting, cleanliness, lifting and carrying, machine guarding, toxic substances, radiation, cloakrooms, toilets and canteens. The protective standards that are operating in the workplace, or the lack of them, are now coming under much closer scrutiny than hitherto.

## Workplace inspections

**J3008** Arrangements for three-monthly and other more frequent inspections and reinspections should be by joint arrangement. The TUC advises that the issues to be discussed with the employer can include:

- more frequent inspections of high risk or rapidly changing areas of work activity;
- the precise timing and notice to be given for formal inspections by safety representatives;
- the number of safety representatives taking part in any one formal inspection;
- the breaking-up of plant-wide formal inspections into smaller, more manageable inspections;
- provision for different groups of safety representatives to carry out inspections of different parts of the workplace;
- the kind of inspections to be carried out, e.g. safety tours, safety sampling or safety surveys;
- the calling in of independent technical advisers by the safety representatives.

Although formal inspections are not intended to be a substitute for day-to-day observation, they have on a number of occasions provided an opportunity to carry out a full-scale examination of all or part of the workplace and for discussion with employers' representatives about remedial action. They can also provide an opportunity to inspect documents required under health and safety legislation, e.g. certificates concerning the testing of equipment. It should be emphasised that, during inspections following reportable accidents or dangerous occurrences, employers are not required to be present when the safety representative talks with its members. In workplaces where more than one union is recognised, agreements with employers about inspections should involve all the unions concerned. It is generally agreed that safety representatives are also allowed under the Regulations to investigate the following:

- potential hazards;
- dangerous occurrences;
- the causes of accidents;
- complaints from their members.

This means that imminent risks, or hazards which may affect their members, can be investigated right away by safety representatives without waiting for formal joint inspections. Following an investigation of a serious mishap, safety representatives are advised to complete a hazard report form, one copy being sent to the employer and one copy retained by the safety representative.

## Rights and duties of safety representatives

### Legal immunity

**J3009** Ever since the *Trade Disputes Act 1906*, trade unions (and employers' associations) have enjoyed immunity from liability in tort for industrial action, taken or threatened, in contemplation or furtherance of a trade dispute (although such freedom of action was subsequently curtailed by the *Trade Union and Labour Relations (Consolidation) Act 1992, s 20*). Not surprisingly, perhaps, this immunity extends to their representatives acting in a lawful capacity. Thus, the *Safety Representatives Regulations (SI 1977 No 500)* state that none of the functions of a safety representative confers legal duties or responsibilities [*Safety Representatives Regulations (SI 1977 No 500), Reg 4(1)*]. As safety representatives are not legally responsible for health, safety

**J3010** *Joint Consultation in Safety – Safety Representatives, Safety Committees, Collective Agreements and Works Councils*

or welfare at work, they cannot be liable under either the criminal or civil law for anything they may do, or fail to do, as a safety representative under the *Safety Representatives Regulations*. This protection against criminal or civil liability does not, however, remove a safety representative's legal responsibility as an employee. Safety representatives must carry out their responsibilities under the *HSWA 1974*, s 7 if they are not to be liable for criminal prosecution by an HSE inspector. These duties as an employee are to take reasonable care for the health and safety of one's self and others, and to co-operate with one's employer as far as is necessary to enable him to carry out his statutory duties on health and safety.

## Time off with pay

### General right

**J3010** Under the *Safety Representatives Regulations* (*SI 1977 No 500*), safety representatives are entitled to take such paid time off during working hours as is necessary to perform their statutory functions, and reasonable time to undergo training in accordance with a Code of Practice approved by HSC.

### Definition of 'time off'

**J3011** The *Safety Representatives Regulations* (*SI 1977 No 500*), Reg 4(2) provides that the employer must provide the safety representative with such time off with pay during the employee's working hours as shall be necessary for the purposes of:

- performing his statutory functions; and

- undergoing such training in aspects of those functions as may be reasonable in all the circumstances.

Further details of these requirements are outlined in the Code of Practice attached to the *Safety Representatives Regulations* and the HSC Approved Code of Practice on time off for training. The Code of Practice is for guidance purposes only – its contents are recommendations rather than requirements. However, it is guidance that an employment tribunal can and will take into account if a complaint is lodged in relation to an employer's unreasonable failure to allow time off.

The combined effect of the ACAS Code No 3: '*Time Off for Trade Union Duties and Activities*' (1991) carried over into the *Trade Union and Labour Relations (Consolidation) Act 1992* (and reissued in 1998) and the HSC Approved Code of Practice on time off is that shop stewards who have also been appointed as safety representatives are to be given time off by their employers to carry out both their industrial relations duties and their safety functions, and also paid leave to attend separate training courses on industrial relations and on health and safety at work – this includes a TUC course on *COSHH* (*Gallagher v The Drum Engineering Co Ltd, COIT 1330/89*).

An employee is not entitled to be paid for time taken off in lieu of the time he had spent on a course. This was held in *Hairsine v Hull City Council [1992] IRLR 211* when a shift worker, whose shift ran from 3 pm to 11 pm, attended a trade union course from 9 am to 4 pm and then carried out his duties until 7 pm. He was paid from 3 pm to 7 pm and he could claim no more. However, where more safety representatives have been appointed than there are sections of the workforce for which safety representatives could be responsible, it is not unreasonable for an employer to deny some safety representatives time off for fulfilling safety functions (*Howard and Peet v Volex plc (HSIB 181)*).

A decision of the EAT seems to favour jointly sponsored in-house courses, except as regards the representational aspects of the functions of safety representatives, where the training is to be provided exclusively by the union (*White v Pressed Steel Fisher [1980] IRLR 176*). Moreover, one course per union per year is too rigid an approach (*Waugh v London Borough of Sutton (1983) HSIB 86*).

## Definition of 'pay'

**J3012** The amount of pay to which the safety representative is entitled is contained in the *Safety Representatives Regulations (SI 1977 No 500), Sch 2*.

## Recourse for the safety representative

**J3013** Where the employer's refusal to allow paid time off is unreasonable, he must reimburse the employee for the time taken to attend. [*Safety Representatives Regulations (SI 1977 No 500), Reg 4(2), Sch 2*]. In the case of *Scarth v East Herts DC (HSIB 181)*: the test of reasonableness is to be judged at the time of the decision to refuse training.

Safety representatives who are refused time off to perform their functions or who are not paid for such time off are able to make a complaint to an employment tribunal. [*Safety Representatives Regulations (SI 1977 No 500), Reg 11*].

## Facilities to be provided by employer

**J3014** The type and number of facilities that employers are obliged to provide for safety representatives are not spelled out in the Regulations, Code of Practice or guidance notes, other than a general requirement in the *Safety Representatives Regulations (SI 1977 No 500), Reg 5(3)* which states, *inter alia*, that 'the employer shall provide such facilities and assistance as the safety representatives shall require for the purposes of carrying out their functions'. Formerly, the requirement to provide facilities and assistance related only to inspections.

Trade unions consider that the phrase 'facilities and assistance' includes the right to request the presence of an independent technical adviser or trade union official during an inspection, and for safety representatives to take samples of substances used at work for analysis outside the workplace. The TUC has recommended that the following facilities be made available to safety representatives:

- a room and desk at the workplace;
- facilities for storing correspondence;
- inspection reports and other papers;
- ready access to internal and external telephones;
- access to typing and duplicating facilities;
- provision of notice boards;
- use of a suitable room for reporting back to and consulting with members;
- other facilities should include copies of all relevant statutes, regulations, Approved Codes of Practice and HSC guidance notes; and copies of all legal or international standards which are relevant to the workplace.

## Disclosure of information

**J3015** Employers are required by the *Safety Representatives Regulations* to disclose information to safety representatives which is necessary for them to carry out their functions. [*Safety Representatives Regulations (SI 1977 No 500), Reg 7(1)*]. A parallel provision exists under the *Management of Health and Safety at Work Regulations 1999, Reg 10(2)* in relation to the information that is to be provided to the parents of a child to be employed by an employer.

*Regulation 7* is consolidated by paragraph 6 of the Code of Practice which details the health and safety information 'within the employer's knowledge' that should be made available to safety representatives. This should include:

- plans and performance and any changes proposed which may affect health and safety;
- technical information about hazards and precautions necessary, including information provided by manufacturers, suppliers and so on;
- information and statistical records on accidents, dangerous occurrences and notifiable industrial diseases; and
- other information such as measures to check the effectiveness of health and safety arrangements and information on articles and substances issued to homeworkers.

The exceptions to this requirement are where disclosure of such information would be 'against the interests of national security'; where it would contravene a prohibition imposed by law; any information relating to an individual (unless consent has been given); information that would damage the employer's undertaking; and information obtained for the sole purpose of bringing, prosecuting or defending legal proceedings. [*Safety Representatives Regulations* (SI 1977 No 500), Reg 7(2)].

However, the decision in *Waugh v British Railways Board [1979] 2 All ER 1169* established that where an employer seeks, on grounds of privilege, to withhold a report made following an accident, he can only do so if its dominant purpose is related to actual or potential hostile legal proceedings. In this particular case, a report was commissioned for two purposes following the death of an employee: (*a*) to recommend improvements in safety measures, and (*b*) to gather material for the employer's defence. It was held that the report was not privileged.

This was followed in *Lask v Gloucester Health Authority [1986] HSIB 123* where a circular 'Reporting Accidents in Hospitals' had to be discovered by order after an injury to an employee whilst he was walking along a path.

Where differences of opinion arise as to the evaluation or interpretation of technical aspects of safety information or health data, unions are advised to contact the local offices of HSE, because of HSE expertise and access to research.

## Technical information

**J3016**   HSE inspectors are also obliged under the *HSWA 1974, s 28(8)*, to supply safety representatives with technical information – factual information obtained during their visits (i.e. any measurements, testing and results of sampling and monitoring), notices of prosecution, copies of correspondence and copies of any improvement or prohibition notices issued to their employer. The latter places an absolute duty on an inspector to disclose specific kinds of information to workers or their representatives concerning health, safety and welfare at work. This can also involve personal discussions between the HSE inspector and the safety representative. The inspector must also tell the representative what action he proposes to take as a result of his visit. Where local authority health inspectors are acting under powers granted by the *HSWA 1974* (see ENFORCEMENT), they are also required to provide appropriate information to safety representatives.

# Safety committees

**J3017**   There is a duty on every employer, in cases where it is prescribed (see below), to establish a safety committee if requested to do so by safety representatives. The committee's purpose is to monitor health and safety measures at work. [*HSWA 1974, s 2(7)*]. Such cases are prescribed by the *Safety Representatives Regulations* (SI 1977 No 500) and limit the duty to appoint a committee to requests made by trade union safety representatives.

## Establishment of a safety committee

**J3018** If requested by at least two safety representatives in writing, the employer must establish a safety committee. [*Safety Representatives Regulations (SI 1977 No 500), Reg 9(1)*].

When setting up a safety committee, the employer must:

- consult with both:
    - the safety representatives who make the request; and
    - the representatives of recognised trade unions whose members work in any workplace where it is proposed that the committee will function;
- post a notice, stating the composition of the committee and the workplace(s) to be covered by it, in a place where it can easily be read by employees;
- establish the committee within three months after the request for it was made.

[*Safety Representatives Regulations (SI 1977 No 500), Reg 9(2)*].

## Function of safety committees

**J3019** In practical terms, trade union appointed safety representatives are now using the medium of safety committees to examine the implications of hazard report forms arising from inspections, and the results of investigations into accidents and dangerous occurrences, together with the remedial action required. A similar procedure exists with respect to representatives for tests and measurements of noise, toxic substances or other harmful effects on the working environment.

Trade unions regard the function of safety committees as a forum for the discussion and resolution of problems that have failed to be solved initially through the intervention of the safety representative in discussion with line management. There is, therefore, from the trade unions' viewpoint, a large measure of negotiation with its consequent effect on collective bargaining agreements.

If safety representatives are unable to resolve a problem with management through the safety committee, or with HSE, they can approach their own union for assistance – a number of unions have their own health and safety officers who can, and do, provide an extensive range of information on occupational health and safety matters. The unions, in turn, can refer to the TUC for further advice.

The 'Brown Book', which contains the *Safety Representatives Regulations*, Code of Practice and guidance, was revised in 1996 to include the amendments made in 1993 by the *Management of Health and Safety at Work Regulations 1992* (now superceded by the *Management of Health and Safety at Work Regulations 1999 (SI 1999 No 3242)*) (see above) and the *Consultation with Employees Regulations* (see J3021).

## Non-unionised workforce – consultation obligations

**J3020** A representative of employee safety is an elected representative of a non-unionised workforce who is assigned with broadly the same rights and obligations as a safety representative in a unionised workforce.

## General duty

**J3021** The *Consultation with Employees Regulations (SI 1996 No 1513)* introduced a new duty to consult any employees who are not members of a group covered by safety representatives (appointed under the *Safety Representatives Regulations*). Employers therefore have the choice of consulting their employees either directly or by way of an appointed representative of employee safety. The obligation is to consult those employees in good time on matters relating to their health and safety at work.

## Number of representatives for workforce

**J3022** Guidance notes on the *Consultation with Employees Regulations* (*SI 1996 No 1513*) state that the number of safety representatives who can be appointed depends on the size of the workforce and workplace, whether there are different sites, the variety of different occupations, the operation of shift systems and the type and risks of work activity. A DTI Workplace Survey has concluded that in a non-unionised workplace which has appointed worker representatives, it is usual for there to be several representatives, with the median being three. The survey estimated that there are approximately 218,000 representatives across all British workplaces with 25 or more employees.

## Role of safety representatives

**J3023** The functions of representatives of employee safety are:

- to make representations to the employer of potential hazards and dangerous occurrences at the workplace which affect or could affect the group of employees he represents;

- to make representations to the employer on general matters affecting the health and safety at work of the group of employees he represents, and in particular on such matters as he has been consulted about by the employer under the *Consultation with Employees Regulations*; and

- to represent that group of employees in consultations at the workplace with inspectors appointed under the *HSWA 1974*.

## Rights and duties of representatives of employee safety
### Time off with pay

**J3024** Under the *Consultation with Employees Regulations* (*SI 1996 No 1513*), Reg 7(1)(b), the right that a representative of employee safety has to take time off with pay is generally the same as that for safety representatives.

### Definition of 'time off'

**J3025** An employer is under an obligation to permit a representative of employee safety to take such time off with pay during working hours as shall be necessary for:

- performing his functions; and
- undergoing such training as is reasonable in all the circumstances.

A candidate standing for election as a representative of employee safety is also allowed reasonable time off with pay during working hours in order to perform his functions as a candidate *SI 1996 No 1513, Reg 7(2)*].

### Definition of 'pay'

**J3026** The *Consultation with Employees Regulations* (*SI 1996 No 1513*), *Sch 1* deal with the definition of pay, and generally the definition is the same as that for union safety representatives.

## Provision of information

**J3027** The employer must provide such information as is necessary to enable the employees or representatives of employee safety to participate fully and effectively in the consultation. In the case of representatives of employee safety, the information must also be sufficient to enable them to carry out their functions under the *Consultation with Employees Regulations* (*SI 1996 No 1513*).

Information provided to representatives must also include information which is contained in any record which the employer is required to keep under *RIDDOR 1995* (*SI 1995 No 3163*) and which relates to the workplace or the group of employees represented by the representatives. Note that there are exceptions to the requirement to disclose information similar to those under the *Safety Representatives Regulations* (*SI 1977 No 500*), Reg 7 (see J3015 above).

## Relevant training

**J3028** Representatives of employee safety must be provided with reasonable training in respect of their functions under the *Consultation with Employees Regulations* (*SI 1996 No 1513*), for which the employer must pay.

## Remedies for failure to provide time off or pay for time off

**J3029** A representative of employee safety, or candidate standing for election as such, who is denied time off or who fails to receive payment for time off, may make an application to an employment tribunal for a declaration and/or compensation. As in the case of safety representatives, the remedies obtainable (set out in the *Consultation with Employees Regulations* (*SI 1996 No 1513*), Sch 2 are similar to those granted to complainants under the *Trade Union and Labour Relations (Consolidation) Act 1992, s 168* (time off for union duties).

# Recourse for safety representatives

**J3030** Safety representatives (whether they are appointed under the *Safety Representatives Regulations* (*SI 1977 No 500*) or the *Consultation with Employees Regulations* (*SI 1996 No 1513*) are provided with statutory protection for the proper execution of their duties.

An employee who is:

- designated by his employer to carry out a health and safety related function,
- a representative of employee safety, or
- a candidate standing for election as such,

has the right not to be subjected to any detriment or unfairly dismissed on the grounds that:

— having been designated by the employer to carry out a health and safety related function, he carried out, or proposed to carry out, the function;

— he undertook, or proposed to undertake, any function(s) consistent with being a safety representative or member of a safety committee;

— he took part in, or proposed to take part in, consultation with the employer; or

— he took part in an election of representatives of employee safety.

In relation to a detriment claim, where the employer infringes any of these rights, the employee has the right to make a complaint to an employment tribunal under the *Employment Rights Act 1996, s 44(1)* (as amended by the *Consultation with Employees Regulations* (*SI 1996 No 1513*),

**J3031** *Joint Consultation in Safety – Safety Representatives, Safety Committees, Collective Agreements and Works Councils*

*Reg 8*. An employment tribunal can make a declaration and also award compensation. [*Employment Rights Act 1996, ss 48, 49*]. There is no minimum qualifying period of service nor upper age limit for bringing such a claim.

If the employer unfairly dismisses such an employee for one of the above reasons, or where it is the principal reason for the dismissal, that dismissal shall be deemed automatically unfair. [*Employment Rights Act 1996, s 100*, as amended by the *Consultation with Employees Regulations (SI 1996 No 1513), Reg 8*]. It will also be an automatically unfair dismissal to select a representative or candidate for redundancy for such a reason. [*Employment Rights Act 1996, s 105*]. The normal minimum qualifying period of service, the normal upper age limit and the cap on the compensatory award for unfair dismissal claims do not apply. [*Employment Rights Act 1996, ss 108, 109 and 124(1A)* as amended by the *Employment Relations Act 1999, s 37(1)*)].

Under the *Employment Rights Act 1996, s 103A* a right to automatically claim unfair dismissal would exist for a worker who is dismissed for making a protected disclosure. A right also exists under the *Employment Rights Act 1996, s 47B* for workers to claim that they have been subjected to a detriment for making a protected disclosure. 'Protected disclosures' in this regard are capable of covering the situation where the health and safety of an individual has been or is likely to be endangered. [*Employment Rights Act 1996, s 43B(1)(d)*].

# European developments in health and safety

## Introduction

**J3031** Health and safety law will continue to be subject to change in the future with the implementation of further European directives and HSC programme of modifying and simplifying health and safety law. At a European level the most important legislation is the directives made under Art 138 of the Treaty of Rome (as amended). The Framework Directive, from which the *Consultation with Employees Regulations (SI 1996 No 1513)* were derived, will continue to drive forward developments in UK health and safety law.

Further developments have occurred under European law which are having an impact at national level these being in the form of the Working Time Directive (93/104) and the European Works Councils Directive (94/45). These Directives have been implemented into national law under the *Working Time Regulations 1998 (SI 1998 No 1833)* (as amended) and the *Works Councils Regulations (SI 1999 No 3323)* respectively.

## The Working Time Regulations 1998 (as amended)

**J3032** A full discussion of the impact of the *Working Time Regulations (SI 1998 No 1833)* is beyond the scope of this chapter – readers are referred to WORKING TIME for a discussion of the impact of the *Working Time Regulations* generally.

That said, the *Working Time Regulations* have introduced a joint consultation function into the operation of the Regulations by use of collective, workforce and relevant agreements. The Regulations provide that it is possible to vary the extent to which the Regulations must be strictly complied with through the use of these devices. The various types of agreements can be described as follows:

- Collective agreements – these are defined by *s 178 of the Trade Union and Labour Relations (Consolidation) Act 1992* as being agreements between independent trade unions and employers;

- Workforce agreements – these were created by the Regulations and are defined in *the Working Time Regulations (SI 1998 No 1833), Reg 2 and Sch 1*. They amount to agreements between an employer and either duly elected worker representatives of the employer or, in the case of an employer employing less than 20 workers, a majority of the individual workers themselves, where the agreement concluded:

- is in writing;
- has effect for a specified period not exceeding five years;
- applies to either:
  (i) all of the relevant members of the workforce, or
  (ii) all of the relevant members of the workforce who belong to a particular sub-group;
- is signed:
  (i) by the worker representatives or by the particular group of workers, or
  (ii) in the case of an employer having less than 20 employees on the date on which the agreement is first concluded, either by appropriate representatives or by a majority of the workers working for the employer; and
- before being made available for signature, copies of the agreement were provided to all of the workers to whom the agreement was intended to apply, together with such guidance as the workers might reasonably require in order to understand the draft agreement;

- Relevant agreements – these are workforce agreements that cover a worker, any provision of a collective agreement that is individually incorporated into the contract of employment of a worker, or any other agreement in writing between a worker and his employer that is legally enforceable (e.g. a staff handbook).

By *Regulation 23(a)* of the *Working Time Regulations (SI 1998 No 1833)*, it is possible to modify the provisions relating to:

- the length of night work (see *Working Time Regulations (SI 1998 No 1833), Reg 6)*; and
- the minimum daily and weekly rest periods and rest breaks (see *Working Time Regulations (SI 1998 No 1833), Regs 10–12* respectively).

Modification must be by way of a collective or workforce agreement and such an agreement must make provision for a compensatory rest period of equivalent length. [*Working Time Regulations (SI 1998 No 1833), Reg 24*].

Further, by *Regulation 23(b)* of the *Working Time Regulations (SI 1998 No 1833)*, it is possible for an employer and its workers to agree by collective or workforce agreement to vary the reference period for calculating the maximum working week from the usual 17 weeks to a 52-week period if there are objective or technical reasons relating to the organisation which justify such a change.

The variation provisions allow a degree of flexibility where it is necessary for the interests of an employer's business to effect such change for operational reasons whilst still ensuring the protection of the health and safety of the workforce. The provisions also ensure that any change that is to be made must survive collective scrutiny of the employer's workforce.

## European Works Councils and Information and Consultation Procedures

J3033     The provisions relating to European Works Councils are at present confined to large pan-European entities.

As stated at J3001, Directive 2002/14/EC will make 'mini works councils' obligatory for any employer employing more than 50 workers. Transitional provisions will apply initially limiting the impact of Directive 2002/14/EC to employers employing more than 150 employees. All employers in the United Kingdom employing 50 or more employees will have to comply with the Directive by 23 March 2007. National legislation to implement Directive 2002/14/EC is not expected until 2004.

## J3033 Joint Consultation in Safety – Safety Representatives, Safety Committees, Collective Agreements and Works Councils

The European Work Council Directive (94/45) was incorporated into national law by the *Works Councils Regulations* (*SI 1999 No 3323*). A full discussion of the operation of the *Works Councils Regulations* is beyond the scope of this chapter, which instead focuses upon the health and safety aspect of the Regulations.

The *Works Councils Regulations* govern employers employing a total of 1,000 or more workers where at least 150 workers are so employed in each of two or more member states.

Their main purpose is procedural. *Part IV* of the *Works Council Regulations* (*SI 1999 No 3323*) creates machinery between workers and their employer for the purpose of establishing either a European Works Council ('EWC') or an Information and Consultation Procedure ('ICP'). *Regulation 17(1)* of the *Works Councils Regulations* (*SI 1999 No 3323*) provides that the central management of the employer and a special negotiating body (defined in *Part III* of the Regulations) are bound to:

'... negotiate in a spirit of co-operation with a view to reaching a written agreement on the detailed arrangements for the information and consultation of employees in a Community-scale undertaking or Community-scale group of undertakings'.

*Regulation 17(3)* of the *Works Councils Regulations* (*SI 1999 No 3323*) leaves the choice of whether to proceed with an EWC or an ICP to the parties.

The parties are free to include in the agreement reference to whatever matters are likely to affect the workers of the employer at a trans-national level. Health and safety is clearly such an issue. If:

- the parties fail to agree the content of the agreement; or
- within six months of a valid request being made to the central management of an employer the employer fails to negotiate so as to create either an EWC or ICP; or
- after three years of negotiation to produce an agreement for an EWC or ICP the parties cannot agree as to the constitution,

default machinery is provided by the *Schedule* to the *Works Councils Regulations* [*Works Councils Regulations* (*SI 1999 No 3323*), *Reg 18(1)*]. Paragraph 6 of the *Schedule* provides in relation to EWCs that:

'The competence of the European Works Council shall be limited to information and consultation on the matters which concern the Community-scale undertaking or Community-scale group of undertakings as a whole or at least two of its establishments or group undertakings situated in different Member States'.

In relation to ICPs, *para 7(3)* of the *Schedule* to the *Works Councils Regulations* (*SI 1999 No 3323*) provides that meetings of ICPs:

'... shall relate in particular to the structure, economic and financial situation, the probable development of the business and of production and sales, the situation and probable trend of employment, investments, and substantial changes concerning organisation, [and] introduction of new working methods or production processes ...'

Although not expressly providing for discussion of health and safety issues, the provisions relating to both EWCs and ICPs will, by implication, include debate of health and safety matters.

# Lifting Operations

## Introduction

**L3001** General requirements for the provision and maintenance of safe plant and equipment set out in the *Health and Safety at Work etc. Act 1974 section 2(2)(a)* and *(b)* apply.

More specific requirements applying to all workplaces came into force on 5 December 1998 in the *Lifting Operations and Lifting Equipment Regulations 1998* (*LOLER*) (*SI 1998 No 2307*) which encompasses all lifting equipment for use at work.

LOLER demands that four main principals are complied with:

- the initial integrity of the equipment is suitable for its intended use;
- the lifting operation is correctly planned and assessed;
- the operation is undertaken in a safe manner; and
- thereafter, that the equipments safe integrity is maintained.

There is a general duty on the employer to select suitable lifting equipment based on a fitness for purpose criteria in both *Regulation 4* of the *Provision and Use of Work Equipment Regulations 1998* (*SI 1992 No 2932*) and its sister legislation, LOLER.

LOLER does introduce some new terminology namely:

| LOLER | Previous Terminology |
|---|---|
| Accessories | Lifting tackle |
| Carrier | Lift, personnel carrier, working platform, etc. |
| Safety Co-efficient | Factor of safety |
| Rated capacity indicator | Automatic safe load indicator |

Accessories for lifting are also lifting equipment. This encompasses and adds to the formerly termed 'Lifting tackle' and includes: chains, ropes, slings and components kept for attaching loads to machinery for lifting i.e. hooks, jigs, eyebolts etc.

## Statutory requirements relating to the manufacture of lifting equipment

**L3002** The statutes that are applied to the manufacture of lifting equipment are:

- *Lifts Regulations 1997* (*SI 1997 No 831*); and
- the Machinery Directive which is implemented into UK law by the *Supply of Machinery (Safety) Regulations 1992* (*SI 1992 No 3073* as amended by *SI 1994 No 2063*).

The relationship between the manufacturers and the owner/users duties are shown in Figure 1

**L3002** *Lifting Operations*

Figure 1 Relationship between the manufacturers and the owner/users

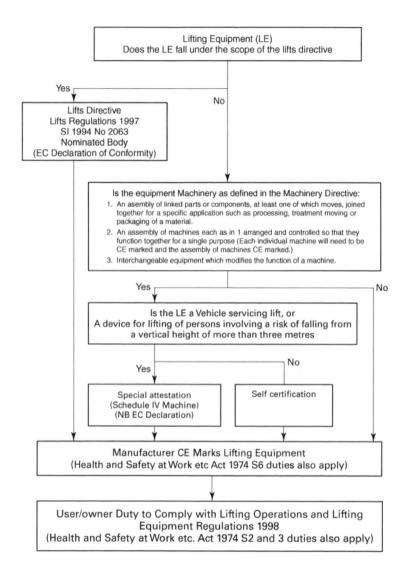

In addition to the specific requirements above, there is also a necessity for lifting equipment controlled by electrical equipment to comply with:

- *Electrical Equipment (Safety) Regulations 1994* (*SI 1994 No 3260*) implementing the Low Voltage Directive; and

- *Electromagnetic Compatibility Regulations 1992* (*SI 1992 No 2372* as amended by *SI 1994 No 3080*).

# Lifts Regulations 1997 (SI 1997 No 831)

L3003      'Lift' – this means an appliance serving specific levels, having a car moving along guides which are rigid – or along a fixed course even where it does not move along guides which are rigid (for example, a scissor lift) – and inclined at an angle of more than 15 degrees to the horizontal and intended for the transport of (there is difference in terminology when LOLER refers to Lifts as Carriers.):

- persons;
- persons and goods; or
- goods alone if the car is accessible, that is to say, a person may enter it without difficulty, and fitted with controls situated inside the car or within reach of a person inside.

## The Regulations

L3004      These Regulations apply to lifts permanently serving buildings or constructions; and safety components for use in such lifts. [*Lifts Regulations 1997 (SI 1997 No 831), Reg 3*].

These Regulations do not apply to:

(*a*)    the following lifts – and safety components for such lifts:

— cableways, including funicular railways, for the public or private transportation of persons,

— lifts specially designed and constructed for military or police purposes,

— mine winding gear,

— theatre elevators,

— lifts fitted in means of transport,

— lifts connected to machinery and intended exclusively for access to the workplace,

— rack and pinion trains,

— construction-site hoists intended for lifting persons or persons and goods;

[*SI 1997 No 831, Reg 4, Sch 14*].

(*b*)    any lift or safety component which is placed on the market (i.e. when the installer first makes the lift available to the user, but see L3009 below) and put into service before 1 July 1997 [*SI 1997 No 831, Reg 5*];

(*c*)    any lift or safety component placed on the market and put into service on or before 30 June 1999 which complies with any health and safety provisions with which it would have been required to comply if it was to have been placed on the market and put into service in the United Kingdom on 29 June 1995.

This exclusion does not apply in the case of a lift or a safety component which;

— unless required to bear the CE marking pursuant to any other Community obligation, bears the CE marking or an inscription liable to be confused with it; or

— bears or is accompanied by any other indication, howsoever expressed, that it complies with the Lifts Directive.

[*SI 1997 No 831, Reg 6*]; and

(*d*)    any lift insofar as and to the extent that the relevant essential health and safety requirements relate to risks wholly or partly covered by other Community directives applicable to that lift [*SI 1997 No 831, Reg 7*].

**L3005** *Lifting Operations*

## General requirements

### General duty relating to the placing on the market and putting into service of lifts

**L3005**

(i) Subject to *Lifts Regulations 1997* (*SI 1997 No 831*), *Reg 12* (see L3010 below), no person who is a responsible person shall place on the market and put into service any lift unless the requirements of paragraph (ii) below have been complied with in relation to it [*SI 1997 No 831, Reg 8(1)*].

(ii) *SI 1997 No 831, Reg 8(2)* provides that the requirements in respect of any lift are that:

— it satisfies the relevant essential health and safety requirements and, for the purpose of satisfying those requirements:

— where a transposed harmonised standard covers one or more of the relevant essential health and safety requirements, any lift constructed in accordance with that transposed harmonised standard shall be presumed to comply with that (or those) essential health and safety requirement(s);

— by calculation, or on the basis of design plans, it is permitted to demonstrate the similarity of a range of equipment to satisfy the essential safety requirements;

— the appropriate conformity assessment procedure in respect of the lift has been carried out in accordance with *SI 1997 No 831, Reg 13(1)* (see L3011 below);

— the CE marking has been affixed to it by the installer of the lift in accordance with *SI 1997 No 831, Sch 3*;

— a declaration of conformity has been drawn up in respect of it by the installer of the lift; and

— it is in fact safe.

*Note:* *SI 1997 No 831, Reg 18(1)* provides that a lift which bears the CE marking, and is accompanied by an EC declaration of conformity in accordance with *SI 1997 No 831, Reg 8(2)*, is taken to conform with all the requirements of the *Lifts Regulations 1997* (*SI 1997 No 831*), including *SI 1997 No 831, Reg 13* (see L3011 below), unless reasonable grounds exist for suspecting that it does not so conform.

(iii) Any technical documentation or other information in relation to a lift required to be retained under the conformity assessment procedure used must be retained by the person specified in that respect in that conformity assessment procedure for any period specified in that procedure [*SI 1997 No 831, Reg 8(3)*].

In these Regulations, 'responsible person' means:

— in the case of a lift, the installer of the lift;

— in the case of a safety component, the manufacturer of the component or his authorised representative established in the Community; or

— where neither the installer of the lift nor the manufacturer of the safety component nor the latter's authorised representative established in the Community, as the case may be, have fulfilled the requirements of *SI 1997 No 831, Reg 8(2)* (see above) or *SI 1997 No 831, Reg 9(2)* (see L3006 below), the person who places the lift or safety component on the market.

## General duty relating to the placing on the market and putting into service of safety components

**L3006** Safety components include devices for locking landing doors; devices to prevent falls; overspeed limitation devices; and electric safety switches.

(i) Subject to *Lifts Regulations 1997 (SI 1997 No 831), Reg 12* (see L3010 below), no person who is a responsible person shall place on the market and put into service any safety component unless the requirements of paragraph (ii) below have been complied with in relation to it [*SI 1997 No 831, Reg 9(1)*].

(ii) *SI 1997 No 831, Reg 9(2)* provides that the requirements in respect of any safety component are that:

— it satisfies the relevant essential health and safety requirements and for the purpose of satisfying those requirements where a transposed harmonised standard covers one or more of the relevant essential health and safety requirements, any safety component constructed in accordance with that transposed harmonised standard shall be presumed to be suitable to enable a lift on which it is correctly installed to comply with that (or those) essential health and safety requirement(s);

— the appropriate conformity assessment procedure in respect of the safety component has been carried out in accordance with *SI 1997 No 831, Reg 13(1)* (see L3011 below);

— the CE marking has been affixed to it, or on a label inseparably attached to the safety component, by the manufacturer of that safety component, or his authorised representative established in the Community, in accordance with *SI 1997 No 831, Sch 3* (which specifies the requirements for CE marking);

— a declaration of conformity has been drawn up in respect of it by the manufacturer of the safety component or his authorised representative established in the Community; and

— it is in fact safe.

*Note: SI 1997 No 831, Reg 18(1)* provides that a safety component – or its label – which bears the CE marking, and is accompanied by an EC declaration of conformity in accordance with *SI 1997 No 831, Reg 9(2)*, is taken to conform with all the requirements of the *Lifts Regulations 1997*, including *SI 1997 No 831, Reg 13* (see L3011 below), unless reasonable grounds exist for suspecting that it does not so conform.

(iii) Any technical documentation or other information in relation to a safety component required to be retained under the conformity assessment procedure used must be retained by the person specified in that respect in that conformity assessment procedure for any period specified in that procedure [*SI 1997 No 831, Reg 9(3)*].

## General duty relating to the supply of a lift or safety component

**L3007** Subject to *Lifts Regulations 1997 (SI 1997 No 831), Reg 12* (see L3010 below) any person who supplies any lift or safety component but who is not a person to whom *SI 1997 No 831, Reg 8* or *9* applies (see L3005 and L3006 above) must ensure that that lift or safety component is safe [*SI 1997 No 831, Reg 10*].

### Penalties for breach of Regs 8, 9 or 10

**L3008** A person who is convicted of an offence under *Lifts Regulations 1997 (SI 1997 No 831), Reg 8, 9* or *10* above is liable to imprisonment (not exceeding three months) or to a fine not exceeding level 5 on the standard scale, or both. *SI 1997 No 831, Reg 22*, however, provides a defence of due diligence: i.e. the defendant must show that he took all reasonable steps and exercised all due diligence to avoid committing the offence.

## Specific duties relating to the supply of information, freedom from obstruction of lift shafts and retention of documents

**L3009**

(i) The person responsible for work on the building or construction where a lift is to be installed and the installer of the lift must keep each other informed of the facts necessary for, and take the appropriate steps to ensure, the proper operation and safe use of the lift. Shafts intended for lifts must not contain any piping or wiring or fittings other than that which is necessary for the operation and safety of that lift.

(ii) Where, in the case of a lift, for the purposes of *Lifts Regulations 1997* (*SI 1997 No 831*), *Reg 8(2)* (see L3005 above) the appropriate conformity assessment procedure is one of the procedures set out in *SI 1997 No 831*, *Reg 13(2)(a)*, *(b)* or *(c)* (see L3011 below), the person responsible for the design of the lift must supply to the person responsible for the construction, installation and testing all necessary documents and information for the latter person to be able to operate in absolute security.

A person who is convicted of an offence under (i) or (ii) above is liable to imprisonment (not exceeding three months) or to a fine not exceeding level 5 on the standard scale, or both. *SI 1997 No 831*, *Reg 22*, however, provides a defence of due diligence: i.e. the defendant must show that he took all reasonable steps and exercised all due diligence to avoid committing the offence.

(iii) A copy of the declaration of conformity mentioned in *SI 1997 No 831*, *Reg 8(2)* or *9(2)* must:

— in the case of a lift, be supplied to the EC Commission, the member states and any other notified bodies, on request, by the installer of the lift together with a copy of the reports of the tests involved in the final inspection to be carried out as part of the appropriate conformity assessment procedure referred to in *SI 1997 No 831*, *Reg 8(2)*; and

— be retained, by the person who draws up that declaration, for a period of ten years – in the case of a lift, from the date on which the lift was placed on the market; and in the case of a safety component, from the date on which safety components of that type were last manufactured by that person.

A person who fails to supply or keep a copy of the declaration of conformity, as required above, is liable on summary conviction to a fine not exceeding level 5 on the standard scale. *SI 1997 No 831*, *Reg 22*, however, provides a defence of due diligence: i.e. the defendant must show that he took all reasonable steps and exercised all due diligence to avoid committing the offence.

*[SI 1997 No 831, Reg 11].*

## Exceptions to placing on the market or supply in respect of certain lifts and safety components

**L3010** For the purposes of *Lifts Regulations 1997* (*SI 1997 No 831*), *Reg 8, 9* or *10*, a lift or a safety component is not regarded as being placed on the market or supplied:

(i) where that lift or safety component will be put into service in a country outside the Community; or is imported into the Community for re-export to a country outside the Community – but this paragraph does not apply if the CE marking, or any inscription liable to be confused with such a marking, is affixed to the lift or safety component or, in the case of a safety component, to its label; or

(ii) by the exhibition at trade fairs and exhibitions of that lift or safety component, in respect of which the provisions of these Regulations are not satisfied, if:

— a notice is displayed in relation to the lift or safety component in question to the effect that it does not satisfy those provisions; and that it may not be placed on the market or supplied until those provisions are satisfied; and

— adequate safety measures are taken to ensure the safety of persons.

*[SI 1997 No 831, Reg 12].*

## Conformity assessment procedures

**L3011** For the purposes of *Lifts Regulations 1997 (SI 1997 No 831), Reg 8(2)* or *9(2)* (see L3005 and L3006 above), the appropriate conformity assessment procedure is as follows:

*For lifts – one of the following procedures:*

— if the lift was designed in accordance with a lift having undergone an EC type-examination as referred to in *SI 1997 No 831, Sch 5*, it must be constructed, installed and tested by implementing:

— the final inspection referred to in *SI 1997 No 831, Sch 6*; or

— the quality assurance system referred to in *SI 1997 No 831, Sch 11* or *Sch 13*; and

the procedures for the design and construction stages, on the one hand, and the installation and testing stages, on the other, may be carried out on the same lift;

— if the lift was designed in accordance with a model lift having undergone an EC type-examination as referred to in *SI 1997 No 831, Sch 5*, it must be constructed, installed and tested by implementing:

— the final inspection referred to in *SI 1997 No 831, Sch 6*; or

— one of the quality assurance systems referred to in *SI 1997 No 831, Sch 11* or *Sch 13*; and

all permitted variations between a model lift and the lifts forming part of the lifts derived from that model lift must be clearly specified (with maximum and minimum values) in the technical dossier required as part of the appropriate conformity assessment procedure;

— if the lift was designed in accordance with a lift for which a quality assurance system pursuant to *SI 1997 No 831, Sch 12* was implemented, supplemented by an examination of the design if the latter is not wholly in accordance with the harmonised standards, it must be installed and constructed and tested by implementing, in addition, the final inspection referred to in *SI 1997 No 831, Sch 6* or one of the quality assurance systems referred to in *SI 1997 No 831, Sch 11* or *Sch 13*;

— the unit verification procedure, referred to in *SI 1997 No 831, Sch 9*, by a notified body (see L3013 below); or

— the quality assurance system in accordance with *SI 1997 No 831, Sch 12*, supplemented by an examination of the design if the latter is not wholly in accordance with the transposed harmonised standards.

*For safety components – one of the following procedures:*

— submit the model of the safety component for EC type-examination in accordance with *SI 1997 No 831, Sch 5* and for production checks by a notified body (see L3013 below) in accordance with *SI 1997 No 831, Sch 10*;

— submit the model of the safety component for EC type-examination in accordance with *SI 1997 No 831, Sch 5* and operate a quality assurance system in accordance with *SI 1997 No 831, Sch 7* for checking production; or

— operate a full quality assurance system in accordance with *SI 1997 No 831, Sch 8*.

[*SI 1997 No 831, Reg 13*].

## Requirements fulfilled by the person who places a lift or safety component on the market

L3012   Where in the case of a lift or a safety component, any of the requirements of *Lifts Regulations 1997* (*SI 1997 No 831*), *Regs 8, 9, 11* and *13* to be fulfilled by the installer of the lift or the manufacturer of the safety component or, in the case of the latter, his authorised representative established in the Community, have not been so fulfilled such requirements may be fulfilled by the person who places that lift or safety component on the market [*SI 1997 No 831, Reg 14*].

This provision, however, does not affect the power of an enforcement authority to take action in respect of the installer of the lift, the manufacturer of the safety component or, in the case of the latter, his authorised representative established in the Community in respect of a contravention of or a failure to comply with any of those requirements.

## Notified bodies

L3013   For the purposes of these Regulations, a notified body is a body which has been appointed to carry out one or more of the conformity assessment procedures referred to in *Lifts Regulations 1997* (*SI 1997 No 831*), *Reg 13* which has been appointed as a notified body by the Secretary of State in the United Kingdom or by a member State.

[*SI 1997 No 831, Reg 15*].

## Lifting Operations and Lifting Equipment Regulations 1998 (LOLER) (SI 1998 No 2307)

L3014   The *Lifting Operations and Lifting Equipment Regulations 1998* (*LOLER*) (*SI 1998 No 2307*) uses the term employer rather than duty-holder and duties specifically assigned to the employer can be assumed to apply to the duty holder, if they have any control over lifting operations.

Where a company provides personnel to undertake work which will involve the use of lifting equipment then that company is regarded as an employer and they have a duty under LOLER to provide persons competent to undertake the work.

Duties under LOLER apply to anyone with control, to any extent, of:

— lifting equipment;

— a person at work who uses or supervises or manages the use of lifting equipment; or

— the way in which lifting equipment is used,

and to the extent of his control.

The following general definitions apply to LOLER:

- 'lifting equipment' – work equipment for lifting or lowering loads and includes its attachments for anchoring, fixing or supporting it;

- 'accessory for lifting' – work equipment for attaching loads to machinery for lifting (pendant, sling, shackle, etc.);

- 'load' – includes material or people lifted by the lifting equipment;
- 'examination scheme' – suitable scheme drawn up by a competent person for such thorough examination of lifting equipment at such intervals as may be appropriate for the purpose described in *SI 1998 No 2307, Reg 9*; and
- 'thorough examination' – means a thorough examination by a competent person including such testing as is appropriate for the purpose.

The lifting equipment that could fall under the range of LOLER is all encompassing and the following list of equipment is covered by the regulations is by no means exhaustive:

— pedestal cranes;

— mobile cranes;

— overhead gantry cranes;

— lifting accessories – slings, shackles, eyebolts, jigs, etc;

— lifts for the lifting of persons or goods;

— abseiling equipment;

— patient-hoists;

— vehicle lifting tables;

— vacuum hoists; and

— vehicle tail-lifts.

Some of these specific items of lifting equipment are looked at in more detail in L3024 below onwards.

## General requirements

### Strength and stability

**L3015** Every employer must ensure that lifting equipment is of adequate strength and stability for each load, having regard in particular to the stress induced at its mounting or fixing point – and that every part of a load and anything attached to it and used in lifting it is of adequate strength [*Lifting Operations and Lifting Equipment Regulations 1998* (*LOLER*) (*SI 1998 No 2307*), *Reg 4*].

To ensure adequate strength and stability the employer must:

- take account of the combination of forces which the equipment may be subjected to;
- assess the implication of the weight of any accessories;
- ensure that the equipment is not susceptible to in-service failure modes (fracture, wear or fatigue);
- have an appropriate factor of safety against foreseeable failure modes;
- take account of any combination of destabilising forces; and
- provide resistance to overturning.

Although the load does not fall within LOLER it is incumbent upon the employer to ensure that any lifting points on the load are of adequate strength.

## Lifting equipment for lifting persons

**L3016** *Lifting Operations and Lifting Equipment Regulations 1998 (LOLER) (SI 1998 No 2307), Reg 5(1)(b)* does contain the exception in the applied duty under LOLER, in this case the standard is qualified **by so far as is reasonably practicable (SFAIRP)**, in that the employer shall SFAIRP prevent a person being injured whilst carrying out activities from a carrier e.g. mobile elevated work platform, suspended cradle etc. All other parts of this regulation demand an **absolute duty**.

Every employer must ensure that lifting equipment for lifting persons:

(i) is such as to prevent a person using it being crushed, trapped or struck or falling from the carrier;

(ii) is such as to prevent so far as is reasonably practicable a person using it, while carrying out activities from the carrier, being crushed, trapped or struck or falling from the carrier;

(iii) has suitable devices to prevent the risk of a carrier falling, and, if the risk cannot be prevented for reasons inherent in the site and height differences, the employer must ensure that the carrier has an enhanced safety coefficient suspension rope or chain which is inspected by a competent person every working day; and

(iv) is such that a person trapped in any carrier is not thereby exposed to danger and can be freed.

[*SI 1998 No 2307, Reg 5*].

Common examples of equipment NOT designed for lifting persons but which are used are forklift trucks and telescopic handlers. It should be emphasising that this type of lifting equipment is not primarily designed for lifting persons and should only be used in exceptional circumstances.

HSE has alerted owners of Mobile Elevating Work Platforms ('MEWP') to an incident in which a work basket became detached from its supporting boom. No one was injured during this incident, but it has raised concerns about the safety of other MEWPs of the same construction and make. The MEWP was made by Access Machines Ltd. It was a Micro 95 trailer mount, class A1, made in 1998, with a Safe Working Load of 120 Kg.

The basket of the MEWP that failed was bolted to a round plate by four bolts and in turn the plate was welded to a pin or stub axle on the main boom. In this instance the pin to plate weld failed allowing the basket to fall off the end of the boom. There was some evidence of fatigue in the weld but HSE is concerned that poor quality welding may have contributed to the failure.

This is possibly a one off incident but nevertheless HSE strongly recommends that owners/operators of these and similar machines have this pin/plate connection thoroughly examined by a competent person at the earliest possible opportunity.

## Positioning and installation

**L3017** Employers must ensure that lifting equipment is positioned or installed in such a way as to reduce to as low as is reasonably practicable the risk of the lifting equipment or a load striking a person, or the risk from a load:

— drifting;

— falling freely; or

— being released unintentionally,

and that otherwise it is safe. Further consideration should be given to:

• the need to lift loads over people, and were this unavoidable minimise the risk;

- crushing is prevented at extreme operating positions;
- loads moving along a fixed path are suitably protected to minimise the risk of the load or equipment striking a person; and
- trapping points are prevented or access limited on travelling or slewing equipment.

Employers must also ensure that there are suitable devices for preventing anyone from falling down a shaft or hoistway and that lifting equipment cannot be unintentionally released during a loss of power to the lifting equipment or through the collision of equipment or their loads.

The use of hooks with safety catches, motion limiting devices and safe systems of work are possible means of minimising these risks. [*Lifting Operations and Lifting Equipment Regulations 1998* (*LOLER*) (*SI 1998 No 2307*), Reg 6].

The HSE has warned of the need for an effective risk assessment for all persons involved in the operation of cranes after the prosecution of the Port of Felixstowe.

The warning followed the death of 51-year-old Dennis Burman from Brantham, Essex. Mr Burman was crushed between railings as he moved between a fixed and moveable walkway on a crane's platform during a dock familiarisation course on 17 June 2003. He then fell approximately 120 feet to the ground.

The Felixstowe Dock and Railway Company Ltd were fined £250,000 with £27,288 costs at Ipswich Crown Court on Wednesday 8 September 2004 after pleading guilty at an earlier hearing. The firm admitted failing to ensure that Mr Burman and other workers were not exposed to risks to their safety and thereby breaching *section 2 (1) of the Health and Safety at Work etc Act* (*HSWA*) *1974*.

HSE Inspector, David Gregory, who investigated the death, said:

'In this instance the crane driver intentionally moved the cab. However his view of the cross-over point between the moving and fixed access walkway was obscured. In any event he was not able to observe the cross-over point whilst at the same time watching where he was driving the crane.

A suitable and sufficient risk assessment would have identified the potential for a fatal or serious injury at the cross-over point. Preventing access to all personnel excluding the driver whilst the crane was in operation would have prevented this incident. Alternatively, interlocking the access gates on the walkways to the movement of the cab would have achieved the same result. I hope this prosecution will remind companies who operate this type of machinery that safety must be the foremost consideration.

## Marking of lifting equipment

L3018     Employers must ensure that:

(i) machinery and accessories for lifting loads are clearly marked to indicate their safe working loads;

(ii) where the safe working load of machinery for lifting loads depends on its configuration, either the machinery is clearly marked to indicate its safe working load for each configuration, or information which clearly indicates its safe working load for each configuration is kept with the machinery;

(iii) accessories for lifting are also marked in such a way that it is possible to identify the characteristics necessary for their safe use;

(iv) lifting equipment which is designed for lifting persons is appropriately and clearly marked to this effect; and

(v) lifting equipment which is not designed for lifting persons but which might be mistakenly so used is appropriately and clearly marked to the effect that it is not designed for lifting persons.

[*Lifting Operations and Lifting Equipment Regulations 1998* (*LOLER*) (*SI 1998 No 2307*), *Reg 7*].

## Organisation of lifting operations

**L3019** Every employer must ensure that any lifting or lowering of a load which involves lifting equipment is properly planned by a competent person, appropriately supervised and carried out in a safe manner.

The *Lifting Operations and Lifting Equipment Regulations 1998* (*LOLER*) (*SI 1998 No 2307*) demands as a minimum standard that all lifting operations are carried out safely under adequate supervision and following a lifting plan. The competent person planning the operation should have adequate practical and theoretical knowledge and experience of planning lifting operations.

The plan will need to address the risks identified during a risk assessment and should identify all resources, procedures and responsibilities necessary for safe operation. The degree of planning will vary considerably depending on the type of lifting equipment and complexity of the lifting operation and degree of risk involved. This regulation does allow for routinely repeated lifting operations to be assessed in the first instance and thereafter as appropriate.

There are two elements to the plan: the suitability of the lifting equipment as per *Regulation 4* of the *Provision and Use of Work Equipment Regulations 1998* (*PUWER*) (*SI 1998 No 2306*) and the individual lifting operation to be performed.

*SI 1998 No 2307, Reg 8* requires that the employer or controller of lifting operations ensures that suitable persons are appointed for planning and supervising of such operations.

For any lifting operation it is necessary to:

(a) carry out a risk assessment under the *Management of Health and Safety at Work Regulations 1999* (*SI 1999 No 3242*);

(b) select suitable equipment for the range of tasks; and

(c) plan the individual lifting operation.

The term 'Competent Person' is not prescriptively described in LOLER and is used to identify a number of different roles under the regulations. In practical terms the competency of a person may be confirmed by formal, vocational qualification or through first hand knowledge of planning or supervising the lifting operations.

**The guidance to the Regulations/ACoP does state that the competent person carrying out the planning task is unlikely to be the same person as the competent person undertaking the thorough examination and testing.** This planning must take into account the location, the load to be lifted, the duration and the specific operation to be carried out.

Reference should be made to the following publications for more explicit guidance on the safe use of particular items of lifting equipment.

*Regulation 4* of (*PUWER*) (*SI 1998 No 2306*) requires suitable work equipment to be provided for the task. There is therefore a close link between this regulation and the requirement for planning. Factors to be considered when selecting lifting equipment so that it is suitable for the proposed task should include:

- the load to be lifted;
- its weight, shape and centre of gravity;
- positioning of the load before and after lifting;

- how often the lifting equipment will be used to carry out the task;
- where the lifting equipment will be used; and
- the personnel available and their knowledge, training and experience (i.e. competence).

## Planning of individual operations

**L3020** Planning for routine operations this will usually be a matter for the persons undertaking the task, such as a trained and certified a forklift operator. The competent person carrying out this exercise will have to have appropriate knowledge and experience. The main elements that would need to be considered in any lifting plan could include:

— assessment of the weight of the load, including any lifting accessories;

— establish that the crane and accessories are in date and covered by a current thorough examination;

— choice of the correct slings and other accessories dependant upon the local environment and conditions;

— check the anticipated travel path;

— prepare a place to set down the load;

— establish the centre of gravity and that this is compensated for if off centre;

— make a trial lift to check the lifting equipments safety devices (as appropriate);

— use tag lines if necessary to prevent the load swinging;

— lift the load and move to the setting down position;

— release the load; and

— clear the work area.

For routine operations the initial plan will only be required for the first lifting operation. This will however need to be reviewed to ensure that nothing has changed and that the plan has remained valid.

Examples where generic lifting plans could be provided for routine lifting operations may include:

(i) a passenger lift in an office block;

(ii) fork lift trucks used in a warehouse operation;

(iii) teagle hoist for unloading HGVs;

(iv) construction site hoist;

(v) mobile elevated work platforms (MEWP's) used for general maintenance; and

(vi) suspended cradle used for window cleaning.

[*Lifting Operations and Lifting Equipment Regulations 1998* (*LOLER*) (*SI 1998 No 2307, Reg 8*].

## Thorough examination and inspection

**L3021** An employer is under a duty to ensure:

(i) before lifting equipment is put into service for the first time by him, that it is thoroughly examined for any defect unless:

— the lifting equipment has not been used before; and

- in the case of lifting equipment for which an EC declaration of conformity could or (in the case of a declaration under the *Lifts Regulations 1997* (*SI 1997 No 831*) should have been drawn up, the employer has received such declaration made not more than twelve months before the lifting equipment is put into service; or

- if obtained from the undertaking of another person, it is accompanied by physical evidence referred to in (iv) below;

[*Lifting Operations and Lifting Equipment Regulations 1998* (*LOLER*) (*SI 1998 No 2307, Reg 9(1)*)].

(ii) where the safety of lifting equipment depends on the installation conditions, that it is thoroughly examined – after installation and before being put into service for the first time and after assembly and before being put into service at a new site or in a new location, to ensure that it has been correctly installed and is safe to operate [*SI 1998 No 2307, Reg 9(2)*];

(iii) that lifting equipment which is exposed to conditions causing deterioration which is liable to result in dangerous situations is:

thoroughly examined;

- at least every six months, in the case of lifting equipment for lifting persons or an accessory for lifting;
- at least every twelve months, in the case of other lifting equipment; or
- in either case, in accordance with an examination scheme; and
- whenever exceptional circumstances which are liable to jeopardise the safety of the lifting equipment have occurred, and

if appropriate for the purpose, is inspected by a competent person at suitable intervals between thorough examinations,

to ensure that health and safety conditions are maintained and that any deterioration can be detected and remedied in good time;

[*SI 1998 No 2307, Reg 9(3)*].

(iv) that no lifting equipment leaves his undertaking; or, if obtained from the undertaking of another person, is used in his undertaking, unless it is accompanied by physical evidence that the last thorough examination required to be carried out under this regulation has been carried out.

[*SI 1998 No 2307, Reg 9(4)*].

## Competent person

**L3022** The competent person responsible for undertaking the thorough examination must remain objective in their approach to this activity and be totally independent of any commercial or other conflicts of interest in the item of lifting equipment that they are to examine.

External competent persons e.g. Engineering Insurance Surveyors are normally accredited by the United Kingdom Accreditation Service (UKAS) to their RG 6 standard.

## Thorough examination

**L3023** The employer (owner of the equipment) must determine the level of thorough examination based on manufacturer's information and other factors. The employer may need to seek specialist advice to comply with this requirement. Factors that must be taken into account by the employer should include the work being carried out, any specific risks at the location that may affect the condition of the equipment, and the intensity of use of the equipment.

A thorough examination may include visual examination, a strip down of the equipment and functional tests. Advice should be sought from manufacturer's instructions, and a competent person for guidance on what an inspection should include for each piece of equipment.

A thorough examination must be carried out under the following circumstances.

- When the equipment is put into service for the first time. If it is new equipment that has not been used before there should be a declaration of conformity, which confirms that the equipment has undergone a thorough examination.

- Where safety depends upon the conditions after installation and before being used for the first time.

If the equipment is obtained from another undertaking, as in the case of hired equipment, then a copy of the previous certificate of thorough examination must accompany the equipment.

Installed equipment applies to lifting equipment erected or built on site, such as tower cranes, hoists or gantry cranes i.e. lifting equipment which is intended to be there for a period of time. It would not apply to equipment such as a mobile crane as it is not 'installed'. For equipment such as a mobile crane, there must be a copy of the previous thorough examination certificate.

## Thorough examination periods

**L3024**

(i) Lifting equipment for lifting persons must be thoroughly examined at least every six months.

(ii) Lifting accessories must be thoroughly examined at least every six months.

(iii) All other lifting equipment must be thoroughly examined at least every twelve months

(iv) Each time that exceptional circumstances that are liable to jeopardise the safety of the lifting equipment have occurred, a thorough examination is required. These may include:

— any accident involving lifting equipment, e.g. collapse of a crane boom;

— after extended periods without proper servicing or maintenance; and

— after long periods out of use after substantial modifications and repairs.

In general **below the hook** items are examined at six monthly intervals with **above the hook** items examined at twelve monthly intervals unless the equipment is used for lifting persons in which case the lifting equipment defaults to a six monthly interval.

The employer may decide to examine specific items of lifting equipment at different intervals in accordance with an examination scheme approved by a competent person e.g. an occasionally used overhead crane could be examined depending on the number of lifting operations, stipulated in the scheme of examination which be may be greater than the previous frequency.

## Reports and defects

**L3025**

(i) A person making a thorough examination for an employer under *Lifting Operations and Lifting Equipment Regulations 1998* (*LOLER*) (*SI 1998 No 2307*), *Reg 9* must:

— immediately notify the employer of any defect in the lifting equipment which in his opinion is or could become a danger to anyone;

- write a report of the thorough examination (see L3022 below) to the employer and any person from whom the lifting equipment has been hired or leased;
- where there is in his opinion a defect in the lifting equipment involving an existing or imminent risk of serious personal injury send a copy of the report to the relevant enforcing authority ('relevant enforcing authority' means, where the defective lifting equipment has been hired or leased by the employer, the Health and Safety Executive – and otherwise, the enforcing authority for the premises in which the defective lifting equipment was thoroughly examined).

(ii) A person making an inspection for an employer under *SI 1998 No 2307, Reg 9* must immediately notify the employer of any defect in the lifting equipment which in his opinion is or could become a danger to anyone, and make a written record of the inspection.

(iii) Every employer who has been notified under (i) above must ensure that the lifting equipment is not used before the defect is rectified; or, in the case of a defect which is not yet but could become a danger to persons, after it could become such a danger.

*[SI 1998 No 2307, Reg 10].*

## Prescribed information

**L3026** *Lifting Operations and Lifting Equipment Regulations 1998* (*LOLER*) (*SI 1998 No 2307*), *Sch 1* specifies the information that must be contained in a report of a thorough examination, made under *SI 1998 No 2307, Reg 10* (see L3021 above). This information can be stored electronically in a secure manner e.g. read only with write access protected, and available only to the competent person.

The information within the report of thorough examination must include the following.

(i) The name and address of the employer for whom the thorough examination was made.

(ii) The address of the premises at which the thorough examination was made.

(iii) Particulars sufficient to identify the lifting equipment including its date of manufacture, if known.

(iv) The date of the last thorough examination.

(v) The safe working load of the lifting equipment or, where its safe working load depends on the configuration of the lifting equipment, its safe working load for the last configuration in which it was thoroughly examined.

(vi) In respect of the first thorough examination of lifting equipment after installation or after assembly at a new site or in a new location:
- that it is such thorough examination; and
- if in fact this is so, that it has been installed correctly and would be safe to operate.

(vii) In respect of all thorough examinations of lifting equipment which do not fall within (vi) above:

(*a*) whether it is a thorough examination under *SI 1998 No 2307, Reg 9(3)*:
- within an interval of six months;
- within an interval of twelve months;
- in accordance with an examination scheme; or
- after the occurrence of exceptional circumstances;

(*b*) if in fact this is so, that the lifting equipment would be safe to operate.

(viii) In respect of every thorough examination of lifting equipment:
- (a) identification of any part found to have a defect which is or could become a danger to anyone, and a description of the defect;
- (b) particulars of any repair, renewal or alteration required to remedy a defect found to be a danger to anyone;
- (c) in the case of a defect which is not yet but could become a danger to anyone:
  — the time by which it could become such a danger;
  — particulars of any repair, renewal or alteration required to remedy the defect;
- (d) the latest date by which the next thorough examination must be carried out;
- (e) particulars of any test, if applicable; and
- (f) the date of the thorough examination.

(ix) The name, address and qualifications of the person making the report; that he is self-employed or, if employed, the name and address of his employer.

(x) The name and address of a person signing or authenticating the report on behalf of its author.

(xi) The date of the report.

Alternative 'multiple exception report' forms are utilised by most competent persons for lifting accessories and simple items of lifting equipment. This method of reporting satisfies the requirements above, but is restricted to multiple items examined during the one visit.

The report clearly identifies the equipment being examined and lists the repairs if necessary as either an A or B defect.

A being the most serious type with an immediate repair required, whereas the B defect is usually time stipulated.

## Keeping of information

**L3027** An employer who obtains lifting equipment to which the 1998 Regulations apply, and who receives an EC declaration of conformity relating to it, must keep the declaration for so long as he operates the lifting equipment [*Lifting Operations and Lifting Equipment Regulations 1998 (LOLER) (SI 1998 No 2307), Reg 11(1)*].

*SI 1998 No 2307, Reg 11(2)(a)* provides that the employer must ensure that the information contained in every report made to him under *SI 1998 No 2307, Reg 10(1)* (see L3025 above) is kept available for inspection:

— in the case of a thorough examination under *SI 1998 No 2307, Reg 9(1)* (see L3021 above) of lifting equipment other than an accessory for lifting, until he stops using the lifting equipment;

— in the case of a thorough examination under *SI 1998 No 2307, Reg 9(1)* (see L3021 above) of an accessory for lifting, for two years after the report is made;

— in the case of a thorough examination under *SI 1998 No 2307, Reg 9(2)* (see L3021 above), until he stops using the lifting equipment at the place it was installed or assembled;

— in the case of a thorough examination under *SI 1998 No 2307, Reg 9(3)* (see L3021 above), until the next report is made under that paragraph or the expiration of two years, whichever is later.

The employer must ensure that every record made under *SI 1998 No 2307, Reg 10(2)* (see L3025 above) is kept available until the next such record is made [*SI 1998 No 2307, Reg 11(2)(b)*].

# Lifting operations and equipment failure

## General

**L3028** Lifting equipment and accessories are not subjected to frequent failure but should failure occur it is often catastrophic. Lifting equipment is subject to statutory inspection to ensure that a fitness for purpose examination, this examination must be current throughout the equipments working life. *Lifting Operations and Lifting Equipment Regulations 1998* (*LOLER*) (*SI 1998 No 2307*) has placed further demands than those that previously existed under repealed lifting legislation in that *SI 1998 No 2307*, *Reg 9(3)(b)* requires that if appropriate, that a competent person undertakes an inspection between thorough examinations (See L3021 above). The specific requirements of this intermediate inspection are based on risk exposure and subsequent determination of the equipment.

Some failures do however still occur and they are often attributed to a combination of procedural and hardware faults.

Examples of specific causes of failure are examined in the following chapters. These examples look at possible causestions, but they are by no means exhaustive and human error will almost certainly be a factor in most failures.

## Cranes

**L3029** Cranes are widely used in lifting/lowering operations in construction, dock and shipbuilding works. The main hazard, generally associated with overloading or incorrect slewing, is collapse or overturning. Contact with overhead power lines is also a danger. In such cases, the operator should normally remain inside the cab and not allow anyone to touch the crane or load; the superintending engineer should immediately be informed. (For reportable dangerous occurrences in connection with cranes etc. see ACCIDENT REPORTING at A3027.)

There are a number of different types of cranes available to perform varying tasks and while there are variants they generally fall into the following categories.

(i)   Fixed crane.

(ii)  Tower crane.

(iii) Mobile crane.

(iv)  Overhead travelling crane.

There are in addition, on construction sites, rough terrain cranes as well as crawler and wheeled cranes. Other variants include ship to shore cranes (portainers) in use at container terminals.

## Safe lifting operations by mobile cranes

**L3030** A mobile crane is a crane which is capable of travelling under its own power.

Safe lifting operations – as per BS 7121 'Safe use of cranes' – depend on co-operation between supervisor (or appointed person), slinger (and/or signaller) and crane driver.

(a)   *Appointed person*

Overall control of lifting operations rests with an 'appointed person', who can, where necessary, stop the operation. Failing this, control of operations will be in the hands of the supervisor (who, in some cases, may be the slinger). The appointed (and competent) person must ensure that:

(i)   lifting operations are carefully planned and executed;

(ii)  weights and heights are accurate;

(iii) suitable cranes are provided;

(iv) the ground is suitable;

(v) suitable precautions are taken, if necessary, regarding gas, water and electricity either above or below ground;

(vi) personnel involved in lifting/lowering are trained and competent; and

(vii) access within the vicinity at where the lifting operation is undertaking is minimised.

(b) *Supervisor*

Supervisors must:

(i) direct the crane driver where to position the crane;

(ii) provide sufficient personnel to carry out the operation;

(iii) check the site conditions;

(iv) report back to the appointed person in the event of problems;

(v) supervise and direct the slinger, signaller and crane driver; and

(vi) stop the operation if there is a safety risk.

(c) *Crane driver*

Crane drivers must:

(i) erect/dismantle and operate the crane as per manufacturer's instructions;

(ii) set the crane level before lifting and ensure that it remains level;

(iii) decide which signalling system is to apply;

(iv) inform the supervisor in the event of problems; and

(v) carry out inspections/weekly maintenance relating to:

— defects in crane structure, fittings, jibs, ropes, hooks, shackles,

— correct functioning of automatic safe load indicator, over hoist and derrick limit switches.

Drivers should always carry out operations as per speeds, weights, heights and wind speeds specified by the manufacturer, mindful that the weight of slings/lifting gear is part of the load. Such information should be clearly displayed in the cab and not obscured or removed. Windows and windscreens should be kept clear and free from stickers containing operational data. Any handrails, stops, machinery guards fitted to the crane for safe access should always be replaced following removal for maintenance; and tools, jib sections and lifting tackle properly secured when not in use.

After a load has been attached to the crane hook by the lifting hook, tension should be taken up slowly, as per the slinger's instructions, the latter being in continuous communication with the driver. In the case of unbalanced loads, drivers/slingers should be familiar with a load's centre of gravity – particularly if the load is irregularly shaped. Once in operation, crane hooks should be positioned directly over the load, the latter not remaining suspended for longer than necessary. Moreover, suspended loads should not be directed over people or occupied buildings.

(d) *Slingers*

Slingers must:

(i) attach/detach a load to/from the crane;

(ii) use correct lifting appliances;

(iii) direct movement of a load by correct signals. Any part of a load likely to shift during lifting/lowering must be adequately secured by the slinger beforehand. Spillage or discharge of loose loads (e.g. scaffolding) can be a problem, and such loads must be properly secured/fastened. Nets are useful for covering palletised loads (e.g. bricks).

(e) *Signallers*

Quite frequently, signallers are responsible for signalling in lieu of slingers. Failing this, their remit is to transmit instructions from slinger to crane driver, when the former cannot see the load.

## Crane failure

**L3031** Crane failure can be attributed to:

(a) failure to lift vertically, e.g. dragging a load sideways along the ground before lifting;

(b) 'snatching' loads, i.e. not lifting slowly and smoothly;

(c) exceeding the maximum permitted moment, i.e. the product of the load and the radius of operation;

(d) excessive wind loading, resulting in crane instability;

(e) defects in the fabrication of the crane, e.g. badly welded joints;

(f) incorrect crane assembly in the case of tower cranes;

(g) brake failure (rail-mounted cranes); or

(h) in the case of mobile cranes:

   (i) failure to use outriggers,

   (ii) lifting on soft or uneven ground, and

   (iii) incorrect tyre pressures.

# Hoists and lifts

**L3032** The safe use and maintenance of hoists and lifts is governed primarily by the *Lifting Operations and Lifting Equipment Regulations 1998 (SI 1998 No 2307)*. Although these regulations would not apply where the lift is not in a workplace and is predominantly used by members of the public e.g. within a shopping precinct.

Examples include: goods lifts; man hoists, scissors lifts and passenger lifts.

Powered working platforms are commonly used for fast and safe access to overhead machinery/ plant, stored products, lighting equipment and electrical installations as well as for enabling maintenance operations to be carried out on high-rise buildings. Their height, reach and mobility give them distinct advantages over scaffolding, boatswain's chairs and platforms attached to fork lift trucks. Typical operations are characterised by self-propelled hydraulic booms, semi-mechanised articulated booms and self-propelled scissors lifts.

Platforms should always be sited on firm level working surfaces and their presence indicated by traffic cones and barriers. Location should be away from overhead power lines – but, if this is not practicable the safe system of work in operation needs to control this additional risk. A key danger arises from overturning as a consequence of overloading the platform. Maximum lifting capacity should, therefore, be clearly indicated on the platform as well as in the manufacturer's instructions, and it is inadvisable to use working platforms in high winds (i.e. above Force 4 or 16 mph). Powered working platforms should be regularly maintained and only operated by trained personnel.

### Hoist and lift failure

L3033   Hoist and lift failure can be attributed to:

(a)   excessive wear in wire ropes;

(b)   excessive broken wires in ropes;

(c)   mechanical failure (e.g. worm and gearing);

(d)   failure of the overload protection device;

(e)   failure of the overrun device;

(f)   failure of the speed governor;

(g)   safety gear failure; or

(h)   failure of the landing door interlocks.

## Fork lift trucks

L3034   Fork lift trucks are the most widely used item of mobile mechanical handling equipment. There are several varieties which are as follows:

(1)   *Pedestrian-operated stackers – manually-operated and power-operated*

Manually-operated stackers are usually limited in operation, for example, for moving post pallets, and cannot pick up directly from the floor. Whereas power-operated stackers are pedestrian-operated or rider-controlled, operate vertically and horizontally and can lift pallets directly from the floor.

(2)   *Reach trucks*

Reach trucks enable loads to be retracted within their wheel base. There are two kinds, namely, (a) moving mast reach trucks, and (b) pantograph reach trucks. Moving mast reach trucks are rider-operated, with forward-mounted load wheels enabling carriage to move within the wheel base – mast, forks and load moving together. Pantograph reach trucks are also rider-operated, reach movement being by pantograph mechanism, with forks and load moving away from static mast.

(3)   *Counterbalance trucks*

Counterbalance trucks carry loads in front counterbalanced to the weight of the vehicle over the rear wheels. Such trucks are lightweight pedestrian-controlled, lightweight rider-controlled or heavyweight rider-controlled.

(4)   *Narrow aisle trucks*

With narrow aisle trucks the base of the truck does not turn within the aisle in order to deposit/retrieve load. There are two types, namely, side loaders for use on long runs down narrow aisles, and counterbalance rotating load turret trucks, having a rigid mast with telescopic sections, which can move sideways in order to collect/deposit loads.

(5)   *Order pickers*

Order pickers have a protected working platform attached to the lift forks, enabling the driver to deposit/retrieve objects in or from a racking system. Conventional or purpose-designed, they are commonly used in racked storage areas and operate well in narrow aisles.

Forklift trucks are involved in far too many incidents although manufacturers are increasing the number of safety devices an attempt to limit human error, these include:

- speed limiters with the forks elevated;
- pressure sensitive seats, that cut off the fuel supply when the driver disembarks;
- TV monitors when the truck has obscured vision;
- intelligent start up systems, swipe cards or similar;
- lap straps; and
- falling object protection.

Forklift trucks are mobile work equipment and with the exception of rough terrain vehicles all are protected from rolling over beyond 90 degrees should an overturn occur by their mast's. Lap straps should be used when fitted although the requirement for their use is risk based.

### Forklift truck failure

L3035  Forklift truck failure can be attributed to:

(a) uneven floors, steeply inclined ramps or gradients, i.e. in excess of 1:10 gradient;

(b) inadequate room to manoeuvre;

(c) inadequate or poor maintenance of lifting gear;

(d) the practice of driving forwards down a gradient with the load preceding the truck;

(e) load movement in transit;

(f) sudden or fast braking;

(g) poor stacking of goods being moved;

(h) speeding;

(i) turning corners too sharply;

(j) not securing load sufficiently e.g. pallets stacked poorly;

(k) hidden obstructions in the path of the truck;

(l) use of the forward tilt mechanism with a raised load; or

(m) generally bad driving, including driving too fast, taking corners too fast, striking overhead obstructions, particularly when reversing and excessive use of the brakes.

## Patient/bath hoist

L3036  Patient/bath hoists are used for the lifting of persons and require a six-month thorough examination.

Patient hoist failure:

(a) mechanical damage to pulleys sheaves and drums;

(b) anchor point failure;

(c) breaking mechanism failure;

(d) boom arm distortion or corrosion;
(e) electrical faults (ere fitted); and
(f) hook catch failure.

## Vehicle lifting table

**L3037** The primary function is as a vehicle lift but personnel are at risk when the vehicle is descending, by definition a twelve monthly thorough examination would normally satisfy the inspection frequency laid down in the *Lifting Operations and Lifting Equipment Regulations 1998 (SI 1998 No 2307)* most of industry have adopted a six monthly examination frequency. Lifting tables are normally either wire suspension type or hydraulic actuated. BS EN 1433:1999 refers.

Vehicle lifting table failure:

(a) mechanical damage to pulleys sheaves and wires;
(b) anchor point failure;
(c) breaking mechanism failure;
(d) electrical faults;
(e) hydraulic failure;
(f) foundation bolt failure;
(g) catching device failure causing vehicles to fall from table; and
(h) carrying arm failure.

## Lifting accessories

**L3038** 'Accessory for lifting' – work equipment for attaching loads to machinery for lifting (pendant, sling, shackle, etc.).

It has become common practice within industry to use a simple colour coded cable tie to identify that the accessory is within its inspection cycle. This colour tie will normally be displayed at the lifting equipment store for ease of recognition.

Lifting accessories refer to:

(a) ropes;
(b) chains;
(c) slings;
(d) shackles act.

### Ropes

**L3039** Ropes are used quite widely throughout industry in lifting/lowering operations, the main hazard being breakage through overloading and/or natural wear and tear. Ropes are either of natural (e.g. cotton, hemp) or man-made fibre (e.g. nylon, terylene). Natural fibre ropes, if they become wet or damp, should be allowed to dry naturally and kept in a well-ventilated room. They do not require certification prior to service, but six months' examination thereafter is compulsory (all lifting equipment accessories require a six monthly examination to meet the requirements under the *Lifting Operations and Lifting Equipment Regulations 1998*). If the specified safe working load (SWL) cannot be maintained by a rope, it should forthwith be withdrawn from service. Of the two, man-made fibre ropes have greater tensile strength, are not subject so much to risk from wear and tear, are more acid/corrosion-resistant and can absorb shock loading better.

## Fibre rope failure

**L3040** Fibre rope failure can be attributed to:

(a) bad storage in wet or damp conditions resulting in rot and mildew;

(b) inadequate protection of the rope when lifting loads with sharp edges;

(c) exposure to direct heat to dry, as opposed to gradual drying in air;

(d) chemical reaction; or

(e) Ultra Violet degradation.

## Wire (steel) rope

**L3041** Wire ropes are much in use in cranes, lifts, hoists, elevators etc. and as such become part of the lifting equipment, they are also commonly utilised as accessories and they need to be lubricant-impregnated to minimise corrosion and reduce wear and tear unless they are manufactured from corrosive resistive materials for specialist applications. BS EN 12385 refers.

## Wire rope failure

**L3042**

Wire rope failure can be attributed to:

(a) excessive broken wires;

(b) failure to lubricate regularly;

(c) frequent knotting or kinking of the rope; or

(d) bad storage in wet or damp conditions which promotes rust.

## Chain

**L3043** A chain is a classic example of a lifting accessory, in spite of an increase in use of wire ropes. There are several varieties, including mild, high-tensile, and alloy steel chains. The principal risk is breakage, usually occurring in consequence of a production defect in a link, or through application of excessive loads. A crucial part of the thorough examination is to measure the chain wear for which a simple gauge is used which allows measurement of elongation and link/pin wear.

## Chain failure

**L3044**

Chain failure can be attributed to:

(a) mechanical defects in individual links;

(b) application of a static in excess of the breaking load;

(c) snatch loading; and

(d) hydrogen embrittlement (manufacturing welding defects refer HSE publication PM 39, Hydrogen cracking of grade T(8) chain and components)).

*Lifting Operations* **L3047**

## Slings

**L3045** Slings are manufactured in natural or man-made fibre or chain. The safe working load of any sling varies according to the angle formed between the legs of the sling.

The relationship between sling angle and the distance between the legs of the sling is also important. (See Table 1 below.)

| Table 1 | |
|---|---|
| **Safe working load for slings** | |
| *Sling Angle* | *Distance between legs* |
| 30° | ½ leg length |
| 60° | 1 leg length |
| 90° | 1⅓ leg length |
| 120° | 1⅔ leg length |

For a one tonne load, the tension in the leg increases as shown in Table 2 below.

| Table 2 | |
|---|---|
| **Safe working load for slings – increased tension** | |
| *Sling leg angle* | *Tension in leg (tonnes)* |
| 90° | 0.7 |
| 120° | 1.0 |
| 151° | 2.0 |
| 171° | 6.0 |

Other causes of failure in slings are:

(*a*) cuts, excessive wear, kinking and general distortion of the sling legs;

(*b*) failure to lubricate wire slings;

(*c*) failure to pack sharp corners of a load, resulting in sharp bends in the sling and the possibility of cuts or damage to it; and

(*d*) unequal distribution of the load between the legs of a multi-leg sling.

## Eyebolts

**L3046** Eyebolts are mainly for lifting heavy concentrated loads; there being several types, namely, dynamo, collar and eyebolt incorporating link.

### Eyebolt failure

**L3047**

Eyebolt failure can be attributed to:

(*a*) mechanical defects;

(*b*) incorrect thread (metric and imperial threads, cross threaded);

(*c*) dynamo eyebolt used where collar eyebolt was required;

(d) eyebolt not hardened down to the shoulder; or

(e) oil remaining in the tapped hole receiving the eyebolt, thus destroying the receiving housing under hydraulic pressure.

## Shackles

**L3048** Shackles come in two main types 'Dee' shackle and 'Bow' so named because of there shapes. The bow shackle has the capability to accept a larger number of ropes, slings etc. The shackle and pin are matched pairs and should not be separated and mixed up.

### Shackle failure

**L3049**

Shackle failure can be attributed to:

(a) mechanical defects;

(b) interchange of pin and shackle;

(c) nut and bolt used instead of the shackle pin; or

(d) wear on the shackle.

## Once only use accessories

**L3050** Also included are 'one off' accessories which are disposed of after there use and examples of these are:

- industrial bulk carriers (IBC's) bags that are commonly used in the building and chemical industry where it would not be possible certify integrity following there first use;

- health care, under body sling used with patient hoists, the problems associated with sterilising such equipment through autoclaving when it would deteriorate and the cost of examination by a competent person have made it both a practical solution and an economic proposition to discard such equipment after use; and

- dynamo eyebolt, are designed for a single lift only.

## Mobile lifting equipment

**L3051** Mobile work equipment is any equipment which carries out work whilst travelling. It may be self-propelled, towed or remotely controlled and it may incorporate attachments. An example of mobile lifting equipment is an excavator involved in digging tasks.

The risk of mobile lifting equipment overturning is another cause for concern, and the problem was addressed by the additional requirements introduced in the *Provision and Use of Work Equipment Regulations 1998* (*SI 1998 No 2306*). These implement into UK legislation the amending directive to the *Use of Work Equipment Directive* (*AUWED*). Under this legislation, workers must be protected from falling out of the equipment and from unexpected movement e.g. overturning.

To prevent such equipment overturning, the following action should be taken:

— increase stability when reasonably practicable (this may be done by reducing the equipment's speed)

— fit stabilisers (e.g. outriggers or counterbalance weights);

- ensure that roll over protection structure is installed which ensures that it does no more than fall on its side;
- ensure the structure gives sufficient clearance to anyone being carried if it overturns more than on its side;
- provision of harnesses or a suitable restraining system to protect against being crushed;
- use only on firm ground; and
- avoid excessive gradients.

# Lighting

## Introduction

**L5001** Increasingly, over the last decade or so, employers have come to appreciate that indifferent lighting is both bad economics and bad ergonomics, not to mention potentially bad industrial relations; and conversely, good lighting uses energy efficiently and contributes to general workforce morale and profitability – that is, operating costs fall whilst productivity and quality improve. Alternatively, poor lighting reduces efficiency, thereby increasing the risk of stress, denting workforce morale, promoting absenteeism and leading to accidents, injuries and even deaths at work.

Ideally, good lighting should 'guarantee' (*a*) employee safety, (*b*) acceptable job performance and (*c*) good workplace atmosphere, comfort and appearance. This is not just a matter of maintenance of correct lighting levels. Ergonomically relevant are:

(*a*)  horizontal illuminance;

(*b*)  uniformity of illuminance over the job area;

(*c*)  colour appearance;

(*d*)  colour rendering;

(*e*)  glare and discomfort;

(*f*)  ceiling, wall, floor reflectances;

(*g*)  job/environment illuminance ratios;

(*h*)  job and environment reflectances;

(*j*)  vertical illuminance.

## Statutory lighting requirements
### General

**L5002** Adequate standards of lighting in all workplaces can be enforced under the general duties of the *Health and Safety at Work etc. Act 1974*, which require provision by an employer of a safe and healthy working environment (see EMPLOYERS' DUTIES TO THEIR EMPLOYEES). *Regulation 8* of the *Workplace (Health, Safety and Welfare) Regulations 1992 (SI 1992 No 3004)* requires that all workplaces have suitable and sufficient lighting and that, so far as is reasonably practicable, the lighting should be natural light.

More particularly, however, people should be able to work and move about without suffering eye strain (visual fatigue) and having to avoid shadows. Local lighting may be necessary at individual workstations and places of particular risk. Outdoor traffic routes used by pedestrians should be adequately lit after dark. Lights and light fittings should avoid dazzle and glare, and be so positioned that they do not cause hazards, whether fire, radiation or electrical. Switches should be easily accessible and lights should be replaced, repaired or cleaned before lighting becomes insufficient. Moreover, where persons are particularly exposed to danger in the event of failure of artificial lighting, emergency lighting must be provided (*Workplace (Health, Safety and Welfare) Regulations 1992 (SI 1992 No 3004), Reg 8(3)* (and ACOP)).

To accommodate these 'requirements', refresh rates (flickering), a combination of general and localised lighting (not to be confused with 'local' lighting, e.g. desk light) is often necessary. Moreover, state of the art visual display units have thrown up some occupational health problems addressed by the *Health and Safety (Display Screen Equipment) Regulations 1992 (SI 1992 No 2792)as amended by the Health and Safety (Miscellaneous) Regulations 2002* (see O5005 *et seq.* OFFICES AND SHOPS).

The Chartered Institute of Building Services Engineers (CIBSE) produces a wide range of technical and industry-specific guidance on design of lighting in various environments, commissioning and installation, etc. As an example, a new edition of *Lighting Guide 7: office lighting* covers all aspects of the lighting of offices, from board room through general offices to the post room.

The Society of Light and Lighting acts as the professional body for lighting in the UK. It operates worldwide and offers a range of publications and advice on specific areas of lighting.

## Specific processes

### Lighting and VDUs

**L5003** Annex A to BS EN 924-6-1999 provides guidance on illuminance levels, specifically in relation to use with display screen equipment (BS EN 9241 – Ergonomic requirements for office work with visual display terminals (VDTs); Part 6 – Guidance on work environment).

### Machinery and work equipment – Provision and Use of Work Equipment Regulations 1998 (SI 1998 No 2306)

**L5004** Lighting requirements, applicable to woodworking machines in the *Woodworking Machines Regulations 1974 (SI 1974 No 903), Reg 43* have now been revoked. This currently comes under the *Provision and Use of Work Equipment Regulations 1998 (SI 1998 No 2306), Reg 21*. This Regulation is very broad and merely requires that suitable and sufficient lighting is provided, taking into account the operations to be carried out. To ascertain what is required to meet this duty, the HSE guidance has been produced. This refers to occasions when additional or localised lighting may be needed such as:

- when the task requires a high perception of detail;
- when there is a dangerous process;
- to reduce visual fatigue;
- during maintenance operations.

Additional guidance is provided in the HSE Guidance Note HS(G) 38, *'Lighting at Work.*

### Lighting requirements on construction sites – Construction (Health, Safety and Welfare) Regulations 1996 (SI 1996 No 1592)

**L5005** Every workplace, approach to the workplace and traffic route in and around the workplace shall have suitable and sufficient lighting. Where reasonably practicable, this lighting shall be by natural light. The colour of every artificial light must not adversely affect or change the perception of any sign or signal provided for health and safety reasons. Suitable and sufficient secondary lighting must be provided in places where there is risk to health and safety in the event of the failure of the primary lighting (*SI 1996 No 1592, Reg 25*).

## Lighting requirements for electrical equipment – Electricity at Work Regulations 1989 (SI 1989 No 635)

**L5006** So as to prevent injury, adequate lighting must be provided at all electrical equipment on which or near which work is being done in circumstances that may give rise to danger (*Electricity at Work Regulations 1989* (*SI 1989 No 635*), *Reg 15*).

# Sources of light

## Natural lighting

**L5007** Daylight is the natural, and cheapest, form of lighting, but it has only limited application to places of work where production is required beyond the hours of daylight, at all seasons, and where daytime visibility is restricted by climatic conditions. But however good the outside daylight, windows can rarely provide adequate lighting alone for the interior of large floor areas. Single storey buildings can, of course, make use of insulated opaque roofing materials, but the most common provision of daylight is by side windows. There is also the fact that the larger the glazing area of the building, the more other factors such as noise, heat loss in winter and unsatisfactory thermal conditions in summer must be considered.

Modern conditions, where the creation of pleasant building interior environment requires the balanced integration of lighting, heating, air conditioning, acoustic treatment, etc., are such that lighting cannot be considered in isolation. At the very least, natural lighting will have to be supplemented for most of the time with artificial lighting, the most common source for which is electric lighting.

## Electric lighting

**L5008** Capital costs, running costs and replacement costs of various types of electric lighting have a direct bearing on the selection of the sources of electric lighting for particular application. Such costs are as important a consideration as the size, heat and colour effects required of the lighting. The efficiency of any type of lamp used for lighting is measured as light output, in lumens, per watt of electricity. Typical values for various types of lamp are as follows (the term 'lumen' is explained in the discussion of standards of illuminance in L5009 below).

| Type of lamp | Lumens per watt |
|---|---|
| Incandescent lamps | 10 to 18 |
| Tungsten halogen | 22 |
| High pressure mercury | 25 to 55 |
| Tubular fluorescent | 30 to 80 (depending on colour) |
| Mercury halide | 60 to 80 |
| High pressure sodium | 100 |

In general, the common incandescent lamps (coiled filament lamps, the temperature of which is raised to white heat by the passage of current, thus giving out light) are relatively cheap to install but have relatively expensive running costs. A discharge or fluorescent lighting scheme (which works on the principle of electric current passing through certain gases and thereby producing an emission of light) has higher capital costs but higher running efficiency, lower running costs and longer lamp life. In larger places of work the choice is often between discharge and fluorescent lamps. The normal mercury discharge lamp and the low pressure sodium discharge lamp have restricted colour performance, although newly developed high pressure sodium discharge lamps and colour corrected mercury lamps do not suffer from this disadvantage.

## Standards of lighting or illuminance

### The technical measurement of illuminance

**L5009** The standard of illuminance (the amount of light) required for a given location or activity depends on a number of variables, including general comfort considerations and the visual efficiency required. The unit of illuminance is the 'lux', which equals one lumen per square metre; this unit has now replaced the 'foot candle' which was the number of lumens per square foot. The term 'lumen' is the unit of luminous flux, describing the quantity of light received by a surface or emitted by a source of light.

### Light measuring instruments

**L5010** For accurate measurement of the degree of illuminance at a particular working point, a reliable instrument is required. Such an instrument, suitable for most measurements, is a pocket lightmeter; this incorporates the principle of the photo-electric cell, which generates a tiny electric current in proportion to the light at the point of measurement. This current deflects a pointer on a graduated scale measured in lux. Manufacturers' instructions should, of course, be followed in the care and use of such instruments.

### Average illuminance and minimum measured illuminance

**L5011** HSE Guidance Note HS(G) 38 'Lighting at work' (1998) relates illuminance levels to the degree or extent of detail which needs to be seen in a particular task or situation. Recommended illuminances are shown in Table 1 at L5012 below.

This guidance note makes recommendations both for average illuminance for the work area as a whole and for minimum measured illuminance at any position within it. As the illuminance produced by any lighting installation is rarely uniform, the use of the average illuminance figure alone could result in the presence of a few positions with much lower illuminance, which pose a threat to health and safety. The minimum measured illuminance is therefore the lowest illuminance permitted in the work area taking health and safety requirements into account.

The planes on which the illuminances should be provided depend on the layout of the task. If predominantly on one plane, for example horizontal, as with an office desk, or vertical, as in a warehouse, the recommended illuminances are recommended for that plane. Where there is either no well-defined plane or more than one, the recommended illuminances should be provided on the horizontal plane and care taken to ensure that the reflectances of surfaces in working areas are high.

### Illuminance ratios

**L5012** The relationship between the lighting of the work area and adjacent areas is significant. Large differences in illuminance between these areas may cause visual discomfort or even affect safety levels where there is frequent movement, for example forklift trucks. This problem arises most often where local or localised lighting in an interior exposes a person to a range of illuminance for a long period, or where there is movement between interior and exterior working areas exposing a person to a sudden change of illuminance. To reduce hazards and possible discomfort, specific recommendations shown in Table 1 below should be followed.

Where there is conflict between the recommended average illuminances shown in Table 1 and maximum illuminance ratios shown in Table 2, the higher value should be taken.

### Table 1

**Average illuminances and minimum measured illuminances for different types of work**

| General activity | Typical locations/types of work | Average illuminance (Lx) | Minimum measured illuminance (Lx) |
| --- | --- | --- | --- |
| Movement of people, machines and vehicles* | Lorry parks, corridors, circulation routes | 20 | 5 |
| Movement of people, machines and vehicles in hazardous areas; rough work not requiring any perception of detail | Construction site clearance, excavation and soil work, docks, loading bays, bottling and canning plants | 50 | 20 |
| Work requiring limited perception of detail** | Kitchens, factories, assembling large components, potteries | 100 | 50 |
| Work requiring perception of detail | Offices, sheet metal work, bookbinding | 200 | 100 |
| Work requiring perception of fine detail | Drawing offices, factories assembling electronic components, textile production | 500 | 200 |

*Notes*

\*  Only safety has been considered, because no perception of detail is needed and visual fatigue is unlikely. However, where it is necessary to see detail to recognise a hazard or where error in performing the task could put someone else at risk, for safety purposes as well as to avoid visual fatigue the figure should be increased to that for work requiring the perception of detail.

\*\*  The purpose is to avoid visual fatigue: the illuminances will be adequate for safety purposes.

## Table 2
### Maximum ratios of illuminance for adjacent areas

| Situations to which recommendation applies | Typical location | Maximum ratio of illuminances | | |
|---|---|---|---|---|
| | | Working area | | Adjacent area |
| Where each task is individually lit and the area around the task is lit to a lower illuminance | Local lighting in an office | 5 | : | 1 |
| Where two working areas are adjacent but one is lit to a lower illuminance than the other | Localised lighting in a works store | 5 | : | 1 |
| Where two working areas are lit to different illuminances and are separated by a barrier but there is frequent movement between them | A storage area inside a factory and a loading bay outside | 10 | : | 1 |

## Maintenance of light fitments

**L5013** The lighting output of a given lamp will reduce gradually in the course of its life but an improvement can be obtained by regular cleaning and maintenance, not only of the lamp itself, but also of the reflectors, diffusers and other parts of the luminaire. A sensible and economic lamp replacement policy is called for (e.g. it may be more economical, in labour cost terms, to change a batch of lamps than deal with them singly as they wear out).

# Qualitative aspects of lighting and lighting design

**L5014** While the quantity of lighting afforded to a particular location or task in terms of standard service illuminance is an important feature of lighting design, it is also necessary to consider the qualitative aspects of lighting, which have both direct and indirect effects on the way people perceive their work activities and dangers that may be present. The quality of lighting is affected by the presence or absence of glare, the distribution of the light, brightness, diffusion and colour rendition.

## Glare

**L5015** This is the effect of light which causes impaired vision or discomfort experienced when parts of the visual field are excessively bright compared with the general surroundings. It may be experienced in three different forms:

(a) disability glare – the visually disabling effect caused by bright bare lamps directly in the line of vision;

(b) discomfort glare – caused by too much contrast of brightness between an object and its background, and frequently associated with poor lighting design. It can cause discomfort without necessarily impairing the ability to see detail. Over a period it can cause visual fatigue, headaches and general fatigue;

(c) reflected glare – the reflection of bright light sources on shiny or wet work surfaces, such as plated metal or glass, which can almost entirely conceal the detail in or behind the object that is glinting.

N.B. The Illuminating Engineering Society (IES) publishes a Limiting Glare Index for each of the effects in (a) and (b) above. This is an index representing the degree of discomfort glare which will be just tolerable in the process or location under consideration. If exceeded, occupants may suffer eye strain or headaches or both.

## Distribution

**L5016** Distribution is concerned with the way light is spread. The British Zonal Method classifies luminaires (light-fittings) according to the way they distribute light from BZ1 (all light downwards in a narrow column) to BZ10 (light in all directions). However, a fitting with a low BZ number does not necessarily imply less glare. Its positioning, the shape of the room and the reflective surfaces present are also significant.

The actual spacing of luminaires is also important when considering good lighting distribution. To ensure evenness of illuminance at operating positions, the ratio between the height of the luminaire and the spacing of it must be considered. The IES spacing: height ratio provides a basic guide to such arrangements. Under normal circumstances, for example offices, workshops and stores, this ratio should be between 1½:1 and 1:1 according to the type of luminaire.

## Brightness

**L5017** Brightness or 'luminosity' is very much a subjective sensation and, therefore, cannot be measured. However, it is possible to consider a brightness ratio, which is the ratio of apparent luminosity between a task object and its surroundings. To ensure the correct brightness ratio, the reflectance (the ability of a surface to reflect light) of all surfaces in the working area should be well maintained and consideration given to reflectance values in the design of interiors. Given a task illuminance factor (the recommended illuminance level for a particular task) of 1, the effective reflectance values should be 0.6 for ceilings, 0.3 to 0.8 for walls and 0.2 to 0.3 for floors.

## Diffusion

**L5018** This is the projection of light in all directions with no predominant direction. The directional flow of light can often determine the density of shadows, which may prejudice safety standards or reduce lighting efficiency. Diffused lighting will reduce the amount of glare experienced from bare luminaires.

## Colour rendition

**L5019** Colour rendition refers to the appearance of an object under a specific light source compared to its colour under a reference illuminant, for example natural light. Good standards of colour rendition allow the colour appearance of an object to be properly perceived. Generally, the colour rendering properties of luminaires should not clash with those of natural light, and should be just as effective at night when there is no daylight contribution to the total illumination of the working area.

## Stroboscopic effect

**L5020** One aspect of lighting quality that formerly gave trouble was the stroboscopic effect of fluorescent tubes, which gave the illusion of motion or even the illusion that a rotating part of machinery was stationary. With modern designs of fluorescent tubes, this effect has largely been eliminated.

# Machinery Safety

## Introduction

**M1001** The introduction of the harmonised European Union ('EU') regulations and standards in health and safety has changed the approach to the prevention of accidents and ill health at work. In the place of a prescriptive set of standards and regulations, the harmonised regulations and standards represent a remarkable breakthrough in the identification, assessment and control of machinery and work equipment risks.

As far as machinery safety is concerned, the introduction of the EU concept for the proactive 'goal-setting' approach has fundamentally changed the approach to the prevention of accidents and ill health at work. In the place of prescriptive legislation and standards, the new risk-based approach to machinery and work equipment safety has fundamentally changed the way for the identification, assessment and control of machinery hazards. The new approach is essentially based on structured and systematic assessment of risks.

This chapter aims to review this approach, which is based on the EU Machinery and Use of Work Equipment Directives as well as the Transposed Harmonised EU Machinery Safety Standards. It aims to give guidance and examples on the link between the structured approach for hazard identification and analysis as well as risk evaluation. This helps in developing action (health and safety) plans for the implementation of a package of measures for machinery and work equipment accident/ill health prevention in a practical way.

Currently, statutory duties are dually (but separately) laid on both manufacturers of machinery for use at work and employers and users of such machinery and equipment, prior compliance, on the part of manufacturers leading to compliance on the part of employers/users. General and specific duties are imposed on manufacturers by the *Supply of Machinery (Safety) Regulations 1992 (SI 1992 No 3073)* as amended by the *Supply of Machinery (Safety) (Amendment) Regulations 1994 (SI 1994 No 2063)*.

Duties of employers are incorporated into the *Provision and Use of Work Equipment Regulations 1998 (SI 1998 No 2306)*.

## Legal requirements relating to machinery

**M1002** The introduction of the *Health and Safety at Work etc. Act* ('*HSWA 1974*') in 1974, in addition to placing general duties on employers under *section 2* of the Act, has placed the responsibility of machinery safety on designers, manufacturers and suppliers. This is still applicable only in the UK, in addition to the EU requirements.

The original version of the text in *section 6(1)* of *HSWA 1974* placed the following duties on designers, manufacturers and suppliers:

'It shall be the duty of any person who designs/supplies ... any article for use at work to ensure, so far as is reasonably practicable, that the article is so designed ... as to be safe and without risks to health ... when properly used'.

The original *section 6* concentrated on protection against physical injuries during normal use of machinery and placed obligations on the user to follow the manufacturer instructions. The term 'so far as is reasonably practicable' has been interpreted by the courts as to indicate that in deciding an adequate level of safety, both the risk and cost of dealing with must be taken into account. This is a difficult concept for the majority of machine designers who understandably are cost driven.

The introduction of the European Directive on product liability, which was implemented in the UK under the *Consumer Protection Act* in 1987, in addition to the deficiencies shown by several court cases, have resulted in the revision of *section 6* in 1988.

The current *section 6(1)* of the 1974 Act requires that:

> 'It shall be the duty of any person who designs/ supplies ... any article for use at work to ensure, so far as is reasonably practicable, that the article is so designed ... that it will be safe without risk to health at all times: when being set, used, cleaned or maintained by a person at work'.

It should be noted that the term 'when properly used' had disappeared, as it is expected that designers should take account of normal behaviour of operators and maintenance personnel and the machine design should accommodate foreseeable misuse.

## Machinery safety – the risk based approach

M1003    In May 1985 the EU Ministers agreed to a New Approach to Technical Harmonisation and Standards to overcome the problem of trade between European partners. The Machinery Safety Directive (89/392/EEC), subsequently amended, is one of the Product Safety Directives and sets out the essential health and safety requirements ('EH&SR's') for machinery which must be met before machinery is placed on the market anywhere within the EU.

EH&SRs are expressed in general terms and it is intended that the European Harmonised Standards should fill in the detail so that machinery designers and suppliers have clear guidance on how to achieve conformity with the Directive.

The Use of Work Equipment Directive (89/655/EEC) ('UWED') was also introduced to outline the responsibilities of management for the protection of workers from machinery and work equipment. In the UK this became the *Provision and Use of Work Equipment Regulations 1992* (*SI 1992 No 2932*) ('*PUWER'92*'). The amending EU Directive on the Use of Work Equipment Directive (95/63/EC) has been implemented as the *Provision and Use of Work Equipment Regulations 1998* (*SI 1998 No 2306*) ('*PUWER '98*').

## The Provision and Use of Work Equipment Regulations 1998 (PUWER'98)

M1004    The first EU Directive aimed at the use of work equipment was the Use of Work Equipment by Workers at Work Directive (89/655/EEC) ('UWED'). This Directive required all member states of the EC to have the same minimum requirements for the selection and use of work equipment, and was implemented in the UK by *PUWER'92* (*SI 1992 No 2932*). The Directive was subsequently amended and the non-lifting aspects of the amendments by Directive 95/63/EC ('AUWED') were implemented in the UK by *PUWER'98* (*SI 1998 No 2306*). These Regulations came into force on 5 December 1998 and on the same date *PUWER'92* was revoked. The lifting aspects of AUWED have been implemented by the *Lifting Operations and Lifting Equipment Regulations 1998* (*SI 1998 No 2307*) ('*LOLER*').

The 1998 Regulations are aimed at employers, and are designed to implement EU Directive 89/655/EEC, made under Article 118A of the *Single European Act*. This is aimed specifically at protecting the health and safety of workers in the Community. PUWER is also part of the so-called '6 pack Regulations'. They came into force on 1 January 1993 and applied immediately to equipment supplied after that date. *Regulations 1–10* apply to existing items of equipment but employers have until 1 January 1997 to ensure existing equipment meets the 'hardware' requirements in *Regulations 11–24*. Until such time as they do, existing equipment in factories will remain subject to the requirements of *sections 12–17* of the *Factories Act 1961*, which *PUWER* repealed.

*PUWER'98*, was made under the *HSWA 1974*. HSC states that the primary objective of *PUWER'98* is to ensure that work equipment should not result in health and safety risks, regardless of its age, condition or origin.

## Structure of PUWER'98

**M1005**   The Regulations are structured in five main parts. These are set out as follows:

*Part I: Regulations 1–3*

This includes interpretation, definitions and application of *PUWER'98*.

*Part II: Regulations 4–10 (General –software)*

This sets out the 'management' duties of *PUWER'98* covering: selection of suitable equipment, maintenance, inspection, specific risks, information, instruction and training. *Regulation 10* covers the conformity of work equipment with the requirements of EU Directives on product safety.

*Part II: Regulations 11–24 (General – hardware)*

This deals with the hardware aspects of *PUWER'98*. It covers the guarding of dangerous parts of work equipment, the provision of appropriate stop and emergency stop controls, stability, lighting and suitable warning or devices.

*Part III: Regulations 25–30 (Mobile work equipment)*

This deals with specific risks associated with the use of self-propelled, towed and remote controlled mobile work equipment.

*Part IV: Regulations 31–35 (Power presses)*

This deals with the management requirements for the safe use and thorough examination of power presses. (Readers are advised to refer to a HSE publication 'Power presses: Maintenance and thorough examination' (HSG236), which provides practical advice on what, when and how to maintain presses.)

*Part V: Regulations 36–39*

Cover transitional provisions, repeal of Acts and revocation of instruments.

## Risk assessment and PUWER'98

**M1006**   There is no requirement in *PUWER'98* (*SI 1998 No 2306*) to carry out a risk assessment. The general risk assessment requirement is made under *Regulation 3(1)* of the *Management of Health and Safety at Work Regulations 1999* (*SI 1999 No 3242*), which is explained in general terms in the associated HSC publications.

The HSC publication, 'Safe use of work equipment: Approved Code of Practice and Guidance' (L22), contains more specific guidance on risk assessment in relation to work equipment. This publication indicates that the factors to be considered in a risk assessment to meet the requirements of *PUWER'98* should include: type of work equipment, substances and electrical or mechanical hazards to which people may be exposed. Action to eliminate/control any risk might include, for example, during maintenance: disconnection of power supply, supporting parts of the work equipment which could fall, securing mobile equipment so that it cannot move and depressurising pressurised systems. Reference is also made to the use of the HSE publication '5 steps to risk assessment' (INDG163) (Revised 1998).

**M1007** *Machinery Safety*

In dealing with risk assessment the HSC publication L22 refers to 'significant risk' which is defined by in that document as 'one which could foreseeably result in a major injury or worse'. This definition implies that if the consequences of exposure to a hazard could result in a minor injury, there is no need to comply with many of *PUWER'98* requirements.

'Risk' is defined as the combination of the probability (or chance) of a harm being realised coupled with the consequences (severity) as a result of exposure to the harm. Depending on which criteria is used to evaluate risks, if the chance of a minor injury is likely, then this would constitute a significant risk. The lack of guidance on agreed criteria for the evaluation of health and safety risks makes the task of deciding whether risks are tolerable somewhat subjective and arbitrary. The HSE's '5 steps to risk assessment' may be too generic to result in adequate hazards identification and analysis or in the evaluation of risks, particularly in relation to machinery.

## Summary of main requirements of PUWER'98
## Part II: General (Regulations 4–24)

**M1007** Part II of *PUWER'98* (*SI 1998 No 2306*) applies to all work equipment, including machinery; mobile plant and work equipment.

### Regulation 4 – Suitability

Equipment must be suitable, by design, construction or adaptation, for the actual work it is provided to do. In selecting the work equipment, the employer must take account of the environment in which the equipment is to be used and must ensure that the equipment is used only for the operations, and under the conditions, for which it is suitable.

### Regulation 5 – Maintenance

Equipment must be maintained in an efficient state, in efficient working order and in good repair. (The term 'efficient' relates to how the condition of the equipment might affect health and safety, not productivity). There is no requirement to keep a maintenance log, but where one exists (and this is good management practice) it must be kept up to date.

### Regulation 6 – Inspection

*Regulation 6* of *PUWER'98* (*SI 1998 No 2306*) includes a requirement for the inspection of work equipment. The purpose of an inspection is to identify whether the work equipment can be operated, adjusted and maintained safely and that any deterioration, e.g. defects, damage or wear can be detected and repaired before it results in unacceptable risks.

The guidance to *PUWER'98* states that inspection is only necessary where there is a significant risk resulting from incorrect installation or re-installation, deterioration or as a result of exceptional circumstances which could affect the safe operation of the work equipment. Examples given in the guidance of work equipment unlikely to need an inspection include hand tools, non-powered machinery and powered machinery such as a reciprocating fixed blade metal cutting saw!

The extent of the inspection required will depend on the level of risk associated with the work equipment, and should include, where appropriate visual and functional checks to the more comprehensive inspection which may require some dismantling and/or testing. The frequency of inspections made under *Regulation 6* is to be decided by a 'competent person'.

A fundamental requirement under *Regulation 6* is that every employer must ensure that the result of an inspection is recorded and kept until the next inspection is recorded. Although records do not have to be kept in a particular form, the example shown in FIGURE 1 below identifies the main items to be recorded on an inspection record sheet.

Components, which must receive particular attention as part of an inspection, are the safety-related and safety-critical parts, for example interlocking devices, safeguards, protection devices, controls and control systems.

*Regulation 6* also states that no work equipment should leave an undertaking, or be obtained from an undertaking of another person (for example by hiring), unless it is accompanied by physical evidence that the last inspection has been carried out.

Figure 1: Example of an inspection record sheet

## INSPECTION RECORD

**Type and model of equipment**: Press Brake-Komato ....................
**Serial number/Identification mark**: KP/012 ..............
**Normal location of equipment**: Press Shop ..................
**Date inspection carried out**: 08/09/2002 ......................

**Type of Inspection**:
    *Visual check:*
    *Functional check:*
    *Dismantling/testing:* ✓

**Who carried out inspection**: John Smith ......................

### Items and safety-related parts inspected

1. Guard-Interlocking Device ..................................
2. Photo-electric Trip Device ..................................
3. Fixed and Interlocking Safeguards .....................
4. Pneumatic Clutch and Brakes Systems .............
5. Machine Control and Electrical Systems ...........

### Faults and defects found:

1. Small air leak near clutch ..........................
2. Interlocking device wiring loose ................
    ............................................................

### Action/corrective measures taken:

1. Seals replaced for pneumatic system .........
2. Re-wiring of interlocking system ................
    ............................................................

**To whom faults have been reported**: Kevin Taylor ............

**Date repairs/actions were carried out**: 12/09/2003 ...........

## Regulation 7 – Specific risks

*Regulation 7 of PUWER'98 (SI 1998 No 2306)* requires employers to ensure that, where the use of work equipment is likely to involve a specific risk, the use, repairs, modifications, maintenance or servicing of such equipment is restricted to those persons who have been specifically designated to perform operations of that description. It is further required that employers must provide adequate training to perform such tasks safely.

The Approved Code of Practice ('ACoP') and Guidance to *PUWER'98* did not define what is meant by 'specific risks', or what makes such risks 'specific'. The examples given in the Guidance L22 on machinery which poses specific risks, e.g. a platen printing machine or a drop-forging machine did not make reference to which specific risks are associated with these machines.

The hazards associated with the use, maintenance, repair etc. of the above example of machines should be identified if a suitable and sufficient risk assessment is carried out. For example, hazards associated with a drop forge might include: noise, vibration, crushing, shearing, heat, splashes of molten metal, toxic fumes, etc. It may not be impractical to provide adequate protection against all these hazards by means of physical safeguards. As a result, significant 'residual' risks remain which may require the use of adequate personal protective equipment and safe systems of work. Are these meant to be 'specific' risks?

The ACoP relating to *Regulation 7* requires employers to ensure that risks are always controlled by (in the order given):

(*a*)  eliminating the risks, or if that is not possible;

(*b*)  taking 'hardware' (physical) measures to control risks such as the provision of guards; but if risks cannot be adequately controlled;

(*c*)  taking appropriate 'software' measures to deal with the residual risks, such as following safe systems of work and the provision of information, instruction and training.

## Regulation 8 – Information and instructions

Managers, supervisors and users of work equipment must be provided with adequate health and safety information and, where appropriate, written instructions relating to work equipment. This must include the conditions in which, and the methods by which, the equipment may be used, the limitations on its use together with any foreseeable difficulties that might arise and the required actions.

## Regulation 9 – Training

Managers, supervisors and all users of work equipment must receive adequate health and safety training in the methods, which may be adopted when using equipment, any risks, which may arise from such use and precautions to be taken. The level of training will depend on the circumstances of use and the competence of the employee. Other regulations may contain specific training requirements. Particular attention must be paid to the training and supervision of young persons.

## Regulation 10 – Conformity with Community requirements

*Regulation 10 of PUWER'98 (SI 1998 No 2306)* (amended by the *Health and Safety (Miscellaneous Amendments) Regulations 2002 (SI 2002 No 2174)*), applies to items of work equipment provided for use in the premises or undertaking of the employer for the first time after 31 December 1992. The amended *Regulation 10(1)* requires employers to ensure that work equipment complies at all times with any EH&SR's that applied to that kind of work equipment at the time of its *first supply* or when it is *first put into service*. It places a duty on the employer that complements those on manufacturers and suppliers in other jurisdictions regarding the initial integrity of equipment.

The amended *Regulation 10(2)* defines the term 'essential requirements' – which are the essential health and safety requirements listed in the Regulations implementing the relevant directives in the UK (which are listed in *Schedule 1* of the 1998 Regulations).

The amended 1998 Regulations no longer disapply *Regulations 11–19* and *22–29*. This means that inspectors are now free to use these particular provisions if they feel that these Regulations are more appropriate than using the more generally worded.

The amendment is designed to deal with the situation where a piece of work equipment is supplied in compliance with the relevant supply legislation when it is first supplied/put into service, but the employer then alters the equipment such that it no longer complies. In this case *Regulations 11–19* and *22–29* can be used to ensure that the employer returns the equipment to a state of safety and compliance.

The amended Regulations also allow the use of *Regulations 11–19* and *22–29* for equipment that was not in compliance with the supply legislation at the time of supply/putting into service. In this situation inspectors can use *Regulations 11–19* and *22–29* to ensure that employers take the necessary action to bring this equipment into compliance, if appropriate.

The employer can demonstrate compliance with *Regulation 10* if the following conditions are satisfied:

- that the employer where appropriate checks to see that the equipment bears a CE marking and asks for a copy of the EU Declaration of Conformity;
- the employer also needs to check that adequate operating instructions have been provided and that there is information about residual hazards such as noise and vibration;
- more importantly, the employer should check the equipment for obvious faults.

The guidance offers additional advice in the form of HSE booklet 'Buying new machinery'.

If however, work equipment does not comply with relevant product legislation, for example a new machine has been supplied without suitable guards, *Regulation 11* of *PUWER'98* (*SI 1998 No 2306*) could be used to ensure the user does provide adequate guards.

The HSE booklet 'Buying new machinery' (INDG271) is written in easy to understand language and contains two checklists: Checklist A – What should I talk to a supplier about? and Checklist B – What do I do when I have bought new machinery? In both checklists, the employer relies on guidance given by the machine designers/suppliers. This guidance is aimed at small/medium-size enterprises and the checklists are somewhat superficial and over simplified. Basic requirements such as the format of Declarations of Conformity/Incorporation, or the structure/relevance of the Transposed Harmonised EU Standards are not discussed.

## Regulation 11 – Dangerous parts of machinery

*Regulation 11(1)* of *PUWER'98* (*SI 1998 No 2306*) requires employers to ensure that access to dangerous parts of the machine is prevented, or if access is needed to ensure that the machine is stopped before any part of the body reaches the danger zone. *Regulation 11(2)* sets out the following hierarchy of preventive measures (see also M1027 onwards):

(*a*)   fixed enclosing guards;

(*b*)   other guards or protective devices;

(*c*)   protection appliances, e.g. jigs, holders, push sticks, etc.

(*d*)   the provision of information, instruction, training and supervision.

This hierarchy follows the traditional preference of fixed guards and is open to criticism. If an individual needs frequent access into the machine danger zone, a fixed guard could be the worst option, as the machine can operate without it.

The European Commission did express concern that the original *Regulation 11(2)(d)* – the provision of information, instruction, training and supervision, could be used as the sole solution to working with the unguarded parts of dangerous machinery.

To address this and other concerns, *Regulation 11* was amended so that the provision of information, instruction, training and supervision is now seen as an additional requirement in each case and not as part of the hierarchy of measures.

*Regulation 11(3)* describes the need for the provision of safeguards/safety devices which are robust, difficult to defeat, well maintained and meet the requirements of relevant EN standards. Illustrated examples of the options given in *Regulation 11* are shown in the guidance to *PUWER'98*.

## Regulation 12 – Protection against specified hazards

Appropriate measures must be taken to prevent or adequately control exposure to 'specified hazards' arising from the use of work equipment. These measures should be other than the provision of personal protective equipment or of information, instruction, training and supervision, so far as is reasonably practicable. The 'specified hazards' are: falling or ejected articles or substances; rupture or disintegration of parts of equipment; work equipment catching fire or overheating; and unintended or premature discharges or explosions.

## Regulation 13 – High or very low temperature

Work equipment, parts of work equipment and any article or substance produced, used or stored in work equipment must be protected so as to prevent burns, scalds or sears.

## Regulations 14 to 18 – Controls and control systems

Work equipment must be provided with:

- One or more controls for starting the equipment or for controlling the operating conditions (speed, pressure etc.), where the risk after the change is greater than or of a different nature from the risks beforehand.

- One or more readily accessible stop controls that will bring the equipment to a safe condition in a safe manner and which operate in priority to any control that starts or changes the operating conditions of the equipment.

- One or more readily accessible emergency stop controls – unless by the nature of the hazard, or by the adequacy of the normal stop controls (above), this is unnecessary. Emergency stop controls must operate in priority to other stop controls.

- Controls should be clearly visible and identifiable (including appropriate marking). The position of controls should be such that the operator can establish that no person is at risk as a result of the operation of the controls. Where this is not possible, safe systems of work must be established to ensure that no person is in danger as a result of equipment starting or where this is not possible, there should be an audible, visible or other suitable warning whenever work equipment is about to start. Sufficient time and means shall be given to a person to avoid any risks due to the starting or stopping of equipment.

- Control systems should not create any increased risk, which ensure that any faults or damage in the control systems or losses of energy supply do not result in additional or increased risk, and which do not impede the operation of any stop or emergency stop control.

## Regulation 18 – Control systems

*Regulation 18* of *PUWER'98* (*SI 1998 No 2306*) relates to work equipment control systems. This requires that:

'(1)  Every employer shall—

    (a)    ensure, so far as is reasonably practicable, that all control systems of work equipment are safe; and

    (b)    are chosen making due allowance for failures, faults and constraints to be expected in the planned circumstances of use.'

A control system is defined as: a system which responds to input signals and generates an output signal which causes the equipment to perform controlled functions.

The input signals may be made by the operator (manual control), or automatically controlled by the equipment through sensors or protection devices, e.g. guard interlocking devices, photoelectric device, emergency stop or speed limiters.

The objective of *Regulation 18* is to prevent or reduce the likelihood that a single failure in the interlocking device, wiring, contactors, hardware, software could cause injury or ill health.

The subject of control systems generally and the safety related functions of control systems in particular are complex, even for engineers. Although the Harmonised Standard BS EN 954-1: 1996 'Control systems with safety-related functions' provides guidance for the design and integrity of control systems based on risk assessment.

The complexity of this subject is further compounded by the use of programmable electronic systems in the control system. Guidance on the use of programmable electronic systems for safety related applications are not covered by the EU Harmonised Standards. However, the International Electro-technical Standard IEC 61508 is generally used to identify, based on risk assessment the relevant Safety Integrity Level (SIL) required for the control system.

## Regulation 19 – Isolation from sources of energy

Equipment must be provided, where appropriate, with identifiable and readily accessible means of isolating it from all its sources of energy. Reconnection of any source of energy must not expose anyone to any health and safety risk.

## Regulation 20 – Stability

All work equipment must be stabilised where necessary – by clamping, tying or fastening.

## Regulation 21 – Lighting

Suitable and sufficient lighting, taking account of the operations carried out, must be provided. This may entail additional lighting where the ambient lighting is insufficient for the required tasks.

## Regulation 22 – Maintenance operations

Work equipment must be constructed or adapted in such a way that maintenance operations, which involve a risk to health or safety, can be carried out while the equipment is shut down. If this is not possible, the work should be carried out in such a way as to prevent exposure to risk and appropriate measures taken to protect those carrying out the maintenance work. Such measures might include, for instance, functions designed into the equipment that limit the power, speed or range of movement of dangerous parts during maintenance.

## Regulation 23 – Markings

Equipment must be marked for the purposes of ensuring health and safety with clearly visible markings.

## Regulation 24 – Warnings

Equipment must incorporate appropriate warnings or warning devices, which must be unambiguous, easily perceived and easily understood. These requirements are nothing new since they already exist in some other part of legislation or at least constitute accepted good practice. However, the terminology of *PUWER'98* (*SI 1998 No 2306*) is very prescriptive when compared to many of the other harmonised 'goal-setting' regulations.

## Second-hand machinery

Second-hand machinery supplied within the EU is not required to be 'CE' marked, nor issued with a Declaration of Conformity/Incorporation. These however should subject to a suitable and sufficient risk assessment and should comply with *Regulations 11* to *24* of *PUWER'98* (*SI 1998 No 2306*).

Second-hand machinery supplied from outside the EU should be treated as new machinery and will be subject to CE marking requirements.

It is the management responsibility to ensure that second-hand machinery and modification work do comply with the requirements of the *PUWER'98*.

## *Part III: Mobile work equipment*

**M1008** *Part III* of *PUWER'98* (*SI 1998 No 2306*) added new requirements for the management of mobile equipment.

Mobile equipment is defined as any work equipment, which carries out work while it is travelling or which travels between different locations where it is used to carry out work. Mobile work equipment may be self-propelled, towed or remote controlled and may incorporate attachments. *Regulations 25* to *30*, in addition to other requirements of *PUWER'98* (e.g. training, guarding and inspection), apply to all work equipment including mobile work equipment. New mobile equipment taken into use from 5th December 1998 should comply with all requirements of *PUWER'98*.

## Regulation 25 – Employees carried on mobile work equipment

*Regulation 25* of *PUWER'98* (*SI 1998 No 2306*) contains a general requirement relating to the risks to people (drivers, operators and passengers) carried by mobile equipment, when it is travelling. This includes risks of people falling from the equipment or from unexpected movement while it is in motion or stopping.

This Regulation requires every employer to ensure that no employee is carried by mobile work equipment unless:

(a)  it is suitable for carrying persons; and

(b)  it incorporates features for reducing to as low as is reasonably practicable risks to their safety, including risks from wheels or tracks.

Mobile work equipment can be made suitable for carrying persons by means of operator stations, seats or work platforms to provide a secure place.

Features for reducing risks may include the following:

- Falling object protective structures ('FOPS'). This may be achieved by a strong safety cab or protective cage.
- Restraining systems – this can be full-body seat belts, lap belts or purpose-designed restraining system. This should provide adequate protection against injury through contact with or being flung from the mobile equipment if it comes to a sudden stop moves unexpectedly or rolls over.
- Speed adjustment – when carrying people, mobile equipment should be driven within safe speed limits for stability.
- Guards and barriers fitted to mobile work equipment to prevent contact with wheels and tracks.

## Regulation 26 – Rolling over of mobile work equipment

In addition to the main general requirements of *Regulation 25*, *Regulation 26* of *PUWER'98* (*SI 1998 No 2306*) covers the measures necessary to protect employees where there are risks from roll-over while travelling. If this risk is significant, then the following measures should be considered:

- Stabilisation – this might include fitting appropriate counterbalance weights, wider wheels and locking devices.
- Structures to prevent rolling over by more than 90°, e.g. boom of a hydraulic excavator, when positioned in its recommended travel position.
- Roll-over protective structures ('ROPS') – normally fitted on mobile equipment to withstand the force for roll over through 180°.

## Regulation 27 – Overturning of forklift trucks

*Regulation 27* of *PUWER'98* (*SI 1998 No 2306*) applies to FLT's fitted with vertical masts, which effectively protect seated operators from being crushed between the FLT and the ground in the event of roll-over. FLT's capable of rolling over 180° should be fitted with ROPS.

FLT's with a seated ride-on operator can roll-over in use. There is also a history of accidents on counterbalanced, centre control, high lift trucks. Restraining systems will normally be required on these trucks.

## Regulation 28 – Self-propelled work equipment

- Preventing unauthorised start-up – this could be achieved by a starter key or device, which is issued only to authorised personnel.
- Minimise consequences of a collision of rail-mounted work equipment – this can be achieved by introducing safe systems of work, buffers or automatic means to prevent contact.
- Devices for stopping and braking – all self-propelled work equipment should have brakes to enable it to slow down and stop in a safe distance and park safely.
- Emergency braking and stopping facilities – a secondary braking system may be required if risk is significant.
- Driver's field of vision – where the driver's direct field of vision is impaired then mirrors or more sophisticated visual or sensing facilities may be necessary.
- Lighting for use in the dark – to be fitted to equipment if outside lighting is inadequate.

## Regulation 29 – Remote-controlled, self-propelled work equipment

- This is self-propelled work equipment that is operated by controls which have no physical link with it, for example radio control.
- Alarms or flashing lights – to be considered for the prevention of collision with others.
- Sensing or contact devices – to prevent or reduce risk of injury following contact.
- Hold-to-run control devices – so that any hazardous movement can stop when the controls are released.
- Control range – once equipment leaves its control range should stop and remain in a safe state.

## Regulation 30 – Drive shafts

- A 'drive-shaft' is a device, which conveys the power from the mobile work equipment to any work equipment connected to it. In agriculture these devices are known as power take-off shafts.
- Shaft seizure – if seizure can result in significant risk, for example the ejection of parts due to equipment break-up, guards should be fitted in accordance with *Regulation 12* of *PUWER'98* (*SI 1998 No 2306*).
- Isolation from source of energy – to reduce the risk associated with drive shaft stalling.
- Support on cradle – when drive shaft and its guards are not in use, the drive shaft should be supported on cradle or equivalent for protection against damage.

# The Supply of Machinery (Safety) Regulations 1992 (as amended)

**M1009**  General and specific duties relating to machinery safety are laid on designers and manufacturers of machinery for use at work by the *Supply of Machinery (Safety) Regulations 1992* (*SI 1992 No 3073*), as amended by the *Supply of Machinery (Safety) (Amendment) Regulations 1994* (*SI 1994 No 2063*).

Supply of safe machinery (including safety components, roll-over protective structures and industrial trucks) is governed by these Regulations. The 1994 Regulations extend to machinery for lifting people (e.g. elevating work platforms). [*SI 1994 No 2063, Reg 4, Sch 2, para 10*]. The combined effect of the two sets of Regulations requires that, when machinery or components are properly installed, maintained and used for their intended purposes, there is no risk (except a minimal one) of their being a cause or occasion of death or injury to persons, or damage to property. [*SI 1992 No 3073, Reg 2(2)*, as amended by *SI 1994 No 2063, Reg 4, Sch 2, para 5*].

The EU Machinery Directive (89/392/EEC) as amended is aimed at machinery designers and suppliers. This is a so-called 'Product Directive' made under Article 100A of the *Single European Act* to ensure fair competition by eliminating barriers to trade. The relevant products have to meet certain essential health and safety requirements ('EH&SRs') before they can be placed on the market within the Community.

The manufacturer affixing a CE-Mark to the machinery claims compliance with all relevant EU Directives. The EU Machinery Directive was implemented in the UK under the *Supply of Machinery (Safety) Regulations 1992*.

## EU Machinery Directive

**M1010** The EU Machinery Directive requires:

- The clear allocation of responsibilities to machinery suppliers.
- Type approval: 'attestation' via a technical construction file and a Declaration of Conformity before the CE mark is affixed to the machine.
- Intended use of machinery has to be clearly defined.
- Principles of Safety Integration. Manufacturers must demonstrate that the following order of options for machinery safety are followed:
    (i) elimination or reduction of risks by seeking to design inherently safe machinery;
    (ii) necessary protection measures taken (e.g. safeguarding or safety devices) to deal with the risks which cannot be eliminated;
    (iii) users of residual risks are informed.

The majority of the requirements of the Directive are couched as 'objects to be achieved'.

The Directive and EN standards show that while traditional mechanical hazards continue to merit attention, manufacturers must also consider: discomfort, fatigue and psychological stress; electricity, including static; sources of energy supply other than electricity; errors of fitting; temperature extremes; fire and explosion; noise and vibration; radiation; laser equipment; and emissions of gases, etc.

## Definition of a machine

**M1011** A 'machine' is defined by the EU Machinery Directive as 'an assembly of linked parts or components, at least one of which moves, with the appropriate actuators, control and power circuits, joined together for a specific application, in particular for the processing, treatment, moving or packaging of a material'. The definition includes an assembly of machines which functions as an integral whole as well as interchangeable equipment (not being a spare part or tool) which can be assembled with a machine by the operator – for instance, an item of agricultural equipment attached to a tractor.

The main elements in the framework for demonstrating compliance are summarised as follows.

## Essential health and safety requirements ('EH&SR')

**M1012** The route to demonstrating compliance starts with the identification of relevant EH&SRs relating to the design and construction of machinery. These are set out in Annex I of the EU Machinery Directive, and include the following:

- Principles of safety integration.
- Materials and products health and safety.
- Lighting.
- Design of the machine to facilitate its handling.
- Safety and reliability of the control system.
- Control devices.
- Starting, change-over and stopping devices.
- Emergency stops.
- Mode selection.

- Power supply failure.
- Failure of the control system, including software.
- Stability.
- Risk of break-up during operation.
- Risks due to falling or ejected objects.
- Sharp surfaces, edges or angles.
- Hazards associated with moving parts.
- Required characteristics of guards and protection devices.
- Fire, explosion, noise, vibration, radiation, laser equipment, etc.

Suppliers should identify relevant EH&SRs applicable to the machine and show how these have been considered.

## *Conformity assessment procedure carried out*

**M1013** The responsibility for demonstrating that the machinery satisfies EH&SR's rests with the manufacturer, supplier or the importer into the EU ('the responsible person').

The EU Machinery Directive sets out various ways in which a supplier can demonstrate compliance with EH&SR's. This is known as 'conformity assessment procedure'. There are three different conformity assessment procedures, detailed in *Regulations 13, 14* and *15* of the *Supply of Machinery (Safety) Regulations 1992 (SI 1992 No 3073)*. These are:

- for most machinery other than that listed in *Schedule 4* of *SI 1992 No 3073* (Annex IV of the Directive);
- Schedule 4 (Annex IV) machinery manufactured in accordance with the EU Transposed Harmonised Standards; or
- Schedule 4 machinery not manufactured in accordance with the EU Harmonised Standards or where they do not exist.

Schedule 4 machinery lists categories of machines that pose special risks. These include mechanical presses, press brakes, injection and compression plastics moulding machines.

For machinery listed in *Schedule* 4 of the *Supply of Machinery (Safety) Regulations 1992 (SI 1992 No 3073)* that is manufactured in conformity with Harmonised Transposed Standards, the machinery technical file is submitted to an approved body for EC type-examination. The approved body may simply verify that the Transposed Harmonised Standards have been correctly applied and draw up a certificate of adequacy.

## *The technical file*

**M1014** The conformity procedure requires the supplier to assemble records, which technically describe the rationale behind the approach to ensuring that the machine satisfies EH&SR's. These records are referred to as the 'technical file'. The technical file should include a risk assessment (if applicable) and other technical information which includes: technical drawings, test results, a list of the EH&SR's, Harmonised Standards, other standards and technical specifications that were used in the design of the machinery. It should also include a description of the methods adopted to eliminate hazards, other relevant technical information and a copy of the instructions for the machinery.

For machinery posing special hazards (listed in Annex IV of the EU Machinery Directive), further action is necessary which usually involves verification by an approved body that either recognised standards have been correctly applied or an EU type-examination of the machine shows that it satisfies the relevant provisions of the Directive.

The supplier must retain the technical file and all documentation relating to particular machinery for a minimum of 10 years after the production of the last unit of that machinery.

## EU Declaration of Conformity/Incorporation

**M1015**   An EU Declaration of Conformity must be drawn up which states that the relevant machinery complies with all the EH&SR's that apply to it.

Such a declaration must include a description of the machinery (with make, type and serial number), indicate all relevant provisions with which the machinery complies, give details of any EU-type examination and specify any standards and technical specifications which have been used. Reference should also be made to all other relevant EU Product Directives, e.g. Low Voltage, EMC and Simple Vessel Directives.

The machine supplier prior to the commissioning or operation of the machine should issue the Declaration of Conformity. A typical example of a Declaration of Conformity is shown in FIGURE 2 below.

The three most relevant sections on this form are:

1. Relevant EU Directives applicable to the product;
2. The Transposed Harmonised Standards used; and
3. Other standards and technical specifications relevant to the machinery.

The declaration must identify European Transposed Harmonised Standards applicable to the machine design and construction. Other national and international standards relevant to the machine, which are not covered by relevant European standards should also be identified on the certificate. For example IEC 61508, which is an international electro-technical standard for the design and selection of programmable electronic systems with safety related applications. The Declaration of Conformity should also make reference to all relevant EU product directives, e.g. Low Voltage and Electro-magnetic Compatibility Directives.

A Declaration of Incorporation is required where the machinery is intended for incorporation into another machinery or assembly with other machinery to constitute machinery covered by the Directive. The Declaration of Incorporation is similar to the declaration of conformity where relevant Harmonised Standards are listed. The only difference is that a 'CE' mark is not required, and a condition included in the declaration, which states that: 'the equipment must not be put into service until the machinery into which it is incorporated has been declared in conformity with the provision of the Machinery Directive'. See Annex IIB of the Directive for more detail.

## 'CE' marking

**M1016**   A 'CE' marking (meaning European Conformity) on a product or system indicates a legal declaration by the supplier that the product complies with EH&SRs of the EU Machinery Directive and all other EU directives relevant to it.

The CE mark is regarded as properly affixed only if: (a) an EU declaration of conformity has been issued, (b) it is affixed in a distinct, visible, legible and indelible manner, and (c) the machinery complies with the requirements of other relevant directives. It is an offence to affix the CE mark to machinery, which does not satisfy the EH&SRs, or to machinery that is not safe. The CE mark is described in the Machinery Directive and consists of a symbol 'CE' and the last two figures of the year in which the mark is affixed.

Figure 2: Example of a Declaration of Conformity

---

**EU DECLARATION OF CONFORMITY**

In accordance with:

**The Supply of Machinery
(Safety) Regulations 1992**

**WE DECLARE THAT THIS MACHINE CONFORMS WITH THE
ESSENTIAL HEALTH AND SAFETY REQUIREMENTS**

Description of the machine, type and serial number:


Business name and full address of manufacturer:

All relevant product directives complied with:
  **Machinery Directive (89/392/EEC) and all amendments thereto.
  Low Voltage, CE marking and EMC Directives, ...**

The machine complies with the following transposed harmonised standards:
  **EN 292-1, EN 292-2, EN 294, EN 60204-1, EN 954-1, EN 1050,
  EN 1088**

Other standards and technical specifications used:
  **IEC 1508**

IDENTIFICATION of the person
empowered to sign on behalf of                    Usually Managing Director
the Company:

I certify that on 14th of January 1996, the above machine conformed
with the EH&SR's of all the relevant Product Directives.

Signature: ..................                     Date: ...............

---

## Requirements of other EU directives

**M1017**   Machinery suppliers need to identify, apart from the Machinery Directive, all other EU Directives relevant to their products. It is then necessary to identify specific legislation, which implement them. Usually more than one EU directive may apply to particular machinery. For example, in addition to the Machinery Directive, an item of electrically driven machine would be subject to the following Directives:

- Electromagnetic Compatibility Directive (89/336/EEC) as amended. This is implemented in the UK as the *Electromagnetic Compatibility Regulations 1992* (*SI 1992 No 2732*).

- Low Voltage Directive (72/23/EEC) implemented in the UK under the *Electrical Equipment (Safety) Regulations 1994* (*SI 1994 No 3260*).

- Other EU directives, which may be relevant to some machinery include Simple Pressure Vessel Directive (87/404/EEC) as amended, Potential Explosive Atmosphere Directive (94/9/EEC) and Telecom Terminal Equipment Directive (91/263/EEC).

## The machine is in fact 'safe'

M1018   The key objective of the EU Machinery Directive is that the onus is placed on suppliers to demonstrate that a machine complies with relevant EH&SRs before it is placed for sale within the market place of the EU. The requirement for the machine to be in fact safe is an UK rather than EU requirement.

'Safe' is defined by the *Supply of Machinery (Safety) Regulations 1992 (SI 1992 No 3073)* as, 'when the machinery is properly installed and maintained and used for the purpose for which it is intended, there is no risk (apart from one reduced to a minimum) of it being the cause or occasion of death or injury to persons or, where appropriate, to domestic animals or damage to property. The practicability at the time of manufacture, of reducing the risk may be taken into account when considering whether or not the risk has been reduced to a minimum'.

Failure to comply with relevant EU requirements will mean that the machinery cannot legally be supplied in the UK and could result in prosecution and penalties on conviction of a fine up to £5,000 or in some cases, of imprisonment for up to three months, or of both.

The same rules apply everywhere in the EU, so machinery complying with the EU regime may be supplied in any Member State.

# The Transposed Harmonised European Machinery Safety Standards

M1019   Application of the European Transposed Harmonised Machinery Safety Standards is fundamental to the risk-based approach. These standards have now replaced corresponding British Standards relating to machinery safety, as well as other EU member countries national standards.

A process for incorporating the European Machinery Safety Standards into international standards has been adopted by ISO/IEC since 1990. The American National Standards Institute (ANSI) is also considering the EU risk-based approach to machinery safety.

## Status and scope of the Harmonised Standards

M1020   The role and status of the European Harmonised Standards has been frequently misunderstood, mainly due to the previously prescriptive nature of the health and safety legislation and standards. The 'new approach' is based on the legal status of the EH&SRs as an objective, but the Harmonised Standards represent ways for fulfilling the details of the individual EH&SRs. The Harmonised Standards are not of the same legal status as the EH&SRs, but do reflect recognised practice and should be the first consideration.

EH&SRs are expressed in general terms and it is intended that the European Harmonised Standards should fill in the detail so that designers have clear guidance on how to achieve conformity with the directive".

The Machinery Directive states 'objectives to be achieved' where the ES&HRs constitute the legal objective requiring to be complied with and the Harmonised Standards are the current recognised methods, and therefore do not constitute the legal duty as to the method of compliance.

## Structure of the Harmonised Standards

**M1021** The European Standards start by listing the machine or process hazards that are considered relevant within its scope and use. In addition, each health and safety requirement should include an explicit safety aim, and to assess whether a risk has been reduced to an acceptable level.

An overview of the structure for the Harmonised European Standards is shown in FIGURE 3 below.

These Standards are introduced in three levels: A, B and C. 'A' type Standards provide general requirements applicable to all machines. The key standard relating to the EU Machinery Directive is EN 292: 1991 Machinery safety: Basic concepts, general principles for design. This Standard is introduced in two parts:

- BS EN 292: 1991 – Part 1: Machinery safety Basic concepts and methodology.
- BS EN 292: 1991 – Part 2 and Part 2/A1: Technical principles and specifications.

BS EN 1050: 1997 – Machinery safety: Principle or risk assessment is largely regarded as an 'A' type Standard. BS EN 1050 provides a framework for risk assessment and the range of hazards to be considered in the assessment.

There is no standard risk assessment technique, as this will depend on the nature of machine hazards and the degree of interactions with humans. Risk assessment however should be suitable and sufficient. In other words, the degree of detail and techniques used are a function of the risk level and the complexity of the situation under study.

'B' type Standards are standards on techniques or components, which could be applicable to a large number of machines. 'B1' Standards provide generic guidance and information, e.g. EN 294: 1992 'minimum height and distance from the hazard'; whereas 'B2' Standards outline generic safety standards for 'hardware', e.g. electro-sensitive safety systems, two-hand controls, control systems.). The B type or horizontal group safety standards therefore deal with only one safety aspect or one safety related device at the time.

'C' type Standards are drawn up to cover a particular type or class of machines. Examples of C Standards include EN 289 1994 – Compression moulding machines and EN 692 for mechanical presses. A large number of 'C' Standards are yet to be produced or introduced in provisional form. These may be classed as vertical standards. In following a particular 'C' Standard, cross-reference is extensively made to the relevant sections of 'A' and 'B' type Standards.

It is important to stress that the use of European Harmonised Standards is not mandatory. However, compliance with relevant EN standards is one of the most effective means for demonstrating that relevant EH&SRs are met. The other alternative would be the application of a robust risk assessment, which considers all relevant hazards and hazardous situations listed in EN 292 and EN 1050.

# Machinery risk assessment

**M1022** The risk-based approach involves a structured and systematic risk assessment. Risk assessment is essentially a proactive means to foresee how accidents can happen and to decide what corrective and preventive actions are needed in advance to prevent such accidents.

Risk assessment is defined as ' ... A structured and systematic procedure for *identifying hazards* and *evaluating risks* in order to prioritise decisions to reduce risks to a *tolerable* level'.

This section will provide guidance on the following key elements of a structured and systematic risk assessment:

- techniques for hazard identification and analysis;

Figure 3: Structure of the Harmonised European Standards

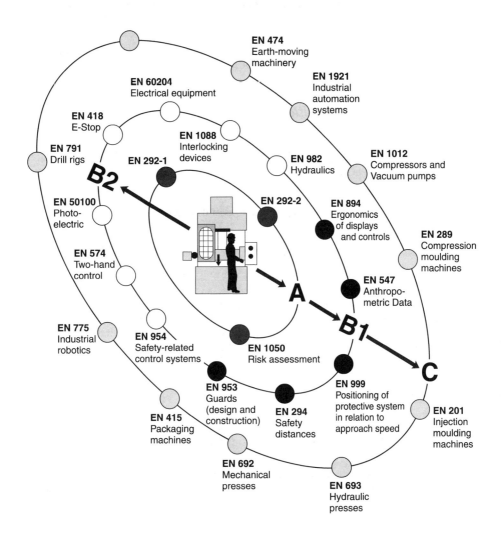

- criteria for the estimation and evaluation of risks;
- the concept of 'ALARP' and risk tolerability, and
- the preferred hierarchy for risk control options.

In addition to demonstrating compliance with relevant legislation/standards, risk assessment is now crucial as it aids in the consistency of decision making and the cost effectiveness in allocation of resources for health, safety and environmental issues.

Risk assessment is vital for two reasons: the level of risk determines priority which should be accorded to and the selection of appropriate health, safety and environmental control measures to deal with different hazards, and the standard of integrity of corrective/preventive measures will depend on the level of risk.

Risk assessment techniques are complementary to the more pragmatic ways of problem identification and assessment. They highlight systematically how hazards can occur and provide a clearer understanding of their nature and possible consequences, thereby improving the decision-making process for the most effective way to prevent injury and damage to health/environment.

The techniques range from relatively simple qualitative methods of hazard identification and analysis to the advanced quantitative methods for risk assessment in which numerical values of risk frequency or probability are derived.

## Hazard and risk – definitions

**M1023**  A *hazard* is defined as 'a potential to cause harm'.

*Harm* is defined as:

- injury or damage to health;
- damage to the environment;
- economic losses (interruption to production/asset damage).

*Risk* is defined as:

- 'the chance (probability) of the harm being realised, combined with its consequences; or
- Risk = Chance of exposure to the hazard x Consequences (severity).

FIGURE 4 below illustrates the concept of risk. If for example it was estimated that the chance of an accident involving the collision between a fork-lift truck and an operator as 'remote', and the resulting injury severity as fatal 'catastrophic', then the Risk Matrix shows this as risk level 'A' or high, which should warrant some urgent attention. If on the other hand it was estimated that the chance of slipping on a wet surface in part of the workplace as 'occasional', which would result in 'minor' injury. The risk matrix shows this to be risk level 'C' or low, which is broadly acceptable.

## Risk assessment framework

**M1024**    Hazard analysis is an essential part of the overall risk assessment framework. The flow-chart shown in FIGURE 5 (based on EN 1050) outlines elements of the overall framework. The effectiveness of hazard analysis and risk evaluation processes involves a number of practical considerations.

Before commencing the study, its objectives and scope must be defined explicitly. The purpose of the study and the fundamental assumptions made must be clearly stated. In order to limit the analysis, the human/system boundaries must also be defined, as well as its intended design, use, operation and layout.

The risk assessment procedure contains two essential elements:

1. *Risk analysis.* Where hazards and hazardous situations are systematically identified and their consequences are analysed. The level of risk associated with each hazardous situation is estimated or measured.
2. *Risk evaluation.* In which a judgement is made as to whether the level of risk is acceptable/tolerable or whether some corrective/preventive actions are needed.

Following any modifications in the plant/process design or operation/maintenance procedures, there is a need to verify that the original hazards/hazardous situations are controlled, and that new hazards are not introduced as a result of these modifications. It is crucial that the risk assessment is updated as a result of design/procedural evolution. The safety management system developed by the employer as a result of risk assessment is also a living system, capable of learning from experience as the people who run it, and should reflect what is actually taking place, and not what should happen.

Figure 4: The Risk Matrix

| Severity | Probability | | | | |
|---|---|---|---|---|---|
| | A | B | C | D | E |
| | Improbable 1 in 100,000 years | Remote 1 in 10,000 years | Occasional 1 in 1,000 years | Probable 1 in 100 years | Frequent 1 in 10 years |
| 5. Catastrophic | | | | HIGH RISK A | |
| 4. Severe | | | | | |
| 3. Critical | | | MEDIUM RISK (ALARP) B | | |
| 2. Marginal | | | | | |
| 1. Negligible | LOW RISK C | | | | |

The main elements of risk assessment, ideally, should include the following:

## Define the system/activity

The first step in any risk assessment is to provide terms of reference to the detailed description of the process/activity being studied. This clear description is needed at this stage in order to avoid confusion at a later stage and to define the boundary and interfaces with other processes/activities. This step also involves the identification of the design objectives, material being handled/processed and exposure patterns for individuals involved in operation and maintenance.

## Define the study objective

Because risk assessment is used as means to demonstrating compliance with a number of codes/regulations, it is important to limit the study to the range of hazards and potential consequences in relation to specific requirements, e.g. COSHH i.e. control of substances hazardous to health', noise at work, the environment, machinery safety or economic risks.

## Identify the hazards

Having defined the scope and objectives, the identification of hazards is the third, but the most important step in any risk assessment study because any hazard omitted at this stage will result in the associated risk not to be assessed.

It is important in this respect to distinguish between continuing hazards (those inherent in the work activity/equipment/machinery/substances under normal conditions), e.g. noise, toxic

Figure 5: Risk assessment framework

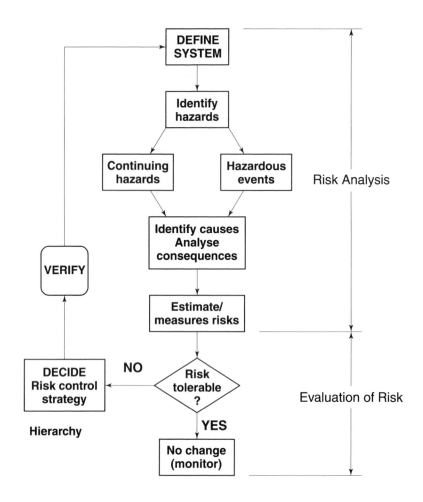

substances, mechanical hazards, and hazards which can result from failures/error, e.g. hardware/software failures as well as foreseeable human error.

The checklist shown in TABLE 1 at M1025 may be used to identify continuing hazards.

The range of continuing hazards outlined in the checklist is based on the EU-Machinery Directive and is intended to be used as an aid-memoire. This approach to hazard identification will focus attention on parts of the work activity/equipment/machinery to identify the sources of each relevant hazard, who may be exposed to them, e.g. the operator, maintenance, and the task involved, e.g. loading, removing, tool setting, cleaning, adjusting.

Hazards resulting from equipment failure and human error require an open-ended type of analysis, based on brain storming. This approach normally involves the following:

1. A detailed hazard identification to be carried out for all stages of the process life-cycle. This ideally should include the identification of contributory causes for each hazards.

2. Analysis of systems of work and established procedures to identify, who might be exposed to a hazard and when, e.g. operator under normal conditions, maintenance/tool fitter/cleaner, etc.

Structured and systematic techniques which could assist in the identification of hazardous events identified by BS EN 1050: 1997 include the following:

- *Hazard and Operability Study* ('HAZOP') – a qualitative technique to identify hazards resulting from hardware failures and human error. This involves potential causes and consequences
- *Failure Modes and Effects Analysis* ('FM&EA') – an inductive technique to identify and analyse hardware failures. This technique can be used to quantify risks.
- *Task Analysis* – an inductive technique to identify the opportunity for human error. An approach, based on task-based hazards identification and analysis is introduced in this document.

## Machinery hazards identification

M1025  A person may be injured at machinery as a result of one or more of the following:

(*a*) contact or entanglement with the machinery;

(*b*) crushing between a moving part of the machine and fixed structures;

(*c*) being struck by ejected parts of the machinery;

(*d*) being struck by material ejected from the machinery.

These are regarded as mechanical hazards. The British Standard 5304: 1988 'Safety of machinery' classified mechanical hazards in the form of type of injury caused by exposure to the hazard.

The main types of mechanical hazards shown in FIGURE 6 are described as:

- **Crushing**: Occurs when part of the body is caught between a moving part of a machine against a fixed object, e.g. underneath scissor lift or between the tools of a press.
- **Shearing**: Parts of the body may be sheared by scissor action caused by parts of the machine, e.g. mechanism of scissor lift or oscillating pendulum.
- **Cutting and severing**: Cutting hazards include contact with circular saws, guillotine knife, rotary knives or moving sheet metal.
- **Entanglement**: Occurs as a result of clothing or hair contact with rotating objects or catching on projections or in gaps, e.g. drills, rotating workpiece or belt fasteners.
- **Drawing-in or trapping**: Occurs when part of the body is caught between two counter-rotating parts, e.g. gears, mixing mills or between belt and pulley or chain and chain wheel.
- **Impact**: Impact hazards are caused by a moving object striking the body without penetrating it. Examples include: being hit by a robot arm or by moving traffic.
- **Stabbing and puncture**: flying objects, swarf or rapid moving parts, e.g. sewing machine.
- **Friction and abrasion**: contact with moving rough or abrasive surfaces, e.g. abrasive wheels.
- **High pressure fluid injection**: sudden release of fluid under pressure can cause tissue damage similar to crushing. Examples include: water jetting, compressed air jets and high pressure hydraulic systems.

Since a hazard is defined as 'a potential to cause injury or damage to health', it is therefore important to include health hazards in addition to the mechanical hazards which are aimed at physical injury. Other hazards considered by BS EN 1050 include the following:

Figure 6: Examples of mechanical hazards

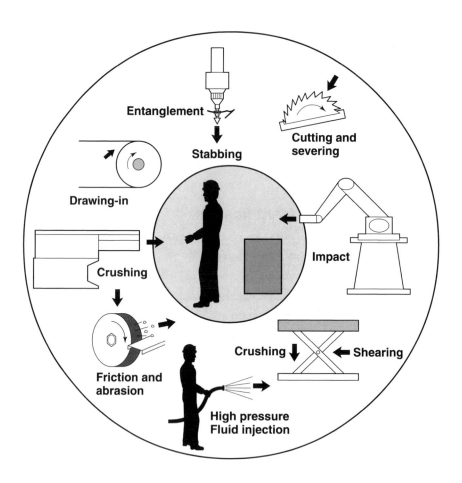

- Electrical hazards – to include direct and indirect contact with live electricity and electro-static phenomena.

- Radiation hazards – to include ionising and non-ionising sources as well as electro-magnetic effect.

- Work environment hazards – to include hot/cold ambient conditions, noise, vibration, humidity and poor lighting.

- Hazardous materials and substances – to include toxic, flammable and explosive substances.

- Work activity hazards – to include highly repetitive actions, stressful posture, lifting/handling heavy items, visual fatigue and poor workplace design.

- The neglect of ergonomic principles.

Table 1: Hazards identification checklist

| Type of hazard | Source | Task involved (Who is exposed and when?) |
|---|---|---|
| **1. Mechanical hazards**<br><br>1.1 Crushing<br>1.2 Shearing<br>1.3 Cutting/severing<br>1.4 Entanglement<br>1.5 Drawing-in/trapping<br>1.6 Impact<br>1.7 Stabbing/puncture<br>1.8 Friction/abrasion<br>1.9 High pressure fluid injection<br>1.10 Slips/trips/falls<br>1.11 Falling objects<br>1.12 Other mechanical hazards | | |
| **2. Electrical hazards**<br><br>2.1 Direct contact<br>2.2 Indirect contact<br>2.3 Electrostatic phenomena<br>2.4 Short circuit/overload<br>2.5 Source of ignition<br>2.6 Other electrical hazards | | |
| **3. Radiation hazards**<br><br>3.1 Lasers<br>3.2 Electro-magnetic effects<br>3.3 Ionising/non-ion radiation<br>3.4 Other radiation hazards | | |
| **4. Hazardous substances**<br><br>4.1 Toxic fluids<br>4.2 Toxic gas/mist/fumes/dust<br>4.3 Flammable fluids<br>4.4 Flammable gas/mist/fumes/dust<br>4.5 Explosive substances<br>4.6 Biological substances<br>4.7 Other hazardous substances | | |
| **5. Work activities hazards**<br><br>5.1 Highly repetitive actions<br>5.2 Stressful posture<br>5.3 Lifting/handling heavy items<br>5.4 Mental overload/stress<br>5.5 Visual fatigue<br>5.6 Poor workplace design<br>5.7 Other workplace hazards | | |

M1026   *Machinery Safety*

| 6. Work environment hazards<br><br>6.1 Localised hot surfaces<br>6.2 Localised cold surfaces<br>6.3 Significant noise<br>6.4 Significant vibration<br>6.5 Poor lighting<br>6.6 Hot/cold ambient temperature<br>6.7 Other work environment hazards | | |
|---|---|---|

## Options for machinery risk reduction

M1026    The most effective risk control measures are those implemented at the machine/work equipment design stage. It is a legal requirement in the EU that all machinery supplied after January 1995 should carry a 'CE' mark. This demonstrates that the machine complies with all relevant EU machinery directives, which include meeting the ES&HRs. Fundamental to this approach, manufacturers and suppliers need to carry out a risk assessment and demonstrate that all risks are adequately controlled by design, rather than by procedures. The following risk control options reflect the preferred order of priority:

### *Technical measures to be taken at the design stage*

1. To make the machinery inherently safer (design out hazards) as a priority:
    — to eliminate hazards at the design stage;
    — to substitute hazardous substances used by the machine or process by less hazardous ones.
2. To provide protection from the hazards by the machine design, e.g. to avoid the need for access to hazardous parts of the machine.
3. To improve machinery reliability by reducing the chance for fail-to-danger of critical components/systems.
4. To provide protection from the hazards by adequate safeguarding.
5. To reduce ease of access to danger zones (e.g. safeguards or safety devices which allow safe access, but prevent access at other times).
6. To reduce the chance of potential human error:
    — by improving the ergonomics of machine control layout to reduce the chance for unintended errors;
    — by reducing the chance for violations by making maintenance, cleaning, adjustments, etc. tasks convenient, easy and safe.

### *Procedural measures (make work tasks safer)*

These may be achieved by the following measures:

- Planned preventive maintenance and regular inspection of machines and safeguards/safety devices.
- Safe systems of work (which would minimise the needs for access into the danger zone).

- Permit-to-work procedures (to formalise precautions in the face of a hazard).
- Adequate personal protection equipment ('PPE') and planning for emergencies.

### Behavioural measures (develop safer people)

This involves the following:

- Selection and certification of personnel against specific work tasks.
- Training: basic skills; systems and procedures as well as knowledge of hazards.
- Adequate instructions, warnings and supervision.
- Improve safety culture within the business.
- Ask: why should safeguards/safety devices be violated/defeated? *

    * Note that violations are prevented if the perceived inconvenience of defeating a safeguard/safety device exceeds the benefit of so doing.

## Types of safeguards and safety devices

M1027   The main types of safeguards and safety devices can be classified as follow:

1. Fixed guards.
2. Fixed guards with adjustable element.
3. Automatic guards.
4. Interlocked guards.
5. Safety devices.

    These are described as:

    5.1  Trip devices, e.g. photoelectric light curtains, pressure sensitive devices.

    5.2  Two-hand control devices.

### Machinery safeguards – general considerations

M1028   It is a general requirement that all machinery safeguards and safety devices must have the following characteristics (adapted from BS EN 292: 1991 Part 2 and EH&SR's) outlined below.

Safeguards and safety devices should:

- Be of robust construction: strength, stiffness and durability to prevent ejected parts of the machine/components or material penetrating the guard.
- Not give rise to additional hazards.
- Not be easy to bypass or render non-operational.
- Be located at an adequate distance from the danger zone.
- Cause minimum obstruction of view for machine operators.
- Enable essential work (e.g. maintenance, cleaning) to be done without safeguard removal.

The two most applicable EU Standards relevant to safeguards are BS EN 953: 1994 (design and construction), and BS EN 294: 1994 (safety distances).

Figure 7: Design/selection of safeguards

## Fixed guards

M1029    BS EN 953: 1998 'Safety of machinery: Guards: General requirements for the design and construction of fixed and movable guards' defines a fixed guard as 'a guard kept in place (i.e. closed)', either permanently or by means of fasteners (screws, nuts) making removal/opening impossible without using tools.

A fixed guard is a guard attached to a machine by a fixing method, which is not linked to, or associated with, the controls, motion, or hazardous condition of a machine. Therefore, a machine can operate normally with the guard removed.

A fixed guard should when fitted be incapable of being displaced casually, so the method of fixing is important. The ideal method of fixing the guard would be of a captive type, which would make unofficial access difficult.

It is clear from experience that if the need for access into hazardous parts of a machine fitted with a fixed guard, there will be a tendency not to replace the guard back onto the machine. Therefore an interlocked guard is more suited in this situation.

## *Adjustable guards*

M1030    Adjustable guards comprise a fixed guard with adjustable elements that the setter or operator has to position to suit the job being worked on. They are widely used for woodworking

Figure 8: Example of a fixed guard

and tool-room machines. Where adjustable guards are used, the operators should be familiar and trained in how to adjust them so that full protection can be obtained.

FIGURE 9 shows the application of a telescopic guard to a heavy duty-drilling machine. The idea here is that as the drill descends into the workpiece, the fixed guard will always enclose the drill. The approach to selecting the most appropriate safeguard should start by hazard identification. Hazards associated with the use of the drilling machine include the following:

- entanglement with the drill;
- stabbing/puncture by the swarf;
- ejection of broken drill bit;
- stabbing by the drill bit; and
- possible ejection of the work-piece.

It is noted that the telescopic guard should provide protection against most of the above hazards apart from stabbing/puncture by the drill and ejection of the work-piece.

**Figure 9: Fixed guard with adjustable element**

Self-adjusting guards are also classified as a fixed guard with adjustable element. This guard prevents access to the hazard except when the guard is forced open by the passage of the work. They usually incorporate a spring-loaded pivoted element. These types of guards are mainly used on woodworking machinery.

## Automatic guards

M1031   An automatic guard is a guard, which is moved into position automatically by the machine, thereby removing any part of a person from the hazardous area of the machine. This is sometimes known as 'sweep away' guard and is used on some paper guillotines and large mechanical or hydraulic presses.

There are however many factors which could limit the effectiveness of these types of guards. The example of the automatic guard shown in FIGURE 10 is used to illustrate how this type of guard can cause injury.

Some possible hazards associated with this automatic guard include the following:

**Figure 10: Example of automatic guard**

- shearing hazard between the moving and fixed parts of the guard;
- impact with the moving part of the guard;
- crushing hazard against fixed structures nearby as well as the ergonomics of the guard, e.g. height of guard in relation to operators and the size of guard gaps.

## *Interlocking guards*

M1032    Interlocking guards are usually movable, e.g. they could be hinged, sliding or removable. These guards are used, where frequent access to hazardous parts of a machine may be needed, and are connected to the machine controls by means of 'position sensors'. These interlocking elements could be electrical, mechanical, magnetic, hydraulic or pneumatic. The position sensors interlock the guard with the power source of the hazard. When the guard is open, the power is isolated thus allowing safe access into the relevant part of the machine.

The British Standard BS EN 1088: 1997 'Interlocking devices associated with guards' provides detailed guidance on the design and selection of the appropriate type of devices. Risk assessment plays a vital role in the selection of type and level of integrity on interlocking devices.

The integrity of the interlocking mechanism must ensure that the device is reliable, capable of resisting interference, difficult to defeat and the system should not, as far as possible, fail-to-danger. The term 'fail-safe' is no longer used, as it can give the impression that if a component fails in a certain mode this will always result in the machine stopping even when the safeguard is closed. Fail-to-danger of an interlocking device means that the machine can be operated when the guard is open. Therefore to minimise the probability of failure-to-danger is a preferred option.

Interlocking guards are convenient, as they give ready access into the danger zone, while ensuring the safety of the operator. However, some of these guards may be defeated to carry out necessary tasks, e.g. maintenance, setting, fault finding etc. Therefore, additional measures such as locking off or isolation may need to be introduced under a safe system of work. It is however a requirement that the machine designer should consider how these tasks can be safely performed at the design stage.

**Figure 11: Main types of electrical interlocking switches**

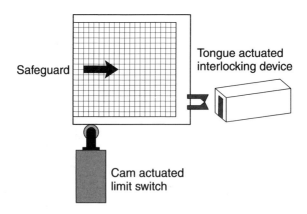

The choice of interlocking method must be compatible with the level of risk and should take into account the probability of failure of the interlocking element(s), proportion of time a person is exposed to the hazard and the potential consequences (e.g. injury severity and/or ill health).

BS EN 1088: 1997 classifies interlocking guards in two broad categories:

- Guard is interlocked with the source of power.
- Guard is interlocked with respect to the motion itself.

There are clear advantages for interlocking the guard with the moving parts rather than the source of power, as the power might fail-to-danger due to wiring/earth faults or failures of the interlocking switch. The main requirements for the movable interlocking guards are:

(a) the hazardous machine functions covered by the guard cannot operate until the guard is closed;

(b) if the guard is opened while hazardous machine functions are operating, a stop instruction is given;

(c) when the guard is closed, the hazardous machine functions covered by the guard can operate, but the closure of the guard does not by itself initiate their operation.

Interlocking guards may be fitted with a locking device so that the guard remains closed and locked until any risk of injury from the hazardous machine functions has passed.

## Types of interlocking devices

M1033    Some common types of interlocking elements include the following:

- *Cam-operated limit switch interlocks*: are one of most popular types of interlocks used generally as they are versatile, effective and easy to install and use. Another type of mechanically actuated device is the 'tongue' actuator, which opens or closes the contacts of a switch. Both types of interlocks are shown in FIGURE 14.

- *Trapped key interlocks*: a master key, which controls the power supply through a switch at the master key box, has to be turned OFF before the keys for individual guards can be released. The master switch cannot be turned to ON until all the individual keys are replaced in the master box. This type of interlock is commonly used in electrical isolation.

- *Captive-key interlocking*: involves a combination of an electrical switch and a mechanical lock in a single assembly.

- *Direct manual switch or valve interlocks*: where a switch or valve controlling power source cannot be operated until a guard is closed, and the guard cannot be opened at any time the switch is in run position (position sensors).

- *Mechanical interlocks*: provide a direct linkage from guard to power or transmission control. In other words the switch cannot be reached when the guard is open.

- *Magnetic switches* with the actuating 'coded' magnet attached to the guard: have the disadvantage that they can be defeated easily. However, types using a shaped magnet have been used with removal guards on screw conveyors. Normal reed switches are not acceptable unless they incorporate special current limiting features. Reed switches are often encapsulated and have application in flammable atmospheres. A similar type of switch relies on inductive circuits.

- *Electro-mechanical device*: can provide a time delay where the first movement of the bolt trips the machine, but the bolt has to be unscrewed a considerable distance before the guard is released. Time delay arrangements are necessary when the machine being guarded has a long run down time. A solenoid-operated bolt can also be used in conjunction with a time delay circuit.

- *Mechanical restraints (scotches)*: are used as a back-up to other forms of interlocks on certain types of presses to protect the operator against crushing injury between platens.

## Trapped key interlocking (Power isolation)

**M1034** The electrical interlocking switches interrupt the power source of the hazard by switching of a circuit, which controls the power-switching device. This is known as control interlocking.

Power interlocking on the other hand involves the direct switching of the power supply to the hazard, e.g. isolation of power. The most practical means of power interlocking is a trapped key system.

An example of such a system is illustrated in FIGURE 12. The power supply isolation switch is operated by key 'A' which is trapped in position while the isolation switch is in the ON position. When the key is turned and released, the isolation switch contacts are locked open thus isolating the power supply. The machine safeguard door in the meantime is locked closed. To open the guard, the isolation key 'A' is inserted into the guard-locking unit and turned to release the door. The key is then trapped in position and cannot be removed until the guard is closed and locked again.

One of the drawbacks of this system is that unknown to others, a person may still be inside the machine while another person decides to start-up the machine by closing and locking the guard. To overcome this foreseeable problem, the use of a personal key 'B' which is also trapped in the guard locking unit and is released when the guard is open. Key 'A' cannot be released from the guard unless key 'B' is inserted and turned into the guard locking unit. Key 'B' can also be used for other tasks, e.g. inch mode for presses or programming mode for robotics.

This type of system is reliable and does not require electrical wiring to the guard. The main disadvantage is that because if requires the exchange of the keys every time, it may not be suitable if access is frequently required. Strict control of the keys is essential.

## Mechanical interlocking

**M1035** The system shown in FIGURE 13 is known as guard inhibited interlocking. This is achieved in the above example by a sliding guard, when open the guard retains the lever of a

Figure 12: Trapped key system for power interlocking

Figure 13: Example of mechanical interlocking system

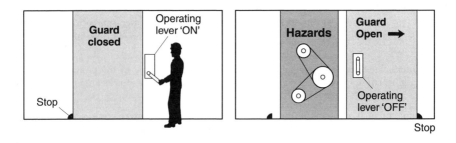

switch or a control valve in the power OFF position. When the guard has been closed, the lever or switch can be moved to the ON position, thus inhibiting the guard from being open.

This system however is easy to defeat by removing one of the door-stops, or even by removing the guard.

Other forms of mechanical interlocks include mechanical restraints (scotches) to prevent gravity fall of vertical machinery, e.g. scissors lifts and presses.

## Safety devices

M1036   A safety device is a protective appliance, other than a guard, which eliminates or reduces risk, alone or associated with a guard. There are many forms of safety device available.

(a)   *Trip device*

A trip device is one which causes a machine or machine elements to stop (or ensures an otherwise safe condition) when a person or a part of his body goes beyond a safe limit (EN

292–1: 1991). Trip devices take a number of forms – for example, mechanical, electro-sensitive safety systems and pressure-sensitive mat systems.

Trip devices may be:

(i) mechanically actuated – e.g. trip wires, telescopic probes, pressure-sensitive devices; or

(ii) non-mechanically actuated e.g. photo-electric devices, or devices using capacitive or ultrasonic means to achieve detection.

(b) *Enabling (control) device*

This is an additional manually operated control device used in conjunction with a start control and which, when continuously actuated, allows a machine to function.

(c) *Hold-to-run control device*

This device initiates and maintains the operation of machine elements for only as long as the manual control (actuator) is actuated. The manual control returns automatically to the stop position when released.

(d) *Two-hand control device*

A hold-to-run control device which requires at least simultaneous actuation by the use of both hands in order to initiate and to maintain, whilst hazardous condition exists, any operation of a machine thus affording a measure of protection only for the person who actuates it. EN 574: 1996 lays down specific recommendations relating to the design of such devices.

(e) *Limiting device*

Such a device prevents a machine or machine elements from exceeding a designed limit (e.g. space limit, pressure limit).

(f) *Limited movement control device*

The actuation of this sort of device permits only a limited amount of travel of a machine element, thus minimising risk as much as possible; further movement is precluded until there is a subsequent and separate actuation of the control.

(g) *Deterring/impeding device*

This comprises any physical obstacle which, without totally preventing access to a danger zone, reduces the probability of access to this zone by preventing free access.

# Managing Absence

## Introduction

M1501    It is advisable for all employers, irrespective of their size or industry sector, to take active steps to manage employee absence, both long and short-term. The causes of absences should be recorded and analysed so that the employer can review whether any patterns exist or whether factors in the workplace are causing or contributing to employee absence. Records should also be kept of all employees' sickness absences when they occur. In this way, employers may be able to minimise the incidence of employee absence, especially short-term absence.

Where, despite good management, employees are nevertheless absent from work, the employer should adopt good practice techniques to manage those absences. This chapter aims to help managers to understand the various employment laws that impact on employee sickness absence and to apply sound management techniques to the management of both long-term and short-term absences.

## Managing long-term sickness absence

M1502    The management of long-term sickness absence requires a pro-active approach from management in order that:

(a)    the employee's absence can be regularly reviewed;

(b)    appropriate steps can be taken to support the absent employee;

(c)    actions be taken to facilitate the employee's return to work;

(d)    the employee's return to work can be effectively managed; and

(e)    where the employee does not recover sufficiently to resume working, the employer can fairly dismiss the employee.

### Reviewing the employee's absence

M1503    Employers should adopt a 'case management' approach when dealing with employees on long-term sickness absence. This means regularly reviewing the employee's absence and state of health/fitness to see whether there is any improvement, and whether the employer can do anything to facilitate the employee's recovery and return to work.

Case reviews would normally be held quarterly or even monthly, and these would involve the employee's line manager, health and safety personnel, HR department, medical specialists and of course the employee who is absent.

### Keeping in touch with an employee who is absent

**M1504** Where an employee is absent from work due to a long-term condition, it is good practice for the employer to nominate an appropriate person to keep in touch with them. This has a number of advantages:

- the employee who is absent will feel reassured of their employer's ongoing support and is less likely to feel 'abandoned' or gain the impression that their employer does not care about their welfare;
- the employer will be able to keep up-to-date on the employee's state of health and progress and the likelihood of a return to work;
- the employer will be able to organise temporary cover more effectively;
- the employer can keep the employee informed of any developments in the business and the work of their department; and
- the employee is more likely to return to work sooner, if they feel confident that their employer adopts a supportive attitude towards them.

Naturally, managers should take care to ensure that contact is not viewed as unwelcome or intrusive. Equally, it is important to make sure that no unfair pressure is put on the employee through such contact. It is, however, a common assumption that contacting an employee who is off sick will be inappropriate, whilst in fact many employees who are absent long-term would welcome regular contact. The absent employee may feel very isolated and excluded whilst off sick and regular contact may help to reduce these feelings and make the employee feel that they are still viewed as part of the team even though they are not able to work.

The type of contact and its frequency should be agreed directly with the employee. Contact may be by telephone, email and/or through visits to the employee's home at agreed times. The employee's views on how contact should be made should always be sought and respected.

The employee's line manager should take responsibility for arranging and maintaining such contact, but he or she will not always be the best person to maintain the contact. It may be that a close colleague of the absent employee can be nominated instead, in particular where visits to the employee's home are agreed.

### Employment laws that impact on long-term sickness absence

**M1505** There are a number of laws that impact on long-term sickness absence:

1. *Disability Discrimination Act 1995*;
2. *Access to Medical Reports Act 1988*;
3. *Employment Rights Act 1996, s 1* (the duty on employers to provide employees with a written statement of employment particulars);
4. *Social Security Contributions and Benefits Act 1992, Pt XI* (the duty to pay statutory sick pay); and
5. *Employment Act 2002, Pt 3* (dispute resolution).

### When an employee's condition may amount to a disability

**M1506** The Disability Rights Commission estimated in 2004 that around 9.8 million people in Britain would qualify as disabled under the terms of the *Disability Discrimination Act 1995*. The definition of disability in the Act is very wide and includes both physical and mental impairments.

## Definition of disability

**M1507** Specifically, a person is disabled if he or she have a physical or mental impairment that has a substantial and adverse long-term effect on his or her ability to carry out normal day-to-day activities. 'Long-term' in this context means twelve months or more. There is no need for an employee to be 'registered' as disabled to qualify for protection under the *Disability Discrimination Act 1995*.

The concept of 'normal day-to-day activities' is defined further as meaning an effect on one of the following: mobility; manual dexterity; physical co-ordination; continence; ability to lift, carry or otherwise move everyday objects; speech, hearing and eyesight; memory and ability to concentrate, learn or understand; and perception of the risk of physical danger. An individual who has a serious disfigurement may also be disabled under the Act, depending on the severity and positioning of the disfigurement.

'Normal day-to-day activities' stands to be interpreted according to the lay-person's understanding of the phrase, in other words any activity that people do on a reasonably regular basis in their every-day lives (as opposed to what they do in their jobs) will be a normal day-to-day activity for the purpose of the Act.

## Scope of the Act

**M1508** All workers are protected by the *Disability Discrimination Act 1995* and all employers, with the exception of the armed forces, are under a duty to comply with the Act's provisions.

There is no qualifying period of employment for an employee to bring a claim of disability discrimination to an employment tribunal, and no ceiling on the amount of compensation that can be awarded if a claim is successful.

The two main duties under the Act are:

(*a*) the duty not to discriminate, i.e. the duty not to treat a disabled worker (or job applicant) unfavourably directly on the grounds of their disability, or for a reason related to their disability; and

(*b*) the duty to make reasonable adjustments to any provision, criterion or practice, or to physical features of premises, that may place a disabled worker at a substantial disadvantage.

Each of these is considered separately below.

## Conditions that may amount to a disability

**M1509** A wide range of physical and mental conditions and illnesses may amount to disabilities, depending always on whether the effect of the condition on the person is substantial and long-term. Conditions that are intermittent, or that fluctuate in their effects, will entitle the person to protection under the *Disability Discrimination Act 1995* at all times (provided the condition is likely to recur), even if at a particular point in time the condition is in remission. A further important point is that a condition will amount to a disability even if the person, as a result of medication or other form of support, experiences no adverse effects on a day-to-day basis. The question that determines whether an employee is disabled is how the condition would affect them if they did not take their medication or use the support.

Progressive conditions such as cancer, multiple sclerosis and AIDS are covered under the definition of disability as soon as the condition is diagnosed and the person's normal day-to-day activities are affected in any way, whether substantial or not. As from October 2005 (when changes to the Act will be implemented), an employee diagnosed with one of these three conditions will automatically be deemed disabled whether or not they are experiencing any symptoms at the time in question and whether or not it is likely that the effects of the condition will, in time, become substantial.

The cause of an illness or other condition is irrelevant to the assessment of whether it amounts to a disability.

There are some exceptions to the Act, i.e. conditions that are specifically deemed not to be disabilities. These include alcohol addiction, addiction to illegal drugs and hay fever. A small number of mental conditions are also excluded.

## The duty not to discriminate

**M1510**  Employers are under a duty not to treat a disabled employee unfavourably:

1. directly on the grounds that they have a particular disability; and
2. on grounds related to disability, unless justified.

Direct disability discrimination, which can never be justified, is where an employer treats a disabled employee or job applicant unfavourably just because they have a disability. An example could be the automatic rejection of a job applicant just because they had disclosed that they had a history of mental illness.

In contrast, it is possible for an employer to justify disability-related discrimination against an employee, for example if an existing employee had, following an accident, become completely unable to perform the job for which he or she was employed. In these circumstances it might be justifiable to terminate the employee's employment, provided the employer had first explored whether any reasonable adjustments could be made (see M1511 below), for example whether the employee could reasonably be transferred to different job duties. The reasons put forward for justification must always be both material and substantial.

Unfavourable treatment can take many forms and may include:

- rejection for employment or promotion;
- exclusion from training opportunities;
- being afforded less favourable terms and conditions;
- removal of job duties or transfer to less interesting or less responsible work;
- exclusion from contractual benefits or non-contractual perks;
- harassment, which can be physical, verbal or non-verbal; or
- dismissal or selection for redundancy.

The list is not exhaustive.

## The duty to make reasonable adjustments

**M1511**  The *Disability Discrimination Act 1995* places a duty on employers to make reasonable adjustments to any provision, criterion or practice and to physical features of premises in order to accommodate the needs of disabled workers and job applicants. This duty arises whenever any aspect of the employer's working practices or premises puts a disabled person at a substantial disadvantage. An employer who fails to comply with this duty will be acting in contravention of the *Disability Discrimination Act 1995* unless they can show that it was not reasonable for them to make adjustments generally, or to make a particular adjustment.

An important point to bear in mind is that the duty to make reasonable adjustments places the employer under a positive duty to take the initiative and consider what adjustments would be possible and practicable for a particular disabled employee (or job applicant). In *Mid Staffordshire*

*General Hospitals NHS Trust v Cambridge [2003] IRLR 566*, the EAT held that employers must, if they are to fulfil their duty to make reasonable adjustments, make a proper assessment of a disabled employee's condition and prognosis, the effects of the condition on their ability to perform relevant job duties, the effect of the physical features of the workplace and the steps that the employer could potentially take to reduce or remove the disadvantages that the employee is experiencing. An employer who fails to obtain and act on a full and proper assessment of the employee's condition will have failed in their duty to make reasonable adjustments, as an assessment is a necessary pre-condition to the fulfillment of that duty.

Although the duty to make reasonable adjustments is on the employer, it is nevertheless advisable for managers to consult the employee concerned about possible adjustments as the employee will, in any event, know more about their condition and its effects than the manager, and will often be able to suggest adjustments that would be helpful for them. Managers should be open-minded and willing to take on board any reasonable suggestions from the employee for adjustments.

Adjustments can include:

(a) transferring the employee to another job, for example lighter work, provided the employee consents to such a move;

(b) adjusting the duties of the job, for example exempting an employee with a back condition from doing any heavy physical work;

(c) changing the method of doing the job, for example allowing an employee who cannot drive on account of a medical condition to use taxis to travel on business;

(d) adjusting working hours, for example allowing someone with a mental illness a later or flexible start time, or more frequent rest breaks, or exempting an employee with a debilitating condition from a requirement to work overtime;

(e) changing the place of work, for example moving someone with limited mobility to a ground floor location or allowing home-working for part of the working week;

(f) adjusting procedural requirements, for example allowing an employee with a recurring illness to take more time off work than would normally be acceptable;

(g) providing additional or tailored training, coaching or mentoring, for example for someone with learning difficulties;

(h) providing special equipment, for example voice activated software for someone with a visual impairment;

(i) modifying instructions or reference manuals, for example providing them in Braille;

(j) providing a reader or interpreter, for example for someone who is profoundly deaf;

(k) providing additional or special supervision; and

(l) modifying premises, for example widening a doorway or relocating door handles or shelves for someone who has difficulty reaching.

Unless an employer has done all they reasonably can to make adjustments for a disabled employee, any subsequent unfavourable treatment (for example dismissal) will always be discriminatory and unlawful.

## Access to medical reports

**M1512** The *Access to Medical Reports Act 1988* places certain restrictions on organisations that wish to obtain medical information about someone from their doctor. The Act also gives individuals a variety of rights in relation to any such medical reports.

The Act applies whenever a medical report is 'prepared by a medical practitioner who is or has been responsible for the clinical care of the individual'. The effect of this is that the Act applies to medical reports prepared by an employee's own doctor, specialist or consultant, but would not normally apply to reports prepared by an occupational doctor.

Employees have a number of rights under the Act:

- to be informed in writing if the employer wishes to seek their consent to contact their doctor for a medical report;
- (at the same time) to be informed by the employer of their rights under the Act;
- to decline to give their consent for the employer to apply to their doctor for a medical report;
- (if they have given consent) to ask their doctor for a copy of the report once it has been prepared;
- to ask the doctor to amend the report, if, in their opinion, it contains anything inaccurate or misleading; and
- to refuse to allow the report to be released to the employer.

The employee's right to refuse to give consent for their employer to apply to their doctor for a medical report and their right to refuse to allow a report, once prepared, to be released to the employer, are statutory rights. This means that the employer cannot override these rights by inserting a clause in employment contracts to the effect that employees are required to give such consent. Any contractual term to this effect will be unenforceable.

Since the Act does not normally apply to reports prepared by occupational doctors, it is open to an employer to include clauses in employees' contracts to the effect that they must agree, on request by the employer, to be examined by an occupational doctor, and allow the doctor to provide a report to the employer.

## *Employees' contractual rights*

**M1513** Under of the *Employment Rights Act 1996, s 1* employers are under a duty to provide a written statement of key terms and conditions of employment to all employees whose employment is to continue for more than one month. The 'written statement' as it is usually known must include information on a number of defined issues, including any terms and conditions relating to 'incapacity for work due to sickness or injury, including any provision for sick pay'.

There is no duty on an employer to pay an employee 'occupational sick pay', i.e. their normal wages or salary, if they are absent from work due to sickness or injury. That is a matter for each employer to decide for themselves. Employers must, however, inform each employee in writing (as part of the written statement) what their entitlement to sick pay will be in the event of sickness absence, and of any conditions attached to the payment of sick pay.

Many employers pay occupational sick pay for defined periods of time, whilst others make the payment of sick pay discretionary. Where the employer wishes to retain discretion over the payment of sick pay, it is advisable to devise clear management guidelines on how discretion should be exercised to ensure consistency and fairness across different departments. In *Clark v Nomura International plc [2000] IRLR 766*, the High Court held that management discretion in relation to pay must not be exercised perversely or irrationally.

It would also be normal practice to make the payment of occupational sick pay dependent on certain conditions being fulfilled, for example to require employees to:

1. telephone a designated person as soon the employee knows that they will be unable to attend work;
2. provide a self-certificate for the first week of absence and doctors' certificates for all periods of absence thereafter;

3. maintain contact with the employer at regular intervals (for example weekly) with a view to advising the employer how long they expect to be absent;
4. agree to be examined by an occupational doctor on request and to the doctor providing a report to the employer; and
5. attend a 'return to work' interview with a designated person (usually the line manager) on return to work.

These conditions allow the employer to manage absence and sick pay effectively.

Statutory sick pay (SSP) is an entirely different matter and is dealt with at M1514 below.

## Statutory sick pay

M1514    Although the employer may decline to make payment of wages or salary during sickness absence, they cannot contract out of their obligation to pay SSP.

There is no length of service required for an employee to qualify for SSP, although complex provisions are in place with regard to when payment is due and record keeping. Both full-time and part-time employees will qualify, but an employee who earns less than the equivalent of the lower limit for the purposes of national insurance contributions (£82.00 per week as from April 2005) are excluded. Where an employee is paid occupational sick pay, SSP must be offset against this, i.e. SSP cannot be paid over and above normal wages or salary.

To qualify for SSP, an employee must comply with their employer's conditions for notification of absence and provide suitable evidence of incapacity.

SSP is a fixed weekly rate, payable to an employee who is absent from work due to personal sickness for up to 28 weeks during any 'qualifying period'. The first three days of absence during any qualifying period are 'waiting days' and no SSP is payable for these days.

When the rules on SSP were first introduced, employers were able to recoup amounts paid out by offsetting payments against their national insurance contributions. These provisions were, however, subsequently abolished, which means that employers now foot the bill for SSP themselves.

## Rehabilitation of an employee who has been absent from work

M1515    Where an employee has been absent from work for a lengthy period of time, it will be essential to manage their return to work carefully. The employee's line manager should take responsibility for discussing the employee's return to work directly with them and for agreeing a detailed plan to support the employee on their return.

Many employees will feel anxious about the prospect of returning after a lengthy period away from the workplace and they may be worried about how they will cope with being back at work, or concerned about their colleagues' reactions.

The first action point will be for the manager to speak to the employee (preferably prior to their return if that is possible, but if not on the employee's first day back) to discuss the details of their return. One way of dealing with this would be to suggest to the employee that they might make a social visit to the workplace shortly before their agreed return date to allow a meeting with their line manager and/or HR department and also allow them to chat informally with their colleagues about any ongoing relevant matters.

The discussion with the line manager should include:

(a) the employee's opinion about their capabilities, for example whether the employee is confident that they are capable of full job performance or only partial performance;

(b) whether the employee's return should be to full-time duties or whether a phased return would be beneficial;

(c) whether the employee will still be taking any medication after their return to work that might have side effects, for example tiredness;

(d) any special arrangements, additional support or adjustments to the employee's duties or environment that would help the employee to reintegrate into the workplace; and

(e) whether an induction programme is desirable or necessary, for example if a number of organisational or procedural changes have taken place during the employee's absence, a 'mini-induction' may be very helpful.

The employer should also arrange for the employee to be medically examined by an occupational doctor to confirm that they are genuinely capable – both physically and mentally – of returning to work. The doctor should also be asked to comment on whether the employee should be exempted initially from any particular job duties and if any special arrangements are necessary or advisable.

After the employee's return, the line manager should:

- monitor the situation over the first few weeks to ensure the employee is coping with their work and the day-to-day pressures of working life;
- make sure the employee is not 'thrown in the deep end', for example by requiring them to deal with a huge backlog of work caused by their absence; and
- take all reasonable steps to ensure the employee's reintegration into the workplace.

## When dismissal on the grounds of long-term absence can be fair

**M1516** Long-term sickness absence can be a fair reason for dismissal for the purposes of the *Employment Rights Act 1996, s 98* under the heading of 'capability'. 'Capability' is defined as having reference to 'skill, aptitude, health or any other physical or mental quality'. For a dismissal to be fair, however, the employer would have to show that the employee's long-term absence was sufficient to justify dismissal and that they had acted reasonably in dismissing the employee for this reason.

In the event of a claim to tribunal, the burden of proving the reason for the dismissal is on the employer.

All employees are entitled to bring a claim for unfair dismissal to an employment tribunal provided that they:

1. are employed under a contract of employment;
2. have a minimum of one year's continuous service with the employer; and
3. are under their employer's normal retirement age at the time of their dismissal, or, if there is no normal or contractual retirement age, under age 65.

Before contemplating dismissal, the employer should review the circumstances to establish whether grounds for dismissal exist. In doing so, the employer should consider:

(a) the effect of the employee's past and likely future absences on the organisation, for example to what extent colleagues are being burdened by the employee's absence;

(b) the degree to which it is possible (if at all) to cover the absent employee's work, for example by using temporary agency workers;

(c) how similar situations have been handled in the past;

(d) medical advice;

(e) the likelihood of the employee being able to return to work in the reasonably near future; and

(f) the availability of suitable alternative work which the employee would be capable of doing, i.e. whether there is any alternative to dismissal.

## *Procedure for ensuring a dismissal is fair*

**M1517** Over and above having a potentially fair reason for dismissal, an employer who is challenged in an employment tribunal in a claim for unfair dismissal, must also be able to demonstrate that they:

- acted reasonably towards the employee in an overall sense; and
- followed the statutory dismissal and disciplinary procedure.

Courts and tribunals have, over the course of many years, developed tests of fairness in cases of dismissal. In relation to dismissals on the grounds of long-term ill-health absence, the employer should, as a minimum, ensure that they:

1. review the employee's absence record to assess whether it is sufficient to justify dismissal;
2. set time limits for review and inform the employee of these;
3. consult the employee regularly;
4. obtain up-to-date medical advice;
5. advise the employee as soon as it is established that termination of employment has become a possibility;
6. review whether there are any other jobs that the employee could do prior to taking any decision on whether to dismiss; and
7. act reasonably throughout.

There is no time limit in law after which it is fair to dismiss an employee who is absent from work. The key question is whether in all the circumstances the employer can reasonably be expected, in light of the requirements of their business, to wait any longer for the employee to recover and return to work. This will depend, amongst other things, on the size and resources of the business and the degree of disruption or difficulty that the employee's long-term absence is causing.

The statutory dismissal and disciplinary procedure (DDP) was implemented to UK law in October 2004 and requires an employer to take certain procedural steps when dismissing an employee. The procedure applies to capability dismissals, including those by reason of long-term sickness absence. It is important to note that any failure by the employer to adhere to the DDP will make the dismissal automatically unfair, irrespective of the reason for non-compliance. In such circumstances, the employee's compensation for unfair dismissal stands to be uplifted by between ten and fifty per cent according to the tribunal's discretion.

The following lists the DDP's three compulsory steps.

(a) The employer must set out in writing the circumstances that may give rise to the termination of the employee's employment and give or send a copy to the employee. This 'step one letter' must give adequate information to the employee without giving the impression that the employer has made up their mind already about what the outcome will be. The letter should also invite the employee to attend a meeting (step two).

(b) The employer must hold a meeting with the employee at which the employee must be permitted to discuss the circumstances and make representations about their ongoing employment.

(c) Once notice of termination has been given, the employee must be allowed the right of appeal against the decision to dismiss, which must be to a more senior manager than the person who took the decision to dismiss.

If the employee is not well enough to attend a meeting, then the employer may put forward a defence that it was not practicable for them to comply with the procedure. It is advisable, however, to retain evidence that step one (writing to the employee) was complied with and that the failure to hold a meeting was caused by factors outside the employer's control.

# Managing short-term absences

**M1518** Frequent, persistent short-term absences can be extremely disruptive for an employer. Where an employee is absent long-term, the employer can at least make plans, as they will have some idea about how long the employee is likely to be absent. The nature of short-term absenteeism, however, is that it tends to be unpredictable and uncertain.

Whilst a certain level of absenteeism is inevitable due to minor illnesses, there may also be occasions when employees take time off work for reasons associated with factors in the workplace. Research conducted by the Health and Safety Commission indicates that in a single year, an estimated 2.3 million people in Britain suffered from an illness that they believed was caused by or made worse by their current job or a previous job.

Furthermore, the CBI has estimated that one third of all sickness absence is due to mental ill-health. Statistics show that absences on account of stress-related conditions are increasing and stress is now thought to be the single most common cause of employee absence, having overtaken the traditional 'bad back' in many sectors of industry.

Employees' rights in relation to short-term absence will be determined according to the terms of their contracts of employment. Contracts or supplementary documentation should state clearly what employees' obligations and entitlements are when they are unable to attend work, for example what they will be paid when they are absent and how they should notify the employer.

To manage short-term absenteeism effectively, employers should:

- require employees who are unable to come to work to notify a designated person of their absence as soon as possible;
- require employees to 'self-certify' for all periods of absence of up to one week and to provide doctors' certificates for all periods of absence beyond the first week;
- conduct regular 'return-to-work interviews' after every period of absence, however short;
- investigate the possible causes of frequent absenteeism;
- look for patterns in absenteeism to see whether there is, for example, a problem in a particular department;
- implement a policy with defined trigger points which, when reached, activate a formal review of the employee's attendance;
- set targets and time limits for improvement; and
- keep detailed records of all employee absences.

These issues are explored further below.

## Sound management techniques for dealing with short-term absenteeism

**M1519** There is much an employer can do to manage short-term absenteeism effectively with a view to reducing the frequency of its occurrence.

## Requirement to notify

**M1520** Employers should have in place a set procedure under which employees are required to notify a designated person (usually their line manager) as soon as they know that they will be unable to come to work. The line manager should record the date and time of the call, the reason given by the employee for inability to attend work and the employee's view of how long they expect to be absent. This creates the start of an absence record. A pre-printed questionnaire could be devised for this purpose, thus ensuring consistency in approach. The questionnaire could also be used as a means of instigating payment of statutory sick pay.

## Certification

**M1521** Employers should routinely require employees who are absent from work due to sickness to complete a self-certification form when they return to work. The employee should be asked to complete and sign the form in front of their manager who should counter-sign the form, and the obligation to do so should apply to every absence of one day or more. The form should require the employee to write down the reason for their absence, the dates and whether they consulted a doctor, rather than just tick a series of boxes. This makes the process of self-certification a more active process.

It is often a sound idea to make the payment of sick pay dependent on the completion of a self-certification form (and on the provision of doctors' certificates in the case of longer periods of absence).

## Return to work interviews

**M1522** As well as requiring the employee to complete a self-certification form, employers should adopt a policy whereby their managers routinely conduct 'return-to-work interviews' every time an employee has been absent from work due to sickness.

The interview should be:

1. informal (but should be more than just a 'corridor chat')
2. private and confidential;
3. structured and factual;
4. carried out in a positive and supportive way; and
5. recorded.

The dual purpose of such an interview will be to establish the reason for the employee's absence (without requiring disclosure of any medical details) and to assess whether the employer can do anything to help or support the employee. It is important to note that this type of interview should not be confused with a disciplinary interview and that fact should be made clear to the employee.

Adopting this procedure will let employees know that management is serious about monitoring absences and may therefore deter casual absences.

## Considering the possible causes of frequent absenteeism

**M1523** Most employers have one or more employees who are absent from work on a regular basis. If the employee has an underlying medical condition that is causing these absences, then the employer will know what they are dealing with and medical advice can be sought to support any employment decisions that the employer might contemplate making. The employer should also consider whether the employee might qualify as disabled under the *Disability Discrimination Act 1995* (see M1507 above) and if so review what reasonable adjustments might be made to the employee's job or working environment.

There are, however, many other possible causes of frequent absenteeism. Apart from ill-health, the employee may:

(a) have personal or family difficulties, for example childcare problems;

(b) have another job which is causing a conflict of interest;

(c) be de-motivated and disinclined to attend work; or

(d) have a specific problem in the workplace (see M1524 below) that is causing or contributing to their frequent absences.

It will be essential for the manager to remain open-minded and to refrain from jumping to the conclusion that the employee is taking time off work without good reason. It may be that the cause of the employee's absences is something outside their control. The manager should speak to the employee to establish whether there is a discrete underlying cause and whether anything can be done to support the employee. Understanding the root cause of the absence problem is important because, until the cause is correctly pinpointed, it will not be possible to identify an appropriate course of action to remedy it.

## Reviewing whether factors in the workplace are causing absence

**M1524** Managers should be constantly alert to the possibility that an employee's absences might be caused by, or exacerbated by, factors in the workplace. For example, the problem may be related to:

- the volume of work being too much for the employee to cope with;
- deadline dates or other pressures that the employee feels unable to face;
- unsatisfactory working relationships or outright conflict with colleagues;
- bullying or harassment;
- de-motivation caused by ineffective management or an authoritarian management style;
- inability to cope with change;
- a feeling of inadequacy or loss of control over the job; or
- other factors at work causing dissatisfaction, for example ineffective procedures or equipment, having no clear goals or targets, etc.

Issues such as those identified above may well lead to absences and managers should be alert to signals that the employee may be suffering from stress to an extent that he or she is not coping. Medical certificates that state 'stress', 'depression' or 'anxiety' should put the manager on notice that there may be a workplace problem that needs to be addressed.

Line managers should always be alert to such issues and should make enquiries of any employee who has frequent short-term absences as to whether there is anything in the workplace causing or contributing to their absences. Such enquiries should be made sympathetically, with the manager reassuring the employee that if he or she is experiencing problems at work, it would be their genuine wish to provide support with a view to resolving those problems.

If a workplace problem is identified, the manager should of course take steps to remove or reduce the factor that is causing the problem, if that is at all possible. Once the cause of the employee's absences is removed, the employee's attendance may well improve. A failure to take steps to support an employee who is experiencing health problems as a result of factors in the workplace may have serious consequences as the employer could be held liable in law if the employee subsequently had a mental breakdown as a result of factors in the workplace and the breakdown was reasonably foreseeable.

## Looking for patterns

**M1525** Employers who are experiencing high levels of short-term absenteeism should use the records created by line managers about individuals' absences to review the absenteeism, and whether any patterns exist, for example:

1. whether absenteeism is higher than average in any particular department or section;
2. whether the rates of absence of employees performing any particular type of work are disproportionate when compared to other employees; or
3. whether there is any particularly common cause of absence (for example back pain).

In each case, the employer should investigate in order to try to establish the root cause of the particular pattern. Following identification of the likely cause, steps can be taken to remedy the situation. For example if absenteeism is unusually high in a particular manager's section, it could possibly be that the manager is the cause. Well-designed training in supervisory skills or inter-personal skills, or individual coaching may improve the manager's people-management skills and subsequently lead to a reduction in levels of absenteeism in his or her department.

On a more local level, line managers should, when reviewing individual employees' absence records, look to see whether there is a pattern amongst the absences. Examples could include the frequent Monday morning absentee or the employee whose absences tend to occur at particular times, for example just before a detailed month-end report is due.

If such a pattern is identified, the manager should speak to the employee about it and ask the employee if they can explain the pattern. Managers should, of course, not make assumptions, nor blame or accuse the employee, but instead should remain open-minded. The main aim in speaking to the employee about a pattern of absences will, once again, be to try to establish the underlying reasons(s) for the frequent absences. It is only when the underlying cause is identified that it will be possible to decide what to do. For example, if absence is due to tiredness and the reason the employee is tired is because they have a second job, then the manager might proceed by explaining to the employee that the situation is unsatisfactory and seek to reach a compromise on how many hours the employee works in the other job.

Even if the employee is unable to put forward any explanation, such a meeting will have the advantage of alerting the employee to the fact that the line manager has noticed the pattern and may deter further casual absences.

## Defined trigger points

**M1526** It can be very effective for employers to devise and implement an attendance procedure with defined trigger points that act to instigate a formal review of the employee's level of absence and its causes. Trigger points are usually defined by reference to:

(*a*) the total number of days' absence the employee has had in a defined period (for example in any twelve-month period); or

(*b*) the number of occasions when the employee has been absent from work during the same period.

A typical example of trigger points could be three separate absences (whatever their length) or absences totalling 15 days in any twelve-month period.

## Setting targets and time limits for improvement

**M1527** Where an employee's level of short-term absences has reached a stage where their level of attendance is unsatisfactory, the line manager should seek to agree with the employee reasonable targets and time limits for improvement in attendance. The manager should try to ensure that the employee is committed to achieving the agreed level of improvement. The manager should of course follow up at the agreed time and review the employee's level of attendance again. If it has not improved to a satisfactory level, formal action may have to be instigated according to the employer's procedures.

## Keeping records

**M1528** Full records should always be kept of employees' absences and of all discussions held with the employee about absence and attendance. Self-certificates and medical certificates should also be retained. Such records should, of course be held confidentially, preferably by the organisation's HR department (if they have one). The terms of the *Data Protection Act 1998* will have to be adhered to in holding such records.

## Giving formal warnings for unsatisfactory attendance

**M1529** Although it is appropriate for managers to be supportive in the first instance towards employees who, for genuine reasons, have frequent absences from work, managers also need to ensure the work of their department is done efficiently and effectively. Even where the reasons for an employee's absences is undisputedly genuine, that does not change the fact that the employee's work still needs to be done on a regular and reliable basis. If, therefore, informal measures have not led to an improvement in the employee's attendance, it may be that formal procedures need to be instigated. This will be appropriate when the employee's absences have become excessive or where they are beginning to cause serious disruption or dissatisfaction.

Some employers use their standard disciplinary procedure to deal with this type of situation, whilst others have separate capability procedures which follow a similar format. The key action steps of any such procedure will be that the employer would:

- write to the employee, explaining that their level of absence has reached a level that is causing concern, enclosing a note detailing the dates of all absences over the last six or twelve-month period (as appropriate);

- invite the employee to attend a meeting to discuss the matter, informing them that they have the right to bring a colleague or trade union official along, either to represent them or simply to provide moral support;

- hold the meeting, and explain to the employee that their absence has reached a level that is considered unsatisfactory and the reasons why this is the case;

- give the employee a full and fair opportunity once more to explain their absences and put forward any mitigating factors or other representations;

- decide after the meeting whether it is appropriate to issue a formal warning, following the terms of the employer's own procedure;

- make it clear in any warning that if the employee's level of attendance does not improve to a defined standard within a defined time period, further formal action will be taken and that the situation could, eventually, lead to dismissal; and

- set down a date for a further review, typically in three or six months time.

This type of procedure may be instigated irrespective of whether or not the employee's absences are for genuine reasons. If some of the absences are not genuine, the action taken would of course fall under the banner of discipline. Even where the employee's absences are for genuine reasons (for example genuine ill-health, family problems, etc.), formal action can be justified under the banner of 'capability' or 'attendance'.

## Fair dismissal on the grounds of short-term absence

**M1530** If, following formal action, the employee's level of attendance does not improve to a satisfactory degree within the defined time-scale, the employer would move to the next stage of the disciplinary or capability procedure. It is usual for two (or three) warnings to be given following which the employee may ultimately be dismissed.

The reason for dismissal will be either 'lack of capability' (i.e. ill-health that has led to the employee being unable to perform their job to a satisfactory standard) or 'some other substantial reason' (i.e. unsatisfactory attendance, whatever the causes). Both of these are stated in the *Employment Rights Act 1996* as potentially fair reasons for dismissal.

For a dismissal to be fair, however, the employer also has to show that the employee's level of absence was sufficient to justify dismissal and that they had acted reasonably in dismissing the employee for this reason.

When the last stage of the procedure has been reached, the employer must take care also to follow the DDP that was implemented into UK law in October 2004. Details of this procedure are given above under 'Procedure for Ensuring a Dismissal is Fair'.

## Conclusion

**M1531** Many employers are wary of tackling employee absence and uncertain as to what action they can lawfully take. The basic underlying principle is that, if an employee is incapable of performing their job, whether due to genuine ill-health or not, the employer may take action up to and including dismissal. Long before the dismissal stage is reached, however, active management intervention can often help to:

1. identify the cause(s) of an employee's absence;
2. provide support to the employee, if appropriate;
3. deter casual absences;
4. identify whether there are any problems inherent in the workplace that are contributing to rates of absenteeism, for example bullying and ensure these are addressed;
5. establish whether or not an employee's level of attendance is likely to improve within a reasonable time-frame;
6. improve morale and motivation; and
7. lead to a reduction in rates of absenteeism within the organisation and an associated reduction in costs.

# Managing Health and Safety

## Introduction

M2001    There are many reasons why employers need to manage health and safety well. These are usually grouped under three main headings:

- *Humanitarian*

    In a typical year in the UK around 300 people are killed in work-related accidents, 30,000 people suffer major injuries (fractured limbs, amputations etc), over 2,000 people die as a result of occupational diseases and at least 10,000 other deaths are partly attributable to occupational ill health. Employers would obviously not wish to be involved in these accidents and cases of ill health. The psychological effects of individual incidents on employers, employees and others are sometimes quite traumatic.

- *Legal*

    Society has been increasingly unwilling to accept such a toll of accidents and ill health. In many respects the UK has led the world in health and safety law and now European directives result in a steady flow of health and safety legislation. Many regulations (particularly the *Management of Health and Safety at Work Regulations1999* (*SI 1999 No 3242*) – see M2005 below), specifically require the application of appropriate management techniques.

- *Financial*

    Accidents and ill-health can have a direct financial impact on a business through fines and compensation claims. However, there may also be significant indirect costs associated with the unavailability of injured employees or damaged equipment, or due to the impact of legal sanctions such as prohibition notices. Loss of reputation following serious accidents has had a significant adverse impact on a number of companies, particularly in the rail sector, and supply-chain pressure can result in those whose health and safety standards are found lacking, finding it increasingly difficult to gain business.

    Insurance will only protect employers against a limited range of direct and indirect costs. In 1993 the HSE published the results of a detailed study of 'The costs of accidents at work'. This was based on five individual case studies and demonstrated that the ratio of 'insured' costs to 'uninsured' costs could be as high as 1:36. The ways in which such costs may be incurred are described in FIGURE 1.

(Expanded version of diagram available in HSE booklet HSG 96.)

## Legal requirements

M2002    Almost all recent health and safety legislation contains provisions relating to effective management, often requiring the application of risk assessment and control techniques. Set out below are the requirements of *sections 2* and *3* of the *Health and Safety at Work etc. Act 1974* ('*HSWA 1974*'), together with those contained in the *Management of Health and Safety at Work Regulations 1999* (*SI 1999 No 3242*) (the '*Management Regulations*').

Figure 1: The costs of accidents

**INSURED**

DIRECT
- Employers' liability claims*
- Public liability claims*
- Major damage*
- External legal costs*

- Major business interruption*
- Product liability*

INDIRECT

- Fines*
- Sick Pay*
- Minor damage
- Lost or damaged product
- Clear-up costs

- Minor business interruption
- Prohibition notice interruption
- Customer dissatisfaction (product delay)
- First aid and medical attention
- Investigation time and internal administration
- Diversion of attention/spectating time
- Loss of key staff
- Hiring/training replacement staff
- Hiring replacement equipment
- Lowered employee morale
- Damage to external image

**UNINSURED**

\* Only these costs can be quantified without considerable effort

## Duties of employers to their employees

**M2003** Section 2 of *HSWA 1974* sets out a number of general duties that employers hold towards their employees. *Subsection (1)* of *section 2* contains an all-embracing requirement:

'It shall be the duty of every employer to ensure, so far as is reasonably practicable, the health, safety and welfare of all his employees'.

*Subsection (2)* of *section 2* contains a number of more specific requirements (again qualified by the phrase 'so far as is reasonably practicable') relating to:

- provision and maintenance of plant and systems of work;
- use, handling, storage and transport of articles and substances;
- provision of information, instruction, training and supervision;
- places of work and means of access and egress;
- the working environment, facilities and welfare arrangements.

It is inconceivable that an employer could achieve compliance with these extremely wide-ranging duties without the application of effective management principles.

In addition, *subsection (3)* of *section 2* requires employers to prepare and revise a written statement of their policy with respect to the health and safety at work of his employees and the organisation and arrangements for carrying out that policy. This duty is examined in greater detail at M2025 below and in the chapter entitled STATEMENTS OF HEALTH AND SAFETY POLICY.

## Duties of employers to others

**M2004**  Section 3(1) of *HSWA 1974* places a general duty on employers in respect of persons other than their employees:

> 'It shall be the duty of every employer to conduct his undertaking in such a way as to ensure, so far as is reasonably practicable, that persons not in his employment who may be affected thereby are not exposed to risks to their health or safety'.

Persons protected by this requirement include contractors (and their employees), visitors, customers, members of the emergency services, neighbours, passers-by and other members of the general public – including trespassers, although the 'so far as is reasonably practicable' qualification is of particular relevance to them. (*Subsection (2) of section 3* places a similar duty on the self-employed.)

Once again, effective health and safety management provides the only way of complying with this duty.

# The Management Regulations

**M2005**  The *Management Regulations (SI 1999 No 3242)* came into operation on 29 December 1999. They modified the 1992 Management Regulations which in turn had developed many of the principles already established by the *HSWA* in 1974. The Regulations make it quite clear that health and safety must be managed systematically, like any other aspect of an organisation's affairs. Some of the main requirements of the Regulations relate to:

- *Risk assessments* – to identify risks and related precautions (*Regulation 3*).
- *Management systems* – to ensure that precautions are implemented (*Regulation 5*).
- *A competent source of health and safety advice* being appointed (*Regulation 7*).
- *Emergency procedures* being developed (*Regulation 8*).

HSE booklet L21 contains the Regulations in full together with an Approved Code of Practice (ACOP) and related guidance. The 1999 Regulations incorporated previous amendments to the 1992 Regulations, relating to young persons, new or expectant mothers and fire risks.

The requirements of a number of the regulations (e.g. risk assessment) are dealt with more fully elsewhere in this publication (see RISK ASSESSMENT) and therefore receive only limited attention here.

## Risk assessment (Regulation 3)

**M2006**  Employers (and the self-employed) must make 'suitable and sufficient' assessments of the risks to their employees and others who may be affected by their work activities, in order to identify the measures necessary to comply with the law. The 'significant findings' of the assessment must be recorded by those with five or more employees.

Given the widely drawn nature of the requirements of *HSWA 1974* and other legislation, all risks must be included in this exercise. There is no need to repeat other 'assessments' required under more specific regulations, provided these are still valid. An assessment of fire risks and precautions must be carried out, whether or not the workplace already has a fire certificate under the *Fire Precautions Act 1971*. Both the risks and the related precautions must be identified.

The *Management Regulations (SI 1999 No 3242)* require special consideration to be given to risks to young persons (under 18s) (*Reg 3(4)* and *(5)*) and to new and expectant mothers and their babies by virtue of *Regulation 16*.

## Principles of prevention to be applied (Regulation 4)

**M2007** Preventive and protective measures must be implemented on the basis of principles specified in *Schedule 1* to the *Management Regulations (SI 1999 No 3242)*. These include:

- avoiding risks;
- evaluating risks which cannot be avoided;
- combating risks at source;
- replacing the dangerous by the non-dangerous or less dangerous;
- giving collective protective measures priority over individual measures, i.e. following good principles of health and safety management.

## Health and safety arrangements (Regulation 5)

**M2008** Employers must make and give effect to appropriate 'arrangements' for the effective planning, organisation, control, monitoring and review of preventive and protective measures. Records should be kept of these arrangements by those with 5 or more employees.

While *Regulation 3* of the *Management Regulations (SI 1999 No 3242)* requires precautions to be identified, *Regulation 5* requires them to be put into practice effectively, utilising a systematic management cycle (see FIGURE 2).

Figure 2

The risk assessment cannot remain a meaningless piece of paper – *Regulation 5* of the *Management Regulations (SI 1999 No 3242)* requires it to be put into effect. Much of the remainder of this chapter is concerned with the practical application of these management principles.

## Health surveillance (Regulation 6)

**M2009** Employers must ensure that employees are provided with appropriate health surveillance where this is identified by the risk assessment as being necessary. In most cases surveillance should already have been identified as being necessary to comply with other specific regulations (e.g. COSHH, asbestos). However, there may be other types of health risk not adequately covered by other legislation, e.g. colour blindness for train drivers or electricians; risks of vibration white finger.

## Health and safety assistance (Regulation 7)

**M2010** Employers must appoint one or more 'competent' persons to assist them in complying with the law. The numbers of such persons and the time and resources available to them must be adequate for the size of and the risks present in the undertaking.

Competence in relation to this requirement might involve:

- knowledge and understanding of the work activities involved;
- knowledge of the principles of risk assessment and prevention;

- understanding of relevant legislation and current best practice;
- the capability to apply this in a practical environment;
- an awareness of one's own limitations (in experience or knowledge);
- the willingness and ability to learn.

Although formal health and safety qualifications give a good guide to competence, they are not an automatic requirement. Small employers are allowed to appoint themselves under this Regulation, providing they are competent. The *Management Regulations* (*SI 1999 No 3242*) state (in *Regulation 7(8)*) that there is a preference for the competent person to be an employee rather than someone from outside, e.g. a consultant.

## Procedures for serious and imminent danger and for danger areas (Regulation 8)

M2011    Procedures must be established 'to be followed in the event of serious and imminent danger to persons at work'. These procedures must be implemented where appropriate. Access must also be restricted to areas which are known to be dangerous.

Most employers are familiar with the process of preparing a fire evacuation procedure. Similar procedures must be established for dealing with other types of emergency which may have been identified through the process of risk assessment as being foreseeable. Such emergencies might include bomb threats, leaks or discharges of hazardous substances, runaway processes, power failure, violence, animal escape, severe weather etc. Competent persons must be appointed to implement evacuation where necessary.

Areas identified as being dangerous, e.g. because of the presence of electrical risks, chemical risks, potentially dangerous animals (or humans), must have access to them restricted to those who have received adequate instruction on the risks present and the precautions to be taken.

## Contacts with external services (Regulation 9)

M2012    Necessary contacts must be arranged with emergency services (and others if appropriate) regarding first aid, emergency medical care and rescue work, particularly in respect of the types of emergencies referred to in M2011 above.

## Information for employees (Regulation 10)

M2013    Employers must provide employees with comprehensible and relevant information on:

- risks to their health and safety identified by risk assessment;
- the related preventive and protective measures;
- emergency procedures (and evacuation co-ordinators);
- risks notified to them by other employers in shared workplaces.

*Regulation 10(2)* of the *Management Regulations* (*SI 1999 No 3242*) requires parents or guardians of children (under the minimum school leaving age) to be provided with information on risks the child may be exposed to and the precautions identified as necessary by the risk assessment. This is of relevance in relation to school work experience programmes and part-time work by schoolchildren.

## Co-operation and co-ordination (Regulation 11)

M2014    Where two or more employers share a workplace (whether temporarily or permanently) each employer must:

- co-operate with the other employers where necessary for them to comply with the law;
- take all reasonable steps to co-ordinate their precautions with others;
- take all reasonable steps to inform the others of risks arising out of their work.

(Self-employed persons are also covered by this requirement.)

This type of co-operation and co-ordination was in effect already required under *sections 2* and *3* of *HSWA 1974*. In many cases a main employer may already control the worksite with the other employers assisting. However, in some circumstances it may be necessary to identify an individual or organisation to carry out a co-ordinating role. Even those without employees in the workplace (e.g. landlords) may still be required to co-operate in order to meet their obligations under *section 4* of *HSWA 1974*.

## Persons working in host employers' or self-employed persons' undertakings (Regulation 12)

M2015    Host employers must ensure that employers whose employees are working in their undertaking are provided with comprehensible information on:

- risks arising out of or in connection with the host's activities; and
- measures being taken by the host to comply with the law.

Hosts have an additional duty to ensure that visiting employees are given instructions and information on the risks present and any emergency or evacuation procedures.

(Under *Regulation 12* of the *Management Regulations* (*SI 1999 No 3242*) the self-employed are treated as both employers and employees).

Over and above the requirements of *Regulation 11* of the *Management Regulations* (*SI 1999 No 3242*) and *HSWA 1974*, hosts are given twin duties of informing visiting employers and of *ensuring* that visiting employees are instructed and informed. Whether they do this themselves or require others to do it on their behalf will depend on local circumstances. *Regulation 12* and *Regulation 11* inter-relate closely with the requirements of the *Construction, Design and Management Regulations 1994* (*SI 1994 No 3140*) (*CDM*).

## Capabilities and training (Regulation 13)

M2016    Employers must take into account employees' capabilities as regards health and safety in entrusting tasks to them. Employees' intellectual and physical capabilities must be taken into account as well as the demands of the task and the knowledge and experience required to perform it. Adequate health and safety training must be provided to employees:

- on recruitment into the undertaking;
- on exposure to new or increased risks because of:
  - transfer or a change in responsibilities;
  - new or changed work equipment;
  - new technology;
  - new or changed systems of work.

Training must:

- be repeated periodically where appropriate;
- be adapted to take account of new or changed risks;
- take place during working hours.

The requirements on training develop on the base provided by *section 2* of *HSWA 1974*. Many other sets of regulations also have specific training requirements. Even where employees have been trained in the same way their capabilities may differ and *Regulation 13* of the *Management Regulations* (*SI 1999 No 3242*) requires employers to take ability into account. It also identifies the possible need for refresher training.

## Employees' duties (Regulation 14)

M2017  Employees must use machinery, equipment, dangerous substances etc in accordance with the training and instructions they have been provided with. They must also inform their employer (or his representative) of situations representing serious and immediate danger and of any shortcomings in the employer's protection arrangements. To a large extent, this requirement duplicates those already contained in *section 7* of *HSWA 1974* and other legislation. A failure to report a dangerous situation is likely to be considered a serious 'omission' under *section 7*.

## Temporary workers (Regulation 15)

M2018  *Regulation 15* of the *Management Regulations* (*SI 1999 No 3242*) creates several detailed requirements in relation to the use of fixed-term contract or agency staff. It should always be remembered that temporary workers are still affected by most of the other requirements of these Regulations.

## Risk assessment in respect of new or expectant mothers (Regulations 16, 17 and 18)

M2019  Where there are:

- women workers of childbearing age, and
- the work could involve risk to the mother or baby, e.g. processes, working conditions, physical, biological or chemical agents (see Annexes I and II of EC Directive 92/85/EEC),

the risk assessments made under *Regulation 3(1)* of the *Management Regulations* (*SI 1999 No 3242*) must take account of such risks.

For individual employees the employer must, *if necessary*:

- alter working conditions or hours of work, or *if necessary*,
- suspend the employee from work,

in order to avoid such risks.

*Regulation 17* of the *Management Regulations* (*SI 1999 No 3242*) allows doctors (or midwives) to require employers to suspend women from night work where necessary for health and safety reasons. *Regulation 18* of the 1999 Regulations states that employers need not take action under *Regulations 16* and *17* in respect of individual employees unless the employee has notified them in writing of her pregnancy or that she has given birth within the previous 6 months or is breastfeeding. (The employer may request confirmation of pregnancy through a certificate from a doctor or midwife.) Issues relating to new and expectant mothers are dealt with in more detail in the chapter entitled VULNERABLE PERSONS.

## Protection of young persons (Regulation 19)

**M2020** Employers must ensure young persons are protected from risks which are a consequence of their:

- lack of experience;
- absence of awareness of risks;
- lack of maturity,

e.g. by appropriate training, supervision, restrictions (see also *Regulation 3* of the *Management Regulations* (*SI 1999 No 3242*)).

Young persons may not be employed for work:

- beyond their physical or psychological capacity;
- involving harmful exposure to toxic, carcinogenic, genetically damaging agents etc;
- involving harmful exposure to radiation;
- involving risks they cannot recognise or avoid (due to insufficient attention, lack of experience or training);
- in which there are health risks from extreme cold or heat, noise or vibration.

Such work may be carried out if it is necessary for the training of young persons who are not children, providing they will be supervised by a competent person and risks are reduced to the lowest level reasonably practicable.

This topic is dealt with more fully in the chapter VULNERABLE PERSONS.

## Other requirements

**M2021** The remaining requirements under the *Management Regulations 1999* (*SI 1999 No 3242*) cover the following:

- *Regulation 20 – Exemption certificates*

   (For the armed forces in the interests of national security)

- *Regulation 21 – Provisions as to liability*

   (Of employers in relation to criminal proceedings)

- *Regulation 22 – Restriction of civil liability for breach of statutory duty*

   (A breach of a duty imposed on an employer by these Regulations does not confer a right of action in civil proceedings insofar as that duty applies for the protection of persons not in his employment. The previous exclusion of right of action under the Regulations by employees was removed in 2003.)

- *Regulation 23 – Extension outside Great Britain*

   (The Regulations apply to various offshore activities)

- *Regulations 23–27*

   (Contain minor amendments in respect of first aid, offshore installations and pipeline works, mines and construction)

- *Regulations 28–30*

   (Deal with administrative matters, revocations and transitional arrangements)

## Accident causation

**M2022** The domino model of accident or loss causation has been used by many to demonstrate how accidents result from failures in health and safety management. The model is based on five dominos standing on their edges in a line:

- The *fifth and final domino* in the line represents all the *potential losses* which can result from an accident or incident – personal injury, damage to equipment, materials, premises, disruption of business etc;
- The *fourth domino* in the line represents the specific *accident or incident* which occurred – for example a failure of an item of lifting equipment;
- This failure could have a number of *immediate causes* which are represented by the *third domino* – overloading or misuse of the lifting equipment, an undetected defect etc;
- Behind these immediate causes are a variety of *basic or underlying causes*, represented by the *second domino* – an absence of training, a lack of supervision, failure to conduct regular inspections and thorough examinations of the lifting equipment;
- These in turn are indicative of a *lack of management control* which is represented by the *first domino* – in a well-managed workplace appropriate training is provided, there is effective supervision, and inspection and thorough examination procedures are in place.

If there is effective management control then all the dominos stay in position. However, if management control breaks down then the other dominos will topple eventually causing an accident or incident and the resultant losses.

## Health and safety management systems

**M2023** The vast majority of occupational safety and health ('OSH') management systems (and environmental management systems) follow the guidance given in HSE's publication 'Successful health and safety management', (HSG65) and embellished in BS 8800:1996, Guide to Occupational Health and Safety Management Systems, which outlines a system based on:

- establishing a policy with targets and goals;
- organising to implement it,
- setting forth practical plans to achieve the targets/goals;
- measuring performance against targets/goals; and
- reviewing performance.

The whole process is overlaid by auditing. This outline may be remembered via the use of the mnemonic: POPIMAR:

**Policy**

**Organising**

**Planning**

**Implementing**

**Measuring/monitoring**

**Auditing**

**Reviewing**

These key elements are linked in the form of a continual improvement loop (see FIGURE 3).

**Figure 3: Continual improvement loop**

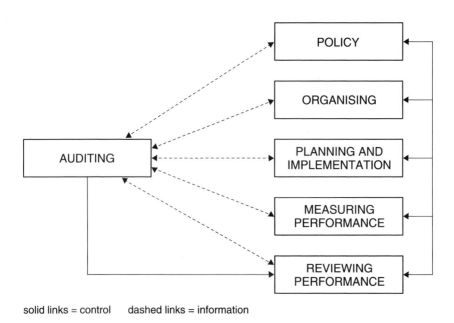

solid links = control    dashed links = information

The more times organisations go round the loop, the better will be their OSH management performance.

Within BS 8800, two alternative approaches are suggested:

- the HSG65 approach;
- the ISO 14001 approach (BSI BS EN ISO 14001: 1996, Environmental Management Systems – Specification with Guidance for Use).

BS 8800 itself is merely a guide; it is *not* a certifiable standard; whereas ISO 14001 *is* a certifiable standard. To date, there is no equivalent in health and safety to the ISO 9000 standard for quality systems or the ISO 14001 standard for environmental management systems. OHSAS 18001 is not a recognised certifiable ISO standard.

BS 8800 builds on the HSG65 approach (outlined above) but starts off the loop with an initial status review in order to establish 'Where are we now?' BS 8800 also includes an alternative approach aimed at those organisations wishing to base their OSH management system not on HSG65 but on BS EN ISO 14001, the environmental management systems standards which again incorporates an initial status review.

## *A review checklist*

**M2024**  In order to undertake an initial status review, use may be made of the following checklist:

- Does a current OSH policy exist for the organisation/location?
- Is the policy up to date (i.e. not more the three years old)?

- Has the policy been signed (and dated) by a director/senior manager who has (site) responsibility for OSH?
- Does the policy recognise that OSH is an integral and critical part of the business performance?
- Does the policy commit the organisation/location to achieve a high level of OSH performance with legal compliance seen as a minimum standard?
- Does this commitment include the concept of continual improvement?
- Does the policy state that adequate and appropriate resources, i.e. time, money, people, will be provided to ensure effective policy implementation?
- Does the policy allow for the setting and publishing of OSH objectives for the organisation/location and/or for individual directors, managers and supervisors?
- Does the policy clearly place the prime responsibility for the management of OSH on to line management, from the most senior executive to first line supervision?
- Has the policy been effectively brought to the attention of *all* employees/agency staff/temporary employees?
- Are copies of the policy on display throughout the organisation/location?
- Is the policy understood, implemented and maintained at all levels within the organisation/location via the use of suitable arrangements?
- Does the policy ensure that employee involvement and consultation takes place in order to gain their commitment to it and its implementation?
- Does the policy require that it gets periodically reviewed (at least every three years) in order to ensure that management and compliance audit systems are in place?
- Does the policy require that all employees at all levels, including agency staff and temporary employees, receive appropriate training to ensure that they are competent to carry out their duties?
- Does the policy contain a section dealing with the organisational framework – people and their duties – so as to facilitate effective implementation?
- Does the policy contain a section dealing with the need for risk assessments to be undertaken – both general and specific – their significant findings acted upon, and an assessment record-keeping system maintained?
- Does the policy contain a section dealing with the arrangements – systems and procedures – by which the policy will be implemented on a day-to-day basis?
- Does the policy contain a section dealing with the monitoring/measurement of OSH performance?
- Does the policy contain a section dealing with planning for and reviewing the organisation's policy implementation and overall OSH performance?

(N.B. Again this may/should be in the arrangements section.)

# Policy

## Requirements under HSWA 1974

**M2025**  *Section 2(3)* of *HSWA 1974* states:

'Except in such cases as may be prescribed, it shall be the duty of every employer to prepare and as often as may be appropriate revise a written statement of his general policy with respect to the health and safety at work of his employees and the organisation and arrangements for the time being in force for carrying out that policy, and to bring the statement and any revision of it to the notice of his employees.'

The *Employers' Health and Safety Policy Statements (Exception) Regulations 1975 (SI 1975 No 1584)* except from these provisions any employer who carried on an undertaking in which for the time being he employs less than five employees. In this respect regard is to be given only to employees present on the premises at the same time (*Osborne v Bill Taylor of Huyton Ltd [1982] IRLR 17*).

## Contents of policy statements

M2026    The policy is the first stage in the POPIMAR management system and the statement must reflect that. Reference to its contents was made in the checklist provided in M2024 above. Some contents will be common to all organisations, whatever their size or work activity. Typical of these are the following:

- A commitment to achieving high standards of health and safety (not only in respect of employees but also others who may be affected by the organisation's activities).
- Reference to the importance of people to the organisation and the organisation's moral obligations.
- A statement that compliance with legal obligations must be achieved (although these should be regarded as a minimum standard).
- Effective management of health and safety must be a key management objective (it must continue right through the line of management responsibility).
- Reference to the provision of resources necessary to implement the policy.
- A commitment to consulting employees on health and safety matters.

Where the organisation is fully committed to the type of management system represented by POPIMAR this should be reflected in the policy statement by:

- A reference to achieving continuous improvement in health and safety standards.
- Reference to the setting of health and safety objectives and the allocation of responsibility for their implementation (whether on an annual basis or otherwise).

The statement must be signed by an appropriate director or senior manager to reflect a high level commitment to the policy and must be dated. Review of the policy as a whole, where appropriate, is a legal requirement, as well as a key component of the POPIMAR system. An out of date policy statement signed by someone who has long since left the organisation is not indicative of a real commitment to effective health and safety management.

# Organisation

M2027    The HSE booklet 'Successful health and safety management' (HSG65) refers to the need to establish a 'health and safety culture' which it defines as 'the product of individual and group values, attitudes, perceptions, competencies and patterns of behaviour that determine the commitment to, and the style and proficiency of, an organisation's health and safety management.' Put in other words it is 'the way we do things here'. In order to achieve such a culture, activities are necessary on four fronts (often called the four Cs):

- establishing and maintaining *control* in respect of health and safety matters;
- securing *co-operation* between individuals and groups;
- achieving effective *communication*;

- ensuring the *competence* of individuals to make their contribution;

In effect the four Cs encompass both the organisation and arrangements which are required to implement the health and safety policy statement.

## Control

M2028  Management must take responsibility and provide clear direction in respect of all matters relating to health and safety. Key elements of control include:

- *Clear definitions of responsibilities*

    Responsibilities must be allocated to all levels of line management as well as persons with specialist health and safety roles. These responsibilities should be set out in the health and safety policy (see the chapter STATEMENTS OF HEALTH AND SAFETY POLICY) and also included in individuals' job descriptions.

- *Nomination of a senior person to control and monitor policy implementation*

    This should ensure that health and safety is seen to be important to the organisation. The senior figure must be involved in the development of plans and particularly in monitoring their implementation. They may also be involved in formal means of consultation with the workforce such as health and safety committees.

- *Establishment of performance standards*

    The old adage of 'what gets measured gets done' applies just as much to health and safety as to other aspects of management. Performance standards must be incorporated into health and safety plans and also into health and safety related procedures and arrangements. Standards must specify:

    — what needs to be done;

    — who is responsible for carrying it out;

    — when (or how often) the task must be carried out;

    — what is the expected result,

    For example, an investigation must be made into all accidents requiring first aid treatment. This must be carried out by the injured employee's supervisor and a report submitted on the company report form within 24 hours.

- *Adequate levels of supervision*

    The precise level of supervision will depend on the risks involved in the work and the competence of those carrying it out. Supervisors are key players in informing, instructing and, to an extent, training the members of their teams. They also act as an important point of reference on health and safety matters, particularly on whether it is safe to continue with a task.

## Co-operation

M2029  Health and safety must become everyone's business – there must be real commitment to health and safety throughout the organisation. Commitment comes where there is a genuine consultation and involvement. This can partially be achieved through formal means of consultation, such as safety representatives and health and safety committees. Hazard report books and suggestion schemes can also play their part. However, much can also be achieved informally by line managers and health and safety specialists by simply 'walking the job' and talking to those at work about the health and safety aspects of their job.

## Communication

**M2030** Good communication is a key element of all aspects of running any organisation. Employees need to be made aware of what is being done to improve health and safety standards (and why), and also of their own role in achieving this improvement. Information needs to move in various directions.

- *Incoming information*

    The organisation must keep abreast of developments in the outside world, e.g:

    — potential and actual changes in legal requirements;

    — technical developments in respect of control measures;

    — new risks coming to light;

    — good practices in health and safety management.

- *Information circulation within the organisation*

    Information will need to be passed down from management levels but there must also be means of passing information up the line and across the organisation. Important information to circulate includes:

    — the health and safety policy document itself;

    — senior management's commitment to implementing it;

    — current plans for achieving improvement;

    — progress reports on previous plans;

    — other performance reports, e.g. accident frequency rates;

    — proposed changes to health and safety procedures and arrangements;

    — employees' views on such proposals;

    — comments and ideas for improvement;

    — details of accidents and incidents, together with lessons learned as a result.

    Information can be communicated in written form (e.g. policy documents, booklets, newsletters, notices, e-mails etc.) but face to face communication is also important. This can include health and safety committee meetings, health and safety discussions or briefings at other meetings, tool box talks. Once again, 'walking the job' by management can be invaluable in achieving two way communication in an informal setting.

- *Outgoing information*

    Health and safety related information may also need to be communicated outside the organisation, for example:

    — formal reports to enforcing authorities (e.g. as required by RIDDOR);

    — information about products supplied or manufactured;

    — information to other organisations involved in the same business (about potential risks or successful control measures);

    — emergency planning information (to the emergency services, local authorities and neighbours who may be involved).

## Competence

M2031   All employees need to be capable of carrying out their jobs safely and effectively. What is required to achieve this will vary depending upon the individual's role in the organisation. Important elements include:

- *Recruitment and appointment arrangements*

    Employees must have appropriate levels of intellectual and physical abilities to carry out their jobs. Whilst knowledge and experience can be acquired subsequently, certain levels may be a pre-requisite for some posts.

- *Training*

    Systems should be in place to ensure that staff receive adequate training for their work, particularly following their recruitment, transfer or promotion. Arrangements will also be necessary to train those given particular responsibilities, e.g. first aiders, fire wardens. Other training needs may also be identified as a result of audits, job appraisals, changes in technology, legislative changes etc. Refresher training may also be necessary for some staff.

- *Succession and contingency planning*

    Account must be taken of the need to have sufficient competent staff available at all times in some safety-critical roles. Staff may be absent for short periods, e.g. holidays or sickness, for longer periods due to serious injury or illness, or leave permanently, whether for other jobs or on retirement.

The chapter TRAINING AND COMPETENCE IN OCCUPATIONAL SAFETY AND HEALTH provides more information on this subject.

## Planning

M2032

'Failure to plan is planning to fail'.

Effective planning results in an occupational safety and health management system ('OSHMS') which controls risks, reacts to changing demands and sustains a positive OSH culture. Such planning involves designing, developing and installing risk control systems ('RCSs') and workplace precautions ('WPs') commensurate to the risk profile of the organisation. It should also involve the operation, maintenance and improvements to the OSHMS to suit changing needs, hazards and risks.

Planning is the first stage of the PDCA cycle: Plan, Do, Check, Act, which is the cornerstone of the vast majority of management systems.

| | |
|---|---|
| Plan: | Policies, organisation, hazard identification, risk assessments, change management |
| Do (Implement): | The implementation of the management plans and processes |
| | Worker consultation is a crucial aspect of planning and doing which helps to create a positive culture |
| Check (Monitor): | Inspections, audits, exposure monitoring, health surveillance, attendance/absence monitoring, accident reporting and investigation, statistical/causal analyses, testing of emergency arrangements/procedures |
| Act (Audit/Review): | Take control action based on monitoring activities, keep track of developments/corrective measures, continually improve processes and plans |

## The planning process

**M2033** The planning process involves:

- Setting objectives.
- Risk control systems.
- Legal and other requirements.
- OSH management arrangements.
- Emergency plans and procedures.

### Setting objectives

This relies on having an annual plan for the organisation with clearly outlined individual and collective 'SMART' OSH targets. SMART refers to the need for the targets to be: specific, measurable, achievable, realistic, trackable.

The annual OSH plan should:

- outline individual and collective action plans for the current year;
- target continual improvement in OSH performance;
- fix accountabilities via SMART targets;
- be linked to job descriptions/performance appraisals;
- be adequately resourced – time, money, people.

### Risk control systems

These should ideally evolve from an ongoing review of all general and specific risk assessments in place within the organisation. Indeed, as part of the annual plan, all current risk assessments should be reviewed, updated, and recommunicated to all concerned.

Risk should be reduced to tolerable or acceptable levels via an ongoing process of hazard identification and risk assessment within the working environment. Suitable RCSs should be developed and implemented with the goal of hazard elimination and risk minimisation. As part of the process all significant findings/risk assessments should be recorded in writing.

### Legal and other requirements

These need to be incorporated into the annual plan in order to establish and maintain arrangements which ensure that all current and emerging OSH legal and other requirements are complied with and understood throughout the organisation.

Bench-marking within the organisation's sector can ensure that best practice and performance measures are adopted and communicated.

### OSH management arrangements

These should reflect both the needs of the business and the overall risk profile and should cover all elements of the POPIMAR model. Specific arrangements should include:

- plans, objectives, personnel, resources;
- operational plans to control risks;
- contingency plans for emergencies;

- planning for organisational arrangements;
- plans covering change management (positive OSH culture);
- plans for interaction with third parties (e.g. contractors);
- planning for performance measurement, audits and reviews;
- implementing corrective actions.

## Emergency plans and procedures

These should be in place and reviewed as part of the annual planning processes. Specific serious and imminent dangers should have been flagged up via previous risk assessments in accordance with the *Management Regulations (SI 1999 No 3242), Reg 8*.

These emergency plans should ensure that controls are in place, are in writing and have been communicated to all concerned in connection with all foreseeable disasters, including fire, flood, structural collapse/subsidence, chemical spillage, transport accident, bomb scare, explosion, plane crash etc.

# Implementation

M2034    In order to ensure smooth implementation of the OSH policy and management system, it is imperative to have sufficient written records and documentation as this is the key element in organisational communication and continual improvement.

It is therefore important to assemble and retain OSH knowledge and competence within the organisation. However, in order to ensure efficiency and effectiveness of the OSHMS, documentation should be kept to the minimum and should be tailored to suit organisational needs. The detail should be in proportion to the level and complexity of the associated hazards and risks.

Specifically, the key risk areas requiring implementation of commensurate risk control systems may be grouped as follows:

- selection and training;
- general/specific risk assessments;
- safe systems of work/permit to work systems;
- management of third parties on site;
- workplace/work equipment safety and health;
- fire precautions;
- first aid requirements;
- use of personal protective equipment.

These are generally referred to as 'arrangements' in most organisational OSH policies.

As outlined in the section on Planning above at M2032, this 'doing' section requires the use of action plans and SMART targets in order to ensure that the written down standards and procedures are actually achieved in practice on a continual basis. The phrase 'Mind the gap(s)' springs to mind!

Most organisations split their arrangements into general and specific. General arrangements require implementation throughout the organisation whereas specific will only operate at certain locations or for certain tasks.

Within the implementation section of the OSHMS, the risk control systems relevant to the organisation should be briefly described, together with who is responsible for their implementation and who should be involved in the implementation process.

In some organisations the detail may well be consigned to an OSH manual which is referred to but is not part of the OSH policy.

## Monitoring

### 'What gets measured gets done!

**M2035** Monitoring of the OSHMS is vital to ensure continual improvement. Monitoring procedures fall into two distinct categories:

- proactive;
- reactive.

*Proactive monitoring* takes place before the event (accident/disease) and includes audits, inspections and specific checks.

*Reactive monitoring* is always after the event and includes accident reporting and investigation, numerical/causal analysis of accidents/diseases, compilation of accident statistics, and adherence to the RIDDOR '95 requirements on accident notification.

- *OSH inspections – proactive monitoring*

    This should involve a competent inspection team going around the workplace on a regular (ideally monthly) basis in order to identify hazards – unsafe acts as well as unsafe conditions – with a view to their rectification.

    Such inspections should be joint – both management and workforce representatives – and should always involve talking to the people at risk in the workplace being inspected. This may help to flag up unsafe practices, near misses and shortcomings in the agreed systems and procedures. In most cases use will be made of a specific inspection checklist which ideally has been compiled by the people undertaking the inspection.

    In all cases there needs to be a post-inspection action plan which clearly indicates who is going to take control action and by when. There should also be full communication of the findings and recommendations from all such inspections.

- *Specific checks – proactive monitoring*

    Such checks in the workplace may involve: housekeeping inspections; checks on adherence to agreed safe systems of work, permit to work systems, site safety rules; utilisation of personal protective equipment; documentary checks – risk assessments, statutory inspection records, maintenance records and training records.

- *Accident reporting and investigation systems – reactive monitoring*

    This should encompass all injury, damage and near-miss accidents and occupational ill-health in order to ensure RIDDOR'95 compliance and an effective accident database with which to plan further accident reduction strategies.

    All reported accidents, including occupational diseases, should be promptly investigated in order to identify causes and to implement commensurate control measures so as to prevent a recurrence. Most organisations will have set up an internal accident reporting and investigation system to capture these accident data. Any investigation culture should be positive, i.e. *not* blame-apportioning or fault-finding.

    The main purposes of accident investigation are therefore to:

    — discover the immediate and underlying causes (note: plural!);

- prevent recurrences;
- minimise legal liability;
- collect sufficient data on which to base future plans;
- identify accident/illness/absence trends;
- ensure RIDDOR'95 compliance;
- maintain records for organisational purposes;
- continually improve the OSHMS performance.

## Audit

**M2036**  Formal OSH audits should be undertaken at regular intervals, ideally annually, by trained and competent auditors, sometimes external to the organisation.

All audit findings (successes as well as shortcomings) should be highlighted and communicated to all concerned. Any recommendations for improvement should be prioritised and allocated to named individuals for action within agreed, finite time-scales. The progress on action completion should also be followed up and communicated.

The main targets of any audit are to reduce risks, with avoidance being the best strategy, and to ensure continual improvement of the OSHMS.

The audit needs to compare actual OSH performance in the workplace against agreed standards – mind the gaps! – such as:

- legislative compliance;
- best practice (bench-marking);
- national/international standards.

Essentially the OSH audit, as with financial audits, is a deep and critical appraisal of all elements (POPIMAR) of the OSHMS and should be seen as being additional to the routine monitoring and reviews. Ideally, the auditors need to be independent of the area/activity/location being audited and the audit checklist should be tailored to suit the needs/hazards/risks of the organisation.

The audit scope should address key areas such as:

- is the overall OSHMS capable of achieving and maintaining the required, agreed standards?
- is the organisation fulfilling all OSH obligations?
- what are the strengths and weaknesses of the OSHMS?
- is the organisation actually doing and achieving what it claims?
- are the audit results acted upon, communicated and followed up?
- are the audit reports used in management reviews?

## Review

**M2037**  Reviews of the performance of the OSHMS generally fall into two categories:

- periodic status review ('PSR');
- management review.

The main purposes of the PSR are to make judgements on the adequacy of performance and to ensure that the right decisions are made and taken in connection with the nature and timings of the remedial actions.

Specifically, the PSR should consider:

- overall OSHMS performance;
- performance of individual elements;
- audit findings;
- internal and external factors such as changes in organisational structure, production cycles, pending legislation, introduction of new technology;
- anticipated future changes;
- information gathered from proactive and reactive monitoring.

The PSR should be a continual process which includes management and supervisory responses to system shortcomings such as non-implementation of agreed risk control systems, substandard performances, or failure to assess the impact of forward plans and objectives on their part of the organisation. Successes should be highlighted.

The effectiveness of the PSR can be enhanced by clearly establishing responsibilities for implementing the remedial actions identified and by setting deadlines for action completion – by whom? by when? The PSR needs to be auditable, trackable, fully communicated and followed up.

The management review should be undertaken annually by the board/senior management. It should ensure that the OSHMS is suitable, adequate, effective and efficient, and that it satisfies and achieves the aims and objectives as stated in the OSH policy.

The review should be based on the audit results and on the periodic status review reports. Specifically it should:

- establish new OSH objectives for continual improvement;
- develop OSH annual plans;
- consider changes to OSHMS arrangements;
- review findings which should then be documented, communicated to stakeholders via annual reports and followed up to ensure implementation and continual improvement.

# Continual improvement

M2038   Continual improvement is at the heart of the OSHMS. It essentially is a fundamental commitment to better manage OSH risks proactively so that accidents and ill-health are reduced (system effectiveness) and/or the system achieves its desired aims using less resource (system efficiency).

The International Labour Office Guidelines on Occupational Safety and Health Management Systems (ILO, 2001) includes the following model which clearly illustrates the importance of the goal of continual improvement.

The route to continual improvement has been highlighted throughout the course of this chapter. It involves the use of SMART targets, the setting of OSH key performance indicators ('KPIs'), the use of performance appraisals which include individual and collective OSH performance reviews, and accountability fixing for OSH responsibilities at all levels within the organisation.

The goal of continual improvement should manifest itself in a number of ways:

- better year on year results – less injuries, diseases, damage, near misses;
- steady, improved results using less resources;

**Figure 4**

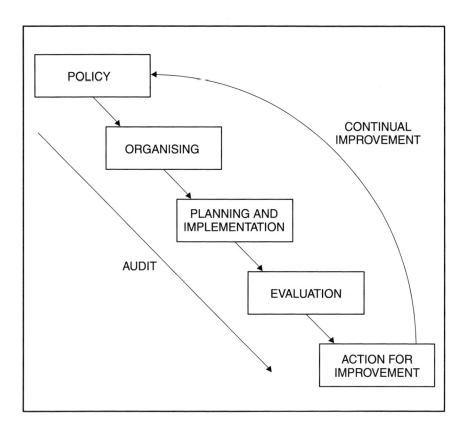

- better targeted and more efficient and effective OSHMS;
- indications of breakthrough performances;
- more positive, proactive OSH culture;
- OSHMS improvements: more comprehensive, easier to understand, better all round, i.e. 'we improve as we get better!'

In order to achieve continual improvement, the following aspects should be taken into account:

- regular OSH audits;
- statistical improvements;
- bench-marking;
- industry/sector guidelines;
- ownership of work processes:
  — managers, supervisors;
  — workers representatives;
- proactive monitoring and targets;

- effective and efficient operation via workforce involvement;
- capture and evaluate ideas for improvement, ensure regular feedback to originators;
- resourced, implemented and monitored action plans;
- SMART improvement action plans;
- OSHMS training;
- OSHMS documentation/reference manual.

# POPIMAR in action

**M2039** The example below shows how the POPIMAR principles can be followed in practice. It relates to the introduction into an organisation of a permit to work ('PTW') system to control certain high risk activities.

## Policy

The PTW system will become part of the 'arrangements' for implementing the health and safety policy statement. The PTW procedure (see below) is likely to contain a general statement about the importance of ensuring that appropriate controls are in place to minimise the risks associated with potentially high risk activities.

## Organisation

This will primarily centre around the development of a formal PTW procedure which will involve:

- *Control*

    The procedure must identify:
    - what activities will be controlled by the PTW;
    - who will issue (and cancel) PTWs;
    - the timing sequence for issuing and cancelling individual PTWs;
    - the format of the PTW form.

- *Co-operation*

    The draft procedure must be reviewed with those likely to be involved in issuing and working under PTWs. This could involve an ad hoc consultative group in addition to safety representatives and the health and safety committee.

- *Communication*

    A wider section of the workforce must be informed about the proposed procedure and the reasons for its introduction.

- *Competence*

    Those who are to issue the PTWs must receive formal training in the procedure and use of the PTW form. There should also be a formal appointment process to ensure that they have successfully assimilated the training and can relate it to their practical work situations. Arrangements for this should be referred to in the procedure.

## Planning and implementation

Once the procedure has been finalised the timescales and responsibilities must be identified for achieving the various component parts of its implementation. This will include:

- delivery of training to PTW issuers;
- formal appointment of PTW issuers (after an appropriate test process);
- printing and distributing the PTW forms;
- obtaining any necessary equipment (e.g. isolation locks, testing equipment);
- announcement of the formal introduction of the PTW procedure.

## Measuring performance

Monitoring of the PTW can be carried out in various ways:

- a survey soon after the procedure's introduction;
- ongoing monitoring during formal health and safety inspections;
- periodic mini-audits of the procedure (e.g. by health and safety specialists or managers).

## Auditing

The operation of the PTW procedure would be included in any general OSH auditing activities.

## Reviewing performance

Both monitoring and audit findings must be reviewed by appropriate individuals and groups such as:

- health and safety specialists;
- the health and safety committee;
- senior management;
- ad hoc review teams.

Where satisfactory standards are not being achieved, the reviewers must consider how the POPIMAR sequence can best be applied to taking remedial action.

# Manual Handling

## Introduction

**M3001** The *Manual Handling Operations Regulations 1992* (*SI 1992 No 2973* as amended by *SI 2002 No 2174*) came into operation on 1 January 1993 as part of the so-called 'six pack' of new Regulations. Although the Regulations implemented European Directive 90/269/EEC, the HSE had previously attempted to introduce Regulations on manual handling in an attempt to reduce the huge toll of accidents from this source. At the time of the introduction of the Regulations more than a quarter of the accidents reported each year to the enforcing authorities (the HSE and local authorities) were associated with manual handling. In 1990/91 65 per cent of handling accidents resulted in sprains and strains, whilst 5 per cent were fractures. Many manual handling injuries have cumulative effects eventually resulting in physical impairment or even permanent disability.

There is already evidence of the beneficial effects of the Regulations, particularly in respect of the increasing use of mechanical handling aids and improved manual handling training. By 1996/97 the proportion of handling accidents causing fractures had been reduced to 3.3 per cent. However, the battle was far from over with 54 per cent of RIDDOR reportable accidents in the health and social work sector still attributable to handling. Several other sectors had over 30 per cent of accidents involving handling, including food and drink, textiles, construction, wholesale and retail, transport/storage and education.

## What the Regulations require

**M3002** Various terms used in the *Manual Handling Operations Regulations 1992* (*SI 1992 No 2973*) are defined in *Regulation 2* which states:

> 'manual handling operations' means any transporting or supporting of a load (including the lifting, putting down, pushing, pulling, carrying or moving thereof) by hand or by bodily force.
>
> 'Injury' does not include injury caused by any toxic or corrosive substance which:
>
> (a)  has leaked or spilled from a load;
>
> (b)  is present on the surface of a load but has not leaked or spilled from it; or
>
> (c)  is a constituent part of a load;
>
> and 'injured' shall be construed accordingly;'

(This in effect establishes a demarcation with the *COSHH Regulations* (*SI 2002 No 2677*) – where there is oil on the surface of a load making it difficult to handle that is a matter for the *Manual Handling Operations Regulations*, whereas any risk of dermatitis is covered by the *COSHH Regulations*).

> 'Load includes any person and any animal'.

(This is of great importance to a number of work sectors including health and social care, agriculture and veterinary work).

HSE guidance accompanying the Regulations states that an implement, tool or machinery being used for its intended purpose is not considered to constitute a load – presumably establishing a demarcation with the requirements of PUWER 98 (*SI 1998 No 2306*). However, when such equipment is being moved before or after use it will undoubtedly be a load for the purpose of these Regulations.

*Regulation 2(2)* imposes the duties of an employer to his employees under the Regulations on a self-employed person in respect of himself. *Regulation 3* excludes the normal ship-board activities of a ship's crew under the direction of the master from the application of the Regulations (these are subject to separate merchant shipping legislation).

The duties of employers under the Regulations are all contained in *Regulation 4* which (as in the *COSHH Regulations*) establishes a hierarchy of measures that employers must take. *Regulation 4* states:

'(1) Each employer shall –

 (a) so far as is reasonably practicable, avoid the need for his employees to undertake any manual handling operations at work which involve a risk of their being injured; or

 (b) where it is not reasonably practicable to avoid the need for his employees to undertake any manual handling operations at work which involve a risk of their being injured–

  (i) make a suitable and sufficient assessment of all such manual handling operations to be undertaken by them, having regard to the factors which are specified in column 1 of Schedule 1 to these Regulations and considering the questions which are specified in the corresponding entry in column 2 of that Schedule,

  (ii) take appropriate steps to reduce the risk of injury to those employees arising out of their undertaking any such manual handling operations to the lowest level reasonably practicable, and

  (iii) take appropriate steps to provide any of those employees who are undertaking any such manual handling operations with general indications and, where it is reasonably practicable to do so, precise information on –

   (aa) the weight of each load, and

   (bb) the heaviest side of any load whose centre of gravity is not positioned centrally.

(2) Any assessment such as is referred to in paragraph (1)(b)(i) of this Regulation shall be reviewed by the employer who made it if –

 (a) there is reason to suspect that it is no longer valid; or

 (b) there has been a significant change in the manual handling operations to which it relates,

and where as a result of any such review changes to an assessment are required, the relevant employer shall make them.'

As a result of the *Health and Safety (Miscellaneous Amendments) Regulations 2002* (*SI 2002 No 2174*), a further paragraph was added to *Regulation 4*. This states:

'(3) In determining for the purpose of this regulation whether manual handling operations at work involve a risk of injury and in determining the appropriate steps to reduce that risk regard shall be had in particular to—

(a) the physical suitability of the employee to carry out the operations;

(b) the clothing, footwear or other personal effects he is wearing;

(c) his knowledge and training;

(d) the results of any relevant risk assessment carried out pursuant to regulation 3 of the Management of Health and Safety at Work Regulations 1999;

(e) whether the employee is within a group of employees identified by that assessment as being especially at risk; and

(f) the results of any health surveillance provided pursuant to regulation 6 of the Management of Health and Safety Regulations 1999.'

The hierarchy of measures which must be taken by employers where there is a risk of injury can be summarised as follows:

- Avoid the operation (if reasonably practicable);
- Assess the remaining operations;
- Reduce the risk (to the lowest level reasonably practicable);
- Inform employees about weights.

Reference will be made later in the chapter to a variety of measures which can be taken to avoid or reduce risks. The factors to which regard must be given during the assessment (as specified in *Schedule 1* to the Regulations) are the *tasks, loads, working environment* and *individual capability* together with other factors referred to in *Regulation 4(3)*. These factors and the questions specified by *Schedule 1* are also covered later in the chapter.

*Regulation 5* places a duty on employees. It states:

'Each employee while at work shall make full and proper use of any system of work provided for his use by his employer in compliance with Regulation 4(1)(b)(ii) of the Regulations.'

*Regulation 6* provides for the Secretary of State for Defence to make exemptions from the requirements of the Regulations in respect of home forces and visiting forces and their headquarters.

*Regulation 7* extends the Regulations to apply to offshore activities such as oil and gas installations, including diving and other support vessels. Previous provisions relating to manual handling were replaced or revoked by *Regulation 8*. The Regulations themselves are extremely brief but they are accompanied by considerable HSE guidance, much of which will be referred to during this chapter.

# Risk of injury from manual handling

## Assessment guidelines

**M3003** Appendix 1 of the guidance accompanying the *Manual Handling Operations Regulations 1992* (*SI 1992 No 2973*) (HSE booklet L23 *Manual Handling Operations 1992: Guidance on the Regulations*) provides assessment guidelines which can be used to filter out manual handling operations involving little or no risk and help to identify where a detailed risk assessment is necessary. However, it must be stressed that *these are not weight limits* – they must not be regarded as safe weights nor must they be treated as thresholds which must not be exceeded. The HSE guidance states that application of the guidelines will provide a reasonable level of protection to around 95 per cent of working men and women. Some workpeople will require additional protection (e.g. pregnant women, frail or elderly workers) and this should be considered in relation to the assessment of individual capability.

In the latter part of 2003 the HSE published a free leaflet, Manual Handling Assessment Charts (MAC) (INDG 383), to provide a tool to assist in identifying high risk manual handling operations during assessments. It uses numerical scores and a green, amber, red, purple colour coding system to categorise risks as low, medium, high and very high respectively. Risks from lifting, carrying and team handling are covered, although not from pushing and pulling. The leaflet also contains guidance on assessment technique. This should only be used as a preliminary screening tool to identify workplace manual handling activities which pose a significant risk. The risk associated with these activities should then be assessed as outlined in the Manual Handling Operations Regulations 1992 (as amended). The user of the MAC tool should also be aware that the graphs presenting the risks associated with frequently handled loads are based on one form of experimental data which is recognised to overestimate the individual's handling capacity when loads are handled frequently. The graphs should be interpreted with caution.

## Lifting and lowering

The guidelines for lifting and lowering operations as set out in the Manual Handling Operations Regulations 1992 (as amended) are shown in the accompanying diagram (below). They assume that the load can be easily grasped with both hands and that the operation takes place in reasonable working conditions, using a stable body position. Where the hands enter more than one box, the lowest weights apply. A detailed assessment should be made if the weight guidelines are exceeded or if the hands move beyond the box zones.

These guideline weights apply up to approximately 30 operations per hour. They should be reduced for more frequent operations:

- by 30 per cent for 1 or 2 operations per minute;
- by 50 per cent for 5 to 8 operations per minute;
- by 80 per cent for more than about 12 operations per minute.

Other factors likely to require a more detailed assessment are where:

- workers do not control the pace of work, e.g. on assembly lines;
- pauses for rest are inadequate;
- there is no change of activity allowing different muscles to be used;
- the load must be supported for any length of time.

## Carrying

Similar guidelines figures can be applied for carrying where the load is held against the body and carried up to 10 metres without resting. For longer distances or loads held below knuckle height a more detailed assessment should be made. The guideline figures can be applied for carrying loads on the shoulder for distances in excess of 10 metres but an assessment may be required for the lifting of the load onto and off the shoulder.

## Pushing and pulling

Where loads are slid, rolled or supported on wheels the guideline figures assume that force is applied with the hands between knuckle and shoulder height. The guideline figure for starting or stopping the load is a force of about 25kg (about 250 Newtons) for men and 10kg (about 100 Newtons) for keeping the load in motion (these figures reduce to 16kg and 7kg respectively for women).

In practice it is extremely difficult to make a judgement as to whether forces of these magnitude are being applied. Many employers are likely to make assessments of significant pushing or pulling activities anyway. There are no guideline limits on distances over which loads may be pushed or pulled, providing there is sufficient opportunity for rest.

## Handling when seated

The accompanying diagram illustrates the guideline figures for handling whilst seated. Any handling activities outside the specified box zones should be assessed in any case.

Fig. 1: Lifting and lowering

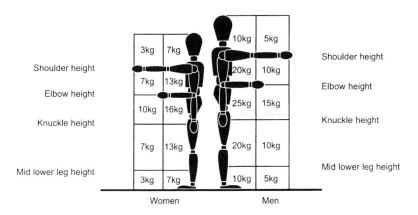

Fig. 2: Handling while seated

*Schedule 1* to the *Manual Handling Operations Regulations 1992* (*SI 1992 No 2973*) contains a list of questions which must be considered during assessments of manual handling operations. All of these indicate factors which may increase the risk of injury. (Most of these factors have been included in the checklist for assessments provided later in the chapter).

## Task related factors

### Holding or manipulating loads at a distance from the trunk

Stress on the lower back increases as the load is moved away from the trunk. Holding a load at arms length imposes five times the stress as the same load very close to the trunk.

### Unsatisfactory body movement or posture

The risk of injury increases with poor feet or hand placings, e.g. feet too close together, body weight forward on the toes, heels off the ground.

*Twisting the trunk* whilst supporting a load increases stress on the lower back – the principle should be to move the feet, not twist the body.

*Stooping* also increases the stress on the lower back whilst *reaching upwards* places more stress on the arms and back and lessens control on the load.

### Excessive movement of loads

Risks increase when further loads must be *lifted or lowered*, especially when lifting from floor level. Movements involving a *change of grip* are particularly risky. Excessive *carrying* of loads increases fatigue – hence the guidance figure of about 10 metres carrying distance before a more detailed assessment is made.

### Excessive pushing or pulling

Here the risks relate to the forces which must be exerted, particularly when starting to move the load, and the quality of the grip of the handler's feet on the floor. Risks also increase when pushing or pulling below knuckle height or above shoulder height.

### Risk of sudden movement

Freeing a box jammed on a shelf or a jammed machine component can impose unpredictable stresses on the body, especially if the handler's posture is unstable. Unexpected movement from a client in the health care sector or an animal in agricultural work can produce similar effects.

### Frequent or prolonged physical effort

Risk of injury increases when the body is allowed to become tired, particularly where the work involves constant repetition or a relatively fixed posture. The periodic inter-changing of tasks within a work group can do much to reduce such risks.

### Clothing, footwear etc

The effects of any PPE or other types of clothing, footwear etc. required for the task may make manual handling more difficult or increase the possibilities of fatigue, dehydration etc.

### Insufficient rest and recovery

The opportunity for rest and recovery will also help to reduce risks, especially in relation to tasks which are physically demanding.

## Rate of work imposed by a process

In such situations workers are often not able to take even short breaks, or to stretch or exercise different muscle groups. More detailed assessments should be made of assembly line or production line activities. Again periodic role changes within a work team may provide a solution.

## Load related factors
### Is the load too heavy?

**M3005**  The guideline figures provided earlier are relevant here but only form part of the picture. Task-related factors (e.g. twisting or repetition) may also come into play or the load may be handled by two or more people. Size, shape or rigidity of the load may also be important.

### Is the load bulky or unwieldy?

Large loads will hinder close approach for lifting, be more difficult to get a good grip of or restrict vision whilst moving. The HSE guidance suggests that any dimension exceeding 75cm will increase risks and the risks will be even greater if it is exceeded in more than one dimension. The positioning of suitable handholds on the load may somewhat reduce the risks.

Other factors to consider are the effects of wind on a large load, the possibility of the load hitting obstructions or loads with offset centres of gravity (not immediately apparent if they are in sealed packages).

### Is the load difficult to grasp?

Where loads are large, rounded, smooth, wet or greasy, inefficient grip positions are likely to be necessary, requiring additional strength of grip. Apart from the possibility of the grip slipping there are likely to be inadvertent changes of grip posture, both of which could result in loss of control of the load.

### Is the load unstable?

The handling of people or animals was referred to earlier. Not only do such loads lack rigidity, but they may also be unpredictable and protection of the load as well as the handler is of course an important consideration. Other potentially unstable loads are those where the load itself or its packaging may disintegrate under its own weight or where the contents may shift suddenly, e.g. a stack of books inside a partially full box.

### Is the load sharp, hot etc?

Loads may have sharp edges or rough surfaces or be extremely hot or cold. Direct injury may be prevented by the use of protective gloves or clothing but the possibility of indirect risks (e.g. where sharp edges encourage an unsuitable grip position or cold objects are held away from the body) must not be overlooked.

## Factors relating to the working environment
### Space constraints

**M3006**  Areas of restricted headroom (often the case in maintenance work) or low work surfaces will force stooping during manual handling and obstructions (e.g. in front of shelves) will result in other unsatisfactory postures. Restricted working areas or the lack of clear gangways will increase risks in moving loads, especially heavier or bulkier items.

## Uneven, slippery or unstable floors

All of the above can increase the risks of slips, trips and falls. The availability of a firm footing is a major factor in good handling technique – such a footing will be a rarity on a muddy construction site. Moving workplaces (e.g. trains, boats, elevating work platforms) will also introduce unpredictability in footing. (The guideline weights should be reduced in such situations.)

## Variation in levels of floors or work surfaces

Risks will increase where loads have to be handled on slopes, steps or ladders, particularly if these are steep. Any slipperiness of their surface, e.g. due to ice, rain, mud will create further risks. The need to maintain a firm handhold on ladders or steep stairways is another factor to consider. Movement of loads between surfaces or shelving may also need to be considered, especially if there is considerable height change, e.g. floor level to above shoulder height.

## Extremes of temperature or humidity

High temperatures or humidity will increase the risk of fatigue, with perspiration possibly affecting the grip. Work at low temperatures (e.g. in cold stores or cold weather) is likely to result in reduced flexibility and dexterity. The need for gloves and bad weather clothing is another factor to consider.

## Ventilation problems or gusts of wind

Inadequate ventilation may increase fatigue whilst strong gusts of wind may cause considerable danger with larger loads, e.g. panels being moved on building sites.

## Poor lighting

Lack of adequate lighting may cause poor posture or an increased risk of tripping. It can also prevent workers identifying risks associated with individual loads, e.g. sharp edges or corners, offset centres of gravity.

### Capability of the individual

**M3007** Manual handling capabilities will vary significantly between individual workers. Reference was made earlier in the chapter to the assessment guideline weights providing a reasonable level of protection to around 95 per cent of workers and those guideline figures differed for men and women. *Regulation 4(3)* of the *Manual Handling Operations Regulations 1992 (SI 1992 No 2973)* (as amended) now specifically refers to factors which must be taken into account in respect of individual workers. Individual factors include:

## Gender

Whilst there is overlap between the capabilities of men and women, generally the lifting strength of women is less than men.

## Age

The bodies of younger workers will not have matured to full strength, whilst with older workers there will be a gradual decline in their strength and stamina, particularly from the mid 40's onwards. Such persons will be the types of groups of employees referred to in *Regulation 4(3)(e)* of the *Manual Handling Operations Regulations 1992 (SI 1992 No 2973)* (as amended).

### Experience

More mature workers may be better able to recognise their own capabilities and pace themselves accordingly, whilst also having acquired better handling techniques.

### Pregnancy

The need to protect new or expectant mothers is emphasised in the chapter RISK ASSESSMENT.

### Previous injury or ill-health

Workers with long term musculoskeletal problems, a history of injuries or ill-health or short term injuries (including those from non-work related causes) will all justify additional protection. Such problems should be detected by the health surveillance required by *Regulation 6* of the *Management of the Health and Safety at Work Regulations 1999 (SI 1999 No 3242)*. (See M3015 later in the chapter.)

### Physique and stature

These factors will vary considerably within any workforce. Whilst bigger will often mean stronger, it must not be overlooked that some tasks or work locations may force taller workers to stoop or adopt other unsuitable handling postures.

Generally the objective should be to ensure that all manual handling operations can be performed satisfactorily by most reasonably fit and healthy employees, although a minority of workers may justify restrictions on their handling activities (see later in the chapter).

During the assessment process it is important that special note is taken of any manual handling operations which:

- Require unusual strength, height etc, e.g. reels of wrapping paper can only be safely positioned in a machine by persons above a certain height.

- Create additional risk to workers who are pregnant, disabled or have health problems.

    Such tasks may need to be reassessed to reduce risks or it may be necessary to restrict some individuals from carrying them out.

- Require special information or training.

    The health care sector provides many examples of situations where both the worker and the client (the load) may be at risk if correct technique is not used. Many other tasks are likely to require specific techniques to be adopted – whether in manual handling itself or the use of handling aids.

### Clothing, footwear etc

Some individuals may be affected more than others by the use of PPE or clothing, e.g. gloves, protective suits, breathing apparatus. These may significantly affect the user's mobility or dexterity in respect of manual handling. Similarly uniforms or costumes (e.g. in the entertainment industry) may also be a factor to be taken into account.

## Avoiding or reducing risks

**M3008**   The *Manual Handling Operations Regulations 1992 (SI 1992 No 2973)* require employers to avoid the need for employees to undertake manual handling operations involving a risk of injury, so far as is reasonably practicable. Measures for avoiding such risks can be categorised as:

- Elimination of handling.
- Automation or mechanisation.
- Reduction of the load.

## Elimination of handling

**M3009**   It may be possible for employers to eliminate manual handling altogether (or to eliminate handling by their own employees) by means such as:

### Redesigning processes or activities

For example, so that activities such as machining or wrapping are carried out in situ, rather than a product being manually handled to a position where the activity take place; a treatment is taken to a patient rather than vice versa (this may have other benefits to the patient).

### Using transport better

For example, allowing maintenance staff to drive vehicles carrying tools or equipment up to where they are working, rather than manually handling them into place.

### Requiring direct deliveries

For example, requiring suppliers to make deliveries directly into a store rather than leaving items in a reception area (such suppliers may be better equipped in terms of handling aids and their staff may be better trained in respect of manual handling).

## Automation or mechanisation

**M3010**   The automation or mechanisation of a work activity is best considered at the design stage although there is no reason in principle why such changes cannot be made later, e.g. as the result of a risk assessment. In making changes it is important to avoid creating additional risks, e.g. the use of fork lift trucks for a task in an already cramped workplace may not be the best solution. However, as employers have become more aware of manual handling risks (many prompted by the advent of the Regulations) a much greater range of mechanical handling solutions to manual handling problems have become available.

These include:

### Mechanical lifting devices

Fork lift trucks, mobile cranes, lorry mounted cranes, vacuum devices and other powered handling equipment are commonly used to eliminate or greatly reduce the manual handling of loads.

### Manually-operated lifting devices

Pallet and stacker trucks, manually operated chain blocks or lever hoists can all be used to move heavy loads using very limited physical force.

### Powered conveyors

As well as fixed conveyors (both belt and driven roller types), mobile conveyors are being used increasingly, particularly for the loading and unloading of vehicles, e.g. stacking of bundles of newspapers in delivery vans.

## Non-powered conveyors, chutes etc

Free-running roller conveyors, chutes or floor-mounted trolleys allow loads to move under the effects of gravity or to be moved manually with little effort. Such devices may operate between different levels or be set into the workplace floor.

## Trolleys and trucks

Trolleys and trucks can be used to greatly reduce the manual handling effort required in transporting loads. Some types also incorporate manual or mechanically-powered lifting mechanisms. Specialist trolleys are designed for carrying drums or other containers whilst an ingenious triple wheel system can be lifted to some trucks to aid them in climbing or descending stairs.

## Lifting tools

Special lifting tools are available to reduce the manual handling effort required for lifting or lowering certain loads, e.g. paving slabs, manhole covers, drums or logs.

All of these and many other types of devices are illustrated in HSE booklet HSG 115 *Manual handling: Solutions you can handle*.

## Reduction of the load

**M3011**   There is considerable scope to avoid the risk of injury by reducing the sizes of the loads which have to be handled. In some cases the initiative for this has come from suppliers of a particular product whereas in others the impetus has come from customers or from users of equipment, often as a result of their manual handling assessments. Examples include:

- Packaged building materials (e.g. cement) reduced to 25kg.

- Photocopying paper now supplied in 5 ream boxes (approximately 12kg) – 10 ream boxes were previously commonplace.

- Newspaper bundle sizes reduced below 20kg (printers reduce the number of copies in a bundle as the number of pages in the paper increases).

- Customers specifying maximum container weights they will accept from suppliers, e.g. weights of brochures from commercial printers.

- Equipment being separated into component parts and assembled where it is to be used, e.g. emergency equipment used by fire services.

Where it is not reasonably practicable to avoid a risk of injury from an manual handling operation, the employer must carry out an assessment of such operations with the aim of reducing the risks to the lowest level reasonably practicable . The four main factors to be taken into account during the assessment are *the task, the load, the working environment* and *individual capability*. Possible risk reduction measures related to each of these factors are described below.

(Later in the chapter these are summarised in an assessment checklist – see M3020 below).

## Task-related measures

### Mechanical assistance

**M3012**   The types of mechanical assistance described earlier can be used to reduce the risk of injury, even if such risks cannot be avoided entirely. The use of roller conveyors, trolleys, trucks or levers can reduce considerably the amount of force required in manual handling tasks.

## Task layout and design

The layout of tasks, storage areas etc should be designed so that the body can be used in its more efficient modes, e.g:

- heavier items stored and moved between shoulder and mid-lower leg height;
- allowing loads to be held close to the body;
- avoiding reaching movements, e.g. over obstacles or into deep bins;
- avoiding twisting movements or handling in stooped positions, e.g. by repositioning work surfaces, storage areas or machinery;
- providing resting places to aid grip changes whilst handling;
- reducing lifting and carrying of loads, e.g. by pushing, pulling, sliding or rolling techniques;
- utilising the powerful leg muscles rather than arms or shoulders;
- avoiding the need for sustaining fixed postures, e.g. in holding or supporting a load;
- avoiding the need for handling in seated positions.

## Work routines

Work routines may need to be altered to reduce the risk of injury using such measures as:

- limiting the frequency of handling loads (especially those that are heavier or bulkier);
- ensuring that workers are able to take rest breaks (either formal breaks or informal rest periods as and when needed);
- introducing job rotation within work teams (allowing muscle groups to recover whilst other muscles are used or lighter tasks undertaken).

## Team handling

Tasks which might be unsafe for one person might be successfully carried out by two or more. HSE guidance indicates that the capability of a two-person team is approximately two thirds of the total of their individual capabilities; and of a three-person team approximately one half of their total. Other factors to take into account in team handling are:

- the availability of suitable handholds for all the team;
- whether steps or slopes have to be negotiated;
- possibilities of the team impeding each other's vision or movement;
- availability of sufficient space for all to operate effectively;
- the relative sizes and capabilities of team members.

## **Load-related measures**

### Weight or size reduction

M3013     The types of measures described earlier in the chapter might be adopted to reduce the risk of injury even if it is not possible to avoid the risk entirely. However, reducing the individual load size will increase the number of movements necessary to handle a large total load. This could result in a different type of fatigue and also the possibility of corners being cut to save time. Sizes of loads may also be reduced to make them easier to hold or bring them closer to the handler's body.

## Making the load easier to grasp

Handles, grips or indents can all make loads easier to handle (as demonstrated by the handholds provided in office record storage boxes). It will often be easier handling a load placed in a container with good handholds than handling it alone without any secure grip points. The positioning of handles or handholds on a load can also influence the handling technique used, e.g. help avoid stooping. Handholds should be wide enough to accommodate the palm and deep enough to accommodate the knuckles (including gloves where these are worn).

The use of slings, carrying harnesses or bags can all assist in gaining a secure grip on loads and carrying them in a more comfortable and efficient position.

## Making the load more stable

Loads which lack rigidity themselves or are in insecure packaging materials may need to be stabilised for handling purposes. Use of carrying slings, supporting boards or trays may be appropriate. Partially full containers may need to have their contents wedged into position to prevent them moving during handling.

## Reducing other risks

Loads may need to be cleaned of dirt, oil, water or corrosive materials for safe handling to take place. Sharp corners or edges or rough surfaces may have to be removed or covered over. There may also be risks from very hot or very cold loads. In all cases the use of containers (possibly insulated) other handling aids or suitable gloves or other PPE may need to be considered.

## **Improving the work environment**

### Providing clear handling space

**M3014** The provision of adequate gangways and sufficient space for handling activities to take place will be linked with general workplace safety issues. Low headroom, narrow doorways and congestion caused by equipment and stored materials are all to be avoided. Good housekeeping standards are essential to safe manual handling.

### Floor condition and design

Even in temporary workplaces such as construction sites, every effort should be made to ensure that that firm, even floors are provided where manual handling is to take place. Special measures may be necessary to allow water to drain away or to clear promptly any potentially slippery materials (e.g. food scraps in kitchens). Where risks are greater special slip-resistant surfaces may need to be considered. In outdoor workplaces the routine application of salt or sand to surfaces made slippery by ice or snow may need to be introduced.

### Differing work levels

Measures may be necessary to reduce the risks caused by different work levels. Slopes may need to be made more gradual, steps may need to be provided or existing steps or stairways made wider to accommodate handling activities. Work benches may need to be modified to provide a uniform and convenient height.

## Thermal environment and ventilation

Unsatisfactory temperatures, high humidity or poor ventilation may need to be overcome by improved environmental control measures or transferring work to a more suitable area. For work close to hot processes or equipment, in very cold conditions (e.g. refrigerated storage areas) or carried on out of doors the use of suitable PPE may be necessary.

## Gusts of wind

Where gusts of wind (or powerful ventilation systems) could affect larger loads extra precautions, in addition to those normally taken, may be necessary e.g:

- using handling aids (e.g. trolleys);
- utilising team handling techniques;
- adopting a different work position;
- following an alternative transportation route.

## Lighting

Suitable lighting must be provided to permit handlers to see the load and the layout of the workplace and to make accurate judgements about distance, position and the condition of the load.

## **Individual capability**

M3015  In ensuring that individuals are capable of carrying out manual handling activities in the course of their work, account must be taken both of factors relating to the individual and those relating to the activities they are to perform. Steps which may be appropriate include:

## Medical screening

Workers may be screened both prior to being offered employment and periodically thereafter to check whether they are physically capable of carrying out the full range of manual handling tasks in the workplace. In some extreme situations (e.g. the emergency services) an individual may be deemed unsuitable for employment if they are not sufficiently fit. However, legislation prevents employers discriminating against those with disabilities *without good reason*. In most workplaces the results of such screening may simply result in restrictions on the range of tasks individuals are allowed to perform.

## Long-term restrictions

Long-term restrictions may be appropriate where workers are identified through screening programmes as not being capable of carrying out certain manual handling activities without undue risk, or they become incapable due to some long term injury or the effects of the ageing process. It may be necessary to place long term restrictions on young workers until they attain an appropriate age at which their capabilities can be re-assessed.

## Short-term restrictions

Relatively short-term restrictions on manual handling activities may be necessary because of:

- pregnancy (or a particular stage of pregnancy)
- recently having given birth

- injury or illness of short duration.

### Fitness programmes

An increasing number of employers actively encourage employers to be physically fit, thus reducing their risk of injury during manual handling activities. A number of Fire Services have particularly pro-active programmes.

### General manual handling training

Many employers offer general training in manual handling, either to all employees or to all those employees expected to engage in significant manual handling work. Aspects to cover in such training include:

- recognition of potentially hazardous operations;
- avoiding manual handling hazards;
- dealing with unfamiliar handling operations;
- correct use of handling aids;
- proper use of PPE;
- working environment factors;
- importance of good housekeeping;
- factors affecting individual capability (knowing one's limitations);
- good handling technique.

### Task-specific training and instruction

Training and instruction will often need to relate to the safety of specific manual handling tasks, e.g:

- how to use specific handling aids;
- specific PPE requirements;
- specific handling techniques.

General principles of handling technique are summarised in HSE leaflet INDG 143 *Getting to grips with manual handling: A short guide for employers*. Many other HSE publications deal with specific occupational areas and often contain guidance on training (see 'HSE Guidance' at M3019). Regulation 4(3)(c) of the *Manual Handling Operations Regulations 1992 (SI 1992 No 2973)* (as amended) now requires employees' level of knowledge and training to be considered during the risk assessment.

## Planning and preparation

M3016    The principles of planning and preparation for manual handling assessments are no different from those for any other types of risk assessments.

### Who will carry out the assessments?

M3017    HSE guidance contained in the main HSE booklet on the *Manual Handling Operations Regulations 1992* (L 23) states that while one individual may be able to carry out an assessment in relatively straightforward cases, in others it may be more appropriate to establish an assessment team. It also refers to employers and managers with a practical understanding of the manual handling tasks to be performed, the loads to be handled, and the working environment being better able to conduct assessments than someone outside the organisation (although it is not stated in the guidance that in-house personnel will also usually be more aware of the capabilities of employees carrying out manual handling tasks).

The guidance refers to the individual or team performing the assessments needing to possess the following knowledge and expertise:

- the requirements of the Regulations;
- the nature of the handling operations;
- a basic understanding of human capabilities;
- awareness of high risk activities;
- practical means of reducing risk.

The HSE also acknowledges that there will be situations where external assistance may be required. These might include:

- training of in-house assessors;
- assessing risks which are unusual or difficult to assess;
- re-designing equipment or layouts to reduce risks;
- training of staff in handling techniques.

## How will the assessments be arranged?

M3018    As for general risk assessments, the workplace or work activities will often need to be divided into manageable assessment units. Depending on how work is organised, these might be based on:

- Departments or sections.
- Buildings or rooms.
- Parts of processes.
- Product lines.
- Work stations.
- Services provided.
- Job titles.

Members of the assessment team can then be allocated to the assessment units where they are best able to make a contribution.

## Gathering information

M3019    It is important to gather information prior to the assessment in order to identify potentially high risk manual handling activities and the precautions which ought to be in place to reduce the risks. Relevant sources might include:

### Accident investigation reports

Dependent upon the requirements of the organisation's accident investigation procedure, these may only include the more serious accidents resulting from manual handling operations.

## Ill health records

These may reveal short absences or other incidents due to manual handling which may have escaped the accident reporting system. Enquiries may need to be made in respect of the health surveillance records of individuals (see M3015 above).

## Accident treatment records

Accident books or other treatment records may identify regularly recurring minor accidents which do not necessarily result in any absences from work, e g cuts or abrasions from sharp or rough loads.

## Operating procedures etc

Reference to manual handling activities and precautions which should be adopted may be contained in operating procedures, safety handbooks etc.

## Work sector information

Trade associations and similar bodies publish information on manual handling risks and how to overcome them.

## HSE Guidance

Much guidance is available from the HSE – some of the more important publications were referred to earlier in the chapter. The table below gives details of additional guidance relating to specific work sectors.

**HSE publications on manual handling in specific work sectors**

|         | |
|---------|---|
|         | Manual handling in the health service (1998) |
|         | Getting to grips with handling problems: Worked examples of assessment and reduction of risk in the health service (1993) |
| HSG 119 | Manual handling in drinks delivery (1994) |
| HSG 196 | Moving food and drink: Manual handling solutions for the food and drink industries (2000) |
| HSG 171 | Well handled: Offshore manual handling solutions (1997) |
|         | Manual handling in paper mills (1998) |
|         | Picking up the pieces: Prevention of musculoskeletal disorders in the ceramics industry (1996) |
| AS 23   | Manual handling solutions for farms (2000) |
| INDG 125 | Handling and stacking bales in agriculture (1998) – free leaflet. |
| IACL 105 | Handling the news: Advice for employers on manual handling of bundles (1999) – free leaflet. |
| IACL 106 | Handling the news: Advice for newsagents and employees on safe handling of bundles (1999) – free leaflet. |
| IACL 103 | Manual handling in the textile industry (1998) |
|         | Manual handling in the rubber industry (1999) |

| APIS 2 | Manual handling operations: Baggage handling at airports |
| INDG 332 | Manual packing in the brick industry (2000) – free leaflet. |
| INDG 326 | Manual handling in the railway industry: Advice for employers (2000) – free leaflet. |
| HSG 149 | Backs for the future: Safe manual handling in construction |
| CIS 37 | Handling heavy building blocks |
| CAIS | Manual handling in the catering industry |
| INDG 318 | Manual handling solutions in woodworking |

## Making the assessment

**M3020** As with other types of assessments, visits to the workplace are an essential part of manual handling assessments. Manual handling operations need to be observed (sometimes in some detail) and discussions with those carrying out manual handling work, their safety representatives and their supervisors or managers need to take place.

The assessment checklist provided on the following pages provides guidance on the factors which may need to be taken into account during the observations and discussions and also on possible measures to reduce the risk. The checklist utilises the four main factors specified in the Regulations – task, load, working environment and individual capability. In practice each of these will have greater or lesser importance, depending upon the manual handling activity.

### Observations

**M3021** Sufficient time should be spent in the workplace to observe a sufficient range of the manual handling activities taking place and to take account of possible variations in the loads handled (raw materials, finished product, equipment etc). Aspects to particularly be considered include:

- comparison of actual methods used with those specified in operating procedures, industry standards etc;
- the level and manner of use of handling aids and their effectiveness;
- handling techniques adopted;
- physical conditions, e.g. access, housekeeping, floor surfaces, lighting;
- variations between employees, e.g. size, strength, technique adopted.

### Discussions

**M3022** In discussing manual handling issues with workpeople an enquiring approach should be taken, utilising open-ended questions wherever possible. Aspects which might be discussed include:

- whether normal conditions are being observed;
- their awareness of procedures, rules etc;
- their training in the use of handling aids or handling techniques;
- possible variations in work activities, e.g. product changes, seasonal variations, differences between day and night shift;
- what happens if handling equipment or handling aids break down or are unavailable;

## MANUAL HANDLING ASSESSMENT CHECKLIST
## THE TASK

| ASSESSMENT FACTORS | REDUCING THE RISK |
|---|---|
| DISTANCE OF THE LOAD FROM THE TRUNK | TASK LAYOUT |
| BODY MOVEMENT / POSTURE | USE THE BODY MORE EFFECTIVELY |
| • Twisting<br>• Reaching<br>• Stooping<br>• Sitting | eg SLIDING OR ROLLING THE LOAD |
|  | SPECIAL SEATS |
| DISTANCE OF MOVEMENT | RESTING PLACE / TECHNIQUE |
| • Height (? grip change)<br>• Carrying (? over 10m)<br>• Pushing or pulling | USE OF TROLLEYS etc<br>TECHNIQUE / FLOOR SURFACE |
| RISK OF SUDDEN MOVEMENT | AWARENESS / TRAINING |
| FREQUENT / PROLONGED EFFORT | IMPROVED WORK ROUTINE |
|  | eg Job rotation |
| CLOTHING / FOOTWEAR ISSUES |  |
| RATE OF WORK IMPOSED BY A PROCESS | ADEQUATE REST OR RECOVERY PERIODS |

## THE LOAD

| ASSESSMENT FACTORS | REDUCING THE RISK |
|---|---|
| HEAVY | MAKE THE LOAD LIGHTER<br>TEAM LIFTING<br>MECHANICAL AIDS |
| BULKY<br>(Any dimensions above 75cm) | MAKE THE LOAD SMALLER |
| UNWIELDY<br>(Offset centre of gravity) | INFORMATION / MARKING |
| DIFFICULT TO GRASP<br>(Large, rounded, smooth, wet, greasy) | MAKE IT EASIER TO GRASP |
|  | • Handles<br>• Handgrips<br>• Indents<br>• Slings<br>• Carrying devices<br>• Clean the load |
| UNSTABLE<br>• Contents liable to shift<br>• Lacking rigidity | WELL-FILLED CONTAINERS<br>USE OF PACKING MATERIALS<br>SLINGS / CARRYING AIDS |
| SHARP EDGES / ROUGH SURFACES | AVOID OR REDUCE THEM<br>GLOVES OR OTHER PPE |
| HOT OR COLD | CONTAINERS (? insulated) |

## THE WORKING ENVIRONMENT

| ASSESSMENT FACTORS | REDUCING THE RISK |
|---|---|
| SPACE CONSTRAINTS AFFECTING POSTURE | ADEQUATE GANGWAYS, FLOORSPACE, HEADROOM |
| UNEVEN, SLIPPERY OR UNSTABLE FLOORS | WELL MAINTAINED SURFACES PROVISION OF DRAINAGE SLIP-RESISTANT SURFACING GOOD HOUSEKEEPING PROMPT SPILLAGE CLEARANCE |
| VARIATION IN WORK SURFACE LEVEL<br>eg steps, slopes, benches | STEPS NOT TOO STEEP GENTLE SLOPES UNIFORM BENCH HEIGHT |
| HIGH OR LOW TEMPERATURES HIGH HUMIDITY POOR VENTILATION | BETTER ENVIRONMENTAL CONTROL RELOCATING THE WORK SUITABLE CLOTHING |
| STRONG WINDS OR OTHER AIR MOVEMENT | RELOCATING THE WORK ALTERNATIVE ROUTE HANDLING AIDS TEAM HANDLING |
| LIGHTING | SUFFICIENT WELL-DIRECTED LIGHT |
| MOVING WORKPLACE<br>• Boat<br>• Train<br>• Vehicle | |

## INDIVIDUAL CAPABILITY

| ASSESSMENT FACTORS | REDUCING THE RISK |
|---|---|
| *INDIVIDUAL FACTORS*<br>STRENGTH<br>HEIGHT<br>FLEXIBILITY<br>STAMINA | BALANCED WORK TEAMS 'SELF SELECTION' MAKE ASSISTANCE AVAILABLE ENCOURAGE FITNESS |
| PROBLEMS WITH CLOTHING ETC.<br>PREGNANCY<br>INJURY / HEALTH PROBLEM<br>• Back trouble<br>• Hernia<br>• Temporary injury | SPECIAL ASSESSMENTS FORMAL RESTRICTIONS CLEAR INSTRUCTIONS |

| TASK FACTORS OPERATIONS OR SITUATIONS REQUIRING PARTICULAR | FORMALISED WORK PROCEDURES CLEAR INSTRUCTIONS ADEQUATE TRAINING |
|---|---|
| • Awareness of risks<br>• Knowledge<br>• Method of approach<br>• Technique | • Recognising potentially hazardous operations<br>• Dealing with unfamiliar operations<br>• Correct use of handling aids<br>• Proper use of PPE<br>• Environmental factors<br>• Importance of housekeeping<br>• Knowing one's own limitations<br>• Good technique |

- whether assistance is available if required;
- what manual handling problems have been experienced?;
- have there been any injuries with manual handling?;
- reasons precautions are not being taken.

## Notes

**M3023** Notes should be taken during the assessment of anything which might to be of relevance such as:

- further detail about manual handling risks found;
- handling risks which are discounted as insignificant;
- handling aids available;
- other control measures in place and appearing effective;
- problems identified or concerns expressed;
- alternative precautions worthy of consideration;
- related procedures, records or other documents.

These rough notes will eventually form the basis of the assessment.

# After the assessment

## Assessment records

**M3024** HSE guidance states that the significant findings of the assessment must be recorded unless:

- the assessment could very easily be repeated and explained at any time because it is so simple and obvious; or
- the manual handling operations are quite straightforward, of low risk, are going to last only a very short time, and the time taken to record them would be disproportionate.

As should be the case in relation to other types of assessment, it is better to make a simple record if there is any doubt.

There is no standard format for recording an assessment. The main HSE Guidance booklet (L 23) includes a sample record format together with a useful worked example. This record format includes a checklist of assessment factors under the four main headings (task, load, working environment, capability) in a similar listing to the assessment checklist we have provided. Alongside each checklist item the assessor has space to:

- identify whether the risk from this factor is low, medium or high;
- provide a more detailed description of the problems;
- identify possible remedial action.

Some will no doubt find this checklist format meets their needs. However, many assessment factors often have little or no relevance to a specific manual handling operation. As a result, the eventual record contains a lot of blank paper, with only occasional comments. Consequently it is often preferable to keep the checklist separate and only include within the record a reference to the risks, precautions and recommended improvements which are relevant to the operation being assessed. Two worked examples of this type of record are included below.

| MANUAL HANDLING ASSESSMENT | | BROADACRES DEVELOPMENT | DATE: 27 January 2004 | ASSESSMENT BY: A Backworth |
|---|---|---|---|---|
| No. | ACTIVITY | RISKS | EXISTING PRECAUTIONS | RECOMMENDATIONS |
| 1 | HANDLING OF STATIONERY AND PRINTED MATERIALS | Heaviest stationery items are 5 ream boxes of paper (approximately 12 kg). Some packs of printed promotional material weigh over 20 kg. | Suppliers deliver items directly to main store. | Require printing suppliers to supply materials in packs weighing no more than 12 kg. |
| | | Items are moved from main store on the ground floor to smaller stores on other floors. | Trolley and lift used for transporting between main store and smaller stores. | Ensure paper and printed materials are stored on shelves between knee and shoulder height. |
| 2 | HANDLING OF ARCHIVED RECORDS | Records are normally stored in file boxes weighing no more than 20 kg. They are moved into and out of the record store on the third floor. | File boxes are kept below shoulder height on substantial racking. They are moved around using the trolley and lift. | Access to the Accounts section of the archive store racking was obstructed by various items which must be removed. |
| | | (Older records are removed for storage off-site by a specialist company.) | Movement to off-site storage is carried out by Facilities staff who have had manual handling training. | Some records were stored in oversize boxes — this practice must cease. |
| 3 | MOVEMENT OF EQUIPMENT PC monitors, TV sets, projection equipment etc. | These items (of varying weights) are moved between the equipment store on the first floor and locations where they are required. | The trolley and lift are used where appropriate. IT staff have been trained in handling techniques. | Training section assistants should also receive handling technique training. Provide dust coats (so that staff can carry dirty items close to their bodies). |
| | | | Some racking is provided in the equipment store but this is inadequate and many items are stored on the floor. | Install additional racking in the equipment store and keep heavier items on more accessible shelves. |
| 4 | MOVEMENT OF FURNITURE Desks, seating etc. | Desks and seating are moved as required for individual workstations. | Facilities section staff have had manual handling training. | Training section assistants require training (see above). |
| | | (Major moves are carried out by contractors.) | Team lifting is used for larger items, e.g. desk tops. The lift is used as appropriate. | Suitable racking must be provided in the equipment store for desk and table tops. |
| | | Seating and tables are moved when required for major meetings and training events. | Desk and table tops are kept in the equipment store. | |

| GOLD STAR ENGINEERING | MANUAL HANDLING ASSESSMENT | |
|---|---|---|
| MAINTENANCE STORES | Assessment by: T Wrightson | Date: 28 January 2004 |
| ACTIVITIES/ RISKS | EXISTING PRECAUTIONS | RECOMMENDATIONS |
| Handling activities involve the movement of equipment and materials within the stores building and the yard outside, together with two storage containers and a flammable liquid store situated within the yard. | A fork lift truck is used to handle heavier items within the yard. A hand-operated pallet truck and a lightweight lifting truck are available. Several trucks and trolleys are provided including a stair-climbing truck, gas cylinder and fire extinguisher trolleys and a drum transporter. Skates, bogie units and dollies are also available for transporting heavy items. | |
| Both equipment and materials must also be moved to locations within the factory. | | |
| There is considerable variation in the sizes and weights of items handled. Some items are palletised, some are supplied in boxes and some are kept in bins and cages. | Most internal and external floor surfaces are in good condition. | A section of the floor just inside the main stores entrance requires repair. |
| | There is a gradual ramp into the main stores building but not into the storage containers or the flammable liquid store (the latter having a sill at its entrance). | Suitable permanent or removable ramps should be provided to allow handling aids to be used in the storage containers and flammable liquid store. |
| Drums (up to 100 litres capacity), gas cylinders and large fire extinguishers must also be moved. | Lighting standards are good in the main stores building, the flammable liquid store and the yard, but there are no lights in the storage containers. | Lighting must be provided within the storage containers. |
| Racking and cupboards are provided within the building and inside the outside storage containers. | Heavier items are generally stored either on the floor or on accessible levels on racking or in cupboards. | Ensure walkways between racking are kept clear. |
| Some items to be handled have sharp or rough edges. | Latex coated knitted gloves (for grip) and gloves with chrome leather palms are available (as are gloves for chemical risks). | |
| | Most stores' staff have attended a one-day course in manual handling techniques tailored to their needs. | Develop a formal training programme to ensure that all stores staff receive training in manual handling technique and the correct use of the various handling aids available. |

## Review and implementation of recommendations

**M3025** As for other types of risk assessments, there will be a need to review recommendations and prepare action plans for their implementation. Recommendations will need to be followed up in order to ensure that they have actually been implemented and also to assess any new risks which may have been introduced. This latter point is particularly relevant to manual handling as changes in working practice may have resulted in different techniques being adopted (with their own attendant risks) or the introduction of handling aids may have created additional training needs.

## Assessment review

**M3026** *Paragraph 4(2)* of the *Manual Handling Operations Regulations 1992* requires a manual handling assessment to be reviewed if:

'(a) there is reason to suspect that it is no longer valid; or

(b) there has been a significant change in the manual handling operations to which it relates;'

As is the case for other assessments, a review would not necessarily result in the assessment being revised, although a revision may be deemed to be necessary. A significant manual handling related injury should automatically result in a review of the relevant assessment. It is recommended that assessments are reviewed periodically in any case in order to take account of small changes in work practices which will eventually have a cumulative effect. The periods for such reviews might vary between two and five years – depending upon the degree of risk involved and the potential for gradual changes to take place.

## Training

**M3027** The need for specific manual handling technique training may frequently be one of the recommendations made as a result of a manual handling assessment. With manual handling activities featuring in a wide range of work sectors many employers have chosen to provide all or a significant proportion of their workforce with general training in good handling technique. Such training is often provided as part of a more comprehensive health and safety induction programme. Guidance on the content of such general training is contained in paragraphs 173 and 174 of HSE Guidance booklet (L 23) and also in other HSE publications dealing with specific work sectors (see the table earlier in the chapter).

# Noise at Work

## Introduction

**N3001** This chapter is aimed at company directors, legal advisors, health and safety officers, staff representatives and others who need a comprehensive, concise reference source on the issue of noise at work.

It will be helpful to anyone who needs to commission a workplace noise assessment or to undertake a noise control project. It will also assist those who wish to devise long-term noise reduction and equipment-buying strategies.

Despite decades of warnings and legislation, noise-induced hearing loss is still a huge but underrated problem, with a slow, cumulative and irreversible effect. As a result, legislation is in the process of being tightened, so that many more workplaces are about to fall into the net (N3002).

This chapter discusses the scale of problems arising from workplace noise exposure, the legal obligations on employers to reduce the risk of noise-induced hearing damage both in the short term and as a long-term obligation (N3012); and the risks in terms of compensation claims (N3022). An overview of social security benefits is also provided (N3018).

The chapter gives a short introduction to the perception of sound and the various indexes used to measure noise (N3004). It provides general guidance on methods of assessing and reducing noise (see N3029 and N3035), and suggests sources of more detailed guidance on these matters (see N3023).

## Scale of the problem and future developments

**N3002** A Medical Research Council survey in 1997–98 estimated that there were 507,000 people in Great Britain suffering from hearing difficulties as a result of exposure to noise at work. This is much higher than HSE's own estimate of 87,000 based on its 2001/02 self-reported work-related Illness (SWI) survey. The large difference is probably due to the survey methods, and suggests that even people who suffer hearing difficulties are often unaware that their noisy workplace has contributed to their problems.

The Department of Social Security (DSS) pays disablement benefit to people who suffer at least 50 decibels (dB) of noise-induced hearing loss in both ears. This level of hearing loss is roughly equivalent to listening to the television or a conversation through a substantial brick wall. Moreover, claimants must have been employed for at least ten years in a specified noisy occupation. Despite these restrictions, the number of new awards appears to have stuck at just over 260 per year, after a significant fall during the 1980s and 1990s from around 1,500 a year in 1984–85 to a low of 226 in 2000. There are currently about 14,000 people – 99 per cent of them men, receiving the benefit from DSS.

The number of claimants with between 35 and 49 decibels of hearing loss (still a severe disability, equivalent to listening through a substantial partition wall in a house) shows an improving trend. According to statistics from audiological examinations, the number of claimants has fallen from 1200 in 1995 to 710 in 1999.

It is likely that some of this improvement is a result of the contraction of traditionally noisy industries such as ship-building and coal mining, with the armed forces now showing some of the highest rates of incidence.

However, there is concern about noise exposure in some of the newer occupations, such as workers in pubs and clubs where noise levels often exceed the second action level of the *Noise at Work Regulations 1989 (SI 1989 No 1790)*.

The EU's Physical Agents (Noise) Directive (2003/10/EC) requires changes to be made to the existing UK regulations (i.e. Noise At Work Regulations) and HSE are currently consulting on new Control of Noise at Work Regulations. The new regulations will reduce the action levels by 5 dB, requiring exposure to be assessed when it exceeds 80 dB(A) averaged over 8 hours (in some cases, averaged over 5 days each of 8 hours). Hearing protection must be made available at this exposure, as well as information and training. Health surveillance must be made available where exposure reaches 85 dB(A) and hearing protectors must be worn. There will be a limit on exposure of 87 dB(A), the exposure to be calculated to include the effect of any hearing protection that is provided. Corresponding changes are made to action levels in relation to peak sound pressures. This proposal is far more practicable and realistic than in the original EU proposals, following representations from Britain (amongst others) that it was preferable to ensure greater compliance with current limits than to introduce more stringent limits that are not rigorously enforced.

## Hearing damage

N3003 We all lose hearing acuity with age, and this loss is accelerated and worsened by exposure to excessive noise. Fortunately, in early life, we have much greater hearing acuity than modern life demands, and a small loss is readily compensated by turning up the volume of the radio or television.

Loud noises cause permanent damage to the nerve cells of the inner ear in such a way that a hearing aid is ineffective. At first, the damage occurs at frequencies above normal speech range, so that the sufferer may have no inkling of the problem, although it could be identified easily by an audiogram, which measures the sensitivity of the ear at a number of frequencies across its normal range. As exposure continues, the region of damage progresses to higher and lower frequencies. Damage starts to extend into the speech range, making it difficult to distinguish consonants, so that words start to sound the same. Eventually, speech becomes a muffled jumble of sounds.

Some people are much more susceptible to hearing damage than others. A temporary dullness of hearing or tinnitus (a ringing or whistling sound in the ears) when emerging from a noisy place are both indicative of damage to the nerve cells of the inner ear. Because these symptoms tend to disappear with continued exposure, they are often misinterpreted as the ear becoming 'hardened', whereas in fact it is losing its ability to respond to the noise. Even when hearing has apparently returned to normal, a little of the sensitivity is likely to have been lost. Therefore, anyone who experiences dullness of hearing or tinnitus after noise exposure must take special care to avoid further exposure. They may also be advised to seek medical advice and an audiogram.

Apart from catastrophic exposures (usually to explosive sounds), the risk of hearing loss is closely dependent on the 'noise dose' received, and especially on the cumulative effect over a period of time. British Standard BS5330 provides a procedure for estimating the risk of 'hearing handicap' due to noise exposure. Handicap is there defined as a hearing loss of 30 dB, which is sufficiently severe to impair the understanding of conversational speech or the appreciation of music. This may be compared with the loss of 50 dB required for DSS disability benefit.

## The perception of sound

N3004 Sound is caused by a rapid fluctuation in air pressure. The human ear can hear a sensation of sound when the fluctuations occur between 20 times a second and 20,000 times a second. The rate at which the air pressure fluctuates is called the frequency of the sound and is measured in Hertz (Hz). The loudness of the sound depends on the amount of fluctuation in the air pressure. Typically, the quietest sound that can be heard (the threshold of hearing) is zero decibels (0 dB) and the sound becomes painful at 120 dB. Surprisingly perhaps, zero decibels is not zero sound. The sensitivity and frequency range of the ear vary somewhat from person to person and deteriorate with age and exposure to loud sounds.

The human ear is not equally sensitive to sounds of different frequencies (i.e. pitch): it tends to be more sensitive in the frequency range of the human voice than at higher or lower frequencies (peaking at about 4 kHz). When measuring sound, compensation for these effects can be made by applying a frequency weighting, usually the so-called 'A' weighting, although other weightings are sometimes used for special purposes.

The ear has an approximately logarithmic response to sound: for example, every doubling or halving in sound pressure gives an apparently equal step increase or decrease in loudness. In measuring environmental noise, sound pressure levels are therefore usually quoted in terms of a logarithmic unit known as a decibel (dB). To signify that the 'A' weighting has been applied, the symbol of dB(A) is often used. However, current practice tends to prefer the weighting letter to be included in the name of the measurement index. For example, dB(A) $L_{eq}$ and dB $L_{Aeq}$ both refer to the 'A' weighted equivalent sound level in decibels.

Depending upon the method of presentation of two sounds, the human ear may detect differences as small as 0.5 dB(A). However, for general environmental noise the detectable difference is usually taken to be between 1 and 3 dB(A), depending on how quickly the change occurs. A 10 dB(A) change in sound pressure level corresponds, subjectively, to an approximate doubling or halving in loudness. Similarly, a subjective quadrupling of loudness corresponds to a 20 dB(A) increase in sound pressure level (SPL). When two sounds of the same SPL are added together, the resultant SPL is approximately 3 dB(A) higher than each of the individual sounds. It would require approximately *nine* equal sources to be added to an original source before the subjective loudness is doubled.

# Noise indices

**N3005** The sound pressure level of industrial and environmental sound fluctuates continuously. A number of measurement indices have been proposed to describe the human response to these varying sounds. It is possible to measure the physical characteristics of sound with considerable accuracy and to predict the physical human response to characteristics such as loudness, pitch and audibility. However, it is not possible to predict *subjective* characteristics such as annoyance with certainty. This should not be surprising: one would not expect the light meter on a camera to be able to indicate whether one was taking a good or a bad photograph, although one would expect it to get the physical exposure correct. Strictly speaking, therefore, a meter can only measure sound and not noise (which is often defined as sound unwanted by the recipient): nevertheless, in practice, the terms are usually interchangeable.

## *Equivalent continuous 'A' weighted sound pressure level, $L_{Aeq}$*

**N3006** This unit takes into account fluctuations in sound pressure levels. It can be applied to all types of noise, whether continuous, intermittent or impulsive. $L_{Aeq}$ is defined as the steady, continuous sound pressure level which contains the same energy as the actual, fluctuating sound pressure level. In effect, it is the energy-average of the sound pressure level over a period which must be stated, e.g. $L_{Aeq, (8-hour)}$.

This unit is now being put forward as a universal noise index, because it can be used to measure all types of noise, although it has yet to supplant older units in certain cases, particularly for the assessment of road traffic noise where calculation techniques and regulatory criteria have not been updated.

Levels measured in $L_{Aeq}$ can be added using the rules mentioned earlier. There is also a time trade-off: if a sound is made for half the measurement period, followed by silence, the $L_{Aeq}$ over the whole measurement period will be 3 dB less than during the noisy half of the period. If the sound is present for one-tenth of the measurement period, the $L_{Aeq}$ over the whole measurement period will be 10 dB less than during the noisy tenth of the measurement period. This is a cause of some criticism of $L_{Aeq}$: for discontinuous noise, such as may arise in industry, it does not limit the maximum noise level, so it may be necessary to specify this as well.

### Daily personal noise exposure, $L_{EP,d}$

N3007   This is used in the *Noise at Work Regulations 1989* (*SI 1989 No 1790*) as a measure of the total sound exposure a person receives during the day. It is formally defined in a *Schedule* to the Regulations.

$L_{EP,d}$ is the energy-average sound level ($L_{Aeq}$) to which a person is exposed over a working day, disregarding the effect of any ear protection which may be worn, adjusted to an 8–hour period. Thus, if a person is exposed to 90 dB $L_{Aeq}$ for a 4-hour working day, their $L_{EP,d}$ is 87 dB, but if they work a 12–hour shift, the same sound level would give them an $L_{EP,d}$ of 91.8 dB.

### Maximum 'A' weighted sound pressure level, $L_{Amax,T}$

N3008   The maximum 'A'-weighted root-mean-square (rms) sound pressure level during the measurement period is designated $L_{Amax}$. Sound level meters indicate the rms sound pressure level averaged over a finite period of time. Two averaging periods ($T$) are defined, $S$ (slow) and $F$ (fast), having averaging times of 1 second and 1/8th second. It is necessary to state the averaging period. Maximum SPL should not be confused with peak SPL which is a measure of the instantaneous peak pressure.

### Peak sound pressure

N3009   Instantaneous peak sound pressure can only be measured with specialist instruments. It is used in the assessment of explosive sounds, such as from gunshots and blasting, and in hearing damage assessments of this type of sound. Where the maximum sound level exceeds 125 dB $L_{Amax,F}$, then an accurate measurement of peak pressure is advisable to check for compliance with the peak action level of the *Noise at Work Regulations 1989* (*SI 1989 No 1790*).

### Background noise level, $L_{A90}$

N3010   $L_{A90}$ is the level of sound exceeded for 90 per cent of the measurement period. It is therefore a measure of the background noise level, in other words, the sound drops below this level only infrequently. (The term 'background' should not be confused with 'ambient', which refers to *all* the sound present in a given situation at a given time).

### Sound power level, $L_{WA}$

N3011   The sound output of an item of plant or equipment is frequently specified in terms of its sound power level. This is measured in decibels relative to a reference power of 1 pico-Watt (dB re $10^{-12}$ W), but it must not be confused with sound pressure level. The sound pressure level at a particular position can be calculated from a knowledge of the sound power level of the source, provided the acoustical characteristics of the surrounding and intervening space are known. As a crude analogy, the power of a lamp bulb gives an indication of its light output, but the illumination of a surface depends on its distance and orientation from the source, and the presence of reflecting and obstructing objects in the surroundings.

To give a rough idea of the relationship between sound power level and sound pressure level, then for a noise source which is emitting sound uniformly in all directions close above a hard surface in an open space, the sound pressure level 10m from the source would be 28 dB below its sound power level.

## General legal requirements

**N3012** Statutory requirements relating to noise at work *generally* are contained in the *Noise at Work Regulations 1989 (SI 1989 No 1790)*. Replacing the previous Department of Employment (voluntary) Code of Practice on Noise (1972), these Regulations require employers to take reasonably practicable measures, on a long-term on-going basis, to reduce employees' exposure to noise at work to the lowest possible level, and to lower noise exposure where employees are exposed to levels of 90 dB(A) or above, or to peak action level or above (200 Pascals). In addition, ear protectors must be provided and worn and ear protection zones designated. Estimates throughout industry overall suggest that about 1.3 million workers may be exposed above 85 dB(A), the first action level.

There are now specific provisions relating to vibration exposure which are covered in the VIBRATION chapter. This is addition to the provision contained in the *Social Security (Industrial Injuries) (Prescribed Diseases) Regulations 1985 (SI 1985 No 967)* and the *Reporting of Injuries, Diseases and Dangerous Occurrences Regulations 1985 (SI 1985 No 2023)*. Occupational deafness and certain forms of vibration-induced conditions, i.e. vibration-induced white finger, are prescribed industrial diseases for which disablement benefit is payable (though the 14 per cent disablement rule will obviously limit the number of successful claimants (see C6011 COMPENSATION FOR WORK INJURIES/DISEASES). Damages may also be awarded against the employer (see further N3022 below).

### *Noise At Work Regulations 1989 (SI 1989 No 1790)*

**N3013** The following duties are laid on employers.

1. To make (and update where necessary) a formal noise assessment, where employees are likely to be exposed to:

    (*a*) first action level or above (85 dB(A)),

    (*b*) peak action level or above (200 Pascals).

    [*SI 1989 No 1790, Regs 2(1), 4(1)*].

    Such assessment should be made by a competent person and adequately:

    (i) identify which employees are exposed, and

    (ii) provide the employer with such information as will enable him to carry out his statutory duties, and

    (iii) when there is reason to suppose that the assessment is no longer valid, or when there has been a significant change in the work to which the assessment relates, review noise levels and make any changes recommended by the review.

    [*SI 1989 No 1790, Reg 4(2)*].

2. To keep an adequate record of such assessment until a further assessment is made. [*SI 1989 No 1790, Reg 5*].

3. (As a long-term strategy, and on an on-going basis), to reduce the risk of damage to the hearing of their employees from exposure to noise to the lowest level reasonably practicable. [*SI 1989 No 1790, Reg 6*].

4. To reduce, so far as is reasonably practicable, the exposure to noise of employees (other than by provision of personal ear protectors), where employees are likely to be exposed to (a) 90 dB(A) or above or (b) peak action level (200 Pascals) or above. [*SI 1989 No 1790, Reg 7*].

5. To provide, at the request of an employee, suitable and efficient personal ear protectors where employees are likely to be exposed to 85 dB(A) or above but less than 90 dB(A). [*SI 1989 No 1790, Reg 8(1)*].

6. To designate ear protection zones, indicating:

    (a) that it is an ear protection zone, and

    (b) the need for employees to wear personal ear protectors whilst in such zone where any employee is likely to be exposed to 90 dB(A) or above, or to peak action level or above. Moreover, no employee should enter such zone unless he is wearing personal ear protectors.

    [*SI 1989 No 1790, Reg 9*].

    Ear protection so provided must be maintained in an efficient state and employees must report any defects in it to the employer and see that it is fully and properly used. [*SI 1989 No 1790, Reg 10*].

7. To provide employees, likely to be exposed to 85 dB(A) or above, or to peak action level or above, with adequate information, instruction and training with regard to:

    (a) risk of damage to that employee's hearing,

    (b) steps the employee can take to minimise the risk,

    (c) the requirement on employees to obtain personal ear protectors from the employer, and

    (d) the employee's duties under the regulations.

    [*SI 1989 No 1790, Reg 11*].

    In addition, there are specific legal requirements applying to tractor cabs and offshore installations and construction sites (see below).

(From the *Health and Safety (Safety Signs and Signals) Regulations 1996 (SI 1996 No 341)*)

## Legal obligations of designers, manufacturers, importers and suppliers of plant and machinery

**N3014** The *Supply of Machinery (Safety) Regulations 1992 (SI 1992 No 3073* as amended) require manufacturers and suppliers of noisy machinery to design and construct such machinery so that the risks from noise emissions are reduced to the lowest level taking account of technical progress. Information on noise emissions must be provided when specified levels are reached.

If a machine is likely to cause people at work to receive a daily personal noise exposure exceeding the first or peak action levels, adequate information on noise must be provided. If the second or peak action levels are likely to be exceeded, this should include a permanent sign or label, or if the machine may be noisy in certain types of use, an instruction label which could be removed following noise testing.

Figure 1: Sign for informing that ear protectors must be worn (white on a circular blue background)

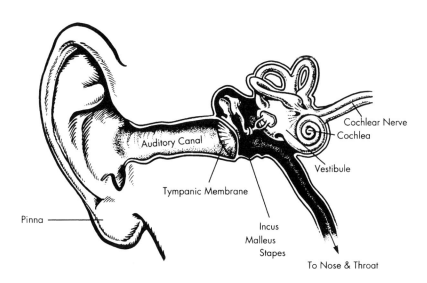

## Specific legal requirements

### Agriculture

**N3015** The *Agriculture (Tractor Cabs) Regulations 1974* (*SI 1974 No 2034*) (as amended by *SI 1990 No 1075*) provide that noise levels in tractor cabs must not exceed 90 dB(A) or 86 dB depending on which annex is relevant in the certificate under Directive 77/311/EEC. [*SI 1974 No 2034, Reg 3(3)*].

### Equipment for construction sites and other outdoor uses

**N3016** An EC Directive relating to the noise emission in the environment by equipment for use outdoors (Directive 2000/14/EC) was issued in 2000 to consolidate and update 23 earlier directives relating to the noise emission of construction plant and certain other equipment. The Directive was implemented by the *Noise Emission in the Environment by Equipment for Use Outdoors Regulations 2001* (*SI 2001 No 1701*). The earlier statutory instruments were revoked from 3 January 2002 and replaced by the requirements of the new statutory instrument.

The 2001 Regulations apply to a very wide range of powered equipment used outdoors. They do not apply to most non-powered equipment, equipment for the transport of goods or persons by road, rail, air or on waterways, or equipment for use by military, police or emergency services.

Each type of equipment is subject to a maximum permissible sound power level, dependent on the power of the equipment's engine or other drive system. The permissible sound power level will be reduced in January 2006 by 2 or 3 dB, relative to the level permitted in 2002.

Examples of the type of equipment covered include: cranes, hoists, excavators, dozers, dump trucks, compressors, generators, pumps, power saws, concrete breakers, compactors, glass recycling containers, lawnmowers and other gardening equipment.

The Regulations define the noise testing methods in great detail and also require a standardised marking of conformity indicating the guaranteed sound power level. Interestingly, it has been noted that many manufacturers simply use the maximum permitted level as the guaranteed level, which would seem to indicate that they see no marketing advantage in promoting quieter products.

*The Supply of Machinery (Safety) Regulations, 1992 (SI 1992 No 3073* as amended) are also relevant to noise emissions from machinery.

## Code of practice for noise and vibration control

N3017   The code of practice for basic information and procedures for noise and vibration control, BS 5228: Parts 1 to 5, gives detailed guidance on the assessment of noise and vibration from construction sites, open-cast coal extraction, piling operations, and surface mineral extraction. Part 1, revised in 1997, gives detailed noise calculation and assessment procedures. Its predecessor (published in 1984) was an approved code of practice under the *Control of Pollution Act, 1974*. Part 1 of the code of practice is principally concerned with environmental noise and vibration, but also briefly recites the *Noise at Work Regulations 1989 (SI 1989 No 1790)*.

# Compensation for occupational deafness
## Social Security

N3018   The most common condition associated with exposure to noise is occupational deafness. Deafness is prescribed occupational disease A10 (see OCCUPATIONAL HEALTH AND DISEASES). Prescription rules for occupational deafness have been extended three times, in 1980, 1983 and 1994. It is defined as: 'sensorineural hearing loss amounting to at least 50 dB in each ear being the average of hearing losses are 1, 2 and 3 kHz frequencies, and being due, in the case of at least one ear, to occupational noise'. [*Social Security (Industrial Injuries) (Prescribed Diseases) Amendment Regulations 1989 (SI 1989 No 1207), Reg 4(5)*]. Thus, the former requirement for hearing loss to be measured by pure tone audiometry no longer applies. Extensions of benefit criteria relating to occupational deafness are contained in the *Social Security (Industrial Injuries) (Prescribed Diseases) Regulations 1985 (SI 1985 No 967)*, which are amended, as regards assessment of disablement for benefit purposes, by the *Social Security (Industrial Injuries) (Prescribed Diseases) Amendment Regulations 1994 (SI 1994 No 2343)*.

### Conditions for which deafness is prescribed

N3019   Occupational deafness is prescribed for a wide range of occupations involving working with, or in the immediate vicinity of:

- mechanically-powered grinding and percussive tools used on metal; on rock and stone in quarries or in mines, in sinking shafts and in tunnelling;
- sawing masonry blocks, jet channelling of stone and masonry, vibrating metal moulding boxes in the concrete products industry;
- forging, cutting, burning and casting metal;
- plasma spray guns engaged in the deposition of metal;
- air arc gouging;
- high pressure jets of water or a mixture of water and abrasive material in the water jetting industry (including work under water);
- textile manufacturing where the work is undertaken in rooms or sheds in which there are machines engaged in weaving man-made or natural (including mineral) fibres or in the high speed false twisting of fibres; or mechanical bobbin cleaning;

- machines for automatic moulding, pressing or forming of glass hollow ware and continuous glass toughening furnaces; and spinning machines using compressed air to produce glass wool or mineral wool;
- a wide range of woodworking and metalworking machines;
- chain saws in forestry;
- work in ships' engine rooms;
- work on gas turbines in connection with performance testing on a test bed, installation testing of replacement engines in aircraft, and acceptance testing in armed service fixed wing combat planes.

The Regulations are detailed and should be consulted for precise definitions of these occupations. [*Social Security (Industrial Injuries) (Prescribed Diseases) Regulations 1985 (SI 1985 No 967)*; *Social Security (Industrial Injuries) (Prescribed Diseases) Amendment No 2 Regulations 1987 (SI 1987 No 2112)*; *Social Security (Industrial Injuries) (Prescribed Diseases) Amendment Regulations 1994 (SI 1994 No 2343)*].

'Any occupation' covers activities in which an employee is engaged under his contract of employment. The fact that the work-force is designated, classified of graded by reference to function, training or skills (e.g. labourer, hot examiner, salvage and forge examiner) does not of itself justify a conclusion that each separate designation, classification or grading involves a separate occupation (*Decision of the Commissioner No R(I) 3/78*).

'Assistance in the use' of tools qualifies the actual use of tools, not the process in the course of which tools are employed. Thus, a crane driver who positions bogies to enable riveters to do work on them and then goes away, assists in the process of getting bogies repaired, which requires use of pneumatic tools – is not assistance in the actual use of tools, for the purposes of disablement benefit. The position is otherwise when a crane holds a bogie in suspension to enable riveters to work safely on them. Here the crane driver assists in the actual use of pneumatic percussive tools (*Decision of the Commissioner No R(I) 4/82*).

## *Conditions under which benefit is payable*

N3020    For a claimant to be entitled to disablement benefit for occupational deafness, the following conditions under the *Social Security (Industrial Injuries) (Prescribed Diseases) Regulations 1985 (SI 1985 No 967)* currently apply:

- they must have been employed:
    (i)    at any time on or after 5 July 1948, and
    (ii)   for a period or periods amounting (in the aggregate) to at least ten years.
- there must be permanent sensorineural hearing loss, and loss in each ear must be at least 50 dB; and
- at least loss of 50 dB in one ear must be attributable to noise at work [*SI 1985 No 967, Sch 1, Pt 1*].

    (There is a presumption that occupational deafness is due to the nature of employment [*SI 1985 No 967, Reg 4(5)*]);

- the claim must be made within five years of the last date when the claimant worked in an occupation prescribed for deafness [*SI 1985 No 967, Reg 25(2)*];
- any assessment of disablement at less than 20 per cent is final [*SI 1985 No 967, Reg 33*].

A person, whose claim for benefit is turned down because he/she had not worked during the five years before the claim in one of the listed occupations, may claim if he/she continued to work and later met the time conditions. If a claim is turned down as the disability is less than 20 per cent, the claimant must wait three years before re-applying. If, by waiting three years, it would be more than five years since the applicant worked in one of the listed occupations, the three-year limit is waived.

## Assessment of disablement benefit for social security purposes

N3021    The extent of disablement is the percentage calculated by:

(a)    determining the average total hearing loss due to all causes for each ear at 1, 2 and 3 kHz frequencies; and

(b)    determining the percentage degree of disablement for each ear; and then

(c)    determining the average percentage degree of binaural disablement.

[Social Security (Industrial Injuries) (Prescribed Diseases) Amendment Regulations 1989, Reg 4(2)].

The following chart (TABLE 1 below), shows the scale for all claims made on/after 3 September 1979.

Table 1

Percentage degree of disablement in relation to hearing loss

| Hearing loss | Percentage degree of disablement |
| --- | --- |
| 50–53 dB | 20 |
| 54–60 dB | 30 |
| 61–66 dB | 40 |
| 67–72 dB | 50 |
| 73–79 dB | 60 |
| 80–86 dB | 70 |
| 87–95 dB | 80 |
| 96–105 dB | 90 |
| 106 dB or more | 100 |

[*Social Security (Industrial Injuries) (Prescribed Diseases) Regulations 1985, Reg 34, Sch 3, Pt II,* as amended]

Any degree of disablement, due to deafness at work, assessed at less than 20 per cent, must be disregarded for benefit purposes. [*Social Security (Industrial Injuries) (Prescribed Diseases) Amendment Regulations 1990 (SI 1990 No 2269)*].

## Action against employer at common law

N3022    There is no separate action for noise at common law; liability comes under the general heading of negligence. Indeed, it was not until as late as 1972 that employers were made liable for deafness negligently caused to employees (*Berry v Stone Manganese Marine Ltd [1972] 1 Lloyd's Rep 182*). Absence of a previous general statutory requirement on employers regarding exposure of employees to noise sometimes led to the law being strained to meet the facts (*Carragher v Singer Manufacturing Co Ltd 1974 SLT (Notes) 28* relating to the *Factories Act 1961, s 29*: 'every place of work must, so far as is reasonably practicable, be made and kept safe for any person working there', to the effect that this is wide enough to provide protection against noise). Admittedly, it is proper to regard noise as an aspect of the working environment (*McCafferty v*

*Metropolitan Police District Receiver [1977] 2 All ER 756* where an employee, the plaintiff, who was a ballistics expert, suffered ringing in the ears as a result of the sounds of ammunition being fired from different guns in the course of his work. When he complained about ringing in the ears – the ballistics room had no sound-absorbent material on the walls and he had not been supplied with ear protectors – he was advised to use cotton wool, which was useless. It was held that his employer was liable since it was highly foreseeable that the employee would suffer hearing injury if no steps were taken to protect his ears, cotton wool being useless). Moreover, although there are specific statutory requirements to minimise exposure to noise (in agriculture and offshore operations and construction operations, see N3015, N3016 above), these have generated little or no case law.

The main points established at common law are as follows.

(*a*) As from 1963, the publication date by the (then) Factory Inspectorate of 'Noise and the Worker', employers have been 'on notice' of the dangers to hearing of their employees arising from over-exposure to noise (*McGuinness v Kirkstall Forge Engineering Ltd 1979, unreported*). Hence, consistent with their common law duty to take reasonable care for the health and safety of their employees, employers should 'provide and maintain' a sufficient stock of ear muffs.

This was confirmed in *Thompson v Smiths, etc.* (see (*d*) below). However, more recently, an employer was held liable for an employee's noise-induced deafness, even though the latter's exposure to noise, working in shipbuilding, had occurred *entirely before* 1963. The grounds were that the employer had done virtually nothing to combat the *known* noise hazard from 1954–1963 (apart from making earplugs available) (*Baxter v Harland & Wolff plc, Northern Ireland Court of Appeal 1990 (unreported)*).

This means that, as far as Northern Ireland is concerned, employers are liable at common law for noise-induced deafness as from 1 January 1954 – the earliest actionable date. Limitation statutes preclude employees suing prior to that date (*Arnold v Central Electricity Generating Board [1988] AC 228*).

(*b*) Because the true nature of deafness as a disability has not always been appreciated, damages have traditionally not been high (*Berry v Stone Manganese Marine Ltd [1972]* – £2,500 (halved because of time limitation obstacles); *Heslop v Metalock (Great Britain) Ltd (1981)* – £7,750; *O'Shea v Kimberley-Clark Ltd (1982)* – £7,490 (tinnitus); *Tripp v Ministry of Defence [1982] CLY 1017* – £7,500).

(*c*) Damages will be awarded for exposure to noise, even though the resultant deafness is not great, as in tinnitus (*O'Shea v Kimberley-Clark Ltd, The Guardian, 8 October 1982*).

(*d*) Originally the last employer of a succession of employers (for whom an employee had worked in noisy occupations) was exclusively liable for damages for deafness, even though damage (i.e. actual hearing loss) occurs in the early years of exposure (for which earlier employers would have been responsible) (*Heslop v Metalock (Great Britain) Ltd, The Observer, 29 November 1981*). More recently, however, the tendency is to *apportion* liability between offending employers (*Thompson, Gray, Nicholson v Smiths Ship Repairers (North Shields) Ltd; Blacklock, Waggott v Swan Hunter Shipbuilders Ltd; Mitchell v Vickers Armstrong Ltd [1984] IRLR 93*). This is patently fairer because some blame is then shared by the original employer(s), whose negligence would have been responsible for the actual hearing loss.

(*e*) Because of the current tendency to apportion liability, even in the case of pre-1963 employers (see *McGuinness v Kirkstall Forge Engineering Ltd* above), contribution will take place between earlier and later insurers.

(*f*) Although judges are generally reluctant to be swayed by scientific/statistical evidence, the trio of shipbuilding cases (see (*d*) above) demonstrates, at least in the case of occupational deafness, that this trend is being reversed (see Table 2 below, the 'Coles-Worgan classification'); in particular, it is relevant to consider the 'dose response' relationship published by the National Physical Laboratory (NPL), which relates long-term continuous noise exposure to expected resultant hearing loss. This graph always shows a rapid increase in the early years of noise exposure, followed by a trailing off.

(g) Current judicial wisdom identifies three separate evolutionary aspects of deafness, i.e. (i) hearing loss (measured in decibels at various frequencies); (ii) disability (i.e. difficulty/inability to receive everyday sounds); (iii) social handicap (attending musical concerts/meetings etc.). That social handicap is a genuine basis on which damages can be (*inter alia*) awarded, was reaffirmed in the case of *Bixby, Case, Fry and Elliott v Ford Motor Group (1990, unreported)*.

# General guidance on noise at work

## Guidance on the Noise at Work Regulations 1989

N3023    HSE has revised and updated its Noise Guides which are regarded as providing authoritative and detailed guidance on the Noise at Work Regulations. These are now published in a single volume called *Reducing noise at work Guidance on the Noise at Work Regulations 1989* (ISBN 0 7176 1511 1). It covers the following areas:

- legal duties of employers to prevent damage to hearing;
- duties of designers, manufacturers, importers and suppliers;
- how to choose a competent person – advice for employers;
- how to carry out a noise assessment – advice for the competent person;
- control of noise exposure – advice for employers and engineers;
- selection and use of personal ear protection – advice for employers.

## Guidance on specific types of equipment

N3024    HSE has also issued a number of guidance documents for specific types of machine or activity, including:

- Hazards arising from ultrasonic processes (Local Authority Circular 59/1, 2000).
- Metal cutting circular saws (Noise Reduction (Local Authority Circular 59/2, 2000).
- The reduction of noise from pneumatic breakers, hammers, drills, etc (Local Authority Circular 59/4, 2000).
- Reducing noise from CNC presses (Engineering Information Sheet No. 39, 2002).

## Other guidance

N3025    Some working environments carry particular difficulties in the assessment and control of noise exposure, mainly because of the variability of the noise levels and the length of exposure. The following specific guidance may be of assistance:

- *A guide to reducing the exposure of construction workers to noise*, R A Waller, Construction Industry Research and Information Association, CIRIA Report 120 (1990).
- *Offshore installations: Guidance on design and construction, Part II, Section 5*, Department of Energy (1977).
- *Guide to health, safety and welfare at pop concerts and similar events*, Health and Safety Commission, Home Office and the Scottish Office, HMSO (1993).

- *Practical solutions to noise problems in agriculture*, HSE Research Report 212 (2004).

# Reducing noise in specific working environments

N3026    Although many noisy industries are in decline, noise is now becoming an issue in some newer industries, as described below.

## Discos and night clubs

N3027    These venues can be a problem because operators have two conflicting needs: they need to provide loud music for their clients, but they must provide a safe working environment for staff. The music levels on the dance floor are often over 100 dB(A) to provide the buzz that clients expect. It is, of course, an interesting point as to whether clients will in the future be looking for compensation for noise-induced hearing loss.

A number of different techniques may need to be used in order to achieve adequate control of the noise exposure of staff. Loudspeakers should be kept away from the bar area. The bar itself can be designed to be partially screened from the loudspeakers and with sound absorbent to reduce noise levels in the locality. A high-quality sound system with low distortion levels can produce the bass frequencies that clients want, with less high-frequency distortion. This results in lower overall sound pressure levels than a poorer system would produce. Such a system can have the additional advantage of reducing environmental noise levels, which are often problematic. Some clubs provide a quiet 'chill-out' room to which staff and members can retreat.

Ear plugs can be used by staff patrolling the dance floor and seating areas, and jobs can be rotated between staff so that the average noise exposure is reduced. Most ear plugs will reduce sound at speech frequencies by more than at the bass frequencies usual in disco music and this may reduce the ability of staff to understand speech. With the anticipated new Control of Noise at Work Regulations, it will be important to ensure that ear plugs give adequate protection at the important bass frequencies.

## Call centres

N3028    Call centres can be very noisy simply because of the large number of people talking on the telephone. If poorly-designed, each operator will inevitably raise their voice so as to be heard above the noise produced by their neighbours. This becomes self-defeating as others then raise their voices further. They may also need to raise the volume of their earpiece, and this could lead to excessive sound exposure. There have been successful cases of compensation for deafness (sometimes only in one ear) suffered by people using telephone and radio headsets, and employers should ensure that staff do not adjust them to be excessively loud.

A number of techniques must be used to achieve a satisfactory acoustic environment. Firstly the basic acoustics of the call centre must be satisfactory. Working areas should be as sound-absorbent as feasible, with acoustic ceiling tiles and a thick, sound absorbent carpet. Acoustic screens should be used between workstations unless they are widely-spaced. The screens need to have thick absorbent surfaces, and should be high enough to screen operators when they are sitting at their workstations.

Staff training must include discussion of appropriate voice and earpiece levels, and the risks of excessive levels. They need to be encouraged to keep their voices down: it should be explained that it is not necessary for them to raise their voices in order for the customer to hear them clearly.

## Workplace noise assessments

### Competent person

**N3029**     The *Noise at Work Regulations 1989* (*SI 1989 No 1790*) require surveys to be made by a 'competent person'. This is someone who understands the requirements of the Regulations, is experienced in using sound level meters and in making workplace noise assessments. The Regulations do not stipulate a particular form of certification for such people, although the Institute of Acoustics and other organisations do provide formal training and certification that is intended to meet these requirements.

### Sound level meters

**N3030**     It is usually more economical to employ an acoustics consultant for workplace noise assessments than to send a member of staff on a training course and to hire a sound level meter to do the measurements in-house. It may be helpful if the report is clearly independent should an employee make a claim. Suitable sound level meters are expensive. A precision meter is required and these cost some thousands of pounds, depending on their capabilities, and they must be sent away for laboratory calibration at regular intervals. Those meters that can be bought cheaply from electronics suppliers are not suitable: they could be misleading even as a 'screening' tool. This is because cheap meters do not have an accurate response to rapid or impulsive sounds, and usually do not contain an $L_{Aeq}$ function as this requires an on-board computer chip for the complex averaging process.

### Assessing the daily personal noise exposure

**N3031**     It is important to appreciate that noise assessments must determine the daily personal noise exposure of each individual member of staff – it is not sufficient simply to measure the noise level in various parts of the workplace and to mark the noisy areas as 'hearing protection zones'.

In some workplaces, this is not particularly difficult to do: if the employee has a fixed workstation and fixed working routine, it will only be necessary to measure at a location representative of their head position, for a sufficient period of time to cover a few work cycles – maybe five minutes or so.

However, it is quite common for a worker to use a variety of machines or to move between different parts of the works during a normal day. Some workers, such as crane or lorry drivers, may have a work environment where noise levels vary a great deal and it is not practicable for the noise surveyor to accompany them throughout a representative period.

These considerations mean that a workplace noise assessment needs planning before it is made. The first item is to establish the different types of work that are being done on the site: this will need a meeting with the site manager to determine how many staff there are, the jobs that they do, shift-working patterns, and the hours of work, including when and where meal-breaks are taken. It is also important to consider the sensitivities of the workforce: it is helpful to explain in brief terms what you are doing and to confirm what they do. They will also want to know the results, but the surveyor should explain that this cannot be provided immediately as the daily personal noise exposure needs to be calculated. In some workplaces, it may be necessary to ensure that union representatives are consulted.

A scale plan of the workplace will be needed so that measurement locations, noise sources and noisy areas can be recorded for the report. The surveyor will wish to ensure that all machinery and plant is tested, and will want to note anything that is out of use during the survey.

### Dosimeters

**N3032**     Where there are people such as lorry drivers, whom the surveyor cannot accompany, then it will probably be best to ask them to wear a dosimeter. This is a small electronic device about the size of a mobile phone that clips into the top pocket, with a microphone that clips to the lapel or some convenient part of the clothing near to the ear. The dosimeter does not record the actual sound, only the sound level. (It is not a 'spy in the cab'.) Dosimeters usually contain some protection to deter tampering or falsification and to prevent the worker obtaining the readout, although the surveyor will need to be alert to the risks.

Where workers move between various workstations during their working day, then it is usually acceptable to measure the noise level at each workstation and to calculate the daily personal noise exposure from an estimate of the proportion of the day that each worker typically spends at each one.

Although it may seem that this is less accurate than using a dosimeter, it is important to recognise that each of the work-cycles will usually produce a slightly different noise exposure, as there are so many things that affect the noise levels. Moreover, it is necessary to obtain actual workstation noise levels in order to identify any ear protection zones that need to be marked out. In some borderline cases, maybe dosimetry might settle the matter, but it would be better to look at ways of reducing noise levels, since the effect on any one individual's hearing is difficult to determine, even though the Regulations define action levels precisely.

### Hazardous environments

**N3033** Special care is required in hazardous working environments, especially those with an explosion risk (such as gas and petroleum processing plants). Many sound level meters use a high 'polarisation voltage' for the microphone, which could produce a spark. 'Intrinsically safe' meters can be obtained for such instances, although they can be less convenient for ordinary use.

### Requirements for the report

**N3034** A written report of the survey must be produced and kept as a record of the survey at least until the next survey is made. The report should state all the information used in the survey, and should contain a plan showing the areas surveyed and hearing protection zones that were identified. It must identify individual staff or jobs where noise exposure exceeds any of the action levels. The report should also contain a statement of the employer's duties resulting from the findings of the survey. However, it is not usual for the report to include detailed specifications for any noise remediation that might be needed. This is usually a separate study, as the issues cannot be known prior to the survey and there will usually be a number of options which will require detailed discussion with the employer.

## Noise reduction

**N3035** It is the duty of every employer to reduce the risk of hearing damage from noise at work as far as reasonably practicable. This is an on-going and long-term obligation. Unfortunately, it is often difficult to reduce the noise of many items of equipment, particularly hand-held tools. Remarkably, it is still the case that quietness does not seem to be a primary selling point for tools and equipment, and it should not be assumed that newer equipment will necessarily be quieter. Moreover, although many items of equipment must be labelled with their sound power, the label often shows only the permitted maximum level rather than the actual level, so this is of little help in choosing the quietest items. Furthermore, it should not be assumed that smaller items of plant will be quieter than larger items. This is because larger items often have more scope for noise control in the form of extra casings, insulation and silencers, as size and weight are less of a limitation.

With the advent of the proposed new Control of Noise At Work Regulations, the buying policy will become a very important aspect of noise reduction, since it is often difficult, inconvenient and expensive to retrospectively fit noise control equipment. A noise target level should be set and then equipment bought to comply with this, taking into account the combined effect of many items operating simultaneously, and the effect of reverberant build-up of sound in the workplace.

When designing noise reduction measures, it is essential to identify the major sources of noise first, since no amount of work in reducing minor sources will have any effect on the overall noise level. Noise is usually created by a moving or vibrating part of the equipment inducing vibrations in the surrounding air. Noise can be reduced by preventing such vibrations from reaching the operator of the equipment: the most effective solution is to block 'air-paths' with a solid, heavy casing. Even tiny air-holes can allow the sound to escape, which is a problem when a flow of cooling air is needed, or where there must be continuous access to feed work-pieces.

A different approach is to fit 'damping material' to vibrating surfaces, to reduce the amplitude of vibration. Damping material is especially effective on thin sheet metal, such as equipment casings, hoppers, containers and ductwork. This is because even a small amount of vibration can cause their surfaces to 'radiate' a large amount of noise energy. The vibrational energy is absorbed in the damping material, rather than being radiated. There are a number of ways of applying damping material: a bitumastic or rubber sheet can be glued to the surface, a second sheet of metal can be glued by an elastic adhesive to the surface, or the item can be made from a laminate composed of two sheets of metal glued with an elastic adhesive.

It should be noted that sound absorptive material (mineral fibre, etc) is not used as a damping material. However, it can be used to provide a vibration break when cladding a vibrating surface. For example, if noise is breaking out from a duct, it can be wrapped in mineral fibre and then clad with a thin metal sheet. The mineral fibre separates the inner and outer cladding, thereby preventing much of the vibration in the wall of the duct from reaching the cladding, which therefore radiates much less noise.

Sometimes, a less 'springy' material than steel can be used. At one time lead was popular for being both heavy and floppy, but is now considered too hazardous for general use. However, plasterboard is a popular acoustic material, as it is reasonably heavy, has good natural damping and is inexpensive. It can often be used to prevent noise breakout, as long as there is a vibration break between it and the radiating surface.

Ventilation and cooling is often a problem, since enclosing a piece of equipment can prevent 'natural' cooling. Moreover, fans used to provide a flow of cooling air can be very noisy in themselves, not counting the noise that escapes along the ductwork. The best solution in such cases is to place the fan on the inside of the casing and to fit an attenuator ('silencer') on the external side of the fan. Attenuators are essentially rectangular or circular ducts with an acoustic absorbent lining.

There are many different types of fan and it is important to select one that is appropriate for the task in hand, taking account of the volume of air that needs to be moved and the pressure (resistance) that the fan must work against. Small, fast-running fans tend to be noisier than larger slower-running fans for a given duty. The amount of noise produced by a fan and the amount of attenuation provided by a silencer are well-characterised and so ventilation systems can be designed with confidence over their performance, although this is a specialist job.

Sometimes, developments intended for another purpose can have the additional benefit of reducing noise. For example, there was a need to develop a circular saw blade that gave better cutting accuracy. It was found that vibration of the saw blade caused irregularities in the cut, so slots were cut in the surface of the blade and these were filled with a resin to maintain the strength of the blade. This extra damping was effective in reducing vibration, thereby improving both the cut and the noise levels.

Despite the technology, it is often difficult to obtain adequate noise reduction. This is because of the huge range of the hearing mechanism. A 10 dB reduction requires the escaping noise energy to be reduced by 90 per cent, a 20 dB reduction requires the escaping noise energy to be reduced by 99 per cent, and a 30 dB reduction requires 99.9 per cent of the energy to be removed. Because of these considerations, it may be advisable to employ a specialist who can design a solution with a minimum of trial and error.

## Ear protection

N3036   There are two main categories of ear protectors:

(i) circumaural protectors (ear muffs) which fit over and surround the ears, and which seal to the head by cushions filled with soft plastic foam or a viscous liquid; and

(ii) ear plugs, which fit into the ear canal.

Ear protectors will only be effective if they are in good condition, suit the individual and are worn properly. Ear protectors can be uncomfortable, especially if they press too firmly on the head or ear canal, or cause too much sweating. Some users may be tempted to bend the head-band so that the muffs do not press so tightly, which reduces the effectiveness of the seal. Ear muffs are also less effective for people with thick spectacle frames, long hair or beards that prevent the muff from sealing fully against the head.

The amount of protection differs between different designs, and usually ear protectors give more protection at middle frequencies and less protection at low frequencies. Where the noise has strong tones (often described by words such as rumble, drone, hum, whine, screech) then particular care is needed to ensure that the protectors give adequate attenuation at those frequencies. However, it can be counterproductive to choose ear protectors that give far more attenuation than necessary, since these will be heavier, with greater head-band or insert pressure and so less comfortable to wear. And an ear protector that is not worn does not give any protection at all. Ear plugs should not be used by people with ear infections and certain other ear conditions, and it may be desirable for such people to obtain medical advice.

Because individuals differ greatly in their preference, the employer should select more than one type of suitable protector and offer the user a personal choice where possible. British Standard BS EN 485: 1994 gives guidance and recommendations for the selection, use, care and maintenance of hearing protectors.

Ear protectors need to be maintained in a clean and hygienic condition. Insert types should not be shared between people, and muffs should not usually be shared. With ear muffs, it is necessary to check that the sealing cushion is not damaged, as this will seriously reduce its effectiveness. On some makes it is possible to change a damaged cushion. The ear cups have a foam lining which must not be discarded. If it gets dirty, it can usually be taken out and washed in detergent, then dried and replaced.

If the headband tension becomes weak, the muffs should be discarded.

Some earplugs are designed for one-off use, but others are reusable. These need to be kept in a protective container when not in use, and regularly washed according to the manufacturer's instructions. If the seals become damaged or hardened, the plugs must be replaced.

The employer needs to ensure that staff know how to check the condition of their ear protection and how to obtain replacements as soon as the need occurs.

## Acoustic booths

N3037   In some workplaces, it may be possible for operators to be located in a control booth from which they can monitor the plant, only occasionally entering the noisy area. They may still need put on ear defenders when leaving the acoustic booth, since it may require only a few seconds or minutes to exceed the allowable daily noise dose.

Even where it is not practical to have a separate control booth, it may be possible to use an acoustic refuge – a small cabin with an open side – that will give some respite from the noise. However, this should not be expected to give more than a nominal sound reduction in most cases.

## Work rotation

N3038  Work rotation can sometimes be a method of reducing the daily personal noise exposure. The staff work part of the day in a noisy area and the rest of the day in a quiet area. If a person works for half their day in a place with a noise level of 93 dB $L_{Aeq}$ (4 hours) and the other half of their day in a place with a noise level of 70 dB $L_{Aeq}$ (4 hours) they would just meet a noise exposure of 90 dB. However, this is not a very satisfactory method as it can easily go wrong. They may spend longer than expected in the noisy workplace, or there may be a higher than expected noise level in the quiet area, for example. And again, whilst this approach could satisfy the legal requirements, some people may suffer hearing loss from shorter exposure to loud noise.

## Noise barriers

N3039  Noise barriers often give very limited noise reduction within buildings. This is because sound can be reflected over the barrier from the ceiling, pipe-work, and other overhead fittings. It is usually necessary to use noise barriers in conjunction with sound baffles suspended from the ceiling. Suspended sound baffles on their own have very limited application. They can sometimes be helpful in very reverberant work-spaces where the staff are not exposed to the direct sound of noisy machines.

## PA and music systems

N3040  Some workplaces use public address and music systems. One study found that they were a major source of excessive noise in lorry cabs. Care must be taken to ensure that in noisy workplaces these system are not so loud that they create excessive noise in themselves. This also applies to people who wear radio or telephone headsets.

## Sound systems for emergency purposes

N3041  Voice messages can be superior to bells or sirens to convey warnings and instructions in emergencies. However, these must be properly audible in noisy places. British Standard BS 7443: 1991 gives specifications for sound systems for emergency purposes.

# Occupational Health and Diseases – An Overview

## What is occupational health?

O1001 Occupational health is concerned with the prevention, identification and management of health issues relating to employment. It is very broad and covers not only the fact that work and the conditions of work may cause ill-health, but also includes the management of employees who have a health problem which may impact on their work or their employment. One definition of occupational health is that it concerns *the effects of work on health and the effects of health on work*. The primary aim of good occupational health practice should always be to prevent ill health. Occupational health practitioners (see O1010 below) are health professionals who are additionally trained to understand the complex relationships between work and the impact this may have on the health of the individual. Occupational ill-health is caused by exposure to a wide variety of chemical, physical and organisational aspects of the work and it's environment and to be effective an occupational health practitioner must have a good understanding of how the environment at work (in it's broadest context) might affect a person's health.

## Why is it important?

### Sickness absence

O1002 Occupational health is an important, but often neglected, aspect of an organisation's management. It is important to the organisation for a number of reasons. Employees are the most important, and often the most costly, asset any organisation has. Sickness absence can be very costly to an organisation. Of *short term* sickness, 40–50% is potentially avoidable and can be put down to a variety of reasons ranging from drug and alcohol abuse to lack of commitment and interest in the job and poor management. Of absence, 20% is *long term* – employees who are absent for weeks or months with a condition that renders them unable to travel to work and/or to carry out their expected duties. This 20% of long-term sickness absence accounts for 70% of the costs of sickness absence. A good occupational health service should be able to considerably reduce the level of *avoidable* sickness absence and hasten the successful return to work of those who are absent long term, with significant financial savings for the organisation (see R2001 REHABILITATION).

### Personnel support

O1003 An occupational health service can be an invaluable support to the Human Resources/Personnel function in managing the many aspects of employment that relate to an individual's medical condition and their consequent fitness to work. As well as managing sickness absence (see O1002 above), the occupational health service can advise on fitness to work at the point of recruitment or placement (pre-employment or pre-placement health screening). This will reduce the risks to the organisation of recruiting someone to a position for which an existing medical condition renders them vulnerable to further illness or injury or for which they may be simply unfit to undertake the proposed work. It may be necessary at the point of recruitment to consider the implications of the *Disability Discrimination Act 1995*, and to determine what restrictions may have to be placed on an employee's work or what reasonable alterations may have to be made to the work or work-place in order to avoid discrimination. Other medical aspects of employment on which the occupational health service can advise include retirement on grounds of ill-health, temporary restrictions on work or working times to aid recovery or rehabilitation, keeping at-risk employees under health surveillance, and advice on managing medical conditions which are impacting on the capacity of the employee to undertake expected duties.

## Legal duties

**O1004** Another important reason for having access to occupational health advice is to ensure that statutory and civil legal obligations are being met. Several regulations place a legal requirement on the employer to place at-risk employees under health surveillance or to have regular medical examinations (usually by a doctor 'appointed' or 'approved' for the particular requirements of the regulation). The *Ionising Radiations Regulations 1999 (SI 1999 No 3232)*, the *Control of Lead at Work Regulations 2002 (SI 2002 No 2676)* and the *Control of Asbestos at Work Regulations 2002 (SI 2002 No 2675)* are examples of regulations which may require medical surveillance by a relevant doctor. In addition, other areas of work, such as working with respirators and working offshore, may require medical examinations either under statute or under industry standards.

## Risk assessment

**O1005** The occupational health department may also need to be involved in risk assessments where these involve health considerations. Under the *Management of Health and Safety at Work Regulations 1999 (SI 1999 No 3242)* there is a general requirement to undertake an assessment of the risk to health and safety of employees. In some circumstances the risk may be to the employee's health (a classic example being the risk assessment for work-related stress (see O1054 below) and the occupational health department can play an important role in helping to determine those aspects of the job that may be stressful. In many instances the risk assessment will have to take account of an individual's vulnerability (perhaps because of age, pregnancy, or previous history of injury or illness) and the occupational health department is in a unique position to translate a previous medical history (or current condition) into the additional risks that this may pose to the employee. As importantly, occupational health can also advise on what needs to be done to avert the risk either by additional protection or modifications to the job or work routine. Examples of where there may be a need to take account of individual vulnerability, in addition to pregnancy, include a history of upper limb disorder, a history or earlier episode of back pain, earlier periods of stress-related disorder, a history of a skin allergy and existing respiratory disorder (such as asthma). There are of course many other examples, but it is extremely important to take such vulnerabilities fully into account in the risk assessment; failure to do so may increase the chances of successful litigation for damages for failure to exercise an appropriate level of duty of care. Other regulations such as the *Control of Substances Hazardous to Health Regulations 2002 (SI 2002 No 2677)*, the *Health and Safety (Display Screen Equipment) Regulations 1992 (SI 1992 No 2792)*, and the *Manual Handling Operations Regulations 1992 (SI 1992 No 2793)*, may require an occupational health input into the risk assessment and may require employees to be placed under health surveillance.

# Relationships between occupational health and safety

**O1006** Generally, it is only the larger organisations which have an occupational health department, although many smaller employers have access to occasional occupational health or medical advice through buying in a service from a local provider or from the local general practice. Most organisations, though, will have some sort of safety provision, either a dedicated safety department (again mostly in the larger organisations) down to a safety person who fits safety in with their main job. Occupational health and safety are two different, but complementary, means of achieving the same goal – ensuring the well being of both the organisation and its employees.

# Prevention is better than cure

## Being reactive

O1007   Inevitably there are cases of employees becoming ill, either as a result of work or for reasons completely unrelated to work. Either way, the role of the occupational health service is to advise management on the employment issues associated with the individual. If the illness is work related or thought to be so, then the advice relates to what can be done to ensure that there is no exacerbation of the condition when the employee returns to work. If it is unrelated to work then the advice may be concerned with how a person's newly acquired health condition may impact on work, either by affecting the way he or she can perform the expected duties or the way in which the medical condition may now render that person more susceptible to the conditions of work. The employer may require advice on employees who are taking frequent short term sick leave or a period of long term sick leave and seek advice on how to manage the sickness absence or manage the rehabilitation and return to work. They may need advice on medical retirement on grounds of permanent ill health. All of these are examples of occupational health, albeit playing an important role, but being *reactive* – reacting to the needs of the organisation in managing the consequences of staff being unfit for work.

A good occupational health service should also be *proactive* – prevention being better than cure. Being proactive involves the anticipation of the possibility of ill health and putting into place measures, which are aimed at preventing or minimising the risk of this occurring. Prevention starts with the pre-employment or pre-placement health assessment (see O1003 above). This is aimed at preventing employees being recruited to jobs for which they are medically unfit or making sure that where preventative measures need to be taken (e.g. vaccinations) then these are undertaken before commencement of work. During employment, prevention is then exercised through a number of initiatives that a good occupational health service will be engaged in. Health education and health promotion programmes are a good way of improving and maintaining the health of the workforce. This may be strictly confined to those health issues related to the work, (for example how to look after your skin, how to prevent hearing damage, or how to prevent musculoskeletal disorders) or it may be concerned with general health promotion such as smoking cessation, women's or men's health, look after your heart campaigns and exercise, diet and obesity. It is obviously good if the organisation is actively involved in promoting health generally and maintaining a healthy workforce will bring benefits and this fits with the Government's initiative in 'Securing Health Together' (see O1020 below). However, not all organisations will see general health promotion as being their responsibility and may not engage in this on account of cost and lack of resources.

## Prevention and risk assessment

O1008   Prevention is also achieved by involving occupational health in the risk assessment process (see O1005 above). This ensures that relevant health issues are fully included the risk assessment process. As discussed at O1005 above, the occupational health practitioner can help ensure that the medical aspects of the risk assessment are fully included and that individuals' vulnerabilities, where relevant, are taken into account. In addition the occupational health input can ensure that, where required, appropriate health surveillance is set up and carried out as a means of early detection and hence prevention of work-related disorders.

## Health surveillance

**O1009** Health surveillance is another important aspect of prevention in which the occupational health department has an important role to play. *Regulation 6* of the *Management of Health and Safety at Work Regulations* (*SI 1999 No 3242*) requires that:

> 'every employer shall ensure that his employees are provided with such health surveillance as is appropriate having regard to the risks to their health and safety which are identified by the assessment'.

The Guidance and Approved Code of Practice to the Regulations (HSE L21), provides the criteria where health surveillance is considered to be necessary. These are where:

(*a*) there is an identifiable disease or adverse health condition related to the work concerned;

(*b*) valid techniques are available to detect indications of the disease or condition;

(*c*) there is reasonable likelihood that the disease or condition may occur under the particular conditions of work; and

(*d*) surveillance is likely to further the protection of the employees to be covered.

Health surveillance can be carried out by anyone competent to do so; it does not necessarily have to be an occupational health professional. Simple surveillance can be done by a trained supervisor (checking for early signs of dermatitis for example), or even by self-surveillance if the individuals at risk know what to look for and who to report concerns to. However, where surveillance includes physiological or biochemical measurements, such as hearing function, lung function, or urine or blood tests, then an occupational health doctor or nurse would usually be necessary. As mentioned at O1004 above, under some regulations it is mandatory for the health surveillance to be carried out by an approved or appointed doctor. What is most important, however the health surveillance is carried out, is that there is a recognised means of referral for expert help if the health surveillance reveals someone with early signs or symptoms of a work-related condition.

In summary, whilst occupational health has an important role to play in investigating and managing illness and the consequences of ill health that occur during the course of employment, the core of occupational health practice should be to maintain a healthy workforce and prevent as many work-related conditions as possible from arising in the first place.

# The occupational health team

**O1010** The occupational health 'team' varies enormously from one organisation to another depending on the complexity of the working environment and the perceived needs of the organisation. It can vary from no provision at all up to a multi-professional team. The majority of organisations who make provision for occupational health, do so with the core service being provided by an occupational health nurse who may be full or part time depending on the size of the organisation. Occupational health nurses, working on their own, will usually have a network of external support, including a physician. The external support should include other occupational health nurses who can participate in clinical audit.

## Occupational health nurses

**O1011** *Occupational health nurses* (also called occupational health advisors) are registered nurses whose names appear on the statutory register to practice held by the Nursing and Midwifery Council (NMC). In addition, an occupational health nurse will hold a qualification in occupational health; i.e. OHN Certificate, Post Graduate Diploma or Degree and their names will be recorded on the specialist register of the Nursing and Midwifery Council.

Experienced occupational health nurses can undertake a wide range of duties including determining fitness for work by pre-employment health assessment, contributing health considerations to the risk assessment process, undertaking health surveillance, and advising the organisation on health issues relating to individuals, such as sickness absence management and return to work. Occupational health nurses also usually play a key role in preventing work-related illness and promoting good heath among the workforce. Many occupational health nurses are also competent to undertake environmental assessments such as simple measurements of noise, lighting etc., and workstation assessments.

The competent occupational health nurse will recognise his or her limitations and be prepared to refer an individual for a medical opinion from a doctor or refer a problem that it outside of his or her competence for expert help.

Where other advice or practical help is required this may be from any one of a number of sources:

- occupational health physicians;
- occupational hygienists;
- ergonomists;
- physiotherapists; and
- counsellors.

## Occupational physicians

O1012    *Occupational physicians* are doctors who have specialised in occupational medicine. To become a specialist in occupational medicine the doctor must first of all undertake a programme of study and examination to acquire Associateship of the Faculty of Occupational Medicine (AFOM). This qualification is an indication of the doctor's expertise in occupational medicine but they are not regarded as accredited specialists (with consultant status in NHS terms) until they have gained full membership of the faculty (MFOM). Full members are admitted to the accreditation register held by the faculty. Some doctors also hold an MSc in Occupational Medicine or a Diploma in Industrial Health (DIH). These are academic qualifications in occupational medicine but holders of these qualifications may or may not be accredited specialists.

The role of the occupational physician is to undertake medical examinations and to diagnose work-related ill health, to determine the work-related causes of ill-health and to advise on the management of health problems which impact on the person's ability to work. Pension funds also often require a doctor to make the decision regarding permanent ill health retirement or admission to pension schemes, although they do not necessarily have to be specialist occupational physicians.

## Occupational hygienists

O1013    Occupational hygienists are employed to measure chemical and physical hazards in the workplace. Information on levels of exposure to chemicals, dusts, fumes, noise etc., is often required in order to determine the risks to health. There is also sometimes a statutory need to determine exposure levels and to keep records of individual exposures. Occupational hygienists are also expert in advising on protective equipment such as the suitability of hearing defenders and dust masks.

## Ergonomists

O1014    *Ergonomists* have specialist skills, which enable them to design work systems that optimise the match between the person and their work and work environment. The important skill of the ergonomist is to understand the mental and physical capabilities of people and the wide range of human variability, and to design work systems which impact least on the person's health.

### Treatment services

**O1015** *Physiotherapists, chiropractors and counsellors.* Occupational health is primarily concerned with the prevention of work-related disorders and the management of health problems that affect employees at work and the organisation employing them. Although it is not primarily the function of occupational health to provide treatment, many organisations consider it a worthwhile investment to provide some types of treatment in order to assist the employee in recovering and returning to work more quickly. Such treatment services typically involve those aimed at recovery from musculoskeletal disorders and psychological illness which together form the two most frequent reasons why employees are off sick. Physiotherapy (and/or chiropractic) and counselling are therefore sometimes provided by the occupational health service, or organised with external providers through the occupational health service.

## Health records

### Health records and confidentiality

**O1016** It is sometimes necessary to obtain the medical records of an employee or a prospective employee (at the time of recruitment), in order to make a judgement about the current state of the person's health and their fitness to work. If this is done then it is important that these are obtained by a heath professional employed by the company (or engaged by the company under contract), who can take responsibility for the confidentiality of the records. Medical records should always be kept confidentially under the guardianship of a registered medical practitioner or registered nurse, and should not form part of the person's personnel file.

### Obtaining records and reports

**O1017** Requesting medical records or a medical report requires the informed consent of the individual and is subject to the legal requirement of the *Access to Medical Reports Act 1988* (in the case of medical *reports*) and the *Data Protection Act 1998*, (in the case of health and medical *records*). The right of personal access to health records used to be under the *Access to Health Records Act 1990*, but this has now been repealed by the *Data Protection Act 1998*. The only provision now contained in the *Access to Health Records Act 1999* is where access is requested to health records pertaining to the deceased. Under the *Access to Medical Reports Act 1990* an individual has the right of access to any medical report relating to him or her which is to be, or has been, supplied by a medical practitioner for employment purposes or insurance purposes.

Subject access to health records under the *Data Protection Act 1998* is much wider than under the *Access to Medical Reports Act 1988*. It applies to any health record kept by any health professional including doctors, nurses, midwives, dentists, chiropodists, physiotherapists and clinical psychologists, who are, or have been, responsible for the care of the individual. An applicant for access to a health record can be the patient or another person that the patient has authorised in writing to make application on his behalf. In the case of an employer seeking access to an employee's health records, the employer is acting on behalf of the patient (his employee) and therefore must seek written informed consent to make an application. Under this act the individual does not have the automatic right to see the record before it is released to the applicant, but does have the right to subsequently request that any misleading or inaccurate entry in the health records is corrected.

The normal way in which an employer seeks to obtain a medical report or access to health records is to explain to the employee what the purpose is in requesting a medical report or copies of health records and asking the employee to sign a consent form. The consent form must clearly spell out the legal rights of the individual under the *Access to Medical Reports Act 1988* and *Data Protection Act 1998*.

## Keeping and making use of the occupational health records.

**O1018**     Some occupational health records have to be kept for a statutory period of time. Under the *Control of Asbestos at Work Regulations 2002* (*SI 2002 No 2675*), for example, the health records of employees exposed to above the action level have to be kept for a minimum of 40 years. Under the *Control of Substances Hazardous to Health Regulations 2002* (*SI 2002 No 2677*), health records pertaining to individuals must be kept for a minimum of 30 years.

Health records are kept for management purposes and these obviously relate to individuals and must be kept confidentially. These records are used firstly to provide a record of any health surveillance, referrals for medical opinion and the opinions received, the results of tests and vaccination history etc. These form an important record of the duty of care afforded by the employer and might be necessary to defend possible claims for damages. Health records are also kept for health management purposes such as recalling employees for routine health checks and tests or making sure that vaccinations are kept up-to-date.

Health records on individuals are confidential and that confidentiality must be preserved. However, if the records are kept in such a way as to allow analysis of collective and anonymised data then they can be a useful means of detecting trends in illness or detecting the emergence of a new illness or condition that has not previously been recognised in the organisation. For example, by analysing the health surveillance records of all the employees working in a particular department (perhaps by producing monthly statistics), it might become apparent that since the introduction a new solvent cleaner that there has been an increase in the incidence of dermatitis. Health records should therefore not simply be a passive record of what was done and was found, but be used positively to promote the health of the individual and employees collectively and to protect the liability of the organisation.

## Buying services and priorities

**O1019**     The priority for any manager contemplating buying in an occupational health service is to determine exactly what the organisation needs and what it wants from its occupational health provider. The needs are determined firstly as a 'must have' list and then a 'nice to have' list. 'Must have' would include any statutory requirement for medical examination or health surveillance and whatever else the organisation determines it must have to manage the business effectively and support the personnel function. This would include pre-employment screening, advice on the management of sickness absence and rehabilitation, and any requirement of the pension fund regarding ill-health retirement (see O1003 above). The nice to have list would build on this, depending on the organisation's resources and intention to promote health and be proactive. This list would include non-statutory health surveillance, occupational health involvement in risk assessment, and health promotion and health education activities.

Organisations intending to buy services from an occupational health provider, can locate one near to them by undertaking an internet search The key words 'occupational health provider' (restricted to a UK search) will identify many such independent providers. The National Health Service offers commercial occupational health provision in many areas, through NHS Plus (http://www.nhsplus.nhs.uk). It is important to ensure that the provider is able to offer competent and properly qualified personnel for the work expected from them. Ask to see CV's of the personnel being put forward and seek confirmation of the professional registration of the medical and nursing staff.

## Initiatives and targets

**O1020**     The Government's 'Securing Health Together' initiative (securing health together – a long-term occupational health strategy for England, Scotland and Wales HSC MISC225) lays out a number of strategies and targets for occupational health. The four main aims are:

- to reduce ill-health in workers and the public caused or made worse by work;

- to help people who have been ill, whether caused by work or not, to return to work;
- to improve work opportunities for people currently not in employment due to ill-health or disability; and
- to use the work environment to help people maintain or improve their health.

The targets are:

- a 20% reduction in the incidence of work-related ill health;
- a 20% reduction in ill-health to members of the public caused by work activity;
- a 30% reduction in the number of days lost due to work-related ill-health; and
- everyone currently in employment but off work due to ill health or disability to be made aware of opportunities for rehabilitation.

The aim is that by 2010 everyone has access to appropriate occupational health support. This is currently being developed by the HSE as a series of pilots (Workplace Health Direct) in which small and medium sized enterprises will be able to get limited occupational health advice (including work-site visits) to kick-start their own occupational health provision.

# Occupational diseases and disorders

## Noise and occupational hearing loss

### What is it?

O1021    Noise is defined as unwanted sound. Sound is created when an object that is vibrating or impacting causes a wave of pressure to move through the air. The magnitude of the air pressure wave is a measure of the intensity of the sound which our ears and brain detect as volume or sound level. The number of times the wave repeats itself in a given time is the frequency of the sound. Again the ear and the brain interpret this as the pitch of the sound. By utilising combinations of intensity and pitch we can interpret the meaning of sounds such as speech or music for example.

Sounds from the environment are an important way for us to communicate and be aware of dangers. However, both the ear and the brain can be affected by sounds that are harmful. Sound can be harmful in two ways:

- when the sound level is sufficiently high to cause damage to the mechanism of the ear; and
- when the level and pitch of the sound is such that it interferes with concentration, communication or the peaceful enjoyment of our surroundings.

Noise is therefore a level of sound sufficient to cause hearing damage or sound of sufficient intensity or frequency to interfere with expected levels of communication or mental well being. The latter effect of noise can have very important consequences at work when, for example, noise in a busy office causes annoyance, distraction, loss of concentration and stress. These psychological effects are important to note but will not be dealt with in further detail here. This section looks in particular at the effects of noise on hearing loss.

### How is it caused?

O1022    The ear receives the changes in air pressure, by a vibration of the eardrum (tympanic membrane). The vibrations of the eardrum are passed through a chain of small bones (the ossicles), and transmitted into the inner ear (the cochlea). In the cochlea the vibrations distort microscopic hairs attached to hair cells, and it is this distortion of the hairs that triggers nerve impulses which pass to the brain where frequency and sound level are interpreted as recognisable sounds. The cochlea is able to separate the sounds into individual frequency and sound level is interpreted by the magnitude of the distortion of the hair cells. Hearing loss is caused when the sound levels are so great that the hairs of the hair cells are distorted so much that they fracture. Once damaged in this way the hair cells do not recover. Also, because the cochlea separates frequencies, the part of the cochlea that is damaged will depend on the frequency of the sound that has caused the damage. It is for this reason that hearing loss is sometimes confined to a small range of frequencies (see figure 1 at O1023 below). The young human ear can normally detect frequencies in the range of about 20 to 20,000Hz (1Hz (Hertz)) is one cycle per second), although this ability does decline significantly with age and the upper limit may fall to 16,000Hz or less as we get older. The ear can detect sound pressure from as little as $2 \times 10^{-5}$ Newtons per square metre ($N.m^{-2}$), (the threshold of hearing), and is most sensitive at around 2 to 4kHz (kilo Hertz) which roughly corresponds with human speech frequencies.

Under most circumstances, hearing loss is gradual with the damage being cumulative over many years. Hearing loss, even in the early stages, may also be accompanied by tinnitus.

## How is it detected?

O1023    Sound level is measured using a sound level meter which measures the sound pressure level according to an agreed international standard. Conventionally the sound intensity is measured using a logarithmic scale, the decibel (dB) scale. The International and British Standards use reference levels that define 0dB as the threshold of hearing. On this scale 0dB is the average threshold of hearing and 130dB is the pain threshold; this roughly defines the range of human hearing ability (although not safe levels of exposure). One other characteristic of sound level meters is the convention to relate the measured sound to the sensitivity of the ear at different frequencies. The A weighted scale is most often used (although others exist), as this is weighted to correlate most closely between measured and perceived sound. It also correlates best with the ability of sound to cause hearing damage. When the A weighted scale is used, the letter A is placed in brackets e.g. 60dB(A).

Hearing loss is measured using an audiometer; the resulting graphical expression of the ability to perceive sounds is the audiogram. The audiometer measures the sound level at which the individual can just perceive the sound, at a range of individual frequencies, usually between 125Hz to 8,000 or 12,000Hz. As this uses the dB(A) scale, a normal audiogram would show a fairly straight line across this frequency range at around 0dB(A) (defined as the threshold of hearing). If there is hearing loss, greater levels of sound will be required to elicit hearing. Hearing loss may occur at any frequency or across the whole range of measured frequencies, but most often starts with a loss at around 3–4,000Hz, the frequency at which the ear is most sensitive.

Figure 1 (a) shows a normal audiogram with the responses close to 0dB across the frequency range. Figure 1 (b) shows the typical drop in hearing ability, in this case of between 50 and 60dB, in the frequencies around 4,000Hz. This is a case of occupational hearing loss.

## How is it managed?

O1024    Hearing loss (occupational deafness) is a serious risk in many industries which involve the use of machinery, or physical percussion or impact. To ensure that the risk of hearing loss is minimised (and legal requirements are being met) the following steps should be taken.

- Identify any areas or work tasks, which may exceed the legal threshold for noise at work (see O1025 below).
- Identify individuals who may be exposed to these noisy areas or types of work.
- Obtain a baseline audiogram on any exposed individuals. (Audiometry should be undertaken in an approved soundproof booth to avoid erroneous results due to interference with the test from background sounds).

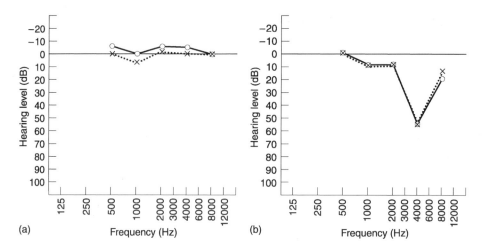

- Have a noise assessment undertaken on the noisy areas by a competent person and act on the report and advice given (see note under O1025 below).
- Ensure that every reasonable effort is taken to reduce the noise at source. Protect the hearing of exposed employees by issuing recommended hearing protection.
- Ensure relevant employees are properly trained and informed in the measures to protect themselves from the damaging effects of noise and how to properly wear and maintain the hearing protection provided.
- Have periodic audiometry undertaken and keep the records.

## What are the legal requirements?

O1025     The primary legal requirement is contained in the *Noise at Work Regulations 1989* (*SI 1989 No 1790*) as amended by the *Quarries Regulations SI 1999 No 2024*. The following are the key duties under these regulations.

(a)   Employers must survey the workplace for areas likely to come under the Regulations. Any area likely to be affected must have a formal 'noise assessment' carried out by a competent person (see note below).

(b)   If the sound level is above 85dB(A), ('the first action level') hearing protection must be provided. If it is above 90dB(A), ('the second action level'), it must not only be provided, but it must be worn (see note below).

(c)   The peak action level of 200 pascals must not be exceeded.

(d)   Risk of hearing damage must be reduced to the lowest reasonably practicable level.

(e)   Noise control measures must be fully maintained and properly used.

(f)   Education and training – employees must know the risks from noise and how to properly use the means to protect themselves from these risks.

(g)   Part of the premises where noise could be damaging must be demarcated and marked as 'Ear Protection Zones'.

Note. Under these regulations it is necessary to assess the cumulative dose of noise during an eight-hour working day. This is done by measurement of the sound levels in the work area (or personal noise dose), and by extrapolation to derive a daily personal exposure over an eight-hour working day (the $L_{EP,d}$). The values of 85 and 90dB(A) quoted above are $L_{EP,d}$ values. It is a legal requirement to have the noise assessment performed by a competent person, which usually means an occupational hygienist with experience of noise assessment work (see O1013 above).

Health surveillance may be considered appropriate under the *Management of Health and Safety at Work Regulations 1999 (SI 1999 No 3242)*. The choice, issue and maintenance of hearing protection is also covered by the *Personal Protective Equipment at Work Regulations 1992 (SI 1992 No 2966)* as amended by the *Health and Safety (Miscellaneous Amendments) Regulations 2002 (SI 2002 No 2174)*.

### The law is set to change

O1026    The EU Physical Agents (Noise) Directive (EU 2003/10/EC) is due to be enacted in this Country by the 15 February 2006. This will reduce the first action level to 80dB(A) and the second action level to 85dB(A). There will be a limit of exposure of 87dB(A), which will take account of the effect of any hearing protection worn.

## Hand arm vibration syndrome (havs) and whole body vibration

### What is it?

O1027    Hand arm vibration syndrome (HAVS) is a set of symptoms suffered by people whose work exposes them to vibrating hand tools. These include drills, de-scalers, grinders, jackhammers, riveting guns, chain saws, brush cutters etc. Whole body vibration causes a different set of symptoms and is potentially caused by driving or being passenger on a variety of vehicles or work platforms, especially those driven off-road and over rough terrain.

Vibration describes the oscillatory movements of an object. Vibration can be characterised by its magnitude (the amount of displacement measured as the distance the object moves during the course of its motion) or by the speed it moves or the rate of change of speed (the acceleration). Vibration, either of the hand-arm or whole body, is also characterised by the direction of movement. This can be in the fore-aft direction, or the horizontal or vertical plane or a complex combination of these axes of movement. Although all of these characteristics can be measured, the value that most closely correlates with the health risk is the acceleration. The acceleration, measured in the unit of metres per second per second (m.s$^2$) is the most usual way of describing the magnitude of the vibration in terms of the likely health risk. It is also usual to measure the acceleration in each of the three axes of movement (fore-aft, vertical and horizontal).

The common symptoms of HAVS are:

- painful finger blanching attacks triggered by cold or wet conditions (vibration white finger);
- loss of sense of touch and temperature;
- numbness and tingling;
- loss of grip strength; and
- loss of manual dexterity.

HAVS is a progressive condition, which starts with minor sensory changes to the fingers and progresses either slowly or rapidly to become a serious and debilitating condition. It may cause the sufferer to be unable to continue working.

Whole body vibration causes a number of symptoms, which commonly include:

- back pain, displacement of intervertebral discs, degeneration of the vertebrae, osteoarthritis;

- abdominal pain, digestive disorders;
- increased urinary frequency, prostatitis;
- visual disturbance;
- sleeplessness; and
- headache.

## How is it caused?

O1028    The causes of HAVS and symptoms of whole body vibration are not fully understood. HAVS can be due to changes in the blood vessels of the arm and hand, the nerves supplying the blood vessels or in the muscles and bones. The most likely cause of the condition in the majority of cases is damage to the nerves supplying the blood vessels caused by the repeated trauma of the shaking of the upper limb. Vibration white finger is caused by a severe reduction in the blood supply to the fingers but may actually be due to damage of the nerves controlling the blood vessels. Once damaged, the nerves do not recover. HAVS is often triggered, and the condition worsened, by exposure to the cold at the same time as using the vibrating tool or equipment.

The symptoms of whole body vibration is also caused by the trauma of repeatedly being shaken and may involve more generalised symptoms such as fatigue, headache and sleeplessness.

## How is it detected?

O1029    HAVS and symptoms of whole body vibration will need to be diagnosed by an occupational physician. Anyone exposed to vibrating equipment should be kept under health surveillance. Health surveillance does not need to be done by an occupational health professional, but the person responsible for undertaking the surveillance (including the exposed individuals if self-surveillance is used) must know what to look for and there must be in place a means of referral for expert occupational health advice for anyone suspected of showing early signs or symptoms of the condition.

Vibration white finger is usually quantified and recorded using the Stockholm Scale.

The Stockholm Scale

| Stage | Grade | Description |
| --- | --- | --- |
| 0 |  | No attacks |
| 1 | Mild | Occasional attacks only affecting the tips of one or more fingers |
| 2 | Moderate | Occasional attacks affecting the distal and middle phalanges of one or more fingers |
| 3 | Severe | Frequent attacks affecting all the phalanges of most fingers |
| 4 | Very severe | As in stage 3 but with trophic skin changes at the finger tips |

## How is it managed?

O1030    The health risks from vibration exposure, especially hand arm vibration, are serious and debilitating. The risk of these conditions must be effectively managed and competent advice taken if there is thought to be a need for exposure assessment, diagnosis or management of the health effects.

Jobs, which expose people to the risk of vibration illness, and the individuals who are potentially affected, must be identified. A competent risk assessment must be made and the control measure identified by the risk assessment must be put in place. This may include measures such as job rotation to reduce the risk, improved maintenance of equipment to reduce vibration, replacement of old equipment by new, wearing vibration absorbing gloves, making sure that the hands are protected from the cold, and placing at-risk individuals under health surveillance.

Individuals identified by the health surveillance as having early signs or symptoms of the condition must be referred for expert occupational health advice.

### What are the legal requirements?

O1031    The EU Physical Hazards (Vibration) Directive (2002/44/EC) has given rise to the *Control of Vibration at Work Regulations 2005* (*SI 2005 No 1093*). These come into force on 6 July 2005. Under these regulations there is a requirement to carry out a health risk assessment and to eliminate or control the risks to health arising from vibration. For hand arm vibration there will be a new daily exposure *limit value* of $5 m.s^{-2}$ and an *action level* of $2.5 m.s^{-2}$. For whole body vibration the daily exposure *limit* is set at $1.15 m.s^{-2}$ and the daily exposure *action value* is $0.5 m.s^{-2}$. These values all refer to the exposure during a working day normalised to an eight-hour day. Whereas the older guidance set limits for vibration of $2.8 m.s^{-2}$, in the dominant axis of movement only, the new regulations require the level of vibration to be determined as the sum of each of the three axes of movement (see O1027 above). The measurement and calculation of daily dose is therefore technically quite complex and will need to be done by a competent person.

The provision of personal protection, which may include gloves to reduce exposure to the cold or to reduce the transmission of vibration, fall under the *Personal Protective Equipment at Work Regulations 1992* (*SI 1992 No 2966*), as amended by the *Health and Safety* (*Miscellaneous Amendments*) *Regulation 2002* (*SI 2002 No 2174*).

In addition the *Provision and Use of Work Equipment Regulations 1998* (*SI 1998 No 2306*), apply insofar as equipment must be suitable for the work, and must be properly maintained in good repair. *SI 1998 No 2306, Reg 7* of these regulations requires 'specific risks' in the use of the equipment to be eliminated or adequately controlled. The regulations also require users of equipment to receive adequate health and safety training. Under the *Supply of Machinery* (*Safety*) *Regulations 1992* (*SI 1992 No 3073*), as amended by the *Supply of Machinery* (*Safety*) *Amendment Regulations 1994* (*SI 1994 No 2063*), manufacturers and suppliers of machinery are obliged to reduce risks to a minimum, to provide data on vibration, and information on risks to health and their control.

## Diseases and disorders of the eye

### What are they?

O1032    Diseases and disorders of the eye relate to:

- disorders of the eye, including injuries, which are *caused* by the conditions of work; and
- conditions of the eye, such as poor visual acuity or defects in colour vision which affect safety at work or affect the health of the worker.

### How are they caused?

O1033    **Eye injuries**. The eye is prone to injury and to disorders at work. Injuries are caused by a number of hazards such as:

- physical penetration of the eye with sharp objects (metal chips or swarf, wood splinters etc.);

- splashes with acid, alkali or other caustic substances; and
- exposure to high flux of UV light, typically in arc welding operations (arc eye).

**Irritation.** The surface of the eye (the conjunctiva) can be irritated by exposure to irritant fumes or chemicals (such as formaldehyde, some pesticides, dusts etc., causing conjunctivitis. Conjunctivitis can also be caused by *allergy* to particles in the atmosphere including animal dander (in veterinary work or animal husbandry), plant material including pollens (in horticultural and agricultural workers). Frequently, allergic conjunctivitis is accompanied by irritation of the nasal passages and the respiratory system generally in a condition commonly called hay fever. Employers have a particular responsibility if the employee's work brings them in contact with known allergens, which cause eye or respiratory allergy. However, these are common conditions in the population and many employees will, at certain times of the year, have to work whilst suffering severe irritation of the eye or have to take time off sick if the condition prevents them doing their work or driving to work. In some cases the employer may have to determine if the employee is safe to continue working if, for example, they work at heights, drive on the roads or drive forklift trucks. The medication taken to alleviate these conditions may itself cause a safety issue.

**Dryness and low humidity.** The surface of the eye is also made sore by low humidity. This causes the surface of the eye to dry and become irritant. This is often seen in air-conditioned buildings where the air has low humidity or in cases of sick building syndrome.

**Photokeratitis and photoconjunctivitis.** Exposure to UV light in particular can cause damage to the tissues of the cornea (photokeratitis) and to the conjunctiva (photoconjunctivitis). Photokeratitis (arc eye) is now rare among arc welders but is, nevertheless, a risk if proper precautions are not observed. Other causes are where there is high reflectivity (snow blindness), in the use of sun beds and in laboratories using UV light. The possibility that with increased outdoor UV radiation in recent years, normal outdoor exposure in the summer may now be sufficient to cause photokeratitis and photoconjunctivitis remains uncertain, but there is sufficient evidence that it may be a risk to suggest that outdoor workers should protect their eyes with suitable sun-glasses.

**Cataract.** Cataracts are caused by the lens of the eye becoming opaque. This may be a natural age-related change or may be caused or hastened by exposures at work. Cataracts may be caused by excessive exposures, or exposures over a period of time, to ultraviolet light, infra-red light or microwave radiation. Sun lamps, sun beds, welding arcs, lasers and furnaces (including glass blowing), are sources of these radiations which may give rise to the risk of cataract. With increasing levels of natural UV in sunlight, there is growing concern that workers who spend long periods of time outdoors may also be at risk. There is sufficient evidence that this may be the case to suggest wearing suitable sunglasses as a precaution.

**Visual acuity.** The other eye disorders, which have great importance in occupational health, relate to visual acuity. Visual acuity is the ability of the eye to sharply focus on objects from near to far distance. The young eye can accommodate (i.e. change its focal length) from about 10cms up to far distance. However, as we get older the ability to focus on near objects diminishes, so that by the age of 60, the near point of vision has receded to about 83cms. Visual acuity is very important in some jobs where the ability to see accurately may have important implications for safety. Driving, for example, has standards determined by the Driver and Vehicle Licensing Authority. For car drivers the basic standard is the ability to read a car number plate at 20.5 metres (67 feet), although there are also restrictions if the visual field is affected. Group 2 (LGV and PCV drivers) have more stringent standards for visual acuity and for binocular vision.

Visual acuity is important in display screen workers because inability to focus sharply at the average distance of the computer screen is sometimes not corrected by glasses or contact lenses, which correct for near or far sight visual defects. Correction for the intermediate distance of the display screen equipment (DSE) screen may require specific lenses. It is for this reason that there is provision in the *Health and Safety (Display Screen Equipment) Regulations 1992 (SI 1992 No 2792)* for eyesight and vision tests and for the employer to supply suitable corrective lenses (see O1036 below).

Colour vision. In certain jobs accurate colour perception may be a safety issue. About 8% of males and 0.4% of females are colour blind. The most common form is red-green colour blindness, which makes the distinction between these colours difficult. People working in jobs that require accurate colour perception, or where safety is dependent on colour perception, should have their colour vision tested.

## How is it detected?

O1034 Defects in visual acuity are measured by a number of standard tests. The basic test is the Snellen's Chart which requires a person to read letters of diminishing size from a distance of 6 meters (20 feet). The normal visual acuity is quoted as 6/6 (or 20/20 in imperial) although many people have better acuity than this, for example, 6/4.5 (20/15). The UK driving standard is equivalent to a Snellen's score of 6/10.

Other tests may measure binocular vision, the ability of the two eyes to work together to give proper depth and distance perception. Vision tests should normally be carried out by a registered optician or ophthalmic practitioner. However, screening tests, using a vision-screening instrument, are often carried out by occupational health advisors to detect individuals who may have visual difficulties. Anyone with a suspected visual impairment would be referred to an optician for full examination.

Tests for colour blindness can be carried out by occupational health advisors using a vision screening instrument or standard colour charts (Ishihara charts).

Other eye conditions or suspected eye conditions, such as injuries to the eye, allergic inflammation, cataracts, photokeratits or photoconjunctivitis must be referred to a doctor or medically qualified ophthalmic practitioner for diagnosis and treatment.

## How is it managed?

O1035 The risk of eye injury must always be anticipated in those work environments where there are hazards such as metal swarf, wood dusts or splinters, mineral and cement, brick and concrete dusts or chips, or when handling chemicals. A risk assessment (see O1036 below) will reveal where such risks may exist and appropriate controls must be put in place. These normally involve the establishment of systems of work (ways of working and adequate supervision), which should prevent such injuries occurring. However good the system of work is, there is always likely to be a residual risk and further safeguards will be necessary. These include the provision and use of eye protection (safety glasses and goggles), and suitable first aid facilities (eye wash bottles) and first aid provision.

Jobs in which visual acuity or colour perception is critical should have eye tests on recruitment and periodically thereafter. The frequency of these will depend on the condition, its likely rate of change and the age of the individual. The occupational health advisor or optician (if referred to one) will normally advise on the frequency of tests. DSE users should have their visual acuity measured at the normal viewing distance of the display screen and, if necessary, be provided with suitable corrective lenses (see O1036 below).

## What are the legal requirements?

O1036 There is a duty under the *Management of Health and Safety at Work Regulations 1999 (SI 1999 No 3242)*, (as amended by the *Management of Health and Safety at Work and Fire Precautions (Workplace) (Amendment) Regulations 2003 (SI 2003 No 2457)*), to undertake a risk assessment wherever the risk of eye injury or eye disease may be a possibility. The risk assessment should identify who is at risk (taking into account any existing vulnerability), what the possible eye injury or condition might be and what controls need to be put in place to prevent or minimise the risk.

Where personal eye protection is considered necessary then the *Personal Protective Equipment at Work Regulations 1992* (*SI 1992 No 2966*), as amended by the *Health and Safety* (*Miscellaneous Amendments*) *Regulations 2002* (*SI 2002 No 2174*), apply.

The *Health and Safety* (*Display Screen Equipment*) *Regulations 1992* (*SI 1992 No 2792*), as amended by *SI 2002 No 2174*, gives employees who are designated as DSE 'users' the right to have an eye and eyesight test carried out by a competent person. Employers have a legal duty to supply corrective lenses, if needed, for the work at the display screen.

# Diseases of the skin

## What are they?

O1037    About 90–95% of diseases of the skin related to work are *dermatitis*. Because these conditions are caused by contact with substances in the workplace they are referred to as *contact dermatitis*. Contact dermatitis is further classified as:

- irritant contact dermatitis; and
- allergic contact dermatitis.

In addition to contact dermatitis, the other, but much less common diseases of the skin are oil folliculitis, chloracne, leucoderma, ulcerations and skin cancers.

## How are they caused?

O1038    *Irritant contact dermatitis*. Contact with a wide variety of agents that are irritant to the skin can cause this disease. The hands are usually affected because they are most likely to be in contact with the irritant agent. These are either chemical substances or physical agents that cause damage to the cells of the outer layer of the skin. Well known highly irritant substances (strong acids and alkalis for example) are normally well protected against and are not the usual causes of contact irritant dermatitis, although accidental splashes may cause skin burns. It is usually the milder substances, which may not cause an immediate reaction, that are the cause. Irritant contact dermatitis often occurs after a lengthy period of contact with agents that do not immediately cause any problem. The onset is therefore insidious and not usually predictable in any particular individual. There is a large individual variation in susceptibility and the only probable risk factor is a history of eczema as a child affecting the hands.

The agents most frequently encountered as causing irritant contact dermatitis are soaps, detergents, mild acids and alkalis, cutting oils, organic solvents, animal and plant products, and oxidising and reducing agents. Physical agents include friction, and low humidity. It is often the case that the dermatitis occurs after exposure to more than one irritant source.

*Allergic contact dermatitis*. This disease is caused by contact with materials, which can penetrate the skin and provoke an immunological response in the skin cells. The type of response, which characterises dermatitis, is called a type IV allergy. The allergic reaction occurs after a period of sensitisation, which may occur after repeated exposure over a prolonged period (many years in some cases) or as little as a single previous contact. Allergic contact dermatitis is much less easy to control than irritant contact dermatitis because once sensitised the person will react to minute quantities of the substance. Substances known to be capable of causing contact allergy are called contact allergens. Hundreds of contact allergens are now known including chromates, methacrylates, formaldehyde, dyes, epoxy resins, and some woods.

*Urticaria*. An alternative allergic reaction is called urticaria (nettle rash). This is a typical wheal and flare reaction, which occurs within 20–30 minutes of contact with the allergen. This is not classified as an allergic contact dermatitis because it involves a different immune response (a type I reaction) and may involve a whole body response, including asthma and anaphylaxis (whole body shock). Reactions to substances causing urticaria may therefore be more serious and even life-threatening. The most common substance at work causing urticaria is latex rubber. Other allergens include proteins from animal skin and hair, and foodstuffs. People working in the food industry, with animals and in the health care industry (using latex gloves) may be at risk.

*Other conditions*. Oil folliculitis (oil acne), chloracne, leucoderma, ulcerations and skin cancers are caused by contact with (or ingestion of) a variety of substances found in the workplace. **Oil folliculitis** is an irritation of the hair follicles caused usually by contact with petroleum oils. **Chloracne** is a serious skin condition caused by ingestion of certain polychlorinated aromatic hydrocarbons. **Leucoderma** is a chemical depigmentation of the skin caused by exposure to a number of chemicals including phenols, catechols and hydroquinones. Such chemicals are commonly found in products such as adhesives. The skin depigmentation may occur at sites remote from the site of skin contact. **Skin ulcers** are usually caused by contact with wet cement or chromium compounds (chrome ulcers). Occupational **skin cancers** are now rare although they were once a common industrial disease from exposure to coal tar and machine oils. Outdoor exposure to sunlight may now be the most likely cause of occupational skin cancers.

## How is it detected?

O1039     Skin conditions, of whatever type, are easily detected by visual examination. Workers identified by the risk assessment as being at risk of skin disease (i.e. because they are working with known irritants or allergens), should be placed under health surveillance. It is important that the person undertaking the periodic skin inspection (and the people themselves who are at risk) should know exactly what to look for and be able to recognise early signs of a skin complaint. Equally important is to have a system in place for rapid referral to a skin specialist or occupational physician. Although abnormal reactions in the skin are easily detected, the diagnosis of the precise nature of the disease, what is causing it and how to manage it will require expert advice from an occupational physician or dermatologist experienced in occupational skin diseases.

## How is it managed?

O1040     Occupational skin disease can be a very debilitating condition, which may render the individual unable to continue their work. Allergic contact dermatitis in particular, is difficult to manage once someone has become sensitised because it may be difficult to avoid the very small levels of exposure needed to bring about a reaction. Prevention is therefore always better than cure. The risk assessment should be able to identify those employees who are at risk from exposure to substances and physical agents, which could possibly cause skin conditions. Avoidance of contact with such agents will prevent irritant contact dermatitis and will greatly reduce the possibility of sensitisation and prevent allergic contact dermatitis. Persons with a history of childhood eczema or a past work history of dermatitis should be carefully screened if they are working with possible irritants or contact allergens. If the job is known to expose people to such substances then a pre-placement or pre-employment health screen should identify those at risk.

Protection against exposure should ideally start with attempting to procure engineering solutions. Enclosure of processes, local exhaust ventilation, and remote working which avoids handling irritant substances are possible solutions which should be considered. Personal hygiene is also very important. Staff must be encouraged to wash their hands after handling irritant material and be provided with easily accessible means to do so. Remember though that soaps and detergents may themselves be a cause of dermatitis. Protection may require the use of gloves (but again remembering that latex gloves themselves may be a risk), or barrier creams. Barrier creams may have a place if gloves are difficult to wear but do not afford the same level of protection.

## What are the legal requirements?

O1041     Since the major causes of dermatitis and other skin complaints are chemical substances, the primary legal obligation falls under the *Control of Substances Hazardous to Health Regulations (COSHH) (SI 2002 No 2677)*, as amended by the *Control of Substances Hazardous to*

Health (*Amendment*) *Regulations 2003* (*SI 2003 No 978*), and the *Control of Substances Hazardous to Health* (*Amendment*) (*COSHH*) *Regulations 2004* (*SI 2004 No 3386*). Under these regulations there is a requirement to undertake a risk assessment of the health effects and to devise and implement effective control measures including, if necessary, personal protection. The COSHH regulations also require health surveillance under certain circumstances – these are the same criteria as given by the Management Regulations (see O1009 above). In the case of skin disease these criteria are met, so health surveillance is mandatory for people at risk. Health records pertaining to individuals have to be kept for 30 years (see O1016 above). Under the COSHH regulations there is requirement for information, instruction and training.

The use of personal protection falls under the *Personal Protective Equipment at Work Regulations 1992* (*SI 1992 No 2966*), as amended by the *Health and Safety* (*Miscellaneous Amendments*) *Regulation 2002* (*SI 2002 No 2174*).

Physical risks, such as friction or low humidity are subject to the general requirements for risk assessment, control, health surveillance, information, instruction and training of the *Management of Health and Safety at Work Regulations 1999* (*SI 1999 No 3242*), as amended by the *Management of Health and Safety at Work and Fire Precautions* (*Workplace*) *Regulations 2003* (*SI 2003 No 2457*).

# Occupational respiratory diseases

## What are they?

**O1042** Respiratory diseases are those that affect the respiratory tract from the nose down to, and including, the lung. The majority of occupational diseases of the respiratory system involve the lower respiratory tract within the lung. Historically, respiratory diseases have been a major cause of industrial illness and death. Diseases associated with dust inhalation in the coalmines (coal worker's pneumoconiosis), in the metal and pottery industries (siderosis and silicosis) and in the cotton industry (byssinosis) were scourges of the industrial revolution. Collectively, diseases of the lung associated with dust inhalation are called the pneumoconioses. Some of these diseases are also associated with development of cancer of the lung. Chronic bronchitis and emphysema (chronic obstructive pulmonary disease – COPD) was also very prevalent, although this was related as much to cigarette smoking as to industrial air pollution. Later, diseases associate with asbestos, (asbestosis, mesothelioma, other asbestos related lung cancers and pleural thickening), became the respiratory lung disease of major concern. Today, most of the older diseases associated with industrial pollution are rare, but have not completely gone way. Asbestos-related lung cancers can have a very long lag time between exposure and the onset of disease. About 3,500 people a year still die from asbestos-related lung cancer from past exposure. This figure is predicted to keep rising until about the year 2015. Asbestos was extensively used in building construction, shipbuilding, locomotive construction and in lagging industrial plant, between the 1950's and the1980's. It is therefore still a considerable risk, especially to those who work in asbestos clearance and in demolition. Builders, plumbers, electricians and anyone working in the refurbishment of older buildings may still be at risk.

Notwithstanding the continuing importance of asbestos-related cancers, today's most common work-related respiratory diseases are allergies of the upper respiratory tract (allergic rhinitis), and the lung (extrinsic allergic alveolitis and asthma). *Extrinsic allergic alveolitis* is an inflammatory immune response in the lung to the inhalation of a wide variety of both organic and inorganic materials. Examples include bird fancier's lung (from inhaling proteins in avian excreta), farmer's lung (from mould spores in mouldy hay or grain), and mushroom worker's lung (from mushroom spores). Some inorganic substances of industrial importance such as those used in spray painting and polyurethane foam, (diisocyanates and diphenyl methane) can cause the same reaction. Asthma is caused by narrowing of the airways, which is reversible usually over a short period of time. *Occupational asthma* is defined as attacks, which are initiated by inhalation of substances at work. Initiators can include irritants (irritant-induced asthma) or agents which bring about an immune response (hypersensitivity-induced asthma).

*Tuberculosis* is worth noting also as an occupational respiratory disease of some importance. The prevalence of tuberculosis in the community is rising and although it is rarely caused by work, it has great importance for those working in the health care industry and others, who may come in close contact with vulnerable groups (community social workers, teachers etc.). There is a risk in these cases for infected staff to pass on the infection to those in their care. Other infectious diseases of the lung have occupational importance (see O1043 below).

## How are they caused?

O1043    Respiratory diseases are normally caused for one of three main reasons:

- inhalation and retention in the lung of dust and particles;
- inhalation of substances that bring about an allergic reaction; and
- inhalation of infectious organisms.

The resulting pathology is, however, very complex. Inhaled and deposited dusts and particles will have different effects depending on the chemical properties of the substance. Some will be fairly inert and be retained in the lung as patches of dust accumulation. Simple coal worker's pneumoconiosis (the early stages of the disease) is an example of this. Some will trigger the formation of protective fibrous capsules, for example iron dust (siderosis), and some will stimulate wider fibrous changes in the lung leading to serious impairment of lung function and eventually death. Silicosis is an example. Some substances are carcinogenic and their retention in the lung significantly increases the risk of lung cancer. Asbestos has been mentioned as one such substance, others include, arsenic, beryllium and cadmium compounds, silica, tars, soots, diesel-engine exhaust particulates, welding fumes, formaldehyde, vinyl chloride monomer and radioactive gases and particles (including radon). This list is of course not exhaustive. Exposure to inhaled carcinogens not only cause cancer in the lung they may also affect the nose and upper airway.

*Allergic responses* in the lung occur after a period of sensitisation. The allergic reaction releases substances into the lung which either cause inflammation in the lowest parts of the lung (the bronchioles and alveoli) or may cause constriction of the slightly higher parts of the airway. The inflammatory reaction, which also causes fluid build-up in the lower parts of the lung, is the cause of extrinsic allergic alveolitis. The bronchial constriction results in an asthmatic attack. Asthma may also be triggered by irritant substances which do not involve an allergic reaction.

*Infectious diseases* of the lung are not common but can have serious occupational implications. Tuberculosis may be caught at work, but the occupational health risk is far more likely to be that an infected employee transmits the infection to someone else. Animal workers are also at risk from catching a variety of animal-borne lung infections (zoonoses). These include brucellosis, psittacosis and Q fever, which are pneumonia-like illnesses. Legionnaire's disease is a serous type of pneumonia caused by the legionella bacterium infecting water systems. To cause infection the water has to be released as fine spray capable of being inhaled. Although it remains an ever-present risk it is now uncommon with elimination of water cooling towers and much better management of building water systems.

## How is it detected?

O1044    This depends on the particular disease. It is important for employers to identify individuals who may be at risk from any of the occupational lung diseases. Employees must be made aware of the risks and what symptoms might arise if they become affected. They should be encouraged to report any symptoms such as chest tightness, breathlessness, cough or wheeze and the organisation must have a system in place for immediate referral to an occupational health doctor or nurse (or advise the employee to go to their own doctor with a note of what they are exposed to at work). It may be necessary to place some employees under health surveillance and this typically involves routine lung function testing. Lung function tests are performed using a spirometer, which measures a number of functional capacities. The most useful predictors of early changes that might indicate the onset of lung disease are the forced vital capacity (FVC) and the forced expiratory volume in one second ($FEV_1$). Monitoring the health of someone with asthma is best done with another measure, the peak flow (PF). Modern electronic spirometers calculate the lung function values and indicate abnormal values by comparison with 'predicted' values for the person's age, sex and height. However, this cannot by itself be relied on to show early significant changes in lung function. It is better to compare individuals with themselves over time. Although modern equipment can be used by anyone with minimum training, it is best to have lung function tests done by an occupational health practitioner who can interpret the findings and ask other relevant questions about the individual's respiratory health at the same time.

Other tests for respiratory diseases include chest x-rays, but they are no longer done routinely unless a physician specifically indicates them.

## How is it managed?

**O1045** Lung disease is usually serious, debilitating and sometimes life-threatening. As with any occupational disease every effort should be put into prevention. This starts with the risk assessment which, by law, must be thorough, systematic and take into account all the relevant factors. Anyone working with any substance or potentially exposed to infectious organisms, capable of causing respiratory illness, must have a risk assessment undertaken on their work. If it is reasonably practicable, then prevention must first of all prevent exposure at source. This may mean suitable engineering solutions (such as enclosure or local exhaust extraction), to prevent dust, particulates or fumes from escaping into the work area. It may be necessary, because of the nature of the work, to use respiratory protection, but this should only be used if engineering solutions are not possible or do not provide complete protection. A wide variety of respiratory protection is available, from simple dust masks to respirators. The correct type and specification must be provided for the type of substance it is intended to protect against. It may be necessary to seek the advice of an occupational hygienist.

Individuals at risk of respiratory illness should be placed under health surveillance. The criteria for health surveillance under the *Management of Health and Safety at Work Regulations 1999* (*SI 1999 No 3242*) and the *Control of Substances Hazardous to Health Regulations 2002* (*SI 2002 No 2677*) are probably met in most cases which means that health surveillance is a legal requirement (see O1009 above). Health surveillance might include self-surveillance for symptoms, respiratory health questionnaires and lung function tests (see O1044 above), and should usually be devised and carried out by an occupational health advisor.

## What are the legal implications?

**O1046** Occupational lung disease usually results from inhalation of 'substances' at work and so the primary legal duty falls under the *Control of Substances Hazardous to Health Regulations* (*COSHH*) (*SI 2002 No 2677*), as amended by the *Control of Substances Hazardous to Health (Amendment) Regulations 2003* (*SI 2003 No 978*), and the *Control of Substances Hazardous to Health (Amendment) Regulations 2004* (*SI 2004 No 3386*). Infectious organisms are also defined as 'substances' under these regulations. Under these regulations there may be a requirement to monitor the levels of exposure to ensure that they do not fall over any prescribed limits (these limits are published in EH40 2002 (HSE) with amended guidance 2005).

If the risk assessment under the COSHH regulations determines a need for personal respiratory protection then the choice of this, it's suitability, issue, maintenance and instruction and training all fall within the *Personal Protective Equipment at Work Regulations 1992* (*SI 1992 No 2966*), as amended by the *Health and Safety (Miscellaneous Amendments) Regulations* (*SI 2002 No 2174*).

If the respiratory risk specifically arises from working with asbestos or potentially coming into contact with asbestos, then the *Control of Asbestos at Work Regulations 2002* (*SI 2002 No 2675*) applies. It is possible also that the intended work with asbestos may require a licence from the

HSE, under the *Asbestos (Licensing) Regulations 1983* (*SI 1983 No 1649*), as amended by the *Asbestos (Licensing) (Amendment) Regulations 1998* (*SI 1998 No 3233*).

# Musculoskeletal disorders

## What is it?

O1047    The term musculoskeletal disorders (MSDs) refer to a range of conditions which affect the skeletal system and its associated muscles, tendons and ligaments. These conditions are usually separated into those affecting the upper limb (from the shoulder to the tips of the fingers), the lower limb (from the hip to the toes), and the back. The most common conditions are those affecting the upper limb (work-related upper limb disorders or WRULDS) and the back (backache, acute and chronic back pain). Conditions which affect the lower limb are less common but in some industries (such as carpet layers) the knee can be damaged by repeated kneeling, causing a condition called beat knee (bursitis). The other part of the body often affected is the neck, especially in sedentary jobs, (at computer workstations for example) where postural problems result in neck pain.

The common conditions affect the tendons of the arm and hand. These tendons run from the muscles of the forearm through the wrists and hand to insert on the individual bones of the wrist and fingers. In places, as they pass across the wrist and the back and front of the hand, they pass through lubricated sheaths (tendon sheaths). Repetitive and forceful hand-arm movements can damage almost any part of this system.

Of the working population, 2.6% suffer from some form of musculoskeletal disorder resulting in 11.8 million lost working days (2003/04). On average each affected person takes 19.4 days off in a 12-month period.

The following is a range of musculoskeletal disorders.

**Upper limb disorders.** These are commonly called repetitive strain injury (RSI). However, RSI is not a medical condition by itself and the courts have rejected claims for damages for RSI on the grounds that it is not a recognised disease. It is usual practice now to refer to these conditions collectively as 'work-related upper disorders' (WRULDS) and then to diagnose one of the recognised conditions that make up this collection of complaints. In practice, however, the precise nature of these complaints is uncertain and drawing a diagnostic distinction between them is difficult.

- *Tenosynovitis.* An inflammatory and painful condition of the tendons as they pass through the tendon sheaths of the wrist and in the hand. It may also involve inflammation and dryness of the tendon sheath itself.

- *Tendinitis.* Inflammation of the tendons but not involving the tendon sheaths. Tendinitis can occur in the wrists and hand and also in the upper arm and shoulder.

- *Peritendinitis.* Inflammation at the point where the tendon joins the muscle in the arm.

- *De Quervain's disease.* Tenosynovitis with pain and swelling at the back of the wrist, particularly involving the tendon to the thumb.

- *Epicondylitis.* Pain at the point where the muscles of the arm join the bone at the elbow (the epicondyles). On the inner side of the elbow this is called medial epicondylitis (golfers elbow). On the outer side this is called lateral epicondylitis (tennis elbow).

- *Carpal tunnel syndrome.* Entrapment or compression of the median nerve as it passes through the front of the wrist (the carpal tunnel). This causes numbness and tingling in the part of the hand supplied by the median nerve – the palm side of the thumb, the second and third fingers and half the fourth (ring) finger.

- *Chronic (non-specific) upper limb pain.* A chronic widespread pain in the upper limb with tenderness to which no certain diagnosis can be made. Linked with repetitive use of the hand and arm at work. Some doctors will diagnose this as RSI, despite it not being a clinical entity, simply because no clearer diagnosis can be made.
- *Beat hand and beat elbow.* Repeated use of picks, shovels and hand tools in labourers, quarrymen and other manual jobs, can cause chronic pain, tenderness, redness and swelling of the hand and the elbow.
- *Cramp.* A condition that appears to affect people in occupations that involve repetitive movements of the hand and arm. There is some dispute about its true nature but it is a prescribed occupational disease. It causes spasm, tremor and pain in the affected muscles when the familiar action is attempted. Typified by writer's cramp.

**Back pain.** Most back pain is caused by mechanical damage to the spine and associated muscles, ligaments and tendons. *Acute* (sudden onset) pain is usually associated with a torn ligament or muscle around the spine caused by sudden forceful lifting or movement associated with awkward posture. In 90% of cases, such episodes usually recover within six weeks. However, these conditions are cumulative and repeated episodes may take longer to recover and recovery may be progressively less complete. The condition therefore becomes *recurrent*. Back pain can also be caused by rupture (prolapse or herniation) of the intervertebral disc. These soft tissue discs act as 'shock absorbers' between each of the vertebral bones. Severe force on the spine can cause them to rupture and compress the nerve roots as they leave the spine. The most common part of the spine for this to occur is in the lumbar region where the nerves leave to supply the leg – compression of these nerves causes pain to be referred down the leg (sciatica). *Chronic* back pain occurs where there is incomplete recovery from repeated trauma. Disc rupture is often associated with subsequent chronic back pain.

**Osteoarthritis.** Osteoarthritis is a very common condition in the population as whole, but there is evidence that its prevalence is higher in some occupations. Osteoarthritis is a degeneration and loss of the articular cartilage in some joints particularly of the hand, knee and hip. Articular cartilage forms a smooth lubricated surface between the joint surfaces – its loss means that bone rubs against bone causing severe damage to the joint. Osteoarthritis of the knee is more common in men involved with heavy manual labour and especially if the work involves kneeling. Osteoarthritis of the hip is less clearly associated with occupation although prolonged heavy lifting may be a risk factor.

## What causes it?

The following are risk factors associated with most of the musculoskeletal disorders.

- **Repetition.** Frequent repeated movements over a relatively long period of time.
- **Force.** The amount of force put into the movement such as pulling, pushing or lifting. The greater the force required, the greater the risk.
- **Duration.** The longer someone is engaged in the same work, the greater is the risk of cumulative damage.
- **Lack of rest.** The risk of musculoskeletal damage is increased if the work is carried out for long periods of time without rest or time for recovery.
- **Posture.** Work carried out in awkward postures greatly increase the risk. This is true of sedentary jobs such as typing or checkout work, or in manual jobs such as lifting.
- **Working in adverse conditions.** Manual work where the hands are cold, the floor uneven or slippery, or where loads are difficult to handle, increase the risk. Working in uncomfortable conditions at computer workstations increases the risk of upper and lower limb disorders.

- **Stress.** Jobs carried out whilst under stress from time pressures etc. increase the risk of musculoskeletal disorder.
- **Individual factors.** Factors such as age and sex, height, weight and strength, a history of such conditions (or a pre-existing condition), and pregnancy will have an influence on the risk of musculoskeletal injury.

## How is it detected?

O1049  Musculoskeletal disorders can normally only be detected by self-reporting or by causes of sickness absence. Employees in any at-risk job should be encouraged to report symptoms. Intervention at an early stage can often result in preventing the problem becoming worse. Acute conditions will result in sudden disabling pain and disability, which will of course require medical intervention. The onset of the chronic condition may be insidious and over a period of time. However, during this time there will be early symptoms, which should not be ignored. With upper limb conditions or slowly developing back pain there may be episodes of acute pain or numbness, which quickly resolve. Recurrent episodes or early signs of disorder such as pain, numbness or tingling, should be referred to an occupational physician (or the person's GP) for diagnosis.

## How is it managed?

O1050  Almost all jobs carry a risk of musculoskeletal disorder. Heavy manual work may carry a more obvious risk but sedentary jobs are no less likely to be associated with problems. In an office the unplanned task of carrying or moving something heavy may be all that is required to cause an acute back injury. All jobs should be considered as inherently capable of giving rise to a musculoskeletal condition, although of course some jobs, which carry the known risk factors (see O1048 above), will be much more at risk. All jobs should, therefore, have a risk assessment for MSD that takes into account all of the known and likely factors associated with the job and any individual vulnerabilities. This must be done by a competent person who recognises all the risk factors and has a good understanding of what control measure should be put in place. Having recognised the risks and implemented the necessary controls it is important that these are adhered to – this has to be done partly by effective supervision (e.g. if rest pauses are built into the job make sure they are taken) and partly by providing good information, education and training of the persons involved.

Employees must be made aware of the symptoms to report and to whom they should be reported. A system of referral to an occupational health nurse or doctor (or if that is not available to the person's GP) must be put in place. Referral to the health professional should ask for advice on what needs to done in the workplace or to the job to ensure that the condition does not worsen. Referral to an occupational health professional is much more likely to result in practical advice, which many GPs may not have the experience to offer.

Conditions that have become chronic or disabling to the point where the individual is not able to carry out their expected duties must be referred to an occupational physician with the view that permanent ill-health retirement may be necessary. However, the *Disability Discrimination Act 1995* (see O1003 above) will almost certainly apply to chronic disabling musculoskeletal disorder.

## What are the legal requirements?

O1051  General duties of risk assessment, risk management, information, education and training, competency of those undertaking risk assessment and health surveillance come under the *Management of Health and Safety at Work Regulations 1999 (SI 1999 No 3242)*, as amended by the *Management of Health and Safety at Work and Fire Precautions (Workplace) Regulations 2003 (SI 2003 No 2457)*. These regulations apply entirely where the risk arises from adverse conditions at work such as slipping and tripping hazards, poor environmental conditions (heat, wet and cold), where the primary hazard is poor or awkward posture and where stress may be a contributing factor. Where the risk arises specifically from manual handling (lifting, pushing or pulling loads) the *Manual Handling Operations Regulations 1992 (SI 1992 No 2793)* as amended by the *Health and Safety (Miscellaneous Amendments) Regulations (SI 2002 No 2174)* applies. Working with display screens (including computer workstations, and video display screens) falls within the *Health and Safety (Display Screen Equipment) Regulations 1992 (SI 1992 No 2792)* as amended by *SI 2002 No 2174*.

# Psychological disorders (stress, anxiety and depression)

## What is it?

**O1052** The HSE defines stress as 'the adverse reaction people have to excessive pressure or other types of demand placed on them'. Pressure is a normal part of work, indeed a normal part of life, but stress occurs when the pressures are perceived by the individual as being greater then their ability to cope. Stress is therefore a very individual phenomenon; people vary greatly both in their perception of the amount of pressure on them and their abilty to cope. Stress often leads to anxiety and depression, but anxiety and depression can occur for many reasons without the underlying cause being stress. Stress also occurs in our everyday lives outside of work and it is important for the employer to understand what contribution, if any, the work has in causing an individual's stress or other psychological problem. An employer can only be responsible for that part of an individual's health complaint that is related to work, but of course most employers will be concerned with an employee who is suffering a psychological illness, whatever the cause, because it is likely to have a major impact on the person's performance at work.

Stress is not by itself an illness, but a set of symptoms which result from the mental and physiological reaction to an environment in which the perceived demands exceed the perceived ability to cope. The *resulting* psychological and physiological illness may be classified as diseases. It is important to recognise this distinction because the *Disability Discrimination Act 1995* only applies to diseases and conditions which are able to be classified as illnesses under the WHO International Classification of Diseases (ICD10). For example, depression is able to be classified under mood (affective) disorders, F30–39 and anxiety states under F40–48. Stress does not have a classiication.

Stress may not only cause psychological illness but may be related to mild to serious physical illness. In 2003/04, 66,000 people reported heart disease related to the stress of work. Stress may cause high blood pressure, gastrointestinal complaints, disturbed sleep and fatigue and a number of other physical illnesses.

Someone suffering from stress may exhibit any of the following signs, some of which will be obvious to work colleagues and management and should trigger action to discover, and hopefully remedy, the cause.

- Irritability with colleagues.
- Expressed feeling of inability to cope.
- Frequent periods of sickness absence.
- Difficulty concentrating.
- Tearfulness.
- Mood changes.
- Undue tiredness/lethargy.
- Headaches.

- Insomnia.
- Palpitations.
- Appetite change either loss of appetite or over-indulgence.
- Abdominal and chest pains.
- Diarrhoea or constipation.

It is possible too that individuals under stress will drink more alcohol, smoke more cigarettes and indulge in 'comfort' eating. These are attempts at reducing the symptoms of stress but will, particularly in the long term, have serious adverse health effects.

One in five of all working people think their job is very or extremely stressful. Stress, depression and anxiety accounted for 13 million working days lost in 2003/04, with an average of 28.5 working days lost per person affected.

## What causes it?

O1053    There are very many factors causing stress at work. Stress may be caused, or the response to it influenced, by any of the following.

- **The work itself and the work environment.** This includes a huge number of possible factors including the workload, time pressures, and complexity (and adequate training for the work expected). It includes also managerial and organisational aspects; how well supported the individual feels, how valued and rewarded they feel, how managers and work colleagues treat them. The physical environment can also be stressful – noise which distracts and interferes with communication and concentration, high or low temperatures which cause physical discomfort, poor lighting etc. Unreliable equipment, frequent system failures, and the pressures brought about by modern technology (including emails) are 'technological' causes of stress in the modern workplace. Bullying, harassment and violence at work are also a cause of stress for some. In some jobs stress arises because of the nature of the job – working in the emergency services, in health care and social work brings emotional stress including traumatic stress. Individual factors that are capable of causing stress are called 'stressors'; *stress* is the adverse reaction people may have to exposure to these stressors. Stress is often caused by an additive effect when several of these stressors are present together.

- **Modifying factors.** The way individuals respond to stressors will depend on a number of modifying factors. These include personality, the support they receive (both at work and outside of work) and their ability to learn successful coping strategies. The effect stressors have will also depend on the person's previous history of stress and their 'stress proneness'. It should usually be taken as true that someone with a history of stress-related disorder is likely to be more vulnerable to continuing or additional levels of stress. Stress frequently arises outside of work and one important modifying factor is the reduced ability to cope with the stressors of work if the person is already suffering stress from non work-related causes.

The HSE has categorised a number of causes of stress which form part of their stress management standards.

- **Demands.** Issues like workload, work pattern and the work environment.
- **Control.** How much say the person has in the way they do their work.
- **Support.** The encouragement and support provided to employees by management and colleagues.
- **Relationships.** Conflict and unacceptable behaviour.

- **Role.** People should understand their role within the organisation and the organisation should avoid creating conflicting roles.
- **Change.** How organisational change is managed and communicated.

### How is it managed?

O1054   As with any occupational health issue the primary aim is to prevent stress happening. However, individuals will become stressed, perhaps mainly for reasons beyond the employer's control (see modifying factors at O1053 above). The organisation, usually with professional support, will need to carefully manage the individual's recovery and return to productive working and address any work-related factors which may have caused or contributed to the illness.

The key to prevention is to undertake a stress risk assessment. Identify any areas of the work or workplace which might place undue pressures on people. Identify aspects of the organisation which might create stress. The list of factors identified in the HSE management standards (at O1053 above) serves as a good check list for undertaking the risk assessment although there may be other factors particular to the organisation. The risk assessment will have to be kept under constant review as things within the organisation change. In particular, it is important to reconsider the risk assessment for an individual who has been diagnosed with 'stress' or a stress-related disorder. Failure to take account of an individual's known vulnerability to stress has led to several succesful civil claims for damages.

Consider having a stress management policy. This would detail the way in which stress will be identified in the organisation. It will also detail who has responsibility for what and what support services are available to employees. An important aspect of a stress management policy is that it shows a concern by the management and a committment to address the issues, but only if it is seen to be working.

Even with the best stress management policy there will be individuals who will become stressed and it is important to have a support system in place to deal quickly and effectively with these cases. If the cause is clearly rooted within the organisation (excessive workload, poor working relationships, bullying, harrassment etc.) then it is a management problem that will need to be addressed and resolved. Individuals may need support during this time and the organisation should have available to it a professional support service (occupational health or a counsellor experienced in stress counselling). If the causes are partly or wholly outside the organisation, but it is nevertheless impacting on the work (causing excessive sickness absence for example), then it is also important for the company to provide the necessary support.

### What are the legal requirements?

O1055   The legal requirement falls under the *Management of Health and Safety at Work Regulations 1999 (SI 1999 3242)*, as amended by the *Management of Health and Safety at Work and Fire Precautions (Workplace) (Amendment) Regulations 2003 (SI 2003 No 2457)*. Under these regulations there is a need to assess the risks of stress occurring and the source of the stress (the stressors). The risk assessment should identify who is at risk (taking into account any existing vulnerability), what the possible adverse outcome might be and what controls need to be put in place to prevent or minimise the risk. Whilst there is not a legal obligation to do so, the HSE Stress Management Guidelines and the associated Indicator Tool are a useful guide to undertaking a risk assessment in this difficult area (http://www.hse.gov.uk/stress/index.htm).

## Other occupational health concerns

O1056   This short review has focussed on the main occupational health issues that confront today's workforce and employers. Inevitably it is not complete and many additional occupational health issues will from time to time arise. Some of these will involve potentially serious health problems and must be addressed pro-actively through the usual practice of risk assessment and risk management. The legal requirement for any of these is much the same – assessment of the risk, devising and implementing suitable controls, education and training, monitoring and review, and keeping at-risk individuals under health surveillance.

Other conditions range for the serious and life threatening to those that affect the well being and comfort of the employee. The more serious (as well as cancers of the lung and skin already mentioned) included cancers of the mouth, the nose, the liver, the bladder and blood cancers. A very large number of chemicals and other agents (such as radiation) used in industry and in manufacturing are known or suspected carcinogens and anyone working with these should be well protected. Other conditions of note are disorders of the nervous system including the brain. Many commonly used substances such as solvents, metals, insecticides, styrene and carbon monoxide are neurotoxic and capable of causing nerve damage. Radiations pose a special risk. Ionising radiations are well known hazards potentially capable of causing cancers and must be well controlled. Protection against the risk of exposure to ionising radiation comes under the *Ionising Radiations Regulations 1999* (*SI 1999 No 3232*). However, the risk from non-ionising radiations, such as extra low frequency (from mains electricity), radio frequency and microwaves is more controversial. High levels of exposure such as that encountered in some occupations do carry a risk of body heating, shock and burn and must be well protected against. However, the possible cancer and other health risks from low levels of non-ionising radiation from such things as hand held transmitters (including mobile phones) are controversial at the present time. If there is a risk it appears to be of very low magnitude. At the less serious end of the spectrum of occupational health concerns are issues such as thermal comfort, low humidity causing eye, nose and throat dryness, postural issues causing aches and pains, and headache and eyestrain caused by poor lighting and glare. Whilst these issues may not cause serious health complaints, they do, nevertheless, impact significantly on the employee's ability to work efficiently and may contribute to stress.

The review is not a complete description of the many and varied diseases and disorders that may be caused or worsened by the conditions of work, nor has it dealt in detail with the various legal requirements. Readers who believe they may have a health issue in their workplace should seek advice on what they may need to do to manage such problems and avoid the inevitable subsequent cost and disruption to work. Occupational health problems rarely go away on their own; many are cumulative and will progress to more serious conditions the longer they are ignored. The key to good occupational health management is to anticipate what might cause ill health and to avert the risk of serious harm to the employee and to the organisation.

Many occupational health conditions will require reporting to the HSE under the *Reporting of Injuries, Diseases and Dangerous Occurrences Regulations 1995* (*RIDDOR*) (*SI 1995 No 3163*). Under these regulations many of the diseases reviewed here such as dermatitis, oil folliculitis, skin cancer, other types of occupational cancer, occupational hearing loss, most work-related lung diseases, hand arm vibration syndrome, certain musculoskeletal disorders and work-related infections will need to be reported.

The legal duties cited in this review apply mainly to regulations which come under the *Health and Safety at Work etc. Act 1974*. These should be interpreted together with the general provisions of the Act itself.

# Occupiers' Liability

## Introduction

**O3001** Inevitably, by far the greater part of this work details the duties of employers (and employees) both at common law, and by virtue of the *Health and Safety at Work etc Act 1974* (*HSWA 1974*) and other kindred legislation. This notwithstanding, duties are additionally laid on persons (including companies and local authorities) who merely *occupy* premises which other persons either visit or carry out work activities upon, e.g. repair work or servicing. More particularly, persons in control of premises (see O3004 below for meaning of 'control'), that is, occupiers of premises and employers, where others work (*although not their employees*), have a duty under *HSWA 1974, s 4* to take reasonable care towards such persons working on the premises. Failure to comply with this duty can lead to prosecution and a fine on conviction (see further ENFORCEMENT). This duty applies to premises not *exclusively* used for private residence, e.g. lifts/electrical installations in the *common parts* of a block of flats, and exists for the benefit of workmen repairing/servicing them (*Westminster City Council v Select Managements Ltd [1985] 1 All ER 897*). Moreover, a person who is injured while working on or visiting premises, may be able to sue the occupier for damages, even though the injured person is not an employee. Statute law relating to this branch of civil liability (i.e. occupiers' liability) is to be found in the *Occupiers' Liability Act 1957* (*OLA 1957*) and, as far as trespassers are concerned, in the *Occupiers' Liability Act 1984* (*OLA 1984*). (In Scotland, the law is to be found in the *Occupiers' Liability (Scotland) Act 1960*.)

## Duties owed under the Occupiers' Liability Act 1957

**O3002** An occupier of premises owes the same duty, the 'common duty of care', to all his lawful visitors. [*OLA 1957, s 2(1)*]. 'The common duty of care is a duty to take such care as in all the circumstances of the case is reasonable to see that the visitor will be reasonably safe in using the premises for the purposes for which he is invited or permitted by the occupier to be there.' [*OLA 1957, s 2(2)*]. Thus, a local authority which failed, in severe winter weather, to see that a path in school grounds was swept free of snow and treated with salt and was not in a slippery condition, was in breach of *s 2*, when a schoolteacher fell at 8.30 a.m. and was injured (*Murphy v Bradford Metropolitan Council [1992] 2 All ER 908*).

The duty extends only to requiring the occupier to do what is 'reasonable' in all the circumstances to ensure that the visitor will be safe. Some injuries occur where no one is obviously to blame. For example, in *Graney v Liverpool County Council (1995) (unreported)* an employee who slipped on icy paving slabs could not recover damages against his employer. The Court of Appeal held that many people slip on icy surfaces in cold weather – it did not follow that in this particular case the employer was to blame.

### Nature of the duty

**O3003** The *OLA 1957* is concerned only with civil liability. 'The rules so enacted in relation to an occupier of premises and his visitors shall also apply, in like manner and to like extent as the principles applicable at common law to an occupier of premises and his invitees or licensees would apply ... '. [*OLA 1957, s 1(3)*]. This contrasts with the *HSWA 1974*, in which the obligations are predominantly penal measures enforced by the HSE (or some other 'enforcing authority'). The *OLA 1957* cannot be enforced by a state agency – nor does it give rise to criminal liability. Action under the *OLA 1957* must be brought in a private suit between parties.

The liability of an occupier towards lawful visitors at common law was generally based on negligence; so, too, is the liability under the *OLA 1957*. It is never strict. In *Neame v Johnson [1993] PIQR 100*, a case concerning an ambulance man who was injured in the defendant's house when carrying him unconscious in a chair in poor lighting conditions, the ambulance man knocked over a pile of books stacked by a wall on a landing, and consequently slipped on a book, injuring himself. It was held that the pile of books did not create a 'reasonably foreseeable risk of injury', and so there was no liability. Nor does the duty of the occupier extend to replacing glass panels in an ageing building. In *McGivney v Golderslea Ltd (1997) (unreported)* the defendant owned a block of flats which had been built in 1955 – the glass in the door at the foot of the communal stairs had been installed at the same time. When the plaintiff, who was visiting a friend on the third floor, came down the stairs, he slipped and his hand went through the glass panel on the front door resulting in personal injuries. The plaintiff claimed that the defendant was in breach of the *OLA 1957, s 2(2)*. However, at the time the flats were built the glass satisfied the building regulations then in force. Since 1955 building regulations have been updated and, in such a building being constructed today, stronger glass must be used such that on the balance of probabilities the plaintiff's accident would not have occurred. The Court of Appeal found that there was no duty on the defendant to replace the glass and the plaintiff's claim therefore failed.

The relationship of occupier and visitor is less immediate than that of employer and employee, and is not, as far as occupiers' liability is concerned, underwritten by compulsory insurance (but by public liability insurance, which is not obligatory but advisable).

## Who is an 'occupier'?

O3004    'Occupation' is not defined in the Act, which merely states that the 'rules regulate the nature of the duty imposed by law in consequence of a person's occupation or control of premises ... but they (shall) not alter the rules of the common law as to the persons on whom a duty is so imposed or to whom it is owed ... .'. [*OLA 1957, s 1(2)*]. The meaning of 'occupation' must, therefore, be gleaned from the rules of common law. Where premises, including factory premises, are leased or subleased, control may be shared by lessor and lessee or by sublessor and sublessee. 'Wherever a person has a sufficient degree of control over premises that he ought to realise that any failure on his part to use care may result in injury to a person coming lawfully there, then he is an "occupier" and the person coming lawfully there is his "visitor" and the "occupier" is under a duty to his "visitor" to use reasonable care. In order to be an occupier it is not necessary for a person to have entire control over the premises. He need not have exclusive occupation. Suffice it that he has some degree of control with others' (per Lord Denning in *Wheat v E Lacon & Co Ltd [1965] 2 All ER 700*). In *Jordan v Achara (1988) 20 HLR 607*, the plaintiff, who was a meter reader, was injured when he fell down stairs in a basement of a house. The defendant landlord, who was the owner of the house, had divided it into flats. Because of arrears of payment, the electricity supply had been disconnected. The local authority, having arranged for it to be reconnected for the tenants, recovered payment by way of rents paid directly to the authority. It was held that the landlord was liable for the injury, under the *OLA 1957*, since he was the occupier of the staircase and the passageway (where the injury occurred) and his duties continued in spite of the local authority being in receipt of rents.

Liability associated with dangers arising from maintenance and repair of premises will be that of the person responsible, under the lease or sublease, for maintenance and/or repair. 'The duty of the defendants here arose not out of contract, but because they, as the requisitioning authority, were in law in possession of the house and were in practice responsible for repairs ... and this control imposed upon them a duty to every person lawfully on the premises to take reasonable care to prevent damage through want of repair' (per Denning LJ in *Greene v Chelsea BC [1954] 2 All ER 318*, concerning a defective ceiling which collapsed, injuring the appellant, a licensee).

Significantly also, managerial control constitutes 'occupation' (*Wheat v Lacon*, see above, concerning an injury to a customer at a public house owned by a brewery and managed by a manager – both were held to be 'in control'). Moreover, if a landlord leases part of a building but retains other parts, e.g. roof, common staircase, lifts, he remains liable for that part of the premises (*Moloney v Lambeth BC (1966) 64 LGR 440*, concerning a guest injured on a defective common staircase in a block of council flats). The Court of Appeal has now held (*Ribee v Norrie, The Times 22 November 2000*), that the proper question was not what actual control the landlord did or did not exercise but what power of control he had, i.e. whether or not he had the authority to act and was in a position where he could have taken steps to prevent certain risks.

However, it is not the occupier's duty to make sure that the visitor is *completely safe*, but only reasonably so. For example, in *Berryman v London Borough of Hounslow, The Times, 18 December 1996*, the plaintiff injured her back when she was forced to carry heavy shopping up several flights of stairs because the lift in the block of flats maintained by the landlord was broken. The landlord's obligation to take care to ensure that the lift he provided was reasonably safe was discharged by employing competent contractors. There was no causal connection between the plaintiff's injury on the stairs and the absence of a lift service. In other words, the injury to the plaintiff was not a 'foreseeable consequence' of the broken lift.

## Premises

**O3005** The *OLA* regulates the nature of the duty imposed by law in consequence of a person's occupation of premises. [*OLA 1957, s 1(2)*]. This means that the duties are not *personal* duties but depend on occupation of *premises*; and extend to a 'person occupying, or having control over, any fixed or movable structure, including any vessel, vehicle or aircraft' [*OLA 1957, s 1(3)(a)*] (e.g. a car, *Houweling v Wesseler [1963] 40 DLR(2d) 956-Canada* or, more recently, a sea wall in *Staples v West Dorset District Council (1995) 93 LGR 536* (see O3006 below)). In *Bunker v Charles Brand & Son Ltd [1969] 2 All ER 59* the defendants were contractors digging a tunnel for the construction of the Victoria Line of London Underground. To this end they used a large digging machine which moved forward on rollers. The plaintiff was injured when he slipped on the rollers. The defendants were held to be occupiers of the tunnel, even though it was owned by London Transport.

## To whom is the duty owed?

### Visitors

**O3006** Visitors to premises entitled to protection under the Act are both (*a*) invitees and (*b*) licensees. This means that protection is afforded to all lawful visitors, whether the visitors enter for the occupier's benefit (clients or customers) or for their own benefit (factory inspectors, policemen), though not to persons exercising a public or private way over premises. [*OLA 1957, s 2(6)*]. (See below O3007 *Countryside and Rights of Way Act 2000*.)

Nevertheless, occupiers are under a duty to erect a notice warning visitors of the immediacy of a danger (*Rae (Geoffrey) v Mars (UK) [1990] 3 EG 80* where a deep pit was situated very close to the entrance of a dark shed, in which there was no artificial lighting, into which a visiting surveyor fell, sustaining injury. It was held that the occupier should have erected a warning). In some instances it may be necessary not only to provide proper warning notices but also to arrange for the supervision of visitors at the occupier's premises (*Farrant v Thanet District Council (1996) (unreported)*). However, there is no duty on an employer to light premises at night, which are infrequently used by day and not occupied at night. It is sufficient to provide a torch (*Capitano v Leeds Eastern Health Authority, Current Law, October 1989* where a security officer was injured when he fell down a flight of stairs at night, while checking the premises following the sounding of a burglar alarm. The steps were formerly part of a fire escape route but were now infrequently used). However, landowners need not warn against obvious dangers (see below O3017).

## Trespassers

**O3007**

'... a trespasser is not necessarily a bad man. A burglar is a trespasser; but so too is a law-abiding citizen who unhindered strolls across an open field. The statement that a trespasser comes upon land at his own risk has been treated as applying to all who trespass, to those who come for nefarious purposes and those who merely bruise the grass, to those who know their presence is resented and those who have no reason to think so.' (*Commissioner for Railways (NSW) v Cardy [1960] 34 ALJR 134*).

Common law defines a trespasser as a person who:

(*a*)    goes onto premises without invitation or permission; or

(*b*)    although invited or permitted to be on premises, goes to a part of the premises to which the invitation or permission does not extend; or

(*c*)    remains on premises after the invitation or permission to be there has expired; or

(*d*)    deposits goods on premises when not authorised to do so.

The *Countryside and Rights of Way Act 2000* (*CRWA 2000*) now gives the public the right to roam across the open countryside in England and Wales. *Section 13* of *CRWA 2000* amends *s 1(4)* of the *OLA 1957* so that: *'A person entering any premises in exercise of [the right to roam] is not, for the purposes of this Act, a visitor of the occupier of the premises.'* Therefore, ramblers under the *CRWA 2000* are also classed as 'trespassers'. Indeed, *CRWA 2000* amends *s 1* of the *OLA 1984* so that no duty of care is owed to a person exercising the right to roam under the Act in respect of: *'a risk resulting from the existence of any natural feature of the landscape, or any river, stream, ditch or pond whether or not a natural feature ... or a risk of that person suffering injury when passing over, under or through any wall, fence or gate, except by proper use of the gate of or a style.'* The 2000 Act therefore explicitly states what the position is in law – whilst 'visitors' falling under the 1957 Act must refer to case law.

## Duty owed to trespassers, at common law and under the Occupiers' Liability Act 1984

### Common law

**O3008**    The position under common law used to be that an occupier was not liable for injury caused to a trespasser, unless the injury was either intentional or done with reckless disregard for the trespasser's presence (*R Addie & Sons (Collieries) Ltd v Dumbreck [1929] AC 358*). In more recent times, however, the common law has adopted an attitude of humane conscientiousness towards simple (as distinct from aggravated) trespassers. '... the question whether an occupier is liable in respect of an accident to a trespasser on his land would depend on whether a conscientious, humane man with his knowledge, skill and resources could reasonably have been expected to have done, or refrained from doing, before the accident, something which would have avoided it. If he knew before the accident that there was a substantial probability that trespassers would come I think that most people would regard as culpable failure to give any thought to their safety' (*Herrington v British Railways Board [1972] 1 All ER 749*). The effect of this House of Lords case was that it became possible for a trespasser to bring a successful action in negligence under the common law. However, the precise nature of the occupier's duty to the trespasser remained unclear, leading to Parliamentary intervention culminating in the *OLA 1984*.

### Occupiers' Liability Act 1984

**O3009**    The *OLA 1984* introduced a duty on an occupier in respect of trespassers, that is persons whether they have 'lawful authority to be in the vicinity or not' who may be at risk of injury on his premises. [*OLA 1984, s 1*]. Thus, an occupier owes a duty of care to a trespasser in respect of any injury suffered on the premises (either because of any danger due to the state of the premises, or things done or omitted to be done) [*OLA 1984, s 1(3)(a)*], in the following circumstances:

(a)   if he was aware of the danger or had reasonable grounds to believe that it exists;

(b)   if he knows or has reasonable grounds to believe that a trespasser is in the vicinity of the danger concerned, or that he may come into the vicinity of the danger; and

(c)   if the risk is one against which, in all the circumstances of the case, he may reasonably be expected to offer some protection.

Where under the *OLA 1984*, the occupier is under a duty of care to the trespasser in respect of the risk, the standard of care is 'to take such care as is reasonable in all the circumstances of the case to see that he does not suffer injury on the premises by reason of the danger concerned'. [*OLA 1984, s 1(4)*]. The duty can be discharged by issuing a warning, e.g. posting notices warning of hazards; these, however, must be explicit and not merely vague. Thus, 'Danger' might not be sufficient whereas 'Highly Flammable Liquid Vapours – No Smoking' would be [*OLA 1984, s 1(5)*] (see O3015 below). Moreover, under *OLA 1984* there is no duty to persons who willingly accept risks (see O3016 below). However, the fact that an occupier has taken precautions to prevent persons going on to his land, where there is a danger, does not mean that the occupier has reason to believe that someone would be likely to come into the vicinity of the danger, thereby owing a duty to the trespasser, under the *OLA 1984, s 1(4)* (*White v St Albans City and District Council, The Times, 12 March 1990*). In *Revil v Newbury, The Times, 3 November 1995*, the court considered the duty of an occupier to an intruder. It found that the intruder could recover damages despite being engaged in unlawful activities at the time the injury occurred. The defendant was held to have injured the intruder by using greater force than was justified (by shooting the intruder through a locked door) in protecting himself and his property. The doctrine of *ex turpi causa non oritur actio* (i.e. no cause of action may be founded on an immoral or illegal act) cannot, without more, be relied upon by the defendant to escape liability. Violence may be returned with necessary violence, but the force used must not exceed the limits of what is reasonable in all the circumstances. For the extent of the intruder's contributory negligence, see O3011 below.

## Children

**O3010**   'An occupier must be prepared for children to be less careful than adults.' [*OLA 1957, s 2(3)(a)*]. Where an adult would be regarded as a trespasser, a child is likely to qualify as an implied licensee, and this in spite of the stricture that 'it is hard to see how infantile temptations can give rights however much they excuse peccadilloes' (per Hamilton LJ in *Latham v Johnson & Nephew Ltd [1913] 1 KB 398*). If there is something or some state of affairs on the premises (e.g. machinery, a boat, a pond, bright berries, a motor car, forklift truck, scaffolding), this may constitute a 'trap' to a child. If the child is then injured by the 'trap', the occupier will often be liable. Though sometimes the presence of a parent may be treated as an implied condition of the permission to enter premises (e.g. when children go on to premises at dusk – *Phipps v Rochester Corporation [1955] 1 All ER 129*). Perhaps the current common law position regarding 'child-trespass' was best put as follows: 'The doctrine that a trespasser, however innocent, enters land at his own risk, that in no circumstances is he owed a duty of reasonable or any care by the owners or occupiers of the land, however conscious they may be of the likelihood of his presence and of the grave risk of terrible injury to which he will probably be exposed, may have been all very well when rights of property, particularly in land, were regarded as more sacrosanct than any other human right ... It is difficult to see why today this doctrine should not be buried' (per Salmon LJ in *Herrington v British Railways Board*, concerning a six-year-old boy electrocuted on the defendant's electrified line) (but see O3017 below, *Titchener v British Railways Board*). Even though the children had been warned of the danger, the defendant company was still held liable in *Southern Portland Cement v Cooper [1974] AC 623*.

The case of *Jolley v London Borough of Sutton*, The Times, May 24 2000, concerned an old wooden boat that had been dumped on a grassed area belonging to a block of local authority flats. It became rotten and decayed, but was not covered or fenced off. The council placed a danger sticker on the boat, warning people not to touch it. This did not deter two teenage boys, who decided to repair it so they could take it sailing. Using a car jack and some wood from home, they jacked up the front of the boat so that they could repair the hull. One of the boys was lying on the ground underneath the boat when the prop gave way. The boat fell on him and caused very serious spinal injuries.

The Court of Appeal ruled that although it was foreseeable that children who were attracted by the boat might climb on it and be injured by the rotten planking giving way, it was not reasonably foreseeable that an accident could occur as a result of the boys deciding to work under a propped-up boat. However, the House of Lords decided this was wrong and that the council was liable; the trial judge was right in saying that the wider risk was that children would meddle with the boat at the risk of some physical injury, and what the boys did was not so different from normal play. The council should have removed the boat.

## Contributory negligence

O3011     In a number of cases brought under the *OLA 1957*, the question of contributory negligence is often considered and impacts considerably on the amount of damages eventually awarded. This may be because the duty on the occupier is only to take such care as is reasonable to see that the visitor will be reasonably safe in using his premises. Therefore, it is possible to imagine that, even where an occupier is found to have breached his duty of care, the injured party may also, by his own conduct, have contributed to the circumstances giving rise to the injury and may have exacerbated the severity of the injuries sustained. The *Law Reform (Contributory Negligence) Act 1945, s 1(1)* provides that 'where any person suffers damage as the result partly of his own fault and partly of the fault of another person, a claim in respect of that damage shall not be defeated by reason of the fault of the person suffering the damage, but the damage recoverable in respect thereof shall be reduced to such extent as the court thinks just having regard to the claimant's share in the responsibility of the damage'. There is nothing to suggest that the above provision does not apply to children or trespassers. However, Lord Denning in *Gough v Thorn [1966] 3 All ER 398* held that 'a very young child cannot be guilty of contributory negligence. An older child may be. But it depends on the circumstances'. In *Revil v Newbury* (see O3009 above), although the defendant's conduct was not reasonable, the plaintiff intruder was found to be two-thirds to blame for the injuries he suffered and his damages were reduced accordingly. A claimant cannot be guilty of 100 per cent contributory negligence (*Anderson v Newham College of Further Education [2002] EWCA Civ 505*) if evidence showed that the entire fault was to lie with the claimant then there could be no liability on the defendant.

## Dangers to guard against

O3012     The duty owed by an occupier to his lawful visitors is a 'common duty of care', so called since the duty is owed to both invitees and licensees, i.e. those having an interest in common with the occupier (e.g. business associates, customers, clients, salesmen) and those permitted by regulation/statute to be on the premises, e.g. factory inspectors/policemen. That duty requires that the dangers against which the occupier must guard are twofold: (*a*) structural defects in the premises; and (*b*) dangers associated with works/operations carried out for the occupier on the premises.

### Structural defects

O3013     As regards structural defects in premises, the occupier will only be liable if either he actually knew of a defect or foreseeably had reason to believe that there was a defect in the premises. Simply put, the occupier would not incur liability for the existence of latent defects causing injury or damage, unless he had *special* knowledge in that regard, e.g. a faulty electrical circuit, unless he were an electrician; whereas, he would be liable for patent (i.e. obvious) structural defects, e.g. an unlit hole in the road. This duty now extends to 'uninvited entrants', e.g. trespassers, under the *OLA 1984, s 1*.

## Workmen on occupier's premises

**O3014**  'An occupier may expect that a person, in the exercise of his calling, will appreciate and guard against any special risks ordinarily incident to it, so far as the occupier leaves him free to do so.' [*OLA 1957, s 2(3)(b)*]. This has generally been taken to imply that risks associated with system or method of work on customer premises are the exclusive responsibility of the employer (not the customer or occupier) (*General Cleaning Contractors Ltd v Christmas* (see W9040 WORK AT HEIGHTS)), though how far this rule now applies specifically to window cleaners themselves is doubtful (see *King v Smith [1995] ICR 339* at W9040 WORK AT HEIGHTS). This means that risks associated with the system or method of work on third party premises are the responsibility of the employer not the occupier (*General Cleaning Contractors Ltd v Christmas [1952] 2 All ER 1110*, concerning a window cleaner who failed to take proper precautions in respect of a defective sash window: it was held that there was no liability on the part of the occupier, but liability on the part of the employer). The occupier will only incur liability for a structural defect in premises which the oncoming workman would not normally guard against as part of a safe system of doing his job, i.e. against 'unusual' dangers. 'And with respect to such a visitor at least, we consider it settled law that he, using reasonable care on his part for his own safety, is entitled to expect that the occupier shall on his part use reasonable care to prevent damage from unusual danger, which he knows or ought to know' (per Willes J in *Indermaur v Dames [1866] LR 1 CP 274*, concerning a gasfitter testing gas burners in a sugar refinery who fell into an unfenced shaft and was injured).

The law is not entirely settled in the case of firemen (*Sibbald v Sher Bros, The Times, 1 February 1981*): 'It (is) ... very unlikely that the duty of care owed by the occupier to workers was the same as that owed to firemen,' (per Lord Fraser of Tullybelton), (but) it is arguable that a 'fireman (is) a "neighbour" of the occupier in the sense of Lord Atkin's famous dictum in *Donoghue v Stevenson [1932] AC 562*, so that the occupier owes him some duty of care, as for instance, to warn firemen of an unexpected danger or trap of which he knew or ought to know' (per Waller LJ in *Hartley v British Railways Board (1981) SJ 125*). The recent case of *Simpson v A I Dairies Farms Limited [2001] LTL January 12 2001* reinforces this as a farmer who failed to warn a fireman of the presence of an uneven drain in a yard which was currently concealed by water was deemed liable when the fireman was injured.

One's neighbour was defined in law as follows: 'persons who are so closely and directly affected by my act that I ought reasonably to have them in contemplation as being so affected when I am directing my mind to the acts or omissions which are called in question' (per Lord Atkin in *Donoghue v Stevenson [1932] AC 562*).

In summary, an occupier will be liable:

(*a*)   if he exposes a fireman to the risk of injury/death over and above the normal risks (see further F5053 (*a*) and (*b*) FIRE AND FIRE PRECAUTIONS); and

(*b*)   in accordance with the general principles of negligence, in non-emergency situations (as per *Ogwo v Taylor [1987] 3 All ER 961*, applying the 'neighbour' principle of *Donoghue v Stevenson*).

In both cases, the basis of action is negligence, either at common law or under the *OLA 1957* or both.

## Dangers associated with works being done on premises

**O3015**   Where work is being done on premises by a contractor, the occupier is not liable if he:

(a) took care in selecting a competent contractor; and

(b) satisfied himself that the work was being properly done by the contractor.

[*OLA 1957, s 2(4)(b)*].

Liability for a person injured as the result of work being carried out by an independent contractor tends also to remain the responsibility of the occupier as it is s/he who has 'control' of the land. The occupier must (1) act reasonably in selecting and entrusting work to the independent contractor; (2) take reasonable steps to supervise the work; and (3) use reasonable care to check that the work has been properly done. In *Jackson v Goodwood Estates [1999] LTL April 5 1999* a visitor to Goodwood Park tripped over a cable which was lying across the gateway. The owner had given instructions to the contractor that all cables be buried. However, the court found that the owner had not made sure that those instructions were implemented and had no system in place for checking, and was therefore liable.

As regards the checking of works, it may be highly desirable (indeed necessary) for an occupier to delegate the 'duty of satisfaction', especially where complicated building/engineering operations are being carried out, to a specialist, e.g. an architect, geotechnical engineer. Not to do so, in the interests of safety of visitors, is probably negligent. 'In the case of the construction of a substantial building, or of a ship, I should have thought that the building owner, if he is to escape subsequent tortious liability for faulty construction, should not only take care to contract with a competent contractor ... but also cause that work to be supervised by a properly qualified professional ... such as an architect or surveyor ... I cannot think that different principles can apply to precautions during the course of construction, if the building owner is going to invite a third party to bring valuable property on to the site during construction' (per Mocatta J in *AMF International Ltd v Magnet Bowling Ltd [1968] 2 All ER 789*).

## Waiver of duty and the Unfair Contract Terms Act 1977 (UCTA)

O3016    Where damage was caused to a visitor by a danger of which he had been warned by the occupier (e.g. by notice), an explicit notice, e.g. 'Highly Flammable Liquid Vapours – No Smoking' (as distinct from a vague notice such as 'Fire Hazards') used to absolve an occupier from liability. Now, however, such notices are ineffective (except in the case of trespassers, see O3008 above) and do not exonerate occupiers. Thus: 'a person cannot by reference to any contract term or to a notice given to persons generally or to particular persons exclude or restrict his liability for death or personal injury resulting from negligence'. [*UCTA 1977, s 2*]. Such explicit notices are, however, a defence to an occupier when sued for negligent injury by a simple trespasser under the *OLA 1984*. As regards negligent damage to property, a person can restrict or exclude his liability by a notice or contract term, but such notice or contract term must be 'reasonable', and it is incumbent on the occupier to prove that it is in fact reasonable. [*UCTA 1977, ss 2(2), 11*].

The question of whether the defendant did what was reasonable fairly to bring to the notice of the plaintiff the existence of a particularly onerous condition, regarding his statutory rights, was considered in *Interfoto Picture Library Ltd v Stiletto Visual Programmes Ltd [1989] QB 433*. It was held that for a condition seeking to restrict statutory rights (e.g. under the *OLA 1957*) to be effective it must have been fairly brought to the attention of a party to the contract by way of some clear indication which would lead an ordinary sensible person to realise, at or before the time of making the contract, that such a term relating to personal injury was to be included in the contract.

## Risks willingly accepted – 'volenti non fit injuria'

O3017    'The common duty of care does not impose on an occupier any obligation to a visitor in respect of risks willingly accepted as his by the visitor'. [*OLA 1957, s 2(5)*]. However, this 'defence' has generally not succeeded in industrial injuries claims. This is similarly the case with occupiers' liability claims (*Burnett v British Waterways Board [1973] 2 All ER 631* where a lighterman was held not to be bound by the terms of a notice erected by the respondent, even though he had seen it many times and understood it). This decision has since been reinforced by the *Unfair Contract Terms Act 1977*. 'Where a contract term or notice purports to exclude or restrict liability for negligence a person's agreement to or awareness of it is not of itself to be taken as indicating his voluntary acceptance of any risk.' [*UCTA 1977, s 2(3)*]. The *OLA 1957* makes clear that there may be circumstances in which even an explicit warning will not absolve the occupier from liability. Nevertheless, the occupier is entitled to expect a reasonable person to appreciate certain obvious dangers and take appropriate action to ensure his safety. In those circumstances no warning would appear to be required. It was not necessary, for example, to warn an adult of sound mind that it was dangerous to go near the edge of a cliff in *Cotton v Derbyshire Dales District Council, The Times, 20 June 1994*. In *Darby v The National Trust [2000] EWCA Civ 189* it was held there was no duty on the defendant to warn against the risk of swimming in a pond as the dangers of drowning were no more then an adult should have foreseen. The risk to a competer swimmer was an obvious one (*Staples v West Dorset District Council [1995] PIQR P439*). There was therefore no relative causative risk; a sign would have told the deceased no more than he already knew.

This is not, however, the position with respect to trespassers. The *OLA 1984, s 1(6)* states, 'No duty is owed ... to any person in respect of risks willingly accepted as his by that person ... '. This applies even if the trespasser is a child (see O3010 above) (*Titchener v British Railways Board 1984 SLT 192* where a 15-year-old girl and her boyfriend aged 16 had been struck by a train. The boy was killed and the girl suffered serious injuries. They had squeezed through a gap in a fence to cross the line as a short-cut to a disused brickworks which was regularly used. It was held by the House of Lords that the respondent did not owe a duty to the girl to do more than they had done to maintain the fence. It would have been 'quite unreasonable' for the respondent to maintain an impenetrable and unclimbable fence. But the 'duty (to maintain fences) will tend to be higher with a very young or a very old person than with a normally active and intelligent adult or adolescent' (per Lord Fraser)). *Scott and Swainger v Associated British Ports and British Railways Board [2000] All ER (D) 1937*, concerned schoolboys who on separate occasions were both severely injured whilst attempting to 'train surf' i.e. hitch rides on slow passing freight trains. The case was decided on the narrow issue of causation. The judge found as a question of fact that the provision of a fence would not have deterred them. In this situation, the court seemed to conclude that defendants may not do more than fence in dangers in order to protect children and that by scaling a fence the claimants would have willingly consented to the risk of injury.

This position might well be otherwise if anti-trespasser devices were not satisfactorily maintained (*Adams v Southern Electricity Board, The Times, 21 October 1993* where the electricity board was held to be liable to a teenage trespasser for injuries sustained when he climbed on to apparatus by means of a defective anti-climbing device). Also, see *Tomlinson v Congleton Borough Council [2002] EWCA Civ 309 CA* in which a lake had warning notices of danger and prohibiting swimming. It was held that by ignoring these notices the claimant entered the lake as a trespasser. The claimant was aware of the danger of swimming and *Darby v The National Trust* was referred to. However as the borough council were aware that notices were being ignored and that injuries had already been caused, and were also aware of further steps which could alleviate the danger but which had not been implemented, the Court of Appeal held by a majority judgement that the council were liable with 2/3 contributory negligence for the claimant.

In the curious case of *Arthur v Anka [1996] 2 WLR 602* a car was parked without permission in a private car park. It was clamped but the driver removed his vehicle with the clamp still attached. The owner of the car park sued the driver for the return of the wheel clamp and payment of the fine for illegal parking. The question of *volenti* was considered in detail. The judge held that the owner parked his car in full knowledge that he was not entitled to do so and that he was therefore consenting to the consequences of his action (payment of the fine); he could not complain after the event. The effect of this consent was to render conduct lawful which would otherwise have been tortious. By voluntarily accepting the risk that his car might be clamped, the driver accepted the risk that the car would remain clamped until he paid the reasonable cost of clamping and de-clamping.

## Actions against factory occupiers

O3018      The *Workplace (Health, Safety and Welfare) Regulations 1992 (SI 1992 No 3004* as amended by *SI 2002 No 2174*) which came into force, in so far as existing workplaces are concerned, on 1 January 1996 now apply with respect to the health, safety and welfare of persons in a defined workplace. The application of the Regulations is not confined to factories and offices – they also affect public buildings such as hospitals and schools, but they do not apply to, for example, construction sites and quarries. Employers, persons who have control of the workplace and occupiers are all subject to the Regulations, the provisions of which address the environmental management and condition of buildings, such as the day-to-day considerations of ventilation, temperature control and lighting and potentially dangerous activities such as window cleaning. In many ways the Regulations reflect the obligations imposed by the old *Factories Act 1961*. For example, *Regulation 12(3)* requires floors and traffic routes, 'so far as is reasonably practicable', to be kept free from obstacles or substances which may cause a person to 'slip, trip or fall'. These Regulations also closely mirror the general obligation placed on employers by *HSWA 1974, s 2* to ensure so far as is reasonably practicable, the health, safety and welfare of their employees whilst at work. Breach of these Regulations gives rise to civil liability by virtue of *HSWA 1974, s 47(2)* (for criminal liability, see O3019 below and ENFORCEMENT).

An approved Code of Practice and Guidance to the Regulations is available from the Health and Safety Commission.

## Occupier's duties under HSWA 1974

O3019      In addition to civil liabilities under the *OLA 1957* and at common law, occupiers of buildings also have duties, the failure of which to carry out can lead to criminal liability under the *HSWA 1974*. More particularly, 'each person who has, to any extent, control of premises (i.e. "non-domestic" premises) or the means of access thereto or egress therefrom or of any plant or substance in such premises' must do what is reasonably practicable to see that the premises, means of access and egress and plant/substances on the premises, are safe and without health risks. [*HSWA 1974, s 4(2)*]. This section applies in the case of (*a*) non-employees and (*b*) non-domestic premises. [*HSWA 1974, s 4(1)*]. In other words, it places health and safety duties on persons and companies letting or sub-letting premises for work purposes, even though the persons working in those premises are not employees of the lessor/sublessor. The duty also extends to non-working persons, e.g. children at a play centre (*Moualem v Carlisle City Council (1994) 158 JPN 786*). As with most other forms of leasehold tenure, the person who is responsible for maintenance and repairs of the leased premises is the person who has 'control' (see, by way of analogy, O3004 above). [*HSWA 1974, s 4(3)*]. Included as 'premises' are common parts of a block of flats. These are 'non-domestic' premises (*Westminster City Council v Select Managements Ltd [1985] 1 All ER 897* which held that being a 'place' or 'installation on land', such areas are 'premises'; and they are not 'domestic', since they are in common use by the occupants of more than one private dwelling).

Moreover, the reasonableness of the measures which a person is required to take to ensure the safety of those premises is to be determined in the light of his knowledge of the expected use for which the premises have been made available and of the extent of his control and knowledge, if any, of the use thereafter. More particularly, if premises were not a reasonably foreseeable cause of danger to anyone acting in a way a person might reasonably be expected to act, in circumstances that might reasonably be expected to occur during the carrying out of the work, further measures would not be required against unknown and unexpected events (*Mailer v Austin Rover Group [1989] 2 All ER 1087* where an employee of a firm of cleaning contractors, whilst cleaning one of the appellant's paint spray booths and the sump underneath it, was killed by escaping fumes. The contractors had been instructed by the appellants not to use paint thinners from a pipe in the booth (which the appellants had turned off but not capped) and only to enter the sump (where the ventilator would have been turned off) with an approved safety lamp and when no one was working above. Contrary to those instructions, an employee used thinners from the pipe, which had then entered the sump below, where the deceased was working with a non-approved lamp, and an explosion occurred. It was held that the appellant was not liable for breach of the *HSWA 1974, s 4(2)*).

When considering criminal liability of an occupier under the *HSWA 1974* in relation to risks to outside contractors, *s 4* is likely to be of less significance in the future because of the decision of the House of Lords in *R v Associated Octel [1994] ICR 281*, a decision which in effect enables the *HSWA 1974, s 3* to be used against occupiers. *Section 3(1)* provides that 'It shall be the duty of every employer to conduct his undertaking in such a way as to ensure, so far as is reasonably practicable, that persons not in his employment who may be affected thereby are not thereby exposed to risks to their health or safety'. The decision establishes that an occupier can still be conducting his undertaking by engaging contractors to do work, even where the activities of the contractor are separate and not under the occupier's control. *Section 3* therefore requires the occupier to take steps to ensure the protection of contractors' employees from risks not merely arising from the physical state of the premises but also from the process of carrying out the work itself.

However, the judgment of *Makepiece v Evans Brothers (Reading) (A firm) (Court of Appeal, 23 May 2000)* should be noted in that it was held that an occupier would not usually be liable to an employee of a contractor employed to carry out work at the occupiers' premises if the employee was injured as a result of any unsafe system of work used by the employee and the contractor. It was not generally reasonable to expect an occupier of premises, having engaged a contractor whom he had reasonable grounds to regard as competent, to supervise the contractor's activities in order to ensure that the was discharging his duties to ensure a safe systems of work for his employees. Even if the occupier may have reason to suspect that the contractors may have been using an unsafe system of work, this would not, in itself, be enough to impose upon him liability under the act or in negligence.

# Offshore Operations

## Introduction

O7001    'Offshore operations' take place, broadly, in three specific stages:

(a)    the initial stage, when the platform is under construction;

(b)    the fitting out or 'hook-up' stage, when plant and equipment are being installed; and

(c)    following commissioning, the operational stage of oil and gas production.

Activities connected with petroleum exploration and production have a history of at least eighty years when, by an enactment of 1918, the searching for petroleum onshore was made the subject of Government licence. This was intended to be a temporary measure, but the *Petroleum (Production) Act 1934* (now repealed) vested property rights in the Crown, and gave new powers to the Board of Trade to make regulations and impose conditions based on 'model clauses' which were incorporated into exploration/exploitation licences.

Model clauses were drafted in conjunction with the Institute of Petroleum, and matters of safety were covered by the phrase 'licensees must execute all operations in accordance with good oilfield practice' and comply with any instructions which might be given by the Board. Good oilfield practice was not, however, defined. The relevant provisions of earlier legislation governing the issue of licences have been amended by the *Petroleum Act 1998*.

In the 1950s petroleum exploration activities were being undertaken at sea. The rights of coastal countries to exploit the natural resources of their continental shelves were first formally recognised in international law by the Geneva Convention on the Continental Shelf 1958. The Convention was ratified by the United Kingdom in 1964 by the *Continental Shelf Act 1964* (*CSA 1964*) which vested in the Crown all rights exercisable in designated areas outside territorial waters in relation to the sea bed and subsoil in those areas. The *CSA 1964* also provided that the model clauses for offshore petroleum licences should make provision for the health, safety and welfare of employees. Enforcement of the Act and subordinate legislation continued to be the responsibility of the Department of Energy under the Safety Directorate of the Petroleum Engineering Division. However, the only sanction available for the enforcement of the safety procedures was withdrawal of a licence.

Following an accident in 1965 to the exploration rig Sea Gem, which capsized and sank with the loss of 13 lives, the Government accepted recommendations for strengthening statutory provisions and the *Mineral Workings (Offshore Installations) Act 1971* (*MWA 1971*) was enacted specifically to regulate offshore safety matters.

Today, most of the provisions of *MWA 1971* have been repealed and replaced by health and safety law. The provisions of the 1934 Act have been re-enacted with modifications by the *Petroleum Act 1998* (*PA 1998*) to cover:

- petroleum (rights vested in Her Majesty, licensing, development fees, mines and workings):
    - 'petroleum' is defined to include any mineral oil or relative hydrocarbon and natural gas petroleum existing in its natural condition in strata. It does not include coal or bituminous shales or other stratified deposits from which oil can be extracted by destructive distillation;

**O7002**   *Offshore Operations*

- offshore activities:
    - application of criminal and civil law etc.;
    - prosecutions for offences with respect to carriage of passengers and goods by helicopter (*Civil Aviation Act 1982 – in part*); infringement of safety zones (*Petroleum Act 1987, s 23*); any offence from remaining provisions of the *Mineral Workings (Offshore Installations) Act 1971*;
- submarine pipelines (see also O7034 below):
    - construction and use of pipelines; authorisations; compulsory modifications; rights to use pipelines; and
    - pipeline inspectors and procedures;
- abandonment of offshore installations (see also O7014 below):
    - abandonment programmes – duties and default procedures.

In addition, the *Petroleum Act 1998* makes amendments to the following statutes associated with offshore (safety) operations:

- *Continental Shelf Act 1964*;
- *Prevention of Oil Pollution Act 1971*;
- *Offshore Petroleum Development (Scotland) Act 1975*;
- *Sex Discrimination Act 1975*;
- *Fatal Accidents and Sudden Deaths Inquiry (Scotland) Act 1976*;
- *Race Relations Act 1976*;
- *Energy Act 1976*;
- *Telecommunications Act 1984*;
- *Food and Environment Protection Act 1985*;
- *Oil and Pipelines Act 1985*;
- *Gas Act 1986*;
- *Petroleum Act 1987*;
- *Territorial Sea Act 1987*;
- *Food Safety Act 1990*;
- *Aviation and Maritime Security Act 1990*;
- *Social Security Contributions and Benefits Act 1992*;
- *Offshore Safety Act 1992*;
- *Trade Union and Labour Relations (Consolidation) Act 1992*;
- *Merchant Shipping Act 1995*;
- *Employment Rights Act 1996*.

## Application of criminal and civil law to offshore installations

**O7002**    The *Petroleum Act 1998* provides that any act or omission constituting an offence under United Kingdom law which takes place on, under or above an offshore installation in United Kingdom waters or within 500 metres of such an installation (the safety zone) shall be treated as taking place in the United Kingdom. Police powers are granted accordingly. The application of criminal law to offshore installations also extends to corporate bodies, including the provisions relating to '… the consent or connivance of, or to be attributable to any neglect on the part of, any director, manager …' (*HSWA 1974, s 37*).

Matters which may be heard under civil law regarding offshore activities are determined in accordance with the law in force in such part of the United Kingdom as may be specified in an Order in Council. Any jurisdiction conferred on a court under this provision cannot prejudice any jurisdiction exercisable by any other court. The application of civil law also extends to mobile installations in transit.

## The Cullen Report

**O7003**    *MWA 1971* and the regulations made under it remained in force until 1992 when, following the Piper Alpha tragedy in 1988 (167 lives were lost), the fatal accident inquiry (the Cullen Report) made 76 recommendations regarding offshore health and safety. During the inquiry Lord Cullen made reference to the Robens Report (1972) and the Burgoyne Report (1980) and in particular referred to the significance of the following:

— the Robens Report (which led to the enactment of *HSWA 1974*) had implications for safety in onshore petroleum exploitation but it did not explicitly consider offshore safety;

— the decision of the Robens Committee to exclude *MWA 1971* because of the similarities in fact and in relevant (marine) law between offshore installations and ships which are not covered by *HSWA 1974*;

— the subsequent decision of the Government to extend *HSWA 1974* to cover oil and gas workers offshore [*Health and Safety at Work etc. Act 1974 (Application outside Great Britain) Order 1977 (SI 1977 No 1232)* – now revoked] but that its enforcement should remain the responsibility of inspectors appointed by the Department of Energy under the powers provided under *MWA 1971*;

— the ineffectiveness of inspections as compared with those of HSE inspectors regulating similar major hazards onshore, under the *Control of Industrial Major Accident Hazards Regulations 1984 (CIMAH 1984) (SI 1984 No 1902 as amended)*.

Regarding *CIMAH 1984* Lord Cullen, although influenced by the HSC/HSE enforcement policy, recognised that due to the close proximity of working and living areas and the impracticability of rapid evacuation from offshore installations in an emergency, the following additional provisions were required which have no direct counterpart in *CIMAH 1984*:

(a)    that the 'duty holder's' standards for management of health and safety and the control of major hazards must be demonstrated by the submission of a safety case and subject to formal acceptance by the regulatory body. Any installation without an accepted safety case will not be allowed to operate;

(b)    that measures to protect the workforce must include arrangements for temporary refuge from fire, explosion and associated hazards until evacuation can be effected from an installation;

(c)    that suitable use should be made of quantitative risk assessment (QRA) techniques as an aid to demonstrating the adequacy of the preventive and protective measures;

(d)    that arrangements should be put in place relating to safety management and audit.

## Repeal and revocation of outdated legislation

**O7004**    Today, as Lord Cullen recommended, offshore health, safety and welfare is regulated by a specialist division of the HSE – the Offshore Safety Division (HSE-OSD) – and the new regime promotes risk-based and goal-setting legislation made under *HSWA 1974*. The change came about following the enactment of the *Offshore Safety Act 1992 (OSA 1992)* and the

**O7005**  *Offshore Operations*

*Offshore Safety (Repeals and Modifications) Regulations 1993* (*SI 1993 No 1823*). The principal change was to expressly extend the application of *Part I* (*General Duties*) of *HSWA 1974* to include:

(a) securing the safety, health and welfare of persons on offshore installations or engaged on pipe-line works;

(b) securing the safety of such installations and preventing accidents on or near them;

(c) securing the proper construction and safe operation of pipe-lines and preventing damage to them; and

(d) securing the safe dismantling, removal and disposal of offshore installations and pipe-lines.

*[OSA 1992, s 1(1)].*

At the same time the following statutes were expressed to come within the ambit of the enforcement provisions of *HSWA 1974*:

— the Pipe-lines Act 1962;

— the Mineral Workings (Offshore Installations) Act 1971;

— the Petroleum and Submarine Pipe-lines Act 1975 (now repealed);

— the Petroleum Act 1987.

Subsequent regulations made under *HSWA 1974* have repealed other sections of *MWA 1971* and its subordinate legislation, leaving only a few provisions of the early offshore legislation still extant. These include:

— *Offshore Installations (Logbooks and Registration of Death) Regulations 1972* (*SI 1972 No 1542*);

— Offshore Installations (Inspectors and Casualties) Regulations 1973 (SI 1973 No 1842);

— Submarine Pipe-lines (Inspectors etc.) Regulations 1977 (SI 1977 No 835).

## What is an offshore installation?

**O7005**   The *Offshore Installations and Pipeline Works (Management and Administration) Regulations 1995* (*SI 1995 No 738 as amended*) specify that for a structure to be classed as a 'fixed installation' (as opposed to a 'mobile installation') it must be:

— undertaking an 'offshore activity'; and

— 'standing or stationed' in 'relevant waters' (see O7009 below).

Because of the complexity of defining offshore installations, appropriate advice should be sought in any case of doubt.

## Parts of offshore installations

**O7006**   The *Offshore Installations and Pipeline Works (Management and Administration) Regulations 1995, Reg 3(3)* states that the following offshore equipment is deemed to be part of an offshore installation:

— wells, if they are connected by a pipeline or cable;

— Christmas trees or blowout preventers;

— pipelines, apparatus or works, attached to the offshore installation and within the 'safety zone' of that installation.

## Safety zones

**O7007** The threat of collision with an offshore installation was first controlled by the *Continental Shelf Act 1964* and later extended by the *Petroleum Act 1987*. Sec 21 of the *Petroleum Act 1987* provides for the automatic establishment of a 500-metre exclusion zone around every offshore installation 'on station' whilst permitting authorised vessels or helicopters to attend the installation.

The prohibition on vessels entering or remaining in a safety zone does not apply to a vessel:

— in connection with pipe laying or inspection activities;

— providing services for transportation of goods or persons or those under the authority of a Government required to inspect an installation;

— performing duties relating to the safety of navigation;

— in connection with the saving or attempted saving of life or property;

— entering because of distress or adverse weather.

*[Offshore (Safety Zone) Regulations 1987 (SI 1987 No 1331)].*

Permission to enter a safety zone is given by the installation manager.

## Temporary exclusion zones

**O7008** In addition to the set safety zones around each offshore installation or sub-sea installation, the Secretary of State for Transport is empowered under marine law to establish temporary exclusion zones around a wrecked ship or structure. In determining whether such a casualty, including those in distress, requires the establishment of an exclusion zone, the Secretary of State will take into account the 'significant harm' factor, that is, in terms of pollution or damage to persons or property. [*Merchant Shipping Act 1995, s 100A* – as introduced by the *Merchant Shipping and Maritime Security Act 1997*].

## Which health and safety law applies to offshore installations?

**O7009** The key as to whether regulations made under *HSWA 1974* are enforceable on offshore installations is the word 'apply'. The *Health and Safety at Work etc. Act 1974 (Application outside Great Britain) Order 1977 (SI 1977 No 1232)*, now revoked, had the effect of extending the application of most of the provisions of *HSWA 1974*, and certain selected regulations made under *HSWA 1974*, to offshore sites. The following regulations were reapplied to offshore installations in their entirety by virtue of the *Health and Safety at Work etc. Act 1974 (Application outside Great Britain) Order 1989 (SI 1989 No 840)*:

— Control of Lead at Work Regulations 1980 (SI 1980 No 1248);

— Asbestos (Licensing) Regulations 1983 (SI 1983 No 1649);

— Freight Containers (Safety Convention) Regulations 1984 (SI 1984 No 1890);

— Ionising Radiations Regulations 1985 (SI 1985 No 1333);

— Control of Asbestos at Work Regulations 1987 (SI 1987 No 2115).

The *Health and Safety at Work etc. Act 1974 (Application outside Great Britain) Order 1989 (SI 1989 No 840)* was revoked by the *Health and Safety at Work etc. Act 1974 (Application outside Great Britain) Order 1995 (SI 1995 No 263)*, which re-enacted its provisions with modifications. Existing regulations which refer to the 1989 Application Order (*SI 1989 No 840*) (or to a previous Order) should be construed as referring to the appropriate Articles of the current Order.

## O7009 Offshore Operations

Determining whether or not health and safety regulations apply to offshore operations can be complex. Advice should be sought from company legal representatives or the HSE-OSD in any cases of doubt. The following comments, however, may be useful:

Firstly, regulations made under *HSWA 1974* apply offshore only if they contain a clause in the current Application Order which explicitly states that they do apply. Each set of new regulations issued should detail the extent of their application by referring to the relevant Articles in the Application Order.

Secondly, regulations which contain an 'Application outside Great Britain' clause but which do not refer to any specific Article in the Application Order should be taken as applying to all the categories of offshore descriptions and activities specified in the Application Order.

To understand how a particular set of regulations is applied offshore, it is necessary to take account of several legally defined terms and expressions which may be included in the 'Application outside Great Britain' clause:

— *territorial waters* – i.e. United Kingdom territorial waters adjacent to Great Britain.

  [*The Health and Safety at Work etc. Act 1974 (Application outside Great Britain) Order 1995* (SI 1995 No 263), Reg 2(1)].

— *within territorial waters* includes on, over or under United Kingdom territorial waters adjacent to Great Britain.

  [*The Health and Safety at Work etc. Act 1974 (Application outside Great Britain) Order 1995* (SI 1995 No 263), Reg 2(1)].

— *territorial sea* – i.e. the breadth of the territorial sea adjacent to the United Kingdom (the 12-mile limit).

  [*Territorial Sea Act 1987, s 1*].

— *relevant waters* – i.e. (a) tidal waters and parts of the sea in or adjacent to Great Britain up to the seaward limits of 'territorial waters', and (b) any area designated by Order under the Continental Shelf Act 1964, s 1(7).

  [*The Offshore Installations and Pipeline Works (Management and Administration) Regulations 1995* (SI 1995 No 738 as amended), Reg 2(1)].

— *associated structure* – i.e. in relation to an offshore installation a 'vessel', aircraft or hovercraft attendant on the installation or any floating structure used in connection with the installation.

  [*The Offshore Installations and Pipeline Works (Management and Administration) Regulations 1995* (SI 1995 No 738 as amended), Reg 2(1)].

— *vessel* – i.e. a hovercraft and any floating structure which is capable of being staffed.

— *jurisdiction* – i.e. (a) the law in force in England and Wales, (b) the law in force in Scotland, (c) the law in force in Northern Ireland,

  and the authority given to the High Court for the determination of any questions of law arising in a designated area made under the *Continental Shelf Act 1964, s 1(7)*, and in particular the application of law as it would apply in those countries by virtue of the *Civil Jurisdiction (Offshore Activities) Order 1987* (SI 1987 No 2197) and the *Criminal Jurisdiction (Offshore Activities) Order 1987* (SI 1987 No 2198).

— *fixed installation* – i.e. an offshore installation other than a 'mobile installation' which is standing in water on station, and which is specified as an Article 4 application. The health and safety legislation so applied relates to most offshore operations and activities.

- *mobile installation* – i.e. an offshore installation (other than a floating production platform) which can be moved from place to place without major dismantling or modification, whether or not it has its own motive power, moves out of station ('relevant waters') to a UK dock etc. ('within territorial waters') for activities to be carried out by shore-based workers ('Article 8 activities').

- *activities within territorial waters* – Article 8 specifies the following activities which may be carried out by non-offshore workers in accordance with the provisions of *HSWA 1974* and its subordinate legislation:

    (*a*)   the construction, reconstruction, alteration, repair, maintenance, cleaning, demolition and dismantling (or any preparatory activities) of any building or other structure (other than a vessel);

    (*b*)   loading, unloading, fuelling or provisioning of a vessel (which includes a 'stacked' installation, i.e. one which is not in use or intended for use at that time);

    (*c*)   the construction, reconstruction, finishing, refitting, repair, maintenance, cleaning or breaking up of a vessel (including a 'stacked' installation) except when carried out by the master or any officer or member of the crew of that vessel;

    (*d*)   diving operations;

    (*e*)   maintaining a 'stacked' installation on station.

The current Application Order is the *Health and Safety at Work etc. Act 1974 (Application outside Great Britain) Order 1995 (SI 1995 No 263)*.

In addition to the necessary compliance with the law in accordance with Article 8, the owner of the mobile installation (also the 'duty holder') would not wish his safety case to be jeopardised, and therefore the legislation outlined below, and specific to offshore installations, must also be taken into consideration.

Generally, if the most onerous of the regulations were complied with then the requirements of other health and safety legislation may be deemed to be satisfied.

# Offshore Installations (Safety Case) Regulations 1992 (SI 1992 No 2885)

**O7010**   As Lord Cullen recommended, new offshore safety regulations, the *Offshore Installations (Safety Case) Regulations 1992* (the *Safety Case Regulations 1992*), were introduced within the *HSWA 1974* framework.

One of the main changes brought in by the new regime (known as the 'safety case regime'), and in line with the principles underlying *HSWA 1974*, is that the responsibility placed on managers (including the manager of an offshore installation) may not extend beyond matters which are in practice within their control, and authority to act must have been explicitly delegated by the employer. The employer remains responsible for ensuring that managers and others who exercise discretion and judgement are competent to do so and have clear guidelines.

Previously, under *Sec 5* of the now defunct *Mineral Workings (Offshore Installations) Act 1971*, a general duty was placed on the manager of an offshore installation for all matters affecting safety, health or welfare on his offshore installation. This duty also included the maintenance of order and discipline over all people on or about the installation.

In drawing up the *Safety Case Regulations 1992* reference was made to the guidance on the effective management of safety and health contained in the following publications:

- HS(G) 65 *Successful Health and Safety Management*; and

— *HS(G) 48 Human Factors in Industrial Safety*: an examination of the roles of organisations, jobs and individuals in industrial safety and a practical guide to control.

## Regulations to underpin the Safety Case Regulations

O7011   In addition to the *Safety Case Regulations 1992*, three further goal-setting regulations have been enacted to implement the recommendations made by Lord Cullen; they complete offshore implementation of the Extractive Industries (Boreholes) Directive and support and complement the objectives of the *Safety Case Regulations 1992*:

— *Offshore Installations and Pipeline Works* (*Management and Administration*) *Regulations 1995* (*SI 1995 No 738*);

— *Offshore Installations* (*Prevention of Fire and Explosion, and Emergency Response*) *Regulations 1995* (*SI 1995 No 743*);

— *Offshore Installations and Wells* (*Design and Construction etc.*) *Regulations 1996* (*SI 1996 No 913*).

The primary aim of these regulations is to reduce risks to the health and safety of the workforce employed on offshore installations or in connected activities. The principal matters to be demonstrated in safety cases and submitted to the HSE in compliance with the *Safety Case Regulations 1992*, whilst taking account of the requirements of the above regulations, are that:

(*a*)   the management system is adequate to ensure compliance with all statutory health and safety requirements;

(*b*)   adequate arrangements have been made for audit, and the preparation of audit reports; and

(*c*)   all hazards with the potential to cause a 'major accident' have been identified, their risks evaluated, and measures taken to reduce risks to persons to as low as reasonably practicable.

The definition of 'major accident' embraces the full range of such accidents which can occur offshore and not just those arising from fire and explosion hazards. The case for safety is required to relate to all activities carried out on, or in connection with, an installation including, for example, relevant activities involving divers, pipelaying or the movement of vessels or helicopters. The various people concerned with these activities are under a duty to co-operate with 'duty holders' in the preparation of installation Safety Cases.

For convenience the expression 'duty holder' is used as a shorthand way of referring to the person (whether owner or operator of an installation) on whom duties are placed by the Regulations which underpin the *Safety Case Regulations 1992*.

In addition to the Safety Case, the *Safety Case Regulations 1992* require the operator to submit prior written notice to the HSE when the following hazardous activities are to take place:

— well operations – describing the location and environmental conditions, type of well operation, the hazards, well control procedures and the proposed programme of work;

— construction activities – describing the location, the type of activity, the hazards, the proposed programme of work, and identifying any installation or heavy lift vessel involved in the described activity.

As Lord Cullen recommended, the *Safety Case Regulations 1992* require Safety Cases for mobile as well as fixed installations, and for both new and existing installations. They are to be submitted at various stages in the 'life cycle' of the installation. The Safety Case is to be seen as a 'living document' to be updated to take account of changing activities, technology or other circumstances.

## Verification of safety-critical elements of installations

**O7012** A major amendment to the *Safety Case Regulations 1992* by the *Offshore Installations and Wells (Design and Construction etc.) Regulations 1996 (SI 1996 No 913)* (the *Design and Construction Regulations 1996*) imposes a duty on the duty holder to set up arrangements for the scrutiny of safety-critical elements of the offshore installation to be verified as suitable by an independent and competent person. This requirement replaces the certification regime established by *the Offshore Installations (Construction and Survey) Regulations 1974 (SI 1974 No 289)* which have been revoked by the *Design and Construction Regulations 1996*.

The purpose of such scrutiny is to obtain assurance of the achievement of the satisfactory condition of the safety critical elements of the installation. The scheme is to be carried out throughout the installation's life cycle (from design through fabrication, construction, hook-up and commissioning, and the whole operating life to decommissioning and dismantling). For the duty holder, the scheme will contribute to the duty holder's ability to demonstrate that the installation is 'fit for purpose'. It is therefore appropriate to include a summary of the scheme in the installation's Safety Case.

## Before commencing operations
**O7013**

(a) Fixed installations

*Reg 4(1)* requires the operators of new fixed installations to submit a Safety Case for such installations by a date which is early enough to allow them to take account in the design of any matters raised by the HSE within three months of the submission.

HSE will indicate in writing any changes to the design or any considerations which it would wish to be taken into account at detailed design, construction or commissioning stages, and which would be likely in due course to assist in acceptance of the full operational Safety Case required under *Reg 4(2)*.

No operations may be commenced until six months have elapsed since the operator sent the Safety Case to the HSE and it has been accepted by the HSE. Commencement of an operation includes the first well drilling operation which may require the release of hydrocarbons beneath the sea bed or when hydrocarbons are brought on to the site for the first time.

(b) Mobile installations

Operators of mobile installations must not commence operations before a Safety Case has been sent to the HSE at least three months before commencing the movement of the installation in the waters, and has been accepted by the HSE. [*Reg 5*].

(c) Combined operations

Operators of fixed installations must not engage in a combined operation with a mobile installation, and vice versa, unless:

(i) the joint operators have prepared a Safety Case and sent it to HSE

— at least four weeks before operations commence where a mobile installation is to carry out an operation on a well connected to a fixed installation (but not a well beneath or immediately adjacent to the installation),

— at least six weeks before operations commence in all other cases; and

(ii) HSE has accepted the Safety Case.

[Reg 6].

The intention of *Reg 6* is that a combined operations Safety Case should be required where two or more installations (each subject to the submission of a Safety Case) interact. Examples include:

— a mobile installation providing accommodation to another installation;

— a mobile drilling unit cantilevered over a fixed wellhead platform;

— a mobile drilling unit working on a subsea well linked back to a fixed production platform.

The duty holder has a general responsibility to revise the Safety Case as appropriate, and for this purpose will need to keep abreast of changes in technology and of other developments which might affect the validity of the original Safety Case. [*Reg 9(1)*].

*Reg 9(4)* reinforces this general obligation by requiring that a suitably revised version be submitted for acceptance at least once every three years. A revised Safety Case should incorporate all previous revisions and should be submitted at least three months before the three-year period expires.

## *Decommissioning fixed installations*

**O7014** Operators of a fixed installation must not decommission an installation unless they have:

(a) prepared a Safety Case;

(b) sent it to the HSE at least six months before the decommissioning starts; and

(c) the HSE has accepted the Safety Case.

*[Reg 7].*

The submission required by *Reg 7* is intended to ensure that the method and arrangements for abandonment of a fixed installation provide for adequate control of risks to persons on the installation or engaged in connected activities. It is an offence to begin decommissioning before the Safety Case has been accepted.

HSE's acceptance of a Safety Case for abandonment of an installation is required, and will be given, independently of any other approval such as the obligation under the *Petroleum Act 1998* to submit an abandonment programme.

## *The abandonment programme*

**O7015** The process of abandonment of a fixed offshore installation or submarine pipeline is subject to the acceptance of a detailed formal document known as the abandonment programme by the Secretary of State which must:

(a) contain an estimate of the cost of the measures proposed in the programme;

(b) specify the periods/times when the programme is to be put into effect, or provide information relevant to determining how the specified times are to be determined;

(c) include provisions regarding the continuing maintenance requirements of installations or pipelines (including those not fully demolished) which are to be left in position.

*Part IV* of the *Petroleum Act 1998* provides that when an abandonment programme has been approved by the Secretary of State, it becomes the duty of the persons who submitted the programme to ensure that it is carried out and that any conditions are complied with.

## Safety Case specifications

O7016    The particulars to be included in all Safety Cases are required to demonstrate that:

(a)    there is a management system that will make sure that the statutory provisions relating to the installation are complied with;

(b)    adequate arrangements have been made for audit and reporting of the management system;

(c)    all hazards that could cause a major accident have been identified;

(d)    risks have been evaluated and measures will be taken to keep these risks to the lowest level that is reasonably practicable.

*[Reg 8]*.

The submission should show among other matters that the duty holder's safety management system is adequate to ensure that the design objectives and arrangements for construction, hook-up of installation facilities and commissioning are fully realised and can be carried out safely.

Matters to be considered include:

(i)    *Organisation and arrangements for the detailed design*

— intended use of contractors or other outside organisations and criteria for their selection;

— means by which the duty holder will supervise, monitor and audit the detailed design (whether prepared internally or externally); and

— arrangements for recording modifications to the design, to facilitate subsequent audit.

(ii)    *Co-operation*

— the contractual arrangements to ensure adequate co-operation between the duty holder, design contractor(s), supplier(s) of plant, fabricators and others involved at any stage from the commencement of detailed design work through to offshore construction and commissioning.

(iii)    *Standards, methodologies and codes of practice*

(iv)    Quality control

— the arrangements for quality assurance at all stages of detailed design, construction (on and offshore) and commissioning to ensure conformity with the design and that the initial testing and proving of the plant is satisfactory.

## Verification of safety-critical elements

O7017    The duty holder (fixed and mobile installations) must ensure that the installation is not operated unless a suitable verification scheme covering the 'safety-critical elements' of the installation has been put in place. These elements include such parts of an installation and of its plant (including computer programs) the failure of which could cause or contribute substantially to a major accident.

To be suitable, the written scheme, prepared by a competent person, must demonstrate that the safety-critical elements are in good repair and condition and are verified as such by competent persons carrying out the following actions:

— examination, including testing where appropriate;

— examination of any design, specification, certificate, CE marking or other document, marking or standard relating to those elements;

— examination of work in progress;

— the taking of appropriate action following reports by such persons.

Competent persons must be independent, that is, their functions will not involve the consideration of elements for which they bear or have borne such responsibility as might compromise objectivity. In addition, it must be shown that they are sufficiently independent of a management system to ensure that they will be objective in discharging their functions.

## Review, revision and recording of verification schemes

O7018    The duty holder must ensure that as often as may be appropriate:

— the verification scheme is reviewed and, where necessary, revised or replaced by or in consultation with an independent and competent person; and

— a note is made of any reservation expressed by such person in the course of drawing it up;

— verification schemes must be kept (at an address notified to the HSE) for a minimum of six months following cessation of the scheme.

[Regs 15A-15D as inserted by SI 1996 No 913].

## Notification of hazardous activities

O7019    Duty holders must ensure that the following operations are not commenced until the HSE has been notified. However, unlike a Safety Case, notification does not require acceptance by HSE before the activities can be commenced. Should the information provided give cause for concern, HSE may intervene by, for example, requesting further information, inspecting, or by issuing an enforcement notice.

When considering whether intervention is appropriate HSE will take account of the following factors (among others):

— the major accident potential of the operations being undertaken;

— the assessed adequacy of the management and other risk control systems;

— whether the operations will take place in conditions approaching the operating limits described in the installation Safety Case (for example, extreme or unusual environmental conditions or well operations).

## Well operations

O7020    Operations involving the drilling of a well require at least 21 days' notice containing the particulars specified in *Sch 6 and 6A*;

Where operations in relation to a well are to be carried out from a mobile installation, sufficient particulars must be provided which demonstrate that the arrangements and procedures of the operator and owner of the installation for managing the operations are co-ordinated to reduce risks from a major accident to the lowest level that is reasonably practicable.

The Safety Case for a fixed installation should have already described (i) the wells proposed to be connected to the installation, and (ii) the well operations to be carried out in so far as these are foreseeable. Similarly, the Safety Case for the mobile installation to be used in the operation will contain the full range of well operations for which it has been designed to operate safely, including any limiting conditions such as maximum well pressures or temperatures.

In both cases, the notifications required by *Reg 11* complement the Safety Case (which will already have been accepted) by providing the additional information which the HSE needs to assess whether the particular well operation described will be performed in a way which reduces risks to as low a level as is reasonably practicable.

In particular, the HSE will wish to be satisfied of the management arrangements of the two installations including:

— the arrangements made for communication between employees both on and offshore;

— the arrangements for co-ordinating, where necessary, systems and procedures (for example, well control procedures, permit to work systems, barrier policies) in order to minimise the potential for any misunderstandings during operations.

If the planned programme of work of the design of the well is changed materially after notification, the Regulations require a further notification which identifies the changes.

## Construction activities

O7021  Similar notification is required at least 28 days before commencement in respect of any of the following activities. The particulars to be notified are contained in *Sch 7*:

— the construction of the installation at the place where it is to be operated;

— the dismantling or demolition of the installation at the place where it was operated; or

— any activity on or in connection with the installation which involves the use of a heavy lift vessel.

## Summary of contents of a Safety Case

O7022  The operational Safety Case which is required to be submitted under *Reg 4(2)* must include a comprehensive account of the following matters:

— management of health and safety, and control of major hazards;

— all foreseeable activities which are intended to be undertaken, or which may need to be undertaken, during the operating lifetime of the installation;

— any occasional activities such as major maintenance projects or diving work and any planned construction or alteration projects;

— activities which may involve other vessels (for example, nearby diving support vessels, supply vessels, floating storage units), aircraft or other installations should be addressed as far as they are foreseeable (*note:* where not foreseeable, further details may need to be provided in a revised Safety Case (under *Reg 9(2)*) or a combined operations Safety Case (as required by *Reg 6*);

— well operations so far as they are foreseeable and, in particular, a description of the intended number of wells and the purpose of each (that is, whether for production, gas injection or water injection);

— the drilling facilities and the general drilling programme;

— the expected performance characteristics of the reservoir and individual wells (including well drive mechanisms);

— well control procedures;

— any simultaneous drilling and production operations from the installation and the arrangements for controlling such activities;

— procedures for the suspension or abandonment of any well;

— the safety management system for the operation, control and emergency procedures of the pipelines connected to the installation;

— a summary of the design philosophy for ensuring the safe operation of the installation.

### Record keeping

**O7023** Copies of the Safety Case and any revision of it must be kept at an address in Great Britain and on the installation for at least three years. Documents relating to the verification scheme (see O7018 above) must be kept for six months.

Reports must be made, and the actions recorded, in respect of audits carried out under *Reg 8(1)(b)*. These requirements are intended to ensure, among other matters, that the reports and action records will be conveniently available for examination by the HSE during the auditing of the duty holder's safety management and audit system.

Documents required to be kept on the installation are normally kept by the installation manager. Arrangements for keeping the documents and revising them as required should be considered as part of the safety management system.

## Offshore Installations and Wells (Design and Construction, etc.) Regulations 1996 (SI 1996 No 913)

**O7024** The *Offshore Installations and Wells (Design and Construction, etc.) Regulations 1996* aim to ensure (*a*) the structural integrity of offshore installations throughout their normal life cycle; (*b*) that risks to installation workers from structural failure or loss of stability are minimised to reasonably practicable levels; and (*c*) to secure the safe condition of wells at all stages of their life cycle.

### Integrity of installations – duties of dutyholders

**O7025** So far as reasonably practicable, duty holders must ensure that:

(*a*) installations are able to withstand such adverse forces as are reasonably foreseeable [*Reg 5(1)(a)*];

(*b*) the structural integrity of an installation is not prejudiced by (i) its layout and configuration, and (ii) its fabrication, transportation, construction, operation, modification, maintenance, repair [*Reg 5(1)(b)(c)*];

(*c*) in the event of reasonably foreseeable damage, the installation will retain sufficient integrity to enable action to be taken to safeguard the health and safety of personnel on or near the installation [*Reg 5(1)(e)*];

(*d*) installations are capable of being decommissioned and dismantled safely [*Regs 5(1)(d), 10*];

(*e*) installations are composed of materials sufficiently proof against and/or protected from anything likely to prejudice their integrity [*Reg 5(2)*];

(*f*) work to an installation does not impair its integrity [*Reg 6*];

(*g*) installations are (i) operated within appropriate limits, and (ii) that environmental conditions in which they may safely operate have been recorded. Such records must be kept on the installation and readily available for inspection by personnel [*Reg 7*];

(*h*) installations are subject to periodic maintenance, and arrangements made for (i) periodic assessment, and (ii) execution of any remedial work [*Reg 8*];

(*j*) within ten days after the appearance of a significant threat to the integrity of an installation, a report is sent to HSE, identifying the threat and specifying any remedial action [*Reg 9*];

(k) every helicopter landing area is large enough for landing and departure purposes and is of adequate design and construction [*Reg 11*].

### Integrity of wells – duties of well-operators

O7026  Well-operators must ensure that:

(a) wells are designed, constructed and operated so that (i) so far as reasonably practicable, there is no unplanned escape of fluids, and (ii) risks to the health and safety of persons are as low as reasonably practicable [*Reg 13*];

(b) prior to design of a well, assessments have been made of geological strata and formations and fluids therein, including any hazards [*Reg 14*];

(c) wells are designed and constructed so that, so far as reasonably practicable, (i) they can be suspended or abandoned safely, and (ii) following suspension or abandonment, there is no unplanned escape of fluids [*Reg 15*];

(d) prior to commencement or adoption of design of a well, written arrangements are made for examination, by independent and competent persons, of a well or parts thereof [*Reg 18*];

(e) in the case of (i) escape of fluids, (ii) completion of a well, (iii) abandonment operations, (iv) drilling operations, and (v) workover operations, a report is sent to the HSE at such intervals as are agreed (or otherwise at weekly intervals following commencement), containing:

— identifying number of well;

— name of installation;

— summary of activity;

— diameter and true vertical and measured depths of (a) any hole drilled, and (b) any casing installed;

— drilling fluid density;

— in the case of an existing well, its current operational state.

[*Reg 19*].

### Duties of other personnel

O7027  Every person concerned (or to be concerned) in whatever capacity, in an operation in relation to a well, must co-operate with the well-operator, so far as is necessary, to enable him to discharge his duties. [*Reg 20*].

## Offshore Installations (Prevention of Fire and Explosion, and Emergency Response) Regulations 1995 (SI 1995 No 743)

O7028  These regulations, applicable to both fixed and mobile installations, require operators, owners and duty holders to:

(a) protect persons on the installation from fire and explosion, and secure effective emergency response [*Reg 4*];

(b) carry out, update, where necessary, and keep a record of an assessment of measures (including performance standards) for effective (i) evacuation, (ii) escape, (iii) recovery and (iv) rescue [*Reg 5*];

(c) establish the organisation/arrangements to be implemented in an emergency, and ensure provision of:
  (i) training/instruction on necessary action, and
  (ii) written information on use of emergency plant
  [Reg 6];
(d) ensure availability of suitable equipment in the event of an accident involving a helicopter [Reg 7];
(e) prepare and update the emergency response plan:
  (i) ensure its availability and that its contents are known, and
  (ii) ensure that it is tested from time to time
  [Reg 8];
(f) take effective measures to prevent fire/explosion, e.g.
  (i) identifying/designating areas where there is a risk of occurrence of flammable/explosive atmospheres,
  (ii) controlling the carrying on of hazardous activities there,
  (iii) prohibiting/limiting use of plant (in such areas) unless suitable for use, and
  (iv) controlling placement (in such areas) of electrical fixtures/sources of ignition
  [Reg 9];
(g) take steps to detect fire and communicate information to places where control action can be taken [Reg 10];
(h) provide suitable emergency response arrangements:
  (i) ensure a warning is given of an emergency (see below), and
  (ii) communicate the purpose of the emergency response.
  These arrangements should ideally remain effective during an emergency.
  [Reg 11(1)].
  *Warnings* take the form of (A) illuminated signs and (B) acoustic signals.
  (A) Illuminated signs
    These are:
    — in the case of warning of toxic gas, a red flashing sign, and
    — in other cases, a yellow flashing sign.
  (B) Acoustic signals
    These are:
    — in the case of warning of evacuation, a continuous signal of variable frequency,
    — in the case of warning of toxic gas, a continuous signal of constant frequency, and
    — in other cases, an intermittent signal of constant frequency.
    [Reg 11];

(*j*) control/limit extent of emergencies (including fire and explosions), e.g. by provision for remote operation of plant and ensuring that it is capable of remaining effective in an emergency [*Reg 12*];

(*k*) protect persons from the effects of fire and explosion [*Reg 13*];

(*l*) provide muster areas and:

    (i) provide safe egress from accommodation and work areas, and

    (ii) provide safe access to muster areas/evacuation and escape points which must be kept unobstructed, adequately lit with emergency lighting, marked with suitable signs and, so far as is reasonably practicable, remain passable in an emergency.

[*Reg 14(1), (2)*];

*Notes:*

- Doors for use in an emergency must open in the appropriate direction, or, if this is not possible, be sliding doors, and must not be fastened.

- Accommodation areas must be provided with at least two means of egress situated suitably apart at each level [*Reg 14(3)*].

- Each person on an installation must be assigned to a muster area; and muster lists must be kept up-to-date and clearly displayed [*Reg 14(4)*];

(*m*) arrange for safe evacuation of persons and their conveyance to a place of safety [*Reg 15*];

(*n*) ensure provision of means of escape, where arrangements for systematic evacuation fail [*Reg 16*];

(*o*) arrange for:

    (i) recovery of persons following evacuation/escape,

    (ii) rescue of persons near the installation, and

    (iii) conveyance of rescued/recovered persons to a place of safety

[*Reg 17*];

(*p*) provide emergency protective equipment and prepare and operate a written scheme for its examination and testing by a competent person [*Reg 18*] (see further PERSONAL PROTECTIVE EQUIPMENT);

(*q*) provide suitable emergency plant and equipment and ensure that it is maintained in good working order by a competent and independent person under a written scheme [*Reg 19*];

(*r*) provide life-saving appliances, e.g. survival craft, life-rafts, life-buoys, life-jackets, which are:

    (i) of conspicuous colour,

    (ii) suitably equipped (where necessary), and

    (iii) kept available in sufficient numbers

[*Reg 20*];

(*s*) provide information to all persons on the installation, in relation to:

    (i) areas in which there is a risk of occurrence of flammable/explosive atmospheres,

    (ii) non-automatic plant for fire-fighting purposes,

    (iii) plant connected with personal protective equipment and life-saving appliances

[*Reg 21*].

In addition to the 'on-board' fire-fighting facilities, the duty holder may elect to utilise the services of a fire brigade from relevant local fire authorities – see *Fire Services Act 1947, s 3(1)(dd)*, as inserted by the *Merchant Shipping and Maritime Security Act 1997*.

The Health & Safety Commission (HSC) has published a revised Approved Code of Practice (ACOP) and guidance* to accompany the *Offshore Installations (Prevention of Fire and Explosion, and Emergency Response) Regulations 1995 (SI 1995 No 743)*. The revised ACOP came into force on 1 June 1997.

The main changes are to paragraphs 167 and 170 of the 1995 ACOP and deal with standby vessels. The revised ACOP sets out the factors to be considered when selecting a standby vessel to provide rescue and recovery services at offshore installations. The factors, such as manoeuvrability, visibility, lighting and communication, are based mainly on Lord Cullen's recommendations following the 1988 Piper Alpha disaster.

\* The revised ACOP, '*Prevention of fire and explosion, and emergency response on offshore installations: Approved Code of Practice and guidance*' (ref L65) is available from HSE Books.

# Offshore Installations and Pipeline Works (Management and Administration) Regulations 1995 (SI 1995 No 738)

O7029    These regulations impose managerial and administrative duties on:

(a)    operators of fixed installations, and

(b)    owners of mobile installations,

not on concession owners or managers (except the duty to co-operate). The installation manager is the delegate of such duties and as such has a limited liability, whilst ultimate residual liability rests with the duty holder (see *McDermid v Nash Dredging and Reclamation Co Ltd [1987] 2 AER 878*).

*Chain of Command*

Concession owner

Primary duty holder

Installation manager

## Duties on operators of fixed installations and owners of mobile installations – duty holders

O7030    Duty holders must:

(1)    notify the HSE in writing of the date of entry/departure of an installation into/out of waters. Where there is a change of duty holder, an installation must not be operated until the HSE has been notified of (i) the date of change, and (ii) the name and address of the new duty holder [*Reg 5*];

(2)    appoint a competent installation manager, who must be known or readily ascertainable by everyone on the installation [*Reg 6*];

(3)    keep both an offshore and onshore record of details of all persons on and working on the installation [*Reg 9*];

(4)    where necessary, introduce a permit to work system [*Reg 10*];

(5)    issue written instructions on procedures to be observed, bringing them to the attention of all personnel [*Reg 11*];

(6) establish a sufficient system for communicating health and safety arrangements between installation and shore, vessels and aircraft [*Reg 12*];

(7) appoint a competent helicopter landing officer to ensure safe helicopter landings and take-offs [*Reg 13*];

(8) collect and keep meteorological/oceanographic data and information concerned with the movement of the installation [*Reg 14*];

(9) ensure that the address and telephone number of the local HSE is known or readily ascertainable by personnel on the installation [*Reg 15*];

(10) provide employees with health surveillance [*Reg 16*];

(11) ensure an adequate supply of clean wholesome drinking water at suitable locations [*Reg 17*];

(12) ensure that all food is fit for human consumption [*Reg 18*]; and

(13) ensure that the installation is readily identifiable by sea or from the air [*Reg 19*].

Certificates of exemption may be issued by the HSE in respect of these duties.

### Duty of installation manager

O7031 Where necessary, in the interests of health and safety, an installation manager can restrain a person and put him ashore. [*Reg 7*].

### Duties of all persons on the installation

O7032 Everyone on the installation must co-operate with the installation manager to enable him to discharge his statutory safety duties and functions; similarly, with the helicopter landing officer. [*Reg 8*].

## Hazardous operations with the potential to cause a major accident

O7033 Lord Cullen recommended that, in parallel with the move to the Safety Case regime, the existing offshore legislation should be comprehensively reviewed with a view to its progressive replacement by a modernised and rationalised structure of Regulations.

This recommendation took account of experience of similar reforms onshore, developed by the Health and Safety Commission using its powers under *HSWA 1974*. Following enactment of the *Offshore Safety Act 1992* (which brought the existing offshore legislation within the scope of the 1974 Act) the Commission published, in August 1992, its outline plans for the offshore reform programme.

The definition of 'major accident' in the *Offshore Installations (Safety Case) Regulations 1992* embraces the full range of such accidents which can occur offshore and not just those arising from fire and explosion hazards. The Safety Case is required to relate to all activities carried out on, or in connection with, an installation including, for example, relevant activities involving divers, pipelaying or the movement of vessels or helicopters. The Regulations place a duty on the various people concerned with these activities, to co-operate with duty holders in the preparation of installation Safety Cases.

## Construction and use of submarine pipelines

O7034 The *Petroleum Act 1998* prohibits the construction of a pipeline (or use of a pipeline constructed after 1975) without the specific written authorisation of the Secretary of State. Once authorised, the pipeline becomes a *'controlled pipeline'* and any work affecting the design, route and boundaries is subject to any terms and conditions specified by the Secretary of State. In particular, it is necessary to ensure that sufficient funds are available regarding liability cover in the event of damage attributable to the release or escape from the pipeline of 'permitted substances', that is, anything or any substance authorised to be transported in that pipeline.

The Secretary of State is granted power to order the compulsory modifications of a pipeline for the purposes of improving flow through that pipeline. A person other than the owner of the pipeline is permitted to make an application to the Secretary of State for a modification order. [*PA 1998, s 16*].

## Pipelines Safety Regulations 1996 (SI 1996 No 825)

O7035     The *Pipelines Safety Regulations 1996* aim to secure the initial and continuing integrity of pipelines throughout their life cycle. To this end, pipelines must be:

(a)     designed so as to be safe within range of all operating conditions [*Reg 5*];

(b)     constructed and installed as designed [*Reg 9*];

(c)     operated within safe operating limits [*Reg 11*];

(d)     abandoned at the end of their life cycle so as not to become a source of danger [*Reg 14*].

Moreover, in the case of pipelines conveying dangerous fluids (e.g. flammable in air, very toxic, oxidising, reacting violently with water):

(i)     details relating to construction/fluid to be conveyed must be notified to the HSE at least six months prior to operation/construction [*Reg 20*];

(ii)    a major accident prevention policy must be prepared (and revised) prior to completion of design [*Reg 23*];

(iii)   details of emergency procedures must be finalised [*Reg 24*]; and

(iv)   local authorities, in whose area the pipeline is to pass, must be notified that construction is to take place, and then prepare and update an emergency plan, for which it may charge a fee [*Reg 26*].

The HSE has published guidance* on the regulations. The guidance is designed to give pipeline operators and others involved with pipeline activities or who may be affected by the regulations, an understanding of what is required.

The regulations provide for high standards to be maintained for all pipelines onshore and offshore while enabling outdated legislation to be removed; and disentangle legislation dealing with health and safety from the consents and authorisations regime operated by the Department of Trade and Industry. The guidance provides detailed guidance on the requirements of the regulations, for example, the general duties which include design, construction, installation and work on a pipeline; arrangements for incidents and emergencies; maintenance and decommissioning. The guidance also sets out the requirements of the regulations for major accident hazard pipelines, covering notification to the HSE at key stages throughout the life of the pipeline, the preparation of major accident prevention documents and emergency procedures by the pipeline operators, and the requirements for local authorities to prepare pipeline emergency plans.

* *'A Guide to the Pipelines Safety Regulations 1996'* (Ref: L82) is available from HSE Books.

## The Diving at Work Regulations 1997

O7036     The *Diving at Work Regulations 1997* (*SI 1997 No 2776*), which revoked the *Diving Operations at Work Regulations 1981*, impose requirements and prohibitions relating to health and safety upon persons who work as divers. Under the 1997 Regulations, a person 'dives' if he enters water or any other liquid; or a chamber in which he is subject to pressure greater than 100 millibars above atmospheric pressure and in order to survive in such an environment he breathes in air or other gas at a pressure greater than atmospheric pressure.

Environments such as scientific clean rooms or submersible craft subjected to a pressure less than 100 millibars above atmospheric pressure are not covered by the Regulations.

Every person who is engaged in a diving project or is responsible for, or has control over, such a project must take such measures as it is reasonable for a person in his position to take to ensure that the provisions of the 1997 Regulations are complied with. [*Reg 4*].

There could be a number of diving projects taking place on one site. Each of these projects could be separate from each other and may have different 'diving contractors' in charge of them.

## *The diving contractor*

O7037    The diving contractor must ensure, so far as is reasonably practicable, that the diving project is planned, managed and conducted in a manner which protects the health and safety of all persons taking part in that project. This requires the preparation of a diving project plan which must be updated as necessary during the continuance of the project.

Other duties placed on the diving contractor include:

— appointing a person to supervise that operation;

— making a written record of that appointment;

— ensuring that the person appointed is supplied with a copy of any part of the diving project plan which relates to that operation;

— ensuring that there are sufficient people with suitable competence to carry out safely and without risk to health both the diving project and any action (including the giving of first-aid) which may be necessary in the event of a reasonably foreseeable emergency connected with the diving project;

— ensuring that suitable and sufficient plant is available whenever needed to carry out safely and without risk to health both the diving project and any action (including the giving of first-aid) which may be necessary in the event of a reasonably foreseeable emergency connected with the diving project;

— ensuring that all diving equipment is maintained in a safe working condition;

— ensuring, so far as reasonably practicable, that any person taking part in the diving project complies with the requirements and prohibitions imposed on him by or under the relevant statutory provisions and observes the provisions of the diving project plan;

— ensuring that a record containing the required particulars is kept for each diving operation and retained for at least two years after the date of the last entry in it.

*[Regs 5–9].*

Although the main duty for the whole project is placed on the diving contractor, supervisors also have a duty to manage the diving operation safely. Supervisors should not participate in a diving operation which they consider to be unsafe because insufficient supervisors have been appointed.

## *Dive planning and risk assessment*

O7038    Similar duties regarding the safe operation of diving projects are also placed on 'clients' who raise diving contracts. They should be involved in the risk assessment elements of the 'diving project plan' required by *Reg 8*.

## Other offshore specific legislation

O7039  The following Regulations were enacted under the *Mineral Workings (Offshore Installations) Act 1971* and apply only to offshore operations:

— Offshore Installations (*Logbooks and Registration of Death*) Regulations 1972 (*SI 1972 No 1542*);

— Offshore Installations (Inspectors and Casualties) Regulations 1973 (SI 1973 No 1842);

— Offshore Installations (Safety Representatives and Safety Committees) Regulations 1989 (SI 1989 No 971);

— Offshore Installations and Pipeline Works (First-Aid) Regulations 1989 (SI 1989 No 1671);

— Offshore Electricity and Noise Regulations 1997 (SI 1997 No 1993).

## Offshore Installations (Logbooks and Registration of Death) Regulations 1972

O7040  These Regulations (*SI 1972 No 1542*) have largely been revoked, leaving only those provisions which cover the registration of deaths and of persons lost from offshore installations, and the offences and penalties for failure to comply with the reporting requirements.

Duties are imposed on the installation manager and the owner of an offshore installation to submit a notification form to the Department of Transport in cases where a person dies or is lost overboard from an offshore installation. [*Regs 8–11 and Sch*]. The notification to the Department of Transport is additional to the reporting requirements of the *Reporting of Injuries, Diseases and Dangerous Occurrences Regulations 1995* (*SI 1995 No 3163*).

## Offshore Installations (Inspectors and Casualties) Regulations 1973

O7041  These Regulations (*SI 1973 No 1842*), which have largely been revoked, cover duties regarding the inspection of offshore installations. The Regulations should be read in conjunction with the *Health and Safety at Work etc. Act 1974, Secs 19–28* and the *Provision and Use of Work Equipment Regulations 1992* (*SI 1992 No 2932*).

The following terms are defined in the Offshore Installations (Inspectors and Casualties) Regulations 1973:

*casualty* means a casualty or other accident involving loss of life or danger to life suffered by a person –

(*a*)  employed on, or working from, an offshore installation; or

(*b*)  employed on, or working from, an attendant vessel, in the course of any operation undertaken on or in connection with an offshore installation;

*equipment* means any plant, machinery, apparatus or system used, formerly used or intended to be used (whether on or from an offshore installation or on or from an attendant vessel) in the assembly, reconstruction, repair, dismantlement, operation, movement or inspection of an offshore installation or the inspection of the sea bed under or near an offshore installation.

## Inspection of offshore installations

### Functions and powers of inspectors

**O7042** In addition to the powers granted by the *Health and Safety at Work etc. Act 1974, s 19*, an inspector may at any time:

— board an offshore installation;

— inspect an offshore installation and any equipment;

— inspect the sea bed and subsoil under or near an offshore installation;

— inspect and take copies from any certificate of insurance or from other records relating to the operation or safety of an offshore installation or of any equipment;

— test any equipment;

— dismantle any equipment or test to destruction or take possession of any equipment, where a casualty has occurred or is apprehended;

— require the owner or manager or any person on board or near an offshore installation to do or to refrain from doing any act as appears to the inspector to be necessary or expedient for the purpose of averting a casualty (whether the danger is immediate or not) or minimising the consequences of a casualty.

*[Reg 2].*

### Offshore equipment and materials

**O7043** Regarding articles, plant and equipment, materials and substances, and any documents relating to the operation of the installation, an inspector also has powers to:

(*a*) make such requirements of any person (including the owner and manager of the installation) as appear to the inspector to be required for the performance of his functions whether by himself or any other person acting at the direction of the Executive, provided that the inspector consults the owner or manager with a view to maintaining safety and to minimising interference with the operation of the installation;

(*b*) require any person to produce to him any article to which this Regulation applies and which is in his possession or custody;

(*c*) make notes, take measurements, make drawings and take photographs of an offshore installation and of any article to which this Regulation applies;

(*d*) require the owner or manager of the installation to furnish him with any article to which this Regulation applies (other than a document) or, in the case of any article on any vessel, may so require the master, captain or person in charge of the vessel;

(*e*) require the owner or manager of an offshore installation or any person employed on or in connection with the installation or equipment to carry out or to assist in carrying out any inspection, test or dismantlement of the offshore installation or of any equipment;

(*f*) require the owner or manager of an offshore installation or the concession owner concerned to assist him in carrying out an inspection of the sea bed or subsoil under or near the installation; and

(*g*) require the owner or manager to provide at any reasonable time conveyance to or from the installation of the inspector, any other person acting at the direction of the Executive, any equipment required by the inspector for testing and any article of which he has taken possession pursuant to these Regulations.

Should the inspector require further information, this would normally be provided in writing. If the information is furnished orally, the person providing the information may do so in the presence of another person, including the installation manager if it is practicable to do so and if the person concerned so wishes.

[Reg 4].

### Duties to assist an inspector

O7044    The owner of the offshore installation must provide an inspector with reasonable accommodation and afford such facilities and assistance as the inspector may reasonably require to enable him to carry out his functions. [Reg 5]. The owner is also obliged to provide (at his expense) inspectors with any necessary helicopter transportation.

### Disclosure of information

O7045    The Regulations prohibit any person acting at the discretion of the HSE (persons holding Government office excepted) from disclosing any information obtained during an inspection of an offshore installation without the consent of:

— the owner of the installation;
— the person who furnished the information; or
— the Health and Safety Executive.

# Offshore Installations (Safety Representatives and Safety Committees) Regulations 1989

O7046    Following the Piper Alpha disaster the present Regulations, the *Offshore Installations (Safety Representatives and Safety Committees) Regulations 1989* (*SI 1989 No 971*) were issued by the then enforcing authority, the Department of Energy. The purpose of these Regulations, which were made under the authority of the *Mineral Workings (Offshore Installations) Act 1971* and not the *Health and Safety at Work etc. Act 1974*, is to ensure that the whole workforce (not just individuals who are trade unionists) is formally involved in promoting health and safety through freely elected safety representatives and the safety committee.

Since the publication of the Cullen Report in 1990, much has been written regarding the identification and control of major accident hazards and protection of the workforce from such risks. The Cullen inquiry made it quite clear that high standards of safety and health performance, necessary on an offshore installation, cannot be achieved without the positive and informed commitment of the workforce. Everyone involved with offshore operations must have a safety management system which describes the arrangements for securing this commitment – the provision of information and advanced consultation – whenever it is appropriate to do so.

Information must be given in a form and manner which ensures that it is both comprehensible and relevant to those who are to receive it.

The HSE have published guidance to the Regulations: '*A Guide to the Offshore Installations (Safety Representatives and Safety Committees) Regulations 1989*' (Ref: L110).

### Establishment of constituencies

O7047    The first stage in the appointment of safety representatives is the establishment of a system of constituencies by the installation manager in consultation with the safety committee. Thus, unlike onshore safety representatives appointed by recognised trade unions, offshore safety representatives are elected by all the workers in a constituency; in other words, by voluntary decision of the workforce. Their principal role is to act as a conduit on safety from workforce to management. Constituencies will take account of

(a)  areas of the offshore installation;

(b)  activities undertaken on or from the installation;

(c)  employees of the workforce;

(d)  other objective criteria which appear to the installation manager to be appropriate to the circumstances of the installation.

There must be, at least, two constituencies and every worker can be assigned to one, but subject to a maximum of forty and a minimum of three. It is the duty of the installation manager to post particulars of a constituency in suitable places on the installation and, if necessary, in appropriate languages. This done, the workers comprising the constituency can elect a safety representative. Any worker can stand for election to represent his constituency as a safety representative, provided that he

(i)  is a member of that constituency;

(ii)  is willing to stand as a candidate;

(iii)  has been nominated by a second member; and

(iv)  his nomination is seconded by a third member.

A list of duly nominated candidates must be posted within a week after nominations have expired.

[Regs 5–10].

The purpose of dividing an installation into constituencies is to provide for appropriate groupings of the workforce from which safety representatives can be elected. Each installation, no matter how small the permanent workforce, must have at least two constituencies. If an installation manager wishes to create new constituencies, his proposals must be agreed by the safety committee, or both safety committees if two installations are temporarily bridge-linked (for example, production and accommodation platforms). Every new arrival to an installation must be given written notification and details of his constituency. Exceptions to this rule may be permitted where a person is not expected to remain on the installation for longer than 48 hours.

Each installation manager should actively encourage the filling of a constituency vacancy. HSE inspectors may wish to discuss such matters with other safety representatives during visits to the installation.

[L110, paras 12–33].

## Functions of offshore safety representatives

**O7048** Safety representatives on offshore installations have similar functions to safety representatives generally. More particularly, they can

(a)  investigate potential hazards/dangerous occurrences and examine causes of accidents;

(b)  investigate *bona fide* health and safety complaints from their members;

(c)  draw matters arising from (a) and (b) to the attention of the installation manager and any employer;

(d)  approach the installation manager/any employer about general health and safety matters;

(e)  attend meetings of the safety committee;

(f)  represent members in consultation with the Inspectorate;

(g)  consult members on any part of the above matters;

(h)  inspect accident sites/equipment;

(j)  alert the HSE if an imminent risk of serious injury is considered to exist;

(k)  carry out site inspections;

(l)  consult with HSE inspectors;

(m)  receive copies of statutory health and safety documentation.

Exercise of these functions cannot give rise to criminal and/or civil liability on the part of a safety representative.

*[Reg 16].*

It is important to distinguish between functions and duties – safety representatives cannot be held to account for not carrying out their functions. However, since the functions are the principal basis of the safety representative system, safety representatives should endeavour to carry them out as fully as possible. Although not normally a formal part of the installation's management team, safety representatives should be able to take matters of safety to the installation manager without delay and without the need for a formal written approach. [*L110, paras 52–55*].

## Inspection of equipment

**O7049**  A safety representative can inspect any part of the installation or its equipment, provided that part of the installation or equipment has not been inspected in the previous three months, and that he has given reasonable notice in writing to the manager and, if his employer is not the duty holder, also his employer.

More frequent inspections can take place, by agreement with the manager.

*[Reg 17].*

## Inspection of safety documents and Safety Case

**O7050**  A safety representative is entitled to see occupational health and safety at work documents which are required by law to be kept on the installation. This entitlement does not extend to a document consisting of or relating to any health record of an identifiable individual.

Regarding the Safety Case for the installation, safety representatives must be supplied with a written summary of the main features of the Safety Case including any particulars concerning remedial work and the time by which it will be done. Appropriate facilities must be made available on the installation for reading the information if it is kept on film or in electronic form.

*[Regs 18,18A].*

The *Health and Safety at Work etc. Act 1974, s 28(8)* requires inspectors appointed under that Act to give certain factual and other information concerning their actions to employees or their representatives. Information is normally provided in the form of letters from inspectors, and the Regulations allow for the safety representatives to receive such information. It is recommended that the safety representatives agree that one of their number will act as a contact point with inspectors for this purpose. This representative can then inform his fellow safety representatives, but such arrangements should not preclude any of the representatives establishing working relationships with inspectors and meeting them when they visit.

One of the functions of safety representatives is the carrying out of formal inspections, both on a regular basis and following an incident. When determining the cause of a 'notifiable incident', safety representatives are empowered to inspect relevant parts of the installation and equipment, provided that the installation manager is notified and confirms:

(a)  there has been a notifiable incident;

(b) it is safe for an inspection; and

(c) the interests of his members might be involved.

A 'notifiable incident' is any death, injury, disease or dangerous occurrence which is required to be reported under the *Reporting of Injuries, Diseases or Dangerous Occurrences Regulations 1995*. In addition, a safety representative can inspect any part of the installation or its equipment, provided that the relevant part of the installation or equipment has not been inspected in the previous three months, and that he has given reasonable notice in writing to the manager and, if his employer is not the duty holder, also his employer.

More frequent inspections can take place by agreement with the manager. [*Reg 17*].

The main purpose of the examination should be to determine the cause so that the possibility of action to prevent a recurrence can be considered. For this reason it is important that the approach to the problem should be a joint one by the installation manager or his representative, relevant employers and the safety representatives. Whilst it may be necessary, following a notifiable incident, for the installation manager to take immediate steps to safeguard against further hazards he should involve the safety representatives at the earliest opportunity. If this is not possible he should notify the safety representatives of the action he has taken and confirm this in writing.

Examinations may include visual inspection, photography (subject to any work permit requirements) and discussions with persons who are likely to be in the possession of relevant information and knowledge regarding the circumstances of the notifiable incident. The examination must not, however, include interference with any evidence or the testing of any machinery, plant, equipment or substance which could disturb or destroy the factual evidence before a HSE inspector has had the opportunity to investigate in accordance with his powers.

The number of safety representatives taking part in a formal inspection should be a matter for agreement with the installation manager. It may be appropriate for an installation safety officer or a specialist to provide advice during an inspection. Where any remedial action has been taken, safety representatives should be given the opportunity to make a reinspection to satisfy themselves in order to report to the safety committee. [*L110, paras 63–70*].

## Risk of serious personal injury

**O7051**     If two or more safety representatives think that there is an 'imminent risk of serious personal injury', arising from an installation activity, they can:

(a) make representations to the manager, who must send a written report to an inspector, and

(b) themselves send a written report to an inspector.

*[Reg 17(4)]*.

Safety representatives are not directly empowered to stop activities which they consider involve imminent risk of serious personal injury. However, where two or more of them consider that there is such imminent risk, they are required by *Reg 17(4)(a)* to make representations on the matter to the installation manager. The manager is then required to report the matter in writing to an inspector as soon as is reasonably practicable.

In addition, the safety representatives may make their own written report on the matter to an inspector, exercising good judgment in invoking this function which should be reserved for genuinely imminent danger. In exercising his powers a safety representative is entitled to seek advice and guidance from persons on the installation or elsewhere. This may usually be informally sought, for example from management, the safety officer, inspectors or from his trade union health and safety adviser. However, if a safety representative considers that there is a need for formal external independent advice, for example a noise survey or a hazard analysis, it is recommended that the matter be raised in the safety committee which may direct the request to the owners. [*L110, paras 71–73*].

## Safety committees

**O7052** Owners of offshore installations, with one or more safety representatives, must establish a safety committee. [*Reg 19*].

This must consist of:

(a)   the installation manager (as chairman),

(b)   one other person (to be appointed by the owner or manager),

(c)   all safety representatives, and

(d)   persons who may be co-opted by unanimous vote, e.g. safety officer(s).

[*Reg 20*].

The person who may be appointed to the committee should preferably be a representative of shore management not directly involved in the day-to-day operation of the installation. Such an appointment would ensure a direct and continuous link with onshore management for the purpose of pursuing any improvements which cannot be made with the resources directly at the disposal of the installation manager. It would also enable solutions to common problems encountered on other installations owned by the same company to be relayed directly to the committee, ensuring consistency of approach.

The Regulations also allow for the safety committee, by unanimous vote, to co-opt on to the committee such additional persons as they wish, but these co-opted individuals are not entitled to vote on the co-option of any further individuals. It is expected that one of them should normally be the installation safety adviser, whilst others might include specialists in particular activities on the installation. Co-option of a minutes secretary may assist the committee. [*L110, paras 78–83*].

## Safety committee meetings

**O7053**   A safety committee must be convened by the chairman (normally the installation manager) within six weeks of the date of its establishment, and thereafter at least once every three months. If the committee wishes to meet more frequently, it may establish a policy on the circumstances under which such meetings may be called, for example to consider an urgent unforeseen matter. A safety representative who is unable, for whatever reason, to attend a meeting can nominate another member of his constituency to attend in his stead. The nominee should act on his representative's behalf at the meeting but has no other functions or powers under the Regulations.

Generally, safety committee representatives should cover certain basic items, for example:

—   all incidents involving the installation since the committee last met;

—   any feedback from onshore management including any incidents on other installations which may have implications for their installation;

and any other matter which could possibly affect the health and safety of people working on the installation. Safety committees on bridge-linked installations may decide to co-ordinate their business but it is a legal requirement for each installation to have a safety committee which must meet at least every three months. [*L110, paras 84–89*].

## Functions of safety committees

**O7054**   Safety committees have the following functions:

(i)   to monitor health and safety at the workplace;

(ii)   to keep under review the constituency system;

(iii)   to monitor arrangements for the training of safety representatives;

(iv) to monitor the frequency of safety committee meetings;

(v) to consider representations from any member of a safety committee;

(vi) to consider the causes of accidents, dangerous occurrences and cases of ill- health;

(vii) to consider any statutory documents relating to occupational health and safety, e.g. registers (but not medical records);

(viii) to prepare and maintain a record of its business, with a copy being kept for one year from the date of the meeting; and

(ix) to review safety representative training arrangements.

[Reg 22].

Installation owners/managers must co-operate with safety committees and safety representatives to enable them to perform their functions and they must disclose relevant information to safety representatives. [Regs 23, 24].

The general philosophy behind the Regulations is that the safety committee will promote co-operation on all matters affecting occupational health and safety between all parties on the installation and will seek to promote and develop measures to ensure the occupational health and safety of the workforce. However, the committee should not attempt to replace the normal day-to-day channels of communication on specific health and safety issues between the individual and his immediate supervisor and ultimately the installation manager, but should serve as a back-up in the event of continuing concern.

The safety committee decides upon the format of its meetings. For example, the installation manager may be required to:

— report on any feedback from shore management on previous meetings;

— make a report on all notifiable incidents involving the installation since the committee last met;

— report any incidents or hazards to health which happened on other installations operated by the same operator which may have repercussions for his own installation (for this he would need briefing from shore);

— report any major changes in the equipment or method of operation planned for the installations; and

— report on anything else which could possibly affect the health and safety of persons working on the installation.

The importance of consultation to satisfy the requirements of *Reg 23* cannot be overemphasised. The installation manager should keep the safety committee informed of progress on agreed actions to be taken. [*L110, paras 90–96*].

## Training

O7055    Employers must permit safety representatives time off from work, without loss of pay, to enable them:

(a)    to perform their functions, and

(b)    to undertake training,

and employers must meet the cost of training, travel and subsistence expenses. [*Regs 26–27*].

Safety representatives must be permitted time off without loss of pay (including any appropriate offshore allowances) during normal working hours for basic training. This should be as soon as possible after their election. The length of training required is not prescribed but basic training should take into account:

— the functions of safety representatives;

— the training necessary for providing an understanding of the role of the safety representatives and safety committees;

— the legal requirements relating to the health and safety of persons on the installation, particularly the constituency members they directly represent;

— the nature and extent of workplace hazards, and the measures necessary to eliminate or minimise them; and

— the health and safety policy of the employer and the organisations and arrangements for fulfilling that policy.

A major element of safety representative training should be devoted to his representational role and the desirability that the representative should be looked upon as a part-time safety officer. Training should also include inspection and investigation techniques, an appreciation of the appropriate level of response to problems and how to pursue any action with management.

The Offshore Petroleum Industry Training Organisation (OPITO) produces and updates training guidelines for offshore safety representative courses. The guidelines may be adopted by any organisation capable of providing suitable facilities and resources for such training. Although the ultimate choice of courses rests with the employer, it could also be a function of the safety committee.

Further training will be needed where the safety representative has special responsibilities, or in order to meet changes in circumstances or relevant legislation.

Installation owners and managers may make allowances in contractual arrangements in order that contractors' employees may be elected as safety representatives. The responsibility for training then rests with the contractor who must ensure not only that training is provided to any employee who is a safety representative, but also that any such employee is given the necessary time off both for training and for the performance of his role and duties as safety representative.

One of the functions of the safety committee is to keep the training requirements of safety representatives under review. All training providers and courses should be considered in order to ensure that safety representatives have been provided with suitable opportunities to become competent for their role, including the requisite interpersonal skills of communication, presentation and representation. Follow-up training may need to be arranged according to developments affecting the installation, new legislation, changing technologies or differing working practices. [*L110, paras 103–109*].

## Offences/defences

O7056

(a) *Offence* – If a duty holder, manager or employer fails to comply with any of his statutory duties, he commits an offence (punishable as with mainline *HSWA 1974* offences – see ENFORCEMENT). [*Reg 28(1)*].

(b) *Defence* – It is a defence for the accused to prove:

   (i) that he exercised all due diligence to prevent the commission of the offence, and

(ii) that the relevant failure to comply was committed without his consent, connivance or wilful default.

*[Reg 28(2)].*

Employment protection for offshore safety representatives is provided for by the *Offshore Safety (Protection Against Victimisation) Act 1992* and the *Employment Rights Act 1996*. Under the 1996 Act, all employees (both onshore and offshore) irrespective of length of service or hours of work, are entitled to complain to an employment tribunal if they are dismissed, selected for redundancy or subjected to any other detriment by their employer. So far as offshore workers are concerned, this means that they can expect not to be disciplined, dismissed, 'blacklisted' or 'not required back' (NRB), transferred to other duties or to another installation, or disadvantaged in any way, because of any actions they take on health and safety grounds. [*L110, App 1*].

## First-aid – the Offshore Installations and Pipeline Works (First-Aid) Regulations 1989

### Medical personnel

**O7057** Installation managers/owners and those in control of pipeline works must make adequate arrangements for first-aid for persons at work. This includes provision of suitably trained personnel and advice or attendance of a registered medical practitioner when needed, and employees must be informed of these arrangements. [*Offshore Installations and Pipeline Works (First-Aid) Regulations 1989 (SI 1989 No 1671), Reg 5(1)* and accompanying Code of Practice (ACOP)].

The Regulations do not apply to diving operations offshore, which are subject to the requirements of the *Diving at Work Regulations 1997*. However, when a diving operation ceases, any members of a diving team who remain present on an installation or barge will be covered by the duty of care which the 1989 Regulations impose upon the person in control.

### Numbers and types of 'suitable persons'

**O7058** The 'suitable persons' required by *Reg 5(1)(a) and (b)* may be offshore medics or offshore first-aiders. These are defined in the Code of Practice as follows:

*offshore medic* means a person who holds a current Offshore Medic Certificate issued by a body approved by the HSE to train, examine and certify offshore medics;

*offshore first-aider* means a person who holds a current Offshore First-Aid Certificate issued by a body approved by the HSE to train, examine and certify offshore first-aiders.

In most circumstances, the minimum number of offshore first-aiders and offshore medics who should be available at all times where people are at work on an offshore installation or a pipelaying barge or a barge, such as a crane barge used in offshore construction, reconstruction, alteration, repair, maintenance, cleaning, demolition or dismantling activities, or in providing accommodation, should be as follows:

| Number of people at work | Number of offshore first-aiders | Number of offshore Medics |
|---|---|---|
| Up to 25 | 2 | – |
| 26 to 50 | 3 | 1 |
| 51 to 100 | 5 | 1 |
| 101 to 150 | 7 | 1 |
| 151 to 200 | 9 | 1 |

| | |
|---|---|
| 201 and above | Additional personnel, if required, to ensure reasonable access by all offshore workers |

In determining whether there is a need for a medic, the number of people likely to be regularly at work at one time on the installation should be considered. In certain circumstances, such as relatively low hazard installations or when the installation is within easy reach of onshore medical services, a medic should be available only where the number of people regularly at work exceeds 50.

## Manned installations – sick bays

O7059    In particular, manned offshore installations and all pipelaying barges and barges used in offshore construction, repair, maintenance, cleaning, demolition or dismantling activities must have a sick bay. The sick bay should:

(*a*)    be clearly identifiable;

(*b*)    be in the charge of a suitable person (e.g. offshore medic or first-aider) who should always be on call;

(*c*)    be available at all times and not used for other purposes;

(*d*)    be kept locked at all times when not in use (though suitable arrangements should be made for immediate access in cases of emergency);

(*e*)    have attached to the door a notice clearly showing names and locations of:

   (i)    offshore medics, or

   (ii)   offshore first-aiders

(who should be easily identifiable, e.g. by arm bands) and means of contacting them (see further Duty of installation manager, O7062 below);

(*f*)    contain suitable furniture, medications and equipment. Medications should be kept in a container and locked, the keys at all times being held by:

   (i)    the installation manager, or

   (ii)   the master, or

   (iii)  the offshore medic;

(*g*)    have facilities for effective two-way communication with onshore medical services.

*[ACOP, para 6].*

## First-aid kits in sick bays

O7060    Sick bays should always contain a first-aid kit; but, in addition, offshore first-aiders must be provided with a first-aid kit readily available for use in emergency and located at the first-aider's place of work or around the installation.

First-aid kits in sick bays should contain:

(i)    12 individually wrapped sterile triangular bandages (90 cm × 127 cm);

(ii)   20 packs of sterile gauze pads, each containing five pads (7.5 cm × 7.5 cm);

(iii)  a suitable device for facilitating mouth to mouth resuscitation.

*[ACOP, paras 3, 4].*

## Contents of sick bay

**O7061** Sick bays on offshore installations should contain the following items:

(a) medications which an offshore medic can administer without the direction of a doctor, e.g. aluminium hydroxide tablets for indigestion, calamine lotion for irritating rashes;

(b) standing orders relating to medications which can only be administered under the direction of a doctor, e.g. ampicillin capsules for infections, diazepam tablets for sedation;

(c) dressings and bandages;

(d) instruments and appliances, e.g. resuscitation devices, mouth gag, face masks;

(e) furnishings and equipment, e.g. alarm bell system, telephone, armchair, accident record book, daily treatment record book.

*[ACOP, paras 31–36, Appendix 2, Parts 1, 2, 3].*

## Duty of installation manager

**O7062** The installation manager must see that all employees (and self-employed persons) are informed of the location of first-aid and medical equipment and how workers can contact the offshore medic or first-aiders rapidly in case of emergency. New workers, in particular, should be told this on first coming aboard. [*Reg 5(1)*].

To this end, notices should be put up (if necessary in various languages) in conspicuous positions (including sick bays), giving locations of first-aid and medical equipment and names and, if possible, locations of medic and first-aider. Offshore medics and first-aiders must be regularly supervised by fully registered medical practitioners and, where necessary, given medical advice [*Reg 5(1)(c)*], and written instructions should be drawn up, setting out arrangements for liaison with the medical practitioner. Copies of such instructions should be displayed in:

(a) the sick bay;

(b) the radio operator's room; and

(c) the master or installation manager's office offshore.

*[ACOP, paras 23–30].*

# The Offshore Electricity and Noise Regulations 1997

**O7063** These Regulations (*SI 1997 No 1993*) amend the following:

— *Electricity at Work Regulations 1989* (*SI 1989 No 635*) (see E3001 ELECTRICITY),

— *Noise at Work Regulations 1989* (*SI 1989 No 1790*) (see N3001 NOISE AND VIBRATION),

by applying them in their entirety to offshore installations.

The duties imposed by these Regulations do not extend to:

(a) the master or crew of a sea-going ship or to the employer of such persons, in relation to the normal ship-board activities of a ship's crew under the direction of the master; or

(b) the crew of any aircraft or hovercraft which is moving under its own power or any other person on board any such aircraft or hovercraft who is at work in connection with its operation.

The Regulations amend the *Offshore Installations and Wells (Design and Construction, etc.) Regulations 1996* (*SI 1996 No 913*) (see O7024 above) by removing references to electrical or noise provisions.

# Legislation for other hazardous offshore activities

O7064      Generally, an installation's Safety Case will not be considered for acceptance unless it contains all the particulars specified in the *Safety Case Regulations 1992, Reg 8* and the relevant *Schedule*. In some cases, with respect to hazardous activities carried out on offshore installations, there will be a degree of overlap with the particulars required in accordance with the provisions of associated legislation.

Set out below are brief references to various regulations which relate to hazards that potentially could cause a major accident.

## Working in confined spaces

O7065      The *Confined Spaces Regulations 1997 (SI 1997 No 1713)* impose requirements and prohibitions with respect to the health and safety of persons carrying out work in confined spaces.

Although offshore installations are not specifically covered by the regulations, the regulations would apply to certain activities aboard installations which are 'stacked' out of use in territorial waters (see O7009 above) and which are not currently classed as offshore installations, such as the activities of shore-based workers undertaking repair, maintenance or cleaning.

A 'confined space' is defined in *Reg 1(2)* as 'any place, including any chamber, tank, vat, silo, pit, trench, pipe, sewer, flue, well or other similar space in which, by virtue of its enclosed nature, there arises a reasonably foreseeable specified risk'.

Under these regulations a confined space has two defining features. Firstly, it is a place which is substantially (though not always entirely) enclosed and, secondly, there will be a reasonably foreseeable risk of serious injury from hazardous substances or conditions within the space or nearby.

## Radiation hazards

O7066      From the site radiography techniques used to assess the integrity of pipeline joints, one can infer the need for an exceptionally high level of radiological protection for operators. There must be total compliance with the *Ionising Radiations Regulations 1985 (SI 1985 No 1333)* in terms of the designation of competent persons, authorised persons and classified workers, medical supervision, personal monitoring procedures following excessive exposure, control of sealed sources of radiation, equipment tests and the maintenance of records (see DANGEROUS SUBSTANCES I).

## Oil pollution

O7067      In addition to the environmental aspects, oil and volatile hydrocarbons spilled into the sea in the vicinity of an offshore installation pose a real threat of fire to that installation. In this context the principal legislation in respect of the control of oil pollution is as follows:

— *Merchant Shipping Act 1995 (as amended);*

— Merchant Shipping (Prevention of Pollution) Order 1983 (SI 1983 No 1106 as amended);

— Merchant Shipping (Prevention of Pollution) (Law of the Sea Convention) Order 1996 (SI 1996 No 282);

— Merchant Shipping (Prevention of Pollution) Regulations 1996 (SI 1996 No 2154).

In the *Merchant Shipping (Prevention of Pollution) Regulations 1996*, the legal definition of 'discharge' expressly excludes the release of harmful substances directly arising from the exploration, exploitation and associated offshore processing of sea-bed mineral resources. [*Reg 1(2)*].

However, there is a prohibition on the discharge into the sea of other harmful substances, or effluents containing such substances, whether such discharge occurred as a result of escaping, disposing, spilling, leaking, pumping and emitting or emptying. In respect of offshore installations, when engaged in the exploration, exploitation and associated offshore processing activities, the Regulations provide that they must be equipped with suitable oil filtering and oil discharge monitoring and control systems.

Each installation is also required to have a shipboard oil pollution emergency plan and an up-to-date Oil Record Book.

### Carriage of dangerous substances

O7068　　In addition to the Regulations relating to the labelling, packaging and transportation of dangerous goods by road or rail (see DANGEROUS SUBSTANCES II), carriage to the installation must comply with the following legislation:

— *Merchant Shipping* (*Dangerous Goods and Marine Pollutants*) *Regulations 1997* (*SI 1997 No 2367*);

— Air Navigation (Dangerous Goods) Regulations 1994 (SI 1994 No 3187).

Reference to these regulations is given for guidance only and must not be regarded as comprising a complete list.

## Written procedures and training programmes for dealing with hazards

O7069　　Detailed plans, procedures, safe systems of work and written instructions, including those covering the operation of permit to work systems, are a standard feature of offshore safety operations. Such procedures and safe systems of work may include the use of access equipment, e.g. scaffolding, boatswain's chairs, ladders, etc., work over the sea, the safe use of lifting equipment, helicopter operations and general fire and emergency procedures.

High standards of operator training, prior to arrival on a working platform and at frequent intervals during their employment, are crucial to ensure safe operation.

Induction training should incorporate the following aspects:

(*a*)　*General safety* – dealing with the hazards of offshore work and the personal precautions necessary, in particular the correct use of all forms of personal protective equipment, i.e. foul weather protection, safety helmets, safety harnesses.

(*b*)　*Safe working procedures* – the use of permit to work systems and safe systems of work covering electricity, welding and ionising radiation in particular; the use of scaffolding, working platforms, the use of plant and equipment, maintenance and inspection procedures and record keeping (see further 43 INTRODUCTION).

(*c*)　*Health precautions* – including precautions against toxic dusts, fumes, gases and vapours, asbestos, the risk of asphyxiation and anoxia, and physical hazards associated with noise, electricity and radiation.

(*d*)　*Survival* – instruction in the use of life jackets, survival capsules, life boats and rafts; emergency platform evacuation procedures, particularly in the event of fire or explosion.

(*e*)　*Helicopter operations* – procedures for safe movement of personnel by helicopter to or from the platform.

(*f*)　*Fire protection* – the correct use of fire appliances, platform evacuation procedure in the event of fire, platform fire alarm system.

(*g*)　*First-aid* – elementary first-aid procedures.

## Hazards in offshore activities

### Cranes and lifting equipment

O7070 The potential for crane and lifting equipment failures, including the use of helicopters in lifting operations during the initial work stage, represents a serious hazard in offshore work. All lifting operations and equipment must comply with the *Lifting Operations and Lifting Equipment Regulations 1998* (see L3023 LIFTING MACHINERY AND EQUIPMENT).

### Burns and eye injuries

O7071 These are the two most common forms of injury to operators, in many cases associated with a failure to wear personal protective equipment provided.

### Access hazards

O7072 Falls from scaffolds and ladders, particularly during the initial work stage, have been common in the past, emphasising the need for good systems of scaffold inspection and maintenance in particular.

### Gassing accidents and incidents

O7073 Precautions against asphyxiation, oxygen enrichment and various fumes, dusts, gases and vapours are now a standard feature of the industry.

# Personal Protective Equipment

## Introduction

**P3001** Conventional wisdom suggests that 'safe place strategies' are more effective in combating health and safety risks than 'safe person strategies'. Safe systems of work, and control/prevention measures serve to protect everyone at work, whilst the advantages of personal protective equipment are limited to the individual(s) concerned. Given, however, the fallibility of any state of the art technology in endeavouring to achieve total protection, some level of personal protective equipment is inevitable in view of the obvious (and not so obvious) risks to head, face, neck, eyes, ears, lungs, skin, arms, hands and feet.

Current statutory requirements for employers to provide and maintain suitable personal protective equipment are contained in the *Personal Protective Equipment at Work Regulations 1992 (SI 1992 No 2966* as amended by *the Health and Safety (Miscellaneous Amendments) Regulations 2002 SI 2002 No 2174*), enjoining them to carry out an assessment to determine what personal protective equipment is needed and to consider employee needs in the selection process. Adjunct to this, common law insists, not only that employers have requisite safety equipment at hand, or available in an accessible place, but also that management ensure that operators use it (see *Bux v Slough Metals Ltd [1974] 1 All ER 262*). That the hallowed duty to 'provide and maintain' has been getting progressively stricter is evidenced by *Crouch v British Rail Engineering Ltd [1988] IRLR 404* to the extent that employers could be in breach of either statutory or common law duty (or both) in the case of injury/disease to a member of the employee's immediate family involved, say, in cleaning protective clothing – but the injury/disease must have been foreseeable (*Hewett v Alf Brown's Transport Ltd [1992] ICR 530*).

This section deals with the requirements of the *Personal Protective Equipment at Work Regulations 1992*, the *Personal Protective Equipment (EC Directive) Regulations 1992 (SI 1992 No 3139)* (now revoked and consolidated by *SI 2002 No 1144*), the duties imposed on manufacturers and suppliers of personal protective equipment, the common law duty on employers to provide suitable personal protective equipment, and the main types of personal protective equipment and clothing and some relevant British Standards.

## The Personal Protective Equipment at Work Regulations 1992 (SI 1992 No 2966 as amended by SI 2002 No 2174)

**P3002** With the advent of the *Personal Protective Equipment at Work Regulations 1992*, all employers must make a formal assessment of the personal protective equipment needs of employees and provide ergonomically suitable equipment in relation to foreseeable risks at work (see further P3006 below). However, employers cannot be expected to comply with their new statutory duties, unless manufacturers of personal protective equipment have complied with theirs under the requirements of the *Personal Protective Equipment (EC Directive) Regulations 1992 (SI 1992 No 3139)* and *SI 1994 No 2326* (now revoked and consolidated by *SI 2002 No 1144*) – that is, had their products independently certified for EU accreditation purposes. Although imposing a considerable remit on manufacturers, EU accreditation is an indispensable condition precedent to sale and commercial circulation (see P3012 below).

It is important that personal protective clothing and equipment should be seen as 'last resort' protection – its use should only be prescribed when engineering and management solutions and other safe systems of work do not effectively protect the worker from the danger, and where risk of injury is still foreseeable after other controls have been implemented (*Health and Safety (Miscellaneous Amendments) Regulations 2002 (SI 2002 No 2174)*). (*Gerrard v Staffordshire Potteries [1995] PIQR P169.*)

Employees must be made aware of the purpose of personal protective equipment, its limitations and the need for on-going maintenance. Thus, when assessing the need for, say, eye protection, employers should first identify the existence of workplace hazards (e.g. airborne dust, projectiles, liquid splashes, slippery floors, inclement weather in the case of outside work) and then the extent of danger (e.g. frequency/velocity of projectiles, frequency/severity of splashes). Selection can (and, indeed, should) then be made from the variety of CE-marked equipment available, in respect of which manufacturers must ensure that such equipment provides protection, and suppliers ascertain that it meets such requirements/standards [*HSWA s 6*] (see further PRODUCT SAFETY). Typically, most of the risks will have already been logged, located and quantified in a routine risk/safety audit (see RISK ASSESSMENT), and classified according to whether they are physical/chemical/biological in relation to the part(s) of the body affected (e.g. eyes, ears, skin).

Selection of personal protective equipment is a first stage in an on-going routine, followed by proper use and maintenance of equipment (on the part of both employers and employees) as well as training and supervision in personal protection techniques. Maintenance presupposes a stock of renewable spare parts coupled with regular inspection, testing, examination, repair, cleaning and disinfection schedules as well as keeping appropriate records. Depending on the particular equipment, some will require regular testing and examination (e.g. respiratory equipment), whilst others merely inspection (e.g. gloves, goggles). Generally, manufacturers' maintenance schedules should be followed.

Equipment must be appropriate for the risk or risks involved, the conditions at the place where exposure to the risk may occur, and the period for which it is worn. It must also take into account the ergonomic requirements and the state of health of the person or persons who may wear it, and of the characteristics of the workstation of each such person.

Suitable accommodation must be provided for protective equipment in order to minimise loss or damage and prevent exposure to cold, damp or bright sunlight, e.g. pegs for helmets, pegs and lockers for clothing, spectacle cases for safety glasses.

On-going safety training, often carried out by manufacturers for the benefit of users should combine both theory and practice. Where it is appropriate, and at suitable intervals, the employer should organise demonstrations in the wearing of personal protective equipment. (*Health and Safety (Miscellaneous Amendments) Regulations 2002 (SI 2002 No 2174).*)

## Work activities/processes requiring personal protective equipment

P3003    Examples abound of processes/activities of which personal protective equipment is a prerequisite, from construction work and mining, through work with ionising radiations, to work with lifting plant, cranes, as well as handling chemicals, tree felling and working from heights. Similarly, blasting operations, work in furnaces and drop forging all require a degree of personal protection.

## Statutory requirements in connection with personal protective equipment

P3004    General statutory requirements relating to personal protective equipment are contained in of the *Health and Safety at Work etc Act 1974, ss 2, 9*, the *Personal Protective*

*Equipment at Work Regulations 1992*, and the *Health and Safety (Miscellaneous Amendments) Regulations 2002 (SI 2002 No 2174)* 'Every employer shall ensure that suitable personal protective equipment is provided to his employees who may be exposed to a risk to their health or safety at work except where, and to the extent that such risk has been adequately controlled by other means which are equally or more effective'. [*Reg 4(1)*]. More specific statutory requirements concerning personal protective equipment exist in the *Personal Protective Equipment at Work Regulations 1992* (in tandem with the *Personal Protective Equipment (EC Directive) Regulations 1992 (SI 1992 No 3139)* and *SI 1994 No 2326*) (now revoked and consolidated by *SI 2002 No 1144*)), and sundry other recent regulations applicable to particular industries/processes, e.g. asbestos, noise, construction (see P3008 below). Where personal protective equipment is a necessary control measure to meet a specific statutory requirement or following a risk assessment, this must be provided free of charge [*HWSA* s 9]. However, where it is optional or non essential, there is nothing to stop the employer from seeking a financial contribution.

## General statutory duties – HSWA 1974

P3005    Employers are under a general duty to ensure, so far as reasonably practicable, the health, safety and welfare at work of their employees – a duty which clearly implies provision/ maintenance of personal protective equipment. [*HSWA s 2(1)*].

## Specific statutory duties – Personal Protective Equipment at Work Regulations 1992 (SI 1992 No 2966 as amended by SI 2002 No 2174)

### Duties of employers

P3006    Employers – and self-employed persons in cases (*a*), (*b*), (*c*), (*d*) and (*e*) below, must undertake the following:

(*a*)    Formally assess (and review periodically) provision and suitability of personal protective equipment. [*Reg 6(1)(3)*].

The assessment should include:

(i)   risks to health and safety not avoided by other means;

(ii)  reference to characteristics which personal protective equipment must have in relation to risks identified in (i); and

(iii) a comparison of the characteristics of personal protective equipment having the characteristics identified in (ii).

[*Reg 6(2)*].

The aim of the assessment is to ensure that an employer knows which personal protective equipment to choose; it constitutes the first stage in a continuing programme, concerned also with proper use and maintenance of personal protective equipment and training and supervision of employees.

(*b*)    Provide suitable personal protective equipment to his employees, who may be exposed to health and safety risks while at work, except where the risk has either been adequately controlled by other equally or more effective means. [*Reg 4(1)*].

Personal protective equipment is not suitable unless:

(i)   it is appropriate for risks involved and conditions at the place of exposure;

(ii)  it takes account of ergonomic requirements, the state of health of the person who wears it and the duration of its use;

(iii) it is capable of fitting the wearer correctly; and

(iv) so far as reasonably practicable (for meaning, see ENFORCEMENT), it is effective to prevent or adequately control risks involved without increasing the overall risk.

[Reg 4(3)].

Where it is necessary to ensure that personal protective equipment is hygienic and otherwise free of risk to health, every employer and every self-employed person shall ensure that personal protective equipment provided is provided to a person for use only by him. [Reg 4]

(c) Provide compatible personal protective equipment – that is, that the use of more than one item of personal protective equipment is compatible with other personal protective equipment. [Reg 5].

(d) Maintain (as well as replace and clean) any personal protective equipment in an efficient state, efficient working order and in good repair. [Reg 7].

(e) Provide suitable accommodation for personal protective equipment when not being used. [Reg 8].

(f) Provide employees with information, instruction and training to enable them to know:

   (i) the risks which personal protective equipment will avoid or minimise;

   (ii) the purpose for which and manner in which personal protective equipment is to be used;

   (iii) any action which the employee might take to ensure that personal protective equipment remains efficient.

   [Reg 9].

(g) Ensure, taking all reasonable steps, that personal protective equipment is properly used. [Reg 10].

## Summary of employer's duties

- duty of assessment;
- duty to provide suitable PPE;
- duty to provide compatible PPE;
- duty to maintain and replace PPE;
- duty to provide suitable accommodation for PPE;
- duty to provide information and training;
- duty to see that PPE is correctly used.

## Duties of employees

**P3007**   Every employee must:

- use personal protective equipment in accordance with training and instructions [Reg 10(2)];
- return all personal protective equipment to the appropriate accommodation provided after use [Reg 10(4)]; and
- report forthwith any defect or loss in the equipment to the employer [Reg 11].

## Specific requirements for particular industries and processes

**P3008** In addition to the general remit of the *Personal Protective Equipment at Work Regulations 1992*, the following regulations impose specific requirements on employers to provide personal protective equipment up to the EU standard.

- *Control of Lead at Work Regulations 2002* (*SI 2002 No 2676*), Regs 6, 8 – supply of suitable respiratory equipment and where exposure is significant, suitable protective clothing and equipment (see APPENDIX 4);

- *Control of Asbestos at Work Regulations 2002* (*SI 2002 No 2675*), Reg 10 – supply of respiratory protective equipment/suitable protective clothing (see APPENDIX 4);

- *Control of Substances Hazardous to Health Regulations 2002* (*SI 2002 No 2677* as amended by *SI 2003 No 978*) (*COSHH*), Regs 7, 8, 9 – supply of suitable protective equipment where employees are foreseeably exposed to substances hazardous to health;

- *Noise at Work Regulations 1989* (*SI 1989 No 1790*), Reg 8 – personal ear protectors;

- *Construction (Head Protection) Regulations 1989* (*SI 1989 No 2209*), Reg 3 – suitable head protection (for the position of Sikhs, see C8115 CONSTRUCTION AND BUILDING OPERATIONS);

- *Ionising Radiations Regulations 1999* (*SI 1999 No 3232*), Reg 9 – supply of suitable personal protective equipment and respiratory protective equipment;

- *Shipbuilding and Ship-repairing Regulations 1960* (*SI 1960 No 1932*), Regs 50, 51 – supply of suitable breathing apparatus, belts, eye protectors, gloves and gauntlets (amended in 1994).

[*Miscellaneous Factories (Transitional Provisions) Regulations 1993* (*SI 1993 No 2482*)].

## Increased importance of uniform European standards

**P3009** To date, manufacturers of products, including personal protective equipment (PPE), have sought endorsement or approval for products, prior to commercial circulation, through reference to British Standards (BS), HSE or European standards, an example of the former being Kitemark and an example of the latter being CEN or CENELEC. Both are examples of *voluntary* national and international schemes that can be entered into by manufacturers and customers, under which both sides are 'advantaged' by conformity with such standards. Conformity is normally achieved following a level of testing appropriate to the level of protection offered by the product. With the introduction of the *Personal Protective Equipment (EC Directive) Regulations 1992* as amended by *SI 1994 No 2326* (now revoked and consolidated by *SI 2002 No 1144*) (see P3012 below), this voluntary system of approval was replaced by a statutory certification procedure.

### Towards certification

**P3010** Certification as a condition of sale is probably some time away. Transition from adherence based on a wide range of voluntary national (and international) standards to a compulsory universal EU standard takes time and specialist expertise. Already, however, British Standards Institution (BSI) has adopted the form BS EN as a British version of a uniform European standard. But, generally speaking, uniformity of approach to design and testing, on the part of manufacturers, represents a considerable remit and so statutory compliance with tougher new European standards is probably not likely to happen immediately. Here, too, however, there is evidence of some progress, e.g. the EN 45000 series of standards on how test houses and certification bodies are to be established and independently accredited. In addition, a range of European standards on PPE have been implemented in BS EN format, for example, on eye protection, fall arrest systems and safety footwear.

## Interim measures

**P3011** Most manufacturers have elected to comply with the current CEN standard when seeking product certification. Alternatively, should a manufacturer so wish, the inspection body can verify by way of another route to certification – this latter might well be the case with an innovative product where standards did not exist. Generally, however, compliance with current CEN (European Standardisation Committee) or, alternatively, ISO 9000: Quality Assurance has been the well-trodden route to certification. To date, several UK organisations, including manufacturers and independent bodies, have established PPE test houses, which are independently accredited by the National Measurement Accreditation Service (NAMAS). Such test houses will be open to all comers for verification of performance levels.

## Personal Protective Equipment (EC Directive) Regulations 1992 (SI 1992 No 3139 as amended by SI 1994 No 2326)

**P3012** The *Personal Protective Equipment (EC Directive) Regulations 1992* (as amended by *SI 1994 No 2326*) required most types of PPE (for exceptions, see P3019 below) to satisfy specified certification procedures, pass EU type-examination (i.e. official inspection by an approved inspection body) and carry a 'CE' mark both on the product itself and its packaging before being put into commercial circulation. Indeed, failure on the part of a manufacturer to obtain affixation of a 'CE' mark on his product before putting it into circulation, is a criminal offence under the *Health and Safety at Work etc Act 1974*, s 6 carrying a maximum fine, on summary conviction, of £20,000 (see ENFORCEMENT). Enforcement is through trading standards officers.

The *Personal Protective Equipment Regulations 2002 (SI 2002 No 1144)* now revoke and consolidate the *Personal Protective Equipment (EC Directive) Regulations 1992*. The 2002 Regulations implement the Product Safety Directive 89/696/EEC, on the approximation of the laws of the Member States relating to personal protective equipment. The new Regulations came into force on 15 May 2002.

The 2002 Regulations place a duty on responsible persons who put personal protective equipment (PPE) on the market to comply with the following requirements:

- the PPE must satisfy the basic health and safety requirements that are applicable to that type or class of PPE;
- the appropriate conformity assessment procedure must be carried out;
- a CE mark must be affixed on the PPE.

The Regulations will be enforced by the Weights and Measures Authorities (District Councils in Northern Ireland).

## CE mark of conformity

**P3013** The *Personal Protective Equipment (EC Directive) Regulations 1992* as amended by the *Personal Protective Equipment (EC Directive) (Amendment) Regulations 1994 (SI 1994 No 2326)* (now revoked and consolidated by *SI 2002 No 1144*) specify procedures and criteria with which manufacturers must comply in order to be able to obtain certification. Essentially this involves incorporation into design and production basic health and safety requirements. Hence, by compliance with health and safety criteria, affixation of the 'CE' mark becomes a condition of sale and approval (see *fig. 1* below). Moreover, compliance, on the part of manufacturers, with these regulations, will enable employers to comply with their duties under the *Personal Protective Equipment at Work Regulations 1992*. Reg 4(3)(e) of the *Personal Protective Equipment at Work Regulations 1992* requires the PPE provided by the employer to comply with this and any later EU directive requirements.

CE accreditation is not a synonym for compliance with the CEN standard, since a manufacturer could opt for certification via an alternative route. Neither is it an approvals mark. It is rather a quality mark, since all products covered by the regulations are required to meet formal quality control or quality of production criteria in their certification procedures. [*Arts 8.4, 11, 89/686/EEC*].

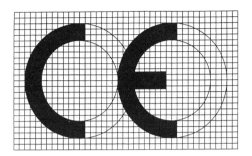

## Certification procedures

**P3014** The *Personal Protective Equipment (EC Directive) Regulations 1992* as amended by *SI 1994 No 2326* (now revoked and consolidated by *SI 2002 No 1144*), identify three categories of PPE:

- PPE of simple design;
- PPE of non-simple design;
- PPE of complex design.

Each of these have different requirements for manufacturers to follow.

In order to obtain certification, manufacturers must submit to an approved body the following documentation (except in the case of PPE of simple design, where risks are minimal).

(1) *PPE – simple design*

This includes use of PPE where risks are minimal e.g. gardening gloves, helmets, aprons, thimbles. This type of equipment must be subject to a manufacturers declaration of conformity.

(2) *PPE – non-simple design*

In order to obtain certification, manufacturers must submit to an approved body the following documentation:

(a) Technical file, i.e.

(i) overall and detailed plans, accompanied by calculation notes/results of prototype tests; and

(ii) an exhaustive list of basic health and safety requirements and harmonised standards taken into account in the model's design.

(b) Description of control and test facilities used to check compliance with harmonised standards.

(c) Copy of information relating to:

(i) storage, use, cleaning, maintenance, servicing and disinfection;

(ii) performance recorded during technical tests to monitor levels of protection;

(iii) suitable PPE accessories;

(iv) classes of protection appropriate to different levels of risk and limits of use;

(v) obsolescence deadline;

(vi) type of packaging suitable for transport;

(vii) significance of markings.

[*Art 1.4, Annex II; Art 8.1, Annex III*].

This must be provided in the official language of the member state of destination.

(3) *PPE – complex design*

This includes use of PPE for protection against mortal/serious dangers and requires compliance with a quality control system.

This category of PPE, including:

(*a*) filtering, respiratory devices for protection against solid and liquid aerosols/irritant, dangerous, toxic or radiotoxic gases;

(*b*) respiratory protection devices providing full insulation from the atmosphere, and for use in diving;

(*c*) protection against chemical attack or ionising radiation;

(*d*) emergency equipment for use in high temperatures, whether with or without infrared radiation, flames or large amounts of molten metal (100°C or more);

(*e*) emergency equipment for use in low temperatures (–50°C or less);

(*f*) protection against falls from heights;

(*g*) protection against electrical risks; and

(*h*) motor cycle helmets and visors;

must satisfy either

(i) — an EU quality control system for the final product, or

— a system for ensuring EU quality of production by means of monitoring;

and in either case,

(ii) an EU declaration of conformity.

[*Art 8.4*].

## Basic health and safety requirements applicable to all PPE

**P3015** All PPE must:

- be ergonomically suitable, that is, be able to perform a risk-related activity whilst providing the user with the highest possible level of protection;

- preclude risks and inherent nuisance factors, such as roughness, sharp edges and projections and must not cause movements endangering the user;

- be kept in clean and hygienic condition and be of a design that enables good hygienic standards, unless it is disposable;

- provide comfort and efficiency by facilitating correct positioning on the user and remaining in place for the foreseeable period of use; and by being as light as possible without undermining design strength and efficiency; and
- be accompanied by necessary information – that is, the name and address of the manufacturer and technical file (see P3014 above).

[*Annex II*].

## General purpose PPE – specific to several types of PPE
P3016

- PPE incorporating adjustment systems must not become incorrectly adjusted without the user's knowledge;
- PPE enclosing parts of the body must be sufficiently ventilated to limit perspiration or to absorb perspiration;
- PPE for the face, eyes and respiratory tracts must minimise risks to same and, if necessary, contain facilities to prevent moisture formation and be compatible with wearing spectacles or contact lenses;
- PPE subject to ageing must contain date of manufacture, the date of obsolescence and be indelibly inscribed if possible. If the useful life of a product is not known, accompanying information must enable the user to establish a reasonable obsolescence date. In addition, the number of times it can be cleaned before being inspected or discarded must (if possible) be affixed to the product; or, failing this, be indicated in accompanying literature;
- PPE which may be caught up during use by a moving object must have a suitable resistance threshold above which a constituent part will break and eliminate danger;
- PPE for use in explosive atmospheres must not be likely to cause an explosive mixture to ignite;
- PPE for emergency use or rapid installation/removal must minimise the time required for attachment and/or removal;
- PPE for use in very dangerous situations (see P3015 above) must be accompanied with data for exclusive use of competent trained individuals, and describe procedure to be followed to ensure that it is correctly adjusted and functional when worn;
- PPE incorporating components which are adjustable or removable by user, must facilitate adjustment, attachment and removal without tools;
- PPE for connection to an external complementary device must be mountable only on appropriate equipment;
- PPE incorporating a fluid circulation system must permit adequate fluid renewal in the vicinity of the entire part of the body to be protected;
- PPE bearing one or more identification or recognition marks relating to health and safety must preferably carry harmonised pictograms/ideograms which remain perfectly legible throughout the foreseeable useful life of the product;
- PPE in the form of clothing capable of signalling the user's presence visually must have one or more means of emitting direct or reflected visible radiation;
- multi-risk PPE must satisfy basic requirements specific to each risk.

[*Art 2, Annex II*].

## Additional requirements for specific PPE for particular risks

P3017   There are additional requirements for PPE designed for certain particular risks as follows.

*PPE protection against*

- *Mechanical impact risks*

    (i) impact caused by falling/projecting objects must be sufficiently shock-absorbent to prevent injury from crushing or penetration of the protected part of the body;

    (ii) falls. In the case of falls due to slipping, outsoles for footwear must ensure satisfactory adhesion by grip and friction, given the state and nature of the surface. In the case of falls from a height, PPE must incorporate a body harness and attachment system connectable to a reliable anchorage point. The vertical drop of the user must be minimised to prevent collision with obstacles and the braking force injuring, tearing or causing the operator to fall;

    (iii) mechanical vibration. PPE must be capable of ensuring adequate attenuation of harmful vibration components for the part of the body at risk.

- (*Static*) *compression of part of the body* – must be able to attenuate its effects so as to prevent serious injury of chronic complaints.

- *Physical injury* – must be able to protect all or part of the body against superficial injury by machinery, e.g. abrasion, perforation, cuts or bites.

- *Prevention of drowning* – (lifejackets, armbands etc.) must be capable of returning to the surface a user who is exhausted or unconscious, without danger to his health. Such PPE can be wholly or partially inherently buoyant or inflatable either by gas or orally. It should be able to withstand impact with liquid, and, if inflatable, able to inflate rapidly and fully.

- *Harmful effects of noise* – must protect against exposure levels of the *Noise at Work Regulations 1989* (i.e. 85 and 90 dB(A)) and must indicate noise attenuation level.

- *Heat and/or fire* – must possess sufficient thermal insulation capacity to retain most of the stored heat until after the user has left the danger area and removed PPE. Moreover, constituent materials which could be splashed by large amounts of hot product must possess sufficient mechanical-impact absorbency. In addition, materials which might accidentally come into contact with flame as well as being used in the manufacture of fire-fighting equipment, must possess a degree of non-flammability proportionate to the risk foreseeably arising during use and must not melt when exposed to flame or contribute to flame spread. When ready for use, PPE must be such that the quantity of heat transmitted by PPE to the user must not cause pain or health impairment. Second, ready-to-use PPE must prevent liquid or steam penetration and not cause burns. If PPE incorporates a breathing device, it must adequately protect the user. Accompanying manufacturers' notes must provide all relevant data for determination of maximum permissible user exposure to heat transmitted by equipment.

- *Cold* – must possess sufficient thermal insulating capacity and retain the necessary flexibility for gestures and postures. In particular, PPE must protect tips of fingers and toes from pain or health impairment and prevent penetration by rain water. Manufacturers' accompanying notes must provide all relevant data concerning maximum permissible user exposure to cold transmitted by equipment.

- *Electric shock* – must be sufficiently insulated against voltages to which the user is likely to be exposed under the most foreseeably adverse conditions. In particular, PPE for use during work with electrical installations (together with packaging), which may be under tension, must carry markings indicating either protection class and/or corresponding operating voltage, serial number and date of manufacture; in addition, date of entry into service must be inscribed as well as of periodic tests or inspections.

- *Radiation*

    (a) Non-ionising radiation – must prevent acute or chronic eye damage from non-ionising radiation and be able to absorb/reflect the majority of energy radiated in harmful wavelengths without unduly affecting transmission of the innocuous part of the visible spectrum, perception of contrasts and distinguishment of colours. Thus, protective glasses must possess a spectral transmission factor so as to minimise radiant-energy illumination density capable of reaching the user's eye through the filter and ensure that it does not exceed permissible exposure value. Accompanying notes must indicate transmission curves, making selection of the most suitable PPE possible. The relevant protection-factor number must be marked on all specimens of filtering glasses.

    (b) Ionising radiation – must protect against external radioactive contamination. Thus, PPE should prevent penetration of radioactive dust, gases, liquids or mixtures under foreseeable use conditions. Moreover, PPE designed to provide complete user protection against external irradiation must be able to counter only weak electron or weak photon radiation.

- *Dangerous substances and infective agents*

    (a) respiratory protection – must be able to supply the user with breathable air when exposed to polluted atmosphere or an atmosphere with inadequate oxygen concentration. Leak-tightness of the facepiece and pressure drop on inspiration (breath intake), as well as in the case of filtering devices purification capacity, must keep the contaminant penetration from the polluted atmosphere sufficiently low to avoid endangering health of the user. Instructions for use must enable a trained user to use the equipment correctly;

    (b) cutaneous and ocular contact (skin and eyes) – must be able to prevent penetration or diffusion of dangerous substances and infective agents. Therefore, PPE must be completely leak-tight but allow prolonged daily use or, failing that, of limited leak-tightness restricting the period of wear. In the case of certain dangerous substances/infective agents possessing high penetrative power and which limit duration of protection, such PPE must be tested for classification on the basis of efficiency. PPE conforming with test specifications must carry a mark indicating names or codes of substances used in tests and standard period of protection. Manufacturers' notes must contain an explanation of codes, detailed description of standard tests and refer to the maximum permissible period of wear under different foreseeable conditions of use.

- *Safety devices for diving equipment* – must be able to supply the user with a breathable gaseous mixture, taking into account the maximum depth of immersion. If necessary, equipment must consist of

    (a) a suit to protect the user against pressure;

    (b) an alarm to give the user prompt warning of approaching failure in supply of breathable gaseous mixture;

    (c) life-saving suit to enable the user to return to the surface.

[*Art 3, Annex II*].

## Excluded PPE

**P3018**

- PPE designed and manufactured for use specifically by the armed forces or in the maintenance of law and order (helmets, shields);
- PPE for self-defence (aerosol canisters, personal deterrent weapons);
- PPE designed and manufactured for private use against:
    (i) adverse weather (headgear, seasonal clothing, footwear, umbrellas),
    (ii) damp and water (dish-washing gloves),
    (iii) heat (e.g. gloves);
- PPE intended for the protection or rescue of persons on vessels or aircraft, not worn all the time.

[*Annex I*].

# Main types of personal protection

The main types of personal protection covered by the *Personal Protective Equipment at Work Regulations 1992* are (*a*) head protection, (*b*) eye protection, (*c*) hand/arm protection, (*d*) foot protection and (*e*) whole body protection.

In particular, these regulations do not cover ear protectors and most respiratory protective equipment – these areas are covered by other existing regulations and guidance, for example the *Noise at Work Regulations 1989* and HSE's guidance booklet HS(G) 53, 'Respiratory protective equipment: a practical guide for users'.

## Head protection

**P3019**  This takes the form of:

- crash, cycling, riding and climbing helmets;
- industrial safety helmets – to protect against falling objects;
- scalp protectors (bump caps) – to protect against striking fixed obstacles;
- caps/hairnets – to protect against scalping,

and is particularly suitable for the following activities:

(i) building work – particularly on scaffolds;
(ii) civil engineering projects;
(iii) blasting operations;
(iv) work in pits and trenches;
(v) work near hoists/lifting plant;
(vi) work in blast furnaces;
(vii) work in industrial furnaces;
(viii) ship-repairing;
(ix) railway shunting;
(x) slaughterhouses;

## Eye protection

**P3020** This takes the following forms:

- safety spectacles – these are the same as prescription spectacles but incorporating optional sideshields; lenses are made from tough plastic, such as polycarbonate – provide lateral protection;
- eyeshields – these are heavier than safety spectacles and designed with a frameless one-piece moulded lens; can be worn over prescription spectacles;
- safety goggles – these are heavier than spectacles or eye shields; they are made with flexible plastic frames and one-piece lens and have an elastic headband – they afford total eye protection; and
- faceshields – these are heavier than other eye protectors but comfortable if fitted with an adjustable head harness – faceshields protect the face but not fully the eyes and so are no protection against dusts, mist or gases.

Eye protectors are suitable for working with:

(i) chemicals;
(ii) power driven tools;
(iii) molten metal;
(iv) welding;
(v) radiation;
(vi) gases/vapours under pressure;
(vii) woodworking.

## Hand/arm protection

**P3021** Gloves provide protection against:

- cuts and abrasions;
- extremes of temperature;
- skin irritation/dermatitis;
- contact with toxic/corrosive liquids,

and are particularly useful in connection with the following activities/processes:

(i) manual handling;
(ii) vibration;
(iii) construction;
(iv) hot and cold materials;
(v) electricity;
(vi) chemicals;
(vii) radioactivity.

## Foot protection

P3022    This takes the form of:
- safety shoes/boots;
- foundry boots;
- clogs;
- wellington boots;
- anti-static footwear;
- conductive footwear,

and is particularly useful for the following activities:
  (i) construction;
  (ii) mechanical/manual handling;
  (iii) electrical processes;
  (iv) working in cold conditions (thermal footwear);
  (v) chemical processes;
  (vi) forestry;
  (vii) molten substances.

## Respiratory protective equipment

P3023    This takes the form of:
- half-face respirators,
- full-face respirators,
- air-supply respirators,
- self-contained breathing apparatus,

and is particularly useful for protecting against harmful:
  (i) gases;
  (ii) vapours;
  (iii) dusts;
  (iv) fumes;
  (v) smoke;
  (vi) aerosols.

## Whole body protection

P3024    This takes the form of:
- coveralls, overalls, aprons;
- outfits to protect against cold and heat;
- protection against machinery and chainsaws;
- high visibility clothing;

- life-jackets,

and is particularly useful in connection with the following activities:

(i) laboratory work;

(ii) construction;

(iii) forestry;

(iv) work in cold-stores;

(v) highway and road works;

(vi) food processing;

(vii) welding;

(viii) fire-fighting;

(ix) foundry work;

(x) spraying pesticides.

## Some relevant British Standards for protective clothing and equipment

**P3025** Although not, strictly speaking, a condition of sale or approval, for the purposes of the *Personal Protective Equipment (EC Directive) (Amendment) Regulations 1994 (SI 1994 No 2326)* (now revoked and consolidated by *SI 2002 No 1144*), compliance with British Standards (e.g. ISO 9000: 'Quality Assurance') may well become one of the well-tried verification routes to obtaining a 'CE' mark. For that reason, if for no other, the following non-exhaustive list of British Standards on protective clothing and equipment should be of practical value to both manufacturers and users.

| Table 1 | |
|---|---|
| **Protective clothing** | |
| BSEN 60903: 1986 | Specification for rubber gloves for electrical purposes |
| BS 3314: (1998) | Specification for protective aprons for wet work |
| BS 5426: 1993 | Specification for workwear and career wear |
| BS 5438: 1976 | Methods of test for flammability of vertically oriented textile fabrics and fabric assemblies subjected to a small igniting flame |
| 1989 (1995) | Methods of test for flammability of textile fabrics when subjected to a small igniting flame applied to the face or bottom edge of vertically oriented specimens |
| BS 6408: (1990) | Specification for clothing made from coated fabrics for protection against wet weather |
| BS EN 340: 1993 | Protective clothing. General Requirements |
| BS EN 341: (1999) | Personal Protective Equipment against falls from a height. Descender devices |

| | |
|---|---|
| BS EN 348: 1992 | Protective clothing. Determination of behaviour of materials on impact of small splashes of molten metal |
| BS EN 353: 1993 | Personal protective equipment against falls from a height. Guided type of fall arresters |
| BS EN 354: 1993 | Personal protective equipment against falls from a height. Lanyards |
| BS EN 355: 1993 | Personal protective equipment against falls from a height. Energy Absorbers |
| BS EN 360: 1993 | Personal protective equipment against falls from a height. Retractable type fall arresters |
| BS EN 361: 1993 | Personal protective equipment against falls from a height. Full body harnesses |
| BS EN 362: 1993 | Personal protective equipment against falls from a height. Connectors |
| BS EN 363: 1993 | Personal protective equipment against falls from a height. Fall arrest systems |
| BS EN 364: 1993 (1999) | Personal protective equipment against falls from a height. Test methods |
| BS EN 365: 1993 | Personal protective equipment against falls from a height. General Requirements for instructions for use and for marking |
| BS EN 366: 1993 | Protective clothing. Protection against heat and fire |
| BS EN 367: 1992 | Protective clothing. Protection against heat and fires |
| BS EN 368: 1993 | Protective clothing. Protection against liquid chemicals |
| BS EN 369: 1993 | Protective clothing. Protection against liquid chemicals |
| BS EN 373: 1993 | Protective clothing: Assessment of resistance of materials to molten metal splash |
| BS EN 381 Part 1: 1993 | Protective clothing for users of hand-held chain saws |
| BS EN 463: 1995 | Protective clothing. Protection against liquid chemicals. Test method |
| BS EN 464: 1994 | Protective clothing. Protection against liquid and gaseous chemicals including liquid aerosols and solid particles. Test method |
| BS EN 465: 1995 | Protective clothing: Protection against liquid chemicals. Performance requirements for chemical protective clothing |
| BS EN 4661: 1995 | Protective clothing: Protection against liquid chemicals. Performance requirements for chemical protective clothing |
| BS EN 467: 1995 | Protective Clothing: protection against liquid chemicals. |

| | |
|---|---|
| BS EN 468: 1995 | Protective clothing for use against liquid chemicals. Test method |
| BS EN 469: 1995 | Protective clothing for firefighters |
| BS EN 470 | Protective clothing for use in welding and allied processes |
| Part 1: 1995 | General requirements |
| BS EN 471: 1994 | Specification for high visibility warning clothing |
| BS EN 510: 1993 | Specification for protective clothing for use where there is a risk of entanglement with moving parts |
| BS EN 530: 1995 | Abrasion resistance of protective clothing material |
| BS EN 531: 1995 | Protective clothing for workers exposed to heat |
| BS EN 532: 1995 | Protective clothing. Protection against heat and flame |
| BS EN 533: 1997 | Clothing for protection against heat or flame |

**Protective footwear**

| | |
|---|---|
| BS 2723: 1956 (1995) | Specification for fireman's leather boots |
| BS 4676: 1983 (1996) | Specification for gaiters and footwear for protection against burns and impact risks in foundries |
| BS 5145: 1989 | Specification for lined industrial vulcanized rubber boots |
| BS 6159: | Polyvinyl chloride boots |
| Part 1: 1987 | Specification for general and industrial lined or unlined boots |
| BS EN | Safety, protective and occupational footwear for professional use |
| BS EN 345 | Safety footwear for professional use |
| BS EN 346 | Protective footwear for professional use |
| BS EN 347 | Occupational footwear for professional use |
| BS EN 381 | Protective clothing for users of hand-held chain saws |

**Head protection**

| | |
|---|---|
| BS 6658: 1985 (1995) | Specification for protective helmets for vehicle users |
| BS 6473: 1984 | Specification for protective hats for horse and pony riders |
| BS 3864: 1989 | Specification for protective helmets for firefighters |
| BS 4033: 1978 | Specification for industrial scalp protectors (light duty) |
| BS 4423: 1969 | Specification for climbers helmets |

| | |
|---|---|
| BS 4472: 1988 | Specification for protective skull caps for jockeys |

**Face and eye protection**

| | |
|---|---|
| BS 679: 1989 | Specification for filters, cover lenses and backing lenses for use during welding and similar industrial operations |
| BS 1542: 1982 (1995) | Specification for eye, face and neck protection against non-ionising radiation arising during welding and similar operations |
| BS 2092: 1987 | Specification for eye protectors for industrial and non-industrial users |
| BS 2724: 1987 (1995) | Specification for sunglare eye protectors for general use |
| BS 4110: 1999 | Specification for eye-protectors for vehicle users |
| BS 7028: 1999 | |
| BS EN 167: 1995 | Personal eye protection. Optical test methods |
| BS EN 168: 1995 | Personal eye protection. Non-optical test methods |
| BS EN 169: 1992 | Specification for filters for personal eye-protection equipment used in welding and similar operations |
| BS EN 170: 1992 | Specification for filters for ultraviolet filters used in personal eye-protection equipment |
| BS EN 171: 1992 | Specification for infra-red filters used in personal eye protective equipment |
| BS EN 172: (2000) | Specification for sunglare filters used in personal eye-protectors equipment for industrial use |

**Respiratory protection**

| | |
|---|---|
| BS 4001 Part 1 (1998) | Care and maintenance of underwater breathing apparatus |
| Part 2 (1995) | Standard diving equipment |
| BS 4275: 1997 | Guide to implementing an effective respiratory protective device programme |
| BS 4400: 1969 (1995) | Method for sodium chloride particulate test for respirator filters |
| BS EN 132: 1999 | Respiratory protective devices. Definitions of terms and pictograms |
| BS EN 133: 1991 | Respiratory protective devices. Classification |
| BS EN 134: 1998 | Respiratory protective devices. Nomenclature of compounds |
| BS EN 135: 1999 | Respiratory protective devices. List of equivalent terms |
| BS EN 136: 1998 | Respiratory protective devices. Full face mask requirements for testing and marking |

| | |
|---|---|
| BS EN 137: 1993 | Specification for respiratory protective devices self-contained open-circuit compressed air breathing apparatus |
| BS EN 138: 2000 | Respiratory protective devices. Specification for fresh air hose breathing apparatus for use with full face mask, half mask or mouthpiece assembly |
| BS EN 139: 1995 | Respiratory protective devices. Compressed air line breathing apparatus for use with a full face mask, half mask or a mouthpiece assembly. Requirements, testing, marking |
| BS EN 141: 1991 | Specification for gas filters and combined filters used in respiratory protective equipment |
| BS EN 143: 1991 | Specification for particle filters used in respiratory protective equipment |
| BS EN 144 | Respiratory protective devices. Gas cylinder valves |
| BS EN 145: 1998 | Respiratory protective devices. Self-contained closed-circuit compressed oxygen breathing apparatus |
| BS EN 146: 1992 | Respiratory protection devices. Specification for powered particle filtering devices incorporating helmets or hoods |
| BS EN 147: 1992 | Respiratory protection devices. Specification for power assisted particle filtering devices incorporating full face masks, half masks or quarter masks |
| BS EN 148 | Specification for thread connection |
| BS EN 149: 1992 | Specification for filtering half masks to protect against particles |
| BS EN 371: 1992 | Specification for AX gas filters and combined filters against low boiling organic compounds used in respiratory protective equipment |
| BS EN 372: 1992 | Specification for SX gas filters and combined filters against specific named compounds used in respiratory protective equipment |

### Radiation protection

| | |
|---|---|
| BS 1542: 1982 (1995) | Specification for equipment for eye, face and neck protection against non-ionising radiation arising during welding and similar operations |
| BS 3664: 1963 | Specification for film badges for personnel radiation monitoring |
| BS EN 269: 1995(2000) | Respiratory protective devices. Specification for powered fresh air hose breathing apparatus incorporating a hood. |
| BS EN 270: 1995 | Respiratory protective devices. Compressed air line breathing apparatus incorporating a hood. Requirements, testing, marking |

### Hearing protection

| | |
|---|---|
| BS EN 352: | Hearing protectors. Safety requirements and testing |
| Part 1 1993 | Ear muffs |
| Part 2 1993 | Ear plugs |
| Part 3 1997 | Ear muffs attached to an industrial safety helmet |

### Hand protection

| | |
|---|---|
| BS EN 374: | Protective gloves against chemicals and microorganisms |
| Part 1: 1994 | Terminology and performance requirements |
| Part 2: 1994 | Determination of resistance to penetration |
| Part 3: 1994 | Determination of resistance to permeation by chemicals |
| BS EN 511: 1994 | Specification for protective gloves against cold |
| BS EN 421: 1994 | Protective gloves against ionising radiation and radioactive contamination |
| BS EN 407: 1994 | Protective gloves against thermal risks (heat and/or fire) |
| BS EN 388: 1994 | Protective gloves against mechanical risks |

## Common law requirements

**P3026** In addition to general and/or specific statutory requirements, the residual combined duty, on employers at common law, to provide and maintain a safe system of work, including appropriate supervision of safety duties, still obtains, extending, where necessary, to protection against foreseeable risk of eye injury (*Bux v Slough Metals Ltd [1974] 1 All ER 262*) and dermatitis/facial eczema (*Pape v Cumbria County Council [1991] IRLR 463*). *Pape* concerned an office cleaner who developed dermatitis and facial eczema after using Vim, Flash and polish. The employer was held liable for damages at common law, since he had not instructed her in the dangers of using chemical materials with unprotected hands and had not made her wear rubber gloves.

### Consequences of breach

**P3027** Employers who fail to provide suitable personal protective equipment, commit a criminal offence, under the *Health and Safety at Work etc Act 1974* and the *Personal Protective Equipment at Work Regulations 1992* and sundry other regulations (see P3006–P3008 above). In addition, if, as a result of failure to provide suitable equipment, an employee suffers foreseeable injury and/or disease, the employer will be liable to the employee for damages for negligence. Conversely, if, after instruction and, where necessary, training, an employee fails or refuses to wear and maintain suitable personal protective equipment, he too can be prosecuted and/or dismissed and, if he is injured or suffers a disease in consequence, will probably lose all or, certainly, part of his damages.

# Pressure Systems

## Introduction

**P7001** Legislation concerning the safety of pressure systems applied in its infancy to steam generating boilers and is traceable back to the original *Boiler Explosions Act 1882*. At this time statutory requirements were placed on the users and owners of the equipment for the safety of the boiler itself rather than the system incorporating it. Periodic examination of boilers became mandatory following the *Factories and Workshops Act 1901* when requirements were also extended to include other items of pressure equipment such as air receivers.

The legislation governing the construction and use of pressure equipment is divided into two areas where the initial duty falls onto those either manufacturing, installing or supplying pressure equipment and thereafter to users or owners of the equipment.

There are three principal statutes that cover the manufacture, installation and operation of pressure equipment:

- *Simple Pressure Vessel Regulations 1991* (*SI 1991 No 2749*) amended by *Simple Pressure Vessels (Safety Amendment) Regulations 1994* (*SI 1994 No 3098*);

- *Pressure Equipment Regulations 1999* (*SI 1999 No 2001*) as amended by the *Pressure Equipment (Amendment) Regulations 2002* (*SI 2002 No 1267*);

- *Pressure Systems Safety Regulations 2000* (*SI 2000 No 12*).

These regulations provide a synergistic approach to the management of pressure systems based on a cradle to grave concept.

The *Pressure Equipment Directive 87/404/EEC* (PED) and the *Simple Pressure Vessel Directive* (SPVD) (see Figure 1) as implemented by the *Pressure Equipment Regulations 1999* (*SI 1999 No 2001*) (PER) as amended by the *Pressure Equipment (Amendment) Regulations 2002* (*SI 2002 No 1267*) and the *Simple Pressure Vessel Regulations 1991* (*SI 1991 No 2749*) as amended by the *Simple Pressure Vessels (Safety Amendment) Regulations 1994* (*SI 1994 No 3098*) (SPVR). This amendment was needed to take account of the *CE Marking Directive* (93/68/EEC) which required all simple pressure vessels to be CE Marked (see P7004 and 7005).

These Regulations set minimum standards of safety and quality for the manufacture of new pressure systems and equipment in the European Union so that non tariff trade barriers are removed and they present a 'level playing field' for all member states. The Department of Trade and Industry implement these requirements in the UK.

The *Pressure Systems Safety Regulations 2000* (*SI 2000 No 128*) are mainly concerned with ensuring that minimum safety standards are satisfied with due regard for in service pressure systems and these are specific to the UK.

Provisions relating to transportable gas containers (receptacles) have now been removed from the governing legislation (*Pressure Systems Safety Regulations 2000* (*SI 2000 No 128*)). These Regulations replaced the *Pressure Systems and Transportable Gas Container Regulations 1989* (*SI 1989 No 2196*) (*PSTGCR*).

**P7001**  *Pressure Systems*

The introduction of the *PSTGCR* in 1989 was a major step forward in the control over pressure systems. They introduced the important concepts of goal setting legislation, the written scheme of examination, and the role of the competent person who was charged within certain limitations of determining the scope and frequency of the thorough examination. Prior to the PSTGCR, pressure systems were covered under the *Factories Act 1961*, supported by prescriptive requirements for their examination.

Transportable gas containers that were previously considered in the PSTGCR have been divorced and are now covered under *the Carriage of Dangerous Goods and Use of Transportable Pressure Equipment Regulations 2004* (*SI 2004 No 568*), these regulations replaced the earlier *Carriage of Dangerous Goods (Classification, Packaging and Labelling) and Use of Transportable Pressure Receptacles Regulations 1996* (*SI 1996 No 2092*) which was in turn replaced by the short lived and now revoked *Transportable Pressure Vessel Regulations 2001* (*SI 2001 No 1426*).

Figure 1. Relationship between Manufacturers' Duties

## Simple pressure vessels

**P7002** Simple pressure vessels are those that are constructed under the criteria of the *Simple Pressure Vessel Regulations 1991* (*SI 1991 No 2749*) amended by the *Simple Pressure Vessels (Safety Amendment) Regulations 1994* (*SI 1994 No 3098*).

A simple pressure vessel is a welded vessel intended to contain air or nitrogen at a gauge pressure greater than 0.5 bars but one that is not intended for exposure to a flame.

## Simple Pressure Vessels (Safety) Regulations (SPVSR) 1991

**P7003** The basic design criteria that must be fulfilled before SPVSR can be applied are that:

- The vessel incorporates circular cross section that is closed at each end by either convexed dished ends or flat end plates.

- The maximum working pressure is not more than 30 bar, and the pressure multiplied by the volume in bar litres is not more than 10,000 bar/litres.

- The minimum working temperature is not lower than minus 50 degrees C, and the maximum working temperature is not higher than 300 degrees C in the case of steel vessels, and 100 degrees C in the case of aluminium or aluminium alloy vessels.

The categories of vessels relevant for the purposes of the SPVSR that are defined in the Regulations are—

- *Category A*: Vessels with a PS/V of more than 50 bar-litres, this category being subdivided into:

    — *Category A.1*: Vessels with a PS/V of more than 3,000 bar-litres;

    — *Category A.2*: Vessels with a PS/V of more than 200 but not more than 3,000 bar-litres; and

    — *Category A.3*: Vessels with a PS/V of more than 50 but not more than 200 bar-litres;

- *Category B:* Vessels with a PS/V of 50 bar-litres or less.

## EC Certificate of Adequacy

**P7004** An application for an EC certificate of adequacy must be accompanied by the design and manufacturing schedule. The approved bodies to which such an application can be made are available from the Department of Trade and Industry.

The schedule must contain all the required information; and additionally provide evidence that the vessel has been manufactured in accordance with the schedule and that it conforms to a relevant national standard e.g. BSI EN 13445.

A certificate of adequacy can then be issued by the approved body which allows the simple pressure vessel to be placed on the market as being "safe".

If the approved body is not so satisfied and refuses to issue a certificate of adequacy it must inform the applicant in writing the reasons for the refusal.

## EC Type-Examination Certificate

**P7005** EC type-examination is the process whereby an approved body ascertains, and certifies by means of an EC type-examination certificate, that a prototype representative of the production envisaged satisfies the requirements of the Directive.

The prototype may be representative of a family of vessels in which case the Approved Body is required to examine the prototype, and perform such tests as it considers are appropriate, and once it is satisfied that the prototype is manufactured in conformity with the schedule, and meets the essential safety requirements specified in Schedule 1, declare that it is safe.

## Pressure Equipment Regulations 1999 (SI 1999 No 2001) and the EU Pressure Equipment Directive (97/23/EC)

**P7006** The Directive was taken into UK legislation as the *Pressure Systems Safety Regulations 1999 (SI 1000 No 2001)* and covers pressure equipment and assemblies with a maximum allowable pressure greater than 0.5 bar. Pressure equipment includes vessels, piping, safety accessories and pressure accessories. Assemblies are several pieces of pressure equipment assembled to form an integrated, functional whole.

The intention of legislation is to remove technical barriers to trade by harmonising national laws of legislation regarding the design, manufacture, marking and conformity assessment of pressure equipment. It does not deal with in-use requirements, which may be necessary to ensure the continued safe use of pressure equipment.

The Directive covers a wide range of equipment such as, reaction vessels, pressurised storage containers, heat exchangers, shell and water tube boilers, industrial pipework, safety devices and pressure accessories. Such equipment is widely used throughout industry.

There are some specific exclusions such as: equipment within the scope of other directives (e.g. simple pressure vessels, aerosols, parts for the functioning of vehicles); and equipment covered by the international conventions for the transport of dangerous goods and equipment for which the pressure hazards are low and are adequately covered by other directives (such as the Machinery Directive).

## Essential safety requirements

**P7007** The principle adopted in the *PER Pressure Equipment Regulations 1999 (SI 1999 No 2001)* is similar to other European Directives, and requires the manufacturer to consider hazards associated with pressure equipment and accordingly take these hazards into consideration when considering the design and construction of the equipment.

The main areas for assessment are:

- General requirements.
- Design requirements:
    - Adequate strength;
    - Experimental design methods;
    - Provisions to ensure safe handling and operation;
    - Means of examination;
    - Means of draining and ventilating;
    - Corrosion;

- — Wear;
- — Assemblies;
- — Provisions for filling and discharge;
- — Protection against exceeding the allowable limits of pressure equipment;
- — Safety accessories;
- — External firing.
- Manufacturing:
  - — Manufacturing procedures;
  - — Final assessment;
  - — Marking and labelling;
  - — Operating instructions.
- Materials.
- Fired or Heated Pressure equipment with a Risk of Overheating.
- Piping.
- Specific Quantitative Requirements for Certain Pressure Equipment:
  - — Allowable stresses;
  - — Joint coefficients;
  - — Pressure limiting devices, particularly for pressure vessels;
  - — Hydrostatic test pressure;
  - — Material characteristics.

## Conformity assessment procedure

**P7008**  Schedule IV of the *Pressure Equipment Regulations 1999* (*SI 1999 No 2001*), or annex III of the *Pressure Equipment Directive 97/23/EC* detail the procedures and responsibilities for the contributing parties. This is a detailed document and it consists of a product classification table and nine classification charts.

In order to declare pressure equipment to be safe the manufacturer or their authorised representative must before supplying any equipment satisfy the certification procedure which is a similar but slightly more detailed procedure to that applied to conformity assessment for verification under the *Simple Pressure Vessel Directive 87/404/EEC*.

- Draw up a technical document which will allow an assessment to be made of the conformity of the pressure equipment, containing:
  - — General description;
  - — Conceptual design and manufacturing drawings;
  - — Descriptions and explanations necessary to understand the drawings;
  - — A list of harmonised standards applied and the solutions adopted to meet the essential safety requirements were harmonised standards have not been applied;
  - — Results of design calculations made;
  - — Test reports.
- Ensure manufacturing process complies with technical documentation.

- Affix CE Mark.
- Draw up written declaration of conformity.
- Retain declaration of conformity and technical documentation for 10 years.

There are a number of different conformity assessment modules available for the different categories of equipment:

E.g. the boiler manufacturer's codes:

- EN 12592 Water Tube Boilers;
- EN 12593 Shell Boilers.

The unfired pressure vessel manufacturer's new code:

- EN 13445 Unfired pressure vessels.

Manufacturers will by following a code, where it exists, have satisfied the requirements of the (*Pressure Equipment Regulations 1999* (*SI 1999 No 2001*).

## Notified bodies

P7009    The Secretary of State for the Department of Trade and Industry (DTI) is responsible for the appointment of recognised third-party notified bodies, following an assessment against criteria laid down in the directive.

## The Pressure Equipment (Amendment) Regulations 2002 (SI 2002 No 1267)

P7010    The *Pressure Equipment (Amendment) Regulations 2002* (*SI 2002 No 1267*) exempt an item of pressure equipment or an assembly of exempting pressure equipment that is placed on the market from the requirements of the 1999 Regulations, provided it has not been used in the course of business at all times since its manufacture or import and that prior to its being made so available it has been used otherwise than in the course of business at all times since its manufacture or import.

There have also been some amendments to the penalties available to the courts where a person guilty of an offence under regulation 25(a) shall be liable:

(a) On summary conviction, to a fine not exceeding the statutory maximum or to imprisonment not exceeding three months or to both;

(b) On conviction on indictment, to a fine or to imprisonment for a term not exceeding two years or to both.

## Pressure Systems Safety Regulations 2000 (SI 2000 No 128)

### Duty holders

P7011    The *Pressure Systems Safety Regulations 2000* impose statutory duties on individuals and organisations connected with pressure systems.

The *Pressure Systems Safety Regulations 2000* ('the Regulations') came into force on 14th February 2000. [*Pressure Systems Safety Regulations 2000* (*SI 2000 No 128*), *Reg 1*]. They are the principal UK statutory controls over the operation and safety of in-service pressure systems. The aim of the Regulations is to prevent serious injury from the hazard of stored energy as a result of failure of the system or one of its component parts.

- **Users and owners** of pressure systems have primary and wide responsibilities for ensuring the safety of their systems.

- **'Competent persons'** are responsible for drawing up or certifying suitable written schemes of examination, for the examination in accordance with the written scheme, and for action in the case of imminent danger.

- **Designers, manufacturers, importers, suppliers and installers** have responsibilities relating to design, materials and construction, and information and marking.

- Employers of **persons, who install**, modify or repair pressure systems have duties to ensure that their work does not gives rise to danger.

- **The enforcing authority** (Health and Safety Executive or Environmental Health Department of the local authority), has powers under the Regulations to enforce and exempt the application of the Regulations.

In cases where pressure systems are supplied by way of lease, hire or other arrangements, the supplier is allowed to discharge the duties of the user with regard to examinations, operation, maintenance and record keeping. [*Pressure Systems Safety Regulations 2000* (*SI 2000 No 128*), *Reg 3(5) and Sch 2*].

A duty holder charged with an offence under the Regulations is allowed a defence if the offence was due to an act or default of another person (not being an employee) and that all reasonable precautions and due diligence had been taken to avoid the offence taking place. [*Pressure Systems Safety Regulations 2000* (*SI 2000 No 128*), *Reg 16*].

## Systems and equipment covered by the regulations

P7012     The Regulations are concerned with systems containing relevant fluids under pressure (See Figure 2). A relevant fluid is:

- Steam (above atmospheric pressure);

- A gas or a mixture of gases;

- A liquid which would turn into a gas if system failure occurred.

The Regulations therefore cover systems containing compressed air, other compressed gases (e.g. nitrogen, oxygen or acetylene), hot water contained above its normal boiling point, and a gas dissolved under pressure in a solvent contained in a porous substance.

Except in the case of steam (which is always a relevant fluid when stored above atmospheric pressure), a fluid is relevant if it exerts a pressure or vapour pressure greater than 0.5 bar above atmospheric pressure.

The Regulations cover both installed (fixed) pressure systems and mobile systems that can be readily removed between and used in different locations. Transportable pressure receptacles such as a tanker used for carrying dangerous goods by rail or road, are covered under the *Carriage of Dangerous Goods and Use of Transportable Pressure Equipment Regulations 2004* (if these Regulations cease to apply and a relevant fluid is carried, then the *Pressure Systems Safety Regulations 2000* will apply). Steam locomotives, including model steam engines, are classed as installed systems for the purposes of the Regulations. Within the Regulations, an installed system is the responsibility of the user, while the owner has responsibility for a mobile system.

Within the Regulations, a pressure system comprises of one or more pressure vessels of rigid construction, any associated pipework, and protective devices. Pipework used in connection with transportable pressure receptacles and pipelines with their associated protective devices are also pressure systems. The terms 'pipework', 'pipeline' and 'protective devices' are defined by the

**P7013** *Pressure Systems*

Regulations and cover all items of pressure equipment including valves, pumps, hoses, compressors, bursting discs and pressure relief devices. [*Pressure Systems Safety Regulations 2000 (SI 2000 No 128), Reg 2*].

Users and owners should decide, seeking suitable advice if required, whether equipment for which they are responsible is covered by these Regulations. Examples of equipment for which the Regulations are likely to apply to include:

- Steam boilers and associated pipework and protective devices.
- Compressed air systems with a receiver of a stored capacity exceeding 250 Bars/litres
- Portable hot water steam cleaning units fitted with a pressure vessel.
- Steam sterilising autoclaves and associated pipework and protective devices.
- Steam pressure cookers.
- Gas loaded hydraulic accumulators if the free gas capacity exceeds 250 Bars/litres
- Steam locomotives including model and narrow gauge steam engines.
- Vapour compression refrigeration systems where the power exceeds 25KW.
- Fixed liquid petroleum gas (LPG) storage systems e.g. supplying fuel for heating in a workplace.

## Systems and equipment excepted from some or all the regulations

**P7013** Some types of pressure systems are excepted from certain Regulations [*Pressure Systems Safety Regulations 2000 (SI 2000 No 128), Sch 1, Part II*]. These exceptions are primarily to avoid overlap with other regulations and apply to the following.

(i) Pressure systems to which the *Medical Devices Regulations 1994 (SI 1994 No 3017)* apply, other than those which contain or are liable to contain steam, are excepted from regulations on design, construction and provisions for information and marking.

(ii) Pressure equipment subject to the *Pressure Equipment Regulations 1999 (SI 1999 No 2001)* is excepted from regulations on design, construction and provisions for information and marking.

(iii) Pressure systems containing a relevant fluid (other than steam) are excepted from provisions on marking, examination and record keeping if the product of the pressure and internal volume control in each pressure vessel is less than 250 bar litres. Where the relevant fluid is steam, all the Regulations apply regardless of vessel pressure and size.

(iv) Tank containers subject to the *Carriage of Dangerous Goods and Use of Transportable Pressure Equipment Regulations 2004 (SI 2004 No 568)* and examined under the provisions of those Regulations, are not subject to the corresponding provisions of the *Pressure Systems Safety Regulations 2000* when they have been removed from the vehicle.

Some twenty five categories of pressure systems are excepted from all the Regulations [*Pressure Systems and Safety Regulations 2000 (SI 2000 No 128), Sch 1, Part I*] and include:

- Any system that forms part of a ship, hovercraft, aircraft or spacecraft.
- Any system that forms part of a weapon system.
- Any system which is only pressurised by a leak test or unintentionally.
- Some systems which form part of a research experiment.
- Any low pressure pipelines where the pressure does not exceed 2 bars.

Figure 2. Pressure Systems Safety Regulations 2000 – Duties Decision Flow Chart

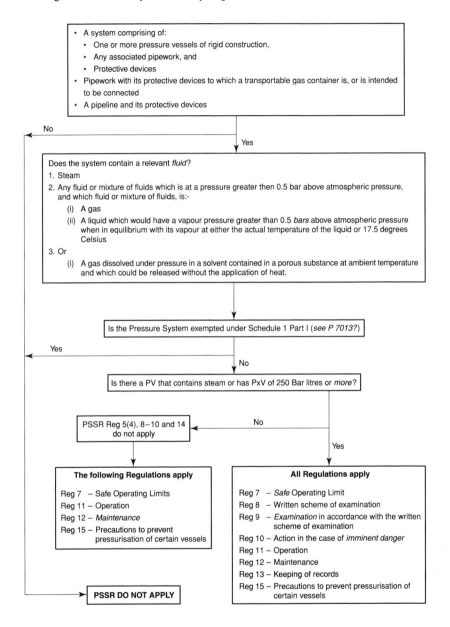

- Any equipment regulated by the *Diving at Work Regulations 1997* (*SI 1997 No 2776*), *Reg 6(3)(b)*.
- A chamber, tunnel or airlock which fall under the *Work in Compressed Air Regulations 1996* (*SI 1996 No 1656*).

## P7014  Pressure Systems

- A tank to which the *Carriage of Dangerous Goods by Rail Regulations 1996* (*SI 1996 No 2089*) apply.
- Any system to which the *Carriage of Dangerous Goods by Road Regulations 1996* (*SI 1996 No 2095*) apply.
- Any system being carried in a vehicle engaged international transport.
- A system that forms part of the braking, control or suspension system of a vehicle.
- Any water cooling system on an international combustion engine or compressor.
- Any tyre used or intended to be used on a vehicle.
- Any system fitted to a vehicle within the meaning of the *Road Traffic Act 1998, s 185*.
- Any portable fire extinguisher with a working pressure of less than 25 bars at 60°C.
- Any part of a tool or appliance designed to be held in the hand.

Some of these categories may, however, be subject to other regulations.

## HSC approved code of practice and guidance

**P7014**  The Regulations are largely non-prescriptive goal-setting regulations. For example, the Regulations state that pressure systems 'shall be properly designed and properly constructed from suitable material so as to prevent danger'. The goal is to prevent danger; what is 'proper' and 'sufficient' to achieve this goal is left to duty holders and ultimately the courts to decide.

The Health and Safety Commission's (HSC) Approved Code of Practice ('the ACoP') to the *Pressure Systems and Safety Regulations 2000* (*SI 2000 No 128*) includes guidance and a guide to the Regulations ('*Safety of Pressure Systems, Approved Code of Practice*', L122, ISBN 07176 1767 X, published by HSE Books). This document is essential reading for duty holders. It has special legal status in relation to compliance with the Regulations.

The ACoP gives practical advice on how to comply with the Regulations. Following the advice in the ACoP should be enough to ensure compliance with the law on specific matters, although alternative methods may be used. If the relevant provisions of the code are not followed, then it is necessary, in any prosecution for breach of the law, to show compliance with the law in other ways.

The additional guidance material is not compulsory but illustrates good practice. The guide sets out the law and explains what it means in simple terms.

## Interpretation

**P7015**  The following words and phrases have a specific defined meaning within the context of the Regulations:

- Competent person.
- Danger.
- Data.
- Examination.
- The Executive.
- Installed system.
- Mobile system.
- Owner.
- Pipeline.

- Pipework.
- Pressure system.
- Protective devices.
- Relevant fluid.
- Safe operating limits.
- Scheme of examination.
- Ship.
- System failure.
- Transportable pressure receptacle.
- User.

[*Pressure Systems Safety Regulations 2000* (*SI 2000 No 128*), *Reg 2*].

## Design, construction and installation

**P7016** The following persons have duties for ensuring that the design and construction of pressure systems and component parts meet the requirements of the Regulations:

(i) Designers.

(ii) Manufacturers.

(iii) Importers.

(iv) Suppliers.

[*Pressure Systems Safety Regulations 2000* (*SI 2000 No 128*), *Reg 4*].

The Regulations require that pressure systems shall be properly designed and properly constructed from suitable material so as to prevent danger. The design and construction must enable necessary examinations to be carried out for preventing danger from defects etc., and that where access to the inside of a pressure system is provided, it can be gained without danger to the inspector. Pressure systems must be provided with suitable protective devices (safety relief valves, bursting discs etc.) to prevent danger, and such devices must be capable of releasing the contents safely.

Pressure systems supplied with a CE marking under the *Pressure Equipment Regulations 1999* (*SI 1999 No 2001*) are presumed to have complied with this Regulation (see P7008.)

Employers of persons installing a pressure system have a duty to ensure that there is nothing about the way it is installed that could give rise to danger from the system or otherwise impair the operation of protective devices or the facility to examine the equipment. [The ACoP draws attention to a number of items that the employer should consider relating to the siting and access to and around the pressure system and the means and manner of installation. Special consideration needs to be given by the manufacturer to any in-service examinations that may require access to constricted areas and the specific requirements of *the Confined Space Regulations 1997* (*SI 1997 No 1713*).

## Documentation and marking

**P7017** The following information and documentation is required for each pressure system.

(*a*) *Information concerning the design, construction, examination, operation and maintenance*

**P7018** *Pressure Systems*

The Regulations require that pressure systems (or any component parts) are provided with written information concerning the design, construction, examination, operation and maintenance. Responsibility for providing this information falls on the person who designs or supplies (whether as manufacturer, importer or in some other capacity) the system. When a pressure system is modified or repaired, the responsibility to provide information about the modification or repair falls on the employer of the person who carried out the work or any self employed individual. [*Pressure Systems Safety Regulations 2000* (*SI 2000 No 128*), *Reg 5*].

The Regulations do not specify the information to be provided, but set the objective that it must be sufficient enough to comply with their provisions. The information must be provided with the design or with the supplied pressure system. When systems are modified or repaired, the information must be provided immediately afterwards. Although it is not possible to give a complete list of all the information which might be needed, the ACoP highlights items that should be considered.

(b) *A plate on each pressure vessel with the information given in Schedule 3 of the Regulations*

Manufacturers of pressure vessels must ensure that before the vessel is supplied, it is marked or has a plate attached containing the information specified in the *Pressure Systems Safety Regulations 2000* (*SI 2000 No 128*), *Sch 3*. This information includes the manufacturer's name, standard of construction, and the design pressure(s) and temperature(s). Similarly, no one may import a vessel unless it is marked or has a plate containing this information. The Regulations prohibit the removal or falsification of information contained on the plate or on any other mark on the vessel.

(c) *A statement of the safe operating limits*

The user of an installed pressure system or the owner of a mobile system will have established *safe operating limits* of the system before the system is operated. The nature of the safe operating limits is not specified by the Regulations and will depend on the system. They should, however, be such as to ensure that there is a suitable margin of safety before failure is liable to occur. Typically, operating limits may be set on pressure, temperature, number of cycles, or time in operation, but it is for the duty holder to determine what is appropriate.

In the case of mobile systems, the owner, if not the user, must supply the user with a written statement specifying the safe operating limits or ensure that the system is marked with this information in a legible, durable and visible way.

(d) *A written scheme of examination* (see P7010 below);

(e) *A report on each examination carried out under the written scheme* (see P7015 below);

(f) *Instructions for safe operation and action in emergencies* (see P7017 below).

Under certain circumstances, the following additional documentation is required:

(g) *An agreement to postpone the date of examination;*

(h) *A report in case of imminent danger* (see P7016 below);

(i) *Information required by the Pressure Equipment Regulations 1999* (*SI 1999 No 2001*);

(j) *A certificate exempting application of the requirements/prohibitions of the Regulations.*

## Competent persons

**P7018** Whilst users and owners have responsibility for the management of their pressure systems, the Regulations require and define the role and responsibilities of a 'competent person'.

The role of the competent person is primarily in connection with the examination and action in the case of imminent danger. The term 'competent person' is used in connection with two distinct functions:

- drawing up or certifying schemes of examination;
- carrying out examinations and making the examination report.

A single individual may carry out these functions, but equally, separate individuals can also carry them out. The user or owner may, however, also use the competent person in an advisory capacity in conjunction with other matters relating to the Regulations. For example, the competent person may assist users/owners in deciding the scope of the examination.

The level of expertise needed by the competent person depends on the size and complexity of the pressure system in question. The ACoP distinguishes between minor, intermediate and major systems and the attributes required of the competent person and back up specialist services and organisation required in each case. In all but minor systems, the competent person must be a chartered engineer.

For larger and more complex systems, it is unlikely that one individual will have sufficient knowledge and expertise. In this case the competent person should be part of a corporate body with a team of employees which have the necessary expertise and experience. The requirements for 'competence' then applies to the company and not to individual employees.

The following bodies may provide competent person services:

- A user company with its own in-house inspection department.
- An inspection organisation providing such services to clients.
- A self employed person.

The Competent Person should be either an individual or company with sufficient knowledge and experience to undertake the work. As a minimum standard the Competent Person should be able to demonstrate sufficient knowledge and experience to undertake the work required. A national certification scheme has been developed for bodies that provide competent person services known as the RG2, UKAS – Accreditation for In-Service Inspection of Pressure Systems/Equipment who will accredit competent persons to EN 45004:1995.

# Written scheme of examination

## Scope

**P7019** At the heart of the Regulations is the requirement for users of installed pressure systems or owners of mobile systems to have a 'suitable' written scheme of examination for every pressure system that they use or own. The user or the owner must ensure that the scheme is drawn up or certified as suitable by a competent person. The written scheme must specify:

- the parts to be examined (the scope);
- the maximum intervals between examinations (the frequency);
- the types of examinations necessary (the nature);
- any special measures necessary to prepare the system for safe examination.

Written schemes of examination may be held as electronic or other forms of data providing that the data is capable of being printed as a written copy and is kept secure from loss or unauthorised interference.

Responsibility for defining the scope of the written scheme of examination rests with the user or owner. The user or owner is likely to be most conversant with the equipment and relevant boundaries between systems. The competent person may be useful in assisting the user or owner to define the scope. The written scheme of examination must cover the relevant parts of the system which are specified by the Regulations as:

(a) All protective devices.

(b) Every pressure vessel and every pipeline in which a defect may give rise to danger.

(c) Those parts of pipework where a defect may give rise to danger.

The user or owner should first identify and establish which parts of the pressure system are pressure vessels, protective devices or pipework as defined by the Regulations. The user or owner will then decide which parts of the system should be included for examination. Where the decision is made to exclude parts from regular examination, users and owners should take advice from a person with appropriate and relevant technical expertise and experience (who need not be the competent person) and be able to justify the decision.

A written scheme of examination must be prepared before the pressure system is operated for the first time. It will specify the examination necessary before the pressure system enters service. The user or owner with the competent person will take account of fabrication and installation quality and inspection reports.

The user or owner must ensure that the scheme is reviewed at appropriate intervals by a competent person to determine whether it is still suitable for the current conditions of use of the system. This normally occurs after every examination. If, as a result of the review, the competent person recommends modifications to the content of the scheme, then the user or owner must ensure that the scheme is modified.

Schemes of examination must be suitable for the purposes of preventing danger arising from the system. Whilst it is for the duty holders to decide how to meet the goal of suitability and whether the scheme achieves it, the ACoP gives detailed advice about the factors that should be considered.

Further explanation regarding the preparation of written schemes of examination can be found in the Safety Assessment Federation (SAFed) Guidelines for the Production of Written Schemes of Examination and the Examination of Pressure Vessels Incorporating Openings to Facilitate Ready Internal Access Ref: PSG4

## *Frequency of examination*

P7020 The written scheme will specify a suitable frequency of examination i.e. the maximum intervals between successive examinations. The ACoP and Guidance give advice about the factors that should be taken into account when deciding the frequency between examinations. Separate considerations apply to the initial examination before entry to service, the first examination after entering service, and the regular examinations thereafter.

The Regulations do not prescribe intervals between examinations and therefore allow flexibility and freedom to exercise good engineering judgement in each case. Different parts of the system may be examined at different intervals depending on the degree of risk associated with each part. Much will depend on the extent and degree of confidence in the knowledge that is available about the predicted condition of the equipment, degradation mechanisms and rates, the operating conditions and operating history.

Advice on the frequency of examinations is available. The *Factories Act 1961* sets limits on the intervals between examinations of boilers and air receivers. These limits are still relevant good practice. The Institute of Petroleum Model Code of Practice (Parts 1 and 2) provides advice on examination periods for chemical pressure vessels, heat exchangers and other equipment depending on the results of earlier examinations. The Safety Assessment Federation (SAFed), which represents some of the major engineering insurers, has published recommendations on the periodicity of examination of pressure equipment (*'Pressure Systems Guidelines on the Periodicity of Examinations'* (PSG1), available from the Safety Assessment Federation, ISBN 1901212106).

## Protective device

Means a device that protects control or measuring equipment which is essential to prevent a dangerous situation from arising. Protective devices are often categorised as:

- Category 1: Devices designed to protect the system against system failure e.g. safety valve;
- Category 2: Devices that warn that a system failure could occur e.g. pressure gauge.

Protective devices in use on Georgia-Pacific GB and Ireland pressure systems are likely to include:

### For steam systems
- Safety (or relief) valves;
- Pressure gauges;
- Water level gauges;
- Low water level devices;
- Pressure limiting devices;
- High pressure cut out devices;
- Flame failure protection systems;
- Pressure reducing vales.

### For compressed air systems
- Safety (or relief) valves;
- Fusible plugs;
- Bursting discs;
- Pressure gauges;
- Pressure limiting or unloading devices.

### For bulk LPG installations
- Safety (or relief) valves;
- Pressure gauges;
- Bursting discs.

The HSE Guidance to the Regulations (L1222) deals with the scheduling of examinations of several identical vessels on the same site. The overriding principle is that each vessel must be examined within the interval specified by the written scheme for that vessel. Within this limit, a form of staged examination may be used.

## Nature of examination

**P7021**   The written scheme must specify the nature of the examination. This may vary, depending on the system, from a simple visual examination to a requirement for the use of non-destructive testing methods. Pressure vessels within a system will need to be examined from the inside when their accessibility permits this. The Safety Assessment Federation (SAFed), have a number of publications covering specialist examinations that provide the explanation and guidance that the competent person needs to follow for:

**P7022** *Pressure Systems*

- Guidelines for the examination of shell-to-endplate and furnace-to-endplate welded joints Ref: SBG1 (ISBN 1 901212 05);
- Guidelines for the examination of longitudinal seams of shell boilers Ref: SBG2 (ISBN 1 901212 30 0).

Some pressure systems will require an examination both with the system stripped down and out of service and with the system running under normal operating conditions so that the systems integrity and protective devices can be verified as operational to protect the equipment under fault conditions, e.g. Pressure relief valve lifts under an overpressure condition.

Other guidance documents produced by the Safety Assessment Federation (SAFed) that cover these areas of the thorough examination include:

- Pressure Systems: Guidelines on Periodicity of Examinations Ref: PSG1 (ISBN 1 901212 10 6);
- Guidelines for the Operation of Steam Boilers Ref: PSG2.

## Examination

**P7022** The Regulations require users and owners to ensure that their pressure systems are examined by a competent person as specified by the written scheme of examination. [*Pressure Systems Safety Regulations 2000* (*SI 2000 No 128*), *Reg 9*]. The examination must take place within the maximum time interval specified by the scheme. There is a clear duty on users/owners to ensure that the equipment is not operated/supplied beyond the date specified in the last examination report until it has been examined again.

Users and owners must ensure that systems are adequately prepared in time for examination and that appropriate precautions for safety are taken. Adequate advance notice should be given to the competent person.

There is provision to postpone the due date of examination once, by agreement in writing between the user or owner and the competent person. The date of examination may be postponed to a later date provided that the following conditions are satisfied:

(*a*)   The postponement does not give rise to danger.

(*b*)   Only one postponement is made for any one examination.

(*c*)   Such a postponement is notified in writing by the user or owner to the enforcing authority before the date set for the next examination in the previous report.

Where the user or owner is the competent person, such an agreement is unnecessary, but the notification to the enforcing authority must contain a declaration that postponement will not give rise to danger. Any postponement must be agreed by the Competent Person who undertook the previous examination. No periods of deferment are quoted within the Regulations, ACoP and Guidance (L122) but when justified the Competent Persons are unlikely to agree more than 25% of the frequency period.

The competent person must take responsibility for all examinations. Often the competent person will carry out the examination, but a user or owner and the competent person may agree that some examinations are carried out by specialist non-destructive testing technicians who are not classed as competent persons within the meaning of the Regulations. The competent person will accept responsibility for ensuring that the examinations are carried out properly according to the scheme and for the results, interpretation and reporting.

## Fitness for service

**P7023** After obtaining the results of the examination, the competent person will make an assessment of the continued fitness for service of the system. If the competent person is not satisfied about continued fitness for service under the current condition or safe operating limits of the system, then s/he will specify in the report any repairs, modifications, or changes to the safe operating limits deemed necessary.

The competent person will then, having regard to the assessment of continued fitness for service, specify the next date, within the limits set by the written scheme, beyond which the system may not be operated without a further examination. The competent person will also consider whether the written scheme remains suitable or whether it should be modified.

## Examination report

**P7024** The Regulations require the competent person to make a written report of the examination and to sign or otherwise authenticate it. [*Pressure Systems Safety Regulations 2000 (SI 2000 No 128), Reg 9*]. The competent person must send the report to the user or owner as soon as is practicable after completing the examination (or for integrated systems, the last examination in the series). In any event, the report must arrive within 28 days of the examination (or the last in the series). The report must also arrive before the date specified by the previous examination report. Where the user or owner is also the competent person, the report must also be made by these times.

The examination report must contain the following information:

(a) A statement of the parts examined, the condition of those parts and the results of the examination.

(b) Specification of any repairs, modifications or changes to the operating limits necessary to prevent danger or ensure the continued effective working of protective devices. The date by which any repairs, modifications or changes are to completed must also be specified. The user or owner must ensure that the system is not operated after the date specified for completion of the repairs, modifications or changes until these have been made.

(c) The next date (within the limits of the scheme of examination) after which the pressure system may not be operated without a further examination under the scheme. The owner of a mobile system must ensure that the date of the next examination is legibly and durably marked on the system and is clearly visible.

(d) A statement of whether the current written scheme of examination is suitable for preventing danger or whether it should be modified (with reasons).

## Imminent danger

**P7025** After the examination of a pressure system and assessment of fitness for service, the competent person must consider whether the condition of the system, or any part thereof, is such that continued operation in its current condition would give rise to imminent danger. If this is the case, then the competent person will immediately make a written report. [*Pressure Systems Safety Regulations 2000 (SI 2000 No 128), Reg 10*]. The report will specify the system and the parts in dangerous condition, and the repairs, modifications or changes to the operating conditions needed in order to make the system safe from imminent danger. A pressure system in a state of imminent danger usually points to wider failings in general management of the user's/owner's organisation.

The competent person will give the report to the user of an installed system or the owner and user of a mobile system. In addition, within 14 days of the examination, the competent person will send a report to the enforcing authority for the place where the pressure system is situated. This will normally be the Health and Safety Executive or the Environmental Health Department of the local authority.

On receipt of this report, the user of an installed system must ensure that the system (or the part effected) is not operated until the repairs, modifications or changes have been carried out. Owners of mobile systems must take all practicable steps to ensure that the system is not operated. Where the competent person is the user or owner, s/he must take responsibility to prepare the report, inform the enforcing authority and ensure that the system is not operated until the repairs, modifications or changes have been carried out.

## Operations, maintenance, modification and repair

### Safe operation

P7026    Users and owners have a duty to provide adequate and suitable operating instructions for the safe operation of the system and the action to be taken in an emergency. Users must ensure that their pressure systems are not operated except in accordance with the instructions.

Users of pressure systems must provide adequate training to persons who operate the systems as required by the *Provision and Use of Work Equipment Regulations 1998* (*SI 1998 No 2306*), Reg 9.

### Precautions to prevent overpressure

P7027    Users have a duty to take precautions to prevent overpressure to vessels. [*Pressure Systems Safety Regulations 2000* (*SI 2000 No 128, Reg 15*]. The Regulation applies to vessels constructed with a permanent outlet to the atmosphere (or to a space where the pressure does not exceed atmospheric pressure) which could become pressurised if the outlet were obstructed. In this case, the user must ensure that the outlet is kept open at all times and is free from obstruction when the vessel is in use.

### Maintenance

P7028    Users and owners must ensure that their systems are properly maintained so as to prevent danger. [*Pressure Systems Safety Regulations 2000* (*SI 2000 No 128, Reg 12*]. This should not be confused with the requirement for examinations under the written scheme. Maintenance might include the periodic replacement of parts, flange tightening, and checks on vulnerable areas of pipework systems not included for examination under the written scheme. Advice on maintenance programmes is given in the ACoP.

### Systems and equipment covered by the regulations

The HSE has warned of the need for effective operation, inspection and maintenance regimes to ensure the safe use of high-pressure equipment, following the prosecution of a chemical manufacturer.

The warning comes after Giancarlo Coletti, 48, a worker at Clariant Life Science Molecules (UK) Limited's plant in Sandycroft, Flintshire, sustained serious injuries to his right arm in October 2003. The accident happened when the clamping system on a pressure vessel lid he was operating failed, causing the lid to fly off and hit him.

Clariant Life Science Molecules (UK) Limited was fined £100,000, split equally between *Regulations 11* and *12* of the *Pressure Systems Safety Regulations 2000* (*SI 2000 No 128*), with £24,272 474 costs, at Mold Crown Court on 17 December 2004 after pleading guilty at an earlier hearing. The company admitted failing to provide workers operating the pressure vessel with adequate and suitable instructions for its safe operation and also failing to ensure the vessel was adequately maintained.

HSE Inspector, Dr Stuart Robinson, who investigated the incident, said:

> '*Our investigation revealed serious deficiencies in Clariant's safe systems of work. Opening and closing of the lid was regarded as a simple process, carried out three times a day, but the hazards had been overlooked. Although the clamping system was designed with a substantial margin of safety, it had been allowed to deteriorate to such an extent that the risk of injury became unacceptably high.*'.

In particular, the HSE's investigation found that Clariant had failed to put in place adequate operating procedures to ensure the system was used correctly. For example, clamps were regularly over-tightened, occasionally causing them to break and the system was allowed to operate with less than its full complement of eight clamps.

In addition, at the time of the accident, one of the clamps was missing and others showed excessive wear and tear, or inadequate repair. Furthermore, it had become common practice for leaks to be nipped up with the system under pressure because the operating procedures failed to state that the system should be depressurised first.

Robinson added:

*'The investigation also revealed that although Clariant had arranged for an independent competent person to examine the pressure system periodically, this was insufficient due to the frequent operation of the clamps, and the high level of wear and tear they showed. Instead, Instead, the firm should have introduced a more frequent system of inspection and maintenance.'.*

### Modification and repair

**P7029** Employers of persons who modify or repair pressure systems have to duty to ensure that the manner in which modifications or repairs are carried out does not give rise to danger, or otherwise impair the operation of any protective devices or inspection facility. [*Pressure Systems Safety Regulations 2000 (SI 2000 No 128), Reg 13*)].

## Record keeping

**P7030** The user of an installed system or the owner of a mobile system must keep the originals or copies of the following documents:

(a) The report of the last examination made by the competent person.

(b) Reports of previous examinations if they contain information that can assist in assessing whether the system is safe to operate or whether any repairs can be carried out safely.

(c) Any documents where reports of the last or previous examinations are recorded as data.

(d) Any agreement, declaration or notification made in relation to the postponement of the date of the next examination.

[*Pressure Systems Safety Regulations 2000 (SI 2000 No 128), Reg 14*].

For an installed system, these documents must be kept at the premises where the system is installed or at other premises approved by the enforcing authority (normally the Health and Safety Executive). For a mobile system, the documents must be kept at the premises from which the deployment of the system is controlled.

Where the user or owner of a pressure system changes, the previous user or owner must give to the new user or owner all originals or copies of documents relating to the system as soon as practicable.

## Related health and safety legislation

**P7031** The Regulations are concerned with the hazard of stored energy. The control of hazards associated with the release of hazardous substances, for example toxic or flammable materials, from containers and systems falls under the *Control of Major Accident Hazards Regulations 1999 (COMAH) (SI 1999 No 743)* as amended by *The Control of Major Accident Hazards (Amendment) Regulations 2005 (SI 2005 No 1088)*. Controls to prevent the release of hazardous materials cover the management of storage systems, containers and tanks.

Other legislation relating to pressure equipment includes:

- *The Carriage of Dangerous Goods and Use of Transportable Pressure Equipment Regulations 2004*
- *Pipelines Safety Regulations 1996 (SI 1996 No 825)*.

The main legislation covering health and safety at work in general is:

- *Health and Safety at Work etc Act 1974*;
- *Workplace (Health and Safety) Regulations 1992 (SI 1992 No 3004)*;
- *Management of Health and Safety at Work Regulations 1999 (SI 1999 No 3242)*;
- Provision and Use of Work Equipment Regulations 1998 (SI 1998 No 2306).

# Product Safety

## Introduction

**P9001** With the expansion of EU directives and other legislation aimed at manufacturers, designers, importers and suppliers, product safety emerges as a fast growth area – a development expedited by the introduction of the *General Product Safety Regulations 1994* (*SI 1994 No 2328*). Indeed, there is an identifiable trend towards placing responsibilities on producers and those involved in commercial circulation as well as on employers and occupiers (see MACHINERY SAFETY and PERSONAL PROTECTIVE EQUIPMENT). Although there is no question of removal of duties from employers and occupiers and users of industrial plant, machinery and products, there is a realisation that the *sine qua non* of compliance on the part of employers and users with their safety duties is compliance with essential health and safety requirements on the part of designers and manufacturers. This trend is likely to continue with more products being brought within the scope of the EU's 'new approach' regime which promotes supply of products throughout the single market provided they meet essential requirements which can be demonstrated by compliance with harmonised European standards. As far as civil liability is concerned, this trend may well result in a long-term shift in the balance of liability for injury away from employers and occupiers towards designers, manufacturers and other suppliers.

It was thalidomide that first focused serious attention on the legal control of product safety, leading to the Medicinal Products Directive (65/65/EEC) and the *Medicines Act 1968*, the first comprehensive regulatory system of product testing, licensing and vigilance. Since then a host of regulations has appeared, covering areas as diverse as pencils, aerosols, cosmetics and electrical products. The General Product Safety Directive (92/59/EEC), implemented by the *General Product Safety Regulations 1994* (*SI 1994 No 2328*), requires in a general way that producers of consumer goods place only safe products on the market and undertake appropriate post-marketing surveillance. The significance of these Regulations lies not in their substantive requirements (which are not at all detailed or especially onerous). Their importance lies more in the completion of an all-embracing consumer product safety regime across the European Union. Whilst specific regulatory requirements will take precedence (e.g. duties specified in sectoral directives or directives dealing with particular risks such as the Electromagnetic Compatibility Directive (89/336/EEC)) aspects of the *General Product Safety Regulations 1994* are meant to cover any gaps in the consumer protection network left by existing regulatory requirements.

The General Product Safety Directive was recently amended and re-issued as Directive 2001/95/EC to incorporate even more post-marketing obligations and powers for enforcing authorities which will apply to consumer products, even those already covered by sector-specific regulatory requirements. At the time of going to print the UK implementing legislation had not been adopted. Member States have until 15 January 2004 to implement the Directive.

The regulation of products for use in the workplace has developed along separate lines (although there is an increasing overlap by virtue of the 'new approach' directives, such as those for machinery or electromagnetic compatibility, which apply uniform requirements for consumer and workplace products albeit with separate enforcement regimes). Under the *Health and Safety at Work etc. Act 1974* (*HSWA 1974*), s 6 all articles and substances for use at work are required to be as safe as is reasonably practicable, and manufacturers and other suppliers are under obligation to provide information and warnings for safety in use. Various product-specific regulations have also been made under *HSWA 1974*, for example the *Chemicals (Hazard Information and Packaging for Supply) Regulations 1994 (SI 1994 No 3247)* (now revoked and re-enacted with amendments by SI 2002 No 1689) and the *Asbestos (Prohibitions) Regulations 1992 (SI 1992 No 3067)*).

As far as commercial considerations are concerned, quality and fitness for purpose of products has been an implied statutory term in contracts for sale of goods in legislation stemming from the late nineteenth century and which is now found in the *Sale of Goods Act 1979*, as amended by the *Sale and Supply of Goods Act 1994* and the *Sale of Goods (Amendment) Act 1995*. These 1994 amendments included a more focused definition of the standard of *satisfactory quality* which a supplier is obliged by contract to provide, the criteria for which now explicitly includes safety, durability, and freedom from minor defects. The ability to exclude these statutory terms, or insist that buyers waive the rights they have, is now strictly curtailed by the *Unfair Contract Terms Act 1977* (banning exclusion of the most significant rights in consumer contracts, and subjecting exclusions in standard form commercial contracts to a statutory test of 'reasonableness'). In addition, and based on the Directive on Unfair Terms and Consumer Contracts (93/13/EEC), there are now other restrictions contained in the *Unfair Terms in Consumer Contracts Regulations 1999 (SI 1999 No 2083)* imposing a requirement of fairness on most terms in consumer contracts which are not individually negotiated. These Regulations also lay down a requirement for suppliers to ensure that terms are expressed in plain, intelligible language. The most important restriction of all for present purposes in these two sets of Regulations is the effective ban on notices and contract terms seeking to exclude liability for death or personal injury resulting from negligence, a prohibition which is reinforced in relation to defective products by the *Consumer Protection Act 1987, s 7*.

This section deals with criminal and civil liabilities for unsafe products for consumer and industrial use as well as the duties imposed on sellers and suppliers of consumer products and after-sales personnel as far as contract law is concerned.

# Consumer products

## General Product Safety Regulations 1994

### Duties of producers and distributors

**P9002**     The *General Product Safety Regulations 1994 (SI 1994 No 2328)* lay down general requirements concerning the safety of products intended for or likely to be used by consumers. The Regulations will be superseded by UK legislation implementing the General Product Safety Directive 2001/95/EC which is required by the Directive to be brought into force by 15 January 2004. Directive 2001/95/EC replaces in its entirety the 1992 General Product Safety Directive 92/59/EEC and significantly increases the obligations on producers and distributors and has other important implications. In the meantime, the *General Product Safety Regulations 1994* remain in full force until that date. See P9008 for details of the revised Directive.

Products used solely in the workplace are not covered by the 1994 Regulations, but the fact that a product has a wider industrial application will not prevent it being a consumer product for the purposes of these Regulations if it is supplied to consumers. All manner of products are potentially caught, including clothing, primary agricultural products, DIY equipment and motor vehicles. What can be considered to be a consumer product will be extended even further by the new Directive (see P9008).

The Regulations do not apply to:

— second-hand products which are antiques;

— products supplied for repair or reconditioning before use, provided that the purchaser is clearly informed accordingly;

— products exported direct by the UK manufacturer to a country outside the European Union.

The Regulations are disapplied in relation to any product where there are specific provisions in European law governing *all aspects* of the safety of a product. Where there are product-specific provisions in European law which do not cover all aspects, the 1994 Regulations apply to the extent that specific provision is not made by European law. [*General Product Safety Regulations 1994 (SI 1994 No 2328), Reg 3(c)* and *Reg 4*].

The main requirements of the Regulations are that:

(a) no producer may place a product on the market unless it is safe [*General Product Safety Regulations 1994 (SI 1994 No 2328), Reg 7*];

(b) producers provide consumers with relevant information, so as to enable them (i.e. consumers) to assess risks inherent in products throughout the normal or foreseeable period of use, where such risks are not immediately obvious without adequate warnings, and to take precautions against those risks [*General Product Safety Regulations 1994 (SI 1994 No 2328), Reg 8(1)(a)*];

(c) producers update themselves regarding risks presented by their products and take appropriate action (if necessary recall and withdrawal), for example by:

   (i) identifying products/batches of products by marking,

   (ii) sample testing,

   (iii) investigating and following up *bona fide* complaints,

   (iv) keeping distributors informed accordingly

   [*General Product Safety Regulations 1994 (SI 1994 No 2328), Reg 8(1)(b), (2)*];

(d) in order to enable producers to comply with their duties, distributors must act with due care and in particular:

   (i) must not supply products to any person which are known or presumed (on the basis of information in their possession and as professionals) to be dangerous products, and

   (ii) must participate in monitoring safety of products placed on the market – in particular, by passing on information on product risks and co-operating in action to avoid risks.

   [*General Product Safety Regulations 1994 (SI 1994 No 2328), Reg 9(a)(b)*].

## *Presumption of conformity with safety requirements*

**P9003** A 'safe' product is a product which, under normal or reasonably foreseeable conditions of use (as well as duration), does not present any risk (or only minimal risks) compatible with the product's use, considered as acceptable and consistent with a high level of protection for the safety and health of consumers, with reference to:

(a) product characteristics (e.g. composition, packaging, assembly and maintenance instructions);

(b) the effect on other products, where use with other products is reasonably foreseeable;

(c) product presentation (i.e. labelling, instructions for use and disposal); and

(d) categories of consumers at serious risk – in particular, children.

*[General Product Safety Regulations 1994 (SI 1994 No 2328), Reg 2(1)].*

In this connection, where a product conforms with specific UK safety requirements (e.g. the *Plugs and Sockets etc (Safety) Regulations 1994 (SI 1994 No 1768)*) there is a presumption that the product is safe, until the contrary is proved. In the absence of specific regulations governing the health and safety aspects of a product, the assessment of safety is to be made by taking into account:

(i) voluntary national standards implementing European standards;

(ii) EU technical specifications, or, failing these,

— standards drawn up in the UK, or

— codes of practice relating to the product, or

— state of the art and technology

and the safety that consumers may reasonably expect.

*[General Product Safety Regulations 1994 (SI 1994 No 2328), Reg 10].*

## Offences, penalties and defences

**P9004**  Contravention of *Regs 7–9(a)* (above) is an offence, carrying, on summary conviction, a maximum penalty of:

(a) imprisonment for up to three months, or

(b) a fine of £5,000, or

(c) both.

*[General Product Safety Regulations 1994 (SI 1994 No 2328), Reg 12].*

## Preparatory acts by producers and distributors

**P9005**  No producer or distributor may:

(a) offer or agree to place on the market any dangerous product, or expose or possess such product for placing on the market, or

(b) offer or agree to supply any dangerous product, or expose or possess same for supply.

Contravention is an offence, carrying the same maximum penalties as breach of *Reg 12* (see P9004 above).

*[General Product Safety Regulations 1994 (SI 1994 No 2328), Reg 13].*

It is worth noting that it may be easier for an enforcement officer to bring a prosecution for the above offences than for the offence under *Reg 9(a)* for supplying products which are known or presumed (on the basis of information in their possession and as professionals) to be dangerous products. In the latter case, the prosecution has the burden of proving knowledge of the defect and also that the distributor did not act with due care. Trading Standards Officers may bring a prosecution on the basis of these offences even in a case where there has been an actual supply and there is evidence to suggest that this is already occurring.

## Offence by another person

**P9006**  As with breaches of *HSWA* and similarly-oriented legislation, where an offence is committed by 'another person' in the course of his commercial activity, that person can be charged, whether or not the principal offender is prosecuted. Similarly, where commission of an offence is consented to or connived at, or attributable to neglect, on the part of a director, manager or secretary, such persons can be charged in addition to or in lieu of the body corporate. *[General Product Safety Regulations 1994 (SI 1994 No 2328), Reg 15].*

## Defence of 'due diligence' – and exceptions

**P9007** It is a defence for a person charged under these Regulations to show that he took all reasonable steps and exercised all due diligence to avoid committing the offence. [*General Product Safety Regulations 1994 (SI 1994 No 2328), Reg 14(1)*]. (See further ENFORCEMENT.)

The exceptions to the above are:

(*a*) Where, allegedly, commission of the offence was due to:

  (i) the act or default of another, or

  (ii) reliance on information given by another,

  a person so charged cannot, without leave of the court, rely on the defence of 'due diligence', unless he has served a notice, within, at least, seven days before the hearing, identifying the person responsible for the commission or default. [*General Product Safety Regulations 1994 (SI 1994 No 2328), Reg 14(2)(3)*].

(*b*) A person so charged cannot rely on the defence of 'information supplied by another', unless he shows that it was reasonable in all the circumstances for him to have relied on the information. In particular, did he take steps to verify the information or did he have any reason to disbelieve the information? [*General Product Safety Regulations 1994 (SI 1994 No 2328), Reg 14(4)*].

(*c*) A distributor charged with an offence cannot rely on the defence where he has contravened *Reg 9(b)* concerning passing on information and co-operating in action to avoid product risks. [*General Product Safety Regulations 1994 (SI 1994 No 2328), Reg 14(5)*].

A mere recommendation on the part of an importer that labels should be attached to boxes by retailers does not constitute 'due diligence' for the purposes of *Reg 14(1)*. (In *Coventry County Council v Ackerman Group plc [1995] Crim LR 140*, an egg boiler imported by the defendant failed to contain instructions that eggs should be broken into the container and yolks pricked before being microwaved. The defendant had learned of the problem and had printed instructions which were sent to all retailers, recommending that they be fixed to the boxes.)

Compliance with recognised standards (such as British Standards) will not amount to the defence of due diligence if a product is nevertheless unsafe for the user. In *Whirlpool (UK) Ltd and Magnet Ltd v Gloucestershire County Council (1993, unreported)* cooker hoods which were intrinsically safe and which met applicable standards for the purposes of the *Low Voltage Electrical Equipment (Safety) Regulations 1989* failed to meet the general safety requirement contained in the then applicable consumer protection legislation because they were liable to result in fires when used in conjunction with certain gas hobs.

# General Product Safety Directive

**P9008** On 15 January 2004, UK legislation implementing the General Product Safety Directive 2001/95/EC will replace the *General Product Safety Regulations 1994 (SI 1994 No 2328)*. The following paragraphs explain the new obligations and other changes that the new Directive will bring.

## Duties of producers and distributors

**P9009** The Directive increases the scope of consumer products covered by it. The new definition (with new wording in italics) is as follows:

' ... any product – *including in the context of providing a service* – which is intended for consumers or likely, *under reasonably foreseeable conditions*, to be used by consumers even if not intended for them, and is supplied or made available, whether for consideration or not, in the course of a commercial activity, and whether new, used or reconditioned.'

This definition shall not apply to second-hand products supplied as antiques or as products to be repaired or reconditioned prior to being used, provided that the supplier clearly informs the person to whom he supplies the product to that effect'.

The scope of the Directive is therefore extended as follows:

1. It encompasses products which are made available to consumers, as well as supplied to them. This would include many products provided in beauty centres, theme parks or playgrounds or in connection with transport, sporting events or health treatments.
2. It includes products supplied or made available in the context of providing a service. This would include repairs and after-sales service, maintenance and cleaning work, hotel and restaurant services, health treatment, provision of gas and electricity etc or a tyre inflation gauge and compressed air pump at a garage.

In addition, new recital 10 makes clear that, although business products are not included *per se*, the new Directive does include products which are designed exclusively for professional use but have subsequently migrated to the consumer market, such as laser pens.

The new Directive includes a provision to prohibit the export from the European Community any products which have been the subject of a European Commission-initiative decision (after consulting the Member States and, whenever it proves necessary also a Community scientific committee) that requires Member States to take measures in relation to that particular product on the grounds that it presents a serious risk (unless the decision specifically provides that the products may still be exported outside the Community).

A number of significant obligations are added to those referred to above in P9002 and these are that:

(*a*) producers keep a register of complaints, if necessary.

   [*General Product Safety Directive 2001/95/EC, Article 5.1*]

(*b*) producers and distributors notify the competent authorities immediately if they know or ought to know, on the basis of information in their possession and as professionals, that a product they have placed on the market poses risks to the consumer that are compatible with the general safety requirement, i.e. that it is dangerous, and of any action taken to prevent such risks [*General Product Safety Directive 2001/95/EC, Article 5.3*];

(*c*) producers and distributors co-operate with the competent authorities, at the request of the latter, on action taken to avoid risks posed by products that they supply or have supplied.

   [*General Product Safety Directive 2001/95/EC, Article 5.4*]

(*d*) producers (as part of their duty to update themselves regarding risks and take appropriate action) adequately and effectively warn consumers or recall from consumers, if necessary (thus the recall obligation is extended beyond the chain of distribution and includes products already supplied or made available to the consumer/user).

   [*General Product Safety Directive 2001/95/EC, Article 5.1*]

(*e*) distributors keep and provide the documentation necessary for tracing the origin of products.

   [*General Product Safety Directive 2001/95/EC, Article 5.2*]

There has been no change in the policy that specific General Product Safety Directive provisions do apply to products regulated by other directives where such provisions are not covered by the particular sectoral directive applicable to the product. These new post-marketing obligations would, therefore, generally apply to most consumer products even if subject also to regulation under sectoral directives, but guidance is awaited from the European Commission to clarify exactly which provisions do not apply to specific-product sectors. It is intended that this guidance will be made available before the date for implementation of the new Directive into national law, so that Member States can incorporate its principles clearly into their national legislation. It may be that sectoral directives will in due course be amended to provide for similar obligations.

The European Commission is to produce a guide containing simple and clear criteria on notifications required in (b) above, together with the content and a standard form for notifications. These notification provisions constitute a major increase in regulation for producers and distributors. They are based on reporting requirements which apply in the United States of America and have arisen out of some frustration that Member States and the Commission have not been informed about product withdrawals in Europe until US authorities inform them that the product has been withdrawn from the US market.

The notification provisions will require producers and distributors to put in place sophisticated procedures to capture and assess safety information. Deciding whether new information means that a product is no longer legally safe can be a major issue in practice and is best undertaken against a background of previous safety data and an assessment that the product's safety is acceptable. Companies' systems may need to provide for obtaining independent expert advice on these issues from scientists and lawyers. For certain products, at least, it will be necessary to undertake a documented risk analysis before a product is placed on the market, and subsequently to keep it up to date, so as to define the level of anticipated risks with a product's use, to determine that these are acceptable and that the general safety requirement is met, and to provide a background and baseline statement of acceptable risks against which any increase in risk may be assessed.

Recalls in (d) above are only required as a last resort where other measures would not suffice to prevent the risks involved, where the producer considers it necessary or where they are obliged to do so by a Member State. Practical issues which may arise include: the ability of a system adequately to contact users over safety issues; to measure the number and percentage of products returned in a recall; and decide what level of returns would be acceptable. UK guidance may assist to an extent (see *Consumer Product Recall: A Good Practice Guide* (1999) Department of Trade and Industry).

Complementary to the obligation to recall products from consumers in (d) above is a new power for Member States to order or organise the immediate withdrawal of any dangerous product already on the market and alert consumers to the risks it presents and to order or co-ordinate or, if appropriate, to organise with producers and distributors its recall from consumers and its destruction in suitable conditions. [*General Product Safety Directive 2001/95/EC, Article 8.1(f)*].

## *Presumption of conformity with safety requirements*

**P9010**  What will be considered to be a safe product is slightly different in the new Directive. In the factors to be taken into account when considering the safety of a product, the new Directive requires that: the instructions for assembly and, where applicable, for installation and maintenance are taken into account in the characteristics of the product; and also that the elderly are a particular category of consumer to consider when taking into account the categories of consumers at risk when using the product, along with children. The factors to be taken into account are in particular:

(a)  product characteristics (e.g. composition, packaging, instructions for assembly and, where applicable, for installation and maintenance);

(b)  the effect on other products, where use with other products is reasonably foreseeable;

(c)  product presentation (i.e. labeling, warnings, instructions for use and disposal, any other indication of information regarding the product); and

(d)  categories of consumers at serious risk – in particular, children and the elderly.

*[General Product Safety Directive 2001/95/EC, Article 2(b)].*

There is a new provision in the new Directive introducing a system whereby European harmonised standards can be mandated under the Directive and will confer a presumption of conformity with the Directive's safety obligations, in the same way as applies under many other product regulatory directives. Reference to such standards will be published in the *Official Journal of the European Communities* with particular reference to this Directive.

A product will still be deemed safe if it conforms with specific national safety requirements, unless the contrary is proved. In the absence of specific harmonised standards mandated and listed in the Official Journal or specific regulations governing the health and safety aspects of a product, the assessment of safety is to be made by taking into account:

(i) voluntary national standards implementing European standards other than those listed in the Official Journal with reference to the General Product Safety Directive; or

(ii) standards drawn up in the Member State in which the product is marketed; or

(iii) European Commission recommendations setting guidelines on product safety assessment; or

(iv) product safety codes of good practice in the relevant sector; or

(v) state of the art and technology; or

(vi) the safety that consumers may reasonably expect.

*[General Product Safety Directive 2001/95/EC, Article 3.3].*

## Market surveillance and enforcement

**P9011** The Directive's market surveillance and enforcement provisions have been strengthened and there are increased obligations on Member States. The European Commission also has an obligation to promote and take part in the operation of a European network of authorities of the Member States.

The obligations and enforcement powers of Member States have been expanded to include the following elements:

(*a*) a detailed definition of the tasks, powers, organisation and cooperation arrangements for market surveillance of competent authorities;

(*b*) establishment of sectoral surveillance programmes by categories of products or risks and the monitoring of surveillance activities, findings and results;

(*c*) the follow-up and updating of scientific and technical knowledge concerning the safety of products;

(*d*) periodic review and assessment of the functioning of the control activities and, if necessary, revision of the approach and organisation of surveillance;

(*e*) procedures to receive and follow-up complaints from consumers and others on product safety, surveillance or control activities;

(*f*) exchange of information between Member States on risk assessment, dangerous products, test methods and test results, recent scientific developments and other aspects relevant for control activities;

(*g*) joint surveillance and testing projects between surveillance authorities;

(*h*) exchange of expertise and best practices, and cooperation in training activities;

(*i*) improved collaboration at Community level on tracing, withdrawal and recall of dangerous products.

*[General Product Safety Directive 2001/95/EC, Articles 6–10]*

An extended list of enforcement powers is included in the new Directive and includes the power for a Member State to order or organise the issuance of warnings about, or recall of, dangerous products. [*General Product Safety Directive 2001/95/EC, Article 8.1*].

## Other significant changes

**P9012** Other significant changes can be summarised as follows:

1. Member States are required to take due account of the precautionary principle when they are taking enforcement measures. The precautionary principle is that Member States should use a precautionary approach to risks where there is scientific uncertainty as to the level of risk.

   [*General Product Safety Directive 2001/95/EC, Article 8.2*]

2. Information which is available to the authorities relating to consumer health and safety, in particular on product identification, the nature of any risk and on measures taken, shall in general be available to the public, in accordance with the requirements of transparency. However, information which is obtained by the authorities shall not be disclosed if the information, by its nature, is covered by professional secrecy in duly justified cases.

   [*General Product Safety Directive 2001/95/EC, Article 16.1*].

# Industrial products

**P9013** The *HSWA 1974* was the first Act to place a *general* duty on designers and manufacturers of industrial products to design and produce articles and substances that are safe and without health risks when used at work. Prior to this date legislation had tended to avoid this approach, e.g. the *Factories Act 1961*, on the premise that machinery could not be made design safe (see M1001 MACHINERY SAFETY). Statutory requirements had tended to concentrate on the duty to guard and fence machinery, and with the placement of a duty upon the user/employer to inspect and test inward products for safety. There is a general residual duty on the employers/users of industrial products to inspect and test them for safety under *HSWA 1974, s 2*.

These duties notwithstanding, the trend of legislation in recent years has been towards safer design and manufacture of products for industrial and domestic use. Thus, *HSWA 1974, s 6* as updated by the *Consumer Protection Act 1987, Sch 3* imposes general duties on designers, manufacturers, importers and suppliers of products ('articles and substances') for use at work, to their immediate users. Contravention of s 6 carries with it a maximum fine, on summary conviction, of £20,000 or an unlimited fine in the Crown Court (see ENFORCEMENT). In addition, a separate duty is laid on installers of industrial plant and machinery. However, because of their involvement in the key areas of design and manufacture, more onerous duties are placed upon designers and manufacturers of articles and substances for use at work than upon importers and suppliers, who are essentially concerned with distribution and retail of industrial products. However, under *HSWA 1974, s 6(8A)* importers are made liable for the first time for the faults of foreign designers/manufacturers.

There has been a recent interesting development in case law regarding the interaction between *HSWA 1974*, European directives concerning the placing on the market and putting into service of certain product types, and the UK legislation which implements the European directives. In the case of *R (on the application of Junttan OY) v Bristol Magistrates' Court [2002] EWHC 566 (Admin)*, concerning a judicial review of a decision of the Bristol Magistrates' Court involving a prosecution brought by the HSE based on offences under *sections 3* and *6* of *HSWA 1974* regarding machinery, the court held that it was not open to the HSE to bring proceedings under *HSWA 1974, s 6*. The court ruled that the HSE should have brought proceedings under the UK Regulations implementing the European directive applicable to such machinery. The basis of the decision is that, where there is a specific statutory offence under regulations implementing a

European directive applicable to the product in question and where the offence under the regulations covers exactly the same ground as *HSWA 1974, s 6*, the offence under the regulations is the offence which gives effect to the directive and proceedings should be brought under the regulations and not *HSWA 1974, s 6*. Fines are lower under the regulations.

## Regulations made under the Health and Safety at Work etc Act 1974

**P9014** Various safety regulations are made under the umbrella of *HSWA 1974* and these often implement European health and safety directives. The *Provision and Use of Work Equipment Regulations 1998 (SI 1998 No 2306)* are such Regulations which implemented European Directive 89/655/EEC.

The 1998 Regulations place obligations on employers and on certain persons having control of work equipment, or of persons who use or supervise or manage the use of work equipment or of the way in which the equipment is used. The key requirement is to ensure that work equipment is maintained in an efficient state, in efficient working order and in good repair. [*Provision and Use of Work Equipment Regulations 1998 (SI 1998 No 2306), Reg 5 (1)*]. It was recently confirmed that this imposes strict liability on the employer (*Stark v The Post Office, The Times, 29 March 2000*) and that, where a workplace product injures someone, the employer will be liable even in situations where through wear and tear the product has become dangerous. The employer's duty to maintain translates into a duty to ensure that there are no latent defects.

## Criminal liability for breach of statutory duties

**P9015** This refers to duties laid down in *HSWA 1974* as revised by the *Consumer Protection Act 1987, Sch 3*. The duties exist in relation to articles and substances for use at work and fairground equipment. There are no civil claims rights available to employees under these provisions except for breach of safety regulations. (For the specific requirements now applicable to machinery for use at work, see MACHINERY SAFETY.)

### Definition of articles and substances

**P9016** An article for use at work means:

(*a*) any plant designed for use or operation (whether exclusively or not) by persons at work; and

(*b*) any article designed for use as a component in any such plant.

*[HSWA 1974, s 53(1)]*.

A substance for use at work means 'any natural or artificial substance (including microorganisms), whether in solid or liquid form or in the form of a gas or vapour'. [*HSWA 1974, s 53(1)*].

An article upon which first trials/demonstrations are carried out is not an article for use at work, but rather an article which *might* be used at work. The purpose of trial/demonstration was to determine whether the article could safely be later used at work (*McKay v Unwin Pyrotechnics Ltd, The Times, 5 March 1991* where a dummy mine exploded, causing the operator injury, when being tested to see if it would explode when hit by a flail attached to a vehicle. It was held that there was no breach of *HSWA 1974, s 6(1)(a)*).

## Duties in respect of articles and substances for use at work

**P9017**  HSWA 1974, s 6 (as amended by the *Consumer Protection Act 1987, Sch 3*) places duties upon manufacturers and designers, as well as importers and suppliers, of (*a*) articles and (*b*) substances for use at work, whether used exclusively at work or not (e.g. lawnmower, hair dryer).

### Articles for use at work

**P9018**  Any person who designs, manufactures, imports or supplies any article for use at work (or any article of fairground equipment) must:

(*a*)  ensure, so far as is reasonably practicable (for the meaning of this expression see E15031 ENFORCEMENT), that the article is so designed and constructed that it will be safe and without risks to health at all times when it is being (i) set, (ii) used, (iii) cleaned or (iv) maintained by a person at work;

(*b*)  carry out or arrange for the carrying out of such testing and examination as may be necessary for the performance of the above duty;

(*c*)  take such steps as are necessary to secure that persons supplied by that person with the article are provided with adequate information about the use for which the article is designed or has been tested and about any conditions necessary to ensure that it will be safe and without risks to health at all such times of (i) setting, (ii) using, (iii) cleaning, (iv) maintaining *and* when being (v) dismantled, or (vi) disposed of; and

(*d*)  take such steps as are necessary to secure, so far as is reasonably practicable, that persons so supplied are provided with all such revisions of information as are necessary by reason of it becoming known that anything gives rise to a serious risk to health or safety.

*[HSWA 1974, s 6(1)(a)–(d) as amended by the Consumer Protection Act 1987, Sch 3].*

(See P9020 and P9021 below for further duties relevant to articles for use at work.)

In the case of an article for use at work which is likely to cause an employee to be exposed to 85 dB(A) or above, or to peak action level (200 pascals) or above, adequate information must be provided about noise likely to be generated by that article. [*Noise at Work Regulations 1989 (SI 1989 No 1790), Reg 12*] (see further NOISE AND VIBRATION).

### Substances for use at work

**P9019**  Every person who manufactures, imports or supplies any substance must:

(*a*)  ensure, so far as is reasonably practicable, that the substance will be safe and without risks to health at all times when it is being (i) used, (ii) handled, (iii) processed, (iv) stored, or (v) transported by any person at work or in premises where substances are being installed;

(*b*)  carry out or arrange for the carrying out of such testing and examination as may be necessary for the performance of the duty in (*a*);

(*c*)  take such steps as are necessary to secure that persons supplied by that person with the substance are provided with adequate information about:

    (i)  any risks to health or safety to which the inherent properties of the substance may give rise;

    (ii)  the results of any relevant tests which have been carried out on or in connection with the substance; and

    (iii)  any conditions necessary to ensure that the substance will be safe and without risks to health at all times when it is being (*a*) used, (*b*) handled, (*c*) processed, (*d*) stored, (*e*) transported and (*f*) disposed of; and

**P9020** *Product Safety*

(d) take such steps as are necessary to secure, so far as is reasonably practicable, that persons so supplied are provided with all such revisions of information as are necessary by reason of it becoming known that anything gives rise to a serious risk to health or safety.

[*HSWA 1974, s 6(4)* as amended by the *Consumer Protection Act 1987, Sch 3*].

## Additional duty on designers and manufacturers to carry out research

**P9020** Any person who undertakes the design or manufacture of any article for use at work must carry out, or arrange for the carrying out, of any necessary research with a view to the discovery and, so far as is reasonably practicable, the elimination or minimisation of any health or safety risks to which the design or article may give rise. [*HSWA 1974, s 6(2)*].

## Duties on installers of articles for use at work

**P9021** Any person who erects or installs any article for use at work in any premises where the article is to be used by persons at work, must ensure, so far as is reasonably practicable, that nothing about the way in which the article is erected or installed makes it unsafe or a risk to health when it is being (*a*) set, (*b*) used, (*c*) cleaned, or (*d*) maintained by someone at work. [*HSWA 1974, s 6(3)* as amended by the *Consumer Protection Act 1987, Sch 3*].

## Additional duty on manufacturers of substances to carry out research

**P9022** Any person who manufactures any substance must carry out, or arrange for the carrying out, of any necessary research with a view to the discovery and, so far as is reasonably practicable, the elimination or minimisation of any health/safety risks at all times when the substance is being (*a*) used, (*b*) handled, (*c*) processed, (*d*) stored, or (*e*) transported by someone at work. [*HSWA 1974, s 6(5)* as amended by the *Consumer Protection Act 1987, Sch 3*].

## No duty on suppliers of industrial articles and substances to research

**P9023** It is not necessary to repeat any testing, examination or research which has been carried out by designers and manufacturers of industrial products, on the part of importers and suppliers, in so far as it is reasonable to rely on the results. [*HSWA 1974, s 6(6)*].

## Custom built articles

**P9024** Where a person designs, manufactures, imports or supplies an article for or to another person on the basis of a written undertaking by that other to ensure that the article will be safe and without health risks when being (*a*) set, (*b*) used, (*c*) cleaned, or (*d*) maintained by a person at work, the undertaking will relieve the designer/manufacturer etc. from the duty specified in *HSWA 1974, s 6(1)(a)* (see P9017 above), to such extent as is reasonable, having regard to the terms of the undertaking. [*HSWA 1974, s 6(8)* as amended by the *Consumer Protection Act 1987, Sch 3*].

## Importers liable for offences of foreign manufacturers/designers

**P9025** In order to give added protection to industrial users from unsafe imported products, the *Consumer Protection Act 1987, Sch 3* has introduced a new subsection (*HSWA 1974, s 6(8A)*) which, in effect makes importers of unsafe products liable for the acts/omissions of foreign designers and manufacturers. *Section 6(8A)* states that nothing in (*inter alia*) *s 6(8)* is to relieve an importer of an article/substance from any of his duties, as regards anything done (or not done) or within the control of:

(*a*) a foreign designer; or

(*b*) a foreign manufacturer of an article/substance.

*[HSWA 1974, s 6(8A)]*.

## Proper use

**P9026** The original wording of *HSWA 1974, s 6(1)(a), 6(4)(a)* and *6(10)* concerning 'proper use' excluded 'foreseeable user error' as a defence, which had the consequence of favouring the supplier. Thus, if a supplier could demonstrate a degree of operator misuse or error, however reasonably foreseeable, the question of initial product safety was side stepped. Moreover, 'when properly used' implied, as construed, that there could only be a breach of *s 6* once a product had actually been *used*. This was contrary to the principle that safety should be built into design/production, rather than relying on warnings and disclaimers. The wording in *s 6* is now amended so that only *unforeseeable* user/operator error will relieve the supplier from liability; he will no longer be able to rely on the strict letter of his operating instructions. [*Consumer Protection Act 1987, Sch 3*].

## Powers to deal with unsafe imported goods

**P9027** *HSWA 1974* does not empower enforcing authorities to stop the supply of unsafe products at source or prevent the sale of products by foreign producers after they have been found to be unsafe, but enforcement officers have the power to act at the point of entry or anywhere else along the distribution chain to stop unsafe articles/substances being imported by serving prohibition notices (see ENFORCEMENT).

Customs officers can seize any imported article/substance, which is considered to be unsafe, and detain it for up to two (working) days. [*HSWA s 25A* (incorporated by *Schedule 3* to the *Consumer Protection Act 1987*)].

In addition, customs officers can transmit information relating to unsafe imported products to HSE inspectors. [*HSWA 1974, s 27A* (incorporated by the *Consumer Protection Act 1987, Sch 3*)].

# Civil liability for unsafe products – historical background

**P9028** Originally at common law where defective products caused injury, damage and/or death, redress depended on whether the injured user had a contract with the seller or hirer of the product. This was often not the case, and in consequence many persons, including employees repairing and/or servicing products, were without remedy. This rule, emanating from the decision of *Winterbottom v Wright (1842) 10 M & W 109*, remained unchanged until 1932, when *Donoghue v Stevenson [1932] AC 562* was decided by the House of Lords. This case was important because it established that manufacturers were liable in negligence (i.e. tort) if they failed to take reasonable care in the manufacturing and marketing of their products, and in consequence a user suffered injury when using the product in a reasonably foreseeable way. More particularly, a 'manufacturer of products, which he sells in such a form as to show that he intends them to reach the ultimate consumer in the form in which they left him with no reasonable possibility of intermediate inspection, and with the knowledge that the absence of reasonable care in the preparation or putting up of the products will result in an injury to the consumer's life or property, owes a duty to the consumer to take reasonable care' (per Lord Atkin). In this way, manufacturers of products which were defective were liable in negligence to users and consumers of their products, including those who as intermediaries, repair, maintain and service industrial products, it being irrelevant whether there was a contract between manufacturer and user (which normally there was not).

*Donoghue v Stevenson* is a case of enormous historical importance in the context of liability of manufacturers, but although the principle has become well established and is still widely applied it did not give a reliable remedy to injured persons, who still had to satisfy the legal burden of proving that the manufacturer had not exercised reasonable care, a serious obstacle to overcome in cases involving technically complex products. In some instances it became possible to avoid this obstacle, by establishing liability on other bases.

## Defective equipment supplied to employees

**P9029** At common law an employer obtaining equipment from a reputable supplier was unlikely to be found liable to an employee if the equipment turned out to be defective and injured him (*Davie v New Merton Board Mills Ltd [1959] AC 604*). Although it represented no bar to employees suing manufacturers direct for negligence, the effect of this decision was reversed by the *Employers' Liability* (*Defective Equipment*) *Act 1969* rendering employers strictly liable (irrespective of negligence) for any defects in equipment causing injury. (See E11004–E11005 EMPLOYERS' DUTIES TO THEIR EMPLOYEES). In such circumstances the employer would be able to claim indemnity for breach of contract by the supplier of the equipment.

## Breach of statutory duty

**P9030** Whilst legislation and regulations dealing with domestic and industrial safety are principally penal and enforceable by state agencies (trading standards officers and health and safety inspectors) if injury, death or damage occurs in consequence of a breach of such statutory duty, it may be possible to use this breach as the basis of a civil liability claim. Here there may be strict liability if there are absolute requirements, or the duty may be defined in terms of what it is practicable or reasonably practicable to do (see E15031 ENFORCEMENT).

As far as workplace products are concerned, *HSWA 1974, s 47* bars a right of action in civil proceedings in respect of any failure to comply with *HSWA 1974, s 6* (or the other general duties under the Act). However a breach of a duty imposed by health and safety regulations will generally be actionable in this way, unless the particular regulations in question contain a proviso to the contrary.

The position is the same in relation to consumer safety regulations made under the *Consumer Protection Act 1987* (*Consumer Protection Act 1987, s 41(1)*). No provision is made in the *General Product Safety Regulations 1994* (*SI 1994 No 2328*) for any breach thereof to give rise to civil liability for the benefit of an injured person. Given that the Regulations stem from European law it is unlikely that they would be construed in such a way as to give more extensive rights than those contained in the Product Liability Directive (85/374/EEC) (see P9031).

# Consumer Protection Act 1987

**P9031** At European level it was deemed necessary to introduce a degree of harmonisation of product liability principles between Member States, and at the same time to reduce the importance of fault and negligence concepts in favour of liability being determined by reference to 'defects' in a product. Thus the nature of the product itself would become the key issue, not the conduct of the manufacturer. After protracted debate the Product Liability Directive (85/374/EEC) was adopted.

## Introduction of strict product liability

**P9032** The introduction of strict product liability is enshrined in Britain within the *Consumer Protection Act 1987, s 2(1)*. Thus, where any damage is *caused* wholly or partly by a defect (see P9035 below) in a product (e.g. goods, electricity, a component product or raw materials), the following may be liable for damages (irrespective of negligence):

(a) the producer;

(b) any person who, by putting his name on the product or using a trade mark (or other distinguishing mark) has held himself out as the producer;

(c) any person who has imported the product into a Member State from outside the EU, in the course of trade/business.

*[Consumer Protection Act 1987, s 2(1)(2)].*

## Producers

**P9033** Producers are variously defined as:

(a) the person who manufactured a product;

(b) in the case of a substance which has not been manufactured, but rather won or abstracted, the person who won or abstracted it;

(c) in the case of a product not manufactured, won or abstracted, but whose essential characteristics are attributable to an industrial process or agricultural process, the person who carried out the process.

*[Consumer Protection Act 1987, s 1(2)(c)].*

## Liability of suppliers

**P9034** Although producers are principally liable, intermediate suppliers can also be liable in certain circumstances. Thus, any person who supplied the product is liable for damages if:

(a) the injured person requests the supplier to identify one (or more) of the following:

   (i) the producer,

   (ii) the person who put his trade mark on the product,

   (iii) the importer of the product into the EU; and

(b) the request is made within a reasonable time after damage/injury has occurred *and* it is not reasonably practicable for the requestor to identify the above three persons; and

(c) within a reasonable time after receiving the request, the supplier fails either:

   (i) to comply with the request, or

   (ii) identify his supplier.

*[Consumer Protection Act 1987, s 2(3)].*

Importers of products into the EU and persons applying their name, brand or trade mark will also be directly liable as if they were original manufacturers. *[Consumer Protection Act 1987, s 2(2)].*

## Defect – key to liability

**P9035** Liability presupposes that there is a defect in the product, and indeed, existence of a defect is the key to liability. Defect is defined in terms of the absence of safety in the product.

More particularly, there is a 'defect in a product ... if the safety of the product is not such as persons generally are entitled to expect' (including products comprised in that product). [*Consumer Protection Act 1987, s 3(1)*].

Defect can arise in one of three ways and is related to:

(a) construction, manufacture, sub-manufacture, assembly;

(b) absence or inadequacy of suitable warnings, or existence of misleading warnings or precautions;

(c) design.

The definition of 'defect' implies an entitlement to an expectation of safety on the part of the consumer, judged by reference to *general* consumer expectations not individual subjective ones. (The American case of *Webster v Blue Ship Tea Room Inc 347 Mass 421, 198 NE 2d 309 (1964)* is particularly instructive here. The claimant sued in a product liability action for a bone which had stuck in her throat, as a result of eating a fish chowder in the defendant's restaurant. It was held that there was no liability. Whatever her own expectations may have been, fish chowder would not be fish chowder without some bones and this is a general expectation.)

The court will require a claimant to be specific and be particular about exactly what the alleged defect is (*Paul Sayers and Others v SmithKline Beecham plc and Others (MMR/MR litigation) [1999] MLC 0117*).

## Consumer expectation of safety – criteria

**P9036** The general consumer expectation of safety must be judged in relation to:

(a) the marketing of the product, i.e.:

  (i) the manner in which; and

  (ii) the purposes for which the product has been marketed;

  (iii) any instructions/warnings against doing anything with the product; and

(b) by what might reasonably be expected to be done with or in relation to the product (e.g. the expectation that a sharp knife will be handled with care); and

(c) the time when the product was supplied (e.g. a product's shelf-life).

A defect cannot arise retrospectively by virtue of the fact that, subsequently, a safer product is made and put into circulation. [*Consumer Protection Act 1987, s 3(2)*].

## Time of supply

**P9037** Liability attaches to the *supply* of a product (see P9041 below). More particularly, the producer will be liable for any defects in the product existing at the time of supply (see P9045(d) below); and where two or more persons collaborate in the manufacture of a product, say by submanufacture, either and both may be liable, that is, severally and jointly (see P9043 below).

## Contributory negligence

**P9038** A person who is careless for his own safety is probably guilty of contributory negligence and, thus, will risk a reduction in damages. [*Consumer Protection Act 1987, s 6(4)*]. (See also E11010 EMPLOYERS' DUTIES TO THEIR EMPLOYEES.) However, in the product liability context, carelessness of the user may mean that there is no liability at all on the part of the producer. If, for example, clear instructions and warnings provided with the product had been disregarded, when compliance would have avoided the accident, it is highly unlikely that the product would be found to be defective for the purposes of the Act. No off-setting of damages for contributory fault of the claimant would arise.

## Absence or inadequacy of suitable/misleading warnings

**P9039** The common law required that the vendor of a product should point out any latent dangers in a product which he either knew about or ought to have known about. Misleading terminology/labelling on a product or product container could result in liability for negligence (*Vacwell Engineering Ltd v BDH Chemicals Ltd [1969] 3 All ER 1681* where ampoules containing boron tribromide, which carried the warning 'Harmful Vapours', exploded on contact with water, killing two scientists. It was held that this consequence was reasonably foreseeable and accordingly the defendants should have researched their product more thoroughly). In a similar product liability action today, the manufacturers would be strictly liable (subject to statutory defences) and, if injury/damage followed the failure to issue a written/pictorial warning, as required by law (e.g. the *Chemicals* (*Hazard Information and Packaging for Supply*) *Regulations 2002* (*SI 2002 No 1689*)), there would be liability.

The duty, on manufacturers, to research the safety of their products, before putting them into circulation (see *Vacwell Engineering Ltd v BDH Chemicals Ltd*) is even more necessary and compulsory now, given the introduction of strict product liability. This includes safety in connection with directions for use on a product. A warning refers to something that can go wrong with the product; directions for use relate to the best results that can be obtained from products, if the directions are followed. In the absence of case law on the point it is reasonable to assume that in order to avoid actions for product liability, manufacturers should provide both warnings, indicating the worst results and dangers, and directions for use, indicating the best results; the warning, in effect, identifying the worst consequences that could follow if directions for use were not complied with.

In the case of *Worsley v Tambrands Ltd (No 2) [2000] MCR 0280*, it was held that a tampon manufacturer had done what was reasonable in all the circumstances to warn a woman about the risk of toxic shock syndrome from tampon use. They had placed a clear legible warning on the outside of the box directing the user to the leaflet. The leaflet was legible, literate and unambiguous and contained all the material necessary to convey both the warning signs and the action required if any risk were present. The manufacturer could not cater for lost leaflets or for those who chose not to replace them. This give valuable guidance as to the extent that manufacturers are expected to warn users of their products of the risks associated with their products.

## Defect must exist when the product left the producer's possession

**P9040** This situation tends to be spotlighted by alteration of, modification to or interference with a product on the part of an intermediary, for instance, a dealer or agent. If a product leaves an assembly line in accordance with its intended design, but is subsequently altered, modified or generally interfered with by an intermediary, in a manner outside the product's specification, the manufacturer is probably not liable for any injury so caused. In *Sabloff v Yamaha Motor Co 113 NJ Super 279, 273 A 2d 606 (1971)*, the claimant was injured when the wheel of his motor-cycle locked, causing it to skid, then crash. The manufacturer's specification stipulated that the dealer attach the wheel of the motor-cycle to the front fork with a nut and bolt, and this had not been done properly. It was held that the dealer was liable for the motor-cyclist's injury (as well as the assembler, since the latter had delegated the function of tightening the nut to the dealer and it had not been properly carried out).

### Role of intermediaries

**P9041**  If a defect in a product is foreseeably detectable by a legitimate intermediary (e.g. a retailer in the case of a domestic product or an employer in the case of an industrial product), liability used to rest with the intermediary rather than the manufacturer, when liability was referable to negligence (*Donoghue v Stevenson [1932] AC 562*). This position does not duplicate under the *Consumer Protection Act 1987*, since the main object of the legislation is to fix producers with strict liability for injury-causing product defects to users. Nevertheless, there are common law and statutory duties on employers to inspect/test inward plant and machinery for use at the workplace, and failure to comply with these duties may make an employer liable. In addition, employers, in such circumstances, can incur liability under the *Employers' Liability (Defective Equipment) Act 1969* (see E11005 EMPLOYERS' DUTIES TO THEIR EMPLOYEES) and so may seek to exercise contractual indemnity against manufacturers.

### Comparison with negligence

**P9042**  The similarity between product liability and negligence lies in causation. Defect must be the material *cause* of injury. The main arguments against this are likely to be along the lines of misuse of a product by a user (e.g. knowingly driving a car with defective brakes), or ignoring warnings (a two-pronged defence, since it also denies there was a 'defect'), or – as is common in chemicals and pesticides cases – a defence based on alternative theories of causation of the claimant's injuries.

Product liability differs from negligence in that it is no longer necessary for injured users to prove absence of reasonable care on the part of manufacturers. All that is now necessary is proof of (*a*) defect (see P9035 above) and (*b*) that the defect caused the injury. It will, therefore, be no good for manufacturers to point to an unblemished safety record and/or excellent quality assurance programmes, or the lack of foreseeability of the accident, since the user is not trying to establish negligence. How or why a defect arose is immaterial; what is important is the fact that it exists.

Liability in negligence still has a role to play in cases where liability under the *Consumer Protection Act 1987* cannot be established because, for example, the defendant is not a 'producer' as defined, or because the statutory defence or time limit would bar a strict liability claim (see P9044 and P9048). There is greater scope under the law of negligence for liability to be established against distributors and retailers who may be held responsible for certain defects, especially where inadequate warnings and instructions have been provided (e.g. *Goodchild v Vaclight, The Times, 22 May 1965*).

### Joint and several liability

**P9043**  If two or more persons/companies are liable for the same damage, the liability is joint and several. This can, for example, refer to the situation where a product (e.g. an aircraft) is made partly in one country (e.g. England) and partly in another (e.g. France). Here both partners are liable (joint liability) but in the event of one party not being able to pay, the other can be made to pay all the compensation (several liability). [*Consumer Protection Act 1987, s 2(5)*].

### Parameters of liability

**P9044**  For certain types of damage including (*a*) death, (*b*) personal injury and (*c*) loss or damage to property liability is included and relevant to private use, occupation and consumption by consumers. [*Consumer Protection Act 1987, s 5(1)*]. However, producers and others will not be liable for:

(*a*)  damage/loss to the defective product itself, or any product supplied with the defective product [*Consumer Protection Act 1987, s 5(2)*];

(b) damage to property not 'ordinarily intended for private use, occupation or consumption' e.g. car/van used for business purposes [*Consumer Protection Act 1987, s 5(3)*]; and

(c) damage amounting to less than £275 (to be determined as early as possible after loss) [*Consumer Protection Act 1987, s 5(4)*].

## Defences

**P9045** The following statutory defences are open to producers:

(a) the defect was attributable to compliance with any requirement imposed by law/regulation or a European Union rule/regulation [*Consumer Protection Act 1987, s 4(1)(a)*];

(b) the defendant did not supply the product to another (i.e. did not sell/hire/lend/exchange for money/give goods as a prize etc. (see below 'supply')) [*Consumer Protection Act 1987, s 4(1)(b)*];

(c) the supply to another person was not in the course of that supplier's business [*Consumer Protection Act 1987, s 4(1)(c)*];

(d) the defect did not exist in the product at the relevant time (i.e. it came into existence after the product had left the possession of the defendant). This principally refers to the situation where for example a retailer fails to follow the instructions of the manufacturer for storage or assembly [*Consumer Protection Act 1987, s 4(1)(d)*];

(e) that the state of scientific and technical knowledge at the relevant time was not such that a producer 'might be expected to have discovered the defect if it had existed in his products while they were under his control' (i.e. development risk) [*Consumer Protection Act 1987, s 4(1)(e)*] (see P9049 below);

(f) that the defect was:

　(i) a defect in a subsequent product (in which the product in question was comprised); and

　(ii) was wholly attributable to:

　　(A) design of the subsequent product; or

　　(B) compliance by the producer with the instructions of the producer of the subsequent product.

[*Consumer Protection Act 1987, s 4(1)(f)*]. This is known as the component manufacturer's defence.

## The meaning of 'supply'

**P9046** Before strict liability can be established under the *Consumer Protection Act 1987*, a product must have been '*supplied*'. This is defined as follows:

(a) selling, hiring out or lending goods;

(b) entering into a hire-purchase agreement to furnish goods;

(c) performance of any contract for work and materials to furnish goods (e.g. making/repairing teeth);

(d) providing goods in exchange for a consideration other than money (e.g. trading stamps);

(e) providing goods in or in connection with the performance of any statutory function/duty (e.g. supply of gas/electricity by public utilities);

(f) giving the goods as a prize or otherwise making a gift of the goods.

*[Consumer Protection Act 1987, s 46(1)]*.

Moreover, in the case of hire-purchase agreements/credit sales the effective supplier (i.e. the dealer), and not the ostensible supplier (i.e. the finance company),is the 'supplier' for the purposes of strict liability. *[Consumer Protection Act 1987, s 46(2)]*.

Building work is only to be treated as a supply of goods in so far as it involves provision of any goods to any person by means of their incorporation into the building/structure, e.g. glass for windows. *[Consumer Protection Act 1987, s 46(3)]*.

## No contracting out of strict liability

**P9047** The liability to person who has suffered injury/damage under the *Consumer Protection Act 1987*, cannot be (a) limited or (b) excluded:

(a) by any contract term; or

(b) by any notice or other provision.

*[Consumer Protection Act 1987, s 7]*.

## Time limits for bringing product liability actions

**P9048** No action can be brought under the *Consumer Protection Act 1987, Part I* (i.e. product liability actions) after the expiry of ten years from the time when the product was first put into circulation (i.e. the particular item in question was supplied in the course of business/trade etc.). *[Limitation Act 1980, s 11A(3); Consumer Protection Act 1987, Sch 1]*. In other words, ten years is the cut-off point for liability. An action can still be brought for common law negligence after this time.

A recent case has serious implications for this cut-off point. In *SmithKline Beecham plc and Another v Horne-Roberts [2001] EWCA CIV 2006*, the court allowed a claimant to substitute a new defendant for an existing one in a strict liability claim despite the fact that the substitution was outside the ten year cut-off period.

However all actions for personal injury caused by product defects must be initiated within three years of whichever event occurs later, namely:

(a) the date when the cause of action accrued (i.e. injury occurred); or

(b) the date when the injured person had the requisite knowledge of his injury/damage to property.

*[Limitation Act 1980, s 11A(4); Consumer Protection Act 1987, Sch 1]*.

But if during that period the injured person died, his personal representative has a further three years from his death to bring the action. (This coincides with actions for personal injuries against employers, except of course in that case there is no overall cut-off period of ten years.) *[Limitation Act 1980, s 11A(5)]*.

## Development risk

**P9049** This will probably emerge as the most important defence to product liability actions. Manufacturers have argued that it would be wrong to hold them responsible for the consequences of defects which they could not reasonably have known about or discovered. The absence of this defence would have the effect of increasing the cost of product liability insurance and stifle the development of new products. On the other hand, consumers maintain that the existence of this defence threatens the whole basis of strict liability and allows manufacturers to escape liability by, in effect, pleading a defence associated with the lack of negligence. For this reason, not all EU states have allowed this defence; the states in favour of its retention are the United Kingdom, Germany, Denmark, Italy and the Netherlands. The burden of proving development risk lies on the producer and it seems likely that he will have to show that no producer of a product of that sort could be expected to have discovered the existence of the defect. 'It will not necessarily be enough to show that he (the producer) has done as many tests as his competitor, nor that he did all the tests required of him by a government regulation setting a minimum standard.' (Explanatory memorandum of EC Directive on Product Liability, Department of Trade and Industry, November 1985).

Additionally, the fact that judgments in product liability cases are 'transportable' could have serious implications for the retention of development risk in the United Kingdom (see P9050 below).

### Transportability of judgments

P9050 The so-called 'Brussels Regulation' (Regulation 44/2001 of 22 December 2000 on *Jurisdiction and the Recognition and Enforcement of Judgments in Civil and Commercial Matters*) requires judgments given in one of the Member States (except Denmark which will continue to apply the 1968 Brussels Convention) to be enforced in another. The United Kingdom is bound by this Regulation as of 1 March 2002, which, by virtue of the *Civil Jurisdiction and Judgments Order 2001* (*SI 2001 No 3929*) (which amends the *Civil Jurisdiction and Judgments Act 1982*), is part of UK law. Where product liability actions are concerned, litigation can be initiated in the state where the defendant is based or where injury occurred and the judgment of that court 'transported' to another Member State. This could pose a threat to retention of development risk in the United Kingdom and other states in favour of it from a state against it, e.g. France, Belgium, Luxembourg.

## Contractual liability for substandard products

P9051 Contractual liability is concerned with defective products which are substandard (though not necessarily dangerous) regarding quality, reliability and/or durability. Liability is predominantly determined by contractual terms implied by the *Sale of Goods Act 1979* and the *Sale and Supply of Goods Act 1994*, in the case of goods sold; the *Supply of Goods* (*Implied Terms*) *Act 1973*, where goods are the subject of hire purchase and conditional and/or credit sale; and the *Supply of Goods and Services Act 1982*, where goods are supplied but not sold as such, primarily as a supply of goods with services; hire and leasing contracts are subject to the 1982 Act as well.

Exemption or exclusion clauses in such contracts may be invalid by virtue of the *Unfair Contract Terms Act 1977*, which also applies to such transactions. Moreover, 'standard form' contracts with consumers where the terms have not been individually negotiated, which contain 'unfair terms' – that is, terms detrimental to the consumer – will have such terms excised, if necessary, by the Director General of Fair Trading, under the *Unfair Terms in Consumer Contracts Regulations 1999* (*SI 1999 No 2083*). These Regulations extend to most consumer contracts between a seller or supplier and a consumer. [*Unfair Terms in Consumer Contracts Regulations 1999*, (*SI 1999 No 2083*), *Reg 4(1)*]. (See P9057 below.)

Contractual liability is strict. It is not necessary that negligence be established (*Frost v Aylesbury Dairy Co Ltd [1905] 1 KB 608* where the defendant supplied typhoid-infected milk to the claimant, who, after its consumption, became ill and required medical treatment. It was held that the defendant was liable, irrespective of the absence of negligence on his part).

### Sale of Goods Acts

P9052 Conditions and warranties as to fitness for purpose, quality and merchantability were originally implied into contracts for the sale of goods at common law. Those terms were then codified in the *Sale of Goods Act 1893*. However, this legislation did not provide a blanket consumer protection measure, since sellers were still allowed to exclude liability by suitably worded exemption clauses in the contract. This practice was finally outlawed, at least as far as consumer contracts were concerned, by the *Supply of Goods (Implied Terms) Act 1973* and later still by the *Unfair Contract Terms Act 1977*, the current statute prohibiting contracting out of contractual liability and negligence. Indeed, consumer protection has reached a height with the *Unfair Terms in Consumer Contracts Regulations 1999* (*SI 1999 No 2083*), invalidating 'unfair terms' in most consumer contracts of a standard form nature (see P9062 below).

More recently, the law relating to sale and supply of goods has been updated by the *Sale of Goods Act 1979* and the *Sale and Supply of Goods Act 1994*, the latter replacing the condition of 'merchantable quality' with 'satisfactory quality'. The difference between the 'merchantability' requirement, under the *1979 Act*, and its replacement 'satisfactory quality', under the *1994 Act*, is that, under the former Act, products were 'usable' (or, in the case of food, 'edible'), even if they had defects which ruined their appearance; now they must be free from minor defects as well as being safe and durable. Further, under the previous law, a right of refund disappeared after goods had been kept for a reasonable time; under current law, there is a right of examination for a reasonable time after buying.

## Sale of Goods Act 1979

**P9053** In 1979 a consolidated *Sale of Goods Act* was passed and current law on quality and fitness of products is contained in that Act. Another equally important development has been the extension of implied terms, relating to quality and fitness of products, to contracts other than those for the sale of goods, that is, to hire purchase contracts by the *Supply of Goods (Implied Terms) Act 1973*, and to straight hire contracts by the *Supply of Goods and Services Act 1982*. In addition, where services are performed under a contract, that is, a contract for work and materials, there is a statutory duty on the contractor to perform them with reasonable care and skill. In other words, in the case of services liability is not strict, but it is strict for the supply of products. This is laid down in the *Supply of Goods and Services Act 1982, s 4*. This applies whether products are simultaneously but separately supplied under any contract, e.g. after-sales service, say, on a car or a contract to repair a window by a carpenter, in which latter case service is rendered irrespective of product supplied.

## Products to be of satisfactory quality – sellers/suppliers

**P9054** The *Sale of Goods Act 1979* (as amended) writes two quality conditions into all contracts for the sale of products, the first with regard to satisfactory quality, the second with regard to fitness for purpose.

Where a seller sells goods in the course of business, there is an implied term that the goods supplied under the contract are of satisfactory quality, according to the standards of the reasonable person, by reference to description, price etc. [*Sale of Goods Act 1979, s 14(2)* as substituted by the *Sale and Supply of Goods Act 1994, s 1(2)*]. The 'satisfactory' (or otherwise) quality of goods can be determined from:

(*a*) their state and condition;

(*b*) their fitness for purpose (see P9057 below);

(*c*) their appearance and finish;

(*d*) their freedom from minor defects;

(*e*) their safety; and

(*f*) their durability.

However, the implied term of 'satisfactory quality' does not apply to situations where:

(i) the unsatisfactory nature of goods is specifically drawn to the buyer's attention prior to contract; or

(ii) the buyer examined the goods prior to contract and the matter in question ought to have been revealed by that examination.

[*Sale of Goods Act 1979, s 14(2)* as substituted by the *Sale and Supply of Goods Act 1994, s 1(2A), (2B)* and *(2C)*].

## Sale by sample

**P9055**     In the case of a contract for sale by sample, there is an implied condition that the goods will be free from any defect making their quality unsatisfactory, which would not be apparent on reasonable examination of the sample. [*Sale of Goods Act 1979, s 15(2)* as substituted by the *Sale and Supply of Goods Act 1994, s 1(2)*].

## Conditions implied into sale – sales by a dealer

**P9056**     The condition of satisfactory quality only arises in the case of sales by a dealer to a consumer, not in the case of private sales. The *Sale of Goods Act 1979, s 14(2)* also applies to second-hand as well as new products. It is not necessary, as it is with the 'fitness for purpose' condition (see P9058 below), for the buyer in any way to rely on the skill and judgment of the seller in selecting his stock, in order to invoke *s 14(2)*. However, if the buyer has examined the products, then the seller will not be liable for any defects which the examination should have disclosed. Originally this applied if the buyer had been given opportunity to examine but had not, or only partially, exercised it. In *Thornett & Fehr v Beers & Son [1919] 1 KB 486*, a buyer of glue examined only the outside of some barrels of glue. The glue was defective. It was held that he had examined the glue and so was without redress.

## Products to be reasonably fit for purpose

**P9057**     'Where the seller sells goods in the course of a business and the buyer, expressly or by implication, makes known

(*a*)    to the seller, or

(*b*)    where the purchase price or part of it is payable by instalments and the goods were previously sold by a credit-broker to the seller, to that credit-broker,

any particular purpose for which the goods are being bought, there is an implied condition that the goods supplied under the contract are reasonably fit for that purpose, whether or not that is a purpose for which such goods are commonly supplied, except where the circumstances show that the buyer does not rely or that it is unreasonable for him to rely, on the skill or judgment of the seller or credit-broker.' [*Sale of Goods Act 1979, s 14(3)*].

Reliance on the skill/judgment of the seller will generally be inferred from the buyer's conduct. The reliance will seldom be express: it will usually arise by implication from the circumstances; thus to take a case of a purchase from a retailer, the reliance will be in general inferred from the fact that a buyer goes to the shop in the confidence that the tradesman has selected his stock with skill and judgment (*Grant v Australian Knitting Mills Ltd [1936] AC 85*). Moreover, it is enough if the buyer relies partially on the seller's skill and judgment. However, there may be no reliance where the seller can only sell goods of a particular brand. A claimant bought beer in a public house which he knew was a tied house. He later became ill as a result of drinking it. It was held that there was no reliance on the seller's skill and so no liability on the part of the seller (*Wren v Holt [1903] 1 KB 610*).

Even though products can only be used normally for one purpose, they will have to be reasonably fit for that particular purpose. A claimant bought a hot water bottle and was later scalded when using it because of its defective condition. It was held that the seller was liable because the hot water bottle was not suitable for its normal purpose (*Priest v Last [1903] 2 KB 148*). But, on the other hand, the buyer must not be hypersensitive to the effects of the product. A claimant bought a Harris Tweed coat from the defendants. She later contracted dermatitis from wearing it. Evidence showed that she had an exceptionally sensitive skin. It was held that the coat was reasonably fit for the purpose when worn by a person with an average skin (*Griffiths v Peter Conway Ltd [1939] 1 All ER 685*).

Like *s 14(2)*, *s 14(3)* extends beyond the actual products themselves to their containers and labelling. A claimant was injured by a defective bottle containing mineral water, which she had purchased from the defendant, a retailer. The bottle remained the property of the seller because the claimant had paid the seller a deposit on the bottle, which would be returned to her, on return of the empty bottle. It was held that, although the bottle was the property of the seller, the seller was liable for the injury caused to the claimant by the defective container (*Geddling v Marsh [1920] 1 KB 668*).

Like *s 14(2)* (above), *s 14(3)* does not apply to private sales.

## Strict liability under the Sale of Goods Act 1979, s 14

**P9058** Liability arising under the *Sale of Goods Act, s 14* is strict and does not depend on proof of negligence by the purchaser against the seller (*Frost v Aylesbury Dairy Co Ltd* (see P9051 above)). This fact was stressed as follows in the case of *Kendall v Lillico [1969] 2 AC 31*: 'If the law were always logical one would suppose that a buyer who has obtained a right to rely on the seller's skill and judgment, would only obtain thereby an assurance that proper skill and judgment had been exercised, and would only be entitled to a remedy if a defect in the goods was due to failure to exercise such skill and judgment. But the law has always gone further than that. By getting the seller to undertake his skill and judgment the buyer gets ... an assurance that the goods will be reasonably fit for his purpose and that covers not only defects which the seller ought to have detected but also defects which are latent in the sense that even the utmost skill and judgment on the part of the seller would not have detected them' (per Lord Reid).

## Dangerous products

**P9059** As distinct from applying to merely substandard products, both *s 14(2)* and *(3)* of the *Sale of Goods Act 1979* can be invoked where a product is so defective as to be unsafe, but the injury/damage must be a reasonably foreseeable consequence of breach of the implied condition. If, for instance, therefore, the chain of causation is broken by negligence on the part of the user, in using a product knowing it to be defective, there will be no liability. In *Lambert v Lewis [1982] AC 225*, manufacturers had made a defective towing coupling which was sold by retailers to a farmer. The farmer continued to use the coupling knowing that it was unsafe. As a result, an employee was injured and the farmer had to pay damages. He sought to recover these against the retailer for breach of *s 14(3)*. It was held that he could not do so.

## Credit sale and supply of products

**P9060** Broadly similar terms to those under the *Sale of Goods Act 1979, s 14* exist, in the case of hire purchase, credit sale, conditional sale and hire or lease contracts, by virtue of the Acts described above at P9051 having been modified by the *Sale and Supply of Goods Act 1994*.

# Unfair Contract Terms Act 1977 (UCTA)

**P9061** In spite of its name, this Act is not directly concerned with 'unfairness' in contracts, nor is it confined to the regulation of contractual relations. The main provisions are as follows:

(a) Liability for death or personal injury resulting from negligence cannot be excluded by warning notices or contractual terms. [*UCTA 1977, s 2*]. (This does not necessarily prevent an indemnity of any liability to an injured person being agreed between two other contracting parties: if A hires plant to B on terms that B indemnifies A in respect of any claims by any person for injury, the clause may be enforceable (see *Thompson v T Lohan (Plant Hire) Ltd [1987] 2 All ER 631)*.)

(b) In the case of other loss or damage, liability for negligence cannot be excluded by contract terms or warning notices unless these satisfy the requirement of reasonableness.

(c) Where one contracting party is a consumer, or when one of the contracting parties is using written standard terms of business, exclusions or restrictions of liability for breach of contract will be permissible only in so far as the term satisfies the requirement of reasonableness [*UCTA 1977, s 3*]; as far as contracts with consumers are concerned, it is not possible to exclude or restrict liability for the implied undertakings as to quality and fitness for purpose contained in the *Sale of Goods Act 1979* (as amended) or in the equivalent provisions relating to hire purchase and other forms of supply of goods [*UCTA 1977, s 6(2)*]. Separate provisions make it an offence to include this type of exclusion clause in a consumer contract [*Consumer Transactions (Restrictions on Statements) Order 1976 (SI 1976 No 1813)*].

(d) In all cases where the reasonableness test applies, the burden of establishing that a clause or other provision is reasonable will lie with the person who is trying to rely on the term in question. The consequence of contravention of the Act is, however, limited to the offending term being treated as being ineffective; the courts will give effect to the remainder of the contract so far as it is possible to do so.

## Unfair Terms in Consumer Contracts Regulations 1999 (SI 1999 No 2083)

**P9062** Consumers faced with standard form contracts are given more ammunition to combat terminological obscurity, legalese and inequality of bargaining power, by the *Unfair Terms in Consumer Contracts Regulations 1999 (SI 1999 No 2083)*. These Regulations have the effect of invalidating any 'unfair terms' in consumer contracts involving products and services, e.g. sale, supply, servicing agreements, insurance and, as a final resort, empowering the Director General of Fair Trading to scrutinise 'standard form' terms, with a view to recommending, where necessary, their discontinued use. Contracts affected by the Act will remain enforceable minus the unfair terms (which will be struck out), in so far as this result is possible. [*Unfair Terms in Consumer Contracts Regulations 1999 (SI 1999 No 2083), Reg 8(2)*].

The contracts affected are only 'standard form' ones, that is, contracts whose terms have not been 'individually negotiated' between seller/supplier and consumer [*Unfair Terms in Consumer Contracts Regulations 1999 (SI 1999 No 2083), Reg 5*]; and a term is taken not to have been 'individually negotiated', where it has been drafted in advance and the consumer has not been able to influence the substance of the term [*Unfair Terms in Consumer Contracts Regulations 1999 (SI 1999 No 2083), Reg 5(2)*]. Significantly, it is incumbent on the seller/supplier, who claims that a term was 'individually negotiated' (i.e. not unfair) to prove that it was so. [*Unfair Terms in Consumer Contracts Regulations 1999 (SI 1999 No 2083), Reg 5(4)*]. However, if the terms of the contract are in plain, intelligible language, fairness, relating to subject matter or price/remuneration, cannot be questioned [*Unfair Terms in Consumer Contracts Regulations 1999 (SI 1999 No 2083), Reg 6(2)*].

The key provisions are:

(a) an 'unfair term' is one which has not been individually negotiated and causes a significant imbalance in the parties' rights and obligations, detrimentally to the consumer, contrary to the underlying tenet of good faith.

[*Unfair Terms in Consumer Contracts Regulations 1999* (*SI 1999 No 2083*), Reg 5(*1*)]

(*b*) an 'unfair term' is not binding on the consumer.

[*Unfair Terms in Consumer Contracts Regulations 1999* (*SI 1999 No 2083*), Reg 8(*1*)]

(*c*) terms of standard form contracts must be expressed in plain, intelligible language and, if there is doubt as to the meaning of a term, it must be interpreted in favour of the consumer except in proceedings brought under *Reg 12* (see (*e*) below).

[*Unfair Terms in Consumer Contracts Regulations 1999* (*SI 1999 No 2083*), Reg 7]

(*d*) complaints about 'unfair terms' in standard forms are to be considered by the Director General of Fair Trading, who may bring proceedings for an injunction to prevent use of the terms in future if the proponent of the unfair terms does not agree to desist from using them.

[*Unfair Terms in Consumer Contracts Regulations 1999* (*SI 1999 No 2083*), Reg 8]

(*e*) instead of the Director General of Fair Trading, a qualifying body named in *Sch 1* (statutory regulators, trading standards departments and the Consumers' Association) may notify the Director that it agrees to consider the complaint and may bring proceedings for an injunction to prevent the continued use of an unfair contract term.

[*Unfair Terms in Consumer Contracts Regulations 1999* (*SI 1999 No 2083*), Regs 10 and 12].

## *Examples of 'unfair terms'*

**P9063** Terms which have the object or effect of:

(*a*) excluding or limiting the legal liability of a seller or supplier in the event of the death of a consumer or personal injury to the latter resulting from an act or omission of that seller or supplier;

(*b*) inappropriately excluding or limiting the legal rights of the consumer vis-à-vis the seller or supplier or another party in the event of total or partial non-performance or inadequate performance by the seller or supplier of any of the contractual obligations, including the option of offsetting a debt owed to the seller or supplier against any claim which the consumer may have against him;

(*c*) making an agreement binding on the consumer whereas provision of services by the seller or supplier is subject to a condition whose realisation depends on his own will alone;

(*d*) permitting the seller or supplier to retain sums paid by the consumer where the latter decides not to conclude or perform the contract, without providing for the consumer to receive compensation of an equivalent amount from the seller or supplier where the latter is the party cancelling the contract;

(*e*) requiring any consumer who fails to fulfil his obligation to pay a disproportionately high sum in compensation;

(*f*) authorising the seller or supplier to dissolve the contract on a discretionary basis where the same facility is not granted to the consumer, or permitting the seller or supplier to retain the sums paid for services not yet supplied by him where it is the seller or supplier himself who dissolves the contract;

(*g*) enabling the seller or supplier to terminate a contract of indeterminate duration without reasonable notice except where there are serious grounds for doing so;

(*h*) automatically extending a contract of fixed duration where the consumer does not indicate otherwise, when the deadline fixed for the consumer to express this desire not to extend the contract is unreasonably early;

(i) irrevocably binding the consumer to terms with which he had no real opportunity of becoming acquainted before the conclusion of the contract;

(j) enabling the seller or supplier to alter the terms of the contract unilaterally without a valid reason which is specified in the contract;

(k) enabling the seller or supplier to alter unilaterally without a valid reason any characteristics of the product or service to be provided;

(l) providing for the price of goods to be determined at the time of delivery or allowing a seller of goods or supplier of services to increase their price without in both cases giving the consumer the corresponding right to cancel the contract if the final price is too high in relation to the price agreed when the contract was concluded;

(m) giving the seller or supplier the right to determine whether the goods or services supplied are in conformity with the contract, or giving him the exclusive right to interpret any term of the contract;

(n) limiting the seller's or supplier's obligation to respect commitments undertaken by his agents or making his commitments subject to compliance with a particular formality;

(o) obliging the consumer to fulfil all his obligations where the seller or supplier does not perform his;

(p) giving the seller or supplier the possibility of transferring his rights and obligations under the contract, where this may serve to reduce the guarantees for the consumer, without the latter's agreement;

(q) excluding or hindering the consumer's right to take legal action or exercise any other legal remedy, particularly by requiring the consumer to take disputes exclusively to arbitration not covered by legal provisions, unduly restricting the evidence available to him or imposing on him a burden of proof which, according to the applicable law, should lie with another party to the contract.

*[Unfair Terms in Consumer Contracts Regulations 1999 (SI 1999 No 2083), Sch 2].*

# Radiation

## Non-ionising and ionising radiation

### Non-Ionising Radiation – What is it?

**R1001** Ionising radiation has sufficient energy to displace tightly bound electrons from atoms, causing the atom to become electrically charged or *ionised*. These ions, although short-lived, can be very reactive, causing tissue damage and changes in DNA, which lead to cancer (see R2001–R2011). *Non-ionising radiation* is that part of the electromagnetic spectrum which has insufficient energy to cause ionisation. Its interaction with body tissues, and hence the potential harm it can cause, is quite different, therefore, from ionising radiation.

All radiation can be thought of as waves of energy radiating away from the source that produces it. In the case of non-ionising radiation, this energy can be characterised by both an *electrical* field (the E-field) and a *magnetic* field (the B or H field – see R1002). Both of these characteristics can be measured and can be used to quantify the strength of the field. In addition, the energy wave has a *frequency*: the number of times it repeats itself in a given time (a frequency of 1 cycle per second is 1 Hertz, or Hz). The *wavelength* (which is directly related to the frequency) is the distance between the waves measured in the unit of metres (or down to nanometres (nm) for the very short wavelengths in the visible and ultraviolet region). The frequency is an important way to characterise non-ionising radiation, since the effects it can have on the body are determined as much by frequency as by the strength of the field. The frequency of non-ionising radiation extends from just above zero Hz (if it was zero and therefore had no frequency, it would be a *static* electrical or magnetic field) to between $10^{15}$ and $10^{16}$ Hz in the ultraviolet region. At this frequency the non-ionising region gives way to the ionising region. The complete range of energies is divided into bands, which are described by the electromagnetic spectrum. (TABLE 1).

# Radiation

Table 1. The electromagnetic spectrum

| | | Region | Wavelength (metres) | Approximate frequency range (Hz)* |
|---|---|---|---|---|
| | | *Static* | *0* | *0* |
| Non – Ionising Radiation | Radiofrequencies | ELF | $10^8$–$10^5$ | 0.3 Hz to 3 kHz |
| | | VLF | $10^5$–$10^4$ | 3 kHz to 30 kHz |
| | | LF | $10^4$–$10^3$ | 30 kHz to 300 kHz |
| | | MF | $10^3$–$10^2$ | 300 kHz to 3 MHz |
| | | HF | $10^2$–10 | 3 MHz to 30 MHz) |
| | | VHF | 10–1 | 30 MHz to 300 MHz) |
| | Micro-waves | UHF | 1–$10^{-1}$ | 300 MHz to 3GHz) |
| | | SHF | $10^{-1}$–$10^{-2}$ | 3 GHz to 30 GHz) |
| | | EHF | $10^{-2}$–$10^{-3}$ | 30 GHz to 300 GHz) |
| | Infrared | IR | $10^{-3}$–$10^{-6}$ | $10^{11}$–$10^{14}$ |
| | Visible ** | Red | 740–625 nm | $10^{14}$–$10^{15}$ |
| | | Orange | 625–590 nm | |
| | | Yellow | 590–565 nm | |
| | | Green | 565–520 nm | |
| | | Cyan | 520–500 nm | |
| | | Blue | 500–435 nm | |
| | | Violet | 435–400 nm | |
| | Ultraviolet | UVA | 400–315 nm | $10^{15}$–$10^{17}$ |
| | | UVB | 315–280 nm | |
| Ionising Radiation | | UVC | 280–100 nm | |
| | X ray | X-ray | $10^{-8}$–$10^{-11}$ | $10^{17}$–$10^{20}$ |
| | Gamma ray | $\gamma$-ray | $10^{-11}$–$10^{-14}$ | $10^{20}$–$10^{22}$ |

\*      1 Hz is 1 cycle per second: kHz = one thousand Hertz (kilo Hertz); MHz = one million Hertz (Mega Hertz); GHz = one thousand million Hertz (Giga Hertz)

\*\*    The visible band is a small range of frequencies that the retina of the eye is able to detect and which gives us the sense of vision. The wavelength of the visible part of the spectrum is usually expressed in terms of nanometres (nm; 1 nm is $10^9$ of a metre). Visible light extends from 740 nm (perceived as red) to 400 nm (perceived as violet). Above the violet part of the visible spectrum is the ultraviolet region, which is broken down into three bands (A, B and C). The transition from non-ionising radiation to ionising radiation occurs around the wavelength of UVC.

**R1002**     Electromagnetic fields can be characterised by their frequency and/or wavelength (see TABLE 1). The frequency is measured in the unit of Hertz (Hz). One Hz equals one cycle per second. The wavelength is measured in the unit of a metre (or for very high frequency radiations, down to nanometres (nm)). One nm is equal to one thousand-millionth of a metre ($10^{-9}$ m). The strength of the field is measured by either the strength of the electrical component of the field or the strength of the magnetic component. The electrical field (the E-field) is measured as volts per metre (v.m$^{-1}$). The magnetic field can be measured as the magnetic field strength in units of Amps per metre (A.m$^{-1}$; the H-field) or the magnetic flux density in units of Tesla (or more usually as micro Tesla – $\mu$T; the B field). 1 $\mu$T is one-millionth of a Tesla. In addition, electromagnetic fields above 100kHz in the radio frequency part of the spectrum have the ability to heat body tissues, but the risk depends on the amount of radiation received. This is measured as the amount of power falling on a given area of the body surface in the units of Watts per square metre (W.m$^{-2}$). This measure is the power density. The rate at which energy is absorbed in the body tissues is measured in the units of Watts per kilogram (W.kg$^{-1}$). This measure, the Specific Absorption Rate (SAR), is a fundamental measure of absorbed dose, which relates to the health and safety risk and is used as the basis for guidelines to ensure safe limits of exposure (see R1003).

## How Does Non-ionising Radiation Interact with the Body?

**R1003** Non-ionising radiation poses a *known* risk and an *uncertain* risk to health. The *known* risks relate to the way in which the radiation is absorbed by the body and causes the heating of body tissues, possible electrical shock and skin burn. The *uncertain* health effects relate to a possible risk of cancers and other diseases that may be linked to exposure to non-ionising radiation. Another way to distinguish these two types of risk is to consider the known effects as being *acute* risks (namely, occurring after a very short period of exposure) and the uncertain health effects as being *chronic* risks (those requiring long and/or intermittent periods of exposure).

## The Known Health Risks

**R1004** The known health effects are separated into direct effects and indirect effects. Direct effects are those that result from a direct interaction between the body and the electromagnetic field to which the body is exposed. Indirect effects occur as a result of an interaction between the electromagnetic field, an external object (usually a metallic structure) and the human body.

## Direct Effects

**R1005** The direct effects are usually considered differently for frequencies above and below 100kHz. Below 100 kHz the hazard is one of induced electrical current in the body. Induced electrical charge on the surface of the body can be detected by a small proportion of people as a tingling sensation or as a perception of the raising or the vibration of the skin hairs. About 10 per cent of people can detect surface charges in this way, but only when the electrical field to which they are exposed is in the order of 10–15kV.m$^{-1}$ (at ELF frequencies – see TABLE 1; 50Hz is the frequency of the electrical mains distribution and would be the usual source of exposure to fields causing this effect). This is not a health hazard but is annoying. A lot of research has focused on what physiological effects induced currents may have on the body. These have included effects on the nervous control of breathing and the heart, the effects of induced electrical signals in the brain and spinal cord, effects on performance and behaviour, effects on vision and hearing, and effects on the blood and on circadian rhythms (the body's daily cycle). The results of much of this work is equivocal and not reproducible. Conservative analysis of the data that is available has, however, allowed limits of exposure to be set, which should prevent adverse effects (see R1009). For frequencies above 100 kHz, the major effects are body heating. Healthy people can tolerate short-term rises in body temperature of about 1 °C without significant adverse effects. However, a rise of 1 °C will cause increased sweating, increased heart rate and a fall in blood pressure. A body temperature rise of 1 °C may also be associated with changes in circulating hormone levels (such as thyroxin and adrenaline), and rises of more than this are considered potentially hazardous. The limits on exposure are set so that the SAR is restricted to a value that will ensure that the rise in body temperature is under 1 °C (see R1009). There is sufficient margin built into these restrictions on exposure to take account of work rate, environmental temperature and clothing, which in themselves will also influence body temperature.

As well as the hazard of whole body heating, frequencies above 100 kHz can cause localised heating, which may present a risk of damage to particular tissues. The lens of the eye, the testes and the embryo and foetus are the tissues most often considered to be at particular risk. These localised effects are most likely to be a problem with microwave radiations, but in the case of damage to the lens of the eye, infrared and ultraviolet radiations are also a significant risk (see TABLE 1). In the case of the eye, localised heating may cause cataracts, an opacity in the lens causing visual impairment and blindness. The testes may be vulnerable because they are usually maintained at a few degrees below body temperature, so that a rise in testicular temperature can cause reduced sperm counts and transient infertility. The embryo and foetus may be particularly sensitive to heating, which has been linked to birth defects.

## *Indirect Effects*

R1006    Indirect effects result from an accumulation of charge in the body, which is then subsequently discharged to a conductive (usually metallic) object. The discharge current may be sufficient to cause significant electric shock, and can have serious consequences, such as paralysis of the respiratory muscles and ventricular fibrillation (an erratic beating of the heart, causing collapse and death). At lower levels of current the discharge may cause pain at the point of exit of the discharge (the point where the body is in contact with the metallic object) and sufficient localised heating to result in burns to the skin. Electrical discharges may be sudden or may be continuous. Sudden discharges occur where the body is effectively insulated from the ground, allowing the accumulation of charge, which is then suddenly discharged when contact is made with a metallic object. This is most likely to happen if the body is in an electromagnetic field of less than 100 kHz but is also extremely common with static electrical fields. This is a common occurrence in indoor environments and in cars where static electricity causes an accumulation of charge sufficient to cause a discharge on subsequent contact with a metallic object. The discharge current in such circumstances will not be sufficient to cause serious effects but will result in pain and discomfort, and is particularly distressing if repeated frequently. Alternatively, discharge may occur if the person is in continuous contact with the metallic object (or other grounded conductor) during the time of exposure to the electromagnetic field. The person effectively becomes a conduit for the electrical current to pass to earth. This is most likely to occur when working with fields of below 100 kHz and at the same time in contact with a grounded conductor. The resulting current may be sufficient to cause electric shock if the field strength is sufficient. The limitations on exposure (see R1009) are set at a value that should limit the contact current to below 1 milli Amp (1 mA), which has been found to be able to be tolerated by adults without adverse effect.

## *Implanted Medical Devices*

R1007    Both direct and indirect exposure to electromagnetic field sources may interfere with implanted devices such as cardiac pacemakers, automatic implantable cardioverter defibrillators (AICDs) and cochlea implants. It is impossible to be precise about what levels of exposure are safe for persons fitted with such devices, because each device varies, but there is a risk that induced currents will interfere with their function. In the case of pacemakers and AICDs, the consequence of malfunction could be serious. Generally, the levels of exposure that might give rise to concern will be lower than those recommended in guidelines for general exposure (see R1009). Anyone newly fitted with an implantable electronic medical device (or newly employed with one fitted) who is working with sources of electromagnetic radiation, or working in an area where exposure to significant sources may be a problem, must have a competent risk assessment undertaken (see R1010). It is also possible that individuals fitted with metallic implants, such as hip or knee replacements, will experience localised heating in certain circumstances. Such problems will again need to be addressed through a risk assessment.

## Uncertain Health Effects

**R1008** These effects result from chronic (namely, long-term) exposure to electromagnetic fields, and may be caused by relatively low levels of exposure. However, what the exact risks are, and indeed if there are any risks at all, is most contentious. The risks cited mainly include cancers of various types, including cancers of the breast, brain and nervous system, leukaemia, degenerative brain and spinal cord conditions such as Amyotrophic Lateral Sclerosis (ALS), and behavioural conditions. The cancer risk is the most frequently cited and the most researched. The possibility of a cancer risk from chronic exposure to electromagnetic fields first appeared in 1979 (Wertheimer and Leeper, 1979). This study, from America, demonstrated a two- to three-fold risk in childhood leukaemia, lymphomas and nervous system tumours in houses that had a high level of magnetic fields from the use of mains electricity (in the USA these are 60 Hz magnetic fields). Since then, scores of epidemiological studies have been done to attempt to confirm these findings. Epidemiological studies attempt to look for statistical differences in disease prevalence in people exposed to a particular factor, compared with people who are not (or are less) exposed. In these studies the cancer prevalence rates are compared for people living near, for example, power transmission lines, compared to people who have never lived near a power line. The difficulty encountered by such studies is that the numbers of people exposed to power lines is fairly small and the number of those with relevant cancers is also small. The statistical power of many of the studies is therefore poor. One of biggest studies ever carried out, the UK Childhood Cancer Study, failed to find evidence that exposure to magnetic fields associated with electricity supply and transmission increases the risk of childhood cancers (UKCC, 1999). However, other studies with similar statistical validity have concluded that there is a risk between two to three times higher when the mean exposure exceeds about 0.2–0.3 µT. (see, for example, Savitz et al, 1988; Feychting and Ahlbom, 1993). The current conclusion of the Health Protection Agency is that for children exposed to above 0.4 µT, there is a doubling of the risk of leukaemia. This has to be taken against their estimate that 99.6 per cent of children live in homes that do not expose them to more than 0.4 µT and that the incidence rate for childhood leukaemia is 1:20,000 per annum (HPA, 2005).

Much of the research over the last 25 years, especially on leukaemia and nervous system tumours, has been concerned with exposure to electrical power fields at 50/60 Hz. However, considerable work has also been done on exposures to radiofrequency fields. To some extent this has been because of concerns about living near to radio or television transmitter masts but probably in greater part because of the concern that the transmitted power from mobile phones may be a health risk. There is also concern that living near to mobile phone base stations poses a health risk.

In the United Kingdom AM radio uses frequencies between about 180 kHz and 1.6 MHz; FM radio between 88 and 108 MHz;and television from 470 to 854 MHz. Mobile phone networks operate between 872 and 960 MHz or 1710 and 1875 MHz. Third generation mobile communication (3G) will utilise frequencies between 1.88 and 2.01 GHz and 2.11 and 2.20 GHz (see TABLE 1). In addition, other frequencies, particularly in the microwave region, are used by the emergency services, for civilian and military radar, for satellite communications and for digital radio and television transmission. There have been a number of epidemiological studies to investigate the health risk of living near to television and radio transmitter sites with almost the same equivocal results as exist for living near to power lines. One UK study found that there was no relationship between living within two kilometres of TV and radio transmitter sites and the prevalence of leukaemia and lymphoma, although the prevalence did decline with increasing distance from the sites. (Dolk et al, 1997a). In addition, the same authors had shown that the incidence of leukaemia and lymphoma around a single transmitter site in the Midlands was increased by a factor of 1.8. (Dolk et al, 1997b). Other studies in Sydney and Hawaii have shown an increased risk of leukaemia to those living close to television and radio masts, (Maskarinec et al, 1994 and Hocking et al, 1996), whilst a study in San Francisco around a single microwave transmitter failed to find any excess incidence of disease (Selvin et al, 1992). These findings are at present too inconclusive to confirm that exposure to radiofrequency radiation is a cancer risk.

Much the same conclusion can be reached when looking at the research on mobile phone cancer risks. Animal and cellular studies suggest that exposure to the power levels received by the brain during a mobile phone conversation can cause changes in the DNA of cells, which, theoretically, could be associated with cancer (Lai and Singh, 1996). However, there is no evidence so far to suggest that this is a real risk in humans during the normal use of mobile phone technology. A study of the mortality rate and leukaemia and brain tumour prevalence among 250,000 mobile phone users did not show any increased risk. (Rothman et al, 1996). The 'Stewart Report' of 2000 (IEGMP, 2000) and the subsequent update by the NRPB (NRPB, 2004) provide a wide-ranging and detailed review of the evidence linking mobile phone technology to health risks.

Much research has focused on the risk of leukaemia in children exposed to power frequency fields. With radiofrequency fields, research has concentrated on the risks of living close to transmission sources. In the occupational health context much of this work (especially that concerning childhood disease) is irrelevant and can only be used to determine what might be the general risks of working with power frequency electrical or magnetic fields or sources of radiofrequency transmission. However, several studies have looked specifically at the risks of working with significant sources of electromagnetic energy. Such jobs are found in electronics and electrical engineering, engineering (welding), in the electrical supply industry, on the railways, in radio and television transmission and transmitter maintenance, and in the military. Some of these studies on occupational exposures have concluded that there is a significant risk of brain cancers, leukaemias and breast cancer (in both men and women) in those occupations in which there is significant exposure to electromagnetic radiation (see, for example, Wilkins and Koutras, 1988; Savitz and Loomis, 1995; Guenel et al, 1993; Floderus et al, 1993; Loomis et al, 1994). In most cases the excess risk is between two and four times higher. However, there are also studies in which no association has been found. (see, for example, Tynes et al, 1992; Sahl et al, 1993; Rosenbaum et al, 1994). At least six studies have looked at paternal occupational exposure to electromagnetic fields and the risk of brain tumour in their offspring. Three of these showed an increased risk of brain tumour (relative risk between 2 and 3.6) in the children of exposed fathers (Spitz and Johnson, 1985; Wilkins and Koutras, 1988; Johnson and Spitz, 1989), whilst the other three failed to show any relationship (Nasca et al, 1988; Wilkins and Hundley, 1990; Bunin et al, 1990).

In conclusion, health risks from chronic exposure either to community or occupational sources of low level electromagnetic radiation remain uncertain. The vast amount of research in this area has failed to uncover any consistent or reliable evidence that there is a health risk. Many of the studies are too weak to reach definite conclusions on a health risk, but the underlying body of evidence that there might be a health risk cannot be totally ignored. If there is a risk, it appears to be a very small one, perhaps doubling the incidence of diseases that are in any case rare. Brain tumours, leukaemias and lymphomas are the most consistent health effects noted. In the absence of conclusive evidence for a health risk, the suggested method of risk management is to apply the precautionary principle ('prudent avoidance').

## Guidelines on Exposure

**R1009** The National Radiological Protection Board (NRPB, now the Radiation Protection Division of the Health Promotion Agency) has published guidelines on exposure to electromagnetic fields (NRPB, 1993). These provide guidance on exposure only to prevent the known health risks of acute exposure (electric shock, body heating and heat damage and burn; see R1004). They do not provide guidance to avoid the uncertain health risks discussed in R1008; if such risks exist, they will occur at much lower levels of exposure. Neither does the guidance ensure that there is no risk of electromagnetic interference with implanted medical devices (R1007); again, these risks may be at much lower levels of exposure than the guidance indicates.

In addition to the NRPB (HPA) guidelines, there are guidelines from various other bodies, including, notably, the International Commission on Non Ionising Radiation Protection (ICNIRP). The impending importance of the ICNIRP guidelines is that they will become the basis of the legal standard in the UK, probably after 2008, when the EC Physical Agents (Electromagnetic Fields) Directive becomes law (see R1010). Other countries may use their own guidelines or will have adopted the ICNIRP guidelines. If a company has staff working in countries other than the UK, it will be necessary to check what standards are applied in that country; in the UK it will usually be satisfactory to work to the NRPB (HPA) guidelines. However, with a new legal requirement pending within the next three years, it may be prudent to start now to adopt the ICNIRP guidelines (ICNIRP, 1988 and NRPB, 1999).

The principle of the guidelines are firstly that standards for frequencies above 100 kHz are set to limit the rate of absorbed radiation (the SAR; see R1002). Remember that these guidelines relate to thresholds (with a considerable margin for uncertainties) to protect against the well-known direct and indirect effects of acute exposure. In the frequency range from 0 to 100 kHz the basic restriction is set as a magnetic field or magnetic flux density value (see R1002).

Above 100 kHz the basic restriction is $0.4$ $W.kg^{-1}$ for the whole body, averaged over any 15-minute period. The basic restriction on SAR for the head, foetus, neck and trunk and the limbs alone, have different (higher) values but are averaged over any 6-minute period. Above 10 GHz the basic restriction is expressed as a power density and is $100$ $W.m^{-2}$.

Although the basic restrictions form the basis of the guidelines, they are not very useful in practice, as it is virtually impossible to measure the SAR under field conditions. The NRPB (HPA) have therefore set levels of exposure measured as magnetic field strength, magnetic flux density, electric field strength or power density. A variety of instruments exist for measuring field strength or power density and it is therefore much more practical to ensure compliance with the basic restrictions by measuring these exposure levels rather than by attempting to directly measure the SAR. NRPB (HPA) call these exposure values *investigation levels* (the ICNIRP calls these *reference levels*). Investigation levels are set at values of exposure that should not result in the SAR being exceeded and therefore should not result in the adverse effects of electric shock or body heating. The investigation levels are set with a considerable margin of safety in extrapolating from the SAR to values of field strength, and worse case situations are taken into account. For example, at frequencies above 10 MHz, children may be more at risk because of their body size, and the investigation levels at these frequencies are given for situations where children might be exposed. If it can be certain that no children will be exposed, the investigation levels can be relaxed.

It is beyond the scope of this chapter to reproduce the guidelines contained in the NRPB (HPA) statement or the ICNIRP guidelines. Readers who need to undertake a risk assessment must refer to the guidelines set out in the Board Statement or the ICNIRP guidelines, and be prepared to understand the complexities and nuances of how they should be interpreted. One particular complexity occurs where exposures may be to more than one frequency simultaneously. There may, under these conditions, be additive effects or combined effects, which cannot easily be related to the guideline investigation levels. It may, therefore, be necessary to seek expert advice on how to determine levels of exposure, how to interpret the guidelines and how to devise suitable means of risk control. Although instruments to measure electric and magnetic fields can be hired, it may be necessary to engage someone with experience of undertaking such measurements in order to determine actual levels of exposure.

In summary, restrictions on exposure:

- Are set to ensure that there is no risk from acute exposure to electromagnetic fields (mainly the risks of electric shock and body heating).
- Do not necessarily ensure that there no risk for those wearing implanted electro-medical devices (such as pacemakers). Individual risk assessments must be undertaken for such individuals.
- Do not consider the risk from chronic (long-term) exposure. These risks are very uncertain but cannot categorically be ruled out, so a precautionary approach should be applied in such circumstances.

- Are set to limit the absorbed dose of radiation (the Specific Absorption Rate – SAR). This is the *basic restriction*.
- Are, in practice, applied as restrictions (limitations) on the levels of exposure measured as electric or magnetic field strengths or power density. These are set out in the NRPB (HPA) guidelines as *investigation levels* or in the ICNIRP guidelines as *reference levels*.
- Can be complex to interpret, and expert help may be needed to determine levels of exposure and interpret compliance with the basic restrictions.

## Legal Requirements

**R1010** There is no specific legislation dealing with non-ionising radiation. The European Commission has now issued a Directive 'On the Minimum Health and Safety Requirements Regarding Exposure of Workers to the Risks Arising from Physical Agents (Electromagnetic Fields)' (OJ L159 30.4.2004; and a corrigendum in L184/1 24.5.2004). The basic restrictions and guidelines on exposure (called reference levels in the Directive) follow those of the International Commission on Non Ionising Radiation (ICNIRP), and these will be adopted in the UK, probably becoming law in 2008. In the meantime the legal requirement is that contained in s 2 of the Health and Safety at Work Act, 1974, which places a general duty on the employer 'to ensure, so far as is reasonably practicable, the health, safety and welfare at work of all his employees'. The requirement to undertake a risk assessment for those employees for whom exposure to electromagnetic radiation may be significantly encountered in their work is contained in Regulation 3 of the Management of Health and Safety at Work Regulations 1999 (SI 1999 No 3242). These Regulations also provide a legal requirement to identify and implement the necessary precautions to avert the risk. Regulation 7 requires that anyone charged with assisting in the management of risk must be competent. It is therefore a legal requirement to ensure that whoever is charged with measuring electromagnetic field strengths, interpreting exposure against accepted guidelines and devising suitable means of controlling the risk must be competent. It is the employer's legal duty to ensure competence in anyone required to undertake such work.

Regulation 5 of the Management Regulations may require exposed individuals to be placed under health surveillance. The need for this should be determined by the risk assessment, in consultation, if necessary, with an experienced Occupational Health Physician.

## Ionising Radiation – What is it?

**R1011** Ionising radiation is radiation with sufficient energy to electrically charge or *ionise* material that the radiation strikes. Ionising radiation arises from a wide range of natural and man-made radioactive sources, and its properties have been harnessed extensively for industrial and medical use. Ionising radiation can, however, adversely influence living tissue and its use must always be assessed and monitored carefully to prevent harmful effects. Types of radiation that cause ionisation include:

- $\alpha$ (alpha) particles;
- $\beta$ (beta) radiation;
- $\gamma$ (gamma) radiation;
- x rays;
- neutrons.

## Units of Measurement: Radioactivity and Radiation Dose

**R1013** Radiation is emitted from radioactive sources and this radioactivity is measured in units known as *bequerels*. Radiation dose to humans is defined as the amount of energy absorbed by the body and the *gray* (*Gy*) is the unit of dose. Different types of radiation have, however, different effects on tissue, and in order to take account of these differences, a further unit of dose, the *sievert*, has been developed.

The *sievert* (*Sv*) is the most commonly used measure of radiation in occupational and environmental exposures. Doses, however, are usually measured in smaller fractions of the sievert, either *millisieverts* or *mSv* (one-thousandth of a sievert), or *microsieverts* or *μSv* (one-thousandth of a millisievert). Natural radiation

**R1013** On earth there are two main sources of natural or background radiation: cosmic radiation from space and natural radiation from the ground. Most cosmic radiation is filtered by the upper layers of the atmosphere, but ionising radiation from space can be significant for those in the space and aviation industry. A typical radiation dose for a flight from the UK to Spain would be about 10 μSv (NRPB, 1994).

Certain rock types, such as granite, emit weak ionising radiation but the most significant source of background radiation from the ground in the UK occurs from radon, a naturally occurring gas, which largely accounts for the difference in background radiation across the UK. Typical annual doses from background radiation in the UK range from 2.1 mSv in London, where Radon levels are low, to 7.8 mSv in Cornwall, where they are much higher (NRPB, 1994).

## Man-Made Radiation

**R1014** The development of ionising radiation for industrial and medical purposes has produced significant benefits. Major uses include:

- Medicine – Ionising radiation has a number of medical applications including the diagnosis of disease, using techniques such as computed tomography (CT) scanning or X ray radiography, and in the treatment of cancer.

- Nuclear Power – In UK in 2004, about 20% electricity was generated by nuclear energy.

- Radiography in industry – Radiographic techniques are used in monitoring the integrity of metal structures such as pipelines and bridges and X ray machines are used widely at ports and airports for baggage security.

- Consumer products – Radioactive materials are used in a number of consumer products including smoke detectors and in luminous dials.

## Harmful Effects

**R1015** Human beings, like all organisms on earth, are made up of a structure of *cells*, the 'building blocks' of living tissue. Ionising radiation can damage essential proteins in these cells, and this damage can give rise to both short and long term health effects.

- Short-term effects are generally related to the capacity of ionising radiation to kill cells, and this is proportional to the dose of radiation. Doses above 3 Gy to the skin will produce redness brought about by the death of skin tissue. In humans whole body doses of greater than about 4–5 Gy can be fatal.

- Lower radiation doses can damage but not kill living cells. The long-term effects of ionising radiation arise as a result of imperfect repair of damage to cells leading to abnormal cell growth. In turn, this abnormal cell growth can lead to cancer. These effects are random, but in a working population the possibility of developing cancer is *increased* by about 4 per cent for every sievert of radiation dose received (NRPB, 1995).

## Irradiation and Contamination

**R1016** In considering the possible harmful effects of radiation, it is useful to differentiate between *irradiation* and *contamination*. Exposure to penetrating radiation from an external source, *irradiation*, may cause an individual to sustain damage to their body depending upon the dose, but the individual would not remain radioactive and would not become a hazard to others. Once the source of radiation is removed, the individual would not receive a further dose. An example of this would be a chest x-ray.

By contrast, if an individual has radioactive material on them or in their body, they are *contaminated*. As the *contamination* continues to emit radiation, they will continue to receive a radiation dose until the contamination is removed or declines naturally, even if they leave the area where the contamination occurred. They may also transfer the contaminated material to others, who may become at risk.

## Protection against Radiation

**R1017** The key to radiation protection lies in understanding the nature of the hazard. Different types of radiation can be blocked or *shielded* by different types of material. For instance, $\alpha$ (alpha) particles can be stopped by a sheet of paper, whereas $\beta$ (beta) radiation requires a sheet of metal and $\gamma$ (gamma) radiation a thickness of lead.

Other properties of radiation can be used when considering protection. The effect of radiation, for example, is reduced by distance from the source, and if the distance is doubled, the dose is reduced to a quarter. This is the *inverse square* phenomenon. The duration of exposure is also important. The less time an individual is exposed to a source, the less the received radiation dose will be.

In UK, background information and advice on radiation protection is available from the Radiation Protection Division at the Centre for Radiation, Chemical and Environmental Hazards (formerly known as the National Radiological Protection Board). The Centre for Radiation, Chemical and Environmental Hazards is the responsibility of the Health Protection Agency (HPA) and access to radiation protection information is via the HPA web site www.hpa.org.uk Legal Requirements

**R1018** In the UK work with ionising radiation is covered by the *Ionising Radiations Regulations 1999 (SI 1999/3232 (IRR99))*. Other legislation covering ionising radiation can be accessed via the UK Government Cabinet Office website www.opsi.gov.uk, and includes:

- The *Radioactive Substances Act 1993*, the legislation by which the Environment Agency regulates the use and disposal of radioactive materials.

- The *Radiation (Emergency Preparedness and Public Information) Regulations 2001 (SI 2001/2975)*, which regulate emergency planning and information if the public is likely to be involved.

- The *Radioactive Material (Road Transport) Regulations 2002 (SI 2002/1093)*, which regulate the package and transport of radioactive materials.

- The *Justification of Practice Involving Ionising Radiation Regulations 2004 (SI 2004/1769)*, by which the Department of Environment, Food and Rural Affairs assesses new processes involving ionising radiation.

Medical exposures to ionising radiation are governed by the Department of Health through:

- The *Ionising Radiation (Medical Exposure) Regulations 2000 (SI 2000/1059)*.

- The *Medicines (Administration of Radioactive Substances) (Amendment) Regulations 1995 (SI 1995/2147)*.

In relation to work with ionising radiation, health and safety managers may also have to consider the *Management of Health and Safety at Work Regulations 1999 (SI 1999/3242)* (MHSWR) and the *Personal Protective Equipment Regulations 2002 (SI 2002/1144)*.

## Key Features of the Ionising Radiations Regulations 1999

**R1019** Employers with staff working with ionising radiation, or those who are considering work with ionising radiation, should study carefully the *Ionising Radiations Regulations 1999*, together with the Approved Code of Practice and Guidance (ACOP) (ISBN 0 7176 1746 7). Key sections for the non-specialist include:

- Risk assessment – Regulation 7 requires employers commencing any new activity with ionising radiation to undertake a 'suitable and sufficient' risk assessment. The ACOP provides useful general advice on risk assessment and lists specific items to be considered.

- Personal protective equipment – Regulations 8 and 9 make reference to the use of personal protective equipment. Employers must ensure that such equipment is appropriate, properly used and maintained in accordance with the *Personal Protective Equipment Regulations 2002*.

- Management of radiation protection – Regulation 13 requires employers to consult a suitable Radiation Protection Advisor (RPA) to advise on the Regulations. Employers must ensure that the RPA fulfils the HSE's criteria of competence but will find that an RPA will be particularly useful in providing practical guidance in applying the Regulations. The HSE website, www.hse.gov.uk/radiation/ionising, is helpful for employers new to work with ionising radiation, and the site has lists of organisations with HSE-approved radiation protection advisors.

- Co-operation between employers – Regulation 15 requires employers sharing the same workplace to co-operate with each other to co-ordinate the measures they take to comply with statutory duties. This reflects the general obligation of Regulation 11 of the MHSWR of employers sharing the same workplace to co-operate.

- Designated areas – Regulations 16 – 19 cover the requirements of employers to designate and monitor areas in which employees may receive a significant radiation dose. These areas may be designated as *controlled* or *supervised*, and employers are obliged to have in place written *local rules* to identify key working instructions to restrict radiation exposure in these areas and the steps to be taken in the event of a radiation accident. Employers are also required to appoint *Radiation Protection Supervisors* to ensure compliance with the Regulations and local rules.

- Designation of classified persons – Regulation 20 requires employers to designate as *classified persons* (or classified workers) employees who are likely to receive a radiation dose greater than 6 mSv per year or an equivalent dose which exceeds three-tenths of any dose limit. In practice many employees become classified workers because they are required to enter designated controlled areas and, frequently, actual radiation doses may be substantially less than 6 mSv per year. Classified persons must be over 18 years of age and under medical surveillance by an appointed doctor (see below).

- Dose assessment and recording – Regulation 21 sets out the requirement for employers to appoint an *approved dosimetry service* (ADS) to monitor the radiation dose and to keep the dose records. Dosimetry services have to be approved by the HSE for certain functions, and employers should confirm that the chosen dosimetry service has an appropriate certificate of approval. The HSE website, www.hse.gov.uk/radiation/ionising, lists dosimetry service providers with HSE approval.

- Medical surveillance – Regulation 24 sets out the requirement for employers who employ classified persons to make arrangements for their medical surveillance. The medical surveillance of classified persons must be undertaken by a registered medical practitioner who has been appointed in writing by the HSE in accordance with the Regulations. Lists of names of *appointed doctors* are held by local HSE offices, but employers should note that an HSE Form FOD MS38A (available from local HSE offices) needs to be completed for each organisation for whom an appointed doctor acts. A *health record* has to be maintained to record the outcome of annual medical surveillance for each classified person, and the requirements of this record are set out in Schedule 7 of the ACOP. Form F2067 is the standard form available from the HSE. Note that this health record is separate from the employee's medical file and should be kept available for scrutiny by a Factory Inspector. Initial health surveillance should include a physical examination of the employee by the appointed doctor. Subsequent annual surveillance may be by review of sickness absence and radiation dose only ('a paper review') but it is good practice for employees to be examined every three or four years. Certain high-risk groups of employees, such as industrial radiographers, must, however, be examined annually.

- Appeals – In the event that an appointed doctor assesses an employee as unfit, the employee has the right to apply to the HSE for a review of that decision. The appeal must be made in writing or by completing the form IRR (Med) 1, available from the local HSE office.

- Female employees – The IRR make special provision for female employees, and Regulation 8(5) sets out the requirement for an assessment of risk for them. The external radiation dose to the abdomen of woman of reproductive capacity should not exceed 13 mSv in any consecutive three-month period (Schedule 4 of IRR), and the employer should confirm this annually before medical surveillance.

- Pregnant and breast-feeding employees – Regulation 8 requires employers, on being notified of pregnancy in an employee, to restrict the dose to the foetus to 1 mSv for the remainder of the pregnancy. In employees who are breast-feeding, the employer must ensure that significant radiological contamination is avoided.

- Outside workers and passbooks – *Outside workers* are employees designated as classified persons who carry out services in a controlled area designated by another employer. A key feature of outside workers is their requirement to hold a *radiation passbook* (Regulation 21(5)), which is provided by the ADS.

- Overexposures – Regulations 25 and 26 detail the action to be taken in the event of an overexposure. The employer must notify the HSE and the appointed doctor as soon as practicable. Care should also be taken to notify other employers in the event that their employees are involved. A special medical examination should be undertaken by the appointed doctor in the event that the effective dose of radiation to an employee is greater than 100 mSv in a year or an equivalent dose of at least twice any relevant annual dose limit. In practice it is recommended that employees are seen by the RPA and the appointed doctor at an early opportunity and provided with as much information and support as possible. Anxiety and the psychological aspects of overexposures are usually much more significant to the employee than the physical factors.

## Key Features of Other Legislation relating to work with Ionising Radiation

**R1020** Much other legislation relating to work with ionising radiation is specialised and beyond the scope of this overview. In the event, however, that an employer believes that any of this legislation might apply, the advice of an appropriate RPA should be sought at an early opportunity.

## Emergencies and Unforeseen Incidents

**R1021** The *Ionising Radiations Regulations 1999* place an obligation on the employer to prepare a contingency plan if the risk assessment shows that an accident is foreseeable (Regulation 12). The *Radiation (Emergency Preparedness and Public Information) Regulations 2001* set out the requirements for emergency planning and the obligation for information to be made available to the public in the event that the public could be affected.

The UK has a scheme, the National Arrangements for Incidents involving Radioactivity (NAIR). NAIR is a set of national arrangements that provide a 'long-stop' to other emergency plans. NAIR has been designed to provide advice and assistance to the police in incidents involving radioactivity where members of the public may require protection. The arrangements have been devised around assistance to the police, since they will normally be among the first informed of any incident in a public place. In an incident the police check whether there are plans to deal with the event. If no suitable plans exist, or if plans fail, NAIR assistance should be sought. NAIR assistance may be obtained by means of a 24-hour national notification telephone number (0800 834 153). This connects to the United Kingdom Atomic Energy Authority Constabulary, Force Communications Centre, who will take details of the incident and contact the nearest respondent (©Health Protection Agency – Radiation Protection Division).

## References

**R1022** Bunin G R, Ward E, Kramer S, et al (1990) Neuroblastoma and parental occupation, Am.J. Epidemiol. 1990; 131:776–780.

Dolk H., Elliott P, Shaddick G, Walls P, and Thakrar B (1997a) Cancer incidence near radio and television transmitters in Great Britain. 2. All high powered transmitters, Am.J. Epidemiol. 1997; 145:10.

Dolk H, Shaddick G, Walls P, Grundy C, Thakrar B, Kleinschmidt L, and Elliott P (1997b) Cancer incidence near radio and television transmitters in Great Britain. 1. Sutton Coldfield Transmitter, Am.J. Epidemiol. 1997; 145:1–9.

Feychting M and Ahlbom A (1993) Magnetic fields and cancer in children residing near Swedish high-voltage power lines, Am.J. Epidemiol. 1993; 138:467–81.

Ioderus B, Persson T, Stenland C et al (1993) Occupational exposure to electromagnetic fields in relation to leukaemia and brain tumours: A case control study in Sweden. Cancer Causes Control 1993; 4:465–476.

Guenel P, Raskmark P, Andersen J B, and Lynge E (1993) Incidence of cancer in persons with occupational exposure to electromagnetic fields in Denmark, Br.J.Ind.Med 1993; 50:758–764.

Hocking B, Gordon I R, Grain H L, Hatfield G E (1996) Cancer incidence and mortality and proximity to TV towers, Med J.Aust 1996; 165:601.

HPA (2005). Health Protection Agency, http://www.hpa.org.uk/radiation/understand/information_sheets/magnetic_fields.htm

ICNIRP (1988), Guidelines for limiting exposure to time-varying electric and magnetic fields (up to 300 GHz) Health Physics 1998; 75(4):494–552.

IEGMP (2000), Independent Expert Group on Mobile Phones. Chairman Sir William Stewart. Mobile Phones and Health. (2000), NRPB Chilton, Didcot, Oxon.

Johnson C C and Spitz M R (1989) Childhood nervous system tumours: An assessment of risk associated with paternal occupations involving use, repair or manufacture of electrical and electronic equipment, Int.J. Epidemiol. 1989; 18:756–762.

Lai H and Singh N P (1996) Single and double-strand DNA breaks in rat brain cells after acute exposure to radiofrequency microwave radiation, Int.J.,Radioat. Biol. 1996; 69:513.

Loomis D P, Savitz D A, and Ananth C V (1994) Breast cancer mortality among female electrical workers in the United States, J. Natl Cancer Inst. 1994; 86:921–925.

Maskarinec G, Cooper J and Swygert L (1994) Investigation of increased incidence in childhood leukaemia near radio towers in Hawaii; preliminary observations. (1994), J.Environ. Pathol. Toxicol. Oncol. 1994; 13:33.

Nasca P C, Baptiste M S, MacCubbin P A et al (1988) An epidemiologic case-control study of central nervous system tumours in children and paternal occupational exposures, Am.J. Epidemiol. 1988; 128:1256–1265.

NRPB (1993) Board Statement on Restrictions on Human Exposure to Static and Time Varying Electromagnetic Fields and Radiation, Documents of the NRPB, Volume 4, No.5. 1993.

NRPB (1994) Radiation Doses – Maps and Magnitudes 'At-a-Glance' series of leaflets published by NRPB.

NRPB (1995) Risk of Radiation-induced Cancer at Low Doses and Low Dose Rates for Radiation Protection Purposes, Documents of the NRPB Vol 6 No 1 1995.

NRPB (1999) Board Statement: Advice on the 1998 ICNIRP Guidelines for Time-varying Electric, Magnetic and Electromagnetic Fields (up to 300 GHz), Documents of the NRPB Vol. 10 No. 2 1999.

NRPB (2004). Mobile Phones and Health 2004: Report by the Board of the NRPB. Documents of the NRPB Vol 15 No 5, NRPB Chilton, Didcot, Oxon.

Rosenbaum P F, Vena J E, Zielezny M A and Michalek A M (1994) Occupational exposures associated with male breast cancer, Am.J. Epidemiol 1994; 139:30–36.

Rothman K J, Loughlin J E., Funch D P, and Dreyer N A (1996) Overall mortality of cellular telephone customers, Epidemiology 1996; 7:303.

Sahl J D, Kelsh M A, and Greenland S (1993) Cohort and nested case-control studies of haematopoietic cancers and brain cancers among electric utility workers, Epidemiology 1993; 4:104–114.

Savitz D A and Loomis D P (1995) Magnetic field exposure in relation to leukaemia and brain cancer mortality among electric utility workers, Am.J. Epidemiol. 1995; 141:123–134.

Savitz D A, Wachtel H A, Barnes F, John E M, Tvrdik J G (1988) Case control study of childhood cancer and exposure to 60 Hertz magnetic fields, Am.J. Epidemiol. 1988; 128:21–38.

Selvin S, Schulman J and Merrill D W (1992) Distance and risk measures for the analysis of spatial data: a study of childhood cancers, Soc.Sci.Med 1992; 34:769.

Spitz M R, and Johnson C C (1985) Neuroblastoma and paternal occupation: A case-control study, Am.J.Epidemiol. 1985; 121:924–929.

Tynes T, Jynge H, and Vistnes A (1994) Leukaemia and brain tumours in Norwegian railway workers: A nested case-control study, Am.J. Epidemiol. 1994; 139:645–653.

UKCC (1999) United Kingdom Childhood Cancer Study. Exposure to power-frequency magnetic fields and the risk of childhood cancer, Lancet 1999; 354: 1925–1931.

Wertheimer N and Leeper E (1979) Electrical wiring configuration and childhood cancer, Am.J. Epidemiol. 1979; 109:273–84.

Wilkins J R 3rd and Hundley V.D (1990) Paternal occupational exposure to electromagnetic fields and neuroblastoma in offspring, Am.J. Epidemiol. 1990; 131:995–1008.

Wilkins J R 3rd, and Koutras R.A. (1988) Paternal occupation and brain cancer in offspring, Am.J.Ind. Med. 1988; 14:299–318.

# Rehabilitation

## What is Rehabilitation?

R2001    The term vocational rehabilitation is now widely used to describe the process of broadly getting people back to work. However different stakeholders have different views of what is meant by vocational rehabilitation. One view is that it refers to getting people into employment when for some reason they may have been out of work for a long period of time or perhaps have never worked, and indeed this is one important aspect of rehabilitation. The reasons why people may not have worked for a long time, or indeed not have worked at all, are complex and only in some cases may this be due to a disability or long term health impairment. The other view is that vocational rehabilitation relates to the successful return to work of an employee who has been ill or has been injured or who is for some other reason, taking significant periods of absence. The employee may eventually return to his or her original job or may return to a modified job or even a new job within the organisation. Either way this is good for the employer who will reduce the costs of absence and will retain a valued employee. The two views therefore refer in the one instance to getting people into employment who are currently unemployed and in the other instance of getting people who do have a job back into the workplace usually following a period of sickness or injury. It is this latter which will be explored in more detail here since it concerns the relationships between employers and their employees and concerns the successful management of the organisation.

## Why should we be concerned?
### The business argument

R2002    All organisations should be concerned to maximise the use of resources and minimise their operational costs. People are usually an organisation's most valuable and most costly asset and it makes good business sense therefore, to ensure as far as possible, that their employees are absent as little as possible. On average in the UK about 2 to 3% of the workforce are absent from work every day because of illness or injury. In some organisations the sickness absence rate can be considerably higher than this. About 40 million days were lost through sickness absence in 2001–2002 at an estimated cost of between £11 billion and £12.9 billion. Clearly a proportion of this is unavoidable and much of it will be short, self-limiting periods of absence with common causes such as colds, headaches and stomach upsets. About 80% of sickness absence is short term, Of this 80% about 50–60% is due to genuine illness and accident. This is often referred to as unavoidable sickness absence and has to be accepted as an inevitable part of employment. However the remaining 40–50% of short term sickness is not due to genuine unavoidable sickness and may be put down to a variety of reasons ranging from drug and alcohol abuse to lack of commitment and interest in the job and poor management. The remaining 20% of sickness absence is long term – employees who are absent for weeks or months with a condition that renders them unable to travel to work and/or to carry out their expected duties. This 20% of long-term sickness absence accounts for 70% of the costs of sickness absence. From a business point of view therefore:

- reducing the impact of long term sickness absence;
- controlling the non-genuine reasons for absence;

- controlling frequent short-term absences,

are essential in reducing costs. The processes of rehabilitation can play an important role in helping an organisation to reduce the impact of these types of absence.

There is no agreed definition of short term and long term. Often short term refers to absences of 7 days or less which coincides with the period most organisations will accept self-certification. However, long term is not normally taken to be more than 7 days and more usually is defined as three weeks or more. Most organisations will have their own definition of short and long term absence and will have these contained within their absence policy.

In addition to single short term absences, there will be people who take very frequent one or two day absences which over the course of a year add up to a significant amount of lost time. Frequent short term absence has different causes from continuous long term absence but nevertheless the problem needs to be addressed.

Absence costs a great deal of money both in terms of direct salary loss, loss of business or business inefficiency and replacement costs and needs to be managed like any other aspects of an organisation. It is important to focus attention to where efforts to manage absence are most likely to pay dividends. It is pointless devoting energy to unavoidable short term absence but having control over frequent short term absence, avoidable short term absence and long term absence can be very cost effective. Rehabilitation can be seen to be the process by which people can be enabled back to work or to a more reliable work pattern (see R2007 below).

## The legal argument

**R2003** Although it is important to get people back to work as soon as possible and to limit the amount of sickness absence time taken, there is no legal duty to do so. The Department of Work and Pensions has recently produced 'A plan for getting people back to work if they are ill or have hurt themselves', but this is guidance for employers and contains no legal obligation. Rehabilitation often involves the intervention of Occupational Health Services but again many organisations will not have access to Occupational Health Provision and in the UK there is no legal obligation to do so (see R2009 below). Rehabilitation therefore is something employers will want to provide because of the benefits to the business rather than any direct legal necessity for it.

## Common law

**R2004** Despite there being no legal obligation to provide for rehabilitation, there are nevertheless a number of important legal ramifications. In the case of *Young v The Post Office [2002] IRLR 660*, the Court of Appeal ruled that employers were in breach of their duty of care in failing to ensure that arrangements made for the claimant's return to work were adhered to. In this case the claimant had returned after an absence of four months due to a nervous breakdown brought on by stress at work. The arrangements for his return for not adhered to with the result that, within seven weeks of returning to work, the claimant suffered a recurrence of his psychiatric illness. An important aspect of this judgement is that the court rejected the suggestion that it was up to the claimant to speak out if he felt that he was under stress. This was a stress case, but the principle would equally apply to return to work following other illnesses or injury. The employer is liable for the subsequent injury to the employee if adequate arrangements are not made for the return to work and/ or if the arrangements made are not adhered to. It is up to the employer to monitor the employee and not to rely on his or her reporting of any worsening of their condition.

This case and others like it (see *Walker v Northumberland County Council [1995] 1 All ER 737*), have established an important principle. If an employer knows that an employee has been off sick and will be returning with a vulnerability, then the employer has a legal duty to ensure that the employee's residual condition is not gong to be made worse on return to work. However the employer's liability will be subject to the usual principles of law. These are as follows.

| | |
|---|---|
| Foreseeability | The employee must show that his employer should have been able to reasonably foresee that the conditions of work on his return would have made his condition worse or prevented his recovery. Foreseeability will depend on what an employer knows or ought reasonably to know about the employee's condition. For an employee who has been off sick for reasons that are well documented by sick notes then it would be difficult for an employer to argue that he did not know of the employee's condition. If the employer has any doubts about the vulnerability the employee may suffer on his return to work and how this vulnerability might be affected by the conditions of work, then it would considered reasonable for the employer to have taken steps to find out. This would normally mean referral to an Occupational Health Professional and a request for an opinion on what adjustments might need to be made to ensure any risks inherent in the employee's return to work were minimised (see R2009 below). |
| Breach of duty | An employer is in breach of his duty of care to his employee if he has failed to take reasonable steps to prevent the employee being harmed. What is reasonable will depend on a number of factors including the size of the undertaking, it's resources, and what is feasible. It may eventually be necessary for an employer to argue in court that what he did, or failed to do, was reasonable. It is important to note that breaches of statutory regulations can often now be used to show breach of common law duty. The most significant change has occurred with The *Management of Health and Safety at Work and Fire Precautions (Workplace) (Amendment) Regulations 2003, (SI 2003 No 2457)* which has amended the *Management of Health and Safety at Work Regulations 1999 (SI 1999 No 3242)*. It is now possible for employees to bring a civil claim for damages as a result of the failure of an employer to comply with *SI 1999 No 3242*. In terms of rehabilitation the most likely implication is that employers will now have to ensure that risk assessments and risk controls (including the need for health surveillance under regulation 6) required under *SI 1999 No 3242* are competently carried out and amended to include any factors which arise as a result of the employee's altered state of heath on return to work. |
| Causation | The employee will need to show that the breach of duty has caused, or has materially contributed, to the harm that he has suffered. It would not be sufficient to show that the work by itself caused an exacerbation of the residual illness or injury someone returned to work with. It must be shown that the employer did not do all that was reasonably practicable (and/or did not comply with relevant statutory duties), and that this breach of duty was responsible for the harm caused. |
| Apportionment and quantification | Where a worsening of a condition on return to work may have more than one cause, then the courts may have to decide on apportioning the damages. The courts may also have to decide on any contribution the claimant may have made to the worsening of his condition (by not complying with the risk controls put in place by the employer for example) and may also have to decide whether the deterioration in a person's condition may have been going to happen anyway (because for example it is a progressive condition). The principle is that the employer should only pay for that proportion of the damage caused by his breach of duty. |

## The Disability Discrimination Act 1995 and The Disability Discrimination Act 1995 (Amendment) Regulations 2003 (SI 2003 No 1673)

**R2005** Another aspect of the legal argument for ensuring that rehabilitation is firmly on an organisation's agenda, is the requirements of the *Disability Discrimination Act 1995* (*DDA 1995*). This Act requires employers to treat people with disabilities no differently from those without that disability and to make reasonable adjustments to the work or the workplace to enable someone with a disability to work. *DDA 1995* applies to all aspects of employment including at the point of recruitment, but in the context of rehabilitation its importance is that it may need to be applied when someone returns to work following an illness or injury. However, much depends on the definition of disability and what adjustments might be considered reasonable under the Act. If an employee claims breach of *DDA 1995*, redress has to be taken through an employment tribunal not the courts.

Under the Act a disability is regarded as any mental or physical impairment that has a substantial and long-term adverse effect on a person's ability to carry out their normal day-to-day activities. Therefore if an employee returns to work with a residual condition, following an illness, injury or operation, and the condition is likely only to be temporary before the person returns to normal full capacity, then that would not fall within the requirements of *DDA 1995*. However if the condition is expected to be long term (the act defines this as lasting at least twelve months) then adjustments have to made. Long term may also be defined as recurring, if it is likely that the condition (for example epilepsy) may recur at least once in twelve months.

*DDA 1995* gives examples of adjustments that may be necessary and which would be considered reasonable. These include making adjustments to premises, re-deploying someone to a different job or different job location, altering the person's work hours, allocating some of the person's job to another person or allowing time off for rehabilitation or treatment. This is not an exhaustive list and many other types of adjustment such as providing new work equipment or modifying existing equipment would need to be considered.

## Health and safety and the Disability Discrimination Act 1995

**R2006** There is a potential conflict between an employer's duties under health and safety law and disability discrimination law. This arises when an employer may decide that it is unsafe to employ a person with a disability or, when considering rehabilitation, that it is unsafe to retain someone in the workplace or allow them back to work. The employer may come to the decision that a person has become a danger to themselves or to other people in the workplace as a result of a disability and uses health and safety law to justify not allowing someone back to work. The Health and Safety Executive's report on this (HSE Research Report No 167 – HSE Books), suggest that in 40% of cases heard at industrial tribunals this justification was not upheld and the tribunal ruled that adequate adjustments were possible but had not been made. What happens though if an employer decides that the it is unsafe for an individual to continue in employment and no reasonable adjustments are possible, but the employee wants to continue working? Case law on this is confused. In the case of *Withers v Perry Chain Co Ltd. [1961] 3 AER*, it was concluded that there was no common law duty to dismiss in these cases. Although this precedes *DDA 1995*, a similar judgement was made in the Appeal Court in the case of *Hatton v Sutherland and three other appeals [2002] PIQR P21*. However, in *Coxall v Goodyear Great Britain Ltd. [2002] IRLR 742*, Mr Coxall brought a claim for damages against his employer for allowing him to continue to work in an environment which had caused his asthma, even though he had wanted to continue working. The Court of Appeal held that an employer may be under a duty to stop an employee from working if he is put at risk by continuing to do so, even though it was the employees wish to continue.

When an employee has been off sick with an illness (whether physical or mental) or an injury or develops a chronic illness and when the individual appears ready to return to work, then the implications of *DDA 1995* and health and safety law have to be weighed up very carefully. If there is a residual disability which appears to be both long term and has substantial adverse effects on the person's ability to carry out day to day activities, then *DDA 1995* applies and the question of reasonable adjustment to avoid discrimination have to be considered. If there appears to be a health and safety reason why the person should not continue in that employment then it may be necessary to cease the person's employment. However, this step should only be made after thorough examination of the possibility of adjustment including additional health and safety precautions have been made.

## Managing sickness absence

**R2007** A number of terms are used to describe the systems that should be in place to reduce absence through sickness. These include:

- absence management;
- sickness absence management;
- attendance management.

Absence management addresses the whole problem of non-attendance at work, some of which include:

- genuine absence for sickness and injury,

but may also include:

- absence for other reasons such as routine dental and doctors appointments;
- staying at home with sick children;
- compassionate leave;
- unauthorised absence.

An organisation will normally want to have a policy on such issues and rules by which it's employees will be expected to conform as part of their contract of employment. When referred to specifically, sickness absence means absence from work because of mental or physical ill-health, injury or recovery from treatments such as surgery, chemotherapy or radio-therapy. Within the boundaries of sickness absence lies the problematic question of absence for non-genuine sickness. Forty to fifty per cent of short term sickness absence may be due to non-genuine reasons (see R2002 above).

Attendance management is not dissimilar to absence management but places the emphasis on the more positive aspects of encouraging people to stay at work or return to work rather than the sometimes negative and heavy-handed methods employed to discourage sickness absence. There is a recent drive in the public services to reduce the level of short-term sickness absence. This has translated into policies to discourage employees from taking sick leave even though the reasons may be genuine. For example some employers now require a return to work interview with the line manager after just one day's sick leave, others require staff who are off sick to report their intended absence on a daily basis to an occupational health nurse, others have penalised sickness absence by withdrawing privileges such as the right to work overtime.

The reverse of absenteeism is presenteeism. Presenteeism is the trend, encouraged by stigmatising absenteeism, of people being at work who should be off for reasons of ill health. Kivimaki et al (2005) have found that there is twice the risk of coronary heart disease episodes (non-fatal myocardial infarction and fatal coronary heart disease) in men who took no sickness absence despite being ill, compared with those who took moderate levels of sickness absence. This seems to suggest that encouraging people to remain at work even though they are ill, has damaging health effects.

The conditions that people often endure and continue working include:

- asthma;
- migraine;
- allergies;
- back pain;
- other musculosketal disorders;
- depression;
- irritable bowel syndrome.

Consider too that there is a huge cost of having people at work who are not well. There could be safety implications, increased risk of accident and reductions in efficiency and productivity. Two studies by Stewart and his colleagues in America have shown that presenteeism with depression costs US Companies $31 billion (£16.5 billion) per year and with common pain conditions $61 billion (£32.5 billion) per year. Pain conditions include headache, back pain, arthritis and other musculoskeletal complaints. The extremely common complaint of headache loses companies on average 3.5 hours per affected employee per week (Stewart et al 2003). Sickness absence management therefore has to take a very careful line between exerting too much pressure on genuinely sick people to remain at work and being too lax in allowing those who are not genuinely ill to take time off without question.

The key to managing sickness absence is to have a well developed policy which is written, transparent and made available to everyone The *Employment Rights Act 1996*, requires employers to provide staff with information on 'any terms and conditions relating to incapacity for work due to sickness or injury, including any provision for sick pay'. The Chartered Institute of Personnel and Development (CIPD) (http://www.cipd.co.uk), suggest that a policy should include:

- details of contractual sick pay terms and its relationship with statutory sick pay;
- the process employees must follow if taking time off sick-covering when and whom employees should notify if they are not able to attend work;
- when (after how many days) employees need a self-certificate form;
- when they require a medical certificate (sick-note) from their doctor to certify their absence;
- mention that the organisation reserves the right to require employees to attend an examination by a company doctor and (with the worker's consent) to request a report from the employee's doctor;
- provisions for return-to-work interviews as these have been identified as the most effective intervention to manage short-term absence.

To control sickness absence it is essential to have an effective system of recording absence (of all types) and to have clear policies of when interventions are required. In larger organisations this will require the full co-operation of line managers and supervisors who will have to ensure that absence is recorded in a central location (usually in HR). Employers must be careful not to breach the *Data Protection Act 1998*, when they store information on employees' health. The Information Commissioner's Office has issued guidance on storing and using data on workers health. (The Employment Practices. Data Protection Code. Part 4. Information about Workers Health). Local managers will also have to be able to give return-to-work interviews and might need to be trained to do this sensitively, confidentially and be able to recognise when a case that gives concern (for example when a work-related illness or injury appears to be developing or when there appears to non-genuine causes or persistent short term absences). There should be clear guidance to managers on who to report problem cases to and this is normally a more senior manager or human resources. In addition to this, someone should have responsibility for monitoring the central absence record to pick up particular patterns of absence (e.g. the persistent Monday absence), to identify particular departments or sections of the workforce with high absence rates, and to detect when someone has reached the threshold for long term absence (usually three weeks). Managers should be given regular absence information for their department. Appropriate interventions for long-term absence will then need to be put in place, which might include the need to consider rehabilitation when the individual returns to work.

According to the CIPD fewer than half of employers monitor the cost of absence or produce data on their absence rates. Sometime this is because sickness absence is not recorded (or not recorded reliably) but sometimes there is just no interest in making use of sickness absence data despite the considerable cost of it to the organisation (see R2002 above). The most common measure of absence is the lost-time rate. This expresses the percentage of the total time available which has been lost to absence in a given period (usually an annual lost time rate).

It is calculated from:

Total Absence (hours or days) / Possible total hours or days worked x 100.

Thus if someone has 20 days absence in a year and could possibly have worked 220 days then the annual lost time rate is:

$20/220 \times 100 = 9.1\%$

It is very useful as part of an absence policy to have a target for lost time rate and some organisations have set this as low as 2%.

The CIPD suggest a number of useful interventions in managing short term absence these include:

- return-to-work interviews;
- disciplinary procedures for unacceptable absence;
- use of trigger mechanisms to review attendance;
- involving line managers in absence management;
- providing sickness absence information to line managers;
- restricting sick pay;
- managers are trained in absence handling;
- occupational health professionals involvement.

Their suggested interventions for long term absence include:

- occupational health involvement;
- line management involvement as part of the absence management programme;
- restricting sick pay;
- changes to work patterns or environment;
- return-to-work interviews;
- rehabilitation programme;
- managers are trained in absence handling.

## Long term absence and rehabilitation

**R2008** Long term absence may need a carefully thought through plan for return to work. The potential legal pitfalls of not having a comprehensive return-to-work plan, or of not adhering to it, were discussed earlier (see R2003 above). The longer an employee is off sick the more likely that a plan for rehabilitation will be needed. However the need for a return-to-work plan will also depend on the nature of the illness or injury for which the person has been off sick and may be essential even if they have only been absent for a relatively short time. This will be important if the employee has a condition such as stress or a musculoskeletal injury and particularly if the condition may have been caused by the conditions of work. Indeed any condition which may be work-related, such as a developed asthma or dermatitis, (neither of which are likely to entail excessively long-term absence), will need to be carefully managed on the person's return to work.

It is important to keep in touch with the long-term absent employee on a regular basis during their absence. This can be done by, say, a weekly telephone call from HR or the immediate line manager to discuss how they are progressing and, at an appropriate time, to start to discuss when they might return to work and what the return-to-work plan might consist of and what adjustments might need to be made.

Return to work options include the following.

- A phased return. A gradual re-introduction to the workplace starting with part-time and building up over a planned number of weeks to full-time. A phased return is an extremely important way of getting someone back to work but will be resisted by some managers in some occupations as being too disruptive and difficult to manage. However, a relatively short period of difficulty will be likely to ensure that the person will successfully return to full productivity; without it they may never return.

- Altering the employee's hours, either temporarily or permanently. Moving the employee from shift work to permanent days. Allowing flexi-working.

- Moving the employee to another job or another department in the organisation. This might be necessary to avoid lifting and carrying following a back injury, to avoid exposure to a chemical that has caused an allergy, for example, or if someone is stressed because of harassment or bullying. It might also be necessary if the job entails driving and if the employee has developed a condition which would now preclude them from holding a licence.

- Considering what adjustments might need to be made to accommodate a new condition that the employee now suffers. This might be, for example, epilepsy following brain surgery, partial paralysis or a speech defect following a stroke, or reduced physical capacity following a heart attack. In these cases there may be no need to redeploy the person; they may be able to carry out their normal work, but with some adjustments made to their workplace, work equipment or working practices. Remember that if the condition falls within the meaning of the *Disability Discrimination Act 1995*, then there is a legal requirement to make reasonable adjustments so as not to discriminate against the person (see *DDA 1995* at R2005 above).

Whatever the agreed return-to-work plan it is essential that someone, probably the line manager, is made responsible to ensure that it is adhered to, that it is reviewed if necessary and that the employee is kept fully consulted. The legal consequences of not seeing through a plan were highlighted in the case of *Young v The Post Office [2002] IRLR 660*, (see R2004 above).

## Using professional advice

R2009    It was mentioned earlier (at R2003 above), that there is no legal requirement for an occupational health service, and indeed many organisations in the UK do not have access to occupational health provision. However, a good occupational health practitioner will be invaluable in helping to manage absence generally and ensuring that rehabilitation is successful. Alternatively, and it sometimes becomes the case, an occupational health physician may need to give advice on the necessity for retirement on the grounds of ill-health.

Many different practitioners contribute to the overall practice of occupational health but the two most relevant are occupational health nurses and occupational health physicians. Having said that however, it is possible that in individual circumstances the advice or input of an ergonomist, an occupational hygienist, physiotherapist or safety practitioner may be required. However the occupational health nurse or occupational health physician will be able to advise on the need for additional expert help and one of these two should always be the first point of reference.

Occupational health nurses (also called occupational health advisors), are qualified nurses who have specialised in occupational health. The occupational health nurse is likely to be the key professional advising on the rehabilitation of employees and advising on suitable return to work options and any necessary work adjustments. The OH nurse would also play an important role in monitoring the progress on return to work.

Occupational health nurses are registered nurses whose names appear on the statutory register to practice held by the Nursing and Midwifery Council (NMC). In addition, an occupational health nurse will hold a qualification in occupational health, i.e. Occupational Health Nursing Certificate (OHNC), post-graduate diploma or degree and their names will be recorded on the specialist register of the Nursing and Midwifery Council. Confirmation of a nurse's general registration and their entry to the specialist practitioner register can be made to the:

Registration Department

Nursing and Midwifery Council

23 Portland Place

London

W1B 1PZ

Tel: 020 7333 9333

Occupational physicians are doctors who have specialised in occupational medicine. An occupational health physician will have at the very least, an AFOM (Associate of the Faculty of Occupational Medicine), although they would not be regarded as accredited specialists unless they have full membership or fellowship. (MFOM/FFOM). Full members are admitted to the accreditation register held by the faculty. Some doctors also hold an MSc in Occupational Medicine or a Diploma in Industrial Health (DIH). These are academic qualifications in occupational medicine but holders of these qualifications may or may not be accredited specialists. Confirmation of a doctors accreditation can be obtained from:

The Registration Department

The General Medical Council

178 Great Portland Street

London

W1N 6JE

Tel: 020 7915 3630

Fax: 020 7915 3558

Organisations who do not have their own occupational health provision can buy into a group practice such as that run by the Robens Centre for Occupational Health and Safety (http://www.rcohs.com) or can buy services from NHS Plus (http://www.nhsplus.nhs.uk).

## Further reading

R2010    Building Capacity for Work. A UK Framework for Vocational Rehabilitation. The Department of Work and Pensions. 27 October 2004.

Tolley's Guide to Employee Rehabilitation. Lexis Nexis Tolley. 2004

Kivimaki M., Head J., Ferrie J.E., Hemingway H., Shipley M.J., Vahtera J., and Marmot M.G. (2005) Working While Ill as a risk factor for Serious Coronary Events: The Whitehall II Study. American Journal of Public Health, 95 (1) 98–102.

Stewart W.F., Ricci J.A., Chee E., Hahn S.R and Morganstein D. (2003) Cost of Lost Productive Work Time Among US Workers with Depression. JAMA 289: 3135–3144

Stewart W.F., Ricci J.A., Chee E., Morganstein D. and Lipton R. (2003). Lost Productive Time and Cost Due to Common Pain Conditions in the US Workforce. JAMA 290: 2443–2454

# Risk Assessment

## Introduction

**R3001** Although the term 'risk assessment' probably first came into common use as a result of the *Control of Substances Hazardous to Health Regulations 1988* (commonly known as the '*COSHH Regulations*' and revised several times since), similar requirements had actually previously been contained in both the *Control of Lead at Work Regulations 1980* and the *Control of Asbestos at Work Regulations 1987*.

In practice a type of risk assessment had already been necessary for some years particularly as a result of the use of the qualifying clause 'so far as is reasonably practicable' in a number of the sections of the *Health and Safety at Work etc Act 1974* ('*HSWA 1974*').

## HSWA 1974 requirements

**R3002** *HSWA 1974, s 2* contains the general duties of employers to their employees with the most general contained within *s 2(1)*:

'It shall be the duty of every employer to ensure, so far as is reasonably practicable, the health, safety and welfare at work of all employees.'

Other more specific requirements are contained in *s 2(2)* and these are also qualified by the term 'reasonably practicable'.

*HSWA 1974, s 3* places general duties on both employers and the self-employed in respect of persons other than their employees. *Section 3(1)* states:

'It shall be the duty of every employer to conduct his undertaking in such a way as to ensure, so far as is reasonably practicable, that persons not in his employment who may be affected thereby are not exposed to risks to their health or safety.'

Employers thus have duties to contractors (and their employees), visitors, customers, members of the emergency services, neighbours, passers-by and the public at large. This may to a certain point extend to include trespassers. Self-employed persons are put under a similar duty by virtue of *HSWA 1974, s 3(2)* and must also take care of themselves. In each case these duties are subject to the 'reasonably practicable' qualification.

*HSWA 1974, s 4* places duties on each person who has, to any extent, control of non-domestic premises used for work purposes in respect of those who are not their employees. *HSWA 1974, s 6* places a number of duties on those who design, manufacture, import or supply articles for use at work, or articles of fairground equipment and those who manufacture, import or supply substances. Many of these obligations also contain the 'reasonably practicable' qualification.

### What is reasonably practicable?

**R3003** The phrase 'reasonably practicable' is not just included within the key sections of *HSWA 1974* but is also contained in a wide variety of regulations. Lord Justice Asquith provided a definition in his judgment in the case of *Edwards v National Coal Board [1949] 1 All ER 743* in which he stated:

'Reasonably practicable' is a narrower term than "physically possible" and seems to me to imply that a computation must be made by the owner in which the quantum of risk placed on one scale and the sacrifice involved in the measures necessary for averting risk (whether in money, time or trouble) is placed in the other, and that, if it be shown that there is a gross disproportion between them – the risk being insignificant in relation to the sacrifice – the defendants discharge the onus on them. Moreover, this computation falls to be made by the owner at a point in time anterior to the accident.'

*HSWA 1974, s 40* places the burden of proof in respect of what was or was not 'reasonably practicable' (or 'practicable', see R3004 below) on the person charged with failure to comply with a duty or requirement. Employers and other duty holders must establish the level of risk involved in their activities and consider the various precautions available in order to determine what is reasonably practicable i.e. they must carry out a form of risk assessment.

## Practicable and absolute requirements

**R3004** Not all health and safety law is qualified by the phrase 'reasonably practicable'. Some requirements must be carried out 'so far as is practicable'. 'Practicable' is a tougher standard to meet than 'reasonably practicable' – the precautions must be possible in the light of current knowledge and invention (*Adsett v K and L Steelfounders & Engineers Ltd [1953] 1 WLR 773*). Once a precaution is practicable it must be taken even if it is inconvenient or expensive. However, it is not practicable to take precautions against a danger which is not yet known to exist (*Edwards v National Coal Board [1949] 1 All ER 743*), although it may be practicable once the danger is recognised.

Many health and safety duties are subject to neither 'practicable' nor 'reasonably practicable' qualifications. These absolute requirements usually state that something 'shall' or 'shall not' be done. However, such duties often contain other words which are subject to a certain amount of interpretation e.g. 'suitable', sufficient', 'adequate', 'efficient', 'appropriate' etc.

In order to determine whether requirements have been met 'so far as is practicable' or whether the precise wording of an absolute requirement has been complied with, a proper evaluation of the risks and the effectiveness of the precautions must be made.

# The Management of Health and Safety at Work Regulations 1999

**R3005** The Management Regulations were introduced in 1992 and revised by the 1999 Regulations. They were intended to implement the European Framework Directive (89/391) on the introduction of measures to encourage improvements in the safety and health of workers at work. The *Management of Health and Safety at Work Regulations 1999 (SI 1999 No 3242), Reg 3* require employers and the self-employed to make a suitable and sufficient assessment of the risks to both employees and persons not in their employment. The purpose of the assessment is to identify the measures needed 'to comply with the requirements and prohibitions imposed ... by or under the relevant statutory provisions ... ' i.e. identifying what is needed to comply with the law.

Given the extremely broad obligations contained in *HSWA 1974, ss 2, 3, 4* and *6*, all risks arising from work activities should be considered as part of the risk assessment process (although some risks may be dismissed as being insignificant). Compliance with more specific requirements of regulations must also be assessed, whether these are absolute obligations or subject to 'practicable' or 'reasonably practicable' qualifications. The requirement for risk assessment introduced in the Management Regulations simply formalised what employers (and others) should have been doing all along i.e. identifying what precautions they needed to take to comply with the law. More detailed requirements of the Management Regulations are covered later in the chapter.

Any such additional precautions must then be implemented (the *Management of Health and Safety at Work Regulations 1999* (*SI 1999 No 3242*), *Reg 5* contains requirements relating to the effective implementation of precautions).

## Common regulations requiring risk assessment

R3006 An increasing number of codes of regulations contain requirements for risk assessments. Several of these regulations are of significance to a wide range of work activities and are dealt with in more detail elsewhere in the looseleaf. The main regulations are as follows.

### Control of Substances Hazardous to Health Regulations 2002 (COSHH)

R3007 The *Control of Substances Hazardous to Health Regulations 2002* (*SI 2002 No 2677*), *Reg 6* require employers to make a suitable and sufficient assessment of the risks created by work liable to expose any employees to any substance hazardous to health and of the steps that need to be taken to meet the requirements of the regulations. See HAZARDOUS SUBSTANCES IN THE WORKPLACE, which provides more details of the COSHH requirements.

The Health and Safety Executive ('HSE') booklet L5, contains both the regulations and the associated Approved Code of Practice. There are many other relevant HSE publications including HSG97 '*A step by step guide to COSHH assessment*'.

### Noise at Work Regulations 1989

R3008 The *Noise at Work Regulations 1989* (*1989 No 1790*), *Reg 4* require employers to make a noise assessment which is adequate for the purposes of:

- identifying which employees are exposed to noise above defined action levels;
- providing information to comply with other duties under the Regulations (e.g. reduction of noise exposure, provision of ear protection, establishment of ear protection zones and informing employees).

See NOISE AND VIBRATION which provides more details of the requirements of the *Noise at Work Regulations 1989*. HSE booklet L108 '*Reducing Noise at Work*' contains guidance both on the regulations and on the assessment process.

### Manual Handling Operations Regulations 1992

R3009 Employers are required by the *Manual Handling Operations Regulations 1992* (*SI 1992 No 2793*), *Reg 4* to make a suitable and sufficient assessment of all manual handling operations at work which involve a risk of employees being injured, and to take appropriate steps to reduce the risk to the lowest level reasonably practicable (They must avoid such manual handling operations if it is reasonably practicable to do so).

See MANUAL HANDLING which provides further details of the Regulations and the carrying out of assessments. HSE booklet L23, '*Manual Handling: Manual Handling Operations Regulations 1992 – guidance on regulations*' provides detailed guidance on the Regulations and manual handling assessments.

### Health and Safety (Display Screen Equipment) Regulations 1992

R3010 The *Health and Safety (Display Screen Equipment) Regulations 1992* (*SI 1992 No 2792*), *Reg 2* require employers to perform a suitable and sufficient analysis of display screen equipment (DSE) workstations for the purpose of assessing risks to 'users' or 'operators' as defined in the Regulations. Risks identified in the assessment must be reduced to the lowest extent reasonably practicable.

The HSE booklet L26 *'Work with display screen equipment'* contains guidance on the Regulations and on workstation assessments.

## Personal Protective Equipment at Work Regulations 1992

**R3011** Under the *Personal Protective Equipment at Work Regulations 1992 (SI 1992 No 2966), Reg 6* employers must ensure that an assessment is made to determine risks which have not been avoided by other means and identify personal protective equipment ('PPE') which will be effective against these risks. The Regulations also contain other requirements relating to the provision of PPE; its maintenance and replacement; information, instruction and training; and the steps which must be taken to ensure its proper use.

See PERSONAL PROTECTIVE EQUIPMENT which provides further details of the requirements under the Regulations. HSE booklet L25, *'Personal protective equipment at work'* provides detailed guidance on the Regulations and the assessment of PPE needs.

## Fire Precautions (Workplace) Regulations 1997

**R3012** The *Fire Precautions (Workplace) Regulations 1997 (SI 1997 No 1840)* together with the *Management of Health and Safety at work Regulations 1999 (SI 1999 No 3242)*, make it quite explicit that employers must carry out an assessment of fire risks and fire precautions. *Part II* of the 1997 Regulations contains specific requirements relating to fire safety which must be included in the assessment.

Amendments to the 1997 Regulations in 1999 removed the exemption from risk assessment requirements originally given to holders of fire certificates (under the *Fire Precautions Act 1971* and other legislation). They too are required to carry out fire risk assessments.

See FIRE AND FIRE PRECAUTIONS which contains further details both on the Regulations and on fire precautions generally. The HSE/Home Office publication *'Fire Safety: An employer's guide'* provides extensive guidance on factors to be taken into account during fire risk assessments.

## Control of Asbestos at Work Regulations 2002

**R3013** The *Control of Asbestos at Work Regulations 2002 (SI 2002 No 2675)* continue the requirement contained in the 1987 Regulations of the same title, for employers to make a risk assessment before carrying out work which is liable to expose employees to asbestos. *Regulation 6* contains several specific requirements on what the risk assessment must involve and other regulations detail many precautions which must be taken in asbestos work. Some types of asbestos work can only be done by companies licensed by the HSE. There are several important HSE reference publications on work with asbestos:

- L11 A guide to the Asbestos (Licensing) Regulations 1983 as amended;
- L27 Work with asbestos which does not normally require a licence;
- L28 Work with asbestos insulation, asbestos coating and asbestos insulating board.

The 2002 Regulations introduced new requirements for 'The management of asbestos in non-domestic premises'. These are contained in *Regulation 4* which does not come into operation until 21 May 2004. Duty holders under this regulation are defined within it and will normally be the owner or leaseholder, the employer occupying the premises or a combination of these (depending upon the terms of any lease or contract or who controls the premises).

Duty holders must carry out an assessment of their premises in order to identify any asbestos-containing material ('ACM') and assess its condition. They must then prepare a written plan for managing the risks from ACM, including such measures as regular inspections of confirmed and presumed ACM, restricting access to or maintenance in ACM areas and, where necessary, protection, sealing or removal of ACM.

Further details on these requirements are contained in ASBESTOS and also in HAZARDOUS SUBSTANCES IN THE WORKPLACE.

Important HSE references worth referring to are:

- L127 The management of asbestos in non-domestic premises.
- HSG227 A comprehensive guide to managing asbestos in premises.

### Dangerous Substances and Explosive Atmospheres Regulations 2002

R3014    These Regulations (known sometimes as 'DSEAR') replace many older pieces of legislation including the *Highly Flammable Liquids and Liquefied Petroleum Gases Regulations 1972* and many of the licensing requirements of the *Petroleum (Consolidation) Act 1928*. They are concerned with protecting against risks from fire and explosion. The definition of 'dangerous substances' contained within the regulations includes such substances as petrol, many solvents, some types of paint, LPG etc. and also potentially explosive dusts.

Employers are required by the *DSEAR (SI 2002 No 2776), Reg 5* to carry out a risk assessment of work involving dangerous substances. The *DSEAR (SI 2002 No 2776), Reg 6* require a range of specified control measures to be implemented in order to eliminate or reduce risks as far as reasonably practicable. Places where explosive atmospheres may occur must be classified into zones and equipment and protective systems in these zones must meet appropriate standards. [*DSEAR (SI 2002 No 2776), Reg 7*]. The *DSEAR (SI 2002 No 2776), Reg 8* require equipment and procedures to deal with accidents and emergencies whilst the *DSEAR (SI 2002 No 2776), Reg 9* states that employees must be provided with information, instruction and training. Further details are contained in HAZARDOUS SUBSTANCES IN THE WORKPLACE.

## Specialist regulations requiring risk assessment

R3015    Some codes of regulations requiring risk assessments are of rather more specialist application, although some are still covered in some detail elsewhere in the publication. References to relevant chapters of the looseleaf and to key HSE publications of relevance are included below. These regulations include:

- *Control of Lead at Work Regulations 2002 (SI 2002 No 2676)*

    See HAZARDOUS SUBSTANCES IN THE WORKPLACE.

    L132 Control of lead at work.

- *Supply of Machinery (Safety) Regulations 1992 (SI 1992 No 3073)*

    The Regulations include a variety of procedures which must be followed in assessing conformity of machinery with essential health and safety requirements set out in the Machinery Directive.

    See MACHINERY SAFETY.

    INDG270 Supplying new machinery: Advice to suppliers, free leaflet.

    INDG271 Buying new machinery: A short guide to the law, free leaflet.

- *Control of Major Accident Hazard Regulations 1999 (SI 1999 No 743) (COMAH)*

The Regulations only apply to sites containing specified quantities of dangerous substances.

See DISASTER AND EMERGENCY MANAGEMENT SYSTEMS (DEMS).

L111 A guide to the Control of Major Accident Hazard Regulations 1999.

HSG 190 Preparing safety reports: Control of Major Accident Hazard Regulations 1999.

HSG 191 Emergency planning for major accidents: Control of Major Accident Hazard Regulations.

- *Ionising Radiations Regulations 1999 (SI 1999 No 3232)*

    See HAZARDOUS SUBSTANCES IN THE WORKPLACE.

    L121 Work With Ionising Radiation: Ionising Radiations Regulations 1999 – approved code of practice and guidance.

# Related health and safety concepts

**R3016** Risk assessment techniques are an essential part of other health and safety management concepts.

- *Safe systems of work*

    Employers are required under *HSWA 1974*, s *2(2)(a)* to provide and maintain 'systems of work that are, so far as is reasonably practicable, safe and without risks to health'.

    Both the *Confined Spaces Regulations 1997 (SI 1997 No 1713)* and the *Lifting Operations and Lifting Equipment Regulations 1998 (LOLER) (SI 1998 No 2307)* contain similar requirements for safe systems of work. A safe system of work can only established through a process of risk assessment.

- *'Dynamic risk assessment'*

    The term 'dynamic risk assessment' is often used to describe the day to day judgements that employees are expected to make in respect of health and safety. However, employers must ensure that employees have the necessary knowledge and experience to make such judgements. The employer's 'generic risk assessments' must have identified the types of risks which might be present in the work activities, established a framework of precautions (procedures, equipment etc.) which are likely to be necessary and provided guidance on which precautions are appropriate for which situations.

- *Permits to work*

    A permit to work system is a formalised method for identifying a safe system of work (usually for a high risk activity) and ensuring that this system is followed. The permit issuer is expected to carry out a dynamic risk assessment of the work activity and should be more competent in identifying the risks and the relevant precautions than those carrying out the work.

- *CDM health and safety plans*

    A key component of the *Construction (Design and Management) Regulations 1994 (CDM) (SI 1994 No 3140)* is the requirement for a health and safety plan. Essentially this process requires an assessment of risks involved in the project and the identification and eventual implementation of appropriate precautions. (The requirements of the CDM Regulations are covered in more detail in CONSTRUCTION AND BUILDING OPERATIONS.)

- *Method statements*

    Method statements usually involve a description of how a particular task or operation is to be carried out and should identify all the components of a safe system of work arrived at through a process of risk assessment.

## Management Regulations requirements

R3017     The general requirement concerning risk assessment is contained in the *Management of Health and Safety at Work Regulations 1999* (*'the Management Regulations'*) (*SI 1999 No 3242*), *Reg 3*. Changes to the original regulations passed in 1992 mean that the risk assessment must now include fire risks and precautions (see R3012 above) and also risks to both young persons (under 18's) (see R3024 below) and new and expectant mothers (see R3025 below). Other regulations require more specific types of risk assessment e.g. of hazardous substances (COSHH), noise, manual handling operations, display screen equipment and personal protective equipment (see R3006–R3015 above).

*Regulation 3(1)* of the Management Regulations states:

'every employer shall make a suitable and sufficient assessment of:

(a)    the risks to the health and safety of his employees to which they are exposed whilst they are at work; and

(b)    the risks to the health and safety of persons not in his employment arising out of or in connection with the conduct by him of his undertaking,

for the purpose of identifying the measures he needs to take to comply with the requirements or prohibitions imposed upon him by or under the relevant statutory provisions and by Part II of the Fire Precautions (Workplace) Regulations 1997.'

*Regulation 3(2)* imposes similar requirements on self-employed persons.

*Regulation 3(3)* requires a risk assessment to be reviewed if:

- there is reason to suspect that it is no longer valid; or
- there has been a significant change in the matters to which it relates.

*Regulation 3(4)* requires a risk assessment to be made or reviewed before an employer employs a young person, and *Regulation 3(5)* identifies particular issues which must be taken into account in respect of young persons (especially their inexperience, lack of awareness of risks and immaturity). Further requirements in respect of young persons are contained in *Regulation 19*. *Regulation 16* contains specific requirements concerning the factors which must be taken into account in risk assessments in relation to new and expectant mothers. These relate to processes, working conditions and physical, biological or chemical agents. Assessments in respect of young persons and new or expectant mothers are dealt within more detail later in the chapter.

*Regulation 3(6)* requires employers who employ five or more employees to record:

—    the significant findings of their risk assessments; and

—    any group of employees identified as being especially at risk.

Methods of recording assessments are described later in this chapter (see R3041 below).

HSE booklet L21, *'Management of health and safety at work'*, contains the Management Regulations in full, the associated Approved Code of Practice ('ACoP') and Guidance on the Regulations.

## Hazards and risks

**R3018** The ACoP to the Management Regulations (*SI 1999 No 3242*) provides definitions of both hazard and risk.

A *hazard* is something with the potential to cause harm.

A *risk* is the likelihood of potential harm from that hazard being realised.

The *extent of the risk* will depend on:

- the likelihood of that harm occurring;
- the potential severity of that harm (resultant injury or adverse health effect);
- the population which might be affected by the hazard i.e. the number of people who might be exposed.

As an illustration, work at heights involves a hazard of those below being struck by falling objects. The extent of the risk might depend on factors such as the nature of the work being carried out, the weight of objects which might fall, the distance to the ground below, the numbers of people in the area etc.

## Evaluation of precautions

**R3019** The ACoP to the Management Regulations (*SI 1999 No 3242*) clearly states that risk assessment involves 'identifying the hazards present and evaluating the extent of the risks involved, taking into account existing precautions and their effectiveness.'

The evaluation of the effectiveness of precautions is an integral part of the risk assessment process. This is overlooked by some organisations who concentrate on the identification (and often the quantification) of risks without checking whether the intended precautions are actually being taken in the workplace and whether these precautions are proving effective.

Using the illustration in the previous paragraph, the risk assessment must take account of what precautions are required, such as:

- use of barriers and/or warning signs at ground level;
- use of tool belts by those working at heights;
- provision of edge protection on working platforms;
- use of head protection by those at ground level.

## 'Suitable and sufficient'

**R3020** Risk assessments under the Management Regulations (*SI 1999 No 3242*) (and several other regulations) must be 'suitable and sufficient', but the phrase is not defined in the Regulations themselves. However, the ACoP to the Regulations states that 'The level of risk arising from the work activity should determine the degree of sophistication of the risk assessment.' The ACoP also states that insignificant risks can usually be ignored, as can risks arising from routine activities associated with life in general ('unless the work activity compounds or significantly alters those risks').

In practice, a risk can only be concluded to be insignificant if some attention is paid to it during the risk assessment process and, if there is any scope for doubt, it is prudent to state in the risk assessment record which risks are considered insignificant.

Winter weather (with its attendant rain, ice, snow or wind) may be considered to pose a routine risk to life. However, driving a fork lift truck in an icy yard or carrying out agricultural or construction work in a remote location may involve a far greater level of risk than normal and require additional precautions to be taken.

The risk assessment must take into account both workers and other persons who might be affected by the undertaking.

For example, a construction company would need to consider risks to (and from) their employees, sub contractors, visitors to their sites, delivery drivers, passers by and even possible trespassers on their sites.

Similarly a residential care home should take into account risks to (and from) their staff, visiting medical specialists, visiting contractors, residents and visitors to residents.

## Reviewing risk assessments

**R3021** The *Management of Health and Safety at Work Regulations 1999* (*SI 1999 No 3242*), *Reg 3(3)* require a risk assessment to be reviewed if:

'(a) there is reason to suspect it is no longer valid; or

(b) there has been a significant change in the matters to which it relates ; and where as a result of any such review changes to an assessment are required, the employer or self-employed person concerned shall make them.'

The ACoP to the Management Regulations states that those carrying out risk assessments 'would not be expected to anticipate risks that were not foreseeable.' However, what is foreseeable can be changed by subsequent events. An accident, a non-injury incident or a case of ill-health may highlight the need for a risk assessment to be reviewed because:

- a previously unforeseen possibility has now occurred;

- the risk of something happening (or the extent of its consequences) is greater than previously thought;

- precautions prove to be less effective than anticipated.

A review of the risk assessment may also be required because of significant changes in the work activity e.g. changes to the equipment or materials used, the environment where the activity takes place, the system of work used or to the numbers or types of people carrying out the activity. The need for a risk assessment review may be identified from elsewhere e.g. from others involved in the same work activity, through trade or specialist health and safety journals, from the suppliers of equipment or materials or from the HSE or other specialist bodies. Routine monitoring activity (inspections, audits etc.) or consultation with employees may also identify the need for an assessment to be reviewed. A review of the risk assessment does not necessarily require a repeat of the whole risk assessment process but it is quite likely to identify the need for increased or changed precautions.

In practice workplaces and the activities within them are constantly subject to gradual changes and the ACoP states 'it is prudent to plan to review risk assessments at regular intervals'. The frequency of such reviews should depend on the extent and nature of the risks involved and the degree of change likely. It is advisable that all risk assessments should be reviewed at least every five years.

There are many activities where the nature of the work or the workplace itself changes constantly. Examples of such situations are construction work or peripatetic maintenance or repair work. Here it is possible to carry out 'generic' assessments of the types of risks involved and the types of precautions which should be taken. However, some reliance must be placed upon workers themselves to identify what precautions are appropriate for a given set of circumstances or to deal with unexpected situations. Such workers must be well informed and well trained in order for them to be competent to make what are often called 'dynamic' risk assessments (see R3016 above).

## Related requirements of the Management Regulations

**R3022** A number of other requirements within the Management Regulations (*SI 1999 No 3242*) are closely related to the risk assessment process.

- *Regulation 4: Principles of prevention to be applied*

    The *Management of Health and Safety at Work Regulations 1999* (*SI 1999 No 3242*), Reg 4 and *Sch 1* require preventive and protective measures to be implemented on the basis of specified principles.

- *Regulation 5: Health and safety arrangements*

    Under the *Management of Health and Safety at Work Regulations 1999* (*SI 1999 No 3242*), Reg 5, employers are required to have appropriate arrangements for the effective planning, organisation, control, monitoring and review of preventive and protective measures. Those employers with five or more employees must record these arrangements. This application of the 'management cycle' to health and safety matters is dealt with in greater detail in R3023 below.

- *Regulation 6: Health surveillance*

    Under the *Management of Health and Safety at Work Regulations 1999* (*SI 1999 No 3242*), Reg 6, where risks to employees are identified through a risk assessment, they must be 'provided with such health surveillance as is appropriate.' (Surveillance is also likely to be necessary to comply with the requirements of more specific regulations e.g. *COSHH* (*SI 2002 No 2677*), *Control of Asbestos at Work Regulations 2002* (*SI 2002 No 2675*), *Ionising Radiations Regulations 1999* (*SI 1999 No 3242*)). Surveillance may be appropriate to deal with risks such as colour blindness and other vision defects (e.g. in electricians and train or vehicle drivers) or blackouts or epilepsy (e.g. for drivers, operators of machinery and those working at heights).

    HSE guidance is available on this important area (see HSG61 *'Health surveillance at work'*).

- *Regulation 8: Procedures for serious and imminent danger and for danger areas*

    The *Management of Health and Safety at Work Regulations 1999* (*SI 1999 No 3242*), Reg 8 require every employer to 'establish and where necessary give effect to appropriate procedures to be followed in the event of serious and imminent danger to persons at work in his undertaking.' It also refers to the possible need to restrict access to areas 'on grounds of health and safety unless the employee concerned has received adequate health and safety instruction.'

    The need for emergency procedures or restricted areas should of course be identified through the process of risk assessment. Situations 'of serious and imminent danger' might be due to fires, bomb threats, escape or release of hazardous substances, out of control processes, personal attack, escape of animals etc. Areas may justify access to them being restricted because of the presence of hazardous substances, unprotected electrical conductors (particularly high voltage), potentially dangerous animals or people etc.

## Application of the management cycle

**R3023** The time spent in carrying out risk assessments will be wasted unless those precautions identified as being necessary are actually implemented. The *Management of Health and Safety at Work Regulations 1999* (*SI 1999 No 3242*), Reg 5 require the application of a five stage 'management cycle' to this process.

- Planning

    The employer must have a planned approach to health and safety involving such measures as:

- a health and safety policy statement;
- annual health and safety plans;
- development of performance standards.

- Organisation

  An organisation must be put in place to deliver these good intentions e.g:

  - individuals allocated responsibility for implementing the health and safety policy or achieving elements of the health and safety plan;
  - arrangements for communicating with and consulting employees (using such measures as health and safety committees, newsletters, noticeboards);
  - the provision of competent advice and information.

- Control

  Detailed health and safety control arrangements must be established. The extent of such arrangements will reflect the size of the organisation and the degree of risk involved in its activities but they are likely to involve:

  - formal procedures and systems e.g. for accident and incident investigation, fire and other emergencies, the selection and management of contractors;
  - the provision of health and safety related training e.g. at induction, for the operation of equipment or certain activities, for supervisors and managers;
  - provision of appropriate levels of supervision.

- Monitoring

  The effectiveness of health and safety arrangements must be monitored using such methods as:

  - health and safety inspections (both formal and informal);
  - health and safety audits;
  - accident and incident investigations.

- Review

  The findings of monitoring activity must be reviewed and, where necessary, the management cycle applied once again to rectify shortcomings in health and safety arrangements.

  The review process might involve:

  - joint health and safety committees;
  - management health and safety meetings;
  - activities of health and safety specialists.

The management cycle can also be applied to ensure the effective implementation of individual health and safety management procedures (e.g. for the investigation and reporting of accidents and non-injury incidents) or to specific types of health and safety precautions (e.g. guarding of machinery or precautions for working at heights).

## Children and young persons

**R3024** Previous Acts and regulations identified many types of equipment which children and young persons were not allowed to use or processes or activities that they must not be involved in. Many of these 'prohibitions' were revoked by the *Health and Safety (Young Persons) Regulations 1997 (SI 1997 No 135)*, since incorporated into the *Management of Health and Safety*

**R3025** *Risk Assessment*

at *Work Regulations 1999 (SI 1999 No 3242)*. (A few 'prohibitions' still remain). The emphasis has now changed to restrictions on the work which children and young persons are allowed to do, based upon the employer's risk assessment.

The *Health and Safety (Training for Employment) Regulations 1990 (SI 1990 No 1380)* have the effect of giving students on work experience training programmes and trainees on training for employment programmes the status of 'employees'. The immediate provider of their training is treated as the 'employer'. (There are exceptions for courses at educational establishments, i.e. universities, colleges, schools etc.). Therefore employers have duties in respect of all children and young persons at work in their undertaking: full-time employees, part-time and temporary employees and also students or trainees on work placement with them.

The term 'child' and 'young person' are defined in the *Management of Health and Safety at Work Regulations 1999 (SI 1999 No 3242), Reg 1(2)*.

- *Child* is defined as a person not over compulsory school age in accordance with:
  - the *Education Act 1996, s 8* (for England and Wales); and
  - the *Education (Scotland) Act 1980, s 31* (for Scotland).

  (In practice this is just under or just over the age of sixteen)

- *Young Person* is defined as 'any person who has not attained the age of eighteen'.

Some of the prohibitions remaining from older health and safety regulations use different cut off ages.

See CHILDREN AND YOUNG PERSONS which provides much more detail on the requirements of the Management Regulations and the factors to be taken into account when carrying out risk assessments in respect of children and young persons. These factors centre around their:

- lack of experience;
- lack of awareness of existing or potential risks; and
- immaturity (in both the physical and psychological sense).

That chapter also contains in APPENDIX C a checklist of 'Work presenting increased risks for children and young persons'.

## New or expectant mothers

**R3025** Amendments made in 1994 to the previous Management Regulations implemented the European Directive on Pregnant Workers, requiring employers in their risk assessments to consider risks to new or expectant mothers. These amendments were subsequently incorporated into the 1999 Management Regulations. The *Management of Health and Safety at Work Regulations 1999 (SI 1999 No 3242), Reg 1* contains two relevant definitions:

- '*New or expectant mother*' means an employee who is pregnant; who has given birth within the previous six months; or who is breastfeeding.
- '*Given birth*' means 'delivered a living child or, after twenty-four weeks of pregnancy, a stillborn child.'

The requirements for 'risk assessment in respect of new and expectant mothers' are contained in the *Management of Health and Safety at Work Regulations 1999 (SI 1999 No 3242), Reg 16* which states in *paragraph (1)*:

'(1) Where —

  (a) the persons working in an undertaking include women of child-bearing age; and

(b) the work is of a kind which could involve risk, by reason of her condition, to the health and safety of a new or expectant mothers, or to that of her baby, from any processes or working conditions, or physical, biological or chemical agents, including those specified in Annexes I and II of Council Directive 92/85/EEC on the introduction of measures to encourage improvements in the safety and health at work of pregnant workers and workers who have recently given birth or are breastfeeding,

the assessment required by regulation 3(1) shall also include an assessment of such risk.'

*Regulation 16(4)* states that in relation to risks from infectious or contagious diseases an assessment must only be made if the level of risk is in addition to the level of exposure outside the workplace. (The types of risk which are more likely to affect new or expectant mothers are described in R3026 below). *Regulation 16(2)* and *(3)* set out the actions employers are required to take if these risks cannot be avoided. *Paragraph (2)* states:

'Where, in the case of an individual employee, the taking of any other action the employer is required to take under the relevant statutory provisions would not avoid the risk referred to in paragraph (1) the employer shall, if it is reasonable to do so, and would avoid such risks, alter her working conditions or hours of work.'

Consequently where the risk assessment required under *Reg 16(1)* shows that control measures would not sufficiently avoid the risks to new or expectant mothers or their babies, the employer must make reasonable alterations to their working conditions or hours of work.

In some cases restrictions may still allow the employee to substantially continue with her normal work but in others it may be more appropriate to offer her suitable alternative work.

Any alternative work must be:

— suitable and appropriate for the employee to do in the circumstances;

— on terms and conditions which are no less favourable.

*Regulation 16(3)* states:

'If it is not reasonable to alter the working conditions or hours of work, or if it would not avoid such risk, the employer shall, subject to section 67 of the 1996 Act, suspend the employee from work for so long as is necessary to avoid such risk.'

(The 1996 Act referred to is the *Employment Rights Act 1996* which provides that any such suspension from work on the above grounds is on full pay. However, payment might not be made if the employee has unreasonably refused an offer of suitable alternative work.)

The *Management of Health and Safety at Work Regulations 1999 (SI 1999 No 3242), Reg 17* deals specifically with night work by new or expectant mothers and states:

'Where —

(a) a new or expectant mother works at night; and

(b) a certificate from a registered medical practitioner or a registered midwife shows that it is necessary for her health or safety that she should not be at work for any period of such work identified in the certificate,

the employer shall, subject to section 46 of the 1978 Act, suspend her from work for so long as is necessary for her health or safety.'

Such suspension (on the same basis as described above) is only necessary if there are risks arising from work. The HSE do not consider there are any risks to pregnant or breastfeeding workers or their children working at night per se. They suggest that any claim from an employee that she cannot work nights should be referred to an occupational health specialist. The HSE's own Employment Medical Advisory Service are likely to have a role to play in such cases.

The requirements placed on employers in respect of altered working conditions or hours of work and suspensions from work only take effect when the employee has formally notified the employer of her condition. The *Management of Health and Safety at Work Regulations 1999 (SI 1999 No 3242), Reg 18* states:

> 'Nothing in paragraph (2) or (3) of regulation 16 shall require the employer to take any action in relation to an employee until she has notified the employer in writing that she is pregnant, has given birth within the previous six months, or is breastfeeding.'

*Regulation 18(2)* states that the employer is not required to maintain action taken in relation to an employee once the employer knows that she is no longer a new or expectant mother; or cannot establish whether this is the case.

## Risks to new or expectant mothers

**R3026** The HSE booklet HSG122 *'New and expectant mothers at work: A guide for employers'* provides considerable guidance on those risks which may be of particular relevance to new or expectant mothers, including those listed in the EC Directive on Pregnant Workers (92/85/EEC). These risks might include:

*Physical agents*

- Manual handling.
- Ionising radiation.
- Work in compressed air.
- Diving work.
- Shock, vibration etc.
- Movement and posture.
- Physical and mental pressure.
- Extreme heat.

*Biological agents*

Many biological agents in hazard groups 2, 3 and 4 (as categorised by the Advisory Committee on Dangerous Pathogens) can affect the unborn child should the mother be infected during pregnancy.

*Chemical agents*

- Substances labelled with certain risk phrases.
- Mercury and mercury derivatives.
- Antimitotic (cytotoxic) drugs.
- Agents absorbed through the skin.
- Carbon Monoxide.
- Lead and lead derivatives.

*Working conditions*

> The HSE's frequently stated position in respect of work with display screen equipment (DSE) is that radiation from DSE is well below the levels set out in international recommendations, and that scientific studies taken as a whole do not demonstrate any link between this work and miscarriages or birth defects. There is no need for pregnant women

to cease working with DSE. However, the HSE recommend that, to avoid problems from stress or anxiety, women are given the opportunity to discuss any concerns with someone who is well informed on the subject.

Other factors associated with pregnancy may also need to be taken into account e.g. morning sickness, backache, difficulty in standing for extended periods or increasing size. These may necessitate changes to the working environment or work patterns.

## Other vulnerable persons

**R3027** As well as children and young persons and new or expectant mothers, there are other vulnerable members of the workforce who merit special consideration during the risk assessment process. Such persons include:

- *Lone workers*

    Some staff members may work alone continuously or only part of the time. They may be working on the employer's own premises or outside, within the community. The level of risk may relate to the worker's location, the activities they are involved in or the type of people they might encounter.

    Amongst lone workers will be:

    — security staff;
    — cleaners;
    — isolated receptionists or enquiry staff;
    — some retail staff;
    — some maintenance workers;
    — staff working late, at nights or at weekends;
    — keyholders (particularly at risk if called to suspected break-ins);
    — delivery staff;
    — sales and technical representatives;
    — property surveyors etc.

- *Disabled persons*

    Legislation concerning disability discrimination will affect general access requirements within work premises. The *Health and Safety (Miscellaneous Amendments) Regulations 2002 (SI 2002 No 2174)* introduced an additional requirement into *Regulation 25* of the *Workplace (Health, Safety and Welfare) Regulations 1992 (SI 1992 No 3004)*:

    'Where necessary, those parts of a workplace (including in particular doors, passageways, stairs, showers, washbasins, lavatories and workstations) used or occupied directly by disabled persons at work shall be organised to take account of such persons.'

    In considering access in general and also specific risks which may affect disabled persons, account must be taken of all possible types of disability:

    — persons lacking mobility (including wheelchair users);
    — those lacking physical strength (particularly in relation to manual handling);
    — persons with relevant health conditions (especially on COSHH-related matters);
    — blind or visually impaired persons;
    — those who have hearing difficulties;

- persons with special educational needs.

Whilst employers have particular duties towards their own employees, the risk assessment must also consider risks to disabled persons who are:

- on work placements;
- visitors;
- service users;
- neighbours or passers-by.

All types of risks must be considered but particular note must be taken of:

- access posing particular dangers to disabled persons;
- the presence of hazardous substances;
- potentially dangerous work equipment;
- hot items or surfaces;
- the need for special emergency arrangements for disabled persons.

- *Inexperienced workers*

Some older persons may (like young people) be lacking in experience of some types of work or lack awareness of existing or potential risks of particular work activities. Some may also lack a suitable degree of maturity, particularly those with special needs. This must be taken into account during the risk assessment process, especially in situations such as:

- training or retraining programmes;
- work activities with rapid staff turnover;
- activities involving people with special needs.

## Who should carry out the risk assessment?

R3028    The *Management of Health and Safety at Work Regulations 1999 (SI 1999 No 3242)*, *Reg 7* require employers to appoint competent health and safety assistance. Such persons must have sufficient training and experience or knowledge together with other qualities, in order to be able to identify risks and evaluate the effectiveness of precautions to control those risks. (*Regulation 7(8)* expresses a preference for such persons to be employees as opposed to others e.g. consultants).

The HSE guidance on risk assessment states that in small businesses the employer or a senior manager may be quite capable of carrying out the risk assessment. Larger employers may create risk assessment teams who might be drawn from managers, engineers and other specialists, supervisors or team leaders, health and safety specialists, safety representatives and other employees. Many employers continue to utilise consultants to co-ordinate or carry out their risk assessments. Whether the assessments are to be carried out by an individual or by a team, others will need to be involved during the process. Managers, supervisors, employees, specialists etc. will all need to be consulted about the risks involved in their work and the precautions that are (or should be) taken.

## Assessment units

R3029    In a small workplace it may be possible to carry out a risk assessment as a single exercise but in larger organisations it will usually be necessary to split the assessment up into manageable units. If done correctly this should mean that assessment of each unit should not take an inordinate amount of time and also allows the selection of the people best able to assess an individual unit. Division of work activities into assessment units might be by department or sections, buildings or rooms, processes or product lines, or services provided.

As an illustration a garage might be divided into:

- Servicing and repair workshop.
- Body repair shop.
- Parts department.
- Car sales and administration.
- Petrol and retail sales.

## Relevant sources of information

R3030    There is a wide range of documents which might be of value during the risk assessment process including:

- *Previous risk assessments*

    (including risk assessments carried out to comply with specific regulations e.g. COSHH or Noise regulations).

- *Operating procedures*

    (where these include health and safety information).

- *Safety handbooks etc.*
- *Training programmes and records*
- *Accident and Incident records*
- *Health and safety inspection or audit reports*
- *Relevant Regulations and Approved Codes of Practice*

    (Risk assessment involves an evaluation of compliance with legal requirements and therefore an awareness of the legislation applying to the workplace in question is essential.)

- *Relevant HSE publications*

    The HSE publishes a wide range of booklets and leaflets providing guidance on health and safety topics. Some of these relate to the specific requirements of regulations, others deal with specific types of risks whilst some publications deal with sectors of work activity.

    All of these can be of considerable value in identifying which risks the HSE regard as significant and in providing benchmarks against which precautions can be measured. The guidance on specific types of workplaces (which includes engineering workshops, motor vehicle repair, warehousing, kitchens and food preparation, golf courses, horse riding establishments and many others) should form an essential basis for those carrying out assessments in those sectors.

    HSE Books regularly publishes a detailed catalogue of HSE publications. Their booklet '*Essentials of Health and Safety at Work*' provides an excellent starting point for those in small businesses needing guidance or carrying out risk assessment. The booklet also contains a useful reference section to other HSE publications which may be of relevance. To obtain these publications contact HSE Books, PO Box 1999, Sudbury, Suffolk CO10 2WA (tel: 01787 881165; fax: 01787 313995; website: www.hsebooks.co.uk).

- *HSE website*

    The HSE website at www.hse.gov.uk contains an ever-growing resource of guidance material. Some of this is available for specific work areas, e.g. agriculture, construction, engineering, food manufacture, haulage, motor vehicle repair, offshore oil and gas, railways, textiles and footwear. It is also possible to search for information on a wide range of particular risk topics. Much of this information (particularly free HSE publications) can be downloaded directly from the website. A subscription service (HSE Direct) also allows access to and downloading of priced HSE publications.

- *Trade Association codes of practice and guidance*
- *Information from manufacturers and suppliers*

    (Equipment handbooks or substance data sheets).

- *General health and safety reference books*

## Consider who might be at risk

R3031    The risk assessment process must take account of all those who may be at risk from the work activities i.e. both employees and others. It is important that all of these are identified.

- *Employees*

    Different categories of employees to take into account might include:

    — production workers;

    — maintenance workers;

    — administrative staff;

    — security officers;

    — cleaners;

    — delivery drivers;

    — sales representatives;

    — others working away from the premises;

    — temporary employees.

    Some of these employees may merit special considerations:

    — children and young persons (see R3024);

    — women of childbearing age (i.e. potential new or expectant mothers) (see R3025 and R3026);

    — other vulnerable persons (see R3027);

- *Contractors and their staff*

    Contracted services might involve:

    — construction or engineering projects;

    — routine maintenance or repair;

    — hire of plant and operators;

    — support services e.g. catering, cleaning, security, transport;

    — professional services e.g. architects, engineers, trainers;

    — supply of temporary staff.

    (Contractors also have duties to carry out risk assessments in respect of their own staff.)

- *Others at risk*

    The types of people might be put at risk by the organisation's activities will depend upon the nature and location of those activities. Groups of people to be considered include:

    — volunteer workers;
    — co-occupants of premises;
    — occupants of neighbouring premises;
    — drivers making deliveries;
    — visitors (both individuals and groups);
    — residents e.g. in the care or hospitality sectors;
    — passers-by;
    — users of neighbouring roads;
    — trespassers;
    — customers or service users.

## Identify the issues to be addressed

**R3032** The issues which will need to be addressed during the risk assessment process should be identified. This may be done in respect of the assessment overall or separately for each of the assessment units and will be based upon the knowledge and experience of those carrying out the assessment and the information gathered together from the sources described earlier. This will create an initial list of headings and sub-headings for the eventual record of the risk assessment, although in practice this list is likely to be amended along the way. The sample assessment records contained later in the chapter (see R3049 and R3050) will demonstrate how such a list can be built up for typical workplaces.

These headings are likely to consist of the more common types of risk e.g. fire, vehicles, work at heights, together with some specialised types of risk associated with the work activities such as violence, lasers or working in remote locations. A checklist of possible risks to be considered during risk assessments is provided in R3034 below. This includes some of the risks which have specific regulations associated with them.

In some situations particular notes may also be made to check on the effectiveness of the precautions which should be in place to control the risks e.g. standards of machine guarding, compliance with personal protective equipment (PPE) requirements or the quality and extent of training.

## Variations in work practices

**R3033** Consideration should also be given at this stage to possible variations in work activities which may create new risks or increase existing risks. Such variations might involve:

- Fluctuations in production or workload demands.
- Reallocation of staff to meet changing workloads.
- Seasonal variations in work activities.
- Abnormal weather conditions.
- Alternative work practices forced by equipment breakdown/unavailability.
- Urgent or 'one-off' repair work.

- Work carried out in unusual locations.
- Difference in work between days, nights or weekends.

Further variations may emerge later on in the assessment process.

## Checklist of possible risks

**R3034** This checklist is intended to assist in identifying which issues need to be addressed during risk assessments. In some cases the headings and/or sub-headings might be used in the form shown, in other cases it may be more appropriate to combine them or modify the titles.

*Work Equipment*
- Process machinery
- Other machines
- Powered tools
- Handtools
- Knives/blades
- Fork-lift trucks
- Cranes
- Lifts
- Hoists
- Lifting equipment
- Vehicles

*Services/power sources*
- Electrical installation
- Compressed air
- Steam
- Hydraulics
- Other pressure systems
- Buried services
- Overhead services

*Access*
- Vehicle routes
- Rail traffic
- Pedestrian access
- Work at heights
- Ladders and stepladders
- Scaffolding
- Mobile elevating work platforms
- Falling objects
- Glazing

*Storage*
- Shelving and racking
- Stacking
- Silos and tanks
- Waste

*Work activity*
- Burning or welding
- Entry into confined spaces
- Electrical work
- Excessive fatigue or stress
- Handling cash/valuables
- Use of compressed gases
- Molten metal

*Fire and explosion prevention*
- Flammable liquids *
- Flammable gases *
- Storage of flammables *
- Hot work

* DSEAR requirements

*External factors*
- Violence or aggression
- Robbery

*Work locations*
- Heat
- Cold

- Large crowds
- Animals
- Clients' activities

*Other factors*
- Vibration
- Lasers
- Ultra violet/infra red radiation
- Work related upper limb disorder
- Severe weather
- Deep water
- Tides
- Remote locations
- Work alone
- Homeworking
- Poor hygiene
- Infestations
- Work abroad
- Work in domestic property
- Clients' premises
- Site security

| Risks/issues subject to separate assessment requirements | |
|---|---|
| Hazardous substances (COSHH) | Lead |
| Noise | Asbestos work |
| Manual handling | Asbestos in premises |
| Display screen equipment workstations | Conformity of machinery |
| PPE needs | Major accident hazards (COMAH) |
| Fire precautions | Ionising radiation |
| Dangerous Substances and Explosive Atmospheres (DSEAR) | |

## Making the risk assessment

**R3035** Good planning and preparation can reduce the time spent in actually making the risk assessment as well as enabling that time to be used much more productively. However, it is essential that time is spent in work locations, seeing how work is actually carried out (as opposed to how it should be carried out).

### Observation

**R3036** Observation of the work location, work equipment and work practices is an essential part of the risk assessment process. Where there are known to be variations in work activities a sufficient range of these should be observed to be able to form a judgement on the extent of the risks and the adequacy of precautions. Evaluations can be made of the effectiveness of fixed guards, the suitability and condition of access equipment, compliance with PPE requirements, the observance of specified operating procedures or working practices and many other aspects. It should also be borne in mind that work practices may change once workers realise they are under observation. Initial or undetected observation of working practices may be the most revealing.

### Discussions

**R3037** Discussions with people carrying out work activities, their safety representatives and those responsible for supervising or managing them are also an essential part of risk assessment. Amongst aspects of the work that might be discussed are possible variations in the work activities, problems that workers encounter, workers' views of the effectiveness of the precautions available, the reasons some precautions are not utilised and their suggestions for improving health and safety standards.

### Tests

R3038    In some situations it may be appropriate to test the effectiveness of safety precautions e.g. the efficiency of interlocked guards or trip devices, the audibility of alarms or warning devices or the suitability of access to remote workplaces e.g. crane cabs, roofs. When carrying out such tests care must be taken by those carrying out the assessment not to endanger themselves or others, nor to disrupt normal activities.

### Further investigations

R3039    Frequently further investigations will need to be made before the assessment can be concluded. Such investigations may involve detailed checks on standards or records, or enquiries into how non-routine situations are dealt with. Examples of further checks or enquiries which might be appropriate are:

- Detailed requirements of published standards e.g. design of guards, thickness of glass.
- Contents of operating procedures.
- Maintenance or test records.
- Contents of training programmes or training records.

Once again the potential list is endless although lines of further enquiry should be indicated by the observations and discussions during the initial phase of the assessment.

### Notes

R3040    Rough notes should be made throughout the assessment process. It will be on these notes that the eventual assessment record will be based. It will seldom be possible to complete an assessment record 'on the run' during the assessment itself. The notes should relate to anything likely to be of relevance, such as:

- Risks discounted as insignificant.
- Further detail on risks or additional risks identified during the assessment.
- Risks which are being controlled effectively.
- Descriptions of precautions which are in place and effective.
- Precautions which do not appear to be effective.
- Alternative precautions which might be considered.
- Problems identified or concerns expressed by others.
- Related procedures, records or other documents.

## Assessment records

R3041    This phase of the risk assessment process concludes with the preparation of the assessment records. It may be preferable to record the findings in draft form initially, with a revised version being produced once the assessment has been reviewed more widely and/or recommended actions have been completed.

- *Decide on the record format*

    Details of the content of assessment records and examples of one assessment record format are provided later in the chapter (see R3049 and R3050) but many alternatives are available. Different types of records may be appropriate for different departments, sections or activities. Use of a 'model' assessment might be relevant for similar workplaces or activities. Some (or all) of the assessment record might be integrated into documented operating procedures.

- *Identify the section headings to be used*

    During the preparatory phase of the assessment a list of risks and other sources to be addressed in each assessment unit was prepared as an aide memoire. This list will now need to be converted into section headings for the assessment records. As a result of the assessment some of the headings may have been sub-divided into different headings whilst others may have been merged.

- *Prepare the assessment records*

    The rough notes made during assessment must then be converted into formal assessment records. Preparing the records will normally require at least half of the time that was spent in the workplace carrying out the assessment and sometimes might even take longer. It is wise not to allow too long to elapse between assessing in the workplace and preparation of the assessment record. Notes will seem much more intelligible and memory will often be able to 'colour in' between the notes.

- *Identify the recommendations*

    The assessment will almost inevitably result in recommendations for improvements and these must be identified. As can be seen later, some assessment record formats incorporate sections in which recommendations can be included but in other cases separate lists will need to be prepared. The use of risk rating matrices (see below) may assist in the prioritisation of recommendations.

## *Risk rating matrices*

**R3042** The ACoP accompanying the *Management of Health and Safety at Work Regulations 1999 (SI 1999 No 3242)* describes how the risk assessment process needs to be more sophisticated in larger and more hazardous sites. Nevertheless it suggests that quantification of risk will only be appropriate in a minority of situations. Some organisations use risk-rating systems to assist in the identification of priorities. Most involve matrices, utilising a simple combination of the likelihood of a hazard having an adverse effect and the severity of the consequences if it did. Some use numbers to produce a risk rating, as in the example below:

| Risk Assessment Matrix | | | Likelihood of adverse effect | | |
|---|---|---|---|---|---|
| | | | Unlikely | Possible | Frequent |
| | | | 1 | 2 | 3 |
| Severity of consequences | Minor | 1 | 1 | 2 | 3 |
| | Moderate | 2 | 2 | 4 | 6 |
| | Severe | 3 | 3 | 6 | 9 |

The numbers can be replaced by descriptions of the level of risk as shown in the next example:

| Risk Assessment Matrix | | Likelihood of adverse effect | | |
|---|---|---|---|---|
| | | Unlikely | Possible | Frequent |
| Severity of consequences | Minor | Low | Low | Medium |
| | Moderate | Low | Medium | High |
| | Severe | Medium | High | Very high |

The author's view is generally against the use of such matrices on the basis that time is often spent considering risk values at the expense of evaluating the effectiveness of the controls which ought to be in place.

## After the assessment

**R3043** Whilst the recording of the risk assessment is an important legal requirement, it is even more important that the recommendations for improvement identified during the assessment are actually implemented. This is likely to involve several stages.

### Review and implementation of the recommendations

**R3044** There may be a need to involve others outside (and probably senior to) the risk assessment team. The reasoning behind the recommendations can be explained and various alternative ways of controlling risks can be evaluated. Changes to the assessment findings or recommendations may be made at this stage but the risk assessment team should not allow themselves to be browbeaten into making alterations that they do not feel can be justified. Similarly they should not hold back on making recommendations they consider are necessary just because they believe that senior management will not implement them. The assessment team should carry out their duties to the best of their abilities in identifying what precautions are necessary in order to comply with the law – the responsibility for achieving compliance rests with their employer.

Some recommendations may need to be costed in respect of the capital expenditure or staff time required to implement them. It is unlikely that all recommendations will be able to be implemented immediately; there may be a significant lead time for the delivery of materials or the provision of specialist services from external sources. Once costings and prioritisation have been agreed, the recommendations should be converted into an action plan with individuals clearly allocated responsibility for each element of the plan, within a defined timescale. Some members of the risk assessment team (particularly health and safety specialists) may have responsibility for implementing parts of the plan or providing guidance to others.

### Recommendation follow-up

**R3045** Even in well-intentioned organisations, recommendations for improvement that have been fully justified and accepted are often still not implemented. It is essential that the risk assessment process includes a follow-up of the recommendations made. The recording format provided later in the chapter includes reference to this. As well as establishing that the improvements have actually been carried out, consideration should also be given to whether any unexpected risks have inadvertently been created.

Once the follow-up has been carried out, the assessment record should be annotated or revised to take account of the changes made. If recommendations have not been implemented there is a clear need for the situation to be referred back to senior management for them to take action to overcome whatever are the obstacles to progress.

## Assessment review

**R3046** The *Management of Health and Safety at Work Regulations 1999 (SI 1999 No 3242)* and the associated ACoP state that assessments must be reviewed in certain circumstances (see R3021 above). The ACoP also states that 'it is prudent to plan to review risk assessments at regular intervals.' The frequency for reviews should be established which relates to the extent and nature of the risks involved and the likelihood of creeping changes (as opposed to a major change which would automatically justify a review).

The review may, however, conclude that the risks are unchanged, the precautions are still effective and that no revision of the assessments is necessary. The process of conducting regular reviews of assessments and, where appropriate, making revisions may be aided by the application of document control systems of the type used to achieve compliance with ISO 9000 and similar standards.

## Content of assessment records

**R3047** The *Management of Health and Safety at Work Regulations 1999 (SI 1999 No 3242)*, *Reg 3(6)* states:

'Where the employer employs five or more employees, he shall record —

(a) the significant findings of the assessment; and

(b) any group of his employees identified by it as being especially at risk.'

The accompanying ACoP refers to the record as representing 'an effective statement of hazards and risks which then leads management to take the relevant actions to protect health and safety.' It goes on to state that the record must be retrievable for use by management, safety representatives, other employee representatives or visiting inspectors. The need for linkages between the risk assessment, the record of health and safety arrangements (required by *Reg 5* of the Management Regulations) and the health and safety policy is also identified. The ACoP allows for assessment records to be kept electronically as an alternative to being in written form.

The essential content of any risk assessment record should be:

- Hazards or risks associated with the work activity.
- Any employees identified as especially at risk.
- Precautions which are (or should be) in place to control the risks (with comments on their effectiveness).
- Improvements identified as being necessary to comply with the law.

Other important details to include are:

- Name of the employer.
- Address of the work location or base.
- Names and signatures of those carrying out the assessment.
- Date of the assessment.
- Date for next review of the assessment.

(These might be provided as a introductory sheet.)

## Illustrative assessment records

**R3048** In the final pages of this chapter the risk assessment methodology described earlier is used to provide illustrations of how the process can be applied in two different types of workplace. In each case, relevant risks or issues to be addressed are listed, one of those risks or issues is selected, relevant regulations, references and other key assessment points relating to that risk or issue are identified and an illustration of how the completed assessment record might look is provided. The content of the form used is similar to that provided in the HSE's *'Five steps to risk assessment'* leaflet but the layout is felt to be more user-friendly.

The illustrations used are for a supermarket and a newspaper publisher.

## A supermarket

R3049   Relevant risk topics are likely to include:

- Food processing machinery – in the delicatessen.
- Knives – delicatessen and butchery.
- Fork-lift trucks – warehousing areas.
- Vehicle traffic – delivery vehicles, customer and staff vehicles.
- Vehicle unloading – in the goods inward area.
- Pedestrian access – external to and inside the store.
- Glazed areas – store windows, display cabinets.
- Electrical installation – the power system within the store.
- Electrical equipment – including tills, cleaning equipment, office equipment.
- Shelving – in the store.
- Racking – in warehouse areas.
- Refrigerators – in the store and warehouse.
- Fire – general risks only, precautions must take account of customers.
- Aggression – e.g. from unhappy customers.
- Possible robbery – the supermarket will hold large quantities of cash.
- WRULD – for checkout operators.
- Hazardous substances – cleaning materials, office supplies.
- Manual handling – e.g. shelf stacking, movement of trolleys.
- Display screen equipment – used at workstations in the office.
- PPE requirements – throughout the supermarket.
- Asbestos – the possible presence of asbestos-containing materials within the premises.
- Use of contractors – for maintenance and repair work.

A sample risk assessment record for **vehicle traffic** is provided below.

In conducting this risk assessment:

- The requirements of the *Workplace (Health, Safety and Welfare) Regulations 1992* must be complied with.
- HSE booklet HS(G)136 *'Workplace Transport Safety'* is likely to be relevant.
- Particular attention should be paid to:

- signage and road markings;
- observation of vehicle movements and speeds;
- conditions during busy periods and hours of darkness.

**ABC Supermarkets, Newtown**

## Risk Assessment

| Reference number: 4 | Risk topic/issue: Vehicle traffic | Sheet 1 of 1 |
|---|---|---|
| Cross references: HSE booklet IND(G) 136 Risk assessments 5 (vehicle unloading) and 6 (pedestrian access) | | |
| **Risks identified** | **Precautions in place** | **Recommended improvements** |
| Vehicles making deliveries to the 'Goods inward' bay present risks to each other and to any pedestrians on the access road | Prominent 10 mph are in place on the roadway. The speed limit is enforced effectively. Signs prohibit use of the road by pedestrians and customer or staff vehicles. | |
| There are also risks to other road users as they leave and rejoin the main road. | There are give way signs and road markings at the junction with the main road. Vehicles are parked in a holding area prior to backing up to the 'Goods Inward' bay. Movement of vehicles is controlled by a designated member of the supermarket staff. The area is well lit by roadside lamps and floodlights on the side of the building. | Reposition the advertising sign which partly blocks visibility on rejoining the main road Provide this designated staff member with a high visibility waterproof jacket. |
| Customer and staff vehicles circulating in the car park area present risks to each other and to pedestrians in the area. | The access road around the car park is one way and well indicated by signs. The entrance and exit are well separated from each other and the good access road. | |
| There are also risks to other road users at the entrance from and access back into the main road. | There are prominent 15 mph signs around the roadways (some vehicles exceed this speed). There are also some give way markings and signs at all roadway junctions. | Provide clearly marked speed ramps at suitable locations. |

| | |
|---|---|
| Parking bays are well marked. | The surface of some bays in the south-west corner of the car park should be repaired. |
| Pedestrian crossing points are clearly marked and signed. | |
| The area is well lit by lighting towers which are protected at the base. | |
| Security staff inspect, and where necessary, salt roadways in icy or snowy weather | Include the goods access road, Goods inward bay and car park in routine safety inspections. |
| **Signature(s)** A Smith, B Jones | **Name(s)** A Smith, B Jones | **Date** 11/12/00 |
| **Dates for** | **Recommendation follow up** March 2001 | **Next routine review** December 2003 |

## A newspaper publisher

**R3050** A large workplace like this would need to be divided into assessment units (see R3029 above), which might consist of:

- Common facilities, services etc. – e.g. fire, electrical supply, lifts, vehicle traffic, presence of asbestos.
- Reel handling and stands – e.g. supply of reels of newsprint to the press area.
- Platemaking – e.g. equipment and chemicals used to produce printing plates.
- Printing press – e.g. press machinery, solvents, noise etc.
- Despatch – e.g. inserting equipment, newspaper stacking, strapping and loading.
- Circulation and transport – e.g. distribution and other vehicles, fuel, waste disposal.
- Maintenance – e.g. workshops, garage, maintenance activities.
- Offices – e.g. editorial and administrative areas.

Selecting the **reel handling and stands** unit, risk topics are likely to include:

- Fork-lift trucks – used to unload, transport and stack reels.
- Reel storage – stability of stacks, access issues.
- Reel handling equipment – hoists, conveyors and the reel stands feeding the press.
- Wrappings and waste – removal of wrappings, storage and disposal of waste paper etc.
- Noise and dust – from the operation of the reel stands and nearby press.

A sample assessment record for **reel handling equipment** is provided below.

In conducting this risk assessment:

- The requirements of the *Provision and Use of Work Equipment Regulations 1998* (*PUWER*), the *Lifting Operation and Lifting Equipment Regulations 1998* and the *Manual Handling Operations Regulations 1992* must be complied with.
- Reference may need to be made to BS 5304:1988 'Safety of machinery' or to other standards for conveyors or specialist handling equipment.
- Particular attention should be paid to:
  — guarding standards and the possible presence of unguarded dangerous parts;
  — any need for manual handling of the reels;
  — training issues relating to the above;
  — statutory examination records (for the hoist).

Newtown News

## Risk Assessment

| Reference number: B3 | Risk topic/issue: **Reel handling and stands – reel handling equipment** | Sheet 1 of 1 |
|---|---|---|
| Cross references: Risk assessments B1 (Fork-lift trucks), B2 (Reel storage), B5 (Noise and dust) | | |
| Risks identified | Precautions in place | Recommended improvements |
| The equipment below presents risks to all staff working in the area. | | |
| *Reel hoist* – Carries reel down from the reels store to the reel stand basement. It is fed by fork-lift trucks and feeds onto the roller conveyor system. | Slow moving hoist protected by a substantial mesh guard.<br><br>No need for access within the hoist enclosure.<br><br>Sign states 'Do not ride on hoist'.<br><br>Gap between base of hoist and roller conveyor (no shear trap).<br><br>Statutory examinations by Insurance Engineers (kept by Work Engineer). | Replace this sign by one complying with the *Safety Signs Regulations*. |
| *Roller conveyor* – This is in a T formation and consists of powered and free running rollers with a turntable at the junction. Reels are transferred to holding bays or floor-based trolleys by fork-lift truck. | All drives for the conveyor are fully enclosed and there are no in-running nips. The conveyor is protected by kerbs from fork-lift damage.<br><br>Signs prohibit climbing on the conveyor system. | Replace these signs as above. |

## Risk Assessment

| | | |
|---|---|---|
| *Floor-based trolleys* – This system carries reels right up to the transfer carriages which load them onto the reel stands. | Reels can be easily moved by one person using the trolley system. Reels can be safely rolled across the floor onto the trolleys if the correct technique is used. | Include formal training on correct techniques in the induction programme for new staff in the area. |
| Some reels must be transferred manually from the holding bays onto the trolleys. | Two persons are required to move the transfer carriages up to the reel stands. | |
| *Reel stands* – The reel stands rotate mechanically ensuring a constant web feed to the press. | There are no accessible in-running nips. Operatives do not need to approach the rotating reels. | |
| The rotating drive shaft for the reel stands is approximately 4 metres above the ground level. | This shaft is considered to be 'safe by position' for normal operating purposes. | Ensure a safe system of work for maintenance work near the shaft, e.g. by using a permit to work procedure. |
| **Signature(s)** J Young, K Old | **Names(s)** J Young, K Old | **Date** 26/3/00 |
| **Dates for Recommendation follow up** June 2000 | | **Next routine review** March 2002 |

# Safe Systems of Work

## Introduction

**S3001** This chapter examines the need for safe systems of work ('SSWs') within the framework of an occupational safety and health management system ('OSHMS').

In essence SSWs outline the control measures needed to avoid or reduce risk in the workplace and are usually one of the main results of a suitable and sufficient risk assessment.

This chapter considers:

- the components of a safe system of work – both legal and others (S3002–S3006);
- appropriate safe system of work required for level of risk involved (S3007);
- development of SSWs, including job safety analysis; preparation of job safety instruction and safe operating procedures (S3008–S3010);
- permit to work systems (S3011–S3022);
- isolation procedures (S3023).

This particular chapter will be of use to guide those managers and supervisors who have the responsibility to develop and utilise SSWs, including permit to work systems, within their areas of control. It will also be useful to occupational safety and health ('OSH') practitioners who have to advise line management on practical development and implementation of such systems.

Essentially, a safe system of work is written down, and chronologically lists the methods of doing a particular job in such a way so as to avoid or minimise risk within the workplace. The safe system of work is therefore a very important workplace precaution or control measure which should be known, understood, adhered to and rigorously enforced within the workplace.

All employers should therefore utilise suitable and sufficient safe systems of work commensurate with the degree of risk highlighted in a risk assessment, especially when the legal requirement for SSWs are taken into account (see S3003 and S3004).

The pitfalls of not utilising SSWs within the risk control system include:

- an increase in the number of accidents/injuries/diseases;
- the increased risk of prosecution;
- increased employers' liability insurance premiums;
- a more dangerous workplace.

Remember – SAFE SYSTEMS OF WORK MEAN FEWER ACCIDENTS!

## Components of a safe system of work

**S3002** The components of a safe system of work comprise of legal elements – common law (S3003) and statute law (S3004) – and other components which explain where SSWs fit into the overall framework of an occupational safety and health management system ('OSHMS') (S3006). Indeed SSWs are considered to be the cornerstone of all successful OSHMSs.

## Legal components – common law

**S3003** The need for SSWs can be traced back to 1905 when Lord McLaren in *Bett v Dalmey Oil Co (1905)7F (Ct of Sess) 787* judged that:

> 'the obligation is threefold: the provision of a competent staff of men, adequate material, and a proper system and effective supervision.'

In 1938, the common law duty of care was highlighted in the case of *Wilsons and Clyde Coal Co Ltd v English [1938 AC 57]* where Lord Wright said that:

> 'the whole course of authority consistently recognises a duty which rests on the employer, and which is personal to the employer, to take reasonable care for the safety of his workmen, whether the employer be an individual, a firm or a company, and whether or not the employer takes any share in the conduct of the operations.'

This duty of care implies the need for SSWs and also that delegation to an employee does not remove the employer's duty to provide a safe system of work.

The question of what constitutes a safe system of work was further considered by the Court of Appeal in 1943 by Lord Greene, Master of the Rolls, in the case of *Speed v Thomas Swift Co Ltd [1943] 1 All ER 539*, who said:

> 'I do not venture to suggest a definition of what is meant by a system. But it includes, or may include according to circumstances, such matters as the physical layout of the job, the setting of the stage, the sequence in which the work is to be carried out, the provision of warnings and notices, and the issue of special instructions.'

He also said that: 'a system may be adequate for the whole course of the job, or it may have to be modified or improved to meet circumstances which arise: such modifications or improvements appear to me equally to fall under the heading of system.'

He added that: 'the safety of a system must be considered in relation to the particular circumstances of each particular job.'

The conclusion here is that the safe system of work must be tailored to the job to which it applies by co-ordinating the work, planning the arrangements and layout, stating the agreed method of work, providing the appropriate equipment, and giving employees the necessary information, instruction, training and supervision. All the above are designed to prevent harm to persons at work.

## Legal components – statute law

**S3004** The above common law precedents (at S3003) were taken into account during the drafting of the *Health and Safety at Work etc. Act 1974* ('*HSWA 1974*').

*Section 2(2)(a)* of *HSWA 1974* requires 'the provision and maintenance of plant *and systems of work* that are, so far as is reasonably practicable, safe and without risks to health'.

Furthermore, *section 2(3)* of *HSWA 1974* requires employers having five or more employees to prepare a *written* health and safety policy statement, together with the organisation and *arrangements* for carrying it out.

These 'written arrangements' should include a safe system of work.

This general duty has been tightened up and more clearly defined since the enactment of the *Management of Health and Safety at Work Regulations1999 (SI 1999 No 3242)* ('*MHSWR*').

Specifically, *Regulation 3* of *MHSWR (SI 1999 No 3242)* requires 'suitable and sufficient' risk assessments. A safe system of work should be the end result of the vast majority of risk assessments.

*Regulation 4* of *MHSWR* (*SI 1999 No 3242*) requires the 'Principles of Protection' to be applied in connection with the introduction of 'any preventive and protective measures'. These measures are control measures and are 'arrangements' within the written health and safety policy. These principles of protection are listed in *Schedule 1* to the Regulations and should be referred to when developing SSWs.

Further clarification of 'arrangements' is given in *Regulation 5* of *MHSWR* (*SI 1999 No 3242*): 'arrangements for the effective planning, organisation, control, monitoring and review of the preventive and protective measures'. This infers that as SSWs are part of the preventive and protective (i.e. control) measures, then they should be planned, monitored and reviewed.

## Summary of legal components of SSWs

S3005    From the common and statute law references above (S3004), it may be seen that the key elements of SSWs are:

- adequate plant and equipment:
    - the right tools for the job;
    - plant designed for safe access and isolation.
- competent staff:
    - properly trained and experienced in the work they do;
    - instructed clearly in the work to be done;
    - provided with the necessary information on substances, safe use of equipment etc.
- proper supervision:
    - to monitor the work as it progresses;
    - to ensure the safe system of work is adhered to.

## Other components

S3006    The bulk of the SSWs component parts are derived from the common and statute law considerations above.

The safe system of work may be therefore seen as the centre piece of a jigsaw which links together all the other component parts including:

- competent staff;
- premises;
- safe design;
- plant safeguards;
- safe installation;
- safe access/egress;
- tools and equipment;
- instruction;
- supervision;
- training;
- information;
- safety rules;

- planned maintenance;
- monitoring;
- PPE;
- environment.

Hence SSWs need to take account of all of the above components, and they need to be in writing.

The relative importance of each of the components will be task-dependent. Some tasks will be heavily dependent on the intrinsic safety built into the plant at the design stage. Other jobs will rely on the skills and expertise of the competent employee(s).

It may be of use therefore to consider the SSWs components under the sub-headings: hardware and software.

The 'hardware' includes: design, installation, premises/plant, tools and equipment, and working environment – heating, lighting, ventilation, noise control.

The 'software' comprises: planned maintenance, competent employees, adequate supervision, safety rules, training and information, correct use of PPE, and correct use of tools and equipment.

## Which type of safe system of work is appropriate for the level of risk?

S3007    Different jobs will require different systems, depending on the levels of risk highlighted via the risk assessment process.

A very low risk job may require adherence to safety rules or a previously agreed guide, whereas a very high risk job may well require a formal written permit to work system. Both types qualify as safe systems of work.

The following matrix may prove useful:

| Risk level | Type of SSW |
| --- | --- |
| Very high | Permit to Work |
| High | Permit (or written SSW) |
| Medium | Written SSW |
| Low | Written SSW |
| Very low | Verbal (with written back-up, such as safety rules) |

Once the SSWs are in place, there is a need to ensure that they are reviewed at least annually, so as to ensure:

- continued legislative compliance (N.B. any new legislation);
- continued compliance with most recent risk assessment;
- SSWs still work in practice;
- any plant modifications are incorporated;
- substituted (safer) substances are allowed for;
- new work methods are incorporated;
- advances in technology are exploited;
- control measures are improved in the light of accident experience; and
- continued involvement in, and awareness of the importance of, all relevant SSWs.

It is vital that regular feedback is needed to all concerned following any updates of existing SSWs.

In essence, SSWs are one of the most important foundations of any OSHMS. They need to be:
- known, i.e. brought to the attention of all relevant employees and third parties;
- well understood by all concerned;
- adhered to by all concerned; and
- enforced by management and supervision.

## Development of safe systems of work

S3008   The steps in the development of SSWs of all types are as follows:
- Hazard identification.
- Risk assessment.
- Define safety operating procedures/methods ('SOP's).
- Implement agreed system.
- Monitor system performance.
- Review system (at least annually).
- Provide feedback on system updates.

### Job safety analysis ('JSA')

S3009   JSA is also known as job hazard analysis and/or task analysis. It is a technique that has evolved from the work study techniques of method study and work measurement which, in turn, formed part of the scientific management approach in the early twentieth century.

The work study engineer analysed jobs in order to improve methods of production by eliminating unnecessary job steps or changing those steps that were inefficient.

From an occupational safety and health ('OSH') viewpoint, JSA assists in the elimination of hazardous steps, thus eliminating or avoiding risky operations.

In essence, JSA is a useful risk assessment tool as it identifies hazards, describes risks and assists in the development of commensurate control measures.

The original work study approach made use of the 'SREDIM' principle:

| | | |
|---|---|---|
| Select | = | work to be studied |
| Record | = | how work is done |
| Examine | = | the total job |
| Develop | = | the best method for doing the work |
| Install | = | the preferred method into the company's way of working |
| Maintain | = | the agreed method via training, supervision etc. |

Work study is used to break the total job down into its component parts and, by measuring (work measurement/time and motion) the quantity of work done in each of the component parts, the total job time could be established. If any of the component parts could be modified to make them more efficient or productive usually by saving time these improvements would be made as part of the overall study.

From measuring a number of similar tasks undertaken by different employees, standard times for each job component – and hence the total job-were derived. These were subsequently used in the development of payment and bonus systems.

JSA uses the SREDIM principle, but measures/assesses the risk content, as opposed to the work content, in each of the component parts of the job. From this detailed breakdown a safe system of work or safe operating procedure can then be compiled.

The basic JSA procedure is as follows:

1. Select the job to be analysed (Select).
2. Break the job down into its component parts in an orderly and chronological sequence of job steps (Record).
3. Critically observe and examine each job step/component part to determine whether there is any risk involved (Examine).
4. Develop control measures to eliminate or reduce the risk (Develop).
5. Formulate written SSWs and job safety instructions ('JSIs') for the job (Install).
6. Review SSWs at regular intervals to ensure continued utilisation (Maintain)

Use may be made of the following three-column layout in order to record the JSA:

| Sequence of job steps | Risk factors | Controls advised |
|---|---|---|
|  |  |  |

It is important to work vertically down the columns in their entirety, rather than to list one or two job steps and then work horizontally across the rows. Experience has shown that the former approach is much less time-consuming!

Once the job to be analysed has been selected, the next stage is to break the job down into its component parts or job steps. On average there should be approximately ten to twenty job steps. If there are more than twenty, too large a job for analysis has probably been chosen and one should therefore split it into two separate jobs. If there are less than ten job steps, then combine two or more jobs together.

The job steps should follow an ordered and chronological approach. Consider the following job – *changing a car wheel*:

The scenario is that you have come out of work and gone to your car where you left it on the works car park. It has a flat tyre. You are not allowed to call for assistance but you have to replace the wheel yourself! The sequence of job steps will be as follows:

1. Check handbrake on and check wheels.
2. Remove spare tyre from boot, check tyre pressure.
3. Remove hub cap/wheel trim.
4. Ensure jack is suitable and located on solid ground.
5. Ensure jacking point is sound.
6. Jack car up part way, but not so wheels leave ground.
7. Loosen wheel nuts.
8. Jack car up fully.
9. Remove nuts, then wheel.
10. Fit spare.
11. Replace nuts and tighten up.
12. Lower car.

13. Remove jack and store in boot with replaced wheel.

14. Re-tighten wheel nuts.

15. Replace trim/hub cap.

16. Ensure wheel is secure before driving off.

Once column 1 has been agreed and completed, then commence listing the associated risk factors by completing column 2. Thereafter, assign commensurate control measures in column 3.

The complete JSA should look like this:

| *Sequence of job steps* | *Risk factors* | *Controls advised* |
|---|---|---|
| 1. Check handbrake, check wheels | Strain to wrist/arm | Avoid snatching, rapid movements |
| 2. Remove spare tyre, check tyre pressure | Strain to back | Use kinetic handling techniques |
| 3. Remove hub cap/trim | Strain, abrasion to hand | Ensure correct lever used |
| 4. Ensure jack on solid ground | Vehicle slipping, jack sinking into ground | Check jack and ground |
| 5. Ensure jacking point is sound | Vehicle collapse | Place spare wheel under car floor as secondary support |
| 6. Jack car up part way, wheels still on ground | Strain, bumping hands on car/jack | Avoid snatching, rapid movements. Use gloves |
| 7. Loosen wheel nuts | Hands slipping-bruised knuckles, strain | Ensure spanner/brace in good order; avoid snatching, rapid movements. Use gloves |
| 8. Jack car up fully | Strain, bumping hands on car/jack | Avoid snatching, rapid movements |
| 9. Remove wheel | Strain to back, may drop onto foot | Use kinetic handling techniques. Use gloves to improve grip |
| 10. Fit spare | Strain to back | Use kinetic handling techniques |
| 11. Tighten nuts | Hands slipping-bruised knuckles, strain | Use gloves, avoid snatching, rapid movements |
| 12. Lower car | Strain, bumping hands on jack/car | Avoid snatching, rapid movements |
| 13. Remove wheel and store with jack in boot | Strain to back | Use kinetic handling techniques |
| 14. Re-tighten nuts | Hands slipping-bruised knuckles | Use gloves, avoid snatching rapid movements |
| 15. Replace hub cap | Abrasion to hand | Use gloves |

| 16. Ensure wheel secure | Vehicle collapse, loose wheel | Check wheel and area around car before driving off. |

The above example serves to illustrate what a typical JSA should look like. Any job/task within the workplace should be able to be recorded in this way and, ideally, displayed in close proximity to where the job is being undertaken.

The ultimate aim must be to undertake JSA on all jobs within the workplace and to communicate the findings and resultant control measures to all concerned. However, to assist in prioritising which jobs to analyse initially, the following criteria should be taken into account:

- past accident/loss experience;
- worst possible outcome/maximum potential loss ('MPL');
- probability of recurrence;
- specific legal requirements;
- newness of the job;
- number of employees at risk.

JSA may be activity-based or job-based.

Activity-based JSAs may cover:

- all work carried out above two metres;
- all driving activities:
    — internal (works transport);
    — external (driving on public highways);
- loading and unloading of vehicles.

Job-based JSAs may cover:

- the activities of a maintenance engineer repairing a roof;
- the activities of a chemical process worker obtaining samples for analysis.

The third column 'Controls advised' is in essence the job safety instructions ('JSI') and forms the basis of the written safe system of work.

## Preparation of job safety instruction and safe operating procedures

S3010     The purpose of job safety instructions ('JSIs') and/or safe operating procedures ('SOPs') is to communicate the safe system of work to all concerned – supervisors, employees and contractors.

For each job step (column 1 in table in S3009) there should be a corresponding control action (column 3) designed to reduce or eliminate the risk factor (column 2) associated with the job step. The chronological ordering of the control measures listed in column 3 becomes the JSI for that particular job.

Such JSIs should be utilised in OSH training and communication, both formal (classroom style) and informal (on the job contact sessions).

All managers and supervisors should be fully knowledgeable and aware of all JSIs and SSWs that are operational within their areas of control and they should ensure that they are adhered to at all times.

JSIs may be:

- verbal (for very low risk tasks) – a word in the ear from a supervisor reminding the employee of the do's and don'ts;
- written (for low, medium and some high risk tasks) – as part of an agreed safe system of work;
- written and incorporated into safe operating procedures (usually for medium and some high risk tasks); or
- as part of an all singing, all dancing formal permit to work system (usually for 'high' or 'very high' risk tasks).

From a practical viewpoint, to aid communication and use of JSIs and SSWs, they should be listed on encapsulated cards and posted in the areas where the jobs are to be carried out. They should also be individually issued to all relevant employees, who should receive training in their application and use. All such training and communication – whether formal or informal should be recorded on individual training records.

The use of the three-column format to communicate JSIs and SSWs is well proven and is an excellent method of revitalising what in most organisations has become a rather moribund set of documents. As stated above, all jobs/tasks should be able to be fitted onto one or two sheets of A4 and use of the technique is a good way to regenerate interest in SSWs via a review.

Most organisations only review SSWs when things have gone wrong, e.g. as part of an accident investigation. This is negative and a form of reactive monitoring.

It is much better to be proactive and to review SSWs at least annually. This can be achieved by setting targets for managers and supervisors to review one safe system of work per month, together with relevant safety representatives, employees and/or contractors. Any findings or changes to the existing SSWs must rapidly be fed back to all concerned, so as to ensure continual improvement in OSH performance.

## Permit to work systems

S3011

'A permit to work system (PTW) is a formal written system of work used to control certain types of work that are potentially hazardous. A permit to work is a document which specifies the work to be done and the precautions to be taken. Permits to work form an essential part of a safe system of work for many maintenance activities. They allow work to start only after safe procedures have been defined and they provide a clear record that all foreseeable hazards have been considered. A permit is needed when maintenance work can only be carried out if normal safeguards are dropped or when new hazards are introduced by the work. Examples are: entry into vessels, hot work, and pipeline breaking.' (HSE, 1997)

### When may PTW systems be required?

S3012    A PTW system may be required in the following instances:

- where the risk assessment indicates 'very high' or 'high' risk;
- entry into confined spaces;
- working at heights;
- working over water;
- asbestos removal;
- high voltage electrical work;

- complex maintenance work – usually involving mechanical/electrical/chemical isolation;
- demolition work;
- work in environments which present considerable health hazards, such as:
  — radiation work;
  — thermal stress (work in hot or cold environments);
  — toxic dusts, gases, vapours;
  — oxygen enrichment or deficiency;
  — flammable atmospheres;
- lone working;
- work on or near overhead travelling cranes (7 metre rule);
- work involving contractors/third parties on site.

## Essential features of PTW systems

S3013    HSE Guidance (1997) states that employers should incorporate the following essential features into bespoke PTW systems:

- provision of information;
- selection and training;
- competency-use of competent persons;
- description of work to be undertaken;
- hazards and precautions;
- procedures.

## Provision of information in PTW systems

S3014    The PTW system should clearly state how the system works in practice, the job(s) it is to be used for, the responsibilities and training of those involved, and how to check that it is working effectively.

There must be clear identification in the system of those persons who may authorise particular jobs and also a note as to the limits of their competency.

There should also be a clear identification of who is responsible and competent for specifying the necessary precautions, e.g. isolation, use of RPE/PPE, emergency arrangements etc.

The PTW system paperwork should be clear, unambiguous and not misleading.

The system should be so designed that it can be used in unusual circumstances, especially to cover contractors.

## Selection and training for PTW systems

S3015    Those persons authorised to issue permits must be sufficiently knowledgeable concerning the hazards and precautions associated with the plant/equipment/substances and the proposed work. They should have sufficient imagination and experience to ask enough 'what if' questions to enable them to identify all potential hazards. Lateral thinking is required here, especially if more than one trade is involved; or when own employees are working alongside contractors.

Ideally formal training should be required for all:

- permit issuers/authorisers;
- permit receivers/users;
- contractors' supervision;
- line managers/supervisors.

All staff and contractors should fully understand the importance and assurance of safety associated with PTW systems and should be 'competent persons' within the framework of the system.

## What is a 'competent person'?

**S3016**   Recent legislation, codes of practice and guidance refer to a competent person in connection with permit to work systems in particular, and the OSHMS in general.

*Regulation 3* of the *MHSWR* (*SI 1999 No 3242*) requires employers to undertake suitable and sufficient risk assessments and *Regulation 7* of the same Regulations requires employers to appoint one or more competent persons to assist them in undertaking such assessments.

*MHSWR '99* defines 'competent' as follows: 'a person shall be regarded as competent for the purposes of these Regulations where he has sufficient training and experience or knowledge and other qualities to enable him properly to assist in undertaking the measures referred to in these Regulations.'

This is especially important in connection with the authorisation and receipt of permit to work documentation. Specifically, permit authorisers and receivers need:

- a knowledge and understanding of:
— the work involved;
— the principles of risk assessments;
— the practical operation of commensurate control measures;
— current OSH best practice.
- the ability to:
— identify OSH problems;
— implement prompt solutions;
— evaluate the PTW system effectiveness;
— promote and communicate the workings of the PTWS to all involved/concerned.

## PTW system: work description

**S3017**   The permit documentation should clearly identify the work to be done, together with the associated hazards. Plans and line diagrams may be used to assist in the description, location and limitations of the work to be done.

All plant and equipment should be clearly tagged/identified/numbered, so as to assist permit issuers and users in selecting and/or isolating the correct piece of kit. All isolation switchgear should be similarly numbered or tagged. An asset register listing all discrete numbering should be maintained.

The permit should also include reference to a detailed method statement for the more complicated tasks. This should form part of the overall PTW system.

## Hazards and precautions

**S3018** The PTW system should require the removal of hazards – risk avoidance being the best strategy. Where this cannot be achieved then effective control measures designed to minimise the risk should be incorporated.

Any control measures should ensure legislative compliance is achieved as an absolute minimum standard. Specific regulations that may apply include: *COSHH 2002 (SI 2002 No 2677)*, *Confined Spaces Regulations 1997 (SI 1997 No 1713)*, *Control of Asbestos at Work Regulations 2002 (SI 2002 No 2675)*. Permit authorisers and users should be aware of all relevant legislation.

The permit should state those precautions that have already been taken before work commences, as well as those that are needed whilst work is in progress. This may include what physical, chemical, mechanical and electrical isolations are required and how these may be achieved and tested. Also, the need for different types of PPE and/or RPE should be specified.

Reference should also be made to any residual hazards that might remain, together with any hazards that might be introduced during the work – e.g. welding fume, vapour from cleaning solvents.

## PTW system procedures

**S3019** The permit should contain clear rules about how the work should be controlled or abandoned in the case of an emergency.

The permit should have a hand-back procedure which incorporates statements that the maintenance work has finished and that the plant has been made safe prior to being returned to production staff.

Time limitations and shift changeovers should be built into the PTW system.

There should be clear procedures to be followed if the work has to be suspended for any reason.

It is vitally important that there is a method for cross-referencing when two or more jobs (at least one of which is subject to a PTW system) may adversely affect each other.

A copy of the permit must be displayed as close as possible to where the work is being undertaken.

All jobs subject to a PTW system should be checked at regular intervals during the duration of the job, so as to ensure that the system is being followed, is still relevant and is working properly.

## General application of PTW systems

**S3020** In order for the PTW system to be applied to risk reduction in the workplace there needs to be a well understood communication system in place that everyone concerned is aware of and becomes involved in.

Most organisations make use of PTW system forms which have been designed to take account of individual site conditions and requirements. In some cases there will be general permits; in other cases special permits will be required for specific jobs, e.g. hot work, entry into confined spaces. This enables sufficient emphasis to be given to the particular hazards present and the precautions advised.

## Outline of a PTW system

**S3021** There are many and varied PTW system documents available in the OSH literature. The key headings/subject areas on PTW system documentation should include the following as a minimum requirement:

- permit title;

- permit number;
- date and time of issue;
- reference to other permits;
- job location;
- plant/equipment identification;
- description of work to be done;
- hazard identification;
- precautions necessary – before, during and after job/work;
- protective equipment;
- authorisation;
- acceptance;
- shift handover procedures;
- hand back;
- permit cancellation – work satisfactorily and safety completed.

## *Operating principles*

S3022    In operating a PTW system the following principles should be observed:

- Title of permit.
- Permit number (including reference to other relevant permits or isolation certificates).
- Precise, detailed and accurate information concerning:
  — job location;
  — plant identification;
- Clear description of work to be done and its limitations.
- Hazard identification – including residual hazards and hazards introduced by the work.
- Specification of plant already made safe, together with an outline of what precautions have been taken, e.g. isolation.
- Specification of what further precautions are necessary prior to the commencing of work–e.g. rpe/ppe
- Specification of what control measures should be in place for the duration of the work.
- Clear statement of when the PTW system comes into effect, and for how long it remains in effect. A re-issue or extension should take place if the work is not completed within the allocated time. Generally, the PTW system should only be valid for the time the authoriser is on site.
- The permit document should be recognised as the master instruction which, unless it is cancelled, overrides all other instructions.
- The authorised/competent person issuing the permit must assure himself/herself – and the persons undertaking the work – that all the precautions/controls specified as necessary to make the job, plant and working environment safe and healthy have in fact been taken. This inevitably should involve an inspection of the job location by the authoriser. The authoriser's signature on the permit document gives this assurance of safety to the people doing the work.

- The person who accepts the permit becomes responsible for ensuring that all specified safety and health precautions continue in force and that only permitted work is undertaken within the location specified on the permit. At this time, confirmation that all permit information has been communicated to all concerned should be obtained. A signature of the accepting person should also be on the permit document.
- A copy of the permit should be clearly displayed in the work area.
- If the permit needs to be extended beyond the initially agreed duration, e.g. because of shift changeover, then signatures of oncoming authorised/competent persons – together with that of the original authoriser – should be added to the appropriate section of the permit to confirm that checks have been made – again involving a visual walk-round inspection to ensure that the plant/equipment remains safe to be worked on. The new acceptor and all involved are then made fully aware of all hazards and associated precautions. A new expiry time is agreed. This should be built into relevant shift hand-over procedures.
- The hand-back procedure should involve the acceptor signing that the work has been satisfactorily completed. The authoriser of the permit also signs off certifying that the work is complete and the plant/equipment is ready for testing and recommissioning.
- The permit is then cancelled, certifying that the work has been tested and the plant/equipment has been satisfactorily completed. The copy permit should also be removed from the work location once the work has been completed.
- It is advisable to keep copies of each PTW system for a period of five years as these constitute maintenance records.

## Isolation procedures

S3023    SSWs and PTW systems frequently require the effective isolation of plant and associated services, including:

- electrical power;
- pneumatic/hydraulic pressure;
- mechanical power;
- steam;
- chemical lines.

The principles of effective isolation require four steps to be taken:

1. Switch off — at the appropriate operating control panels. It is a good idea to post warning signs on the panel to alert other workers that it is switched off.

2. Isolate — by creating a physical break or gap between the source of energy and the part to be worked on. This can be achieved by operating an electrical isolator, removing a section of pipe or blanking off with a plate. N.B. Removing pipe sections is preferable to blanking off.

Reliance on control valves alone has been a frequent cause of accidents and is not considered to be effective isolation. From a mechanical isolation viewpoint, drive belts and linkages should be physically removed.

3. Secure — the means of isolation by locking off. Multiple hasp padlock devices (e.g. Isolok) can be used when several people are working on the same piece of kit. Each puts his own padlock on the isolator so that the power cannot be reinstated until all personal padlocks are removed. It is therefore imperative to have a suite of different keys/padlocks.

4. Check the isolation is effective by trying all available operating controls.

Many maintenance procedures will justify either a written safe systems of work or a formal PTW system; both can have the isolation procedure incorporated into the documentation.

## Further reading

S3024    The following publications are all available from HSE Books.

- Permit to work systems, INDG98 (rev) (1997).
- Safe work in confined spaces: Confined Spaces Regulations 1997: Approved Code of Practice, Regulations and guidance, L101 (1997).
- Safe work in confined spaces, INDG258 (1999).
- Management of health and safety at work: Management of Health and Safety at Work Regulations 1999: Approved Code of Practice and guidance, L21 (rev) (2000).
- Emergency isolation of process plant in the chemical industry, CHIS 2 (1999).

# Statements of Health and Safety Policy

## Introduction

S7001    Employers have a statutory duty under s 2(3) of the *Health and Safety at Work etc. Act 1974* (*HSWA 1974*) to prepare a written statement of their health and safety policy, including the organisation and arrangements for carrying it out, and to bring the statement to the attention of their employees. Employers with less than five employees are excepted from this requirement.

This section sets out the legal requirements for a policy, explains the essential ingredients that a policy should contain and provides checklists to assist in the preparation of policies.

## The legal requirements

S7002    *Section 2(3) of the HSWA 1974* states:

'Except in such cases as may be prescribed, it shall be the duty of every employer to prepare and as often as may be appropriate revise a written statement of his general policy with respect to the health and safety at work of his employees and the organisation and arrangements for the time being in force for carrying out that policy, and to bring the statement and any revision of it to the notice of his employees.'

The *Employers' Health and Safety Policy Statements (Exception) Regulations 1975* (*SI 1975 No 1584*) except from these provisions any employer who carries on an undertaking in which for the time being he employs less than five employees. In this respect regard is to be given only to employees present on the premises at the same time *(Osborne v Bill Taylor of Huyton Ltd [1982] IRLR 17).*

Directors, managers and company secretaries can be personally liable for a failure to prepare or to implement a health and safety policy. (see further *Armour v Skeen* at E15040 ENFORCEMENT).

## Content of the policy statement

S7003    The policy statement should have three main parts:

— *The Statement of Intent*

This should involve a statement of the organisation's overall commitment to good standards of health and safety and usually includes a reference to compliance with relevant legislation. Whilst *s 2(3)* of the *HSWA 1974* only requires the statement to relate to employees, many organisations also make reference to others who may be affected by their activities e.g. contractors, clients, members of the public (This may be of particular relevance if the policy is being reviewed by other parties such as clients). In order to demonstrate that there is commitment at a high level, the statement should preferably be signed by the chairman, chief executive or someone in a similar position of seniority.

— *Organisation*

It is vitally important that the responsibilities for putting the good intentions into practice are clearly identified. This may be relatively simple in a small organisation but larger employers are likely to need to identify the responsibilities held by those at different levels in the management structure as well as those for staff in more specialised roles e.g. health and safety officers. In all cases the competent person who is to assist in complying with health and safety requirements should be identified *[Management of Health and Safety at Work Regulations 1999 (SI 1999 No 3242), Reg 7]*.

— *Arrangements*

The practical arrangements for implementing the policy (e.g. emergency procedures, consultation mechanisms, health and safety inspections) should be identified in this section. It may not be practicable to detail all of the arrangements in the policy document itself but the policy should identify where they can be found e.g. in a separate health and safety manual or within procedure systems.

More detailed guidance on what should be included in each of these parts is contained later together with guidance on two other important issues:

- Communication of the policy to employees.
- Revision of the policy.

## The statement of intent

**S7004** The statement should emphasise the organisation's commitment to health and safety. The detailed content must reflect the philosophical approach within the organisation and also the context of its work activities. Typically it might include:

— a commitment to achieving high standards of health and safety in respect of its employees;

— a similar commitment to others involved or affected by its activities (possibly specifying them e.g. contractors, visitors, clients, tenants, members of the public);

— a recognition of its legal obligations under the *HSWA 1974* and related legislation and a commitment to complying with these obligations,

— references to specific obligations e.g. to provide safe and healthy working conditions, equipment and systems of work;

— reference to the importance of health and safety as a management objective;

— a commitment to consultation with its employees.

Most employers keep the statement of intent relatively short – less than a single sheet of paper. This allows it to be prominently displayed (e.g. in reception areas or on noticeboards), or to be inserted easily into other documents such as employee handbooks.

### Example:

The Midshires Development Agency (MDA) is committed to achieving high standards of health and safety not only in respect of its own employees but also in relation to tenants of the Agency's property, contractors working on the Agency's property or projects, visitors and members of the community who may be affected by the Agency's activities.

The Agency is well aware of its obligations under the Health and Safety at Work Act and related legislation and is fully committed to meeting those obligations. The successful management of health and safety is a key management objective.

MDA supports the concept of consultation with its staff on health and safety matters and has established a Health and Safety Committee to provide a forum for such consultation. This policy will be distributed to all staff and will be reviewed on an annual basis.

A Champion

Chief Executive

August 2000

## *Organisation*

S7005    The statement of intent is of little value unless the organisation for implementing these good intentions is clearly established. This usually involves identifying the responsibilities of people at different levels in the management organisation and also those with specific functional roles e.g. the health and safety officer. How these responsibilities are allocated will depend upon the structure and size of the organisation. The responsibilities of the managing director of a major company will be quite different from those of a managing director in a small business. The latter will quite often have responsibilities held by a supervisor in a larger organisation e.g. ensuring staff comply with PPE requirements.

Most organisations prefer to identify the responsibilities through job titles rather than by name as this avoids the need to revise the document when individuals change jobs or leave. Responsibilities might be allocated to levels in the organisation e.g.

— Managing Director/Chief Executive

— Senior Managers

— Junior Managers

— Supervisors/Foremen/Chargehands

— Employees

and also to those with specialist roles e.g.

— Health and Safety Officer

— Occupational Hygienist

— Occupational Health staff

— Personnel staff

— Maintenance Engineer

Where health and safety advice is provided from outside (e.g. a consultant or a specialist at a different location), this should be stated and the means of contacting this person for advice should be identified.

The following examples are provided as an illustration of how responsibilities might be allocated in a medium sized organisation. The allocation will undoubtedly be different for businesses of greater or lesser size.

## *Examples:*

- Managing Director

    The Managing Director has overall responsibility for health and safety and in particular for:

    — ensuring that adequate resources are available to implement the health and safety policy;

- ensuring health and safety performance is regularly reviewed at board level;
- monitoring the effectiveness of the health and safety policy;
- reviewing the policy annually.

- **Works Manager**

  The Works Manager is primarily responsible for the effective management of health and safety within the factory. In particular this includes;
  - delegating specific health and safety responsibilities to others;
  - monitoring their effectiveness in carrying out those responsibilities;
  - ensuring the company has access to adequate competent health and safety advice;
  - ensuring that safe systems of work are established within the factory;
  - ensuring that premises and equipment are adequately maintained;
  - ensuring that risk assessments are carried out;
  - ensuring that adequate health and safety training is provided;
  - chairing the health and safety committee.

- **Supervisors**

  Each Supervisor is responsible for the effective management of health and safety within his or her own area or function. In particular this includes:
  - ensuring that safe systems of work are implemented;
  - enforcing PPE requirements;
  - ensuring that staff are adequately trained for the tasks they perform;
  - monitoring premises and work equipment, reporting faults where necessary;
  - identifying and reporting health and safety related problems and issues;
  - identifying training needs;
  - investigating and reporting on accidents and incidents;
  - participating in the risk assessment programme;
  - setting a good example on health and safety matters.

- **All employees**

  All employees have a legal obligation to take reasonable care for their own health and safety and for that of others who may be affected by their actions e.g. colleagues, contractors, visitors, delivery staff. Employees are responsible for:
  - complying with company procedures and health and safety rules;
  - complying with PPE requirements;
  - behaving in a responsible manner;
  - identifying and reporting defects and other health and safety concerns;
  - reporting accidents and near miss incidents to their supervisor;
  - suggesting improvements to procedures or systems of work;
  - co-operating with the company on health and safety matters.

- Health and Safety Officer

  The Health and Safety Officer is responsible for co-ordinating many health and safety activities and for acting as the primary source of health and safety advice within the company. These responsibilities specifically include:

  — co-ordinating the company's risk assessment programme;
  — acting as secretary to the Health and Safety Committee;
  — administering the accident investigation and reporting procedure;
  — liaising with the HSE, the company's insurers and other external bodies;
  — submitting reports as required by RIDDOR;
  — co-ordinating the health and safety inspection programme;
  — identifying health and safety training needs;
  — providing or sourcing health and safety training;
  — providing health and safety induction training to new staff;
  — identifying the implications of changes in legislation or HSE guidance;
  — preparing and submitting progress reports on an annual health and safety action programme;

  sourcing additional specialist health and safety assistance when necessary.

## Arrangements

S7006    The duty to provide information on the arrangements for carrying out the health and safety policy overlaps with the duty imposed by *Regs 3* and *5* of the *Management of Health and Safety at Work Regulations* (*SI 1999 No 3242*) to record the results of risk assessments and of arrangements 'for the effective planning, organisation, control, monitoring and review of the preventive and protective measures'. Rather than providing all of the detail on arrangements within the health and safety policy many employers prefer to provide information on where these details can be found, for example in:

— risk assessment records;
— health and safety manuals or handbooks;
— formal health and safety procedures (e.g. within an ISO 9000 controlled system).

This has the benefit of keeping the policy document itself relatively short – an aid in the effective communication of the policy to employees. Some key aspects of health and safety arrangements which should be included or referred to in this section of the policy are those for:

— conducting and recording risk assessments;
— specialised risk assessments e.g. COSHH, Noise, Manual Handling, DSE workstations;
— assessing and controlling risks to young persons and new or expectant mothers;
— establishing PPE standards and the provision of PPE;
— operational procedures (where these relate to health and safety);
— permit to work systems;
— routine training e.g. induction, operator training;
— specialised training e.g. first aiders, fork lift drivers;
— health screening e.g. pre-employment medicals, regular monitoring, DSE user eye tests;

- statutory inspections;
- consultation with employees e.g. health and safety committees, staff meetings, shift briefings;
- health and safety inspections;
- auditing health and safety arrangements;
- investigation and reporting of accidents and incidents;
- selection and management of contractors;
- control of visitors;
- fire prevention, control and evacuation;
- dealing with other emergencies e.g. bomb threats, chemical leaks;
- first-aid;
- maintaining and repairing premises and work equipment.

## Communication of the policy

**S7007**   Employers must bring the policy (and any revision of it) to the notice of their employees. The most important time for doing this is when new employees join the organisation and communication of the policy should be an integral part of any induction programme. Young people in particular need to be reminded of their personal responsibilities at this stage, as well as any restrictions on what they can and can't do. Even experienced workers may need to be made aware that they are joining an organisation that takes health and safety seriously, as well as knowing what the PPE rules are and where they can obtain the PPE from.

Many employers provide their employees with a personal copy of the policy whilst others include it in a general employee handbook or display it prominently e.g. on noticeboards. In all cases communication will be aided by keeping the policy itself relatively brief and providing the detail (particularly in respect of 'arrangements') elsewhere.

## Review

**S7008**   The policy must be revised 'as often as may be appropriate' but the need for revision can only be identified through a review. In their free guidance leaflet for small firms on preparing a health and safety policy document (IND(G) 324 *'Stating your business'*) the HSE recommend an annual review and this should not be too onerous even for small employers, for whom it should be a fairly simple affair.

In larger organisations the policy could be subjected to an annual review by incorporating it within an ISO 9000 document control system or making the review an annual task of the health and safety committee. Alternatively the review could be made the responsibility of an individual (e.g. the Managing Director or the Health and Safety Officer), either formally through specifying this within the organisational responsibilities or informally through a simple diary entry.

The review may well conclude that there is no need for change but it may identify the need for revision such as:

- A new Managing Director should make his personal commitment to the statement of intent by signing and re-issuing it.
- Changes in the management structure necessitate a reallocation of responsibilities for health and safety.

— New 'arrangements' for health and safety have been established (or existing ones have been altered) and the policy needs to be amended to match this.

A policy which is clearly well out of date does not reflect well on the organisation's commitment to health and safety, nor on the effectiveness of its health and safety management.

# Stress at Work

## Introduction

**S11001** Stress is becoming an increasing concern for employers and employees alike. More seems to be demanded by organisations in less time. At work, life is becoming more competitive, with more people chasing the same goals.

People are very complex and the way in which they react to events within their lives varies considerably. Sometimes, a single event may cause a crisis in someone's life. How they respond will depend upon a number of factors, including other circumstances that may exist within their life, their own reaction to the event and the length of time it takes to deal with the event. There are people who, when stressed, will make sure that everyone around them knows it. They will become withdrawn, and internalise their stress. Physically retreating from stressful situations, such as by taking frequent smoking breaks, is one way that some people cope. Others will obviously be angry, they care little about anyone except themselves, and their tension is obvious. This may result in behaviour that appears to be aggressive or threatening. Indeed, in some cases the stress may build to an extent that someone does resort to expressing their feelings by using violent behaviour. For others, stress will be a result of them being exposed to violent or threatening situations. Examples where this may occur include: care workers, retailers, police and other emergency support workers, and call centre staff.

## Definition of stress

**S11002** Taken in its simplest form, the Oxford English dictionary defines stress as 'demand upon physical or mental energy'. The HSE's definition of stress is given as 'the adverse reaction people have to excessive pressures or other types of demand placed upon them'. This has been seen by some as a rather negative view, as it implies that stress produces only adverse effects.

Stress often occurs when the pressures placed upon a person exceed their capacity to cope. This can be negative or positive. For some people, the pressure of a tight deadline will 'positively stress' them, with the end result being that it enhances their performance. This may be due to their fear of failure or the consequences of not meeting the deadline. This is particularly true of people working in competitive sport and those people who set themselves challenges e.g. to increase sales figures by 50 per cent in the next 12 months. In these cases, the pressures placed upon them have a positive effect in that it produces drive, challenge and excitement.

## Understanding the stress response

**S11003** People's ability to respond to difficult or stressful situations has evolved over many million years. The mind recognises a particular need to respond to an external source (both physical and mental). The alert message is sent to the brain which then initiates dramatic changes in the hypolamus. This is the control centre which integrates our reflex actions and coordinates the different activities within our bodies. When the hypothalamus is activated, the muscles, the brain lungs and the heart are given priority over any other bodily activity. The chemical adrenaline is produced and the message of a threat is passed on to other parts of the body which prepares itself for vigorous physical action – known as the fight or flight instinct. When the threat has abated, the parasympathetic nervous system takes over again and the body returns to a state of equilibrium, (known as homeostasis) where all our functions are in balance. This is demonstrated in the stress response curve.

**Figure 1: Stress Response Curve**

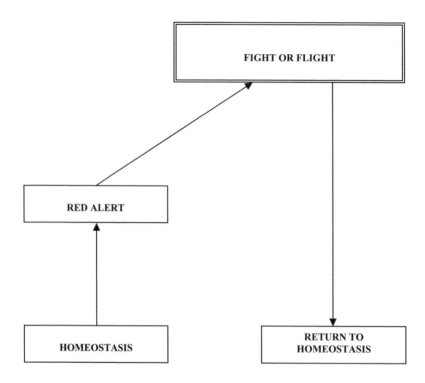

To summarise, given below are the physical responses to the 'fight or flight' mechanism.

- brain goes on red alert and stimulates hormonal changes including the production of adrenalin and noradrenaline (stress hormones).
- muscles tense, ready for action.
- eye pupils dilate to enable the danger to be seen more clearly.
- the heart beat increases to provide more blood to the muscles, therefore blood pressure increases.
- breathing increases to provide extra oxygen for the tensed muscles.
- liver releases glucose to provide extra energy for the muscles.
- digestive system shut down, making the mouth go dry and the sphincters close.
- sweating occurs in anticipation of expending extra energy.
- immune system slows down.

*Stress at Work* S11009

These reactions are highly complex, but demonstrate the increased demands that are placed upon the body when feelings of stress occur. It is easy to see, therefore, how this response can have a detrimental effect upon well being and health if the stress is prolonged.

## Why manage stress?

S11004 The following paragraphs explain the importance of managing stress.

### Legal issues

S11005 There is no legislation that relates specifically to stress but the following Regulations touch upon the issue.

### Health and Safety at Work etc Act 1974

S11006 Under the *Health and Safety at Work etc Act 1974*, s 2(1) employers have a general duty to ensure, so far as is reasonably practicable, the health, safety and welfare of employees at work. This implies that measures should be in place to ensure that employees do not suffer stress-related illness as a result of their work.

### Management of Health and Safety at Work Regulations 1999

S11007 Under the *Management of Health and Safety at Work Regulations 1999* (*SI 1999 No 3242*), *Reg 3* employers must also take into consideration the risk of stress-related ill health or unsafe working practices that may be caused or exacerbated by stress. Stress needs to be considered when undertaking general risk assessments (e.g. a risk assessment of the work environment, or of a particular task) as people's responses may differ when they are feeling the symptoms of stress, or overload. Some organisations may decide to undertake a specific assessment of occupational stress within their workplace, i.e. factors associated with the work that are known to affect the way people feel and respond.

### Working Time Regulations 1998

S11008 The *Working Time Regulations 1998* (*SI 1998 No 1833*) seek to reduce the effects on employees who are subjected to working long hours, or who have insufficient rest periods during and between work shifts (these issues are known to be occupational stressors). The Regulations are detailed, and cover minimum rest times during a shift (20 minutes every six hours), minimum rest time between shifts (11 hours) and maximum number of hours that can be worked within a weekly period (48 hours). The maximum number of hours may be calculated over a 17-week reference period.

There are a number of exclusions that apply to certain categories of personnel under the Regulations, and these are under review in the European Courts at present.

Detailed requirements apply to employers to maintain a record of hours worked for each employee who is covered by this legislation.

For further information see WORKING TIME.

### Criminal and civil liability

S11009 Whilst there is potential for prosecution in criminal law for failure to meet the above duties, the likelihood is that compliance will be driven by civil litigation. To date, there have been no prosecutions under criminal law for failure to manage stress at work. Enforcement of the general duty of care under health and safety legislation has been due to successful claims being made for compensation following allegations by an employee that they have suffered illness or ill-health as a result of being exposed to unacceptable levels of stress at work.

Indeed, there have been some large insurance settlements for employees who have successfully claimed for stress at work. The most notable was *Walker v Northumberland County Council [1995] IRLR 35*. In this case, Mr Walker, a senior social worker, had been employed by the council for 17 years. During the 1980s his workload had increased, and as a result, he suffered a nervous breakdown. Upon his return to work he told his employers that his workload must be reduced. This was not done, and Mr. Walker suffered another nervous breakdown which resulted in him being dismissed for permanent ill health. He claimed compensation from his employer, and was successful on the grounds that the employer had been aware that he was under extreme pressure at work. A key question in the case was whether Mr Walker's risk of mental illness was materially higher than that which would normally affect a senior social worker with a heavy workload.

As regards the second breakdown, the employer should have foreseen a risk that Mr Walker's career would end and they should have appreciated that Mr Walker was more vulnerable to damage than he had been before his first breakdown. Because of this, it was deemed reasonable for Mr Walker to have received assistance, but in not providing such assistance, the council was in breach of their common law duty of care.

In February 2002, a landmark decision by the Court of Appeal in *Hatton v Sutherland and Others [2002] 2 AER 1* overturned three out of four stress claims and quashed the awards for damages initially given to the workers in the High Court. The Appeal Court ruling stated that for a compensation claim to succeed it must first be 'reasonably foreseeable' to the employer that the employee concerned would suffer a psychiatric injury as a result of stress in the workplace. When deciding whether a risk is 'foreseeable' the following needs to be taken into account:

- nature and extent of the work undertaken by the employee (e.g. is the workload more than is normal for that particular job?).
- signs from the employee themselves of impending harm to their health as a result of work-related stress (e.g. has an employee already suffered harm to their health from work-related stress?).

The Appeal Court ruling further stated that employers are entitled to assume that an employee can withstand the normal pressures of a job, except where they are aware of a particular problem or vulnerability (e.g. the employee has warned the employer that they are suffering from work-related stress). In addition, employers are entitled to take at face value what they have been told by their employees about the effects of work-related stress on their mental health. For example, if an employee has had sickness absence due to stress but does not inform their employer of the cause of the absence, then it would be reasonable for the employer to expect that the worker would be fit to return to their normal work.

The Court of Appeal set out a number of factors that must be taken into account by the courts once it has been established that it was reasonably foreseeable that an employee would suffer harm as a result of work-related stress. These include:

- an employer is only in breach of their duty of care if they fail to take reasonable steps to prevent the employee from suffering as a result of stress.
- indications of impending harm to health must be plain enough for any reasonable employer to realise.
- when considering what actions are reasonable for an employer to take to reduce risk of harm from stress, the size and resources of the employer's operation may be taken into account.
- it is not enough for a claimant to show they are suffering harm from a result of occupational stress – they must demonstrate that their 'injury' was caused by the employer's breach of duty of care by failing to take reasonable steps to protect the worker.

- an employer will not be deemed to be in breach of their duty of care if they allow a willing employee to continue with their job, despite them stating they have been suffering from stress, if the only reasonable alternative is to dismiss the employee.
- if an employer offers a confidential staff counselling service for those who experience stress at work with referral to treatment services, they will be unlikely to be found in breach of their duty of care.

The court ruling places more responsibility on the employee to let an employer know if they are suffering from symptoms of occupational stress, rather than leaving it totally to the employer to identify a 'stressed' employee. In practice, therefore, an employee is unlikely to have a valid claim unless they can demonstrate that they have told their employer that they are suffering from work-related stress. Note that this does not take away the employer's duty to be proactive and seek to identify stress risk factors in the workplace, and to implement suitable precautions to minimise these risks.

See also OCCUPATIONAL HEALTH AND DISEASES O1052.

## Occupational stress risk factors

S11010    The following paragraphs discuss in detail the most common occupational stress risk factors.

### The physical environment

S11011    The physical environment can be described as the general conditions in which people have to work. The work environment is often seen as a contributory cause for stress at work, i.e. something that compounds a problem that is being experienced at work. For others the work environment may be the main cause of stress amongst the workforce, especially if the conditions prevent work targets from being achieved, or detract workers from undertaking their work effectively.

Our general work conditions have improved drastically over the past century, but so have our expectations. The environments in which people work will vary considerably, and will be dependant upon the type of work that is being undertaken. For example, a die casting factory is traditionally a hot, dirty, and malodorous environment. Any investments are usually spent upon the equipment to produce the product rather than to enhance the work environment, and managers may be reluctant to spend a heavy outlay on a workplace that is likely to become dusty and dirty within a short time. These types of environment often have less time and effort spent on them in maintaining high standards of cleanliness, good housekeeping or ensuring that lighting remains in good order throughout the premises.

This can, however, have a knock-on effect for staff, both in their mental attitude towards the workplace, in their general conduct and in their general well being. It has been found, following several studies of the workplace, that an environment that is poorly maintained produces a negative attitude amongst the staff – 'if management don't care then why should we' type approach. If the workplace is allowed to become dirty, untidy or cramped, then workers will respond accordingly with the 'why should we tidy up – management don't seem to mind' type attitude. A workplace that is dull, untidy, badly designed or scruffy is likely to make employees feel negative about coming to work, and they are more likely to respond negatively to issues that arise at work.

### The condition of the workplace

S11012    The condition of the workplace can be described as its general standard of maintenance and decoration, both internal and external. Housekeeping standards within the workplace could be included under this heading. To understand how a workplace condition can affect how a person feels and responds, consider the following examples.

- *Scenario 1*

    An old hospital building was first constructed during the early 1900s. The hospital management has struggled to keep up with the increase in usage over the years and the structural changes that have been necessary to accommodate more patients and more services. The hospital is crowded and cluttered. Government financial cutbacks have been a common setback and funding has focussed on provision of medical care, rather than on maintenance.

    The walls are grey and flooring is mainly grey concrete. It is cracked in many areas, causing ruts and trip hazards. Lighting is poor and several overhead tubes are not working. Paint on the wall is dirty, marked in places and chipped in others. Some of the windows are cracked.

    Doors are damaged due to the constant throughfare of hospital trolleys being pushed against them.

    Sacks of dirty laundry, and black sacks containing rubbish, seem to be constantly accumulating in corners awaiting collection.

- *Scenario 2*

    This scene depicts another hospital that was constructed in the early 1900s. Similar challenges have been faced within the hospital to provide increased services and a higher level of care. Budgets restraints also present a challenge.

    The walls are still grey but have been painted within the last few years. However, different parts of the hospital have received a splash of colour by way of painting the support pillars, door surrounds etc with bright contrasting colours. The doors are less damaged as they have metal plates fitted to reduce damage from the trolleys. Windows are cracked in some places, but have been cleaned, which has resulted in more light coming into the hospital.

    Rubbish bags and dirty laundry bags are out of sight. The floor looks cleaner and there are less pot holes and cracks in the flooring.

- *Scenario 3*

    This scene depicts a hospital that was also built in the early 1900s. However, despite the hospital facing the same challenges as the hospitals in the first two scenarios, the management team have adopted a much more proactive approach to general decoration. Staff have been encouraged to participate in choosing colour schemes when new services have been introduced and the patients' views have been taken into account when planning any refurbishment. This has resulted in local colleges being invited to practise their artistic skills in main areas at little or no cost to the hospital. Waiting rooms have brightly coloured murals along the walls. Corridors are coloured differently for each separate unit (e.g. cardiology, rehabilitation) and collages and other pictures have been displayed along the walls. These have resulted in interesting schemes for people to enjoy as they walk through the corridors.

    Some extra money has been made available each year for ongoing repairs which has resulted in a higher standard of maintenance. A proactive maintenance schedule is in place whereby all lighting tubes are replaced annually, thereby significantly minimising lighting failure. Staff have been encouraged to share their requirements or

> preferences for lighting in different areas, and for reporting to management when problems are occurring. Windows are kept clean.
>
> Wards are brightly coloured, and curtains have been fitted at windows which are pleasant to look at. Plants are situated around the hospital and a water feature has been installed in the main reception waiting area.
>
> Waste collection points have been clearly defined. Laundry collection/delivery areas are clearly marked and waste is not allowed to accumulate.

An employee working in any one of these environments, doing work which is known to be emotionally and physically demanding, would obviously prefer to be working in an environment depicted in scenario 3.

People feel better and therefore respond more positively when their environments are pleasant. This positive factor is increased when staff feel that they have had some level of control over their own environment, even if it is something as basic as deciding on the best areas for storage or waste collection points.

## *The workplace layout*

S11013    The workplace layout can make a significant impact upon a worker's efficiency, safety and general well being. This is where the study of ergonomics can have a major input into reducing stress levels at work. Ergonomics is a specialist area, which looks at the interaction of the worker, the work equipment and the work environment. The use of an expert may be needed at some stage in workplace layout, but the first step should always be to discuss employee requirements and to see how these can be accommodated within current restraints. See ERGONOMICS E17021.

Simple changes to a work layout can make a drastic improvement in a person's efficiency, safety and general well being, especially when the idea has emanated from the workforce itself. The worker is the expert in the particular task that is being undertaken, and whilst an employer does not want to provide a 'wish list' to the workforce, valuable information can be ascertained merely by finding out if and what causes frustration within a work layout.

An employer might consider asking its staff the following questions.

| Question | Yes/No | Comments |
|---|---|---|
| Are worksurfaces large enough for the work and equipment that are needed? | | |
| Are worksurfaces of a suitable height (too high or too low) for the worker(s)? | | |
| Is there sufficient space for the worker to get to, from and around their own work area? | | |
| Are storage facilities in close proximity and easy to access? | | |
| Is there sufficient storage? | | |
| Is storage being used to full effect? | | |
| Do workers have somewhere to store their own personal items? | | |

| | | |
|---|---|---|
| Are seats or suitable rest areas provided, at or near to the workstations? | | |
| Are walkways wide enough to enable people and equipment to pass through (e.g. trolleys)? | | |
| Are work areas in large areas clearly identifiable, signed and labelled? | | |
| Is the layout designed to minimise environmental hazards (e.g. glare from windows, noise from nearby machines)? | | |

Poorly designed layouts can induce other health conditions such as musculo-skeletal disorders e.g. back problems from having to stoop, stretch or crouch to undertake a task. Stress creates physical tension which can contribute to these health conditions.

## Staff facilities

S11014    Everyone occasionally feels a need to take a break from their work, and this may be more of a need than a desire in some work situations that are known to incur high levels of stress e.g. dealing with the public, working in the care profession. Having nowhere to go to take a break or let off steam may result in stress levels increasing to a point where a worker becomes distressed. This will be compounded when other aspects of the work are problematic, for example, the shift patterns are badly designed or working hours are excessive.

Staff facilities should include basic facilities such as sanitary conveniences and washing facilities, and somewhere from where drinks can be prepared or obtained. Employers should ensure that such facilities:

- accommodate the numbers of people using them.

- take into consideration the need for privacy (especially if workers need to change clothes or shower before or after work).

- are maintained to a good state of hygiene and cleanliness.

- provide basic equipment such as soap, drying equipment, toilet paper and sanitary bins.

- enable workers to obtain both hot and cold drinks, including provision to clean spillages or clean utensils if drinks are being prepared by the workers.

- provide a ready supply of hot/warm and cold water, and potable water for drinking purposes.

The *Workplace (Health, Safety and Welfare) Regulations 1992 (SI 1992 No 3004)* and supporting guidance contain detailed minimum standards for sanitary conveniences, washing facilities, drinking water facilities for changing clothes.

These Regulations also contain provisions for rest facilities, and require that somewhere that is suitable and sufficient is made available for people to eat meals where meals are regularly eaten in the workplace. Resting and eating facilities must ensure that people who do not smoke are protected from the effects of tobacco smoke. Therefore the onus of protection is for the non-smoker rather than the rights of the smoker. Implementation of these requirements will vary considerably, according to the company's activities and the resources available. Separate smoking and non-smoking facilities would be ideal but are not practicable in many workplaces. Some workplaces offer rest rooms for smokers, whilst others require that smokers take their 'smoke break' outside of the workplace.

This legislation also requires the provision of rest facilities for workers who are pregnant, or who are nursing mothers. Some employers have expressed concern at the practicalities of meeting this requirement. However, the female workforce is increasing as more mothers are returning to work earlier and employers have a legal as well as moral duty to ensure that these provisions are available. These types of issue do have an impact upon the stress levels of employees.

For further information see WORKPLACES – HEALTH, SAFETY AND WELFARE W11001.

## Environmental conditions

**S11015**  When considering stress induced by environmental conditions, an employer should consider issues such as lighting, heating, ventilation, humidity and space.

In some cases, the environmental conditions may be the sole or primary cause of stress, in others it may be a factor that exacerbates other stressful conditions within the workplace. Poor environmental conditions are the cause of other workplace issues such as sick building syndrome.

Several regulations contain specific duties for provision of environmental standards. These include the *Workplace (Health, Safety and Welfare) Regulations 1992 (SI 1992 No 3004)*, the *Schedule* to the *Health and Safety (Display Screen Equipment) Regulations 1992 (SI 1992 No 2792)* and the *Provision and Use of Workplace Regulations 1998 (SI 1998 No 2306)*.

An employer should take the following into account when deciding on the standards that should be achieved in the workplace.

- legal requirements contained in various statutes.
- official guidance issued by the HSE, trade associations or other recognised bodies such as the Chartered Institute of Building Service Engineers.
- the nature of the work (work requiring intricate detail will need a higher level of lighting than one undertaking general tasks).
- the amount of physical effort in a task (a cooler background temperature will be needed where workers exert physical effort).
- the need to combine workers' needs with other stakeholders (e.g. customers, patients).
- the level of humidity in the environment (work in a humid environment will reduce a worker's ability to control their own internal temperature by sweating, thus increasing the risk of heat exhaustion).
- the amount of space for the work components, the work equipment, and the worker's need for movement.

There are several sources of guidance which will help an employer to interpret conditions that will be deemed as 'reasonable' to meet statutory requirements for a work environment. A good starting point, other than a specific trade association, is the Chartered Institute of Building Service Engineers. This institute can provide detailed guidance on recommended environmental standards, and should be referred to when deciding optimum work conditions for different types of work activity.

Employers should take into account, however, that any such guidance should not be used in isolation. The most effective way of reducing physical and emotional stress from work environments is to involve the staff. It is not possible to please everyone, but by obtaining a general consensus an employer has an opportunity of ascertaining if any problem exists, where the problem might be, and the extent that the environment may be cause or adding to a problem.

Another important key to reducing stress induced by environmental conditions is to provide as much control as possible over the local environment. For example, it is usual in large open workspaces for the lights to be controlled by only a few switches. The lighting circuits could be changed so that a smaller area is controlled by each switch, enabling more control over an employee's immediate working vicinity. The strength and colour of the tubes could also be altered to further meet individual preferences.

## The work

**S11016** There are many aspects of work that may induce pressure. Pressure at work is not necessarily bad as it helps to generate enthusiasm and can produce energy. However, prolonged pressure is likely to create a state of nervous tension as the person has no respite from the constant rush of adrenaline that occurs when stress is induced. Unless this is relieved, the end result can lead to health problems.

The trick is to try to achieve a balance between providing a challenge to staff, and overloading them.

The following areas are known to be occupational stressors, and should be reviewed when any stress management strategy is being considered.

### Targets deadlines and standards

**S11017** Targets are necessary at work. This is because an employee who is not given a target will set their own target which may not be in line with organisational expectations. However, problems arise when the standard that has been set by the employer exceeds the individual's ability to meet it.

When a deadline or target is unachievable, a sense of helplessness may be experienced by the person. Similar emotions may be experienced when each time the target is met the employer increases the next target. Whilst for some employees this will represent a sense of excitement, there will inevitably be a time when an employee begins to feel defeated, undermined or useless. This problem becomes more serious when there is no mechanism for the employee to let the employer know how they are feeling, or when there is no system for monitoring staff behaviour and performance once the target has been set. If the employee feels that their job depends upon meeting the target then the feeling of pressure is increased, as the consequences of failure are more significant to that individual.

There are various reasons why a deadline or target may be unachievable. Managers should consider the following questions, and perhaps agree with the employee what is reasonable when all the restraints have been taken into consideration. This will help the employee to perceive a sense of involvement and control over the amount of pressure being placed upon them.

| Assessment of targets | | | |
|---|---|---|---|
| Question | Yes/ No | Comments | Agreed action |
| Is the employee sufficiently trained/experienced to meet the standard? | | | |
| Are there adequate tools/equipment to meet the deadline/target? | | | |
| What other demands are being made which may affect the standard/deadline being met? | | | |
| Is there any mechanism for the employee to report early if he/she feels they are not likely to meet the standard? | | | |

| | | | |
|---|---|---|---|
| To what extent can machine failure etc affect the target being met? | | | |
| Is the time frame for the amount of work being set reasonable (i.e. is this being achieved reasonably by others)? | | | |
| What are the consequences of meeting or failing the target? | | | |
| Are there any other factors that the employee feels may detrimentally affect the target/deadline being achieved? | | | |

People who have experienced continual pressure from unrealistic deadlines or targets often explain their feelings with comments such as not being able to see the light at the end of the tunnel and being on a treadmill that they cannot get off. When these feelings become excessive, a common response is for the employee to take time off for sickness absence, or even to leave the company. Early warning signs that an employee may be feeling excessive pressure may include erratic behaviour or mood swings, reduced or erratic work performance, looking worn out or tired or even compulsive behaviour (increased smoking, drinking or eating). Some people, of course, will respond with the opposite effects, i.e. by losing their appetite, becoming very quiet or seeming to become particularly conscientious.

## *Time demands*

**S11018** Stress from conflicting demands on time may be a sign of 'role conflict' or 'role ambiguity', i.e. which task is more important. It may be that the time available is insufficient to deal with the extent of work being required. If time demands being made upon the employer are such that everything is expected to be undertaken at the same time, then the worker will suffer.

Different people have different ways of handling their work. Some people prefer to complete one task before giving any attention to another, whilst others appear to thrive on dealing with numerous matters all at the same time. Problems arise when the manager operates to one of these principles whilst the subordinates preferred style is totally different. The end results can be disastrous, with neither party feeling that a job is being completed satisfactorily and both parties feeling that they are being misunderstood. Two-way dialogue between management and employees is needed to ensure that each party understands the other's needs, and to agree parameters for working.

A commonly reported cause of stress is knowing that several other tasks are building up whilst only one task is being addressed. The employee will either rush through all the work, to try to reduce the mountain that is piling, or select one task that they feel is the most important at the expense of others which are perceived to be less important. Neither response is likely to result in staff feeling satisfied that they are coping and doing a good job. A key to managing this type of problem is to:

- review work loads.
- take into account deadlines.
- agree priorities for each task that is building up.

- ensure the most suitable type of person is allocated (some people deal better with a high throughput that requires low levels of accuracy whilst others prefer work that require less output but a higher level of attention).
- encourage and develop self-organisation for the individual to manage their work load.

## *Working hours and shift patterns*

**S11019** Management of working hours is a key feature in reducing stress levels at work. When considering working hours, the employer should take into consideration:

- the length of the shift (how many hours worked each day).
- the shift pattern, especially if rotating shifts are used.
- rest times between shifts.
- rest and break times during the shift.

As mentioned previously (see S11008) the *Working Time Regulations 1998 (SI 1998 No 1833)* contain requirements on the number of hours that should be worked each week, the minimum rest time between shifts and the minimum rest breaks during shifts. Regardless of this whether an employee is excluded from the Regulations or chooses to 'opt out', the employer has a duty of care to ensure that health, safety and welfare are not detrimentally affected by the work.

Problems with working hours seem to crystalize when the time spent at work impinges on employees' home and social activities. Someone who habitually works long hours often feels too tired to enjoy their private time. An employee will be prepared to 'trade off' a well-designed shift pattern in order meet social demands being made at home. It is clear, therefore, that a balance must be struck between shift patterns that have been deemed to be the most effective, against the social needs of the workers.

The most frequently reported source of stress among shift workers who work night shifts is the interruption in their sleep, or difficulty in being able to sleep during the day. Possible sources of interruptions are numerous and include doorbells, telephones, increased traffic noise, family, pets and neighbours. Workers who operate to rotating shifts are more likely to feel the effects of sleep disturbance.

The need for variety during night shifts in terms of the range of tasks, and the mental and physical output necessary, is more important for night workers. The attention span will reduce, typically between 3.00 a.m. – 4.00 a.m. and this will be exacerbated if the work is monotonous, requiring high levels of concentration or is physically demanding. Fatigue is unavoidable during night work, but will be reduced if the interest of the shift worker is maximised.

## *Travel during work*

**S11020** Many employees enjoy travelling as part of their work as it provides a sense of variety, change in environment and a level of autonomy about their work pattern (e.g. by not feeling that they are being watched). Problems arise when the amount of time travelling exceeds the amount of time that can be spent in the workplace, or impinges upon the time available for the remainder of work that needs to be completed. Travelling is also exhausting, both physically and mentally, and can reduce a person's work performance. Other problems associated with travelling include feelings of isolation from colleagues, family and friends, especially when extended stays away from home occur.

Apart from drivers of heavy goods or public services vehicles, there are no maximum times set in legislation that a worker may travel on the road without taking a break. Therefore an employer should undertake a risk assessment, and agree with the workers what travel arrangements are considered reasonable. Factors that need to be taken into account include:

- preferred method of travel.
- manual handling issues of the traveller (cases, files etc).
- the length of the journey (including times for delays).
- how much time the employee has been given to reach their destination (i.e. work time versus private travelling).
- travel routes (especially if using the road).
- the need for regular two-way communication with the worker whilst away from the workplace.
- suitability of overnight accommodation, including location of accommodation.

Organisations such as ROSPA (Royal Society for the Prevention of Accidents) have produced detailed guidance and literature detailing results of their research into occupational road risk, and this may be a useful source of reference when considering the physical and psychological risks associated with travelling.

## Technology and work equipment

S11021   There is nothing more frustrating that having a mismatch between the tools necessary to complete a task and the task itself. If the problem is simply being given the wrong tool to complete the task, then this can quickly be rectified (provided the employer is informed). However, stress levels increase when equipment constantly fails to meet its design criteria, breaks down due to overload or poor maintenance, or when there is a skill's shortage amongst the workers.

To overcome the problem the employer should undertake a task analysis to identify what tools and equipment are needed, and to ensure that they are provided. Ongoing maintenance may be needed, including an emergency call-out service where the job is dependant upon the work being completed within a certain time. This needs to be followed by a training needs' analysis, to identify if and what type of training is required to ensure workers can use the equipment effectively and safely.

Most people are instinctively sociable, and respond to both verbal and non-verbal behaviour and communication whenever they meet. Someone making a light-hearted comment can lighten a dull or pressurised meeting, a smile will indicate whether an ambiguous comment is meant as a joke. This type of behaviour is missing from computerised communication i.e. e-mail or even telephone. Messages received without the additional benefit of non-verbal communication may come across as blunt, aggressive or hostile and may make the recipient feel threatened when the opposite was the intention. Stress will result from the perception of aggression or threat, especially if that person feels helpless to respond or defend themselves. This needs to be taken into account when transmitting difficult messages, to avoid inducing unnecessary anxiety. In extreme cases, e-mail communication has been known to cause severe disruption to workplace harmony and cause individuals to become insecure or even sick.

## Role conflict or ambiguity

S11022   People can work much more effectively if they know what they are supposed to do, how it is supposed to be undertaken, why it needs to be undertaken in a particular way and how their role fits into the corporate objective.

When job roles conflict or are ambiguous the end result is confusion and frustration for everyone involved. If this becomes a frequent event then those feelings of frustration may increase into anger and resentment. This may be compounded when the person's level of authority is limited and they have no access or authority to change what is causing the problem. The following scenarios may be indicative of conflicting or ambiguous roles.

> - *Scenario 1*
>
>   A secretary is told by her manager to manage the diary and to keep all the clients satisfied when they ring. Several potential clients have telephoned to book meetings which the secretary has entered into the book, but her boss is very displeased when he finds out and tells her to cancel them. Nearly all other bookings made into the diary by the secretary get changed or altered, causing the secretary additional work and upset. When challenged by the manager about why all these bookings have been made, the secretary defends herself by stating that she is doing what she was told to do. The boss responds by saying that nothing should be booked without checking with him first.

> - *Scenario 2*
>
>   A departmental manager has responsibility for reducing expenditure in his area of authority. However, he has also been given responsibility for managing health and safety. The health and safety standards in the department are poor. The manager identifies a need for health and safety training which will improve standards but this will take him over the agreed spending threshold which will directly affect profitability and his requirement to reduce expenditure.

## *Clerical/administration support*

S11023    Many employees now undertake their own basic administration tasks, using basic computer software packages. This is acceptable provided they have the basic skills to use the equipment, and the time allocated within their work patterns to complete the administration. Common indicators that completion of paperwork is becoming problematic may include:

- workers regularly taking work home in the evenings/weekends.
- employees not using their holiday entitlement.
- increase in errors within written correspondence (typographical, grammar or content).
- tasks being half completed.
- paperwork piling up on workstations.
- messages not being answered.
- comments from workers about bureaucracy.

Employers should ensure that time is made available to enable workers to complete any necessary paperwork, and that the equipment is accessible. This is particularly relevant where the worker is based in a non-office environment, and would not automatically have access to a desk, paper, filing system etc readily to hand.

Most people do not like excessive paper-work, and some do not like any paper-work. Employers should therefore seek to streamline their systems so that administration is minimised, and simplified.

## *Cover during staff absence*

S11024    Administration often becomes a stress issue when staff are constantly required to cover for colleagues. The additional pressures this causes are unlikely to be problematic if it is for only short duration, e.g. sickness or holiday. However, staff who feel that they are frequently undertaking additional work to cover for shortages will begin to feel the effects of the burden.

Early symptoms to look for may include reduced efficiency, increased error rates, lower patience or attention thresh-hold, or increased sickness absence.

Managers should seek to avoid relying on employee goodwill as the sole method for dealing with the workload when a colleague is absent. Use of temporary staff should only be used as a stopgap. Often their effectiveness is limited as they are unfamiliar with the work and work systems. It also takes time to train a temporary worker, and if this becomes a regular exercise then the use of temporary workers will be seen as a hindrance rather than a help.

Employers should therefore adopt a more proactive stance, and consider what alternatives would be suitable to deal with staff shortages. Examples may include use of multi-skilled workers who could be moved around, or 'floating' staff who are employed specifically to cover during other staff absences.

## Company structure and job organisation

S11025    The following paragraphs discuss the relevance of company structure and job organisation on stress levels.

### Hierarchy of control

S11026    Ambiguity over job roles and the level of control over a task is a major stressor at work.

Problems occur when employees are given responsibility for a task but have no authority to implement necessary systems, training etc to effectively deal with the task. Colleagues and fellow team members will judge both the corporate reaction and the reaction of the particular manager to any decisions that are made within an organisation, and will build their own attitudes as a result when showing commitment back to the organisation. It is therefore very important that everyone knows what their level of responsibility is, and the authority that they have to deal with a given situation. When a challenge is presented at work the employee's physiological state will alter to deal with the event, (increased blood pressure, etc). Any nervousness or heightened state of alertness to deal with the situation will be better controlled by the individual if they know exactly how they are supposed to respond, and the parameters in which they can deal with the given situation.

The interface between two overlapping parts of an organisation can also be a cause of stress – each will have their own priorities and their own way of working. Good communication and clearly defined lines of authority are vital to identify the essential patterns of overlap and to enhance the organisation's efficiency.

To overcome such problems the employer should review where any overlap in processes or responsibilities may occur, and where any gaps or areas that have not been clearly defined may be present. These should be cross-checked against the recipients (i.e. employee's) perceptions and interpretations of the responsibilities to ensure that misunderstandings do not occur.

### Organisational changes and arrangements

S11027    People generally do not like change. It upsets the person's comfort factor and invariably leads to a period of uncertainty whilst they are adapting to the changes. Some people adapt better to changes in their lives than others. Having an awareness that any change is likely to cause stress is the first step to managing the problem.

It is necessary to communicate as clearly as possible the nature of any changes, and the consequences of these to the workers who are affected. Even bad news can be dealt with positively if it is communicated honestly, openly and with sensitively.

## Responsibility

S11028     Once a responsibility has been allocated to an individual, there should be sufficient scope allocated to enable the person to carry out the task. This does not mean that the employee is then left to deal with a given responsibility without any mechanism for support or assistance if the responsibility becomes more extensive or complex than first anticipated. Being given a responsibility without any access to support when needed will inevitably lead to stress.

The feeling of a lack of appreciation, through non-realisation of an individual's potential, often arises from being given insufficient responsibility. Many people, especially those in jobs which do not enjoy high status in their organisation, feel frustration through lack of challenge, or feel frustrated when they can see things going on around them when they do not have an opportunity to be involved in making a contribution. Giving people autonomy, i.e. increasing the degree of self-control over their work, will help to counteract this.

Developing a system for employee recognition and staff development will help to identify where an employee needs more or less responsibility within their work. This will also provide a mechanism for a manager to explain why more responsibility has not been given to an individual who feels they want or need it, and for managers and employees to work together to reach a level which is mutually agreeable.

## Individual perception of company

S11029     If staff have a negative view of the company, they will respond accordingly. Employees who feel that the company is going nowhere, or does not care about its image, will not care about the company's future or its image. This will give them no sense of achievement or pride in their employment. The end result will be a higher staff turnover and the associated problems that go with this. Output will not be as effective as a motivated team of workers, and standards of work are likely to be lower.

Involving staff in the company's vision, or business plan will help employees to see what the company is hoping to achieve and the goals that have been set. Even if the employees do not fully understand or agree with the vision they will at least have an understanding of what is happening when changes begin to occur.

## Perception of one's self–esteem and role within the company

S11030     Everyone needs to feel that they have a valid role to play, in their life and in their work; everyone needs to feel valued and needed. When this does not occur in the workplace, the result is an employee who feels worthless, undervalued and misunderstood. The problem will become compounded if the individual concerned is someone who is ambitious or is trying to achieve promotion or a more fulfilling role.

## Previous experience

S11031     Someone who has experience of the environment will be more aware of the penalties and rewards associated with the work. They are likely to have been exposed to any associated work stressors before and may be more adept at managing them. However, this can sometimes have the opposite effect, in that if someone has been exposed to high levels of stress in a similar type job previously then this may well have reduced their tolerance to pressures.

# Personal factors

S11032     Everyone responds differently to events that occur in their lives. This very much depends upon their basic personality traits, their upbringing and their experiences in life.

Various studies have been undertaken in behaviour patterns. The research undertaken by Friedman and Rosenman *Type 'A' behaviour and your heart* (1975) (ISBN 0704501589) gave some insight into why some people are more prone to stress-related disease. In these studies, two main risk behaviours were identified, which were called Type A for high risk, and Type B for low risk. In general, Type B behaviours are the opposite of Type A behaviours. These categories should only be considered as a general yardstick as there are no absolutes, and most people will fall somewhere between the two extremes identified.

Type A people were said to be:

- autonomous, dominant and self-confident.
- very self-driven, busy and self-disciplined.
- aggressive, hostile and impatient.
- ambitious and striving for upward social mobility.
- preoccupied with competitive activities, enjoying the challenge of responsibility.
- believing that their behaviour pattern is responsible for their own success.

Type A people were said to work longer hours and spend less time in relaxation and recreational activities. They received less sleep, communicated less with their partners and tended to derive less pleasure from socialising.

In contrast, type B people tend to:

- find it easier and are more effective at delegating.
- work well in a team.
- allow subordinates some flexibility in how a task is undertaken.
- avoid, or do not set, unrealistic targets.
- keep a sense of balance in the events that occur in their lives.
- be more accepting of a given situation.

Type B people are often more secure within themselves and more accepting of their strengths and limitations. A very simplistic summary of this type of character could be that they are simply more 'laid back'.

## The level of responsibility and control over work

S11033     These are major factors as to whether someone is likely to suffer the effects of stress. Karasek and Theorell *Healthy Work* (1996), a comprehensive study of job demand and autonomy, showed that stress is a function of the combination of psychological demand and 'decision latitude', or autonomy.

The term 'executive stress' has been widely used, insinuating that perhaps people with high levels of responsibility are more likely to suffer from occupational stress. Research has shown, in fact, that this is not true. Whilst the level of responsibility increases with 'white collar' roles, so does the level of autonomy about how the work is to be carried out. Material rewards are usually higher, which helps to enhance self-esteem within the worker.

Karasek and Theorell's studies showed that the greatest risks are among those occupations with least autonomy and of a lower status. Examples include bus drivers, production line workers and care staff. These are classic examples of the conclusion reached by Karasek and Theorell that it 'is the bossed, and not the bosses' who are most likely to suffer from stress. They often have little or no control over how the work is to be undertaken but have to deal with the problems and difficulties that arise when a job is badly designed or managed. They also have less material reward for the work undertaken.

## Qualifications and competencies

**S11034** There is a rather cynical saying that people are promoted to their level of incompetence. Whilst this may be an exaggerated view there is some truth in the concept that people who are newly promoted may not immediately be competent. The same could apply to almost any role at work, if there is a mismatch between the task that is required and the skills that are possessed by the people undertaking the work. People need to grow and develop into their roles at work, to a point that they feel comfortable and competent in what they do without becoming bored or complacent. An employee who works beyond their competence is likely to cause problems for the organisation in that work standards may reduce and production may be affected. From the workers' point of view they are not likely to feel confident or able that they can achieve what is required, and this will have a knock-on effect upon their self-esteem. The end result is likely to be that such an employee would feel frustrated and dissatisfied, and maybe very unhappy. The manager is likely to feel equally frustrated and dissatisfied because the work is not being completed satisfactorily.

## Personal issues

**S11035** Separating personal problems from work issues is difficult for some people, and impossible for others. Both work and personal issues have an indirect effect upon each other and if an employee is experiencing a stressful situation in their home life then this may impact upon work performance.

An employer cannot be responsible for personal issues in an employee's life, but an awareness that a problem exists for the individual is useful knowledge for the employer. Systems should be developed that enable a manager to offer support and sympathy whilst the difficulty exists. For example, if a family member is in hospital perhaps working hours could be adapted slightly to fit in with the hospital's visiting times. The reaction to an individual who is going through a crisis, in terms of the level of support and flexibility shown by the manager, can stay with that person for a long time. Genuine help being offered by the manager can be an investment which promotes commitment from that person at a later stage.

A pre-agreed workload level may be acceptable in normal circumstances, but intolerable when a long term problem is occurring in someone's personal life. Therefore, the events going on in the employees' lives outside work will, at some stage, have an effect on their ability to cope with work. If these events are positive then the ability to cope will improve, with the opposite effect happening when the personal events are negative.

Managers need to take a balanced approach when responding to an individual's personal issues. An employee who is considered to have a good 'credit' balance in terms of their commitment and enthusiasm to the organisation may be treated more sympathetically by the organisation if a crisis develops in their lives. However, this could in itself have a significant knock-on effect to other employees who feel that they have not been treated so considerately, and this can induce feelings of being undervalued, misinterpreted or simply overlooked by those colleagues.

# Stress management techniques

**S11036** The following paragraphs provide detailed guidance and advice on managing stress.

## Stress auditing/surveys

**S11037** Stress audits or surveys are a good starting point, as this will give the employer an indication of whether a stress issue exists in the workplace, and the areas where this may be most problematic.

There are several approaches to audits and surveys, ranging from a simple checklist that employees are asked to complete, to detailed analyses with scoring systems that are undertaken by specialists. The type of audit that is used will depend upon the organisational culture and the extent to which an employer feels that a stress problem exists.

A survey in its simplest form may merely be a short questionnaire that asks employees to list their three most positive and three most negative aspects about their work. Guided headings may be included, perhaps with a tick box and a space for the employee to write comments explaining why it is seen as a positive/negative aspect of work. Headings that are used could include areas that the employer suspects are causing pressures upon the employees. This type of approach is often used as a pre-audit review, to enable employers to gain a feel as to any patterns which might be obvious to the workers.

Stress assessments are more detailed, and require planning, thought and sensitivity. They usually take the form of a guided list of questions or areas that the assessor wishes to cover about the work. They can include all members of staff, or selected members of staff from certain job titles to act as a representation. The interview needs to be undertaken by someone who is sufficiently competent to complete such a task. Competencies necessary to achieve this may include:

- counselling skills.
- polished interview techniques.
- knowledge of the company and its management structure.
- knowledge of procedures and practices.
- understanding of health and safety issues, including stressors at work.
- understanding knowledge of human behaviour (i.e. human factors at work).
- the ability to gain the confidence of the interviewee.
- the ability to listen.

---

Points to remember:

- Reassure staff that all comments will be confidential.
- Respect the confidentiality of your staff.
- Tell your staff what you are planning to do, what you hope to achieve, why you are doing it, and the benefits that may be in it for them.
- Provide a mechanism by which anyone can discuss any queries or concerns about the audit process beforehand.
- Involve *all* levels of employment, from senior board members to junior staff.
- Consider issuing a pre-interview questionnaire which you can review before your interview (e.g. three best and three worst aspects of their job).
- Design a simple framework around which you can develop your audit/survey format.
- Ask simple, open questions.
- Give people time to consider questions asked, or points that have been raised.
- Allow plenty of time for the interview, and sufficient time afterwards to write up the key findings.
- Tell your staff what you plan to do with any information collected.
- Involve them as much as possible in any further decisions or developments arising from the survey.

After the audit:

- Inform staff and/or representatives about the key findings and what the company plans to do in response.
- Provide (where possible) an indication of time frames as to when and how these changes can expect to occur.
- Involve staff and/or representatives.
- Explain why any suggestions may not be taken forward by the company, don't just refuse.
- Follow up any changes you make to ensure they are having the effect you intend.
- Keep staff and representatives involved at all times during the design and implementation of any systems or procedures resulting from the audit.
- Lead by example – managers can communicate important signals about the importance of managing stress.

## Stress management policy

S11038     A proactive management strategy for occupational stress should include development of a policy. This should highlight the company's philosophy and recognition of the issue, and clearly state the procedures in place to eliminate or manage stress associated with work.

The policy should include:

- the company's definition of occupational stress.
- recognition that stress is a work issue that will be taken seriously and sympathetically by management.
- acknowledgement that stress is not an illness or a weakness.
- explanation of the procedures the company will implement to identify occupational stressors and implement risk reduction control systems.
- details of personnel who have been allocated with responsibilities for implementing and monitoring risk reduction measures for stress.
- explanation of employees' responsibilities and procedures for reporting to management when they feel they are suffering from stress.
- training and support packages, including any counselling services that may be available.

## On-going monitoring

S11039     Like any other health and safety system it is necessary to monitor systems and procedures to ensure they are effective. Monitoring will have the added benefit of helping to identify a stressed employee who is unwilling to report they have a problem. Monitoring can be undertaken in many ways including:

- monitoring during team meetings, by encouraging a general discussion about positive and negative aspects of work that have occurred.
- monitoring during appraisals and other one-to-one meetings with staff.
- implementing an anonymous stress report box in which people who are finding a work situation to be continuously problematic can report this without any fear of recrimination.
- circulating staff attitude or opinion surveys which will encourage feedback from workers.

- observation of workers' behaviour patterns and relationships with colleagues (looking for any obvious differences from the norm, or spotting relationships/atmospheres that appear to be tense).
- monitoring accident and incident statistics, and sickness absence records.

## Information and training

**S11040** Provision of information and training is a general duty under the *Health and Safety at Work etc Act 1974*, s 2(2)(c). This will include information and training in occupational stress. Information should be given to employees to enable them to understand the organisation's views on occupational stress, its policies and its procedures for dealing with stress and the part that employees can play in managing occupational stress for themselves.

Training can be used to underpin any findings from surveys or assessments that may have been undertaken, and to help the policy on stress to be understood and effectively implemented. Training should be provided to managers and to staff in general. Managers will need to know:

- What stress is and it contrasts with or effects occupational stress.
- Symptoms of stress and how to recognise someone who may be suffering from stress.
- Health risks associated with stress.
- Occupational factors which cause or contribute towards stress.
- Identifying, separating and managing personal and management stressors.
- Outline of company policy and procedures on stress.
- Identifying measures to identify stress.
- Understanding of measures to manage/reduce occupational stress.

Employees will need to understand for themselves exactly what stress is, and how to recognise for themselves when they or a colleague may be experiencing symptoms. The training will enable the organisation to demonstrate that it is a caring employer.

The Chartered Institute of Environmental Health (CIEH) has developed a one day accredited course called 'Stress Awareness'. This provides a detailed programme that covers the key issues associated with occupational stress. The fact that is accredited by a recognised body will help to increase its profile to employees who may be more cynical about the subject itself, or the organisation's commitment to employee welfare.

# Training and Competence in Occupational Safety and Health

## Introduction

**T7001** Although the term 'incompetent' is widely recognised by people in the world at large and can quite easily be defined as not qualified or able to undertake a defined action or task, when looking at the definition of competent or competence, the boundaries are less easy to define. Someone may be qualified or able to undertake a task to a certain level or in a specified area, so they could be considered competent either generally at a lower level or in part. This particular dilemma is one that has faced most professions and occupations during their development stages, and has led to the formation of professional bodies and trade associations, who as one of their major reasons for existence have set standards of practice for their own particular disciplines. This is perhaps where competent performance related to health and safety is particularly different to other professions and trades, because everyone, whether at work or not, needs to have knowledge of how to remain safe and healthy.

This becomes particularly relevant in workplaces where both employers and employees have moral and legal responsibilities both for themselves and also for others, in terms of maintaining a safe working environment. Everyone needs to have a level of competence in terms of occupational health and safety, from employees who need to ensure that their actions do not endanger either themselves or their colleagues, to directors of the company who are responsible for the health and safety of the workforce and need to know what this entails. Additionally, specialist advisers in health and safety will need to have much higher and broader levels of competence in occupational health and safety issues, so that the advice they give to both the management and the workforce in organisations is clear and of a standard expected from a professional practitioner.

The Health and Safety Commission ('HSC'), made a clear statement regarding health and safety competence in their 'Strategy on Health and Safety Training', which entailed the aim that: 'Everyone at work should be competent to fulfil their roles in controlling risk.' It also makes a statement about the training required to do this:

'Health and safety training covers all training and developmental activities aimed at providing worker, including safety representatives, and managers with:

(a)   greater awareness of health and safety issues;

(b)   specific skills in risk assessment and risk management;

(c)   skills relating to the hazards of particular tasks and occupations; and

(d)   a range of other skills, including those relating to job specification and design, contract management, ergonomics, occupational health etc.

The specific aims of the strategy are to:

- raise awareness of the importance of health and safety training;
- bring about a substantial improvement in the quality and quantity of health and safety training;
- promote an awareness of the importance of competence in controlling risk;

- influence providers of the education system to provide the necessary framework of basic knowledge and skills.

With the objectives of the strategy being to ensure competence by:

- encouraging employers and trade unions to recognise the need to provide health and safety training of good quality;
- engaging HSE and Local Authority (LA) inspectors to assess the competence of workers and managers and the adequacy of training provided by employers. This forms an important part of inspection, investigation and enforcement activities;
- continuing to seek partnerships elsewhere to provide training of appropriate quality and quantity;
- influencing other government departments to reflect the need for health and safety training provision in their responsibility; and
- ensuring that all parts of the education system provide a foundation of knowledge upon which health and safety training can be built.'

This chapter focuses on:

- the components of competence and the acquisition of this;
- how it may be attained and assessed for all the sectors of the workforce;
- the legal and managerial requirements for competence; and
- the various organisations that fulfil roles in the process.

In all cases the chapter recognises that any level of competence is time sensitive. Skills and knowledge become outdated very quickly in the light of new technologies and practices, and it is a particular aspect that needs to be considered in maintaining the health and safety of the workforce.

## Recognising competence

**T7002** It is worth having an in-depth look at some of the more common definitions of competence before exploring the ways and means of attaining specific and relevant levels of health and safety competence. These definitions, although not in actual conflict with each other, do have slightly different slants on the concept, which very much depends on the source of the definition. Clearly people from outside the health and safety world may well view competence in a slightly different way to those more familiar with the legal concepts. This may well lead to the position where employers believe they have an appropriate level of competent advice (as described in the next section at T7002), but may in fact be leaving themselves exposed to both criminal and civil actions in the event of an accident, where it can be proved that the appropriate level of advice was not available. When either selecting individuals or training programmes for employees, employers should be advised to seek appropriate advice to ensure that they are legally and morally covered in terms of health and safety; good advice can also have a considerable effect on a company's profit margins!

Looking at the concept of competence from a plain English perspective, *The New Oxford Dictionary of English* defines 'competence' as the ability to do something successfully or efficiently, and 'competent' as having the necessary ability, knowledge or skill to do something successfully.

The National Vocational Standards define competence as 'the ability to perform to the standards required in employment across a range of circumstances and to meet changing demands'.

In case law a competent person is viewed as 'one who is a practical and reasonable man who knows what to look for and how to recognise it when he sees it' (*Gibson v Skibs A/S Marina and Orkla*

*Grobe A/B and Smith and Coggins Ltd [1966] 2 All ER 476*). It could also be added that they should also know what to do with the acquired knowledge once they have found it.

All of these definitions basically amount to the same thing but with a slightly different slant: that competence is a combination of knowledge, experience and skills. It should be recognised that competence itself is based on the outcomes of this combination of requirements, and it is a demonstration that an individual can perform to specified standards rather than simply a record of a person's qualifications and experience – with which it is often confused. It is this simple definition of knowledge, skills and experience which is used throughout this chapter, taking into account that the actual determination of an individual's level of competence does require a further stage of assessment by an employer. Only the training and education leading to the development of competence can be identified here.

## Legal obligations for training and competence
### Health and Safety at Work etc. Act 1974

T7003    The *Health and Safety at Work etc. Act 1974* ('*HSWA 1974*') is the principal enabling legislation for the UK. It imposes duties on employers, employees, bodies corporate, manufacturers and others, as well as establishing the HSC that recommends policy regarding health and safety, and the Health and Safety Executive ('HSE') that enforces this policy and serves a range of legal notices to employers.

Training is specifically mentioned within section 2 of the *HSWA 1974*, which prescribes the general duties of employers. Section 2(2) states that in addition to other requirements, it is an employer's duty to ensure 'the provision of such information, instruction, training and supervision as is necessary to ensure, so far as is reasonably practicable, the health and safety at work of his employees'. This in fact does not place an absolute duty on an employer to provide training, but recognises that having carried out generic and specific risk assessments of the workplace, an employer may well believe that training would be an effective measure of control. Many employers take this approach in addition to other risk control methods.

Competence is not specifically identified within the *HSWA 1974*, other than an allusion in section 19(1) that refers to the appointment of inspectors. Here it relates to 'such persons having suitable qualifications as it thinks necessary for carrying into effect the relevant statutory provisions within its field of responsibility'.

Perhaps the most important part of the *HSWA 1974* in relation to training and competence is in section 15, which relates to the power to make health and safety regulations. The first set of these regulations as relevant to training and competence are the *Safety Representatives and Safety Committees Regulations 1977 (SI 1977 No 500)*, which require an employer to allow a trade union-appointed safety representative to take time off 'to undergo such training as may be reasonable in the circumstances'.

### Management Regulations

T7004    Major changes to the requirement for competence on a general basis came about as a result of the UK's enactment of a series of European directives on health and safety via a series of regulations. The *Management of Health and Safety at Work Regulations 1992 (SI 1992 No 2051)* (now revoked) was a set of generic regulations relating to the overall management of health and safety, and worked in conjunction with already existing regulations such as the Control of Substances Hazardous to Health Regulations ('COSHH') and the Noise at Work Regulations. The 1992 Management Regulations actually introduced the concept of a 'competent person'. The Regulations were amended in 1999, with *Regulation 6* becoming *Regulation 7* in the *Management of Health and Safety at Work Regulations 1999 (SI 1999 No 3242)* ('*MHSWR 1999*'). These Regulations are the main focus when considering competence in health and safety.

The *MHSWR 1999 (SI 1999 No 3242), Reg 7(1)* states that:

> 'Every employer shall, appoint one or more competent persons to assist him in undertaking the measures he needs to take to comply with the requirements and prohibitions imposed upon him by or under the relevant statutory provisions.'

The MHSWR 1999 (SI 1999 No 3242), Reg 7(5) states that:

> 'A person shall be regarded as competent for the purposes of paragraph (1) where he has sufficient training and experience or knowledge and other qualities to enable him to assist in undertaking the measures referred to in that paragraph.'

The Approved Code of Practice that accompanies the *MHSWR 1999 (SI 1999 No 3242)* makes some specific conditions regarding the competent person role in these Regulations. In summary, these are:

- Employers are solely responsible for ensuring that the person appointed to fulfil the 'competent person' role is capable of applying the principles of risk assessment and prevention, together with a current understanding of legislation and health and safety standards.

- Where there is more than one person appointed to fulfil the role, or parts of it, then these people must co-operate.

- Competence in the Regulations does not necessarily depend on the particular possession of specific skills or qualifications, and simple situations may require an understanding of relevant current best practices and an awareness of the limitations of one's own knowledge whilst having the ability to seek external help where necessary.

- More complex situations will require a person with a higher level of knowledge and experience, and in these cases employers are advised to check the appropriateness of health and safety qualifications and/or membership of professional bodies. The competence-based qualifications accredited by the Qualification Curriculum Authority (QCA) can act as a guide.

The *MHSWR 1999 (SI 1999 No 3242), Reg 8(1)(b)* imposes a duty on employers to nominate competent persons who have the ability to carry out evacuations of an establishment in the event of serious or imminent danger. *Regulation 8(3)* goes on to say that such a person will have sufficient training and experience or knowledge and other qualities, defining competence in a similar way to the *MHSWR 1999 (SI 1999 No 3242), Reg 7*. Also, in *Regulation 15* of the 1999 Regulations, which relates to temporary workers, an employer is again required to determine if a temporary employee has the competence to carry out the job in a healthy and safe manner.

The *MHSWR 1999 (SI 1999 No 3242)* make several statements particularly in relation to the requirement for training. *Regulation 3(5)* relates specifically to young people and *Regulation 13* specifically relates to capabilities and training for all employees. *Regulation 13(2)* particularly states that every employer shall ensure that his employees are provided with adequate health and safety training, and very specifically relates to the various times that this should be given, namely:

- on recruitment;

- on being exposed to new or increased risks because of being transferred or having a change of responsibility;

- on the introduction of new work equipment or a significant change to current equipment;

- on the introduction of a new system of work or a change to an existing system.

The *MHSWR 1999 (SI 1999 No 3242), Reg 13* also says that this training should be repeated periodically and be adapted to take account of any changes or new risks, and should take place during working hours.

Overall the *MHSWR 1999* (*SI 1999 No 3242*) imposes a highly comprehensive requirement for both training and the recognition of competent advice in the workplace. They have highlighted and made a legal requirement of what has been good practice in workplaces for many years. In addition to these Regulations, there are several more hazard-specific pieces of legislation, which also require training to be delivered and competence to be demonstrated.

The policy unit of HSE has published an outline map on competence, training and certification. The map gives an overview for competence and training as described in legislation, and can be found on the HSE website. The mapping process identifies that there are four different groups in which training and competence requirements can be identified:

- *Group 1*

    General goal-setting requirements where there is no precise detail of how the requirement should be met – there are 37 pieces of legislation that have this approach;

- *Group 2*

    General goal-setting requirements qualified by some specific requirements for people or bodies with particular responsibilities – there are eight instances of this in legislation;

- *Group 3*

    General goal-setting requirements qualified by particularly strong specific guidance, sometimes agreed with the industry/sector – there are six instances of this in legislation;

- *Group 4*

    Specific requirements for competence or training, certification/qualifications and approval, including statutory requirements to meet a performance standard – there are 40 specific requirements of this nature in legislation.

Within group 1 there are five high-risk areas where an employer is required to have arrangements for training and levels of competence written into plans or reports. These are:

- construction (site supervisor's plan);
- control of major hazards (in accordance with the *Control of Major Accident Hazards Regulations 1999* (*SI 1999 No 3242*);
- offshore safety cases;
- railway safety cases (in accordance with the *Railways (Safety Case) Regulations 2000* (*SI 2000 No 2688*);
- nuclear site licences.

There are specific requirements for people or bodies within group 2. This includes medical personnel responsible for medical surveillance under the *Control of Substances Hazardous to Health Regulations 2002* (*SI 2002 No 2677* as amended by *SI 2003 No 978*) and the requirement of specific training within the *Provision and Use of Work Equipment Regulations 1998* (*SI 1998 No 2306* as amended by *SI 2002 No 2174*).

In group 3 there is strong specific guidance advocating the use of particular programmes of training. Regulations included in this group are the *Noise at Work Regulations 1989* (*SI 1989 No 1790*), the *Manual Handling Operations Regulations 1992* (*SI 1992 No 2793* as amended by *SI 2002 No 2174*) and the *Health and Safety (Display Screen Equipment) Regulations 1992* (*SI 1992 No 2792* as amended by *SI 2002 No 2174*).

Group 4 lists those areas where there are specific legal requirements for named training programmes. Activities worth noting in particular in this area are radiation protection advisors, gas fitting (CORGI), adventure activity licensing, mining, the carriage of dangerous goods and diving.

Employers should make themselves aware of any specifically cited requirements in their sphere of business in addition to the more general management requirements described in the *MHSWR 1999* (*SI 1999 No 3242*). They should give the necessary training required that will lead their workforce to perform competently with regards to health and safety.

## Setting national standards in health and safety

**T7005** During the 1980s it was identified by a Government Working Party that there was no effective system of vocational qualifications within the UK. What did exist had evolved from individual employment sectors, and as such there was no commonality of level and the standards varied. This was a difficult situation for employers, who were unable to determine the validity of qualifications; whilst some sectors had highly respected qualifications, others had none. A system was required that would recognise the skills people needed and was reliable, consistent and well structured. In 1986 the Government established the National Council for Vocational Qualifications ('NCVQ') to set up a comprehensive framework of vocational standards covering all occupations and industries. One notable fact was that all the standards were required to have a unit of health and safety within their structure.

The standards were developed by organisations know as 'Lead Bodies'. These were independent groups of professionals representing employers, trade unions, government departments, local authorities and other relevant groups. In the health and safety area the organisation setting the standards was known as the Occupational Health and Safety Lead Body ('OHSLB'). The HSE formed the secretariat for the Lead Body, and included representatives from the CBI, TUC the Department for Employment and Local Authorities. They developed the national standards for Occupational Health and Safety Practice, Regulation and Radiation Protection. These standards were launched in 1995 and became available as the National and Scottish Vocational Qualifications ('NVQ/SVQ').

The HSE adopted the OHSLB Enforcement Standards, in addition to its established academic training programme, in the Guidance Note attached to the *Management of Health and Safety at Work Regulations 1992* – recognising the Practice Standards as representing levels of competence for those giving advice on health and safety. The Institution of Occupational Safety and Health ('IOSH'), the chartered professional body for practitioners in health and safety, also recognised the standards as underpinning their competence categories of membership: level 4 at their full member category and level 3 at technician level (see T7006 below for clarification of levels).

The ownership of the standards passed from the OHSLB to a new organisation, the Employment National Training Organisation ('ENTO'), in 1998, following a government rethink of the structure of how the national vocational standards should be organised. ENTO's immediate priority was to cross-functionally map the constituent standards it had inherited. As well as occupational health and safety, this included the personnel, training and development and trade union standards. Once this was completed, its next task was to revise the existing standards, which commenced in 2000 with a consultation process.

The new standards in Occupational Health and Safety Practice ('OHS') were launched on 13 August 2002 in conjunction with the leading players in the health and safety world: IOSH, the Royal Society for the Prevention of Accidents (RoSPA), the National Examining Board for Occupational Safety and Health ('NEBOSH') and the British Safety Council. The OHS standards themselves had undergone a fundamental revision during this process and now became available at levels 4 and 5 – the latter reflecting the increasing requirement for higher standards of competence for those who strategically manage or advise on health and safety issues. The HSE also adopted the revised level 5 standards for enforcement.

An earlier initiative by ENTO had identified that the mandatory health and safety units included in the wider range of vocational standards were numerous (well over 400) and variable. It was its belief, as the training organisation for people who work with people at work, that a collection of stand-alone units for people who needed the competence to fulfil certain health and safety responsibilities, but who were not health and safety practitioners, was necessary to bring a level of consistency in the competence requirements across the whole range of vocational standards. These were produced and launched in 1999, and as such they have no particular level and can be fitted into other sets of vocational standards as appropriate. The eight separate units known as 'Health and Safety for People at Work' can be used to demonstrate a level of health and safety competence for the whole array of employees, from workers to managers and directors. They cover the range of knowledge, skills and practical experience that a person has to have to be able to do a particular task in a safe manner, and are designed to allow people to show that they are competent in the safety issues described.

In conclusion, there are now standards available within all the UK national frameworks to recognise the level of competence required by all people in the workforce. These standards, as well being capable of being used as the standards against which the NVQ/SVQ qualifications are assessed, also form the basis for other types of qualification in the health and safety area.

## Education and training towards national standards of competence

**T7006** This section will focus in more detail on the required competence levels in health and safety across the whole spectrum of the workforce. It will include strategic health and safety practitioners or managers, health and safety advisers and practitioners, health and safety technicians, managers and supervisors with line responsibilities for health and safety, and last but by no means least, the workforce itself. Some of the qualifications traditionally associated with health and safety, such as those from the British Safety Council or NEBOSH, will be equated to the national frameworks to give an indication of the parity of the different routes available.

The framework of competence standards and associated vocational-related qualifications ('VRQs') is organised in the constituent bodies by the nationally associated bodies highlighted in T7005 above. For brevity they will be referred to using the English body QCA. However, this should be taken as meaning all the bodies that do in fact have a working association.

In addition to these frameworks there is another UK qualification framework specifically for the higher education sector. This is known as the Quality Assurance Agency for Higher Education, and is referred to as 'QAA'. For a clearer understanding, the relationship between qualifications on the QAA framework and the QCA framework are shown in the following chart:

| QCA | QAA |
| --- | --- |
| *Level 5* | Post-graduate Diploma, Masters Degree |
| *Level 4* | Higher Education Diploma, Ordinary Degree and Honours Degree |
| *Level 3* | 'A' Level (access to higher education) |
| *Level 2* | — |
| *Level 1* | — |

This chart is only a rough guide to be used for the purposes of this chapter. It has no legal standing as to the equivalence of level.

## Competence requirements for occupational health and safety practitioners

**T7007** The IOSH is the leading professional body in the UK for those involved in occupational health and safety practice. It was founded in 1945, incorporated by Royal Charter in 2003 and currently has over 27,000 subscribers, including over 12,000 in competent practice categories of membership. The IOSH, as the chartered professional body, sets the standards for those in the practice of health and safety. However, since their inception, the IOSH has taken an active part in the development of the national vocational standards for the OHS and Health and Safety for People at Work, and now uses these national vocational standards as part of the definition of its membership criteria.

The national vocational standards at level 4 are recognised by the profession as representing the core competence requirements for those in health and safety practice. There are, however, many ways of achieving these standards through the various routes of qualification, together with the development of skills and experience. The recently revised standards, which are loosely based on the safety management model developed in the HSE publication *Successful health and safety management* (HSG65), have moved away from the approach taken by the original level 4 standards. These original standards focused on the level of risk faced by an organisation and the competence a person would need to develop to ensure that the advice given to control these risks was given in an effective manner. For the level 4 standards this was deemed to be for all industry sectors, which included complex, co-existing high risks where failure to control them could be life or organisation threatening. The level 3 standards introduced at the same time were for those in lower risk workplaces, where control is well documented, or for those who formed part of a team in higher risk areas reporting to a level 4 standard competent person. Level 3 Practice Standards do not appear in the revised suite but have been replaced by a different level 3 in health and safety, based on the generic standards Health and Safety for People at Work. The new standards, which are all at level 4, form the basis of competent practice, irrespective of the organisation in which a person operates.

The titles of the revised OHS Standards, which are available on a CD-ROM from ENTO, are as follows:

G3 Evaluate and develop own practice

H2 Promote a positive health and safety culture

H3 Develop and implement a health and safety policy

H4 Develop and implement effective communication systems for health and safety information

H5 Develop and maintain individual and organisational competence in health and safety matters

H6 Identify and evaluate health and safety hazards

H7 Assess health and safety risks

H8 Determine and implement health and safety risk control measures

H9 Develop and implement active monitoring systems for health and safety

H10 Develop and implement reactive monitoring systems for health and safety

H11 Develop, implement and test health and safety emergency response systems and procedures

H12 Develop and implement health and safety review systems

H13 Develop and implement health and safety audit systems

The NVQ qualifications are actually assessed against these standards. This type of qualification is an assessment of the ability of a person to perform to the laid-down standards and involves an assessor who will monitor and lead the candidate through the process of developing a practice portfolio. For an NVQ at this higher level it is also necessary for the assessor, backed up by the verification processes required by the awarding bodies, to determine that a candidate has the requisite domain knowledge to carry out the competence-based tasks required.

This type of qualification has proved extremely popular in occupational health and safety practice, as it is particularly suited to those who have many years of experience in giving health and safety advice based on sound experience, but who are not happy or comfortable with the traditional exam-based type of qualifications available. In fact, in IOSH's subscription year ending on 31 March 2005 25 per cent of all those entering the Member (corporate) category of IOSH held a qualification of this type.

The national vocational standards are also used by the QCA to assess the relevance of vocationally-related qualifications to the overall national framework. Organisations applying to the QCA to be placed on this national framework must show that they have robust systems of quality control in place and have an examination process totally divorced from the delivery of their courses. This is known as part A accreditation. They must also show that their syllabus delivers the underpinning knowledge requirements of the national vocational standards, known as part B accreditation. NEBOSH is one organisation that follows this route, and the long-established NEBOSH Diploma (Dip2OSH) is currently being revised to meet the new OHS Standards at level 4. The NEBOSH Diploma, which is divided into part 1 and part 2, was structured to meet the original Practice Standards at level 3 and 4. However, as the level 3 Practice Standards have been withdrawn, this has lead to the reversion back to a one-part NEBOSH Diploma. The national qualifications framework for related vocational qualifications was revised to reflect a better equivalence to qualifications from the higher education sector and came into effect on 1 September 2004. This means that the one-part NEBOSH Diploma, which has been marketed as NEBOSH Diploma level 4, will in due course have its name changed to the NEBOSH National Diploma in Occupational Health and Safety, although there is in fact no change to the syllabus. In a similar market the British Safety Council have undertaken a thorough review of their existing qualifications and have produced a level 6 Diploma in Occupational Safety and Health. This has been accredited by IOSH and QCA, and have been available since September 2005. Although other organisations, such as the Chartered Institute of Environmental Health (CIEH) and the Royal Society for the Promotion of Health (RSPH), have qualifications on the framework, as yet they have nothing above level 3.

The QCA accreditation verifies that the qualifications on its framework are robust and have fair levels of achievement recognised at the specified levels. Universities offering qualifications in occupational health and safety also have to undergo similar processes. Universities are themselves responsible for awarding qualifications within the QAA frameworks, and have introduced stringent and searching quality control procedures internally by a system of internal validation and external examination by appointed examiners. The QAA holds the overarching quality control on the qualifications offered. Furthermore, it is normal for universities offering vocationally-related degrees to seek external accreditation from the professional body in that area. IOSH undertakes this accreditation process for occupational health and safety degrees and post-graduate qualifications, and published its own criteria for the conditions of accreditation and syllabus content in 1997. These have currently been revised to fall in line with the national vocational standards of competence, and to maintain the IOSH accreditation universities will be expected to adjust their syllabuses to meet the knowledge requirements within the next few years.

There are a range of different levels of qualification available from the higher education ('HE') sector that form the Higher Education Diploma and Foundation degrees, which roughly equate to level 4 on the QCA framework to the more academically challenging Bachelors and Post-Graduate degree programmes. There are currently 28 universities and related colleges in the UK offering these types of programme, which are proving very popular. In fact this is now by far the biggest route of entry into the professional membership category of IOSH, amounting to 52 per cent of all entrants in the subscription year ending on 31 March 2003.

It should be clarified that academic qualifications such as those from the HE sector, NEBOSH or other similar bodies are not assessments of competence in the same way as an NVQ qualification. They are a demonstration that a programme of academic work leading to the development of knowledge and some skills has been undertaken and formally assessed by examination and/or other suitable form of academic assessment. This by itself would not be sufficient to demonstrate competence, but with suitable experience and further development of skills in a practical setting, the information gained can form part of the knowledge, skills and experience equation. However, it should also be recognised that although the achievement of an NVQ actually demonstrates competence, this is only in the workplace setting in which the person undertaking the qualification is employed. This means that the broader knowledge base required of a competent practitioner has not necessarily been covered or assessed to the same extent to a person who has taken the academic qualifications.

There are clearly pros and cons for each type of qualification and this is a debate that continues throughout the safety profession and the deliverers of training and education. This is not helpful to employers and those needing to employ competent health and safety practitioners who need to be able to assess the similarities and differences between the varying types of initial qualification. Health and safety is a diverse profession and attracts people from many different backgrounds, so it is advantageous and necessary to have the variety of different ways to qualification currently available. IOSH as the professional body currently accepts all the different qualification routes but recognises that all have particular strengths and weaknesses. In the revised IOSH membership structure that begins to operate in November 2005, further assessment in addition to initial qualifications will be necessary to show overall competence . This will be tailored to the initial qualification, so that those with an NVQ will need to undertake some further formal assessment of knowledge, whilst those with knowledge-based qualifications will need to demonstrate the acquisition of skills. Potential employers will be able to use the IOSH membership structure and particularly the requirements for its chartered category membership as a yardstick of competence for safety practitioners, irrespective of whether they are a member of the professional body.

## Competence in health and safety practice at strategic management level

**T7008** In the current set of national vocational standards, a level 5 set of competencies in OHS was introduced. This reflects the fact that many more health and safety professionals are now moving into more strategic management roles within organisations and that many who would not consider themselves to be health and safety professionals also operate at this strategic management level role. Although there are a plethora of health and safety training courses specifically targeted at managers, very few of them are actually pitched at this higher level of management expertise. This is perhaps not surprising, as most managers would incorporate health and safety within their workplace role on a day-to-day basis, implementing the rules and policies that have already been developed, using their personal knowledge of the processes of the organisation for which they work.

It has been a long-standing request from the health and safety community that health and safety should be included in MBA programmes so that potential future leaders of industry should have health and safety included within their academic programmes. As yet this has not borne much fruit, but one or two organisations do now do this. Perhaps of more potential is the fact that the HSC included in its *Revitalising Health and Safety* document that the education for those in safety-critical professions should include health and safety specifically.

The titles of standards for OHS at level 5 are listed as follows:

| | |
|---|---|
| G3 | Evaluate and develop own practice |
| H1 | Develop and review the organisation's health and safety policy |
| H2 | Promote a positive health and safety culture |
| H3 | Develop and implement the health and safety policy |
| P11 | Develop a strategy and plan for people resourcing change management |
| H15 | Influence and keep pace with improvements in health safety practice |
| L2 | Identify organisational learning and development needs |
| B3 (METO) | Manage and use financial resources |

B4  Secure financial resources for your organisation
(METO)

These standards are reflective of health and safety strategies but are in reality management standards. Some of the standards are actually taken from other sets of standards, such as those from the management-training organisation (METO). How much of these standards will be used as NVQs is as yet undetermined, as they are still relatively new and form part of the new suite of standards launched by ENTO during 2002.

## Competence in health and safety for managers and supervisors

T7009  Although those in safety practice can set the standards and form the systems to ensure that health and safety risks are controlled, it is actually those in everyday contact with the people and work interface who can actually make the most difference to the overall health and safety performance. The competence of this group can be achieved by a variety of ways and means, and it is this group who have the knowledge of the day-to-day workings of an organisation and therefore will be able to have the most effect in the workplace.

The ENTO suite of standards, Health and Safety for People at Work, were designed as competence standards for all people in the workplace, and many of them were targeted at the managerial and supervisory groups. Initially it was anticipated that managers and supervisors would make their own choice from the standards depending on their own role in the organisation for which they are employed. However, at the recent revision it was suggested that the complete suite, with the exception of Unit , which is for placement review officers, would form a level 3 NVQ qualification in its own right A new standard, I, is currently in development; this has been designed for those who undertake supervision of placement students in the workplace, and augments Unit D. The titles of the standards are given below:

H&SA  Ensure your own actions reduce risks to health and safety

H&SB  Monitor procedures to control risks to health and safety

H&SC  Develop procedures to control risks to health and safety

H&SE  Promote a health and safety culture within the workplace

H&SF  Investigate and evaluate incidents and complaints in the workplace

H&SG  Conduct an assessment of risks within the workplace

H&SH  Ensure your own actions aim to protect the environment

H&SI  Supervise the health, safety and Welfare of a learner in the workplace

These standards can be assessed as competence via the NVQ route, but at this level the majority of industries still favour the more traditional training approach to providing knowledge and skills. There are numerous programmes that take this approach. Some are assessments of both knowledge and skills, whilst others programmes take a more traditional knowledge approach, assuming that the skills can be gained in the actual workplace. This section highlights a few of these programmes but it is not exhaustive, and normal training selection criteria should be applied when selecting a programme.

The NEBOSH National General Certificate in Occupational Safety and Health is an examined-basis qualification designed to help non-specialists to discharge their duties or functions in workplace health and safety. Such people include managers, supervisors and employee representatives. This certificate covers much of the underpinning knowledge requirements for the ENTO level 3 Health and Safety for People at Work standards. It is normally taught over 80–100 hours by all modes of delivery, including distance learning. NEBOSH makes it clear in its literature that being awarded a certificate does not imply competence but the underpinning knowledge towards competence. Although not specifically designed as an introduction to safety practice, this qualification is often used as a starter for people moving into the profession who would then progress to the NEBOSH Diploma. It is accredited by IOSH as meeting the academic requirements for a Technician member in the IOSH (Tech IOSH) membership category (from November 2005).

The Chartered Institute of Environmental Health ('CIEH') Advanced Health and Safety Certificate is also classified as a level 3 qualification on the QCA framework. However, it is a much shorter programme than the NEBOSH Certificate. It is at a similar level but does not have the full breadth of syllabus found in the NEBOSH Certificate.

The Royal Society for the Promotion of Health ('RSPH') Advanced Diploma in Health and Safety in the Workplace is again a QCA level 3 course. It has 36 guided learning hours and assessments are carried out by coursework assignments. The title is a little misleading, as this is a basic health and safety qualification. The QCA framework does not specify the titles of programmes, only the levels.

The British Safety Council Level 3 Certificate in Occupational Safety and Health is a single unit qualification with two elements, 'Principles of Health and Safety' and 'Applied Health and Safety', and is intended for those with some responsibility for health and safety in the workplace. It is assessed by two three-hour examinations, including a risk assessment exercise, and is approximately 60 guided learning hours, with self-study time required as well. It is accredited by IOSH as meeting the academic requirements for Tech IOSH membership. (from November 2005).

The IOSH Managing Safety course has been specifically written for managers and focuses on two to three units of the ENTO Health and Safety for People at Work standards. The course particularly concentrates on the risk management process and how to use it in day-to-day workplace activities. This course does not appear on the QCA framework, as it is run as a collaboration between IOSH and its training delivering members, and as such does not have the separation of marking from delivery that is required for this accreditation. The course is flexible and can be tailored to meet the specific requirements of organisations.

In addition to these qualifications run by health and safety bodies, most of the large awarding bodies also offer health and safety programmes. Both OCR and City and Guilds have qualifications at QCA level 2, as well as offering the level 3 NVQ Health and Safety for People at Work.

The IOSH also offers a Safety for Senior Executives briefing programme, which is tailored to give executives, directors and the most senior managers in organisations the knowledge and information they require to run them in a safe and healthy manner. This group is vital to the facilitation of good health and safety standards in the workplace, and they must show a commitment towards making sure standards are maintained and where possible improved on a continuing and regular basis. Training at this level needs to specifically recognise the requirements of this group of people, and routine training hammering out day-to-day hazards and specific details of the legislative requirements is not appropriate. More in-depth consideration and exploration of the financial and moral aspects of health and safety is required, together with an understanding of the legal responsibilities and possible penalties for failure to ensure health and safety in the organisations for which they are responsible.

Most of the organisations also offer courses of a slightly shorter nature for those in supervisory roles. These are shortened versions of the manager programmes. It can be argued, however, that supervisors are the real front-line for safety and that their training should be exactly the same as for managers, but the reality is that financial constraints often mean that these shorter programmes are used.

## Competence in health and safety for safety representatives

T7010    Many people who perform the function of safety representatives in the workplace, as defined by the *Safety Representatives and Safety Committees Regulations 1977* (*SI 1977 No 500*), actually train in health and safety in a similar way to their colleagues undertaking comparable courses to those described for managers and supervisors. There is, however, a specific programme run by the TUC under the auspices of the National Open College Network (NOCN), which has been specifically designed to meet the requirements of this group. This programme, the TUC Certificate in Occupational Health and Safety, is a level 3 programme that builds on the TUC basic stage 1 and stage 2 health and safety courses or other unions' equivalent courses. The assessment of the course is by a competence-based type of assessment similar to the NVQ qualifications, with the addition of workplace-biased projects of practical value to the candidate's workplace and union.

The course is available for union representatives on either a part-time day release basis or by distance learning. The course additionally gives access to higher education for those who wish to progress further.

The TUC Education Service normally pays for the course, provided that the candidate is nominated by a trade union, and is a valuable contribution towards competent safety advice in the workplace. This is recognised by the IOSH, who currently offer TechSP membership for those with this qualification and able to demonstrate at least two years pro-rata full-time equivalent experience.

## Competent performance by the workforce

**T7011** Last but by no means least in the range of the competence chain that leads to better health and safety performance in the workplace is the workforce itself. The *HSWA 1974* makes it quite clear that employers have a legal duty to ensure, as far as is reasonably practicable, the health, safety and welfare at work of the people for whom they are responsible and the people who may be affected by the work that they do. The 1974 Act further requires employers to provide information, instruction training and supervision to ensure, as far as reasonably practicable, the health and safety at work of employees.

It is the training aspect of these requirements, which has a huge variance across organisations, on which this section focuses.

The ENTO recognised the fact that there should be a standard available for everyone in the different suites of occupational standards, namely the Health and Safety for People at Work suite of standards. Within this suite there is a unit specifically targeted at everyone in a working environment, regardless of position or the number of hours worked. Unit A in the suite titled 'Ensure that your own actions reduce risks to health and safety' is about making sure that risks to health and safety are not created or ignored. It is divided into two elements:

A1 Identify the hazards and evaluate the risks in your workplace

A2 Reduce the risks to health and safety in your workplace

The elements then go on to identify what a person should be able to do to act in a competent manner towards health and safety issues in the workplace. Clearly this is competence that is developed whilst undertaking an employment role, but it will be necessary to include training that delivers the knowledge required to perform to the appropriate level. This is normally carried out at an early stage in a person's employment, normally during induction training on starting a job.

These standards are straightforward and form a good series of competence requirements for a workforce. The series of performance criteria for identifying hazards and evaluating risks include such things as:

- correctly naming and locating the persons responsible for health and safety in the workplace;
- identifying workplace policies that are relevant to working practices;
- identifying working practices in any part of a job role which could harm yourself or other people;

- identifying those aspects of the workplace which could harm yourself or other people;
- evaluating which of the potentially harmful working practices and the potentially harmful aspects of the workplace are those with the highest risk to yourself or to others;
- reporting those hazards that present a high risk to the persons responsible for health and safety in the workplace;
- dealing with hazards with low risks in accordance with workplace policies and legal requirements.

The attainment of this level of competence would be assessed for an NVQ across a wide range of workplace policies, including methods of working, equipment, hazardous substances, smoking, eating, drinking and drugs and what to do in the event of an emergency. Similar sets of competence requirements are also required for A2 but these are based on reducing the risks to health and safety in the workplace in addition to the recognition required in unit A1.

These standards can be used both to plan health and safety training and also to analyse where training is actually needed, by comparing existing skills and competence of the workforce with the standards and identifying if there are any gaps, which then can be covered by training. This can be done on an individual or group basis, and the training offered as a result of this analysis can be offered in a specifically targeted way.

There are a variety of training programmes at this primary level which develop the knowledge requirements for health and safety in the workplace. Many are quality controlled by either awarding bodies or bodies with a health and safety remit. Some of the well-known programmes are the IOSH Working Safely, CIEH Basic Health and Safety Certificate, RSPH Foundation Certificate in Health and Safety, and the Workplace awards from City and Guilds, OCR and Edexcel. Many organisations seek this external verification to show that they have carried out their training requirements with regard to their duties under the *HSWA 1974*. However, there are many perfectly good induction programmes run by organisations themselves which allow this knowledge part of the competence requirements to be achieved. It is the fact that good training is available that is of vital importance, and the national standards of competence give an indication of what needs to be achieved from any training programme.

## Competence for contractors and safety passports

**T7012** Leading on from the requirement of the development of a workforce that is competent to perform its role in a healthy and safe manner is the requirement to look at all those people who also may be present in a workplace.

Although organisations have the ability to control the training and competence of individuals they have directly employed, by developing programmes that meet their specific requirements, there is one area that they need to, but do not necessarily have, complete control over: outside contractors coming into their workplaces. This is particularly relevant in industrial sectors such as construction and related industries, where the use of contracted labour is extensive. However, at some point all employing organisations may have people other than their own employees within their workplaces. Many organisations control this situation by having 'permit to work' systems to ensure that anyone entering a workplace has sufficient knowledge of the health and safety arrangements that are in place.

However, within the last few years there has been a move to a more transferable skills type of scheme that can allow those who move between workplaces to show that they have sufficient-knowledge of health and safety matters to allow briefer induction programmes to be given when moving between different workplaces and sites. These are the safety passport schemes.

These passports, available from several industry-related schemes, allow employers to determine who may be given access to their workplaces. The employing organisation makes their workplace a passport-controlled environment. The passports themselves generally take the form of a credit card-sized plastic card, containing the name, photograph and other relevant details of an individual; they also verify what training a person has undertaken towards receiving this passport. The passport usually also has an expiry date.

The advantages of passport schemes are:

- they can save time and money through reduced induction training, as they are the starting point for workers training;
- they demonstrate an organisation's commitment to a safe and healthy working environment;
- they can create good relationships with the supply chain between organisations and their suppliers;
- they can be used by employees of companies as well as sub-contractors;
  - contractors can move between companies and demonstrate that they have the necessary health and safety awareness;
  - they can be used to control access to worksites;
  - probably most importantly, they can increase a positive safety culture within organisations, which can lead to a reduction in accidents or ill health.

However, it does need to be borne in mind that a passport can neither identify whether a worker is competent nor be an effective substitute for workplace management including risk assessment and site-specific information.

Many large organisations are now taking up this approach as part of their safety management systems; this in turn will lead to smaller organisations, who are often suppliers to or sub-contractors for the larger organisations, to also adopt the schemes as a business decision.

The largest of the schemes currently available is the Construction Skills Certification Scheme ('CSCS'), which to date has issued over 250,000 passports. This organisation has a management board that includes members from the Construction Confederation, Federation of Master Builders, GMB Trade Union, National Specialist Contractors Council, Transport and General Workers Union and the Union of Construction Allied Trades and Technicians. It also includes observer members from relevant government departments and the Confederation of Construction Clients.

The Construction Industry Training Board (CITB) manages the operation of the CSCS and delivers training through its series of regional offices. To obtain a passport the candidates have to sit a 40-question multiple choice test through the same centres as used for driving tests, although this is expected to be expanded to other centres using an on-line technique in the near future.

The Client/Contractor National Safety Group ('CCNSG') is one of the oldest schemes and was developed for contractor site personnel. Representatives from major clients are nominated by local safety groups to serve on the CCNSG and trade unions; contractors and safety groups also have places. The training is a basic two-day training programme. The Engineering Construction Industry Training Board (ECITB) manages the scheme. Around 160,000 people currently hold this passport.

Other passport initiatives are: Airport Construction Training Alliance (ACTA), Gas and Water Industry, Register of Energy Sector Engineers, Technologist and Support Staff (RESETS), Engineering Services SKILL Card, Sentinel for Railway Track Workers, Federation of Bakers' Contractors Passport (FBCP), IOSH in partnership with other organisations, National Microelectronics Institute (NMI), Metal Industry Skills and Performance Institute (MetSkill), Safety Certification for Contractors (SCC), Safety Pass Alliance and the Port Skills and Safety Limited (PSSL).

In addition to this list there are some other initiatives within specific industries. Clearly some rationalisation of the schemes would be useful to improve the currency of the passports, and this may well be a factor in this market over the next few years. To accommodate such a rationalisation approach the HSE have produced a leaflet INDG381 Passport Schemes – A good practice guide. This leaflet covers:

- What health, safety and environment passports are
- How Passport schemes work
- Benefits and Advantages of Passport Schemes
- How Passports help employers and workers stay within the law
- How a worker can get a Passport
- Withdrawal arrangements for training, refresher training and monitoring

Helpfully, the HSE has also produced a core syllabus containing all the items that the schemes should include within their training and assessment programmes.

## Maintaining competence

T7013    As we have seen, competence is the development of sufficient knowledge, skills and experience to be able to perform a task to the laid down standards in a workplace. These standards may be prescribed by a legislative requirement, national standards or by best practice in the workplace. However competence is prescribed or developed, there is a cycle that needs to be recognised in the attainment and then in the maintenance of the desired level.

Competence is actually a developed attribute and passes through several phases. These can generally be recognised as distinct stages:

- *unconscious incompetence* – a person is unaware of the requirements to perform in a competent manner;
- *conscious incompetence* – a person begins to undertake training and becomes aware of what they do not know;
- *conscious competence* – a person has undergone sufficient training to be able to complete a task in a competent manner and is aware of this;
- *unconscious competence* – a person now performs tasks in a fully competent manner and has now developed to such an extent as to become unaware of this, as it is part of their behavioural patterns.

At this point in the cycle there can be two pathways: a person can continue to perform in a competent manner or they can unconsciously develop bad habits, which can lead to a diminishment in their performance.

The final bullet point indicates that there is a need for those responsible for the workforce to revisit both their own current knowledge and that of their employees on a regular basis, to ensure that performance relating to health and safety is in fact to the level they believe it to be.

There can be a number of triggers which may identify a possible need for refresher courses or further training. This is not necessarily simply because of a time lapse since the initial training took place – other issues such as the introduction of new equipment, changes to methods of use, the re-site of equipment, changes of shift patterns, etc. should be considered. Safety audits may also identify that there have been behavioural changes that need addressing.

Many of the external agencies, which play a role in the quality control of training programmes, will specify an expiry date for their accreditation. The safety passport schemes have particular expiry dates, usually three to five years, built into the actual passport, whilst organisations such as the IOSH have a three-year expiry term on their Managing and Working series of programmes.

For professionals who practice in health and safety there is a particular need for the maintenance and development of their skills. The area of health and safety is constantly changing, both in terms of legislative requirements and also in the knowledge of best practice in health and safety issues. For Chartered Members and Fellows of the IOSH (CMIOSH/CFIOSH), newly developed categories of IOSH membership, recently approved by the Privy Council and as such acknowledged as competent in the practice of health and safety, there is a mandatory requirement to complete Continuing Professional Development ('CPD') to a prescribed level, roughly equating to five days of activity per year. All those in the Chartered categories of the IOSH are currently required to produce a log book with details of the activities that show how both the maintenance and development of health and safety have been achieved. Employers looking to recruit either an employee or consultant at this level are advised that they should ask to view this log as an indication of continuing competence. This is one particular aspect that the IOSH have reviewed, and they are currently incorporating a modern, more reflective approach to this activity, in which the professional develops plans for both development and maintenance of skills and after a predetermined period, usually three years, reflects back to determine their achievement or, sometimes, lack or deficiency of achievement over this period. This is an extension of the concept of life-long learning that is now being incorporated into the initial training programmes for those in occupational health and safety practice and in fact features within the national vocational standards. The IOSH CPD scheme will offer on-line learning from late 2005, so that members can maintain their CPD as efficiently as possible without mounds of paperwork. As part of this revision the competent practitioner also takes overall ownership of their own CPD requirements and allows them to be more flexible in the type of activities that they can use to demonstrate CPD. Although there are a number of well designed CPD-taught courses available, these form only part of the requirement for CPD, and practitioners are being encouraged to think beyond attendance at events. The CPD scheme will also be extended to Tech IOSH members towards the end of 2005, as it is critical that all those involved in health and safety are maintaining their competence.

To sum up, it is critical for good performance in health and safety in the workplace that the initially developed knowledge and skills are constantly maintained. This can only be achieved by regular checking of the competence levels of the entire workforce in organisations. Some industry sectors have formalised this approach within their safety management systems and in this instancecompetence frameworks are developed and monitored on a formal basis. This is good practice and should be considered by those who have not yet taken this approach.

## Useful contacts

T7014     The following is a list of useful contacts for organisations that have been mentioned in this chapter:

**Airport Construction Training Alliance**
Website: www.acta.easitrack.com

**British Safety Council**
70 Chancellors Road
London
W6 9RS
Tel: 020 8741 1231
Fax: 020 8741 4555
Website: www.britsafe.org

**Chartered Institute of Environmental Health (CIEH)**
Chadwick Court
15 Hatfields
London
SE1 8DJ
Tel: 020 7928 6006
Fax: 020 7828 5866
Website: www.cieh.org

**City and Guilds**
1 Giltspur Street
London
EC1A 9DD
Tel: 020 7294 2468
Fax: 020 7294 2400
Website: www.city-and-guilds.co.uk

**Client Contractor National Safety Group (CCNSG)/Engineering Construction Industry Training Board**
Blue Court
Church Lane
Kings Langley
Hertfordshire
WD4 8JP
Tel: 01502 712 329
Website: www.ecitb.org.ik/safetypassport/safetypassport.html

**Construction Skills Certification Scheme (CSCS)**
Tel: 01485 578 777
Website: www.cscs.uk.com

**Employment National Training Organisation (ENTO)**
Kimberley House
47 Vaughan Way
Leicester
LE1 4SG
Tel: 0116 251 7979
Fax: 0116 251 1464
Website: www.empnto.co.uk

**Engineering Services Skill Card**
Old Mansion House
Eamont Bridge
Penrith
Cumbria
CA10 2BX
Tel: 01768 860 406
Fax: 01768 860 401
Website: www.skillcard.org.uk

**Federation of Bakers (FBCP)**
6 Catherine Street
London
WC2B 5JW
Tel: 020 7420 7190
Website: www.bakersfederation.org.uk

**Gas and Water Industries National Training Organisation (GWINTO)**
The Business Centre
Edward Street
Redditch
Worcestershire
B97 6HA
Tel: 01527 584 848
Website: www.gwinto.co.uk

**Health and Safety Commission (HSC)** (Contact HSE)

**Health and Safety Executive (HSE)**
Rose Court
2 Southwark Bridge
London
SE1 9HS
Tel: 020 7717 6000
Fax: 020 7717 6996
Website: www.hse.gov.uk

**Institution of Occupational Safety and Health (IOSH)**
The Grange
Highfield Drive
Wigston
Leicester
LE18 1NN
Tel: 0116 257 3100
Fax: 0116 257 3101
Website: www.iosh.co.uk

**Metal Industries Skills and Performance (MetSkill)**
5–6 Meadowcroft
Amos Road
Sheffield
S9 1BX
Website: www.sinto.co.uk

**National Examination Board in Occupational Safety and Health (NEBOSH)**
5 Dominus Way
Meridian Business Park
Leicester
LE19 1QW
Tel: 0116 263 4700
Fax: 0116 282 4000
Website: www.nebosh.org.uk

**National Microelectronics Institute (NMI)**
1 Michaelson Square
Kirkton Campus
Livingston
EH54 7DP
Tel: 0131 449 8507
Website: www.nmi.org.uk

**OCR Examinations**
Westwood Way
Coventry
CV4 8JQ
Tel: 0247 647 0033
Fax: 0247 642 1944
Website: www.meg.org.uk

**Port Skills and Safety Ltd**
Africa House
64–78 Kingsway
London
WC2B 6AH
Tel: 020 7242 3538
Website: www.bpit.co.uk

**Qualification Assurance Agency for Higher Education (QAA)**
Southgate House
Southgate Street
Gloucester
GL1 1UB
Tel: 01452557000
Fax: 01452 557 070
Website: www.qca.ac.uk

**Qualification Curriculum Authority (QCA)**
83 Piccadilly
London
W1J 8QA
Tel: 020 7509 5555
Fax: 020 7509 6666
Website: www.qca.org.uk

**Royal Society for the Promotion of Health**
38A St George's Drive
London
SW1V 4BH
Tel: 020 7630 0121
Fax: 020 7976 6847
Website: www.rsph.org

**Register of Energy Sector Engineers, Technologists and Support Staff (RESETS)**
Operations Enterprise House
Cherry Orchard Lane
Salisbury
SP2 7LD
Tel: 01722 427 226
Fax: 01722 414 165
Website: www.resets.org

**Safety Pass Alliance**
P.O. Box 7772
Braunstone
Leicester
LE3 3XY
Tel: 0116 2221228
Fax 0116 2220329
Website: www.safetypassalliance.co.uk

**Trade Union Congress (TUC)**
Congress House
Great Russell Street
London
WC1B 3LS
Tel: 020 7636 4030
Website: www.tuc.org.uk

# Ventilation

## Introduction

**V3001** Air being invisible and intangible, there is a dangerous tendency to take it for granted. As breathing is predominantly an involuntary activity, it underlines the need to be selective about what we breathe, with whom we do so, and in what conditions. The twin objectives of good ventilation are the elimination or dispersal of particulate matter (see V3010 below) and the replacement of stale, hot or humid air (often associated with production processes or equipment) with fresh air free from impurity, the aim being to induce a sense of individual comfort in employees. The *Workplace (Health, Safety and Welfare) Regulations 1992* guidance states that the fresh air supply rate to workplaces should not normally fall below 5 to 8 litres per second, per occupant. When considering the appropriate fresh air supply rate, the factors that should be considered include: the amount of floor space available for each occupant; the work actives being carried out; whether there is smoking allowed within the work areas; and whether there are other sources of airborne contamination, such as that arising from work activities, process machinery, heaters, furniture, furnishings, etc. In addition to air changes, the movement of air across the work area/occupant should also be taken into consideration. The CIBSE Guide *'Volume A: Environmental design'* recommends that 'At normal temperatures the air flow velocity should be between 0.1 to 0.15 metres per second and up to 0.25 metres per second during summer.' Draughts should be kept to a minimum. The composition of pure dry air is 20.94% oxygen, 0.03% carbon dioxide and 79.03% nitrogen and inert gases.

Satisfactory ventilation is normally achievable by means of windows (or other similar apertures), as well as by recycled air, particularly in warm weather or hot workplaces. Air recirculated via mechanical ventilation or air conditioning systems should be adequately filtered and impurities removed (see 'Legionella' at D1032 DANGEROUS SUBSTANCES I), and be impregnated with fresh air prior to recirculation. To that end, ventilation systems should be designed with fresh air inlets and kept open, regularly and properly cleaned, tested and maintained. In order to avoid uncomfortable draughts the velocity of ventilation systems may have to be controlled. Employers who fail to ventilate effectively and follow relevant official guidance on medical surveillance of workers can be liable at common law if an employee becomes sensitised to a respiratory sensitiser in the workplace, even in the absence of evidence that the relevant occupational exposure limits have been exceeded (*Douglas Reilly v Robert Kellie & Sons Ltd (1989) HSIB 166* where an employee contracted occupational asthma whilst working with low levels of isocyanate fumes. It was held that the employer was liable at common law for breach of the *Factories Act 1961, s 63* (now repealed), since the hazards associated with isocyanate fumes had been known about since the early 1980s).

This section deals with current statutory requirements, contained in the *Health and Safety at Work etc Act 1974* (general) (see V3004 below), the *Workplace (Health, Safety and Welfare) Regulations 1992* (*SI 1992 No 3004*) (specific) (see V3005 below), and certain specific processes where adherence to ventilation requirements is paramount (see V3006 below). (For requirements relating to ventilation on construction sites see C8047 CONSTRUCTION AND BUILDING OPERATIONS.) Origins of pollution of the working atmosphere, specific pollutants and their insidious effect on health, along with strategies for controlling them, are also covered.

## Pollution of the working environment and dangers of exposure to dust and fumes

**V3002** Pollution of the working environment can occur through the generation of airborne particulates, such as dusts, fumes, mists and vapours (as defined in V3003 below). Of these, probably dust and fumes pose the principal hazard to workers, some of the situations being well documented, e.g. tungsten and silicon processes resulting in pneumonia followed by progressive fibrosis, milling cotton resulting in byssinosis, weaving hemp in chronic bronchitis, underground mining in emphysema and copper welding/electroplating with cadmium also causing emphysema (see further O1042 OCCUPATIONAL HEALTH AND DISEASES). Dusts can be fibrogenic, that is they can lead to fibrotic changes to lung tissue, or, alternatively be toxic, eventually leading to poisoning of body systems. Examples of fibrogenic dusts are: silica, cement dust and certain metals such as tungsten; toxic dusts include: arsenic, lead, mercury and beryllium. More particularly, dusts are classified according to the response on the worker, as follows:

(a) benign pneumoconiosis, such as siderosis, associated with work with iron particles; here there is no permanent lung disorder;

(b) pneumonitis or acute inflammation of lung tissue, caused by inhalation of metallic fumes, e.g. zinc oxide fumes; this can result in death;

(c) extrinsic allergic alveolitis, e.g. farmer's lung; this can eventuate into a disabling condition;

(d) tumour-forming, e.g. asbestosis, plural mesothelioma (caused by work with crocidolite); characterised by a high mortality rate;

(e) nuisance particulates, e.g. dust from combustion of solid fuels; however causing no permanent lung damage.

Dust at the workplace has a variety of origins ranging from:

(i) dust arising in connection with cleaning and treatment of raw materials, e.g. sandblasting in foundries;

(ii) dust emitted in operations such as refining, grinding, milling, cutting, sanding; and

(iii) dusts manufactured for specific treatments or dressings; to

(iv) general environmental dust, e.g. caused by sweeping factory floors or by fuel combustion.

Fumes are formed through vaporisation or oxidation of metals, e.g. typically lead and welding fumes. Lead processes emitting dust and fume should be enclosed and maintained under negative pressure by an enclosing hood. Moreover, the fume should be treated before any escape into the atmosphere. Welding fume, caused by action of heat and ultraviolet light, produces carbon monoxide and ozone and is potentially harmful, leading to 'welder's lung'. For that reason, welding should only be carried out in ventilated areas. Welding workshops should be equipped with adequate mechanical ventilation, and local exhaust ventilation should be incorporated at the point of fume emission as a supplement to general ventilation. If welding is carried out in a confined space (see the *Confined Spaces Regulations 1997* (*SI 1997 No 1713*)), a permit to work system should be in force. Welders should of necessity be familiar with various relevant forms of respiratory protection (see further PERSONAL PROTECTIVE EQUIPMENT).

### Polluting agents

**V3003** Pollution of the workplace environment may take place through the generation of airborne particulates, such as dusts, fumes, mists and vapours. These are defined below.

*Particulate*: a collection of solid particles, each of which is an aggregation of many molecules.

*Dust*: an aerosol composed of solid inanimate particles. (Standard ILO definition).

*Fumes*: airborne fine solid particulates formed from a gaseous state, usually by vaporisation or oxidation of metals.

*Mist*: airborne liquid droplets.

*Vapour*: a substance in the form of a mist, fume or smoke emitted from a liquid.

## Statutory requirements
### General – HSWA 1974

**V3004** The general duty upon employers under the *Health and Safety at Work etc Act 1974* (*HSWA 1974*) to 'ensure, so far as reasonably practicable, the health, safety and welfare at work of all their employees' [*HSWA 1974, s 2(1)*], must be regarded as including a duty to provide employees with an adequate and renewable supply of pure and uncontaminated air. Employers must also provide and maintain a safe working environment. [*HSWA 1974, s 2(2)(e)*]. In consequence, employers may be in breach if the workplace is not adequately ventilated and specific dust and fume hazards not controlled. In addition, all employers must inform, instruct and train their employees in health and safety procedures. This means that they, or their representatives, must know how to use, test and maintain equipment for ensuring air purity and ventilation as well as dust control (*R v Swan Hunter Shipbuilders Ltd [1981] ICR 831*, where the deaths of the subcontractor employees might have been avoided, had they been properly instructed and trained in the use of respirators). (See further C8134 CONSTRUCTION AND BUILDING OPERATIONS.)

Facilities managers (i.e. those responsible for the well-being of occupants and visitors) are required to maintain a clean and healthy ventilation system in the workplace. This may mean the procurement of ventilation hygiene services from a specialist contractor which often demands a level of technical expertise outside the scope of the majority.

'*General ventilation guidance for employers*', HSG202, is published by HSE Books.

The Heating and Ventilating Contractors' Association (HVCA) have published a Guide to Good Practice, '*Cleanliness of Ventilation Systems – TR/17*'. This Guide and others on ventilation are available from HVCA Publications (tel: 01768 864771).

Responding to industry needs, the Building Services Research and Information Association has developed a Standard Specification for Ventilation Hygiene\* to allow competitive tenders to be provided by contractors on an equal basis. The client thereby achieves optimum value for money by securing complete but not excessive action by the contractor. Correspondingly, contractors face less ambiguity in performance expectations.

The Standard Specification is in two parts, the basic specification with accompanying detailed guidance. Supported by the Department of Environment Transport and Regions, this is the first of a range of specifications written for facilities managers. It provides straightforward, easy to use clauses of industry-standard performance requirements. The guidance document explains in detail the scope of work required to fulfil a ventilation hygiene contract.

\* '*Guidance and Standard Specification for Ventilation Hygiene*', price £40, is available from BSRIA publications sales, tel: 01344 426511.

### Specific – Workplace (Health, Safety and Welfare) Regulations 1992 (SI 1992 No 3004)

**V3005** Effective and suitable provision must be made to ensure that every enclosed workplace is ventilated by a sufficient quantity of fresh or purified air. Plant designed and used for such purposes must be accompanied with visible and/or audible means of warning of any failure, which might affect health or safety. [*Workplace (Health, Safety and Welfare) Regulations 1992 (SI 1992 No 3004), Reg 6*].

# Specific processes or industries

## Highly flammable liquids and liquefied petroleum gases

**V3006** Where a dangerous concentration of vapours from a highly flammable liquid may be expected, the process must be carried on within a cabinet or other enclosure, which is:

(a) effective to prevent the escape of such vapours into the general atmosphere;

(b) adequately ventilated; and

(c) fire-resistant.

[*Highly Flammable Liquids and Liquefied Petroleum Gases Regulations 1972* (*SI 1972 No 917*), *Reg 10(2)*].

Where compliance with this requirement is not reasonably practicable the workroom must have exhaust ventilation adequate to remove vapours from the workroom. [*Highly Flammable Liquids and Liquefied Petroleum Gases Regulations 1972* (*SI 1972 No 917*), *Reg 10(3)*].

## Lead processes

**V3007** The *Control of Lead at Work Regulations 1998* (*SI 1998 No 543*) came into force on 1 April 1998. They were introduced to ensure that work involving lead might be covered by a single set of regulations.

Regulations revoked include the *Control of Lead at Work Regulations 1980* (*SI 1980 No 1248*). See the *Control of Lead at Work Regulations 1998, Sch 3* for a full list of revocations. The *Factories Act 1961, ss 74, 128, 131 and 132* have been repealed by *Reg 14(1)* of the 1998 Regulations.

The aims of the new Regulations and Approved Code of Practice, Regulations and Guidance, (COP2) are to:

(a) protect the health of people at work by preventing or, where this is not practicable, adequately controlling their exposure to lead;

(b) monitor the amount of lead that employees absorb so that individuals whose work involves significant exposure (as defined by *Control of Lead at Work Regulations 1998* (*SI 1998 No 543*), *Reg 2*) to lead at work can be taken off such work before their health is affected.

The Regulations apply to all work which exposes employees to lead in any form in which it may be inhaled, ingested or absorbed through the skin, for example as dust, fume or vapour.

The duties imposed by these Regulations extend not only to employers but also to self-employed persons. [*Control of Lead at Work Regulations 1998* (*SI 1998 No 543*), *Reg 3(2)*].

## Control of Lead at Work Regulations 1998 (SI 1998 No 543)

**V3008** An employer who is working with lead, or a substance or material containing it has a duty to ensure that his employees are not exposed to lead, or, where this is not reasonably practicable (see E15039 ENFORCEMENT), to ensure that appropriate control measures are adequately controlling the exposure to lead of his employees and of anyone else who may be affected by their work with lead and/or lead-based products – such persons may include:

— persons working for other employers, such as maintenance staff or cleaners;

— visitors to the work area; and

— families of those exposed to lead at work who may be affected by lead carried home unintentionally on clothing and/or footwear.

Exposure to lead covers all routes of possible exposure, i.e. inhalation, absorption through the skin and ingestion.

Without prejudice to the *Management of Health and Safety at Work Regulations 1999* (*SI 1999 No 3242, Reg 3*), employers must not carry on any work which is liable to expose any employees to lead at work unless they have first made an assessment as to whether the exposure of any employees to lead is liable to be significant. [*Control of Lead at Work Regulations* (*SI 1998 No 543*), *Reg 5*].

Measures for controlling exposure to lead may include one or more of the following:

(a) using substitutes, i.e. lead-free or low-solubility lead compounds;

(b) using lead or lead compounds in emulsion or paste form to minimise the formation of dust;

(c) using temperature controls to keep the temperature of molten lead below 500°C, the level above which fume emission becomes significant, though the formation of lead oxide and the emission of dust is still possible below this temperature;

(d) the containment of lead, lead materials, compounds, fumes or dust in totally enclosed plant and in enclosed containers such as drums and bags. The container must be so designed that no lead is allowed to leak out. Where it is necessary to open such containers, this should be carried out under exhaust ventilation conditions, if reasonably practicable;

(e) where total enclosure is not reasonably practicable, an effective ventilation system must be in operation before work is allowed to commence. This may consist of:

   (i) partial enclosure such as booths designed to prevent the lead escaping – these should be fitted with exhaust ventilation;

   (ii) various types of exhaust hoods which should be as close to the lead source as is reasonably practicable so that they may take the lead dust, fume or vapour away from the employee's breathing zone;

   (iii) an extract ductwork system that is adequate to remove the dust, fume or vapour from the source area;

   (iv) a dust and/or fume collection unit with a sound and adequate filtration system that will both remove the lead source from the workplace and prevent it from re-entering;

   (v) fans of a suitable type placed in the system after the collection and filtration units so that the units are kept under negative pressure, thus ensuring that any escape of lead is minimised;

(f) wet methods which include:

   (i) the wetting of lead and lead materials, e.g. wet grinding and pasting processes. Wet methods should be used during rubbing and/or scraping lead-painted surfaces;

   (ii) the wetting of floors and work benches whilst certain types of work are being carried out, e.g. work with dry lead compounds and pasting processes in the manufacture of batteries.

Wetting should be sufficiently thorough to prevent dust forming, and the wetted materials or surfaces should not be allowed to dry out since this can create dry lead dust which then is liable to be hazardous if it becomes airborne. Water sprays are not a fully effective method of controlling airborne dust.

Wetting methods should not be used when they are liable to be unsafe, such as:

   (i) at furnaces where they could cause an explosion;

   (ii) when lead materials containing arsenides or antimonides could, on contact with water, produce highly toxic arsine or stibine gases;

(g) providing and maintaining a high standard of cleanliness.

Special care and attention should be given to the design of plant and systems to eliminate possible areas which might increase the risk to the employee of exposure from lead.

Before any plant, equipment or systems are used in lead or lead compound work, there should be an adequate checking procedure in place to ensure that such plant, equipment or systems are as designed and that they meet the standard required under the *Control of Lead at Work Regulations 1998* for the safety of all those who could be affected by lead.

## Asbestos

**V3009**   No process must be carried on in any factory unless exhaust equipment is provided, maintained and used, which prevents the entry into air of asbestos dust. [*Control of Asbestos at Work Regulations 1987* (*SI 1987 No 2115*), *Reg 13(2)*].

In addition, where it is not reasonably practicable to reduce exposure of employees to below the 'control limits', employers must provide employees concerned with suitable respiratory protective equipment. [*Control of Asbestos at Work Regulations 1987* (*SI 1987 No 2115*), *Reg 8(2)*].

Moreover, where the concentration of asbestos is likely to exceed any 'control limit', the employer must designate that area a 'respirator zone' and ensure that only permitted employees enter/remain in that zone. [*Control of Asbestos at Work Regulations 1987* (*SI 1987 No 2115*), *Reg 14(2)*]. (See further ASBESTOS.)

## Control of airborne particulates

**V3010**   Airborne particulates may be controlled in the following ways.

### (a) Substitution replacement

The use of a less harmful toxic substance or modification to a process may minimise or totally eliminate the hazard. For example, substitution of soap solutions for organic solvents is sometimes possible in cleaning operations.

### (b) Suppression

The use of a wet process for handling powders or other particulates is an effective form of control. In the cleaning process, it may be possible to damp down floors prior to removal of dust sooner than resorting to dry sweeping.

### (c) Isolation

This form of control entails enclosure of all or part of a process or the actual point of dust production and may be incorporated in machinery/plant, in which case it is necessary to ensure seals are maintained. Total enclosure of large processes involving grinding of metals or other materials, linked to an extract ventilation system, is an effective method of isolating dust from the operator's breathing zone.

### (d) Extract/exhaust ventilation

Removal of dust or fume at the point of emission by entraining it in a path of fresh air and taking it to an extract hood or other collection device is the role of extract/exhaust ventilation (see the HSE's guidance booklet HSG54: '*The maintenance, examination and testing of local exhaust ventilation*'). The air velocity required to provide this movement depends upon the type of material, varying from about 0.5 metres per second for gases to 10 metres per second or above for some dusts. In fact for large dense dust particles from grinding or cutting operations, it may be necessary to arrange for the trajectory of the emitted particles to be encompassed by the exhaust hood so that the material is literally thrown into the hood by its own energy.

Extract ventilation systems generally take three distinct forms:

(i) *Receptor systems*: the contaminant enters the system without inducement, and is transported from the hood through ducting to a collection point by the use of a fan.

(ii) *Captor systems*: in this system moving air captures the contaminant at some point outside the hood and induces its flow into it. The rate of air flow must be sufficient to capture the contaminant at the furthest point of origin, and the air velocity induced at this point must be high enough to overcome the effects of cross currents created by open doors, windows or moving parts of machinery.

(iii) *High velocity low volume systems*: dusts from high speed grinding machines in particular require very high capture velocities. With an HVLV system high velocities at the source are created by extracting from small apertures very close to the source of the contaminant. These high velocities can be achieved with quite low air flow rates.

## (e) Dilution ventilation

In certain cases, it may not be possible to extract particulates close to their point of origin. Where the quantity of contaminant is small, uniformly evolved and of low toxicity, it may be possible to dilute it by inducing large volumes of air to flow through the contaminated region. Dilution ventilation is most successfully used to control vapours from low toxicity solvents, but is seldom satisfactory in the control of dust and fumes.

## (f) Cleaning procedures

Whilst the above methods may be effective in preventing environmental contamination of the workplace, there will inevitably be a need for efficient cleaning procedures wherever dusty processes are operated. Hand sweeping should be replaced by the use of industrial vacuum cleaning equipment, or in situ (ring main) systems which incorporate hand-held suction devices connected via ducting to a central collection point.

## (g) Air cleaning

Air cleaning is often employed, either to prevent emission of noxious substances into the atmosphere or to enable some of the air to be recirculated during winter months, thus reducing heated air costs and lowering fuel bills. For particulates such as dusts and grit, air cleaning may involve some form of inertial separator, such as settling chambers or cyclones, usually followed by bag filters. For very fine dusts and fumes, bag filters are often used as the pre-filter, followed by absolute or electrostatic filters. For gases and vapours, the cleaning is usually achieved by wet scrubbers (device where the air is passed in close contact with a liquid) to take the gas into solution or react chemically with it. Scrubbers can also be used for particulate material. The selection of air cleaning types will depend upon the properties of the materials emitted and the size ranges of the particulates.

To generate air flow in the duct, various types of fans are available. The type of fan must be suitable for the system in which it is installed. This requires a knowledge of the system resistance and the fan characteristics in order to generate the desired air flow with minimum noise and power consumption.

For systems with little ducting and airflow resistance, axial fans may be used to generate high air flows. For ducted systems, centrifugal fans are often used, creating generally lower air flows per size, but at the higher air pressures required to overcome the resistances imposed by ducting and air cleaners. The fan should not be used outside its duty range. It should be sited either outside the building or as near to the discharge as possible, so that air cleaners and all ducting within the building are on the negative static pressure side of the fan, thus preventing leakage of the contaminant through cracks or defects in the duct.

New guidance* on the use of the dust lamp for observing the presence of airborne particles in the workplace has been published by the Health & Safety Executive (HSE) in the series 'Method for the Determination of Hazardous Substances' (MDHS). The guidance has been written for occupational hygienists, ventilation engineers, and health and safety practitioners. It briefly explains the principle of the dust lamp, its use in observing the presence of airborne particles and identifies its advantages and limitations. Despite its name, the dust lamp can be used to reveal the presence of many different types of airborne particulates both solids (dusts, fumes, fibres) and liquids (organic or inorganic mists). The dust lamp can be used to gauge the size and direction of movement of a particle cloud, but it does not give a quantitative measure of either concentration or particle size.

* '*MDHS 82: The dust lamp: A simple tool for observing the presence of airborne particles*', price £15.00, is available from HSE Books.

## Respiratory protection

V3011    Respiratory protection implies the provision and use of equipment such as dust masks, general purpose dust respirators, positive pressure powered dust respirators, self-contained breathing apparatus or other forms of such protection. Where dust emission is intermittent or other controls are not available (as above), then resort may be made to respiratory protection. The work and care required in the selection and establishment of a respiratory protection programme, including the training of operators in the correct use of the equipment, is likely to be at least as great as any other control system. The protection must be selected to give adequate cover and minimum discomfort and the need for other personal protective devices should not be ignored. (See also PERSONAL PROTECTIVE EQUIPMENT.)

# Vibration

## The scope of this chapter

**V5001** This chapter deals with the effects of vibration on people and the obligations of employers to assess, monitor and reduce the risks of adverse health effects of hand-arm and whole-body vibration on their workforce. This area has emerged from relative obscurity following an EC Directive (2002/44/EC) called the Physical Agents (Vibration) Directive. The Health and Safety Commission (HSC) is currently preparing new regulations on the Control of Vibration at Work, which are expected to come into operation in July 2005.

This chapter is intended to provide some basic information on vibration and to help companies to get a feel for the likely scope and requirements of the new regulations, and to consider the implications for their businesses.

## An introduction to the effects of vibration on people

**V5002** The effects of vibration on people can be divided into three broad classes:

- building vibration perceptible to occupiers;
- whole-body vibration of people in vehicles etc, and industrial situations;
- hand-arm vibration of people operating certain tools or machines.

It is convenient to consider these classes separately, as the cause, effect and method of control is different in each case.

### Hand-arm vibration and vibration-induced white finger (VWF)

**V5003** The effects of industrial vibration on people received little attention until an award of damages to seven ex-British Coal miners in July 1998, in a landmark Court of Appeal decision (*Armstrong and Others v British Coal Corporation (1998) (unreported)*). The miners received compensation ranging from £5,000 to £50,000 for the effects of vibration white finger ('VWF'), a form of hand-arm vibration syndrome characterised by the fingers becoming numb and turning white. In its early stages, the disease is reversible, but continued exposure leads to permanent damage and even gangrene, resulting in the loss of fingers or even a complete hand. The decision was important not only for the size of the awards, but also in setting out the terms in which British Coal were negligent and the standards of exposure that would be reasonable.

The Government set up a compensation scheme for miners (now closed to new claimants) to avoid the need for individual litigation, which is eventually expected to award 40,000 miners a total of £500 million. However, by January 2001, only £212 million had been paid.

This landmark award is inevitably being followed by many others. For example, in August 2000, eight workers of North-West Water received a £1.2 million settlement. They contracted VWF whilst using jackhammers and whackers when breaking and re-instating concrete surfaces.

A Medical Research Council survey in 1997–98 gave an estimate of over 300,000 sufferers from VWF in Great Britain – more than eight times the number of self-reported sufferers identified in HSE's self-reported work-related illness surveys in 1995.

The number of new cases of VWF assessed for disablement benefit dropped to 2,428 in 2001/02 from a peak of 5,403 in 1990/91. The majority of claims are from current or former coal-miners and water supply workers, probably because of the higher awareness in these industries, due to the legal actions.

In recent years, VWF has become the most common prescribed disease under the Industrial Injuries Scheme. Approximately 40 per cent of those reporting VWF in the 1995 survey also reported work-related deafness or other ear problems, indicating that work which exposes people to hand-arm vibration is often noisy.

## Carpal Tunnel Syndrome

**V5004** Use, either frequent or intermittent, of hand-held vibratory tools, can result in injury to the wrist: Carpal tunnel syndrome ('CTS') is one example of this. It is thought to arise in part from trapping or compression of nerves in the wrist. It can arise from repetitive twisting or gripping movements of the hand as well as from the use of vibrating tools. CTS arising from the use of vibrating tools was prescribed occupational disease A12 in April 1993. [*Social Security (Industrial Injuries) (Prescribed Diseases) Amendment Regulations 1993* (*SI 1993 No 862*), *Reg 6(2)*]. The number of new cases assessed for benefit continues to rise, with 797 cases in 2001/02.

## Whole-body vibration

**V5005** The four principal effects of whole-body vibration ('WBV') are considered to be:

- degraded health;
- impaired ability to perform activities;
- impaired comfort;
- motion sickness;

Exposure to whole-body vibration causes a complex distribution of oscillatory motions and forces within the body. These may cause unpleasant sensations giving rise to discomfort or annoyance, resulting in impaired performance (e.g. loss of balance, degraded vision) or present a health risk (e.g. tissue damage or deleterious physiological change). However, there is little evidence that vibration directly affects thought processes. Many factors influence human response to vibration, including the type and direction of vibration and the type of person involved. The present state of knowledge does not permit a definitive dose-effect relationship between whole-body vibration and injury or health.

People can overcome moderate effects of vibration on task performance by making greater effort, and this may initially improve performance, although at the cost of greater fatigue, with ultimate degradation of performance.

Effects on comfort depend greatly on the circumstances, for example whether the person expects to be able to read or write in the prevailing conditions.

Low frequency oscillation of the body can cause the motion sickness syndrome (kinetosis) characterised principally by pallor, sweating, nausea and vomiting.

Musculoskeletal disorders may result from WBV, a common example being back pain in drivers of lorries, tractors and other off-road vehicles. However, it is not easy to assess the degree to which WBV is the cause of back problems, as these can also be caused by lifting goods, climbing when entering or leaving vehicles, and even from prolonged sitting in the driving position.

## Buildings and structures

**V5006** Vibration can affect buildings and structures, but a detailed treatment of this is beyond the scope of the present chapter. People are more sensitive to vibration than are buildings or structures, so vibration within a building is likely to become unacceptable to the occupants at values well below those which pose a threat to a structurally sound building. For detailed technical guidance on the measurement and evaluation of the effects of vibration on buildings, British Standard BS 7385: Part 1: 1990 [ISO 4866: 1990] and BS 7385: Part 2: 1993 may be consulted.

## Vibration measurement

### Measurement units

**V5007** Vibration is the oscillatory motion of an object about a given position. The rate at which the object vibrates (i.e. the number of complete oscillations per second) is called the frequency of the vibration and is measured in Hertz (Hz). The frequency range of principal interest in vibration is from about 0.5 Hz to 100 Hz, i.e. below the range of principal interest in noise control.

The magnitude of the vibration is now generally measured in terms of the acceleration of the object, in metres per second squared (i.e. metres per second, per second) denoted $m.s^{-2}$ or $m/s^2$. (In mathematical notation, the minus sign in front of the index, such as in $s^{-2}$, means that $s^2$ is on the bottom line. Thus in $m.s^{-2}$, m is 'divided by' $s^2$. In vibration work, other values of index may be encountered, such as 1 and 1.75.)

Vibration magnitude can also be measured in terms of peak particle velocity, in metres per second (denoted $m.s^{-1}$ or $m/s$), or in terms of maximum displacement (in metres or millimetres).

For simple vibratory motion, it is possible to convert a measurement made in any one of these terms to either of the other terms, provided the frequency of the vibration is known, so the choice of measurement term is to some extent arbitrary. However, *acceleration* is now the preferred measurement term because modern electronic instruments generally employ an *accelerometer* to detect the vibration. As the name suggests, this responds to the acceleration of the vibrating object and hence this characteristic can be measured directly.

### Direction and frequency

**V5008** The human body has different sensitivity to vibration in the head to foot direction, the side to side direction, and in the front to back direction, and this sensitivity varies according to whether the person is standing, sitting or lying down. In order to assess the effect of vibration, it is necessary to measure its characteristics in each of the three directions and to take account of the recipient's posture. The sensitivity to vibration is also highly frequency-dependent. BS 6841: 1987 'Measurement and evaluation of human exposure to whole-body mechanical vibration and repeated shock' provides a set of six frequency-weighting curves for use in a variety of situations. The curves may be considered analogous to the 'A'-weighting used for noise measurement. However, few instruments have these weightings built into them.

### Vibration dose value

**V5009** In most situations, vibration magnitudes do not remain constant, and the concept of vibration dose value ('VDV') has been developed to deal with such cases. This is analogous to the concept of noise dose used in the *Noise at Work Regulations 1989* (*SI 1989 No 1790*). Again, few instruments are capable of measuring VDV. By applying the frequency-weighting system, it is possible to introduce a single-number vibration rating. The proposed Control of Vibration at Work Regulations are likely to set an exposure limit value for whole-body vibration of 9.1 $m.s^{-1.75}$.

Structural vibration of buildings can be felt by the occupants at low vibration magnitudes. It can affect their comfort, quality of life and working efficiency. Low levels of vibration may provoke adverse comments, but certain types of highly sensitive equipment (e.g. electron microscopes) or delicate tasks may require even more stringent criteria.

Adverse comment regarding vibration is likely when the vibration magnitude is only slightly in excess of the threshold of perception and, in general, criteria for the acceptability of vibration in buildings are dependent on the degree of adverse comment rather than other considerations such as short-term health hazard or working efficiency.

Vibration levels greater than the usual threshold may be tolerable for temporary or infrequent events of short duration, especially when the risk of a startle effect is reduced by a warning signal and a proper programme of public information.

Detailed guidance on this subject may be found in British Standard BS 6472: 1992 'Evaluation of human exposure to vibration in buildings (1 Hz to 80 Hz)'.

## Causes and effects of hand-arm vibration exposure

V5010     Intense vibration can be transmitted to the hands and arms of operators from vibrating tools, machinery or workpieces. Examples include the use of pneumatic, electric, hydraulic or engine-driven chain-saws, percussive tools or grinders. The vibration may affect one or both arms, and may be transmitted through the hand and arm to the shoulder.

The vibration may be a source of discomfort, and possibly reduced proficiency. Habitual exposure to hand-arm vibration has been found to be linked with various diseases affecting the blood vessels, nerves, bones, joints, muscles and connective tissues of the hand and forearm, most commonly VWF.

VWF arises from progressive loss of blood circulation in the hand and fingers, sometimes resulting in necrosis (death of tissue) and gangrene for which the only solution may be amputation of the affected areas or complete hand. Initial signs are mild tingling and numbness of the fingers. Further exposure results in blanching of the fingers, particularly in cold weather and early in the morning. The condition is progressive to the base of the fingers, sensitivity to attacks is reduced and the fingers take on a blue-black appearance. The development of the condition may take up to five years according to the degree of exposure to vibration and the duration of such exposure.

## Vibration-induced white finger as an occupational disease

V5011     Vibration white finger ('VWF') is reportable under the *Reporting of Injuries, Diseases and Dangerous Occurrences Regulations 1995* (*RIDDOR*) (*SI 1995 No 3163*).

VWF is prescribed occupational disease A11. It is described as: 'episodic blanching, occurring throughout the year, affecting the middle or proximate phalanges or in the case of a thumb the proximal phalanx, of:

(*a*)   in the case of a person with five fingers (including thumb) on one hand, any three of those fingers;

(*b*)   in the case of a person with only four such fingers, any two of those fingers; or

(*c*)   in the case of a person with less than four such fingers, any one of those fingers or ... the remaining one finger.'

[*Social Security (Industrial Injuries) (Prescribed Diseases) Regulations 1985* (*SI 1985 No 967*)].

## Action at common law

**V5012** In July 1998, the Court of Appeal upheld an award of damages to seven employees of British Coal claiming damages for VWF. The court determined that after January 1976, British Coal should have implemented a range of precautions, including training, warnings, surveillance and job rotation where exposure to vibration was significant.

The court went on to consider what degree of exposure would be reasonable. They supported the standards set out in the HSE Guidance booklet *Hand-arm vibration* published in 1994, which establishes limits for vibration dose measured as an eight-hour average or A8. The court considered that an exposure of 2.8 m/sec$^2$ A8 would be an appropriate level for prudent employers to use. This level of exposure would lead to a 10 per cent risk of developing finger blanching (the first reversible stage of VWF) within eight years.

The court also suggested a form of wording to warn employees: 'If you are working with vibrating tools and you notice that you are getting some whitening or discolouration of any of your fingers then, in your own interests, you should report this as quickly as possible. If you do nothing, you could end up with some very nasty problems in both hands.'

## Stages of vibration white finger

**V5013** Damages are awarded according to the stage of the disease at the time of the action from the date the employer should have known of the risk. The Taylor-Pelmear Scale is usually used to describe these stages as follows.

| Taylor-Pelmear Scale System | | |
|---|---|---|
| *Stage* | *Grade* | *Description* |
| 0 | — | No attacks |
| 1 | Mild | Occasional attacks affecting the tips of one or more fingers |
| 2 | Moderate | Occasional attacks affecting the tips and the middle of the fingers (rarely the base of the fingers) on one or more fingers |
| 3 | Severe | Frequent attacks affecting the entire length of most fingers |
| 4 | Very severe | As in stage 3, with damaged skin and possible gangrene in finger tips |

## Prescription of hand-arm vibration syndrome

**V5014** Foreseeably, prescription may extend to hand-arm vibration syndrome ('HAVS') instead of VWF. The former would cover recognised neurological effects as well as the currently recognised vascular effects of vibration. Neurological effects will include numbness, tingling in the fingers and reduced sensibility. Moreover, the current list of occupations, for which VWF is prescribed, may be replaced by a comprehensive list of tools/rigid materials against which such tools are held, including:

- percussive metal-working tools (e.g. fettling tools, riveting tools, drilling tools, pneumatic hammers, impact screwdrivers);
- grinders/rotary tools;
- stone working, mining, road construction and road repair tools;

- forest, garden and wood-working machinery (e.g. chain saws, electrical screwdrivers, mowers/shears, hedge trimmers, circular saws); and
- miscellaneous process tools (e.g. drain suction machines, jigsaws, pounding-up machines, vibratory rollers, concrete levelling vibratibles).

## Physical Agents (Vibration) Directive (2002/44/EC) and proposed Control of Vibration at Work Regulations

**V5015** After ten years of debate, in 2002 the European Commission issued the Physical Agents (Vibration) Directive 2002/44/EC on the protection of workers from exposure to vibration. Member states have until July 2005 to transpose the directive into national law and accordingly, HSC have produced and consulted on draft Control of Vibration at Work Regulations. These recognise that the health risks and methods of control of hand-arm vibration and of whole-body vibration are very different. The Regulations will apply to work activities on mainland Britain, on offshore workplaces and in aircraft in flight over Britain, although exemption certificates may be granted to emergency services, air transport and the Ministry of Defence. Separate Vibration Regulations will be introduced in Northern Ireland and for sea transport.

For *hand-arm vibration*, there will be an exposure action value of 2.5 m/s$^2$, at which specified action is to be taken to reduce risk, and an exposure limit value at 5 m/s$^2$ daily personal exposure which must not be exceeded except in certain pre-permitted circumstances. The daily personal exposure values are to be adjusted (normalised) to an 8-hour reference period.

For *whole-body vibration*, there will be an exposure action value at 0.5 m/s$^2$ and an exposure limit value of 1.15 m/s$^2$ daily personal exposure, offending work activities must be notified. (HSC are consulting on whether vibration dose value should be used instead of acceleration for the action value.)

A long transitional period is proposed. The limit values will not come into effect until July 2007 for new equipment or July 2010 for existing equipment, or until July 2014 for existing equipment used by employees in the agricultural and forestry sectors.

Employers must undertake a health risk assessment if any of their employees are likely to be exposed to a vibration risk. This includes consideration of people who may be particularly susceptible to vibration, the maintenance of equipment and the availability of replacement equipment that reduces vibration exposure.

Consideration must be given to aggravating circumstances, such as exposure to cold, awkward working conditions and workplace exposure extending beyond normal working hours, such as when rest areas subject to vibration, as often encountered in offshore installations.

Health surveillance is required in circumstances where there is a link between the level of exposure and known adverse health effects, and there are a range of obligations when an adverse effect is detected.

Information, instruction and training is to be provided to persons who may be exposed to a risk from vibration.

### Impact on employers

**V5016** The new Regulations will impose a new discipline on employers in an unfamiliar area. Few employers or health and safety officers will have any feeling for the degree of vibration risk that their processes entail, even though there have been some specific obligations on employers for some years, in addition to the general duty of care under the *Health and Safety at Work etc. Act 1974*.

Such obligations include, amongst others, those under:

- the *Reporting of Injuries, Diseases and Dangerous Occurrences Regulations 1995* (*SI 1995 No 3163*);
- the *Provision and Use of Work Equipment Regulations 1998* (*SI 1998 No 2306*);
- the *Supply of Machinery* (*Safety*) *Regulations 1992* (*SI 1992 No 3073* as amended by *SI 2004 No 2063*).

## Advice and obligations in respect of hand-arm vibration

### Vibration magnitudes

**V5017** The energy level of the hand tool is significant. Percussive action tools, such as compressed air pneumatic hammers, operate within a frequency range of 33–50 Hz. These cause considerable damage whereas rotary hand tools, which operate within the frequency range 40–125 Hz, are less dangerous.

The Health and Safety Executive (HSE) intends to publish guidance on the vibration magnitudes of a variety of tools in a selection of typical working conditions. Vibration can vary not only with the type and design of the tool, but also with other things such as the task, the operator's technique and the material being worked.

For example, electric impact drills (hammer drills) create vibration of around 10 m/s$^2$ on average, but can go as high as 25 m/s$^2$. At 10 m/s$^2$, the exposure action value would be reached after about 30 minutes use and the exposure limit value would be reached after about 2 hours use.

Angle grinders produce a similar range of exposure values. Older models of needle scalers produce average vibration magnitudes in the range of 8 to 18 m/s$^2$, but newer vibration-reduced models produce vibration magnitudes between 4 and 7 m/s$^2$. It is clear from this data that most of these tools could not be used for a whole working day without exceeding the vibration limit value, and none could be used without exceeding the exposure action value. Indeed, it appears that most currently-available vibrating hand tools will exceed the vibration exposure action value, and are very likely to exceed the exposure limit value if used for a whole working day.

It would seem that all employers whose work processes use vibratory tools will be advised to undertake a vibration risk assessment.

### Managing vibration exposure

**V5018** Where a tool is used for a well-defined process, it might be helpful to control the vibration exposure by limiting the number of at-risk tasks in a day, rather than the number of hours use. For example, it may be found that an operator can drill 50 to 60 holes 100 mm deep in concrete before reaching the exposure action value, and 200 to 230 such before reaching the exposure limit (this is a theoretical example).

Many work processes that involve hand-held power tools will require employees to undertake a variety of tasks, possibly using different tools, each with different vibration exposures. In this case, it could be more difficult to set a limit on the number of operations that are allowable. An alternative is to use a 'points' system.

For example, an employer might allocate points to vibration exposure such that 100 points per day is equal to the exposure action value. The employer assesses the vibration 'dose' of an angle grinder as 30 points per hour, and of an impact drill as 120 points per hour. This means that the grinder gives an exposure of 5 points every 10 minutes and the impact drill gives 20 points for every 10 minutes use. Thus 20 minutes use of the drill plus 120 minutes use of the grinder would cause the exposure action value to be reached.

The allocation of points to each tool will have to be calculated using a formula which depends on the vibration magnitude of the tool and process. Note that the system described here is only intended to be indicative, as vibration dose does not strictly obey the rules of simple arithmetic.

## Personal protection against hand-arm vibration

**V5019** Workers can reduce the risk of injury from hand-arm vibration by keeping the blood flowing while working, for example by keeping warm, not smoking especially whilst working, and exercising hands and fingers. Gloves and other warm clothing should be provided when working in cold conditions.

Tools should be designed for the job, both to lessen vibration and to reduce the strength of grip and amount of force needed. The equipment should be used in short bursts rather than long sessions. Symptoms should not be ignored

There are a number of 'anti-vibration' gloves on the market that conform with the relevant standard BS EN ISO 101819:1997. However, these may seriously limit dexterity and can in some circumstances actually increase the risk of vibration injury. The standard is under revision, but at present, anti-vibration gloves should not be relied upon to reduce vibration exposure significantly. Very often, good quality industrial gloves will provide equal levels of protection.

### *Manufacturers' warning of risk*

**V5020** The *Supply of Machinery (Safety) Regulations 1992 (SI 1992 No 3073* as amended by *SI 2004 No 2063*), establish essential health and safety requirements for machinery supplied in the European Economic Area. This includes information on vibration from hand-held, hand guided and mobile machinery. Machinery manufacturers and suppliers have a duty to warn whenever a machine carries a vibration risk, and to provide information for hand-held or hand-guided machines where vibration emissions exceed 2.5 m/s$^2$. This is the vibration emission of the machine, not the daily exposure of the user, which will depend on the amount of use of this and other machines. It should be noted that 2.5 m/s$^2$ is not to be regarded as an acceptable target for the amount of vibration a machine may emit – there is an overriding duty to reduce vibration emission as far as possible within the current state of the art.

Manufacturers are required to test vibration emissions in accordance with test procedures aimed at obtaining a reproducible value, and they must declare both the typical measured value and the amount of uncertainty (variation) of measurements from the typical value. Thus, the typical value plus the uncertainty value should give the 'worst case', but it may still not represent the total amount of vibration in real use. This can complicate the choice of tool, to the extent that sometimes a tool that gives slightly more vibration than another tool may still be better if it is more comfortable, appropriate or quicker for the task.

Moreover, tools that are incorrectly assembled, used wrongly, with worn or loose parts can give many times the amount of vibration than normal.

### *An action plan*

**V5021** Although there is a long transitional period, employers should commence an action plan as soon as possible, starting with a risk assessment. All hand-held vibrating tools, and especially those using some sort of hammer action, have a risk: rotary tools usually have lower vibration emissions, but can still present a risk when used for long periods.

Employers should also check any warnings that the equipment suppliers provide, and whether there is any generally-known vibration problem in their industry. They should also ask their employees whether any of them have any problems that could be symptomatic of HAVS. If any employee is diagnosed with HAVS or vibration-related carpal tunnel syndrome, the employer must report this to the local health and safety authority (either HSE or the local authority depending on the industry.)

## Formal assessments

**V5022** It is quite difficult to measure the vibration emission of equipment in actual use, as the measuring device must be attached to the equipment in such a place and in such a manner as to represent the vibration being transmitted to the operator. The measuring system must be capable of determining the amount of vibration over a range of frequencies and in different directions. Hence, it is usually necessary to rely on general advice.

The first part of the plan should therefore be to try to find ways to eliminate the use of vibrating tools, to use the most appropriate equipment to get the job done quickly and safely, and to make sure that the tool is correctly assembled and used, with the correct cutting bit kept sharp.

# Advice and obligations in respect of whole-body vibration

## Incidence of whole-body vibration

**V5023** Although some millions of people are exposed to whole-body vibration ('WBV') at work, most of them are users of road vehicles and are not likely to be at risk of injury from WBV. Nevertheless, more than 1.3 million workers are thought to be exposed to vibration above the exposure action value and more than 20,000 are thought to be exposed to more than the WBV limit value.

WBV can cause or worsen back pain, and the effect is aggravated when the vibration contains severe shocks and jolts. These conditions occur particularly in off-road driving, such as in farming, construction and quarrying, but can also occur in small, fast boats at sea and in some helicopters. Old railway vehicles running on trackwork in poor condition can also give rise to high levels of WBV.

Back pain can arise from many causes other than WBV, and can be aggravated by poor driving posture, sitting for long periods without being able to change position, awkwardly placed controls that require the operator to stretch or twist to reach them, manual lifting, and repeatedly climbing into or jumping out of a high cab.

Jobs which combine two or more of these factors can increase the risk further, for example, driving over rough ground whilst twisting round to check on the operation of equipment being operated behind the vehicle.

## Assessing the risk of WBV problems

**V5024** Road transport drivers are not usually at risk unless there are other aggravating factors such as long hours of driving, or use of unmade or poor road surfaces. However, if there is a history of back pain in the job, this could be indicative of a problem, and it may be wise to consider the regulations and guidance on manual handling of materials in addition to the vibration regulations.

Operators of off-road vehicles are at greater risk. Indicators of risk include the operator being thrown around in the cab, or receiving shocks and jolts through the seat, or if the manufacturer warns of a risk. A history of back pain is again an indicative factor.

## Keeping records

**V5025** Risk assessments should be recorded. These should show those jobs or processes identified as being at risk from WBV; an estimate of the amount of exposure; comparison of exposure with the action and limit values; the available risk controls; the plans to control and monitor risks; and the effectiveness of the control measures.

## Reducing the risk of WBV problems

**V5026** When a risk is identified the obvious initial actions are to ensure that the machine is suitable for the job, is used correctly and driven at safe speeds for the ground conditions, and that seats are correctly adjusted for the driver's size and weight. Access routes on work sites should be maintained in good condition, for example by using a scalping blade and roller.

## Exposure assessments

**V5027** An exposure assessment should be done by a competent person, but an initial assessment will usually be made from published data and does not need to be highly accurate, so it can often be done by company staff who have had some training. If the initial assessment shows that exposure is approaching or in excess of the exposure limit value, a measurement of vibration levels may be needed: this is more complex and is likely to require a specialist.

## Health surveillance

**V5028** HSE currently considers that because there is no way in which specific occurrences of back pain can be attributed to WBV, it is not appropriate to undertake health surveillance. However, it suggest that 'health monitoring' is advisable, though not obligatory. Health monitoring is a system for reporting and recording symptoms that could be attributable to WBV, manual handling or posture, so that action can be taken if there is a high incidence of problems.

# HSE guidance on hand-arm and whole-body vibration

**V5029** HSE publishes a full guide on *Hand-arm vibration* (HS(G)88), ISBN 0 7176 0743 7. It also publishes a book of 51 case studies dealing with vibration, titled *Vibration solutions: Practical ways to reduce the risk of hand-arm vibration injury'*, ISBN 0 7176 0954 5. An HSE video pack *Hard to handle* includes a book and a leaflet, along with a 15-minute video giving an introduction to the risks of hand-arm vibration and how to manage them. A CD-ROM *The successful management of hand-arm vibration* (1999), for managers and safety specialists, has also been published by HSE.

In addition to the above, HSE publishes a number of leaflets, including:

- *Hand-arm vibration: Advice for employees and the self-employed* (1999), ISBN 0 7176 1554 5;
- *Health risks from hand-arm vibration: Advice for employers* (1998), ISBN 0 7176 1553 7;
- *Reducing the risk of hand-arm vibration injury among stonemasons* (1998), MISC 112.

A whole new series of guidance documents are in preparation for use with the proposed new regulations.

# Violence in the Workplace

## Introduction

**V8001** The Health and Safety Executive (HSE) have identified work related violence to be an emerging problem of significant magnitude and wide-ranging consequences. Whereas violence was once seen exclusively as a problem for occupations such as police officers and prison staff, it is now widely acknowledged that almost anyone can be exposed to it. At particular risk of exposure are those who come into contact with the public, those who handle money, those who, on occasions must become a service denier rather a service provider, and those who deal with people in pain and distress or under the influence of alcohol and other drugs. There are many jobs to which some, or all, of these dimensions apply, e.g. teachers, nurses, taxi drivers, transport staff, shop assistants, public house staff, social workers and 'helpline' staff.

HSE emphasise that the process of managing the risk of violence is the same as for any other health and safety risk. They acknowledge, however, that violence is a problem within society and the risk of violence in the workplace depends on the interaction of a range of factors some of which lie outside of the employer's ability to control. There are no easy solutions or short cuts and there are no universal blueprints for dealing with the risk of violence. What works for one organisation may not work for another. When developing a workplace violence prevention programme it is imperative that organisations make discernible judgements based on the peculiarities of their particular working environments.

HSE guidelines suggest the adoption of a risk management framework with a recursive cycle of activities to ensure continuous improvement in the assessment and management of risk, beginning with an appraisal of the nature and extent of the problem, and followed by the design, implementation and evaluation of appropriate preventive measures.

## The nature and extent of the problem

**V8002** There is very little empirical data in the United Kingdom from which we can gain a broad picture of the extent of the problem. Fortunately, the British Crime Survey (BCS) has been collecting data about workplace violence for several years.

The BCS provides a count of crime across in England and Wales and includes crimes not reported to or recorded by the police. It has been conducted bi-annually since 1986 and, since 2000, is now conducted on a yearly basis.

The definition of workplace violence used in the survey is:

> 'All assaults or threats which occurred while the victim was working and were perpetrated by a member of the public'.

In 1999 HSE commissioned the Home Office to carry out a retrospective analysis of data from the BCS, relating to violence at work covering the period from 1991 to 1997. Between 1991 and 1997 the estimated number of workers who had been a victim of violence at work rose by 30%, from 423,000 to 551,000 victims. The number of workers who had been a victim of assaults rose by a fifth (19%) over this period while the number who had been a victim of threats increased by two-fifths (39%). Since 1997 the number of victims of violence at work has fallen by 32% to 376,000; although the 2002/03 figure is below that in 1991 the difference is not statistically significant.

Between the 2001/02 BCS and the 2002/03 BCS the number of victims of all violence fell by 3%, largely due to a decrease in the number of victims of threats at work, however this fall was not statistically significant.

## Who is at risk?

**V8003** The 2002/03 BCS provided information on which the following table is based, showing the occupations that carry an above average risk of assault or threat of assault from members of the public. The data is based upon the combined data from 2001/02 and 2002/03 BCS.

Occupations with above average risks of violence while working, 1994/1996/1998 British Crime Survey (based on working adults of working age)

| Occupation | High risk of assaults<br>Average risk = 0.9% | High risk of threats<br>Average risk = 1.0% |
| --- | --- | --- |
| Protective service occupations | 12.6% | 3% |
| Health and social welfare professionals | 3.3% | 2.3% |
| Public transport | 1.9% | 1.3% |
| Managers and proprietors in agriculture | 1.8% | 2.2% |
| Health professionals | 1.4% | 2.3% |
| Caring personal services | 1.3% | – |
| Leisure services | 1.1% | 1.5% |
| Teaching and research | 1.0% | 1.2% |
| Elementary administration | 0.9% | – |
| Corporate managers | 0.8% | 4.0% |
| Sales occupations | – | 1.3% |
| Business and public services | – | 1.0% |

In summary the findings from the survey show those people who work directly with the public continue to have an increased risk of violence. Those in protective service occupations, for example, police officers, fire service officers and prison service officers are most at risk of violence. Also at high risk are health and welfare workers, including nurses, paramedics and welfare officers; health professionals such as medical and dental practitioners; transport workers; teachers; workers in the leisure and hospitality industry; and corporate managers (including senior officials in national and local government) who come into contact with the public.

The survey found that 41% of assaults at work and 34% of threats happened after 6 pm. This suggests that the risk of violence at work is higher for people working in the evening or night, than it is for those who work during the daytime.

Forty-one per cent of physical assaults at work involved offenders aged 25 to 39, and a further 30% involved offenders aged 16 to 24. Threats at work were also most likely to involve offenders aged 25 to 39 (42%), but were more likely than assaults to involve offenders aged 40 and over. Offenders in this age group were responsible for 22% of threats at work, compared with 15% of assaults, although this difference was not significant.

Incidents of threats at work were less likely to involve younger offenders than assaults; 7% of threats and 14% of assaults were said to involve offenders under the age of 16.

Previous findings from the BCS indicated that victims of violent incidents involving offenders under the age of 16 tended to be in teaching occupations, although many were also in protective service occupations (Budd, 1999).

The survey found that 16% of workers who had some form of contact with members of the public were worried about being assaulted. Unsurprisingly, those in occupations with a higher risk of violence, such as public transport workers, nurses and teachers, were most concerned.

To put all this in perspective, although the total number of reported incidents seems to be high, the estimated risk of an employee being assaulted or threatened in a given year are actually relatively low. The 2002/03 BCS indicates that 1.7% of working adults were the victim of at least one violent incident at work. This is of course cold comfort to the actual people who represent those statistics.

## Work-related violence – the legislation

**V8004** In common with most countries, in the UK there is little new legislation or regulation designed to deal specifically with the issues of work-related violence. Instead, we need to look to several existing laws that provide for regulation, enforcement and remedy in such cases.

In simple terms, the law provides:

- Regulation to ensure that employers provide a working environment in which the risks of being subjected to violence are minimised and the fears of employees are taken into account. Employees also owe a duty of care to themselves and to others who may be affected by their acts or omissions.

- Enforcement to provide sanctions if employers fail to comply with the regulation and punishment of individuals who act violently towards another person in a workplace context.

- Remedy for employees who become victims of workplace violence. This will usually be in the form of compensation through an industrial tribunal, civil court, or criminal injuries compensation board.

The legal responsibilities and rights relating to workplace violence are contained in a range of provisions that include:

- Health and safety legislation.
- Employment law.
- Criminal law.
- Civil law.

### Statutory health and safety requirements
### Health and Safety at Work etc Act 1974

**V8005** Employers have a legal duty under the *Health and Safety at Work etc Act 1974* (*HSWA 1974*) to ensure, so far as is reasonably practicable, the health, safety and welfare at work of their employees. [*HSWA 1974, s 2(1)*].

**V8006** *Violence in the Workplace*

Besides a common law 'duty of care' to others, employers have a statutory duty to do 'everything reasonable and practicable' to 'eradicate or minimise' the risk of harm from all hazards to health – including violence.

Employers are also required to conduct their undertaking in such a way as to ensure, so far as reasonably practicable, the safety of other people who are not their employees, and to whom the premises have been made available.

The 1974 Act is considered as the most important piece of legislation dealing with health and safety at work. Although it was implemented to cater for the risks associated with industries such as chemicals, mining and construction work, the basic principles are translated relatively easily within the context of the protection required against the risks of workplace violence. Lord Skelmersdale, who chaired the DHSS Advisory Committee on Violence to Staff, wrote in the 1988 report:

> 'Where violent incidents are foreseeable employers have a duty under section 2 [of the Health and Safety at Work etc Act 1974] to identify the nature and the extent of the risk and to devise measures which provide a safe workplace and a safe system of work.'

The Health and Safety Executive publication, *Successful Health and Safety Management*, (HSG65), says that accidents and ill health are seldom random events. They generally arise from the failure of control and involve multiple contributory causes.

The duty of care also embraces employees. Employees are required under *section 7* of the *HSWA 1974 to* 'take reasonable care for the health and safety of himself and of other persons who may be affected by his acts or omissions at work'. The duties placed on the employee do not reduce the responsibility of the employer to comply with his health and safety duties under the 1974 Act.

## Management of Health and Safety at Work Regulations 1999

**V8006** The *Management of Health and Safety at Work Regulations 1999* (*MHSWR*) (*SI 1999 No 3242*) requires employers to undertake a 'suitable and sufficient' assessment of the risks to which employees are exposed while they are at work and put in place measures to eliminate, minimise or control the risks. In many workplaces the potential for violence is a significant hazard and thus the duty on employers extends to the risk of violence.

Employers must:

- Establish how significant these risks are.
- Identify what can be done to prevent or control the risks.
- Produce a clear management plan to achieve this.

To satisfy the requirements employers must put into place a comprehensive policy for dealing with work-related violence. This will include effective measures for identifying and assessing risk, clear guidelines and procedures for dealing with violent incidents and their aftermath, and the provision of appropriate training and equipment.

## Reporting of Injuries, Diseases and Dangerous Occurrences Regulations 1995

**V8007** Employers must notify their enforcing authority in the event of an accident at work to any employee resulting in death, major injury or incapacity to work for three or more days. This includes any act of non-consensual physical violence inflicted on a person at work. [*Reporting of Injuries, Diseases and Dangerous Occurrences Regulations 1995* (*RIDDOR*) (*SI 1995 No 3163*)].

## Safety Representatives and Safety Committees Regulations 1977 and the Health and Safety (Consultation with Employees) Regulations 1996

**V8008** Employers must inform and consult with employees in good time on matters relating to their health and safety. Employee representatives either:

- appointed by recognised trade unions under the *Safety Representatives and Safety Committees Regulations 1977* (*SI 1977 No 500*); or

- elected under the *Health and Safety (Consultation with Employees) Regulations 1996* (*SI 1996 No 1513*),

may make representations to their employer on matters affecting the health and safety of those they represent.

This places a responsibility on the employer to consult and inform on issues of violence at work.

## Employment law

### The Employment Rights Act 1996

**V8009** Section 44 of the *Employment Rights Act 1996* (*ERA 1996*) relates to 'Health and safety cases' and reminds us that responsibility for safety at work is something shared between employers and employees. The section covers:

- Employees statutory entitlement to a safe way of working, and to be able to fulfil their responsibility under the *HSWA 1974, s 7* to take care of themselves and others, without fear of recriminations ('any detriment') from their employer for doing so.

- The circumstances in which an employee should withdraw from danger (see V8014 below).

- Employees duty and right to be able to draw attention to safety deficiencies that may exist and to take 'appropriate action' to withdraw/remove themselves from 'serious and imminent' danger that they would be unable to avert.

- Warning to employees that if they continue to knowingly and recklessly undertake work that is unsafe and get injured they may not be able to make a claim against their employer for liability.

### Contract law

**V8010** A contract of employment imposes two obligations on the part of the employer:

- The employer must provide a workplace in which employees are subjected to minimal exposure to risk.

- The employer must provide 'trust and support' to the employee in carrying out their role.

It is well established that workplace violence should be regarded as a risk in this context. The employer is expected to provide 'trust and support' to ensure that the employee is working under minimum stress and that the employer responds to issues of perceived risk as well as real risk.

Failure to provide appropriate support might reach a point where an individual can no longer tolerate the working conditions and decides to leave. This could be construed as constructive dismissal where the situation is treated as if the employee had been dismissed. The individual will usually seek a remedy through an industrial tribunal or a county court.

The obligations of an employer are unambiguous. There is a clear requirement to assess the possible risk of an employee being subjected to violence and to provide appropriate measures which will minimise that risk. There is also a requirement to recognise and respond to the perceived fears of employees by providing appropriate support in relation to those fears.

## Criminal law

**V8011** The laws in relation to violence, disorder, assault and threats are well established and cover almost every incident of work-related violence. Recent legislation offers greater protection against harassment and more serious penalties for racist incidents. Thankfully, the incidence of serious assault is comparatively rare but this brings with it a difficulty in relation to prosecuting the offender. In general, it requires the police to pursue the matter to prosecute the offender and many different issues have to be addressed when making the decision.

Quite often the nature of the assault is not serious enough for a criminal charge to be pursued or, for a variety of reasons, the circumstances do not warrant it. It is unlikely that an assault that results in only minor bruising, for example, would find its way into a court. Whilst these decisions often make economic, procedural or legal sense, they leave the victims feeling frustrated and unsupported. A common cause of dissatisfaction felt by employees who have been assaulted is related to the apparent lack of action that follows the incident. Organisations need to be clear as to their stance on prosecution and their preparedness to support a private/civil action.

## Civil Law

### Negligence

**V8012** The tort of negligence imposes a duty to take reasonable care to avoid acts or omissions that may cause reasonably foreseeable harm to others. This common law duty of care runs in parallel to statutory health and safety requirements.

In cases where the victim has suffered serious injury, either mental or physical, they will undoubtedly look for substantial compensation for things like their loss of earnings, pain, suffering, and inconvenience. Although it is clear that the person who actually committed the act of violence is directly at fault, it is unlikely that the individual will have the funds to provide for any substantial compensation award. Consequently, the victim will be more likely to seek redress through his or her employer by claiming that the employer was negligent in providing appropriate measures to prevent the incident from happening, and/or in failing to provide adequate support post-incident.

In such cases, key questions will be asked:

- Was violence in the workplace something that ought to have been assessed?
- If so, what were the risks of violence that would have been identified in the risk assessment?
- What reasonably practicable control measures would the risk assessment have identified should it have been put in place?

An employer who has failed to assess the risk of violence will have great difficulty in defending a legal action for injury caused through employer negligence.

### Vicarious liability

**V8013** In recent years the boundaries have been extended even further to cover a company's responsibility for its employees' actions. 'Vicarious liability' is the legal concept whereby a company can be liable for the negligent conduct of its employees. A claim for compensation for personal injury will be successful if the conduct occurred during the course of employment.

### Withdrawing services

**V8014** At what point is it appropriate for an employee to say that they are not prepared to continue because they feel they are in danger? This is a question which is faced from time to time where an employee feels that they are in danger of being physically hurt and they are not prepared to continue to deal with a situation. Legislation makes clear the duties upon employers to protect staff, and *s 7* of the *HSWA 1974*, *Reg 8* of the *MHSWR 1999* (*SI 1999 No 3242*) and *s 44(1)(d)* of the *ERA 1996* also place clear requirements upon employees to take action if their safety is compromised.

### The law relating to the use of force and self defence

**V8015** The use of force is considered under both common law and statute law and it is important that those employees likely to need to use force are aware of the law relevant to their circumstances. Although the powers are drawn from various sources, most specify that force can only be used where it is absolutely necessary, and that the amount of force used must be reasonable in the circumstances. The *Human Rights Act 1998* adds the term 'proportionate'.

## Developing and implementing policy

**V8016** Effective policies translate priorities into action and set a clear direction for the organisation to follow with defined accountabilities. In any organisation that has workers at risk from violence, a comprehensive and workable policy is the cornerstone for building an effective response. Effective policy development requires the active commitment of senior management and the involvement of workers affected by violent incidents. Unless an implementation plan is put in place supported by adequate resources to drive and embed change the policy is likely to remain as words on a page. Once implemented, the processes, procedures, equipment, training and incidence of violence need to be monitored, analysed and evaluated to ensure a proactive and problem-solving approach to the ever-changing risks associated with work-related violence and aggression.

### Problem recognition

**V8017** The most important first step developing policy is to define what 'workplace violence' means within the organisation. Most organisations use the definition offered by the HSE:

> 'Any incident in which a person is abused, threatened, or assaulted in circumstances relating to their work'.

This definition encompasses a range of unacceptable behaviour likely to cause physical or psychological harm. Importantly, it includes incidents where someone uses abusive language or behaviour that does not amount to a threat of or actual physical assault. It also covers incidents that might occur when people are not actually at work, for example, being followed home or confronted when arriving at or leaving work.

### Conducting a review

**V8018** Having defined violence the next step in developing an effective policy is to understand the nature and extent of the problem facing the organisation and the effectiveness, or otherwise, of existing policy, procedures and guidance. This is best achieved through a systematic review of existing arrangements. The review should include 'top down' and 'bottom up' elements of assessment to ensure that problems facing front line workers are understood and a 'reality check' made on the effectiveness of existing policy and procedures.

The purpose of such a review is to establish:

- the nature and extent of the problem;
- the specific roles that are at risk;
- the nature of the risks;
- the solutions to minimise the risks identified.

The effectiveness or otherwise of existing policy and procedures

## Collaboration

**V8019** An important element in the eventual success of such a review will be the approach adopted from the outset. It is relatively easy to secure employees' collaboration because the intention of the review is primarily to ensure they operate in an environment where risks from work-related violence are minimised.

## Sources of information

**V8020** To carry out an effective review, there are a variety of sources from which data and information can be gathered. These include:

- Generic violence risk assessment of specific roles.
- Interviews, observation, focus groups and questionnaires.
- Analysis of data.

## Generic violence risk assessment of specific roles

**V8021** The people who face the customers, clients, patients and public every day are best placed to identify and understand the risks they face from violence and aggression, and risk assessment of roles is a statutory requirement under health and safety legislation.

## Interviews, observation, focus groups and questionnaires

**V8022** It is important to gather views from front line staff about the effects that the risks of workplace violence have upon them.

The information can be gathered through informal interviews with affected staff, focus groups of staff from a particular role or through a questionnaire which is designed to extract the appropriate information. Often, a combination of these techniques is used.

There are issues around confidentiality and individuals often fear their views will 'get back to management'. Confidentiality must be guaranteed, be evident in the procedure and be completely respected in practice.

## Analysis of data

**V8023** The data available will vary according to the nature of the organisation and the sophistication of the reporting and recording procedures available. In the best case, a dedicated reporting system will inform on not only the number of incidents of violence and aggression but also information about times of day, locations, nature of injuries, frequency and types of injury, roles most at risk, and so on.

# Designing the policy

## Consultation and collaboration

**V8024** Once the nature and extent of the problem of violence is fully understood consideration can be given to designing policy that will provide direction and focus for action.

It is important that a policy development group is formed and made up of people who understand the issues and practical implications of the decisions that are made. In practice, such a group will probably exist for the general health and safety issues in the form of a health and safety committee or group (see figure 1 below).

Figure 1: Model for effective policy development group

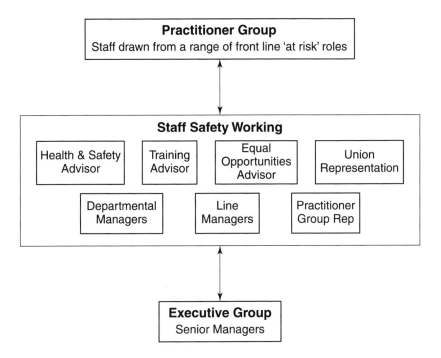

In this model, the practitioner group drives the policy formation and therefore the front line staff have ownership of the issues from the beginning and have been fully involved in the creation of the solution.

## Policy content

**V8025** These are the basic elements of an effective policy:

- Purpose.
- Scope and definitions.
- Who the policy affects and their specific responsibilities.
- Risk assessment, reporting and review process.
- Risk reduction and training.
- Response to incidents:
  — Incident management.
  — Post-incident management.
- Reporting.
- Involvement of other agencies and sanctions
- Maintenance.
- Communication strategy.

- Policy review and revision.

## Purpose

**V8026** This part of the policy sets out the context and aims, and expresses the corporate values that support the existence of the policy.

## Scope and definitions

**V8027** This part of the policy defines what is meant by 'workplace violence'. The HSE definition is recommended. This section should state clearly what is included within the policy and what is not.

## Who the policy affects

**V8028** This part of the policy considers who is affected by the policy and identifies these individuals' specific roles and responsibilities in relation to the policy.

## Risk assessment, reporting and review process

**V8029** This part of the policy outlines the risk assessment process and requirements, together with clear expectations of managers and staff with regard to recording, classification and monitoring of incidents.

## Risk reduction and training

**V8030** Risk reduction is at the heart of the policy and underpins the whole point of its existence. This is the part of the policy that outlines the measures that are to be taken to minimise the risks identified.

The key elements of risk reduction will be:

- advice, guidance, and procedures aimed at safer working practices;
- equipment and design of the working environment;
- improved service delivery and reduction in triggers of violence;
- training and development in handling conflict and potentially violent situations.

## Response to incidents

**V8031** The response will typically cover two key areas:

1. Incident management

    This part of the policy outlines the procedures that managers and staff are required to follow should a violent incident occur, and provides clarity on roles, responsibilities and communications. It also includes communications with other agencies such as police, security and media. It should state the position with regard to use of physical intervention and the provision of appropriate guidance and training.

2. Post-incident management

    This part of the policy clearly defines post-incident management procedures include reporting requirements, operational support, and immediate and longer-term staff support. It should also state the responsibilities of line managers and their health and safety and occupational health functions.

### Involvement of other agencies and sanctions

**V8032** This part of the organisation policy will outline guidance about the involvement of other agencies such as the police and local authorities. It should also outline the stance on prosecution and the options available including legal/financial support for employees and, where appropriate, the policy and guidance concerning exclusion and withdrawal of services.

### Maintenance

**V8033** Performance indicators need to be established and managers held accountable for ensuring that breaches of policy and guidance are addressed.

### Communication strategy

**V8034** This part of the policy states the corporate position with regard to internal and external publicity, and identifies key messages/statements to be promoted. It should also include a strategy for managing media attention following an incident.

### Policy review and revision

**V8035** This part of the policy establishes a monitoring and review process that allows periodic re-evaluation of the strategy and policy.

## *Implementation plan*

**V8036** Implementation is about making the policy happen. It includes educating people about new processes and procedures, ensuring managers understand and monitor new practices, introducing and using new equipment, and setting up and running a programme of training and development. It is important that the implementation plan is adequately resourced to ensure that change is driven forward and embedded in the organisation.

## *Communication*

**V8037** There are two aspects to this – internal communication to the people who work within the organisation – and external communication to the people who make up the clients, service users and the general public.

### Internal communication

**V8038** As always this will depend upon the structure, size and nature of the particular organisation. If it has a department that deals with communication and the media, their help will be invaluable. There are some factors that will help to make the communication process successful.

- Plan

    Thought needs to be given to the target audience, the communication structures in the organisation and how to make the information accessible.

- Policy launch

    It is much more positive if people recognise the start of a new process.

- Workplace briefings

    Front line managers should be given a full understanding of the policy first, and then should communicate this to their staff.

- Easy access 'help desk'

  A system is needed where people can quickly have their queries, worries and doubts dealt with.

- Reinforcement

  Reinforcement of the messages with a second wave of communication needs to be made some weeks later by checking out how well things are going.

### External communication

**V8039** The media will naturally look for a story or a slant and care must be taken that the original good intentions are not skewed to provide an interesting angle – at the expense of the launch.

### New systems, procedures, practices and equipment

**V8040** The implementation of a policy will involve the introduction of new systems of working, new procedures, practices, and perhaps equipment, to learn about and to follow or use.

### Implementing the policy

**V8041** If the organisation has developed an effective strategy and plan, the actual implementation should be relatively straightforward.

To safeguard the organisation it is necessary to show a training and development plan that clearly outlines the phases of training being undertaken and the thinking that has gone into the development of those phases and the sequence of training for different roles.

## Evaluating the policy

**V8042** It is very likely that polices and practices which made perfect sense in theory will run into trouble when applied in the workplace. It is important therefore to monitor and evaluate the implementation of the policy (see figure 2 below).

In this model there are two feedback loops.

- An 'immediate loop' where the quality of the measure is evaluated when first received. This will include information about the initial perceptions of the staff about quality of the training, or the usefulness of the equipment, or the workability of a particular policy or procedure.
- The second feedback loop is through the reporting and recording procedure and the quality of this will be defined by the quality of the information and the information gathering techniques.

### Co-ordination and revision of policy

**V8043** It is advisable to have a single point of reference for monitoring the effectiveness of the policy and its measures. All the immediate feedback and information from the reporting and recording process should go through this point of reference.

# Risk assessment for workplace violence

**V8044** Risk assessment lies at the heart of an effective organisational response to workplace violence. Without a deep understanding of the nature and extent of the risk it is impossible to develop a targeted preventative strategy that will successfully address the hazards.

Figure 2: A model for monitoring and evaluating policy

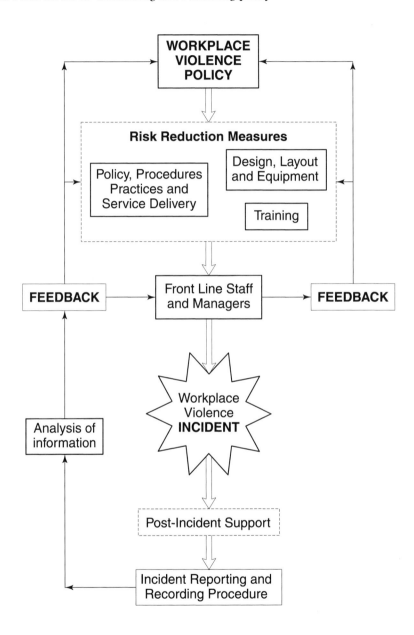

The legal requirement for employers to carry out risk assessments is contained the *Management of Health and Safety at Work Regulations 1999* (*SI 1999 No 3242*), *Reg 3* which states:

'Every employer shall make a suitable and sufficient' assessment of ... the risks to health and safety of his employees to which they are exposed whilst at work ...'

'Suitable and sufficient' is not defined however detailed guidance is set out in the Management of Health and Safety at Work: Approved Code of Practice & Guidance L 21 (ACOP). National Occupational Standards also re-enforce the importance of the risk assessments in managing work related violence.

The above Regulations, Code of Practice and standards also require employers to review, and if necessary, modify their risk assessments following any adverse event. Assessment is not to be regarded as a once-and-for-all activity. This idea links together the elements, risk identification, risk assessment, risk control, incident report, incident investigation and risk review into a continuous improvement cycle.

There are three basic ways in which the risk of violence can be assessed:

- Generic violence risk assessment of role.
- Risk assessment of pre-planned event.
- Dynamic Risk Assessment.

## Generic violence risk assessment of role

**V8045** This is perhaps the most recognised form of risk assessment. It involves an analysis of all the activities commonly associated with each job role that carry a level of risk from work-related violence. Each activity is examined, the possible risks are identified, and a judgement is made about the likelihood of the perceived risk and the severity of the consequences.

A good example to illustrate the point is any role that has an element of working alone and visiting clients or service users in their homes.

Some of the risks inherent in such a role are:

- Being targeted by criminals when travelling and in car parks, isolated locations, entrances to flats and apartments.
- Violence from a service user or other individual inside the person's home.
- Being held against will in premises.
- Being injured in an isolated location without access to help.

Measures to eliminate, minimise or control could include:

- Procedures in relation to callout, providing details of location, client visits, times of appointments to a central source and call in when complete.
- Personal safety guidance to reduce vulnerability when travelling and visiting.
- Training in Dynamic Risk Assessment, conflict resolution, communication skills etc.
- Provision of equipment where appropriate such as a mobile phone or attack alarm.

Each role should be examined in this way to determine all the risks and measures appropriate for that role.

Once this assessment and the resulting control measures have been established for a role, then everyone performing that role should be required to have the appropriate equipment, conform to the procedures and undertake the training required.

Management should ensure that the generic violence risk assessment is carried out on all roles where there is a potential danger and that the control measures are clearly stated and understood by everyone in each role.

## Violence risk assessment – pre-planned event

**V8046** The generic violence risk assessment caters well for the main activities of a role. However, in every job there are events and activities that are out of the ordinary or, perhaps, done only rarely. It is quite possible that the generic risk assessment will not cover every circumstance and, in such cases, it is important to assess the risk involved in an special event or activity as part of the planning process. The risk assessment will identify if there is an increased risk of potential for violence and a need for extra measures in these circumstances.

The risk assessment should be recorded and included in the planning documentation for the event itself.

## Dynamic Risk Assessment

**V8047** There are many occasions when front line staff are confronted with situations which are unique and not catered for in a generic risk assessment or assessment of a planned event. These are often the situations where people get hurt because they don't have a way of assessing the risks in the situation they are confronted with, and do not respond appropriately.

'Dynamic Risk Assessment' is a process which helps an individual to effectively assess a situation from a personal safety perspective, as it is unfolding. The person can continuously assess the circumstances and adjust his or her response to meet the risk presented moment by moment. Training which develops understanding and skills in Dynamic Risk Assessment can help an employee to assess situations more effectively and respond in the most appropriate way.

## Risk assessment – who should do it?

**V8048** There is little doubt that the individual who is working at the customer interface is best placed to be able to identify the risks faced. For generic violence risk assessment of roles and pre-planned events, the process needs to include someone who is trained and experienced in identifying risks and who can spot the hidden dangers in workplace environments, situations and incidents. The most effective solution is to combine the two – a person who is trained and competent in risk assessment and has experience in the management of violence, should examine the roles alongside the people who carry out those roles on a day-to-day basis.

## Risk assessment – how should it be approached?

**V8049** It is recommended that risk assessment be approached on a role-specific basis rather than purely on a geographical area or departmental basis. By definition, the risk of workplace violence will always involve some sort of interaction with another person. This unique feature means the risks of work-related violence are much more difficult to predict and to control because it is hard to predict the range of responses that someone might use in a situation involving conflict.

For each role the following main areas need to be assessed:

- What contact is made with the clients, patients, service users or members of the public? This will usually be described in the job description. For example, enforcing role, caring profession, advising or dealing with complaints.
- What sort of people generally make up this contact? For example, travelling public, offenders, patients, clients or relatives.
- Where does the contact take place? For example, interview rooms, reception, platform, home address, classroom or on the streets.
- What sort of 'state' will the client/patient etc be in? For example, will they be influenced by drink or drugs, anxious, frustrated or hostile.

- What specific tasks are performed? For example, handling complaints, delivering 'bad news', denying benefits, evicting or arresting, or even treating an individual.

## Content of the risk assessment

**V8050** The first question that should be asked is 'Is there any likelihood that someone doing this role will be abused, threatened or assaulted in circumstances relating to their work?' If the answer is 'yes', then a risk assessment for workplace violence needs to be carried out.

Figure 3 below outlines five steps that provide a basic risk assessment model for both the generic risk assessment and for the risk assessment of a pre-planned event.

Figure 3: Risk assessment model

### Step 1 – Identify the risks inherent in the role and plan consultation

**V8051** An examination of the job description should show the potential situations where the jobholder will be at risk. However, few job descriptions will capture all the possible scenarios that might arise and other sources of information should be used.

It is important to research other potentially useful sources of information such as HSE and sector-specific guidance and research.

Consultation with workers needs to be planned to ensure this covers a cross section of staff performing the roles being risk assessed. The consultation can take the form of a combination of the following:

- Structured and informal interviews.
- Small focus groups (typically 3–6 staff).
- Questionnaires.
- Workplace observation.

## Step 2 – Research incidents and identify who may be harmed

**V8052** An examination and analysis of past-incident reports, local log books etc associated with workplace violence will provide invaluable data in a risk assessment. Unfortunately violence is under-reported in most organisations so it is important to try to establish the true extent of the problem by consulting with managers and staff, and gathering anecdotal evidence of occurrences. Interviews, workplace observations, discussions and surveys should be carried out.

## Step 3 – Examine risks relating to the specific activities and tasks performed

**V8053** Assessing the level of risk has its difficulties. By reviewing the actual activities and tasks performed it is possible to get a realistic measure of the risks involved and to ensure controls are both relevant and proportionate.

This stage involves an analysis if the data obtained in Step 2 and further consultation to establish:

- The risks associated with specific activities and tasks performed

    For example, delivering bad news, cash handling, or fare evaders. The risks may be compounded at certain times and locations, or by factors relating to drink or drugs.

- Practical risk reduction and support measures

    Those performing the job will often have the best ideas for reducing risk.

When considering the risks present in job activities and tasks, it is helpful to identify the risk level based on two elements:

1. the *likelihood* of harm actually being caused, and
2. the likely impact or *severity* of the harm caused.

The *likelihood* can be assessed on a sliding scale, such as:

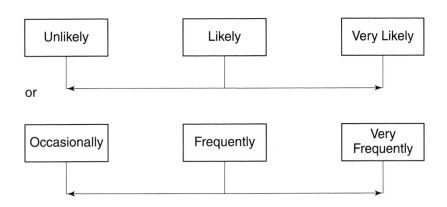

The *severity* of harm can be assessed on a similar scale such as:

From this, a 'risk rating' can be established which identifies the risk as being low, medium or high. When considering the severity of an incident there are number of considerations including:

- Physical harm to those involved.
- Emotional impact on staff.
- Impact on the perpetrator and other service users.
- Numbers of people involved.
- Duration of the incident.
- Effect on operations and service delivery.
- Financial impact on both individuals and organisation.

The likelihood/severity approach to risk assessment is long established and is helpful in providing a broad assessment of risk. However, it does not take into account the fear of violence experienced by the person doing the job. An employer has a duty to include this in the risk assessment and respond to this fear, for example through rationalising fears in training and introducing guidance to further minimise the risk.

## Step 4 – Identify and examine existing risk reduction measures

**V8054** There will be risk reduction measures already in place. These existing measures need to be compared with the risks identified in the assessment and an evaluation made about their effectiveness and how relative they are to the current risks.

## Step 5 – Identify and implement new risk reduction measures

**V8055** Effective measures for reducing the assessed risks should be identified and implemented. These may be a combination of safer working practices and procedures, policy, training and equipment.

## Continuous review

**V8056** The risk assessment for each role should be a 'live' document. Staff should have easy access to it and should be encouraged to involve themselves in the process of keeping the assessment live and up-to-date.

# Reporting, recording and monitoring system

**V8057** Central to an effective strategy for combating workplace violence is an understanding of the nature and extent of the problem. Integral to this is a system that monitors and provides data and information about the trends surrounding violence in the workplace for a particular organisation.

Figure 4 below shows a typical model of an effective reporting and monitoring system for workplace violence.

## Designing the system

**V8058** The outcomes required of the system will be specific to the needs of a particular organisation but the essential elements of a system are:

- Information gathered through a dedicated workplace violence incident report form.
- An investigation process that will examine the circumstances of an incident and provide recommendations for improving risk reduction.
- An information database and analysis process.
- A workplace violence co-ordinator.
- A communication process.

### Workplace violence incident report form

**V8059** The design of the form will depend on many factors, some of which will be dictated by the way that the information is to be recorded and analysed.

The following details are suggested as a minimum whether using a dedicated violence report, or a combined use report form:

- Name and contact details of person making the report.
- Date, time and location of incident.
- Description of the incident and any injuries sustained.
- Details of any other staff involved.
- Details of the assailant.
- Details of any possible witnesses.

Further information will be required from the incident form, but this will differ between organisations based upon the management data required.

The report form can also provide the basis for recording post-incident action by the line manager, including staff support, the investigation process and subsequent recommendations and actions.

### The investigation process

**V8060** The investigation process should intend to establish the facts of the incident and make recommendations that will help to reduce the risk of this type of incident happening again.

The recommendations from this investigation will be fed directly to the person responsible for co-ordinating issues of workplace violence for a decision as to how they should be taken forward.

### Information database and analysis

**V8061** The information gathered from all the incidents of workplace violence can provide an invaluable insight into the trends, hotspots and risks that are not obvious in the day-to-day management of such incidents.

The database also provides a powerful tool for monitoring the effectiveness of risk reduction measures by monitoring over a similar period to see if the measures have impacted upon the identified risks.

Figure 4: A model for reporting and monitoring incidents of workplace violence

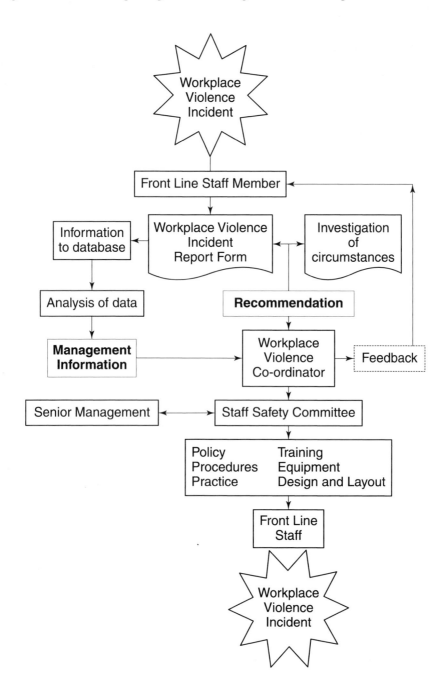

## Workplace violence co-ordinator

**V8062**  This person (or department) will be responsible for:

- Maintaining the reporting and recording procedure.
- Monitoring the quality of the content and the investigation process.
- Monitoring the recommendations from the reporting and investigation process.
- Searching and monitoring the database for useful management information, trends, and hotspots that need further investigation.
- Informing the Staff Safety Committee of recommendations and work-related violence issues for discussion and further action.
- Monitoring the effectiveness of risk reduction measures.
- Liaising with the police and other relevant agencies.
- Administering or assisting in the process of litigation or action against the attacker.

## A communication process

**V8063**  A clear communication process should be built into the system so that the front line staff are kept fully aware of the most up-to-date policy, procedure and practice for their role. A common complaint is that individuals complete incident forms and then hear nothing more about the outcomes.

## Implementation

**V8064**  The following areas are essential for the successful implementation of a reporting process:

- Management involvement and training.
- Staff training.
- Monitoring the completion of reports.
- Provision of feedback to those affected and information to the wider organisation.

## Manager involvement and training

**V8065**  Managers at every level need to understand and appreciate the complete reporting, recording and monitoring process.

## Staff training

**V8066**  Training helps to establish best practice and a sound consistent approach. It ensures that individuals understand the importance of the process and that completion is part of their duty of care.

## Monitoring the completion of reports

**V8067**  Four fundamental questions need to be asked on a routine basis:

- Are all relevant incidents being recorded?
- Are reports correctly entered?
- Are accounts of incidents adequate and professional?

- Is anything actually being done as a consequence?

## Feedback and regular provision of information across the organisation

**V8068** Following the introduction of a reporting process, particular effort must be made to provide immediate and ongoing feedback in relation to reported incidents. This applies not only to the affected individual, but also across the whole organisation.

A *formal review* of the reporting system should be undertaken some months after initial implementation to establish whether the system is being used, whether it can be improved and what is being done with the information generated.

## Measurable outcomes

**V8069** Measurable outcomes using information from the reporting process may include:

- The number of injuries to staff.
- The number of injuries to service users.
- The extent and type of injuries suffered.
- The number of working days lost due to assaults on staff.
- The number of prosecutions taken against those assaulting staff.
- Extent to which safety procedures have been followed.
- Which techniques taught in training have been used and their effectiveness.
- Incident and assault patterns.
- The groups at most risk.
- The highest risk tasks and activities.
- Profiles of perpetrators of violence.

The more sophisticated the database, the more comparisons and measures available.

## Under-reporting

**V8070** It has been mentioned several times that most organisations suffer from under-reporting of workplace violence-related incidents. There are many reasons for this, for example, over complicated and lengthy reporting forms and processes, fear of criticism after having brought an incident to light.

The organisation should work to eliminate as many barriers to reporting as possible.

## Disclosure rules

**V8071** Written records of any incident that results in a court case may be called upon as evidence. This would include any internal incident reports, notes, debriefing notes, statements or correspondence.

The purpose of such 'disclosure' is to ensure that all potential sources of evidence are made available to defendants and the courts. In criminal cases the rules are drawn from the *Police and Criminal Evidence Act 1984* and for civil cases they are drawn from the *Civil Disclosure Rules* and *Civil Procedure Rules*.

From a practical point of view the purpose of the reporting process is to ensure that reports are essentially factual and as free as possible from conjecture and opinion.

## Risk reduction measures

**V8072** The *Management of Health and Safety at Work Regulations 1999* (*SI 1999 No 3242*), *Reg 4* and *Sch 1* of require employers to design and implement protective and preventive measures on the basis of hierarchy of general preventative principles. The associated Approved Code of Practice & Guidance L 21 (ACOP) contains further guidance. The principles may be summarised as follows:

- Eliminate risk by avoiding or removing the hazard
- Substitute the process etc by one that carries less risk
- Control the risk at source
- Devise safe systems of work
- Provide adequate training and instructions
- Ensure adequate supervision is provided
- Provide personal protective equipment as a last resort

Most risk reduction measures fall under one of the following broad categories:

- Training and Information
- Work environment
- Job design

It may not be possible to completely eradicate violence in a workplace but it is possible to minimise the risk of violence and it's impact by introducing appropriate control measures based upon an understanding of the particular problems. Effective risk reduction is a blend of measures that embrace all the areas of policy, effective risk assessment, practice and procedures, training, equipment, design and layout.

### Risk reduction – achieving a balance

**V8073** Appropriate risk reduction measures are a balance between understanding the identified risk and providing a balanced response in terms of acceptability, cost and relative effectiveness. This is not an easy equation to balance and it requires some thought before a reduction measure is introduced. The appropriate balance is, in the end, a matter of judgement, which can only be made by the organisation itself. The cost of a glass screen may be relatively modest and the simple balance may suggest it should be installed. However, other aspects can be added to the equation; is a glass screen acceptable in terms of the organisation's ethos and approach? Does it, in fact, reduce the level of violence – or does it provide a trigger for the level of aggression to increase? On the other hand, if there is a high risk of serious violence involving weapons this may outweigh other considerations and it may be imperative that bullet-proof glass screen is in place.

### Service delivery

**V8074** Workplace conflict can occur for many different reasons. Sometimes organisations and staff can create or exacerbate the environment within which a conflict develops and increase the risk of violence by the way they deliver services or approach their work.

## Working practice

### Staffing levels

**V8075** Queuing and apparent 'queue jumping', long waiting times, cancellations, missed connections and flights, poorly informed staff and rude or abrupt service are all the common sparks which can eventually ignite violence. Organisations should ensure that the ratios of staff to customers, clients or service users are realistic and allow for the cost effective provision of the best service.

### Safe practice guidance

**V8076** Some roles have inherent risks associated with them and require specific guidance. For example, safe practice and communication systems need to be put in place for workers working alone, and for those working out in the community who can be isolated and vulnerable to crime.

Staff are more likely to adopt guidance that they have actually been involved in developing.

## Systems and procedures

**V8077** Effective risk assessment will anticipate and deal proactively with situations likely to cause conflict. High demands on service, running out of stock, maintenance work likely to cause delays and particularly unpalatable news are some of the situations that can be anticipated and contingencies for minimising and resolving conflict put in place.

Those working in certain urban areas may be more likely to be exposed to theft, robbery, aggressive begging, prostitution, public disorder, drug dealing and drug abuse. Rural areas present different challenges, as staff can be a long way from help.

Some roles, particularly in health care, necessarily involve close contact with a person to carry out checks, tests and medical procedures. Some workers deal with aggressive people in confined spaces where there are risks to other people and little prospect of immediate help from the police or security staff.

Each risk should be examined and appropriate guidance and practice designed which will minimise the risks involved in the working practice or situation.

## Communications

**V8078** Poor information and communication often causes or contributes to conflict. Effective communication can quickly calm and resolve potential conflict. Many people will accept the inevitability of a delay or cancellation – as long as they know how long it will be or what alternatives are available.

## Design and layout

**V8079** Designers are becoming more aware of the issue of work-related violence and significant developments are being made in this area as it becomes clear that personal safety and good service delivery go hand in hand. Some pub chains for example use clever layouts to reduce potential frustrations and conflict flash points, thereby 'enhancing the customer experience', selling more drinks and reducing the risk of violence and aggression.

'Queue jumping' is a very common cause of conflict and effective signage and queue management can reduce this.

Risk can be further reduced with the introduction of wider desks, raised floors, and areas offering access to those with special needs or requiring greater privacy. Staff need to have access to a secure place should a serious incident occur, and private areas need to be clearly identified and secured.

## Proactive service delivery

**V8080** The quality of service delivery has a major bearing on the incidence of violence in the workplace. A climate for trouble is created when people are frustrated and their expectations are not met and staff respond unprofessionally to their concerns.

Sometimes, staff cannot meet the customers' expectations because of circumstances beyond their control. The response should be to influence and adjust the customer expectation. The development of basic interpersonal skills and a service oriented culture lays the foundations for a safer environment. This area is probably the most under-rated as a control measure.

Simple inexpensive gestures like providing refreshments, upgrades, free use of a telephone or concessions can help to reduce tension, and it is important that staff are empowered to offer these, with appropriate support guidance.

## Warning 'flags'

**V8081** A contentious issue for many organisations is the placing of warning 'flags' against an individual service user or address. Many organisations cite client/patient confidentiality, data protection and human rights as reasons for not placing warnings. Some also argue that if staff know someone could be a problem then this will affect the way they deal with them and risk a self fulfilling prophecy. Although each of these issues is valid and needs to be considered, employees also have rights and a balance must be sought in order to ensure staff safety.

Data protection agencies provide guidance on the use of such warning systems and their management.

## Security responses

**V8082** Security responses available to organisations break down into three areas:

- Security personnel.
- Security procedures.
- Physical measures.

## Security personnel

**V8083** Some organisations use external security contractors, others choose to employ an in-house team or mix the two. There are benefits and drawbacks to each.

If considering an external provider it is important to establish the level of investment that the provider makes in staff training and support. In a traditionally low paid area of work with a high turnover of staff, contractors are reluctant to invest in more than the bare minimum required for training.

More control over selection, training and performance is achieved through developing an in-house security team.

Whichever option is chosen it is important to remember that security staff need to be actively involved in the initiatives, so that they can win the trust and support of other staff and made to feel part of the team.

## Selection

**V8084** Security staff, by the very nature of their role, will deal with conflict on a regular basis. It is important to select staff who demonstrate the attitudes and behaviours that help to deal positively with situations where conflict is inevitable.

## Training

**V8085** Security staff play a key role in service delivery and are often the first and last point of contact for a customer or service user. It is therefore a worthwhile investment to develop their communication and interpersonal skills.

Security staff are often expected to confront potential criminals, respond to violent situations, and to protect staff and service users. They need to be equipped with the necessary knowledge and skills to manage conflict and violent behaviour.

## *Security procedures*

**V8086** Security staff face very difficult and sometimes dangerous situations where their actions may be subjected to detailed scrutiny. It is important therefore that staff are clear as to the expectations upon them for preventing and responding to violent incidents. Clear procedures need to be put in place that outline the action to be taken in areas such as incident management, refusing entry, searching and evicting people and making an arrest.

Other employees will also need guidance as to what they can expect of their security colleagues, and how they can help security staff when the need arises.

## *Physical measures*

**V8087** There is a vast range of physical measures that can be introduced to reduce the risk of physical injury through a workplace violence incident.

Some of these measures can be extremely expensive and it is worth emphasising here that a thorough risk assessment needs to be undertaken before committing to any expenditure on specialised equipment.

## Access control

**V8088** There are many variations of access control ranging from barriers to open plan receptions. However, funding and existing building design are major influences on this. A well-designed and managed reception can contribute greatly to access control, yet be seen as offering a friendly and helpful service rather than looking intimidating and overtly tight on security.

## Closed circuit television (CCTV)

**V8089** Although the outlay on CCTV can be high, it is one of the most popular security responses available for deterring and managing crime and violence. However, organisations need to be realistic about its purpose and use. A recent Home Office commissioned report suggests that CCTV is more effective as a detection tool than as a deterrent. CCTV needs to be monitored and used proactively to provide effective prevention.

## Alarms

**V8090** There are many types of alarm system, each designed for a different purpose. The key alarm systems involved with enhancing personal safety are:

- Intruder alarms – commonly used to protect buildings after hours, or to protect restricted areas.
- Panic alarms – increasingly used in reception areas, interview rooms and isolated areas.
- Personal alarms – carried on the person as a means of attracting attention and temporarily distracting an assailant.

Before purchasing an alarm, be sure of the organisation's needs and consider them in the overall strategy for tackling violence.

### Radios, paging and public address systems

**V8091** These communication systems have advanced greatly in recent years and can be an asset to staff safety. Some personal radios and pagers have built in panic alarms and sophisticated tracking facilities that pinpoint the location of the unit.

### Mobile phones

**V8092** Mobile phones can provide a means of communication and comfort to lone workers. They are useful for keeping colleagues informed of movements and of unforeseen travel/schedule difficulties.

### Protective vests

**V8093** Often referred to as 'body armour' these vests can be designed to protect against the threats presented by firearms and edged weapons.

Any organisation can purchase protective vests as no specific authority or licence is required, and they can be a necessary risk reduction measure in some areas of work. It is however important that the decision to adopt protective vests is not taken lightly, as it can be a sensitive issue for employees, service providers, and the public generally. Great care needs to be taken when considering purchasing this expensive equipment and effective risk assessment, incident reporting and monitoring processes should be carefully scrutinised to help make any decision.

### Restraints

**V8094** Restraints are common in policing, custodial services, airlines and some areas of healthcare. Although the concept of using restraints will understandably concern most employers, they may be a necessary measure to prevent a disruptive passenger putting an aircraft at risk, or to prevent a patient harming themselves or others. Use of such equipment needs to be justified by the role requirements and risk assessments, and staff need comprehensive training and guidance in its use.

Advice should be sought before adopting handcuffs and other restraints to ensure an appropriate and lawful choice is made. New restraints such as the emergency response belt are becoming popular within some areas of work, as they allow a great deal of control without pain.

### Batons

**V8095** Batons are deemed offensive weapons and can only be used by the police.

## Training

**V8096** Providing effective training is an essential element of an employer's duty of care to employees. Through training employees can identify ways in which they can reduce risk and develop the skills they need to deal with potentially violent situations. Appropriate training for workplace violence can take many forms and can be sourced internally or externally. For training to be effective it is essential that it is not viewed as a 'stand alone' solution but integrated into a wider learning programme that is actively supported by the organisation.

## A training development model

**V8097**   A well designed training or development activity is developed through a three-stage process (see figure 5 below):

1.   Training Needs Analysis.
2.   Training design, development and testing.
3.   Implementing, monitoring and evaluation.

Each stage of this model is essential to the development of an effective training solution and should be evident in any training programme being offered whether that be internal or external.

**Figure 5: General training and development model**

## Training Needs Analysis

**V8098**   The first step in a Training Needs Analysis is to identify the gap between how things are now and how things ought to be.

Any training or development solution should begin by asking two questions:

1.   How has the need for the proposed training been identified?
2.   What are the intended learning outcomes of the proposed training?

## Identifying the need

**V8099** The variety of ways that a training need can be identified are shown in figure 6 below.

A complete review of the risk of conflict and workplace violence across an organisation will identify a range of needs. Amongst these will be specific training and development requirements for staff performing specific roles and for line managers who are responsible for reducing risk and managing the issues that result from workplace incidents.

Figure 6: Identifying the need

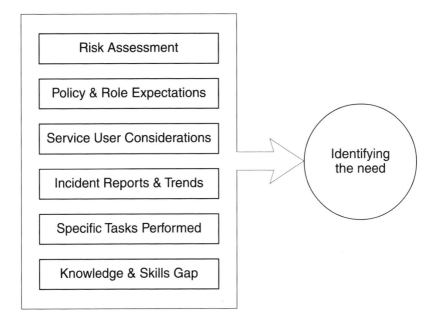

## Defining the learning outcomes

**V8100** The outcomes should link directly to the identified needs. A good way to start is by saying: 'At the end of the training the delegates will ... ' – and then describe the outcome. The outcome will describe what individuals will be able to do after they have attended the programme.

Here is a simple example. A review has shown that the workplace violence reporting form is not being filled in correctly in 80% of cases. A training need has been identified. The training outcome will be: 'At the end of the training the delegates will be able to correctly complete the reporting form.'

The above example also highlights the importance of ensuring that the need is correctly identified – in this case it may well be the form that is ambiguous, unclear or difficult to follow.

## Checklist

*Training Needs Analysis*

- How has the need for the training been identified?

  For example, through risk assessment; staff consultation; statistical data showing an increase in assaults.

- Who needs the training and why do they need it?

  For example, specific individuals, specific roles, specific departments, lone workers.

- What are the specific roles – when, where and how are they exposed to the risk of conflict or violence?

- What is the level of risk in terms of frequency and seriousness faced in each role and what are the expectations on staff in dealing with this?

- What challenges do service users and the environment present?

- What statistical data are available to help in the analysis of the risk?

  For example, data from reporting system on types of assaults, times of day, locations.

- What research is required to investigate and pinpoint the specific issues relating to each role?

  For example, interviews, focus groups.

- Learning outcomes

  For example, what do should each role to be able to understand, appreciate, do, or do differently, at the end of the proposed training?

## Training design, development and testing

**V8102** The aims, objectives and outcomes of the training should be clearly identified from the desired learning outcomes. Decisions then need to be made about the learning methodologies to be used and the models, tutorials, activities and assessment criteria required to match the specification.

The design of a learning activity, course or programme is a specialist task and should be undertaken by the chosen training provider. It is essential to work closely with them to ensure that the design meets all the requirements.

Generic and 'off-the-shelf' courses are unlikely to meet the needs of staff working in different roles and who have different problems and concerns.

Modern technology provides a great deal of choice in methods of delivering training including distance learning, video, audio and computer-based learning, and virtual classrooms as well as traditional classrooms, workshop and workplace-based learning such as coaching and mentoring. This choice can be somewhat bewildering and it must be recognised that not all of these methods suit all occasions. Distance learning, in particular, can be attractive as a cheap alternative to other methods but it will fail to deliver some important areas. Generally, distance learning is effective as a support element to a training programme, particularly for the 'knowledge' aspects. The classroom is good for practising skills, discussing issues and experiences and problem solving. The workplace is good for coaching and applying the knowledge and skills.

## Soft skills and physical intervention training

**V8103**   Managing conflict demands a unique set of skills which fall into two distinct areas:

1. *Soft skills* – communication and interpersonal skills that can be used to calm and control situations.
2. *Physical intervention skills* – used when it is necessary to engage in physical contact with another person.

## Core content – soft skills

**V8104**   The Training Needs Analysis should identify the learning needs of people who perform a particular role. In general the following constitute the core ingredients of any training solution which is intended to meet soft skills needs.

- *Organisational policy and values* – core values which underpin the organisation's approach and the policy on workplace violence.
- *Definitions of workplace violence* – what the organisation defines as 'violence' and what type of incidents should be considered for reporting.
- *Risk assessment* – dynamic assessment of a situation to assess the risk and the consequent appropriate action.
- *Risk reduction and safety systems* – safe working practice, security procedures, the location, testing and use of alarms, panic buttons and CCTV equipment.
- *Theoretical models of aggression, violence and conflict management* – understanding the physiological and psychological processes that occur when people become angry, frightened or aggressive.
- *Triggers and escalation* – recognising the triggers and signs of increasing aggression so that employees can anticipate and defuse situations before they become more serious.
- *Verbal and non-verbal communication skills* – understanding and practising the specific skills of communication and controlling personal space.
- *Special groups of people* – eg drugs, alcohol, mental illness, learning disabilities, cultural differences and the elderly.
- *De-escalation and calming skills and exit strategies* – de-escalation and calming through empathy, problem solving, win-win thinking and strategies for getting out of difficult situations.
- *Legal issues, self defence and use of force* – employees' understanding of their rights and responsibilities when confronted with aggressive and violent behaviour.
- *Support for staff* – post-incident – what support is available through the organisation from line manger support to specialist counselling.
- *Post-incident reporting and debriefing* – understanding how to properly report an incident and be aware of the reasons why it this is important.

## Core content – physical intervention skills

**V8105**   It is more difficult to identify the core content for physical skills, as it will vary a great deal across different sectors. The Training Needs Analysis and risk assessment will identify whether there is a specific need for physical intervention in a particular role. Some self-protection systems include strikes and locks to joints in order to gain compliance through pain. Others are based on methods that do not employ aggressive techniques or use pain as a way of gaining compliance. Most organisations requiring physical intervention training are well advised to opt for an effective non-aggressive system, as this will reduce the likelihood of an escalation and the risk of injury to both staff and the assailant.

The following are the key pre-requisites to physical skills training:

- *Core soft skills* – core content described for soft skills should be regarded as an essential pre-requisite of any course involving physical intervention skills.
- *Skills for protecting oneself or another from unlawful assault* – commonly referred to as 'breakaway' or 'disengagement' techniques; skills to protect against strikes and skills to release holds and remove oneself or another from danger.
- *Interventions that are used to hold and restrain another person* – often referred to as 'control and restraint' or 'holding skills'; used to prevent someone from escaping lawful arrest or detention, or preventing them harming themselves or others.
- *Legal issues* – employees must be critically aware of their powers in relation to detaining someone.

Additional physical skills can be added to these core areas, including safer approaches to day-to-day tasks such as escorting or guiding.

Specialist skills can also be added to the highest-level training such as the use of restraint equipment.

## Reviewing the training content

**V8106** It is important at this stage to ensure that the solution will satisfy three basic criteria:

1. Relevance

    Is the training content based upon a thorough Training Needs Analysis that considers the role and tasks performed and the risks associated with these?

2. Legal review

    Is the content of the course legally correct? Will it stand up to examination in legal proceedings if tested?

3. Medical review

    Are the learning methods and content safe? Will this stand up to examination in legal proceedings if tested?

The legal and medical basis for the training provided may be challenged in any legal proceedings and it is important that the organisation is confident that the training being provided will stand up to such scrutiny.

It is the responsibility of the training provider to show the appropriate legal and medical basis that underpins the learning solution being designed.

## Testing the solution

**V8107** Before implementing the course or programme, it is a good idea to test it out using a pilot course or courses. The pilot course should be delivered under the same conditions that the fully implemented course will be delivered and feedback about the programme should be sought from delegates.

## Implementation

**V8108** The rollout of the programme needs consideration, with some thought about who should receive training first. There are no hard and fast rules but it is important to develop a cohesive, prioritised strategy for implementation.

## Checklist

**V8109**

> *Design and Implementation*
> 
> - What are the aims and objectives of the training programme?
> - How does the content of the programme achieve the aims and objectives?
> - What areas of knowledge and understanding are required?
>   For example, theory, models, legislation, policies and procedures.
> - What skills are needed by staff?
>   For example, communication, problem solving, disengagement, escorting, restraint, detaining.
> - What are the appropriate attitudes and behaviours required?
>   For example, avoiding physical contact, assertiveness.
> - What learning methodologies are being used to achieve the development and how appropriate are they to these circumstances?
>   For example, distance learning, group discussion, practical role play, practice physical skills.
> - Has the training been tailored to the role and tasks performed?
> - Have staff been consulted in the design process?
> - Can any of the training be delivered in the workplace for realism and problem-solving opportunities?
> - How is the content reviewed to ensure it is tactically effective, legally correct and medically safe?
> - How will the programme be piloted and who should take part in the pilot?
> - How will the pilot be evaluated?
> - Who needs to know about the programme and how will it be communicated?
> - What backing and support (and from whom) is available at senior level?
> - How will the programme be rolled out? Who will receive it first – or will it be a mixed rollout? Can some groups, roles, individuals be trained together?

## Monitoring and evaluating the solution

**V8110** The monitoring and evaluation of the programme needs to fit with the existing infrastructure of the organisation.

Figure 7 below shows a typical process that combines the individual development needs of the delegate with the feedback required to monitor and evaluate the programme. From an individual point of view, the delegate should define his or her personal learning objectives before the course, preferably involving his or her line manager. A short time after the course they should review the objectives and establish the learning outcomes achieved.

From an organisational point of view, delegates can complete an immediate post-course evaluation, which will establish reactions to, and satisfaction with, the programme. About four to six weeks later, delegates should provide further feedback about the impact that the programme has had in they way they perform in the workplace. This is then fed into the process of Training Needs Analysis and design for continuous improvements to the programme.

Figure 7: A typical delegate feedback process combining individual and organisational needs

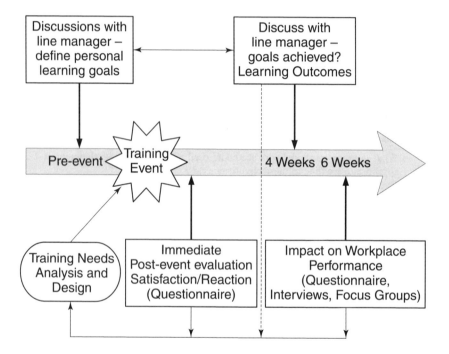

## Internal or external training providers

**V8111** The decision about whether to use an internal training department or an external training provider is quite a difficult one and needs some careful consideration.

### Internal provider

**V8112** The most compelling reason for choosing an internal training department is usually that it costs less. However, there are hidden costs involved which include specialist research into the subject of violence in the workplace, design of the programme, and specific training for trainers necessary to deliver an effective conflict management training programme.

If an organisation's workplace violence issues require physical intervention training, then it must be remembered that the risks to the organisation increase with the level of physical intervention required. The internal provider must be able to satisfy managers that the interventions are both appropriate and medically and legally defensible.

Most internal training departments attract 'can do' people who are willing and eager to help and enthusiastic about new challenges. Care is needed to ensure that they can provide training specific to requirements – not what they may think is needed. The following checklist will help in this regard.

---

*Internal provider*

- How well is the department equipped to do a Training Needs Analysis on the requirements for dealing with workplace violence?
- Can they design a properly researched solution that will provide effective development for the employees?
- Are the trainers suitably experienced, qualified and credible to provide specialised training in managing conflict and workplace violence?
- Does the trainer development programme include coaching and ongoing support?
- Does the Training Needs Analysis indicate a requirement for skills in physical intervention?
- How has the training solution been reviewed to ensure that it is tactically appropriate, legally correct and medically safe?
- How will the training be monitored and evaluated to ensure it is effective and meeting the current needs of the workplace?

---

## External provider

**V8113** The extra investment of engaging an external provider needs to be justified and the most compelling justification is that the provider has specialist knowledge, experience and expertise in the field of managing conflict and workplace violence. Of course, most external providers will claim to have such expertise in abundance and the difficulty lies in making the right choice from the plethora of providers who claim to have just the solution being sought.

In simple terms the client will need to find out from the provider if:

- Their solution will equip the individuals at risk with the knowledge, skills, attitudes and behaviour to deal effectively with the incidents of workplace violence they may face.
- Their organisation has processes which ensure that the training need is correctly identified, the solution is properly designed, and that there is an evaluation process that provides feedback to validate and improve the product.
- Their training is properly designed to meet specific needs and is sufficiently robust to be medically and legally defensible if required.
- The people used to deliver the training are properly trained and experienced in delivering this type of training and understand the particular problems of the client.

The following checklist provides the basic questions that need to be asked.

---

*External providers*

- What evidence can they provide of their expertise in the field of managing conflict and workplace violence?
- What experience do they have in the organisation's specific sector and work?

- What do the other organisations they have worked with say about them?
- What evidence can they provide as to the effectiveness of their training in the workplace?
- What methodology will they use to identify the training needs of the organisation in the area of conflict management and workplace violence?
- How will they develop their understanding of the business and organisation?
- What learning methods are they proposing to use? Are they practically-based and in line with the training needs identified?
- How do they ensure that the content of their programmes is appropriate and legally correct?
- How do they ensure that the physical interventions used are medically safe?
- How do they conduct their training to ensure the delegates are trained in a safe environment?
- How do they develop their trainers?
- What level of resources will they commit to the programme and how will they deal with short-term issues such as a trainer becoming ill?
- What level of programme evaluation do they offer and how will it be fed back to the organisation?
- Are they likely to provide credible expert witness support if required?

## Combining internal and external provision

**V8114** Combining the two options can provide a compromise and solution to this problem. It can be achieved by engaging an external provider to undertake a Training Needs Analysis and design a solution. The provider could be asked to deliver training directly to the higher risk and more demanding staff groups, and they could train internal trainers to deliver the bulk of the remaining programmes.

This solution is attractive because the visible costs are kept relatively low and the benefits of using an external provider are achieved to some degree.

*Combine internal and external providers*

- What experience can the external provider evidence to show they are capable of trainer training?
- Will the training enable internal trainers to provide the programme in a way that is professional and credible?
- What level of ongoing trainer coaching and support is offered?
- Does the training require physical interventions? If so, review the decision and be sure the trainers can teach these safely and effectively.

## Training the managers

**V8115** Managers are often forgotten when it comes to violence management training. The needs of managers, particularly line managers, are two fold.

Firstly, they are often called into disputes involving customers or members of the public which have escalated and become too difficult for the member of staff to deal with. They may have already made the situation worse. Additionally, they are often dealing with situations when things have gone wrong and the potential for conflict is high. Consequently, there is a clear need for line managers in these situations to be personally skilled in conflict resolution as part of their role. They also need to be aware of the techniques and skills in this area provided to the people they manage.

Secondly, they have a vital role to play in the management of incidents of workplace violence and the provision of front line support both during and after the process. They are key players in ensuring that such incidents are properly recorded and reported and that staff carry out their roles in a way which achieves the organisation's goals in a safe and customer-friendly way. Staff can be affected by incidents in many different ways and need an individual response from their line manager.

The second point above is particularly important. Line managers have a great deal of influence in relation to the approach to dealing with conflict and violence in the workplace. Their individual attitude towards the whole issue will greatly influence how seriously the people who work for them approach the subject. This influence will affect attitudes towards risk assessment, approach to conflict situations, reporting and aftercare. Some people are badly affected by incidents that other members of staff might take in their stride. Sometimes the after effects of an incident do not set in until several hours or several days later and managers need to be aware of and look for the signs which indicate that the incident remains unresolved. The way in which the manager deals with the individual will have an influence upon the speed that he or she will recover from the incident.

## Incident management

**V8116** Until this point the focus has been on preventing violence or at least reducing it. This section focuses on the effective management of an incident that does occur, and how its impact can be reduced. It will then consider the steps that should be taken following the incident to support staff and learn from what has happened.

Although every eventuality cannot be planned for, it is important to plan and practise the response to both the most common and most serious scenarios that are likely to be experienced. Even though events will not always 'go to plan', the rigour of the planning process will help ensure that the systems are in place and that staff and managers are better prepared to deal with the situations that arise.

It is particularly important to prepare and practise a response to violent incidents, as during these highly emotive times employees will find it difficult to think clearly and objectively.

Figure 8 reinforces the importance of undertaking a violence risk assessment to establish key risk areas, and putting in place measures to reduce these risks. This process will help with the identification of key incident scenarios that the organisation can then plan and train for.

### Key scenario's

**V8117** The risk assessment process will have highlighted key areas of risk and control measures will be put in place to tackle these. In preparing for the management of incidents it is important to use the risk assessment findings as the basis for the scenarios to be planned for and practised.

### Roles and responsibilities

**V8118** It is important to be clear as to roles and responsibilities prior to an incident occurring to prevent a breakdown in communication when under pressure. This extends further than a specific job role but to the wider team working in an area, for example, what part will the reception, security, domestic and office staff play, when an incident occurs in their area? Who

**Figure 8: Incident management – planning and practice**

takes control, how are they supported, who calls for help, who takes down descriptions – are all questions that need to be addressed. Everyone needs to be involved and has a part to play – even if it is in the background. In lone working environments communications and expectations need to be clear between lone workers, office colleagues, control centre or switchboard colleagues.

## Response teams

**V8119** Some organisation such as hospitals operate incident response teams. These multi-disciplinary teams follow a similar concept to 'crash teams' that respond to cardiac arrest. There are pro's and con's to this approach, some organisation feel the response team is an intrusion and can damage relationships between local staff and patients/service users. If approached well staff can however benefit from the knowledge that they will get competent help quickly. Careful selection of response teams and a high level of training are vital if they are to operate safely and earn the respect of staff.

## Leadership

**V8120** The most critical role in incident management as with any crises, is leadership. Supervisors and managers need to have the confidence and skills to take control of an incident and its immediate aftermath, and staff will look to them for direction and example. Managers often miss out on staff conflict management training, yet they need these valuable skills and the additional knowledge in how to manage an incident and aftermath. Some managers will find difficulty in adjusting from their preferred leadership style to a more directive style required in a crisis. The concept of situational leadership is useful in training, to help managers understand that they need to adapt their style to the situation and that staff will want clear direction and decisions. Senior management needs to provide clarity as to the level of authority given to managers and staff during an incident and as to how and when decisions should be escalated to a senior level.

## Procedures

**V8121**  The importance of clear role expectations and communications has already been discussed, and these need to be written down in the form of clear guidance and procedures. Flow charts will provide a useful means of simplifying and communicating responsibilities. Most organisations will have clear fire procedures in place and the same approach should be taken with violent incidents

One key area requiring guidance and procedure is that of *preserving evidence*.

Violent incidents often result in the prosecution of the perpetrator. The success of the prosecution will depend on the quality of the evidence that supports it, and unfortunately many cases collapse at court due to simple errors made at the time of the incident in gathering and protecting evidence. Things that may seem trivial here, can have a dramatic impact in court as the defence seeks to discredit prosecution evidence.

The rules of disclosure highlight the duty of the prosecution to disclose all evidence, whether it will stand for or against the defendant. Organisations must therefore realise that anything that could be deemed as evidence relating to the case, such as any reports or records, will form part of the legal process. There are limited exceptions such as medical records.

Some basic tips and common sense will help to strengthen a prosecution and to ensure justice is done. Guidance should cover:

- Importance of securing witnesses.
- Identification evidence.
- Use of Closed Circuit Television (CCTV) as evidence.
- Preserving a scene.

## Communications

**V8122**  Another important consideration in incident management is communications with other agencies such as the police, specialist consultants and the media. It is important to be proactive in responding to media interest following a violent incident. Considerable harm can be done to both the individuals involved and the reputation of the organisation if this is not handled well.

## Environment and equipment

**V8123**  Design, layout, security and equipment will be put to the test should an incident occur, and it is wise to ensure that these are considered at the earliest opportunity and tested regularly. Panic alarms, for example, are all to often found to be disconnected or out of order.

## Sanctions

**V8124**  Many airlines and hospital trusts operate warning systems, for example they issue a letter to a disruptive individual or a tallow or red warning card. Clear sanctions play an important part in preventing and responding to violent behaviour. However, it must be recognised that the confronting of individual/s could be the trigger of an assault. It is essential that staff are taught how to accurately assess people and situations and are provided with training in when, where and how to confront a violent person. Some organisation are quick to launch these schemes but do not consider the position of the manager or member of staff who has to carry out these difficult and high risk tasks.

## Training

**V8125** Training will focus firstly on raising safety awareness and reducing risk. This is largely *knowledge* based and staff also need to develop the *skills* for dealing with the incident that does occur. If staff are to develop confidence in managing violence the training will need to be dynamic and realistic, and cover the key scenarios identified.

Teamwork will be tested during stressful incidents and this should also be a key consideration in staff development to ensure staff communicate effectively, understand their responsibilities and respect each other. This should also extend to the wider team operating in a certain area, as communication often breaks down across roles or functions.

Physical fitness is particularly important and vital in some roles where staff are expected to respond to violence. Looking and feeling fit and behaving professionally will help to earn respect and deter assault. Fitness will also play a vital part when restraining a violent person or even running away from one.

## Rehearsal and testing

**V8126** Without doubt rehearsal and testing is one of the most under-rated and least performed aspects of incident management. With the exception of some areas of the emergency services, psychiatric care and the armed forces, organisations rarely practise their response to violent incidents. This is somewhat strange as fire drills to test communications and procedures are regularly conducted within organisations.

In sport coaches are often heard telling athletes 'you play as you train' and the same applies to incident management. Realistic practise and scenario-based training will make a big difference to the effectiveness of the incident response. Unforeseen problems with communications, equipment and procedures will be highlighted through a rehearsal and staff can be actively involved in solving these.

In some areas of work it will be beneficial to set up multi-agency exercises involving police and other agencies. This helps to clarify expectations and iron out any problems.

## Review and improvement

**V8127** Training, rehearsal and real incidents provide valuable feedback, and it is important that procedures are continuously reviewed to respond to this.

## Checklist

**V8128**

- Assess the risk – does the organisation face a foreseeable risk of serious incidents of violence towards staff?
- Are roles, responsibilities and communication lines clearly identified for the management of an incident of workplace violence?
- Are the individuals and departments concerned aware of the roles, responsibilities and communication lines?
- Is the risk to staff and other service users great enough to warrant the development of specially trained response teams?
- Are the leadership roles and decision-making levels clear and have the appropriate people received training for taking a lead role in such incidents?

- Are there clear procedures outlined for managing the different aspects of an incident? In particular, for serious incidents are there clear procedures for securing and preserving evidence?
- Is there a clear strategy for managing communications and the media in the event of an incident?
- Has specialist equipment been identified and have staff been trained in its use in the event of an incident?
- Have the all the people who might be called upon to respond to an incident received appropriate training?
- Is there a robust and regular method in place for rehearsing and testing the complete response to an incident?
- Is there a review and evaluation process in place to provide feedback about the effectiveness of the incident management either after a test or a real incident?

# Post-incident management

## Investigation

V8129     The post incident management process includes investigating the circumstances of the incident, attending to the needs of the victim and, where appropriate, dealing with the perpetrator.

Investigating incidents is an integral element of the risk management process. By establishing the factual circumstances surrounding an incident of workplace violence and gaining an understanding of why and how the incident took place it may be possible to take remedial action to prevent a similar happening in the future. As with all other phases of the risk management process it is important that employees are consulted and their suggestions taken into consideration.

An internal investigation should:

- collect facts on who, what, when, where and how the incident occurred;
- record information;
- identify contributing causes;
- encourage appropriate follow-up; and
- consider changes in controls, procedures and policy.

Incident investigation forms part of the wider legal requirement under the *Management of Health and Safety at Work Regulations* (*SI 1999 No 3242*), *Reg 5* for an employer to monitor and review preventive and protective measures. The accompanying Approved Codes of Practice (ACOP L21), HSE management guidance (HSG 65) and National Occupational Standards for managing work related violence provide further guidance. The guidance acknowledges that the amount of time and effort put into the investigation will vary dependent upon the nature and severity of the event.

HSE research clearly demonstrates the link between workplace violence and abuse and stress. Insensitive questioning of the victim during the early stages of an investigation may compound the traumatic effects of the incident itself leading to longer-term psychological damage. Those responsible for conducting the investigation should be aware of this dynamic and carefully balance the priority of providing support to the victim against the requirement to investigate, question and learn from what happened to prevent recurrence.

## How people are affected by workplace violence

**V8130** Perhaps the most important thing to recognise is that everyone has a different way of responding to and dealing with the aftermath of a violent or aggressive incident. There is no 'right' or 'wrong' way to react and people must be allowed to deal with it in their own way.

Being the victim of violence is particularly traumatic because it involves an interaction with another person at a very personal level and this can produces some difficult and complex emotional reactions. These reactions will vary over the short, mid and longer term and it is important that support is provided at each stage, where problems persist, professional support should be provided (see figure 9 below). It is important that those involved recognise that these reactions are quite 'normal' following an abnormal event, also that no one says they have to experience them.

Figure 9: Timescale of reactions to workplace violence

| Short Term | Medium Term | Long Term |
|---|---|---|
| 24 hours | 1–3 days | Weeks, months – possibly years |

## Common reactions to an abnormal event such as an assault

### Short-term reactions

**V8131** In the first few hours following the incident, the victim will have some initial reactions to the aggression and violence inflicted upon them. These reactions are predominantly emotional and are a direct response to the incident. Many factors will influence the severity of the reaction, not least of which is the individual's level of resilience towards traumatic situations.

The level of aggression, suddenness of the confrontation and physical injury sustained are also some of the factors which will influence how the victim will react. The following are the most likely reactions:

- Shock, confusion, disbelief, fear, helplessness.
- Anger, embarrassment, feeling of violation.

Many of these initial reactions will begin to lessen as the victim moves into the next phase.

### Medium-term reactions

**V8132** The short-term reactions are characterised by their 'immediate' nature, formed before the victim has had any time to think about, and begin to rationalise, what has happened. The medium-term reactions begin to appear when the victim has had a chance to consider the incident, to work though what happened and to think about the consequences, near misses and alternatives. This will be around 24 hours after the incident. Reactions can include:

- Feelings of loss, guilt, shame, embarrassment, humiliation.
- Exhaustion and tiredness, lack of sleep.
- Denial of effects, ready to get back to work.

- Anger, frustration and resentment.
- Lack of confidence, anxiety about similar situations or meeting the aggressor.

Moving successfully through this medium-term phase is often the key to recovery. Once the victim has acknowledged what has happened and come to terms with it then he or she can then move back towards a normal life. Line managers can provide vital support in this phase and are pivotal in the successful recovery of most of the victims.

### Long-term reactions

**V8133** Generally, reactions that persist beyond a couple of weeks after the incident are indicative that the victim is finding difficulty in coming to terms with the incident and that he or she probably needs professional specialist help. Examples include:

- Persistent tiredness, exhaustion, depression, bouts of anxiety.
- Excessive drinking and smoking, anti-social behaviour, irritable and aggressive behaviour.
- Nightmares, flashbacks, headaches, nausea, difficulty in eating and sleeping.

A victim who displays these long-term reactions clearly needs specialised help, and an organisation which wants to provide a complete response to the range of issues that result from workplace violence, will need to set up a procedure to facilitate this.

## *Returning to normal*

**V8134** Many factors will influence the speed of recovery including the victim's life circumstances and their emotional resilience at the time of the incident. The turning point for most victims is the acceptance of what has happened to them. Once they accept the incident as a reality, they stop going through the 'if only ... ' scenarios and stop blaming themselves for what took place.

Most people reach a point where they can move on from the event and get back to their normal daily lives. They achieve this when they regain confidence and self esteem and recognise that, although life has changed in some aspects as a result of what occurred, they can become positive about being back in their working environment.

Although these various reactions have been described in discrete stages it will be rare for anyone to pass smoothly through them all. In reality, many things will cause a victim to progress and regress in the move towards normality. The return to work, for example, can be quite difficult and bring back feelings of insecurity and fear. Having to appear in court as a witness or learning of a colleague who has been involved in a similar incident may well trigger a reoccurrence of one or more of the reactions previously experienced.

For a few people, a return to 'normal' is virtually impossible, particularly if they have been permanently disabled by physical or mental injury. In such cases, the victim will need the most specialised care and support in trying to come to terms with their circumstances.

Thankfully, for most people the support of family, friends, colleagues and managers will be enough to help them recover from the trauma of the incident and return to a normal working life.

## *Providing post-incident support*

**V8135** An appropriate post-incident support system should be an integral part of an organisation's overall response to workplace violence (see figure 10 below). The sophistication of the support system will depend upon the level of risk to which the staff are being exposed.

Staff working in the emergency services can often experience violent and other traumatic events, and therefore require a high level of post-incident support. This will range from line manager support, formal debriefing processes, to occupational health services, which will include access to specialist help for the most serious consequences of work-related violence. Some organisations adopt formal psychological debriefing processes, sometimes referred to a 'critical incident stress debriefing'. Although these approaches have considerable support in some areas, there is ongoing debate about their value, and concern at the potential risks associated with victims 're-living' the traumatic experience in the debriefing process. Whether or not the organisation adopts a formal debriefing process, it is essential that support mechanisms and procedures are put in place for managers and staff.

Even where the frequency of serious incidents is low, the organisation should have a post-incident support procedure and line mangers should be trained in the skills appropriate to helping a victim through the first stages of coping with an incident of work-related violence. It should also be possible to access specialist help if necessary.

**Figure 10: Supporting the victim**

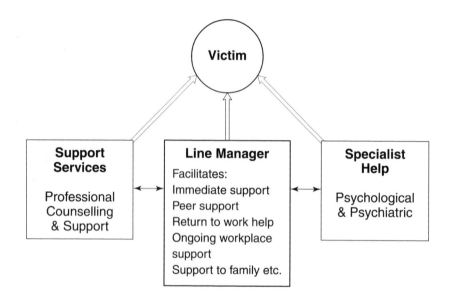

## *The role of the line manager*

**V8136** In the vast majority of cases the support for the victim will be provided through his or her line manager. The short and medium-term reactions following an incident have been described earlier in this chapter and the focus of the line manager's support will be concentrated on helping the victim work though those reactions and in facilitating further support where necessary. Successfully dealing with these phases is vital in providing the optimum conditions for the individual to recover and return to a normal working life.

There are three points at which this support is crucial:

- Immediately after the incident has happened.

- During any absence from work.

- Preparing for and returning to work.

Organisations should provide guidance and training for managers to help them to perform their key role sensitively and effectively.

## Other immediate post-incident considerations

**V8137** There will be a need to complete the workplace violence report form, which will require a formal narrative of the incident. For evidential reasons, this should wherever possible be completed personally by the victim as soon as practicable after the incident. Every effort should be made to get the victim to complete at least the basic requirements whilst it is fresh in his or her memory.

It is tempting to leave this task but it can have far reaching consequences if the incident becomes the subject of criminal or civil proceedings. An accurate and early account of the incident can make all the difference to the outcome of a case.

## Support services

**V8138** There is a point where some victims need more help than the line manager can be expected to offer. If an individual is suffering from the sort of persistent reactions already discussed in the medium to long term following the incident then they should be referred to more professional help. In this context it will probably consist of professional counselling.

Occasionally, a victim may find it impossible to move on from the incident. Some may develop a condition called post-traumatic stress disorder or exhibit some of the symptoms of it. In all these cases, the victim will need specialist psychological and psychiatric help. Remember the reactions outlined earlier that may indicate that an individual is in need of specialist help. Sometimes colleagues will bring attention to the fact that the individual is simply not 'his normal self'.

## Preparing for the worst

**V8139** Organisations need to be in a position to quickly access expert management advice following incidents that are likely to have a severe impact on individuals and/or the organisation. It is important to plan this in advance as poor quality decisions can easily be made when emotions are high and everyone is under pressure. Plans should include access to advice and support concerning:

- Professional help for employees, service users, families and partners.
- Advice on legal issues.
- Advice on managing media attention.

## The police investigation and court case

**V8140** Most victims want to see their aggressor brought to justice. However, it may prove to be quite a daunting process and it can help to be aware of how the process works and what might be expected. There are several prosecution routes including:

- Prosecution through police.
- Private prosecution (a similar process but instigated through a solicitor, not police).
- Civil prosecution (instigated through a solicitor in a civil court).

Prosecution through police is the most common route. However employers should have clear policy and guidance on other avenues should the police or prosecution service decide not to progress the matter. This should include the degree of support they can provide to the victim, such as funding and providing time off for the legal process in preparing and giving evidence.

Criminal and civil courts provide further protection from harassment and violence in a number of ways, including:

- Injunctions.
- Restraining orders (e.g. preventing convicted persons contacting the victim/s).
- Anti-social behaviour orders.

When a court order is in place, the police are in a strong position to take action should it be breached.

Note: Offences can carry extra penalties if proven to be racially aggravated, and such offences will be recorded as crimes by the police and can constitute a more serious offence under the *Crime and Disorder Act 1998*.

Giving evidence in court will understandably cause anxiety and it is especially difficult for the victim of violence 're-living' the trauma of what happened. It is vital that support is provided through this process by the organisation.

Victim Support (tel: 020 7735 9166) is a national organisation with a great deal of experience in helping people to cope with being the victim of a crime, regardless of whether it has been reported to the police. Victim Support also run the Witness Support Service.

They provide confidential, practical and emotional support through a network of local schemes across England, Wales and Northern Ireland. Sister organisations exists for Scotland (tel: 0131 668 4486) and the Republic of Ireland (tel: 00 353 1878 0870).

They also produce a helpful handbook 'Helping People Cope with Crime' (ISBN 0 340 78049 5).

## Support during the process of the investigation and court case

V8141    For anyone who has to undergo the process of a court case it can prove to be a very difficult time.

As mentioned earlier, Victim Support is an organisation with many years of experience in helping individuals to cope with all the different aspects of surviving the trauma of being the victim of a crime. They have developed a Witness Support service, which provides practical help and support for anyone who has to attend a Crown Court as a witness.

# Sources of information and support

V8142    There are a number of sources of information and support that deal with workplace violence. Readers are advised to consult the website: www.workplaceviolence.co.uk, which is a one-stop access point to other useful sites.

## Maybo Risk Management Model

V8143

### Acknowledgement

V8144    This section on work related violence was written by Maybo Limited, the UK's leading specialist consultancy in managing work related conflict and violence.

Figure 11: The process of prosecution

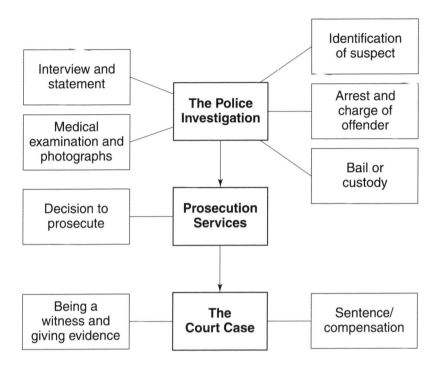

**V8144** *Violence in the Workplace*

Figure 12: Maybo Risk Management Model

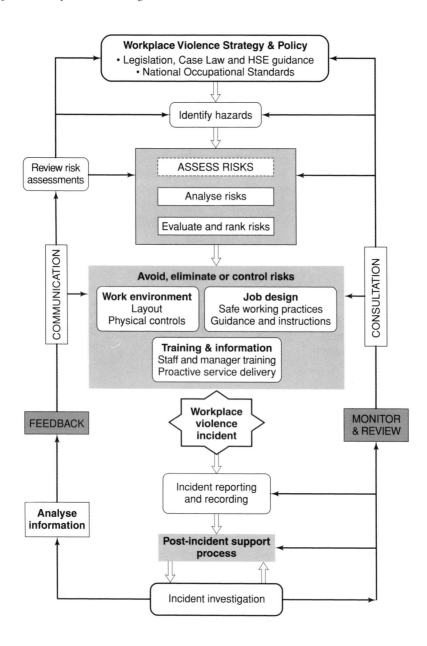

# Vulnerable Persons

## Introduction

V12001 This chapter deals with groups of people who, for one reason or another, are more vulnerable to health and safety risks. Such people include:

- children and young persons;
- new or expectant mothers;
- lone workers;
- disabled persons; and
- inexperienced workers.

Where these persons are at work they are subject to the protection of the full range of health and safety legislation – the *Health and Safety at Work etc. Act 1974* ('*HSWA 1974*') and the various regulations made under it. Of particular relevance are the *Management of Health and Safety at Work Regulations 1999* (*SI 1999 No 3242*) ('the *Management Regulations*') and their requirements for risk assessments. The potential vulnerability of some members of the workforce must be taken into account during the risk assessment process. In addition, the *Management Regulations* contain specific requirements in respect of children and young persons and also new or expectant mothers. These will be explained later in the chapter.

However, *HSWA 1974, s 3* places duties on employers and the self-employed in relation to persons not in their employment such as visitors, occupants of premises, service users, neighbours or passers-by. Risks to such persons must also be considered during risk assessments with appropriate attention given to those who are particularly vulnerable. The general principles of risk assessment are described in the chapter dealing with that subject.

This chapter concentrates on legislation of particular relevance to vulnerable persons and many of the practical considerations to be taken into account when carrying out risk assessments relating to such people.

## Relevant legislation

V12002 Whilst the *Management Regulations* (*SI 1999 No 3242*) impose specific requirements in relation to children and young persons, together with new or expectant mothers, there are other regulations where the vulnerability of some people is either recognised directly or must be taken into account indirectly through some form of risk assessment.

### Management Regulations requirements

V12003 The terms 'child' and 'young person' are defined in *Regulation 1(2)* of the *Management Regulations* (*SI 1999 No 3242*), whilst *subsections (4)* and *(5)* of *Regulation 3* contain requirements relating to risk assessments and young people. *Regulation 19* of the Regulations deals with precautions which must be taken to ensure the 'Protection of Young Persons'.

## V12004 Vulnerable Persons

'New or expectant mother' is also defined in *Regulation 1(2)* whilst there are three regulations relating directly to them:

- *Regulation 16* – deals with risk assessments;
- *Regulation 17* – covers possible suspension from night work;
- *Regulation 18* – contains requirements in respect of notification to the employer of pregnancy or having given birth.

All these definitions and requirements are explained in full later in the chapter.

## Other regulations directly affecting vulnerable persons
V12004

- *Children and young persons*

    Several sets of regulations contain age-related prohibitions or restrictions on specified activities. These are set out and explained in TABLE A later in the chapter (see V12015 below).

    Also of relevance are the *Health and Safety (Training for Employment) Regulations 1990 (SI 1990 No 1380)* which give students on work experience programmes and trainees on training for employment programmes the status of 'employees'. The immediate provider of the training is treated as their 'employer'. Whilst this particularly affects children and young persons, it also applies to adult participants in training programmes. (Courses at schools, colleges and universities etc. are not covered by these requirements.)

- *New or expectant mothers*

    Some regulations contain requirements for employees engaged in certain activities to be free from medical or physical conditions which make them unfit for that activity. Typical of these are the *Work in Compressed Air Regulations 1996 (SI 1996 No 1656)* and the *Diving at Work Regulations 1997 (SI 1997 No 2776)*.

    Both the *Ionising Radiations Regulations 1999 (SI 1999 No 3232)* and the *Control of Lead at Work Regulations 2002 (SI 2002 No 2676)* impose tighter exposure standards on *all* women of reproductive capacity whilst the latter Regulations would almost certainly result in the suspension from lead work of a woman once her pregnancy was notified. (These requirements are explained in more detail later in the chapter – see V12018 below.)

## Regulations with indirect implications relating to vulnerable persons
V12005

- *Construction (Design and Management) Regulations 1994 (SI 1994 No 3140) (CDM)*

    When construction work is to be carried out in or close to premises occupied by children (or those with special educational needs) the CDM Health and Safety Plan must take particular account of their lack of awareness of risks and general curiosity. If normal access routes in premises are affected this may create risks or problems for pregnant women, disabled persons or the elderly.

- *Manual Handling Operations Regulations 1992 (SI 1992 No 2793)*

    The limited manual handling capabilities of, and potential risks to, both children and young persons, and new or expectant mothers must be taken into account during risk assessments. However, risks to lone workers (with no source of assistance), persons with temporary or permanent physical disabilities and inexperienced workers (who may be unaware of relevant handling techniques) must also be considered.

- *Control of Substances Hazardous to Health Regulations 2002 (SI 2002 No 2677) (COSHH)*

    Children, persons with special educational needs or those with fading mental faculties may all be at extra risk from hazardous substances as they may not be able to read or understand warnings or instructions. COSHH assessments should take account of the potential presence of such vulnerable persons and ensure the secure storage of hazardous substances where this is appropriate, e.g. by locking cleaners' cupboards.

- *Personal Protective Equipment at Work Regulations 1992 (SI 1992 No 2966)*

    Use of certain types of personal protective equipment ('PPE') may present difficulties for pregnant women or workers with physical difficulties. Young and inexperienced workers are likely to need more detailed explanations of the purposes of different types of PPE and how and when they are to be used, whilst those with limited intellectual capabilities will need even closer attention.

- *Health and Safety (Display Screen Equipment) Regulations 1992 (SI 1992 No 2792)*

    The suitability of a display screen equipment ('DSE') workstation for a pregnant woman will need to be assessed, particularly as her size increases. Disabled workers or other staff with physical problems (e.g. past injuries) are likely to require special attention during workstation assessments, especially in relation to the suitability of chairs and other ergonomic aspects.

## Children and young persons

V12006   For many years work done by children and young persons was subject to an often bewildering array of prohibitions – equipment that children and young persons were not allowed to use or processes or activities that they must not be involved in. However, since 1997 the emphasis has switched from such prohibitions towards restrictions, based upon a process of risk assessment.

Many of the previous age-related prohibitions were revoked but some still remain. The more important of these are listed in TABLE A at V12015 below. However, the ongoing programme of modernisation of health and safety legislation is continually sweeping away many of these residual prohibitions. (Any cases of doubt should be referred to the HSE.)

### Definitions

V12007   The terms 'child' and 'young person' are defined in *Regulation 1(2)* of the *Management Regulations (SI 1999 No 3242)*.

*Child* is defined as under the minimum school leaving age ('MSLA') in accordance with:

- *section 8* of the *Education Act 1996* (for England and Wales);

- *section 31* of the *Education (Scotland) Act 1980* (for Scotland).

(The MSLA is just before or just after the age of sixteen)

*Young Person* is defined as 'any person who has not attained the age of eighteen'.

Some of the prohibitions remaining from older health and safety regulations use different cut off ages – where relevant these are referred to in TABLE A at V12015 below. Further detail is also provided in the HSE guidance booklet HSG 165 'Young people at work'.

## Risk assessment requirements

**V12008** *Regulation (3)* of the *Management Regulations (SI 1999 No 3242)* contains important requirements (first introduced in 1997) in relation to risk assessments and young persons. These are:

'(4) An employer shall not employ a young person unless he has, in relation to risks to the health and safety of young persons, made or reviewed an assessment in accordance with paragraphs (1) and (5).

(5) In making or reviewing the assessment, an employer who employs or is to employ a young person shall take particular account of—

    (a) the inexperience, lack of awareness of risks and immaturity of young persons;

    (b) the fitting-out and layout of the workplace and the workstation;

    (c) the nature, degree and duration of exposure to physical, biological and chemical agents;

    (d) the form, range and use of work equipment and the way in which it is handled;

    (e) the organisation of processes and activities;

    (f) the extent of the health and safety training provided or to be provided to young persons; and

    (g) risks from agents, processes and work listed in the Annex to Council Directive 94/33/EC(b) on the protection of young persons at work.'

(Note: *Paragraph (1)* of the regulation contains the basic requirement for employers to carry out risk assessments.)

## Protection of young persons

**V12009** In assessing the risks to young persons, the employer must take particular note of *Regulation 19* of the *Management Regulations (SI 1999 No 3242)* which states:

'(1) Every employer shall ensure that young persons employed by him are protected at work from any risks to their health or safety which are a consequence of their lack of experience, of absence of awareness of existing or potential risks or the fact that young persons have not yet fully matured.

(2) Subject to paragraph (3), no employer shall employ a young person for work—

    (a) which is beyond his physical or psychological capacity;

    (b) involving harmful exposure to agents which are toxic or carcinogenic, cause heritable genetic damage or harm to the unborn child or which in any way chronically affect human health;

    (c) involving harmful exposure to radiation;

    (d) involving the risk of accidents which it may reasonably be assumed cannot be recognised or avoided by young persons owing to their insufficient attention to safety or lack of experience or training; or

    (e) in which there is a risk to health from—

        (i) extreme cold or heat;

        (ii) noise; or

        (iii) vibration,

and in determining whether work will involve harm or risk for the purposes of this paragraph, regard shall be had to the results of the assessment.

(3) Nothing in paragraph (2) shall prevent the employment of a young person who is no longer a child for work—

(a) where it is necessary for his training;

(b) where the young person will be supervised by a competent person; and

(c) where any risk will be reduced to the lowest level that is reasonably practicable.'

Some of the practical implications of these provisions are considered later in the chapter.

## Information on risks to children

**V12010** *Regulation 10* of the *Management Regulations* (*SI 1999 No 3242*) deals with 'information for employees' and contains two paragraphs relating to the employment of children. These state:

'(2) Every employer shall, before employing a child, provide a parent of the child with comprehensible and relevant information on—

(a) the risks to his health and safety identified by the assessment;

(b) the preventive and protective measures; and

(c) the risks notified to him in accordance with Regulation 11(1)(c).

(3) The reference in paragraph (2) to a parent of the child includes—

(a) in England and Wales, a person who has parental responsibility, within the meaning of section 3 of the Children Act 1989, for him; and

(b) in Scotland, a person who has parental responsibility, within the meaning of section 8 of the Law Reform (Parent and Child) (Scotland) Act 1986, for him.'

This requirement to provide information to parents includes situations where children are on work experience programmes (where they have the status of employees by virtue of the *Health and Safety (Training for Employment) Regulations 1990* (*SI 1990 No 1380*)) and also includes part time or temporary work.

Some of the practical implications of this requirement are considered later in this chapter.

## Disapplication of requirements

**V12011** The wide-ranging impact of these provisions is lessened to a limited extent by *paragraph (2)* in R*egulation 2* of the *Management Regulations* (*SI 1999 No 3242*) – 'Disapplication of these Regulations'.

'(2) Regulations 3(4), (5), and 10(2) and 19 shall not apply to occasional work or short-term work involving—

(a) domestic service in a private household; or

(b) work regarded as not being harmful, damaging or dangerous to young people in a family undertaking.'

However, the term 'family undertaking' is not defined in the Regulations. HSE guidance indicates that this should be interpreted as meaning a firm, owned by, and employing members of the same family, i.e. husbands, wives, fathers, mothers, grandfathers, grandmothers, stepfathers, stepmothers, sons, daughters, grandsons, granddaughters, stepsons, stepdaughters, brothers, sisters, half-brothers and half-sisters.

As far as is known this narrow interpretation of 'family undertaking' has not been tested in the courts. Common usage of the term would suggest small and medium-sized businesses controlled and managed by members of the same family but not necessarily only employing family members, as implied by the words used by the HSE.

## Capabilities and training

**V12012** Regulation 13 of the *Management Regulations* (*SI 1999 No 3242*) states in paragraph (*1*) that 'Every employer shall, in entrusting tasks to his employees, take into account their capabilities as regards health and safety'. Quite clearly the capabilities of children and young persons will be somewhat different from those of more experienced and mature employees. (This requirement is also of relevance to other vulnerable persons e.g. disabled or inexperienced workers.)

## Risk assessment in practice

**V12013** The requirements of *Regulations 3(4)*, *(5)* and *19* of the *Management Regulations* (*SI 1999 No 3242*) are based upon the contents of a European Directive (94/33/EC) and it is not particularly easy for the employer to identify exactly what he or she must (or must not) do.

A good starting point is to consider the three characteristics associated with young people which are mentioned in both *Regulations 3(5)* and *19(1)*:

- lack of experience;
- lack of awareness of existing or potential risks; and
- immaturity (in both the physical and psychological sense).

All young people share these characteristics but to differing extents – for example, one would have different expectations of a school leaver who had already been playing a prominent role in a family business, e.g. a farm, as opposed to a work experience student with no previous exposure to the world of work. Employers must also be aware that the degrees of physical and psychological maturity of young people vary hugely.

These three characteristics must then be considered in respect of the risks involved in the employer's work activities and particularly those identified in *Regulation 3(5)(b)–(g)* and *Regulation 19(2)*.

The contents of those Regulations (including those risks listed in the Annex to Council Directive 94/33/EC) have been consolidated into a single checklist for employers which is contained in TABLE B.

The purpose of all risk assessments is to identify what measures the employer needs to take in order to comply with the law. Additional measures which the employer must consider in order to provide adequate protection for young persons are:

- not exposing the young person to the risk at all;
- providing additional training;
- providing close supervision by a competent person;
- carrying out additional health surveillance (as required by *Regulation 6* of the *Management Regulations* (*SI 1999 No 3242*) or other regulations e.g. COSHH (*SI 2002 No 2677*));
- taking other additional precautions.

*Regulation 19(3)* allows more latitude in respect of young persons who are no longer children in relation to the requirement to completely exclude them from the types of risks specified in *Regulation 19(2)*.

In deciding what precautions are required, the employer must consider both young people generally and the characteristics of individual young persons. Additional training and/or supervision may be necessary in respect of young people with 'special needs'.

Young people should gradually acquire more experience, awareness of risks and maturity, particularly as they pass through formal training programmes within the NVQ system. As this occurs, restrictions on their activities may be progressively removed.

## Provision of information

V12014    *Regulation 10(1)* of the *Management Regulations (SI 1999 No 3242)* requires employers to provide all employees with comprehensible and relevant information on:

- risks to their health and safety identified by risk assessments (or notified by other employers);
- preventive and protective measures (i.e. appropriate precautions);
- emergency procedures and arrangements.

This requirement is of particular importance in relation to young persons. *Regulation 10(2)* requires employers *also* to provide information to parents of children (under the MSLA) on the risks the children will be exposed to and the precautions which are in place.

The information may be provided either orally or in writing or possibly both. Key items (e.g. critical restrictions or prohibitions) should be recorded. Means of providing information might include:

- induction training programmes;
- employee handbooks or rulebooks;
- job descriptions;
- formal operating procedures;
- trainee agreement forms (increasingly common for young persons on formal training programmes e.g. modern apprenticeships);
- information forms for parents of work experience students.*

    * Further guidance on work experience is provided in the HSE booklet HSG 199 'Managing health and safety on work experience: A guide for organisers'.

The information must be comprehensible – special arrangements may be necessary for young people whose command of English is poor or for those with special needs. The type of information required to be provided will obviously relate to the work activities and the risks involved. The content must be relevant – both to the workplace and the young person. It might include the following types of information:

- general risks present in the workplace

    e.g. fork lift trucks are widely used in the warehouse.

- general precautions taken in respect of those risks

    e.g. all fork lift drivers are trained to the standard required by the Approved Code of Practice ('ACoP').

- specific precautions in respect of the young person

    e.g. the induction tour includes identification of areas where fork lift trucks operate and indication of warning signs.

- restrictions or prohibitions on the young person

**V12015** *Vulnerable Persons*

      e.g. X will not be allowed to drive fork lift trucks or any other vehicles (he will be considered for fork lift truck training after attaining the age of 17).

- supervision arrangements

      e.g. X will be supervised by the warehouse foreman (or other persons designated by him).

- PPE requirements

      e.g. safety footwear must be worn by all employees working in the warehouse (this is supplied by the company).

Where restrictions or prohibitions are removed (e.g. after successful completion of training programmes) an appropriate record should be made, either on the original restriction/prohibition or within the individual's training record.

**V12015**

| Table A |
|---|
| *Prohibitions on children and young persons* |
| Outright prohibitions on work by children and young persons continue to be revoked, and to be replaced by a risk assessment approach. Prohibitions known still to be in place at the time of publication in 2000 of the HSE booklet 'Young people at work: A guide for employers' (HSG 165) were: |
| **Carriage of explosives and dangerous goods** |
| Under-18s may not be employed as a driver or attendant of an explosives vehicle, be responsible for the security of the explosives and may only enter the vehicle under the direct supervision of someone over the age of 18. (There are some exceptions where the risks are low). [*Carriage of Explosives by Road Regulations 1996 (SI 1996 No 2093)*]. Under-18s may not supervise road tankers or vehicles carrying dangerous goods nor supervise the unloading of petrol from a road tanker at a petrol filling station. [*Carriage of Dangerous Goods by Road Regulations 1996 (SI 1996 No 2095* as amended]. |
| **Agriculture** |
| Under-13s may not ride on vehicles and machines including tractors, trailers etc. [*Prevention of Accidents to Children in Agriculture Regulations 1998 (SI 1998 No 3262)*]. HSE guidance also states that children (under the minimum school leaving age) should not operate certain machines and tractors carrying out certain operations. |
| **Ionising radiation** |
| Under-18s may not be designated as 'classified persons'. Dose exposure limits are lower for under-18s. [*Ionising Radiations Regulations 1999 SI 1999 No 3232*]. |
| **Lead** |
| Under-18s may not be employed in certain lead processes – lead smelting and refining, lead-acid battery manufacturing – or to clean places where such processes are carried out. |
| **Mines and quarries** |
| Various restrictions exist for under-18s and under-16s relating to use of winding and rope haulage equipment, conveyors at work faces, locomotives, shunting and shot firing. Some restrictions also apply to under-21s and under-22s. |

| Shipbuilding and shiprepairing |
|---|
| Under-18s, until they have been employed in a shipyard for six months, may not be employed on staging or in any part of a ship where they are liable to fall more than two metres or into water where there is a risk of drowning. [*Shipbuilding and Shiprepairing Regulations 1960 (SI 1960 No 1932)*]. |
| Docks |
| Under-18s may not operate powered lifting appliances in dock operations unless undergoing a suitable course of training under proper supervision of a competent person (serving members of HM Forces are exempt). [*Docks Regulations 1988 (SI 1988 No 1655)*]. |

## *Work presenting increased risks for children and young persons*

**V12016** The checklist below (TABLE B) is based upon *Regulations 3(5)* and *19(2)* of the *Management Regulations (SI 1999 No 3242)* and the Annex to the European Council Directive 94/33/EC. It is intended to assist employers conducting risk assessments in respect of work by children and young persons.

These types of work or situations of exposure to risk are not necessarily prohibited, although the requirements of *Regulation 19(2)* must be taken into account. However, restrictions could be required for young persons (particularly children) and additional precautions may be required to provide them with adequate protection from risk. Many situations present no greater risk to young persons than adults and restrictions (e.g. close supervision) may only be necessary until the employer is sure that the young person is fully aware of the risks and is capable of taking the necessary precautions.

| Table B |
|---|
| Excessively physically demanding work |
| • Manual handling operations where the force required or the repetitive nature could injure someone whose body is still developing (including production line work); |
| • Certain types of piece work (particularly if peer pressure may result in them tackling tasks or working at speeds that are too much for them). |
| Excessively psychologically demanding work |
| • Work with difficult clients or situations where there is a possibility of violence or aggression; |
| • Difficult emotional situations e.g. dealing with death, serious illness or injury; |
| • Decision making under stress. |
| Harmful exposure to physical agents |
| • Ionising radiation (separate exposure limits apply to young persons); |
| • Non-ionising electromagnetic radiation, e.g. lasers, UV from electric arc welding or lengthy exposure to sunlight, infra red from furnaces or burning/welding; |
| • Risks to health from extreme cold or heat; |
| • Excessive noise; |
| • Hand-arm vibration, e.g. from portable tools; |

- Whole-body vibration, e.g. from off-road vehicles;
- Work in pressurised atmospheres and diving work.

**Harmful exposure to biological or chemical agents**

- Very toxic, toxic, harmful, corrosive and irritant substances;
- Substances causing heritable genetic damage or harming the unborn child;
- Carcinogenic substances;
- Asbestos (including asbestos-containing materials);
- Lead and lead compounds;
- Biological risks, e.g. legionella, leptospirosis (Weil's disease), zoonoses.

**Work equipment**

Where there is an increased risk of injury due to the complexity of precautions required or the level of skill required for safe operation e.g:

- Woodworking machines (particularly saws, surface planing and vertical spindle moulding machines);
- Food slicers and other food processing machinery;
- Certain types of portable tools such as chainsaws;
- Power presses;
- Vehicles such as fork lift trucks, mobile cranes, construction vehicles;
- Other cranes and lifting hoists;
- Firearms.

Some young people may not have the physical size or strength necessary to operate equipment which has been designed for adults.

**Dangerous materials or activities**

- Work with explosives, including fireworks;
- Work with fierce or poisonous animals, e.g. on farms, in zoos or veterinary work;
- Certain types of electrical work, e.g. exposure to high voltage or live electrical equipment;
- Handling of flammable liquids or gases, e.g. petrol, acetylene, butane, propane;
- Work with pressurised gases;
- Work in large slaughterhouses (risks from animal handling, use of stunning equipment);
- Holding large quantities of cash or valuables.

**Dangerous workplaces or workstations**

- Work at heights, e.g. on high ladders or other unprotected forms of access
- Work in confined spaces, particularly where the risks specified in the *Confined Spaces Regulations 1997 (SI 1997 No 1713)* are present;
- Work where there is a risk of structural collapse, e.g. in construction or demolition activities or inside old buildings.

Further guidance on risks to young people and appropriate precautions is contained in HSE booklet HSG 165 'Young people at work. A guide for employers.'

## New or expectant mothers

**V12017** Amendments made in 1994 to the previous *Management Regulations* (*SI 1992 No 2051* now revoked) implemented the European Directive on Pregnant Workers (92/85/EEC), requiring employers in their risk assessments to consider risks to new or expectant mothers. These amendments were subsequently incorporated into the *Management Regulations 1999* (*SI 1999 No 3242*). *Regulation 1* of the *Management Regulations* contains two relevant definitions:

'New or expectant mother' means an employee who is pregnant; who has given birth within the previous six months; or who is breastfeeding.

'Given birth' means 'delivered a living child or, after twenty-four weeks of pregnancy, a stillborn child'.

### Requirements of the Management Regulations

**V12018** The requirements for 'Risk assessment in respect of new and expectant mothers' are contained in *Regulation 16* of the *Management Regulations 1999* (*SI 1999 No 3242*) which states in *paragraph (1)*:

'Where—

(a) the persons working in an undertaking include women of child-bearing age; and

(b) the work is of a kind which could involve risk, by reason of her condition, to the health and safety of a new or expectant mothers, or to that of her baby, from any processes or working conditions, or physical, biological or chemical agents, including those specified in Annexes I and II of Council Directive 92/85/EEC on the introduction of measures to encourage improvements in the safety and health at work of pregnant workers and workers who have recently given birth or are breastfeeding,

the assessment required by Regulation 3(1) shall also include an assessment of such risk.'

*Paragraph (4)* of the Regulation states that in relation to risks from infectious or contagious diseases an assessment must only be made if the level of risk is in addition to the level of exposure outside the workplace.

The types of risk which are more likely to affect new or expectant mothers are described later in the chapter. *Paragraphs (2)* and *(3)* of *Regulation 16* set out the actions employers are required to take if these risks cannot be avoided. *Paragraph (2)* states :

'Where, in the case of an individual employee, the taking of any other action the employer is required to take under the relevant statutory provisions would not avoid the risk referred to in paragraph (1) the employer shall, if it is reasonable to do so, and would avoid such risks, alter her working conditions or hours of work.'

Consequently where the risk assessment required under *Regulation 16(1)* shows that control measures would not sufficiently avoid the risks to new or expectant mothers or their babies, the employer must make reasonable alterations to their working conditions or hours of work.

In some cases restrictions may still allow the employee to substantially continue with her normal work but in others it may be more appropriate to offer her suitable alternative work.

Any alternative work must be:

- suitable and appropriate for the employee to do in the circumstances;

- on terms and conditions which are no less favourable.

*Paragraph (3) of Regulation 16* states:

'If it is not reasonable to alter the working conditions or hours of work, or if it would not avoid such risk, the employer shall, subject to section 67 of the 1996 Act, suspend the employee from work for so long as is necessary to avoid such risk.'

The 1996 Act referred to here is the *Employment Rights Act 1996* which provides that any such suspension from work on the above grounds is on full pay. However, payment might not be made if the employee has unreasonably refused an offer of suitable alternative work. Employment continues during such a suspension, counting as continuous employment in respect of seniority, pension rights etc. Contractual benefits other than pay do not necessarily continue during the suspension. These are a matter for negotiation and agreement between the employer and employee, although employers should not act unlawfully under the *Equal Pay Act 1970* and the *Sex Discrimination Act 1975*. Enforcement of employment rights is through employment tribunals.

*Regulation 17* of the *Management Regulations 1999 (SI 1999 No 3242)* deals specifically with night work by new or expectant mothers and states:

'Where—

(a) a new or expectant mother works at night; and

(b) a certificate from a registered medical practitioner or a registered midwife shows that it is necessary for her health or safety that she should not be at work for any period of such work identified in the certificate,

the employer shall, subject to section 46 of the 1978 Act, suspend her from work for so long as is necessary for her health or safety.'

Such suspension (on the same basis as described above) is only necessary if there are risks arising from work. The HSE does not consider there are any risks to pregnant or breastfeeding workers or their children from working at night per se. It suggests that any claim from an employee that she cannot work nights should be referred to an occupational health specialist. The HSE's own Employment Medical Advisory Service are likely to have a role to play in such cases.

The requirements placed on employers in respect of altered working conditions or hours of work and suspensions from work only take effect when the employee has formally notified the employer of her condition. *Regulation 18* of the *Management Regulations 1999 (SI 1999 No 3242)* states:

'(1) Nothing in paragraph (2) or (3) of regulation 16 shall require the employer to take any action in relation to an employee until she has notified the employer in writing that she is pregnant, has given birth within the previous six months, or is breastfeeding.

(2) Nothing in paragraph (2) or (3) of regulation 16 or in regulation 17 shall require the employer to maintain action taken in relation to an employee—

   (a) in a case—

      (i) to which regulation 16(2) or (3) relates; and

      (ii) where the employee has notified her employer that she is pregnant, where she has failed, within a reasonable time of being requested to do so in writing by her employer, to produce for the employer's inspection a certificate from a registered medical practitioner or a registered midwife showing that she is pregnant;

   (b) once the employer knows that she is no longer a new or expectant mother; or

   (c) if the employer cannot establish whether she remains a new or expectant mother.'

## Risks to new or expectant mothers

**V12019** The HSE booklet HSG 122 'New and expectant mothers at work: A guide for employers' provides considerable guidance on those risks which may be of particular relevance to new or expectant mothers, including those listed in the EC Directive on Pregnant Workers (92/85/EEC). This guidance is both summarised and augmented below.

## Physical agents
**V12020**

*Manual handling*

Pregnant women are particularly susceptible to risk from manual handling activities as also are those who have recently given birth, especially after a caesarean section. Manual handling assessments (as required by the *Manual Handling Operations Regulations 1992* (*SI 1992 No 2793*) are dealt with in the chapter MANUAL HANDLING in this publication.

*Noise*

The HSE does not consider that there are any specific risks from noise for new or expectant mothers. Compliance with the requirements of the *Noise at Work Regulations 1989* (*SI 1989 No 1790*) should provide them with sufficient protection.

*Ionising radiation*

The foetus may be harmed by exposure to ionising radiation, including that from radioactive materials inhaled or ingested by the mother. The *Ionising Radiations Regulations 1999* (*SI 1999 No 3232*) set an external radiation dose limit for the abdomen of any woman of reproductive capacity and also contain a specific requirement to provide information to female employees who may become pregnant or start breast-feeding. Systems of work should be such as to keep exposure of pregnant women to radiation from all sources as low as reasonably practicable. Contamination of a nursing mother's skin with radioactive substances can create risks for the child and special precautions may be necessary to avoid such a possibility.

Several HSE publications provide detailed guidance on work involving ionising radiation. (See L121 'Work with ionising radiation: Ionising Radiations Regulations 1999: Approved Code of Practice and Guidance' (2000); and INDG334 'Working safely with radiation: Guidelines for expectant and breast-feeding mothers' (2001), in particular.)

*Other electromagnetic radiation*

The HSE does not consider that new or expectant mothers are at any greater risk from other types of radiation, with the possible exception of over exposure to radio-frequency radiation which could raise the body temperature to harmful levels. Compliance with the exposure standards for electric and magnetic fields published by the National Radiological Protection Board should provide adequate protection.

*Work in compressed air*

Although pregnant women may not be at greater risk of developing the 'bends', potentially the foetus could be seriously harmed by gas bubbles in the circulation should this condition arise. There is also evidence that women who have recently given birth have an increased risk of the bends.

The *Work in Compressed Air Regulations 1996* (*SI 1996 No 1656*), Reg 16(2) states:

' ... the compressed air contractor shall ensure that no person works in compressed air where the compressed air contractor has reason to believe that person to be subject to any medical of physical condition which is likely to render that person unfit or unsuitable for such work.'

This would appear to prohibit such work by pregnant women or those who have recently given birth. In its booklet HS(G)122 the HSE states that there is no physiological reason why a breastfeeding mother should not work in compressed air although they point out that practical difficulties would exist. HSE booklet L96 'A guide to the Work in Compressed Air Regulations 1996' (1996) provides detailed guidance on the requirements of the *Work in Compressed Air Regulations*.

*Diving work*

The HSE draws attention to the possible effects of pressure on the foetus during underwater diving by pregnant women and states that they should not dive at all.

Under the *Diving at Work Regulations 1997 (SI 1997 No 2776)*, *Reg 15* divers must have a certificate of medical fitness to dive and the HSE guidance to doctors issuing such certificates advises that pregnant workers should not dive.

*Regulation 13(1)* of the *Diving at Work Regulations 1997 (SI 1997 No 2776)* states that 'No person shall dive in a diving project ... if he knows of anything (including any illness or medical condition) which makes him unfit to dive.'

A series of HSE booklets provide general guidance on different types of diving projects.

*Shock, vibration etc.*

Major physical shocks or regular exposure to lesser shocks or low frequency vibration may increase the risk of a miscarriage. Activities involving such risks (e.g. the use of vehicles off road) should be avoided by pregnant women.

*Movement and posture*

Fatigue from standing and other physical work has been associated with miscarriage, premature birth and low birth weight. Ergonomic considerations will increase as the pregnancy advances and these could affect display screen equipment workstations, work in restricted spaces (e.g. for some maintenance or cleaning activities) or work on ladders or platforms. Underground mining work is likely to involve movement and posture problems and will also be subject to some of the other 'physical agents' described in this section. Driving for extended periods or travel by air may also present postural problems.

Pregnant women should be allowed to pace their work appropriately, taking longer and more frequent breaks. They may need to be restricted from carrying out certain tasks. Seating may need to be provided for work that is normally done standing and adjustments to display screen equipment and other workstations may be necessary.

*Physical and mental pressure*

Excessive physical or mental pressure could cause stress and lead to anxiety and raised blood pressure. Workplace stress is a complex issue which has recently been receiving increased attention. Fatigue issues are referred to above but other possible causes of stress may also need to be considered. These might be associated with the workload of individual pregnant employees (e.g. for those in management or administrative roles), the pressure of decision-making (e.g. in the health care or financial sectors) or the trauma of potential work situations (e.g. serious accidents dealt with by the emergency services).

*Extreme cold or heat*

Pregnant women are less tolerant of heat and may be more prone to fainting or heat stress. Although the risk is likely to reduce after birth, dehydration may impair breastfeeding. Exposure to prolonged heat at work, e.g. at furnaces or ovens, should be avoided.

Maintenance or cleaning work in hot situations should also be avoided, particularly if this involves use of less secure forms of access such as ladders, where fainting could result in a serious fall.

The HSE does not consider that there are any specific problems from working in extreme cold although obviously appropriate precautions should be taken as for other workers e.g. the provision of warm clothing.

## Biological agents

**V12021** Many biological agents in hazard groups 2, 3 and 4 (as categorised by the Advisory Committee on Dangerous Pathogens) can affect the unborn child should the mother be infected during pregnancy. Some agents can cause abortion of the foetus and others can cause physical or neurological damage. Infections may also be passed on to the child during or after birth, e.g. while breastfeeding.

Agents presenting risks to children include hepatitis B, HIV, herpes, TB, syphilis, chickenpox, typhoid, rubella (german measles), cytomegalovirus and chlamydia in sheep. Most women will be at no more risk from these agents at work than living within the community but the risks are likely to be higher in some work sectors, e.g. laboratories, health care, the emergency services and those working with animals or animal products.

Details of appropriate control measures for biological agents are contained in *Schedule 3* and *Appendix 2* of the ACoP booklet to the *COSHH Regulations 2002 (SI 2002 No 2677)*. There is a separate HSE publication on 'Infections in the workplace to new and expectant mothers' (1997). When carrying out a risk assessment in respect of new or expectant mothers, normal containment or hygiene measures may be considered sufficient, but there may be the need for special precautions such as use of vaccines. Where there is a high risk of exposure to a highly infectious biological agent it may be necessary to remove the worker entirely from the high risk environment.

## Chemical agents

### V12022

*Substances labelled with certain risk phrases*

The *Chemicals (Hazard Information and Packaging for Supply) Regulations 2002 (SI 2002 No 1689)* ('*CHIP*') require many types of hazardous substances to be labelled with specified risk phrases. Several of those are of relevance to new or expectant mothers:

R40: possible risk of irreversible effects;

R45: may cause cancer;

R46: may cause heritable genetic damage;

R49: may cause cancer by inhalation;

R61: may cause harm to the unborn child;

R63: possible risk of harm to the unborn child;

R64: may cause harm to breastfed babies;

R68: possible risk of irreversible effects.

The *CHIP Regulations 2002* are regularly subject to changes and employers should be alert for other risk phrases which indicate risks in respect of new or expectant mothers.

Control of such substances at work is already required by the *COSHH Regulations 2002 (SI 2002 No 2677)* or the separate regulations governing lead and asbestos. Risk assessments in relation to pregnant women or those who have recently given birth may indicate that normal control measures are adequate to protect them also. However, additional precautions may be necessary, e.g. improved hygiene procedures, additional PPE or even restriction from work involving certain substances.

The ACoPs relating to the *COSHH Regulations 2002* (*SI 2002 No 2677*) (contained in HSE booklet L5) together with HSG 193 'COSHH Essentials: Easy steps to control chemicals' provide further details of the types of precautions which may be appropriate.

*Mercury and mercury derivatives*

The HSE states that exposure to organic mercury compounds can slow the growth of the unborn baby, disrupt the nervous system and cause the mother to be poisoned. It considers that there is no clear evidence of adverse effects on the foetus from mercury itself and inorganic mercury compounds. Mercury and its derivatives are subject to the *COSHH Regulations 2002* (*SI 2002 No 2677*) and normal control measures (see above) may be adequate to protect new or expectant mothers.

(See HSE publications L5 and HSG193 as mentioned above; EH17 'Mercury and its inorganic divalent compounds' (1996); and MS12 'Mercury: Medical guidance notes' (1996).)

*Antimitotic (cytotoxic) drugs*

These drugs (which may be inhaled or absorbed through the skin) can cause genetic damage to sperm and eggs and some can cause cancer. Those workers involved in preparation or administration of such drugs and disposal of chemical or human waste are at greatest risk, e.g. pharmacists, nurses and other health care workers. Antimitotic drugs can present a significant risk to those of either gender who are trying to conceive a child as well as to new or expectant mothers, and all those working with them should be made aware of the hazards.

The *COSHH Regulations 2002* (*SI 2002 No 2677*) again apply to the control of these substances (see above). Since there is no known threshold limit for them, exposure must be reduced to as low a level as is reasonably practicable.

*Agents absorbed through the skin*

Various chemicals including some pesticides may be absorbed through the skin causing adverse effects. These substances are identified in the tables of occupational exposure limits in HSE booklet EH40 'Occupational exposure limits' (an updated version is published annually), in which some substances are accompanied by an annotation 'Sk'. Many such agents (particularly pesticides which are subject to the *Control of Pesticides Regulations 1986* (*SI 1986 No 1510*)) will also be identified by product labels.

Effective control of these chemicals is obviously important in respect of all employees (the *COSHH Regulations 2002* (*SI 2002 No 2677*) applying once again) although risk assessments may reveal the need for additional precautions in respect of new or expectant mothers, e.g. modified handling methods, additional PPE, or even their restriction from activities involving exposure to such substances.

HSE booklet L9 'Safe use of pesticides for non-agricultural purposes' contains a further ACoP in relation to COSHH obligations.

*Carbon monoxide*

Exposure of pregnant women to carbon monoxide can cause the foetus to be starved of oxygen with the level and duration of exposure both being important factors. There are not felt to be any additional risks from carbon monoxide to mothers who have recently given birth or to breast-fed babies.

Once again it is important to protect all members of the workforce from high levels of carbon monoxide (the *COSHH Regulations 2002* (*SI 2002 No 2677*) require it) but it may also be necessary to ensure that pregnant women are not regularly exposed to carbon monoxide at lower levels, e.g. from use of gas-fired equipment or other processes or activities. HSE Guidance Note EH43 provides general guidance on carbon monoxide.

*Lead and lead derivatives*

High occupational exposure to lead has historically been linked with high incidence of spontaneous abortion, stillbirth and infertility. Decreases in the intellectual performance of children have more recently been attributed to exposure of their mothers to lead. Since lead can enter breast milk there are potential risks to the child if breastfeeding mothers are exposed to lead.

All women of reproductive capacity are prohibited from working in many lead processing activities. All work involving exposure to lead is subject to the *Control of Lead at Work Regulations 2002* (*SI 2002 No 2676*). Even where women of reproductive capacity are allowed to work with lead or its compounds the blood-lead concentrations contained within the Regulations as an 'action level' and a 'suspension level' are set at half the figures for adult males. This is intended to ensure that women who may become pregnant already have low blood lead levels.

Once pregnancy is confirmed, the doctor carrying out medical surveillance (as required by *Regulation 10* of *SI 2002 No 2676*) would normally be expected to suspend the woman from work involving significant exposure to lead. There is, however, no specific requirement in the Regulations themselves that this must happen. Detailed guidance on the Regulations is contained in HSE booklet L132 'Control of lead at work' (2002).

## Working conditions

**V12023** HSE guidance in HS(G)122 in respect of the 'working conditions' referred to in the *Management Regulations* (*SI 1999 No 3242*), *Reg 16(1)* repeats the HSE's stated position on work with DSE contained in Appendix 2 of the HSE booklet L26 'Work with display screen equipment'. Its view is that radiation from DSE is well below the levels set out in international recommendations and that scientific studies taken as a whole do not demonstrate any link between this work and miscarriages or birth defects.

There is no need for pregnant women to cease working with DSE. However, the HSE recommends that, to avoid problems from stress or anxiety, women are given the opportunity to discuss any concerns with someone who is well informed on the subject.

## Other aspects of pregnancy

**V12024** Although the HSE booklet on 'New and expectant mothers at work' (HSG 122) draws attention to other features of pregnancy which employers may wish to take into account, it suggests that employers have no legal obligation to do so. However, this view has not been tested by the courts and an argument could be advanced that some of these represented 'working conditions' which involve risk to the mother or her baby within the terms of the *Management Regulations* (*SI 1999 No 3242*), *Reg 16(1)(b)*.

The impact of these aspects will vary during the course of the pregnancy and employers are likely to need to keep the situation under review. A modified version of the appendix from the HSE booklet dealing with these 'other aspects' is provided in TABLE C below.

**Aspects of pregnancy that may affect work**

| Table C | |
|---|---|
| *Aspects of pregnancy* | *Factors in work* |
| Morning sickness | Early shift work<br>Exposure to nauseating smells<br>Availability for early meetings<br>Difficulty in leaving job |

| | |
|---|---|
| Frequent visits to toilet | Difficulty in leaving job/site of work |
| Breastfeeding | Access to private area<br>Use of secure, clean refrigerators to store milk |
| Tiredness | Overtime/long working hours<br>Evening work<br>More frequent breaks<br>Private area to sit or lie down |
| Increasing size | Difficulty in using protective clothing<br>Work in confined areas<br>Manual handling difficulties<br>Posture at DSE workstations<br>(Dexterity, agility, co-ordination, reach and speed of movement may also be impaired) |
| Comfort | Problems working in confined or congested workspaces |
| Backache | Standing for extended periods<br>Posture for some activities<br>Manual handling |
| Travel<br>i.e. fatigue, stress, static posture | Reduce need for lengthy journeys<br>Change travel methods |
| Varicose veins | Standing/sitting for long periods |
| Haemorrhoids | Working in hot conditions/sitting |
| Vulnerability to passive smoking | Effective smoking policy (priority to the needs of non-smokers) |
| Vulnerability to stress | Modified duties<br>Adjustments to working conditions or hours |
| Vulnerability to falls | Restrictions on work at heights or on potentially slippery surfaces |
| Vulnerability to violence<br>(e.g. from contact with the public) | Changes to workplace layout or staffing<br>Modified duties |
| Lone working | Effective communication and supervision<br>Possible need for improved emergency arrangements |

## *A practical approach to assessing risks*

**V12025** Essentially risk assessment in respect of new or expectant mothers must be carried out by an employer if:

- there are women of childbearing age; and
- their work could involve risks to new or expectant mothers or their babies.

Most organisations employ women of childbearing age and, particularly in the case of large businesses, there may be a number of work activities that could create relevant risks. In some cases (e.g. those involving exposure to hazardous substances or radiation) it may be appropriate to stipulate that women are not allowed to work in certain activities, processes or departments once their pregnancy has been notified and/or for a finite period after they return to work after giving birth.

However, in many workplaces the issues will be far from clear cut – it would be neither necessary nor practical to prohibit pregnant workers from carrying out *any* manual handling or from standing up, sitting down or climbing up ladders. Also it is difficult for any employer to identify in advance exactly what steps need to be taken to alter working conditions or hours of work so that risks can be avoided for every female employee who may become pregnant.

A much more practical approach is for the employer to develop a checklist similar to the sample provided at the end of this section. Such a checklist can be prepared by carrying out a review of the organisation's activities in order to identify risks which are present which *may* be of relevance in relation to new or expectant mothers. The risks described in the previous part of this chapter provide a good starting point from which a workplace-specific checklist can be developed. In some workplaces it may be appropriate to develop more than one such checklist (e.g. for production, maintenance and administration) because the profiles of risks are different.

Such a checklist should be completed by a suitable employer's representative (a personnel or health and safety specialist or the pregnant woman's manager perhaps) together with the pregnant woman herself. The checklist is intended to provoke discussion about the employee's possible exposure to the risks that the employer has identified so that any necessary additional precautions (or changes to work practices) can be agreed. There is the opportunity to identify whether any further review of the situation is necessary. The process would be repeated once the mother returned to work after giving birth.

The sample checklist provided below is for a bakery. Female staff are likely to be involved in production work where there is significant potential for risk to new or expectant mothers, and they may also be employed in laboratory or maintenance work. Risks for staff working in administrative activities must also be considered, e.g. those associated with DSE workstations. However, the greatest problems could be for female staff delivering to retail outlets. As well as risks from manual handling, falls and possibly excessive driving, the difficulties for such staff in accessing suitable rest and toilet facilities must also be taken into account.

# Vulnerable Persons

| BUNN THE BAKERS | NEW OR EXPECTANT MOTHERS | | RISK ASSESSMENT CHECKLIST | |
|---|---|---|---|---|
| Name of Employee: | | | Work location: | |
| | Possible risks to consider | | Precautions/changes agreed | |
| Bakery and delivery work | Manual handling (production, maintenance, deliveries) <br> Excessive standing (production lines) <br> Awkward sitting (production lines) <br> Difficult access (cleaning, maintenance) <br> Excessive heat (near to ovens) <br> Hazardous substances (e.g. cleaning materials, maintenance, laboratory) <br> Significant risk of falls (e.g. slippy floors, maintenance work, trips on delivery work) <br> Excessive driving (deliveries) <br> Other | | | |
| Administrative activities | DSE workstation layout <br> Unsuitable meeting times <br> Manual handling (e.g. stationery, records) <br> Significant travel (especially by car or air) <br> Other | | | |
| All activities | Ability to leave workplace temporarily <br> Availability of rest area/toilets <br> Major pressure or stress situations <br> Possible effects of fatigue <br> Other | | | |
| Signature (for Bunn the Bakers): | | | Signature (new/expectant mother): | |
| Date: | | | Date for further review (if any): | |
| This form should be completed when the pregnancy is first notified *and* when the new mother returns to work. | | | | |

## Risks to non-employees

**V12026** As part of their general risk assessments, employers must also consider risks to new or expectant mothers who are not part of their own workforce. Such persons may be employed by another organisation or be self-employed but they may also be service users or members of the public. Even though other employers have duties to assess risks to their staff, 'host' employers still have a statutory duty (and a 'duty of care') to other people's employees, particularly in respect of risks of which they may be aware, but which might not be immediately apparent to others. In this context, risks to pregnant women especially could include:

- slips, trips and falls, e.g. due to slippy or uneven surfaces;
- accidental impact from persons, e.g. running children or accidental contacts related to sports activities;
- accidental collision with mobile equipment;
- use of off-road transport;
- assault, e.g. on police or prison premises;
- ionising radiation;
- hazardous substances (see the earlier reference to risk phrases at V12022 above).

# Lone workers

**V12027** There are no general legal restrictions on working alone – indeed many work activities would be extremely impractical if there were. However, the activities of lone workers must be subject to the same sort of risk assessment process as any other type of work. Even those regulations imposing the greatest restrictions on lone working (i.e. *Electricity at Work Regulations 1989* (*SI 1989 No 635*) and the *Confined Spaces Regulations 1997* (*SI 1997 No 1713*) still require the element of risk assessment to be applied in identifying suitable safe systems of work.

## The various types of lone workers

**V12028** There are many different types of lone working. These can be divided into three categories:

*People working alone in fixed workplaces*

This category includes:

- workers in small shops and kiosks;
- people working in more remote parts of larger premises, e.g. isolated reception or enquiry desks, private interview rooms, security gatehouses;
- laboratories, stores;
- staff working late, at nights or at weekends;
- staff responding to alarm calls etc, e.g. 'keyholders';
- staff working at home.

*Staff working on a variety of premises belonging to other employers*

Such people will be working independently of their colleagues but may have others working around them for at least part of the time. They include:

- persons carrying out construction or plant installation work;
- maintenance and repair workers;

**V12029** *Vulnerable Persons*

- cleaners, painters, decorators etc;
- other types of contracting staff.

*Persons working within the wider community*

These staff may be working in urban, rural or remote locations, sometimes in places which are accessible to the public but also on private property (often in domestic premises). This includes:

- mobile security staff;
- domestic installation, maintenance and repair staff;
- delivery staff (including postal workers);
- sales and technical representatives;
- architects, property surveyors and estate agents;
- rent collectors and meter readers;
- social workers, home carers and district nurses;
- police, traffic wardens and probation staff;
- agricultural and forestry workers;
- utility workers (electricity, gas, water, sewage, telecommunications etc.);
- vehicle recovery staff;

and many other similar types of work.

## Risks for lone workers

**V12029** In some situations lone workers may be at risk because they are not physically capable of carrying out certain activities safely on their own. These may relate to:

- *Manual handling activities*

  – where team handling is identified as being necessary.

- *Use of certain types of equipment*

  – particularly heavy and bulky equipment, or where another person needs to be available to carry out adjustments.

- *Operation of controls etc.*

  – controls may be physically separate from each other or from related instrumentation requiring a second person to simultaneously activate controls or relay information (e.g. where a banksman may be required for lifting operations).

- *Achieving safe access*

  – a second person may be necessary to foot a ladder or ensure safe use of other forms of temporary access equipment.

Some work activities may involve such a high degree of risk that another person needs to be present to assist in the identification of risks or act as a deterrent to such risks. These situations include:

- *Some types of maintenance and repair work*

  – ensuring that equipment has been correctly isolated and other precautions taken (two heads are better than one!).

- *Live electrical work*

    – ensuring that the person carrying out the work is fully aware of what is live and what is not.

- *Work involving potential risks of aggression or even violence*

    – handling complaints, passing on unwelcome news, dealing with persons with aggressive or violent histories, working in high risk areas of the community, handling cash or valuables.

- *Activities where staff feel highly vulnerable*

    – one to one encounters, possibly with unknown persons, particularly in remote locations, empty premises or domestic property (female staff often feel more vulnerable but males should also be made aware of risks of unwarranted accusations against them)

There may also be risks to lone workers where a second person may be needed to take appropriate emergency actions such as:

- *Isolating electrical supplies*

    – particularly where live work is being carried out.

- *Rescuing a worker from inside a confined space*

    – e.g. by use of a rescue line or wearing suitable respiratory protective equipment ('RPE') and PPE.

- *Giving immediate first aid attention*

    – particularly important for electrical accidents or persons seriously affected by hazardous substances, or asphyxiated.

- *Safely recovering a worker who has fallen whilst wearing a safety harness*

The general potential for accidents and illness to lone workers from less foreseeable causes must also be considered, even though these may not justify the immediate presence of another person.

The above examples are of the more common types of risks that lone workers may encounter. Employers will no doubt be able to add to the list, particularly after consulting those who carry out lone work and their representatives.

## Controlling risks to lone workers

V12030     The previous section identified the types of risks that lone workers may be subject to. Set out below is a variety of measures which may be necessary to control risks. Whilst some measures will be relevant to most, if not all risk situations, others may only have limited application. It is in effect a menu from which to choose the appropriate precautions.

- *General precautions*

    Lone workers must be made aware of the types of risks that they may encounter and the precautions which must be taken to control those risks. This will require information, instruction and training, supported by effective ongoing supervision.

    Some types of lone working may justify the creation of formal written procedures or rules whilst in others the lone workers themselves may be expected to carry out a degree of dynamic risk assessment (see R3016 in RISK ASSESSMENT).

    As a result of the employer's risk assessment some activities may be identified as being ones which workers must never carry out alone. These must be communicated effectively to the lone worker, who must also be given guidance on where the necessary assistance can be obtained. This may be achieved by the temporary presence of work colleagues or assistance may be forthcoming from the customer's personnel or from others working in the vicinity.

- *Communication*

    Good communications are essential to ensuring the safety of lone workers. This is likely to involve:

    — provision of suitable communication equipment, e.g. mobile phones, portable radios, direct radio links, monitored CCTV surveillance;

    — availability of emergency alarms, e.g. fixed alarm buttons, portable attack alarms (some types of alarms detect an absence of movement by the alarm user or require a regular response from the user);

    — availability of information on lone workers' whereabouts to others, e.g. work colleagues, control centres, security staff, partners, friends etc. (this may involve the use of location boards, computerised or desk diaries, itinerary sheets or just word of mouth);

    — ongoing communication from the lone worker, e.g. about delays, problems, changes in plans;

    — ongoing checks on the lone worker's wellbeing;

    — pre-emptive contacts from lone workers encountering unexpected risks (explaining where they are, who they're with or what they're doing and requesting a return call in an agreed period of time);

    — checks to ensure that the lone worker has safely returned to base (or home).

- *Specific precautions*

    There are many specific precautions which can be taken to reduce the risks to lone workers. The nature of these will depend upon the location of the lone worker and the type of activity involved. Examples of these include:

    — physical separation of the lone worker from people they must deal with, e.g. by use of screens or wide desks;

    — ensuring that assistance is readily available to the lone worker when required, e.g. for certain manual handling operations or live electrical work;

    — placing lone workers where assistance from members of the public is at hand, e.g. arranging encounters, particularly with higher risk contacts, in public buildings or areas, including hotels, cafes etc.

- *Emergency arrangements*

    Here too, the nature of the arrangements will vary according to the location and activity of the lone worker. Arrangements are likely to include:

    — response to alarms – who is expected to respond and what actions should they take;

    — follow-up of lone workers who do not return when expected or do not respond to calls;

    — provision of basic first aid equipment for lone workers, so that they can carry out simple first aid for themselves or receive treatment from others coming to their aid.

## Monitoring of lone workers

**V12031** Those responsible for managing and supervising lone workers must ensure that their activities are monitored effectively. Since only a limited amount of observation of their work in the field is likely to be practicable, regular consultation with the staff involved will be important. Aspects to monitor for are:

- new risks which have become apparent;
- existing risks which may have increased;
- the effectiveness of existing precautions;
- workers' views on new precautions which may be appropriate.

It will also be necessary to actively monitor the effectiveness of precautions through actions such as:

- checking the effectiveness of communication equipment;
- testing alarms;
- enquiring as to the current whereabouts of lone workers;
- testing the awareness of staff on how they should react to emergency situations.

# Disabled persons

## Types of disability

**V12032** Disabilities come in many different forms. HSE guidance to local authorities on the *Disability Discrimination Act 1995* refers to impairments affecting the ability to perform normal day to day activities affecting at least one of the following:

- mobility;
- manual dexterity;
- physical co-ordination;
- continence;
- ability to lift, carry or otherwise move everyday objects;
- speech, hearing or eyesight;
- memory or ability to concentrate, learn or understand; or
- perception of the risk of physical danger.

The last two on this list have a close relationship with the consideration of risks to young persons referred to earlier in the chapter.

Although not included in the list above, account must also be taken of those with relevant health conditions and allergies, particularly when assessing risks relating to hazardous substances.

## Disabled persons at risk

**V12033** Employers in their risk assessments must take account of all disabled persons who may be affected by their activities, including:

- permanent and temporary employees;
- persons on training or work experience placements (who have the status of employees);
- visitors;

- customers or service users;
- neighbours and passers-by.

## Access for disabled persons

**V12034** The *Health and Safety (Miscellaneous Amendments) Regulations 2002 (SI 2002 No 2174)* introduced an additional requirement into *Regulation 25* of the *Workplace (Health, Safety and Welfare) Regulations 1992 (SI 1992 No 3004)*:

'Where necessary, those parts of a workplace (including in particular doors, passageways, stairs, showers, washbasins, lavatories and workstations) used or occupied directly by disabled persons at work shall be organised to take account of such persons.'

Factors which must be taken into account in complying with the above requirement and in preventing accidents to disabled persons are:

- the locations where disabled persons work or must make access to;
- the width of access routes;
- maintaining access routes free from obstructions and slipping or tripping hazards;
- standards of guardrails and handrails;
- heights of door handles, light swtiches, alarms, controls etc.

(It is not the intention of this chapter to consider the general accessibility of work premises to disabled members of the public.)

## Other issues of relevance

**V12035** There are a number of other issues which must be considered when carrying out risk assessments in respect of disabled persons, whether members of the workforce or not, including:

- hazardous substances:
  - their potential effects on those with relevant health conditions or allergies;
  - the abilities of exposed persons to understand the risks or implement precautions.
- work equipment:
  - physical ability to handle the equipment;
  - intellectual capability to understand risks and implement necessary precautions.
- hot (or very cold) items or surfaces:
  - the ability to identify such items or surfaces;
  - a perception of the risks involved;
  - the ability to keep clear of such risks.
- workstation layout:
  - ergonomically suited to the needs of the disabled worker;
  - provision of suitable seating is of particular importance.
- emergency arrangements:
  - the ability to hear or see alarm signals;
  - the physical abilty to react to the alarm with the necessary speed;

— the potential for panic amongst those with some types of disability.

## Controlling risks to disabled persons

**V12036** As a result of the risk assessment process employers must control risks to the standards required by legislation – either the general standards required by the *HSWA 1974* or the more specific requirements of individual sets of regulations. In respect of employees they must also comply with the requirements of *Regulation 13(1)* of the *Management Regulations* (*SI 1999 No 3242*) which states:

> 'Every employer shall, in entrusting tasks to his employees, take into account their capabilities as regards health and safety.'

This will require employers to:

- Identify the disabilities of individual workers (and others) via:
    - application forms and questionnaires;
    - interviews;
    - medical examinations.
- Identify situations where persons with such disabilities may be at risk:
    - the types of issues referred to above.

In order to control risks to the standards required, it may be necessary for the employer to:

- modify existing equipment;
- provide alternative equipment;
- make modifications to the workplace;
- provide a work location suited to the individual's disability;
- provide additional or modified training (making special arrangements for those with learning difficulties);
- ensure emergency arrangements take account of disabled persons;
- restrict the activities and/or locations of disabled persons.

Employers have a duty under the *Disability Discrimination Act 1995* ('DDA') to make 'reasonable adjustment' to the workplace or working arrangements of disabled persons.

## Avoiding discrimination

**V12037** Employers must find a balance between avoiding exposing disabled persons to risk and discriminating against them unfairly. HSE guidance (Local Authority Circular 63/3) states that:

> 'The DDA does not prevent employers from continuing to stipulate essential health requirements for particular types of employment but they must be able to justify these and show that it would not be reasonable to waive them in any individual case.'

Employers may enquire about disability or ask a disabled person to undergo a medical examination but this must be justified by the relevance of the disability to the job in question.

There are concerns that employers may use health and safety reasons as an excuse for discriminating against disabled persons. Some of these concerns were found to be justified by research carried out on behalf of the HSE entitled 'Extent of use of health and safety requirements as a false excuse for not employing sick or disabled persons'. Although this showed that health and safety was frequently being used as the rationale for non-recruitment or dismissal, it found considerably more evidence of employers overcoming health and safety difficulties to retain workers with a disability, health condition or injury.

There are many practical actions which can be taken to accommodate disabled workers safely in the workplace, as described in the section above. Even where restrictions must be placed on the disabled, these can often relate to a very limited range of activities or locations, rather than preventing their employment altogether.

## Inexperienced workers

**V12038** Workers of all ages may (like young persons) be lacking in experience of some types of work or lack awareness of the risks associated with particular work activities. *Regulation 13* of the *Management Regulations* (*SI 1999 No 3242*) as well as requiring employers to take into account the capabilities of employees in entrusting tasks to them (see above), also contains several requirements about training.

*Paragraph (2)* of the *Regulation 13* states:

> 'Every employer shall ensure that his employees are provided with adequate health and safety training – on their being recruited into the employer's undertaking; and on their being exposed to new or increased risks ... '

Such new or increased risks are specified as:

- transfer or change of responsibilities;
- due to new or changed work equipment;
- because of new or changed systems of work.

### Training and supervision of inexperienced workers

**V12039** General health and safety induction training and job specific training is necessary in all workplaces but particular account of how to cope with these training needs must be taken in:

- work activities with rapid staff turnover;
- employment utilising large numbers of temporary workers, e.g. to meet seasonal needs or special orders (special arrangements will be necessary for workers without a good command of English);
- workplaces providing placements for those on government-funded training and re-training schemes;
- work activities involving workers with special educational needs.

Even when inexperienced workers have received appropriate training they will still require ongoing supervision, in many respects similar to that required for young persons. The imposition of restrictions as described earlier in this section (in relation to children and young persons) may also be appropriate.

# Work at Heights

## Introduction

**W9001** There are many circumstances in working life where people are substantially above ground or floor level (or near the edge of an opening or excavation) and at risk, however remote, of falling. Such risk also arises where access is necessary to roofs clad in material which is either inherently fragile, or has become weakened through age. Then there are situations where those working at a height may cause articles to fall on those below. Two-fifths of all reported major injuries are caused by falls from a height; as are half the fatal injuries to construction workers. The management of such risks is clearly a priority in many situations, particularly in construction, facilities management and plant maintenance.

People may fall, or cause things to fall, while going to or leaving an elevated workplace; or such an accident may happen while they are there and preoccupied with what they are doing. In the one case there will be a need to provide safe stairways or ladders etc., and in the other, the proper protection of edges from which persons or articles might fall, or where appropriate, the provision of nets or harnesses. But it must be realised that since the introduction of the *Management of Health and Safety at Work Regulations 1992*, (*SI 1992 No 2051*), and particularly since the amended *Management of Health and Safety at Work Regulations 1999* (*SI 1999 No 3242*), such customary approaches to the management of the risk are not sufficient – once, that is, a risk requiring a remedy has been identified. The strategy (principles of prevention and protection) set out in *SI 1999 No 3242, Reg 4* and *Sch 1* requires that the actual need to be at such heights, or to be there *so often*, must be questioned *before* the appropriate precautions are considered. This has been made explicit in the present context by *Regulation 6* of the *Work at Heights Regulations 2005* (*SI 2005 No 735*) (these cover working at heights across all industries including construction, ship repairing etc.).

This is not to say, of course, that where well recognised and *reliable* means of access and workplace protection have been provided, there is any pressing need to redesign the job to avoid work at heights – but when changes are contemplated (and change occurs all the time in construction, for instance) this approach must be properly considered. Indeed where, say, the existing precautions entail the use of portable ladders, there will be an immediate need to consider if this is appropriate.

The approach of Health and Safety Executive (HSE) inspectors when encountering portable ladders on sites and premises is not now just to check the angle at which the ladder is set and how well it is footed or secured etc., but to ask whether its use can be justified at all in the circumstances found (*Regulation 8(e)* and *Schedule 6* of the *Work at Heights Regulations 2005* (*SI 2005 No 735*). This questioning of the need to be at heights at all and the assumption that something so commonplace as an ordinary portable ladder may well be illegitimate could cause some to lose patience with the argument as an example of nanny state extremism. What needs to be considered, however, is that health and safety law is driven by national statistics, and what might appear locally as a remote possibility, when repeated daily across the country, may produce a very objectionable number of deaths at the end of the year. Furthermore, the particular strategic drive to avoid or reduce risk in the first place – rather than manage it later – rests on the sound perception that human failure is a fact of existence. Precautions that rely on the constant repetition of correct actions by people will simply not be as reliable as the consequences of reduced inherent risk. If a job can be re arranged so that it can be done from the ground nobody can fall; and a permanent ladder will not slip.

However, the view of the Health and Safety Commission (HSC) that the *Work at Height Regulations*, as originally prepared for submission to Parliament, should have applied to those at work giving leadership or training to recreational climbers and cavers has called into question the universal applicability of the strategy (see W9007 below). An exception from the substantive regulations was given for these activities.

This strategic approach to the management of risk extends to designers within the meaning of the *Construction (Design and Management) Regulations 1994 (SI 1994 No 3140)* as amended by *SI 2000 No 2380*, ('CDM Regulations'), Reg 13 (see C8029). This duty, to have *adequate regard* to the need to avoid or reduce risk applies not just to risks to those constructing a building, but to risks to those maintaining it and specifically (through inclusion of the defined term *cleaning work*) to the protection of window cleaners. Further, work at heights in and on the completed structure is one of the matters to be covered in the health and safety file required under these Regulations (see C8038 CONSTRUCTION AND BUILDING OPERATIONS). It is no longer the case that an architect or engineer can design a structure without careful thought as to how it is to be maintained and how the risks of that can be minimised. So increasingly in the case of new buildings, one can expect there to be somewhat less need to be above floor level for foreseeable maintenance purposes and that where that remains necessary, as it often will be, there should be properly thought-through provision for the work to be done safely. It should be noted that HSC proposals for revision of the CDM Regulations (see W9002 below) would extend this duty to designing for safety in use, which might have a considerable long-term impact on the risks of falling from height.

## Risk management of working at heights – employers' legal duties

W9002   Before describing and commenting on the legal duties concerning the management of the risks of working at heights, it will be helpful to consider how employers may incur those duties. The following cases are not exhaustive but cover the more common situations:

- as the employer of those working at height in that employer's premises;
- as an employer bringing in contractors to those premises while work continues there;
- as a contractor whether or not the client's workforce is present;
- as an employer who brings in a contractor to carry out construction work (as defined in the *Construction (Design and Management) Regulations 1994 (SI 1994 No 3140)*) for the purposes of those and other regulations) while normal work continues; and
- as an employer who engages a contractor to carry out construction work on empty premises or a green or brown-field site.

The first three of these cases are either clear or fairly straightforward. For instance the general duties of an employer to contractors carrying out work in the employer's premises are clear – specific requirements about organisation and the provision of information are in the *Management of Health and Safety at Work Regulations 1999 (SI 1999 No 3242)*. All these duties, insofar as they relate to work at heights, will be discussed in the following paragraphs of this chapter. In the case of contractors engaged to undertake construction work, as defined in *SI 1994 No 3140* as amended by *SI 2000 No 2380* (see C8031 CONSTRUCTION AND BUILDING OPERATIONS), however, the situation is more complicated and not without doubt. Since construction contributes so largely to the to the risk of falling, it is necessary to refer to a significant change in the official view of the extent of the duties of the construction client.

There is no need to repeat here all the considerations set out in C8028 CONSTRUCTION AND BUILDING OPERATIONS, which can readily be referred to, but it should be noted that the in the current edition of the Approved Code of Practice to *SI 1994 No 3140* the HSC expresses the view that construction clients should take a more active role in relation to the health and safety management of the construction work itself, than just following the duties of a client as set out in *SI 1994 No 3140*. This view is reiterated in the Consultative Document (*CD 200*) on the proposed revision of *SI 1994 No 3140* and remaining *Construction (Health Safety and Welfare) Regulations 1996 (SI 1996 No 1592)*.

HSC apparently believes that *SI 1999 No 3242* applies directly to the employer's acts as a client over and above any duties they might have in relation to the interaction of their own staff and those of the contractor. This would mean, for instance, that an employer having work done to the roof of an empty building, with none of their own workers even remotely present, would have to monitor the observance of the necessary precautions by the contractor. According to HSC, if the employer lacks the technical knowledge to do this effectively they must engage advice under *SI 1999 No 3242, Reg 7*; whereas, previously it was thought that the need for such advice would relate only to the specifics of his own operations. This might very well mark a very substantial increase in an employer's responsibilities in relation to working at heights. As stated in C8028 CONSTRUCTION AND BUILDING OPERATIONS, it is arguable that HSC are not correct in their opinion, but employers having construction work done must be aware of this official view and it would seem wise for occupiers of premises etc. to pay rather more attention in what follows to matters that ordinarily are the duty of contractors. The changes proposed in the health and safety law relating to construction would seem, if adopted, to put the matter beyond doubt.

If an occupier extended their interest to the point of controlling the way that construction work is done they could well acquire typical and specific contractor duties, including those relating to the prevention of falls. [*SI 1996 No 1592, Reg 4* and the *Work at Heights Regulations 2005 (SI 2005 No 735), Regulation 3 ((2)(b)*)]

Sometimes work at heights is undertaken in a crisis situation. For instance, a fault may have developed on an item of plant which needs immediate attention for production to continue, or the factory roof may be leaking resulting in damage to stored products below. In both cases, there may be insufficient time to properly consider the precautions necessary to ensure safe working at height. Procedures should be planned and in force to deal with this sort of contingency. The danger of being at a height when an emergency occurs is addressed in the *SI 2005 No 735*. It should be noted that the special position of the emergency services is recognised (in relation to the weather for instance).

## The Work at Heights Regulations 2005

W9003      The specific statutory requirements relevant to risk from working at heights are set out in the *Work at Heights Regulations 2005 (SI 2005 No 735)*, under the umbrella of the *Health and Safety at Work etc. Act 1974 (HSWA 1974)* and the *Management of Heath and Safety at Work Regulations 1999 (SI 1999 No 3242)*, particularly *SI 1999 No 3242, Regs 3–4*. The substantive regulations in *SI 2005 No 735, Regs 6–7* call up the risk assessment required under *SI 1999 No 3242, Reg 3* and articulate, for working at heights, the hierarchy of *SI 1999 No 3242, Reg 4* and *Sch 1*. The immediate consequence of this is that people cannot be required to work at heights from which they may be injured or killed until this assessment has taken place. Then there is a clear duty, according to the degree of risk revealed, to avoid the risk altogether or to reduce the extent or potentially dangerous nature of the work, wherever reasonable, before adopting the specific precautions referred to later in *SI 2005 No 735*. Paragraphs 30 and 31 of the Approved Code of Practice (ACoP) to *SI 1999 No 3242* explain this approach. As already seen, in practical terms this means asking at the outset 'do we have to do it at all?', 'do we only have to do so much of it?' or 'do we have to do it in that way?'.

Some specific requirements relating to windows and window cleaning are contained in the *Workplace (Health, Safety and Welfare) Regulations 1992 (SI 1992 No 3004* as amended by *SI 2002*

## W9003  Work at Heights

*No 2174), Regs 14–16* and the present Regulations. We have already referred to the designers' duty to take into consideration window cleaning along with the risk implications of their designs for construction and maintenance workers. [*Construction (Design and Management) Regulations 1994, (SI 1994 No 3140), Reg 4* as amended by *SI 2000 No 2380)*].

*SI 2005 No 735* are thus embedded in the framework of the strategic management of risk. Starting with a critical examination of what is existing, or proposed in the course of a business or undertaking, the aim is to ensure that everything is likely to be done with due regard to health and safety and that nothing entails difficulties of execution and management that might be avoided or at least reduced.

Thus the need to work at heights must be anticipated, and questioned to see that it is justified, either altogether of to the extent that seems desirable. What has to be done at height must be properly planned and the most suitable means of reaching and carrying out the work selected – strictly following the doctrine that collective means – such as platforms with guard rails etc. – are preferable to individual means such as safety harnesses, since the latter rely on constant attention to safe practice by the individual. Once the means of carrying out the work have been decided, the health and safety competence of those concerned can be established and trained for.

*SI 2005 No 735* applied specific requirements for work at heights across (nearly) all industries and work activities for the first time. However, it will be seen from the above that, practically speaking, there is nothing much that they require that ought not to have been done simply by following the general provisions of *HSWA 1974* and *SI 1999 No 3242*. However, *SI 2005 No 735* makes it easier to focus on the specific steps that have to be taken in respect of a widespread risk and they were required following he adoption of the second amendment to the Use of Work Equipment Directive 89/655/EEC in 2001 – the *Temporary Work at Heights Directive*.

The text of *SI 2005 No 735* may be downloaded from the HMSO website (Statutory Instrument 735 of 2005). Guidance has been produced: INDG 401 *The Work at Heights Regulations 2005 – A brief guide*, which is not much more than a convenient paraphrase of *SI 2005 No 735* under topic headings, and a *Question and answer brief for the construction industry on the Work at Heights Regulations 2005*. This is more helpful and its value is not confined to construction. Both may be downloaded from the HSE website or obtained free for single copies from HSE Books. A very helpful guide to the practicalities, *'Height Safe'* may be similarly obtained (see W9006 below).

The wording of the regulations themselves is perhaps rather more difficult to understand for non-lawyers than is usually the case and this is particularly true of the critical *SI 2005 No 735, Reg 6* referred to already above. It is tempting to blame Directive 89/655/EEC for this, but it would be unfair: its wording is in fact fairly straight forward. It refers simply to *temporary* work at heights, whereas the UK regulations cover any place where there is a risk of falling sufficiently far to cause injury – and so define *work at height*. This is not a meaning the words ordinarily carry. *Work at height* includes work at any place where there is an opening in the floor though which a person might fall and be injured – perhaps going down only a short distance. Likewise an *existing workplace* – introduced in *SI 2005 No 735, Reg 6* as the first preference – once the need to work at height has been established – is a workplace protected against falls, or a place where such a risk dies not exist at all. Again, this is not simply what the words mean in ordinary usage. This term seems to suggest a permanent work place, which one might expect to be governed more logically by the *Workplace (Health Safety and Welfare) Regulations 2002 (SI 2002 No 2174)*. On the contrary, former *SI 2002 No 2174, Regs 13(1)–(4)*, which dealt with the risk of falling, was repealed in favour of *SI 2005 NO 735*.

Thus, *SI 2005 No 735, Reg 6(4)* in its preference for doing work at height from an *existing workplace* and reaching it via existing means of access, leads back to standards formerly required under *SI 2002 No 2174* by referring to standards set out in *SI 2005 No 735, Sch 1*. This might be perplexing because one ordinarily thinks of work at height as work above the level of such a normal workplace! The wording of the Brief Guide at paragraphs 21 and 22 is not very helpful in this case, being *over* brief and in paragraph 21 couched in legalese. Fortunately the pyramid diagram on page 5 is very clear overview and the construction *Question and Answer Brief*, at questions 7–12 explains *SI 2005 No 735, Regs 6–7* in plain English, giving helpful examples for clarity. At question 9 an *existing place of work* is shown to be a legal concept meaning an existing safe place of work. Such a workplace ceases the moment it does not satisfy *SI 2005 No 735, Sch 1*. It may thus be transient in its legal sense while remaining much as before – at first sight.

One advantage of the notion of an *existing place of work* as adopted in the Regulations is that it is perfectly general, applying to plant and vehicles as well as buildings and fixed structures. Thus, if the catwalk at the top of a tanker does not satisfy *SI 2005 No 735, Sch 1*, other precautions in the hierarchy set out in *SI 2005 No 735* may be necessary (in the light of the risk assessment) though its mere existence might suggest that people would be expected to use it without further precautions (in fact the official guidance, *Avoiding falls from vehicles* (see W9006 below) is not very specific on this practical point).

## Summary of Regulations 6–7

**W9004**     In plain terms *SI 2005 No 735, Regs 6–7* say:

- risk assessment under *SI 1999 No 3242* must have been carried out;
- work at height is not justifiable if it is reasonably practicable to do it otherwise;
- where work requires people to be at a sufficient height above a usually safe place to cause injury should they fall, appropriate means must be provided for them to get to and work at that height;
- work at height cannot be avoided by requiring people to over-reach themselves or to work in awkward positions – this of course is particularly important when handling heavy articles (*appropriate ergonomic conditions*);
- in choosing equipment for access to and working at height, preference must be given to collective over individual means, bearing in mind the particular circumstances (platforms with toe boards and guard rails rather than rope positioning systems, for instance);
- thought must be given to what to do if an emergency occurs while work is going on at height; and
- where the precautions do not eliminate the risk, equipment is needed to limit the distance and/or consequences of a fall (fall arrest devices, nets air bags etc.).

*SI 2005 No 735, Reg 4* contains a separate duty for the work to be properly planned and supervised and an overriding requirement for it to be carried out in a manner which is, so far as is reasonably practicable, safe. It is made clear that the planning includes the actual selection of the equipment to be used and extends to emergencies that might occur. These will include emergencies affecting those at height, say how to get them down in a fire; and emergencies occurring in the course of the work at height, for instance what to do if an individually protected worker falls and is hanging in the safety harness. The HSE does not believe that reliance should be placed on the emergency services in such a case. See question 5 of INDG 401 *The Work at Heights Regulations 2005 – A brief guide*.

A factor to be considered in risk assessment and consequent planning is the weather. *SI 2005 No 735, Reg 4* prohibits work at height where the health or safety of those at work would be jeopardised. This is a considerably stronger specific requirement than *Construction (Health Safety and Welfare) Regulations 1996 (SI 1996 No 1592, Reg 24 (2)* (which remains in force in that industry).

With the work method and equipment established, the competencies necessary for the work to be done safely can be identified; but *SI 2005 No 735, Reg 5* does not just cover the competence of workers, but that of those responsible for the planning and organisation as well; question 6 gives advice on this subject.

**W9005** Work at Heights

The *Question and Answer Brief* explains the hierarchy of preventive measures and at pages 9 and 10 gives some illustrations to clarify what it means in practical terms. Risk is avoided by designing so that it is not necessary to run it, or so that the need to run it is reduced, In construction the erection of pre-assembled roof trusses is a long established example adopted before the health and safety duty existed under the *Construction (Design and Management) Regulations 1994 (SI 1994 No 3140)*, no doubt for reasons of economy and productivity.

Falls are prevented by properly protected access systems. The document explains that they may also be prevented by the use of personal protective equipment (work restraint system) that does not allow workers to approach, say, an edge from which they might fall, although this as an individual measure may be used only after the possibilities of collective means of protection have been exhausted. It appears to be a safer use of PPE than PPE as a means of limiting falls in the sense that it would not allow a fall at all. However, it is open to exactly the same objection: that it is as reliable as the discipline of the wearer.

The hierarchy of measures continues with examples of how the risk might be mitigated, for example by nets or air bags as close as possible below the place from which the worker might fall or, failing that, by fall arrest equipment (as required by *SI 2005 No 735, Reg 6(5)*). The place of PPE as a means of *prevention* (work restraint system) above that of collective means of *mitigation* seems to raise a doubt and arguably, might be thought a contradiction of the major principle that this excellent document exists to promote. Intuitively, one would surely be more comfortable with collective means (nets, air bags) where these could be provided just a short distance below any place from which a person might fall.

At the bottom of the hierarchy comes training and instruction seen as a means of mitigation. Clearly, for instance, where portable ladders may be legitimately used, training is the essential element in ensuring safe use.

The question arises as to what is *actually* the height at which work has to be avoided, if possible, or protected under the remainder of the hierarchy of measures. The answer, in the definitions contained in *SI 2005 No 735, Reg 2*, is *a distance liable to cause injury*. This marks a change from the standard of 2m formerly appearing in construction regulations in relation to the provision of guard-rails and toe boards. Proper risk assessment will reveal that one may be seriously injured by falling less than this distance and indeed, examination of HSE statistics has shown that a significant proportion of the serious injuries from falls from a height are caused in this way. Question 2 of the *Question and Answer Brief* gives examples of work at height. Question 13 addresses the issue of falls below 2m. An example given is the use of edge protection on bricklayers' bandstands and this advice obviously will apply to similar low working platforms in other industries.

Fears were expressed that the omission of the 2m standard would result in weaker protection for the worker. It will be seen that the HSC view was that a specific figure should not be mentioned in order to *increase* protection and this view has prevailed in the Regulation. The Regulation must not be read in isolation but, as ever, in conjunction with *SI 1999 No 3242, Reg 3*.

## Legal requirements

**W9005** *SI 2005 No 735, Reg 8* brings in the various Schedules to the Regulations setting standards for the different kinds of equipment that may be used in accordance with the hierarchy.

*SI 2005 No 735, Sch 2* sets out the standards for guard-rails, toe-boards etc. which will be familiar, but for slight differences in dimensions, from previous construction regulations. The height of the top guard-rail has to be 950mm, though rails fixed at the time the Regulations came into force might remain at 910mm. No gap between the intermediate rail and the upper rail or toe-board should exceed 470mm.

*SI 2005 No 735, Sch 3* likewise repeats familiar requirements for working platforms. Part two of the Schedule contains additional requirements for scaffolding, some of which are new to UK legislation relating to this equipment. Paragraph 7 in this part requires strength and stability calculations, unless previous relevant calculations are available, or the scaffolding has been erected according to a generally recognised standard configuration, which usually will be the case. Those in industries other than construction need to be aware that the standards for scaffolding are given in EN 12811 (BS 5973). A guide to this published by the National Access and Scaffolding Contractors (NASC) is accepted as authoritative by the HSE. Plans for assembling and dismantling the scaffold are also required.

*SI 2005 No 735, Sch 4* is concerned with collective means for arresting falls nets, air bags etc. Note that the devices may be used only if the work to be done does not compromise their effectiveness. This is of course after the possible use of such devices can be considered in the first place (*SI 2005 No 735, Sch 4, para 2(c)*). Note also the condition that a sufficient number of people must have received appropriate training, particularly in rescue procedures. The favouring of these devices above personal fall arrest systems is a reversal of the order of precautions formerly set out in the repealed part of *SI 1996 No 159*.

*SI 2005 No 735, Sch 5* covers personal, rope based, access or protection systems.

*SI 2005 No 735, Sch 5, Pt 1* applies to all personal fall protection systems. Note again that the conditions for use, of compatibility with the work; inability to use safer systems; and appropriate training of all concerned including of enough people to deal with the situation created by such devices' coming into play. There are separate requirements for each of the categories of equipment. Sketches and brief explanations are given in the *Question and Answer Brief* to help distinguish these.

*SI 2005 No 735, Sch 5, Pt 2* applies to work positioning systems. The system is required to have a backup system unless this is not reasonably practicable. Question 21 illustrates the principles of such a system and says that a bo'suns chair is an example.

*SI 2005 No 735, Sch 5, Pt 3* applies to rope access and positioning techniques, which are dealt with in Question 22. These methods are commonly known as abseiling and have particular benefits where examination and other short-term activities are necessary, particularly where cradles are impracticable. These are to be two line systems, the working line having a self-locking device (to prevent the operator losing control) and a mobile fall-protection system on the safety line. There is an exemption from the safety line where risk assessment shows more risk in rigging one than proceeding without one.

*SI 2005 No 735, Sch 5, Pt 4* deals with fall arrest systems and draws attention to the need to consider the zone through which the worker might fall and the effect of the likely swinging. Fall arrest systems have to be able to absorb the resultant shock. Question 23 gives advice and illustrates the preferred positioning of the anchor-above the head: good; at chest height: adequate; at foot level: poor.

*SI 2005 No 735, Sch 5, Pt 5* on work restraint systems is little more than a definition. Question 20 illustrates what is required.

*SI 2005 No 735, Sch 6* sets out the conditions on which ladders may be used. *SI 2005 No 735, Sch 6, para 1* requires that ladders may be used only if the risk assessment has demonstrated that the use of more suitable means is not justified because of the low risk and short duration of use, or because of site features which the employer cannot alter. The other conditions are of the kinds that have been required in construction for many years.

*SI 2005 No 735, Sch 6, para 5* merits some comment. A portable ladder is to be prevented from slipping by securing the stiles either at top or bottom, or by an effective anti-slip device or securing device, or any other arrangement that gives equivalent protection. It is perhaps worth mentioning that research carried out for the HSE has revealed that footing a ladder is not a very effective means of preventing it from slipping. Merely holding the stiles while standing on the ground is not effective at all. Standing on the bottom rung will prevent the ladder cocking over, but may do little to prevent actual slipping. The guidance produced for window cleaning will be helpful in understanding what is required (see W9006 below).

*SI 2005 No 735, Sch 6, para 10* is important in that it allows the use of stepladders where the worker has both hands engaged in carrying an article. Justification by risk assessment is necessary.

By August 2005, HSE should have published comprehensive guidance on the use of ladders.

*SI 2005 No 735, Reg 9* applies to fragile surfaces. Similar requirements previously existed for construction, but the main risk arises from going on roofs in existing buildings, which are either inherently lacking in the strength necessary to bear a person's weight, or have become so through age and weathering. Much of this approach to danger is not in the course of construction work as defied and this requirement has considerably increased the protection given to workers by specific legislation. Those familiar with the previous construction regulation in *SI 1996 No 1592* will note a significant change from 'materials' to 'surfaces' This is to make it clear that a roof sheet previously of adequate strength which has now been weakened by damage or age will be subject to this control.

*SI 2005 No 735, Regs 9–10* deal with managing the risk of falling objects by prevention altogether or by taking common sense precautions, or failing that, by keeping people away from areas where they might be struck. As the *Question and Answer Brief* makes clear, special vigilance is necessary when work is undertaken at heights adjacent to places where the public have access and the duties here will give shape to the general duty under *HSWA 1974, s 3*.

*SI 2005 No 735, Reg 12* requires initial inspection of equipment and further inspection at intervals. These requirements are common sense and largely familiar from the former relevant parts of *SI 1996 No 1592*. It should be noted that to preserve the specific standard of those regulations as they applied to working platforms, an interval of seven days is prescribed where a person might fall 2m or more. On the other hand, the exemption from making a written report where that distance was less then 2m no longer appears. *SI 2005 No 735, Sch 7* gives the particulars to be recorded. It should be noted that details are required of the action taken to remedy defects found. *SI 2005 No 735, Reg 13* contains a requirement particularly important in relation to the maintenance of existing buildings. The employer must check surfaces, parapets and permanent rails etc. rather than assuming that they will give protection.

These duties to inspect are reinforced by a specific duty on workers to report defects among their general duties to co-operate etc. in *SI 2005 No 735, Reg 15*.

**Practical guidance**

**W9006**  In contrast to the rather sparse official general guidance for those seeking to understand the *Work at Height Regulations 2005* (*SI 2005 No 735*), there is a good deal of more practical advice for those translating their legal duties, once understood, into action. Apart from the *Question and Answer Brief* already mentioned, all of this, at the time of writing, lacks reference to these Regulations specifically, and some of it may contain references to measures since repealed. But this scarcely matters in this context and indeed in some cases the text was prepared after the *Temporary Work at Heights Directive* (Directive 89/655/EEC) was adopted and the general shape of the Regulations was known.

The main practical guidance for the construction industry is a useful general resource, with its many references to the methods of managing the risk of falls. There are also a number of information sheets for the industry that may be downloaded from HSE's website.

The main construction guidance is:

(a)  Health and Safety in Construction, HSG 150;

(b)  Protecting the Public – your next move, HSG 151; and

(c)  Health and Safety in Roof Work, HSG 333.

An illustrated general guide is: *Height Safe – absolutely essential health and safety information for people who work at height.*

*Avoiding falls from vehicles,* INDG 395, is a shorter all-industry leaflet.

Information sheet MISC 614, *Preventing falls from boom-type mobile elevating work platforms,* ('cherry pickers') gives advice on these access devices generally. It is a facile assumption on the part of building designers and others that these provide a ready made access solution to maintenance and cleaning operations at heights.

The special problems of window cleaning are well covered in a series of information sheets, which may be downloaded from the HSE website at: http://www.hse.gov.uk/pubns/fallindex.htm.

MISC 611, *Safety in window cleaning using suspended and powered access equipment* is comprehensive. It guides the cleaning business through all the safety considerations relevant to using this equipment, The checks necessary to ensure that the building occupier has properly maintained the devices and assessed their safe use under *Health and Safety at Work etc. Act 1974, s 3* and the *Management of Heath and Safety at Work Regulations 1999 (SI 1999 No 3242), Reg 3* are set out together with the visual checks by the cleaning business itself. Access to the equipment is an important consideration.

MISC 612, *Safety in window cleaning using rope access techniques* includes training objectives for rope access managers supervisors and operatives. Rope access is being used increasingly by window cleaning firms and HSE are concerned that the complexities are sometimes not properly understood.

MISC 613 *Safety in window cleaning using portable ladders*, was produced in collaboration with the National Federation of Master Window and General Cleaners. It is in relation to houses and smaller business premises that the most difficult problems in window cleaning may arise. Little can be expected of domestic customers, even in the way of understanding the difficulties. And smaller businesses may be reluctant to co-operate, particularly when similar structures – with windows at no greater heights than in a house – are involved. The guidance points to common risks, for instance, the dormer window, which can be reached from the outside only from sloping tiles. Even where these may be reached from the inside, many occupiers are naturally reluctant to have window cleaners wandering, as they would see it, through their bedrooms. HSE puts forward the opinion that some windows should not be cleaned at all. Comprehensive advice is given on the safe use of portable ladders and this will be of benefit elsewhere – at least until HSE's general guidance on ladders appears.

Also available at the above web address is guidance for teachers and classroom assistants avoiding the risk of falls and no doubt other sectoral guidance will appear.

### Recreational climbing

**W9007** *Work at Height Regulations 2005 (SI 2005 No 735), Reg 3(4)(d)* exempts instruction or leadership in climbing or caving for sport, recreation, team building or similar activities. This is the result of a campaign by organisations concerned with outdoor or adventure activities. It is understood that those who were most concerned had subscribed to a declaration by the British Mountaineering Council (BMC) that:

> 'the BMC recognises that climbing and mountaineering are activities with a danger of personal injury or death. Participants in these activities should be aware of this and accept these risks and be responsible for their own actions and involvement'.

This does not sit well with the principles of prevention and protection as in *Management of Heath and Safety at Work Regulations 1999 (SI 1999 No 3242), Sch 1,* nor with these as now spelled out in the *Working at Height Regulations (SI 2005 No 735), Reg 6,* since risk is actually courted – if in a responsible fashion – in these activities. However, if the explicit duty to avoid work at height was objectionable, it was the extent of the measures that would have had to be observed failing this, that was the main trouble. These measures are clearly aimed at a low level of risk indeed and if adopted, would render the commercial provision of climbing instruction on actual mountainsides impracticable – even ridiculous – the plastering of Snowdonia with warning signs was feared! That is not to say that much might not be done in the controlled environment of a training centre, but at some stage instruction must extend to where the techniques being taught are to be practised. It seems, therefore, that a successful case was made for both workers in a particular activity (the instructors) and those protected via *HSWA 1974, s 3* (the would be mountaineers) to be exposed to what ordinarily would be unacceptable risk (the HSC maintain that Directive 89/655/EEC applies to such activities and that special regulations for them will be introduced). Furthermore, the argument has been advanced that there is a human right to at least somewhat risky recreation, together with, one deduces a practical corollary that commercial support is necessary, if this is not to be stultified.

In fact of course, the general statement of the strategy of risk avoidance in *SI 1999 No 3242, Reg 4* and *Sch 1* still applies. It is perhaps, nor altogether surprising, that this seemingly abstract requirement has nor drawn adverse comment, but the exemption granted suggests that the principle of the primary avoidance of risk might not have universal application – or rather be universally fair.

There is perhaps a further aspect here. It was hard not to be struck, while reading the original consultative document, with the thought that, when the Directive was being negotiated, climbers on mountain sides were not often in mind, if at all. The extension of orthodox (industrially derived) health and safety thinking to other areas has received hostile comment in relation to the HSE's role in the regulation of the railways, and this was endorsed by no less an authority than the House of Commons Select Committee on Transport. In that case the HSE's role is to be transferred to the Office of the Railway Regulator.

# Working Time

## Introduction

**W10001** The *Working Time Regulations 1998* (*SI 1998 No 1833*) ('the Regulations') came into force on 1 October 1998 and were the first piece of English law which set out specific rules governing working hours, rest breaks and holiday entitlement for the majority of workers (as opposed to those in specialised industries). The Regulations were introduced in order to implement the provisions of the Working Time Directive (93/104/EC) and the Young Workers Directive (94/33/EC). This chapter deals only with the Regulations as they govern adult workers, i.e. those aged 18 or over.

The Working Time Directive ('the Directive') was adopted on 23 November 1993 as a health and safety measure under Article 138 (formerly Article 118a) of the Treaty of Rome 1957, thereby requiring only a qualified majority, rather than a unanimous, vote. As such, the UK Government, which was opposed to the Directive, was unable to avoid its impact. The Government challenged the Directive on the basis that it was not truly a health and safety measure and should, in fact, have been introduced under Article 100 as a social measure which would have required a unanimous vote. The European Court of Justice (ECJ) has nevertheless upheld the status of the Directive as a health and safety measure.

The stated purpose of the Working Time Directive (acknowledged by the ECJ) is to lay down a minimum requirement for health and safety as regards the organisation of working time. This is, of course, significant to the extent that the English courts and tribunals will interpret the Regulations in accordance with the Directive by adopting the 'purposive' approach, which is now accepted under English law.

When they were first introduced, the Regulations (and accompanying guidance produced by the DTI) were met with vociferous criticism from the business world. Just a year later the Government sought to address these initial concerns. Following a limited consultation in July of 1999, amendments were tabled to two key areas of the Regulations (see W10013 and W10033 below). In addition in March 2000, the DTI issued a second (more concise) version of its guidance, which contains many changes from the original. It is still unclear how much weight tribunals will attach to this publication, which does not have the force of law.

Despite all the controversy regarding the Regulations and the implementation of the Directive it would appear that the Regulations have had little impact on the long-hours culture within the UK. This may of course change if the European Commission removes the 48-hour limit 'opt out' when it is reviewed towards the end of 2003 (see W10013).

The structure of the Regulations is to prescribe various limits and entitlements for workers and then to set out a sequence of exceptions and derogations from these provisions. The structure of this chapter will roughly follow that format.

## Definitions

**W10002** The *Working Time Regulations 1998* (*SI 1998 No 1833*), *Reg 2* sets out a number of basic concepts (some more familiar than others), which recur consistently and which underpin the legislation. These are as follows.

## Working time

**W10003**  Working time is defined, in relation to a worker, as:

(a) any period during which he is working, *at his employer's disposal* and carrying out his activity or duties;

(b) any period during which he is receiving relevant training; and

(c) any *additional* period which is to be treated as working time for the purposes of the Regulations under a relevant agreement (see W10006 below).

This definition initially raised numerous uncertainties, in particular, in relation to workers who are 'on call' or who operate under flexible working arrangements.

Despite this initial confusion, the ECJ decision in the case of *SIMAP v Conselleria de Sandidad y Consumo de la Generalitat Valencia (C303/98)* confirmed that time spent on call (in this case by doctors at health centres) is 'working time' provided the doctors are available and physically present. In addition, periods during which doctors are subject to the 'on-call' system (namely being available to come in to work while being outside the health centre is working time to the extent that the doctor spends it effectively exercising his professional responsibility). A similar approach may now be taken in other sectors.

The DTI has given examples of what it considers constitutes working time. In its revised guidance it states that 'time spent travelling outside normal working time' is not working time. This seems to address the difficult question of non-routine travel. For example, it has never been clear whether someone who is required to travel from London to Edinburgh on a Sunday, to be at a meeting in Edinburgh at 9 a.m. on Monday morning, can count that time as working time. The DTI's guidance would suggest that it would not be working time. In practice, however, it seems that it would depend on the facts.

It should be noted that the definition of working time can be clarified in a relevant agreement between the parties, but only by specifying any *additional* period which can be treated as working time. The relevant agreement cannot alter, and in particular cannot narrow, the absolute definition of working time.

## Worker

**W10004**  A worker is any individual who has entered into or works under (or where the employment has ceased), worked under:

(a) a contract of employment; or

(b) any other contract, whether express or implied and (if express) whether oral or in writing, whereby the individual undertakes to do or perform personally any work or services for another party to the contract whose status is not by virtue of the contract that of a client or customer of any profession or business undertaking carried on by the individual.

This definition is wider than just employees and covers any individuals who are carrying out work for an employer, unless they are genuinely self-employed, in that the work amounts to a business activity carried out on their own account (see also the reference to agency workers at W10005). When the Regulations were first introduced, the definition of 'worker' was considered to be a novel concept. It has now become a familiar one and appears in several other contexts such as the *Part Time Workers (Prevention of Less Favourable Treatment) Regulations 2000 (SI 2000 No 1551)* and the *National Minimum Wage Act 1998*.

In *Byrne Bros (Foamwork) Ltd v Baird & Others [2002] ICR 667* the Employment Appeal Tribunal considered whether carpenters and labourers who had been offered work on a building site were 'workers' for the purposes of the Regulations. One significant factor was that the tradesmen were entitled to appoint a substitute to carry out the work but the EAT concluded that a limited power to appoint substitutes was inconsistent with an obligation to provide work on a

## Agency workers

**W10005** The *Working Time Regulations 1998* (*SI 1998 No 1833*), *Reg 36* makes specific provision in relation to agency workers who do not otherwise fall into the general definition of workers. This provides that where (i) any individual is employed to work for a principal under an arrangement made between an agent and that principal, and (ii) the individual is not a worker because of the absence of a contract between the individual and the agent or principal, then the Regulations will apply as if that individual were a worker employed by whichever of the agent or principal is responsible for paying or actually pays the worker in respect of the work. Again, individuals who are genuinely self-employed are excluded from this definition.

## Collective, workforce and relevant agreements

**W10006** These play a significant role in the Regulations as employers and employees can, by entering into such agreements (where the Regulations so allow), effectively supplement or derogate from the strict application of the Regulations. Employers and employees who need a certain amount of flexibility in their working arrangements may well find one or other of these agreements will facilitate compliance with the Regulations.

### Collective agreement

**W10007** This is an agreement with an independent trade union within the meaning of the *Trade Union and Labour Relations* (*Consolidation*) *Act 1992, s 178*. The definition in *s 178* is wide enough to cover agreements reached between trade unions and employers under the statutory recognition process.

### Workforce agreement

**W10008** This is a new concept under English law and since its introduction by the Regulations has not been widely utilised. It is an agreement between an employer and the duly elected representatives of its employees or, in the case of small employers (i.e. those with 20 or fewer employees), potentially the employees themselves. In order for a workforce agreement to be valid, it must comply with the conditions set out in the *Working Time Regulations 1998* (*SI 1998 No 1833*), *Sch 1*, which provides that a workforce agreement must:

(*a*) be in writing;

(*b*) have effect for a specified period not exceeding five years;

(*c*) apply either to

　(i) all of the relevant members of the workforce, or

　(ii) all of the relevant members of the workforce who belong to a particular group;

(*d*) be signed by

　(i) the representatives of the workforce or of the particular group of workers, or

　(ii) where an employer employs 20 or fewer workers on the date on which the agreement is first made available for signature, either appropriate representatives or by a majority of the workers employed by him;

(e) before the agreement was made available for signature, the employer must have provided all the workers to whom it was intended to apply with copies of the text of the agreement and such guidance as they might reasonably require in order to understand it fully.

'Relevant members of the workforce' are defined as all of the workers employed by a particular employer (excluding any worker whose terms and conditions of employment are provided for wholly or in part in a collective agreement). Therefore, as soon as a collective agreement is in force in respect of any worker, the provisions of any workforce agreement in respect of that worker would cease to apply.

*Paragraph 3* of *Schedule 1* to the Regulations sets out the requirements relating to the election of workforce representatives. These are as follows:

(a) the number of representatives to be elected shall be determined by the employer;

(b) candidates for election as representatives for the workforce must be relevant members of the workforce, and the candidates for election as representatives of a particular group must be members of that group;

(c) no worker who is eligible to be a candidate can be unreasonably excluded from standing for election;

(d) all the relevant members of the workforce must be entitled to vote for representatives of the workforce and all the members of a particular group must be entitled to vote for representatives of that group;

(e) the workers must be entitled to vote for as many candidates as there are representatives to be elected;

(f) the election must be conducted so as to secure that (i) so far as is reasonably practicable those voting do so in secret, and (ii) the votes given at the election are fairly and accurately counted.

### Relevant agreement

**W10009** This is an 'umbrella provision' and means a workforce agreement which applies to a worker, any provision of a collective agreement which forms part of a contract between a worker and his employer, or any other agreement in writing which is legally enforceable as between the worker and his employer. A relevant agreement could, of course, include the written terms of a contract of employment. It would only include the provisions of staff handbooks, policies etc where it could be shown that these were 'legally enforceable' as between the parties.

## Maximum weekly working time

### 48-hour working week

**W10010** It is, of course, the 48-hour working week which initially caused so much controversy. The *Working Time Regulations 1998* (*SI 1998 No 1833*), *Reg 4* provides that an employer shall take all reasonable steps, in keeping with the need to protect the health and safety of workers, to ensure that a worker's *average* working time (including overtime) shall not exceed 48 hours for each 7-day period.

*Regulation 4(6)* provides a specific formula for calculating a worker's average working time over a reference period, which the Regulations have determined as 17 weeks (for exceptions, see W10012). This is as follows:

$$\frac{a+b}{c}$$

where

*a*   is the total number of hours worked during the reference period;

*b*   is the total number of hours worked during the period which

   (i)   begins immediately after the reference period, and

   (ii)  consists of the number of working days equivalent to the number of 'excluded days' during the reference period; and

*c*   is the number of weeks in the reference period.

'Excluded days' are days comprised of annual leave, sick leave or maternity leave and any period in respect of which an individual has opted out of the 48-hour limit (see W10013 below).

When the case of *Barber and Others v RJB Mining (UK) Ltd [1999] IRLR 308* was first reported it was initially thought that this decision would add considerably to the importance of this provision. The facts of this case were as follows. For the 17 weeks from 1 October 1998 (the date the Regulations came into force), the plaintiffs worked more than the average of 48 hours per week. Letters seeking the workers' agreement to opt out of the 48-hour limit were sent out on 7 December 1998. Each of the plaintiffs refused to sign the opt-out. On 25 January 1999, each plaintiff refused to carry out further work until his average working hours fell to within the specified limit. Each was required to continue working and did so 'under protest' and without prejudice to their rights in the proceedings.

In his judgement, Mr Justice Gage said:

> 'It seems to me clear that Parliament intended that all contracts of employment should be read so as to provide that an employee should work no more than an average of 48 hours in any week during the reference period. In my judgement, this is a mandatory requirement which must apply to all contracts of employment.'

He stated that although *Regulation 4(1)* does not prohibit an employer from requiring his employees to work longer hours, it does not preclude this interpretation as the obligation is in keeping with the stated objective of the Directive, of providing for the health and safety of employees.

The plaintiffs had clearly worked more than their 48-hour average. The judge continued:

> 'Having held that para(1) of Reg 4 provides free-standing legal rights and obligations under their contract of employment, it must follow that to require the plaintiffs to continue to work before sufficient time has elapsed to bring the weekly average below 48 hours is a breach of Reg 4(1).'

In theory, the ruling could add an extra string to the worker's bow in the form of a claim for breach of contract if their employer fails to comply with their obligations under *Regulation 4(1)*. Workers may be able to claim damages, a declaration or possibly an injunction in the event of a breach. However, it should be remembered that the case is a first instance decision and unfortunately, due to the 'politics' of the circumstances surrounding the cases will not be appealed. Indeed, there have been no other reported decisions where a worker has succeeded in bringing a claim for breach of contract where their employer has failed to comply with its obligations under *Regulation 4(1)*. It is, of course, worth bearing in mind that the steps being taken by the European Commission in relation to the potential infringement proceedings against the UK Government for failure to implement the Directive correctly may give further guidance on this point in due course.

The issue of an employer's obligations under the Regulations where they are not the only employer of the worker is problematic. *Regulation 4(2)* states 'an employer shall take all reasonable steps, in keeping with the need to protect the health and safety of workers, to ensure that the limit specified in *paragraph (1)* is complied with in the case of each worker employed by him in relation to whom it applies'. The DTI, in its guidance, recommend that an employer should agree an opt-out with a worker (see below) if the employer knows that the worker has a second job, if the total time worked is in excess of 48 hours a week. They also suggest that employers may wish to make an enquiry of their workforce about any additional employment. If a worker does not tell an employer about other employment and the employer has no reason to suspect that the worker has another job, it is extremely unlikely, the DTI suggest, that the employer would be found not to have complied.

In *Brown v Controlled Packaging Services Ltd (1999) (unreported)*, a worker who had another job in a bar on Tuesday and Thursday evenings was told at his interview that his basic working week would be 40 hours but that on occasion he would be required to work overtime. He soon discovered that the amount of overtime expected of him conflicted with the bar job, as on a number of occasions he was required to start work at 6 a.m. instead of 8 a.m. and he did not want to start work early on Wednesdays or Fridays. After the Regulations came into force, the worker was asked to sign an opt-out agreement which was viewed by his manager as a formality to ensure that existing overtime arrangements complied with the Regulations. Brown refused to sign the agreement because long hours would be incompatible with his evening job. Brown left and claimed constructive dismissal. The Employment Tribunal found that the dismissal was automatically unfair under the *Employment Rights Act 1996, s 101A* – i.e. for a reason connected with the employee exercising his rights under the Regulations.

## Reference period

**W10011** The crucial issue in calculating the average number of hours worked is the question of when the reference period starts. The reference period is any period of 17 weeks in the course of a worker's employment, unless a relevant agreement provides for the application of successive 17-week periods. [*Working Time Regulations 1998 (SI 1998 No 1833), Reg 4(3)*]. This means that unless the parties specify that the reference period is a defined period of 17 weeks, followed by a successive period of 17 weeks, then the reference period will become a 'rolling' 17-week period. This could be significant where there is a marked variation in the hours that an individual works in a particular period, from week to week. In such a situation an employer would be recommended to include provision in a relevant agreement for successive 17-week reference periods to ensure that it can comply with the maximum weekly working requirements. Such a provision may be desirable in any event, for ease of administration.

For the first 17 weeks of employment, the average is calculated by reference to the number of weeks actually worked. [*Working Time Regulations 1998 (SI 1998 No 1833), Reg 4(4)*].

## Exceptions and derogations to Regulation 4

**W10012** There are currently various exceptions to the maximum working week.

(a) Regulation 5 provides that an individual can agree with his employer to opt out of the 48-hour week (see W10013 below).

(b) The 48-hour limit does not apply at all in the case of certain workers whose working time is unmeasured; in respect of other workers, part of whose working time is unmeasured, it only applies to that part of their working time which is measured. [*Working Time Regulations 1998 (SI 1998 No 1833), Reg 20* – see W10033 below].

(c) In special cases (as described in *Reg 21* – see W10034 below), the 17-week reference period over which the 48 hours are averaged will be automatically extended to 26 weeks. [*Working Time Regulations 1998 (SI 1998 No 1833), Reg 4(5)*].

(d) A collective or workforce agreement can extend the 17-week reference period to a period not exceeding 52 weeks, if there are objective or technical reasons concerning the organisation of work. [*Working Time Regulations 1998 (SI 1998 No 1833), Reg 23*].

## Agreement to exclude 48-hour working week

**W10013** The 48-hour limit on weekly working time will not apply where a worker agrees with his employer in writing that the maximum 48-hour working week should not apply in the individual's case, provided that certain requirements are satisfied. These are:

(a) the agreement must be in writing;

(b) the agreement may specify its duration or be of an indefinite length – however, it is always open to the worker to terminate the agreement by giving notice, which will be the length specified in the agreement (subject to a maximum of 3 months), or, if not specified, 7 days; and

(c) the employer must keep up-to-date records of all workers who have signed such an agreement.

Until December 1999, employers were under controversial and onerous record keeping requirements in relation to workers who had signed opt out agreements. Since December 1999, employers only need to keep records of who has opted out.

A worker cannot be forced to sign an opt-out agreement. Moreover, the Regulations provide protection where an employee has suffered any detriment on the grounds that he has refused to waive any benefit conferred on him/her by the Regulations. [*Employment Rights Act 1996, s 45A(1)(b)*, inserted by the *Working Time Regulations 1998 (SI 1998 No 1833), Reg 31(1)(b)* – see W10041 below]. Furthermore, a worker will always have the right to terminate the opt-out agreement so that the 48-hour limit applies, on giving, at most, three months' notice.

On a practical note, the agreement by a worker to exclude the 48-hour working week must be a written agreement between the employer and the individual worker. It cannot take the form of or be incorporated in a collective or workforce agreement. However, it could, of course, be part of the contract of employment, (although it would be advisable for an employer to clearly delineate the agreement to work more than 48 hours from the rest of the contract of employment). It seems that employers' record keeping obligations in this regard would be satisfied by keeping an up-to-date list of workers who have signed opt-out agreements.

The ability of employers to seek the agreement of workers to 'opt out' of the 48-hour week (which is specifically permitted by the Directive) arguably runs a 'coach and horses' through the spirit of the legislation. Severe criticism has been levelled at the UK which is the only country which allows workers generally to opt-out of the maximum working week. The use of the opt-out is due to be reviewed by the EU Council by 23 November 2003 and it will be interesting to see if the long-hours culture which seems to prevail in the UK will be affected should a decision be taken to remove the 'opt-out'.

# Rest periods and breaks

## Daily rest period

**W10014** It should be noted that currently an employer's only obligation is to ensure that its employees are allowed to take their rest breaks – they are not required to compel employees to take them. This is one area which is currently being reviewed by the European Commission as part of the first steps in a potential infringement action against the UK Government for failing to implement the Directive properly. Adult workers are entitled to an uninterrupted rest period of not less than 11 consecutive hours in each 24-hour period. [*Working Time Regulations 1998 (SI 1998 No 1833), Reg 10*].

## Exceptions and derogations

**W10015** The provisions regarding daily rest periods do not apply where a worker's working time is unmeasured. [*Working Time Regulations 1998 (SI 1998 No 1833), Reg 20(1)*].

Derogations from the rule may be made with regard to shift work [*Reg 22*], or by means of collective or workforce agreements [*Reg 23*] or where there are special categories of workers [*Reg 21*]. However, in all cases (except for workers with unmeasured working time under *Reg 20(1)* – see W10033 below) compensatory rest must be provided [*Reg 24*].

## Weekly rest period

**W10016** Adult workers are entitled to an uninterrupted rest period of not less than 24 hours in each seven-day period. [*Working Time Regulations 1998 (SI 1998 No 1833), Reg 11*]. However, if his employer so determines, an adult worker will be entitled to either two uninterrupted rest periods (each of not less than 24 hours) in each 14-day period or one uninterrupted rest period of not less than 48 hours in each 14-day period.

The entitlement to weekly rest is in addition to the 11-hour daily rest entitlement which must be provided by virtue of *Regulation 10*, except where objective or technical reasons concerning the organisation of work would justify incorporating all or part of that daily rest into the weekly rest period.

For the purposes of calculating the entitlement, the 7 or 14-day period will start immediately after midnight between Sunday and Monday, unless a relevant agreement provides otherwise. The Regulations do not require Sunday to be included in the minimum weekly rest period.

Derogations from the entitlement to weekly rest periods are the same as those for daily rest.

## Rest breaks

**W10017** By virtue of the *Working Time Regulations 1998 (SI 1998 No 1833), Reg 12*, adult workers are entitled to a rest break where their daily working time is more than six hours. Details of this rest break, including duration and the terms on which it is granted, can be regulated by a collective or workforce agreement. If no such agreement is in place, the rest break will be for an uninterrupted period of not less than 20 minutes and the worker will be entitled to spend that break away from his workstation, if he has one. In its latest guidance the DTI states that the break should be taken during the six-hour period and not at the beginning or end of it, at a time to be determined by the employer.

Derogations from the entitlement to rest breaks include cases specified in the *Working Time Regulations 1998 (SI 1998 No 1833), Reg 21* (see above), where working time is unmeasured and cannot be predetermined [*Reg 20(1)*], and also by means of a collective or workforce agreement [*Reg 23*]. As before, compensatory rest must be provided other than where *Regulation 20* applies [*Reg 24*].

## Monotonous work

**W10018** The *Working Time Regulations 1998 (SI 1998 No 1833), Reg 8* provides that where the pattern according to which an employer organises work is such as to put the health and safety of a worker employed by him at risk, in particular because the work is monotonous or the work rate is predetermined, the employer shall ensure that the employee is given adequate rest breaks. This Regulation is phrased in virtually identical terms to Article 13 of the Working Time Directive and unfortunately, its incorporation into the Regulations has not clarified its meaning in any way!

It is not clear how rest breaks in *Regulation 8* would differ from those under *Regulation 12*. It may be that in the case of monotonous work, an employer might have to consider giving employees shorter breaks more frequently as opposed to one longer continuous break. This was the suggestion in the Government's consultative document.

No derogations from these provisions apply except in the case of domestic workers.

# Night work

## Definitions

**W10019** Before considering the detailed provisions in relation to night work contained in *Regulation 6*, an understanding of the definitions of 'night time' and 'night worker' contained in the *Working Time Regulations 1998* (*SI 1998 No 1833*), *Reg 2*, is necessary. These are as follows.

*Night time* in relation to a worker means a period:

(*a*)  the duration of which is not less than 7 hours; and

(*b*)  which includes the period between midnight and 5 a.m.,

which is determined for the purposes of the Regulations by a relevant agreement or, in the absence of such an agreement, the period between 11 p.m. and 6 a.m.

*Night worker* means a worker:

(i)  who *as a normal course* works at least three hours of his daily working time during night time (for the purpose of this definition, it is stated in the Regulations that (without prejudice to the generality of that expression) a person works hours 'as a normal course' if he works such hours on a majority of days on which he works – a person who performs night work as part of a rotating shift pattern may also be covered); or

(ii)  who is likely during night time to work at least such proportion of his annual working time as may be specified for the purposes of the Regulations in a collective or workforce agreement.

The question of when someone works at least 3 hours of his daily working time 'as a normal course' was addressed in the case of *R v Attorney General for Northern Ireland ex parte Burns [1999] IRLR 315*. In this case the Northern Ireland High Court held that a worker who spent one week in each 3-week cycle working at least 3 hours during the night was a 'night worker'. The High Court said that 'as a normal course' meant nothing more than as a regular feature. The DTI in its guidance refers to this case and says that it expects further clarification from the European Court in due course. It states that occasional and or ad hoc work at night does not make you a 'night worker'.

The question is certainly open, following the *Burns* case, of how frequent something has to be in order to become 'regular'.

## Length of night work

**W10020** If a worker falls within these definitions, then he or she is a night worker for the purposes of the Regulations. The *Working Time Regulations 1998* (*SI 1998 No 1833*), *Reg 6* then goes on to provide that an employer shall take all reasonable steps to ensure that the normal working hours of a night worker do not exceed an average of 8 hours in any 24-hour period. This is averaged over a 17-week reference period which is calculated in the same way as in *Regulation 4* (see W10011 above).

As with the maximum working week, there is a formula for calculating a night worker's average normal hours for each 24-hour period as follows:

$$\frac{a}{b-c}$$

where:

*a*  is the normal (not actual) working hours during the reference period;

*b*  is the number of 24-hour periods during the applicable reference period; and *c* is the number of hours during that period which comprise or are included in weekly rest periods under *Regulation 11*, which is then divided by 24.

The Regulations exclude night shift overtime hours from those which count towards 'normal' hours. This is currently being considered by the European Commission as part of the preliminary steps in a potential infringement procedure against the UK Government for not implementing the Directive correctly.

If the night work involves 'special hazards or heavy physical or mental strain', then a strict eight-hour time limit is imposed on working time in each 24-hour period and no averaging is allowed over a reference period. The identification of night work with such characteristics is by means of either a collective or workforce agreement which takes account of the specific effects and hazards of night work, or by the risk assessment which all employers are required to carry out under the *Management of Health and Safety at Work Regulations 1999 (SI 1999 No 3242)* (see RISK ASSESSMENT).

The derogations in the *Working Time Regulations 1998 (SI 1998 No 1833), Reg 21* (special categories of workers) apply to the provisions on length of night work. Workers whose working time is unmeasured or cannot be predetermined (and where it is only partly unmeasured or predetermined, to the extent that it is unmeasured) are also excluded. [*Working Time Regulations 1998 (SI 1998 No 1833), Reg 20*]. Other exemptions may be made by means of collective or workforce agreement. [*Working Time Regulations 1998 (SI 1998 No 1833), Reg 23*].

### Health assessment and transfer of night workers to day work

**W10021** An employer must, before assigning a worker to night work, provide him with the opportunity to have a free health assessment. [*Working Time Regulations 1998 (SI 1998 No 1833), Reg 7*]. The purpose of the assessment is to determine whether the worker is fit to undertake the night work. While there is no reliable evidence as to any specific health factor which rules out night work, a number of medical conditions could arise or could be made worse by working at night, such as diabetes, cardiovascular conditions or gastric intestinal disorders.

Employers are under a further duty to ensure that each night worker has the opportunity to have such health assessments 'at regular intervals of whatever duration may be appropriate in his case'. [*Working Time Regulations 1998 (SI 1998 No 1833), Reg 7(1)(b)*].

The Regulations do not specify the way in which the health assessment must be carried out, nor is there specific reference to medical assessments, so that strictly speaking, such assessments could be carried out by qualified health professionals rather than by a medical practitioner. The latest DTI guidance contains a sample health questionnaire.

Contrast this with the position as regards the transfer from night to day work. If a night worker is found to be suffering from health problems that are recognised as being connected with night work, he or she is entitled to be transferred to suitable day work 'where it is possible'. In such a situation, the Regulations provide that a 'registered medical practitioner' must have advised the employer that the worker is suffering from health problems which the practitioner considers to be connected with the performance by that night worker of night work.

## Annual leave

### Entitlement to annual leave

**W10022** The right under the *Working Time Regulations 1998 (SI 1998 No 1833), Reg 13* to 4 weeks' paid holiday each year is the first time under English law that workers have had a statutory right to paid holiday. The right was originally available to all workers within the scope of the Regulations, providing that they had been continuously employed for 13 weeks (this requirement is satisfied if the worker's relations with his employer have been governed by a contract during the whole or part of each of those weeks). The 13-week qualifying period was challenged by BECTU in the case *R v Secretary of State for Trade and Industry ex parte BECTU Case–173/99 (25 June 2001) (ECJ)* and the ECJ concluded that the 13 week qualifying period was unlawful.

The Regulations were amended by the *Working Time (Amendment) Regulations 2001 (SI 2001 No 3256)* and now provide for the entitlement to paid annual leave to begin on the first day of employment.

Where a worker began work after 25 October 2001, the employer now has the option to use an accrual system during the first year of employment. The amount of leave which the worker may take builds up monthly in advance at the rate of one twelfth of the annual entitlement each month. If the calculation does not produce a whole number the accrued leave entitlement is rounded up either to a half day (if the calculation is less than a half day) or a whole day (if the calculation exceeds a half day). Any rounded up element of leave taken would, of course, be deducted from the worker's annual entitlement to paid leave.

For the purposes of *Regulation 13*, a worker's leave year begins on any day provided for in a relevant agreement, or, where there is no such provision in a relevant agreement, 1 October 1998 or the anniversary of the date on which the worker began employment (whichever is the later). In the majority of cases, written statements of terms and conditions contain a reference to the holiday or leave year. If they do not, employers are recommended to ensure that they do so. Failure in this regard could result in administrative confusion as each employee could have a different holiday year for the purposes of calculations under the Regulations.

The statutory leave entitlement may be taken in instalments, but it can only be taken in the leave year to which it relates and a payment in lieu of the statutory entitlement cannot be made except where the worker's employment is terminated (but see comments at W10024 below). It should be noted that this relates only to the entitlement under the Regulations, so that if an employer provides for annual leave over and above the statutory entitlement, the enhanced element of the holiday can be carried forward or paid for in lieu as agreed between the parties. In *Miah v La Gondola Ltd (1999) (unreported)*, the Employment Tribunal found that a worker who was paid an additional week's pay instead of taking a week's holiday (due under the Regulations) during his employment, was entitled, on the termination of that contract, to payment for that proportion of annual leave which was due to him in accordance with *Regulation 14(3)(b)* and the payment paid to him during his employment in lieu of the week's statutory holiday had to be disregarded.

For many workers (and employers), the introduction of rights to paid annual leave was significant. It is understood that before the Regulations came into force some 2.5 million workers had no right to a holiday at all. Employment Tribunals have had to resolve many disputes arising from this right.

## Compensation related to entitlement to annual leave

**W10023** Where an employee has outstanding leave due to him when the employment relationship ends, the *Working Time Regulations 1998 (SI 1998 No 1833), Reg 14* specifically provides that an allowance is payable in lieu. It states that in the absence of any relevant agreement to the contrary, the amount of such allowance will be determined by the formula

$(a \times b) - c$

where:

| | |
|---|---|
| $a$ | is the period of leave to which the worker is entitled under the Regulations; |
| $b$ | is the proportion of the worker's leave year which expired before the effective date of termination; |
| $c$ | is the period of leave taken by the worker between the start of the leave year and the effective date of termination (*Reg 14(3)*). |

*Regulation 14(4)* provides that where there is a relevant agreement which so provides, an employee shall compensate his employer – whether by way of payment, additional work or otherwise – in relation to any holiday entitlement taken in excess of the statutory entitlement.

Interestingly in the case of *Witley & District Mens Club v Mackay [2001] IRLR 595* the EAT held that where a relevant agreement (in this case a collective agreement) purported to allow an employer to dismiss an employee for gross misconduct without payment for any accrued holiday pay this provision would be void as the *Working Time Regulations 1998 (SI 1998 No 1833), Reg 35* specifically provides for a sum to be paid. It is open to debate whether a provision of a relevant agreement which permitted the payment of a notional sum (say, £1) on termination in the event of gross misconduct would be valid.

## Payment, and notice requirements, for annual leave

**W10024** The *Working Time Regulations 1998 (SI 1998 No 1833), Reg 16* specifies the way in which a worker is paid in respect of any period of annual leave: this is at the rate of a week's pay in respect of each week of leave.

In order to calculate the amount of a 'week's pay', employers are referred to *s 221–224* of the *Employment Rights Act 1996* (although the calculation date is to be treated as the first day of the period of leave in question, and the relevant references to *ss 227 and 228* do not apply). The application of these provisions can lead to a disproportionately high level of holiday pay for workers paid on a commission only basis. This is because *s 223(2)* stipulates that if, during any of the 12 weeks preceding the worker's holiday, no remuneration was payable by the employer to the worker concerned, account should be taken of remuneration in earlier weeks so as to bring the number of weeks of which account is taken up to 12. In other words, it seems that the calculation of a 'commission only' worker's holiday entitlement will be based solely on the worker's earnings for any weeks (up to a maximum of 12) in which they received commission payments – any slow, unproductive weeks will be disregarded.

In the case of *Smith & Others v Chubb Security Personnel Ltd (1999) (unreported)*, employees who were obliged to work such hours as were detailed in their duty roster were paid holiday pay on the basis of a 40-hour week when in fact they worked a 3-week cycle of fourteen 12 hour days which averaged out at 56 hours per week. The Tribunal decided that their pay should be calculated on the basis of a 56-hour week.

The *Working Time Regulations 1998 (SI 1998 No 1833), Reg 15* deals with the dates on which the leave entitlement can be taken, and sets out detailed requirements as to notice. This Regulation is extremely complicated and before going into the details, it should be noted that alternative provisions concerning the notice requirements can be contained in a relevant agreement. Bearing in mind the detailed nature of the provisions, it is recommended that employers consider specifying such notice requirements in their contracts of employment, thereby avoiding the need for confusion at a later stage.

*Regulation 15* provides that a worker must give his employer notice equivalent to twice the amount of leave he is proposing to take. An employer can then prevent the worker from taking the leave on a particular date by giving notice equivalent to the number of days' leave which the employer wishes to prohibit. An employer can also, by giving notice equivalent to twice the number of days' leave in question, require an employee to take all or part of his leave on certain dates. This would, of course, be useful with regard to seasonal shutdowns over summer and Christmas holidays etc. An employee has no right to serve a counter notice and so, provided the employer has given proper notice, the employee must take the holiday as required. The Regulations do not require any notice to be given in writing. However, given the level of detail that must be included in any notice, employers would be advised to give the notice in writing and require their employees to do the same. This would also ensure that employers have accurate records of required holiday periods.

Employers who give more than 4 weeks' holiday and who are contractually obliged to pay their workers more for the statutory 4 weeks than they pay for the remaining leave will be interested in *Barton & Others v SCC Ltd (1999) (unreported)*. In this case, an employer sought to argue that monies paid by way of holiday pay for leave in excess of the statutory minimum leave under the Regulations can go towards the employer's liability to pay a week's pay for each week of leave taken under the Regulations (calculated under the *Working Time Regulations 1998* (*SI 1998 No 1833*), *Reg 16*). In addition, the employer argued that they had the right to chose which days off would be paid at the statutory rate and which at the contractual rate. The Employment Tribunal found in favour of the workers on both issues. *Regulations 16(1)* and *(5)* make it clear that the Employment Tribunal have to focus on the individual holiday week in question and the payment made for that individual week. With regard to which holiday is paid at which rate, the Employment Tribunal said that the right to holiday pay is an entitlement and it is a matter for the worker to choose which holiday weeks he or she seeks to take pursuant to his or her statutory entitlement.

The right to paid leave together with the relevant rate of pay have resulted in a number of controversial and sometimes conflicting decisions. In particular, there have been conflicting decisions as to whether it is possible to 'roll-up' holiday pay in an hourly rate and whether an employee is entitled to paid holiday whilst on sick leave.

There have been conflicting decisions as to whether it is permissible to roll-up holiday pay into an hourly rate. In the first case of *Johnson v Northbrook College (unreported) (Case No. 3102727/99)*, the Employment Tribunal expressly found that if an hourly rate incorporated an element of holiday pay, the employer's duties under the Regulations would have been discharged. (On the facts, as the employee had not agreed to the purported variation of her contract to include holiday pay in an hourly rate, the employer was found to be in breach of the Regulations.) This decision is, of course, not binding on subsequent Employment Tribunals and indeed, is difficult to reconcile with the case of *Barton & Others v SCC Ltd* where the Employment Tribunal considered that it was necessary to look at the week in which the leave was actually taken and to consider what payments, if any, were made during that week. In the *Johnson* case, no payment was actually made when leave was taken as the holiday pay was rolled up into the hourly rate.

The same issue arose again at EAT level in the case of *Gridquest Ltd t/a Select Employment v Blackburn [2002] IRLR 168* where the EAT concluded that it was permissible for employers to 'roll-up' an element of holiday pay into the overall hourly rate paid to workers. In direct contrast, a later decision of the EAT in a Scottish case, *MPB Structures v Munro (EAT/1257/01)* (which was argued under *Reg 35* (Restrictions on Contracting Out) of the 1998 Regulations and did not (at EAT level) refer to either *Reg 16(5)* or indeed the *Gridquest case*), held that adding an allowance of 8 per cent to each pay packet to cover the cost of holidays did not comply with the requirement of the Regulations to ensure that workers are entitled to paid leave. The EAT in the *Munro* case took the view that the aim of the Regulations was to ensure that workers were able to take their leave and that this could only be achieved where a worker was paid at the time when he took the leave.

The *Gridquest* case was subsequently appealed to the Court of Appeal and it was hoped that their decision would clarify the state of the law. The Court of Appeal concluded that, on the facts, there was no agreement between the workers and their employer regarding rolling up holiday pay into the hourly rate and accordingly it declined to comment on whether any such agreement would be lawful under the Regulations. The *Munro* decision was also appealed and the Scottish Court of Session concluded that rolling up holiday pay was not lawful under the Regulations. The decision of the Scottish Court of Session is not, strictly speaking, binding on English tribunals as was evidenced by the subsequent English EAT decision in *Marshalls Clay Products v Caulfield (as yet unreported)*. At the time of going to print, only a brief summary of this decision was available. However, it appears to suggest that employers can roll up holiday pay into an hourly rate provided that this is done with the express agreement of the worker. Hopefully, when the case is fully reported, employers will be given additional guidance as to when it is permissible to roll up holiday pay.

In addition, until recently there have been conflicting decisions on the question of a worker's entitlement to annual leave while on sick leave. The issue has now been clarified by the EAT decision in *Kigass Aero Components Ltd v Brown (EAT/481/00)* where the EAT held that workers are entitled to take their paid annual leave under the Regulations even if they are on sick leave during that period provided that they comply with the notification requirements under the Regulations. The EAT acknowledged that this was a somewhat surprising result which does not reflect the underlying health and safety purpose of the Regulations.

A further case which has given employers some cause for concern is the decision in *List Design Group v Catley (EAT/0966/00)*. In this particular case, the EAT held that a worker could mount a claim for a series of failures to make payments in respect of holiday pay where the claim was based on the unlawful deduction from wages provisions in the *Employment Rights Act 1996*. (See also EMPLOYMENT PROTECTION.) Normally, under the Regulations a worker would only be able to make a claim in respect of untaken holiday where the claim is brought within 3 months of the failure to allow the worker to take paid holiday. The Regulations do not permit a worker to bring a claim in respect of untaken holiday in previous holiday years. In contrast, under the provisions for unlawful deductions from wages workers can, in certain circumstances, bring a claim for serial failures to pay 'wages' (which for these purposes includes holiday pay due under the Regulations). Therefore, by bringing a claim under the deduction from wages provisions rather than the Regulations a worker may be able to bring a claim covering several holiday years where holiday was neither taken nor paid for. Accordingly as Mr Catley's case was based on a series of deductions from wages, he was able to claim 'back holiday pay' to the date when the Regulations came into force.

# Records

**W10025** The publicity and controversy surrounding the introduction of the Regulations related primarily to the maximum working week and the annual leave entitlement. However, in practice the provisions relating to record keeping proved very significant for employers, in particular those with workers who had signed opt-outs, where the rigorous record keeping requirements arguably undermined the benefit to an employer of the freedom given by the opt-out. Accordingly, the UK Government significantly watered down the requirements for employers *vis a vis* opted out workers, and the current position is outlined below.

The *Working Time Regulations 1998 (SI 1998 No 1833)*, *Reg 9* introduces an obligation on employers to keep specific records of working hours. All employers (except in relation to workers serving in the armed forces – *Reg 25(1)*) are under a duty to keep records which are adequate to show that certain specified limits are being complied with in the case of each entitled worker employed by him. These limits are:

(*a*) the maximum working week (see W10010 above);

(*b*) the length of night work (including night work which is hazardous or subject to heavy mental strain) (see W10020 above); and

(*c*) the requirement to provide health assessments for night workers (see W10021 above).

These records are required to be maintained for two years from the date on which they were made.

In addition, where a worker has agreed to exclude the maximum working week under the *Working Time Regulations 1998 (SI 1998 No 1833)*, *Reg 5* (see W10013 above), an employer is required to keep up-to-date records of all workers who have opted out. Note that, in contrast with the obligations under *Regulation 9*, no time limit is imposed for the maintenance of such records.

Note that the obligation on employers to keep records may increase as a result of the steps taken by the European Commission as part of a potential infringement action against the UK government for failing to implement the Directive correctly.

## Excluded sectors

**W10026** Prior to 1 August 2003, the following sectors of activities (listed below) were totally excluded from the terms of the Regulations by *Regulation 18*:

(*a*)   air, rail, road, sea, inland waterway and lake transport;

(*b*)   sea fishing;

(*c*)   other work at sea;

(*d*)   the activities of doctors in training.

## Following the adoption of five Directives aimed at extending the working time provisions to previously excluded sectors, this is no longer the case and most of the above sectors have gained some protection under the Regulations. The Directives are:

- Horizontal Amending Directive.
- Road Transport Directive.
- Seafarers' Directive (outside the scope of this chapter).
- Seafarers' Enforcement Directive (outside the scope of this chapter).
- Aviation Directive (outside the scope of this chapter).

The Horizontal Amending Directive was implemented by the UK with effect from 1 August 2003 (see the *Working Time (Amendment) Regulations 2003* (*SI 2003 No 1684*).

### Application of the new directives to previously excluded sectors

#### Mobile and non-mobile workers

**W10027** A mobile worker is defined in the *Working Time (Amendment) Regulations 2003* (*SI 2003 No 1684*) as: 'any worker employed as a member of travelling or flying personnel by an undertaking which operates transport services for passengers or goods by road or air'. Non-mobile workers are undefined and are therefore any worker who is not a mobile worker, for example, all office based staff, warehouse staff etc. Unfortunately the Amendment Regulations do not address the situation where a worker carries out a combination of mobile and non-mobile activities (for example, a worker who spends part of his time based in a depot loading goods onto lorries and the remainder of his time driving an HGV delivering the goods.) It is likely that future cases will provide further guidance on this point.

#### Road transport sector

##### Non-mobile workers

**W10028** Following the implementation of the *Working Time (Amendment) Regulations 2003* (*SI 2003 No 1684*) on 1 August 2003, the Regulations are extended in full to non-mobile workers in the road transport sector.

##### Mobile workers

**W10029** As of 1 August 2003, mobile workers who are *not* covered by the Road Transport Directive (such as drivers of vehicles of less than 3.5 tonnes) will benefit from the average 48-hour working week, four weeks' paid holiday, appropriate health checks if they are night workers and provision for adequate rest (unless there are exceptional circumstances as provided for in *Regulation 21(e)* of the *Working Time Regulations 1998* (*SI 1998 No 1833*). They are not entitled to the daily rest requirements set out in *Regulation 10(1)* or the weekly rest periods set out in *Regulation 11(1)* and *11(2)* of the 1998 Regulations. In addition, they are not entitled to the rest breaks set out in *Regulation 12(1)*. Further, the exemptions for special categories of workers may apply (see W10034 below).

As of 1 August 2003, mobile workers who are covered by the Road Transport Directive (mainly drivers of vehicles of over 3.5 tonnes) will be entitled to 4 weeks' paid holiday and appropriate health checks if they are night workers.

In addition, mobile workers who are covered by the Road Transport Directive, will gain additional protection once the Directive has been implemented. The Road Transport Directive specifies a maximum working week of 60 hours (provided always that the average working week over a 4-month period does not exceed 48 hours). Night work is restricted to 10 hours and self-employed drivers are excluded from the ambit of the Directive. It is due to be implemented by 23 March 2005.

## Rail sector

**W10030** Following the implementation of the *Working Time (Amendment) Regulations 2003* (*SI 2003 No 1684*) on 1 August 2003, the Regulations are, in principle, extended in full to all workers in the rail transport sector.

It is, however, possible to derogate from the Regulations in respect of railway workers whose activities are intermittent, whose working time is spent on board trains or whose activities 'are linked to transport timetables and to ensuring the continuity and regularity of traffic'. For more information on these partial exemptions see W10034 below.

## Other sectors

**W10031** It should be noted that the Horizontal Amending Directive together with the other sector-specific directives detailed above provide different entitlements for mobile workers in the sea-fishing sector, aviation sector, inland waterway and lake transport sectors and also in relation to offshore working. In addition, in relation to doctors in training the Horizontal Amending Directive provides for a gradual phasing in of the Regulations: initial implementation must be completed by 1 August 2004 and full implementation of the provisions by 1 August 2012 at the latest.

## **Partial exemptions**
### Domestic service

**W10032** The following provisions of the Regulations do not apply in relation to workers employed as domestic servants in private households:

(*a*)   the 48-hour working week (see W10010 above);

(*b*)   length of night work (see W10020 above);

(*c*)   health assessments for night workers (see W10021 above);

(*d*)   monotonous work (see W10018 above).

## Unmeasured working time

**W10033** The requirements of the Regulations listed below are, by virtue of the *Working Time Regulations 1998* (*SI 1998 No 1833*), *Reg 20(1)*, not applicable to workers where, on account of the specific characteristics of the activity in which they are engaged, the duration of their working time is not measured or pre-determined or can be determined by the workers themselves. The provisions excluded are:

(a) the 48-hour working week (see W10010 above);

(b) minimum daily and weekly rest periods and rest breaks (see W10014 above);

(c) length of night work (see W10020 above).

This would leave such workers with requirements relating to monotonous work, the requirement for health assessments for night workers, the requirement to keep records and annual leave. It should also be noted that there is no requirement to provide such workers with compensatory rest where the requirements of the Regulations are not complied with.

This provision is often thought to exclude the Regulations in relation to 'managing executives or other persons with autonomous decision-making powers'. However, it should be noted that this is simply *one* of the three examples given in *Regulation 20(1)*, as regards the type of worker with unmeasured working time. The application of the Regulation could be much wider than this, and will depend on whether the worker in question is genuinely able to control his work, to the extent of being able to determine how many hours he works.

The UK Government made an amendment to *Regulation 20* in December 1999 so that the working time of workers does not, effectively count towards their weekly working time, to the extent that it is unmeasured, even if their time is not wholly unmeasured. Prior to the amendment, workers were either within the *Regulation 20* 'unmeasured' exemption or they were outside it. The exemption provides a 'half way house' for workers who have some part of their working time set in advance (e.g. under their contract of employment) but choose on their own account to work longer hours. It is as yet unclear how widely this provision will be interpreted i.e. when extra hours can genuinely be said not to be 'required' by the employer.

## Special categories of workers

**W10034** The *Working Time Regulations 1998* (*SI 1998 No 1833*), *Reg 21* provides that in the case of certain categories of employees, the following Regulations do not apply:

(a) length of night work (see W10020 above);

(b) minimum daily rest and weekly rest periods and rest breaks (see W10014 above).

This leaves the 48-hour working week, requirements relating to monotonous work, the requirement for health assessments for night workers, the requirement to keep records and annual leave. However, it should be noted that where any of the allowable derogations are utilised, an employer will have to provide compensatory rest periods (see W10037 below).

Further, for these categories of workers, the reference period over which the 48-hour working week is averaged is 26 weeks and not 17. [*Working Time Regulations 1998* (*SI 1998 No 1833*), *Reg 4(5)*].

The special categories of worker are as follows:

(i) where the worker's activities mean that his place of work and place of residence are distant from one another, or the worker has different places of work which are distant from one another (including cases where the worker carries out offshore work);

(ii) workers engaged in security or surveillance activities, which require a permanent presence in order to protect property and persons – examples given are security guards and caretakers;

- (iii) where the worker's activities involve the need for continuity of service or production – specific examples are:
    - (A) services relating to the reception, treatment or care provided by hospitals or similar establishments (including the activities of doctors in training), residential institutions and prisons;
    - (B) workers at docks or airports;
    - (C) press, radio, television, cinematographic production, postal and telecommunications services, and civil protection services;
    - (D) gas, water and electricity production, transmission and distribution, household refuse collection and incineration;
    - (E) industries in which work cannot be interrupted on technical grounds;
    - (F) research and development activities;
    - (G) agriculture;
- (iv) any industry where there is a foreseeable surge of activity – specific cases suggested are:
    - (A) agriculture;
    - (B) tourism;
    - (C) postal services;
- (v) where the worker's activities are affected by:
    - (A) unusual and unforeseeable circumstances beyond the control of the employer;
    - (B) exceptional events which could not be avoided even with the exercise of all due care by the employer;
    - (C) an accident or the imminent risk of an accident;
- (vi) where the worker works in railway transport and:
    - (A) his activities are intermittent; or
    - (B) he works on board trains; or
    - (C) his activities are linked to transport timetables and to ensuring the continuity and regularity of traffic.

## Shift workers

**W10035** Shift workers are defined by the *Working Time Regulations 1998* (*SI 1998 No 1833*), *Reg 22* as workers who work in a system whereby they succeed each other at the same work station according to a certain pattern, including a rotating pattern which may be continuous or discontinuous, entailing the need for workers to work at different times over a given period of days or weeks.

In the case of shift workers the provisions for daily rest periods (see W10014 above) and weekly rest periods (see W10016 above) can be excluded in order to facilitate the changing of shifts, on the understanding that compensatory rest must be provided (see W10037 below). In addition, those provisions also do not apply to workers engaged in activities involving periods of work split up over the day – the example given is that of cleaning staff.

## Collective and workforce agreements

**W10036** By virtue of the *Working Time Regulations 1998* (*SI 1998 No 1833*), *Reg 23(a)*, employers and employees have the power to exclude or modify the following provisions:

(a) length of night work (see W10020 above);

(b) minimum daily rest and weekly rest periods and rest breaks (see W10014 above),

by way of collective or workforce agreements. Compensatory rest must be provided (see W10032 below).

In addition, the *Working Time Regulations 1998* (*SI 1998 No 1833*), *Reg 23(b)* allows the reference period for calculating the maximum working week to be extended to up to 52 weeks, if there are objective or technical reasons concerning the organisation of work to justify this.

This Regulation gives the parties a good deal of flexibility (on the understanding that they are prepared to consent with each other) to opt out of significant provisions contained in the Regulations.

## Compensatory rest

**W10037** The *Working Time Regulations 1998* (*SI 1998 No 1833*), *Reg 24* provides that where a worker is not strictly governed by the working time rules because of:

(a) a derogation under *Regulation 21* (see W10034 above); or

(b) the application of a collective or workforce agreement under *Regulation 23(a)* (see W10036 above); or

(c) the special shift work rules under *Regulation 22* (see W10035 above);

and as a result, the worker is required by his employer to work during what would otherwise be a rest period or rest break, then the employer must wherever possible allow him to take an equivalent period of compensatory rest. In exceptional cases, in which it is not possible for objective reasons to grant such a rest period, the employer must afford the employee such protection as may be appropriate in order to safeguard his health and safety.

# Enforcement

**W10038** The Regulations divide the enforcement responsibilities between the Health and Safety Executive and the Employment Tribunals. In essence, the working time limits are enforced by the Health and Safety Executive and local authority environmental health departments. The entitlements to rest and holiday are enforced through Employment Tribunals.

Interestingly, despite the assertions that the Directive is a health and safety measure, there do not appear to be any successful prosecutions by the Health and Safety Executive and there has been only one successful prosecution by a local authority under the Regulations in the past two years. In contrast, there have been numerous cases lodged at the Employment Tribunal or High Court since the implementation of the Regulations.

## Health and safety offences

**W10039** The *Working Time Regulations 1998* (*SI 1998 No 1833*), *Reg 28* provides that certain provisions of the Regulations (referred to as 'the relevant requirements') will be enforced by the Health and Safety Executive (except to the extent that a local authority may be responsible for their enforcement by virtue of *Regulation 28(3)*). The relevant requirements are:

(a) the 48-hour working week;

(b) length of night work;

(c) health assessment and transfers from night work;

(d) monotonous work;

(e) record keeping; and

(f) failure to provide compensatory rest, where the provision concerning the length of night work is modified or excluded.

Any employer who fails to comply with any of the relevant requirements will be guilty of an offence and shall be liable on summary conviction (in the magistrates' court) to a fine not exceeding the statutory maximum and on conviction on indictment (in the Crown Court) to a fine.

In addition, the Health and Safety Executive can take enforcement proceedings utilising certain provisions of the *Health and Safety at Work etc Act 1974*, as set out in *Regulations 28* and *29*. Employers may face criminal liability under these provisions, the sanctions for which range, according to the offence, from a fine to two years' imprisonment. (For enforcement of health and safety legislation generally, see ENFORCEMENT.)

## Employment tribunals

### Enforcement of the Regulations

**W10040** By virtue of the *Working Time Regulations 1998* (*SI 1998 No 1833*), *Reg 30*, certain provisions of the Regulations may be enforced by a worker presenting a claim to an Employment Tribunal where an employer has refused to permit him to exercise such rights. These are:

(a) daily rest period;

(b) weekly rest period;

(c) rest break;

(d) annual leave;

(e) failure to provide compensatory rest, insofar as it relates to situations where daily or weekly rest breaks are modified or excluded;

(f) the failure to pay the whole or any part of the amount relating to paid annual leave, or payment on termination in lieu of accrued but untaken holiday.

A complaint must be made within (i) three months (other than in the case of members of the armed forces – see below) of the act or omission complained of (or in the case of a rest period or leave extending over more than one day, of the date on which it should have been permitted to begin) or (ii) such further period as the tribunal considers reasonable, where it is satisfied that it was not reasonably practicable for the complaint to be presented within that time. (The time limit in respect of members of the armed forces is six months.)

Where an Employment Tribunal decides that a complaint is well-founded, it must make a declaration to that effect and can award compensation to be paid by the employer to the worker. This shall be such amount as the Employment Tribunal considers just and equitable in all the circumstances, having regard to (i) the employer's default in refusing to permit the worker to exercise his right and (ii) any loss sustained by the worker which is attributable to the matters complained of. With regard to complaints relating to holiday pay or payment in lieu of accrued holiday on termination, the tribunal can also order the employer to pay the worker the amount which it finds properly due.

There is no qualifying service period with regard to such complaints being presented to a tribunal.

## Protection against detriment, and against unfair dismissal

**W10041** The *Working Time Regulations 1998 (SI 1998 No 1833), Reg 31* inserts a new *section 45A* into the *Employment Rights Act 1996*, so as to provide protection for a worker against being subjected to any detriment, where the worker has:

(a) refused (or proposed to refuse) to comply with a requirement which the employer imposed (or proposed to impose) in contravention of the Regulations;

(b) refused (or proposed to refuse) to forgo a right conferred on him by the Regulations;

(c) failed to sign a workforce agreement or make any other agreement provided for under the Regulations, such as an individual opt-out from the maximum weekly working time limit;

(d) been a candidate in an election of work place representatives or, having been elected, carries out any activities as such a representative or candidate; or

(e) in good faith, (i) made an allegation that the employer has contravened a right under the Regulations, or (ii) brought proceedings under the Regulations.

The right to bring these claims would (as with discrimination claims) allow an individual to pursue a claim relating to a breach of the Regulations whilst continuing in employment.

By virtue of the inserted *section 101A* of the *Employment Rights Act 1996*, the dismissal of an employee on all but ground (e) above is automatically unfair, although this will only apply to employees and not to the wider definition of worker (for the position of workers who are not employees, see below). The compensation available would be subject to any cap on unfair dismissal compensation (the current cap on unfair dismissal compensatory awards is £52,600). Employees whose contracts were terminated on ground (e) above would be protected from dismissal for assertion of a statutory right, by virtue of *section 104* of the 1996 Act.

An employee would also be protected if selected for redundancy on any of the grounds listed above. This would be automatically unfair selection, by virtue of *section 105* of the 1996 Act.

Workers (i.e. those who are not employees) whose contracts are terminated for any of the grounds set out above can claim that they have suffered a detriment, and they may claim compensation, which would be capped in the same way as an award for unfair dismissal.

## Breach of contract claims

**W10042** Following the *RJB Mining case* (see W10010), there is an argument that workers may also be able to claim for breach of contract, if their employer breaches the obligations in the *Working Time Regulations 1998 (SI 1998 No 1833), Reg 4(1)*. (In practice, however, this type of claim has not arisen since the determination of this case.) If the claim is for damages and is outstanding at the termination of the employment, then the worker could in principle take their case to the Employment Tribunal (within 3 months of the effective date of termination of the contract giving rise to the claim). If the claim is for a declaration or an injunction, then the worker will have to apply to the High Court, where special rules apply.

## Contracting out of the Regulations

**W10043** Under the *Working Time Regulations 1998 (SI 1998 No 1833), Reg 35*, an agreement to contract out of the provisions of the Regulations can be made via a conciliation officer or by means of a compromise agreement, and the provisions are similar and consistent with the current provisions in the *Employment Rights Act 1996, s 203* (as amended by the *Employment Rights (Dispute Resolution) Act 1998*).

## Conclusion

**W10044**  The *Working Time Regulations 1998* (*SI 1998 No 1833*) broke new ground in English law. Despite the adverse publicity surrounding them their introduction does not appear to have unduly changed current industrial practice and does not appear to have led to a reduction in the long hours culture which is prevalent within the UK.

The amendments to the Directive which extend the Regulations to both non-mobile and mobile workers within the previously excluded sectors are, however, likely to have a significant impact on some sectors, most notably the road transport sector. In addition, it is arguable that the impact of the Regulations will be increased if the European Commission's review of the use of opt-outs in 2003 results in the removal of the opt-out provisions.

Finally, it should be noted that the European Commission has taken the first steps which could lead to infringement proceedings against the UK government for failing to implement the Directive properly. There are three grounds of complaint: the lack of a requirement to measure time worked voluntarily over normal working time, the exclusion of night shift overtime hours from those which count towards normal hours and the lack of a duty on employers to ensure that workers take breaks and holidays. We shall have to wait and see whether this results in further changes to the Regulations.

# Workplaces – Health, Safety and Welfare

## Introduction

**W11001** The *Workplace (Health, Safety and Welfare) Regulations 1992* (*SI 1992 No 3004* as amended by *SI 2002 No 2174*) set out a wide range of basic health, safety and welfare standards applying to most places of work. The Workplace Regulations cover not only offices, shops and factories covered by earlier legislation in this area, but apply to a much wider range of workplaces, including schools, hospitals, theatres, cinemas and hotels, for example.

The *Health and Safety (Miscellaneous) Amendments Regulations 2002* (*SI 2002 No 2174*) mean that the 1992 Regulations now apply to every workplace, as previously excluded workplaces such as factories and mines are now included. In addition, a new *Regulation 25A* states that employers must consider the needs of disabled workers.

The Health and Safety Executive (HSE) reviewed the Workplace Regulations for the European Commission in 2003 and concluded that they were working well.

The Workplace Regulations set down detailed standards for premises used as places of work in the following areas:

(*a*)   maintenance [*SI 1992 No 3004, Reg 5*] (see W11003);

(*b*)   ventilation [*SI 1992 No 3004, Reg 6*] (see VENTILATION);

(*c*)   temperature [*SI 1992 No 3004, Reg 7*] (see W11005);

(*d*)   lighting [*SI 1992 No 3004, Reg 8*] (see LIGHTING);

(*e*)   cleanliness and waste storage [*SI 1992 No 3004, Reg 9*] (see W11007);

(*f*)   room dimensions and space [*SI 1992 No 3004, Reg 10*] (see W11008);

(*g*)   workstations and seating [*SI 1992 No 3004, Reg 11*] (see W11009);

(*h*)   conditions of floors and traffic routes [*SI 1992 No 3004, Reg 12*] (see W11010);

(*j*)   freedom from falls and falling objects [*SI 1992 No 3004, Reg 13*] (see WORK AT HEIGHTS);

(*k*)   windows and transparent or translucent doors, gates and walls (see W11012);

(*l*)   windows, skylights and ventilators [*SI 1992 No 3004, Reg 18*] (see W11013);

(*m*)   window cleaning [*SI 1992 No 3004, Regs 14—16*] (see WORK AT HEIGHTS);

(*n*)   organisation of traffic routes [*SI 1992 No 3004, Reg 17*] (see W11015);

(*o*)   doors and gates [*SI 1992 No 3004, Reg 18*] (see W11016);

(*p*)   escalators and travelators [*SI 1992 No 3004, Reg 19*] (see W11017);

(*q*)   sanitary conveniences/washing facilities [*SI 1992 No 3004, Regs 20, 21*] (see W11019–W11023);

(*r*)   drinking water [*SI 1992 No 3004, Reg 22*] (see W11024);

(s)  clothing accommodation and facilities for changing clothing [*SI 1992 No 3004, Reg 23*] (see W11026–W11027); and

(t)  rest/meal facilities [*SI 1992 No 3004, Reg 25*] (see W11027).

In addition to these Regulations, there is other legislation which contain health, safety and welfare standards for particularly high risk industries and workplaces, including construction sites, mines, quarries and the railway industry. This legislation is dealt with in other chapters.

This chapter looks at the provisions of the *Workplace (Health, Safety and Welfare) Regulations 1992*, together with guidance provided in the associated approved code of practice (ACoP). It outlines the requirements of the *Building Regulations 1991 (SI 1991 No 2768)*, which apply where workplaces are being built, extended or modified, and looks at the *Disability Discrimination Act 1995, s 6*, which requires that reasonable adjustments be made to workplaces where necessary to ensure that disabled workers are not put at a substantial disadvantage.

It also sets out the provisions of the *Health and Safety (Safety Signs and Signals) Regulations 1996 (SI 1996 No 341)*.

# Workplace (Health, Safety and Welfare) Regulations 1992 (SI 1992 No 3004)

## Definitions

**W11002**  A 'workplace' is any non-domestic premises available to any person as a place of work, including:

(a)  canteens, toilets;

(b)  parts of a workroom or workplace (e.g. corridor, staircase, or other means of access/egress other than a public road);

(c)  a completed modification, extension, or conversion of an original workplace;

but excluding

(i)  boats, ships, hovercraft, trains and road vehicles (although the requirements in *regulation 13*, which deal with falls and falling objects, apply when aircraft, trains and road vehicles are stationary inside a workplace);

(ii)  building operations/works of engineering construction;

(iii)  mining activities.

[*Workplace (Health, Safety and Welfare) Regulations 1992 (SI 1992 No 3004), Regs 2, 3* and *4*].

The definition of 'work' includes work carried out by employees and self-employed people and the definition of 'premises' includes outdoor places. The Regulations do not apply to private dwellings, but they do apply to hotels, nursing homes and to parts of premises where domestic staff are employed, for example in the kitchens of hostels.

*Regulations 20* to *25* (which deal with toilets, washing, changing facilities, clothing accommodation, drinking water and eating and rest facilities) apply to temporary work sites, but only so far as is reasonably practicable.

## General maintenance of the workplace

**W11003**  All workplaces, equipment and devices should be maintained

(a)  in an efficient state,

(b)  in an efficient working order, and

(c) in a good state of repair.

[*Workplace (Health, Safety and Welfare) Regulations 1992 (SI 1992 No 3004), Reg 5*].

Dangerous defects should be reported and acted on as a matter of good housekeeping (and to avoid possible subsequent civil liability). Defects resulting in equipment/plant becoming unsuitable for use, though not necessarily dangerous, should lead to decommissioning of plant until repaired – or, if this might lead to the number of facilities being less than required by statute, repaired forthwith (e.g. a defective toilet).

To this end, a suitable maintenance programme must be instituted, including:

(a) regular maintenance (inspection, testing, adjustment, lubrication, cleaning);

(b) rectification of potentially dangerous defects and the prevention of access to defective equipment;

(c) record of maintenance/servicing.

There are more detailed regulations dealing with plant and equipment used at work, including the *Provision and Use of Work Equipment Regulations 1998 (SI 1998 No 2306* as amended by *SI 2002 No 2174*), which are examined in detail in MACHINERY SAFETY.

## *Ventilation*

**W11004** *Workplace (Health, Safety and Welfare) Regulations 1992 (SI 1992 No 3004), Reg 6* deals with the provision of sufficient ventilation in enclosed workplaces. There must be effective and suitable ventilation in order to supply a sufficient quantity of fresh or purified air. If ventilation plant is necessary for health and safety reasons, it must give warning of failure.

The ACoP to the Regulations sets out that ventilation should not cause uncomfortable draughts, and that it should be sufficient to provide fresh air for the occupants to breathe, to dilute any contaminants and to reduce odour.

The HSE publication, *General ventilation in the workplace: Guidance for employers*, provides detailed guidance in this area. It sets out that the fresh air supply rate to workplaces should not normally fall below 5 to 8 litres per second, per occupant, and that in deciding the appropriate rate, employers should consider factors including:

- The amount of floor space available to each worker;
- The type of work being carried out;
- Whether people smoke in the workplace; and
- What contaminants are being discharged by, for example, any process machinery, heaters, furniture and furnishings.

This guidance also lists sources of further information and help, from organisations including the Chartered Institution of Building Services Engineers (CIBSE) and the Heating and Ventilating Contractors Association (HVCA) and is available from HSE Books price £4.00.

Ventilation is dealt with in detail in VENTILATION.

## *Temperature*

**W11005** The temperature in all workplaces inside buildings should be reasonable during working hours. [*Workplace (Health, Safety and Welfare) Regulations 1992 (SI 1992 No 3004), Reg 7*]. Workroom temperatures should enable people to work (and visit sanitary conveniences) in reasonable comfort, without the need for extra or special clothing. Although the Regulations themselves do not specify a maximum or minimum indoor workplace temperature, the approved code of practice (ACoP) sets out that the minimum acceptable temperature is 16°C at the workstation, except where work involves considerable physical effort, when it reduces to 13°C (dry bulb thermometer reading). Space heating of the average workplace should be 16°C, and this should be maintained throughout the remainder of the working day. However, this is a minimum temperature. The method of heating or cooling should not result in dangerous or offensive gases or fumes entering the workplace.

The following temperatures for different types of work are recommended by the Chartered Institute of Building Services Engineers (CIBSE):

(a) heavy work in factories 13°C;

(b) light work in factories 16°C;

(c) hospital wards and shops 18°C; and

(d) office and dining rooms 20°C.

Maintenance of such temperatures may not always be feasible, as, for instance, where hot/cold production/storage processes are involved, or where food has to be stored. In such cases, an approximate temperature should be maintained. With cold storage, this may be achievable by keeping a small chilling area separate or by product insulation; whereas, in the case of hot processes, insulation of hot plant or pipes, provision of cooling plant, window shading and positioning of workstations away from radiant heat should be considered in order to achieve a reasonably comfortable temperature. Moreover, where it is necessary from time to time to work in rooms normally unoccupied (e.g. storerooms), temporary heating should be installed. Thermometers must be provided so that workers can periodically check the temperatures.

Where, despite the provision of local heating or cooling, temperatures are still not reasonably comfortable, suitable protective clothing or rest facilities should be provided, or there should be systems of work in place, such as job rotation, to minimise the length of time workers are exposed to uncomfortable temperatures.

HSE guidance on the Regulations, *Workplace health, safety and welfare: A short guide for managers*, sets out how to carry out an assessment of the risk to workers' health from working in either a hot or cold environment. This advises that employers should look at personal factors, such as body activity, the amount and type of clothing and duration of exposure, together with environmental factors, including the ambient temperature and radiant heat, and if the work is outdoor, sunlight, wind velocity and the presence of rain or snow.

It sets out that any assessment needs to consider:

- Measures to control the workplace environment, particularly heat sources;

- Restriction of exposure by, for example, reorganising tasks to build in rest periods or other breaks from work;

- Medical pre-selection of employees to ensure that they are fit to work in these environments;

- Use of suitable clothing;

- Acclimatisation of workers to the working environment;

- Training in the precautions to be taken; and

- Supervision to ensure that the precautions the assessment identifies are taken.

HSE produces further guidance on temperature at work, *Thermal comfort in the workplace: Guidance for employers*. This is available from HSE Books, price £3.50.

For employers with outdoor workers there is a free leaflet, *Keep your top on*, which contains guidance on working outdoors in sunny weather, also available from HSE Books.

## Lighting

**W11006** Every workplace must be provided with suitable and sufficient lighting. [*Workplace (Health, Safety and Welfare) Regulations 1992 (SI 1992 No 3004), Reg 8*]. This should be natural lighting, so far as is reasonably practicable. There should also be suitable and sufficient emergency lighting where necessary. The ACoP sets out that in order to be suitable and sufficient, lighting must enable people to work and move about safely. HSE guidance on the Regulations advises that where necessary, local lighting should be provided at individual workstations, and at places of particular risk such as crossing points on traffic routes.

An HSE booklet, *Lighting at work*, gives detailed guidance in this area and is available from HSE Books, price £9.25. LIGHTING also deals with this area in greater detail.

## General cleanliness

**W11007** All furniture and fittings of every workplace must be kept sufficiently clean. Surfaces of floors, walls and ceilings must be capable of being kept sufficiently clean and waste materials must not accumulate other than in waste receptacles. [*Workplace (Health, Safety and Welfare) Regulations 1992 (SI 1992 No 3004), Reg 9*].

The level and frequency of cleanliness will vary according to workplace use and purpose. Obviously a factory canteen should be cleaner than a factory floor. Floors and indoor traffic routes should be cleaned at least once a week, though dirt and refuse not in suitable receptacles should be removed at least daily, particularly in hot atmospheres or hot weather. Interior walls, ceilings and work surfaces should be cleaned at suitable intervals and ceilings and interior walls painted and/or tiled so that they can be kept clean. Surface treatment should be renewed when it can no longer be cleaned properly. In addition, cleaning will be necessary to remove spillages and waste matter from drains or sanitary conveniences. Methods of cleaning, however, should not expose anyone to substantial amounts of dust, and absorbent floors likely to be contaminated by oil or other substances difficult to remove, should be sealed or coated, say, with non-slip floor paint (not covered with carpet!).

## Workroom dimensions/space

**W11008** Every room in which people work should have sufficient

(a)  floor area,

(b)  height, and

(c)  unoccupied space

for health, safety and welfare purposes. [*Workplace (Health, Safety and Welfare) Regulations 1992 (SI 1992 No 3004), Reg 10*].

Workrooms should have enough uncluttered space to allow people to go to and from workstations with relative ease. The number of people who may work in any particular room at any time will depend not only on the size of the room but also on the space given over to furniture, fittings, equipment and general room layout. Workrooms should be of sufficient height to afford staff safe access to workstations. If, however, the workroom is in an old building, say, with low beams or other possible obstructions, this should be clearly marked, e.g. 'Low beams, mind your head'.

The total volume of the room (when empty), divided by the number of people normally working there, should be 11 cubic metres (minimum) per person, although this does not apply to:

(i)  retail sales kiosks, attendants' shelters etc.,

(ii)  lecture/meeting rooms etc.

[*Workplace (Health, Safety and Welfare) Regulations 1992 (SI 1992 No 3004), Sch 1*].

In making this calculation, any part of a room which is higher than 3 metres is counted as being 3 metres high.

Where furniture occupies a considerable part of the room, 11 metres may not be sufficient space per person. Here more careful planning and general room layout is required. Similarly, rooms may need to be larger or have fewer people working in them depending on the contents and layout of the room and the nature of the work.

## Workstations and seating

W11009   The Regulations stipulate the following.

(a)   *Workstations*

Every workstation must be so arranged that:

(i)   it is suitable for

(a)   any person at work who is likely to work at the workstation, and

(b)   any work likely to be done there;

(ii)   so far as reasonably practicable, it provides protection from adverse weather;

(iii)   it enables a person to leave it swiftly or to be assisted in an emergency;

(iv)   it ensures any person is not likely to slip or fall.

[*Workplace (Health, Safety and Welfare) Regulations 1992 (SI 1992 No 3004), Reg 11(1)(2)*].

(b)   *Workstation seating*

A suitable seat must be provided for each person at work whose work (or a substantial part of it) can or must be done seated. The seat should be suitable for:

(i)   the person doing the work, and

(ii)   the work to be done.

Where necessary, a suitable footrest should be provided.

[*Workplace (Health, Safety and Welfare) Regulations 1992 (SI 1992 No 3004), Reg 11(3)(4)*].

It should be possible to carry out work safely and comfortably. Work materials and equipment in frequent use (or controls) should always be within easy reach, so that people do not have to bend or stretch unduly, and the worker should be at a suitable height in relation to the work surface. Workstations, including seating and access, should be suitable for special needs, for instance, disabled workers. The workstation should allow people likely to have to do work there adequate freedom of movement and ability to stand upright, thereby avoiding the need to work in cramped conditions. More particularly, seating should be suitable, providing adequate support for the lower back and a footrest provided, if feet cannot be put comfortably flat on the floor.

Workstations with visual display units (VDUs) are subject to the *Health and Safety (Display Screen Equipment) Regulations 1992 (SI 1992 No 2792* as amended by *SI 2002 No 2174*).

The Health and Safety Executive (HSE) revised its guidance on seating in 1998, and advises employers to use risk assessments in order to ensure that safe seating is provided. *Seating at Work*, HS(G)57 is available, price £5.95, from HSE Books.

## Condition of floors and traffic routes

W11010   The principal dangers connected with industrial and commercial floors are slipping, tripping and falling. Slip, trip and fall resistance are a combination of the right floor surface and the appropriate type of footwear. Employers should ensure that level changes, multiple changes of floor surfaces, steps and ramps etc. are clearly indicated. Safety underfoot is at bottom a trade-off between slip resistance and ease of cleaning. Floors with rough surfaces tend to be more slip-resistant than floors with smooth surfaces, especially when wet; by contrast, smooth surfaces are much easier to clean but less slip-resistant. Use of vinyl flooring in public areas – basically slip-resistant – is on the increase. Vinyl floors should be periodically stripped, degreased and resealed with slip-resistant finish; linoleum floors similarly.

Apart from being safe, floors must also be hygienically clean. In this connection, quarry tiles have long been 'firm favourites' in commercial kitchens, hospital kitchens etc. but can be hygienically deceptive. In particular, grouted joints can trap bacteria as well as presenting endless practical cleaning problems. Hence the gradual transition to seamless floors in hygiene-critical areas. Whichever floor surface is appropriate and whichever treatment is suitable, underfoot safety depends on workplace activity (office or factory), variety of spillages (food, water, oil, chemicals), nature of traffic (pedestrian, cars, trucks).

Thus, floors in workplaces must:

(a) be constructed so as to be suitable for use. [*Workplace (Health, Safety and Welfare) Regulations 1992 (SI 1992 No 3004), Reg 12(1)*].

They should always be of sound construction and adequate strength and stability to sustain loads and passing internal traffic; they should never be overloaded (see *Greaves v Baynham Meikle [1975] 3 All ER 99* for possible consequences in civil law).

(b) (i) not have holes or slopes, or

(ii) not be uneven or slippery

so as to expose a person to risk of injury. [*Workplace (Health, Safety and Welfare) Regulations 1992 (SI 1992 No 3004), Reg 12(2)(a)*].

The surfaces of floors and traffic routes should be even and free from holes, bumps and slipping hazards that could cause a person to slip, trip or fall, or drop or lose control of something being lifted or carried; or cause instability or loss of control of a vehicle.

Holes, bumps or uneven surfaces or areas resulting from damage or wear and tear should be made good and, pending this, barriers should be erected or locations conspicuously marked. Temporary holes, following, say, removal of floorboards, should be adequately guarded. Special needs should be catered for, for instance, disabled walkers or those with impaired sight. (Deep holes are governed by *Workplace (Health, Safety and Welfare) Regulations 1992 (SI 1992 No 3004), Reg 13* (see W9013 WORK AT HEIGHTS).) Where possible, steep slopes should be avoided, and otherwise provided with a secure handrail. Ramps used by disabled persons should also have handrails.

(c) be kept free from

(i) obstructions, and

(ii) articles/substances likely to cause persons to slip, trip or fall

so far as reasonably practicable. [*Workplace (Health, Safety and Welfare) Regulations 1992 (SI 1992 No 3004), Reg 12(3)*].

Floors should be kept free of obstructions impeding access or presenting hazards, particularly near or on steps, stairs, escalators and moving walkways, on emergency routes or outlets, in or near doorways or gangways or by corners or junctions. Where temporary obstructions are unavoidable, access should be prevented and people warned of the possible hazard. Furniture being moved should not be left in a place where it can cause a hazard.

(d) have effective drainage. [*Workplace (Health, Safety and Welfare) Regulations 1992 (SI 1992 No 3004), Reg 12(2)(b)*].

Where floors are likely to get wet, effective drainage (without drains becoming contaminated with toxic, corrosive substances) should drain it away, e.g. in laundries, potteries and food processing plants. Drains and channels should be situated so as to reduce the area of wet floor and the floor should slope slightly towards the drain and ideally have covers flush with the floor surface. Processes and plant which cause discharges or leaks of liquids should be enclosed and leaks from taps caught and drained away. In food processing and preparation plants, work surfaces should be arranged so as to minimise the likelihood of spillage. Where a leak or spillage occurs, it should be fenced off or mopped up immediately.

Staircases should be provided with a handrail. Any open side of a staircase should have minimum fencing of an upper rail at 900mm or higher, and a lower rail.

It is important also to consider the dangers posed by snow and ice upon, for example, external fire escapes.

## Falls and falling objects

**W11011** *Workplace (Health, Safety and Welfare) Regulations 1992 (SI 1992 No 3004), Reg 13* deals with falls and falling objects, which are covered in detail in WORK AT HEIGHTS.

It also requires that tanks, pits and other structures containing dangerous substances are securely covered or fenced where there is a risk of a person falling. Traffic routes over such open structures should also be securely fenced.

## Windows and transparent or translucent doors, gates and walls

**W11012** Transparent or translucent surfaces in windows, doors, gates, walls and partitions should be constructed of safety material or be adequately protected against breakage, where necessary for health and safety reasons, where:

(*a*)   any part is at shoulder level or below in doors and gates; or

(*b*)   any part is at waist level or below in windows, walls and partitions, with the exception of glass houses.

Screens or barriers can be used as an alternative to the use of safety materials. Narrow panels of up to 250mm width are excluded from the requirement.

Transparent or translucent surfaces should be marked to make them apparent where this is necessary for health and safety reasons.

[*Workplace (Health, Safety and Welfare) Regulations 1992 (SI 1992 No 3004), Reg 14*].

## Windows, skylights and ventilators

**W11013** Openable windows, skylights and ventilators must be capable of being opened, closed and adjusted safely. They must not be positioned so as to pose a risk when open.

They should be capable of being reached and operated safely, with window poles or similar equipment, or stable platforms, made available where necessary. Where there is the danger of falling from a height, devices should be provided to prevent this by ensuring the window cannot open too far. They should not cause a hazard by projecting into an area where people are likely to collide with them when open. The bottom edge of opening windows should normally be at least 800mm above floor level, unless there is a barrier to prevent falls.

[*Workplace (Health, Safety and Welfare) Regulations 1992 (SI 1992 No 3004), Reg 15*].

## Ability to clean windows etc. safely

**W11014** *Workplace (Health, Safety and Welfare) Regulations 1992 (SI 1992 No 3004), Reg 16* deals with the safe cleaning of windows and skylights where these cannot be cleaned from the ground or other suitable surface. This is dealt with in detail in WORK AT HEIGHTS.

## Organisation of traffic routes

**W11015** Traffic routes in workplaces should allow pedestrians and vehicles to circulate safely, be safely constructed, be suitably indicated where necessary for health and safety reasons, and be kept clear of obstructions.

They should be planned to give the safest route, wide enough for the safe movement of the largest vehicle permitted to use them, and they should avoid vulnerable items like fuel or chemical plants or pipes, and open and unprotected edges.

There should be safe areas for loading and unloading. Sharp or blind bends should be avoided where possible, and if they cannot be avoided, one-way systems or mirrors to improve visibility should be used. Sensible speed limits should be set and enforced. There should be prominent warning of any limited headroom or potentially dangerous obstructions such as overhead electric cables. Routes should be marked where necessary and there should be suitable and sufficient parking areas in safe locations.

Traffic routes should keep vehicles and pedestrians apart and there should be pedestrian crossing points on vehicle routes. Traffic routes and parking and loading areas should be soundly constructed on level ground. Health and Safety Executive (HSE) guidance in this area can be found in the publication, *Workplace transport safety – guidance for employers*, HS(G)136, price £7.50, available from HSE Books.

[*Workplace (Health, Safety and Welfare) Regulations 1992 (SI 1992 No 3004), Reg 17*].

This area is dealt with in detail in ACCESS, TRAFFIC ROUTES AND VEHICLES.

## Doors and gates

**W11016** Doors and gates must be suitably constructed and fitted with safety devices. In particular,

(i) a sliding door/gate must have a device to prevent it coming off its track during use;

(ii) an upward opening door/gate must have a device to prevent its falling back;

(iii) a powered door/gate must

    (*a*) have features preventing it causing injury by trapping a person (e.g. accessible emergency stop controls),

    (*b*) be able to be operated manually unless it opens automatically if the power fails;

(iv) a door/gate capable of opening, by being pushed from either side, must provide a clear view of the space close to both sides.

[*Workplace (Health, Safety and Welfare) Regulations 1992 (SI 1992 No 3004), Reg 18*].

Doors and gates that swing in both directions should have a transparent panel, unless they are low enough to see over.

## Escalators and travelators

**W11017** Escalators and travelators must:

(*a*) function safely;

(*b*) be equipped with safety devices;

(*c*) be fitted with emergency stop controls.

[*Workplace (Health, Safety and Welfare) Regulations 1992 (SI 1992 No 3004), Reg 19*].

## Welfare facilities

**W11018** 'Welfare facilities' is a wide term, embracing both sanitary and washing accommodation at workplaces, provision of drinking water, clothing accommodation (including facilities for changing clothes) and facilities for rest and eating meals (see W11027 below). The need for sufficient suitable hygienic lavatory and washing facilities in all workplaces is obvious. Sufficient facilities must be provided to enable everyone at work to use them without undue delay. They do not have to be in the actual workplace but ideally should be situated in the building(s) containing them and they should provide protection from the weather, be well-ventilated, well-lit and enjoy a reasonable temperature. Where disabled workers are employed, special provision should be made for their sanitary and washing requirements. Wash basins should allow washing of hands, face and forearms and, where work is particularly strenuous, dirty, or results in skin contamination (e.g. molten metal work), showers or baths should be provided. In the case of showers, they should be fed by hot and cold water and fitted with a thermostatic mixer valve. Washing facilities should ensure privacy for the user and be separate from the water closet, with a door that can be secured from the inside. It should not be possible to see urinals or the communal shower from outside the facilities when the entrance/exit door opens. Entrance/exit doors should be fitted to both washing and sanitary facilities (unless there are other means of ensuring privacy). Windows to sanitary accommodation, showers/bathrooms should be obscured either by being frosted, or by blinds or curtains (unless it is impossible to see into them from outside).

This section examines current statutory requirements in all workplaces. For requirements relating to sanitary conveniences and washing facilities on construction sites see C8056 CONSTRUCTION AND BUILDING OPERATIONS.

## Sanitary conveniences in all workplaces

**W11019** Suitable and sufficient sanitary conveniences must be provided at readily accessible places. In particular,

(*a*) the rooms containing them must be adequately ventilated and lit;

(*b*) they (and the rooms in which they are situated) must be kept clean and in an orderly condition;

(*c*) separate rooms containing conveniences must be provided for men and women except where the convenience is in a separate room which can be locked from the inside.

[*Workplace (Health, Safety and Welfare) Regulations 1992 (SI 1992 No 3004), Reg 20*].

## Washing facilities in all workplaces

**W11020** Suitable and sufficient washing facilities (including showers where necessary (see W11018 above)), must be provided at readily accessible places or points. In particular, facilities must:

(*a*) be provided in the immediate vicinity of every sanitary convenience (whether or not provided elsewhere);

(*b*) be provided in the vicinity of any changing rooms – whether or not provided elsewhere;

(*c*) include a supply of clean hot and cold or warm water (if possible, running water);

(*d*) include soap (or something similar);

(*e*) include towels (or the equivalent);

(*f*) be in rooms sufficiently well-ventilated and well-lit;

(*g*) be kept clean and in an orderly condition (including rooms in which they are situate);

(h) be separate for men and women, except where they are provided in a lockable room intended to be used by one person at a time, or where they are provided for the purposes of washing hands, forearms and face only, where separate provision is not necessary.

[*Workplace (Health, Safety and Welfare) Regulations 1992 (SI 1992 No 3004), Reg 21*].

## *Minimum number of facilities – sanitary conveniences and washing facilities*

(a) People at work

**W11021**

| Number of people at work | Number of WCs | Number of wash stations |
|---|---|---|
| 1 to 5 | 1 | 1 |
| 6 to 25 | 2 | 2 |
| 26 to 50 | 3 | 3 |
| 51 to 75 | 4 | 4 |
| 76 to 100 | 5 | 5 |

(b) Men at work

| Number of men at work | Number of WCs | Number of urinals |
|---|---|---|
| 1 to 15 | 1 | 1 |
| 16 to 30 | 2 | 1 |
| 31 to 45 | 2 | 2 |
| 46 to 60 | 3 | 2 |
| 61 to 75 | 3 | 3 |
| 76 to 90 | 4 | 3 |
| 91 to 100 | 4 | 4 |

For every 25 people above 100 an additional WC and wash station should be provided; in the case of WCs used only by *men*, an additional WC per every 50 men above 100 is sufficient (provided that at least an equal number of additional urinals is provided). [*Workplace (Health, Safety and Welfare) Regulations 1992 (SI 1992 No 3004), Sch 1, Part II*].

## *Particularly dirty work etc.*

**W11022** Where work results in heavy soiling of hands, arms and forearms, there should be one wash station for every 10 people at work up to 50 people; and one extra for every additional 20 people. And where sanitary and wash facilities are also used by members of the public, the number of conveniences and facilities should be increased so that workers can use them without undue delay.

## *Temporary work sites*

**W11023** At temporary work sites suitable and sufficient sanitary conveniences and washing facilities should be provided so far as is reasonably practicable (see E15039 ENFORCEMENT for meaning). If possible, these should incorporate flushing sanitary conveniences and washing facilities with running water.

## Drinking water

**W11024** An adequate supply of wholesome drinking water must be provided for all persons at work in the workplace. It must be readily accessible at suitable places and conspicuously marked, unless non-drinkable cold water supplies are clearly marked. In addition, there must be provided a sufficient number of suitable cups (or other drinking vessels), unless the water supply is in a jet. [*Workplace (Health, Safety and Welfare) Regulations 1992 (SI 1992 No 3004), Reg 22*].

Where water cannot be obtained from the mains supply, it should only be provided in refillable containers. The containers should be enclosed to prevent contamination and refilled at least daily. So far as reasonably practicable, drinking water taps should not be installed in sanitary accommodation, or in places where contamination is likely, for instance, in a workshop containing lead processes.

## Clothing accommodation

**W11025** Suitable and sufficient accommodation must be provided for:

(*a*) any person at work's own clothing which is not worn during working hours; and

(*b*) special clothing which is worn by any person at work but which is not taken home, for example, overalls, uniforms and thermal clothing.

[*Workplace (Health, Safety and Welfare) Regulations 1992 (SI 1992 No 3004), Reg 23(1)*].

Accommodation is not suitable unless it:

(i) provides suitable security for the person's own clothing where changing facilities are required;

(ii) includes separate accommodation for clothing worn at work and for other clothing, where necessary to avoid risks to health or damage to clothing; and

(iii) is in a suitable location.

[*Workplace (Health, Safety and Welfare) Regulations 1992 (SI 1992 No 3004), Reg 23(2)*].

Work clothing is overalls, uniforms, thermal clothing and hats worn for hygiene purposes. Workers' own clothing should be able to hang in a clean, warm, dry, well-ventilated place. If this is not possible in the workroom, then it should be put elsewhere. Accommodation should take the form of a separate hook or peg. Clothing which is dirty, damp or contaminated owing to work should be accommodated separately from the worker's own clothes.

## Facilities for changing clothing

**W11026** Suitable and sufficient facilities must be provided for any person at work in the workplace to change clothing where:

(*a*) the person has to wear special clothing for work, and

(*b*) the person cannot be expected to change in another room.

Facilities are not suitable unless they include:

(i) separate facilities for men and women, or

(ii) separate use of facilities by men and women.

[*Workplace (Health, Safety and Welfare) Regulations 1992 (SI 1992 No 3004), Reg 24*].

Changing rooms (or room) should be provided for workers who change into special work clothing and where they remove more than outer clothing; also where it is necessary to prevent workers' own clothes being contaminated by a harmful substance. Changing facilities should be easily accessible from workrooms and eating places. They should contain adequate seating and clothing accommodation, and showers or baths if these are provided (see W11018 above). Privacy of user should be ensured. The facilities should be large enough to cater for the maximum number of persons at work expected to use them at any one time without overcrowding or undue delay.

*Post Office v Footitt [2000] IRLR 243*, involved an employers' appeal against an improvement notice requiring the construction of a separate changing room for women postal workers to change into and out of their uniforms, and looked at the definition of 'special clothing' and at the concept of propriety.

An environmental health officer had served an improvement notice under the *Workplace (Health, Safety and Welfare) Regulations 1992 (SI 1992 No 3004), Reg 24* as she had found that any female employees wishing to change their clothing could only do so in the general area of the women's toilet facilities.

In the High Court, the judge held that the uniform worn by postal workers was 'special clothing' for the purposes of *Regulation 24*. It was held that 'special clothing' is not merely limited to clothing that is worn only at work. Therefore the fact that postal workers wear their uniform to and from work does not prevent it from being 'special clothing'. The changing facilities for women provided by the Post Office were therefore not 'suitable and sufficient' within the meaning of the Regulation.

The court also held that the fact that the changing facilities for men and women were separated was not in itself enough to satisfy the concept of propriety referred to in *Regulation 24(2)*. There is no reason why requiring one female to undress in the presence of another cannot be said to offend against the principles of propriety. The fact that many people would have no objection to changing in the company of others of the same sex does not absolve the employer from providing facilities for those who may prefer privacy.

## *Rest and eating facilities*

**W11027** Suitable and sufficient rest facilities must be provided at readily accessible places. [*Workplace (Health, Safety and Welfare) Regulations 1992 (SI 1992 No 3004), Reg 25(1)*].

### (a) Rest facilities

A rest facility is:

(i)   in the case of a new workplace, extension or conversion – a rest room (or rooms);

(ii)  in other cases, a rest room (or rooms) or rest area; including

(iii) (in both cases):

— appropriate facilities for eating meals where food eaten in the workplace would otherwise be likely to become contaminated;

— suitable arrangements for protecting non-smokers from tobacco smoke. The ACoP advises that this can be achieved by providing separate areas for smokers and non-smokers, or by prohibiting smoke in rest areas;

— a facility for a pregnant or nursing mother to rest in.

Canteens or restaurants may be used as rest rooms provided that there is no obligation to buy food there (ACoP). [*Workplace (Health, Safety and Welfare) Regulations 1992 (SI 1992 No 3004), Regs 25(2)–(4)*].

### (b) Eating facilities

Where workers regularly eat meals at work, facilities must be provided for them to do so. [*Workplace (Health, Safety and Welfare) Regulations 1992 (SI 1992 No 3004), Reg 25(5)*].

In offices and other workplaces where there is no risk of contamination, seats in the work area are sufficient, although workers should not be interrupted excessively during breaks, for example, by the public. In other cases, rest areas or rooms should be provided and in the case of new workplaces, this should be a separate rest room. Rest facilities should be large enough, and have enough seats with backrests and tables, for the number of workers likely to use them at one time.

Where workers regularly eat meals at work, there should be suitable and sufficient facilities. These should be provided where food would otherwise be contaminated, by dust or water for example. Seats in work areas can be suitable eating facilities, provided the work area is clean. There should be a means to prepare or obtain a hot drink, and where persons work during hours or at places where hot food cannot be readily obtained, there should be the means for heating their own food. Eating facilities should be kept clean.

## *Smoking*

**W11028** Providing protection to non-smokers from tobacco smoke in rest areas is the only specific legal requirement concerning smoking at work, although the general duty to ensure the health, safety and welfare of the employees under the *Health and Safety at Work etc. Act 1974 (HSWA 1974)*, s 2 applies. However there have been a number of legal cases, and compensation awards have been made to employees who have claimed their health has been affected by breathing in tobacco smoke at work.

Stockport Metropolitan Council has made two out of court settlements of £25,000 and £15,000 to employees who claimed that their health was damaged as a result of passive smoking. In addition, an Employment Appeal Tribunal (EAT) case, *Dryden v Greater Glasgow Health Board [1992] IRLR 469*, held that a change to a complete smoking ban, which meant an employee who smoked had to leave her job, did not amount to constructive dismissal. In this case, the employer had consulted workers about the introduction of the ban.

In another EAT case, *Walton and Morse v Dorrington [1997]*, a tribunal decision that an employee had been constructively dismissed when her employer failed to provide her with a smoke free environment or deal with her problems relating to passive smoking, was upheld.

In the first passive smoking case to reach the courts, in May 1998, a nurse lost her action for damages against her employer. Silvia Sparrow claimed that she had developed asthma as a result of exposure to environmental tobacco smoke in a residential care home for elderly people. But the court said she had failed to prove that her former employers, St Andrew's Homes Ltd, were negligent so as to cause injury to her.

The Health and Safety Commission (HSC) has recommended that an approved code of practice (ACoP) on passive smoking should be introduced. This would give authoritative guidance on the employer's legal obligation to protect their employees from exposure to environmental tobacco smoke (ETS).

As with other ACoPs, failing to follow the code would not in itself be an offence, but the employer would have to demonstrate that equally effective methods have been adopted to signal compliance with the law.

Under the code, employers will have to determine the most reasonably practical way of controlling ETS. This could involve:

- Banning smoking in the workplace completely or partially;
- Physically segregating non-smokers from tobacco smoke;
- Providing adequate ventilation; or
- Implementing a system of work to reduce the time an employee is exposed to ETS.

There will, however, be a two-year exemption for parts of the hospitality industry, such as bars, clubs and restaurants. They will need to instead comply with the Public Places Charter during this period.

The code is currently awaiting ministerial approval.

The current HSE publication, *Passive smoking at work*, IND(G)63(L), recommends that all employers should introduce a policy to control smoking in the workplace in full consultation with employees.

The guidance sets out that environmental tobacco smoke contains carbon dioxide, hydrogen cyanide and ammonia and that passive smoking has irritant effects on the eyes, throat and respiratory tract. In addition to aggravating asthma, research indicates that it can increase the risk of lung cancer and may increase the risk of heart disease.

The HSE agrees with the advice of the Independent Scientific Committee which says that employers should regard non-smoking as the norm in enclosed workplaces and make provision for smoking, rather than vice versa, and that smokers should be segregated from non-smokers.

The guidance advises employers that they should have a specific written policy on smoking in the workplace which gives priority to the needs of non-smokers. Any policy should be introduced with proper consultation of employees and their representatives and a minimum of three months' notice of its introduction should be given. It also advises that may employers provide help to smokers to reduce or give up smoking.

It outlines that the only effective ways to achieve a smoke-free environment for non-smokers are to:

- Introduce a complete ban on indoor smoking; or
- Ban smoking in all parts of the building with the exception of designated, enclosed smoking areas.

A less effective method, which employers may have to resort to if the constraints of the workplace do not allow the above, is to segregate smokers and non-smokers in separate rooms and ban smoking in common areas.

The guidance advises employers in buildings with mechanical ventilation to consider discharging air from smoking areas separately, and if this is not reasonably practicable, decontamination systems should be used to bring the re-circulated air up to an appropriate standard.

### Civil liability

W11029    There is no specific reference to civil liability in the Regulations. However, safety regulations are actionable, even if silent (as here), and, if a person suffered injury/damage as a result of breach by an employer, he could sue. Certainly, there is civil liability for breach of the *Building Regulations 1991 (SI 1991 No 2768)* (see W11044 below).

## Safety signs at work – Health and Safety (Safety Signs and Signals) Regulations 1996 (SI 1996 No 341)

W11030    Traditionally, safety signs, communications and warnings have played a residual role in reducing the risk of injury or damage at work, the need for them generally having been engineered out or accommodated in the system of work – a situation unaffected by these Regulations.

### Types of signs

W11031    Safety signs and signals can be of the following types:

(*a*)    permanent (e.g. signboards);

(*b*)    occasional (e.g. acoustic signals or verbal communications – acoustic signals should be avoided where there is considerable ambient noise).

## Interchanging and combining signs

**W11032** Examples of interchanging and combining signs are:

(a) a safety colour (see W11033 below) or signboard to mark places where there is an obstacle;

(b) illuminated signs, acoustic signals or verbal communication; and

(c) hand signals or verbal communication.

[*Health and Safety (Safety Signs and Signals) Regulations 1996 (SI 1996 No 341), Sch 1, Part I, para 3*].

## Safety colours

**W11033**

| Colour | Meaning or purpose | Instructions and information |
|---|---|---|
| Red | Prohibition sign | Dangerous behaviour |
| | Danger | Stop, shutdown, emergency cut-out services |
| | | Evacuate |
| | Fire-fighting equipment | Identification and location |
| Yellow or Amber | Warning sign | Be careful, take precautions |
| | | Examine |
| Blue | Mandatory sign | Specific behaviour or action |
| | | Wear personal protective equipment |
| Green | Emergency escape, first-aid sign | Doors, exits, routes, equipment and facilities |
| | No danger | Return to normal |

[*Health and Safety (Safety Signs and Signals) Regulations 1996 (SI 1996 No 341), Sch 1, Part I, para 4*].

## Varieties of safety signs and signals

**W11034** Safety signs and signals include, comprehensively:

(a) safety signs – providing information about health and safety at work by means of a signboard, safety colour, illuminated sign, acoustic signal, hand signal or verbal communication;

(b) signboards – signs giving information by way of a simple pictogram, lighting intensity providing visibility (these should be weather-resistant and easily seen);

(c) mandatory signs – signs prescribing behaviour (e.g. safety boots must be worn);

(d) prohibition signs – signs prohibiting behaviour likely to cause a health and safety risk (e.g. no smoking);

(e) hand signals – movement or position of arms/hands for guiding persons carrying out operations that could endanger employees;

(*f*) verbal communications – predetermined spoken messages communicated by human or artificial voice, preferably short, simple and as clear as possible.

[*Health and Safety (Safety Signs and Signals) Regulations 1996 (SI 1996 No 341), Reg 2*].

## Duty of employer

**W11035** It is only where a risk assessment carried out under the *Management of Health and Safety at Work Regulations 1999 (SI 1999 No 3242)* indicates that a risk cannot be avoided, engineered out or reduced significantly by way of a system of work that resort to signs and signals becomes necessary. In these circumstances, all employers (including offshore employers) must:

(*a*) provide and maintain any appropriate safety sign(s) (see W11037–W11041 below) (including fire safety signals) but not a hand signal or verbal communication;

(*b*) so far as is reasonably practicable, ensure that correct hand signals or verbal communications are used;

(*c*) provide and maintain any necessary road traffic sign (where there is a risk to employees in connection with traffic); and

(*d*) provide employees with comprehensible and relevant information, training and instruction and measures to be taken in connection with safety signs.

[*Health and Safety (Safety Signs and Signals) Regulations 1996 (SI 1996 No 341), Regs 4, 5*].

*Schedule 1* to the Regulations sets out the minimum requirements concerning safety signs and signals with regard to the type of signs to be used in particular circumstances, interchanging and combining signs, signboards, signs on containers and pipes, the identification and location of fire-fighting equipment, signs used for obstacles and dangerous locations, and for marking traffic routes, illuminated signs, acoustic signals, verbal communication and hand signals.

## Exclusions

**W11036** Excluded from the operation of these Regulations are:

(*a*) the supply of dangerous substances or products;

(*b*) the transportation of dangerous goods;

(*c*) road traffic signs (except where there is a particular risk to employees. Where there is a risk arising from the movement of traffic and the risk is addressed by a sign stipulated in the *Road Traffic Regulations Act 1984* (e.g. speed restriction sign), these signs must be used, whether or not the Act applies to that place of work. In effect this means that where road speed and other signs are needed on a company's road, these must replicate the signs used for the purpose on public roads; and

(*d*) activities on board ship.

[*Health and Safety (Safety Signs and Signals) Regulations 1996 (SI 1996 No 341), Reg 3(1)*].

## Examples of safety signs

### Prohibitory signs

**W11037** Intrinsic features:

— round shape

— black pictogram on white background, red edging and diagonal line (the red part to take up at least 35% of the sign area).

## fig. 1 Safety signs (prohibitory)

No smoking

Smoking and naked flames forbidden

No access for pedestrians

Do not extinguish with water

Not drinkable

No access for unauthorised persons

No access for industrial vehicles

Do not touch

## Warning signs

**W11038**  Intrinsic features:

- triangular shape
- black pictogram on a yellow background with black edging (the yellow part to take up at least 50% of the area of the sign).

## fig. 2 Safety signs (warning)

Flammable material or high temperature

Explosive material

Toxic material

*Workplaces – Health, Safety and Welfare* **W11039**

Corrosive material

Radioactive material

Overhead load

Industrial vehicles

Danger: electricity

General danger

Laser beam

Oxidant material

Non-ionizing radiation

Strong magnetic field

Obstacles

Drop

Biological risk

Low temperature

Harmful or irritant material

## *Mandatory signs*

**W11039**  Intrinsic features:

- round shape
- white pictogram on a blue background (the blue part to take up at least 50% of the area of the sign).

fig. 3 Safety signs (mandatory)

## Emergency escape or first-aid signs

**W11040**   Intrinsic features:

- rectangular or square shape
- white pictogram on a green background (the green part to take up at least 50% of the area of the sign).

**fig. 4 Safety signs (emergency escape or first-aid)**

Emergency exit/escape route

This way
(supplementary information sign)

First-aid post   Stretcher   Safety shower   Eyewash

## Fire-fighting signs

**W11041**   Intrinsic features:

- rectangular or square shape
- white pictogram on a red background (the red part to take up at least 50% of the area of the sign).

**W11042**  *Workplaces – Health, Safety and Welfare*

Emergency telephone for first-aid or escape

fig. 5 Safety signs (fire-fighting)

    Fire hose           Ladder           Fire          Emergency fire
                                     extinguisher    telephone

This way
(supplementary information sign)

## Examples of hand signals
W11042

*Meaning*                *Description*                *Illustration*

**A. General signals**

fig. 6 Hand signals

| Meaning | Description | Illustration |
|---|---|---|
| START<br>Attention<br>Start of<br>Command | both arms are extended horizontally with the palms facing forwards. |  |

Workplaces – Health, Safety and Welfare  **W11042**

| | | |
|---|---|---|
| STOP<br>Interruption<br>End of movement | the right arm points upwards with the palm facing forwards. |  |
| END<br>of the operation | both hands are clasped at chest height. |  |

## B. Vertical movements

| | | |
|---|---|---|
| RAISE | the right arm points upwards with the palm facing forward and slowly makes a circle. |  |
| LOWER | the right arm points downwards with the palm facing inwards and slowly makes a circle. |  |
| VERTICAL DISTANCE | the hands indicate the relevant distance. |  |

## C. Horizontal movements

| | | |
|---|---|---|
| MOVE FORWARDS | both arms are bent with the palms facing upwards, and the forearms make slow movements towards the body. |  |

W110/23

| | | |
|---|---|---|
| MOVE BACKWARDS | both arms are bent with the palms facing downwards, and the forearms make slow movements away from the body. |  |
| RIGHT to the signalman's | the right arm is extended more or less horizontally with the palm facing downwards and slowly makes small movements to the right. |  |
| LEFT to the signalman's | the left arm is extended more or less horizontally with the palm facing downwards and slowly makes small movements to the left. | |
| HORIZONTAL DISTANCE | the hands indicate the relevant distance. |  |

### D. Danger

| | | |
|---|---|---|
| DANGER Emergency stop | both arms point upwards with the palms facing forwards. |  |
| QUICK | all movements faster. | |
| SLOW | all movements slower. | |

## Disability Discrimination Act 1995

**W11043** The *Disability Discrimination Act 1995* (*DDA 1995*) makes it unlawful for employers (who employ 15 or more people) to treat a disabled person less favourably, without a justifiable reason.

In order to gain protection under the *DDA 1995*, a person must have a substantial and long-term disability, affecting them (or which could affect them) for more than 12 months, impacting on their ability to carry out normal day-to-day activities, and affecting their mobility, dexterity, co-ordination, continence or memory. It covers mental as well as physical impairments, and applies to recruitment, selection, promotion and redundancy. It applies to all workers, including the self-employed, contract workers and agency workers, as well as employees. It does not currently apply to the police, prison and fire services nor to the armed forces.

The *DDA 1995, s 6* requires employers to make reasonable adjustments to working conditions or to the workplace to avoid putting disabled workers at a substantial disadvantage, and could therefore involve making adjustments to the physical features of workplace premises.

The Act gives examples of adjustments to the workplace that could be required, including access for wheelchairs and acquiring or modifying wheelchairs.

There is a provision in the Act to enable employers who lease their premises to obtain their landlord's permission to make alterations in order to make any necessary reasonable adjustments to the workplace.

The *Disability Discrimination (Employment Relations) Regulations 1996 (SI 1996 No 1456)* set out the duty to make adjustments in more detail. The Regulations specify that physical features of the employer's premises are:

(a) any feature arising from the design or construction of a building on the premises or any approach to, exit from or access to such a building;

(b) any fixtures, fittings, furniture, equipment or materials in or on the premises; and

(c) any other physical element or quality of land included in the premises.

Complaints about disability discrimination are heard in employment tribunals. In addition, the Disability Rights Commission, which came into effect in April 2000, can carry out formal investigations where it believes there is discrimination and can enter into legally binding agreements with employers where the employer can choose to comply with it's recommendations. It can also assist individuals taking cases to tribunals.

In *Tawling v Wisdom Toothbrushes Ltd*, a 1997 tribunal case, a woman who had a club foot and was experiencing sciatica and other pain and discomfort as a result of standing for long periods had to take increasing amounts of sickness absence.

Her employers sought advice from an organisation providing financial support for people with disabilities, the Shaw Trust. It recommended the purchase of one of two specific types of chair costing £500 and around £1000 respectively, to which the employer would have had to fund 20 per cent of the cost. The employer did not take the advice and instead provided her with a series of ordinary chairs, which did not meet her needs and then dismissed her for poor performance.

The tribunal found that the woman had been unfairly dismissed because the employer had failed to make reasonable adjustments in not providing a suitable chair for her.

## Building Regulations 1991 (SI 1991 No 2768)

W11044    Where workplaces are built, extended or altered, or where fittings such as drains, heat-producing appliances, washing and sanitary facilities or hot water storage are provided, the *Building Regulations 1991 (SI 1991 No 2768)* normally apply.

The Regulations set down minimum standards of design and building work for the construction of buildings. They contain requirements designed to ensure the health and safety of people in and around buildings, to provide for energy conservation and to provide access for disabled people.

The requirements, which are set out in *Schedule 1* of the Regulations, concern:

- *Structure*

    Buildings must be constructed to withstand loads and ground movement. Buildings of five or more storeys must be constructed in such a way that in the event of an accident it will not suffer collapse to an extent disproportionate to the cause. Where parts of a public building, shop or shopping mall have a roof with a clear span exceeding nine metres between supports, these should be constructed so that in the event of the failure of any part of the roof, the building will not suffer collapse to an extent disproportionate to that failure;

- *Fire safety*

    Buildings must be designed and constructed with a means of escape, in the case of fire, from the building to a place of safety outside the building. Internal linings of the building must inhibit the spread of fire and the building must be designed and constructed to maintain stability for a period of time in the event of fire. Party walls should resist the spread of fire between buildings and the spread of fire within buildings should be inhibited by sub-division with fire-resisting construction as appropriate to the size and intended use of the building. The building should be designed and constructed so that the unseen spread of fire and smoke within concealed spaces in the structure and fabric of the building is inhibited. The external walls and the roof of the building should resist the spread of fire, and the building should be designed and constructed to provide facilities for fire fighters and to enable fire fighting appliances to gain access to the building. Fire safety in workplaces is dealt with in detail in FIRE PREVENTION AND CONTROL.

- *Site preparation and resistance to moisture*

    The ground to be covered by the building should be reasonably free from vegetable matter, and precautions should be taken to avoid risks to health and safety from any dangerous or offensive substances found on or in this ground, and subsoil drainage should be provided where necessary to avoid damage to the fabric of the building or the passage of ground moisture to the interior of the building. The walls, floors and roof of the building should also prevent damp penetration.

- *Toxic substances*

    If cavity wall insulation is used, precautions should be taken to prevent any subsequent release of toxic fumes into any occupied parts of the building.

- *Resistance to the passage of sound*

    This part applies only to dwellings.

- *Ventilation*

    Rooms containing sanitary conveniences must be adequately ventilated, and provision must be made to prevent excessive condensation in a roof or a roof void above an insulated ceiling.

- *Hygiene*

    Adequate sanitary conveniences shall be provided in rooms provided for that purpose and they must be separated from places where food is prepared. Adequate wash basins must be provided in or next to rooms containing wcs, and again must be separated from areas where food is being prepared. There should be suitable installation for the provision of hot and cold water to washbasins, and wcs and washbasins should be designed and installed to allow them to be effectively cleaned. There are also requirements concerning the installation of hot water storage systems.

- *Drainage and waste disposal*

    There must be a waste water drainage system which takes foul water from toilets, washbasins and sinks to a sewer, or properly constructed cesspool or septic or settlement tank. There should be adequate rainwater drainage and adequate means of storing solid waste.

- *Heat producing appliances*

    Heat producing appliances designed to burn solid fuel, oil or gas and incinerators should have adequate air supply, and adequate provision for the discharge of combustion products to the outside air. Appliances should be installed, and fireplaces and chimneys constructed to reduce the risk of the building catching fire as a consequence of their use.

- *Protection from falling, collision and impact*

    Stairs, ladders and ramps should allow users to move safely between different levels of the building, and there should be barriers where necessary to protect people from falling. Vehicle ramps and floors and roofs with vehicular access should also be guarded with barriers where necessary to protect people in and around the building.

- *Conservation of fuel and power*

    Reasonable provision should be made for the conservation of fuel and power in buildings where the floor area exceeds 30 metres squared.

- *Access and facilities for disabled people*

    Reasonable provision should be made for disabled people to gain access to and use the building; and to use any sanitary conveniences provided. This applies to people who have an impairment which limits their ability to walk or requires them to use a wheelchair for mobility, or who have impaired hearing or sight. It excludes extensions which do not include a ground storey; material alterations and parts of a building used solely for the maintenance, inspection or repair of fittings or services.

- *Glazing*

    Glazing which people are likely to come into contact with must break on impact in a way unlikely to cause injury, or resist impact without breaking or be shielded or protected from impact. Transparent glazing with which people are likely to collide should be made apparent; and

- *Materials and workmanship*

    Any building work should be carried out with proper materials and in a workmanlike manner.

There are approved documents setting out detailed practical and technical guidance on how these requirements are to be met, although the requirements can be met in alternative ways if these are available.

The *Building Regulations 1991* (*SI 1991 No 2768*), *Reg 11* requires that anyone intending to carry out building work or make a material change of use must give a building notice to, or deposit full plans with the local authority. This does not have to be done where gas appliances are being installed by, or are under the supervision of people approved under the Gas Safety Regulations. Full plans are only required where buildings are to be put to a use designated under the *Fire Precautions Act 1971*.

An approved inspector can be chosen to ensure compliance with the Regulations rather than the local authority, in which case a building notice or deposit of full plans is not required. The list of approved inspectors is available from the Association of Corporate Approved Inspectors (see website: http://www.acai.org.uk).

The Regulations and approved documents can be purchased from the Stationary Office on 0870 600 5522.

# Factories Act 1961

**W11045** Although large parts of the *Factories Act 1961* have now been repealed and replaced by more recent legislation, some provisions are still in force, although the Health and Safety Commission (HSC) is considering complete repeal. Additionally, civil actions for injury, relating to breach of health and safety provisions of the *Factories Act* (though not welfare) may well continue for some time, since actions for personal injury can be initiated for up to three years after injury/disease has occurred. [*Limitation Act 1980, s 11*].

## Residual application of the Factories Act 1961

**W11046** The *Factories Act 1961* applies to factories, as defined in *s 175*, including 'factories belonging to or in the occupation of the Crown, to building operations and works of engineering construction undertaken by or on behalf of the Crown, and to employment by or under the Crown of persons in painting buildings', e.g. hospital painters. [*Factories Act 1961, s 173(1)*].

## Enforcement of the Factories Act 1961 and regulations

**W11047** Offences under the Act are normally committed by occupiers rather than owners of factories. Unless they happen to occupy a factory as well, the owners of a factory would not normally be charged. Offences therefore relate to physical occupation or control of a factory (for an extended meaning of 'occupier', see OCCUPIERS' LIABILITY). Hence the person or persons or body corporate having managerial responsibility in respect of a factory are those who commit an offence under *s 155(1)*. This will generally be the managing director and board of directors and/or individual executive directors. Moreover, if a company is in liquidation and the receiver is in control, he is the person who will be prosecuted and this has in fact happened (*Meigh v Wickenden [1942] 2 KB 160; Lord Advocate v Aero Technologies 1991 SLT 134* where the receiver was 'in occupation' and so under a duty to prevent 'accidents by fire or explosion', for the purposes of the *Explosives Act 1875, s 23*).

## Defence of factory occupier

**W11048** The main defence open to a factory occupier charged with breach of the *Factories Act 1961* is that the Act itself, or more likely regulations made under it, placed the statutory duty on some person other than the occupier. Thus, where there is a contravention by any person of any regulation or order under the *Factories Act 1961*, 'that person shall be guilty of an offence and the occupier or owner ... shall not be guilty of an offence, by reason only of the contravention of the provision ... unless it is proved that he failed to take all reasonable steps to prevent the contravention ...'. [*Factories Act 1961, s 155(2)*].

Before this defence can be invoked by a factory occupier or company, it is necessary to show that:

(*a*) a statutory duty had been laid on someone other than the factory occupier by a regulation or order passed under the Act;

(*b*) the factory occupier took all reasonable steps to prevent the contravention (a difficult test to satisfy).

*NB*. This statutory defence is not open to a building contractor (in his capacity as a notional factory occupier).

## Effect on possible civil liability

**W11049** Whether conviction of an employee under the *Factories Act 1961, s 155(2)* would prejudice a subsequent claim for damages by him against a factory occupier, must be regarded as an open question. Thus, in *Potts v Reid [1942] 2 All ER 161* the court said 'Criminal and civil liability are two separate things ... The legislation (the *Factories Act 1937*) might well be unwilling to convict an owner who failed to carry out a statutory duty of a crime with which he was not himself directly concerned, but still be ready to leave the civil liability untouched'. Similarly in *Boyle v Kodak Ltd [1969] 2 All ER 439* it was said, 'When considering the civil liability engrafted by judicial decision upon the criminal liability which has been imposed by statute, it is no good looking to the statute and seeing from it where the criminal liability would lie, for we are concerned only with civil liability. We must look to the cases' (per Lord Diplock). Moreover, a breach of general duties of *HSWA 1974* gives rise only to civil liability at common law and not under statute. (Though this is not the position where there is a breach of a specific regulation under *HSWA 1974*.) On the other hand, there is at least one isolated instance of an employee being denied damages where he was in breach of specific regulations (*ICI Ltd v Shatwell [1964] 2 All ER 999*). It is thought, however, that this decision would not apply in the case of breach of a *general* statutory duty, such as *s 155(2)*.

# Appendix 1
# Associations, organisations and departments connected with health and safety

**Advisory, Conciliation and Arbitration Service (ACAS)**
Brandon House
180 Borough High Street
London
SE1 1LW
Telephone: (020) 7210 3613

**Association of British Insurers (ABI)**
Head Office
51 Gresham Street
London
EC2V 7HQ
Telephone: (020) 7600 3333

**Building Research Establishment**
Bucknalls Lane
Garston
Watford
Hertfordshire
WD25 9XX
Telephone: (01923) 664000

**British Chiropractic Association (BCA)**
Blagrave House
17 Blagrave Street
Reading
Berkshire
RG1 1QB
Telephone: (0118) 950 5950

**British Industrial Truck Association (BITA)**
5–7 High Street
Sunnighill
Ascot
Berkshire
SL5 9NQ
Telephone: (01344) 623 800

**British Occupational Hygiene Society**
Suite 2
Georgian House
Great Northern Road
Derby
DE1 1LT
Telephone: (01332) 298 101

*Appendix*

**British Safety Council (BSC)**
National Safety Centre
70 Chancellors Road
London
W6 9RS
Telephone: (020) 8741 1231

**British Standards Institution (BSI)**
389 Chiswick High Road
London
W4 4AL
Telephone: (020) 8996 9000

**Chartered Institute of Environmental Health (CIEH)**
Chadwick Court
15 Hatfields
London
SE1 8DJ
Telephone: (020) 7928 6006

**Construction Industry Training Board (CITB)**
Bircham Newton
Kings Lynn
Norfolk
PE31 6RH
Telephone: (01485) 577 577
Telephone: (0845) 609 9960

**Department for Transport and the Office of the Deputy Prime Minister**
Public Enquiries
Ashdown House
123 Victoria Street
London
SW1E 6DE
Telephone: (020) 7890 3000

**Department of Health (DoH)**
Public Enquiries
Richmond House
79 Whitehall
London
SW1A 2NS
Telephone: (020) 7210 4850

**Department for Work and Pensions (DWP)**
Correspondence Unit
Room 540
The Adelphi
1–11 John Adam Street
London
WC2N 6HT
Telephone: (020) 7712 2171

**Department of Trade and Industry (DTI)**
Public Enquiries
1 Victoria Street
London
SW1H 0ET
Telephone: (020) 7215 5000

*Appendix*

**Environment Agency**
Apollo Court
2 Bishops Square Business Park
St Albans Road West
Hatfield
Hertfordshire
AL10 9EX
Telephone: (08708) 506 506

**Fire Protection Association (FPA)**
Bastille Court
2 Paris Garden
London
SE1 8ND
Telephone: (0160) 881 2500

**Health and Safety Commission (HSC)**
Rose Court
2 Southwark Bridge
London
SE1 9HS
Telephone: (020) 7717 6000

**Health and Safety Executive (HSE)**
Information Centre & Public Enquiry Point
Caerphilly Business Park
Caerphilly
CF83 3GG
Telephone: (08701) 545 500

**Health and Safety Publications**
HSE Books
PO Box 1999
Sudbury
Suffolk
CO10 2WA
Telephone: (01787) 881 165

**Independent Training Standards Scheme and Register**
Second Floor Suite
Armstrong House
28 Broad Street
Wokingham
RG40 1AB
Telephone: (0118) 989 3229

**British Institute of Occupational Hygienists**
Suite 2
Georgian House
Great Northern Road
Derby
DE1 1LT
Telephone: (01332) 298 087

*Appendix*

**Institution of Occupational Safety and Health (IOSH)**
The Grange
Highfield Drive
Wigston
Leicestershire
LE18 1NN
Telephone: (0116) 257 3100

**Lantra**
Lantra House
Stoneleigh Park
Coventry
CV8 2LG
Telephone: (024) 7669 6996

**Loss Prevention Council**
(*see* Fire Protection Association above)
Qualifications and Curriculum Authority (QCA)
83 Piccadilly
London
W1J 8QA
Telephone: (020) 7509 5555

**Skills for Logistics**
14 Warren Yard
Warren Farm Office Village
Stratford Road
Milton Keynes
MK12 5NW
Telephone: (01908) 313 360

**RTITB Ltd**
Ercall House
8 Pearson Road
Central Park
Telford
TF2 9TX
Telephone: (01952) 520 204

**Royal Society for the Prevention of Accidents (RoSPA)**
Edgbaston Park
353 Bristol Road
Edgbaston
Birmingham
B5 7ST
Telephone: (0121) 248 2000

**Occupational & Environmental Diseases Association (OEDA)**
PO Box 26
Enfield
Middlesex
EN1 2NT

**Storage Equipment Manufacturers' Association (SEMA)**
6th Floor
MacLaren Buildings
35 Dale End
Birmingham
B4 7LN
Telephone: (0121) 200 2100

# Appendix 2
# Current HSE Publications

## Guidance Notes

There are 6 principal series of guidance notes available. These are: Chemical Safety (CS); Environmental Hygiene (EH); General Series (GS); Medical Series (MS); Plant and Machinery (PM), and Legal Series (L).

The following list indicates publications currently in print available from HSE Books.

### Chemical safety

| | |
|---|---|
| CS 3 | Storage and use of sodium chlorate and other similar strong oxidants. 1998 |
| CS 15 | Cleaning and gas freeing of tanks containing flammable residues. 1985 |
| CS 21 | Storage and handling of organic peroxides. 1991 |
| CS 22 | Fumigation. 1996 |
| CS 23 | Disposal of waste explosives. 1999 |
| CS 24 | The interpretation and use of flashpoint information. 1999 |

### Environmental hygiene

| | |
|---|---|
| EH 1 | Cadmium: health and safety precautions. 1995 |
| EH 2 | Chromium and its inorganic compounds: health and precautionary measures. 1998 |
| EH 10 | Asbestos: exposure limits and measurement of airborne dust concentrations. 2001 |
| EH 13 | Beryllium: health and safety precautions. 1995 |
| EH 16 | Isocyanates: health hazards and precautionary measures. 1999 |
| EH 17 | Mercury and its inorganic divalent compounds. 1996 |
| EH 19 | Antimony and its compounds: health hazards and precautionary measures. 1997 |
| EH 38 | Ozone: health hazards and precautionary measures. 1996 |
| EH 40/02 | Occupational exposure limits. 2001 |
| EH 40/03 | Occupational exposure limits. 2003 |
| EH 40 | Occupational exposure limits 2002/2003. 2003 |
| EH 43 | Carbon monoxide: health hazards and precautionary measures. 1998 |
| EH 44 | Dust: general principles of protection. 1997 |
| EH 46 | Man-made mineral fibres. 1990 |
| EH 47 | Provision, use and maintenance of hygiene facilities for work with asbestos insulation, asbestos coatings and asbestos insulating board. 2002 |
| EH 50 | Training operatives and supervisors for work with asbestos insulation and coatings. 1988 |

*Appendix*

| | |
|---|---|
| EH 51 | Enclosures provided for work with asbestos. 1999 |
| EH 57 | The problems of asbestos removal at high temperatures. 1990 |
| EH 59 | Respirable crystalline silica. 1997 |
| EH 60 | Nickel and its inorganic compounds: health hazards and precautionary measures. 1997 |
| EH 63 | Vinyl chloride: toxic hazards and precautions. 1992 |
| EH 64 | Summary criteria for occupational exposure limits. 2001 |
| EH 64supp | Summary criteria for occupational exposure limits. 2002 |
| EH 65/1 | Trimethylbenzenes – criteria document for an OEL. 1992 |
| EH 65/2 | Pulverised fuel ash – criteria document for an OEL. 1992 |
| EH 65/3 | N,N-Dimethylacetamide – criteria document for an OEL. 1992 |
| EH 65/4 | 1,2-dichloroethane – criteria document for an OEL. 1993 |
| EH 65/5 | 4.4-Methylene dianiline – criteria document for an OEL. 1993 |
| EH 65/6 | Epichlorohydrin – criteria document for an OEL. 1993 |
| EH 65/7 | Chlorodifluoromethane – criteria document for an OEL. 1994 |
| EH 65/8 | Cumene – criteria document for an OEL. 1994 |
| EH 65/9 | 1,4-dichlorobenzene – criteria document for an OEL. 1994 |
| EH 65/10 | Carbon tetrachloride – criteria document for an OEL. 1994 |
| EH 65/11 | Chloroform – criteria document for an OEL. 1994 |
| EH 65/12 | Portland cement dust – criteria document for an OEL. 1994 |
| EH 65/13 | Kaolin – criteria document for an OEL. 1994 |
| EH 65/14 | Paracetamol – criteria document for an OEL. 1994 |
| EH 65/15 | 1,1,1,2-Tetrafluoroethane HFC 134a – criteria document for an OEL. 1995 |
| EH 65/16 | Methyl methacrylate – criteria document for an OEL. 1995 |
| EH 65/17 | p-Aramid respirable fibres – criteria document for an OEL. 1995 |
| EH 65/18 | Propranolol – criteria document for an OEL. 1995 |
| EH 65/19 | Mercury and its inorganic divalent compounds – criteria document for an OEL. 1995 |
| EH 65/20 | Ortho-toluidine – criteria document for an OEL. 1996 |
| EH 65/21 | Propylene oxide – criteria document for an OEL. 1996 |
| EH 65/22 | Softwood dust – criteria document for an OEL. 1996 |
| EH 65/23 | Antimony and its compounds – criteria document for an OEL. 1996 |
| EH 65/24 | Platinum metal and soluble platinum salts – criteria document for an OEL. 1996 |
| EH 65/25 | Iodomethane – criteria document for an OEL. 1996 |
| EH 65/26 | Azodicarbonamide – criteria document for an OEL. 1996 |
| EH 65/27 | Dimethyl and diethyl sulphates – criteria document for an OEL. 1996 |
| EH 65/28 | Hydrazine – criteria document for an OEL. 1996 |
| EH 65/29 | Acid anhydrides – criteria document for an OEL. 1996 |
| EH 65/30 | Review of fibre toxicology – criteria document for an OEL. 1996 |
| EH 65/31 | Rosin-based solder flux fume – criteria document for an OEL. 1997 |
| EH 65/32 | Glutaraldehyde – criteria document for an OEL. 1997 |

| | |
|---|---|
| EH 66 | Grain dust. 1998 |
| EH 67 | Grain dust in maltings (maximum exposure limits). 1993 |
| EH 68 | Cobalt: health and safety precautions. 1995 |
| EH 70 | Control of fire-water run-off from CIMAH sites to prevent environmental damage. 1995 |
| EH 72/1 | Phenylhydrazine – risk assessment document. 1997 |
| EH 72/2 | Dimethylaminoethanol – risk assessment document. 1997 |
| EH 72/3 | Bromoethane – risk assessment document. 1997 |
| EH 72/4 | 3-Chloropropene – risk assessment document. 1997 |
| EH 72/5 | Chlorotoluene – risk assessment document. 1997 |
| EH 72/6 | 2-Furaldehyde. 1997 |
| EH 72/7 | 1,2-Diaminoethane (Ethylenediamine (EDA)) – risk assessment document. 1997 |
| EH 72/8 | Aniline – risk assessment document. 1998 |
| EH 72/9 | Barium sulphate – risk assessment document. 1998 |
| EH 72/10 | N-Methyl-2-Pyrrolidone – risk assessment document. 1998 |
| EH 72/11 | Flour dust – risk assessment document. 1999 |
| EH 72/12 | Bromochlromethane – risk assessment document. 2000 |
| EH 72/13 | Methyl and ethyl cyanoacrylate – risk assessment document. 2000 |
| EH 72/14 | Chlorine dixide. 2000 |
| EH 72/15 | Vanadium and its inorganic compounds – risk assessment document. 2002 |
| EH 72/16 | Acetic anhydrate – risk assessment document. 2002 |
| EH 73 | Arsenic and its compounds: Health hazards and precautionary measures. 1997 |
| EH 74/1 | Exposure assessment: Dichloromethane. 1998 |
| EH 72/2 | Respirable crystalline silica. 1999 |
| EH 74/3 | Dermal exposure to non-agricultural pesticides. 1999 |
| EH 74/4 | Metalworking fluids: Exposure assessment document. 2000 |
| EH 75/1 | Medium density fibreboard: MDF. 1999 |
| EH 75/2 | Occupational exposure limits for hyperbaric conditions. 2000 |
| EH 75/3 | N-propyl bromide – hazard assessment document. 2002 |
| EH 75/4 | Respirable crystalline silica: phase 1 – hazard assessment document. 2002 |
| EH 75/5 | Respirable crystalline silica: phase 2 carcinogenicity – hazard assessment document. 2003 |
| EH 76 | Control of laboratory animal allergy. 2002 |

## General series

| | |
|---|---|
| GS 4 | Safety in pressure testing. 1998 |
| GS 6 | Avoidance of danger from overhead electrical lines. 1997 |
| GS 6W | Avoidance of danger from overhead electrical lines – Welsh version. 1999 |
| GS 28/3 | Safe erection of structures – part 3 working places and access. 1986 |
| GS 32 | Health and safety in shoe repair premises. 1984 |
| GS 38 | Electrical test equipment for use by electricians. 1995 |

*Appendix*

| | |
|---|---|
| GS 46 | In-situ timber treatment using timber preservatives: health, safety and environmental precautions. 1989 |
| GS 49 | Pre-stressed concrete. 1991 |
| GS 50 | Electrical safety at places of entertainment. 1997 |
| GS 51 | Façade retention. 1992 |
| GS 53 | Single-flue steel industrial chimneys: inspection and maintenance. 1997 |

## Medical series

| | |
|---|---|
| MS 7 | Colour vision. 1987 |
| MS 12 | Mercury: medical guidance notes. 1996 |
| MS 13 | Asbestos: medical guidance notes. 1999 |
| MS 17 | Medical aspects of work-related exposures to organophosphates. 2000 |
| MS 24 | Medical aspects of occupational skin disease. 1998 |
| MS 25 | Medical aspects of occupational asthma. 1998 |
| MS 26 | A guide to audiometric testing programmes. 1995 |

## Plant and machinery

| | |
|---|---|
| PM 4 | High temperature textile dyeing machines. 1997 |
| PM 5 | Automatically controlled steam and hot water boilers. 1989 |
| PM 15 | Safety in the use of pallets. 1998 |
| PM 16 | Eyebolts. 1978 |
| PM 17 | Pneumatic nailing and stapling tools. 1979 |
| PM 24 | Safety at rack and pinion hoists. 1981 |
| PM 29 | Electrical risks from steam/water pressure cleaners. 1995 |
| PM 33 | Reducing bandsaw accidents in the food industry. 2000 |
| PM 38 | Selection and use of electric handlamps. 1992 |
| PM 39 | Hydrogen cracking of grade T(8) chain and components. 1998 |
| PM 48 | Safe operation of passenger carrying amusement devices: the octopus. 1985 |
| PM 55 | Safe working with overhead travelling cranes. 1985 |
| PM 56 | Noise from pneumatic systems. 1985 |
| PM 57 | Safe operation of passenger carrying amusement devices: the big wheel. 1986 |
| PM 59 | Safe operation of passenger carrying amusement devices: the paratrooper. 1986 |
| PM 60 | Steam boiler blowdown systems. 1998 |
| PM 61 | Safe operation of passenger carrying amusement devices: the chair-o-plane. 1986 |
| PM 65 | Worker protection at crocodile (alligator) shears. 1986 |
| PM 66 | Scrap baling machines. 1986 |
| PM 69 | Safety in the use of freight containers. 1987 |
| PM 70 | Safe operation of passenger carrying amusement devices: ark/speedway. 1988 |
| PM 71 | Safe operation of passenger carrying amusement devices: water chutes. 1989 |
| PM 72 | Safe operation of passenger carrying amusement devices: the trabant. 1990 |

| | |
|---|---|
| PM 73 | Safety at autoclaves. 1998 |
| PM 74 | Forced air filtration units for agricultural vehicles. 1991 |
| PM 75 | Glass reinforced plastic vessels and tanks: advice to users. 1991 |
| PM 77 | Fitness of equipment used for medical exposure to ionising radiation. 1998 |
| PM 78 | Passenger carrying aerial ropeways. 1994 |
| PM 81 | Safe management of ammonia refrigeration systems: food and other workplaces. 1995 |
| PM 82 | The selection, installation and maintenance of electrical apparatus for use in and around buildings containing explosives. 1997 |
| PM 83 | Drilling machines: guarding of spindles and attachments. 1998 |
| PM 84 | Control of safety risks at gas turbines used for power generation. 2003 |

## Health and Safety: Guidance Booklets

The purpose of this series is to provide guidance for those who have duties under *HSWA* and other relevant legislation. It gives guidance on the practical application of regulations made under *HSWA*, but should not be regarded as an authoritative interpretation of the law.

| | |
|---|---|
| HS(G) 6 | Safety in working with lift trucks. 2000 |
| HS(G) 17 | Safety in the use of abrasive wheels. 2000 |
| HS(G) 28 | Safety advice for bulk chlorine installations. 1999 |
| HS(G) 31 | Pie and tart machines. 1986. |
| HS(G) 33 | Health and safety in roof work. 1998 |
| HS(G) 37 | An introduction to local exhaust ventilation. 1993 |
| HS(G) 38 | Lighting at work. 1998 |
| HS(G) 39 | Compressed air safety. 1998 |
| HS(G) 40 | Safe handling of Chlorine from drums and cylinders. 1999 |
| HS(G) 42 | Safety in the use of metal cutting guillotines and shears. 1988 |
| HS(G) 43 | Industrial robot safety. 2000 |
| HS(G) 45 | Safety in meat preparation: guidance for butchers. 1988 |
| HS(G) 47 | Avoiding danger from underground services. 2000 |
| HS(G) 48 | Reducing error and influencing behaviour. 1999 |
| HS(G) 51 | The storage of flammable liquids in containers. 1998 |
| HS(G) 53 | Selection, use and maintenance of respiratory protective equipment. 1998 |
| HS(G) 54 | Maintenance, examination and testing of local exhaust ventilation. 1998 |
| HS(G) 57 | Seating at work. 1997 |
| HS(G) 60 | Upper limb disorders in the workplace: a guide to prevention. 2002 |
| HS(G) 61 | Health Surveillance at Work. 1999 |
| HS(G) 62 | Health and safety in tyre and exhaust fitting premises. 1991 |
| HS(G) 65 | Successful health and safety management. 1997 |
| HS(G) 66 | Protection of workers and the general public during the development of contaminated land. 1991 |
| HS(G) 67 | Health and safety in motor vehicle repair. 1991 |
| HS(G) 71 | Chemical warehousing: storage of packaged dangerous substances. 1998 |
| HS(G) 72 | Control of respirable silica dust in heavy clay and refractory processes. 1992 |

*Appendix*

| | |
|---|---|
| HS(G) 73 | Control of respirable crystalline silica in quarries. 1992 |
| HS(G) 78 | Dangerous goods in cargo transport units: packing and carriage for transport by sea. 1998 |
| HS(G) 79 | Health and safety in golf course management and maintenance. 1994 |
| HS(G) 85 | Electricity at work: safe working practices. 2003 |
| HS(G) 87 | Safety in the remote diagnosis of manufacturing plant and equipment. 1995 |
| HS(G) 88 | Hand-arm vibration. 1994 |
| HS(G) 89 | Safeguarding agricultural machinery: advice for designers, manufacturers, suppliers and users. 1998 |
| HS(G) 90 | The law on VDUs: an easy guide. 2003 |
| HS(G) 92 | Safe use and storage of cellular plastics. 1996 |
| HS(G) 93 | The assessment of pressure vessels operating at low temperature. 1993 |
| HS(G) 94 | Safety in the design and use of gamma and electron irradiation facilities. 1998 |
| HS(G) 95 | The radiation safety of lasers used for display purposes. 1996 |
| HS(G) 96 | The costs of accidents at work. 1997 |
| HS(G) 97 | A step-by-step guide to COSHH assessment. 2004 |
| HS(G) 100 | Prevention of violence to staff in banks and building societies. 1993 |
| HS(G) 101 | The costs to Britain of workplace accidents and work-related ill health in 1995/96. 1999 |
| HS(G) 103 | Safe handling of combustible dusts: precautions against explosions. 2004 |
| HS(G) 107 | Maintaining portable and transportable electrical equipment. 1994 |
| HS(G) 109 | Control of noise in quarries. 1993 |
| HS(G) 110 | Seven steps to successful substitution of hazardous substances. 1994 |
| HS(G) 110W | Seven steps to successful substitution of hazardous substances. (Welsh version) 1996 |
| HS(G) 112 | Health and safety at motor sports events: a guide for employers and organisers. 1999 |
| HS(G) 113 | Lift trucks in potentially flammable atmospheres. 1996 |
| HS(G) 114 | Conditions for the authorisation of explosives in Great Britain. 1994 |
| HS(G) 115 | Manual handling: solutions you can handle. 1994 |
| HS(G) 117 | Making sense of NONS: a guide to the Notification of New Substances Regulations 1993. 1994 |
| HS(G) 118 | Electrical safety in arc welding. 1994 |
| HS(G) 119 | Manual handling for drinks delivery. 1994 |
| HS(G) 121 | A pain in your workplace? Ergonomic problems and solutions. 1994 |
| HS(G) 122 | New and expectant mothers at work: a guide for employers. 2002 |
| HS(G) 123 | Working together on firework displays: a guide to safety for firework display organisers and operators. 1999 |
| HS(G) 124 | Giving your own firework display: how to run and fire it safely. 1995 |
| HS(G) 125 | A brief guide on COSHH for the offshore oil and gas industry. 1994 |
| HS(G) 129 | Health and safety in engineering workshops. 1999 |
| HS(G) 131 | Energetic and spontaneously combustible substances: identification and safe handling. 1995 |

*Appendix*

| | |
|---|---|
| HS(G) 132 | How to deal with sick building syndrome: guidance for employers, building owners and building managers. 1995 |
| HS(G) 133 | Preventing violence to retail staff. 1995 |
| HS(G) 135 | Storage and handling of industrial nitrocellulose. 1995 |
| HS(G) 136 | Workplace transport safety: guidance for employers. 1995 |
| HS(G) 137 | Health risk management: a practical guide for managers in small and medium-sized enterprises. 1995 |
| HS(G) 137W | Health risk management: a practical guide for managers in small and medium-sized enterprises. (Welsh version). 1996 |
| HS(G) 138 | Sound solutions: techniques to reduce noise at work. 1995 |
| HS(G) 139 | The safe use of compressed gases in welding, flame cutting and allied processes. 1997 |
| HS(G) 140 | Safe use and handling of flammable liquids. 1996 |
| HS(G) 141 | Electrical safety on construction sites. 1995 |
| HS(G) 142 | Dealing with offshore emergencies: a guide to the roles and responsibilities of the government departments and agencies that may become involved in emergencies in the offshore oil and gas industry. 2003 |
| HS(G) 143 | Designing and operating safe chemical reaction processes. 2000 |
| HS(G) 144 | Safe use of vehicles on construction sites: a guide for clients, designers, contractors, mangers and workers involved with construction transport. 1998 |
| HS(G) 146 | Dispensing petrol: assessing and controlling the risk of fire and explosion at sites where petrol is stored and dispensed as a fuel. 1996 |
| HS(G) 149 | Backs for the future: safe manual handling in construction. 2000 |
| HS(G) 150 | Health and safety in construction. 2001 |
| HS(G) 151 | Protecting the public – your next move. 1997 |
| HS(G) 153/1 | Railway safety principles and guidance: part 1. 1996 |
| HS(G) 153/3 | Railway safety principles and guidance: part 2 section B. Guidance on stations. 1996 |
| HS(G) 153/4 | Railway safety principles and guidance: part 2 section C. Guidance on electric traction systems. 1996 |
| HS(G) 153/5 | Railway safety principles and guidance: part 2 section D. Guidance on signalling. 1996 |
| HS(G) 153/6 | Railway safety principles and guidance: part 2 section E. Guidance on level crossings. 1996 |
| HS(G) 153/7 | Railway safety principles and guidance: part 2 section F. Guidance on trains. 1996 |
| HS(G) 153/8 | Railway safety principles and guidance: part 2 section G. Guidance on tramways. 1997 |
| HS(G) 154 | Managing crowds safely: a guide for organisers at events and venues. 2000 |
| HS(G) 155 | Slips and trips: guidance for employers on identifying hazards and controlling risks. 1996 |
| HS(G) 156 | Slips and trips: guidance for the food processing industry. 1996 |
| HS(G) 158 | Flame arresters: preventing the spread of fires and explosions in equipment that contains flammable gases and vapours. 1997 |
| HS(G) 159 | Managing contractors: a guide for employers. 1997 |

*Appendix*

| | |
|---|---|
| HS(G) 165 | Young people at work: a guide for employers. 2000 |
| HS(G) 166 | Formula for health and safety: guidance for small to medium-sized firms in the chemical manufacturing industry. 1997 |
| HS(G) 167 | Biological monitoring in the workplace: a guide to its practical application to chemical exposure. 1997 |
| HS(G) 168 | Fire safety in construction: guidance for clients, designers and those managing and carrying out construction work involving significant fire risks. 1997 |
| HS(G) 169 | Camera operations on location: guidance for managers and camera crews for work in news gathering, current affairs and factual programming. 1997 |
| HS(G) 170 | Vibration solutions: practical ways to reduce hand-arm vibration injury. 1997 |
| HS(G) 171 | Well handled: offshore manual handling solutions. 1997 |
| HS(G) 172 | Health and safety in sawmilling: a run-of-the-mill-business? 1997 |
| HS(G) 173 | Monitoring strategies for toxic substances. 1997 |
| HS(G) 174 | Anthrax: safe working and the prevention of infection. 1997 |
| HS(G) 175 | Fairgrounds and amusement parks: guidance on safe practice: practical guidance on the management of health and safety for those involved in the fairgrounds industry. 1997 |
| HS(G) 176 | The storage of flammable liquids in tanks. 1998 |
| HS(G) 177 | Managing health and safety in dockwork. 2002 |
| HS(G) 178 | The spraying of flammable liquids. 1998 |
| HS(G) 179 | Managing health and safety in swimming pools. 2003 |
| HS(G) 180 | Application of electro-sensitive protective equipment using light curtains and light beam devices to machinery. 1999 |
| HS(G) 181 | Assessment principles for offshore safety cases. 1998 |
| HS(G) 182 | Sound solutions offshore: practical examples of noise reduction. 1998 |
| HS(G) 183 | Five steps to risk assessment: case studies. 1998 |
| HS(G) 184 | Guidance on the handling, storage and transport of airbags and seat pretensioners. 1998 |
| HS(G) 185 | Health and safety in excavations: be safe and shore. 1999 |
| HS(G) 186 | The bulk transfer of dangerous liquids and gasses between ship and shore. 1999 |
| HS(G) 187 | Control of diesel engine exhaust emissions in the workplace. 1999 |
| HS(G) 188 | Health risk management: a guide to working with solvents. 1999 |
| HS(G) 189/1 | Controlled asbestos stripping techniques for work requiring a licence. 1999 |
| HS(G) 189/2 | Working with asbestos cement. 1999 |
| HS(G) 190 | Preparing safety reports: Control of Major Accident Hazard Regulations 1999 (COMAH). 1999 |
| HS(G) 191 | Emergency planning for major accidents: Control of Major Accident Hazard Regulations 1999 (COMAH). 1999 |
| HS(G) 192 | Charity and voluntary workers: a guide to health and safety at work. 1999 |
| HS(G) 193 | COSHH essentials: easy steps to control chemicals: Control of Substances Hazardous to Health Regulations. 2003 |
| HS(G) 194 | Thermal comfort in the workplace: guidance for employers. 1999 |

| | |
|---|---|
| HS(G) 195 | The event safety guide: a guide to health, safety and welfare at music and similar events. 1999 |
| HS(G) 196 | Moving food and drink: Manual handling solutions for the food and drink industries. 2000 |
| HS(G) 197 | Railway safety principles and guidance: part 3 section A. Developing and maintaining staff competence. 2002 |
| HS(G) 198 | The transitional arrangements for the Biocidal Products Regulations: a guide for importers and suppliers of biocides. 2001 |
| HS(G) 199 | Managing health and safety on work experience: a guide for organisers. 2000 |
| HS(G) 200 | Go-karts: guidance on safe operation and use. 2000 |
| HS(G) 201 | Controlling exposure to stonemasonry dust: guidance for employers. 2001 |
| HS(G) 202 | General ventilation in the workplace: Guidance for employers. 2000 |
| HS(G) 203 | Controlling exposure to coating powders. 2000 |
| HS(G) 204 | Health and safety in arc welding. 2000 |
| HS(G) 205 | Assessing and managing risks at work from skin exposure to chemical agents: Guidance for employers and health and safety specialists. 2001 |
| HS(G) 206 | Cost and effectiveness of chemical protective gloves for the workplace: guidance for employers and health and safety specialists. 2001 |
| HS(G) 207 | Choice of skin care products for the workplace: Guidance for employers and health and safety specialists. 2001 |
| HS(G) 208 | A guide to the Biocidal Products Regulations for importers and suppliers of biocides. 2001 |
| HS(G) 209 | Aircraft turnaround: management responsibilities at airports. 2000 |
| HS(G) 210 | Asbestos essential tasks manual: task guidance sheets for the building maintenance and allied trades. 2001 |
| HS(G) 212 | The training of first-aid at work: a guide to gaining and maintaining HSE approval. 2000 |
| HS(G) 213 | Introduction to asbestos essentials: comprehensive guidance on working with asbestos in the building maintenance and allied trades. 2001 |
| HS(G) 215 | A guide to the Biocidal Products Regulations for users of biocidal products. 2001 |
| HS(G) 216 | Passenger-carrying miniature railways: guidance on safe practice. 2001 |
| HS(G) 217 | Involving employees in health and safety: forming partnerships in the chemical industry. 2001 |
| HS(G) 218 | Tackling work-related stress: a managers' guide to improving and maintaining employee health and well-being. 2001 |
| HS(G) 220 | Health and safety in care homes. 2001 |
| HS(G) 221 | Technical guidance on the safe use of lifting equipment offshore. 2002 |
| HS(G) 222 | Effective health and safety training: a trainers' resource pack. 2001 |
| HS(G) 223 | The regulatory requirements for medical exposure to ionising radiation: an employer's overview. 2001 |
| HS(G) 224 | Managing health and safety in construction: Construction (Design and Management) Regulations 1994 – approved code of practice and guidance. 2001 |
| HS(G) 225 | Handling homecare: achieving safe efficient and positive outcomes for care workers and clients. 2002 |

*Appendix*

| | |
|---|---|
| HS(G) 227 | A comprehensive guide to managing asbestos in premises. 2002 |
| HS(G) 228 | CHIP for everyone. 2002 |
| HS(G) 229 | Work-related violence: case studies: managing the risk in smaller businesses. 2002 |
| HS(G) 230 | Keeping electrical switchgear safe. 2002 |
| HS(G) 231 | Health and Safety with metalworking fluids: good practice manual. 2002 |
| HS(G) 232 | Sound solutions for the food and drink industries: reducing noise in food and drink manufactoring. 2002 |
| HS(G) 233 | A bakers dozen: thirteen essentials for a safer bakery. 2003 |
| HS(G) 234 | Caring for cleaners: guidance and case studies on how to prevent musculoskeletal disorders. 2003 |
| HS(G) 235 | Bulk storage of acids: guidance on the storage of hydrochloric acid and nitric acid in tanks. 2003 |
| HS(G) 236 | Power presses: maintenance and thorough examination. 2003 |
| HS(G) 238 | Out of control: why control systems go wrong. 2003 |
| HS(G) 240 | Managing health and safety at recreational dive sites. 2003 |
| HS(G) 242 | Railway safety principles and guidance: part 3 section B. Safe movement of trains. 2003 |
| HS(G) 243 | Fuel cells: understand the hazards, control the risks. 2004 |

# Legal – Approved Codes of Practice (COP series), HS(R) series and new L series

*Code numbers in the HS(R) series and the COP series are gradually being superseded by the L series. Publications in the L series contain guidance on Regulations and Approved Codes of Practice.*

## Approved Codes of Practice

| | |
|---|---|
| COP 2 | Control of lead at work (in support of SI 1980 No 1248) – approved code of practice. 1998 |
| COP 15 | Zoos: safety, health and welfare standards for employers and persons at work – approved code of practice and guidance notes. 1985 |
| COP 20 | Standards of training in safe gas installation – approved code of practice. 1987 |
| COP 25 | Safety in docks: Docks Regulations 1988 – approved code of practice with regulations and guidance. 1988 |
| COP 28 | Safety of exit from mines underground workings: Mines (Safety of Exit) Regulations 1988 – approved code of practice. 1988 |
| COP 35 | The use of electricity at quarries: Electricity at Work Regulations 1989 – approved code of practice. 1989 |

## HS(R) series

| | |
|---|---|
| HS(R) 17 | A guide to the Classification and Labelling of Explosives Regulations 1983. 1983 |
| HS(R) 25 | Memorandum of guidance on the Electricity at Work Regulations 1989. 1989 |

*Appendix*

| | |
|---|---|
| HS(R) 27 | A guide to the Dangerous Substances in Harbour Areas Regulations 1987. 1988 |
| HS(R) 28 | Guide to the Loading and Unloading of Fishing Vessels Regulations 1988. 1988 |
| HS(R) 29 | Notification and marking of sites: The Dangerous Substances (Notification and Marking of Sites) Regulations 1990 – guidance on regulations. 1990 |

## L series

| | |
|---|---|
| L 5 | Control of substances hazardous to health: The Control of Substances Hazardous to Health Regulations 2002 – approved code of practice. 2002 |
| L 8 | Legionnaires' disease: The control of legionella bacteria in water systems – approved code of practice. 2000 |
| L 9 | Safe use of pesticides for non-agricultural purposes: Control of Substances Hazardous to Health Regulations 1994 – approved code of practice. 1991 |
| L 10 | A guide to the Control of Explosives Regulations 1991 – guidance on regulations. 1991 |
| L 11 | A guide to the Asbestos (Licensing) Regulations 1983 as amended. The Asbestos (Licensing) Regulations 1983 – guidance on regulations. 1999 |
| L 13 | A guide to the Packaging of Explosives for Carriage Regulations 1991 – guidance on regulations. 1991 |
| L 21 | Management of health and safety at work. Management of Health and Safety at Work Regulations 1999 – approved code of practice and guidance. 2000 |
| L 22 | Safe use of work equipment: Provision and Use of Work Equipment Regulations 1998 – approved code of practice and guidance. 1998 |
| L 23 | Manual handling: Manual Handling Operations Regulations 1992 (as amended) – guidance on regulations. 1998 |
| L 24 | Workplace health, safety and welfare. Workplace (Health, Safety and Welfare) Regulations 1992 – approved code of practice and guidance. 1992 |
| L 25 | Personal protective equipment at work: Personal Protective Equipment at Work Regulations 1992 – guidance on regulations. 1992 |
| L 26 | Work with display screen equipment: Health and Safety (Display Screen Equipment) Regulations 1992 as amended by the Health and Safety (Miscellaneous Amendments) Regulations 2002 – guidance on regulations. 2002 |
| L 27 | Work with asbestos which does not normally require a licence: Control of Asbestos at Work Regulations 2002 – approved code of practice and guidance. 2002 |
| L 28 | Work with asbestos insulation, asbestos coating and asbestos insulating board: Control of Asbestos at Work Regulations 1987 – approved code of practice. 2002 |
| L 29 | A guide to the Genetically Modified Organisms (Contained Use) Regulations 1992. 2000 |
| L 30 | A guide to the Offshore Installations (Safety Case) Regulations 1992. 1998 |
| L 31 | A guide to the Public Information for Radiation Emergencies Regulations 1992 – guidance on regulations. 1993 |
| L 42 | Shafts and winding in mines: Mines (Shafts and Winding) Regulations 1993 – approved code of practice. 1993 |

*Appendix*

| | |
|---|---|
| L 43 | First-aid at mines: Health and Safety (First-Aid) Regulations 1981 – approved code of practice. 1993 |
| L 44 | The management and administration of safety and health at mines: Management and Administration of Safety and Health at Mines Regulations 1993 – approved code of practice. 1993 |
| L 45 | Explosives at coal and other safety-lamp mines: Coal and Other Safety-lamp Mines (Explosives) Regulations 1993 – approved code of practice. 1993 |
| L 46 | The prevention of inrushes in mines – approved code of practice. 1993 |
| L 47 | The Coal Mines (Owners Operating Rules) Regulations 1993 – guidance on Regulations. 1993 |
| L 50 | Railways Safety Critical Work: Railways (Safety Critical Work) Regulations 1994 – approved code of practice and guidance. 1996 |
| L 52 | Railways (Safety Case) Regulations 2000 including 2001 and 2003 amendments – guidance on regulations. 2000 |
| L 55 | Preventing asthma at work: How to control respiratory sensitisers. 1994 |
| L 56 | Safety in the installation and use of gas systems and appliances: The Gas Safety (Installation and Use) Regulations 1998 – approved code of practice and guidance. 1998 |
| L 59 | A guide to the approval of railway works, plant and equipment. 1994 |
| L 60 | Control of substances hazardous to health in the production of pottery: The Control of Substances Hazardous to Health Regulations 1994: The Control of Lead at Work Regulations 1980 – approved code of practice. 1995 |
| L 64 | Safety signs and signals: The Health and Safety (Safety Signs and Signals) Regulations 1996 – guidance on regulations. 1997 |
| L 65 | Prevention of fire and explosion, and emergency response on offshore installations: Offshore Installations (Prevention of Fire and Explosion and Emergency Response) Regulations 1995 – approved code of practice and guidance. 1997 |
| L 66 | A guide to the Placing on the Market and Supervision of Transfers of Explosives Regulations 1993 (POMSTER) 1993 – guidance on regulations. 1995 |
| L 70 | A guide to the Offshore Installations and Pipeline Works (Management and Administration) Regulations 1995 – guidance on regulations. 2002 |
| L 71 | Escape and Rescue from Mines. Escape and Rescue from Mines Regulations 1995 – approved code of practice. 1995 |
| L 72 | A guide to Borehole Sites and Operations Regulations 1995 – guidance on regulations. 1995 |
| L 73 | A guide to the Reporting of Injuries, Diseases and Dangerous Occurrences Regulations 1995 – guidance on regulations. 1999 |
| L 74 | First aid at work: The Health and Safety (First-Aid) Regulations 1981 – approved code of practice and guidance. 1997 |
| L 75 | Guidance for railways, tramways, trolley vehicle systems and other guided transport systems on Reporting of Injuries, Diseases and Dangerous Occurrences Regulations 1995 – guidance on regulations. 1996 |
| L 77 | Guidance to the licensing authority on the Adventure Activities Licensing Regulations 1996: The Activity Centres (Young Persons Safety) Act 1995 – guidance on regulations. 1996 |

*Appendix*

| | |
|---|---|
| L 80 | A guide to the Gas Safety (Management) Regulations 1996 – guidance on regulations. 1996 |
| L 81 | Design, construction and installation of gas service pipes: Pipelines Safety Regulations 1996 – approved code of practice and guidance. 1996 |
| L 82 | A guide to the Pipelines Safety Regulations 1996 – guidance on regulations. 1996 |
| L 84 | A guide to the well aspects of amendments of the Offshore Installations and Wells (Design and Construction etc) Regulations 1996 – guidance on regulations. 1996 |
| L 85 | A guide to the integrity, workplace environment and miscellaneous aspects of the Offshore Installations and Wells (Design and Construction etc) Regulations 1996 – guidance on regulations. 1996 |
| L86 | Control of Substances Hazardous to health in fumigation operations: Control of Substances Hazardous to Health Regulations 1994 – approved code of practice. 1996 |
| L87 | Safety representatives and safety committees (the brown book) – approved code of practice and guidance on the regulations. 1996 |
| L 88 | Approved requirements and test methods for the classification and packaging of dangerous goods for carriage: Carriage of Dangerous Goods (Classification, Packaging and Labelling) and Use of Transportable Pressure Receptacles Regulations 1996 – approved requirements. 1996 |
| L 93 | Approved tank requirements: the provisions for bottom loading and vapour recovery systems of mobile containers carrying petrol: Carriage of Dangerous Goods by Road Regulations 1996. Carriage of Dangerous Goods by Rail Regulations 1996. 1996 |
| L 95 | A guide to the Health and Safety (Consultation with Employees) Regulations 1996 – guidance on regulations. 1996 |
| L 96 | A guide to the Work in Compressed Air Regulations 1996. 1996 |
| L 97 | A guide to the Level Crossings Regulations 1997. 1997 |
| L 98 | Railway safety miscellaneous provisions. Railway Safety (Miscellaneous Provisions) Regulations 1997 – guidance on regulations. 1997 |
| L 101 | Safe work in confined spaces. Confined Spaces Regulations 1997 – approved code of practice, regulations and guidance. 1997 |
| L 102 | A guide to the Construction (Head Protection) Regulations 1989. 1998 |
| L 103 | Commercial diving projects offshore. Diving at Work Regulations 1997 – approved code of practice. 1998 |
| L 104 | Commercial diving projects inland/inshore: Diving at Work Regulations 1997 – approved code of practice. 1998 |
| L 105 | Recreational diving projects: Diving at Work Regulations 1997 – approved code of practice. 1998 |
| L 106 | Media diving projects: Diving at Work Regulations 1997 – approved code of practice. 1998 |
| L 107 | Scientific and archaeological diving projects: Diving at Work Regulations 1997 – approved code of practice. 1998 |
| L 108 | Reducing Noise at Work: Guidance on the Noise at Work Regulations 1989. 1998 |
| L 110 | A guide to the Offshore Installations (Safety Representatives and Safety Committees) Regulations 1989 – guidance on regulations. 1998 |

*Appendix*

| | |
|---|---|
| L 111 | A guide to the Control of Major Accident Hazard Regulations (COMAH) – guidance on regulations. 1999 |
| L 112 | Safe use of power presses: Provision and Use of Work Equipment Regulations 1998 as applied to power presses – approved code of practice and guidance. 1998 |
| L 113 | Safe use of lifting equipment: Lifting Operations and Lifting Equipment Regulations 1998 – approved code of practice and guidance. 1998 |
| L 114 | Safe use of woodworking machinery: Provision and Use of Work Equipment Regulations 1998 as applied to woodworking machinery – approved code of practice and guidance. 1998 |
| L 116 | Preventing accidents to children in agriculture – approved code of practice. 1999 |
| L 117 | Rider-operated lift trucks: operator training – approved code of practice and guidance. 1999 |
| L 118 | Health and safety at quarries: Quarries Regulations 1999 – approved code of practice. 1999 |
| L 119 | The control of ground movement in mines: The Mines (Control of Ground Movement) Regulations 1999 – approved code of practice and guidance. 1999 |
| L 120 | Train protection systems and mark 1 rolling stock: Railway Safety Regulations 1999 – guidance on regulations. 1999 |
| L 121 | Work with ionising radiation: Ionising Radiations Regulations 1999 – approved code of practice and guidance. 2000 |
| L 122 | Safety of pressure systems: Pressure Systems Safety Regulations 2000 – approved code of practice. 2000 |
| L 123 | Health care and first-aid on offshore installations and pipeline works: Offshore Installations and Pipeline Works (First-aid) Regulations 1989 – approved code of practice and guidance. 2000 |
| L 126 | A guide to the Radiation (Emergency Preparedness and Public Information) Regulations 2001 – guidance on regulations. 2002 |
| L 127 | The management of asbestos in non-domestic premises: Regulation 4 of the Control of Asbestos at Work Regulations 2002 – approved code of practice and guidance. 2002 |
| L 128 | The use of electricity in mines: Electricity at Work Regulations 1989 – approved code of practice. 2001 |
| L 129 | Approved Supply List (Seventh edition): Information approved for the classification and labelling of substances and preparations dangerous for supply Chemicals (Hazard Information and Packaging for Supply) Regulation 2002 – approved list. 2002 |
| L 130 | The compilation of safety data sheets (Third edition): Chemicals (Hazard Information and Packaging for Supply) Regulations 2002 – approved code of practice. 2002 |
| L 131 | Approved classification and labelling guide (Fifth edition): Chemicals (Hazard Information and Packaging for Supply) Regulations 2002 – guidance on regulations. 2002 |
| L 132 | Control of Lead at Work: Control of lead at work Regulations 2002 – approved code of practice and guidance. 2002 |

*Appendix*

| | |
|---|---|
| L 133 | Unloading petrol from road tankers: Dangerous Substances and Explosive Atmospheres Regulations 2002 – approved code of practice and guidance. 2003 |
| L 134 | Design of plant, equipment and workplaces: Dangerous Substances and Explosive Atmospheres Regulations 2002 – approved code of practice and guidance. 2003 |
| L 135 | Storage of dangerous substances: Dangerous Substances and Explosive Atmospheres Regulations 2002 – approved code of practice and guidance. 2003 |
| L 136 | Control and mitigation measures: Dangerous Substances and Explosive Atmospheres Regulations 2002 – approved code of practice and guidance. 2003 |
| L 137 | Safe maintenance, repair and cleaning procedures: Dangerous Substances and Explosive Atmospheres Regulations 2002 – approved code of practice and guidance. 2003 |
| L 138 | Dangerous Substances and Explosive Atmospheres: Dangerous Substances and Explosive Atmospheres Regulations – approved code of practice and guidance. 2003 |

## Other Publications

### Accidents and emergencies

| | |
|---|---|
| – | Basic advice on first aid at work (rigid PVC poster). 2002 |
| – | Basic advice on first aid at work (encapsulated poster). 2002 |
| – | Reducing at-work road traffic incidents. 2001 |
| BI 510 | Accident book. 2003 |
| HSG212 | The training of first aid at work: a guide to gaining and maintaining HSE approval. 2000 |
| HSR29 | Notification and marking of sites: The Dangerous Substances (Notification and Marking of Sites) Regulations 1990 – guidance on regulations. 1990 |
| L73 | A guide to the Reporting of Injuries, Diseases and Dangerous Occurrences Regulations. Revised 1999 |
| L74 | First aid at work. 1997 |
| HSE31 | RIDDOR explained (available in Welsh). 1999 |
| HSE32 | RIDDOR information for doctors. 1996 |
| HSE33 | RIDDOR Offshore. Revised 1999 |
| HSIS1 | RIDDOR 95 Guidance for employers in the healthcare sector. 1996 |
| INDG62 | Protecting your health at work. 1996 |
| INDG116 | Your work and your health: what your doctor needs to know. 1992 |
| INDG214 | First aid at work – your questions answered (available in Welsh). 1996 |
| INDG347 | Basic advice on first aid at work 2002 |
| MISC471 | Jobs to die for. 2002 |

### Asbestos

| | |
|---|---|
| – | Asbestos – Effects on health of exposure to asbestos. R Doll and J Peto. 1996 |
| – | How are you today? The risks of asbestos in buildings. 2001 |

*Appendix*

| | |
|---|---|
| MDHS39/4 | Asbestos fibres in the air: sampling and evaluation by phase contrast microscopy (PCM) under the Control of Asbestos at Work Regulations. 1995 |
| MDHS77 | Asbestos in bulk materials: sampling and identification by polarised light microscopy (PLM). 1994 |
| MDHS100 | Surveying, sampling and assessment of asbestos-containing material. 2001 |
| HSG189/1 | Controlled asbestos stripping techniques for work requiring a licence. 1999 |
| HSG189/2 | Working with asbestos cement. 1999 |
| HSG210 | Asbestos essentials task manual: task guidance sheets for the building maintenance and allied trades. 2001 |
| HSG213 | Introduction to asbestos essentials: comprehensive guidance on working with asbestos in the building maintenance and allied trades. 2001 |
| HSG227 | A comprehensive guide to managing asbestos in premises. 2002 |
| L11 | A guide to the Asbestos (Licensing) Regulations 1983 as amended: The Asbestos (Licensing) Regulations 1983 – guidance on regulations. 1999 |
| L27 | Work with asbestos which does not normally require a licence: Control of Asbestos at Work Regulations 2002 – approved code of practice and guidance. 2002 |
| L28 | Work with asbestos insulation, asbestos coating and asbestos insulating board: Control of Asbestos at Work Regulations 2002 – approved code of practice and guidance. 2002 |
| L127 | The management of asbestos in non-domestic premises: Regulation 4 of the Control of Asbestos at Work Regulations 2002 – approved code of practice and guidance. 2002 |
| INDG188 | Asbestos alert for building maintenance, repair and refurbishment workers. 1995 |
| INDG223 | Managing asbestos in workplace buildings. Revised 1999 |
| INDG255 | Asbestos dust kills. 1999 |
| INDG288 | Selection of suitable respiratory protective equipment for work with asbestos. 2003 |
| INDG289 | Working with asbestos in buildings. 1999 |

## Chemical Industry

| | |
|---|---|
| – | Handle with care – assessing musculoskeletal risk in the chemical industry. 2000 |
| – | Control of exothermic chemical reactions. 1993 |
| – | Fill safe … Safe filling of LPG cylinders. 1994 |
| HSG71 | Chemical warehousing – the storage of packaged dangerous substances. Revised 1998 |
| HSG143 | Designing and operating safe chemical reaction processes. 2000 |
| HSG159 | Managing contractors – a guide for employers. 1997 |
| HSG166 | Formula for health and safety: guidance for small and medium-sized firms in the chemical industry. 1997 |
| HSG217 | Involving employees in health and safety – forming partnerships in the chemical industry. 2001 |
| CHIS2 | Emergency isolation of process plant in the chemical industry. 1999 |

| | |
|---|---|
| CHIS3 | Major accident prevention policies for lower-tier COMAH establishments. 1999 |
| CHIS4 | Use of LPG in cylinders. 1999 |
| CHIS5 | Small-scale use of LPG in cylinders. 1999 |
| CHIS6 | Better alarm handling in the chemical and allied industries. 2000 |
| INDG98 | Permit to work systems. Revised 1997 |
| INDG243 | Computer control – a question of safety. 1997 |
| INDG245 | Biological monitoring for chemicals in the workplace. 1997 |
| INDG246L | Prepared for emergency. 1997 |
| INDG254 | Chemical reaction hazards and the risk of thermal runaway. 1997 |
| INDG260 | Emergency action for burns. Revised 1997 |
| INDG352 | Rread the label – how to find out if chemicals are dangerous. Revised 2002 |

## Construction

| | |
|---|---|
| HSG33 | Health and safety in roof work. Revised 1998 |
| HSG47 | Avoiding danger from underground services. Revised 2000 |
| HSG141 | Electrical safety on construction sites. 1995 |
| HSG144 | Safe use of vehicles on construction sites – a guide for clients, designers, contractors, managers and workers involved in construction transport. 1998 |
| HSG149 | Backs for the future: safe manual handling in construction. 2000 |
| HSG150 | Health and safety in construction. Revised 2001. |
| HSG151 | Protecting the public: your next move. 1997 |
| HSG168 | Fire safety in construction work – guidance for clients, designers and those managing and carrying out construction work involving significant fire risks. 1997 |
| HSG185 | Health and safety in excavations: be safe and shore. 1999 |
| HSG201 | Controlling exposure to stonemasonry dust – guidance for employers. 2001 |
| HSG224 | Managing health and safety in construction: Construction (Design and Management) Regulations 1994 – approved code of practice and guidance. 2001 |
| L96 | A guide to the Work in Compressed Air Regulations 1996 – guidance on regulations. Rvised 1996 |
| L102 | Construction (Head Protection) Regulations 1989 – guidance on regulations. 1998 |
| SIR58 | Safety of construction transport: a survey of standards and accidents associated with construction vehicles. 2001 |
| SIR59 | Issues surrounding the failure of an energy absorbing lanyard. 2001 |
| CIS8 | Safety in excavations. Revised 1997 |
| CIS10 | Tower scaffolds. Revised 1997 |
| CIS18 | The provision of welfare facilities at fixed construction sites. Revised 1996 |
| CIS24 | Chemical cleaners. Revised 1998 |
| CIS27 | Solvents. Revised 1998 |
| CIS36 | Silica. 1993 |
| CIS37 | Handling heavy building blocks. 1999 |

*Appendix*

| | |
|---|---|
| CIS39 | Construction (Design and Management) Regulations 1994: the role of the client. 1995 |
| CIS41 | Construction (Design and Management) Regulations 1994: the role of the designer. Revised 1995 |
| CIS43 | Construction (Design and Management) Regulations 1994: the health and safety plan during the construction phase. 1995 |
| CIS44 | CDM Regulations 1994: the health and safety file. 1995 |
| CIS45 | Establishing exclusion zones when using explosives in demolition. 1995 |
| CIS46 | Provision of welfare facilities at transient construction sites. 1997 |
| CIS47 | Inspections and reports. 1997 |
| CIS49 | General access to scaffolds and ladders. Revised 1997 |
| CIS50 | Personal Protective Equipment (PPE): safety helmets. 1997 |
| CIS51 | Construction fire safety. 1997 |
| CIS52 | Construction site transport safety: safe use of compact dumpers. 2000 |
| CIS53 | Crossing high-speed roads on foot during temporary traffic management works. 2000 |
| CIS54 | Dust control on concrete cutting saws used in the construction industry. 2000 |
| CIS56 | Safe erection, use and dismantling of falsework. 2003 |
| IRIS3 | Portable nuclear moisture/density gauges in the construction industry. Revised 2002 |
| INDG127 | Noise in construction: further guidance on the Noise at Work Regulations 1989. Revised 1994 |
| INDG212 | Workplace health and safety – Glazing: guidance on glazing for employers and people in control of workplaces. 1996. |
| INDG220 | A guide to the Construction (Health, Safety and Welfare) Regulations 1996. 1996 |
| INDG242 | In the driving seat: advice to employers on reducing back pain in drivers and machinery operators. 1997 |
| INDG258 | Safe work in confined spaces. (Available in Welsh) 1999 |
| INDG284 | Working on roofs. (Available in Welsh) 1999 |
| INDG315 | Stone dust and you – guidance for stonemasons. 2001 |
| INDG332 | Manual packing in the brick industry. 2000 |
| INDG344 | The absolutely essential health and safety toolkit for the smaller construction contractor. 2001 |
| INDG367 | Inspecting fall arrest equipment made from webbing or rope. 2002 |
| INDG368 | Use of contractors – a joint responsibility. 2002 |
| INDG384 | The high 5: Five ways to reduce risk on site. 2003 |
| MISC112 | Reducing the risk of hand-arm vibration injury among stone masons. 1998 |
| MISC193 | Having Construction Work Done? – duties of clients under the Construction (Design and Management) Regulations 1994. 1999 |

## Dangerous substances – general

| | |
|---|---|
| – | Major hazard aspects of the transport of dangerous substances. 1991 |

*Appendix*

| | |
|---|---|
| – | Major hazard aspects of the transport of dangerous substances: report and appendices. 1991 |
| – | The technical basis for COSHH essentials: easy steps to control chemicals. 1999 |
| HSR29 | Notification and marking of sites: The Dangerous Substances (Notification and Marking of Sites) Regulations 1990 – guidance on regulations. 1990 |
| HSG97 | A step by step guide to COSHH assessment. Revised 2004 |
| HSG110 | Seven steps to successful substitution of hazardous substances. 1994 |
| HSG117 | Making sense of NONS: a guide to the Notification of New Substances Regulations. 1993 |
| HSG173 | Monitoring strategies for toxic substances. 1997 |
| HSG188 | Health risk management: a guide to working with solvents. 1998 |
| HSG193 | COSHH essentials: easy steps to control chemicals. Control of Substances Hazardous to Health Regulations. Revised 2003 |
| HSG205 | Assessing and managing risks at work from skin exposure to chemical agents – guidance for employers and health and safety specialists. 2001 |
| HSG228 | CHIP for everyone. 2002 |
| L5 | Control of substances hazardous to health: the Control of Substances Hazardous to Health Regulations 2002 – approved code of practice and guidance. 2002 |
| L13 | A guide to the Packaging of Explosives for Carriage Regulations 1991 – guidance on regulations. 1991 |
| L55 | Preventing asthma at work: how to control respiratory sensitisers. 1994 |
| L86 | Control of Substances Hazardous to Health in fumigation operations: Control of Substances Hazardous to Health Regulations 1994 – approved code of practice. 1996 |
| L88 | Approved requirements and test methods for the classification and packaging of dangerous goods for carriage: Carriage of Dangerous Goods (Classification, Packaging and Labelling) and Use of Transportable Pressure Receptacles Regulations 1996 – approved requirements 1996 |
| L129 | Approved supply list (seventh edition): information approved for the classification and labelling of substances and preparations dangerous for supply. Chemicals (Hazard Information and Packaging for Supply) Regulations 2002 – approved list. Revised 2002 |
| L130 | The compilation of safety data sheets (third edition): Chemicals (Hazard Information and Packaging for Supply) Regulations 2002 – approved code of practice. Revised 2002 |
| L131 | Approved classification and labelling guide (fifth edition): Chemicals (Hazard Information and Packaging for Supply) Regulations 2002 – guidance on regulations. 2002 |
| INDG95 | Respiratory sensitisers and COSHH: breath freely – an employers' leaflet on preventing occupational asthma. Revised 1995 |
| INDG136 | COSHH: a brief guide to the Regulations. Revised 1999 |
| INDG172 | Breath freely: a workers' information card on respiratory sensitisers. 1994 |
| INDG233 | Preventing dermatitis at work: advice for employers and employees. 1996 |
| INDG273 | Working safely with solvents: a guide to safe working practices. 1998 |

*Appendix*

| | |
|---|---|
| INDG350 | The idiot's guide to CHIP 3: Chemicals (Hazard Information and Packaging for Supply) Regulations 2002. Revised 2002 |
| INDG353 | Why do I need a safety data sheet? CHIP 3. 2002 |

## Dangerous substances – by type

| | |
|---|---|
| – | Asthmagen: critical assessments of the evidence for agents implicated in occupational asthma. 1997 |
| – | Recommendations for training users of non-agricultural pesticides. 1990 |
| – | Anaesthetic agents: controlling exposure under COSHH. 1996 |
| HSG28 | Safety advice for bulk chlorine installations. Revised 1999 |
| HSG40 | Safe handling of chlorine from drums and cylinders. Revised 1999 |
| HSG187 | Control of diesel engine exhaust emissions in the workplace. 1999 |
| HSG198 | The transitional arrangements for the Biocidal Products Regulations. 2001 |
| HSG203 | Controlling exposure to coating powders. 2000 |
| HSG215 | A guide to the Biocidal Products Regulations for users of biocidal products. 2001 |
| HSG235 | Bulk storage of acids: guidance on the storage of hydrochloric acid and nitric in tanks. 2003 |
| L9 | The safe use of pesticides for non-agricultural purposes: Control of Substances Hazardous to Health Regulations 1994 – approved code of practice. Revised 1995 |
| L132 | Control of lead at work – approved code of practice and guidance. Revised 2002 |
| INDG140 | Grain dust in non-agricultural workplaces. 1993 |
| INDG197 | Working with sewage: the health hazards – a guide for employees. 1995. |
| INDG198 | Working with sewage: the health hazards – a guide for employers. 1995 |
| INDG230 | Storage and safe handling of ammonium nitrate. 1996 |
| INDG257 | Pesticides: use them safely. 1997 |
| INDG276 | Feral honey bees. 1999 |
| INDG286 | Diesel engine exhaust emissions. 1999 |
| INDG305 | Lead and you – a guide to working safely with lead. Revised 1998 |
| INDG307 | Hydrofluoric acid poisoning: recommendations on first aid procedures. 1999 |
| INDG309 | Safe waters: using antifouling paints safely – a guide for boat owners. 2000 |
| INDG310 | Drycleaners: are you in control? Health and safety drycleaning. 2000 |
| INDG311 | Beryllium and you: working with beryllium: are you at risk? 1995 |
| INDG319 | Working safely with coating powders. 2000 |
| INDG321 | A simple guide to the Biocidal Products Regulations. 2001 |
| INDG329 | Benzine and you: working with benzene: are you at risk? 2000 |
| INDG346 | Chromium and you. 1992 |
| INDG351 | Nickel and you. 1997 |
| INDG391 | Cadmium and you: working with cadmium: are you at risk? Revised 1995 |
| MSA8 | Arsenic and you: arsenic is poisonous: are you at risk? 1996 |
| MSA17 | Cobalt and you: working with cobalt: are you at risk. 1995 |

*Appendix*

| | |
|---|---|
| MSA19 | PCBs and you – do you know how to work safely with PCBs? 1995 |
| MSA21 | MbOCA and you: do you use MbOCA? 1996 |
| MSB4 | Skin cancer by pitch and tar. 1996 |

## Diving

| | |
|---|---|
| HSG240 | Managing health and safety at recreational diving sites. 2003 |
| L103 | Commercial diving projects offshore: Diving at Work Regulations 1997 – approved code of practice. 1998 |
| L104 | Commercial diving projects inland/offshore: Diving at Work Regulations 1997 – approved code of practice. 1998 |
| L105 | Recreational diving projects: Diving at Work Regulations 1997 – approved code of practice. 1998 |
| L106 | Media diving projects: Diving at Work Regulations 1997 – approved code of practice. 1998 |
| L107 | Scientific and archaeological diving projects: Diving at Work Regulations 1997 – approved code of practice. 1998 |
| INDG266 | Are you involved in a diving project? 1998 |
| DVIS1 | General hazards. 1998 |
| DVIS2 | Diving system winches. 1998 |
| DVIS3 | Breathing gas management. 1998 |
| DVIS4 | Compression chambers. 1998 |
| DVIS5 | Exposure limits for air diving operations. 1998 |
| DVIS6 | Maintenance of diving bell hoists. 1998 |
| DVIS7 | Bell run and bell lock-out times in relation to habitats. 1998 |
| DVIS8 | Diving in benign conditions and in pools, tanks, aquariums and helicopter underwater escape training. 1999 |

## Electricity and electrical systems

| | |
|---|---|
| – | Electrical incidents in Great Britain: statistical summary. 1997 |
| – | Live wires: what to look for when inspecting portable electrical equipment. 1994 |
| – | Shock horror: the dangers of electrocution by overhead power lines. 1993 |
| GS6 | Avoidance of danger from overhead electric power lines. Revised 1997 |
| GS6W | Avoidance of danger from overhead electric power lines. (Available in Welsh) Revised 1999 |
| GS38 | Electrical test equipment for use by electricians. Revised 1995 |
| GS50 | Electrical safety at places of entertainment. Revised 1997 |
| HSG85 | Electricity at work: safe working practices. 2003 |
| HSG107 | Maintaining portable and transportable electrical equipment. 1994 |
| HSG118 | Electrical safety in arc welding. 1994 |
| HSG180 | Application of electro-sensitive protective equipment using light curtains and light beam devices to machinery. 1999 |
| HSG230 | Keeping electrical switchgear safe. 2002 |

*Appendix*

| | |
|---|---|
| HSR25 | Memorandum of guidance on the Electricity at Work Regulations 1989 – guidance on regulations. 1989 |
| PM29 | Electrical risks from steam/water pressure cleaners. Revised 1995 |
| INDG68 | Do you use a steam/water pressure cleaner? You could be in for a shock. Revised 1997 |
| INDG139 | Electric storage batteries: you could be in for a shock! Revised 1997 |
| INDG231 | Electrical safety and you. (Available in Welsh) 1996 |
| INDG236 | Maintaining portable electrical equipment in offices and other low-risk environments. 1996 |
| INDG237 | Maintaining portable electrical equipment in hotels and tourist accommodation. 1996 |
| INDG247 | Electrical safety for entertainers. 1997 |
| INDG354 | Safety in electrical testing at work – general guidance. 2002 |
| INDG372 | Electrical switchgear and safety: a concise guide for users. 2003 |
| INDG389 | Shock horror: safe working near overhead power lines in agriculture. 2003 |

## Engineering

| | |
|---|---|
| – | Health and safety in motor vehicle repair. 1991 |
| – | One careful owner: dangers in the metal recycling industry. 1996 |
| – | Power presses: parts 1 and 2. 1992 |
| EH74/4 | Metal working fluids. 2000 |
| HSG17 | Safety in the use of abrasive wheels. 2000 |
| HSG42 | Safety in the use of metal cutting guillotines and shears. 1998 |
| HSG62 | Health and safety in tyre and exhaust fitting premises. 1991 |
| HSG67 | Health and safety in motor vehicle repair. 1991 |
| HSG87 | Safety in the remote diagnosis of manufacturing and plant equipment. 1995 |
| HSG118 | Electrical safety in arc welding. 1994 |
| HSG129 | Health and safety in engineering workshops. 1999 |
| HSG139 | The safe use of compressed gases in welding, flame cutting and allied processes. 1997 |
| HSG204 | Health and safety in ark welding. 2000 |
| HSG231 | Working safely with metal working fluids: good practice manual. 2002<br>Working safely with metal working fluids: eight task sheets. 2002<br>Working safely with metal working fluids: pack. 2002<br>1 x HSG213 8 x task sheets 10 x INDG365 1 x A2 wallchart |
| MDHS95/2 | Measurement of personal exposure of metalworking machine operators to airborne water-mix metal working fluids: elemental marker method using flame atomic absorption spectrometry or inductively coupled plasma-atomic emission spectrometry. 2003 |
| EIS1 | Hot work on vehicle wheels. 1992 |
| EIS2 | Accidents at metalworking lathes using emery cloth. 1993 |
| EIS3 | Monitoring requirements in the electroplating industry including electrolytic chromium processes. Revised 1998 |
| EIS4 | Workplace welfare in the electroplating industry. Revised 1998 |
| EIS5 | Health surveillance requirements in the electroplating industry. Revised 1998 |

| | |
|---|---|
| EIS6 | Electrical systems in the electroplating industry. Revised 1998 |
| EIS7 | Safeguarding 3 roll bending machines. Revised 1998 |
| EIS12 | Safety at manually-fed pivoting-head metal-cutting circular saws. Revised 1998 |
| EIS13 | Safeguarding of combination metalworking machines. Revised 2000 |
| EIS15 | Control of exposure to triglycidyl isocyanurate (TGIC) in coating powders. Revised 2003 |
| EIS16 | Preventing injuries from the manual handling of sharp edges in the engineering industry. 1997 |
| EIS19 | Engineering machine tools: retrofitting CNC. 1997 |
| EIS20 | Maintenance and cleaning of solvent degreasing tanks. Revised 1998 |
| EIS21 | Immersion and cold cleaning of engineering components. 1998 |
| EIS26 | Noise in engineering. 1998 |
| EIS27 | Control of noise at metal-cutting saws. 1998 |
| EIS28 | Safeguarding at horizontal boring machines. 1998 |
| EIS29 | Control of noise at power presses. 1998 |
| EIS30 | Safety in the use of hand- and foot-operated presses. 1999 |
| EIS31 | Cadmium in silver soldering or brazing. 1999 |
| EIS32 | Chromate primer paints. 1999 |
| EIS33 | CNC turning machines: controlling risks from ejected parts. 2001 |
| EIS34 | Surface cleaning: solvent update including the reclassification of trichloroethylene. 2002 |
| EIS35 | Safety in electrical testing: servicing and repair of domestic appliances. 2002 |
| EIS36 | Safety in electrical testing: servicing and repair of audio, TV and computer equipment. 2002 |
| EIS37 | Safety in electrical testing: switchgear and control gear. 2002 |
| EIS38 | Safety in electrical testing: products on production lines. 2002 |
| EIS39 | Reducing noise from CNC punch presses. 2002 |
| EIS40 | Safe use of solvent degreasing plant. 2003 |
| INDG248 | Solder fume and you. Revised 2001 |
| INDG249 | Controlling health risks from rosin (colophony) based solder fluxes. 1997 |
| INDG280 | A guide to handling and storage of air bags and seatbelt pretensioners at garages and motor vehicle repair shops. 1998 |
| INDG297 | Safety in gas welding, cutting and similar processes. 1999 |
| INDG313 | Safe unloading of steel stock. 2000 |
| INDG314 | Hot work on small tanks and drums. 2000 |
| INDG327 | Take care with acetylene. 2000 |
| INDG349 | Safe working with vehicle air-conditioning systems – guidance for employers, self employed people and supervisors. 2002 |
| INDG356 | Reducing ill health and accidents in motor vehicle repair. 2002 |
| INDG365 | Working safely with metalworking fluids – a guide for employees. 2002 |
| INDG388 | Safety in motor vehicle repair: working with 2-pack isocyanate paints. 2003 |
| INDG390 | Choosing a welding set? Make sure you can handle it. 2003 |
| WGIS1 | Supply of welding consumables. 1999 |

*Appendix*

## Flammable and explosive substances

| | |
|---|---|
| – | Acquisition and use of explosives by historical societies. 1998 |
| – | Selection and use of explosion effects and consequences of models for explosives. 2000 |
| – | Fire Safety: an employers guide. 1999 |
| – | List of classified and authorised explosives 1994 (LOCAE): Explosives Acts, 1875 and 1923 The Classification and Labelling of Explosives Regulations 1983 (CLER) – approved list. 1994 |
| – | List of classified and authorised explosives 1994 (LOCAE): Explosives Acts, 1875 and 1923 The Classification and Labelling of Explosives Regulations 1983 (CLER) Supplement No 1. 1996 |
| – | List of classified and authorised fireworks (LOCAE): Explosives Acts, 1875 and 1923 The Classification and Labelling of Explosives Regulations 1983 (CLER). 1990 |
| – | Guide to general fire precautions in explosives factories and magazines: Fire Certificate (Special Premises) Regulations 1976. 1990 |
| COP6 | Plastic containers with normal capacities up to 5 litres for petroleum spirit: Requirements for testing and marking or labelling in support of SI 1982 No 830 – approved code of practice. 1982 |
| CS23 | Disposal of waste explosives. 1999 |
| HSG51 | The storage of flammable liquids in containers. Revised 1998 |
| HSG71 | Chemical warehousing. 1998 |
| HSG103 | Safe handling of combustible dusts: precautions against explosions. 2003 |
| HSG114 | Conditions for the authorisation of explosives in Great Britain. 1994 |
| HSG131 | Energetic and spontaneously combustible substances: identification and safe handling. 1995 |
| HSG135 | Storage and handling of industrial nitrocellulose. 1995 |
| HSG140 | Safe use and handling of flammable liquids. 1996 |
| HSG146 | Dispensing petrol: assessing and controlling the risk of fire and explosion at sites where petrol is stored and dispensed as a fuel. 1996 |
| HSG158 | Flame arresters: preventing the spread of fires and explosions in equipment that contains flammable gases and vapours. 1997 |
| HSG176 | The storage of flammable liquids in tanks. 1998 |
| HSG178 | The spraying of flammable liquids. 1998 |
| HSR17 | A guide to the Classification and Labelling of Explosive Regulations 1983. 1983 |
| L10 | A guide to the Control of Explosives Regulations 1991 – guidance on regulations. 1991 |
| L13 | A guide to the Packaging of Explosives for Carriage Regulations 1991 – guidance on regulations. 1991 |
| L66 | A guide to the Placing on the Market and Supervision of Transfers of Explosives Regulations (POMSTER) 1993. 1995 |
| L133 | Unloading petrol from road tankers: Dangerous Substances and Explosive Atmospheres Regulations 2002 – approved code of practice and guidance. 2003 |

| | |
|---|---|
| L134 | Design of plant, equipment and workplaces: Dangerous Substances and Explosives Atmospheres Regulations 2002 – approved code of practice and guidance. 2003 |
| L135 | Storage of dangerous substances: Dangerous Substances and Explosive Atmospheres Regulations 2002 – approved code of practice and guidance. 2003 |
| L136 | Control and mitigation measure: Dangerous Substances and Explosive Atmospheres Regulations 2002 – approved code of practice and guidance. 2003 |
| L137 | Safe maintenance, repair and cleaning procedures: Dangerous Substances and Explosive Atmospheres Regulations 2002 – approved code of practice and guidance. 2003 |
| L138 | Dangerous Substances and Explosive Atmospheres: Dangerous Substances and Explosive Atmospheres Regulations – approved code of practice and guidance. 2003 |
| HSE8 | Take care with oxygen: fire and explosion hazards in the use of oxygen. 1999 |
| INDG216 | Dispensing petrol as a fuel: health and safety guidance for employees. 1996 |
| INDG227 | Safe working with flammable substances. 1996 |
| INDG314 | Hot work on small tanks and drums. 2000 |
| INDG331 | Safe use of petrol in garages. 2000 |
| INDG335 | Is it explosive? Dangers of explosives in metal recycling. 2002 |
| INDG370 | Fire and explosion: how safe is your workplace? 2002 |
| INDG387 | Safe working with LPG-fuelled motor vehicles. 2003 |

## Foundries

| | |
|---|---|
| – | Hand-arm vibration in foundries. 2001 |
| – | The selection, use and maintenance of molten metal protective clothing. 1996 |
| – | Noise control at foundry shakeouts. 1995 |
| FNIS2 | Foundry machine guarding: introductory sheet. 1995 |
| FNIS3 | Foundry machine guarding: mould and core-making machinery. 1995 |
| FNIS4 | Foundry machine guarding: sand handling equipment. 1995 |
| FNIS5 | Foundry machine guarding: shakeouts, sand mixers and shotblasts. 1995 |
| FNIS6 | Hazards associated with foundry processes: fettling – noise hazards. 1995 |
| FNIS8 | Hazards associated with foundry processes: hand-arm vibration – the current picture. 1996 |
| FNIS9 | Hazards associated with foundry processes: hand-arm vibration – symptoms and solutions. 1996 |
| FNIS10 | Hazards associated with foundry processes: hand arm vibration – assessing the need for action. 1999 |
| FNIS11 | Hand-arm vibration in foundries: furnace an ladle relining operations. 2002 |
| FNIS12 | A purchasing policy for vibration-reduced tools in foundries. 2002 |
| IACL83 | Hearing protection in foundries: advice for employees. 1994 |
| IACL104 | Health surveillance in the foundry industry. 1998 |

*Appendix*

## Health issues

| | |
|---|---|
| – | A matter of life and breath: occupational asthma – the causes, the effects and how to prevent it. 2001 |
| – | Real situations, real people: a managers' guide to tackling work-related stress. 2003 |
| HSIS1 | The Reporting of Injuries, Diseases and Dangerous Occurrences Regulations 1995 – guidance for employers in the healthcare sector. 1998 |
| HSG61 | Health surveillance at work. Revised 1999 |
| HSG122 | New and expectant mothers at work: a guide for employers. Revised 2002 |
| – | Infections in the workplace to new and expectant mothers. 1997 |
| INDG63 | Passive smoking at work. (Available in Welsh) Revised 1997 |
| INDG91 | Drug misuse at work: a guide for employers. 1998 |
| INDG240 | Don't mix it! A guide for employers on alcohol at work. 1996 |
| INDG281 | Help on work-related stress. 1999 |
| INDG304 | Understanding health surveillance at work: an introduction for employers. 1999 |
| INDG341 | Tackling work-related stress: a guide to employees. 2001 |
| INDG373 | A guide to new and expectant mothers who work. (Available in Welsh) Revised 2003 |
| GEIS1 | Heat stress in the workplace: what you need to know as an employer. 2003 |

## Health services

| | |
|---|---|
| – | Getting to grips with handling problems: worked examples of assessment and reduction of risk in the health services. 1994 |
| – | The management of occupational health services for healthcare staff. 1993 |
| – | Safe disposal of clinical waste. Revised 1999 |
| – | Safe working and the prevention of infection in clinical laboratories and similar facilities. Revised 2003 |
| – | Safe working and the prevention of infection in the mortuary and post-mortem room. Revised 2003 |
| – | Violence and aggression to staff in the health services: guidance on assessment and management. 1997 |
| – | Manual handling in the health services. Revised 1998 |
| HSG220 | Health and safety in care homes. 2001 |
| HSG225 | Handling home care: achieving safe efficient and positive outcomes for care workers and clients. 2002 |
| HSIS1 | The Reporting of Injuries, Diseases and Dangerous Occurrences Regulations 1995 – guidance for employers in the healthcare sector. 1998 |
| HSIS2 | Slips and trips in the health services. 2003 |
| MISC186 | Safe use of pneumatic air tube transport systems for pathology specimens. 1999 |

## Infection risks

| | |
|---|---|
| – | The occupational zoonoses. 1992 |
| – | Working safely at Containment Level 3. 1997 |

| | |
|---|---|
| – | Microbiological risks when working with sewage and sludge. 2001 |
| HSG174 | Anthrax: safe working and the prevention of infection. 1997 |
| INDG184 | Leptospirosis: are you at risk? 1990 |
| INDG85 | BSE and carcase disposal. 2003 |
| INDG376 | Legionnaires' disease: essential information for providers of residential accommodation. 2003 |
| IAC27 | Legionnaires' disease: a guide for employers. Revised 2000 |
| L8 | Legionnaires' disease: the control of legionella bacteria in water systems – approved code of practice and guidance. Revised 2000 |
| – | Legionnaires' disease: control of legionella bacteria in water systems – audit checklist. 2003 |

## Legislation and enforcement

| | |
|---|---|
| HSE40 | Employers' Liability (Compulsory Insurance) Act 1969: a guide for employers. Revised 1998 |
| HSE5 | An Employment Medical Advisory Service and you. 2000 |
| HSE39 | Employers' Liability (Compulsory Insurance) Act 1969: a guide for employers and their representatives. 2002 |
| HSC13 | Health and safety regulation: a short guide (available in Welsh). Revised 1995 |
| HSC14 | What to expect when a health and safety inspector calls: a breif guide for businesses, employees and their representatives. 1998 |
| HSC15 | HSC's enforcement policy statement (available in Welsh). 2002 |
| INDG184 | Signpost to the Health and Safety (Safety Signs and Signals) Regulations 1996: guidance on the regulations. 1996 |

## Major hazards

| | |
|---|---|
| – | Fire safety: an employer's guide. Revised 1999 |
| – | Risk criteria for land-use planning in the vicinity of major industrial hazards. 1989 |
| – | Quantified risk assessment. It's input to decision making. 1989 |
| – | Hazard analysis report. Canvey. A second report – summary of a review of potential hazards from operations in the Caney Island/Thurrock area three years after the publication of the Canvey report. 1981 |
| – | Hazard analysis report. Canvey. A second report – a review of potential hazards operations in the Canvey Island/Thurock area three years after the publication of the Canvey report. 1981 |
| – | Hazard analysis report. PWR. A report by HSE to the Secretary of State for Energy on a review of the generic safety issues of pressurised water reactors. 1979 |
| – | Further guidance on emergency plans for major accident hazard pipelines: The Pipelines Safety Regulations 1996. 1997 |
| – | Toxicology of substances in relation to major hazards: acrylonitrile. 1989 |
| – | Toxicology of substances in relation to major hazards: ammonia. 1990 |
| – | Toxicology of substances in relation to major hazards: chlorine. 1990 |
| – | Toxicology of substances in relation to major hazards: hydrogen fluoride. 1993 |

*Appendix*

| | |
|---|---|
| – | Toxicology of substances in relation to major hazards: hydrogen sulphide. 1990 |
| – | Toxicology of substances in relation to major hazards: sulphuric acid mist. 1992 |
| HSG66 | Protection of workers and the general public during the development of contaminated land. 1991 |
| HSG190 | Preparing safety reports: Control of Major Accident Hazards Regulations (COMAH) 1999. 1999 |
| HSG191 | Emergency planning for major accidents: Control of Major Accident Hazards Regulations 1999. 1999 |
| HSG238 | Out of control: why control systems go wrong and how to prevent failure. 2003 |
| L111 | A guide to the Control of Major Accident Hazards regulations 1999 (COMAH) – guide on regulations. 1999 |
| INDG196 | Safety reports: how HSE assesses them. Control of Industrial Major Accident Hazards Regulations 1984. 1995 |

## Mines

| | |
|---|---|
| – | Supplementary guidance on use of flexible bolts in reinforcement systems for coal mines. 2000 |
| – | The safety of free steered vehicle operations below ground in British coal mines. 1992 |
| – | Safe use of bolt conveyors in mines. 1993 |
| – | Guidance on the design and construction of safety critical electric systems at mines. 2001 |
| – | Guidance on the use of cablebolts to support roadways in coal mines. 1996 |
| – | Guidance on the repair and overhaul of apparatus intended for use in coal mines susceptible to fire damp. 1997 |
| – | Workmen at miscellaneous mines: Book 6. 1997 |
| – | The law relating safety and health in mines and quarries: part 1 acts 1972. 1997 |
| – | The law relating safety and health in mines and quarries: part 2 extracts relating to mines of coal, stratified ironstone, shale and fireclay – section A subject headings A–H. 1979 |
| – | The law relating safety and health in mines and quarries: part 2 extracts relating to mines of coal, stratified ironstone, shale and fireclay – section B subject headings I–S. 1979 |
| – | The law relating safety and health in mines and quarries: part 2 extracts relating to mines of coal, stratified ironstone, shale and fireclay – section C subject headings T–Z. 1979 |
| – | The law relating safety and health in mines and quarries: part 4 and supplement number 1 – quarries. 1972 |
| – | Guidance on the support of salvage faces in coal mines. 1997 |
| – | Safe manriding in mines. 2001 |
| – | Guidance and information on escape from mines. 2001 |

*Appendix*

| | |
|---|---|
| – | Extensive fall of roof at Bilsthorpe Colliery: a report of HSE's investigation into the extensive fall of roof at Bilsthorpe Colliery, Nottinghamshire on 18 August 1993. 1994 |
| – | Guidance on the design, installation and use of free standing support systems (including powered supports) in coal mines. 2002 |
| COP28 | Safety of exit of mines underground workings: Mines (Safety of Exit) Regulations 1988 – approved code of practice. 1988 |
| L42 | Shafts and winding in mines: Mines (Shafts and Winding) Regulations 1993 – approved code of practice. 1993 |
| L43 | First aid at mines: Health and Safety (First Aid) Regulations 1981 – approved code of practice. 1993 |
| L44 | The management and administration of safety and health at mines: Management and Administration of Safety and Health at Mines Regulations 1993 – approved code of practice. 1993 |
| L45 | Explosives at coal and other safety-lamp mines: Coal and Other Safety-lamp Mines (Explosives) Regulations 1993 – approved code of practice. 1993 |
| L46 | The prevention of inrushes in mines – approved code of practice. 1993 |
| L47 | The Coal Mines (Owners' Operating Rules) Regulations 1993: guidance on regulations. 1993 |
| L71 | Escape and Rescue from Mines: Escape and Rescue from Mines Regulations 1995 – approved code of practice. 1995 |
| L119 | The control of ground movement in mines – approved code of practice and guidance. 1999 |
| L128 | The use of electricity in mines: Electricity at Work regulations 1989 – approved code of practice 2001 edition. 2001 |
| QIS1 | Managers' guide to safe coal cleaning and the control of pedestrians at opencast coal sites. 1994 |
| TOP7 | Improving visibility on underground free steered vehicles. 1996. |

## Noise

| | |
|---|---|
| HSG138 | Sound solutions: techniques to reduce noise at work. 1995 |
| INDG362 | Noise at work: advice for employers. 2002 |
| INDG363 | Protect your hearing or lose it! 2002 |
| L108 | Reducing noise at work: guidance on the Noise at Work Regulations 1989. 1998 |

## Nuclear

| | |
|---|---|
| – | Arrangements for responding to nuclear emergencies. 1994 |
| – | Safety assessment principles for nuclear plants. 1992 |
| – | The tolerability of risk from nuclear power stations. 1992 |
| – | ACSNI Human Factors Study Group: third report – organising for safety. 1993 |
| – | Structures and principles of the regulations of civil nuclear licensed sites. 2003 |
| L31 | A guide to the Public Information for Radiation Emergencies Regulations 1992 – guidance on regulations. 1993 |

*Appendix*

| | |
|---|---|
| INDG206 | Wear your dosemeter. Revised 1995 |
| INDG207C | Wear your dosemeter. Revised 2000 |
| IRIS2 | Radiation doses: assessment and recording. Revised 2000 |
| MISC434 | Safety audit of Dounreay 1998 – Final Report 2001. 2001 |
| HSG120 | Nuclear site licences under the Nuclear Installations Act 1965: (as amended) notes for applicants. 1994 |

## Oil Industry

| | |
|---|---|
| – | Guidance on multi-skilling in the petroleum industry. Revised 1998 |
| – | Guidance on permit-to-work systems in the petroleum industry. 1997 |
| – | Management of occupational health risks in the offshore oil and gas industry. 1996 |
| – | The safe isolation of plant and equipment. 1997 |
| – | Play your part: how offshore workers can help improve health and safety. 1994 |
| – | Rash decisions offshore: HSE's video on work-related dermititus – its causes, effects and prevention. 1996 |
| – | Lifeboat on-load release gear: safe inspection and maintainance. 1997 |
| – | To cap it all: the capping of stranded wire ropes with polyester resin. 1994 |
| HSG125 | A brief guide on COSHH for the offshore oil and gas industry. 1994 |
| HSG142 | Dealing with offshore emergencies: a guide to the roles and responsibilities of the government departments and agencies that may become involved in emergencies in the offshore oil and gas industry. Revised 2003 |
| HSG171 | Well handled: offshore manual handling solutions. 1997 |
| HSG181 | Assessment principles for offshore safety cases. 1998 |
| HSG221 | Technical guidance on the safe use of lifting equipment offshore. 2002 |
| L30 | A guide to the Offshore Installations (Safety Case) Regulations 1992 – guidance on regulations. Revised 1998 |
| L65 | Prevention of fire and explosion and emergency response on offshore installations: Offshore Installations (Prevention of Fire and Explosion and Emergency Response) Regulations 1995 – approved code of practise and guidance. 1997 |
| L70 | A guide to the Offshore Installations and Pipeline Works (Management and Administration) Regulations 1995 – guidance on regulations. Revised 2002 |
| L72 | A guide to Borehole Sites and Operations Regulations 1995 – guidance on regulations. 1995 |
| L82 | A guide to the Pipelines Safety Regulations 1996 – guidance on regulations 1996 |
| L84 | A guide to the well aspects of amendments of the Offshore Installations and Wells (Design and Construction etc) Regulations 1996 – guidance on regulations. 1996 |
| L85 | A guide to the integrity, workplace environment and miscellaneous aspects of the Offshore Installations and Wells (Design and Construction etc) Regulations 1996 – guidance on regulations. 1996 |
| L110 | A guide to the Offshore Installations (Safety Representatives and Safety Committees) Regulations 1989. Revised 1998 |

| | |
|---|---|
| L123 | Health care and first aid on offshore installations and pipeline works. Offshore Installations and Pipeline Works (First Aid) Regulations 1989 – approved code of practise and guidance. 2000 |
| INDG119 | Safety representatives and safety committees on offshore installations: a brief guide for the workforce. Revised 1999 |
| INDG219 | How offshore helicopter travel is regulated. 1996 |
| INDG221 | Technical guidance on the safe use of lifting equipment offshore. 2002 |
| INDG239 | Play your part! How you can help improve health and safety offshore. 1996 |
| INDG250 | How HSE assesses offshore safety cases. 1997 |
| INDG277 | Leadership for the offshore industry. 1997 |
| INDG361 | Regulating health and safety in the UK offshore oil and gas fields: who does what? 2002 |

## Personal Protective Equipment (PPE)

| | |
|---|---|
| HSG53 | The selection, use and maintenance of respiratory protective equipment. 1998 |
| HSG206 | Cost and effectiveness of chemical protective gloves for the workplace: guidance for employers and health and safety specialists. 2001 |
| HSG207 | Choice of skincare products for the workplace guidance for employers and health and safety specialists. 2001 |
| L25 | Personal protective equipment at work: Personal Protective Equipment at Work Regulations 1992 – guidance on regulations. 1992 |
| APIS1 | PPE: High visibility clothing for airport workers. 1994 |
| INDG147 | Keep your top on: health risks from working in the sun. (Available in Welsh) Revised 1998 |
| INDG174 | A short guide to the Personal Protective Equipment at Work Regulations 1992. (Available in Welsh) 1995 |
| INDG288 | Selection of suitable respiratory protective equipment for work with asbestos. (Available in Welsh) Revised 2003 |
| INDG330 | Selecting protective gloves for work with chemicals: guidance for employers and health and safety specialists. 2000 |
| INDG337 | Sun protection: advice for employers of outdoor workers. 2001 |

## Quarries

| | |
|---|---|
| – | Dust in quarries. 1992 |
| COP35 | The use of electricity at quarries: Electricity at Work Regulations 1989 – approved code of practice. 1989 |
| HSG73 | Control of respirable crystalline silica in quarries. 1992 |
| HSG109 | Control of noise in quarries. 1993 |
| INDG303 | Do you work in a quarry? 1999 |
| L118 | Health and safety at quarries: Quarries Regulations 1999 – approved code of practice. 1999 |

## Vibration

| | |
|---|---|
| – | The successful management of hand-arm vibration. 2000 |

*Appendix*

| | |
|---|---|
| – | Hard to handle: hand-arm vibration – manage the risk. 1998 |
| HSG88 | Hand-arm vibration. 1994 |
| INDG126 | Health risks from vibration white finger: advice for employees and the self employed. Revised 1998 |
| INDG175 | Health risks from hand arm vibration: advice for employers. Revised 1998 |
| INDG242 | In the driving seat: advice to employers on reducing back pain in drivers and machinery operators. 1997 |
| INDG296P | Hand-arm vibration syndrome: pocket card for employees. 1999 |
| INDG338 | Power tools: How to reduce vibration health risks. 2001 |
| MISC493 | Hand-arm vibration in the cast stone industry: reducing the risk. 2003 |

## Woodworking

| | |
|---|---|
| – | Woodworking posters. 1999 |
| – | The cutting edge II: health and safety in woodworking. Part 1 managing for safety. 1998 |
| – | The cutting edge II: health and safety in woodworking. Part 2 a safe machine safety. 1998 |
| – | The cutting edge II: health and safety in woodworking. Part 3 a healthy business. 1998 |
| – | The cutting edge II (value pack, contains the three videos above). 1998 |
| HSG172 | Health and safety in sawmilling: a run-of-the-mill business? 1997 |
| L114 | Safe use of woodworking machinery: Provision and Use of Work Equipment Regulations 1998: approved code of practice and guidance. 1998 |
| INDG318 | Manual handling solutions in woodworking. 2000 |
| TOP06 | Supervising for safety in woodworking: are you a good supervisor? 1994 |
| WIS1 | Wood dust: hazards and precautions. Revised 1997 |
| WIS2 | Safe stacking of sawn timber and board materials. Revised 2000 |
| WIS3 | Beam panel saws. 1990 |
| WIS4 | Noise reduction at band re-saws. 1990 |
| WIS5 | Noise enclosure at band re-saws. 1990 |
| WIS6 | COSHH and the woodworking industries. Revised 1997 |
| WIS7 | Accidents at woodworking machines. Revised 1999 |
| WIS12 | Assessment and control of wood dust: use of the dust lamp. 1991 |
| WIS13 | Noise at woodworking machines. 1991 |
| WIS14 | Selection of respiratory protective equipment suitable for use with wood dust. 1991 |
| WIS15 | Safe working at woodworking machines. 1992 |
| WIS16 | Circular saw benches: safe working practices. Revised 1999 |
| WIS17 | Safe use of hand-fed planing machines. Revised 2000 |
| WIS18 | Safe use of vertical spindle moulding machines. Revised 2001 |
| WIS19 | Health risks during furniture stripping using dichloromethane (DCM). 1993 |
| WIS20 | Safeguarding veneer guillotines. 1992 |
| WIS23 | LEV: general principles of system design. 1992 |

*Appendix*

| | |
|---|---|
| WIS24 | LEV: dust capture at sawing machines. 1992 |
| WIS25 | LEV dust capture at fixed belt sanding machines. 1992 |
| WIS26 | LEV dust capture at fixed drum and disc sanding machines. 1992 |
| WIS27 | Safeguarding board-edge processing machinery. 1994 |
| WIS28 | Rotary knife wood turning lathes: safeguarding and reducing risks to health. 1994 |
| WIS29 | Occupational hygiene and health surveillance at industrial timber treatment plants. Revised 2002 |
| WIS30 | Toxic woods. 1995 |
| WIS31 | Safety in the use of narrow band saws. 1995 |
| WIS32 | Safe collection in woodwaste: prevention of fire and explosion. 1997 |
| WIS33 | Health surveillance and wood dust. 1997 |
| WIS34 | Health and safety priorities for the woodworking industry. 1997 |
| WIS35 | Safe use of power-operated cross-cut saws. 1998 |
| WIS36 | Safe use of manually operated cross-cut saws. 1998 |
| WIS37 | PUWER 98: selection of tooling for use with hand-fed woodworking machines. 1998 |
| WIS38 | PUWER 98: retrofitting of braking to woodworking machines. 1998 |
| WIS39 | Safe use of single-end tenoning machines. 2000 |

## Work Equipment

| | |
|---|---|
| – | PUWER 1998: Provision and Use of Work Equipment Regulations 1998 – open learning guidance. 1999 |
| – | LOLER 1998: Lifting Operations and Lifting Equipment Regulations 1998 – open learning guidance. 1999 |
| – | The use of computers in safety critical application. 1998 |
| HSG43 | Industrial robot safety: your guide to the safeguarding of industrial robots. Revised 2000 |
| L22 | Safe use of work equipment: Provision and Use of Work Equipment Regulations 1998 – approved code of practice and guidance. Revised 1998 |
| L113 | Safe use of lifting equipment: Lifting Operations and Lifting Equipment Regulations 1998 – approved code of practice and guidance. 1998 |
| L117 | Rider operated-lift trucks: operating training – approved code of practice and guidance. 1999 |
| INDG229 | Using work equipment safely. Revised 2002 |
| INDG270 | Supplying new machinery – a short guide to the law. 1998 |
| INDG290 | Simple guide to LOLER. 1999 |
| INDG291 | Simple guide to PUWER. 1999 |
| INDG317 | Chainsaws at work. 2000 |
| INDG339 | Thorough examination and testing of lifts – simple guidance for lift owners. 2001 |
| MISC156 | Hiring and leasing out of plant – application of PUWER 98 Regulations 26 and. 1998 |
| MISC175 | Retrofitting of roll over protective structures, restraining systems and their attachment points to mobile work equipment. 1999 |

*Appendix*

| | |
|---|---|
| MISC241 | Fitting and use of restraining systems on lift trucks. 2000 |
| MISC611 | Safety in window cleaning using suspended and powered access equipment. 2003 |
| MISC612 | Safety in window cleaning using rope access techniques. 2003 |
| MISC613 | Safety in window cleaning using portable ladders. 2003 |
| MISC614 | Preventing falls from boom-type elevating work platforms. 2003 |

## Workplace

| | |
|---|---|
| – | Reducing at-work road traffic incidents. 2001 |
| – | Stop slips – managing slips to reduce injuries and costs. 2000 |
| – | Mental well-being in the workplace – a resource pack for management training and development. 1998 |
| – | An easy guide to the VDU Regulations (checklist). 1994 |
| – | Improving maintenance – a guide to reducing human error. 2000 |
| – | Change in manufacturing: How to manage stress-related risks – case studies in practitioner format. 2001 |
| – | Danger – vehicles at work. 1995 |
| L24 | Workplace health, safety and welfare: Workplace (Health, Safety and Welfare) Regulations 1992 – approved Code of Practice and guidance. 1992 |
| L26 | Work with display screen equipment: Health and Safety (Display Screen Equipment) Regulations 1992 as amended by the Health and Safety (Miscellaneous Amendments) Regulations 2002 – guidance on regulations. Revised 2003 |
| HSG37 | An introduction to local exhaust ventilation. Revised 1993 |
| HSG38 | Lighting at work. Revised 1998 |
| HSG54 | Maintenance, examination and testing of local exhaust ventilation. Revised 1998 |
| HSG57 | Seating at work. Revised 1997 |
| HSG90 | The law on VDUs – an easy guide. Revised 2003 |
| HSG132 | How to deal with sick building syndrome – guidance for employers, building owners and building managers. 1995 |
| HSG165 | Young people at work – a guide for employers. Revised 2000 |
| HSG194 | Thermal comfort in the workplace – guidance for employers. 1999 |
| HSG199 | Managing health and safety on work experience – a guide for organisers. 2000 |
| HSG202 | General ventilation in the workplace – guidance for employers. 2000 |
| HSG218 | Tackling work-related stress – a managers' guide to improving and maintaining employee health and well-being. 2001 |
| L8 | Legionnaires' disease: the control of legionella bacteria in water systems – approved code of practice and guidance. Revised 2000 |
| L101 | Safe work in confined spaces: Confined Spaces Regulations 1997 – approved code of practice. 1997 |
| IACL27 | Legionnaires' disease – a guide for employers. Revised 2000 |
| INDG36 | Working with VDUs. Revised 1998 (Also available in Welsh) |
| INDG73 | Working alone in safety. Revised 1998 |

| | |
|---|---|
| INDG90 | Understanding ergonomics at work: reduce accidents and ill-health and increase productivity by fitting the task to the worker. Revised 2003 |
| INDG173 | Officewise. (Also available in Welsh). 1994 |
| INDG225 | Preventing slips, trips and falls at work. Revised 2003 |
| INDG226 | Homeworking: guidance for employers and employees on health and safety. 1996 |
| INDG244 | Workplace health, safety and welfare – a short guide. Revised 1997 |
| INDG258 | Safe work in confined spaces. 1999 |
| INDG281 | Work related stress – a short guide. Revised 2001 |
| INDG293 | Welfare at work – guidance for employers on welfare provision. 1999 |
| INDG333 | Back in work: managing back pain in the workplace – a leaflet for employers and workers in small businesses. 2000 |
| INDG341 | Tackling work related stress – a guide for employees. 2001 |
| INDG364 | The right start: work experience for young people. 2002 |
| OSR1 | Office, Shop and Railway Premises Act. (Also available in Welsh). 1994 |

# Appendix 3
# Relevant British Standards

| | |
|---|---|
| Access equipment | |
|     permanently installed, suspended | BS 6037 |
|     rope access methods for industrial purposes | BS 7985 |
| Agricultural machinery: | |
|     combine and forage harvesters | BS EN 632 |
|     silage cutters | BS EN 703 |
| Airborne noise emission | |
|     earth-moving machinery | BS 6812, BS ISO 6393, 6394 |
|     hydraulic fluid power systems and components | BS 5944 |
|     portable chain saws | BS 6916–6 |
| Ambient air: determination of asbestos fibres – direct-transfer transmission electron microscopy method | BS ISO 13794 |
| Anchorages | |
|     industrial safety harnesses | BS EN 795 |
|     self-locking, industrial | BS EN 353, 355, 360, 362, 365 |
| Arc welding equipment | BS 638 |
| Artificial daylight lamps | |
|     colour assessment | BS 950 |
|     for sensitometry | BS 1380–4 |
| Artificial lighting | BS 8206–1 |
| Barriers, in and about buildings | BS 6180 |
| Bromochlorodifluoromethane | |
|     fire extinguishing systems | BS 5306–5-2, BS EN 27201–1 |
|     fire extinguishers | BS EN 27201 |
| Carbon steel welded horizontal cylindrical storage tanks | BS 2594 |
| Carpet cleaners, electric, industrial use | BS EN 60335–2–67, BS EN 60335–2–72 |
| Cellulose fibres | BS 1771–2 |
| Chain lever hoists | BS 4898 |
| Chain pulley blocks, hand-operated | BS 3243 |
| Chain slings | BS EN 816–3, 816–5, 816–6 |
| Chairs | |
|     office furniture, performance | BS 5459–2 |
|     office furniture, ergonomic design | BS 3044 |
| Chemical protective clothing | |
|     against gases and vapours | BS EN 464 |

*Appendix*

| | |
|---|---|
| liquid chemicals | BS EN 466, 467 |
| Circular saws | |
|     hand-held electric | BS EN 50144–2–5 |
|     woodworking | BS 411 |
| Cleaning and surface repair of buildings | BS 6270, 8211 |
| Closed circuit escape breathing apparatus | BS EN 400 |
| Clothing for protection against heat and fire | BS EN 366, 367, 702 |
| Concrete cladding | BS 8297 |
| Construction equipment | |
|     hoists | BS 7212 |
|     suspended safety chairs, cradles | BS 2830 |
| Control of noise (construction and open sites) | BS 5228 |
| Cranes, safe use | BS 5744, 7121 |
| Disabled people, means of escape | BS 5588–8 |
| Drill Rigs | BS EN 791 |
| Dust | |
|     high efficiency respirators | BS EN 136, 143 |
|     particulate emission | BS 3405 |
| Ear protectors, acoustics | BS EN 24869, BS EN ISO 4869 |
| Earphones | |
|     audiometry, calibration, acoustic couplers | BS EN 60318–2, 60318–3 |
|     audiometry, calibration, artificial ears | BS 4669 |
| Earthing | BS 7430 |
| Earth moving equipment | BS 6912, BS EN 4741 |
| Electrical equipment | |
|     explosive atmospheres | BS 4683, 5501, 6941, BS EN 50014, 50015, 50017–50019, 50021, 60079 |
|     fire hazard testing | BS 6458, BS EN 60695, BS IEC 60695 |
|     guidance to wiring regulations | BS 7671 |
| Electrical resistance materials | |
|     bare fine resistance wires | BS 1117 |
|     earth-leakage circuit-breakers | |
|         — AC voltage operated | BS 842 |
|         — current-operated | BS 4293 |
|         — portable RCDs | BS 7071 |
|     electric shock protection, construction of electrical equipment | BS 2754 |

*Appendix*

| | |
|---|---|
| enclosures for high-voltage cable terminations, transformers and reactors | BS 6435 |
| fans for general purposes | BS 848 |
| industrial electric plugs | BS EN 60309 |
| industrial machines | BS 2771 |
| marking for low-voltage industrial switchgear/controlgear | BS 6272 |
| metallic | BS 115 |
| resistivity measurement | BS 5714 |
| static electricity | BS 5958 |
| switchgear | BS 5486 |
| test for resistance per unit length | BS 3466 |
| Emergency exits | BS EN 179, 1125 |
| Emergency lighting | BS 5266, BS EN 1838 |
| Environmental management systems | BS EN ISO 14001 |
| Ergonomics of the thermal environment | BS ISO 10551, BS EN ISO 7726 |
| Ergonomic requirements for office work with VDUs | BS EN 29241 |
| Eye protection | |
| equipment for eye and face protection during welding operations | BS EN 175 |
| glossary of terms | BS EN 165 |
| specification for sunglare filters used in personal eye protectors for industrial use | BS EN 172 |
| Eye protectors | BS EN 165–168, 170–172, 174, 175, 205, 207, 379, 1731, 1836, 1938 |
| Fabrics, curtains and drapes | BS 5867 |
| Falling-object protective structures | BS EN 13627, BS EN ISO 3164, 6055 |
| Falsework, code of practice | BS 5975 |
| Filters | |
| specification for infra-red filters used in personal eye protection equipment | BS EN 171 |
| specification for personal eye protection equipment in welding | BS EN 169, 379 |
| specification for protection against industrial sunglare | BS EN 172 |
| specification for protection against laser radiation | BS EN 207 |
| specification for ultra-violet filters used in personal eye protection equipment | BS EN 170 |
| Fire blankets | BS EN 1869 |
| Fire classification | BS EN 2 |
| Fire detection/alarm systems | BS 5839, 7807, BS EN 54, BS ISO 1338-7-7 |

*Appendix*

| | |
|---|---|
| Fire door assemblies | BS 8214 |
| Fire extinguishers | |
|     disposable aerosol type | BS 6165 |
|     media | BS 6535, BS EN 1568, 27201 |
|     on premises | BS 5306 |
|     portable | BS EN 3, 6643, 7867, 7937 |
| Fire hose reels | BS 5306–1, BS EN 671–1, 671–3 |
| Fire point determination, petroleum products | |
|     Able closed up method | BS EN ISO 13736 |
|     Cleveland open cup method | BS EN ISO 22592 |
|     Equilibrium method | BS 6664–6664-4 |
|     Pensky-Martens apparatus method | BS 2000–35, BS EN 22719 |
| Fire precautions in design/construction and use of buildings | BS 5588 |
| Fire protection measures, code of practice for operation | BS 7273 |
| Fire safety signs | BS 5499 |
| Fire terms | BS 4422 |
| Fire tests | BS 476 |
| Fire tests for furniture | BS 5852 |
| Firefighters gloves | BS EN 659 |
| First-aid reel/hoses | BS 3169 |
| Flameproof industrial clothing | BS EN 469, 1486 |
| Footwear | |
|     footwear fitted with toe caps | BS EN 345, 346, 12568 |
|     methods of test for safety | BS EN 344–1, 345 |
|     occupational footwear for professional use | BS EN 347 |
|     protective clothing for users of hand-held chain saws | BS EN 381 |
|     requirements/test methods for safety protective and occupational footwear for professional use | BS EN 344 |
| Freight containers | BS 3951, BS ISO 1496 |
| Gaiters and footwear for protection against burns and impact risks in foundries | BS 4676 |
| Gas detector tubes | BS 5343 |
| Gas fired hot water boilers | BS 5871, 5978, 5986, 6644, 6798, BS EN 297, 677 |
| Gas welding equipment | BS 1453, 1821, 2640, BS EN 731, 962, BS EN ISO 9692–3 |

App/43

*Appendix*

| | |
|---|---|
| Glazing | BS 6262 |
| Gloves: medical gloves for single use | BS EN 455 |
| Gloves: rubber gloves for electrical purposes | BS 697 |
| Goggles, industrial/non-industrial use | BS EN 166, 167, 168 |
| Grinding machines | |
|     hand-held electric | BS 2769-2-3 |
|     pneumatic, portable | BS 4390 |
|     spindle noses | BS 1089 |
| Head protection — fire fighters | BS EN 443 |
| Headforms for use in testing protective helmets | BS EN 960 |
| Hearing protectors | BS EN 352, 458, 24869, BS EN ISO 4869 |
| High visibility warning clothing | BS EN 471, 1150 |
| Hoists | |
|     construction, safe use | BS 7212 |
|     electric, building sites | BS 4465 |
| Hose reels with semi-rigid hose | BS EN 671-1 |
| Hose systems with lay-flat hose cloth | BS EN 671-2 |
| Hot environments | |
|     estimation of heat stress on the working man | BS EN 27243 |
| Household and similar electrical appliances | BS EN 60335 |
| Industrial gloves | BS EN 374, 381-7, 388, 407, 420, 697, 6093 |
| Industrial safety helmets, firemen's | BS EN 443 |
| Industrial trucks | |
|     hand-operated stillage trucks, dimensions | BS 4337 |
|     pallet trucks, dimensions | BS ISO 509 |
|     pedals, construction/layout | BS 7178 |
| Insulating material | BS 2844, 5691, 7737, 7822, 7831, BS EN 60383-2, |
| Ionising radiation | |
|     recommendation for data on shielding from ionizing radiation | BS 4094 |
|     units of measurement | BS 5775 |
| Jib cranes | |
|     high pedestal and portal | BS 2452 |
|     power-driven | BS 357, 1757 |
| Ladders | |
|     code of practice | BS 5395 |
|     permanent for chimneys, high structures | BS 4211 |

*Appendix*

| | |
|---|---|
| portable | BS 1129, 2037, BS EN 131, 1147 |
| Life jackets | BS EN 394, 395, 396, 399 |
| Lifting chains | BS 4942, 5594, 6521, 7255, BS EN 818, BS ISO 1834 |
| Lifting slings | BS 6166 |
|     chain, welded | BS 6304, BS EN 818–2, 818–4 |
|     textile | BS 6668, BS EN 1492 |
|     wire rope | BS 1290, 6210 |
| Lighting systems: automatic change-over contractors for emergency lighting | BS 764 |
| Machine guards | |
|     chain saws | BS 6916–3, 6916–4 |
|     conveyors and elevators | BS 5667–18, 5667–19 |
|     earth-moving equipment | BS EN ISO 3457 |
| Machine tools | |
|     emergency stop equipment, functional aspects | BS EN 418 |
| Machinery, safety of | |
|     drafting and presentation | BS EN 414 |
|     ergonomic design principles | BS EN 614–1 |
|     hazardous substances emitted by machines | BS EN 626 |
|     indication, marking and actuation, requirements for visual, auditory and tactile signals | BS EN 61310–1 |
|     indication, marking and actuation, requirements for marking | BS EN 61310–2 |
|     minimum gaps to avoid crushing parts of body | BS EN 349 |
|     principles and specifications for machinery manufacturers | BS EN 626–1 |
| Machines, vibration | BS 4675, 7854, BS EN 12786 |
| Materials handling | |
|     conveyor belts | BS 5767 |
|     freight containers | BS 3951 |
| Mobile cranes | BS 1757 |
| Mobile road construction machinery | BS EN 500 |
| Mortising machines, single chain | BS 4361–16 |
| Natural fibre ropes | BS EN 698, 701, 1261 |
|     cords, lines, twines | BS 6125 |
| Nets, safety | BS EN 1263 |
| Noise | |

*Appendix*

| | |
|---|---|
| airborne, chain saws | BS 6916–6 |
| airborne, earth-moving equipment | BS 6812, BS ISO 6393, 6394 |
| airborne, hydraulic transmission systems | BS 5944 |
| effects on hearing handicap | BS 5330 |
| industrial noise, method for rating | BS 4142 |
| sound exposure meters | BS EN 61252 |
| Noise induced hearing loss | |
| effects of noise exposure | BS 5330 |
| pure tone air conduction threshold audiometry | BS 6655 |
| Occupational safety and health management systems | BS 8800 |
| Office buildings, fire precautions | BS 5588–11 |
| Office furniture, ergonomic design and selection | BS 3044 |
| Office machines | |
| electrically energised, safety | BS EN 60950 |
| glossary | BS 4277 |
| keyboard layout | BS 5448 |
| noise measurement | BS 7135 |
| Oil burning equipment | BS 799 |
| Oil firing code of practice | BS 5410 |
| Open bar gratings – specification | BS 4592–1 |
| Overhead travelling cranes | |
| power-driven | BS 466 |
| safe use | BS 5744 |
| Packaging | |
| pictorial marking for handling of goods | BS EN ISO 780 |
| Particulate air pollutants | |
| in effluent gases, measurement | BS 3405 |
| Passenger hoists | |
| electric, building sites | BS 4465 |
| vehicular | BS 6109–2 |
| Patent glazing | BS 5516 |
| Pedestrian guardrails (metal) | BS 7818 |
| Performance of windows | BS 6375 |
| Personal eye protection | |
| filters for welding and related techniques | BS EN 169 |
| infrared filters | BS EN 171 |
| non-optical test methods | BS EN 168 |
| optical test methods | BS EN 167 |
| specifications | BS EN 166 |
| ultraviolet filters | BS EN 170 |
| vocabulary | BS EN 165 |

| | |
|---|---|
| Pipelines, identification marking | BS 1710 |
| Pneumatic tools | |
|     portable grinding machines | BS 4390 |
| Portable fire extinguishers | BS 7863, BS EN 3, 6643, 7867, 7937 |
| Portable tools | |
|     pneumatic grinding machines | BS 4390 |
| Powder fire extinguishers | |
|     disposable, aerosol type | BS 6165 |
|     extinguishing powders for | BS EN 615 |
|     fixed fire fighting systems | BS EN 12416 |
|     on premises | BS 5306 |
|     portable, recharging | BS 6643 |
| Power take-off | |
|     agricultural tractors, front-mounted | BS 6818 |
|     agricultural tractors hydraulic equipment | BS 4742–4742-8 |
|     agricultural tractors, rear-mounted | BS 5861 |
| Powered industrial trucks | |
|     driverless, safety | BS EN 1525 |
|     safety signs and hazard pictorials | BS ISO 15870 |
| Process control – safety of analyser houses | BS EN 61285 |
| Protective barriers | BS 2655–10, 6180, 6779, 7818, BS EN 1317, 1930 |
| Protective cabs | |
|     agricultural tractors | BS 5947 |
|     controls for external equipment | BS 5731 |
|     sound insulation performance | BS EN ISO 11957 |
| Protective clothing | |
|     against cold weather | DD ENV 342 |
|     against foul weather | DD ENV 343 |
|     against heat and fire | BS EN 366 |
|     against heat and flame | BS EN 532, 702, BS ISO 3150, 11612, 15025, 17943 |
|     against molten metal splash | BS EN 373 |
|     against risk of being caught up in moving parts | BS EN 510 |
|     eye, face and neck protection, welding | BS EN 175 |
|     flameproof | BS EN 469, 531 |
|     for firefighters | BS EN 469, 1486 |
|     for industrial workers exposed to heat | BS EN 531 |
|     for use where there is risk of entanglement | BS EN 510 |
|     for users of hand-held chain saws | BS EN 381 |

| | |
|---|---|
| for welders | BS EN 470 |
| for workers exposed to heat | BS EN 531 |
| gaiters for foundries | BS 4676 |
| general requirements | BS EN 340 |
| mechanical properties | BS EN 863 |
| protection against intense heat | BS EN 366, 367 |
| protection against liquid chemicals | BS EN 369, 466, 467 |
| welding | BS EN 175, 470 |
| Protective equipment | |
| against falls from a height | BS EN 341 |
| against falls from a height – guided fall type arresters | BS EN 353 |
| Protective footwear | |
| vulcanised rubber | BS 5145, 7193 |
| firemen's leather boots | BS 2723 |
| for foundries | BS 4676 |
| lined industrialised rubber boots | BS 5145, 7193 |
| polyvinyl chloride boots | BS 6159 |
| professional use | BS EN 346 |
| Protective gloves | |
| against chemicals and micro-organisms | BS EN 374 |
| against cold | BS EN 511 |
| against ionising radiation | BS EN 421 |
| against mechanical risks | BS EN 388, BS EN ISO 237 |
| against thermal hazards | BS EN 407 |
| for users of hand-held chain saws | BS EN 381–7 |
| general requirements | BS EN 420 |
| mechanical test methods | BS EN 388 |
| vibration transmissibility | BS EN ISO 10819 |
| Protective helmets | BS EN 397 |
| Quality control | BS 5750, 7679, BS EN 295–2, 450, 12306, BS EN ISO 9000–9004 |
| Radiation measures | |
| detectors, nuclear reactors | BS 5548 |
| electroscope, exposure meters | BS 3385 |
| film badges | BS 3664 |
| neutron detectors | BS 5552 |
| personal photographic dosemeters | BS 6090 |
| portable equipment | BS 7515, 7516 |
| test tube for | BS 5542 |

| | |
|---|---|
| Radiation protection, | |
|     area radiation monitors, X-ray and gamma radiation | BS 5566 |
| Refrigeration systems | BS 1586, 1608, 2626, 7005, BS EN 378, 1861 |
| Resistance to ignition of upholstered furniture for non domestic seating | BS 7176, 13136, 13215 |
| Respirators | |
|     full masks for respiratory protective devices | BS EN 136 |
| Respiratory protective devices | BS EN 138, 148, 269, 270, 271 |
| Roll-over protective structures | |
|     agricultural tractors | BS 5947 |
|     earth moving equipment | BS EN 13510, BS EN ISO 6683 |
|     industrial trucks, stacking with masted tilt forward | BS 5778 |
|     pallet stackers/high lift platform trucks | BS 5777 |
|     reach and straddle fork trucks | BS 4436 |
| Rope pulley blocks | |
|     gin blocks | BS 1692 |
|     synthetic fibre | BS 4344 |
|     wire | BS 4018, 4536 |
| Rope slings | |
|     fibre rope slings | BS 6668–1 |
|     wire rope slings | BS 1290, 6210 |
| Rubber/plastics injection moulding machines | BS EN 201 |
| Safety anchorages | |
|     industrial safety harnesses | BS 6858, BS EN 795 |
| Safety distances to prevent danger zones being reached by upper limbs | BS EN 294, 349 |
| Safety harnesses, industrial | BS 6858, BS EN 354, 355, 358, 361–365 |
| Safety helmets | BS EN 397, 403, 12941 |
| Sampling methods | |
|     airborne radioactive materials | BS 5243 |
|     particulate emissions | BS 3405 |
| Scaffolds, code of practice | BS 5973, 5974 |
| Scalp protectors | BS EN 812 |
| Shaft construction and descent | BS 8008 |
| Specification for artificial daylight for colour assessment | BS 950 |
| Staging, portable | BS 1129, 2037 |
| Stairs, ladders, walkways | BS 5395 |

*Appendix*

| | |
|---|---|
| Steam boilers | |
|     electric boilers | BS 1894 |
|     oil fuelled boilers, code of practice | BS 5410–2 |
|     safety valves for | BS 6759–1 |
|     welded steel low pressure boilers | BS 855 |
| Step ladders | |
|     portable | BS 1129, 2037 |
|     safety of access to machinery | BS EN ISO 14122–3 |
| Storage tanks | |
|     carbon steel welded horizontal cylindrical | BS 2594 |
|     vertical steel welded non-refrigerated butt-welded shells | BS 2654 |
|     welded | BS 7122, BS EN 12493, 12573 |
| Suspended access equipment | BS 6037, BS EN 1808 |
| Suspended safety chairs | BS 2830 |
| Suspended scaffolds, temporarily installed | BS 5974 |
| Tables, office furniture | BS 3044, 5459–1, BS EN 527–1 |
| Textile floor coverings | BS 3655, 5287, 5325, 5808, 5921 |
| Textile machinery, safety requirements | BS EN ISO 11111 |
| Transportable gas containers | |
|     acetylene | BS 6071, BS EN 1800, 1801, 12755 |
|     periodic inspection, testing and maintenance | BS 5430 |
|     welded steel tanks for road transport of liquefiable gases | BS 7122 |
| Travelling cranes | BS 357, 3037, 5744 |
| Vertical steel welded non-refrigerated storage tanks, manufacture of | BS 2654 |
| Vibration measurement | |
|     chain saws | BS 6916–8 |
|     rotating shafts | BS ISO 7919, 10817 |
| Water absorption and translucency of china or porcelain | BS 5416 |
| Water services, installation, testing and maintenance | BS 6700 |
| Welders, protective clothing | BS EN 470 |
| Window cleaning | BS 8213–1 |
| Windows, performance of | BS 6375 |
| Woodworking noise | BS 7140 |
| Wool and wool blends | BS 1771–1 |
| Working platforms | |
|     mobile, elevating | BS 7171, BS EN 280 |

| | |
|---|---|
| permanent, suspended access | BS 6037 |
| Workplace atmospheres | |
| performance of procedures for measurement of chemical agents | BS EN 482 |
| size definitions for measurement of airborne particulates | BS EN 481 |
| Workwear and career wear | BS 5426 |

# Appendix 4
# A Summary of Risk Assessment Requirements

## Control of Asbestos at Work Regulations 2002

### On whom duties placed
1. Employers.
2. Self-employed persons.

### Identify of those assessed and geographical location
1. Employers.
2. Any other person, whether at work or not, who may be affected by work activity carried out by employer/self-employed person.

### Hazards or risks to be assessed
Risks created by exposure of employees and other relevant persons to asbestos.

### Purpose of assessment
(a) Identify type of asbestos.
(b) Determine degree and nature of exposure which may occur.
(c) Consider the effect of control measures.
(d) Set out steps to be taken to prevent that exposure or reduce it to lowest level reasonably practicable.
(e) Consider results of relevant medical surveillance.

### Qualification of duty
Suitable and sufficient. In so far as the duty extends to any other person, it is also subject to the 'so far as is reasonably practicable' qualification.

### When the assessment has to be made
Before work is commenced.

### Record provision
Employer/self-employed person must record the significant findings of the risk assessment as soon as is practicable after it is made.

*Appendix*

### Review provision
Risk assessment to be reviewed 'regularly and forthwith' if:
(a) reason to suspect existing risk assessment is no longer valid;
(b) significant change in the work to which the risk assessment relates; or
(c) results of any monitoring carried out in accordance with Regulations show it to be necessary.

### Action on review
Where, as a result of the risk assessment review, changes to it are required, such changes shall be made.

## Control of Asbestos at Work Regulations 2002 (the new duty, under Reg 4 to manage the risk from asbestos in non-domestic premises, coming into force on 21 May 2004)

### On whom duties placed
1. Persons having, by reason of contract or tenancy, a maintenance/repair obligation (to any extent) for non-domestic premises.
2. Persons who, where there is no such contract or tenancy, have (to any extent) control of premises or part thereof.

### Identity of those assessed and geographical location
Non-domestic premises.

### Hazards or risks to be assessed
Risks from asbestos.

### Purpose of assessment
(a) To determine whether asbestos is, or is likely to be, present in the premises.
(b) To consider condition of any asbestos which is, or has been assumed to be, present in the premises.

### Qualification of duty
Suitable and sufficient.

### When assessment to be made
By 21 May 2004.

### Record provision
Duty-holder must record conclusions of the risk assessment and every review.

*Appendix*

### Review provision

Risk assessment to be reviewed forthwith if:

(a) reason to suspect assessment is no longer valid; or

(b) significant change in the premises to which assessment relates.

### Action on review

Where review reveals changes to risk assessment necessary, such changes shall be made.

## Control of Lead at Work Regulations 2002

### On whom duties placed

1. Employers.
2. Self-employed persons.

### Identity of those assessed and geographical location

1. Employees
2. Any other person, whether at work or not, who may be affected by the work carried on by employer or self-employed person.

*Note*: Young persons and women of 'reproductive capacity' cannot be employed in any of the lead-working activities set out in *Schedule 1* of the Regulations.

### Hazards or risks to be assessed

The risks to health created by working with lead.

### Purpose of assessment

To allow employer/self-employed person to make a valid decision about whether the work concerned is likely to result in the exposure of any employees and other persons not employees to lead being 'significant', and to identify the measures needed to prevent or adequately control exposure.

### Qualification of duty

Suitable and sufficient. In so far as the duty extends to persons not employees it is also subject to the 'so far as is reasonably practicable' qualification.

### When assessment has to be made

Before work is commenced.

### Record provision

Employer/self-employed person employing 5 or more employees must record:

(a) significant findings of the risk assessment as soon as is practicable after the assessment is made; and

(b) the steps taken to prevent or control exposure to lead; and

(c)  the significant findings of any risk assessment review.

## Review provision

The record of the assessment should be a living document, which must be revisited to ensure that it is kept up-to-date. The employer/self-employed person should make arrangements to ensure assessment is reviewed regularly. The date of the first review and the length of time between successive reviews will depend on the type of risk, the work and the judgment of the employer/self-employed person on the likelihood of changes occurring. In particular, the risk assessment should be reviewed when there is:

(a)  evidence to suggest that it is no longer valid;

(b)  new information on health risks;

(c)  significant change in the circumstances of work.

## Action on review

Implementation of any necessary changes identified by review.

# Control of Substances Hazardous to Health Regulations 2002 (as amended by SI 2003 No 278)

## On whom duties placed

1. Employers.
2. Self-employed persons.

## Identify of those assessed and geographical location

1. Employers liable to be exposed to substances hazardous to health by work.
2. Any other person, whether at work or not, who may be affected by work carried out by employer or self-employed person.

## Hazards or risks to be assessed

Those likely to arise from exposure to substances hazardous to health.

## Purpose of assessment

To enable employer/self-employed person to make a valid decision about the measures necessary to prevent or adequately control the exposure of their employees/other relevant persons, as above, to substances hazardous to health arising from work.

## Qualification of duty

Suitable and sufficient. In so far as the duty extends to any other person, it is also subject to the 'so far as is reasonably practicable' qualification.

When the assessment has to be made

Before work is commenced.

*Appendix*

### Record provision

Where employer/self-employed person employs 5 or more employees, obligation to record:

(a) the significant findings of the risk assessment as soon as practicable after it is made; and

(b) the steps taken to prevent or control exposure to substances hazardous to health.

### Review provision

Risk assessment to be reviewed 'regularly and forthwith' if:

(a) reason to suspect that it is no longer valid;

(b) there has been a significant change in the work to which the assessment relates; or

(c) results of any monitoring carried out in accordance with Regulations show it to be necessary.

### Action on review

Where, as a result of the risk assessment review, changes to it are required, such changes shall be made.

## Dangerous Substances and Explosive Atmospheres Regulations 2002

### On whom duties placed

1. Employers.
2. Self-employed persons.

### Identity of those assessed and geographical location

1. Employees.
2. Any other person, whether at work or not, who may be affected by the work of the employer.

### Purpose of assessment

(a) The elimination or reduction of risks to employees and others from dangerous substances present or liable to be present in the workplace.

(b) The classification of places at the workplace where explosive atmospheres may occur into hazardous or non-hazardous places so that the appropriate steps be taken in accordance with the regulatory requirements.

(c) The putting in place of appropriate arrangements to deal with accidents, incidents or emergencies related to the presence of a dangerous substance at the workplace.

### Qualification of duty

Suitable and sufficient.

### When assessment has to be made

When a dangerous substance is or is liable to be present at the workplace.

## Record provision

The 'significant' findings of the risk assessment, including the matters specified in the Regulations.

## Review provision

Review assessment regularly and if:

(a) there is reason to suspect it is no longer valid; or

(b) there has been a significant change in the matters to which the assessment relates.

## Action on review

Changes to assessment to be made where required.

*Note*: The record-keeping requirements only apply to those employing five or more employees.

# Genetically Modified Organisms (Contained Use) Regulations 2000 (as amended by SI 2002 No 63)

## On whom duties placed

Any person undertaking any activity involving genetic modification of micro-organisms and organisms other than micro-organisms.

## Identity of those assessed and geographical location

Persons and the environment put at risk by the above activity.

## Hazards or risks to be assessed

As above i.e. the risks created by the activity to human health and the environment.

## Purpose of assessment

To identify those risks.

## Qualification of duty

Suitable and Sufficient.

## When assessment has to be made

Prior to undertaking any such activity.

## Record provision

(a) A record of the assessment relating to the activity, and any review of that assessment.

(b) At least 10 years from the date of cessation of the activity.

## Review provision

Review assessment if:

*Appendix*

(a) reason to suspect it is no longer valid,

(b) there has been a significant change in the activity to which assessment relates.

### Action on review

Risk assessment to be modified, as necessary.

## Health and Safety (Display Screen Equipment) Regulations 1992 (as amended by SI 2002 No 2174)

### On whom duties placed

Employers.

### Identity of those assessed and geographical location

1. Users (employees who habitually use display screen equipment as a significant part of their normal work).
2. Operators (self-employed persons who habitually use display screen equipment as a significant part of their normal work).

### Hazards or risks to be assessed

Workstations.

### Purpose of assessment

To assess workstations for health and safety risks to which users/operators are exposed.

### Qualification of duty

Suitable and sufficient.

### Review provision

Review assessment if:

(a) reason to suspect it is no longer valid

(b) there has been a significant change.

### Action on review

Changes to assessment to be made where required.

### Details

Employers shall reduce risks identified by assessment to lowest extent reasonably practicable.

## Ionising Radiations Regulations 1999

### On whom duties placed

1. Employers.

2. Self-employed persons.
3. Mine managers – in so far as the duties relate to the mine or part thereof which he is manager/operator and to matters within his control.
4. Quarry operators – in so far as the duties relate to the quarry or part thereof which he is manager/operator and to matters within his control.
5. The holder of a nuclear site licence – in so far as the duties relate to the licensed site.

## Identity of those assessed and geographical location

1. Employees.
2. Persons other than an employer's employees, only in so far as the exposure of these persons to ionising radiation arises from work with ionising radiation undertaken by that employer.
3. Self-employed persons.
4. Trainees, as defined.

## Hazards or risks to be assessed

Any new activity involving work with ionising radiation – the 'Prior risk assessment'.

## Purpose of assessment

To identify the measures needed to be taken to restrict the exposure of employees or other persons to ionising radiation.

## Qualification of duty

Suitable and Sufficient.

## When assessment has to be made

Before a duty holder commences a new activity involving working with ionising radiation.

## Record provision

(a) No stipulation in the Regulations but ACoP and guidance says that it 'makes sense' to record the significant findings of the prior risk assessment where employers have five or more employees and/or where any group of employees is especially at risk.

(b) The significant findings.

## Review provision

Review prior risk assessment if:

(a) reason to suspect it is no longer valid (or can be improved),
(b) there has been a significant change.

## Action on review

Prior risk assessment to be modified, as necessary.

*Appendix*

# Management of Health and Safety at Work Regulations 1999

## On whom duties placed

1. Employers.
2. Self-employed persons.

## Identity of those assessed and geographical location

1. Employees (at work).
2. Persons not in the employer's employment who may face risks arising out of or in connection with the conduct by him of his undertaking.
3. New and expectant mothers.
4. Young persons.

The Management Regulations' ACoP recommends that employers should identify groups of workers particularly at risk such as night workers, homeworkers, those who work alone and disabled staff.

## Hazards or risks to be assessed

(a) All risks.
(b) A risk assessment carried out by a self-employed person in circumstances where he or she does not employ others does not have to take into account duties arising under Part II of the Fire Precautions (Workplace) Regulations 1999, as amended by the Fire Precautions (Workplace) (Amendment) Regulations 1999 (SI 1999 No 1877).
(c) Physical, biological and chemical agents.

## Purpose of assessment

To identify measures needed to be taken to comply with the requirements and prohibitions imposed by or under the relevant statutory provisions and by Part II of the Fire Precautions (Workplace) Regulations 1999, as amended (see above).

## Qualification of duty

Suitable and Sufficient.

## Record provision

(a) Five or more employees.
(b) Significant findings of the assessment and any group of employees identified by it as being especially at risk.

## Review provision

Review assessment if:

(a) reason to suspect it is no longer valid (or can be improved),
(b) there has been a significant change.

### Action on review

Risk assessment to be modified, as necessary.

## Manual Handling Operations Regulations 1992 (as amended by SI 2002 No 2174)

### On whom duties placed

1. Employers.
2. Self-employed persons.

### Identity of those assessed and geographical location

Employees (at work).

The HSE's Manual Handling Regulations Guidance emphasises that employers must make allowance for those who might be pregnant or have a disability or health problem.

### Hazards or risks to be assessed

All manual handling operations (with regard to factors – task, load, working environment, individual capability, other factors – listed in Schedule 1).

### Purpose of assessment

To consider the questions set out in column 2 of Schedule 1 pertaining to the factors listed in that Schedule (see above).

### Qualification of duty

Suitable and Sufficient.

### When assessment has to be made

Make assessment where it is not reasonably practicable to avoid the need for employees to undertake any manual handling operations which involve risk of injury.

### Review provision

Review assessment if:

(a) reason to suspect it is no longer valid,

(b) there has been a significant change.

### Action on review

Changes to assessment to be made where required.

### Details

Employers shall:

(a) take steps to reduce risk of injury to lowest level reasonably practicable,

*Appendix*

(b) take steps to provide (to employees) general indications, and, where reasonably practicable, precise information on:
    (i) weight of load,
    (ii) heaviest side of load where centre of gravity is not central.

# Noise at Work Regulations 1989 (most recently amended by SI 1999 No 2024)

## On whom duties placed

Employers and self-employed persons to ensure competent person makes assessment.

## Identity of those assessed and geographical location

Employees (at work).

## Hazards or risks to be assessed

Noise.

## Purpose of assessment

(a) To identify which employees are exposed.
(b) To provide such information, with regard to the noise, as will facilitate compliance with Regs 7,8,9 and 11.

## Qualification of duty

Adequate.

## When assessment has to be made

Assessment to be made when any employee is likely to be exposed to the first action level or above or to the peak action level or above.

## Record provision

Record to be kept until a further assessment is made.

## Review provision

Review assessment if:
(a) reason to suspect that it is no longer valid,
(b) there has been a significant change.

## Action on review

Changes to assessment to be made where required.

# Personal Protective Equipment at Work Regulations 1992 (as amended most recently by SI 2002 No 2174)

## On whom duties placed

1. Employers.
2. Self-employed persons.

## Identity of those assessed and geographical location

Employees (at work).

## Hazards or risks to be assessed

Any risks which have not been adequately controlled by other means.

## Purpose of assessment

To assess risks to health and safety which have not been avoided by other means.

## When assessment has to be made

Assessment to be made *before* choosing any personal protective equipment (PPE).

## Review provision

Review assessment if:

(a) reason to suspect it is no longer valid,

(b) there has been a significant change.

## Action on review

Changes to assessment to be made where required.

## Details

Assessment to include:

(a) definition of the characteristics which PPE must have in order to be effective against the risks (taking into account any risks which the equipment itself may create),

(b) comparison of the characteristics of the PPE available to the required characteristics.

# Appendix 5
# A Summary of Requirements for the Provision of Information

## Carriage of Dangerous Goods by Road Regulations 1996 (most recently amended by SI 1999 No 303)

Consignor of dangerous substances to supply sufficient accurate written information to enable the operator to:

(a) comply with his duties under these Regulations; and

(b) be aware of the hazards created by the substance to health and safety,

(c) adequate information about identity, quantity and nature of hazards created by substance information to operator. This must be accurate, sufficient and so far as possible in writing.

Operator to keep information for at least three months after completion of journey and must have information before carriage.

## Carriage of Explosives by Road Regulations 1996 (as amended by SI 1999 No 303)

Employer/self-employed person to provide for other employers, self-employed persons/others

The following information must be provided in writing at the start of the journey:

(a) in the case of classified explosives the Division and Compatibility Group of each type of explosive carried,

(b) the net mass of each type of explosive carried,

(c) whether, in the case of explosives in Compatibility Group C, D or G, the explosives carried are explosive substances or explosive articles,

(d) name and address of consignor, operator and consignee,

(e) information to enable driver and attendant to know nature of danger and action to be taken in an emergency.

## Chemicals (Hazard Information and Packaging for Supply) Regulations 2002

Employer/self-employed person to provide for other employers, self-employed persons/others

Supplier of dangerous chemicals to communicate with recipients information on the hazards presented.

*Appendix*

# Construction (Design and Management) Regulations 1994 (as amended by SI 2000 No 2380)

## Principal contractor

To provide:

— so far is reasonably practicable, every contractor with comprehensible information on the risks to the health or safety of that contractor or of any other employees or other persons under the control of that contractor arising out of or in connection with the construction work.

To ensure:

— so far is reasonably practicable, that every contractor who is an employer provides his employees carrying out the construction work with the information the employer must provide to those employees under the Management of Health and Safety at Work Regulations 1999, namely: the risks identified by the risk assessment; the preventative and protective measures thereby made necessary; the procedures in event of serious and imminent danger; the 'competent persons' as regards evacuation; and, the risks notified by other employers.

# Control of Asbestos at Work Regulations 2002

## Employer/self-employed person to provide own employees/persons not employees on the premises where the work is being carried out with adequate information

If such persons are, or are liable to be, exposed to asbestos, or if they supervise such persons who are, so that they are aware of:

(a)   significant findings of the risk assessment;

(b)   risks to health from asbestos;

(c)   precautions which should be observed;

(d)   relevant control limit and action level, in order to safeguard themselves and other relevant persons.

The above information must be:

(a)   given at regular intervals;

(b)   adapted to take account of significant changes in the type of work carried out by, or methods of work used by employer/self-employed person; and

(c)   provided in a manner appropriate to the nature and degree of exposure identified by the risk assessment.

*Note*: The duty to provide adequate information to persons not employees is subject to the 'so far as is reasonably practicable' qualification.

# Control of Lead at Work Regulations 2002

## Employer/self-employed person to provide own employees/persons other than employees on the premises where the work is being carried out with suitable and sufficient instruction and training including:

(a)   details of the form of lead to which employee/other relevant person is liable to be exposed;

*Appendix*

(b) the significant findings of the risk assessment;

(c) the appropriate precautions and actions to be taken by employees and other persons, as above, coming within the duty, in order to safeguard both themselves and those other persons;

(d) the results of any monitoring of exposure to lead carried out in accordance with the Regulations; and

(e) the collective results of any medical surveillance undertaken in accordance with the Regulations.

Such instruction and training must be adapted to take account of significant changes in the type of work or methods of work used by the employer/self-employed person; and, provided in a manner appropriate to the level, type and duration of exposure identified by the risk assessment.

In addition, every employer/self-employed person must ensure that *any* person carrying out work in connection with the employer's/self-employed person's duties under the Regulations has sufficient instruction and training.

*Note*: The duty to provide suitable and sufficient instruction and training to persons other than employees is subject to the 'so far as is reasonably practicable' qualification.

## Control of Substances Hazardous to Health Regulations 2002 (as amended by SI 2002 No 278)

**Employer/self-employed person to provide own employees/persons not employees on the premises where the work is being carried out with suitable and sufficient information, including:**

(a) details, as specified, of the substances hazardous to health to which employees/other relevant persons are liable to be exposed;

(b) the significant findings of the risk assessment;

(c) the appropriate precautions and actions to be taken by employees/other relevant persons in order to safeguard both themselves and those other persons at the workplace;

(d) the results of any monitoring of exposure in accordance with the Regulations;

(e) the collective results of any health surveillance undertaken in accordance with the Regulations; and

(f) where employees and other relevant persons are working with a Group 4 biological agent (or material that may contain such an agent) the provision of written instruction and, if appropriate, the display of notices outlining procedure for handling such an agent or material.

Such information must be adapted to take account of significant changes in type of work carried out or methods of work used by employer/self-employed person; and, be provided in a manner appropriate to the level, type and duration of exposure identified by the risk assessment.

*Note*: The duty to provide suitable and sufficient information to persons other than employees is subject to the 'so far as is reasonably practicable' qualification.

## Dangerous Substances in Harbour Areas Regulations 1987 (as amended by SI 1998 No 2885)
### Employer to provide for own employees

Employer to provide information necessary to ensure his health and enable him to perform any operations in which he is involved with due regard to the health and safety of others.

### Employer/self-employed person to provide for other workers

Self-employed person to ensure that he has information to ensure his and others' health and safety.

### Employer/self-employed person to provide for other employers, self-employed persons/others

Operator to provide persons present on the berth with information to ensure their and others' health and safety.

## Health and Safety (Display Screen Equipment) Regulations 1992 (as amended by SI 2002 No 2174)
### Employer to provide for own employees

For his employees who are 'users' employer must provide information on:

(a) All aspects of health and safety relating to their workstation;

(b) Measures taken to comply with Regulations 2 (risk assessment), 3 (workstations), 4 (breaks), 5 (eyes and eyesight) and 6 (training).

### Employer/self-employed person to provide for other workers

For 'users' employed by other employers, and 'operators', at work in his undertaking the employer must provide information on:

(a) All aspects of health and safety relating to their workstation;

(b) Measures taken to comply with Regulations 2 and 3 for 'users' only, Regulations 4 and 6(2) (training when workstation modified).

## Health and Safety (First-aid) Regulations 1981 (most recently amended by SI 2002 No 2174)
### Employer to provide for own employees

Inform his employees of the arrangements that have been made in connection with the provision of first-aid, including the location of equipment, facilities and personnel.

*Appendix*

# Health and Safety Information for Employees Regulations 1989 (as amended by SI 1995 No 2923)

## Employer to provide for own employees

Employer to ensure that the approved poster is kept displayed in a readable condition at an accessible place to the employee while he is at work and must provide the employee with the approved leaflet. Employer has to have information of EMAS office and enforcing authority on poster.

# Ionising Radiations Regulations 1999

## Employer/self-employed person to provide for own employees

Such information as is suitable for them to know:

(a) the health risks created by exposure to ionising radiation;

(b) the precautions which should be taken; and

(c) the importance of complying with the medical, technical and administrative requirements of the Regulations.

## Employer/self-employed person to ensure

Adequate information is given to persons other than employees who are directly concerned with ionising radiation work carried on by the employer/self-employed person to ensure, so far is reasonably practicable their health and safety.

# Management of Health and Safety at Work Regulations 1999

## Employer/self-employed person to provide for own employees

(a) Risks identified by the assessment.

(b) Preventive and protective measures as per Schedule 1.

(c) Procedures in event of serious and imminent danger.

(d) Competent persons re evacuation.

(e) Risks notified by other employers.

## Employer/self-employed person to provide for other workers

(a) Risks to health and safety arising from the conduct of the undertaking.

(b) To enable them to identify competent persons re evacuation.

(c) Any special occupational qualifications or skills needed to work safely.

(d) Any health surveillance required.

## Employer/self-employed person to provide for other employers, self employed persons/others

(a) Risks to health and safety arising from the conduct of the undertaking.

(b) Measures taken in compliance.

(c) To enable them to identify competent persons re evacuation.

*Appendix*

(d) Any special occupational qualifications or skills needed to work safely.

(e) Specific features of jobs in relation to health and safety.

## Manual Handling Operations Regulations 1992 (as amended by SI 2002 No 2174)

### Employer/self-employed person to provide for own employees

Those undertaking manual handling operations.

(a) General indications, and where reasonably practicable, precise information on:
  (i) Weight of each load,
  (ii) Heaviest side.

## Noise at Work Regulations 1989 (most recently amended by SI 1999 No 2024)

### Employer/self-employed person to provide for own employees

Every employer shall provide any of his employees, who are likely to be exposed to the first action level or above, information on:

(a) Risk of damage to hearing.

(b) What steps can be taken to minimise risk.

(c) Steps that the employee must take in order to obtain personal ear protection.

(d) Employees' obligations.

## Offshore Installations and Wells (Design and Construction, etc) Regulations 1996

### Well-operator in installation duty holder

Provision of appropriate information so that those carrying well operations can carry them out competently.

## Packaging, Labelling and Carriage of Radioactive Material by Rail Regulations 2002

'Relevant' employers must provide all employees with responsibilities relating to the carriage of radioactive material with information appropriate to those responsibilities so that they can understand:

(a) the nature of the dangers of the material being carried;

(b) the precautions the employee should take to ensure that:
  (i) his exposure to radiation; and
  (ii) the exposure of others who may be affected by the employee's actions, are restricted;

(c) what action an employee should take in an emergency concerning such radioactive material; and

(d) the regulatory requirements relating to the carriage of radioactive material by rail.

*Appendix*

# Packaging, Labelling and Carriage of Radioactive Material by Road Regulations 2002

'Relevant' employers must provide all employees with responsibilities relating to the carriage of radioactive material with such information as is appropriate to those responsibilities so that they can understand such matters as are mentioned in APPENDIX 5 in connection with these Regulations in so far as they relate to instruction and training.

# Personal Protective Equipment at Work Regulations 1992 (as amended by SI 2002 No 2174)

## Employer to provide for own employees

Such information as is adequate and appropriate on:

(a) Risks PPE will avoid/limit.

(b) Purpose/manner of use.

(c) Action to be taken by employee.

# Pressure Systems Safety Regulations 2000

## Employer to provide for own employees

(a) Any person who:

  (i) designs for another any pressure system or any article which is intended to be a component or part thereof; or

  (ii) supplies (whether as manufacturer or importer or in any other capacity) any pressure system or any such article, shall provide sufficient information concerning its design, construction, examination, operation and maintenance as may reasonably foreseeably be needed to comply with the Regulations.

(b) The employer of a person who modifies or repairs any pressure system must provide sufficient information concerning the modification or repair as may reasonably foreseeably be needed to comply with the Regulations.

# Provision and Use of Work Equipment Regulations 1998 (most recently amended by SI 1999 No 2001) (new provisions replacing the revoked Power Press Regulations are incorporated in Part IV of these Regulations)

## Employer to provide for own employees

Every employer must ensure that all employees using work equipment are provided with adequate health and safety information and, where appropriate, written instructions about the use of such equipment. The employer must also ensure that any employee supervising or managing the use of work equipment has available adequate health and safety information and, as above, where appropriate, written instructions. The information and instructions required above must include information and, where appropriate, written instructions on:

(a) All health and safety aspects arising from the use of the work equipment including conditions in which and methods by which it can be used.

(b) Any limitations on these uses.

(c) Any foreseeable difficulties that could arise.
(d) The methods to deal with them.
(e) Any conclusions to be drawn from experience in using the work equipment.

### Employer/self-employed person to provide for other workers

Employers and the self-employed are under a duty to provide information etc. to other workers only in so far as they have control, to *any* extent, of

(a) Work equipment.
(b) A person at work who uses or supervises or manages the use of work equipment.
(c) The way in which such equipment is used at work to the extent of that control. *Note:* the Regulations do *not* apply to a person who has supplied work equipment by way of sale, agreement for sale, or hire purchase agreement.

### Employer/self-employed person to provide for other employers, self-employed persons/others

See above.

## Safety Representatives and Safety Committees Regulations 1977 (most recently amended by SI 1999 No 2024)

### Employer to provide for own employees

To allow safety representatives to carry out inspections to allow access to relevant documents to make available all the information necessary to allow fulfilment of functions.

### Employer/self-employed person to provide for other employers, self-employed persons/others

Safety representatives to receive information from inspectors.

## Work in Compressed Air Regulations 1996 (as amended by SI 1997 No 2776)

### Employer to provide for own employees

The 'compressed air contractor', as defined, is to ensure that adequate information etc. is given to *all* persons working in compressed air on the risks arising from such work and the precautions to be taken. The HSE Guidance to the 1996 Regulations stresses that the provision of information etc. in such circumstances may be carried out by anyone competent to do so, but the compressed air contractor is responsible for ensuring that it is carried out to standard.

### Employer/self-employed person to provide for other employers, self-employed persons/others

See above.

*Appendix*

**Employer/self-employed person to provide for other workers**

See above.

# Appendix 6
# Summary of the Requirements for Training/Instruction and Consultation

## Carriage of Dangerous Goods by Road (Driver Training) Regulations 1996 (as amended by SI 1999 No 303)

### Employer provides training/instruction for specified employees/others

Operator must ensure driver of vehicle has received adequate instruction and training. This includes understanding his duties under other relevant health and safety legislation.

## Confined Spaces Regulations 1997

### Employer/self-employed person to ensure

Persons likely to be involved in any emergency rescue to be trained for that purpose. The ACoP and guidance specifies the content of that training, where appropriate.

## Construction (Health, Safety and Welfare) Regulations 1996 (most recently amended by SI 1998 No 2306)

### Employer provides training/instruction for employees

Adequate training of employees to avoid acts causing injury or damage.

## Control of Asbestos at Work Regulations 2002

### Employer/self-employed person to provide own employees/persons not employees on the premises where the work is being carried out with adequate instruction and training

If such persons are, or are liable to be, exposed to asbestos, or if they supervise such persons who are, so that they are aware of:

(a) significant findings of the risk assessment;
(b) risks to health from asbestos;
(c) precautions which should be observed;
(d) relevant control limit and action level, in order to safeguard themselves and other relevant person.

The above information must be:

(a) given at regular intervals;
(b) adapted to take account of significant changes in the type of work carried out by or methods of work used by employer/self-employed person; and

*Appendix*

(c) provided in a manner appropriate to the nature and degree of exposure identified by the risk assessment.

*Note*: The duty to provide adequate information to persons not employees is subject to the 'as far as is reasonably practicable' qualification.

## Control of Lead at Work Regulations 2002

**Employer/self-employed person to provide own employees/persons other than employees on the premises where the work is being carried out with suitable and sufficient information, including:**

(a) details of the form of lead which employee/other relevant person is liable to be exposed;

(b) the significant findings of the risk assessment;

(c) the appropriate precautions and actions to be taken by employees and other persons, as above, coming within the duty, in order to safeguard both themselves and those other persons;

(d) the results of any monitoring of exposure of lead carried out in accordance with the Regulations; and

(e) the collective results of any medical surveillance undertaken in accordance with the Regulations.

Such information must be adapted to take account of significant changes in the type of work or methods of work used by the employer/self-employer persons; and, provided in a manner appropriate to the level, type and duration of exposure identified by the risk assessment.

In addition, every employer/self-employed person must ensure that *any* person carrying out work in connection with the employer's/self-employed person's duties under the Regulations has suitable and sufficient information.

*Note*: The duty to provide suitable and sufficient information to persons other than employees is subject to the 'so far as is reasonably practicable' qualification.

## Control of Substances Hazardous to Health Regulations 2002 (as amended by SI 2002 No 278)

**Employer/self-employed person to provide own employees/persons not employees on the premises where the work is being carried out with suitable and sufficient instruction and training, including:**

(a) details, as specified, of the substances hazardous to health to which employees/other relevant persons are liable to be exposed;

(b) the significant findings of the risk assessment;

(c) the appropriate precautions and actions to be taken by employees/other relevant persons in order to safeguard both themselves and those other persons at the workplace;

(d) the results of any monitoring of exposure in accordance with the Regulations; and

(e) the collective results of any health surveillance undertaken in accordance with the Regulations.

Such information must be adapted to take account of significant changes in type of work carried out or methods of work used by employer/self-employed person; and, be provided in a manner appropriate to the level, type and duration of exposure identified by the risk assessment.

*Note*: The duty to provide such suitable and sufficient instruction and training to persons other than employers is subject to the 'so far as is reasonably practicable' qualification.

## Dangerous Substances and Explosive Atmospheres Regulations 2002

Employer/self-employed person to provide own employees/persons other than employees at the workplace where the work is being carried out with suitable and sufficient information, including:

(a) details as specified of any dangerous substances present at the workplace;

(b) significant findings of the risk assessment; and

(c) suitable and sufficient information on the appropriate precautions and actions to be taken by employees and other persons, as above, coming within the duty, in order to safeguard both themselves and those other persons.

Such information must be adapted to take account of significant changes in type of work or methods of work used by employer/self-employed person; and, be provided in a manner appropriate to the risk identified in the risk assessment.

*Note*: The duty to provide suitable and sufficient information to persons other than employees is subject to the 'so far as is reasonably practicable' qualification.

## Dangerous Substances in Harbour Areas Regulations 1987 (as amended by SI 1998 No 2885)

### Employer provides training/instruction for employees

Employer and self-employed person to provide instruction, training and supervision necessary to ensure his health and safety and to enable him to perform any operations in which he is involved with due regard to the health and safety of others.

### Employer provides training/instruction for specified employees/others

Operator to provide instruction, as necessary, to persons present on the berth to ensure their health and safety.

## Health and Safety (Consultation with Employees) Regulations 1996 (as amended by SI 2002 No 2174)

### Employer provides training/instruction for employees

Employer to consult either:

(a) with employees directly, or

(b) with representatives of employee safety in non-unionised workforces,

and to allow time-off for training.

*Appendix*

# Health and Safety (Display Screen Equipment) Regulations 1992

## Employer provides training/instruction for employees

Employer shall provide adequate health and safety training in use of workstation to:

(a) his employees who are 'users' on date of coming into force of the Regulations.

(b) his employees about to become users. Employers also shall provide such training to employee 'users' when the organisation of their workstation is substantially modified.

## Employer provides training/instruction for specified employees/others

Current users of DSE. Adequate health and safety training in the use of workstation, and whenever organisation of workstation modified.

# Ionising Radiations Regulations 1999

## Employer/self-employed person to ensure

Those of his employees engaged in work with ionising radiation are given appropriate training in the field of radiation protection and receive instruction as is suitable and sufficient for them to know:

(a) the risks to health created by exposure to ionising radiation;

(b) the precautions which should be taken; and

(c) the importance of complying with the medical, technical and administrative requirements of the Regulations.

# Ionising Radiation (Medical Exposure) Regulations 2000

## Employer provides training/instruction for 'practitioners or operators', as defined

Steps must be taken to ensure that every 'practitioner or operator' engaged by the employer to carry out medical exposures involving ionising radiation undertakes continuing education and training after qualification, as specified (where the employer is concurrently 'practitioner or operator' he must ensure that he undertakes such continuing education and training as may be 'appropriate').

# Management of Health and Safety at Work Regulations 1999

## Employer provides training/instruction for employees

Adequate health and safety training: on recruitment on being exposed to new/increased risks e.g. on transfer/change of responsibilities; new work equipment; new technology; new system of work.

## Employer provides training/instruction for specified employees/others

Employees who have access to any area (where such access is subject to restriction): adequate instruction for evacuation procedures.

## Noise at Work Regulations 1989 (most recently amended by SI 1999 No 2024)

### Employer provides training/instruction for employees

Every employer shall provide each of his employees, who is likely to be exposed to the first action level or above, instruction and training on:

(a) risk of damage to hearing

(b) what steps can be taken to minimise risk

(c) steps that employee must take in order to obtain personal ear protectors

(d) employees' obligations.

## Offshore Installations and Wells (Design and Construction, etc) Regulations 1996

### Well-operator or installation duty holder to ensure

Well operations are not carried out unless persons carrying out the operation have received appropriate instruction and training.

## Packaging, Labelling and Carriage of Radioactive Material by Rail Regulations 2002 (SI 2002 No 2099)

'Relevant' employers must provide all employees with responsibilities relating to the carriage of radioactive material with training and instruction appropriate to those responsibilities so that they can understand:

(a) the nature of the dangers of the material being carried;

(b) the precautions the employee should take to ensure that:

    (i) his exposure to radiation, and

    (ii) the exposure of others who may be affected by the employee's actions,

    are restricted;

(c) what action an employee should take in an emergency concerning such radioactive material; and

(d) The regulatory requirements relating to the carriage of radioactive material by rail.

## Personal Protective Equipment at Work Regulations 1992 (as amended most recently by SI 2002 No 2174)

### Employer provides training/instruction for employees

Adequate and appropriate training about risks the PPE will avoid/limit; purposes of PPE; action to be taken by employee.

*Appendix*

# Provision and Use of Work Equipment Regulations 1998 (most recently amended by SI 2002 No 2174)

## Employer provides training/instruction for employees

All persons who use work equipment must be provided with:

adequate training for purposes of health and safety including training in the methods of use, risks and precautions. For employers' obligations regarding instruction see Provision of Information above.

## Employer provides training/instruction for 'practitioners or operators', as defined

Obligations as above. The HSC's ACoP and Guidance on the Regulations emphasises that 'training and proper supervision of "young people" is particularly important because of their relative immaturity and unfamiliarity with the working environment'.

# Quarries Regulations 1999 (as amended by SI 2002 No 2174)

## Quarry operator must

Ensure that no person undertakes any work in a quarry unless: that person is either competent to do the work or does so under the instruction and supervision of some other person who is competent to give instruction in and to supervise the doing of that work for the purpose of training him.

# Safety Representatives and Safety Committees Regulations 1977 (most recently amended by SI 1999 No 2024)

## Employer provides training/instruction for specified employees/others

Allow time off with pay for training of safety representatives (as may be reasonable); consult with safety representatives and with other trade unions representatives over establishment of safety committees; consult safety representatives on:

(a)  instruction of measures which may affect health and safety of employees.

(b)  arrangements for appointing those who give health and safety assistance.

(c)  health and safety information provided to workforce.

# Work in Compressed Air Regulations 1996 (as amended by SI 1997 No 2776)

## Employer provides training/instruction for employees

The 'compressed air contractor', as defined, to provide adequate training on risks and precautions. See entry for Regulations under Provision of Information.

# Table of Cases

This table is referenced to paragraph numbers in the work.

| | |
|---|---|
| Ackerman and Another v Stratton (1999) (unreported) | W10022 |
| Addie (R) & Sons (Collieries) Ltd v Dumbreck [1929] AC 358 | O3008 |
| Adsett v K and L Steelfounders & Engineers Ltd [1953] 1 WLR 773 | R3004 |
| Alcock v Chief Constable of South Yorkshire Police [1992] 1 AC 310; [1991] 3 WLR 1057 | C6033 |
| Alexander v Midland Bank [1999] IRLR 723 | O1038 |
| Allen and Others v BREL and RFS (23 February 2001) (unreported) | O1027 |
| Aluminium Wire and Cable Co Ltd v Allstate Insurance Co Ltd [1985] 2 Lloyd's Rep 280 | E13020 |
| AM v WC & SPV [1999] IRLR 410 | H1722 |
| AMF International Ltd v Magnet Bowling Ltd and Trentham [1968] 2 All ER 789 | O3015 |
| Amosu and Others v Financial Times (August 1998) (unreported) | O1038 |
| Anderson v Newham College of Further Education [2002] EWCA Civ 500 | O3011 |
| Anyanwu v South Bank Student Union [2001] IRLR 305 | H1722 |
| Aparau v Iceland Frozen Foods plc [1996] IRLR 119 | E14004 |
| Armour v Skeen (Procurator Fiscal, Glasgow) [1977] IRLR 310 | E15032 |
| Armstrong & Others v British Coal Corporation, The Times, 6 December 1996 | N3033, O1027 |
| Armstrong & Others v British Coal Corporation (1998) (unreported) | V5003 |
| Arnold v Central Electricity Generating Board [1988] AC 228 | N3022 |
| Arthur v Anka [1996] 2 WLR 602 | O3017 |
| Aspden v Webbs Poultry and Meat Group (Holdings) Ltd [1996] IRLR 521 | E14004 |
| Attia v British Gas plc [1988] QB 304 | C6033 |
| | |
| Babcock Int Ltd v National Grid Co plc, CA, December 2001 | O1009 |
| Bailey v Armes [1999] PLSCS 34 | O3004 |
| Baker v TE Hopkins & Son Ltd [1959] 3 All ER 225 | A1022, C8072 |
| Balgobin v London Borough of Tower Hamlets [1987] IRLR 401 | H1708, H1715, H1721, H1728, H1745 |
| Bamsey v Albon Engineering and Manufacturing plc (Judgment 25 March 2004) | W10024 |
| Barber v RJB Mining (UK) Ltd [1999] IRLR 308 | E14002, W10010, W10042 |
| Barber v Somerset County Council [2004] UKHL 13 | O1035 |
| Barclays Bank plc v Fairclough Building Ltd (No 2) [1995] 1 All ER 289, [1995] IRLR 605 | C8001, C8004, C8133 |
| Barclays Bank plc v Kapur [1989] ICR 753, CA | H1710 |
| Barclays Bank plc v Kapur (No 2) [1995] IRLR 87 | H1710 |
| Barker v Sr Gobain Pipelines plc [2004] EWCA Civ 545 | O1009 |
| Barton v Wandsworth Council (1995) IDS Brief 549 | E14009 |
| Barton and Others v SCC Limited (1999) (unreported) | W10024 |
| Baxter v Ford Motor Co (1932) 168 Wash 456, 12 P 2d 409 | P9028 |
| Baxter v Harland & Wolff plc, Northern Ireland Court of Appeal (1990) (unreported) | N3022 |
| Beattie v Department of Social Security [2001] EWCA Civ 498 | C6041 |
| Belhaven Brewery Co Ltd v McLean [1975] IRLR 370 | E15017 |
| Berry v Stone Manganese Marine Ltd [1972] 1 Lloyd's Rep 182 | N3022 |
| Berryman v London Borough of Hounslow, The Times, 18 December 1996 | O3004 |
| Bett v Dalmey Oil Co (1905)7F (Ct of Sess) 787 | S3003 |
| Bews v Scottish Hydro-Electric plc (1992) SLT 749 | C6053 |
| Biesheuval v Birrell [1999] PIQR Q40 | C6023, C6024 |
| Bixby, Case, Fry and Elliott v Ford Motor Group (1990) (unreported) | N3022 |

*Table of Cases*

| | |
|---|---|
| Blacklock, Waggott v Swan Hunter Shipbuilders Ltd [1984] IRLR 93–116 | Intro 17, N3022 |
| Bourhill v Young [1943] AC 92 | C6033 |
| Bowden & Others v Tuffnell Parcels Express Ltd (6 April 2000) (EAT 622/99) | W10026 |
| Boyle v Kodak Ltd [1969] 2 All ER 439 | W9023, W11049 |
| Bracebridge Engineering Ltd v Darby [1990] IRLR 3 | H1714, H1715, H1728, H1731 |
| Bradburn v Great Western Railway Co (1874) LR 10 Exch 1 | C6053, E13007 |
| Bradley v Eagle Star Insurance Co Ltd [1989] 1 All ER 961 | E13002 |
| Brice v Brown [1984] 1 All ER 997 | C6033 |
| British Airways Ltd v Moore [2000] IRLR 296 | E14016 |
| British Home Stores v Burchell [1978] IRLR 379 | E14013, H1708 |
| British Transport Commission v Gourley [1955] 3 All ER 796; [1956] AC 185 | C6021, C6022, C6026 |
| Brown v Controlled Packaging Services Ltd (1999) (unreported) | W10010 |
| Brown v Kigass Aero Components Ltd (2000) (unreported) | W10024 |
| Bunker v Charles Brand & Son Ltd [1969] 2 All ER 59 | O3005 |
| Burnett v British Waterways Board [1973] 1 WLR 700; 2 All ER 631 | O3017 |
| Burton v De Vere Hotels [1986] IRLR 596 | H1715 |
| Bux v Slough Metals Ltd [1974] 1 All ER 262 | P3001, P3026 |
| Byrne Bros (Foamwork) Ltd v Baird and Others [2002] ICR 667 | W10004 |
| | |
| Campbell v Mylchreest [1999] PIQR Q17 | C6041 |
| Campion DJM and AJ v Hughes (HM Inspector of Factories) [1975] IRLR 291 | E15024 |
| Canterbury City Council v Howletts and Port Lompyne Estates Limited, The Times, 13 December 1996 | E15022 |
| Capitano v Leeds Eastern Health Authority, Current Law, October 1989 | O3006 |
| Cappoci v Bloomsbury Health Authority (21 January 2000) | C6023 |
| Carragher v Singer Manufacturing Co Ltd 1974 SLT (Notes) 28 | N3022 |
| Castellain v Preston [1883] 11 QBD 380 | E13007 |
| Century Insurance Co Ltd v Northern Ireland Road Transport Board [1942] 1 All ER 491 | A1020 |
| Cerberus Software Ltd v Rowley [1999] IRLR 690 | E14018 |
| Chadwick v British Railways Board [1967] 1 WLR 912 | C6033 |
| Chapman v Eurostaff Personnel Ltd (1999) (unreported) | W10022 |
| Chessington World of Adventures v Reed [1997] IRLR 556 | H1713, H1715 |
| Chief Adjudication Officer v Faulds [2000] 1 WLR 1035 | C6005 |
| Chief Adjudication Officer v Rhodes [1999] ICR 178 | C6006 |
| Chief Constable of the Lincolnshire Police v Stubbs [1999] IRLR 81 | H1714, H1720, H1732 |
| Chrysler (UK) Ltd v McCarthy [1978] ICR 939 | E15009, E15016 |
| Clark v Novacold Ltd [1999] IRLR 318 | H1709 |
| Clemisnon v Ashley (1999) LTL 6 August 1999 | |
| Codling v Paglia (1973) 32 New York 2d 330 | P9028 |
| Collins v Wilcock [1984] 3 All ER 374 | H1731, H1738 |
| Commissioner of Police for the Metropolis v Harley [2001] IRLR 190 | H1728 |
| Commissioner for Railways (NSW) v Cardy [1960] 34 ALJR 134 | O3007 |
| Condogianis v Guardian Assurance Co Ltd [1912] 2 AC 125 | E13005 |
| Cotton v Derbyshire Dales District Council, The Times, 20 June 1994 | O3017 |
| Courtaulds Northern Spinning Ltd v Sibson [1988] IRLR 305; [1988] ICR 451 | E14004 |
| Coventry County Council v Ackerman Group plc [1995] Crim LR 140 | P9007 |
| Cox v Angus [1981] ICR 683 | A1021 |
| Cox v Hockenhull [1999] 3 All ER 577 | C6037 |
| Crawley v Mercer, The Times, 9 March 1984 | C6052 |
| Cresswell v Eaton [1991] 1 All ER 484, [1991] 1 WRL 1113 | C6039 |
| Crouch v British Rail Engineering Ltd [1988] IRLR 404 (CA) | P3001 |
| Cunningham v Harrison [1973] 3 All ER 463 B942 | C6053 |

*Table of Cases*

| | |
|---|---|
| Cunningham v Wheeler [1994] 113 DLR | C6053 |
| | |
| Darby v National Trust [2000] EWCA Civ 189 | O3017 |
| Dashiell v Luttitt [2000] 3 QR 4 | C6024 |
| Davie v New Merton Board Mills Ltd [1959] AC 604 | P9029 |
| Davies v Taylor [1974] AC 207 | C6036 |
| Davies and Others v MJ Wyatt (Decorators) Ltd (1999) (unreported) | W10022 |
| Day v T Pickles Farms Ltd [1999] IRLR 217 | E14019, E14025 |
| Deary v Mansion Hide Upholstery Ltd [1983] ICR 610 | E15021, E15031 |
| Deeley v Effer (1977) COIT No 1/72, Case No 25354/77 | E15023 |
| Derby Special Fabrication Ltd v Burton [2001] IRLR 69 | H1728 |
| Dewhurst (JH) Ltd v Coventry Corporation [1969] 3 All ER 1225 | E3007 |
| Dhanjal (SS) v British Steel General Steels (Case No 50740/91) | C8050, C8115 |
| DJM & AJ Campion v Hughes (HM Inspector of Factories) [1975] IRLR 291 | E15024 |
| Doherty & Ors v Rugby Joinery (UK) Ltd (2004) EWCA Civ 147 | O1027 |
| Doleman v Deakin, The Times, 30 January 1990, (1990) 87 (13) LSG 43 | C6035 |
| Donoghue v Stevenson [1932] AC 562 | H1732, O3014, P9028, P9040, P9041 |
| Driskel v Peninsula Business Services [2000] IRLR 151 | H1708, H1711, H1713 |
| Dryden v Greater Glasgow Health Board [1992] IRLR 469 | W11028 |
| Dunn v British Coal Corporation [1993] ICR 591 | C6023 |
| Dunnachie (Christopher) v Kingston Upon Hull City Council. HL (15 July 2004) | O1035 |
| Dyson v John Watson Field & Others, CA, December 2001 | O1009 |
| | |
| Eastwood & Williams v Magnox Electric plc, HL (15 July 2004) | O1035 |
| Edwards v National Coal Board [1949] 1 All ER 743 | E15015, R3003, R3004 |
| | |
| Fairchild v Glenhaven Funeral Services Ltd, [2002] UKHL 22, [2002] All ER (D) 139 | O1004, O1009 |
| Fairhurst v St Helens and Knowsley Health Authority [1995] PIQR 41 | C6029 |
| Farrant v Thanet District Council (1996) (unreported) | O3006 |
| Faulkner v Chief Adjudication Officer, [1994] PIQR P244 | C6006 |
| Ferguson v John Dawson & Partners (Contractors) Ltd [1976] 1 WLR 1213; IRLR 346 | C8021 |
| Fish v British Tissues (Sheffield County Court, 29 November, 1994) | C6058 |
| Flanagan v Watts Blake Bearne & Co plc [1992] PIQR Q144 | C6052 |
| Fraser v State Hospital Board of Scotland, The Times, 11 September 2000 | O1035 |
| Fraser v The State Hospital Board of Scotland (11 January 2000) (2000 Rep LR 94) | E14003 |
| Frost v Aylesbury Dairy Co Ltd [1905] 1 KB 608 | P9051, P9058 |
| | |
| Gabriel v Nuclear Electric [1996] PIQR Q1 | C6030 |
| Gaca v Pirelli General plc (Judgment 26 March 2004) | C6053 |
| Gallagher v The Drum Engineering Co Ltd, COIT 1330/89 | J3011 |
| Galt v British Railways Board (1983) 133 NLJ 870 | C6033 |
| Garrett v London Borough of Camden [2001] EWCA Civ 395 | O1035 |
| Gaugh v Thorn [1966] 3 All ER 398 | O3011 |
| Geddling v Marsh [1920] 1 KB 668 | P9052 |
| General Cleaning Contractors Ltd v Christmas [1952] 2 All ER 1110 | O3014, W9040 |
| Gentle v Perkins Group Ltd EAT/670/99 | E14022 |
| Gibson v Skibsn A/C Marina and Orkla Grpbe A/B and Smith & Coggins Ltd [1996] 2 All ER 476 | T7002 |
| Gill v Donald Humberstone & Co Ltd [1963] 3 All ER 180 | C8041 |
| Goodchild v Vaclight, The Times, 22 May 1965 | P9042 |
| Goold (WA) (Pearmak) Ltd v McConnell [1995] IRLR 516 | E14008 |
| Gough v Thorn [1966] 3 All ER 398 | O3011 |

*Table of Cases*

| | |
|---|---|
| Graney v Liverpool County Council (1995) (unreported) | O3002 |
| Grant v Australian Knitting Mills Ltd [1936] AC 85 | P9052 |
| Grant v South-West Trains Ltd [1998] CMLR 993 | H1708 |
| Greaves & Co (Contractors) Ltd v Baynham Meikle & Partners [1975] 3 All ER 99 | W11009 |
| Greene v Chelsea BC [1954] 2 All ER 318 | O3004 |
| Gridquest Ltd t/a Select Employment v Blackburn [2002] IRLR 168 | W10024 |
| Griffiths v British Coal Corporation (unreported) | O1038 |
| Griffiths v Peter Conway Ltd [1939] 1 All ER 685 | P9057 |
| Grovehurst Energy Ltd v Strawson, HM Inspector, COIT No 5035/90 | E15024 |
| Gunion v Roche Products Ltd, The Times, 4 November 1994 | A1001 |
| | |
| Hadley v Hancox (1987) 85 LGR 402 | E15011 |
| Hairsine v Hull City Council [1992] IRLR 211 | J3011 |
| Harris v Evans and Another [1998] 3 All ER 522 | E15028 |
| Harrison (Newcastle-under-Lyme) Ltd v Ramsay (HM Inspector) [1976] IRLR 135 | E15024 |
| Hartley v British Railways Board (1981) SJ 125 | O3014 |
| Hartley v Provincial Insurance Co Ltd [1957] Lloyd's Rep 121 | E13020 |
| Hartley v Sandholme Iron Co [1975] QB 600 | C6052 |
| Harvest Press Ltd v McCaffrey [1999] IRLR 778 | E14009 |
| Hassall and Pether v Secretary of State for Social Security [1995] 1 WLR 812 | C6050 |
| Hatton v Sutherland & Others [2002] 2 All ER 1 | O1035, S11011 |
| Heal v Garringtons Ltd (1984) (unreported) | N3030 |
| Healy and Others v Corporation of London (24 June 1999) (unreported) | E14005 |
| Heil v Rankin [2001] QB 272 | C6033 |
| Henningsen v Bloomfield Motors Inc & Chrysler Corp (1960 32 NJ 358, 161 A 2d 69) | P9023 |
| Herrington v British Railways Board [1972] 1 All ER 749 | F5079, O3008, O3010 |
| Hewett v Alf Brown's Transport Ltd [1992] ICR 530 | P3001 |
| Hewson v Downs [1970] 1 QB 73 | C6053 |
| Heslop v Metalock (Great Britain) Ltd, The Observer, 29 November 1981 | N3022 |
| Hinz v Berry [1970] 1 All ER 1074; 2 QB 40 | C6033 |
| Holtby v Brigham & Cowan (Hull) Ltd (6 April 2000) (unreported) | O1009 |
| Hounslow London Borough Council v Twickenham Garden Developments Ltd [1970] 3 All ER 326 | C8141 |
| Houweling v Wesseler [1963] 40 DLR (2d) 956 – Canada | O3005 |
| Howard and Peet v Volex plc (HSIB 181) | J3011 |
| HSE v George Tancocks Garage (Exeter) (1993) COD 284 | E15014 |
| Hunt v Severs [1994] AC 350 | C6028 |
| Hunter v British Coal Corporation, [1999] QB 140, [1998] 2 All ER 97 | C6033 |
| Hurditch v Sheffield Health Authority [1989] QB 562 | C6040 |
| Hussain v New Taplow Paper Mills [1988] AC 514 | C6052 |
| | |
| Imperial Chemical Industries Ltd v Shatwell [1964] 2 All ER 999; [1965] AC 656 | W11051 |
| Indermaur v Dames [1866] LR 1 CP 274 | O3014 |
| Insitu Cleaning Co Ltd v Heads [1995] IRLR 4 | H1714 |
| Interfoto Picture Library Ltd v Stiletto Visual Programmes Ltd [1989] QB 433 | O3016 |
| IPC Magazines Ltd v Clements, EAT/456/99 | E14022 |
| | |
| Jackson v Goodwood Estates [1999] LTL 5 April 1999 | O3015 |
| Jameson and Another v Central Electricity Generating Board and Another [2000] AC 455, [1999] 2 WLR 141 | C6037 |
| Joel v Law Union and Crown Insurance Co [1908] 2 KB 863 | E13004 |
| Johnson v Coventry Churchill International Ltd [1992] 3 All ER 14 | E13010 |

*Table of Cases*

| | |
|---|---|
| Johnson v Northbroke College (1999) (unreported) | W10024 |
| Johnson v Unisys Ltd [2001] IRLR 279, [2001] 2 WLR 1076 | H1729, H1736, O1035 |
| Johnstone v Bloomsbury Health Authority [1992] QB 333; [1991] 2 All ER 293; [1991] 2 WLR 1362; [1991] IRLR 118 | E14002, E14003 |
| Jolley v London Borough of Sutton, [2000] 1 WLR 1082 | O3010 |
| Jones v Sainsbury's Supermarkets Ltd (24 July 2000) (unreported) | E14013 |
| JJones (Olwen) v Sandwell Metropolitan BC | O1035 |
| Jordan v Achara (1988) 20 HLR 607 | O3004 |
| | |
| Kendall v Lillico [1969] 2 AC 31 | P9058 |
| Kerr v Nathan's Wastesavers Ltd (1995) IDS Briel 548 | E14009 |
| Kigass Aero Components Ltd (EAT/481/00) | W10024 |
| King v Smith [1995] ICR 339 | W9029, W9036, W9040 |
| | |
| Lambert v Lewis [1982] AC 225 | P9059 |
| Larby v Thurgood [1993] ICR 66, PIQR 219 | C6026 |
| Lask v Gloucester Health Authority [1986] HSIB 123 | J3015 |
| Latham v Johnson & Nephew Ltd [1913] 1 KB 398 | O3010 |
| Lea v British Aerospace plc [1991] 2 WLR 27 | C6040 |
| Ledeatte v Tower Hamlets London Borough Council (26 June 2000) EAT | H1716 |
| Liffen v Watson [1940] 1 KB 556 | C6054 |
| Lim Poh Choo v Camden and Islington Area Health Authority [1980] AC 174 | C6027, C6052 |
| Lindop v Goodwin Steel Castings Ltd, The Times, 19 June 1990 | C6043 |
| List Design Group v Catley (EAT/0966/00) | W10024 |
| Livingstone v Rawyards Coal Co (1880) 5 App Cas 25 | H1733 |
| Longden v British Coal Corporation [1998] AC 653 | C6053 |
| Lord Advocate v Aero Technologies 1991 SLT 134 | W11047 |
| | |
| McCabe (Robert) v Cornwall County Council and the Governing Body of Mounts Bay School, HL (15 July 2004) | O1035 |
| McCafferty v Metropolitan Police District Receiver [1977] 2 All ER 756 | N3022 |
| McCamley v Cammell Laird Shipbuilders Ltd [1990] 1 WLR 963 | C6044, C6053 |
| McClanahan v California Spray Chemical Corporation 75 SE 2d 712 (Va 1953) | P9038 |
| McLoughlin v O'Brian [1983] 1 AC 410 | C6033 |
| McDermid v Nash Dredging & Reclamation Co Ltd [1987] 2 All ER 878 | C8150 |
| McDonald v Ministry of Defence [2000] IRLR 748; [2001] IRLR 431 | H1705, H1709 |
| McFarlane v EE Caledonia [1994] 2 All ER 1 | C6033 |
| McFaul v Garringtons Ltd (1985) HSIB 113 | N3030 |
| McGeown v Northern Ireland Housing Executive [1995] 1 AC 233 | O3006 |
| McGivney v Golderslea Ltd (1997) (unreported) | O3003 |
| McGuinness v Kirkstall Forge Engineering Ltd (1979) (unreported) | N3022 |
| McIlgrew v Devon County Council [1995] PIQR 66 | C6027 |
| McKay v Unwin Pyrotechnics Ltd, The Times, 5 March 1991 | P9016 |
| McLoughlin v O'Brian [1982] 2 All ER 298; [1983] AC 410 | C6032 |
| McNulty v Marshall's Food Group Ltd, The Times, 17 January 1999 | C6027 |
| Maguire (Teresa) v Harland & Wolff plc & Ors (2004) EWHC 577 | O1009 |
| Mailer v Austin Rover Group [1989] 2 All ER 1087 | C8147, O3019 |
| Majrowski v Guy's & St Thomas NHS Trust (2004) | O1034 |
| Makepeace v Evans Brothers (Reading) (a firm) and Alfred Macalpine Construction Ltd, (23 May 2000) (unreported) | O3014 |
| Malcolm v Broadbent [1970] 3 All ER 508 | H1732 |
| Malik v BCCI [1997] IRLR 462 | H1736 |
| Marshalls Clay Products v Caulfield (EAT) (unreported) | W10024 |
| Martin v Owen, [1992] PIQR Q151 | C6036 |

TC/5

*Table of Cases*

| | |
|---|---|
| Matthews (Edwin) v Associated Portland Cement Manufacturers (1978) Ltd British Uralsite plc, CA, December 2001 | O1009 |
| Mattis v Pollock (t/a Flamingo's Nightclub) [2003] EWCA Civ 887 | E13001 |
| Meigh v Wickenden [1942] 2 KB 160 | W11047 |
| Miah v La Gondola Ltd (1999) (unreported) | W10022 |
| Mills v Hassal [1983] ICR 330 | C6053 |
| Mitchell v Vickers Armstrong Ltd [1984] IRLR 93 | N3022 |
| Moloney v Lambeth BC (1966) 64 LGR 440 | O3004 |
| Moorcock, The (1889) 14 PD 64 | E14004 |
| Morris v Ford Motor Co Ltd [1973] 2 All ER 1084 | E13007 |
| Mosiak v City Restaurants (UK) Ltd [1999] IRLR 780 | E14009 |
| Moualem v Carlisle City Council (1994) 158 JPN 786 | O3019 |
| MPB Structures v Munro (EAT/1257/01) | W10024 |
| Mughal v Reuters [1993] IRLR 571 | O1037, O1038 |
| Muir (William) (Bond 9) Ltd v Lamb [1985] IRLR 95 | E14013 |
| Murphy v Bradford Metropolitan Council [1992] 2 All ER 908 | O3002 |
| | |
| Nagarajan v London Regional Transport [199] IRLR 572 | H1716, H1744 |
| National Trust v Haden Young Ltd, The Times, 11 August 1994 | F5029, F5089 |
| Neame v Johnson [1993] PIQR 100 | O3003 |
| Newsholme Brothers v Road Transport & General Insurance Co Ltd [1929] 2 KB 356 | E13003 |
| Nico Manufacturing Co Ltd v Hendry [1975] IRLR 225 | E15024 |
| | |
| Oakley v Walker (1977) 121 SJ 619 | C6026 |
| Ogwo v Taylor [1987] 3 All ER 961 | O3014 |
| Osborne v Bill Taylor of Huyton Ltd [1982] IRLR 17 | M2025, S7002 |
| O'Shea v Kimberley-Clark Ltd, The Guardian, 8 October 1982 | N3022 |
| | |
| Page v Freight Hire (Tank Haulage) Ltd [1981] IRLR 13 | E14023 |
| Page v Sheerness Steel plc [1996] PIQR 26 | C6052 |
| Page v Smith [1995] 2 WLR 644; [1995] 2 All ER 736, [1996] 1 WLR 855 | C6033, O1035 |
| Palfrey v GLC [1985] IRLR 437 | C6044 |
| Pape v Cumbria County Council [1991] IRLR 463 | P3026 |
| Parry v Cleaver [1970] AC 1, [1969] 1 All ER 555 | C6044 |
| PCW Syndicates v TCW Reinsurers [1996] 1 Lloyd's Rep 241 | E13004 |
| Pearce v Governing Body of Mayfield Secondary School [2001] IRLR 669 | H1709 |
| Pendleton Stone v Webster Eng Ltd & Others, CA, December 2001 | O1009 |
| Philips v Syndicate 922 (14 May 2003) | O1009 |
| Phipps v Rochester Corporation [1955] 1 All ER 129 | O3010 |
| Pickford v Imperial Chemical Industries plc [1998] 3 All ER 462; [1998] 1 WLR 1189 | C6058, O1038 |
| Ping v Essette-Letraset Ltd [1992] PIQR P74 | C6058 |
| Polkey v AE Dayton Services Ltd [1988] AC 344; [1987] 3 All ER 974; [1987] 3 WLR 1153; [1987] IRLR 503 | E14013 |
| Porcelli v Strathclyde Regional Council [1986] IRLR 134 | H1702, H1708, H1709, H1710, H1713, H1731 |
| Post Office v Footitt [2000] IRLR 243 | C8056, W11026 |
| Potter v Arafa [1995] IRLR 316 | C6027 |
| Potts v Reid [1942] 2 All ER 161 | W11049 |
| Priest v Last [1903] 2 KB 148 | P9057 |
| | |
| Quinn v McGinty 1999 SLT (ShCt) 27 | E13001 |

*Table of Cases*

| | |
|---|---|
| R v Associated Octel Co Ltd [1996] 4 All ER 846; [1996] 1 WLR 1543 | Intro 17, C8021, C8082, C8147, E15039, O3019 |
| R v Attorney General for Northern Ireland ex parte Burns [1999] IRLR 315 | W10019 |
| R v Boal [1992] QB 591 | E15040 |
| R v Board of Trustees of the Science Museum [1994] IRLR 25 | C8120, C8147 |
| R v British Steel plc [1994] IRLR 540 | E15039, E15039 |
| R v British Steel plc [1995] ICR 586 | C8002 |
| R v Chapman (1992) (unreported) | E15040 |
| R v Colthorp Board Mill Ltd, CA (January 2002) (unreported) | Intro 15, E15036 |
| R v D'Albuquerque, ex parte Bresnahan [1966] 1 Lloyd's Rep 69 | C6007 |
| R v F Howe & Son (Engineering) Ltd [1999] 2 All ER 249, | Intro 15, E15036, E15044 |
| R v Friskies Petcare (UK) Ltd (CA) [2000] 2 CrAppR (S) 401 | Intro 15, E15036 |
| R v Gateway Foodmarkets Ltd, The Times, 2 January 1997 | E15039 |
| R v Hill (1996) (unreported) | E15036 |
| R v Kerr; R v Barker (1996) (unreported) | E15017 |
| R v Lambert [2001] 3 WLR 206 | E15031 |
| R v Mara [1987] 1 WLR 87 | W9036 |
| R v National Insurance Commissioner, Ex parte Michael [1977] 1 WLR 109 | C6006 |
| R v Neath Port Talbot Council | C8061 |
| R v Nelson Group Services (Maintenance) Limited [1999] IRLR 646 | E15039 |
| R (on the application of Junttan OY) v Bristol Magistrates' Court [2002] EWHC 566 (Admin) | P9009 |
| R v Paul Wurth SA, The Times, 29 March 2000 | C8029 |
| R v Rhone-Poulenc Rorer Ltd [1996] ICR 1054 | C8146 |
| R v Rollco Screw and Rivet Co and Others [1998] HSE E198: 98 | Intro 15, C8133, E15043 |
| R v Secretary for Trade and Industry ex parte BECTU (1999) (unreported) | W10022 |
| R v Swan Hunter Shipbuilders Ltd and another [1981] ICR 831; [1982] 1 All ER 264 | A1022, C8142, C8145, C8147, V3004 |
| Rae (Geoffrey) v Mars (UK) [1990] 3 EG 80 | O3006 |
| Read v J Lyons & Co Ltd [1947] AC 156 | G1001 |
| Readmans v Leeds CC [1992] COD 419 | E15015 |
| (1) Reed (2) Bull Information Systems Ltd v Steadman [1999] IRLR 299 | E14020, H1702, H1710, H1711, H1712, H1713, H1714 |
| Reid v Rush Tompkins Group plc [1989] 3 All ER 228 | E13008 |
| Reilly v Merseyside Regional Health Authority [1995] 6 Med LR 246 | C6032 |
| Reilly (Douglas) v Robert Kellie & Sons Ltd (1989) HSIB 166 | V3001 |
| Revil v Newbury, The Times, 3 November 1995 | O3009, O3011 |
| Ribee v Norrie, The Times, 22 November 2000 | O3004 |
| Richardson v Pitt-Stanley [1995] 1 All ER 460 | E13001 |
| Richardson v West Lindsey District Council and others [1990] 1 All ER 296 | W11045 |
| RMC Roadstone Products Ltd v Jester, [1994] 4 All ER 1037 | C8147 |
| Roberts v Day (1977) COIT No 1/133, Case No 3053/77 | E15022 |
| Roberts v Dorothea Slate Quarries Co Ltd (No 1) [1948] 2 All ER 201 | C6005 |
| Robertson v Forth Bridge Joint Board [1995] IRLR 251 | C6033 |
| Robinson v Harman (1848) 1 Exch 850 | H1736 |
| Rose v Plenty [1976] 1 All ER 97 | A1020 |
| Rowe v Herman and Others, The Times, 9 June, 1997 | C8140 |
| Rylands v Fletcher [1861–73] All ER Rep 1; (1868) LR 3 HL 330 | G1001 |
| | |
| Sabloff v Yamaha Motor Co 113 NJ Super 279, 273 A 2d 606 (1971) | P9040 |
| Salquiero da Silva Mouta v Portugal 33290/96 21 Dec 1999 EAT H1709 | |
| Sayers (Paul) v Smithkline Beecham plc (MMR/MR litigation) [1999] MLC 0117 | P9035 |
| Scarth v East Hertfordshire District Council (HSIB 181) | J3013 |
| Scott and Swainger v Associated British Ports and British Railways Board [2000] All ER (D) 1937 | O3017 |

*Table of Cases*

Secretary of State for Employment v Associated Society of Locomotive Engineers and
    Firemen (No 2) [1972] 2 QB 455; [1972] 2 All ER 949; [1972] 2 WLR 1370     E14007
Shirlaw v Southern Foundries Lrd [1939] 2 KB 206     E14004
Sibbald v Sher Bros, The Times, 1 February 1981     O3014
SIMAP v Conselleria de Sandidad y Consumo de la Generalitat Valencia (Case No
    C303/98)     W10003
Simpson v AI Dairies Farm Ltd [2001] LTL (12 January 2001)     O3014
Skinner v HMA [1994] SCCR 316
Smith v Gardner Merchant Limited [1998] IRLR 520     H1705, H1709, H1728
Smith v Leech Brain & Co Ltd [1961] 3 All ER 1159; [1962] 2 QB 405     C6022
Smith v Manchester City Council (1974) 118 SJ 597     C6026
Smith & Grady v UK [1999] IRLR 734     H1709
Smith & Others v Chubb Security Personnel Ltd (1999) (unreported)     W10024
SmithKline Beecham plc v Home-Roberts [2001] EWCA CIV 2006     P9048
Smoker v London Fire and Civil Defence Authority [1991] 2 AC 502;
    [1991] 2 All ER 449     C6058
Snowball v Gardner Merchant Ltd [1987] IRLR 397     H1723
South Surbiton Co-operative Society Ltd v Wilcox [1975] IRLR 292     E15025
Southampton and South West Hampshire AHA V Marshall [1993] IRLR 445     H1729
Southern Portland Cement v Cooper [1974] AC 623     O3010
Speed v Thomas Swift Co Ltd [1943] 1 All ER 539     S3003
Staples v West Dorset District Council (1995) 93 LGR 536 [1995] PIQR 439     O3017
Stark v The Post Office, The Times, 29 March 2000     P9008
Stewart v Cleveland Guest (Engineering) ltd [1994] IRLR 440     H1708
Stokes v Guest, Keen & Nettlefold (Bolts and Nuts) Ltd [1968] 1 WLR 1776     O1018,
    O1035
Stringman v McArdle [1994] 1 WLR 1653     C6041
Sutherland v Hatton [2002] EWCA Civ 76 (CA)     M2008

Tawling v Wisdom Toothbrushes (1997) (unreported)     W11043
Taylor v Furness, Withy & Co Ltd (1969) 6 KIR 488; [1969] 1 Lloyd's Rep 324     E14001
Taylor v Somerset Health Authority [1993] 4 Med LR 34     C6033
Tesco Stores Ltd v Kippax COIT No 7605-6/90     E15016
Tesco Supermarkets Ltd v Nattrass [1972] AC 153, 2 All ER 127     E15039
Thomas v Bunn [1991] 1 AC 362, [1991] 2 WLR 27     C6043
Thompson v T Lehan (Plant Hire) Ltd [1987] 2 All ER 631     P9061
Thompson v Smiths Shiprepairers [1984] QB 405     N3022
Thompson, Gray, Nicholson v Smiths Shiprepairers (North Shields) Ltd; Blacklock,
    Waggott v Swan Hunter Shipbuilders Ltd; Mitchell v Vickers Armstrong Ltd
    [1984] IRLR 93; [1984] QB 405; [1984] 1 All ER 881     N3022, O1018
Thornett & Fehr v Beers & Son [1919] 1 KB 486     P9056
Titchener v British Railways Board 1984 SLT 192     O3010, O3017
Tomlinson v Congleton Borough Council [2000] EWCA Civ 309 (CA)     O3017
Tower Boot Co Ltd v Jones [1995] IRLR 529     H1713, H1731
Trim Joint District School Board v Kelly [1914] AC 667     C6005
Tripp v Ministry of Defence [1982] CLY 1017     N3022
TSB Bank v Harris [2000] IRLR 157     E14020
Twine v Bean's Express Ltd [1946] 1 All ER 202     A1020

Vacwell Engineering Co Ltd v BDH Chemicals Ltd [1969] 3 All ER 1681     P9039
Vernon v Bosley [1997] 1 All ER 577, 614     C6033

Walker v Northumberland County Council [1995] IRLR 35; [1995] 1 All ER 737;
    [1995] ICR 702     Intro 14, E14003, H1732, H1736, O1035, S11011
Walter Lilly & Co Ltd v Westminster CC, The Times, 1 March 1994     N3021

| | |
|---|---|
| Walton & Morse v Dorrington [1997] IRLR 488 | E14003, E14019, W11027 |
| Warnes v Situsec Contractors Ltd (1999) (unreported) | W10024 |
| Waters v Commissioner of Police of the Metropolis [1997] IRLR 589 | H1720 |
| Waters v Commissioner of Police for the Metropolis (No 2) [2000] IRLR 720 | H1720, O1035 |
| Watson v Willmot [1991] 1 QB 140 | C6039 |
| Waugh v British Railways Board [1979] 2 All ER 1169 | A3029, J3015 |
| Waugh v London Borough of Sutton (1983) HSIB 86 | J3011, J3015 |
| Wayne Coxall v Goodyear GB Ltd [2003] 1 WLR 5365 (CA) | O1035 |
| Webster v Blue Ship Tea Room Inc 347 Mass 421, 198 NE 2d 309 (1964) | P9035 |
| West (H) & Son Ltd v Shephard [1964] AC 326 | C6030 |
| West Cumberland By Product Ltd v DPP, The Times, 12 November 1987 | E15042 |
| West Midlands Co-operative Society Ltd v Tipton [1986] AC 536; [1986] 1 All ER 513; [1986] 2 WLR 306; [1986] IRLR 112 | E14013 |
| Western Excavating (ECC) Limited v Sharp [1978] IRLR 27 | H1728 |
| Westminster City Council v Select Managements Ltd [1985] 1 All ER 897 | O3001, O3019 |
| Westwood v The Post Office [1973] 3 All ER 184; [1974] AC 1 | A1001 |
| Wheat v E Lacon & Co Ltd [1966] 1 All ER 582; [1965] 2 All ER 700 | O3004 |
| Whirlpool (UK) Ltd and Magnet Ltd v Gloucestershire County Council (1993) (unreported) | P9007 |
| White v Chief Constable of South Yorkshire Police [1999] 1 All ER 1 | C6033 |
| White v Holbrook Precision Castings Ltd [1985] IRLR 215 (CA) | O1027 |
| White v Pressed Steel Fisher [1980] IRLR 176 | J3011 |
| White v St Albans City and District Council, The Times, 12 March 1990 | O3009 |
| Wieland v Cyril Lord Carpets Ltd [1969] 3 All ER 1006 | C6022 |
| Wileman v Minilec Engineering Ltd [1988] IRLR 144 | H1723 |
| Williams v Gloucestershire County Council (10 September 1999) | C6023 |
| Willson v Ministry of Defence [1991] 1 All ER 638 | C6040 |
| Wilson v Graham [1991] 2 WLR 27 | C6043 |
| Wilson v National Coal Board 1981 SLT 67 | C6053 |
| Wilsons and Clyde Coal Co Ltd v English [1938] AC 57 | S3003 |
| Winterbottom v Wright (1842) 10 M & W 109 | P9028 |
| Witley & District Mens Club v Mackay [2001] IRLR 595 | W10023 |
| Woking Borough Council v BHS, (1995) 93 LGR 396 | A3007 |
| Woolfall and Rimmer Ltd v Moyle and Another [1941] 3 All ER 304 | E13020 |
| Woolmington v DPP [1935] AC 462 | E15031 |
| Workvale (No 2), Re [1992] 2 All ER 627 | E13002 |
| Worsley v Tambrands Ltd (No 2) [2000] MCR 0280 | P9039 |
| Wren v Holt [1903] 1 KB 610 | P9057 |
| Wright v British Railways Board [1983] 2 AC 773 | C6043 |
| Young v The Post Office [2002] EWCA Civ 661 | O1035 |

# Table of Statutes

| | | | |
|---|---|---|---|
| Access to Justice Act 1999 | C6021 | s 2(5) | P9043 |
| Administration of Justice Act 1982 | | s 3(1) | P9035 |
| s 1(1) | C6023, C6032, C6037 | (2) | P9031 |
| (2) | C6037 | s 4(1) | P9045 |
| s 3(1) | C6037 | s 5(1)–(4) | P9044 |
| s 5 | C6047 | s 6(4) | P9038 |
| s 6 | C6052 | s 7 | P9001, P9047 |
| (1) | C6040 | s 11 | P9011 |
| (2) | C6040 | s 41 | P9030 |
| s 74 | C6043 | s 46(1)–(3) | P9046 |
| Sch 1 | C6043 | Sch 1 | P9048 |
| | | Sch 3 | E15016, E15027, P9009, P9010, P9017–P9019, P9022, P9024–P9027 |
| Building Act 1984 | F5031, F5049, G1026 | | |
| s 38(3) | W11045 | Consumer Safety Act 1978 | A5030 |
| s 72 | F5049 | Consumer Safety Act 1980 | P9001 |
| | | Control of Pollution Act 1974 | N3017 |
| Children Act 1989 | | s 60 | N3021–N3023 |
| s 3 | V12010 | s 61 | N3022, N3023 |
| Civil Defence Act 1948 | D6006 | ss 63, 64 | N3001, N3024 |
| s 3 | D6007 | s 65 | N3001, N3024, N3025 |
| Civil Jurisdiction and Judgments | | s 66 | N3001, N3025, N3026 |
| Act 1982 | P9050 | s 67 | N3001, N3026 |
| Civil Protection in Peacetime Act 1986 | | s 68 | N3001, N3027 |
| | D6009 | ss 69–72 | N3001 |
| Companies Act 1985 | | s 71 | C8137 |
| s 310(3) | E15040 | Countryside and Rights of Way | |
| s 651 | E13002 | Act 2000 | O3007 |
| (5), (6) | E13002 | s 13 | O3007 |
| s 722 | E13013 | Crime and Disorder Act 1998 | H1737, H1741 |
| s 723 | E13013 | | |
| Companies Act 1989 | | Criminal Justice Act 1982 | F5040 |
| s 141 | E13002 | s 37(2) | E13014, E13015 |
| Company Directors' Disqualification | | Criminal Justice Act 1991 | |
| Act 1986 | | s 17(1) | E13014, E13015, F5040, F5068 |
| s 2(1) | E15040 | Criminal Justice and Public Order | |
| Competition and Service (Utilities) | | Act 1994 | E15026 |
| Act 1992 | G1002 | | |
| ss 21–23 | E3013 | Damages Act 1996 | C6001, C6022 |
| Consumer Credit Act 1974 | P9001, P9009, P9046, P9048, P9049, P9056 | s 1(1) | C6027 |
| | | s 2 | C6021 |
| | | Data Protection Act 1988 | A3025 |
| Consumer Protection Acts 1961/1971 | P9001 | Disability Discrimination Act 1995 | E14002, E14020, E14021, E14022, H1701, H1707, H1708, V12036, W11043 |
| Consumer Protection Act 1987 | Intro 24, E3027, G1026, M1002, P9008, P9013, P9015, P9016, P9018–P9021, P9026, P9041, P9042, S7012 | | |
| | | s 4(2) | H1728 |
| | | s 5(1) | H1725 |
| s 1(2)(c) | E3011, P9025, P9033 | s 6 | W11043 |
| s 2(1) | P9032 | s 55 | H1716 |
| (2) | P9032 | s 58 | H1721, H1722 |
| (3) | P9034 | | |

# Table of Statutes

| | | | |
|---|---|---|---|
| Education Act 1996 | | s 100(1)(ba) | J3029 |
| s 8 | R3024, R3025, V12007 | s 100(1)(e) | E14010 |
| Education (Scotland) Act 1980 | | s 101A | W10041 |
| s 31 | R3022, V12007 | s 103A | J3030 |
| Emergency Powers Act 1920 | D6010 | s 104 | E14011, W10041 |
| Emergency Powers Act 1964 | D6010 | s 105 | J3029, W10041 |
| Employers' Liability (Compulsory Insurance) Act 1969 | Intro 24, C6001, C8121, E13011, E13021 | s 108 | J3029 |
| | | (3)(c) | E14010 |
| | | s 109 | J3029 |
| s 1 | E13001 | (2)(c) | E14010 |
| (1) | E13008 | ss 113–118 | H1729 |
| s 2(2)(a) | E13010 | s 124(1A) | J3030 |
| s 3 | E13012 | s 203 | W10041 |
| s 4(1) | E13013 | ss 221–224 | W10024 |
| s 5 | E13001 | ss 227–228 | W10024 |
| Employers' Liability (Defective Equipment) Act 1969 | C6001, P9029, P9041 | Employment Rights (Dispute Resolution) Act 1998 | W10041 |
| | | Environment Act 1995 | A5030, D6021, D6024, E15001 |
| Employment Act 1989 | | | |
| ss 11, 12 | C8115 | ss 14–18 | D6015 |
| Employment Act 2002 | Intro 18 | s 95(2) | E15040 |
| Employment Relations Act 1999 | E14013, E14019, J3005 | (3) | E15040 |
| | | (4) | E15040 |
| ss 10–15 | E14008 | s 108 | E15027 |
| s 37(1) | J3030 | s 109 | E15027 |
| Sch A1 | J3005 | Environment and Safety Information Act 1988 | |
| Employment Rights Act 1996 | E14001, E14002, E14011, E14017, V12018 | s 1 | E15029, F5049 |
| | | s 2 | E15029 |
| | | (2) | E15029 |
| s 1 | E14006 | s 3 | E15029, F5049 |
| s 13 | E14014 | (3) | E15029 |
| s 43B | J3030 | Sch | E15029, F5021 |
| s 44 | E14009, E14010, E14011 | Environmental Protection Act 1990 | A5030, C8004, C8133, C10001, E15001, E15031, E15040, N3008, P9062, R3012 (App B), S7012 |
| (1)(ba) | J3029 | | |
| (e) | E14010 | | |
| (2) | E14009 | | |
| (3) | E14009 | | |
| s 45A | W10041 | | |
| s 45(1)(b) | W10013 | ss 33–46 | C8118 |
| s 47B | J3030 | s 79 | C8118, N3001 |
| s 48 | E14009, J3030 | s 157 | E15040 |
| s 49 | J3030 | Equal Pay Act 1970 | E14002, H1749, V12018 |
| (2) | E14009 | | |
| s 64 | E14016 | Explosives Acts 1875 & 1923 | E15012 |
| s 67 | E14015, E14025, R3023, V12018 | s 23 | W11049 |
| s 68 | E14016 | | |
| s 70(1) | E14016 | Factories Act 1937 | E13020, O1003, O1009, W11049 |
| (4) | E14015 | | |
| s 72(1) | E14028 | s 14(3) | E13020 |
| s 86 | E14002, E14006, 14011, E14018 | Factories Act 1961 | C8001, E15001, E15012, E15015, F5031, F5051, L3013, O1001, O1003, O1009, O3018, P9013, W11001, W11045 |
| s 95 | E14019 | | |
| (1)(c) | H1727, H1728 | | |
| s 98 | E14020 | | |
| (4) | E14021 | | |
| s 100 | E14009, E14010, E14011, E14022 | s 12 | M1004, O3003 |

| | |
|---|---|
| s 13 | E15015, M1004 |
| s 14 | M1004 |
| (1) | E13001 |
| ss 15–17 | M1004 |
| s 26(3) | L3007 |
| s 29 | N3022 |
| (1) | A1001 |
| s 30 | A1023 |
| (6) | P3008 |
| s 31(3)(4) | G1034 |
| s 63 | V3001 |
| s 123 | E3011, G1010 |
| ss 124–126 | G1010 |
| s 155(1) | W11047 |
| (2) | W11048, W11049 |
| s 173(1) | W11046 |
| s 175 | G1010 |
| Fatal Accidents Act 1976 | C6034, C6037, C6040 |
| s 1A | C6032 |
| s 3 | C6034, C6036 |
| s 4 | C6033 |
| s 5 | C6034 |
| Financial Services and Markets Act 2000 | E13024 |
| Fire Precautions Act 1971 (FPA) | E15029, F5030, F5031, F5032, F5033–F5035, F5052, F5059, M2006, R3012 |
| s 1 | F5053 |
| s 5 | F5035, F5009 |
| s 6(1)(2) | F5039 |
| (8) | F5066 |
| s 7 | F5040 |
| (1), (5) | F5040, F5068 |
| (6) | F5066 |
| s 8 | F5040, F5068 |
| (1), (2) | F5067 |
| s 9D | F5046 |
| (3) | F5047 |
| s 10 | E15029, F5032, F5038 |
| (2) | F5038 |
| ss 10A, 10B | F5038 |
| s 12 | F5032, F5051 |
| s 19(1)(c) | F5032 |
| (6) | F5040, F5068 |
| s 22(1) | F5040, F5068 |
| s 23 | E15040 |
| s 40 | F5077 |
| (2) | F5077 |
| Sch 2 Part II | F5037 |
| Fire Safety and Safety of Places of Sport Act 1987 | F5031, F5032, F5034 |
| s 13 | F5049 |
| Fire Services Act 1947 | F5031, F5055 |
| s 1 | F5055 |
| Fires Prevention (Metropolis) Act 1774 | F5031, F5078 |
| s 83 | F5078 |
| s 86 | F5078 |
| Food and Environment Protection Act 1985 | |
| s 19 | E15029 |
| Food Safety Act 1990 | F9001–F9004, F9006, F9013, F9022, F9023, F9070, F9072, F9078, P9062 |
| s 1 | F9007 |
| s 2 | F9077 |
| s 3 | F9007 |
| ss 4–6 | F9007 |
| s 7 | F9008–F9010 |
| s 8 | F9008–F9010, F9011 |
| s 9 | F9008–F9010, F9013, F9014 |
| s 10 | F9008–F9010, F9013, F9015 |
| s 11 | F9008–F9010, F9013, F9016–F9018 |
| s 12 | F9008–F9010, F9013, F9018 |
| s 13 | F9008–F9010, F9019 |
| s 14 | F9008, F9011, F9024 |
| s 15 | F9008, F9012 |
| ss 16–18 | F9008, F9020 |
| s 19 | F9008, F9021 |
| s 20 | F9008 |
| s 21 | F9008, F9023, F9070 |
| s 22 | F9008 |
| ss 29–31 | F9013 |
| s 32 | F9013 |
| s 33 | F9013 |
| s 35 | F9024 |
| s 37 | F9015 |
| s 39 | F9015 |
| s 40 | F9013, F9025 |
| s 54(2) | E15044 |
| s 254(2) | E15045 |
| Food Standards Act 1999 | F9005 |
| Foreign Judgments (Reciprocal Enforcement) Act 1933 | P9045 |
| Gaming Act 1968 | F5072 |
| Gas Act 1986 (as amended) | G1001, G1002 |
| s 4A | G1002 |
| s 7 | G1002, G1005 |
| ss 8–11 | G1002 |
| s 18 | G1002 |
| s 18A | G1002 |
| ss 21, 22, 22A, 23 | G1002 |
| s 33 | G1002 |
| Gas Act 1995 | G1001 |

*Table of Statutes*

| | |
|---|---|
| Health and Safety at Work etc. Act 1974 (HSWA) | Intro 7, Intro 21, Intro 22, Intro 23, Intro 24, A1001, A5030, C6001, C8001, C8002, C8006, C8009, C8021, C8029, C8056, C8069, C8074, C8115, C8123, C8124, C8128, D0407, D6007, D6012, D8203, E3001, E3004, E3005, E3012, E13015, E13020, E14003, E14023, E14027, E15001, E15013, E15026, E15028, E15037, E17060, F5031, G1001, G1026, G1029, J3016, J3023, L5002, M1002, M1004, M2006, M2015, O1001, O3001, O3003, O3019, P3005, P3027, P9001, P9006, P9013–P9015, P9022, R3001, R3003, S3004, S7004, S11006, T7024, T7025, V3001, V3004, V12001, V12036, W11001, W11039 |
| s 1 | E15012 |
| (1) | Intro 17, C8002, C8013, C8109, C8132–C8134, C8137, D6012, G1028, O3019 |
| (2) | C8002, C8013, G1027 |
| s 2 | A1009, C8002, C8021, C8069, C8104, C8119, C8120, C8145, D6022, E14001, E14006, E14012, E14023, E15012, E15022, E15036, E15039, L3052, M2002, M2003, M2014, M2016, , O3018, P3004, P9013, R3002, R3005, S7012, S11042, S11045, T7003 |
| (1) | Intro 21, F7001, M1002, M2003, P3005, S11006, V3004, W9034, W9036 |
| (2) | Intro 21, C8134, M2003, N3001, N3008, R3001, R3016, T7003, V3004, W9034, W9036, W9040 |
| (a) | C8145, R3001, S3004 |
| (c) | C8145 |
| (3) | Intro 21, Intro 27, C8120, E15004, M2003, M2025, O1040, R3001, S3004, S7001–S7003 |
| (6) | J3002 |
| (7) | J3017 |
| s 3 | Intro 21, C8026, C8029, C8119, C8120, C8128, C8144, C8146, C8147 D6022, E15012, E15027, E15028, E15031, E15036, E15039, M2002, M2014, O3019, R3002, V12001 |
| (1) | Intro 21, C8021, C8028, C8120, C8143, C8144, C8145, M2004, W9036, W9038, W9039 |
| (2) | C8021, M2004, W9034, W9036 |
| s 4 | Intro 21, C8021, C8069, C8116, C8128, E15012, E15036, E15039, M2014, O3001, R3002, R3005 |
| (1) | C8002, O3019, C8021 |
| (2) | C8002, C8021, C8143, O3019 |
| (3) | O3019 |
| s 5 | E15012, E15036, E15039 |
| s 6 | Intro 21, C8029, C8069, C8131, E3008, E15009, E15012, E15036, E15039, M1002, P3002, P3012, P9001, P9013, R3002, R3005 |
| (1) | M1002, P9011, P9013, P9019, P9021 |
| (a–d) | P9012 |
| (1A) | P9010 |
| (2) | P9020 |
| (3) | P9021 |
| (4) | P9014, P9026 |
| (5) | P9022 |
| (6) | P9023 |
| (8) | P9024 |
| (8A) | P9025 |
| (10) | P9026 |
| s 7 | Intro 21, A1020, C8002, C8020, C8068, E15012, E15036, J3009, M2017 |
| s 8 | Intro 21, E15012, E15034 |
| s 9 | Intro 21, E15012, E15034, P3004 |
| s 10 | E15012 |
| (2), (3) | E15002 |
| s 11 | E15012 |
| (2) | E15003 |
| (4)(a) | E15003 |
| s 12 | E15012 |
| s 13 | E15003, E15012 |
| s 14 | E15003, E15012, E15033 |
| (1) | D6006, D6012 |
| s 15 | E15012 |
| (3)(c) | E15004 |
| s 16 | E15004, E15012 |
| s 17 | E15012 |
| s 18 | E15012 |
| (1) | E15005, E15008 |
| (7)(a) | E15008 |
| s 19 | E15012 |
| (1) | E15007, T7003 |
| s 20 | E15012, E15026, E15033, G1033 |
| (1)(j) | E15026 |

*Table of Statutes*

| | |
|---|---|
| s 20(7) | E15026 |
| s 21 | E15012, G1033, J3003 |
| s 22 | E15012, E15022, E15015 |
|   (1), (2), (3) | E15015 |
|   (4) | E15016 |
| s 23 | C8108, E15012 |
| s 24 | E15012, E15024 |
|   (2) | E15019 |
|   (3) | E15019 |
| s 25 | E15012, E15020, E15034 |
|   (1), (2), (3) | E15027 |
| s 25A | E15027, P9027 |
| s 26 | E15012, E15027, E15028, H3012 |
| s 27 | E15012 |
| s 27A | E15027, P9027 |
| s 28 | E15012 |
|   (8) | J3016 |
| ss 29–32 | E15012 |
| s 33 | C10019, D6008, D6014 |
|   (1)(c) | E15034 |
|     (g) | E15017, E15031 |
|   (1A) | E15036 |
|   (2) | E15036 |
|   (2A) | E15017, E15036 |
|   (3) | E15036 |
|     (b)(i) | E15017 |
| s 34 | D6014, E15012 |
|   (1) | E15034 |
| s 35 | D6014, E15012 |
| s 36 | D6014, E15012, E15042, P7030 |
|   (1) | E15042 |
| s 37 | D6014, E15012, E15040, E15041 |
|   (1) | E15040 |
| ss 38, 39 | D6014, E15012 |
| s 40 | D6014, E15012, E15031, R3003 |
| s 41 | D6014, E15012 |
| s 42 | D6014, E15012, E15034, E15036 |
| ss 43–46 | E15012 |
| s 47 | E15012 |
|   (2) | O3018, P9030 |
| s 48 | E15012 |
|   (1)(2) | E15045 |
| ss 49–52 | E15012 |
| s 53 | E15012 |
|   (1) | E15012, E15027, P9016 |
| s 82(1)(c) | E15019 |
| Sch 1 | E15002 |
| Highways Act 1980 | |
| s 168 | C8118 |
| s 169 | C8118 |
| s 174 | C8118 |
| Human Rights Act 1988 | E15031, H1709, H1723 |
| Arts 8, 14 | H1709 |
| Sch 1 | H1709 |
| Industrial Diseases (Notification) Act 1981 | A3010 |
| Judgments Act 1838 | |
| s 17 | C6043 |
| Law Reform (Contributory Negligence) Act 1945 | |
| s 1(1) | O3011 |
| Law Reform (Miscellaneous Provisions) Act 1934 | C6036, C6038 |
| Law Reform (Parent and Child) (Scotland) Act 1986 | |
| s 8 | V12010 |
| Law Reform (Personal Injuries) Act 1948 | C6001 |
| Limitation Act 1980 | E13002 |
| s 11 | W11045 |
| s 11A(3), (4), (5) | P9048 |
| s 30 | E13002 |
| Local Government Act 1972 | |
| s 138 | D6008 |
| Local Government (Miscellaneous Provisions) Act 1982 | |
| Sch 1 | F5073 |
| Magistrates' Courts Act 1980 | |
| s 127(1) | E15034 |
| Medicines Act 1968 | P9001 |
| Merchandise Marks Act 1887 | F9023 |
| Mineral Workings (Offshore Installations) Act 1971 | E15012 |
| Mines and Quarries Act 1954 | E15002, G1010 |
| s 69(4) | A3035 |
| s 79 | A3035 |
| Mines and Quarries (Tips) Act 1969 | A3035 |
| Misuse of Drugs Act 1971 | |
| s 28(3) | E15031 |
| National Health Service and Community Care Act 1990 | |
| s 60 | E15044 |
| National Minimum Wage Act 1998 | W0004 |
| New Roads and Street Works Act 1991 | |
| s 50 | C8118 |
| Noise and Statutory Nuisance Act 1993 | C8118, N3001 |
| Occupiers' Liability Act 1957 | Intro 24, C8116, C8119, C8129, O3001, |

*Table of Statutes*

| | | | |
|---|---|---|---|
| | O3002, O3003, O3004, O3005, O3006, O3011, O3016, O3017, O3019 | Race Relations Act 1976 | E14002, H1701, H1708, H1719 |
| s 1(2) | O3004, O3005 | s 1(1) | H1725 |
| (3) | O3003 | s 4(2) | H1707, H1725 |
| (a) | O3005 | s 32 | H1719, H1721, H1722 |
| (4) | O3007 | s 33 | H1722 |
| s 2(1) | O3002 | Radioactive Material (Road Transport) Act 1991 | E15001, P9062 |
| (2) | O3002, O3003 | s 5 | E15027 |
| (3) | O3010, O3014 | Radioactive Substances Act 1993 | E15001, E15029 |
| (4) | O3015 | | |
| (5) | O3017 | Regulatory Reform Act 2001 | F5031, F9015 |
| (6) | O3006 | | |
| Occupiers' Liability Act 1984 | Intro 24, C8119, F5079, O3001, O3008, O3009, O3016 | Road Traffic Act 1988 | |
| | | s 145 | E13001 |
| s 1 | O3007, O3009, O3013 | Road Traffic Regulation Act 1984 | A1005 |
| (3), (4), (5) | O3009 | | |
| (6) | O3017 | Safety of Sports Grounds Act 1975 | F5032 |
| Occupiers' Liability (Scotland) Act 1960 | O3001 | s 10 | E15029 |
| Offices, Shops and Railway Premises Act 1963 (OSRPA) | E15001, E15012, E15025, F5051, O1001, V3001, W11001 | Sale and Supply of Goods Act 1994 | P9001, P9051, P9052, P9055, P9059 |
| | | s 1(2) | P9054, P9055 |
| s 16(1) | E15023 | (2A)–(2C) | P9054 |
| Offshore Safety Act 1992 | E15012, E15036 | Sale of Goods Act 1893 | P9052 |
| | | Sale of Goods Act 1979 | P9001, P9051–P9054, P9055, P9061 |
| s 1 | E15004 (Table 17) | s 14 | P9058–P9060 |
| | | (2) | P9054, P9056, P9052, P9054 |
| Petroleum Spirit (Consolidation) Act 1928 | F5031, F5060, H2144, R3014 | (3) | P9057 |
| | | s 15(2) | P9055 |
| Pneumoconiosis etc. (Workers' Compensation) Act 1979 | C6059, C6061 | Sale of Goods (Amendment) Act 1995 | P9001 |
| Police and Criminal Evidence Act 1984 | E15026 | Sex Discrimination Act 1975 (SDA) | E14002, E14023, H1701, H1702, H1705, H1707, H1708, H1710, H1716, H1722, H1733, H1749, V12018 |
| Policyholders Protection Board Act 1997 | E13024 | | |
| Prices Act 1974 | F9004 | | |
| Private Places of Entertainment (Licensing) Act 1967 | F5073 | s 1 | H1716 |
| Protection from Harassment Act 1997 | H1733, H1734, H1737, H1739, O1034 | (1)(a) | H1702, H1725 |
| | | s 3 | H1716 |
| | | s 4 | H1716 |
| s 1(2) | H1739 | s 6(2)(b) | H1707, H1710 |
| s 3 | H1744 | (c) | H1725 |
| s 7 | H1744 | s 41(1) | H1722, H1732 |
| (4) | H1734, H1744 | (3) | H1721, H1722 |
| Public Health Act 1936 | F5031 | s 42 | H1722 |
| s 205 | E14028 | (1) | H1719 |
| Public Health Act 1961 | E15012 | s 51(1) | E14023 |
| Public Interest Disclosure Act 1998 | E14001, E14009 | s 82 (1A) | H1728 |
| | | Sex Discrimination Act 1986 | E14002 |
| Public Order Act 1986 | H1737 | Shops and Railway Premises Act 1963 | E15018 |
| s 4A | H1740 | | |

*Table of Statutes*

| | | | |
|---|---|---|---|
| Social Security Act 1989 | | Supply of Goods (Implied Terms) | |
| s 22(4) | C6036 | Act 1973 | P9051–P9053 |
| Social Security Act 1998 | | Supreme Court Act 1981 | |
| s 29(4) | C6012 | s 35A | C6043 |
| Social Security Administration | | | |
| Act 1992 | C6001, C6041 | Theatres Act 1968 | |
| Social Security (Contributions and | | s 12(1) | F5071 |
| Benefits) Act 1992 | C6001, C6059 | Third Parties (Rights Against Insurers) | |
| s 2 | C6008 | Act 1930 | |
| s 44 | C6012 | s 1(1) | E13008 |
| s 94 | C6004 | (b) | E13002 |
| s 95 | C6008 | s 2(2)(a) | E13010 |
| s 97 | C6007 | Trade Descriptions Act 1968 | E15039, |
| s 98 | C6007 | | F9003, F9012 |
| s 99 | C6007 | Trade Disputes Act 1906 | J3009 |
| s 100 | C6007 | Trade Union and Labour Relations | |
| s 101 | C6007 | (Consolidation) Act 1992 | E14011, J3005, |
| s 103 | C6011 | | J3011 |
| s 110 | C6060 | s 5 | J3005 |
| s 171(A)–(G) | C6002 | s 6 | J3005 |
| Sch 6 | C6011 | s 8 | J3005 |
| Social Security (Incapacity for Work) | | s 20 | J3009 |
| Act 1994 | C6001, C6002 | s 168 | J3029 |
| s 5 | C6002 | s 178 | J3032, W10007 |
| Social Security (Recoupment of | | (3) | J3005 |
| Benefits) Act 1997 | C6001 | s 179(1) | E14005 |
| Social Security (Recovery of Benefits) | | | |
| Act 1997 | C6044 | Unfair Contract Terms Act 1977 | |
| s 4 | C6045 | (UCTA) | O3016, O3017, P9001, P9046, |
| s 5 | C6045 | | P9047, P9061, W9039 |
| s 6 | C6042 | s 2 | E14002, O3016, P9056 |
| s 8 | C6044 | (2) | O3016 |
| s 9 | C6045 | (3) | O3017 |
| s 24 | C6041 | s 3 | P9061 |
| Statutory Sick Pay Act 1994 | C6001 | s 6(2) | P9056 |
| Supply of Goods and Services | | s 11 | O3016 |
| Act 1982 | P9046, P9053 | | |
| s 4 | P9048 | Weights and Measures Act 1985 | F9004 |

# Table of Statutory Instruments

Aerated Water Regulations 1921 (SR & O 1921 No 1932)   P3001
Agriculture (Tractor Cabs) Regulations 1974 (SI 1974 No 2034 as amended by SI 1990
    No 1075)   E15005, N3015
Asbestos Industry Regulations 1931 (SI 1931 No 1140)   O1003
Asbestos (Licensing) Regulations 1983 (SI 1983 No 1649)   A5005, A5006, A5012, A5029,
    A5030, C8022, C8133, E15005 H2138, O1003,
    R3013 R3031
Asbestos (Licensing) (Amendment) Regulations 1998 (SI 1998 No 3233)   A5029, A5030, O1003
Asbestos Products (Safety) Regulations 1985 (SI 1985 No 2042)   A3050
Asbestos (Prohibitions) Regulations 1992 (SI 1992 No 3067 as amended)   A5004, A5005, A3050,
    O1003, P9001
Asbestos (Prohibitions) (Amendment) Regulations 1999 (SI 1999 No 2373)   O1003
Asbestos (Prohibitions) (Amendment) Regulations 1999 (SI 1999 No 2977)   O1003
Asbestos (Prohibitions) (Amendment) Regulations 2003 (SI 2003 No 1889)   Intro 18, O1003

Biocidal Products (Amendments) Regulations 2003 (SI 2003 No 429)   Intro 18
Building (Amendment) Regulations 2003 (SI 2003 No 2692)   Intro 18
Building Regulations 1991 (SI 1991 No 2768) (Revoked)   A1001, G1026, W11001, W11029,
    W11044
Building Regulations 2000 (SI 2000 No 2531)   F5031, F5046
Building (Safety, Health and Welfare) Regulations 1948 (SI 1948 No 1145)   W9023

Carriage of Dangerous Goods (Amendment) Regulations 1998 (SI 1998 No 2885)   D0407,
    App 5, App 6
Carriage of Dangerous Goods (Amendment) Regulations 1999 (SI 1999 No 303)   D0407
Carriage of Dangerous Goods and Transportable Pressure Vessels (Amendment)
    Regulations 2003 (SI 2003 No 1431)   D0407
Carriage of Dangerous Goods and Use of Transportable Pressure Equipment
    Regulations 2004 (SI 2004 No 568)   Intro 18, A3050, D0406–D0410, D0428, D0432,
    D0434, D0436, D0440, D0441, D0444–D0447,
    D6001, D6019, D6020, E15005, App 5, App 6
Carriage of Dangerous Goods by Rail Regulations 1996 (SI 1996 No 2089)   A3050, P7031
Carriage of Dangerous Goods by Road (Driver Training) Regulations 1996 (SI 1996
    No 2094) (Revoked)   D0407, E15005
Carriage of Dangerous Goods by Road Regulations 1996 (SI 1996 No 2095) (Partially
    revoked)   A3035, D0407, D0439, D6001, D6019, V12015
Carriage of Dangerous Goods (Classification, Packaging and Labelling) and Use of
    Transportable Pressure Receptacles Regulations 1996 (SI 1996 No 2092) (Revoked)   A3035,
    D0407
Carriage of Explosives by Road Regulations 1996 (SI 1996 No 2093) (Revoked)   D0407, V12015
Celluloid (Manufacture, etc.) Regulations 1921 (SR & O 1921 No 1825)   F5063
Chemicals (Hazard Information and Packaging for Supply) Regulations 2002 (SI 2002
    No 1689)   Intro 18, A3035, C10003, D0402, H2101, H2104, H2117, H2137, P9039,
    V12022, App 5
Chemicals (Hazard Information and Packaging for Supply) Regulations 1994 (SI 1994
    No 3247) (CHIP 2) (Revoked)   A3035, C8128, P9001, P9039
Children's Homes Regulations 1991 (SI 1991 No 1506)   F5075
Cinematograph Film Stripping Regulations 1939 (SR&O 1939 No 571)   F5063
Cinematograph (Safety) Regulations 1955 (SI 1955 No 1129)   F5070
Civil Defence (General Local Authority Functions) Regulations 1993 (SI 1993 No 1812)   D6010
Civil Defence (Grant) Regulations 1953 (SI 1953 No 1777)   D6007

*Table of Statutory Instruments*

Civil Defence (Scotland) Regulations 2001 (SI 2001 No 139)     D6010
Civil Procedure Rules (SI 1998 No 3132)     C6040, C6041
Classification and Labelling of Explosives Regulations 1983 (SI 1983 No 1140)     A3035
Coal and Other Mines (Fire and Rescue) Regulations 1956 (SI 1956 No 1768)     A3035
Community Homes Regulations 1972 (SI 1972 No 319)     F5075
Conduct of Employment Agencies and Employment Businesses Regulations 2003 (SI 2003 No 3319)     Intro 18
Confined Spaces Regulations 1997 (SI 1997 No 1713)     A1023, C8022, C8029, C8030, C8069, C8070, C8072, C8073, C8079, C8084, C8128, D6018, R3016, S3018, V3002, V12016, V12027, App 6
Construction (Design and Management) Regulations 1994 (CDM) (SI 1994 No 3140)     Intro 17, Intro 23, A3050, C8001, C8003, C8004, C8005, C8006, C8009, C8017, C8019, C8020, C8021, C8024, C8025–C8035, C8039, C8040, C8111, C8122, C8129, E3004, E3007, G1010, M2015, R3016, V12005, W9003, W9004, W9006, W9007, W9014, W9028, W9042, W9050, App 5
Construction (Design and Management) (Amendment) Regulations 2000 (SI 2000 No 2380)     C8001, C8003, C8004, C8005, C8009, C8017, C8019, C8020, C8021, C8022, C8024, C8025–C8035, C8039, C8040, C8111, C8122, C81129, W9003, W9004, W9006, W9007, W9014, W9028, W9042, W9050, App 5
Construction (General Provisions) Regulations 1961 (SI 1961 No 1580)     E3004, L5004, V3009
Construction (Head Protection) Regulations 1989 (SI 1989 No 2209)     C8020, C8031, C8067, C8108–C8109, C8112, C8114, C8104, P3008
Construction (Health, Safety and Welfare) Regulations 1996 (SI 1996 No 1592)     C8001, C8007, C8020, C8040, C8043–C8066, D6017, F5052, F5063, L5003, W9002, W9004, W9007, W9008, W9014–W9027, App 6
Construction (Lifting Operations) Regulations 1961 (SI 1961 Nos 1580, 1581)     L3013
Construction Plant and Equipment (Harmonisation of Noise Emission Standards) Regulations 1988 (SI 1988 No 36 as amended)     C8137
Construction Products Regulations 1991 (SI 1991 No 1620)     C8117, P9062
Construction Regulations 1961–1966 (SIs 1961 Nos 1580, 1581, SIs 1966 Nos 94, 95)     C8001, C8030
Construction (Working Places) Regulations 1966 (SI 1966 No 94)     C8021, C8146
Consumer Transactions (Restrictions on Statements) Order 1976 (SI 1976 No 1813)     P9061
Control of Asbestos at Work Regulations 1987 (SI 1987 No 2115) (Revoked)     A5005, O1003, R3001
Control of Asbestos at Work Regulations 2002 (SI 2002 No 2675)     Intro 18, A5003, A5005, A5006, A5007, A5009, A5011, A5012, A5020, A5022, A5025, A5029, A5030, C8019, C8022, C8068, C8133, H2101, H2138, O1003, O1041, P3008, R3013, R3022, S3018, V3009, App 4, App 5, App 6
Control of Asbestos in the Air Regulations 1990 (SI 1990 No 556)     A5030
Control of Explosives Regulations 1991 (SI 1991 No 1531)     C8128
Control of Industrial Major Accident Hazards Regulations 1984 (SI 1984 No 1902) (CIMAH) (Revoked)     C8147, C10001, C10009, C10011, C10013
Control of Lead at Work Regulations 1980 (SI 1980 No 1248) (Revoked)     R3001
Control of Lead at Work Regulations 1998 (SI 1998 No 543) (Revoked)     V3007
Control of Lead at Work Regulations 2002 (SI 2002 No 2676)     Intro 18, E14002, E15005, H2101, H2142, O1041, P3008, V3007, V3008, V12004, V12022, App 4, App 5, App 6

*Table of Statutory Instruments*

Control of Major Accident Hazards (Amendment) Regulations 2005     C10001
Control of Major Accident Hazards Regulations 1999 (COMAH) (SI 1999 No 743)     Intro 14,
         A3002, C8120, C8147, C10001–C10019,
         D6001, D6003, D6014, D6048, D6049, D6057,
         D6063, D6070, F5063, H2101, H2142, R3015
Control of Noise (Codes of Practice for Construction and Open Sites) Order 1984
    (SI 1984 No 1992)     C8137
Control of Noise (Codes of Practice for Construction and Open Sites) Order 1987
    (SI 1987 No 1730)     C8003, C8137
Control of Pesticides Regulations 1986 (SI 1986 No 1510)     C8003, V12022
Control of Substances Hazardous to Health Regulations 1988 (SI 1988 No 1657)
    (COSHH) (Revoked)     Intro 16, C8128, O1051, R3001
Control of Substances Hazardous to Health Regulations 2002 (SI 2002 No 2677)
    (COSHH)     Intro 4, Intro 18, Intro 23, C8128, E15037, H2101, H2104–H2107, H2109–
         H2115, H2125–H2138, J3011, M3002, O1041,
         P3008, R3007, R3022–R3025, R3028, R3042,
         R3046, S3018, V12005, V12013, V12021,
         V12022, App 4, App 5, App 6
Control of Substances Hazardous to Health (Amendment) Regulations 2003 (SI 2003
    No 978)     Intro 18, Intro 23, App 4, App 5, App 6
Control of Waste (Registration of Carriers and Seizure of Vehicles) Regulations 1991
    (SI 1991 No 1624 as amended)     C8118
Controlled Waste (Registration of Carriers and Seizure of Vehicles) Regulations 1991
    (SI 1991 No 1624)     A5030
Controlled Waste Regulations 1992 (SI 1992 No 588)     A5030

Damages for Bereavement (Variation of Sum) (England and Wales) Order 1990 (SI 1990
    No 2575)     C6033
Dangerous Machines (Training of Young Persons) Order 1954 (SI 1954 No 921)     E3025,
         H2101, H2144–H2146, R3014
Dangerous Substances and Explosive Atmospheres Regulations 2002 (SI 2002 No 2776)
         Intro 18, C8008, C8018, C8053, F5031, F5063,
         V12020, App 4, App 6
Dangerous Substances in Harbour Areas Regulations 1987 (SI 1987 No 37)     E15005, F5001,
         F5063, App 5, App 6
Dangerous Substances (Notification and Marking of Sites) Regulations (SI 1990 No 304)
         H2101, H2147
Disability Discrimination Act 1995 (Amendment) Regulations 2003 (SI 2003 No 1673)     Intro 18
Disability Discrimination (Employment Relations) Regulations 1996 (SI 1996 No 1456)
         W11043
Diving at Work Regulations 1997 (SI 1997 No 2776)     V12004, App 5
Diving Operations at Work (Amendment) Regulations 1990 (SI 1990 No 996)     F7014
Diving Operations at Work (Amendment) Regulations 1992 (SI 1992 No 608)     F7013
Diving Operations at Work Regulations 1981 (SI 1981 No 399) (Now Revoked)     C8050
Docks Regulations 1988 (SI 1988 No 1655)     F5052, V12015

Electrical Equipment (Safety) Regulations 1994 (SI 1994 No 3260)     M1017
Electrical Equipment for Explosive Atmospheres (Certification) Regulations 1990
    (SI 1990 No 13) (Revoked)     E3001, E3004, E3010, E3025, E3026
Electrical Equipment (Safety) Regulations 1994 (SI 1994 No 3260) (Revoked in part)     E3012,
         E3013–E3023, L3001
Electricity at Work Regulations 1989 (SI 1989 No 635)     Intro 23, C8128, E3004, E3006, E3008,
         E3009, E3010, E3027, F5063, L5006, R3005,
         R3008, T7020, V12027
Electricity Safety, Quality and Continuity Regulations 2002 (SI 2002 No 2665)     Intro 18, E3029

*Table of Statutory Instruments*

Electricity (Standards of Performance) Regulations 1993 (SI 1993 No 1193)    E3028
Electricity (Standards of Performance) Regulations 2001 (SI 2001 No 3265) (Revoked)    E3028
Electromagnetic Compatibility Regulations 1992 (SI 1992 No 2372; 1994 No 3080)
   (Revoked in part)    L3001, M1017
Employers' Health and Safety Policy Statements (Exception) Regulations 1975 (SI 1975
   No 1584)    E15004, M2025, S7002
Employer's Liability (Compulsory Insurance) Regulations 1998 (SI 1998 No 2573)    E13001,
    E13008, E13011, E13012, E13013, E13017–E13020
Employment Equality (Religion or Belief) Regulations 2003 (SI 2003 No 1660)    E14002
Employment Equality (Sexual Orientation) Regulations 2003 (SI 2003 No 1661)    E14002
Employment Rights (Increase of Limits) Order 2003 (SI 2003 No 3038)    Intro 18
Employment Tribunals (Constitution and Rules of Procedure) Regulations 2001 (SI 2001
   No 1171)    E15019
Environmental Protection (Duty of Care) Regulations 1991 (SI 1991 No 2839)    A5027, A5030
Environmental Protection (Prescribed Processes and Substances) Regulations 1991
   (SI 1991 No 472)    A5030
Equipment and Protective Systems Intended for Use in Potentially Explosive
   Atmospheres Regulations 1996 (SI 1996 No 192)    C8090, C8091, E3004, E3025

Factories (Testing of Aircraft Engines and Accessories) Special Regulations 1952
   (SI 1952 No 1689)    F5063
Fertilisers Regulations 1991 (SI 1991 No 2197)    C10003
Fire Certificates (Special Premises) Regulations 1976 (SI 1976 No 2003)    F5031, F5060, F5062, F5063, F5064
Fire Precautions (Application for Certificate) Regulations 1989 (SI 1989 No 77)    F5036
Fire Precautions (Factories, Offices, Shops and Railway Premises) Order 1989 (SI 1989
   No 76)    F5032, F5034
Fire Precautions (Hotels and Boarding Houses) Order 1972 (SI 1972 No 238)    F5032
Fire Precautions (Workplace) Regulations 1997 (SI 1997 No 1840)    F5030, F5031, F5032, F5051, F5052, F5069, F5074–F5077, R3003, R3004, T7015
Fire Precautions (Workplace) (Amendment) Regulations 1999 (SI 1999 No 1877)    F5030, F5031, F5032, F5051, F5052, T7015
Fishing Vessels (Working Time: Sea-fishermen) Regulations 2004 (SI 2004 No 1713)    Intro 18
Fixed-term Workers (Prevention of Less Favourable Treatment) Regulations 2002
   (SI 2002 No 2034)    E14002
Flexible Working (Eligibility, Complaints and Remedies) Regulations 2002 (SI 2002
   No 3236)    Intro 18
Flexible Working (Procedural Requirements) Regulations 2002 (SI 2002 No 3207)    Intro 18
Food Labelling Regulations 1996 (SI 1996 No 1499)    F9003, F9012
Food Labelling (Amendment) Regulations 1998 (SI 1998 No 1398)    F9003
Food Premises (Registration) Regulations 1991 (SI 1991 No 2825, as amended by
   SI 1993 No 2022 and SI 1997 No 723)    F9004, F9021, F9027
Food Premises (Registration) Amendment Regulations 1993 (SI 1993 No 2022)    F9031
Food Premises (Registration) Amendment Regulations 1997 (SI 1997 No 723)    F9031
Food Safety (General Food Hygiene) Regulations 1995 (SI 1995 No 1763)    F9002, F9021, F9025, F9034–F9038, F9046, F9049, F9050
Food Safety (Temperature Control) Regulations 1995 (SI 1995 No 2200)    F9002, F9034, F9050, F9052, F9053, F9058, F9063, F9064, F9066
Freight Containers (Safety Convention) Regulations 1984 (SI 1984 No 1890)    A3035, E15005

Gas Appliances (Safety) Regulations 1992 (SI 1992 No 711)    G1015
Gas Appliances (Safety) Regulations 1995 (SI 1995 No 1629)    G1026
Gas Cylinders (Pattern Approval) Regulations 1987 (SI 1987 No 1160)    D0407

*Table of Statutory Instruments*

Gas Safety (Installation and Use) Regulations 1994 (SI 1994 No 1886)    G1028, G1029, G1031, G1033
Gas Safety (Installation and Use) Regulations 1998 (SI 1998 No 2451)    G1001, G1010–G1026, G1035
Gas Safety (Installation and Use) (Amendment) Regulations 1996 (SI 1996 No 550)    C8128
Gas Safety (Management) Regulations 1996 (SI 1996 No 551)    C8128, G1003–G1005, G1009
Gas Safety (Rights of Entry) Regulations 1996 (SI 1996 No 2535)    G1001, G1006–G1008
General Product Safety Regulations 1994 (SI 1994 No 2328)    P9030
Genetically Modified Organisms (Contained Use) Regulations 2000 (SI 2000 No 2831)    E15005, App 4
Genetically Modified Organisms (Contained Use) Regulations 2002 (SI 2002 No 63)    Intro 18, App 4

Health and Safety (Consultation with Employees) Regulations 1996 (SI 1996 No 1513)
    Intro 23, A3028, C8022, E14009, J3002, J3005, J3019, J3021, J3022, J3024, J3026, J3028–J3030, J3031, R3008, App 6
Health and Safety (Display Screen Equipment) Regulations 1992 (SI 1992 No 2792)    Intro 23, D8201–D8211, D8213, D8219, E17044, E17052, E17063, L5002, O1039, O1041, R3010, S11017, V12005, W11009, App 4, App 5, App 6
Health and Safety (Emissions into the Atmosphere) Regulations 1983 (SI 1983 No 943)    A3050
Health and Safety (Enforcing Authority) Regulations 1998 (SI 1998 No 494)    C8022, E15001, E15008, E15009, E15010, E15011
Health and Safety (Fees) Regulations 2004 (SI 2004 No 456)    Intro 18, A3050, E15005
Health and Safety (First-Aid) Regulations 1981 (SI 1981 No 917)    Intro 23, C8003, C8053, D6016, E15005, F5031, F7001, App 5
Health and Safety Information for Employees Regulations 1989 (SI 1989 No 682)    S7012, App 5
Health and Safety Information for Employees (Modifications and Repeals) Regulations 1995 (SI 1995 No 2923)    App 5
Health and Safety (Miscellaneous Amendments) Regulations 2002 (SI 2002 No 2174)    Intro 18, Intro 23, D8201, M1007, M3001, M3002, O1039, R3027, V12034, W9006, W9007, W9008, W9009, W9013, W9030–W9032, W9034, W9036, W9038–W9040, App 4, App 5, App 6
Health and Safety (Safety Signs and Signals) Regulations 1996 (SI 1996 No 341)    Intro 23, C8128, E3009, F5018, F5032, F5078, N3013, W11001, W11031–W11035
Health and Safety (Training for Employment) Regulations 1990 (SI 1990 No 1380)    Intro 23, R3024, V12004, V12010
Health and Safety (Young Persons) Regulations 1997 (SI 1997 No 135)    C8003, R3024
Highly Flammable Liquids and Liquefied Petroleum Gases Regulations 1972 (SI 1972 No 917) (Revoked)    C8008, F5063, H2144, P9062, R3014

Income Support (General) Regulations 1987 (SI 1987 No 1967)    C6041
Industrial Tribunals (Constitution of Rules of Procedure) Regulations 1993 (SI 1993 No 2678)    E15021
Insurance Companies (Amendment) Regulations 1992 (SI 1992 No 2890)    E13008
Ionising Radiation (Medical Exposure) Regulations 2000 (SI 2000 No 1059)    App 6
Ionising Radiations Regulations 1985 (SI 1985 No 1333)    Intro 14, E13020, E15012
Ionising Radiations Regulations 1999 (SI 1999 No 3232)    C8069, C8128, E15005, H2101, H2148, P3008, R3015, R3022, V12004, V12015, V12020, App 4, App 5, App 6

*Table of Statutory Instruments*

Level Crossings Regulations 1997 (SI 1997 No 487)        Intro 14
Lifting Operations and Lifting Equipment Regulations 1998 (SI 1998 No 2307)    C8020, C8022,
                 L3001, L3002, L3004, L3010, L3013, L3024–
                 L3033, M1004, R3016, R3050
Lifting Plant and Equipment (Records of Test and Examination etc.) Regulations 1992
   (SI 1992 No 195)        L3013
Lifts Regulations 1997 (SI 1997 No 831)        L3001, L3013–L3024
Low Voltage Electrical Equipment (Safety) Regulations 1989 (SI 1989 No 728)      P9007

Magnesium (Grinding of Castings and Other Articles) Special Regulations 1946 (SR&O
   1946 No 2017)        F5063
Management of Health and Safety at Work and Fire Precautions (Workplace)
   (Amendment) Regulations 2003 (SI 2003 No 2457)        App 4, App 5
Management of Health and Safety at Work Regulations 1992 (SI 1992 No 2051)
   (Revoked)    Intro 12, Intro 14, C8001, C8002, , F5051, J3002, J3007, J3019, R3023, T7015,
                 V12017
Management of Health and Safety at Work Regulations 1999 (SI 1999 No 3242)       Intro 5,
                 Intro 16, Intro 23, Intro 24, Intro 27, Intro 28,
                 A1010, A1013, A1015, A1023, A3005, A3030,
                 A5030, C8001, C8002, C8003, C8008, C8010,
                 C8014, C8019, C8020, C8021, C8026, C8028,
                 C8029, C8032, C8034, C8047, C8053, C8069,
                 C8116, C8120, C8127, C8128, D6007, D6013,
                 D6033, D6034, D6062, E3004, E3006, E13006,
                 E14023–E14027, E17061, F5022, F5051,
                 F5082, F7004, G1004, G1026, H2115, H2127,
                 H2135, J3002, J3007, J3015, M1006, M2001,
                 M2002, M2005–M2021, O1034, O1039, O1041,
                 R3005, R3012, R3017–R3024, R3039, R3045,
                 S3004, S11007, S11043, T7004–T7009, V3008,
                 V12001–V12003, V12007–V12014, V12016–
                 V12018, V12023, V12024, V12036, V12038,
                 W9001, W9004, W9007, W9008, W9009,
                 W9014, W9015, W9028, W9042, W11035,
                 App 4, App 5, App 6
Manual Handling Operations Regulations 1992 (SI 1992 No 2793)     Intro 23, C8022, E17044,
                 E17052, E17062, M3001–M3003, M3007,
                 M3008, M3015, M3026, O1041, P9062, R3009,
                 R3047, S7003, S7006, T7007, V12005, V12020,
                 App 4, App 5
Manufacture of Cinematograph Film Regulations 1928 (SR&O 1928 No 82)       F5063
Maternity and Parental Leave Regulations 1999 (SI 1999 No 3312)       E14002
Mines (Safety of Exit) Regulations 1988 (SI 1988 No 1729)       A3035
Miscellaneous Factories (Transitional Provisions) Regulations 1993 (SI 1993 No 2482)       P3008
Motor Vehicles (Compulsory Insurance) Regulations 1992 (SI 1992 No 3036)       E13001

National Assistance (Conduct of Homes) Regulations 1962 (SI 1962 No 2000)       F5076
Noise at Work Regulations 1989 (SI 1989 No 1790)     Intro 23, C8022, C8126, N3002, N3007,
                 N3009, N3012, N3013, N3017, N3023, N3029,
                 O1015, O1041, P3008, P3018, P3019, P9018,
                 R3005, R3008, T7004, T7007, V5009, V12020,
                 App 4, App 5, App 6
Noise Emission in the Environment by Equipment for use Outdoors Regulations 2001
   (SI 2001 No 1701)        N3016

*Table of Statutory Instruments*

Notification of Cooling Towers and Evaporative Condensers Regulations 1992 (SI 1992
No 2225) — C8131
Notification of New Substances (Amendment) Regulations 2002 (SI 2002 No 2176) — Intro 18
Notification of New Substances Regulations 1993 (SI 1993 No 3050 as amended) — E15005
Nursing Homes and Mental Nursing Homes Regulations 1981 (SI 1981 No 932) — F5076

Offices, Shops and Railway Premises (Hoists and Lifts) Regulations 1968 (SI 1968
No 849 as amended by SI 1974 No 1943) — L3013
Offshore Installations and Pipeline Works (First-Aid) Regulations 1989 (SI 1989
No 1671) — E15005
Offshore Installations and Pipeline Works (Management and Administration)
Regulations 1995 (SI 1995 No 738 as amended) — F5052
Offshore Installations and Wells (Design and Construction, etc.) Regulations 1996
(SI 1996 No 913) — C8137, App 5, App 6
Offshore Installations (Safety Representatives and Safety Committees) Regulations 1989
(SI 1989 No 971) — J3002

Packaging of Explosives for Carriage Regulations 1991 (SI 1991 No 2097) — D0407
Part Time Workers (Prevention of Less Favourable Treatment) Regulations 2000
(SI 2000 No 1551) — W10004
Personal Protective Equipment Regulations 2002 (SI 2002 No 1144) — Intro 18
Personal Protective Equipment at Work Regulations 1992 (SI 1992 No 2966) — Intro 23, A3050,
C8022, C8069, C8101, C8114, E17065, O1041,
O1049, P3001, P3002, P3004, P3006–P3008,
P3013, P3027, R3011, T7010, V12005, App 4,
App 5
Personal Protective Equipment (EC Directive) (Amendment) Regulations 1993 (SI 1993
No 3074) — P3012
Personal Protective Equipment (EC Directive) (Amendment) Regulations 1994 (SI 1994
No 2326) — P3001, P3002, P3004, P3009, P3012, P3013, P3025
Personal Protective Equipment (EC Directive) Regulations 1992 (SI 1992 No 3139) — P3001,
P3002, P3004, P3009, P3012–P3019
Personal Protective Equipment Regulations 2002 (SI 2002 No 1144) — P3001, P3002, P3004,
P3009, P3012–P3014, P3025
Pipelines Safety Regulations 1996 (SI 1996 No 825) — C8128, G1001, G1009
Pipelines Safety (Amendment) Regulations 2003 (SI 2003 No 2563) — Intro 18
Planning (Control of Major Accident Hazards) Regulations 1999 (SI 1999 No 981) — C10001
Plugs and Sockets etc. (Safety) Regulations 1994 (SI 1994 No 1768) — P9003
Pneumoconiosis etc. (Workers' Compensation) (Determination of Claims)
Regulations 1985 (SI 1985 No 1645) — C6061, C6062
Pneumoconiosis etc. (Workers' Compensation) (Specified Diseases) Order 1985 (SI 1985
No 2034) — C6049
Potentially Explosive Atmospheres Regulations 1996 (SI 1996 No 192) — C8078, C8079
Power Presses Regulations 1965 (SI 1965 No 1441 as amended by SI 1972 No 1512) — App 5
Pressure Equipment Regulations 1999 (SI 1999 No 2001) — App 5
Pressure Systems Safety Regulations 2000 (SI 2000 No 128) — C8128, G1026, App 5
Pressure Vessels (Verification) Regulations 1988 (SI 1988 No 896) — D0407
Prevention of Accidents to Children in Agriculture Regulations 1998 (SI 1998 No 3262)
V12015
Provision and Use of Work Equipment Regulations 1992 (SI 1992 No 2932) (Revoked) — A1001,
A3050, E15022, E17064, L5003, M1003,
M1004, P9067, S7012
Provision and Use of Work Equipment Regulations 1998 (SI 1998 No 2306) — Intro 23, A1009–
A1012, A1015, A1020, C8015, C8022, C8069,
E3004, E3010, E3024, G1026, L3043, M1001,

*Table of Statutory Instruments*

Public Information for Radiation Emergencies Regulations 1992 (SI 1992 No 2997)   M1003–M1008, M3002, P9008, P9014, R3047, S1101, V5016, W11003, App 5, App 6  C10020, C10021
Public Works Contracts Regulations 1991 (SI 1991 No 2680)   C8001

Quarries Regulations 1999 (SI 1999 No 2024) A3035, App 4, App 5, App 6

Radiation (Emergency Preparedness and Public Information) Regulations 2001 (SI 2001 No 2975)   C10020, C10021
Radioactive Material (Road Transport) (Amendment) Regulations 2003 (SI 2003 No 1867)   D0407
Radioactive Material (Road Transport) Regulations 2002 (SI 2002 No 1093)   D0407, D0412, D0413, D0436
Railway Safety (Miscellaneous Provisions) Regulations 1997 (SI 1997 No 553)   Intro 14
Railways (Safety Case) Regulations 2000 (SI 2000 No 2688)   T7007
Registration of Births and Deaths Regulations 1987 (SI 1987 No 2088)   A3010
Regulation of Investigatory Powers (Communications Data) Order 2003 (SI 2003 No 3172)   Intro 18
Reporting of Injuries, Diseases and Dangerous Occurrences Regulations 1985 (SI 1985 No 2023) (RIDDOR) (Revoked)   C8003, C8023, N3012, S7012
Reporting of Injuries, Diseases and Dangerous Occurrences Regulations 1995 (SI 1995 No 3163) (RIDDOR)   Intro 23, A3001, A3002, A3007–A3016, A3018–A3025, A3029, A3030, A3036, A5030, C6004, C8022, C8032, C8128, C10010, D6003, D6009, D6034, E17066, G1005, J3027, R3010, V5011, V5016
Road Traffic (Carriage of Dangerous Substances in Packages, etc.) Regulations 1992 (SI 1992 No 742) (Revoked)   P9062
Road Traffic (Carriage of Dangerous Substances in Road Tankers and Tank Containers) Regulations 1992 (SI 1992 No 743) (Revoked)   P7036, P9062
Road Traffic (Carriage of Explosives) Regulations 1989 (SI 1989 No 615) (Revoked)   P9062

Safety Representatives and Safety Committees Regulations 1977 (SI 1977 No 500)   Intro 23, A3028, C8022, J3004–J3007, J3015, J3018, J3019–J3015, J3018, J3019, R3007, T7003, T7016, T7017, T7023, App 5, App 6
Sex Discrimination (Gender Reassignment) Regulations 1999 (SI 1999 No 1102)   H1709
Shipbuilding and Ship-repairing Regulations 1960 (SI 1960 No 1932)   O1003, P3008, V12015
Social Security Benefits Up-rating Order 2003 (SI 2003 No 526)   C6017
Social Security (Claims and Payments) Regulations 1979 (SI 1979 No 628)   A3025
Social Security (Claims and Payments) Regulations 1987 (SI 1987 No 1968)   C6012, F7019
Social Security (Employed Earners' Employments for Industrial Injuries Purposes) Regulations 1975 (SI 1975 No 467)   C6008
Social Security (General Benefit) Regulations 1982 (SI 1982 No 1408)   C6011
Social Security (Incapacity Benefit) (Transitional) Regulations 1995 (SI 1995 No 310)   C6002
Social Security (Incapacity for Work) (General) Regulations 1995 (SI 1995 No 311)   C6002
Social Security (Incapacity for Work) (General) Regulations 1996 (SI 1996 No 484)   C6002
Social Security (Industrial Injuries and Adjudication) Regulations 1993 (SI 1993 No 861)   O1053
Social Security (Industrial Injuries and Diseases) (Miscellaneous Amendments) Regulations 1996 (SI 1996 No 425)   O1053
Social Security (Industrial Injuries) (Prescribed Diseases) Regulations 1985 (SI 1985 No 967)   C6001, C6045, C6061, N3012, N3018–N3020, O1001, O1053, V5011
Social Security (Industrial Injuries) (Prescribed Diseases) Amendment Regulations 1987 (SI 1987 No 335)   O1053

*Table of Statutory Instruments*

Social Security (Industrial Injuries) (Prescribed Diseases) (Amendment No 2) Regulations 1987 (SI 1987 No 2112) — N3019, O1053
Social Security (Industrial Injuries) (Prescribed Diseases) Amendment Regulations 1989 (SI 1989 No 1207) — N3018, N3021, O1053
Social Security (Industrial Injuries) (Prescribed Diseases) Amendment Regulations 1990 (SI 1990 No 2269) — N3021, O1053
Social Security (Industrial Injuries) (Prescribed Diseases) Amendment Regulations 1991 (SI 1991 No 1938) — O1053
Social Security (Industrial Injuries) (Prescribed Diseases) Amendment Regulations 1993 (SI 1993 No 862) — O1053, V5004
Social Security (Industrial Injuries) (Prescribed Diseases) Amendment (No 2) Regulations 1993 (SI 1993 No 1985) — O1053
Social Security (Industrial Injuries) (Prescribed Diseases) Amendment Regulations 1994 (SI 1994 No 2343) — N3018, N3019, O1053
Social Security (Industrial Injuries) (Prescribed Diseases) Amendment Regulations 2000 (SI 2000 No 1588) — O1053
Social Security (Industrial Injuries) (Prescribed Diseases) Amendment Regulations 2002 (SI 2002 No 1717) — O1053
Social Security (Industrial Injuries) (Prescribed Diseases) Amendment Regulations 2003 (SI 2003 No 270) — O1053
Social Security (Industrial Injuries) (Prescribed Diseases) Amendment (No 2) Regulations 2003 (SI 2003 No 2190) — O1053
Social Security (Recovery of Benefits) Regulations 1997 (SI 1997 No 2205) — C60445, C6048, C6049
Special Waste Regulations 1996 (SI 1996 No 972) — A5027, C8118
Standards for School Premises Regulations 1972 (SI 1972 No 2051) — F5074
Supply of Machinery (Safety) (Amendment) Regulations 1994 (SI 1994 No 2063) — M1001, M1009, P9062, V5016, V5020
Supply of Machinery (Safety) Regulations 1992 (SI 1992 No 3073 as amended) — L3001, M1001, M1009, M1012, M1018, M7010, N3014, N3016, P9067, R3015, V5016, V5020

Transnational Information and Consultation of Employees Regulations 1999 (SI 1999 No 3323) — J3001, J3002
Transport of Dangerous Goods (Safety Advisers) Regulations 1999 (SI 1999 No 257) — D0407, E15005
Transportable Pressure Vessels Regulations 2001 (SI 2001 No 1427) — D0407

Unfair Terms in Consumer Contracts Regulations 1994 (SI 1994 No 3159) — P9062
Unfair Terms in Consumer Contracts Regulations 1999 (SI 1999 No 2083) — E13004, P9001, P9051, P9052, P9062, P9063

Waste Management Licensing Regulations 1994 (SI 1994 No 1056) — A5030, C8118
Woodworking Machines Regulations 1974 (SI 1974 No 903 as amended by SI 1978 No 1126) — L5003, P3001
Work in Compressed Air Regulations 1996 (SI 1996 No 1656) — C8067, C8092–C8104, C8107, C8096, C8128, F5063, P7031, V12004, V12020, App 5, App 6
Work in Compressed Air Special Regulations 1958 (SI 1958 No 61) (Revoked) — C8081
Working Time (Amendment) Regulations 2001 (SI 2001 No 3256) — W10022
Working Time Regulations 1998 (SI 1998 No 1833) — Intro 14, E14001, E14002, E14011, J3001, J3002, J3031, J3032, O1034, S11008, S11021, W10029–W10044
Working Time Regulations 1999 (SI 1999 No 3372) — J3001, J3002, J3031, J3032, O1034, W10001

*Table of Statutory Instruments*

Working Time (Amendment) Regulations 2003 (SI 2003 No 1684)   Intro 18, W10026–W10028, W10030
Workplace (Health, Safety and Welfare) Regulations 1992 (SI 1992 No 3004)   Intro 23, A1001, A1002, A1003, A1004, A1015, A3050, C8021, C8022, C8040, D8203, E15012, E15023, E17029, L5002, O1039, O1041, O3018, R3027, R3049, S11016, S11017, V3001, V3005, V12034, W3001–W3013, W9006, W9007, W9008, W9009, W9013, W9030–W9032, W9034, W9036, W9038–W9040, W11004–W11012, W11024–W11027
Works Council Regulations 1999 (SI 1999 No 3323)   J3033

# Index

References are to sections and paragraph numbers of this book.

ACAS,
  code on disciplinary procedures in
    industry,   E14007, E14013,
      E14014, E14021

ACCESS,
  access/egress points,
    employers to provide and
      maintain,   A1001
    hazards,   A1016–A1019
  safe workplace transport
    systems,   A1003
  unreasonable means of,   A1001
  business continuity planning,   B8008
  common law duty of
    care,   A1001, A1021
  confined spaces, *see* Confined Spaces
  construction work,   V12005
  disabled persons, facilities
    for,   A1001, V12005,
      V12034, W11043
  elderly,   V12005
  employer's statutory duty,   A1001
  hazards,   A1016–A1019
  heights, working at,   W9001
  lone workers,   V12029
  new or expectant mothers,   V12005
  operations with electricity,   A1025
  parking spaces,   A1004
  pedestrians,   A1007
  safe workplace transport
    systems,   A1003
  separation of vehicles and
    pedestrians,   A1004
  speed limits regarding,   A1005
  suitability of traffic
    routes,   A1004, A1005
  traffic routes, *see* Traffic Routes
  unreasonable means of,   A1001
  vehicles,   A1003–A1019
    access to,   A1015
    contractors',   A1014
    drivers, training of,   A1013, A1014
    hazardous operations
      with,   A1016–A1019
    maintenance,   A1012
    provision of,   A1011
    statutory requirements
      concerning,   A1009
    subcontractors',   A1014
    unauthorised lifts in,   A1020
  windows and window
    cleaning,   W9045–W9050
  workplace transport systems,
    organisation of safe,   A1003
    parking areas,   A1004
    pedestrians,   A1004, A1007
    safe traffic routes,   A1006, A1007
    suitability of traffic routes,   A1004
    vehicles,   A1005

ACCIDENT
  INVESTIGATIONS,   A3030–A3034
  allocation of blame and
    liability,   A3030
  causes, analysis of,   A3033
  disciplinary procedures,   A3030
  elements of, main,   A3031–A3034
  evidence, collating,   A3032, A3033
  guidance,   A3030
  Health and Safety Executive,   A3031
  information,   A3030, A3033
  interviewing witnesses,   A3032
  lessons, learning,   A030
  management of health and
    safety,   M2035
  manual handling,   M3019
  monitoring,   A3030
  performance,   A3030
  rationale for,   A3030
  reassurance and explanation,   A3030
  reports,   A3031, A3033–
    A3034, M3019
  risk assessment,   A3030
  witnesses,   A3032

ACCIDENT LINE,   C6021–C6062

ACCIDENT REPORTING,
  accident book,   A3002
  action to be taken by
    employers,   A3017
  basic rules,   A3001
  biological agents,   A3035
  breathing apparatus,   A3035
  building work,   A3035
  causes of accidents,   A3001
  coverage,   A3006
  dangerous occurrences,   A3009
    list of reportable,   A3035, A3036
  death of employee, reporting,   A3018
  defence for breach of
    Regulations,   A3023
  definition of terms,   D6003

# Index

| | |
|---|---|
| disclosure of accident data, | |
| generally, | A3027 |
| legal representatives, | A3029 |
| safety representatives, | A3027 |
| disease, | |
| duty to keep records of, | A3016 |
| prescribed form, | A3037 |
| reportable, list of, | A3002, A3010, A3035 |
| diving operations, | A3005, A3035 |
| electricity, | A3035 |
| employees, duties on, | A3024–A3026 |
| enforcing authority, | A3003 |
| ergonomics and, | E17066 |
| exceptions to notification and reporting, | A3013 |
| explosion, | A3035 |
| explosives, | A3035 |
| fairgrounds, | A3035 |
| fines, | A3022 |
| fire, | A3035 |
| flammable substance, escape of, | A3035 |
| freight containers, | A3035 |
| gas fitters, | A3021 |
| gas incidents, | A3002, A3020, A3021 |
| generally, | A3001 |
| Health and Safety Executive Regional Offices, | A3039 |
| incapacity for more than three days, | A3008 |
| incident reporting centre, | A3004 |
| injuries and dangerous occurrences duty to keep records of, | A3015, A3031 |
| major, | A3006 |
| legal representatives, | A3029 |
| level crossings, | A3035 |
| lifting machinery, | A3035 |
| major accidents | C10010 |
| major injuries or conditions, | A3007, A3011 |
| management of health and safety, | M2035 |
| manual handling, | M3001, M3019 |
| members of public, | A3001 |
| mines, | A3005, A3035 |
| notify, duty to, | A3011 |
| objectives, | A3026 |
| offshore operations, | A3005, A3035 |
| overhead electric lines, | A3035 |
| penalties, | A3022 |
| persons responsible for notification and reporting, | A3005 |
| pipelines, | A3005, A3035 |
| prescribed form, | |
| dangerous occurrences, | A3036 |
| diseases, | A3037 |
| injuries, | A3036 |
| pressure systems, | A3035 |
| quarries, | A3005, A3035 |
| radiation, | A3035 |
| records, duty to keep, | A3014–A3017 |
| action to be taken by employers, | A3017 |
| disease, | A3016 |
| duration, | A3014 |
| general rule, | A3014 |
| injuries and dangerous occurrences, | A3015, A3036 |
| regulations, | Intro 23 |
| relevant enforcing authority, | A3003 |
| report, duty to, | A3012 |
| requirements, | A3001 |
| responsible persons, | A3002, A3005, A3011 |
| RIDDOR, | A3002–A3023 |
| accident book, | A3002 |
| dangerous occurrences, | A3009, A3029 |
| death of employee, reporting, | A3018 |
| defences, | A3023 |
| duties under, | A3002 |
| enforcing authority, | A3003 |
| exceptions to notification and reporting, | A3013 |
| gas incidents, | A3020, A3021 |
| generally, | C6004 |
| incapacity for more than three days, injuries, | A3008 |
| major injuries or conditions, | A3007 |
| notify, duty to, | A3011 |
| penalties, | A3022 |
| persons responsible for notification and reporting, | A3005 |
| records, duty to keep, *see* records, duty to keep *above* | |
| relevant enforcing authority, | A3003 |
| report, duty to, | A3012 |
| responsible persons, | A3002, A3005 |
| road accidents, | A3002, A3019 |
| scope of regulations, | A3002, A3006 |
| specified dangerous occurrences, | A3009 |

| | |
|---|---|
| three days, injuries or incapacity for more than, | A3008 |
| road accidents, | A3019, A3035 |
| safety representatives, | A3027, A3028 |
| scaffolding, | A3035 |
| scope of duty, | A3001, A3002 |
| scope of provisions, | A3002, A3006 |
| specified dangerous occurrences, | A3009 |
| statistics, | A3001 |
| substances, escape of, | A3035 |
| three days, injuries or incapacity for more than, | A3009 |
| trains, | A3035 |
| transport systems, | A3035 |
| transportation of dangerous substances, | A3035 |
| visitors, | A3001 |
| wells, | A3005, A3035 |
| **ACCOMPANIED, RIGHT TO BE,** | E14001 |
| **ACOPS,** *see* Approved codes of practice | |
| **ACOUSTIC BOOTHS,** | N3037 |
| **ACOUSTIC SIGNALS,** | |
| safety signs regulations, | W11030 |
| **ACTORS,** | |
| safety representatives, | J3005 |
| **ACTUAL DISMISSAL,** | |
| case law illustrating, | E7025 |
| test of dismissal, | E7017 |
| **ACTUARY,** | |
| role of, | D6002 |
| **ADR AGREEMENT (EUROPEAN AGREEMENT CONCERNING THE INTERNATIONAL CARRIAGE OF DANGEROUS GOODS)** | D0404–D0406, D0408–D0409, D0411–D0414 |
| Approved List, | D0416, D0419 |
| concessions, | D0431 |
| consignment duties, | D0409 |
| documentation, | D0429 |
| exemptions, | D0431 |
| identification, | D0419 |
| loading, | D0440 |
| markings, | D0426, D0436–D0437 |
| placarding, | D0436–D0437 |
| radioactive materials, | D0412, D0414 |
| segregation, | D0442 |
| selection, | D0421 |
| stowage, | D0441 |
| structure of, | D0411 |
| UN Economic Commission for Europe, | D0404 |
| vehicle requirements, | D0434, D0436–D0437, D0440–D0442 |
| **ADVICE, SOURCES OF,** | Intro 28 |
| **ADVISORY, CONCILIATION & ARBITRATION SERVICE,** | |
| Code of Practice, disciplinary procedures, | E14007, E14013, E14014, E14021 |
| **AGE** | |
| capability, | M3007 |
| children and young persons, | V12004, V12006–V12007 |
| discrimination, | E14001 |
| manual handling, | M3007 |
| **AGENCY WORKERS,** | W10005 |
| **AGGREGATES LEVY,** | E16006 |
| **AGRICULTURE,** | |
| children and young persons, | V12015 |
| **AIR,** *see also* Air pollution, | |
| Compressed air on construction sites, work in | |
| airborne particles, control of, | V3010 |
| cleaning, types of, | V3010 |
| construction and building works, | C8057 |
| control of substances hazardous to health, | H2132 |
| general duty on employers, provision of pure and uncontaminated air, | V3004 |
| specific processes, control of dust and fumes, | V3006 |
| ventilation of excavations and pits, | V3009 |
| **AIR POLLUTION,** | |
| emissions trading, | E16006 |
| polluting agents, | V3003 |
| **AIRCRAFT,** | |
| occupier's liability, | O3005 |
| **ALARMS,** | |
| British Standard, | F5017 |
| fire and fire precautions, | F5017, F5022 |
| soundings, | F5022 |
| **ALCOHOL,** | |
| compressed air on construction sites, | C8103 |

IND/3

*Index*

ALLERGY,
  food, in connection
    with, F9011, F9075
AMENITY, DAMAGES FOR LOSS
  OF, C6030, C6034,
    C6044
ANCHORAGE, W9006, W9037–
    W9039, W9042, W9048
ANNUAL LEAVE,
  compensation regarding entitlement
    to, W10023
  entitlement, W10022
  instalments, W10022
  notice, W10024
  payment requirements, W10024
  qualifying period, W10022
  records, W10025
  rolling up, W10024
  sick leave, W10024
  working time,
ANTHROPOMETRIC
  DATA, E17007, E17010
APPEALS,
  improvement notice,
    against, E15001, E15019–
    E15025
  prohibition notice,
    against, E15001, E15019–
    E15025
APPOINTED PERSON,
  cranes, L3006
APPROVED CODES OF
  PRACTICE, H3007, App 2
  advice, sources of, Intro 28
  asbestos, A5005, A5012,
    A5022, A5030, C8068,
    H2139
  confined spaces, C8069
  construction and building
    works, C8005–C8006,
    C8028, C8029, C8032
  control of substances hazardous to
    health, H2105, H2113,
    H2122, H2125, H2129,
    H2132, H2135, V12022
  EU Law Intro 10
  heights, working at, W9004, W9007,
    W9009
  lead, H2143
  machinery safety, M1007
ARCHITECTURE,
  disaster and emergency management
    and, D6025
ARMED FORCES, M2021, M3002

ARRANGEMENTS,
  POLICY, Intro 27
ASBESTOS, A5001–A5030
  action levels, A5003
  Approved Code of
    Practice, A5005, A5012,
    A5022, A5030, C8068,
    H2139
  asbestos-containing
    material, A5011–A5012,
    H2139–H2141
    assessing the condition of, H2140
    knowledge of, A5013
    management, A5013
  asbestos-related
    diseases, O1001, O1003–
    O1009
  asbestos zones, A5003
  asbestosis, A5002, O1003,
    O1005
  buildings
    maintenance workers, exposure
      of, A5006
    removal from, A5005
    surveys, A5024–A5026
  causation, O1009
  civil liability, O1009
  clearance indicators, A5005
  compliance strategy, development
    of, A5016
  construction and building
    works, C8019, C8068
  contractors, A5026
  contractual agreements, A5021
  control measures
    domestic premises, repair and
      maintenance of, O1003
    historical A5003
    seven step
      programme, A5013–A5019
  Control of Asbestos at Work
    Regulations 2002, A5006–A5022
    contractual and tenancy
      agreements, A5021
    exceptions, A5007
    criminal offences, Intro 15, C8133
    definition, A5001
    demolition, C8133
  diffuse pleural
    thickening, O1003, O1007
  diseases, A5002, O1001,
    O1003–O1009
  disposal, C8133
  domestic premises, repair and
    maintenance of, O1003

IND/4

| | |
|---|---|
| duty holders, | A5007–A5019 |
| identification of, | A5009 |
| third parties, | A5020 |
| employers, | |
| duties, | A5007–A5019 |
| liability insurance, | E1301 |
| exhaust equipment, | O1003, V3009 |
| exhaust ventilation equipment, | |
| maintenance and inspection, | |
| pathological effects, | O1003 |
| exposure of building maintenance | |
| workers, | A5006 |
| *Fairchild* test, | O1009 |
| generic term for a variety of | |
| minerals, | O1003 |
| historical exposures, | A5003 |
| HSE guidance | |
| notes, | A5030, C8019, H2139, R3013 |
| identification of exposure, | A5006 |
| inspection, | A5015, R3013 |
| insurance, | E1301 |
| laboratories, | A5012, A5025 |
| leases, | A5021 |
| legislation, | A5030–A5031 |
| licensing of works, | A5029 |
| long tail liability, | O1001 |
| lung cancer, | A5002, A5003, O1003, O1006 |
| maintenance, | |
| building maintenance workers, | |
| exposure of, | A5006 |
| control measures, | O1003 |
| plans, | A5006, A5012 |
| records, | A5011 |
| management, | |
| asbestos-containing materials, | A5013 |
| duty as to, | C8019, C8068, H2139, R3013 |
| leases, | A5021 |
| long-term plans, | A5018 |
| plans, | A5012, A5018–A5019, A5022 |
| third parties, | A5020 |
| mesothelioma, | A5002, A5003, O1001, O1004 |
| minerals, generic term for a variety of, | O1003 |
| minimum requirements, | A5022 |
| notification of work with, | H2138 |
| occupational diseases, | O1003–O1009 |
| pathological effects of exposure to, | O1003 |
| plans, | |
| long-term, | A5018 |
| monitoring, | A5019 |
| review of, | A5019 |
| risk | |
| management, | A5012, A5018–A5019, A5022, H2141 |
| pleural plaques, | O1003, O1008 |
| precautions, | C8133, R3013 |
| premises, | |
| affected, | A5008 |
| control of work on, | A5014 |
| guidance, | A5005 |
| management of, | C8019, C8068, R3013 |
| survey of the, | H2140 |
| priorities, | A5017 |
| prohibitions, | A5004 |
| raw asbestos, storage, distribution and labelling of, | H2138 |
| records, | A5011, H2140 |
| register, | A5011, A5022 |
| regulations on | C8068, H2138–H2141 |
| removal work, | A5005–A5006 |
| licensed contractors, | A5026 |
| qualifications, | C8133 |
| specifications, | A5026 |
| surveys, | A5026 |
| waste, | A5027 |
| repairs, | |
| domestic premises, | O1003 |
| leases, | A5021 |
| risk assessment, | A5003, A5010, A5017, H2138, R3013, R3015 |
| risk management, | A5012, H2141 |
| samples, | A5011, A5025 |
| self-employment, period of, | A5007, O1009 |
| social security, | C6010 |
| surveys, | A5024–A5026, H2140 |
| removal specifications, | A5026 |
| risk assessments, | A5025 |
| type 1, | A5024 |
| type 2, | A5025 |
| type 3, | A5026 |
| tenancy agreements, | A5021 |
| third parties, | A5020 |
| types of, | A5001 |
| UK Accreditation Service | |
| laboratories, | A5012 |
| ventilation, | V3009 |
| visitors and the public, protection of, | C8133 |

## Index

| | |
|---|---|
| waste | A5027 |
| removal, | A5027 |
| special, | A5027 |
| storage, distribution and labelling of, | H2138 |
| work clothes, spouse's exposure to | O1009 |
| works, licensing of, | A5029–A5030 |
| **ASSAULT,** | |
| tort of, | H1731 |
| **ASSOCIATION OF CHARTERED CERTIFIED ACCOUNTANTS AWARDS,** | E16024 |
| **ASTHMA,** | |
| damages, | C6057 |
| disablement benefit, | C6057 |
| occupational disease, | O1051 |
| social security, | C6057 |
| **ATMOSPHERIC HAZARDS,** | |
| confined spaces, | A1022 |
| **ATTENDANCE ALLOWANCE,** | C6014, C6020 |
| **AUDITING,** | |
| business continuity planning, | B8008, B8034 |
| DEMS process, | D6061, D6062 |
| environmental management systems, | E16025 |
| management of health and safety, | M2036, M2039 |
| **AUTOMATION** | M3010 |
| **AVIATION,** | |
| major incident, | D6064, D6067 |
| **BACK,** | |
| display screen equipment, | D8206 |
| pain, | V5005, V5023–V5024, V5028 |
| vibration, | V5005, V5023–V5024, V5028 |
| whole-body vibration, | V5023–V5024, V5028 |
| **BAD NEWS, RECEIVING,** | C6033 |
| **BANKRUPTCY,** | |
| employer's liability insurance, | E13002 |
| **BANKS,** | |
| disasters or emergencies, help with, | D6030 |
| **BARRIERS,** | C8122 |
| **BATHROOM,** | |
| gas appliances in, | G1018 |
| **BATTERY,** | |
| offence of, | H1737, H1738 |
| **BCF EXTINGUISHERS,** | F5014 |
| **BENEFIT RATES,** | C6012 |
| **BENZENE POISONING,** | |
| prescribed occupational disease, | O1051 |
| **BEREAVEMENT DAMAGES,** | C6035 |
| **BHOPAL DISASTER,** | D6025, D6030 |
| **BIOLOGICAL AGENTS,** | |
| accident reporting, | A3035 |
| children and young persons, | V12016 |
| control of substances hazardous to health, | H2113 |
| new or expectant mothers, | V12021 |
| **BREACH OF CONTRACT,** | |
| damages, | H1736 |
| harassment, | H1730, H1735, H1736 |
| remedies for, | H1736 |
| working time, | W10010, W10042 |
| **BREACH OF STATUTORY DUTY,** | Intro 24, E15023–E15035 |
| damages, | C6001 |
| windows and window cleaning, | W9033, W9036 |
| **BREAST FEEDING MOTHERS,** | |
| risk assessment, | E14024, E14025, R3025, R3026 |
| **BREATHING APPARATUS,** | |
| accident reporting, | A3035 |
| confined spaces, | C8071 |
| **BRITISH SAFETY COUNCIL,** | T7013, T7020 |
| **BRONCHITIS,** | C6060 |
| **BSE,** | F9020 |
| **BUILDING,** see Construction and Building Operations | |
| **BUILDING REGULATIONS,** | |
| civil liability for breach of, | W11044 |
| health and safety regulations, | W11030–W11042 |
| **BURDEN OF PROOF,** | |
| human rights legislation, | F15031 |
| **BURGOYNE REPORT,** | O7003 |
| **BUSINESS ASSOCIATIONS,** | |
| major incidents, provision of information for, | D6030 |
| **BUSINESS CHARTER FOR SUSTAINABLE DEVELOPMENT,** | E16011 |
| **BUSINESS CONTINUITY PLANNING,** | B8001–B042 |
| access, effects of denial of, | B8008 |
| after the crisis, communication, | B8042 |
| audits, | B8008, B8034 |

| | |
|---|---|
| British Standard, | B8006 |
| business case for, | B8010 |
| Business Continuity Institute, | B8001 |
| business impact analysis, | B8011, B8016, B8018 |
| buy-in, securing, | B8027 |
| Cadbury guidance, | B8006 |
| call deluge, | B8040 |
| commercial drivers, | B8005 |
| command and control team, | B8024 |
| communication, | B8038–B8042 |
| component approach, | B8014, B8022 |
| component testing, | B8036 |
| context of, | B8002 |
| creditworthiness, loss of, | B8017 |
| crisis communication, | B8038–B8042 |
| critical business approach, | B8015, B8018, B8022, B8037 |
| customer base, erosion of, | B8017 |
| damage assessment team, | B8024 |
| departmental teams, | B8024 |
| desk check/unit, | B8034 |
| development of, | B8011, B8020, B8025 |
| dilemma, the planning, | B8021 |
| disasters, | |
| imported, | B8008 |
| natural, effects of, | B8008 |
| public relations, | B8009, B8038 |
| recovery, | B8001–B8003, B8023 |
| drivers, | B8004–B8009 |
| during crisis, communications, | B8041 |
| emergency response, | B8023 |
| fallback procedures, | B8023 |
| financial drivers, | B8007 |
| full simulation, | B8037 |
| hierarchical structure, | B8024 |
| implementation, | B8011, B8026–B8032 |
| importance of, | B8003 |
| individual responsibilities, | B8030 |
| insurance, | B8007 |
| key staff, loss of, | B8017 |
| maintenance, | B8011, B8033–B8037 |
| media training, | B8040 |
| model, | B8011, B8015 |
| objectives, | B8001 |
| outputs, | B8018 |
| phased roll-out, | B8031 |
| policy, | B8028 |
| pre-crisis measures, | B8040 |
| press statements, | B8040 |
| project authorisation, | B8029 |
| public relations, | B8038–B8042 |
| disasters, | B8009, B8038 |
| quality drivers, | B8006 |
| quantification of business impact, | B8017 |
| regulatory and contracted impact, | B8017 |
| regulatory drivers, | B8006 |
| reputation, loss of, | B8017 |
| resumption of business, | B8023 |
| risk assessment, | B8012–B8018 |
| risk management, | B8007 |
| simulation, | B8037 |
| specialist services, | B8032 |
| standards, | B8006 |
| strategy, | B8011, B8019 |
| testing, | B8011, B8033–B8037 |
| Turnbull Committee, | B8001, B8006 |
| understanding your business, | B8011, B8012 |
| walkthrough test, | B8035 |
| **BUSINESS IN THE COMMUNITY,** | E16002 |
| **BUSINESS MANAGEMENT,** | D6002 |
| **BYSSINOSIS,** | C6060, O1051 |
| **CADBURY GUIDANCE,** | B8006 |
| **CALL CENTRES, NOISE AND,** | N3028 |
| **CALL DELUGE,** | B8040 |
| **CALL, WORKERS ON,** | W10003 |
| **CAPABILITY,** | |
| children and young persons, | V12012 |
| manual handling, | M3007, M3015 |
| **CARBON DIOXIDE,** | |
| extinguishers, | F5015 |
| new or expectant mothers, | V12022 |
| **CARCINOGENS,** | |
| control of substances hazardous to health, | C10001, H2113 |
| directive on | C10001 |
| **CARE,** | |
| damages, | C6028 |
| voluntary, | C6028 |
| **CARPAL TUNNEL SYNDROME,** | V5004 |
| **CARRIAGE OF DANGEROUS GOODS,** see Dangerous goods systems | |
| **CARS,** see also Vehicles | |
| taxation, | E16006 |
| **CATCHMENT PLATFORMS,** | C8013 |
| **CATERING PREMISES,** | |
| registrations, | F9027 |
| **CE MARKS,** | C8117, E3025, M1009, M1015–M1016, M1026 |

*Index*

CERTIFICATE OF RECOVERABLE BENEFITS, C6045–C6050
CHAIN SLINGS, L3010
CHALLENGER SPACE SHUTTLE, D6024
CHANGING CLOTHES,
  facilities for, W11026
CHARGES, Intro 21
CHARTERED INSTITUTE OF ENVIRONMENTAL HEALTH (CIEH), T7018
CHEMICAL HAZARDS,
  confined spaces, A1022
  regulations, D0402
  transport, D0402
CHEMICAL PLANT,
  major incident, D6064, D6068
  visitors and the public, protection of, C8120
CHEMICAL SUBSTANCES,
  children and young persons, V12016
  information, H2137
  new or expectant mothers, V12022
  packaging, H2137
CHERNOBYL DISASTER, D6001, D6026, D6030
CHILDREN AND YOUNG PERSONS, V12006–V12016
  age, V12004, V12006–V12007
  agriculture, V12015
  biological agents, V12016
  capability, M3007, V12012
  characteristics of, V12013
  checklist, V12016
  chemical agents, V12016
  children's homes, F5075
  construction work, V12005
  control of substances hazardous to health, V12005
  dangerous goods, carriage of, V12015
  dangerous materials or activities, V12016
  dangerous workplaces or workstations, V12016
  definition of child and young person, V12003, V12007
  docks, V12015
  domestic service, V12011
  employees, V12004
  equipment, V12016
  European Union Directive on, W10001
  excessively physically demanding work, V12016
  excessively psychologically demanding work, V12016
  experience, V12013
  explosives, carriage of, V12015
  family undertakings, V12011
  fire and fire precautions, F5075
  immaturity, V12013
  increased risks, work presenting, V12016
  information on risks to, V12010, V12014
  ionising radiation, V12015
  lead processes, E14002, V12015
  management of health and safety, M2013, M2020
  Management of Health and Safety at Work Regulations 1999, V12002–V12003, V12008–V12016
    disapplication of requirements, V12011
  manual handling, M3007, V12005
  mines, V12015
  occupier's liability, O3010
    assessment of, R3024, R3031
  parents, information for, V12010, V12014
  physical agents, harmful exposure to, V12016
  precautions, V12016
  prohibition and restrictions, V12004, V12006, V12014–V12016
  protection of, V12009
  quarries, V12015
  risk assessment, V12006, V12008, V12013, V12016
  shipbuilding and repairing, V12015
  training, V12004, V12010, V12012
  work experience, V12010
  young persons,
    definition, V12003, V12007
    training and competence, T7006
    working time, W10001
CHURCH HALL,
  consumption of food in, F9027
CHUTES, M3010
CINEMAS, FIRE PRECAUTIONS AND, F5070
CITY AND GUILDS, T7022

# Index

**CIVIL DEFENCE,**
disaster and emergency management
  systems, D6006, D6007,
  D6009–D6011
**CIVIL LIABILITY**, *see also* Damages,
  Negligence Intro 24
  asbestos, O1009
  breach of statutory
    duty, Intro 24, W11045
  construction and building
    works, C8040, C8041,
      C8067, C8149–C8151
  contractors, C8149
  director, E15041
  employment protection, Intro 24
  experts, Intro 24
  Factories Act, effect of conviction
    under, W11049
  fire and fire precautions, F5078
  pre-action protocol, Intro 24
  procedure, Intro 24
  settlements, Intro 24
  time limits, Intro 24
  voluntary assumption of risk, Intro 24
  windows and window
    cleaning, W9033
  Woolf reforms, Intro 24
**CLEANING,**
  cleaners, R3031
  construction and building
    works, C8062
  control of airborne particles, V3010
  risk assessment, R3031
  windows, *see* Windows and window
    cleaning
  workplaces, level and frequency
    of, W11007
**CLIENT/CONTRACTOR
  NATIONAL SAFETY
  GROUP,** T7027
**CLIMATE CHANGE LEVY,** E16006
**CLIMBERS, INSTRUCTION
  OF,** W9008, W9051
**CLOTHING,**
  accommodation, C8056, S11016,
    W11025, W1102
  asbestos, O1009
  capability, M3007
  changing, facilities for, W11026
  construction and building
    works, C8056
  manual handling, M3004, M3007
  workplaces, S11016, W11025,
    W11026
**CLUBS, NOISE AND,** N3002, N3027

**COAL MINERS,**
  compensation, vibration white finger
    for, V5003, V5012
  pneumoconiosis, C6061
  vibration, V5003, V5012
**COCKLE PICKERS TRAGEDY,
  MORECAMBE BAY,** intro 26
**CODES OF PRACTICE,** *see also*
  Approved Codes of Practice
  employers' liability insurance, E13001
  noise, N3017
  Safety Case Regulations, O7016
  vibration, N3017
**CODEX ALIMENTARIUS,** F9038
**COFFERDAMS AND
  CAISSONS,** C8048, C8064
**COLLECTIVE
  AGREEMENTS,** E14004, W10007,
    W10013, W10036
**COMBUSTIBILITY,**
  method of test for, H7024 (Table 2)
**COMMERCIAL PREMISES,**
  health and safety law, enforcement
    of, E15007
**COMMUNITY HOMES,** F5075
**COMMUNITY RESPONSE,**
  disaster or emergency, to, D6030
**COMPANY,**
  Companies Commission, E16005
  corporate social
    responsibility, E16002, E16005,
      E16024
  director, *see* Directors
  health and safety
    offences, E15039–E15042
  Hermes Principles, E16007
  occupier's liability, O3001
  operating and financial
    review, E16005
**COMPENSATION,** C6001–C6062
  coal miners, vibration white finger
    and, V5003
  Compensation Recovery Unit, C6050
  construction and building
    works, C8041
  damages, *see* Damages
  harassment, for, H1726, H1729,
    H1733
  unfair dismissal, for,
    additional award, E14022
    basic award, E14022
    compensatory award, E14022
    constructive dismissal, H1729
    formula for, E14017
    special award, E14022

IND/9

*Index*

| | |
|---|---|
| vibration white finger, | V5003 |
| working time, | W10040, W10041 |

**COMPENSATION RECOVERY UNIT,** C6050

**COMPETENT PERSONS,** *see also*
| | |
|---|---|
| Training and competence | Intro 28 |
| compressed air on construction sites, | C8095, C8106 |
| control of substances hazardous to health, | H2115 |
| management of health and safety, | M2010 |
| noise, | N3029 |
| permit to work systems, | S3016 |

**COMPRESSED AIR ON CONSTRUCTION SITES, WORK IN,**
| | |
|---|---|
| alcohol, prohibition against, | C8103 |
| compression and decompression procedures, | C8097 |
| competent persons, | C8095, C8106 |
| construction and building works, | C8092–C8107 |
| contractors' duties of, | C8093 |
| criminal offences, | C8107 |
| drugs, prohibition against, | C8103 |
| emergencies, | C8099 |
| employees' duties, | C8104 |
| enforcement, | C8107 |
| equipment, | C8096 |
| fire precautions, | C8100 |
| fitness for work, | C8102 |
| information, | C8101 |
| instructions, | C8101 |
| medical advisers, contract, | C8105 |
| medical facilities, provision and maintenance of, | C8098 |
| new or expectant mothers, | V12004, V12020 |
| notification, | C8094–C8103 |
| plant and equipment, | C8096 |
| training, | C8101 |

**COMPUTER,**
ergonomic design, *see* Ergonomic Design
| | |
|---|---|
| laptop, | E17017, E17063 |

**CONDITIONAL FEE ARRANGEMENTS,** C6021

**CONDITIONS,** *see* Working environment

**CONFINED SPACES,**
| | |
|---|---|
| access to, | A1001, A1022 |
| approved codes of practice, | C8069 |
| breathing apparatus, | C8071 |
| checklist for entry to, | A1023 |
| communications, | C8081 |
| competence for, | C8080 |
| construction and building works, | C8069–C8071 |
| control of substances hazardous to health, | C8069 |
| definition of, | A1023 |
| designing safe system of work, | A1023 |
| duties and responsibilities, | C8072–C8074 |
| emergencies, | C8081, D6012 |
| arrangements and procedures, | C8084–C8088 |
| breathing apparatus, | C8071 |
| employee involvement, | C8077 |
| engine driven equipment, | C8081 |
| entry or exit, to or from, | A1001, C8069–C8070 |
| equipment, | C8074, C8088–C8091 |
| explosive atmospheres, in, | C8090 |
| fire extinguishers, | C8087 |
| maintenance of safety and rescue, | C8089 |
| rescue, | C8088–C8089 |
| selection and use of, | C8091 |
| examples of, | C8070 |
| exemption certificates, | A1023 |
| explosions, | C8082, C8090 |
| fatal accidents, | C8072 |
| fire prevention and procedures, | C8082, C8087 |
| first aid, | C8081 |
| fumes, | C8081 |
| gas purging, | C8081 |
| hazards, | |
| atmospheric, | A1022 |
| chemical, | A1022 |
| physical, | A1022 |
| isolation requirements, | C8076 |
| lighting, | C8081 |
| lone workers, | V12029 |
| openings, size of, | C8085 |
| permit-to-work procedure, use of, | A1022, C8076 |
| personal protective equipment, | C8074 |
| planning entries into confined spaces, | C8079–C8082 |
| plant and equipment, | C8074 |
| prevent entry, duty to, | C8073, C8075 |
| procedures, | C8081 |
| prohibition, | A1023 |
| public emergency services, | C8086 |
| radon gas, | C8069 |

*Index*

| | | | |
|---|---|---|---|
| rescues, | C8084–C8088, V12029 | compensation, | C8041 |
| respiratory protective equipment, | C8071, C8074 | compressed air, work in, | C8092–C8107 |
| risk assessment, | A1023, C8073, C8075–C8080, C8083 | confined spaces and harmful atmospheres, | C8069–C8091 |
| safe system of work, | A1023, C8075–C8081 | Construction (Design and Management) Regulations 1994 | Intro 23, C8023–C8040 |
| smoking, fire and, | C8082 | exclusions from, | C8025 |
| specified risk, | A1023 | Construction (Health, Safety and Welfare) | |
| static electricity, fire and, | C8082 | Regulations 1996 | C8041–C8067 |
| statutory requirements relating to work in, | A1023 | duties under the, | C8042–C8066 |
| supervision, | C8083 | construction work, definition of, | C8020 |
| testing and monitoring the atmosphere, | C8081, C8089 | contractors, | |
| toxic gases, | C8081 | civil liability, | C8149 |
| training, | C8083, C8087 | duties of, | C8032 |
| ventilation, | C8081 | principal, | C8028, C8031, C8035 |
| water surges, | C8081 | responsibility for, | C8140–C8141, C8146 |
| working time, | C8081 | | |
| written instructions, | C8081 | | |
| **CONSTANT ATTENDANCE ALLOWANCE,** | C6014, C6020 | control of substances hazardous to the health, | C8001 |
| **CONSTRUCTION AND BUILDING OPERATIONS,** | C8001–C8151 | co-operation, duty of, | C8003 |
| | | cranes, | C8020 |
| access, | | criminal offences, | C8041, C8142–C8148 |
| disabled persons, | W11043 | | |
| routes, | V12005 | dangerous substances | C8008, C8053 |
| accident reporting, | A3035 | delegation, | C8150 |
| advice, duty to provide, | C8006 | demolition, | C8017, C8045, C8133 |
| air, supply of fresh, | C8057 | | |
| Approved Code of Practice, | C8005–C8006, C8028, C8029, C8032 | design, | C8004, C8005, C8032, F5027 |
| arbitration, | C8140 | designers, | C8029 |
| asbestos, | A5006, C8019, C8068, C8133 | appointment of, co-operation between, | C8029 |
| surveys, | A5023–A5026 | ensuring, | C8030 |
| Associated Octel case, | C8147 | duties of, | C8029, C8030 |
| CE marks, | C8117 | planning supervisors, | C8030 |
| children and young persons, | V12005 | roofs, | W9028 |
| civil liability, | C8040, C8041, C8067, C8149–C8151, W11044 | disabled persons, | V12005, W11043 |
| | | dismantling, | C8045 |
| | | doors, | C8052 |
| cleanliness, | C8062 | drinking water, supply of, | C8056 |
| clients, | | drivers, | C8009 |
| definition of, | C8002 | drowning, prevention of, | C8049 |
| duties of, | C8004–C8005, C8027–C8028 | Egan Report, | C8001 |
| | | elderly, | V12005 |
| clothing, accommodation for, | C8056 | emergency procedures, | D6011, C8054–C8055, F5026 |
| cofferdams and caissons, | C8048, C8064 | emergency routes and exits, | C8054 |

IND/11

*Index*

| | | | |
|---|---|---|---|
| employers' liability, | C8141–C8151 | licensee and licensor, relationship of, | C8141 |
| enforcement, | C8066 | lifting operations, | C8013, C8020, L3011 |
| equipment, | C8015, C8061, C8064 | lighting, | C8015, C8060, L5004 |
| European Union law, | C8001–C8003, C8005–C8007, C8023, C8043 | manslaughter by gross negligence, | C8143 |
| excavations, | C8014, C8047, C8064 | materials falling, | C8043 |
| explosions, | C8008, C8046, C8053, C8147 | McDermid Nash case, | C8150 |
| Factories Act, relating to, | W11045–W11049 | mineral resources, exploration and extraction of, | C8020 |
| falls, | C8043, C8146, W9014–W9015, W9019–W9022 | mobile sites, | C8002 |
| | | multiple occupation of sites, | C8141 |
| | | new or expectant mothers, | V12005 |
| | | notification of projects, | C8024 |
| fatal accidents, | C8001, C8029 | occupier's liability, | C8141–C8151, O3015 |
| fire, | C8018, C8145, F5024–F5030, F5062 | personal suspension equipment, | C8064 |
| equipment, | C8053 | planning supervisor, | C8004, C8025 |
| risks, prevention of, | C8053 | appointment of, | C8005, C8024, C8027 |
| fragile roofing materials, | C8043, C8146 | designers, | C8030 |
| Framework Directive, | C8001 | duties of, | C8030 |
| gates, | C8052 | health and safety plans, | C8035 |
| good order, keeping the site in, | C8062 | importance of, | C8005 |
| hand tools and equipment, powered, | C8015 | plant and equipment, | C8061, F5028 |
| head protection, | C8108–C8116 | Port Ramsgate ferry walkway case, | C8148 |
| Health and Safety Commission, | C8001, C8028 | portable tools, | C8015 |
| | | precautions, | C8041 |
| Health and Safety Executive, | C8001, C8028 | principal contractors, | C8027 |
| | | appointment of, | C8028 |
| health and safety files, | C8038 | duties of, | C8031 |
| health and safety law, | C8020–C8023 | health and safety plans, | C8035 |
| health and safety plans, | C8004, C8025, C8027, C8034–C8037, C8040 | products, composition of construction, | C8117 |
| | | prosecution, | C8039–C8040 |
| heights, working at | C8007, C8011–C8012, W9002–W9004, W9007–W9008, W9014–W9029 | public, protection of the, | C8118–C8139 |
| | | reports, | C8065 |
| | | rest facilities, | C8056 |
| highways, hazards left on, | C8151 | Rhone-Poulenc Rorer case, | C8146 |
| HSE Construction Division Inspectorate, | C8001 | RIBA arbitration, | C8140 |
| | | risk assessment, | C8021, C8029, C8030, R3002, R3031 |
| HSE guidance, | C8029 | | |
| independent contractors, | C8021, C8151 | risk management, | C8001, C8010, C8025 |
| inspection, | C8064, C8065 | risk reduction, | C8001–C8002, C8005 |
| insurance, | C8141 | | |
| JCT forms of contract, | C8140–C8141, F5029 | roof work, | W9028 |
| | | roofing materials, | C8043, C8146 |
| | | safe place of work, | C8043 |
| ladders, | C8011 | safe system of work, | C8145, C8150 |

## Index

safety criteria, C8009–C8019
sanitary conveniences, C8056, W11019
self-employment, C8001, C8021
signs, C8062
sites,
  mobile, C8002
  temporary, C8002
  transport, C8016
stability of structures, C8044
standard forms of contract, C8140–C8141
subcontracting, C8001, C8021
  criminal liability, C8145
  responsibility for, C8140–C8141
supervision of projects, C8002–C8005
Swan Hunter case, C8145
temperature, C8058
temporary sites, C8002, F5062
traffic routes, C8050
training, C8005, C8063, T7026–T7027
trenches, C8014
vehicles, C8051
ventilation, C8056
vibration, C8015
vicarious liability, C8150
visitors and the public, protection of, C8118–C8139
walkways, C8148
washing facilities, C8056
weather protection, C8059
welfare facilities, C8056
working platforms, C8012–C8013, C8064
working time, W10004

**CONSTRUCTION INDUSTRY TRAINING BOARD,** T7027

**CONSTRUCTION SKILLS CERTIFICATION SCHEME,** T7027

**CONSTRUCTIVE DISMISSAL,** E14014, E14019, H1727, H1728, H1733

**CONSULTANTS,** Intro 28

**CONSULTATION, JOINT,**
  generally, J3001
  regulatory framework, J3002
  safety committees, *see also* Safety Committees
  safety representatives, *see also* Safety Representatives

**CONSUMER CONTRACTS,**
  unfair terms, P9060

**CONTINUING PROFESSIONAL DEVELOPMENT,** T7028

**CONTRACT CLEANING COMPANIES,** W9008, W9036

**CONTRACT OF EMPLOYMENT,**
disciplinary procedures, *see* Disciplinary Procedures
new or expectant mothers, V12018
other contracts contrasted, E14001
terms of,
  additional duties, as to, E14001
  collective agreements, terms from, E14004
  disciplinary procedures, E14005, E14007
  express terms, E14002
  grievance procedures, E14005, E14007
  implied duties, E14006, E14008
  implied terms, E14003
  statement of terms and conditions, E14005
  statutory minimum notice, E14002, E14018
  working hours, E14002, E14005
  works rules, E14006, E14007

**CONTRACTORS,**
asbestos removal, A5026
civil liability, C8149
compressed air on construction sites, C8093
construction and building works, C8028, C8031–C8032, C8035, C8140–C8141, C8146, C8149
independent, C8021, C8151
licensing, A5026
local authorities, Intro 17
major incidents, D6027
permit to work systems, S3015
principal, C8027–C8028, C8031, C8035
risk assessments, Intro 17
standards, Intro 17
subcontractors, C8001, C8021, C8140–C8141, C8145
training and competence, T7026–T7027
visitors and the public, protection of, C8127

**CONTRIBUTORY NEGLIGENCE,** Intro 24, P9033
damages, C6037, C6046
occupier's liability, O3011
product liability, in cases of, P9033

**CONTROL OF LEAD AT WORK,** *see* Lead

*Index*

| | |
|---|---|
| **CONTROL OF MAJOR ACCIDENT HAZARDS REGULATIONS,** *see* Major Accident Hazards, | |
| **CONTROL OF SUBSTANCES HAZARDOUS TO HEALTH,** | H2101, H2104–H2136 |
| accidents, | H2135 |
| adequate control, | H2111 |
| air quality, | H2132 |
| approved codes of practice, | H2105, H2113, H2122, H2125, H2129, H2131, H2135, V12022 |
| biological agents, | H2113 |
| carcinogens, | H2113 |
| children and young persons, | V12005 |
| competent persons, | H2115 |
| confined spaces, | C8069 |
| construction and building works, | C8001 |
| control measures, | H2108–H2111, H2127–H2131 |
| maintenance, examination and testing of, | H2128–H2132 |
| definition of substances hazardous to health, | H2104 |
| emergencies, | H2135 |
| employees, discussions with, | H2119 |
| health surveillance, | H2133 |
| HSE guidance, | H2105, H2113, H2120, H2133, H2135 |
| incidents, | H2135 |
| information, | H2117, H2134 |
| instructions, | H2134 |
| investigations, | H2121 |
| local exhaust ventilation plant, | H2130 |
| Maximum Exposure Limit, | H2111, H2134 |
| mental disabilities, | V12005 |
| monitoring exposure, | H2132 |
| new or expectant mothers, | V12022 |
| observations in the workplace, | H2118 |
| Occupational Exposure Standard, | H2111 |
| personal protective equipment, | H2110, H2117, H2128, H2131 |
| prevention of control or exposure, | H2107–H2108 |
| reasonably practicable, | H2108 |
| recommendations, review and implementation of, | H2123 |
| records, | H2122–H2124 |
| examples of, | H2124 |
| preparation of, | H2122 |
| regulations, | Intro 23, H2101, H2104–H2136 |
| duties under the, | H2105 |
| respiratory protective equipment, | H2128, H2131 |
| risk assessments, | H2105–H2125 |
| emergencies, | H2135 |
| organisation of, | H2116 |
| records, | H2122–H2124 |
| reviews, | H2125 |
| suitable and sufficient, | H2114 |
| self-employed, | H2105 |
| training, | H2134 |
| **CONVEYORS,** | M3010 |
| **COOLING,** | F5005, N3035 |
| **COOLING TOWERS,** | C8120, C8131 |
| **CORPORATE GOVERNANCE,** | Intro 3, Intro 5 |
| **CORPORATE MANSLAUGHTER,** | Intro 20, E15030 |
| construction and building operations, | C8143 |
| corporate, | Intro 20, |
| gross negligence, by, | C8143 |
| *Herald of Free Enterprise*, | D6022 |
| high profile cases, | D6022 |
| *Involuntary Homicide Bill*, | D6022 |
| Law Commission Report, | D6022 |
| Lyme Bay canoeing disaster, | E15030 |
| proposals for the offence of, | D6016 |
| unlawful act manslaughter, | D6022 |
| workplace, in, | E15038 |
| **CORPORATE SOCIAL RESPONSIBILITY,** | |
| AA 1000 Assurance Standard, | E16024 |
| benchmark index, | E16002 |
| Business in the Community, | E16002 |
| definition, | E16005 |
| environmental management, | E16002, E16005, E16024 |
| European Business Campaign, | E16002 |
| European Commission Green Paper, | E16005 |
| matrix, | E16002 |
| multi-stakeholder forum, | E16005 |
| performance, | Intro 5 |
| **COSTS,** | |
| director's liability, | E15041 |
| **COUNCIL OF REGISTERED GAS INSTALLERS,** | G1013 |

IND/14

COUNSELLING,
  major incident, after,    D6052
CRANES,
  construction and building
    works,    C8020
  crawler,    L3005
  failure, causes of,    L3033, L3034
  fixed,    L3005
  hazards,    L3005
  mobile,    L3005, L3034
    appointed person,    L3006
    crane driver,    L3006
    meaning,    L3006
    safe lifting operations by,    L3006
    signallers,    L3006
    slingers,    L3006
    supervisor,    L3006
  rough-terrain,    L3005
  signallers,    L3005, L3006
  tower,    L3005
  types,    L3005
  wheeled,    L3005
CRIMINAL LIABILITY, *see also*
  Enforcement, Corporate
    manslaughter    Intro 20
  asbestos,    Intro 15, C8133
  breach of statutory duty,
    for,    E15023–E15035
  burden of proof,    Intro 20
  compressed air on construction
    sites,    C8107
  construction and building
    works,    C8041, C8142–C8148
  Crown Court,    Intro 20
  employers' liability insurance,    E13001
  fire and fire
    precautions,    F5040, F5043, F5050, F5058, F5068
  fire certificates,    F5040
  machinery safety,    M1018
  magistrates,    Intro 20
  management of health and
    safety,    M2021
  prohibition notices,    F5050
  prosecutions,    Intro 15, Intro 20, C8039–C8040
  subcontractors,    C8140–C8141
  working time,    W10039
CRISIS
  COMMUNICATION,    B8038–B8042
CROWN COURT,    Intro 20
CROWN, THE,
  fire and fire
    precautions,    F5059, F5077

  health and safety law and,    E15045
  immunity,    E15044
CRUSHING,    M1025
CULLEN REPORT,    D6021
CUSTODIAL SENTENCES,    Intro 15
CUSTOM BUILT ARTICLES,    P9019
CUSTOMER BASE, EROSION
  OF,    B8017
CUTTING AND SEVERING,    M1025

DAMAGES,    C6021–C6062
  Accident Line,    C6021
  appeals against certificate of
    recoverable benefits,    C6050
  asthma,    C6057
  bad news, receiving,    C6033
  basis of claims,    C6022–C6023
  benevolence exception,    C6053
  bereavement,    C6035
  breach of contract, for,    H1736
  breach of statutory duty,    C6001
  capitalisation of future losses,    C6027
  care,    C6028
  certificate of recoverable benefits,
    applications for,    C6045–C6050
  coal miners,    V5003, V5012
  Compensation Recovery Unit,    C6050
  complex cases,    C6048
  conditional fee arrangements,    C6021
  contributory
    negligence,    C6037, C6046
  deafness,    C6056
  deductions,    C6027, C6036, C6051–C6054
  dependants,    C6037, C6039, C6040
  deterioration,    C6040
  director's liability,    E15041
  expenses,    C6024, C6028, C6029, C6054
  fatal accidents,    C6035–C6039
  foreseeability,    C6022
  general,    C6021, C6044
  home care,    C6028
  index-linked government
    securities,    C6021
  inflation,    C6027, C6043
  institutional care,    C6028
  insurance,    C6053
  interest on,    C6043
  interim awards,    C6041
  investments,    C6027
  legal aid,    C6021
  Legal Services Commission,    C6021
  life expectancy,    C6029, C6034

*Index*

| | |
|---|---|
| loss of amenity, | C6030, C6034, C6044 |
| loss of earnings, | C6026, C6043 |
| loss of future earnings, | C6027 |
| lost years, | C6029 |
| lump sum, | C6021, C6026, C6040, C6048 |
| mental distress, | C6032 |
| multiplicand, | C6027 |
| multipliers, | C6027 |
| nervous shock, | C6032–C6033 |
| noise, | N3012 |
| non-pecuniary losses, | C6023, C6030–C6035 |
| Ogden Tables, | C6027 |
| pain and suffering, | C6032, C6044 |
| part 36 payments and offers, | C6042 |
| payments into court, | C6042 |
| pecuniary losses, | C6023, C6026 |
| pneumoconiosis, | C6059–C6062 |
| prescribed diseases, | C6055–C6058 |
| product liability, | P9032, P9038 |
| provisional awards, | C6040 |
| psychiatric harm, | C6032–C6033, O1035 |
| quantum of, | C6023–C6024, C6026–C6027, C6039, C6058 |
| repetitive strain injury, | C6058 |
| rescuers, nervous shock and, | C6033 |
| sexual harassment, for, | H1723, H1733, H1734, H1736 |
| social security, | |
|   certificate of recoverable benefits, | |
|     applications for, | C6045–C6049 |
|     deductions, | C6036, C6051–C6054 |
|     information, | C6049 |
|   prescribed diseases, | C6055–C6058 |
|   recovery of, | C6044–C6050 |
| special, | C6021, C6024 |
| stress, | O1035 |
| structured settlements, | C6021, C6025, C6039, C6047 |
| survival of actions, | C6036, C6038 |
| vibration white finger, | V5003, V5012, V5013 |
| voluntary care, | C6028 |
| windows and window cleaning, | W9036 |
| wrongful dismissal, for, | E14017 |
| **DAMPERS,** | F5007 |

| | |
|---|---|
| **DANCING PREMISES, FIRE PRECAUTIONS FOR,** | F5073 |
| **DANGER,** | |
| danger areas, | M2011 |
| dangerous goods, *see* Dangerous goods systems | |
| dangerous occurrences, *see* Dangerous occurrences | |
| dangerous substances, *see* Dangerous substances | |
| management of health and safety, | M2011 |
| serious and imminent, | M2011 |
| **DANGEROUS GOODS SYSTEMS,** | D0401–D0447 |
| accident reporting, | A3035 |
| acute effects, | D0402 |
| ADR Agreement (European Agreement concerning the International Carriage of Dangerous Goods) | D0404–D0406, D0408–D0409, D0411–D0414 |
| Approved List, | D0416, D0419 |
| concessions, | D0431 |
| documentation, | D0429 |
| exemptions, | D0431 |
| identification, | D0419 |
| markings, | D0426 |
| selection, | D0421 |
| structure of, | D0411 |
| Approved List, | D0408, D0416, D0419, D0421 |
| aquatic pollutants, | D0416 |
| bulk containers, | D0422 |
| bulky loads, | D0402, D0422 |
| carriage duties, | D0410 |
| Carriage of Dangerous Goods and Use of Transportable Pressure Equipment Regulations 2004, | D0406–D0407, D0428 |
| categories, | D0402 |
| children and young persons, | V12015 |
| CHIP Regulations, | D0402 |
| chronic effects, | D0402 |
| classification, | D0402–D0403, D0408–D0409, D0411, D0415–D0417 |
|   basic principles, | D0415 |
|   training, | D0432 |
| common dangerous goods, | D0401 |
| compliance thresholds, | D0430 |

*Index*

concessions,
  ADR, D0431
  UK special, D0405
consignment duties, D0409
containment, D0420–D0422
  principles of, D0420
control systems, fundamentals
  of, D0402–D0403
Dangerous Goods Safety
  Advisor, D0408
documentation
  requirements, D0427–D0429
Emergency Action Code
  system, D0406, D0425, D0428
European Union, D0404, D0406
exemptions, D0431
flammable liquids, D0417
gases, D0417
groupings of hazards, D0403
identification of dangerous
  goods, D0418–D0419
interface of different modal
  controls, D0405
intermediate bulk containers, D04022
labelling, D0423–D0426
legal framework, D0404–D0410
marking, labelling and
  placarding, D0423–D0426
  basic principles, D0423
  exemptions and
    concessions, D0431
motor vehicles, see Vehicle
  requirements, dangerous
  substances and
Orange Book, D0403–D0404
packaging, D0403, D0415, D0417, D0420–D0421
packing groups
  classification, D0403, D0415, D0417
placarding, D0423–D0426
prohibited items, D0402
Proper Shipping Names
  (PSNs) D0418–D0419, D0404, D0431
purpose of measures on, D0401
radioactive materials, D0412–D0414
receptacle thresholds, D0430
regulations, D0406–D0407, D0428, D0432
regulatory principles, D0401–D0403
revocations, D0407
standards, D0420

structure of UK Carriage of
  Dangerous Goods
  Regulations, D0406
tank systems, D0422
terminology, D0423
training, D0432
transport, D0401
Transportable Pressure Equipment
  Directive, D0406
types of hazards, D0402
UN Committee of Experts, D0415
UN dangerous goods classification
  system, D0403–D0405, D0417–D0418, D0420–D0421
UN Numbers, D0418, D0428
vehicles, see Vehicle requirements,
  dangerous substances and
warning labels, D04024–D0425

**DANGEROUS MACHINERY**, see
  Machinery Safety

**DANGEROUS OCCURRENCES**,
children and young persons, V12016
list of reportable, A3035, A3037

**DANGEROUS SUBSTANCES**, see
  also Asbestos, Dangerous goods
  system. Hazardous Substances;
  Major Accident Hazards
aggregation, C10005
ammonium nitrate, C10003
breach of the implied
  condition, P9055
carriage by road, D6001, D6019
children and young persons, V12016
classification, C10003
construction and building
  works, C8008, C8053
definition, H2145, R3014
emergency services, H2147
enforcement of legislation, E15001
EU law, Intro 11
explosions, H2144–H2146
fire and fire
  precautions, F5031, F5063
fire authorities, H2147
liability for, P9054
list, C10003
notification, H2147
regulations, H2144–H2146
risk assessment, R3014
site marking, H2147
threshold levels, C10003
total quantity present at
  establishment, C10003

IND/17

*Index*

transport, *see* Dangerous goods systems
DATA,
  ergonomic design and, E17056, E17057
DEAFNESS, *see also* Hearing loss O1010–O1015
  damages, C6056
  noise-induced, O1010–O1017
  occupational, compensation for, C6056, O1018
  social security, C6056
DEATH, *see* Fatal Injuries
DECLARATION, remedy as, H1726
DEFECTIVE PRODUCTS, *see* Product Liability
DEFENCES,
  factory occupier, W11048
  food legislation, under, F9022, F9023, F9070
  HSWA, under, E15029, E15031
  occupier's liability, O3017
  offshore operations, O7056
DEMOLITION,
  asbestos, C8133
  construction and building works, C8017, C8045
DEMS, *see* Disaster and Emergency Management Systems
DEPARTMENT OF HEALTH,
  food, policy on, F9005
DEPARTMENT OF WORK AND PENSIONS,
  Health and Safety Commission, Intro 7, Intro 12
DEPENDENTS, C6015, C6037, C6039, C6040
DEREGULATION, Intro 12
DERMATITIS, M3002
DESIGN,
  construction and building works, C8004, C8005, C8032
  designer, *see* Designer
  disaster and emergency management and, D6025
  ergonomic, *see* Ergonomic Design
  fire and fire precautions, F5027
  floors, M3006, M3014
  guards, M1028
  lighting, E17031
  machinery safety, M1002, M1009, M1026
  manual handling, M3009, M3012
  role of, E17001, E17002
  roofs, W9028
  windows and window cleaning, W9032, W9046, W9050
  work environment, M3014
DESIGNER,
  construction and building works, C8029–C8030
  product safety, as to, *see* Product Safety
DIFFUSE PLEURAL THICKENING, O1003, O1007
DILUTION VENTILATION, V3011
DIPLOMA IN SAFETY MANAGEMENT, T7013
DIRECTIVES, EU, *see* European Union law
DIRECTORS,
  civil liability, E15041
  criminal liability, E15040
  damages, E15041
  health and safety offences, E15040
  insurance, E15041
  liability insurance, E13023, E15041
  offences committed by, E15032
DISABILITY DISCRIMINATION,
  assault and battery claims, H1731
  breach of contract, action for, H1735
  compensation, H1726, H1729
  criminal law, H1737
  damages for, H1736
  definition, H1707, H1708, H1710, H1711
  employer's liability, H1718, H1721–H1723
  future legal developments, H1749
  guidance, HSE, V12037
  harassment policy, H1742–H1746
  legal actions and remedies, H1717, H1725–H1741
  monitoring and review, H1748
  statutory claims, H1734, H1739, H1740
  tort, actions in, H1730–H1736
  training of workforce, H1747
  unfair dismissal, E1402–E14022, H1727–H1729
  unlawful discrimination, H1725
DISABILITY LIVING ALLOWANCE, C6002, C6020

**DISABLED PERSONS,** see also
  Disability
    discrimination           V12032–V12037
  access facilities for,       A1001, V12005,
                          V12034, W11043
  construction work,             V12005
  day to day activities,           V12032
  disablement benefit, see Disablement
    Benefit
  display screen equipment,      V12005
  harassment,                H1735, H1736
  manual handling,              V12005
  platform lifts,                 W11043
  risk assessment,            R3027, R3031,
                    V12033, V12035–V12036
  sanitary conveniences, situation
    of,                           W11043
  tax credit,                C6003, C6016
  types of disability,            V12032

**DISABLEMENT**
  **BENEFIT,**              C6002, C6011–
                      C6012, C6056–C6062
  occupational deafness, for
    assessment of benefit,        N3021
    conditions under which
      payable,                     N3020
    entitlement to,          N3002, N3018
    vibration white finger,         N3012

**DISASTER AND EMERGENCY**
  **MANAGEMENT SYSTEMS,**
  adequacy of resources,         D6052
  audit and review,          D6061–D6063
  business continuity,      D6052, D6054
  case studies,            D6064–D6067
  civil defence
    legislation,            D6006, D6007,
                            D6009–D6011
  classification of incidents,       D6003
  code of conduct,              D6050
  community, effective means
    of,                     D6037, D6038
  competence,              D6037, D6040
  confined spaces regulations,     D6018
  control,                 D6037, D6041
  co-operation,             D6037, D6039
  co-ordination,                  D6050
  corporate reasons for,      D6004, D6024
  counselling support,             D6052
  damage assessment,             D6052
  dangerous goods etc, carriage
    of,                  D6019, D6020
  definition of terms,              D6003
  DEM policy,
    activation,                    D6050

  awareness of issues, raised
    by,                     D6037–D6041
  establishment of,         D6032–D6036
  environmental reasons
    for,                    D6004, D6026
  explosives, carriage of,         D6020
  external factors
    influencing,           D6029–D6030
  flood defences,                D6021
  grants and loans for,           D6008
  health and safety
    legislation,            D6012, D6013
  historic development of,      D6001
  humanitarian reasons for,    D6027
  insurance reasons
    for,                    D6002, D6004,
                          D6023, D6030
  internal factors
    affecting,             D6029, D6031
  investigation of
    incident,              D6052, D6056
  involuntary manslaughter and,  D6022
  legal reasons for,        D6004–D6022,
                                D6030
  live exercises,                  D6048
  major accident hazard
    regulations,                 D6014
  major disasters,               E17002
  major incidents: definition,    D6003
  media, role of,                D6001,
                    D6027 D6030, D6037
  medical assistance,       D6027, D6030
  methods used,                D6002
  monitoring,              D6060, D6069
  need for,                D6004–D6027,
                                D6069
  off-site emergency plan,      D6049
  on-site emergency plan,      D6049
  origin of, as a modern
    discipline                   D6002
  outline of,                     D6028
  planning,
    after the event,          D6042, D6051–
                                D6059
    before the event,        D6042–D6049
    contingency planning,        D6043
    decisional planning,          D6045
    definition,                     D6042
    factors relevant during
      event,                  D6042, D6050
    monitoring of plan,            D6060
  reporting of injuries etc,       D6015
  reviews,                 D6036, D6058,
                                D6063
  seminar exercises,              D6048

*Index*

| | | | | |
|---|---|---|---|---|
| societal reasons for, | D6004, D6025, D6030 | | asbestos-related, | O1001, O1003–O1009 |
| state of emergency, | D6010, D6011 | | list of reportable, | A3035 |
| synthetic simulation, | D6048 | | lung cancer, | O1003, O1006 |
| table-top exercises, | D6048 | | mesothelioma, | C6010, O1001, O1004 |
| testing, | D6046 | | | |
| training and competence, | D6040, D6047, D6056, D6059 | | occupational, *see* Occupational Diseases | |
| | | | pneumoconiosis, *see* Pneumoconiosis | |
| | | | record keeping duty, | A3016 |

**DISASTERS**, *see also* **DISASTER AND EMERGENCY MANAGEMENT SYSTEMS**

| | | | | |
|---|---|---|---|---|
| business continuity planning, | B8001–B8003, B8008, B8023 | | **DISMISSAL**, civil actions, | Intro 24 |
| | | | constructive, | E14019, H1727, H1728, H1733 |
| fines, | Intro 15 | | statutory dismissal and disciplinary procedure, | E14008 |
| Morecambe Bay cockle pickers tragedy, | Intro 26 | | unfair, *see* Unfair Dismissal | |
| natural, | B8008 | | wrongful, | E14017, E14018 |
| public relations, | B8009 | | **DISPLAY SCREEN EQUIPMENT**, | D8201–D8222 |
| recovery, | B8001 | | back complaints, | D8205 |

**DISCIPLINARY PROCEDURES**,

| | | | | |
|---|---|---|---|---|
| | | | breaks, | D8206 |
| ACAS procedures, | E14013, E14014 | | checklists, | D8212 |
| accident investigations, | A3030 | | daily work routine, | D8206 |
| accompanied, right to be, | E14001 | | definitions, | D8202 |
| contract of employment, in, | E14005, E14007 | | desk and chair, | D8216 |
| | | | disabled persons, | V12005 |
| essential elements, | E14013 | | discomfort glare, | L5014 |
| appeal, | E14013 | | discussions, | D8215 |
| hearing, | E14013 | | ergonomic approach, | E17044 |
| investigation, | E14013 | | eye tests, | D8201, D8207 |
| rules, scope of, | E14012 | | exclusions, | D8203 |
| sanctions, | | | guidance, HSE, | D8201–D8202, D8205–D8208, D8211–D8213, D8216–D8220 |
| demotion, | E14014 | | | |
| dismissal, *see* Dismissal | | | | |
| fines or deductions, | E14014 | | homeworkers, | D8213 |
| suspension, | E14014, E14015 | | hot desking, | D8218 |
| warning, | E14014 | | information, | D8209 |
| statutory dismissal and disciplinary procedure, | E14008 | | keyboards, | D8216 |
| | | | laptops, | D8219 |
| **DISCOS**, | N3027 | | lighting, | D8216 |
| | | | mobile phones, | D8219 |

**DISCRIMINATION**, *see* Disability Discrimination

| | | | | |
|---|---|---|---|---|
| | | | new or expectant mothers, | V12005 |
| age, | E14001 | | operators, discussions with, | D8215 |
| EU law, | E14001 | | personal organisers, | D8219 |
| harassment, *see* Harassment | | | pointing devices, | D8220 |
| racial, *see* Racial Harassment | | | portable display screen equipments, | D8219 |
| religion or beliefs, | E14001 | | posture, | D8214 |
| sex, | E14002, E14023 | | re-assessments, | D8222 |
| sexual, *see* Sexual Harassment | | | records, | D8216 |
| sexual orientation, | E14001 | | reflected glare, | L5015 |

**DISEASES**,

| | | | | |
|---|---|---|---|---|
| accident reporting, | A3016, A3035 | | regulations, | Intro 23, D8201–D8222 |
| prescribed form, | A3038 | | repetitive strain injury, | O1027–O1029 |

review and implementation of
    recommendations, D8221–D8222
risk assessment, D8201, D8204,
    D8210–D8222, R3010
risk reduction, D8204
screens, D8216
self-assessments, D8212, D8215–D8216
shared workstations, D8218
teleworkers, D8213
training, D8208
users
    discussions with, D8215
    identification of, D8211
vision, D8214
work environment, D8216
workstations
    observations at the, D8214
    requirements for, D8205
    shared, D8218

**DIVING OPERATIONS,**
accident reporting, A3035
new or expectant mothers, V12020
safety devices for, P3018

**DOCKS,**
children and young persons, V12015

**DOCTORS,** W10003, W10026

**DOMESTIC SERVICE**
children and young persons, V12011
working time and, W10032

**DOORS AND GATES,**
construction and building
    works, C8052
workplaces, safety
    regulations, W11016

**DRAWING-IN OR TRAPPING,** M1025

**DRINKING WATER,**
construction and building
    works, C8056
provision of, C8056, S11016, W11024

**DRIVE SHAFTS,** M1008

**DRIVERS,**
access, when using vehicles, *see* Access
construction and building
    works, C8009
training and
    competence, A1013, A1014
whole-body
    vibration, V5023–V5024, V5026

**DROWNING,** C8049

**DRUGS,**
compressed air on construction
    sites, C8103
new or expectant mothers, V12022

**DRY POWDER EXTINGUISHERS,** F5013

**DUCTS,** F5007

**DUST,**
dangers of exposure to, V3002
polluting agent, as, V3003
ventilation, *see* Ventilation,
visitors and the public, protection
    of, C8132
workplace, origins of, V3002

**DUTY OF CARE,** Intro 24

**EARNINGS, DAMAGES FOR LOSS OF,** C6026, C6027, C6043

**EATING FACILITIES,**
provision of, W11027

**EC law,** *see* European Union (EU) laws,

**EDUCATION,** *see also* Training and competence
Safety Representatives'
    Charter, J3002

**ELDERLY,**
access, V12005
construction work, V12005

**ELECTRIC SHOCK,**
causes of, E3002
current flow, calculations, E3002
effect on the body, E3002

**ELECTRICAL HAZARDS,**
causes of main, E3002
design/construction of
    system, E3007, E3010
electric shock, E3002
equipment in explosive
    atmospheres, E3024
fire and fire precautions, F5004
general duties under HSWA, E3005
legal requirements, E3004
lone workers, V12029
machinery safety, M1025
maintenance of system, E3010
preventative action, E3003
product liability, E3026

**ELECTRICITY,**
access, means of, E3022
accident reporting, A3035
capability of equipment, E3012
CE marks, E3025
competent staff, E3021, E3023

IND/21

*Index*

conductors,
  insulation, protection and placement, E3014
  work on or near live, E3020
conformity assessment, E3025
construction of system, E3007
cutting off supply, E3018
design of system, E3008, E3010
earthing, E3015
employees, records of, E3023
equipment, lighting requirements, L5006
equipment, E3025
  conformity assessment, E3025
  inspection and certification of, E3025
  precautions as to, E3019
excess current protection, E3017
experience statements, E3023
explosive atmospheres, E3024, E3025
guidance notes and publications, E3001, E3012
hazards as to, *see* Electrical Hazards
HSE guidance, E3009–E3012
information on safety, E3001
Institute of Electrical Engineers website, E3008
integrity of referenced connectors, E3016
isolation of electrical equipment, E3018
lighting for equipment, E3022
maintenance, regular, E3010
manufacturers, E3025
non-statutory standards and codes, E3027
offshore operations, O7063
overhead lines, E3001
portable electrical equipment, E3011
product liability, E3026
prosecutions, E3025
qualifications of employees, E3023
records, E3023
references, E3023
regulations as to, Intro 23, E3006–E3008, E3026, E3028
responsible person, duties imposed on, E3025
safe system of work, general requirements, E3009
shock and burns, preventative action, E3003
siting of equipment, E3013
standards of performance, E3027, E3028

static, C8082
strength of equipment, E3012
working space, E3022
**EMERGENCIES**, *see also* Disaster and Emergency Management Systems, Emergency services
breathing apparatus, C8071
business continuity planning, B8023
compressed air on construction sites, C8099
confined spaces, C8071, C8081, C8084–C8088, D6012
construction and building works, C8011, C8054–C8055, F5026
control of substances hazardous to health, H2135
dangerous substances, transport of, D0406, D0425, D0428, D0445–D0446
exits, C8054
explosives, R3014
fire and fire precautions, F5026, F5055
heights, working at, W9005
lone workers, V12030
major accident hazards plans, H2142
  implementation, C10015
  off-site, C10014
  on-site, C10013, D6008
  radiation emergencies, C10020
  review, C10015, D6008
  testing, C10015
noise, N3041
radiation emergencies, C10020–C10021, D6001
risk assessment, H2135
routes, C8054
sound systems, N3041
training, C8087
**EMERGENCY SERVICES**,
confined spaces, C8086
contacts with, M2012
dangerous substances, H2147
fire brigade, delay in calling, F5008
management of health and safety, M2012
risk assessment, R3002
visitors and the public, protection of, C8125
**EMISSIONS TRADING**, E16006
**EMPHYSEMA**, C6060
**EMPLOYEES**,
accident reporting, A3024–A2036

## Index

| | | | |
|---|---|---|---|
| agency workers, | W10005 | construction and building | |
| call, workers on, | W10003 | works, | C8141–C8151 |
| children and young persons, | V12004 | disciplinary procedures, see | |
| civil liability, | Intro 24 | Disciplinary Procedures | |
| compressed air on construction | | duties to employees, see Employers' | |
| sites, | C8104 | duties to employees | |
| confined spaces, | C8077 | enforcement of health and safety | |
| consultation with, see Consultation, | | laws against, see Enforcement | |
| Joint; Safety Committees; | | head protection, | C8109 |
| Safety Representatives | | social security, | C6002 |
| contract, see Contract of | | vibration, | V5016 |
| Employment | | vicarious liability, see Vicarious | |
| control of substances hazardous to | | liability | |
| health, | H2133 | visitors and the public, protection | |
| disciplinary procedures, see | | of, | C8123–C8128 |
| Disciplinary Procedures | | **EMPLOYERS' DUTIES TO** | |
| dismissal, | Intro 24 | **EMPLOYEES,** | |
| duties of, | Intro 21 | asbestos, | A5007–A5019 |
| implied, | E14003 | employees' safety, | E14008 |
| obey lawful reasonable orders, | | enforcement of safety rules by | |
| to, | E14008 | employer, | E14012, E14021 |
| work with reasonable care and | | health and safety law, | Intro 21 |
| skill, to, | E14008 | implied duties, | E14003 |
| electricity, records of people working | | management of health and | |
| with, | E3023 | safety, | M2003, M2017 |
| enforcement of health and safety | | manual handling, | M3002 |
| laws against, | E15001, E15040– | occupational diseases, as to, see | |
| | E15042 | Occupational Diseases | |
| experience statements, | E3023 | safe system of work, | S3003 |
| head protection, | C8112 | statutory duties, | |
| implied duties, | E14003 | civil liability for breach, | W11045 |
| negligence, see Negligence | | work-related upper limb disorders, | |
| qualifications, | E3023 | as to, | O1039 |
| redundancies, | Intro 24 | **EMPLOYERS' LIABILITY** | |
| references, | E3023 | **INSURANCE,** | |
| safety rights, | | assessment of premium, | E13022 |
| not to be dismissed on certain | | bankruptcy of company, | E13002 |
| grounds, | E14010, E14011, | certificates of insurance, | |
| | E14022 | display, | E13013 |
| not to suffer a detriment, see also | | issue, | E13013 |
| Employment Protection, | E14009 | retention, | E13013 |
| sexual harassment, liability | | code of practice, | E13001 |
| for, | H1722 | conditions of cover, | E13019 |
| suspension from work, see | | prohibition of certain | |
| Suspension from Work | | conditions, | E13020 |
| training and | | cover provided by | |
| competence, | T7024–T7025, | policy, | E13016–E13019 |
| | V12004 | conditions, | E13019 |
| windows and window | | geographical limits, | E13018 |
| cleaning, | W9034 | necessary, | E13011 |
| women workers, see Women Workers | | persons, | E13016 |
| workers contracted, | E14001 | scope of cover, | E13017 |
| **EMPLOYERS,** | | criminal offences | E13001 |
| asbestos, | E1301 | degree of cover necessary, | E13011 |
| | | directors' personal liability, | E13023 |

IND/23

*Index*

| | |
|---|---|
| disclosure, duty of, | E13004 |
| display of certificate, | E13013 |
|   failure, | E13015 |
| duty to take out, | E13008 |
| employees, | |
|   covered by Act, | E13009 |
|   not covered by Act, | E13010 |
| exempted employers, | E13012 |
| extension of cover, | E13023 |
| failure, | |
|   display, to, | E13015 |
|   insure, to, | E13001, E13014 |
|   maintain insurance, to | E13014 |
| first-aid, | F7012 |
| generally, | E13001 |
| geographical limits, | E13018 |
| importance of, | E13001 |
| insolvency, | E13002, E13024 |
| insurance contracts, law relating to, | E13003–E13006 |
|   disclosure, duty of, | E13004 |
|   loss mitigation, | E13006 |
|   mitigation of loss, | E13006 |
|   proposal form, filling in, | E13005 |
| issue of certificate, | E13013 |
| legal liability policy, as, | E13002 |
| limitation periods, | E13002 |
| loss mitigation, | E13006 |
| maintain, duty to, | E13008, E13014 |
| measure of risk, | E13022 |
| minimum cover, | E13001 |
| offshore operations, | E13001 |
| penalties, | E13014, E13015 |
|   display, failure to, | E13015 |
|   failure to display, | E13015 |
|   failure to insure, | E13014 |
|   failure to maintain, | E13014 |
|   insure, failure to, | E13014 |
|   maintain, failure to, | E13014 |
| persons covered by policy, | E13016 |
| premiums, | E13001, E13022 |
| proposal form, filling in, | E13005 |
| public liability policy, | E13002, E13016 |
| purpose, | E13002 |
| renewal notifications, code of practice on, | E13001 |
| retention of certificate, | E13013 |
| risk, measure of, | E13022 |
| scope of cover, | E13017 |
| statutory duty to take out, | E13001 |
| subrogation, | E13007 |
| third party claims, | E13002 |
| trade endorsements for certain types of work, | E13021 |
| websites, | E13001 |
| **EMPLOYMENT NATIONAL TRAINING ORGANISATION (ENTO),** | T7008–T7009, T7015–T7017, T7021, T7024 |
| **EMPLOYMENT PROTECTION,** | E14001–E14028 |
| accompanied, right to be, | E14001 |
| constructive dismissal, | E14014, E14019, H1727, H1728, H1733 |
| contractual terms of employment, *see* Contract of Employment | |
| employees' rights, | E14009–E14011 |
|   disclosures, as to, | E14009 |
| dismissal, in respect of, *see* Dismissal | |
| health and safety cases, | E14009, E14010 |
|   scope of, | E14009–E14011 |
|   safety representatives, for | J3030 |
| EU law, | E14001 |
| implied duties, | E14003 |
| suspension from work, | E14015, E14016 |
| women workers, *see* Women Workers | |
| **EMPLOYMENT TRIBUNALS,** | |
| complaints to, | E14009, E14011 |
| working time, | W10040 |
| **ENFORCEMENT,** | |
| act of 'another person', offences due to, | E15034 |
| breach of an absolute duty, | E15018 |
| compressed air on construction sites, | C8107 |
| construction and building works, | C8066 |
| Crown, position of, | E15001, E15045 |
| dangerous substances legislation, | E15001 |
| directors' insurance, | E15033 |
| employee, against, | E15001, E15040–E15042 |
| employers, against civil action, by, generally, | E15001 |
| offences, in cases of *see* offences below | |
| employer's safety rules, of, | E14012–E14022 |
|   demotion, | E14014 |
|   dismissal, | E14017–E14022 |
|   fines or deductions, | E14014 |
|   procedural fairness, | E14012, E14013 |

IND/24

| | | | |
|---|---|---|---|
| procedures, essential elements of, | E14013 | directors, | E15040–E15041 |
| suspension without pay, | E14014 | due to act of another, | E15042 |
| warnings, | E14014 | main offences, | E15032–E15035 |
| enforcing authority, *see* Enforcing authority | | manslaughter, | E15038 |
| | | method of trial, | E15032 |
| Environment Agency, role of, | E15001 | penalties for, | E15036 |
| environmental health officers, role of, | E15001 | prosecution, *see* prosecution, below | |
| Factories Act, of, | W11049 | publication of conviction for, | E15044 |
| fire and fire precautions, | F5057 | sentencing guidelines, | E15043 |
| food safety legislation, *see* Food Safety and Standards | | prohibition notice, *see* Prohibition Notice | |
| Health and Safety Commission, | | prosecution, | |
| constitution of, | E15002 | breach of statutory provisions, | E15030 |
| role of, | E15001–E15004 | burden of proof, | E15031 |
| transfer of responsibility by, | E15011 | offences, *see* offences, above | |
| Health and Safety Executive, | | publication of convictions, | E15044 |
| constitution of, | E15002 | sanction, as, | E15001 |
| enforcing authority, as, | E15007, E15008, E15009 | service of notice, coupled with, | E15018 |
| fees, | E15005 | public register of notices, | E15001, E15029 |
| inspectors' powers, | E15001 | 'relevant statutory provisions', | |
| publications, | E15006 | meaning, | E15001 |
| role of, | E15001, E15002, E15005 | breach of, prosecution for, | E15030 |
| | | execution of, | E15028 |
| transfer of responsibility by, | E15011 | scope of, | E15012 |
| improvement notice, *see* Improvement Notice | | sentencing guidelines, | E15001, E15043 |
| inspectors' powers | | training and competence, | T7008 |
| improvement notice, as to, *see* Improvement Notice | | working time, | W10038–W10043 |
| indemnification as to, | E15028 | **ENFORCING AUTHORITY,** | |
| investigation, as to, | E15001, E15026 | activities for which local authorities are the, | E15004 (Table 18) |
| prohibition notice, as to, *see* Prohibition Notice | | activities for which the HSE is the, | E15004 (Table 17) |
| search and seizure, of, | E15001, E15027 | appropriate, | E15008 |
| | | commercial premises, for | E15007 |
| ladders, | W9002 | indemnification by, | E15001, E15028 |
| local authority, | | industrial premises, for | E15007 |
| enforcing authority, as, | E15007, E15008, E15010 | responsibilities of, | E15001, E15007–E15011 |
| role of, | E15001, E15007 | **ENGINE DRIVEN EQUIPMENT,** | C8081 |
| transfer of responsibility by, | E15011 | **ENTANGLEMENT,** | M1025 |
| notices, | F5057 | **ENTERTAINMENT,** | |
| offences, | | food safety and standards, | F9007 |
| company officers, | E15040 | **ENTRY TO PREMISES,** | |
| corporate offences, | E15039 | gas safety, as to, *see* Gas Safety | |
| defences, | E15037, E15039 | hazardous substances, | H2102 |
| | | **ENVIRONMENT,** | |
| | | dangerous substances, vehicles and, | D0446 |

*Index*

directive on substances dangerous to the C10001
Environment Agency, *see* Environment Agency
Environmental Health Authority, enforcement powers of E15001, F9005
environmental management, *see* Environmental Management
noise, *see* Environmental noise
work environment, *see* Work environment

**ENVIRONMENT AGENCY,**
major accident hazards, role as to, C10001, C10002
notice-serving powers, E15001
training and competence, E17022

**ENVIRONMENTAL HEALTH AUTHORITY,**
enforcement powers, E15001, F9005

**ENVIRONMENTAL MANAGEMENT,**
AA 1000 Assurance Standard, E16024
aggregates levy, E16006
auditing,
  environmental reporting, E16024
  systems, of, E16025
benefits of, E16026
climate change levy, E16006
communications, E16022
corporate social responsibility, E16002, E16005, E16024
disaster and emergency management, D6004, D6016, D6027
effective management, E16026
emissions trading, E16006
ergonomic approach, *see* Ergonomics
ergonomic design, *see* Ergonomic Design
external pressures,
  business
    community, E16001, E16011
  community pressure
    groups, E16009
  customers, E16012
  environmental crises, E16010
  environmental pressure
    groups, E16001, E16009
  financial
    community, E16001, E16007
  legislation, E16001, E16005
  market
    mechanisms, E16001, E16006
  role of, E16001
  supply chain, E16008
  Exxonmobil, E16009
future of, E16027
Global Reporting Initiative, E16024
Greenpeace Stopesso campaign, E16009
guidelines,
  Eco-Management and Audit Scheme (EMAS), E16013, E16014, E16016
  International Standards, E16013, E16015, E16016
standards for environmental management, E16013
Hermes principle, E16007
implementation, E16017–E16025
information management, E16024
insurers' interest, E16007
internal drivers,
  corporate governance, E16001, E16002
  employees, E16004
  management information needs, E16003
  investors, and, E16001, E16007
landfill tax, E16006
management information needs, E16003
meaning, E16001
objectives, E16020
operational controls, E16023
Operating and Financial Review, E16005
policy as to, E16019
practical requirements, E16017
public reporting, E16024
responsibilities, E16021
review, E16017, E16024
shareholders and, E16001, E16007
stress and, S11012–S11017
sustainability reporting, E16024
training, E16022
vehicle and fuel levy, E16006
World Summit on Sustainable Development, E16027

**ENVIRONMENTAL NOISE,**

**EQUAL OPPORTUNITIES COMMISSION,**
equal opportunity policy, H1741, H1742, H1744–H1749

| | | | |
|---|---|---|---|
| sexual harassment, recommendations as to, | H1749 | equipment, controls, design of, | E17013, E17015 |
| White Paper, | H1748 | design of, generally, | E17011 |
| **EQUAL PAY LEGISLATION,** | | display screen equipment, | E17044, E17063 |
| employment contract and, | E14002 | displays, design of, | E17013–E17015 |
| **EQUIPMENT,** *see also*, Personal protective equipment | | hand tool design, | E17012 |
| asbestos, | O1003, V3009 | information instruction, | E170116 |
| buying policy, noise and, | N3035 | labelling, | E17016 |
| children and young persons, | V12016 | legislation as to, | E17060, E17064 |
| compressed air on construction sites, | C8096 | personal protective equipment, | E17065 |
| confined spaces, | C8074, C8088–C8091 | provision of, | E17042 |
| construction and building works, | C8015, C8061, C8084 | software design, | E17017 |
| | | suitability of, | E17064 |
| dangerous substances, vehicles and, | D0444 | user trial, | E17055 |
| display screen, | | guidelines as to, | E17003, E17005–E17043, E17063, E17067, E17068 |
| ergonomic approach, | E17044 | | |
| regulations, | Intro 23 | job design | |
| risk assessment, | R3010 | enlargement of job, | E17036, E17038 |
| engine driven, | C8081 | enrichment of job, | E17036, E17039 |
| EU law, | M1003–M1004 | rotation of work, | E17036, E17037 |
| exhaust, | O1003, V3009 | organisational support and, | E17043 |
| explosive atmospheres, | C8090 | scheduling of work, | E17036, E17040 |
| falls, | W9024 | | |
| fire, | C8053, C8087, F5001, F5006, F5009–F5017 | shift-work, | E17036, E17041 |
| | | legislation promoting, | E17003, E17059–E17066 |
| extinguishers, | C8087, F5009–F5017 | | |
| hand tools, | C8015 | machinery, *see* equipment *above* | |
| heights, working at, | W9008, W9024 | musculoskeletal disorders and, *see* Musculoskeletal disorders | |
| lone workers, | V12029 | | |
| machinery safety, | M1007 | people, for, | E17005 |
| maintenance, | W9024 | physical work environment | |
| mobile work, | M1007 | lighting and, | E17031 |
| noise, | N3016, N3035 | noise and, | E17032 |
| outdoors, noise and equipment used, | N3016 | physical hazards, | E17035 |
| | | role of, | E17030 |
| permit to work systems, | S3017 | thermal environment | E17034 |
| portable tools, | C8015 | vibration and, | E17032, E17033 |
| regulations, | Intro 23 | population stereotypes, | E17008 |
| rescue, | C8088–C8089 | posture, role of, | E17006, E17010, E17021, E17031 |
| Use of Work Equipment Directive, | M1003–M1004 | risk assessment, | E17061, E17063, E17068 |
| windows and window cleaning, | W9024 | | |
| **ERGONOMIC DESIGN,** | | task, allocation of, | E17009 |
| anthropometric data, | E17007, E17010 | designing, | E17010 |
| capabilities, | | tools assisting, | E17003, E17051–E17058 |
| machinery, of, | E17009 | | |
| person, of, | E17009 | | |

*Index*

workstation,
  factors to be considered,    E17018–E17029
  arrangement of items on,    E17024
  characteristics of,    E17020
  height of,    E17019
  layout of,    E17021–E17024
  lighting,    E17031, E17035
  normal working area    E17023
  sitting versus standing,    E17029
  user trial,    E17055
  visual considerations,    E17025–E17028
  zone of convenient reach,    E17022

**ERGONOMICS,**
  aim of,    E17001
  application of,    E17002
  core disciplines,    E17003
  data,
    collection of,    E17056
    existing, use of,    E17057
  design, *see* Ergonomic Design
  employee involvement,    E17058
  ergonomic approach,
    illustration of,    E17004
    overview of,    E17003, E17004
  frequency analysis,    E17055
  legislation as to,    E17003, E17059–E17066
  origin of term,    E17001
  risk assessment,    E17052, E17061, E17063, E17065
  role of,    E17001
  task analysis,    E17053
  tools of,    E17051–E17058
  use of term,    E17001
  workflow analysis    E17054

**ESCALATORS,**
  workplaces, safety regulations,    W11017

**ESCAPE,**
  definition of,    F5045
  fire and fire precautions,    F5007, F5020–F5021, F5020, F5044–F5047, F5049, F5062
  improvement notices,    F5046–F5047
  local authorities,    F5049
  prohibition notices,    F5049
  routes,    F5007, F5020

**EUROPEAN AGREEMENT CONCERNING THE INTERNATIONAL CARRIAGE OF DANGEROUS GOODS,**
  *see* ADR Agreement

**EUROPEAN BUSINESS CAMPAIGN,**    E16002

**EUROPEAN COMMISSION,**
  sexual harassment,
    Code of Practice,    H1701–H1703, H1705, H1706
    recommendation as to,    H1701, H1704, H1706

**EUROPEAN UNION (EU) LAWS,**
  accession states,    intro 9
  age discrimination,    E14001
  Approved Codes of Practice,    Intro 10
  carcinogens,    C10001
  climbers and hill walkers,    W9051
  Constitution,    Intro 9
  construction and building works,    C8001–C8003, C8005–C8007, C8023, C8043
  corporate social responsibility,    E16005
  current initiatives,    Intro 11
  dangerous substances,    Intro 11, D0404, D0406
  directives,    Intro 9, J3002
    carcinogens    C10001
    dangerous substances, transport of,    D0405
    environment, substances dangerous to the    C10001
    Framework Directive,    Intro 9
    health and safety, as to,    E15004
    heights, working at,    W9008, W9051
    implementation,    Intro 10–11
    Machinery Safety Directive,    M1003, M1009–M1018
    Physical Agents (Vibration) Directive,    V5001, V5015–V5016
    Product Liability Directive,    M1002, P9001, P9008–P9012
    Seveso II Directive    C10001, D6014
    vibration,    V5001, V5015–V5016

*Index*

working time, as to, E14002, W10001, W10010, W10026–W10031, W10044
Young Workers Directive, W10001
discrimination, E14001
Eco-label award scheme, E16008
employment protection, E14001
enlargement, intro 9
environment, substances dangerous to the, C10001
fire and fire precautions, F5004
food legislation, F9001, F9002, F9020, F9052
Framework Directive, Intro 9
hand arm vibration, Intro 11
Health and Safety Commission, Intro 10
heights, working at, W9008
Information and Consultation Procedures (ICPs) J3032
machinery safety, M1001–M1004, M1007, M1009–M1020, M1026
Machinery Safety Directive, M1003, M1009–M1018
manual handling, M3001
noise, N3002, N3016
outdoors, noise and equipment for use, N3016
Physical Agents (Vibration) Directive, N3002, V5001
Product Liability Directive, M1002
product safety, P9001, P9008, P9026, P9044
religion or belief, discrimination on the grounds of, E14001
role of, Intro 9–11
Seveso II Directive C10001, D6014
sexual orientation, discrimination on the grounds of, E14001
Social Charter, Intro 9
Transportable Pressure Equipment Directive, D0406
Use of Work Equipment Directive, W9008, W9051
vibration, V5001, V5015–V5016
whole body vibration, Intro 11
working time, W10001, W10010, W10026–W10031, W10044
Young Workers Directive, W10001

**EVACUATION,**
fire and fire precautions, F5022

safety sign regulations, W11033, W11040
training and competence, T7005
**EVIDENCE,**
accident investigations, A3032, A3033
major incident, as to, D6028
**EXCAVATIONS,**
construction and building works, C8014, C8047, C8064
HSE Guidance, C8047
**EXCEPTIONALLY SEVERE DISABLEMENT ALLOWANCE,** C6014
**EXECUTIVES, WORKING TIME AND,** W10033
**EXHAUST EQUIPMENT,** V3009
**EXPECTANT MOTHERS,** *see* New or expectant mothers
**EXPENSES,** Intro 26
damages, C6024, C6028, C6029, C6054
**EXPERIENCE,** *see* Inexperienced workers
**EXPERTS,** Intro 24
**EXPLOSIONS,** *see also* Explosives
construction and building works, C8008, C8046, C8053, C8147
equipment, C8082, C8090
**EXPLOSIVES,** *see also* Explosions
accident reporting, A3035
carriage by road, D6014
children and young persons, V12015
classification, R3014
dangerous substances, H2144–H2146
emergencies, R3014
explosive atmospheres, definition of, H2146
information sheets, C8008
major accident hazards, C10004
noise and, E17032
protection of, P3020
risk assessment, R3014, R3015 (App B)
zoning of workplaces, C8008, C8053
**EXTERNAL SERVICES, CONTACTS WITH,** M2012
**EXX OMMOBIL,** E16009
**EYE TESTS,** D8201, D8207

**FACILITIES,**
disabled person, for, W11043
generally, *see* Welfare Facilities

IND/29

*Index*

**FACTORIES,**
  absence of overcrowding, W11008
  access for disabled persons, W11043
  changing clothes facilities, W11026
  clothing accommodation, W11025
  doors and gates, W11016
  drinking water, supply of, W11024
  eating facilities, W11027
  employees, medical examinations of, O1041
  FA 1961
    defence of factory occupier, W11048, W11049
    enforcement of, W11047
    possible civil liability, W11049
    replacement of, W11045
    residual application of, W11046
  floors, condition of, safety underfoot, W11010
  gas safety requirements, G1034
  general cleanliness, W11007
  occupiers, actions against, O3018
  occupiers, defences of, W11048
  occupier's liability, O3018
  pedestrian traffic routes, W11015
  rest facilities, W11027
  sanitary conveniences, W11019–W11021
  sedentary comfort, W11009
  temperature control, W11005
  washing facilities, W11020
  work clothing, W11025
  workroom dimensions/space, W11008
  workstations, W11009

**FAIR DISMISSAL,**
  health and safety, on grounds of, E14010

**FAIRGROUNDS,**
  accident reporting, A3035

**FALLING OBJECTS,** W11011, W9026

**FALLS,** W11001
  construction and building works, C8043, C8148, W9014–W9015, W9019–W9022
  equipment, maintenance of, W9024
  fencing, W9010
  heights, working at, W9001, W9007, W9008–W9010
  ladders, from, W9041
  personal suspension equipment, W9019–W9020
  windows and window cleaning, W9031, W9033, W9040–W9041
  construction sites, W9014–W9015, W9019–W9022
  equipment, maintenance of, W9024
  ladders, from, W9041
  personal suspension equipment, W9019–W9020

**FAMILY UNDERTAKINGS,**
  children and young persons, V12011
  definition, V12011
  guidance, HSE, V12011

**FATAL INJURIES,**
  confined spaces, C8072
  construction and building works, C8001, C8029
  corporate manslaughter, Intro 20, D6016, E15030
  damages, C6035–C6039
  head protection, C8108
  management of health and safety, M2001
  Morecambe Bay cockle pickers tragedy, intro 26
  reporting, A3018
  statistics, intro 26
  windows and window cleaning, W9041

**FENCING,**
  heights, working at, W9010
  visitors and the public, protection of, C8122

**FINES,** Intro 15
  accident reporting, A3022

**FIRE AND FIRE PRECAUTIONS,** F5001–F5080
  active fire protection measures, F5005, F5007
  appeals, certificates, F5065
  improvement notices, F5048
  BCF extinguishers, F5014
  British Standards, F5016–F5017, F5047
  carbon dioxide extinguishers, F5015
  certificates, F5031–F5035
    alterations, notification of, F5040, F5067
    appeals, F5065
    applications for, F5035–F5037, F5064
    change of conditions, F5042
    construction work, F5062
    contents of, F5033, F5039

IND/30

## Index

criminal offences, F5040
dangerous processes, F5063
escape, F5020, F5044, F5062
exemptions, F5041–F5044
form of applications for, F5036
Health and Safety
  Executive, F5061
hospitals, F5035
keep on premises, duty to, F5066
local authorities, F5033–F5035
special premises, F5060–F5064
children's and community
  homes, F5075
cinemas, F5070
civil liability, F5078
classification, F5003
common causes, F5002
compressed air on construction
  sites, C8100
confined spaces, C8082, C8087
construction sites, C8018, C8145, C8053, F5024–F5030
  certificates, F5062
  designing out fire, F5027
  emergency procedure, F5026
  failure to comply with the code on, F5029
  fire extinguishers, F5028
  fire safety co-ordinator, role of site, F5025
  JCT standard form contract, breach of, F5029
  temporary buildings, F5062
control, F5001
cooling, F5005
co-ordinator, role of fire safety, F5025
criminal offences, F5040, F5043, F5058, F5068
  prohibition notices, F5050
Crown, F5059, F5077
dampers, F5007
dancing, premises for, F5073
dangerous and hazardous
  material, F5031, H2147
  certificates, F5063
delay in calling the fire
  brigade, F5008
designing out fire, F5027
detection, F5054
drills, F5022–F5023
  notices, typical, F5023
dry powder extinguishers, F5013
ducts, F5007
electrical fires, F5004
emergency procedures, F5026, F5055

enforcement notices, F5057
equipment, C8053, C8087, F5001, F5006, F5009–F5017
escape,
  certified
    premises, F5020, F5044, F5062
  definition of, F5045
  improvement
    notices, F5046–F5047
  local authorities, F5049
  means of, F5007, F5020–F5021
  prohibition notices, F5049
  routes, F5007, F5020
  unsatisfactory means of, F5021
EU law, F5004
evacuation, F5022, M2011
extinction, F5005
fire alarms, F5017
  British Standard, F5017
  certified premises, in, F5017
  soundings, F5022
fire brigades, calling, F5008
fire evacuation, F5022, M2011
fire extinguishers, C8087, F5009–F5017
  British Standard, F5016
  colour coding, F5016
  confined spaces, C8087
  distribution of, F5016
  portable, F5016, F5028
  recommendations, F5009
  types of, F5010–F5015
fire fighting, F5054
fire resistant compartments, F5007
foam extinguishers, F5012
fuel, F5001
gaming houses, F5072
good housekeeping, F5018, F5022
Health and Safety Executive, F5061
hospitals, F5035
HSE guidance, C8053
improvement notices, F5046–F5048
inhalation, C8136
insurance, F5080
legislation, F5030–F5045, F5051–F5059
licensed premises, F5069–F5075
local authorities,
  certificates, F5033–F5035
  escapes, F5049
  prohibition notices, F5049
maintenance, F5056
management, F5030, M2011
music, premises for, F5073
nursing homes, F5076

IND/31

*Index*

| | |
|---|---|
| occupier's liability, | F5079 |
| passive fire protection, | F5007 |
| petroleum spirit, storage of, | F5031 |
| plant on construction sites, | F5028 |
| pre-planning, | F5019–F5021 |
| prevention, | F5001 |
| procedures, | F5008 |
| prohibition notices, | F5049–F5050 |
| property risk, | F5006 |
| residential homes, | F5076 |
| risk assessment, | F5006, F5017, F5035, F5051, F5053, R3012, R3032 |
| risk management, | F5001 |
| risk premises, categories of, | F5038 |
| routines, | F5001, F5008 |
| schools, | F5074 |
| sentencing, | F5058, F5068 |
| signs, | F5032, M11040 |
| smoke inhalation, | C8136 |
| smoking, | F5018 |
| smothering, | F5005 |
| sprinkler systems, | F5007 |
| stairways, | F5007 |
| starvation, | F5005 |
| static electricity, | C8082 |
| telephone equipment rooms, | F5006 |
| temporary buildings, | F5062 |
| theatres, | F5071 |
| training, | F5001 |
| triangle of fire, | F5001 |
| ultra-hazardous premises, certificates for, | F5060 |
| visitors and the public, protection of, | C8122, C8136 |
| wardens, | F5022 |
| water, | F5005 |
| water extinguishers, | F5011 |
| Workplace Regulations, | F5051–F5059 |
| exceptions, | F5052 |

**FIRST-AID,**

| | |
|---|---|
| absences of first-aiders or appointed persons, | F7002, F7011 |
| annual leave, | F7011 |
| appointed persons, | F7019 |
| absence, | F7002, F7011 |
| assessment of first-aid needs, | F7003–F7013 |
| absences of first-aiders or appointed persons, | F7011, F7002 |
| aim, | F7003 |
| annual leave, | F7011 |
| basic rule, | F7003 |
| checklist, | F7025 (App A) |
| distance to emergency medical services, | F7008 |
| distant workers, | F7009 |
| doctors, qualified medical, | F7003 |
| factors affecting, | F7003 |
| hazards and risk in the workplace, | F7004 |
| history of accidents in organisation, | F7006 |
| insurance, employers' liability, | F7012 |
| lone workers, | F7009 |
| multi-occupied sites, | F7010 |
| nature and distribution of the workforce, | F7007 |
| no need to appoint first-aiders, where, | F7003 |
| nurses, registered, | F7003 |
| occupational health service advice, | F7003 |
| public, members of the, | F7012 |
| reassessment of needs, | F7013 |
| requirements, | F7003 |
| shared sites, employees on, | F7010 |
| size of organisation, | F7005 |
| trainees, | F7012 |
| travelling workers, | F7009 |
| workforce, nature and distribution of the, | F7007 |
| certificates, | F7018 |
| confined spaces, | C8081 |
| containers, | F7022 |
| courses, | F7018 |
| disaster situations, | D6016 |
| distance to emergency medical services, | F7008 |
| distant workers, | F7009 |
| doctors, qualified medical, | F7003 |
| emergency situations, | D6015, D6026 |
| employer's duty, | |
| informing employees of arrangements, | F7014 |
| provision for first-aid, | |
| assessment of first-aid needs. *See* assessment of first-aid needs *above* | |
| first-aiders, provision of, | F7002 |
| requirements, | F7002 |
| scope of duty, | F7002 |
| eye irrigation, | F7023 |
| first-aiders, provision of, | F7002, F7016–F7018 |
| abilities, | F7017 |
| absence of first-aider, | F7002, F7011 |

*Index*

| | | | |
|---|---|---|---|
| appointed persons, | F7019 | specially trained staff, action by, | F7001 |
| certificates, | F7018 | summary of requirements, | F7001 |
| competencies, | F7025 (App B) | trainees, | F7012 |
| courses, | F7018 | training of first-aid personnel, | F7018 |
| generally, | F7016 | travelling, kits for, | F7024 |
| no need to appoint first-aiders, where, | F7003 | travelling workers, | F7009 |
| numbers required, | F7017, F7016 | workforce, nature and distribution of the, | F7007 |
| offshore operations, | O7058 | **FITNESS PROGRAMMES,** | M3015 |
| position in company, | F7017 | **FIXBOROUGH DISASTER,** | D6068 |
| qualifications, | F7018 | **FLAMMABLE SUBSTANCES,** | |
| refresher training, | F7018 | accident reporting, | A3035 |
| selection, | F7017 | escape of, | A3035 |
| special training, | F7018 | transport, | D0417 |
| standard, | F7017 | **FLEXIBLE WORKING,** | Intro 6, W10003, W10006 |
| training, | F7018 | | |
| unusual risks/hazards, | F7018 | **FLOOD,** | |
| updating skills, | F7018 | defence committees, | D6021, D6030 |
| generally, | D6009, F7001 | **FLOORS,** | |
| hazards and risk in the workplace, assessment of, | F7004 | condition of, in workplaces, | M3006, M3014, W11010 |
| history of accidents in organisation, | F7006 | design, | M3006, M3014 |
| inform employees of arrangements, employer's duty to | | free from obstructions, | W11010 |
| insurance, employers' liability, | F7012 | hygienically clean, | W11010 |
| lone workers, | F7009, V12029 | manual handling, | M3006, M3014 |
| multi-occupied sites, | F7010 | uneven, slippery or unstable, | M3006 |
| nature and distribution of the workforce, | F7007 | **FLUORESCENT TUBES,** | |
| nurses, registered, | F7003 | stroboscopic effect, | L5019 |
| public, members of the, | F7012 | **FOAM EXTINGUISHERS,** | F5012 |
| qualifications of first-aid personnel, | F7018 | **FOIE GRAS,** | F9077 |
| reassessment of needs, | F7013 | **FOOD SAFETY AND STANDARDS,** | |
| record keeping, | F7020 | administrative regulations, | F9004 |
| records, | F7020 | adulteration of food, | F9001 |
| refresher training, | F7018 | allergies, | F9001, F9075 |
| regulations, | Intro 23 | BSE, | F9020 |
| resources, first-aid, | F7021–F7024 | canned foods, | F9056 |
| additional resources, | F7023 | codes of practice, | F9025 |
| containers, | F7022 | complaints, | F9080 |
| eye irrigation, | F7023 | composition of food, | F9002, F9078 |
| materials, | F7022 | consumer concerns, | F9077 |
| protective equipment, | F7023 | contamination, prevention of, | F9038, F9047 |
| provision, | F7021 | dietary preferences, | F9076 |
| stock of first-aid materials, | F7022 | enforcement, | |
| travelling, kits for, | F7024 | bodies responsible, | F9005 |
| rooms, first-aid, | F7025 | detention and seizure of food, | F9014 |
| selection of first-aiders, | F7017 | emergency control order, | F9019 |
| self-employment, | F7015 | emergency prohibition, | F9015, F9018 |
| shared sites, employees on, | F7010 | improvement notice, use of, | F9015 |
| signs, *see* Safety Signs | | | |
| size of organisation, assessment of, | F7005 | | |

IND/33

# Index

| | | | |
|---|---|---|---|
| licensing, by, | F9021 | HACCP system, | F9021, F9035–F9038 |
| powers as to, | F9013 | industry guides, | F9049 |
| officers for, | F9005 | meaning of, | F9035 |
| prohibition order, use of, | F9016, F9017 | meat, | F9005 |
| | | personal hygiene, | F9038, F9046 |
| registration, by, | F9021 | 'pre-requisites', | F9038, F9039 |
| regulations as to, | F9004 | regulations as to, | F9002, F9034, F9035 |
| sample taking, | F9013 | temporary businesses, | F9038, F9041 |
| environmental health officer, role of, | F9005 | transport of food, | F9038, F9042 |
| European Union, | F9001, F9002, F9020, F9052 | vending machines, | F9038, F9041 |
| | | water supply, | F9038, F9045 |
| falsely presenting food, | F9008, F9012 | 'industry' guides, | F9025, F9049 |
| foie gras, | F9077 | inspectors, | F9004 |
| food, definition of, | F9007 | instruction of staff, | F9081 |
| food analysts, | F9004 | labelling of food, regulations as to, | F9003, F9012, F9075, F9078 |
| food directives, | F9001, F9002 | | |
| food handler, | | licensing, | F9021 |
| policy as to, | F9074 | mail order foods, | F9054 |
| supervision, | F9038, F9048, F9081 | Ministry of Agriculture, role of, | F9005 |
| training, | F9035, F9038, F9048, F9081 | premises, registration of, | |
| food hazard warnings, | F9019 | exemption from, | F9028–F9033 |
| food premises, registration of, | F9004 | scope of regulations, | F9026, F9027 |
| food safety legislation, | | price of food, display of, | F9004 |
| codes of practice, | F9025 | product recall, | F9079 |
| defences against charges, | F9022, F9023, F9070 | registration, | |
| definitions, | F9007 | FSA, under, | F9021 |
| enforcement, *see* enforcement, *above* | | premises regulations, under, | F9026–F9033 |
| main provisions of, | F9008–F9025 | rendering food injurious to health, | F9008, F9009 |
| offences, | F9008–F9012 | sale of food, | |
| penalties under, | F9024 | meaning, | F9007 |
| scope of, | F9001–F9006 | non-compliance with safety requirements, | F9009, F9011 |
| Food Safety Manual | | | |
| HACCP, | F9073, F9074 | not of nature etc demanded, | F9008, F9011 |
| organisation, | F9072 | | |
| policy statement, | F9070–F9072 | Scotland, in | F9001, F9005, F9050 |
| procedures, | F9071–F9081 | supervision of staff, | F9081 |
| use of, | F9070, F9082 | temperature control, | |
| Food Standards Agency, role of, | F9005, F9019 | canned foods, | F9056 |
| | | chilled storage, | F9053 |
| genetically modified foods, | F9020, F9077 | cool storage, | F9064 |
| | | cooling, | F9062, F9066 |
| government policy, | F9005 | England and Wales, | F9052–F9062 |
| health, protection of, | F9001 | exemptions, | F9060, F9067 |
| hygiene, | | four hour rule, | F9059 |
| contamination, | F9036, F9038, F9047 | gelatine, | F9069 |
| cooking, | F9036 | hot food, | F9061, F9065 |
| documentation and records, | F9037 | mail order foods, | F9054 |
| equipment, | F9038, F9043 | raw food for processing, | F9057 |
| food room, in, | F9038, F9040 | regulations as to, | F9002, F9050, F9053 |
| food handler, | F9038, F9048 | reheating, | F9068 |
| food waste, | F9038, F9044 | | |

*Index*

| | | | |
|---|---|---|---|
| short 'shelf life' products, | F9055 | **GAS SAFETY,** | |
| time control and, | F9036, F9051 | appliances, | |
| variations from 8°C storage, | F9058 | CE mark etc, | G1015 |
| themes, | F9001 | installation of, | G1017 |
| trading standards officer, role | | room-sealed, | G1018 |
| of, | F9005 | unsafe, | G1017, G1033 |
| training of staff, | F9081 | competent person, | |
| veal production, | F9077 | qualifications, | G1014 |
| vegetarians, | F9076 | quality control, | G1013 |
| weights and measures, | F9004, F9005 | supervision, | G1014 |
| **FOOTWEAR,** | | confined spaces, | C8081 |
| capability, | M3007 | Council of Registered Gas | |
| manual handling, | M3004, M3007 | Installers, | G1013 |
| personal protective | | court cases, | G1027–G1033 |
| equipment, | P3022, P3025 | definition of gas, | G1001 |
| **FORK-LIFT TRUCKS,** | | Director General of Gas | |
| counterbalance trucks, | L3003 | Supply, | G1002 |
| failure, common causes of, | L3041 | employer's duties, | G1001, G1011 |
| generally, | L3001 | escape of gas, | |
| machinery safety, | L3045 | reporting of, | G1005 |
| narrow aisle trucks, | L3003 | responsible person's | |
| operators, | | duties, | G1005, G1012 |
| safety requirements, | L3047 | strict product liability, | G1001 |
| training, | L3050 | factories, in, | G1034 |
| order pickers, | L3003 | Gas Consumers' Council, | G1002 |
| overturning, | M1008 | generally, | G1001 |
| pedestrian-operated | | health and safety manager's | |
| stackers, | L3003, L3051 | duties, | G1001 |
| reach trucks, | L3003 | installation and use, | |
| safe operation | | appliances, | G1010, G1026 |
| requirements, | L3046–L3049 | competent persons, | G1013, G1014 |
| safety measures, | L3045 | court cases, | G1027–G1033 |
| driver, | L3047 | dampers, | G1019 |
| system of work, | L3049 | duties as to, | G1011 |
| truck, | L3048 | emergency controls, | G1021 |
| training of operators, | L3050 | escape of gas, | G1012 |
| types, | L3003 | flues, | G1019 |
| **FRAGILE ROOFING** | | gas fittings, | G1020 |
| **MATERIALS,** | C8043, C8146 | inspection, | G1025 |
| **FREIGHT CONTAINERS,** | | maintenance | |
| accident reporting, | A3035 | requirements, | G1024, G1025 |
| **FRICTION AND** | | materials, | G1015 |
| **ABRASION,** | M1025 | meters, | G1022 |
| **FUMES,** | | pipework, | G1023 |
| confined spaces, | C8081 | pressure fluctuations, | G1035 |
| polluting agent, as, | V3003 | regulations as to, | G1026 |
| ventilation, | V3002 | regulators, | G1022 |
| visitors and the public, protection | | responsibilities as to, | G1011 |
| of, | C8134 | room-sealed appliances, | G1018 |
| welding, | V3002 | safety precautions, | G1016 |
| | | testing, | G1024 |
| **GAMING HOUSES, FIRE** | | work, scope of, | G1010 |
| **PRECAUTIONS AND,** | F5072 | workmanship, | G1015 |
| **GANGMASTERS, LICENSING** | | landlords' duties, | G1001, G1010, |
| **OF,** | intro 26 | | G1024, G1025, G1029 |

*Index*

| | |
|---|---|
| leaks, notification of, | G1005, G1026 |
| legislation, | G1001 |
| maintenance | |
|    requirements, | G1024, G1026 |
| management, | G1028 |
| pipelines, | G1009 |
| reconnection, prohibition of, | G1008 |
| reporting gas | |
|    incidents, | A3002, A3020, A3021 |
| rights of entry, | |
|    disconnection, for | G1007 |
|    generally, | G1006 |
|    inspection, for, | G1007 |
|    reconnection, prohibition of, | G1008 |
|    testing, for, | G1007 |
| Scotland, in, | G1011 |
| self-employed, duties of, | G1011 |
| supply of gas, | |
|    domestic customer, responsibilities of, | G1002 |
|    incidents, reporting of, | G1005 |
|    management of, | G1003–G1005 |
|    owner's duties, | G1002 |
|    relevant authority, role of, | G1002 |
|    scope of legislation, | G1002 |
|    training and competence, | G1014 |
| transport, | D0417 |
| **GATES,** | |
| construction and building works, | C8052 |
| **GENDER,** | |
| capability, | M3007 |
| manual handling, | M3007 |
| **GENETICALLY MODIFIED FOODS,** | F9020, F9077 |
| **GENETICALLY MODIFIED ORGANISMS,** | R3015 |
| **GEOGRAPHY,** | |
| disaster site, around, | D6029 |
| **GLOBAL LEGAL FRAMEWORK,** | Intro 8 |
| **GLOBAL REPORTING INITIATIVE,** | E16024 |
| **GLOVES,** | V5019 |
| **GOVERNMENT POLICY,** | |
| disaster and emergency, as to, | D6029 |
| **GREENPEACE STOPESSO CAMPAIGN,** | E16009 |
| **GRIEVANCE PROCEDURES,** | |
| accompanied, right to be, | E14001 |
| contract of employment in, | E14005, E14007 |
| statutory procedure, | E14008 |

| | |
|---|---|
| **GUARDS,** | M1027–M1035 |
| adjustable, | M1030 |
| automatic, | M1031 |
| design, | M1028 |
| drills, | M1030 |
| electrical interlocking switches, | M1032, M1034 |
| fixed, | M1007, M1029–M1030 |
| heights, working at, | W9017 |
| interlocking, | M1032–M1035 |
| movable, | M1032 |
| selection, | M1028 |
| self-adjusting, | M1030 |
| standards, | M1028, M1032 |
| telescopic, | M1030 |
| trapped key interlocking, | M1033–M1034 |
| **GUIDANCE NOTES, HSE,** | App 2 |
| asbestos, | A5030, C8019, H2139, R3013 |
| construction and building operations, | C8029 |
| control of substances hazardous to health, | H2105, H2113, H2120, H2133, H2135 |
| dangerous substances, | C10005 |
| disability discrimination, | V12037 |
| display screen equipment, | D8201–D8202, D8205–D8208, D8211–D8213, D8216–D8220 |
| electricity, | E3009–E3012 |
| family undertakings, | V12011 |
| fire, | C8053 |
| hazardous substances, | H2149 |
| HSE Direct, | R3030 |
| lead, | H2143 |
| machinery safety, | M1007 |
| management of health and safety, | M2023, M2027 |
| manual handling, | M3002, M3012, M3017, M3019 |
| new or expectant mothers, | V12019, V12020–V12024 |
| permit to work systems, | S3013 |
| risk assessment, | R3030 |
| stress, | S11009 |
| training and competence, | T7007 |
| vehicles, | C8051 |
| visitors and the public, protection of, | C8118 |
| website, downloading from the, | R3030 |

| | |
|---|---|
| windows and window cleaning, | W9042 |

**HAND ARM VIBRATION SYNDROME (HAVS),**
| | |
|---|---|
| action plans, | V5021 |
| advice, | V5017–V5018 |
| aggravating circumstances, | V5015 |
| carpal tunnel syndrome, | V5004 |
| causes, | V5010 |
| civil liability, | O1027 |
| daily dose, | O1023, V5015 |
| diseases caused by, | V5010 |
| effects, | V5010, V5014 |
| emissions, testing, | V5020 |
| ergonomic approach to, | E17012, E17033 |
| EU law, | Intro 11, V5015 |
| examples of causes of, | V5010 |
| exposure action value, | V5015 |
| frequency, | O1022, V5017 |
| generally, | O1019 |
| gloves, | V5019 |
| guidance, | V5017 |
| health surveillance, | V5015 |
| magnitude, | O1021, V5017 |
| managing exposure, | V5018 |
| measurement factors, | O1020 |
| noise, | V5003 |
| occupations, examples of, | V5014 |
| personal protection against, | V5019–V5022 |
| Physical Agents (Vibration) Directive, | V5015 |
| prescription of, | V5014 |
| preventative steps, | O1026 |
| risk assessment, | V5015, V5021–V5022 |
| symptoms, | O1025 |
| tools, vibration magnitudes of, | O1021, V5017, V5020 |
| vibration, | |
| daily dose, | O1023 |
| excessive effect of, | O1024 |
| frequency, | O1022 |
| magnitude, | O1021 |
| measurement factors, | O1020 |
| warnings, | V5020–V5021 |

**HAND SIGNALS,** W11032, W11042

**HAND TOOLS,**
| | |
|---|---|
| construction and building works, | C8015 |

**HARASSMENT,**
| | |
|---|---|
| criminal offence, | H1734, H1739 |
| disability, on grounds of, | H1735, H1736, H1742–H1746 |
| EOC Code of Practice, | H1741–H1750 |
| future developments, | H1749, H1750 |
| general law as to, | H1701 |
| racial, *see* Racial Harassment | |
| sexual, *see* Sexual Harassment | |
| tort of, | H1734, H1739 |
| training and competence, | H1747 |
| victimisation, | H1716–H1717 |

**HARBOUR AREAS,**
| | |
|---|---|
| liquefied gases in, | H7023 |

**HAZARDOUS SUBSTANCES,** *see also* Control of substances
| | |
|---|---|
| hazardous to health, Dangerous substances | H2101–H2150 |
| asbestos, *see* Asbestos, | |
| biological problems, | H2103 |
| chemical substances, | H2137 |
| control regulations, | |
| biological agents, | H2113 |
| carcinogens, | H2113 |
| entry, routes of, | H2102 |
| harm caused by, | H2102–H2103 |
| HSE guidance, | H2149 |
| ingestion, | H2102 |
| inhalation, | H2102 |
| ionising radiation, | H2148 |
| lead, | H2143 |
| machinery safety, | M1025 |
| major accident hazards, control of, | H2142 |
| occupational ill-health, examples of, | H2103 |
| personal protective equipment, | P3017 |
| poisoning, | H2103 |
| references, | H2149 |
| regulations, | H2136–H2148 |
| respiratory problems, | H2103 |
| risk assessments, | R3002, R3003 |
| skin, | H2102, H2103 |
| visitors and the public, protection of, | C8135 |

**HEAD PROTECTION,**
| | |
|---|---|
| British Standard, | C8113, P3025 |
| construction and building works, | C8108–C8116 |
| control of sites, persons in, | C8110–C8111 |
| employees, duties of, | C8112 |
| employers, duties of, | C8109 |
| ensuring the wearing of, | C8109 |
| exceptions, | C8115 |

*Index*

| | |
|---|---|
| fatal accidents, | C8108 |
| maintenance, | C8109 |
| personal protective equipment, | C8112 |
| procedures and rule-making, | C8111 |
| provision of, | C8109 |
| purpose of, | C8108 |
| self-employed, | C8114 |
| standards, | C8113, P3025 |
| suitable, | C8113 |
| supervision, | C8111 |
| visitors, | C8116 |

**HEADROOM,** M3006

**HEATHROW RAIL TUNNEL COLLAPSE,** Intro 15

**HEALTH AND SAFETY COMMISSION,**
| | |
|---|---|
| communications, | Intro 6 |
| composition of, | E15002 |
| functions of, | E15001–E15004 |
| construction and building works, | C8001, C8028 |
| Department of Work and Pensions, | Intro 7, Intro 12 |
| European Union, | Intro 10 |
| functions of, | E15001–E15004 |
| heights, working at, | W9004 |
| performance review of, | intro 6 |
| priorities, | intro 6 |
| resources of, | Intro 6, Intro 7 |
| *Revitalising* targets | Intro 6, Intro 7 |
| safety management, assistance with, | Intro 6 |
| strategy statement, | Intro 6, Intro 7 |
| training and competence, | T7001, T7008 |

**HEALTH AND SAFETY EXECUTIVE,**
| | |
|---|---|
| accident investigations, | A3031 |
| Area Offices, | A3037 |
| communications, | intro 6 |
| composition of, | E15002 |
| construction and building works, | C8001, C8028 |
| Construction Division Inspectorate, | C8001 |
| enforcing authority, as, | E15007 |
| fees payable to, | E1505 |
| fire and fire precautions, | F5061 |
| fire certificates, | F5061 |
| functions of, | E15001, E15002, E15005 |
| guidance, *see* Guidance Notes, HSE | |
| Infoline, | Intro 28 |
| inspectors' powers, | E15001 |

| | |
|---|---|
| partnerships, development of closer, | Intro 6 |
| performance review of, | Intro 6 |
| Policy Section, | Intro 13 |
| Priority Programmes, | Intro 6, Intro 13 |
| publications, | E1506, App 2 |
| regional offices, | A3039 |
| resources of, | Intro 6, Intro 7 |
| *Revitalising* targets | Intro 6, Intro 7 |
| safety management, assistance with, | Intro 6 |
| strategy statement, | Intro 6 |
| stress, | Intro 13, S11009, S11010 |
| training and competence, | T7007, T7008 |
| transfer of responsibilities, | E15011 |
| working time, | W10038–W10039 |

**HEALTH AND SAFETY LAW,**
| | |
|---|---|
| absolute duty, | Intro 22 |
| charge, duty not to, | Intro 21 |
| construction and building works, | C8020–C8023 |
| employees, duties of, | Intro 21 |
| employers, duties of, | Intro 21 |
| Health and Safety at Work Act 1974, | Intro 21–22 |
| application, | Intro 21, D6006 |
| duties under, | D6006, D6032, D6033 |
| improvement notice, | |
| contents of, | E15014 |
| contravention of, | E15001, E15015, E15017, E15030, E15036 |
| Crown, position of, | E15045 |
| prohibition notice contrasted, | E15015, E15016 |
| prosecution, coupled with, | E15018 |
| public register, | E15029 |
| service of, | E15014 |
| use of, | E15001, E15013 |
| inference and misuse, | Intro 21 |
| levels of duties, | Intro 22 |
| management of health and safety, | |
| arrangements, | D6007 |
| risk assessment, | D6007, R3002–R3004 |
| manufacturers, duties of, | Intro 21 |
| policy statements, | Intro 21 |
| practicable, meaning of, | Intro 22 |
| precautions, | Intro 22 |
| premises, duties relating to, | Intro 21 |
| reasonably practicable, meaning of, | Intro 22 |

*Index*

| | | | |
|---|---|---|---|
| Regulations, | Intro 23, D6007 | falls, | W9001, W9007, W9008–W9010 |
| risk assessment, | Intro 22 | construction | |
| self-employed, | Intro 21 | sites, | W9014–W9015, W9019–W9022 |
| stress-related illnesses, | S11006, S11007 | equipment, maintenance of, | W9024 |
| suppliers, duties of, | Intro 21 | personal suspension | |

**HEALTH AND SAFETY POLICIES,** Intro 27

equipment, W9019–W9020
fencing, W9010

**HEALTH AND SAFETY POLICY STATEMENT,**

fragile materials, W9008, W9025, W9028

arrangements to be considered, S7003, S7006
guard rails, W9017
format, S7003

Health and Safety Commission, W9004

organisation, S7003, S7005
high risk situations, W9015
policy, S7008
hill walkers, W9051

*Strategy for Workplace Health and Safety,* Intro 6

information, W9004
inspections, W9027

written statements of, E14012, S7001–S7008
kiers, W9013
ladders, C8011, W9002, W9023

**HEALTH SURVEILLANCE,**

construction and building operations, C8011

control of substances hazardous to health, H2133
enforcement campaign on, W9002

hand-arm vibration syndrome, V5015
fixed, W9012, W9013

management of health and safety, M2009
mobile tower scaffolds, W9027

personal suspension equipment, W9019–W9020, W9022

manual handling, M3015
periodic examinations, O1049
inspection of, W9027
pre-employment, O1049
pits, W9013
risk assessment, R3022
precautions, W9005, W9008–W9009, W9015
training and competence, T7007
whole-body vibration, V5028
collective, W9016

**HEARING LOSS,** *see also* Deafness
review of, W9028
noise-induced, N3002, O1010–O1017
reasonably practicable, so far as, W9008, W9051
recreation, W9008

**HEIGHT,**
regulations on, W9008, W9019

capability, M3007
risk assessment, W9007, W9014, W9028

manual handling, M3007

**HEIGHTS, WORKING AT,** *see also* Windows and window cleaning W9001–W9051
risk management, W9002–W9003
roof work, W9028

access, W9001
rope working, single or double, W9019

approved code of practice, W9004, W9007, W9009
safety harnesses, W9014
safety nets, W9028

climbing instruction, W9008
scaffolds, installation and erection of, W9022, W9027

construction sites, C8007, C8011–C8012, W9002–W9004, W9007–W9008, W9014–W9029
statutory requirements, W9007–W9013
surfaces, fragile, W9008

covers, W9011
tanks, W9013

emergencies, W9005
toe caps, W9017

equipment, W9008, W9024

EU law, W9008, W9051

falling objects, W9026

IND/39

*Index*

| | |
|---|---|
| Use of Work Equipment Directive, | W9008 |
| vats, | W9013 |
| working platforms, | C8012–C8013, W9008, W9014–W9020 |
|   catchment platforms, | C8013 |
|   construction and building operations, | C8012–C8013, C8064 |
|   definition, | W9016 |
|   guard rails, | W9017 |
|   height to base ratios, | C8012 |
|   housekeeping, | C8013 |
|   inspection | W9027 |
|   lifting operations, | C8013 |
|   load, | W9018 |
|   materials, | C8012 |
|   mobile, | C8012, W9015 |
|   stability, | W9018 |
|   toe caps, | W9017 |
| **HERALD OF FREE ENTERPRISE,** | D6016 |
| **HERMES PRINCIPLES,** | E16007 |
| **HIGH PRESSURE FLUID INJECTIONS,** | M1025 |
| **HIGHLY FLAMMABLE LIQUIDS,** | |
|   ventilation regulations, | V3006 |
| **HIGHWAYS,** | |
|   construction and building operations, | C8151 |
| **HILL WALKERS,** | W9051 |
| **HIRE, CONTRACT OF,** | |
|   product liability, | P9048, P9055 |
| **HIRE PURCHASE,** | |
|   product liability, | P9048, P9055 |
| **HOISTS AND LIFTING MACHINERY,** *see* Lifting | |
| **HOLIDAYS,** *see* Annual Leave | |
| **HOME CARE,** | |
|   violence and, | S11001 |
| **HOMEWORKERS,** | |
|   display screen equipment, | D8213 |
| **HOMOSEXUALS,** | |
|   harassment of, | H1705 |
| **HOSPITAL,** | |
|   fire and fire precautions, | F5035 |
|   occupier's liability, | O3018 |
|   social security, | C6019 |
| **HOST EMPLOYERS,** | |
|   management of health and safety, | M2015 |
| **HOT DESKING,** | D8218 |
| **HOURS OF WORK,** *see* Working Time | |
| **HUMAN COST OF DISASTERS,** | D6004, D6021 |
| **HUMAN ERRORS,** | |
|   major incidents, contribution to, | D6025 |
| **HUMAN RIGHTS,** | |
|   burden of proof and, | E15031 |
| **HUMIDITY,** | M3006 |
| **HYGIENE,** | |
|   contamination, | F9036, F9038, F9047 |
|   cooking, | F9036 |
|   documentation and records, | F9037 |
|   equipment, | F9038, F9043 |
|   floors, | W11010 |
|   food, | F9002, F9005, F9021, F9034–F9049 |
|   food room, in, | F9038, F9040 |
|   food handler, | F9038, F9048 |
|   food waste, | F9038, F9044 |
|   HACCP system, | F9021, F9035–F9038 |
|   industry guides, | F9049 |
|   meaning of, | F9035 |
|   meat, | F9005 |
|   personal hygiene, | F9038, F9046 |
|   'pre-requisites', | F9038, F9039 |
|   regulations as to, | F9002, F9034, F9035 |
|   temporary businesses, | F9038, F9041 |
|   transport of food, | F9038, F9042 |
|   vending machines, | F9038, F9041 |
|   water supply, | F9038, F9045 |
| **ILLUMINANCE,** | |
|   average, | L5010 (Table 26) |
|   light measuring instruments, | L5009 |
|   lighting, *see* Lighting, | |
|   maximum ratios, adjacent areas, | L5011 (Table 27) |
|   minimum measured, | L5010 |
|   ratios, | L5011 |
|   safety signs regulations, | W11034 |
|   technical measurement of, | L5008 |
| **ILLUMINATING ENGINEERING SOCIETY,** | |
|   limited glare index, | L5014 |
| **IMPROVEMENT NOTICES,** | |
|   appeal against, | E15001, E15019–E15025 |
|   enforcement powers of inspectors, | E15006, E15007, E15011 |
|   fire and fire precautions, | F5046–F5048 |
|   public register, | E15022 |
| **INCAPACITY BENEFIT,** | C6002, C6015 |

| | |
|---|---|
| INCIDENT CONTACT CENTRE, | A3004 |
| INCOME SUPPORT, | C6015 |
| INDEMNITY, | |
| directors, for, | E15041 |
| inspectors, for, | E15028 |
| INDEPENDENT CONTRACTORS, | |
| construction and building operations, | C8021, C8151 |
| INDEX-LINKED GOVERNMENT SECURITIES, | C6021 |
| INDUCTION PROGRAMMES, | T7025 |
| INDUSTRIAL ATMOSPHERIC POLLUTION, *see* Air pollution | |
| INDUSTRIAL INJURIES BENEFIT, | C6002, C6003, C6008, C6015 |
| INDUSTRIAL PREMISES, | |
| health and safety law enforcement of, | E15007 |
| INDUSTRIAL RELATIONS, | |
| policy and procedure, | E17043 |
| INEXPERIENCED WORKERS, | V12038–V12039 |
| capability, | M3007 |
| children and young persons, | V12013 |
| manual handling, | M3007, V12005 |
| statements, | E3023 |
| supervision, | V12039 |
| training, | V12038–V12039 |
| INFLATION, | C6027, C6043 |
| INFORMATION PROVISION, | App 5 |
| accident investigations, | A3030, A3033 |
| children and young persons, | V12010, V12014 |
| disaster or emergencies, | D6030, D6031 |
| display screen equipment, | D8209 |
| machinery safety, | M1007 |
| major accident hazards, | C10017 |
| management of health and safety, | M2013, M2030 |
| manual handling, | M3019 |
| permit to work systems, | S3014 |
| radiation emergencies, | C10021 |
| INGESTION AND INHALATION OF HAZARDOUS SUBSTANCES, | H2102 |
| INJUNCTION, | |
| breach of contract, in respect of, | H1736 |
| tort, remedy in, | H1733, H1736 |
| INLAND REVENUE, | |
| disclosures to, | E14009 |

| | |
|---|---|
| INSOLVENCY, | |
| employers' liability insurance, | E13002, E13024 |
| INSPECTION, | |
| asbestos, | A5015, R3013 |
| construction and building operations, | C8064, C8065 |
| machinery safety, | M1007 |
| management of health and safety, | M2035 |
| personal suspension equipment, | W9027 |
| windows and window cleaning, | W9027 |
| INSTITUTE OF ELECTRICAL ENGINEERS, | E3001 |
| INSTITUTION OF OCCUPATIONAL HEALTH AND SAFETY (IOHS), | T7011–T7014, T7021–T7022, T7027–T7028 |
| INSTITUTIONAL CARE, | C6028 |
| INSURANCE, | |
| asbestos, | E1301 |
| business continuity planning, | B8007 |
| civil liability, | Intro 24 |
| construction and building operations, | C8141 |
| damages, | C6053 |
| disaster and emergency management, | D6002, D6004, D6023, D6030 |
| employers' liability, *see* Employers' Liability Insurance, | |
| environmental management and, | E17043 |
| fire and fire precautions, | F5080 |
| management of health and safety, | M2001 |
| loss adjusters, | D6002, D6034, D6052 |
| public liability insurance, | O3003 |
| visitors and the public, protection of, | C8121 |
| INTENT, STATEMENTS OF, | Intro 27 |
| INTERCEPTION OF COMMUNICATIONS, | |
| regulation of, | E14002 |
| INTERNATIONAL ATOMIC ENERGY AGENCY, | D0412 |
| INTERNATIONAL LABOUR ORGANISATION, | Intro 8 |

*Index*

INVESTIGATIONS, *see* Accident investigations
INVESTMENT,
  damages, C6027
  socially responsible (SRI), E16001, E16007
IONISING RADIATION,
  children and young persons, V12015
  new or expectant mothers, V12004, V12020
  risk assessment, H2148, R3015
ISOLATION PROCEDURES, M1007, S3023

JCT FORMS OF CONTRACTS, C8140–C8141, F5029
JOB MARKET, CHANGES IN THE, Intro 6
JOB SAFETY ANALYSIS, S3009
JOB SAFETY INSTRUCTIONS, S3010
JOINT CONSULTATION IN SAFETY,
  generally, *see* Consultation, Joint
  safety committees, *see* Safety Committees
  safety representatives, *see* Safety Representatives

KEGWORTH AIR DISASTER, D6067
KIERS, W9013
KILLING, *see also* Corporate manslaughter
  corporate killing, D6022
  gross carelessness, by, D6022
  reckless killing, D6022
KINGS CROSS DISASTER, D6030

LABELLING,
  chemicals, H2137
  dangerous substances, transport of, D0423–D0426, D0437
  food, F9003, F9012, F9075, F9078
  performance, Intro 4
LADBOKE GROVE RAIL DISASTER, D6027, E17002
LADDERS,
  construction and building operations, C8011
  enforcement campaign on, W9002
  falls, W9041
  fixed, W9012, W9013
  portable, W9047
  precautions, W9002, W9047
  step, W9029
  travelling, W9042, W9047
  windows and window cleaning, C8011, W9002, W9023, W9041, W9042, W9047
  enforcement campaign on, W9002
  falls, W9041
  fixed, W9012, W9013
  portable, W9047
  precautions, W9002, W9047
  travelling, W9042, W9047
LANDFILL TAX, E16006
LANDLORD,
  gas safety requirements, G1001, G1010, G1024, G1025, G1029
LAPTOP COMPUTERS, D8129, E17063
LEAD,
  approved codes of practice, H2143
  children and young persons, V12015
  employers' duties, V3008
  exhaust ventilation system, V3008
  fumes, control of, V3007
  HSE guidance, H2143
  new or expectant mothers, V12004, V12022
  occupational disease, O1043
  poisoning, O1043
  risk assessment, R3015
  women, employment of, E14002
  young persons and, E14002
LEASES,
  asbestos, A5021
LEAVE, *see* Annual Leave
LEGAL ADVISER,
  disclosures to, E14009
LEGAL AID, C6021
LEGAL FRAMEWORK FOR HEALTH AND SAFETY, Intro 7–11, Intro 19–24
LEGAL REPRESENTATIVES,
  accident reporting, A3029
LEGAL SERVICES COMMISSION, C6021
LEVEL CROSSINGS,
  accident reporting, A3035
LICENSING,
  asbestos, A5026, E15005

*Index*

| | | | |
|---|---|---|---|
| construction and building operations, | C8141 | pulley blocks, | L3010 |
| contractors, asbestos removal and, | A5026 | ropes, | L3009, L3036, L3037 |
| gangmasters, | intro 26 | slings, | |
| licensed premises, fire precautions and, | F5069–F5075 | chain, | L3010 |
| | | generally, | L3010 |
| Morecambe Bay cockle pickers tragedy, | Intro 26 | rope, | L3010 |
| | | safe working load, | L3039 |

**LIFE EXPECTANCY, DAMAGES AND,** C6029, C6034

**LIFTING,**
- accessory for lifting, L3012
- accident reporting, A3035
- categories of machinery, L3001
- chains, L3008
  - failure, causes of, L3038
- construction sites, lifting operations on, C8013, C8020, L3011
- cranes, *see* Cranes
- defects, L3030
- examination, thorough, L3029
- eyebolts, L3010
- failure, causes of,
  - chains, L3038
  - cranes, L3034
  - fork-lift trucks, L3041
  - hoists and lifts, L3035
  - hooks, L3040
  - ropes, L3036, L3037
  - slings, L3039
- fork-lift trucks, *see* Fork-lift Trucks
- generally, L3001, L3004, L3023
- hoists and lifts,
  - examples, L3002
  - failure, causes of, L3033, L3035
  - meaning, L3002
- hooks, L3010
  - failure, causes of, L3040
- information,
  - keeping of, L3032
  - prescribed, L3031
- inspection, L3029
- installation, L3026
- keeping information, L3032
- lifting tackle,
  - chain slings, L3010
  - chains, L3008, L3038
  - examples, L3007
  - eyebolts, L3010
  - failure, causes of, L3039
  - hooks, L3010, L3040
  - meaning, L3007

- lifts, *see* Lifts
- machinery safety, M1004
- manual handling, M3003, M3010
- manually operated lifting devices, M3010
- marking, L3027
- mechanical lifting devices, M3010
- organisation of operations, L3028
- person, lifting equipment for lifting, L3025
- positioning, L3026
- prescribed information, L3031
- pulley blocks, L3010
- reports, L3030
- ropes,
  - failure, causes of, L3036, L3037
  - fibre, L3036
  - generally, L3009
  - wire, L3037
- safe stacking of materials, L3042, L3045
- self-employment, L3023
- slings,
  - chain, L3010
  - generally, L3010
  - rope, L3010
  - safe working load, L3039
- stability, L3024
- statutory requirements, L3012
- strength, L3024
- thorough examination, L3029
- tools, M3011
- working platforms, C8013

**LIFTS,**
- *see also* Lifting
- breach of regulations, L3017
- conformity assessment procedures, L3020
- documents, retention of, L3018
- exceptions from regulations, L3013, L3019
- freedom from obstruction of lift shafts, L3018
- general requirements, L3014–L3016
- generally, L3001
- information, supply of, L3018
- meaning, L3012

IND/43

| | | | |
|---|---|---|---|
| notified bodies, | L3022 | maximum measured, | L5012 (Table 2) |
| penalties for breach of | | natural, | L5002, L5007 |
| regulations, | L3017 | qualitative aspects, | L5014 |
| placing on the | | reflected glare, | L5015 |
| market, | L3014, L3019, L3021 | specific processes, statutory requirements, | L5003–L5006 |
| putting into service, | L3014 | standards of, | L5009–L5013 |
| regulation, | L3013 | statutory requirements, | |
| requirements fulfilled by person placing lift/safety component on market, | L3021 | general, | L5002 |
| | | stroboscopic effect, | L5020 |
| | | tripping, | M3006 |
| retention of documents, | L3018 | VDUs, | L5003 |
| safety components for use in, | L3013, L3015 | workplace, | W11006 |
| | | **LIMITATION OF ACTIONS,** | L5003 |
| shafts, freedom from obstruction of, | L3018 | **LIQUEFIED FLAMMABLE GASES,** | |
| statutory requirements, | L3012 | definition, | H7008 |
| supply of lift or safety component, | L3016, L3019 | **LIQUEFIED PETROLEUM GASES,** | |
| | | ventilation regulations, | V3006 |
| **LIGHTING,** | | **LOADING,** | |
| brightness, | L5017 | bays, | A1001 |
| colour rendition, | L5019 | vehicles, dangerous substances and, | D0440 |
| confined spaces, | C8081 | | |
| construction and building operations, | C8015, C8060, L5004, L5005 | workplace access, generally, | A1001 |
| | | **LOADING VEHICLES** | A1006 |
| design, | L5014 | **LOBBY GROUPS,** | Intro 14 |
| diffusion, | L5018 | **LOCAL AUTHORITY,** | |
| disability glare, | L5015 | contractors, | Intro 17 |
| discomfort glare, | L5015 | disaster and emergency plans, | D6030 |
| display screen equipment, | D8216 | enforcement of legislation by, *see* Enforcement | |
| distribution, British Zonal Method, | L5016 | | |
| economic lamp replacement policy, | L5013 | escapes, | F5049 |
| | | fire and fire precautions, | F5033–F5035, F5059 |
| electric, | L5008 | | |
| electrical equipment, | E3022, L5006 | occupier's liability, | O3001 |
| ergonomic design and, | E17004, E17031, E17035 | prohibition notices, | F5049 |
| | | standards, | Intro 17 |
| fitments, maintenance of, | L5013 | **LOCAL EXHAUST VENTILATION PLANT,** | H2130 |
| fluorescent tubes, | L5008, L5020 | | |
| glare, | L5015 | **LONE WORKERS,** | V12027–V12031 |
| illuminance, | | access, | V12029 |
| average, | L5011 (Table 1) | categories of, | V12027–V12028 |
| light measuring instruments, | L5010 | communication, | V12030 |
| | | confined space, rescues from, | V12029 |
| minimum measured, | L5011 (Table 1) | controls, operation of, | V12029 |
| | | electrical work, live, | V12029 |
| ratios, | L5012 | emergency arrangements, | V12030 |
| technical measurement of, | L5009 | equipment, | V12029 |
| lumen, meaning, | L5009 | first aid, | F7009, V12029 |
| machinery safety, | L5004, M1007 | fixed workplaces, in, | V12028 |
| maintenance of fitments, | L5013 | maintenance, | V12029 |
| manual handling, | M3006, M3014 | manual handling, | V12005, V12029 |
| | | monitoring, | V12031 |

| | |
|---|---|
| other employers, premises belonging to, | V12028 |
| precautions, | V12030–V12031 |
| repairs, | V12029 |
| rescues, | V12029 |
| risk assessment, | R3027, V12027, V12030 |
| risks for, | V12029 |
| controlling, | V12030 |
| safety harnesses, rescuing people who have fallen whilst wearing, | V12029 |
| violence or aggression, risks of, | V12029 |
| vulnerable, where staff feel highly, | V12029 |
| wider community, workers in the, | V12028 |
| **LOSS ADJUSTERS**, | D6002, D6034, D6052 |
| **LOSS OF AMENITY, DAMAGES FOR**, | C6030, C6034, C6044 |
| **LOSS OF EARNINGS, DAMAGES FOR**, | C6026, C6043 |
| **LOSS OF FUTURE EARNINGS, DAMAGES FOR**, | C6027 |
| **LOST YEARS, DAMAGES FOR**, | C6029 |
| **LUMP SUMS**, | C6021, C6026, C6040, C6048 |
| **LUNG CANCER**, | A5002–A5003, O1003, O1006 |
| **MACHINERY SAFETY**, | M1001–M1036 |
| abrasion, | M1025 |
| Approved Code of Practice, | M1007 |
| behavioural measures, | M1026 |
| buying machinery, guidance on, | M1007 |
| CE marks, | M1009, M1015–M1016, M1026 |
| checklist, | M1025 |
| conformity assessment procedure, | M1013–M1014 |
| continuing hazards, | M1024–M1025 |
| control systems, | M1007 |
| criminal offences, | M1018 |
| crushing, | M1025 |
| cutting, | M1025 |
| dangerous parts of machinery, | M1007 |
| declarations of conformity, | M1015–M1016 |
| definitions, | |
| harm, | M1023 |
| hazard, | M1023 |
| machine, | M1011 |
| design, | M1002, M1009, M1026, M1028 |
| deterring/impeding device, | M1036 |
| drawing-in, | M1025 |
| electrical hazards, | M1025 |
| enabling (control) device, | M1036 |
| entanglement, | M1025 |
| equipment, | M1007 |
| essential health and safety requirements, | M1003, M1007, M1009, M1012–M1015, M1018, M1020 |
| EU law, | M1001–M1004, M1007, M1009–M1020, M1026 |
| examples of hazards, | M1025 |
| foreseeable misuse, | M1002 |
| fork-lift trucks, | M1008 |
| friction, | M1025 |
| guards, *see* Guards | M1007 |
| guidance, HSE, | M1007 |
| harm, definition of, | M1023 |
| hazard analysis, | M1024 |
| hazardous substances, | M1025 |
| Health and Safety at Work etc Act 1974, | M1002 |
| high/low temperature, | M1007 |
| hold-to-run control device, | M1036 |
| identification of hazards, | M1024 |
| impact hazards, | M1025 |
| incorporation, declarations of, | M1015 |
| information, | M1007 |
| inspection, | M1007 |
| instructions, | M1007 |
| isolation from energy sources, | M1007, M1034 |
| legal requirements, | M1001–M1021 |
| lifting/moving persons, | M1004 |
| lighting, | M1007 |
| limited movement control device, | M1036 |
| limiting device, | M1036 |
| machine, definition of a, | M1011 |
| Machinery Safety Directive, | M1003, M1009–M1018 |
| maintenance, | M1007 |
| manufacturers, | M1001–M1002, M1009 |
| markings, | M1007 |
| mechanical hazards, | M1025 |

*Index*

| | |
|---|---|
| mechanical interlocking, | M1035 |
| mobile work equipment | M1008 |
| procedural measures, | M1026 |
| Product Liability Directive, | M1002 |
| programmable electronic systems, | M1007 |
| puncture, | M1025 |
| PUWER '92, | M1003–M1004 |
| PUWER '98, | M1003–M1008 |
| risk assessment, | M1006 |
| structure of, | M1005 |
| summary of, | M1007 |
| radiation hazards, | M1025 |
| reasonably practicable, | M1002 |
| risk assessment, | M1001, M1006, M1007, M1020, M1022–M1024, M1026 |
| risk-based approach, | M1003, M1019–M1020 |
| risk, definition of, | M1006, M1023 |
| risk reduction, options for, | M1026 |
| safe, definition of, | M1018 |
| safeguards and safety devices, types of, | M1027–M1035 |
| second-hand machinery, | M1007 |
| severing, | M1025 |
| shearing, | M1025 |
| significant risk, definition of, | M1006 |
| specific risks, | M1007 |
| stabbing, | M1025 |
| stability, | M1007 |
| standards, | M1001, M1003, M1007, M1010, M1012–M1015, M1019–M1020 |
| suitability, | M1007 |
| suppliers, | M1002, M1009–M1018 |
| Supply of Machinery (Safety) Regulations 1992, | M1009–M1018 |
| technical file, | M1014 |
| technical measures to be taken at the design stage, | M1026 |
| temperature, | M1007 |
| training, | M1007 |
| trapped key interlocking, | M1034 |
| traps, | M1025 |
| trip device, | M1036 |
| two-hand control device, | M1036 |
| types of safeguards and safety devices, | M1027–M1035 |
| Use of Work Equipment Directive, | M1003–M1004 |
| warnings, | M1007 |
| work activities hazards, | M1025 |
| work equipment, | M1007 |
| **MAGISTRATES,** | Intro 20 |
| **MAIL ORDER FOOD,** | |
| temperature control, | F9054 |
| **MAINTENANCE,** | |
| asbestos, | |
| building maintenance workers, exposure of, | A5006 |
| control measures, | 01003 |
| plans, | A5006, A5012 |
| records, | A5011 |
| business continuity planning, | B8011, B8033–B8037 |
| equipment, | W9024 |
| fire and fire precautions, | F5066 |
| head protection, | C8109 |
| hoists and lifts, | L3005 |
| lifts for carrying persons, | L3018 |
| lone workers, | V12029 |
| machinery safety, | M1007 |
| permit to work systems, | S3011 |
| records, | A5011 |
| training and competence, | T7028 |
| vehicles, | A1011 |
| **MAJOR ACCIDENT HAZARDS,** | C10001–C10022 |
| aggregation, | C10005 |
| ammonium nitrate, | C10003 |
| application of Regulations, | C10003, D6008 |
| background, | C10001 |
| breach of regulations, | D6008 |
| charging regime, | C10002 |
| classification of dangerous substances, | C10003 |
| competent authority, | C10002 |
| enforcement, | C10019 |
| inspections, | C10002 |
| powers to prohibit use, | C10018 |
| dangerous substances, | C10003 |
| aggregation, | C10005 |
| ammonium nitrate, | C10003 |
| classification, | C10003 |
| list, | C10003, D6014 |
| meaning, | C10003 |
| threshold levels, | C10003, D6014 |
| total quantity present at establishment, | C10003 |
| definition of terms, | C10002, D6003 |
| DEMS, *see* Disaster and Emergency Management Systems | |
| emergency plans, | H2142 |
| implementation, | C10015 |
| local authority charges, | C10016 |

IND/46

| | | | |
|---|---|---|---|
| off-site, | C10014, D6014, D6049 | top tier, | C10002 |
| | | fees payable by, | C10002 |
| on-site, | C10013, D6014, D6049 | meaning, | C10002 |
| | | powers to prohibit use, | C10018 |
| review, | C10015, D6014, D6049 | prohibit use, powers to, | C10018 |
| | | provision of information to public, | C10017 |
| testing, | C10015, D6036, D6059, D6063 | radiation emergencies. *See* Radiation Emergencies | |
| enforcement of Regulations, | C10019 | | |
| Environment Agency, role of, | C10001, C10002 | regime, safety, | C10002 |
| | | responsibility for provision, | C10002 |
| establishment, | C10003 | review, | |
| exemptions, | C10006 | emergency plans, | C10015 |
| explosives, | C10004 | safety report, | C10012 |
| fees, | C10002 | revision of safety report, | C10012 |
| hazardous substances, | H2142 | risk assessment, | R3015 |
| HSE, | | safety report, | |
| guide, | C10002 | before start-up, | C10011 |
| responsibilities of, | C10001, C10002 | competent authority, sent to, | C10011 |
| implementation of emergency plans, | C10015 | contents, | C10011 |
| | | demonstration of incapacity to create major accident, | C10011 |
| industrial chemical process, | C10003 | | |
| information provision to public, | C10017 | examination by competent authority, | C10011 |
| legislative history of control, | C10001 | existing establishments, | C10011 |
| local authority charges for review of emergency plan, | C10016 | information to be included, | C10011 |
| | | operator duties, | C10011 |
| lower tier duties, | C10002, C10007–C10010 | purpose, | C10011 |
| | | reference to other reports, | C10011 |
| best practice, | C10007 | requirement to produce, | C10011 |
| general duty, | C10007, D6008 | review, | C10012 |
| Major Accident Prevention Policy (MAPP), | C10008 | revision, | C10012 |
| | | time of preparation, | C10011 |
| notifications, | C10009 | Scottish Protection Agency, role of, | C10001 |
| report major accident, duty to, | C10010 | | |
| standard of care, | C10007 | Seveso I Directive, | D6014 |
| Major Accident Prevention Policy (MAPP), | C10008 | Seveso II Directive, | C10001 |
| | | testing emergency plans, | C10015 |
| management, *see* Disaster and Emergency Management Systems | | top tier duties, | C10002, C10011–C10016 |
| | | emergency plans, | |
| meaning of major accident, | C10002 | implementation, | C10015 |
| notifications, | | local authority charges, | C10016 |
| lower tier duties, | C10009 | off-site, | C10014, D6014, D6049 |
| objective of provision, | C10002 | | |
| off-site emergency plan, | C10014, D6014, D6049 | on-site, | C10013, D6014, D6049 |
| | | review, | C10015, D6008 |
| on-site emergency plan, | C10013, D6014, D6049 | testing, | C10015 |
| | | off-site emergency plan, | C10014 |
| | | on-site emergency plan, | C10013 |
| operator, duties, | | safety report. *See* safety report *above* | |
| lower tier, | C10002 | typical major accidents, | C10002 |

IND/47

*Index*

| | |
|---|---|
| **MANAGEMENT OF HEALTH** | |
| **AND SAFETY,** | M2001–M2039 |
| accident causation, | M2022 |
| accident investigations, | M2035 |
| accident reporting, | M2035 |
| advice, sources of, | Intro 28 |
| armed forces, national security and, | M2021 |
| arrangements, | M2008 |
| asbestos, | |
|   asbestos-containing materials, | A5013 |
|   duty as to, | C8019, C8068, H2139, R3013 |
|   leases, | A5021 |
|   long-term plans, | A5018 |
|   plans, | A5012, A5018–A5019, A5022 |
|   third parties, | A5020 |
| assistance with health and safety, | M2010 |
| audits, | M2036, M2039 |
| breach of statutory duty, | M2021 |
| British Standard, | M2023 |
| capabilities, | M2016 |
| causation, | M2022 |
| checklist, | M2024 |
| children, | M2013 |
| communication, | M2030, M2039 |
| competence, | M2010, M2031, M2039 |
| competent persons, | M2010, M2031 |
| contingency planning, | M2031 |
| continual improvement, | M2038 |
|   loop, | M2023 |
| control, | M2028, M2039 |
| co-operation, | M2014, M2029, M2039 |
| co-ordination, | M2014 |
| criminal liability, | M2021 |
| danger | |
|   areas, | M2011 |
|   procedure for serious and imminent, | M2011 |
| documentation, | M2034 |
| emergency plans and procedures, | M2033 |
| emergency services, contacts with, | M2012 |
| employees' duties, | M2017 |
| employers' duties, | M2003–M2004 |
| exemption certificates | M2021 |
| external services, contacts with, | M2012 |
| fatal accidents, | M2001 |
| financial reasons for, | M2001 |
| fire, | F5030 |
|   evacuation, | M2011 |
|   precautions, | F5030 |
| Health and Safety Commission/Executive, strategy of, | Intro 6 |
| health and safety management systems, | M2023–M2024 |
| health surveillance, | M2009 |
| HSE guidance, | M2023, M2027 |
| host employers undertakings, persons working in | M2015 |
| humanitarian reasons for, | M2001 |
| implementation | M2034, M2039 |
| improvement, continual, | M2038 |
| information, | M2013, M2030 |
| inspections, | M2035 |
| insurance, | M2001 |
| leases, asbestos and, | A5021 |
| legal reasons for, | M2001 |
| legal requirements, | M2002–M2004, M2033 |
| monitoring, | M2035 |
| new or expectant mothers, | M2019 |
| objectives, setting, | M2033 |
| offshore activities, | M2021 |
| organisation, | M2027–M2031, M2039 |
| performance measurement, | M2039 |
| permit-to-work system, | M2039 |
| planning, | M2031–M2033, M2039 |
| policy statement, | M2003, M2025–M2026 |
| POPIMAR, | M2023, M2026, M2033, M2038 |
| precautions, | M2008 |
| prevention, application of principles to, | M2007 |
| records, | M2008, M2034 |
| recruitment and appointment arrangements, | M2031 |
| regulations, | M2005–M2021 |
| responsibilities, clear definitions of, | M2028 |
| review, | M2024, M2037, M2039 |
| risk assessment, | M2002, M2006, M2008–M2009, M2011, M2019 |
| risk control systems, | M2032–M2034 |
| self-employed persons' undertakings, persons working in | M2015 |
| senior person, nomination of, | M2028 |

*Index*

| | |
|---|---|
| SMART, | M2033, M2034, M2038 |
| standards, | M2023, M2028, M2036 |
| succession and contingency planning, | M2031 |
| supervision, adequate levels of, | M2028 |
| systems, | M2023–M2024, M2032 |
| temporary workers, | M2018 |
| third parties | |
|   asbestos, | A5020 |
|   duties owed to, | M2004 |
| training, | M2016, M2020, M2031, T7007, T7015–T7022 |
| written statements, | M2003, M2025–M2026 |
| young persons, | M2013, M2020 |
| **MANDATORY SIGNS,** | W11034, W11039 |
| **MANSLAUGHTER,** *see* Corporate manslaughter | |
| **MANUAL HANDLING OPERATIONS,** | M3001–M3027 |
| accident reporting, | M3001 |
| accident investigation reports, | M3019 |
| accident statistics, | M3001 |
| accident treatment records, | M3019 |
| age, | M3007 |
| armed forces, | M3002 |
| automation, | M3010 |
| avoiding risk, | M3008–M3015 |
| bulky or unwieldy loads, | M3005 |
| capability of the individual, | M3007, M3015 |
| carrying, | M3003 |
| charts, | M3003 |
| checklist, | M3020, M3024 |
| children or young persons, | V12005 |
| chutes, | M3010 |
| clothing, | M3004, M3007 |
| control of substances hazardous to health, | M3002 |
| conveyors, powered and non-powered, | M3010 |
| deliveries, requiring direct, | M3009 |
| dermatitis, | M3002 |
| design, | |
|   floors, | M3014 |
|   processes or activities, | M3009, M3012 |
|   revision of, | M3009 |
| disabled persons, | V12005 |
| discussions, | M3022 |
| elimination of handling, | M3009 |
| employers' duties, | M3002 |
| ergonomic approach, | E17044, E17052, E17062 |
| European Union, | M3001 |
| excessive movement of loads, | M3004 |
| experience, | M3007, V12005 |
| fitness experience, | M3015 |
| floors | |
|   condition and design, | M3006, M3014 |
|   uneven, slippery or unstable, | M3006 |
| footwear, | M3004, M3007 |
| frequent physical effort, | M3004 |
| gender, | M3007 |
| grasp, loads difficult to, | M3005, M3013 |
| guidance, | |
|   checklist, | M3020 |
|   HSE, | M3002, M3012, M3017, M3019 |
|   risk assessment, | M3003, M3017, M3024 |
|   team handling, | M3012 |
| headroom, restricted, | M3006 |
| heaviness, | M3005, M3013 |
| hierarchy of measures, | M3002 |
| humidity, | M3006 |
| ill-health records, | M3019 |
| individual capability, | M3007, M3015 |
| information gathering, | M3019 |
| instability, | |
|   floors, | M3006 |
|   loads, | M3005, M3013 |
| layout, | M3012 |
| levels, differing work, | M3006, M3014 |
| lifting and lowering, | M3003, M3010 |
| lighting, | M3006, M3014 |
| load, | |
|   factors related to, | M3005 |
|   reduction of the, | M3011 |
|   related measures, | M3013 |
| lone workers, | V12005, V12029 |
| long-term restrictions, | M3015 |
| Manual Handling Operations Regulations 1992, | M3001–M3004, M3008 |
| manually-operated lifting devices, | M3010 |
| mechanical lifting devices, | M3010 |
| mechanisation, | M3010, M3012 |
| medical screening, | M3015 |

IND/49

*Index*

| | |
|---|---|
| movement or posture, unsatisfactory, | M3004 |
| new or expectant mothers, | V12005, V12020, V12025 |
| notes, | M3023 |
| observations, | M3021 |
| obstruction, | M3006 |
| offshore activities, | M3002 |
| oil, | M3002 |
| operating procedures, | M3019 |
| physique, | M3007 |
| planning and preparation, | M3016–M3019 |
| pregnant workers, | M3007 |
| previous ill-health or injury, | M3007 |
| prolonged physical effort, | M3004 |
| pushing and pulling, | M3003, M3004 |
| rate of work imposed by a process, | M3004 |
| reasonably practicable, | M3002 |
| records, | M3019, M3024 |
| redesigning processes or activities, | M3009 |
| reducing the risk, | M3008–M3015 |
| reduction of the load, | M3011 |
| regulations, | Intro 23, M3001–M3004 |
| rest and recovery, insufficient, | M3004 |
| restricted working areas, | M3006 |
| risk assessment, | M3002–M3007, M3020–M3024, R3009 |
| after, | M3024 |
| arrangement of, | M3017 |
| automation or mechanisation, | M3010 |
| checklist, | M3020, M3024 |
| discussions, | M3022 |
| HSE guidance, | M3003, M3017, M3024 |
| load, reduction of the, | M3011 |
| notes, | M3023 |
| observations, | M3021 |
| persons carrying out, | M3017 |
| visits, | M3020 |
| risk reduction, | M3008–M3015 |
| routines, | M3012 |
| seated, handling when, | M3003 |
| sharpness of loads, | M3005 |
| short-term restrictions, | M3015 |
| slipping, | M3005, M3006 |
| space | M3006 |
| providing clear handling, | M3014 |
| statistics, | M3001 |
| stature, | M3007 |
| stopping, | M3006 |
| sudden movement, risk of, | M3004 |
| task related factors, | M3004 |
| task related measures, | M3012 |
| team handling, | M3012 |
| temperature, | M3014 |
| extremes, of, | M3006 |
| loads, of, | M3005 |
| training and competence, | M3015 |
| transport, use of, | M3009 |
| tripping, | M3006 |
| trolleys, | M3010 |
| trucks, | M3010 |
| trunk, holding or manipulating loads at a distance from the, | M3004 |
| ventilation, | M3006, M3014 |
| weight reduction, | M3013 |
| whole-body vibration, | V5024 |
| wind, gusts of, | M3006, M3014 |
| work surfaces, levels of | M3006 |
| working environment, | M3006, M3014 |
| young workers, | M3007 |
| **MANUFACTURER,** | |
| duties of, | Intro 21 |
| electricity, | E3025 |
| hand-arm vibration syndrome, | V5020–V5021 |
| machinery safety, | M1001–M1002, M1009 |
| product liability, *see* Product Liability; Product Safety | |
| **MARKET,** | |
| registration for sale of food, | F9027 |
| **MARKINGS,** | |
| CE marks, | C8117, E3025, M1009, M1015–M1016, M1026 |
| dangerous substances, | D0423–D0426, D0431, D0435–D0347 |
| machinery safety, | M1007 |
| **MATERNITY LEAVE,** | |
| regulations regarding, | E14002, E14028 |
| **MEALS,** | |
| facilities for, | W11026 |
| **MEAT,** | |
| licensing of premises, | F9021 |
| Meat Hygiene Service, | F9005 |
| **MECHANISATION,** | M3010 |
| **MEDIA,** | |
| business continuity planning, | B8009, B8038–B8042 |
| crisis communication, | B8038–B8042 |

| | |
|---|---|
| disaster and emergency situations, | D6001, D6027, D6030, D6037 |
| **MEDICAL SERVICES** | |
| compressed air on construction sites, | C8098, C8105 |
| disasters or emergencies, for, | D6030 |
| **MEDICAL SURVEILLANCE**, *see* Health surveillance | |
| **MENTAL DISABILITIES AND LEARNING DIFFICULTIES**, | |
| control of substances hazardous to health, | V12005 |
| personal protective equipment, | V12005 |
| **MENTAL DISTRESS, DAMAGES FOR**, | C6021, C6026, C6040, C6048 |
| **MERCURY**, | |
| new or expectant mothers, | V12022 |
| occupational disease, | O1051 |
| poisoning, | O1051 |
| **MESOTHELIOMA**, | A5002–A5003, C6010, O1001, O1004 |
| **MINERAL RESOURCES**, | |
| construction and building operations, | C8020 |
| **MINER'S NYSTAGMUS**, | O1051 |
| **MINES**, | |
| accident reporting, | A3005, A3035 |
| children and young persons, | V12015 |
| **MINIMUM SPACE PER PERSON**, | |
| factories, offices and shops, | W11008 |
| **MINISTRY OF AGRICULTURE, FISHERIES AND FOOD**, | F9005 |
| **MISTS**, | C8134 |
| **MOBILE CONSTRUCTION SITES**, | C8002 |
| **MOBILE CRANES**, | L3005, L3006, L3034 |
| **MOBILE FOOD PREMISES**, | |
| registration, | F9027 |
| **MOBILE PHONES**, | D8219 |
| **MOBILE SAFETY ANCHORS**, | W9042 |
| **MOBILE TOWER SCAFFOLDS**, | W9027 |
| **MOBILE WORK EQUIPMENT**, | M1007 |
| **MOBILE WORKERS**, | W10027–W10029 |
| **MONOTONOUS WORK**, | W10018, W10033, W10034 |

| | |
|---|---|
| **MOORGATE UNDERGROUND DISASTER**, | D6066 |
| **MORECAMBE BAY COCKLE PICKERS TRAGEDY**, | Intro 26 |
| **MULTI-OCCUPATION**, | |
| construction and building operations, | C8141 |
| **MUSCULOSKELETAL DISORDERS**, | |
| accident reporting, | E17066 |
| back disorders, | E17045 |
| clinical effects, | E17044 |
| guidance as to, | E17067, E17068 |
| individual differences, | E17049 |
| legislation as to, | E17044 |
| litigation, | E17044 |
| lower limb discomfort, | E17047 |
| prevention of, | E17050 |
| psychosocial risk factors, | E17048 |
| scope of, | E17003, E17044–E17050 |
| statistics as to, | E17002 |
| task design and, | E17010 |
| upper limb disorders, | E17046, E17052, E17067 |
| use of term, | E17044 |
| vibration, exposure to, | E17033, V5003 |
| **MUSIC PREMISES, FIRE PRECAUTIONS AND**, | F5073 |
| **MUSICIANS**, | |
| safety representatives, | J3005 |
| **NATIONAL COUNCIL FOR VOCATIONAL QUALIFICATIONS**, | T7008 |
| **NATIONAL EXAMINATION BOARD IN OCCUPATIONAL HEALTH AND SAFETY (NEBOSH)**, | T7013, T7017–T7018 |
| **NATIONAL GENERAL CERTIFICATE IN OCCUPATIONAL HEALTH AND SAFETY**, | T7017 |
| **NATIONAL OPEN COLLEGE NETWORK**, | T7023 |
| **NATIONAL VOCATIONAL QUALIFICATIONS (NVQs)**, | T7012–T7013, T7015, T7016, T7022, T7024 |
| **NATURAL LIGHTING**, | |
| source of lighting, | L5006 |

*Index*

| | |
|---|---|
| **NEGLIGENCE**, *see also* Contributory negligence | |
| construction and building operations, | C8143 |
| corporate manslaughter, | C8143 |
| damages, assessment of, | Intro 24 |
| defences, | Intro 24 |
| duty of care, | Intro 24 |
| foreseeability, | Intro 24 |
| gross, manslaughter by, | C8143 |
| harassment cases, | H1732 |
| noise, | N3022 |
| occupier's liability, and, | O3003 |
| product liability, | |
| comparison with, | P9042 |
| contributory negligence, | P9038 |
| vicarious liability, | Intro 24 |
| *volenti non fit injuria*, | O3017 |
| **NERVOUS SHOCK, DAMAGES FOR,** | C6032–C6033 |
| **NEW OR EXPECTANT MOTHERS,** | V12017–V12025 |
| access, | V12005 |
| antimitotic drugs, | V12022 |
| biological agents, | V12021 |
| breast feeding mothers, | E14024, E14025, R3025, R3026 |
| capability, | M3007 |
| carbon monoxide, | V12022 |
| checklist, | V12025 |
| chemical agents, | V12022 |
| compressed air, | V12004, V12020 |
| conditions of work, | V12023–V12025 |
| construction work, | V12005 |
| contracts of employment, | V12018 |
| control of substances hazardous to health, | V12022 |
| cytotoxic drugs, | V12022 |
| definition, | V12003, V12017 |
| display screen equipment, | V12005 |
| diving work, | V12020 |
| electromagnetic radiation, | V12020 |
| EU law, | V12019 |
| guidance, HSE, | V12019, V12020–V12024 |
| ionising radiation, | V12004, V12020 |
| labelling, | V12022 |
| lead, | V12004, V12022 |
| management of health and safety, | M2019 |
| Management of Health and Safety at Work Regulations 1999, | V12002–V12003, V12017–V12018 |
| manual handling, | M3007, V12005, V12020, V12025 |
| maternity leave, | E14002, E14028 |
| medical certificates, | V12018 |
| mental pressure, | V12020 |
| mercury and mercury derivatives, | V12022 |
| movement, | V12020 |
| musculoskeletal disorders, | E17041, E17049 |
| night work, | M2019 |
| noise, | V12020 |
| non-employees, risks to, | V12026 |
| notification of pregnancy, birth or breastfeeding, | E14027 |
| personal protective equipment, | V12005 |
| physical agents, | V12020 |
| physical pressure, | V12020 |
| posture, | V12020 |
| Pregnant Workers Directive, | V12019 |
| restrictions, | V12004, V12018 |
| risk assessment, | E14024–E14025, R3025, R3026, V12017–V12018, V12025–V12026 |
| risks to, | V12019 |
| safety provisions, | E14024, E14027 |
| shock, | V12020 |
| skin, agents absorbed through the, | V12022 |
| social security, | C6002 |
| suitable alternative employment, | V12018 |
| suspension, | V12018 |
| temperatures, | V12020 |
| vibration, | V12020 |
| working conditions, | V12023–V12025 |
| working time, | V12018 |
| **NEWSPAPER PUBLISHER,** | |
| risk assessment, | R3051 |
| **NIGHT CLUBS, NOISE AND,** | N3002, N3027 |
| **NIGHT WORK,** | |
| day work, transfer to, | W10021 |
| definitions, | W10019 |
| health assessments, | W10021 |
| length of, | W10020 |
| limits on, | E14002 |
| new or expectant mothers, | M2019 |
| overtime, | W10020 |
| risk assessment, | W10020 |
| stress, | E17041 |
| working time, | W10020–W10021 |
| **NOISE,** | |
| acoustic booths, | N3037 |

*Index*

| | |
|---|---|
| acoustic signals | W11030 |
| action against employer at common law, | N3022 |
| agriculture, specific legal requirements, | N3015 |
| assessments, formal, | N3013, N3029–N3034 |
| background noise level, | N3010 |
| barriers to noise, | N3039 |
| British Standard, | N3003, N3036, N3041 |
| buying policy for equipment, | N3035 |
| call centres, | N3028 |
| clubs, | N3002, N3027 |
| code of practice, | N3017 |
| competent person, | N3029 |
| conditions when deafness prescribed, | N3019 |
| construction and building operations, | N3016–N3017 |
| construction sites, specific legal requirements, | N3016, N3017 |
| continuous, | N3005–N3006 |
| contribution, | N3022 |
| cooling, | N3035 |
| daily personal noise exposure, | N3007, N3031, O1011 |
| damage occurs, how, | N3003 |
| damping material, | N3035 |
| deafness, noise-induced hearing loss, | N3003, O1010–O1015 |
| statistics, | N3002 |
| designers of machinery, duties of, | N3014 |
| disablement benefit, payment of, | N3002 |
| assessment of benefit, | N3021 |
| claimants, number of, | N3002 |
| conditions under which payable, | N3020 |
| entitlement to, | N3002, N3018 |
| vibration white finger, | N3012 |
| discos, | N3027 |
| dosimeters, | N3003, N3032 |
| ear muffs, | N3036 |
| ear plugs, | N3027, N3036 |
| ear protection, | N3002, N3012, N3013, N3022, N3027, N3036–N3041, P3018 |
| ear protection zones, | N3012, N3014 |

| | |
|---|---|
| emergency purposes, sound systems for, | N3041 |
| employers' duties, | N3013 |
| equipment buying policy, | N3035 |
| ergonomic approach to, | E17004, E17032 |
| EU law, | N3002, N3016 |
| evidence, | N3022 |
| general legal requirements, | N3012–N3017, N3023–N3024 |
| hazardous environments, | N3033 |
| health surveillance, | N3002 |
| hearing disability through ageing, | N3003 |
| hearing loss, | N3003, O1010–O1015 |
| civil liability, | O1015 |
| control measures, | O1012 |
| daily dose, | O1011 |
| disablement benefit, | N3002 |
| frequency, | O1011 |
| intensity, | O1011 |
| tinnitus, | O1014 |
| hearing protection zones, | N3031, N3034 |
| high noise levels at work, | O1011–O1012 |
| HSE guidance, | N3023–N3025 |
| human ear, effect of sound on, | N3004, O1013 |
| human responses to | N3005 |
| importers of machinery, duties of, | N3014 |
| indices, | N3005–N3011 |
| legal requirements relating to, | N3012–N3017 |
| liability of employer at common law, | N3022 |
| manufacturers' duties, | N3014 |
| maximum sound pressure level, | N3008 |
| measurement of sound, | N3005–N3011, N3029–N3030, O1011 |
| meters, | N3029–N3030, N3033 |
| music systems, | N3037, N3040 |
| negligence, | N3022 |
| new or expectant mothers, | V12020 |
| night clubs, | N3002, N3027 |
| Northern Ireland, | N3022 |
| occupational, statutory controls over, | O1012 |

# Index

occupational deafness,
  assessment of disablement
    benefit, N3021
  compensation for, N3018–N3021
  conditions for benefit to be
    payable, N3019
  damages for, N3022
  hearing loss, degree of
    disablement, N3021 (Table 29)
  physical causes, O1010–O1011
  outdoors, equipment used, N3016
  PA and music systems, N3040
  peak sound pressure, N3009
  personal exposure, N3007
  Physical Agents (Noise)
    Directive, N3002
  pubs, N3002
  reducing noise N3035
  specific environments,
    in, N3026–N3028
  regulations, Intro 23, N3001,
    N3013, N3023, P3018
  reports, N3034
  risk assessment, R3008
  rotation of work, N3038
  scale of the problem, N3002
  sound level meters, N3029–N3030,
    N3033
  sound, perception of N3004
  sound power level, N3011
  sound systems, N3027, N3040–
    N3041
  statistical evidence, N3022
  statutory controls
    over, N3012–N3013,
    O1012
  successive employers, N3022
  survey of, N3002
  tinnitus, O1017
  tractor cabs, N3015
  ventilation, N3035
  vibration, N3012, N3017
  vibration white
    finger, E17032, N3012,
    V5003
  visitors and the public, protection
    of, C8137
  weighted sound pressure
    level, N3006, N3008

**NOISE-INDUCED HEARING**
  **LOSS,** O1010–O1015
  civil liability, O1015
  control measures, O1012
  daily dose, O1011
  frequency, O1011
  intensity, O1011
  tinnitus, O1014

**NORTHERN IRELAND,**
  noise, N3022

**NOTIFICATION,**
  see also Accident Reporting
  asbestos, H2138
  compressed air on construction
    sites, C8094–C8103
  construction and building
    operations, C8024
  dangerous substances, H2147
  major accident hazards, C10009

**NURSING HOMES, FIRE**
  **PRECAUTIONS AND,** F5076

**OCCUPATIONAL DEAFNESS,** see
  Deafness

**OCCUPATIONAL DISEASES,**
  asbestos, C1003–C1009
  causation, O1001, O1003
  changing pattern of, O1028
  compensation, and, N3018–N3021,
    O1001, O1008
  current list of, O1054
  employers' duties, O1037–O1051
    ceasing exposure, O1042
    enclosure, O1044, O1045
    information, O1037, O1050
    mechanisation, O1046
    medical surveillance, O1051
    minimising
      exposure, O1043, O1047
    monitoring exposure, O1048
    personal protective
      equipment, O1049
    risk assessments, O1037, O1041
    statement of policy, O1038
    training, O1037, O1050
    warnings, O1037
  extent, O1001
  list of prescribed, O1053
  long-tail diseases, O1001, O1002–
    O1028
    asbestos-related
      diseases, O1003–O1009
    asbestosis, O1003, O1005
    diffuse pleural
      thickening, O1003, O1007
    lung cancer, O1003, O1006
    mesothelioma, O1001, O1004
    noise-induced hearing
      loss, O1010–O1017
    pleural plaques, O1003, O1008
    tinnitus, O1017

| | | | |
|---|---|---|---|
| vibration white finger, | O1001, O1019–O1023 | duties under, HSWA, | O3019 |
| 'neighbourhood exposure', | O1001 | OLA 1957, | O3002 |
| occupational deafness, see Deafness | | factory occupiers, actions against, | O3018 |
| pleural plaques, | O1003, O1008 | fire, as to, | F5079, O3014 |
| pneumoconiosis, | C6059–C6062, O1033 | intruders, | O3009 |
| prescribed diseases, | C6055–C6058 | maintenance and repair of premises, | O3004 |
| range of, | O1001 | nature of duty, | O3003 |
| risk assessment, | O1033 | occupier, definition, | O3004 |
| short-tail diseases, | O1001 | premises, | |
| social security, | C6055–C6058 | definition, | O3005 |
| stress, see Stress | | maintenance and repair, | O3004 |
| tuberculosis, | C6060 | visitors to, | O3006 |
| vibration white finger, | O1001, O1016–O1021, V5011–V5014 | work being done, dangers of, | O3015 |
| | | public right of way, | O3006 |
| work-related upper limb disorders (WRULDs), see Work-Related Upper Limb Disorders (WRULDS) | | risks willingly accepted, | O3017 |
| | | scope of duty, | O3004 |
| | | Scotland, in, | O3001 |
| | | structural defects, | O3013 |
| **OCCUPATIONAL NOISE**, see Noise at work | | supervision, | O3006 |
| | | trespassers, | |
| **OCCUPATIONAL SAFETY AND HEALTH**, | | common law definition, | O3007 |
| | | common law duty to, | O3008 |
| disaster and emergency management, | D6002 | duty to, under OLA, | O3009 |
| | | unfair contract terms, | O3016 |
| ergonomics and, see Ergonomic Design ; Ergonomics management systems, | Intro 25 | visitors, | C8119, C8125 |
| | | invitees, | O3006 |
| | | licensees, | O3006 |
| musculoskeletal disorders, see Musculoskeletal Disorders | | waiver of duty, | O3016 |
| | | warning notices, | O3006, O3009, O3016 |
| national occupational health support programme, | Intro 7 | window cleaners, to, | O3014 |
| pilot studies, | Intro 7 | workmen on premises, | O3014, O3015 |
| **OCCUPIER**, | | **OFFICES**, | |
| dangers to guard against, | O3012 | cleanliness, level and frequency of, | W11006 |
| definition, | O3004 | | |
| duty to provide, | | **OFFSHORE OPERATIONS**, | |
| sanitary conveniences, | W11019 | accident reporting, | A3035 |
| washing facilities, | W11020 | employers' liability insurance, | E13001 |
| factory, actions against, | O3018 | management of health and safety, | M2021 |
| liability, see Occupiers' Liability | | manual handling, | M3002 |
| **OCCUPIERS' LIABILITY**, | | *Piper Alpha* disaster, | D6001, D6022, D6024, D6027, D6065 |
| child-trespass, | O3010 | | |
| children, | O3010 | | |
| common law duty of care, | Intro 24, O3002 | safety representatives, | J3002 |
| | | **OGDEN TABLES**, | C6027 |
| construction and building operations, | C8141–C8151, O3015 | **OIL**, | M3002 |
| | | **OLDER WORKERS**, | |
| contributory negligence, | O3011 | risk assessment, | R3027 |
| dangers to guard against, | O3012 | | |

*Index*

| | |
|---|---|
| ORANGE BOOK (UN DANGEROUS GOODS CLASSIFICATION SYSTEM), | D0403–D0405 |
| ORGANISATIONAL RESPONSIBILITIES, | Intro 27 |
| OUTDOORS, EQUIPMENT USED, | N3016 |
| OVERCROWDING, | W11008 |
| OVERHEAD ELECTRIC LINES, | |
| accident reporting, | A3035 |
| cranes, and, | L3005 |
| OVERTIME, | W10010 |
| OVERTURNING OF MOBILE WORK EQUIPMENT, | M1008 |
| PA SYSTEMS, | N3040 |
| PACKAGING DANGEROUS SUBSTANCES, | D0403, D0415, D0417, D0420–D0421, D0442 |
| PAIN AND SUFFERING, DAMAGES FOR, | C6032, C6044 |
| PARENTAL LEAVE, | E14002 |
| PARKING AREAS, | |
| workplace access, statutory duties, | A1009 |
| PART 36 OFFERS AND PAYMENTS, | C6042 |
| PART-TIME WORKERS, | |
| treatment of, | E14002 |
| PAYMENTS INTO COURT, | C6042 |
| PEDESTRIANS, | |
| traffic routes, | W11015 |
| workplace access, statutory duties, | A1007 |
| PENALTIES AND PROSECUTIONS, | Intro 15, Intro 20 |
| PENSION, | |
| employee, for, | E14005 |
| SERPS, | E14005 |
| PERFORMANCE, | |
| accident investigations, | A3030 |
| compliance, | Intro 5 |
| corporate governance, | Intro 3, Intro 5 |
| corporate social responsibility, | Intro 5 |
| drivers for, | Intro 3–5 |
| Health and Safety Commission/Executive, review of, | Intro 6 |
| labelling, | Intro 4 |
| large companies, | Intro 3, Intro 5 |
| management of health and safety, | M2028, M2039 |
| public sector, | Intro 5 |
| risk management, large companies and, | Intro 5 |
| safety data sheets, | Intro 4 |
| small businesses, | Intro 3–4 |
| social norms, | Intro 3 |
| sources of information, | Intro 4 |
| suppliers, | Intro 4 |
| training and competence, | T7024–T7025 |
| voluntary sector, | Intro 5 |
| PERMIT-TO-WORK SYSTEM, | S3011–S3022 |
| application, general, | S3020 |
| competent persons, | S3016 |
| confined spaces, | C8076 |
| contractors, | S3015 |
| definition, | S3011 |
| description of work, | S3017 |
| documentation, | S3017 |
| equipment, | S3017 |
| essential features of, | S3013–S3019 |
| hazards, | S3018 |
| HSE guidance, | S3013 |
| information, provision of, | S3014 |
| isolation procedures, | S3023 |
| maintenance, | S3011 |
| management of health and safety, | M2039 |
| operating principles, | S3022 |
| outline of, | S3021 |
| plant and equipment, | S3017 |
| POPIMAR, | M2039 |
| precautions, | S3018 |
| procedures, | S3019 |
| risk assessments, | S3016 |
| selection, | S3015 |
| time for, | S3012 |
| training, | S3015, T7026 |
| written, | S3007 |
| PERSONAL HYGIENE, | |
| food handlers, and, | F9033 |
| PERSONAL ORGANISERS, | D8219 |
| PERSONAL PROTECTIVE EQUIPMENT, *see also* Respiratory protection | |
| British Standards, | P3025 (Table 1) |
| CE mark of conformity, | P3012 |
| certification procedures, | P3010–P3014 |
| clothing, | P3025 (Table 1) |
| common law requirements, | P3026 |
| confined spaces, | C8074 |
| consequences of breach, | P3027 |

IND/56

| | | | |
|---|---|---|---|
| control of substances hazardous to health, | H2110, H2117, H2126, H2131 | work related upper limb disorders and, | O1049 |
| employees' duty to wear, | P3007 | PETROLEUM, | |
| equipment bearing CE mark, | P3012 | fire and fire precautions, | F5031 |
| ergonomic requirements | E17035, E17065 | storage, | F5031 |
| | | PHYSICAL AGENTS, | |
| exclusions, | P3018 | children and young persons, | V12016 |
| eye protection, | P3020, P3025 | new or expectant mothers, | V12020 |
| face protection, | P3020, P3025 | PHYSICAL HAZARDS, | |
| floor protection, | P3022 | confined spaces, | A1020 |
| foot protection, | P3022, P3025 | PHYSIQUE AND STATURE, | |
| general purpose, | P3016 | capability, | M3007 |
| general requirements, | O1041, P3016 | manual handling, | M3007 |
| general statutory duties under HSWA, | P3005 | PIPELINES, accident reporting, | A3005, A3035 |
| hand protection, | P3025 | gas safety, | G1009 |
| handwear protection, | P3026 | PIPER ALPHA | |
| hand/arm protection, | P3021 | DISASTER, | D6001, D6022, D6024, D6027, D6065 |
| hazardous substances, | P3017 | PITS, | W9013 |
| head protection, | C8112, P3019 | PLACARDING, | D0423–D0426, D0435–D0347 |
| health and safety requirements, | P3015, P3025 | PLANT, | |
| hearing protection, | N3012, N3013, N3022, P3025 | compressed air on construction sites, | C8096 |
| main types of protection, | P3018–P3025 | confined spaces, construction and building | C8074 |
| mechanical impact risks, | P3017 | operations, | C8161, F5028 |
| mental disabilities, | V12005 | fire and fire precautions, | F5028 |
| mortal/serious dangers, protection against, | P3014 | permit to work systems, PLEURAL PLAQUES, | S3017 O1003, O1008 |
| new or expectant mothers, | V12005 | PNEUMOCONIOSIS, | |
| particular industries, | P3008 | damages, | C6059–C6062 |
| particular risks, additional requirements, | P3017 | social security, specified disease, as a, | C6059–C6062 O1033 |
| quality control system, | P3014 | POINTING DEVICES, | D8220 |
| radiation protection, | P3017, P3025 | POISONING, | H2103 |
| regulations, | Intro 23 | POLICE OFFICERS | |
| respiratory protection, | P3023, P3025 | violence and, | S11001 |
| risk assessment, | O1041, R3011 | POLICIES, see also Statements of | |
| specific statutory duties, | | Health and Safety Policy | Intro 27 |
| duties of employees, | P3007 | POPIMAR, | M2023, M2026, M2033, M2039 |
| duties of employers, | P3006 | | |
| summary of employers' duties, | P3006 | PORTABLE DISPLAY SCREENS, | D8219 |
| statutory requirements, | P3004 | PORTABLE TOOLS, | C8015 |
| suspension equipment, | C8064, W9019–W9020, W9022, W9027 | PRE-ACTION PROTOCOLS, PRE-EXISTING INJURIES OR | Intro 24 |
| testing of, | P3010 | ILL-HEALTH, | |
| uniform European standards, increased importance of, | P3009 | capability, manual handling, | M3007 M3007 |
| whole body protection, | P3024 | PRECAUTIONS, see also Fire and | |
| work activities requiring, | P3003 | Fire precautions | |
| work regulations, | P3002 | asbestos, | C8133, R3013 |

*Index*

| | |
|---|---|
| collective, | W9016 |
| construction and building operations, | C8141 |
| criminal offences, | C8041 |
| fumes, mists and vapours, | C8134 |
| heights, working at, | W9005, W9008–W9009, 39015–W9016 |
| ladders, | W9002, W9047 |
| lone workers, | V12030–V12031 |
| management of health and safety, | M2008 |
| permit to work systems, | S3018 |
| risk assessment, | Intro 16 |
| roof work, | W9028 |
| windows and window cleaning, | W9002, W9028–W9029, W9036, W9042, W9047 |

**PREGNANT WORKERS AND NEW MOTHERS**, *see* New or expectant mothers

**PREMISES**,

| | |
|---|---|
| asbestos, | C8019, C8068, H2140, R3013 |
| health and safety law, | Intro 21 |
| occupiers' liability, | O3005 |

**PRESSURE SYSTEMS**,

| | |
|---|---|
| accident reporting, | A3035 |

**PRIORITY PROGRAMMES**, Intro 13

**PRODUCT LIABILITY**, *see also* Product Safety

| | |
|---|---|
| action, time limit for bringing, | P9048 |
| consumer protection, | P9031–P9049 |
| damages, | |
| contributory negligence, | P9038 |
| persons liable for, | P9032 |
| dangerous products, for, | P9059 |
| defect in product, | |
| definition of, | P9035 |
| key to liability, as, | P9035 |
| material time for existence of, | P9040 |
| development risk as a defence, | P9049 |
| EC Directive, | P9008–P9012, P9031, P9049 |
| hire contract, | P9053, P9060 |
| hire purchase, where, | P9053, P9060 |
| intermediaries, where, | P9041 |
| joint and several liability, | P9043 |
| negligence, | |
| comparison with, | P9042 |
| contributory negligence, | P9038 |
| parameters of, | P9044 |
| producers, | |
| damages, liability for, | P9032 |
| defences open to, | P9045, P9049 |
| definition of, | P9033 |
| duties of, | P9002–P9009 |
| sale of goods legislation, *see* Sale of Goods Acts | |
| strict liability, | |
| introduction of, | P9032 |
| no contracting out of, | P9047 |
| substandard products, contractual liability for, | P9051 |
| suppliers' liability, | P9034 |
| supply, | |
| attachment to, | P9037 |
| definition of, | P9046 |
| time of, | P9037 |
| unfair contract terms, | P9061–P9063 |
| warnings, absence or inadequacy of, | P9039 |

**PRODUCT SAFETY**, *see also* Product Liability

| | |
|---|---|
| articles for use at work, | |
| definition of, | P9015 |
| designers' duties, | P9017, P9018, P9020 |
| importers' duties, | P9017, P9018 |
| installers' duties, | P9021 |
| manufacturers' duties, | P9017, P9019, P9020 |
| suppliers' duties, | P9017–P9020 |
| civil liability, | |
| breach of statutory duty, | P9030 |
| historical background, | P9028 |
| conformity with safety requirements, | P9010 |
| consumer expectation of, | P9036 |
| consumer products, | P9002–P9007 |
| criminal liability for breach of statutory duty, | P9015–P9027 |
| custom built articles, | P9024 |
| dangerous products, | P9059 |
| defective equipment supplied to employees, | P9029 |
| designers, | |
| duties of, | P9017, P9018, P9020 |
| foreign, offences by, | P9025 |
| distributors, duties of, | P9002 |
| EC Directives, | P9008–P9012, P9031, P9049 |
| enforcement, | P9011 |
| general regulations, | |
| conformity with, presumption of, | P9003 |
| defences under, | P9004, P9007 |
| distributors' duties, | P9002 |
| producers' duties, | P9002 |

| | |
|---|---|
| offences and penalties under, | P9004–P9006 |
| scope of, | P9002 |
| importers, duties of, | P9017–P9019 |
| liabilities of, for foreigner's offences, | P9025 |
| industrial products, | P9013, P9023 |
| manufacturers, duties of, | P9017–P9020, P9022 |
| foreign, offences by, | P9025 |
| negligence, comparison with, | P9042–P9009 |
| producers, duties of, | P9002 |
| regulations EC Directive and, | P9008–P9012 |
| general regulations, *see* general regulations, *above* | |
| HSWA 1974, under, | P9014 |
| substance for use at work, definition of, | P9015 |
| designer's duties, | P9016 |
| importer's duties, | P9016, P9019 |
| manufacturer's duties, | P9016, P9019 |
| suppliers, duties of, | P9014–P9019 |
| industrial products, of, | P9023 |
| liability of, | P9034 |
| sale of goods legislation, *see* Sale of Goods Acts | |
| unsafe imported products, powers to deal with, | P9027 |
| unsafe products, civil liability for, | P9028 |
| voluntary national standards, | P9003 |
| **PROHIBITION NOTICES,** | |
| appeal against, | E15001, E15019–E15025 |
| contravention of, | E15001, E15015, E15017, E15030 |
| Crown, position of, | E15045 |
| enforcement powers of inspectors, | E15001, E15006, E15008–E15011 |
| fire and fire precautions, | F5049–F5050 |
| improvement notice, contrasted, | E15015, E15016 |
| prosecution, coupled with, | E15018 |
| public register, | E15029 |
| **PROHIBITORY SIGNS,** | W11036 |
| use of, | E15001, E15013, E15015 |
| **PROSECUTIONS,** | Intro 15, Intro 20 |
| construction and building operations, | C8039–C8040 |
| **PSYCHIATRIC HARM,** | |
| damages, | C6032–C6033, O1035 |
| stress | O1035 |
| **PUBLIC LIABILITY INSURANCE,** | O3003 |
| **PUBLIC OPINION,** | Intro 14 |
| **PUBLIC, PROTECTION OF THE,** *see* Visitors and the public, protection of | |
| **PUBLIC RELATIONS,** *see* Media | |
| **PUBLIC SECTOR,** | Intro 5 |
| **PUBS, NOISE AND,** | N3002 |
| **PUNCTURE,** | M1025 |
| **QUALIFICATIONS AND CURRICULUM AUTHORITY (QCA),** | T7010, T7013, T7018–T7019, T7021–T7022 |
| **QUALITY ASSURANCE FOR HIGHER EDICATION,** | T7010, T7013 |
| **QUARRIES,** | |
| accident reporting, | A3005, A3035 |
| children and young persons, | V12015 |
| **RACIAL HARASSMENT,** | |
| assault, action for, | H1731 |
| battery, | H1731, H1738 |
| breach of contract, action for, | H1735 |
| compensation, | H1726, H1729 |
| criminal law, | H1737–H1741 |
| damages for, | H1736 |
| definition, | H1707, H1708, H1710, H1711 |
| employee's liability, | H1722 |
| employer's liability, | H1718–H1721, H1723 |
| examples of, | H1713 |
| future legal developments, | H1749 |
| general law, | H1701 |
| harassment policy, | H1742–H1746 |
| injunction for, | H1736 |
| monitoring, | H1748 |
| remedies for, | H1717, H1725–H1741 |
| statutory claims, | H1734, H1739, H1740 |
| tort, actions in, | H1730–H1736 |
| training of workforce, | H1747 |

*Index*

unfair dismissal
  claims, H1727–H1729
  victimisation, H1716, H1717, H1725–H1741
**RADIATION,**
  accident reporting, A3035
  emergencies, D6001, C10020, C10021
  exposure to, protection from, P3025
  incidents involving, D6001
  machinery safety, M1025
  new or expectant mothers, V12020
  offshore operations, O7066
  planning aspects, C10020
  public information, C10021
  visitors and the public, protection of, C8135
**RADIOACTIVE SUBSTANCES,**
  ADR Agreement, D0412, D0414
  International Atomic Energy Agency, D0412
  ionising radiation, H2148
  regulations on, D0412–D0413
  transport, D0412–D0414
**RADON,** C8069
**RAILWAYS,**
  Heathrow rail tunnel, collapse of, Intro 15
  lobby groups, Intro 14
  major incidents, D6027, D6064, D6066
  public opinion, Intro 14
  Southall rail crash, Intro 15
  underground, D6064, D6066
  working time, W10030
**RAMSGATE PEDESTRIAN WALKAWAY COLLAPSE,** Intro 15, Intro 17
**RECKLESS KILLING,** D6022
**RECORDS,**
  accidents, as to, *see* Accident Reporting
  asbestos, A5011, H2140
  control of substances hazardous to health, H2122–H2124,
  display screen equipment, D8216
  electricity, employees working with, E3023
  first-aid, F7020
  maintenance, A5011
  management of health and safety, M2008, M2034
  manual handling, M3019

risk assessment,
  control of substances hazardous to health, H2122–H2124
  content of, R3047
  display screen equipment, D8216
  illustrative, R3048–R3050
  manual handling, M3024
  risk rating matrices, R3042
  training etc, for, R3039
  Safety Case Regulations, O7023
  whole-body vibration, V5025
  working time, W10013, W10025
**RECREATION,**
  climbers, instructing, W9008
  heights, working at, W9008
**RECRUITMENT,**
  management of health and safety, M2031
**RED CROSS,** D6021, D6025, D6028
**REDUNDANCY,**
  civil liability, Intro 24
  selection for, E14010, W10041
  working time, W10041
**REFERENCES,** E3023
**REGISTER OF SAFETY PRACTITIONERS,** T7028
**REGULATION,**
  balancing business with, Intro 12
**RELIGION, DISCRIMINATION ON THE GROUNDS OF,** E14001
**REMOTE-CONTROLLED, SELF-PROPELLED WORK EQUIPMENT,** M1008
**REPAIRS,**
  asbestos, O1003
  lone workers, V12029
**REPETITIVE STRAIN INJURY (RSI),** O1033–O1035
  damages, C6058
  social security, C6058
**REPORTABLE DANGEROUS OCCURRENCES,** *see* Accident reporting
**REPORTABLE OCCUPATIONAL DISEASES,**
  accident reporting, A3015, A3029
  vibration white finger, V5011, V5012
**REPUTATION, LOSS OF,** B8017
**RESCUE,**
  confined spaces, C8084–C8089, V12029
  damages, C6033
  equipment, C8088–C8089
  lone workers, V12029

| | |
|---|---|
| nervous shock, | C6033 |
| safety harnesses, | V12029 |
| training, | C8087 |

**RESIDENTIAL HOMES, FIRE PRECAUTIONS AND,** F5076

**RESPIRATORY PROTECTION,**
| | |
|---|---|
| breathing apparatus, | C8071 |
| confined spaces, | C8071, C8074 |
| control of substances hazardous to health, | H21218, H2131 |
| dust respirators, | V3011 |
| generally, | V3011 |
| personal protective equipment/ clothing, | P3023, P3025 |

**REST FACILITIES,**
| | |
|---|---|
| construction and building operations, | C8056 |
| provision of, | W11027 |

**REST PERIODS AND BREAKS,**
| | |
|---|---|
| duration, | W10014, W10017 |
| EU law, | W10014 |
| exceptions and derogations, | W10015 |
| monotonous work, | W10018 |
| shift workers, | W10035 |
| weekly rest period, | W10016 |
| working time, | E17040, J3002, J3032, W10014–W10018, W10035 |

| | |
|---|---|
| **REVERSING,** | A1015 |

**RIDDOR,** *see* Accident Reporting

**RISK ASSESSMENT,** Intro 16
| | |
|---|---|
| absolute requirements, | R3004 |
| accident investigations, | A3030 |
| asbestos, | A5003, A5010, A5017, A5025, H2138, R3013, R3015 |
| assessment unit, | R3029 |
| business continuity planning, | B8012–B8018 |
| CDM health and safety plans, | R3016 |
| checklist, | R3034 |
| children and young persons, | R3024, R3031, V12006, V12008, V12013, V12016 |
| cleaners, | R3031 |
| confined spaces, | A1023, C8073, C8075–C8080, C8083 |
| construction and building operations, | C8021, C8029, C8030, R3002, R3031 |
| contractors, | Intro 17, R3002, R3031 |
| control, | R3023 |

| | |
|---|---|
| control of substances hazardous to health, | H2105–H2125 |
| customers, | R3002 |
| danger areas, | R3022 |
| dangerous substances, | R3014 |
| disabled persons, | R3027, V12033, V12035–V12036 |
| discussions, | R3037 |
| display screen equipment, | D8201, D8204, D8210–D8222, R3010 |
| dynamic risk assessment, | R3016 |
| emergencies, | H2135, R3002 |
| employees, | R3002, R3031 |
| ergonomic tool, as, | E17052, E17061, E17063, E17068 |
| expectant mothers, | R3025, R3026 |
| explosive atmospheres, | R3013 |
| fire and precautions, | F5006, F5017, F5035, F5051, F5053, R3012, R3031, R3032 |
| genetically modified organisms, | R3015 |
| hand-arm vibration syndrome, | V5015, V5021–V5022 |
| hazardous substances, | R3002, R3003, R3007 |
| health and safety at work requirements, | D6007, R3002–R3004 |
| health surveillance, | R3022 |
| heights, working at, | W9007, W9014, W9028 |
| HSE Guidance, | R3030 |
| identification of issues, | R3032 |
| implementation of recommendations, | R3043 |
| information required, | R3030 |
| ionising radiation, | H2148, R3015 |
| lead at work, | R3015 |
| lone workers, | R3027 |
| machinery safety, | M1001, M1006, M1007, M1020, M1022–M1024, M1026 |
| major incident potential, | D6044, D6060, R3015 |
| making the assessment, | R3035–R3041 |
| management of health and safety, | M2002, M2005, M2006, M2008–M2009, M2011, M2019 |

*Index*

| | | | |
|---|---|---|---|
| management regulations, | | control of substances hazardous to | |
| ACOP to, | R3018, R3019, R3042 | health, | H2122–H2124 |
| application of management | | content of, | R3047 |
| cycle, | R3023 | display screen equipment, | D8216 |
| children and young persons, | R3024 | illustrative, | R3048–R3050 |
| evaluation of precautions, | R3019 | manual handling, | M3024 |
| hazards and risks, | R3018 | risk rating matrices, | R3042 |
| new or expectant | | training etc, for, | R3039 |
| mothers, | R3025, R3026 | regulations, | |
| scope of, | R3005, R3017, R3022 | common regulations, | R3006–R3012 |
| 'suitable and sufficient', | R3020 | management regulations, *see* Management Regulations, *above* | |
| manual handling | | specialist regulations, | R3015 |
| operations, | M3002–M3007, M3020–M3024, R3009 | review, | R3021, R3023, R3044, R3046 |
| after, | M3024 | safe systems of work, | R3016, S3001 |
| arrangement of, | M3017 | sales representatives, | R3031 |
| automation or | | self-employment, | R3002 |
| mechanisation, | M3010 | serious and imminent danger, | |
| checklist, | M3020, M3024 | where, | R3022 |
| discussions, | M3022 | source of risk, | R3034 |
| HSE guidance, | M3003, M3017, M3024 | substances hazardous to health, | R3007 |
| load, reduction of the, | M3011 | supermarket, | R3049 |
| notes, | M3023 | temporary workers, | R3031 |
| observations, | M3021 | tests, | R3038 |
| persons carrying out, | M3017 | training and competence, | T7003 |
| visits, | M3020 | variations in work practices, | R3033 |
| method statements, | R3016 | VDUs | R3010 |
| monitoring, | R3023 | visitors, | C8119–C8122, C8129, R3002, R3031 |
| new or expectant | | | |
| mothers, | M2019, R3025, R3024, V12018, V12025 | volunteer workers, | R3031 |
| | | vulnerable persons, | R3027 |
| newspaper publisher, | R3050 | who might be at risk, | R3031 |
| noise, | R3008 | who should carry out, | R3028 |
| notes, | R3040 | whole-body | |
| observation, | R3036 | vibration, | V5024–V5025, V5027 |
| occupational diseases, and, | O1033 | | |
| occupational stress, | S11048–S11052 | work activities examined under, | R3034 |
| older people, | R3027 | work related stress, | S11048–S11052 |
| organisation, | R3023 | work related upper limb disorder | |
| permit to work, | R3016 | and, | O1041 |
| personal protective | | young persons, | R3025, R3031 |
| equipment, | O1049, R3013 | **RISK CONTROL** | |
| personnel, as to, | R3002, R3031 | **SYSTEMS,** | M2032–M2034 |
| planning, | R3023, R3035 | **RISK MANAGEMENT**, | |
| precautions, | Intro 16 | asbestos, | A5012, A5018–A5019, A5022, H2141 |
| preventative measures, | R3022 | | |
| preventative principles, | R3022 | business continuity planning, | B8007 |
| public at large, | R3002 | construction and building | |
| recommendations, | R3044, R3045 | operations, | C8001, C8010, C8025 |
| records, | | | |
| assessment, for, | R3041 | corporate governance, | Intro 5 |

IND/62

| | | | |
|---|---|---|---|
| fire and fire precautions, | F5001 | communication, | S3010 |
| heights, working at, | W9002–W9003 | components of, | S3002–S3006 |
| large companies, | Intro 5 | confined spaces, | A1023, C8075– C8081 |
| media, | Intro 5 | | |
| performance, | Intro 5 | construction and building operations, | C8146, C8150 |

ROAD,
- accident reporting, A3010, A3035
- dangerous substances, transportation of, *see* Dangerous goods systems
- visitors and the public, protection of, C8118
- working time, W10028–W10029

ROAM, RIGHT TO, O3007

ROBENS COMMITTEE,
- enforcement powers of inspectors, E15013

ROLLING OVER, M1008

ROOFS,
- construction sites, C8043, C8146, W9028
- design, W9028
- fragile materials, C8043, C8146
- heights, working at, W9028
- precautions, W9028
- safety nets, W9028

ROPE SLINGS, L3010, L3039
- failure, causes of,
  - fibre ropes, L3036
  - wire ropes, L3037

ROPES,
- fibre, L3036
- heights, working at, W9019
- lifting, generally, L3009
- wire, L3037

ROUTINES,
- display screen equipment, D8206
- manual handling, M3012

ROYAL SOCIETY FOR THE PROMOTION OF HEALTH ADVANCED DIPLOMA IN HEALTH AND SAFETY IN THE WORKPLACE, T7019

SAFE OPERATING PROCEDURES, S3010

SAFE PLACE OF WORK,
- construction and building operations, C8043

SAFE STACKING OF MATERIALS,
- forklift trucks, L3045
- generally, L3042

SAFE SYSTEMS OF WORK, S3001–S3024
- common law, S3003, S3005–S3006
- development of, S3008–S3010
- employers' duty, S3003
- hardware, S3006
- isolation procedures, S3023
- job safety analysis, S3009
- job safety instructions, preparation of, S3010
- legal components, S3003–S3006
- matrix, S3007
- occupational health and safety practitioners, S3001
- permit to work systems, *see* Permit to work systems
- pitfalls of not utilising, S3001
- policy statements, S3004
- publications, S3024
- review, S3007
- risk assessment, S3001
- safe operating procedures, preparation of, S3010
- software, S3006
- statute law, S3004–S3006
- windows and window cleaning, W9040
- written arrangements, S3004, S3007

SAFE WORKING LOAD, L3033
- slings, L3039

SAFETY ADVISERS, R3008
- hazardous substances, H2102, H2103

SAFETY COLOURS, W11032, W11033

SAFETY COMMITTEES,
- 'Brown Book', J3019
- consultation on establishment, J3018
- duty to establish, J3017
- establishment, J3017, J3018
- function, J3019
- risk assessment, involvement in, R3007
- role, J3019

SAFETY DATA SHEETS, Intro 4

SAFETY DEVICES,
- nets, W9028

SAFETY FOR PEOPLE AT WORK STANDARDS, T7021

SAFETY HARNESSES, V12029, W9037– W9039, W9042, W9048

IND/63

*Index*

| | |
|---|---|
| **SAFETY HELMETS,** | |
| personal protective equipment, | P3029, P3025 |
| **SAFETY NETS,** | |
| roofs, | W9028 |
| **SAFETY PASSPORTS,** | T7026–T7028 |
| **SAFETY REPORTS,** | C10011–C10012 |

**SAFETY REPRESENTATIVES,** *see also* Safety Committees

| | |
|---|---|
| accident reporting, | A3027, A3028 |
| actors, | J3005 |
| appointment, | |
| unionised workforce, | J3005 |
| armed forces, | J3002 |
| 'Brown Book', | J3019 |
| disclosure of information, | J3015 |
| duties of, | J3007 |
| education, | J3002 |
| employment protection legislation, overlap with, | J3002 |
| European developments, | J3030, J3031 |
| European Works Councils | J3032 |
| extension of duty to consult, | J3002 |
| facilities to be provided by employer, | J3014 |
| general duty, | |
| non-unionised workforce, | J3021 |
| unionised workforce, | J3004 |
| independent, | J3005 |
| information disclosure, | |
| non-unionised workforce, | J3027 |
| unionised workforce, | J3015 |
| musicians, | J3005 |
| non-unionised workforce, | |
| general duty, | J3021 |
| generally, | J3020 |
| information provision, | J3002, J3027 |
| number of representatives for workforce, | J3022 |
| recourse for representative, | J3030 |
| rights and duties of representatives, | J3024–J3029 |
| role of safety representatives, | J3023 |
| time off with pay, | J3024–J3026 |
| training, relevant, | J3028 |
| number of representatives for workforce, | |
| non-unionised workforce, | J3022 |
| unionised workforce, | J3006 |
| obligation to consult, | J3001, J3002 |
| offshore operations, | J3002 |
| prosecutions for non–compliance, | J3002 |
| regulatory framework, | J3002 |
| rights and duties of safety representatives, | |
| non-unionised workforce, | J3024–J3030 |
| unionised workforce, | J3010–J3013 |
| risk assessment, involvement in, | R3008 |
| role, | |
| non-unionised workforce, | J3023 |
| unionised workforce, | J3007 |
| Safety Representatives' Charter, | J3002 |
| statutory protection for, | J3030 |
| technical information, | J3016 |
| time off with pay, | J3010–J3013, J3029 |
| non-unionised workforce, | J3024–J3026 |
| unionised workforce, | J3010–J3013 |
| training and competence, | E14001, J3028, T7023 |
| unionised workforce, | |
| appointment of safety representatives, | J3005 |
| disclosure of information, | J3015 |
| duty to consult, | J3003, J3004 |
| facilities to be provided by employer, | J3014 |
| functions of safety representatives, | J3007 |
| general duty, | J3004 |
| generally, | J3003 |
| immunity, legal, | J3009 |
| 'in good time', consultation, | J3007 |
| information disclosure, | J3015 |
| number of representatives for workforce, | J3006 |
| rights and duties of safety representatives, | J3001, J3010–J3013 |
| role of safety representatives, | J3007 |
| technical information, | J3016 |
| terms of reference, | J3007 |
| time off with pay, | J3010–J3013 |
| workplace inspections, | J3008 |
| workplace inspections, | J3008 |

**SAFETY SIGNS,**

| | |
|---|---|
| colour, | W11033 |
| construction and building operations, | C8062 |
| emergency escape, | W11040 |
| employers' duties, | W11035 |
| exclusions, | W11036 |
| fire and fire precautions, | F5032, W11040 |
| fire-fighting signs, | W11041 |

IND/64

*Index*

| | | | |
|---|---|---|---|
| hand signals, | W11032, W11034, W11042 | SECURITY, | |
| | | emergency situations, | D6002 |
| interchanging and combining, | W11032 | risk assessment, | R3015 (App B) |
| | | visitors and the public, protection of, | C8122 |
| mandatory signs, | W11034, W11039 | SEDENTARY COMFORT, *see also* Workplaces | W11001, W11009 |
| occupiers' liability and, | O3006, O3009, O3016 | SELF-EMPLOYMENT, | |
| prohibitory signs, | W11034, W11037 | asbestos, | A5007, O1009 |
| regulations, | Intro 23 | construction and building operations, | C8001, C8021 |
| types of, | W11031 | control of substances hazardous to health, | H2105 |
| variety of, | W11034 | first-aid, | F7015 |
| warning signs | W11038 | gas safety, duties as to, | G1011 |
| SALE OF GOODS ACTS, | | head protection, | C8114 |
| consolidation, | P9053 | management of health and safety, | M2015 |
| dangerous products, | P9059 | risk assessment, | R3002 |
| dealer, sales by, | P9056 | social security, | C6004 |
| fitness for purpose, | P9052–P9054, P9057 | windows and window cleaning, | W9034, W9036–W9039 |
| history of, | P9052 | working time, | W10004 |
| implied terms, | P9052–P9057 | SELF-PROPELLED WORK EQUIPMENT, | M1008 |
| quality of product, | P9052–P9054 | | |
| sample, sale by, | P9055 | SELF-REGULATION, | |
| strict liability under s 14, | P9058 | 'so far as is reasonably practicable', | E15017 |
| SANITARY CONVENIENCES, | | | |
| construction and building operations, | C8156, W11019 | SENTENCING, | |
| | | fines, | Intro 15 |
| factories, | W11017–W11020 | fire and fire precautions, | F5058, F5068 |
| minimum number of facilities, | W11020 | health and safety offences, | Intro 15, E15043 |
| occupier's duty to provide, | W11018 | working time, | W10039 |
| offices/shops, | W11019 | SERVICE INDUSTRIES, | Intro 6 |
| provision of, generally, | S11016 | SETTLEMENTS, | Intro 24 |
| temporary work sites, | W11022 | SEX DISCRIMINATION, *see also* Sexual harassment | E14002, E14023 |
| SCAFFOLDING, | | | |
| accident reporting, | A3029 | SEXUAL HARASSMENT, | |
| heights, working at, | W9022, W9027 | complaints procedures, | H1742 |
| mobile towers, | W9027 | confidentiality, | H1742, H1744 |
| suspended, | W9042 | course of conduct cases, | H1712, H1713 |
| windows and window cleaning, | W9027, W9042 | criminal law, | H1724, H1737–H1740 |
| SCHOOLS, | | damages, | H1723, H1733, H1734, H1736 |
| fire and fire precautions, | F5074 | | |
| occupier's liability, | O3018 | definition, | H1702, H1707–H1711 |
| SCOTLAND, | | | |
| food policy in, | F9001, F9063–P9069 | direct, | H1716, H1725 |
| Food Standards Agency in, | F9005 | EC Code of Practice, content of, | H1705 |
| gas safety, | G1011 | | |
| occupiers' liability, | O3001 | | |
| Scottish Protection Agency, role of, | C10001, C10002 | | |
| SCREENS, COMPUTER, *see* Visual Display Units | | | |
| SEATING, *see* Workstations | | legal effect of, | H1706 |
| SECOND-HAND MACHINERY, | M1007 | | |

IND/65

*Index*

| | | | |
|---|---|---|---|
| publication of, | H1701–H1703 | monitoring, | H1747 |
| EC Recommendation, | | review of policy as to, | H1747 |
| content of, | H1704 | single incident cases, | H1712, H1714 |
| legal effect of, | H1706 | three stage test for, | |
| publication of, | H1701–H1703 | detriment, | H1707, H1710– |
| employee's liability, | H1722 | | H1711 |
| employer, | | differential treatment on grounds | |
| defence for, | H1702, H1721 | of sex, | H1707, H1709 |
| harassment policy, | H1742–H1746 | less favourable | |
| liability of, | H1718–H1720, | treatment, | H1707, H1708 |
| | H1722–H1723, H1732 | training of staff, | H1747 |
| EOC Code of | | vicarious liability, | H1718–H1720, |
| Practice, | H1742–H1750 | | H1722, H1732 |
| examples of, | H1712–H1715 | victimisation, | H1716, H1717 |
| failure to investigate or protect | | **SEXUAL ORIENTATION,** | |
| cases, | H1712, H1715 | **DISCRIMINATION ON THE** | |
| formal action as to, | H1746 | **GROUNDS OF,** | E14001 |
| future developments in | | **SHEARING,** | M1025 |
| law, | H1749, H1750 | **SHIFT WORKERS,** | |
| general law, | H1701 | ergonomics and, | E17036, E17041 |
| history of, | H1702 | stress and, | S11008, S11021 |
| homosexuals, | H1705 | working time, | W10035 |
| indirect, | H1716 | **SHIPBUILDING AND** | |
| informal action as to, | H1742, H1745 | **SHIPREPAIRING,** | |
| legal action and remedies, | | children and young persons, | V12015 |
| assault, for, | H1731 | **SHOPS,** | |
| battery, for, | H1731, H1737, | absence of overcrowding, | W11008 |
| | H1738 | access for disabled persons, | W11043 |
| breach of contract, | | changing clothes, facilities | |
| for, | H1730, H1735, | for, | W11026 |
| | H1736 | doors and gates, | W11016 |
| civil law, under, | H1724–H1736 | drinking water, supply of, | W11024 |
| compensation, | H1726, H1729, | eating facilities, | W11027 |
| | H1733 | escalators, | W11017 |
| criminal law, | H1724, H1737– | first-aid, | F7017 |
| | H1740 | floors, safety regulations, | W11010 |
| damages, | H1733, H1734, | pedestrian traffic routes, | W11015 |
| | H1736 | rest facilities, | W11027 |
| declaration, use of, | H1726 | sanitary | |
| injunction, | H1733, H1736 | conveniences, | W11018–W11021 |
| negligence claim, | H1732 | temperature control, | W11005 |
| Protection from Harassment Act, | | washing facilities, | W11020 |
| under, | H1734, H1737, | **SHOWER ROOM,** | |
| | H1739 | gas appliances in, | G1018 |
| Public Order Act 1986, | | **SIGNALS,** *see* Safety Signs | |
| under, | H1737, H1740 | **SIGNS,** *see* Safety Signs | |
| recommendation, use of, | H1726 | **SMALL BUSINESSES,** | |
| re-engagement, order for, | H1729 | definition, | Intro 3 |
| reinstatement, order for, | H1719 | deregulation, | Intro 12 |
| sex discrimination legislation, | | increase in, | Intro 6 |
| under, | H1725 | performance, | Intro 3–4 |
| unfair dismissal, claim | | time off, | Intro 26 |
| for, | H1727–H1729 | **SMART,** | M2033–M2034, |
| victimisation, for, | H1717 | | M2038 |
| marital, | H1716 | **SMOKE INHALATION,** | C8136 |

| | |
|---|---|
| **SMOKING,** | |
| confined spaces, | C8082 |
| fire and fire precautions, | F5018 |
| passive, | W11028 |
| **SMOTHERING FIRE,** | F5005 |
| **SOCIAL CHARTER,** | Intro 9 |
| **SOCIAL RESPONSIBILITY,** *see* Corporate social responsibility | |
| **SOCIAL SECURITY,** | C6001–C6020 |
| accidents, | C6004–C6010 |
| appeals, | C6018 |
| asbestos, | C6010 |
| asthma, | C6057 |
| attendance allowance, | C6014, C6020 |
| bronchitis, | C6060 |
| byssinosis, | C6060 |
| certificate of recoverable benefits, applications for, | C6045–C6049 |
| change of circumstances, | C6019 |
| claims in accident cases, | C6012 |
| constant attendance allowance, | C6014, C6020 |
| course of employment, accidents arising out of, | C6006–C6007 |
| damages, | |
| certificate of recoverable benefits, applications for, | C6045–C6049 |
| deductions, | C6036, C6051–C6054 |
| information, | C6049 |
| prescribed diseases, | C6055–C6058 |
| recovery of, | C6044–C6050 |
| deafness, | C6056 |
| dependents, | C6015 |
| deductions, | C6036, C6051–C6054 |
| disability living allowance, | C6002, C6020 |
| disabled person's tax credit, | C6003 |
| disability tax credit, | C6016 |
| disablement benefit, | C6002, C6011–C6013, C6056–C6062 |
| emphysema, | C6060 |
| employers, persons treated as, | C6009 |
| exceptionally severe disablement allowance, | C6014 |
| hospital, time spent in, | C6019 |
| incapacity benefit, | C6002, C6015 |
| income support, | C6015 |
| industrial injuries benefit, | C6002, C6003, C6008, C6015 |
| mesothelioma, | C6010 |
| overlapping benefits, | C6020 |
| payment, | C6017 |
| personal injuries, | C6004–C6010 |
| pneumoconiosis, | C6059–C6062 |
| pregnant workers, | C6002 |
| prescribed diseases, | C6010, C6015, C6055–C6058 |
| rate of industrial injuries disablement benefit, | C6012 |
| relevant employer, | C6008 |
| repetitive strain injuries, | C6058 |
| self-employed, | C6004 |
| supplementary rules, | C6007 |
| tax, | C6019 |
| tuberculosis, | C6060 |
| voluntary work, | C6002 |
| working tax credit, | C6016 |
| **SOUND SYSTEMS,** | N3027, N3040–N3041 |
| **SOUTHALL RAIL CRASH,** | Intro 15 |
| **SOURCES OF ADVICE,** | Intro 28 |
| **SPACE,** | |
| manual handling, | M3014 |
| work environment, | M3014 |
| **SPRINKLER SYSTEMS,** | F5007 |
| **ST JOHN'S AMBULANCE,** | D6030 |
| **STABBING,** | M1025 |
| **STABILITY,** | C8044 |
| **STAIRS,** | |
| fire and fire precautions, | F5007 |
| workplace, safety regulations on, | W11010 |
| **STALKING,** | H1734 |
| **STANDARDS,** | |
| AA 1000 Assurance Standard, | E16024 |
| business continuity planning, | B8006 |
| CE marks, | C8117, E3025 |
| contractors, | Intro 17 |
| dangerous substances, transport of, | D0420 |
| electricity, | E3025 |
| environmental management, | E16024 |
| fire alarms, | F5017 |
| fire extinguishers, | F5018 |
| fire and fire precautions, | F5016–F5018, F5047 |
| guards, | M1028, M1032 |
| head protection, | C8113, P3025 |
| local authorities, | Intro 17 |
| machinery safety, | M1001, M1003, M1007, M1010, M1012–M1015, M1019–M1020 |
| management of health and safety, | M2023, M2028, M2036 |
| noise, | N3003, N3036 |

IND/67

*Index*

stress, intro 13, S11010
  training and
    competence, T7008–T7027
STATE EARNINGS RELATED
  PENSION SCHEME,
  contracting out of, E14005
STATEMENTS OF
  INTENT, Intro 27
STATEMENT OF TERMS AND
  CONDITIONS OF
  EMPLOYMENT, E14005
STATEMENTS OF HEALTH AND
  SAFETY POLICY, *see* Health
  and Safety Policy Statements
STATURE,
  capability, M3007
  manual handling, M3007
STEP LADDERS, W9029
STREET VENDOR,
  food safety and standards, F9007
STRESS,
  administrative support
    and, S11025, S11026
  ambiguity, role of, S11022
  auditing of stress, S11039
  causation, O1032
  civil liability, O1033–O1035, S11011
  clerical support and, S11025, S11026
  company structure, relevance
    of, S11028–S11033
  conflict, role of, S11024
  cover during staff absence, S11026
  criminal liability, S11011
  damages, O1035
  deadlines, role of, S11019
  definition, S11002
  environmental
    conditions, S11012–S11017
  ergonomics, role of, E17002
  excessive pressure, S11002
  extent of problem, S11001
  guidance, S11009
  growth areas, O1029
  HSE guidance, Intro 13, S11009, S11010
  information and training, S11042
  job organisation, relevance
    of, S11028–S11033
  legal issues, S11005–S11008, S11043
  management standards, Intro 13
  management techniques
    auditing of stress, S11039
    information and training, S11042
  management policy, S11040
  on-going monitoring S11041
  role of, S11004–S11011
  surveys and, S11039
  on-going monitoring S11041
  personal factors, S11034–S11037
  physical environment and, S11010
  pressure, role of, S11018
  preventative steps, O1036
  psychiatric injury, damages
    for, O1035
  reasonable and prudent
    employer, O1035
  research, S11010
  response to, understanding, S11003
  risk factors, S11012–S11017
  shift patterns, S11021
  standard, Intro 13, S11010, S11019
  surveys and, S11039
  targets, role of, S11019
  technology, role of, S11023
  time demands, S11020
  training, S11042
  travel and, S11022
  unfair dismissal, O1035
  upper limb disorders, *see*
    Work-Related Upper Limb
    Disorders (WRULDS)
  use of term, O1031
  violence,
    definition of, S11044
    hazard, as, S11043
    #legal issue, S11045
    processing information as
      to, S11047
    recognition of problem, S11046
    risk management
      procedures, S11048–S11052
    useful publications, S11053
  work equipment, role of, S11023
  working time and, S11010, S11021
STRUCTURED
  SETTLEMENTS, C6021, C6025, C6039, C6047
SUBCONTRACTORS,
  construction and building
    operations, C8001, C8021
SUBROGATION,
  employer's liability insurance, E13007
SUBSTANCES, ESCAPE OF,
  accident reporting, A3029
SUBSTANCES HAZARDOUS TO
  HEALTH, *see* Control of
  Substances Hazardous to Health

# Index

**SUBSTANDARD PRODUCTS,**
  contractual liability for, P9050
**SUITABLE ALTERNATIVE EMPLOYMENT,**
  new or expectant mothers, V12018
**SUPERMARKET,**
  risk assessment, R3049
**SUPPLIERS,**
  duties of, Intro 21
  machinery safety, M1002, M1009–M1018
**SUPPLY-CHAIN PRESSURE,** Intro 17
**SURVEILLANCE,** *see* Health surveillance
**SURVEYS,**
  asbestos, A5023–A5026, H2140
**SURVIVAL OF ACTIONS,** C6036, C6038
**SUSPENSION EQUIPMENT,**
  construction and building operations, C8064
  heights, working at, W9019–W9020, W9022, W9027
  inspections, W9027
**SUSPENSION FROM WORK,**
  maternity grounds, E14015, E14016, E14025–E14027, V12018
  remuneration, E14016
  suitable alternative work, E14015
  without pay, E14014
**SUSTAINABLE DEVELOPMENT,**
  Global Reporting Initiative, E16024
  Sustainable Consumption and Production strategy, E16024
  World Summit, E16027

**TANKS,** D0422, W9013
**TAX,**
  disability tax credit, C6016
  disabled person's tax credit, C6003
  landfill, E16006
  social security, C6019
**TEAM HANDLING,**
  manual handling, M3012
**TECHNOLOGICAL EQUIPMENT,**
  disaster or emergencies, for, D6030
  stress and, S11023
**TELECOMMUNICATIONS,**
  interception of, E14002
**TELEPHONE EQUIPMENT ROOMS, FIRE PRECAUTIONS AND,** F5006
**TELEWORKERS,**
  display screen equipment, D8213
**TEMPERATURE,**
  construction and building operations, C8058
  extremes of, M3006
  food, *see* Food Safety and Standards
  machinery safety, M1007
  manual handling, M3006, M3014
  new or expectant mothers, V12020
  workplaces, W11005
**TEMPORARY SITES,**
  construction and building operations, C8002, F5062
  fire and fire precautions, F5062
**TEMPORARY WORKERS,** M2018, T7005
**TENANCY AGREEMENTS,**
  asbestos, A5021
**THALIDOMIDE,**
  product safety and product liability, P9001
**THEATRES, FIRE PRECAUTIONS AND,** F5071
**THREE MILE ISLAND DISASTER,** E17002
**TIME,** *see also* Working time
  limits, Intro 24
  off, Intro 26
**TOE CAPS,** W9017
**TINNITUS,** O1017
**TRACTOR CABS,**
  noise, N3015
**TRADE SECRETS,**
  disclosure of, F9013
**TRADE UNIONS,**
  civil actions, Intro 14
  collective agreements, W10007, E14004
  recognition of, E14001
  training and competence, T7023
**TRADING STANDARDS OFFICER,**
  role of, F9005
**TRAFFIC ROUTES,**
  checklist,
    pedestrians, A1007
    vehicles, A1006
  construction and building operations, C8050
  internal traffic, A1005
  markings, A1005
  number of, A1004
  obstructions, A1005
  pedestrians, A1007

*Index*

| | | | |
|---|---|---|---|
| positions, | A1004 | Diploma in Safety | |
| safe access, | A1004 | Management, | T7013 |
| speed limits, | A1005 | disaster and emergency management | |
| speed ramps, | A1005 | systems, | D6040, D6047, D6056 |
| statutory duties, | A1004 | | |
| sufficiency, | A1005 | display screen equipment, | D8208 |
| suitability, | A1004 | doctors, | W10026 |
| vehicles, for, | A1005 | employees, of, | T7024–T7025 |

**TRAFFIC SYSTEMS,**
see also Access,
- common law duty of care, A1021
- organisation of safe, A1003
- safe traffic routes, checklist, A1006
- traffic routes, suitability of, A1004
- vehicles allowed to circulate safely, A1005

**TRAINING AND COMPETENCE,** T7001–T7029, App 5
- advice, sources of, Intro 28
- British Safety Council (BSC), T7013, T7020
- business continuity planning, B8040
- Chartered Institute of Environmental Health (CIEH), T7018
- children and young persons, V12004, V12010, V12012
- City and Guilds, T7022
- Client/Contractor National Safety Group, T7027
- competent persons, T7004–T7005
- compressed air on construction sites, C8101
- confined spaces, C8083, C8087
- construction industry, C8005, C8063, T7026–T7027
- Construction Industry Training Board, T7027
- Construction Skills Certification Scheme, T7027
- contacts, useful, T7029
- continuing professional development, T7028
- contractors, competence for, T7026–T7027
- control of substances hazardous to health, H2134
- coverage, T7001
- dangerous substances, transport of, D0432, D0447
- definition of competence, T7001–T7002

Employment National Training Organisation (ENTO), T7008–T7009, T7015–T7017, T7021, T7024
Enforcement Standards, T7008
environmental management, as to, E17022
evacuations, T7005
fire and fire precautions, F5001
food handlers, of, F9035, F9038, F9048, F9081
gas installer, G1014
harassment, as to, H1747
Health and Safety Commission, T7003
Enforcement Standards, T7008
strategy on, T7001
Health and Safety Executive, T7007, T7008
health surveillance, T7007
heights, working at, W9008
higher education, T7013
HSE guidance, T7007
induction programmes, T7025
inexperienced workers, V12038–V12039
Institution of Occupational Health and Safety (IOSH), T7011–T7014, T7021–T7022, T7028
Safety Pass Alliance Scheme, T7027
lack of, E17002
lead bodies, T7008
legal obligations, T7003–T7009
machinery safety, M1007
maintenance of competence, T7028
management, M2016, M2031, T7007, T7015–T7022
manual handling, M3015, M3027
MBA programmes, inclusion in, T7015
National Council for Vocational Qualifications (NCVQ), T7008

IND/70

*Index*

| | | | |
|---|---|---|---|
| National Examination Board in Occupational Health and Safety (NEBOSH), | T7013, T7017–T7018 | temporary workers, | T7005 |
| | | TUC Certificate in Occupational Health and Safety, | T7023 |
| National General Certificate in Occupational Health and Safety, | T7017 | universities, | T7013 |
| | | vehicles, dangerous substances and, | D0447 |
| National Open College Network, | T7023 | vocational qualifications, | T7008, T7010 |
| | | work experience, | V12010 |
| National Vocational Qualifications (NVQs), | T7012–T7013, T7015–T7016, T7022, T7024 | working time, | W10026 |
| | | WRULDS, as to, | O1050 |
| | | young persons, | T7006 |
| need for, assessment of, | E17010 | **TRAINS**, *see* Railways | |
| Occupational Health and Safety Lead Body, | T7008 | **TRANSPORT**, | |
| | | accident reporting, | A3035 |
| Occupational Health and Safety Practice standards, | T7009 | dangerous goods, *see* Dangerous goods systems | |
| OCR qualifications | T7022 | manual handling, | M3009 |
| outline map, | T7007 | systems, *see also* Dangerous goods systems | A3035 |
| performance, | T7024–T7025 | **TRANSPORTATION OF FOOD**, | F9038, F9042 |
| permit to work systems, | S3015, T7026 | **TRAPPING**, | M1025 |
| Practice Standards, | T7012 | **TRAVELATORS**, | |
| Qualifications and Curriculum Authority (QCA), | T7010, T7013, T7018–T7019, T7021–T7022 | workplaces, safety regulations, | W11017 |
| | | **TRENCHES**, | C8014 |
| | | **TRESPASS**, | |
| Quality Assurance Agency for Higher Education (QAA), | T7010, T7013 | occupiers' liability, | |
| | | common law definition, | O3007 |
| | | common law duty to, | O3008 |
| recognising competence, | T7002 | duty to, under OLA, | O3009 |
| Register of Safety Practitioners, | T7028 | person, to the, | H1731 |
| | | risk assessment, | R3031 |
| regulations, | Intro 23, T7003–T7007 | **TROLLEYS**, | M3010 |
| | | **TRUCKS**, | M3010 |
| resources, | E17042 | **TUBE TRAIN**, | |
| risk assessment, | T7003 | Moorgate disaster, | D6066 |
| Royal Society for the Promotion of Health (RSPH) Advanced Diploma in Health and Safety in the Workplace, | T7019 | **TUBERCULOSIS**, | C6060 |
| | | **TURNBULL COMMITTEE**, | B8001, B8006 |
| Safety for People at Work standards, | T7021 | **UN DANGEROUS GOODS CLASSIFICATION SYSTEM**, | D0403–D0405, D0417, D0420–D0421 |
| safety passports, | T7026–T7027 | | |
| examples of, | T7026 | UN Committee of Experts, | D0415 |
| safety representatives, of, | E14001, J3028, T7023 | UN Numbers, | D0418, D0428 |
| | | **UNFAIR CONTRACT TERMS**, | |
| standards, | T7008–T7027 | occupiers' liability, | O3016 |
| strategic management level, competence at a, | T7015 | product liability and, | P9061–P9063 |
| | | **UNFAIR DISMISSAL**, *see also* Enforcement | |
| supervisors, competence for, | T7016–T7022 | assertion of statutory right, for, | E14011 |

IND/71

*Index*

| | | | |
|---|---|---|---|
| compensation for, | E14022 | equipment, | D0443 |
| additional award, | E14022 | filling, | D0440 |
| basic award, | E14022 | hazard identification numbers, | D0437 |
| compensatory award, | E14022 | labels, | D0437 |
| constructive dismissal, | H1729 | loading, | D0440 |
| formula for, | E14017 | marking, | D0435–D0437 |
| health and safety cases, | E14010 | operational | |
| special award, | E14022 | procedures, | D0443–D0447 |
| constructive | | packaging, | D0442 |
| dismissal, | E14014, E14019, | placarding, | D0435–D0437 |
| | H1727, H1728, H1733 | regulated loads, | D0438–D0439 |
| harassment cases, | E14022, H1727– | safety advisors, | D0445 |
| | H1729 | safety kit, | D0444 |
| health and safety grounds, | | segregation, | D0442 |
| on, | E14010, E14011, | statutory | |
| | E14020 | requirements, | A1009–A1014, |
| procedural fairness, | E14012, E14013 | | W11001 |
| protected disclosures, | J3030 | stowage, | D0440–D0441 |
| protection against, | E14017, E14019 | traffic routes for, *see* Traffic Routes | |
| qualifying period for, | E14019 | training, | D0447 |
| reasons for dismissal, | E14020 | UK requirements, | D0434, D0436, |
| reasonableness of employer's | | | D0444 |
| action, | E14021 | vocational training certificates, | D0447 |
| remedies for, | | **VEHICLES,** | |
| compensation, | E14022, H1729 | access to, | A1015 |
| re-engagement, | E14022, H1729 | construction and building | |
| reinstatement, | E14022, H1729 | operations, | C8051 |
| stress, | O1035 | contractors, | A1014 |
| working time, | W10041 | co-operation requirements, | A1015 |
| wrongful dismissal and, | E14017 | dangerous products, *see* Vehicle | |
| **UNIVERSITIES,** | T7013 | requirements, dangerous | |
| **UPPER LIMB DISORDERS,** *see* | | substances and | |
| Work Related Upper Limb | | deaths and injuries caused by, | A1008 |
| Disorders (WRULDS) | | fall prevention, | A1015 |
| | | giving unauthorised lifts, | A1020 |
| **VAPOURS,** | C8134 | hazardous operations, *see* potentially | |
| **VATS,** | W9013 | hazardous operations *below* | |
| **VDUs,** *see* Display screen equipment | | HSE guidance, | C8051 |
| **VEAL PRODUCTION,** | F9077 | loading, | A1018 |
| **VEGETARIANS,** | F9076 | maintenance, | A1012 |
| **VEHICLE REQUIREMENTS,** | | off-road, whole-body vibration | |
| **DANGEROUS** | | and, | V5024 |
| **SUBSTANCES AND,** | | parking areas, | A1009 |
| access to and from workplaces, *see* | | potentially hazardous | |
| Access | | operations, | A1016–A1019 |
| ADR requirements, | D0434, D0436– | generally, | A1016 |
| | D0437, D0440–D0442 | loading, | A1018 |
| checks, | D0434 | reversing, | A1017 |
| concessions, | D0439 | tipping, | A1019 |
| Dangerous Goods Safety | | unloading, | A1018 |
| Advisor, | D0445 | private vehicles, | A1009 |
| Emergency Action Code, | D0436 | provision, | A1011 |
| emergency response | | reversing, | A1017 |
| requirements, | D0445–D0446 | safety of, | A1008–A1010 |
| environmental protection, | D0446 | speed limits, | A1009 |

| | |
|---|---|
| statutory requirements, | A1009 |
| subcontractors, | A1014 |
| tipping, | A1019 |
| traffic routes, *see* Traffic Routes | |
| training of drivers, | A1013 |
| unauthorised lifts, | A1020 |
| unloading, | A1018 |
| vehicle and fuel duty, | E16006 |
| work equipment, | A1009 |
| work vehicles, | A1010–A1015 |
|    access to, | A1015 |
|    contractors, | A1014 |
|    co-operation requirements, | A1015 |
|    fall prevention, | A1015 |
|    generally, | A1010 |
|    maintenance, | A1012 |
|    provision, | A1011 |
|    subcontractors, | A1014 |
|    training of drivers, | A1013 |

**VEHICLES CARRYING DANGEROUS PRODUCTS**, *see* Vehicle requirements, dangerous substances and

**VENTILATION**,
| | |
|---|---|
| air cleaning, | V3010 |
| air purity, | V3001 |
| airborne particles, control of, | V3010 |
| asbestos, | V3009 |
| confined spaces, | C8081 |
| construction and building operations, | C8156 |
| dust and fumes, danger of exposure to, | V3002 |
| employer's duty, | V3004 |
| exposure to dust and fumes, | V3002 |
| extract/exhaust, | V3010 |
| health and safety procedures, | V3004 |
| highly flammable liquids, | V3006 |
| lead processes, *see* lead | |
| liquefied petroleum gases, | V3006 |
| local exhaust ventilation plant, | H2130 |
| manual handling, | M3006, M3014 |
| noise, | N3035 |
| objectives of good, | V3001 |
| opening etc. of ventilators, | W11013 |
| polluting agents, | V3003 |
| pollution of working environment, | V3002 |
| respirators, use of, | V3004 |
| respiratory protection, | V3011 |
| safe working environment, | V3004, V3005 |
| sanitary conveniences, offices/shops, | W11019 |

| | |
|---|---|
| specific processes or industries, | V3006–V3009 |
| statutory requirements, | V3004, V3005 |
| temperature control, | W11005 |
| welding fumes, | V3002 |

**VIBRATION**, V5001–V5029
| | |
|---|---|
| back pain, | V5005 |
| body, | E17033 |
| buildings and structures, | V5006, V5009 |
| carpal tunnel syndrome, | O1051 |
| code of practice, noise and, | N3017 |
| compensation scheme for minors, government, | V5003 |
| construction site, on, | C8015, N3017 |
| damages, | V5003 |
| direction and frequency, | V5008 |
| dose value, | V5009 |
| effects on people of, | V5002–V5006 |
| employers, impact on, | V5016 |
| EU law, | Intro 11, V5001, V5015–V5016 |
| frequency of, | V5008 |
| Hand Arm Vibration Syndrome (HAVS), *see* Hand Arm Vibration Syndrome (HAVS) | |
| hand-held vibratory tools, | O1020, O1051 |
| measurement, | V5007–V5009 |
|    direction and frequency, | V5008 |
|    dose value, | V5009 |
|    units, | V5007 |
| musculoskeletal disorders, | V5005 |
| new or expectant mothers, | V12020 |
| noise and, | E17032, N3012, V5003 |
| occupational disease, cause of, | O1019–O1022 |
| Physical Agents (Vibration) Directive, | V5001, V5015–V5016 |
| regulations relating to, | N3012, V5001, V5015 |
| Vibration White Finger (VWF), *see* Vibration White Finger (VRF) | |
| Whole-body vibration, *see* Whole-body vibration | |

**VIBRATION WHITE FINGER (VRF)** O1019–O1027
| | |
|---|---|
| civil liability, | O1027 |
| coal miners, | V5003, V5012 |
| common law, action at, | V5012 |
| damages, | V5003, V5012, V5013 |

*Index*

| | |
|---|---|
| definition, | V5011 |
| disablement benefit, | N3012 |
| dosage, | V5012 |
| guidance, | V5012 |
| knowledge, | O1027 |
| noise, | N3012 |
| number of sufferers of, | V5003 |
| occupational disease, | |
| as, | V5011–V5014 |
| preventative steps, | O1026 |
| reporting, | V5011–V5012 |
| stages of, | V5013 |
| symptoms, | O1024, O1025 |
| Taylor-Pelmear Scale, | V5013 |
| warnings, wording of, | V5012 |
| water supply workers, | V5003 |
| whole-body vibration, | V5005 |
| woodworking industry, guilty knowledge in the, | O1027 |
| **VICARIOUS LIABILITY,** | Intro 24 |
| construction and building operations, | C8150 |
| sexual harassment, for, | H1718–H1720, H1722, H1733 |
| **VICTIMISATION,** | |
| definition, | H1716 |
| protection from, scope of, | H1716 |
| remedies for, | H1717 |
| **VIOLENCE,** | |
| definition of, | S11044 |
| hazard, as, | S11043 |
| legal issue, | S11045 |
| lone workers, | V12029 |
| processing information as to, | S11047 |
| recognition of problem, | S11046 |
| risk management procedures, | S11048–S11052 |
| stress and, | S11043–S11053 |
| useful publications, | S11053 |
| **VISITORS AND THE PUBLIC, PROTECTION OF,** | |
| accident reporting, | A3001 |
| asbestos, | C8133 |
| barriers, | C8122 |
| chemical installations, | C8120 |
| children, | C8124 |
| construction and building operations, | C8118–C8139 |
| contractors, | |
| cooperation with, | C8122 |
| duties of, | C8127 |
| control measures, | C8125, C8128 |
| control of premises, | C8119 |
| cooling towers, | C8120, C8131 |
| dust and fibres, | C8132 |
| emergency services, | C8125 |
| employers' liability, | C8123–C8128 |
| fencing, | C8122 |
| fire and smoke inhalation, | C8136 |
| fumes, mists and vapours, | C8134 |
| general construction operations, | C8138 |
| hazards, identifying, | C8119–C8122 |
| head protection, | C8116 |
| HSE guidance, | C8118 |
| insurance, | C8121 |
| layout, | C8122 |
| new or expectant mothers, | V12026 |
| noise control, | C8137 |
| occupiers' liability, | C8119, C8125, O3003, O3006 |
| publications, | C8139 |
| radiation and hazardous waste, | C8135 |
| references, | C8139 |
| risk assessment, | C8119–C8122, C8129, R3002, R3031 |
| roads and streets, | C8118 |
| safety policies, | C8120 |
| security, | C8122 |
| signing in and out, | C8126 |
| site planning, | C8122 |
| vulnerability of the public, | C8130–C8132 |
| waste, | C8118, C8135 |
| written arrangements, | C8120 |
| **VISUAL DISPLAY UNITS,** *see* Display screen equipment | |
| **VOCATIONAL QUALIFICATIONS,** | T7008, T7010 |
| **VOLENTI NON FIT INJURIA,** | Intro 24 |
| occupiers' liability claims, | O3017 |
| **VOLUNTARY CARE,** | C6028 |
| **VOLUNTARY SECTOR,** | intro 5 |
| social security, | C6002 |
| **VULNERABLE PERSONS,** *see also* Children and young persons, Disabled persons, Lone workers, New or expectant mothers | V12001–V12039 |
| elderly, | V12005 |
| Health and Safety at Work etc Act 1974, | V12001 |
| indirect implications, regulations with, | V12005 |
| inexperienced workers, | V12005, V12013, V12038–V12039 |
| legislation, | V12001–V12005 |

IND/74

*Index*

Management of Health and Safety at Work Regulations 1999, V12001–V12004
mental disabilities, V12005
older workers, R3027, V12005
risk assessment, R3027, V12001

**WALKWAYS,**
construction and building operations, C8148
windows and window cleaning, W9048, W9050

**WARNINGS,**
dangerous substances, D0425–D0426
hand-arm vibration syndrome, V5020–V5021
machinery safety, M1007
offshore operations, O7028
vibration white finger, V5012

**WARNING SIGNS,** *see also* Safety Signs W11037

**WASHING FACILITIES,**
construction and building operations, C8056
minimum number of facilities required, W11021
occupiers' duties to provide, W11020
provision of, generally, S11016

**WASTE,**
asbestos, A5027
visitors and the public, protection of, C1118, C8135

**WATER SUPPLY,**
confined spaces, C8081
drinking, C8056, S11016, W11024
extinguishers, F5005
fire and fire precautions, F5005
food hygiene, for, F9038, F9045
surges, C8081
vibration white finger, V5003

**WEATHER,**
construction and building operations, C8059
disaster or emergency, effect on, D6024

**WEIGHTS AND MEASURES,**
trading standards officer, F9005

**WELDING FUMES,**
ventilation requirements, V3002

**WELFARE FACILITIES,**
clothing accommodation, W11025
construction and building operations, C8056

current specific statutory requirements, W11018–W11029
drinking water, provision of adequate supply, W11024
facilities for changing clothes, W11026
minimum number of facilities, W11021
occupier's duties,
 sanitary conveniences, W11019
 washing facilities, W11020
particularly dirty work, wash stations required, W11022
rest and eating facilities, W11027
temporary work sites, W11023

**WELLS,**
accident reporting, A3005, A3035

**WHITE FINGER,**
vibration induced, O1019–O1029

**WHOLE-BODY VIBRATION,** *see also* Vibration Intro 11
advice, V5023–V5028
back pain, V5023–V5024, V5028
driving, V5023–V5026
effects of, V5005
exposure assessments, V5027
guidance, V5029
health surveillance, V5028
HSE guidance, V5029
incidence of, V5023
manual handling, V5024
off-road vehicles, V5024
Physical Agents (Vibration) Directive, V5015
record-keeping, V5025
risk assessment, V5024–V5025, V5027
risk reduction V5026

**WIND, GUSTS OF,** M3006, M3014

**WINDOWS AND WINDOW CLEANING,** W9006, W9007, W9029–W9050
access for cleaning, W9045–W9050
anchorage points, W9006, W9037–W9039, W9042, W9048
breach of statutory duty, W9033, W9036
causes of accidents, W9041
civil liability, W9033
common law liability, W9040
contract cleaning companies, W9006, W9036
damages, W9036

IND/75

*Index*

design and
  construction, W9032, W9046, W9050
employees, window cleaners as, W9034
falls, W9031, W9033, W9040–W9041
  construction sites, W9014–W9015, W9019–W9022
  equipment, maintenance of, W9024
  ladders, from, W9041
  personal suspension equipment, W9019–W9020
fatal accidents, W9041
footholds and handholds, W9050
fragile materials, W9025, W9028, W9041
frequency of cleaning, W9044
general safety requirements, W11001, W11014
glazing, W9030, W9043, W9050
high risk commercial buildings, W9006, W9029
  contract firms, W9036–W9039
  mobile safety anchors, W9042
  responsibility for protection of window cleaners and, W9035–W9039
  self-employed contractors and, W9036–W9039
HSE guidance, W9042
inside, cleaning from the, W9029, W9049
inspections, W9027
ladders, C8011, W9002, W9023, W9042, W9047
  enforcement campaign on, W9002
  falls, W9041
  fixed, W9012, W9013
  portable, W9047
  precautions, W9002, W9047
  travelling, W9042, W9047
methods of cleaning, W9043
mixed system of cleaning, W9050
mobile safety anchors, W9042
mobile tower scaffolds, W9027
occupiers' liability, O3014
opening, W9031
position and use of, W11012–W11013, W9031
precautions, W9029, W9036, W9042
  ladders, W9002, W9047
  review of, W9028
  protection methods, W9042–W9043
  safe system of work, W9040
  safety harnesses, W9037–W9039, W9042, W9048
scaffolds,
  mobile tower, W9027
  suspended, W9042
self-employed contractors, window cleaners as, W9034, W9036–W9039
step ladders, W9029
statutory requirements, W9030–W9034
walkways, W9046, W9050
**WITNESSES,**
accident investigations, A3032
interviewing, A3032
**WOMEN WORKERS,** *see also* New or expectant mothers
general statutory requirements, E14023
generally, E14023
HGV driving, E14023
hours of work, E14002
lead processes, E14002
night work, E14002, E14026
risk assessment, E14025, R3031
sex discrimination legislation, effect of, E14002, E14023
statutory requirements, E14002
suspension from work, health and safety reasons, E14015
unfair dismissal, qualifying period for, E14019
unlawful sex discrimination, E14023
**WOODWORKING INDUSTRY,**
knowledge, guilty, 01027
lighting requirements, L5004
machines, L5004
vibration white finger, 01027
**WOOLF REFORMS,** Intro 24
**WORK ENVIRONMENT,**
display screen equipment, D8216
ergonomic design,
  lighting and, E17031
  noise and, E17032
  physical hazards, E17035
  role of, E17030
  thermal environment E17034
  vibration and, E17032, E17033
improvements, M3014
lighting and, E17031
machinery safety, M1025
manual handling, M3014

| | |
|---|---|
| new or expectant mothers, | V12023–V12025 |
| noise and, | E17032 |
| physical hazards, | E17035 |
| role of, | E17030 |
| space, | M301 |
| thermal environment | E17034 |
| vibration and, | E17032, E17033 |

**WORK EQUIPMENT,** *see* Equipment

**WORK RELATED UPPER LIMB DISORDERS (WRULDS),**

| | |
|---|---|
| civil liability, | O1038 |
| definition, | O1037 |
| employer's duties, | O1039 |
| enclosing, | |
|   operator, of, | O1045 |
|   process, of, | O1044 |
| exposure to, | |
|   cessation of, | O1042 |
|   minimising, | O1043, O1047 |
|   monitoring, | O1048 |
| future trends, | O1052 |
| information as to, | O1050 |
| medical surveillance, | O1051 |
| musculoskeletal disorder, as, *see* Musculoskeletal Disorder | |
| personal protective equipment, | O1049 |
| prescribed occupational disease, | O1023 |
| preventative management, | O1040–O1052 |
| process, automating or mechanising, | O1046 |
| risk assessment, | O1041 |
| safety policy, | O1040 |
| training and competence, | O1050 |

**WORK STATION,**

design of, *see* Ergonomic Design
musculoskeletal disorders, *see* Musculoskeletal Disorders

| | |
|---|---|
| rest breaks, | E17040 |
| user trial, | E17055 |

**WORK VEHICLES,** *see also* Vehicles

| | |
|---|---|
| access to, | A1015 |
| drivers of, | A1013 |
| loading, | A1018 |
| maintenance, | A1012 |
| provision of, | A1011 |
| statutory requirements relating to, | A1009 |
| tipping, | A1019 |
| training of drivers, | A1013 |
| reversing, | A1017 |
| unauthorised lifts, giving of, | A1020 |

**WORKERS,**

| | |
|---|---|
| definition of, | E14001 |
| part-time, | E14002 |
| women, *see* Women Workers | |

**WORKFORCE AGREEMENTS,** W10008–W10009, W10013, W10036

**WORKING AT HEIGHTS,** *see* Heights, working at

**WORKING PLATFORMS,** C8012–C8013, W9008, W9014–W9020

| | |
|---|---|
| catchment platforms, | C8013 |
| construction and building operations, | C8012–C8013, C8064 |
| definition, | W9016 |
| guard rails, | W9017 |
| height to base ratios, | C8012 |
| housekeeping, | C8013 |
| inspection | W9027 |
| lifting operations, | C8013 |
| load, | W9018 |
| materials, | C8012 |
| mobile, | C8012, W9015 |
| stability, | W9018 |
| toe caps, | W9017 |

**WORKING TAX CREDIT,** C6016

**WORKING TIME,** W10001–W10044

| | |
|---|---|
| 48 hour week, | W10010–W10013 |
| agency workers, | W10005 |
| agreements on, | W10003, W10006, W10013 |
| annual leave, | W10022–W10024 |
| breach of contract, claims for, | W10010, W10042 |
| calculation of, | W10010, W10036 |
| call, workers on, | W10003 |
| collective agreements, | W10007, W10013, W10036 |
| compensation, | W10040, W10041 |
| compensatory rest, | W10037 |
| confined spaces, | C8081 |
| construction work, | W10004 |
| contracting out, | W10043 |
| criminal offences, | W10039 |
| definitions, | W10002–W10004 |
| detriment, protection against, | W10041 |
| doctors, | |
|   call, on, | W10003 |
|   training, | W10026 |
| domestic services, | W10032 |
| employment tribunals, | W10040 |

IND/77

*Index*

| | |
|---|---|
| enforcement, | W10038–W10043 |
| exceptions and derogations, | W10012, W10032–W10036 |
| excluded days, | W10010 |
| excluded sectors, | W10026–W10031 |
| executives, | W10033 |
| flexible working, | W10003, W10006 |
| Health and Safety Executive, | W10038–W10039 |
| long hours culture, | W10001 |
| maximum weekly working time, | W10010–W10013 |
| mobile and non-mobile workers, | W10027–W10029 |
| monotonous work, | W10018, W10033, W10034 |
| night work, | W10020–W10021 |
| opt-out, | W10001, W10010, W10012, W10013, W10044 |
| overtime, | W10010 |
| partial exemptions, | W10032–W10036 |
| rail sector, | W10030 |
| records, | W10013, W10025 |
| redundancy, selection for, | W10041 |
| reference period, | W10011 |
| regulations, | W10001 |
| application to previously excluded sectors, | W10027–W10031 |
| rest periods, | W10014–W10018, W10035 |
| breaks, | E17040, J3002, J3032, W10017 |
| daily, | J3002, J3032, W10014–W10015, W10035 |
| weekly, | J3002, J3032, W10016, W10035 |
| road transport workers, | W10028–W10029 |
| self-employed, | W10004 |
| sentencing, | W10039 |
| shift workers, | W10035 |
| special categories of workers, | W10033–W10036 |
| stress and, | S11008, S11021 |
| substitutes, appointment of, | W10004 |
| training, | W10026 |
| transport workers, | W10027–W10030 |
| travel, | |
| non-routine, | W10003 |
| unfair dismissal, | W10041 |
| unmeasured, | W10033 |
| workers, definition of, | W10004 |
| workforce agreements, | W10008–W10009, W10013, W10036 |
| workforce representatives, | W10008 |
| Working Time Directive, | W10001 |
| Horizontal Amending Directive, | W10026–W10031, W10044 |
| implementation of, | W10001, W10010 |
| status of, | W10001 |
| Young Workers Directive, | W10001 |

**WORKPLACES,**

| | |
|---|---|
| access, | |
| parking areas, | A1009 |
| pedestrians, | A1007 |
| safe traffic routes, *see* Traffic Routes | |
| safe transport systems, | A1003 |
| traffic routes, *see* Traffic Routes | |
| building work, controls over, | W11042 |
| civil liability, | W11029, W11043, W11049 |
| clothing accommodation, | W11025, W11026 |
| conditions of, | S11016, W11010 |
| definition of, | W11002 |
| dimensions/space, | W11008 |
| doors and gates, | W11012, W11016 |
| drinking water, provision of, | S11016, W11024 |
| eating facilities, | W11027 |
| enforcement of legislation as to, *see* Enforcement | |
| escalators and travelators, | W11017 |
| factories, *see* Factories, | |
| falls from a height, safety measures, | W11011 |
| fire precautions, *see* Fire and fire precautions | |
| floor coverings, | W11010 |
| gates and walls, | W11012 |
| general cleanliness, | W11001, W11007 |
| general maintenance of, | W11003 |
| general requirements, | W11001 |
| harassment in, *see* Harassment | |
| inspections by safety representatives, | J3007 |
| layout, | S11015 |
| lighting, *see also* Lighting, | W11001, W11006 |
| loading bays, statutory duties, | A1009 |
| maintenance requirements, | W11001, W11003 |
| minimal standards, | W11001 |

| | |
|---|---|
| organisation of safe, | A1004 |
| overcrowding, | W11008 |
| parking areas, statutory duties, | A1009 |
| pedestrian access, | A1007 |
| pedestrian traffic routes, | W11001, W11015 |
| regulations, | Intro 23 |
| rest facilities, | W11027 |
| safety signs, *see* Safety Signs | |
| sanitary conveniences, | S11016, W11001, W11019, W11021 |
| seating, | W11009 |
| skylights, | W11013 |
| smoking, | W11028 |
| staff facilities, | S11016 |
| steep slopes provided with handrail, | W11010 |
| temperature controls, | W11001, W11005 |
| traffic routes, *see* Traffic Routes | |
| vehicle traffic routes, *see also* Traffic Routes, | W11001 |
| ventilation of, *see also* Ventilation, | W11001 |
| washing facilities, | S11016, W11020–W11022 |
| welfare facilities, | W11018 |
| windows, | W11012–W11014 |
| workstations, | W11009 |

**WORKS RULES,**
| | |
|---|---|
| incorporation into contract of employment, | E14006, E14007 |

**WORKSTATIONS,**
| | |
|---|---|
| adequate freedom of movement, | W11009 |
| sedentary comfort, | W11009 |
| suitable seating, | W11009 |

**WORLD HEALTH ORGANISATION,** Intro 8

**WRITTEN POLICY STATEMENT,**
| | |
|---|---|
| arrangements, | S7006 |
| basic objectives, | S7004 |
| communication, | S7007 |
| content, | S7003 |
| essential format, | S7003 |
| exceptions, | S7003 |
| people and their duties, | S7005 |
| requirements to prepare, | S7002 |
| review, | S7008 |
| statutory duty to prepare, | S7001 |
| systems and procedures, | S7007 |

**WRONGFUL DISMISSAL,** E14017, E14018

**YOUNG PERSON,** *see* Children and Young Persons.